ANIMAL BEHAVIOR DESK REFERENCE

*A Dictionary of Animal Behavior,
Ecology, and Evolution*

Second Edition

ANIMAL BEHAVIOR DESK REFERENCE

*A Dictionary of Animal Behavior,
Ecology, and Evolution*

Second Edition

Edward M. Barrows

CRC Press
Boca Raton   London   New York   Washington, D.C.

Library of Congress Cataloging-in-Publication Data

Barrows, Edward M.
    Animal behavior desk reference ; a dictionary of animal behavior, ecology, and
evolution / by Edward M. Barrows. —2nd ed.
        p. cm.
    Includes bibliographical references.
    ISBN 0-8493-2005-4 (alk. paper)
    1. Animal behavior—Dictionaries.   2. Ecology—Dictionaries.   3. Evolution
(Biology)—Dictionaries.   I. Title.

    QL750.3 .B37 2000
    591.5¢03—dc21

                                                                                    00-051850

This book contains information obtained from authentic and highly regarded sources. Reprinted material
is quoted with permission, and sources are indicated. A wide variety of references are listed. Reasonable
efforts have been made to publish reliable data and information, but the author and the publisher cannot
assume responsibility for the validity of all materials or for the consequences of their use.

Visit the CRC Press Web site at www.crcpress.com

DEDICATION

To my mother, M.L. Barrows, and to the memory of my father, S.E. Barrows, biophiles and bibliophiles who encouraged my interests in nature and my keeping insect pets (even in our living room!).

INTRODUCTION

Aim of this Book

Animal-Behavior Desk Reference, A Dictionary of Animal Behavior, Ecology, and Evolution is an annotated dictionary intended to help improve scientific communication, in particular in the fields of animal behavior, ecology, evolution, and related branches of biology. I created this book for students, teachers, researchers, writers, editors, and others active in science. This second edition contains more than 1200 additional terms and definitions and thousands of additions and improvements on material presented in the first edition.

A book such as this one can never be complete. Therefore, I certainly welcome constructive criticisms and further information that should be included for its improvement, such as previously published definitions, new definitions of existing terms, your view of definitions already in this book, and new terms with their definitions. Your e-mails (barrowse@georgetown.edu), letters, and annotated reprints regarding these subjects are especially welcome. If you have coined a term(s) included in this book, and I do not attribute this properly, please let me know.

Many references are secondary ones — *e.g.,* Smith (1946 in Jones 1990) — because I wanted to indicate the source of an author's interpreted information, I could not examine certain primary references, or both. If I have cited your work secondarily, I would be very pleased to receive a copy of your publication (with relevant parts highlighted) to facilitate quoting your work directly in the future. Further, if I have misinterpreted your writing with regard to a definition(s) or other information, please let me know.

The Need for Precise Terminology

Clearly defined biological terminology aids the advancement of basic, theoretical, and applied biology. Phenomena are often further elucidated when they are placed in proper order with useful names (Brosius and Gould 1992, 10709). For effective, efficient communication, people should have, or at least understand, the same precise terminology. Dawkins (1986, 110–111) discusses how definitions have had an enormous impact on the way people perceive, for example, in communication. "They have affected what we regard as 'communication' and 'signals' and whether we see communication as involving any sort of transfer of information. ...Much of the confusion over 'information transfer,' too, appears to have been largely brought about by the way it is defined. Arguments which appear to be over something profound about animal communication (such as whether information is being transferred or not) often turn out to be nothing more than whether a protagonist is adopting the technical or the more everyday meaning of the term."

To promote the understanding of terminology, some societies (*e.g.,* the Animal Behavior Society and American Society of Parasitologists) have, or have had, committees dedicated to this subject. Useful dictionaries and encyclopedias that define biological terms (meaning biological in the broad sense) include Campbell and Lack 1985; Chapin 1968; Collocot and Dobson 1974; Cumming 1972; Drever 1974; Eysenck et al. 1979; Goldenson 1984; Harré and Lamb 1983; Heymer 1977; Hinsie and Campbell 1976; Hurnik et al. 1995; Immelmann 1977; Immelmann and Beer 1989; Keller and Lloyd 1992; King and Stansfield 1985; Lincoln et al. 1985; Medawar and Medawar 1983; Roe and Frederick 1981; Seymour-Smith 1986; Steen 1973; Storz 1973; and Wolman 1973.

Confusion About Meanings of Terms

Biologists in general strive to obtain a superior terminology, unfortunately not yet with total success. Confusion and controversy about meanings are common, as I indicate in many entries of this book. Many terms, not to mention the concepts that they represent, are obviously in need of further study of their already-designated meanings. The confusion and controversies arise for many reasons, including:

- A satisfactory definition of a term might not yet exist due to our present state of knowledge (*e.g.,* see aggression, awareness, and consciousness).
- The name of a term is a nontechnical English word which already has a common dictionary definition (*e.g.,* see altruism, selfishness, and spite). Such nontechnical words often "have a variety of meanings and are surrounded by a dense atmosphere of values and associations"

(Wilson and Dugatkin 1992, 29). Therefore, such words can confound communication even if they are precisely defined in scientific literature. Keller (1992a) describes how "competition" with its established colloquial meaning permits the simultaneous transfer and denial of its colloquial connotations when it is used technically. Colloquial connotations can lead plausibly to one set of inferences about a concept and close off others. "Eugenics" was originally defined technically and nonpolitically, yet some people now find it difficult to think of eugenics without negative political connotations.

- The name of a term is a foreign word that may be hard to remember and interpret accurately (*e.g.,* see *umwelt* and *zeitgeber*).
- The name of a term is a technical word that may be difficult to remember and understand without an explicit definition (*e.g.,* see apotreptic behavior and epidiectic display).
- The term is only implicitly defined with regard to all, or some, of its meanings in the scientific literature. To my dismay, I have omitted some terms and some definitions from this book because I could not formulate a satisfactory definition from information provided in written material.
- The term is explicitly, but not correctly, defined in all scientific literature, textbooks, or both. For example, consider "inclusive fitness." Grafen (1982) lists 14 commonly used animal-behavior textbooks that defined this term erroneously or inadequately. Dawkins (1986, 44) later advises, "It is best not to read too much about inclusive fitness as much of what has been written is misleading."
- The term has more than one technical definition that can vary in their degrees of difference, from one word (such as partially, possibly, or potentially) to many words (*e.g.,* see species). In some cases, the same term can even have opposite meanings (*e.g.,* see true altruism and overdispersed distribution).
- Many terms have evolved in their meanings as scientists learned more about the concepts that they represent, research focus shifted, or both. "Isolation" was originally used for both geographical isolation and reproductive isolation; "variety" was used, for instance, by Darwin, for a single individual or a population; and "teleological" was used for four distinct phenomena (Mayr 1982, 43) (*e.g.,* also see communication, evolution, fitness, gene, homology, species, and territory).
- The term may have one to many synonyms, sometimes including a perplexing one that is also the same word(s) used for a very different concept (*e.g.,* see adaptation and homology). In some cases, imprecise language led to virtual synonymies of important terms that should retain distinct usages (*e.g.,* these terms and their associated ones: coevolution, guild, and keystone species) (Fauth et al. 1996, 282).
- The term has more than one spelling, capitalization, or hyphenation style (*e.g.,* see homeotherm, endosymbiosis hypothesis, the Modern Synthesis, and molecular-clock hypothesis).
- The same concept is referred to using two or more of these words: hypothesis, law, rule, and theory. Examples are locale-odor hypothesis and Allen's rule.
- Authors often shorten a concept's name; for example, they use "adaptation" to mean a specific kind of adaptation, such as physiological adaptation, and "selection" to mean a specific kind of selection, such as artificial selection.
- Many scientific words are long. Especially long words are the chemical names for deoxyribonucleic acid (207,000 letters), tryptophan synthetase (1900), tobacco mosaic virus (1185), pneumonoultramicroscopicsilicovolcanoconiosis (45, a lung disease caused by breathing fine dust), and hepaticocholangiocholecystenterostomies (39, a surgical operation that creates channels of communication between gall bladders and hepatic ducts or intestines) (Ash 1996, 108).
- Different dictionaries define the term differently (*e.g.,* atavism) (Hrdy 1996, 851; Thornhill and Gangestad 1996, 853).

Many concerned scientists have tried to rescue us from terminological pitfalls. I applaud them. For instance, Hailman (1976) insightfully suggests how to improve behavioral terminology by emphasizing logic, informational content, and efficiency, rather than historical origins of stability. Brown (1987, 297) lists very helpful guidelines for using biological terms. Fitch (1976) even organized others to petition against a confusing definition of *homology* that refers to similarity of nucleotide sequences (= sequence homology) (Lewin 1987a).

How To Use This Book

Main format. This book uses a dictionary format to present definitions in a standard, quick-to-use way and to help especially users whose first language is not English. The body of the book contains terms that are not organism names. Organism names can be found in Appendix 1, and selected organizations are provided in Appendix 2.

Type of terms emphasized. I emphasize conceptual terms, not anatomical parts or taxonomic names, in the main body of the book; however, it does include a small sampling of these parts, including some organs and glands. It also defines a few of the millions of taxonomic groups in Appendix 1. Biology texts, taxonomic catalogs, and primary literature have thousands of definitions of anatomical parts and taxonomic names that I do not include. This book concentrates on definitions of nouns or noun forms, rather than adjectives, adverbs, and verbs, due to space limitations.

New terms. A term within quotation marks in an entry line is one that my correspondents or I coined as part of producing this book.

Superscripted entries. When needed for clarity, a term with two or more distinct meanings is listed as more than one superscripted entry (*e.g.,* ^{1}sensation and ^{2}sensation).

Term hierarchies. As conceptual structures, relationships among many terms are summarized in tables, and many terms with their definitions are organized in hierarchical clusters either by conceptual relationships or under key words, combining forms, and groups of letters that they have in common, or both. A term that is not part of a hierarchy or one that is the first term in a hierarchy (= a first-level term) is preceded by a diamond (◆). Second- through fifth-level terms are increasingly indented. Second- and fourth-level terms are not preceded by symbols; a third-level term is preceded by an arrowhead (▶); and a fifth-level term is set in sans-serif type, as shown below.

♦ **chemical-releasing stimulus, CRS** *n.* A chemical that…
  semiochemical, semiochemic *n.* A chemical that…
   ▶ **allelochemic** *n.* …
    1. A chemical that…
    allomone *n.* A chemical substance, …
     allelopathic substance *n.* A waste product,…

Definition sources. Most references do not give definitions in the exact form used in this book; therefore, almost all definitions are paraphrased for conciseness and to conform with the style of this book. This includes changes from British to American spellings, where appropriate. Author-date citations indicate references on which I have based definitions. Many definitions are based on ideas combined from more than one source. In some cases, definitions are my perceptions of how authors *seem* to define terms because they did not present their formal definitions. I often indicate this by stating "inferred from… ."

See directives. "See…" directs the reader to where a concept is defined under a synonymous name, is part of a term hierarchy in which it is defined, is defined as part of the definition of another term, or is otherwise characterized (*e.g.,* in a table). For example, "prophototaxis" is a sub-subentry under "phototaxis," which is a subentry under "taxis." The main listing for "prophototaxis," then, reads as follows:

♦ **prophototaxis** See taxis: phototaxis: prophototaxis.

Note that colons separate main entries, subentries, and sub-subentries in the context of a cross-reference. The first term listed is always a first-level entry; reference to a single term (*e.g.,* See behavior) directs the reader to the main, alphabetical listing for the term (in this case, in the "b" section)

Multiple definitions. For many terms, I document some of the diversity of meaning. When I give more than one definition for a term, I list the definitions in *chronological* order with regard to citation date. When I can, I indicate which definition seems appropriate for wide use today. One meaning of a term can often be better understood after its other definitions(s), as well as those of associated terms, are examined. Some terms (*e.g.,* evolution and heterochrony) have been modified through descent by "conceptually selective forces" so that they now have families of definitions, even some with opposite meanings (Gould 1992, 158; Richard 1992, 95). Further, one meaning of a term (*e.g.,* eugenics) can even affect another meaning of it in a negative way, making it a "dirty word" (Kelves 1992, 94).

Nontechnical and obsolete definitions. For many terms, especially ones originally developed to refer to Humans and then extrapolated to nonhumans, I first give nontechnical English definitions, and then I list obsolete definitions when I think they are helpful in understanding the evolution of meaning of a term. An older meaning(s) of a term (*e.g.,* evolution) can lie below the surface of a newer meaning and affect the term's significance (Richards 1992, 95).

Pronunciations. This book lists pronunciations for some of the terms that it defines, particularly for some acronyms and terms with words that are not normally found in standard English dictionaries. For words with more than one syllable, I indicate the stressed syllable by capitalizing it — *e.g.,* **apoptosis** (ap POE toe sis; the second p is silent).

Common-denominator entries. Many terms are grouped as divisions under a main entry (a key word, combining form, or terminal group of letters) that they have in common. This results in many divisions that do not have definitions that relate directly, or at all, to definition(s) in the main entry that precedes a cluster. Also, because authors often did not specify which main-entry definition relates to a division, I often could not specify this information.

The common denominators and main words are that are listed as primary entries (those preceded by diamonds) are: abundance, acme, -acmic, acoustics, action, activity, adaptation, affinity, aggression, agonist, allele, altruism, analysis, angel, animal, animal names, animal sounds, anthropic, attachment, attention, -auxesis, awareness, bee, behavior, beneficiary, benefit, benthic, biont (bion), -bios, biota, -biotic, biotope, bite, bond (bonding), bout, breeding, brood, buffering, calling, camouflage, canalization, cannibalism, care, caste, castration, cause (causation), cell, ceremony, chain, character, cheat (cheater), chemical-releasing stimulus, chimera (chimaera), -chore, chromosome, chronology, -cial, -cide, cline, clock, clone (clon), coefficient, -cole, colonial, coloration, communication, community, companion, competition, complementation, conflict, consciousness, consumer, copulation (copulatory behavior), correlation, cost, courtship (courtship behavior), -cron, cross, crossing over, crypsis, cue, curve, cycle, cyclic, dance, Darwinism, datum, definition, deme, density, dialect, discontinuity, dispersal, display, distribution, diurnal, diversity, doctrine, dog, dominance, domatium, dormancy, dress, drift, drive, -dromous, drug, dynamic, effect, efficiency, effort, electrophoresis, elimination, emotion, endemic, energy, environment, equilibrium, error, estrus (oestrous), eugenics, evolution, evolutionism, evolutionist, experiment, experimental design, exploration, extinction, facial expression, facilitation, factor, fauna, fallacy, family, fauna, fear, feedback, feeding -ference, fight, fitness, flight, food, forager, foraging (foraging behavior), fossil, fostering, frequency, function, game, gamete, -gametic, -gamety, -gamy, gene, gene flow, -genesis, -genetic, genetic load, -genic, -genous, -geny, gland, -grade, gradualism, -gram, -graph, graph, -graphy, greeting, gregarious, grooming, group, growth, guild, -gyny, -haline, heritability, hermaphrodite, hermaphroditism, heterochrony, hierarchy, homeostasis, homolog (homologue), homology, hormone, host, hybrid, hypothesis, illusion, inheritance, inhibition, insemination, instar, instinct, intelligence, investment, isolation, -karyon, kin, -kinesis, kinesis, kiss, Lamarckian, language, law, learning, -lectic, lek, link, locomotion, male, manipulation, map, mate choice, mating, mating system, maze, mechanism, mechanist, meiosis, memory, metamorphosis, method, -metrics, -metrosis, -metry, migration, mimicry, mitosis, -mixis, modality, model, molt, -morph-, -morphic, -morphism, morphology, -morphosis, mortality, mother, movement, mutation, need, nest, nesting, nestling, neurotransmitter (neural transmitter), niche, nucleic acid, object, observation, odor, -oecism, optimality, order, organ, organism, -orial, ovular, ovulator, p (P, *p*, *P*), -paedium, pair, parasite, parasitism, parent, parity, parthenogenesis, -patric, pattern, -pelagic, perception, period, -phage, -phagia, -phagy, phase, -phile, phenomenon, -phobe, pigeon, -planetic, plankton, plasticity, -plasia, play, -ploidy, -plont, pollination, polyethism, population, posture, potential, predation, predator, preference, pregnancy, principle, probability, procedure, process, production (rate of production, primary production, productivity, basic productivity, primary productivity), program, provisioning, puzzle box, race, range, reaction, receptor, recognition, reductionism, reflex, regulation, reinforcement, relationship, reliability, repertory, replication, representation, reproduction, resource, response, revolution, rhythm, ritual, ritualization, role, rule, sample, sampling technique, scale, scientist, scratching, secretion, sex, sex termination, -sexual, sexual reproduction, sexuality, shaping, signal, sister, skewness, sleep, sociality, society, -somatic, -somy, song, -sound, speciation, species, -spermy, -sphere, stage, stasis, state, statistical test, stimulus, strategy, stress, study of, substance, success, succession, suckling, swarming, swimming, symbiont, symbiosis, synapse, syndrome, synethogametism, system, taxis, teleology, territory, test, theory, -therm, time, -toky, -topy, tradition, transport, -troph, trophallaxis, -trophy, -tropic, -tropism, -tropous, -trosis, -type, value, variable, variance, variation, vitalism, -voltine, -vore, -welt, -xene, -xenia, -xenic, -xeny, yawning, -zoan, -zoic, -zoite, zone, -zoon, -zygosity, -zygote, and -zygous.

Under "study of," the reader will find terms grouped by acoustics, biology, botany, chemistry, chronology, climatology, dynamics, ecology, endocrinology, ethology, evolution, genetics, geography, ichnology, limnology, -metry, morphology, ornithology, paleontology, palynology, pathology, pharmacology, phenology, phycology, phylogeny, physics, physiology, psychology, semiotics, sociality, systematics, taxonomy, toxicology, and zoology.

"In" and "For example, in" statements. I indicate when phenomena relate to particular organism groups, or taxa, by writing phrases such as "In the Honey Bee:..." or "For example, in some carpenter-bee and bird species:... ." When I use "In" alone, it means to my knowledge that the concept relates only to the taxon (taxa) listed. "For example, in..." followed by a list of organisms indicates that the phenomenon relates to these organisms and others that I have not listed.

***cf.* directives.** "*cf.* ..." directs the reader to a related term (sometimes an antonym), related information, or both.

Synonyms. I list synonyms known to me and suggest how each is used when I can. When it is obvious where a synonym is published from the body of a term entry, this synonym occurs in bold type right after a main entry term. Authors have sometimes used the same word for terms with very different meanings, and synonyms of many terms often do not have exactly the same meanings. Thus, synonyms should be used with care.

Notes and comments. *Note,* or *Notes,* refers to a particular definition of a term, while *Comment,* or *Comments,* refers to an entire entry.

Hyphens and one-em dashes. Scientific writers follow different "rules" when they use hyphens and one-em dashes, resulting in the presence, absence, or different combinations of these marks in the same term. In this book, I attempt to standardize use of these marks by using a hyphen(s) in an adjectival phrase that precedes the nouns that it modifies. I use a one-em dash(es) to separate hyphenated terms within adjectival phrases. This avoids the use of a forward slash (/) for a hyphen, or one-em dash, except when I am true to original published information. This book does not hyphenate names that are capitalized in the literature, although such names are grouped under key words (*e.g.,* under "project," one finds "Biological Dynamics of Forest Fragment Project" and "Deep Green Project").

Gender-neutral writing. With all due respect for treating female and male Humans and other organisms equitably, for the reason of more direct writing, this book uses "he" to mean he or she, "him" to mean him or her, and so forth in many entries. My referring only to he, him, or his in many definitions from past times is done to reflect English style of these times.

Other information. For selected terms, this book gives further information such as etymologies and related facts. When they are known by me, I include term originators. This book includes some euphemistic and obscene terms and indicates them as such at the request some foreign scientists.

PLEASE CHOOSE DEFINITIONS AND SYNONYMS CAREFULLY

You should be very careful in selecting which definition to use for a term. A term can have different shades of meaning to different people and in different contexts, and, as discussed above, its meaning can change as we discover more about the parts of our natural, or hypothetical, world represented by the term. The same phenomenon can have different names, depending on a person's frame of reference. One definition of a term allows us to think of certain related concepts, but not others. For example, the environmental concept of niche allows us to conceive of an empty niche, while the population definition of niche does not (Colwell 1992). Further, perfect definitions for many terms (*e.g.,* aggression or consciousness) may not yet exist or be possible to formulate. Many of the definitions in this book are controversial (Verplanck 1957) and should be used *only* as initial attempts to understand concepts they represent. These definitions should not necessarily be regarded as the final, or best, ones that can be formulated. If you consult references to definitions listed in this book, you will often find more information that relates to their meanings and may discover useful denotations and connotations that I have not included. I indicate areas of controversy regarding terms when they are known to me. I reiterate that definitions within an entry are given in *chronological* order. Many definitions are not necessarily widely used or considered to be correct today, but I include them because they help to understand earlier work, subsequent definitions, or both.

Some of the listed synonyms may be used by only one, or a few, authors, and their uses may be debatable. Sometimes synonyms are used for terms that have very different meanings (*e.g.,* "true altruism" for altruism involving helping only nonkin or for altruism involving helping only kin). I indicate this kind of information when it is known to me. Some entries have more than one list of synonyms because different lists refer to different meanings of a term.

PLEASE COIN TERMS AND DEFINE CONCEPTS CAREFULLY

Coining new terms and defining them deserves special attention for enhanced communication. These are rules of thumb to consider: (1) thoroughly research your concept and related concepts and their names, (2) choose the best possible word(s) for the name of the term, and (3) write an explicit, full definition in your publication. Full dictionary-style definitions are seldom given in biological papers, yet they could certainly enlighten and reduce confusion. I urge editors to allow authors to include such definitions in their papers when they are appropriate.

Key to Abbreviations and Conventions

abbr.	abbreviation
adj.	adjective
ant.	antonym
Brit.	British spelling
cf.	compare [Latin *confer*]; indicates related terms that are useful to examine in understanding the term in question
Comment(s)	information that refers to an entire entry
def.	definition.
e.g.	for example
i.e.	that is
n.	noun
Note(s)	information that relates to a particular part of an entry
pl.	plural
q.v.	which see [Latin *quod vide*]
See	indicates where to find a definition of a particular term given under a synonym or with associated terms elsewhere
sing.	singular
syn.	synonym
v.i.	intransitive verb
v.t.	transitive verb

AUTHOR

Edd Barrows is a professor of biology at Georgetown University, Washington, D.C. He has had a lifetime interest in the natural world, in particular, insects and their kin, plants, and geology. Childhood years spent roaming the countrysides of southern Michigan and central Florida fortified his earlier interests. He earned his B.S. in biology at the University of Michigan and his Ph.D. in entomology at the University of Kansas. Since 1975, Edd has been engaged in biological research and teaching at Georgetown University. His scientific publications primarily relate to insect reproductive strategies, ecology, and pollination. Currently, besides scientific communication, he is studying the biodiversity of a freshwater tidal marsh and adjacent habitats; effects of a pesticide on beneficial, nontarget insects; and the evolution of an insect mimicry complex. He is a member of many scientific and conservation societies and is a former president of the Entomological Society of Washington. This book originated in 1984 as a glossary compiled as a reference for teaching and research.

ACKNOWLEDGMENTS

In addition to people I thanked in the first edition, I would like to thank, or thank again, the Barrows family; many Georgetown students who have helped with library research; people at CRC Press who directly helped to produce this book (Carolyn Lea, Patricia Roberson, John Sulzycki, and Susan Zeitz); Patrick O'Brien-McGinty, who helped with word processing; people who sent me specific information for this edition; and many others (whose names can be found in the text) who have discussed terms and definitions with me and contributed definitions and other information. The cover of the second edition celebrates the school colors of Georgetown University, which kindly awarded me a summer research grant to finish this second edition and supported it in other ways.

a

- **A-chromosome** See chromosome: A-chromosome.
- **a-rates** See [2]evolution: tachytely.
- **a selection** See selection: a selection.
- **ABA** See hormone: abscissic acid.
- **abandonment call** See call: abandonment call.
- **abasement** *n.*
  1. A person's lowering, casting down, or humbling another person, in rank or character; humiliation (*Oxford English Dictionary* 1972, entries from 1561); one's lowering in position, rank, prestige, or estimation (Michaelis 1963).
     syn. cast downness, humbleness
  2. A person's being debased, humiliated, degraded (*Oxford English Dictionary* 1972, entries from 1611).
  3. A person's passive submission to external force; acceptance of injury, blame, criticism, punishment; surrender; resignation to fate; admittance of inferiority, error, wrong-doing, or defeat; confession and atonement; self-blame, belittling, or mutilation; seeking and enjoying pain, punishment, illness, and misfortune (Murray 1938, 1961–1962).
  4. A person's behavior indicative of submission to another person (*e.g.,* aggression or punishment) (Cumming 1972, 1).
  5. A person's degradation of himself; excessive complying, surrendering, accepting punishment (Wolman, 1973, 1).
  v.t. abase
  syn. degradation, disgrace, dishonor, mortification, shame
  cf. aggression; appeasement; need: abasement
- **abdominal bursting** See behavior: defensive behavior: autothysis.
- **abdominal up-and-down movement** See [2]movement: abdominal up-and-down movement.

- **abient** *adj.* Referring to an animal's avoiding, or moving away, from a stimulation source (Hurnik et al. 1995).
  ant. adient
- **abiogenesis** See -genesis: abiogenesis.
- **abiosis** See -biosis: anabiosis.
- **abiotic pollination** See pollination: abiotic pollination.
- **ablation** *n.*
  1. Surgical removal of an organism's tissue or organ (*Oxford English Dictionary* 1972, entries from 1846).
  2. An investigator's deliberate destruction of an animal's neural tissue for experimental purposes (*e.g.,* by surgically severing a nerve tract in its brain) (Immelmann and Beer 1989, 175).
  syn. lesion (Immelmann and Beer 1989, 175)
  cf. extirpation
- **abnormal behavior** See behavior: abnormal behavior.
- **abnormal psychology** See study of: psychology: abnormal psychology.
- **abominable mystery** *n.* How and when flowering plants (Angiosperms) first evolved on Earth (Charles Darwin in Weiss 1999, A1).
  cf. Amborella (Appendix 1)
- **abortion** *n.*
  1. An artificially produced human miscarriage, including an illegal one (Michaelis 1963).
  2. A mammal's expelling its fetus prematurely; a miscarriage (Michaelis 1963).
  3. The defective result of a premature birth; a monstrosity (Michaelis 1963).
  4. An embryo's partial through complete arrest of development in its early stages (Michaelis 1963).
  5. Anything's failure during its progress and before its maturity (Michaelis 1963).
  [Latin *abortio, -onis < aboriri,* to miscarry]

selective fruit abortion *n.* A plant's dropping a particular fruit before it matures (Niesenbaum 1999, 261).
Comment: Mirabilis jalapa, the Four-O'Clock, tends to produce mature fruits from flowers that had higher and more diverse pollen loads on their stigmas compared to flowers that had lower and less diverse pollen loads (Niesenbaum 1999, 261).

♦ **abscissa** *n.* The horizonal, or *x*-axis, of a graph (Lincoln et al. 1985).
cf. ordinate

♦ **abscissic acid** See hormone: abscissic acid.

♦ **absconding** *n.* For example, in Honey Bees: departure of an entire colony for a new nest site (Michener 1974, 371).

♦ **absenteeism** *n.* In Tree Shrews: parents' nesting away from their offspring and visiting them from time to time only to provide them with food and a minimum of other parental care (Wilson 1975, 577).

♦ **absolute dominancehierarchy** See hierarchy: dominance hierarchy: absolute dominance hierarchy.

♦ **absolute experiment** See experiment: mensurative experiment.

♦ **abstract homotypy** See homotypy: abstract homotypy.

♦ **abundance, absolute abundance** *n.* The total number of individuals of a taxon, or taxa, in an area, volume, population, or community, often measured as cover in plants (Lincoln et al. 1985).
relative abundance *n.* The total number of individuals of one taxon compared to the total number of individuals of all other taxa combined per unit area, volume, or community (Lincoln et al. 1985).

♦ **abundant** See frequency (table).

♦ **Ac-Ds** See transposable element: *Ac-Ds.*

♦ **acarodomatium** See *domatium: acarodomatium.*

♦ **acarophile, acarophilous, acarophily** See [1]-phile: acarophile.

♦ **acceleration** See [2]heterochrony: acceleration.

♦ **accessory chromosome** See chromosome: accessory chromosome.

♦ **accessory pigment** See molecule: accessory pigment.

♦ **accidental osmallaxis** See osmallaxis: accidental osmallaxis.

♦ **accidental parasitism** See parasitism: accidental parasitism.

♦ **acclimation** See [3]adaptation: [1]physiological adaptation: acclimation; [2]physiological adaptation (comment).

♦ **acclimatization** See [3]adaptation: [1]physiological adaptation: acclimatization; [2]physiological adaptation (comment).

♦ **acentric society** See [2]society: acentric society.

♦ **acetogen** See -troph-: acetogen.

♦ **acetylcholine** See neurotransmitter: acetylcholine.

♦ **Acheulean Tool Industry** See tool industry: Acheulean Tool Industry.

♦ **achievement** See need: achievement.

♦ **acid** *n.*
1. A compound that contains hydrogen in which all or a part of the hydrogen may be exchanged for a metal, or basic, radical, forming a salt (Michaelis 1963)
2. A compound that yields hydrogen ions when dissolved in an ionizing solvent (Michaelis 1963).
cf. molecule
See nucleic acid.
Comment: Aqueous solutions of acids are sour and redden vegetable substances such as litmus (Michaelis 1963).
[Latin *acidus*, acid]

amino acid *n.* A molecule composed of an amino group and a carboxyl radical; a building block of a protein (Michaelis 1963).
Comment: There are 20 kinds of amino acids that differ from one another only by their inside chains and are universally found in proteins (King and Stansfield 1985).

♦ **acid deposition** *n.*
1. A mixture of sulfuric acid and nitric acid that is washed out of the atmosphere by precipitation and is caused by sulfer and nitrogen emissions from volcanoes (Lean et al. 1990, 85).
2. A mixture of sulfuric acid, nitric acid, and sometimes hydrochloric acid that is washed out of the atmosphere by precipitation and is caused mainly by sulfer and nitrogen emmissions from the burning of fossil fuels such as coal and oil in power plants, industrial boilers, and car engines (Lean et al. 1990, 85).
cf. precipitation: acid precipitation
syn. acid rain (Lean et al. 1990, 85)
Comments: Acid deposition decreases the pH of freshwater streams, ponds, and lakes and sometimes soils (Lean et al. 1990, 85). Many bodies of freshwater are now so acidic that fish cannot live in them. Acid depositions may be causing *Waldsterben* (tree death) in Germany. Volcanic acid deposition is slight compared to artificially produced acid deposition. Rain has a pH around 5.6; acid deposition is often from pH 3.5 to 4.3. Acid deposition increases levels of toxic heavy metals in water which kill wildlife. This deposition also greatly alters the nutrient content in forest soils.

♦ **acid rain** See acid deposition.

♦ **acidophile** See [1]-phile: acidophile.

♦ **acidosis** *n*. Too much carbon dioxide in an animal's blood that causes excess carbonic acid, which can kill the animal (Browne 1996b, C10).
cf. hypothesis: soda-water hypothesis; hypercapnia

♦ **acidotroph** See -troph-: acidotroph.

♦ **acme** *n*.
1. A period of maximum phylogenetic vigor in a taxon (Lincoln et al. 1985).
2. The highest point attained in a taxon's phylogenetic, or ontogenetic, development (Lincoln et al. 1985).

epacme *n*. The period in a group's, or organism's, phylogenetic, or ontogenetic, development just prior to its point of maximum vigor or adulthood (Lincoln et al. 1985).

heteracme See -gamy: dichogamy.

paracme, phylogerontic period *n*. The period of decline after the point of maximum vigor, or peak of development, in the phylogeny of a group (Lincoln et al. 1985).

♦ **-acmic** (suffix)

diacmic *adj*. For example, in *Cladocera:* referring to a taxon that exhibits two abundance peaks per year (Lincoln et al. 1985).

monacmic *adj.* For example, in *Cladocera:* referring to a taxon that exhibits one abundance peak per year (Lincoln et al. 1985).

polyacmic *adj*. Referring to a taxon that exhibits more than two abundance peaks per year (Lincoln et al. 1985).

♦ **acoustic eucrypsis** See mimicry: acoustic eucrypsis.

♦ **acoustic window** *n*. The sound frequency range to which an animal's environment is most permeable (Immelmann and Beer 1989, 182).
cf. melotope

♦ **acoustics** See study of: acoustics.

♦ **acquired behavior** See behavior: acquired behavior.

♦ **acquired character, acquired trait** See character: acquired character; inheritance: inheritance of acquired characters.

♦ **acquired hereditary immunity** See hypothesis: hypothesis of directed mutation.

♦ **acquired releasing mechanism (ARM)** See mechanism: releasing mechanism.

♦ **acquired trait** See character: acquired character.

♦ **acquisition** *n*.
1. "Progressive increments in [an animal's] response-strength observed over a series of occasions on which the response is measured" (Verplanck 1957).
2. "A modification of [an animal's] behavior in which a response changes in strength

or topography, or occurs in new environments" (Verplanck 1957).

♦ **acrasin** See chemical-releasing stimulus: semiochemical: pheromone: acrasin.

♦ **acridophage** See -phage: acridophage.

♦ **acrocentric chromosome** See chromosome: acrocentric chromosome.

♦ **acrodendrophile** See [1]-phile: acrodendrophile.

♦ **acrophobia** See phobia (table).

♦ **acrophytia** See [2]community: acrophytia.

♦ **acrophytism** See symbiosis: acrophytism.

♦ **act** See behavior, behavior act, law.

♦ **ACTH** See hormone: adrenocorticotropic hormone.

♦ **action** *n*.
1. A domesticated animal's trained body, or limb, movement (*Oxford English Dictionary* 1972, entries from 1599).
Note: "Action" now also refers to a movement in any kind of animal.
2. An organism organ's performing its proper function (Michaelis 1963).
3. An animal's "behaviour performed in a particular situation" (Maynard Smith 1982, 204).
cf. behavior; [3]theory: game theory

transitional action *n*. An animal's multipurpose movement, *q.v.,* that occurs in the same, or similar, form in two of its different functional systems and whose performance can lead to a switch in motivation from one of these systems to another (Immelmann and Beer 1989, 196, 317).

♦ **action chain** See chain: action chain.

♦ **action pattern** See pattern: fixed-action pattern.

♦ **action potential** See potential: action potential.

♦ **action-specific energy (ASE)** See energy: action-specific energy.

♦ **action-specific exhaustibility, action-specific exhaustion, behavioral fatigue** *n*. A particular behavior's being refractory to a particular stimulus (quality and intensity) (Immelmann and Beer 1989, 3).
cf. effect: Coolidge effect; fatigue; learning: habituation

♦ **action system** See behavior pattern; [2]system: action system.

♦ **actions** *pl. n*. A person's habitual behavior; conduct (Michaelis 1963).

♦ **actium** See [2]community: actium.

♦ **active** See dominant (genetic).

♦ **active avoidance learning** See learning: avoidance learning: active avoidance learning.

♦ **active competition** See competition: contest competition.

♦ **active ingredient (ai)** *n*. The active ingredient in a pesticide formulation, expressed as an amount of active ingredient per unit area; *e.g.*, 140 g (ai) ha⁻¹ (Johansen 1977, 184; Harrahy et al. 1993, 2192).
syn. AI (Barrows et al. 1994); a.i. (Harrahy et al. 1994, 521)

♦ **active nuclear species** See ²species: active nuclear species.

♦ **active process** *n*. A system, or process, that requires metabolic-energy expenditure (Lincoln et al. 1985).
cf. passive process

♦ **active replicator** See replicator (comment).

♦ **active sensory system** See ²system: active sensory system.

♦ **active sleep** See sleep: stage-1 sleep.

♦ **active space** *n*. The space within which the concentration of a pheromone, or any other behaviorally active chemical substance, is at, or above, threshold concentration; the chemical signal itself (Wilson 1975, 185, 577).

♦ **activity** *n*.
1. A person's physical exercise, gymnastics, athletics (*Oxford English Dictionary* 1972, entries from 1552).
Note: This is an obsolete definition.
2. An individual animal's general, or specific, movement (Immelmann and Beer 1989, 3).

instinctive activity *n*. An ordered sequence of fixed-action patterns (Immelmann and Beer 1989, 151).
cf. pattern: fixed-action pattern

overflow activity See activity: vacuum activity.

spontaneous activity, resting-level activity *n*. Continuous firing of nerve cells without apparent external stimulation (*e.g.*, in cells of olfactory bulbs and vertebrate eyes) (Adrian 1950, Leyhausen 1954 in Heymer 1977, 49).

structural activities *n*. A group of heterogeneous behaviors, including especially grooming, scratching, and feeding, that accompany and surround the appetence for rest and require no other specific context for their occurrences (Hassenberg 1965 in Heymer 1977, 138).

vacuum activity *n*.
1. For example, in some bird species: an instinctive behavior that does not occur for an extended period and eventually occurs without a demonstrable eliciting stimulus (Lorenz 1937 in Heymer 1977, 110); *e.g.*, a well-fed Starling, which has had no opportunity to catch flies for some time, suddenly goes through all the movements of searching for a fly, catching it, and killing it, although no

fly is present (Lorenz 1935 in Hinde 1970, 312).
Notes: Lorenz (1981, 127) indicates that "vacuum activity" cannot be exactly defined. Morris and Moynihan (1953 in Heymer 1977) suggest that "vacuum activity" be replaced with "overflow activity" because one cannot be certain that no stimulus is eliciting the behavior.
2. A fixed-action pattern that occurs in the absence of any stimulus, because, according to the energy model of behavior, its action-specific energy is very great (Dewsbury 1978, 19).
3. A behavior that is normally dependent on external stimulation but occurs spontaneously due to its extreme threshold change, *e.g.*, a weaver bird's showing complex nest-building behavior without nest material or even a substitute object (Immelmann and Beer 1989, 290, 323).
cf. behavior: spontaneous behavior

♦ **activity drive** See drive: activity drive.

♦ **activity period** See period: activity period.

♦ **activity range** See ³range: activity range.

♦ **actophile** See ¹-phile: actophile.

♦ **actualism** *n*. The principle "that the same causes (physical laws) have operated throughout geological time, since the immanent characteristics of the world have always remained the same" (Mayr 1982, 377).
cf. ³theory: uniformitarianism
Comment: According to Lyell (1830–1833), it is legitimate to "attempt to explain the former changes of the Earth's surface by reference to causes now in operation" (Mayr 1982, 377).

♦ **acyclic** See cyclic: acyclic.

♦ ***ad hoc* adaptations** See habit (def. 3).

♦ ***ad hoc* mimicry** See mimicry: *ad hoc* mimicry.

♦ ***ad libitum* sampling** See sampling technique: *ad libitum* sampling.

♦ **adaptability** See ³adaptation: ²physiological adaptation.

♦ ¹**adaptation** See habituation; learning: habituation.

♦ ²**adaptation** *n*. "Reproduction of anything modified to suit new uses" (*Oxford English Dictionary* 1972, entries starting with Darwin 1859).
[Latin *adaptare*, to fit]

♦ ³**adaptation** *n*. *Comments:* "Adaptation" with, or without, adjectives, is used to refer to both physiological and evolutionary processes as well as the results (organism characters) of these processes; in addition, "adaptation" is a kind of learning. Confusion

A Classification of ³Adaptation

I. ¹evolutionary adaptation
   (*syn.* adaptation, adaption, genotypic adaptation, postadaptation, postadaption)
   A. coadaptation (*syn.* coadaption, internal balance)
   B. idioadaptation
   C. iterative adaptation (*syn.* iterative evolution)
II. ²evolutionary adaptation
   (*syn.* adaptation, adaption, adaptive character, genotypic adaptation, phylogenetic adaptation, postadaptation, postadaption)
   A. alternative adaptation
   B. exaptation
   C. juvenile adaptation
   D. preadaptation
III. ¹physiological adaptation (*syn.* adaptation)
   A. acclimation
   B. acclimatization
   C. dark adaptation
   D. light adaptation
   E. resistance adaptation
   F. sensory adaptation (*syn.* neuronal adaptation)
   G. stimulus adaptation (*syn.* habituation)
IV. ²physiological adaptation
   (*syn.* adaptation, adaptability, phenotypic adaptation)
V. Some of the other kinds of adaptation (only some defined in this book)
   A. *ad hoc* adaptation
   B. aquatic adaptation
   C. behavioral adaptation
   D. counteradaptation
   E. cultural adaptation
   F. defensive adaptation
   G. environmental adaptation
   H. heritable adaptation (*syn.* genetic adaptation)
   I. intentional adaptation
   J. learning adaptation
   K. local adaptation
   L. long-term adaptation
   M. nongenetic adaptation
   N. ontogenetic adaptation
   O. sensitive adaptation
   P. short-term adaptation
   Q. terrestrial adaptation
   R. transitory adaptation

arises because many of these concepts are often simply called "adaptation" without modifying adjectives and some of the concepts have several synonyms.
See ²evolution: adaptation.
cf. study of: teleonomy; trend

¹evolutionary adaptation, adaptation *n.*

1. An evolutionary process that involves a genetic change, creation of a new phenotypic character from this change, and increased genetic fitness of the individual that possesses this character (inferred from Brown 1975, 268; Wilson 1975, 577; Bateson 1979, 241).
   Notes: This phenotypic character is often termed an "adaptation." It would be clearer to call it an "evolutionary adaptation" to distinguish it from a "physiological adaptation" or other kinds of adaptation.
2. "The production of a superior variant" in a particular environment ("adaptation" in Wilson 1975, 67).
3. Phyletic evolution ("adaptation" in Mayr 1982, 358).
4. Transgenerational alterations of features and capacities of organisms in a lineage that enable them to "solve (or improve on previous solutions of) problems" posed by their environment (*e.g.,* those of internal integration and reproducing) (Burian 1992, 7).
5. "The evolutionary modification of a character under selection for efficient or advantageous (fitness-enhancing) functioning in a particular context or set of contexts" (West-Eberhard 1992, 13).

syn. adaption, genotypic adaptation, postadaptation, postadaption (Lincoln et al. 1985)

▶ **coadaptation** *n.*
1. The evolution of mutually advantageous adaptations in two or more interactive species (Lincoln et al. 1985).
2. Evolution resulting in the accumulation of harmoniously interacting genes in a population's gene pool (Lincoln et al. 1985).

syn. coadaption, internal balance (Lincoln et al. 1985)

▶ **iterative adaptation** See ²evolution: iterative evolution.

²evolutionary adaptation, adaptation *n.*
Note: These definitions relate to the result, or presumed result (one or more characters), of the process known as evolutionary adaptation.

1. A distinctive "helpful" character favored by natural selection (Darwin 1859 in Ruse 1984, 102); a character that is "good for" its owner because it enables it to fit better with its environment (Lloyd 1992, 338).
   Note: Lloyd (1992, 338) calls this the "engineering" definition of adaptation.

a — c

2. A species', or subspecies', genetically fixed condition or evolution that favors its survival in a particular total environment, *q.v.* (given as "genotypic adaptation" in Bligh and Johnson 1973, 943).
3. A character that is a result of natural selection (Dawkins 1982, 292; Burian 1992, 7).
   Note: Lloyd (1992, 338) calls this the "product-of-selection" definition of adaptation.
4. A character that increases an organism's relative fitness in its natural environment (Wilson 1975, 577; Burian 1992, 8).
   Notes: This is a widely used concept of "adaptation" in the sense of "evolutionary adaptation." Williams (1966, 4) indicates that "adaptation is a special and onerous concept that should be used only where it is really necessary." "Adaptation" should be used only as a last resort when less onerous principles, such as those of physics and chemistry or that of unspecific cause and effect, are not sufficient for a complete explanation (Williams 1966, 11).
   syn. adaptive character (Wilson 1975, 21)
5. A character that is the product of selection *for* the character in question (Sober 1984 in Burian 1992, 8).
   syn. adaption, phylogenetic adaptation (Immelmann and Beer 1989, 4), genotypic adaptation (McFarland 1984, 4), postadaptation (Lincoln et al. 1985)
6. An individual organism's "'propensity to survive and reproduce' in a particular environment" (West-Eberhard 1992, 13).
   syn. adaptedness (Mayr 1986 in West-Eberhard 1992, 13); general adaptation (West-Eberhard 1992, 13)
   cf. [3]adaptation: [2]evolutionary adaptation: exaptation, preadaptation; [1]physiological adaptation: sensory adaptation; [2]evolution: adaptation; postadaptation
   Comments: Types of characters resulting from "evolutionary adaptation" include behaviors, developmental traits, physiological processes, and structures. Bateson (1979, 241) uses "adaptation" to refer to an organism's character that *seemingly* enables it to fit better into its environment and way of life. Fitzgerald (1995, 32) metaphorically uses "evolutionary wisdom" for a behavioral adaptation. Burian (1992) and West-Eberhard (1992) discuss the evidence used to determine whether a character is truly an adaptation.

▸ **alternative adaptation** *n.* One phenotypic character, of a group of different adaptive phenotypic characters, that is maintained in the same life stage of a population but not necessarily simultaneously expressed in the same individual (West-Eberhard 1986, 1388).
cf. hypothesis: alternative-adaptation hypothesis

▸ **exaptation** *n.* "Any organ not evolved under natural selection for its current use — either because it performed a different function in ancestors (classical preadaptation) or because it represented a nonfunctional part available for later co-optation."
cf. [3]adaptation: [2]evolutionary adaptation: preadaptation (comment)
[coined by Gould and Vrba 1981 in Gould 1985b, 18]

▸ **juvenile adaptation** *n.* The interpolation of new features into early ontogenetic stages for their own immediate evolutionary utility (Haeckel 1905 in Gould 1992, 160).
cf. -genesis: cenogenesis

▸ **preadaptation** *n.*
1. A previously existing structure, physiological process, or behavioral pattern that "is already functional in another context and available as a stepping stone to the attainment of a new adaptation" (Wilson 1975, 34).
   Note: "Preadaptation" happens, for example, when a population with a certain song dialect (which originally arose through geographical isolation but later acts as a means of reproductive isolation) becomes sympatric with a conspecific population with a different dialect (Immelmann and Beer 1989, 229).
2. "An ancestral characteristic or trait that predisposes a directional evolutionary change to proceed in one direction rather than in another" (Wittenberger 1981, 620).
syn. classical preadaptation (Gould 1985b, 18)
Comment: Gould and Vrba (1981) indicate that "preadaptation" should be dropped in favor of a more inclusive term — "exaptation," *q.v.*, because "preadaptation" seems to imply that certain features of organisms are predestined to become adaptations (arise with a different, future use in view), and it does not cover the important class of features that arise without functions (*e.g.*, as developmental consequences of other primary adaptations) but remain available for later coaptation. Campbell (1990, 493) indicates that "preaptation" is gradually replacing "preadaptation."

[1]**physiological adaptation, adaptation** *n.* The process of an individual organism's changing one or more of its

characters due to its adjusting to its environment but not changing genetically (Bligh and Johnson 1973, 942; inferred from Brown 1975, 268).

▸ **acclimation** *n*. Physiological adaptation involving an organism's altering its tolerance of a single environmental factor (*e.g.*, temperature) under laboratory conditions; contrasted with acclimatization (Bligh and Johnson 1973, 942; McFarland 1985, 287).

Comment: "Acclimation" and "acclimatization" are etymologically indistinguishable; both terms have several and different meanings, and it is not always clear how an author is using these terms (Bligh and Johnson 1973, 942).

▸ **acclimatization** *n*. Physiological adaptation involving an organism's altering its tolerance of environmental factors under natural conditions; contrasted with acclimation (Bligh and Johnson 1973, 942; McFarland 1985, 287).

cf. anticipatory adjustment

▸ **dark adaptation** *n*. A person's gradual improvement in visual perception after moving from a bright area to a dark one due to pupil enlargement and retinal chemical changes; contrasted with light adaptation (Storz 1973, 63).

▸ **light adaptation** *n*. A person's gradual improvement in perception after moving from a dark area to a light one; contrasted with dark adaptation (Storz 1973, 148).

▸ **resistance adaptation** *n*. An organism's phenotypic adaptation at its tolerance limit (McFarland 1985, 286).

▸ **sensory adaptation** *n*.

1. "Change, incremental or decremental, in response or response-strength that has been experimentally demonstrated to depend solely upon changes in the state of a receptor organ produced by protracted or repetitive stimulation of, or by recovery from, such stimulation of that organ" (Verplanck 1957); *e.g.*, a person's strong sense of pressure from gloves fades with time while his gloves remain on his hands (Storz 1973, 251).

2. Physiological adaptation in which an animal stops responding to a kind of stimulus because its sense organ is no longer responding to the stimulus; *e.g.*, some kinds of flies stop feeding when their chemoreceptors become sensorially adapted (Hinde 1970, 288–290).

3. A cell's, or tissue's, showing a response that diminishes to a relatively low level from an initial response due to a regularly repeated, or continuous, stimulus (Brown, 1975, 268).

Note: Cells of the yeast *Saccharomyces cerevisiae* adapt to (= recover from) growth arrest induced by their mating pheromones. Their adaptation is, therefore, their ceasing to respond to the pheromone (Grishin et al. 1994, 1081).

syn. neuronal adaptation (Brown 1975, 268)

cf. learning: habituation

²**physiological adaptation, adaptation** *n*.

1. An organism's change that reduces the physiological strain produced by a stressful component of its total environment, *q.v.* (given as "phenotypic adaptation" in Bligh and Johnson 1973, 942).

2. A result (organism attribute, or attributes) of the process called physiological adaptation, *q.v.* (inferred from Brown 1975, 268); a short-term physiological adjustment by a phenotypically plastic individual (West-Eberhard 1992, 13).

3. An organism's receptor cell's adjusting its responsiveness to a stimulation change, either very quickly as in phasic cells or very slowly as in tonic cells (Gould 1982, 84).

Note: "Physiological adaptation" often also involves a group of cells.

syn. adaptability (Brown 1975, 268)

Comment: Physiological adaptations include the results of acclimation, acclimatization, dark adaptation, learning, light adaptation, resistance adaptation, and sensory adaptation, as well as a person's breathing faster at higher altitudes, developing calluses on feet and hands in response to wear, and skin tanning due to sunlight (Brown 1975, 268).

preadaptation See ¹adaptation: ²evolutionary adaptation: preadaptation.

♦ **adaptationist program** *n*.

1. A way of thinking that "regards natural selection as so powerful and the constraints upon it so few that direct production of adaptation through its operation becomes the primary cause of nearly all organic form, function, and behaviour" [coined by Gould and Lewontin 1979, 584–585].

Note: The adaptationist program does recognize constraints on natural selection, but it usually dismisses them as unimportant or simply acknowledges them (Gould and Lewontin 1979, 585).

syn. adaptationist programme, Panglossian paradigm

[coined by Gould and Lewontin 1979, 584]

2. A research paradigm that tries to envision a biological phenomenon as an aspect of adaptation, "some sort of biological machinery or process shaped by

natural selection to help solve one or more problems faced by the organism" (Gould and Lewontin 1979, interpreted by Williams and Nesse, 1991, 1–2).
Note: The adaptationist program can predict otherwise unsuspected facets of human biology and provides new insights into the causes of medical disorders (Williams and Nesse 1991, 1–2).

♦ **adapted system** See [1]system: adapted system.

♦ **adaptedness** See [3]adaptation: [2]evolutionary adaptation.

♦ **adaptive** *adj.* Pertaining to an adaptation, *q.v.*

♦ **adaptive correlation** *n.* Association between an organism's form and function (Immelmann and Beer 1989, 5).
See study of: adaptive correlation.

♦ **adaptive focus** See adaptive peak.

♦ **adaptive landscape** *n.*
1. Fitness of one or more kinds of organisms graphed as an ordinate vs. phenotype (Wilson 1975, 24, illustration; Sewell Wright in Campbell 1990, 473, illustration).
   cf. zone: adaptive zone
2. "A graphical 'map' on which the altitude at any given location indicates Darwinian fitness for a specified combination of alleles or linkage groups" (Wittenberger 1981, 612).

♦ **adaptive mutation** See [2]mutation: directed mutation.

♦ **adaptive neutrality** See evolutionary synthesis; [2]evolution: neo-Darwinian evolution; [3]theory: neutral theory.

♦ **adaptive peak, Wright's principle of maximization of W, adaptive focus** *n.* A small region of high fitness in an adaptive landscape (Futuyma 1979, 329).

♦ **adaptive radiation** *n.* For example, in Hawaiian fruit flies, East African cichlid fish, Darwin's Finches, Hawaiian Honeycreepers, Australian marsupials: divergent evolution of members of a single phylogenetic line into a variety of different adaptive forms, usually with reference to their diversification in their use of resources, habitats, or both (Futuyma 1986, 550).
cf. [2]evolution

♦ **adaptive significance, adaptively significant** See statistically significant.

♦ **adaptive significance, biological function, survival value** *n.* The value of an organism's trait to its survival and fitness (Immelmann and Beer 1989, 300).
cf. adaptive value

♦ **adaptive strategy** See strategy: adaptive strategy.

♦ **adaptive trait** See character: adaptive character.

♦ **adaptive value** *n.*
1. "A measure of the tendency for a particular allele or linkage group to change in frequency within a gene pool" (Wittenberger 1981, 612).
2. The comparative fitness of different genotypes in a given environment (Lincoln et al. 1985).
3. The survival and reproductive value of one genotype compared to another in a population (Lincoln et al. 1985).
cf. fitness: genetic fitness; selective advantage

♦ **adaptive zone** See zone: adaptive zone.

♦ **adaptor** See communication: nonverbal communication.

♦ **additive-genetic variance** See variance: additive-genetic variance.

♦ **adelphogamy** See mating: sib mating.

♦ **adelphoparasitism** See parasitism: adelphoparasitism.

♦ **adenosine diphosphate** See molecule: adenosine phosphate: adenosine diphosphate.

♦ **adenosine monophosphate** See molecule: adenosine phosphate: adenosine monophosphate.

♦ **adenosine phosphate** See molecule: adenosine phosphate.

♦ **adenosine triphosphate** See molecule: adenosine phosphate: adenosine triphosphate.

♦ **adequate stimulus** See [2]stimulus: adequate stimulus.

♦ **adient** *adj.* Referring to an animal's seeking, or moving toward, a stimulation source (Hurnik et al. 1995).
ant. abient

♦ **adjunctive behavior** See behavior: displacement behavior.

♦ **adjustable mimicry** See mimicry: adjustable mimicry.

♦ **adolescent** See animal names: adolescent.

♦ **adoption** *n.* For example, in some ant species; bird species with creches; the African Wild Dog; Chimpanzees; the Coati; the Hamadryas Baboon and other baboons; Human; Langurs; Macaques: an animal's giving essentially full-time parental care to an orphaned conspecific young; the "parent" varies in genetic relationship to the young (Estes and Goddard 1967, Lawick-Goodall 1968, Kummer 1986 in Wilson 1975, 352, 512; Immelmann and Beer 1989, 6).
Comment: In Hamadryas Baboons, a young adult male abducts sexually immature females from the female group of an older male; this starts the young male's female group (Immelmann and Beer 1989, 6).

◆ **adoption substance** See substance: adoption substance.

◆ **adrenal corticoid** See hormone: adrenal corticoid.

◆ **adrenal gland** See gland: adrenal gland.

◆ **adrenalin** See hormone: adrenalin; toxin: adrenalin.

◆ **adrenergic synapse** See synapse: adrenergic synapse.

◆ **adrenocorticotropin (ACTH)** See hormone: adrenocorticotropin.

◆ **adrenogenital syndrome** See syndrome: adrenogenital syndrome.

◆ **adsere** See ²succession: adsere.

◆ **adult** See animal names: adult.

◆ **adult transport** See transport: adult transport.

◆ **adultery** *n.* A married person's voluntary sexual intercourse with a person of the opposite sex who is not the person's spouse and is either married or unmarried; "violation of a marriage bed" (*Oxford English Dictionary* entries from 1366).
See copulation: extrapair copulation.
[Latin *adulter*]

◆ **advanced character** See character: derived character.

◆ **advanced species** See ²species: advanced species.

◆ **advantageous altruism** See altruism (def. 6).

◆ **adventitious mimicry** See mimicry: adventitious mimicry.

◆ **adventive** *adj.* Referring to a group, or species, that has arrived in an area previously unoccupied by it, whether by its own means or due to Humans (Frank and McCoy 1990, 4).
cf. autochthonous, ¹endemic, immigrant, indigenous, introduced, precinctive
Comment: "Immigrant" and "introduced species" are divisions of "adventive species" (Frank and McCoy 1990, 4).
[Latin *advenire*, to arrive]

◆ **advergence** See deceptive advergence.

◆ **advertisement call** See animal sounds: call: advertisement call.

◆ **advertising, advertisement, advertisement behavior** See display: advertisement.

◆ **advertising dress** See dress: advertising dress.

◆ **advertising mimicry** See mimicry: advertising mimicry.

◆ **advocacy method of developing science** See method: advocacy method of developing science.

◆ **aedeagus** See organ: copulatory organ: aedeagus.

◆ **aeitiology** See study of: etiology.

◆ **aerial copulation** See copulation: aerial copulation.

◆ **aerial display** See display: aerial display.

◆ **aerial respiration** See breathing: air breathing.

◆ **aerie** See ²group: aerie.

◆ **aerobe** *n.* A prokaryote that can live in the presence of free oxygen; contrasted with anaerobe (Michaelis 1963).
[Greek *aēro*, air + *bios*, life]
anaerobe *n.*
1. A microorganism that flourishes without free oxygen; distinguished from an aerobe (Michaelis 1963).
2. Either a facultative anaerobe or an obligate anaerobe (Campbell 1996, 506).
[New Latin < Greek *an*, without + *aēro*, air + *bios*, life] *n.* A prokaryote that can use oxygen in energy production if it is present and can also grow by fermentation in an anaerobic environment (Campbell 1996, 506).
▸ **obligate anaerobe** *n.* A prokaryote that cannot use oxygen and is poisoned by it (Campbell 1996, 506).

◆ **aerobic** *adj.*
1. Referring to an organism that grows or occurs only in the presence of molecular oxygen (Lincoln et al. 1985).
syn. oxybiotic, oxybiontic (Lincoln et al. 1985)
2. Referring to an environment whose oxygen partial pressure is similar to normal atmospheric levels; oxygenated (Lincoln et al. 1985).

◆ **aerobiology** See study of: biology: aerobiology.

◆ **aerochore, aerochorous, aerochory** See -chore: aerochore.

◆ **aerohygrophile** See ¹-phile: aerohygrophile.

◆ **aerophile** See ¹-phile: aerophile.

◆ **aerophobia** See phobia (table).

◆ **aerophytobiota** See biota: aerophytobiota.

◆ **aeroplankton** See plankton: aeroplankton.

◆ **aeroscepsy** *n.* An organism's perception of airborne sound or chemical stimuli (Lincoln et al. 1985).

◆ **aestarifruticeta** See ²community: aestarifruticeta.

◆ **aestarisilune** See ²community: aestarisilune.

◆ **aestidurilignosa** See ²community: forest: mixed forest.

◆ **aestilignosa** See ²community: aestilignosa.

◆ **aestival, estival** *adj.*
1. Of, or belonging to, summer or the summer solstice (*Oxford English Dictionary* 1972, entries from 1386).

2. Appearing, or produced, in summer (*Oxford English Dictionary* 1972, entries from 1682).
cf. dormancy: aestivation
♦ **aestivation** See estivation.
♦ **aethogametism** See synethogametism.
♦ **affect gesture** See communication: nonverbal communication.
♦ **affectional drive** See drive: affectional drive.
♦ **afference** See -ference: afference.
♦ **afferent neuron** See cell: neuron: afferent neuron.
♦ **affiliation** *n*.
1. A person's need to draw near and enjoyably cooperate with another person, to form friendships and remain loyal, to please and win affection of important others (Wolman 1973). See need: affiliation.
syn. affiliative need (Wolman 1973)
2. A form of social behavior involving an individual animal's tending to approach and remain near conspecifics (Dewsbury 1978, 107).
♦ **affiliative drive** See drive: affiliative drive.
♦ **affiliative need** See affiliation.
♦ **affinity** *n*.
1. A person's relationship by marriage, as opposed to consanguinity (*e.g.,* a relationship between a man and his wife's blood relatives) (*Oxford English Dictionary* 1972, entries from 1303).
2. A person's relationship, or kinship, generally between individuals or races (*Oxford English Dictionary* 1972, entries from 1382).
3. Structural resemblance between two organisms that suggests gradual modification from a common ancestor (*Oxford English Dictionary* 1972, entries from 1794).
Comment: Lincoln et al. (1985) also define "absolute affinity" and "relative affinity."
cladistic affinity *n*. The degree of recency of common ancestry to two or more organisms (Lincoln et al. 1985).
cf. affinity: patristic affinity
patristic affinity, patristic similarity *n*. Organisms' degree of similarity due to their common ancestry, including parallel, but not convergent, evolution (Lincoln et al. 1985).
cf. affinity: cladistic affinity
♦ **African-origin hypothesis (of human mitochondrial DNA evolution)** See hypothesis: African-origin hypothesis of human mitochondrial DNA evolution.
♦ **after-discharge** *n*.
1. "That part of a response that occurs after the termination of the stimulus that elicited it or set the occasion for its occur-

rence. [Uncommon usage.]" (Verplanck 1957).
2. "Discharges of nerve impulses in efferent neurons that persist in time after the stimulus that set up the discharge of which they are part is terminated. [This observation forms part of the empirical basis of the concept of synapse.]" (Verplanck 1957).
♦ **agamete** See gamete: agamete.
♦ **agametic** *adj*. Referring to reproduction without gametes (Bell 1982, 501).
♦ **agamodeme** See deme: agamodeme.
♦ **agamont** *n*. An individual organism that gives rise to agametes (Bell 1982, 501).
♦ **agaricole** See -cole: agaricole.
♦ **age**
1. *n*. The period that a thing (*e.g.,* an individual, group of individuals, or species) has existed (*Oxford English Dictionary* 1972, entries from 1325).
2. *v.t.* To become old or attain maturity (*Oxford English Dictionary* 1972, entries from 1398).
♦ ***age*-1 gene** See gene: *age*-1 gene.
♦ **age-and-area hypothesis** See hypothesis: age-and-area hypothesis.
♦ **age dimorphism** See -morphism: dimorphism: age dimorphism.
♦ **age polyethism** See polyethism: age polyethism.
♦ **age polymorphism** See -morphism: polymorphism: age polymorphism.
♦ **age-specific selection** See selection (table).
♦ **agent** *n*. Any force, organism, or substance that causes a material change (Michaelis 1963).
[Latin *agens, agentis*, present participle of *agere*, to do]
oxidizing agent *n*. In chemistry, in an oxidation-reduction reaction, the atom, or compound, that accepts an electron(s); contrasted with reducing agent (Campbell 1987, 184).
Comment: Oxygen is one kind of oxidizing agent. It has a strong affinity for electrons; that is, it is strongly electronegative.
reducing agent *n*. In chemistry, in an oxidation-reduction reaction, the atom, or compound, that donates an electron(s) and thereby *reduces* the charge of the electron-accepting element or compound; contrasted with oxidizing agent (Campbell 1987, 184).
♦ **aggregate character** See character (comments).
♦ **aggregated distribution** See distribution: aggregated distribution.
♦ **aggregation** See ²group: aggregation; ²society: modular society: discontinuous modular society.

♦ **aggregation-attachment pheromone**
See chemical-releasing stimulus: semiochemical: pheromone: aggregation-attachment pheromone.

♦ **aggregation pheromone** See chemical-releasing stimulus: semiochemical: pheromone: aggregation pheromone.

♦ **aggregational hierarchy** See hierarchy: aggregational hierarchy.

♦ **aggression** *n*.

1. A person's "unprovoked attack" on another person; "the first attack in a quarrel; an assault, an inroad" (*Oxford English Dictionary* 1972, first entry 1611).
2. A person's "setting upon any one;" one's "making of an attack or assault" (*Oxford English Dictionary* 1972, first entry 1704).
3. An animal's "tendency to initiate a vigorous fight" (King and Gurney 1954, 326).
4. In animals, "a broad class of behavior that includes both threat and attack behavior" (Verplanck 1957).
5. An organism's "response that delivers noxious stimuli to another organism" (Buss 1961 in Dewsbury 1978, 104).
   Note: Dewsbury (1978, 104) discusses the problems with this definition.
6. An animal's "fighting" and "the act of initiation of an attack" (Scott 1975, 1).
7. One animal's "physical act or threat of action that reduces the freedom or genetic fitness of another" (Wilson 1975, 242, 577).
   Note: Aggression serves very diverse functions in different species. Wilson (1975, 242–243) lists eight categories of aggression: antipredatory, dominance, moralistic, parental, predatory, sexual, territorial, and weaning aggression.
8. "A vague term used to designate an array of behaviors, with various functions, that we intuitively feel resemble human aggression" (Wilson 1975, 22).
9. In population biology, behavior actively directed against competitors that decreases the reproductive prospects of an attacked animal relative to those of an attacker (*e.g.,* by denying access to certain resources) (Markl 1976 in Immelmann and Beer 1989, 9).
10. An animal's "overt behavioral act or threat directed at harming another individual with the intent of gaining some advantage" (Wittenberger 1981, 612).
11. An animal's making offensive acts such as threat to and encroachment, or attack, upon another animal (Hand 1986, 203).
syn. aggressive behavior
cf. aggressiveness; behavior: agonistic behavior; meddling; need: aggression; violence

Comments: Aggression "is virtually universal in some form or another in almost every animal which has the necessary motor apparatus to fight or inflict injury" (Southwick 1970, 2).
"The term aggression has so many meanings and connotations that in effect it has lost its meaning ... it is not a simple, unitary concept and therefore cannot be defined as such. There is no single kind of behavior which can be called 'aggression' nor is there any single process which represents 'aggression.' ...Where only modest levels of precision are necessary, it may be useful to adopt the term *agonistic behavior* as a desirable alternative to 'aggressive behavior' (Johnson 1972, 8). "In attempting to define aggression ... one of the difficulties is that aggression may be situationally determined so that a particular behavior may be considered aggressive in one instance and not in another" (Johnson 1972, 17). "The word 'aggression' is widely used and misused in a variety of contexts" (Scott 1975, 1). Dewsbury (1978, 104) states, "The solution to the problem of 'aggression' is simply to treat the word as a convenient, loosely defined aid to communication and organization, recognizing that we cannot provide an adequate definition and that we are probably lumping together a number of diverse phenomena." Moyer (1968, 1976 in Dewsbury 1978, 104) lists eight kinds of aggression: "predatory aggression," "inter-male aggression," "fear-induced aggression," "irritable aggression," "territorial-defense aggression," "maternal aggression," "instrumental aggression," and "sex-related aggression."

antipredatory aggression *n*. Aggression that is purely defensive behavior that can be escalated into a full-fledged attack on a predator (*e.g.,* mobbing) (Wilson 1975, 343).

aversion-induced aggression *n*. In psychology, an animal's attack due to an aversive stimulus (*e.g.,* pain), which may be directed at any bystander or inanimate object that is within its reach (Immelmann and Beer 1989, 24).
cf. aggression: frustration-induced aggression, redirected aggression

behavioral aggression *n*. Aggression that enables animals "to survive in competitive situations" (Southwick 1970, 2).
cf. aggression: ecological aggression

defensive aggression *n*. In some prey-animal species: an individual's using aggression, often as a last resort, to defend itself or its offspring from predators (Wittenberger 1981, 614).

direct aggression *n.* Aggression in which animals confront one another with physical interactions; *e.g.,* a barnacle's displacing another, inter- and intraspecific fights between ants, and cannibalism in some kinds of predaceous insects (Wilson 1975, 244–246).
cf. competition: interference competition

displaced aggression, displacement *n.* Psychologically speaking, human behavior somewhat analogous to redirected aggression; contrasted with displacement behavior, *q.v.* (Immelmann and Beer 1989, 247).

dominance aggression *n.* Aggression involving displays and attacks mounted by dominant animals against fellow group members used primarily to prevent subordinates from performing actions for which the dominant animal claims priority (Wilson 1975, 242).

ecological aggression *n.* Aggression that enables animals to invade and colonize new areas and exploit new habitats (Southwick 1970, 2).
cf. aggression: behavioral aggression

frustration-induced aggression *n.* Aggression due to frustration, *q.v.*; *e.g.,* when an investigator switches a pigeon from constant to partial reinforcement in an operant situation, the bird becomes aggressive and pecks bystanders within range (Immelmann and Beer 1989, 114).
cf. aggression: aversion-induced aggression, redirected aggression

interspecific aggression *n.* Aggression that is frequently physically injurious between, or among, members of different species (Immelmann and Beer 1989, 9).

intrasexual aggression *n.* "Aggression directed at like-sexed conspecifics of the same age class or reproductive class" (Wittenberger 1981, 617).

intraspecific aggression *n.* Aggression that is often ritualized fighting between, or among, members of the same species (Immelmann and Beer 1989, 9).

moralistic aggression *n.* Human aggression used to enforce reciprocation; this "aggression is manifested in countless forms of religious and ideological evangelism, enforced conformity of group standards, and codes of punishment for transgressors" (Wilson 1975, 243).
cf. morality

parental aggression *n.* "Aggression directed at progeny by a parent" (Wittenberger 1981, 619).

parental disciplinary aggression *n.* In some mammal species: aggression used by parents; *e.g.,* to keep offspring close at hand, urge them into motion, to break up fighting, and to terminate unwelcome suckling (Wilson 1975, 243).

predatory aggression *n.* A predator's aggression toward its prey (Davis 1964 in Wilson 1975, 243).

redirected aggression, redirection *n.* An animal's action that it deflects from an object that arouses it toward a neutral (or substitute) object, *e.g.,* a lower-ranking conspecific, a stone, or a clump of grass which it might pull (grass pulling) (Dewsbury 1978, 19; Immelmann and Beer 1989, 246).
cf. aggression: displaced aggression; behavior: displacement behavior
Comments: "Redirected aggression" involves an animal's switching its target, while "displacement behavior" involves its changing its kind of behavior. This concept is comparable to the psychological term "aversion-induced aggression" (Immelmann and Beer 1989, 24) and the psychoanalytic term "displaced aggression" (Dewsbury 1978, 19).

sexual aggression *n.* A male's aggression involving threats and attacks directed toward a female that forces her into a more prolonged sexual alliance with him (Wilson 1975, 242).
Comment: The ultimate development of sexual aggression in higher vertebrates might be found in Hamadryas Baboons (Wilson 1975, 242).

territorial aggression *n.* An animal's aggression used in defending its territory, often involving dramatic signaling behavior to repulse intruders and with escalated fighting used as a last resort (Wilson 1975, 242).

weaning aggression *n.* In some mammal species: aggression of parents that involves threatening and even gently attacking their own offspring at weaning time, when the young continue to beg for food beyond that age when it is necessary to do so (Wilson 1975, 243).
cf. weaning

♦ **aggression-frustration hypothesis** See hypothesis: aggression-frustration hypothesis.

♦ **aggressive call** See animal sounds: call: territorial call.

♦ **aggressive cryptic mimesis** See mimicry: mimesis: cryptic mimesis: aggressive cryptic mimesis.

♦ **aggressive mimicry** See mimicry: aggressive mimicry.

♦ **aggressive neglect** *n.* A reduction in an animal's parental care that results from its spending time and energy on aggressive interactions, such as territoriality, and reduces its number of courtships, successful copulations, offspring surviving to independence, or a combination of these things (Wilson 1975, 269, Wittenberger 1981, 612).

♦ **aggressive reproductive mimicry**
See mimicry: aggressive reproductive mimicry.

♦ **aggressive signal** See signal: aggressive signal.

♦ **aggressiveness** *n.*
1. A person's disposition to attack others (*Oxford English Dictionary* 1972, entries from 1859).
2. An individual animal's "tendency to attack" (Scott and Fredericson 1951, 273); an individual's "level of preparedness to fight," contrasted with the act of aggression (Wilson 1975, 251); typical tendency of a species, or individual, to attack or an individual's specific tendency to attack on a particular occasion or under a particular circumstance (Immelmann and Beer 1989, 9).
3. An individual animal's behavior shown by threatening or fighting or, in a more subtle form, expressed simply by independence of action (Collias 1944, 83).
4. A human neurosis brought out by abnormal circumstances and, hence, by implication, nonadaptive for the individual (Montagu, 1968 in Wilson 1975, 255).
cf. aggression

♦ **agium** See ²community: agium.

♦ **agonist** *n.*
1. A human contender for a prize (*Oxford English Dictionary* 1972, entries from 1626).
2. An individual organism involved in agonistic behavior, *q.v.*
3. A drug that can substitute for a hormone, or neurochemical substance, at a neural-receptor binding site and produce the same effects as a natural substance or enhance its effectiveness (Immelmann and Beer 1989, 10).

antagonist *n.* A drug that can substitute for a hormone, or neurochemical substance, at neural receptor binding sites and blocks the effect of a natural substance (Immelmann and Beer 1989, 10).

♦ **agonistic** *adj.* "Referring to any activity related to fighting, whether aggression or conciliation and retreat" (Scott and Fredericson 1951 in Wilson 1975, 242, 578).
cf. behavior: agonistic behavior

♦ **agonistic behavior** See behavior: agonistic behavior.

♦ **agonistic buffer** *n.* An infant used in agonistic buffering, *q.v.*

♦ **agonistic buffering** See buffering: agonistic buffering.

♦ **agoraphobia** See phobia (table).

♦ **agrium** See ²community: agrium.

♦ **agrophile** See ¹-phile: agrophile.

♦ **agrostology** See study of: graminology.

♦ **aha experience** *n.* Possibly in Chimpanzees; in Humans: an individual's sudden recognition of the relationships, causes, and effects of a particular contingency or occurrence which is apparently based on familiarity with the separate parts necessary to solve a problem (Köhler 1963 in Heymer 1977, 25, 60).
cf. learning: insight learning
Comment: According to Büchler (1922), an "aha experience" is part of insight learning.

♦ **ai** See active ingredient.

♦ **aichinophobia** See phobia (table).

♦ **aigialophile** See ¹-phile: aigialophile.

♦ **aigicole** See -cole: aigicole.

♦ **ailurophobia** See phobia (table).

♦ **aiphyllophile** See ¹-phile: aiphyllophile.

♦ **aithallium** See ²community: aithallium.

♦ **aithalophile** See ¹-phile: aithalophile.

♦ **aitionomic, aitionomous** See -genic: aitiogenic.

♦ **"Akilia Island Rocks"** See formation: "Akilia Island Rocks."

♦ **akinesis** See behavior: defensive behavior: playing dead.

♦ **aktology** See study of: aktology.

♦ **alarm** *n.* An alert that an organism communicates to its group mates (Wilson 1975, 211).
v.t. alarm
cf. behavior: alarm behavior

♦ **alarm behavior** See behavior: alarm behavior.

♦ **alarm call** See animal sounds: call: alarm call.

♦ **alarm-defense system** See ¹system: alarm-defense system.

♦ **alarm pheromone** See chemical-releasing stimulus: semiochemical: pheromone: alarm pheromone.

♦ **alarm reaction** See reaction: alarm reaction.

♦ **alarm-recruitment system** See ¹system: alarm-recruitment system.

♦ **alarm signal** See signal: alarm signal.

♦ **ALAS** See Arthropods of La Selva.

♦ **Albert-Lasker Award** See award: Albert-Lasker Award.

♦ **albinism** *n.* An organism's state, or condition, of being an albino (*Oxford English Dictionary* 1972, entries from 1836).

♦ **albino** *n.*
1. A person who has a partial to total absence of pigment in his skin, hair, and eyes so that his skin and hair can be as light as white and his eyes are pink

(*Oxford English Dictionary* 1972, entries from 1777).

2. A nonhuman animal with lighter than usual pigmentation (*e.g.,* a white rat, rabbit, cat, or elephant) (Darwin 1859 in the *Oxford English Dictionary* 1972).

3. An individual plant that lacks chlorophyll in its leaves but belongs to a species whose members usually have such chlorophyll (*Oxford English Dictionary* 1972, entries from 1879).

[Spanish or Portuguese *albino*, white, originally referring to white Negroes on the coast of Africa]

♦ **alertness** See facial expression: alertness.

♦ **aletophile** See ¹-phile: aletophile.

♦ **alimentary egg** See gamete: trophic egg.

♦ **algeny** See genetic engineering.

♦ **algicole** See -cole: algicole.

♦ **algology** See study of: phycology.

♦ **algophage** See -phage: algophage.

♦ **alimentation** *n.*
 1. A person's taking in nourishment (*Oxford English Dictionary* 1972, entries from 1605).
 2. An individual organism's taking in nourishment (Lincoln et al. 1985).
 syn. feeding

♦ **alkaliphile** See ¹-phile: alkaliphile.

♦ **all-occurrences sampling** See sampling technique: all-occurrences sampling.

♦ **all-or-none character** See character: binary character.

♦ **all-or-none law** See law: all-or-none law.

♦ **allaesthetic selection** See selection (table).

♦ **allautogamy** *n.* See -gamy: autoallogamy.

♦ **Allee effect, Allee principle, Allee law** See law: Allee effect.

♦ **allelarkean society** See ²society: allelarkean society.

♦ **allele** *n.*
 1. One of the two or more alternate conditions of a particular trait (Mayr 1982, 733).
 Note: "Allele" was originally coined as "allelomorph" by Bateson (1901), who did not specify whether it referred to a phenotypic or genotypic character (Mayr 1982, 733). However, prior to about 1910, there was the silent, almost universal assumption that there is a 1:1 relation between a genetic factor (gene) and a phenotypic character (Mayr 1982, 733, 736, 754).
 2. "A particular form of a gene, distinguishable from other forms or alleles of the same gene" (Wilson 1975, 578).
 Note: This is a widely used modern definition.
 3. Alternative form of a gene found in a population that occupies the same locus

on a particular chromosome (Dawkins 1982, 283).

4. One of a series of possible alternative forms of a particular gene (cistron), differing in DNA sequence and affecting the functioning of a single product (RNA, protein, or both) (King and Stansfield 1985).

syn. allelomorph (Mayr 1982, 733)

cf. gene

[Greek *allelon,* of one another]

amorphic allele *n.* An inoperative allele that does not influence a phenotype (Lincoln et al. 1985).

dominant allele *n.* In diploid organisms: an allele that is phenotypically manifested when it is homozygous or heterozygous (present with its corresponding recessive allele) (King and Stansfield 1985).

syn. protogene (Lincoln et al. 1985)

cf. allele: recessive allele; dominance (genetic dominance).

hypermorphic allele, hypermorph *n.* A mutant allele with an effect greater than that of its corresponding wild-type allele (Lincoln et al. 1985).

hypomorphic allele, leaky gene, hypomorph *n.* A mutant allele with a lesser effect than that of its corresponding wild-type allele (Lincoln et al. 1985).

isoalleles *pl. n.* Alleles that are so similar in their phenotypic effects that special techniques are required to distinguish them (Lincoln et al. 1985).

meiotic-drive locus *n.* A locus, or allele, that drastically modifies the Mendelian segregation ratio in heterozygotes (Sandler and Novitski 1957 in Rieger et al. 1991).

Comments: Meiotic-drive loci are often found in more than 90% of gametes produced by heterozygotes (Rieger et al. 1991). These loci can severely reduce fitnesses of their bearers.

multiple alleles *pl. n.* Two or more different forms of a gene occupying a specific locus (Lincoln et al. 1985).

neomorphic allele, neomorph *n.* A mutant allele that produces a phenotypic effect that is quantitatively different from that of its corresponding wild-type allele (Lincoln et al. 1985).

neutral allele *n.*
 1. An allele upon which selection does not act (Wilson 1975, 72).
 2. A gene that differs chemically from another one at the same locus but that has indistinguishable effects on its bearer's phenotype (Bell 1982, 510).

syn. neutral mutation

neutral mutation See allele: neutral allele.

null allele *n.* An allele that produces no functional product and, therefore, usually behaves as a recessive allele (King and Stansfield 1985).
cf. allele: silent allele
Comment: Strickberger (1990, 173) suggests that a null allele can also produce a partly functional product or even no product.

pseudoallele *n.* One of two or more mutations that have the same effect but are not structurally allelic (Lincoln et al. 1985).

recessive allele *n.* In diploid organisms: an allele that is not phenotypically manifest when it is present with its corresponding dominant allele (Mayr 1982, 715; King and Stansfield 1985).
syn. allogene (Lincoln et al. 1985), latent allele (de Vries in Mayr 1982, 715)
cf. allele: dominant allele; recessiveness
[Mendel probably introduced the term "recessiv" (now used as recessive) independently of others such as Martinin and Sageret (Mayr 1982, 715).]

recognition allele *n.* A hypothesized allele that encodes the production of a recognition cue and simultaneously an organism's ability to recognize this cue in conspecifics, leading to its discriminating kin from nonkin (Wilson 1987, 12).
cf. phenotype matching
Comment: Such recognition is metaphorically called the "green-beard effect," *q.v.* Existence of recognition alleles has not been demonstrated (Wilson 1987, 12).

revertant *n.* "An allele that undergoes back mutation" (Lincoln et al. 1985). See revertant.

segregation-distorter allele *n.* A meiotic-drive locus in *Drosophila* (Rieger et al. 1991).

silent allele, silent mutant, silent section (of a chromosome) *n.* "An allele that manifests no detectable phenotypic effect" (Lincoln et al. 1985); "an allele that makes no detectable product" (King and Stansfield 1985).
cf. allele: null allele; gene: pseudogene

t allele *n.* A meiotic-drive locus in mice (Rieger et al. 1991).

wild-type allele *n.* The natural, or typical, form of an allele that is arbitrarily designated as standard, or normal, for comparison with mutant, or aberrant, alleles (Lincoln et al. 1985).
Comment: Wild-type alleles are often selected to be dominant in diploid organisms because they produce advantageous products in the presence of other alleles with less advantageous, or even deleterious, products (Strickberger 1990, 173).

♦ **allelic-replacement hypothesis** See hypothesis: allelic-replacement hypothesis.

♦ **allelic-switch-alternative phenotype** See -type, type: phenotype: allelic-switch-alternative phenotype.

♦ **allelo-** *prefix* One another (Lincoln et al. 1985).

♦ **allelochemic** See chemical-releasing stimulus: semiochemical: allelochemic.

♦ **allelogenic** See -genic: allelogenic.

♦ **allelomimetic behavior** See behavior: allelomimetic behavior.

♦ **allelopathic substance** See chemical-releasing stimulus: semiochemical: allelochemic: allomone: allelopathic substance.

♦ **allelopathy** *n.* An organism's release of a chemical (= allelopathic substance) that is an excretory product, metabolite, or waste product that regulates, or inhibits, the growth of another organism (Lincoln et al. 1985).
Comments: Allelopathic substances include antibiotics, germination inhibitors, growth inhibitors, stimulants, and toxins (International Society of Chemical Ecology 1999, www.isce.ucr.edu/society/). Hundreds of plant species including Arizona Fescue (grass), Australian *Eucalyptus*, Black Walnut, Broomsedge Grass, Cherry-Bark Red Oak, Pine-Muhly Grass, Sugar Maple seedlings, Tree of Heaven, Western Bracken Fern, and Yellow Birch show allelopathy (Walker 1990, 180–181). Root microorganisms of Peach trees have allelopathic effects on these trees. The Second World Congress on Allelopathy met at Lakehead University, Canada, in 1999 (www.lakeheadu.ca/~allelo99/).

♦ **allelothetic orientation** See taxis: allelothetic orientation.

♦ **Allen's law** See law: Allen's law.

♦ **allesthetic character** See character: allesthetic character.

♦ **alliesthesia** See -thesia: alliesthesia.

♦ **allo-** *combining form*
  1. Other; alien (Michaelis 1963).
  2. In biology, extraneousness; different from or in opposition to the normal (Michaelis 1963).

♦ **allobiosphere** See -sphere: biosphere: eubiosphere: allobiosphere.

♦ **allocation, principle of** See principle: principle of allocation.

♦ **allochroic** *adj.* Referring to an organism that shows color variation or has the ability to change color (Lincoln et al. 1985).

♦ **allochronic** See chronic: allochronic.

♦ **allochronic speciation** See speciation: allochronic speciation.

♦ **allochronic species** See ²species: allochronic species.

♦ **allochthonous** See -genous: exogenous.

a – c

♦ **allochthonous behavior** See behavior: allochthonous behavior, displacement behavior.

♦ **allochthonous drive** See drive: allochthonous drive.

♦ **allocryptic** See crypsis: allocryptic.

♦ **allodiploidy** See -ploidy: allodiploidy.

♦ **alloethism** *n*. In ants, the regular and disproportionate change in a particular behavior category as a function of worker size (Hölldobler and Wilson 1990, 635).

♦ **allofeeding** See feeding: allofeeding.

♦ **allofusion** See chimera: genetic mosaic.

♦ **allogamy** See -gamy: allogamy.

♦ **allogene** See gene: allogene.

♦ **allogenic** See -genic: allogenic.

♦ **allogenous** See -genous: allogenous.

♦ **allogrooming** See grooming: allogrooming.

♦ **allohospitic parasite** See parasite: allohospitic parasite.

♦ **allohyperparasitism** See parasitism: hyperparasitism: interspecific tertiary hyperparasitism.

♦ **allokinesis** *n*. "Passive or involuntary movement; drifting; planktonic transport" (Lincoln et al. 1985).
cf. kinesis

♦ **allomarking** See behavior: marking behavior.

♦ **allomaternal** See allomother.

♦ **allometry** See auxesis: heterauxesis.

♦ **allomimetic behavior** See behavior: mood induction.

♦ **allomixis** *n*. Cross fertilization (Lincoln et al. 1985).

♦ **allomone** See chemical-releasing stimulus: semiochemical: allelochemic: allomone.

♦ **allomorphosis** See auxesis: heterauxesis: allomorphosis; evolution: allomorphosis.

♦ **allomother** See mother: allomother.

♦ **alloparasitism** See parasitism: alloparasitism.

♦ **alloparent** See parent: alloparent.

♦ **alloparental care** See care: alloparental care.

♦ **allopaternal** *n*. Referring to a male alloparent (Lincoln et al. 1985).
cf. allomaternal

♦ **allopatric** See -patric: allopatric.

♦ **allopatric speciation, allopatric theory of speciation** See speciation: allopatric speciation.

♦ **allophile** See ²-phile: allophile.

♦ **alloploidy** See -ploidy: alloploidy.

♦ **allopreening** See grooming: allopreening.

♦ **allosematic** *adj*. Referring to coloration or markings in one species that imitate warning patterns of another typically noxious, or dangerous, species (Lincoln et al. 1985).
cf. behavior: aposematism; mimicry

♦ **allospecies** See ²species: allospecies.

♦ **allosuckling** See ²suckling: allosuckling.

♦ **allotetraploidy** See -ploid: allotetraploidy.

♦ **allotherm** See -therm: allotherm.

♦ **allotopic population** See ¹population: microdichopatric population.

♦ **allotrophic** See -trophic: allotrophic.

♦ **allotropic** See -tropic: allotropic.

♦ **allotropous** See -tropous: allotropous.

♦ **alloxenic** See -xenic: alloxenic.

♦ **allozygous** See -zygous: allozygous.

♦ **almost certain** See frequency: almost certain.

♦ **alpha, α** *n*.
1. For example, in the Domestic Chicken, Wolves: the highest-ranking individual within a dominance hierarchy (Wilson 1975, 578).
See p: "stipulated p."
cf. beta, omega
2. In statistics, p, p level, *q.v.*
adj. alpha
[α, the first letter of the Greek alphabet]

♦ **α-diversity** See diversity: α-diversity.

♦ **α-diversity index** See index: species-diversity index: α-diversity index.

♦ **alpha selection** See selection (table).

♦ **alpha taxonomy** See taxonomy: alpha taxonomy.

♦ **alsocole** See -cole: alsocole.

♦ **alsophile** See ¹-phile: alsophile.

♦ **alternating scratching** See scratching: run scratching.

♦ **alternation learning** See learning: alternation learning.

♦ **alternation of generations** See -genesis: metagenesis.

♦ **alternative adaptation** See ³adaptation: ²evolutionary adaptation: alternative adaptation.

♦ **alternative-adaptation hypothesis** See hypothesis: alternative-adaptation hypothesis.

♦ **alternative phenotype** See -type, type: phenotype: alternative phenotype.

♦ **alternative strategy** See strategy: alternative strategy.

♦ **altophobia** See phobia (table).

♦ **altricial** See -cial: altricial.

♦ **altruism, altruistic behavior** *n*.
1. A person's regard for the well-being of others; benevolence (*Oxford English Dictionary* 1972, entries from 1853 as the adjective altruistic).
syn. motivated altruism (Rosenberg 1992, 18)
2. Evolutionarily, an individual animal's decreasing its own direct fitness while simultaneously helping a relative in a way that increases the relative's fitness (Haldane 1932, 131, 208–209; Hamilton

A CLASSIFICATION OF ALTRUISM

I. The actor and recipient are relatives:
  A. advantageous altruism
  B. biological altruism (*syn.* cooperative altruism, nepotistic altruism, regular altruism, social donorism, true altruism)
  C. extreme altruism
  D. imposed altruism
  E. low-cost altruism
  F. parental altruism
  G. reciprocal altruism
  H. reproductive altruism
  I. weak altruism
II. The actor and recipient are conspecific nonrelatives:
  A. altruistic altruism
  B. imposed altruism
  C. low-cost altruism
  D. reciprocal altruism
  E. weak altruism

1963, 1964, 13; Alcock 1979, 400; Wittenberger 1981, 60).

Note: Some Humans and workers of ants, bees, Naked Mole Rats, Termites, and wasps show this kind of altruism.

syn. biological altruism (for this definition and others that do not involve intentional motivation to be altruistic, Rosenberg 1992, 18), cooperative altruism, nepotistic altruism (Axelrod and Hamilton 1981, 1391), regular altruism, social donorism (Wilson 1975), true altruism (Axelrod and Hamilton 1981, 1391)

cf. selection: kin selection

Notes: This definition is commonly used or implied. The helper is expected to benefit by its "altruism" because it increases its indirect fitness, but not its direct fitness. "Altruism" in this sense was possibly first suggested by Haldane (1932, 131, 208–209).

3. Sociobiologically, an individual animal's "self-denying or self-destructive behavior performed for the benefit of others" (Wilson 1975, 578; 1987, 10).

Note: Wilson (1987, 10) indicates that some workers are starting to drop this term in favor of expressions such as "nepotism" and "reciprocation," but he prefers to retain it.

4. Sociobiologically, an individual animal's helping a conspecific nonrelative in a way that increases the nonrelative's fitness but not the helper's fitness (Alcock 1979, 11; Barash 1982, 108).

Note: This definition is related to Wynne-Edwards' (1962) concept of nonkin group

selection. This kind of altruism occurs in Humans but is probably, at best, rare in nonhumans.

syn. altruistic altruism (Donald Carr in Alcock 1979), true altruism

cf. altruism: reciprocal altruism

5. Any behavior that benefits another animal while harming or imperiling the individual that performs it (Wittenberger (1981, 613).

Note: Benefit and harm are measured in terms of direct fitness (Wittenberger 1981, 613).

6. For example, in social animal species, behavior of a focal animal that increases its indirect fitness (West Eberhard 1975, 7).

syn. advantageous altruism (West Eberhard 1975, 7)

cf. cooperation

7. One individual organism's "promoting the welfare of another entity, at the expense of its own welfare" (Dawkins 1982, 284).

Note: Dawkins (1982, 284) indicates that "various shades of meaning of 'altruism' result from various interpretations of 'welfare.'"

adj. altruistic

n. altruist

cf. behavior: selfish behavior; donor; investment; nepotism; selfishness; spite

Comments: Wilson and Dugatkin (1992, 31) explain how the same behavior can be labeled selfish, cooperative, or altruistic depending on one's frame of reference. Some workers consider altruism to be a helper's promoting the fitness of a helped conspecific that may be either the helper's relative or nonrelative. McFarland (1985, chap. 9) includes kin selection, parental care, and cooperation, *q.v.*, between symbionts as altruism (see symbiosis); he also uses "altruism" to refer to an animal's helping a heterospecific individual; other workers do not agree with his classification. Fletcher (1987, 22) indicates that use of the terms "reciprocity," "nepotism," and "parental manipulation" instead of "altruism" can help to solve our semantic problem with "altruism." Intentional motivation is not generally ascribed to altruism of nonhuman animals (Rosenberg 1992, 18). However, the evolutionary definitions of "altruism" appear to rely on something similar to motivation, not of the altruistic animal, but of the evolutionist as he calculates what evolves in an altruistic situation (Wilson and Dugatkin 1992, 32). Uyenoyama and Feldman (1992) give two definitions of altruism based on the functions $f_A(x)$ and $f_S(x)$, which denote the

probabilities of survival to reproductive age of altruistic (A) and selfish (S) members of a group of N individuals of which x are altruists and $N - x$ are nonaltruists.
[French *altruiste* (adjectival form of *altruisme* < Latin *alter*, other)]

advantageous altruism See altruism (def. 6).

altruistic altruism See altruism (def. 4).

biological altruism See altruism (def. 2).

cooperative altruism See altruism (def. 2).

extreme altruism *n.* Altruism (def. 2) of animals that cannot have direct fitness, *e.g.,* altruism of sterile workers in some insect species (West Eberhard 1975, 17).

helping behavior, help, helping See behavior: helping behavior.

human reciprocity See altruism: reciprocal altruism.

imposed altruism *n.* Altruism (def. 2 or 4) that is forced upon an individual by a conspecific, *e.g.,* manipulation of an animal's behavior by an adult that adopts it and manipulation of behavior and resources of others by dominant animals (coined by West Eberhard 1975, 17).

kin altruism *n.* An organism's altruism towards only its kin (Fletcher 1987, 23). *Comment:* Fletcher (1987, 23) differentiates this term from "nepotism," *q.v.*

kin selection See selection: kin selection.

low-cost altruism *n.* Altruism (def. 2) that causes a small reduction in an altruistic organism's indirect fitness, *q.v.* (*e.g.,* alarm calling and social grooming in some circumstances) (West Eberhard 1975, 11).

motivated altruism See altruism (def. 1).

nepotistic altruism See altruism (def. 2).

parental altruism *n.* An organism's investment in its offspring (Fletcher 1987, 23). *Comment:* Fletcher (1987, 23) differentiates "parental altruism" from "nepotism," *q.v.*

parental care See care: parental care.

reciprocal altruism *n.* In the Human, Olive Baboon, and Vervet Monkey: altruism in which an individual helps a relative, or nonrelative, with the expectation that the helped individual will return aid to the helper in the near to distant future (Trivers 1971, 35, 39; Packer 1977, 441; Seyfarth and Cheney 1984, 541; Immelmann and Beer 1989, 246). *syn.* reciprocally altruistic behavior; reciprocation, reciprocity (Brown 1987, 305); temporary altruism

cf. cheater, reciprocity
Comments: Trivers (1985 in Connor 1986, 1562) presents "reciprocity" as a special case of the more general category "return-benefit altruism." Brown (1987, 305–306) discusses confusion about "reciprocal altruism" and the relationships between this term and "byproduct mutualism" and "score-keeping mutualism" (coined by Trivers 1971 in Brown 1987, 305).

▸ **pseudo-reciprocity** *n.* An interaction between two individuals in which the actor (A) performs a beneficent act toward a receiver (B) that is a byproduct, or incident effect, of A's egoistic behavior; thus, B is not considered to be a cheater if B does not perform a beneficent act toward A (coined by Connor 1986, 1562).

regular altruism See altruism (def. 2).

reproductive altruism *n.* An individual animal's forgoing its reproduction entirely and helping other individuals to reproduce (Honeycutt 1992, 43).

return-benefit altruism *n.* All acts of social beneficence in which the return benefit to the initial actor (A) comes from the initial recipient (B) but the effect of the return benefit on B is not specified (Trivers 1985 in Connor 1986, 1562).

social donorism See altruism (def. 2).

temporary altruism See altruism: reciprocal altruism.

true altruism See altruism (def. 2 and 4).

truly altruistic behavior. See altruism (def. 4).

weak altruism *n.* An interaction between an actor and recipient in which (1) the actor has reduced fitness and the recipient has no fitness change, or (2) that actor has no fitness change and the recipient has increased fitness (Gadagkar 1993, 233).

♦ **altruism investment** See investment: altruism investment.

♦ **altruistic altruism** See altruism (def. 4).

♦ **altruistic behavior** See altruism.

♦ **altruistic character** See gene: altruistic gene.

♦ **altruistic genes** See character: altruistic traits.

♦ ***Alu*** See transposable element: *Alu*.

♦ **amanthicole** See -cole: amanthicole.

♦ **amanthium** See ²community: amanthium.

♦ **amatophile** See ¹-phile: amatophile.

♦ **amatory behavior** See behavior: amatory behavior.

♦ **amber** See fossil: amber.

♦ **ambient temperature (t_a)** See temperature: ambient temperature.

♦ **ambiguity** *n.* A receiver's potential of giving more than one response to a signal due to "noise" in its communication system (Wilson 1975, 195).

♦ **ambisexual mate desertion** See mate desertion: ambisexual mate desertion.

♦ **ambivalent behavior** See behavior: ambivalent behavior.

♦ **ambivalent movements and postures** See behavior: ambivalent movements and postures.

♦ **ambivore** See -vore: ambivore.

♦ **ambush predator** See predator: ambush predator.

♦ **ameiotic parthenogenesis** See parthenogenesis: ameiotic parthenogenesis.

♦ **ameiotic thelytoky** See parthenogenesis: apomixis.

♦ **American behaviorism** See study of: psychology: classical animal psychology.

♦ **American Sign Language** See language: Ameslan.

♦ **Ames room** *n.* A distorted room (with trapezoidal walls, a ceiling, floor, and back wall that slope away from an observer but are shaded to appear rectangular) designed to test human visual perception (Storz 1973, 13).

♦ **Ameslan** See language: Ameslan.

♦ **ametabolous metamorphosis** See metamorphosis: ametabolous metamorphosis.

♦ **amicable behavior** See behavior: amicable behavior.

♦ **amino** *combining form* In chemistry, of, or pertaining to, the NH_2 group combined with a nonacid radical (Michaelis 1963).

♦ **amino acid** See acid: amino acid.

♦ **amixis** See mixis: amixis.

♦ **ammochthium** See ²community: ammochthium.

♦ **ammochthocole** See -cole: ammochthocole.

♦ **ammocole** See -cole: ammocole.

♦ **ammophile** See ¹-phile: ammophile.

♦ **amnicole** See -cole: amnicole.

♦ **amorphic** See -morphic: amorphic.

♦ **amorphic allele** See allele: amorphic allele.

♦ **AMPA receptor** See ²receptor: AMPA receptor.

♦ **ampherotoky** See parthenogenesis: amphitoky.

♦ **amphi-** *prefix*
1. On both, or all, sides; at both ends (Michaelis 1963).
2. Around (Michaelis 1963).
3. "Of both kinds; in two ways" (Michaelis 1963)

[Greek *amphi*, around]

♦ **amphiclexis** See selection: epigamic selection.

♦ **amphigamy** See -gamy: amphigamy.

♦ **amphigenesis** See -genesis: amphigenesis.

♦ **amphigony** *n.* "Sexual reproduction involving cross-fertilization" (Lincoln et al. 1985).

♦ **amphimict** *n.* An organism that reproduces by amphimixis (sexual reproduction) (Lincoln et al. 1985).

♦ **amphimixis** See -mixis: amphimixis.

♦ **amphioecism, amphioecious, amphitopic** See -oecism: amphioecism.

♦ **amphiplexus** *n.* "A sexual embrace" (Lincoln et al. 1985).
cf. amplexus

♦ **amphitoky** See parthenogenesis: amphitoky.

♦ **amphitroph** See -troph-: amphitroph.

♦ **amphogenic** See -genic: amphogenic.

♦ **amphoterosynhesmia** See ²group: amphoterosynhesmia.

♦ **amphoterotoky** See parthenogenesis: amphitoky.

♦ **amplexus** *n.*
1. In many amphibian species: "Sexual embrace" (Oliver 1955, 329).
2. In most Frog and Toad species and many Salamander species: a mating position in which a male clasps a female with one or both pairs of his legs before she releases her eggs and he fertilizes them (Dewsbury 1978, 77, 209; Halliday and Adler 1986, 146; Massey 1988, 205).
syn. clasping hold (Heymer 1977, 48)
cf. amphiplexus; reflex: grasping reflex
[Latin *amplecti*, to embrace, clasp]

♦ **amplitude modulation** See modulation: amplitude modulation.

♦ **amygdala** See organ: brain: amygdala.

♦ **amyloplast** See organelle: amyloplast.

♦ **anabiosis** See -biosis: anabiosis.

♦ **anabolism** See metabolism: anabolism.

♦ **anachoresis** *n.* An organism's living in a fissure, hole, or crevice (Lincoln et al. 1985).

♦ **anadamide** *n.* A compound produced by human brains that activates the same brain area as marijuana does (Associated Press 1996b, D18).
Comments: Anadamide that occurs naturally in chocolate might alter human mood (Associated Press 1996b, D18).

♦ **anadromous, anadromesis, anadromy** See -dromous: anadromous.

♦ **anaerobe** See aerobe: anaerobe.

♦ **anagenesis** See evolution²: anagenesis; -genesis: anagenesis.

♦ **anal feeding** See trophallaxis: anal trophallaxis.

♦ **anal gland** See gland: anal gland.

♦ **anal trophallaxis** See trophallaxis: anal trophallaxis.

◆ **analog, analogue** *n.*

1. A trait of one animal species that is mistaken for a homolog, *q.v.,* of another (Owen 1843 in Hailman 1976, 181).
2. A part, or organ, in one animal species that has the same function as another part, or organ, in a different animal species regardless of ancestry (Owen 1848 in Mayr 1982, 464; Donoghue 1992, 170). *syn.* analogy
3. A structure, physiological process, or behavior in one species that is similar to one in another species owing to convergent evolution as opposed to common ancestry (Wilson 1975, 578); *e.g.,* dorsal fins of Dolphins and Sharks, lens eyes of Cephalopods and Vertebrates, echolocation of bats and Oilbirds, or distraction displays of Plover and some ducks (Immelmann and Beer 1989, 14). *Note:* This is a commonly used definition. *syn.* analogy
4. A species that has a similar habitat, or distribution, as another species (Lincoln et al. 1985).

adj. analogous, nonhomologous
cf. homogeny, homolog, homoplasy
Comment: Hailman (1976) gives more definitions of "analog" and discusses how authors have confounded the definitions of analog, homogeny, homolog, and homoplasy.
[French < Greek *analogon,* originally neutral singular of *analogos,* proportionate, conformable < *ana-,* according to + *logos,* proportion]

◆ **analog signal** See signal: analog signal.

◆ **analogy** See analog.

◆ **analysis** *n.*

1. A method used to determine, or describe, the nature of a thing by separating it into its parts (Michaelis 1963).
2. The results of using the method in def. 1 (Michaelis 1963). See analysis: psychoanalysis.

form analysis *n.* An investigator's close comparison of behaviors within or among species to elucidate evolutionary origins of behaviors (Immelmann and Beer 1989, 111).
cf. method: comparative method; analysis: sequence analysis, situation analysis

message-meaning analysis *n.* Analysis of what a signal expresses about its signaler (the message) and what the signal, in context, conveys to a recipient (the meaning) (Smith 1965 in Immelmann and Beer 1989, 184).

motivational analysis *n.* Analysis of regularities in the occurrences of instinctive behaviors, dealing with the number and type of drive factors mediating a particular behavior pattern (Tinbergen 1939, etc., in Heymer 1977, 119).

multivariate analysis *n.*

1. In statistics, "a method whose theoretical framework allows for the simultaneous considerations of several dependent variables (Kleinbaum and Kupper 1978, 14).
Note: Examples of multivariate analyses include multivariate ANOVA and multivariate multiple regression.
2. In biomedical and health research, "any statistical technique involving several variables, even if only one dependent variable is considered at a time" (Kleinbaum and Kupper 1978, 14).
Note: Kleinbaum and Kupper (1978, 14) prefer "multivariable analysis" instead of "multivariate analysis" for this concept.

cf. analysis: univariate analysis

psychoanalysis *n.*

1. A group of doctrines concerned with the study and interpretation of human mental states in terms of the dynamic interplay of conflicting drives and processes originating in one's unconscious (Michaelis 1963).
2. A psychotherapy system introduced by Freud that seeks to alleviate human neuroses and other mental disorders by the systematic technical analysis of unconscious factors that are revealed, for example, in dreams, free association, and memory lapses (Michaelis 1963; Collocot and Dobson 1974).

syn. analysis (Michaelis 1963)

sequence analysis *n.* An investigator's recording the order of behaviors and subsequently examining it to determine whether the order is random or structured (Immelmann and Beer 1989, 265).
cf. analysis: form analysis, situation analysis

situation analysis *n.* An investigator's noting the circumstances in which displays are performed to try to elucidate their evolutionary derivations (Immelmann and Beer 1989, 271).
cf. analysis: form analysis, sequence analysis

stochastic analysis *n.* An investigator's computing the transition probabilities among items (in mutually exclusive and collectively exhaustive sets) in different sequence combinations and determining whether they depart from randomness (Immelmann and Beer 1989, 265, 293–295).
Comment: Communicatory behaviors are often examined by this analysis (Immelmann and Beer 1989, 265, 293–295).

univariate analysis *n.* In statistics, a method whose theoretical framework allows for the simultaneous considerations of only one dependent variable (Kleinbaum and Kupper 1978, 14).
cf. analysis: multivariate analysis

♦ **analysis method** For some specific kinds of analysis methods, see procedure and program.

♦ **analysis of variance (ANOVA)** See statistical test: analysis of variance.

♦ **anamorphosis** See ²evolution: anamorphosis; metamorphosis: ametabolous metamorphosis.

♦ **anaplasia** See -plasia: anaplasia.

♦ **anautogenous** See -genous: anautogenous.

♦ **anauxotrophic** See -trophic: anauxotrophic.

♦ **ancestor, archetype** *n.* A progenitor of a more recent descendant taxon, group, or individual; any preceding member of a lineage (Lincoln et al. 1985).
cf. -type: archetype
cenancestor *n.* The most recent common ancestor of taxa in question (Hasegawa and Fitch 1996, 1750).
syn. last common ancestor (Vigilant et al. 1991, 1504)
common ancestor *n.* A species that gave rise to one or more new species (Mayr 1982, 466).
Comment: The common-ancestor concept replaced the concept of "morphological type" or "archetype" as understanding of evolution improved (Mayr 1982, 466).

♦ **ancestral** *adj.*
1. Referring to the earliest stage in ontogeny, or development, of an organ or system (Lincoln et al. 1985).
   syn. original, primordial, primary, primitive (Lincoln et al. 1985)
2. "Early; simple; poorly developed; unspecialized" (Lincoln et al. 1985).
   syn. plesiomorphic, primitive, primordial (in some cases), protomorphic (Lincoln et al. 1985)
   cf. derived, morph
3. Referring to a taxon with relatively many ancestral characters.
syn. lower, primitive
cf. character: ancestral character; Darwin's admonition; derived; genetic: palingenetic; homotenous; primitive
Comments: Some workers indicate that it is better to use "ancestral" and "derived" to describe organisms, rather than "higher" and "lower" or "primitive" and "modern." "Higher" and "lower" unscientifically can suggest better and inferior. "Primitive" and "modern" are troublesome terms because

they are sometimes applied to extant species, which, because they are alive now, are thus technically "recent," or "modern." Smith (1984) makes a case for designating vertebrates as "ectotherm vertebrates" and "endotherm vertebrates" rather than "higher" and "lower vertebrates."

♦ **"ancestral-brain hypothesis"** See hypothesis: "ancestral-brain hypothesis."

♦ **ancestral character, ancestral trait** See character: ancestral character.

♦ **ancestral mitochondrial genome** See genome: ancestral mitochondrial genome.

♦ **ancium** See ²community: ancium.

♦ **ancocole** See -cole: ancocole.

♦ **ancophile** See ¹-phile: ancophile.

♦ **andr-** *combining affix* Male, testis, stamen.

♦ **andrenarche** *n.* Turning on of the andrenal androgen dihydroespiandrosterone in a Human which usually occurs around 6 years of age.
cf. puberty

♦ **andric** *adj.* Male.
ant. gynic

♦ **androchore** See -chore: androchore.

♦ **androchromatypic mimicry** See mimicry: androchromatypic mimicry.

♦ **androcyclic parthenogenesis** See parthenogenesis: androcyclic parthenogenesis.

♦ **androgamy** See -gamy: androgamy.

♦ **androgen** See hormone: androgen.

♦ **androgenesis** See parthenogenesis: androgenesis.

♦ **androgenous** See -genous: androgenous.

♦ **androgyne** See hermaphrodite.

♦ **androgynous, androgynal, androgynic** *adj.*
1. Referring to an individual organism that has characteristics of both males and females (*Oxford English Dictionary* 1972, entries from 1651).
2. Referring to an individual plant, or flower, with both stamens and pistils (*Oxford English Dictionary* 1972, entries from 1793).
3. Referring to an individual organism that has both male and female reproductive organs (Lincoln et al. 1985).
syn. hermaphroditic
cf. hermaphrodite
[Latin *androgynus* < Greek *androgynos*, hermaphrodite < *aner, andros,* man + *gynē,* woman]

♦ **andromorphic** See -morphic: andromorphic.

♦ **androphile** See ¹-phile: androphile.

♦ **androphobia** See phobia (table).

♦ **androsterone** See chemical-releasing stimulus: semiochemical: pheromone: androsterone.

♦ **anemochore** See -chore: anemochore.

♦ **anemochorous** See -chorous: anemochorous.

♦ **anemodium** See [2]community: anemodium.

♦ **anemophile** See [1]-phile: anemophile.

♦ **anemophobia** See phobia (table).

♦ **anemotaxis** See taxis: anemotaxis.

♦ **anestrus** See estrus: anestrus.

♦ **angel, radar angel** *n.* One of a group of discrete radar echoes that are often observed in profusion on radars and that were at first thought to be echoes from a heterogeneous atmosphere but were later identified as flying birds (Eastwood 1967, 60; D.B. Quine, personal communication).

ring angel *n.* A radar angel that starts as a bright blob and then becomes a diffuse ring, or a series of concentric rings (Eastwood 1967, 166).

Comment: Ring angels were first postulated to be attributable to reflections from consecutive wave fronts of a shear-gravity wave moving through the atmosphere but were later found to be birds leaving their roosts (Eastwood 1967, 166; D.B. Quine, personal communication) [coined by Elder 1957 in Eastwood 1967].

♦ **anger** See facial expression: anger.

♦ **angiotensin** See hormone: angiotensin.

♦ **anheliophile** See [1]-phile: anheliophile.

♦ **anholocyclic parthenogenesis** See parthenogenesis: anholocyclic parthenogenesis.

♦ **animal** See Appendix 1.

♦ **animal behavior** See behavior: animal behavior; study of: animal behavior.

♦ **animal communication** See communication.

♦ **animal group** See [2]group.

♦ **animal hypnosis** See playing dead.

♦ **animal language** See language.

♦ **animal names** *n.* The specific names given to males, females, and young of particular kinds of animals.

cf. [2]group: brood, fry, immature

Comments: Some workers informally refer to immatures of any kind of organism, especially mammals and birds, as "babies" (*e.g.,* hummingbird babies) (Gould 1982, 188). Some animal names are given in the table below. More complete lists of animal names are in Barnes (1974).

adolescent *n.* A person who is approaching adulthood (Michaelis 1963).

Comments: "Adolescent" can also refer to nonhuman animals. Elephant adolescents are 10.1 to 15 years old (Lee 1987, 278).

adult *n.*

1. A "fully grown, sexually mature (where appropriate) animal" (Immelmann and Beer 1989, 7).

Note: Elephant adults are individuals over 15 years old (Lee 1987, 278).

2. A group of coral polyps >4 centimeters in diameter; contrasted with juvenile (Edmunds 1996, 95).

3. A plant that is old enough to reproduce (*e.g.,* an adult fig tree) (inferred from Nason et al. 1998, 685).

▸ **subadult** *n.* A lion that is 25 through 48 months old (Schaller 1972 in Handby and Bygott 1987, 161).

buck *n.* An adult male of deer, elk, goat, hare, kangaroo, moose, rabbit, or rat.

cf. animal names: doe

calf *n.* An immature elephant, hoofed animal, or pinniped (Sea Elephant, Seal, Walrus).

Comment: Elephant calves are individuals 0 to 24 months old (Lee 1987, 279).

cheekpadder *n.* A male Orangutan with fleshy checks (Galdikas 1995, 15).

doe *n.* An adult female of a deer, elk, goat, hare, kangaroo, moose, or rabbit.

cf. animal names: buck

fry *n., pl.* **fry**

1. A very young fish (Michaelis 1963).

Note: Many researchers call a very young postembryonic fish a larva (Balon 1975b). Embryos of many fish species swim.

2. A small adult fish when in the company of a large number of other such fish (Michaelis 1963).

3. The young of certain animals (*e.g.,* frogs, when produced in very large numbers) (Michaelis 1963).

4. Young children, or other young animals (Michaelis 1963).

cf. animal names: small fry

5. A fish larva or juvenile (Balon 1975b; Blumer 1982, 3).

[Old Norse *frið,* seed]

gilt *n.* A female pig that has not yet had piglets; a young sow (Michaelis 1963).

hart *n.* A male Red Deer (Lipton 1968, 13).

hind *n.* A female Red Deer (Lipton 1968, 13).

jimmy *n., pl.* **jimmies** An adult male Blue Crab; contrasted with sook (Phillips 1992).

juvenile *n.*

1. A young person; a youth (*Oxford English Dictionary* 1972, entries from 1733).

2. A subadult mammal between an infant and adult, not yet sexually mature (Immelmann and Beer 1989, 161).

cf. animal names (table), hebetic, immature

3. In Dragonflies: a young adult (Dunkle 1989, 10).

Notes: In some Dragonflies, juveniles change color as they mature. Parts of their abdomens, wings, or both, become covered by a waxy powder (pruinescense) which is usually white or pale blue, but in the slaty skimmer is black (Dunkle 1989, 10). Elephant young juveniles are 25 to 60 months old; old juveniles are 61 to 120 months old (Lee 1987, 279).

4. A group of Coral polyps < 4 centimeters in diameter; contrasted with adult (Edmunds 1996, 95).

larva *n., pl.* **larvae**

1. In many aquatic and marine invertebrates, insects, Frogs, and Toads: an immature stage that is radically different in form from an adult (Michaelis 1963; Wilson 1975, 587).
   Notes: "Larva" is used to refer to immature ametabolous, paurometabolous, hemimetabolous, and holometabolous insects. See table for more specific larval names.
2. In termites: an immature individual without any external trace of wing buds or soldier characteristics (Wilson 1975, 587).
3. In fish: a postembryonic stage that is younger than a juvenile fish (Balon 1975, 829, etc.).

adj. larval
cf. animal names: juvenile, pupa
Comment: "Larvas" is the anglicized plural (Buchsbaum et al. 1987, 502).
[Latin *larva*, ghost, spectre, hobgoblin, mask]

planula *n., pl.* **planulae** [PLAN new la, PLAN new LEE] The freely moving, ciliated embryo of certain cnidarians, including hydroids (Michaelis 1963).
Note: A planula might have evolved into the first Porifera, Cnidaria, and Platyhelminthes. See planula.
adj. planular, planulate
[New Latin < Latin, diminutive of *planus*, flat]

pupa *n., pl.* **pupae**

1. In holometabolous insects: an individual's usually nonmotile, developmental stage during which it completes its development into adulthood (*Oxford English Dictionary* 1972, entries from 1815); also, an insect in its pupal stage.
2. In cirriped and holothurian species: a developmental stage (*Oxford English Dictionary* 1972, entries from 1877).

n. pupation
v.i. pupate
cf. animal names: larva, nymph (table)
[New Latin < Latin *pupa*, girl, doll, puppet]

queen *n.* A female cat (*Oxford English Dictionary* 1972, entry from 1898).

small fry *n., pl.* **small fry**

1. A small, young fish (Michaelis 1963).
2. A young child (Michaelis 1963).
3. A small, or insignificant, person or thing (Michaelis 1963).

cf. animal names: fry

sook *n.* An adult female Blue Crab; contrasted with jimmy (Phillips 1992).

spawn *n.*

1. The minute eggs of fishes and various other oviparous animals, chiefly aquatic or amphibian, usually laid in large numbers and forming a more or less coherent gelatinous mass; also, the newly hatched young from such eggs (*Oxford English Dictionary* 1972, entries from 1491).
2. A fish egg; an undeveloped fish (*Oxford English Dictionary* 1972, entries from 1563).

See spawn.
v.i., v.t. spawn
cf. group: spawn; gamete: egg

♦ **animal physiology** See study of: physiology: animal physiology.

♦ "**animal piercers**" See [2]group: functional feeding group: animal piercers.

♦ **animal psychology** See study of: psychology.

♦ **animal sounds** This group of terms includes sounds produced by animals' vocalizations, wing movements, and other movements. Only some of the many kinds of animal sounds are included.
cf. behavior: rapping behavior, verbal behavior; carnival; ceremony; communication; jargon; language; message; mocking; panting; repertory; signal; yawn
Comment: Bird sounds are described in many books including Peterson (1947), Chandler et al. (1966), Peterson et al. (1967), and Peterson and Chalif (1973).

baa *n.* A Sheep's, or lamb's, cry; a bleat (*Oxford English Dictionary* 1972, entries from 1589).
Comment: Some frog and toad species make bleating sounds (Conant 1958).
v.i., v.t. baa

bark *n.*

1. In Dogs: a sharp, explosive cry (*Oxford English Dictionary* 1972, entries from 1562).
2. For example, in some toad, gecko, and squirrel species, the Barking Frog, the Barking Tree Frog, Foxes, and Prairie Dogs: a vocalization similar to a Dog's bark (*Oxford English Dictionary* 1972, entries from 1562; Conant 1958; Wilson 1975, 473; Grady and Hoogland 1986, 108).

v.i., v.t. bark

booming *n.* A male Prairie Chicken's making a hollow three-syllabled *oo-loo-woo* that suggests the sound made by blowing across the mouth of a soda bottle (Peterson 1947, 76).
cf. animal sounds: drumming

a – c

NAMES OF SEXES AND LIFE STAGES OF SELECTED ANIMAL TAXA

ADULT MALE	ADULT FEMALE	OFFSPRING	TAXA
		Sponges[a]	
male	female	egg, amphiblastula, or parenchymula larva (depending on species)	
		Cnidarians (Coelenterates)	
male medusa	female medusa	egg, planula	Jellyfish[b]
male polyp	female polyp	egg, planula	Coral, Sea Anemones, Hydra
male	female	egg, planula, scyphistoma, strobila, young medusa (= ephyra), adult medusa	*Aurelia* Jellyfish
		Molluscs	
male	female	trochophore, larva	Chitons[c]
male	female	trochophore, veliger	Bivalves[c]
male	female	trochophore	Archaeogastropods
male	female	veliger	nonarchaeogastropods[d]
male	female	juvenile	freshwater and land snails[e] (pulmonate gastropods)
male	female	juvenile	Cephalopods[b]
		Flatworms	
male	female	egg, larva	Monogenetic Trematodes
male	female	egg, miracidium, sporocyst, redia, cercaria, metacercaria	Digenetic Trematodes
male	female	egg, oncospore, cysticercus, procercoid, plerocercoid	Tapeworms
		Nematodes[c]	
male	female	egg, young, or larva	
		Annelids	
male	female	egg, trochophore, larva	Polychaetes[c]
male	female	eggs, young worm	Oligocheates[c]
male	female	eggs, young leech	Oligocheates[c]
		Spiders	
male	female	egg, spiderling	
		Crustaceans[f]	
male (some species hermaphroditic)	female	egg, nauplius, species	many marine crustacean species
male	female	egg, zoea, postlarva	more derived malacostracans
		Insects[g,h]	
male[a]	female	egg, larva	Collembolans, Thysanurans, Proturans, Diplurans
male	female	egg, larva or nymph, pupa	Orthopteroids, Hemipteroids

Names of Sexes and Life Stages of Selected Animal Taxa (cont.)

Adult Male	Adult Female	Offspring	Taxa
male	female	egg, larva or naiad, pupa	Odonatans, Plecopterans
male	female	egg, larva, pupa	many neuropteroids
male	female	egg, larva or maggot, pupa	many fly and beetle species
male	female	egg, larva or caterpillar, pupa	Moths, Sawflies
male	female	egg, larva or caterpillar, chrysalid, chrysalis, or pupa	Butterflies
king, soldier, worker	queen, soldier, worker	egg, larva or nymph, subadult	Termites
drone	queen, worker	egg; larva, grub, or maggot; prepupa; pupa	Honey Bees

Echinoderms

Adult Male	Adult Female	Offspring	Taxa
male	female	egg, bipinnaria larva, brachiolaria larva	Starfish[c]
male	female	egg, echinopluteus, or larva	Brittle-Stars[c]
male	female	egg, echinopluteus, or larva	Sea Urchins
male	female	egg, auricularia, doliolaria, vitellaria (some species), pentacula larva	Sea-Cucumbers[c]
male	female	egg, vitellaria, pentacrinoid	Sea-Lilies, Feather-Stars

Hemichordates

Adult Male	Adult Female	Offspring	Taxa
male	female	egg, tornaria larva	Acorn worms

Tunicates

Adult Male	Adult Female	Offspring	Taxa
male	female	egg, appendicularia (= tadpole)	Sea-Squirts[a]

Arrow worms

Adult Male	Adult Female	Offspring	Taxa
hermaphroditic		egg, larvae	

Fish

Adult Male	Adult Female	Offspring	Taxa
male	female	egg, larva (*syn.* fry), juvenile	many fish species[c]
male	female	egg, pup	Sharks

Amphibians

Adult Male	Adult Female	Offspring	Taxa
male	female	egg, tadpole (sometimes = fry), froglet (in frogs)	Frogs, Toads
male	female	egg, larva, eft	Newts

Reptiles

Adult Male	Adult Female	Offspring	Taxa
male	female	egg, young, or immature	Lizards, Snakes, Turtles, Tortoises

Birds

Adult Male	Adult Female	Offspring	Taxa
gander	goose	egg, gosling	Geese
cob	pen	egg, cygnet	Swans
rooster, cock	hen	egg, chick	Chickens, Pheasants

a — c

NAMES OF SEXES AND LIFE STAGES OF SELECTED ANIMAL TAXA (CONT.)

ADULT MALE	ADULT FEMALE	OFFSPRING	TAXA
drake	female	egg, duckling	Ducks
male	reeve	egg, chick	Ruffs
male	female	egg, nestling, fledgling, juvenile	many bird species

Mammals[i,j]

ADULT MALE	ADULT FEMALE	OFFSPRING	TAXA
buck	doe	joey	Kangaroos
male	female	infant, cub	Koala
boar	sow	cub (= whelp)	Bears
male	female	cub (= whelp)[j]	Coati, Wolves
dog fox	vixen	cub, kit	Foxes
male	female	kit	Beaver
male	bitch	puppy, pup	Dogs
tom	queen	kitten, kit	Cats
lion	lioness	cub, subadult	Lions
tiger	tigress	cub	Tigers
male	female	pup	Gerbils, other rodents, Bats
buck	female	pup	Rats
buck	doe	kit	Rabbits
buck	doe	leveret	Hares
bull	cow	calf	Giraffes, Sea Lions, Walruses, Whales, Dolphins, Porpoises, Cattle, Oxen
bull	cow	calf, juvenile, adolescent	Elephants
bull	cow	calf, pup	Seals
man, "jack"	woman, "dame"	child, infant, toddler, tot, wean, kid, juvenile, subadult, "buck," "colt," "cub," "small fry"	Humans
silverback, male	female	young	Gorilla
stag	doe	fawn	Red Deer, other large deer
buck	doe	fawn	Deer, Antelopes
hart	hind	fawn	Red Deer
buck	cow, "doe"	young	Elk, Moose
buck, billy	"nanny"	kid	Goats
ram	ewe	lamb	Sheep
male, stallion	mare	foal, colt	Horses, Zebras
jack, jackass	mare, sheass	colt	Donkey (= Ass)
boar	gilt, sow	piglet	Pigs, Hogs, Peccaries

[a] Most species are hermaphrodites.
[b] A few species are hermaphrodites.
[c] Some species are hermaphrodites.
[d] Many species are hermaphrodites.
[e] All species are simultaneous hermaphrodites.
[f] Depending on the species, one or more larval stages may be suppressed (Barnes 1974, 521). The postlarva of a crab is called a "megalops;" of a sergestid prawn, an "acanthosoma." The zoea of lobsters is called a "mysis." In some species, later nauplius instars are called "metanauplii," and the prezoeal instar is called a "protozoea."
[g] Some species are parthenogenetic.
[h] An adult insect is also called an "imago."
[i] Not included in this table is "suckling," a young mammal that is still feeding at its mother's breast, and "weanling," a newly weaned young.
[j] Names in quotes are less preferred, or more specifically, less used ones.

call *n.*
1. A person's loud vocal utterance or speech, a shout, a cry; loud vocal address or supplication (*Oxford English Dictionary* 1972, entries from *ca.* 1300).
2. In some animal species, especially birds: a cry (*Oxford English Dictionary* 1972, verb entries from 1486).
3. In the Honey Bee: an individual's buzzing before swarming (*Oxford English Dictionary* 1972, verb entry from 1609).
4. In some animal species: an individual's vocalization that involves short, rather simple signals that are generally uttered by either sex at any time of the year and with little individual variation (Dewsbury 1978, 168).

v.i., v.t. call
cf. calling: female calling; animal sounds: song: calling song
Comments: "Calls" are sometimes difficult to distinguish from "songs" and have different functions (Immelmann and Beer 1989, 38). "Call" now sometimes refers to a pheromone release; see "calling." Some of the many kinds of calls are defined below.

▶ **advertisement call** *n.* In many frog species: a call used to indicate a male's motivational state and location to a potential mate or rival (Halliday and Adler 1986, 58).
▶ **aggressive call** See animal sounds: call: territorial call.
▶ **alarm call, protective call** *n.* For example, in the Belding's Ground Squirrel and Scrub Jay: a call made when danger threatens but is still far away (Lincoln et al. 1985).
▶ **coquette call** *n.* In some duck species: an epigamic vocalization (Lorenz 1941 in Heymer 1977, 99).
▶ **distress call** *n.*
1. In many kinds of vertebrates including Humans: a vocalization made by a young animal that is separated from its parent, surrogate parent, or sibling that can cause these animals to come back in contact with the young (Lorenz 1935 in Heymer 1977, 149).
   syn. lost call
2. In many frog species: a frog's call given when it is grasped by a predator (Halliday and Adler 1986, 58).
   Note: A distress call is given with an open mouth and sounds like human screaming in some species (Halliday and Adler 1986, 58).
   syn. fright call, mercy call

3. In some bird species: a nidifugous chick's call that helps its parent locate it (Brémond and Aubin 1990, 504).
   syn. lost call
4. In many bird species, including Corvids, Gulls, Sparrows, Starlings, and other small passerines: a bird's call that it produces when it is struggling with a predator or being handled or forcibly restrained by a person (Brémond and Aubin 1990, 504).
   syn. fright call, mercy call
   cf. animal sounds: gecker
   Comment: Avian distress calls are loud, harsh calls of long duration that cover a wide frequency range that make them easy to locate (Stefanski and Falls 1972 in Hill 1986, 590).
▶ **encounter call** See animal sounds: call: territorial call.
▶ **food call, food calling** *n.* In the Domestic Fowl: a cluck-cluck sound emitted during ground pecking by a hen that can result in her young's starting to feed or by a cock that can result in a hen's starting to feed (Schenkel 1956, 1958 in Heymer 1977, 72).
   v.i. food call
▶ **fright call, lost call, mercy call** See animal sounds: call: distress call.
▶ **gargle call** *n.* In the Black-Capped Chickadee: a vocalization used in agonistic encounters (Miyasato and Baker 1999, 1311).
▶ **mew call** *n.* In Herring Gulls: a long, drawn-out tone given with a bird's neck stretched forward and down and with its bill open wide; possibly used for attracting other gulls or expressing antagonism (Tinbergen 1958 in Heymer 1977, 96).
   v.i. mew call
▶ **milk call** *n.* In mammals: a female's vocalization that summons young prior to feeding (Dunbar 1985).
▶ **predator warning call** *n.* A call note used to communicate a predatory threat and with properties making it difficult for a predator to locate its sender (Marler 1957 in Wilson 1975, 236).
▶ **protective call** See animal sounds: call: alarm call.
▶ **release call** *n.*
1. In many frog species: a call given by a male that is grasped by another male (Halliday and Adler 1986, 58).
2. In many frog species: a call given by an unreceptive female that is grasped by a male (Halliday and Adler 1986, 58).
▶ **territorial call** *n.* In many frog species: a male's call that appears to inform

a – c

a rival male that he may attack; or, in the early stage of a conflict, may mainly disrupt the calling of a competitor Halliday and Adler 1986, 58).

syn. aggressive call, encounter call

▶ **trumpet call** *n.* In Elephants: a call directed at a smaller opponent (Wilson 1975, 495).

cf. animal sounds: trumpet

▶ **vocalization** *n.*

1. A person's vocalizing; uttering with his voice; or the fact of being vocalized (*Oxford English Dictionary* 1972, entries from 1842).
2. An individual animal's song, call, and other sound produced by its vibrating membranes in its respiratory tracts (*e.g.,* vocal cords in bird syrinxes and mammal larynxes) (Immelmann and Beer 1989, 324).

Comment: Apart from human speech, the more complex vocalizations are songs of some songbird and whale species, Howler Monkeys, and the Gibbon (Immelmann and Beer 1989, 324).

▶ **wiggling call** *n.* In the House Mouse: a pup's low-frequency call that is emitted when it is struggling in its nest, mainly when pushing for a teat during suckling (Ehret and Bernecker 1986, 821).

call note *n.* A bird's vocalization that consists of one or a few short bursts of sound; much simpler in structure than a song (Wilson 1975, 236).

cf. animal sounds: call, song

Comments: Call notes are used in situations of alarm, mobbing, distress, contact-maintenance, flight intention, and releasing following in, for example, chicks or goslings. They include fear thrills and predator warning calls, *q.v.* (Wilson 1975, 236).

crepitation *n.* In many species of oedipodine grasshoppers: a peculiar sound made by males by snapping their hind wings while they make display flights that females appear to be watching from the ground (Otte 1970 in Wilson 1975, 228). See crepitation.

croak *n.* For example, in male Frogs and Toads when mate calling, young Gharials during hatching from eggs: a hoarse, low-pitched vocalization (Michaelis 1963).

v.i. croak

cuckoo *n.* A Cuckoo's cry (Michaelis 1963). See organism: cuckoo.

[Old French *cucu, coucou,* imitative word]

drone *n.* A continued deep, monotonous sound of humming, or buzzing, of Bees, some fly species, and the bass of a bagpipe (*Oxford English Dictionary* 1972, entries from 1500). See bee: drone.

Note: Some beetle species also make droning sounds when they fly.

v.i. drone

drumming *n.*

1. A male Ruffed Grouse's beating his wings and making a booming sound during a mating display (Wing 1956, 508).

Note: This drumming sounds like a distant outboard motor starting up on a distant lake. It starts off slowly and gains speed until it ends in a whir: "*bup ... bup ... bup ... bup . . bup . bup . up . rrrrr*" (Peterson 1947, 75).

syn. booming (Peterson 1947, 75)

2. A woodpecker's hitting wood with its bill many times (Wing 1956, 508).

syn. tattooing (Wing 1956, 508)

3. In some mammal species: an individual's thumping, or tapping, its paws against an object (Heymer 1977).

v.i. drum

cf. animal sounds: booming

▶ **footdrumming** *n.* In several rodent species, including the Banner-Tailed Kangaroo Rat and Mongolian Gerbil: an individual's tapping its hind feet on the ground which produces a sound (Eisenberg 1963, etc., in Randall 1989, 620).

Comment: Banner-Tailed Kangaroo Rats have individual footdrumming signatures used in communication (Randall 1989, 620).

▶ **marking drumming** *n.* In Hamsters and Mice: an animal's drumming involving its hind paws that it rubbed across its scent glands (Heymer 1977, 129).

cf. scent mark

▶ **paw drumming** *n.* In many rodents: an animal's drumming on the ground during a threat display (Heymer 1977, 129).

fear trill *n.* In some bird species: a call note indicating fright (Wilson 1975, 236).

gecker *n.* In Hamadryas Baboons: a distress sound (Sigg and Falett 1985, 980).

v.i., v.t. gecker

growl *n.*

1. For example, in Lions, Tigers, Wolves, Dogs, Bears, and some bird species: an individual's deep, "angry," guttural sound (*Oxford English Dictionary* 1972, entries from 1727).
2. A person's utterance of anger or dissatisfaction (*Oxford English Dictionary* 1972, entries from 1821).

v.i., v.t. growl

grunt *n.*

1. A person's deep, guttural sound, sometimes expressive of approbation, or the opposite (*Oxford English Dictionary* 1972, entries from 1553).

2. In Hogs, some Domestic Dogs: a deep, guttural sound (*Oxford English Dictionary* 1972, entries from 1615).

cf. grunt

▶ **baby-grunt, cough-grunt, chortle, gurgle** *n*. For example, in the Rhesus Monkey: a vocalization that is given specifically to newborns (de Waal 1989, 107).
[chortle, blend of chuckle and snort; coined by Lewis Carrol in *Through the Looking-Glass*]

grunt whistle *n*. For example, in drakes (Anatidae): a sharp whistling sound followed by a grunt, both caused by the release of compressed air during neck bending while the animal is rising out of water (Lorenz 1941 in Heymer 1977, 79).

hoot *n*.
1. An owl's cry (Michaelis 1963).
2. An animal species' cry that is similar to an owl's cry.
3. An animal's, or other thing's, making a sound like, or similar to, an owl's cry (Michaelis 1963).
4. A person's jeer or mocking with derisive cries (Michaelis 1963).
5. A person's driving off another person with hoots (Michaelis 1963).
6. A person's expressing (disapproval, scorn, etc.) by hooting (Michaelis 1963).
cf. animal sounds: yowl
Comments: Chimpanzees give a long-distance call which is a slowly swelling hoot (de Waal and Lanting 1997, 24)
[Scandinavian; *cf.* Swedish *huta*]

howl *n*.
1. In some Domestic Dogs, Wolves, the Coyote: a prolonged, mournful cry (*Oxford English Dictionary* 1972, entries from 1605).
2. A person's loud wail or outcry of anguish; savage yell of rage or disappointment (*Oxford English Dictionary* 1972, entries from 1599).
Note: Persons, especially adolescent males in the U.S., sometimes howl, or bark (woof), when viewing a member of the opposite sex who is attractive, or unattractive, to them.
cf. animal sounds: yowl

kiss-squeak *n*. An Orangutan vocalization given by an irritated animal (Galdikas 1995, 15).

laugh, laughing, laughter *n*.
1. A person's characteristic explosive or inarticulate sounds, facial expressions, and other physical manifestations expressive of merriment, elation, amusement, satisfaction, etc. (*Oxford English Dictionary* 1971, entries from 611).

Notes: During playing, human laughing may indicate friendly intention or mild aggression (Heymer 1977, 109). Laughing (involving particular facial expressions and sounds) occurs in Bonobos (de Waal 1989, 219, 225). Human laughing may be a form of ritualized aggression (Eibl-Eibesfeldt 1972, 1973 in Heymer 1977, 109).
2. In nonhuman animals: an animal's sound that resembles or is suggestive of human laughter — *e.g.,* in the Laughing Bird (= Green Woodpecker), Laughing Crow, Laughing Dove, Laughing Goose (= White-Fronted Goose), Laughing Gull, Laughing Jackass (= Kookaburra Bird), Laughing Owl, Laughing Thrush, and Spotted Hyena (= Laughing Hyena) (Michaelis 1963; *Oxford English Dictionary* 1989); an alarm call; "a hoarse, rhythmic *hahaha!-hahahaha!*" in Herring Gulls (Tinbergen 1958 in Heymer 1977, 109; Tinbergen 1961, 11).
[Old English *hliehhan, hlæhhan*]

lip smacking, lipsmacking *n*.
1. In more derived primates: an all-purpose conciliatory greeting that consists of rapidly repeated sucking movements, typified by the Yellow Baboon, *Papio cynocephalus* (Wilson 1975, 227).
2. For example, in the Rhesus Monkey, Stump-Tailed Monkey: an individual's series of rapid lip and tongue movements accompanied with brief glances at its partner (de Waal 1989, 106).
Comment: Rhythmic lip smacking commonly occurs during grooming but also occurs when primates are apart from one another. Lip smacking with raised eyebrows signals "friendly intentions" (de Waal 1989, 106).

murmuration *n., pl.* **murmurations** Humans' murmuring; utterance of low continuous sounds; grumbling, complaining (*Oxford English Dictionary* 1972, entries from 1386). See ²group: murmuration.

piping *n*. A sound made by new queens within the same Honey Bee colony (Wilson 1975, 141).

quack, quacking *n*.
1. A person's state of hoarseness, or croaking, in his throat; a now rarely used word (*Oxford English Dictionary* 1972, entries from 1386).
2. In some duck species: a characteristic harsh vocalization; a sound that resembles or imitates this (*Oxford English Dictionary* 1972, verb entries from 1617).
3. In some animals species: a harsh sound; a noisy outcry (*Oxford English Dictionary* 1972, entries from 1624).

a – c

4. In some frog species, the Raven: a croak (*Oxford English Dictionary* 1972, entries from 1727 as a verb).

quacking. *n.* A sound exchanged between rival Honey Bee queens in the same colony (Wilson 1975, 141).

scream *n.*

1. An animal's shrill cry or any similar sound (*Oxford English Dictionary* 1972, entries from 1513); *e.g.,* in the Hamadryas Baboon and Humans, a high-pitched distress sound (Sigg and Falett 1985, 980).

2. A person's shrill, piercing cry, usually expressive of pain, alarm, or other sudden emotion (*Oxford English Dictionary* 1972, entries from 1605).

v.i., v.t. scream

cf. animal sounds: call: yowl

screech *n.*

1. A person's loud, shrill cry, usually expressive of violent and uncontrollable pain or alarm (*Oxford English Dictionary* 1972, entries from 1560).

2. In Screech Owls: a shrill, harsh cry; "shriek" (Michaelis 1963).

   Note: Other birds named for their screeching include Screech Thrush, Screech-Martin, and Screech-Hawk (*Oxford English Dictionary* 1972).

v.i., v.t. screech

cf. animal sounds: call, scream

snort *n.* For example, in the White-Tailed Deer: an individual's relatively high-pitched auditory signal in which it forcibly discharges air through its mouth and nostrils (Richardson et al. 1983 in LaGory 1987, 20).

Comment: Snorting is part of a deer's alarm behavior and indicates to a predator the deer has detected it (LaGory 1987, 20). White-Tailed Deer snort at Humans in their ranges (personal observation).

song *n.*

1. A person's act, or art, of singing; the result, or effect, of this; vocal music; that which is sung, in a general or collective sense; occasionally poetry (*Oxford English Dictionary* 1972, entries from *ca.* 1063).

2. In some bird species: a "musical utterance;" also the cry of the Sea-Hawk and Eagle (*Oxford English Dictionary* 1972, entries from *ca.* 1000).

3. An animal's elaborate vocal signal (Lincoln et al. 1985).

4. In many insect, fish, and bird species; Frogs; Toads; primates, including the Gibbons, the Howler Monkey, and Humans: an individual's sound production (vocal and other) of length and complexity that is distinguished from a call, which is shorter and simpler (Immelmann and Beer 1989, 281).

cf. animal sounds: call

Comments: The song of the humpback whale, *Megaptera novaeangliae,* is possibly "the most elaborate *single* display known in any animal species" (Wilson 1975, 220, 478, based on Schevill and Watkins 1962, Schevill 1964, Payne and McVay 1971; D.B. Quine, personal communication). "Song" and "call" overlap and can be difficult to differentiate, at least morphologically (Immelmann and Beer 1989, 281).

▶ **antiphonal song, antiphonal singing** *n.* In some cricket and bird species and some mammal species (Gibbons, Humans, some squirrels): a song in which two individuals sing parts alternately (Immelmann and Beer 1989, 17).

cf. animal sounds: song: duet

▶ **calling song** *n.* For example, in Crickets that produce sound by wing rubbing: a male's song that attracts females (McFarland 1985, 44).

cf. animal sounds: call

▶ **chorus** *n.* The sound of a group of calling insects, frogs, or toads (Wilson 1975, 443, 580; Brush and Narins, 1989, 33).

cf. animal sounds: song: communal song; ²group: chorus

v. chorus.

dawn chorus *n.* In many passerine bird species: an early-morning peak in singing (Mace 1986, 621).

▶ **communal song** *n.*

1. A number of conspecific animals' singing together (Immelmann and Beer 1989, 50).

2. In many frog species, "The collective production of trains of calls" (Immelmann and Beer 1989, 50).

syn. group song, chorusing (Immelmann and Beer 1989, 50); group calling (Schwartz 1991, 565)

cf. animal sounds: song: chorus; behavior: marking behavior

▶ **courtship song** *n.* For example, in many orthopteran, homopteran, *Drosophila,* and bird species (for example, Estrildine Finches, Bullfinches, Grassquits): a male's song used only for attracting or stimulating a female (Crossley 1986, 1146; Ewing and Miyan 1986, 421; Immelmann and Beer 1989, 64).

▶ **duet, duett** *n.*

1. A musical composition for two human voices or two performers (*Oxford English Dictionary* 1972, entries from 1740); performance of a duet.

2. For example, in some orthopteran-insect and bird species; *Hyla* frogs: calling in which two animals alternate sounds (Goin 1949 in Wilson 1975, 443).

3. For example, in some katydid species, Barbets, Cranes, the Cuckoo Shrike, Field Crickets, Geese, Greves, Kingfishers, Megapode Scrub Hens, *Melidectes* Honeyeaters, Quail, the Sea Eagle, Siamangs, Woodpeckers: "The rapid and precise exchange of notes between two individuals, especially mated birds" (Thorpe 1963a,b, Diamond and Terborgh 1968 in Wilson 1975, 222–223, 583; Wickler 1980 in Immelmann and Beer 1989, 80).

v.t. duet

syn. antiphonal song, duetting, duet singing (Immelmann and Beer 1989, 80)

cf. animal sounds: call, duetting; animal sounds: song: chorus, quartet, trio

Comment: Some researchers believe that duetting keeps mates in contact with one another over long periods (Wilson 1975, 222–223, 583).

counter singing, song dueling *n.* In some cricket and bird species: a duet of an individual and its territorial neighbor (Immelmann and Beer 1989, 81).

pair duet *n.* In some bird species: a highly elaborate duet of a mated pair (Immelmann and Beer 1989, 81).

▸ **functional song, motif song** *n.* In many bird species: the usual song of an adult male (Immelmann and Beer 1989, 298).

cf. animal sounds: song: subsong

▸ **group song** See animal sounds: song: communal song.

▸ **juvenile song** *n.* In many bird species: a young male's subsong given prior to his first utterances of functional song (Immelmann and Beer 1989, 299).

▸ **luring song, soliciting song** *n.* In some insect species: a male's female-attracting song (Immelmann and Beer 1989, 64).

▸ **soliciting song** See animal sounds: song: luring song.

▸ **subsong** *n.* In many bird species: a male's song that is of lower intensity, more variable, and of wider frequency range with longer components and a virtual absence of consistent motifs compared to functional song (Immelmann and Beer 1989, 298–299).

Comment: Depending on species, a male might utter subsongs prior to functional songs or utter them side by side. Subsongs commonly occur in spring and during a short period of resurgent gonad activity

in autumn (autumn song) (Immelmann and Beer 1989).

▸ **territorial song** *n.* Song that advertises an animal's possession of a territory to the same sex (Immelmann and Beer 1989, 309).

▸ **unison-bout singing** *n.* For example, in many frog and insect species: collective calling by groups of males punctuated by variable periods of relative quiet (Schwartz 1991, 565).

Comment: Males might use unison-bout singing to save energy (Schwartz 1991, 565).

stridulation *n.* For example, in many ant, beetle, cricket, grasshopper, mutillid-wasp species: an individual's producing sound by rubbing one part of its body surface against another, which is a common form of communication (Wilson 1975, 595; Borror et al. 1989, 82–83).

v.i. stridulate

Comments: Different species of arthropods rub different anatomical parts together in stridulating; *e.g.,* Leafcutter Ants scrape a ridge on their third abdominal segments against a row of finer ridges on their fourth segments, probably using this sound as a distress signal (Wilson 1975, 211). Males of many species of crickets and grasshoppers stridulate to attract mates. Lewis and Cane (1990, 1003) discuss "aposematic stridulation" and "disturbance stridulation" (= "noncommunicative stridulation").

▸ **substratum stridulation** *n.* For example, in a sawfly species: a larva's scratching the end of its abdomen, which has fine teeth, over a leaf surface and making a sound (Hograefe 1984, 234).

Comment: This sound, audible to Humans, aids in keeping a group of larvae together and in their locating fresh food (Hograefe 1984, 234).

trumpet *n.*

1. For example, in some crane and pelican species, elephants, the Trumpeter Swan: an individual's loud, penetrating sound reminiscent of one from a trumpet, especially the vocalization of an "enraged" or excited elephant (*Oxford English Dictionary* 1972, entries from 1850).

2. In some Gnat and Mosquito species: a shrill hum (made when an individual is about to bite) (*Oxford English Dictionary* 1972, verb entry from 1900).

v.i., v.t. trumpet

cf. animal sounds: call: trumpet call

whistle *n.*

1. A clear, shrill sound that one makes by blowing a whistle or whistling with one's lips (Michaelis 1963).

a – c

2. Any whistling sound, from a missile, the wind, etc. (Michaelis 1963).

3. In *Opsanus tau* (Toadfish): a high-pitched sound made by a male's vibrating its swim bladder (Weiss 1996b, A3). *Notes:* A male uses his whistle as a mate-attraction call. The muscle used to vibrate his swim bladder contracts up to 200 times per second at 60 to 75°F and is the fastest vertebrate muscle known (Weiss 1996b, A3). The fastest vertebrate locomotory muscle (used by a sprinting lizard) vibrates at 35 times per second. A rattlesnake's rattle muscle vibrates at 90 times per second at 95°F and 30 times per second at 60 to 75°F.

v.i., v.t. whistle
cf. animal sounds: grunt whistle

yowl, yawl *n.*

1. A person's crying out loudly (from pain, grief, or distress); a cat's "wauling;" a dog's howling; a Peacock's screaming (*Oxford English Dictionary* 1972, entries from *ca.* 1225).

2. Formerly, an owl's hooting; a dove's cooing (*Oxford English Dictionary*).

v.i., v.t. yowl, yawl
cf. animal sounds: howl
Comment: "Yowling" refers especially to a prolonged, wailing cry (*Oxford English Dictionary* 1972, entries from 1225).

♦ **animal speech** See language.

♦ **animal suffering, suffering** *n.*

1. A person's undergoing, or bearing, pain, distress, or tribulation (*Oxford English Dictionary* 1971, entries from 1597).

2. A person's painful condition; pain suffered (*Oxford English Dictionary* 1971, entries from 1392).

3. In many animal species, including Humans: an individual's wide range of unpleasant "emotional states," *e.g.,* fear, pain, frustration, exhaustion, and loss of a companion (Dawkins 1980, 25). *Notes:* "A major difficulty with any definition of suffering is to decide how much (that is, how intense, or how prolonged) of an unpleasant emotional state constitutes 'suffering'" (Dawkins 1980, 25). The Littlewood Committee (1965 in Rowan 1986, 82) indicates that suffering includes pain, discomfort (indicated, for example, by signs such as torpor or poor condition), and stress (*i.e.,* a condition of tension or anxiety). The Brambell Committee (1965 in Rowan 1986, 82) includes fear, pain, frustration, and exhaustion as some examples of "suffering."

4. A person's "unpleasant emotional response to more than minimal pain and distress" (Kitchen et al. 1987 in DeGrazia

and Rowan; DeGrazia, personal communication).

syn. pain (by some), distress (by some, DeGrazia, personal communication)
cf. anxiety; fear; general-emergency reaction; pain; pleasure; stress; syndrome: fight-or-flight syndrome, general-adaptation syndrome

♦ **animal welfare** *n.* Physical and mental "well-being" of a nonhuman animal while it is alive; some people feel that nonhuman animal welfare does not involve killing a nonhuman animal in question for any reason (Dawkins 1980, 7).

♦ **aniso-** *prefix* Unequal.

♦ **anisogamete** See gamete: anisogamete.

♦ **anisogamy** See -gamy: anisogamy.

♦ **anisohologamy** See -gamy: anisohologamy.

♦ **anisomerogamy** See -gamy: anisomerogamy.

♦ **anisosymbiosis** See symbiosis: anisosymbiosis.

♦ **Anlagen** *n.* An element of a trait that remains discrete in hybrids and separates again in the formation of germ cells of these hybrids (Mendel 1866 in Mayr 1982, 722).

♦ **annual** *n.*

1. A plant species with individuals that last only one season (*Oxford English Dictionary* 1972, entries from 1710).

2. An organism species with a life cycle that lasts one growing season or year (Wilson 1975, 578).

adj. annual
cf. biennial, perennial

♦ **annual periodicity, annual rhythm** See rhythm: annual rhythm.

♦ **annuation** *n.* Fluctuations in an organism's abundance, or behavior, resulting from annual changes in environmental factors (Lincoln et al. 1985).

♦ **anoestrus** See estrus: anestrus.

♦ **anomaly** *n.*

1. Unevenness, inequality, of condition, motion, etc. (*Oxford English Dictionary* 1972, entries from 1571).

2. Irregularity, deviation from the common order, exceptional condition or circumstance; also, a thing that exhibits such irregularity; an anomalous thing or being (*Oxford English Dictionary* 1972, entries from 1664).

3. Deviation from natural order, *e.g.,* an animal with an unusual character such as a bird that cannot fly (*Oxford English Dictionary* 1972, entries from 1646).

4. An observed fact that is difficult to explain in terms of an existing conceptual framework (Lightman and Gingerich 1992, 690).

Note: Anomalies often point to the inadequacy of a current theory and herald a new one (Lightman and Gingerich 1992, 690). [Latin *anomalia* < Greek *anomalia* < *anomalos* < *an-* not + *homalos* even < *homos*, same]

♦ **anonymous DNA** See nucleic acid: deoxyribonucleic acid: anonymous DNA.

♦ **anonymous group** See ²society: open society: anonymous group.

♦ **anonymous monogamy** See mating system: monogamy: anonymous monogamy.

♦ **anonymous society** See ²society: open society: anonymous society.

♦ **anorthogenesis** See -genesis: anorthogenesis.

♦ **anosmatic, anosmic** *adj.* Referring to an animal that has no sense of smell (Lincoln et al. 1985).

♦ **anosmize** *v.t.* To cause an animal (*e.g.,* a lab rat) to lose its olfaction temporarily or permanently by an investigator [coined by Cheal 1975, 3, personal communication]. *syn.* make anosmic [Greek *anōsmē*, without smell + ize, to render

♦ **ANOVA** See statistical test: analysis of variance.

♦ **ant cattle, ant cow** *n.* An insect that is a mutualist with ants and tended by them (*e.g.,* an aphid) (Wilson 1975, 354) or a Caterpillar (Ross 1985). *cf.* honey dew

♦ **ant pollination** See pollination: ant pollination.

♦ **ant rain** *n.*
1. The dropping of newly fertilized queen ants (*e.g., Formica* and *Lasius*) from the sky after their nuptial flights (Haemig 1997, 89).
2. A continuous drift of free-falling ants (*e.g., Formica aquilonia*) from various parts of trees to the ground (Haemig 1997, 89). *Note:* Ant rain in this species increases with the amount of bird foraging in the trees where the ants occur (Haemig 1997, 89). By raining, an ant can quickly move away from a predator in a tree and more quickly move from a foraging site in a tree to the ground.

♦ **ant worker** See caste: worker: ant worker.

♦ **antagonism** *n.*
1. The interference of one chemical (*e.g.,* antiandrogen) with the action of another inside an animal (Immelmann and Beer 1989, 16).
2. One muscle's having the opposite effect (flexing or extending) of its complementary muscle (Immelmann and Beer 1989, 16).

3. An individual animal's opposition between two incompatible behavioral tendencies, such as attacking and fleeing (Immelmann and Beer 1989, 16).

♦ **antagonist** See agonist: antagonist.

♦ **ante-** *prefix*
1. Before in time or order (Michaelis 1963).
2. Before in position; in front of (Michaelis 1963).
[Latin *ante*, before]

♦ **antebrachial gland** See gland: antebrachial gland.

♦ **antenatal** *n.* "Before birth; during gestation" (Lincoln et al. 1985).

♦ **antepisematic character** See character: episematic character: antepisematic character.

♦ **antennation** *n.* In arthropods: an individual's touching another animal or some object with its antennae (Wilson 1975, 578). *Comment:* This behavior can serve as a sensory probe, or as a tactile signal, to an antennated arthropod (Wilson 1975, 578).

♦ **antepisematic character** See character: episematic character.

♦ **anthecology** See study of: ecology: anthecology.

♦ **anthogenesis** See -genesis: anthogenesis.

♦ **anthophile** See ¹-phile: anthophile.

♦ **anthrophobia** See phobia (table).

♦ **anthophyte hypothesis** See hypothesis: angiosperm-origin hypotheses: anthophyte hypothesis.

♦ **anthropic** *adj.* Referring to Humans' influence on something (Lincoln et al. 1985).
eusynanthropic *adj.* Referring to an organism that lives on, or in, human habitations (Lincoln et al. 1985).
exanthropic *adj.* Referring to an organism that lives far from human habitations (Lincoln et al. 1985).
synanthropic *adj.* Referring to an organism that lives close to human habitations (Lincoln et al. 1985).

♦ **anthropocentricity** *n.*
1. A person's regarding Humans as a central fact or aim of the universe, or other system, to which all surrounding facts have reference (*Oxford English Dictionary* 1972, entries from 1863; Michaelis 1963).
2. A person's comparing something else with Humans (Michaelis 1963).
3. A person's interpreting the activities of other organisms with respect to human values (Lincoln et al. 1985).
adj. anthropocentric
cf. zoocentric

♦ **anthropochore** See -chore: anthropochore.

♦ **anthropochorous** See -chorous: anthropochorous.

♦ **anthropogenesis** See -genesis: anthropogenesis.

♦ **anthropogenic** See -genic: anthropogenic.

♦ **anthropology** See study of: anthropology.

♦ **anthropomorphism** See -morphism: anthropomorphism.

♦ **anthropophagite** See cannibal.

♦ **anthropophile** See ¹-phile: anthropophile.

♦ **anthropophily** *n.* A insect's biting a Human or sucking his blood (Lincoln et al. 1985).
cf. ¹-phile: androphile

♦ **anti-** *prefix*
1. Against; opposed to (Michaelis 1963).
2. Opposite to; reverse (Michaelis 1963).
3. Rivaling; spurious (Michaelis 1963).
4. In medicine, counteracting, curative, neutralizing. Usually "ant-" before words beginning with a vowel; sometimes "anth-" before the aspirate in words of Greek formation or analogy (Michaelis 1963).
[Greek *anti*, against]

♦ **antiaphrodisiac** See chemical-releasing stimulus: semiochemical: pheromone: aphrodisiac: antiaphrodisiac.

♦ **anticipatory adjustment** *n.* An animal's response to some aspect of its environment or to its biological clock that predisposes it to climatic changes (McFarland 1985, 288).
cf. ³adaptation: ¹physiological adaptation: acclimation; ²physiological adaptation (comment)

♦ **anticipatory visual illusion** See illusion: anticipatory visual illusion.

♦ **anticryptic** See cryptic: anticryptic.

♦ **antidisplay** See display: antidisplay.

♦ **antidiuretic hormone** See hormone: antidiuretic hormone.

♦ **antigenicity** *n.* Exhibition of sexual dimorphism (Lincoln et al. 1985).

♦ **antigeny** See morphism: dimorphism.

♦ **antihormone** See hormone: antihormone.

♦ **antimone** See chemical-releasing stimulus: semiochemical: allelochemic: antimone.

♦ **antimonotony principle** See principle: antimonotony principle.

♦ **antimorph** See gene: antimorph.

♦ **anting** *n.*
1. A bird's "bathing" in ant nests or swarms, dressing its plumage with crushed ants, placing ants among its feathers, and all apparent substitutes for these actions (Stresemann 1935 in McAtee 1938, 98).
Note: McAtee (1938, 98) suggests that "all apparent substitutes for these actions" should be omitted from this definition.

An early report of anting is by Frazar (1876 in McAtee 1938, 102).
2. In many bird species, including Scarlet Tanagers and Starlings: an individual's seizing one or more ants and placing it in its feathers under its wing or elsewhere; crushing the ant with its bill and rubbing its juices on its feathers; dusting itself in an ant hill; or a combination of these activities (Groskin 1945, 55; 1950, 201).
cf. pharmacognosy: zoopharmacognosy
Comments: Groskin (1950) lists many of the bird and ant species that are involved in anting. Hypotheses for the function of anting include: Anting disinfects a bird with ant exudates that contain acid, the exudates help to rid ectoparasites, and irritants in the exudate produce an "exquisite sensation" on the bird's skin (Groskin 1943, 1950; Immelmann and Beer 1989, 17). Over 200 bird species show anting, but its function is still largely a mystery and a source of controversy. Some birds might ant to prepare ants as food by causing them to discharge their formic acid before they are ingested. Some birds also rub fruit peels, flowers, mothballs, and other items, many of which have antiparasitic properties, on their feathers (Clayton and Wolfe 1993).

♦ **antipathetic symbiosis** See symbiosis: antipathetic symbiosis.

♦ **antiphonal singing, antiphonal song** See animal sounds: song: antiphonal song.

♦ **antipredator aggression** See aggression: antipredator aggression.

♦ **antisocial factor** *n.* "Any selection pressure that tends to inhibit or reverse social evolution" (Wilson 1975, 578).

♦ **antithesis, principle of** See principle: principle of antithesis.

♦ **anxiety** *n.*
1. A person's uneasiness, or trouble of mind, about some uncertain event; solicitude; concern (*Oxford English Dictionary* 1972, entries from 1525).
cf. fear
2. Pathological, a person's condition of agitation and depression with a sensation of tightness and distress in his precordal region (*Oxford English Dictionary* 1972, entries from 1661).
3. A person's strained or solicitous desire (*Oxford English Dictionary* 1972 entries from 1769).
4. A person's diffuse state of tension "…which magnifies and even causes the illusion of an outer danger, without pointing to appropriate avenues of defense or mastery" (Erickson 1950 in DeGrazia and Rowan, personal communication).

syn. fear (used by some authors)
cf. fear, suffering
Comment: Behavior and physiological changes in nonhuman animals suggest that they are anxious (Gray 1982 in DeGrazia and Rowan, personal communication).

♦ **apandry** *n.* An organism's lacking, or having nonfunctional, male reproductive organs (Lincoln et al. 1985).
adj. apandrous

♦ **apeirophobia** See phobia (table).

♦ **aphagia** See -phagia: aphagia.

♦ **aphanic species** See ²species: sibling species.

♦ **aphatetic coloration** See coloration: aphatetic coloration.

♦ **aphidicole** See -cole: aphidicole.

♦ **aphidivore** See -vore: aphidivore.

♦ **aphotometric** *adj.* Referring to organisms, or structures, not affected by light (Lincoln et al. 1985).

♦ **aphototaxis** See taxis: phototaxis: aphototaxis.

♦ **aphrodisiac** *n.* For example, in Humans: a drug, food, etc. that arouses, or increases, sexual desire or potency (Michaelis 1963). See chemical-releasing stimulus: semiochemical: pheromone: aphrodisiac.

♦ **antilogy** See contranym.

♦ **aphydrotaxis** See taxis: hydrotaxis: aphydrotaxis.

♦ **aphyletic trogloxene** See -xene: trogloxene: aphyletic trogloxene.

♦ **apiculture** See study of: apiculture.

♦ **apidology** See study of: mellitology.

♦ **aping** See learning: social-imitative learning.

♦ **apiophobia** See phobia (table).

♦ **apiphobia** See phobia (table).

♦ **apivore** See -vore: apivore.

♦ **aplanogamete** See gamete: aplanogamete.

♦ **apneumone** See chemical-releasing stimulus: apneumone.

♦ **Appalachian Trail** See place: Appalachian Trail.

♦ **apparent competition** See competition: apparent competition.

♦ **apo-** *prefix* Off; from; away (Michaelis 1963). Also, ap- before vowels; aph- before an aspirate.
[Greek *apo*, from, off]

♦ **apocrine sweat gland** See gland: apocrine sweat gland.

♦ **apoendemic** See ¹endemic: apoendemic.

♦ **apogamy** See -gamy: apogamy.

♦ **apolegamy** See -gamy: apolegamy.

♦ **apomict** *adj.* Referring to apomixis, *q.v.*

♦ **apomixis** See reproduction: parthenogenetic reproduction.

♦ **apomorph, apomorphy** See morph: apomorph.

♦ **apomorphic** See -morphic: apomorphic.

♦ **apoptosis** See death: apoptosis.

♦ **aposematic behavior, aposematism** See behavior: aposematic behavior.

♦ **aposematic coloration** See coloration: aposematic coloration.

♦ **aposematic stridulation** See animal sounds: stridulation.

♦ **apostatic selection** See selection: apostatic selection.

♦ **aposymbiotic** *adj.* Referring to a normally symbiotic organism that has been deprived experimentally of its partner (Lincoln et al. 1985).
syn. symbiont-free
cf. symbiosis

♦ **apothermotaxis** See taxis: apothermotaxis.

♦ **apotreptic behavior** See behavior: threat behavior.

♦ **appeasement, appeasement behavior** See behavior: appeasement behavior.

♦ **appeasement display** See display: appeasement display.

♦ **appeasement gland** See gland: appeasement gland.

♦ **appeasement substance** See substance: appeasement substance.

♦ **appeasing signal** See behavior: appeasement.

♦ **appetitive behavior** See behavior: appetitive behavior.

♦ **applied ethology** See study of: ³ethology: applied ethology.

♦ **applied mathematics** See study of: mathematics: applied mathematics.

♦ **applied physics** See study of: physics: applied physics.

♦ **applied psychology** See study of: psychology: applied psychology.

♦ **appraisor** *n.* A component of a signal that allows a responder to react more to one object (or signaler) than to another (Marler 1961 in Wilson 1975, 217).
Comment: Marler's classification of signal components includes "appraisor," "designator," "identifior," and "prescriptor," *q.v.* (Marler 1961 in Wilson 1975, 217).

♦ **apprehending error** See error: observer error.

♦ **approach-approach conflict** See ²conflict: approach-approach conflict.

♦ **approach-avoidance conflict** See ²conflict: approach-avoidance conflict.

♦ **approach-withdrawal theory** See ³theory: approach-withdrawal theory.

♦ **approximation conditioning** See learning: conditioned learning: approximation conditioning.

♦ **aptosochromatosis** *n*. In some bird species: color change without molting (Lincoln et al. 1985).

♦ **apyrene sperm** See gamete: sperm: apyrene sperm.

♦ **Aqua-lung, aqualung** See self-contained underwater breathing apparatus.

♦ **aquatic respiration** See breathing: water breathing.

♦ **aquatic-surface respiration** See respiration: aquatic-surface respiration.

♦ **aquaticole** See -cole: aquaticole.

♦ **aquatosere** See ²succession: aquatosere.

♦ **aquiherbosa** See ²community: aquiherbosa.

♦ **arachneophobia** See phobia (table).

♦ **arachnophobia** See phobia (table).

♦ **Arber's law** See law: Arber's law.

♦ **arboreal** *adj*. Referring to an organism's living in trees; "adapted for life in trees" (Lincoln et al. 1985).

♦ **arboricole** See -cole: arboricole.

♦ **arbovirus** See virus: arbovirus.

♦ **arbusticole** See -cole: arbusticole.

♦ **archegenesis** See -genesis: abiogenesis.

♦ **archetype, archtype** See -type, type: archetype.

♦ **archigenesis, archigony** See -genesis: archegenesis.

♦ **arching** See display: wing display: arching.

♦ **ardium** See ²community: ardium.

♦ **area effect** See effect: area effect.

♦ **area-species curve** See curve: species-area curve.

♦ **arena, arena display** See ²lek.

♦ **arenicole** See -cole: arenicole.

♦ **argillicole** See -cole: argillicole.

♦ **argillophile** See ¹-phile: argillophile.

♦ **argotaxis** See taxis: argotaxis.

♦ **argument from design** *n*. Attributing the creation of an organism's complex character (*e.g.*, behavioral pattern) to a designer (God, according to Paley 1802, or natural selection, according to Darwinists) because it seems ideally suited for the function in question (D.M. Lambert, personal communication).
cf. teleology
Comments: To structuralists, the designer is the laws of form and generative mechanisms (D.M. Lambert, personal communication). Another view is that the argument from design is strictly a creationist concept and the core of creationist polemics (C.K. Starr, personal communication).

♦ **aristogenesis** *n*. One of the autogenetic theories, *q.v.* (Osborn 1934; Simpson 1953).
syn. orthogenesis (Lincoln et al. 1985)
cf. -genesis

♦ **aristogenic** See -genic: aristogenic.

♦ **Aristotelian, Aristotelean** *n*.
1. An adherent of Aristotle's teachings; one who tends to be empirical, or scientific, in his method rather than speculative or metaphysical (*Oxford English Dictionary* 1972, entries from 1607; Michaelis 1963).
2. A maligning term used for botanists in the 18th century that implied that they used a deductive approach and a blind reliance on tradition and authority (Mayr 1982, 165).

♦ **Aristotelian mimicry** See mimicry: Aristotelian mimicry.

♦ **Aristotle's theory of inheritance** See *eidos*.

♦ **arithmetic average, arithmetic mean** See mean.

♦ **arithmetic mimicry** See mimicry: arithmetic mimicry.

♦ **ARM** See mechanism: releasing mechanism.

♦ **arms race** See ²race: arms race.

♦ **aromorphosis** See ²evolution: aromorphosis.

♦ **arousal** *n*.
1. A person's stirring up another person from sleep or inactivity (*Oxford English Dictionary* 1972, entries from 1593).
2. A person's awakening himself from sleep (*Oxford English Dictionary* 1972, entry from 1822).
3. An individual animal's general state of excitability or activation (Immelmann and Beer 1989, 20).
syn. drive (sometimes)
cf. drive
Comment: Specific examples of arousal include: an animal's transition from sleep to wakefulness; its level of responsiveness, as indicated by stimulus intensity necessary to elicit its reaction; its activation level, as indicated by its kind of exhibited behavior (*e.g.*, relaxed grooming, frantic fleeing, etc.); its physiological indicators, such as heat rate and skin conductances, as measured in lie-detector tests; and its attentiveness to sensory stimuli (Immelmann and Beer 1989, 20).

♦ **arquetype** See -type: archetype.

♦ **arrestant** See chemical-releasing stimulus: arrestant.

♦ **arrested evolution** See ²evolution: arrested evolution.

♦ **arrhenogenic** See -genic: arrhenogenic.

♦ **arrhenotoky** See parthenogenesis: arrhenotoky.

♦ **arrhythmic** *adj*. Displaying no diurnal periodicity.
syn. nonrhythmic
cf. rhythm

♦ **arrow greeting** See greeting: arrow greeting.

♦ **Arrow's impossibility theorem** *n.*
Optimum socioeconomic systems can never
be perfect (Arrow in Wilson 1975, 575).

♦ **Artenkreis, Artenkreise** See ²species:
superspecies.

♦ **arthrogenous evolution** See ²evolution: arthrogenous evolution.

♦ **Arthropods of La Selva (ALAS)** *n.* An
effort to inventory all of the arthropods of
the La Selva Biological Station, Costa Rica
(Yoon 1995, C4).

♦ **artificial chromosome** See vector: artificial chromosome.

♦ **artificial intelligence** See intelligence:
artificial intelligence.

♦ **artificial parthenogenesis** See parthenogenesis: artificial parthenogenesis.

♦ **artificial selection** See selection: artificial selection.

♦ **AS** See sleep: stage-1 sleep.

♦ **ascaphus reflex** See thanatosis.

♦ **Aschoff's rule** See rule: Aschoff's rule.

♦ **asexual** See sexual: asexual.

♦ **asexual meiotic parthenogenesis**
See parthenogenesis: asexual meiotic parthenogenesis.

♦ **asexual reproduction** See parthenogenesis: asexual reproduction.

♦ **aspect** *n.* The mean direction that a sloping land area (*e.g.,* a watershed) faces
(inferred from Edwards 1995, 7).

♦ **ASR** See respiration: aquatic-surface respiration.

♦ **assembly** *n.* One or more member's calling together other members of its society
for any general communal activity (Wilson
1975, 211, 415, 579).
cf. ¹,²recruitment

♦ **assessment signal** See signal: assessment signal.

♦ **associate** See helper.

♦ **association** *n.* An animal's forming a
learned connection between a stimulus and
a response or between two stimuli (Immelmann and Beer 1989, 21).
See ²group: association.
cf. learning: associative learning

♦ **association element** See ²species: association element.

♦ **associationism** *n.* The view that organisms do not have any freedom of will, and
behavior cannot necessarily be accounted
for in physical or physiological terms
(McFarland 1985, 362).
cf. instinct, materialism, rationalism, *tabula
rasa*
Comment: The empiricists John Locke (1700)
and David Hume (1739) held the view that
human behavior develops entirely through
experience, according to laws of association
(McFarland 1985, 362).

♦ **associative learning** See learning: associative learning.

♦ **assortative mating** See mating: assortative mating.

♦ **assortative mating by size** See mating: assortative mating: assortative mating
by size.

♦ **assortative pair formation** See mating: assortative mating.

♦ **asteroid** *n.* A small rocky object in our
Solar System (Mitton 1993).
Comments: Asteroids may have formed 4.6
billion years ago when our Solar System
formed, and they might be parent bodies of
meteorites (Mitton 1993). Astronomers classify them based on their reflected spectra.

♦ **astraphobia** See phobia (table).

♦ **astrophobia** See phobia (table).

♦ **asymmetric game** See game: asymmetric game.

♦ **asymptotic species-accumulation
curve** See curve: species-accumulation
curve: asymptotic species-accumulation curve.

♦ **asynethogametism** See synethogametism: asynethogametism.

♦ **atavism** *n.*
1. A person's resembling his grandparents,
   or more remote ancestors, rather than
   parents (*Oxford English Dictionary* 1972,
   entry from 1833).
2. The sporadic occurrence of individuals of
   a species with characteristics of phylogenetically ancestral forms; *e.g.,* a second or
   third hoof on the leg of a perissodactyl, a
   second pair of wings on a true fly (Diptera),
   or display of behaviors of ancestral species in species hybrids of some ducks and
   some parrots (*Oxford English Dictionary*
   1972, entry from 1872; Immelmann and
   Beer 1989, 22).
3. The reappearance of a characteristic in
   an organism after several generations of
   absence, usually caused by the chance
   recombination of genes (American Heritage Dictionary 1992 in Hrdy 1996, 851).
See gene conversion.
syn. reversion (Lincoln et al. 1985)
cf. ²evolution: regressive evolution; -gen-
esis: biogenesis; relict; speciation: saltational
speciation

♦ **atokous** *adj.* Without offspring; nonreproductive; vegetative (Lincoln et al. 1985).
cf. -otoky

♦ **atomistic mechanist** See mechanist:
atomistic mechanist.

♦ **atrophic** See -trophic: atrophic.

♦ **attachment** *n.*
1. The fact, or condition, of a person's being
   attached to another person by sympathy;
   "affection, devotion, fidelity" (*Oxford English Dictionary* 1972, entries from 1704).

2. Relationships between one person and another perceived as stronger, or wiser, than himself, irrespective of their age [coined by Bowlby 1958, 1969 in Ainsworth and Bell 1970, 50; Hinde 1982, 267].

3. An affectional tie that one animal forms between itself and another specific one which binds them together in space and endures over time (Ainsworth and Bell 1970, 50).

   Notes: The behavioral hallmark of attachment is seeking to gain and to maintain a certain degree of proximity to the object of attachment, which ranges from close physical contact under some circumstances to interaction or communication across some distance under other circumstances (Ainsworth and Bell 1970, 50). An attachment relationship "is developed and mediated through an organized hierarchical behavioral system, the attachment system, which encompasses infant attachment behaviors and maternal behaviors, both designed to promote mother-infant proximity" (Mann 1991, 1).

4. An aspect of the enduring relationship between a human infant and its attachment figure (Hinde 1982, 228).

5. "A tie between an animal and some place, object, or social companion;" *e.g.,* a retreat, a nest site, a breeding territory, a specific habitat used seasonally (Immelmann and Beer 1989, 22).

cf. behavior: attachment behavior; learning: imprinting: place imprinting

mother-infant attachment *n.* In mammals, most markedly in primates: an especially strong bond between a mother and her young (Immelmann and Beer 1989, 190).

cf. affiliation

♦ **attachment behavior** See behavior: attachment behavior.

♦ **attachment-behavior system** See [1]system: attachment-behavior system.

♦ **attack behavior** See behavior: attack behavior.

♦ **attack-cone avoidance** See behavior: inspection behavior: attack-cone avoidance.

♦ **attend** *v.t.* "To give any response whatsoever to a stimulus [this term defines what is probably the broadest possible class of behavior]" (Verplanck 1957).

♦ **attendant** *n.* A bird that takes care of young in a nest, including the young's parents and helpers (Dow 1980 in Brown 1987, 297).

♦ **attendant species** See [2]species: attendant species.

♦ **attention** *n.*

1. A person's act, or state, of heeding, directing his mind toward, or regarding something (*Oxford English Dictionary* 1972, entries from 1374).

2. An individual animal's "reification, as faculty or process, of attending" (Verplanck 1957).

3. An individual animal's directing its interest, or concern, toward a particular object at a particular moment (Storz 1973, 29; Immelmann and Beer 1989, 22–23).

selective attention *n.* Usually, an organism's limiting its responsiveness such that some stimuli control its behavior over a considerable period and do so to the exclusion of other stimuli that simultaneously impinge on it (Hinde 1970, 125).

♦ **attention structure** *n.*

1. A conceptualization of individual social fields, *q.v.,* as they relate to a whole society (Chance 1967, Chance and Jolly 1970 in Wilson 1975, 517).

2. The directions of regard among animals in a group in relation to their social statuses and spatial positions (Immelmann and Beer 1989, 23).

cf. [2]society: acentric society, centripetal society

♦ **attractant** See chemical-releasing stimulus: attractant.

♦ **attractivity, female attractivity** *n.*

1. A female's stimulus value in evoking a male's sexual responses (Beach 1976 in Dewsbury 1978, 237).

2. The full range of stimuli (behavioral as well as nonbehavioral) produced by a female, from those that attract males from a distance to those that promote ejaculation (Hinde 1982, 155).

cf. proceptivity, receptivity

♦ **aucuparious** *adj.* "Attractive to birds" (Lincoln et al. 1985).

♦ **audience-effected social facilitation** See [2]facilitation: social facilitation: audience-effected social facilitation.

♦ **audiogenic** See -genic: audiogenic.

♦ **audition** See modality: audition.

♦ **aufwuchs** See [2]community: periphyton.

♦ **August-Krogh principle** See principle: August-Krogh principle.

♦ **aunt, auntie** *n.*

1. The sister of a person's father or mother (*Oxford English Dictionary* 1972, entries from 1297).

2. A woman who does benevolent, practical things for her acquaintances (*Oxford English Dictionary* 1972, entries from 1861).

3. A nonreproducing female Eider Duck that joins another female with her brood (Darling 1938, 5).

4. In Rhesus Monkeys: a female, other than an infant's mother, that helps to care for the infant (Rowell et al. 1964).

5. In some primate species and Elephants: any female that assists a parent in caring for conspecific young (Wilson 1975, 125, 349, 491, 579).
See mother: play mother.
cf. parent: alloparent; uncle
♦ **aural** *adj.* Referring to hearing or ears (Lincoln et al. 1985).
♦ **Aurignacian Tool Industry** See tool industry: Aurignacian Tool Industry.
♦ **Außendienst** *n.* In social insects: an individual's work outside of her nest (Wilson 1975, 412).
cf. Innendienst
[German *außen*, outside; *dienst*, duty]
♦ **autallogamy** See -gamy: autallogamy.
♦ **autapomorph** See character: apomorph: autapomorph; -morph: autapomorph.
♦ **autarkean society** See ²society: autarkean society.
♦ **autecology** See study of: ecology: autecology.
♦ **author** *n.*
1. "The original writer of a book, treatise, etc." (Michaelis 1963).
2. "One who makes literary composition his profession" (Michaelis 1963).
3. The name of a person who is responsible both for the name of a taxon and for its diagnostic description (Mayr 1969, 317, 363, who discusses exceptions to this kind of authorship).
4. A person who created a work, scientific name, nomenclatural act, or a nominal taxon (Lincoln et al. 1985).
5. A person who contributes to research that is reported in a scientific paper (Higley and Stanley-Samuelson 1993, 74).
Note: Higley and Stanley-Samuelson (1993, 75) discuss trading authorship, authorship as a payoff for resources, coerced authorship, adding a luminary to the author list to enhance chances of publication, and using an authorship to buy a competing group out of a research area. They conclude that "authorship is an exercise in interpretation, reflection, creativity, written artistry, and, perhaps most importantly, responsibility" (p. 75).
[Anglo-French *autour*, Old French, *autor* < Latin *auctor*, originator, producer < *augere*, to increase]
♦ **autism** *n.*
1. A person's "tendency toward day-dreaming or introspection in which external reality is unduly modified by wishful thinking" (Michaelis 1963).
Note: This definition is not widely used today.
2. A child's active withdrawal from relationships with other people, such as occurs in schizophrenia (Clayman 1989, 145).

Note: This definition is not widely used today (Clayman 1989, 145).
3. A child's condition that involves failure to form relationships with other people and other traits (Clayman 1989, 145).
Notes: Autism is evident before the age of 30 months and is usually apparent within a person's first year of life (Clayman 1989, 145). Many autism symptoms appear throughout much of a person's life. These symptoms include avoiding eye-to-eye contact; becoming attached to unusual objects or collections; being aloof from parents and others; being hyperactive; being hypersensitive to touch; being indifferent to feelings of others and to social conventions; being obsessed with one particular topic or idea; delaying speaking as a toddler; developing rituals in play; displaying sudden screaming fits; echoing spoken words; extremely resisting change of any kind; failing to form relationships with other Humans; flicking or twiddling fingers for hours on end; having a heightened power of concentration; having a heightened visual perception; having an isolated special skill such as drawing, rote memory, or musical ability; having a normal appearance and muscular coordination; making up words; preferring to play alone; self-injuring oneself; showing difficulty in learning manual tasks; showing unusual fears; resisting being cuddled with screaming to be put down when picked up, even if one is hurt or tired; responding to sounds inappropriately; rocking; using immature, unimaginative speech; and walking on tiptoes. Some victims of autism are unable to speak or function in usual ways (Raver 1997, C1). Temple Grandin, an animal behaviorist, uses her autism-affected perceptions in designing livestock facilities and making livestock slaughter less traumatic for the animals (Raver 1997, C1).
4. A person's being shut off to varying extents from social contact with other people and showing symptoms such as persistent stereotyped movements and reduction of sensory responsiveness (Immelmann and Beer 1989, 23).
Note: Captive animals sometimes display behavior similar to autism (Immelmann and Beer 1989, 23).
♦ **auto-** *combining form*
1. Arising from some process, or action, within an object; not induced by any stimulus from without (Michaelis 1963).
2. Acting, acted, or directed upon the self. Before vowels as aut- (Michaelis 1963).
[Greek *autos*, self]

♦ **autoallogamy** See -gamy: autoallogamy.
♦ **autobiology** See study of: biology: idiobiology.
♦ **autobiosphere** See -sphere: biosphere: eubiosphere: autobiosphere.
♦ **autocatalysis** *n.* Any process whose rate is increased by its own products (Wilson 1975, 567, 579).
Comment: Thus, autocatalytic reactions, fed by positive feedback, tend to accelerate until their ingredients are exhausted or some external constraint is imposed (Wilson 1975, 567, 579).
♦ **autocatalysis model** See [4]model: autocatalysis model.
♦ **autochore** See -chore: autochore.
♦ **autochthonous** *adj.* Referring to a group, or species, that is native, or aboriginal, to a particular geographic area; contrasted with a group, or species, considered to have immigrated from outside the area (Tillyard 1926, etc. in Frank and McCoy 1990, 2).
syn. indigenous, native (Frank and McCoy 1990, 4)
cf. adventive, [2]endemic, immigrant, indigenous, introduced, precinctive
♦ **autochthonous behavior, autochthonous activity** See behavior: autochthonous behavior.
♦ **autochthonous drive** See drive: autochthonous drive.
♦ **autocoprophagous, autocoprophagy, autocoprophagous** See -phagy: autocoprophagy.
♦ **autodeme** See deme: autodeme.
♦ **autodiploidy** See -ploidy: autodiploidy.
♦ **autoecology** See study of: ecology: autecology.
♦ **autogamy** See -mixis: automixis
♦ **autogenesis** See -genesis: abiogenesis.
♦ **autogenetic theories** See [3]theory: autogenetic theories.
♦ **autogenous** See -genous: autogenous.
♦ **autogeny** See -genesis: autogenesis.
♦ **autogony** See -genous: autogenous.
♦ **autogrooming** See grooming: autogrooming.
♦ **autohyperparasitism** See parasitism: hyperparasitism: intraspecific tertiary hyperparasitism.
♦ **autoinstruction** See learning: programmed learning.
♦ **automarking** See behavior: marking behavior.
♦ **automated teaching** See learning: programmed learning.
♦ **automatic ovulator** See ovulator: spontaneous ovulator.
♦ **automatic selection** See selection (table).
♦ **automictic parthenogenesis** See parthenogenesis.

♦ **automimicry** See mimicry: automimicry.
♦ **automixis** See -mixis: automixis.
♦ **autonomic nervous system** See [2]system: nervous system: autonomic nervous system.
♦ **autonomic temperature regulation** See temperature regulation: autonomic temperature regulation.
♦ **autonomous behavior** See behavior: autonomous spontaneous behavior.
♦ **autonomy** See need: autonomy.
♦ **autoparasitism** See parasitism: autoparasitism.
♦ **autoparthenogenesis** See parthenogenesis: autoparthenogenesis.
♦ **autophage** See -phage: autophage.
♦ **autophagous, autophagy** See -phagy: autophagy.
♦ **autophile** See [1]-phile: autophile.
♦ **autophobia** See phobia (table).
♦ **autoploidy** See -ploidy: autoploidy.
♦ **autopoiesis** *n.* A cell's, or organism's, metabolic self-maintenance with the use of carbon and energy sources (Margulis et al. 1985, 69).
adj. autopoietic
Comment: "Autopoiesis" is characteristic of all cells but not of viruses or plasmids (Varela and Maturana 1974 in Margulis et al. 1985, 69).
♦ **autopolyploidy** See -ploidy: autopolyploidy.
♦ **autoradiograph** See graph: autoradiograph.
♦ **autoselection** See selection: autoselection.
♦ **autosex** *v.t.* A person's using sex-linked morphological characters to identify the sex of an immature organism before the development of adult sexual dimorphism (Lincoln et al. 1985).
♦ **autoshaping** See shaping: autoshaping.
♦ **autosome** See chromosome: autosome.
♦ **autotherm** See -therm: autotherm.
♦ **autothysis** See behavior: defensive behavior: autothysis.
♦ **autotilly** See autotomy.
♦ **autotomy** *n.* An animal's, or colony's, shedding part of its whole; *e.g.,* papillae of Nudibranch Molluscs; legs of Daddylonglegs, Crabs, and Spiders; arms of Starfish; tails of some lizard and snake species, and zooids of colonial organisms (Fabre 1928, etc. in Formanowicz 1990, 400, who lists review articles on autotomy in some taxa; Matthews and Matthews 1978, 341).
syn. autotilly (Lincoln et al. 1985), autotomization (Halliday and Adler 1986, 89)
v.t. automize
Comments: In some crab species, individuals autotomize a particular leg if it is pulled by the individual itself or an outside force;

in other species, an autotomizer muscle can shed a leg without its being pulled (Barnes 1974, 615).
[*auto* + Greek *tomē*, a cutting < *temnein*, to cut]

♦ **autotroph** See -troph-: autotroph.

♦ **autozygous** See -zygous: autozygous.

♦ **autumn song** See animal sounds: song: subsong.

♦ **autumnal equinox** See equinox: autumnal equinox.

♦ **auxesis** *n.* Growth, sometimes used in a restricted sense to refer to the special case of "isauxesis" (Lincoln et al. 1985). See auxesis: heterauxesis: isauxesis; growth.

heterauxesis *n.*

1. The differential growth of body parts (x and y), expressed by the equation $y = bx^a$, where a and b are fitted constants (Huxley 1932 in Futuyma 1986, 413). *Note:* Heterauxesis is important in the differentiation of castes of the social insects, especially ants (Wilson 1975, 578). "Large ants (but small humans) tend to have relatively large heads; the head grows at a different rate from the body as a whole" (Dawkins 1982, 283). *syn.* allometry, canalization (sometimes), developmental canalization (sometimes), differential relative growth, heterogony (sometimes)

2. Change of shape, or proportion, with increase in size (Lincoln et al. 1985). *syn.* allometry, heterogony

   ▸ **allomorphosis** *n.* Individual, population, specific, or phylogenetic heterauxesis (Lincoln et al. 1985).

   ▸ **bradyauxesis** *n.* Heterauxesis in which a, from the equation $y = bx^a$, is less than unity so that a given structure is relatively smaller in large individuals than in small ones (Lincoln et al. 1985).

   ▸ **isauxesis, isometric growth, isometry, auxesis** *n.* Heterauxesis in which a, from the equation $y = bx^a$, equals unity and in which no change in shape occurs with increasing size (Lincoln et al. 1985).

   ▸ **monophasic allometry** *n.* "Polymorphism in which the allometric regression line has a single slope" (Wilson 1975, 589). *Comment:* In ants, "monophasic allometry" also implies that the relation of some of the body parts measured is nonisometric (Wilson 1975, 589).

   ▸ **tachyauxesis** *n.* Heterauxesis in which a, from the equation $y = bx^a$, is greater than unity so that a given structure is relatively larger in large individuals than in small ones (Lincoln et al. 1985).

♦ **auxiliary** *n.* In social insects, especially in Bees, Wasps, and Ants: a female that associates with other females of the same generation and becomes a worker (Wilson 1975, 579). See helper. *cf.* caste: worker [Latin *auxiliarius* < *auxilium*, help < *augere*, to increase]

♦ **auxiliary character** See character: auxiliary character.

♦ **auxin** See hormone: auxin.

♦ **auxins** See hormone: auxins

♦ **average** See mean.

♦ **aversion-induced aggression** See aggression: aversion-induced aggression.

♦ **avian Davian behavior** See behavior: Davian behavior: avian Davian behavior.

♦ **aviary** See ²group: aviary.

♦ **Avida** *n.* A flexible computer platform for research on artificial organisms (Adami 1998 in Lenski et al. 1999, 661; http://www.krl.caltech.edu/avida/).

♦ **avifauna** See fauna: avifauna.

♦ **aviphobia** See phobia (table).

♦ **avoidance, gradient of** *n.* An animal's strength is inversely proportional to the animal's distance from its goal box; its strength is measured as the animal's running, or pulling, away from its goal box and is shown when an investigator places the animal in a series of positions in a runway after the animal has been repeatedly presented with a strongly aversive stimulus (*e.g.,* intense electric shock) in the goal box (Verplanck 1957).

♦ **avoidance-avoidance conflict** See ²conflict: avoidance-avoidance conflict.

♦ **avoidance conditioning, avoidance learning** See learning: avoidance conditioning.

♦ **avoidance learning** See learning: avoidance learning.

♦ **avoiding reaction** See kinesis: klinokinesis.

♦ **award** *n.* A badge, citation, medal, money, etc. that a person receives by winning a contest or passing an examination (Michaelis 1963). *Comment:* Below, I define some of the many prizes given for promise, work, or both in science and related fields.

Lasker Medical Research Awards *pl. n.* Awards, established by Albert and Mary Woodard Lasker in 1945, made to physicians, public servants, and scientists whose accomplishments have made major advances in the understanding, diagnosis, prevention, treatment, and even cure of many of the great crippling and killing diseases (www.laskerfoundation.org/intro, 2000).

syn. America's Nobels (www.laskerfoun dation.org/intro, 2000)

Comments: The Laskers sought to raise public awareness of the value of biomedical research to a healthy society (www. laskerfoundation.org/intro, 2000). An award recipient receives an honorarium, a citation highlighting achievements, and an inscribed statuette of the Winged Victory of Samothrace, the Foundation's symbol of victory over disability, disease, and death.

Japan Prize   *n.* A prize, first awarded in 1985, given to scientists and researchers who have made original and outstanding achievements in science and technology (www. meshnpt.or.ip/jst/ prize).

Comments: The Japan Prize includes 50 million yen, a certificate, and a medal (www. meshnpt.or.ip/jst/prize). Among the biologists who have won this award are Robert C. Gallo, Luc Mastagnier, Kary B. Mullis, and Frank Sherwood Rowland.

MacArthur Prize   *n.* A prize, first awarded in 1981 and given by the MacArthur Fellows Program of the John D. and Catherine T. MacArthur Foundation to exceptionally creative individuals, regardless of of their fields of endeavor (www. macfdn.org).

syn. genius grant, MacArthur-Foundation "Genius" Award (Rimer 1996, B7)

Comments: The MacArthur Prize is an unrestricted 5-year grant of up to $350,000, which an individual winner may spend in any way that he chooses (www. macfdn. org). Biologists who have won this prize include Eric L. Charnov, Philip James DeVries, Jared M. Diamond, Paul R. Ehrlich, Michael Ghiselin, Stephen Jay Gould, Daniel H. Janzen, Richard E. Lenski, Jane Lubchenco, Barbara McClintock, Nancy A. Moran, George F. Oster, Roger S. Payne, Naomi Pierce, Margie Profet, Peter H. Raven, Alan Walker, Carl. R. Woese, and Richard Walter Wrangham.

National Medal of Science *n.* An award, enacted by the U.S. Congress in 1959 and given to scientists and engineers who have made a large impact on biology, behavior or social science, engineering, mathematics, or chemistry (www. asee.org/ hstmf/html/medals, 1998).

Comments: This medal is the highest honor bestowed by the President of the U.S. to a leading scientist. Biologists who have won this prize include Seymour Benzer, Carl Djerassi, Theodosius Dobzhansky, Thomas Eisner, Harry F. Harlow, G. Evelyn Hutchinson, Edwin H. Land, Ernst Mayr, Barbara McClintock, Ruth Patrick, Linus Pauling, Wendell L. Roelofs, James D. Watson, and Edward O. Wilson.

National Medal of Technology   *n.* An award, enacted by the U.S. Congress in 1980 and first awarded in 1985, whose primary purpose is to recognize technological innovators who have made lasting contributions to enhancing the U.S.'s competitiveness and standard of living (www. asee.org/hstmf/html/medals, 1998).

Comments: This award "highlights the national importance of fostering technological innovation based upon solid science, resulting in commercially successful products and services" (www. asee.org/hstmf/ html/medals, 1998). This medal is the highest honor bestowed by the President of the U.S. to a leading innovator. The President awards one annually "to individuals, teams, or companies for accomplishments in the innovation, development, commercialization, and management of technology, as evidenced by the establishment of new or significantly improved products, processes, or services." Awardees include William H. Gates III and Edwin H. Land.

Nobel Prize   *n.* A prize, first awarded in 1901, given for exceptional research in the fields of chemistry, literature, peace, physics, and physiology or medicine by the Nobel Foundation (www. nobel.se, 1998).

Comments: Karl von Frisch, Konrad Lorenz, and Nikolass Tinbergen shared the Nobel Prize in 1973 in Physiology or Medicine for their discoveries concerning organization and elicitation of individual and social behavior. Other biologists who have won this prize include Francis Harry Compton Crick, Karl von Frisch, Walter Rudolf Hess, Schack August Steenberg Krogh, Konrad Lorenz, Barbara McClintock, Thomas Hunt Morgan, Ivan Petrovitch Pavlov, Nikolass Tinbergen, and James Dewey Watson (www.almaz.com/nobel/medicine/, 2000).

Pulitzer Prize   *n.* A prize, established by Joseph Pulitzer and first awarded in 1917, given for excellence in education, journalism, or letters and drama or for a traveling scholarship (www. pulitzer.org).

Comments: Joseph Pulitzer, a newspaper publisher, established this prize in his will. In letters, prizes were for a history of public service by the press, American biographies, American novels, books on U.S. history, and original plays performed in New York (www. pulitzer.org). He established an overseer advisory board and willed it "power in its discretion to suspend or to change any subject or subjects, substituting, however, others in their places, if in the judgment of the board such suspension, changes, or substitutions shall be conducive to the public good or rendered advisable by public necessities, or by reason of change of time." The board later added music, non-fiction books, photography, and poetry as award subjects. In 1997, the board started recognizing online journalism. In 1998, the board began including American music in addition to classical music. Biologists who have won this prize include Jared Diamond, Bert Hölldobler, and Edward O. Wilson.

Tyler Prize *n.* A premier award, established by John and Alice Tyler in 1973, that honors significant achievements in environmental science and environmental protection (www. usc.edu/admin/provost/tylerprize, 1998).

Comments: The prize is $200,000 and a gold medallion (www.usc.edu/admin/provost/tylerprize, 1998). Biologists who have won this prize include Herbert Bormann, Anne and Paul Ehrlich, Thomas Eisner, Charles Elton, Birute Galdikas, Jane Goodall, G. Evelyn Hutchinson, Gene E. Likens, Eugene P. Odum, Peter Raven, George Schaller, and Edward O. Wilson.

◆ **awareness** *n.*

1. A person's being watchful, vigilant, cautious (*Oxford English Dictionary* 1972, entries from 1095).
   Note: One can parsimoniously apply this definition to nonhuman animals, as well.
2. A person's being informed cognizant, conscious, sensible (*Oxford English Dictionary* 1972, entries from 1205); consciousness, *q.v.*; cognizance (Michaelis 1963).
3. A person's "whole set of interrelated mental images of the flow of events; they may be close at hand in time and space, like a toothache, or enormously remote, as in an astronomer's concept of stellar evolution" (Griffin 1976, 5).
4. A person's "experiencing of interrelated mental images" (Griffin 1981, 12, in Burghardt 1985, 908).
5. A person's behavior that "entails the experiencing of thinking about something" including some "evaluation or recognition of relationships between the objects and events about which we are thinking" (Griffin 1982, 4, in Burghardt 1985, 908).
6. An animal's appearing to know something or responding to stimuli (Burghardt 1985, 908).
   Note: "Awareness" is currently preferred to the term "consciousness" (Burghardt 1985, 909).

syn. conscious awareness (Griffin 1981 in Burghardt 1985, 908), consciousness (Burghardt 1985, 905)
cf. ³consciousness

Comment: Definitions and synonymies for these terms are often confusing: awareness, conscious awareness, consciousness, mental experience, self-awareness, self-consciousness, and true consciousness, *q.v.* [Old English *gewaer*, watchful]

conscious awareness *n.* Human behavior with five levels: proprioceptive body awareness, awareness of situational contingency, awareness of agency, awareness of social agency, and linguistic self-consciousness (Crook 1983).

Comments: "Proprioceptive body awareness" is closed-loop systems' monitoring the interdependence of bodily processes involving a continuous attention to inputs from many sources (*e.g.,* bodily positioning in aerobatic flight or in echolocational pursuit of prey). "Awareness of situational contingency" is an awareness of the relations between bodily positioning and environmental situations at times of intense activity, possibly being concomitant with movement and providing an implicit reference to one's position. "Awareness of agency" is a continuous monitoring of body positions in time and space, suggesting an implicit recognition of one's body as an agent in the tactical expression of a foraging strategy. "Awareness of social agency" is a continuous monitoring of the relations between self and others in complex mammalian societies, being highly likely in the context of collaborative, manipulative, or cheating behaviors. "Linguistic self-consciousness" is one's insight relating inner feeling and agency in behavior, leading to clear symbolization of self as an agent and using pronouns in language. As language develops, using metaphors in expressions of knowledge means that consciousness becomes primarily concerned with meanings and constrained by words.

self-awareness *n.*

1. A person's perception of himself as distinct from other selves (Hubbard 1975 in Dawkins 1980, 24; Gallup 1979, 418); a person's "capacity to think in terms of one's self-concept, focusing attention, processing information, and acting with regard to this concept" (Harré and Lamb 1983).
2. An individual animal's being aware of its own existence, with self-recognition as one criterion (Gallup 1975); an individual animal's being aware of itself as different from all other entities (Bunge 1984, 186–187).
3. An individual animal's "ability to abstract and to form a conceptual framework of its environment so that it can perceive itself and its actions in relation

to the environment" (*e.g.*, in great apes) (Universities Federation of Animal Welfare workshop 1980 in Crook 1983, 11); an animal's ability to become the subject of its own attention (Gallup 1982, 243). *cf.* awareness; awareness: conscious awareness; [3]consciousness: self-consciousness; self *Comments:* Epstein et al. (1981) question an animal's using a mirror to examine parts of its body as a test of self-awareness. They report how Pigeons learn to use a mirror to locate spots on their bodies that they cannot see directly and how their behavior is not likely to be due to self-awareness.

▶ **objective self-awareness** *n.* A person's evaluating himself and attempting to attain correctness and consistency in his beliefs and behaviors (Wicklund and Duval 1971); a person's focusing on himself as an object to be evaluated (Liebling and Shaver 1973, 298). *Comment:* This is very close to "self-consciousness" in everyday usage and is very similar to Wine's concept of "self-attention" (Liebling and Shaver 1973, 298).

▶ **subjective self-awareness** *n.* A person's focusing his attention outward; being concerned with other persons, objects, and events and not being cognizant of himself "as a distinct entity to be compared against standards" (Liebling and Shaver 1973, 298).

triadic awareness *n.* For example, in the Rhesus Monkey: a focal animal's taking into account the interrelationships of two conspecific individuals when it interacts with each of these individuals; *e.g.,* animal A is likely to be friendly with B in the presence of C if B and C are allies (de Waal 1989, 107).

◆ **axenic** See -xenic: axenic.

◆ **axiom** *n.*
1. A self-evident or universally recognized truth (Michaelis 1963).
2. An established principle or rule (Michaelis 1963).
3. A self-evident proposition of logic or mathematics, accepted as true without proof (Michaelis 1963).
cf. doctrine, dogma, law, principle, rule, theorem, [3]theory, truism
[Latin *axiōma, -atos* < Greek *axioma*, a thing thought worthy, a self-evident thing]

◆ **azygote** See zygote: azygote.

♦ **B-chromosome** See chromosome: B-chromosome.

♦ **b-rates** See [2]evolution: horotely.

♦ **b selection** See selection: b selection.

♦ **baa** See animal sounds: baa.

♦ **Babinski reflex** See reflex: Babinski reflex.

♦ **baby coloration** See coloration: baby coloration.

♦ **baby grunt** See animal sounds: grunt: baby grunt.

♦ **baby schema** *n.* A combination of stimuli (including a relatively large head, forehead arch, and eyes found in young of some bird and mammal species) that releases epimeletic behavior of some primate species including Humans (Lorenz 1943 in Heymer 1977, 97).

♦ **baby-sitter** See parent: alloparent.

♦ **bachelor group** See [2]group: bachelor group.

♦ **back mutation** See [4]mutation: back mutation.

♦ **backcross** See cross: backcross.

♦ **background genotype** See -type, type: genotype: background genotype.

♦ **background imitation** See camouflage.

♦ **background phenotype** See -type: phenotype: background phenotype.

♦ **backward conditioning** See learning: conditioned learning (comments).

♦ **Baconian inductivism** See inductivism.

♦ **bacterial pyrogen** See pyrogen: bacterial pyrogen.

♦ **bacteriology** See study of: bacteriology.

♦ **bacteriophage, bacteriophagous** See -phage; -phage: bacteriophage.

♦ **badge of status** *n.* An animal's features, or appearance, that indicate its ability to maintain control of a resource, such as a territory; *e.g.,* the area of black feathers below the bill of a male Harris' Sparrow or the red epaulet of a Red-Winged Blackbird [coined by Dawkins and Krebs 1978 in Immelmann and Beer 1989, 26].
syn. badge (Hansen and Rohwer 1986, 69)

♦ **Baer's laws** See law: von Baer's laws.

♦ **Bakerian mimicry** See mimicry: Bakerian mimicry.

♦ **Baker's law** See law: Baker's law.

♦ **balance of nature** *n.* The state of equilibrium in nature that results from the constant interaction between organisms and their environments (Lincoln et al. 1985).

♦ **balance-shift theory** See [3]theory: balance-shift theory.

♦ **balance theories** See [3]theory: balance theories.

♦ **balanced genetic polymorphism** See -morphism: polymorphism: balanced polymorphism.

♦ **balanced hermaphrodite population** See [1]population: balanced hermaphrodite population.

♦ **balanced incomplete block** See experimental design: incomplete block: balanced incomplete block.

♦ **balanced load** See genetic load: segregational load.

♦ **balanced polymorphism** See morphism: polymorphism.

♦ **balancing selection** See selection (table).

♦ **Baldwin-Waddington effect** See effect: Baldwin-Waddington effect.

♦ **ballistrophobia** See phobia (table).

♦ **ballooning** *n.* For example, in some spider species, larvae of the Evergreen Bagworm Moth and Gypsy Moth: an individual's being carried through the air by a strand of silk (a gossamer) that it produces (Haseman 1912; Borror et al. 1989, 112).
syn. parachuting (Borror et al. 1989, 112)
Comments: Populations disperse by ballooning. A population of the Gypsy Moth can move up to many kilometers per year by this mechanism.

♦ **band** *n.* In genetics, one of the many light and dark patterns on chromosomes that

results from certain artificial chemical treatment of them; depending on the staining technique used, one sees C, G, Q, R, or T bands (Mayr 1982, 775).
See ²group: band.

♦ **barachore** See -chore: barachore.

♦ **bared-teeth display** See display: bared-teeth display.

♦ **bared-teeth-scream display** See display: bared-teeth-scream display.

♦ **baren, barren** See ²group: baren, barren.

♦ **bark** See animal sounds: bark.

♦ **barokinesis** See kinesis: barokinesis.

♦ **barophile** See ¹-phile: barophile.

♦ **barotaxis** See taxis: barotaxis.

♦ **Barro Colorado Island** See place: Barro Colorado Island.

♦ **basal ganglia** See organ: brain: basal ganglia.

♦ **basic productivity** See production.

♦ **basic rank** See ¹rank: basic rank.

♦ **basifuge** See -phobe: calciphobe.

♦ **basophile** See ¹-phile: basophile.

♦ **bat dectector** See instrument: bat detector.

♦ **Bateman's principle** See principle: Bateman's principle.

♦ **Batesian mimicry, Bate's theory of mimicry** See mimicry; mimicry: Batesian mimicry, deceptive-floral mimicry.

♦ **Batesian-mimicry ring** See mimicry ring: Batesian-mimicry ring.

♦ **Batesian-Poultonian mimicry** See mimicry: Batesian-Poultonian mimicry.

♦ **Batesian-Wallacian mimicry** See mimicry: Batesian-Wallacian mimicry.

♦ **bathophile, bathyphile** See ¹-phile: bathophile.

♦ **bathophobia** See phobia (table).

♦ **batrachophobia** See phobia (table).

♦ **battered-Earth hypothesis** See hypothesis: battered-Earth hypothesis.

♦ **battery** See ²group: battery.

♦ **Bauplan** *n., pl.* **Baupläne** An organism's basic arrangement of its parts that gives it a unity (von Uexküll 1899 in Gherardi 1984, 373).
See -type: archetype.
[German *Bauplan*, ground plan]

♦ **bay** See habitat: bay.

♦ **Bayesian forager** See forager: Bayesian forager.

♦ **Bayesian statistics** See study of: statistics: Bayesian statistics.

♦ **BCI** See place: Barro Colorado Island.

♦ **BDFFP** See project: Biological Dynamics of Forest Fragments Project.

♦ **beacon hypothesis** See hypothesis: firefly-flash-synchronization hypothesis: beacon hypothesis.

♦ **Beagle** See Dog: Beagle.

♦ **bean-bag genetics** See study of: genetics: bean-bag genetics.

♦ **bearing-and-caring** *n.* Informally, the sex that invests more in offspring shows more "concern" for its offspring (Dawkins 1976, 117, 186).

♦ **beater effect** See effect: beater effect.

♦ **Beau Geste effect, Beau Geste hypothesis** See effect: Beau Geste effect.

♦ **becoming accustomed** See learning: becoming accustomed.

♦ **bed load** *n.*
1. The part of a stream's transported materials that move on, or immediately above, its bed, such as the larger, heavier particles (boulders, gravel, pebbles) rolled along its bottom (Bates and Jackson 1984).
2. The part of a stream's transported particles that is not continuously in solution or suspension (Bates and Jackson 1984).
syn. bottom load, traction load (Bates and Jackson 1984)
Comment: When researchers quantify bed load, they trap it in a sediment box placed in a stream, remove it, and weigh it.

♦ **bee bread** *n.* The vernacular name for pollen combined with nectar, or honey, and beneficial microbes and stored in open hexagonal comb cells by Honey Bees (Buchmann and Nabhan 1996, 242).

♦ **bee dance** See dance: bee dance.

♦ **bee-language hypothesis** See hypothesis: dance-"language" hypothesis.

♦ **beetlephilia** See -philia: beetlephilia.

♦ **behavior** (*Brit.* **behaviour**) *n.* The manner in which a person, substance, machine, or other thing acts under specified conditions, or circumstances, or in relation to other things (*Oxford English Dictionary* 1971, entries from 1490; Michaelis 1963).
See behavior: animal behavior, organism behavior.
cf. act; action; activity; aggression; animal sounds; behavior: human behavior; behavior, classification of; behavior pattern; behavioral act; behavioral type; brooding; care; coenotrope; copulation; courtship; dating; display; episode; flehmen; flight foraging; greeting; habit; head-up/tail-up behavior; homing; hunting; kidnaping; kinesis: klinokinesis; language; learning; ligulation; mating; mouse pounce; movement; must; panting; play; posture; pouting; quivering; recruitment; reflex; response; scratching; submission; suckling; swimming; symbiosis: phoresy; temperature regulation, territoriality; treading; vigilance; wing fanning
v.i. behave
adj. behavioral (*Brit.* behavioural)

Comments: Some workers use "behavior" as both a singular and plural term; other workers use "behavior" as singular and "behaviors" as plural (Edwards 1986). Only some of the many kinds of named behavior are listed below.

[Formed on BEHAVE by form analogy with HAVOUR; BEHAVE; behavior < Middle English *be-*, thoroughly + *have*, to hold oneself, act < Old English *behabben*, had < Old High German *bihabên* < *be-*, about + *habban*, to hold, HAVE, in senses "encompass, contain, detain" (*Oxford English Dictionary*)].

abnormal behavior *n*. In many animal species: a very large mixed category of behavior that includes persistent, undesirable actions shown by a minority of individuals of a population, is not due to any obvious damage to an animal's nervous system, and is generalized (not confined to any situation that originally elicited it) (Broadhurst 1960 in Dawkins 1980, 77).

syn. behavioral anomaly (Immelmann and Beer 1989, 1)

cf. behavioral deficit; captivity degeneration; hospitalism; hypersexuality; hypertrophy; Kasper-Hauser; learning: imprinting: erroneous imprinting; mounting attempt; syndrome: deprivation syndrome, separation syndrome; stereotypy

Comments: Abnormal behavior is damaging or maladaptive to an animal, according to Fraser (1968 in Dawkins 1980, 77) but not according to Meyer-Holzapfel (1968 in Dawkins 1980, 77). "Abnormal behavior" may overlap broadly with "fear" and "conflict behavior" (Dawkins 1980, 77). Examples of abnormal behavior include a monkey's biting itself so hard that it screams with pain, a cockatoo's pulling out all of its feathers except a single one on the top of its head, and pigs' biting each other's ears and tails (references in Dawkins 1980, 77).

▸ **deprivation syndrome** See syndrome: deprivation syndrome.

▸ **erroneous imprinting** See learning: imprinting.

▸ **separation syndrome, separation trauma** See syndrome: separation syndrome.

▸ **stereotypy** See stereotypy.

acquired behavior *n*. An individual's "behavior that has been experimentally demonstrated, in either its topography or stimulus control, or both, to be dependent in part upon the operation of variables encountered in conditioning and learning, such as the occurrence of reinforcing stimuli..." (Verplanck 1957).

ant. inborn behavior, unlearned behavior (Verplanck 1957)

cf. learning: conditioning

Comment: All animal species might have acquired behavior.

adjunctive behavior See behavior: displacement behavior.

advertising, advertisement (behavior) See display: advertisement.

aggression, aggressive behavior See aggression.

agonistic behavior *n*.
1. In many animal species: a behavior complex that includes all attack, threat, appeasement, and flight behavior (Scott and Fredericson 1951 in Hinde 1970, 336; Verplanck 1957).
2. Animal fighting and competitive behavior that includes threats and offensive attacks as well as defensive fighting (Johnson 1972, 8).

syn. aggression (excluding fleeing, but a clear distinction between "agonistic behavior" and "aggression" is preferred; Immelmann and Beer 1989, 10)

cf. aggression

alarm behavior *n*.
1. In some species of mud snails, sea slugs, sea urchins, and anuran amphibians; many fish species: an individual's retreating in response to chemicals discharged from a conspecific individual that is injured, alarmed, or both (Frisch 1938, etc. in Hews 1988, 125; Maschwitz 1964).
2. An animal's behavior shown in response to a predator's appearance that alerts conspecifics of a threat, alerts a predator that an alarmer is well aware of its presence, or both; *e.g.*, acoustic warning signals in many bird and some mammal species, ground thumping by some mammal species, releasing alarm pheromones in some insect species, white-tail flashes of some deer and rabbit species, touching in some fish species, and mobbing in some bird species (LaGory 1987, 20; Immelmann and Beer 1989, 326).

syn. alarm response (LaGory 1987, 20); alarm signaling (Maschwitz 1964, 268); recruitment (sometimes, Wilson 1965, 1065); warning behavior

cf. chemical releasing stimulus: semiochemical: pheromone: alarm pheromone; enemy specification; signal: alarm signal

allelomimetic behavior *n*.
1. Any behavior in which animals "do the same thing with some degree of mutual stimulation and consequent coordination" (Scott 1956, 215).
   syn. mood induction (Scott 1946)
2. Two or more animals' imitative behavior that results from mutual stimulation (Lincoln et al. 1985).

3. Mimicry of a single animal's behavior by others in its group, *e.g.*, llamas that perform excretion in a localized area, lining up at the area as soon as one member urinates or defecates (Altmann 1969 in Heymer 1977, 168).

cf. effect: group effect

Comment: Immelmann and Beer (1989, 275) distinguish between "mood induction" and "social facilitation," *q.v.* In "mood induction," an animal switches to a new mood (*e.g.*, from fleeing to preening) due to observing its group conspecifics' showing the behavior to which it switches.

allochthonous behavior *n.* An animal's focal behavior that is energized by the drive built up by other activities besides the focal behavior (Kortlandt 1940 in McFarland 1985, 383)

See behavior: displacement behavior.

cf. behavior: autochthonous behavior

altruistic behavior See altruism.

amatory behavior *n.* "All the minor and less stimulating attentions, such as licking and nuzzling," that may keep two animals together (Southern 1948, 179).

ambivalent behavior *n.*

1. An animal's alternately showing two different reactions, or their incipient forms, because two different sign stimuli are present, each eliciting a different behavior (*e.g.*, the zig-zag dance of a male stickleback) (Tinbergen 1951, 50).
2. An animal's tending to behave in two or more incompatible ways at the same time (Hinde 1982, 277).

cf. emotional ambivalence

ambivalent movements and postures *pl. n.* In animals in which two or more conflicting drives are weakly activated: a combination of intention movements of introductory appetitive behavior that belongs to the drives involved and, in some cases, integrated movements that typically occur later in the chain from introductory appetitive to consummatory behavior (Bastock et al. 1953, 67).

cf. behavior: ambivalent behavior, compromise behavior, displacement behavior; conflict

Comments: These behaviors occur because the individual in question responds to all activated drives simultaneously. "Although ambivalent combinations of movements may visually resemble various displacement activities, they differ basically from displacement activities in that they are autochthonous" (Bastock et al. 1953, 67). An example of an ambivalent posture is found in a duck that has approached a person for bread and then cranes its neck forward to reach the bread and simultaneously turns its body away from the person (McFarland 1985, 401).

amicable behavior *n.* In rats: all specific forms of behavior that involve the association together of two or more individuals and do not involve conflict, including sexual behavior and care of young (Barnett 1958, 117).

animal behavior *n.*

1. "The total of movements made by the intact animal" (Tinbergen 1951, 2).
2. All observable, recordable, or measurable activities of a living animal, including muscular contractions that cause movement, organelle (*e.g.*, cilium) movement, color changes, and glandular secretion, due to the animal's interactions with, or reactions to, its environment (Collocot and Dobson 1974); "what an animal does" (Lehner 1979, 8).
3. One or more behaviors of one animal, species, or other grouping of animals.

See study of: behavior.

syn. behavior

antipredatory behavior *n.* An organism's behavior that helps to minimize its being attacked and consumed by a predator (Ylönen and Magnhagen 1992, 179).

cf. aggression: antipredator aggression

aposematic behavior, aposematism *n.* In many animal species with defenses (bites, sprays, stings, toxicity) and either warning coloration, *q.v.*, or warning structures, or both: an individual's acting as if it is dangerous to a potential predator (*e.g.*, by resting in a conspicuous place or walking slowly in the open) (Wilson 1975, 579; Matthews and Matthews 1978, 312, 323).

syn. warning (behavior)

cf. allosematic; coloration: aposematic coloration

apotreptic behavior See behavior: threat behavior.

appeasement, appeasement behavior *n.* In many animal species: an individual's behavior that occurs in agonistic contexts and commonly signifies lack of hostility towards, or intent to attack, another conspecific animal and is often directly opposite in form from threats, *q.v.*; *e.g.*, a wolf that is losing a fight throws up its head and turns the ventral surface of its neck toward the jaws of the winner (Darwin 1872; Verplanck 1957; Sparks 1965; Hand 1986, 216).

See display: appeasement display.

syn. appeasement display (in some contexts, Wilson 1975), appeasing signal (in some contexts, Hand 1986, 216), submission (Brown 1985) *cf.* submission

Comments: Appeasement functions either to halt aggression or facilitate conciliation; "appeasement" does not necessarily indicate either submission or subordination (Hand 1986, 216). "Appeasement," defined as behavior that activates behavior of another animal that is incompatible with aggression, is sometimes distinguished from "submissive behavior," defined as the cessation of aggressive signaling. "Appeasement" commonly includes the connotation of "submission" (Immelmann and Beer 1989, 18).

▸ **trophallactic appeasement** *n*. In ants: a worker's use of a liquid food offering to appease a potentially hostile worker (Hölldobler and Wilson 1990, 644).

appetitive behavior *n*.

1. "A term applied to characterize, in terms of an inferred or anticipated (by the ethologist) consummatory act, the behavior of an animal that is not at rest or 'doing nothing'..." (Verplanck 1957).
2. An animal's "variable, agitated behavior that is terminated by the occurrence of a stimulus releasing a consummatory act" (W. Craig in Dewsbury 1978, 14).
3. Behavior that enables an animal to perform a specific kind of action or to attain a specific state of being, *e.g.,* satiety, stimulus reception, arrival at a suitable nesting place, or contact with particular individuals (Immelmann and Beer 1989, 18).
4. An animal's orientation and adjustment movements that are more or less reflexes (Immelmann and Beer 1989, 19).

cf. behavior: consummatory behavior

Comments: Appetitive behavior is usually comparatively plastic with a learned component (Immelmann and Beer 1989, 19). [coined by Sherrington 1906 in Hailman 1984]

attachment behavior *n*.

1. A number of evolved response systems (behaviors) that maintain proximity between a human mother and her infant which is necessary for the infant's survival (Bowlby 1969, 1973 in Hinde 1974, 189).
2. The number of different types of behavior contributing to maintaining proximity, or contact, between a human parent and his, or her, young; these include proximity- and contact-seeking behavior, such as approaching, following, and clinging, and signaling behaviors, such as smiling, crying, and calling of infants (Ainsworth and Bell 1970, 50; Hinde 1982, 228).

Note: Attachment figures can be non-parents or even other children (Harré and Lamb 1983).

3. Any behavior that results in a person's obtaining, or retaining, proximity to some other differentiated and preferred individual, usually conceived as stronger, wiser, or both (Bowlby 1975 in Goldenson 1984).

cf. attachment

Comment: Human infant attachment behaviors include signals (*e.g.,* crying, smiling, vocalizing) and active proximity seeking (approaching, following, and clinging) to an attachment figure, typically the infant's mother (Mann 1991, 1).

attack behavior *n*. In many animal species: "a broad response class including those behaviors of an animal that, when carried to completion, bring to bear one or more of the animal's effectors on the body surface of a second animal in such a way that injury and possibly death of the second animal will occur if the behavior continues. [In a given species, more precise specification is possible in terms of biting, clawing, hitting with the wings, and the like.]" (Verplanck 1957).

autochthonous behavior, autochthonous activity *n*. A behavior caused by its proper (own) drive (Tinbergen 1952, 6).

cf. behavior: displacement behavior; drive; instinct

autonomous behavior See behavior: autonomous spontaneous behavior.

beneficent behavior *n*. An organism's behavior that raises the fitness of a conspecific regardless of its evolutionary basis (West Eberhard 1975, 7).

cf. altruism, cooperation, mutualism

bonding behavior *n*. For example, in a crustacean, some mammal, and many bird species: an individual's behavior that establishes, or maintains, a pair bond; *e.g.,* sexual behavior, duet singing, courtship feeding, and allogrooming (Immelmann and Beer 1989, 34, 212).

caching, caching behavior *n*.

1. A person's concealing, or storing, something in the earth or other place (Michaelis 1963).
2. A person's hiding an object in a secret place (Michaelis 1963).
3. An animal's storing food in a particular place.
4. An individual's storing food (such as seeds and succulent fruits and leaves) for long periods (Janzen 1977, Smith and Reichman 1984 in Reichman 1988, 1535; Balda and Kamil, 1989, 486).

5. An animal's handling food in a way that conserves it for future use (Vander Wall 1990, 1).

syn. food hoarding, food storing (Van der Wall 1990, 1)

Notes: In this definition, an animal's handling food involves its deterring other organisms from consuming the food and one or more of the following: food concealment, placement, preparation, and transportation (Van der Wall 1990, 1–2). This definition excludes an animal's accumulation of food in its crop or other diverticular of its digestive tract as occurs in birds including finches, grouse, and hummingbirds. The time of hoarding food varies from a few minutes (*e.g.,* in Barbados Green Monkeys) to 2 years (*e.g.,* in Red Squirrels).

v.t. cach, cached

cf. behavior: hoarding behavior: larder hoarding, scatter hoarding; cache

Comments: Food-hoarding animals can control the availability of their food in space and time, which gives them a marked advantage over nonhoarders (Vander Wall 1990, 1). Caching behavior occurs in many species, including the Clark's Nutcracker, Eastern Woodrat, European Jay, Fox Squirrel, Gray Squirrel, Kangaroo Rat, Lewis Woodpecker, Pinyon Jay, Pocket Mouse, and Red Squirrel; Argiope spiders; harvester ants; spider wasps; and many bee species.

care-giving behavior See behavior: epimeletic behavior.

cf. care

catasematic behavior See submission.

chemical self-marking See behavior: marking behavior: automarking.

chin raising *n.* In some duck species: an individual's raising its bill, without just previously lowering it, while its body moves lower in water, as part of courtship behavior (Lorenz 1941 in Heymer 1977, 98).

v.t. chin raise

chinning *n.* For example, in the European Rabbit, Mongolian Gerbil: an individual's rubbing the lower surface of its chin against objects on the ground, resulting in the deposit of pheromones (Wilson 1975, 280).

v.i., v.t. chin

coenotrope *n.* Behavior common to all members of a group or species (Lincoln et al. 1985).

comfort behavior *n.*

1. In many kinds of animals: animals' body-care behaviors (*e.g.,* preening, grooming, scratching, shaking, rubbing, washing, and dustbathing) (Heymer 1977, 100; Immelmann and Beer 1989, 49).

2. An animal's behaviors that relieve tensions arising from metabolic functions (*e.g.,* defecating, urinating, stretching, and yawning) (Heymer 1977, 100; Immelmann and Beer 1989, 49).

syn. comfort search (Kortlandt 1940 in Heymer 1977, 100), comfort movements (Baerends 1950 in Heymer 1977, 100)

cf. stretching syndrome, yawning

compromise behavior *n.* Behavior that consists of separate components of conflicting tendencies (McFarland 1985, 401).

conflict behavior *n.* An animal's tension (often leading to displacement behavior) shown when simultaneous releasers activate contradictory activities (*e.g.,* attack and flee) (Heymer 1977, 101).

cf. behavior: displacement behavior; conflict; [3]theory: conflict theory of display

Comment: "Conflict behavior" includes "ambivalent behavior," "displacement behavior," and "redirection" (Immelmann and Beer 1989, 57).

conscious behavior, consciousness *n.* A person's behavior that involves empathy, analogical sympathy, and concordance (Crook 1983, 12).

See [3]consciousness.

Comment: Similar behavior is likely to occur in other vertebrates (Crook 1983, 13).

consummatory act, consummatory behavior *n.*

1. In many organism species: the last phase of an individual's appetitive-behavior sequence that ends an elicited series of actions and leads under normal conditions to drive reduction (*e.g.,* food ingestion or copulatory behavior) (Craig 1918 in Heymer 1977, 61).

Note: In a behavioral sequence, an act can simultaneously represent appetitive and consummatory behavior (Craig 1918 in Heymer 1977, 61). Some kinds of appetitive behavior can reduce an animal's likelihood of immediate repetition or continuation of a sequence in which they occur; this blurs the distinction between appetitive and consummatory behavior (Immelmann and Beer 1989, 86).

2. An organism's act that "constitutes the termination of a given instinctive behavior pattern or sequence" (Thorpe 1951, 4).

3. One of a group of an organism's acts that "complete a reaction chain and give very plain evidence of release of appetitive tension" (Thorpe 1956, 26).

4. An organism's behavior that "constitutes the terminal phase of an instinctive behavior sequence" (Lincoln et al. 1985).

syn. end act (Immelmann and Beer 1989, 86)

cf. behavior: appetitive behavior, comfort behavior; pattern: fixed-action pattern [Coined by Sherrington 1906 in Hailman 1984; linking "consuming" and "consummatory" is an unintended pun. Consuming comes from Latin *consumere*, to devour or destroy; consummatory comes from Latin *summa*, a total, or sum (Immelmann and Beer 1989, 86).]

contact behavior, contactual behavior *n.*

1. In some wren, swift, and penguin species: two or more usually conspecific individuals' maintaining bodily contact (Scott 1956, 214); in many cases, the birds position themselves so as to have as much of their body surfaces as possible in contract with another (Immelmann and Beer 1989, 58, 136). For example, in Emperor Penguins, during severe storms, members of a group gather together, forming a sort of roof by holding their heads together; other animals move toward the group's center and inner ones are pushed outward (Prevost 1961 in Heymer 1977, 156).
   syn. huddling, huddling behavior
   cf. animals: contact animals
2. For example, in clinging young of primates: a young's holding onto its parent, usually its mother (Immelmann and Beer 1989, 58).
3. In social animals: signal exchange (Immelmann and Beer 1989, 59).

contact-seeking behavior *n.* Contact-oriented display and movements, especially in small children (Heymer 1977, 92).
cf. baby schema, hand patting

conventional behavior *n.*

1. An animal's "voluntarily agreeing" to curtail reproduction when it realizes that its population density is rising due to its perceiving ritualized displays from a conspecific(s) (Allee et al. 1949; Kalela 1954; Wynne-Edwards 1962 in Wilson 1975, 87).
2. "Any behavior by which members of a population reveal their presence and allow others to assess the density of the population," according to Wynne-Edwards' hypothesis (Wilson 1975, 581).
syn. epideictic display (Lincoln et al. 1985)
cf. display; selection: group selection
Comment: "Conventional behavior" is a controversial concept.

copulatory behavior See copulation.
   ▸ **postcopulatory behavior** *n.* Behavior that occurs immediately after copulation (*e.g.,* the postcopulatory display of Uganda

Kob or mate guarding of Japanese Beetles) (Barrows and Gordh 1978).
   syn. copulatory afterplay, postcopulatory play (Immelmann and Beer 1989, 229, who do not recommend these synonyms because "postcopulatory behavior" does not appear to be "play" in its strict sense)
   cf. behavior: mate guarding
   ▸ **precopulatory behavior** See courtship.
   ▸ **pseudocopulatory behavior** See copulation: pseudocopulation.

courtship behavior See courtship.

cryptic behavior *n.*

1. In many animal species: an individual's behavior that enables it to be concealed, or concealed better, in its environment, including altering its coloration (flatfish), sticking pieces of its environment onto itself (spider crabs, caterpillars of some geometrid moth species), concealing itself with feces (tortoise beetles), and resting in its environment in a concealing way (flatfish) (Wickler 1968, 54).
   cf. camouflage
2. A behavior that a species has in its repertoire but is never shown by certain individuals (White et al. 2000, 885).

curiosity behavior See behavior: exploratory behavior.

Davian behavior *n.* For example, in the Richardson's Ground Squirrel: a male animal's copulating with a conspecific dead female (Dickerman 1960 in Lehner 1988). [named after a ribald Limerick about necrophilia]
   ▸ **avian Davian behavior** *n.* In Mallard Ducks: a male bird's copulating with a dead conspecific female [coined by Lehner 1988].
   Comment: Male birds also copulate with dead heterospecific females (Lehner 1988).

defensive behavior *n.*

1. In many animal species: the measures that an individual takes to avoid being eaten as prey (Immelmann and Beer 1989, 71).
2. An animal's self-protective behavior shown in agonistic encounters with conspecifics (Immelmann and Beer 1989, 71).
Comment: Examples of defensive behaviors include an animal's keeping away from a predator; rapidly retreating from an aggressive conspecific; withdrawing into a burrow, crevice, or shell; submerging into water; displaying protean behavior; displaying thanatosis; and hiding (Immelmann and Beer 1989, 90).
cf. behavior: threat
Comment: Hermann and Blum (1981, 77–197) describe many of the defensive

behaviors used by social Hymenoptera, including abdominal bursting; ant-wasp symbioses; architectural defense; attack (stinging and mobbing); biting; defensive immobility; defensive smearing; mimicry; stridulation; territoriality; use of defensive alarm pheromones, allomones, mandibular-gland secretions, phragmotic heads, protective body spines and other integumental modifications, and sticky fluids; and venom spraying.

▸ **autothysis** *n*. In some *Camponotus*-ant species: a worker's increasing the hydrostatic pressure inside her gaster until it bursts (Maschwitz and Maschiwitz 1974 in Hermann and Blum 1981, 79; 86, illustration).

syn. abdominal bursting (Hermann and Blum 1981, 79)

Comment: An ant's bursting releases a large quantity of a sticky fluid that immobilizes an ant of another species that is attacking (Hermann and Blum 1981, 79).

▸ **cycloalexy** *n*. For example, in some chrysomelid-beetle and tenthredinid-sawfly species, a ceratopogonid species: larvae's resting in a "tight circle" with their heads or abdominal ends at its periphery, surrounding other larvae inside the circle, and using coordinated threatening postures, regurgitation, and biting to repel predators or parasites (Lewis 1836, Vesconcellos-Neto and Jolivet 1988, etc. in Jolivet et al. 1990, 133, 139).

syn. circular defense, circular-defense strategy (Jolivet et al. 1990, 133–134)

Comment: Similar behavior occurs in a saturniid moth species, penguins, the Muskox, the Oryx, and other vertebrates (Jolivet et al. 1990, 133, 139).

[Greek *kyklos*, circle; *alexo*, defend, avert, ward off]

▸ **deimatic behavior** *n*.
1. For example, in large-cat species: an animal's adopting an intimidating posture that frightens another animal (Lincoln et al. 1985).
2. In animals with eyespots (*e.g.*, some mantid, moth, and butterfly species): an individual's frightening a predator by suddenly exposing an eyespot, *q.v.* (Immelmann and Beer 1989, 71).

syn. dymantic behavior

cf. display: deimatic display

▸ **enteric discharge** *n*. Regurgitation (*e.g.*, in some grasshoppers that produce "tobacco juice" and in some sawfly larvae) or defecation that is used as predator repellent (Matthews and Matthews 1978, 335).

cf. elimination: egestion

▸ **"heat defense"** *n*. In Japanese Honey Bees: worker's covering a Japanese hor-

net and killing it by baking it with their body heat (Masato Ono in Anonymous 1997a, 16, illustration).

Comments: The bees raise the temperature to 116°F by vibrating their wing muscles (Anonymous 1997a, 16, illustration). They can tolerate a temperature of up to 122°F, but the hornets perish at 114°F.

▸ **playing dead** *n*. An animal's defensive behavior in which it becomes essentially motionless and, thus, acts as if it were not alive (Maier and Schneirla 135, 222; Wilson 1975, 363).

syn. self-mimesis (a kind of) (Pasteur 1982, 184); akinesis, animal hypnosis, catalepsy, trance reaction (Hermann and Blum 1981, 104); ascaphus reflex, death feigning, death-feigning behavior, death feinting, feigning death, freezing, hypnotic reflex, playing O'possum, thanatosis, tonic immobility (Lincoln et al. 1985)

cf. behavior: defensive behavior; kinesis: akinesis; hypnosis; response: immobility response

Comment: Playing dead occurs in some anole-lizard, ant, bat, bee, beetle, butterfly, frog, hog-nose snake, sawfly, spider, and wasp species; the African Ground Squirrel, Domestic Chicken, Gundi (rodent), Human, O'possum, and Turkey Vulture (Maier and Schneirla 135, 222; Wilson 1975, 363; Burghardt and Greene 1988, 1842; Allen 1996, 639; D.R. Smith, personal communication).

▸ **primary-defensive behavior** *n*. An animal's defensive behavior it continuously shows, including living in burrows, constructed tubes, or shells; camouflaging behavior, aposematic advertisement, and acting like a Batesian model (Immelmann and Beer 1989, 71).

▸ **secondary-defensive behavior** *n*. An animal's defensive behavior shown when it is faced with a predator, including escaping, thanatosis, protean behavior, deimatic behavior, attack deflection, and various forms of defensive attack (*e.g.*, mobbing) (Immelmann and Beer 1989, 71).

detour behavior *n*. An animal's immediately following an alternative route with success after its path is blocked (Immelmann and Beer 1989, 151).

developmentally fixed behavior *n*. In all animal species: an individual's behavior that arises from interactions of its environment and genetics during its ontogeny (Dawkins 1986, 62).

Comment: "Developmentally fixed" is not synonymous with "genetic" (inherited) (Dawkins 1986, 62).

displacement behavior, displacement activity *n.*

1. For example, in stickleback fish, some bird species, Humans: behavior that appears irrelevant to the situation in which it appears (Huxley 1914, etc. in McFarland 1985, 381).
2. Behavior that results from the activation of the specific action potential of one instinct, or of the action pattern belonging to another instinct, and appears when a charged instinct is denied the opportunity for adequate discharge through its own consummatory acts (Kortlandt 1940 in Dawkins 1986, 79; Tinbergen 1951, 113; Lincoln et al. 1985).
3. "Behavior that is not activated by its own drive" (Verplanck 1957).
4. "Behavior that is activated as a consequence of the frustration of behavior activated by some drive other than that which most often controls it" (Verplanck 1957).
5. "A seemingly irrelevant display of behavior given in conflict situations where it has no direct functional significance;" displacement behavior "presumably does have communicatory significance" (Wittenberger 1981, 614).
6. An activity that seems irrelevant to its behavioral context to a human observer and often occurs when its performer is in a conflict situation, *e.g.,* a chaffinch's suddenly breaking off an encounter and wiping its bill while it is engaged with a rival (Hinde 1982, 66) or a person's fidgeting with cigarettes or jewelry in a tense circumstance.

syn. allochthonous behavior (Verplanck 1957), allochthonous reaction (Lorenz 1957, 296), ambivalence (Lincoln et al. 1985), displacement activity (Tinbergen and van Iersel 1947 in Tinbergen 1951, 114), irrelevant behavior, irrelevant movements (Rand 1943 in Tinbergen 1951, 114), redirected behavior (in the broad sense according to some authors), sparking-over movement (Makkink 1936 in Tinbergen 1951, 114), substitute activity (Kirkman 1937, Tinbergen 1951, 114)

cf. autochthonous behavior; inhibition: behavioral inhibition; redirected behavior (in the strict sense)

Comments: Dawkins (1986, 79–80) indicates that because displacement behavior relates to the abandoned Lorenz's hydraulic model of behavior, *q.v.,* this term should be abandoned. The exact cause of displacement behavior is not known; it might have more than one cause.

[coined by Armstrong 1947 and Tinbergen and van Iersel 1947 in McFarland 1985, 382 < energy displaced from two mutually antagonistic behaviors into a third behavior]

▸ **adjunctive behavior** *n.* In experimental psychology, apparently irrelevant rat behavior that occurs in learning situations, such as grooming or drinking, by a hungry individual that is pressing a bar for food (Falk 1971 in Hinde 1982, 66).

▸ **displacement feeding** *n.* In some gallinaceous bird species: displacement behavior involving a bird's pecking the ground but not feeding (Heymer 1977, 183).

▸ **displacement preening** *n.* For example, in some duck species: displacement behavior involving preening (Heymer 1977, 183).

▸ **displacement sleeping** *n.* For example, in limicolans: displacement behavior involving sleeping in place (Heymer 1977, 183).

▸ **grass pulling** *n.* For example, in Herring Gulls: an individual's probable displacement behavior consisting of its pulling up tufts of grass which it does not use for nest building while it is confronting another conspecific individual (Goethe 1956, Tinbergen 1958 in Heymer 1977, 78).

▸ **redirected behavior, redirected activity** *n.*

1. In some bird species, Humans: irrelevant behavior that involves an individual's directing a behavior, such as an act of aggression, away from its primary target and toward another, less appropriate object (Wilson 1975, 225, 594).
2. A behavior shown toward a seemingly less appropriate conspecific, *e.g.,* an animal experiencing the conflict of attacking or not attacking another animal might attack another (less threatening) animal rather than its opponent (Dewsbury 1978, 19); redirection activity (Dewsbury 1978, 262).

cf. aggression: redirected aggression, displaced aggression

▸ **sham pecking** *n.* In some bird species: an individual's pecking at the ground but not consuming food (Heinroth 1911, etc. in Heymer 1977, 154).

▸ **sham preening** *n.* In some bird species: a male's preening his wings during courtship, apparently not because they require cleaning (Heymer 1977, 155).

cf. preening

▸ **sublimation** *n.* In Freudian psychoanalysis, human behavior similar to redirected behavior (Immelmann and Beer 1989, 75).

duration-meaningful behavior See behavior state.

egg-rolling behavior *n*. In many ground-nesting birds, including the Emu and Greylag Goose: a bird's painstakingly rolling an egg back into its nest by pulling it with the underside of its bill and making sideways correction movements; classified as a fixed-action pattern, *q.v.*, with a taxis component by Lorenz and Tinbergen (1970 in Gould 1982, 36; Immelmann and Beer 1989, 84).
cf. retrieving

eliminative behavior, elimination *n*. For example, in mammals: an individual's behavior associated with urination and defecation (Scott 1950 in Lehner 1979, 64).
cf. behavior: comfort behavior; elimination; enurination

emotional behavior *n*. For example, in the Domestic Pigeon, rats: "an arbitrary class of responses that is defined differently for different species and that is based on the covariation in response-strength of the behaviors as a function of certain often ill-defined independent variables" (Verplanck 1957).
cf. emotion, emotionality
Comments: "Despite the rather fuzzy origins of this class, once a particular response has been placed in the class, the term emotional can be consistently applied to it; and a concept that is initially theoretical seems to acquire some empirical status" (Verplanck 1957). In rats, emotional behaviors include urination, defecation, freezing, vocalizing, and trembling, when two or more of them occur together. In Domestic Pigeons, emotional behavior includes cooing and wing-beating (Verplanck 1957).

end act See behavior: consummatory behavior.

epideictic behavior See display: epideictic display.

epigamic behavior *n*. In many animal species: an individual's expressive activities that are directly related to its reproduction (Heymer 1977, 61).
cf. display: epigamic display
▸ **nonepigamic behavior** *n*. In many animal species: an individual's behavior that normally occurs apart from its reproduction (Heymer 1977, 61).
cf. behavior: epigamic behavior

epimeletic behavior *n*. In many animal species: an individual's giving care and attention (Scott 1950 in Lehner 1979, 64), including attentive behavior and nurturance (Scott 1956, 214), guarding, protecting, and allogrooming (Heymer 1977, 129).

syn. care-giving behavior (Heymer 1977, 129)
cf. baby schema, infantilism
▸ **et-epimeletic behavior** *n*. In some young and adult animal species: an animal's soliciting care and attention from another animal (Scott 1950 in Lehner 1979, 64; Scott 1956, 214).

epitreptic behavior See behavior: treptic behavior: epitreptic behavior.

equilibratory behavior *n*. An animal's response to potential, or overt, threat which occurs between attacking and fleeing (Chance 1956).

escape behavior *n*. An animal's rapid locomotion from any given location that occurs after specific stimuli are presented there (Verplanck 1957).
syn. flight, flight behavior, flight response (LaGory 1987, 20)
Comment: "The stimuli effective in producing escape behavior are almost identical with those identified as negative reinforcing stimuli or aversive stimuli in conditioning situations" (Verplanck 1957).
▸ **protean behavior** *n*. A pursued prey animal's escape behavior in which it switches unpredictably between (among) different fleeing or avoidance behaviors, *e.g.,* erratic flying of a moth under attack by a bat or an insect's simulating a frightening predator (Chance and Russell 1959 in Heymer 1977, 133; Immelmann and Beer 1989, 236).
cf. display: protean display
[named after the minor deity Proteus, the Greek sea god, who escaped his pursuers by continually changing his form]

exploratory behavior *n*.
1. In many vertebrate (*e.g.,* rats) and invertebrate species: an animal's species-typical orientation in time and space necessary for its effective learning and resulting in remembering its nest site or surroundings (Heymer 1977, 62).
2. In many bird and mammal species, some fish and reptile species: an animal's searching for and active investigation of novel situations in the absence of a pressing physiological need (Immelmann and Beer 1989, 95).
syn. curiosity behavior, exploration
Comment: Exploratory behavior is more conspicuous in young than adults, but it persists throughout life in raptors, parrots, rodents, and primates (Immelmann and Beer 1989).

finning *n*. A whale's slapping the surface of water with its long flipper, instead of its flukes (Baker and Herman 1985, 55).
v.t. fin
cf. behavior: lobtail

fixed-reaction behavior *n.* A human infant's quiet orientation toward sound with his eyes, *e.g.,* in congenitally blind infants when they are spoken to by their mothers or in infants with normal vision, especially during nursing, probably being used in memorizing their mothers' faces (Heymer 1977, 68).

flight behavior See behavior: escape behavior.

following behavior *n.*

1. In nidifugous birds: a bird's innate disposition to approach and follow a moving object shortly after hatching (Hess 1958, 1959a in McFarland 1985, 369; Heymer 1977, 121).
2. In species with parental care (*e.g.,* some fish, precocial-bird, ungulate, and mammal species): a young animal's attempts to stay close to one or both of its parents (Immelmann and Beer 1989, 109).

cf. learning: imprinting: object imprinting

foraging, foraging behavior *n.*

1. A person's searching about, or rummaging around, for something, especially food or supplies (given as a verb in Michaelis 1963).
2. A person's making a raid to find, or capture, supplies (given as a verb in Michaelis 1963).
3. A person's obtaining (food, supplies) by rummaging about (given as a verb in Michaelis 1963).
4. An organism's searching for or obtaining, or both, a resource (*e.g.,* food or water).

[French *fourrage* < Old French *feurre*, fodder < Germanic]

▶ **cooperative foraging** *n.* Foraging, in which animals show at least some temporary altruistic restraint, group members' behaviors are often diversified, and communication modes are typically complex (Wilson 1975, 51).

▶ **handling, handling behavior** *n.* A predator's behavior after it catches a prey, including its physically manipulating the prey for consumption, ingestion, and sometimes digestion (Stephens and Krebs 1986).

▶ **imitative foraging** *n.* An animal's simply going where its social group goes and eating what it eats (Wilson 1975, 51).

▶ **percussive foraging** *n.* An Aye-Aye's gentle tapping on wood surfaces that enables it to locate cavities in wood that may contain prey (Erickson 1991, 793). *Comment:* This primate might use echolocation, or a cutaneous sense in its extremely narrow middle digit, to find cavities and detect insect movement in them (Erickson 1991, 793).

SOME CLASSIFICATIONS OF FORAGING BEHAVIOR

I. Classification of behavior of an individual forager
   A. handling
   B. pursuing (*syn.* pursuit)
   C. searching (*syn.* search)
      1. detecting (*syn.* detection)
      2. encountering (*syn.* encounter)
   D. traveling
II. Classification by kinds of foraging behavior
   A. central-place foraging
   B. cooperative foraging
   C. imitative foraging
   D. percussive foraging
   E. risk-sensative foraging

▶ **risk-sensitive foraging behavior** *n.* An organism's foraging in which it responds to risk in food rewards as well as mean food reward (Stephens and Paton 1986, 1659).

grooming behavior See grooming.

guarding behavior *n.* A parent's or other older animal's watching over conspecifics other than mates (Immelmann and Beer 1989, 124).

cf. behavior: mate-guarding behavior

head lunge *n.* For example, in Humpback Whales: a kind of aggressive behavior (Baker and Herman 1985, 55).

head-neck dipping *n.* For example, in pelicans and swans: an individual's dipping its head into water in a rhythmic, snake-like fashion during bathing (Heymer 1977, 103).

v.i. head lunge

head nodding *n.*

1. In male blennid fishes: an individual's moving its head up and down while in a tube-like enclosure and encountering a rival (Heymer 1977, 103).
2. In many human cultures: one's moving one's head up and down to signify agreement (Heymer 1977, 103).

cf. head shaking

head scratching *n.*

1. In birds: an individual's scratching its head with its claws (Heymer 1977, 103).
2. In the Chimpanzee, Humans: a displacement activity made during concentration or while being perplexed (Heymer 1977, 103).

head shaking *n.*

1. For example, in ducks, swans: an individual's moving its head from side to side (Heymer 1977, 103).

2. A person's moving his head from side to side, signifying disagreement (Heymer 1977, 103).

cf. head nodding

head-up–tail-up behavior *n.* In some duck species: a drake's simultaneously pulling up his head sharply, whistling, bending his chin down towards his breast, and bending his tail region up steeply (Heymer 1977, 107).

helping behavior, help, helping *n.*
1. For example, in Long-Tailed Tits (birds), some species of jays: an individual's assisting a conspecific breeding pair (often its relatives) in raising its young but not reproducing itself (Wilson 1975, 448, 451).
2. For example, in Gray-Crowned Babblers (in usually nonbreeding helpers): an individual's showing parent-like behavior toward young not its offspring (Brown et al. 1982).

cf. altruism

homing behavior See homing.

homosexual behavior *n.* In many animal species including the Bonobo, Human, Japanese Beetle: sexual behavior that is directed toward a conspecific member of the same sex (*e.g.,* pseudocopulation) (Barrows and Gordh 1978, 341; de Waal 1989, 201; de Waal and Lanting 1997, 23, illustration).

cf. copulation: pseudocopulation: homosexual pseudocopulation; genital-genital rubbing; mating: homosexual mating; transvestism; sexual: homosexual

Comments: White et al. (2000, 892) consider "true homosexuality" to involve a male's actively seeking another male for sexual purposes, which has not been found in Fruit Flies (Tephritidae).

huddling behavior See behavior: contact behavior.

human behavior *n.*
1. A person's manner of conducting himself in external relations of life (*Oxford English Dictionary* 1971, entries from 1490).
   syn. bearing, demeanor, deportment, manners
2. "Conduct, general practice, course of life; course of actions *towards* or *to* others, treatment of others" (*Oxford English Dictionary* 1971, entries from 1754).
3. In psychology, the form of glandular and muscular activity characteristic of an individual in relation to internal, or external, stimuli, with special reference to emotional, linguistic, and other responses (Michaelis 1963).
   syn. behavior

hygienic behavior *n.* In the Honey Bee: a worker's removing a diseased, or dead,

immature bee from its cell and her nest (Rothenbuler 1964 in McFarland 1985, 39).

cf. behavior: nest-cleaning behavior

imitative behavior See learning: social imitative learning.

inborn behavior See innate (note).

infantile behavior, infantilism, infant-like behavior *n.* An adult animal's returning to child-like behavior, especially in connection with courtship or submissive behavior; *e.g.,* mature male mammals' imitating juvenile vocalizations during courtship or a human male's speaking in a distinctly childlike fashion using diminutives which may be accompanied by sucking breasts (Heymer 1977, 92; Immelmann and Beer 1989, 146).

cf. mouth-to-mouth feeding

ingestive behavior *n.* An animal's eating, drinking, or both (Scott 1950 in Lehner 1979, 64).

innate behavior *n.*
1. An animal's "behaviour that has not been changed by learning processes" (Tinbergen 1951, 2).
2. An animal's inherited behavior that is independent of its experience (early ethologists in McFarland 1985, 36).
3. An animal's "behavior that develops without obvious environmental influence" (McFarland 1985, 37).

cf. behavior: instinctive behavior

Comment: Some workers synonymize "innate behavior" and "instinctive behavior" in certain contexts; other's distinguish between these two behaviors.

innovative behavior, innovation *n.* In some reptile, bird, and mammal species, especially primates: an individual's new behavioral combination that arises neither from trial and error nor from imitation, but occurs spontaneously when the animal encounters a new situation and occurs as though the animal "anticipates" a new circumstance and produces the "correct" behavioral response for the first time (Immelmann and Beer 1989, 150); *e.g.,* a Chimpanzee's spontaneously putting fitted sticks together and reaching food with its new long stick or an animal's following a successful alternative route after its original path is blocked (Immelmann and Beer 1989, 150).

syn. insight learning (which is little used today because it carries a connotation of consciousness) (Immelmann and Beer 1989, 150)

cf. learning: insight learning

inspection behavior *n.* In some fish species: an individual's, or small group's, approaching a predator and then retreating

(George 1960 and Curio et al. 1983 in Magurran and Seghers 1990, 443).

cf. attack-cone avoidance

▸ **attack-cone avoidance** *n*. In some fish species: inspection behavior in which an individual selectively avoids a predatory fish's anterior portion when it approaches the predator (George 1960, etc. in Magurran and Seghers 1990, 443).

Comment: This behavior enables an inspecting fish to obtain information about a predator (*e.g.,* its size and motivation) while avoiding the dangerous zone in front of its mouth (attack cone) (George 1960, etc. in Magurran and Seghers 1990, 443).

instinct, instinctive behavior *n*.

1. An animal's complex reflex made up of units compatible with inheritance mechanisms and, thus, a product of natural selection that evolved with other aspects of an animal's life (Darwin 1859 in McFarland 1985, 363).

2. An animal's "purposeful action without "foresight of the ends and without previous education in the performance" (James 1890 in Immelmann and Beer 1989, 151).

3. An animal's behavior that is entirely genetically based ("innate behavior" of Lorenz 1932, 1937 in Dawkins 1986, 66).

   Notes: Lorenz considered all environmental effects on behavior under the category of learning; however, some workers point out that environmental factors (apart from those that promote learning) can determine whether a behavior develops in a usual way or if it is shown in the first place. Further, Dawkins (1986, 57) discusses the problems with equating "innate" with "inherited."

4. An animal's fixed-action pattern, *q.v.* (Lorenz 1937 in McFarland 1985, 363).

5. An animal's "behavior that has been experimentally demonstrated, in both its stimulus control and topography, to be independent of and unmodified by the operation of variables encountered in conditioning and learning, such as the occurrence of reinforcing stimuli" (Verplanck 1957).

   syn. inborn behavior

6. A behavioral difference between two individual animals, or two species, that is based, at least in part, on a genetic difference (Wilson 1975).

7. An organism's behavior that either is subject to relatively little modification in an individual's lifetime or varies very little throughout the population, or (preferably) both (Wilson 1975, 26).

Note: "This definition can never be precise, and it really has informational content only when applied to the extreme cases" (Wilson 1975, 26). "Closed instinct" is considered to be an unmodified instinct, or instinct in the strict sense, and "open instinct" is considered to be a kind of learning (Alcock 1979).

cf. learning

8. An animal's behavior that is highly stereotyped, more complex than the simplest reflexes, and usually directed at particular objects in its environment (Wilson 1975, 587).

   Note: Learning may or may not be involved in the development of instinctive behavior; "the important point is that the behavior develops toward a narrow, predictable end product" (Wilson 1975, 587).

9. An animal's behavior that is not learned, or acquired, during ontogeny (Heymer 1977, 28).

10. An animal's behavior that is present at birth, but not necessarily functional, and may appear only in the course of ontogenetic maturation (Heymer 1977, 28).

11. An animal's "highly stereotyped, species-specific behavior that is more complex than a simple reflex and is attributable largely to genetic influences" (Wittenberger 1981, 617).

Note: Dominance, subordinance, intolerance of outsiders, aggression, altruism, and the need to form amicable groups might be human instincts (Eibl-Eibesfeldt 1973 in Heymer 1977, 194).

syn. preprogrammed behavior, programmed behavior, species-specific behavior (Heinroth 1911 in Heymer 1977, 178)

cf. behavior: innate behavior, instinct, instinctive behavior pattern, learning

Comments: Dawkins (1986, chap. 5) explains why an "innate character" is not synonymous with a "developmentally fixed, or genetic, character" and why "instinctive behavior" is not synonymous with "innate behavior;" however, based on certain subsets of criteria, some authors seem to synonymize "innate" and "instinctive behavior." McFarland (1985, 363–364) explains problems with early definitions of "instinct." McFarland (1987, 310) explains that modern research regards previously designated features of "instinct" (genetically determined parts, reflexes, and motivational aspects) as separate issues. Historically, "instinct" and "innate behavior" have been used to indicate separate concepts; these terms are sometimes used interchangeably today (Immelmann and Beer 1989, 151).

[Latin *instinctus*, past participle of *instinguere*, to impel < root *-stig*]

▶ **primary instinct** *n.* Instinct that directly results from natural selection (Romanes 1882 in Gherardi 1984, 381).

▶ **secondary instinct** *n.* Animal habits that are unconsciously formed and directly inherited in some unknown manner (Romanes 1882 in Gherardi 1984, 381).

intention movement *n.*

1. For example, in some bird, fish, and mammal species, including Humans, etc.: the beginning of an innate behavior pattern that communicates an individual's readiness to perform a specific behavior and indicates its "mood" (Heinroth 1911 in Tinbergen 1951, 141).

2. An animal's preparatory motions shown prior to a complete behavioral response (*e.g.,* a crouch before a leap or a snarl before a bite) (Daanje 1950, Andrew 1956 in Wilson 1975, 225; Wilson 1975, 560, 587).

cf. behavior: displacement behavior

Comments: An intention movement is not necessarily followed by the behavior that it seems to indicate. Some intention movements have been ritualized into communicative signals (Wilson 1975, 560). "Intention" in this term does not imply subjective intent (Immelmann and Beer 1989, 155).

▶ **fight-intention movement** *n.* An intention movement that communicates combat readiness to a rival (Heymer 1977, 96).

cf. irrelevant behavior

▶ **flight-intention movement** *n.* An intention movement that communicates fleeing readiness to a rival (Heymer 1977, 96).

intentional behavior *n.*

1. A person's behavior that involves a mental representation of a goal that is instrumental in guiding his behavior; this representation is not necessarily conscious (McFarland 1985, 500).

2. A person's having a mental image, or images, of future events in which he "pictures" himself as a participant and makes a choice as to which image he will try to bring to reality (Griffin 1976 in McFarland 1985, 526).

cf. intent, intentionality

intraspecific behavior *n.* An animal's behavior directed toward a conspecific (Immelmann and Beer 1989, 158).

cf. behavior: social behavior

investigative behavior *n.* An organism's exploring its social, biotic, and abiotic environment (Scott 1950 in Lehner 1979, 64).

cf. behavior: exploratory behavior

joking behavior *n.* A person's teasing, mocking, and kidding around that might serve to divert aggression, educate, or both (Heinz 1967, Eibl-Eibesfeldt 1972 in Heymer 1977, 155).

v.i. joke

juvenile behavior *n.* Behavior shown by a juvenile, or immature, animal that is never shown during its later life, *e.g.,* food begging by juvenile swallows, alarm behavior of Herring Gull chicks, or larval behavior of Lepidoptera (McFarland 1985, 36).

killing behavior, killing *n.* An animal's causing the death of another animal, *e.g.,* a predator's killing a prey or an animal's committing infanticide or siblicide, *q.v.* (Immelmann and Beer 1989, 164).

cf. cannibalism; -cide: infanticide

Comments: Wilson (1975, 246–248, 504) uses "murder" for "killing behavior" and indicates that it is done without reference to morality of a killing act; de Waal (1989, 69) suggests that "murder" implies intent to kill in Chimpanzees which cannot be proved. Many animal species kill conspecifics, including praying mantids, Honey Bees, Lions, Chimpanzees, and Humans (de Waal 1989, 6). Many animal species kill heterospecifics in predating or fighting with them over resources (*e.g.,* ants or stingless bees).

learned behavior See learning.

leisure behavior, luxury *n.* A behavior that completely disappears from an animal's activities when it has to ration its time (McFarland 1985, 452).

ligulation See ligulation.

lobtail *n.* A whale's slapping its tail flukes against the surface of water (Baker and Herman 1985, 55).

v.i, v.t. lobtail

cf. fin

marking behavior, marking *n.*

1. In some bee and desert-woodlice species, many mammal species: an individual's making, or strengthening, its scent mark (urine, feces, saliva, pheromones, or a combination of these things, depending on the species) (Immelmann and Beer 1989, 258).

syn. scent marking, scent-marking behavior (Clapperton 1989, 436)

2. An animal's making, or strengthening, any kind of a mark (*e.g.,* scent mark or acoustic, or optical, signal) (Immelmann and Beer 1989, 179).

cf. scent mark

Comment: As a type of bookkeeping system, a red fox scent-marks a place where food, or food odor, remains after it has eaten most or all of the food in that particular location (Henry 1993, 90).

▶ **allomarking** *n.* An animal's marking a conspecific (Immelmann and Beer 1989, 259).
cf. behavior: marking behavior: automarking

▶ **automarking** *n.* In some primate species; many deer species; the Hedgehog, Muskox: an individual's applying an odorous substance to its own body and using the odor in communication (Immelmann and Beer 1989, 259, 262); *e.g.,* a deer's automarking its belly and forelegs with sprayed urine, resulting in almost black staining of these areas (Heymer 1977, 83).
syn. chemical self-marking (Immelmann and Beer 1989, 259), self-marking

▶ **marking drumming** *n.* In hamsters and mice: an individual's marking the ground with its hind paws after rubbing them across its scent glands (Heymer 1977, 129).

▶ **self-marking** See behavior: marking: automarking.

▶ **territory marking, territory signaling** *n.* An animal's delineating its territory with signals or markings; *e.g.,* many dragonfly and fish species mark with special conspicuous movements; electric fish may mark with weak electric fields; many mammal species scent mark with gland secretions, urine, or feces; many bird species and the howler monkey mark with songs or other vocalizations (Heymer 1977, 147).

▶ **urine marking** *n.* In many mammal species: an individual's marking a landmark with its urine.

▶ **urine spraying** *n.*
1. In some mammal species, especially rodents: an animal's spraying a partner with urine (Altmann 1969 in Heymer 1977, 83).
2. In some rodent species, hares: an individual's (usually a male) forceful ejection of urine used to mark inanimate objects in its territory or conspecifics (rodents, hares) (Immelmann and Beer 1989, 322).
syn. urine marking

▶ **urine washing** *n.* In several species of less-derived and more-derived primates: an individual's moistening its palms with urine and wiping it on its foot soles with rapid swipes; this leaves an odor trail as the animal moves about (Hill 1938 and Eibl-Eibesfeldt 1953 in Heymer 1977, 83).

mashing, scamming, scrumping, shacking *n.* U.S. slang, a person's having sexual relations with another person (Gabriel, 1997, 22).

Comments: These are terms used by U.S. people (especially students) in the 1990s that suggest a lack of emotional and romantic content (Gabriel, 1997, 22).

mate guarding, mate-guarding behavior *n.* In many species of invertebrates and vertebrates: an individual's staying near before or after copulation, or both, and stopping others from copulating with its mate (Wilson 1975, 322; Barrows and Gordh 1978, 341; Convey, 1989, 56; Immelmann and Beer 1989, 124; Dunham and Hurshman 1990, 976).
syn. "wife watching" (Immelmann and Beer 1989)
cf. escort
v.t. mate guard

▶ **postcopulatory mate guarding** *n.* For example, in some cricket, damselfly, and dragonfly species; the Cactus Fly, Dung Fly, Human, and Japanese Beetle: a male's mate-guarding behavior shown after initially, or completely, copulating with his mate, depending on the species (Jacobs 1955, Parker 1970, Mangan 1979 in Parker 1984, 17; Barrows and Gordh 1978, 341; Waage 1984, 253; Smith 1984, 636; Sakaluk 1991, 207).
Comment: In decorated crickets, mate guarding deters rivals from courting a recently mated female (Sakaluk's 1991, 207).

▶ **precopulatory mate guarding** *n.* For example, in some insect species; many crustacean and spider species; *Bonellia* echiuroid worms, Humans: a male's staying in close association with a female before copulation; this guarding can involve physical attachment of the pair in some species (Austad 1984, 245; Parker 1984, 19; Dunham et al., 1986, 1680).
syn. suitor phenomenon (Robinson and Robinson 1980 in Austad 1984, 245; Smith 1984, 636)
Comment: In Humans, a female may also guard a male mate. Humans mate guard in many ways, including the use of societal rules and laws and male anticuckoldry techniques (Dickeman 1979, 1981 in Smith 1984).

maternal behavior See care: parental care: maternal care.

mating behavior See mating.

mimetic behavior *n.* One animal's making a species-specific response after perceiving a conspecific second animal's making the same response, under conditions where it can be demonstrated that the latter animal's response is the stimulus for the former animal and that no opportunity for discrimination and differentiation training has been given to either animal (Verplanck 1957).

cf. behavior: imitative behavior; learning: social imitative learning

momentary behavior See behavior event.

mounting, mounting behavior *n.* A male animal's adopting a position that facilitates copulation (Immelmann and Beer 1989, 194–195).

syn. sexual mounting

cf. bout: mount bout; copulation: pseudocopulation

Comment: In mammals, mounting usually consists of a male's standing on his hindlegs and grasping a female's flanks from behind (Immelmann and Beer 1989, 194–195).

necrophoric behavior *n.* In ants, parents of some bird species, elephants, mothers of some primate species, the Honey Bee: an individual's carrying the corpse of a dead conspecific (Immelmann and Beer 1989, 198).

nest-cleaning behavior *n.* In social-insect and many songbird species: an individual's removing fecal matter, debris, remains of dead animals, or some other object from its nest (Immelmann and Beer 1989, 200).

cf. behavior: hygienic behavior

nonepigamic behavior See behavior: epigamic behavior: nonepigamic behavior.

nurturance, nurturing behavior *n.* An animal's behavior that is essential for ensuring the survival of relatively helpless offspring (Morris et al. 1995, 1697).

cf. care

Comment: There has been considerable ambiguity in researchers' use of the term "nurturance;" they have used broad definitions, often including any form of cooperative behavior (Morris et al. 1995, 1699). A more fine-grained definition of nurturance would be useful.

[Old French *norriture, norreture* < Late Latin *nutrire,* to nourish]

operant behavior *n.* An animal's totality of operant responses (Verplanck 1957).

organism behavior *n.*
1. Any observable action, or response, of an organism (Lincoln et al. 1985).
2. The response(s) of a single organism, group, species, or other taxon, to environmental factors (Lincoln et al. 1985).

syn. behavior

pairing behavior See copulation, courtship.

parental behavior See care: parental care.

paternal behavior See care: parental care: paternal care.

phoretic behavior See symbiosis: phoresy.

play behavior See play.

playing dead, playing O'possum See behavior: defensive behavior: playing dead.

postcopulatory behavior See behavior: copulatory behavior.

pouting, pouting behavior *n.* A person's hanging his head, turning his eyes away, and making a long face (*e.g.,* in children who have just lost an argument) (Heymer 1977, 157).

preceptive behavior See behavior: soliciting behavior.

precoital behavior, precopulatory behavior See courtship.

preprogrammed behavior See behavior: instinct.

prey-catching behavior *n.* A predator's behavior that leads to prey capture (Heymer 1977, 42).

proceptive behavior *n.*
1. A female's reactions that constitute her initiative in establishing and maintaining a sexual interaction with a male (Beach 1976 in Dewsbury 1978, 237), including genital presentation in some primate species (Immelmann and Beer 1989, 233).
2. Attraction of estrous females to males (Hinde 1982, 155).

syn. proceptivity

cf. attractivity, receptivity

programmed behavior See behavior: instinct.

protean behavior See behavior: escape behavior: protean behavior.

pseudosexual behavior See behavior: sexual behavior: pseudosexual behavior.

puddling *n.* Butterflies' congregating at puddles where they imbibe salts and other nutrients that are found in the evaporating water (Anderson 1995, 88).

v.i. puddle (Anderson 1995, 88)

Comments: Butterflies and moths also congregate at excrement, rotting fruit, urine, and other nutritious substances (Anderson 1995, 88; personal observation). Bees also congregate at salt sources (Barrows 1974, 189).

[Middle English *podel,* diminutive of Old English *pudd,* ditch]

purposive behavior *n.* Behavior that is considered with respect to its goals (Verplanck 1957).

cf. teleology.

Comment: Verplanck (1957) discusses problems with "purposive behavior" in view of the fact that the definition of "goal" varies. "... purposive behavior is not a very useful concept, for it is neither empirical fish nor theoretical fowl."

rapping behavior *n.* In Hermit Crabs: a short, rapid tapping that an initiator in shell-exchange interaction makes against a noninitiator's shell (Hazlett 1975 in Hazlett 1987, 218).
Comment: A noninitiator can obtain information about an initiator's shell size from the rapping (Hazlett 1987, 218).
recruitment behavior See recruitment.
reflex behavior See reflex.
relict behavior *n.*

1. A behavior that has become functionless (vestigial) yet persists, usually in a fragmentary or greatly reduced form; *e.g.,* performance of balancing movements with their tails which are no longer appropriate for balancing in some macaque species or egg-rolling behaviors typical of their ground-nesting ancestors shown by many pigeon species and tree-nesting rails (Immelmann and Beer 1989, 252).
   syn. evolutionary vestige (Immelmann and Beer 1989, 252)
2. A species' behavior that remains essentially intact long after an environmental stimulus that influenced its evolution vanished (Yoon 1996d, C1).

Notes: California Ground Squirrels from populations that have been free of snakes for 30,000–70,000 years still clearly recognize rattlesnakes and exhibit stereotypic anti-rattlesnake behavior (Coss et al. in Yoon 1996d, C1). Stickleback fish from a population that has long been free of sculpin (a dangerous predator of these fish) immediately engage in stereotypic antisculpin behavior when confronted with a sculpin (Foster et al. in Yoon 1996d, C6). Human females report that they are fearful about being attacked from below, and males report that they are fearful about being attacked from the side. This behavior might have evolved when female hominids spent more time in trees than males, and males spent more time on the ground the females. A possible relict behavior is the fast running of the North American Pronghorn Antelope, which can run nearly 60 miles per hour (Byers in Yoon 1996d, C1). This running ability might have evolved in response to now extinct predators, including Jaguars, Lions, Long-Legged Hyenas, North American Cheetahs, and Short-Faced Bears.
cf. organ: relict
reproductive behavior *n.* Any behavior relating to breeding, including pair formation and mating; some investigators include territoriality, nest building, brooding, and parental care in reproductive behavior (Immelmann and Beer 1989, 253).
cf. copulation, mating

respectful behavior, respect of possession *n.* In Hamadryas Baboons: an individual's inhibition of taking possessions of conspecific animals (*e.g.,* males' not taking over mates or food of others) (Kummer et al. 1973 in Sigg and Falett 1985, 978).
cf. showing respect
respondent behavior *n.* An animal's totality of behavior respondents (Verplanck 1957).
risk-sensitive behavior *n.* An animal's showing different behavioral options based on variance in its reward or punishment (Brown 1987a, 306).
ritualized behavior *n.*

1. An animal's behavior that appears in most members of its species, is relatively invariant in its topography, and is typically a social releaser (Verplanck 1957).
2. An animal's behavior that appears in repertoires of members of several related species and usually varies in stereotypy among species (Verplanck 1957).

syn. ritualized response
cf. emancipation; grooming: ritualized grooming; ritual; ritualization
Comments: "Ritualized responses occur in sharply restricted situations" (*e.g.,* grass pulling during Herring Gull threat behavior); theoretically, they have become specialized through evolution, most often as a social releaser, and are often associated with a special marking of fur or feathers (*e.g.,* ritualized preening in the courtship of some anatid duck species that reveals a usually hidden, light-colored patch) (Verplanck 1957). "Ritualized behavior" has been traced evolutionarily to intention movement, ambivalent behavior, displacement activities, or redirection, augmented by additional secondary signal characteristics in most cases studied (Immelmann and Beer 1989, 255).
rutting behavior *n.* Animal sexual behavior, especially in deer and other ruminants (Michaelis 1963; Immelmann and Beer 1989, 257).
syn. estrus (Michaelis 1963)
cf. rut
saliva spreading *n.* An animal's spreading its saliva on its body surface, often for cooling purposes (Bligh and Johnson 1973, 954).
cf. temperature regulation: behavioral temperature regulation
self-medicating behavior, self-medication See pharmacognosy: zoopharmacognosy.
selfish behavior, selfishness *n.*

1. A person's devotion to, or concern with, his own advantage, or welfare, to the exclusion of regard for others (*Oxford*

English Dictionary 1971, entries regarding selfish from 1640).

2. Sociobiologically, an individual organism's doing something that benefits its own genetic fitness at the expense of the genetic fitness of conspecifics (Wilson 1975, 594).

3. Sociobiologically, an individual organism's doing something that benefits itself while harming or imperiling others that it affects (Wittenberger 1981, 621). *Note:* Benefit and harm can be measured in terms of direct fitness [to the performer] (Wittenberger 1981, 621).

cf. altruism; behavior: spiteful behavior; cooperation

▶ **weak selfishness** *n.* An interaction between an actor and recipient in which the actor has a fitness increase and the recipient has no fitness change (Gadagkar 1993, 233).

sentinel behavior *n.* For example, in two baboon species; the Pigtail Macaque and Verbet Monkey: an alert nonforaging individual's stationing itself in a prominent place while members of its group forage nearby (Horrocks and Hunte 1986, 1566).

sexual behavior *n.* An organism's courtship and copulatory behavior (Scott 1950 in Lehner 1979, 64).

cf. courtship, copulation

▶ **pseudosexual behavior** *n.* In some lizard species (whiptails, racerunners): a reproductively active female's pursuing, mounting, and gripping another such female with her limbs and jaws (Halliday and Adler 1986, 93).

cf. copulation: pseudocopulation
Comment: This behavior might synchronize ovulation and oviposition in a female group (Halliday and Adler 1986, 93).

shell-searching behavior *n.* In pagurid crabs (Hermit Crabs): an individual's searching for a new shell when it outgrows its present one or when its shell is artificially removed (Heymer 1977, 151).

shelter-building behavior *n.* An animal's using parts of its environment, body products, or both for producing a shelter for itself, other organisms, or both (inferred from Fitzgerald 1995, 30).

Comments: In Lepidoptera (butterflies and moths), larvae of many species make shelters from plant parts, silk, or both. Some species roll leaves; some cut and fold over leaves; some fasten leaves and twigs together with silk; and so forth (Fitzgerald 1995).

shelter-seeking behavior *n.* An organism's seeking out and coming to rest in a favorable part of its environment (Scott 1950 in Lehner 1979, 64).

showing-the-nest behavior *n.* In many bird and fish species: a male's drawing a female toward his nest, or nesting hole, which may be followed by his showing her nesting behavior (broodiness) and her occupying his nest (Heymer 1977, 122).

signaling behavior See display.

snoring *n.* A sleeping animal's breathing through its nose and open mouth with a hoarse, rough noise and rattling vibrations of its soft palate (Michaelis 1963).

n. snorer
v.i. snore
Comments: Snoring occurs in mammals including the Domestic Dog, Human, and Orangutan (Galdikas 1995, 11).

[Middle English *snoren*; akin to Middle Low German *snorren*, to hum, drone]

social behavior See sociality.

▶ **selfish social behavior** *n.* Behavior of a focal organism that includes overt selfishness (promoting its direct fitness), such as aggressiveness and territoriality, and quasi-altruistic selfishness, such as cooperation and reciprocal (temporary) altruism, *q.v.* (West Eberhard 1975, 7).

soliciting behavior *n.* For example, in primates: a female's movements and postures that invite a male to copulate with her (Immelmann and Beer 1989, 281).

syn. preceptive behavior

▶ **presenting** *n.* In some species of Old World Monkeys: Soliciting behavior in which a female shows her swollen red sex skin to a male(s) (Dewsbury 1978, 87).

syn. genital presentation (Immelmann and Beer 1989, 281)
v.i. present

spacing behavior *n.* In many animal species, including social insects, mockingbirds, Humans: behavior (including territoriality and individual-distance behavior) that maintains an appropriate distance between individual organisms or groups (Heymer 1977, 52).

cf. dispersion

species-specific behavior *n.* Behavior that is unique to one species (Brown 1975, 400).

See instinct.
syn. species-characteristic behavior, species-typical behavior (G.M. Burghardt, personal communication)
cf. behavior; behavior: innate behavior

spinning *n.* A worker Honey Bee's rotating her body and moving erratically due to pesticide poisoning (Robinson and Johansen 1978, 16).

spite, spiteful behavior *n.* "Behavior that lowers the genetic fitnesses of both the perpetrator and the individual toward which the behavior is directed" (Wilson 1975, 595).

cf. altruism; behavior: selfish behavior

Comment: "Harm is measured in terms of direct fitness" (Wittenberger 1981, 622).

▸ **weak spite** *n.* An interaction between an actor and recipient in which the actor has no fitness change and the recipient has decreased fitness (Gadagkar 1993, 233).

sponge carrying *n.* A Bottlenose Dolphin's transporting a sponge on its rostrum (Smolker et al., 1997, 454).

Comment: Available data indicate that sponge carrying is probably involved in foraging and a likely case of tool use, *q.v.* (Smolker et al. 1997, 454).

spontaneous behavior *n.*

1. Behavior that occurs in the ostensible absence of any stimuli that can be shown to elicit, release, or set the occasion for its occurrence (Verplanck 1957). *Notes:* To call a behavior "spontaneous" indicates our current ignorance of the events that control it. Spontaneous behavior is not capricious. "This term would be synonymous with operant behavior, except that operants that have come under the control of discriminative stimuli are no longer spontaneous" (Verplanck 1957).

2. Internally caused and controlled behavior that is neither released nor maintained by external or peripheral stimulation, *e.g.,* some vocalizations of turkeys and crickets, initiation of appetitive behavior, and some embryo movements (Immelmann and Beer 1989, 289).

syn. autonomous behavior (Immelmann and Beer 1989, 289)

cf. activity: vacuum activity

stereotype, stereotyped-motor acts, stereotypic behavior See stereotypy.

submissive behavior See submission.

superstitious behavior *n.* For example, in Domestic Pigeons: an individual's repeating certain behavior that it associates with a reward because it once showed the behavior by chance when a reward was given (*e.g.,* a pigeon's turning quickly in a small circle or tic-like head movements) (Skinner 1948 and Herrnstein 1966 in Hinde 1970, 449).

symbolic behavior *n.*

1. An animal's intention movement that became ritualized and is used for intraspecific communication (Lorenz 1941 in Immelmann and Beer 1989, 301).

2. An animal's display (Immelmann and Beer 1989, 301).

syn. symbolic action, symbolic movement

3. A person's verbal behavior (Verplanck 1957).

4. A hypothesized class of behavior that cannot necessarily be directly observed (Verplanck 1957).

5. The observed behavior from which symbolic behavior (def. 4) is inferred (Verplanck 1957).

6. Loose and colloquially, thinking (Verplanck 1957).

cf. thinking

Comment: This term was used in earlier literature, but without consistency; it is seldom used today (Immelmann and Beer 1989, 301).

territorial behavior See territoriality.

threat, threat behavior *n.*

1. An animal's behavior that produces flight behavior of some strength in another, usually conspecific animal when it occurs in its presence; threat behavior most often elicits other agonistic behavior in a threatened animal (Verplanck 1957).

syn. apotreptic behavior (Barnett 1981 in Immelmann and Beer 1989, 18)

2. In many animal species: avoidance behavior with aggressive elements, *e.g.,* an intimidation display with aspects of threat (hair bristling, feather ruffling, raising skin folds and crests, teeth displaying, horn displaying, making sounds, etc.); threat always contains components of attacking and fleeing, and it can express an animal's readiness to fight (Heymer 1977, 55; Immelmann and Beer 1989, 311).

3. Behavior that occurs in agonistic contexts and commonly signifies hostility toward, or intent to attack, another animal (Hand 1986).

cf. behavior: appeasement behavior, treptic behavior: epitreptic behavior; display: bared-teeth display, bared-teeth-scream display, challenge display; facial expression: ear-flap threat, open-mouth threat [apotreptic behavior coined by Barnett 1981 in Immelmann and Beer 1989, 18]

▸ **canine-tooth threat** *n.* In several deer species: a male's drawing back his upper lip along his entire upper jaw, exposing his rudimentary canine teeth in the corner of his mouth, making his eyes appear white to his rival due to his standing diagonal to his rival with only the side of his eyeball exposed, and erecting his neck and back hair (Bützler 1974 in Heymer 1977, 58).

Comment: This threat is associated with specific types of confrontations (*e.g.,* rank changing) (Heymer 1977, 58).

▶ **defensive threat** *n*. An animal's showing a threat while it withdraws from a combatant during an agonistic encounter (Eibl-Eibesfeldt 1957 in Heymer 1977, 50).

▶ **offensive threat** *n*. An animal's showing a threat while it approaches a combatant during an agonistic encounter (Eibl-Eibesfeldt 1957 in Heymer 1977, 50).

▶ **phallic threat** See display: genital display: genital presentation.

▶ **protected threat** *n*. In baboons: a tactic used by females of a one-male group ("harem") that involves confronting a rival only when the group's male is nearby (Wilson 1975, 534).

▶ **teeth grinding** *n*. In rodents and ruminants that use their teeth as weapons: an individual's noisily rubbing its teeth together by sideways mandibular movement (Heymer 1977, 204).

tongue-lashing behavior See ligulation.

trekking *n*.
1. In South Africa, a person's traveling by ox wagon (Michaelis 1963).
2. Traveling, especially slowly or arduously (Michaelis 1963).
3. In South Africa, an ox's drawing a vehicle or load (Michaelis 1963).
4. In Yanomami, an entire community's packing up its possessions, abandoning its communal shelter, and going into the forest in one to several groups to hunt and gather wild foods (Good 1995, 57).

n. trek
v. trekked
v.i. trekking
[Dutch *trekken*, to draw, travel]

treptic behavior, treptics *n*. Animal social interactions of approach and withdrawal during interactions.
[coined by Barnett 1981 in Immelmann and Beer 1989, 18, who indicate that this term and apotreptic and epitreptic behavior have not caught on in the literature]
cf. intimidation

▶ **apotreptic behavior** See behavior: threat.

▶ **epitreptic behavior** *n*. An animal's behavior that tends to cause a conspecific to approach.
[coined by Barnett 1981 in Immelmann and Beer 1989, 18]

trial-and-error behavior See kinesis: klinokinesis.

triggered behavior *n*. Behavior that continues independently of external stimulation after its first elicitation by a stimulus (*e.g.*, a fixed-action pattern) (Dewsbury 1978, 15).

unlearned behavior See instinct.
cf. behavior: innate behavior; innate

vacuum behavior See activity: vacuum activity.

verbal behavior *n*.
1. A person's vocalization, writing words, response to written words, or a combination of these activities (Verplanck 1957).
2. A person's behavior (including vocalization, writing, gesturing, and other forms of communication) whose reinforcement is contingent upon stimulation of and response by another individual (Verplanck 1957).

vestigial behavior See behavior: relict.

vigilant behavior See vigilance.

wallowing *n*.
1. An animal's rolling about (Michaelis 1963).
2. A animal's actively being immersed in mud, sand, snow, etc. (Michaelis 1963). *Note*: This is often done with pleasure in Humans (Michaelis 1963).
3. A animal's trashing about, floundering (Michaelis 1963).
4. A person's living self-indulgently: to wallow in sensuality (Michaelis 1963).
5. An animal's spreading fluid (*e.g.*, mud, water, or urine) on its body surface, resulting in evaporative heat loss (Bligh and Johnson 1973, 958).
cf. temperature regulation: behavioral temperature regulation
n. wallower
v.i. wallow
[Old English *wealwian*]

warning behavior See behavior: alarm behavior.

♦ **behavior chain** See chain.

♦ **behavior-chain diagram** See -gram: behavior-chain diagram.

♦ **behavior event, behavioral event** *n*. An animal's change in behavior states that approaches being an instantaneous occurrence (*e.g.*, a bird's taking flight) (Altmann 1974 in Lehner 1979, 69).
syn. momentary behavior (Sackett 1978 in Lehner 1979, 69)
cf. behavior state

♦ **behavior hypothesis** See hypothesis: population-limiting hypotheses: behavior hypothesis.

♦ **behavior modification, behavioral modification** *n*. A technique that seeks to modify animal (including human) behavior, in which rewards and reinforcements, or punishments, are used to establish desired habits or patterns of behavior (Michaelis 1963).

♦ **behavior pattern** See classification of behavior (table); pattern: behavior pattern.

♦ **behavior-sequence diagram** See behavior-chain diagram.

behavior state, behavioral state *n.*
The behavior that one or more animals is
engaged in; an ongoing behavior (*e.g.,* a
bird's flying) (Altmann 1974 in Lehner 1979,
69).
syn. duration-meaningful behavior (Sackett
1978 in Lehner 1979, 69)
cf. behavior event

behavior stereotype See stereotypy.

behavior syndrome *n.* A behavioral
category comprised of several smaller behavioral categories (Immelmann and Beer
1989, 303).

behavior type *n.* One kind of behavior
in Scott's (1950 in Lehner 1979, 64) classification: ingestive, investigative, shelter-seeking, eliminative, sexual, epimeletic, et-
epimeletic, allelomimetic, and agonistic behavior.

behavioral act, act *n.* Part of a behavior
pattern, *e.g.,* parts of flight: taking off, flying,
and landing (Lehner 1979, 65).
See classification of behavior (table).
cf. behavior pattern; behavior type; behavioral act: component part of a behavioral act;
episode

**component part of a behavioral
act** *n.* Part of a behavior act (*e.g.,* parts of
the act of taking off), movements of particular body parts, and neurological activity (Lehner 1979, 65).
cf. behavioral act, behavior pattern, behavior type

behavioral biology See study of: biology: behavioral biology.

behavioral catalog See -gram: ethogram.

behavioral characteristic See character: behavioral characteristic.

behavioral deficit *n.* An animal's abnormally low production of a particular
behavior (Immelmann and Beer 1989, 137).
cf. behavior: abnormal behavior; hypertrophy

behavioral dichotomy See courtship:
behavioral dichotomy.

behavioral dimorphism See polyethism.

behavioral ecology See study of: ecology: behavioral ecology.

behavioral embryology See study of:
embryology: behavioral embryology.

behavioral endocrinology See study of:
endocrinology: behavioral endocrinology.

behavioral event See behavior event.

behavioral genetics See study of: genetics: behavioral genetics.

behavioral inventory See -gram: ethogram.

behavioral-isolating mechanism See
mechanism: isolating mechanism: sexual
isolating mechanism.

behavioral mimicry See mimicry: behavioral mimicry.

behavioral modification See behavior modification.

**behavioral ontogenesis, behavioral
ontogeny** See -genesis: ethogenesis.

behavioral phylogeny See study of:
phylogeny: behavioral phylogeny.

behavioral physiology See study of:
physiology: behavioral physiology.

behavioral polymorphism See polyethism.

behavioral program See program: behavioral program.

**behavioral repertoire, behavioral
repertory** See -gram: ethogram.

behavioral resilience *n.* A measure of
the extent to which any of an animal's
behaviors can be reduced in duration by its
other behaviors (McFarland 1985, 452).
cf. behavior: leisure behavior
Comment: Low-resilience behaviors are reduced in duration rather than high-resil-
ience ones when an animal faces a time
constraint (McFarland 1985, 452).

behavioral scaling *n.*
1. Variation in the magnitude, or in the
qualitative state, of an animal's behavior
that is correlated with stages of its life
cycle, population density, or certain parameters of its environment (Wilson 1975,
20).
Note: Behavioral scaling is likely to be
adaptive (Wilson 1975, 20).
2. "The range of forms and intensities of a
behavior that can be expressed in an
adaptive fashion by the same society or
individual organism" (Wilson 1975, 579).
Note: For example, a society may be organized into individual territories at low
densities but shift to a dominance system
at high densities (Wilson 1975, 579).
syn. behavior scaling

behavioral sciences See study of: behavioral sciences.

behavioral silence See learning: behaviorally silent learning.

behavioral state See behavior state.

behavioral system See cycle: functional
cycle.

behavioral temperature regulation
See temperature regulation: behavioral temperature regulation.

behavioral tendency See potential: specific-action potential(ity).

behavioral unit *n.* The kind of behavior
that one measures in a study (*e.g.,* fights,
songs, journeys, blows, notes, or footsteps)
(Lehner 1979, 63).
Note: Lehner (1979, 64–67) lists classifications of behavioral units.

a – c

♦ **behavioral vestige** See relict.
♦ **behaviorism** (*Brit.* **behaviourism**) *n.*
1. A theory and method of psychological investigation that is based on the objective, experimental study and analysis of behavior (*Oxford English Dictionary* 1989, entries from Watson 1913).
2. The psychological theory that the behavior of Humans and nonhuman animals is determined by measurable external and internal stimuli acting independently (Michaelis 1963).
3. The view that "psychology studies behavior itself, rather than mental events" (McFarland 1985, 8), such as sensation, emotion, mind, consciousness, will, and imagery (Collocot and Dobson 1974).
syn. behaviorist school of psychology
cf. study of: behavior, behavioral biology; psychology: behaviorism
Comments: "In its strong form, behaviorism rejects all reference to inner processes in the explanation of behavior." Behaviorism was initiated by John B. Watson in 1913 (McFarland 1985, 8). Eysenck et al. (1979) describe "Watsonian behaviorism," "Tolman's purposive behaviorism," etc.

American behaviorism See study of: psychology: classical animal psychology.

descriptive behaviorism *n.*
1. The learning theory of B.F. Skinner which explains behavior largely in terms of linguistically formulated observations and their interrelationships (Eysenck et al. 1979, 121).
2. "A collective term for the techniques of operant conditioning by which operants, or emitted responses, are examined without necessary reference to originating rather than reinforcing stimuli" (Eysenck et al. 1979, 121).

ethical behaviorism *n.* One of several theories of human ethical behavior that proposes that "moral commitment is entirely learned, with operant conditioning's being the dominant mechanism; in other words, children simply internalize the behavioral norms of the society" (Scott 1971 in Wilson 1975, 562–563).
See ethical institutionism.
Comment: This theory is opposed by the "developmental-genetic concept" (Scott 1971 in Wilson 1975, 562–563).

logical behaviorism *n.* Behaviorism that has the view that any statements about mental events are meaningless (Dawkins 1980, 12).

metaphysical behaviorism, radical behaviorism *n.* Behaviorism that denies the existence of mind; asserts that conscious, or mental, processes are beyond the realm of scientific inquiry; and emphasizes observability, operationalism, falsifiability, experimentation, and replication (Blackburn 1984, 291–292).
Comments: Skinner's "metaphysical behaviorism" is often exaggerated by his followers; Skinner makes several references to the essential study of thoughts in psychological studies (Blackburn 1984, 291–292).

methodological behaviorism *n.* Behaviorism that acknowledges that subjective experiences may exist, but it is no concern of science to study them (Dawkins 1980, 12).

molar behaviorism, molarism See behaviorism: purposive behaviorism.

molecular behaviorism *n.* Behaviorism that seeks to explain all behavior as built from the smallest possible units (Eysenck et al. 1979, 121).

purposive behavior, purposive behaviorism *n.*
1. The concept that behavior is not a mechanical succession of causes and effects but is a set of actions strictly linked and oriented towards an organism's achieving a goal (Tolman 1932 in Gherardi 1984, 371).
cf. teleology
2. The learning theory of E.C. Tolman, who does not start from the smallest possible elements of behavior but instead considers the holistic, purposive units or species (Eysenck et al. 1979, 121).
syn. molar behaviorism, molarism (Eysenck et al. 1979, 121)

radical behaviorism See behaviorism: metaphysical behaviorism.

♦ **behaviorist** *n.* A proponent, or practitioner, of behaviorism, *q.v.* (McKechnie 1979; McFarland 1985, 9).
adj. behaviorist, behavioristic
♦ **behaviorist school of psychology** See behaviorism.
♦ **behavioristic method** See method: behavioristic method.
♦ **Behrens-Fisher t test** See statistical test: multiple-comparisons test: planned-multiple-comparisons procedure: Behrens-Fisher *t* test.
♦ **Bekner-Marshall hypothesis** See hypothesis: hypotheses regarding the origin of the Cambrian explosion: Bekner-Marshall hypothesis.
♦ **bell-shaped distribution** See distribution: normal distribution.
♦ **bellowing** See ²group: bellowing.
♦ **beneficence** See symbiosis: beneficence.
♦ **beneficent behavior** See behavior: beneficent behavior.
♦ **beneficiary** *n.* A person who receives benefits or favors (*Oxford English Dictionary* 1971, entries from 1662).

super-beneficiary *n.* An organism that is a very efficient at being a recipient of help from conspecific workers, or subordinates (*e.g.,* queens of some kinds of social insects) (West Eberhard 1975, 14).
cf. super-donor

◆ **benefit** *n.*
1. Any consequence of a phenotypic trait that increases Darwinian fitness (Wittenberger 1981, 613).
   Note: Sometimes benefits are operationally defined in terms of direct and indirect fitness, especially in empirical studies or theoretical analyses, pertaining to them (Wittenberger 1981, 613).
2. An organism's gain in viable offspring (Dawkins 1986, 31).
3. An organism's energy intake (Dawkins 1986, 31).
   cf. cost, utility

benefits of sex *pl. n.* Benefits including assembly of better genotypes (Willson 1983, 47).

expected benefit *n.* Benefit, measured as fitness, that an individual can expect to receive at some future time by performing a particular behavior rather than some alternative behavior; "benefit multiplied by the probability of realizing the benefit" (Wittenberger 1981, 615).

  ▸ **net expected benefit** *n.* "The average net increase in fitness that an individual receives by choosing one of several possible behavioral options in a given context;" expected benefit minus expected cost (Wittenberger 1981, 618).

◆ **¹benthic** *n.* A stickleback species that always feeds on the bottom of a lake; contrasted with limnetic (Weiner 1995, 31).

◆ **²benthic, benthal, benthonic** *adj.* Referring to a sea bed, river bed, or lake floor (Lincoln et al. 1985); contrasted with pelagic.
See benthos.

eurybenthic *adj.* Referring to an organism that lives on a sea, or lake, bed and is tolerant of a wide depth range; contrasted with stenobenthic (Lincoln et al. 1985).

holobenthic *adj.* Referring to an organism that is confined to a benthic existence throughout its life cycle (Lincoln et al. 1985).

nectobenthic, nektobenthic *adj.* Referring to an organism that swims off a sea bed (Lincoln et al. 1985).

stenobenthic *adj.* Referring to an organism that lives on a sea, or lake, bed and is tolerant of a narrow depth range; contrasted with eurybenthic (Lincoln et al. 1985).

◆ **benthon** See community: benthos.

◆ **benthopotamous** *adj.* Referring to an organism that lives on the bed of a river or stream (Lincoln et al. 1985).

◆ **benthos** (*adj.* benthic) See ²community: benthos.

◆ **Bergmann's law, Bergmann's rule** See rule: Bergmann's rule.

◆ **Bernard's dictum, Bernard's principle** See principle: Bernard's principle.

◆ **berried** See gravid.

◆ **bestiality** See sodomy.

◆ **best-man hypothesis** See hypothesis: hypotheses of the evolution and maintenance of sex: best-man hypothesis.

◆ **beta, β** *n.* The second ranking animal in a dominance hierarchy.
cf. alpha, omega
[β, the second letter of the Greek alphabet]

◆ **β-diversity** See diversity: β-diversity.

◆ **β-diversity index** See index: species-diversity index: β-diversity index.

◆ **beta karyology** See karyology (comment).

◆ **beta taxonomy** See taxonomy: beta taxonomy.

◆ **beta waves** See brain waves: beta waves.

◆ **bevy** See ²group: bevy.

◆ **bew** See ²group: bew.

◆ **bi-, bin** *prefix* Twice; doubly; two; especially, occurring twice or having two. Also *bis-* before the letters *c* or *s* (Michaelis 1963).
[Latin *bi-* < *bis*, twice]

◆ **bias** *n.* In statistics, the difference between an expectation (= estimation) of a quantity and the actual magnitude of that quantity (Efron and Tibshirani 1991,124).

◆ **bicentric** *adj.*
1. Referring to a taxon that has two centers of distribution or evolution (Lincoln et al. 1985).
2. Referring to an organism's distributional range with two zones of concentration separated by an impoverished region (Lincoln et al. 1985).

◆ **biennial** *adj.*
1. Referring to an event that occurs every 2 years (Michaelis 1963).
2. A plant that produces flowers and fruit in its second year, then dies (Michaelis 1963).
   Note: Some plant species contain both biennial and annual individuals.
3. Referring to an organism that requires two years to complete its life cycle (Lincoln et al. 1985).
   cf. annual, perennial

◆ **big bang** *n.* A virtually simultaneous radiation at the base of the eukaryotic evolutionary tree that involves the appearance of almost all extant eukaryotic phyla (Gray et al. 1991, 1980).
See Cambrian explosion.

a–c

◆ **big-bang hypothesis** See hypothesis: hypotheses of the beginning of the Universe: big-bang hypothesis.

◆ **big-bang reproduction strategy** See parity: semelparity.

◆ **Big Chill** *n.* The ending of our Universe if its present expansion continues unabated and the entire Universe dilutes itself to zero density and temperature [coined by David N. Schramm, physicist (1979 in Hitt 1992)].
cf. Big Crunch

◆ **Big Crunch** *n.* The catastrophic ending of our Universe if its present expansion stops and the whole universe collapses on itself [coined by David N. Schramm, Beatrice Tinsley, James Gunn, and Richard Gott, physicists (1973 in Hitt 1992)].
cf. Big Chill

◆ **Big Science** See science: Big Science.

◆ **bike** See ²group: bike.

◆ **bilateral-primary-linkage principle** See principle: bilateral-primary-linkage principle.

◆ **billing** *n.* In birds that maintain permanent pair bonds: two bird's mutual beak contact in which their bills are either crossed or the bill of an arriving partner is grasped by the other; billing sometimes involves actual food transfer (Gwinner 1964, Wickler 1969 in Heymer 1977, 158).
cf. food begging, feeding: courtship feeding
Comment: Billing without food transfer may be symbolic feeding (Heymer 1977, 158).

◆ **bimodal distribution** See distribution: bimodal distribution.

◆ **binary digit** See bit.

◆ **binary fission** *n.* In prokaryotes: a kind of cell division in a parent cell splits transversely into approximately equal-sized daughter cells that each receives a copy of a single parental chromosome (King and Stansfield 1985; Campbell 1990, G-2).
cf. mitosis

◆ **binaural hearing** See modality: mechanoreception: audition: binaural hearing.

◆ **binomial distribution** See distribution: binomial distribution.

◆ **bio-** *combining form* "Life" (Michaelis 1963).
[Greek *bios*, life]

◆ **bioacoustics** See study of: acoustics.

◆ **biocenose, biocoenose, biocoenosis** *n.* An organism community, or natural assemblage, excluding abiotic aspects (Lincoln et al. 1985).
syn. life assemblage
cf. biotope, ²community

◆ **biocenotics** See study of: biocoenology.

◆ **biochemical pathway** *n.* A directional series of chemical reactions within a living organism (inferred from Strickberger 1996, 515).

Comments: Biochemical pathways are linear or cyclical, make molecules more or less complex, and consume and produce energy. Organisms have hundreds of biochemical pathways.

C_3 photosynthesis *n.* Plant photosynthesis that fixes carbon via ribulose bisphosphate carboxylase (= rubisco, RUBP carboxylase), the Calvin-cycle enzyme that adds CO_2 to birubulose bisphosphate (RuBP); contrasted with C_4 photosynthesis (Campbell et al. 1999, 182).
cf. Kingdom Plantae: C_3 plant (Appendix 1)
[C_3, after the fact that the first organic product of carbon fixation is a three-carbon compound, 3-phosphoglycerate]

C_4 photosynthesis *n.* Plant photosynthesis that prefaces its Calvin cycle with a mode of carbon fixation that forms the four-carbon compound malate as its first product; contrasted with C_3 photosynthesis (Campbell et al. 1999, 182).
cf. Kingdom Plantae: C_4 plant (Appendix 1)
[C_4, after the fact that this process makes the four-carbon compound malate as its first product]

Calvin cycle *n.* A cyclic biochemical pathway that incorporates CO_2 from air into sugar (as the process of carbon fixation) (Campbell 1990, 209).
Comments: The Calvin cycle occurs in the stroma of chloroplasts. In most plants, it occurs during daylight, but because it does not require light directly, this cycle is sometimes referred to as "dark reactions" (Campbell 1990, 209).
[after M. Calvin, who begin to elucidate the steps along with his colleagues in the late 1940s]

citric-acid cycle *n.* A biochemical cycle that uses a two-carbon acetyl fragment from pyruvic acid which is combined with the four-carbon oxaloacetic acid to make citric acid and other compounds and releases a net of 2 ATPs.
syn. tricarboxylic-acid cycle, TCA cycle, Krebs cycle (Campbell 1987, 191)
Comments: This cycle produces NADH and $FADH_2$, which provide electrons used in the electron transport chain, *q.v.* (Campbell 1987, 193).
[Krebs cycle, after British scientist Hans Krebs, who was largely responsible for elucidating the pathway in the 1930s (Campbell 1987, 191)]

fermentation *n.* An anaerobic metabolic pathway that converts sugar into a waste product and energy (Campbell 1987, 187).
syn. anaerobic glycolysis (Strickberger 1996, 151)
cf. biochemical pathway: glycolysis

Comments: Fermentation includes "glycolysis" and the steps necessary to regenerate NAD⁺ (Campbell 1987, 187–188). Types of fermentation include "alcohol fermentation" and "lactic acid fermentation." Living organisms use part to all of a fermentation pathway, depending on the taxon.

glycolysis *n*. An anaerobic metabolic pathway (= sequence of reactions) that converts one glucose molecule into two pyruvic acid molecules and releases a net of two ATPs (Alberts et al. 1989, 12).
cf. fermentation
Comments: Glycolysis probably occurred in early life because it occurs in cytoplasm and thus does not require membrane-bound organelles; it is widespread in living organisms, probably being in almost all living cells; and different sections of the glycolytic pathway are thought to occur in all living species (Alberts et al. 1989, 12; Strickberger 1990, 139). Glycolysis (as fermentation) is the first pathway in energy production of present-day aerobes (Campbell 1987, 187). Some organisms change amino acids, fats, or sugars into products that enter glycolysis (King and Stansfield 1985; Strickberger 1996, 151).
[Greek sugar splitting < *glukus*, sweet; *lysis*, loosening]

glycolytic pathway *n*. Glycolysis that starts with almost any organic material; *e.g.,* sugars, fats, or amino acids (Strickberger 1990, 139).
Comments: Strickberger (1990, 139) does not clearly indicate the different between "glycolytic pathway" and "glycolysis." Some researchers are likely to make them synonymous.

Krebs cycle *n*. A biochemical cycle that is enzymatically controlled, uses a two-carbon acetyl fragment from pyruvic acid (from glycolysis) which is combined with the four-carbon oxaloacetic acid to make citric acid and other compounds, and releases a net of 2 ATPs (Campbell 1990, 188).
syn. citric-acid cycle, tricarboxylic-acid cycle, TCA cycle (Campbell 1987, 191), citrate cycle (King and Stansfield 1985)
Comments: Each turn of this cycle produces 3 NADH and 1 FADH$_2$, which provide high-energy electrons used in the electron transport chain, *q.v.* (Campbell 1990, 189, illustration).
[Krebs cycle, after British scientist Hans Krebs who was largely responsible for elucidating the pathway in the 1930s (Campbell 1987, 191); citric-acid cycle, after citric acid which is formed by the condensation of acetyl-coenzyme A with the four-carbon oxaloacetic acid]

photosynthesis *n*. The enzymatic conversion of light energy into chemical energy resulting in the formation of carbohydrates and oxygen from carbon dioxide and water (King and Stansfield 1985).
Comments: Photosynthesis is found in anaer-obic photosynthetic bacteria, Cyanobacteria, green plants, many protist species including algae, and sulfur bacteria (Campbell 1990, 204; Strickberger 1990, 237).
[Greek synthesis by light < *photos*, light; *syn*, together; *tithenai*, to place]

"photosystem-I photosynthesis" *n*. Photosynthesis that uses only photosystem-I and obtains protons and electrons from ammonia, hydrogen sulfide, ferrous ions, or some other chemical, but not water (Campbell 1990, 529; Strickberger 1990, 146).
Comments: "Photosystem-I photosynthesis" is found in anaerobic photosynthetic bacteria: the Green Sulfur Bacteria and Purple Sulfur Bacteria. The chemical used as a source of protons and electrons depends on the taxon.

"photosystem-II photosynthesis" *n*. Photosynthesis that uses both photosystem-I and photosystem-II and obtains protons and electrons from water (Strickberger 1990, 146).

♦ **biochronology** See study of: biochronology.

♦ **bioclimatology** See study of: bioclimatology.

♦ **biocoenology** See study of: biocoenology.

♦ **biodeterioration zone** See zone: biodeterioration zone.

♦ **biodiversity** *n*. The total spectrum of living variability from gene through species through higher taxa, including ecological interactions, communities, and populations (Buchmann and Nabhan 1996, 242).
cf. species
Comments: Earth is currently experiencing the greatest die-off of species since the end of the Cretaceous Period, 65 Ma (Turning Point Project, A14).

♦ **biodiversity hotspot** *n*. A region that has an unusually high number of life forms, is threatened, and deserves special protection by Humans (Myers 1999, 35).
Comments: Norman Myers, Oxford ecologist, formulated the concept of biodiversity hotspot (Myers 1999, 35). His current list contains 25 biodiversity hotspots. They occupy 2% of Earth's land yet contain more the 50% of its terrestrial species.

♦ **bioecology** See study of: ecology.

♦ **bioenergetics** See study of: energetics.

♦ **biogenesis, biogenic law** See law: biogenetic law.

♦ **biogenic social regulation** See regulation: social regulation: biogenic social regulation.

♦ **biogeny** See study of: biogeny.

♦ **biogeochemical cycle** See cycle: biogeochemical cycle.

♦ **biogeochemistry** See study of: chemistry: biogeochemistry.

♦ **biogeography** See geography: biogeography.

♦ **biogeosphere** See -sphere: biosphere: eubiosphere.

♦ **biogram** See -gram: biogram.

♦ **biohydrology** See study of: biohydrology.

♦ **bioinert body** See ¹system: ecosystem.

♦ **bioinvasion** *n.* The entry of one or more species from distant ecosystems into a focal ecosystem (Turning Point Project 1999, A14). *cf.* Kingdom Plantae: invasive plant (Appendix 1).

Comments: Bioinvaders alter vegetation, compete with native species, prey upon native species, and sometimes bring in new diseases that affect native organisms, as well as Humans (Turning Point Project, 1999, A14). About 50,000 bioinvaders have entered U.S. ecosystems. More than 200 plant bioinvaders occur in eastern U.S. deciduous forests. The damage of bioinvaders to the U.S. is about $138 billion annually. Pathogens that arrive in raw logs cause about $2.1 billion loss in forest products each year. Bioinvaders include, in Australia, Black-Striped Mussel, feral Domestic Cats; in Europe, Rootworm; in Guam, Brown Tree Snake; in Hawai'i, the Central American tree *Miconia calvescens*, Giant African Snail, feral Domestic Goats, Lantana, feral Domestic Pigs, rats, Rosy Wolf Snail; in Lake Victoria, Africa, Nile Perch; in South Africa, pines, Varroa Mite; in Tahiti, *Miconia calvescens*; in Tierra del Fuego, South America, the Beaver; in the Mediterranean, *Caulerpa taxifolia;* and in Mainland U.S., the Asian Longhorned Beetle, Asiatic Clematis, Bottlebrush Tree, Baby's Breath (plant), Brazilian Pepper, Brown Trout, Eurasian Cheatgrass, Eurasian Milfoil, Giant Cane (*Arundo donax*), Hydrilla, Japanese Knotweed, Kudzu Vine, Leafy Spurge, Mosquito Fish, Norway Maple, Purple Loosestrife, Russian Olive, Russian Thistle (= Tumbleweed), Tall Fescue, *Tamarisk* Tree, Varroa Mite, Water Hyacinth, Witchweed, Yellow Star Thistle, and Zebra Mussel (Kaiser 1999, 1839–1841; Malakoff 1999, 1841–1843; Raver 1999, D1; Stone 1999, 1837; Turning Point Project 1999, A14). Nurseries continue to sell many species of invasive plants.

♦ **biological** *adj.*
1. Referring to biology, *q.v.*
2. Colloquially, referring to the genetic relationship between relatives (Alexander 1987, 8).

♦ **biological altruism** See altruism (def. 2).

♦ **biological clock** See clock: biological clock.

♦ **biological complementariness, complementariness, principle of biological complementariness** *n.* The phenomenon that organisms are not self-sufficient, in that they require stimuli from their environments to mature normally (Brownlee 1981, iv).
cf. -genesis: epigenesis

♦ **Biological Dynamics of Forest Fragments Project** See project: Biological Dynamics of Forest Fragments Project.

♦ **biological efficiency** See ¹efficiency: ecological efficiency.

♦ **biological function** See adaptive significance.

♦ **biological mimicry** See mimicry.

♦ **biological race** See ¹race: biological race.

♦ **biological-rank ordering** See hierarchy: dominance hierarchy: biological-rank ordering.

♦ **biological rhythm** See rhythm: biological rhythm.

♦ **biological species** See ²species: biological species.

♦ **biological theory of evolution** See evolutionary synthesis.

♦ **biology** *n.* Colloquially, genetic variations (Gould 1984 in Alexander 1987, 7).
See study of: biology.

♦ **bioluminescence** *n.* In many organism species: an organism's emission of light due to an energy-yielding chemical reaction in which luciferin undergoes oxidation, catalyzed by luciferase (Aristotle, Boyle 1672, etc. in Johnson 1966, 3; Barnes 1974, 141, 277, 531; Hastings 1989).
cf. communication: flash communication; phosphorescence

Comments: Bioluminescence is a kind of chemiluminescence (Hastings 1989, 545) found in many organisms including a limpet, acorn-worm, earthworm, sea-pansy, polycheate-worm, and midge (fly) species; some bacterium, fungus, dinoflagellate, sponge, jellyfish, brittle-star, tunicate, ostracod-crustacean, millipede, and centipede species; and many ctenophore, beetle (including some firefly, elaterid, phengotid), and fish species. Luminescent bacteria emit a continuous blue-green light (Hastings 1989, 547). A species of deep-sea octopus squirts a luminescent fluid instead of ink, which would be ineffective in total darkness (Hastings 1989, 548). A marine crustacean, *Cypridina*, squirts luciferin and luciferase into water; this light in the water might divert, or confuse, predators (Hastings 1989, 547). Fireflies use their lights to attract mates and prey, depending on the species (White 1983, 190). Firefly lights are coppery yellow, greenish yellow, orangish yellow, or yellow, depending on the species (McDermott 1948, 13). Flashlight Fish appear to use their luminescence (from symbiotic luminous bacteria in the fish's eye pouches) for assisting in predation and predator escape and for intraspecific communication (Hastings 1989, 548). The Pony Fish has a diffuse glow from symbiotic bacteria over its entire ventral surface; this light might camouflage it against a light background seen when an animal looks up at it (Hastings 1989, 548). Some angler-fish species attract prey by dangling luminescent bodies that they hold in front of their mouths; a *Neosopelus* fish has a photophore on its tongue that lures prey (Hastings 1989, 548).

♦ **biomass** *n*. The weight of a set of organisms (Wilson 1975, 579).
Comment: The set is chosen for convenience; it can be, for example, a colony of insects, a population of wolves, or an entire forest (Wilson 1975, 579).

♦ **biome** *n*.
1. "A biogeographical region or formation" (Lincoln et al. 1985).
2. A major regional community characterized by distinctive life forms and principal plant (terrestrial biomes) or animal (marine biomes) species (Lincoln et al. 1985).
cf. habitat
Comment: Terrestrial biomes are chaparral, desert, mountains, northern coniferous forest, temperate deciduous forest, temperate rain forest, temperate grassland, tropical deciduous forest, tropical grassland and savanna, tropical rain forest, tropical scrub forest, and tundra.

♦ **biomechanics** See study of: biomechanics.

♦ **biometeorology** See study of: meteorology: biometeorology.

♦ **biometrician** *n*.
1. An evolutionist who is interested in population phenomena and holistic interpretations (Mayr 1982, 778–779).
   Note: Biometricians opposed the "Mendelians," *q.v.*, from about 1894 to 1906 (Mayr 1982, 778–779).
2. A biostatistician.
cf. study of: biometry

♦ **biometrics** See study of: -metry: biometry.
♦ **biometry** See study of: -metry: biometry.
♦ **BioMOO Center, MOO** *n*. A computer-produced science facility with a simulated, large growing laboratory and an office complex used by hundreds of biologists in different parts of the world and in different fields who are communicating, collaborating, and designing electronic tools for scientific work (Anderson 1994, 900).
adj. MOO (Anderson 1994, 900)
Comments: The BioMOO Center was founded by Gustavo Glusman in 1993 (Anderson 1994, 900, who includes an illustration). It is the first major effort to use a virtual environment to perform day-to-day science and scientific communication. In using the bioMOO center, a scientist can walk down a virtual hall and into a lab, strike up a conversation with one of its workers (communicating in real time, almost instantaneously), use a computer program in the lab, inspect and manipulate a research model, and participate in a journal club. Ecologists are creating EcoMOO, and astronomers have made a prototype MOO center called AstroVR.
[*bio*, life + MOO, multiple-user dimension, object oriented; the M originally stood for multiple-user dungeon, referring to the computer version of the game "Dungeons and Dragons," from which the BioMOO Center evolved (Anderson 1994, 900)]

♦ **bion** See biont.

♦ **bionomic axes, resource axes** *pl. n*. Axes that relate to resources, such as food and space, for which there may be competition with neighboring species in the context of a multidimensional hyperspace (niche) (Lincoln et al. 1985).
cf. scenopoetic axes

♦ **biont, bion** *n*.
1. An elementary, difficult concept that expresses the extent to which one living thing is separate from, or independent of, other living things (Bell 1982, 508).
   syn. individual (Bell 1982, 508)
2. An individual organism (Lincoln et al. 1985).
3. A physiological individual (*e.g.*, a Portuguese Man-of-War); contrasted with morphont (Haeckel in Hull 1992, 184).
See individual: biont.
adj. biontic, biotic
cf. -cole, ¹-phile, -vore
Comment: "Biont" and "morphont" are problematical terms because "morphology and physiology do not provide sufficiently well articulated theoretical contexts" (Hull 1992, 184).

chorotobiont See -cole: graminicole.

coprobiont *n.* An animal (coprozoite) or a plant (coprophyte) that lives, or feeds, on dung (Lincoln et al. 1985).

diplobiont, diphophalontic organism *n.* An organism with a regular alternation of haploid and diploid generations during its life cycle (Lincoln et al. 1985).

endobiont *n.* An organism that lives within a substratum (Lincoln et al. 1985).

eobiont *n.* One of the early living organisms that developed from prebiotic macromolecular precursors (Lincoln et al. 1985).

epibiont See -cole: epicole.

eremobiont deserticole *n.* An organism that lives in desert regions (Lincoln et al. 1985).
See -cole: deserticole.

eurybiont *n.* An organism that tolerates a wide range of a particular environmental factor; contrasted with stenobiont (Lincoln et al. 1985).

eutroglobiont See [1]-phile: troglophile.

exobiont *n.* An organism that lives on the outer surface of another organism or on the surface of a substratum (Lincoln et al. 1985).

halobiont, halobion *n.*
1. A marine organism (Lincoln et al. 1985).
2. An organism that lives in a saline habitat (Lincoln et al. 1985).

haplobiont *n.*
1. An organism that does not have a regular alternation of haploid and diploid generations during its life cycle (Lincoln et al. 1985).
   cf. biont: diplobiont
2. A plant that flowers once per season (Lincoln et al. 1985).

geobiont *n.*
1. An organism that spends its whole life in soil (Lincoln et al. 1985).
2. A member of the permanent soil fauna (Lincoln et al. 1985).
syn. terricole

limnobiont, limnobion *n.* A freshwater organism (Lincoln et al. 1985).

mycobiont *n.* The fungal partner of an algal-fungal symbiosis of a lichen; contrasted with phycobiont (Lincoln et al. 1985).

neobiont *n.* A primordial life form that arises independently from nonliving matter (Lincoln et al. 1985).

oxybiont, aerobe *n.* An organism that grows or occurs only in the presence of molecular oxygen (Lincoln et al. 1985).

patabiont, patobiont *n.* An organism that permanently inhabits forest litter (Lincoln et al. 1985).

petrobiont *n.* An organism that lives on, or among, rocks or stones (Lincoln et al. 1985).

phycobiont *n.* The algal partner of an algal-fungal symbiosis of a lichen; contrasted with mycobiont (Lincoln et al. 1985).

phytobiont *n.* An organism that spends most of its active life on, or within, plants (Lincoln et al. 1985).
See -cole: planticole.

planktobiont *n.* An organism that lives solely in plankton (Lincoln et al. 1985).
adj. holoplanktonic, planktobiontic

polyoxybiont *n.* An organism that requires abundant free oxygen (Lincoln et al. 1985).

protobiont *n.*
1. "A hypothetical prebiotic complex of proteins and nucleic acids that eventually gave rise to self-replicating, and hence living, organisms" (Wittenberger 1981, 620).
2. A protistan; *adj.* protistan.
[Greek *prōtos*, first + *bios*, life]

psammobiont *n.* An organism that lives interstitially between, or attached to, sand particles (Lincoln et al. 1985).
See -cole: arenicole.

saprobiont, saprophage *n.* An organism that feeds on dead, or decaying, organic matter (Lincoln et al. 1985).
adj. saprobiontic, saprophagic

stenobiont *n.* An organism that requires a stable uniform habitat; contrasted with eurybiont (Lincoln et al. 1985).

trophobiont *n.* An organism that provides food in trophobiosis, *q.v.* (Wilson 1975, 354, 597).

xenobiont *n.* A species that lives with its host species in xenobiosis, *q.v.* (Wilson 1975, 362).

♦ **biophage** See -phage: biophage.

♦ **biophasic processes** See [3]theory: approach-withdrawal theory.

♦ **biophile** See [1]phile and [3]phile: biophile.

♦ **biophilia** See -philia: biophilia.

♦ **biophysics** See study of: physics: biophysics.

♦ **biopiracy** *n.* A person's obtaining biological material from a country without reimbursing that country for it (Faiola 1999, A21).
Comment: Tropical countries are trying to stop biopiracy (Faiola 1999, A21).

♦ **biopoiesis** *n.* "The origin of life, including the abiotic synthesis of macromolecular systems and the transformation (eobiogenesis) of these systems into the first living organisms (eobionts)" (Lincoln et al. 1985).

♦ **biorealm** *n.* "A group of similar biomes" (Lincoln et al. 1985).

♦ **-bios** See biota.
cf. biome, biorealm, biotope, [2]community, flora, fauna

geobios *n.*
1. Total life on land (Lincoln et al. 1985).
2. The part of the Earth's surface that is occupied by terrestrial organisms (Lincoln et al. 1985).

halibios, halobios *n.*
1. The total life of the sea (Lincoln et al. 1985).
2. The part of the surface of the Earth occupied by marine organisms (Lincoln et al. 1985).

heteroplanobios *n.* Organisms that are passively transported by flood water (Lincoln et al. 1985).

hydrobios *n.*
1. All aquatic life (Lincoln et al. 1985).
2. The part of the Earth's surface that is occupied by aquatic organisms (Lincoln et al. 1985).

hygropetrobios *n.*
1. "Hygropetric fauna and flora" (Lincoln et al. 1985).
2. "Hygropetrical fauna" (Lincoln et al. 1985).

limnobios *n.*
1. All life of fresh waters (Lincoln et al. 1985).
2. The part of the Earth's surface occupied by freshwater organisms (Lincoln et al. 1985).

prebios *n.* Conditions existing before the origin of life on Earth (Lincoln et al. 1985).
adj. prebiological, prebiotic

protobios *n.* "All ultramicroscopic life forms" (Lincoln et al. 1985).

saproxylobios *pl. n.* Organisms that live in, or on, rotting wood (Lincoln et al. 1985).

skatobios *n.* Organisms that inhabit detritus or fecal matter (Lincoln et al. 1985).

♦ **bioseries** *n.* "The evolutionary sequence of changes in any heritable character" (Lincoln et al. 1985).

♦ **-biosis** *combining form* "Manner of living" (Michaelis 1963).
cf. symbiosis.
[Greek *biōsis* < *bios*, life]

anabiosis *n.* An organism's state of greatly reduced metabolic activity assumed during unfavorable environmental conditions (Lincoln et al. 1985).
syn. cryptobiosis (Lincoln et al. 1985)
cf. estivation; -biosis: cryptobiosis; dormancy; hibernation; suspended animation; viable lifelessness; viability

cleptobiosis See symbiosis: cleptobiosis.

cryptobiosis *n.* A dormant organism's showing no external signs of metabolic activity (Lincoln et al. 1985).
cf. -biosis: hypobiosis

ecotrophobiosis See trophallaxis.

epibiosis See symbiosis: epibiosis.

hamabiosis *n.* "Symbiosis without obvious advantage to either symbiont" (Lincoln et al. 1985).

hypobiosis *n.* A dormant organism's showing only minimal outward signs of metabolic activity (Lincoln et al. 1985).
cf. -biosis: cryptobiosis; dormancy.

kleptobiosis See symbiosis: cleptobiosis.

lestobiosis See symbiosis: lestobiosis.

metabiosis See symbiosis: metabiosis.

oecotrophobiosis See trophallaxis.

parabiosis *n.* An organism's temporary suspension of physiological activity (Lincoln et al. 1985).
See symbiosis: mutualism.

phylacobiosis See -biosis: parabiosis.

plesiobiosis *n.*
1. In social animals: the close association of nests of two or more species, accompanied by no mixing of individuals and little, or no, benefit to any of the species (*e.g.,* some ant species) (Wilson 1975, 354, 591).
2. Organisms' living close to one another (Lincoln et al. 1985).
[Greek *plēsios*, near + *biōsis* < *bios* life]

symbiosis See symbiosis.

synclerobiosis See symbiosis: synclerobiosis.

trophobiosis See symbiosis: mutualism.

♦ **biospace** See niche: realized niche.

♦ **biospecies** See ²species: biological species.

♦ **biospeleology, biospeology** See study of: biospeleology.

♦ **biosphere** See -sphere: biosphere.

♦ **Biosphere 2** *n.* A large, artificial, enclosed habitat designed to study closed, large-scale, integrated ecosystems (Dempster 1997, 1247–1248; Nelson 1997, 1248–1249; Allen 1997, 1249).

♦ **biospherian** *n.* A person who was enclosed in Biosphere 2 for a period as part of a study of large-scale, artificial ecosystems (Dempster 1997, 1247–1248).

♦ **biostratigraphy** See study of: paleontology: stratigraphic paleontology.

♦ **biostratinomy** See study of: biostratinomy.

♦ **biosystem** See ¹system: ecosystem.

♦ **biosystematics** See study of: systematics.

♦ **biosystematy** See study of: systematics.

♦ **biota** *n.*
1. Flora and fauna of a region or geological period (*e.g.,* forest biota or pond biota) (Michaelis 1963).
2. Organisms of a region or geological period.
cf. biome, biorealm, -bios, biotope, community, flora, fauna

aerophytobiota *n.* "Aerobic soil flora" (Lincoln et al. 1985).

▸ **anaerophytobiota** *n*. "Anaerobic soil flora" (Lincoln et al. 1985).

macrobiota *n*.
1. Soil organisms longer than 1 mm (Lincoln et al. 1985).
2. Larger soil organisms, or their parts, that one can readily remove with one's hands from a soil sample, including particular burrowing vertebrates (*e.g.*, moles or rabbits) and tree roots (Allaby 1994).

mesobiota *n*. Soil organisms of intermediate size, from about 40 mm long to a size just visible with the use of a hand lens (Lincoln et al. 1985).
Comments: Mesobiota includes the mesofauna (*e.g.*, annelids, arthropods, molluscs, and nematodes) and the mesoflora, *q.v.*

microbiota *n*. Soil organisms that are too small to be seen with a hand lens, including algae, bacteria, fungi, and protozoa (Lincoln et al. 1985; Allaby 1994).

mycobiota *n*. Fungi of an area or habitat (Lincoln et al. 1985).

A CLASSIFICATION OF BIOTA BASED ON ORGANISM SIZE

I.   macrobiota (includes macroflora, megafauna)
II. mesobiota (includes macroflora, megafauna)
III. microbiota (includes meiofauna, microfauna, microflora)

♦ **biotaxis** See taxis.

♦ **biotechnology** *n*.
1. Human use of organisms to make specific industrial products (Ereky 1917 in Bugos 1993, 121).
   [After *Biotechnologie* coined by Ereky (1917) as part of a campaign to revolutionize Hungarian agriculture]
2. In a very broad sense, every possible human use and manipulation of life (Bugos 1993, 121).
Comments: The concept of biotechnology has its roots in bacteriology (Bugos 1993). Zymotechnology was the bridge between biotechnology's ancient roots and its modern association with chemical engineering. Scientists gave "biotechnology" meanings that conveyed their specific hopes for industrial uses of life. Biotechnology and genetic engineering were wed in 1974.

♦ **biotelemetry** See -metry: biotelemetry.

♦ **bioterrorism** *n*. A person's effort to spread germs to destroy a country's crops, livestock, or both for financial gain, political purposes, or both (Miller 1999, A1, A25).

♦ **biotic** Also see -biosis.
cryptobiotic *adj*. Referring to an organism that is typically hidden, or concealed, in crevices or under stones (Lincoln et al. 1985).
geobiotic *adj*. Referring to a terrestrial organism (Lincoln et al. 1985).
hemiendobiotic *adj*. Referring to an organism that is usually found inside its host (Lincoln et al. 1985).
hypobiotic *adj*. Referring to an organism that lives in a sheltered microhabitat (Lincoln et al. 1985).
macrobiotic *adj*. Referring to a long-lived organism (Lincoln et al. 1985).
oxybiotic, oxybiontic See aerobic.
prebiotic *n*. Referring to conditions on Earth that existed before the beginning of life (Lincoln et al. 1985).
syn. prebiological
n. prebios
zoobiotic *adj*. Referring to an organism that lives as a parasite on an animal (Lincoln et al. 1985).

♦ **biotic pollination** See pollination: biotic pollination.

♦ **biotic potential** See rate: *r*.

♦ **biotic succession** See ²succession.

♦ **biotically sympatric population** See ¹population: biotically sympatric population.

♦ **biotope** *n*.
1. The smallest geographical unit of the biosphere, or a habitat, that can be delimited by convenient boundaries, is characterized by its biota, and is labeled by its predominant vegetation type (Lincoln et al. 1985; Immelmann and Beer 1989, 33).
   Note: Organisms of a particular biotope are its ecological community (= biocenose).
   syn. habitat (according to some authors)
2. A parasite's location within its host's body (Lincoln et al. 1985).
cf. habitat
crenon *n*. "The spring-water biotope" (Lincoln et al. 1985).
stygon *n*. "The groundwater biotope" (Lincoln et al. 1985).
thalasson *n*. "The marine biotope" (Lincoln et al. 1985).
troglon *n*. The biotope comprised of subterranean water bodies in caves and subterranean passages (Lincoln et al. 1985).

♦ **biotrophic** See -trophic: biotrophic.

♦ **biotrophic symbiosis** See symbiosis: biotrophic symbiosis.

♦ **biparous** See -parous: biparous.

♦ **biped** *n*. For example, in many dinosaur and primate species, birds: an animal with two feet (*Oxford English Dictionary* 1972, entries from 1607).

adj. biped, bipedal
[Latin *bipes, bipedis* two-footed]
♦ **bipolar mating system** See mating system: bipolar mating system.
♦ **bipolar mimicry system** See ¹system: mimicry system: bipolar mimicry system.
♦ **bird-watching** *n.*
  1. A person's observing, or identifying, wild birds in their natural habitats (Michaelis 1963).
  syn. birding
  2. Derogatorily, a person's observing scientific phenomena, making little or no relationships of these phenomena to scientific principles. (Thomson 1985, 570).
  v.t. bird-watch
  cf. stamp-collecting
♦ **birth rate** See natality.
♦ **bisexual** See sexual: bisexual.
♦ **bit** *n.* "The basic quantitative unit of information; specifically, the amount of information required to control, without error, which of two equiprobable alternatives is to be chosen by the receiver" (Wilson 1975, 194, 579).
  [*b*inary + dig*it*]
♦ **bite** *n.*
  1. An animal's seizing, tearing, or wounding another animal with its teeth (Michaelis 1963).
  2. An animal's tearing of something with its teeth (Michaelis 1963).
  3. An animal's puncturing the skin of another animal with its sting or fangs (Michaelis 1963).
  See ²group: bite.
  v.t., v.i. bite
  neck bite *n.*
   1. In many bird and carnivorous-mammal species: a male's bite given on his mate's neck during copulation (Heymer 1977, 122).
    syn. pairing bite (Immelmann and Beer 1989, 197)
   2. A bite given by a predator on a prey's neck in contrast to shaking a prey to death (Heymer 1977, 122; Immelmann and Beer 1989, 197).
   3. In felid carnivores and some rodent species: neck grasping of a young by its parent when carrying it (Heymer 1977, 197).
  cf. posture: limp posture
  repeated bites *n.* In many carnivore species: a predator's biting into its prey, shaking it, relaxing its jaws, and biting again (Eibl-Eibesfeldt 1956, Leyhausen 1965 in Heymer 1977, 121).
  tail-bite *n.* In crowded, captive pigs: a bite given by one pig to another pig's tail, followed by other pigs' doing likewise and

the possible death of the victim due to the biting (Colyer 1970 in Dawkins 1980, 77). *Comment:* "Tail-biting" is often accompanied by "ear-biting" (Colyer 1970 in Dawkins 1980, 77).
♦ **bivoltine** See voltine: bivoltine.
♦ **bivouac** *n.*
  1. In Army Ants: the mass of workers within which the queen and brood find refuge (Schneirla 1933–1971 in Wilson 1975, 425).
  2. The site of a bivouac (def. 1) (Wilson 1975, 579).
♦ **black-box view of behavior, whole-animal view of behavior** *n.* The claim that to understand behavior, one should study the behavior of whole, intact animals without examining their insides (Dawkins 1986, 84).
  cf. identified-neurone chauvinist
♦ **black hole** *n.* A predicted region of space where the gravitational force is so strong that not even light can escape from it (Mitton 1993).
  Comments: Black holes might suck up nearby objects, and material from them might reappear somewhere else in our Universe, in another universe, or both (Sawyer 1997, A3). They may be plentiful, important players in the evolution of the Universe. In the two-star system V404Cyg, in the constellation Cygnus, a dense object believed to be a black hole is sucking material from a companion star.
  galactic black hole *n.* A black hole in the central area of a galaxy (Sawyer 1997, A3). *Comments:* Almost all galaxies might have central black holes which may have played key roles in galaxy formation (Sawyer 1997, A3). A galactic black hole might have the mass of three billion suns, all squeezed into an area no larger than our Solar System (Wilford 1997, C7). There is evidence for a black hole in the middle of the giant galaxy M87, the Milky Way, and other galaxies based on a sharp rise in star velocities near the centers of galaxies.
  stellar black hole *n.* A black hole believed to be the remnant of a supernova of a star that was 3 or more solar masses (Mitton 1993). *Comments:* Stellar black holes are about 10 km or less in diameter. We observe black holes indirectly by their gravitational effects and their X-ray emission. Their X-rays result from the energy released when matter streams into them. Scientists found strong evidence for the existence of an event horizon, the rim of a black hole (Sawyer 1997, A3). A black-hole candidate seems to be swallowing nearly 100 times as much energy as it is radiating.

a – c

♦ **blade** See tool: blade.
♦ **blastochore** See -chore: blastochore.
♦ **blastogenesis** See -genesis: blastogenesis.
♦ **bleat** See animal sounds: baa.
♦ **blending** *n.*
1. A gradual return to a parental type of an artificially bred line of organisms (Mayr 1982, 740).
2. An intermediate appearance of phenotypes, particularly in species crosses (Mayr 1982, 781).
   Comment: "Blending" does not necessarily make any commitment with regard to the behavior of genetic material (Mayr 1982, 781).
♦ **blending concept of heredity, blending inheritance** See inheritance: blending inheritance.
♦ **blind sight** *n.* A brain-damaged person's not being able to name objects presented to him in certain parts of his visual field, although he might be able to point to such an object even though he claims he cannot see it (Weiskrantz 1980 in McFarland 1985, 523).
♦ **bloat** See ²group: bloat.
♦ **block** *n.* A hypothetical state of the pathways between two centers of an instinct that stops an interaction between centers (Verplanck 1957).
   Comment: This state may be terminated, or reduced, by the action of an innate releasing mechanism that has been activated by a sign stimulus. After such nullification of the state, "motivational impulses" can flow from a higher to a lower center, activating the latter and, hence, yielding a response. There is no direct physiological evidence for such a state (Verplanck 1957).
♦ **blocking** *n.* The phenomenon in which stimulus A stops ("blocks") an animal from learning that stimulus B is correlated with an unexpected, or surprising, occurrence because the animal has already experienced the occurrence in stimulus A's presence (Kamin 1969 in McFarland 1985, 352).
♦ **blood theory of inheritance** See inheritance: blending inheritance.
♦ **blue moon** *n.*
1. The second full moon in the same month (Mitton 1993; Anonymous 1996h, A7).
   Note: Blue moons occur every 2.7 years and are the result of human-made calendars which do not coincide with our moon's 29-day cycles (Anonymous 1996h, A7). Further, atmospheric effects occasionally make the moon appear blue, possibly due to dust in the upper atmosphere from forest fires, volcanoes, or both (Mitton 1993).
2. The fourth full moon in a season (Anonymous 1999, B4).

Note: The editors of *Sky and Telescope* say they incorrectly defined "blue moon" 53 years ago and that def. 2 is correct, not def. 1 (Anonymous 1999, B4).
[possibly from Old English *belewe*, to betray, referring to a blue moon's betraying the rule of one full moon per month]
♦ **body** *n.*
1. The entire physical part of an organism (Michaelis 1963).
2. A colony of genes' using cells as convenient working units for their chemical industries (Dawkins 1977).
♦ **body-brooding** See brooding: body-brooding.
♦ **bog** See habitat: bog.
♦ **bog forest** See habitat: bog forest.
♦ **bombykol** See chemical-releasing stimulus: semiochemical: pheromone: bombykol.
♦ **bonanza strategy** See strategy: bonanza strategy.
♦ **bond, bonding, social attachment** *n.* A specific dependence between, or among, individual animals; *e.g.*, mated pairs, young animals and their mother or both parents, and groups, especially individualized groups (Immelmann and Beer 1989, 34).
   cf. attachment; behavior: bonding behavior; harem
   life-long pair bond *n.* For example, in some raptor species, the Gray Goose: a pair bond, *q.v.*, that endures during much, or all, of the lives of partners (Immelmann and Beer 1989, 176).
   long-term pair bond *n.* In some bird species, many mammal species: a pair bond, *q.v.*, that transcends a species' single reproductive period (Immelmann and Beer 1989, 176).
   pair bond, pair bonding *n.* For example, in some insect and mammal species, many bird species: a close association formed between a conspecific male and female which can last until the end of a breeding season or longer (Wilson 1975, 327; Immelmann and Beer 1989, 211).
   See marriage.
   syn. pair formation (according to some authors), marriage, sexual bonding (Wilson 1975, 315)
   cf. pair formation
   Comment: In mammals, pair bonding serves primarily for cooperative rearing of young (Wilson 1975, 327, 590).
   partner bonding *n.* Individuals' of a mated pair maintaining continuous visual contact, auditory contact, or both, with one another (Heymer 1977, 128).
   spatial bonding *n.* Individuals' of a mated pair maintaining close physical proximity with one another [coined as

Paarsitizen (spatial bond) by Seibt and Wickler (1972, 128)].

♦ **bonding behavior** See behavior: bonding behavior.

♦ **bonding drive** See drive: bonding drive.

♦ **Bonnet's chain of being** *n.* Bonnet's (1769) idea that there is a gradual, unbroken transition from inanimate matter to the most perfect organic being (Mayr 1982, 350).
cf. scala naturae

♦ **booming** See animal sounds: booming, drumming.

♦ **booming field** See ²lek: booming field.

♦ **Boorman-Levitt model of group selection** See ⁴model: Boorman-Levitt model of group selection.

♦ **boot**
1. *n.* An often boot-shaped nesting cavity made in a Saguaro Cactus by a Gila Woodpecker which may be used by other species of birds (*e.g.,* Screech Owls and Elf Owls) that do not make their own cavities (Scott 1974, 49; Venning 1974, 36; Shelton 1985, 67).
   Comment: Isolated boots can be found on the ground after surrounding Saguaro tissue disappears (personal observation). Pima Indians used them as jugs (Venning 1974, 36).
2. *v.t.* To start up a computer from a set of core instructions (Efron and Tibshirani 1993, 5).
syn. bootstrap (Efron and Tibshirani 1993, 5)
cf. statistical test: bootstrap
[after one's pulling oneself up by one's own bootstraps in the *Adventures of Baron Munchausen*]

♦ **bootstrap** See boot; statistical test: bootstrap.

♦ **botany** See study of: botany.

♦ **botrology** See study of: botrology.

♦ **bottleneck** *n.* A single episode of small population size of a particular population that changes its size through time (Hartl and Clark 1997, 291).

♦ **bottleneck effect, bottlenecking** See effect: bottleneck effect.

♦ **bottom-up control** See ³control: bottom-up control.

♦ **bounded rationality model of human behavior** See ⁴model: bounded rationality model of human behavior.

♦ **bouquet** See ²group: bouquet.

♦ **bout** *n.*
1. An animal's repetitive display of the same behavior (*e.g.,* a bout of pecking by a bird) (Lehner 1979, 70).
2. An animal's relatively stereotyped sequence of behaviors that occur in a burst (*e.g.,* a courtship-display bout) (Lehner 1979, 70).

Comment: Sibly et al. (1990) indicate how to divide behavior into bouts.

mount bout *n.* In mammals: a male's series of mountings, with or without pelvic thrusting and intromission, that occurs before he can ejaculate (Immelmann and Beer 1989, 194).

superbout *n.* A cluster of bouts of behavior, *q.v.,* which can be revealed by long-term records of behavior (Machlis 1977 in Lehner 1979, 73).
cf. bout

♦ **bow** See ²group: bow.

♦ **bow-coo display** See display: bow-coo display.

♦ **bower** See territory: display territory.

♦ **brace** See ²group: brace.

♦ **brachial gland** See gland: brachial gland.

♦ **brachiation** *n.* In some arboreal primates: locomotion by swinging among tree limbs with use of "hands" and forelegs (Michaelis 1963).
[Latin *brachiatus*, having arms < *brachium*, arm]

♦ **bradyauxesis** See auxesis: heterauxesis: bradyauxesis.

♦ **bradygenesis, bradytely** See ²evolution: bradygenesis.

♦ **bradymetabolism** See metabolism: bradymetabolism.

♦ **bradys** *combining form*
[Greek *bradys*, slow]

♦ **brain** *n.* Figuratively, intellectual power, intellect, sense, thought, imagination; often used as a plural (brains) (*Oxford English Dictionary* 1971, entries from 1393).
See organ: brain.

♦ **"brain-first hypothesis"** See hypothesis: "brain-first hypothesis."

♦ **brain stem** See organ: brain: brain stem.

♦ **brain stimulation** See method: brain stimulation.

♦ **branchicole** See -cole: branchiocole.

♦ **breakage-fusion theory** See ³theory: breakage-fusion theory.

♦ **breaking dance** See dance: bee dance: buzzing run.

♦ **breathing** *n.*
1. An animal's inhaling and exhaling air (Michaelis 1963).
   syn. respiring (Michaelis 1963)
   cf. respiration
2. A vertebrate's inhaling and expelling air from its lungs (Michaelis 1963).
   syn. respiring (Michaelis 1963)
   cf. respiration
3. An organism's being alive (Michaelis 1963).
4. An animal's pausing for breath, resting (Michaelis 1963).
5. A person's murmuring; whispering (Michaelis 1963).

6. An animal's exhaling something (*e.g.*, an odor) (Michaelis 1963).

v.i., *v.t.* breathe

[Middle English *brethen* < *breth*, breath]

air breathing *n.* For example, in 34 fish families, terrestrial animals: an individual's extracting oxygen from air; contrasted with bimodal breathing and water breathing (Kramer and Mehegan 1981, 299; Kramer 1983, 145).

syn. aerial respiration (Kramer 1983, 145)

bimodal breathing *n.* For example, in some fish species: an individual's extracting oxygen from both air and water; contrasted with air breathing and water breathing (Kramer 1983, 145).

Comment: Fish species with bimodal breathing vary from having almost complete dependence on air breathing to complete dependence on water breathing except when dissolved oxygen is extremely limited (Kramer 1983, 145).

water breathing *n.* For example, in fish: an individual's extracting dissolved oxygen from the water in which it lives; contrasted with air breathing and bimodal breathing (Kramer 1983, 145).

syn. aquatic respiration (Kramer 1983, 145)

♦ **breed** *n.*

1. "A group of organisms related by descent; an artificial mating group having a common ancestor," especially with regard to genetic studies of domesticated species (Lincoln et al. 1985).

   cf. copulation

2. *v.i.* In organisms: to reproduce.

3. *v.t.* In Humans: "to propagate organisms under controlled conditions" (Lincoln et al. 1985).

♦ **breed true** *v.t.* To produce offspring phenotypically identical to parents (Lincoln et al. 1985).

♦ **breeding** *n.* Reproducing (Lincoln et al. 1985).

communal breeding *n.* For example, in the Groove-Billed Ani, Mexican Jay: a breeding system in which helpers are normally present at some, or all, nests, often resulting in three or more birds' attending the young of a nest (Brown 1987a, 298).

syn. cooperative breeding (which Brown 1987a, 298, indicates is not preferable to "communal breeding")

cf. mating system: polygynandry

Comments: By tradition, "communal breeding" excludes "brood capture," *q.v.*, and cases in which individuals care for young not their own as a result of deception through intraspecific, or interspecific, brood parasitism (Brown 1987a, 298). "Group living" should not be synonymized with "communal breeding" (Brown 1987a, 300).

cooperative breeding See breeding: communal breeding.

crossbreeding See breeding: outbreeding; crossing.

explosive breeding *n.* Simultaneous breeding in a large group of conspecific animals (Halliday and Adler 1986, 146).

inbreeding *n.*

1. Mating of kin (Wilson 1975, 586); mating or crossing of individuals that are more closely related than average pairs in a population (Lincoln et al. 1985).

2. Sexual reproduction in which there is a greater frequency of mating between related, or relatively closely related, individuals than would occur by chance alone (random mating) (Lincoln et al. 1985).

syn. endogamy, endokaryogamy

cf. breeding: outbreeding

Comment: The degree of inbreeding is measured by the fraction of genes that are identical owing to common descent (Wilson 1975, 586).

▸ **optimal inbreeding** See breeding: outbreeding: optimal outbreeding.

interbreeding *n.* Mating, or hybridization, between different individuals, populations, varieties, races, or species (Lincoln et al. 1985).

cf. hybrid

outbreeding *n.*

1. Sexual reproduction between individuals that are not closely related (Lincoln et al. 1985).

   syn. exogamy

2. Mating, or crossing, of conspecific organisms that are either less closely related than average pairs in their population or from different populations (Lincoln et al. 1985).

   syn. outcrossing, open-pollination, crossbreeding

   cf. breeding: inbreeding

▸ **optimal outbreeding** *n.* An individual's breeding with conspecifics of particular genetic relationships so that there will be the least possible build-up of deleterious homozygous genes in offspring due to inbreeding and, simultaneously, the least possible loss of adaptive genetic complexes due to too much outbreeding; balancing the costs of inbreeding and outbreeding (Bateson 1978 in Keane 1990, 264; Bateson 1983, 257–277).

syn. optimal inbreeding (Shields 1982)

cf. hypothesis: optimal-outbreeding hypothesis; law: Knight-Darwin law

plural breeding *n.* In many bird species: a social system in which two or more monogamous or nonmonogamous con-

specific females breed in the same communal social unit (Brown 1987a, 21–22, 305).
cf. breeding: singular breeding

selective breeding See selection: artificial selection.

singular breeding *n.* In many bird species: a communal social system in which no more than one conspecific female breeds in each social unit (Brown 1987a, 18–21, 306).
cf. breeding: plural breeding

♦ **breeding colony** See colony: breeding colony.

♦ **breeding system** See ¹system: breeding system.

♦ **broad-sense heritability** See heritability: heritability in the broad sense.

♦ **broadcast promiscuity** See mating system: polybrachygamy: broadcast promiscuity.

♦ **Brock-Riffenburgh model** See ⁴model: Brock-Riffenburgh model.

♦ **broken-wing display** See display: broken-wing display.

♦ **brontophobia** See phobia (table).

♦ **brood capture** *n.* In some waterfowl species: a parent's taking over the care of young that are not its own by forcing their real parents to abandon them (Brown 1987a, 298).
cf. parent: alloparent

♦ **brood cell** *n.* A special chamber, or pocket, that an insect builds that houses its immature stages (Wilson 1975, 579).

♦ **brood parasitism** See parasitism: brood parasitism.

♦ **brood patch** *n.* For example, in gulls, songbirds: an area, or areas, of an individual's ventral skin that becomes defeathered, vascularized, and edematous during egg laying and stays in this condition during incubation (Immelmann and Beer 1989, 36).

♦ **brood provisioning** See provisioning: brood provisioning.

♦ **brood raiding** See parasitism: social parasitism: slavery.

♦ **brooding** *n.*
1. In many egg-laying animal species: an adult's incubating eggs (*Oxford English Dictionary* 1972, entries from 1440).
2. In many bird species: an adult's protecting young by covering with its wings (Michaelis 1963).
3. In many bird species: a parent's caring for brood, including incubating its eggs (Immelmann and Beer 1989, 36).
4. In many bird species: a parent's covering and warming, or cooling, hatched young by sitting on them, squatting over them, crouching on them, or a combination of these behaviors (Immelmann and Beer 1989, 36, 142).

See ²group: brood.
n. brooder
v.t. brood
syn. incubating (not a preferred synonym, Immelmann and Beer 1989, 142)
cf. care: parental care; incubation; parity
Comments: Balon (1975, 821, etc.) classifies a major division of ecoethological guilds of fish as "bearers." This division includes the "external bearers" (forehead brooders, gill-chamber brooders, mouth brooders, pouch brooders, skin brooders, and transfer brooders) and the "internal bearers" (oviovoviviparous, ovoviviparous, and viviparous fish).

body-brooding *n.* In some frog species, Giant Water Bugs: an animal's taking care of eggs that are on its own body (Townsend et al. 1984).

egg-brooding *n.* In some stink-bug species, many social-insect and bird species, a tropical-frog species, the Gharial: an animal's taking care of eggs by being near them and protecting them from egg predators (Townsend et al. 1984).

forehead-brooding *n.* In *Kurtus*-fish species: an adult male's taking care of eggs on his superoccipital hook (Balon 1975, 852, illustration, 853).

gastric-brooding See brooding: stomach-brooding.

gill-brooding *n.* In four genera of North American cavefishes: a mother's taking care of her eggs in her gill cavities (Balon 1975, 852, illustration, 853).

mouth-brooding *n.* In many fish taxa, including in some cichlid- and belontiid-fish species: a mother's (or father's) taking care of eggs in her (or his) mouth (buccal cavity) until they hatch (Wickler 1968, 221; Balon 1975, 852).
syn. oral brooding (Blumer 1982, 3)
cf. brooding: open-brooding; mouth breeder
Comments: The fry are typically well developed when they leave their parent's mouth (Blumer 1982, 3). In some fish species, fry return to their parent's mouth when they sense danger (Immelmann and Beer 1989, 195).

nest-brooding *n.* For example, in carrion beetles, social insects, many bird species, alligators: an animal's taking of eggs, or young, that are in its nest.

open-brooding *n.* An animal's taking care of eggs that are in its environment and not on its body or in its mouth (Wickler 1968, 221).

pouch-brooding *n.* In a group of South American catfishes; some pipefish species, including seahorses; female marsupials:

an adult's taking care of young in its body pocket (= marsupium) (Balon 1975, 854). *Comment:* In fish, either a female or a male pouch broods, depending on the species (Balon 1975, 854).

skin-brooding *n.* In some pipefish species, a group of South American catfishes: an adult female's taking care of eggs attached to the ventral surface of her body before depositing them (Balon 1975, 852, illustration, 853).
cf. care: parental care: ectodermal feeding
Comments: In the catfish, a stalked, vascularized cup envelops each egg (Balon 1975, 854). Skin-brooding also occurs in males of some fish species.

stomach-brooding, gastric-brooding *n.* An adult's taking care of its young in its stomach, *e.g.,* an Australian frog "that swallows its fertilized eggs, broods its tadpoles in its stomach, and gives birth to young frogs through its mouth" (Gould 1985a, 12).

transfer-brooding *n.* In some fish species: an adult female's taking care of her eggs by carrying them in her fins for a certain time before depositing them (Balon 1975, 852, illustration, 853).

vocal-pouch-brooding *n.* An adult's taking care of young in its vocal pouch; *e.g.,* male *Rhinoderma darwini* (frogs) which take advanced eggs into their pouches where they hatch and from which they emerge in 52 days as froglets (Gould 1985a, 16).

♦ **Brook's law** See law: Dyar's law.

♦ **brotherhood** *n.* For example, in Turkeys: a pair of brothers that assist each other in fierce competition for mates (Watts and Stokes 1971 in Wilson 1975, 125).

♦ **broticole** See -cole: broticole.

♦ **Browerian mimicry** See mimicry: Browerian mimicry.

♦ **browse** *n.*
1. "The edible plant material within the reach of browsing animals" (Lincoln et al. 1985).
2. *v.t.* In many vertebrate species: "To feed on parts of plants" (Lincoln et al. 1985).

♦ **browse line** *n.* The height of a plant to which grazing animals can reach and eat plant parts (*e.g.,* acacia trees fed on by giraffes) (Lincoln et al. 1985).

♦ **Bruce effect** See effect: Bruce effect.

♦ **bryocole** See -cole: bryocole.

♦ **bryology** See study of: bryology.

♦ **bryophile** See [1]-phile: bryophile.

♦ **bubble nest** See nest: bubble nest.

♦ **bubble net** *n.* In some whale species: a perimeter of bubbles that an individual produces around its prey that might con-

fuse them and improve the whale's feeding efficiency (Baker and Herman 1985, 55).

♦ **bubbling** *n.* In Humpback Whales: an individual's releasing air underwater during male-male competitions (Baker and Herman 1985, 54).
v.i. bubble

♦ **budding** *n.*
1. For example, in hydras: reproduction by the direct growth of a new individual from the body of an old one (Barnes 1974, 94; Wilson 1975, 390, 580).
2. In Florida scrub jays: a means of forming a new social unit (mated pair with helpers) (Wilson 1975, 455).
See swarming: budding.

♦ **budget** *n.* A person's, group's, institution's, or country's summary of probable income and expenditures for a given period; also, its plan for adjusting expenditures to income (*Oxford English Dictionary* 1972, entries from 1733; Michaelis 1963).

resource budget *n.* The total amount of resources (*e.g.,* time and energy) that an organism has at a particular time (Willson 1983, 6).

time budget *n.* The total amount of time that an organism has for a particular activity.

♦ **buffer species** See [2]species: buffer species.

♦ **buffered populations** See [1]population: buffered populations.

♦ **buffering** *n.*
1. Stabilization of population size fluctuation (Kluijver and Tinbergen 1953 in Wilson 1975, 274).
2. An animal's using another animal as part of its social communication with a third animal.

agonistic buffering *n.*
1. For example, in some baboon species, multimale groups of the Barbary macaque: one male's using an infant to inhibit aggression by another conspecific male animal (Deag and Crook 1971, 195–196; Wilson 1975, 352, 578; Immelmann and Beer 1989, 10).
2. For example, in Australian Aborigines, Waika Indians: a person's use of a child to signal friendly intention (Immelmann and Beer 1989, 10).
cf. buffering: social buffering; care
[coined by Deag and Crook 1971, 195–196]

social buffering *n.* For example, in Barbary Macaques: an individual's using another conspecific "individual to regulate, in whatever context, its relations with a third party" [coined by Deag and Crook 1971, 196].
cf. care

Comment: "Agonistic buffering," *q.v.*, may be a category of "social buffering."

♦ **buffering gene** See gene: buffering gene.

♦ **bufotenin** See toxin: bufotenin.

♦ **bufotoxin** See toxin: bufotoxin.

♦ **bug perceiver** *n*. The convex-edge detector in a frog's retina which facilitates its seeing flying insects (Dewsbury 1978, 181).

♦ **building** See ²group: building.

♦ **bull** See animal names (table).

♦ **bunch** See ²group: bunch.

♦ **bunt order** See hierarchy: dominance hierarchy: bunt order.

♦ **bursa copulatrix** See organ: copulatory organ: bursa copulatrix.

♦ **bury** See ²group: bury.

♦ **bush** See habitat: bush; Plantae: bush, shrub (Appendix 1).

♦ **busyness** See ²group: busyness.

♦ **Butler's aphorism** *n*. The chicken is only an egg's way of making another egg (Samuel Butler in Wilson 1975, 3).

♦ **buzz pollination** See pollination: buzz pollination.

♦ **buzzing run** See dance: bee dance: buzzing run.

♦ **byproduct mutualism** See symbiosis: mutualism: byproduct mutualism.

a–c

C

♦ **C$_3$ photosynthesis** See biochemical pathway: C$_3$ photosynthesis.

♦ **C$_4$ photosynthesis** See biochemical pathway: C$_4$ photosynthesis.

♦ **C-selected species, C strategist, C-strategist** See ²species: c-selected species. *cf.* ²species: C-selected species, r-selected species, s-selected species

♦ **cache** *n.*
1. A place where a person stores equipment, provisions, or other things (Michaelis 1963).
2. A place where an animal stores food (Vander Wall 1990, 3; Clarke and Kramer 1994, 299).
cf. behavior: hoarding behavior; larder [French < *cacher*, to hide]

♦ **caching, caching behavior** See behavior: caching.

♦ **cacogenesis** See -genesis: cacogenesis.

♦ **cadavericole** See -cole: cadavericole.

♦ **caecotroph** See food: caecotroph; -troph-: caecotroph.

♦ **caenogenesis, caenogenetic** See -genetic: caenogenetic.

♦ **caenomorphic, caenomorphism** See -morphism: caenomorphism.

♦ **caespiticole** See -cole: caespiticole.

♦ **cage effect** See effect: cage effect.

♦ **Cainism** See -cide: fratricide.

♦ **Cairnsian mutation** See mutation: Cairnsian mutation.

♦ **calcicole** See -cole: calcicole.

♦ **calcifuge** See -phobe: calciphobe.

♦ **calciphile** See ¹-phile: calciphile.

♦ **calciphobe** See -phobe: calciphobe.

♦ **calcosaxicole** See -cole: calcosaxicole.

♦ **calf** See animal names: calf.

♦ **caliology** See study of: caliology.

♦ **call** See animal sounds: call.

♦ **calling** *n.* An animal's making a call, *q.v.*
  female calling, calling *n.* For example, in some ant and moth species: a female's releasing sex pheromone and standing in one place, thereby "calling" males to her with her pheromone (Hölldobler and Wilson 1990, 638).
  tandem calling *n.* In ants: a leader's pheromone release that recruits a nest mate for tandem running (Hölldobler and Wilson 1990, 644).

♦ **callow** *adj.*
1. Referring to an unfledged bird, without feathers (*Oxford English Dictionary* 1972, entries from 1603).
2. Referring to an inexperienced person; immature person; a callow youth (person) (Michaelis 1963).
3. Referring to an individual arthropod that is newly emerged from its last larval skin, chrysalis, or pupa (Torre-Bueno 1978).
syn. teneral (Torre-Bueno 1978) [Old English *calu* bare, bald]

♦ **callow worker** See caste: worker: callow worker.

♦ **calobiosis** See symbiosis: calobiosis.

♦ **Calvin cycle** See biochemical pathway: Calvin cycle.

♦ **Cambrian explosion** *n.* The rapid diversification of marine invertebrates during the Cambrian Period, resulting in the many now extinct taxa and extant animal phyla and classes (Bowring et al. 1993, 1293; Strickberger 1996, 308).
syn. big bang (Kerr 1993a, 1274)
Comments: Researchers date the Cambrian explosion from *ca.* 570 to 530 Ma ago (Kerr 1993a, 1274), 530 to 520 Ma ago (Bowring et al. 1993, 1297), and 530 to 525 Ma ago (Erwin et al. 1997, 132). The first known fossils of Annelida, Arthropoda, Brachiopoda, Cheatognatha, Echinodermata, Hemichordata, Mollusca, Onycophora, Pogonophora, Porifera, and Priapulida are from the Cambrian (Erwin et al. 1997, 132). Sites for these fossils include the Burgess Shale and

Chengjiang. Multicellular organisms first appeared about 570 Ma ago (Graham Logan in Oliwenstein 1996b, 43). Jeffrey Levinton, Gregory Wray, and Leo Shaprio compared seven genes from annelids, arthropods, chordates, echinoderms, and molluscs (Zimmer 1996c, 52). Their data suggest that chordates and echinoderms diverged from the other phyla about 1.2 Ga ago and that chordates and echinoderms diverged about 1 Ga ago, long before the Cambrian explosion.

♦ **camnium** See ²succession: camnium.

♦ **camouflage** *n.*

1. Human military measures, or material, used to conceal, or misrepresent, the identity of, *e.g.*, installations, ships, or persons (Michaelis 1963).
   Note: Game hunters also use camouflage.
2. In many animal species, including *Aeshna* dragonfly larvae, ptarmigans, the Snowshoe Hare, and Ermine: an individual's resembling its environmental background coloration (Bruns 1958 in Heymer 1977, 186).
3. An individual organism's imitation of "certain environmental background features," involving "at least shape, color, and color pattern, and sometimes scent and sound;" *e.g.*, a swallowtail butterfly caterpillar's looking like a bird dropping or a katydid's resembling a leaf (Matthews and Matthews 1978, 310).

v.i., v.t. camouflage
syn. crypsis (Immelmann and Beer 1989, 39)
cf. coloration, mimicry, mimesis
[French *camoufler*, to disguise]

background imitation *n.* An organism's resembling its background.

▸ **countershading** *n.* For example, in caterpillars of some moth species, some fish species, many mammal species: an individual's coloration that decreases, or eliminates, its three-dimensional appearance when it is in a certain position with regard to light direction; for example, a caterpillar's having coloration that becomes lighter from its dorsum to its venter so that when it rests and the greater amount of light strikes its dorsum, it tends to look two-rather than three-dimensional from its side (Matthews and Matthews 1978, 315).
syn. obliterative coloration, obliterative shading (Lincoln et al. 1985)
cf. hypothesis: Thayer's countershading hypothesis

▸ **disruptive coloration** *n.* In some arthropod, fish, and mammal species: an individual's coloration that breaks up its outline "so that parts of it appear to fade separately into its background" (Matthews and Matthews 1978, 314).

▸ **dorsoventral flattening** *n.* In some arthropod species: an animal's body morphology that involves its body's being flattened from its venter to its dorsum; dorsoventral flattening may be accompanied by lateral flaps, various irregular body protuberances, or both, which bridge the gap between its body and its substrate (Matthews and Matthews 1978, 314).

▸ **hunting camouflage** *n.* For example, in the Black-Headed Gull: coloration (*e.g.*, white underparts) that make an individual inconspicuous to its aquatic prey when viewed against a bright sky (Götmark 1987, 1786).

▸ **mimesis** See mimesis, mimicry.

▸ **transparency** *n.* For example, in some butterfly, crustacean, and fish species: an individual's having a mostly clear, or partially clear, body that allows its background to show through (Immelmann and Beer 1989, 39).

eye camouflage *n.* In several coral-fish species: obscuring of eyes by black bars (Heymer 1977, 35).

♦ **canalization** *n.* The buffering of an organism's developmental pathways that tends to produce a standard phenotype despite environmental fluctuations and underlying genetic variability (Lincoln et al. 1985).
See canalization: genetic canalization.
syn. developmental homeostasis (Lincoln et al. 1985), allometry (sometimes), developmental canalization, developmental flexibility, phenotypic flexibility

environmental canalization *n.* An organism's having the same phenotype despite changes in its environment (Stearns 1989, 436).
syn. autonomous development (Schmalhausen 1949 in Stearns 1989, 436)
cf. auxesis: heterauxesis; plasticity: phenotypic plasticity; reaction norm

genetic canalization *n.* An organism's having the same phenotype despite changes in its genotypes (due to mutation or recombination) (Stearns 1989, 436).
syn. canalization (Waddington 1942 in Stearns 1989, 436)

♦ **cancer phobia** See phobia (table).

♦ **canine-tooth threat** See behavior: threat behavior: canine-tooth threat.

♦ **cannibal** *n.*

1. A person, especially a savage, that eats human flesh; a man eater; an anthropophagite (*Oxford English Dictionary* 1971, entries from 1153).
2. In over 1300 animal species, including the Chimpanzee, Guppy, Human, Hyena, Lion, Mongolian Gerbil; many termite species;

some ant, bird, langur, praying mantid, salamander, spider species: an animal that eats members of its own species (*Oxford English Dictionary* 1971, entries from 1796; Ehrlich and Raven 1964, 598; Wilson 1975, 84–85, 246; Jones 1982 in Picman and Belles-Isles 1987, 236; Goodall in Gould 1989, 29; Immelmann and Beer 1989, 39).
Comments: An individual that eats its own mother is a "gerontophage," *q.v.* Cannibalism might have been widespread in *Homo neanderthalensis* (T. White in Gugliotta 1999, A3).
[Spanish *Canibales*, var. of *Caribes*, Caribs, a fierce nation of the West Indies reported to eat human flesh]
♦ **cannibalism** *n.* An animal's being a cannibal, *q.v.*
syn. metasitism (Lincoln et al. 1985)
cf. -cide
cronism *n.* For example, in birds (some duck, eagle, gull, hawk, jay, owl, shrike, stork, and tern species): a parent's actual, or attempted, swallowing of its dead or sickly young (Welty 1975, 349; Schüz 1957 in Campbell and Lack, 1985).
syn. Kronismus (German)
cf. cannibal; -cide: infanticide
Comments: Immelmann and Beer (1989, 145) do not indicate that cronism involves dead or sickly young. Captive fish (*e.g.*, cichlids, labyrinth fish, and live-bearing tooth carps) commonly eat their young (Gilbert 1976, 145). Laboratory gerbils also eat their own young, especially when their young are sick or when a person disturbs the gerbils too often (personal observation). Under disturbed conditions, mother tree shrews cannibalize their young (Immelmann 1977, 567).
[term proposed by Schüz (1957) in Campbell and Lack (1985) after the Greek mythological Cronus, the son of Uranus (Heaven) and Gaea (Earth), who swallowed his sons, except for Zeus, as they were born (Duckworth and Rose 1989)]
egg cannibalism *n.*
1. For example, in the ant *Leptothorax acervorum*: a queen's consuming eggs of other queens and workers in her nest (Bourke 1991, 295).
Note: This egg cannibalism appears to be part of reproductive conflict among queens and workers in this ant (Bourke 1991, 295).
2. For example, in stickleback fish: intraspecific egg consumption by groups composed of females, non-nesting males, and occasionally nesting males (Woriskey 1991, 989).
syn. oophagy

filial cannibalism *n.* In some frog species: cannibalism in which parents eat their own eggs (Rohwer 1978 in Townsend et al. 1984, 422).
syn. filial ovicide (Picman and Belles-Isles 1987, 35)
gerontophagy See -phage: gerontophage.
heterocannibalism *n.* In some frog species: cannibalism in which animals eat conspecific eggs that are not their own (Rohwer 1978 in Townsend et al. 1984, 422).
oophagy See cannibalism: egg cannibalism; -phagy: oophagy.
sexual cannibalism *n.* In some arachnid and mantid species: a female's eating her potential mate before copulation or her mate during or after copulation (Barrows 1982, 16; Elgar and Nash 1988, 1511; Birkhead et al. 1988).
Comment: Sexual cannibalism appears to be the exception, rather than the rule, in mantids (Barrows 1982, 16; Brown 1986, 421).
♦ **canopy** *n.*
1. Part of forest community that is formed by trees (Allaby 1994, 66).
Note: Some forests have more than one canopy layer. A complex tropical rain forest can have an upper emergent zone, middle zone, and lower zone (Allaby 1994, 66).
2. The upper layer of scrub, or shrub communities, or any terrestrial plant community in which a distinctive habitat is formed in the upper, denser regions of its taller plants (Allaby 1994, 66).
[Middle English *canape* < Medieval Latin *canapeum*, mosquito net < Lain *conopeum* < Greek *kōnōpion* < *Kanōpos, Canopus*, a city of ancient Egypt]
♦ **cantharidin** See toxin: cantharidin.
♦ **cantharophile** See ²-phile: cantharophile.
♦ **capacity laws** See law: capacity laws.
♦ **Cape Floral Kingdom** See ²community: fynbos.
♦ **caprification** See pollination: caprification.
♦ **captivity degeneration** *n.* All of an animal's disturbances in mental and physical performance that result from captivity (Fox 1968 in Heymer 1977, 95).
cf. behavior: abnormal behavior; stereotypy
♦ **capture hypothesis** See hypothesis: Moon-origin hypotheses: capture hypothesis.
♦ **"carbon-dioxide hypothesis"** See hypothesis: hypotheses regarding the Permian mass extinction: "carbon-dioxide hypothesis."
♦ **carbonicole** See -cole: carbonicole.
♦ **carboxyphile** See ¹-phile: carboxyphile.
♦ **carcinology** See study of: carcinology.
♦ **care** *n.*
1. A person's charge; oversight with a view to protection, preservation, or guidance

(*Oxford English Dictionary* 1972, entries from *ca.* 1400).

2. In many animal species: interactions between, or among, individuals (immature, adult, or both) that, depending on the species, may involve behaviors such as protection from dangerous situations caused by the physical environment and conspecific and heterospecific animals, holding, and grooming (or preening) (Deag and Crook 1971, 195).

v.i., v.t. care

cf. behavior: care-giving behavior; buffering: agonistic buffering, social buffering

alloparental care *n.* A nonparent's assisting in care of offspring which may be shown either by females (allomaternal care) or by males (allopaternal care) (Wilson 1975, 349, 578).

cf. parent: alloparent

female care *n.* For example, in the Barbary Macaque: care given to baby monkeys by females other than their mothers [coined by Deag and Crook 1971, 196].

cf. aunt; care: maternal care

male care *n.* In some primate species, including marmosets; the Barbary Macaque: care given to baby monkeys by males of unknown relationship to the young [coined by Deag and Crook 1971, 195].

cf. care: female care, maternal care, paternal care; uncle

Comment: Because it is difficult to distinguish clearly between a male's "paternal care" and other kinds of behavior, "male care" is sometimes extended to mean the totality of behaviors between mature males and conspecific young (Immelmann and Beer 1989, 179, 214).

parental care *n.* In some myriapod species; many amphibian, bird, cephalopod, crustacean, insect, fish, leech, mammal, and spider species: all forms of assistance (*e.g.*, protecting, nourishing, and nurturing young) provided by parents to their progeny following their birth and prior to their independence or maturity; all parental investments other than a parent's investment in its gametes (Wilson 1975, chap. 16; Wittenberger 1981, 619; Kutschera and Wirtz 1986, 941; Nafus and Schreiner 1988, 1425; Immelmann and Beer 1989, 214).

syn. parental behavior

cf. brooding; ²facilitation: parental facilitation; -parity; provisioning: brood provisioning

Comments: Most arthropod, fish, amphibian, and reptile species have no parental care; most bird species have parental care by both parents; and most mammal species have only maternal care (Maynard Smith 1978b in McFarland 1985, 131; Halliday and

Adler 1986). About 89 of the approximately 422 families of bony fishes exhibit parent care (Blumer 1982, 1–2). Forms of parental care in fishes include "brood-pouch egg carrying" (pouch-brooding), "cleaning," "coiling," "ectodermal feeding" (skin-brooding), "egg burying," "external-egg carrying," "fanning," "internal gestation," "guarding," "moving," "nest building and maintenance," "removal," "splashing," and "substrate cleaning." Major categories of "parental care" in frogs are "egg attendance," "tadpole attendance," "egg transport," and "tadpole transport" (Townsend et al. 1984, 421). McFarland (1985, 132) classifies parental care as a kind of altruism; his classification is controversial.

▸ **brood-pouch egg carrying** See brooding: pouch-brooding.

▸ **cleaning** *n.* A fish's taking an egg into its mouth, manipulating it inside its buccal cavity, and returning it to its original site (Blumer 1982, 3).

▸ **coiling** *n.* A parent fish's coiling its body around its egg mss while guarding it (Blumer 1982, 3).

Comment: This guarding posture reduces the eggs' exposure to air at low tide in intertidal oviposition sites (Blumer 1982, 3).

▸ **ectodermal feeding** *n.* A parent fish's feeding its young fry with a specialized mucous produced on the parent's body surface (Blumer 1982, 3).

▸ **egg burying** *n.* A fish's depositing eggs beneath a substrate surface, or covering its eggs with substrate material (Blumer 1982, 3).

▸ **external egg carrying** *n.* A fish's carrying an egg(s) outside of its body until the egg hatches (Blumer 1982, 3).

▸ **fanning** *n.* A fish's moving its anal, caudal, pectoral, pelvic, or a combination of these fins over its egg mass or fry (Blumer 1982, 3).

Comments: This behavior aerates the young and removes sediment from their area. Some fish perform these acts by forcing water through their gill cavities or mouths which flows over their young (Blumer 1982, 3).

▸ **internal gestation** *n.* A mother fish's retaining her egg(s) inside her oviducts, or ovaries, while it develops (Blumer 1982, 3).

cf. -parity: ovoviviparity, oviparity, ovoviparity, viviparity

▸ **guarding** *n.* A fish's actively chasing, displaying toward, or both, a conspecific or heterospecific individual that approaches its eggs, fry, or the site where they reside (Blumer 1982, 2).

a–c

▶ **maternal care** *n.* For example, in a sawfly species; some spider, burying-beetle, crustacean, frog, and reptile species (including alligators); many bee, wasp, bird, and mammal species; scorpions; burying beetles: care given by a mother to her young (Deag and Crook 1971, 196; Fink 1986, 34; Immelmann and Beer 1989, 178; Polis 1989; Scott and Traniello 1989; D.R. Smith, personal communication).
syn. female-parental care, maternal behavior (Fink 1986, 34)
cf. care: female care, male care

▶ **moving** *n.* A fish's taking eggs, or fry, by its mouth from one location to another, often from one nest to another (Blumer 1982, 3).

▶ **nest building and maintenance** *n.* A fish's making a structure used to hold its eggs, fry, or both (Blumer 1982, 2).
Comment: Depending on the species, this parental care includes one or more of the following behaviors: assembling a cup, or tube, with pieces of vegetation; blowing mucus-covered bubbles that form a floating mass; digging a burrow, or tube, into a substrate; and making an elevated mound with substrate materials (Blumer 1982, 2).

▶ **oral-brooding** See brooding: mouth-brooding.

▶ **paternal care** *n.* In some insect, seahorse, and other fish species; many frog, bird, and mammal species: care given by a father to his young (Townsend et al. 1984; Bisazza and Marconato 1988, 1352; Immelmann and Beer 1989, 178; Scott and Traniello 1989).
syn. male-parental care, paternal care, paternal behavior
cf. care: male care

▶ **removal** *n.* A fish's removing dead or diseased eggs from its egg mass with its mouth (Blumer 1982, 3).

▶ **retrieval** *n.* A fish's taking an egg, or fry, that falls, or strays, from its nest, or school, into its mouth and returning it to its nest, or school (Blumer 1982, 3).
cf. retrieving

▶ **splashing** *n.* A fish's splashing water upon its eggs that are deposited out of water or upon eggs exposed to air during low tide (Blumer 1982, 3).

▶ **substrate cleaning** *n.* A fish's removing an alga, animal, detritus, or combination of these items from a site where she will deposit her eggs (Blumer 1982, 2).

♦ **care-giving behavior** See behavior: epimeletic behavior.

♦ **carnival** *n.*
1. In the Human: "a season or course of feasting, riotous revelry, or indulgence" (*Oxford English Dictionary* 1972, entries from 1598); any gay festival, wild revel, or merrymaking (Michaelis 1963).
2. In the Chimpanzee: a display produced by troop members that involves unleashing of "a deafening outburst of noise — shouting at maximum volume, drumming trunks and buttresses of trees with their hands, and shaking branches, all the while running rapidly over the ground or brachiating from branch to branch" (Savage 1844, Sugiyama 1972, Reynolds and Reynolds 1965 in Wilson 1975, 222, 542).
[Italian *carnevale* < Medieval Latin *carnelevarium* < Latin *caro, carnis*, flesh + *levare*, to remove]

♦ **Carolina bay** See habitat: bay.

♦ **carpogenous** See -genous: carpogenous.

♦ **carpophage** See -phage: caropophage.

♦ **carrying capacity (K)** *n.*
1. The number of reindeer stock that a range can support without injury to the range (Hadwen and Palmer 1922, who evidently first introduced this term, according to Pulliam and Haddad 1994, 141).
2. The number of animals that a particular range can support (Leopold 1933, 50–51, in Pulliam and Haddad 1994, 141).
Notes: The carrying capacities of a particular animal species is habitat dependent (Pulliam and Haddad 1994, 141–144). "Carrying capacity is a useful concept only when used in the original sense of a limit set to population size by the availability of resources; carrying capacity can vary in the same location from year to year."
3. A "threshold of security" that is reached when all available cover is saturated and mortality increases rapidly because "surplus animals" become especially vulnerable to predation (Errignton 1934 in Pulliam and Haddad 1994, 141).
4. Parameter K of the logistic growth equation; the "upper bound beyond which no major increase in an animal's population size can occur (assuming no major environmental changes)" (Odum 1953, 122, in Pulliam and Haddad 1994, 142).
Note: Odum assumed that carrying capacity and saturation point are the same concept (Pulliam and Haddad 1994, 142).
5. "A measure of the amount of renewable resources in the environment in units of the number of organisms these resources can support" (Roughgarden 1979 in Pulliam and Haddad 1994, 142).
6. The largest number of individuals of a particular species that can be maintained indefinitely in a given part of their environment (Wilson 1975, 81, 580).

7. The numbers of individuals of different species that can live in an area without harming its resource base (Odum et al. 1988, 25).

cf. saturation point

Comment: Dhont (1988 in Pulliam and Haddad 1994, 141) reviewed the concept of carrying capacity. Textbooks most frequently use Odum's (1953) definition of carrying capacity.

human carrying capacity *n.*

1. The maximum number of persons that can be supported in perpetuity in an area with a given technology and set of consumptive habits and without causing environmental degradation (Allan 1965 in Pulliam and Haddad 1994, 150).

   Note: Pulliam and Haddad (1994, 150) chose to omit "without causing environmental degradation" in defining human carrying capacity.

2. The maximum number of Humans that an environment can sustain without reference to the quality of life of the people (Hardin 1986; Pulliam and Haddad 1994, 154).

Comments: Human carrying capacity is more complicated than that of other organisms because it includes, in addition to biophysical components, social and cultural components (Daily and Ehrlich 1992 in Pulliam and Haddad 1994, 150). Human carrying capacity changes through time due to human innovation and technology. The number of people, their consumptive habits, and their damage to ecosystems determine human carrying capacity. A human population that leads to irreparable degradation of its life-support mechanisms is, by definition, above human carrying capacity (Pulliam and Haddad 1994, 154).

♦ **carrying in** *n.* In rodents and carnivores: an individual's transporting objects (*e.g.*, nest material or winter provender) or young to its nest, den, or burrow (Immelmann and Beer 1989, 40).

cf. transport of young

♦ **carsinomaphobia** See phobia (table).

♦ **carsinomatophobia** See phobia (table).

♦ **carsinophobia** See phobia (table).

♦ **carton** *n.* "The chewed vegetable fibers used by many kinds of ants, wasps, and other insects to construct nests" (Wilson 1975, 580).

♦ **-caryo** See -karyo.

♦ **caryogamy** See -gamy: karyogamy.

♦ **caryotype** See -type, type: karyotype.

♦ **cascade** See trophic cascade.

♦ **cast downness** See abasement.

♦ **caste** *n.*

1. "A hereditary group, endogamously breeding, occupied by persons belonging to the same rank, economic position, or occupation, and defined by mores that differ from those of other castes" (*Oxford English Dictionary* 1971, entries from 1613; Wilson 1975, 299).

2. In eusocial bees and wasps: a functional group of females, either workers or queens (Michener 1974, 371).

   Note: A caste may be differentiated by behavior and physiology alone; it may also be morphologically distinct (Michener 1974, 371).

3. In ants: the workers, soldiers, ergatogynes, dichthadiiform ergatogynes, or queen(s) in a colony (Wilson 1971, 136).

   Note: The three castes — queens, soldiers, and workers — are found in only a minority of ant species; males comprise a caste only in the very loosest sense (Wilson 1971, 136).

4. In termites: the larvae (apterous nymphs), nymphs (brachypterous nymphs), workers, pseudergates, soldiers, primary reproductives (first-form reproductives, imagos), supplementary reproductives, replacement reproductives, neoteinic reproductives, adultoid reproductives, nymphoid reproductives (second-form reproductives, secondary reproductives, brachypterous neoteinics), and ergatoid reproductives (third-form reproductives, tertiary reproductives, apterous neoteinics) (Wilson 1971, 184).

   Note: The castes present depend on species of termite and colony development.

5. In social insects, the Naked Mole Rat: any set of individuals of a particular morphological type, age group, or both, that performs specialized labor within its colony (Wilson 1975, 299; Jarvis 1981, 571).

6. In social insects: "any set of individuals in a given colony that are both morphologically distinct from other individuals and specialized in behavior" (Wilson 1975, 299, 580).

7. In sponges, coelenterates, ectoprocts, and tunicates: a set of cells that has a particular role in social cell groups (Wilson 1975, 315).

See ²group: brace; imago.

cf. auxiliary, division of labor, king, role

Comment: Castes can be classified by type of polyethism (physiological and temporal) and by kind of individual (worker, soldier, and reproductive).

[Portuguese *casta*, unmixed breed < Latin *castus*, pure]

dichthadiigyne *n*. In some ant species: a large female with a huge gaster (Wilson 1971, 70, 136).
syn. dichthadiiform ergatogyne (Wilson 1971, 136)

ergatandromorph *n*. In some ant species: A form that is morphologically intermediate between a worker and male, having normal male genitalia and a worker-like body (Wilson 1971, 138).
adj. ergatandromorphic
syn. eratomorphic male, ergatoid male
cf. male

ergatogyne *n*. In some social-insect species: any form that is morphologically intermediate between a worker and a queen (Wilson 1971, 138; 1975, 371, 583).
adj. ergatogynous
syn. queen-worker intercaste

gyne *n*.
1. In ants: a sexual female that is not socially a functional reproductive (Brian 1957 in Wilson 1971, 138).
2. In bees: a potential, or actual, queen (Michener 1974, 372).
3. In ants: a female reproductive-caste member that may, or may not, be functioning as a reproductive at a particular time; broadly, a queen (Hölldobler and Wilson 1990, 638).
syn. queen (Wheeler 1907 in Wilson 1971, 138)
cf. queen

gynergate *n*. In ants: a female with patches of both queen and worker tissue (Hölldobler and Wilson 1990, 638).

king *n*. In termites: the male that accompanies a queen (egg-laying female) and inseminates her from time to time (Wilson 1975, 435, 587).

larva See caste: reproductive: nymph.

microgyne *n*. In an ant species with two sizes of queens: a queen of the smaller form (Hölldobler and Wilson 1990, 640).
cf. female: macrogyne

neoteinic *n*. A supplementary reproductive termite (Wilson 1975, 589).
syn. neotene, neoteinic (Lincoln et al. 1985)

nymph, nympha See caste: reproductive: nymph.

pseudergate *n*. In less-derived termites: a caste comprised of individuals that either have regressed from nymphal stages by molts that reduced, or eliminated, their wing buds or else were derived from larvae by undergoing nondifferentiating molts (Wilson 1975, 435, 593).
Comment: Pseudergates serve as the principal elements of the worker caste but remain capable of developing into other castes (*e.g.*, secondary neoteinic, *q.v.*) by further molting (Wilson 1975, 435, 593).

pseudocaste *n*. A group of social symbionts in its host species' colony (Wilson 1975, 353).

pupa See caste: reproductive: nymph.

queen *n*.
1. In ants, termites, and some species of bees and wasps: "a perfect female" (*Oxford English Dictionary* 1972, entries from 1609); a female member of a reproductive caste (Wilson 1975, 435, 593).
2. In the Naked Mole Rat: a reproductive female (Jarvis 1981).
v.i., v.t. queen
syn. gyne (Wheeler 1907 in Wilson 1971, 138)
cf. caste: gyne; female
Comments: In social animals, the existence of a queen presupposes the existence of a worker caste at some stage of a colony's life cycle. Queens may, or may not, be morphologically different from workers (Wilson 1975, 435, 593)

▸ **foundress queen** *n*. In social halictine bees: a principal egg layer that never functioned as a worker (Eickwort and Kukuk 1987 in Michener 1988a, 77).

▸ **macrogyne** *n*. In an ant species with two sizes of queens: a queen of the larger form (Hölldobler and Wilson 1990, 640).

▸ **microgyne** *n*. In an ant species with two sizes of queens: a queen of the smaller form (Hölldobler and Wilson 1990, 640).

▸ **replacement queen** *n*.
1. In highly eusocial bees: a queen that a colony produces that will replace a queen that has died or became lost (Michener 1974, 374).
2. In social halictine bees: a principal egg layer that is a daughter, with a high probability of originally being a worker, of a disappeared queen (Eickwort and Kukuk 1987 in Michener 1988a, 77).

▸ **supersedure queen** *n*. In highly eusocial bees: a queen that a colony produces while an aging queen is still present and that will replace her (Michener 1974, 374).

reproductive *n*. An individual that has offspring. See imago.

▸ **primary reproductive** *n*. In termites: a colony-founding type of queen, or male, derived from a winged adult (Wilson 1975, 592).

▸ **neoteinic** *n*. In termites: a supplementary reproductive (Wilson 1975, 589).

▸ **nymph** *n*. In termites: an immature individual that possesses external wing buds and enlarged gonads and is capable of developing into a functional reproductive by further molting (Wilson 1975, 590).

syn. larva (confusing synonym), nympha, pupa (uncommon synonym, *Oxford English Dictionary* 1972)

▶ **secondary neoteinic** *n.* In some termite species: a male, or female, that replaces a king or queen, respectively, that is lost; a secondary neoteinic transforms in one molt from a worker-like pseudergate (Wilson 1975, 435).

nasute soldier *n.* A termite soldier with a nasus (Wilson 1975, 589).

scout *n.* In ants: a worker that searches outside her nest for food or, in a slave-making species, a worker that searches for a host colony suitable for raiding (Hölldobler and Wilson 1990, 642).
cf. caste: worker

soldier *n.*
1. A nonreproductive worker ant, or termite, that is specialized for colony defense (Wilson 1975, 595).
2. A morphologically distinct, second-instar aphid that defends its area against other insects (Aoki et al. 1977 in Aoki et al. 1991).
Note: Soldier aphids do not molt into third instars and they do not reproduce (Aoki et al. 1991).

worker *n.*
1. A "neuter or undeveloped female" of certain social Hymenoptera (*e.g.*, bees and ants) that "supplies food and provides other services for the community" (*Oxford English Dictionary* 1972, entries from 1747).
2. In most ant species: an ordinarily nonreproductive (sterile) female that has reduced ovarioles and a greatly simplified thorax with nota that are typically represented by no more than one sclerite each (Tulloch 1935 in Wilson 1971, 138).
Note: In ants, workers are minors (minima), medias, majors (maxima, soldiers).
syn. ergate (Torre-Bueno 1978)
[Greek *ergatēs*, workman]
3. In semisocial and eusocial insect species, the Naked Mole Rat: a member of the nonreproductive, laboring caste (Wilson 1975, 598; Jarvis 1981).
4. In termites, usually termitid species: a male, or female, individual that undertakes usual colony chores; *e.g.*, constructing a nest, cleaning, nursing, and foraging (Wilson 1971, 184).
Note: In termitid termites: a worker lacks wings; has a reduced pterothorax; has reduced, or no, eyes; and has a rudimentary genital apparatus (Wilson 1971, 184; 1975, 598).

5. In Bumble Bees, Honey Bees, Stingless Bees, and other kinds of social bees: a female that lays no eggs, or few eggs relative to a queen, and performs tasks in her nest (Michener 1974, 374).
See worker.
cf. scout
Comment: The existence of a worker caste presupposes the existence of a royal (reproductive) caste in a species.

▶ **callow worker** *n.* In social insects: a teneral adult worker; a newly emerged adult worker, whose exoskeleton is still relatively soft and lightly pigmented (Wilson 1975, 426, 580).

▶ **frequent worker** *n.* In the Naked Mole Rat: an animal that frequently performs tasks associated with nest building and foraging (Jarvis 1981, 571); contrasted with infrequent worker.

▶ **gamergate** *n.* In ants: "a mated, egg-laying worker" (Hölldobler and Wilson 1990, 638).

▶ **infrequent worker** *n.* In the Naked Mole Rat: an animal that shows role overlap with frequent workers but performs tasks in its colony at less than half the rate of frequent workers (Jarvis 1981, 571); contrasted with frequent worker.

▶ **major worker** *n.* In ants and termites: "a member of the largest-worker subcaste, especially in ants" (Wilson 1975, 588).
cf. caste: worker: media worker, minor worker
Comment: In ants, this subcaste is usually specialized for defense, and an adult belonging to it is often also referred to as a soldier (Wilson 1975, 588).

▶ **media worker** *n.* In polymorphic ant series involving three or more worker subcastes: an individual belonging to the medium-sized subcaste(s) (Wilson 1975, 588).
cf. caste: worker: minor worker, major worker

▶ **minima** See caste: worker: minor worker.

▶ **minor worker** *n.*
1. In eusocial insects, especially in ants: "A member of the smallest worker subcaste" (Wilson 1975, 589).
2. In eusocial insects: a sterile but not necessarily nonreproductive caste that performs most energy acquisition, colony maintenance, and brood care functions (Wittenberger 1981, 618).
syn. minima (sing., Wilson 1971, 589)
cf. caste: worker: major worker
Comment: Minor workers may be differentiated into several functionally different castes on the basis of age, ontogeny, or both (Wittenberger 1981, 618).

▶ **nanitic worker** *n.* In ants: a dwarf worker that is produced from either the first, or late, ant brood of a colony and is small due to her being starved (Hölldobler and Wilson 1990, 640).

▶ **nonworker** *n.* In the Naked Mole Rat: a larger male, or female, that cares for young and very rarely digs or transports materials (Jarvis 1981, 571).
Comment: A male nonworker is likely to mate with the sole reproductive female of his colony (Jarvis 1981, 571).

▶ **replete** *n.* For example, in the False Honey Ant, Honeypot Ant: a worker whose crop is greatly distended with liquid food, to the extent that her abdominal segments are pulled apart and her intersegmental membranes are stretched tight (Talbot 1943, 32; Wilson 1975, 594).
Comment: Repletes usually serve as living reservoirs that regurgitate food on demand to their nest mates (Wilson 1975, 594).

▶ **temperature messenger** *n.* In some ant species: a worker that runs along her nest surface in winter and tests external temperature; the warmer it becomes, the longer the messenger remains near the surface and the more heat she transfers to her nest mates when she returns (Wilson 1971, 1973 in Heymer 1977, 173).

◆ **caste polyethism** See polyethism.

◆ **casting** *n.* An object that has been cast off, or voided, by an organism (*e.g.*, a fecal pellet or worm casting) (Lincoln et al. 1985).
cf. exuviae

◆ **Castles' law** See law: Hardy-Weinberg law.

◆ **Castles' theory** See ³theory: Castles' theory.

◆ **castration** *n.*
1. A person's removal of, or interference with the function of, testes (gelding, emasculation) or ovaries (spaying) of another animal, including a Human (*Oxford English Dictionary* 1972, entries from 1420; Lincoln et al. 1985).
2. In the Bonobo, Chimpanzee, Rhesus Monkey: one individual's removing the testicles of another (de Waal 1989, 73–74, 221).
syn. gonadectomy
[Latin *castratus*, past participle of *castrare*, to castrate, prune, expurgate, deprive of vigor]

◆ **emasculation** *n.* A person's removal of male reproductive organs, or inhibition of male reproductive capacity, of another animal (Lincoln et al. 1985).
syn. castration, orchidectomy, orchiectomy (Immelmann and Beer 1989, 120)

ovariectomy *n.* A person's removal of another animal's ovaries (Immelmann and Beer 1989, 120).

parasitic castration *n.* A host's reproductive death due to a parasitic infection (Lincoln et al. 1985).

psychological castration *n.* In some primate species: the phenomenon that the mere presence of higher-ranking rivals suppresses the breeding capability of subordinate males (Immelmann and Beer 1989, 120).

◆ **casual society, casual group** See ²society: casual society.

◆ **cata-, kata-** *prefix*
1. Down; against; upon (Michaelis 1963).
2. Back; over (Michaelis 1963).
Before vowels, cat-.
[Greek *kata-* < *kata* down, against, back]

◆ **catabolism** See metabolism: catabolism.

◆ **cataclysmic evolution** See ²evolution: cataclysmic evolution.

◆ **catadromous** See -dromous: catadromous.

◆ **catagenesis** See -genesis: catagenesis.

◆ **catalepsy** See playing dead.

◆ **catalog** See -gram: ethogram.

◆ **cataplasia** See -plasia: cataplasia.

◆ **catastrophic selection** See selection: catastrophic selection.

◆ **catastrophism** See hypothesis: catastrophism.

◆ **catastrophist** See directionist.

◆ **catch-mark-recatch method** See Lincoln index.

◆ **catecholamine** See hormone: catecholamine.

◆ **category** *n.* The rank, or level, in a hierarchic classification; *e.g.*, the Dog (*Canis familiaris*) and House Fly (*Musca domestica*) are in the species category (Mayr 1982, 207).
cf. taxon

◆ **cause, causation** *n.* An agent, or force, that produces an effect; *e.g.*, a person, occasion, or condition that gives rise to a result or action (Michaelis 1963).
See etiology.

configurational causes *pl. n.* Factors in different constellations that can have drastically different results [coined by Simpson 1970 in Mayr 1982, 377].

proximate cause, proximate causation *n.*
1. "The immediate cause underlying expression of a phenotypic trait" (Spencer, Romanes, Baker 1938 in Mayr 1982, 68; Wittenberger 1981, 620).
2. A factor (*e.g.*, external stimuli or an animal's motivational state) that governs the immediate causation of the animal's physiological process or behavior; for example, in many birds, the

tactile stimuli from sitting on eggs leads to cessation of egg laying (Immelmann and Beer 1989, 236).
syn. proximate factor
cf. causation: ultimate causation; function: proximate function of behavior

▶ **proximate causation of behavior** *n.* A condition(s) in an organism's external or internal environment (physiology, development) that triggers the organism's response (Wilson 1975, 593), *e.g.*, increase in day length that triggers the beginning of a bird's breeding cycle (Immelmann and Beer 1989, 321); contrasted with ultimate causation of behavior.
syn. proximal causation, proximate cause
cf. cause: ultimate causation: ultimate causation of behavior

ultimate causation *n.* All environmental conditions that lead to an animal's longer life expectancy and greater reproductive success (Immelmann and Beer 1989, 321).

▶ **ultimate causation of behavior** *n.*
1. Past environmental forces (*e.g.*, food supply) that influenced the evolution of behavioral adaptations (Wilson 1975, 23, 593); for example, environmental factors that affect bird breeding time over evolutionary time (Immelmann and Beer 1989, 320); contrasted with proximate causation of behavior.
2. Both evolutionary and ecological reasons for why an organism exhibits particular behavior (Alcock 1975 in Dewsbury 1978, 6).
3. Adaptive consequences of behavior presumed to have had selective significance during the history of a species; why a behavior evolved (Baker 1938 in Mayr 1982, 68; Daly and Wilson 1983).
4. In experimental psychology, function of behavior that is not necessarily the result of natural selection (Daly and Wilson 1983).
5. In physiology, an utterly proximate stimulus for behavior (Daly and Wilson 1983).
syn. ultimate factor, ultimate function, selective factor (Immelmann and Beer 1989, 320)
cf. cause: proximate cause: proximate causation of behavior

♦ **cavalry** See ²group: cavalry.
♦ **cavernarious** *adj.* Living in caves.
syn. cavernicolous
♦ **cavernicole** See -cole: cavernicole.
♦ **CC** See coefficient: coefficient of community.
♦ **cDNA** see nucleic acid: deoxyribonucleic acid: copy DNA.
♦ **cDNA library** See library: cDNA library.

♦ **celestial-body taxis** See taxis: celestial-body taxis.
♦ **cell** *n.*
1. "A minute portion of protoplasm, enclosed usually in a membranous investment" (*Oxford English Dictionary* 1971, entries from 1672).
  Note: This definition is misleading because all living cells are thought to have peripheral limiting membranes (G.B. Chapman, personal communication).
2. The structural unit of organisms which is surrounded by a membrane and composed of cytoplasm, one or more nuclei, and other structures (Curtis 1983, 1090).
3. The smallest, membrane-bound protoplasmic body that is capable of independent reproduction (King and Stansfield 1985).
4. A mass of protoplasm limited in space by a plasma membrane (= cell membrane, cytoplasmic membrane) and containing one or more nuclei (or nucleoids) and various organelles and inclusions (G.B. Chapman, personal communication).
See nest: wasp nest: cell.
cf. -cyte; organized structure: quasi-cell
Comments: In most plants, fungi, and bacteria, a cell wall surrounds a cell's membrane (Curtis 1983, 1090). "Cell" is also used to refer to a dead cell, such as wood cell, which may have lost its contents after death. Only a few of the hundreds of kinds of cells are defined below.
[Latin *cella*, chamber]

echo detector *n.* In some bat species: one of a class of neurons that responds to a faint, bat-like echo preceded by a loud, bat-like sound (Feng et al. 1978 in Dawkins 1986, 89).

germ cell *n.*
1. A gamete, or cell, that gives rise to a gamete(s) (Mayr 1982, 658).
2. An ancestral (primitive) male or female element (Mayr 1982, 658).
syn. agamete, gamete, germinal cell
cf. gamete
Comment: The concept of "germ cell" arose in the second half of the 18th century (Mayr 1982, 658).

glial cell, supporting cell [GLEE al] *n.* A cell in a vertebrate's brain and spinal cord (Campbell et al. 1999, 964).
Comment: Glial cells give nervous systems structural integrity and aid in normal neuron functioning (Campbell 1999, 964). The brain of a human baby's brain contains about 1 trillion glial cells at birth which form a network that protects and nourishes neurons (Nash 1997, 50).
[Greek *glia*, glue]

neuroendocrine cell *n*. A cell that delivers its neurohormone(s) into an animal's bloodstream (Immelmann and Beer 1989, 202).

neuron *n*. A nerve cell (Immelmann and Beer 1989, 202).

Comment: The brain of a human baby, at birth, contains about 100 billion neurons (Nash 1997, 50).

▸ **afferent neuron** *n*. A neuron that transmits an impulse from an animal's receptor to its central nervous system (Michaelis 1963).

▸ **combination-sensitive neuron** *n*. For example, in the Mustached Bat: a neuron in the auditory cortex that shows a facilitative response to a combination of different frequencies in a bat's sound pulse and its echo (Fitzpatrick et al. 1993, 931). *cf.* cell: echo detector

▸ **efferent neuron** *n*. A neuron that transmits an impulse from an animal's central nervous system to another part of its body (*e.g.*, a muscle) (Michaelis 1963).

▸ **interneuron** *n*. A neuron that intervenes between an afferent and efferent neuron in a reflex chain of three neurons (= disynaptic reflex) (Immelmann and Beer 1989, 248).
syn. intercalated association, intercalated association neuron, intercalated commissural neuron, intercalated neuron, internuncial neuron (G.B. Chapman, personal communication)

protocell *n*. A hypothetical structure that is composed of proteins and other kinds of organic compounds, interacts as a unit with its environment, and reproduces (Strickberger 1990, 120–121).

Comments: One kind of protocell might have had ancestral heredity (without nucleic acids) that enabled it to pass on some of its metabolic and enzymatic properties. This kind of protocell possibly evolved into the first cell (the first organism) which had heredity based on nucleic acids (Strickberger 1990, 120–123). Inorganic selection (= chemical selection) evolved into organic selection (= biological selection) with the appearance of the first cell. Protocells might have evolved from organized structures.

quasi-cell See organized structure: quasi-cell.

sense cell See ²receptor.

♦ **cell assembly** *n*. The basic unit of storage in brains which is a group of neurons connected in a ring or "closed circuit" (Hebb 1949 in Immelmann and Beer 1989, 41).
cf. engram

♦ **cell biology** See study of: biology: cell biology.

♦ **cell theory (of multicellarity)** See ³theory: cell theory (of multicellarity).

♦ **cenancestor** See ancestor: cenancestor.

♦ **cenobiology** See study of: biocoenology.

♦ **cenogenesis** See -genesis: cenogenesis.

♦ **cenogenous** See -genous: coenogenous.

♦ **cenophobia** See phobia (table).

♦ **cenospecies** See ²species: ceonospecies.

♦ **cenothermy** See -thermy: cenothermy.

♦ **cenozoology** See study of: zoology: ceno-zoology.

♦ **centeener** See ²group: centeener.

♦ **center** *n*.
1. A locus in an animal's nervous system characterized by the presence of a number of cell bodies and synapses, whose excitation by appropriately specified electrical stimulation may yield discrete motor, or autonomic, behavior patterns, and whose experimental destruction is followed by the disappearance, or gross modification, of the same, or similar, discrete motor or autonomic behavior patterns (Verplanck 1957).
2. A functionally coordinated, but not necessarily localized, group of neural structures that have the properties of a center (def. 1) (Verplanck 1957).
3. A hypothetical neural structure, place, or set of places in an animal's central nervous system, of unspecified anatomical properties, presumed to act as a unit upon excitation by another such place, or other such places, by sending nerve impulses that govern the occurrence of some innate response (Verplanck 1957).
syn. Erbkoördinationen
Comments: A center is the theoretical neural correlate of a species-specific response. Theory endows this correlate with all the properties of a physiologist's center. These centers may become empirical concepts, with specified anatomical loci and properties (Verplanck 1957).

♦ **center of creation** *n*. A place where a species is thought to originate according to a creationist; a center of creation varies from a single pair of organisms in one particular location to groups of organisms all over its present range, depending on the theorist (Mayr 1982, 440).

♦ **centimorgan (cM)** See genetic map (comments).

♦ **central dogma, central dogma of genetics** *n*.
1. During protein production within a cell or virus, amino acids and proteins are translated from nucleic acids and not the other way around (Mayr 1982, 574).

Note: "This discovery supplied the last and most conclusive proof for the impossibility of an inheritance of acquired characters" (Mayr 1982, 574). The term "acquired characters" refers to those that theoretically could be induced in a cell nucleus by environmental protein; however, newly acquired cytoplasmic DNA and nuclear DNA that are incorporated by a cell may be transmitted to offspring in some bacteria.

2. "The dogma that genes exert an influence over the form of a body, but the form of a body is never translated back into genetic code; acquired characteristics are not inherited" (Dawkins 1982, 285).

3. In molecular biology, "the dogma that nucleic acids act as templates for the synthesis of proteins, but never the reverse" (Dawkins 1982, 285).

4. Acquired characters are not inherited and only changes in germ plasm are transmitted from generation to generation (Lincoln et al. 1985).

syn. Weismannism (Lincoln et al. 1985)
cf. Lamarckism, transposable element

♦ **central dogma of evolutionary theory** See [3]theory: central dogma of evolutionary theory.

♦ **central excitatory mechanism (CEM)** See mechanism: central excitatory mechanism.

♦ **central excitatory state, central motive state** See state: central excitatory state.

♦ **central executive** See working memory (comments).

♦ **central grouping tendency** *n.* The array of the median numbers in each sex-age category calculated from a sample of societies (Wilson 1975, 518).

♦ **central hierarchy** See hierarchy: central hierarchy.

♦ **central-place system** See [1]system: central-place system.

♦ **central motive state** See state: central excitatory state.

♦ **central nervous system** See [2]system: nervous system: central nervous system.

♦ **centrifugal selection** See selection: disruptive selection.

♦ **centripetal selection** See selection: stabilizing selection.

♦ **centripetal society** See [2]society: centripetal society.

♦ **cephalization** [SEF ah leh zaye shun, SEF a lye zaye shun] *n.* Evolutionary concentration and localization of functions, powers, parts, or a combination of these things in, or toward, an animal's head; the cephalization of a vertebrate nervous system (Michaelis 1963).

[French *céphalique*, ultimately from Greek *kephalilos < kephalē*, head]

♦ **cerebellum** See organ: brain: cerebellum.

♦ **ceremony** *n.*
1. A person's formal act, or observance, expressive of difference, or respect, to superiors in rank, or established by custom in social intercourse; a usage of courtesy, politeness, or civility (*Oxford English Dictionary* 1972, entries from *ca.* 1386).

2. In many animal species: "a highly evolved and complex display used to conciliate others and to establish and maintain social bonds" (Wilson 1975, 224, 491, 560, 580).

3. A complex social behavior sequence that is distinguished by a high degree of form constancy (Immelmann and Beer 1989, 42).

cf. display
Comment: "Ceremony" is more commonly used in earlier, rather than later, ethological literature (Immelmann and Beer 1989).
[Old French *cerymonie < Latin caerimonia*, awe, veneration]

courtship feeding See feeding: courtship feeding.

defecation ceremony *n.* In Dikdiks: sequential defecation by a pair of animals, often involving more than one defecation bout of each partner (Pilters 1956, Altmann 1969, Hendrichs 1971 in Heymer 1977, 105).

feeding ceremony *n.* For example, in some species of empidid flies and terns: a ceremony involving one individual's giving food to another (Heymer 1977, 73).

cf. feeding: courtship feeding

greeting ceremony *n.* Behavior displayed when conspecific animals meet (Wilson 1975, 193, 227, 495 509).

Comments: When African Wild Dogs and Timber Wolves have a greeting ceremony, subordinate individuals approach higher ranked pack members in a groveling posture and "enthusiastically" nip and lick the latter's mouth areas; elephants show a similar greeting ceremony (Wilson 1975, 193, 227, 495 509).

relief ceremony *n.* Special behaviors that an animal shows just prior to nest relief, *q.v.* (Immelmann and Beer 1989, 201).

triumph ceremony *n.* In the Greylag Goose: a ceremony in which a male makes a sham attack toward objects that are normally avoided (including a rival male), returns to his intended or actual mate, and gives a "triumph" call and threat displays toward a point near her (Lorenz 1943 in Heymer 1977, 180; Immelmann and Beer 1989, 319).

a – c

Comment: Similar ceremonies occur in other goose species (Immelmann and Beer 1989, 319).

urination ceremony *n.* For example, in some rodent species, Hottentots: urination on one animal by another (Heymer 1977, 84).

cf. enurination; odor: group odor

Comment: In Hottentots, a circumciser may urinate upon a newly circumcised youth; a marriage official may urinate upon a bride, bridegroom, and bystanders; or a man may urinate upon a successful hunter (Eibl-Eibesfeldt 1970 in Heymer 1977, 84).

♦ **cerophage** See -phage: cerophage.

♦ **cespiticole** See -cole: cespiticole.

♦ **cete** See ²group: cete.

♦ **cetology** See study of: cetology.

♦ **CFC** See molecule: chlorofluorocarbon.

♦ **CFIRMS** See instrument: C/N-continuous-flow-isotope-ratio mass spectrometer (CFIRMS).

♦ **CGH** See congenital generalized hypertrichosis.

♦ **CG–mass spec, GC–mass-spectrometric analytical method (CG–MS)** See method: gas-chromatography–mass-spectrometric analytical method.

♦ **chain** *n.* "Any connected series; a succession: a chain of events" (Michaelis 1963).

action chain *n.*

1. A preprogrammed response sequence often shown by animals in social interactions, *e.g.*, in courtship, copulation, ritualized fights, food searching, and prey catching (Szymanski 1913, Meisenheimer 1921, etc. in Heymer 1977, 142).
2. Two animals' stereotyped sequence in which one's response is the stimulus for the other's next act, which in turn stimulates the next act of the first animal, etc. (Immelmann and Beer 1989, 2).

syn. reaction chain (Immelmann and Beer 1989, 2)

cf. graph: behavior-chain diagram; reflex: chain reflex

behavior chain *n.* A sequence of stimuli and an animal's responses that can be observed repeatedly with only minor variations in the ordinal position of each stimulus and response (Verplanck 1957).

cf. pattern: behavior pattern

Comment: In many chaining cases, one cannot identify all of the stimuli of a chain, but only the responses as they occur in order. In other cases, an animal's response may move it in such a way that it is confronted with a new set of stimuli that release, or elicit, the next response, and so on (Verplanck 1957).

courtship chain *n.* An action chain of a pair of courting animals.

electron-transport chain *n.* A system of electron carriers embedded in the inner membrane of a mitochondrion that produces a net of 32 ATPs from the electrons obtained from one glucose molecule that is involved in glycolysis and the Krebs cycle (Campbell 1987, 194).

Comment: The electrons are carried by NADH and $FADH_2$.

food chain *n.*

1. Part of a food web, *q.v.*, that is most frequently a simple sequence of prey species and the predators that consume them (Wilson 1975, 584).
2. The energy transfer from primary producers through a series of organisms that eat and are eaten, assuming that each organism feeds on only one other type of organism (*e.g.*, caterpillar → sparrow → Sparrow Hawk) (Allaby 1994, 159).

Comments: Much energy is lost at each stage, a fact that usually limits the number of steps (trophic levels) in the chain to four or five (Allaby 1994, 159). The two basic food-chain types are grazing and detrital pathways. They often interact, giving a complex food web.

Markoff chain *n.* A series of items in which the different-from-chance transition probability between two focal items is independent from one to all preceding items in the sequence (Immelmann and Beer 1989, 295).

▶ **first-order Markoff chain** *n.* A Markoff chain in which the different-from-chance transition probability between two focal items is independent of all preceding items in the sequence (Immelmann and Beer 1989, 295).

▶ **second-order Markoff chain** *n.* A Markoff chain in which the different-from-chance transition probability between two focal items is independent of one preceding item in the sequence (Immelmann and Beer 1989, 295).

reaction chain See chain: action chain.

♦ **chain of being** See *scala naturae*.

♦ **chain reaction** See reaction: chain reaction.

♦ **chain reflex** See reflex: chain reflex.

♦ **chain-reflex theory** See ³theory: chain-reflex theory.

♦ **chain transport** See transport: chain transport.

♦ **chaining** *n.* An animal's linking together its responses in a learning test; each response gives rise to a new stimulus that the animal associates with its next response in a series (Storz 1973, 41).

See chain: behavior.

♦ **chalcone synthase** See enzyme: chalcone synthase.

♦ **challenge display** See display: challenge display.

♦ **challenge ritual** See ritual: challenge ritual.

♦ **change in motivation** See emancipation.

♦ **change of function** *n.* Evolution of a new function by an organ or behavior, *e.g.*, the food calls' of certain phasianid birds becoming mate-attracting calls (Schenkel 1956, 1958 in Heymer 1977, 72; Immelmann and Beer 1989, 42).
cf. ritualization

♦ **change of mood** *n.* A change from one motivational state, or mood, to another (von Holst and von St. Paul 1960 in Heymer 1977, 185).

♦ **channel** *n.* A pathway through which a signal travels in communication (Dewsbury 1978, 99).
See ¹system: communication system.

♦ **character** *n.*
1. A distinguishing feature of a species or genus (Chambers 1727 in *Oxford English Dictionary* 1972).
2. A taxonomic trait (Wilson 1975, 580); an organism's reliably recognizable attribute, or distinguishing mark (Wickler and Seibt 1977 in Immelmann and Beer 1989, 28).
   Notes: Mendel referred to traits (*Merkmale*) in his genetic work and characters (*Charaktere*) as essentially phenotypic traits (Mendel 1865 in Mayr 1982, 735). A character can also be (1) an organism's correlated set of observable features caused by a single developmental, or ecological, process (*e.g.*, a group of features that are due to paedogenesis), or (2) a historic event in the evolution of a feature (*e.g.*, an unobservable ancestral character that gave rise to observable characters) (Fristrup 1992, 45–51).
3. A property (of a structure or behavior) of an organism, or population, chosen for study (Hinde 1982, 277; Darden 1992, 41).
4. Any feature, or trait, transmitted from parent to offspring (Lincoln et al. 1985).
5. Part of an individual: *the cat's blue eyes* (Colless 1985 in Fristrup 1992, 46).
   syn. character part (Colless 1985 in Fristrup 1992, 46)
   Note: Use of this kind of character is problematic (Fristrup 1992, 46).
6. A basis of comparison between taxa: *the cat's eye color is blue* (Colless 1985 in Fristrup 1992, 46).
   syn. character-variable, *fundmentum divisionis* (Colless 1985 in Fristrup 1992,

46), variable (Sokal and Rohlf 1981, etc. in Fristrup 1992, 47)
7. An attribute of an individual: *the cat is blue-eyed* (Colless 1985 in Fristrup 1992, 46).
   syn. character-attribute (Colless 1985 in Fristrup 1992, 46)
8. An organism's feature that can be evaluated as a variable with two or more mutually exclusive and ordered states (Pimentel and Riggins, 1987, 201, in Fristrup 1992, 45).

See variable.
syn. characteristic (according to some authors), trait (many authors), variable (statistics)
cf. characteristic
Comments: Many authors use "character," "characteristic," and "trait" interchangeably; I have not listed synonyms that differ only by these words. Character is "not an unproblematic, easily operationalized concept" (Darden 1992, 41). Fristrup (1992) also discusses "aggregate character," "conditional character," "emergent character," and "extra-individual character" and sources of confusion regarding the concept of character.

acquired character *n.* A character resulting from direct environmental influences (Lincoln et al. 1985).
See inheritance: inheritance of acquired characteristics; Lamarckism.

adaptation, adaptive character See ³adaptation: ²evolutionary adaptation (def. 2).

advanced character See character: derived character.

aggregate character See character (comments).

allesthetic character *n.* An animal's character that evolves via its effects on a potential mate's senses, perceptions, and emotions (Huxley 1914 in Gherardi 1984, 388).
cf. selection: mutual sexual selection

all-or-none character See character: binary character.

altruistic character See gene: altruistic gene.

ancestral character *n.*
1. A character that appeared first in evolution and gave rise to other, more derived characters (Wilson 1975, 593).
2. A character that is similar to an ancestral character (Wilson 1975, 424).

syn. nonderived character, primitive character, plesiomorph, *q.v.*
cf. ancestral; character: plesiomorph, specialized character; Darwin's admonition
Comment: Ancestral characters are often, but not always, less complex than derived ones (Wilson 1975, 593).

antepisematic character See character: episematic character: antepisematic character.

apomorph, apomorphy, apomorph character *n*. A character that is unique and derived [coined by Hennig in Mayr 1982, 227].

syn. derived character, modern character; advanced character, apotypic character, apotypy, apomorphous character, derived character, specialized character (Wiley 1981, 122)

cf. character: derived character, plesiomorph [Greek *apo*, after + *morph*, form]

▸ **autapomorph** *n*. An apomorph that has evolved in only one of two sister groups (Hennig in Mayr 1982, 230, 869).

▸ **nonhomologous apomorph, pseudoapomorph** *n*. An apomorph that is similar to another one with which it is compared because of convergent evolution (Mayr 1982, 228).

▸ **synapomorph** *n*. An apomorph that is shared in both sister groups.

cf. -typy: synapotypy

homologous synapomorph, genuine synapomorph *n*. An apomorph that is actually, or inferred to be, derived from the nearest common ancestor of two taxa (Wiley 1981, 9–10; Mayr 1982, 228).

cf. character: apomorph, nonhomologous apomorph, pseudoapomorph; homolog

auxiliary character, confirmatory character *n*. A character that is used to confirm a taxon's identification (Lincoln et al. 1985).

cf. character: diagnostic character

behavioral character *n*.

1. An aspect of an animal's behavior that can be reliably measured, is reasonably stable over an animal's life, and, thus, can be studied as a single phenomenon in relation to, for example, its heredity (Collocot and Dobson 1974).

2. Any character connected with behavior, *e.g.*, a movement pattern, scent signal, sensitive period, response to a sign stimulus, or vocalization (Immelmann and Beer 1989, 28).

binary character, all-or-none character, two-state character *n*. A character that has only two states (Lincoln et al. 1985).

cf. character: multistate character

character-attribute See character.

character part See character.

character-variable See character.

complex character *n*. A phenotypic character that is determined by more than

one gene and, consequently, is not transmitted to offspring as a simple unit (Lincoln et al. 1985).

conditional character See character (comments).

confirmatory character See character: auxiliary character.

continuous character *n*. A character for which there is a continuum of possible phenotypes (*e.g.*, growth rate, height, milk yield, and weight); contrasted with meristic character (Hartl 1987, 216).

See character: continuous character.

demonstrative character *n*. An anatomical character, or behavioral character (locomotory pattern, distraction display), that an animal presents during a demonstrative movement, *q.v.* (Heymer 1977, 50).

dependent character *n*. A character that is applicable to an organism only if another character is already present in a particular state; *e.g.*, hairy legs occur only in an organism with legs (Lincoln et al. 1985).

derived character *n*. A character that is more recently evolved than one with which it is compared (*e.g.*, feathers of birds which have evolved from scales of reptiles) (Lincoln et al. 1985).

syn. apomorph character, advanced character, modern character

cf. character: apomorph, ancestral character, plesiomorph, stasimorphic character; Darwin's admonition

diagnostic character *n*.

1. A character that is particularly important in a group of organisms evolving in a new direction; *e.g.*, feathers were a key character in evolution of birds from reptiles (given as "key character" in Mayr 1982, 613–614).

2. A character, or character state, that unambiguously differentiates one taxon from others (Lincoln et al. 1985).

syn. key character, peculiar character (Lincoln et al. 1985)

dichotomous character *n*. A character that exists in two states only (Lincoln et al. 1985).

dimorphism *n*. A character with two alternative states (Lincoln et al. 1985).

discrete character *n*. A character that is either present or absent in any one individual of a taxon (Hartl and Clark 1997, 399).

See character: qualitative character.

domestication character *n*. A character that distinguishes a domestic strain of a species from its wild ancestral types (Immelmann and Beer 1989, 78, 262).

Comment: Domestication characters include increase (hypertrophy) in sexual behavior, decrease (hypotrophy) in aggressiveness and parental care, body-size changes, shape changes, color changes (albinism, piebalding), brain-weight decrease, hair changes (Angora hair), and feather changes (curling feathers) (Immelmann and Beer 1989, 78, 262).

emergent character See character (comments).

epigamic character *n.* An organism's trait (other than one that is essential or copulation behavior) that attracts, or stimulates, individuals of the opposite sex during courtship (Lincoln et al. 1985).

episematic character *n.* A character (episeme; *e.g.,* some color markings) that aids in one animal's recognizing another (Lincoln et al. 1985).

▶ **antepisematic character** *n.* An episematic character that implies a threat (Lincoln et al. 1985).

▶ **proepisematic character** *n.* An episematic character that aids social recognition (Lincoln et al. 1985).

▶ **pseudepisematic character, pseudoepisematic character, pseudosematic character** *n.* A character that aids an animal in recognition of another animal but involves deception (Lincoln et al. 1985).
cf. cheat

extra-individual character See character (comments).

extrinsic character *n.* A trait that exists, or has its origins, outside an individual organism, group, or system (Lincoln et al. 1985).
syn. extraneous character
cf. character: intrinsic character

fundmentum divisionis See character.

heterogynistic character, heterogynism *n.* A character that is more strongly marked in females than males (Lincoln et al. 1985).

holandric character, holandry *n.*
1. A character that occurs only in males (Lincoln et al. 1985).
2. A character carried on the Y-chromosome in the heterogametic male (Lincoln et al. 1985).
cf. character: hologynic character

hologynic character, hologenous character, hologyny *n.*
1. A character found only in females (Lincoln et al. 1985).
2. A sex-linked character in a species with heterogametic females.
cf. character: holandric character

homolog, homologue, homology See homolog.

homologous synapomorph See character: apomorph: synapomorph: homologous synapomorph.

ideal character *n.* An organism's character that has the best features for the organism's present environment (Wilson 1975, 23).

independent character *n.* A character that is not linked to a second character of the same individual organism because it is part of a different ontogenetic pathway and is free to respond to evolution differently than the second one (Wiley 1981, 117).
Comments: Independent characters of the same organism are not absolutely independent of one another because they are both of this same organism. Examples of independent characters are a gar's number of teeth that border its snout and its number of caudal-fin rays (Wiley 1981, 117).

intrinsic character *n.* A character that originates, or occurs, within an individual, group, or system (Lincoln et al. 1985).
cf. character: extrinsic character

juvenile characteristic *n.* One of a set of characters that distinguishes a young animal from a conspecific infant or adult (*e.g.,* juvenile plumage of birds or juvenile behavior, including vocalization) (Immelmann and Beer 1989, 161–162).
cf. coloration: baby coloration; dress: juvenile dress

key character See character: diagnostic character.

lethal character *n.* Any inherited character that causes an organism's death before it can reproduce (Lincoln et al. 1985).

linked character See character: dependent character.

Mendelian character *n.* Any character that is inherited in accordance with Mendel's law of independent assortment, *q.v.* (Lincoln et al. 1985).

meristic character *n.* A character for which the phenotype is expressed in discrete, integral classes, *e.g.,* bristle number on a fruit fly, ear number on a corn plant, litter size, and petal number; contrasted with continuous character (Hartl 1987, 216–217).
See character: quantitative character.

modern character See character: apomorph, derived character.

multistate character *n.* A character that has three or more different states (Lincoln et al. 1985).
cf. character: binary character

nonadaptive character *n.* An "abnormal character that reduces the fitness of individuals that consistently manifest it under environmental circumstances that are usual for their species" (Wilson 1975, 21).

numerical character See character: polygenic character.

optimum-permissible character *n.* An organism's character that has the best possible features given the genetic and environmental constraints upon it and that is less than its ideal conceivable condition (Wilson 1975, 24).

paratelic character *n.* One of two or more superficially similar characters that result from convergent evolution (Lincoln et al. 1985).

patristic character *n.* A character that is inherited by all members of a group from their most recent common ancestor (Lincoln et al. 1985).

peculiar character See character: diagnostic character.

phene *n.* A phenotypic character that is genetically determined (Lincoln et al. 1985).

plesiomorph, plesiomorphy, plesiomorph character, ancestral character, primitive character *n.* An ancestral character or character state (Lincoln et al. 1985).
syn. generalized character, plesiotypic character, plesiotypy, plesiomorphous character (Wiley 1981, 122)
cf. character: apomorph
[Greek *plesios*, near + *morph*, form]
▸ **symplesiomorph** *n.* A plesiomorph shared by two or more taxa (Lincoln et al. 1985).
cf. -typy: symplesiotypy

polygenic character *n.* A character that is controlled by the integrated action of multiple independent genes (Lincoln et al. 1985; Hartl 1987, 215); contrasted with digenic character, monogenic character, oligogenic character, trigenic character.
syn. polygenetic character, polyergistic character, polyfactorial character, multigenic character (Lincoln et al. 1985); multifactorial trait, quantitative trait (Hartl 1987, 215)
Comment: An organism's external environmental influences polygenic characters to varying degrees (Hartl 1987, 215).
[multifactorial, referring to the fact that multiple genetic and environmental factors affect many polygenetic characters (Hartl 1987, 215)]
[polygenic, referring to the fact that multiple genetic factors affect many polygenetic characters (Hartl 1987, 215)]
[quantitative, referring to the fact that combined effects of many quantities (environ-

mental and genetic) affect many polygenic characters (Hartl 1987, 216)].

polymorphic character *n.* A character with two or more alternative states (suggested by Campbell 1990, 446); contrasted with dimorphism (under character).
cf. -morphism: dimorphism, polymorphism

primitive character See character: ancestral character, plesiomorph.

pseudepisematic character, pseudoepisematic character, pseudosematic character See character: episematic character: pseudepisematic character.

qualitative character, discrete character *n.* A character that has discrete states that are not numerical or morphometric (Lincoln et al. 1985).
See character: discrete character.
▸ **qualitative-multistate character** *n.* A qualitative character that exhibits several states that cannot be arranged linearly along a single axis (Lincoln et al. 1985); contrasted with quantitative-multistate character.

quantitative character *n.*
1. A character based on counts, measurements, ratios, or other numerical values (Lincoln et al. 1985).
   syn. meristic character, *q.v.* (Hartl and Clark 1997, 398), numerical character (Lincoln et al. 1985)
   See character: polygenic character.
2. A character with continuous variation and controlled by several interacting genes (Lincoln et al. 1985; Hartl and Clark 1997, 398).
   syn. continuous character (Hartl and Clark 1997, 398)
3. A character that results from combined environmental and genetic effects
   cf. study of: quantitative genetics

A CLASSIFICATION OF QUANTITATIVE CHARACTERS

I.   continuous character
     (*syn.* quantitative character)
II. discrete character
III. meristic character (*syn.* numerical character, quantitative character)
IV. polygenic character (*syn.* polygenetic character, polyergistic character, polyfactorial character, multigenic character multifactorial trait, quantitative trait)
V.   quantitative-multistate character

▶ **quantitative-multistate character** *n.*
A quantitative character that exhibits several states that can be arranged linearly along a single axis (Lincoln et al. 1985); contrasted with qualitative-multistate character.

redundant character *n.* "Any character that does not contribute useful information to an analysis" (Lincoln et al. 1985).

regressive character *n.* A character that shows a gradual loss of differentiation due to aging or evolution (Lincoln et al. 1985).

rudimentary character *n.*
1. An imperfectly developed character (Lincoln et al. 1985).
2. A character that is arrested at an early stage of its development or evolution (Lincoln et al. 1985).
cf. character: stasimorphic character
Comment: "Rudimentary character" is sometimes incorrectly synonymized with "vestigial character" (Lincoln et al. 1985).

semantid *n.* "Any biochemical character" (Lincoln et al. 1985).

sex-limited character *n.* "A character expressed in only one sex" (Lincoln et al. 1985).

sexual character *n.* In sexually dimorphic organisms: a character in which sexes differ (Immelmann and Beer 1989, 266).

▶ **primary-sexual character** *n.*
1. "Any character that is directly involved in reproduction or parental care" (Wittenberger 1981, 620).
2. A difference between the sexes of a species that is related to reproductive organs and gametes (Lincoln et al. 1985).

▶ **secondary-sexual character** *n.* A difference between the sexes of a species that is related to characters other than reproductive organs and gametes (Lincoln et al. 1985).
cf. character: epigamic character

somatic character *n.* A character that results from an organism's bodily change during its lifetime due to practice or environmental impact (Bateson 1979, 245).

specialized character *n.*
1. A character that is highly modified from its ancestral condition and can no longer carry out its ancestral function (Lincoln et al. 1985).
2. A character that is adapted to perform a particular function (Lincoln et al. 1985).

species character, taxonomic character *n.* A character that does not vary within a species but does vary among species (Mayr 1963, 1969 in Fristrup 1992, 48).

species-specific character *n.*
1. A character that is specific only to a particular species or part of a species, *e.g.*, innate-movement patterns or learning dispositions (Immelmann and Beer 1989, 287).
Note: Some workers use "species-characteristic" and "species-typical" for characters that are found in more than one species and to distinguish these characters from "species-specific characters" (def. 1) (Immelmann and Beer 1989, 287).
2. A character that is specific to a group of species (Immelmann and Beer 1989, 287).
cf. character: specific character

specific character *n.* A character that is diagnostic of a species (Lincoln et al. 1985).
cf. character: species-specific character

stasimorphic character, stasimorph, stasimorphy *n.* A character that retains an ancestral condition (Lincoln et al. 1985).

synapomorph See character: apomorph: synapomorph.

taxonomic character See character: species character.

threshold character *n.*
1. A character that is either present, or absent, in any one individual of a population (Hartl 1987, 217).
Note: Hartl and Clark (1997, 399) call this kind of trait a "discrete trait," *q.v.*
2. A character that an individual expresses when its liability value is greater than a particular threshold, or triggering, level (*e.g.*, human diabetes and schizophrenia).

two-state character See character: binary character.

vestigial, vestige *n.* A character, or function, that has become diminished, or reduced, through regressive evolution or ontogeny (Lincoln et al. 1985).
cf. relict
Comment: "Vestigial character" is not synonymous with "rudimentary character" (Lincoln et al. 1985).

♦ **character attribute** See character.

♦ **character convergence** *n.* Two newly evolved species' interacting in such a way that one, or both, converges in one or more characters toward the other; contrasted with character displacement (Moynihan 1968, Cody 1969 in Wilson 1975, 277, 580).
syn. theory of character convergence

♦ **character-data methods** See method: character-data methods.

♦ **character displacement** *n.* For example, in some frog and Darwin's Finch species: the phenomenon in which the range of closely related species with similar ecological niches comes to overlap, and the differences in some traits of these species evolve to be more pronounced compared to areas where the species do not overlap; contrasted with character convergence (Murray 1971 in Wilson 1975, 277, 580; Immelmann and Beer 1989, 43; D.M. Lambert, personal communication). *syn.* character divergence (Darwin 1859 in Price 1996, 78); theory of character displacement.

Comment: Since the 1950s, researchers broadened "character displacement" to include character divergence, convergence, and parallel characteristic shifts in zones of sympatry (Taper and Case 1992 in Price 1996, 79). [This term was introduced by Brown and Wilson (1956) according to Futuyma and Slatkin (1983, 4). The idea of character displacement can be traced to Lack (1947 in Wilson 1992, 367).]

♦ **character enhancement** *n.* A character's further differentiation caused by selection on sympatric animals (Heymer 1977, 102).

Comment: Characters involved include grasshopper vocalizations and firefly flashing (Heymer 1977, 102).

♦ **character gradient** See cline.

♦ **character part** See character.

♦ **character release** *n.* An increase in the variation of certain phenotypic characters associated with ecological release of a species (Lincoln et al. 1985).

♦ **character space** *n.* "The hypothetical space occupied by a taxon, having as many dimensions as characters used to define the taxon" (Lincoln et al. 1985).

♦ **character state** *n.*
1. A particular character possessed by one individual and not another, or by one species and not another (Wilson 1975, 580).
2. "Any of the range of values, conditions, or expressions of a particular taxonomic character" (Lincoln et al. 1985).

♦ **character-variable** See character.

♦ **characteristic** *n.* A distinguishing feature, often used loosely as a synonym of character although, more precisely, it refers to the distinctive state, or expression, of that character (Lincoln et al. 1985).

♦ **characteristic species** See ²species: characteristic species.

♦ *Charaktere* See character (def. 2).

♦ **charge-strike** *n.*
1. A kind of aggressive behavior (*e.g.*, in Humpback Whales) (Baker and Herman 1985, 55).
2. *v.t.* To show a charge-strike.
cf. head lunge

♦ **charm** *n.*
1. Any fascinating or alluring human quality or feature that excites love, admiration, or both (*Oxford English Dictionary* 1972, entries from 1598).
2. In lekking birds: a male's attribute that attracts females as mates and is expected to be correlated with genetic quality of the male (Darwin in Zuk 1984).
See ²group: charm.
[French *charme* < *carmen*, song, incantation]

♦ **chasmophile** See ¹-phile: chasmophile.

♦ **Chauvet Grotto** See place: Chauvet Grotto.

♦ **cheater, cheat** *n.*
1. A person who deals fraudulently; deceives; swindles (*Oxford English Dictionary* 1972, entries from 1607).
2. In organisms: an individual that does not reciprocate, or reciprocate fully, in a reciprocal-altruism interaction with another organism (West Eberhard 1975, 21; Axelrod and Hamilton 1981).
cf. robber, thief
3. An individual organism that deceives a conspecific organism in a way that is likely to increase the deceiver's fitness (Trivers 1976, 139).
v.i. cheat
cf. deceit; signaler: deceitful signaler

gross cheater *n.*
1. A cheater that does not reciprocate in an interaction with an altruist (Trivers 1971, 46).
2. A cheater that reciprocates so little, if at all, that the altruist receives less benefit from the gross cheater than the cost of the altruist's acts of altruism to the cheater [coined by Trivers 1971, 46].

subtle cheater *n.* A cheater that reciprocates but always attempts to give less than it was given, "or, more precisely, to give less than the partner would give if the situation were reversed" [coined by Trivers 1971, 46].

♦ **cheekpadder** See animal names: cheekpadder.

♦ **cheiropterophile** See ¹-phile: cheiropterophile.

♦ **chemical communication** See communication: chemical communication.

♦ **chemical-degradation method** See method: chemical-degradation method.

♦ **chemical dialect** See dialect: chemical dialect.

♦ **chemical ecology** See study of: ecology: chemical ecology.

♦ **chemical evolution** See ¹evolution: chemical evolution.

♦ **chemical releaser** See chemical-releasing stimulus: semiochemical: pheromone.

♦ **chemical-releasing stimulus (CRS)**
n. A chemical that an organism makes, bears, or both, that may, or actually does, influence the behavior of one or more other organisms (Weldon 1980).
cf. odor
Comment: The same chemical may function in more than one different type of organism interaction (Nordlund and Lewis 1976, 212).

A Classification of Chemical-Releasing Stimuli

I. apneumone
II. arrestant
III. attractant
IV. deterrent
V. semiochemical (*syn.* chemosignal, semiochemic)
  A. allelochemic (*syn.* "allelochemical," xenomone)
    1. allomone (*syn.* false pheromone)
      a. allelopathic substance
      b. defense allomone
        (1) plant-derived allomone
        (2) plant-produced allomone
        (3) predator-released allomone
        (4) chemiocryptic allomone (*syn.* mimetic allomone)
        (5) herbivore-released allomone
        (6) plant-manipulation allomone
        (7) predator-defense allomone
        (8) prey-attraction allomone
        (9) prey-disruption allomone
        (10) prey-subduction allomone
      c. exocrine substance (*syn.* ectocrine substance, environmental hormone)
      d. propaganda substance
    2. antimone
    3. kairomone
      a. herbivore-released kairomone
      b. predator-released kairomone
      c. trail kairomone
    4. synomone
      a. plant-produced synomones
        (1) extrafloral-nectar synomone
        (2) floral-scent synomone
        (3) food-body synomone
        (4) guardian synomone
      b. herbivore-released synomone
      c. predator-released synomone
      d. vegetative synomone
  B. pheromone (*syn.* ectohormone, social hormone, etc.)

1. acrasin
2. aggregation pheromone
3. aggregation-attachment pheromone
4. alarm pheromone
5. androstenone
6. antiaphrodisiac
7. aphrodisiac
8. bombykol
9. copulin
10. death pheromone (*syn.* funeral pheromone)
11. dispersant pheromone
12. epideictic pheromone
13. human-female "pheromone"
14. human-male "pheromone"
15. incitin
16. individual pheromone
17. inhibitory pheromone
18. mating pheromone
19. matrone
20. primer pheromone
21. propaganda pheromone
22. queen pheromone
23. recognition pheromone
24. recruitment pheromone
25. releaser pheromone
26. reproductive pheromone
27. sex pheromone
28. superpheromone
29. surface pheromone
30. synchronization pheromone
31. territory pheromone
32. trail pheromone (*syn.* trail-marking pheromone, trail-following pheromone)
VI. stimulant
  A. feeding stimulant (*syn.* phagostimulant)
  B. locomotory stimulant
  C. mating stimulant
  D. ovipositional stimulant

apneumone *n.* A chemical that is emitted by nonliving material and evokes a behavioral, or physiological, reaction that is adaptively favorable to a receiving organism but detrimental to a heterospecific organism that may be found in, or on, the nonliving material; *e.g.*, the odor of a host's food (oatmeal) that attracts an ichneumonid parasitoid to this food [coined by Nordlund and Lewis 1976, 216, from Greek *ā-pneum*, breathless or lifeless]

arrestant *n.* A chemical that causes organisms to aggregate in contact with it, the mechanism of aggregation being kinetic or having a kinetic component (Dethier et al. 1960 in Nordlund et al. 1981, 22).
Comment: An arrestant may slow the linear progression of organisms by reducing their

actual speed of locomotion or by increasing their turning rate (Dethier et al. 1960 in Nordlund et al. 1981, 22). Arrestants, attractants, deterrents, and stimulants might be non-semiochemicals or semiochemicals, depending on their origins.

attractant *n.* "A chemical that causes an organism to make oriented movements toward its source" (Dethier et al. 1960 in Nordlund et al. 1981, 22).

deterrent *n.* "A chemical that inhibits feeding, mating, or oviposition when in a place where an organism would, in its absence, feed, mate, or oviposit" (Dethier et al. 1960 in Nordlund et al. 1981, 22).

semiochemical, semiochemic *n.* A chemical that mediates interactions between (among) organisms [coined by Law and Regnier 1971; from Greek *semeon*, a mark or signal + CHEMICAL].

syn. chemosignal (Houck and Reagan 1990, 729)

Comment: A substance produced by a species can have different simultaneous functions; *e.g.*, a pine terpenoid can act as an allomone, kairomone, and synomone (Whitman 1988, 13).

▶ **allelochemic** *n.*

1. A chemical that mediates interspecific interactions and is significant to organisms of a species different from its source for reasons other than indicating food (Whittaker 1970a,b in Nordlund and Lewis 1976, 213).

2. A secondary substance produced by an organism that modifies the growth, behavior, or population dynamics of other species, often having an inhibitory or regulatory effect (Lincoln et al. 1985).
   syn. allelochemical (Weldon 1985, personal communication, suggests this word should not be used as a noun), xenomone (Chernin 1970 in Nordlund and Lewis 1976, 213)

Some Allelochemics and Their Characteristics

	Beneficial to		Example of Allelochemic/ Receiver
	Sender	Receiver	
allomone	yes	no	venom of a snake/person
antimone	no	no	chemicals of a pathogen/host
kairomone	no	yes	honey-bee wax/wax moth
synomone	yes	yes	floral scent/ pollinator

cf. chemical-releasing stimulus: semiochemical: allelochemic: allomone substance [coined by Whittaker 1970a in Nordlund and Lewis 1976, 213]

3. *adj.* Referring to the general phenomena of chemical interactions between and among species (Whittaker and Feeny 1971).

allomone *n.* A chemical substance, produced or acquired by an emitter organism, that contacts a heterospecific receiver organism in nature and evokes the receiver's behavioral, or physiological, response that is adaptively favorable to the emitter but not the receiver (Brown 1968 in Nordlund and Lewis 1976, 214; Nordlund et al. 1981, 17); *e.g.*, venoms used in defense, chemicals used as repellents, or plant defense chemicals used against herbivores and that reduce competition from other plants.

syn. false pheromone (Immelmann and Beer 1989, 156); heterotelergone (Kirschenblatt 1958 in Maschwitz 1964, 268); pseudopheromone

cf. chemical-releasing stimulus: semiochemical: allelochemic: kairomone; -cide: siblicide

Comment: Whitman (1988) describes "defense allomone," "plant-derived allomone," "plant-produced allomone," "predator-released allomone," "chemiocryptic (= mimetic) allomone," "herbivore-released allomone," "plant-manipulation allomone," "prey-disruption allomone," "prey-attraction allomone," "prey-subduction allomone," and "predator-defense allomone."
[derived from alloiohormone; coined by Beth 1932 in Nordlund and Lewis 1976, 214]

allelopathic substance *n.* A waste product, excretory product, or metabolite produced by one organism that inhibits or regulates other organisms (Lincoln et al. 1985).

cf. chemical-releasing stimulus: semiochemical: allelochemic

exocrine substance, ectocrine substance, environmental hormone *n.* An organism's externally secreted substance that excites or inhibits a biological system or process (Lincoln et al. 1985).

propaganda substance *n.* In slave-making ants: an odor deposited during a raid that disperses defenders — slave ants [coined by Regnier and Wilson 1971, 267].

cf. chemical-releasing substance: semiochemical: pheromone: propaganda pheromone; gland: Dufour's gland

antimone *n.* A chemical substance, produced or acquired by an emitter organism, that contacts a heterospecific receiver organism in nature and evokes the receiver's behavioral or physiological response that is maladaptive to both emitter and receiver; *e.g.*, a substance released by a pathogenic microorganism that can cause abnormal behavior, or premature death, of its host [coined by Whitman 1988, 13, 18].

false pheromone See chemical-releasing stimulus: semiochemical: allelochemic: allomone.

kairomone *n.* A chemical substance, produced or acquired by an emitter organism, that contacts a heterospecific receiver organism in nature and evokes the receiver's behavioral or physiological response that is adaptively favorable to the receiver, not the emitter (Brown et al. 1970 in Nordlund and Lewis 1976, 214; Nordlund et al. 1981, 18–19); *e.g.*, a host's odors that a parasite uses to find the host (Conte et al. 1989).

Comments: Some workers contend that "allomone" should be used in a broader sense and the term "kairomone" should be dropped, but Weldon (1980) defends its use. Kairomones include "plant-produced attractants," "oviposition stimulants," and "phagostimulants" of arthropods and "herbivore-released kairomones" and "predator-released kairomones" (Whitman 1988). Blum (1974 in Nordlund and Lewis 1976, 214) suggests that the so-called kairomones appear to be pheromones and allomones that have "evolutionarily backfired" and thus "do not represent a class of chemical signals different from allomones and pheromones." Instead of "kairomone," Blum uses the term "kairomonal effect." Establishing whether a sender, or receiver, benefits from a kairomone is difficult and occasionally impossible. [Greek *kairos*, opportunistic; coined by Brown et al. 1970 in Nordlund and Lewis 1976, 214).

trail kairomone *n.* A trail pheromone that is deposited by one species that is responded to by another species; *e.g.*, an ant trail odor used by inquiline cockroaches (Moser 1964).

pseudopheromone See chemical-releasing stimulus: semiochemical: allelochemic: allomone.

synomone *n.* A chemical substance, produced or acquired by an emitter organism, that contacts a heterospecific receiver organism in nature and evokes the receiver's behavioral or physiological response that is adaptively favorable to both emitter and receiver [coined by Nordlund and Lewis 1976, 215; Nordlund et al. 1981, 20; from Greek *syn*, with jointly + *horman*, to excite or stimulate].

Comments: "Synomones" have been regarded as "allomones," as either "allomones" or "kairomones," or as "allomone-kairomones;" *e.g.*, the molting hormone (= ecdysone) of a wood-eating cockroach that induces the sexual cycle of some of its symbiotic protozoa or pheromones produced by two species of *Ips* beetles that also serve as substances that reduce competition between the two species (Brown et al. 1970 and Whittaker and Feeny 1971 in Nordlund and Lewis 1976, 215). Whitman (1988) describes many kinds of plant-produced synomones, including "extrafloral-nectar synomone," "floral-scent synomone," "food-body synomone," "guardian synomone," "herbivore-released synomone," "predator-released synomone," and "vegetative synomone."

▶ **pheromone** *n.*

1. "A substance secreted by an animal to the outside that causes a specific reaction" by one or more conspecifics (Nordlund and Lewis 1976, 213).

2. A substance, usually glandular, secreted by an organism outside its body that causes a specific reaction in a conspecific receiving organism that smells or tastes it (Wilson 1975, 414, 591).

3. A chemical used for communication among conspecific animals (Nordlund et al. 1981, 15).

syn. chemical releaser, ectohormone (Karlson and Butenandt 1959, 39); ectoincretion (Karlson and Butenandt 1959, 39); homotelergone (Kirschenblatt 1958 in Maschwitz 1964, 268); social hormone, sociohormone

cf. odor: colony odor

Comments: "Pheromone" has essentially replaced all of its synonyms. "Ectohormone" is a self-contradictory term (Wilson 1965, 1064). A pheromone is usually comprised of more than one type of chemical substance (Nordlund et al. 1981, 15–16). The same pheromone may elicit different behavior in different contexts in the same species. A pheromone may be autostimulatory; that is, it may directly affect the behavior of its producer (Bradshaw et al. 1986, 234). Some of the many kinds of pheromones are defined below.

[Greek *phereum*, to carry, *horman*, to excite or stimulate; coined by Karlson and Butenandt 1959, 39; Karlson and Luscher 1959 in Nordlund and Lewis 1976, 213]

acrasin *n.* In slime molds: an aggregation pheromone which is cyclic AMP in *Dictyostelium discoideum* (Konijn et al. 1967 in Wilson 1975, 235, 391).

aggregation-attachment pheromone *n.* In a tick species: a male-produced pheromone that causes unfed adult ticks to aggregate and attach themselves to a particular cattle individual (Norval et al. 1989, 364).

aggregation pheromone *n.* A pheromone that brings two or more members of a group together (*e.g.*, for feeding in scolytid beetles) (Pitman et al. 1968; Pitman and Vité 1968; Borden 1984, 123).

syn. attractant, population-aggregating pheromone

alarm pheromone *n.* For example, in some aphid, termite, ant, wasp, and fish species; the Honey Bee: a pheromone that induces a state of alertness, or alarm, due to a common threat (Butler 1609; Frisch 1942; Buren 1958, 121; Moore, 1968, 33; Moser et al. 1968, 529; Crewe and Blum 1970, 141; Regnier and Wilson 1971; Wilson 1975, 578).

syn. alarm releaser (Wilson 1958 in Ghent and Gary 1962, 1), alarm substance, sting pheromone (Wilson 1971, 234); *Schreckstoff* (sing., Frisch 1942); warning substance (Karlson and Butenandt 1959, 40)

cf. behavior: alarm behavior

Comments: The same substance can be an alarm pheromone as well as serve other functions, most alarm substances are not species specific, and many known and probable alarm substances have been chemically identified (Wilson 1971, chap. 12). Some frog and snail species have chemical-alarm systems (Maschwitz 1964, 267).

androstenone *n.* In the Domestic Pig: a presumed pheromone that induces lordosis in sows (Forsyth 1985, 30).

cf. chemical-releasing stimulus: semiochemical: pheromone: human-male "pheromone"

Comments: This compound is in hog saliva and is given off by truffles. Androstenone is used in Boar Mate® (Forsyth 1985, 30).

antiaphrodisiac *n.*

1. For example, in a heliconiine butterfly, the beetle *Tenebrio molitor*, a sweat-bee species: a pheromone that is pro-

duced by a male, is placed on a conspecific female by him, counteracts an aphrodisiac, and, thus, tends to stop other males from mating with the female (Gilbert, 1976; Kukuk 1983).

2. A repellant pheromone that a male puts on his mate that decreases the chances that other males will try to mate with her (Thornhill and Alcock 1983, 336).

aphrodisiac *n.*

1. For example, in a sweat-bee species: a female pheromone that releases copulatory behavior (Barrows 1975b).

2. In a plethodontid-salamander species: a pheromone produced by a male and secreted during courtship that increases a female's receptivity (Houck and Reagan 1990, 729).

See aphrodisiac in section "a".

syn. courtship pheromone (Houck and Reagan 1990, 729)

cf. chemical-releasing stimulus: semiochemical: pheromone: antiaphrodisiac

[Greek *aphrodisiakos*, *Aphrodite*, goddess of love]

bombykol *n.* A sex attractant, *trans*-10-*cis*-12-hexadecadienol, produced by female silk moths, *Bombyx mori.*

caste pheromone *n.* In the termite *Kalotermes flavicollis:* one of several pheromones that control reproductive caste formation (Lüscher 1961 in Wilson 1965, 1067).

Comments: In *Kalotermes flavicollis*, a king and queen produce pheromones that inhibit the development of pseudergates into their own royal castes, a king produces a pheromone that stimulates female pseudergates to become reproductives, a king produces a pheromone by which other reproductive males recognize them, and a queen produces a pheromone by which other reproductive females recognize her (Lüscher 1961 in Wilson 1965, 1067).

copulin *n.* A possible pheromone comprised of fatty acids in human vaginal secretions (Forsyth 1985, 30).

cf. odor: human vaginal odor

death pheromone *n.* In some ant, bee, and wasp species: a pheromone that causes insects to drag their dead from their nests and abandon them (Matthews and Matthews 1978, 212).

syn. funeral pheromone

epideictic pheromone *n.*

1. A pheromone that indicates a receiver's population density and serves in its population-size regulation (Corbet 1971).

2. A pheromone that elicits organism dispersal from currently, or potentially, scarce or overcrowded resources (*e.g.*, food, space, and refugia) and thereby acts to partition intraspecific foraging activities (Prokopy 1980 in Nordlund 1981, 496).

false pheromone See chemical-releasing substance: semiochemical: allelochemic: allomone.

funeral pheromone See chemical-releasing substance: semiochemical: pheromone: death pheromone.

group pheromone See odor: clan-specific odor.

human-female "pheromone" *n.*

1. A possible pheromone produced by women that synchronizes their menstrual cycles (McClintock 1971).
   Notes: Underarm secretions from women in the early (= follicular) phase of their menstrual cycles can shorten the menstrual cycles of women exposed to the secretions (McClintock and Stern 1998 in Angier 1998, A22). Underarm secretions from women in their midcyles (= times of ovulation) can lengthen the menstrual cycles of exposed women. In theory, ancestral women might have benefited from synchronized menstrual cycling, group mothering, and the ability to wet-nurse one another's offspring. Also, competent ovulators might have helped their female kin to become more regular cyclers and ovulators by pheromonal communication.
2. A possible pheromone that is a "fishy" primary odor in menstrual blood and sweat and might serve as an indicator of estrus and menstruation (Hirth et al. 1986, 4).

cf. chemical-releasing stimulus: semiochemical: pheromone: human-female "pheromone;" odor: human vaginal odor

human-male "pheromone" *n.*

1. A possible pheromone, 5-alpha-androst-16-en-3-alpha-ol and 5-alpha-androst-16-en-3-one, in men's sweat (secreted by apocrine glands), smegma, and urine which may increase feelings of submissiveness, but not sexual feelings, of women in the middle of their menstrual cycles (Benton 1982 in Benton and Wastell 1986, 141).
2. A possible pheromone in men's pubic sweat and semen that may be an aphrodisiac, mating marker, response substance, and territorial marker (Hirth et al. 1986, 4).

cf. chemical-releasing substance: semiochemical: pheromone: androstenone
Comments: MHC genes, *q.v.*, code for individual odors of Humans (Claus Wedekind in Richardson 1996, 26). Women find the perspiration odors of individual men to vary from pleasant to unpleasant, and they prefer the odors from men whose MHC genes were dissimilar to theirs. Pregnant women prefer the odors of relatives compared to non-relatives. Humans who have MHC-dissimilar mates might have increased fertility, hardier offspring, and less risk of genetic disease. Androstadienone is a human-male steroid and estratetraene is a human-female steroid that subliminally alter mood (M.K. McClintock and S. Jacob in Holden 1999b). These steroids make women feel more happy and energized; men, tired and less elated.

individual pheromone *n.* For example, in some halictid-bee and ant species, many mammal species: a pheromone that is unique to one individual of a group and that is used by other members of its group to recognize it as unique (Bowers and Alexander 1967; Müller-Schwarze 1974, 319; Halpin 1974 in Newman and Halpin 1988, 1779; Jessen and Maschiwitz 1985, 1987, etc. in Hölldobler and Wilson 1990, 277; Hefetz et al. 1986, 197).
Comment: "Individual odor" rather than "individual pheromone" is used when an investigator does not know if the odor meets the criteria of being a pheromone.

inhibitory pheromone *n.* In some species of insects and mammals: a pheromone that stops something, such as aggression, movement, or the production of reproductives within a colony (Wilson 1975, 413).

mating pheromone *n.*

1. For example, in many insect species: a pheromone that attracts members of the opposite sex prior to mating.
2. In the yeast *Saccharomyces cerevisiae*: a pheromone that promotes mating by arresting growth that would stop yeast cells from mating (Grishin et al. 1994, 1081; Telford et al. 1997, 228).

cf. chemical-releasing substance: semiochemical: pheromone: sex pheromone

matrone *n.* In *Aedes* mosquitoes: a postulated pheromone secreted by a male's accessory glands that reduces female receptiveness to other males (Craig 1967 in Wilson 1975, 321).

cf. chemical-releasing substance: semiochemical: pheromone: antiaphrodisiac

plethodontid receptivity factor (PRF)
n. A 22-kilodalton cytokine protein from the submandibular (mental) gland of male *Plethodon jordani* (salamander) that increases female receptivity during courtship (Rollmann et al. 1999, 1907).

primer pheromone *n.* A pheromone that physiologically alters an animal's endocrine and reproductive system, reprogramming it for an altered response pattern (Matthews and Matthews 1978, 189); *e.g.*, a pheromone that causes desert locusts to mature more quickly (Richards and Mangoury 1968).
cf. chemical-releasing substance: semiochemical: pheromone: releaser pheromone

propaganda pheromone *n.* A pheromone that is part of the propaganda substance, *q.v.*, of some ant species that attracts slave-making ants to slave ants (Regnier and Wilson 1971, 267; Hölldobler and Wilson 1990, 394).

pseudopheromone See chemical-releasing substance: semiochemical: allelochemic: allomone.

queen pheromone *n.*
1. For example, in Honey Bees and termites: a pheromone produced by a queen of a colony that mediates worker behavior (Morse 1963; Gary 1974).
   syn. queen substance
2. For example, in Honey Bees: a pheromone produced by a queen that attracts males (Morse et al. 1962).
   syn. queen-mating attractant, queen substance
3. In Honey Bees: a pheromone produced by a queen that prevents the development of other queens in her colony (Heymer 1977, 129).
   syn. royal substance

recognition pheromone, discriminator *n.* A pheromone (genetically determined chemical cue) that enables an animal to classify conspecifics as kin or nonkin (Wilson 1987, 10).

recruitment pheromone *n.* In some ant species: a pheromone that elicits tandem running in a recipient (Heymer 1977, 145).

releaser pheromone *n.* A pheromone that stimulates an immediate behavioral response mediated wholly by an animal's nervous system (Matthews and Matthews 1987, 189).
cf. chemical-releasing stimulus: semiochemical: pheromone: primer pheromone

reproductive pheromone *n.* In many animal species: a pheromone that can alter a receiver's reproductive behavior or physiology (Houck and Reagan 1990, 729).

sex pheromone *n.* In many kinds of organisms including some alga, ciliate, fungus, mammal, nematode, and snake species; many insect species; *Streptococcus faecalis*: a pheromone used in communication between members of the opposite sex in a mating context (Barnes 1974, 221; Dunny et al. 1978, Ensign 1978, etc. in Clewell 1985, 13; Harborne 1982, 182–192; Mason et al. 1989).
cf. chemical-releasing substance: semiochemical: pheromone: mating pheromone
Comments: Some species (*e.g.*, *Drosophila melanogaster*) have both male and female sex pheromones (Scott et al. 1988, 1164). Sex pheromones include sex attractants and aphrodisiacs produced by either sex and "antiaphrodisiacs" and "incitins" (Wilson 1975, 235; Immelmann and Beer 1989, 266).

sodefrin *n.* In the Red-Bellied Newt (*Cynops pyrrhogaster*): a sex attractant released by a male from his cloacal gland (Kikuyama et al. 1995 in Sawyer, 1995, A2).
Comment: This is the first pheromone identified from an amphibian (Wabnitz et al. 1999, 444).
[Japanese *sodefuri*, soliciting (Sawyer 1995, A2)]

splendiferin *n.* In the Magnificent Tree Frog (*Litoria splendida*): a male sex pheromone (Wabnitz et al. 1999: 444).
Comments: This is the first sex pheromone reported in an anuran (Wabnitz et al. 1999, 444). This chemical is comprised of 24 amino-acid residues, is waterborne, and is produced by males' parotoid and rostral glands.
[coined by Wabnitz et al. 1999, 444]

superpheromone *n.* A pheromone that is strong and long lasting (*e.g.*, alarm pheromones in slave-making ants) [coined by Regnier and Wilson 1971, 269].

surface pheromone *n.* A pheromone whose active space is restricted so close to the body of a sending organism that direct contact, or something approaching it, must be made with the sender's body in order for the receiver to perceive the pheromone (*e.g.*, colony odors of many species of social insects) (Wilson 1975, 596).

territory pheromone *n.* For example, in ants: a pheromone that an individual deposits on or around its nest, is colony or species specific, and aids in excluding alien colonies (Nordlund and Lewis 1976, 213, Hölldobler and Wilson 1990, 644).

syn. territorial-demarcation substance (Karlson and Butenandt 1959, 40)

trail pheromone *n.* For example, in many termite and ant species: a pheromone deposited by an individual that is followed by and oriented to by others, (Wilson 1959; Moser and Blum 1963, 1228; and Bossert and Wilson 1963 in Farkas and Shorey 1974, 82).

syn. footprint pheromone, trail odor, trail-following pheromone (Nordlund and Lewis 1976, 212), trail-marking pheromone or substance, trail substance (Hölldobler and Wilson 1990, 644)

cf. odor trail, [2]recruitment

Comment: Trail pheromones are deposited on substrates or in the air.

stimulant *n.* A chemical that increases the probability that an organism will show a particular behavior (Dethier et al. 1960 in Nordlund et al. 1981, 22).

▸ **feeding stimulant** *n.* A chemical that elicits an organism's feeding (Dethier et al. 1960 in Nordlund et al. 1981, 22).
syn. phagostimulant
[coined by Thorsteinson 1953, 1955 in Nordlund et al. 1981].

▸ **locomotory stimulant** *n.* A chemical that elicits locomotion in an organism (Dethier et al. 1960 in Nordlund et al. 1981, 22).

▸ **mating stimulant** *n.* A chemical that elicits mating in an organism (Dethier et al. 1960 in Nordlund et al. 1981, 22).

▸ **ovipositional stimulant** *n.* A chemical that elicits egg laying in an arthropod (Dethier et al. 1960 in Nordlund et al. 1981, 22).

▸ **phagostimulant** See chemical-releasing stimulus: stimulant: feeding stimulant.

♦ **chemical self-marking** See behavior: marking behavior: automarking.

♦ **chemical temperature regulation** See temperature regulation: chemical temperature regulation.

♦ **chemiotaxis** See taxis: chemotaxis.

♦ **chemistry** See study of: chemistry.

♦ **chemoautotroph** See -troph-: autotroph: chemoautotroph.

♦ **chemoheterotroph, chemoheterotrophic** See -troph-: heterotroph, chemoheterotroph.

♦ **chemokinesis** See kinesis: chemokinesis.

♦ **chemolithotroph** See -troph-: chemolithotroph.

♦ **chemoorganotroph** See -troph-: chemoorganotroph.

♦ **chemoreception** See modality: chemoreception.

♦ **chemoreceptor** See receptor: chemoreceptor.

♦ **chemosignal** See chemical-releasing stimulus: semiochemical.

♦ **chemotaxis** See taxis: chemotaxis.

♦ **chemotaxonomy** See study of: taxonomy: chemotaxonomy.

♦ **chemotroph** See -troph-: chemoautotroph.

♦ **cheradium** See [2]community: cheradium.

♦ **cheradophile** See [1]-phile: cheradophile.

♦ **chersium** See [2]community: chersium.

♦ **chersophile** See [1]-phile: chersophile.

♦ **chianophile** See [1]-phile: chianophile.

♦ **chianophobe** See -phobe: chianophobe.

♦ **chiasma** *n., pl.* **chiasmata** The place where, during the later stages of prophase-1 of meiosis, two homologous chromosomes establish close contact, usually where an exchange of homologous parts between non-sister chromatids (crossing over) takes place (Mayr 1982, 766, 957).

Comment: Janssens correctly postulated crossing over at chiasmata in 1901 (Mayr 1982, 766).

♦ *chico* See gene: *chico*.

♦ **Chiczulib Crater** See place: Chiczulib Crater.

♦ **child companion** See companion: child companion.

♦ **child labor** *n.* In Kalotermitid and Rhinotermitid Termites: immature stages' performing a substantial amount of work in their nest (Wilson 1975, 345, 413).

♦ **child schema** *n.* An animal's combination of bodily and facial features that arouses a person's sentiments typically experienced toward young children; this schema includes large roundish eyes, a proportionately large head, small nose and chin, and rounded, chubby cheeks (Immelmann and Beer 1989, 44).

Comment: Some animals (*e.g.*, Pekinese Dogs) elicit child-schema sentiments from persons.

♦ **chimera, chimaera** *n.*
1. A novel or unusual organism that has tissue characteristic of two, or more types; especially, a hybrid of mixed characteristics produced by grafting (Michaelis 1963).
2. A DNA molecule that is formed from DNA fragments from different sources that are spliced together (Curtis 1983, 321).
3. An organism comprised of tissues of two or more genetic types (Lincoln et al. 1985).
4. A plant that is a comprised of two or more genotypes, in cell layers that make up its shoot apex (Geneve 1991, 33).
adj. chimeric

syn. mosaic (Lincoln et al. 1985)

cf. -morph: gynandromorph

Comments: According to Geneve (1991), Winkler selected "chimera" in 1907 as a name for a "plant mosaic." He dropped the "a" from "chimaera" to reduce the confusion of a mosaic with the fish genus *Chimaera.* [Latin *chimaera* < Greek *chimaira,* she-goat; a mythological fire-breathing monster with a lion's head, goat's body, and dragon's tail]

genetic mosaic *n.* A fusion of more than one conspecific individual that results in a new "individual" with parts that are genetically different, *e.g.,* probably certain species of strangler fig trees found to be comprised of multiple genotypes (Thomson et al. 1991, 1214).

syn. allofusion (Putz and Holbrook 1986 in Thomson et al. 1991, 1215)

graft chimera, graft hybrid *n.* A mosaic plant comprised of cells from two species and that results from artificial grafting (Geneve 1991, 33–34).

Comments: +*Laburnocytisus adamii* is a graft chimera of *Cytisus purpureus* and *Laburnum anagyroides.* The plus sign and combining the two parent-plant genus names is the designation for a graft chimera as outlined by the International Code of Botanical Nomenclature (Geneve 1991, 34). [coined by Winkler 1907 in Geneve 1991, 34]

periclinal chimera *n.* A plant with a thin outer layer of cells that differ genetically from its inner cells (*e.g.,* +*Laburnocytisus adamii*) [coined by Bauer 1909 in Geneve 1991, 34].

cf. chimera: variegated-leaf chimera

Comment: Periclinal chimeras include those with variegated leaves and flower petals, changed fruit-skin color and fuzziness, and loss of stem thorniness (Geneve 1991).

sectorial chimera *n.* A plant chimera with one third to one half of all three layers of its entire shoot apex comprised only of mutant cells, resulting in a stem that is part white, or yellow, and part green (*e.g.,* in *Kerria japonica* 'Kin Kan') (Bauer 1909 in Geneve 1991, 35).

variegated-leaf chimera *n.* A plant that is a periclinal chimera and has white, yellow, or other color markings on its green leaves or entire leaf-color change, due to different leaf-cell layers' being white or green (Geneve 1991, 35–36).

Comment: For example, a variegated *Hosta* has light leaf areas due to an outer layer of white cells and an inner layer of green cells (Geneve 1991, 35). Bracts of pink *Poinsettias* have inner and middle red cells and outer white cells, compared to red *Poinsettias* that are not chimeras and have all three layers comprised of red cells.

♦ **chimonophile** See [1]-phile: chimonophile.

♦ **chin raising** See behavior: chin raising.

♦ **chinning** See behavior: chinning.

♦ **chinophobia** See phobia (table).

♦ **chionium** See [2]community: chionium.

♦ **chionophile** See [1]-phile: chionophile.

♦ **chionophobe** See -phobe: chionophobe.

♦ **chirm** See [2]group: chirm.

♦ **chiropterophile** See [2]-phile: chiropterophile.

♦ **chitin** See molecule: chitin.

♦ **Chitty hypothesis** See hypothesis: population-limiting hypotheses: Chitty hypothesis.

♦ **chledium** See [2]community: chledium.

♦ **chledophile** See [1]-phile: chledophile.

♦ **chlorofluorocarbon** See molecule: chlorofluorocarbon.

♦ **chlorophyll** See molecule: accessory pigment: chlorophyll.

♦ **chlorophyll *a*** See molecule: accessory pigment: chlorophyll *a.*

♦ **chlorophyll *b*** See molecule: accessory pigment: chlorophyll *b.*

♦ **chloroplast** See organelle: chloroplast.

♦ **chloroplast DNA** See nucleic acid: deoxyribonucleic acid: chloroplast DNA.

♦ **choice point** *n.* The position in a T- or other kind of maze, or on a discrimination apparatus, from which it is possible for an animal to give only one of two or more alternative responses; *e.g.,* move down only one of two or more runways or jump to one of two or more doors.

Comment: In a T-maze, a choice point is the place at which the base of the T touches its cross-arm (Verplanck 1957).

♦ **choice test** See test: choice test.

♦ **choking** *n.* For example, in the Herring Gull: a disturbed bird's emitting a deep *huoh-huoh-huoh-huoh* sound while moving its breast in a conspicuous manner (Heymer 1977, 115).

syn. threat choking (Tinbergen 1958 in Heymer 1977, 115)

♦ **cholinergic synapse** See synapse: cholinergic synapse.

♦ **-chore** *neutral suffix* Spread, used to mean region or agent of dispersal (Lincoln et al. 1985).

adj. -chorous, -choric

n. -chory; the act of being spread by a particular region or agent

aerochore, anemochore, aerophile *n.* An organism that is disseminated by wind (Lincoln et al. 1985).

adj. aerochorous, anemochorous, aerophilous

cf. [1]-phile: anemophile

androchore *n.* An organism that is dispersed by Humans (Lincoln et al. 1985).

anemochore See -chore: aerochore.

anthropochore *n.* An organism whose propagules are dispersed by Humans (Lincoln et al. 1985).

autochore *n.* A plant that disperses its own propagules (Lincoln et al. 1985).

barachore *n.* An organism that has propagules dispersed by their own weight (Lincoln et al. 1985).

blastochore *n.* A plant that is dispersed by offshoots (Lincoln et al. 1985).

bolochore *n.* An organism whose propagules are dispersed by propulsive mechanisms (Lincoln et al. 1985).

brotichore *n.* An organism that lives in close proximity to Humans (*e.g.*, in houses or other buildings) (Lincoln et al. 1985).

clitochore *n.* An organism that is dispersed by gravity (Lincoln et al. 1985).

crystallochore *n.* An organism that is dispersed by glaciers (Lincoln et al. 1985).

endozoochore *n.* An organism that is dispersed by animals after it passes through their guts (Lincoln et al. 1985).
cf. -chore: epizoochore

entomochore *n.* An organism that is dispersed by insects (Lincoln et al. 1985).

epizoochore *n.* An organism that is dispersed by attaching to the surfaces of animals (Lincoln et al. 1985).
cf. -chore: endozoochore

eurychore *n.* A widely dispersed species (Lincoln et al. 1985).
adj. eurychoric, eurychorous

gynochore *n.* An organism that is dispersed by motile females (Lincoln et al. 1985).

hydrochore *n.* An organism that is dispersed by water (Lincoln et al. 1985).

myrmecochore *n.* An organism that is dispersed by ants (Lincoln et al. 1985).

saurochore, saurophile *n.* An organism that is dispersed by lizards or snakes (Lincoln et al. 1985).

stenochore *n.* An organism that has a narrow range of distribution (Lincoln et al. 1985).
adj. stenochoric, stenochorous.

synzoochore, zoochore *n.* An organism that is dispersed by animals (Lincoln et al. 1985).
adj. synzoochorous, zoochoric, zoochorous

zoochore See -chore: synzoochore.

♦ **-choric** See -chore.

♦ **chorology** See study of: geography.

♦ **chorotobiont** See -cole: graminicole.

♦ **-chorous** *combining form* Dispersed by the agency of.
See -chore.

♦ **chortle** See animal sounds: grunt: baby-grunt.

♦ **chorus** See animal sounds: song: chorus; group: chorus.

♦ **-chory** See -chore.

♦ **chrom-** See chromo-.

♦ **chromatid** *n.* One of two longitudinal units of a chromosome resulting from a split in early prophase, becoming a daughter chromosome later in mitosis (Mayr 1982, 957).

♦ **chromatin** *n.* Threadlike parts of a cell nucleus, at a particular stage in the cell's mitosis, that stain heavily and are now known to be genetic material (Mayr 1982, 674). [Flemming (1879) first called these parts chromatin (Mayr 1982, 674)]

♦ **chromesthesia** See -thesia: chromesthesia.

♦ **chromo-, chrom-** *combining form* Color; in, or with, color (Michaelis 1963). [Greek *chrōma*, color]

♦ **chromoplast** See organelle: chromoplast.

♦ **chromosomal dimorphism** See -morphism: dimorphism: chromosomal dimorphism.

♦ **chromosomal inversion** *n.* The 180% rotation of a middle piece of a chromosome that occurs during chiasmata formation (Mayr 1982, 769).
syn. chromosome inversion (Mayr 1982, 769)

♦ **chromosomal mutation** See [4]mutation: chromosomal mutation.

♦ **chromosomal polymorphism** See -morphism: polymorphism: chromosomal polymorphism.

♦ **chromosome** *n.*
1. One of an eukaryote cell's complex, often rodlike structures in its nucleus that consists of chromatin and carries genetic information (genes) arranged in linear sequences (King and Stansfield 1985).
2. A prokaryote cell's circular DNA molecule that contains all genetic instructions essential for the cell's life (King and Stansfield 1985).
syn. karyomite (King and Stansfield 1985)
Comments: In general, within a species individuals all have the same chromosome number. An exception is the Raccoon Dog, whose individuals have from 38 through 56 chromosomes, with the number even varying among cells of the same individual (Doris Wurster-Hill in Anonymous, 1986, A12). Different numbers of B-chromosomes account for this phenomenon.
[term proposed by Waldeyer in 1888 (Mayr 1982, 674)]

A-chromosome *n.* A normal chromosome of a eukaryotic organism; contrasts with allosome (Lincoln et al. 1985).
cf. chromosome: supernumerary chromosome

accessory chromosome *n.* In the hemipteran insect *Pyrrhocoris*: the 12th chromosome, in the set of 12 chromosomes, which is a sex chromosome determining a bug's sex (Stevens 1905, Wilson 1905 in Mayr 1982, 751).
cf. chromosome: B-chromosome, sex chromosome

acrocentric chromosome *n.* A chromosome, or chromosome fragment, with a nearly terminal centromere (Mayr 1982, 769; King and Stansfield 1985).
cf. chromosome: metacentric chromosome, telocentric chromosome

allosome, heterochromosome *n.* A chromosome, or chromosome fragment, that is not an A-chromosome, *q.v.* (Lincoln et al. 1985).

artificial chromosome See vector: artificial chromosome.

autosome *n.* A chromosome other than a sex chromosome (Dawkins 1982, 284; Mayr 1982, 751).

B-chromosome See chromosome: supernumerary chromosome.

driving Y-chromosome *n.* In *Drosophila*: a mutant Y-chromosome that causes males to make only Y-bearing sperm (Hamilton 1967 in Thornhill and Alcock 1983, 7).

heterochromosome See chromosome: allosome.

homeologous chromosome *n.* A chromosome that is only partially homologous to another one with which it exhibits reduced synaptic attraction and which is apparently derived from the same ancestral chromosome (Lincoln et al. 1985).

homologous chromosome, homologue, homolog *n.* A chromosome that is structurally similar (has identical genetic loci in the same sequence) to another chromosome with which it pairs during nuclear division within a cell (Lincoln et al. 1985).

idiochromosome See chromosome: sex chromosome.

metacentric chromosome *n.* A chromosome with its centromere at its center (King and Stansfield 1985).
cf. chromosome: acrocentric chromosome, telocentric chromosome

parasitic chromosome *n.* A chromosome that does not confer a selective advantage to its bearer (Ostergren 1945 in Nur et al. 1988); *e.g.*, the psr chromosome (Nur et al. 1988).

polytene chromosome *n.* In some angiosperm, ciliate, and dipteran species: a giant cable-like chromosome that consists of many identical chromatids lying in par-

allel and in register (King and Stansfield 1985; Jasny 1991).
Comment: Because the chromatids are in register, there is a pattern of bands oriented perpendicularly to the long axis of a chromosome.

sex chromosome, idiochromosome *n.* In eukaryotic organisms: a chromosome, or group of chromosomes, represented differently in the sexes, responsible for the genetic determination of sex (Lincoln et al. 1985).

supernumerary chromosome *n.* In plants and animals: a chromosome, or chromosome fragment, that is devoid of structural genes and differs from normal A-chromosomes in structure, genetic effectiveness, and pairing behavior and never pairs with an A-chromosome (Lincoln et al. 1985; King and Stansfield 1985).
syn. accessory chromosome (King and Stansfield 1985), B-chromosome (Lincoln et al. 1985)
cf. nucleic acid: deoxyribonucleic acid: selfish DNA
Comment: A supernumerary chromosome may be a parasitic chromosome (Nur et al. 1988).

synthetic lethal *n.* A lethal chromosome formed as a result of crossing over between normally viable chromosomes (King and Stansfield 1985).
See gene: synthetic lethal.
Comment: A synthetic lethal can convey superior fitness in some combinations but is lethal in combination with other combinations of chromosomes (Mayr 1982, 580).

telocentric chromosome *n.* A chromosome with its centromere at one end (King and Stansfield 1985).
cf. chromosome: acrocentric chromosome, metacentric chromosome

W-chromosome *n.* In species with female heterogamety (*e.g.*, Lepidoptera): the sex chromosome that is present in females only (*i.e.*, females are WZ) (King and Stansfield 1985).

X-chromosome *n.* In species with male heterogamety (*e.g.*, *Drosophila*, mammals): the sex chromosome that is present in both sexes (*i.e.*, females are XX and males are XY) (King and Stansfield 1985).

Y-chromosome *n.* In species with male heterogamety (*e.g.*, *Drosophila*, mammals): the sex chromosome that is present in males only (*i.e.*, males are XY) (King and Stansfield 1985).

yeast-artificial chromosome (YAC) See vector: yeast-artificial chromosome.

Z-chromosome *n.* In species with female heterogamety (*e.g.*, Lepidoptera): the sex chromosome that is present in both sexes (*i.e.,* males are ZZ and females are WZ) (Lincoln et al. 1985).

♦ **chromosome band** *n.* Alternating light or dark staining sections along a chromosome that are visible by light microscopy after staining (Jasny 1991).

♦ **chromosome map, cytogenetic map** See map: genetic map.

♦ **chromosome ring** *n.* In a species of evening primrose, *Oenothera*: a group of fused chromosomes (Mayr 1982, 731).

♦ **chromosome set** *n.* The number of chromosomes characteristic of a somatic cell of a diplodiploid species (*e.g.,* 4 in *Drosophila*, 23 in *Homo sapiens*) (King and Stansfield 1985).
syn. basic number, chromosome number (Lincoln et al. 1985)

♦ **chromosome theory** See ³theory: Sutton-Boveri chromosome theory.

♦ **chromosome walking** See method: chromosome walking.

♦ **-chronic** *combining form* Referring to time (Michaelis 1963).
allochronic *adj.* Referring to populations, or species, that live, grow, or reproduce during different seasons (Lincoln et al. 1985).
monochronic *adj.* Referring to a phenomenon that happens only once (Lincoln et al. 1985).
synchronic, contemporaneous, synchronous *adj.* Referring to a phenomenon that occurs, or exists, at the same time as another one, or ones (Lincoln et al. 1985).

♦ **chronic-fatigue syndrome** See syndrome: chronic-fatigue syndrome.

♦ **chronistics** See study of: chronistics.

♦ **chronocline** See cline: chronocline.

♦ **chronometry** See -metry: chronometry.

♦ **chronospecies** See ²species: chronospecies.

♦ **chronostratigraphy** See study of: chronostratigraphy.

♦ **chrymosymphile** See ¹-phile: chrymosymphile.

♦ **CHS** See enzyme: chalcone synthase.

♦ **-cial** *combining form*
altricial *adj.*
1. In some animals: referring to a young that is born in a very underdeveloped state and requires significant parental care for a substantial period following its birth (Reingold 1963, 123; Wilson 1975, 578; Eisenberg 1981, 505); often used especially for birds, sometimes Humans.
2. In many kinds of birds, including passerines: referring to young that hatch with

closed eyes and usually no feathers and cannot walk and find their own food (Immelmann and Beer 1989, 12).
3. In mammals including rats, mice, Golden Hamsters, cats, dogs, rabbits: referring to young that are born with closed eyes and ears and an inability to walk, maintain their body temperatures, and eliminate wastes by themselves (McFarland 1987, 363).
cf. -cial: precocial; neoteny; nestling: nidicolous nestling, nidifugous nestling
[New Latin *altricialis* < Latin *altrix, -icis,* a nurse]

precocial, praecocial *adj.*
1. In some bird, fish, and mammal species: referring to young that are able to engage in a wide range of behavior patterns, including locomotion, soon after hatching or birth (Dewsbury 1978, 135).
2. In mammals, including the Hare, Guinea Pig, all herd ungulates (Cow, Deer, Goats, Horses, Sheep, etc.): referring to young that are born with open eyes and ears and ability to walk, maintain their body temperatures, and eliminate wastes by themselves (McFarland 1987).
3. For example, in many anseriform- and galliform-bird species: referring to young that hatch with open eyes; an ability to walk, swim, or both; and an ability or inability to find their own food in foraging areas (Immelmann and Beer 1989, 230).
4. For example, in Army Ants: referring to a species whose workers participate in colony treks and raids very soon after becoming adults (Immelmann and Beer 1989, 230).
n. praecoces
cf. -cial: altricial; nestling: nidicolous nestling, nidiferous nestling; nidifugous; precocious
▸ **semiprecocial** *adj.* In some bird species, including gulls: referring to a young that hatches with open eyes, down feathering, and enough motor control to stand and walk, but depends on its parent(s) for food and warmth (Immelmann and Beer 1989, 203, 230).

♦ **cicada principle** See principle: cicada principle.

♦ **-cide** *combining form*
1. Killer or destroyer of (Michaelis 1963).
2. Murder, or killing, of (Michaelis 1963).
cf. behavior: killing behavior; pesticide.
[Latin *-cida,* killer < *caedere,* to kill]
filial ovicide See cannibalism: filial cannibalism.
fratricide See -cide: siblicide.

infanticide *n.*

1. In over 1300 animal species, including some lizard, falconiform-bird, and social-insect species; many rodent species; the Hanuman Langur, House Sparrow, Human, Lion, Long-Billed Marsh Wren, and White Stork: an individual's killing conspecific young which may, or may not, be its own (Wilson 1975, 84; Fox 1975 and other reviews in Jenssen et al. 1989, 1054; Veiga 1990, 496).

    Note: Some bumble-bee species practice infanticide by oophagy and ejection of living larvae from their brood combs (Free et al. 1969, Pomeroy 1979 in Fisher and Pomeroy 1990, 801).

2. An animal's behavior that makes a direct and significant contribution to the immediate death of an embryo, or newly hatched or born, conspecific individual (Hausfater and Hrdy 1984 in Wasser 1986, 624; Mock 1984 in Immelmann and Beer 1989, 145).

syn. infant killing

cf. cannibalism: cronism; effect: Bruce effect

Comments: Infanticide has been considered pathological, but in the last decade, biologists have found it to be adaptive in many animal species (Jenssen et al. 1989, 1054). Infanticide may also include any deaths resulting from fetal resorption and abortion as well as from nutritional neglect, desertion, or abuse of immature offspring that may be already weaned. The latter end of this continuum is "deferred infanticide" and is especially prevalent in *Homo sapiens*. Some authors refer to young-killing without their consumption as "infanticide" as distinguished from "cannibalism," *q.v.* (Immelmann and Beer 1989, 65). Infanticide by males is known only in a few bird, several mammal, and one spider species (Schneider and Lubin 1997, 305–306).

ovicide *n.* For example, in the White-Winged Chough: a bird's tossing an egg from the nest of another bird (Heinsohn 1991, 1097).

siblicide, sibicide *n.* In many animal species and probably many plant species: a developmentally immature individual's killing its sibling (Bragg 1954 and Hamilton 1979 in Mock et al. 1990, 438; Immelmann and Beer 1989, 113).

syn. Cainism [after the Biblical Cain who slew his brother Abel], fratricide (Immelmann and Beer 1989, 113)

Comments: Siblicide is found in some ant, fish, lacewing, and praying-mantid species; the Black Eagle, Blue-Footed Booby, Cattle Egret, *Dalbergia* sp. (tree), Domestic Pig, Great Egret, Human, Kittiwake (bird), Pronghorn Antelope and Spadefoot Toad (Bragg 1954 and Hamilton 1979 in Mock et al. 1990, 438; Immelmann and Beer 1989, 113; Yoon 1996b, C4). In some species, siblicide may lead to cannibalism (Mock et al. 1990, 438). Siblicide is a form of sibling rivalry in many animals (Yoon 1996b, C1, C4). For example, in the Cattle Egret, a mother often lays three eggs and puts much more testosterone and other androgens into the first two eggs compared to the third one. After they hatch, the parents preferentially feed those young compared to the young that hatches from the third egg. The two older young cow the youngest one, which is a runt that survives when food is plentiful. In Proghorns, embryos literally grow through their siblings in the womb. In sharks, embryos swim about and devour others *in utero*. *Dalbergia* trees produce seeds in pods, which settle to the ground. The seed in a pod that sprouts first makes an exocrine substance that annihilates other seeds (its siblings) in its pod. In Humans, many pregnancies that initiate with twin embryos end with only one baby, and siblings often fight after they are born.

▸ **facultative siblicide** *n.* For example, in some booby, eagle, and pelican species: species-specific siblicide in which one chick usually kills another chick (Kepler 1969, etc. in Mock et al. 1990, 441).

▸ **obligate siblicide** *n.* For example, in some egret species; the Blue-Footed Booby, the Osprey: species-specific siblicide in which one chick often does not kill another chick although it fights with it (Mock 1985, etc. in Mock et al. 1990, 441).

suicide *n.*

1. A person's intentionally killing himself (Michaelis 1963).

2. An individual Pea Aphid's displaying behavior that increases its chance of death (McAllister et al. 1990, 167).

♦ **cin4** See transposable element: cin4.

♦ **circadian mimicry** See mimicry: adjustable mimicry: circadian mimicry.

♦ **circadian rhythm** See rhythm: circadian rhythm.

♦ **circalunadian rhythm** See rhythm: circalunadian rhythm.

♦ **circamonthly rhythm** See rhythm: circamonthly rhythm.

♦ **circannual rhythm** See rhythm: circannual rhythm.

♦ **circasemiannual rhythm** See rhythm: circasemiannual rhythm.

♦ **circular-networks hypothesis** See hypothesis: hypotheses of species richness: circular-networks hypothesis.

♦ **circular statistics** See study of: statistics: circular statistics.

♦ **circumcision** *n.*
1. A person's cutting off all or part of a prepuce, either as a religious rite or as a prophylactic operation; the result of this procedure (Michaelis 1963).
2. A person's cutting off a person's clitoris; the result of this procedure (Michaelis 1963).
   syn. female genital cutting (Dugger 1996, 1)
3. Spiritual purification (Michaelis 1963).
cf. infibulation
Comments: In parts of Africa, it is an ancient religious custom to circumcise a girl and stitch her genital lips together (= infibulation) to preserve her virginity (Dugger 1996, 1, 9). This practice became a federal crime in the U.S. in 1996. For Sierra Leoneans, genital cutting is part of an elaborate, highly secret initiation rite.
[from Old French *circonciser* < Latin *circumcisus*, past participle of *circumcidere*, *circum*, round + *caedere*, to cut]

♦ **circumneutrophile** See [1]-phile: circumneutrophile.

♦ ***cis*-9,10-octadecenoamide** See molecule: *cis*-9,10-octadecenoamide.

♦ **cistron** See gene.

♦ **citadel** See [2]group: citadel.

♦ **citric-acid cycle** See biochemical pathway: citric-acid cycle.

♦ **clade** *n.*
1. A delimitable monophyletic unit [coined by Huxley 1957, 455].
2. A separate evolving line of organisms (Wilson 1975, 26).
3. A species, or species set, that represents a distinct branch in a phylogenetic tree, or cladogram (Wilson 1975, 580).
4. A monophyletic group of taxa that share a closer common ancestry with one another than with members of any other clade (Lincoln et al. 1985).
cf. grade (evolutionary); -gram: cladogram

♦ **cladism** See study of: systematics: cladistic systematics.

♦ **cladistic** *adj.* Referring to a clade or holophyletic group (Lincoln et al. 1985).
stratocladistic *adj.* Referring to phylogenetic relationships inferred from weighted derived similarities between fossil taxa and selected stratigraphic data not contradicted by morphological analysis (Lincoln et al. 1985).

♦ **cladistic affinity** See affinity: cladistic affinity.

♦ **cladistic method** See study of: systematics: cladistic systematics.

♦ **cladistics** See study of: systematics: cladistic systematics.

♦ **clado-, klado** *prefix* Branch, offshoot, stem, twig (Brown 1956; Lincoln et al. 1985).

♦ **cladogenesis** See -genesis: cladogenesis.

♦ **cladogram** See -gram: cladogram.

♦ **clamping reflex** See reflex: grasp reflex.

♦ **clan** See [3]group: [1,2]clan.

♦ **clan-specific odor** See odor: clan-specific odor.

♦ **clandestine** *adj.*
1. Referring to evolutionary changes not evident in an organism's adult stage (Lincoln et al. 1985).
2. Referring to adult characters of a descendant species derived from embryonic characters of an ancestral species (Lincoln et al. 1985).

♦ **clasp reflex** See reflex: grasp reflex.

♦ **clasping hold** See amplexus.

♦ **class A to class E levels of social organization** See social organization.

♦ **classic scientist** See scientist: classic scientist.

♦ **classical animal psychology** See study of: psychology: behaviorism.

♦ **classical Batesian mimicry** See mimicry: Batesian mimicry: classical Batesian mimicry.

♦ **classical conditioning** See learning: conditioning.

♦ **classical ethology** See study of: [3]ethology: classical ethology.

♦ **classical fitness** See fitness (def. 2, 4); fitness: genetic fitness.

♦ **classical gene-chromosome theory** See [3]theory: classical gene-chromosome theory.

♦ **classical genetics** See study of: genetics: Mendelian genetics.

♦ **classical imprinting** See learning: imprinting: classical imprinting.

♦ **classical individual fitness** See fitness (def. 2).

♦ **classical lek** See [2]lek: classical lek.

♦ **classical model of human rationality** See [4]model: classical model of human rationality.

♦ **classical molecular evolution** See study of: evolution: classical molecular evolution.

♦ **classical personal fitness** See fitness (def. 4).

♦ **classical population genetics** See study of: genetics: classical population genetics.

♦ **classical theory** See [3]theory: neutral theory.

♦ **classification** *n.*
1. The process used by systematists for assembling populations and taxa into groups and these, in turn, into even larger groups, using a large number of characters (Mayr 1982, 147).

syn. taxonomy, systematics (Michaelis 1963)

2. The taxonomic scheme that results from classification (def. 1).

cf. ¹identification

♦ **classification of behavior** *n.* A hierarchical determination of kinds of behavior, *e.g.*, Scott's (1950) classification and Delgado and Delgado's (1962) classification.

Comment: See other areas of this book for definitions of kinds of behaviors, especially under "behavior."

DELGADO AND DELGADO'S (1962) BEHAVIOR CLASSIFICATION[a]

  I. simple behavior units
    A. individual
      1. static or postural units
      2. dynamic or gestural units
        a. localized
        b. generalized
    B. social
      1. static
      2. dynamic
  II. complex behavioral units
    A. simultaneous
    B. sequential
    C. syntactic
    D. roles
      1. active
      2. passive

---

[a] Lehner (1979, 64–66)

LEHNER'S BEHAVIOR CLASSIFICATION[a]

---

behavior type
  behavior pattern of a type
    behavioral act(s), or act(s), in a pattern
      component parts of an act

---

[a] Lehner (1979, 65–66)

♦ **claustral colony founding** *n.* In queens of ants and other social hymenopterans, royal pairs in termites: an individual's sealing itself off in a cell and rearing the first generation of workers on nutrients obtained mostly or entirely from its own storage tissues, including fat bodies and histolyzed wing muscles (Wilson 1975, 422, 580).

♦ **claustrophobia** See phobia (table).

♦ **claw-waving mode** See display: claw-waving mode.

♦ **"clay aggregation"** See organized structure: "clay aggregation."

♦ **cleaner** *n.* An animal, often a mutualist, that removes particular things from bodies of hosts; *e.g.*, over 40 decapod-crustacean species; about 40 fish species that remove and eat bacteria, ectoparasites, diseased and damaged tissue, or excess food particles from fish and other marine animals (Wickler 1968, 157; Immelmann and Beer 1989, 46); many bird species that pick parasites off skin of large mammals or reptiles (crocodiles, tortoises, iguanas); a mite that cleans a beetle species; a rotifer that cleans a water flea (Immelmann and Beer 1989, 46).

♦ **cleaner-fish dance** See dance: cleaner-fish dance.

♦ **cleaner mimicry** See mimicry: cleaner mimicry.

♦ **cleaning appetence swimming** See dance: cleaner-fish dance.

♦ **cleaning customer** *n.* An animal cleaned by a cleaner, *q.v.* (Wickler, 157).

♦ **cleaning drive** See drive: cleaning drive.

♦ **cleaning-invitation posture** See posture: cleaning-invitation posture.

♦ **cleaning station** *n.* The home base of a cleaner fish (Heymer 1977, 136).

♦ **cleaning symbiosis** See symbiosis: cleaning symbiosis.

♦ **cleaver** See tool: cleaver.

♦ **cleistogamy** *n.* In some jewelweed, violet species; some pansy cultivars; many other taxa: a plant's having small, closed, self-fertilizing flowers, usually in addition to its regular flowers (Michaelis 1963).

Comment: Cleistogamous flowers are usually on or near the ground (Lawrence 1951, 744).

adj. cleistogamous

[Greek *cleistos*, closed < *kleinein*, to close]

♦ **cleithrophobia** See phobia (table).

♦ **clenched-teeth yawning** See yawning: clenched-teeth yawning.

♦ **cleptobiosis** See -biosis: cleptobiosis; parasitism: cleptoparasitism.

♦ **cleptoparasitism, kleptoparasitism** See parasitism: cleptoparasitism.

♦ **clesiophobia** See phobia (table).

♦ **cliff, clift** See habitat: cliff.

♦ **climatic climax** See ²community: climax: climatic climax.

♦ **climatic hypothesis of human emergence** See hypothesis: climatic hypothesis of human emergence.

♦ **climatic race** See ¹race: climatic race.

♦ **climatic rule** See rule: ecogeographical rule.

♦ **climatic selection** See selection (table).

♦ **climatology** See study of: climatology.

♦ **climax species** See ²species: climax species.

♦ **cline** *n.*

1. A gradient of gene-frequency change, phenotypic-frequency change, or both, in a population from one part of its geographic range to another; *e.g.*, many mammal species show clines of increasing size toward the colder portions of their ranges (Wilson 1975, 580; King and Stansfield 1985).
2. "Continuous variation on the expression of a character through a series of contiguous populations" (Lincoln et al. 1985).

syn. geographic-character gradient
adj. clinal
[Greek *klinein*, to slope, bend]

chronocline *n.* A cline through an extended geological period (Lincoln et al. 1985).

coenocline *n.* A sequence of communities distributed along an environmental gradient (Lincoln et al. 1985).

ecocline *n.*

1. A more or less continuous trait variation in a sequence of populations distributed along an ecological gradient, with each population exhibiting local adaptation to its particular part of the gradient; a gradient of ecotypes (Lincoln et al. 1985).
2. Community-structure differences resulting from changes in slope aspect around a mountain or ridge (Lincoln et al. 1985).

ethocline *n.* A graded series in the expression of a particular behavioral trait within a group of related species (Lincoln et al. 1985) "interpreted to represent stages in an evolutionary trend" (Wilson 1975, 347, 584); *e.g.*, in *Peromyscus* mouse nesting behavior and possibly in garter-snake feeding preferences to chemical cues (Brown 1975, 18).

geocline *n.*

1. A graded sequence of morphological variation through a series of populations, resulting from spatial, or topographical, separation (Lincoln et al. 1985).
2. A cline resulting from hybridization between adjacent, but genetically distinct, populations (Lincoln et al. 1985).

halocline *n.*

1. A discontinuity in salinity (Lincoln et al. 1985).
2. A zone with a marked salinity gradient (Lincoln et al. 1985).

hybrid cline See cline: nothocline.

morphocline, morphological transformation series *n.* "A graded series of character states of a homologous character" (Lincoln et al. 1985).

nothocline *n.* A graded series of characters, or forms, produced by hybridization.
syn. hybrid cline

ontocline *n.* An organism's gradation of phenotypic characters during its ontogenic development (Lincoln et al. 1985).

phenocline *n.* A graded series of phenotype frequencies within a species' geographical range (Lincoln et al. 1985).

ratio cline *n.* In polymorphic species: a cline in which successive populations show progressive change in relative frequency of morphs (Lincoln et al. 1985).

sociocline *n.* A graded series of social organization among related species that is interpreted to represent stages in an evolutionary trend (Lincoln et al. 1985).

topocline *n.* A graded series of forms that occur through the geographical range of a taxon, but not necessarily correlated with an ecological gradient (Lincoln et al. 1985).

♦ **clinging young** *n.* For example, in Bats, Koalas, Primates, Sloths: an offspring that holds firmly to its mother during its first days to weeks postpartum; this offspring is comparable to precocial young in its sensory capacities and to altricial young in its motility (Immelmann and Beer 1989, 47).
cf. reflex: grasp reflex; transport of young

♦ **clinical psychology** See study of: psychology: clinical psychology.

♦ **clinodeme** See deme: clinodeme.

♦ **clinology** See study of: clinology.

♦ **clinotaxis** See taxis: clinotaxis.

♦ **clique** *n.* "A set of species with the property that every pair in the set has some food source in common" (Yodzis 1982 in Hawkins and MacMahon 1989, 434).
cf. guild

♦ **clisere** See succession: clisere.

♦ **clithrohobia** See phobia (table).

♦ **clitochore** See -chore: clitochore.

♦ **cloaca, cloacal gland** See gland: cloacal gland.

♦ **cloacal kiss** See kiss: cloacal kiss.

♦ **clock** *n.* An instrument used for measuring and indicating time (Michaelis 1963).

biological clock *n.* In many animal species: an animal's internal physiological mechanism that measures time or maintains endogenous rhythms (Lincoln et al. 1985; Kolata 1985, 929; Immelmann and Beer 1989, 155).
syn. endogenous clock, escapement clock, internal chronometer, internal clock, physiological clock, rhythm, time sense (McFarland 1985, 290; Immelmann and Beer 1989, 155)
cf. rhythm
Comments: Animals have different means of regulating their master biological clocks (Blakeslee 1998, A1, A2). Fruit flies have timekeeping genes that are active in their

bristles, legs, and wings. Horseshoe crabs have clock sensors on their tails, and swallows have them just inside their skulls. The suprachiasmatic nucleus of a human brain might possess a master clock. Humans have light-sensitive cells on their legs and possibly elsewhere in their skin that affect their biological rhythms. Hemoglobin in these cells might transmit day-length information to a person's brain by carrying nitric oxide, a neurotransmitter. Hemoglobin in human eyes, not particular light-sensitive cells, might function in setting biological clocks. Chlorophyll regulates plant circadian rhythms.

molecular clock, molecular-clock model of evolution, molecular evolutionary clock See hypothesis: molecular-clock hypothesis.

♦ **clone, clon** *n.*
1. "A population of individuals all derived asexually from the same single parent" (Wilson 1975, 581).
2. "A set of organisms all of whose cells are members of the same clone;" *e.g.*, a pair of identical twins are members of the same clone (Dawkins 1982, 285).
3. In cell biology, "a set of genetically identical cells, all derived from the same ancestral cell;" *e.g.*, a human body is a gigantic clone of some 10^{15} cells (Dawkins 1982, 285).
4. A group of bionts that are identical by descent for every allele at every locus (Bell 1982, 504).

cf. genet, ortet, ramet
Comments: Hans Spemann proposed vertebrate cloning in 1938; John Gurdon cloned frogs that lived to the tadpole stage in 1970; Steen Willadsen cloned a sheep from immature sheep embryo cells in 1984; Neal First cloned calves from embryos that had at least 120 cells; and Ian Wilmut produced the first mammal clone, a sheep, from an adult mammal cell in 1996 (Kolata 1997, 1; Specter and Kolata 1997, A1, A20–A23). This sheep grew from a sheep egg from which Wilmut removed the DNA, which he replaced with DNA from a sheep mammary cell.
[Greek *klōn*, sprout, twig]

euclone *n.* A group of conspecific organisms that is comprised of genotypically identical modular units (ramets), each of which can follow an independent existence if separated from its parental organism (given as adjective "euclonal" in Lincoln et al. 1985).

hemiclone *n.* "A lineage of hybrido-genetic animals with a common maternal haploid genome" (Bell 1982, 507).

paraclone *n.* A group of conspecific organisms that is comprised of genotypically identical modular units, each of which cannot follow an independent existence if separated from its parental organism (given as adjective "paraclonal" in Lincoln et al. 1985).

pseudoclone *n.* A group of conspecific organisms that is comprised of nongenotypically identical modular units that are so closely related and coordinated that they can be regarded as ecologically and functionally equivalent to a clone (given as adjective "pseudoclonal" in Lincoln et al. 1985).

♦ **cloning** *n.*
1. An organism's reproducing by producing a clone, *q.v.*
2. A person's reproducing an organism or its gene(s) by causing the organism to clone or the gene to multiply.

gene cloning *n.* A person's multiplying a gene(s) with the use of DNA technology (Campbell et al. 1999, 365).

♦ **cloning vector** See vector: cloning vector.

♦ **clonodeme** See deme: clonodeme.

♦ **closed behavioral program** See program: behavioral program: closed behavioral program.

♦ **closed community** See ²community: closed community.

♦ **closed genetic program** See program: genetic program: closed genetic program.

♦ **closed group** See ²society: closed society.

♦ **closed population** See ¹population: closed population.

♦ **closed society** See ²society: closed society.

♦ **cloud** See ²group: cloud.

♦ **clowder** See ²group: clowder.

♦ **clump** See ²group: clump.

♦ **clumped distribution** See distribution: clumped distribution.

♦ **clusium** See ²community: clusium.

♦ **CLUSTAL W** See program: CRUSTAL W.

♦ **cluster** See ²group: cluster.

♦ **clutch** See ²group: clutch.

♦ **clutter** See ²group: clutter.

♦ **cM** See genetic map (comments).

♦ **C/N-continuous-flow-isotope-ratio mass spectrometer (CFIRMS)** See instrument: C/N-continuous-flow-isotope-ratio mass spectrometer.

♦ **co-accretion hypothesis** See hypothesis: Moon-origin hypotheses: co-accretion hypothesis.

♦ **coacervate** See organized structure: coacervate.

♦ **coaction-effected social facilitation** See ²facilitation: social facilitation: coaction-effected social facilitation.

♦ **coadaptation, coadaption** See [3]adaptation: [1]evolutionary adaptation: coadaptation.

♦ **coarse-grained environment** See environment: coarse-grained environment.

♦ **coarse-grained resource** See resource: coarse-grained resource.

♦ **coarse-grained species** See [2]species: coarse-grained species.

♦ **coarse-grained variation** See variation: coarse-grained variation.

♦ **coarse-particulate-organic matter (CPOM)** See particulate-organic matter: coarse-particulate-organic matter.

♦ **coast** See habitat: coast.

♦ **cob** See animal names: cob.

♦ **Cochran's approximate-*t* test** See statistical test: multiple-comparisons test: planned-multiple-comparisons procedure: Cochran's approximate-*t* test.

♦ **cocoon** *n.*
  1. For example, in Ants, nymphal reproductive termites, and some moth, skipper-butterfly, sphecid-wasp species: the encasement, often of silk, made by a larva in which it pupates (*Oxford English Dictionary* 1972, entries from 1699; Wilson 1975, 139).
  2. The protective envelope for an animal's eggs, larvae, or pupae (*Oxford English Dictionary* 1972, entries from 1699; Immelmann and Beer 1989, 47).
     Note: Animals including bagworm moths, earthworms, and spiders lay their eggs in cocoons (Barnes 1974, 298).
  3. In Mudfish: a clay cell made of an individual fish (Wood 1883 in *Oxford English Dictionary* 1972).
  [French *cocon* < *coque*, shell]

♦ **code** *n.* The complete set of possible signals and contexts in a communicatory situation (Dewsbury 1978, 99).
  cf. system: communication system.

♦ **codemic** *adj.* Referring to organisms that belong to the same deme (Lincoln et al. 1985).

♦ **codominance** See dominance: codominance.

♦ **codon** *n.* A sequence of three nucleotides (= a triplet) on DNA or RNA that specifies either a particular amino acid sequence or the end of a functional unit (*i.e.*, allele) (Wittenberger 1981, 613; Dawkins 1982, 285).

♦ **codon bias** *n.* The nonuniform distribution of codon usages; *e.g.*, the phenomenon that an organism frequently uses a particular codon, rather than its alternative, to specify a given amino acid (Air et al. 1976 in Rieger et al. 1991).

♦ **coefficient** *n.*
  1. Statistically, a constant (Lincoln et al. 1985).

  2. Statistically, a dimensionless description of a distribution or data set (Lincoln et al. 1985).

CC See coefficient: coefficient of community.

coefficient of community (CC) *n.* A coefficient that indicates the degree of similarity of two communities based on the number of species that they have in common (Mueller-Dombois and Ellenberg 1974 in Dubey 1995, 33; Krebes 1985, 447–448); CC = $2c/(a + b)$, where a = the number of taxa in community 1, b = the number of taxa in community 2, and c = the number of taxa both communities have in common.
  syn. index of similarity (Krebs 1985, 447–448)
  Comments: The coefficient of community ranges from 0 to 1.0, indicating no similarity to complete similarity (Krebs 1985, 448).

coefficient of consanguinity See coefficient: coefficient of relatedness.

coefficient of genetic determination See heritability.

coefficient of inbreeding (F, f) *n.* "The probability that both alleles on one locus in a given individual are identical by virtue of identical descent" (Wright 1948 in Wilson 1975, 730).
  syn. inbreeding coefficient
  Comment: "Any value of *f* above zero implies that the individual is inbred to some degree, in the sense that both of its parents share an ancestor in the relatively recent past" (Wright 1948 in Wilson 1975, 73).

coefficient of kinship See coefficient: coefficient of consanguinity.

coefficient of relatedness (r) *n.*
  1. The probability that two individuals share an allele that is identical due to common descent (Wright 1922 in Keane 1990, 265).
  2. "The fraction of genes in two individuals that are identical by descent, averaged over all loci" (Wright 1948; Hamilton 1963, 1964, 9; Wilson 1975, 74, 581).
  3. The conditional probability that a second individual has a given gene if a related individual is known to have the gene (Trivers and Hare 1976).
  4. Coefficient of consanguinity (F_{IJ}, f_{IJ}): "The probability that a pair of alleles drawn at random from the same locus in two individuals will be autozygous;" I and J refer to the individuals that are being compared (Wilson 1975, 73, 581).
  syn. coefficient of kinship, coefficient of parentage, degree of relationship, panmictic index, Wright's inbreeding coefficient
  cf. r

panmictic index See coefficient: coefficient of relatedness.

regression coefficient *n*. In statistics, the rate of change of a dependent random variable with regard to one or more independent variables; *e.g.*, *b* in the regression equation $y = a + bx$ (Sokal and Rohlf 1969, 408).

relative-crowding coefficient (RCC) *n*. A measure of the aggression of one plant species towards another, derived from the results of a replacement-series experiment (Lincoln et al. 1985).

selection coefficient(s) *n*. In population genetics, a measure of natural-selection intensity, calculated as the proportional reduction in gametic contribution of one genotype compared with that of a standard one; $0 < s < 1$ (Futuyma 1986, 152).
syn. coefficient of selection.

Wright's inbreeding coefficient See coefficient: coefficient of relatedness.

◆ **coelom** [SEE lim] *n*. An animal body cavity that is between the body wall and viscera and lined with mesoderm; contrasted with pseudocoelom (Michaelis 1963).
syn. celom, coelome (SEE lome)
Comments: Coelomate phyla include Arthropoda, Echinodermata, and Chordata.
[Greek *koilōma*, cavity < *koilos*, hollow]

pseudocoelom [sue doe SEE lim] *n*. An animal body cavity that is between the body wall and viscera and partly lined with mesoderm; contrasted with coelom.
syn. pseudocelom, pseudocoelome
Comment: Aschelminthes and related phyla are pseudocoelomates.

◆ **coeno-, ceno-, koino-** *combining form* Sharing, in common (Lincoln et al. 1985).
cf. ¹community: phytocoenosis
[Greek *koinos*, common]

◆ **coenobiology** See study of: biocoenology.

◆ **coenocline** See cline: coenocline.

◆ **coenogamodeme** See deme: coenogamodeme.

◆ **coenogenesis** See -genesis: coenogenesis.

◆ **coenogenous** See -genous: coenogenous.

◆ **coenomonoecism** See -oecism: coenomonoecism.

◆ **-coenosis**
cf. ²group
necrocoenosis, liptocoenosis *n*. "An assemblage of dead organisms" (Lincoln et al. 1985).
phytocoenosis *n*. "The total plant life of a given habitat or community" (Lincoln et al. 1985).
cf. ¹community: phytocoenosis; flora
psephonecrocoenosis *n*. A necrocoenosis of dwarf individuals (Lincoln et al. 1985).

◆ **coenosium** See ²community: coenosium.

◆ **coenospecies** See ²species: coenospecies.

◆ **coenotrope** See behavior: coenotrope.

◆ **coevolution** See ²evolution: coevolution.

◆ **cognition** *n*.
1. A person's action, or faculty, of knowing taken in its broad sense, including sensation, perception, conception, etc.; distinguished from feeling and volition; more specifically, one's recognizing an object in perception proper (*Oxford English Dictionary* 1972, entries from 1651).
2. A hypothetical stimulus-stimulus association, or perceptual organization, postulated to account for expectancies (Verplanck 1957).
Note: Cognition is not yet possible to define in other than intuitive terminology, except for trivial cases (Verplanck 1957).
3. "All of the various modes and aspects of knowing, including perceiving, recognizing, remembering, imagining, conceptualizing, judging, and reasoning," potentially including "all inner mediating systems whereby an organism processes stimulus information and construes, represents, organizes, interprets, and responds to ongoing events" (Storz 1973, 49).
Note: In Humans, this includes thought processes involving imagery, language, and symbols (Storz 1973, 49).
4. In Humans: a state that modulates the expression of emotions (Hinde 1985, 987).
Note: It is often useful to consider a continuing cognitive-emotion interaction when studying cognition and emotion (Hinde 1985, 987).
5. Animal mental processes that cannot be observed directly but for which there is scientific evidence (McFarland 1985, 345).
Note: Cognition may involve an animal's having a mental image of a "goal" to be achieved (McFarland 1985, 356).
6. Animal "brain processes through which an organism acquires information about its environment" (Zayan and Duncan 1987 in Dawkins 1988, 316).
7. A general psychological term for a person's mental functioning including perception, memory, and thinking; a person's knowledge of local geography, awareness of self, learning, judgment, and use of language (Immelmann and Beer 1989, 48).
cf. awareness, cognitive map
[Latin *cognitio, -onis*, knowledge < *cognoscere*, to know < *co-* together + (*g*)*noscere*, to know]

visuospatial-construction cognition *n*. A person's ability to take a whole apart mentally and to reconstruct it from its parts (Blakeslee 1996, C3).

cf. syndrome: Williams syndrome

Comment: This process appears to take place in the parietal lobes of the back of a person's brain (Blakeslee 1996, C3).

♦ **cognitive** *adj.* Referring to imparting objective information unrelated to emotion (Sebeok 1962 in Wilson 1975, 217).

Comment: This is one of Sebeok's six basic functions of communication. The other basic functions are conative, emotive, metacommunicative, phatic, and poetic (Sebeok 1962 in Wilson 1975, 217).

♦ **cognitive-developmental concept** See developmental-genetic conception of ethical behavior.

♦ **cognitive ethology** See study of: [3]ethology: cognitive ethology.

♦ **cognitive map** *n.*
  1. An elaboration of a cognition (Verplanck 1957).
  2. In more derived vertebrates, possibly in Honey Bees: an ordered mental "map" that tells an animal how to interpret cues that occur in a particular sequence (Gould 1982, 263; Gould and Gould 1988, 108–109).
  3. An animal's internal representation of the layout of an area in its environment (Immelmann and Beer 1989, 48).

syn. landmark-based mental map (Gould 1986 in Dyer 1991, 239), mental map (Dyer 1991, 239)

Comment: Dyer's (1991) study does not support the existence of cognitive maps in Honey Bees reported by Gould and Towne (1987).

♦ **cognitive psychology** See study of: psychology: cognitive psychology.

♦ **cohesion species concept** See [2]species (def. 32).

♦ **coition, coitus** See copulation.

♦ ***coitus interruptus*** *n.* An individual act of onanism, *q.v.*, with regard to intercourse (Storz 1973, 183).

♦ **cold-blooded animal** See -therm: poikilotherm.

♦ **-cole** *combining form* Inhabitant of, in, or on (Lincoln et al. 1985).
adj. -colous
cf. [1]-phile, -zoite

agaricole *n.* An organism that lives on mushrooms and toadstools (*e.g.*, some springtails [arthropods]) (Lincoln et al. 1985).

aigicole *n.* An organism that lives on beaches (*e.g.*, a tiger-beetle species) (Lincoln et al. 1985).

algicole *n.* An organism that lives on algae (*e.g.*, a periwinkle) (Lincoln et al. 1985).

alsocole *n.* An organism that lives in woody groves (*e.g.*, the Piliated Woodpecker) (Lincoln et al. 1985).

cf. -cole: arboricole, hylacole, nemoricole, silvicole; [1]-phile: acrodendrophile, aiphyllophile, dendrophile, halorgadophile, hylodophile, orgadophile

amanthicole *n.* An organism that lives on sandy plains (*e.g.*, the Pronghorn) (Lincoln et al. 1985).

cf. -cole: aigicole, ammocole, amnicole, arenicole, thinicole; [1]-phile: aigialophile, amathophile, ammochtophile, ammophile, anemophile, cheradophile, enaulophile, psamathophile, psammophile, syrtidophile, thinophile

ammocole *n.* An organism that lives on or in sand (*e.g.*, the Slash Pine) (Lincoln et al. 1985).

cf. -cole: aigicole, amanthicole, amnicole, arenicole, thinicole; [1]- phile: aigialophile, amathophile, ammochtophile, ammophile, anemophile, cheradophile, enaulophile, psamathophile, psammophile, syrtidophile, thinophile

amnicole *n.* An organism that lives on sandy river banks (*e.g.*, a digger-wasp species) (Lincoln et al. 1985).

cf. psammon

ancocole *n.* An organism that lives in canyons (*e.g.*, a hummingbird species) (Lincoln et al. 1985).

cf. [1]-phile: ancophile

aphidicole *n.* An organism that lives among aphid aggregations (*e.g.*, a larva of a syrphid-fly species) (Lincoln et al. 1985).

aquaticole *n.* An organism that lives in water or aquatic vegetation (*e.g.*, the Sunfish) (Lincoln et al. 1985).

arboricole, dendricole, dendrophile *n.* An organism that lives in trees or large woody shrubs (*e.g.*, the Orangutan) (Lincoln et al. 1985).

cf. -cole: alsocole, arboricole, arbusticole, fruticole, hylacole, nemoricole, silvicole, thamnocole; [1]-phile: acrodendrophile, aiphyllophile, dendrophile, halorgadophile, hylodophile, orgadophile

arbusticole *n.* An organism that lives on scattered shrubs and shrub-like perennial herbs (*e.g.*, a grasshopper species) (Lincoln et al. 1985).

cf. -cole: arboricole, fruticole, hylacole, thamnocole

arenicole, psammobiont, sabulicole *n.* An organism that lives in sand (*e.g.*, the Coquina Clam, the grass *Elymus arenicola*, the wasp *Sphex sabulosa*) (Lincoln et al. 1985).

cf. -cole: aigicole, amanthicole, ammocole, amnicole, thinicole; [1]- phile: aigialophile, amathophile, ammochtophile, ammophile, anemophile, cheradophile, enaulophile, psamathophile, psammophile, syrtidophile, thinophile

a – c

argillicole *n*. An organism that lives on or in clay (*e.g.*, the evening primrose *Oenothera argillicola*) (Lincoln et al. 1985).
cf. [1]-phile: argillophile, pelophile, phellophile, spiladophile

branchicole *n*. An organism that lives on gills of fish or other aquatic animals (*e.g.*, a parasitic-crustacean species) (Lincoln et al. 1985).

broticole *n*. An organism that lives in close proximity to Humans, in houses or other buildings (*e.g.*, the American Cockroach) (Lincoln et al. 1985).

bryocole *n*. An organism that lives on or in moss (*e.g.*, a tardigrade species) (Lincoln et al. 1985).
cf. -cole: muscicole, sphagnicole; [1]-phile: bryophile, sphagnophile

cadavericole *n*. An organism that feeds on dead bodies or carrion (*e.g.*, the beetle *Dermestes cadaverinus*) (Lincoln et al. 1985).
cf. saprobe

caespiticole *n*. An organism that lives in grassy turf or pastures (*e.g.*, the Broadleaf Plantain) (Lincoln et al. 1985).
cf. -cole: nomocole; [1]-phile: nomophile

calcicole, calcipete, calciphyte, gypsophyte *n*. A plant that grows in soils rich in calcium salts (Lincoln et al. 1985).
cf. calcifuge; -cole: gelicole, geocole, halicole, humicole, pergelicole, perhalicole, silicole, terricole; [1]-phile: agrophile, geophile, gypsophile, nitrophile

calcosaxicole *n*. An organism that lives in rocky limestone areas (*e.g.*, a fern species) (Lincoln et al. 1985).
cf. -cole: petrimadicole, petrocole, rupicole; [1]-phile: actophile, calicophile, chasmophile, lithophile, petrochthophile, petrodophile, petrophile, phellophile

carbonicole *n*. An organism that lives on burnt, or scorched, substrates (*e.g.*, Fireweed) (Lincoln et al. 1985).

caulicole *n*. An organism that lives on plant stems (*e.g.*, some aphid species) (Lincoln et al. 1985).
cf. -cole: culmicole, folicaulicole

cavernicole *n*. An organism that lives in subterranean cave or passages (*e.g.*, a salamander species) (Lincoln et al. 1985).
cf. -cole: stygobie; [1]-phile: troglophile

cespiticole See -cole: caespiticole.

corticole *n*. An organism that lives on bark (*e.g.*, a lichen species) (Lincoln et al. 1985).

crenicole *n*. An organism that lives in springs or in brook water fed from a spring (*e.g.*, a dragonfly species) (Lincoln et al. 1985).

culmicole *n*. An organism that lives on grass stems (*e.g.*, a planthopper species) (Lincoln et al. 1985).
cf. -cole: caulicole, folicaulicole

dendricole See -cole: arboricole.

deserticole *n*. An organism that lives on open ground in an arid, or desert, region (*e.g.*, a cactus species) (Lincoln et al. 1985).
cf. -cole: siccocole, xerocole; [1]-phile: chersophile, hydrophile, hylodophile, lochnodophile, subxerophile, syrtidophile, xerohylophile, xerophile

domicole *n*. An organism that lives in a tube, nest, or other domicile (*e.g.*, a polychaete-worm species) (Lincoln et al. 1985).
cf. cole: tubicole

epicole, epibiont *n*. An organism that lives attached to the surface of its organism "host" without benefit, or detriment, to its host (*e.g.*, a bromeliad species) (Lincoln et al. 1985).

fimbricole *n*. An organism that grows in or on dung (*e.g.*, a dung-beetle species) (Lincoln et al. 1985).

fimicole, meridicole *n*. An organism that lives in or on dung (*e.g.*, the Dung Fly) (Lincoln et al. 1985).
cf. -zoite: coprozoite

folicaulicole *n*. An organism that lives attached to leaves and stems (*e.g.*, a scale-insect species) (Lincoln et al. 1985).
cf. -cole: folicole, forbicole

folicole *n*. An organism that lives on leaves (*e.g.*, the Poison-Ivy Sawfly) (Lincoln et al. 1985).
cf. -cole: folicaulicole, forbicole; hypophyllus

forbicole, herbicole *n*. An organism that lives on broad-leaved plants (*e.g.*, the Eastern Tent Caterpillar) (Lincoln et al. 1985).
cf. -cole: folicaulicole, folicole

fructicole *n*. An organism that lives on or in fruits (*e.g.*, the Mediterranean Fruit Fly) (Lincoln et al. 1985).
cf. -cole: arboricole, arbusticole, hylacole, thamnocole

fruticole *n*. An organism that lives on, or grows on, shrubs (*e.g.*, a mistletoe species) (Lincoln et al. 1985).

fungicol *n*. An organism that lives in or on fungi (*e.g.*, the Handsome Fungus Beetle) (Lincoln et al. 1985).

gallicole *n*. An organism that lives in galls (*e.g.*, a mite species) (Lincoln et al. 1985).

gelicole *n*. An organism that lives in geloid soils having a crystalloid content between 0.2 and 0.5 parts per thousand (Lincoln et al. 1985).
cf. calcifuge; -cole: calcicole, geocole, halicole, humicole, pergelicole, perhalicole,

silicole, terricole; [1]-phile: agrophile, geophile, gypsophile, nitrophile

geocole *n.* An organism that lives in soil for part of its life cycle (*e.g.*, a 17-year cicada) (Lincoln et al. 1985).
cf. calcifuge; -cole: calcicole, gelicole, halicole, humicole, pergelicole, perhalicole, silicole, terricole; [1]-phile: agrophile, geophile, gypsophile, nitrophile

graminicole *n.*
1. An animal that spends most of its life in a grassy habitat (Lincoln et al. 1985).
2. An organism that grows on grasses (*e.g.*, a grasshopper or the dodder, a parasitic plant) (Lincoln et al. 1985).
syn. chorotobiont

halicole *n.*
1. An organism that lives in haloid soils having a crystalloid content between 0.5 and 2 parts per thousand (Lincoln et al. 1985).
2. An organism that lives in a habitat with a high salt content (*e.g.*, Red Mangrove or *Aster halophilus*) (Lincoln et al. 1985).
cf. calcifuge; -cole: calcicole, gelicole, geocole, humicole, pergelicole, perhalicole, silicole, terricole; [1]-phile: agrophile, geophile, gypsophile, nitrophile

herbicole See -cole: forbicole.

humicole *n.* An organism that lives on, or in, soil (*e.g.*, the Arctic Rockcress, *Arabis humifusa*) (Lincoln et al. 1985).
cf. calcifuge; -cole: calcicole, gelicole, geocole, halicole, pergelicole, perhalicole, silicole, terricole; [1]-phile: agrophile, geophile, gypsophile, nitrophile

hydrocole *n.* An organism that lives in an aquatic habitat (*e.g.*, the beaver) (Lincoln et al. 1985).

hygrocole *n.* An organism that lives in a moist, or damp, habitat (*e.g.*, a slime-mold species) (Lincoln et al. 1985).

hylacole, hylocole *n.* An organism that lives among trees or in woodland understory shrubs or forest (*e.g.*, the Gibbon, *Hylobates agilis,* and the Veery bird, *Hylocichla fuscescens*) (Lincoln et al. 1985).
cf. -cole: alsocole, arboricole, arbusticole, fruticole, hylacole, nemoricole, silvicole, thamnocole; [1]-phile: acrodendrophile, aiphyllophile, dendrophile, halorgadophile, hylodophile, orgadophile

lapidocole, saxicole, saxitile *n.* An organism that lives under, or among, stones (*e.g.*, the lichen *Parmelia saxatilis* or the Pavement Ant) (Lincoln et al. 1985).
cf. hypolictic; [1]-phile: petrochthophile

latebricole *n.* An organism that lives in holes (*e.g.*, the Gopher Tortoise) (Lincoln et al. 1985).

leimocole, leimicole *n.* An organism that lives in moist grassland or meadowland (*e.g.*, False Hellebore) (Lincoln et al. 1985).
cf. -cole: poocole, pratinicole; [1]-phile: coryphile, helolochmophile, poophile, telmatophile

lichenicole *n.* An organism that lives in, or on, a lichen (*e.g.*, a bagworm species) (Lincoln et al. 1985).
cf. [1]-phile: lichenophile

lignicole *n.* An organism that lives on or in wood (*e.g.*, the spider *Selenops lignicolus*) (Lincoln et al. 1985).
cf. [1]-phile: lignophile, proxylophile, xylophile

limicole *n.* An organism that lives in mud shores (*e.g.*, a mudwort, *Limosella aquatica*) (Lincoln et al. 1985).
cf. -cole: luticole; [1]-phile: argillophile, pelochthophile

limnicole, limniicole *n.* An organism that lives in a lake (*e.g.*, the amphipod *Gammarus limnaeus*) (Lincoln et al. 1985).
cf. [1]-phile: limnophile

lochmocole *n.* An organism that lives in thickets (*e.g.*, a blackberry species) (Lincoln et al. 1985).

lucicole *n.* An organism that lives in open habitats with ample light (*e.g.*, the White Oak) (Lincoln et al. 1985).
cf. -cole: umbriticole

luticole *n.* An organism that lives in mud (*e.g.*, the Spatterdock, a water-lily) (Lincoln et al. 1985).
cf. -cole: limicole; [1]-phile: argillophile, pelochthophile

merdicole See -cole: fimbicole.
adj. merdicolous, stercoraceous

monticole *n.* An organism that lives in mountain habitats (*e.g.*, the plant *Paronychia montana*) (Lincoln et al. 1985).

muscicole *n.* An organism that lives on or in mosses or moss-rich communities (*e.g.*, a millipedes species) (Lincoln et al. 1985).
cf. -cole: bryocole, sphagnicole; [1]-phile: bryophile, sphagnophile

myrmecocole *n.* An organism that lives in ant or termite nests (*e.g.*, a butterfly species) (Lincoln et al. 1985).

nemoricole, orgadocole *n.* An organism that lives in open woodlands (*e.g.*, the goldenrod *Solidago nemoralis*) (Lincoln et al. 1985).
adj. nemoricolous, nemeral, nemorose
cf. -cole: alsocole, arboricole, hylacole, silvicole; [1]-phile: acrodendrophile, aiphyllophile, dendrophile, halorgadophile, hylodophile, orgadophile

nervicole *n.* An organism that lives on or in leaf veins (*e.g.*, a scale-insect species) (Lincoln et al. 1985).

nidicole *n*. An organism that lives in a nest (*e.g.*, stingless bees and the Gray Squirrel) (Lincoln et al. 1985).

nivicole *n*. An organism that lives in snow or snow-covered habitats (*e.g.*, the Snowshoe Hare) (Lincoln et al. 1985).

nomocole *n*. An organism that lives in a pasture (*e.g.*, the Bull Thistle) (Lincoln et al. 1985).
cf. -cole: ceaspiticole; [1]-phile: nomophile

omnicole *n*. An organism that lives on many kinds of substrates (*e.g.*, a bacterium species) (Lincoln et al. 1985).

orgadocole See -cole: nemoricole.

ovariicole *n*. An organism that lives in ovaries (Lincoln et al. 1985).

paludicole *n*. An organism that lives in marshy habitats (*e.g.*, the Marshmallow) (Lincoln et al. 1985).
cf. -cole: pontohalicole, telmicole; [1]-phile: helophile, limnodophile

patacole *n*. An organism that lives in forest litter part of its life cycle (*e.g.*, the Mexican Bean Beetle) (Lincoln et al. 1985).

pergelicole *n*. An organism that lives in haloid soils with crystalloid content below 0.2 parts per thousand (Lincoln et al. 1985).
cf. calcifuge; -cole: calcicole, gelicole, geocole, halicole, humicole, perhalicole, silicole, terricole; [1]-phile: agrophile, geophile, gypsophile, nitrophile

perhalicole *n*. An organism that lives in haloid soils with crystalloid content above 0.2 parts per thousand (Lincoln et al. 1985).
cf. calcifuge; -cole: calcicole, gelicole, geocole, halicole, humicole, pergelicole, silicole, terricole; [1]-phile: agrophile, geophile, gypsophile, nitrophile

petrimadicole *n*. An organism that lives in the surface film of water on rocks (*e.g.*, an alga species) (Lincoln et al. 1985).
cf. -cole: calcosaxicole, petrocole, rupicole; [1]-phile: actophile, calicophile, chasmophile, lithophile, petrochthophile, petrodophile, petrophile, phellophile
adj. hygropetric, petrimadicolous

petrocole *n*. An organism that lives in a rocky habitat (*e.g.*, a scorpion species) (Lincoln et al. 1985).
cf. -cole: calcosaxicole, petrimadicole, rupicole; [1]-phile: actophile, calicophile, chasmophile, lithophile, petrochthophile, petrodophile, petrophile, phellophile

phreatocole *n*. An organism that lives in groundwater habitats (*e.g.*, a crustacean species) (Lincoln et al. 1985).
adj. phreatobie, phreatocolous

piscicole *n*. An organism that lives on or within fish (*e.g.*, the Lamprey) (Lincoln et al. 1985).
cf. -zoite: histozoite

planticole, phytobiont, phytophile *n*. An organism that spends most of its active life on or within plants (*e.g.*, the Oleander Aphid) (Lincoln et al. 1985).
cf. -zoite: histozoite

pontohalicole *n*. An organism that lives in a salt marsh (*e.g.*, the Salt-Marsh Skipper butterfly) (Lincoln et al. 1985).
cf. -cole: paludicole, telmicole; [1]-phile: helophile, limnodophile

poocole *n*. An organism that lives in a meadow (*e.g.*, the Meadow Fritillary butterfly) (Lincoln et al. 1985).
cf. -cole: leimocole, pratinicole; [1]-phile: coryphile, helolochmophile, poophile, telmatophile

potamincole *n*. An organism that lives in a river (*e.g.*, the Hippopotamus) (Lincoln et al. 1985).

pratinicole *n*. An organism that lives in a grassland or meadowland (*e.g.*, the spider *Pardosa prativaga* and Timothy Grass [*Phleum pratense*]) (Lincoln et al. 1985).
cf. -cole: leimocole, poocole, pratinicole, psicole; [1]-phile: coryphile, helolochmophile, poophile, psilophile, telmatophile

psicole *n*. An organism that lives in a savanna or prairie (*e.g.*, the Prairie Coneflower) (Lincoln et al. 1985).
cf. -cole: pratinicole; [1]-phile: psilophile

radicicole *n*. An organism that lives on or in roots (*e.g.*, the Carrot Wireworm [beetle]) (Lincoln et al. 1985).

ramicole *n*. An organism that lives on twigs and branches (*e.g.*, a leafhopper species) (Lincoln et al. 1985).

ripicole *n*. An organism that lives on banks of rivers and streams (*e.g.*, otters) (Lincoln et al. 1985).

rubicole *n*. An organism that lives on brambles (*e.g.*, a sawfly species) (Lincoln et al. 1985).

rupicole *n*. An organism that lives on walls or rocks (*e.g.*, a bristletail [insect] species) (Lincoln et al. 1985).
adj. mural, rupestral, rupestrine, rupicolous
cf. -cole: calcosaxicole, petrimadicole, petrocole; [1]-phile: actophile, calicophile, chasmophile, lithophile, petrochthophile, petrodophile, petrophile, phellophile

sabulicole See -cole: arenicole.

sanguicole *n*. An organism that lives in blood (*e.g.*, a malaria *Plasmodium*) (Lincoln et al. 1985).

saxicole, saxitile See -cole: lapidicole.

sepicole *n*. An organism that lives in hedge rows (*e.g.*, Jack-of-the-Hedge, an aroid plant) (Lincoln et al. 1985).

siccocole, siccicole, siccole *n*. An organism that lives in arid habitats (*e.g.*, the Creosote Bush) (Lincoln et al. 1985).

cf. -cole: deserticole, xerocole; [1]-phile: chersophile, hydrophile, hylodophile, lochnodophile, subxerophile, syrtidophile, xerohylophile, xerophile

silicole *n.* An organism that lives in soil rich in silica or silicates or on flints (*e.g.*, a cactus species) (Lincoln et al. 1985).

cf. calcifuge; -cole: calcicole, gelicole, geocole, halicole, humicole, pergelicole, perhalicole, terricole; [1]-phile: agrophile, geophile, gypsophile, nitrophile

silvicole, sylvicole *n.* An organism that lives in woodlands (*e.g.*, the Oven Bird) (Lincoln et al. 1985).

adj. silvicolous, sylvestral, sylvicolous

cf. -cole: alsocole, arboricole, hylacole, nemoricole; [1]-phile: acrodendrophile, aiphyllophile, dendrophile, halorgadophile, hylodophile, orgadophile

sphagnicole *n.* An organism that lives in peat moss (*e.g.*, the Rose-Pogonia Orchid) (Lincoln et al. 1985).

cf. -cole: bryocole, muscicole; [1]-phile: bryophile, sphagnophile

spongicole *n.* An organism that lives on or in sponges (*e.g.*, a starfish species) (Lincoln et al. 1985).

stagnicole *n.* An organism that lives in stagnant water (*e.g.*, the pond snail *Lymnaea stagnalis*) (Lincoln et al. 1985).

stygobie, troglobie *n.* "An obligate cavernicole" (Lincoln et al. 1985).

cf. -cole: cavernicole; [1]-phile: troglophile

subgeocole *n.* An organism that lives underground (*e.g.*, a mole cricket) (Lincoln et al. 1985).

sylvicole See -cole: silvicole.

tegulicole *n.* An organism that lives on tiles (*e.g.*, a lichen species) (Lincoln et al. 1985).

telmicole *n.* An organism that lives in freshwater marshes (*e.g.*, the Marsh Wren *Telmatodytes griseus*) (Lincoln et al. 1985).

cf. -cole: paludicole, pontohalicole; [1]-phile: helophile, limnodophile

termitocole *n.* An organism that lives in termite nests (*e.g.*, a silverfish species) (Lincoln et al. 1985).

terricole *n.* An organism that lives on, or in, soil, and spends most of its life there, *e.g.*, the Naked Mole Rat (Lincoln et al. 1985).

cf. calcifuge; -cole: calcicole, gelicole, geocole, halicole, humicole, pergelicole, perhalicole, silicole; [1]-phile: agrophile, geophile, gypsophile, nitrophile

thamnocole *n.* An organism that lives on or in bushes or shrubs (*e.g.*, the Yellow Warbler [bird]) (Lincoln et al. 1985).

cf. -cole: arboricole, arbusticole, fruticole, hylacole, thamnocole

thinicole *n.* An organism that lives on sand dunes (*e.g.*, a grass species) (Lincoln et al. 1985).

cf. -cole: aigicole, amanthicole, ammocole, amnicole, arenicole; [1]-phile: aigialophile, amathophile, ammochtophile, ammophile, anemophile, cheradophile, enaulophile, psamathophile, psammophile, syrtidophile, thinophile

tiphicole *n.* An organism that lives in ponds (*e.g.*, a bladderwort species) (Lincoln et al. 1985).

torrenticole *n.* An organism that lives in river torrents (*e.g.*, a fly *Charadromyia torrenticola*) (Lincoln et al. 1985).

tubicole *n.* An organism that lives in tubes (*e.g.*, a worm *Tubifex multisetosus*, a mud-dauber *Sceliphron tubifex*) (Lincoln et al. 1985).

cf. -cole: domicole

umbraticole, skiophile *n.* An organism that lives in shaded habitats (*e.g.*, a fern *Pteris umbrosa*) (Lincoln et al. 1985).

cf. -cole: lucicole, umbraticole; [1]-phile: anheliophile, heliophile, helioxeriophile, lygophile, phengophile, skotophile, sciophile

vaginicole *n.* An organism that lives in a secreted sheath or case (*e.g.*, a bagworm-moth larva) (Lincoln et al. 1985).

viticole *n.* An organism that lives on vines (*e.g.*, the Pipevine Swallowtail [butterfly]) (Lincoln et al. 1985).

xerocole *n.* An organism that lives in dry conditions (*e.g.*, an African ground squirrel *Xerus rutilans* and an amaranth *Philoxerus vermicularis*) (Lincoln et al. 1985).

cf. -cole: deserticole, siccocole; [1]-phile: chersophile, hydrophile, hylodophile, lochnodophile, subxerophile, syrtidophile, xerohylophile, xerophile

♦ **Cole's Paradox, Cole's Result** *n.* A semelparous parthenogenetic female should produce only one more offspring or a semelparous sexually reproducing female should produce only two more offspring than an "immortal" iteroparous female in order to match her fitness, even with the extreme and unrealistic condition that the iteroparous female is reproducing at no cost whatever to her reproductive potential (Cole 1954 in Brown 1975, 191; Daly and Wilson 1983, 182).

Comments: Cole's Paradox is unrealistic because juveniles are commonly more susceptible to mortality than are adults and other realistic complications are not included.

♦ **collaplankton** See plankton: kollaplankton.

♦ **collateral kin** See kin.

◆ **collective group** See ²species aggregate.

◆ **collective mimicry** See mimicry: collective mimicry.

◆ **collective species** See ²species: superspecies.

◆ **collector-gatherer** See ²group: functional feeding group: gathering collector.

◆ **colonial** *adj.*
cf. colony

multicolonial *adj.* Referring to a social-insect population that is divided into colonies that recognize nest boundaries (Hölldobler and Wilson 1990, 640).
cf. colonial: unicolonial

unicolonial *adj.* Referring to a social-insect population in which there are no behavioral colony boundaries (Hölldobler and Wilson 1990, 644).
cf. colonial: multicolonial

◆ **colonial mating system** See mating system: colonial mating system.

◆ **colonist** *n.* An organism, typically r-selected, that invades and colonizes a new habitat or territory (Lincoln et al. 1985).

◆ **colony** See ²group: colony; ²society: modular society: continuous-modular society.

◆ **colony fission** See swarming: fission.

◆ **colony odor** See odor: colony odor.

◆ **color change** *n.* An animal's alteration of its body color that can involve color change of its whole body surface or of specific parts or structures; only one or a few color changes during its development, seasonal changes, or irregular changes and color change that lasts from only seconds to up to months (Immelmann and Beer 1989, 49).
cf. metachrosis

◆ **coloration** *n.* Arrangement of colors of an organism's body (Michaelis 1963).
cf. discolorous, metachrosis

apatetic coloration *n.* In many kinds of animals with camouflage: misleading coloration, *e.g.*, coloration resembling physical features of an organism's habitat (Lincoln et al. 1985).
cf. camouflage, crypsis, mimicry

aposematic coloration, aposematism *n.* In some amphibian, fish, insect, and reptile species: conspicuous coloration with bold patterns (often orange, red, white, or yellow, or a combination of these colors, on a black background) that advertises the identity of an animal that is dangerous (ferocious, venomous, foul-tasting); this coloration is likely to have a protective function (Matthews and Matthews 1978, 312, 323).
syn. proaposematic coloration, warning coloration (Matthews and Matthews 1978, 312, 323)
cf. mimicry: Batesian mimicry

▸ **pseudaposematic coloration, pseudoaposematic coloration, pseudoaposematism** *n.* In some Batesian mimics: aposematism that involves misleading potential enemies (Lincoln et al. 1985).

▸ **synaposematic coloration** *n.* In some Müllerian mimicry complexes: aposematic coloration whose warning signal is shared by more than one species (Lincoln et al. 1985).

baby coloration *n.* In many primate species: coloration of a young that is markedly different from adult coloration; *e.g.*, infant black colobus monkeys are almost completely white while adults are predominantly black (Immelmann and Beer 1989, 162).
cf. dress: juvenile dress

cryptic coloration *n.* In camouflaged animals: coloration that resembles an animal's substratum, or surroundings, and aids in concealment (Lincoln et al. 1985).
cf. coloration: phaneric coloration; mimicry

directive coloration *n.* A prey's surface markings that divert its predator's attention, attack, or both to nonvital parts of its body (Lincoln et al. 1985).

fighting coloration *n.* For example, in some cephalopod, cichlid- and blennioid-fish, and some chamaeleonid-lizard species: coloration taken on by dominant individual animals during and after fights (Heymer 1977, 95).
cf. coloration: submissive coloration

flash coloration *n.* For example, in some Australian katydid and underwing-moth species: usually hidden coloration that an animal suddenly exposes when threatened (Matthews and Matthews 1978, 336).

gynochromatypic coloration *n.* For example, in some butterfly and dragonfly species: coloration of a mature adult female with usual color patterns of mature, adult, conspecific females [gynochromatypic coined by Hilton, 1987, 222].
syn. heterochromatic coloration, heterochrome coloration, heteromorphic coloration (Johnson 1964 in Hilton 1987, 221), heteromorphous coloration
cf. mimicry: androchromatypic mimicry

heterochromatic coloration, heterochrome coloration, heteromorphic coloration, heteromorphous coloration See coloration: gynochromatypic coloration.

juvenile coloration See dress: juvenile dress.

nuptial coloration See dress: advertising dress.

obliterative coloration See camouflage: countershading.

phaneric coloration *n.* Conspicuous coloration (Lincoln et al. 1985).
cf. coloration: cryptic coloration

poster coloration *n.* In some coral fishes: bright spotted and banded coloration that is used as an assembling signal (Franzisket 1960 in Wilson 1975, 212).

proaposematic coloration See coloration: aposematic coloration.

protective coloration See coloration: cryptic coloration.

pseudaposematic coloration See coloration: aposematic coloration.

submissive coloration *n.* In some cephalopod, blenniodi-fish, cichlid-fish, chamaeleonid-lizard species: a coloration, after color change, in subordinate individuals (Ohm 1958 in Heymer 1977, 92).
cf. coloration: fighting coloration

synaposematic coloration See coloration: aposematic coloration.

warning coloration See coloration: aposematic coloration.

♦ **-colous** See -cole.

♦ **column raid** *n.* In Army Ants: a raid conducted in branching columns whose termini are each headed by a relatively small group of workers that lay down chemical trails and capture prey (Hölldobler and Wilson 1990, 636).

♦ **comb** *n.*
1. In nests of many species of bees and wasps: a layer of brood cells, or cocoons, clustered together in a regular arrangement (Wilson 1975, 581).
   Note: Honey Bees make combs of wax; some vespid-wasp species make combs of paper.
2. In nests of some termite species: a spongy, dark, reddish-brown material made by workers from excreta that is used for making fungus beds (Torre-Bueno 1978).
   See nest: wasp nest: comb.

♦ **combat dance** See dance: combat dance.

♦ **combined-switch-alternative phenotype** See -type: phenotype: combined-switch-alternative phenotype.

♦ **comfort behavior** See behavior: comfort behavior.

♦ **comfort movement** See ²movement: comfort movement.

♦ **comfort sucking** *n.* A child's sucking a breast, or rubber nipple, to pacify itself, not because it is hungry (Eibl-Eibesfeldt 1972, Heymer 1974 in Heymer 1977, 40).
cf. behavior: comfort behavior

♦ **commensal of civilization** *n.* An animal species that changes its niche by using resources resulting from human activities, *e.g.*, swallows and swifts that originally nested in cliffs and now nest in houses and other buildings (Heymer 1977, 106).

♦ **commensalism** See symbiosis: commensalism.

♦ **commensalist mimicry** See mimicry: commensalist mimicry.

♦ **common** See frequency (table).

♦ **common ancestor** See ancestor: common ancestor.

♦ **common descent** *n.* Part of Darwin's theory of evolution, proposing that all species evolved from a single, or few, common ancestors, originally stated as, "There is a grandeur in this view of life, with its several powers, having been originally breathed into a few forms or into one; and that ... from so simple a beginning endless forms most beautiful and most wonderful have been, and are being, evolved" (Darwin, 1859, last text of book).

♦ **commonality, principle of** See principle: principle of commonality.

♦ **commons** *n.* A resource, or habitat, whose depletion, or deterioration, affects all individuals using it to an similar extent regardless of imbalances in the rate that particular individuals deplete or harm it (Wittenberger 1981, 613).

♦ **communal breeding** See breeding: communal breeding.

♦ **communal courtship** See courtship: communal courtship.

♦ **communal sexual display** See display: communal sexual display.

♦ **communal sociality** See sociality: communal sociality.

♦ **communal song** See animal sounds: song: communal song.

♦ **communal territory** See territory: group territory.

♦ **commune** *n.* A society, or group, of conspecific organisms that have a social structure and consists of repeated members, or modular units, with a high level of coordination, integration, and genotypic relatedness (Lincoln et al. 1985).

♦ **communication** *n.*
1. A person's imparting, conveying, or exchanging ideas, knowledge, or information (*Oxford English Dictionary* 1972, entries from 1690).
2. Any sort of an animal's anticipatory movement that may signalize an activity to another individual, or vocalizations, or gestures, that clearly direct or predicate (Crawford 1939).
   Note: "Communication" lacks definition in its application to animal behavior, for it may be stretched to include too much (Crawford 1939).
3. "All procedures by which one [human] mind may affect another" (Shannon and Weaver 1949, 3, in Burghardt 1970, 7).

4. "The discriminatory response of an organism to a stimulus" (Stevens 1950 in Burghardt 1970, 6–7).
5. "The establishment of a social unit from individuals by the use of language or signs" (Cherry 1957 in Marler 1961).
6. Information transfer between, or among, conspecific organisms (Frings and Frings 1964, Diebold 1968 in Burghardt 1970, 8).
7. The "transformation of information from one carrier to another" (Batteau 1968 in Burghardt 1970, 7).
8. One individual organism's using a specialized stimulus-producing mechanism to emit a chemical, or other physical signal, that influences the behavior of a receiving organism which has specialized receptors and responds in a specific manner; communication involves signal, symbolic, and language levels (Tavolga 1968 in Sebeok 1968a, 271–275).
9. "Any stimulus arising from one animal and eliciting a response in another" (Scott 1968 in Burghardt 1970, 8).
10. A process in which information is exchanged between individual animals to the mutual adaptive advantage of both (Klopfer and Hatch 1968 in Lewis and Gower 1980, 1).
11. One individual organism's producing a "signal that, when responded to by another organism, confers some advantage (or the statistical probability of it) to the signaler or his group" (Burghardt 1970, 16).
12. "Action on the part of one organism (or cell) that alters the probability pattern of behavior in another organism (or cell) in an adaptive fashion" (Wilson 1975, 10, 176, 581).
    Note: This definition has been criticized because it would define such interspecific interactions, such as those of predators and prey, as communicative (Lewis and Gower 1980, 1).
13. Animal "signal transmission and information exchange as a prerequisite of social behaviour" (Heymer 1977, 192).
    syn. understanding
14. "The transmission of a signal or signals between two or more organisms where selection has favoured both the production and reception of the signal(s)" (Lewis and Gower 1980, 2).
    Note: "It is impossible to achieve a definition, at once comprehensive and specific, of a term so widely used (and misused)" (Lewis and Gower 1980, 2).
15. "The imparting of information from one organism to another in a way that evokes a detectable response from the recipient at least some of the time" (Wittenberger 1981, 613).

syn. animal communication, biocommunication (Immelmann and Beer 1989, 51), true communication
cf. animal sounds, kinopsis, ²movement
Comment: Burghardt (1970) discusses problems with many definitions of "communication." McGregor (1997, 754) remarks that it is difficult to find two authors who use the same definition of communication.

chemical communication n. In many animal and some plant species: communication by means of chemical substances (e.g., pheromones, q.v.).

dance language, dance-language communication n. In Honey Bees: an individual's communication of distance and direction of an object (e.g., food or a nesting site) (Heymer 1977, 171) with use of signs (Immelmann and Beer 1989, 170).
cf. dance: bee dance; language
Comment: Many behaviorists do not consider this communication to be a language (Immelmann and Beer 1989, 170).

electrical communication, electrocommunication n. In some catfish species; electric fish, rays, sharks; the Common Eels: communication using low-frequency, feeble voltage gradients (Möhres 1957; Valone 1970; Black-Cleworth 1970; Kalmijn 1971; Bullock 1973 in Wilson 1975, 239–240; Immelmann and Beer 1989, 85).

flash communication n. Communication with light by fireflies and other animals (White 1983, 190).
cf. luminescence

instrumental communication See animal sounds: stridulation.

interspecific communication n. Communication between, or among, heterospecific animals (e.g., mutualists and songbirds) (Immelmann and Beer 1989, 51).

mass communication n. Transfer among groups of individuals of information of a kind that cannot be transmitted from a single individual to another; e.g., spatial organization of Army Ant raids, the regulation of numbers of worker ants on odor trails, and certain aspects of thermoregulation of a nest (Wilson 1962a; 1971a; 1975, 193, 588).

mechanical communication n. Intraspecific communication among woodpeckers using different pecking patterns as signals (Blume 1967 in Heymer 1977, 93).
cf. animal sounds: drumming

metacommunication n.
1. Communication about how further communication should be interpreted, e.g.,

primates' emitting a set of social messages that affect the way in which other social messages are interpreted (Ruesch and Bateson 1951 and Ruesch 1953 in Bateson 1955), or the word "advertisement" on an advertisement page in a magazine (C.K. Starr, personal communication).

2. An animal's provision of information about the frame of reference within which a subsequent message should be viewed (Lewis and Gower 1980, 5).

modulatory communication *n*. In ants: communication that influences a receiver's behavior by not forcing her into a narrowly defined behavioral channel, but by slightly shifting her probability of performing another behavioral act (Hölldobler and Wilson 1990, 640).

nonverbal communication *n*.

1. Human communication without using vocal words (McFarland 1985, 486).
2. In Humans of nonlinguistic means: blushing, crooning, crying, eyebrow flashing, facial expression, gesture, laughing, movement, posture, shouting, smiling, and touching that communicate without words (Immelmann and Beer 1989, 204).

cf. study of: kinesics
Comment: True nonverbal communication does not include systems such as gestural sign languages or whistling languages (Immelmann and Beer 1989, 204).

▸ **adaptor** *n*. An act of nonverbal communication that also has noncommunication function, *e.g.*, grooming, intention movements (Ekman and Friesen 1969 in McFarland 1985, 486).

▸ **affect gestures** *n*. Body movements that communicate emotion (McFarland 1985, 486).

▸ **emblem** *n*. An act of nonverbal communication with a verbal counterpart (*e.g.*, long-distance signaling movements, obscene gestures, or sign language) (Ekman and Friesen 1969 in McFarland 1985, 486).

▸ **illustrator** *n*. An act of nonverbal communication that illustrates points that are also being made verbally (Ekman and Friesen 1969 in McFarland 1985, 486).

▸ **regulator** *n*. A gestures that controls the flow of conversation between two persons (*e.g.*, head nodding, eye-contact movements, and various body-posture shifts) (Ekman and Friesen 1969 in McFarland 1985, 486).

sematectonic communication, stigmergic communication *n*. Communication by means of constructed objects, *e.g.*, sand pyramids of male Ghost Crabs and various portions of the nest structures of social insects (Wilson 1975, 186, 594).
syn. stigmergy, "incited to work," stigmergic communication (Grassé in Wilson 1975, 186)
cf. ³theory: stigmergy theory
[coined by Wilson 1975, according to Downing and Jeanne 1988, 1729; < Greek *sema*, sign, token + *tekton*, craftsman, builder]

trail communication *n*. Communication with a trail pheromone, *e.g.*, in ants (Wilson 1975, 55) and stingless bees (Lindauer and Kerr 1958, 1960; Nedel 1960).

vibrational communication *n*. Communication by substrate vibration, *e.g.*, in larvae of many butterfly species that are symbiotic with ants (DeVries 1990, 1104).

♦ **communication channel** *n*. A means of transmitting information (Wilson 1975, 231–241).

A Classification of Communication Channels

I. sound
II. chemical
   A. taste
   B. smell
   C. toxin
   D. vomeronasal
III. electricity
IV. surface wave
V. touch
VI. sight
   A. movement
   B. posture
   C. color (ultraviolet to red)
   D. poster

♦ **communication system** See ¹system: communication system.

♦ **communication triad** *n*. Communication that includes a signal and a female's and male's response toward it [coined by Morris and Ryan 1996, 1017].
Comments: After studying communication triads in Swordtail Fish, Morris and Ryan (1996) suggest that male response to a male signal (= vertical body bars) coevolved more closely than female response to this signal.

♦ ¹**community** *n*.

1. "A body of men living in the same locality" (*Oxford English Dictionary* 1972, entries from 1600).
2. "The people of a country (or district) as a whole; the general body to which all alike belong, the public" (*Oxford English Dictionary* 1972, entries from 1789).

◆ ²**community** *n.*

1. "An assemblage of populations living in a prescribed area" (Krebs 1985, 458); heterospecific populations that co-inhabit that same site (*e.g.*, a dung pat, forest leaf litter, or a pond) (Taylor 1992, 52).

   Note: Some ecologists claim that a community has one or more of these attributes: species co-occurrence, conspecific-group reoccurrence, and homeostasis or self-regulation (Krebs 1985, 458).

2. Any group of heterospecific organisms that co-occur in the same habitat, or area, and interact through trophic and spatial relationships; typically characterized by reference to one or more dominant species (Lincoln et al. 1985).

3. A group of species that are frequently found together in a geological stratum (Lincoln et al. 1985).

4. A collection of species that occur in the same place at the same time (Fauth et al. 1996, 283).

syn. biological community, biotic community (Lincoln et al. 1985)

cf. biocenose; biome; biorealm; -bios; biota; biotope; -coenosis; -cole; ditch; flora; fauna; group: assemblage, ensemble; group: kingdom; guild; habitat; ¹-phile; ²succession; vegetation; waste ground

Comments: Odum (1969, 245) indicates that "community" is, and should remain, a broad term that may be used to designate natural assemblages of different sizes from the biota of a log to that of a vast forest. The concept of "community" is one of the more important ones in ecology (Odum 1969, 246). Communities are named and classified according to their: (1) major structural features such as dominant species, life forms, or indicators; (2) physical habitats; or (3) functional attributes such as the type of community metabolism (Odum 1969, 253). Functional attributes, rather than structural features, offer a better basis for comparison of all communities according to Odum (1969). Fauth et al. (1996, 285) demonstrate that authors of some more current ecology textbooks define communities by space, time, organism interactions, and taxa, including the first two, three, or all four subjects in their definitions.

With regard to classification of plant communities, for example, Küchler (1964) divides the vegetation of the conterminous U.S. into 116 named kinds; Brush et al. (1976) divide the vegetation of Maryland into tidal marsh and 15 associations; and Olson (1979) classifies the vegetation of New Jersey pine barrens into six major vegetational types, each with one to four divisions.

The organismic school holds that communities are integrated units with discrete boundaries, and the opposing individualistic school holds that communities are not integrated units but collections of heterospecific populations that require the same environmental conditions (Krebs 1985, 458). Data support the individualistic interpretation more than the organismic interpretation. Communities grade continuously in space and time, and species groups are not consistent from place to place. Community classification is for convenience and is not a description of the fundamental structure of nature.

Plant ecology now tends to use "community" descriptively; explanations, especially evolutionary ones, are now advanced in terms of populations or population variants (Taylor 1992, 54). Plant communities in South Africa are grassland, forest, fynbos, nama karoo, savanna, succulent karoo, and thicket (Becker 1997, 10). Many of the large number of kinds of communities that have been named are defined below.

acrophytia, acrophyta *pl. n.* "Plant communities of alpine regions" (Lincoln et al. 1985).

cf. ²community: acrophytia, coryphium, krummholz, orohylile community, orophytium, orothamnic community

actium *n.* A rocky-shore community (Lincoln et al. 1985).

cf. ²community: actium, petrochthium, phellium, promunturium, psamathium

aestatifruticeta *n.* A deciduous-bush community of an aestilignosa (Lincoln et al. 1985).

aestatisilvae *n.* A deciduous-woodland community of the aestilignosa (Lincoln et al. 1985).

cf. ²community: forest

aestidurilignosa See ²community: mixed forest.

aestilignosa *n.* Broadleaf-deciduous bush and woodland vegetation of temperate regions that experiences alternating periods of mild, damp and cold, dry climate (Lincoln et al. 1985).

cf. ²community: forest

agium, aigialium *n.* "A beach community" (Lincoln et al. 1985).

agrium *n.* A community on cultivated land or land subject to human activities (Lincoln et al. 1985).

syn. culture community (Lincoln et al. 1985)

aithallium *n.* An evergreen-thicket community (Lincoln et al. 1985).

cf. ²community: thicket

alsium *n.*

1. A grove community (Lincoln et al. 1985).

2. An area with trees and grasses (Lincoln et al. 1985).

cf. ²community: forest

amanthium, amathium *n.* A sandhill, or plain, community (Lincoln et al. 1985). *cf.* ²community: drimium, Jack Pine plain; habitat: prairie

ammochthium *n.* "A sandbank community" (Lincoln et al. 1985).

ancium *n.* A canyon-forest community (Lincoln et al. 1985). *cf.* ²community: forest

anemodium, anemium *n.* A blowout community (Lincoln et al. 1985).

aquiherbosa *pl. n.* The herbaceous communities of ponds and swamps, subdivided into emersiherbosa, sphagniherbosa, and submersiherbosa (Lincoln et al. 1985).

ardium *n.* A succession that follows irrigation (Lincoln et al. 1985).

aufwuchs See ²community: periphyton.

benthos *pl. n.*
1. The organisms that live on the sea bottom; distinguished from plankton, *q.v.* (Lincoln et al. 1985).
2. Organisms attached to or living on, in, or near a sea bed, river bed, or lake floor (Lincoln et al. 1985).

syn. benthon (Lincoln et al. 1985)
adj. -benthic, benthonic
Comment: Ali and Lord (1980) refer to "benthic invertebrates" which they collected from freshwater ponds.
[Greek *benthos*, depth of the sea]

SOME CLASSIFICATIONS OF BENTHOS

I. Classification by size
  A. macrobenthos
  B. meiobenthos
  C. microbenthos
II. Classification by habitat
  A. endobenthos
  B. epibenthos
  C. haptobenthos
  D. herpobenthos
  E. hyperbenthos (*syn.* suprabenthos)

▸ **endobenthos** *n.* Organisms living within the sediment on a sea bed or lake floor; contrasted with epibenthos, haptobenthos, herpobenthos, and hyperbenthos (Lincoln et al. 1985).
syn. infauna (Lincoln et al. 1985)
Comments: Endobenthos are more common in subtidal and deeper zones in marine habitats; some endobenthos make burrows or tubes (Allaby 1994).

▸ **epibenthos** *n.* Organisms living on the substratum surface of a sea bed or lake floor; contrasted with endobenthos, haptobenthos, herpobenthos, and hyperbenthos (Lincoln et al. 1985).

▸ **geobenthos** *n.* All terrestrial life (Lincoln et al. 1985).
cf. ²community: benthos: phytobenthos; habitat: geobenthos

▸ **haptobenthos** *n.* Benthos that live closely applied to, or growing on, submerged surfaces; contrasted with endobenthos, epibenthos, herpobenthos, and hyperbenthos (Lincoln et al. 1985).

▸ **herpobenthos** *n.* Benthos growing, or moving, through muddy sediments; contrasted with endobenthos, epibenthos, haptobenthos, and hyperbenthos (Lincoln et al.).

▸ **hyperbenthos, suprabenthos** *n.* Benthos that live above but close to a substratum; contrasted with endobenthos, epibenthos, haptobenthos, and herpobenthos (Lincoln et al. 1985).

▸ **macrobenthos** *n.* The larger benthos that are over 1 mm long; contrasted with meiobenthos and microbenthos (Lincoln et al. 1985).

▸ **meiobenthos** *n.* Benthos that pass through a 1-mm mesh sieve but are retained by a 0.1-mm mesh; contrasted with macrobenthos and microbenthos (Lincoln et al. 1985).

▸ **microbenthos** *n.* Microscopic benthos that are less than 0.1 mm long; contrasted with macrobenthos and meiobenthos (Lincoln et al. 1985).

▸ **nectobenthos, nektobenthos** *n.* Organisms typically associated with benthos that swim actively in a water column at certain periods (Lincoln et al. 1985).

▸ **phytobenthos** *n.* "A bottom-living plant community" (Lincoln et al. 1985).
cf. habitat: phytobenthos
syn. phytobenthon (Lincoln et al. 1985)

▸ **rhizobenthos** *n.* The organisms that are rooted in a substratum (Lincoln et al. 1985).

biome See biome.

biotic community See ²community.

biotic succession See ²succession.

carpolochmium See ²community: lochmium

chalicium, chalicodium *n.* A plant community of gravel-slide habitats (Lincoln et al. 1985).

chaparral *n.* Fire-adapted trees (oaks, pines, etc.) and shrubs (*Ceanothus*, Manzanita, Scot's Broom, etc.) usually growing in dense thickets in California and southwest and northern Mexico (Kricher and Morrison 1993, 336–344).
syn. little oak (Spurr and Barnes 1980, 593)
Comments: Fires frequently burn chaparral areas (Walker 1990, 250).
[Mexican Spanish (more specifically, Basque) *chappara*, evergreen scrub oak (Walker 1990, 250)]

cheradium *n.* "A sandbar community" (Lincoln et al. 1985).

chersium *n.* A dry-wasteland plant community (Lincoln et al. 1985).

chionium *n.* A plant community that is associated with snow (Lincoln et al. 1985).

chledium *n.* "A plant community on waste ground or rubbish heaps" (Lincoln et al. 1985).

climax *n.* A more or less stable biotic community that is in equilibrium with existing environmental conditions and represents the terminal stage of an ecological succession (Lincoln et al. 1985).
syn. formation (sometimes) (Lincoln et al. 1985)

▸ **biotic climax** *n.* A plant community that is maintained at climax by a biotic factor (*e.g.*, grazing) (Lincoln et al. 1985).

▸ **climatic climax, regional climax** *n.* A more or less stable community in which the major factors affecting vegetation are climatic, typical of zonal soils (Lincoln et al. 1985).

▸ **consociation** *n.* A small climax community, or vegetative unit, dominated by one particular species (the physiognomic dominant) that has the life form characteristic of its association (Lincoln et al. 1985).

▸ **disclimax** *n.*
1. "A disturbed climax" (Lincoln et al. 1985).
2. A succession that is maintained below climax by rapid expansion of introduced species, climatic instability, fire, grazing, human activities, or a combination of these things (Lincoln et al. 1985).

▸ **edaphic climax, soil climax** *n.* A more or less stable community whose structure and composition are determined largely by soil, or substratum, properties (Lincoln et al. 1985).

▸ **panclimax, panformation** *n.* Two, or more, related climaxes having similar dominants, life forms, and climatic factors (Lincoln et al. 1985).

▸ **paraclimax** See ²community: climax: proclimax, subclimax.

▸ **physiographic climax** *n.* A climax that is determined largely by topographic and edaphic factors (Lincoln et al. 1985).

▸ **plagioclimax** *n.* A climax formed following the deflection of an ecological succession by human activity or influence (Lincoln et al. 1985).

▸ **polyclimax** *n.* A compound climax that occurs within a particular climatic region and is comprised of a number of local edaphic climaxes (Lincoln et al. 1985).

▸ **postclimax** *n.* A climax that ordinarily occurs in more mesic conditions than present ones and considered to be a remnant of an earlier climatic climax (Lincoln et al. 1985).
cf. ²community: climax: preclimax

▸ **preclimax** *n.* A climax occurring in more xeric conditions than present ones and often considered to be a seral stage preceding attainment of a climatic climax (Lincoln et al. 1985).
cf. ²community: climax: postclimax

▸ **prochosium** *n.* An ecological succession that occurs on an alluvial soil (Lincoln et al. 1985).

▸ **proclimax, paraclimax, subclimax** *n.*
1. A stage in an ecological succession that replaces a typical climatic climax (Lincoln et al. 1985).
2. A stable plant community that reflects favorable local variations in edaphic, or biotic, factors (Lincoln et al. 1985).
See ²community: climax: subclimax.

▸ **regional climax** See ²community: climax: climatic climax.

▸ **serclimax** *n.* A stable plant community that persists at a stage before a subclimax (Lincoln et al. 1985).

▸ **socies** *n.* A sere with at least one dominant species (Lincoln et al. 1985).

▸ **soil climax** See ²community: climax: edaphic climax.

▸ **stasium** *n.* "A stagnant-water community" (Lincoln et al. 1985).

▸ **subclimax, paraclimax** *n.* A state in a succession that precedes a final, or climax, community and is prevented from attaining its full development by one or more edaphic, or biotic, factors (arresting factors) (Lincoln et al. 1985).
See ²community: climax: proclimax.

▸ **zootic climax** *n.* A stable climax whose structure and composition depend upon continued grazing or other stress from animal use (Lincoln et al. 1985).

closed community, closed association *n.* A plant community that covers the entire ground area of a habitat and effectively prevents establishment of other species (Lincoln et al. 1985).

clusium, clysium *n.* A succession on flooded soil (Lincoln et al. 1985).

coenosium *n.* "A plant community" (Lincoln et al. 1985).

coniferous forest *n.* A forest dominated by coniferous trees (Voss 1972, 18).
syn. conophorium (Lincoln et al. 1985)
cf. ²community: forest.

conifruticeta *n.* "A forest dominated by coniferous shrubs" (Lincoln et al. 1985).
cf. ²community: forest

conilignosa *n.* "A forest dominated by trees and shrubs with needle-like foliage" (Lincoln et al. 1985).
cf. ²community: forest

conisilva See ²community: coniferous forest.

conodrymium *n.* An evergreen-plant community (Lincoln et al. 1985).

conophorium See ²community: coniferous forest.

consocies *n.*
1. A successional stage (Lincoln et al. 1985).
2. Part of an association that lacks one or more of its dominant species (Lincoln et al. 1985).
3. "A group of mores" (Lincoln et al. 1985).
cf. sere

coryphium *n.* An alpine-meadow community (Lincoln et al. 1985).
cf. ²community: acrophytia, meadow

cremnium, cremnion *n.* A plant community that is associated with cliffs (Lincoln et al. 1985).

crenium *n.* A plant community that is associated with spring water (Lincoln et al. 1985).

crymium, crymophium *n.* A plant community of a polar region (Lincoln et al. 1985).

culture community See ²community: agrium.

dendrium *n.* "An orchard community" (Lincoln et al. 1985).
cf. ²community: orchard

deciduous forest, hardwoods, hemlock-hardwoods *n.* A Michigan forest dominated by American Beech, Sugar Maple, and associated trees and with some coniferous trees, especially Eastern Hemlock (Voss 1972, 18).
cf. ²community: forest, ptenophyllium

dendrium See ²community: orchard.

drimium *n.* An alkaline-plain, or salt-basin, plant community (Lincoln et al. 1985).
cf. ²community: amanthium

driodium See ²community: lochmium: carpolochmium.

drymium *n.* "A woody plant community" (Lincoln et al. 1985).

dyticon *n.* "An ooze-inhabiting community" (Lincoln et al. 1985).

ecological succession See ²succession.

elfin forest *n.* A forest of higher elevations in warm moist regions that has stunted trees with abundant epiphytes (Lincoln et al. 1985).
See ²community: krummholz.
cf. ²community: forest

eluvium *n.* A sand-dune community (Lincoln et al. 1985).

eremium *n.* "A desert community" (Lincoln et al. 1985).

forest *n.* A large group of trees and associated organisms.
syn. hylium, hylion, wood, woods (Michaelis 1963; Voss 1972, 18)
See habitat: forest, grove.
adj. forest, sylvan
cf. ²community: aestilignosa, aestatisilvae, alsium, ancium, bog forest, coniferous forest, conifruticeta, conilignosa, deciduous forest, elfin forest, helodium, helohylium, hylodium, Jack Pine plain, jungle, krummholz, mangrove, mixed forest, northern-evergreen forest, northern hardwoods, orchard, orgadium, orohylile community, pleuriilignosa, ptenophyllium, rain forest, shale barren: shale-barren woodland, swamp, therodrymium, thicket, tropodrymium, woodland, wood, xerohylium
Comments: Forest researchers have devised many classifications of forest types but no universally adopted classification (Spurr and Barnes 1980, 573). Forest names derive from things including dominant-tree species, historic names, latitude, moisture availability, percentage of kind of trees (broadleaf and conifers), physiognomy (tree density), specific geographic location, structure (evergreen, deciduous, thorn), and combinations of these characters. Physiognomic-structural classifications are useful for analysis at different levels. There is and can be no compartmentalized classification scheme of vegetation types that can set up mutually exclusive divisions because each geographical area has its own peculiar history and each species and genus has its own particular distribution.

Some phytogeographers divide Earth's forests into four kingdoms subdivided into provinces, sectors, and districts (Spurr and Barnes 1980, 574). Hammond, Inc. (1978) divides world forests into two types and six subtypes. Spurr and Barnes (1980, 579) divide world forests into two types, six subtypes, and 20 subsubtypes. Kricher and Morrison (1988, 1993) divide forests of the U.S. into nine main types and about 50 subtypes. Strausbaugh and Core (1978) divide forests of West Virginia into three main types and six subtypes. Brown and Brown (1972) divide Maryland forests into three main types, three subtypes, and two subsubtypes. The Council of Europe Commission of the European Communities (1987) describes many kinds of European forests.
[Old French < Medieval Latin (*silva*) *foresta*, an unenclosed (wood) < Latin *foris*, outside]

fynbos *n.* A plant community in South Africa (Becker 1997, 10–11).

a – c

syn. Cape Floral Kingdom (Becker 1997, 10; Boroughs 1999, 17)
Comments: The fynbos is about 3 million years old and comprises about 17,000 square miles of the Southern and Southwestern Cape in the southernmost tip of Africa (Becker 1997, 10–11). It occurs on dry, poor soil and has frequent fires. The fynbos is a world hotspot of biodiversity with its 7000 plant species and the highest known concentration of plant species per unit area, with about 1300 per 4000 square miles. The South American rain forest is the runner up to this record, with 400 plant species per 4000 square miles.
[Dutch *fynbos*, fine bush, fine-leaved plants]

geobenthos See ²community: benthos: geobenthos.

geosere *n.* A series of climax formations in a particular area through geological time (Lincoln et al. 1985).
cf. succession: geosere

halodrymium See ²community: mangrove.

heath barren *n.* A more or less dense growth of ericaceous shrubs and other low growing plants on a mountain top in West Virginia (Strausbaugh and Core 1978, x).
cf. ²community: Jack-Pine plain, orothamnic community, xeropoium; habitat: shale barren

heliophytium *n.* A plant community that thrives in full sunlight (Lincoln et al. 1985).

helium *n.* "A marsh community" (Lincoln et al. 1985).
cf. habitat: marsh

helodium, helorgadium *n.* An open-swampy-woodland community (Lincoln et al. 1985).

helodrium *n.* A swamp-thicket community (Lincoln et al. 1985).

helohylium *n.* A swamp-forest community (Lincoln et al. 1985).
cf. ²community: forest

helolochmium *n.* A meadow-thicket community (Lincoln et al. 1985).
cf. ²community: meadow, thicket

helorgadium See ²community: helodium.

herbosa *pl. n.* "Communities of grasses and herbs" (Lincoln et al. 1985).
syn. herbaceous vegetation (Lincoln et al. 1985)

hydrophytium *n.* A bog, or swamp, plant community (Lincoln et al. 1985).
cf. habitat: bog

hydrothermal-vent community *n.* An oceanic community characterized by large clams, mussels, and vestimentiferan worms (= tube worms) that live in hydrothermal-vent areas and thrive on chemosynthetic microbial production (Grassle 1985).

cf. ²community: thalassium; ¹-phile: hyperthermophile
Comments: The first hydrothermal-vent community was discovered in 1977 off the Galápagos Islands (Broad 1993a, C1, C12). The vents are fissures where the oceanic tectonic plates are moving apart. Other members of this community include several species of corals, galatheid crustaceans, hyperthermophiles, palm worms, sea anemones, and spider crabs.

hygrodrymium See ²community: rain forest.

hygrophorbium *n.* A moist-pasture, fenland, or low-moorland community (Lincoln et al. 1985).

hygrosphagnium *n.* A high-moor community (Lincoln et al. 1985).

hylium, hylion See ²community: forest.

hylodium *n.* A dry-open-woodland community (Lincoln et al. 1985).
cf. ²community: forest

infauna See ²community: benthos: endobenthos.

isocoenoses *pl. n.* Communities of plants and animals that have similar life forms (Lincoln et al. 1985).

isocommunity *n.* "A community that shows a close structural and ecology resemblance to another natural community" (Lincoln et al. 1985).

Jack Pine plain, Jack Pine barren *n.* A Michigan forest dominated by sparse to sometimes dense stands of Jack Pine and sometimes Trembling Aspen (Voss 1972, 19).
cf. ²community: amanthium, forest, heath barren
Comment: Sites of Jack Pine plains were once forested by Red Pine, White Pine, or both (Voss 1972, 19).

jungle *n.*
1. A dense tropical thicket of high grass, reeds, vines, brush, trees, or a combination of these plants, "choked with undergrowth and usually inhabited by wild animals" (Michaelis 1963).
2. "Any similar tangled growth" (Michaelis 1963).
3. Dense seral vegetation especially characteristic of tropics with high precipitation (Lincoln et al. 1985).
cf. ²community: forest, thicket
[Hindustani *jangal*, desert, forest < Sanskrit *jangala*, dry, desert]

krummholz *n.* A discontinuous belt of stunted forest, or scrub, typical of windswept alpine regions close to tree lines (Lincoln et al. 1985).
cf. ²community: acrophytia, forest
Comments: Krummholz (*syn.* elfinwood) can also mean an individual dwarfed tree

(Walker 1990, 137). Cold, hail, ice, lightning, snow, and wind cause krummholz trees. Electrical discharges strike trees causing them to exhibit odd shapes.
[German *Krummholz*, *krumm*, stunted; *holz*, wood; dwarf mountain pine, knee timber, scrub]

limnium *n.* "A lake community" (Lincoln et al. 1985).

lochmium *n.* "A thicket community" (Lincoln et al. 1985).
cf. thicket
▸ **carpolochmium, driodium, lochmodium, xerodrymium** *n.* A dry-thicket community (Lincoln et al. 1985).
cf. ²community: thicket

lochmodium See ²community: lochmium: carpolochmium.

lophium *n.* "A hill-top community" (Lincoln et al. 1985).

major community *n.* A community that is of sufficient size and completeness of organization that it is relatively independent of adjoining communities and needs to receive only sun energy from the outside; contrasted with minor community (Odum 1969, 245).

mangrove, mangle, halodrymium *n.* A tidal-salt-marsh community dominated by trees and shrubs, especially *Rhizophora*, many of which produce adventitious aerial roots (Lincoln et al. 1985).
cf. ²community: forest

maquis *n.* Shrubby, mostly evergreen plants in the Mediterranean region, known as cover for bandits, game, etc. (Michaelis 1963).
See habitat: maquis.
syn. *macchia* (Michaelis 1963); matorral (Council of Europe Commission of the European Communities 1987, 57)
[French < Italian *macchia*, thicket, originally spot < Latin *macula*, spot]

mictium *n.* A heterogenous assemblage of species that often occurs in a transition zone between different vegetation stands and often results from the interspersion of two or more seres and has the dominant species of each equally represented (Lincoln et al. 1985).

minor community *n.* A community that is more or less dependent on neighboring communities for energy; contrasted with major community (Odum 1969, 245).

mixed forest, mixed woods *n.* A forest with substantial proportions of both coniferous and nonconiferous trees (Voss 1972, 18).
syn. aestidurilignosa
cf. ²community: forest

namatium *n.* A brook community; a stream community (Lincoln et al. 1985).

northern-evergreen forest *n.* A high-elevation Appalachian forest that is dominated by Balsam Fir and Red Spruce and is generally at an elevation above 3000 feet in West Virginia (Strausbaugh and Core 1978, viii).
cf. ²community: forest

northern hardwoods *n.* A forest dominated by American Beech, Sugar Maple, and Yellow Birch, present in a zone from about 2500 to 3000 feet of elevation in West Virginia (Strausbaugh and Core 1978, ix).
cf. ²community: forest

ochthium *n.* "A bank community" (Lincoln et al. 1985).

olisthium, olisthion *n.* "A landslip community" (Lincoln et al. 1985).

open community *n.* "A community that is open to invasion by immigrant species" (Lincoln et al. 1985).

opium *n.* "A parasite community" (Lincoln et al. 1985).

orchard *n.* A growth of dwarf trees that are twisted, gnarled, or both, that merges into northern hardwoods and grassy balds in West Virginia (Strausbaugh and Core 1978, x).
cf. ²community: dendrium; habitat: orchard

orgadium *n.* An open-woodland community (Lincoln et al. 1985).
cf. ²community: forest

orohylile community *n.* An alpine-, or subalpine-, forest community or habitat (Lincoln et al. 1985).
cf. ²community: acrophytia, forest

orophytium *n.* A subalpine-plant community (Lincoln et al. 1985).
cf. ²community: acrophytia

orothamnic community *n.* An alpine-heath community (Lincoln et al. 1985).
cf. ²community: acrophytia, heath barren

oryctocoenosis, oryctocoenose *n.*
1. An outcrop community or assemblage of fossils (Lincoln et al. 1985).
2. The part of a thanatocoenosis preserved as fossils (Lincoln et al. 1985).

oxodium, oxodic community, oxodion, oxylium *n.*
1. A humus-marsh community (Lincoln et al. 1985).
2. A peat-bog community (Lincoln et al. 1985).
cf. habitat: bog, marsh

paleobiocoenosis, palaeobiocoenosis, palaeocoenosis, paleocoenosis *n.* A fossil-organism assemblage that once existed as an integrated community (Lincoln et al. 1985).

paleocommunity, palaeocommunity *n.* "All preservable taxa comprising a fossil community" (Lincoln et al. 1985).

paleothanatocoenosis, palaeothana-tocoenosis *n.* A group of organisms buried together in the geological past (Lincoln et al. 1985).

panclimax See ²community: climax.

panformation See ²community: climax: panclimax.

paraclimax See ²community: climax: subclimax.

pelagium *n.* "A sea-surface community" (Lincoln et al. 1985).
cf. ²community: pontium

pelochthium *n.* "A mud-bank community" (Lincoln et al. 1985).

periphyton, aufwuchs *n.* A community of organisms and associated detritus adhering to and forming a surface coating on stones, plants, and other submerged objects (Lincoln et al. 1985).
[German *Aufwuchs*, growth]

petrochthium *n.* "A rock-bank community" (Lincoln et al. 1985).
cf. ²community: actium

petrodium *n.* A boulder-field, or ravine, community (Lincoln et al. 1985).
cf. ²community: actium

phellium *n.* "A rock-field community" (Lincoln et al. 1985).
cf. ²community: actium

phretium *n.* A water-tank community (Lincoln et al. 1985).

physically controlled community *n.* A community formed under high physiological stress and controlled largely by physical fluctuations and unfavorable physical conditions (Lincoln et al. 1985).
cf. hypothesis: stability-time hypothesis

phytobenthos, phytobenthon See ²community: benthos: phytobenthos.

phytocoenosis *n.*
1. A plant community (spelled "phytocenose" in Küchler 1964, 1).
2. All plants of a given habitat or community (Lincoln et al. 1985).
3. In plant communities, a syntaxon comprised of one or more synusiae (Lincoln et al. 1985).
Comments: Küchler 1964 identified 113 plant communities in the conterminous U.S.
[Greek *phyton*, plant + *koinos*, common]

phytoedaphon *n.* "Soil flora" (Lincoln et al. 1985).

phytoma, phytome *n., pl.* **phytomata** A plant community (Lincoln et al. 1985).

pioneer *n.* The first community to colonize, or recolonize, a barren or disturbed area, thereby starting a new succession (Lincoln et al. 1985).
cf. ²species: pioneer

plagioclimax See ²community: climax: plagioclimax.

pleuriilignosa, pluviilignosa *n.* A rain forest or rainbush community (Lincoln et al. 1985).
cf. ²community: forest, rain forest

pluviilignosa See ²community: pleuriilignosa.

pnoium *n.* "A ecological succession on an aeolian soil" (Lincoln et al. 1985).

poium *n.* "A meadow community" (Lincoln et al. 1985).
cf. ²community: coryphium, helolochmium, pratum, telmatium

polyclimax See ²community: climax: polyclimax.

pontium *n.* A deep-sea community (Lincoln et al. 1985).
cf. community: pelagium

postclimax See ²community: climax: postclimax.

potamium *n.* "A river community" (Lincoln et al. 1985).

pratum *n.* A meadow, or grassy-field, community (Lincoln et al. 1985).
cf. ²community: poium; habitat: field, meadow

preclimax See ²community: climax: preclimax.

proclimax See ²community: climax: proclimax.

promunturium *n.* A rocky-seashore community (Lincoln et al. 1985).
cf. ²community: actium

psamathium *n.* "A strandline community of a sandy seashore" (Lincoln et al. 1985).
cf. ²community: actium

psilium *n.* A prairie, or savanna, community (Lincoln et al. 1985).
cf. habitat: prairie, savanna

ptenophyllium *n.* A deciduous-forest community (Lincoln et al. 1985).

ptenothalium *n.* A deciduous-thicket community (Lincoln et al. 1985).
cf. ²community: thicket

pyrium *n.* "An ecological succession following fire or burn-off" (Lincoln et al. 1985).

rain forest, rainforest *n.* A forest with luxuriant tree growth and very high annual precipitation.
▶ **temperate rain forest** *n.* A rain forest in a temperate region; contrasted with tropical rain forest (Smith 1996, 265).
Comments: Rainfall in a temperate rain forest can exceed 600 centimeters per year (Smith 1996, 265). Coniferous trees dominate temperate rain forests. They include forests along the Pacific Coast from southern Alaska and Canada (Tongass National Forest) through Washington State (The Olympic National Forest), and Redwood forests in California.

▶ **tropical rain forest** *n.* A rain forest in a tropical region; contrasted with temperate rain forest, tropical dry forest, and tropical season forest (Smith 1996, 281). *syn.* hygrodrymium (Lincoln et al. 1985), rainforest (Golden Turtle Press 1997), rain forest (Morris 1982)

cf. ²community: forest, pleuriilignosa

Comments: Tropical rain forests have at least 245 cm of rain per year (Lincoln et al. 1985), come in about 40 different types, cover less than 2% of Earth's surface, may have up to 50% of Earth's species (as many as 30 million species), and are disappearing at about 150 acres per minute due to human activities (Smith 1996, 281; Golden Turtle Press 1997, 2). The Amazon rain forest is the largest one. Types of tropical rain forests include equatorial-lowland tropical rain forest, evergreen mountain forest, evergreen savanna forest, monsoon forest, and tropical evergreen-alluvial forest (Smith 1996, 281). Tropical rain forests are biologically characterized by being evergreen, containing many tree species that grow up to 30 meters tall, and having very high species numbers, numerous epiphytic plant species, and thick-stemmed woody vines (lianas) (Allaby 1994). These forests commonly have four intergrading plant strata. [coined by A.F.W. Schimper, German botanist, in 1898 as *tropische Regenwald* (Allaby 1994; Smith 1996, 281)]

repium, rhepium *n.* An ecological succession on soil that is subject to subsidence (Lincoln et al. 1985).

rhepium See ²community: repium.

rhoium *n.* "A creek community" (Lincoln et al. 1985).

rhyacium *n.* "A torrent community" (Lincoln et al. 1985).

rhysium, rhysion *n.* "An ecological succession following volcanic activity" (Lincoln et al. 1985).

saprium *n.* "A saprophyte community" (Lincoln et al. 1985).

sciophytium *n.* "A shade community" (Lincoln et al. 1985).

seminatural community *n.* A community that is modified by human activities (Lincoln et al. 1985).

serclimax See ²community: climax: serclimax.

soil climax See ²community: climax: edaphic climax.

soil climax See ²community: climax: edaphic climax.

sphagniherbosa *n.* A plant community dominated by *Sphagnum* growing on peat (Lincoln et al. 1985).

cf. community: aquiherbosa

steganochamaephytium *n.* "A community of dwarf shrubs under trees" (Lincoln et al. 1985).

sterrhium *n.* "A moorland community" (Lincoln et al. 1985).

subclimax See ²community: climax: subclimax.

suprapelos *n.* Aquatic organisms that swim above soft mud that are dependent upon this substratum as a food source (Lincoln et al. 1985).

suprapsammon *n.* Aquatic organisms that swim above sand and are dependent upon this substratum as a food source (Lincoln et al. 1985).

synusia *n.* A community of species with a similar life form and ecological requirements (Lincoln et al. 1985).

See habitat: synusia.

syrtidium *n.* "A dry sand-bar community" (Lincoln et al. 1985).

taphrium *n.* "A ditch community" (Lincoln et al. 1985).

telmatium, telmathium *n.* A wet-meadow, or marsh, community (Lincoln et al. 1985).

cf. habitat: meadow

thalassium *n.* "A marine-plant community" (Lincoln et al. 1985).

cf. ²community: hydrothermal-vent community, habitat: sandy beach

thermium *n.* A hot-spring community (Lincoln et al. 1985).

therodrymium *n.* "A leafy-forest community" (Lincoln et al. 1985).

cf. ²community: forest

thicket *n.* A dense growth of shrubs, underwood, and small trees; a place where low trees, or bushes, grow thickly together (*Oxford English Dictionary* 1972, entries from *ca.* 1000).

See habitat: thicket.

syn. brake (*Oxford English Dictionary* 1972), coppice (Michaelis 1963)

cf. ²community: aithallium, forest, heloloch-mium, jungle, lochmium, ptenothalium

tiphium *n.* A pond, or pool, community (Lincoln et al. 1985).

tropodrymium *n.* A savanna-forest community (Lincoln et al. 1985).

cf. ²community: forest

varzea *n.* An area of rain forest that is partially submerged for much of the year (*e.g.*, along the Amazon River) (Alexander 1994, 606).

Comments: The varzea between the Japurá and Solimões Rivers in Brazil is inundated up to 10 meters or more for 7 months (Alexander 1994). Seed- and fruit-eating fish invade the varzea during the wet season.

wood *n.* A collection of trees growing more or less thickly together (especially naturally as distinguished from a plantation), of considerable extent, usually larger than a grove or copse (but including them), and smaller than a forest; a piece of ground covered with trees with or without undergrowth (*Oxford English Dictionary* 1972, entries from 825).
See ²community: forest, wood.
syn. grove, woods (Michaelis 1963)
cf. ²community: forest
[Old English *wudu, widu*]

xerodrymium See ²community: lochmium: carpolochmium.

xerohylium *n.* "A dry-forest community" (Lincoln et al. 1985).
cf. ²community: forest

xeropoium *n.* "A heathland community" (Lincoln et al. 1985).
cf. ²community: heath barren

xerothamnium *n.* "A spiny-shrub community" (Lincoln et al. 1985).

xylium *n.* "A woodland community" (Lincoln et al. 1985).

♦ **community psychology** See study of: psychology: community psychology.

♦ **community structure** *n.* The physical and biological structures (combinations of related parts) of a community; the physical structure includes abiotic and physical manifestations of the biotic parts of a community (*e.g.*, rocks, soil, water, tree trunks, logs, leaves, shrub and herb thickets); the biological structure includes species composition and abundance, temporal changes in these factors, and interspecific relationships (Krebs 1985, 462).

♦ **companion** *n.*
1. An organism, or object, in an animal's environment that has specific significance in some, or all, functional systems to the animal (Lorenz 1935 in Heymer 1977, 107).
2. A social partner in a behavior nexus (a term used in earlier ethological literature, Immelmann and Beer 1989, 52).
cf. animal: companion animal

child companion See ²companion: parent companion.

flight companion *n.* A Human that an animal (*e.g.*, Hooded Crow) flies near when the Human moves from place to place (Lorenz 1935 in Heymer 1977, 107).

parent companion *n.* A Human that an animal (*e.g.*, a Jackdaw [bird] or child companion) imprints on when it is young (Lorenz 1935 in Heymer 1977, 107).

sexual companion *n.* A Human upon which a nonhuman animal (*e.g.*, Jackdaw) is imprinted as a sexual partner (Lorenz 1935 in Heymer 1977, 107).

sibling companion *n.* A young animal that is raised with young of another species (Lorenz 1935 in Heymer 1977, 107).

social companion *n.* A Human with which a nonhuman animal (*e.g.*, Hooded Crow) interacts (Lorenz 1935 in Heymer 1977, 107).

♦ **companion species** See ²species: companion species.

♦ **company** See ²group: company.

♦ **comparative ethology** See study of: ³ethology: comparative ethology.

♦ **comparative experiment** See experiment: manipulative experiment.

♦ **comparative method** See method: comparative method.

♦ **comparative psychology** See study of: psychology: comparative psychology.

♦ **compartmentalization** *n.* "The extent to which the subgroups of a society operate as discrete units" (Wilson 1975, 17, 581).

♦ **compass orientation** See taxis: menotaxis.

♦ **compensatory mechanism** See mechanism: compensatory mechanism.

♦ **compensatory mitigation** *n.* A person's attempt to restore, or create, an ecosystem(s) so that a natural one can be ceded to development; contrasted with ecological restoration (Roberts 1993a, 1891).
Comment: Many ecologists do not favor this practice because it usually does not work (Roberts 1993a, 1890; 1993b, 1891).

♦ **compensatory-mortality hypothesis** See hypothesis: hypotheses of species richness: compensatory-mortality hypothesis.

♦ **competition** *n.*
1. "The relation between plants occupying the same area and dependent upon the same supply of physical factors" (Clements 1905, 317, in McIntosh 1992, 62).
Note: This might be the earliest published, formal definition of competition in the biological sense (McIntosh 1992, 62). Clements added elements to this definition in the first extended survey of competition: "when the immediate supply of a single necessary factor falls below the combined demand" of the competitors (Clements et al. 1929, 317 in McIntosh 1992, 62).
2. "A more or less active demand in excess of the immediate supply of material or condition on the part of two or more organisms" (Clements and Shelford 1939, 139 in McIntosh 1992, 63).
3. "The demand, typically at the same time, of more than one organism for the same resources of the environment in excess of immediate supply" (Crombie 1947 in McIntosh 1992, 63).

4. An organism's interfering with a hetero-specific organism whether or not both use the same resources (Birch 1957, 16).
5. Predation within the context of one organism's interfering with another (Birch 1957, 16).
6. Utilization of common resources that are short in supply by conspecific or heterospecific animals or, if resources are not in short supply, harming of one animal by another while seeking resources (Birch 1957 in McIntosh 1992, 64).
   Note: In evolutionary and ecological literature, definitions of competition are converging on the ideas in this definition (McIntosh 1992, 67).
7. "The endeavour of two (or more) animals to gain the same particular thing, or to gain the measure each wants from the supply of a thing when that supply is not sufficient for both (or all)" (Milne 1961, 60 in McIntosh 1992, 65).
8. An interaction between two species in which an increase in population size in either one harms the other; *e.g.*, species A and B fight, A reduces B's food supply, or A, by its own losses, increases B's predators (MacArthur 1972, 256, who wrote that no precise definition of competition can be related and that current precise definitions are premature, in McIntosh 1992, 65).
9. The active demand by two or more individual conspecific, or heterospecific, organisms at the same trophic level for a common resource, or requirement, that is actually or potentially limiting (Wilson 1975, 243).
   Notes: Competition may harm some contestants (Birch 1957, 16; Wilson 1975, 243); resources can include mating partners or rank positions (Immelmann and Beer 1989, 54).
10. Organisms' vying for the same resource(s), which may, or may not, involve direct confrontation (Wilson 1975, 581).
Comments: "Competition" has been described at the individual, population, and species levels. McIntosh (1992) describes Charles R. Darwin's conception of competition although he did not explicitly define it. Birch (1957) reviewed the biological meanings of "competition" and concluded that it had lost its usefulness as a scientific term due to misunderstanding and confusion regarding it (McIntosh 1992, 64). Andrewartha (1961, 174 in McIntosh 1992, 65) proposed that "competition" be added to "a list of abandoned carcasses." McIntosh (1992, 65–66) explains that it is difficult to define "competition" because it is applied to di-verse categories. He includes more definitions of competition and writes, "Although considerable agreement on the definition of competition has been achieved, dispute about its mechanisms and significance for ecological and evolutionary dynamics of population continues." Keller (1992a, 71–73) explains that confusion about "competition" occurs for two main reasons: (1) "competition" is a colloquial as well as a technical word, and as the latter it carries colloquial connotations in the minds of some biologists; and (2) as a technical term, "competition" can denote an operation of comparison between organisms, or species, but only in a biologist's mind. Further, "competition" is sometimes extended to mean "natural selection," although these are separate concepts.

SOME CLASSIFICATIONS OF COMPETITION

I. Classification with regard to species and lower levels
   A. interspecific competition
   B. intraspecific
      1. organismal competition
      2. sexual competition
         (*syn.* mate competition)
      3. sperm competition
II. Classification with regard to behavior
   A. contest competition
      (*syn.* active competition, interference competition)
   B. diffuse competition
      (*syn.* consumptive competition, exploitation, exploitative competition, passive competition, scramble competition)
III. Classification with regard to resource type
   A. mate competition (*syn.* sexual competition)
   B. resource competition
   C. territory competition (which can include III.A and III.B)
IV. Classification with regard to mechanism[a]
   A. apparent competition
   B. chemical competition
   C. consumptive competition
   D. encounter competition
   E. overgrowth competition
   F. preemptive competition
   G. territorial competition

[a] Schoener (1983, in Feller 1987, 1466).

active competition See competition: contest competition.

apparent competition *n.* Competition that occurs when two prey species are limited in numbers by a common predator;

if one prey species increases in numbers, the predator increases in numbers, and this leads in turn to an increase in the mortality rate of the other prey species (Holt 1977 in McIntosh 1992, 67).

contest competition *n.*

1. Competition for a resource in which the winner physically deters another organism from obtaining part, or all, of a fought-over resource (referred to as "interference" by Crombie 1947 in McIntosh 1992, 63; Nicholson 1954 in Wilson 1975, 85, 268).
2. Competition for a resource that is not limited (Lincoln et al. 1985).

syn. active competition (McIntosh 1992, 67), encounter competition (Feller 1987, 1466), interference competition (Park 1954, 178–181 in McIntosh 1992, 64; Feller 1987, 1466)

cf. competition: diffuse competition

diffuse competition *n.*

1. Competition for a resource in which the winner uses all of the resource that it needs without specific behavioral responses to other competitors that may be in the same area (Nicholson 1954 in Wilson 1975, 85).
2. Competition "in which individuals compete by exploiting a resource faster or more efficiently than their competitors without resort to aggression" (Wittenberger 1981, 621).
3. Competition for a resource that is equally partitioned between competitors so that some, or all, of them obtain an insufficient amount of the resource for survival or reproduction (Lincoln et al. 1985).
4. Competition for an actually, or potentially, limited resource (Lincoln et al. 1985).
5. Simultaneous interspecific competition among numerous species, each with a small degree of niche overlap with other species (Lincoln et al. 1985).

syn. consumptive competition, exploitation (Park 1954, 178–181 in McIntosh 1992, 64), exploitative competition (Feller 1987, 1466), passive competition (McIntosh 1992, 67), scramble competition (Nicholson 1954 in McIntosh 1992, 62)

cf. competition: contest competition

direct competition *n.* One individual's, or species', exclusion of another individual, or species, from a resource by direct aggressive behavior, toxins, or both (Lincoln et al. 1985).

cf. competition: contest competition, indirect competition

exploitation, exploitative competition See competition: diffuse competition.

indirect competition *n.* Competition for a resource in which one individual's, or species', exploitation reduces its availability to other individuals or species (Lincoln et al. 1985).

cf. competition: direct competition

intense competition *n.* Simultaneous interspecific competition between a few species each with considerable niche overlap with one another (Lincoln et al. 1985).

cf. competition: diffuse competition

interference competition See competition: contest competition.

interspecific competition *n.* Competition between members of different species for a limited resource.

cf. competition: intraspecific competition (Wittenberger 1981, 617)

intraspecific competition *n.* Competition between, or among, individuals of the same species.

local-mate competition (LMC) *n.* The relative amount of intrasexual competition for mates in a localized area; *e.g.,* in some chalcidoid wasps, individual hosts are isolated (or localized) and new males, which emerge before females, compete for particular groups of females (Hamilton 1967).

mate competition *n.* In many organism species: "Competition among members of one sex for sexual access to members of the opposite sex" (Wittenberger 1981, 618).

syn. sexual competition

motivational competition *n.* Operationally, a behavioral changeover due to competition, "recognized when a change in the level of causal factors for a second-in-priority activity results in an alternation in the temporal position of the occurrence of this activity" (McFarland 1969 in McFarland 1985, 389).

resource competition *n.*

1. Competition for commodities besides mates (Wilson 1975, 243).
2. Competition of two or more populations, or species, for a limited resource (Lincoln et al. 1985).

cf. competition: sexual competition

scramble competition See competition: diffuse competition.

sexual competition See competition: mate competition.

sperm competition *n.* For example, in some mammal and spider-mite species; the Dung Fly and Japanese Beetle: competition among the sperm of more than one male to fertilize a female's eggs which may occur in an external medium following gamete release or within a female's reproductive tract (Wittenberger 1981, 621–622; Schwagmeyer and Foltz 1990, 156).

cf. sperm displacement, sperm precedence

status competition *n.* For example, in young male Humans: competition for a position in a dominance hierarchy (Wilson and Daly 1985, 59).

♦ **competitive exclusion, competitive-exclusion principle** See principle: Gause's principle.

♦ **competitive release** *n.* An organism's expansion of its habitat range and food preferences due to a reduction in intensity of interspecific competition (Lincoln et al. 1985).

♦ **competitive selection** See selection (table).

♦ **competitive species** See ²species: c-selected species.

♦ **compilospecies** See ²species: compilospecies.

♦ **complemental male** See male: parasitic male.

♦ **complementariness** See biological complementariness.

♦ **complementarity** *n.* The distinctness, or dissimilarity, of biota in more than one location (Colwell and Coddington 1994, 103). *Comments:* Complementarity covers distinctness of species composition over a broad spectrum of environmental scales, from small-scale ecological differences (*e.g.,* differences between mite faunas on trunks vs. leaves of the same tree species) through climatically distinct sites in different biomes (Colwell and Coddington 1994, 103). There are scores of published indices that can be used to measure complementarity. A simple one is C_{jk} (Marczewski-Steinhaus distance, M-S distance) which varies from 0 through 1, or 0 through 100%. Colwell and Coddington (1994, 103, 104) use "complementarity" instead of the statistical equivalents of dissimilarity, distance, or distinctness to capture the sense that complementary biotas form parts of a whole, a sense that distinctness, or its equivalent, does not convey. They prefer a single, broad term to a series of more specific, scale- or gradient-dependent concepts.

♦ **complementary gene, complementary factor** See gene: complementary gene.

♦ **complementation** *n.*
1. Genetic masking of deleterious recessive mutations resulting from recombination of homologous alleles in sexual reproduction involving outcrossing (Bernstein et al. 1985, 1278–1279).
2. Cooperative interaction of mutant genes in double mutants, giving a phenotype closer to the wild type than any one of the mutant genes could produce alone (Lincoln et al. 1985).

intercistronic complementation *n.* Complementation (def. 2) of a pair of mutant alleles belonging to different functional units of a chromosome (cistrons) (Lincoln et al. 1985).

intracistronic complementation *n.* Complementation (def. 2) of a pair of mutant alleles within the same functional unit of a chromosome (cistron) (Lincoln et al. 1985).

♦ **complete record** See sampling technique: all-occurrences sampling.

♦ **completely randomized block** See experimental design: completely randomized block.

♦ **complex character** See character: complex character.

♦ **complex digital organism** See program: digital organism: complex digital organism.

♦ **complex system** See ¹system: complex system.

♦ **"complexity-level hypothesis"** See hypothesis: hypotheses regarding the origin of the Cambrian explosion: "complexity-level hypothesis."

♦ **component part of a behavioral act** See behavior, classification of.

♦ **components of fitness** *n.* An organism's "devices of adaptation, together with genetic stability in constant environments" and an organism's ability to generate new genotypes that cope with fluctuating environments (Thoday 1953 in Wilson 1975, 67).

♦ **composite signal** See signal: composite signal.

♦ **composite species** See ²species: composite species.

♦ **composition, principle of** See principle: principle of composition.

♦ **compositional method** See emergence.

♦ **compound nest** See nest: compound nest.

♦ **compression hypothesis of interspecific competition** See hypothesis: compression hypothesis of interspecific competition.

♦ **compromise behavior** See behavior: compromise behavior.

♦ **compromise of evolution** *n.* The concept that the phenotype of an individual organism is a combination of traits that vary in the amount of fine-tuning that has been acquired through evolution (Endler 1978 in Mayr 1982, 589).
syn. optimization process of evolution (Mayr 1982, 589)
Comment: Compromise of evolution occurs because its genotype "is a compromise

between various selection pressures, some of which may be opposed to each other, as, for instance, sexual selection and crypsis" (Endler 1978 in Mayr 1982, 589).

♦ **computational error** See error: observer error.

♦ **computer-compatible data logger** *n.* A machine that transfers behavioral records to a computer which can store and analyze the data (Lehner 1979, 162–169).

♦ **computer vision** See study of: computer vision.

♦ **¹conceive** *v.t.* In Mammals: to become pregnant (with) (Michaelis 1963).

♦ **²conceive** *v.t.* In Humans:
1. To form a concept in one's mind; develop mentally; think of; imagine (*Oxford English Dictionary* 1972, entries from 1340; Michaelis 1963).
2. "To understand; grasp" (*Oxford English Dictionary* 1972, entries from 1362; Michaelis 1963).
3. "To believe or suppose; think" (Michaelis 1963).
4. *v.i.* "To form a mental image; think: with *of*" (Michaelis 1963).
adj. conceivable
adv. conceivably
n. conceivability, conceivableness, conceiver
cf. idea
Comment: We do not know if any nonhuman animals conceive as we do.
[Old French *conceveir* < Latin *concipere* < *com-*, thoroughly + *capere*, to grasp, take]

♦ **concept** *n.*
1. A person's "mental image; especially, a generalized idea formed by combining the elements of a class into the notion of one object;" a thought; an opinion (*Oxford English Dictionary* 1972, entries from 1556; Michaelis 1963).
2. A response, verbal or motor, that is under discriminative control of a broad class of environmental objects or events (Verplanck 1957).
Note: The members of a class may differ from one another in all respects other than a single quantifiable property. Most concepts are statements that refer to a common property (*e.g.*, blue, square, velocity, beauty, length) (Verplanck 1957).
cf. conceive, idea, law, principle, rule

♦ **concept formation** *n.* An animal's ability to from a concept, *q.v.*; *e.g.*, a Civet Cat's and Rhesus Macaque's concept of same and different or a Chimpanzee's and Rhesus Macaque's correlating different kinds of tokens with different amounts of foods (Immelmann and Beer 1989, 54–55).
cf. counting, language

Comment: Concept formation is the highest intellectual achievement of animals (Immelmann and Beer 1989, 54–55).

♦ **concept learning** See learning: concept learning.

♦ **conceptive** *adj.* "Capable of being fertilized" (Lincoln et al. 1985).

♦ **concerted evolution** See ²evolution: concerted evolution.

♦ **conchology** See study of: conchology.

♦ **conchometry** See study of: conchometry.

♦ **concolorous** *adj.* Referring to an object that has uniform coloration (Lincoln et al. 1985).
cf. coloration

♦ **concomitant mortality** See mortality: concomitant mortality.

♦ **Concorde Effect, Concorde Fallacy, Concorde Strategy** See effect: Concorde Effect.

♦ **concrete homotypy** See homotypy: concrete homotypy.

♦ **condensation** *n.* In chemistry, two molecule's joining to form one larger molecule and simultaneously producing one molecule of water (Curtis 1983, 1092).
Comment: Many kinds of monomers condense when they form polymers.

♦ **conditional character** See character (comments).

♦ **conditional phenotype, condition-sensitive phenotype** See -type, type: phenotype: conditional phenotype.

♦ **conditional response, conditioned response (CR)** See response: conditional response.

♦ **conditional stimulus, conditioned stimulus (CS)** See ²stimulus: conditional stimulus.

♦ **conditional strategy** See strategy: conditional strategy.

♦ **conditioned-emotional response** See response: conditioned emotional response.

♦ **conditioned learning** See learning: conditioned learning.

♦ **conditioned-suppression technique** See method: conditioned-suppression technique.

♦ **conditioning** See learning: conditioning.

♦ **confabulation** *n.*
1. In psychology, a person's compensating for memory loss, or impairment, by fabrication of details (Michaelis 1963).
2. A person's finding a reason or purpose for his own behavior or external events, *e.g.*, in split-brain patients who justify why they make incorrect choices in experimental situations (Gould 1982, 492–498).
syn. justification (Gould 1982, 492–498)

♦ **confidence interval** *n.* A quantitative range between and including an upper and lower confidence limit that has a particular probability of containing the mean (whose value is fixed) in question (Sokal and Rohlf 1969, 139).
cf. confidence limit
Comment: For example, a confidence interval that has a 0.95 probability of containing the mean in question is bounded by the 95% confidence limits (Sokal and Rohlf 1969, 139). The upper limit is $\bar{x} + 1.96\ \sigma_{\bar{x}}$ and the lower limit is $\bar{x} - 1.96\ \sigma_{\bar{x}}$, with $\sigma_{\bar{x}} = 1$ standard deviation of the mean.

♦ **confidence-interval statistics** See study of: statistics: confidence-interval statistics.

♦ **confidence limit** *n.* A value that is larger or smaller than a mean and that is the border value of a confidence interval, *q.v.* (Sokal and Rohlf 1969, 139).
cf. confidence interval

♦ **configurational causes** See cause: configurational causes.

♦ **configurational stimulus** See ²stimulus: configurational stimulus.

♦ **confirmatory characters** See character: confirmatory characters.

♦ **¹conflict** *n.*
1. An armed human encounter; a fight, battle; fighting, battling (*Oxford English Dictionary* 1972, entries from *ca.* 1440).
2. In many animal species: individuals' clashing over particular resources (Immelmann and Beer 1989, 57).
v.i. conflict
syn. squabble (used by Mock and Forbes 1992, 409, to avoid the ambiguity of "conflict")
cf. war

social conflict *n.*
1. In many animal species: tension generated between, or among, socially interacting animals when they attempt to carry out mutually incompatible goal-directed behavior (Hand 1986, 204–205).
2. Operationally, "when the behaviors of two (or more) individuals indicate that their motivation priorities are incompatible; they seek the same thing, or different things, and both cannot be satisfied;" *e.g.*, when two pair-bonded birds each try to incubate their eggs simultaneously (Tinbergen 1960, 135, in Hand 1986, 201–204).
[coined by Hand 1986, 203]

♦ **²conflict** *n.*
1. Two tendencies (each of an incompatible behavior type) being simultaneously aroused in an animal; *e.g.*, a bird's simultaneous tendency to continue its feeding or follow its flock when its flock is leaving (Hinde 1970, 361).

Note: "Conflict" is used to refer to incompatible response probabilities, not to a hypothetical, or physiological, state (Hinde 1970, 361).
2. An animal's state when two of its incompatible behaviors are aroused equally, *e.g.*, attacking and fleeing at an animal's own territory boundary (Immelmann and Beer 1989, 57).
syn. motivational conflict (Hinde 1970, 361)
cf. behavior: conflict behavior; motivation: mixed motivation

approach-approach conflict *n.* An animal's simultaneous behavioral tendencies to move toward two different objects that are some distance apart (Hinde 1970, 361–362); *e.g.*, a positively phototactic insect's moving toward, but between, two equally bright lights.

approach-avoidance conflict *n.* An animal's simultaneous behavioral tendencies of moving toward and away from an object, *e.g.*, a male three-spined stickleback's moving both toward and away from a rival male near its territorial boundary (Hinde 1970, 362; McFarland 1985, 399).
cf. behavior: ambivalent movements and postures, compromise behavior

avoidance-avoidance conflict *n.* An animal's simultaneous behavioral tendencies to move away from two different objects (Hinde 1970, 362); *e.g.*, a negatively phototactic fly larva's moving away from both equally bright lights that are on each side of it.

motivational conflict See ²conflict.

♦ **³conflict** *n.* An evolutionary, or genetical, competition between two or more genes within an organism or between two or more conspecific organisms that have nonidentical genetic interests (Brockmann and Grafen 1989, 232).
syn. evolutionary conflict (Brockmann and Grafen 1989, 232)

ecological conflict *n.* A behavioral conflict between, or among, conspecific animals, viewed from an evolutionary, or ultimate, perspective (Selander 1972 in Hand 1986).

intergenerational conflict *n.* For example, in gulls: conflict between a chick that leaves its parents due to inadequate parental care and an unrelated adult that resists adopting the chick (Pierotti and Murphy 1987, 435).

mate conflict *n.* In haplodiploid arthropods: a mother's and father's conflict over the sex of their offspring; *i.e.*, a mother would favor 50% sons and daughters, and a father would favor production of daughters rather than sons (Brockmann and Grafen 1989, 232).
cf. ³conflict: reproductive conflict

parent-offspring conflict (POC) *n.*

1. For example, in some mammal species: a parent's tendency to stop investing resources in an offspring while, at the same time, this offspring continues to try to obtain these resources from its parent (Trivers 1974, 249; Immelmann and Beer 1989, 215–216).
2. The concept that parents and offspring have divergent fitness interests; this divergence creates "disagreements" between these parties over the allocation of parental investment; individual offspring "value" themselves more highly than their siblings (as reflected by coefficients of relatedness), but parents value each offspring equally, and this results in offspring that theoretically should behave more selfishly than their parents "desire" (Mock and Forbes 1992, 409).

Notes: If a parent-offspring conflict exists in a species, both parties affect each other in ways that change their fitnesses. Most studies of avian brood reduction are open to multiple interpretations and thus do not definitely support this concept; some plant species (*e.g., Dalbergia sissoo*) have siblicide, but it is not known if this is parental-offspring conflict (Mock and Forbes 1992, 409, 412).

cf. ³conflict: queen-worker conflict, weaning conflict

Comment: Parent-offspring conflict, queen-worker conflict, and weaning conflict are analyzed with the theory of parent-offspring conflict which has not been rigorously tested in field studies (Mock and Forbes 1992, 409).

queen-worker conflict *n.* In haplodiploid social insects: a queen's and one of her worker's being in competition because the worker could hypothetically maximize her inclusive fitness by channeling 1/4 of her colony's resources into male reproductives and 3/4 of its resources into female reproductives; however, the queen could maximize her inclusive fitness by channeling 1/2 of her colony's resources into reproductive males and 1/2 into reproductive females (Trivers and Hare 1976; Fisher 1987a, 1026).

Comment: Queen-worker conflict is supported by empirical studies (Mock and Forbes 1992, 412–413).

reproductive conflict *n.* In many animal species: a conspecific conflict between males and females, viewed from an evolutionary (or ultimate) perspective (Trivers 1972 and Emlen 1982 in Hand 1986); *e.g.,* a mother's favoring obtaining as much paternal care for her offspring as she can from their father and the father's favoring

seeking additional mates and, thus, decreasing his paternal care.

syn. sexual conflict (Davies 1989, 226)

cf. ³conflict: mate conflict

sexual conflict See ³conflict: reproductive conflict.

weaning conflict *n.* For example, in some mammal species: a parent's tendency to attempt to wean its offspring sooner than the offspring is willing to be weaned (Trivers 1974, 251; Immelmann and Beer 1989, 215–216).

cf. ³conflict: parent-offspring conflict; investment: parental investment

Comment: The offspring might increase its fitness if it weans later, but its parent might lower its own fitness because it invests in this offspring rather than in a new one (Trivers 1974, 251). Most studies of mammalian weaning are open to multiple interpretations and thus do not definitely support this concept as it relates to parent-offspring conflict (def. 2) (Mock and Forbes 1992, 409).

♦ **conflict behavior** See behavior: conflict behavior.

♦ **conflict block** *v.t.* To show an opposing behavior instead of a usual behavior toward a stimulus (Blume 1967 in Heymer 1977, 45).

♦ **conflict hypothesis** See hypothesis: conflict hypothesis.

♦ **conflict of interests** *n.* A behavioral conflict viewed from an evolutionary (or ultimate) perspective (Maynard Smith 1982, 123, in Hand 1986).

♦ **conflict theory of display** See ³theory: conflict theory of display.

♦ **conformer** *n.* An animal with certain aspects of its bodily condition that tend to be similar to its environmental conditions, *e.g.,* its body temperature's being similar to environmental temperature (McFarland 1985, 284).

cf. regulator

♦ **confusion effect** See effect: confusion effect.

♦ **congener** *n.*

1. A member of the same genus (Lincoln et al. 1985).
2. A fellow member of the same genus, class, family, or kind (Michaelis 1963).

♦ **congenetic** *adj.* "Having the same origin" (Lincoln et al. 1985).

cf. -genetic

♦ **congenital** See -genous: geneogenous.

♦ **congested** *adj.* "Crowded together; packed close together" (Lincoln et al. 1985).

♦ **conglobation** *n.* For example, in Armadillos, Chrysidid Wasps; some woodlice and millipede species: an individual's rolling up its body (Lincoln et al. 1985).

v.t. conglobate, conglobe

♦ **congregate** *v.t.* To collect together, or assemble, into a group (Lincoln et al. 1985).

♦ **congregation** See ²group: congregation.

♦ **conidiun** *n., pl.* **conidia** In many fungus species: a nonsexually produced propagative cell, or spore, borne upon special branches of a hypha (Michaelis 1963).
syn. conidiospore ((Michaelis 1963)
[New Latin < Greek *konis*, dust]

♦ **coniferous forest** See ²community: coniferous forest.

♦ **conifruticeta** See ²community: conifruticeta.

♦ **conilignosa** See ²community: conilignosa.

♦ **coniophile** See ¹-phile: coniophile.

♦ **conisilvae** See ²community: conisilvae.

♦ **conjugation** *n.*
  1. Reproductive union, or fusion, of two (apparently) similar cells of a less derived plant or animal (*Oxford English Dictionary* 1972, entries from 1843).
  2. Union of gametes, nuclei, cells, or individuals (Lincoln et al. 1985).
  3. Sexual reproduction in which two cells unite but have only limited genetic exchange (Lincoln et al. 1985).
  4. Direct transfer of DNA from one prokaryote to another (Campbell 1996, 504).
  syn. zygosis (Lincoln et al. 1985)
  cf. transduction, ¹transformation

♦ **conjunct mimicry system** See ¹system: mimicry system: conjunct mimicry system.

♦ **conlocal** *adj.* Referring to a local character species whose geographical range coincides with that of a vegetation unit (Lincoln et al. 1985).

♦ **connascent** *adj.* Referring to organisms that are born in the same clutch (Lincoln et al. 1985).

♦ **connectedness** *n.*
  1. "The number and direction of communication links within and between societies" (Wilson 1975, 581).
  2. "The number and frequency of interactions between the member species of a community" (Lincoln et al. 1985).

♦ **connections, principle of** See principle: principle of connections.

♦ **connubial display** See display: connubial display.

♦ **conodrymium** See ²community: conodrymium.

♦ **conophorium** See ²community: coniferous forest.

♦ **conophorophile** See ¹-phile: conophorophile.

♦ **consanguineous** *adj.* Referring to individuals that share a common recent ancestor (Lincoln et al. 1985).

♦ **consanguinity** *n.* "Relationship by descent from a common ancestor" (Lincoln et al. 1985).

♦ **conscious awareness** See awareness: conscious awareness.

♦ **conscious behavior** See behavior: conscious behavior.

♦ **¹consciousness** *n.* In psychoanalysis, human conscious (Michaelis 1963).

♦ **²consciousness** *n.*
  1. In Humans and possibly some other animals: "the presence of mental images, and their use by an animal to regulate its behavior;" an individual's "ability to think about objects and events, whether or not they are part of the immediate situation" (Griffin 1976, 5).
  2. In animals: an individual's "immediate awareness of things, events, and relations" (Immelmann and Beer 1989, 57).

♦ **³consciousness** *n.*
  1. A person's awareness of himself and surroundings (*Oxford English Dictionary* 1971, entries from 1678; Michaelis 1963).
  2. A person's state, or fact, of being mentally conscious or aware of anything (*Oxford English Dictionary* 1971, entries from 1746).
  3. A person's mental and emotional experience and awareness of another individual or a group: mob *consciousness* (*Oxford English Dictionary* 1971, entries from 1837; Michaelis 1963).
  4. A person's "awareness of some object, influence, etc.; a feeling or conviction" (Michaelis 1963).
  5. A person's attribute with many elements such as self-awareness (knowledge of oneself as distinct from other selves), anticipation of the future, ability to manipulate abstract ideas, and ability to pay and switch attention (Hubbard 1975 in Dawkins 1980, 24).
  6. In Humans and nonhuman animals: awareness "of what one is doing or intending to do, having a purpose and intention in one's actions" (Griffin 1982 in McFarland 1985, 526).
  7. A person's special kind of self-awareness, which is not simple awareness of parts of his body or of processes occurring within his brain; "a propositional awareness that it is *I* who am feeling or thinking, *I* am the animal aware of the circumstances" (McFarland 1985, 523).
  Note: Bateson (1991, 831) uses the phrase "full-blown human-style reflective consciousness" which relates to this definition.
  See ³consciousness: self-consciousness.
  syn. awareness (Michaelis 1963, Griffin 1976), cognizance, intention (according to some workers by not others)

cf. awareness: conscious awareness, self-awareness; cognition; ³consciousness: self-consciousness, true consciousness; intention
Comments: Mayr (1982, 74) says consciousness is impossible to define. Harnad (1985) states, "The only individual one can know is conscious is oneself." According to Watson (1913 in Crook 1983, 11), consciousness is a term irrelevant to nonhuman animals. Definitions and synonyms of consciousness and related terms are very confusing; it is controversial whether nonhuman animals have consciousness (McFarland 1985, 526). Although Griffin has defined consciousness in his writing (in 1987), I heard him say that we do not know enough about consciousness to define it. Despite all of the controversy regarding whether consciousness exists in nonhuman animals, some animal behaviorists now consider nonhuman animal consciousness to be a legitimate area of study. de Waal (1989, 87) suggests that the differences in consciousness between Chimpanzees and most other nonhuman primates seem to be gradual rather than radical. Kinds of consciousness include "objective consciousness" and "subjective consciousness" (Duval and Wickland 1972 in Crook 1983, 12).
[Latin *conscius,* knowing inwardly < *com-,* together + *scire,* to know]

group consciousness *n.* A person's feeling of belonging to a particular group, or organization; group consciousness in a well-integrated group increases aggressiveness toward other groups (Heymer 1977, 80).

self-consciousness *n.*
1. A person's state of being unduly conscious that he is observed by others; embarrassed by inability to forget himself; ill at ease (*Oxford English Dictionary* 1971, entries from 1712; Michaelis 1963).
2. A person's intentionality in performing a behavior (*Oxford English Dictionary* 1971, entries from 1860; Crook 1983, 12).
3. A person's consciousness of his existence (Michaelis 1963).
   syn. consciousness (Gallup 1982, 243)
4. In Humans, possibly in Chimpanzees and Orangutans: an individual's having a consciousness of some of its own past conscious states (Bunge 1984, 186-187; Lewin, 1987b).
cf. awareness: conscious awareness, self-awareness; consciousness; self.

true consciousness *n.* A self-scanning mechanism in an animal's central nervous system (Armstrong 1984).

♦ **consecutive hermaphrodism** See hermaphrodism: consecutive hermaphrodism.
♦ **consecutive sexuality** See hermaphrodite: consecutive hermaphrodite.
♦ **consensus sequence** *n.* A particular (idealized) nucleotide sequence in which each position represents the base most often found when many actual sequences of a given class of genetic elements are compared (Rieger et al. 1991).
Comments: Researchers have defined consensus sequences by comparison of those naturally occurring DNA or RNA sequences that they believe to encode a particular genetic function and by generation of many mutations of an individual genetic element (Rieger et al. 1991).
♦ **conservation** *n.*
1. A person's, or group's, preserving natural resources (such as endangered species, forests, fisheries) for economic use, recreational use, scientific use, scientific interest, aesthetic value, or for a combination of these things (Michaelis 1963).
2. An area preserved for conservation reasons (Michaelis 1963).
cf. preservation
[Latin *conservare*< *com-* thoroughly + *servare,* to keep, save]
♦ **consilience of inductions** *n.* A form of argumentation based on a group of compatible inductions that support a hypothesis (Whewell 1840).
Comment: This kind of reasoning is used to support Darwin's theory of organic evolution (Ruse 1984).
♦ **consilva** See ²community: coniferous forest.
♦ **consociation** See ²community: consociation.
♦ **consociation dominant** See ²species: physiognomic dominant.
♦ **consocie** See ²community: consocies.
♦ **consort pair** See ²pair: consort pair.
♦ **consort relationship** See relationship: consort relationship.
♦ **consortium** *n.* A group of heterospecific individuals, typically of different phyla, that live in close association (Lincoln et al. 1985). See symbiosis: consortism.
♦ **conspecific** *adj.* Referring to organisms that belong to the same species (Wilson 1975, 581).
ant. heterospecific
♦ **constant-environment principle** See principle: constant-environment principle.
♦ **constant extinction, law of** See hypothesis: red-queen hypothesis.
♦ **constant species** See ²species: constant species.

◆ **constipated** *adj.* Referring to objects that are crowded together (Lincoln et al. 1985). *cf.* dispersion; distribution: contagious distribution

◆ **constitutive hierarchy** See hierarchy: constitutive hierarchy.

◆ **constitutive reductionism** See reductionism: constitutive reductionism.

◆ **constraint** *n.*
1. "Confinement; restriction" (Michaelis 1963).
2. The act of constraining; the state of being constrained (Michaelis 1963).
3. Anything that constrains" (Michaelis 1963). [Old French *constreindre* < Latin *constringere* < *com-* together + *stringere*, to bind]

architectural constraint *n.* An organism's morphological restriction that was never an adaptation but rather a necessary consequence of materials and designs used to build its basic *Bauplan* (*e.g.*, the patterns of ridges on shells of some mollusc species) (Seilacher 1970 in Gould and Lewontin 1979, 595).

developmental constraint *n.* The resistance of an organism's developmental pathways to evolutionary change (Gould and Lewontin 1979, 594).
Comments: Developmental constraint is a form of phyletic constraint (Gould and Lewontin 1979, 594).

phyletic constraint See phylogenetic inertia.

◆ **consumer** *n.* An organism that feeds on another organism, or on existing organic matter; consumers include herbivores, carnivores, parasites, and all other saprotrophs and heterotrophs (Lincoln et al. 1985). *cf.* parasite, primary producer, -troph-, -vore

primary consumer *n.* "A heterotrophic organism that feeds directly on a primary producer" (Lincoln et al. 1985).

secondary consumer *n.* A heterotrophic organism that feeds directly on a primary consumer (Lincoln et al. 1985).

◆ **consumer psychology** See study of: psychology: consumer psychology.

◆ **consummatory act, consummatory behavior** See behavior: consummatory behavior.

◆ **contact avoidance** *n.*
1. For example, in Humans, the Orangutan, praying mantids: avoidance of close bodily contact and maintenance of individual distance (Heymer 1977, 102).
2. Human avoidance of social relations (Heymer 1977, 102).
cf. distance: individual distance

◆ **contact behavior, contactual behavior** See behavior: contact behavior.

◆ **contact greeting** See greeting: contact greeting.

◆ **contact pair** See ²pair: contact pair.

◆ **contact-seeking behavior** See behavior: contact-seeking behavior.

◆ **contact withdrawal** *n.* Removal of contact with conspecifics in early childhood that may lead to irreversible damage, *e.g.*, long hospitalization in Humans (hospitalism; Tinbergen and Tinbergen 1973 in Heymer 1977, 101), isolation in monkeys (Harlow and Harlow 1962a,b in Heymer 1977, 101).

◆ **contagious distribution** See distribution: contagious distribution.

◆ **contamination theory** See ³theory: Castle's theory.

◆ **contemporaneous polygamy** See mating system: polygamy: simultaneous polygamy.

◆ **contest** See game.

◆ **contest competition** See competition: contest competition.

◆ **context** *n.* The setting (including recipient's sex, other signals besides the one in question, etc.) in which a signal is emitted and received that can influence its effect on its recipient (Dewsbury 1978, 99; Immelmann and Beer 1989, 59). *cf.* ¹system: communication system; ³theory: communication theory; syntax

◆ **contiguity, law of** See law: law of contiguity.

◆ **contiguous-gene-deletion syndrome** See syndrome: continuous-gene-deletion syndrome.

◆ **continental drift** See hypothesis: continental drift.

◆ **"continental-drift hypothesis"** See hypothesis: hypotheses regarding the origin of the Cambrian explosion: "continental-drift hypothesis."

◆ **contingency** *n.* The simultaneous, or sequential, occurrence of two behaviors, a stimulus and behavior, or behavior and reinforcement that can give rise to persistent linkage (Immelmann and Beer 1989, 59). *cf.* learning: conditional learning, prenatal learning

◆ **continuity of germ plasm, theory of** See ³theory: theory of continuity of germ plasm.

◆ **continuous drift** See drift: genetic drift: continuous drift.

◆ **continuous modular society** See ²society: modular society.

◆ **continuous patrolling** See patrol.

◆ **continuous reinforcement schedule** See reinforcement schedule: continuous reinforcement schedule.

◆ **continuous variable** See variable: continuous variable.

◆ **continuous variation** See variation: continuous variation.

♦ **contranatant** *adj.* Referring to swimming, moving, or migrating against water current (Lincoln et al. 1985).
cf. denatant

♦ **contranym** *n.* A word or phrase that is an antonym of itself through distortion, if not misuse (Safire 1997, 22).
syn. antilogy, Janus word
Comments: "Arguably" is a contranym which can mean certainly or hardly. "True altruism" and "overdispersed distribution" (both in this book) each have opposite meanings, but it is not clear if they meet the criterion(a) of Safire's (1997) definition.
[Janus word, after Janus the Roman god of beginnings; Latin, *ianus*, gateway (Safire 1997, 22). Janus' face was on the gates of his temple in the Roman Forum. The gates were open in war time and closed in peace time, requiring a head with two faces. One sense of Janus-faced is deceitful and two-faced. Another is sensitive to dualities and polarities.]

♦ **contrast** See statistical test: multiple-comparisons test: planned-multiple-comparisons procedure: contrast.

♦ **¹control** *n.*
1. A parallel experiment, or test, carried out to provide a standard against which an experimental result can be evaluated (Lincoln et al. 1985); any treatment against which one or more other treatments is compared (Hurlbert 1984, 191).
2. In experimental design, all of the obligatory design features (*e.g.*, control treatments, randomized assignment of experimental units to treatments, and replication of treatments) that reduce or eliminate sources of confusion (*e.g.*, temporal change in measured units, procedure effects on these units, experimenter bias, and experimenter-generated variability) (Hurlbert 1984, 191).
3. In experimental design, one's regulation of the conditions under which an experiment is conducted, *e.g.*, homogenizing experimental units, making procedures as precise as possible, or, most often, regulating the physical environment in which one conducts an experiment (Hurlbert 1984, 191).

♦ **²control** *n.* For example, in some primate species: the intervention by one or more individuals to reduce or halt aggression between other members of the group (Wilson 1975, 581).

♦ **³control** *n.* Power to regular and direct; verify; rectify (Michaelis 1963).

bottom-up control *n.* Either a primary producer's or a limiting nutrients' regulating an ecosystem's higher food-web components; contrasted with top-down control (Pace et al. 1999, 483).
Comments: A problem with the ideas of bottom-up control and top-down control is that researchers often find them difficult to separate in practice (Pace et al. 1999, 483). Both phenomena occur in many ecosystems. Also, researcher often used these concepts in the context of equilibrium conditions, but most natural food webs are probably rarely near equilibrium.

top-down control *n.* An upper-level predator's regulating an ecosystem's lower food-web components; contrasted with bottom-up control (Pace et al. 1999, 483).

♦ **control theory** See study of: mathematics: cybernetics.

♦ **convenience polyandry** See mating system: polyandry: convenience polyandry.

♦ **Convention on International Trade in Endangered Species of Wild Fauna and Flora (CITES)** See organization: Convention on International Trade in Endangered Species of Wild Fauna and Flora (Appendix 2).

♦ **conventional behavior** See behavior: conventional behavior.

♦ **conventional signal** See signal: conventional signal.

♦ **conventional statistics** See study of: statistics: conventional statistics.

♦ **convergence, convergent evolution** See ²evolution: convergent evolution.

♦ **convulsionism** See ³theory: catastrophism.

♦ **Coolidge effect** See effect: Coolidge effect.

♦ **cooperation** *n.*
1. An organism's showing (sociobiological) altruism toward genetic relatives (given as a verb in Axelrod and Hamilton 1981, 1393).
2. An organism's helping conspecific nonrelatives (*e.g.*, with reciprocal altruism) (given as a verb in Axelrod and Hamilton 1981, 1395).
See symbiosis: mutualism, protocooperation.
v.i. cooperate
syn. protocooperation (Lincoln et al. 1985)
cf. altruism
Comment: "Symbiosis," "protocooperation," "obligacy," "facilitation," and "altruism" are used with partial to total overlap in meaning with cooperation and mutualism (Boucher 1992, 208).

weak cooperation *n.* An interaction between an actor and recipient in which neither has a fitness change (Gadagkar 1993, 233).

♦ **cooperative altruism** See altruism (def. 2).

♦ **cooperative breeding** See breeding: communal breeding.

♦ **cooperative foraging** See foraging: cooperative foraging.

♦ **co-option** *n.* The transfer of an adaptation from an earlier evolutionary situation to a later one, *e.g.*, the transfer of the sucking mouthparts of an insect family from seed plants that occurred before angiosperms to angiosperms when these plants originated on Earth (inferred meaning from Labandeira and Sepkoski 1993, 314).

♦ **coordination** *n.*
1. Interaction among units of a group such that the overall effort of the group is divided among the units without leadership being assumed by any one of them, *e.g.*, formation of a school of fish, trophallaxis between worker ants, and encirclement of prey by a pack of wolves (Wilson 1975, 10, 581).
   Note: Coordination may be influenced by a unit in a higher level of a social hierarchy, but such outside control is not essential (Wilson 1975, 10).
2. The special postural and locomotor behaviors that always occur together in a given context, *e.g.*, in mammalian elimination behavior (Heymer 1977, 79).
syn. basic-movement pattern
cf. taxis

♦ **coordination signal** See signal: coordination signal.

♦ **copal** *n.* A hard, transparent resin that is almost colorless to bright yellowish-brown and is exuded by several species of tropical trees (Michaelis 1963; Preece 1973, vol. 6, 458).
cf. fossil: amber
[Spanish < Nahuatl *copalli*, incense]

♦ **Cope's rule** See law: Cope's rule.

♦ **copia element** See transposable element: copia element.

♦ **coprobiont** See biont: coprobiont.

♦ **coprolite** See fossil: coprolite.

♦ **coprology** See study of: coprology.

♦ **coprophage, coprophagy, coprophagous** See -phage: coprophage.

♦ **coprophile** See ¹-phile: coprophile.

♦ **coprophilia** See -philia: coprophilia.

♦ **coprophilous** See -philous: coprophilous.

♦ **coprozoite** See -zoite: coprozoite.

♦ **copula** See *in copulo*.

♦ **copulation, copulatory behavior** *n.* In many animal species: an animal's engaging in sexual intercourse (Michaelis 1963); animals' "uniting" male and female sex organs which can lead to fertilization; in many bird species: pressing together cloacae during which a male ejaculates sperm into a female's reproductive tract (Heymer 1977, 104); in mammals, a male's mounting a female, usually from behind, intromission (insertion of his penis into her vagina), and

pelvic thrusting and ejaculation (forceful discharge of sperm into her vagina) (Immelmann and Beer 1989, 61).
v.i., v.t. copulate
syn. cloacal kiss (euphemistic, refers to birds), coition, coitus, courting (some authors, seemingly an euphemistic term for copulation), fucking (obscene), mating (used somewhat euphemistically for copulation in scientific literature), making love and pairing (somewhat euphemistic), pairing behavior, treading (in male birds)
cf. behavior: precopulatory behavior, postcopulatory behavior; courtship; *in copulo*; insemination; interest in participation with a copulating pair; mating; mounting; pair; sexual reproduction, sperm plug; stuck; tie
Comments: "Copulation," as "homosexual copulation," is also used when same-sexed individuals engage in "sexual intercourse." There are scores of euphemistic synonyms for "copulation," especially related to Humans (Spears 1981). "Mating" alone does not indicate whether insemination, conception, or both occur. I suggest that one should use "inseminative mating," "inseminative copulation," "conceptional mating," or "conceptual copulation" to clarify matters. "Successful mating" is a confusing term which could be considered synonymous with "inseminative, or conceptional, copulation."
[Latin *copulatus*, past participle of *copulare*, to fasten < *copula*, a link]

aerial copulation *n.* For example, in many insect, some swift (bird) species: copulation in flight (Rothgänger 1973 in Heymer 1977, 112).

bond-oriented copulation *n.* For example, in Humans: copulation that maintains pair bonds, rather than fertilizes ova (Heymer 1977, 104).

extrapair copulation *n.* For example, in many bird species, Humans: copulation involving a pair-bonded individual and an individual outside its pair bond (Wittenberger 1981, 615; McKenney et al. 1984; Westneat 1987, 865; Lank et al. 1989, 86).
syn. adultery (for Humans, sometimes for nonhumans)

forced copulation *n.*
1. An animal's forced coition or insemination (Gladstone 1979; Burns et al. 1980; Gowaty 1982).
2. A male animal's forced insemination of a female that simultaneously enhances his fitness and decreases her fitness (Thornhill 1980; Thornhill and Alcock, 1983).
3. A male animal's forced coition with a resisting female that is willing, or unwilling, to mate (Estep and Bruce 1981).

4. "A copulation forced onto an unwilling female by an aggressive or overpowering male" (Thornhill and Alcock 1983, 272, 275, 404).

cf. rape

Comments: Forced copulation occurs in animals, including some insect species and the Mallard Duck (Gladstone 1979; Burns et al. 1980), Human, and Orangutan (Galdikas 1995, 178). A male Orangutan forced copulation with a female Human (Galdikas 1995, 294). McKinney et al. (1984, 528) indicate that "forced copulation" is shorthand for "forced extrapair copulation" in waterfowl which was reported as early as 1912 by Huxley; female ducks can be exhausted, wounded, or even killed during forced copulation attempts. Many behaviorists prefer to use "forced copulation" rather than "rape" for nonhuman animals because "rape" has so many emotional connotations (Wittenberger 1981, 615; Gowaty 1982).

homosexual copulation *n.* For example, in an acanthocephalan worm, Humans, the Pukeko (bird), *Uta* lizards: "Copulation" between two male animals that may involve one male's ejaculating semen into an orifice of the other (Tinkle 1967 in Wilson 1975, 444; Jamieson and Craig 1987, 1251).

syn. homosexual mating

cf. behavior: homosexual behavior; copulation: pseudocopulation: homosexual pseudocopulation

pseudocopulation *n.*

1. In some bird, fish, and mammal species: Heterosexual and homosexual mounting with, or without, intromission, that is not used for inseminative purposes (Immelmann and Beer 1989, 237).
   syn. pseudocopulatory behavior
   cf. behavior: sexual behavior: pseudosexual behavior
   Note: Pseudocopulation in the parthenogenetic whiptail lizard *Cnemidophorus uniparens* facilitates ovarian growth, much like male courtship behavior stimulates ovarian growth in sexual species (Crews 1989 in Crews and Young 1991, 512).

2. In some species of bees and wasps: a male's making copulatory movements with a flower of some orchid species which resemble conspecific females, but do not offer nectar rewards (Correvon and Pouyanne 1916, Wolff 1950, Kullenberg 1973, etc. in Faegri and van der Pijl 1979, 74).
   Note: The orchids are pollinated by this process; this appears to be a case of a plant's parasitizing an animal.
   cf. mimicry: Pouyannian mimicry

▸ **heterosexual pseudocopulation** *n.* Pseudocopulation between a male and female that may serve to reinforce their pair bonding (Immelmann and Beer 1989, 237).

▸ **homosexual pseudocopulation** *n.*

1. Pseudocopulation between males (Wilson 1975, 22, 281; Immelmann and Beer 1989, 237).
   Notes: In all-male groups, homosexual pseudocopulation may serve to maintain males' reproductive readiness; in some primate species (baboons and macaques), it may be used to express rank among males (Maslow 1936, 1940 in Wilson 1975, 281; Carpenter 1942a in Wilson 1975, 22; Immelmann and Beer 1989, 237). Depending on the situation, dominant males either mount or are mounted by other males. Similar behavior occurs in domestic cattle (Thornhill and Alcock 1983).

2. A male insect's attaining a copulatory position on a conspecific male.
   Note: This occurs in a meloid beetle and may result from female incitation of male aggression (Thornhill and Alcock 1983, 406). When this occurs in Japanese beetles, it may be due to female sex attractant's being on males under laboratory conditions (Barrows and Gordh 1978).
   syn. homosexual mounting
   cf. perversum simplex; copulation: rage copulation

rage copulation *n.* For example, in some baboon species, the House Mouse, Humans, Timber Wolf: an animal's mounting a conspecific of the same sex, representing an aggressive dominance demonstration (Heymer 1977, 203).

cf. homosexual mounting

sneak copulation *n.*

1. For example, in some fish species, a carpenter bee, Humans: a copulation by a male that invades another's territory and copulates with a female in the territory (Barrows 1983, 806; Constanz 1984, 474; Parker 1984, 33).

2. In a bat species: a male's copulation with a sleeping female during her hibernation (Fenton 1984, 581).

♦ **copulator** *n.* An animal engaged in copulation.
syn. mater
cf. copulation

♦ **copulatory afterplay** See behavior: postcopulatory behavior.

♦ **copulatory organ** See organ: copulatory organ.

♦ **copulatory plug** See mating plug.

♦ **copulin** See chemical-releasing stimulus: semiochemical: pheromone: copulin.

♦ *copulo* See *in copulo*.

♦ **copy DNA** See nucleic acid: deoxyribonucleic acid: copy DNA.

♦ **coquette call** See call: coquette call.

♦ **coquette swimming** See swimming: coquette swimming.

♦ **coral bleaching** *n.*
1. A coral's expelling its symbiotic algae due to damage by light at higher than normal temperatures, leaving stark, white skeletons (Pockley 1999, 98; www.wri.org/indictrs/rrbleach, 1999, illustrations, references).
2. The result of the process of coral bleaching as defined in def. 1 above (inferred from Pockley 1999, 98).
cf. habitat: coral reef
Comments: Ove Hoegh-Guldberg predicts that, given global warming, coral bleaching will eliminate most coral reefs by 2100 (Pockley 1999, 98). His prediction is controversial.

♦ **coral reef** See habitat: coral reef.

♦ **corbicula** *n., pl.* **corbiculae** A pollen holding area on the hind tibia of a female apid bee (bumble bee, Honey Bee, orchid bee, or stingless bee) (Michener 1974, 10, illustration; 143; O'Toole and Raw 1991, 185).
syn. pollen basket (O'Toole and Raw 1991, 185)
cf. scopa
Comments: A corbicula is formed by the smooth, slightly concave outer surface of a tibia, fringed by long, stiff hairs, and is used to transport pollen (combined with nectar) and building materials (such as resin) to a bee's nest (Michener 1974, 10; O'Toole and Raw 1991, 185).

♦ **core area** *n.*
1. The area that an animal uses more frequently than other areas in its home range (Kaufmann 1962 in Wilson 1975, 256, 581).
Note: In mammals, conspecifics of the same sex usually exclude others from their core areas (Eisenberg 1981, 506).
2. The central location of a colony, or other central-place system, around an animal (or group of animals) in which it focuses its activities and movements (Wittenberger 1981, 614).
cf. ³range, territory

♦ **core temperature** See temperature: core temperature.

♦ **corkscrew swimming** See swimming: corkscrew swimming.

♦ **cormidium** *n.* In siphonophores: a group of zooids (individual members) of a colony that can separate from the remainder of the colony and live an independent existence (Wilson 1975, 581).
Note: The cormidium is the unit of organization between a zooid and a complete colony (Wilson 1975, 581).

♦ **Cornell-style Malaise trap** See trap: Malaise trap: Cornell-style Malaise trap.

♦ **cornicle** *n.* In some aphid species: an external abdominal organ that produces a quickly hardening wax that is used in defensive behavior (Wilson 1975, 356; Boyle and Barrows 1978, 452).

♦ **corpse** *n.* A dead embryo (*e.g.,* in a *Drosophila* fly) (White et al. 1994, 677).

♦ **corpus-allatum hormones** See hormone: juvenile hormones.

♦ **corpus callosum** See organ: brain: corpus callosum.

♦ **corpuscular theory of the gene** See ³theory: classical-gene-chromosome theory.

♦ **correction procedure** See procedure: correction procedure.

♦ **correlation** See statistical test: correlation.

♦ **correlation of parts, principle of** See principle: principle of correlation of parts.

♦ **Corrositex** See test: Corrositex.

♦ **corticole** See -cole: corticole.

♦ **corticospinal tract** See ²system: pyramidal system.

♦ **corticosteroid** See hormone: corticosteroid.

♦ **corticotropin** See hormone: adrenocorticotropic hormone.

♦ **corticotropin-releasing factor (CRF)** See hormone: corticotropin-releasing factor.

♦ **coryphile** See ¹-phile: coryphile.

♦ **coryphium** See ²community: coryphium.

♦ **coryphophile** See ¹-phile: coryphile.

♦ **cosexual** See hermaphrodite.

♦ **cosmic dust, interplanetary dust** *n.* Micrometeoroids present in the space between planets (Mitton 1993).
Comments: Cosmic dust might originate in collisions between asteroids in the asteroid belt and from the gradual breakup of comets (Mitton 1993). Cosmic dust that settled on Earth might have brought organic compounds important in the origin of life.

♦ **cosmic teleology** See teleology: cosmic teleology.

♦ **cosmology** See study of: philosophy: metaphysics: cosmology.

♦ **cosmopolitan** See ubiquitous.

♦ **Cosquer Grotto** See place: Chauvet Grotto (comments).

♦ **cost** *n.*
1. What a person gives, or surrenders, in order to acquire, produce, accomplish, or maintain something; the price he pays for something (*Oxford English Dictionary* 1972, entries from 1300).

2. A person's "loss; suffering; detriment" (Michaelis 1963).
3. Any consequence of a phenotypic trait that decreases an organism's Darwinian fitness, due to an energy expenditure, loss in viable offspring, etc. (Dawkins 1986, 31).

Note: "Sometimes, costs are operationally defined in terms of direct and indirect fitness, especially during empirical studies or theoretical analyses pertaining to them" (Wittenberger 1981, 614).

cf. benefit, risk, trade-off, utility

cost of anisogamy   *n.* "The twofold reduction in Darwinian fitness that results from producing male progeny in anisogametic, sexually reproducing organisms" (Wittenberger 1981, 614).

cost of males   *n.* The decrease in number of females produced by a sexually reproducing form of a species in subsequent generations compared to an asexually reproducing form of this same species (Maynard Smith 1978 in Willson 1983, 48).

cf. cost: cost of meiosis

cost of mating   *n.*
1. A male's expenditures in time and energy due to factors such as securing mates, courting, growing or securing "ornamentation," producing pheromones, undertaking dangerous travels, and attracting predators and parasites by conspicuous courting (Daly and Wilson 1983, 62).
2. A female's expenditures in time and energy due to factors including the above plus a male's badgering her so that she cannot feed, wounding her in battling for her, and even killing her *in copulo* (Daly and Wilson 1983, 62).

cost of meiosis   *n.*
1. The decrease in the number of genes passed on to future generations by a sexually reproducing form of a species compared to an asexually reproducing form (Williams 1975, 1980 in Willson 1983, 47).
2. "The alleged twofold reduction in Darwinian fitness that results from being only 50% related to each progeny in sexually reproducing organisms, as compared to the 100% relatedness present in asexually reproducing organisms" (Wittenberger 1981, 614).
3. A female sexually reproducing organism's halving her chromosome number but not being able to double her gamete (Dawkins 1986, 135).

syn. cost of sex, meiotic cost
cf. cost: cost of males

cost of outcrossing   *n.* An organism's fitness loss (in the short run) from outbreeding compared to inbreeding (Bernstein et al. 1985, 1279).

Comment: The cost comes from having offspring with poorly adapted gene complexes, etc. (Bernstein et al. 1985, 1279).

cost of parthenogenesis   *n.* An organism's fitness loss due to parthenogenetic reproduction resulting in poor egg hatching success caused by abnormal development, reduced fecundity, etc. (Bernstein et al. 1985, 1980).

cost of recombination   *n.*
1. A parent's cost due to producing offspring with new genetic combinations that might not survive well in the parent's present habitat (Daly and Wilson 1983, 62).
2. In genetics, recombinational load, *q.v.* (Daly and Wilson 1983, 62).

cost of selection   *n.* The accumulating cost to a population, in terms of genetic deaths, or replacing one allele by another during evolution (Lincoln et al. 1985).

syn. cumulative substitutional load (Lincoln et al. 1985)

▶ **substitutional load, transitional load**   *n.* Genetic load due to one allele's being replaced by another during evolutionary change (Lincoln et al. 1985).

Comment: Substitutional load is the expression of the cost of selection, *q.v.*, as a kind of genetic load (Ridley 1996, 160).

costs of sex   *pl. n.* Costs including mating cost and breaking up favorable genetic combinations each generation (Daly and Wilson 1983, 60; Willson 1983, 46).

See cost: cost of meiosis.
cf. benefit: benefits of sex; cost: cost of mating, major costs of sex

energy cost of play   *n.* The "net daily energy expenditure (in excess of resting metabolism) that is due to play, expressed as a percentage" of an individual's total daily energy budget (Martin 1982, 295).

expected cost   *n.* Cost, measured in fitness, that an individual can expect to incur at some future time by performing a particular behavior rather than some alternative behavior (in any behavior context) at a given time; "expected cost equals cost multiplied by the probability of incurring the cost" (Wittenberger 1981, 615, 621).

cf. risk

leverage cost   *n.* Cost, measured in fitness, incurred by the winner of a social conflict in a situation where winning ultimately reduces its fitness; during a social conflict, an animal has a leverage advantage, *q.v.*, if the leverage cost it incurs if it wins is less than the leverage cost its opponent incurs [coined by Hand 1986, 213].

major costs of sex *pl. n.* Cost of mating, males, high recombinational load, and lower genetic relatedness between parents and their offspring (Bernstein et al. 1985, 1279).

meiotic cost See cost: cost of meiosis.

opportunity cost *n.* An organism's cost of performing one activity (equal to the value of some alternative foregone activity) that results from the organism's investing its resources in this former activity instead of the foregone one (Mishan 1975 in Winterhalder 1983, 74).
syn. principle of lost opportunity (Stephens and Krebs 1986, 11)

♦ **cost function** See function: cost function.

♦ **coterie** See ²group: coterie.

♦ **counseling psychology** See study of: psychology: counseling psychology.

♦ **counteracting selection** See selection: counteracting selection.

♦ **counteraction** See need: counteraction.

♦ **counterconditioning** See learning: conditioned learning: counterconditioning.

♦ **counterevolution** See ²evolution: counterevolution.

♦ **counterfeit signal** See signal: counterfeit signal.

♦ **countershading** See camouflage: countershading.

♦ **countersinging** See animal sounds: song: duet: counter singing.

♦ **counting, counting ability** *n.*
1. A person's action, or process, of listing or calling off the units of a group or collection, one by one, to determine the total; numbering; enumerating; calculating; computing; reckoning (*Oxford English Dictionary* 1989, entries from 1330; Michaelis 1963).
2. For example, in dolphins, mice, rats, seals; the Human, Jackdaw, Magpie, Raccoon, Raven, and Rhesus Monkey: an individual's capacity for forming nonverbal counting or number recognition (Koehler 1937–1955, Bucholtz 1973, Rensch 1973 in Heymer 1977, 203; Davis and McIntire 1969 in Davis and Memmott 1983, 95; Davis 1984, 409; Davis and Albert 1986, 57).
Comments: In counting tests, an African Grey Parrot, Magpies, Ravens, and squirrels can "count" up to 6; Budgerigars and Jackdaws, up to 6 to 7; and Domestic Pigeons, up to 8 (Immelmann and Beer 1989, 61). In most tests of "counting," nonhuman animals do not enumerate objects by applying a series of cardinal labels. "Counting" has

been applied to cases of behavior control when all nonnumerical properties of an animal's stimulus environment (*e.g.,* visual, auditory, and temporal) have been controlled (Davis 1984, 409). Davis and Memmott (1983, 99) conclude that after laboratory training nonhuman animals will not continue to count simply because they have been taught to do so; however, they will continue to count if their environment demands this behavior. Fully developed counting includes a person's producing a standard sequence of number words, applying a unique number word to each item of a set that he counts, remembering which items he has already counted, and comprehending that the last used number word indicates how many items he counted (Pepperberg 1987, 37).

♦ **court** *n.*
1. In some species of birds and mammals: an area defended by individual males within a lek or communal display area (Wilson 1975, 331).
   cf. ²lek
2. In social insects, especially Honey Bees: The group of workers in an insect colony that surrounds its queen; a court's worker composition changes constantly (Wilson 1975, 581).
   syn. retinue (Wilson 1975, 581)
3. *v.i., v.t.* To show courtship behavior, *q.v.*
   Note: This definition is preferable to def. 4.
4. To mate or copulate.
   cf. copulation

♦ **courtship, courtship behavior** *n.*
1. A man's action or process of paying court to a woman with a view to marriage; wooing; courting (*Oxford English Dictionary*, entries starting at 1596).
2. Mating, copulating.
   cf. copulation
3. Broadly, precopulatory, pair-forming, and pair-bonding behavior (Immelmann and Beer 1989, 62).
4. Behavior that brings two conspecific animals of different sexes together under conditions where inseminative copulation, or fertilization, is likely to occur (Dewsbury 1978, 71).
   Notes: This definition seems appropriate for wide use.
   Comments: "It may be difficult to designate the point at which courtship ends and mating begins" (Oliver 1955, 218). Halliday and Adler (1986, 146) indicate that courtship can accompany the act of fertilization. Wickman's (1986, 153) "courtship solicitation" in the small heath butterfly is "courtship" based on this

definition. White et al. (2000, 883) divide courtship (in the broad sense) into "courtship" (strict sense) and "copulatory-induction behavior" with regard to Fruit Flies (Tephritidae).

5. All of a male's behavior that he directs towards a female after he detects her and before he attempts to copulate with her (Boake 1983, 29).
See display: sex advertisement.
syn. copulation (some authors), mating (some authors, especially with regard to mammals; Immelmann and Beer 1989), precopulatory or precoital behavior (Immelmann 1977)
cf. behavior: postcopulatory behavior; court; dance: mating dance; display: advertisement: sex advertisement; estrus: heat, rut; play: foreplay
Comments: Armstrong (1965, 22) suggests that "courtship" restricted to prenuptial display and postnuptial display be called "connubial display." "Courtship" helps an animal find a mate, determine that a potential mate is the correct species, form a pair bond, reinforce a pair bond, stimulate a mate, and synchronize sexual arousal of mates (Immelmann 1977). Also "courtship" may allow an animal to choose among potential mates that are competing for it. It is useful for a scientist to state at the beginning of his publication precisely how he is using "courtship" (Immelmann and Beer 1989, 62). Jamieson and Craig (1987) describe male-male and female-female courtship and copulatory behavior in a bird.

behavioral dichotomy *n.* Courtship in which the sexes distinctly differ in their behaviors (Aronson 1949 in Heymer 1977, 198).

communal courtship *n.*
1. Courtship in which a group of conspecific males displays to females (Heymer 1977, 79; Immelmann and Beer 1989, 50).
cf. heterogenous summation
2. For example, in flamingos: courtship in which both males and females display to one another more or less equally (Immelmann and Beer 1989, 50).
syn. group courtship, group mating (Immelmann and Beer 1987, 121)
Comment: Communal courtship is found in lekking species including the Cock-of-the-Rock, Ruff, and Uganda Kob; birds-of-paradise; some bat dance-fly, fruit-fly, and bee species; and in nonlekking species including the Oyster Catcher and some insect species.

homosexual courtship *n.* For example, in the Pukeko (bird): male-male, or female-female, courtship (Jamieson and Craig 1987, 1251).

mutual courtship *n.* Courtship in which both partners participate equally (Huxley 1923, 1938 in Heymer 1977, 198).

unilateral display *n.* Courtship in which only one partner is a courter (Heymer 1977, 198).

♦ **courtship chain** See chain: courtship chain.

♦ **courtship colors** *pl. n.* For example, in male stickleback fish: colors and color patterns that occur only during a species' mating season (Heymer 1977, 37).
cf. coloration

♦ **courtship dance** See dance: courtship dance.

♦ **courtship feeding** See feeding: courtship feeding.

♦ **courtship flight** See flight: courtship flight.

♦ **courtship pheromone** See chemical-releasing stimulus: semiochemical: pheromone: aphrodisiac.

♦ **courtship rippling** *n.* In *Rhagadotarsus* (water-strider insects): propagation of patterned-water-surface waves by a male and female during their courtship (Wilcox 1972 in Wilson 1975, 238).

♦ **courtship ritual** *n.* A genetically determined characteristic behavior involving the production and reception of a complex sequence of visual, auditory, and chemical stimuli by a male and female animal before mating (Lincoln et al. 1985).

♦ **courtship song** See animal sounds: song: courtship song.

♦ **covert** See ²group: covert.

♦ **covey** See ²group: covey.

♦ **cow** See animal names: cow.

♦ **cowardess** See ²group: cowardess.

♦ **coy** *adj.*
1. Quiet, still (*Oxford English Dictionary* 1972, entries from 1330).
2. Referring to a person's being not demonstrative; being shyly reserved or retiring; displaying modest backwardness or shyness (sometimes with emphasis on the displaying); not responding readily to familiar advances; now especially referring to a girl or young woman (*Oxford English Dictionary* 1972, entries from 1386).
3. In sociobiology, referring to an animal that is hesitant to copulate and thus evokes more courtship displays from a potential mate (Wilson 1975, 320).
cf. Lex-Heinze mood
syn. cautious.
Comment: Cronin (1992, 293) explains why "coy" is a terminological oddity: When females choose mates, they are often called "coy" and males "eager, but if males were choosy would they be coy or discriminating,

judicious, and prudent; would females then be eager or wanton, brazen, and flighty? [French *coi* (fem. *coitte*), at rest, still, quiet]

♦ **coyness** *n.* An animal's being coy (Wilson 1975, 320).

♦ **cpDNA** See nucleic acid: deoxyribonucleic acid: chloroplast DNA.

♦ **CPOM** See particulate-organic matter: coarse-particulate-organic matter.

♦ **crash** *n.* A precipitous decline in population size (Lincoln et al. 1985).
See ²group: crash.
ant. flush

♦ **crawling** See creeping

♦ **creaght** See ²group: creaght.

♦ **creation science** *n.*

1. A subject involving "scientific evidence to explain geology by 'catastrophism, including the occurrence of a worldwide flood,' and evidence that showed Earth to be no more than a few thousand years old" (S. Nelson quoting B. Keith in Kamet 1986a).

2. "The scientific evidences for creation and inferences from those scientific evidences" (Kamet 1986a); the evidence involves the abrupt appearance of complex life in the fossil record and the systematic gaps between fossil categories without transitional forms (W.R. Bird in Kamet 1986a).

syn. creationism (Gillam 1999, A13)
cf. special creation; study of: evolution
Comments: It is controversial whether creation science is a science and should be taught in U.S. schools; the U.S. Supreme Court tried to define creation science (Kamet 1986b). In 1982, a federal district court ruled in *McLean v. Arkansas Board of Education* that creation science is in fact religion and that is should not be taught alongside evolution in Arkansas public schools (Schmidt 1996, 422; Gillam 1999, A13). A court also struck down a similar law in Louisiana in 1982. In 1987, the U.S. Supreme Court ruled in *Edwards v. Aguillard* that Louisiana could not mandate that schools give equal time to teaching creation science and evolution (Rosin 1999, A22). In 1999, the Kansas Board of Education voted to eliminate most evolutionary concepts, including common ancestors, evolution as an underlying principle of biology and other sciences, evolution as a way to describe the emergence of new species, natural selection, and the origin of our Universe, in grades kindergarten through 12 (Rosin 1999, A1; Belluck 1999, A1; Gillam 1999, A13; Holden 1999b, 1186). This move did not eliminate teaching evolution in Kansas, however. According to a Gallup Poll, about 44% of U.S. citizens believe in a biblical creationist view that God created

Humans essentially in their present form at one time within the last 10,000 years (Rosin 1999, A22). About 40% believe in "theistic evolution" (= the idea that God oversaw and guided the millions of years of evolution that culminated with Humans). About 10% believed in a strict secular evolutionist perspective.

♦ **creationism** See creation science, special creation.

♦ **creationist** *n.* A person who believes that an intelligent creator somehow brought about the origin of life on Earth (inferred from Todd 1999, 423).
Comment: A creationist, depending on the person, also believes one or more of the following: DNA changes (mutations) give rise to variation in an organism's characteristic(s) that are subject to natural selection, DNA encodes the features of an organism, genetic and phenotypic changes could result in speciation, macroevolution is empirically proven, scientific enquiry should take place for biblical reasons, and the Earth is not older than 10,000 years (Rosen 1999, A22; Todd 1999, 423).

♦ **CREB gene** See gene: CREB gene.

♦ **creche, crèche, crêche** See ²group: creche.

♦ **creeping** *n.* For example, in many protozoan, insect, and mammal species: an individual's moving with its body close to, or touching, its substrate (Michaelis 1963; Ricci 1990, 1052).
syn. crawling.

♦ **cremnium** See ²community: cremnium.

♦ **cremnophile** See ¹-phile: cremnophile.

♦ **crenicole** See -cole: crenicole.

♦ **crenium** See ²community: crenium.

♦ **crenon** See biotope.

♦ **crenophile** See ¹-phile: crenophile.

♦ **creophage, creophagous** See -phage: creophage.

♦ **crepitation** *n.* A defense mechanism involving explosive discharge of fluid (*e.g.*, in insects) (Lincoln et al. 1985).
See animal sounds: crepitation.

♦ **crepuscular, vespertine** *adj.* Referring to a species that is active at twilight hours of dawn, dusk, or both (Lincoln et al. 1985).
cf. diurnal, matinal, nocturnal, pomeridanus

♦ **criterion of minimum specification** *n.* In sociobiology, "the number of individuals which, on the average, must be put together in order to observe the full behavioral repertory of the species" (Wilson 1975, 19).

♦ **critical period** See period: sensitive period.

♦ **croak** See animal sounds: croak.

♦ **cron** See unit of measure cron.

♦ **cronism** See cannibalism: cronism.

♦ **crop** *n.* In many insect species: a dilated posterior portion of an individual's foregut just behind its esophagus (Borror et al. 1989). *Comment:* Ant repletes (workers) store liquid food in their crops (= "social stomachs") and regurgitate it for nest mates (Hölldobler and Wilson 1990, 643).

♦ **crop milk** *n.* In Ring Doves and other dove species: a regurgitated substance produced by the epithelial lining of a bird's crop that is used to feed young (Dewsbury 1978, 86).

♦ **cropper** *n.* An organism that feeds actively on other organisms from its surrounding environment (Lincoln et al. 1985).

♦ **cross** *n.*
1. Interbreeding of specific organisms, *e.g.*, members of two genera, species, populations, varieties, or genetic types.
2. The product of a cross (def. 1); a hybrid (Michaelis 1963).
See hybrid.
cf. crossing, mating

backcross *n.* An organism produced by a hybrid's breeding with either of its two parental species (Mayr 1982, 643; Dewsbury 1978, 118).

diallel cross *n.* An experimental design in which one produces all possible F_1 crosses among three or more inbred strains which suggests which traits are relevant to fitness and estimates a wide range of genetic parameters (Dewsbury 1978, 128).

dihybrid cross *n.* Interbreeding of two individuals that are heterozygous for two pairs of alleles (Lincoln et al. 1985).
cf. cross: monohybrid cross

incross *n.* Interbreeding of homozygotes (Lincoln et al. 1985).
cf. cross: intercross

intercross *n.* Interbreeding of heterozygotes (Lincoln et al. 1985).
cf. cross: incross

Mendelian crosses *n.* Interbreeding between a homozygous recessive parent and a homozygous dominant parent and interbreeding of their F_1 offspring which produces F_2 offspring (Dewsbury 1978, 116–119).

monohybrid cross *n.* Interbreeding of two individuals that are heterozygous for only one pair of alleles (Lincoln et al. 1985).
cf. cross: dihybrid cross

reciprocal crosses See hybrid: reciprocal hybrids.

♦ **cross-cultural comparison** See study of: [3]ethology: human ethology.

♦ **cross-fertilization** See fertilization: cross-fertilization.

♦ **cross-fostering** See fostering: cross-fostering.

♦ **cross-gait** See locomotion: cross-gait.

♦ **cross-pollination** See pollination: cross-pollination.

♦ **cross-validation** See statistical test: cross-validation.

♦ **crossbreeding** See breeding: outbreeding; crossing.

♦ **crossing** *n.* Organisms' making a cross (Michaelis 1963).
syn. hybridizing, crossbreeding (Lincoln et al. 1985)

♦ **crossing over** *n.*
1. "A complicated process whereby chromosomes, while engaged in meiosis, exchange portions of genetic material," resulting in an almost infinite variety of gametes (Dawkins 1982, 285).
2. The reciprocal exchange of homologous parts between two non-sister chromatids at the four-strand stage.
cf. chiasma
Comments: De Vries (1903) originally proposed that crossing over might occur (Mayr 1982, 764, 766, 957). Crossing over rarely occurs during mitosis (Lincoln et al. 1985).

unequal crossing over, oblique crossing over *n.* For example, in bar-eyed and ultrabar *Drosophila:* crossing over of sister chromosomes resulting in one's having a duplicate piece of a chromosome and the latter lacking this piece (Mayr 1982, 769, 798).

♦ **Crow-Kimura hypothesis, Crow-Kimura model** See hypothesis: hypotheses of the evolution and maintenance of sex: Crow-Kimura hypothesis.

♦ **cruel bind** *n.* In animals with both maternal and paternal care: an animal's dilemma involving whether or not to rear its offspring alone or abandon them and attempt to promote its fitness with a new mate and brood, after it has placed much parental investment into them and its mate deserts them (Trivers 1972 in Wilson 1975, 326).

♦ **crustecdysone** See hormone: crustecdysone.

♦ **crymium** See [2]community: crymium.

♦ **crymophile** See [1]-phile: crymophile.

♦ **-cryo** *combining form* "Cold; frost" (Michaelis 1963).
[Greek *kryos*, frost]

♦ **cryophile** See [1]-phile: cryophile.

♦ **cryothermy** See -thermy: cryothermy.

♦ **-crypsis** *adj.* cryptic. See mimicry: crypsis.
cf. coloration: cryptic coloration; mimesis

allocrypsis *n.* An organism's concealing itself under a covering of living, or nonliving, material that it does not produce (Lincoln et al. 1985).

anticrypsis *n.*
1. An organism's coloration that facilitates attack of its enemies (Lincoln et al. 1985).
2. A predator's protective coloration that facilitates attack, or capture, of its prey (Lincoln et al. 1985).
   cf. -crypsis: procrypsis; mimicry: aggressive mimicry

procrypsis *n.* An organism's coloration, or behavior, that affords protection against its enemies (Lincoln et al. 1985).
cf. behavior: cryptic behavior

syncrypsis *n.* Resemblance between unrelated organisms that have similar cryptic coloration (Lincoln et al. 1985).

♦ **crypsis** See camouflage.
♦ **cryptic** See -crypsis.
♦ **cryptic behavior** See behavior: cryptic behavior.
♦ **cryptic mimesis** See mimicry: mimesis: cryptic mimesis.
♦ **cryptic polymorphism** See -morphism: polymorphism: cryptic polymorphism.
♦ **cryptic polyploidy** See -ploidy: cryptic polyploidy.
♦ **cryptic species** See ²species: sibling species.
♦ **cryptobiosis** See -biosis: cryptobiosis; biosis: anabiosis.
♦ **cryptobiotic** See biotic: cryptobiotic.
♦ **cryptofauna** See fauna: cryptofauna.
♦ **cryptogam** *n.*
1. A member of the former division of plants Cryptogamia that has no carpels or stamens, propagates by spores, and includes Algae, Bryophyta, Fungi, Lycophyta, Psilophyta, Pterophyta, Rhiniophyta, and Sphenophyta (Michaelis 1963).
2. A plant that lacks true seeds and flowers; distinguished from phanerogam (Michaelis 1963).
syn. cryptophyte
[French *cryptogame*, ultimately from Greek *kryptos*, hidden; *gamos*, marriage]
♦ **cryptogenic** See -genic: cryptogenic.
♦ **cryptohybrid** See hybrid: cryptohybrid.
♦ **cryptomere** *n.* "A hidden recessive hereditary factor" (Lincoln et al. 1985).
♦ **cryptozoology** See study of: zoology: cryptozoology.
♦ **crystallochore** See -chore: crystallochore.
♦ **C-S-R triangle** *n.* A three-component ecological-strategy system that is conceptualized as a triangle with its three extremes representing competitive species (C-strategists), stress-tolerant species (S-strategists), and ruderal species (R-strategists) (Lincoln et al. 1985).
♦ **cteinotrophic** See -trophic: cteinotrophic.
♦ **ctetology** See study of: ctetology.

♦ **cub** See animal names (table).
♦ **cuckoldry** *n.*
1. A wife's dishonoring her husband by committing adultery with another person; another person's dishonoring a man by committing adultery with his wife (*Oxford English Dictionary* 1972, entries from 1529).
   Note: Due to his wife's adultery, the husband is said to be cuckolded or a cuckold, a state socially disgraceful to him (C.K. Starr, personal communication).
2. In sociobiology, a male animal's taking care of offspring that are not his (Wilson 1975, 327).
syn. kleptogamy (May and Robertson 1980; Gowaty 1981)
Comment: Cuckoldry is expected to lower the male's fitness.
[Old French *cucu*, cuckoo, a parasitic bird]
♦ **cuckoo** *v.t.*
1. To repeat without cessation (Michaelis 1963).
2. To utter, or imitate, a cuckoo's cry (Michaelis 1963).
See animal sounds: cuckoo; Animalia (Appendix 1): bee: cuckoo bee; wasp: cuckoo wasp.
[Old French *cucu*, *coucou*, imitative word]

cue *n.* "A hint or suggestion; reminder" (Michaelis 1963).

extrinsic cue *n.* A label that an animal acquires from its environment that enables a discriminating conspecific to identify it as kin, or nonkin, with a certain probability (Wilson 1987, 10).
cf. odor

intrinsic cue *n.* A phenotypic label of an animal that enables a discriminating conspecific to identify the animal with respect to its genetic relatedness (Wilson 1987, 11).
♦ **cue bearer** *n.* An aspect of an animal's environment that is significant in its life (von Uexküll 1921, 1937 in Heymer 1977, 38).
See individual: recognized individual.
♦ **cue strength** *n.* A single index that represents a combination of additive components of stimulus summation that operates within a relatively restricted range of features, such as egg recognition (McFarland and Houston 1981 in McFarland 1985, 220).
♦ **culmicole** See -cole: culmicole.
♦ **cultivar** See variety (def. 5).
♦ **cultural anthropology** See study of: anthropology: cultural anthropology.
♦ **cultural ethology** See study of: ³ethology: cultural ethology.
♦ **cultural evolution** See ²evolution: cultural evolution.

a – c

♦ **cultural exchange** *n.* Animal's passing information from one generation to the next one by nongenetic means (McFarland 1985, 514).

 cf. learning: social imitation

♦ **cultural fitness** See fitness: cultural fitness.

♦ **cultural heritability** See [2]heritability (comments).

♦ **cultural inheritance** See tradition.

♦ **cultural selection** See selection: cultural selection.

♦ [1]**culture** *n.* A person's cultivation or development of his mind, faculties, manners, etc.; improvement by education or training (*Oxford English Dictionary* 1972, entries from 1510).

♦ [2]**culture** *n.*
 1. Human "culture or civilisation, taken in its wide ethnographic sense, is that complex whole which includes knowledge, belief, art, morals, law, custom, and any other capabilities and habits acquired by man as a member of society" (Tylor 1871 in Seymour-Smith 1986).
 Note: There is no overall consensus as to the precise meaning of "culture" (Seymour-Smith 1986). Kroeber and Kluckhohn (in Seymour-Smith 1986) list nearly 300 definitions of culture.
 2. The sum total of human attainments and learned behavior (including artistic, social, ideological, and religious patterns of behavior and techniques for coping environmentally) of any specific period, race, or people, regarded as expressing a traditional way of life subject to gradual but continuous modification by following generations (Winick 1956; Michaelis 1963).
 3. A nonhuman primate's discovering a new way to do something, other members' of its group using its new method, and persistence of this acquired method over generations (Immelmann and Beer 1989, 65).
 4. Information that an animal acquires from a conspecific individual(s) by learning (*e.g.*, social imitative learning and vocal learning) that may cause variation in this acquiring animal's behavior (Boyd and Richerson 1985 in Whitehead 1998, 1710).
 Note: Some cetacean species have vocal learning and possibly cultural transmission of feeding specializations (Whitehead 1998, 1708).
 cf. selection: cultural selection; tradition

♦ [3]**culture** *n.* The conditions produced by human culture; refinement, enlightenment (Michaelis 1963).

♦ [4]**culture** *n.*
 1. A specific stage, or period, in the development of a civilization (Michaelis 1963).

 2. "An autonomous population unit defined by distinctive cultural characteristics or shared tradition" (Seymour-Smith 1986).

♦ [5]**culture** *n.* A system of human values, ideas, and behaviors that may be associated with one or more social, or national, group (*e.g.*, Black American culture or Western culture) (Seymour-Smith 1986).

 human culture *n.* The highest form of tradition, *q.v.*, in animals (Wilson 1975, 168).

♦ [6]**culture** *n.* The medium in which one cultivates microorganisms or cell types (Immelmann and Beer 1989, 65).

 v.t. culture

♦ **culture community** See [2]community: agrium.

♦ **cumulative distribution** See distribution: cumulative distribution.

♦ **cumulative recorder** *n.* A special type of event recorder (Lehner 1979, 160).

♦ **cumulative-substantial load** See cost: cost of selection.

♦ **cuniculine** *adj.* Referring to an animal that lives in burrows resembling rabbit burrows (Lincoln et al. 1985).

♦ **cupola** See nest: wasp nest: cupola.

♦ **curiosity** See behavior: exploratory behavior.

♦ **curiosity drive** See drive: curiosity drive.

♦ **cursorial** See -orial: cursorial.

♦ **curve** *n.* A diagram that represents variations in a relationship between (among) two (or more) factors by means of a series of connected points, bars, curves, lines, etc.; a curve can have two or more axes (Michaelis 1963).

 syn. graph, plot (Lincoln et al. 1985), model (Colwell and Coddington 1994, 107)

 cf. -gram, graph

 Comment: Curves are used in nonparametric methods for estimating species richness from samples; see method.

 extinction curve *n.* A graph of the total number of species in a particular habitat, or area, plotted against either time or extinction rate (Lincoln et al. 1985).

 frequency curve *n.* A graph of a number of individuals, particular traits, etc. vs. their frequency (Wilson 1975, 584).

 hollow curve *n.* A plot of y vs. x in which y decreases quickly and shows a long right tail as x increases (Krebs 1985, 514).

 Comment: Plots of number of species vs. number of individuals represented in a community are often hollow curves, indicating that rarer species are more abundant in communities than commoner ones (Krebs 1985, 514).

immigration curve *n.* A graph of a species' immigration rate vs. either the number of resident species or time (Lincoln et al. 1985).

learning curve *n.* A graph of an animal's performance in a learning situation (Storz 1973, 144).

linear curve *n.* A straight-line graph of the relationship between two variables, plotted against each other on orthogonal Cartesian coordinates (Bateson 1979, 242). *cf.* curve: nonlinear curve, regression curve: linear-regression curve

logistic-population-growth curve *n.* One of many graphs of population increase described by mathematical equations (*e.g.,* an S-shaped curve) (Wilson 1975, 81).

Lyellian survivorship curve *n.* A graph of the proportion of taxa in any interval of geologic time that are still alive today (Labandeira and Sepkoski 1993, 311, figure 3).

Comment: Lyellian survivorship curves that decay backward rapidly through time represent groups that undergo rapid extinction (*e.g.,* bivalve families); those that decay slowly represent groups that undergo slow extinction (*e.g.,* insect families) (Labandeira and Sepkoski 1993, 312).

nonlinear curve *n.* A graph (which is not a straight line) of the relationship between two variables, plotted against each other on orthogonal Cartesian coordinates. *cf.* curve: linear curve, regression curve: curvilinear-regression curve

normal curve *n.* The typical bell-shaped graph of a normal distribution (Lincoln et al. 1985). *cf.* distribution: normal distribution

regression curve *n.* A graph of one or more dependent variables vs. one or more independent variables (Sokal and Rohlf 1969, chap. 14; Kleinbaum and Kupper 1978, chaps. 9 and 10, who describe many kinds of regression curves).

▸ **curvilinear-regression curve** *n.* A regression curve in which the dependent variable has a curvilinear relationship with one or more independent variables; *e.g.,* one represented by the equation:

$$Y = b_0 + b_1 X_1^{z_1} + \ldots b_n X_n^{z_n}$$

where Y = the dependent variable; b_0 = the Y intercept; $b_1 \ldots b_n$ = the coefficients of X; $X_1 \ldots X_n$ = the independent variables; $z_1 \ldots z_n$ = exponents of the independent variables (Sokal and Rohlf 1969, 468).

▸ **linear-regression curve** *n.* A regression curve in which the dependent variable has a linear relationship with one or more independent variables; *e.g.,* one represented by the equation:

$$Y = b_0 + b_1 X_1 + \cdots b_n X_n$$

where Y = the dependent variable; b_0 = the Y intercept; $b_1 \ldots b_n$ = the coefficients of X; $X_1 \ldots X_n$ = the independent variables (Sokal and Rohlf 1969, 413).

reproduction curve *n.* A curve of the relationship between the number of individuals at a given stage of one generation and the number of individuals at that stage in the previous generation (Lincoln et al. 1985).

sigmoid curve *n.* "An S-shaped curve," *e.g.,* a kind of population growth curve (Lincoln et al. 1985).

species-abundance curve *n.* A graph of species numbers vs. the number of individuals per species (Lincoln et al. 1985).

species-accumulation curve *n.* A plot of the cumulative number of species, $S(n)$, vs. the number of samples pooled (Colwell and Coddington 1994, 104, illustration). *syn.* collector's curve (Colwell and Coddington 1994, 105) *cf.* curve: species-area curve *Comment:* Colwell and Coddington (1994, 105) use species-accumulation curves from an area of a habitat that is roughly homogeneous. They use "species-area curve" to refer to one from a heterogenous area. These authors discuss the pros and cons of using different species-accumulation curves for estimating species richness.

▸ **asymptotic-species-accumulation curve** *n.* A species-accumulation curve that asymptotes (Holdridge et al. 1971 in Colwell and Coddington 1994, 106). *Comments:* An asymptotic species-accumulation curve starts at 0 and reaches an asymptote (Colwell and Coddington 1994, 106). Different functions for these curves are in Holdridge et al. (1971, trees in Costa Rica), de Caprariis et al. (1976, palaeoecology), and Clench (1979, entomology). All of these references are in Colwell and Coddington (1994, 106). The equation of Holdridge et al. (1971) is the same as that for the Michaelis-Menten equation of enzyme kinetics in different notation. A transformation of the equation of Caprariis et al. (1976) is the Eadie-Hofstee equation.

▸ **nonasymptotic-species-accumulation curve** *n.* A late species-accumulation curve that does not asymptote (Colwell and Coddington 1994, 106). *Comments:* The nonasymptotic-species-accumulation curve is a species-accumulation curve that does not reach an asymptote (Palmer 1990 in Colwell and

Coddington 1994, 107). Examples are the log-linear model (Gleason 1922) and log-log model (MacArthur and Wilson 1967).

log-linear curve *n.* A nonasymptotic-species-accumulation curve that assumes $S(n)$ is a linear function of the logarithm of the area (Gleason 1922 in Colwell and Coddington 1994, 107).

log-log curve *n.* A nonasymptotic-species-accumulation curve that assumes the logarithm of $S(n)$ is a linear function of the logarithm of area (MacArthur and Wilson 1967 in Colwell and Coddington 1994, 107).
Comment: This is the standard species-area curve of island biogeography (Colwell and Coddington 1994, 107).

▶ **rank-abundance curve** *n.* A curve of species abundances in large samples (May 1975 and others in Colwell and Coddington 1994, 108).
Comments: Rank-abundance curves include the lognormal curve, log-series curve, Poisson-lognormal curve, and zero-truncated-generalized-inverse-Gaussian-Poisson curve.

log-normal curve *n.* A parametric rank-abundance curve of number of species vs. abundance class on a log scale, using log base 2 so that each class, or octave (= R), involves doubling the size of the population (Preston 1948, 1960, 1962 in Southwood 1979, 424, illustration).

log-series curve *n.* A parametric rank-abundance curve that assumes that the modal class is always the singleton species, regardless of how large the sample becomes (Williams 1964 in Colwell and Coddington 1994, 108).

Poisson-lognormal curve *n.* A parametric rank-abundance curve that does not require the assumption that discrete numbers of individuals approximate a continuous curve (Bulmer 1974 in Colwell and Coddington 1994, 108).
syn. discrete lognormal curve (Colwell and Coddington 1994, 108)

zero-truncated-generalized-inverse-Gaussian-Poisson curve *n.* A parametric rank-abundance curve (Colwell and Coddington 1994, 108).

species-area curve, area-species curve *n.* A curve of the number of species found, $S(n)$, vs. the size of an area surveyed (Lincoln et al. 1985).
cf. curve: species-accumulation curve; [3]theory: theory of island biogeography; hypothesis: Red-Queen hypothesis

survivorship curve *n.* A curve of the number of surviving individuals of a given cohort vs. age (Lincoln et al. 1985).

▶ **Type-1 survivorship curve** *n.* A survivorship curve in which the probability of an organism's survival decreases with age (Lincoln et al. 1985).

▶ **Type-2 survivorship curve** *n.* A survivorship curve in which the probability of an organism's survival is constant with age (Lincoln et al. 1985).

▶ **Type-3 survivorship curve** *n.* A survivorship curve in which the probability of an organism's survival increases with age (Lincoln et al. 1985).

♦ **cut-off** *n.* In social-conflict situations, an animal's means of reducing stimulation that tends to make it distressed or behave in ways contrary to what it is "aiming at;" *e.g.,* an animal's looking away, or averting its eyes, from another animal that is arousing its fear or hostility, or a person's covering his ears to avoid hearing something embarrassing (introduced by Chance 1962 in Immelmann and Beer 1989, 66).

♦ **cut-off posture** See posture: cut-off posture.

♦ **cuticle** *n.*
1. The outer, nonvascular covering of human skin, overlying the corium (Michaelis 1963).
   syn. epidermis, scarfskin Michaelis 1963)
2. The crescent of toughened skin around the base of a human fingernail or toenail (Michaelis 1963).
3. The transparent film that covers a plant's surface, derived from layers of epidermal cells (Michaelis 1963).
4. A thick lining membrane (*e.g.,* the integument of an arthropod) (Michaelis 1963).
Note: In insects, a cuticle is secreted by epidermal cells, and it covers an insect's entire body as well as lining ectodermal invaginations such as the proctodaeum, stomodaeum, and tracheae (Torre Bueno 1989). Kinds of cuticle are endocuticle, epicuticle, exocuticle, and procuticle.
adj. cuticular
syn. cuticula (Michaelis 1963)
[Latin *cuticula*, diminutive of *cutis*, skin]

♦ **cuticula** See cuticle.

♦ **cybernetics** See study of: mathematics: cybernetics.

♦ **cycle** *n.*
1. A recurring period within which particular events, or phenomena, occur and complete themselves in definite sequence: a round of years or ages (Michaelis 1963).
2. "A pattern of regularly recurring events; a series that repeats itself" (Michaelis 1963).

biogeochemical cycle *n.* The cycling and recycling of an inorganic substance (*e.g.,* calcium, carbon, chlorine, cobalt, magnesium, nitrogen, phosphorus, potassium, sodium, sulfur, or water) through the biosphere (Curtis 1983, 986).

▶ **nitrogen cycle** *n.*

1. The process by which nitrogen is cycled and recycled through the Earth's biosphere, with the principal stages of ammonification, nitrification, and assimilation (Curtis 1983, 402, 988).

2. A description of the balance, changes, and nature of nitrogen-containing compounds that circulate among the Earth's atmosphere, soil, and organisms (Allaby 1994).

Comments: Earth's atmosphere is the chief reservoir of nitrogen (Curtis 1983, 402, 988). Most species cannot use elemental atmospheric nitrogen to make nitrogen-containing compounds. A shortage of nitrogen in soil is often a major limiting factor in plant growth. Ammonification is release of ammonia and ammonium primarily by certain soil bacteria and fungi that decompose organic matter. Nitrification is oxidation of ammonia and ammonium by several species of bacteria common in soil that produce nitrite (NO_2^-) and energy from these compounds. Another group of bacteria oxidizes nitrites and produce nitrate (NO_3^-) and energy. Nitrogen moves as nitrate from soil into roots. Assimilation is a plant's reducing nitrate to ammonium, which is then transferred into carbon-containing compounds making amino acids and other required nitrogenous organic compounds.

developmental cycle *n.* In insects: the period from the birth of an egg to the eclosion of an adult (Wilson 1975, 582).

estrous cycle, estrual cycle, oestrous cycle *n.* In mammals: a female's repeated series of changes in her reproductive physiology and behavior that culminates in estrus (Wilson 1975, 583).

cf. estrus

functional cycle *n.*

1. The relation of an animal's organs and behavior to particular aspects of its environment, essentially the part of its surroundings that is filtered out by its sensory receptors and contains certain objects whose recognition and perception is necessary for the animal's survival (Heymer 1977, 71).

Notes: These aspects act as effector cues that control the animal's evolutionary preprogrammed behavior. As soon as such an aspect is perceived, an animal may react toward it in some manner, thereby eliminating it as an effector cue. [coined as *Functionskreis* by von Uexküll

and Kriszat 1934 and von Uexküll 1937 in Heymer 1977, 71]

2. A set of behaviors that serve the same or similar functions (*e.g.*, aggression, courtship, foraging, locomotion, or parental care) (Immelmann and Beer 1989, 115).

syn. behavior system, functional system (Immelmann and Beer 1989, 114)

hydrologic cycle *n.* Water flow (in its different states) through Earth's terrestrial and atmospheric environments (Allaby 1994).

syn. water cycle (Bates and Jackson 1984)

Comments: In the hydrologic cycle, water is stored in the atmosphere, groundwater, ice caps, oceans, and surface water (Allaby 1994). Water is a condensation (clouds), a liquid (precipitation), and a vapor (during evaporation and transpiration).

life cycle *n.*

1. The sequence of events from an organism's origin as a zygote to its death; the stages through which an organism passes between the production of gametes from one generation to the next (Wilson 1975, 588; Lincoln et al. 1985).

2. In social insects: a society's life span from its origination to reproduction time (Wilson 1975, 588).

▶ **dimorphous life cycle** *n.* A life cycle with two different types of individuals, each bearing a different type of reproductive structure; *e.g.*, in plants, alternation of sporophyte and gametophyte generations (Lincoln et al. 1985).

cf. cycle: life cycle: monomorphous life cycle

▶ **diplontic life cycle** *n.* A life cycle characterized by a diploid adult stage that produces haploid gametes by meiosis and zygotes that form by fusion of a pair of gametes (Lincoln et al. 1985).

▶ **direct life cycle** See cycle: life cycle: homogonic life cycle.

▶ **heterogonic life cycle** *n.*

1. A life cycle comprised of alternating parthenogenetic and sexually reproducing phases (Lincoln et al. 1985).

2. A life cycle with an alternation of a parasitic and a free living generation (Lincoln et al. 1985).

syn. indirect life cycle

▶ **homogonic life cycle** *n.* "A life cycle in which all generations are either parasitic or free living" (Lincoln et al. 1985).

syn. direct life cycle (Lincoln et al. 1985)

▶ **indirect life cycle** See cycle: life cycle: heterogonic life cycle.

▶ **monomorphous life cycle** *n.* "A life cycle in which successive generations comprise only one type of individual and a single reproductive strategy" (Lincoln et al. 1985).

cf. cycle: life cycle: dimorphous life cycle

limit cycle *n.* An oscillation that endlessly passes through the same sequence of points and if perturbed returns to this sequence after the removal of the perturbing force (Bell 1982, 508).

sex cycle *n.* A cycle leading to genetic recombination, such as the alternation of karyogamy and meiosis (Lincoln et al. 1985).

taxon cycle *n.* "A cycle in which a species spreads while adapted to one habitat and restricts its range and splits into two or more species while adapting to another habitat" (Wilson 1975, 597).

Comment: Taxon cycles have been especially noted in large systems of islands. Dispersal commonly occurs in open habitats, and restriction and speciation occur in forests (Wilson 1975, 597).

♦ **cycleology** See study of: cycleology.

♦ **cyclic** *adj.*
1. Of, or referring to, a cycle; revolving or recurring in cycles (*Oxford English Dictionary* 1972, entries from 1794).
2. Referring to a flower with parts arranged in whorls (*Oxford English Dictionary* 1972, entries from 1875).

acyclic *adj.*
1. Referring to an organism that reproduces by obligate thelytoky (Bell 1982, 501).
2. Referring to an organism without sexual periods (Bell 1982, 501).
cf. cyclic: holocyclic.

dicyclic *adj.* For example, in *Cladocera, Rotifera:* referring to an organism that has two sexual periods in a single growing season (Bell 1982, 505).

holocyclic *adj.* For example, in aphids: referring to an organism that has sexual periods (Bell 1982, 507).
cf. cyclic: acyclic

monocylic *adj.* Referring to a species that has two or more broods, or generations, per year and that changes its mode of reproduction once a year, typically to produce resting stages (Lincoln et al. 1985).
cf. -voltine: multivoltine

pleiocyclic *adj.* "Existing through more than one cycle or activity" (Lincoln et al. 1985).

polycyclic *adj.* For example, in *Cladocera, Mono-gonota:* referring to an organism that has more than two sexual periods in a single growing season (Bell 1982, 511).

♦ **cyclic evolution** See ²evolution: cyclic evolution.

♦ **cyclic parthenogenesis** See parthenogenesis: cyclic parthenogenesis.

♦ **cyclical selection** See selection: cyclical selection.

♦ **cycloalexy** See behavior: defensive behavior: cycloalexy.

♦ **cyclogeny** See -morphosis: cyclomorphosis.

♦ **cygnet** See animal names (table).

♦ **cynophobia** See phobia (table).

♦ **-cyte**
cf. cell
oocyte *n.* A cell that produces female gametes by meiosis (Lincoln et al. 1985).
spermatocyte *n.* A cell that produces male gametes by meiosis (Lincoln et al. 1985).

♦ **cytocatalytic evolution** See ²evolution: cytocatalytic evolution.

♦ **cytodeme** See deme: cytodeme.

♦ **cytodiaeresis** See mitosis.

♦ **cytogamy** See -gamy: cytogamy.

♦ **cytogenesis** See -genesis: cytogenesis.

♦ **cytogenetics** See study of: genetics.

♦ **cytokinesis** See -kinesis: cytokinesis.

♦ **cytokinins** See hormone: cytokinins.

♦ **cytology** See study of: cytology.

♦ **cytophage, cytophagous** See -phage: cytophage.

♦ **cytoplasm** *n.* The aggregate of organelles and other structures in a cell's fluid. [Kölliker proposed this term, which has generally replaced the term protoplasm, *q.v.* (Mayr 1982, 654)]

♦ **cytoplasmic androgamy** See -gamy: cytoplasmic androgamy.

♦ **cytoplasmic gynogamy** See -gamy: cytoplasmic gynogamy.

♦ **cytoplasmic inheritance** See inheritance: cytoplasmic inheritance.

♦ **cytoplasmic-inheritance theory** See ³theory: cytoplasmic-inheritance theory.

♦ **cytotaxis** *n.*
1. Movement of cells in relation to each other; used to describe cell's separating (negative cytotaxis) or aggregating (positive cytotaxis) (Lincoln et al. 1985).
2. "The arrangement of cells within an organ" (Lincoln et al. 1985).
cf. taxis

♦ **cytotaxonomy** See study of: taxonomy: cytotaxonomy.

♦ **cytotype** See -type, type: cytotype.

♦ **d** See unit of measure: dalton, day.

♦ **D** See index: species-diversity index: Simpson's Index of Diversity.

♦ **dabbling** *n.*

1. A person's moving his feet, or hands, in shallow water, or liquid mud, thus causing some splashing (*Oxford English Dictionary* 1972, entries from 1611).

2. An individual duck's moving its bill in shallow water, or liquid mud, thus causing some splashing (*Oxford English Dictionary* 1972, entries from 1661).

3. A person's playing in shallow water; paddling (*Oxford English Dictionary* 1972, entries from 1611); his playing in a liquid, as with his hands; splashing gently (Michaelis 1963).

4. In surface-feeding ducks: an individual's feeding on aquatic vegetation by moving its bill in water and tipping up its head without completely submerging its body under water (Peterson 1947, 36).

v.i. dabble

cf. dive

♦ **dairy** See ²group: dairy.

♦ **dalton** See unit of measure: dalton.

♦ **dam** *n.* In animal breeding, a female parent (Lincoln et al. 1985).

cf. sire

♦ **damaging fight** See fight: damaging fight.

♦ **damming up** *n.* An animal's being stopped from displaying a particular behavior (*e.g.*, sexual behavior) for a considerable period (Immelmann and Beer 1989, 67).

Comment: "Damming up" is a now rarely used, metaphorical term linked with hydraulic models of motivation (Immelmann and Beer 1989, 67).

♦ **dance** *n.*

1. A person's rhythmical skipping and stepping with regular body and limb turnings and movements, usually to musical accompaniment, either as an expression of joy, exultation, and the like or as an amusement or entertainment; the action, act, or round of dancing (*Oxford English Dictionary* 1972, entries from 1300).

2. An individual animal's highly stylized repetitive movement, often associated with courtship behavior (Lincoln et al. 1985).

bee dance *n.* In the Honey Bee (*Apis mellifera*): a worker's communicative dance.

cf. dance-language communication

▸ **bumping run** *n.* A returning forager's bounding forward onto her comb and trampling over other workers which may cause them to leave their hive (Schmid 1964; Frisch 1965 in Heymer 1977, 149).

▸ **buzzing run, breaking dance, *Schwirrlauf*** *n.* A worker's buzzing her wings and running through her colony or swarm, randomly and energetically butting into hive mates (Frisch 1967, 283).

Comment: A buzzing run increases nestmate activity and leads to flight; it signals the exit of a swarm or induces a swarm to alight (Winston 1987, 162).

▸ **dorsoventral-abdominal-vibrating dance** *n.* A worker's vibrating her body, particularly her metasoma, dorsoventrally while grasping a queen or another worker with her legs (Milum 1955 in Schneider et al. 1986a, 377; Winston 1987, 152).

syn. dorso-ventral abdominal vibration (D-VAV) (Milum 1955 in Frisch 1967); jerking dance (Frisch 1967, 281); DVAV, vibration dance (Schneider et al. 1986a, 377), vibratory dance (Frisch 1967, 284)

Comments: This dance prevents a queen from destroying developing queens in their cells (Fletcher 1975, 722), thus it controls queen emergence and times of swarming during colony reproduction (Winston 1987, 152, 160). This dance also regulates both daily and seasonal foraging (Schneider et al. 1986a, 377), and it may operate as a two-level feedback system that allows a colony to adjust its foraging activity to both long- and short-term fluctuations in food availability (Schneider et al. 1986b, 386).

▶ **DVAV, D-VAV** See dance: bee dance: dorsoventral-abdominal-vibrating dance

▶ **grooming dance** See dance: bee dance: shaking dance.

▶ **jerk dance** *n.* A forager's short abdominal-wagging bouts made between bouts of depositing scopal (pollen-sac) contents into comb cells (Frisch 1965 in Heymer 1977, 149).

▶ **jostling dance** *n.* A returning, successful forager's running into her nest mates and pushing them aside (Frisch 1967, 283; Winston 1987, 162).
syn. jostling run (Frisch 1967, 283)
Comment: A jostling dance might alert nest mates that a dance is about to occur (Winston 1987, 162).

▶ **jostling run** See dance: bee dance: jostling dance.

▶ **night dance** *n.* A worker's sun-oriented dance performed at night when she cannot see the sun's position (Lindauer 1954 in Heymer 1977, 122).

▶ **persistent dance** *n.* A scout's spontaneous sun-oriented dance performed inside her nest without a new view of the sun (Lindauer 1954, 1955; Frisch 1965 in Heymer 1977, 50).

▶ **quiver dance** *n.* A bee's wandering on four legs slowly over her comb, rocking back and forth and left and right in jerky fashion and holding her twitching forelegs in the air (Frisch 1923, 90, in Frisch 1967, 282; Winston 1987, 162).
syn. trembling dance (Frisch 1967, 282)
Comments: A bee shows this dance in response to rough handling by Humans; her getting her legs stuck in viscous sugar water or on a milkweed pollinium; discovering a sugar solution that is suddenly adulterated with salt, quinine, acid, or poisons; crowding at a feeding dish; and hive robbing (Frisch 1967, 282). Winston (1987, 162) says the function of quiver dances is unknown.

▶ **round dance** *n.* A forager's running in circles, going clockwise or counterclockwise for one to two circles before changing directions (Aristotle; Spitzner 1788, 102; Unhock 1823, 115; Park 1923 in Frisch, 1967, 5–6).
Comments: A round dance signifies a rich food source near a hive. Different honeybee subspecies signify different food distances with their round dances. Honey

Bees show a continuum of round dance, sickle dance, and waggle dance depending on food distance and direction (Frisch, 1967). Round dances are used primarily for recruitment to resources (Winston 1987, 151).
[possibly coined by Spitzner (1788), based on translation in Frisch (1967, 6)]

▶ **scout dance** *n.* A scout's dance that indicates the presence of a nesting site for a queen with her swarm of workers (Frisch 1953, 132–133).

▶ **shaking dance** *n.*
1. A worker's shaking her body rapidly from side to side and back and forth (Frisch 1967, 283).
   Note: A shaking dance induces nearby workers to groom her in places that are difficult for her to reach (Frisch 1967, 283).
2. A worker's shaking her abdomen rapidly up and down, frequently in response to a queen bee (Hammann 1957; Allen 1959a,b; Frisch 1965 in Heymer 1977, 150).
syn. grooming dance (Milum 1947, 1955, in Heymer 1977, 137)
[coined by Haydak 1929, 1945, in Heymer 1977, 137]

▶ **sickle dance** *n.* A forager's running in a path that is roughly a bowed figure eight (Frisch 1967, 283).
Comments: This dance is a transitional dance between the round and waggle dances. A sickle dance indicates the direction of food to other bees (Frisch 1967, 283).

▶ **spasmodic dance** *n.* A worker's dance in which her crop-contents distribution into comb cells is interspersed with short tail-wagging movements (Frisch 1967, 283).
Comment: This dance might signify that another kind of dance is about to occur (Frisch 1967, 283).

▶ **trembling dance** See dance: bee dance: quiver dance.

▶ **vibration dance, vibratory dance** See dance: bee dance: dorsoventral abdominal vibrating dance

▶ **waggle dance, wagging dance** *n.* A forager's running a short, straight distance while wagging her body and then making an approximate semicircle while running to the right of her straight run, returning to her starting point, making another straight run in essentially the same place as her first one, and then making an approximate semicircle to the left of her straight run, returning to her starting point, and then repeating these movements (Frisch 1953, 116–117, 1965 in Heymer 1977, 161; Winston 1987, 157).

syn. figure-eight dance (Winston 1987, 154), *Schwänselltänze,* tail-wagging dance (Frisch 1923 in Frisch 1967, 4).

Comments: A raspy sound accompanies the straight run. The waggle dance communicates distance and direction of food to other bees, and the number of waggles per run and dance tempo have information value (Frisch 1953, 116–117, 1965 in Heymer 1977, 161; Winston 1987, 157). The workers use waggle dances when the indicated resources are 15 or more meters from their hive (Michener 1974, 164).

cleaner-fish dance *n.* A set of signals sent to a customer fish by a cleaner fish that communicates it will clean it (Wickler 1963 in Heymer 1977, 136).

syn. cleaning-appetence swimming (Heymer 1977, 136)

combat dance See dance: courtship dance.

courtship dance *n.* In some snake species: two or more individuals' contacting each other with their bodies parallel and partially entwined and struggling, each trying to tighten its body loops around the others' bodies (combat dance of Oliver 1955, 217).

cf. display

Comments: The snakes are usually males and probably interacting aggressively. This is a misnamed phenomenon (Dewsbury 1978, 74; Burghardt, personal communication).

mating dance *n.* For example, in some ciliate protozoan species: a series of rarely performed, periodically recurrent, modular behavior patterns that lead to the fusion of potential partners into conjugating pairs (Ricci 1990, 1059).

cf. courtship

zigzag dance *n.* In male three-spined sticklebacks: a stereotyped swimming display, involving movements toward and away from a gravid female (Tinbergen 1951, 79–80).

◆ **dance language, dance-language communication** See communication: dance language.

◆ **dance-"language" hypothesis** See hypothesis: dance-"language" hypothesis.

◆ **dark adaptation** See [3]adaptation: [1]physiological adaptation: dark adaptation; [2]physiological adaptation (comment).

◆ **Darling effect** See effect: Fraser-Darling effect.

◆ **darwin** See units of measure: darwin.

◆ **darwinize** *v.i.* To theorize wildly (Moorehead 1969, 26).
[derived from speculation about evolution done by Erasmus Darwin, 1731–1802, grandfather of Charles R. Darwin (Moorehead 1969, 26)]
cf. [3]theory

◆ **Darwin's abominable mystery** *n.* The unknown origin and rapid evolution of angiosperms (Darwin's letter to Hooker 1879 in Darwin 1903).
cf. hypothesis: anthophyte hypothesis; Plantae: *Amborella trichopoda* (Appendix 1)
Comment: Doyle (1978, 387) suggests a monophyletic origin of angiosperms by progenetic modification of caytoniaceous, corystospermaceous, or related seed ferns. On the other hand, molecular data suggest that the dicot lineage and gymnosperm lineage split about 290 Ma in the late Carboniferous Period (Kenrick 1999, 359; Qui et al. 1999, 404).

◆ **Darwin's admonition** *n.* "Never say higher or lower!" (Mayr 1982, 367).
Comment: Dawkins (1992) also recommends that evolutionary writers should no longer use the adjectives "higher" and "lower," and he suggests that it would be as sensible to present animal taxa alphabetically as in their presumed phylogenetic order. Futuyma (1986, 8) indicates that Darwin did not always follow his own admonition.

◆ **Darwin's theory of gemmules** See [3]theory: theory of pangenesis.

◆ **Darwinian** *n.*
1. A follower of Charles Robert Darwin; one who accepts part to all of his theory (*Oxford English Dictionary* 1972, entries from 1871).
Note: Which ideas and theories that are concerned with Darwinian thinking vary with the different times of Darwin's career and thereafter.
syn. Darwinist, Darwinite (*Oxford English Dictionary* 1972)
cf. neo-Darwinist
2. *adj.* Of, or referring to, Erasmus Darwin (1731–1802) and his speculations or poetical style (*Oxford English Dictionary* 1972, entries from 1804).
3. *adj.* Of, or referring to, Charles R. Darwin (1809–1882) and his scientific views and observations, especially his theory of evolution (*Oxford English Dictionary* 1972, entries from 1867).

neo-Darwinist *n.* A person holding the beliefs of neo-Darwinism, *q.v.*

◆ **Darwinian evolution** See Darwinism.

◆ **Darwinian fitness** See fitness (def. 2, 4); fitness: genetic fitness.

d – g

♦ **Darwinian sexual selection** See selection: sexual selection.

♦ **Darwinism** *n*.

1. The theory of evolution by natural selection, as originally set forth by Charles R. Darwin in *On the Origin of Species...* (1859) and other works by Darwin and co-thinkers (*Oxford English Dictionary* 1972, entries from 1871).

    Notes: This includes five lines of thought: (1) life evolves rather than being constant; (2) all organisms evolved from only one, or a few, common ancestors; (3) adaptation is the result of natural selection; (4) speciation occurs in populations of organisms; and (5) evolution is usually gradual, rather than saltational (Darwin 1859 in Mayr 1982, 505–510). In the period immediately after 1859, Darwinism "referred most often to the totality of Darwin's thinking, while it strictly means natural selection for the evolutionary biologist of today" (Mayr 1982, 505). Ruse (1992, 75) further elucidates that conception of Darwinism.

    syn. scientific Darwinism (Ruse 1992, 80)

2. Popularly, evolution (Gould 1982, 380).

    Note: Some workers are including evolutionary thinking in economics, psychology, and social sciences (Williams 1997, 29).

3. A world view that "everything, certainly in the material world but usually also in the world of thought and human culture, is in a state of Heraclitean flux or becoming; change is seen as unidirectional, but almost always value-impregnated, *i.e.,* progressive (Ruse 1992, 75).

    Notes: This world view is a metaphysical notion in a nonpejorative sense, a framework for "understanding the world rather than of something read directly from the surface of nature" (Ruse 1992, 75). It is a sort of philosophy even akin to a faith or religion. This concept views Humans as the endpoint of already accomplished change, a mere stepping stone to even better organisms, or a fallback from an endpoint, or peak, in the organic world. This view existed before Darwin wrote and has continued after him.

    syn. metaphysical Darwinism (Ruse 1992, 75)

Neo-Darwinism *n*.

1. A synthesis of Darwinian natural selection theory and the new population genetics that first occurred in the 1920s (Wilson 1975, 63).

2. Darwin's theory of evolution, except for any inheritance of acquired characters (Mayr 1982, 698, 958).

Note: This was Weismann's theory of evolution which Romanes (1896 in Mayr 1982, 698, 958) designated as neo-Darwinism. "Near universal acceptance [of Weismann's theory of evolution] did not occur until the 1930s and '40s, as a result of the evolutionary synthesis (Mayr and Provine, 1980)" (Mayr 1982, 701).

3. The ideas held by a group of evolutionists in the 1880s (Dawkins 1982, 291).

    Note: The purpose of neo-Darwinism "was to emphasize (and in my opinion exaggerate) the distinctness of the modern synthesis of Darwinism and Mendelian genetics, achieved in the 1920s and 1930s, from Darwin's own view of evolution. I think the need for the 'neo' is fading, and Darwin's own approach to 'the economy of nature' now looks very modern" (Dawkins 1982, 291).

syn. adaptive neutrality, neo-Darwinian evolution, neo-Darwinist evolutionary theory, synthetic theory

cf. drift: genetic drift; ^{2}evolution: non-Darwinian evolution; modern synthesis; ^{3}theory: neutral theory

Comment: Neo-Darwinists seem to dismiss three seeds of Darwin's thought: species evolution should be modeled on individual evolution, evolution is progressive, and embryogenesis recapitulates phylogenies (Richards 1992, 105).

Social Darwinism *n*. Praise of the struggle for existence, unmerciful competition, and social bias under the erroneous excuse that this is what Darwin taught (Mayr 1982, 536).

Comments: The concept of social Darwinism was created by, and the term was coined by, Herbert Spencer. Because Spencer is the intellectual father of this concept, "it would be better to call it social Spencerism. Social Darwinism was confused with real Darwinism during the 1880s and 1890s" (Mayr 1982, 536, 598, 883). Ruse (1992, 75) also discusses social Darwinism.

♦ **database** *n*. A data collection.

Comments: Many databases can be obtained from Internet sites. I list and define a few of the many databases below.

GenBank *n*. The genetic sequence database of the U.S. National Institutes of Health, which is an annotated collection of all publicly available DNA sequences (www. ncbi.nim.nig.gov/web/Genebank/index. html).

Comments: Genbank has approximately 3,400,000,000 bases in 4,610,000 sequence records as of August 1999 (GenBank website). GenBank is part of the International Nucleotide Sequence Database Collaboration, comprised of the DNA DataBank of Japan (DDBJ), the European Molecular Biology Laboratory (EMBL), and GenBank at the National Center for Biotechnology Information which exchange data daily.

Organelle Genome Database (GOBASE) *n.* Information on mitochondrial genomes on the Internet (Gray et al. 1991, 1477; www.megasun.bch.umontreal. ca/gobase/).

Protist Image Database *n.* Information on Protista on the Internet (Gray et al. 1991, 1477; www.megasun.bch.umontreal. ca/protists/).

Tree of Life (TOL) *n.* A database with illustrations, dendrograms, and lists of taxa on the Internet (Maddison and Maddison 1998).

Comments: The TOL's basic goals are to increase education about and appreciation of biological diversity, present a modern scientific view of the evolutionary tree that unites all organisms on Earth, and provide a uniform and linked framework within which to publish information electronically about the evolutionary history and characteristics of all groups of organisms, (eventually) a life-wide database and searching system about characteristics of organisms and a means to find taxon-specific information on the Internet, both taxonomic and otherwise (Maddison and Maddison 1998, goals page; www.phylogeny.arizona.edu/ tree/homepages/recent1998.html). David and Wayne Maddison put a prototype of the Tree of Life online in 1994 (Maddison and Maddison 1998, history page). The base pages and others of the tree are in a computer in Tucson, AZ, and other pages are in other computers at other locations. Peers review pages of the TOL before they are available on the Internet. As of January 1996, the TOL had over 120 contributors.

TreeBASE *n.* A relational database being developed to manage and explore information on phylogenetic relationships (www.herbaria.harvard.edu/treebase/).

Comments: TreeBASE's main function is to store published phylogenetic trees and data matrices according to its website.

Vascular Tropicos Nomenclatural Database, VAST Nomenclatural Database *n.* A database produced by the Missouri Botanical Garden which includes information on full plant names and references regarding the plants (http:// mobot.mobot.org/Pick/Search/pick.html). [VAST, acronym for VAScular Tropicos]

◆ **"data-set-choice p"** See p: "data-set-choice p."

◆ **date-rape drug** See drug: gamma hydroxybutyrate.

◆ **dating** *n.* A person's participating in a social appointment or engagement for a specified time, in particular with a person of the opposite sex (inferred from date, Michaelis 1963).

v.i., v.t. dating

pack dating *n.* A person's socializing in an unpartnered group (pack) (Gabriel, 1997, 22).

Comments: U.S. college students frequently pack date, rather than date as conventional couples (Gabriel, 1997, 22). People may pair up during pack dating and have intimate experiences with another individual, often after getting inebriated. The packs give the people a sense of self-assurance and identity but keep them from deeper, more committed relationships.

◆ **dating method** See method: dating method.

◆ **datum** *n., pl.* **data**
1. Something given or granted; something known as, or assumed to be, a fact and used as a the basis of reasoning or calculation; an assumption, or premise, from which one draws inferences (*Oxford English Dictionary* 1972, entries from 1646).
2. Part of the results of an experiment or study (Lincoln et al. 1985); one of a group of data (facts or figures) from which one draws conclusions (Michaelis 1963).

syn. fact, observation (Lincoln et al. 1985)
Comments: Some workers commonly use "data" with a singular article and verb (*e.g.,* this data is interesting). It is preferable to use data with a plural article and verb (*e.g.,* these data are interesting).
[Latin *datus*, past participle of *dare*, to give, grant]

raw data, uncooked data *n.* Collected data that have not been organized numerically or treated statistically (Lincoln et al. 1985).

◆ **"datum-choice p"** See p: "datum-choice p."

◆ **daughter** *n.*
1. A female offspring (Michaelis 1963).
2. Any offspring of a given generation, referring to either sex (Lincoln et al. 1985).

[Old English *dohtor*]

◆ **Davian behavior** See behavior: Davian behavior.

◆ **Dawkins' Aphorism** *n.* An organism is only the genes' way of making more genes (Dawkins 1977).

◆ **dawn chorus** See animal sounds: song: chorus: dawn chorus.

◆ **day** See unit of measure: day.

◆ **d.b.h.** See unit of measure: diameter at breast height.

♦ **DCA** See method: detrended-correlation analysis.

♦ *de novo* *adj.* Arising anew (Michaelis 1963). [Latin]

♦ **de Vries' genetic theory** See [3]theory: de Vries' genetic theory.

♦ **de Vriesianism, mutationism** *n.* The belief that evolution, in general, and speciation, in particular, result from drastic and sudden mutational changes (Lincoln et al. 1985).
See mutationism.

♦ **dead-end replicator** See replicator (comment).

♦ **dead gene** See gene: pseudogene.

♦ **"dead-plant shredders"** See [2]group: functional feeding group: animal piercers.

♦ **deafferentation** *n.* One's surgically severing an animal's afferent nerves to eliminate its external stimulation and allow investigation of the spontaneous component of its behavior (Immelmann and Beer 1989, 68).

♦ **dealate** *n.*
1. In ants and termites: an individual that has shed its wings, usually after mating (Wilson 1975).
2. *adj.* Referring to dealate.

♦ **dear-enemy phenomenon** *n.* An animal's recognizing its territorial neighbors as individuals, with the result that it keeps aggressive interactions with them at a minimum (Fisher 1954 in Getty 1989; Wilson 1975, 273–274, 582).

♦ **death** *n.*
1. The act, or fact, of dying; the end of life; an animal's, or plant's, final cessation of vital functions (*Oxford English Dictionary* 1972, entries from 971).
2. The total and irreversible cessation of all life processes in an organism (Lincoln et al. 1985).

apoptosis [ap POE toe sis; the second p is silent] *n.* Cell death that is controlled genetically during an organism's metamorphosis (White et al. 1994, 677; Campbell 1996, 980).
cf. mutation: repear
Comments: Caenorhabditis elegans shows 31 apoptotic bouts during its normal development (Campbell 1996, 980). Its genome includes two "suicide genes" (including ced-9, named after cell death) that code for proteins that are either directly toxic to a cell or that change other metabolites into toxins. Apoptosis is an essential part of normal development in vertebrates, as well.
[Greek *apo*, from, off; New Latin < Greek *ptōsis*, falling < *piptein*, to fall; referring to dead, or dying, cell's falling away from other tissue in development]

♦ **death feign, death feigning, death feint** See playing dead.

♦ **death pheromone** See chemical-releasing stimulus: semiochemical: pheromone: death pheromone.

♦ **death phobia** See phobia (table).

♦ **death shake** *n.* In many carnivore species, especially canids: a predator's shaking a prey that is held loosely in its teeth, causing respiratory failure, death by neck dislocation, or both in the prey (Leyhausen 1965 in Heymer 1977, 174).

♦ **decay** See error: observer error.

♦ [1]**deceit, evolutionary deceit** See deception.

♦ [2]**deceit** See [2]group: deceit.

♦ **deceitful signal** See signal: deceitful signal.

♦ **deceitful signaler** See signaler: deceitful signaler.

♦ **decent** See [2]group: decent.

♦ **deception** *n.*
1. A person's deceiving or cheating (*Oxford English Dictionary* 1972, entries from 1490).
2. An individual organism's use of its behavior, appearance, signs, signals, or a combination of these things in a way that creates a "false impression" to another organism; misinforming; disinforming; *e.g.,* an *Ophrys* orchid "tricking" a wasp or bee to "pseudocopulate" with it, mate mimicry of heterospecific fireflies by the firefly *Photuris versicolor,* distraction displays of some species of ground-nesting birds, or possibly the Beau Geste effect (Immelmann and Beer 1989, 69).
See deceit.
syn. deceit, evolutionary deceit
cf. intentionality; lie; signal: counterfeit signal, deceitful signal
Comments: Mitchell and Thompson (1986 in Krebs 1986, 1906) define deception in terms of intention and effect. Alexander (1987) discusses human deception from an evolutionary point of view. Whether deception is intentional in some animals (*e.g.,* some bird and nonhuman-primate species) is currently under study (Immelmann and Beer 1989, 69). de Waal (1989, 238–241) describes deception examples in Chimpanzees and Humans.

tactical deception *n.* In Primates: "an individual's capacity to use an 'honest act' from his normal repertoire in a different context such that even familiar individuals are misled" (Byrne and Whiten in Lewin, 1987b).
Comments: "Tactical deception" is compelling evidence that animals understand enough of their own intentions to manipulate those

of others; it suggests that nonhuman animals have self-awareness, *q.v.* (Bateson 1991, 831).

◆ **deception advergence, advergence** *n.* The tendency of a deceptive signaling animal to resemble more closely an "honest" signaling conspecific animal (Brower and Brower 1972; P.J. Weldon, personal communication, 1985).
cf. deception divergence

◆ **deception divergence** *n.* "The [evolutionary] tendency of 'honest' signalers to become noticeably different from 'deceptive' signalers, which might be the mechanism whereby conspicuous epigamic features evolve to extravagance in the runaway process hypothesized by Fisher (1958)" Weldon and Burghardt 1984, 89).
[coined by Weldon and Burghardt 1984]

◆ **deceptive analogies** See mimicry.

◆ **deceptive floral mimicry** See mimicry: deceptive floral mimicry.

◆ **deceptive resemblances** See mimicry.

◆ **deceptive signal** See signal: deceitful signal.

◆ **deciding** *n.* An animal's making a decision, *q.v.*
v.i., v.t. decide
syn. decision making (Krebs and Kacelnik 1991, 105)

◆ **deciduous** *adj.* Referring to an organism's structure that is shed at regular intervals or at a given stage of an organism's development; *e.g.,* antler or leaf (Lincoln et al. 1985).

◆ **deciduous forest** See ²community: deciduous forest, ptenophyllium.

◆ **decision** *n.*
1. A person's act of determining (an issue, question, etc.) (Michaelis 1963).
2. A person's conclusion, or judgment, that he reaches by determining; a verdict (Michaelis 1963).
3. A person's making up his mind (Michaelis 1963).
4. A person's firmness in action, character, or judgment; determination (Michaelis 1963).
5. An animal's performance of one particular behavior out of a range of possible alternatives, with no conscious intent necessarily implied (McFarland 1985, 455; Immelmann and Beer 1989, 69–70; Krebs and Kacelnik 1991, 105).
syn. decision making (Krebs and Kacelnik 1991, 105)
[French *décision* < Latin *decisio, -onis* < *decidere*, to decide]

◆ **decision making** See deciding.
optimal decision making *n.* Decision making that maximizes an organism's fitness, or some fitness index, under prevailing circumstances (McFarland 1985, 471, 479).

◆ **decision rule** See rule: decision rule.

◆ **declarative representation** See representation: declarative representation.

◆ **decoration stealing** *n.* In some bowerbird species: a male's taking a nest ornamentation (*e.g.,* blue poker chip, blue toy, feather, flower, or snail shell) from another male's bower (Borgia and Gore 1986, 727).
Comment: Decoration stealing includes "feather stealing" (Borgia and Gore 1986, 727).

◆ **decremental conduction** *n.* An animal's loss of nerve impulses from a volley due to impulse expenditure in synaptic transmission, *e.g.,* volleys' diminishing with distance from stimulation site in a cnidarian nerve net (Immelmann and Beer 1989, 70).
cf. ¹facilitation

◆ **dedifferentiation** *n.* Evolutionary regression to a less specialized structure or condition (Lincoln et al. 1985).

◆ **deduction** *n.* A person's reasoning from the general to the particular, *e.g.,* deducing facts from a biological concept (Michaelis 1963; Moore 1984).
cf. induction

◆ **deductive method** See method: deductive method.

◆ **deep-body temperature** See temperature: core temperature.

◆ **Deep Green Project** See project: Deep Green Project.

◆ **deep-sea trench** *n.* A deep, narrow, submarine depression found in certain areas of the sea floor where lithosphere descends into asthenosphere (Stanley 1989, 181, 185).

◆ **deep sleep (DS)** *n.* Human sleep in which delta brain waves predominate and periods of a desynchronized EEG reminiscent of wakefulness occurs (Campbell 1996, 1018).

◆ **deep structure** *n.* The origin of a generative process leading to an organism's overt characteristics (D.M. Lambert, personal communication).
deep structure of language *n.* A clear, basic, abstract, cross-cultural core of language (Chomsky in Gould 1982, 538).
cf. innate-language-acquisition device

◆ **defaunation** *n.*
1. Depletion of animals, or symbiotic microorganisms, in a habitat (Lincoln et al. 1985).
2. A person's removing animals from a region, either unintentionally or in the course of experimentation (Buchmann and Nabhan 1996, 245).
adj. defaunated

◆ **defecation** See elimination.

◆ **defecation ceremony** See ceremony: defecation ceremony.

♦ **defended territory** See territory: defended territory.

♦ **defensive aggression** See aggression: defensive aggression.

♦ **defensive behavior** See behavior: defensive behavior.

♦ **defensive threat** See behavior: threat behavior: defensive threat.

♦ **deference** See need: deference.

♦ **definable-model-virtual-model mimicry"** See mimicry: "virtual-model mimicry:" "definable-model-virtual-model mimicry."

♦ **definition** *n.*
1. A precise statement of the nature of a thing; a statement, or form, by which anything is defined (*Oxford English Dictionary* 1972, entries from 1398).
2. A statement of the meaning of a word, phrase, term, etc. (Michaelis 1963).

operational definition *n.*
1. A statement that identifies the conditions under which an empirical test gives a particular result that is related to the occurrence of a particular behavioral phenomenon; if this result is not obtained after testing, then the phenomenon does not occur (McFarland 1985, 390).
2. A definition that is restricted to a worker's observational, measurement, or manipulation procedures, or a combination of these procedures; *e.g.,* a behavioral tendency's being defined in terms of stimulation level necessary or sufficient to elicit it, "with no 'surplus meaning' implied about unobservable states or processes" (Immelmann and Beer 1989, 207).

♦ **definitive host** See [3]host: definitive host.

♦ **deflection display** See display: distraction display.

♦ **degradation** See abasement.

♦ **degree-day** See unit of measure: day: degree day.

♦ **degree of genetic determination** See heritability: heritability in the broad sense.

♦ **degree of polyandry** *n.* The mathematical variance (within a group of females) in the number of males with which individual females have reproductive success (Thornhill and Alcock 1983, 82).
cf. degree of polygyny, mating system
Comment: Only females that have offspring are used in calculating the variance in number of mates.

♦ **degree of polygyny** *n.* The mathematical variance (in a group of males) in number of females with which individual males have had reproductive success (Thornhill and Alcock 1983, 82).

cf. degree of polyandry, mating system
Comment: Only males that have produced offspring are used in calculating the variance in number of mates.

♦ **degree of relatedness** See coefficient of relatedness.

♦ **degree of satisfaction, psi, ψ** *n.* The index of the extent to which a density-dependent effect is limiting (Wilson 1975, 96).
Comment: "At lowest densities, when the control is negligible, psi is equal to one (satisfaction with this aspect of the environment is total). As the population grows dense, and the control becomes severe, psi approaches its minimum value of zero" (Wilson 1975, 96).

♦ **deimatic behavior** See behavior: defensive behavior: deimatic behavior.

♦ **deimatic display** See display: deimatic display.

♦ **delayed-density-dependent mortality** See mortality: density-dependent mortality: delayed-density-dependent mortality.

♦ **delayed dominance** See dominance (genetic): delayed dominance.

♦ **delayed Mendelian inheritance** See inheritance: delayed Mendelian inheritance.

♦ **delayed mortality** See mortality: delayed mortality.

♦ **delayed-reaction method, delayed-response method** See method: delayed-reaction method.

♦ **delayed reflex** See reflex: delayed reflex.

♦ **delayed return benefit** See altruism: reciprocal altruism.

♦ **deleterious** *adj.* Referring to a trait that impairs survivorship or a mutation that reduces fitness (Lincoln et al. 1985).
ant. beneficial

♦ **deletion** *n.* A chromosome's losing a part (Lincoln et al. 1985).

♦ **delousing** *n.* In many primate species: an individual's removing a conspecific's lice during social grooming (Immelmann and Beer 1989, 275).

♦ **δ¹⁵N** *n.* The ratio of the stable nitrogen isotopes $^{15}N/^{14}N$.
Comments: These isotopes are widely used as natural tracers in ecosystem studies (Ehleringer and Rundel 1998 in Ben-David et al. 1998, 48). Pacific Salmon obtain ^{15}N in the ocean and bring this nitrogen into freshwater habitats when they spawn.

♦ **delta karyology** See study of: karyology: delta karyology.

♦ **delta sleep** See sleep: stage-3 and -4 sleep.

♦ **δ¹³C** *n.* The ratio of the stable carbon isotopes $^{13}C/^{12}C$.

Comment: These isotopes are widely used as natural tracers in ecosystem studies (Ehleringer and Rundel 1998 in Ben-David et al. 1998, 48).

♦ **delta waves** See brain waves: delta waves.

♦ **deme** *n.*

1. An undifferentiated aggregation of cells, plastids, or monads (*Oxford English Dictionary* 1972, entry from 1883).

2. The smallest local set of organisms within which interbreeding occurs freely (Wilson 1975, 9, 582).

3. The smallest local set of organisms that is panmictic (has random interbreeding); hence, a deme is the largest population unit that can be analyzed by the simpler models of population genetics (Wilson 1975, 9, 582).
   syn. panmictic population (Wilson 1975, 590)

4. Any local group of individuals of a given species (Lincoln et al. 1985).

5. A local population in which mate choice is random with regard to most gene loci (Brown 1987a, 299).

[Greek *dēmos*, district, township, people]

agamodeme See deme: gamodeme.

autodeme *n.* A deme composed predominantly of self-fertilizing organisms (Lincoln et al. 1985).

clinodeme *n.* A deme that forms part of a graded sequence of demes distributed over a particular geographic area (Lincoln et al. 1985).

clonodeme *n.* A deme composed predominately of vegetatively reproducing organisms (Lincoln et al. 1985).

coenogamodeme See deme: gamodeme.

cytodeme *n.* A deme that differs in cytological characters from other such demes within the same taxon (Lincoln et al. 1985).

ecodeme *n.* A deme occurring in a particular habitat (Lincoln et al. 1985).

endodeme *n.* A deme composed of predominantly inbreeding, but dioecious, individuals (Lincoln et al. 1985).

gamodeme *n.* A relatively isolated deme (Lincoln et al. 1985).

▸ **agamodeme** *n.* A deme composed of predominantly asexual (apomictic) individuals (Lincoln et al. 1985).

▸ **coenogamodeme** *n.* A biosystematic unit comprising all the hologamodemes that is capable of exchanging genes to some extent, but not with freedom, and hybridizing with other coenogamodemes producing sterile offspring (Lincoln et al. 1985).

▸ **hologamodeme** *n.*

1. A deme in which all individuals are able to interbreed with a high level of freedom under a particular set of conditions (Lincoln et al. 1985).

2. A deme capable of hybridizing with other hologamodemes, giving hybrids with some fertility (Lincoln et al. 1985).

▸ **merogamodeme** *n.* Part of a gamodeme; a phenodeme of intrapopulation variants (Lincoln et al. 1985).

▸ **syngamodeme** *n.* A biosystemic unit comprised of all coenogamodemes that are linked by some members' abilities to form viable but sterile hybrids; members of one syngamodeme are not capable of hybridizing with any other syngamodemes (Lincoln et al. 1985).

▸ **topogamodeme** *n.* A gamodeme occupying a specific area (Lincoln et al. 1985).

genodeme *n.* A deme characterized by genotypic features (Lincoln et al. 1985).

genoecodeme *n.* An ecodeme that is characterized by genotypic features (Lincoln et al. 1985).

hologamodeme See deme: gamodeme: hologamodeme.

merogamodeme See deme: gamodeme: merogamodeme.

phenodeme *n.* A deme characterized by observed structural and functional properties (Lincoln et al. 1985).

plastodeme *n.* A deme characterized by environmentally induced phenotypic features (Lincoln et al. 1985).

plastoecodeme *n.* An ecodeme characterized by environmentally induced phenotypic features (Lincoln et al. 1985).

serodeme *n.* A deme characterized by immunological properties (Lincoln et al. 1985).

syngamodeme See deme: gamodeme.

topodeme *n.* A deme occurring in a particular geographic area (Lincoln et al. 1985).

topogamodeme See deme: gamodeme.

xenodeme *n.* A deme of a parasite species that differs from the species' other demes in its host specificity (Lincoln et al. 1985).

♦ **demographic society** See ²society: demographic society.

♦ **demography** See study of: demography.

♦ **demonstration of trust** *n.* A person's display of defenselessness during a greeting gesture; *e.g.,* Masai greet each other by throwing their spears into the ground in front of themselves (Heymer 1977, 193).

♦ **demonstrative character** See character: demonstrative character.

♦ **demonstrative movement** See ²movement: demonstrative movement.

♦ **demophobia** See phobia (table).

♦ **den** See ²group: den.

d – g

♦ **denatant** *adj.* Referring to an organism's swimming, moving, or migrating with water current (Lincoln et al. 1985).
cf. contranatant

♦ **denaturation** *n.*
1. Biochemistry, alteration of a protein by chemical or physical means (Michaelis 1963).
2. Alteration of a DNA molecule by separating its strands by chemicals or heat (Campbell et al. 1999, 368).

♦ **dendricole** See -cole: dendrocole.

♦ **dendritic evolution** See [2]evolution: cladogenesis.

♦ **dendrium** See [2]community: dendrium.

♦ **dendrochronology** See method: dendrochronology.

♦ **dendroclimatology** See study of: climatology: dendroclimatology.

♦ **dendrogram** See -gram: dendrogram.

♦ **dendrology** See study of: dendrology.

♦ **dendrophage** See -phage: dendrophage.

♦ **dendrophile** See [1]-phile: dendrophile.

♦ **dense school** See [2]group: school: dense school.

♦ **density** *n.* The number of specified units per specific space (Michaelis 1963).

population density *n.*
1. The number of conspecific organisms in an area or volume of a habitat (Lincoln et al. 1985).
2. The mean number of individuals of a population per unit area (Immelmann and Beer 1989, 228).
syn. density (Lincoln et al. 1985)

relative density *n.* A measure (usually expressed as a percentage) of one species' density in an area or community (N_a) compared to the total number of individuals (N_{tot}) of all other species within the same unit; relative density = $100(N_a/N_{tot})$.

♦ **density dependence** *n.* A population density's being affected by a density-dependent factor that tends to retard growth (by increasing mortality and decreasing fecundity) as population density increases, or enhance population growth (by decreasing mortality and increasing fecundity) as density decreases (Lincoln et al. 1985).
cf. factor: density-dependent factor, density-independent factor, law; Allee effect

inverse density dependence *n.* A population density's being affected by a density-dependent factor that tends to enhance growth (by decreasing mortality and increasing fecundity) as density increases, or to retard population growth (by increasing mortality and decreasing fecundity) as density decreases (Lincoln et al. 1985).

♦ **density-dependent factor** See factor: density-dependent factor.

♦ **density-dependent mortality** See mortality: density-dependent mortality.

♦ **density-dependent selection** See selection: density-dependent selection.

♦ **density-independent factor** See factor: density-independent factor.

♦ **deoxyribonucleic acid (DNA)** See nucleic acid: deoxyribonucleic acid.

♦ **dependence** See dependent; need: dependence.

♦ **dependent** *adj.*
1. Referring to a variable's being influenced, or controlled, by another variable (Lincoln et al. 1985).
2. Referring to a datum whose value is affected by the magnitude of other variables in its set.
n. dependence
cf. independent

♦ **dependent character** See character: dependent character.

♦ **dependent rank** See [1]rank: dependent rank.

♦ **dependent variable** See variable: dependent variable.

♦ **Depéret's rule** See law: Cope's law.

♦ **depletion effect** See effect: depletion effect.

♦ **deplumation** *n.* In birds: molting (Lincoln et al. 1985).

♦ **Depo-Provera®** See drug: medroxyprogesterone.

♦ **deposit feeder** See -vore: detritivore.

♦ **deposit feeders** See [2]group: functional feeding group: gathering collectors.

♦ **deposit feeding** See feeding: deposit feeding.

♦ **depression** *n.* A predator's lowering its capture rates with prey in its immediate vicinity due to its foraging activities [coined by Charnov et al. 1976, 247].
syn. resource depression (Charnov et al. 1976, 247, 257)
Comment: Depression results from a number of different processes and need not require actual harvesting of any prey items by a predator (Charnov et al. 1976, 247).

behavioral depression *n.* Depression due to a predator's visiting a focal area and causing prey to change their behaviors in ways that make them more difficult to capture by the predator or another one [coined by Charnov et al. 1976, 248].
Comment: Prey behavior changes include becoming more alert, decreasing activities (*e.g.,* advertising, courting, and feeding) that increase predation risk, and flocking differently (Charnov et al. 1976, 248).

exploitation depression *n.* Depression due to a predator's previously harvesting prey from a focal area [coined by Charnov et al. 1976, 247].

microhabitat depression *n.* Depression due to a prey's changing its position in a focal habitat to where it is more difficult for a predator to find and capture it [coined by Charnov et al. 1976, 248].

Comments: "Behavioral depression" and "microhabitat depression" are often not distinct from one another in that a prey's position shift is usually accompanied by a change in its behavior (Charnov et al. 1976, 248). However, if the depression is mostly from a prey's position change, the prey may be still, or more, available to a second predator using a different hunting method.

♦ **depression effect** See effect: depression effect.

♦ **deprivation experiment** See experiment: deprivation experiment.

♦ **deprivation syndrome** See syndrome: deprivation syndrome.

♦ **depth perception** See perception: depth perception.

♦ **derived** *adj.* Referring to an organism with relatively many derived characters (Mayr 1982, 375).
syn. higher, modern (Mayr 1982, 344)
cf. ancestral; character: derived character; Darwin's admonition

♦ **derived character** See character: derived character.

♦ **derived mitochondrial genome** See genome: derived mitochondrial genome.

♦ **derived-primary-social dominance** See dominance: social dominance: primary-social dominance: derived-primary-social dominance.

♦ **derived species** See ²species: advanced species.

♦ **derived trait** See character: derived character.

♦ **-derm** See organ: -derm.

♦ **dermatotroph** See -troph-: dermatotroph.

♦ **dermatozoon** See -zoon: dermatozoon.

♦ **descriptive behaviorism** See behaviorism: descriptive behaviorism.

♦ **descriptive observation** See observation: descriptive observation.

♦ **descriptive statistics** See study of: statistics: descriptive statistics.

♦ **desert** See ²group: desert.

♦ **deserticole** See -cole: deserticole.

♦ **desertion** See mate desertion.

♦ **desiccation** *n.* Water removal; the process of drying (Lincoln et al. 1985).
v.i. desiccate

♦ **desiccation avoidance** *n.* Delaying drying out by an organism that uses different mechanisms that enable it to maintain a favorable tissue water content despite dry air or soil (Lincoln et al. 1985).

♦ **desiccation avoidant** *n.* An organism that performs desiccation avoidance.

♦ **desiccation tolerance** *n.* An organism's capacity to tolerate its protoplasm's drying out without damage (Lincoln et al. 1985).

♦ **design** *n.* The concept that natural selection results in characters that enable an organism to compete for necessary resources (McFarland 1985, 427).
cf. optimality

♦ **designator** *n.* A component of a signal that identifies the nature of the object toward which the attention of the responder is directed (Marler 1961 in Wilson 1975, 217).
cf. appraisor

♦ **despot** *n.* For example, in some lizard species: one individual that dominates all other members of a group with no rank distinctions being made among the subordinates (Wilson 1975, 279).
cf. alpha

♦ **despotism** See hierarchy: despotism.

♦ **desynchronized sleep** See sleep: stage-1 sleep.

♦ **determinants of social organization** *pl. n.* Factors including the demographic parameters (birth rates, death rates, and equilibrium population size), the rates of gene flow, and the coefficients of relationship that influence how the social system of a species is organized (Wilson 1975, 32).

♦ **deterministic** *adj.* In mathematics, "referring to a fixed relationship between two or more variables, without taking into account the effect of chance on the outcome of particular cases" (Wilson 1975, 582).
ant. probabilistic, stochastic

♦ **deterministic model** See ⁴model: deterministic model.

♦ **deterrent** See chemical-releasing stimulus: deterrent.

♦ **detour behavior** See behavior: defensive behavior: detour behavior.

♦ **detrended-correlation analysis** See method: detrended-correlation analysis.

♦ **detriophage** See -phage: detriophage.

♦ **detritivore** See -vore: detritivore.

♦ **detritus** *n.*
1. Loose fragments, or particles, separated from rock masses by erosion, glacial actions, or other forces (Michaelis 1963).
2. "Any mass of disintegrated material; debris" (Michaelis 1963).
3. Fragments of dead organic material, including corpses, feathers, feces, and leaves (Allaby 1994).
syn. organic debris (Lincoln et al. 1985)
[Latin *deterere* < *de-*, away + *terere*, to rub]

♦ **deuterogenesis** See -genesis: deuterogenesis.

d – g

♦ **deuterotoky** See parthenogenesis: deuterotoky.

♦ **development** See study of: embryology.

♦ **development, law of** See law: von Baer's laws.

♦ **developmental canalization** See canalization.

♦ **developmental cycle** See cycle: developmental cycle.

♦ **developmental flexibility** See developmental homeostasis.

♦ **developmental-genetic conception of ethical behavior** *n.* One of several theories of ethical behavior in Humans that opposes ethical behaviorism and claims that moral commitment is not entirely learned, with operant conditioning being the dominant mechanism (Kohlberg 1969 in Wilson 1975, 562).
See behaviorism: ethical behaviorism; ethical institutionism.

♦ **developmental genetics** See genetics: developmental genetics.

♦ **developmental homeostasis** See canalization.

♦ **developmental mimicry** See mimicry: developmental mimicry.

♦ **developmental psychology** See study of: psychology: developmental psychology.

♦ **developmentally fixed behavior** See behavior: developmentally fixed behavior.

♦ **DHEA** See hormone: dihydroepiandrosterone.

♦ **di-** *prefix* "Twice; double" (Michaelis 1963). [Greek *di-* < *dis*, twice]

♦ **diacmic** See -acmic: diacmic.

♦ **diadromous** See -dromous: diadromous.

♦ **diagenesis** See -genesis: diagenesis.

♦ **diagnostic character** See character: diagnostic character.

♦ **diagram** See graph.

♦ **diakineses** *n.* A stage in meiosis, at the end of prophase, during which chromosomes are strongly condensed and chiasmata are particularly visible (Mayr 1982, 957).

♦ **dialect** *n.*
1. In Humans: one of the subordinate forms of a language that arises from local peculiarities of vocabulary, pronunciation, and idiom (*Oxford English Dictionary* 1972, entries from 1577).
2. In some animal species: a population-characteristic vocalization of one of two or more neighboring populations of potentially interbreeding individuals (Marler 1960, Nottebohm 1969 in Conner 1982, 297).
Note: "Dialect" is distinguished from "geographical variation" in vocalization which occurs over long distances and between populations that normally do not come together (Nottebohm 1969 in Conner 1982, 297).

3. "A consistent difference in the predominant song type between one population and another of the same species" (Marler and Tamura 1962 in Conner 1982, 297). *Note:* Conner (1982, 297) indicates how this definition has caused confusion between "dialect" and "geographical variation."
4. In the Honey Bee; some amphibian, bird, cricket, mammal species: a variant of a species' acoustical, or nonacoustical, trait that is restricted to the population of a particular region and differs from those of other regions (Immelmann and Beer 1989, 73).

syn. geographic variation (but Conner 1982, 297, discusses why this is an inappropriate synonymy)

cf. tradition

chemical dialect *n.* A dialect of a species' sex attractant (in some lepidopteran species) or territorial marking pheromone (in some mammal species) (Immelmann and Beer 1989, 73).

instrumental dialect *n.* A dialect of a species' nonvocal sound production, *e.g.,* woodpeckers' drumming or African Clapper Larks' wing clapping (Immelmann and Beer 1989, 73).
cf. dialect: vocal dialect

play dialect *n.* For example, in the American Big-Horned Sheep: a dialect of a species' play behavior (Immelmann and Beer 1989, 74).

regional dialect *n.* A dialect with a large range (Immelmann and Beer 1989, 73).

vocal dialect *n.* A dialect of a species' vocal sound production (Immelmann and Beer 1989, 73).
cf. dialect: instrumental dialect

♦ **diallel cross** See cross: diallel cross.

♦ **diamesogamy** See -gamy: diamesogamy.

♦ **diapause** See dormancy.

♦ **diaphototaxis** See taxis: diaphototaxis.

♦ **dichogamy** See -gamy: dichogamy.

♦ **dichopatric** See -patric: dichopatric.

♦ **dichotomous character** See character: dichotomous character.

♦ **dichthadiigyne, dichthadiiform ergatogyne** See caste: dichthadiigyne.

♦ **diclinous** See sexual: unisexual.

♦ **dictum** See law.

♦ **dicyclic** See cyclic: dicyclic.

♦ **dideoxy-chain-termination method** See method: dideoxy-chain-termination method.

♦ **diecdysis** See ecdysis: diecdysis.

♦ **dieciopolygamy** See -gamy: dieciopolygamy.

♦ **diecodichogamy** See -gamy: diecodichogamy.

♦ **diencephalon** See organ: brain: diencephalon.

♦ **dientiophile** See [2]-phile: dientiophile.

♦ **dientomophile** See [2]-phile: dientomophile.

♦ **diestrous, diestrus** See estrus: diestrus.

♦ **differential fertilizing capacity** *n.*
The relative ability of sperm from rodent
males of different genotypes to gain representation in litters that are the result of
competitive matings, when order of mating,
number of ejaculates, and time of matings
are controlled (Lanier et al. 1979 in Dewsbury
1984, 551).

♦ **differential psychology** See study of:
psychology: differential psychology.

♦ **differential species** See [2]species: differential species.

♦ **diffuse coevolution** See [2]evolution: coevolution: diffuse coevolution.

♦ **diffuse competition** See competition:
exploitative competition.

♦ **digametic** See -gametic: heterogametic.

♦ **digamety** See -gamety: digamety.

♦ **digenesis** See -genesis: digenesis.

♦ **digenetic** See -genetic: digenetic.

♦ **digenetic parasite** See parasite: digenetic parasite.

♦ **digenic** See genic: digenic.

♦ **digenomic** *adj.* Referring to an organism
that contains two independently derived
genomes (Lincoln et al. 1985).

♦ **digenous** See -genous: digenous.

♦ **digeny** See sexual reproduction.

♦ **digital organism** See program: digital
organism.

♦ **digital signal** See signal: digital signal.

♦ **digitigrade** See -grade: digitigrade.

♦ **digitigrade locomotion** See locomotion: digitigrade locomotion.

♦ **digoneutic** See -voltine: bivoltine.

♦ **digonic** *adj.* Referring to an organism that
produces male and female gametes in different gonads (Lincoln et al. 1985).

♦ **dihybrid cross** See cross: dihybrid cross.

♦ **dihydroepiandrosterone** See hormone: dihydroepiandrosterone.

♦ **dilution factor** See factor: dilution factor.

♦ **dimegaly** *n.* Gametes' having marked size
dimorphism (Lincoln et al. 1985).
adj. dimegalic

♦ **dimension of preparedness** *n.* An
organism's "predisposition to learn vitally
relevant things more readily than anything
else" (Immelmann and Beer 1989, 173).
cf. equipotentiality assumption, learning
disposition

♦ **dimorphic** See -morphic: dimorphic.

♦ **dimorphism** See -morphism: dimorphism.

♦ **dimorphous life cycle** See life cycle:
dimorphous life cycle.

♦ **dinosaur-extinction hypothesis** See
hypothesis: dinosaur-extinction hypothesis.

♦ **dioecious** See -oecious: dioecious.

♦ **dioecopolygamy** See -gamy: dioecopolygamy.

♦ **diphenic population** See [1]population:
diphenic population.

♦ **diphygenic** See -genic: diphygenic.

♦ **diphyletic group** See [2]group: phyletic
group: diphyletic group.

♦ **diphyly** See -phyletic: diphyletic.

♦ **diplanetic** See -planetic: diplanetic.

♦ **diplo-** *prefix* Double, twofold (Michaelis
1963).

♦ **diplobiont** See biont: diplobiont.

♦ **diplodiploidy, diplo-diploidy** See
ploidy: diplodiploidy.

♦ **diplogenesis** See -genesis: diplogenesis.

♦ **diplohaplontic organism** See biont:
diplobiont.

♦ **diploid** See -ploid: diploid.

♦ **diploid apogamy** See -gamy: diploid
apogamy.

♦ **diploid effect** See effect: diploid effect.

♦ **diploid parthenogenesis** See -gamy:
apogamy: parthenoapogamy.

♦ **diploidy** See -ploidy: diploidy.

♦ **diplont, diplophase** See -plont: diplont.

♦ **diplontic life cycle** See life cycle: diplontic life cycle.

♦ **diploses** *n.* A cell's doubling its chromosome number; establishing a zygotic chromosome number (Lincoln et al. 1985).

♦ **diplotype** See -type: genoholotype.

♦ **direct aggression** See aggression: direct aggression.

♦ **direct competition** See competition: direct competition.

♦ **direct fitness** See fitness (def. 2, 4).

♦ **direct hyperparasitism** See parasitism:
hyperparasitism: direct hyperparasitism.

♦ **direct life cycle** See life cycle: homogonic
life cycle.

♦ **direct mimicry** See mimicry: direct mimicry.

♦ **direct role** See role: direct role.

♦ **direct scratching under the wing** See
grooming: direct scratching under the wing.

♦ **direct selection** See selection: direct selection.

♦ **direct tradition** See tradition: direct tradition.

♦ **directed movement** See taxis: topotaxis.

♦ **directed mutation** See [2]mutation: directed mutation.

♦ **directed-mutation hypothesis** See
hypothesis: hypothesis of directed mutation.

♦ **directed panspermia** *n.* "The idea that
Earth's original life arrived as microorganisms

dispatched by intelligent beings who chose not to make the long journey themselves" (F. Crick in Gould 1981a).

♦ **directing stimulus** See [2]stimulus: directing stimulus.

♦ **directional dominance** See dominance: directional dominance.

♦ **directional selection** See selection: directional selection.

♦ **directional signal** See signal: directional signal.

♦ **directionism, catastrophism** *n*. The views held by directionists, *q.v.*

♦ **directionist** *n*. A member of a school of geologists who holds that the evolution of and life on the Earth are affected by direct divine intervention, different causes operated in the early history of the Earth compared to the later history, the intensity of causal forces was irregular and varied and steadily decreased in geological time, configurational causes were different in certain former geological periods, many truly cataclysmic changes occurred, and changes occur in a more or less directional manner, being progressive as indicated in the sequence invertebrates-fishes-reptiles-mammals (Mayr 1982, 365, 375–378).
syn. catastrophist (Mayr 1982, 378)
cf. special creation; [3]theory: uniformitarianism
Comments: Cuvier used the milder term "revolution" rather than "catastrophe" to refer to the presumed upheavals. Catastrophic elimination of species is supported by many data (Mayr 1982, 365, 375–378; Kerr 1984).

♦ **directive coloration** See coloration: directive coloration.

♦ **directive selection** See selection: directive selection.

♦ **directive species** See [2]species: directive species.

♦ **directiveness** *n*.
1. The concept that at least some of an organism's behavior has been modified by natural selection in a way that leads the organism towards states that favor survival and reproduction (Thorpe 1963, 4).
2. The concept that animal behavior is "directed to a certain end (goal, or purpose)" (McDougall 1933 and Russell 1934, 1935 in Tinbergen 1951, 3).
cf. purposiveness, teleology

♦ **disaccharide** See molecule: -saccharide: disaccharide.

♦ **disassortative mating** See mating: disassortative mating.

♦ **disclimax** See [2]community: climax: disclimax.

♦ **discolorous** *adj.* An object's having a nonuniform coloration (Lincoln et al. 1985).
cf. coloration

♦ **discontinuity** *n*.
1. A marked gap in an organism's range (Mayr 1982, 542).
2. A marked interruption in an otherwise continuous variation of sequence, populations, objects, or events (Lincoln et al. 1985).
syn. disjunction (Lincoln et al. 1985)
primary discontinuity *n*. A discontinuity in a species' range that originates when colonists reach an isolated area and succeed in establishing a permanent population there; *e.g.*, Scandinavian insects and plants dispersed to Iceland in the post-Pleistocene period, probably moving over a large watergap (Mayr 1982, 452).
secondary discontinuity *n*. A discontinuity in a species' distribution due to the fractionation of its originally continuous range resulting from a geological, climatic, or biotic event; *e.g.*, the Blue Magpie occurs in eastern Asia and in a completely isolated colony in Spain and Portugal (Mayr 1982, 452).

♦ **discontinuous distribution** See distribution: discontinuous distribution.

♦ **discontinuous modular society** See [2]society: discontinuous modular society.

♦ **discontinuous variable** See variable: discontinuous variable.

♦ **discontinuous variation** See variation: discontinuous variation.

♦ **discounting, discounting future events** In the Human, Starling (bird): an individual's trading off the value of future opportunities for immediate rewards (A. Kacelnik in Williams 1997, 30).

♦ **discrete character** See character: discrete character.

♦ **discrete signal** See signal: digital signal.

♦ **discrete variable** See variable: discontinuous variable.

♦ **discriminate analysis** See statistical test: discriminate analysis.

♦ **discriminating individual** See individual: discriminating individual.

♦ **discrimination learning** See learning: discrimination learning.

♦ **discrimination test** See test: discrimination test.

♦ **discriminator, recognition label** *n*. A genetically determined cue that permits an individual to classify another as kin or non-kin (Hölldobler and Wilson 1990, 637).
See chemical-releasing stimulus: semiochemical: pheromone: recognition pheromone.

♦ **disease** *n*. A condition of ill health, or malfunctioning, in a living organism; especially a disordered physical condition, or processing with particular symptoms and affecting part through all of an organism (Michaelis 1963).

Comments: Diseases are important factors that affect organism's behavior, ecology, and evolution. A few of the thousands of organism diseases are provided below. [Anglo-French, Old French *desaise* < *des-*, away (< Latin *dis-*) + *aise* < *adjacens*, close at hand, within easy reach]

allergic diseases *n.* Human diseases including asthma, rhinitis, and sinusitis (American Academy of Allergy, Asthma, and Immunology 1998, S-1).

Comments: Allergic diseases result from overreactions of human immune systems to just about any normally innocuous foreign substance including certain animal dander, cosmetics, drugs, dust, food, mold spores, and pollen (American Academy of Allergy, Asthma, and Immunology 1998, S-1, 10). Ragweed pollen is a major cause of outdoor allergy in the U.S. (S-8).

allergic rhinitis *n.* A human allergic disease that involves inflammation of nasal mucous membranes (American Academy of Allergy, Asthma, and Immunology 1998, S-10).

Alzheimer's disease *n.* A human age-related dementia (Hall 1998, 28).

Comments: Alzheimer's disease affects nearly 4 million U.S. citizens (Hall 1998, 28).

anaphylaxis *n.* A person's acute systemic, allergic reaction (American Academy of Allergy, Asthma, and Immunology 1998, S-10).

Comments: Anaphylaxis occurs after a person is exposed to an antigen (*e.g.,* an allergen) to which he was previously sensitized (American Academy of Allergy, Asthma, and Immunology 1998, S-10).

asthma *n.* A human allergic disease that involves inflammation and narrowing of bronchial tubes (American Academy of Allergy, Asthma, and Immunology 1998, S-21).

Comments: Asthma symptoms include chest tightness, coughing, whizzing while talking, shortness of breath, wheezing, and whistling while talking (American Academy of Allergy, Asthma, and Immunology 1998, S-21). About 15 million U.S. citizens have asthma.

hayfever, seasonal allergic rhinitis *n.* A human allergic disease that includes the following symptoms: hot, red, swollen, and tender tissues, including nasal passages; itchy eyes; reduced general physical energy level; running nose; sneezing; and wheezing (American Academy of Allergy, Asthma, and Immunology 1998, S-9–S-10, personal observation).

Comments: About 35 million U.S. citizens suffer from this disease (American Academy of Allergy, Asthma, and Immunology 1998, S-1). One year recently, about 1.8 million U.S. citizens required emergency-room services for asthma attacks.

intrinsic asthma *n.* Asthma, *q.v.*, that has no apparent external cause and can develop in a person of any age (American Academy of Allergy, Asthma, and Immunology 1998, S-10).

sinusitis *n.* A human allergic disease that involves inflammation of membranes lining facial tissues and is often caused by bacteria, viruses, or both (American Academy of Allergy, Asthma, and Immunology 1998, S-10).

Wilson's disease *n.* A form of human retardation caused by one gene (Dewsbury 1978, 119).

♦ **diserotization** *n.* Interference with, or inhibition of, an organism's copulation by low temperature (Lincoln et al. 1985).

♦ **disgrace, dishonor** See abasement.

♦ **disgust** See facial expression: disgust.

♦ **disgust reaction** See reaction: disgust reaction.

♦ **dishabituation** See learning: habituation.

♦ **disharmony law** See law: disharmony law.

♦ **disinhibition** See inhibition: disinhibition.

♦ **disinhibition hypothesis** See hypothesis: disinhibition hypothesis.

♦ **disintegrative-capture hypothesis** See hypothesis: Moon-origin hypotheses: disintegrative-capture hypothesis.

♦ **disjunct mimicry system** See ¹system: mimicry system: disjunct mimicry system.

♦ **disjunct species** See ²species: disjunct species.

♦ **disorientation** *n.* An organism's temporary, or permanent, inability to maintain a point of reference with respect to its surroundings; disorientation includes spatial and temporal disorientation (Heymer 1977, 52). *cf.* taxis

♦ **dispermy** *n.* Penetration of a single ovum by two spermatozoa at fertilization time (Lincoln et al. 1985).

♦ **dispersal** *n.*
1. Outward spreading of organisms, or propagules, from their point of origin or release (Lincoln et al. 1985).
2. Outward extension of a species' range, often by a chance event (Lincoln et al. 1985). *syn.* accidental migration (Lincoln et al. 1985)
3. An animal's movement (excluding migration) from a source, such as its birth place (Johnston 1961, etc. in Brown 1987a, 299).

natal dispersal *n*. An animal's successful, or nonsuccessful, dispersal from its hatching site to first-breeding site (Brown 1987a, 299).

presaturation dispersal *n*. Emigration of individuals (typically in good condition and of either sex) from a population that remains at low density or is in its early growth phase (Lincoln et al. 1985).

saturation dispersal *n*. Emigration of surplus individuals (typically juvenile, aged, or in poor condition) from a population that is at, or near, its environment's carrying capacity (Lincoln et al. 1985).

♦ **dispersed lek** See ²lek: dispersed lek.

♦ **dispersion** *n*.
1. The distribution pattern of organisms, or populations, in space (Lincoln et al. 1985).
2. Nonaccidental movement of bionts into, or out of, an area or population, often a movement over a relatively short distance and of a more or less regular nature (Lincoln et al. 1985).
   cf. migration
3. Statistically, the distribution, or scatter, of observations, or values, about the mean or central value (Lincoln et al. 1985).
   cf. behavior: spacing behavior; distribution

♦ **displaced aggression, displacement** See aggression: displaced aggression.

♦ **displacement activity, displacement behavior** See behavior: displacement behavior.

♦ **displacement feeding** See behavior: displacement behavior: displacement feeding.

♦ **displacement preening** See behavior: displacement behavior: displacement preening.

♦ **displacement sleeping** See behavior: displacement behavior: displacement sleeping.

♦ **displacement theory** See hypothesis: continental drift.

♦ **display** *n*. A behavior (a kind of signal) modified by evolution (by ritualization) to convey information (Wilson 1975, 560, 582), including all specifically differentiated behavior that serves in intraspecific and sometimes interspecific communication (Immelmann and Beer 1989, 76).
syn. expressive behavior, signaling behavior (Immelmann and Beer 1989, 75)
cf. animal sounds; behavior: treptic behavior; ceremony; dance; eyebrow flash; facial expression; flehmen; posture; presentation; movement; signal
Comments: Some of the many kinds of described displays are given below. White et al. (2000, 891–892) define many kinds of displays given by male Fruit Flies (Tephritidae).

advertising, advertisement *n*. A display, or other kind of signal.
cf. display

▸ **honest advertising** *n*. A signal, or display, that indicates to receivers, or viewers, the true information that is being broadcast (Wilson and Daly 1985, 66); *e.g.*, in Humans, a competitive, risk-taking young male might be communicating to others that he has high-quality genes or other resources.
syn. honest signaling, honest salesmanship (Wilson and Daly 1985, 66)
ant. dishonest advertising, dishonest salesmanship, dishonest signaling

aerial display *n*. In Humpback Whales: an individual's breaching the water surface and exposing itself to aerial viewers (Baker and Herman 1985, 55).
Comment: An aerial display might be part of competitive behavior; communication used when a whale joins, or leaves, a group; or a result of human disturbance (Baker and Herman 1985, 55).

antidisplay *n*. A behavior that is effective in misleading its recipient(s), *e.g.*, the resting postures of many orthopteroid insects that increase their camouflaging (Moynihan 1975 in Baerends et al. 1975, 278).

appeasement display, appeasement, appeasement gesture *n*. A display that typically inhibits aggression from another animal; *e.g.*, some kinds of birds respond to conspecific threats, or attacks, as if they were about to be preened, and this usually stops the attacks (Wilson 1975, 209).
cf. appeasement
Comment: Appeasement displays may also involve an animal's making itself appear as small as possible, flattening its hair or feathers, retracting limbs, withdrawing combat-releaser weapons such as its teeth or bill from its superior adversary, and immobility (Cumming 1982, 73).

bared-teeth display, threat mimic *n*.
1. In most primates: a phylogenetically ancestral social display involving an individual's opening its mouth, drawing back its lips, and showing its upper and lower teeth, commonly without making vocal sounds (van Hooff 1972 in Wilson 1975, 227).
   Notes: This display is often shown when primates are confronted with aversive stimuli and have moderate to strong tendencies to flee. Chimpanzees may use a form of this display in making friendly within-troop contacts. It may be homologous to human smiling and laughing (van Hooff 1972 in Wilson

1975, 227). Some workers indicate that Humans have bared-teeth displays.

2. In some mammals, especially carnivores: a threat display involving an individual's curling back its lips and showing its teeth (Heymer 1977, 204).

cf. behavior: threat: teeth grinding; facial expression: smile

▶ **bared-teeth-scream display** *n.* In many species of primates excluding Humans: a bared-teeth display, *q.v.*, accompanied with screaming; this display indicates extreme fear and submission, as well as readiness to attack if the animal is pressed further (van Hooff 1972 in Wilson 1975, 227).

▶ **silent bared-teeth display, bared-teeth display** *n.* A bared-teeth display with no vocalizations (McFarland 1985, 486).

cf. smiling

bow-coo display *n.* In male Ring Doves: a courtship display involving an individual's approaching a female in a stereotyped manner and emitting a distinctive cooing sound (Lehrman 1964, 1965 in Dewsbury 1978, 86).

broken-wing display See display: distraction display: broken-wing display.

buttocks display *n.* A person's buttocks presentation that represents scorn, or mockery, toward others, depending on the culture (Heymer 1977, 76).

cf. notify

Comments: In Bushmen, buttocks displays may involve grasping sand between pelvic cheeks and releasing it (and often gas) during a deep bow directed toward a mocked person. European children use buttocks displays as a form of protest to their parents. In European adults, buttocks display is sometimes ritualized in dancing (Heymer 1977, 76). Especially during the 1960s in the U.S., some usually young persons of both sexes sometimes performed "mooning" for fun, which is presenting bare buttocks (*e.g.*, out of car windows) to strangers. "Streaking" (running partially to fully naked in front of other people) was also popular in some areas of the U.S. in the 1960s.

challenge display *n.* A high-intensity, aggressive display performed by a male to a conspecific male (Lincoln et al. 1985).

claw-waving modes See display: fiddling display.

communal sexual display *n.* In male courting animals: a group display that can be spectacular, as in synchronously flashing fireflies, mass emergences of cicadas, and battles of mountain sheep and elk (Wilson 1975, 331).

cf. lek

connubial display See courtship.

deflection display See display: distraction display.

deimatic display *n.* A display thought to scare off enemies (*e.g.*, the "snakehead" morphology of a hawk-moth caterpillar) (McFarland 1985, 97) or eyespots on many kinds of moths (Immelmann and Beer 1989, 76).

cf. behavior: deimatic behavior

distraction display *n.*

1. A display used to attract an enemy's attention and draw it away from any object that a displayer is trying to protect (Armstrong 1947 in Wilson 1975, 122).

2. For example, in the three-spined stickleback and many bird and other vertebrate species: a display used by an adult guardian to divert an predator's attention from its nest, eggs, or young (Simmons 1955; etc. in Woriskey 1991, 989).

syn. deflection display, paratrepsis (Lincoln et al. 1985)

cf. display: diversionary display

▶ **injury-feigning distraction display** *n.* In the Sandpipers, the Killdeer, and some other ground-nesting birds: a distraction display in which an individual acts as though it were injured and trails its apparently broken wing to lure a potential predator from its nest (McFarland 1985, 500).

▶ **rodent run** *n.* In some plover species: a distraction display in which a parent mimics the movement of a mouse or rat, thus diverting a predator's attention toward itself (Immelmann and Beer 1989, 77).

diversionary display *n.* A bird's activities that deflect a potential predator from its nest or young (Oliver 1955, 121).

cf. display: distraction display

epideictic display *n.*

1. The most refined form of "conventional behavior" in which members of a population reveal themselves and allow all to assess the density of the population (Allee et al. 1949; Kalela 1954; Wynne-Edwards 1962 in Wilson 1975, 87, 110).

2. A hypothesized social display whose function is to convey information about local population density.

Note: This term is used solely in the context of interdemic selection for population-regulating mechanisms (Wittenberger 1981, 615).

syn. epideictic behavior

Comment: The existence of epideictic displays is controversial.

epigamic display *n.* Courtship behavior that an animal, usually a male, performs in a situation of epigamic selection (*e.g.*, lekking of ruffs and sage grouse) (Immelmann and Beer 1989, 88).

fiddling display *n*. In Fiddler Crabs: claw-waving used as a courtship signal (Heymer 1977, 201).

fin flicking *n*. In the Glowlight Tetra (fish): an individual's very rapid movement of its caudal, dorsal, and pectoral fins, with no resulting change in its body position (Brown and Godin 1999 in Brown et al. 1999, 470).

Comment: An individual shows fin flicking after detecting a conspecific's alarm pheromone, signaling alarm to conspecifics, and possibly reducing its risk to predation (Brown et al. 1999, 469).

▶ **genital display** *n*. In some primate species: an individual's using its genitals for signaling (Immelmann and Beer 1989, 118).

▶ **genital presentation, presentation** *n*. In some primate species, including Humans: a behavior that ranges, depending on species, from simply exposing genitals (which are colorful in some species) to raising a tail and showing a swollen often colorful scrotum to showing an erect penis (phallic threat) (Ploog et al. 1963, etc. in Heymer 1977, 76).

cf. behavior: treptic behavior: apotreptic behavior: phallic threat

▶ **pubic presentation** *n*. In Bushman girls: a frontal genital display, often made before and after vulva presentation, as part of ridiculing another person (Eibl-Eibesfeldt 1972 in Heymer 1977, 152).

cf. display: sticking out one's tongue

▶ **vulva presentation** *n*. In Bushman girls: a deep bow, similar to a buttocks display, in which one exhibits genitals toward a person who is being ridiculed, followed by raucous laughter (Eibl-Eibesfeldt 1972 in Heymer 1977, 151).

cf. display: genital display: genital presentation

hula display *n*. In the Red-Spotted Newt: a male's undulating his body in front of a stationary female as part of his reproductive behavior (Arnold 1972 in Massey 1988, 205).

machismo *n*. Aggressive displays made by male animals during competition for females; *e.g.,* horn fighting of male sheep, deer, and antelopes; spectacular displays of male grouse and other lek birds; or male fights in rhinoceros and stag beetles (Wilson 1975, 243, 320).

[Spanish]

major-domo stroll *n*. In Rhesus Macaques: an individual's display involving leisurely walking with its head and tail up (Wilson 1975, 242).

cf. morphological support, movement

paratrepsis See display: distraction display.

presentation *n*. An animal's displaying or "showing off" specific body parts or appendages to a conspecific, including horn presentation in some antelope species and making a genital display in some primate species (Immelmann and Beer 1989, 232).

cf. behavior: solicitation behavior

pronking See display: stotting.

protean display *n*. The unpredictable random flights of fleeing animals, which may stymie a predator more effectively than a more orderly retreat (*e.g.,* a group of resting butterflies' taking wing) (Humphries and Driver 1967, 1970 in Matthews and Matthews 1978, 351).

pushup *n*. In the lizard *Anolis cristalellus*: an individual's moving its body up and down in a vertical plane by flexion and extension of its legs (Leal 1999, 522).

cf. signal: pursuit-deterrent signal

relaxed-open-mouth display *n*. In more derived primates, including Humans: a display that involves an individual's opening its mouth, without drawing back its lips as much as in a bared-teeth display, often accompanied by a short expirated vocalization and ordinarily associated with play (van Hooff 1972 in Wilson 1975, 227–228).

ritualized display *n*. A display that evolved from a new communicatory function, *e.g.,* courtship displays in Fiddler Crabs that evolved from threat displays (Heymer 1977, 201).

sex advertisement *n*.

1. All of an animal's behavior and appearance that may contribute to its pair formation in nonpair-bonding species; in general, this term refers to an animal, usually a male, that initiates, or takes the more active role in, pair formation (Immelmann and Beer 1989, 265).

2. Courtship behavior; see behavior: courtship.

sticking out one's tongue *n*. A person's gesture of contempt and rejection (Heymer 1977, 207).

cf. display: genital display: pubic presentation, vulva presentation

Comment: This gesture is widespread in human cultures (Heymer 1977, 207).

stotting *n*. In many antelope, bovid, deer, and gazelle species; *Dolichotis patagonium* (rodent): an individual's display involving a conspicuous, stiff-legged, bounding gait, with its tail raised and white rump flashing (Walther 1948, 1966, etc. in Heymer 1977, 132; Estes and Goddard 1967 in Wilson 1975, 124; Caro 1988, 26).

syn. pronking

Comments: Walther (1969 in Caro 1986, 650) gives a more detailed definition of stotting in ungulates. Caro (1986, 649; 1988) suggests that Thomson's Gazelles inform Cheetahs that they see them, and mother Thomson's Gazelles distract cheetahs from their young with stotting.

[coined by Percival 1928 in Caro 1986, 650]

stripes display *n.* In the stenogastrine wasp *Parischnogaster mellyi*: a male's showing the three white stripes on his tergites by fully stretching his abdomen (Beani and Turillazzi 1999, 1233).

Comment: There is a positive relationship between mating and both display frequency and successful aerial duels (Beani and Turillazzi 1999, 1233).

submissive display *n.*

1. In animals: a display of a fight's loser that is demonstrably nonassertive in all contexts (*i.e.,* when it "indicates complete capitulation, Marler, 1956, 47") (Hand 1986, 216).

2. The display of a subordinate animal shown after it meets a dominant animal (inferred from Henry 1993, 51).

   Note: In the Red Fox, a submissive display involves a subordinate fox's acting as if it is a young fox begging for food (Henry 1993, 51). This fox crouches low, whines, and beats its tail quickly in every direction. Then it slowly creeps up to the dominant fox and carefully reaches up and smells and licks the corner of its mouth.

cf. display: threat display; submission

threat display *n.* An animal's intimidation display (inferred from Immelmann and Beer 1989, 311).

cf. behavior: threat behavior; display: submissive display; submission

Comments: In the Red Fox, threat displays involve an animal's arching its back and erecting its body hair which increase its size; two foxes' rising up and pressing their forepaws against each other's shoulders while jaw-gaping and gekkering; and a fox's walking stiff-legged and staring at a subordinate fox (Henry 1993, 53).

threat mimic See display: bared-teeth display.

tidbitting *n.*

1. In some gallinaceous-bird species: a display that consists of an adult's pecking at the ground and calling, which entices chicks to approach and feed near the adult (Immelmann and Beer 1989, 313).

2. In some gallinaceous-bird species: a ritualized display, consisting of a male's pecking at the ground and calling,

which lures a female to him (Immelmann and Beer 1989, 313).

Comment: Analogous behavior is found in other bird species (Immelmann and Beer 1989, 313).

unilateral display See courtship: unilateral display.

upward stretch *n.* In several species of bovids, cervids, and giraffids: an individual's stretching its neck and pointing its nose upward, associated with threat, dominance, or mating depending on the species involved (Walther 1966 in Heymer 1977, 33).

wing display *n.* For example, in Flies (Diptera), a wing movement, apart from one used for flight (White et al. 2000, 906).

Comment: Fruit Flies (Tephritidae) have complex, elaborate wing displays which include "arching," "enantion," "hamation," "lofting," and "supination."

wing fanning *n.* For example, in Fruit Flies (Tephritidae): wing displays that occur during reproductive behavior (White et al. 2000, 907).

♦ **display circling** *n.* In bovids and cervids: two animals' of the opposite sex circling around each other as they are trying to back beside each other within a small area; *e.g.,* in savanna-dwelling animals or around a vegetation island in forest-dwellers (Backhaus 1958; Walther 1958 in Heymer 1977, 127).

cf. witch's circle

♦ **display flight** See flight: courtship flight.

♦ **display territory** See territory: display territory.

♦ **dispute** *n.*

1. "A controversial discussion; debate" (Michaelis 1963).

2. "An altercation; quarrel" (Michaelis 1963).

v.t, v.t. dispute

♦ **disruptive coloration** See camouflage: background imitation: disruptive coloration.

♦ **disruptive selection** See selection: disruptive selection.

escalated-showing-off dispute *n.* Two or more persons trying to best one another in front of witnesses (Wilson and Daly 1985, 64).

♦ **dissimulation** See ²group: dissimulation.

♦ **¹dissociation** *n.* For example, in geese: the process that dissolves connections and associations responsible for integrated mental function, resulting in memory disturbances, poorly coordinated movements, etc. (Heymer 1977, 52).

♦ **²dissociation** *n.* In some mite and pseudoscorpion species: a male's depositing a spermatophore on a substrate irrespective of a female's presence (Alexander 1964 in Thomas and Zeh 1984, 180–182).

◆ **dissogony, dissogeny** *n.* Sexual maturation at two different life-cycle stages, with an intervening period of no gamete production (Lincoln et al. 1985).

◆ **dissonance** *n.* Lack of consistency between, or among, a person's attitudes, beliefs, or both (Hinde 1982, 278).

◆ **distance** *n.* "The extent of spatial separation between things, places, or locations" (Michaelis 1963).

evolutionary distance *n.* The amount of character difference between taxa (*e.g.,* based on DNA sequences or gross morphology) measured as a distance, often on a phylogenetic tree (Niesbach-Klösgen et al. 1987, 219).

flight distance *n.*
1. The distance between an animal and its predator, or rival, which causes the animal to flee (Heymer 1977, 68).
2. The distance an animal will flee due to its being alarmed by a predator or intimidated by a rival's threat; flight distance in hand-reared animals can be very short (Immelmann and Beer 1989, 107).

genetic distance *n.*
1. In population genetics, a quantitative measure of genetic relationship between two individuals, or populations from which an evolutionary tree may be constructed (Rieger et al. 1991).
   Notes: Measurement is in terms of the probability of the common possession of a given gene or character (= coefficient of genetic distance) (Rieger et al. 1991).
2. The cumulative genetic difference between two populations expressed as a summary of their evolutionary history which is proportional to their separation time and inversely related to their intermigration (Rieger et al. 1991).
3. The distance between two gene loci as defined by the average number of crossover points per chromatid (Rieger et al. 1991).

Haldane Map Distance *n.* A distance on a linkage map measured in centimorgans (cM) (Bradshaw et al. 1995, 763).

individual distance *n.* In many animal species, including starlings and swallows: the minimum distance tolerated between conspecific individuals under different social conditions (Hediger 1942, 1950 in Heymer 1977, 91).
syn. social distance (Heymer 1977, 91)
cf. territoriality

peck distance *n.* In colonial nesting birds: the distance that individuals nest from one another that stops them from pecking one another when they remain in their nests (Heymer 1977, 91).

social distance See distance: individual distance.

◆ **distance greeting** See greeting: distance greeting.

◆ **distance methods** See method: distance methods.

◆ **distance-Wagner tree** See -gram: distance-Wagner tree.

◆ *distal-less* See gene: *distal-less.*

◆ **distraction display** See display: distraction display.

◆ **distress call** See animal sounds: call: distress call.

◆ **distribution** *n.*
1. Scattering, or spreading, of infectious agents, seeds, spores, or other propagules (Lincoln et al. 1985).
   syn. dissemination (Lincoln et al. 1985)
2. "The geographical range of a taxon or group" (Lincoln et al. 1985).
3. The spatial pattern, or arrangement, of organisms, organism groups, values, or observations (Lincoln et al. 1985).
4. The geometric plot of a set of data with frequencies of observations per group vs. kind of group.
cf. dispersion

aggregated distribution See distribution: contagious distribution.

bimodal distribution *n.* An extreme platykurtic distribution, *q.v.* (Sokal and Rohlf 1969, 113).

binomial distribution n. A distribution that is described by the formula $\{N!/[x! (N-x)!]\}(P^xQ^{N-x})$; x = 0, 1, 2, ... , N; has values, each of which gives the probability that a particular event will occur; becomes normal when the sample size upon which it is formulated is large; and is used as a model when each individual in a sample may exhibit a character in either one of two alternative states (Siegel 1956; Lincoln et al 1985).

▸ **negative-binomial distribution** *n.* A distribution in which the presence of a datum at any given point increases the probability of another datum's occurring nearby and in which its variance is greater than its mean; this results in an aggregated, or contagious, distribution (Lincoln et al. 1985); *e.g.,* sizes of troops of baboons, gibbons, and langurs during healthy periods (Wilson 1975, 134).

▸ **positive-binomial distribution** *n.*
1. A distribution of frequency of sample vs. kind of sample comprised of individuals that can exhibit a character in one of two character states (Lincoln et al. 1985).

2. A distribution whose variance is less than its mean and is used as an approximate model of a uniformly distributed population in which the presence of one individual at any given place decreases the probability of another one's occurring nearby (Lincoln et al. 1985).

cf. distribution: uniform distribution

clumped distribution See distribution: contagious distribution.

contagious distribution, aggregated distribution, clumped distribution, overdispersed distribution, patchy distribution, superdispersed distribution, underdispersed distribution (sometimes) *n.* A distribution in which values, observations, or individuals are more aggregated, or clustered, than in a random distribution, indicating that the presence of one individual, or value, increases the probability of another's occurring nearby (Lincoln et al. 1985).

cf. constipated

cumulative distribution *n.* A distribution showing the number of observations above, or below, a given value (Lincoln et al. 1985).

discontinuous distribution *n.* A species' occurrence in two or more separate areas, but not in intervening regions (Lincoln et al. 1985).

frequency distribution *n.*
1. "The array of numbers of individuals showing differing values of some variable quantity," *e.g.,* the numbers of animals of different ages or nests containing different numbers of young (Wilson 1975, 584).
2. "The arrangement of data grouped into classes, each with its corresponding frequency of occurrence" (Lincoln et al. 1985).

Gaussian distribution See distribution: normal distribution.

hyperdispersed distribution See distribution: overdispersed distribution.

hypodispersed distribution See distribution: underdispersed distribution (def. 1).

ideal-dominance distribution *n.* A theoretical dispersion pattern in which individuals that occupy intrinsically higher quality habitats achieve higher fitness than those that occupy intrinsically poor-quality habitats (Wittenberger 1981, 616).

cf. distribution: ideal free distribution

ideal free distribution *n.* A theoretical dispersion pattern in which individuals that occupy intrinsically higher quality habitats achieve the same average

fitness as those that occupy intrinsically poorer quality habitats (Wittenberger 1981, 616).

cf. distribution: ideal dominance distribution

infradispersed distribution See distribution: underdispersed distribution.

J-shaped distribution *n.* A distribution with the general shape of a "J" that describes the rapid exponential phase of population growth (Lincoln et al. 1985).

leptokurtic distribution *n.* A distribution with more items near its mean and at its tails and with few items in its intermediate regions relative to a normal distribution with the same mean and variance (Sokal and Rohlf 1969, 113).

cf. distribution: bimodal distribution, mesokurtic distribution, platykurtic distribution

log-normal distribution, logarithmic normal distribution *n.* A distribution of logs of a variable that is normal.

mesokurtic distribution *n.* A distribution that has a close resemblance in peakedness to a normal one (Lincoln et al. 1985).
n. mesokurtosis

normal demographic distribution *n.* The age distribution of the sexes and castes in a population with a high degree of fitness (Wilson 1975, 15).

normal distribution, Gaussian distribution *n.* A probability distribution described by $[(2\pi)^{-1/2}]^{(-x^2/2)}$; if x has a normal distribution then $y = \mu + \sigma(x)$ is said to have a normal distribution with mean μ and a standard deviation σ (Lincoln et al. 1985).

syn. bell-shaped distribution, Gaussian distribution

overdispersed distribution *n.*
1. A spatial dispersion pattern characterized by individuals' being more evenly distributed than would be the case in a random dispersion pattern (Wittenberger 1981, 619).
   syn. hyperdispersed distribution
   cf. distribution: underdispersed distribution, uniform distribution
2. Contagious distribution, *q.v.* (Lincoln et al. 1985).
Comment: This term can have opposite meanings depending on the author.

patchy distribution See distribution: contagious distribution.

platykurtic distribution *n.* A distribution with fewer items near its mean and at its tails and with more items in its intermediate regions relative to a normal distribution with the same mean and variance (Sokal and Rohlf 1969, 113).

d – g

cf. distribution: bimodal distribution, lepto-kurtic distribution, mesokurtic distribution

Poisson distribution *n.* A mathematical distribution used to describe, or test for, a random distribution, or process, and as a model of randomly distributed populations in which an individual's presence at any given point does not increase or decrease another individual's probability of occurring nearby (Sokal and Rohlf 1969, 83); in a Poisson distribution, the variance is approximately equal to the mean.
syn. Poisson series (Lincoln et al. 1985)

random distribution *n.*
1. A distribution whose outcome is a result of chance alone (Lincoln et al. 1985).
2. A distribution in which the presence of one biont has no influence on the distribution of other bionts (Lincoln et al. 1985).
syn. random dispersion pattern (Lincoln et al. 1985)

regular distribution See distribution: uniform distribution.

spatial distribution *n.* "The distribution of organisms in space" (Lincoln et al. 1985).

superdispersed distribution See distribution: contagious distribution.

underdispersed distribution *n.*
1. Uniform distribution, *q.v.*
   syn. hypodispersed distribution
2. Contagious distribution, *q.v.*
syn. infradispersed distribution (Lincoln et al. 1985)
Comment: This term has opposite meanings.

uniform distribution, regular distribution *n.* A distribution (def. 3) with observations more regularly spaced than in a random distribution, indicating that the presence of one observation decreases the probability of another nearby observation (Lincoln et al. 1985).
cf. distribution: positive binomial distribution, underdispersed distribution

zero-truncated-Poisson distribution *n.* A Poisson distribution with groups of one or more individuals; *e.g.,* Howler-Monkey troop sizes following epidemics show this distribution (Wilson 1975, 133).

♦ **disturbance stridulation** See animal sounds: stridulation.

♦ **ditch** *n.* A long, narrow trench or channel dug in the ground, typically used for irrigation or drainage (Michaelis 1963).
Comment: A ditch is usually at least seasonally wet; contrasted with a swale, a natural depression, *q.v.* (Voss 1972, 20).
[Old English *dīc*, related to DIKE]

♦ **ditopogamy** See -gamy: ditopogamy.

♦ **ditotoky** See -toky: ditotoky.

♦ **ditypism** See -morphism: dimorphism.

♦ **diurnae** *adj.* Referring to day-flying insects (Lincoln et al. 1985).
cf. nocturnae

♦ **diurnal** *adj.*
1. Referring to an organism that is active during daylight hours (Lincoln et al. 1985).
2. Referring to a phenomenon that lasts for only one day (Lincoln et al. 1985).
3. Referring to a phenomenon that occurs during the day, as distinct from the night (Bligh and Johnson 1973, 945).
cf. crepuscular, matinal, nocturnal, pomeridanus
[Latin *adj. diurnus < dies,* day]

semidiurnal *adj.* Referring to a phenomenon (*e.g.,* a tide) that has a period of about one half a lunar day (12.42 hours) (Lincoln et al. 1985).

♦ **diurnal rhythm** See rhythm: diurnal rhythm.

♦ **dive, diving** *n.* For example, in diving ducks: an individuals's quick entry followed by moving deeper into water in an arching path to feed on submerged vegetation (Peterson 1947, 36; Welty 1966, 442).
v.i. dive
cf. dabble

neck-over-neck diving *n.* In swans: an epigamic behavior that occurs just before mating when members of a pair, with one's neck over that of the other, dip their bills into water several times (Petzold 1964 in Heymer 1977, 53).

♦ **divergent evolution, divergence** See ²evolution: divergent evolution.

♦ **diversifying selection** See selection: disruptive selection.

♦ **diversionary display** See display: diversionary display.

♦ **diversity** *n.*
1. The total number of species in an assemblage, community, or sample (Lincoln et al. 1985).
   syn. species richness (a confusing synonym) (Lincoln et al. 1985)
   See index: species richness.
2. A measure of the number of species and their relative abundances in a community (Lincoln et al. 1985).
   syn. species diversity
3. The condition of having differences with regard to a given character or trait (Lincoln et al. 1985).
cf. index: diversity index

α-diversity *n.* The diversity of animal species within a community or habitat (Whittaker 1972 in Southwood 1978, 420).
Comments: Researchers describe α-diversity as a single number (a scalar) (Southwood 1978, 420). Researchers use α-diversity with regard to all organism taxa.

β-diversity *n.* A measure of the rate and extent of change in species along a gradient, from one habitat to another (Whittaker 1972 in Southwood 1978, 420).
Comment: Researchers describe -diversity as a single number (a scalar) (Southwood 1978, 420). Researchers use β-diversity with regard to all organism taxa.

τ-diversity *n.* The richness in species of a range of habitats in a geographical area (*e.g.,* an island) (Whittaker 1972 in Southwood 1978, 420).
Comment: τ-diversity is a consequence of α-diversity of habitats together with the extent of β-diversity among them (Southwood 1978, 420).

low diversity *n.* Diversity (def. 2) involving only a few species or species with unequal abundances (Lincoln et al. 1985).

high diversity *n.* Diversity (def. 2) involving many species or species with similar abundances (Lincoln et al. 1985).

hyperdiversity *n.* Diversity in which a certain taxon contains more species, genera, or higher ranked groups than is expected by a null model of random assortment (Dial and Marzluff 1989 in Ehrlich and Wilson 1991, 759).

species diversity See diversity (def. 2).

♦ **diversity index** See index: species-diversity index: diversity index.

♦ **diving** See dive.

♦ **diving reflex** See reflex: diving reflex.

♦ **division of labor** *n.*
1. Organisms' parceling out functions among their body parts, life-history stages, or colony or society members that are at least to some degree specialized for these functions (Immelmann and Beer 1989, 77).
2. In social animals: workers' (of different body types, sex, or age) performing different tasks in their colonies (Lindauer 1952 in Heymer 1977, 31; Immelmann and Beer 1989, 77).
syn. work division
cf. sociality

♦ **divorce** *n.* Breaking a pair bond, *q.v.* (Wilson 1975, 331).

♦ **dixenic** See -xenic: dixenic.

♦ **dixenous** See -xenous: dixenous.

♦ **dixeny** See -xeny: dixeny.

♦ **dizygotic** See -zygotic: dizygotic.

♦ **DNA** See nucleic acid: deoxyribonucleic acid.

♦ **DNA chip** *n.* A device that determines which genes are expressed in a particular tissue (Wade 1998, C1).

♦ **DNA fingerprint, DNA fingerprinting** See method: DNA fingerprinting.

♦ **DNA world** See world: DNA world.

♦ **DNAML Computer Program** See program: DNAML Computer Program.

♦ **doctrine, dogma** *n.*
1. That which a person, or group, presents for acceptance or belief; teachings, as of a religious or political group (Michaelis 1963).
2. A particular principle or tenet that is taught, or a body of such principles or tenets (Michaelis 1963).
See dogma.
cf. axiom, dogma, law, principle, rule, theorem, theory, truism
[Old French < Latin *doctina*, teaching < *docere*, to teach]

doctrine of specific nerve energies *n.* A given nerve always produces the same type of sensation regardless of its mode of stimulation (Müller 1827 in McFarland 1985, 166).

♦ **Dodsonian mimicry** See mimicry: Dodsonian mimicry.

♦ **doe** See animal name: doe.

♦ **dogma, doctrine** *n., pl.* **dogmas, dogmata**
1. A person's, or group's, belief, principle, or tenet more or less formally stated and held to be authoritative (Michaelis 1963).
2. A system of such beliefs or principles: the dogmas of art (Michaelis 1963).
See doctrine.
cf. axiom, doctrine, law, principle, rule, theorem, theory, truism
[Latin < Greek *dogma*, *-atos*, opinion, tenet < *dokeein*, to think, deem right]

♦ **Dollo's law, Dollo's law of irreversibility** See law: Dollo's law.

♦ **domatium** *n., pl.* **domatia**
acarodomatium *n.* A specialized chamber produced by a plant that houses mites that may maintain leaf hygiene (Lundströem 1887 in O'Dowd et al. 1991, 88).
syn. domatium (O'Dowd et al. 1991, 88)
Comment: Domatia and associated mites are found in Eocene fossils (O'Dowd et al. 1991).

myrmecodomatium *n.* A specialized structure (*e.g.,* an inflated stem, inflated leaf base, leaf-blade pouch, swollen node with a cavity, hollow thorn, pseudobulb, or hollow root) that a plant evolved for housing ants (Bequaert 1922, etc. in Hölldobler and Wilson 1990, 536, who describe many kinds of myrmecodomatia).
syn. domatium (Hölldobler and Wilson 1990, 531)

♦ **domestic-bliss strategy** See strategy: domestic-bliss strategy.

♦ **domestication** *n.*
1. A person's accustoming a nonhuman animal to live under human care and near human habitations; taming or bringing (another Human, a nonhuman animal) under control (*Oxford English Dictionary* 1972, verb entries from 1641).

2. A person's more or less controlling the breeding, care, and feeding of a particular kind of animal (*e.g.,* Cattle, Sheep, etc.) (Hale 1969, 22, in Dewsbury 1978, 267).

3. Organisms' adaptation to life in intimate association with Humans (Lincoln et al. 1985).

cf. parasitism: social parasitism: slavery; selection: artificial selection; taming

Comment: Nonhuman animals may undergo behavioral, morphological, physiological, etc. changes due to domestication (Dewsbury 1978, 267).

[*v.t.* domesticate, Medieval Latin *domesticatus*, past participle of *domesticare*, to live in a house < Latin *domus*, house]

◆ **domestication character, domestication trait** See character: domestication character.

◆ **domicile** *n.* A home, nest, burrow, tube, den, or other refuge (Lincoln et al. 1985). See ³range: home range.
cf. hibernaculum

◆ **domicole** See -cole: domicole.

◆ **dominance, ecological dominance** *n.*

1. The extent to which a particular species predominates in a community due to its size, abundance, coverage, or a combination of these traits and affects the fitness of its associated species (Lincoln et al. 1985).

2. The abundance of one taxon compared to others and its overall "ecological and evolutionary impact" on all other organisms (Wilson 1990 in Ruse 1993, 58).

See need: dominance.
cf. dominance: faunal dominance, genetic dominance, social dominance

codominance *n.* A plant community's having two equally dominant species (Lincoln et al. 1985).
n. codominant

faunal dominance *n.* A species' dominating its community (*sensu* dominance, def. 3) or dispersing to another community and dominating it (Michener 1987, 449).
cf. hypothesis: age-and-area hypothesis
Comment: Some biologists deny the existence of "faunal dominance."

interspecific dominance *n.* In some bird, rodent, and ungulate species; Humans: one species' dominance over another during their encounters (Wilson 1975, 296).

relative dominance *n.* A measure of one species' relative importance in a habitat calculated as this species' basal area as a percentage of all species' total basal area (Lincoln et al. 1985).

◆ **dominance, genetic dominance** *n.* An allele's fully manifesting its phenotype

when present in a heterozygous, heterokaryotic, or heterogenotic state (King and Stansfield 1985).
syn. complete dominance
cf. allele: dominant allele; dominance (genetic): incomplete dominance; law: Mendel's laws; pleiotropy; recessiveness; ³theory: Fisher's theory of dominance
Comment: An allele may be dominant for one or more phenotypic characteristic that it influences but recessive for others.
[This term was introduced by Gregor Mendel as *dominierend* (Mayr 1982, 715).]

codominance See dominance (genetic): incomplete dominance.

delayed dominance *n.* An allele's dominance that expresses itself late in an organism's development (*e.g.,* in Huntington's chorea in Humans) (King and Stansfield 1985, 144).

directional dominance *n.* An F_1 character's being closer to one parental value than to the other (Dewsbury 1978, 127).

incomplete dominance *n.* The effect of alleles upon one another resulting in a heterozygote that shows intermediate phenotypic characteristics (Dawkins 1982, 285).
syn. codominance, semidominance (Lincoln et al. 1985)
cf. dominance: complete dominance; recessiveness
Comment: Incomplete dominance was first noted by Mendel and is found in both plants and animals (Mayr 1982, 735, 959).

overdominance *n.* Superior phenotypic expression of a heterozygote over both of its related homozygotes (Lincoln et al. 1985).
cf. heterosis

semidominance See dominance (genetic): incomplete dominance.

◆ **dominance, social dominance** *n.*

1. An animal's showing behavior without reference to similar behavior of a subordinate animal (Maslow 1935, 58).

2. One animal's determining behavior of others by aggressive behavior or other means (Collias 1944, 83).

3. An animal's feeding, sexual, and locomotion priority, and its being superior in aggressiveness in controlling one or more other animals (Wilson 1975, 287).

4. One individual's priority of access to an approach situation, or of leaving an avoidance situation, that it has over another individual (van Kreveld 1970, 146, in Dewsbury 1978, 92, 298).

5. One animal's, or conspecific group's, ruling another animal or group (Morse 1974, Rowell 1974 in Wilson 1975, 279).

6. An animal's chastising "another with impunity" (Klopfer 1974, 154, in Dewsbury 1978, 92).

7. An animal's "consistent winning at points of social conflict, regardless of tactic used" (Hand 1986, 201).

Note: Crawford's (1939) definition is similar to this one.

syn. dominance, primary-social dominance (in many papers; Hand 1986, 206)

cf. dominance (ecological); dominance (genetic); need: dominance; spheres of dominance; submission; subordination

Comments: Hand (1986, 202) indicates that there is no agreement regarding how to define, or measure, social dominance. Kinds of social dominance include primary social dominance and secondary social dominance (Hand 1986, 201).

intergroup dominance *n.* In some social-insect species; the rhesus monkey and common langur: one conspecific group of animals' rule over another such group in a confrontation (Wilson 1975, 295).

monarchistic dominance *n.* One individual's dominance of its intraspecific group that represses the interactions between, or among, other group members (Lincoln et al. 1985).

syn. monarchy (Lincoln et al. 1985)

primary-social dominance *n.* Social dominance that an animal obtains chiefly by using superior force, real or apparent; contrasted with secondary social dominance [coined by Hand 1986, 206].

syn. dominance, social dominance (in many contexts, Hand 1986, 206)

▸ **derived-primary social dominance** *n.* Primary social dominance that "is due to an asymmetry in potential force that is not primarily intrinsic to the individual, although to utilize the asymmetry the animal must usually have reasonably well-developed dominance feelings" (Kawai 1958); *e.g.,* the animal may be supported by allies (Walters 1980) or may use a weapon, or instrument, of intimidation (Lawick-Goodall 1968 in Hand 1986, 206–207).

▸ **intrinsic-primary social dominance** *n.* Primary social dominance that "is due to superiority in an asymmetry of some relevant *personal* trait that determines the amount of force than can or would be exercised (superior size, age, fighting experience, or confidence born of familiarity with the arena; Yasukawa, 1979)" (Hand 1986, 206).

secondary-social dominance *n.* Social dominance that an animal obtains chiefly by using leverage advantage, *q.v.,*

not superior force; contrasted with primary social dominance [coined by Hand 1986, 207].

♦ **dominance aggression** See aggression: dominance aggression.

♦ **dominance hierarchy** See hierarchy: dominance hierarchy.

♦ **dominance hypothesis** See hypothesis: dominance hypothesis.

♦ **dominance order** See hierarchy: dominance hierarchy.

♦ **dominance rank** See [1]rank: dominance rank.

♦ **dominance-subordination social relationship** See relationship: dominance relationship: dominance-subordination social relationship.

♦ **dominance system** See hierarchy: dominance hierarchy.

♦ **dominant** *n.*

1. "The highest ranking animal in a dominance hierarchy" (Lincoln et al. 1985).
   syn. α, alpha (Lincoln et al. 1985)

2. An organism that exerts considerable influence upon a community due to its size, abundance, or coverage (Lincoln et al. 1985).
   cf. allele: dominant allele, dominance: faunal dominance

♦ **dominant allele** See allele: dominant allele.

♦ **dominant species** See [2]species: dominant species.

♦ **dominate** See dominance (social).

♦ **dominion** *n.* "A form of spatial organization in which individuals defend nonexclusive use of their home ranges; the degree to which intruders are excluded usually decreases with increasing distance from a central area of exclusive use" (Wittenberger 1981, 614).
   cf. territory, [3]range

♦ **dominule** *n.* A dominant organism in a microhabitat, or a community, in a serule (Lincoln et al. 1985).

♦ **-domous**

lithodomous, lithotomous *adj.* Referring to an organism that lives in holes in rock, or that bores into rock (Lincoln et al. 1985).

monodomous, monodomic *adj.* In ants: referring to single colonies that occupy one nest (Lincoln et al. 1985).

polydomous *adj.* In ants: referring to single colonies that occupy more than one nest (Wilson 1975, 592).
   syn. polycalic, polydomic (Lincoln et al. 1985)

♦ **donor** *n.* An organism that helps another organism.

super-donor *n.* An organism that is a very efficient helper and may have extreme morphological specializations (*e.g.,* workers of some kinds of social insects) (West Eberhard 1975, 13).

cf. altruism, super-beneficiary

♦ **Doppler effect, Doppler shift** See effect: Doppler effect.

♦ **dormancy** *n.* In animals: a state of relative metabolic quiescence, including anabiosis (crypotobiosis), diapause, estivation (aestivation), hibernation, and hypobiosis (Lincoln et al. 1985).

cf. hypnote

anabiosis See biosis: anabiosis.

diapause *n.* In some annelid and arthropod species: an animal's state of arrested development and reduced metabolic rate, during which its growth, differentiation, and metamorphosis cease (Barnes 1974, 300; Borror et al. 1989, 794).

Comments: Diapause can occur during dry, warm periods or cool periods. It is not necessarily correlated with adverse environmental conditions (Borror et al. 1989, 794). Some workers classify "hibernation" and "estivation" as kinds of "diapause" (Immelmann and Beer 1989, 131).

estivation, aestivation *n.*

1. An organism's passing the summer in a state of torpor or suspended animation (given as the verb "aestivate" in the *Oxford English Dictionary* 1972, entries from 1626).
2. An organism's remaining dormant, or torpid, during a dry season or extreme heat of summer; "summer sleep" (Darwin 1845 in the *Oxford English Dictionary* 1972).
3. For example, in some arthropod, crocodile, fish, and frog species; many desert-animal species: an individual's passing a dry, warm period in a dormancy (Dewsbury 1978, 54).

v.i. estivate, aestivate

cf. aestival; dormancy: diapause, hibernation

hibernation *n.* For example, in some amphibian, bird, mammal, reptile, and terrestrial-mollusc species: an individual's passing an unfavorable period of low temperature in dormancy characterized by a sleep-like state with lowered body temperature, rates of respiration, and heartbeat (McFarland 1985, 294–295).

v.i. hibernate

cf. dormancy: diapause, estivation

[Latin *hibernatus,* past participle of *hibernare,* to pass the winter < *hiems,* winter]

▸ **partial dormancy** *n.* For example, in the European Brown Bear and American Black Bear: an individual's hibernation in which its body temperature falls to about 30°C (McFarland 1985, 295).

▸ **true hibernation** *n.* In some species of small mammals including badgers, raccoons, and woodchucks: an individual's hibernation in which its body temperature falls as low as 2°C (McFarland 1985, 295; Line 1997, C2).

Comments: Some birds show a similar kind of torpor (McFarland 1985, 295). Body temperature of hibernating woodchucks is about 5°C, and if a hibernating woodchuck is not below that frost line it can freeze to death (Line 1997, C2).

hypobiosis See biosis: hypobiosis.

torpidity, torpor *n.* In hummingbirds: dormancy that lasts only a few hours during which an individual's body temperature falls to that of its environment (McFarland 1985, 295).

♦ **dormer** See nest: wasp nest: dormer.

♦ **dorsal-light reaction** See taxis: dorsal-light reaction.

♦ **dorsoventral abdominal vibration** See dance: bee dance: dorsoventral abdominal vibrating dance.

♦ **dorsoventral flattening** See camouflage: background imitation: dorsoventral flattening.

♦ **double autogamy** See -gamy: cytogamy.

♦ **double-blind experiment** See experiment: double-blind experiment.

♦ **double crossover** *n.* The phenomenon in which two crossovers occur on the same chromosome and which simulates a nonoccurrence of crossing over for distant genes (Mayr 1982, 767).

♦ **double helix** *n.* The Watson-Crick model of DNA structure that involves plectonemic coiling of two hydrogen-bonded polynucleotide, antiparallel strands wound into a right-handed spiral; the strands are connected by base pairs (one purine and one pyrimidine per pair) (King and Stansfield 1985).

Comments: In plectonemic coiling, two parallel threads coil in the same direction about one another and cannot be separated unless they are uncoiled. Antiparallel strands are those that have opposite orientations; *i.e.,* one string is oriented from left to right 5′ to 3′, and the other is left to right 3′ to 5′ (King and Stansfield 1985). Watson and Crick's discovery of the double helix in 1953 had a profound impact not only on genetics but also on embryology, evolutionary theory, physiology, and even philosophy (Mayr 1982, 824).

♦ **double selection** See selection: double selection.

♦ **dout** See ²group: dout.

♦ **down** See ²group: down.

♦ **down-up movement** See ²movement: down-up movement.

♦ **downward causation** *n*. The phenomenon that a whole can affect properties of components at lower levels (Campbell 1974, 182 in Mayr 1982, 64).

♦ **downward classification** *n*. Classification by logical division (Mayr 1982, 150). *ant.* upward classification

♦ **doylt** See ²group: doylt.

♦ **Dr. Fox effect** See effect: Dr. Fox effect.

♦ **draught** See ³lactation: milk ejection (comments).

♦ **drave** See ²group: drave.

♦ **dray** See ²group: dray.

♦ **dress** *n*.
  1. In later use, a person's external clothing that serves for adornment as well as covering (*Oxford English Dictionary* 1972, entries from 1868).
  2. An external covering and adornment (*e.g.*, birds' plumage) (*Oxford English Dictionary* 1972, entries from 1713).
  cf. coloration

advertising dress *n*. For example, in males of many animal species, females of the phalarope and seed snipe: sexual dimorphism in which one sex (usually males, sometimes females) is more conspicuous in form, coloration, or both, than the other sex (Immelmann and Beer 1989, 7).
  syn. nuptial dress
  cf. eclipse plumage

juvenile dress, juvenile coloration *n*. In many vertebrate species: coloration of a juvenile that is markedly different than a conspecific adult (Immelmann and Beer 1989, 161).

  ▸ **nuptial dress, nuptial coloration** *n*. For example, in stickleback-fish, salamander, and duck species: advertising dress that occurs only during a species' breeding season (Bakker and Sevenster 1983 in Rowland 1989, 282; Immelmann and Beer 1989, 7).

♦ **drift** *n*.
  1. An object's moving along, or being carried along, in a current of fluid; an object's drifting; a slow course or current (*Oxford English Dictionary* 1972, entries from 1562).
  2. A force, or influence, that drives something along steadily in a given direction (Michaelis 1963).
  3. "The course along which something is directed, or the direction in which it tends; tendency or intent: the *drift* of conversation" (Michaelis 1963).
  See ²group: drift.
  cf. error: observer error

continental drift See hypothesis: continental drift.

continuous drift *n*. Genetic drift in which a population remains small in size

and sampling error is effective in each generation (Wilson 1975, 65).

genetic drift *n*. Alteration of a population's gene frequencies through sampling error (chance processes) alone, not natural selection, mutations, or immigration; its most important effect is the loss of population heterozygosity (Wilson 1975, 64, 585). See drift: random drift.
  syn. drift, Sewall-Wright effect (Lincoln et al. 1985); neutral drift (Ridley 1996, 151); random drift, random genetic drift (Dobzhansky et al. 1977, 108, 568)
  cf. drift: intermittent drift, social drift; ³theory: shifting-balance theory, Sewall-Wright theorem
  Comments: In his first major account, Wright (1931a) expressed himself in a way that sounded as if he were proposing genetic drift as an alternative mechanism to natural selection, and this caused considerable confusion. Also, although Wright meant genetic drift to designate stochastic processes of changes in allele frequencies in small populations, it was misinterpreted by some workers as a steady one-directional drift (Mayr 1982, 555). The bottleneck effect, founder effect, and Hagedoorn effect can be viewed as special cases of genetic drift.
  [coined by Sewall Wright (Mayr 1982)]

intermittent drift *n*. Genetic drift occurring during one generation, not in succession with another generation in which genetic drift occurred (Wilson 1975, 65).

random drift *n*. Fluctuations in the frequency of variations that have no adaptive significance or are otherwise equally fit (concept suggested by Darwin 1859 in Beatty 1992b, 274). See drift: genetic drift.
  syn. genetic drift (in some cases, Dobzhansky et al. 1977, 568)
  Comments: "Random drift is a heterogenous category of evolutionary causes and effects, whose overall significance relative to other modes of evolution (especially evolution by natural selection) has been greatly disputed" (Beatty 1992b, 271). Phenomena called "random drift" usually involve a biological form of random, or indiscriminate, sampling and consequent sampling error. Beatty (1992b) seems to synonymize random drift and genetic drift; depending on the author, phenomena included in the category of random drift are the founder effect, Gulick effect, Hagedoorn effect, and Sewall-Wright effect. Wright broadened "random drift" "to include some cases of natural selection (where selection pressures vary at random)" (Beatty 1992b, 278). The "Gulick effect" is named

after J.T. Gulick. Fisher and Ford (1950 in Beatty 1992, 273) referred to random drift as the "Sewall-Wright effect," named after the first and last names of Wright; perhaps they tried to avoid the self-fulfilling sense conferred by the "Wright effect" (Beatty 1992b, 273).

social drift *n*. "The random divergence in the behavior and mode of organization of societies or groups of societies;" contrasted with tradition drift (Wilson 1975, 13, 595).

Comment: This process is due to behavioral differences that are not the result of adaptation to physical environmental conditions or underlying genetic changes due to genetic drift (Wilson 1975, 13, 595); *e.g.,* in Humans, a new idea that spreads (Wilson 1975, 14).

tradition drift *n*. Social drift that is based purely on differences in experience of individuals and hence is passed on as part of tradition; contrasted with social drift (Burton 1972 in Wilson 1975, 14, 597).

♦ **drifting** *n*. An organism's passive transport by wind or water current; distinguished from locomotion (Immelmann and Beer 1989, 176).
cf. locomotion

♦ **drimium** See ²community: drimium.

♦ **drimophile** See ¹-phile: drimophile.

♦ **drinking** *n*. An animal's taking liquid into its mouth and swallowing it (Michaelis 1963).
cf. ingestion

vacuum drinking *n*. An animal's drinking movements made without accompanying fluid uptake (Heymer 1977, 180).

♦ **driodium** See ²community: driodium.

♦ **¹drive** *n*.
1. An energizing, as opposed to a directing, aspect of motivation (Woodworth 1918 in McFarland 1985, 275).
   Note: Woodworth introduced drive as a alternative to McDougall's concept of instinct (McFarland 1985, 275).
2. "A complex of internal an external states and stimuli leading to a given behavior" (Thorpe 1951, 4).
3. A strong, motivating power or stimulus (Michaelis 1963).
4. A hypothetical physiological variable that in some way reflects an organism's internal state and alters its tendencies to employ certain behaviors (Alcock 1979, 144).
5. An organism's internal mechanism that works over a longer term than a motivation, *q.v.,* and turns on and off, or modulates, behavioral programs and, as a consequence, the organism's responsiveness to stimuli (Gould 1982, 189).
6. In psychology, the mobilization of an animal's action as a consequence of its bodily need (Immelmann and Beer 1989, 80).
7. A quantitative factor (*e.g.,* hours of food deprivation, rate of bar pressing) that relates the dependent and independent variables in an equation (Immelmann and Beer 1989, 80).
   Note: This is an operational definition.
syn. arousal (Dewsbury 1978, 172), concept of drive, instinct (in one sense) motivation, mood (in one sense of drive, Immelmann and Beer 1989, 80), motivation (Hinde 1970, 200), motivational state (Wilson 1975, 153), tendency, variable: intervening variable
cf. arousal, instinct, motivation, motivational state, urge

Comments: "Drive" is a problem term with many definitions. Schöne et al. (1972, etc. in Heymer 1977, 83) suggest that "drive" be used in preference to "motivation" because motivation has diverse psychological and ethological connotations. Some authors state that "drive" is not synonymous with "instinct;" others seem to make them synonymous. McFarland (1985, 275–278) discusses why some animal behaviorists no longer use "drive." Immelmann and Beer (1989, 80) state that "drive" "is now more or less avoided except in contexts that do not pretend to theoretical precision." Some of the hundreds of kinds of named "drives" are defined below.
[Old English *drīfan,* introduced to behavior by Woodworth 1918]

activity drive *n*. In psychology, an animal's "fundamental need to expend physical energy, as expressed in restlessness or in general movement" (Storz 1973, 3).

affectional drive *n*. In psychology, an animal's "urge or need to be close to or in contact with another living being either physically or emotionally through some form of tie or bond" (Storz 1973, 7).

affiliative drive *n*. In psychology, an animal's "urge or need to associate with other living beings in order to form social attachments" (Storz 1973, 7).

allochthonous drive *n*. A drive that causes a displacement behavior; contrasted with autochthonous [introduced by Kortlandt 1940 in Immelmann and Beer 1989, 10].

autochthonous drive *n*. A drive that causes a nondisplaced behavior; contrasted with allochthonous drive [introduced by Kortlandt 1940 in Immelmann and Beer 1989, 11].

bonding drive, gregarious disposition, herd instinct, social drive, social tendency *n*. A drive that draws conspecifics together (Immelmann and Beer 1989, 34–35).

Comment: This is a controversial concept (Immelmann and Beer 1989, 34–35).

cleaning drive *n.* A drive that relates to a bird's having a strong tendency to preen after bathing or a fish's strong tendency to get cleaned at a cleaning station (Heymer 1977, 135).

curiosity drive *n.* In psychology, an animal's "motive or need for stimulation" (Storz 1973, 62).

exploratory drive *n.* In psychology, an animal's "drive related to curiosity" (Storz 1973, 95).

general drive *n.*
1. Possibly in many animals species: an individual's central drive that governs all of its other drives (Thorpe 1963, 8). *Note:* "General drive" is closely linked with the ideas of "expectancy" and "purpose" (Thorpe 1963, 13).
2. An organism's overall tendency to behave (McFarland 1985, 278).

predatory drive *n.* A drive that activates an animal to hunt prey (Heymer 1977, 42). *syn.* predatory motivation

primary drive *n.* A drive resulting from tissue needs; contrasted with secondary drive (Woodworth 1918 in McFarland 1985, 275).

secondary drive *n.* A drive resulting from learned habits; contrasted with primary drive (Woodworth 1918 in McFarland 1985, 275).

sex drive *n.* In psychology, an animal's urge or need to engage in sex (Storz 1973, 252).

social drive See drive: bonding drive.

thirst drive *n.* In psychology, an animal's "need to satisfy thirst" (Storz 1973, 276).

♦ **²drive** *n.* An urgent pressure or rush (Michaelis 1963).

meiotic drive *n.* A meiotic deviation from a Mendelian segregation ratio in heterozygotes that results exclusively from a disturbance in the meiotic mechanism; *e.g.,* segregation distortion (= an exception to the rule of a one-to-one recovery of segregating alleles) (Rieger et al. 1991). *syn.* segregation distortion (Lincoln et al. 1985) *cf.* allele: meiotic-drive locus, segregation-distorter allele, t allele *Comments:* A driven allele, or chromosome (= one that becomes more frequent by meiosis), may increase in frequency even when it is deleterious (Rieger et al. 1991). In some cases, meiotic drive might result from the action of complex genetic systems rather than a single gene (= a meiotic-drive locus). "Drive" refers to the fact that favored genes spread through a population in spite of any deleterious effects they may have on organisms (Dawkins 1982, 290). Meiotic drive is a cause of maintenance of otherwise deleterious genes in populations (Mayr 1982, 768).

♦ **drive model** See ⁴model: drive model.

♦ **drive reduction** *n.* In drive-reduction theory, reinforcement's removal of a need state and hence reduction of the drive associated with it (Immelmann and Beer 1989, 80).

♦ **drive satisfaction** *n.* An organism's cessation of showing a behavior after performing an appropriate consummatory act (Heymer 1977, 177).

♦ **drive-training interlocking** See instinct-training interlocking.

♦ **driving Y-chromosome** See chromosome: driving Y-chromosome.

♦ **-dromous** *combining form* Running (Michaelis 1963). *n.* -dromy [Greek *dromous,* running + -OUS]

anadromous *adj.* Referring to an animal that migrates from salt to fresh water, *e.g.,* fish that migrate from the sea into a river for spawning (Lincoln et al. 1985). *n.* anadromesis, anadromy

catadromous, katadromous *adj.* Referring to an animal that migrates from fresh to salt water, *e.g.,* a fish that moves into the sea for spawning (Lincoln et al. 1985).

diadromous *adj.* Referring to an animal that migrates between salt and fresh water (Lincoln et al. 1985), *e.g.,* anadromous and catadromous fish.

oceanodromous *adj.* Referring to organisms that migrate only within the ocean (Lincoln et al. 1985).

potamodromous *adj.* Referring to organisms that migrate only within fresh water (Lincoln et al. 1985).

♦ **drone** See animal sounds: drone; male: drone; Appendix 1, Part 3, bee: drone.

♦ **drone parasitism** See parasitism: drone parasitism.

♦ **dropping** See feces.

♦ **drosophile** See ²-phile: drosophile.

♦ **drought** *n.* A period of abnormally low precipitation. *syn.* drouth (Michaelis 1963) *Comments:* The worst recent drought in the U.S. occurred from March 1930 through February 1931 and gave rise to the Dust Bowl (Salmon and Bombardieri, 1999, A1). The second worst drought occurred in the Mid-Atlantic and Northeast U.S. from August 1998 through July 1999. [Old English *drūgath;* akin to DRY]

d – g

♦ **drove** See ²group: drove.

♦ **drug** *n.*

1. Any chemical, or biological, substance other than food that a person uses to treat, prevent, or diagnose a disease in a Human or nonhuman animal (Michaelis 1963).
2. A chemical, or biological, substance other than food that a person uses to modify his own, or another animal's, behavior (*e.g.,* a psychedelic drug).

Comment: A few of the thousands of kinds of drugs are below.

ampakine *n.* A drug that enhances memory amplifying signals received by AMPA receptors (Gary Lynch in Hall 1998, 28, 49).

donepezil *n.* A drug, developed in Japan in the early 1980s, used for treating Alzheimer's disease (Hall 1998, 49).
syn. Aricept™ (a brand name) (Hall 1998, 49)

gamma hydroxybutyrate (GHB) *n.* A highly addictive drug that depresses a person's central nervous system and even causes death (Bradsher 1999, A8).
Comments: Very small doses of GHB can cause mild euphoria and sometimes hallucinations (Bradsher 1999, A8). Larger amounts, even tiny doses, cause lassitude, unconsciousness, or even respiratory failure and death. GHB can also stop someone from remembering, and sexual predators have put it in women's drinks, giving it the name "date-rape drug."

Inderal® *n.* A drug, used as a beta blocker for cardiac patients (Hall 1998, 49).

medroxyprogesterone acetate, Depo-Provera® *n.* A drug used for treating male sex offenders (Leary 1998, A11).

Prozac® *n.* A drug that is widely used to treat human depression (Hall 1998, 56).
Comments: Cesare Mondadori, head of research for central-nervous-system drugs at Hoechst Marion Roussel, says that people pay about $1.8 billion per year for Prozac, and one third of the users have no medical need for it (Hall 1998, 56).

psychopharmacological drug *n.* A drug that affects an animal's central nervous system and leads to changes in its behavior, *e.g.,* strychnine, which affects human short-term memory, or certain antibiotics that affect human long-term memory (Immelmann and Beer 1989, 239).

tacrine *n.* A drug, discovered in 1977, used for treating Alzheimer's disease (Hall 1998, 49).
syn. Cognex™ (a brand name) (Hall 1998, 49)

triptorelin *n.* A drug that reduces the blood level of testosterone, pedophilia, and other sexual tendencies in men (Leary 1998, A11).
Comments: Triptorelin is a synthetic hormone that mimics a hormone made by a person's hypothalamus (Leary 1998, A11).

♦ **drumming** See animal sounds: drumming.

♦ **dry-bulb temperature** See temperature: ambient temperature.

♦ **dry prairie** See habitat: prairie: dry prairie.

♦ **dryft** See ²group: dryft.

♦ **drymium** See ²community: drymium.

♦ **dual-inheritance model** See ⁴model: dual-inheritance model.

♦ **dualism** *n.*

1. The theory, or system of thought, that recognizes two independent principles: (1) mind and matter exist as distinct entities, as opposed to idealism and materialism; and (2) there are two independent principles, one good and the other evil (*Oxford English Dictionary* 1972, entries from 1794).
2. The theory that a Human's body and mind are two different entities but are intimately related and interacting (Michaelis 1963).
3. A belief that organisms have separate material and spiritual attributes (Mayr 1982, 357).

cf. ³theory: organismic theory

♦ **duet, duette, duetting** See animal sounds: duetting; animal sounds: song: duet.

♦ **Dufour's gland** See gland: Dufour's gland.

♦ **dule** See ²group: dule.

♦ **dulosis** See parasitism: social parasitism: slavery.

♦ **dummy** See ²model: stimulus model.

♦ **dune slack** See habitat: dune slack.

♦ **Dunnett's test** See statistical test: multiple-comparisons test: Dunnett's test.

♦ **dupe** *n.* An animal enemy, or victim of a mimic, whose senses are receptive to a model's signals and is thus deceived by the mimic's similar signals in a mimicry system, *q.v.* (Pasteur 1972 in Pasteur 1982, 169).
syn. enemy, operator, predator, signal receiver (Pasteur 1982, 171, who discusses problems with these synonyms)
cf. mimicry

♦ **duplicate gene** See gene: duplicate gene.

♦ **duration-meaningful behavior** See behavioral state.

♦ **DVAV, D-VAV** See dance: bee dance: dorsoventral-abdominal-vibrating dance.

♦ **dwarfism** See nanism.

♦ **dyad** *n.* An interrelationship of two individual animals, *e.g.,* often a sexually interacting pair (Hinde 1982, 278).

♦ **dyadic relationship** See relationship: social relationship: dyadic relationship.

♦ **Dyar's rule** See law: Dyar's rule.

♦ **dymantic behavior** See behavior: deimatic behavior.

♦ **dynamic** *adj.* Producing, or involving, changes or action (Michaelis 1963).

Comment: This is one of several definitions of dynamic.

heterodynamic *n.*

1. Referring to an organism that has a life cycle with two or more distinct forms, or stages, with dissimilar biological activity (Lincoln et al. 1985).
2. Referring to different genes that simultaneously effect the same developmental process (Lincoln et al. 1985).

homodynamic *n.*

1. Referring to an organism with a life cycle that progresses from generation to generation without a resting stage with a distinct change of biological activity (Lincoln et al. 1985).
2. Referring to different genes that simultaneously effect different developmental processes (Lincoln et al. 1985).

♦ **dynamic psychology** See study of: psychology: dynamic psychology.

♦ **dynamic selection** See selection: directional selection.

♦ **dynamics** See study of: dynamics; study of: physics: dynamics.

♦ **dysgenesis** See -genesis: dysgenesis.

♦ **dysgenic** See -genic: dysgenic.

♦ **dysmetric load** See genetic load: dysmetric load.

♦ **dysploidy** See -ploidy: dysploidy.

♦ **dysteleology** See teleology: dysteleology.

♦ **dystropic** See -tropic: dystropic.

♦ **dyticon** See ²community: dyticon.

♦ **Dzierzon's rule** See rule: Dzierzon's rule.

d–g

♦ **e** See unit of measure: erg.

♦ **E-species** See ²species (comments).

♦ **eat-and-run hypothesis** See hypothesis: eat-and-run hypothesis.

♦ **eavesdropper** *n.* In some firefly species: an illegitimate receiver of another firefly's signal (Burk 1982, 93).
syn. illegitimate receiver (Burk 1982, 93; Alcock 1983, 449)
v.i. eavesdrop

♦ **ecad, oecad, ecophene, ecophenotype, oecophne, habitat type** *n.* A plant, or animal, form produced in response to particular habitat factors with characteristic adaptations that are not heritable (Lincoln et al. 1985).
See -type, type: ecophene, ecotype, phenotype: ecophenotype.
cf. ²group: variety

♦ **ecdemic species** See ²species: ecdemic species.

♦ **ecdysis** *n.* For example, in arthropods: an individual's shedding, or molting, its external skeleton (= exoskeleton) (Lincoln et al. 1985).
cf. molt
diecdysis *n.* An animal's continuous molting with one molting grading rapidly into the next (Lincoln et al. 1985).

♦ **ecdysteroids** See hormone: ecdysteroids.

♦ **ece** See habitat.

♦ **echo detector** See cell: echo detector.

♦ **echolocation** *n.* In many bat species; possibly Shrews; Dolphins, Porpoises, Whales; the Asiatic Cave Swiftlet, Oilbird: an animal's finding objects in its environment by using sonar (Spallanzani 1793 in Heymer 1977, 57; Wilson 1975, 475; Immelmann and Beer 1989, 82).
v.t., v.i. echolocate
Comments: Many bats, for example, produce ultrasounds whose echoes they hear

with finely tuned ears (Allen 1996, 639). The echoes enable them to locate nearly objects and fly among objects without bumping into them.

♦ **-ecious** See -oecism.

♦ **eclectic taxonomy** See taxonomy: evolutionary taxonomy.

♦ **eclipse plumage** *n.* In many bird species: a male's dull, inconspicuous plumage that alternates with his brighter breeding plumage (Lincoln et al. 1985).
cf. dress: advertising dress

♦ eclosion *n.*
1. In holometabolous insects: the emergence of an adult (imago) insect from its pupa (Torre-Bueno, 1978).
2. In most insect species: an individual's hatching from its egg (Torre-Bueno, 1978).
v.i. eclose

♦ **ecochronology, oekochronology** *n.* A person's dating biological events using paleoecological evidence (Lincoln et al. 1985).

♦ **ecoclimatology** See study of: climatology: ecoclimatology.

♦ **ecocline** See cline: ecocline.

♦ **ecodeme** See deme: ecodeme.

♦ **ecodichogamic** See -gamic: ecodichogamic.

♦ **ecoethology** See study of: ecology: behavioral ecology.

♦ **ecogeographical rule** See rule: ecogeographical rule.

♦ **ecography** See study of: -graphy: ecography.

♦ **ecological aggression** See aggression: ecological aggression.

♦ **ecological biochemistry** See study of: ecology: chemical ecology.

♦ **ecological biogeography** See study of: geography: biogeography: ecological biogeography.

♦ **ecological community** See ²community.

- **ecological conflict** See ³conflict: ecological conflict.
- **ecological death** See mortality: density-dependent mortality.
- **ecological dominance** See dominance (ecological).
- **ecological efficiency** See ¹efficiency: ecological efficiency.
- **ecological efficiency rule** See rule: ecological efficiency rule.
- **ecological environment** See environment: ecological environment.
- **ecological equivalent** See ²species: ecological equivalent.
- **ecological genetics** See study of: genetics: ecological genetics.
- **ecological geography** See study of: geography: ecological geography.
- **ecological isolation** See isolation: ecological isolation.
- **ecological longevity** *n.* A biont's mean life span in a particular population under stated conditions (Lincoln et al. 1985).
- **Ecological Monitoring and Assessment Program** See program: Ecological Monitoring and Assessment Program.
- **ecological niche** See niche.
- **ecological pressure, environmental pressure** *n.* "The set of all environmental influences that constitute the agents of natural selection (Wilson 1975, 32).
- **ecological pyramid** *n.* A graphical representation of the trophic structure and function of an ecosystem (Allaby 1994, 131).
  syn. Eltonian pyramid [after Charles Elton, British ecologist, who devised the ecology pyramid]
  Comments: The first trophic level is the producers (usually plants), which form the base of the pyramid. The second level is the primary consumers; the third, the secondary consumers; the fourth, the tertiary consumers.

  pyramid of biomass *n.* An ecological pyramid of biomass (organism weight) at different trophic levels, usually plotted as dry matter or calorific value per unit area or volume; contrasted with pyramid of energy and pyramid of numbers (Allaby 1994, 321).
  Comments: A "pyramid of biomass" is typically a gradually sloping pyramid, except where organism sizes vary dramatically from one trophic level to another (Allaby 1994, 321). In this case, smaller organisms' higher metabolic rate may result in a greater biomass of consumers than producers, giving an inverted pyramid. Aquatic communities in winter typically show inverted biomass pyramids.

  pyramid of energy *n.* An ecological pyramid of rates of energy flow through different trophic levels of an ecosystem; contrasted with pyramid of biomass and pyramid of numbers (Allaby 1994, 321).
  Comments: A pyramid of energy reflects the rates of photosynthesis, respiration, etc. (and not the standing crop, as in a pyramid of biomass), can never be inverted because energy is dissipated through the ecosystem, and is the most fundamental and most useful of the three ecological pyramids.

  pyramid of numbers *n.* An ecological pyramid of numbers of individual organism present at each trophic level of an ecosystem; contrasted with pyramid of biomass and pyramid of energy (Allaby 1994, 321).
  Comments: A pyramid of numbers is less useful than a pyramid of biomass and a pyramid of energy because it makes no allowance for the organisms' different sizes and metabolic rates (Allaby 1994, 321). A pyramid of numbers typically slopes more steeply than the other pyramids and can be inverted (*e.g.,* when it is based on studies of temperate forests in summer).
- **ecological race** See ¹race: ecological race.
- **ecological release** *n.*
  1. The reduction in interspecific competition that allows a species' trait variation (*e.g.,* in social structure) which seems suboptimal to persist (Wilson 1975, 550).
  2. A species' using a broader range of resources in the absence of a competitor compared to in its presence (Futuyma 1986, 32).
  syn. niche expansion
- **ecological restoration** *n.* A person's effort to bring back a lost ecosystem; contrasted with compensatory mitigation (Roberts 1993b, 1891).
  Comment: Many ecologists favor this practice (Roberts 1993b, 1891).
- **ecological sociology** See study of: synecology.
- **ecological succession** See ²succession.
- **ecological time** See time: ecological time.
- **ecological zoogeography** See study of: geography: zoogeography: ecological zoogeography.
- **ecology** See study of: ecology.
- **economic defensibility** *n.*
  1. An animal's defending the amount of terrain, the defense of which gains more energy than the animal expends (Brown 1964 in Wilson 1975, 268).
  2. The value of an animal's aggressively controlling a particular resource in terms of its inclusive fitness (Wittenberger 1981, 615).

d — g

Note: A resource is economically defensible if the benefits of defense exceed the costs (Wittenberger 1981, 615).
cf. optimality, parsimony, territoriality

♦ **ecoparasitism** See parasitism: ecoparasitism.

♦ **ecophene** See -type, type: ecophene, phenotype.

♦ **ecophenotype** See -type, type: phenotype: ecophenotype.

♦ **ecophysiology** See study of: physiology: ecophysiology.

♦ **ecospace** See niche (def. 7).

♦ **ecospecies, ecological species** See ²species: ecospecies.

♦ **ecosphere** See -sphere: biosphere.

♦ **ecosystem** See ¹system: ecosystem.

♦ **"ecosystem-diversity-productivity hypothesis"** See hypothesis: "ecosystem-diversity-productivity hypothesis."

♦ **ecosystem ecology** See study of: ecology: ecosystem ecology.

♦ **ecosystematics** See systematics.

♦ **ecotone** *n.*
1. A zone of tension between two ecological communities (Clements 1907, 297, in Harris, 1988, 330); the zone wherein two different groups of plants contend for dominance (Michaelis 1963).
2. The boundary, or transitional zone, between adjacent communities or biomes (Lincoln et al. 1985).
syn. tension zone (Lincoln et al. 1985)
cf. effect: edge effect
Comment: Some biologists consider an ecotone to be a distinct community because it is inhabited by distinct species (Johnson 1947 in Harris 1988, 330).
[Greek *oikos*, home + *tonus*, stress, tension]
edge-interior ecotone *n.* An area of a focal habitat from where it has an edge with another habitat through partly within the focal one (Didham 1997, 69).

♦ **ecotope** *n.*
1. A habitat type within a large geographic area (Lincoln et al. 1985).
2. A "species' full range of environmental relationships including both niche and habitat factors (some of which are closely related [Whittaker et al., 1973])" (Whittaker 1977, 27).

♦ **Ecotron** *n.* A greenhouse-like chamber used for experiments on species diversity and productivity (British group, Nature, 1994 in Yoon 1996b, C4).
cf. hypothesis: "ecosystem-diversity-productivity hypothesis"

♦ **ecotrophobios** See trophallaxis.

♦ **ecotype** See -type, type: ecotype.

♦ **ecotypification** *n.* Formation of ecotypes in a particular habitat (Lincoln et al. 1985).

♦ **ectendotroph** See -troph-: ectendotroph.

♦ **ecto-, ect-** *combining form* "Without, outside; external" (Michaelis 1963).
[Greek *ekto-* < *ektos*, outside]

♦ **ectocommensal** See symbiosis: commensalism.

♦ **ectocrine substance** See chemical-releasing stimulus: semiochemical: allelochemic: allomone: exocrine substance.

♦ **ectoderm** See organ: -derm: ectoderm.

♦ **ectogenesis** See -genesis: ectogenesis.

♦ **ectogenous** See -genous: ectogenous.

♦ **ectohormone, ectoincretion** See chemical-releasing stimulus: semiochemical: pheromone.

♦ **ectoophage** See -phage: ectoophage.

♦ **ectoparasite** See parasite: ectoparasite.

♦ **ectophage** See -phage: ectophage.

♦ **ectophagous hyperparasitism** See parasitism: hyperparasitism: ectophagous hyperparasitism.

♦ **ectosite** See parasite: ectoparasite.

♦ **ectosymbiosis** See symbiosis: ectosymbiosis.

♦ **ectotherm** See -therm: ectotherm.

♦ **ectotroph** See -troph-: ectotroph.

♦ **ectozoon** See -zoon: ectozoon.

♦ **ecumene** See -sphere: biosphere.

♦ **ecumenical** See ubiquitous.

♦ **edaphic climax** See ²community: climax: edaphic climax.

♦ **edaphology** See study of: edaphology.

♦ **edge-interior ecotone** See ecotone: edge-interior ecotone.

♦ **edge species** See ²species: edge species.

♦ **education psychology** See study of: psychology: education psychology.

♦ **EEG** See -gram: electroencephalogram.

♦ **effect** *n.*
1. Something brought about by a particular cause, or agency; result; consequence (Michaelis 1963).
2. Something's capacity to produce a particular result; efficacy (Michaelis 1963).
area effect *n.* The effect of an area of a habitat on its organisms, *e.g.,* the size of a patch of forest on the kinds of organisms that live in the patch (Didham 1997, 55).
Baldwin effect, Baldwin-Waddington effect *n.*
1. A phenomenon in which an extreme individual's previously hidden genetic potential (and its associated trait) is exposed by an environmental change, the change favors an increase in the number of individuals with the trait, and finally the population may evolve to a point where most individuals develop the trait spontaneously even if the species' environment returns to its original state (*e.g.,* a population's shifting from

individuals with territories to one with dominance hierarchies) (Baldwin 1896 and Waddington 1953 in Wilson 1975, 72–73).

2. "A largely hypothetical evolutionary process … whereby natural selection can create an illusion of the inheritance of acquired characteristics. Selection in favour of a genetic tendency to acquire a characteristic in response to environmental stimuli, and eventual emancipation from the need for them" (Spalding 1873 in Dawkins 1982, 284).

syn. organic selection (Baldwin 1902 in Bowler, 1992, 192)

cf. genetic assimilation, organic selection [after J.M. Baldwin]

beater effect *n.* An animal's, or human-made vehicle's, flushing up prey by its moving, thus making prey more vulnerable to capture by predators (Wittenberger 1981, 613).

Beau Geste effect, Beau Geste hypothesis *n.* An animal's "pretending" to be several animals at once, *e.g.,* a male bird's "fooling" other birds that the density of males in his area is high with his large song repertoire, thus the other birds do not settle in the area (Immelmann and Beer 1989, 283).

[named from the novel of the same name in which a unit of the French Foreign Legion used a comparable tactic]

bottleneck effect *n.* Genetic drift that occurs in a population that has become reduced in size, typically by a natural disaster, and results in a population that is no longer genetically representative of the original population (Lincoln et al. 1985; Campbell 1980, G-3).

syn. bottlenecking (Campbell 1996, 422)

cf. bottleneck; drift: genetic drift; effect: founder effect

Comments: Early Humans might have witnessed the bottleneck effect when calamities decimated tribes (Campbell 1990, 443). Northern elephant seals witnessed this effect due to overhunting. This effect, caused by the last ice age and overhunting, might have caused low genetic variability in Cheetahs.

[after the metaphor of a population's genes moving through the narrow neck of a bottle (Campbell 1996, 422, illustration)]

Bruce effect *n.* In some strains of mice: a recently impregnated female's failure to implant embryos and rapid return to estrus after she is exposed to a male with an odor sufficiently different from that of her stud (Bruce 1960 in Dewsbury 1978, 185; Whitten and Bronson 1970, Bronson 1971 in Wilson 1975, 154, 321; Forsyth 1985, 31).

cf. infanticide, natural birth control [after Hilde Bruce]

cage effect *n.* The inadvertent differences in organisms in different cages of an experiment that result from different environments within the cages (or other chambers), although a researcher attempts to maintain the cages under the same conditions.

Comments: Rigorous biologists control for cage effects in their experiments. The chambers (cages) include aquaria, terraria, and wire cages. "Plot effects," which are analogous to cage effects, can occur when a researcher uses areas such as cultivated fields, indoor greenhouse areas, lakes, ponds, streams, watersheds, and woodlots. Regarding a cage effect, consider an experiment that tests the H_0 that male longevity = female longevity in adult Mexican Bean Beetles. A researcher tries to make the environment of each cage identical by putting five 1-day-old beetles of the same sex and a flower pot of five bean seedlings of the same age and variety into each of 20 cages and by using a growth chamber designed to have uniform conditions throughout. However, the cage environments will still vary due to differences among the individual plants, different beetle interactions, slight differences in climate in different parts of the growth chamber, and other possible factors. Further, as beetles die, the cages will have different numbers of interacting individuals. To control for a cage effect in this experiment, a researcher uses one mean value of longevity per cage for quantitative analysis. The experiment has 20 longevity means as its data, not 100 individual longevity values, which would be a data set with pseudoreplication, *q.v.* Another way to control for cage effect is to place only one beetle in a cage at the onset of an experiment.

Concorde effect, Concorde fallacy, Concorde strategy *n.* An organism's "deciding" its future behavior on the basis of its past resource investment in "goal A," instead of on possible future benefits that it could obtain from abandoning A and investing in a new goal, B. It may thus continue to invest in A even though its costs will exceed its benefits from this investment (Dawkins and Carlisle 1976 in McFarland 1985, 108). Possibly found in the great golden digger wasp (Dawkins and Brockmann 1980).

[after the Concorde airliner and decisions regarding building it]

confusion effect *n.* Prey animals' grouping that decreases a predator's efficiency in obtaining a prey (*e.g.,* the schooling of fish) (Heymer 1977, 101).

Coolidge effect *n.* In some fish and mammal species: the phenomenon that a male animal can be stimulated to resume copulation (that had just previously waned) by replacing his mate with a new one or reoffering his present mate in a new setting (Immelmann and Beer 1989, 60).

Darling effect See effect: Fraser-Darling effect.

depletion effect *n.* A species' decline in immigration rate to an island as its number of resident species increases (Lincoln et al. 1985).

depression effect *n.* In laboratory rats: one group of test animals that initially received a larger reward for a particular behavior than a second test group shows the behavior at a lower rate, or intensity, than the second group if its reward is decreased to the second group's level (Dewsbury 1978, 354).

diploid effect *n.* A phenotypic character of a sperm that is dictated by its producer's diploid genotype (Sivinski 1984, 95).

Doppler effect, Doppler shift *n.* In physics, the change in frequency of a sound, light, or other wave, caused by movement of its source relative to an organism perceiver (Michaelis 1963).
Comment: Bats can use Doppler shifts in their echoes to determine relative velocities of their prey (D.B. Quine, personal communication).
[after C.J. Doppler]

Dr. Fox effect *n.* Students' possessing the "illusion of having learned" due to a lecturer's personality, presentation style, or both, even if the lecturer gives erroneous information [coined by Naftulin et al. 1973 in Brewer 1981, 121].

edge effect *n.*
1. The increase in species richness and population sizes in edges formed by the junction of two communities, habitat types, or seral stages, compared to the communities, habitats, or stages themselves (Leopold 1933, Odum 1971, Smith 1977, and others in Reese and Ratti 1988, 127).
syn. edge-effect principle (Harris 1988, 330)
Note: These edges are ecotones, *q.v.* (Lincoln et al. 1985). This edge effect occurs in organisms including birds and spiders (Aspey 1976, 42; LaRue et al. 1994), but not in mammals in highly fragmented landscapes (Heske 1995, 562). Wildlife managers often consider edges to be beneficial because they

enhance the amount of big-game forage and provide improved habitats and nesting sites for woodland passerine birds (references in Reese and Ratti 1988, 127). "Standard habitat management guides include the prescription to create as much 'edge' as possible because wildlife is a product of the places where two habitats meet" (Yoakum and Dasmann 1971 in Harris 1988, 330). Wildlife managers and others debate the value of creating edges because of their beneficial and deleterious effects on habitats.
2. The effect of a habitat edge on a species' abundance, behavior, disease incidence, distribution, or other attribute (inferred from Dennis 1984a, 85; 1984b, 157; Alverson et al. 1988, 348; Roland and Kaupp 1995, 1175; Sekgororane and Dilworth 1995, 1432; Oğurlu 1996, 427).
Note: Many butterfly species, primarily in the Pieridae, tend to lay their eggs on host plants at the edges of host-plant patches (Dennis 1984a, 85).
3. The effect of Humans', or other factors', rapid creation of abrupt edges in a large unit of previously undisturbed habitat on the community in this habitat (Lovejoy et al. 1986, Soulé 1986 in Reese and Ratti 1988, 127).
Note: This edge effect can lead to reduced populations of species that are dependent on large blocks of previously undisturbed habitat, large blocks of forest interior, or both (references in Reese and Ratti 1988, 127).

effect of volition See hypothesis: volition.

fixed effect *n.* A main effect "whose levels are the only relevant levels of interest," contrasted with random effect, *q.v.* (given as "fixed factor" in Kleinbaum and Kupper 1978, 246).
Comment: Fixed effects include sex, age, marital status, and education. Locations, treatments, drugs, and tests can be either fixed effects or random effects, depending on the experimental situation.

founder effect *n.*
1. A new population's being started by a small number of individuals that carry only a fraction of the genetic variability of the parental population; hence, the new population differs statistically from its parental population (Mayr 1942, 237, in Beatty 1992b, 274; Wilson 1975, 65, 584).
2. The genetic differentiation of isolated populations that arises because founders of these populations differ genetically from each other and the average members of their parent population (Wittenberger 1981, 615).

3. A cause of genetic drift that is attributable to colonization by a limited number of individuals from a parent population (Campbell 1990, G-10).
4. Genetic drift in a new colony of organisms (Campbell 1990, 444).

Notes: The founder effect probably contributed to divergent evolution of Darwin's finches after the first birds landed on the remote Galápagos Islands from the mainland of South America (Campbell 1990, 444). This effect is also probably responsible for the relatively high frequency of certain inherited disorders found in human populations that were established by small numbers of colonists, *e.g.,* congenital total color blindness in Pingelap Islanders, diastrophic dystrophy in Finns, familial hyperchylo-micronemia in Quebecois, retinitis pigmentosa on Tristan da Cunha, and Tay-Sachs diseases in Ashkenazi Jews (Campbell 1996, 423; Hartl and Clark 1997, 291).

syn. founder principle
cf. drift: genetic drift, random drift; effect: bottleneck effect

Fraser-Darling effect, Fraser-Darling law *n.*
1. For example, in the African-Village Weaverbird, Herring Gull, Red-Winged Blackbird, Tri-Colored Weaverbird, Viellot's Blackweaver: an individual's stimulation, or synchronization, of breeding with conspecifics due to social stimulation beyond its sexual pairing alone (Darling 1938 in Wilson 1975, 41–42, 584; Wittenberger 1981, 615).
2. A breeding group's acceleration and synchronization of its reproductive state due to communal social stimulation (Lincoln et al. 1985).

syn. Darling effect (Lincoln et al. 1985)
cf. courtship: communal courtship; effect: Orians effect
[after Fraser Darling, biologist (Immelmann and Beer 1989, 113)]

green-beard effect See hypothesis: green-beard effect.

greenhouse effect *n.*
1. A greenhouse's warming as a result of sunlight entering it and heating it and the building's glass stopping the heat from radiating out of it (Rensberger 1993a, A1).
Note: A greenhouse also warms due to having still air (Allaby 1994).
2. The warming of Earth's lower atmosphere that results from sunlight passing through its atmosphere as short-wave radiation ($< 4\,\mu m$) and heating it because clouds and gases (including carbon dioxide, chlorofluorocarbons and other halocarbons, methane, and water vapor) absorb the heat and reradiate it as long-wave radiation ($> 4\,\mu m$), much of which does not exit the atmosphere (Rensberger 1993a, A1; Allaby 1994).
Note: The clouds and gases make the atmosphere transparent to incoming short-wave radiation but partially opaque to long-wave radiation.

group effect *n.*
1. An alteration in behavior, or physiology, within a conspecific group caused by signals that are directed in neither space nor time (Wilson 1975, 415, 585).
Note: A simple example is social facilitation, in which there is an increase of an activity merely from the sight or sound (or other form of stimulation) of other individuals engaged in the same activity (Wilson 1975, 415, 585); other examples are eating more by group-housed chicks and other animals compared to solitary animals and facilitation of yawning in a human group (Katzran and Révész 1909 in Tinbergen 1951, 143; Heymer 1977, 168).
syn. partner effect, sympathetic induction (Tinbergen 1951, 143)
2. All the benefits that individual animals accrue from group living, *e.g.,* antipredator vigilance, collective defense, cooperative foraging, and reproductive synchronization (Immelmann and Beer 1989, 121).

cf. effect: Fraser-Darling effect, Orians effect; mood induction; ²facilitation: social facilitation

Gulick effect See drift: genetic drift.

Hagedoorn effect *n.*
1. The phenomenon that much genetic variation in natural populations exists because it is selectively neutral (Fisher 1922, 328, in Mayr 1982, 555).
2. Reduction of genetic variation in a population by means other than natural selection; *e.g.,* a gene can be lost because heterozygous parents may leave homozygous offspring (Beatty 1992b, 274).

cf. drift: genetic drift; ³theory: the neutral theory

Comment: Fisher thought that most allelic polymorphism in populations was due to a superiority of heterozygotes rather than genetic drift (Mayr 1982, 555).
[coined by Ronald A. Fisher (1922, 328, in Mayr 1982, 555) after A.L. and A.C. Hagedoorn, Dutch geneticists who gathered a great deal of evidence in support of a category of random drift]

d – g

Haldane effect *n.* "Sterility or deficiency of the heterogametic sex among offspring in F_1 hybrids" (Scriber 1996, 25).

haploid effect *n.* A phenotypic characteristic of sperm that is produced by expression of its haploid genotype (Thornhill and Alcock 1983, 8).

Comment: Haploid effects are rare in sperm; generally, gamete attributes are controlled by their parents' diploid genotypes (Thornhill and Alcock 1983, 8).

hitchhiking effect *n.* The increase in frequency of a selectively neutral, or nearly neutral, gene that is closely linked to a gene that is favored by selection (introduced by Kojima and Schaffer 1967 in Reiger et al. 1991).

syn. hitchhiking, hitch-hiking effect (Kaplan et al. 1989, 887)

Comments: The hitchhiking effect can cause changes in gene frequency, changes in association of alleles at different loci, and gametic disequilibrium (Reiger et al. 1991).

Lagerstätten effect *n.* One's recording a taxon's apparent, possibly erroneous, time of evolutionary origination based on sampling a single exceptional fossil deposit (Benton 1995, 54).

Comments: The actual origination time of the taxon might be earlier than the one indicated by known fossils (Benton 1995, 54). The actual time of origination of many insect families might be earlier than the date based on their fossils in Baltic amber.

Lansing effect *pl. n.* The phenomenon that a mother's age can influence the life histories of her offspring (Lints 1978 in Willson 1983, 34).

Comment: Lansing effects can be transmissible from one generation to the next and are often cumulative over several generations. A Lansing effect is sometimes reversible (Lints 1978 in Willson 1983, 34).

Lee-Boot effect *n.*
1. In House Mice: loss of regular estrous cycling and display of spontaneous pseudopregnancies of females housed as a group (van der Lee and Boot 1955 in Dewsbury 1978, 185).
2. In female mice: the suppression of estrus and occurrence of pseudopregnancies that occur in an all-female group in the absence of a male (Whitten and Bronson 1970 and Bronson 1971 in Wilson 1975, 154).
[after S. van der Lee and L.M. Boot]

magnet effect *n.* A fish fin's, or pair of fins', maintaining a constant rhythm and inducing other fins to follow the rhythm (von Holst 1937, 1939 in Heymer 1977, 112).
[coined by von Holst]

main effect *n.*
1. In statistics, an independent variable (manipulated, or not, by an investigator) that might, or actually does, affect a dependent variable (Sokal and Rohlf 1969, 187, 305).
   syn. factor (Kleinbaum and Kupper 1978, 246), treatment effect (Sokal and Rohlf 1969, 187), treatment
2. In statistics, the influence of a main effect (def. 1) on a dependent variable (Sokal and Rohlf 1969, 187).
cf. level

Marsden's effect *n.* As subordinate troops of Rhesus Monkeys are confined to smaller spaces due to intergroup dominance, their members fight less among one another; however, within the dominant group which acquires more space, aggressive interactions increase (Marsden 1971 in Wilson 1975, 295–296).

minority effect See rare-male mating advantage.

multiplier effect *n.* Amplification of a small evolutionary change (in the behavior pattern of individuals) into a major social effect by the expanding upward distribution of the effect into multiple facets of social life (Wilson 1975, 11, 589).

Comment: Multiplier effects can speed social evolution (Wilson 1975, 11, 589).

oddity effect *n.* The increased risk of predation that a focal fish witnesses because it is smaller, or larger, than the other fish in its shoal, which has similarly sized fish except for the focal fish (Peuhkuri 1997, 271).

Orians effect *n.* Localized breeding synchrony due to individuals' in the same reproductive state attracting one another to the same area.

cf. courtship: communal courtship; effect: Fraser-Darling effect
[after Gordon H. Orians, biologist (1961 in Immelmann and Beer 1989, 113)]

Petit effect See rare-male mating advantage.

position effect *n.* A change in the phenotypic effect of a gene owing to a change of its position on its chromosome.
[coined by Sturtevant 1925 in Mayr 1982, 798, 805]

Comment: A given gene may have a different effect if it is on the same chromosome as a second gene or on the second gene's homologous chromosome.

random effect *n.* A main effect "whose levels may be regarded as a sample from some large population of levels," contrasted with fixed effect, *q.v.* (given as "random factor" in Kleinbaum and Kupper 1978, 246).

Comment: Random effects include subjects, litters, observers, days, and weeks.

rare-male effect See rare-male mating advantage.

reproductivity effect *n.* In social insects, perhaps other colonial organisms: a colony's productivity rate decreases as its colony size increases (Michener 1964 and Wilson 1971 in Wilson 1975, 36, 594).

Comment: The productivity rate is measured as the number of a colony's new individuals divided by the number of its already-existing ones.

Ropartz effect *n.* A decrease in reproductive capacity of a mouse resulting from its perception of odor of other mice that causes its adrenal glands to grow heavier and increase corticosteroid production (Whitten and Bronson 1970 and Bronson 1971 in Wilson 1975, 154, 247).

runting effect *n.* One human twin's being smaller than another because the latter obtained more nutrients than the former while it was a fetus (Allen 1998, 22).

Sewall-Wright effect See drift: random drift.

Signor-Lipps effect *n.* One's recording a taxon's extinction time that is erroneously too soon because of inadequate sampling up to its actual extinction time (Benton 1995, 54).

sperm effect *n.* In *Drosophila melanogaster:* a female's long-term re-mating inhibition due to stored sperm in her ventral receptacle (Scott, 1987, 142).

Tryon effect *n.* The phenomenon that the F_2 variance of a behavioral trait often fails to exceed the variance of the F_1 or parental strains (Hirsch 1967 in Dewsbury 1978, 123).

Comment: This is probably a result of a behavioral trait's being affected by a large number of genes (Dewsbury 1978, 123).

Wallace effect *n.*

1. The development and reinforcement of reproductive isolation between sexual populations that have reached the level of elementary biological species due to selection against hybrids that have relatively lower fitness than members of these populations (Lincoln et al. 1985).
 syn. Wallace's hypothesis (Lincoln et al. 1985)

2. The situation in which hybrids of two species have low fitness, introgression does not occur, and natural selection then improves reproductive isolating mechanisms so that hybridization diminishes (Price 1996, 79).

Comment: Butlin (1989 in Price 1996, 79) questions whether the Wallace effect occurs.

Whitten effect *n.* In laboratory mice: the more or less simultaneous entrance into the state of sexual receptivity by a group of females (whose estrus cycles have been suppressed by being with other females alone) that results from the females' smelling an odorant found in the urine of a male mouse (Whitten 1956 in Dewsbury 1978, 1985; Whitten and Bronson 1970 and Bronson 1971 in Wilson 1975, 154).

cf. effect: Bruce effect

[after W.K. Whitten]

Zajonc effect *n.* For example, in rats: the phenomenon in which repeated exposure of an animal to a given stimulus object is enough to increase its attraction to the object (Zajonc 1971 in Wilson 1975, 274).

♦ **effect, law of** See law: law of effect.

♦ **effective core temperature** See temperature: effective core temperature.

♦ **effective population number (N_e)** *n.* The number of individuals in an ideal, randomly breeding population with a 1:1 sex ratio that would have the same rate of heterozygosity decrease as the real population under consideration (Wilson 1975, 77, 583).

Comment: Typically, "N_e" is well below the real population number, N(Wilson 1975, 77, 583).

♦ **effector** *n.* A muscle, gland, or color cell that can produce behavior (Immelmann and Beer 1989, 84).

♦ **efference** See -ference: efference.

♦ **efference copy** See *Sollwert.*

♦ **efferent neuron** See cell: neuron: efferent neuron.

♦ **¹efficiency** *n.* The ratio of work done, or energy expended, by an organism or machine to the energy supplied in the form of food or fuel (Michaelis 1963).

ecological efficiency *n.*

1. A food-cycle level's efficiency of production at any level (λ_m) relative to the productivity of any of its previous levels (λ_n), defined as the percentage: $100(\lambda_n)/(\lambda_m)$ (Hutchinson 1942 in Lindeman 1942, 407; Odum 1971, 75; McClendon 1975, 214; Conrad 1977).

2. The ratio of energy per unit time obtained by one population (*e.g.,* a predator population) divided by energy per unit time ingested by another population (*e.g.,* a prey population) (Slobodkin 1960, 222, in Conrad 1977, 99).

3. (Gross) ecological efficiency = ε = Y/I, where Y is the yield (or food extracted from the population) and I is the food (or light energy) available to the population (Phillipson 1966 in Conrad 1977, 101).

syn. biological efficiency, efficiency

cf. ³theory: optimal-diet theory

♦ ²**efficiency** *n*. The relative cost in time, energy, and other resources of collecting data balanced against the strength of the data (Lehner 1979, 83).

♦ **effort** *n*.

1. An organism's expenditure of physical or mental energy to get something done; exertion (Michaelis 1963).
2. "Something produced by exertion" (Lincoln et al. 1985).

reproductive effort *n*.

1. An organism's effort required to reproduce at present, measured in terms of the decrease in its ability to reproduce at later times (Williams 1966, 171–172; Wilson 1975, 95, 317, 594).
2. An organism's effort related to putting its genes into its next generation which is partitioned into mating and parental effort (Low 1978 in Gwynne 1984, 119).
3. The proportion of an organism's total available resources that it uses in reproduction (Thornhill and Alcock 1983, 65–67).

▶ **mating effort** *n*. Reproductive effort (including nuptial gifts) expended by a male in trying to obtain, and actually obtaining, a mate(s) (Alexander and Borgia 1979 in Thornhill and Alcock 1983, 65–67).

nonpromiscuous mating effort *n*. Resources donated by a male to his mate in exchange for copulation (Gwynne in Thornhill and Alcock 1983, 66).

promiscuous mating effort *n*. All activities whereby an organism finds a mate(s) (Thornhill and Alcock 1983, 67).

▶ **parental effort** *n*. The sum of an individual organism's parental investments in its offspring (Gwynne in Thornhill and Alcock 1983, 66).

♦ **egalitarian social relationship** See relationship: social relationship: egalitarian social relationship.

♦ **egalitarian society** See ²society: egalitarian society.

♦ **egestion** See elimination: egestion.

♦ **egg** See gamete: egg, trophic egg.

♦ **egg-brooding** See brooding: egg-brooding.

♦ **egg cannibalism** See cannibalism: egg cannibalism.

♦ **egg dumping** *n*.

1. In some bird species: a female's depositing one or more eggs in a conspecific, or heterospecific, individual's nest (Weller 1959, etc. in Tallamy 1986, 599).
2. In a lace-bug species: a mother's placing her eggs in the egg mass of a neighboring female (Tallamy 1985 in Tallamy 1986, 599).

♦ **egg follicle** *n*. In vertebrates: a ripening ovum's covering which provides it with nourishment and secretes estrogen, thereby influencing behavior (Immelmann and Beer 1989, 210).

syn. Graafian follicle (in mammals) (Immelmann and Beer 1989, 210)

♦ **egg imprinting** See learning: imprinting: egg imprinting.

♦ **egg-laying hormone** See hormone: egg-laying hormone.

♦ **egg mimicry** See mimicry: egg mimicry, Gilbertian mimicry, Wicklerian-Barlowian mimicry.

♦ **egg robbing** *n*. In the dark chub (fish): an individual's eating eggs of another species that are guarded by a male of that species (Baba et al. 1990, 776–777).

cf. robber

♦ **egg-rolling behavior** See behavior: egg-rolling behavior.

♦ *eide* *n*. The limited number of fixed and unchanging forms that comprise the variable world of phenomena (Plato in Mayr 1982, 38).

syn. essences (of the Thomists in the Middle Ages) (Mayr 1982, 38)

♦ **eidetic mental image** *n*. A mental image that has all "characteristics of a percept, especially if it is referred to a sense organ and so seems to come in from the outside" (Bateson 1979, 241).

♦ *eidos* *n*. Something that "orders the raw material into the patterned systems of living beings" (Aristotle in Mayr 1982, 636).

Comments: "*Eidos*" is conceptually virtually identical with the ontogenetic program of developmental physiology (Mayr 1982, 56). Aristotle's theory of inheritance indicates that semen contributed the *eidos* (form-giving principle), and menstrual blood, *catamenia*, is the unformed substance that is shaped by the *eidos* (Mayr 1982, 636).

♦ **eight-months fear** See fear: fear of strangers.

♦ **18S-ribosomal RNA (18S-rRNA)** See nucleic acid: ribonucleic acid: 18S-ribosomal RNA.

♦ **ejaculate, ejaculation** *n*.

1. An animal's, or plant's, sudden and quick discharge of something (*e.g.*, seminal fluid, seeds); emission (*Oxford English Dictionary* 1972, entries from 1603).
2. The sperm and other secretions that an animal discharges at one time (Immelmann and Beer 1989, 289).

v.t. ejaculate

♦ **El Niño** *n., pl.* **Los Niños** Above-average surface warming of water in the South Pacific; contrasted with La Niña (Claiborne 1997a, A14; Mydans 1997b, A6).

Comments: Los Niños occur every 3 to 7 years and are associated with the Southern

Oscillation. They have occurred for several millennia and have major effects on Earth's weather, including delay of onset of monsoon rains in Indonesia; droughts in Borneo, Indonesia, and New Guinea; dry weather in the northern U.S., the South Pacific, and West Africa; above-average rainfall in Asia, Europe, and the southern U.S.; and unseasonable warmth in Southern America and Europe (Mydans 1997b, A6; Suplee 1997b, A1, illustration). Effects on organisms include food shortages and surpluses and changing ocean temperatures that result in more southern animals' occurring in northern waters.
[Spanish *El Niño*, Christ child, named by Spanish sailors, referring to the fact that Los Niños typically show up around Christmas time in Peru (Suplee 1997b, A16).]

♦ **El Niño–Southern Oscillation event (ENSO event)** *n*. The collective effects of El Niño and the Southern Oscillation (Allaby 1994).
Comments: The ENSO event affects climate in many parts of the world (Allaby 1994). A similar phenomenon occurs in the Atlantic. Exceptional ENSO events occurred in 1891, 1925, 1953, 1972–1973, 1982–1983 (strongest one on record), and 1992–1995 (possibly three ENSO events in these 4 years) (Allaby 1994; Suplee 1997b, A16). A major ENSO event occurred in 1997–1998. Effects of ENSO events during the Christmas season are weakening of prevailing trade winds and strengthening of the Equatorial countercurrent in the Pacific.

♦ **elaboration** *n*. Evolution, or production, of complex structures from simpler ones (Lincoln et al. 1985).
cf. reduction

♦ **electrical communication, electrocommunication** See communication: electrical communication.

♦ **electrical reception** See modality: electrical reception.

♦ **electroencephalogram (EEG)** See -gram: electroencephalogram.

♦ **electron transport chain** See chain: electron transport chain.

♦ **electronic synapse** See synapse: electronic synapse.

♦ **electrophoresis** See method: electrophoresis.

♦ **electroporation** See method: electroporation.

♦ **electrotaxis** See taxis: galvanotaxis.

♦ **elfin forest** See ²community: elfin forest.

♦ **elimination** *n*. An organism's expelling something (waste matter, foreign substances, etc.), especially to get rid of it from its tissues (*Oxford English Dictionary* 1972, entries from 1794 as the verb "eliminate").

cf. behavior: eliminative behavior; elimination: egestion; movement

defecation *n*. An animal's excretion of its feces or egesta (Heymer 1977, 105).
syn. poop (euphemistic), shit (obscene)
v.t. defecate, defaecate
Comment: "Defecation" has many euphemistic synonyms, especially related to Humans (Spears 1981).

egestion *n*. An organism's voiding fecal, or regurgitated, matter (Lincoln et al. 1985).
syn. defecation, elimination, and movement (for fecal matter) (Lincoln et al. 1985)
cf. behavior: defensive behavior: enteric discharge; elimination; excretion; ingestion; movement

excretion *n*. An organism's eliminating wastes from its body (Lincoln et al. 1985).
v.t. excrete

urination *n*. An animal's voiding, or passing, urine (Michaelis 1963).
v.t. urinate
syn. pee (euphemistic), piss (obscene)
Comment: "Urination" has scores of other euphemistic synonyms, especially related to Humans (Spears 1981).

♦ **eliminative behavior** See behavior: eliminative behavior.

♦ **elite, élite** *n*.
1. The choicest part, as of a particular social group (Michaelis 1963).
2. An insect colony member that displays greater than average initiative and activity (Wilson 1975, 285).
3. *adj.* Referring to an elite (Wilson 1975, 582).
[French < élire < Latin *eligere*]

♦ **eluvium** See ²community: eluvium.

♦ **emancipation** *n*.
1. The evolutionary process that results in a new signal function's being derived from an animal's pre-existing motor pattern with a different original function (Tinbergen 1952 in Wilson 1975, 225); *e.g.*, genital presentation displays of baboons (Wickler 1966 in Heymer 1977, 119).
2. A derived behavior's disengagement during ritualization, *q.v.*, from its original causal control and its transferral to the control of the causal system in which it functions in communication (Immelmann and Beer 1989, 85).
cf. behavior: ritualized behavior, ritualization

♦ **EMAP** See program: Ecological Monitoring and Assessment Program.

♦ **emasculation** See castration.

♦ **embarrassment** See gesture of embarrassment.

♦ **emblem** See communication: nonverbal communication.

◆ **embryo** *n.*

1. An animal's developmental state that extends from fertilization of an egg to birth or hatching (*Oxford English Dictionary* 1972).
   Note: Older embryos can make distinct and controlled movements, including vocalization in some bird species (Immelmann and Beer 1989, 85).
2. A human fetus *in utero*, less than 4 months old (*Oxford English Dictionary* 1972, entries from 1350).
3. A rudimentary plant contained in a seed (*Oxford English Dictionary* 1972, entries from 1728).

◆ **embryogenesis** See -genesis: embryogenesis.

◆ **embryology** See study of: embryology.

◆ **emergence** *n.* The appearance of previously unsuspected new characteristics, or properties, at higher levels of integration in complex hierarchical systems (Morgan 1894 in Mayr 1982, 63, 131).
syn. compositional method (Simpson 1964b), emergentism (Gherardi 1984, 387), fulguration (Lorenz 1973)
Comment: Wheeler wrote that a novel behavior arises "from the specific interaction or organization of a number of elements, whether inorganic, organic, or mental, which thereby constitute a whole, as distinguished from their mere sum, as 'resultant'" (Gherardi 1984, 387).

◆ **emergent character** See character (comments).

◆ **emergent evolution** See ²evolution: emergent evolution.

◆ **emergentism** See emergence.

◆ **emergy, solar emjoule** *n.* The amount of solar energy (low-quality energy) to make a product of high-quality energy, expressed in emjoules (Odum et al. 1988, 28).
cf. solar transformity
Comment: The "emergy" required to produce 1 emjoule of coal is 40,000 emjoules.

◆ **Emery's rule** See rule: Emery's rule.

◆ **emetophobia** See phobia (table).

◆ **emigration** *n.* An individual's, or society's, movement from one nest site, or other location, to another (Wilson 1975, 583).

◆ **emoticon, smiley** *n.* A combination of typewriter characters used to make symbols that represent a psychological state, a physiological state, or a person; commonly used in computer messages (Garreau 1993, D1, D10).
Comments: Emoticons are turned 90 degrees clockwise.
[emoticon, *emotion* + *icon*; smiley, after the happy face of "Have a nice day"]

SOME EMOTICONS

EMOTICON	MEANING
%−)	befuddlement
:−O	shocked and amazed
:−	a kiss
'·:−(	very hot
% }	very drunk
%−)	one's state after staring at a video monitor for 15 hours straight
%+{	lost a fight
:'−(	a tear of sympathy
:−(	sad
:−C	really unhappy
:.−(	bawling
;−)	I'm saying this with a wink.
;·)	winking
:·}	President Richard Milhouse Nixon
@:−{}	Mrs. Tammy Faye Bakker
−>=:−)X	Zippy the Pinhead
=| :−)	President Abraham Lincoln
=:O]	President Bill Clinton
:(=)	President Jimmy Carter
+O<:−)	The Pope
7:^]	President Ronald Reagan
>−< −	Bill the ant run over by a truck
8(:−)	Micky Mouse
8(:−)8	Ms. Annette Funicello
:−)−!<	Mr. Hank Aaron

◆ **emotion** *n.*

1. That which leads a person's "condition to become so transformed that his judgement is affected, and which is accompanied by pleasure and pain" (Aristotle in Calhoun and Solomon 1984, 44).
2. Any agitation or disturbance of a person's mind, his feeling, passion; "any vehement or excited emotional state" (*Oxford English Dictionary*, entries from 1660).
3. "A person's mental 'feeling' or 'affection' (of pleasure or pain, desire or aversion, surprise, hope or fear, etc.), as distinguished from cognitive or volitional states of consciousness" (*Oxford English Dictionary* 1972, entries from 1808).
4. "'Feeling' as distinguished from other classes of mental phenomena" (*Oxford English Dictionary*, entries from 1808).
5. Modifications of a person's body that increase or decrease his active powers (Spinoza in Calhoun and Solomon 1984, 72).
6. A person's "hereditary 'pattern-reaction' involving profound changes of the bodily mechanism as a whole but particularly of the visceral and glandular systems" (Watson 1919, 195, in Lyons 1980, 18).

7. Broadly, an animal's episodes, or sequences, of overt and incipient somatic adjustment, often loosely patterned and variable, usually with concurrent exciting sensory effects, perhaps also perceptual attitudes characterizable as desirable or undesirable, pleasant or unpleasant, related to the intensity effects or perceptual meaning of a stimulus and synergic with organic changes of approach or withdrawal types (Schneirla 1959, 26).

8. A person's "felt tendency toward an object judged suitable, or away from an object judged unsuitable, reinforced by specific bodily changes according to the type of emotion" (Arnold and Gasson 1968, 203).

9. A person's "acutely disturbed affective state that is psychological in origin and revealed in three aspects as: (1) behavior, (2) conscious experience, (3) bodily processes, especially visceral functioning" (Young, 1961a, 358).

10. A person's "strong surge of feeling marked by an impulse to outward expression and often accompanied by complex bodily reactions; any strong feeling, as love, hate, or joy" (Michaelis 1963).

11. A person's "power of feeling; sensibility" (Michaelis 1963).

12. In Humans: one of "a set of processes that (1) reflect the state of relative organization or disorganization of an ordinarily stable configuration of neural systems; and (2) reflect those mechanisms which operate to redress an imbalance, not through action, but by the regulation of input" (Pribram 1967, 4).
Note: Characteristics of emotion include input from sensory systems and viscera, orienting behavior evoked by this input, arousal or activation, cognitive processes in the form of memory, against which input is matched and which provides direction and course of emotional experience and response (Pribram 1967, 4).

13. "An inferred complex sequence of reactions to a stimulus, and includes cognitive evaluations, subjective changes, autonomic and neural arousal, impulses to action, and behavior designed to have an effect upon the stimulus that initiated the complex sequence" (Plutchnik 1980, 81–83).

14. In nonhuman animals: "An internal motivator that creates a readiness to change behavior to increase adaptation" (Weinrich 1980, 133).
Note: Characteristics of an emotion include: it can be discharged by an act that would (if successful) bring the external world more into line with what would be adaptive for an individual having the emotion, or it can be dissipated by a further change in the individual's environment, or internal cognition, and reduces the value of adaptive action; also, it is the result of a conscious, or unconscious, decision-making process, and it results from an external event's changing what is adaptive to do for the individual feeling the emotion (Weinrich 1980, 133).

15. "One of a set of welfare responses marked by total nervous system action in which that of the vegetative system predominates" (Jacobson 1987, xi).

syn. affect, emotional stimulus (in some cases, Hinde 1985)

cf. behavior: emotional behavior; emotionality; test: open-field test

Comments: Many workers generally do not use subjective connotations of "emotion" with regard to nonhuman animals. For these animals, they use emotion, or kind of emotion, in a sense close to that of motivational state, arousal, or action tendency (Immelmann and Beer 1989, 85–86). Storz (1973, 81) indicates that there is no satisfactorily precise definition of "emotion."

[Latin *emotio, -onis < e-*, out + *movere*, to move]

emotion-as-a-state *n.* An emotional stimulus intervening between another stimulus, or stimuli, and a response (Hinde 1985, 986).
Comment: As a state, emotion may affect behavior of a short to long duration (state-trait continuum); it may carry physiological implications; it may be thought of as energizing behavior directly, as a reinforcing event, as a goal or positive incentive, or as a modulator, or disrupter, of behavior; it may be seen as related to different numbers of behavior from one or more limited groups of behavior to many, or all, types of behavior; and it may, or may not, be regarded as associated with subjective experience (Hinde 1985).

emotion-as-an-output *n.* An animal's behavior occurring in conditions of excitement, danger, joy, etc. (Hinde 1985, 986).
syn. emotional behavior (sometimes), motivated behavior (sometimes), emotional expression (sometimes, Hinde 1985, 986)
Comment: Both an animal's somatic and autonomic nervous systems are usually involved, and its movement involved may, or may not, have been adapted for a signal function. This definition has problems because one cannot clearly distinguish emotional from nonemotional behavior. Emotional behavior may lie along a continuum

d – g

from behavior that is more or less purely expressive to that concerned primarily with negotiation between individuals (Hinde 1985, 989).

♦ **emotional ambivalence** *n.* A person's simultaneously having two incompatible emotions (*e.g.,* love and hate) (Heymer 1977, 22).

♦ **emotional behavior** See behavior: emotional behavior; emotion-as-an-output.

♦ **emotional deprivation** *n.* Removal of certain emotional stimuli by an experimenter from an animal or group of animals (Hinde 1985, 985).

♦ **emotional intelligence** See intelligence: emotional intelligence.

♦ **emotive** *adj.* Referring to emotional responses that are induced in communication (Sebeok 1962 in Wilson 1975, 217).
cf. cognitive

♦ **empathic learning** See learning: empathic learning.

♦ **empiricism** *n.* The doctrine that human experience is the only reliable source of knowledge (*Oxford English Dictionary* 1972, entries from 1803), as exemplified in the philosophies of John Locke, George Berkeley, and David Hume (C.K. Starr, personal communication).
Comments: Empiricism was dominant in biological thinking in Great Britain from 1790 to 1860 (Mayr 1982, 110). The great flowering of empiricism which occurred in England "resulted in an overemphasis on the physical and experimental science. The pursuit of natural history was entirely in the hands of ordained ministers and inevitably led to the belief in the perfect design of a created world, a belief totally incompatible with the concept of evolution" (Mayr 1982, 340).

♦ **empirical taxonomy** See taxonomy: empirical taxonomy.

♦ **empty niche** See niche: vacant niche.

♦ **Emsleyan mimicry** See mimicry: Emsleyan mimicry.

♦ **enantiomer** *n.* One of two forms of a molecule that are mirror images of one another; a mirror-image isomer (King and Stansfield 1985; Bradley 1993, 1117).
syn. enantiomorph (King and Stansfield 1985)
Comment: Different enantiomers of the same compound can have very different effects on an organism; *e.g.,* one enantiomer of the drug thalidomide sedates a person, but the other interferes with fetal development (Bradley 1993, 1117).

♦ **enantiomorph** See enantiomer.

♦ **enantion** See display: wing display: enantion.

♦ **enauophile** See ¹-phile: enauophile.

♦ **encasement theory** See ³theory: encasement theory.

♦ **encephalization quotient** *n.* In mammals: the degree to which an animal's brain size exceeds or falls below the mean brain size for animals of this focal animal's weight (Jerrison 1973 in Eisenberg 1981, 506).

♦ **encounter call** See animal sounds: call: territorial call.

♦ **encounter site** See mating site.

♦ **enculturation** *n.*
1. "The act of learning one culture in all its uniqueness and particularity" (Mead 1963 in Wilson 1975, 159).
2. "The transmission of a particular culture, especially to the young members of the society" (Wilson 1975, 583).

♦ **end act** See behavior: consummatory behavior.

♦ **endangered** *adj.* Referring to a species, or population, that is in danger of becoming extinct (Cooper 1984, 9, in Norden et al. 1984).
cf. extant, extinct, extirpated, threatened

♦ **endangered species** See ²species: endangered species.

♦ **Endangered Species Act (ESA)** See law: Endangered Species Act.

♦ **¹endemic** *n.* An organism that is ³endemic, *q.v.*
Comments: There are two principal schemes for classifying endemic taxa: the cyto-evolutionary scheme and the historical-biogeographic scheme; both approaches have their disadvantages (Keener 1983, 78). Keener (1983) presents a cyto-evolutionary scheme which contains "apoendemic," "paleoendemic," "patroendemic," and "schizoendemic" (including "neoschizoendemic" and "holoschizoendemic").

apoendemic *n.* A geographically restricted neopolyploid taxon that is derived from a widely distributed diploid (*e.g.,* most British endemics) (Favarger and Contandriopoulos 1961, etc. in Keener 1983, 81).
cf. ¹endemic: paleoendemic, patroendemic, schizoendemic

insular endemic *n.* An endemic that evolved in a restricted geographical area and has retained its limited range without spreading to a wider area (Lincoln et al. 1985).

paleoendemic *n.* A generally morphologically uniform, taxonomically isolated ancient relict with an obscure origin and which may now be nearly extinct (Favarger and Contandriopoulos 1961, etc. in Keener 1983, 80).
cf. ¹endemic: apoendemic, patroendemic, schizoendemic

Comment: Paleoendemics are often eco-logical specialists with contracted ranges, may occupy areas remote from their locus of origin, are diploids, or high polyploids, without closely related diploids, and are usually monotypic genera, or sections of genera (Keener 1983, 80).

patroendemic *n.* An ancient element of a restricted, relatively old diploid that is ancestral (at least in part) to widespread polyploids (Favarger and Contandriopoulos 1961, etc. in Keener 1983, 81).
cf. ¹endemic: apoendemic, paleoendemic, schizoendemic

schizoendemic *n.* A diploid, or polyp-loid, species that originated from diploid ancestors by gradual speciation involving a progressive divergence of a taxon through-out its area (Favarger and Contandriopoulos 1961, etc. in Keener 1983, 81).
cf. ¹endemic: apoendemic, paleoendemic, patroendemic

> **holoschizoendemic** *n.* A "mature," sta-bilized, and diversified schizoendemic that occupies a relatively maximum area but is restricted by habitat, geography, or both; contrasted with neoschizoendemic (Favarger and Contandriopoulos 1961, etc. in Keener 1983, 81).

> **neoschizoendemic** *n.* A relatively youthful schizoendemic that is restricted geographically principally because of its age, not ecology nor depauperate bio-types, etc.; contrasted with neoschizo-endemic (Favarger and Contandriopoulos 1961, etc. in Keener 1983, 81).

> **strict endemic** *n.* An endemic that oc-curs only, or preferentially, in a certain habitat (*e.g.,* a bog) (Keener 1983, 82).

♦ ²**endemic** *adj.* Epidemiologically and eco-logically, referring to a non-outbreak size of a population (Lodge 1603, etc. in Frank and McCoy 1990, 4).
ant. epidemic (Frank and McCoy 1990, 4)

♦ ³**endemic** *adj.*
1. Biogeographically, referring to a group, or species, found in a particular area and nowhere else (Darwin 1887, 187, etc. in Frank and McCoy 1990, 3).
Notes: Allaby (1977 in Frank and McCoy 1990, 3) adds "and having originated in the particular area" to this definition. Frank and McCoy (1990, 4) suggest that "precinctive" is a more valid alternative to "endemic" with regard to this definition, and they consider "endemic" to be a subclass of "indigenous."
syn. indigenous (sometimes, Frank and McCoy 1990, 3)
2. Referring to a geographically restricted organism that evolved from a geographi-

cally neighboring species (Ridley 1925 in Keener 193, 78).
3. Referring to a taxon with a relative de-gree of restriction to a single, natural, geographical, or ecological region (Cain 1971, Daubenmire 1978 in Keener 1983, 78).
4. Referring to a taxon that is "restricted, for whatever reason, to a relatively narrowly circumscribed geographic region, al-though an exact operational delimitation of such an area can be problematic" (Keener 1983, 78).
5. Referring to an insect species that is rare or uncommonly abundant (Wallner 1987 in Frank and McCoy 1990, 3).
cf. adventive; autochthonous; endemicity; immigrant; indigenous; introduced; precinctive; ²species: endemic species
[French *endémiqué* < Greek *en,* in + *demos,* population]

♦ **endemic species** See ²species: endemic species

♦ **endemicity** *n.*
1. A trait's being peculiar to a particular country or people (Michaelis 1963).
2. An organism's being native to, and re-stricted to, a particular geographical re-gion (Lincoln et al. 1985).
adj. endemic
syn. endemism (Lincoln et al. 1985)
cf. precinctive
[Greek *endēomos,* native < *en-* in + *dēmos,* people]

♦ **endemism** See endemicity.

♦ **endo-, end-** *combining form* "Within; in-side" (Michaelis 1963).
[Greek < *endon,* within]

♦ **endobenthos** See ²community: benthos: endobenthos.

♦ **endobiont** See biont: endobiont.

♦ **endocrine disrupter** *n.* An artificial en-vironmental pollutant that affects an organism's endocrine systems (Suplee 1996a, A4).
Comments: Endocrine disrupters include breakdown products of certain plastics, polychlorinated biphenyls (PCBs), and some pesticides (Suplee 1996a, A4). Combina-tions of endocrine disrupters might increase the risk of human birth defects, cancers, and other ills.
cf. environmental estrogen

♦ **endocrine gland** See gland: endocrine gland.

♦ **endocrine signaling** See signaling: en-docrine signaling.

♦ **endocrine system** See ²system: endo-crine system.

♦ **endocrinology** See study of: endocri-nology.

♦ **endocylic selection** See selection: cyclical selection.

♦ **endodeme** See deme: endodeme.

♦ **endoderm** See organ: -derm: endoderm.

♦ **endodyogeny** See -geny: endodyogeny.

♦ **endogamy** See -gamy: endogamy.

♦ **endogenous** See -genous: endogenous.

♦ **endogenous clock** See clock: biological clock.

♦ **endogenous pyrogen** See pyrogen: endogenous pyrogen.

♦ **endogenous rhythm** See rhythm: endogenous rhythm.

♦ **endokaryogamy** See -gamy: endogamy.

♦ **endolithophage** See -phage: lithophage (comment).

♦ **endomitosis** See mitosis: endomitosis.

♦ **endomixis** See -mixis: endomixis.

♦ **endomorphology** See morphology: endomorphology.

♦ **endoophage** See -phage: endoophage.

♦ **endoparasitism** See parasitism: endoparasitism.

♦ **endophage** See -phage: entophage.

♦ **endophagous hyperparasitism** See parasitism: hyperparasitism: endophagous hyperparasitism.

♦ **endophenotypic** See -typic: endophenotypic.

♦ **endopolygeny** See -geny: endopolygeny.

♦ **endoreceptor** See ^{2}receptor: interoceptor.

♦ **endorphin** See hormone: endorphin.

♦ **endosite** See parasite: endoparasite.

♦ **endosymbiont** See symbiont: endosymbiont.

♦ **endosymbiont hypothesis** See hypothesis: endosymbiosis hypothesis.

♦ **endosymbiosis** See symbiosis: endosymbiosis.

♦ **endosymbiosis hypothesis, endosymbiotic hypothesis** See hypothesis: endosymbiosis hypothesis.

♦ **endotherm** See -therm: endotherm.

♦ **endotoxic pyrogen** See pyrogen: endotoxic pyrogen.

♦ **endotroph** See -troph-: endotroph.

♦ **endozoochore** See -chore: endozoochore.

♦ **enemy specification** *n.* In some ant species: a colony's exaggerated alarm response to an ant, or other arthropod, species that poses a threat to the colony (Hölldobler and Wilson 1990, 637).
cf. alarm; behavior: alarm behavior

♦ **energetics** See study of: energetics.

♦ **energy** *n.*
1. A person's vigor, or intensity, of action, expression, or vocalization (*Oxford English Dictionary* 1972, entries from 1809).

2. A person's capacity, or tendency, to perform vigorous action (Michaelis 1963).
3. Something's inherent power to produce an effect (Michaelis 1963).

action-specific energy (ASE) *n.* "A hypothetical construct inferred from changes in stimulus threshold, intensity, and rate of occurrence of an [animal's] unlearned response with time and with frequency of occurrence of the response. Its quantity becomes very low when the response is made; it then recovers with time in the absence of response" (Symposia of the Society for Experimental Biology 1950 in Verplanck 1957).
syn. motivation energy (Immelmann and Beer 1989, 192), specific-action energy
cf. instinct; ^{4}model: energy model of behavior
Comments: "This concept is almost identical with Skinner's respondent reflex reserve" (Skinner 1938 in Verplanck 1957), and it dates back to Lorenz (1950) and Tinbergen (1951). "Specific-action potential(ity)," *q.v.,* came to be preferred to "ASE" (Immelmann and Beer 1989, 2).

♦ **energy cost of play** See cost: energy cost of play.

♦ **energy maximizer** See ^{2}species: energy maximizer.

♦ **energy model of behavior** See ^{4}model: energy model of behavior.

♦ **engineering psychology** See study of: psychology: engineering psychology.

♦ **engram** See -gram: engram.

♦ **engulfers** See ^{2}group: functional feeding group: engulfers.

♦ **ENIAC** *n.* The first electronic computer (Levy 1995, B4).
Comments: John V. Atanasoff conceived the idea of an electronic computer in 1937, and J.P. Eckert and J.W. Mauchly, drawing on Atanasoff's ideas and prototype computer, built and patented ENIAC in the mid-1940s (Levy 1995, B4).
[electronic numerical integrator and computer, named by J.P. Eckert and J.W. Mauchly (Levy 1995, B4)]

♦ **enrichment** *n.* An experimenter's introduction of some kind of environmental manipulation, or event, above and beyond an animal's usual early laboratory environment (Dewsbury 1978, 148).

♦ **ENS** See external-nontranscribed spacers.

♦ **ENSO event** See El-Niño–Southern-Oscillation event.

♦ **entelechy** See ^{2}evolution: orthogenesis.

♦ **enteric discharge** See behavior: defensive behavior: enteric discharge.

♦ **enteroreceptor** See ^{2}receptor: interoreceptor.

♦ **entomochore** See -chore: entomochore.

♦ **entomogamous** See [2]-phile: entomophile.

♦ **entomogenous** See -genous: entomogenous.

♦ **entomography** See -graphy: entomography.

♦ **entomology** See study of: entomology.

♦ **entomophage** See -vore: insectivore.

♦ **entomotomy** See study of: entomotomy.

♦ **entoparasite** See parasite: endoparasite.

♦ **entoparasitism** See parasitism: endoparasitism.

♦ **entophage** See -phage: entophage.

♦ **entrainment** *n.* An organism's coupling, or synchronizing, its biological rhythm to an external time source (*zeitgeber*), causing its rhythm to display the *zeitgeber's* frequency (*e.g.,* entraining to lunar periodicity) (Lincoln et al. 1985).
adj. entrained
v.t. entrain

♦ **entropy** *n.* "The degree to which relations between [among] components of any aggregate are mixed up, unsorted, undifferentiated, unpredictable, and random" (Bateson 1979, 242).
ant. negentropy
Comment: "In physics, certain sorts of ordering are related to quantity of available energy" (Bateson 1979, 242).
[first proposed by Clausius, 1865, as *entropie* (German) (*Oxford English Dictionary* 1972]
   negentropy *n.* The degree of ordering, sorting, or predictability in an aggregate (Bateson 1979, 242).
   ant. entropy

♦ **enurination** *n.* In Chinchillas, Guinea Pigs, Porcupines, Rabbits: A male's releasing urine in the direction of a female (Dewsbury 1978, 60).
syn. epuresis (Southern 1948, 177)
cf. behavior: marking behavior
Comment: This behavior may reflect a male's "frustration" or rejection (Ewer 1968 in Dewsbury 1978, 60).

♦ **envelope** See nest: wasp nest: envelope.

♦ **environment** *n.* The complex of biotic, climatic, edaphic, and other conditions that comprise an organism's immediate habitat; the physical, chemical, and biological surroundings of an organism at any particular time (Lincoln et al. 1985), including factors that affect and do not affect the organism (Immelmann and Beer 1989, 87).
cf. environmentalist; -welt: *Umwelt*
Comment: Any to all of these factors can influence behavior, but their effects vary greatly from one individual to another (Storz 1973, 88).

   coarse-grained environment *n.* The environment experienced by an organism (*e.g.,* with poor homeostatic mechanisms or low behavioral plasticity and mobility) as sets of alternative, disconnected conditions; contrasted with fine-grained environment (Lincoln et al. 1985).
   cf. variation: fine-grained variation

   ecological environment *n.* The features of an organism's external environment that affect the organism's contribution to it population's growth; contrasted with external environment and selective environment (Brandon 1992, 82).

   external environment *n.* The sum total of both biotic and abiotic factors that are external to an organism of interest; contrasted with ecological environment and selective environment (Brandon 1992, 81).

   fine-grained environment *n.* The environment experienced by an organism (*e.g.,* with good homeostatic mechanisms or high behavioral plasticity and mobility) as a succession of interconnected conditions; contrasted with coarse-grained environment (Lincoln et al. 1985).
   cf. variation: coarse-grained variation

   patchy environment, environmental patchiness *n.* An environment with resource patches, *q.v.*

   selective environment *n.* An environment that "is measured in terms of the relative actualized fitnesses of different genotypes [of the same species] across time or space;" contrasted with ecological environment and external environment (Brandon 1992, 82).

   total environment *n.* All abiotic and biotic environmental factors that exert an influence on an organism and to which it must be adequately adapted in order to survive (Bligh and Johnson 1973, 958).

♦ **environmental canalization** See canalization: environmental canalization.

♦ **environmental estrogen** *n.* A chemical, such as a pesticide, that has the effect of estrogen on a culture of yeast cells genetically modified to carry the human estrogen receptor (Leary 1996, A15; Suplee 1996a, A4).
cf. endocrine disrupter
Comments: Pairs of the pesticides chlordane, dieldrin, endosufan, and toxaphene synergize and act as environmental estrogens (Leary 1996, A15). These chemicals alone produce no to little estrogen-like effects.

♦ **environmental hormone** See chemical-releasing stimulus: semiochemical: allelochemic: allomone: exocrine substance.

♦ **environmental imprinting** See learning: habitat imprinting.

◆ **environmental odor** See odor: environmental odor.

◆ **environmental patchiness** See environment: patchy environment.

◆ **environmental pressure** See prime mover of social evolution.

◆ **environmental sex determination** See sex determination: phenotypic sex determination.

◆ **environmental tracking** *n.* An organism's changing its foraging gains, or other resource gains, in relation to changes in resource availability (Stephens and Krebs 1986; Kramer and Weary 1991, 443).
syn. tracking (Kramer and Weary 1991, 443)

◆ **environmental variance** (**VE**) See variance: environmental variance.

◆ **environmentalism** *n.*
1. Analysis that stresses the role of environmental influences in the development of an organism's behavioral or other traits (Wilson 1975, 583).
2. Doctrine emphasizing the role of environmental factors rather than hereditary ones in the development of biological traits, especially behaviors (Lincoln et al. 1985).

◆ **environmentalist** *n.* A person who believes that all, or most significant, behavior or knowledge is a product of an animal's experience (Immelmann and Beer 1989, 88).

◆ **enzyme** *n.* A protein produced by a cell that initiates, or accelerates, its specific chemical reactions in metabolism without being more than transiently changed in the process; an organic catalyst; ferment (Michaelis 1963; Immelmann and Beer 1989, 88).
cf. ribozyme
Comment: Some enzymes are named by elaborate standardized nomenclature (Berenbaum 1996, 5).
[Latin < Greek *enzymos*, leavened < *en-*, in + *zyme-*, leaven]

chalcone synthase (CHS) *n.* A key enzyme of the anthocyanin biosynthesis pathway in plants (Niesbach-Klösgen et al. 1987, 213).
Comment: The gene for this enzyme is rather conserved among investigated plants and may be a useful one for exploring the evolutionary relationships among plants (Niesbach-Klösgen et al. 1987, 219).

coenzyme *n.* A nonprotein organic molecule that is necessary for an enzyme to function (Campbell 1990, 106).
Comments: Most vitamins are coenzymes or raw materials from which cells make coenzymes (Campbell 1990, 106). Coenzymes generally act as donors, or acceptors, of groups of atoms that have been added, or removed, from substrates (King

and Stansfield 1985). Coenzymes include ATP, coenzyme A (CoA), coenzyme Q, FAD, FMN, NAD, and NADP.

DNA ligase *n.* An enzyme that seals strands of DNA together by catalyzing the formation of phosphodiester bonds (Campbell et al. 1999, 367).
Comment: This enzyme functions in DNA replication and repair, and making recombinant DNA bonds (Campbell et al. 1999, 367).

flavin adenine dinucleotide (FAD) *n.* A coenzyme that assists enzymes in electron transfer during metabolic redox reactions (Campbell 1990, 184).
Comment: $FADH_2$ is the reduced form of FAD.

housekeeping enzyme *n.* An enzyme that is active in all cell types (inferred from Niesbach-Klösgen et al. 1987, 213).
Comment: Housekeeping enzymes include cytochrome-c, ferredoxin, plastocyanin, and ribulose-1,5-biphosphate carboxylase-oxygenase (Niesbach-Klösgen et al. 1987, 213).

luciferase *n.* An enzyme, in bioluminescent organisms, that catalyzes oxidation of luciferin, *q.v.* (McDermott 1948, 17).
cf. firefly, glowworm (Appendix 1); molecule: luciferin
[named by French physiologist Raphael DuBois (1886) in McDermott 1948, 17]

nicotinamide adenine dinucleotide (NAD⁺) *n.* A coenzyme that assists enzymes in electron transfer during metabolic redox reactions (Campbell 1990, 184–185, illustration).
Comment: NADH is the reduced form of NAD^+.

nicotinamide adenine dinucleotide phosphate (NADP⁺) *n.* A coenzyme that accepts two energized electrons from FAD and one proton from the light reactions of photosynthesis and provides electrons to fix carbon in the Calvin cycle (Campbell 1990, 209–210).
syn. coenzyme 2, TPN
Comment: NADPH is the reduced form of $NADP^+$.

repair enzyme *n.* An enzyme that repairs DNA (Bernstein et al. 1985, 1278).

▸ **excision repair enzyme** *n.* A repair enzyme that removes damages to only one strand of DNA (single-strand damages) (Bernstein et al. 1985, 1278).

restriction enzyme *n.* An enzyme that cuts double-stranded DNA at, or near, specific short nucleotide sequences (Campbell 1990, G-21; Rieger et al. 1991; Futuyma 1998, glossary).
syn. restrictase, restriction endonuclease (Rieger et al. 1991)

Comments: The resulting DNA fragments are called "restriction fragments," and they have discrete molecular weights (Rieger et al. 1991). Variation in the short nucleotide sequences within a population results in variation in DNA sequence lengths after treatment with a restriction enzyme or restriction-fragment-length polymorphism (RFLP). Restriction enzymes are widespread in bacteria and are classified as type-I restrictases, type-II restrictases, and type-III restrictases.

◆ **eobiogenesis** See biopoiesis.

◆ **eobiont** See biont: eobiont.

◆ **eonism** *n.* A male Human's adopting a female's clothing, mannerisms, etc. (Michaelis 1963).
cf. mimicry: automimicry: female mimicry; sex-role reversal; transvestism
[after Chevalier Charles d'Eon, 1728–1810, French diplomat]

◆ **epacme** See acme: epacme.

◆ **epeirophoresis theory** See hypothesis: continental drift.

◆ **ephebic** *adj.*
1. Referring to adult animals' stages between juvenile and old age (Lincoln et al. 1985).
2. Referring to an evolutionary peak or acme (Lincoln et al. 1985).
3. Referring to features that first appeared in evolution in adult stages of organisms (Lincoln et al. 1985).
cf. neanic

◆ **ephemer** *n.*
1. An organism that survives only a short period when introduced into a new area (Lincoln et al. 1985).
2. An organism that is sexually mature for only one day (Lincoln et al. 1985).

◆ **ephemeral, short-lived, transient** *adj.* Referring to an organism, or other phenomenon, that lasts for a relatively short time; *e.g.,* an organism that grows, reproduces, and dies within a few hours or days, or a flower that lasts for 1 day for less (Lincoln et al. 1985).
cf. horary

◆ **ephydrogamous** See -gamous: ephydrogamous.

◆ **epi-** *prefix*
1. Upon; above; among; outside (Michaelis 1963).
2. Besides; over; in addition to (Michaelis 1963).
3. Near; close to; beside (Michaelis 1963). Also ep- before vowels; eph-, before an aspirate: ephemeral.
[Greek *epi-, ep-, eph-* < *epi*, upon, on, besides]

◆ **epibenthos** See ²community: benthos: epibenthos.

◆ **epibiont** See -cole: epicole.

◆ **epibiosis** See -biosis: epibiosis.

◆ **epicole** See -cole: epicole.

◆ **Epicurean** *n.* One who believes that everything is made of unchanging atoms which whirl about and collide at random, all things happen through natural causes, and life is due to the motions of lifeless matter as proposed by Epicurus (342–271 BC) (Mayr 1982, 90).
See atomistic materialist, Stoic.

◆ **epideictic behavior, epideictic display** See display: epideictic display.

◆ **epideictic pheromone** See chemical-releasing stimulus: semiochemical: pheromone: epideictic pheromone

◆ **epidemic** *adj.* Epidemiologically and ecologically, referring to an outbreak size of a population (Lodge 1603, etc. in Frank and McCoy 1990, 4).
ant. ²endemic (Frank and McCoy 1990, 4)
[French *épidémique* < Late Latin *epidemia* < Greek *epi*, on + *demos*, population]

◆ **epidemiology** See study of: epidemiology.

◆ **epidermis** See cuticle.

◆ **epifauna** See fauna: epifauna.

◆ **epigamic behavior** See behavior: epigamic behavior.

◆ **epigamic character** See character: epigamic character.

◆ **epigamic display** See display: epigamic display.

◆ **epigamic selection** See selection: epigamic selection.

◆ **epigamy** See -gamy: epigamy.

◆ **epigenesis** See -genesis: epigenesis.

◆ **epigenetic approach** *n.* An experimental approach to animal development (including behavioral development) that holds that all of an animal's response systems are synthesized during its ontogeny, this synthesis involves the integrative influence of both intraorganic processes and extrinsic stimulative conditions, gene effects are contingent on environmental conditions, an animal's genotype is capable of entering into different relationship classes depending on its prevailing environment, and an animal's environment is not benignly supportive but actively implicated in determining the very structure and organization of each of an animal's response systems (Moltz 1965, 44, in Dewsbury 1978, 132).
cf. -genesis: epigenesis

◆ **epigenetics** See study of: genetics: epigenetics.

◆ **epigenotype** See -type, type: genotype.

◆ **epigenous** See -genous: epigenous.

◆ **epilithophage** See -phage: lithophage (comment).

♦ **epimelectic behavior** See behavior: epimelectic behavior.

♦ **epimorphosis** See morphosis: epimorphosis.

♦ **epinephrine** See hormone: adrenalin.

♦ **epineuston** See neuston: epineuston.

♦ **epiontology** See study of: epiontology.

♦ **epiorganism** See organism: superorganism (Appendix 1, Part 1).

♦ **epiparasite** See parasite: ectoparasite.

♦ **epiphenomenon, Epiphänomen** See phenomenon: epiphenomenon.

♦ **epiphyte** *n.*
1. A plant that grows on another plant (the phorophyte) for support, anchorage, or both, rather than for water or nutrients (Lincoln et al. 1985).
  syn. aerophyte (Lincoln et al. 1985)
2. Any organism that lives on the surface of a plant (Lincoln et al. 1985).
  cf. zoon: epizoon

♦ **episematic character** See character: episematic character.

♦ **episite** See parasite: ectoparasite.

♦ **episodes of proliferation** See ²evolution: explosive evolution.

♦ **episodic memory** See memory: episodic memory.

♦ **episome** See plasmid: episome.

♦ **epistasis** *n.*
1. A nonreciprocal interaction of an organism's nonallelic genes (King and Stansfield 1985).
2. One gene's (the epistatic gene) masking the effect of another gene (hypostatic gene) in an organism (King and Stansfield 1985), *e.g.*, with regard to wing condition in *Drosophila* or coat color in rabbits.
3. In molecular and biochemical genetics, one gene's product's being conditional upon the success, or failure, of another gene that operates at an earlier step in the same pathway (Wade 1992a, 87).
4. In statistical and quantitative genetics, "the between-locus 'nonadditive' component of the genetic variance for a trait" (Wade 1992a, 87).
  Note: This component measures statistical effects of variations in gene combinations between individuals in relation to the total phenotypic variance between individuals in a population (Wade 1992a, 87). Epistasis (def. 1 and 2) could be pervasive, but yet undetectable, at the population level.
  cf. hypostasis
  Comment: "Epistasis" generally refers to genes at different loci, but some authors consider dominance and recessiveness to be special cases of it (Dawkins 1982, 286).
  [coined by Bateson (Mayr 1982, 792)]

♦ **epistatic gene** See gene: epistatic gene.

♦ **epistemology** *n.* "A particular theory of cognition" (Michaelis 1963).
  See philosophy: epistemology; study of: epistemology.

♦ **epitoky** See -toky: epitoky.

♦ **epitreptic behavior** See behavior: treptic behavior: epitreptic behavior.

♦ **epitropism** See tropism: geotropism.

♦ **epizoochore** See -chore: epizoochore.

♦ **epizoon** See -zoon: epizoon.

♦ **epizootic** *n.* In nonhuman animals: an epidemic (Lincoln et al. 1985).

♦ **epuresis** See enurination.

♦ **equal-chance hypothesis** See hypothesis: hypotheses of species richness: equal-chance hypothesis.

♦ **equifinality** *n.* The phenomenon in which a structure, or behavior pattern, develops via different ontogenetic routes (McFarland 1985, 28).

♦ **equilibration** *n.* A threatened animal's adjusting the spatial relation between it and an actual, or potential, threat (Chance 1958, 2).

♦ **equilibratory behavior** See behavior: equilibratory behavior.

♦ **equilibrium** *n.*
1. "Any state of balance, compromise, or adjustment between opposites or opposing forces" (Michaelis 1963).
2. Any state of a system that once attained tends to persist indefinitely (Bell 1982, 505).
  See homeostasis.

Nash equilibrium *n.* In an evolutionarily stable strategy, *q.v.*, an equilibrium in which a player's payoff is reduced if it deviates unilaterally from the equilibrium (Maynard Smith 1978 in Parker 1984, 3).

neutral equilibrium *n.* An equilibrium that conserves its new state after a perturbing force is removed (*e.g.*, a ball's remaining in its position on a plane) (Bell 1982, 505).

species equilibrium *n.* A condition in which the species' extinction rate in a certain place equals the arrival rate of new species by immigration (Lincoln et al. 1985).

stable equilibrium *n.* An equilibrium that moves to its previous state after a perturbing force is removed (*e.g.*, a ball's eventually settling in the base of a trough) (Bell 1982, 505).

unstable equilibrium *n.* An equilibrium that is not restored to its previous state after a perturbing force is removed (*e.g.*, a ball's not remaining on a ridge) (Bell 1982, 505).

♦ **"equilibrium hypothesis of species organization"** See hypothesis: hypotheses of species organization: "equilibrium hypothesis of species organization."

♦ **"equilibrium hypothesis of species richness"** See hypothesis: hypotheses of species richness: "equilibrium hypothesis of species richness."

♦ **equilibrium population** See ¹population: equilibrium population.

♦ **equinophobia** See phobia (table).

♦ **equinox** *n.*
1. One of two opposite points at which the Sun crosses the celestial equator, when day and night are equal (Michaelis 1963).
2. The time when an equinox (def. 1) occurs (Michaelis 1963).
cf. solstice
[French *éqinoxe* < Latin *aequinoctium* < *aequus*, equal + *nox*, night]

autumnal equinox *n.* The second equinox of a year, which occurs on about September 21 in the northern hemisphere; the first day of autumn (Michaelis 1963).
Comment: This is when Earth's axis of tilt is sideways with regard to the Sun (diagrams in Lehr et al. 1975, 48–49).

vernal equinox *n.* The first equinox of a year, which occurs on about March 21 in the northern hemisphere; the first day of spring (Michaelis 1963).
Comment: This is when Earth's axis of tilt is sideways with regard to the Sun (diagrams in Lehr et al. 1975, 48–49).

♦ **equipotentiality assumption** *n.* An assumption implicit in earlier learning theory that one of an organism's stimulus-response associations is as easily formed as another (Immelmann and Beer 1989, 173).
cf. dimension of preparedness, learning disposition

♦ **equivocation** *n.* The potential effectiveness of more than one signal's evoking a particular response (Wilson 1975, 195).

♦ ***Erbkoordination*** *n.*
1. One of two main components of an organism's oriented movement; a more or less fixed pattern that is controlled by external releasing stimuli in cooperation with internal motivating factors and which, once released, is integrated by internal mechanisms only, often quite independently of further external stimulation (Tinbergen 1951, 87).
2. The "aspect of a stereotyped species-characteristic movement that, though elicited by external stimuli, is not subsequently guided by them" (Hinde 1982, 278).
cf. Taxiskomponente
[Possibly coined by Kühn (1919) in Tinbergen (1951, 87). Tinbergen translated this term as "fixed pattern."]

♦ **erection** *n.*
1. An erectile tissue's being filled with blood (Morris 1982).
2. An erect penis (Morris 1982).
Comments: Bonobo males display to one another and to females with erections (de Waal and Lanting 1997, 22–23, illustration).

♦ **eremium** See ²community: eremium.

♦ **eremobiont** See -cole: deserticole.

♦ **eremophile** See ¹-phile: eremophile.

♦ **erg** See unit of measure: erg.

♦ **ergate** See caste: worker.

♦ **ergatandromorph** See morph: ergatandromorph.

♦ **ergatandrous species** See ²species: ergatandrous species.

♦ **ergatogyne** See caste: ergatogyne.

♦ **ergatomorphic** See morphic: ergatomorphic.

♦ **ergatomorphic male, ergatoid male** See morph: ergatandromorph.

♦ **ergonomic stage** See stage: ergonomic stage.

♦ **ergonomics** See study of: ergonomics.

♦ **erroneous imprinting** See behavior: abnormal behavior: erroneous imprinting; learning: imprinting: erroneous imprinting.

♦ **error** *n.*
1. "Something done, said, or believed incorrectly; a mistake" (Michaelis 1963).
2. One's deviating from what is correct, or true, in judgment, belief, or action (Michaelis 1963).
3. In mathematics, the magnitude of difference between an observed value and mean value (Michaelis 1963).
4. In mathematics, any deviation from a true, or mean, value not due to gross blunders of observation or measurement (Michaelis 1963).
Comment: In statistics, "error" is used as a shorthand for many different quantities or concepts, including random and systematic errors introduced by an experimenter, the discrepancy between μ and $\bar{x}$, type-I and type-II errors, variation among replicates, and variation among samples (Hurlbert 1984, 188).

experimental error *n.*
1. One's extraneous, chance variation in measurements of an experiment due to all of its nuisance variables (Lehner 1979, 83).
2. Data variation that is not accounted for by a particular hypothesis (Lincoln et al. 1985).
cf. error: observer error, sampling error

measurement error *n.* One's inaccurate measuring of a variable (Lewis-Beck 1980, 26).
Comment: Measurement error might result from one's making an experimental error,

an observational error, observer error, or a combination of these errors.

observer error *n.* Mistakes made by an investigator in observing behavior, analyzing behavior, or both (Lehner 1979, 128).

▸ **apprehending error** *n.* Observer error due to the physical arrangement of both an animal and its human observer (Lehner 1979, 128).

▸ **computational error** *n.* Observer error due to an investigator's inappropriate choice of a statistical test, its inappropriate performance, or both (Lehner 1979, 129).

▸ **decay** *n.* Observer error due to an observer's instrument (which includes the observer and scoring categories) that drifts beyond the bounds of acceptable measure error (Hollenbeck 1978, 96, in Lehner 1979, 128).

▸ **drift** *n.* Observer error due to movement of an observer in time from a base point in either a positive or negative direction (Hollenbeck 1978, 96, in Lehner 1979, 128).

▸ **recording error** *n.* Observer error due to an observer's inexperience, mental lapses, poor techniques, poor equipment, or a combination of these things (Lehner 1979, 129).

cf. error: recording error

orthographic error *n.* One's unintentional misspelling (Lincoln et al. 1985).

prediction error *n.*

1. In regression analysis, the difference between a predicted $\hat{Y}_i$ and Y_i (Lewis-Beck 1980, 38).

syn. residual (Lewis-Beck 1980, 38)

2. A quantity that measures how well a model predicts a response value of a future observation (Efron and Tibshirani 1993, 237).

cf. statistical test: cross-validation

Comment: Researchers often use prediction error to select models that they will use in their statistical work because a model with a lower prediction error can be more useful than one with a higher one (Efron and Tibshirani 1991, 237).

random error *n.* Any deviation whose magnitude and direction cannot be predicted (Lincoln et al. 1985).

recording error *n.* A person's incorrectly recording a datum or data.

cf. error: observer error: recording error

residual See error: prediction error.

sampling error *n.*

1. "The difference between an observed value of a statistic and the parameter it is intended to estimate" (Lincoln et al. 1985).

2. "The variation between the observations in any one cell of an analysis of variance" (Lincoln et al. 1985).

cf. error: experimental error

specification error *n.* In statistical regression, one's assuming that a relationship between a dependent and independent variable is linear when in fact it is not; excluding a relevant independent variable in a regression analysis; including an irrelevant independent variable in a regression analysis; or a combination of these errors (Lewis-Beck 1980, 26).

type-I error *n.* In statistics, one's rejection of a true null hypothesis; contrasted with type-II error (Siegel 1956, 9).

cf. power of a statistical test

▸ **experimentwise-type-I error** *n.* The probability of one or more type-I errors when all of the possible comparisons in an experiment are made (Day and Quinn 1989, 434).

type-II error *n.* In statistics, one's nonrejection of a false null hypothesis; contrasted with type-I error (Siegel 1956, 9).

♦ **erst** See ²group: erst.

♦ **erucivore** See -vore: erucivore.

♦ **erucivory** See -vore: erucivore.

♦ **eruptive evolution** See ²evolution: eruptive evolution.

♦ **erythrism** *n.* For example, in some bird and mammal species: excessive and abnormal redness in the color of an individual or population (Lincoln et al. 1985).

cf. albinism, melanism

♦ **erythrophobia** See phobia (table).

♦ **ESA** See law: Endangered Species Act.

♦ **escalated-showing-off dispute** See dispute: escalated-showing-off dispute.

♦ **escalation hypothesis** See hypothesis: escalation hypothesis.

♦ **escape behavior** See behavior: escape behavior.

♦ **escape conditioning** See learning: conditioned learning: escape conditioning.

♦ **escape learning** See learning: escape learning.

♦ **escapement clock** See clock: biological clock.

♦ **escargatoire** See ²group: escargatoire.

♦ **escort** *v.t.* For example, in a male locust Leaf-Miner Beetle: to remain with a female mate for a long time after copulating with her and while she feeds, rests, and moves about [coined by Kirkendall 1984, 909].

cf. behavior: mate-guarding behavior

♦ **escort phase** See phase: escort phase.

♦ **Eshkol-Wachmann movement notation** *n.* A notation for recording animal motor patterns, originally devised for ballet (Immelmann and Beer 1989, 194).

◆ *espèces jumelles* See ²species: sibling species.
◆ **ESS** See strategy: evolutionarily stable strategy.
◆ **essences** See *eide*.
◆ **essentialism** *n.*
1. Plato's view that nature is regulated by laws that can be stated in mathematical terms (Mayr 1982, 38, 45).
   Note: Essentialism is similar to the deductive method of Descartes and the mechanistic world picture of Galileo, *q.v.*
2. "All members of a species share the same essence (unaffected by external influences or occasional accidents), the study of nature is simply the study of species" (Mayr 1982, 639).
   Note: Essentialism completely dominated western thought from the Middle Ages to the 19th century (Mayr 1982, 639).
syn. fulguration (Lorenz 1973 in Mayr 1982, 64), typological thinking
cf. population thinking, ²species (def. 4)
◆ **essentialist species concept** See ²species (def. 4).
◆ **essentualism** See essentialism.
◆ **estival** See aestival.
◆ **estivation** See dormancy: estivation.
◆ **estrodiol** See hormone: estrodiol.
◆ **estrogen** See hormone: estrogen.
◆ **estrous cycle** See cycle: estrous cycle.
◆ **estrus, oestrus** *n.*
1. A female mammal's "vehement bodily appetite or passion," specifically, sexual orgasm; rut (*Oxford English Dictionary* 1972, entry from 1890).
2. A female mammal's period of heat, or maximum sexual receptivity; ordinarily, estrus is also the time of egg release (Wilson 1975, 583).
3. A female mammal's relatively delimited period of female sexual receptivity, a stage defined by that receptive behavior rather than by hormonal, or other physiological, factors (Daly and Wilson 1983).
Note: "Estrus," or "heat," is sometimes synonymized with "courtship" in female mammals with the connotation of physiological readiness for mating (Immelmann and Beer 1989, 91).
See behavior: rutting behavior.
adj. estrous, oestrous, estrual
syn. estrum (Michaelis 1963), heat, oestrum, rut (in mammals), season
cf. courtship, rut
[Latin *oestrus* < Greek *oistros*, gad fly, breeze, hence frenzy, mad impulse]
anestrus, anoestrus *n.*
1. A nonbreeding period (Lincoln et al. 1985).

2. In mammals: the phase of a female's reproductive cycle following estrus when she is sexually unreceptive and incapable of being fertilized (Immelmann and Beer 1989, 14).
cf. cycle: estrus cycle
diestrus, dioestrous, diestrous *adj.* In mammals: referring to a female's being between estrous periods (Dewsbury 1978, 60).
monestrus, monoestrous *adj.* Referring to an organism that has a single breeding period in a sexual season.
polyestrus species, polyoestrous species See ²species: polyestrus species.
Rammelzeit, shag time *n.* The reproduction period of hares and rabbits (Heymer 1977, 47).
[German, sexually motivated chasing]
Ranz *n.* The reproduction period of foxes and badgers (Heymer 1977, 47).
[German hunting jargon]
Rauschzeit *n.* The reproduction period of wild boars (Heymer 1977, 47).
Rolligkeit *n.* The reproduction period of cats (felids) (Heymer 1977, 47).
Rossigkeit *n.* The reproduction period of horses (Heymer 1977, 47).
◆ **et-epimeletic behavior** See behavior: epimeletic behavior.
◆ **etheogenesis** See parthenogenesis: etheogenesis.
◆ **ethical behaviorism** See behaviorism: ethical behaviorism.
◆ **ethical institutionism** *n.* One of several theories of ethical behavior in Humans, proposing that a human mind has a direct awareness of true right and wrong that it can formalize by logic and translate into rules of social action (Wilson 1975, 562).
cf. behaviorism: ethical behaviorism; developmental-genetic conception of ethical behavior
◆ **ethnobotany** See study of: botany.
◆ **ethnography** See study of: ethnography.
◆ **ethnology** See study of: ethnology.
◆ **ethnozoology** See study of: zoology.
◆ **ethocline** See -cline: ethocline.
◆ **ethoendocrinology** See study of: endocrinology: behavioral endocrinology.
◆ **ethodynamic pollination** See pollination: ethodynamic pollination.
◆ **ethogenesis** See -genesis: ethogenesis.
◆ **ethogram** See -gram: ethogram.
◆ **ethological** *adj.* Pertaining to behavior, particularly species-specific aspects (Lincoln et al. 1985).
cf. study of: ³ethology
◆ **ethological isolating mechanism** See mechanism: isolating mechanism: sexual isolating mechanism.

d–g

♦ **ethological isolation** See isolation: ethological isolation.

♦ **ethology** See study of: [1,2,3]ethology.

♦ **ethology of domestic animals** See study of [3]ethology: applied ethology.

♦ **ethometry** See -metry: ethometry.

♦ **ethomimicry** See mimicry: behavioral mimicry.

♦ **ethoparasitism** See parasitism: social parasitism: ethoparasitism.

♦ **ethospecies** See [2]species: ethospecies.

♦ **ethosystematics** See study of: systematics: ethosystematics.

♦ **ethotype** See -type: ethotype.

♦ **ethylene** See hormone: ethylene.

♦ **etiology, aetiology** n. The demonstrated cause of a disease or trait (Lincoln et al. 1985).
See study of: etiology.
syn. causation (Lincoln et al. 1985)

♦ **etymology** n.
1. Tracing out and describing the elements of a word with their modifications of form and sense (*Oxford English Dictionary* 1972, entries from 1588); the derivation and meaning of a word (*e.g.,* scientific name or taxonomic epithet) (Lincoln et al. 1985).
2. A statement resulting from def. 1 (Michaelis 1963).
See study of: etymology.

♦ **eu-** *prefix*
1. "Good; well; easy; agreeable" (Michaelis 1963).
2. True, as in eusociality (true sociality).
[Greek *eu*, well < *eus*, good]

♦ **euapogamy** See -gamy: euapogamy.

♦ **eubiosphere** See -sphere: biosphere: eubiosphere.

♦ **euclonal society** See [2]society: euclonal society.

♦ **eucrypsis** See mimicry: eucrypsis.

♦ **eugenic, aristogenic** *adj.* Pertaining to, or having the capacity for, increasing the fitness of a breed of organisms (Lincoln et al. 1985).
cf. dysgenic

♦ **eugenics** n.
1. "The 'science' of improving human stock by giving the 'more suitable races or strains of blood a better chance of prevailing speedily over the less suitable'" (Galton 1883, 24–25, in Kevles 1992, 92); the application of artificial selection to improve *Homo sapiens* biologically (Mayr 1982, 623).
Notes: The idea of eugenics dates back at least to Plato (Kevles 1992, 92). "Eugenics" had many meanings that mirrored a broad range of social attitudes. No political bias was at first attached to eugenics, but in the long run it led to the horrors

of Hitler's holocaust (Mayr 1982, 623). "Eugenics [with implications of social prejudice] remains a dirty word" (Kevles 1992, 94)
[coined by Francis Galton, geneticist (1883 in Kevles 1992, 92)]
2. The science of breeding (Lincoln et al. 1985).
3. Application of genetic principles to improve inherited qualities of a breed (Lincoln et al. 1985).

negative eugenics n.
1. A branch of eugenics that "endeavors to reduce the number of deleterious genes in a population by preventing the reproduction of carriers of dominant genes and by reducing the reproductive rate of heterozygous carriers of recessives (where such heterozygotes can be diagnosed);" contrasted with positive eugenics (Mayr 1982, 623).
2. Improvement of the genetic constitution of a population by preventing reproduction in individuals with undesirable heritable characteristics (Lincoln et al. 1985).
3. Encouraging socially unworthy people to breed less, or better yet, not at all (Kevles 1992, 92).

positive eugenics n.
1. Improvement of the genetic constitution of a population by selective breeding of individuals with desirable heritable characteristics; contrasted with negative eugenics (Lincoln et al. 1985).
2. Fostering a greater representation of "socially valuable" people in society (Kevles 1992, 92).

♦ **euhydrophile** See [1]-phile: euhydrophile.

♦ **eukaryote** See -karyote: eukaryote.

♦ **eukaryotic sex** See sexual reproduction: eukaryotic sex.

♦ **eulectic** See -tropic: eutropic.

♦ **euphile** See [2]-phile: euphile.

♦ **euplankton** See plankton: euplankton.

♦ **euploidy** See ploidy: euploidy.

♦ **eupyrene sperm** See gamete: sperm: eupyrene sperm.

♦ **European ethology** See study of: [3]ethology: classical ethology.

♦ **eurotophile** See [1]-phile: eurotophile.

♦ **eury-** *prefix* Wide, broad (Michaelis 1963).
cf. steno-
[Greek *eurys*, wide]

♦ **euryadaptive parasite** See parasite: euryadaptive parasite.

♦ **eurybaric** *adj.* Referring to an organism that is tolerant of a wide range of atmospheric, or hydrostatic, pressure (Lincoln et al. 1985).
cf. stenobaric

◆ **eurybathic** *adj.* Referring to an organism that is tolerant of a wide range of depth (Lincoln et al. 1985).
  cf. stenobathic
◆ **eurybenthic** See ²benthic: eurybenthic.
◆ **eurybiont** See biont: eurybiont.
◆ **eurychore** See -chore: eurychore.
◆ **eurycladous** See -haline: euryhaline.
◆ **eurycoenose** *adj.* Referring to a widespread organism (Lincoln et al. 1985).
  cf. stenocoenose
◆ **euryhaline** See -haline.
◆ **euryhydric** *adj.* Referring to an organism that is tolerant of a wide range of moisture or humidity (Lincoln et al. 1985).
  cf. stenohydric
◆ **euryionic** *adj.* Referring to an organism that is tolerant of a wide range of pH (Lincoln et al. 1985).
  cf. stenionic
◆ **eurylume, euryphotic** *adj.* Referring to an organism that is tolerant of a wide range of light intensity (Lincoln et al. 1985).
  cf. stenolume
◆ **euryoecism, euryoecious** See -oecism: euryoecism.
◆ **euryphage** See -phage: euryphage.
◆ **euryphotic** See eurylume.
◆ **euryplastic** *adj.* Referring to an organism that shows a broad developmental response (phenotypic variation) to different environmental conditions (Lincoln et al. 1985).
  cf. stenoplastic
◆ **eurysubstratic** *adj.* Referring to an organism that is tolerant of a wide range of substratum types (Lincoln et al. 1985).
  cf. stenosubstratic
◆ **eurytherm** See -therm: eurytherm.
◆ **eurythermophile** See ¹-phile: thermophile: eurythermophile.
◆ **eurytopic, eurytopy** See -topy: eurytopy.
◆ **eurytropism** See -tropism: eurytropism.
◆ **euryxenous** See -xenous: euryxenous.
◆ **euryxeny** See -xeny: euryxeny.
◆ **eusexual** See sexual: eusexual.
◆ **eusexual reproduction** See sexual reproduction.
◆ **eusociality** See sociality: eusociality.
◆ **eusynanthropic** See anthropic: eusynanthropic.
◆ **euthenics** See study of: euthenics.
◆ **euthermy** See -thermy: cenothermy.
◆ **eutroglobiont, eutroglophile** See ¹-phile: troglophile: eutroglophile.
◆ **eutrogloxene** See -xene: eutrogloxene.
◆ **eutrophapsis** *n.* In some social-insect species: a female's feeding her young in a previously prepared nest (Lincoln et al. 1985).

◆ **eutrophy** See -trophy: eutrophy.
◆ **eutropic** See -tropic: eutropic.
◆ **eutropous** See -tropous: eutropous.
◆ **evaporation pan** See instrument: evaporation pan.
◆ **evapotranspiration** See transpiration: evapotranspiration.
◆ **Eve** See mitochondrial Eve.
◆ **Eve hypothesis, Eve theory** See hypothesis: African-origin hypothesis of human mitochondrial DNA evolution.
◆ **event** See behavior event.
◆ **event recorder** *n.* A time-based device used to record occurrences of behavioral events and states (Lehner 1979, 158–161).
◆ **Everglades** See place: Everglades.
◆ **¹evolution** *n.*
  1. The process of opening out, unrolling, expanding (*e.g.,* tissues' developing during embryogenesis) (*Oxford English Dictionary* 1972, entries from 1641).
  2. An organism's process of developing from a less mature to a more mature, or fully mature or complete, state (*Oxford English Dictionary* 1972, entries from 1670; Richards 1992, 95).
  3. The concept that an embryo, or germ, expands from a pre-existing form that contains the rudiments of all the parts of the developed organism, rather than the new organism's arising from a fertilized egg (Bonnet 1762 in the *Oxford English Dictionary* 1972).
    Note: This obsolete definition of evolution is the opposite of ideas under ²evolution below.
  4. The process of change, production, or emission (*e.g.,* in a chemical reaction) (*Oxford English Dictionary* 1972, entries from 1800; Lincoln et al. 1985).
  5. The process of undergoing "gradual directional change" (Lincoln et al. 1985).
  6. The process of undergoing (organic) evolution, *q.v.*
  7. The result of any of these evolutionary processes (*Oxford English Dictionary* 1972).
  v.t., v.i. evolve
  Comment: Richards (1992) discusses the changes in the meaning of "evolution" through time.
  [Latin *ēvolvĕre*, to roll out, unroll]
  chemical evolution *n.* A progressive development of organic molecules in a manner analogous to organic evolution (Immelmann and Beer 1989, 94).
◆ **²evolution, organic evolution** *n.*
  1. Development of a more highly organized biological phenomenon from a more rudimentary one during evolutionary time (*Oxford English Dictionary* 1972, entries from 1832).

2. Species gradual modification from earlier forms (*Oxford English Dictionary* 1972, entries from 1832).

3. Species' origination from earlier forms, rather than by the process of "special creation" (*Oxford English Dictionary* 1972, entries from Lyell 1832).

   syn. doctrine of evolution (*Oxford English Dictionary* 1972)

4. Changes undergone by organisms in their phylogenetic histories (Immelmann and Beer 1989, 93).

5. Organism changes from generation to generation in gene frequencies of their gene pools that generally lead to complex organizations and specialized forms from simpler precursors (Immelmann and Beer 1989, 93–94).

Note: A broad definition of "evolution" should include both "monotypic evolution" (= "transformation," change of gene frequencies in populations) and "polytypic evolution" (diversification, processes that occur simultaneously, such as species multiplication) (Mayr 1982, 400).

See [2]evolution: macroevolution; [3]theory: theory of preformation.

syn. natural selection (a confusing synonym, Endler 1992, 223)

cf. elaboration; [1]evolution: chemical evolution; -genesis: diagenesis; [3]theory; [3]theory: autogenetic theory, central dogma of evolutionary theory, encasement theory, molecular theory of evolution, neutral theory, theory of evolution, uniformitarianism

Comments: "A retrospective survey of the various terms and definitions for evolution proposed since 1800 reveals quite clearly the ambiguities and uncertainties that have bedeviled evolutionists almost up to the present" (Bowler 1975 in Mayr 1982, 400). Richards (1992) provides more information on the evolution of the term "evolution." Broadly speaking, "evolution" can be "genetic evolution" or "nongenetic evolution" (= "cultural evolution"); it is also classified by many other criteria. "Evolution" can refer to a process or the result of an evolutionary process. I list only some of the many kinds of evolution here.

adaptation See [3]adaptation: [1,2]evolutionary adaptation.

allomorphosis, evolutionary allometry, idioadaptation *n.* Evolution by a rapid increase in an organism's structural specialization (Lincoln et al. 1985).

cf. [2]evolution: aromorphosis

anagenesis *n.*

1. An evolutionary advance in organisms' general organization or perfection of some major biological function (Rensch

1954 in Huxley 1957, 454); "progressive evolution toward higher taxonomic levels" (Futuyma 1986, 550).

2. "All types or degrees of biological improvement from detailed adaptation to general organizational advance" (Huxley 1957, 454).

3. Directional evolution within a single lineage (Futuyma 1986, 286).

4. Directional evolution of a feature over an arbitrarily short lineage segment (Futuyma 1986, 550).

See [2]evolution: phyletic evolution.

syn. phyletic evolution (Dobzhansky 1962, 220; Mettler et al. 1988, 267)

cf. [2]evolution: catagenesis, cladogenesis, stasigenesis

aromorphosis *n.* Evolution by an increase in an organism's degree of integration and organization without marked specialization (Lincoln et al. 1985).

cf. [2]evolution: allomorphosis

arthrogenous evolution *n.* A hypothesis that a creative principle in living matter manifests itself in response to environmental stimuli (Lincoln et al. 1985).

bradygenesis, bradytely *n.*

1. Evolution with very slow, or almost no, change; contrasted with horotely and tachytely (Simpson in Boas 1984).

2. A smaller distribution of evolutionary speed within the total horotelic distribution of evolutionary speed, with a distinct central tendency of smaller values than the central tendency of the horotelic distribution (Simpson 1944 in Gould 1994, 6765).

Comment: See [2]evolution: horotely.

cataclysmic evolution See [2]evolution: instantaneous evolution.

catagenesis, katagenesis *n.* Regressive evolution involving an organism's losing its independence from, and control over, its environment (Lincoln et al. 1985).

cf. [2]evolution: anagenesis, stasigenesis

cladogenesis *n.*

1. Evolutionary splitting of organisms, "from subspeciation through adaptive radiation to the divergence of phyla and kingdoms (Huxley 1957, 454, after Rensch 1947).

2. A single species' splitting into two species (Stebbins and Ayala 1985, 79).

3. Splitting and subsequent divergence of populations (Lincoln et al. 1985).

syn. dendritic evolution, evolutionary diversification, kladogenesis (Lincoln et al. 1985)

cf. [2]evolution: anagenesis

coenogenesis, syngenesis *n.* "Descent from a common ancestor" (Lincoln et al. 1985).

cf. -genesis: cenogenesis

coevolution *n.*

1. Evolution in which heterospecific organisms (*e.g.,* parasites and their hosts) affect one another (Mode 1958, 158, who does not formally define coevolution).

2. The interaction between two major organism groups with a close and evident ecological relationship, such as plants and herbivores (Ehrlich and Raven 1964, 586, who do not formally define coevolution).

3. "Development of genetically determined traits in two species to facilitate some interaction, usually mutually beneficial" (Ricklefs 1979 in Schemske 1982, 70, who states two drawbacks of this definition: many coevolution examples involve negative interactions between taxa, *e.g.,* predator vs. prey, and coevolution results from the interaction, not to facilitate it).

4. "The simultaneous evolution of interacting populations" (Roughgarden 1979, 451).
   Note: Schemske (1982, 70) states that this is a clear, and operationally correct, definition, but it "perhaps does not sufficiently emphasize the point made by both Ricklefs and Janzen that coevolution results form interactions between taxa at the level of particular traits."

5. "An evolutionary change in a trait of the individuals in one population in response to the trait of the individuals of a second population, followed by an evolutionary response by the second population to the change in the first" (Janzen 1980, 611).
   Note: Schemske (1982, 70) states that this definition "places unnecessary importance on the notion of reciprocity; coevolution can operate simultaneously or sequentially among interacting taxa."
   cf. ²evolution: coevolution: diffuse coevolution

6. "The joint selective effects on characters of interacting taxa, based on heritable variation in these characters" (Schemske et al. in Schemske 1982, 70, who indicates that this definition includes both positive and negative effects of an interaction and emphasizes that coevolution is dependent upon genetic variation in characters relevant to the interaction, and those characters that are genetically correlated with the selected ones).

7. Species adaptation to features of their biotic environments — features that may remain effectively constant for long periods (Futuyma and Slatkin 1983, 2).

Note: This is a very general definition that is equivalent to "evolution;" therefore, Futuyma and Slatkin (1983, 2) use a more restricted definition of "coevolution."

8. An adaptive response in two or more ecologically interacting species to genetic change in the other(s) (Futuyma and Slatkin 1983, 2).
   Note: Some persons have interpreted this idea as a gene-for-gene coevolutionary reciprocity between heterospecific organisms; this kind of coevolution might not be common (Spencer 1988, 3).

9. "Stepwise, reciprocal evolution of ecologically intimate organisms" (Chu 1985).
   Notes: Thompson (1983, 3), Futuyma (1989, 992), and others caution that evidence for "coevolution" requires evidence for reciprocal evolutionary response among interacting species. Buchmann and Nabhan (1996, 253) give the bee genus *Rediviva* and the plant genus *Diascia* as an example of reciprocal coevolution in which both organisms apparently directed each other's evolution. These organisms probably cannot live long without each other in nature. Roubik (1989, 55) illustrates these organisms.
   See ²evolution: coevolution: diffuse coevolution.
   Comments: "The idea of coevolution is as old as the study of evolution itself" (Futuyma and Slatkin 1983, 3). Janzen (1980) discusses misuses of the concept of coevolution; he suggests removing "coevolution" as a synonym of "interaction," "symbiosis," "mutualism," and "animal-plant interaction." "Applied to plant-herbivore and predator-prey systems, it [coevolution] is often referred to as an evolutionary arms race between two species: Adaptations in one species lead to counteradaptations in the other, which in turn provoke a response by the first, and so on" (Chu 1985).

▶ **diffuse coevolution** *n.*

1. Coevolution that occurs when either, or both, interacting parties are represented by an array of populations that generate a selective pressure as a group (Janzen 1980, 611).

2. Evolution of a particular trait in one or more species in response to a trait, or suite of traits, in several other species; *e.g.,* many plants have evolved chemical and physical defenses against a diverse suite of insects, and many insects have evolved the ability to detoxify a wide range of plant chemicals (Futuyma and Slatkin 1983, 2).

syn. coevolution (Futuyma and Slatkin 1983, 2)

d – g

► **pairwise coevolution** *n*. Coevolution involving only two species, *e.g.*, a plant and its herbivore or a plant and its pollinator (Futuyma and Slatkin 1983, 2).

► **runaway coevolution** *n*. A hypothetical kind of coevolution that has resulted in exaggerated traits, *e.g.*, the bright orange body color of the Guppy or long male eyestalks of the stalk-eyed fly (Dugatkin and Godin 1998, 59).

concerted evolution *n*. The maintenance of a homogeneous nucleotide sequence in members of a gene family that evolves over time (Futuyma 1986, 551).

convergent evolution, convergence *n*.

1. Two or more distantly related organisms' evolving similar morphological, biochemical, behavioral, or other kinds of adaptations due to adapting to the same, or similar, ecological conditions (Wilson 1975, 25).

   Note: Convergent evolution is on a continuum from very shallow to very deep, based on the degree of complexity of adaptation involved and the extent to which a species has organized its way of life around it (Wilson 1975, 25).

2. The evolution of similar nonhomologous apomorph characteristics (nonhomologous apomorphy or pseudoapomorphy) (Mayr 1982, 228); *e.g.*, swimming adaptations in sharks and whales; nectar foraging in Old World nectariniid birds and New World hummingbirds; grazing in Old World ruminants and New World rodents (Heymer 1977, 102); flight in insects, birds, and bats; gliding in certain squirrels; flying lemurs, lizards, and frogs (Brown 1975, 13); egg rolling and distraction displays in many species of ground-nesting birds; mouth breeding in different kinds of fish; sucking drinking by pigeons and some estrildine finches; or hovering in some insect and bird species (Immelmann and Beer 1989, 60).

   syn. evolutionary convergence, (Wilson 1975, 584), syntechny (Lincoln et al. 1985)
   cf. ²evolution: divergent evolution, parallel evolution; mimicry

counterevolution *n*. Evolution of traits in members of a population in response to adverse interactions with another population (Lincoln et al. 1985).

cultural evolution *n*.

1. For example, in Darwin's Medium Ground Finch, Japanese Macaques, Humans: "The patterns and processes involved in changes in behavioural traditions across generations" in a population (Cavalli-Sforza and Feldman 1981 and Boyd and Richardson 1985 in Gibbs 1990, 253).

2. The appearance of a new behavior and its transmission to subsequent generations by tradition (Immelmann and Beer 1989, 94).

syn. nongenetic evolution
cf. tradition

cyclic evolution *n*. A species' evolution involving an initial phase of rapid expansion, followed by a long period of relative stability and a final brief episode of degeneration and overspecialization leading to its extinction (Lincoln et al. 1985).

cytocatalytic evolution *n*. In plants: evolution that is initiated by an abrupt mutation that results in the formation of a polyploid or aneuploid (Lincoln et al. 1985).

dendritic evolution See ²evolution: cladogenesis.

diffuse coevolution See ²evolution: coevolution.

divergent evolution *n*.

1. "An arborescent multiplication of types arising simultaneously in space" (given as "polytypic evolution" in Romanes 1897, 22).

2. Evolution of dissimilar characters by related organisms (Lincoln et al. 1985).

3. Evolution in which a model changes in appearance from its mimic (Brower and Brower 1972, 65).

cf. ²evolution: convergent evolution, parallel evolution, phyletic evolution, polytypic evolution

emergent evolution *n*. Appearance of entirely novel and unpredictable characters by the rearrangement of existing potentialities (Lincoln et al. 1985).

entelechy See -genesis: orthogenesis.

eruptive evolution See ²evolution: explosive evolution.

explosive evolution *n*. For example, in cichlid fish, Hawaiian *Drosophila*, Honeycreepers (birds): a group's, or population's, splitting into numerous lines of descent within a relatively short period of geological time; rapid adaptive radiation (Greenwood 1974 and Mayr 1976, 268–270, in Greenwood, 819; Lincoln et al. 1985).

syn. episodes of proliferation, eruptive evolution (Simpson 1953, 234); explosive radiation (Lincoln et al. 1985); *Virenzperioden* (Rensch 1947 in Simpson 1953, 234); tachytelic evolution (Grant 1963, 555)

Comments: Simpson (1953, 234) coins "episodes of proliferation" and explains why he prefers it to other names for this phenomenon. He says "explosive evolution" is ill chosen because it has a "certain unintended humor." The "explosions" commonly take millions of years and end with a "silent bang." "Eruptive evolution" suggests an inept mental picture to him. He writes that "*Virenzperioden*" might carry some undertone of a life-cycle analogy. About 500 cichlid-fish species apparently evolved in Lake Victoria in about 12,400 years (Yoon 1996c, C1; Seehausen et al. 1997a, 1808). In contrast, about 20 species of Darwin's finches evolved on the Galápagos Islands in about 4 million years. About half of Lake Victoria's original cichlid species are now extinct, probably primarily as a result of its eutrophication, which led to a breakdown of cichlid reproductive barriers and a loss of their diversity (Hanson 1997, 1737), and consumption by Nile Perch (Vick 1999, A28).

Falconer-style evolution *n*. Evolution in which adaptation is a rare event and in which, most of the time, most species do not show structural changes even when these changes would apparently increase their adaptiveness to their changing environments; a species is a "stuttering and sporadic adaptive trajectory" (Bakker 1985, 74).

cf. hypothesis: hypothesis of punctuated equilibrium

Comment: Falconer-style evolution is being accepted by an increasing number of biologists (Bakker 1985, 74).

gradualism, gradual evolution See ²evolution: horotely.

haplogenesis *n*. Evolution of new forms (Lincoln et al. 1985).

hologenesis *n*.
1. The hypothesis that all species arose simultaneously throughout their modern distribution without subsequent dispersal (Lincoln et al. 1985).
2. Origin of species by multiple mutation from a single extinct ancestor (Lincoln et al. 1985).

homeotely *n*. Evolution of homologous structures (Lincoln et al. 1985).

horotely *n*.
1. Evolution with gradual change (Simpson 1944 in Boas 1984).
2. The entire distribution of ordinary evolutionary rates which includes a left region of slower rates with a distinct central tendency (bradytely) and a right region of faster rates with a distinct central tendency (tachytely) (Simpson 1944 in Gould 1994, 6765).

Notes: In specifying bradytely, horotely, and tachytely, Simpson (1944) was trying to identify separate peaks (modes in the statistical sense) in the distribution

of tempos of evolution in order to specify distinct modes (in the ordinary sense) of evolution (Gould 1994, 6766). Many researchers have misunderstood these kinds of evolution.

3. Evolution at a normal, or average, rate (Lincoln et al. 1985).

Comment: Many researchers have misunderstood these kinds of evolution, and definitions 1 and 3 are erroneous (Gould 1994, 6766).

hypertely *n*.
1. Evolutionary overspecialization (Lincoln et al. 1985).
2. An organism's, or structure's, progressive attainment of disproportionate size (Lincoln et al. 1985).

instantaneous evolution *n*. Evolution through polyploidy in which new species arise in as short a time as one generation (Stebbins 1951, 54–55; Stebbins and Ayala 1985, 73).

syn. cataclysmic evolution

cf. speciation: instantaneous speciation

iterative evolution, iterative adaptation *n*. Repeated production of an adaptive type at successive points in an organism's lineage (Lincoln et al. 1985).

kladogenesis See ²evolution: cladogenesis.

lipogenesis *n*. Accelerated evolutionary development due to loss of certain developmental stages (Lincoln et al. 1985).

lipopalingenesis *n*. A descendant's loss of one or more developmental stages that were present in its ancestral form (Lincoln et al. 1985).

cf. ²evolution: regressive evolution

Lizzie-Borden evolution *n*. Evolution in which, first, daughter species bud off parental stock (hypothesis of punctuated equilibrium) and, second, daughter species spread and exterminate their parental stock over most of its range (Bakker 1985, 74).

cf. Schankler immigration event

[After Lizzie Borden, who allegedly killed her stepmother and father with a hatchet]

macroevolution *n*.
1. Evolution above the species level (Rensch 1959 in Futuyma 1986, 397).
   Note: This is a common definition of macroevolution.
2. The origin of higher taxonomic units (Mayr 1982, 607; Lincoln et al. 1985).
   syn. transpecific evolution, macrophylogenesis, megaevolution (Lincoln et al. 1985)
3. Evolution as studied by paleontologists and comparative anatomists (Mayr 1982, 607).

4. Major evolutionary events, or trends, usually viewed through the perspective of evolutionary, or geological, time (Lincoln et al. 1985).

5. A change in species composition within a monophyletic group in space and time (Eldredge and Cracraft 1980).

Note: "If species are discrete reproductive units, microevolutionary processes cannot logically be extrapolated, in a reductionist manner, to explain macroevolutionary patterns" (Eldredge and Cracraft 1980).

See study of: evolution: macroevolution.
syn. evolution (Wilson 1975, 589)
cf. ²evolution: microevolution
Comment: "Macroevolutionary change is usually recognized as change in gross morphology in a series of fossils. There is some controversy over whether macroevolutionary change is fundamentally just cumulated microevolutionary change, or whether the two are 'decoupled' and driven by fundamentally different kinds of process" (Dawkins 1982, 289).

macrophylogenesis See ²evolution: macroevolution.

megaevolution See ²evolution: macroevolution.

microevolution *n.*

1. A change in gene frequency within a population or species; evolution in its slightest, most elemental form (Wilson 1975, 64; Eldredge and Cracraft 1980; Futuyma 1986, 397).

2. "A small amount of evolutionary change, consisting of minor alterations in gene proportions, chromosome structure, or chromosome numbers" (Wilson 1975, 589).

Note: "A larger amount of change would be referred to as macroevolution or simply as evolution" (Wilson 1975, 589).
syn. evolution, microphylogenesis (Lincoln et al. 1985)
cf. ²evolution: macroevolution

monotypic evolution, transformation *n.* Change of gene frequencies in populations (Mayr 1982, 400).
syn. transformation in time (Romanes 1897, 21)
cf. ²evolution: polytypic evolution
[coined by Gulick 1888 in Mayr 1982, 400]

morphogenesis, morphogeny *n.* Evolution of morphological structures (Lincoln et al. 1985).
cf. -genesis: morphogenesis

mosaic evolution, nonharmonious character transformation *n.* Evolution of different adaptive structures, traits, or other components of a phenotype at different times, or different rates, in an evolutionary sequence (Lincoln et al. 1985). See principle: principle of mosaic evolution.

neo-Darwinian evolution See evolutionary synthesis.

non-Darwinian evolution *n.* Genetic change owing to stochastic processes (essentially neutral mutations) with neither the environment (directly or through selection) nor internal factors influencing the direction of variation and evolution (Kimura 1979, 98; Mayr 1982, 361; King and Jukes 1969 and Crow and Kimura 1970 in Mayr 1982, 593).
See ³theory: non-Darwinian evolution.
cf. ³theory: neutral theory

nongenetic evolution See ²evolution: cultural evolution.

orthogenesis, aristogenesis, entelechy *n.* Evolution continuously in one direction over a considerable time, commonly with the implication that the direction is determined by a factor internal to an organism, or at least not determined by natural selection (Lincoln et al. 1985).

parallel evolution *n.*

1. Evolution of and maintenance of similar traits in two organismal lines for a significant period of time (Mayr 1982, 234); *e.g.,* similar adaptations of eutherian mammals in North and South America and marsupial mammals in Australia, possibly longitudinal contraction used for escape by several worm genera (Brown 1975, 14), or flight refinements in paleopterous and neopterous insects (Kukalova-Peck 1978).

2. The independent acquisition in two or more related descendant species of similar derived character states evolved from a common ancestral condition (Lincoln et al. 1985), *e.g.,* similar morphological and behavioral adaptations in African antelope species (Immelmann and Beer 1989, 60).

3. "The maintenance of constant differences in the evolution of characters between two unrelated lines" (Lincoln et al. 1985).

syn. evolutionary parallelism, parallelism (Lincoln et al. 1985)
cf. ²evolution: convergent evolution, divergent evolution

phyletic evolution *n.*

1. A species' evolutionary transformation into another species (Darwin in Mayr 1963, 428; Mayr 1982, 400).

2. Evolution of a population from one state to another resulting in a single new species (Simpson 1944, 198, in Eldredge and Cracraft 1980, 262; Eldredge and Gould 1972, 87).

3. A sequence of changes that occurs more or less uniformly throughout a species' entire geographical range over an extended period, often regarded as a type of speciation (Lincoln et al. 1985).
See ²evolution: anagenesis.
syn. anagenesis (Dobzhansky 1962, 220; Mettler et al. 1988, 267), modification in time, modification by descent (Darwin in Mayr 1963, 428), monotypic evolution (Gulik in Romanes 1897, 21), transformation, transformation in time (Romanes 1897, 21), vertical evolution
cf. ²evolution: divergent evolution; ²extinction: pseudoextinction; speciation: phyletic speciation

polytypic evolution *n.* Evolutionary processes that occur simultaneously, such as species multiplication (Mayr 1982, 400).
See ²evolution: horizontal evolution.
syn. diversification (Mayr 1982, 400), transformation in space (Romanes 1897, 21)
cf. evolution: divergent evolution, monotypic evolution
[coined by Gulick 1888 in Mayr 1982, 400]

punctuational evolution *n.*
1. Periodic and rapid evolution (Simpson in Boas 1984).
2. A model of evolution in which species are relatively stable and long-lived and in which new species appear during concentrated outbursts of rapid speciation, followed by differential success of certain species (Lincoln et al. 1985).
syn. macrogenesis (Mayr 1982, 75)
cf. gradualism: phyletic gradualism; hypothesis: hypothesis of punctuated equilibrium; saltation

quantum evolution *n.*
1. "The relatively rapid shift of a biotic population in disequilibrium to an equilibrium distinctly unlike an ancestral condition" [coined by Simpson 1944, 206, in Mayr 1982, 609]; "the fastest kind of speciation, *e.g.,* bats' originating from an insectivore within a few million years…" (Mayr 1982, 609–610).
2. "Rapid and rare, but efficacious, 'all-or-nothing' transitions from one adaptive zone to another through an inadaptive phase (a process analogized with Wright's model of genetic drift)" (Simpson 1944 in Gould 1994, 6766).
3. A rapid evolutionary shift to a new adaptive zone under strong selective pressure (Lincoln et al. 1985).
syn. saltational evolution (according to some authors)
cf. ²evolution: instantaneous evolution, phyletic evolution, saltational evolution; speciation: quantum speciation; ³theory: catastrophe theory

rectilinear evolution *n.* Continued change in one direction within a line of descent over a considerable time period, with no implication as to how this direction is maintained (Lincoln et al. 1985).
cf. ²evolution: orthogenesis

regressive evolution, regression *n.*
1. Evolution in which an organism changes back to an ancestral condition (*e.g.,* whales' re-evolving the streamlined form of their distant fish-like ancestors).
2. Evolution of increasing structural simplification (Lincoln et al. 1985).
3. Evolution of decreasing specialization (Lincoln et al. 1985).
cf. ²evolution: lipopalingenesis; reduction

reticulate evolution *n.* Evolution that depends on repeated intercrossing between a number of lineages that produces a network of relationships in a series of related allopolyploid species with anastomoses in the network representing hybridization sites (Lincoln et al. 1985).

retrograde evolution, backward evolution *n.* Evolution that lengthens a biochemical pathway when a particular substrate is depleted; contrasted with "nonretrograde evolution" (Horowitz 1945 in Strickberger 1990, 141).
Comment: Retrograde evolution is believed to account for many intermediate steps in biochemical pathways that lead to the synthesis of compounds such as amino acids (Strickberger 1990, 141). A hypothetical example of retrograde evolution originally involves four substrates and one enzyme, enzyme-1. Substrate-D is a precursor of -C, -C is a precursor of -B, and -B is a precursor of -A. Enzyme-1 converts substrate-B to -A. After all substrate-B is depleted in an organism's environment, an enzyme eventually mutates by chance and becomes enzyme-2, which converts substrate-C to -B, which enzyme-1 converts to -A. After all substrate-C is depleted in the organism's environment, an enzyme mutates by chance and becomes enzyme-3, which converts substrate -D to -C, which enzyme-2 converts to -B.
[retrograde < Latin *retrogradus* < *retrogradi* < *retro,-* backward + *gradi,* to walk]

saltational evolution, saltatory evolution, saltation *n.* Possible evolution by major reorganization, rather than by progression through slight intermediate stages; by new features emerging by Lamarckian mechanisms; and by internal, "autogenetic" drives rather than natural selection (Mayr 1982; Futuyma 1986, 397).

See speciation: saltational speciation.
cf. ²evolution: punctuational evolution; saltation
Comment: This is a controversial concept.
speciation See ²evolution: macroevolution.
stasigenesis *n.*
1. "All processes leading to stabilization and persistence of types and of patterns of organization from species up to phyla" [coined by Huxley 1957, 454].
2. Persistence of a form, or group, showing little change over an extended time, irrespective of changing environmental conditions (Lincoln et al. 1985).
3. An organism's tendency to form well-integrated gene pools with well-adapted phenotypic products that persist under stabilizing selection (Lincoln et al. 1985).
cf. ²evolution: anagenesis, catagenesis
tachygenesis *n.* Accelerated ontogenetic, or phylogenetic, development, *e.g.*, omission of certain larval stages during development (Lincoln et al. 1985).
cf. ²evolution: bradygenesis
tachytely *n.*
1. A smaller distribution of evolutionary speed within the total horotelic distribution of evolutionary speed, with a distinct central tendency of larger values than the central tendency of the horotelic distribution; contrasted with bradytely and horotely (Simpson 1944 in Gould 1994, 6765).
2. Relatively fast evolution, *e.g.*, when populations shift from one major adaptive zone to another (Lincoln et al. 1985).
Comment: See ²evolution: horotely.
tandem evolution *n.* The concept that, in evolution, an organism transforms from one homozygous type to another homozygous one (Mayr 1982, 429).
Comment: This concept neglects the fact that, during evolution, genetic change in a species can proceed simultaneously at thousands, if not millions, of loci (Mayr 1982, 429).
transformation See ²evolution: monotypic evolution, phyletic evolution.
transpecific evolution See ²evolution: macroevolution.
zoogenesis *n.*
1. "The origin of animal life on Earth" (Lincoln et al. 1985).
   syn. zoogeogenesis (Lincoln et al. 1985)
2. The evolution and development of an animal species" (Lincoln et al. 1985).
syn. zoogeny (Lincoln et al. 1985)
cf. -genesis
♦ **evolution-by-natural-selection tautology** *n.* "The fitter genotypes are those that leave more descendants, which be-

cause of heredity, resemble the ancestors; and the genotypes that leave more descendants have greater Darwinian fitness" (Wilson 1975, 67).
Note: This is a statement of what has already happened and does not predict which genotypes will be more fit than others in the future (Wilson 1975, 67). Thus, it does not have to be considered a tautology.
♦ **evolution, theory of** See ³theory: theory of evolution.
♦ **evolutional load** See genetic load: substitutional load.
♦ **evolutionarily stable population** See ¹population: evolutionarily stable population.
♦ **evolutionarily stable set of genes** *n.* A gene pool that cannot be invaded by another gene (Dawkins 1977).
♦ **evolutionarily stable strategy (ESS)** See strategy: evolutionarily stable strategy.
♦ **evolutionary adaptation** See ³adaptation: ¹,²evolutionary adaptation.
♦ **evolutionary allometry** See ²evolution: allomorphosis.
♦ **evolutionary biology** See study of: biology: evolutionary biology.
♦ **evolutionary clock** See clock: molecular clock.
♦ **evolutionary conflict** See ³conflict: evolutionary conflict.
♦ **evolutionary conservatism** *n.* Preservation of ancestral similarity in species on diverging evolutionary pathways, due to retention of a high proportion of common ancestral alleles (Lincoln et al. 1985).
♦ **evolutionary convergence** See ²evolution: convergent evolution.
♦ **evolutionary developmental biology** See biology: evolutionary developmental biology.
♦ **evolutionary distance** See distance: evolutionary distance.
♦ **evolutionary diversification** See -genesis: cladogenesis.
♦ **evolutionary genetics** See study of: genetics: evolutionary genetics.
♦ **evolutionary grade** See grade: evolutionary grade.
♦ **evolutionary lability** *n.* The ease and speed with which particular trait categories evolve (Lincoln et al. 1985).
♦ **evolutionary lag, lag load** *n.* A measure of how far from an adaptive peak a species finds itself and therefore a measure of the rate of evolution this species is likely to be undergoing (Stenseth and Maynard Smith 1984, 87; Lewin 1985a).
♦ **evolutionary medicine** See study of: evolutionary medicine.

♦ **evolutionary method** See study of: systematics: evolutionary systematics.

♦ **evolutionary morphology** See study of: morphology: evolutionary morphology.

♦ **evolutionary-morphology school** *n*. A school of evolutionists that starts with a common ancestor of a group of organisms and asks questions such as, "What evolutionary processes were responsible for divergence of descendants?" (Mayr 1982, 468). See study of: morphology: evolutionary morphology.

♦ **evolutionary novelty** *n*. An entirely new organ, structure, physiological capacity, or behavior that occurs in a group of related species (Mayr 1982, 610).

♦ **evolutionary optimum** *n*. A balance that is struck between (among) forces of evolution in a population (Wilson 1975, 131).

♦ **evolutionary pacemaker** *n*. The first trait to change in an interrelated group of traits due to evolutionary pressure (Wilson 1975, 13).
Comment: This idea extends back at least as far as Darwin's 1872 edition of *On the Origin of Species*... . Behavior may often be an evolutionary pacemaker that is followed by associated morphological change (Dewsbury 1978, 258).

♦ **evolutionary physiology** See study of: physiology: evolutionary physiology.

♦ **evolutionary rate** See rate: evolutionary rate.

♦ **evolutionary reductionism** See reductionism: evolutionary reductionism.

♦ **evolutionary response** See response: genetic response.

♦ **evolutionary species** See [2]species: evolutionary species.

♦ **evolutionary synthesis, modern synthesis** *n*.
1. The unified evolutionary theory that asserts that "major evolutionary phenomena such as speciation, evolutionary trends, the origin of evolutionary novelties, and the entire systematic hierarchy could be explained in terms of the genetic theory" (Mayr 1982, 117, 119–120, 567, 569–570; Futuyma 1986, 11).
Note: A refined version of the evolutionary synthesis is the paradigm of today. The evolutionary synthesis developed markedly from 1936 to 1947 (Mayr 1982, 567).
2. "The modern synthetic theory of evolution;" a reformulation of systematics, comparative morphology and physiology, paleontology, cytogenetics, ecology, and ethology in the language of early population genetics from about 1930 to 1960 (Wilson 1975, 4, 63–64).

Note: In this approach, each phenomenon is weighed for its adaptive significance and then related to the basic principles of population genetics (Wilson 1975, 4).
3. An expanded version of Darwinism that started in the 1930s, continues at present, and is expected to continue in the future (Stebbins and Ayala 1985, 72).
syn. adaptive neutrality, biological theory of evolution, the modern synthetic theory of evolution, neo-Darwinian evolution, theory of synthetic evolution (Wilson 1975, 4, 63–64), Neo-Darwinian theory, synthetic theory (Stebbins and Ayala 1985, 72); Neo-Darwinism (Gould 1994, 6765)
cf. [3]theory: theory of monistic evolution
Comments: The evolutionary synthesis now stresses many phenomena: opportunistic, steady evolution through natural selection on variations that arise by chance and are selected in accordance with environmental demands; natural selection acting on genes, heritable units of information governing structure, development, and function of organisms; the importance of population structure and distribution in the development of new species; the biological species concept; mosaic evolution; and molecular evolution. Three main challenges to the synthetic theory are molecular determinism, the neutral theory of evolution, and the hypothesis of punctuated equilibrium. Mayr (1993) further elucidates facets of the evolutionary synthesis, *q.v.*
[coined by Huxley 1942 in Mayr 1982, 567]

♦ **evolutionary systematics** See study of: systematics.

♦ **evolutionary taxonomy** See study of: taxonomy.

♦ **evolutionary time** See time: evolutionary time.

♦ **evolutionary tree** See -gram: phylogenetic tree.

♦ **evolutionary trends** See -genesis: orthogenesis.

♦ **evolutionary vestige** See organ: relict.

♦ **evolutionary wisdom** See [2]evolutionary adaptation (comments).

♦ **evolutionism** *n*.
1. The theory of evolution or development (*Oxford English Dictionary* 1972, entries from 1869).
2. One's belief in the occurrence of organic evolution; evolutionism comprises different groups of ideas that depend on the evolutionist (Mayr 1982, 835).
cf. Darwinism

neo-Lamarckian evolutionism *n*. Evolutionism, which permeated U.S. behaviorists starting in the late 19th century and stressed the roles of animal minds in "satisfying" environmental needs by experience, considered evolution to be an active process performed by individual efforts, and minds to be propulsive forces that yield changes that are transmitted to offspring (Gherardi 1984, 394).

◆ **evolutionist** *n*.
1. An adherent, or proponent, of organic evolution (*Oxford English Dictionary* 1972, entries from 1859).
2. One who advocates the theory of evolution (now known as the theory of preformation) as opposed to epigenesis (*Oxford English Dictionary* 1972, entry from 1875).
cf. Darwinism

macroevolutionist *n*.
1. One who studies macroevolution, *q.v.* (Dawkins 1982, 290).
2. One who is a partisan of the idea that micro- and macroevolution are "decoupled" processes (Dawkins 1982, 290).

◆ **evolve** See [1,2]evolution.
◆ **exafference** See -ference: exafference.
◆ **exagamy** See -gamy: exagamy.
◆ **exaltation** See [2]group: exaltation.
◆ **exanthropic** See -anthropic: exanthropic.
◆ **exaptation** See [3]adaptation: [2]evolutionary adaptation: exaptation.
◆ **exchange and interdependence theories** See [2]theory: exchange and interdependence theories.
◆ **excision repair enzyme** See enzyme: repair enzyme: excision repair enzyme.
◆ **excitability** *n*. In living organisms: a basic property involving elicitation of arousal responses, with inherently defined characteristics, due to milieu or behavioral-state changes (Heymer 1977, 62).
◆ **excitation** *n*.
1. A stimulus' affecting an organ, or tissue, so as to produce or intensify its characteristic activity (*Oxford English Dictionary* 1972, entries from 1831).
2. A stimulus' action on an organism (Michaelis 1963).
3. An animal's immediate, time-limited behavioral response to an external stimulus or internal-milieu change (Heymer 1977, 62).
◆ **exclusion** See competitive exclusion.
◆ **exclusive species** See [2]species: exclusive species.
◆ **exclusive territory** See territory: exclusive territory.
◆ **excreta** *pl. n*. Wastes that are eliminated from an organism's body (*e.g.*, urine and components in sweat) (Michaelis 1963).
◆ **excretion** See elimination: excretion.

◆ **exendotrophic, exendotrophy** See pollination: exendotrophy.
◆ **exhabited symbiont** See symbiont: exhabited symbiont.
◆ **exhabiting symbiont** See symbiont: exhabiting symbiont.
◆ **exhaustive sampling** See sampling: exhaustive sampling.
◆ **exhibition** See need: exhibition.
◆ **existentialism** *n*.
1. A philosophical movement that stresses the active role of human will rather than reason in confronting problems posed by a hostile universe; existentialism regards human nature as consisting of decisive actions rather than an inner, or latent, disposition (Michaelis 1963).
2. A cult of nihilism and pessimism that is supposedly based on Sartre and other existentialist writers (Michaelis 1963).
3. A school of thought that maintains that each Human is at all times faced with the dilemma of making a choice of what actions to pursue when he does not have all of the information relevant to a decision (Storz 1973, 92).
◆ **exo-** *combining form* "Out; outside; external" (Michaelis 1963). Also before vowels, ex-.
[Greek *exo-, ex-* < *exō*, outside]
◆ **exobiology** See study of: biology: exobiology.
◆ **exobiont** See biont: exobiont.
◆ **exocrine gland** See gland: exocrine gland.
◆ **exocrine substance** See chemical-releasing stimulus: semiochemical: allelochemic: allomone: exocrine substance.
◆ **exogamy** See -gamy: exogamy.
◆ **exogenous** See -genous: exogenous.
◆ **exoisogamy** See -gamy: exoisogamy.
◆ **exomixis** See -mixis: exomixis.
◆ **exomorphology** See morphology: exomorphology.
◆ **exon** *n*. In Eukaryotes: a portion of a split gene (comprised of DNA) that is included in the gene's RNA transcript (heterogeneous nuclear RNA) and remains in mature messenger RNA that a cell makes from the heterogeneous nuclear RNA (Curtis 1983, 340; King and Stansfield 1985).
cf. gene; intron; nucleic acid: deoxyribonucleic acid: parasitic DNA
Comment: Archaea have introns and exons in tRNA genes (Strickberger 1990, 155).
[from *expressed* sequences of DNA (Curtis 1983, 338); coined by Walter Gilbert (Travis 1993)]
◆ **exon-shuffling hypothesis** See hypothesis: exon-shuffling hypothesis.
◆ **exoparasite** See parasite: ectoparasite.

♦ **exoparasitism** See parasitism: exoparasitism.

♦ **exophenotypic** See -typic: exophenotypic.

♦ **exoskeleton** *n*. In arthropods: the hardened outer body layer that functions as a protective covering, a skeletal attachment for muscles, site for receptors, etc. (Wilson 1975, 584).

♦ **exosymbiont** See symbiont: ectosymbiont.

♦ **exosymbiosis** See symbiosis: exosymbiosis.

♦ **exotropism** See -tropism: exotropism.

♦ **expansion of function** *n*. An organ's evolution of more than one function; *e.g.*, besides locomotion, fish fins can serve as supporting, or copulatory, organs (Heymer 1977, 71).
syn. extension of function (Immelmann and Beer 1989, 96)

♦ **expected benefit** See benefit: expected benefit.

♦ **expected cost** See cost: expected cost.

♦ **expendable surplus, principle of** See principle: principle of expendable surplus.

♦ **experience** *n*.
1. A person's actually observing facts, or events, considered as a source of knowledge (*Oxford English Dictionary* 1972, entries from 1377).
2. A person's consciously being the subject of a state, or condition, or being affected by an event; also an instance of this phenomenon; a state, or event, viewed subjectively; an event that affects one (*Oxford English Dictionary* 1972, entries from 1382).
3. A person's knowledge obtained from observation or what he has undergone (*Oxford English Dictionary* 1972, entries from 1553).
4. An individual animal's reactions with regard to stimuli directly contacted and learned about, *e.g.*, site-orientation memory, learned behavior sequences, or stimulus-response relationships (Heymer 1977, 62).

♦ **experience deprivation** See experiment: deprivation experiment.

♦ **experiment** *n*.
1. A person's act, or operation, designed to discover, test a hypothesis, or establish, or illustrate, a truth, principle, or effect (*Oxford English Dictionary* 1972, entries from 1362; Michaelis 1963).
2. *v.t.* To make an experiment (*Oxford English Dictionary* 1972, entries from 1787).
v.i. experiment
cf. test

deprivation experiment, experience deprivation, isolation experiment, Kasper-Hauser experiment *n*. An experiment in which an investigator rears an animal in an environment that lacks some of its usual stimuli (*e.g.*, conspecifics and particular physical stimuli) (Dewsbury 1978, 159); used to determine which of an animal's capabilities develop normally, even when apparently relevant experiences are absent, and which deficits are related to which kinds of deprivation (Immelmann and Beer 1989, 159).
cf. Kasper Hauser

double-blind experiment *n*. An experiment in which neither an interrogator nor a subject has access to correct answers to questions prior to and during testing (Gould 1982, 10).

isolation experiment See experiment: deprivation experiment.

Kasper-Hauser experiment See experiment: deprivation experiment.

manipulative experiment *n*. An experiment that involves two or more treatments; attempts to make one or more comparisons; gives different experimental units different treatments; and randomizes, or can randomize, treatments to experimental units (Hurlbert 1984, 190).

masking experiment *n*. A test for time sharing, *q.v.*, involving prevention of the occurrence of ongoing behavior from occurring by an experimentally controlled interruption (McFarland 1985, 466).
syn. comparative experiment (Anscombe 1948 in Hurlbert 1984, 190)
cf. experiment: mensurative experiment

matching experiment *n*. An experiment in which an individual (*e.g.*, a Domestic Pigeon, Dolphin, Chimpanzee, or Human) is first shown a single object (the sample) and then is shown an array of objects that includes the sample; its correct response is choosing the sample from the array (Siegel and Honig 1970, etc. in McFarland 1985, 345).

maze experiment *n*. An experiment involving a simple maze (*e.g.*, T or Y maze), or more complicated maze (*e.g.*, three-dimensional maze) and an animal that learns the correct path at each choice point of the maze, eventually reaching a reward (*e.g.*, food) (Buchholtz 1973, Rensch 1973 in Heymer 1977, 108).

mensurative experiment *n*. An experiment that involves making a measurement at one or more points in space or time, with space and time as the only experimental variables or treatments; that may, or may not, involve a significance test; and that

d − 9

usually does not involve an experimenter's imposition of some external factor(s) on experimental units (Hurlbert 1984, 189).
syn. absolute experiment (Anscombe 1948 in Hurlbert 1984, 190)
cf. experiment: manipulative experiment
Comment: Some workers distinguish between "measurement experiment" and "mensurative experiment;" a mensurative experiment involves one's taking "somewhat elaborate" measurements (Hurlbert 1984, 189).

Miller-Urey experiment *n.* Creation of amino acids in a laboratory from ammonia, hydrogen, methane, and water and energy from a sparking electrode (Miller and Urey 1950s in Orgel 1994, 79).
[after Stanley L. Miller and Harold C. Urey, who performed the experiment]

natural experiment *n.* An experiment whose variables have been "manipulated" by natural phenomena without help from Humans and whose result from the manipulation is interpreted by us (*e.g.,* effects of similar environmental conditions on isolated groups of mammals, such as American eutherians and Australian marsupials).

♦ **experimental design** *n.*
1. A precisely planned procedure for testing hypotheses about the relationships among studied variables; for example, the effects of independent variables (*e.g.,* temperature, light periodicity, learning) on a dependent variable (*e.g.,* feeding) of an animal (Storz 1973, 93; Lehner 1979, 77).
Notes: The independent variables are sometimes called "treatments" whose states, or magnitudes, can be set by natural conditions or manipulation by an investigator. One's knowing the relationship between a dependent variable and one or more independent variables can be useful in predicting the state, or magnitude, of the dependent variable in future studies.
2. The logical structure of an experiment (Fisher 1971, 2, in Hurlbert 1984, 188), including a specification of the nature of experimental units, the number and kinds of treatments to be imposed, the properties or responses (of the experimental units) that will be measured, the manner in which treatments are assigned to the available experimental units, the number of experiment units (replicates) that receive each treatment, the physical arrangement of the experimental units, the temporal sequence in which treatments are applied to the different experimental units, and the measurements made on these units (Hurlbert 1984, 188).
cf. statistical test

completely randomized block *n.* An experimental design used to compare the effects of two or more treatment levels of an independent variable on a dependent variable (Lehner 1979, 77).

incomplete block *n.* An experimental design that is used "when the number of subjects available for study is not large enough to measure each treatment effect for each block" (Lehner 1979, 80).
▸ **balanced incomplete block** *n.* An incomplete block design in which each block "contains the same number of subjects, each treatment level occurs the same number of times, and subjects are assigned to the treatment levels so that each possible pair of treatment levels occurs together within some block an equal number of times" (Lehner 1979, 80).
▸ **partially balanced incomplete block** *n.* An incomplete block design in which "some pairs of treatment levels occur together within the blocks more often than do other pairs" (Lehner 1979, 80).

Latin-square design *n.* A randomized, two-way, complete-block design in which each treatment level is replicated only once in a given row or column (Sokal and Rohlf 1969, 362–363).

randomized block *n.* An experimental design that "attempts to control for additional sample variability (expressed in the error effect) by assigning subjects that are similar in one or more characteristics (or ways treatments are applied to them) to blocks" (Lehner 1979, 79).

sociometric matrix *n.* An experimental design often used to test one-sidedness in dyadic interactions (Lehner 1979, 115).

♦ **experimental error** See error: experimental error.

♦ **experimental hypothesis** See hypothesis: experimental hypothesis.

♦ **experimental manipulation** *n.* A scientific investigation which "...usually consists of making an event occur under known conditions where as many extraneous influences as possible are eliminated and close observation is possible so that relationships between phenomena can be revealed" (Beveridge 1950, 20, in Lehner 1979, 59).
cf. naturalistic observation

♦ **experimental psychology** See study of: psychology: experimental psychology.

♦ **experimentation** *n.* A person's conducting an experiment (Michaelis 1963).

♦ **experimentwise-type-I error** See error: type-I error: experimentwise-type-I error.

♦ **explanatory reductionism** See reductionism: explanatory reductionism.

♦ **exploded lek** See ²lek: dispersed lek.
♦ **exploitably vulnerable species** See ²species: exploitably vulnerable species.
♦ **exploitation** See competition: exploitation.
♦ **exploration** *n.*
1. A person's examining, investigating, or scrutinizing something (*Oxford English Dictionary* 1972, entries from 1543, obsolete definition).
2. A person's exploring, especially unfamiliar, or unknown, regions (Michaelis 1963).
   pure exploration *n.* An animal's learning about a new object or a strange part of its environment (Wilson 1975, 165).
   Comment: An animal's particular investigatory responses are determined by a new object's properties (Hutt 1966 in Wilson 1975, 165).
♦ **exploratory behavior** See behavior: exploratory behavior.
♦ **exploratory drive** See drive: exploratory drive.
♦ **exploratory learning** See learning: exploratory learning.
♦ **exploratory trail** See odor trail: exploratory trail.
♦ **explosive breeding** See breeding: communal breeding: explosive breeding.
♦ **explosive evolution, explosive radiation** See ²evolution: explosive evolution.
♦ **explosive-mating-assemblage polygyny** See mating system: polygyny: scramble-competition polygyny: explosive-mating-assemblage polygyny.
♦ **exponential growth** See growth: exponential growth.
♦ **exponential-growth phase** See phase: log-growth phase.
♦ **expression vector** See vector: expression vector.
♦ **expulsion reflex** See reflex: expulsion reflex.
♦ **extant** *adj.* Referring to a taxon that is still living on Earth (Lincoln et al. 1985).
See recent.
ant. extinct
cf. endangered, extirpated, threatened
♦ **extended family** See ¹family: extended family.
♦ **extended phenotype** See -type, type: phenotype: extended phenotype.
♦ **extension of function** See expansion of function.
♦ **external environment** See environment: external environment.
♦ **external-nontranscribed spacers (ENS)** *n.* Nontranscribed regions of ribosomal-RNA genes (Berenbaum 1996, 5).
syn. intergenic spacers (IGS), nontranslated spacers (NTS) (Berenbaum 1996, 5)

♦ **external stimulus, external stimulation** See ²stimulus: external stimulus.
♦ **external validity** See validity: external validity.
♦ **exteroception** See modality: chemoreception: exteroception.
♦ **extinct** *adj.* Referring to a taxon that is no longer living on Earth (Cooper 1984 in Norden et al. 1984, 9).
ant. extant
cf. endangered; extirpated; threatened; type: necrotype
♦ ¹**extinction** *n.* A dying out of an animal's response to a primary stimulus (unconditioned stimulus) because a secondary stimulus (conditioned stimulus) that it comes to associate with its response is repeated without reinforcement by the primary stimulus (Scott and Fredericson 1951, 286; Dewsbury 1978, 323); contrasted with forgetting (Immelmann and Beer 1989, 97).
cf. learning: conditional learning: Pavlovian conditioning
♦ ²**extinction** *n.*
1. The process of elimination of, *e.g.,* less fit genotypes from a population (Lincoln et al. 1985).
2. The disappearance of all living individuals of a species, or other taxon, from a particular habitat, biota, or entire Earth (Lincoln et al. 1985).
See ²extinction: terminal extinction.
cf. extinct
Comments: Extinction of a population, or taxon, is often considered to be an example of a general biological phenomenon of considerable importance in biology and paleontology. Evidently, most species that have lived on Earth are now extinct (Damuth 1992, 106). Primarily due to anthropogenic causes, Earth is losing about 32,000 organism populations per day (Myers 1999, 32). Some hypotheses related to extinction are defined under hypothesis.
   K extinction, K-extinction *n.* The regular extinction of populations when they have reached, or exceeded, carrying capacity of their environments (Lincoln et al. 1985).
   cf. ²extinction: r extinction
   linked extinction *n.* One, or more, other organism's going extinct due to the extinction of another organisms (*e.g.,* a keystone mutualist, *q.v.*) (inferred from Buchmann and Nabhan 1996, 249).
   mass extinction *n.* An extinction that involves many taxa; a major extinction episode (Raven and Johnson 1989, 428).
   Comments: Some of Earth's many mass extinctions occurred about 505, 438, 360, 243, and 65 million years ago. Mass extinctions might be periodic, occurring an average of

about every 26 million years (Raup and Sepkoski 1986). Benton (1995, 52) found no periodicity in mass extinctions. Seehausen et al. (1997a) state that the dying off of about 200 cichlid species is probably Earth's largest "mass extinction" of contemporary vertebrates.

natural extinction *n.* Extinction due to causes besides Humans; contrasted with "human-caused extinction" (Ehrlich and Wilson 1991, 759).

pseudoextinction *n.* The evolution of one species into another, with the disappearance of the ancestral species (Ehrlich and Wilson 1991, 759).
syn. ²extinction: phyletic extinction; speciation: phyletic speciation (Lincoln et al. 1985)
cf. ²extinction: terminal extinction, true extinction

r extinction, r-extinction *n.* Extinction that occurs when organism colonists fail to establish themselves in a new site (Lincoln et al. 1985).
cf. ²extinction: K extinction

terminal extinction, extinction *n.* The loss of an entire species without the evolution of a descendant lineage (Lincoln et al. 1985).
cf. ²extinction: phyletic extinction

true extinction *n.* Extinction of a species and any species descended from it (altogether called a "clade") (Ehrlich and Wilson 1991, 759).
cf. ²extinction: pseudoextinction

♦ **extinction curve** See curve: extinction curve.

♦ **extinction rate** See rate: extinction rate.

♦ **extirpated** *adj.* Referring to a population of a species that has been exterminated (Cooper 1984 in Norden et al. 1984, 9).
cf. endangered, extant, extinct, threatened

♦ **extirpation** *n.*
1. Removal of an animal's organ or organ part (Immelmann and Beer 1989, 97).
2. A population's elimination by either human or natural causes (Buchmann and Nabhan 1996, 245).
cf. ablation

♦ **extrafloral nectary** See nectary: extrafloral nectary.

♦ **extra-individual character** See character (comments).

♦ **extraneous character** See character: extrinsic character.

♦ **extraocular light perception** See perception: extraocular light perception.

♦ **extrapair copulation** See copulation: extrapair copulation.

♦ **extrapyramidal system** See ²system: nervous system: extrapyramidal system.

♦ **extreme altruism** See altruism: extreme altruism.

♦ **extreme feather ruffling** See feather ruffling: extreme feather ruffling.

♦ **extreme inbreeding** See -mixis: automixis.

♦ **extreme vitalism** See vitalism: extreme vitalism.

♦ **extrinsic character** See character: extrinsic character.

♦ **extrinsic isolating mechanism** See mechanism: isolating mechanism: extrinsic isolating mechanism.

♦ **extrinsic selection pressure** See selection pressure: extrinsic selection pressure.

♦ **exudativore** See -vore: exudativore.

♦ **exuviae** *pl. n.* Skins, shells, etc. cast off, or shed, by animals, often referring to arthropods (Michaelis 1963).
adj. exuvial
syn. exuvia, exuvium (not preferred synonyms, Torre-Bueno 1978)
cf. molt
[Latin < *exuere*, to cast off, undress]

♦ **eye** See ²group: eye.

♦ **eye camouflage** See camouflage: eye camouflage.

♦ **eye contact** *n.* An element of communication in which one person looks into the eyes of another which may be part of flirting, aggression, dominance, or a combination of these behaviors (Heymer 1977, 45).

♦ **eye of a needle** See effect: bottleneck effect.

♦ **eyebrow flash** *n.* A person's quick raising of his eyebrows, often followed by a smile, and being part of flirting, greeting, or both (Heymer 1977, 34).
Comments: Because eyebrow flashes have been found to have wider communicative function than first supposed, Eibl-Eibesfeldt (1972a,b in Heymer 1977, 47) proposed the less-restrictive term "eyebrow raising" for this behavior. Eyebrow flashes have different meanings in different human cultures. [coined by Eibl-Eibesfeldt 1968 in Heymer 1977, 47]

♦ **eyebrow raise** See eyebrow flash.

♦ *eyeless* See gene: *eyeless.*

♦ **eyespot** *n.* A pattern ranging from a simple round black spot, which scarcely contrasts with its background, to a black spot with a light contrasting ring, to startlingly impressive eye markings; *e.g.,* on some caterpillars; some adult butterflies, homopterans, mantids, and moths; fish; Peacock feathers (Wickler 1968, 64–70).
cf. behavior: defensive behavior, deimatic behavior
Comment: Potential predators (*e.g.,* birds) are frightened by eyespots of a butterfly (Heymer 1977, 34).

♦ **eyrar** See ²group: eyrar.

♦ **eyrie** See ²group: aerie.

♦ **F** See index: fixation index.

♦ ***F, f*** See coefficient: coefficient of inbreeding; function.

♦ **FACE System** See system: Free Air CO_2 Enrichment System.

♦ **face-to-face-mating hypothesis** See hypothesis: face-to-face-mating hypothesis.

♦ **facial expression** *n.* In more derived mammals, *e.g.,* canines, felines, many ungulate and primate species, notably Chimpanzees and Humans: movements of facial parts that convey mood, or intention, as intraspecific communicatory signals (Heymer 1977, 116; Immelmann and Beer 1989, 98).

cf. display and threat for other facial expressions, flehmen

anger *n.* A person's facial expression that usually involves withdrawing his lips, baring his teeth, staring with his eyes, and frowning (McFarland 1985, 486).

alertness *n.* A person's facial expression that involves a relatively fixed gaze and a certain tension of facial muscles (McFarland 1985, 486).

copulation grimace *n.* In Rhesus Monkeys: an adult male's withdrawing his lips during copulation; contrasted with ear-flap threat, fear grimace, and open-mouth threat (Hauser 1993, 475).

disgust *n.* A person's ritualized facial expression that involves nose wrinkling, upper-lip raising, eye-screwing, and turning his face away (McFarland 1985, 486).

Comment: These components are derived from protective responses serving to exclude noxious stimuli (McFarland 1985, 486).

ear-flap threat *n.* In Rhesus Macaques: a dominant individual's retracting its ears back against its head when it is attacking, or intimidating, a subordinate; contrasted with copulation grimace, fear grimace, and open-mouth threat (Hauser 1993, 475).

fear *n.* In Humans and many other primates: a facial expression that involves wide open eyes and withdrawn lips (McFarland 1985, 486).

cf. fear

fear grimace *n.* In Rhesus Monkeys: a subordinate individual's withdrawing its lips when it is attacked, or intimidated, by a higher ranking group member; contrasted with copulation grimace, ear-flap threat, and open-mouth threat (Hauser 1993, 475).

flehmen See flehmen.

funny face *n.* In juvenile Bonobos and Chimpanzees: a weird facial expression aimed at no other individual in particular (de Waal 1989, 197–198).

grin *n.*

1. For example, in the Chimpanzee, Rhesus Macaque; Humans: an animal's drawing back its lips and exposing its teeth (*Oxford English Dictionary* 1972, entries from 1000; de Waal 1989, 21), as in a snarl or a grimace of pain, rage, etc. (Michaelis 1963).
2. A person's broad smile (Michaelis 1963).

cf. facial expression: smile

joy *n.* In Humans, possibly some other primates: a facial expression that involves smiling, laughter, or both (McFarland 1985, 486).

open-mouth laughter *n.* In human infants, rarely in adults: laughing with an open mouth and without retracted lips (McFarland 1985, 486).

open-mouth threat *n.* In Rhesus Macaques: a dominant individual's slightly protruding its lips and placing them into an O-shaped configuration when it is attacking, or intimidating, a subordinate; contrasted with copulation grimace, ear-flap threat, and fear grimace (Hauser 1993, 475).

play signal *n.* An animal's facial expression that signifies its readiness for social play or serves to avoid "misunderstandings," especially during aggressive play (Immelmann and Beer 1989, 224–225).

▶ **play face** *n.* In many carnivore and primate species: a play signal (Immelmann and Beer 1989, 225).

pout face *n.* For example, in Bonobos: an animal's extending its lips while its mouth is slightly open (de Waal 1989, 207–208).

relaxed open-mouth display See display: relaxed open-mouth display.

sadness *n.* In the Human, possibly Chimpanzee: a ritualized facial expression that involves one's arched eyebrows, retracted corners of one's mouth that are turned downward, outward curling of ones's lips, and tears (in intense cases) (McFarland 1985, 486).

smile, smiling *n.* A person's raising his lip corners with his mouth open or closed, with or without vocalizing (*Oxford English Dictionary* 1972, entries from 1562).

cf. facial expression: grin

Comment: Smiling can be a low-intensity form of laughter and suggest pleasure, satisfaction, good humor, approval, or amity or can be associated with mild fear, amused contempt, disdain, incredulity, etc. "Smile," "grin," "simper," and "smirk" are distinguished from one another (Michaelis 1963; McFarland 1985, 486).

surprise *n.* A person's facial expression that involves prolonged eyebrow raising, open eyes, and often an open mouth (McFarland 1985, 486).

threat face *n.* In some carnivore, cervid, and primate species: an individual's showing threat behavior with its facial expression (*e.g.*, baring teeth) (Immelmann and Beer 1989, 312); in Humans and Mandrills, exposing canine teeth and lifting lips during anger and offensiveness (Heymer 1977, 54).

threat gape, threat yawn *n.* In the Hippopotamus, some carnivore and primate species: an animal's showing a threat face that is enhanced by its opening its jaws very widely (Heymer 1977, 54; Immelmann and Beer 1989, 312).

cf. yawning

◆ **facial-feature graph** See graph: facial-feature graph.

◆ **facies fossil** See fossil: facies fossil.

◆ **[1]facilitation** *n.* The additive effect that multiple nerve impulses' arriving at a synapse may have in bringing about synaptic transmission (Immelmann and Beer 1989, 98).

spatial facilitation *n.* The summation of results from impulses that arrive via different nerve fibers simultaneously; these impulses are ineffective singly, but together they produce sufficient change to cause a response (Immelmann and Beer 1989, 99, 299).

temporal facilitation *n.* The summation of results from impulses that arrive close together in succession via a single nerve fiber (Immelmann and Beer 1989, 99).

syn. temporal summation (Immelmann and Beer 1989, 99, 299)

◆ **[2]facilitation** *n.*

1. One animal's behavior being initiated, or increased, in pace or frequency by the mere presence, or actions, of another animal that is not necessarily engaged in this behavior (Wilson 1975, 51, 595).

2. In psychology, an animal's enhancing its behavior (*e.g.*, quickening its perception or response) due to contributory internal, or external, factors, or processes, such as attention, arousal, or reinforcement (Immelmann and Beer 1989, 99).

See learning: facilitation.

cf. effect: group effect; [2]facilitation: social facilitation; [1]induction

parental facilitation *n.* A parent's aid to its offspring that improves its offspring's abilities to achieve breeding status; this is not extended parental care in the usual sense, because the usual types of parental care, such as feeding young, are not involved (Brown 1987a, 299).

cf. care: parental care

social facilitation *n.* An animal's taking up the behavior (*e.g.*, feeding or taking flight) of another member of its social group; this tendency can spread so that most to all members of a group show the same behavior (Immelmann and Beer 1989, 275).

cf. imitation, mood induction

▶ **audience-effected social facilitation** *n.* Social facilitation in which the facilitating organisms are passive spectators (Zajonc 1965 in Dewsbury 1978, 105).

▶ **coaction-effected social facilitation** *n.* Social facilitation wherein interacting organisms are engaged in the same behavior (Zajonc 1965 in Dewsbury 1978, 105) (*e.g.*, feeding in rats and chickens or copulating in rats).

◆ **[3]facilitation** *n.* In behavior, the accelerating influence of environmental factors on behavioral development during an animal's ontogeny; distinguished from induction (Immelmann and Beer 1989, 99).

◆ **[4]facilitation** *n.* The positive effect of species-2 on species-1, without any reference to the effect of species-1 on species-2; species-2 thus causes the facilitation of species-1 (Boucher 1992, 208).

cf. cooperation; symbiosis: mutualism; parasitism; predation

♦ **fact** *n.* "Something that has really occurred or is actually the case; something certainly known to be of this character; hence, a particular truth known by actual observation, or authentic testimony, as opposed to what is merely inferred, or to a conjecture or fiction; a datum of experience, as distinguished from the conclusions that may be based upon it" (*Oxford English Dictionary* 1972, entries from 1632).
[Latin *factum* < *facere*, to do]

fact-in-itself *n.* A "fact" that does not have a logical explanation in an existing theoretical framework but is, nevertheless, "unquestioned and ignored, or accepted as a given property of the world, or simply postulated to be true" (Aristotle in Lightman and Gingerich 1992, 694).
syn. to oti (Aristotle), *a quia* (Middle Ages) (Lightman and Gingerich 1992, 694)
cf. phenomenon: retrorecognition phenomenon

reasoned fact *n.* A "fact" that has a reasonable explanation (Aristotle in Lightman and Gingerich 1992, 694).
syn. di oti (Aristotle), *propter quid* (Middle Ages) (Lightman and Gingerich 1992, 694)
cf. phenomenon: retrorecognition phenomenon

♦ **fact-in-itself** See fact: fact-in-itself.
♦ **fact memory** See memory: fact memory.
♦ **factoid** *n.* Any fact that has no existence other than that it once appeared in print as a purported fact and has been treated as a fact ever since (Norman Mailer, novelist, in Hitt 1992).
[FACT + OID]

♦ **factor** *n.*
1. An element, or cause, that contributes to producing a result (Michaelis 1963).
2. In mathematics, one or more quantities that, when multiplied together, produce a given quantity (Michaelis 1963).
3. In physiology, an element that is important in metabolism and nutrition (*e.g.*, an enzyme, hormone, or vitamin) (Michaelis 1963).
See effect: main effect.
cf. gene

density-dependent factor *n.* A factor (*e.g.*, intraspecific competition, parasitism, predation, or disease) whose influence is affected by the size of a population that it affects (Wilson 1975, 82).
syn. endogenous factor (Turchin et al. 1999, 1069)
cf. dependence: density dependence

density-independent factor *n.* A factor (*e.g.*, volcanic eruption, drought, or flood) that is not affected by the size of a population that it influences and influences one or more

of the following: the population's birth, death, and migration rates (Wilson 1975, 82).
syn. exogenous factor (Turchin et al. 1999, 1069)

dilution factor, mean r, r̄ *n.* The fraction of genes in a population, or species, that is unlike that of a focal altruist and that slows down natural selection but does not alter its direction (Hamilton 1970 in West Eberhard 1975, 5).

kappa factor *n.* A microorganism in the cytoplasm of *Paramecium* that affects its phenotypes (Preer et al. 1974 in Mayr 1982, 789).
cf. factor: sex-ratio factor, sterility factor

limiting factor *n.* Any environmental factor, or group of related factors, that exists at a suboptimal level and thereby prevents an organism from reaching its full biotic potential (Lincoln et al. 1985).
▸ **master limiting factor, master factor** *n.* The single limiting factor that prevents a population from realizing its biotic potential (Lincoln et al. 1985).
▸ **paired limiting factor** *n.* One of two or more limiting factors that acts synergistically to prevent a population from realizing its biotic potential (Lincoln et al. 1985).

proximate factor, proximate causation, zeitgeber *n.* An environmental factor that acts as an immediate stimulus for periodic biological activity (Lincoln et al. 1985).
See causation: proximate causation; ^{2}stimulus: *zeitgeber.*
cf. factor: ultimate factor

sex-ratio factor *n.* Microorganisms that are passed to gametes during their formation in *Drosophila* and affect their progeny sex ratio (Mayr 1982, 789).
cf. factor: kappa factor, sterility factor

sterility factor *n.* Microorganisms that are passed to gametes during their formation that cause sterile offspring in *Culex* mosquitoes (Mayr 1982, 789).
cf. factor: kappa factor, sex-ratio factor

ultimate factor *n.* An environmental factor to which the timing of a biological event is ultimately associated; ultimate causation (Lincoln et al. 1985).
See causation: ultimate causation.

♦ **factor analysis** See test: statistical test: factor analysis.
♦ **facultative** *adj.*
1. Referring to an organism that can assume a particular role, or mode, of life but is not restricted to it (Lincoln et al. 1985).
2. Referring to an organism that can live, or is living, under atypical conditions (Lincoln et al. 1985).
cf. obligate

♦ **facultative anaerobe** See aerobe: anaerobe: facultative anaerobe.

♦ **facultative gamete** See gamete: facultative gamete.

♦ **facultative hyperparasitism** See parasitism: hyperparasitism: facultative hyperparasitism.

♦ **facultative learning** See learning: facultative learning.

♦ **facultative parasitism** See parasitism: facultative parasitism.

♦ **facultative parthenogenesis** See parthenogenesis: facultative parthenogenesis.

♦ **facultative polygamy** See mating system: polygamy: facultative polygamy.

♦ **facultative schooling species** See ²species: facultative schooling species.

♦ **facultative siblicide** See -cide: siblicide: facultative siblicide.

♦ **facultative social parasitism** See parasitism: social parasitism: facultative social parasitism.

♦ **facultative thermophile** See ¹-phile: thermophile: facultative thermophile.

♦ **faculty psychology** See study of: psychology: faculty psychology.

♦ **faithful species** See ²species: exclusive species.

♦ **Falconer-style evolution** See ²evolution: Falconer-style evolution.

♦ **fall** See ²group: fall.

♦ **fall song** See animal sounds: song: subsong.

♦ **fallacy** *n.*
1. One's reasoning, argument, etc. that is contrary to the rules of logic (*Oxford English Dictionary* 1972, entries from 1562; Michaelis 1963).
2. One's delusive notion, an error, especially one founded on false reasoning (*Oxford English Dictionary* 1972, entries from 1590).
   Concorde fallacy See effect: Concorde effect.
   fallacy of affirming the consequent
   n. One's constructing "a particular model from a set of postulates, obtaining a result, noting that approximately the predicted result does exist in nature, and concluding thereby that the postulates are true" (Wilson 1975, 29).
   Comment: "The difficulty [with this process] is that a second set of postulates, inspiring a different model, can often lead to the same result" (Wilson 1975, 29).
   naturalistic fallacy *n.* One's "supposition that what is 'natural' is 'good'" (Futuyma 1986, 8); "whatever biology tells us is so is also what ought to be" (David Hume in Alexander 1987, xvi).
   nominal fallacy *n.* The false concept that naming a phenomenon explains it

(*e.g.,* naming a kind of instinct explains it) (Dewsbury 1978, 11).

♦ **false head** *n.* For example, in some membracid-insect, hairstreak-butterfly, and snake species: coloration and morphology at the posterior end of an animal that resembles its true head (Wickler 1968, 70–77).
syn. dummy head (Wickler 1968, 71)
cf. mimicry: head mimicry
Comment: False heads might help to protect their bearers because predators might strike at them instead of true heads, inflicting no, or less significant, damage on the prey (Wickler 1968, 70–77; Robbins 1980).

♦ **false parasitism** See parasitism: pseudoparasitism.

♦ **false pheromone** See chemical-releasing substance: semiochemical: allelochemic: allomone.

♦ **falsifiable hypothesis** See hypothesis: falsifiable hypothesis.

♦ **¹family** *n.* For example, in some invertebrate and fish species, almost all bird species, all mammal species: one or both parents and their offspring, which might be accompanied by other closely associated kin (Wilson 1975, 584).
cf. ²group
Comment: "Two-parent family" is on a continuum with "one-parent family" in terms of the amount of care that parents give their young (Immelmann and Beer 1989, 100).
   extended family *n.* A nuclear family plus other kin; contrasted with nuclear family (Brown 1987a, 299).
   maternal family *n.*
   1. A mother and her offspring (Heymer 1977, 120).
      syn. gynopaedium (Deegener 1918 in Heymer 1977, 120)
   2. In some gallinaceous-bird, hummingbird, insect, fish, mammal, scorpion, spider species: a family in which only a mother cares for her young (Immelmann and Beer 1989, 100).
   nuclear family *n.* A family consisting only of a mother, father, and their offspring; contrasted with extended family (Brown 1987a, 299).
   paternal family *n.* In the Painted Snipe, Phalarope, Shea; labyrinth fish, seahorses, stickleback fish: a family in which only a father cares for his young (Immelmann and Beer 1989, 100).
   two-parent family *n.* In the majority of bird species, some fish species (including many cichlid species) and mammal species (canids, Gibbons, Humans): a family in which both parents make similar investments in their young (Immelmann and Beer 1989, 100).

◆ ²**family** *n.* A taxon comprised of closely related genera, sometimes only one genus (Immelmann and Beer 1989, 99).

◆ ³**family** *n.* A class, or group, of like, or related, things (Michaelis 1963).

gene family *n.* Two, or more, loci with similar nucleotide sequences, that have evolved from a common ancestral sequence (Futuyma 1986, 552).

◆ ⁴**family** See ²group: family.

◆ **fanning** *n.* For example, in male stickle-back fish, some other fish species: rhythmic pectoral-fin movement of water over eggs (van Iersel 1953 in Heymer 1977, 66).

wing fanning *n.* Rhythmic wing movement (*e.g.,* in Honey Bees that are cooling their hive) (Heymer 1977, 66).

◆ **fare** See ²group: fare.

◆ **farming** See husbandry.

◆ **fast *ad hoc* mimicry** See mimicry: *ad hoc* mimicry: fast *ad hoc* mimicry.

◆ **fast-evader mimicry** See mimicry: fast-evader mimicry.

◆ **fatigue** *n.*
1. Physiologically, the loss of energy, lessened activity, and decreased response to stimulation of an organism, or one of its parts, that results from excessive exertion or stimulation (Michaelis 1963).
2. Behaviorally, specific exhaustibility, *e.g.,* the reduced releasibility of a behavior due to its preceding performance (Immelmann and Beer 1989, 101).
cf. learning: habituation

◆ **fauna** *n.*
1. Animal species in a particular location or during a stated period; distinguished from flora (Michaelis 1963).
2. A treatise on these animals (Michaelis 1963).
cf. biota; ²community: benthos; flora
Comment: Macrofauna, megafauna, meiofauna, mesofauna, and microfauna are confusing terms. Someone needs to distinguish them on the basis of animal adult size.
[New Latin from Latin *Fauna*, a rural goddess. Introduced by Linnaeus as a companion to flora.]

A CLASSIFICATION OF FAUNA BASED ON ANIMAL SIZE

I. megafauna
   A. macrofauna
II. mesofauna
III. microfauna
   A. meiofauna (*syn.* interstitial fauna)

avifauna *n.* The birds of an area or period (Lincoln et al. 1985).

▶ **resident avifauna** *n.* The avifauna that live in an area during the season of highest breeding activity (Lincoln et al. 1985).

cryptofauna *n.* Fauna of protected, or concealed, microhabitats (Lincoln et al. 1985).
cf. fauna: phytalfauna

epifauna *n.*
1. The total fauna inhabiting a sediment, or water, surface (Lincoln et al. 1985).
syn. epibenthos (Lincoln et al. 1985)
2. Organisms (characteristic of an intertidal zone) that live on the surface of a seabed, either attached or unattached to objects on its bottom (Allaby 1994).
See ²community: benthos: epibenthos.
cf. fauna: infauna

ichnofauna *n.* An area's animal traces (Lincoln et al. 1985).

ichnyofauna *n.* Fish species of a particular region (Lincoln et al. 1985).

infauna See ²community: benthos: endobenthos.

interstitial fauna See fauna: meiofauna.

macrofauna *n.* Larger soil animals that one can readily remove with one's hands from a soil sample, in particular burrowing vertebrates (*e.g.,* moles or rabbits); contrasted with megafauna, meiofauna, mesofauna, and microfauna (Allaby 1994).
Note: Some researchers include larger earthworms and insects in macrofauna (Allaby 1994).

megafauna *n.*
1. Fauna large enough to be seen with unaided human eyes; contrasted with macrofauna, meiofauna, mesofauna, and microfauna (Lincoln et al. 1985; Allaby 1994).
2. "A widespread group of animals" (Lincoln et al. 1985).
3. Animals of a large uniform habitat (Lincoln et al. 1985).

meiofauna *n.*
1. Small interstitial animals that pass through a 1-mm-mesh sieve but are retained by a 0.1-mm one; contrasted with macrofauna, megafauna, mesofauna, and microfauna (Lincoln et al. 1985).
2. That part of the microfauna that inhabits algae, rock fissures, and the superficial layers of the muddy sea bottom (Allaby 1994).
syn. interstitial fauna

mesofauna *n.* Soil organisms of intermediate size; small invertebrates, characteristically annelids, arthropods, molluscs, nematodes; contrasted with macrofauna, megafauna, meiofauna, and microfauna (Allaby 1994).

d–g

Comment: Mesofauna are readily removed from soil with a Berlese funnel, Tullgren funnel, or other similar device (Allaby 1994).

microfauna *n.*

1. Fauna not visible with unaided human eyes; contrasted with macrofauna, megafauna, meiofauna, and mesofauna (Lincoln et al. 1985).

   Notes: A person with normal 20-20 vision can see most kinds of tiny animals but not many kinds protozoa without magnification (Kingdom Protista). "Microfauna" takes on a different meaning when one places protozoa in Protista and not in Animalia.

2. "A localized group of animals" (Lincoln et al. 1985).

3. Animals of a microhabitat (Lincoln et al. 1985).

phytalfauna *n.* Fauna in protected, or concealed, plant microhabitats (Lincoln et al. 1985).

cf. fauna: cryptofauna

stenohygrobia *n.* A fauna that is tolerant of only a narrow humidity range (Lincoln et al. 1985).

♦ **faunal dominance** See dominance (ecological): faunal dominance.

♦ **fear** *n.*

1. A person's emotion of pain, or uneasiness, caused by his sense of impending danger or by the prospect of some possible evil (*Oxford English Dictionary,* entries from 1155).

   cf. anxiety

2. A person's solicitude, anxiety for a person or thing (*Oxford English Dictionary,* entries from 1490).

3. A person's state of apprehension that focuses on isolated and recognizable dangers; distinguished from anxiety (Erickson 1950 in DeGrazia and Rowan, personal communication).

4. In animals: an emotion that is manifested in outward signs or behavioral tendencies of fleeing, escaping, or showing defensive behavior in a dangerous or threatening situation (Hinde 1970, 349; Immelmann and Beer 1989, 101).

   syn. fear response (Hinde 1970, 349)

5. In animals: a motivation that impels fleeing, escaping, or showing defensive behavior in a dangerous or threatening situation (Immelmann and Beer 1989, 101).

cf. anxiety; facial expression: fear; fright; response: immobility response; suffering

fear of strangers *n.* A child's fear of unfamiliar persons that peaks when he is about 8 months old and is often expressed by turning his head away from a stranger, avoiding eye contact, and protesting if touched (Eibl-Eibesfeldt 1972 and Hassenstein 1972 in Heymer 1977, 69).

syn. eight-months fear (Heymer 1977, 69)

♦ **fear trill** See animal sounds: fear trill.

♦ **feather ruffling** *n.* In birds: feather erection (Immelmann and Beer 1989, 101).

cf. sleeking

extreme feather erection, ruffling *n.* Feather ruffling that causes feather tips to separate from one another, which allows more air to get to a bird's skin and cool it (Immelmann and Beer 1989, 101).

fluffing *n.* Feather ruffling that stops short of separating feather tips from one another, which increases the thickness of a feather insulating layer and increases a bird's heat retention (Immelmann and Beer 1989, 101).

♦ **feather stealing** See decoration stealing.

♦ **"fecal hypothesis"** See hypothesis: hypotheses regarding the origin of the Cambrian explosion: "fecal hypothesis."

♦ **feces** (*Brit.* **faeces**) *n.*

1. Animal excrement; ordure (Michaelis 1963).

2. Any foul refuse matter or sediment (Michaelis 1963).

syn. dropping, issue, poop (colloquial) (Wright 1996, 59), scat (Robinson and Challinor 1995, 224), shit (obscene)

[Latin *faex, faecis,* sediment]

♦ **Fechner's law** See law: Fechner's law.

♦ **fecundity ratio** See ratio: fecundity ratio.

♦ **feed-forward** *n.* An animal's action that results from cues that signify what condition is to be expected, rather than from just its current physiological state and feedback, *e.g.,* an animal's drinking before eating and thus "anticipating" the dehydrating effects of digestion (Immelmann and Beer 1989, 103).

♦ **feedback** *n.*

1. A system's output's influencing its input in a way that modifies and controls its future output (Michaelis 1963; Immelmann and Beer 1989, 102).

2. Information processing in which the result of a certain system component's action serves as an initiator of new action; feedback involves cycling and information transfer (Heymer 1977, 149).

negative feedback *n.* Feedback that negates deviations from a set value, thus it is a means of maintaining stability; *e.g.,* when a homeotherm's body temperature deviates from its normal magnitude, its body takes measures to return it to this magnitude (Immelmann and Beer 1989, 102).

positive feedback *n.* Feedback in which a system's output adds to its cause and results in a progressive increase of output;

e.g., local depolarization of a neuron membrane increases its permeability to sodium ions at a particular point, the consequent sodium-ion influx further depolarizes the membrane, the sodium influx is further increased, etc., until limiting conditions are reached (Immelmann and Beer 1989, 102).

♦ **feedback circuit** *n.* A self-regulating system in which an actual parameter value (*Istwert*) is compared with and corrected towards a predetermined reference value (*Sollwert*) (Heymer 1977, 143).
cf. principle: reafference principle

♦ **feeder** See feeding.

♦ **feeding** *n.*
1. A person's giving, or supplying, food to another person or animal (*Oxford English Dictionary* 1972, verb entries from 950).
   Note: "Feeding" now refers to nonhuman animals as actors.
2. A person's grazing, or pasturing, cattle, sheep, etc. (*Oxford English Dictionary* 1972, entries from 1382).
3. A person's consuming food; eating (*Oxford English Dictionary* 1972, entries from 1387).
   Note: This sense is now used for nonhuman animals as well.
n. feeder
syn. alimentation (Lincoln et al. 1985)
cf. fluid recycling, foraging, guild, hawking, trophallaxis

allofeeding *n.* An animal's feeding young that are not its own (Brown 1987a, 297).
Comment: This is a special case of alloparenting (Brown 1987a, 297).

courtship feeding *n.* In some spider species, members of 10 insect orders (including some dance-fly and scorpionfly species, a pomace-fly species), many bird species: an animal's (usually a male's) actual, or sham, feeding of its mate during courtship (Heymer 1977, 37; Thornhill and Alcock 1983, 417–418; Lifjeld and Slagsvold 1986, 1441; Steele 1986, 1087; Immelmann and Beer 1989, 63).
syn. greeting feeding [because some kinds of birds present food only during, or after, copulation (Wickler 1969 in Heymer 1977, 37)]
cf. billing, ceremony

deposit feeding *n.* For example, in some bivalve-mollusc and polychaete-worm species: feeding on the bottom sediments in which an individual lives (Barnes 1974, 62).
cf. feeding: suspension feeding
Comments: Deposit feeders consume the substratum in which they live and digest organic detritus away from sand grains and other inorganic substances, egesting

the latter. Deposit feeders ingest material directly with their mouths or first gather food with special appendages.

filter feeding *n.* For example, in many bivalve-mollusc species, some crustacean species: feeding by straining suspended particulate organic matter from water; distinguished from suspension feeding (Barnes 1974, 63).
See feeding: suspension feeding.
syn. suspension feeding (a confusing synonym, Lincoln et al. 1985)
cf. feeding: deposit feeding

free feeding *n.* An arthropod's consuming all tissue from a leaf's margin to its midvein, leaving only the larger veins uneaten; contrasted with hole feeding, margin feeding, skeletonization, and window feeding (Coulson and Witter 1984).

greeting feeding *n.* For example, in some bird species: feeding one partner of a pair by the other when they meet after being separated for a while (Wickler 1969 in Heymer 1977, 120).

hawking *n.* A bird's feeding while flying (Lincoln et al. 1985).

hole feeding *n.* An arthropod's consuming all tissue of part of a leaf; contrasted with free feeding, margin feeding, skeletonization, and window feeding (Coulson and Witter 1984).

kiss feeding *n.* For example, in Humans, some other anthropoid-ape species: a ritualized form of mouth-to-mouth feeding during lovemaking or bonding (Eibl-Eibesfeldt 1970, 1972, 1973 in Heymer 1977, 120).

margin feeding *n.* An arthropod's consuming all tissue of a leaf's margin; contrasted with free feeding, hole feeding, skeletonization, and window feeding (Coulson and Witter 1984).

mouth-to-mouth feeding *n.*
1. In many bird species: an individual's inserting food into a young's mouth (Heymer 1977, 120).
2. In Humans (*e.g.*, Ituri Pygmies, Papuans): a woman's inserting prechewed food into the mouth of an infant or young domesticated animal with her tongue (Heymer 1977, 120).

reciprocal feeding See trophallaxis.

skeletonization *n.* An arthropod's consuming all of a leaf's tissue except for many of its veins; contrasted with free feeding, hole feeding, margin feeding, and window feeding (Coulson and Witter 1984).

suspension feeding *n.* Feeding in which an animal traps or collects (but does not filter) organisms, or detritus, that are suspended in surrounding water; distinguished from filter feeding (Barnes 1974, 63).

d – g

See feeding: filter feeding.

syn. ciliary feeding (which may or may not be synonymous), filter feeding (a confusing synonym, Barnes 1994, 63)

window feeding *n*. An arthropod's consuming a layer of a leaf without making a hole through it; contrasted with free feeding, hole feeding, margin feeding, and skeletonization (Coulson and Witter 1984).

♦ **feeding aggregation** See aggregation: feeding aggregation.

♦ **feeding ceremony** See ceremony: feeding ceremony.

♦ **feeding stimulant** See chemical-releasing stimulus: feeding stimulant.

♦ **feeding territory** See territory: feeding territory.

♦ **feeding tradition** *n*. For example, in some bird and mammal species: a regional diet peculiarity that develops by young animals' learning food preferences from their mothers or both parents (Heymer 1977, 73).

♦ **fellatio** *n*. In Bonobos, Humans: an individual's sucking another individual's penis (de Waal 1989, 204–205).

♦ **female** *n*. In bisexual, or dioecious species: an egg-producing form (Lincoln et al. 1985).

cf. "caste" for different kinds of "gynes," hermaphrodite, male

Comments: In dioecious species in which males and females produce very similar gametes, the individuals that produce larger gametes can be considered to be females while those that produce smaller gametes can be considered to be males. Workers of many social insect species are morphologically females and classified as such, although they do not lay eggs.

metafemale *n*. A female with an additional female sex chromosome (Lincoln et al. 1985).

neofemale *n*. A female that arises by sex reversal from a male (Lincoln et al. 1985).

♦ **female calling** See calling: female calling.

♦ **female care** See care: female care.

♦ **female choice, female preference** See mate choice: female mate choice.

♦ **female copying** See mate choice: female copying.

♦ **female-defense polygyny** See mating system: polygyny: female-defense polygyny.

♦ **female mate choice** See mate choice: female mate choice.

♦ **female mimicry** See mimicry: female mimicry.

♦ **female pattern** See female schema.

♦ **female preference** See mate choice: female mate choice.

♦ **female receptivity** See receptivity.

♦ **female schema** *n*. For example, in a dragonfly and flesh-fly species: characteristics of an object that elicit male courtship behavior when they are presented in a particular combination to a male animal (Vogel 1954, 1957, 1958; Buchholtz 1951; and Heymer 1973 in Heymer 1977, 198).

♦ **fen** See habitat: fen.

♦ **-ference**

afference *n*. Excitation of afferent nerves conveyed from an animal's sense organs (including proprioceptors and visceral receptors) to its central nervous system (Immelmann and Beer 1989, 7).

▸ **exafference** *n*. Afference that results from environmental stimuli not caused by a focal animal (*e.g.*, afference caused by a cloud passing over the sun rather than a focal animal's own movements) (Immelmann and Beer 1989, 95).

▸ **reafference** *n*. Afference that results from stimuli caused by a focal animal's own movement and enables the animal to make compensatory movement in its orientation, *q.v.* (Immelmann and Beer 1989, 243). *cf.* principle: reafference principle; *Sollwert*

efference *n*. Nerve impulses from an animal's central nervous system to its effectors via its efferent nerves (Heymer 1977, 58; Immelmann and Beer 1989, 7, 84). *cf.* -ference: afference

♦ **fermentation** See biochemical pathway: fermentation cycle.

♦ **fertility ratio** See ratio: fertility ratio.

♦ **fertility schedule** *n*. The average number of daughters that will be produced by one female at each particular age in a population (Wilson 1975, 90).

♦ **fertilization** *n*.

1. In many organism species: a male's insemination, or fecundation, of a female (*Oxford English Dictionary* 1972, verb entries from 1859).

2. In many organism species: union of the nuclei of an egg and sperm of the same species or similar species (Immelmann and Beer 1989, 103).

v.t. fertilize

syn. copulation (loosely), mating (loosely), pollination (in plants)

cf. copulation, -gamy, insemination, pollination, sex

cross-fertilization, allogamy, allomixis, staurogamy, xenogamy *n*. In many organism species: the union of male and female gametes each from different conspecific individuals (Lincoln et al. 1985).

cf. -gamy: allogamy

plasmagynogamous fertilization *n*. Fertilization in which a zygote's cytoplasm is derived only from the female gamete that formed it (Lincoln et al. 1985).

plasmaheterogamous fertilization *n.* Fertilization in which a zygote's cytoplasm is derived unequally from the male and female gametes that formed it (Lincoln et al. 1985).

plasmaisogamous fertilization *n.* Fertilization in which a zygote's cytoplasm is derived equally from the male and female gametes that formed it (Lincoln et al. 1985).

self-fertilization, idiogamy, mychogamy *n.* In many plant species: the union of male and female gametes that are produced by the same individual organism (Lincoln et al. 1985).

♦ **fesnyng, fesynes** See ²group: fesnyng.
♦ **fiddling display** See display: fiddling display.
♦ **fidelity to place** See ortstreue.
♦ **field** See habitat: field.
♦ **field capacity** *n.* Water that remains in soil after excess moisture has drain freely from it (Allaby 1994).
Note: Researchers usually measure field capacity as a percentage of the weight, or volume, of oven-dried soil (Allaby 1994).
♦ **51 Pegasi** See star: 51 Pegasi.
♦ **fight** *n.*
1. A hostile encounter, or engagement, between opposing human forces (*Oxford English Dictionary* 1972, entries from 893).
2. In animals, including Humans: a combat between two or more individuals (*Oxford English Dictionary* 1972, entries from 1300).
3. A person's power, or disposition, to fight; pugnacity (*Oxford English Dictionary* 1972, entries from 1812).
v.i., v.t. fight
syn. fighting

damaging fight *n.* For example, in some ant species; the Human, Wolf: a fight in which animals severely harm one another, even to the point of killing one another (Heymer 1977, 41).
syn. injurious fight (Immelmann and Beer 1989, 148)
cf. ritualized fight
Comment: Damaging fights usually occur between heterospecific individuals (*e.g.,* species competing for the same resource or a predator and prey). Conspecific-damaging fights usually occur in social species (Immelmann and Beer 1989, 148).

injurious fight See fight: damaging fight.

mouth clashing *n.* For example, in mouth-breeding cichlid species: fishes' butting and striking one another with open mouths (Heymer 1977, 115).
cf. fight: mouth fight

mouth fight, mouth pulling *n.* For example, in a salariine blenniid fish: a fight involving mouthing one another (*e.g.,* in substrate-nesting cichlids that grasp each other lip to lip, wrestle, and shove each other back and forth) (Heymer 1977, 115).
cf. fight: mouth clashing

ritualized fight *n.* For example, in the Ibex, Wolf: an intraspecific fight that proceeds according to set rules (*e.g.,* an animal's not attacking at a crucial time, or at all, due to perceiving certain appeasement signals from an opponent) and that usually results in no or only minor physical injury to combatants (Heymer 1977, 41, 100; Immelmann and Beer 1989, 256).

♦ **fight-intention movement** See behavior: intention movement: flight-intention movement.
♦ **fight-or-flight syndrome** See syndrome: fight or flight.
♦ **fighting coloration** See coloration: fighting coloration.
♦ **fighting instinct** See instinct: fighting instinct.
♦ **F_{ij}, f_{ij}** See coefficient: coefficient of consanguinity.
♦ **filial cannibalism** See cannibalism: filial cannibalism.
♦ **filial imprinting** See learning: imprinting: filial imprinting.
♦ **filial ovicide** See cannibalism: filial cannibalism.
♦ **filial regression** See law: Galton's law of filial regression.
♦ **filter feeding** See feeding: filter feeding.
♦ **"filtering collectors"** See ²group: functional feeding group: filtering collectors.
♦ **fimbricole** See -cole: fimbricole.
♦ **fimicole** See -cole: fimicole.
♦ **fin** See behavior: fin.
♦ **fin flicking, fin-flicking behavior** See display: fin flicking.
♦ **final common path** *n.* The same motor neurons and muscles that are affected by different impulses that arise from any of many sources of reception (Sherrington 1906 in Immelmann and Beer 1989, 104).
♦ **final instar** See instar: last instar.
♦ **fine-grained environment** See environment: fine grained environment.
♦ **fine-grained species** See ²species: fine grained species.
♦ **fine-grained variation** See variation: fine grained variation.
♦ **fine-particulate-organic matter (FPOM)** See particulate-organic matter: fine-particulate-organic matter.
♦ **firefly-flash-synchronization hypothesis** See hypothesis: firefly-flash-synchronization hypothesis.

d – g

♦ **first Darwinian revolution** See revolution: first Darwinian revolution.

♦ **first instar, first larval instar** See instar: first instar.

♦ **first law of thermodynamics** See law: laws of thermodynamics: first law of thermodynamics.

♦ **first order** See order: first order.

♦ **first-order Markoff chain** See chain: Markoff chain.

♦ **first-order stream** See stream: first-order stream.

♦ **Fisher's exact test, Fisher's exact probability test** See statistical test: Fisher's exact test.

♦ **Fisher's fundamental theorem** See ³theory: Fisher's fundamental theorem.

♦ **Fisher's theory of 1:1 sex ratios** See ³theory: Fisher's theory of 1:1 sex ratios.

♦ **fission** *n.* Spontaneous division of a cell, or organism, into new cells or organisms, especially as a mode of reproduction (Michaelis 1963).
See swarming: fission.
syn. cell division (Michaelis 1963)
[Latin *fissio, -onis* < *fussus*, past participle of *findere*, to split]

architomical fission *n.* In Nemerteans and Sponges: fission "in which a large parental individual proliferates vegetatively into a number of smaller progeny" (Bell 1984, 600).

binary fission *n.* In prokaryotes: a kind of cell division in which a parent cell splits transversely into approximately equal sized daughter cells, each of which receives a copy of a single parental chromosome (King and Stansfield 1985; Campbell 1990, G-2).
cf. mitosis

paratomical fission *n.* In Actiniarians, Aelosomatid and Naiad Oligochaetes, Hydras, and Planarians: fission "in which the two products are similar in size and both require only a small amount of growth and differentiation to regain the actively reproducing adult state" (Bell 1984, 600).

♦ **fission-fusion group** See ²group: fission-fusion group.

♦ **fission hypothesis** See hypothesis: Moon-origin hypotheses: fission hypothesis.

♦ **fitness**
Comments: Because of the many kinds of fitness, one general, broad definition of fitness as it relates to all kinds of fitness is not appropriate. Fitness can be defined as genetic or nongenetic and from the vantage points of memes, genes, individual organisms, or particular groups of organisms. The vast literature on fitness is confusing because authors frequently call whatever fitness they are discussing simply "fitness" without informative adjectives, and some fitness terms that some authors synonymize have very different meanings to other authors.

SOME CLASSIFICATIONS OF FITNESS[a]

I. General classification
  A. genetic fitness (*syn.* Darwinian fitness)
  B. nongenetic fitness (*syn.* cultural fitness)
II. Classification by type of genetic "group"
  A. below the individual organism level
    1. single-gene fitness
    2. gene-group fitness
  B. individual organism level
    1. direct fitness
    2. inclusive fitness
    3. indirect fitness
  C. above the individual organism level
    1. mated-pair fitness
    2. family-group fitness
    3. deme fitness
    4. population fitness, etc.
III. Classification by kind of relative
  A. cousin fitness
  B. grandchild fitness (granddaughter fitness, grandson fitness), etc.
  C. grandparent fitness (grandfather fitness, grandmother fitness)
  D. nephew fitness, niece fitness
  E. offspring fitness (daughter fitness, hermaphrodite fitness, son fitness)
  F. parent fitness (father fitness, mother fitness)

[a] These classifications are only some of many that can be made based on the many definitions of the different kinds of fitness.

adaptive value See fitness: genetic fitness.

classical fitness See fitness (def. 2, 4); fitness: genetic fitness.

classical-individual fitness See fitness (def. 2).

classical-personal fitness See fitness (def. 4).

cultural fitness See fitness: nongenetic fitness.

Darwinian fitness See fitness (def. 2, 4); fitness: genetic fitness.

direct fitness See fitness (def. 2, 4).

fitness[1] See fitness (def. 1).

fitness[2] See fitness: genetic fitness.

fitness[3] See fitness (def. 4).

fitness *n.*
1. An animal's capacity to survive and reproduce (Spencer 1864, Wallace 1866, Darwin 1866 in Dawkins 1982, 181).

Notes: In certain contexts, "fitness" refers to some individuals' having better survivorship traits than others (*e.g.,* harder teeth, keener eyes) (Spencer 1864, Wallace 1866, Darwin 1866 in Dawkins 1982, 181). This definition of fitness is not precisely synonymous with "reproductive success." The 19th-century idea of fitness as an animal's being favored physically in its "struggle for existence" is markedly different from some more recent statistical concepts of fitness (Immelmann and Beer 1989, 106). Generally speaking, fitness refers to an organism's state of good physical condition; good health (Michaelis 1963).
syn. classical fitness (Brown 1987a, 299), fitness[1] (Dawkins 1982), classical individual fitness, individual fitness (Brown 1987a, 299), personal fitness

2. The number of offspring reaching adulthood that an individual organism leaves in the next generation in the absence of chance effects (Williams 1966 in West Eberhard 1975, 2).
See fitness: individual fitness.
syn. classical fitness (West Eberhard 1975, 6); Darwinian fitness (Haldane 1932); direct fitness (Brown 1975); individual fitness; personal fitness (West Eberhard 1975, 6)

3. The probability that a lineage will continue for a very long time, such as 10^8 generations, and will be contributed to by such "biotic" factors (Williams 1966) as "genetic flexibility" (Thoday 1953 in Dawkins 1982, 193).
Note: Some of Thoday's basic ideas have been recently recast in Cooper's (1984) interpretation of fitness as "expected time to extinction" (Beatty 1992a, 119).

4. The genes, relative to conspecifics, that an individual organism contributes to its next generation, measured in different ways which vary in accuracy, including number of a male's matings, number of eggs in a female (fecundity), number of eggs a female lays (fecundity), number of eggs that hatch or number of live births (fertility), and number of an individual's offspring that live to maturity.
Notes: This is a commonly used concept of fitness. An individual's fitness is related to different factors including its health, feeding efficiency, resource availability, and survival time (Dawkins 1982, 183). Different fitness measurements suggest an individual's actual, or potential, reproductive success, which is very difficult to measure under field conditions. In the long run, it is an individual's relative contribution to the gene pool of its next generation and future generations that represents the "bottom line" of this individual's fitness (Dewsbury 1978, 295).
syn. classical fitness, classical personal fitness, personal fitness (Wilson 1975), direct fitness (Brown 1975, Brown and Brown 1981 in Dawkins 1982), fitness[3] (Dawkins 1982), genetic fitness (Dewsbury 1978, 43)

5. A verbal trick to make it possible to discuss natural selection at the level of the individual organism instead of at the level of the gene (Dawkins 1982, 179).
Note: Dawkins (1982, 179) contends that "fitness" is a confusing term with many different meanings, and this term probably should not be used (Dawkins 1982, 179).

6. A hypothetical quantity that tends to be maximized by natural selection (Dawkins 1982, 182).
Note: Dawkins (1982, 182), who opposes many concepts represented by the term "fitness," argues that this definition makes "survival of the fittest" into a tautology.

7. "A measure of the ability of genetic material to perpetuate itself in the course of evolution" (McFarland 1985, 78).
Note: McFarland (1985) uses fitness in reference to long-term reproductive success throughout his book.
cf. fitness: genetic fitness, inclusive fitness, indirect fitness
Comments: "'Fitness' is perhaps the most contentious concept in evolutionary biology" (Paul 1992b, 113). Darwin (1959) used "fit" and its forms to refer to adapted and suitable characters; he used "fitness" once (Paul 1992b, 113). Despite Dawkin's (1982) analyses, many biologists consider "fitness" to be a useful concept, as indicated in def. 4. "The precise meaning of 'fitness' has yet to be settled, in spite of the fact — or perhaps because of the fact — that the term is so central to evolutionary thought" (Beatty 1992a, 115).
[from "survival of the fittest" (Spencer 1864 in Dawkins 1982, 179); a synonym for "natural selection" (Paul 1992b, 113)]

genetic fitness *n.*
1. The relative increase, or decrease, of a particular gene in the gene pool of the next generation of a population (Fisher 1930, Haldane 1957 in Mayr 1982, 588); contrasted with nongenetic fitness.

2. *W*, a measure of the relative frequency of a genotype, usually a single locus (Falconer 1960 in Dawkins 1982, 182).

Notes: $W = 1 - s$, where s is the coefficient of selection against this genotype; 1 is an arbitrary fitness value of the more common, or most common, allele at the locus in question; and W can be regarded as the number of offspring, relative to the arbitrary value of 1, that the average individual of a certain genotype produces that live up to their reproductive ages (Falconer 1960 in Dawkins 1982, 182).

syn. fitness, fitness[2] (Dawkins 1982)

3. "A measure of the rate at which allele or linkage-group frequencies change within a gene pool through time" (Wittenberger 1981, 614).

See fitness (def. 4).

syn. adaptive value, selective value, Darwinian fitness (Wittenberger 1981)

cf. fitness (def. 7)

inclusive fitness *n*. "The animal's production of adult offspring ... stripped of all components ... due to the individual's social environment, leaving the fitness he would express if not exposed to any of the harms or benefits of the environment, ... and augmented by certain fractions of the quantities of the harm and benefit, the individual himself causes to the fitnesses of his neighbours. The fractions in question are simply the coefficients of relationship ..." (Hamilton 1964, 14).

syn. neighbor-modulated fitness (Hamilton 1964), personal fitness (not a good synonym as indicated in Orlove 1975, 1979 in Dawkins, 1982)

Comments: Primary literature and textbooks are confusing in their implied and explicit definitions of "inclusive fitness." Different authors (Hamilton 1964; West Eberhard 1975; Grafen 1982, 1984) in defining "inclusive fitness" use "a focal organism's (helper's) number of offspring" with different combinations of the following characteristics: (1) plus any increase, or decrease, in offspring equivalents due to this focal animal's helping them; (2) plus the number of the focal organism's offspring equivalents present, or not present, due to help, or hindrance, from any relatives involved with the helper; (3) plus all offspring equivalents present due to its relatives' reproducing, whether or not the focal organism had anything to do with these offspring equivalents being produced; (4) minus the number of the focal organism's offspring equivalents present, or not present, due to help, or hindrance, from any relatives involved with the helper. These authors' papers should be consulted for further information. The concept of inclusive fitness is now so

generally accepted by biologists that it is implied in most uses of fitness, regardless of whether it differs from classical fitness in the case in point (C.K. Starr, personal communication).

[coined by W.D. Hamilton (1964)]

indirect fitness *n*. "The component of inclusive fitness that arises from [a focal organism's] effects on nondescendent kin and other nondescendent co-gene-carriers" [coined by Brown 1978 in Brown 1987a, 302)].

cf. selection: indirect selection, kin selection

Comment: Brown (1987a, 302–304) discusses the confusion regarding the concepts represented by the terms indirect fitness, indirect selection, kin selection, and kinship component.

individual fitness *n*.

1. "The (average) net contribution of an individual of a particular genotype to the next generation" (Keller 1992b, 120).

2. "The geometric rate of increase of a particular genotype" in a population (Keller 1992b, 120).

See fitness (def. 2).

Comment: There is a chronic confusion in evolutionary literature regarding these two definitions (Keller 1992b, 120).

kinship component See fitness: indirect fitness.

neighbor-modulated fitness See fitness: inclusive fitness.

nongenetic fitness *n*. In animals with culture: an animal's perpetuation of its memes; contrasted with genetic fitness.

syn. cultural fitness

personal fitness See fitness (def. 1, 2, 4); fitness: inclusive fitness.

selective value See fitness: genetic fitness.

♦ **fittest** *n*. Referring to animals that survive in their natural environ ments (Dawkins 1982, 181); animals of a group with high fitness, *q.v.*

See survival of the fittest.

Comment: Using "fittest" to refer to more than one animal of a group can be considered poor English; "fitter" seems preferable.

♦ **fixation** *n*. "The complete prevalence of one gene form (allele), resulting in the complete exclusion of another (Wilson 1975, 584).

random fixation *n*. Fixation due to chance loss of an allele's homologous allele in a population (Lincoln et al. 1985).

♦ **fixed-action pattern (FAP)** See action pattern: fixed-action pattern.

♦ **fixed effect** See effect: main effect: fixed effect.

♦ **fixed factor** See effect: main effect: fixed effect.

♦ **fixed-interval reinforcement schedule** See reinforcement schedule: fixed-interval reinforcement schedule.

♦ **fixed-ratio reinforcement schedule** See reinforcement schedule: fixed-ratio reinforcement schedule.

♦ **fixed-reaction behavior** See behavior: fixed-reaction behavior.

♦ **fixed territory** See territory: territory.

♦ **flash coloration** See coloration: flash coloration.

♦ **flash communication** See communication: flash communication.

♦ **flashbulb memory** See memory: flashbulb memory.

♦ **flehmen,** *Flehmen* n.
1. In males of many ungulate species, felid species, some insectivore and primate species: an individual's extending his neck, closing his nasal openings, and curling his upper lip (lip curling, exposing his upper incisors and possibly gum ridge), especially in his reproductive season after licking a female's urine or vulva, or smelling a female (Schneider 1930 in Heymer 1977, 68; Immelmann and Beer 1989, 107).
Notes: This behavior appears to be more important in facilitating a male's perception of the odor more than being a courtship display (Dewsbury 1978, 76). Flehmen appears to be a ram's olfactory mechanism for confirming the reproductive state of a ewe (Bland and Jubilan 1987, 735).
syn. lip curl (Immelmann and Beer 1989, 107)
cf. facial expression, organ: Jacobson's organ
2. In the Asian Elephant: a female's touching a liquid sample with her trunk tip and moving it and a bit of the liquid to the ducts in the roof of her mouth that lead to her vomeronasal organ (Schulte and Rasmussen 1999, 1269, illustration).
[German *Flehmen*]

♦ **Flemming's aphorism** n. *Omnis nucleus e nucleoa* (the nucleus of a fertilized egg cell — a zygote — never disappears, and there is complete continuity between it and all nuclei on the organism into which it develops) (Mayr 1982, 666, 678).
Comment: By 1880, every cytological investigation confirmed this aphorism (Flemming in Mayr 1982, 666, 678).

♦ **flexibility** See plasticity: phenotypic plasticity: flexibility.

♦ **flick** See ²group: flick.

♦ **flight** n.
1. In the majority of insect species, most bird species, bats: an individual's act, or

manner, of flying (*Oxford English Dictionary* 1972, entries from 900).
2. In Falcons: an individual's pursuit of game; also the pursued quarry (*Oxford English Dictionary* 1972, entries from 1530).
3. In many insect and bird species: a migration or issuing forth of bodies (*Oxford English Dictionary* 1972, entries from 1823).
4. In animals: an individual's fleeing, or running away, from danger; hasty departure or retreat (*Oxford English Dictionary* 1972, entries from 1200).
cf. behavior: escape behavior
5. *v.i.* For example, in wild fowl: to migrate, or move in flights (*Oxford English Dictionary* 1972, entries from 1604).
See ²group: flight.
cf. locomotion

courtship flight n. Flight during which a bird sings (skylarks, nighthawks) or makes noise with its wings (pigeons) or tail feathers (snipe) (Immelmann and Beer 1989, 63).
syn. song flight (Immelmann and Beer 1989, 63)
cf. flight: nuptial flight
Comment: This is a misleading term because most so-called "courtship flights" are not part of mating initiation but serve as territory advertisements. Better terms for this behavior are "advertisement flight," "territorial flight," and "display flight." "Courtship flight" should be reserved for the few bird species whose song flights are used solely for courtship, including some heron species and many hummingbird species (Immelmann and Beer 1989, 63).
▸ **individual courtship flight** n. In fireflies: flight during which solitary males signal to females with light flashes (Immelmann and Beer 1989, 64–65).

nuptial flight n. In many social-insect species: the mating flight of the winged reproductive females and males of a colony (Wilson 1975, 435, 590).
syn. marriage flight (Wheeler 1930, 9), mating flight
cf. flight: courtship flight
Comment: Reproductives' copulations may occur in flight or after landing, depending on species (Immelmann and Beer 1989, 204).
[Latin *nuptialis* < *nuptus,* past participle of *nubere,* to marry]

orientation flight n. For example, in some digger-wasp and sand-wasp species, the Honey Bee: an individual's flying about and learning the landmarks of its environment (Tinbergen 1932, Baerends 1941, Frisch 1965 in Heymer 1977, 126–127).

d – g

song flight See flight: courtship flight.

territorial flight See flight: courtship flight.

♦ **flight behavior** See behavior: flight behavior.

♦ **flight companion** See companion: flight companion.

♦ **flight distance** See distance: flight distance.

♦ **flight-intention movement** See behavior: intention-movement: flight-intention movement.

♦ **flight year** See unit of measure: year: flight year.

♦ **Fligner-Policello test** See statistical test: multiple-comparisons test: planned-multiple-comparisons procedure: Fligner-Policello test.

♦ **fling** See ²group: fling.

♦ **floater** *n.*
1. For example, in some digger-wasp and sand-wasp species, the Honey Bee: an individual animal that is "unable to claim a territory and hence forced to wander through less suitable surrounding areas" (Wilson 1975, 271, 584; Barrows 1983, 808; Arcese 1987, 773; Caro 1989, 54).
syn. nonresident, nonterritorial animal (Caro 1989)
2. A nonbreeding animal that is a potential breeder and not bound to a particular territory (Brown 1987a, 299).
cf. ¹population: floater population; territory: floating territory; resident; territory holder

♦ **floater population** See population: floater population.

♦ **floating territory** See territory: floating territory.

♦ **flock** See ²group: flock.

♦ **flora** *n.*
1. The plant species of and usually peculiar to a particular region or period; distinguished from fauna (Michaelis 1963).
2. A treatise on these plants (Michaelis 1963). [New Latin from Latin *Flora*, goddess of flowers]
ichnoflora *n.* An area's plant flora (Lincoln et al. 1985).

♦ **floral color change** *n.* A natural color change in a part of a flower (Müller 1877 and others in Weiss 1991, 227).
Comments: Flowers in at least 74 diverse angiosperm families undergo dramatic, often localized, color changes (Weiss 1991, 227; 229, illustrations). In many studied plants, floral color change directs pollinators to flowers with nectar and pollen.

♦ **floral mimicry** See mimicry: floral mimicry.

♦ **floral mutualism** See mimicry: floral mimicry: nondeceptive floral mimicry.

♦ **floral nectary** See nectary: floral nectary.

♦ **floral reward** See reward: floral reward.

♦ **floral visitor** *n.* An animal that visits a flower where it finds food, a mate, a resting place, shelter, or a combination of these things (Buchmann and Nabhan 1996, 246).
Comment: A particular floral visitor may, or may not, be a pollinator.

♦ **flote** See ²group: flote.

♦ **flower** See organ: flower.

♦ **flower constancy** *n.*
1. In many bee species: a forager's restricting its visits to a single plant species, or morph within a species, during one foraging bout or a longer period (Plateau 1901, etc. in Waser 1986, 594).
2. In many bees species: a forager's tendency to use the same flower species as a food source, especially a pollen source, on any one foraging trip or during a longer period (Michener 1974, 372).
adj. flower constant

♦ **fluctuating asymmetry** *n.* Asymmetry of the two sides of bilateral characters (*e.g.,* ears, feet, fins, hands, or wings) for which the signed differences between the two sides have a population mean of zero and are normally distributed (Van Valen 1962 in Thornhill et al. 1995, 1602).
Comments: Because the two sides of such characters are not controlled by different genes, some researchers think that fluctuating asymmetry represents imprecise expression of underlying developmental design because of developmental perturbations. Reduced fluctuating asymmetry is associated with male mating success in, for example, the Barn Swallow, *Drosophila*, nonhuman primates, and scorpionflies.

♦ **fluffing** See feather erection.

♦ **fluid recycling** *n.* In many skipper-butterfly species: an adult's extending its proboscis under its thorax and imbibing fluid that it excreted from its abdomen (Atkins 1989, 103).

♦ **flurry** See ²group: flurry.

♦ **flush** See ²group: flush.

♦ **flutter** See ²group: flutter.

♦ **fluviology** See study of: fluviology.

♦ **fly along a mating path** See patrol.

♦ **flying-fish hypothesis of wing origin** See hypothesis: hypothesis of the origin of insect wings: flying-fish hypothesis of wing origin.

♦ **flying-squirrel hypothesis of insect wing origin** See hypothesis: hypotheses of the origin of insect wings: paranotal hypothesis of insect wing origin.

♦ **FMGP** See program: Fungal Mitochondrial Genome Project.

♦ **focal-animal sampling** See sampling technique.

♦ **fold** See ²group: fold.

♦ **folicaulicole** See -cole: folicaulicole.

♦ **folicole** See -cole: folicole.

♦ **folivore** See -vore: folivore.

♦ **follicle-stimulating hormone (FSH)** See hormone: follicle-stimulating hormone.

♦ **follower** *n*. In many ungulate species, including Chamois, Gnus, Goats, Sheep, Wildebeest: a precocial young that, from its birth onwards, tends to follow its mother (or a substitute moving object) and stay near her (Immelmann and Beer 1989, 108).
cf. hider; period: lying-out period
Comment: "Follower" is on a continuum with "hider" (Immelmann and Beer 1989, 108).

♦ **following behavior, following response** See behavior: following behavior.

♦ **following-response imprinting** See learning: imprinting: filial imprinting.

♦ **food** *n*.
1. Matter that a person eats, drinks, or absorbs and uses for his maintenance and growth (*Oxford English Dictionary* 1972, entries from 1000).
   syn. aliment, nourishment, nutriment, provisions (*Oxford English Dictionary* 1972)
2. Matter that a plant absorbs from the Earth and air (*Oxford English Dictionary* 1972, entries from 1759).
   syn. nutriment (*Oxford English Dictionary* 1972)
3. A person's nourishment consumed in more or less solid form vs. liquid form: food and drink (*Oxford English Dictionary* 1972, entries from 1610).
Comment: Definitions 1 and 3 now refer to many species of nonhuman animals as well.

caecotroph *n*. In all lagomorph and rodent species, a prosimian species: material that is higher in raw protein than feces, arises from an animal's caecum, and is eaten by the animal that produces it (Harder 1949, Frank et al. 1951, Hladik et al. 1971 in Heymer 1977, 49).
cf. -troph-: caecotroph

pap *n*. In the Koala: a green fecal fluid that is eaten by a young individual and is composed of partly digested eucalyptus leaves and produced by its mother (Lee and Martin 1990, 41).

proctodeal food *n*. In Kalotermitid and Rhinotermitid Termites: food originating from the hindgut that is exchanged in trophallaxis; this food is emitted from the anus and is different from ordinary feces because it contains symbiotic flagellates and is more watery (Wilson 1975, 207).
cf. food: stomodeal food
[Greek *proktos*, rectum]

stomodeal food *n*. Food exchanged in trophallaxis originating in the salivary glands and crop in Kalotermitid and Rhinotermitid Termites (Wilson 1975, 207).
cf. food: proctodeal food
Comment: This food is the principal source of nutriment for the royal pair and larvae (Wilson 1975, 207).

♦ **food begging** *n*. In many animal species: one individual's soliciting food from a conspecific; for example, a young's soliciting food from its parents (in many animal species), a courtship partner's soliciting food from its mate during courtship feeding (in some bird species), an adult's soliciting food from another adult that is not its mate during food sharing (*e.g.*, in the African Wild Dog and the Honey Bee) (Heymer 1977, 72; Immelmann and Beer 1989, 109).
v.t. food beg
syn. food solicitation, solicitation (C.K. Starr, personal communication)
cf. feeding: courtship feeding, greeting feeding

♦ **food call** See animal sounds: call: food call.

♦ **food chain** See chain: food chain.

♦ **food-deception mimicry** See mimicry: floral mimicry: food-deception mimicry.

♦ **food dispenser** *n*. A device used in a Skinner box that automatically gives out a particular amount of food following a particular performance by an animal in the box (Heymer 1977, 73).

♦ **food hiding** *n*. For example, in Chipmunks, Ravens, Squirrels: an individual's concealing its gathered food in certain places (Eibl-Eibesfeldt 1951, Gwinner 1962 in Heymer 1977, 73).
See hoarding: food hoarding.
cf. behavior: caching; food storage; hoarding: food hoarding

♦ **food hoarding** See hoarding: food hoarding.

♦ **food imprinting** See learning: imprinting: food imprinting.

♦ **food mimicry** See mimicry: food mimicry.

♦ **food pilfering** See parasitism: cleptoparasitism.

♦ **food provisioning** See provisioning: food provisioning.

♦ **food-related patrolling** See patrol.

♦ **food-source mimicry** See mimicry: floral mimicry: deceptive floral mimicry.

♦ **food storage** See hoarding: food hoarding.

♦ **food web** *n*. The complete set of food links among species in a community (Wilson 1975, 584).

See -gram: food web.

cf. food chain

♦ **fool's gold** See mineral: iron pyrite.

♦ **foot preference** See handedness.

♦ **footdrumming** See animal sounds: drumming: footdrumming.

♦ **footedness** See handedness.

♦ **footprint substance** See chemical-releasing stimulus: semiochemical: pheromone.

♦ **forager** *n*. An organism that practices foraging, *q.v.* (*e.g.*, a feeding animal or a social-insect worker that brings food to its nest).

cf. predator

Bayesian forager *n*. A forager that assesses resource-patch quality by using information on the distribution of resources among patches (McNamara and Houston 1980 in Valone 1991, 569).

cruising forager See predator (table).

intensive forager See predator (table).

prescient forager *n*. A forager that assesses resource-patch quality by using information obtained through its modalities and remembering patches that have temporally predictable qualities (Valone and Brown 1989 in Valone 1991, 569).

♦ **foraging, foraging behavior** See behavior: foraging.

♦ **forbicole** See -cole: forbicole.

♦ **forbivore** See -vore: forbivore.

♦ **forced copulation** See copulation: forced copulation.

♦ **forebrain** See organ: brain: forebrain.

♦ **forehead-brooding** See brooding: forehead-brooding.

♦ **forehead gland** See gland: forehead gland.

♦ **forensic psychology** See study of: psychology: forensic psychology.

♦ **foreplay** See play: foreplay.

♦ **forest** See ²community: forest.

♦ **form** *n*. A shallow depression dug by a hare that serves as a camouflage blind from which it monitors its surroundings (Stutz 1987).

♦ **form analysis** See analysis: form analysis.

♦ **form constancy** *n*. An animal's always performing a behavior in a rigidly stereotyped manner (Immelmann and Beer 1989, 111).

cf. pattern: fixed-action pattern, stereotypy

♦ **form species** See ²species: form species.

♦ **formation** *n*. A body of rock strata that consists of a certain lithologic type or combination of types (Bates and Jackson 1984).

Comment: A formation is the fundamental lithostratigraphic unit (Bates and Jackson 1984). Some formations are combined into groups and divided into members.

"Akilia Island Rocks" *n*. A banded-iron formation from Akilia Island, Western Greenland, that is >3800 million years old and has tiny bits of elemental carbon in apatite that may be remains of early life on Earth (Mojzis et al. 1996 in Holland 1997, 38).

cf. life

Comments: This formation has very low $\delta^{13}C$ values, which are typical of organic matter produced by methanotrophs (Holland 1997, 38). Akilia Island Rocks have the oldest possible remains of life known.

Isua Rocks *n*. A banded-iron formation from Isua, Western Greenland, that is 3800 million years old and has carbonaceous residues that may be remains of early life on Earth (Schidlowski et al. 1979 in Holland 1997, 38).

Comments: The Isua rocks may have originated by chemical precipitation in marine basins, and their quartz layers may have initially been chert that precipitated from seawater (Stanley 1989, 251). The Isua rocks have $\delta^{13}C$ values that characterize present-day organisms

Warrawoona Rocks *n*. A formation of sedimentary rocks in Northwestern Australia that are 3450 million years old and contain carbonaceous material that might be remains of early life on Earth (Schopf 1993 in Holland 1997, 38).

syn. Warrawoona Formation (Strickberger 1996, 167, illustration)

♦ **formic acid** See toxin: formic acid.

♦ **formicarium, formicary** See nest: formicarium.

♦ **forward mutation** See ⁴mutation: forward mutation.

♦ ***FOSB*** See gene: FOSB.

♦ **fossil** *n*.

1. Any curious object, either organic or nonorganic, dug from the earth (obsolete definition) (Trowbridge 1962; Olson 1973, 649).

   Note: This early definition included relics of ancient civilizations (Mayr 1985, 9).

2. The preserved remains of an organism dating before historical time, *e.g.*, an altered original soft part, altered original hard part, an unaltered hard part, or an unaltered soft part (Trowbridge 1962; Curtis 1983, 1095).

   Note: This definition dates from the 16th century (Olson 1973, 649). Coal, petroleum, come graphite deposits, and some limestone deposits are technically fossils but are left out of the concept of "fossil" by some researchers.

   syn. necrotype (Lincoln et al. 1985)

3. The trace of an organism's existence (*e.g.*, a boring, a cast, a coprolite, an ichnofossil, a gastrolith, a mold, a track, or a trail) (Michaelis 1963; Lapidus 1987).

4. Any preserved remains, or trace, of an organism from before the Holocene Epoch, which started from 10,000 through 12,000 years ago, depending on the researcher (inferred from Mayr 1985, 9).

5. One of certain inorganic objects, or physical features, that record natural activities or phenomena of past geological times (*e.g.,* solidified ripple marks) (Michaelis 1963).

6. *adj.* Referring to extinction or age (*e.g.,* a fossil fern, fossil rain prints, fossil hail prints) (Lapidus 1987).

7. Referring to a geological form that has become covered or changed: a fossil feature (Lapidus 1987).

See Appendix 1 for entries on particular fossils.

cf. -type: necrotype; stromatolite
[Latin *fossilis,* dug up; *fodere,* to dig]

A CLASSIFICATION OF FOSSILS[a]

I. abiotic fossils
   A. astronomical fossil, including "faint hotspots"
   B. geological fossil, including ripple marks, sea beds, seashores
II. biotic fossils
   A. amber
   B. characteristic fossil (*syn.* diagnostic fossil; index fossil, not preferred)
   C. coprolite
   D. facies fossil
   E. guide fossil
   F. index fossil (*syn.* characteristic fossil, not preferred)
   G. microfossil
   H. molecular fossil
   I. mummy
   J. organism body remains
   K. stromatolite fossil
   L. trace fossil (*syn.* ichnofossil, Lebensspur, trace, vestigiofossil)
     1. boring
     2. caste
     3. ichnite (*syn.* ichnolite)
       a. ornithichnite
     4. gastrolith (gastric mill stone, stomach stone)
     5. mold
     6. scolite
     7. trail

---

[a] "Living fossils," "pseudofossils," and "subfossils," defined below, are not true fossils.

amber *n.* Fossilized tree resin that has become stabile due to loss of its volatile constituents, oxidation, and polymerization during lengthy burial in the ground (Frondel 1973, 717).

Note: Amber is a general name for a group of very heterogeneous resins from different plant species (Frondel 1973, 717). Amber is white ("bone amber") through yellow, reddish yellow, red, and brownish yellow, or rarely bluish; brittle; easily electrified by friction; hard; and translucent. *cf.* copal

Comments: Amber, which often contains fossils (bacteria through vertebrates), is used as jewelry and for pharmaceutical purposes. The natural embalming features of amber are due to inert dehydration, encapsulation of organic material by flowing resin (of some plant species including the legume tree *Hymenaea protera*), and bactericidal action of terpenes, according to Desalle et al. (1992, 1936), but the exact process of fossil preservation in amber is considered to be an enigma (Fischman 1993, 656). Amberization preserves fine features of organisms, even molecules such as DNA. Desalle et al. (1992) sequenced DNA from 18s and 16s ribosomal DNA genes of a 30-million-year-old termite fossil in Dominican amber. The plant that produced Baltic amber is unknown (Poinar and Poinar 1995, 4).

Amber sites include the Bavarian Alps, Germany (Triassic, 225 Ma); Indonesia, Jamaica, Mexico (Chiapas), Morocco (Atlas Mountains), Philippines, and Alberta, Canada (Cretaceous, 80–70 Ma); Danish, Polish, and Russian Baltic Coasts (Eocene and Oligocene, 60–40 Ma); Alaska, New Jersey, and Taimyr Peninsula, Siberia (Cretaceous); Lebanon (Cretaceous, 135–120 Ma); and Dominican Republic (Oligocene, 40–25 Ma); New Zealand (Miocene, 20 Ma) (Poinar and Poinar 1995; Grimaldi 1996, illustrations).

Behavior is sometimes preserved in amber; for example, animals' struggling while caught in the sap that became amber, an ant's struggling while in a spider web, and a spider's attacking a termite (Poinar and Poinar 1995, illustrations, unnumbered central pages), as well as a queen ant's carrying prey, a scale insect (Poinar and Poinar 1995, 180, illustration).
See fossil: succinite.
[Old French *ambre* < Arabic *anbar*, ambergris]

characteristic fossil *n.* A fossil species, or genus, that is often found in a particular stratigraphic unit or time unit; contrasted with guide fossil and index fossil (Bates and Jackson 1984).
See fossil: index fossil.
syn. diagnostic fossil, index fossil (not preferred) (Bates and Jackson 1984)

Comment: A characteristic fossil is either confined to a particular unit or is particularly abundant in it (Bates and Jackson 1984).

coprolite *n.* Fossilized fecal material (Lincoln et al. 1985; Wright 1996, illustrations).
Comment: Early coprolites (probably from arthropods) occur in rocks from the Upper Silurian (412 Myr) and Lower Devonian (390 Myr) (Edwards et al. 1995).

diagnostic fossil See fossil: characteristic fossil.

facies fossil *n.* A fossil that characterizes the environment that prevailed in a particular area of sedimentation (*e.g.,* a coral species) (Mayr 1985, 11).

gastrolith *n.* A highly polished, rounded stone believed to have been a part of the digestive system of certain extinct reptiles, such as plesiosaurs (Lapidus 1987).
syn. gastric mill stone (Norman 1985, 78), stomach stone (Lapidus 1987)
Comment: Gastroliths aid in grinding food in an animal's stomach (Lapidus 1987).

ichnite, ichnolite *n.* A fossilized animal track or footprint (Michaelis 1963).
[Greek ichnos, footprint]

ichnofossil See fossil: trace fossil.

guide fossil *n.*
1. A fossil that has actual, potential, or supposed value in identifying the stratum's age in which it is found or in indicating the conditions under which it lived; contrasted with characteristic fossil and index fossil (Bates and Jackson 1984).
2. A fossil used especially as an index, or guide, in the local correlation of strata (Bates and Jackson 1984).

index fossil *n.* A fossil used to identify and date the stratum in which it occurs; contrasted with characteristic fossil and guide fossil (Bates and Jackson 1984).
See fossil: characteristic fossil.
syn. characteristic fossil (not preferred) (Bates and Jackson 1984)
Comments: Index fossils are usually genera, not species (Bates and Jackson 1984). An index fossil has morphologic distinctiveness; a relatively broad, even worldwide, occurrence; and a short span of geological time and, therefore, a narrow, or restricted stratigraphic range. The better index fossils include swimming, or floating, organisms that evolved rapidly and were distributed widely (*e.g.,* ammonites and graptolites).

Lebensspur See fossil: trace fossil.

living fossil *n.*
1. A relict organism from an ancient group of organisms (*e.g.,* a lungfish) [coined by Charles R. Darwin (1859 in Thomson 1991, 71)].

2. A species that has persisted to the present with little, or no, change over a long period of geological time (Lincoln et al. 1985).
Note: Some people consider a genus that has persisted to the present with little or no change over a long period of geological time to be a living fossil, also.
3. "The living representative of an ancient group of organisms that is expected to be extinct (it may for a long time have been thought to be extinct) but isn't" (Thomson 1991, 71).
Note: "Living fossil" connotes one or more of these characters: rare, or at least uncommon; a restricted geographical range; a member of a taxon formerly widely distributed in time and space as indicated by the fossil record; and very ancestral in comparison to other organism groups, even closely related ones (Thomson 1991, 71–72).
Comments: Scientists debate about the meaning of "living fossil" (Thomson 1991, 71–72). Many biologists do not like this term. It sounds like an oxymoron. A living organism cannot be a fossil; however, a living species can be represented by fossils. Researchers have called the following "living fossils:" the African Lungfish, *Protopterus* sp.; Australian Lungfish, *Neoceratodus forsteri*; bowfin (fish), *Amia calva*; coelacanth, *Latimeria chalumnae*; the Dawn Redwood, *Metasequoia glypto-stroboides*; Horseshoe Crab, *Limulus polyphemus*; Mississippi paddlefish, *Polyodon* sp.; South American lungfish, *Lepidosiren paradoxa*; Virginia Opossum, *Didelphis virginianum*; Gars, *Atractosteus* spp.; and Sturgeons, *Lepisosteus* sp. (Thomson 1991, 71–75). Xiphosurida, which contains *Limulus polyphemus*, is represented by Silurian fossils about 425 million years old, but this species has no fossil record and its genus is known from the Miocene (Gould and Eldredge 1993, 224). Some extant genera have lived for millions of years with little apparent change: *Lingula* (Brachiopoda), 400 million years; *Metasequoia* (Coniferophyta), *ca.* 70 million years (Stewart and Rothwell, 1993, 432–433).

microfossil *n.* A fossil of a unicellular organism (*e.g.,* a dinoflagellate, foraminiferan, or nanoplankter) (Kerr 1997a, 1265).

molecular fossil *n.* A modern biomolecule whose parts appear to be frozen in time, thus preserving remnants of ancient events that forged early living molecules [coined by Alan Weiner and Nancy Maizels in Gibbons 1992c, 31].

mummy *n.* A desiccated, prehistoric mammal corpse preserved in northern permafrost (Guthrie 1990, 22).
See mummy.
Comments: There are many prehistoric mummies (= fossils) (Guthrie 1990).

ornithichnite *n.* A fossilized bird track or footprint (Lincoln et al. 1985).

pseudofossil *n.* An object that looks like a true biotic fossil but is not such a fossil (Mayr 1985, 11).
Comments: Abiotic chemical factors, mechanical factors, physical factors, or a combination of these factors produce pseudofossils (Mayr 1985, 11).

scolite *n.* A fossil worm burrow (Lincoln et al. 1985).

stromatolite fossil *n.* A banded dome of sediment produced by bacteria that is strikingly similar to ones constructed by bacterial mats today (Campbell 1987, 512–513).
Comments: Stromatolite fossils from western Australia and southern Africa are dated at 3.5 billion years old and are the oldest evidence of early life. Many of them may have formed before the evolutionary appearance of grazing animals (Stanley 1989, 96). John Grotzinger and Daniel Rothman reported that abiotic processes can produce formations that resemble stromatolites (Anonymous 1997a, 16–17).

subfossil *n.* An organism's remains, or trace, from the Holocene Epoch (which started *ca.* 10,000–12,000 years ago and continues through today); a "post-Pleistocene fossil" (Lincoln et al. 1985).

succinite *n.* Baltic amber.
See fossil: amber.
[after succinic acid found in this amber]

trace See fossil: trace fossil.

trace fossil *n.* A prehistoric sedimentary structure that resulted from an animal's behavior (*e.g.,* a bore hole of predatory marine snail, rodent burrow, spider web, termite's nest, and worm tube in petrified wood or shale), or preserved feces (= coprolite) or stomach contents (Lincoln et al. 1985; Immelmann and Beer 1989, 111–112).
syn. ichnofossil, lebensspur, trace, vestigiofossil (Lincoln et al. 1985)

vestigiofossil See fossil: trace fossil.

♦ **fossil feature** *n.* A geological form that has become covered or changed (Lapidus 1987).

♦ **Fossil Record 2, The** *n.* A database of fossils that includes all groups of algae, animals, fungi, plants, protists, and microbes, including 7186 families or family-equivalent taxa (Benton 1995, 53).

♦ **fossil sea** *n.* An ancient sea bed (Lapidus 1987).

♦ **fossorial** See -orial: fossorial.

♦ **fostering** *n.* A person's bringing up a child, often a foster child, with parental care (*Oxford English Dictionary* 1972, entries from 1205).
Comment: "Fostering" now refers to nonhuman animals as well.

cross-fostering *n.*
1. An experimental method for controlling for subtle environmental factors that affect observed differences among strains of a species; in mice, this method involves switching newborn litters of pups at birth of two strains differing in adult behavior so that foster mothers rear the pups (Dewsbury 1978, 113).
2. Nonparental rearing of young in humanly contrived situations, thus brood parasitism and adoption are excluded (Immelmann and Beer 1989, 112).
   syn. foster rearing (Immelmann and Beer 1989, 112)
cf. adoption; parasitism: brood parasitism

interspecific fostering *n.* Rearing young animals by heterospecific adults (Immelmann and Beer 1989, 112).
syn. cross-species rearing, cross-fostering (Immelmann and Beer 1989, 112)

▸ **hand-rearing** *n.* A person's rearing a young heterospecific animal (Immelmann and Beer 1989, 112).

♦ **"found p," "calculated p"** See p: "found p."

♦ **founder effect** See effect: founder effect.

♦ **founding stage** See stage: founding stage.

♦ **foundress queen** See caste: queen: foundress queen.

♦ **four why's of animal behavior** See Tinbergian research programme.

♦ **Fourier analysis** *n.* "A technique that attempts to approximate a time series like the [animal] extinction record by a combination of sine and cosine curves" (Kerr 1984).
[after Jean Baptiste Joseph Fourier, French mathematician and physicist]

♦ **FPOM** See particulate-organic matter: fine-particulate-organic matter.

♦ **Fraser-Darling effect, Fraser-Darling law** See effect: Fraser-Darling effect.

♦ **fraternal polyandry** See mating system: polyandry.

♦ **fraternal twin** *n.* See twin: fraternal twin.

♦ **fratricide** See -cide: fratricide.

♦ **Free Air CO$_2$ Enrichment System** See system: Free Air CO$_2$ Enrichment System.

♦ **free-cell-formation theory** See ³theory: free cell-formation theory.

d–g

◆ **free-running period** See period: natural period.

◆ **freehold** See territory.

◆ **French kiss** See kiss: French kiss.

◆ **frequency** *n.*
1. The number of times something occurs within a particular extent of time, a particular group, etc. (Michaelis 1963).
2. The number of times something occurs in relation to its total number of possible occurrences (Michaelis 1963).
cf. abundance, rarity

A CLASSIFICATION OF FREQUENCY IN INCREASING OCCURRENCE

    I.   very infrequent (*syn.* rare)
    II. infrequent (*syn.* occasional, sporadic, somewhat common)
    III. frequent (*syn.* common)
    IV. very frequent (*syn.* abundant)

---

almost certain *n.* An event that occurs 95 out 100 times (Mosteller et al. in Kolata 1986).

rare *n.* An event that occurs less than once out of a thousand times, depending on the kind of event (Mapes in Kolata 1986).
See frequency (table).
cf. rarity

relative frequency *n.*
1. A measure of the occurrence of a species, usually expressed as a percentage, calculated as the ratio of the frequency of a given species to the sum of the frequencies of all species present (Lincoln et al. 1985).
2. The number of times an event occurs in x trials, divided by the number of trials (n) (Lincoln et al. 1985).

very likely *n.* An event that occurs 9 out 10 times (Mosteller et al. in Kolata 1986).

◆ **frequency curve** See curve: frequency curve.

◆ **frequency-dependent selection** See selection: frequency-dependent selection.

◆ **frequency-dependent sexual selection** See selection: frequency-dependent sexual selection.

◆ **frequency distribution** See distribution: frequency distribution.

◆ **frequency modulation** See modulation: frequency modulation.

◆ **frequent worker** See caste: worker: frequent worker.

◆ **freshwater marsh** See habitat: marsh: freshwater marsh.

◆ **friend** *n.*
1. A person joined to another in mutual benevolence and intimacy, not usually applied to relatives or lovers (*Oxford English Dictionary* 1972, entries from 1018).
2. A person who wishes another person, or a cause, well; a sympathizer, favorer, patron, supporter (*Oxford English Dictionary* 1972, entries from 1205).
3. A person who is on good terms with another, not hostile or at variance; one who is on the same side in warfare, politics, etc. (*Oxford English Dictionary* 1972, entries from 1000).
4. In Olive Baboons: a conspecific that an animal is frequently near and with which it frequently engages in grooming; friends can be mothers and adolescent sons, older females and older males, adolescent males and females, or older females and young adult males (Smuts 1987, 39).

◆ **fright** *n.*
1. A person's sudden fear, violent terror, alarm; an instance of this (*Oxford English Dictionary* 1972, entries from 825).
2. An individual animal's feeling of being threatened accompanied with its alertness and acumen (Tembrock 1961 in Heymer 1977, 72).
v.t. fright (poetic, to frighten)
cf. fear
Comment: Unlike "fear," "fright" is object oriented; *i.e.,* it occurs only in response to tangible danger, and it probably does not result in physiological arousal such as fear-induced defecation and urination (Tembrock 1961 in Heymer 1977, 72).

◆ **fright call** See animal sounds: call: fright call.

◆ **fright molt** See molt: fright molt.

◆ **frigofuge** *n.* An organism that is intolerant of cold (Lincoln et al. 1985).
adj. frigofugous

◆ **frigophile** See [1]-phile: frigophile.

◆ **frith** See [2]group: frith.

◆ **frogger** See [2]group: frogger.

◆ **frost** *n.* Temperature cold enough to freeze a particular object (Michaelis 1963).

killing frost *n.*
1. A sharp temperature drop that severely damages and kills a plant (Allaby 1994).
2. A sharp temperature drop that prevents reproduction of an annual, biennial, or ephemeral plant (Allaby 1994).

◆ **fructicole** See -cole: fructicole.

◆ **frugivore** See -vore: frugivore.

◆ ***fruitless*** See gene: *fruitless*.

◆ **frustration** *n.* Failure of an animal's goal-directed behavior to realize its objective through obstruction, or absence, of a requisite condition (Heymer 1977, 70; Immelmann and Beer 1989, 113).
cf. hypothesis: aggression-frustration hypothesis

Comment: Frustration may include an animal's "emotional state" due to its not reaching its goal (Heymer 1977, 70; Immelmann and Beer 1989, 113).

♦ **frustration-induced aggression** See aggression: frustration-induced aggression.

♦ **fruticole** See -cole: fruticole.

♦ **fry** See animal names: fry; ²group: fry.

♦ **fucivore** See -vore: fucivore.

♦ **fugitive species** See ²species: fugitive species.

♦ **fulguration** See essentialism.

♦ **full charge** *n.* The ultimate aggressive act of an African elephant directed only at dangerous predators including Humans (Wilson 1975, 495).

♦ **full phylogenetic tree** See -gram: full phylogenetic tree.

♦ **full sister** See sister: full sister.

♦ **¹function** *n.*
1. Anything's activity that is proper to it; the mode of action by which something fulfills its purpose; contrasted with structure (*Oxford English Dictionary* 1972, entries from 1590).
   Note: Function was first used with regard to animals, then plants.
2. Evolutionary advantage of a trait (Hailman 1967 in Hailman 1976, 184).
   syn. ultimate function (Immelmann and Beer 1989, 114)
3. Ecological adaptiveness (Dewsbury 1978, 6).
   syn. proximate function (Immelmann and Beer 1989, 114)
   cf. four why's of animal behavior, structure
   proximate function of behavior *n.* The immediate effects, or consequences, of an animal's behavior, distinguished from ultimate function of behavior, *q.v.* (Immelmann and Beer 1989, 236), *e.g.*, a male bee's charging after another conspecific male and driving the intruder from his territory.
   ultimate function of behavior *n.* The effects of an animal's behavior on its survivorship and reproduction (genetic fitness), distinguished from proximate function of behavior, *q.v.* (Immelmann and Beer 1989, 236); *e.g.*, a male bee's increasing his genetic fitness by maintaining a territory where he obtains food and mates.

♦ **²function, (F, f)** *n.* In mathematics, a quantity whose value is dependent on the value of some other quantity (Michaelis 1963).
   cost function *n.* An organism's long-term optimization (*e.g.*, energy intake per life time) (McFarland and Houston 1981 in Dawkins 1986, 31).
   cf. ³theory: optimality theory, utility [originally an economics term]

goal function *n.* An organism's short-term optimization (*e.g.*, energy intake per unit time) (McFarland and Houston 1981 in Dawkins 1986, 31).
   cf. ³theory: optimality theory; utility [originally an economics term]

utility function *n.* In economics, a consistent set of values (Tierney 1983).

♦ **functional cycle** See cycle: functional cycle.

♦ **functional feeding group** See ²group: functional feeding group.

♦ **functional fixedness** *n.* An animal's inability to solve an insight problem because it cannot perceive an unusual function for an object (Storz 1973, 103).

♦ **functional group** See guild.

♦ **functional monogyny** See -gyny: monogyny: functional monogyny.

♦ **functional morphology** See morphology: functional morphology.

♦ **functional play** See play: functional play.

♦ **functional psychology** See study of: psychology: functional psychology.

♦ **functional response** See response: functional response.

♦ **functional song** See animal sounds: song: functional song.

♦ **functional system** See cycle: functional cycle.

♦ **functionalism** *n.*
1. The ideas that natural selection for an adaptation results in a function of an organism, and form (structure) is a manifestation (materialization) of this function (Geoffroy; Russell 1916, 77, in Mayr, 1982, 463); function determines structure (Mayr 1982, 463).
   Note: Functionalists include Cuvier, Darwin, and Lamarck (Mayr 1982, 463; Gould 1986, 18).
2. A framework of thinking in which parts of the whole perform functions; these functions represent "biological significance" and within an historical framework lead to the notion of "purpose" (Hughes and Lambert 1984, 787).
See study of: psychology: functionalism.
Comments: Consequently, functionalism represents the view that structures result from "need" (D.M. Lambert, personal communication). Gould (1986, 27) suggests that an adequate theory of evolution must meld certain aspects of both functionalism and structuralism. Lambert and Hughes (1988) argue that there is a conceptual and keyword schism within evolutionary biology and that these represent incommensurable viewpoints.

♦ **fundamental law of biogenesis** See -genesis: biogenesis.

♦ **fundamental niche** See niche (def. 7).

♦ **fundmentum divisionis** See character.

♦ **funeral pheromone** See chemical-releasing stimulus: semiochemical: pheromone: funeral pheromone.

♦ **Fungal Mitochondrial Genome Project** See program: Fungal Mitochondrial Genome Project.

♦ **fungicole** See -cole: fungicole.

♦ **fungivore** See -vore: fungivore.

♦ **funny face** See facial expression: funny face.

♦ **fusion-fission group** See ²group: fission-fusion group.

♦ **Futterparasitismus** See parasitism: social parasitism: trophic parasitism.

♦ **fynbos** See ²community: fynbos.

g

♦ **g** See intelligence: general-cognitive ability.

♦ **Ga** See unit of measure: Ga.

♦ **Gadgil-Bossert model** See [4]model: Gadgil-Bossert model.

♦ **gaggle** See [2]group: gaggle.

♦ **Gaia hypothesis** See hypothesis: Gaia hypothesis.

♦ **galactic black hole** See black hole: galactic black hole.

♦ **gall** *n.* A tumorous plant growth caused by an associated organism, including some bacterium, roundworm, insect, and some mite species (Kritcher and Morrison 1988, 252).

Comment: Many plant species bear galls, often more than one kind. In insects, some aphid, beetle, fly, moth, thrips, and wasp species cause galls. About 800 species (mostly wasp species) produce oak galls (Walker 1990, 201). The gall-producing organisms usually live inside their galls where they obtain food and enjoy protection from enemies.

♦ **gallicole** See -cole: gallicole.

♦ **galliphage, gallivore** See -phage: galliphage.

♦ **Galton's law of ancestral inheritance** See law: Galton's law of ancestral inheritance.

♦ **Galton's law of filial regression** See law: Galton's law of filial regression.

♦ **galvanotaxis** See taxis: galvanotaxis.

♦ **gam** See [2]group: gam.

♦ **game** *n.* An interaction between, or among, organisms that relates to the promotion of their fitnesses (Maynard Smith 1982, 2).

syn. contest (Maynard Smith 1982, 2)

See [2]group: game.

asymmetric game *n.* For example, in the Speckled-Wood Butterfly: a game in which "there is a difference in 'role' between the contestants, of a kind which enables either of them to adopt a strategy

---

SOME KINDS OF GAMES

  I. Classification with regard to the number of players

    A. pairwise

    B. individual vs. conspecific group

 II. Classification with regard to capabilities of players

    A. asymmetric game

    B. symmetric game

III. Classification with regard to game rules

    A. hawk-dove game

    B. hawk-dove-assessor game

    C. hawk-dove-bourgeois game

    D. life-history-strategy game

    E. prisoner's-dilemma game (one-shot, reiterated)

    F. retaliator game

    G. rock-scissors-paper game

    H. territory game

    I tit-for-tat game

    J. war-of-attrition game

'in role 1, do A; in role 2, do B;' "contrasted with symmetric game (Davies 1978; Maynard Smith 1982, 204).

syn. asymmetric contest (Maynard Smith 1982, 204)

symmetric game *n.* A game in which there is no difference in "role" between the contestants; contrasted with asymmetric game (Maynard Smith 1982, 204).

♦ **game theory** See [3]theory: game theory.

♦ **gamergate** See caste: worker: gamergate.

♦ **gamete** *n.* A mature, usually haploid, sexual reproductive cell that fuses with another gamete of the opposite sex and forms a zygote, which is usually diploid (Vines 1886 in the *Oxford English Dictionary* 1972; Lincoln et al. 1985).

Comment: Most animal species have two kinds of gametes: motile sperm and nonmotile eggs; at fertilization, a sperm and an egg fuse to form a zygote (Immelmann and Beer 1989, 116).

[New Latin *gameta* < Greek *gametē*, wife, or *gametēs*, husband]

agamete *n.*
1. A mature reproductive cell that does not fuse with another to form a zygote; a noncopulating germ cell (Bell 1982, 501; Lincoln et al. 1985).
2. A haploid propagule that has the capacity for syngamy but develops without it (Bell 1982, 501).

syn. arygospore (Bell 1982, 501)

aplanogamete *n.* A gamete without cilia (Vines 1886 in the *Oxford English Dictionary* 1972); "a nonmotile gamete" (Lincoln et al. 1985).

cf. gamete: zoogamete

anisogamete See gamete: heterogamete.

egg *n.*
1. A female gamete (Michaelis 1963).
   syn. egg cell, ovum, *pl.* ova (Michaelis 1963); gynogamete (Lincoln et al. 1985)
2. An unfertilized egg cell (Lincoln et al. 1985).
3. An unfertilized fish egg, a fish embryo, or both (Blumer 1982, 3).

cf. animal names: spawn; female; gamete: macrogamete

facultative gamete *n.* "A motile spore (zoospore) that can function as a gamete" (Lincoln et al. 1985).

gynogamete See gamete: egg.

heterogamete *n.*
1. A gamete that is produced by the heterogametic sex (Lincoln et al. 1985).
2. A gamete that belongs to one of two distinguishable types (Lincoln et al. 1985).

syn. anisogamete (Lincoln et al. 1985)
cf. gamete: homogamete

homogamete *n.* A gamete, of only one distinguishable type, produced by the homogametic sex; contrasted with heterogamete (Lincoln et al. 1985).

macrogamete *n.* The larger of two anisogametes, often called an ovum or female gamete; contrasted with microgamete (Lincoln et al. 1985).

syn. megagamete (Lincoln et al. 1985)

megagamete See gamete: macrogamete.

microgamete *n.* The smaller of two anisogametes, often called a sperm or male gamete; contrasted with macrogamate (Lincoln et al. 1985).

obligate gamete *n.* A gamete that cannot develop parthenogenetically (Lincoln et al. 1985).

oosphere *n.* A female gamete before it is fertilized (Lincoln et al. 1985).

oospore *n.* A fertilized female gamete (Lincoln et al. 1985).

ovum *n., pl.* **ova** See gamete: egg.

parthenogamete *n.* A gamete that can develop parthenogenetically (Lincoln et al. 1985).

planogamete See gamete: zoogamete.

planont *n.* A motile gamete, spore, or zygote (Lincoln et al. 1985).

syn. planospore (Lincoln et al. 1985)

progamete *n.* A cell that gives rise directly to gametes, either by a single division or intracellular metamorphosis (Lincoln et al. 1985).

sperm, spermatogamete, spermatosome, spermatozoan, spermatozoid, spermatozooid, spermium *n.* A male gamete (Lincoln et al. 1985).
cf. ejaculate; gamete: microgamete; male
[Greek *sperma*, seed]

▸ **apyrene sperm** *n.* A sperm that lacks a genome (Sivinski 1984, 92).

▸ **eupyrene sperm** *n.* A sperm with a complete haploid genome (Sivinski 1984, 92).

▸ **kamikaze sperm** *n.* A human aberrant sperm that may sacrifice itself to protect normal sperm during fertilization (Baker and Bellis 1988, 938; 1989, 867; Small 1989, 544).

▸ **oligopyrene sperm** *n.* A sperm with only part of its usual chromosomal complement (Sivinski 1984, 92).

▸ **outlaw sperm** *n.* A sperm that competes in a way that lowers the effectiveness of an ejaculate as a whole (Sivinski 1984, 108).

▸ **worker sperm** *n.* An apyrene sperm that aids its ejaculate mates but cannot fertilize an egg (Sivinski 1984, 108).

syngamete See zygote.

zoogamete, zoogygogamete, planogamete *n.* A gamete with cilia (Vines 1886 in the *Oxford English Dictionary* 1972); a motile gamete (Lincoln et al. 1985).
cf. gamete: aplanogamete

♦ **gamete selection** See selection: gamete selection.

♦ **-gametic**
heterogametic, digametic *adj.* Referring to an organism that has two kinds of gametes, one producing males and the other producing females; contrasted with homogametic (Lincoln et al. 1985).

homogametic *adj.* Referring to an organism that has only one kind of gamete; contrasted with heterogametic (Lincoln et al. 1985).

♦ **gametic disequilibrium** See linkage disequilibrium.

♦ **gametic meiosis** See meiosis: gametic meiosis.

♦ **gametic number** *n.* The number of chromosomes in a gamete's nucleus, which is usually half the somatic chromosome number of a particular species (Lincoln et al. 1985).

♦ **gametic pool** *n.* The total potential gamete production of any particular generation of a sexually reproducing population (Lincoln et al. 1985).

♦ **gameto-** *combining form* Gamete (Michaelis 1963).
[Greek *gametēs*, husband < *gamos*, marriage]

♦ **gametogamy** See -gamy: syngamy.

♦ **gametogenesis** See -genesis: gametogenesis.

♦ **gametogenic** See -genic: gametogenic.

♦ **gametogeny** See -geny: gametogeny.

♦ **gametophyte** *n.* In plants with haploid and diploid generations, the haploid phase (IN) which produces gametophytes (Curtis 1983, 424).

♦ **gametotoky** See parthenogenesis: amphitoky.

♦ **-gamety** *combining form*
cf. gameto-

diagamety, heterogamety *n.* An organism's having two kinds of gametes, one producing male and the other producing female offspring (Lincoln et al. 1985).
heterogamety See sex determination: heterogamety.
homogamety *n.* An organism's production of only one kind of gamete from its homogametic sex (Lincoln et al. 1985).
See sex determination: homogamety.
pseudoanisogamety *n.* The occurrence of different-sized gametes whose sizes are not correlated with their genders (Bell 1982, 511).
sex diagamety *n.* The capacity of an individual of an organism's heterogametic sex to produce male- and female-determining gametes through a sex-chromosome mechanism (Lincoln et al. 1985).
sex monogamety *n.* A hermaphrodite's, parthenogenetic individual's, or population's producing gametes of only one kind with regard to sex determination (Lincoln et al. 1985).
sex polygamety, sex plurigamety *n.* An individual organism's "producing gametes of many different types with regard to sex determination through the mechanism of polyfactorial sex determination" (Lincoln et al. 1985).

♦ **gamic** *adj.* "Fertilized" (Lincoln et al. 1985).

♦ **-gamic** See -gamous, -gamy.

♦ **γ-diversity** See diversity: γ-diversity.

♦ **γ-diversity index** See index: species-diversity index: γ-diversity index.

♦ **gamma hydroxybutyrate** See drug: gamma hydroxybutyrate.

♦ **gamma karyology** See study of: karyology: gamma karyology.

♦ **gamma link** See link: gamma link.

♦ **gamma taxonomy** See taxonomy: gamma taxonomy.

♦ **gamodeme** See -deme: gamodeme.

♦ **gamogenesis** See -genesis: gamogenesis.

♦ **gamogenic, gamogonic** See -genetic: gamogenetic.

♦ **gamontogamy** See -gamy: gamontogamy.

♦ **gamophase** See -plont: haplont.

♦ **gamophobia** See phobia (table).

♦ **gamosematic** *adj.* Referring to coloration, markings, or behavior that assists members of a pair in locating each other (Lincoln et al. 1985).

♦ **gamotropism** See -tropism: gamotropism.

♦ **-gamous** See -gamy.
ephydrogamous *adj.* Referring to a plant species that has, or is pollinated by, waterborne pollen that is transported at the water surface (Lincoln et al. 1985).
cf. -gamous: hyphydrogamous
hercogamous, herkogamous *adj.* Referring to a flower whose stamens and stigma are positioned in a way that prevents self-pollination by a stamen's touching the stigma (Lincoln et al. 1985).
n. hercogamy, herkogamy
Comment: The woodland lily is hercogamous, but it also appears that an individual's own pollen cannot fertilize its ovules (Barrows 1979a).
hyphydrogamous, hyphydrogamic, hypohydrogamic *adj.* Referring to a plant species that has waterborne pollen that is transported below the water surface (Lincoln et al. 1985).
cf. -gamous: ephydrogamous
monogamous *adj.* Referring to a population, or species, that has a monogamous mating system (Lincoln et al. 1985).
See mating system: monogamy.
polygamous *n.* Referring to a plant species with hermaphroditic and unisexual flowers on the same, or on different, individuals, *e.g.*, *Ampelopsis cordata* (Grape Family) (Fernald 1950, 1580).
progamous *adj.* Referring to a phenomenon that occurs, or exists, before fertilization (Lincoln et al. 1985).

♦ **-gamy** *combining form* "Marriage, fertilization, sexual union" (Lincoln et al. 1985).
adj. -gamic, -gamous
cf. fertilization, mating system, pollination
[Greek *gamos*, marriage]
adelphogamy See mating: sib mating.
allautogamy See -gamy: autoallogamy.

allogamy *n.* Fertilization of the stigma for one flower with pollen from another flower (Faegri and van der Pijl 1979, 24).
cf. fertilization: cross-fertilization; -gamy: autogamy, gneisiogamy; pollination: cross-pollination

▶ **geitonogamy** *n.* Allogamy that involves two flowers from the same plant (Faegri and van der Pijl 1979, 24).

▶ **xenogamy** *n.* Allogamy that involves two flowers from different plants (Faegri and van der Pijl 1979, 24).

amphigamy *n.* Cross-fertilization; the fusion of male and female gametes (Lincoln et al. 1985).

androgamy *n.* Impregnation of a male gamete by a female gamete (Lincoln et al. 1985).

anisogamy *n.*
1. The condition in which a female sex cell (ovum) is larger than a male sex cell (sperm) (Wilson 1975, 316, 578).
   Note: Anisogamy, not isogamy, is the rule in organisms (Wilson 1975, 316, 578).
2. "A sexual system in which fusion takes place at fertilization between a large (female) and a small (male) gamete" (Dawkins 1982, 284).
See -gamy: heterogamy, isogamy.

anisohologamy *n.* Fusion of hologametic gametes that are not quite identical in some respect (Lincoln et al. 1985).

anisomerogamy See -gamy: anisogamy.

apogamy *n.*
1. A kind of apomixis (reproduction without fertilization) in which meiosis and gamete fusion are partially, or totally, suppressed (Lincoln et al. 1985).
2. Direct production of a plant from a gametophyte by budding with gamete formation (Lincoln et al. 1985).

▶ **euaphogamy diploid apogamy** *n.* Development of a diploid sporophyte from one or more cells of a gametophyte without gamete fusion (Lincoln et al. 1985).

▶ **parthenoapogamy, obligate apogamy, diploid parthenogenesis, somatic parthenogenesis** *n.* Parthenogenesis of diploid individuals in which meiosis has been suppressed so that neither chromosome reduction nor any corresponding phenomenon occurs (Lincoln et al. 1985).
cf. parthenogenesis

▶ **obligate apogamy** See -gamy: apogamy: parthenoapogamy.

apolegamy *n.* Selected breeding (Lincoln et al. 1985).

autoallogamy, allautogamy, autallogamia, autallogamy, homodichogamy *n.* The condition in which some individuals of a species are adapted to cross-fertilization and others to self-fertilization (Lincoln et al. 1985).

autogamy *n.*
1. Self-fertilization or self-pollination (Lincoln et al. 1985).
   syn. orthogamy (Lincoln et al. 1985)
2. Fusion of two reproductive nuclei within a single cell from one parent (Lincoln et al. 1985).
3. "Reproduction in which a single cell undergoes reduction division producing two autogametes which subsequently fuse" (Lincoln et al. 1985).
4. For example, in some protistans: the division of micronuclei to produce eight, or more, nuclei of which two fuse and give rise to a new macronucleus (Lincoln et al. 1985).
See -mixis: automixis.

▶ **paedogamous autogamy** *n.* Autogamy (def. 2) in which only nuclei, not cytoplasm, are involved in zygote formation (Lincoln et al. 1985).

caryogamy See -gamy: karyogamy.

cytogamy *n.*
1. Fusion of two cells (Lincoln et al. 1985).
2. Fusion of both male nuclei of one of two conjugating organisms with the other's female nucleus; double autogamy (Lincoln et al. 1985).

cytoplasmic androgamy *n.* A male gamete's being fertilized by the cytoplasm of a female gamete (Lincoln et al. 1985).

cytoplasmic gynogamy *n.* A female gamete's apparently being fertilized by the cytoplasm of a male gamete (Lincoln et al. 1985).

diamesogamy *n.* "Fertilization through an external agency" (Lincoln et al. 1985).

dichogamy *n.* Maturation of male and female reproductive organs at different times in a flower or hermaphroditic organism which prevents self-fertilization; contrasted with homogamy (Faegri and van der Pijl 1979, 27; Lincoln et al. 1985).
syn. heteracme (Lincoln et al. 1985)

diecodichogamy *n.* Maturation of male flowers first in some plants and female flowers first in other plants in a population of dichogamic plants (Lincoln et al. 1985).

dioeciopolygamy *n.* Both unisexual and hermaphrodite individual's being present in the same species (Lincoln et al. 1985).

ditopogamy See -gamy: heterogamy.

double autogamy See -gamy: cytogamy.

ecodichogamy *n.* The condition in which individuals of a dioecious species have different maturation times (given as the adjective "ecodichogamic" in Lincoln et al. 1985).

endogamy See breeding: inbreeding; mating: assortative mating.

endokaryogamy See -gamy: endogamy.

epigamy *n.* In polychaete worms: modification of an asexual individual (atoke) into a sexual one (epitoke) without any intervening reproductive process; contrasted with schizogamy (Bell 1982, 505).

exogamy See breeding: outbreeding; mating: disassortative mating.

exoisogamy *n.* The condition in which an isogamete undergoes fusion only with an isogamete from a different brood (Lincoln et al. 1985).

gametogamy See -gamy: syngamy.

gamontogamy *n.* "Aggregations or union of gamonts (organisms)" (Margulis et al. 1985, 73).
See -gamy: hologamy.

geitonogamy See -gamy: allogamy.

gneisiogamy, intraspecific zygosis *n.* Cross-fertilization between two conspecific individuals (Lincoln et al. 1985).
cf. -gamy: allogamy

gymnogamy *n.* Fertilization (Lincoln et al. 1985).

hemigamy, semigamy *n.* Activation of an ovum by a male nucleus with fusion (Lincoln et al. 1985).

herkogamy *n.* The spatial separation of anthers and stigma(s) in a single flower (Faegri and van der Pijl 1979, 27).

heterodichogamy *n.* For example, in some maple species: the phenomenon in which all male flowers of a plant finish anthesis before female flowers open, or vice versa (Faegri and van der Pijl 1979, 26).

heterogamy *n.*
1. A species' having two kinds of gametes, one kind producing males and the other producing females; contrasted with homogamy (Lincoln et al. 1985).
   cf. -gametic: homogametic
2. The union of gametes of different shape, size, or behavior (heterogametes) (Lincoln et al. 1985).
   syn. anisogamy, ditopogamy (Lincoln et al. 1985)
   cf. -gamy: anisomerogamy, isogamy
3. For example, in some cecidomyiid midges and aphids: "alternation of two sexual generations, one syngamic and the other parthenogenetic" (Thornhill and Alcock 1983, 35; Lincoln et al. 1985).
See parthenogenesis: heterogony.
adj. heterogametic
syn. heterogony (Lincoln et al. 1985)

hologamy, gamontogamy, macrogamy *n.* A mode of protistan reproduction involving gametes similar in size to vegetative cells (Lincoln et al. 1985).
See -gamy: gamontogamy.

homiogamy, homoiogamy *n.* Union of two gametes of the same sex (Lincoln et al. 1985).

homodichogamy See -gamy: autoallogamy.

homogamy *n.*
1. The condition of having only one kind of gamete (Lincoln et al. 1985).
2. A flower's simultaneously having its own mature pollen and a receptive stigma; contrasted with dichogamy, heterogamy (Faegri and van der Pijl 1979, 27).
See breeding: inbreeding; mating: assortative mating.
adj. homogametic
syn. autogamy (Faegri and van der Pijl 1979, 27)

homoiogamy See -gamy: homiogamy.

hylogamy *n.* The fusion of a haploid nucleus with a diploid one (Lincoln et al. 1985).

hyperanisogamy *n.* An organism's having large, active female gametes (Lincoln et al. 1985).

idiogamy See fertilization: self-fertilization.

isogamy *n.* In a few species of sexually reproducing organisms: "the condition in which the male and female sex cells are the same size" (Wilson 1975, 316, 587) and not differentiated into sperm and ova (Wittenberger 1981, 617); contrasted with anisogamy.
syn. zygogamy

karyallogamy *n.* Reproduction involving the fusion of two morphologically identical gametes (Lincoln et al. 1985).

karyogamy, caryogamy, karyapsis *n.* Fusion of gametic nuclei (Lincoln et al. 1985).

macrogamy See -gamy: hologamy.

merogamy, microgamy, merogony, merogeny *n.* Protistan reproduction involving gametes that are smaller than vegetative cells and are produced by multiple fission; contrasted with hologamy (Lincoln et al. 1985).

metagamy *n.* Reproduction involving cycles that alternate between sexual and asexual phases (Lincoln et al. 1985).

microgamy See -gamy: merogamy; isolation: reproductive isolation.

misogamy See isolation: reproductive isolation.

mychogamy See fertilization: self-fertilization.

neogamy *n.* "Precocious syngamy" (Lincoln et al. 1985).

nothogamy See -gamy: heteromorphic xenogamy.

orthogamy See -gamy: autogamy.

pangamy, panmixis *n.* "Unrestricted random mating" (Lincoln et al. 1985).
See -mixis: panmixis.

paragamy *n.* Zygote nucleus formation by fusion within a syncytium (Lincoln et al. 1985).

parthenogamy See parthenogenesis: automictic parthenogenesis.

parthenokaryogamy, parthenocaryogamy *n.* "The fusion of two female haploid nuclei" (Lincoln et al. 1985).

pedogamy, paedogamy *n.*
1. Zygote formation by fusion of two gametes produced by one individual organism (Lincoln et al. 1985).
2. A form of automixis (Lincoln et al. 1985).

perittogamy *n.* Random cytoplasmic fusion between a gametophyte's undifferentiated cells without nuclear fusion (Lincoln et al. 1985).

phytogamy *n.* Plant cross-fertilization (Lincoln et al. 1985).

plasmagynogamy *n.* Fertilization in which a zygote's cytoplasm is derived from only a female gamete (Lincoln et al. 1985).

plasmaheterogamy *n.* Fertilization in which a zygote's cytoplasm is derived unequally from both a male and female gamete (Lincoln et al. 1985).

plasmaisogamy *n.* Fertilization in which a zygote's cytoplasm is derived equally from both a male and female gamete (Lincoln et al. 1985).

plasmogamy, plasmatogamy *n.* Cytoplasm fusion of two or more cells without nuclear fusion (Lincoln et al. 1985).

plastogamy *n.* Cytoplasm fusion of unicellular organisms to form a plasmodium without nuclear fusion (Lincoln et al. 1985).

protogamy *n.* Fusion of gametes producing a binucleate zygote (Lincoln et al. 1985).

pseudogamy, gynogenesis, pseudomixis, somatogamy *n.* Egg activation by a male gamete that degenerates without its nucleus fusing with that of the egg (Lincoln et al. 1985).

schizogamy *n.* In polychaete worms: creation of sexual individuals by fission; contrasted with epigamy (Bell 1982, 512).

semigamy See -gamy: hemigamy.

somatogamy See -gamy: pseudogamy.

staurogamy See fertilization: cross-fertilization.

syngamy *n.*
1. The final step in fertilization in which the nuclei of sex cells (gametes) meet and fuse, forming a zygote (Wilson 1975, 596).
2. Contact or fusion of gametes (cells or nuclei) (Margulis et al. 1985, 73).

syn. fertilization (Bell 1982, 513), gametogamy, sexual reproduction (Lincoln et al. 1985)
cf. -gamy: neogamy

xenautogamy *n.* A species' condition in which cross-fertilization normally occurs but self-fertilization is possible (Lincoln et al. 1985).

xenogamy See fertilization: cross-fertilization; -gamy: allogamy: xenogamy.

zoidiogamy *n.* Fertilization by motile male gametes (Lincoln et al. 1985).

zoogamy *n.*
1. An animal's reproducing sexually (Lincoln et al. 1985).
2. A plant species' having mobile gametes (Lincoln et al. 1985).

zygogamy See -gamy: isogamy.

♦ **gang** See ²group: gang.

♦ **gaping** *n.*
1. A person's opening his mouth; yawning (*Oxford English Dictionary* 1972, entries from 1660).
cf. yawning
2. An individual bird's holding its bill wide open, upward in young birds and toward parents in older birds (Tinbergen and Kuenen 1939 in Heymer 1977, 166).
v. gape

♦ **gargle call** See animal sounds: call: gargle call.

♦ **gas-chromatography–mass-spectrometric analytical method** See method: gas-chromatography–mass-spectrometric analytical method.

♦ **gaster** *n.* The "terminal major body part behind the 'waist' of Ants and other hymenopterans" (Wilson 1975, 585).
syn. metasoma (Wilson 1975, 585)

♦ **gastric-brooding** See brooding: stomach-brooding.

♦ **Gause's axiom, Gause's hypothesis, Gause's law, Gause's principle, Gause's rule, Gause's exclusion principle, Gause-Volterra principle** See principle: Gause's principle.

♦ **Gaussian distribution** See distribution: normal distribution.

♦ **GC–mass spec, GC–mass-spectrometric analytical method, GC-MS** See method: gas-chromatography–mass-spectrometric analytical method.

♦ **gecker** See animal sounds: gecker.

♦ **geitonogamy** See -gamy: allogamy: geitonogamy.

♦ **gel electrophoresis** See method: electrophoresis: gel electrophoresis.

♦ **Gelber-Jensen controversy** *n.* A debate regarding whether *Paramecium aurelia* showed modified behavior due to learning or environmental modification in

Gelber's investigation (Gelber 1952, etc. in Dewsbury 1978, 343).

♦ **gelicole** See -cole: gelicole.

♦ **geminate species, law of** See law: Jordan's law.

♦ **gemmule** *n.* A hypothetical small hereditary particle that carries information from all parts of an organism's body into its germ cells (Darwin 1868; Dawkins 1982, 287).
Comment: This is a discredited concept (Dawkins 1982, 287).

♦ **geitonogamy** See -gamy: allogamy.

♦ **GenBank** See database: GenBank.

♦ **gender** *n.*
1. Informally, the quality of being of the male or female sex; "sex classification: a humorous use" (Michaelis 1963).
2.. The set of individuals, or gametes, of the same species that are incapable of fertilizing one another, *e.g.,* males and females in gonochores or microgametes and macrogametes in anisogametic amphimicts (Bell 1982, 506).
See sex.

♦ **gene** *n.*
1. A unit of heredity; an "element" (Mendel 1866 in Mayr 1982, 736).
2. An elementary unit of inheritance; pangen, *q.v.* [Darwin's "pangen" renamed "gene" by Johannsen (1909 in Mayr 1982, 736)].
3. Johannsen's concept of gene: a kind of accounting or calculating unit (*Rechnungseinheit*); not a material, morphologically characterized structure in the sense of Darwin's gemmules or Weismann's biophores; a kind of force (Mayr 1982, 736).
Note: Johannsen coined "gene" as a term to be free of any hypothesis about its nature or how it functioned (Maienschein 1992, 123).
4. "Any hereditary information for which there is a favorable or unfavorable selection bias equal to several or many times its rate of endogenous change" (Williams 1966, 25, in Dawkins 1982, 287); any portion of chromosomal material that potentially lasts for enough generations to serve as a unit of natural selection (Dawkins 1977).
5. A heritable unit of information governing structure, development, and function of an organism (Stebbins and Ayala 1985, 72).
6. A hereditary unit that, in the classical sense, occupies a specific position (locus) within an organism's genome or chromosome, has one or more specific effects upon the organism's phenotype, can mutate to various allelic forms, and recombines with other such units (King and Stansfield 1985).

syn. cistron (for a gene that specifies for the formation of a particular polypeptide chain); factor; genetic material; idioplasm (Nägeli in Mayr 1982, 958); inheritance factor

7. Commonly, a shorthand term (metaphor) that indicates that one conspecific organism has a genetic makeup that is related to its showing a particular character (*e.g.,* protein, biochemical pathway, organelle, organ, behavior) and another conspecific organism does not have such a genetic makeup (McFarland 1985, 100; Dawkins 1986, 47).

8. A DNA sequence that is transcribed as a single unit and encodes a functional RNA molecule or one set of closely related polypeptide chains (protein isoforms) (Alberts et al. 1989, 486, 592).
Notes: This gene definition was formulated after the discovery of alternative gene splicing. This definition seems appropriate for wide use.

cf. mutation; nucleic acid: deoxyribonucleic acid; regulon
Comments: "The concept of gene continues to undergo considerable revision and even fragmentation" (Maienschein 1992, 127). Kitcher (1992) discusses what a gene is physically and concludes that a "gene is anything a competent biologist chooses to call a gene" and that evolutionary biology can do without a precise definition of gene. "Gene" in the definitions below refers to different gene definitions. I usually have not tried to indicate which definitions below are relevant to "gene" because the authors of the definitions usually do not stipulate this information. Some gene names suggest that one gene, alone, is responsible for a particular trait (*e.g.,* homosexuality gene, I.Q. gene, novelty-seeking gene). These are metaphorical names because more than one gene is involved in producing these traits (Gould 1999, 62).
[gene < pangen < gemmule (Mayr 1982, 704; Brosius and Gould 1992, 10706)]

*age***-1 gene** *n.* A gene that increases life span of *Caenorhabditis elegans* (Johnson 1990, 908).
Comment: The *age-*1 gene increases life span a mean of 65% and is the first known gene that shortens life in its normal form (Johnson 1990, 908).

aggression gene *n.* A gene for monoamine oxidase-A that possibly underlies the aggressive and sometimes violent behavior displayed in certain male Humans (Brunner et al. in Morell 1993b, 1722).
Comment: Monoamine oxidase-A helps break down several neurotransmitters that

ESTIMATED NUMBERS OF BASE PAIRS AND GENES IN SELECTED TAXA

TAXON	NUMBER OF BASE PAIRS	NUMBER OF GENES	COMMENTS, REFERENCES
Prokaryota	—	up to 8000	a
Eukaryota	—	up to 140,000	b
Arabidopsis thaliana, Mouse-Ear Cress	100,000,000	—	c
Caenorhabditis elegans, roundworm	100,000,000	18,424	d, v
Drosophila melanogaster, fruit fly	180,000,000	13,601	e, v
Escherichia coli, eubacterium	4,700,000	—	f
Fugu rubripes, puffer fish	400,000,000	—	g
HIV-1 virus	5386	—	h
Haemophilus influenze, bacterium	1,830,121	—	i
Helicobacter pylori, bacterium	—	1700	j
Homo sapiens sapiens, Human	3,300,000,000	50,000–140,000	k
Methanobacterium thermoautotrophicum	—	1700	l
Methanoccocus jannaschii, archaean	12,000,000	1738	m
Mus musculus, Domestic Mouse	3,300,000,000	100,000	n
Mycoplasma genitalium, bacterium	—	600	o
Mycoplasma pneumoniae, bacterium	—	800	p
Oryza sativa, Rice	430,000,000	—	q
Saccharomyces cerevisiae, Baker's Yeast	13,700,000	600	r
Synecocystis sp., bacterium	12,068	6000	s
Variola virus, Smallpox Virus	186,000	—	t
Zea mays, corn, Maize	3,300,000,000	—	u

[a] Researchers have primarily sequenced prokaryotes so far. They may have sequenced the genomes of other species but have not published their results.

[b] Researchers sequenced several eukaryotic species.

[c] Hale (1996, 11A). Geneticists expect to sequence the entire genome of this organism by 2004.

[d] Hale (1996, 11A), Wade (1998, C1, 17,000 genes; 1999b, A17, 19,099 genes), Pennisi (2000, *ca.* 18,000 genes). Researchers completed sequencing this species in 1998.

[e] Merriam et al. (1991, 221, 5000–15,000 genes), Pennisi (2000, 2183, *ca.* 13,600 genes). Celera Genomics Systems (a U.S. company) sequenced most of this fly's genome in 2000 (Butler 1999, 729; Pennisi 2000, 2183).

[f] Hale (1996, 11A), Blattner et al. (1997).

[g] Hale (1996, 11A).

[h] This virus has a single strand of RNA bases.

[i] Fleischmann et al. (1995), Venter and Smith (in Sawyer 1995, A2).

[j] Goffeau et al. (1996), Tomb et al. (1997).

[k] Jasney (1991), Lander et al. (1996). Geneticists working with the Public Consortium aim to se-quence the entire human genome by 2005 (Wade 1998, C1). The Institute of Genomic Research, led by J. Craig Venter, plans to complete sequencing the entire human genomes of five people by the end of 2001 (Wade 1999a, D1). Randy Scott (Incyte Pharmaceuticals, Inc.) announced that humans have 140,000 genes (Wade 1999b, A17).

[l] Goffeau et al. (1996).

[m] Bult et al. (1996), Suplee (1996c, A3). Geneticists sequenced some of the genome of this bacterium in 1996 (Suplee 1996c, A3).

[n] Hale (1996, 11A), Wade (1998, C1).

[o] Fraser et al. (1995).

[p] Fraser et al. (1995), Himmelreich (1996). Researchers completely sequenced this species.

[q] Hale (1996, 11A).

[r] Hale (1996, 11A), Goffeau et al. (1996). Researchers completely sequenced this species in 1996.

[s] Fraser et al. (1995), Kaneko (1996).

[t] Hale (1996, 11A).

[u] Hale (1996, 11A).

[v] Wade (2000, D4). Researchers completely sequenced *Drosophila melanogaster* in 2000.

might, if their concentrations build up abnormally, cause a person to respond excessively, and at times even violently, to stress (Morell 1993b).

allele See allele.

allogene See allele: recessive allele.

altruistic gene *n*. A metaphor of a group of genes that underlie altruism (Wilson 1975, 3).

antimorph *n*. A gene that produces the opposite effect of a wild-type gene (Lincoln et al. 1985).

behavior gene *n*. A metaphor for a group of genes (a genetic makeup) that underlies a particular behavior, *e.g.*, a "gene" for sibling care or a "gene" for a dove strategy (McFarland 1985, 100; Dawkins 1986, 47).
Comment: Genes control many behaviors, including movements of paramecia, singing of crickets, human retardation due to phenylketonuria, and mouse maze learning (Dawkins 1986, 51–55).

buffering gene See gene: polygene.

cheap gene *n*. A hypothetical kind of gene that arises from a repeat sequence that was not subject to natural selection (Zuckerkandl et al. 1989, 506).

chico *n*. In *Drosophila:* a gene that affects fly size (Mitchell 1999, 115).
Comment: Flies with mutations in *chico* are about half the size of normal ones (Mitchell 1999, 115).
[Spanish *chico*, small boy]

cistron *n*.
1. A segment of DNA with the property that the phenotype of a double mutation within the segment differs according to whether the mutations occur in the *cis* or *trans* arrangement (Benzer 1957 in Kitcher 1992, 129).
   Note: Thus, "cistron" has "a precise definition in terms of a specific experimental test" (Dawkins 1982, 285).
2. "The gene of function; the functional unit of inheritance" (Mayr 1982, 806, 957).
3. The length of a chromosome responsible for the encoding of one complete, mature tRNA, rRNA, or polypeptide chain; a cistron can include regions that precede and follow the coding region (leader and trailer), introns, and exons (King and Stansfield 1985).
syn. gene (Dawkins 1982), structural gene
cf. gene: muton, recon
[Benzer (1957 in Mayr 1982, 806) coined "cistron" from the *cis-trans* difference in gene position effects.]

clock gene *n*. In *Drosophila melanogaster:* a gene that enables an individual to maintain a circadian rhythm at different environmental temperatures (Sawyer et al. 1997, 2117).

complementary genes, complementary factors *n*. Two, or more, nonallelic factors that are collectively required to produce a single phenotypic trait (Lincoln et al. 1985).
syn. reciprocal genes (Lincoln et al. 1985)

CREB gene *n*. A gene that functions as a kind of master switch of dozens of other genes that function to convert recent experience into long-term memory (Hall 1998, 28).
syn. CREB switch (Tim Tully in Hall 1998, 26)
Comments: This gene makes the molecule CREB, which switches on other genes after it binds with an animal's DNA (Hall 1998, 31). Blocking the action of CREB stops long-term memory in *Aplysia californica* and *Drosophila* (Hall 1998, 32). Mice, deprived of the CREB gene by genetic engineering, showed normal learning and short-term memory but not long-term memory (Alcino Silva in Hall 1998, 32). Genetically engineered *Drosophila* with a huge excess of the activator protein from the CREB gene memorized tasks after only one training session, and researchers dubbed these flies "smart flies" (Tulley et al. in Hall 1998, 33). Further, CREBs are in many body cells besides neurons and might play roles in circadian rhythms, drug addiction, and hormone metabolism (Jerry Yin in Hall 1998, 33).

Distal-less *n*. A regulatory gene that helps generate both arthropod and mammal legs (Erwin et al. 1997, 135).

duplicate gene *n*. One of a set of genes, each of which has an identical but noncumulative effect (Lincoln et al. 1985).

epistatic gene *n*. A gene that inhibits the expression of another gene (hypostatic gene) (King and Stansfield 1985).
syn. masking gene (King and Stansfield 1985)

eyeless *n*. In *Drosophila:* a gene that encodes the development of an eye (Mestel 1996, 110).
Comment: Mestel (1996, 110) includes a photograph of an eyeless fruit fly and fruit flies with several other eye conditions.

FOS B *n*. A kind of immediate-early gene that triggers female and male mice to care for her young (Associated Press 1996a, A2).
syn. good-mother gene (Associated Press 1996a, A2)
Comments: A mother mouse without a *FOSB* gene ignores her young and lets them die (Associated Press 1996a, A2). *FOSB* is probably a regulatory gene that turns on other genes in response to outside cues, probably the sight and smell of baby mice. The function of the human *FOSB* gene is not yet known. In Humans, a group

d–g

of genes and external environmental conditions probably influence a complex behavior such as nurturing (E. Kandel in Anonymous 1996e).

fruitless *n.* A gene that coordinates all sexual behavior of *Drosophila melanogaster* (Weiss 1996c, A2).

syn. fruity (Weiss 1996c, A2)

Comments: Scientists found evidence for this gene in 1963 (Weiss 1996c, A2) and isolated it in 1996. This is an enormous gene, about 70 times larger than the average fruit-fly gene, and is active in just 500 of the fly's 100,000 brain cells. It apparently serves as a "master switch" of other genes that together control a fly's sexual behavior. This gene resembles no known human gene in its DNA structure.

[fruitless, referring to having no offspring, a name change from the offensive "fruity," 1960s U.S. slang for homosexual (Weiss 1996c, A2)]

"gene for a behavior" See gene: behavior gene.

good-mother gene See gene: *FOSB.*

homeobox-containing gene *n.* A gene (*e.g.,* homeotic-selector gene) that contains a homeobox, *q.v.* (Alberts et al. 1989, 937–938).

syn. homeobox gene (Marx 1992, 401)

cf. homeobox; ²mutuation: homeotic mutation

Comments: Homeobox genes each encode a sequence of about 60 amino acids, are evolutionarily conserved in many kinds of animals, are key regulators of embryonic development, and may have helped to generate the enormous diversity of present-day organisms (Marx 1992, 401).

homeotic gene *n.* A gene that appears to control the developmental fate of cell groups (Campbell 1990, 391).

Comments: Homeotic genes were first discovered in *Drosophila* where mutants of such genes produce extra sets of wings, extra or missing body segments, and legs growing from heads instead of antennae (Campbell 1990, 391). *Drosophila* homeotic genes and other developmental genes contain homeoboxes.

homeotic homeobox gene, *Hox* **gene** *n.* In animals, a homeobox-containing gene that is master architect of development (Duboule et al. in Nash 1995, 56).

cf. gene: homeobox-containing gene

Comments: Non-*Hox* homeobox genes and *Hox* genes are not independent agents but members of vast genetic networks that connect hundreds, perhaps thousands, of other genes (Nash 1995, 57). About 38 of the some 50,000 to 100,000 genes in derived vertebrates are *Hox* genes (Nash 1995, 56).

homosexuality gene *n.* A gene (in the Xq28 of DNA on the X-chromosome) that is possibly the basis for some instances of male homosexuality in Humans (Pool 1993, 292; Hamer et al. 1993).

house-keeping gene *n.* A gene that is active in all cell types (Rieger et al. 1991).

syn. housekeeping gene (Rieger et al. 1991)

Comments: House-keeping genes include those that function in DNA recombination, repair, and replication and produce molecular chaperones (Koonin and Galperin 1997, 760). House-keeping genes produce rare through moderately abundant mRNAs, as opposed to tissue-specifically expressed genes. Cells regulate house-keeping genes primarily at the level of mRNA maturation, stability, or both (Rieger et al. 1991).

***Hox* cluster** *n.* A group of *Hox* genes that specifies the developmental fate of a region of a complex organism (Erwin et al. 1997, 133; illustration, 136).

Comments: Erwin et al. (1997) describe the evolution of *Hox* clusters from a cnidarian ancestor through mammals. There is one gene in the *Hox* cluster of sponge and eight in arthropods. Mammals have four *Hox* clusters, with a total of 38 genes, derived from an ancestral one.

***Hox* genes** See gene: homeotic homeobox genes.

hypostatic gene *n.* A gene that is inhibited by another (epistatic gene) (King and Stansfield 1985).

I.Q. gene *n.* A metaphor for a group of genes that influence the learning ability, or I.Q., *q.v.,* of an animal (*e.g.,* mouse or Human) (Lemonick 1999, 56).

Comment: "Intelligence is an array of largely independent and socially defined mental attributes, not a measure of a single something, secreted by one gene measurable as one number and capable of arranging human diversity into one line ordered by relative mental worth" (Gould 1999, 62).

key gene See gene: oligogene.

Lcyc *n.* A gene whose mutant (in this case, silencing due to methylation) causes radial symmetry in the normally bilaterally symmetric flowers of *Linaria vulgaris* (Butter-and-Eggs, Toadflax).

Comments: Lcyc is a homolog of the *cycloidea* gene, which controls dorsoventral asymmetry in flowers of *Antirrhinum*, Snapdragon.

[Lcyc from *Linaria cycloidea*-like gene]

***Lim*1 gene** *n.* An organizer gene that determines whether or not a mouse embryo develops a head (Oliwenstein 1996a, 34, photograph).

major gene See gene: oligogene.

major-histocompatibility-complex genes (MHC genes) *n.* A gene group that varies widely among mice and Humans, encodes for cell-surface proteins that offer a window display of proteins being made within a cell, and encodes individual odors (Claus Wedekind in Richardson 1996, 26).

cf. chemical-releasing stimulus: semiochemical: human-male pheromone

Comment: The displayed proteins help an immune system identify cells that have been invaded by a virus or other pathogen (Richardson 1996, 26).

masking gene See gene: epistatic gene.

mimic *n.* A gene that has phenotypic effects similar to another gene that is not its allele (Lincoln et al. 1985).

cf. genocopy

mobile genetic element See transposable element.

modifier gene *n.*
1. "A gene whose phenotypic effect is to modify the effect of another gene" (Dawkins 1982, 290).
2. "A gene that modifies the phenotypic expression of a nonallelic gene" (King and Stansfield 1985).

syn. modifier, modifying factor, specific modifier (Lincoln et al. 1985)

Comments: "Many (and perhaps most) genes modify the effects of many (and perhaps most) other genes" (Dawkins 1982, 290). Modifier genes readily respond to natural selection and provide populations with flexibility to respond to sudden environmental changes (Mayr 1982, 791). Strickberger (1990, 173) considers a gene to be a modifier gene when its effects on another gene can be measured quantitatively. Some modifier genes have their own phenotypic effects (Rieger et al. 1991). Modifier genes are classified as enhancers (= intensifiers, extension genes) and reducers (= restriction genes).

multigene See gene: polygene.

mutable gene *n.* An unstable gene that mutates spontaneously with relatively high frequency during an organism's development (Lincoln et al. 1985).

muton *n.*
1. "The minimum unit of mutational change;" distinguished from cistron and recon (Dawkins 1982, 290).
2. "The mutational site within a gene" (Lincoln et al. 1985).

3. The smallest unit (a single nucleotide) that, when modified, changes an organism's genetic code (King and Stansfield 1985).

[Benzer (1957) coined this term, which is not widely used (Mayr 1982, 805).]

novelty-seeking gene *n.* A metaphor for a group of genes that underlie an animal's searching for novel experiences (Allen 1998, 24).

syn. novelty gene (Allen 1998, 24)

Comments: Three follow-up studies failed to find a novelty-seeking gene (Allen 1998, 24). According to geneticist David Dox, if there is a novelty-seeking gene, its contribution to human personality is likely to be small. The novelty-seeking gene is also related to a propensity for heroin addiction (Gould 1999, 62).

oligogene, key gene, major gene, switch gene *n.* "A gene with a major obvious phenotypic effect" (Lincoln et al. 1985), *e.g.,* on a kind of behavior.

See gene: switch gene.

operator gene *n.* The gene within an operon that is responsible for switching on, or off, the operon in which it resides (King and Stansfield 1985).

syn. regulator gene (Lincoln et al. 1985, a confusing synonym)

cf. gene: regulatory gene

ortholog *n.* A gene that is related to another particular gene by vertical descent; contrasted with paralog (Koonin and Galperin 1997, 759).

Comment: Orthologs typically retain the same function in the course of evolution, whereas paralogs acquire novel, but usually mechanistically related, functions (Koonin and Galperin 1997, 759).

outlaw gene *n.* A gene that is favored by selection at its own locus, in spite of its deleterious effects on an organism's other genes (*e.g.,* a gene involved in meiotic drive, *q.v.*) (Dawkins 1982, 291).

syn. outlaw concept (Dawkins 1982, 291)

paralog *n.* A gene that is related to another particular gene by gene duplication; contrasted with ortholog (Koonin and Galperin 1997, 759).

***Pax*-6** *n.* A regulatory gene at the top of a gene cascade that produces eyes in mice (Erwin et al. 1997, 135).

plastogene *n.* "A hereditary factor contained within a plastid" (Lincoln et al. 1985).

pleiotropic gene *n.* A gene that has more than one seemingly unrelated phenotypic effects (Lincoln et al. 1985).

cf. gene: polygene; pleiotropy

polygene *n.*
1. One of an integrated group of independent genes that collectively control the
   expression of a character or trait (Lincoln et al. 1985).
   syn. multigene, multiple factors (Lincoln et al. 1985), quantitative-trait locus
   (Strickberger 2000, 219)
2. A gene that controls quantitative characters (Lincoln et al. 1985).
cf. gene: pleiotropic gene

protogene See allele: dominant allele.

pseudogene *n.*
1. A DNA region that closely resembles the
   DNA sequence of a gene but is nonfunctional due to nucleotide additions and
   deletions that prevent normal transcription, translation, or both (King and
   Stansfield 1985; Kimura 1985, 46).
2. A defective copy of a gene that lacks
   introns and is rarely, if ever, expressed
   (Nowak 1994).
syn. dead gene (Kimura 1985, 46)
cf. allele: silent allele

reciprocal genes See gene: complementary genes.

recognition gene *n.* A gene that promotes combining of zooids into colonies
(*e.g.,* in a tunicate species) (Burnet 1971 in
Wilson 1975, 386).

recon *n.* The smallest genetic unit that
can experience recombination, as determined by cross-over location, which is an
adjacent pair of nucleotides in the *cis*
position (King and Stansfield 1985).
cf. gene: cistron, muton
[Benzer (1957) coined "recon," which is not
widely used (Mayr 1982, 805).]

regulator gene See gene: operator gene,
regulatory gene.

regulatory gene *n.*
1. A gene whose primary function is to
   control the synthesis rate of products of
   other distant genes (King and Stansfield
   1985).
   syn. regulator gene (King and Stansfield
   1985)
2. A gene whose product regulates gene
   expression via transcription initiation or
   termination (Jacob and Monod 1961 in
   Rieger et al. 1991).
3. A gene that directs RNA synthesis, which
   can directly control gene expression
   (Rieger et al. 1991).
cf. ²mutation: regulatory mutation
Comments: Regulatory genes are found in
both prokaryotes and eukaryotes (Rieger et
al. 1991). In eukaryotes, they also control
the timing and tissue specificity of the
expression of other genes; some control
only the structural gene that they are linked

to (*cis*-regulation), and others appear to
produce diffusible effector molecules that
result in *trans* regulation. Regulatory genes
may be more important in the evolution of
new adaptations than structural gene
changes. Many different classes of regulatory genes have homeoboxes, *q.v.* (Erwin
et al. 1997, 133).

segregation distorter *n.* "A gene whose
phenotypic effect is to influence meiosis so
that the gene has a greater than 50 percent
chance of ending up in a successful gamete" (Dawkins 1982, 293).
syn. segregation-distorting allele (Dawkins
1982, 293)

self-sterility gene *n.* In monoecious
plants: a gene that prevents self-fertilization by controlling the rate of pollen-tube
growth down a style (Lincoln et al. 1985).

selfish gene *n.* A gene that "does what it
can," through acting on individual organisms that carry it, to make more copies of
itself (Dawkins 1976, 95).
cf. nucleic acid: deoxyribonucleic acid:
selfish DNA
Comments: Dawkins (1976) used this term
in the context that genes, not individual
organisms, are the main units of natural
selection. The idea of a selfish gene is not
universally accepted (Gould 1981a, 13).

**sex-conditioned gene, sex-influ-
enced gene** *n.* A gene whose expression is modified by the sex of the individual
organism bearing it (Lincoln et al. 1985).

sex-limited gene *n.* A gene that is not
necessarily carried on a sex chromosome,
but it is expressed phenotypically in only
one sex (Dawkins 1982, 294).

sex-linked gene *n.* A gene on a sex
chromosome (Lincoln et al. 1985).

specific modifier See gene: modifier
gene.

split gene *n.* In some animal viruses,
Archaebacteria, and Eukaryotes: a gene
with introns between exons (King and
Stansfield 1985; Strickberger 1990, 155).
Comments: Split genes were discovered in
1977 by Phillip Sharp and Richard Roberts,
who won the Nobel Prize in 1993 for their
discovery (Travis 1993). Before this discovery, geneticists firmly believed that genes
were uninterrupted pieces of DNA.

structural gene See gene: cistron.

sublethal gene See ²mutation: subvital
mutation.

supergene *n.* A group of tightly linked
genes that are usually transmitted as a
single unit (Lincoln et al. 1985).

suppressor gene *n.* A gene that stops
phenotypic expression of a gene at a different locus (Lincoln et al. 1985).

switch gene *n*. A gene that activates a group of other genes, *e.g.*, a gene that determines which wing pattern of a mimetic butterfly will occur in an individual (Mayr 1982, 41).
See gene: oligogene.

synthetic lethal *n*. A gene, or chromosome, that conveys superior fitness in some combinations but is lethal in combination with other combinations of chromosomes (Mayr 1982, 580).
Comment: Dobzhansky's research revealed synthetic lethals and "spelled the end of the faith in constant fitness values of genes" (Mayr 1982, 580).

transposon See transposable genetic element: transposon.

♦ **gene cloning** See cloning: gene cloning.

♦ **gene complex** *n*.
1. Two or more genes that have a similar function and are sometimes closely adjacent on the same chromosome (Mayr 1982, 804).
2. An organism's entire system of interacting genetic factors (Lincoln et al. 1985).

♦ **gene-conservation guild** See guild.

♦ **gene conversion** See ³theory: transmutation.

♦ **gene dosage** *n*. The number of alleles present in a given genotype that is dependent upon its ploidy state (Lincoln et al. 1985).

♦ **gene family** See ³family: gene family.

♦ **gene fitness** See fitness: genetic fitness.

♦ **gene flow** *n*.
1. Exchange of genes between different conspecific populations caused by migrants and commonly resulting in simultaneous changes in gene frequencies at many loci in the recipient gene pool (Wilson 1975, 66, 585; King and Stansfield 1985).
Note: Gene flow that occurs between populations, artificially bred strains, and especially species is called "hybridization" (Wilson 1975; King and Stansfield 1985).
2. Genes' spreading through a gene pool due to organisms' dispersal and mating patterns occurring within, between, or among populations (Wittenberger 1981, 615).
syn. migration (Lincoln et al. 1985)
cf. hybridization, migration

island-model migration *n*. "Gene flow by periodic migration between groups of semi-isolated populations" (Lincoln et al. 1985).
cf. gene flow: river-model migration

river-model migration *n*. "Gene flow by migration along a linear series of subpopulations" (Lincoln et al. 1985).
cf. gene flow: island-model migration

♦ **gene imprinting** *n*. The phenomenon that some genes function properly only when they are donated to an offspring by its father and other genes function properly only when they are donated by its mother (Hoffman 1991, 1250).
cf. learning: imprinting

♦ **gene locus** *n., pl.* **gene loci** The specific location on a chromosome where a gene resides (King and Stansfield 1985).
syn. locus (King and Stansfield 1985)
[Latin doublet of LIEU]

quantitative-trait locus (QTL) *n*. The specific location on a chromosome where a gene that influences a quantitative character, *q.v.*, resides (Bradshaw et al. 1995, 762).
See gene: polygene.

♦ **gene machine, survival machine** *n*. Metaphorically, an individual organism used by its genes to replicate themselves (Dawkins 1976, 47).
cf. vehicle

♦ **gene mutation** See ²mutation: gene mutation.

♦ **gene pool** See pool: gene pool.

♦ **gene selfishness** *n*. A fundamental concept of biology: Genes are programmed to survive and reproduce their own kind (Dawkins 1977).
cf. gene: selfish gene
Comment: This concept is controversial.

♦ **gene substitution** *n*. Replacement of one allele by a mutant allele in a population (Lincoln et al. 1985).

♦ **geneagenesis** See parthenogenesis.

♦ **genealogy, geneology** *n*. "An ancestor-descendant lineage" (Lincoln et al. 1985).
See study of: genealogy.

♦ **genecology** See study of: ecology: genecology.

♦ **geneogenous** See -genous: geneogenous.

♦ **geneology** See genealogy.

♦ **genepistasis** *n*. One of two related form's developing further than the other (Lincoln et al. 1985).

♦ **general adaptation** See ³adaptation: ²evolutionary adaptation.

♦ **general adaptation syndrome (GAS)** See syndrome: general-adaptation syndrome.

♦ **general cognitive ability** See intelligence: general cognitive ability.

♦ **general drive** See drive: general drive.

♦ **general emergency reaction** See reaction: general-emergency reaction.

♦ **general physics** See study of: physics: general physics.

♦ **generalist, generalist species** See ²species: generalist.

♦ **generalization** See stimulus generalization.

♦ **generalized** *adj.* Referring to an individual animal, population, or species that eats more than a few species of other organisms as food.
cf. specialized; -topic: stenotopic; ²species: generalist

♦ **generation** *n.* All the individuals produced within a single life cycle of a species (Lincoln et al. 1985).
Comment: "Generation" is sometimes misused to refer to offspring from a single brood-rearing cycle (*e.g.,* in social wasps). The term "brood cohort" (or "brood cycle") is preferred for these offspring (C.K. Starr, personal communication).

♦ **generation time** See time: generation time.

♦ **generational mutualism** See symbiosis: mutualism: generational mutualism.

♦ **generative parthenogenesis** See parthenogenesis: generative parthenogenesis.

♦ **generator potential** See potential: generator potential.

♦ **genesiology** See study of: genesiology.

♦ **-genesis** *combining form* Development, descent, genesis, evolution, origin, formation (Michaelis 1963; Lincoln et al. 1985).
adj. -genetic
n. -geny
[Greek *genēsis*, origin]
abiogenesis *n.*
1. An organism's hypothetical origination, or evolution, from inanimate matter without the action of living parents (Huxley 1870 in the *Oxford English Dictionary* 1972; Mayr 1982, 582, 959).
2. The concept that life can arise spontaneously from nonliving matter by natural processes without the intervention of supernatural powers (Lincoln et al. 1985).
syn. archebiosis, archegenesis, archigenesis, archigony, autogenesis, diagenesis, heterogenesis, nomogenesis, spontaneous generation, xenogenesis (Lincoln et al. 1985)
cf. -genesis: biogenesis, nomogenesis, xenogenesis
Comments: The occurrence of spontaneous generation during human history as a regular form of reproduction was refuted by Spallanzani and Redi (Mayr 1982, 258). Before then it was a widely accepted idea. Many biologists believe that life on Earth originated by spontaneous generation, although it seems unlikely to occur today on Earth.
[coined by Huxley (1870)]
amphigenesis *n.* Fusion of two dissimilar gametes; sexual development (Lincoln et al. 1985).
anagenesis *n.* Regeneration of tissues (Lincoln et al. 1985).
See ²evolution: anagenesis.

androgenesis *n.* See -genesis: patrogenesis.
anorthogenesis *n.* "Changes in direction of evolution based on preadaptation" (Lincoln et al. 1985).
anthogenesis *n.* Production of male and female offspring by asexual forms (Lincoln et al. 1985).
anthropogenesis *n.* The descent of Humans; origin (phylogeny) and development (ontogeny) of Humans (Lincoln et al. 1985).
archegenesis, archigenesis See -genesis: abiogenesis.
aristogenesis See ²evolution: orthogenesis.
autogenesis *n.* Reproduction, or origin, within a given system or individual (Lincoln et al. 1985).
See -genesis: abiogenesis.
syn. autogeny, autogony (Lincoln et al. 1985)
behavioral ontogenesis, behavioral ontogeny See -genesis: ontogenesis: ethogenesis.
biogenesis *n.*
1. Living matter always arises from the agency of pre-existing living matter (Huxley 1870 in *Oxford English Dictionary* 1972).
[coined by Huxley (1870) as the hypothesis of biogenesis]
2. The phenomenon that many kinds of animals show ancestral traits in their embryonic development that are present in their ancestors' adult stages (Rensch 1954 in Immelmann and Beer 1989, 244) or in closely related (often less derived) species (Immelmann and Beer 1989, 244).
Note: "Behavioral biogenesis" occurs in some songbird species (Immelmann and Beer 1989, 244).
See law: biogenic law.
3. The principle that all living organisms have arisen from previously existing living organisms (Lincoln et al. 1985).
cf. -genesis: abiogenesis; law: biogenic law
▸ **neobiogenesis** *n.* The hypothesis that life originated repeatedly from inorganic substrates (Lincoln et al. 1985).
▸ **probiogenesis** *n.* "The origin of life" (Lincoln et al. 1985).
▸ **symbiogenesis** *n.* The evolutionary origins of symbiotic relationships between (among) organisms (Lincoln et al. 1985).
blastogenesis *n.* "Asexual reproduction by budding or gemmation" (Lincoln et al. 1985).
cf. parthenogenesis
bradygenesis See ²evolution: bradygenesis.

cacogenesis, kakogenesis *n*. An organism's being unable to hybridize (Lincoln et al. 1985).

catagenesis See ²evolution: catagenesis.

cenogenesis *n*. The unequal development of different organs in the same organism during its ontogeny; contrasted with palingenesis (Haeckel 1875 in Gould 1992, 160). See ²evolution: coenogenesis.

Comments: Haeckel considered cenogenesis to be a "bad" ontogenetic phenomenon because it did not support his biogenetic law (Gould 1992, 160). He called it *Fälschungsgeschichte* (falsified history) and divided it into (1) juvenile adaptation (the interpolation of new features into early ontogenetic stages for their own immediate evolutionary utility), and (2) changes in time and placement of one organ relative to others in the same body. Category 2 has two subcategories: heterotopy and ²heterochrony.

cladogenesis See coenogenesis; ²evolution: cladogenesis.

coenogenesis See ²evolution: coenogenesis.

cytogenesis *n*. "Cell production and development" (Lincoln et al. 1985).

deuterogenesis *n*. The second phase of embryonic development after gastrulation (Lincoln et al. 1985).
cf. -genesis: protogenesis

diagenesis *n*.
1. The regular alternation of sexual and asexual generations of an organism (Lincoln et al. 1985).
   syn. metagenesis, heterogenesis (Lincoln et al. 1985)
2. A mutant's appearance in a population (Lincoln et al. 1985).
See -genesis: abiogenesis.
cf. ²evolution

diplogenesis *n*.
1. "An abnormal duplication of a structure" (Lincoln et al. 1985).
2. The postulated changes in germ plasm that accompany modifications attributed to use and disuse of body organs and structures (Lincoln et al. 1985).

dysgenesis *n*. The condition of infertility between hybrids that are themselves cross-fertile with parental stocks (Lincoln et al. 1985).
adj. dysgenetic

ectogenesis *n*.
1. Variation generation by extrinsic factors (Lincoln et al. 1985).
2. Development of an embryo *in vitro* (Lincoln et al. 1985).

epigenesis, epigenetics *n*.
1. An embryo's development (ontogenesis) from undifferentiated cells into

organs that are not themselves present at conception (*Oxford English Dictionary* 1972, entries from 1807; Mayr 1982, 106, 958; Strickberger 1990, 12).
Note: Early biologists believed this differentiation of undifferentiated tissues occurred due to mystical, nonphysical forces (Strickberger 1990, 12).
2. The process in which development of each small stage in an organism's growth and differentiation from a zygote to mature adult arises from a preceding stage through the joint action of its genetic and environmental determinants; thus, an organism's phenotype, genotype, and environment "work in dynamic conjunction and progressive sequence" (Immelmann and Beer 1989, 89).
cf. -genesis: biogenesis, ontogenesis; preformationism
Comment: "Epigenesis" contrasts with "preformationism," which views an organism as prepackaged in miniature or in coded information in its zygote (thus, as a "blueprint" rather than a "recipe") (Dawkins 1982, 286)

embryogenesis *n*. Development of an embryo.

ethogenesis *n*.
1. A person's development and maturation of innate behavior during post-embryonic and early-childhood development (Heymer 1977, 64).
2. An individual animal's behavioral development as it grows (Immelmann and Beer 1989, 206).
syn. behavioral ontogeny (Immelmann and Beer 1989, 206), behavioral ontogenesis, ontogenesis of behavior
[coined by Jaisson (1974 in Heymer 1977)]

gametogenesis *n*. "The specialized series of cellular divisions that leads to the production of sex cells (gametes)" (Wilson 1975, 585).
syn. gametogeny (Lincoln et al. 1985)

gamogenesis *n*.
1. Formation of gametes (Lincoln et al. 1985).
2. Reproduction by union of gametes (Lincoln et al. 1985).
adj. -genetic
cf. gamogenetic

geneagenesis See parthenogenesis.

gonogenesis *n*. Formation of germ cells by meiotic division (Lincoln et al. 1985).

gynogenesis See parthenogenesis: gynogenesis.

haematogenesis, haemopoiesis *n*. "The process of blood formation" (Lincoln et al. 1985).

haplogenesis See ²evolution: haplogenesis.

heterogenesis See -genesis: diagenesis.
heteroparthenogenesis See partheno-genesis: heteroparthenogenesis.
histogenesis *n.*
1. Tissue formation and development (Craig 1847 in *Oxford English Dictionary* 1972).
   syn. histogeny (*Oxford English Dictionary* 1972)
2. Reorganization of body tissue during metamorphosis (Lincoln et al. 1985).
hologenesis See ²evolution: hologenesis.
homogenesis, homogeny *n.* An organism's having a succession of morphologically similar generations, not alternation of generations (Lincoln et al. 1985).
cf. -genesis: heterogenesis, hypogenesis
hybridogenesis See parthenogenesis: hybridogenesis.
hypogenesis *n.* An organism's development without alternation of generations (Lincoln et al. 1985).
kakogenesis See -genesis: cacogenesis.
katagenesis See ²evolution: catagenesis.
lipogenesis See ²evolution: lipogenesis.
lipopalingenesis See ²evolution: lipopalingenesis.
macrogenesis See speciation: saltational speciation.
macrophylogenesis See ²evolution: macroevolution.
megagametogenesis *n.* Development of megagametes (Lincoln et al. 1985).
metagenesis See -genesis: diagenesis.
microgametogenesis *n.* Development of microgametes (Lincoln et al. 1985).
monogenesis, unigenesis *n.* The development of an organism from a single cell (Lincoln et al. 1985).
See parthenogenesis: asexual reproduction.
morphogenesis *n.* The total process of embryological development and growth (Lincoln et al. 1985).
See ²evolution.
neobiogenesis See -genesis: biogenesis.
nomogenesis See hypothesis: nomogenesis.
ontogenesis, ontogeny *n.* An individual organism's growth and development from an egg, or other cell, to maturity; contrasted with phylogeny (*Oxford English Dictionary* 1972, entries from 1875).
cf. law: biogenetic law; phylogeny; prochronism
Comment: Some workers include senescence and death in "ontogenesis" (Immelmann and Beer 1989, 206). "Ontogenesis" includes whatever changes environment and habit may impose on an organism during this process (Bateson 1979, 242).

According to the doctrine of the extended phenotype, "ontogenesis" includes "the 'development' of extracorporeal adaptations, for example artifacts like beaver dams" (Dawkins 1982, 291).
[coined by Haeckel (1866 in Heymer 1977, 64)]
ontogenesis of behavior See -genesis: ethogenesis.
oogenesis, female gametogenesis, ovogenesis *n.* "The formation, development, and maturation of female gametes (ova)" (Lincoln et al. 1985).
organogenesis *n.* Organ formation during an organism's development (Lincoln et al. 1985).
orogenesis, orogeny *n.* The process of mountain formation (Lincoln et al. 1985).
orthogenesis *n.*
1. The assertion that a "nonphysical (perhaps even nonmaterial) force" drives the living world upward toward ever greater perfection (Mayr 1982, 50, 361, 883).
   Note: This is one of the autogenetic theories, *q.v.* (Lamarck, Osborn, Teilhard de Chardin in Mayr 1982, 50, 361, 883).
2. The notion that once evolution has begun to proceed in a certain direction it tends to continue in this direction due to its own momentum (Bell 1982, 510).
syn. Cartesian transformations, evolutionary trends, trends
cf. -genesis: anorthogenesis; ²evolution; ²evolution: orthogenesis, nomogenesis; progress
ovogenesis See -genesis: oogenesis.
paedogenesis, pedogenesis *n.* In some beetle and gall-midge species: parthenogenetic reproduction by a larva or pupa (Gould 1976, 24; Thornhill and Alcock 1983, 35; Borror et al. 1989, 3, 57, 538).
See ²heterochrony: progenesis.
syn. progenesis (a confusing synonym)
cf. ²heterochrony: neoteny, merostasis
[New Latin < Greek *pais, paidos*, child + GENESIS]
palaeogenesis See neoteny.
palingenesis *n.*
1. The equal and coordinated acceleration of organ development in the same organism; contrasted with cenogenesis, *q.v.* (Haeckel 1875 in Gould 1992, 160).
   Note: Haeckel considered palingenesis to be a "good" ontogenetic phenomenon, which he called *Auszugsgeschichte* (epitomized history) because it supported his biogenetic law (Gould 1992, 160).
2. Development of individual organisms already preformed within eggs (Lincoln et al. 1985).

3. The true repetition, or recapitulation, of past phylogenetic states in ontogeny of an organism's descendants (Lincoln et al. 1985).

cf. law: biogenetic law; [3]theory: theory of preformation

pangenesis See pangenesis.

paragenesis *n.* "Reproduction induced by a cross between a parent and an otherwise sterile hybrid" (Lincoln et al. 1985).

parthenogenesis See parthenogenesis.

pathogenesis *n.* The course, or development, of a disease (Lincoln et al. 1985).

patrigenesis, androgenesis, male parthenogenesis *n.* Parthenogenetic development of an organism from an enucleated egg following fusion with a normal male gamete (Lincoln et al. 1985).
See parthenogenesis: androgenesis.

pedogenesis See -genesis: paedogenesis.

perigenesis *n.* Theoretically, somatic cells are modified by their environment and activity of organs containing them, and these changes are passed as wave motions to sex cells and influence the course of an organism's heredity (Lincoln et al. 1985).

phenogenesis *n.* Development of genetically determined phenotypic characters or traits (Lincoln et al. 1985).

phylogenesis See phylogeny: phylogenesis.

physiogenesis *n.*
1. "Origin and development of physiological processes" (Lincoln et al. 1985).
2. "Cellular differentiation during ontogeny" (Lincoln et al. 1985).

phytogenesis *n.*
1. "Origin of plants on Earth" (Lincoln et al. 1985).
   syn. phytogeogenesis (Lincoln et al. 1985)
2. "Evolution and development of a plant species" (Lincoln et al. 1985).
syn. phytogeny (Lincoln et al. 1985)

probiogenesis See -genesis: biogenesis.

progenesis See -genesis: paedogenesis; [2]heterochrony: progenesis.

proterogenesis See hypothesis: proterogenesis.

protogenesis *n.*
1. "Early embryonic development to gastrulation" (Lincoln et al. 1985).
2. "Reproduction by budding" (Lincoln et al. 1985)
cf. -genesis: deuterogenesis

psychogenesis *n.* The origin and development of social instincts and behavior (Lincoln et al. 1985).
adj. psychogenetic

saturation mutagenesis *n.* A procedure that is used to make many mutations in a study animal species (*e.g.*, fruit flies or zebra fish), involving treating the animal with a chemical mutagen and searching for embryos that develop abnormally three generations later (Kahn 1994, 904).
Comment: Saturation mutagenesis was pioneered by Christiane Nüsslein-Volhard and Eric Wieschaus (Kahn 1994, 904).

schizogenesis *n.* "Reproduction by multiple fission" (Lincoln et al. 1985).
adj. schizogenetic

sociogenesis *n.* For example, in ants: the collective processes and patterns that lead to a colony's development during its life cycle (Hölldobler and Wilson 1990, 643).

somagenesis *n.* "Development of somatic structure" (Lincoln et al. 1985).

somatogenesis *n.* "The development of somatic structure" (Lincoln et al. 1985).

spermatogenesis *n.* Sperm formation (Lincoln et al. 1985).

spermiogenesis, spermiohistogenesis, spermateleosis, spermioteleosis *n.* The transformation of a spermatid into a spermatozoon (Lincoln et al. 1985).

spermogenesis See -genesis: spermatogenesis.

stasigenesis See [2]evolution: stasigenesis.

strophogenesis *n.* The evolution of alternation of generations in an organism that previously passed through a succession of similar generations (Lincoln et al. 1985).

syngenesis *n.* "Sexual reproduction" (Lincoln et al. 1985).
See [2]evolution: coenogenesis.

tachygenesis See [2]evolution: tachygenesis.

trophogenesis *n.* In the Honey Bee, ants: different caste traits' originating from differential feeding of immatures as opposed to genetic control of castes and blastogenesis (Hölldobler and Wilson 1990, 644).

tychoparthenogenesis See parthenogenesis: tychoparthenogenesis.

typogenesis *n.* A period of explosive evolution in which new forms are produced rapidly (Lincoln et al. 1985).
cf. typolysis; -stasis: typostasis

unigenesis See -genesis: monogenesis.

xenogenesis *n.* "Any unusual method of reproduction" (Lincoln et al. 1985).
See -genesis: abiogenesis.

zoogenesis See [2]evolution: zoogenesis.

zygogenesis *n.* "Reproduction during which male and female nuclei fuse" (Lincoln et al. 1985).

♦ **genet** *n.* A unit, or group, derived by asexual reproduction from a single original zygote, such as a seedling or a clone (Lincoln et al. 1985).
cf. ortet, ramet

♦ ¹**genetic** *adj.*

1. Of, referring, or based on genetics (Michaelis 1963).
2. Of, referring, or produced by genes (Michaelis 1963).
3. Innate (Lorenz 1932, 1937 in Dawkins 1986, 57).

   Note: Dawkins (1986) discusses why "innate" should not be used synonymously with "genetic" or "inherited."

cf. developmentally fixed, innate, instinct, learning

♦ ²**genetic** *adj.* Of, or referring to, the origin, generation, or development of something (Michaelis 1963).

♦ **-genetic** *combining form*
cf. ¹,²genetic

caenogenetic *adj.*

1. Referring to an organism of recent origin (Lincoln et al. 1985).
2. Referring to transitory adaptations developed in an organism's early ontogenetic stages (Lincoln et al. 1985).

digenetic *adj.* Referring to a symbiont that requires two different hosts during its life cycle (Lincoln et al. 1985).
cf. -genetic: monogenetic, trigenetic

gamogenetic, gamogenic, gamogonic, sexual *adj.* Referring to an organism produced from a union of gametes (Lincoln et al. 1985).
cf. -genesis: gamogenesis

gynogenetic *adj.* For example, in some lizard and chalcidoid wasp species, a mantid species, aphids: referring to a strain of females that produces other females without fertilization (Wilson 1975, 286).
cf. parthenogenetic

heterogenetic, heterogen *adj.* Referring to organisms that are derived from different ancestral stocks (Lincoln et al. 1985).
cf. -genetic: homogenetic

homogenetic, homogene *adj.* Referring to organisms that have a common origin or ancestral stock (Lincoln et al. 1985).
cf. -genetic: heterogenetic

karyogenetic *adj.* Referring to a trait that is "heritable, not subjected to direct environmental influences" (Lincoln et al. 1985).

monogenetic *adj.*

1. Referring to a symbiont's having only one host throughout its life cycle (Lincoln et al. 1985).
   cf. -genetic: digenetic, trigenetic
2. Referring to an asexually reproducing organism or asexual reproduction (Lincoln et al. 1985).

syn. monogenous (Lincoln et al. 1985)

morphogenetic *adj.* "Pertaining to the development of anatomical structures dur-

ing the growth of an organism" (Wilson 1975, 589).

paragenetic *adj.* Referring to a chromosomal change that affects a gene's expression rather than its constitution, such as a position effect (Lincoln et al. 1985).

palingenetic, ancestral *adj.* Referring to a trait that is remote or of ancient origin (Lincoln et al. 1985).
cf. ancestral, derived

phylogenetic *adj.* Referring to evolutionary relationships within and between organism groups (Lincoln et al. 1985).

psychogenetic See -genesis: psychogenesis.

trigenetic *adj.* Referring to a symbiont that requires three different hosts during its life cycle (Lincoln et al. 1985).
cf. -genetic: digenetic, monogenetic

♦ **genetic assimilation** *n.* "Reinforcement, by alternation in genetic material, of phenotypic modifications which themselves had no genetic basis" (Lincoln et al. 1985).
cf. effect: Baldwin-Waddington effect

♦ **genetic background** *n.* The part of an organism's genotype other than a gene, or genes, under consideration (Lincoln et al. 1985).

♦ **genetic-behavioral-polymorphism hypothesis** See hypothesis: population-limiting hypotheses: polymorphic-behavior hypothesis.

♦ **genetic-benefit polyandry** See mating system: polyandry: genetic-benefit polyandry.

♦ **genetic canalization** See canalization: genetic canalization.

♦ **genetic code** *n.* The sequence of nucleotide base pairs on an organism's DNA polynucleotide chain that encodes its genetic information (Lincoln et al. 1985).

♦ **genetic damages** *pl. n.* Physical alterations in the structural regularity of DNA such as breaks, depurinations, depyrimidinations, cross-links, thymine dimers, and modified bases (Bernstein et al. 1985, 1278).
Note: Damages usually interfere with replication and transcription, and damages are neither replicated nor inherited (Bernstein et al. 1985, 1278).

♦ **genetic death** *n.*

1. Selective elimination of a genotype carrying mutant alleles that reduce fitness (Lincoln et al. 1985).
   syn. genetic extinction (Lincoln et al. 1985)
2. An individual organism's being unable to reproduce (Lincoln et al. 1985).

cf. genetic load

♦ **genetic determinism of behavior** *n.* The opinion that a Human's basic behavior is genetically determined and should not be changed by cultural means (Mayr 1982, 598).

Comment: "E.O. Wilson and other sociobiologists have been accused of preaching the genetic determinism of behavior. This does not represent their views accurately. All they have said, and one can argue about the validity of this claim, is that much of man's social behavior has a genetic component" (Mayr 1982, 598).

♦ **genetic disharmony** *n.* Incompatibility between parental genes manifested in progeny (Lincoln et al. 1985).

♦ **genetic distance** See distance: genetic distance.

♦ **genetic dominance** See dominance (genetic).

♦ **genetic drift** See drift: genetic drift.

♦ **genetic engineering** *n.*
1. Human experimental manipulation of the genetic composition of an organism or cell (Lincoln et al. 1985).
   syn. algeny
2. Human manipulation of genes for practical purposes (Campbell et al. 1999, 364).

♦ **genetic epistemology** See developmental-genetic conception of ethical behavior.

♦ **genetic equilibrium** See law: Hardy-Weinberg law.

♦ **genetic equivalence of all chromosomes, theory of** See ³theory: theory of genetic equivalence of all chromosomes.

♦ **genetic extinction** See genetic death.

♦ **genetic factor** See gene.

♦ **genetic fitness** See fitness: genetic fitness.

♦ **genetic heritability** See ²heritability (comments).

♦ **genetic homeostasis** See homeostasis: genetic homeostasis.

♦ **genetic load** *n.*
1. Human genes that are deleterious for our welfare (Muller 1950 in Crow 1992, 132). [after "Our Load of Mutations," a 1950 paper by Muller (Crow 1992, 132)]
2. All fitness-reducing genes in any population; those that are maintained by mutation or by other mechanisms (Dobzhansky 1950 in Crow 1992, 132).
   Note: A load is the consequence of normal genetic variability; the less favorable genes and genotypes constitute a load (Crow 1992, 132).
3. The reduction in fitness of a gene locus, chromosome, individual organism, deme, population, or species due to genetic polymorphism (Wilson 1975, 71).
   Note: Genetic loads are static; they refer to a population at equilibrium for the load-determining forces (Crow 1992, 135).
4. "The average loss of genetic fitness (*q.v.*) in an entire population due to the pres-

ence of individuals less fit than others" (Wilson 1975, 585).
5. The average number of lethal equivalents (potential genetic deaths) per individual in a population, viewed as the accumulated depression of fitness from a theoretical optimum, caused by deleterious genes, and comprised of mutational, segregational, and substitutional load (Lincoln et al. 1985).

syn. load (Crow 1992, 132)
Comments: Genetic load can be precisely defined mathematically (Crow 1992, 132). Any factor that leads to variability in fitness can create a genetic load (Crow 1992, 134).

balanced load See genetic load: segregational load.

cumulative substantial load See cost: cost of selection.

drift load *n.* Genetic load caused by random gene-frequency drift in small populations (Crow 1992, 134).

dysmetric load *n.*
1. Genetic load caused from a mismatch between genes and their environment (Crow 1992, 132).
2. Genetic load caused by less than optimum allocation of different genotypes to appropriate environments (Crow 1992, 134).

hidden load *n.* Genetic load caused by a gene whose effect is not manifest (*e.g.,* a recessive gene whose effect is concealed by heterozygosity with a favorable dominant) but might be expressed at a later time or brought out by special techniques such as inbreeding (Crow 1992, 132).

lag load See ²evolutionary lag.

incompatibility load *n.* Genetic load caused by mother-fetus incompatibility (Crow 1992, 134).

meiotic-drive load *n.* Genetic load caused by deleterious genes maintained by meiotic, or gametic, advantage (Crow 1992, 134).

mutational load *n.* Genetic load caused by deleterious mutations in an organism's genome (Haldane 1937 in Crow 1992, 132; Bernstein et al. 1985, 1279).

syn. mutation load (Crow 1992, 132)
Comment: Haldane regarded mutational load as the price a species paid for the privilege of evolution; without mutation, there would be no evolution (Crow 1992, 133).

recombinational load *n.*
1. The loss in an organism's fitness caused by gene randomization from outcrossing (Bernstein et al. 1985, 1280).
2. The genetic load caused by breakup of favorable linked epistatic gene combinations (Crow 1992, 134).

See cost: cost of recombination.

d – g

syn. cost of recombination
cf. genetic load

segregational load *n.* Genetic load due to a particular gene's being favored in a heterozygous, but not homozygous, condition (Lincoln et al. 1985; Crow 1992, 132). *syn.* balanced load [after the balance between mutation and natural selection, Dobzhansky (1955 in Crow 1992, 132)], cost of natural selection (not a preferred synonym; Haldane 1957 in Crow 1992, 135), segregation load (Crow 1992, 134)
Comment: The segregation load is the price a species pays for Mendelian inheritance; an asexual population has no segregation load (Crow 1992, 134).

substitutional load, transitional load *n.* Genetic load due to one allele's being replaced by another during evolutionary change (Lincoln et al. 1985).
Comment: Substitutional load is the expression of the cost of selection, *q.v.,* as a kind of genetic load (Ridley 1996, 160).

total load *n.* Hidden plus expressed genetic loads (Morton et al. 1956 in Crow 1992, 132).

transitional load See genetic load: substitutional load.

♦ **genetic map, genetic linkage map** See map: genetic map.

♦ **genetic material** See gene.

♦ **genetic mosaic** See chimera: genetic mosaic.

♦ **genetic polymorphism** See -morphism: polymorphism: genetic polymorphism.

♦ **genetic program** See program: genetic program.

♦ **genetic psychology** See study of: psychology: genetic psychology.

♦ **genetic recombination** See recombination.

♦ **genetic response** See response: genetic response.

♦ **genetic revolution** See revolution: genetic revolution.

♦ **genetic selection** See selection: genetic selection.

♦ **genetic species concept** See ²species.

♦ **genetic surgery** *n.* A therapeutic alteration of the genotype of a specific tissue to cure a disease (Williams and Nesse 1991, 13).

♦ **genetic swamping** *n.* An influx of genes (= gene flow) from a nearby population adapted to other circumstances than the population receiving the genes (Wilson 1975, 34).

♦ **genetic system** See ¹system: genetic system.

♦ **genetic variance** See variance: genetic variance.

♦ **genetical theory** See altruism.

♦ **genetics** *pl. n.* An organism's inherited traits (Michaelis 1963).
See study of: genetics; -type: genotype.

♦ **genic** *adj.* Referring to, characteristic of, or produced by one or more genes (Michaelis 1963).

♦ **-genic** *combining form* Related to generation or production; producing, produced by (Michaelis 1963; Lincoln et al. 1985).
n. **-geny**

aitiogenic, aitiogenous, aitionomic, aitionomous, paratonic *adj.* Referring to an individual organisms's movement, or reaction, induced by an external stimulus (Lincoln et al. 1985).
See paratonic.

allelogenic, allelogenous *adj.* Referring to an organism that produces offspring in broods that are entirely of one sex (Lincoln et al. 1985).
cf. -genic: amphogenic, arrhenogenic, monogenic, thelygenic

amphogenic *adj.*
1. Referring to an organism that produces sons and daughters, usually in approximately equal numbers (Bell 1982, 501; Lincoln et al. 1985).
   cf. allelogenic, arrhenogenic
2. Referring to cecidomyiids (flies) that produce both thelytokous female larvae and sexual male larvae (Bell 1982, 501).
cf. genic: arrhenogenic, monogenic, thelygenic

anthropogenic *adj.* Referring to something caused, or produced, by Humans (Lincoln et al. 1985).

aristogenic See -genic: eugenic.

arrhenogenic *adj.* Referring to an organism that has offspring that are entirely, or predominantly, male (Lincoln et al. 1985).
cf. genic: amphogenic, monogenic, thelygenic

audiogenic *adj.* Referring to an organism's response that results from an auditory stimulus (*e.g.,* audiogenic seizures) (Dewsbury 1978, 117).

cacogenic See -genic: dysgenic.

carcinogenic *adj.* Cancer producing.

cryptogenic *adj.*
1. Referring to a fossil species of unknown, or obscure, phylogenetic descent from species occurring in earlier geological formations (Lincoln et al. 1985).
2. "Of indeterminate descent" (Lincoln et al. 1985).
cf. phanerogenic

digenic *adj.* Referring to traits controlled by the integrated action of two genes (Lincoln et al. 1985).
syn. digenous
cf. -genic: monogenic, oligogenic, polygenic, trigenic

diphygenic *adj.* Referring to an organism that has two types of development (Lincoln et al. 1985).

dysgenic, cacogenic, kakogenic *adj.* Referring to, or having, the capacity for decreasing the fitness of a race or breed (Lincoln et al. 1985).
cf. -genic: eugenic

eugenic, aristogenic *adj.* Referring to, or having, the capacity for increasing the fitness of a race or breed (Lincoln et al. 1985).
cf. -genic: dysgenic

gametogenic *adj.* Referring to a variation that arises from spontaneous changes in a gametic chromosome (Lincoln et al. 1985).

gamogenic See -genetic: gamogenetic.

heterogenic *adj.* Referring to a population, or gamete, that contains more than one allele at a given locus (Lincoln et al. 1985).

hysterogenic *adj.* Referring to a structure, or trait, that appears late in an organism's development (Lincoln et al. 1985).

intergenic *adj.* Referring to a mutation that involves more than one gene (Lincoln et al. 1985).

intragenic *adj.* Referring to something within the same gene (Lincoln et al. 1985).

isogenic, isogeneic, homozygous, syngeneic, syngenic *adj.* Referring to individuals that have the same set of genes (Lincoln et al. 1985).
cf. allogenic

kakogenic See -genic: dysgenic.

monogenic *adj.*
1. Referring to traits controlled by a single gene (Lincoln et al. 1985).
   syn. monofactorial, unifactorial (Lincoln et al. 1985)
   cf. -genic: digenic, oligogenic, polygenic, trigenic
2. Referring to an organism that produces only sons or only daughters (Lincoln et al. 1985).
   n. monogeny
   cf. -genic: allelogenic, amphogenic, arrhenogenic, thelygenic

multigenic See -genic: polygenic.

nectogenic *adj.* Referring to something derived from nekton (Lincoln et al. 1985).

nosogenic, pathogenic *adj.* Referring to a disease-producing, or pathogenic, agent (Lincoln et al. 1985).

oligogenic *adj.* Referring to traits controlled by only a few genes (Lincoln et al. 1985).
cf. -genic: digenic, monogenic, polygenic, trigenic

oncogenic *adj.* Referring to a tumor-causing agent (Lincoln et al. 1985).

ornithogenic *adj.* Referring to sediments that are rich in bird droppings (Lincoln et al. 1985).

pathogenic See -genic: nosogenic.

phanerogenic *adj.*
1. Referring to a fossil species that has an established phylogeny from species occurring in earlier geological formations (Lincoln et al. 1985).
2. Referring to a biological phenomenon of known descent (Lincoln et al. 1985).
   cf. -genic: cryptogenic

photogenic *adj.* Referring to a light-producing agent (Lincoln et al. 1985).
See bioluminescent.

phytogenic *adj.* Arising from, or caused by, plants (Lincoln et al. 1985).

polygenic, polygenetic, multigenic, polyergistic, polyfactorial *adj.* Referring to traits controlled by the integrated action of multiple independent genes (Lincoln et al. 1985).
cf. -genic: digenic, monogenic, trigenic

protogenic *adj.* Referring to structures that persist from an organism's very early developmental stages (Lincoln et al. 1985).

psychogenic *adj.*
1. Referring to something caused by mental activity (Lincoln et al. 1985).
2. Referring to something originating in a mind (Lincoln et al. 1985).

saprogenic, saprogenous *adj.* Something's causing, or something caused by, the decay of organic matter (Lincoln et al. 1985).
n. saprogen

syngenic See -genic: isogenic.

thelygenic *adj.*
1. Referring to isopod crustaceans that produce only daughters (Bell 1982, 513).
2. Referring to an organism that has offspring that are entirely, or predominantly, female (Lincoln et al. 1985).
   syn. thelygenous (Lincoln et al. 1985)
   cf. -genic: allelogenic, amphogenic, arrhenogenic, monogenic; -genous: gynogenous

trigenic *adj.* Referring to traits controlled by the integrated action of three genes (Lincoln et al. 1985).
cf. -genic: monogenic, digenic, oligogenic, polygenic

zoogenic, zoogenous *adj.* Referring to something produced by, or associated with, the activity of animals (Lincoln et al. 1985).
cf. -morphic: zoomorphic; -morphism: zoomorphism

♦ **genic balance** *n.* The sum of the combined effects of all of an organism's genes that interact to produce a given phenotypic trait (Lincoln et al. 1985).

d – g

♦ **genic selection** See selection: genetic selection.

♦ **genital display, genital presentation** See display: genital display.

♦ **genital gland** See gland: genital gland.

♦ **genital pouch** See organ: copulatory organ: bursa copulatrix.

♦ **genital touching** *n*.

1. In some tribes of New Guinean Humans: one's touching a person's scrotum with an upward movement as part of a greeting (Eibl-Eibesfeldt 1977 in de Waal 1989, 79).

2. In the Chimpanzee: one male's fingering another's scrotum at moments of mild tension (de Waal 1989, 79).

syn. ball bouncing (colloquial) (de Waal 1989, 79)

♦ **genitalic recognition hypothesis** See hypothesis: hypothesis of divergent evolution of animal genitalia: genitalic-recognition hypothesis.

♦ **genito-genital rubbing, GG-rubbing** *n*. In the Bonobo: two females' holding their bellies together and faces close together and rubbing their genital swellings together with rapid sideways movements (de Waal 1989, 201–202).

♦ **genocline** See cline: genocline.

♦ **genocopy** *n*. The phenotypic expression of one gene that resembles that of another (nonallelic) gene (Lincoln et al. 1985).

♦ **genodeme** See deme: genodeme.

♦ **genoecodeme** See deme: genoecodeme.

♦ **genome** *n*.

1. An organism's complete genetic constitution (Wilson 1975, 585).

    syn. total genetic complement (Stebbins and Ayala 1985, 73)

2. "The complete set of genetic materials carried by an organism in its primary sex cell" (Wittenberger 1981, 616).

    Note: Viruses have from 1300–20,000 nucleotide pairs; bacteria, 4 million; fungi, 10–20 million; most animals and plants, several billion; a few groups of higher plants, salamanders, and some primitive fishes, 10^{10}; certain species of amoeba and Psilophyta (plant phylum), 10^{12} (Stebbins and Ayala 1985, 73).

3. The minimum set of nonhomologous chromosomes required for a cell's proper functioning (Lincoln et al. 1985).

4. A particular species' basic (monoploid) set of chromosomes (Lincoln et al. 1985).

syn. gametic chromosome number, genom (Lincoln et al. 1985)

cf. gene

ancestral mitochondrial genome *n*. A mitochondrial genome that has retained clear vestiges of its eubacterial ancestry; contrasted with derived mitochondrial genome (Gray et al. 1999, 1477).

Comments: The contrast between ancestral and derived mitochondrial genomes is not always clear and sharp in some taxa (Gray et al. 1999, 1477). *Reclinomonas americana* is the prototypical example of an ancestral mitochondrial genome, with its 69,034 base pairs of mtDNA (http://megasun.bch. umontreal.ca/ogmp/projects/ramer/ramer. html).

derived mitochondrial genome *n*. A mitochondrial genome that radically departs from an ancestral one, having little or no evidence of retained ancestral traits, and usually being reduced in overall size; contrasted with ancestral mitochondrial genome (Gray et al. 1999, 1477).

Comments: Animal and most fungal mtDNAs, apicomplexa such as *Plasmodium* sp., and the highly atypical mtDNAs of green algae are derived mitochondrial genomes (Gray et al. 1999, 1477).

mitochondrial genome *n*. All of the genes in a mitochondrion (Gray et al. 1999, 1478).

Comments: Phylogenetic evidence derived from both SSU rRNA and protein data and gene arrangements support the hypothesis that all mitochondrial genomes descended from a single protochondrial ancestor (Gray et al. 1999, 1478).

♦ **genome blotting** See method: Southern blotting.

♦ **genome library** See library: genomic library.

♦ **genomic bank** See library: genomic library.

♦ **genomic library** *n*. A set of thousands of DNA segments that are from a genome and are each carried by a plasmid or phage (Campbell 1990, 402-403, G-10).

♦ **genomics** See study of: genomics.

♦ **genomorph** See -morph: genomorph.

♦ **genonomy** See study of: systematics: biosystematics.

♦ **genophenes** See -type, type: phenotype.

♦ **genophyletic** *adj*. Referring to a group of taxa inferred to have greater genotypic similarity to one another than to members of any other groups (Lincoln et al. 1985). *n*. genophyly

♦ **genospecies** See ²species: genospecies.

♦ **genotype** See -type, type: genotype.

♦ **genotypic adaptation** See ³adaptation: [1,2]evolutionary adaptation.

♦ **genotypic distance** *n*. A measure of the disparity between two genotypes in terms of the probability that two individuals differ with respect to a gene at a particular locus (Lincoln et al. 1985).

♦ **genotypic sex determination** See sex determination.

♦ **genotypic sterility** *n*. Hybrid sterility resulting from an imbalance in its zygote's genotype (Lincoln et al. 1985).

♦ **-genous** *suffix to adjectives*
  1. Generating; yielding (Michaelis 1963).
  2. Produced, or generated, by (Michaelis 1963).
  [GEN- (that which is produced < French *-gène* < Greek *-genēs*) + -OUS]

allogenous *adj*. Referring to one species' genetic change produced from its incorporation of another species' genes which leads to phyletic evolution in a lineage (Mayr 1963, 429).
  cf. -genous: autogenous

anautogenous, nonautogenous *adj*. Referring to a female insect that must feed before she can produce mature eggs (Lincoln et al. 1985).

androgenous *adj*. Pertaining to an organism that produces only sons (Lincoln et al. 1985).
  cf. -genous: gynogenous

autogenous *adj*.
  1. Referring to genetic change (mutation, recombination, selection, etc.) that leads to phyletic evolution in a lineage (Mayr 1963, 429).
  cf. -genous: allogenous
  2. Referring to a female insect that does not have to feed in order to facilitate maturation of her eggs (Lincoln et al. 1985).
  cf. -genous: anautogenous

biogenous *adj*. Living on, or in, other organisms (Lincoln et al. 1985).
  syn. parasitic, symbiotic (Lincoln et al. 1985)

carpogenous *adj*. Growing on, or in, fruit (Lincoln et al. 1985).

coenogenous, cenogenous *adj*. Referring to an animal that produces offspring oviparously during one season and ovoviparously during another season of a year (Lincoln et al. 1985).

digenous See -genetic: digenetic; sexual: bisexual.

ectogenous, ectogenesis, ectogenic *adj*. Arising, or originating, outside an organism or system (Lincoln et al. 1985).
  cf. -genous: endogenous

endogenous, endogenetic *adj*.
  1. Referring to something that arises, or originates, within an organism or system (Lincoln et al. 1985), *e.g.*, behavioral causation, or control, that is located, or originates, within an animal's body. "Endogenous" does not necessarily imply a genetic basis of a factor (Immelmann and Beer 1989, 87).

  2. Referring to something that grows inside something else (Lincoln et al. 1985).
  syn. autochthonous, intrinsic (a psychological synonym, Immelmann and Beer 1989, 87)
  cf. ectogenous; exogenous; rhythm: endogenous rhythm

entomogenous *adj*. Living in, or on, insects (Lincoln et al. 1985).

epigenous *adj*. Referring to an organism that grows, or develops, on a surface (Lincoln et al. 1985).
  cf. epiphyte; zoon: epizoon

exogenous *adj*.
  1. Referring to something that originates outside an organism or system (Lincoln et al. 1985).
  2. Referring to something that is due to, or triggered by, external environmental factors (Lincoln et al. 1985).
  syn. allochthonous, ectogenous, xenogenous (Lincoln et al. 1985)
  cf. -genous: endogenous

geneogenous *adj*.
  1. Referring to a character that is present when an organism is born (Lincoln et al. 1985).
  syn. congenital
  2. Referring to an organism's condition that has resulted from an embryonic aberration (Lincoln et al. 1985).

gynogenous *adj*. Referring to an organism that produces only daughters (Lincoln et al. 1985).
  cf. -genic: thelygenic; -genous: androgenous

homogenous *adj*. Referring to organisms that are more or less similar owing to descent from a common stock (Lincoln et al. 1985).

hypogenous *adj*. Referring to an organism that lives, or grows, underneath an object or on its lower surface (Lincoln et al. 1985).

monogenous See -genetic: monogenetic.

necrogenous *adj*.
  1. Referring to an organism that grows on, or inhabits, dead bodies (Lincoln et al. 1985).
  2. Referring to organisms, or other factors, that promote decay (Lincoln et al. 1985).

nonautogenous See -genous: anautogenous.

psammogenous *adj*.
  1. Referring to an agent that produces sand (Lincoln et al. 1985).
  2. Referring to an agent that forms sandy soils (Lincoln et al. 1985).

rhexigenous, rhexogenous *adj*. Produced as a result of breakage or rupture (Lincoln et al. 1985).

saxigenous See -cole: saxicole.

d – g

xenogenous See -genous: exogenous.

♦ **genovariation** See ²mutation: gene mutation.

♦ **gens** *n., pl.* **gentes**
1. In the European cuckoo, *Cuculus canorus*: a group of females within a population that lay their eggs primarily in the nests of a single host species whose eggs theirs mimic (Wilson 1975, 585).
2. "A species group" (Lincoln et al. 1985).
3. "A distinct evolutionary lineage" (Lincoln et al. 1985).

♦ **genuine synapomorphy** See character: homologous synapomorphy.

♦ **-geny** *combining form* Mode of production of; generation, or development, of (Michaelis 1963).
adj. -genic, genous
[French -*génie* < Latin -*genia* < Greek -*geneia* < *gen-*, stem of *gignesthai*, to become]
cyclogeny See -morphosis: cyclomorphosis.
digeny See sexual reproduction.
endodyogeny *n.* A rare type of sexual reproduction shown by some protistans in which two daughter cells develop within a single parent cell that is destroyed in the process (Lincoln et al. 1985).
endopolygeny, endopolygony *n.* Formation of many daughter cells, each surrounded by her own membrane within a mother cell (Lincoln et al. 1985).
gametogeny See -genesis: gametogenesis.
gynecogeny See parthenogenesis.
herpetogeny *n.* "The history of colonization and evolution in establishment of modern amphibian and reptile faunas" (Lincoln et al. 1985).
histogeny See -genesis: histogenesis.
ontogeny See -genesis: ontogenesis.
thelygeny *n.* An organism's producing offspring that are entirely, or predominantly, female (Lincoln et al. 1985).
cf. -genic: allelogenic, amphogenic, arrhenogenic, monogenic

♦ **geo-** *combining form* "Earth; ground; soil" (Michaelis 1963).

♦ **geoaesthesia** See perception: graviperception.

♦ **geobenthos** See ²community: benthos: geobenthos.

♦ **geobiology** See study of: biology: geobiology.

♦ **geobiont** See biont: geobiont.

♦ **geobios** See -bios: geobios.

♦ **geobiotic** See -biotic: geobiotic.

♦ **geobotany** See study of: geography: biogeography: phytogeography.

♦ **geochronology** See study of: geochronology

♦ **geochronometry** See -metry: geochronometry.

♦ **geochrony** See study of: geochronology.

♦ **geocline** See -cline: geocline.

♦ **geocole** See -cole: geocole.

♦ **geocosmology** See study of: geocosmology.

♦ **geodyte** See -cole: terricole.

♦ **geoecology** See study of: geology: geoecology.

♦ **Geoffroyism** See inheritance of acquired characters; hypothesis: Geoffroyism.

♦ **geographic barrier** *n.*
1. A large landmark, such as a mountain, chain of mountains, or body of water, that isolates populations of a species, or different species, from one another (Mayr 1982, 601).
Note: Geographic barriers are important factors that promote allopatric speciation (Mayr 1982, 601).
2. "Any geographical feature that prevents gene flow between populations" (Lincoln et al. 1985).
syn. geographical barrier (Lincoln et al. 1985)

♦ **geographic-character gradient** See cline.

♦ **geographic race** See ²species: subspecies.

♦ **geographic speciation** See speciation: geographic speciation.

♦ **geographic variation** See variation: geographic variation.

♦ **geographical barrier** See geographic barrier.

♦ **geographical ecology** See study of: ecology: geographical ecology.

♦ **geographical equivalents, syngeographs** *n.* Two or more taxa that occupy the same geographical area and have patterns of distribution that are nearly congruent (Lincoln et al. 1985).
See isolation: geographical isolation.

♦ **geographical isolating mechanism** See mechanism: isolating mechanism: geographical isolating mechanism.

♦ **geographical isolation, geographic isolation** See isolation: geographic isolation

♦ **geographical race** See variety.

♦ **geographical speciation** See speciation: allopatric speciation.

♦ **geography** See study of: geography.

♦ **geologic range** See ¹range: geologic range.

♦ **geological chronology** See study of: geochronology.

♦ **geological stratum** *n.* A layer of rocks with diagnostic characteristics including particular fossil species.
See "fossil site" for many of the strata with fossils.

- **geological time** *n*. Time intervals from the Earth's origin (4.6 Byr) to present.
  Comments: The times of the intervals vary among published references. Strickberger (1996) lists two eons, six eras, periods, and epochs.
- **geology** See study of: geology.
- **geonemy** See study of: geography: biogeography.
- **geoperception** See graviperception.
- **geophage** See -phage: geophage.
- **geophile** See -¹phile: geophile.
- **geosere** See ²community: geosere; ²succession: geosere.
- **geotaxis** See taxis: geotaxis.
- **geotectic geology** See study of: geology: structural geology.
- **geotropism** See -tropism: geotropism.
- **geoxene** See -xene: geoxene.
- **geratology** See study of: geratology.
- **germ cell** See cell: germ cell.
- **germ-line** *n*. The part of an organism that "is potentially immortal in the form of reproductive copies: the genetic contents of gametes and of cells that give rise to gametes" (Dawkins 1982, 287).
  cf. soma
- **germ-line replicator** See replicator (comment).
- **germ-line theory** See ³theory: germplasm theory.
- **germ plasm** *n*. The genetic material within a cell (Mayr 1982, 670).
  syn. idioplasm (Nägeli 1884 in Mayr 1982, 670, 958), gonoplasm, *stirp* (Galton 1876 in Mayr 1982, 696)
  [Weismann coined "germplasm" in 1883 (Mayr 1982, 670).]
- **germ-plasm theory** See ³theory: germplasm theory.
- **germinal** *adj*. Referring to, or something's influencing, germ cells (Lincoln et al. 1985).
- **germule** *n*. "A unit of colonization or migration" (Lincoln et al. 1985).
- **gerontic, gerontal** *adj*. Referring to later stages of phylogeny or ontogeny (Lincoln et al. 1985).
- **gerontology** See study of: gerontology.
- **gerontomorphosis** See -morphosis: gerontomorphosis.
- **gerontophage** See -phage: gerontophage.
- **gerontophagy** See -phagy: gerontophagy.
- **gestalt, Gestalt** *n., pl.* **gestalts, Gestalten**
  1. A shape, configuration, or structure that a person perceives as a specific whole, or unity, that one cannot express simply in terms of its parts (*e.g.,* a melody as distinguished from its individual notes) (*Oxford English Dictionary* 1989, entries from 1890).

  2. A functional configuration, or synthesis, of separate elements of emotion, experience, sign stimuli, etc. (Tinbergen 1951, 78) that constitutes more than the mechanical sum of their parts (Michaelis 1963).
  3. A pattern of biological phenomena in which the properties of a functional whole differ from those predicted from the sum of its component parts (Lincoln et al. 1985).
  syn. holism (Lincoln et al. 1985)
  cf. holism; perception: gestalt perception; reductionism
  [German *Gestalt* form, shape]
- **gestalt model** See ⁴model: gestalt model.
- **gestalt odor** See odor: gestalt odor.
- **gestalt perception** See perception: gestalt perception.
- **gestalt psychology** See study of: psychology.
- **gestation** *n*. The duration of development of an embryo within a viviparous animal's uterus, from conception to birth (Lincoln et al. 1985).
  v.t. gestate
- **gesture** *n*.
  1. A person's manner of carrying his body; bearing; carriage; deportment (*Oxford English Dictionary* 1972, entries from 1410).
  2. A person's manner of placing his body; position; posture; attitude (*Oxford English Dictionary* 1972, entries from 1533).
  3. A person's movement of his body, or limbs, as an expression of feeling (*Oxford English Dictionary* 1972, entries from 1545).
  4. A person's movement of his body or any part of it; now in the strict meaning of a movement that expresses thought or feeling (*Oxford English Dictionary* 1972, entries from 1551).
  5. An individual animal's expressive movement, or body posture, that involves its torso, extremities, and appendages (*e.g.,* tail) (Immelmann and Beer 1989, 120).
  cf. display, facial expression
  Comment: In Humans who are hearing and speech impaired, gestures can become conventionalized into a gesture language (Drever 1974).
- **gesture of embarrassment** *n*. Partial covering of one's face in persons who are mildly embarrassed (*e.g.,* in all human cultures) (Heymer 1977, 189).
- **GHB** See drug: gamma hydroxybutyrate.
- **gibberellins** See hormone: gibberellins.
- **gift** See nuptial gift.
- **Giftsterzeln** *n*. In the Honey Bee: a guard's raising her abdomen, exposing her sting, releasing an alarm pheromone, and fanning her wings, which aid in pheromone dissemination (Maschwitz 1964 in Heymer 1977, 78).
  [German]

♦ **gigantism** *n.* An organism's being much larger than normal or exhibiting excessive growth, often associated with polyploidy (*Oxford English Dictionary* 1972, entry from 1885; Lincoln et al. 1985).
adj. gigantic
cf. nanism

♦ **Gilbertian mimicry** See mimicry: Gilbertian mimicry.

♦ **gill-brooding** See brooding: gill-brooding.

♦ **gill-cover hypothesis of insect wing origin** See hypothesis: hypotheses of the origin of insect wings: gill-cover hypothesis of insect wing origin.

♦ **gilt** See animal names: gilt.

♦ **glade** See habitat: glade.

♦ **gland** *n.*
1. In animals: an organ that is composed of nucleated cells, is from simple to complex in structure, and removes certain constituents from blood for body use or ejection (*Oxford English Dictionary* 1972, entries from Ray 1692).
2. In plants: a secreting cell, or group of cells, on the surface of a structure (*Oxford English Dictionary* 1972, entries from Martyn 1785).
3. In animals: an organ that is composed of modified epithelial cells and specialized to produce one or more secretions discharged outside of itself (Michaelis 1963; Curtis 1983, 1096).
Note: This definition seems appropriate for wide use.
4. In animals: a gland-like structure (*e.g.,* lymph gland) (Michaelis 1963).
cf. chemical-releasing stimulus: semiochemical; hormone
Comments: Animal glands (*e.g.,* adrenals, spleen, thymus, and thyroid) have no ducts and hence are called "ductless glands" or "aporic glands." Glands are named for their body position (*e.g.,* cervical or iliac) or for their discoverers (*e.g.,* Dufour's gland). Barnes (1974) describes many kinds of glands found in invertebrates. Some of the many kinds of glands are defined below.
[Greek *glans, glandis,* acorn]

adrenal gland *n.* In vertebrates: an endocrine gland that produces cortisol, aldosterone, and other steroid hormones in its cortex and adrenaline and noradrenaline in its medulla (Curtis 1983, 1086).

anal gland *n.* For example, in the Domestic Dog, European Rabbit (*Oryctolagus cuniculus*), Phalanger (*Petaurus breviceps*), and Wolf; Dolichoderine Ants: a gland near an animal's anus that produces a territory-marking pheromone (Wilson 1975, 205, 231, 509).

antebrachial gland *n.* For example, in male Ring-Tailed Lemurs: a forearm gland that produces a odorous substance used in stink fights, *q.v.* (Wilson 1975, 531).
cf. gland: brachial gland

apocrine gland *n.* In Humans: a gland found especially under arms, on breasts, and in genital regions, that produces chemicals fed on by lipophilic diphtheroid bacteria that transform the chemicals into human-perspiration odor (Curtis 1983, 805; Booth 1990).
syn. apocrine-sweat gland (Curtis 1983, 805), human-scent gland (Booth 1990)
cf. odor: human underarm odor
Comment: Human pubic and axillary hair may function to retain and amplify human odors (Curtis 1983, 805).

appeasement gland *n.* In *Atemeles pubicollis* (beetle): a gland near an individual's abdominal tip that produces a material that seems to calm ants that feed on it (Hölldobler 1967–1971 in Wilson 1975, 375).

brachial gland *n.* In male Ring-Tailed Lemurs: a gland high on the chest that produces an odorous substance (Wilson 1975, 531).

cloacal gland *n., pl.* **cloacae** For example, in the African Toad; most reptile species; many bird species: a male's, or female's, gland that is brought together with one of the opposite sex during mating in a cloacal kiss, *q.v.* (Dewsbury 1978, 77; Curtis 1983, 840).
syn. cloaca (Curtis 1983, 840)
Comment: Depending on the species, a cloaca is an exit chamber from an animal's digestive system and the exit for an animal's reproductive and urinary systems (Curtis 1983, 1091).
[Latin *cloaca,* sewer]

Dufour's gland *n.* In female Hymenoptera: an abdominal gland that produces different substances with markedly different functions, depending on the species (*e.g.,* a sex pheromone in the ant *Xenomyrmex floridanus*); offensive and defensive chemicals, alarm pheromone, and a propaganda substance in the ants *Formica pergandei* and *F. subintegra*; a nest lining in some sweat-bee species; and a polyethylene used in lining brood cells in the bee *Colletes thoracicus* (Dufour 1835, Wheeler 1910, Talbot and Kennedy 1940 in Wilson 1975, 370; Regnier and Wilson 1971, 269; Hölldobler 1971b in Wilson 1975, 141; Hefetz et al. 1979; Barrows et al. 1986).
syn. sebiferous gland, silk gland (Dufour 1835), alkaline gland (Carlet 1884, 1890),

glande appendiculaire, glande de Dufour,
basic gland (Bordas 1894, Semichon 1906)
Comment: Carlet (1884, 1890) states that
Dufour used the names "sebiferous gland"
and "silk gland" for the gland we now call
"Dufour's gland."
[named after L. Dufour, biologist, by later
workers]

endocrine gland *n.* An often ductless
gland that synthesizes, stores, or secretes
hormones, or a combination of these activities (*e.g.,* adrenal, pituitary, and sex glands);
contrasted with exocrine gland (Wilson
1975, 583; Immelmann and Beer 1989, 86).
[Greek *endon,* within + *krinein,* to separate]

exocrine gland *n.* A gland that secretes
to the outside of an animal's body or into its
alimentary tract (*e.g.,* salivary and sweat
glands); contrasted with endocrine gland
(Curtis 1983, 1094).
Comment: Exocrine glands are the most
common sources of pheromones (Wilson
1975, 584).
[Greek *ex,* out of + *krinein,* to separate]

forehead gland *n.* For example, in
Black-Tailed Deer: a pheromone-produc-
ing gland in an individual's forehead (Müller-
Schwarze 1971 in Wilson 1975, 233).

genital gland *n.* In the Wolf: a gland in
the genital region that appears to produce
a pheromone (Wilson 1975, 509).

gland of construction *n.* In many (perhaps all) paper-making wasps: a gland in a
worker's head that produces a fluid that
apparently acts as the adherent in preparation of nest carton from fibers and water
(Evans and West Eberhard 1970, 166).

honeydew gland *n.* For example, in
Ross' Metalmark Butterfly: one of a pair of
glands on the posterior part of a caterpillar
that produces a sweet substance used by
ants (Ross 1985).
cf. honeydew

hypopharyngeal gland *n.* In the Honey
Bee: a gland that secretes royal jelly and
brood food that workers feed to queens
and larvae (Free 1961 in Wilson 1975, 207).

hypophysis See gland: pituitary gland.

hypothalamus See organ: hypothalamus.

interdigital gland *n.* For example, in
the Black-Tailed Deer, Blue Wildebeest: a
pheromone-producing gland found between toes that deposits odor directly on
the ground (Müller-Schwarze 1971 in Wilson 1975, 234, 491).

interrenal gland *n.* In some fish species: a gland that produces corticosteroids
(Wilson 1975, 289).

intersegmental gland *n.* In the Honey
Bee: an abdominal gland that secretes wax
(Wilson 1975, 430).

Koschevnikov's gland *n.* In nest
queens of Honey Bees: a gland at the base
of a queen's sting that produces a volatile
attractant that is one of three pheromones
causing workers to tend to queens (Wilson
1975, 203, 212).

mandibular gland *n.* In hive queens of
Honey Bees: a gland that produces *trans*-
9-keto-2-decenoic acid, an odor which prevents workers from constructing queen
cells in which new queens would be reared
(Wilson 1975, 216).

mental gland *n.* In males of most salamander species: a gland on an individual's
chin that produces a courtship pheromone
(Houck and Reagan 1990, 729).
syn. courtship pheromone (Houck and
Reagan 1990, 729)

metanotal gland *n.* For example, in
Oecanthine Grasshoppers: a gland on the
metanotum of a male that produces a
secretion eaten by a female during or after
copulation (Boake 1984, 696).

metapleural gland *n.* In some ant species: a gland that produces antibiotic substances (phenylacetic acid and other bio-
cidal secretions) used for protection against
fungi and bacteria (Wilson 1975, 211, 422).

metatarsal gland *n.* For example, in the
Black-Tailed Deer: a pheromone gland
found on hind legs that transmits airborne
scents (Müller-Schwarze 1971 in Wilson
1975, 234).

Nasanov gland *n.* In worker Honey Bees:
a gland that produces a pheromone consisting of a mixture of geraniol, nerolic acid,
geranic acid, and citral (Frisch and Rösch
1926 and Butler and Calam 1969 in Wilson
1975).
Comment: This pheromone is used to
attract and assemble workers at a food
source or at the queen during swarming
(Velthuis and van Es 1964, Mautz et al.
1972 in Wilson 1975, 212).

oral-scent gland *n.* In the Columbian
Ground Squirrel: a gland in the oral angle
of an individual that produces a scent that
appears to enable adults to tell familiar
from unfamiliar individuals, as well as
recognize conspecifics as individuals (Harris and Murie 1982, 140).

pituitary gland *n.* In vertebrates: an endocrine gland that hangs from the floor of
an animal's brain stem and produces hormones that affect an animal's body directly
and regulate activity of other endocrine
glands (thyroid, adrenal cortex, gonads)
(Curtis 1983, 1103; Immelmann and Beer
1989, 221).
syn. hypophysis (Immelmann and Beer
1989, 221)

Comment: The anterior lobe of a pituitary produces tropic hormones, growth hormone, and prolactin (Curtis 1983, 1103). [Latin *pituita*, phlegm]

▸ **neurohypophysis** *n.* The posterior part of a pituitary gland which stores and releases oxytocin and antidiuretic hormone (ADH) produced by the hypothalamus (Curtis 1983, 1103).

postcornual-skin gland *n.* For example, in the Chamois: one of a pair of glands located directly behind horns which is most developed during rutting (Heymer 1977, 48).

precaudal gland *n.* In the Wolf: a gland that appears to produce pheromones (Wilson 1975, 509).

sternal gland *n.* In termites: a gland that produces trail pheromone (Wilson 1975, 231).

tarsal organ *n.* For example, in the Black-Tailed Deer: a pheromone-producing gland that produces individual odors (Müller-Schwarze 1969 in Wilson 1975, 205).

temporal gland *n.* In Elephants: a gland located between ears and eyes that secretes a viscous, strong-smelling liquid, especially when an individual is excited or under stress (Wilson 1975, 495).

♦ **glandotrophic hormones, glandotropic hormone** See hormone: glandotropic hormones.

♦ **glial cell** See cell: glial cell.

♦ **global warming** *n.*
1. A period of an increasing yearly average temperature on Earth, not caused by human activities.
2. An increase in the average temperature of Earth's lower atmosphere due to greenhouse gases produced as byproducts of human activities (Allaby 1994).

Comments: Nonanthropogenic global warming and cooling have occurred many times during Earth's history. At present, Earth seems to be in a warming phase with an accompanying sea-level rise (Rensberger 1993b, A1; Stevens 1995, C4; Sawyer and Lee 1995, A3). Scientists disagree regarding whether some of this putative warming is caused by human activities, such as burning fossil fuels. Global warming models disagree regarding the details of how global warming will affect our planet. The amount of greenhouse gas may double in the 21st century, causing an increase in global temperature of 3 to 8°F and increased weather fluctuations. Many Web sites discuss global warming, including the EPA Global Warming Site, www.epa.gov/globalwarming/home.htm; NASA's Global Change Master Directory, http://gcmd.gsfc.nasa.gov; and the Sierra Club Global Warming and Energy Program, www.toowarm.com.

♦ **Gloger's law, Gloger's rule** See law: Gloger's rule.

♦ **glucagon-like factor-1** See hormone: glucagon-like factor-1.

♦ **gluttony principle** See principle: gluttony principle.

♦ **glycolysis** See biochemical pathway: glycolysis cycle.

♦ **glycolytic pathway** See biochemical pathway: glycolytic pathway.

♦ **gnesiogamy** See -gamy: gnesiogamy.

♦ **goal expectation** *n.* An animal's striving for an expected thing, *e.g.,* a Rhesus Monkey's refusing a lettuce leaf after seeing a banana that it might receive instead of the leaf (Heymer 1977, 205).

♦ **goal function** See ²function: goal function.

♦ **goal orientation** See taxis: telotaxis.

♦ **GOBASE** See database: Organelle Genome Database.

♦ **Goldschmidt's speciation hypothesis** See hypothesis: Goldschmidt's speciation hypothesis.

♦ **gonad** *n.* In multicellular-animal species: a gamete-producing organ; ovary or testes (Curtis 1983, 836).
cf. ovariole

♦ **gonadectomy** See castration.

♦ **gonadotropic hormone, gonadotropin** See hormone: gonadotropic hormone.

♦ **gonadotropin-releasing hormone (GNRH)** See hormone: gonadotropin-releasing hormone (GNRH).

♦ **gono-, gon-** *combining form* Generation, offspring, semen, procreative, sexual, etc. (*Oxford English Dictionary* 1972; Michaelis 1963).
[Greek *gonos*, seed]

♦ **gonochorism** *n.*
1. Fusion of male and female gametes produced from separate unisexual individuals (Lincoln et al. 1985).
syn. sexual reproduction (Lincoln et al. 1985)
2. The history, or development, of sex differentiation (Lincoln et al. 1985).
3. A species' having individuals of separate sexes (Lincoln et al. 1985).
syn. dioecism, dioecy, diecism, unisexuality (Lincoln et al. 1985)
cf. sexual: unisexual
4. Male and female individuals' being present in the same population (Lincoln et al. 1985).
adj. gonochoric, gonochoristic
syn. bisexuality (Lincoln et al. 1985)
cf. hermaphroditism

♦ **gonogenesis** See -genesis: gonogenesis.

♦ **good-mother gene** See gene: *FOSB.*
♦ **goosery** See ²group: goosery.
♦ **Graafian follicle** See egg follicle.
♦ **-grade** *combining form*
1. "Progressing or moving: *retrograde*" (Michaelis 1963).
2. Walking in a specified manner (Michaelis 1963).
cf. locomotion
[Latin *-grandus* < *grandi*, to walk]

digitigrade *n.* An animal that locomotes with digits of its foot, but not its entire sole, in contact with its substrate (*e.g.,* dogs or cats) (Michaelis 1963; Eisenberg 1981; 506).
adj. digitigrade

palmigrade See -grade: plantigrade.

plantigrade *n.* An animal that locomotes with the entire soles of its feet (or palms of its hands) in contact with its substrate (*e.g.,* bears or Humans) (Michaelis 1963; Eisenberg 1981, 508).
adj. plantigrade
syn. palmigrade (Lincoln et al. 1985)

pinnigrade *n.* An animal that locomotes utilizing flippers as paddles (*e.g.,* seals or dolphins) (Lincoln et al. 1985).
adj. pinnigrade

saltigrade *n.* An animal that locomotes by leaping or hopping (*e.g.,* grasshoppers or kangaroos) (Lincoln et al. 1985).
adj. saltigrade

syringograde *n.* An animal that locomotes by propelling itself with a jet of water (*e.g.,* dragonfly larvae or squids) (Lincoln et al. 1985).
adj. syringograde

unguligrade *n.* An animal that walks, or runs, on hooves or the modified tips of one or more of its digits (*e.g.,* deer or antelope) (Lincoln et al. 1985).
adj. unguligrade

♦ **grade, evolutionary grade** *n.*
1. A group of animals whose members are presumed to have arisen from the same common ancestor at about the same time; a higher grade is separated from a lower one by a certain "advance in differentiation of structure" [coined by Lankester 1877 in *Oxford English Dictionary* 1972].
2. A distinct stage in a series of stages in morphological, physiological, or behavioral traits that an organism may pass through during its evolution (Lankester 1909 and others in Mayr 1982, 234, Wilson 1975, 25).
Note: A principal characteristic of macroevolution "is the relative rapidity with which shifts into new adaptive zones occur. ...When a phyletic line enters a new adaptive zone, ...it undergoes at first a very rapid morphological reorganiza-tion until it has reached a new level of adaptation. Once it has achieved this new *grade*, it can radiate into all sorts of minor niches without any major modifications of its basic structure" (Bather 1927, Huxley 1958 in Mayr 1982, 614).
3. "The evolutionary level of development in a particular structure, physiological process, or behavior occupied by a species or group of species" (Wilson 1975, 584).
Note: "Evolutionary grade is distinguished from the phylogeny of a group, which is the relationship of species by descent" (Wilson 1975, 584).
4. "A delimitable unit of anagenetic advance or biological improvement" (Lincoln et al. 1985).
cf. clade

intergrade *n.* An individual in an intergradation zone (Lincoln et al. 1985).

♦ **graded potential, graded response** See response: graded response.
♦ **graded signal** See signal: analog signal.
♦ **gradual-change hypothesis** See hypothesis: hypotheses of species richness: gradual-change hypothesis.
♦ **gradualism** *n.* The belief that the Earth's geological changes were and are gradual (Baker 1978 and Alvarez et al. 1980 in Mayr 1982, 377–378).
Comments: Leibniz, Buffon, Lamarck, and most of Darwin's so-called forerunners believed in gradualism; however, their upholding gradualism became more difficult after the discovery of the frequency of stratigraphic breaks. Both Lyell and, later, Darwin were aware of the fact that earthquakes and volcanic eruptions could produce rather drastic effects, but they were smaller by several orders of magnitude than the catastrophes postulated by some geologists. Modern geological research indicates that certain events in the Earth's past history qualify as catastrophes (Baker 1978, Alvarez et al. 1980 in Mayr 1982, 377–378).

phyletic gradualism *n.*
1. A paleontological conception of evolution with these tenets: new species arise by transformation of an ancestral population into modified descendants; the transformation is slow and even and involves large numbers of conspecifics, usually an entire ancestral population; and the transformation occurs over all, or a large part, of the ancestral species' geographic range (Eldredge and Gould 1972, 89).
2. "The doctrine that [organic] evolutionary change is gradual and does not go in jumps" (Dawkins 1982, 288).

d–g

cf. ²evolution: anagenesis, quantum evolution, saltation; hypothesis: hypothesis of punctuated equilibrium; selection: species selection; speciation: allopatric speciation
Comments: "All sane Darwinians are gradualists in the extreme sense that they do not believe in the *de novo* creation of very complex and therefore statistically improbable new adaptations like eyes. This is surely what Darwin understood by the aphorism 'Nature does not make leaps.' But, within the spectre of gradualism in this sense, there is room for disagreement about whether evolutionary change occurs smoothly or in small jerks punctuating long periods of *stasis*. It is this that is the subject of the modern controversy, and it does not remotely bear, one way or the other, on the validity of Darwinism" (Dawkins 1982, 288).

♦ **graft chimera, graft hybrid** See chimera: graft chimera.

♦ **grail number** See number: grail number.

♦ **grain** *n.* The spatial pattern of a resource with respect to size and spacing of its patches (Wittenberger 1981, 616).
Comment: Coarse-grained resources are widely dispersed and patchy. Fine-grained resources are more uniformly distributed (Wittenberger 1981, 616).

♦ **grallitorial** See -orial: grallitorial.

♦ **-gram** *combining form* "Something written or drawn" (Michaelis 1963).
cf. curve, graph
[Greek *gramma*, letter, writing]

bar diagram *n.* A graph of counts of something vs. categories (Sokal and Rohlf 1969, 21, illustration; 29).
Comments: Sides of bars of a bar diagram do not touch. Categories include religion (Protestant, Catholic, or Hindu), collection interval (1st, 2nd, 3rd, etc.), or color. Examples of bar diagrams include: number of bees caught vs. season (spring, summer, or fall) and number of bees vs. color of flower visited (blue, yellow, or white).

behavior-chain diagram *n.* A graph of interactions between two or more organisms (Lehner 1979, 118–120).
syn. behavior-reaction chain (Evans and Matthews 1976, fig. 2)
cf. graph: kinematic graph

biogram *n.* The behaviors and rules by which an organism increases its Darwinian fitness (Count 1958, Tiger and Fox 1971 in Wilson 1975, 548).

cladogram (CLADE o gram) *n.*
1. "A branching diagram that shows how species split and form new species" (Simpson 1961; Mayr 1969; Wilson 1975, 26).

2. "A phylogenetic tree that depicts only the splitting of species and groups of species through evolutionary time" (Wilson 1975, 580).
3. A hypothetical tree that orders organisms according to nested sets of novelties; consequently, the organisms are ordered as well into nested sets — (hypothesized) monophyletic taxa (Eldredge and Cracraft 1980).
4. A branching diagram constructed by cladistic methods (Futuyma 1998, 95).
Note: Cladograms can represent different entities, including nucleic acid sequences, molecules, subspecies, species, and higher taxa.
cf. -gram: dendrogram, phenogram, phylogenetic tree; study of: systematics: cladistic systematics
Comment: Cladistics, initiated by Willi Hennig, utilizes cladograms (Futuyma 1998, 94). Researchers use cladograms as suggestions of evolutionary changes within investigated entities.
See Lincoln et al. (1985) for more definitions.
[Greek *clados*, branch, stem, twig; *gramma*, letter, writing]

dendrogram *n.*
1. "A diagram showing evolutionary change in a biological trait, including the branching of the trait into different forms due to the multiplication of the species possessing it" (a more narrow definition than def. 2, Wilson 1975, 582).
2. "A tree-like diagram of relationship" (Mayr 1982, 957) or resemblance (Lincoln et al. 1985).
syn. tree (Vigilant et al. 1991, 1505)
Comment: Biological dendrograms include cladograms, phenograms, and phylograms, which are often working hypotheses of relationship.
[Greek *dendros*, tree, stem, twig; *gramma*, letter, writing]

distance-Wagner tree *n.* A cladogram that shows the differences in characters among taxa as differences in branch distances (Shaffer et al. 1991, 289).
cf. -gram: Wagner tree

electroencephalogram (EEG) *n.* A recording of an animal's brain electrical activity (Michaelis 1963; Immelmann and Beer 1989, 20).

engram, memory trace *n.* A hypothetical physical manifestation of learning within an animal's central nervous system (Dewsbury 1978, 338).
Comment: The physiologist Karl Lashley spent several decades in the 20th century looking for engrams in Humans (Campbell

1996, 1022). Because partial damage to one area of a person's cerebral cortex does not destroy individual memories, Lashley ultimately concluded that there is no highly localized memory trace in a person's nervous system; rather, one of his memories seems to be stored within a certain association area of his cortex with some redundancy.

ethogram *n.* "A complete inventory of the behaviour patterns of a species" (Tinbergen 1951, 6).

syn. action system (Jennings 1906 in Heymer 1977, 64), behavioral catalog, catalog, behavioral repertoire (Fagen 1978, 26), behavioral repertory, repertoire (Lehner 1979, 49)

cf. repertory

Comments: Different authors give different definitions of "ethogram" (and terms listed here as synonyms), and some of the definitions are contradictory (Brown 1975, 2; Dewsbury 1978, 20; Lehner 1979, 49; Lincoln et al. 1985). Workers might reduce confusion by using "ethogram" and "partial ethogram" only with the meanings given in this book instead of the many synonyms of debatable meanings and the same synonyms for both "partial ethogram" and "ethogram."

▸ **partial ethogram** *n.* Part of an ethogram.

syn. behavioral catalog (Fagen 1978, 26), catalog, ethogram (Lehner 1979, 49)

Comment: It is very difficult to compile an ethogram, rather than a partial ethogram, for a species because it takes a long time to see rare behaviors and many behaviors are variable with regard to complexity.

evolutionary tree See -gram: phylogenetic tree.

food web *n.* A diagram that indicates which organisms are the consumers and which are consumed (Wilson 1975, 584).

cf. chain: food chain

full phylogenetic tree *n.* A dendrogram that "contains the information of the cladogram, plus some measure of the amount of divergence between the branches, plotted against a time scale" (Wilson 1965).

histogram *n.* A graph of counts of something vs. a continuous (= nondiscrete) variable represented as class intervals (Sokal and Rohlf 1969, 25, illustration; 29).

Comments: The width of a bar along the abscissa represents that of a class interval, and sides of bars touch each other to show that the actual class limits are contiguous (except for the left side of the left-most bar and the right side of the right-most bar). The midpoint of an interval corresponds to a class mark. Continuous variables include

human blood pressure, height, or time. Examples of histograms include: mean number of leaves on a group of mangroves vs. time (3-day intervals from June to October) and number of people vs. height (1-cm intervals). Data used in a histogram may also be correctly presented as a frequency polygon.

idiogram See -gram: karyogram.

karyogram *n.* A photographic representation of a cell's karyotype (chromosome complement) (Lincoln et al. 1985).

syn. idiogram, karyotype

kingram *n.* A graphical technique used to evaluate decision rules and corresponding maximum efficiencies in different animal recognition systems (Getz 1981 in Wilson 1987, 11).

nomogram, nomograph *n.* A graph of numerical relations, especially one that has graduated scales for three interrelated variables, arranged in a straight line that joins values of two of the variables and cuts the scale of the third variable at its related value (Michaelis 1963).

phenogram *n.* A dendrogram (def. 2) of overall similarity among species, generated by a numerical algorithm that uses as many traits of the species as possible (Michener and Sokal 1957 in Futuyma 1998, 92).

Comment: A phenogram does not necessarily represent phylogenetic relationships, although a researcher might infer them from a phenogram (Lincoln et al. 1985; Futuyma 1998, 94).

cf. study of: systematics: phenetic systematics

[Greek *pheno* < *phaino*, appear, shine; *gramma*, letter, writing]

phylogenetic tree *n.*

1. A hypothetical taxonomic tree, more detailed than a cladogram, that specifies actual series of ancestral and descendant taxa (Eldredge and Cracraft 1980).

2. A dendrogram (def. 2) representing inferred lines of descent (Lincoln et al. 1985).

syn. evolutionary tree, phyletic tree, phylogeny (*Oxford English Dictionary* 1972, entries from 1870; Wilson 1975, 591), phylogram (Mayr 1982, 233)

[phylogram: Greek *phylo*, race, species, tribe; *gramma*, letter, writing]

rooted-phylogenetic tree *n.* A phylogenetic tree that has a base that represents the ancestor of the organisms in the tree (Niesbach-Klösgen et al. 1987, 219).

Comment: Construction of a rooted-phylogenetic tree requires information concerning the time scale of evolution (Niesbach-Klösgen et al. 1987, 219).

rootless phylogenetic tree *n.* A phylogenetic tree that does not have a base that represents the ancestor of the organisms in the tree (Niesbach-Klösgen et al. 1987, 219). *syn.* undirected tree (Mayr and Ashlock 1991, 292)

scatter plot, scatter diagram *n.* A graph of sets of paired, or multiple, points made to indicate the general nature of the relationship between two or more variables (Lincoln et al. 1985).

sociogram *n.*
1. A complete description of the social behavior of a particular species (Wilson 1975, 16, 595).
2. A graph of social relationships, or interaction frequencies, in an animal group (Immelmann and Beer 1989, 280).
cf. -gram: ethogram

sonagram, sonogram, sound spectrogram *n.* A graph of an animal's vocalizations, showing their amplitudes, frequencies, and sound pressures (Heymer 1977, 99; Lincoln et al. 1985).

undirected tree See -gram: unrooted tree.

vibragram *n.* A graph of robber-fly wingbeat frequency vs. time (Lavigne and Holland 1969, 13).

Wagner tree *n.* A dendrogram based on the Wagner algorithm (Kluge and Farris 1969 in Wiley 1981, 176).
cf. -gram: distance-Wagner tree

♦ **graminicole** See -cole: graminicole.
♦ **graminivore** See -vore: graminivore.
♦ **graminology** See study of: graminology.
♦ **granivore** See -vore: granivore.
♦ **graph** *n.* A diagram that represents variations in the relationship between two or more factors by means of a series of connected points, bars, curves, lines, and so forth (Michaelis 1963)
cf. curve, -gram

behavior-chain diagram See -gram: behavior-chain diagram.

"facial-feature graph" *n.* A type of graph devised by H. Chernoff that encodes data in human facial features to help a viewer detect data patterns, groupings, and correlations (Levine 1990, 457).

frequency polygon *n.* A graph, of counts of something vs. a continuous variable represented as class intervals, which is made up of the lines that connect the dots of y vs. x when $y > 0$ (Sokal and Rohlf 1969, 30, illustration).
Comment: The data of a frequency polygon may also be correctly presented as a histogram.

hydrograph *n.*
1. A graph on which a water course's water flow is plotted against time (Allaby 1994).

2. A graph on which the elevation of groundwater in a borehole above a particular datum point is plotted against time (Allaby 1994).
Comment: Peaks on the hydrograph indicate times of high water level.

kinematic graph *n.* A graph that represents a temporal, or sequential, ordering of behavior states (Sustare 1978, 278).
syn. behavior-sequence diagram
cf. graph: behavior-chain diagram

rose diagram *n.* A circular histogram that displays directional data and their frequency in each class (Allaby 1994).
cf. wind rose

spindle diagram *n.* A graph with geological time on its ordinant that shows the number of focal taxa during geological time as a group of centered, stacked, horizontal bars, each bar being the number of taxa that are know from a time interval (Labandeira and Sepkoski 1993, 311, fig. 2).

wind rose *n.* A rose diagram that represents the relative frequencies of different wind directions and wind speeds at a climatic station over a time interval (Adams et al. 1994, 19, illustration; Allaby 1994).
cf. rose diagram

♦ **-graph** *combining form* "That which writes, portrays, or records" (*Oxford English Dictionary* 1972).
[French -*graphe* < Latin -*graphus* < Greek -*graphos* < *graphein*, to write]

autoradiograph *n.* A sheet of X-ray film that is exposed by radioactive chemicals (Campbell et al. 1999, 369).
Comment: Autoradiographs are used in DNA fingerprinting and other methods that seek to identify particular molecules (Campbell et al. 1999, 369).

Sonagraph® *n.* The brand name of a kind of sound spectrograph (Immelmann and Beer 1989, 284).

sound spectrograph, sonograph *n.* A machine that produces sonagrams, *q.v.* (Lehner 1979, 174–180; Immelmann and Beer 1989, 284).

stenograph *n.* A machine that can be used to record the behavior of two animals simultaneously (Lehner 1979, 161).

♦ **-graphy** *combining form*
1. Denoting a style of writing, drawing, or graphic representation (*Oxford English Dictionary* 1972).
2. Denoting a descriptive science (*Oxford English Dictionary* 1972).
[Greek *graphein*, to write]

ecography *n.* "Descriptive ecology" (Lincoln et al. 1985). See study of: ecology.

entomography *n.* Description of an insect or its life history (Lincoln et al. 1985).

onomatography *n.* The correct writing of plant or animal names (Lincoln et al. 1985).

orthography *n.* "The spelling of words" (Lincoln et al. 1985).

paleontography, palaeontography *n.* "The systematic description of fossil taxa" (Lincoln et al. 1985).

phytography *n.* "Descriptive botany" (Lincoln et al. 1985).
cf. study of: botany

phytotopography See study of: phytogeography: phytotopography.

♦ **grasp reflex, grasping reflex** See reflex: grasp reflex.

♦ **grass bald** See habitat: grass bald.

♦ **grass pulling** See behavior: displacement behavior: grass pulling.

♦ **graveolent** *adj.* Possessing a strong, or offensive, odor (Lincoln et al. 1985).

♦ **gravid** *n.*
1. In mammals, including Humans: a female's being pregnant; heavy with young (*Oxford English Dictionary* 1972, entries from 1597).
2. In nonmammalian animals: referring to an individual that carries eggs or young internally (Lincoln et al. 1985).
Note: Def. 2 is usually used today.
syn. berried, ovigerous (Lincoln et al. 1985), pregnant (in less precise modern writing)
cf. pregnancy

♦ **graviperception, geoaesthesia, geoperception** *n.* Gravity perception (Lincoln et al. 1985).

♦ **gravity pollination** See pollination: gravity pollination.

♦ **gray matter** See organ: brain: gray matter.

♦ **grazers** See ²group: functional feeding group: scrapers.

♦ **grazing** *n.* An animal's feeding on herbage, algae, or phytoplankton by consuming whole food plants or cropping entire surface growth of herbage (Lincoln et al. 1985).
v.t., v.i. graze
cf. feeding, predator

♦ **great bevy** See ²group: great bevy.

♦ **great chain of being** See *scala naturae*.

♦ **green-beard effect** See hypothesis: green-beard effect.

♦ **green development** *n.* A trend in homebuilding that includes greater environmental concern by including wastewater recycling, better building energy management, transportation solutions, wetlands protection, and preservation of wildlife habitats (Lehman 1991, E1).

♦ **greenhouse effect** See effect: greenhouse effect.

♦ **greeting, greeting behavior** *n.*
1. A person's approaching, or addressing, another with expressions of goodwill, or courtesy, upon meeting; a person's offering, in speech or writing, to another person his own, or another person's, friendly or polite regard (*Oxford English Dictionary* 1972, entries from 1000).
2. A person's receiving, or meeting, another person with welcome (*Oxford English Dictionary* 1972, entries from 1605).
See greeting: notifying.
Comment: "Greeting" now refers to nonhuman animals as well.

arrow greeting *n.* For example, in certain Amazonian Indians: an aggressive greeting in which a barrage of arrows is fired just in front of a visitor (Eibl-Eibesfeldt 1970 in Heymer 1977, 128).

contact greeting *n.* Any greeting in which bodily contact occurs, *e.g.,* kissing (Eibl-Eibesfeldt 1970), hand shaking, and nose rubbing in Humans; mutual rubbing against one another in some mammals; billing in some birds; naso-nasal, nasal-anal, or naso-genital greetings in some mammals (Heymer 1977, 102).

notifying *n.* In the Hamadryas Baboon: a type of male-male greeting behavior characterized by a sequence of approach-retreat patterns used by one or both males, swinging-gait locomotion, and no physical contact when the animals come close to each other during an interaction (Kummer 1968; etc. in Colmenares 1991, 49).
syn. greeting (Colmenares 1991, 49)
Comments: In the context of male rivalry over females, "notifying" might be a negotiation strategy whereby harem possessors and their rivals can assess a situation, influence each other's roles in their relationship, and eventually resolve a conflict without resorting to fighting. "Notifying" originally referred to the behavior's presumed function of a male's signaling the direction of a foraging march (Colmenares 1991, 49).

threat greeting *n.* For example, in the Horse, Human, Laughing Gull: greeting behavior that contains some elements of threat (*e.g.,* aggressive facial expression of a horse combined with erected ears which is a nonaggressive position) (Heymer 1977, 54).

♦ **greeting ceremony** See ceremony: greeting ceremony.

♦ **greeting etiquette** *n.* Human greeting responses that often contain components of symbolic submission (such as repeated bowing, lifting a hat and showing a bare head, falling to knees or to the ground in front of rulers or royalty) and may include present giving (Heymer 1977, 80–81).

d – g

Comment: Greeting etiquette may establish, or maintain, bonds and appease aggression (Heymer 1977, 80–81).

♦ **greeting feeding** See feeding: greeting feeding.

♦ **gregarious** *adj.*

1. In classes, or species, of animals: referring to individuals that live in conspecific groups (*Oxford English Dictionary* 1972, entries from 1668).
2. Referring to a person who is "inclined to associate with others, fond of company" (*Oxford English Dictionary* 1972, entries from 1789).
3. Referring to plants that grow in open groups (*Oxford English Dictionary* 1972, entries from 1829).
4. Of, or pertaining to, a group of persons; characteristic, or affecting, persons gathered in crowds (*Oxford English Dictionary* 1972, entries from 1833).
5. Referring to a parasitic wasp species whose females lay more than one egg into or on a host individual (Hooker and Barrows 1992).

cf. ²group: swarm

subgregarious *adj.* Referring to organisms that form loose aggregations or groups (Lincoln et al. 1985).

♦ **gregarious disposition** See drive: bonding drive.

♦ **grin** See facial expression: grin.

♦ **grind** See ²group: grind.

♦ **Grinnell's axiom** See principle: Gause's principle.

♦ **grip** *n.* A person's tight, or strained, hand grasp upon an object (*Oxford English Dictionary* 1972, entries from 1148).

power grip *n.* In primates, including Humans: an individual's strong grasping with its hand (Wilson 1975, 516).

precision grip *n.* In primates: an individual's grasping an object using some amount of separate control in its index finger and thumb, permitting its fine manipulation of food particles and grooming its fur (Wilson 1975, 516).

Comment: Humans have the most refined precision grip of all primates (Wilson 1975, 516).

♦ **grist** See ²group: grist.

♦ **grooming** *n.*

1. A person's carrying, feeding, and generally attending to a horse (*Oxford English Dictionary* 1972, entries from 1809).
2. A person's tending, or attending, to something carefully; giving a neat, tidy, or smart appearance to something (*Oxford English Dictionary* 1972, entries from 1843).
3. An individual animal's cleaning itself of foreign matter and parasites, including insects' mouthing, scraping, and brushing body parts; birds' nibbling their feathers with their bills; and mammals' nibble-grooming their fur with their teeth and licking it with their tongues and scratching with hind legs, rubbing against objects, rolling in dust, and washing.

v.i., v.t. groom

cf. behavior: comfort behavior; displacement behavior: sham preening; ²movement: abdominal-up-and-down movement

Comments: Among primates, grooming serves both to keep pelage clean and in communication (Dewsbury 1978, 59). In Humans, grooming can have a calming effect. Bird grooming is usually called "preening," which can involve cleaning, combing, carding, smoothing feathers, dressing feathers with oil-gland secretions, scratching, head rubbing, and rehooking detached barbules (Immelmann and Beer 1989, 230).

allogrooming *n.* An animal's grooming a conspecific animal (Wilson 1975, 208, 578).

cf. grooming: autogrooming

Comment: "Allopreening" is usually used for birds.

autogrooming *n.* An animal's grooming its own body (Lincoln et al. 1985).

cf. grooming: allogrooming

direct scratching (under the wing) *n.* In many bird species: an individual's raising its leg up without moving its wing and scratching its head (Heymer 1977, 194).

cf. grooming: scratching over the wing

habit preening *n.* In ducks: a ritualized preening movement common in courting individual's involving, for example, an individual's tilting up its enlarged and brightly colored feathers (*e.g.*, in Mandarin Ducks) or an individual's pointing its bill to its bright speculum (Immelmann and Beer 1989, 126).

Lausen *n.* In primates: an individual's searching for lice with its hands (Heymer 1977, 104).

cf. grooming: grooming talk

Comment: "*Lausen*" indicates a groomer's familiarity with a groomee, may strengthen social bonds, and may calm groomees (*e.g.*, in Humans) (Heymer 1977, 104).

[German]

nibble-cleaning, nibble-preening *n.* In birds: grooming with a bill (Meischner 1959, Petzold 1964 in Heymer 1977, 104).

ritualized grooming *n.* A grooming movement, or movements, that an animal uses for communication (McFarland 1985, 392).

cf. behavior: ritualized behavior; ritualization

scratching over the wing *n.* In a few bird species: a bird's raising its ipsilateral leg over and above its dropped wing and scratching its head (Heinroth 1930, Wickler 1970 in Heymer 1977, 87).

cf. grooming: direct scratching

self-grooming See grooming: auto-grooming.

social grooming *n.* Allogrooming and allopreening (Immelmann and Beer 1989, 275).

♦ **grooming dance** See dance: bee dance: shaking dance.

♦ **grooming talk, small talk** *n.* A conversation in which substantive exchange of information does not occur but which contains social information (that a partner is interested in the other person and ready to listen and answer) and may function in group bonding (Morris 1968 in Heymer 1977, 197).

cf. grooming: *Lausen*

♦ **gross cheater** See cheater: gross cheater.

♦ **gross primary productivity, gross primary production** See production: gross primary productivity.

♦ **ground litter** See litter.

♦ **ground ozone** See molecule: ozone.

♦ **ground pecking** *n.* For example, in some gallinaceous bird species: pecking at the ground, which functions in hens' stimulating their chicks to follow them, cocks' holding their groups of hens together, and agonistic behavior (Schenkel 1956, 1958; Wickler 1968; Feekes 1972 in Heymer 1977, 46).

♦ ¹**group** *n.* A set of taxa (*e.g.,* Protostomia or Deuterostomia).

♦ ²**group** *n.*

1. A number of persons, or things, that form a unity because they have a common relationship; a number of such items classed together because they have a certain degree of similarity (*Oxford English Dictionary* 1972, entries from 1809).

2. A set of conspecific organisms that remain together for any period while interacting with one another to a much greater degree than with other conspecific organisms; no specific information regarding the organization of the set is implied (Kummer 1971 in Immelmann and Beer 1989, 122; Wilson 1975, 8–9).

Note: Specific group terms, many that denote particular information about a group, are used for different types of animals; I list many below. Some of the animal groups listed below are further defined under alphabetical listings. Hierarchy of groups may occur, as in the Hamadryas Baboon: "one-male unit," "two-male team," "band," and "troop" (Kummer, 1968).

3. In the Anubis Baboon: the basic unit of society; an assembly of females, offspring, and multiple males (Wilson 1975, 534).

4. A set of related species (*e.g.,* a genus) (Wilson 1975, 585).

See taxon.

cf. biota, -coenosis, community, chorus, fauna, flora, harem, jordanon, ²lek, one-male group, -paedium, social group, ²society, ²species, warren

Comments: Categories designated as the term "group" (which I have not entered) are used at different levels between divisions in this hierarchy. An increasing number of students

A TAXONOMIC CLASSIFICATION OF ANIMAL GROUPS[a]

domain
superkingdom
kingdom[b]
subkingdom
superphylum
phylum[b] (*pl.* phyla)
subphylum
infraphylum
subterphylum
superclass
class[b]
subclass
infraclass
subterclass
supersection
section[b]
subsection
infrasection
supercohort
cohort[b]
subcohort
infracohort
subtercohort
superorder
order[b]
suborder
infraorder
superfamily
family[b]
subfamily
supertribe
tribe[b]
subtribe
genus[b] (*pl.* genera)
subgenus
superspecies
species[b] (*pl.* species)
geographical race (*syn.* race, subspecies)

---

[a] Boudreaux (1979, 83, 141); Lincoln et al. (1985, 279); Borror et al. (1989, 147).

[b] This is a major taxonomic division.

An Ecological and Population-Biology Classification of Animal Groups[a]

biota (flora and fauna)
  biome
    community
      species
        geographical race (= subspecies)
          metapopulation
            population
              deme
                colony
                  troop
                    family
                      pair
                        individual

[a] These categories do not refer to all species.

A Taxonomic Classification of Plants at the Species and Lower Levels[a]

species
  subspecies (*syn.* geographical race, variety)
    variety (*syn.* genetic variant,
      microgeographic race, nongenetic
      variant, subspecies, varietas,
      varietas)
        subvariety (*syn.* subvaritas)
          form (*syn.* form, *forma,*
            race, variety)
              clone

[a] Lawrence (1951), Munson (1981). Different authors (in Munson 1981) proposed different classificational schemes for taxa between the species and clone levels, none of which is widely accepted.

are arguing for abolishing these Linnaean categories as arbitrary and even nonsensical and overtly misleading (Minelli 1996).

aerie, brood, eyry, eyrie, nest *n.* A group of birds of prey, ravens, or children (Sparkes 1975).

aggregation *n.*
1. A temporary group of conspecific, or heterospecific, organisms that forms for nonsocial reasons in response to environmental factors (*e.g.,* a group of insects or tree frogs around a light at night) (Dice 1952, 266; Dewsbury 1978, 92).
   Note: Aggregating individuals do not cooperate and may even be in strong competition with one another (Dice 1952, 266).
2. In some fish species: a group of fish that occurs incidentally, not as a result of

"biosocial attraction" (Breder 1967 in Partridge 1982, 298).
   cf. [2]group: school
3. A conspecific group of individuals, "comprised of more than just a mated pair or a family, gathered in the same place but not internally organized or engaged in cooperative behavior;" distinguished from a true society, *q.v.* (Wilson 1975, 8, 577).
4. A society, or group, of conspecific organisms that have a social structure and consist of repeated members or modular units but with a low level of coordination, integration, genotypic relatedness, or a combination of these attributes (Lincoln et al. 1985).
5. A group of nests of conspecific bees of a social, or nonsocial, species.
   Note: Members of some of these species may attract one another with aggregation pheromones.
   See [2]group: pseudosocial group.
   syn. nest aggregation
See [2]society: modular society: modular discontinuous society.
syn. subsocial grouping (Immelmann and Beer 1989, 8)
cf. [2]group: colony, pseudosocial group; [1]population; school; [2]society: discontinuous modular society

amphoterosynhesmia *n.* A swarm of both sexes of an animal species gathered for reproductive purposes (Deegener 1918 in Wilson 1975, 16).

anonymous group See [2]society: open society.

assemblage *n.* A group of species in the same taxon that live in the same location at the same time (*i.e.,* live in the same community); contrasted with ensemble, [2]community, guild, local guild, and taxon in the Venn diagram of Fauth et al. (1996, 283).

association *n.*
1. A large assemblage of organisms in a particular area, with one or two dominant species (Lincoln et al. 1985).
2. Living together by conspecific, or heterospecific, organisms (Immelmann and Beer 1989, 21).
See association.
▶ **kormogene association** *n.* An organism colony in which different individuals are morphologically attached to each other without organic connection (Lincoln et al. 1985).

aviary *n.* A group of birds in an aviary (Sparkes 1975).

bachelor group *n.* For example, in the Anubis Baboon, Vicuña: a social unit comprised of reproductively mature males without territories (Wilson 1975, 490, 534).

band *n*.
1. A social group of any kind of social mobile animals, composed of two or more individuals; "a social group of indefinite composition" (Dice 1952, 266).
2. A group of social animals (*e.g.,* a band of Coatis or Humans) (Wilson 1975, 456, 579).
3. In the Hamadryas Baboon: a group of a few one-male units (Wilson 1975, 534). See ²group: troop; hierarchy: central hierarchy.
4. A group of clans of human Pygmies (Heymer 1977, 99).

baren, barren *n*. A pack, or herd, of Mules (Sparkes 1975).
[Middle English *berynge*, bearing (Lipton 1968, 53); or from barren, referring to the mule's not being capable of reproducing (Sparkes 1975)]

battery *n*. A group of birds housed together to encourage egg laying (Sparkes 1975).

bellowing *n*. A group of Bullfinches (Sparkes 1975).

bevy *n*. A group of things, *e.g.,* six head of Roe Deer or swans, a bevy of Conies, larks, Otter, quail, ladies, or beauties (attractive human females) (Sparkes 1975; Choate 1985, 13).
cf. ²group: company, covey, flight, flock, great bevy, middle bevy

bew *n*. A flock of Partridges (Sparkes 1975).
syn. bow, flock (Sparkes 1975)

bike *n*. A nest of wild bees, a swarm, or crowd, of people (*e.g.,* a bike of ants, hornets, or wasps) (Sparkes 1975).
syn. swarm

bite *n*. A group of mites (Sparkes 1975).

bloat *n*. A group of Hippopotami (Sparkes 1975).

bouquet *n*. A group of pheasants (Sparkes 1975).
[from the appearance of a bouquet (Lipton 1968, 9)]

bow *n*. A herd of cattle; the cattle on a farm (Sparkes 1975).

brace, cast *n*. A pair of usually hunted animals (*e.g.,* a brace of pike, trout, partridges, pheasants, ducks, bucks, hares, foxes, Greyhounds, hounds, or, rarely, people) (Sparkes 1975).

brood *n*.
1. The offspring of an animal that hatch, or are born alive, contemporaneously (*Oxford English Dictionary* 1972, entries from 1000).
   syn. hatch (*Oxford English Dictionary* 1972)
2. "Any young animals that are being cared for by adults;" in social insects, in particular, the immature members of a colony collectively, including eggs, larvae, and pupae (Wilson 1975, 579).
   Note: In the strict sense, eggs and pupae are not members of a society, but they are, nevertheless, referred to as part of a brood (Wilson 1975, 579).
3. A group of young animals, all from the same parents or same mother (*e.g.,* a brood of eggs, eagles, hawks, chickens, hens, grouse, or kittens) (Sparkes 1975).
syn. aerie, fry, breed (Sparkes 1975)
cf. ²group: aerie, eye, eyrar, fall

building See ²group: rookery.

bunch *n*. A group of conspecific animals that are close together (*e.g.,* a bunch of cattle, ducks, waterfowl, or Widgeon) (Sparkes 1975).

bury *n*. A group of rabbits (Sparkes 1975).

busyness *n*. A group of ferrets (Sparkes 1975).

cast See ²group: brace.

cavalry *n*. A group of horses, or horsemen (Sparkes 1975).

centeener *n*. A large number of organisms that have a common parentage (Sparkes 1975).

cete *n*. An assemblage or company; a group of badgers (Sparkes 1975).

charm *n*. A group of goldfinches (Sparkes 1975).

chirm *n*. A group of finches (Sparkes 1975).
[from chirm, din, or chatter (Sparkes 1975)]

chorus *n*. A group of calling frogs, toads, or insects (Wilson 1975, 443, 580; Brush and Narins, 1989, 33).
cf. animal sounds: song: chorus, communal song
▶ **wood choir** *n*. A bird chorus (Sparkes 1975).

citadel *n*. A group of mole burrows at different levels that are connected by vertical shafts (Sparkes 1975).

¹**clan** *n*.
1. A group of persons who claim descent from a common ancestor and associate together; tribe (*Oxford English Dictionary* 1972, entries from *ca.* 1425); a small, usually exogamous, group of several families that traces its origin from a common ancestor (Dice 1952, 272; Heymer 1977, 99).
2. In Pygmies (Humans): one of several groups, each with an elder head, within a band (Heymer 1977, 99).
   cf. group
3. For example, in African Elephants, Spotted Hyenas: a large social complex which is co-extensive with a local population

(Wilson 1975, 494; van Lawick-Goodall in Heymer 1977, 99).

See ²group: clan.

²clan, clique, sept, society *n.*

1. A group of social animals of common decent (*e.g.,* a clan of African Elephants, Spotted Hyenas, or Humans) (Sparkes 1975).
2. A collection of animals, plants, or lifeless things (Sparkes 1975).

See clan.

cf. ²group: zoo

clatter See ²group: clutter.

cloud *n.* A large number things (*e.g.,* a cloud of locusts, grasshoppers, flies, gnats, locusts, seafowl, or Starlings) (Sparkes 1975).

clowder *n.* A group (kendle or kindle) of cats (Sparkes 1975).

[possibly the same word as "clutter" (Lipton 1968, 53)]

clump, group, thicket *n.* An unshaped mass, a heap (*e.g.,* a clump of plants, reeds, or trees) (Sparkes 1975).

cluster *n.* A number of like things that are growing together, a number of similar things that are collected together (*e.g.,* a cluster of islands, nuts, grapes, spiders, or bees) (Sparkes 1975).

clutch *n.*

1. The number of progeny (eggs or live births) produced by a female at one time (Wilson 1975, 581; Lincoln et al. 1985; Hooker and Barrows, 1989, 460).
   syn. clutch size, litter size
2. A number of chickens (Choate 1985, 13).
3. A brood (group of young) of animals (*e.g.,* wasps) (Hooker and Barrows, 1989, 460).
4. *v.t.* "To hatch (chickens)" (*Oxford English Dictionary* 1972, entry from 1774).

[*cletch* < *cleck* to hatch < Old Norse *klekja* hatch]

clutter, clatter *n.* A confused collection of something (Sparkes 1975).

Comment: Some usages are clutter of cats or spiders (Sparkes 1975).

colony *n.*

1. A human settlement in a new country that remains subject to, or connected with, its parent state (*Oxford English Dictionary* 1972, entries from 1548).
2. An aggregate of individual organisms that forms a physiologically connected structure (*e.g.,* compound ascidians or coral polyps) (*Oxford English Dictionary* 1972, entries from 1872); in social animals, a group of individuals that are highly integrated by physical union of their bodies, by division into specialized zooids or castes, or both (Wilson 1975,

8, 383, 456, 472, 581); *e.g.,* tightly integrated masses of bryozoans, siphonophores, sponges, other "colonial" invertebrates, or societies of social insects. *Note:* In social insects, a "colony" should be distinguished from a "nest" (Michener 1974, 372).

3. An aggregation of conspecific individuals that has a more or less permanent location in its ecological community (Dice 1952, 266); almost any group of organisms, especially if they are fixed in one locality (Wilson 1975); *e.g.,* a group of sedentary spiders, a group of nesting birds, or a cluster of rodents living in dens. No implication about group cohesion is made in this term (Immelmann and Beer 1989, 48).
   cf. ²group: association: kormogene association
4. Mature female bees that work in a nest plus actively cared-for immatures (Michener 1974, 372).
5. In bees: a group of two or more nest-making, or nest-tending, adult females (Michener 1988b).
   cf. ²group: aggregation, hive, nest; ²society: continuous modular society

[Latin *colonia* < *colonus*, farmer]

▶ **breeding colony** *n.* For example, in the Starling, Waxbill; Weaver Finches, Herons: a nesting concentration of communally breeding individuals (Immelmann and Beer 1989, 35).

cf. ²group: heronry, rookery

▶ **mixed colony** *n.* A group of two or more social-insect species whose brood are placed together and tended communally (Wilson 1975, 354).

▶ **supercolony** *n.*

1. In ants: "a unicolonial population, in which workers move freely from one nest to another, so that the entire population is a single colony" (Hölldobler and Wilson 1990, 643).
2. In Fire Ants: a group of workers with more than one egg-laying queen (as "super colony" in Scott 1996, 26).

syn. multiple-queen colony, polygene colony (Scott 1996, 26)

Comment: Fire Ant supercolonies occur in parts of Texas, where an acre can support over 250 mounds (= ant hills), compared to areas with single-queen colonies which have up to about 50 mounds per acre (Norvell 1996, 26).

company, assembly, bevy *n.* A fellowship, band, or retinue (*e.g.,* a company of moles, parrots, or Widgeon) (Sparkes 1975).

congregation *n*. A group of things (*e.g.,* Humans, Plovers, or Rooks) (Sparkes 1975; Choate 1985, 13).

cortège See ²group: parade.

coterie *n*.

1. In the Prairie Dog: the basic society unit, consisting of a small group of individuals (often as many as two males and five females with their offspring) that occupies communal burrows (King 1955 in Wilson 1975, 139, 472, 581).
   Note: More than two coteries comprise a "ward," *q.v.*
2. An animal social group that defends a common territory against members of other coteries (King 1955, 54; Lincoln et al. 1985).
3. A group of people that meet familiarly, usually for social or literary purposes (Sparkes 1975).

syn. clique, set (Sparkes 1975)
cf. ²group: colony, town.
[French earlier, an organization of tenants holding land from the same lord < *citier,* cotter < *cote,* hut]

covert *n*. A flock of birds (*e.g.,* coots) (Sparkes 1975).

syn. covey (Sparkes 1975)

covey *n*.

1. A flock of Bobwhites (quail) or Partridges (Michaelis 1963) or ptarmigans (Sparkes 1975).
2. A brood, or hatch, of birds (Sparkes 1975).
   syn. bevy, company, covert (Sparkes 1975)
3. A sport term for a foraging flock of Partridges, Bobwhites, or pheasants (Choate 1985, 13).

[Middle English *cove, covy* < Latin *cubare,* to be lying down (Lipton 1968, 27)]

cowardess *n*. A group of curs (mongrel dogs) (Lipton 1968).

[from an observer's viewpoint (Lipton 1968)]

crash *n*. A group of rhinoceros (a modern pun) (Sparkes 1975).

See ²group: stubborness.

creaght, booly, bow *n*. A herd of cattle that is driven from place to place for pasturing (Sparkes 1975).

creche, crèche, crêche *n*.

1. A public nursery for human infants (*Oxford English Dictionary* 1972, entries from 1882).
2. In some bird species: a group of young of more than one pair of parents, tended by one or more adults, usually fewer than one parent from each family (Wilson 1975, 504; Sparks 1975; Immelmann and Beer 1989, 64).

3. In precocial birds with young that can forage for themselves (*e.g.,* Eider Ducks): a gathering of young that adults guide and defend against predators (Immelmann and Beer 1989, 64).
4. In altricial birds (*e.g.,* Emperor Penguins, Flamingos, some gull species): an assemblage of young, that have left their nests or brood places, are cared for by a few adults at a time or are left by themselves for short periods, and are periodically fed by their parents (Immelmann and Beer 1989, 64).

syn. nursery group
cf. ²group, kindergarten
Comment: Young lions may form creche-like groups (Wilson 1975, 504).
[French *crèche*, crib, cradle]

cultivar *n*.

1. A horticultural variety of a plant or flower (Michaelis 1963).
2. An assemblage of cultivated individuals that are distinguished by any character (*e.g.,* chemical, cytological, morphological, or physiological) significant for the purpose of agriculture, forestry, or horticulture and that when reproduced (asexually or sexually) retain their distinguishing characters (Gilmer 1969 in Munson 1981, 2).
3. A plant variety produced and maintained by cultivation (Lincoln et al. 1985).

syn. cultigen, *cultivaritas* (Lincoln et al. 1985)
abbr. CV (Lincoln et al. 1985), cv
Comments: A cultivar name is capitalized and placed inside single quotes: *Calluna vulgaris* 'Flore Pleno,' *Cyclamen persicum giganteum* 'All Sorts Mixed,' *Rudbeckia fulgida* 'Goldsturm.' Cultivar names derive from scientific names, nonscientific names, persons' names, and newly invented words. Cultivars can be difficult to identify because their written descriptions are lacking or ambiguous, the same name is used for different clones, different names are used for the same clone, cultivars are distinguished by their fragrance or hardiness, or a combination of these phenomena (Munson 1981, 1). Jeffrey (1968 in Munson 1981, 2) indicates that, according to the International Code of Nomenclature for Cultivated Plants, a cultivar is an atomic unit whose division is not permitted. Munson (1981) provides a key to cultivars of heaths and heathers. People protect some new plant cultivars with plant-variety-protection certificates (Adams 1996, 12). Researchers use plant characters, including genetic fingerprinting as developed by P.B. Cregan, to ascertain whether

a new variety is distinct enough from others to warrant such a certificate. [*cultiv*ated + *var*iety]

dairy *n.* A group of milking cows (Sparkes 1975).

deceit, desert *n.* A flock of Lapwing (Sparkes 1975).
[from an observer's viewpoint, Lipton 1968]

decent *n.* A group of woodpeckers (Sparkes 1975).

den *n.* Animals in a den or lair (*e.g.,* a den of foxes or snakes) (Sparkes 1975).

desert See ²group: deceit.

diphyletic group *n.* A group that is derived from two distinct ancestral lineages; one that is comprised of descendants of two different ancestors (given as the adjective "diphyletic" in Lincoln et al. 1985).
n. diphyly

dissimulation *n.* A flock of birds (Sparkes 1975).

doggery *n.* A group of dogs (Sparkes 1975).

dole, dule *n.* A company of doves (Sparkes 1975).
[French *deuil*, mourning, from doves "mournful" calls (Lipton 1968, 36)]

dout *n.* A group of wild cats (Sparkes 1975).

down *n.* A group hares or sheep (Sparkes 1975).

doylt *n.* A group of swine (Sparkes 1975).

drave *n.* A haul, or shoal, of fish (Sparkes 1975).

dray *n.* A group of squirrels (Sparkes 1975).
[Middle English *dray*, squirrel nest (Lipton 1968, 52)]

drift, drive, creaght *n.* A group of animals that are driven or move along in a body (*e.g.,* a drift of bees, birds, cattle, Fishers, sheep, or swans) (Sparkes 1975).

drove, concourse, drift, flock, mulada *n.* A group of cattle, or other animals, that are driven in a body; a crowd of people that moves in one direction (Sparkes 1975).
Comment: Some usages are drove of Asses, hares, Bullocks, Oxen, sheep, or swine (Sparkes 1975).

dryft *n.* A herd of tame swine (Sparkes 1975).

dule See ²group: dole.

ensemble *n.* A group of species in the same taxon that live in the same location at the same time (*i.e.,* live in the same community) and are in the same guild (*i.e.,* a group of species that exploit the same class of environmental resources in a similar way); *e.g.,* the ensemble of seed-eating insects in the Mojave desert; contrasted with assemblage, ²community, guild, local

guild, and taxon in the Venn diagram of Fauth et al. (1996, 283).

erst *n.* A group of bees (Sparkes 1975).

escargatoire *n.* A nursery of snails (Sparkes 1975).
[French *escargot*, snail]

exaltation *n.* A flock of larks (Sparkes 1975).
[from an observer's viewpoint (Lipton 1968, 9)]

eye, nye *n.* A brood of pheasants or other animals (Sparkes 1975).

eyrar *n.* A brood of swans (Sparkes 1975).

eyrie See ²group: aerie.

fall, brood, cast, clutch *n.* A brood of lambs, spawn, or Woodchucks (Sparkes 1975).
cf. ²group: spawn

family See ¹,²,³family; ²group: family (table).

fare *n.* A group of animals, *e.g.,* a fare of pigs (litter) or a fare of flies (swarm) (Sparkes 1975).

fesnyng, fesynes *n.* A pack of Ferrets (a 15th-century word) (Sparkes 1975).

fission-fusion group *n.* In the Spider Monkey, some other more-derived primate species: a casual social group that changes in size due to joining and leaving of smaller groups of conspecific animals (Kummer 1971 in Wilson 1975, 137, 522).
syn. fusion-fission group (Wilson 1975, 137)

flick *n.* A group of rabbits or hares (Sparkes 1975).

flight, bevy, covey, skein *n.* A group of things that pass through the air together, *e.g.,* flight of birds (young birds that have taken flight together), cormorants, doves, Dunbirds, Dunlin, Goshawks, larks, pigeons, plover, rails, storks, swallows, or widgeon (Sparkes 1975).

fling *n.* A flock of Dunlin or other sandpipers (Sparkes 1975).
Comment: Another usage is fling of Oxbirds (Sparkes 1975)

flock *n.*
1. A group of conspecific animals that feed and travel together, especially birds, goats, and sheep (*Oxford English Dictionary* 1972, entries from 1299).
   Note: This is the most frequently used term for a group of birds (Choate 1985, 13).
2. A number of domestic animals, especially sheep or goats, tended by one or more persons (*Oxford English Dictionary* 1972, entries from 1300).
3. A company of birds or mammals, including people, *e.g.,* a flock of Auks (at sea), birds, Bitterns, Bustards, Camels, Coots, Cranes, Elephants, Fish, Geese, Lice, Lions, Parrots, Seals, Sheep, or Swifts (Sparkes 1975).

syn. bevy, bew, bow, canaille, covy, deceit, dissimulation, drove, exaltation, fling, gaggle, herd, hurtle, line, loft, lute, meinie, meiny, mob, murder, murmuration, raft, rafter, screech, scry, skein, sord, swarm, tiding, trace, trip, volery, wing (Sparkes 1975); many flock "synonyms" are explained in more detail under ²group.

flote *n.* A herd of Cattle; a shoal of fish (Sparkes 1975).

flurry *n.* A fluttering group of things (*e.g.,* a group of birds that flutter around before they set out or when they settle down on a lake or marsh) (Sparkes 1975).

flush *n.* A flock of startled birds (*e.g.,* mallards) (Sparkes 1975).

fluther, smuth, stuck *n.* A group of jellyfish (Sparkes 1975).
cf. ²group: smack

fold *n.* A flock that is enclosed within a fence or shelter (*e.g.,* a fold of sheep) (Sparkes 1975).

form *n.*
1. In plants, the lowest taxonomic rank in classification right below subvariety (Lincoln et al. 1985).
   syn. forma, morph (Lincoln et al. 1985), forma, *forma* (Lawrence 1957, 55)
2. In plants, any minor variant or recognizable subset of a population or species based on characteristics of corollas, fruit, habitat, or other things (Lawrence 1957, 55).
3. In plants, any sporadic variant, irrespective of its morphological variation or constancy, with the significant criterion that this variant has no geographic discontinuity (Lawrence 1957, 55).
syn. race, variety (of many botanists, Lawrence 1957, 55)

forma See ²group: form.

frith *n.* Woods; wooded country (Sparkes 1975).
cf. ²community: xylium

froggery *n.* "A gathering of frogs" (Sparkes 1975).

fry *n.*
1. The young, or hatched brood, of a fish (Sparkes 1975).
2. The young, or brood, of animals such as oysters (young), bees (eggs, larvae, and pupae), or eels (spawns) (Sparkes 1975).
cf. ²group: spawn

functional feeding group *n.* A group of insect species that performs a particular kind of feeding (Cummins and Merritt in Merritt and Cummins 1984, 63).
cf. feeding, guild
Comments: Cummins and Merritt (in Merritt and Cummins 1984, 63) consider a functional feeding group to be analogous to "guild," defined as a group of organisms that use a particular resource class. Aquatic insects in functional feeding groups are usually omnivores because they often ingest algae (including diatoms), bacteria, fungi, microarthropods, and protozoans with their plant food (Merritt and Cummins 1984, 61). Cummins and Merritt (in Merritt and Cummins 1984) indicate which insect orders have dominant representatives in different functional feeding groups.

▸ **"animal piercers"** *n.* A functional feeding group of aquatic insects whose species puncture and suck fluids from animals; contrasted with "plant piercers" and "predators" (Cummins and Merritt in Merritt and Cummins 1984, 63).
Comments: Animal piercers include Backswimmers, Broad-Shouldered Water-Striders, Creeping Water Bugs, Giant Water Bugs, Pigmy Backswimmers, Riffle Bugs, Velvet Water Bugs, Water-Boatmen, Water-Measurers (= Marsh-Treaders), Waterscorpions, Water-Striders, and Water-Treaders (Borror et al. 1989). These animals can also be catergorized as "predators" in the broad sense of this word.

▸ **"dead-plant shredders"** *n.* A functional feeding group of aquatic insects whose species consume dead-plant material that they shred before eating; contrasted with "live-plant shredders" (Cummins and Merritt in Merritt and Cummins 1984, 63).
Comments: "Dead-plant shredders" include chewers of coarse particulate organic material and gougers that consume wood (Cummins and Merritt in Merritt and Cummins 1984, 61). This group includes many crane-fly, mayfly, and stonefly species and is important in recycling a large amount of plant material that falls and washes into bodies of water. Dead-plant shredders are saphrophages.

▸ **"filtering collectors," suspension feeders** *n.* A functional feeding group of aquatic insects whose species obtain food by filtering it from water; contrasted with gathering collectors (Cummins and Merritt in Merritt and Cummins 1984, 63).
Comment: "Filtering collectors" consume fine particulate organic material (Cummins and Merritt in Merritt and Cummins 1984, 61). This group includes net-making caddisfly species.

▸ **gathering collectors, deposit feeders** *n.* A functional feeding group of aquatic insects whose species obtain food from losse surface films and sediments; contrasted with "filtering collectors" (Cummins and Merritt in Merritt and Cummins 1984, 63).

syn. collectors-gatherers (Harper and Stewart in Merritt and Cummins 1984, 226; Harrahy et al. 1994, 517)

Comment: Gathering collectors consume fine particulate organic material in loose surface films and sediments (Cummins and Merritt in Merritt and Cummins 1984, 61).

▶ **"live-plant shredders"** *n.* A functional feeding group of aquatic insects whose species consume living plants whose parts they shred before eating; contrasted with "dead-plant shredders" (Cummins and Merritt in Merritt and Cummins 1984, 63).

Comment: "Live-plant shredders" include "chewers" and "miners" (Cummins and Merritt in Merritt and Cummins 1984, 61). This group includes Nymphuline Pyralid Moths (Borror et al. 1989) and plant parasites and predators in the broad senses of these terms.

▶ **parasites** *n.* A functional feeding group of aquatic parasitic insects (Cummins and Merritt in Merritt and Cummins 1984, 61).
cf. parasite

▶ **"plant piercers"** *n.* A functional feeding group of aquatic insects whose species puncture and suck fluids from plants; contrasted with "animal piercers" (Cummins and Merritt in Merritt and Cummins 1984, 63).

Comment: This group includes some species of mymarid chalcidoid wasps (Borror et al. 1989, 712).

▶ **predators, engulfers** *n.* A functional feeding group of aquatic insects whose species attack prey and ingest their parts or entire bodies; contrasted with animal piercers (Cummins and Merritt in Merritt and Cummins 1984, 61–63).
cf. predator

Comments: Predators includes larvae of Crawling Water Beetles, Damselflies, Dragonflies, Phantom Midges, Water-Scavenger Beetles, and Whirligig Beetles; larvae of some caddisfly, crane-fly, and mosquito species; and adults and larvae of Predaceous Diving Beetles and Trout-Stream Beetles (Borror et al. 1989). "Animal piercers" that kill their hosts by sucking out their fluids are predators in the broad sense of this term.

▶ **scrapers, grazers** *n.* A functional feeding group of aquatic insects whose species obtain food by scraping it off other objects (Cummins and Merritt in Merritt and Cummins 1984, 63).

Comment: Scrapers consume periphyton (attached algae and associated organ-

isms) (Cummins and Merritt in Merritt and Cummins 1984, 61).

gaggle *n.* A flock of geese (Sparkes 1975). [echoic word based on goose vocalizations]

gam *n.* A herd, or school, of whales or porpoises; whaling ships in company (Sparkes 1975).
[from "gam," an exchange of crews via whaleboats when two whale boats encounter each other, which reminded sailors of playful sporting of whales (Lipton 1968, 34)]

game *n.* A group of animals raised and kept for sport or pleasure (*e.g.,* a game of bees, swans, conies, or Red Deer) (Sparkes 1975).

gang *n.* A full set of things, including some animals (*e.g.,* a gang of Dogs, Elk, or Humans) (Sparkes 1975).

goosery *n.* A group of geese (Sparkes 1975).

great bevy *n.* Twelve head of Roe Deer (Sparkes 1975).

grind *n.* A school of Blackfish or Bottlenosed Whales (Sparkes 1975).

grist *n.* A lot of things such as bees or flies (Sparkes 1975).

grove, bosquet *n.* A small group of trees or something similar (Sparkes 1975).
cf. ²community: grove

gulp *n.* A group of cormorants or swallows (Lipton 1968; Sparkes 1975).

haras, harras *n.* A group of breeding mares or wild horses (Lipton 1968; Sparkes 1975).
[Latin *hara*, pigsty, hence any enclosure for animals (Lipton 1968, 56)]

hatch *n.* A group of young oviparous animals.

head *n.*
1. A single animal (*e.g.,* a head of Cattle) (Michaelis 1963).
2. A collection of animals; an indefinite number (*e.g.,* a head of pheasants, Cattle, or hungry Wolves) (Sparkes 1975).

herd *n.* An assemblage of animals, usually large ones (*e.g.,* a herd of Antelopes, Asses, Bison, Boars, Buffalo, Camels, Caribous, Cattle, Chamois, Cranes, Curlew, Deer, Elephants, Giraffes, Goats, Ibex, Moose, Oxen, parasites, ponies, Porpoises, Seals, Swans, Swine, Whales, or Wrens) (Sparkes 1975).
syn. baren, bow, creaght, dryft, flock, flote, mob, pace, rangale, rabble, sounder, zeal (of Zebras) (Sparkes 1975)
cf. ²group: little herd
Comment: Many synonyms of herd are explained in more detail under "group."

heronry *n.* A place where Herons congregate and breed (Michaelis 1963).

hive *n.* A group of oysters or bees (Sparkes 1975).

cf. hive

Comment: Some biologists prefer to use "hive" to mean a bee nest and "colony" to mean the bees themselves, as discussed under ²colony.

hoggery *n.* A group of Hogs (Sparkes 1975).

holophyletic group See ²group: monophyletic group (def. 2).

horde, gang, troop *n.* A large group of gnats, other insects, urchins, or wolves (Sparkes 1975).

host *n.* A large number of things such as daffodils, sparrows, or Humans (Sparkes 1975). See host.

hover *n.* A group of trout waiting for food; a group of crows (Sparkes 1975).

huddle *n.* A group of things crowded together (*e.g.,* a huddle of puppies, Walruses, or Humans) (Sparkes 1975).

hurtle *n.* A flock of sheep (Sparkes 1975).

huske *n.* A down, or group, of Hares (Sparkes 1975).

individualized group *n.* In some bird and mammal species: a group whose members recognize each other as individuals (Immelmann and Beer 1989, 122).

cf. recognition: individual recognition; ²society: open society

ingratitude *n.* A group of children (Sparkes 1975).

Imperium Naturae *n.* The three kingdoms: animals, minerals, and plants (Linnaeus in Minelli 1996, 1193).

intrusion *n.* A group of cockroaches (Lipton 1968).

jug, covey *n.* A group of partridge, quail, or roosting grouse (Sparkes 1975).

[from jug, to nestle or collect in a covey, therefore the covey itself, Sparkes 1975]

juvenile group *n.* For example, in some stink-bug and lepidopteran species: young that remain together after hatching (Heymer 1977, 97).

syn. sympaedium (Deegener 1918 in Heymer 1977)

kennel, crew, gang *n.* A pack of Dogs, hounds, Humans, or other animals (Sparkes 1975).

kindergarten *n.* In many ungulate species with lying out periods: a gathering of young animals (Immelmann and Beer 1989, 64).

cf. ²group: creche

kindle, kendle, kindling, kyndyll *n.* A litter or brood (*e.g.,* a kindle of hares, kittens, Leverets, or rabbits) (Sparkes 1975).

kingdom *n.*

1. A taxonomic group between superkingdom and subkingdom.
   See group above (see ²group: table).

2. One of Earth's six large plant communities (Boroughs 1999, 17)

Comments: These kingdoms include the Antarctic Kingdom, Australasian Kingdom, Boreal Kingdom, Cape Floral Kingdom (*syn.* fynbos, *q.v.*), Neotropical Kingdom, and Paleotropical Kingdom (Boroughs 1999, 17).

kip *n.*

1. A set, or bundle, of hides (Sparkes 1975).

2. Young or small beasts (*e.g.,* a kip of calves or lambs) (Sparkes 1975).

knob *n.* A group of Pochard or Widgeon (on water) (Sparkes 1975).

knot *n.* A small cluster, or group, of things (*e.g.,* a knot of stars, islands, mountains, palm trees, toads, young snakes, or people) (Sparkes 1975).

[from the appearance of a knot (Lipton 1968, 9)]

labor (*Brit.* **labour**) *n.* A group of moles; a 15th-century term (Sparkes 1975).

laughter *n.* A clutch of eggs (Sparkes 1975).

leafag *n.* A group of leaves (Sparkes 1975). *syn.* foliage (Sparkes 1975)

leap *n.* A group of Leopards (Lipton 1968). [after leopard leaping behavior (Lipton 1968)]

lease *n.* A group of three (*e.g.,* a lease of fish or hares) (Sparkes 1975).

cf. ²group: leash

leash, lece *n.* A sporting term for a group of three; a brace and a half; a tiere (*e.g.,* a leash of Trout, Teal, Snipe, Partridges, hawks, bucks, Deer, Hares, Foxes, Greyhounds, or hounds) (Sparkes 1975).

legion *n.* A large group of things (*e.g.,* a legion of whelps) (Sparkes 1975).

lepe *n.* A group of leopards (Sparkes 1975).

line, string *n.* A series, or rank, of things, usually of the same kinds, *e.g.,* a line of geese (flock), ponies, or people (Sparkes 1975).

lineage group *n.* A group of species allied by common descent (Wilson 1975, 588).

linkage group *n.* A group of genes associated to a greater or lesser extent by linkage; the maximum number of linkage groups is equal to the number of chromosomes in a cell (Lincoln et al. 1985).

litter *n.* The young of a viviparous vertebrate that are born at about the same time (*e.g.,* a litter of piglets, kittens, or puppies) (Sparkes 1975).

little herd *n.* Twenty deer (Sparkes 1975).

lodge *n.*

1. A group animals that live close together (*e.g.,* a lodge of Beavers or Otters) (Sparkes 1975).

2. A family unit of four or six people (Sparkes 1975).

loft *n.* A flock of Pigeons (Sparkes 1975).

lot, back, break, sort *n.* A group of things (*e.g.,* a lot of cattle or people) (Sparkes 1975).

lute *n.* A flock of Mallard (Sparkes 1975).

mating group *n.* A group of haploid, or diploid, individuals having genetic or environmental characteristics that favor mating within the group at the expense of mating outside the group (Lincoln et al. 1985).

matriarchy *n.* For example, in Elephants: a social group led by a female (Wilson 1975, 491).

cf. ²group: patriarchy
[a form of matriarch after patriarchy, Latin *mātr(i), māter,* mother + *archein,* to rule]

meiny, meinie *n.* A large group (*e.g.,* a meiny of brooks, Oxen, pilgrims, plants, sheep, or sparrows) (Sparkes 1975).

syn. flock, retinue, train (Sparkes 1975)

menagerie *n.* A collection of wild, or foreign, animals; an aviary (Sparkes 1975).

meute, meuse, mews *n.* A pack of hounds (Sparkes 1975).

cf. ²group: mews

mews *n.* A group of molting hawks or fattening hens or capons (Sparkes 1975).

cf. ²group: meute

middle bevy *n.* Ten Roe Deer or 40 deer in general (Sparkes 1975).

cf. ²group: bevy

migration *n.* A group of organisms that moves from one place to another (*e.g.,* a migration of locusts, butterflies, birds, or mammals, including Humans) (Sparkes 1975).

mixed group *n.* In some fish, bird, and mammal species: a group comprised of more than one species that usually lead similar lives (Heymer 1977, 29; Immelmann and Beer 1989, 187).

syn. mixed-species group (Immelmann and Beer 1989, 122)

mob, canaille, flock, herd *n.* A collection of things (*e.g.,* a mob of Ducks, Horses, Kangaroos, Sheep, Wallabies, Whales, or Humans) (Sparkes 1975).

monophyletic group *n.*

1. A group of taxa that are derived from one or more lineages from one immediately ancestral taxon of the same, or lower, rank [derived from Simpson's (1961, 124) definition of monophyly in Sober (1992a, 203)].
   Note: This is an evolutionary-taxonomist definition of monophyletic group. Reptilia is a monophyletic group based on this definition (Sober, 1992a, 205).

Problems with this definition have caused many evolutionary taxonomists to abandon it (Wiley 1981, 256).

2. "A group of species descended from a single ('stem') species, and which includes all species descended from this stem species" (Hennig 1966, 73, in Wiley 1981, 83, and Sober, 1992a, 203).
   Note: This is a cladist definition of monophyletic group. Reptilia is not a monophyletic group based on this definition (Sober, 1992a, 205).
   syn. holophyletic group (Ashlock 1971 in Sober, 1992a, 207; Lincoln et al. 1985; not a frequently used synonym, C.K. Starr, personal communication)

3. "A group of species in which every species is more closely related to every other species than to any species that is classified outside the group" (Hennig 1966 in Wiley 1981, 84).

4. "A group whose most recent common ancestor is cladistically a member of the group" (Ashlock 1971 in Wiley 1981, 84).

5. A group into which one places all species, or species groups, that are assumed to be descendants of a single hypothetical ancestral species; that is, a complete sister-group system (Nelson 1971 in Wiley 1981, 84).

6. "A group that includes a common ancestor and all of its descendants" (Farris 1974 in Wiley 1981, 84).

7. "A group with unique and unreversed group membership characters; an algorithm definition" (Farris 1974 in Wiley 1981, 84).

8. An assemblage of organisms that are related by descent from a common ancestor (Lincoln et al. 1985).

syn. natural group (Lincoln et al. 1985)
cf. ²group: paraphyletic group, phyletic group, polyphyletic group, sister group
Comments: Wiley (1981) and Sober (1992a) give more details regarding how to distinguish between the cladists' and evolutionary taxonomists' definitions of monophyletic group. Ashlock (1971) has classified both paraphyletic group and monophyletic group (*sensu* Hennig = Ashlock's holophyletic group) as kinds of monophyletic groups; others have objected to this classification (Wiley 1981, 258). Wiley (1981) gives other definitions of "monophyletic group."
["monophyly" coined by Haeckel (1866 in Wiley 1981, 85)]

mores, mune *n.* Groups of organisms that have similar ecological requirements and behavioral attributes (Lincoln et al. 1985).

cf. ²species: ecospecies

mulada See ²group: drove.

multimale group *n.* In many baboon and macaque species: a group comprised of several sexually mature males that live with sexually mature females and young of all ages (Immelmann and Beer 1989, 195).
cf. harem

murder *n.* A flock of crows (Sparkes 1975). [from an observer's viewpoint (Lipton 1968, 9)]

murmuration *n.* In falconry, a flock (of Starlings) (*Oxford English Dictionary* 1972, entries from 1470; Choate 1985, 13). [echoic word < French *murmurer* < Latin *murmurare*, to mummur]

muster, levy *n.* A number of things that are assembled on a particular occasion (*e.g.,* a muster of Peacocks, storks, troops, or other people) (Sparkes 1975).

mutation *n.* A group of molting birds; a mutation of thrushes (Sparkes 1975).
cf. mutation

mute *n.* "A pack of hounds" (Sparkes 1975).

natural group See ²group: monophyletic group.

nest *n.*
1. A number of animals that occupy the same nest or habitation (*Oxford English Dictionary* 1972, entries from 1470), *e.g.,* a nest of caterpillars, bees, wasps, vipers, chickens, toads, crocodiles, Dormice, mice, Hedgehogs, kittens, rabbits, or toads (Sparkes 1975).
2. "An accumulation of similar things" (*e.g.,* a nest of hummocks or low bushes) (Sparkes 1975).
v.i., v.t. nest
syn. aerie, brood, bike, colony, swarm (*Oxford English Dictionary* 1972; Sparkes 1975)
cf. ²group: aerie, brood, bike, colony, hive, swarm
[from the habitat (nest) of the animals involved (Lipton 1968, 9)]

nide, bike, nye *n.* A nest, or brood, of young birds (*e.g.,* a nide of eggs, geese, or pheasants) (Sparkes 1975).
cf. ²group: bike

nursery group See ²group: creche.

obstinacy *n.* A group of Buffaloes (Sparkes 1975).
[from an observer's viewpoint (Lipton 1968, 9)]

open anonymous group See ²society: open society: anonymous group.

ostentation *n.* A group of Peacocks (Sparkes 1975).

out-group *n.* A species, or a higher monophyletic taxon, that is examined in the course of a phylogenetic study to determine which of two homologous char-

acters may be inferred to be apomorphic (Wiley 1981, 7).
syn. out group (Shaffer et al. 1991, 286)
Comments: One to several out-groups may be examined in a study. The most critical out-group comparisons involve the sister group of the taxon studied (Wiley 1981, 7).

pace, herd *n.* A herd of Asses (Sparkes 1975).
[Latin *passus*, step, stride (Lipton 1968, 34)]

pack *n.* A group of animals (*e.g.,* a pack of Dogs, perch, grouse, hounds, Stoats, Weasels, or Wolves) (Sparkes 1975).
syn. baren, crew, fesnyng, fesynes, gang, kennel, mute, rabble, singular (Sparkes 1975)
Comment: Many synonyms of "pack" are explained in more detail under ²group.

paddling *n.* A group of ducks (Sparkes 1975).

parade, cortège, procession, walk *n.* A procession of animals (*e.g.,* a parade of Elephants or Humans) (Sparkes 1975).

paraphyletic group *n.*
1. "A group of species that has no ancestor in common only with them and thus no point of origin in time only to them in the true course of phylogeny" (Hennig 1966 in Wiley 1981, 84).
2. "A group based on symplesiomorphous characteristics" (Hennig 1966 in Wiley 1981, 84).
3. "A group that does not contain all of the descendants of the most recent common ancestor" (Ashlock 1971 in Wiley 1981, 84).
4. "An incomplete sister-group system lacking one species or one monophyletic species group" (Nelson 1971 in Wiley 1981, 84).
5. "A group that includes a common ancestor and some but not all of its descendants" (Farris 1974 in Wiley 1981, 84).
6. "A group with unique but reversed group membership characters" (Farris 1974 in Wiley 1981, 84).
cf. ²group: monophyletic group, phyletic group, polyphyletic group

parcel *n.* A small group of animals (*e.g.,* a parcel of chickens, penguins, crows, hens, horses, sheep, or Humans) (Sparkes 1975).

parliament *n.* A group of owls or fowls (Sparkes 1975).

party *n.*
1. A group of animals (*e.g.,* a party of birds, jays, or people) (Sparkes 1975).
2. A group of people that travel together (Sparkes 1975).

patriarchy *n.* For example, in the Vicuña: a social group led by a male (Wilson 1975, 490).
cf. ²group: matriarchy

[Greek *patriarchia*, office of a patriarch < Old French *patriarche* < Latin *patriarcha* < Greek *patriarchēs*, head of a family, *patria*, family, clan + *archein*, to rule]

peck *n.* A large group of bees (Sparkes 1975).

peep *n.* A brood of chickens (Sparkes 1975).

perversum confusum *n.* A unisexual aggregation of heterospecific individuals during their breeding season (Lincoln et al. 1985).

phalanx *n.*
1. A compact group of animals prepared for attack, defense, or both (*e.g.,* a phalanx of lawyers or troops) (Sparkes 1975).
2. A group of things drawn together in a common purpose (*e.g.,* a phalanx of migrating storks) (Sparkes 1975).

phyletic group *n.* "A group of species related to one another through common descent" (Wilson 1975, 591).
cf. ²group: holophyletic group, natural group, paraphyletic group, polyphyletic group

pinnacles of social evolution *pl. n.* Four animal groups: the more derived colonial invertebrates, the more derived social insects, the more derived social vertebrates, and Humans (Wilson 1975, 379).

plague, swarm *n.* A swarm of locusts (Lipton 1968, 22).
[from a human observer's viewpoint]
cf. ²group: swarm

pleophyletic group See ²group: polyphyletic group.

plump *n.* A compact body of animals or other things (*e.g.,* a plump of trees, wild fowl, ducks, Moorhens, seals, whales, or Humans) (Sparkes 1975).

pod *n.*
1. A school of fish whose bodies touch (Breder 1959 in Wilson 1975, 439, 591).
2. A small group of birds or mammals, for example, a pod of Whiting, birds (*e.g.,* Coots), otters, seals, whales, or porpoises (Sparkes 1975).
cf. ²group: turmoil
[from peas in a pod, a sailor's term (Lipton 1968, 33)]

polyphyletic group *n.*
1. A group of descendants minus their ancestor (inferred from Hennig 1966 by Wiley 1981, 84); a group of taxa derived from two or more distinct ancestral lineages (Lincoln et al. 1985).
   syn. pleophyletic group (inferred from Lincoln et al. 1985 who synonymize pleophyletic and polyphyletic)
2. "A group based on convergent similarity" (Hennig 1966 in Wiley 1981, 84); a group

based on convergence rather than common ancestry (Lincoln et al. 1985).
3. A group of descendants minus their most recent common ancestor (Ashlock 1971 in Wiley 1981, 84).
4. "An incomplete sister-group system lacking two species or monophyletic species groups that together do not form a single monophyletic group" (Nelson 1971 in Wiley 1981, 84).
5. "A group in which the most recent common ancestor is assigned to some other group and not the group itself" (Farris 1974 in Wiley 1981, 84).
6. "A group whose membership characters are not uniquely derived" (Farris 1974 in Wiley 1981, 84).
cf. ²group: monophyletic group, paraphyletic group, phyletic group

population See ²population.

pride *n.* A group of animals (*e.g.,* a pride of Lions or Peacocks) (Sparkes 1975).

procession See ²group: parade.

pseudosocial group, aggregation *n.* A group of animals that gather "by accident" due to individual animals' seeking the same kind of place (Immelmann and Beer 1989, 122).
cf. ²group: aggregation

psittosis *n.* A group of parrots (Sparkes 1975).
Comment: This is a modern pun (Sparkes 1975).

puddling *n.* A group of Mallards (Sparkes 1975).

quartet, quartett *n.*
1. A set of four human singers, or players, who make music (render a quartet); a set of four persons (*Oxford English Dictionary* 1972, entries from 1814).
2. For example, in *Hyla* frogs: a group of four calling individuals that alternate calls (Wilson 1975, 443).
cf. ²group: duet, trio, chorus

rabbitry *n.* A group of hutches of tame rabbits (Sparkes 1975).

rabble, mon, swarm *n.* A pack, string, or swarm of animals, *e.g.,* a rabble of bees, butterflies, flies (including gnats), or other insects (Sparkes 1975).

raft *n.*
1. A large group of things that is taken indiscriminately (*e.g.,* a raft of people) (Sparkes 1975).
2. "A dense flock of swimming birds," *e.g.,* a raft of auks (at sea) (Sparkes 1975) or ducks (Choate 1985, 13).

rafter *n.* A group of Turkeys (Sparkes 1975).

rag, rake *n.* A group of colts (Sparkes 1975).

rake *n.* A group of colts or Mules (Sparkes 1975).

ramage *n.* A group of branches (Sparkes 1975).

ramification *n.* A tree's branches (Sparkes 1975).

rangale *n.* A herd of deer (Sparkes 1975).

range *n.* A series of things; a role, line, or file (*e.g.,* a range of beehives) (Sparkes 1975).

richness *n.* A group of Martins (Lipton 1968).
[from an observer's viewpoint, Lipton 1968, 9]

rookery *n.*
1. A rook colony or breeding place; a group of Rooks (Sparkes 1975).
   syn. building (Sparkes 1975)
2. A breeding place, or large colony, of seabirds or marine mammals (*e.g.,* a rookery of Albatrosses, Herons, penguins, Rooks, or seals) (Sparkes 1975).
cf. ²group: colony

roost *n.* A group of fowl that roost together (Sparkes 1975).

rout, route *n.* A troop, throng, or company (*e.g.,* a rout of snails or Wolves) (Sparkes 1975).
[Old French *route*, troop, throng (Lipton 1968, 55)]

run *n.* A group of migrating aquatic animals (*e.g.,* a run of fish, salmon, or whales) (Sparkes 1975).

rush *n.* A group of animals that moves forward with great speed (*e.g.,* a rush of Dunbirds) (Sparkes 1975).

safe *n.* A group of ducks (Sparkes 1975). See ²group: sord.

sawt *n.* A group of Lions (Sparkes 1975).
[French leap]

school *n.* A group of individuals, many to all typically in the same life stage, that swim together in an organized fashion (Wilson 1975, 594), *e.g.,* a school of fish, Herrings, dolphins, gulls, pheasants, Hippopotami, porpoises, or whales (Sparkes 1975).
v.i. school
cf. ²group: aggregation
Comment: Shaw (1970 in Partridge 1982) and Partridge (1982, 298) review more than a dozen attempts to define "school," scientifically. This is very difficult because of the great variability in the kinds of schools.
[Middle English, *sceald*, shallow; "school" comes from an erroneous transcription of shoal (Lipton 1968, 9, 23)]
▶ **dense school** *n.* A school of poisoned fish of some species that is more tightly grouped than usual (Geiger et al. 1985, 11; R.A. Drummond, personal communication).

Comment: In a dense school, fish can appear "stacked" with all school members facing the same direction (Geiger et al. 1985, 11; R.A. Drummond, personal communication).
▶ **loose school** *n.* A fish school that is less tightly grouped than usual (*e.g.,* a group of poisoned fish compared to a control group) (Geiger et al. 1985, 11; R.A. Drummond, personal communication).

screech *n.* A flock of gulls (Sparkes 1975).

scry *n.* A flock of wild fowl (Sparkes 1975).

sedge *n.*
1. A group of rush-like marsh plants (Sparkes 1975).
2. A group of sea or marsh, birds (*e.g.,* a sedge of Bitterns, Cranes, or Herons) (Sparkes 1975).
syn. sege, siege (of Herons) (Sparkes 1975)

shoal *n.* A school of fish (Lipton 1968, 9).
[from the habitat of a school of fish (Lipton 1968, 9)]

shrewdness *n.* A group of apes (Lipton 1968).

siege See ²group: sedge.

single-male group See harem.

singular *n.* A pack of Boars (Sparkes 1975).

sister group *n.* A species, or a higher monophyletic taxon, hypothesized to be the closest genealogical relative of a given taxon exclusive of the ancestral species of both taxa (Wiley 1981, 7).
syn. sister-species group (Lincoln et al. 1985)
cf. ²group: holophyletic group, natural group, paraphyletic group, phyletic group, polyphyletic group; species: sibling species

sitting *n.* The number of eggs that a fowl covers in a single brooding (Sparkes 1975).

skein *n.*
1. A flight of geese; a line of geese in flight (Michaelis 1963; Choate 1985, 13).
2. A flock of wild fowl, ducks, or geese (Sparkes 1975).

skulk *n.* A group of foxes (Lipton 1968, 9).
[from fox behavior, Lipton 1968, 9]

skull *n.* A shoal of fish (Sparkes 1975).
Comment: This is an obsolete term (Sparkes 1975).

sleuth, slewthe, sloth, slought *n.* A group of bears (Sparkes 1975).

smack *n.* A small group of jellyfish (Sparkes 1975).
cf. ²group: fluther

social group, society See ²society.

sord, sore, safe *n.* A flock of Mallards or other ducks (Sparkes 1975).

sort, sorte *n.* A group of animals (Sparkes 1975).

sounder *n.*
1. A herd of boars, Pigs, or Wild Swine (Sparkes 1975).
2. A group of swans (Sparkes 1975).
[English sounder < Norman French *soundre* < Old English *sunor*, herd, Lipton 1968, 50]

sowse *n.* A group of Lions (Sparkes 1975).

span *n.* A group of Mules (Sparkes 1975).

spawn *n.* A brood; numerous offspring (*Oxford English Dictionary* 1972, entries from 1590).
See spawn.
cf. ²group: fall, fry

species-group *n.*
1. A group of closely related species, usually with partially overlapping ranges (Lincoln et al. 1985).
   syn. superspecies (sometimes) (Lincoln et al. 1985)
   cf. ²species: superspecies
2. In zoology, the categories of species and subspecies (Lincoln et al. 1985).

spring *n.*
1. A group of animals that is flushed from its covert (*e.g.,* a spring of teal) (Sparkes 1975).
2. A group of teal (Choate 1985, 13).
[This term possibly comes from the bird's quick flight as they spring into flight (Choate 1985, 13)]

stable *n.* A group of horses (Sparkes 1975).
See ²group: stud.

stand, set *n.*
1. A hive of bees (Sparkes 1975).
2. A group of animals (*e.g.,* a stand of plovers or flamingos) (Sparkes 1975).
See hive.

string *n.* A line of ponies (Lipton 1968, 27).
See ²group: stud.
cf. ²group: line

stubbornness *n.* A group of Rhinoceros (Sparkes 1975).
See ²group: crash.
[from an observer's viewpoint (Lipton 1968, 9)]

stuck See ²group: smack, smuth.

stud, stable, string *n.* A group of Horses, or other animals, kept for breeding, racing, or riding (*e.g.,* a stud of Dogs, Horses, mares, or racehorses) (Sparkes 1975).

swarm *n.*
1. A group of Honey Bees that, at a particular season, leave their hive and form a dense cluster and may find a new nesting site or be artificially transferred to one (*Oxford English Dictionary* 1972, entries from 725).
   syn. synhesma (Lincoln et al. 1985)

2. A human crowd, throng, or multitude (*Oxford English Dictionary* 1972, entries from 1423).
3. A very large, dense group of insects, other small organisms, or rarely large animals (*Oxford English Dictionary* 1972, entries from 1560).
4. A large number of unicellular organisms, gametes, or small animals, usually in motion (*e.g.,* a swarm of zoospores, Palolo Worms, termites, flies, ants, bees, hornets, other wasps, or other insects) (Michaelis 1963; Barnes 1974, 276–277; Sparkes 1975).
syn. bike, fare, flock, nest, plague, rabble (Sparkes 1975)

sympedium See ²group: juvenile group.

syncheimadium *n.* A hibernating aggregation (Deegener 1918 in Wilson 1975).

taxon See taxon.

taxonomic groups See ²group (table).
▸ **phylum** *n.* The taxon between superphylum and subphylum.
See ²group (table).
syn. phylon (Lincoln et al. 1985)
Comments: Scientists divide animals into up to 40 phyla (Anonymous 1996b, 24). Some phyla are defined in Appendix 1.

team *n.*
1. A group of animals that may be moving together (*e.g.,* a team of ducks, swans, dolphins, or polo horses) (Sparkes 1975).
2. A group of animals that are harnessed together (*e.g.,* a team of carriage horses or Oxen) (Sparkes 1975).
syn. yoke (of Oxen) (Sparkes 1975)

teem *n.* A brood of ducks (Sparkes 1975).
[Anglo Saxon, Sparkes 1975]

tiding *n.* A flock of Magpies (Sparkes 1975).

toft *n.* A small group of trees (Sparkes 1975).

tok *n.* A nesting place of Capercaillies (Great Grouse); an assembly of Capercaillies (Sparkes 1975).

tower *n.* A group of towering things (*e.g.,* a tower of Giraffes) (Sparkes 1975).

town *n.* A local population of Prairie Dogs (Wilson 1975, 472).

trace *n.* A group of hares (Sparkes 1975).
[from hare traces in snow (Sparkes 1975)]

train *n.* A procession of Camels (Sparkes 1975).
syn. line (Sparkes 1975)
cf. ²group: parade

trait group *n.* "A recognizable entity comprised of more than one individual within a population that represents a unit upon which selection can act" (Wittenberger 1981, 622).

Comment: For any individual, a "trait group" is the group of conspecifics the individual interacts with during a specified time interval (Wittenberger 1981, 622).

tribe *n.*

1. Any group of Humans that perceives itself as a distinct group and is so perceived by other Humans outside its group (Wilson 1975, 565).

   Note: The group may be a human "race," a religious sect, a political group, or an occupational group. A tribe follows "a double standard morality — one kind of behavior for in-group relations, another for out-group" (Hardin 1972 in Wilson 1975, 565).

2. A social group that consists of a number of families; any group whose members have a common feature (*e.g.,* a tribe of sparrows, goats, or savages) (Sparkes 1975).

trio *n.* For example, in *Hyla* frogs: three calling individuals that alternate sounds (Wilson 1975, 443).

See animal sounds: trio.

cf. animal sounds: duet; ²group: chorus; quartet

trip *n.* A flock or troop; a brood or litter, *e.g.,* a trip of Dottrel (Migratory Plover), goats, Hares, hippies (Humans), seals, sheep, swine, or Wild Fowl (Sparkes 1975).

syn. trippe (of goats) (Sparkes 1975)

[trip of hippies seems to be a pun related to hippies' "tripping out" (getting high) on drugs and being on a "trip" (being high on drugs)]

troop, troupe *n.*

1. A group of things, *e.g.,* a troop of baboons, Dogfish, Humans (acrobats, actors, dancers, or soldiers), kangaroos, Lions, or monkeys (Sparkes 1975).

2. A conspecific group of apes, Humans, lemurs, or monkeys (Sparkes 1975).

3. In some baboon species: a sleeping unit comprised of as many as 750 individuals in regions where suitable shelters are scarce and as few as 12 where they are common (Wilson 1975, 534).

cf. ²group: band, horde, rout, trip

turmoil *n.* A group of porpoises (Sparkes 1975).

cf. ²group: pod

turn *n.* A group of turtles (Sparkes 1975).

tussock *n.* A tuft or small cluster (*e.g.,* a tussock of grass, hair, or twigs) (Sparkes 1975).

unkindness *n.* A group of Ravens (Sparkes 1975).

[from an observer's viewpoint, 15th-century term (Lipton 1968, 9)]

variety *n.*

1. In horticulture, any variant of a species (Lawrence 1957, 55).

2. In botany, a morphological variant of a species without regard to distribution (Lawrence 1957, 55).

3. In botany, a morphological variant with its own geographical distribution (Lawrence 1957, 55).

4. In botany, a morphological variant that shares an area in common with one or more other conspecific varieties (Lawrence 1957, 55).

5. In botany, a variant that represents a color, or habit, phase (Lawrence 1957, 55).

6. In organisms, an individual or group of individuals that differs from the type species in certain characters and is usually fertile with any other member of the species; a subdivision of species (Michaelis 1963).

7. In botany, a taxonomic rank between species and form (Lincoln et al. 1985).

   cf. ²species: subspecies

8. An infrasubspecific taxon, such as varietas or cultivar (Lincoln et al. 1985).

9. An ambiguous term for any variant group within a species (Lincoln et al. 1985).

10. A group that differs from other varieties of the same subspecies, including nongenetic variants and microgeographic races (Lincoln et al. 1985).

abbr. var.

Comments: Different workers designate the same intraspecific plant group as a subspecies, variety, or form (Lawrence 1957, 55). When more than one variety occurs within a species, specialists follow the trinomial form of nomenclature (Lawrence 1957, 54). For example, *Carex aquatilis* var. *aquatilis* (a sedge) is the typical element, and *Carex aquatilis* var. *altior* is an infraspecific taxon of this element.

volery See ²group: flock.

walk, parade *n.* A procession or a group of animals in procession (*e.g.,* a walk of snails or snipe) (Sparkes 1975).

ward *n.*

1. A group of Prairie Dogs (Wilson 1975, 472).

2. A group of Prairie Dog societies (coteries) separated from others by some kind of physical barrier, such as a band of vegetation, stream, or ridge (Wilson 1975, 472).

cf. ²group: colony, group, town

[Old English *weard*, watching, *weardian*, to watch, guard]

warp *n.* A set of four (*e.g.,* a warp of oysters or Herrings) (Sparkes 1975).

d – g

warren *n.*
1. The inhabitants of a warren (*Oxford English Dictionary* 1972, entries from 1607).
2. Any collection or assemblage of small animals (*Oxford English Dictionary* 1972, entries from 1607).
See warren.

watch *n.* A group of Nightingales (Lipton 1968, 53).

wedge *n.* A group of things in the form of a wedge (*e.g.,* a wedge of swans) (Sparkes 1975).

whisp, wisp *n.*
1. A group of Snipe (Sparkes 1975).
2. A group of shorebirds in flight (Choate 1985, 13).
[This term derives from the shorebird group's looking like a thin smoke column (Choate 1985, 13).]

wing *n.* A flock of plovers (Sparkes 1975).

wisp See ²group: whisp.

zoo *n.* The Zoological Gardens in Regent's Park, London; a similar collection of living animals elsewhere (*Oxford English Dictionary* 1972, entries from *ca.* 1847).
syn. zoological garden, zoological park (Michaelis 1963)
Comment: In the broad sense, some people now think of a zoo as a collection of living animals plus things including a research center, an education building, a miniature tropical rain forest, souvenir shop, and restaurant. The word "zoo" was popularized in the 1860s in the song "Walking the Zoo Is the Thing to Do" (Reuter 1992).

zoological garden, zoological park See ²group: zoo.

♦ **group calling** See animal sounds: song: communal call.

♦ **group cohesion** *n.* In social-animal species, especially those with closed societies, including Humans: a group of adult animal's continuous association, often with their offspring, over time (Eibl-Ebesfeldt 1972, 1973 in Heymer 1977, 80; Immelmann and Beer 1989, 121).

♦ **group consciousness** See ³consciousness: group consciousness.

♦ **group courtship** See courtship: communal courtship.

♦ **group effect** See effect: group effect.

♦ **group-living species** See ²species: group-living species.

♦ **group odor** See odor: group odor.

♦ **group pheromone** See odor: clan-specific odor.

♦ **group predation** See predation: group predation.

♦ **group recognition** See recognition: group recognition.

♦ **group scent** See odor: group odor.

♦ **group selection** See selection: group selection.

♦ **group song** See animal sounds: song: group song.

♦ **group territory** See territory: group territory.

♦ **group transport** See transport: group transport.

♦ **grouping, upward classification** *n.* Classifying organisms into larger hierarchical groups (Mayr 1982, 238).

♦ **grove** See ²community: grove; ²group: grove.

♦ **growing season** See season: growing season.

♦ **growl** See animal sounds: growl.

♦ **growth** *n.*
1. Something's action, process, or manner of growing both in material and immaterial senses; vegetative development; increase (*Oxford English Dictionary* 1972, entries from 1587).
2. An organism's increase in its biomass (Willson 1983, 5).
syn. auxesis, *q.v.*
cf. principle: principle of allocation

exponential growth *n.* "Growth, especially in the number of organisms of a population, that is a simple function of the size of the growing entity; the larger the entity, the faster it grows" (Wilson 1975, 584).

population growth *n.* Change in population size with time as a net result of natality, mortality, immigration, and emigration (Lincoln et al. 1985).

♦ **grunt, stomp** *v.t.* In the Human, some tortoise species: to cause earthworms to come to the surface of the ground by vibrating it (Kaufmann 1989, 10).
cf. animal sounds: grunt
Comment: Some tortoises grunt (for worms) by stomping their feet (Kaufmann 1989, 10).

♦ **grunt whistle** See animal sounds: grunt whistle.

♦ **Gruppenauslese** See selection (table).

♦ G_{ST} See index: fixation index: G_{ST}.

♦ **guano** *n.*
1. An accumulation of sea-bird droppings rich in phosphates and nitrates (*Oxford English Dictionary* 1972, entries from 1604).
2. An accumulation of bat droppings.

♦ **guard bee** See bee: guard bee.

♦ **guarding behavior** See behavior: guarding behavior: mate-guarding behavior.

♦ **guest** *n.*
1. A person who is entertained at the house, or table, of another (*Oxford English Dictionary* 1972, entries from 1800).

2. An animal that lives, breeds, or both within a heterospecific nest, domicile, or colony (Lincoln et al. 1985).

See parasite.

v.i., v.t. guest

cf. inquiline

♦ **guidance procedure** See procedure: guidance procedure.

♦ **guide species** See ²species: characteristic species.

♦ **guild** *n.*

1. An ecological group of plants that is distinguished from other ecological groups by a special mode of life and sometimes similar physiological requirements (*Webster's New International Dictionary of the English Language* 1960 in Balon 1975, 825).

2. A group of plants that is dependent on another plant in some way (*e.g.,* climbing vines, epiphytes, parasites, or saprophytes) (Atsatt and O'Dowd 1976, 24, who indicate that this botanical definition predates def. 2.)

3. "A group of species that exploit the same class of environmental resources in a similar way" [coined by Root 1967 in Hawkins and MacMahon 1989, 423, who indicate that Elton alluded to a "guildlike niche"].

   Notes: Root (1967) emphasized that a guild is (1) an ecologically appropriate context for studying interspecific competition, and (2) a natural ecological unit that recurs across communities. Cummins (1973) emphasized Root's second point and the use of guilds to simplify the complex, numerous interactions among species that characterize real communities and ecosystems. Hawkins and MacMahon (1989, 442) suggest that the concept of guild is more useful if it contains the idea that guild members are co-occurring, interacting species of different taxa.

4. A group of plants that are functionally dependent, or interdependent, with respect to their herbivores (Atsatt and O'Dowd 1976, 24).

   syn. plant-defense guild (Atsatt and O'Dowd 1976, 24)

   Note: This definition does not necessarily imply spatial association. "Although close spatial relationships are often important, functional guild boundaries are in each case defined by herbivore feeding behavior and dispersal capacity" (Atsatt and O'Dowd 1976, 24–25).

5. "All organisms that use the same investigator-defined resource" (Hawkins and MacMahon 1989, 444).

syn. functional group [coined by Cummins (1973 in Hawkins and MacMahon 1989, 425) for aquatic invertebrates]

cf. clique, -vore, parasite, predator

Comments: Different workers have used "guild" in ways that are very different from Root's (1967) original meaning, *e.g.,* reproductive styles, characteristics of a basic body plan, and physiological systems (Hawkins and MacMahon 1989, 423). Guild definitions are "loose and are inconsistent among investigators (or even among studies by the same investigator), and guilds are seldom analyzed before they are accepted as existing in nature;" some workers "view guilds as natural units worthy of detailed study, while others deny that guilds exist except in the minds of ecologists" (Hawkins and MacMahon 1989, 424, 441). Fauth et al. (1996, 283) attempt to clear up definitions of assemblage, ensemble, ²community, local guild, and guild with the use of a Venn diagram. Balon (1975, 827) putatively classifies all living fishes into 32 ecoethological guilds in three divisions each with two subdivisions. Balon (1984, 40–41) classifies reproductive styles into 34 reproductive guilds in three sections. Jaksic (1981 in Hawkins and MacMahon 1989, 426) distinguished between resource-based "true community guilds" and taxonomically based "assemblage guilds." Researchers have described many kinds of guilds, including "dung feeders," "epiphyte grazers," "fungivores," "gall formers," "generalists," "large-particle shredders," "macrophyte piercers," "nectarivores," "omnivores (scavengers)," "parasites," "pit feeders," "predators" (including "engulfers"), "sap feeders" (miners and suckers), "scrapers" of algae and other attached organic material, "seed feeders," "small-particle shredders," "stem borers," "strip feeders" ("chewers"), "tourists," and "wood borers" (references in Hawkins and MacMahon 1989, 430, 432, 440–441). Hubbell et al. (1999, 554) classify light-gap-inhabiting woody plants into three regeneration niche guilds: "intermediate species," "shade-tolerant species," and "strongly-gap-dependent-pioneer species."

local guild *n.* A group of species that live in the same location at the same time (*i.e.,* live in the same community) and are in the same guild (*i.e.,* a group of species that exploit the same class of environmental resources in a similar way); *e.g.,* local bird species of a foliage-gleaning guild; contrasted with assemblage, ensemble, ²community, guild, and taxon in the Venn diagram of Fauth et al. (1996, 283).

management guild *n*. A group of species that respond in a similar way to a variety of changes likely to affect their environment [coined by Verner 1984 in Hawkins and MacMahon 1989, 441].

♦ **Gulick effect** See drift: random drift.

♦ **gumivore** See -vore: gumivore.

♦ **Gunflint Formation** See geological stratum: Gunflint Formation.

♦ **gurgle** See animal sounds: grunt: baby-grunt.

♦ **gustation** See modality: chemoreception: gustation.

♦ **gymnogamy** See -gamy: gymnogamy.

♦ **gynandromorphism** See -morphism: gynandromorphism.

♦ **gyne** See caste: gyne.

♦ **gynecoid** *n*. In ants: referring to a female that is queen-like, specifically a worker with some typically queen characters (*e.g.,* an enlarged abdomen) (Hölldobler and Wilson 1990, 638).

♦ **gynecogeny** See parthenogenesis.

♦ **gynergate** See caste: gynergate.

♦ **gynic** *adj*. Female (Lincoln et al. 1985). *cf*. andric

♦ **-gyno, gyn** *combining affix*
  1. "Woman; female" (Michaelis 1963).
  2. "Female reproductive organ; ovary; pistil" (Michaelis 1963).
  [Greek *gynē*, woman]

♦ **gynochore** See -chore: gynochore.

♦ **gynochromatypic coloration** See coloration: gynochromatypic coloration.

♦ **gynogamete** See gamete: gynogamete.

♦ **gynogenesis** See parthenogenesis: gynogenesis.

♦ **gynogenetic** See study of: genetic: gynogenetic.

♦ **gynogenous** See -genous: gynogenous.

♦ **gynogeny** See -geny: gynogeny.

♦ **gynomorphic** See -morphic: gynomorphic.

♦ **gynopaedium** See -paedium: gynopaedium.

♦ **-gynous** See -gyny.

♦ **-gyny** *combining form* Female.
See mating system.
adj. -genic, -gynous
[Greek *gynē*, woman]

heterogyny *n*. An organism's having two kinds of females (Lincoln et al. 1985).

hologyny See character: hologeny.

metagyny See -gyny: protogyny.

monogyny *n*. In some social-insect species: the existence of only one functional queen in a colony (Wilson 1975, 589).
cf. -gyny: oligyny, polygyny
Comment: Hölldobler and Wilson (1990, 640) distinguish between "primary monogyny" and "secondary monogyny."

▸ **functional monogyny** *n*. In some ant species: the condition in which several mated queens coexist, but only one produces reproductive brood (Hölldobler and Wilson 1990, 638).

oligyny *n*. In ants: two to several functional queens' occurring in the same colony (Hölldobler and Wilson 1990, 641).
cf. -gyny: monogyny, polygyny

poecilogyny *n*. A species' having more than one form of female (Lincoln et al. 1985).
cf. poecilandry

polygyny *n*. In some social-insect species: the coexistence of two or more egg-laying queens in the same colony (Wilson 1975, 592).
See mating system.
adj. multiple queen, polygynous (Bourke 1991, 295)
cf. -gyny: monogyny, oligyny

protogyny *n*. The rare phenomenon in insects in which females tend to emerge from their natal nests sooner than males (*e.g.,* in a tephritid fly and mosquito species, and some solitary bee species) (Thornhill and Alcock 1983, 104).
See hermaphrodism: protogynous hermaphrodism.
adj. protogenous, proterogynous
syn. metagyny (Lincoln et al. 1985)
cf. protandry

pseudopolygyny *n*. In ants: several dealated virgin queens' coexisting with one egg-laying mated queen (Hölldobler and Wilson 1990, 642).

spanogyny *n*. A scarcity, or progressive decrease, in the number of females (Lincoln et al. 1985).

♦ **gypsophile** See -¹phile: gypsophile.

h

♦ **H** See index: species-diversity index: Shannon-Weiner Function.

♦ **habit** *n.*

1. A person's bearing, demeanor, deportment, behavior; posture (*Oxford English Dictionary* 1972, entries from 1413).
   Note: This is an obsolete definition.
2. A person's bodily condition or constitution (*Oxford English Dictionary* 1972, entries from 1576).
3. An individual organism's characteristic mode of growth and general appearance (*Oxford English Dictionary* 1972, entries from Ray 1691); all of a species' adaptations present at a particular time, including its external appearance and growth form (Gregory 1913, 1936 in Mayr 1982, 590; Lincoln et al. 1985).
   syn. ad hoc adaptations, constitutional type, habitus (Lincoln et al. 1985).
4. A person's mental, or moral, constitution (*Oxford English Dictionary* 1972, entries from 1386).
5. A person's settled disposition, or tendency, to act in a certain way, especially one acquired by frequent repetition of the same act until it becomes quite, or almost, involuntary; a settled practice, custom, usage; a customary way, or manner, of acting (*Oxford English Dictionary* 1972, entries from 1581).
6. An individual animal's stimulus-response connection that it acquires through experience (Immelmann and Beer 1989, 125).
7. In animals, especially Humans: an individual's recurrent behavior performed without its thought or attention (Immelmann and Beer 1989, 125).
8. An animal's behavior; *habits* of a certain species (Immelmann and Beer 1989, 125).
See behavior.
v.i., v.t. habit

syn. behavior, behavior pattern, conditioned reflex (Immelmann and Beer 1989, 125)
cf. behavior, heritage

♦ **habit formation** See learning: becoming accustomed.

♦ **habit preening** See grooming: habit preening.

♦ **habitat** *n.*

1. The locality (*e.g.,* chalky hill, rocky cliff, or sea shore) in which an organism normally grows or lives (*Oxford English Dictionary* 1972, entries from 1854).
   Note: "Habitat" sometimes means an organism's geographical range or a special locality to which it is confined; a particular spot where a specimen is found (*Oxford English Dictionary* 1972).
2. "The organisms and physical environment in a particular place" (Wilson 1975, 585).
3. The place in a physical environment (*e.g.,* climate) and biotic community (*e.g.,* vegetation type) where a species lives at a particular time (Wittenberger 1981, 616; Immelmann and Beer 1989, 125).
4. Broadly, an area of Earth's physical environment that is distinct from other areas in a broad range of abiotic and biotic variables, *e.g.,* a coral-reef habitat or a running-water habitat (= lotic habitat) (Kramer et al. 1995, 2).
   syn. ecosystem, habitat type
5. Narrowly, an area within an ecosystem that differs in enough ways to appear qualitatively distinct from other areas, *e.g.,* a pool habitat and a riffle habitat within a stream, or a backreef habitat and reef-crest habitat within a coral reef (Kramer et al. 1995, 2).

syn. biotope (of some authors), habitation (*Oxford English Dictionary* 1972), ece, local environment, oike, oikos (Lincoln et al. 1985)
cf. biome, biotope, ²community

Comments: The same term (*e.g.,* bog, forest, marsh, or prairie) can refer to a habitat or community. Some kinds of habitats are defined below. Those not defined below include coniferous forest, Coral reef, deciduous forest, elfin forest, Jack Pine plain, jungle, krummholz, lake, northern hardwoods, ocean, pond, rain forest, river, spring, stream, tidal zone, and wood. Habitats vary in their extents from those of small ranges (*e.g.,* grass bald or shale barren) to those of large ranges (*e.g.,* forest or ocean).

SOME CLASSIFICATIONS OF HABITATS[a]

I. Classification with regard to amount of homogeneity
    A. synusia
II. Classification with regard to amount of water, kind of water, etc.
    A. aquatic habitat
       1. acidic habitat
          a. dystrophic lake (*syn.* acidic lake)
             (1) bog lake
             (2) brown-water lake
             (3) humic lake
          b. fen
          c. volcanic lake
       2. alkaline habitat
          a. desert-alkali lake (*syn.* soda lake)
          b. fen
          c. volcanic lake
       3. brackish habitat
          a. bay
          b. beach (*syn.* coast, shore)
             (1) back shore
             (2) fore shore
             (3) litteral zone (*syn.* intertidal zone)
             (4) near-subtidal zone
             (5) supralittoral zone
          c. canyon
          d. cave
          e. cliff
          f. creek (*syn.* gut, run, stream)
          g. estuary
          h. fluvial plain
          i. lake
          j. lagoon
          k. marsh
          l. mudflat
          m. near-subtidal zone
          n. pond
             (1) limnetic zone (*syn.* open-water zone, submersed-plant zone)
             (2) littoral zone
                (a) emergent-plant zone
                (b) floating-leaf plant zone
             (3) profundal zone
          o. ridge
          p. river
          q. spring
          r. sublittoral zone (*syn.* subtidal zone)
          s. surface film
          t. swale
          u. swamp
          v. swamp forest
          w. talus slope
          x. terrace
          y. tide pool
          z. wet prairie
       2. freshwater
          a. bay
          b. beach (*syn.* coast, shore)
             (1) back shore
             (2) fore shore
             (3) litteral zone (*syn.* intertidal zone)
             (4) near-subtidal zone
             (5) supralittoral zone
          c. canyon
          d. cave
          e. cave pool
          f. cliff
          g. creek (*syn.* run, stream)
             (1) pool zone
             (2) rapids zone (*syn.* riffles zone)
          h. estuary
          i. everglade
          j. fen
          k. fluvial plain
          l. geobenthos
          m. glade
          n. impoundment
          o. lake
             (1) epilimnion
             (2) euthrophic lake
             (3) hypolimnion
             (4) mud
             (5) oligotrophic lake
             (6) thermocline
             (7) volcanic lake
          p. marsh
          q. mudflat
          r. phytobenthos
          s. pond
             (1) artificial pond
             (2) beaver pond
             (3) bog pond
             (4) cypress pond
             (5) farm pond
             (6) flood-plain pond
             (7) limnetic zone (*syn.* open-water zone, submersed-plant zone)

(8) littoral zone
    (a) emergent-plant zone
    (b) floating-leaf plant
       zone
(9) meadow-stream pond
(10) mill pond
(11) mountain pond
(12) profundal zone
(13) sink-hole pond
(14) temporary pond
  t. ridge
  u. river
  v. seep
  w. swale
  x. spring
  y. swamp
  z. swamp forest
aa. surface film
bb. talus slope
cc. terrace
dd. tide pool
ee. wet prairie
3. saltwater habitat
  a. bay
  b. beach (*syn.* coast, shore)
    (1) back shore
    (2) fore shore
    (3) neritic zone
      (a) littoral zone (*syn.*
        intertidal zone)
      (b) lower neritic zone
      (c) near-subtidal zone
      (d) subtidal zone
    (4) supralittoral zone (*syn.*
      supratidal zone)
  c. cave
  d. canyon
  e. cliff
  f. coral reef[b]
    (1) barren zone
    (2) breaker zone
    (3) deep forereef
    (4) forereef escarpment
    (5) forereef slope
    (6) forereef terrace
    (7) inshore zone
    (8) lagoon
    (9) mixed zone (*syn.*
      buttress zone)
    (10) rear zone
    (11) reef flat (*syn.* rubble
      zone)
  g. estuary
    (1) salt-marsh estuary
  h. desert salt lake
  i. lagoon
  j. marsh
    (1) tide-land marsh
  k. mudflat
  l. ocean
    (1) abyssal plain

(2) abyssopelagic zone
(3) abyssobenthic zone
(4) bathybenthic zone
(5) bathypelagic zone
(6) benthos
(7) mesopelagic zone (*syn.*
    twilight zone)
(8) cave
(9) canyon
(10) cliff
(11) continental rise
(12) continental shelf
(13) continental slope
(14) hadobenthic zone
(15) pelagic zone
    (a) epipelagic zone
(16) ocean rise
(17) ocean floor
(18) outer sublittoral zone
(19) ridge
(20) sea floor
(21) seamont
(22) talus slope
(23) terrace
(24) trench
  l. plain
    (1) fluvial plain
  m. pond
  n. sea
  o. spring
  p. surface film
  q. swale
  r. swamp
    (1) mangrove swamp
  s. swamp forest
  t. talus slope
  u. terrace
  v. tide pool
  w. wet prairie
B. terrestrial habitat
  1. atoll
  2. bay
  3. beach
  4. bush
  5. canyon
  6. cave
  7. cliff
  8. coast (*syn.* shore)
  9. dry prairie
10. dune
11. field
12. forest (*syn.* woods)
13. glade
14. grass bald
15. hammock
16. heath
17. hummock
18. island
19. meadow
20. maqui (*syn.* matorral)
21. orchard

h – m

22. plain
    a. alluvial plain
23. ridge
24. savanna (*syn.* savannah)
25. shale barrens
26. steppe
27. swale
28. talus slope
29. terrace
30. thicket
31. woodland

III. Classification with regard to amount of water movement
  A. lentic freshwater habitat
    1. lake
    2. pond
    3. stagnant pool
  B. lotic freshwater habitat
    1. river
      a. pool
      b. rapids (*syn.* riffle)
    2. stream
      a. pool
      b. rapids (*syn.* riffle)
IV. Classification with regard to complexity and size
  A. habitat (*syn.* ecosystem, habitat type)
  B. microhabitat
V. Classification with regard to supporting a population
  A. sink habitat
  B. source habitat

---

[a] Odum (1969).
[b] Kaplan (1982, 12).

bay

1. A body of water that is partly enclosed by land (*e.g.,* an inlet of the sea) (Michaelis 1963).
2. "A recess of lowland between hills" (Michaelis 1963).
3. In the U.S., land partly surrounded by forest (Michaelis 1963).
4. An often oval, shallow, wetland depression in the Eastern Coastal Plain of the U.S. from New Jersey south into northern Florida (Krajick 1997, 45, illustrations).

Notes: Bays are ringed with ridges of sand, and there are several hypotheses regarding how they formed; they possibly formed due to strong prevailing winds and ocean waves during a time of dropping sea level (Krajick 1997). The oval ones have northwest-southeast axes. Bays were probably important in the evolution of carnivorous plants, including the Venus flytrap, as well as many other organisms. People have drained most of the bays.

syn. Carolina bay (after the fact that bays are more concentrated in the Carolinas than other U.S. states), whale wallow (a term for a bay in the Delmarva Peninsula, after the erroneous hypothesis that a beached whale made the depression in land) (Krajick 1997, 48–49).

[bay, after bays trees, which often grow in these habitats (Krajick 1997, 48)]

[bay, Old French *baie* < Late Latin *baia*]

bog *n.*

1. A piece of wet, spongy ground, consisting chiefly of decaying or decayed moss and other organic matter, that is too soft to bear the weight of any heavy body (*Oxford English Dictionary* 1972, entries from 1505).
   syn. bogland, boggy soil, morass, moss (*Oxford English Dictionary* 1972)
2. A typically zoned habitat with a marsh-like area surrounding a pond through low and high shrubs and tree invaders, to a surrounding swamp forest, usually coniferous (Voss 1972, 17–18).

Notes: The edge of a bog pond is a floating (and hence quaking) mat that borders the open water and is formed largely of interlacing rhizomes and roots, usually with much sphagnum moss that causes the mat waters to be very acidic (Voss 1972, 18). Some bogs have alkaline areas, totally filled in ponds, or both.

See habitat: glade, swamp.

cf. [2]community: hydrophytium; habitat: fen, forest, bog forest, marsh, swamp

[Irish or Gaelic, *bogach,* a bog < *bog,* soft]

▸ **Arctic bog, palsa bog** *n.* A bog with an ice core characteristic of Arctic zones with permafrost that can form a small hillock up to 7 m high (Council of Europe Commission of the European Communities 1987, 198).

▸ **blanket bog** *n.* A bog that forms a continuous carpet on very wet hills in some parts of northern Europe (Council of Europe Commission of the European Communities 1987, 18).
Comment: The peat thickness of a blanket bog is up to 7 meters (Council of Europe Commission of the European Communities 1987, 16, illustration; 18).

▸ **raised bog** *n.* A bog with a convex profile that results from abundant rainfall and a peripheral extension supplied by an aquifer maintained within peat by infiltrating rainfall (Council of Europe Commission of the European Communities 1987, 16, illustration; 18).
Comment: The peat thickness of a raised bog may be up to 7 meters at the center of the bog's bulge (Council of Europe Commission of the European Communities 1987, 18).

▶ **string bog, aapa mire** *n.* A bog that occupies an elongated depression supplied with melt water and may be separated from another string bog by a peat ridge (Council of Europe Commission of the European Communities 1987, 19).

bog forest, coniferous swamp *n.* The area of a bog that is dominated by trees (Voss 1972, 18).
cf. ²community: forest

bogland, boggy soil See habitat: bog.

bush *n.*
1. A clump of shrubs; thicket; undergrowth (Michaelis 1963).
2. Wild, uncleared land covered with scrub; any rural or unsettled area; usually preceded by "the."
3. Wood lot (Canadian).
cf. forb; habitat: forest; shrub; tree

cliff *n.* A high, steep face of rock; a precipice (Michaelis 1963).
obs. clift (Michaelis 1963)
Comment: A similar face of unconsolidated sediment is also called a cliff (*e.g.,* Calvert Cliffs, MD, a rich Miocine fossil deposit primarily in silt and clay).

coast *n.* The land next to the sea; seashore (Michaelis 1963).
syn. shore (Michaelis 1963)
[Old French < *coste* < *costa*, rib, flank]
▶ **rocky coast** *n.* The rocky area, above an intertidal fringe habitat of larger marine algae in Mediterranean areas, which has *Zostera* spp. and *Cymodocea nodosa* (Council of Europe Commission of the European Communities 1987, 13).

coral reef *n.* A ridge of coral skeletons, with or without living coral, formed by the gradual buildup of these skeletons (Michaelis 1963).

dune *n.* A hill of loose sand heaped up by wind (Michaelis 1963).
syn. down (Michaelis 1963)
cf. habitat: dune slack
Comment: The Council of Europe Commission of the European Communities (1987, 14) describes calcareous and noncalcareous dunes.
[French < Middle Dutch akin to DOWN]
▶ **coastal dune** *n.* A dune along a body of water (*e.g.,* along an ocean or lake) (Council of Europe Commission of the European Communities 1987, 14; 15, illustration).
▶ **fixed calcareous dune, gray dune** *n.* An inshore dune that is less exposed to wind than a moving calcareous dune and is stabilized by grasses and other plants (Council of Europe Commission of the European Communities 1987, 14).

▶ **moving calcareous dune, white dune** *n.* A high littoral dune, rich in calcium carbonate and exposed to constant wind impact, colonized by grasses and other plants (Council of Europe Commission of the European Communities 1987, 14).

dune slack, panne, slack *n.* An original European hydrosere ranging from *Molinion* and *Littorellion* groupings to basic and minerotrophic mire communities (Council of Europe Commission of the European Communities 1987, 14; 15, illustration).

everglade *n.* A tract of low, swampy land covered with tall grass (Michaelis 1963).
cf. place: everglades

fen *n.*
1. Low land covered wholly, or partly, with water or subject to frequent inundations; a tract of such land (*Oxford English Dictionary* 1972, entries from Beowulf).
   See community: meadow: sedge meadow.
   syn. marsh (*Oxford English Dictionary* 1972)
2. A hollow where the water table is permanently above ground level (Council of Europe Commission of the European Communities 1987, 18).
   syn. soligenous mire (Council of Europe Commission of the European Communities 1987, 18)
3. A hillside where the water table is permanently above ground level (Council of Europe Commission of the European Communities 1987, 18).
syn. topogenous mire (Council of Europe Commission of the European Communities 1987, 18)
cf. bog, marsh, swamp
Comments: A fen can be acidic or alkaline (Council of Europe Commission of the European Communities 1987, 18). Most of the plant debris of a fen decomposes and only a few decimeters of fibrous peat accumulate from abundant sedges.
[Old English, *fen, fenn*]

field *n.*
1. Open ground as opposed to a forest; a stretch of open land; a plain (*Oxford English Dictionary* 1972, entries from 1050).
   Note: This is an obsolete definition.
2. Land that is tilled or used as a pasture, usually parted off by hedges, fences, boundary stones, and so forth (*Oxford English Dictionary* 1972, entries from 1025).
3. A piece of land with few, or no, trees, covered with grass, weeds, or similar

vegetation that are growing wild (Michaelis 1963).

4. "A piece of cleared land covered with grass, clover or other plants suitable for grazing animals and set aside for use as a pasture" (Michaelis 1963).

5. A piece of cleared, cultivated land on which crops are grown (Michaelis 1963).

6. An area that is, or has been, cultivated or pastured (Voss 1972, 20).

cf. ²community: amanthium; habitat: meadow, waste ground
[Old English *feld*, akin to Old English *folde*, earth]

▸ **old field**   *n.* A field that was cleared by Native Americans (Strausbaugh and Core 1978, x).

forest *n.*

1. A large tract of land covered with trees and undergrowth, sometimes intermingled with pasture; also the trees themselves (*Oxford English Dictionary* 1972, entries from 1300).

2. An area dominated by trees (Voss 1972, 18); contrasted with savanna, woodland.

3. A plant formation composed of trees with touching crowns that form a continuous canopy; contrasted with woodland (Allaby 1994).

4. In Great Britain, from Norman times, a district reserved for deer hunting, often belonging to the sovereign, managed by special officers, and to which special laws applied (Allaby 1994).
   Note: Such forest ranged from being plots of dense trees only to areas with trees and other habitats including bogs, grassland, and heaths (Allaby 1994).

See ²community: forest.
adj. forest, sylvan
syn. wood, woods (Michaelis 1963; Voss 1972, 18)
cf. habitat: bush
[Old French < Medieval Latin (*silva*) *foresta*, an unenclosed (wood) < Latin *foris*, outside]

▸ **broadleaf forest**   *n.* A forest with 81–100% angiosperms and 0–19% gymnosperms; contrasted with broad-leaved-evergreen forest, deciduous forest, evergreen forest, evergreen-mixed forest, mixed forest (inferred from Allaby 1994).

▸ **broad-leaved-evergreen forest**   *n.* A forest dominated by evergreen angiosperms; contrasted with broadleaf forest, deciduous forest, evergreen forest, evergreen-mixed forest, mixed forest (Allaby 1994).
Comment: This kind of forest occurs in the coastal plain of the Gulf of Mexico,

southern Japan, and central China, and south of the tropics, except for Chilean Patagonia (Allaby 1994).

▸ **deciduous forest**   *n.* An angiospermous forest in which most to all of the trees shed their leaves in autumn; contrasted with broadleaf forest, broad-leaved-evergreen forest, evergreen forest, evergreen-mixed forest, mixed forest (Allaby 1994). *syn.* deciduous summer forest (Allaby 1994)
Comments: Leaf loss is probably an adaptation to winter drought when the ground is frozen (Allaby 1994). Some gymnospermous trees (*e.g.,* Larches) are deciduous.

▸ **evergreen forest**   *n.*

1. A forest with 81–100% gymnosperms and 0–19% angiosperms (inferred from Allaby 1994).

2. A forest with no complete, seasonal leaf loss and comprised of angiosperms, gymnosperms, or both; contrasted with broadleaf forest, broad-leaved-evergreen forest, deciduous forest, evergreen-mixed forest, mixed forest (Allaby 1994).
   Note: Evergreen forests include temperate coniferous forests and tropical rain forests.

▸ **evergreen-mixed forest**   *n.* A forest dominated by both evergreen angiosperms and gymnosperms; contrasted with broadleaf forest, broad-leaved-evergreen forest, deciduous forest, evergreen forest, mixed forest (Allaby 1994).
Comments: This kind of forest occurs in particular in the southern hemisphere in Chile, New Zealand, South Africa, and Tasmania, and in the northern hemisphere in eastern Asia and the Mediterranean Basin (Allaby 1994).

▸ **mixed forest**   *n.* A forest with 20–80% angiosperms or 20–80% gymnosperms; contrasted with broadleaf forest, broad-leaved-evergreen forest, deciduous forest, evergreen forest, evergreen-mixed forest (inferred from Allaby 1994, who defines mixed woodland).

▸ **Niagara-Escarpment forest**   *n.* A sparse, old-growth forest on the cliffs of the Niagara Escarpment, Ontario, Canada, dominated by eastern white cedar, some up to 1600 years old (Douglas Larson and Steven Spring in Wheeler 1996, 76–82, illustrations).

geobenthos *n.* That part of the bottom of a stream, or lake, not covered by vegetation (Lincoln et al. 1985).
See ²community: geobenthos; habitat: geobenthos.
cf. ²community: benthos

glade *n.*

1. A clear, open space, or passage, in forest that is either natural or produced by Humans felling trees (*Oxford English Dictionary* 1972, entries from 1529).
2. A poorly drained, treeless area on a mountain in West Virginia that is similar to a more northern bog community in both appearance and floristic composition (Strausbaugh and Core 1978, xii).

syn. bog (Strausbaugh and Core 1978, xii)
[probably akin to *glad* in the obsolete sense of bright, sunny]

grass bald *n.* treeless mountain top in West Virginia covered with a more or less continuous stand of grasses and their kin (Strausbaugh and Core 1978, x).

grove *n.*

1. "A small stand of trees" (Lincoln et al. 1985).
2. A fragment of a former extensive forest (Lincoln et al. 1985).

cf. ²community: forest

hammock *n.* A thickly wooded tract of fertile land (that is often elevated), found in the southern U.S. (Michaelis 1963).

cf. habitat: hummock

Comments: The hammocks in The Everglades are on raised land that is drier than surrounding grasslands and harbor subtropical forest (Kritcher and Morrison 1988, 84–85, plate 20).

[a variation of hummock (origin unknown)]

heath *n.*

1. An area of open land in Britain overgrown with heath (*Erica*) or coarse herbage (Michaelis 1963).
   Note: Williams (1989, 6) discusses heath in the strict sense and broad senses.
2. A habitat dominated by ericaceous plants ("heath" and "heathland" in Council of Europe Commission of the European Communities 1987, 21, which describes many kinds of heathlands).

hummock *n.* A wooded tract of land that rises above an adjacent marsh or swamp (Michaelis 1963).

adj. hummocky

cf. habitat: hammock

[origin unknown]

maquis, matorral *n.* An area of low, sclerophyllous ligneous plants in the Mediterranean Region (Council of Europe Commission of the European Communities 1987, 51, illustration; 57).

See ²community: maquis.

marsh *n.*

1. A tract of low-lying land, flooded in winter and usually more or less watery throughout the year (*Oxford English Dictionary* 1972, entries from *ca.* 725).

2. A wet, treeless area (Voss 1972, 17); contrasted with bog, fen, meadow, swamp. See habitat: swamp.

syn. Marsch and polder (in the North Sea area) (Council of Europe Commission of the European Communities 1987, 14)

cf. ²community: mangrove

[Old English *mersc, merisc,* akin to morass]

▸ **freshwater marsh** *n.* A marsh in fresh water (Council of Europe Commission of the European Communities 1987, 14; 15, illustration)

cf. ²community: marsh

Comment: Plants of Maryland salt marshes are listed by Brown and Brown (1984, xx–xxi).

▸ **salt marsh** *n.* A marsh in salt water.

cf. ²community: marsh

Comment: Plants of Maryland salt marshes are listed by Brown and Brown (1984, xx). A salt marsh along the North Sea is called a "shorre" (Council of Europe Commission of the European Communities 1987, 13; 154, illustration).

▸ **tidal freshwater marsh** *n.* A marsh that occurs along a tidal freshwater river.

Comment: Examples of tidal freshwater marshes occur along rivers in eastern U.S. Dyke Marsh Wildlife Preserve, on the west side of the Potomac River near Washington, D.C., has a tide up to about 1 meter and high organism diversity.

meadow *n.*

1. A piece of land permanently covered with grass that is mown to be used as hay; any piece of grassland, used either for cropping or pasture; in some cases, well-watered ground usually near a river (*Oxford English Dictionary* 1972, entries from 969).
2. In North America, a low, level tract of uncultivated grassland especially along a river or a marshy region near the sea (*Oxford English Dictionary* 1972, entries from 1670).
3. One of a diversity of treeless areas such as a grassy field, a grazed area (= pasture), and a grassy, or sedgy, place this is usually less level (and therefore less wet) than what is generally termed a marsh; a vague term when it is used for so many kinds of habitats (Voss 1972, 20); contrasted with marsh, prairie, swamp.

cf. ²community: poium

[Old English *mǣdwe,* oblique case of *mǣd,* meadow]

▸ **sedge meadow** *n.* A treeless area that is dominated by sedges in West Virginia (Strausbaugh and Core 1978, vii). See fen.

syn. fen (Strausbaugh and Core 1978, vii)

h–m

microhabitat *n.*
1. A small, specialized habitat (Lincoln et al. 1985).
   syn. microenvironment (Lincoln et al. 1985)
2. A narrowly defined environment that a species occupies (Goldschmidt 1996, 248). *Notes:* For example, the cichlid species *Haplocromis argens* occurs almost exclusively in the uppermost 2 meters of water (Goldschmidt 1996, 248).
3. A division of a habitat that is relatively homogeneous and differs primarily in quantitative values of a small number of variables from other divisions of the same habitat, *e.g.,* the deep-water microhabitat and shallow-water microhabitat of a pool, or the live-branched-coral microhabitat and the rubble-substrate habitat within a backreef habitat (Kramer et al. 1995, 2).
Comments: Researchers often use "habitat" for the general type of place where an animal lives and, therefore, apply it to a scale larger than the animal's normal daily range. In contrast, researchers typically use "microhabitat" for a finer division of an animal's space use that occurs within its normal daily range.

morass See habitat: bog.

moss See habitat: bog.

mud flat *n.* A muddy area in a coast area (Council of Europe Commission of the European Communities 1987, 13).
syn. silkke (Council of Europe Commission of the European Communities 1987, 13)
▸ **marisma, sansourire** *n.* A Mediterranean, or thermoatlantic, mudflat (Council of Europe Commission of the European Communities 1987, 13).

orchard *n.*
1. A garden for herbs and fruit trees (obsolete definition) (*Oxford English Dictionary* 1972, entries from *ca.* 1000).
2. An enclosure for fruit-tree culture (*Oxford English Dictionary* 1972, entries from *ca.* 1000).
3. A plantation of trees grown for their products, such as fruit, nuts, or oils (Michaelis 1963).
4. The enclosure, or ground, that contains trees grown for their products (Michaelis 1963).
cf. ²community: dendrium, orchard
[Old English *orceard* < *ort-geard*, garden < *ort* (akin to Latin *hortus*) + *geard*, yard, enclosure]

phytobenthos *n.* The part of a stream, or lake, bottom covered with vegetation (Lincoln et al. 1985).
cf. ²community: phytobenthos

plain *n.* An expanse of low, treeless land; a prairie, *q.v.* (Michaelis 1963).
[Old French < Latin *planus,* flat]
▸ **fluvial plain** *n.* An alluvial plain that experiences periodic flooding (Council of Europe Commission of the European Communities 1987, 17; 16, illustration).

prairie *n.*
1. A treeless, level, or undulating tract of grassland, usually of great extent; applied chiefly to the grassy plains of North America (*Oxford English Dictionary* 1972, entries from *ca.* 1682).
   syn. savanna, steppe (*Oxford English Dictionary* 1972)
2. A naturally treeless area, dominated by grasses and often maintained in this condition by fire (Voss 1972, 20).
cf. ²community: amanthium, plain
[French *prairie,* large meadow < *pré,* meadow < *pratum*]
▸ **dry prairie** *n.* A prairie that is not periodically flooded; contrasted with wet prairie (Voss 1972, 20).
▸ **wet prairie** *n.* A prairie that is periodically flooded; contrasted with dry prairie (Voss 1972, 20).

sandy beach *n.* The sandy-shore habitat of a lake, ocean, pond, river, or stream.
cf. ²community: actium, thalassium

savanna *n.*
1. A treeless plain, properly one found in tropical America (*Oxford English Dictionary* 1972, entries from 1555).
2. A tract of level land covered with low plants (Michaelis 1963).
3. Any large area of tropical, or subtropical, grassland covered in part with trees and spiny shrubs (Michaelis 1963).
See prairie.
syn. savannah
cf. habitat: steppe
[earlier *zavana* < Spanish < Carib]

shale barrens *n.* (construed as singular or plural) A habitat of shale exposures in different stages of disintegration in the eastern region of the Central Appalachian Mountains.
[coined by Steele 1911, 359, after the areas of sparse, or no, vegetation in this habitat]
syn. shale-barrens (Wherry 1930, 43; Henry 1933, 65), shale bank (Artz 1937, 45), shale slope (Wherry 1929, 104)
cf. ²community: heath barren
Comments: "Shale barrens" refers to an area in this habitat from an isolated one to all of this habitat in the U.S. Valleys of shale barrens often have heavy clay soil derived from shale and were formerly covered with forest (Silberhorn 1968, 111–114, personal observation). Declivities and

uplands harbor a low, open oak-pine forest or shrubs and herbs. A shale barrens' kind and density of vegetation varies with the aspect of its slope and the direction that it faces, which affect moisture availability and rate of organic decomposition. Vegetation is absent on some steep, southerly facing slopes of shale barrens. The sparse vegetation of shale barrens grades into dense forest where conditions are moister (Wherry 1930, 43). Shale barrens usually occur on Devonian shale but also occur on Ordovician and Silurian shale (Artz 1937, 45; Platt 1951, 273; Strausbaugh and Core 1978, xviii–xix; Morse 1983). A shale barrens is up to about 20 hectares (Keener 1983, 71), and is not now believed to harbor any plant species restricted only to them.

"shale-barrens forest" *n.* Part of a shale barren that is dominated by more closely spaced trees.

"shale-barrens opening" *n.* Part of a shale barren that has no plants or is dominated by nonwoody plants and has no to only a few low woody plants.

"shale-barrens woodland" *n.* Part of a shale barren that is dominated by more widely spaced trees.

shore *n.* The coast, or land, next to lake, large river, ocean, or sea (Michaelis 1963).
Comment: Shores can be muddy, rocky, or sandy and have particular dominant organisms depending on a shore's location (Council of Europe Commission of the European Communities 1987, 13).

sink habitat *n.* A habitat in which a population has a net size decrease; contrasted with source habitat (Pulliam 1988, 652; Lewin 1989b, 477).
Comments: Pulliam (1988) argues that for many populations, a large fraction of individuals may regularly occur in sink habitats, where within-habitat reproduction is insufficient to balance local mortality; nevertheless, populations may persist in such habitats, being locally maintained by continued immigration from nearby, more-productive source habitats.

source habitat *n.* A habitat that supports a population's net size increase; contrasted with sink habitat (Pulliam 1988, 652; Lewin 1989b, 477).

steppe *n.*
1. One of the vast, comparatively level, treeless plains in Southeastern Europe and Siberia (*Oxford English Dictionary* 1972, entries from 1671).
   Note: Steppe also occurs in Southwestern and Western Asia (Robinson and Challinor 1995, 225).

2. An extensive, usually treeless plain (*Oxford English Dictionary* 1972, entries from 1837).
See habitat: prairie.
Comment: The Council of Europe Commission of the European Communities (1987, 69) describes steppe woodlands.
[Russian *step*]

swale *n.*
1. A hollow, low place; especially in the U.S., a moist, or marshy, depression in a tract of land, especially in the midst of rolling prairie (*Oxford English Dictionary* 1972, entries from 1584).
2. A naturally elongated depression, such as an interdunal trough; contrasted with ditch (Voss 1972, 20).
syn. swail (Michaelis 1963)
[probably from Scandinavian, *cf.* Old Norse *svalr*, cool]

swamp *n.*
1. A tract of low-lying ground in which water collects; a piece of wet spongy ground (*Oxford English Dictionary* 1972, entries from 1624).
   syn. bog, marsh (*Oxford English Dictionary* 1972)
2. A tract of rich soil harboring trees and other vegetation but too moist for cultivation (*Oxford English Dictionary* 1972, entries from 1741).
3. A wet wooded area, generally subjected to periodic flooding, as occurs on river flood plains; swamp forest (Voss 1972, 17).
syn. swampland (Michaelis 1963)
cf. ²community: forest; habitat: bog, marsh
[Lower German *swampen*, to quake (said of a bog), akin to sump]

swampland See ²community: swamp.

synusia *n.* A habitat of characteristic and uniform conditions (Lincoln et al. 1989).
See ²community: synusia.
adj. synusial

thicket *n.* Usually a small, or narrow, area characterized by dense shrubs or small second-growth trees (Voss 1972, 19).
See ²community: thicket.
[Old English *thiccet* < *thicce*, thick]

tundra *n.* A vast cold, dry, fairly level, treeless plain found in Asia, Europe, and North America (Robinson and Challinor 1995, 227).
cf. biome

veld, veldt *n.* An area of open grasslands with scattered shrubs and trees, found in parts of Southern Africa (Robinson and Challinor 1995, 227).

wetland *n.* A low-lying, water-saturated area, such as a bog, marsh, swamp, or swamp forest (Robinson and Challinor 1995, 227).

h–m

woodland *n.*

1. Land covered with wood (= trees); a wooded region or piece of ground (*Oxford English Dictionary* 1972, entries from 869).
2. An area with trees that are less dense than those in a forest (Voss 1972, 19).

cf. ²community: forest, xylium; habitat: bush; shale barren: shale-barren woodland

woods See habitat: forest.

♦ **habitat choice** See habitat selection.

♦ **habitat corridor** *n.* A linear band, or series of "stepping stones," of undisturbed or secondary habitat that connects two protected regions for a particular species (Buchmann and Nabhan 1996, 247).

♦ **habitat form** *n.* The characteristic growth form of an organism under a particular set of environmental conditions (Lincoln et al. 1985).

syn. ecad (Lincoln et al. 1985)

♦ **habitat fragmentation** *n.* The division of natural ecosystems into patchwork habitats as a result of forestry, land conversion for agriculture, and urbanization (Buchmann and Nabhan 1996, 247).

Comments: Habitat fragmentation often stops an area from supporting its original complement of species and can entirely extirpate populations or species (Buchmann and Nabhan 1996, 247).

♦ **habitat imprinting** See learning: imprinting: habitat imprinting.

♦ **habitat preference** *n.* A measure of the degree to which an organism chooses one habitat over another (Johnson 1980 in Kramer et al. 1995, 3).

♦ **habitat saturation** *n.* The point on a continuum when all of a population's suitable territories are occupied and defended (Brown 1987, 300).

♦ **habitat selection** *n.* An animal's nonrandom use of space that results from its voluntary movements in its environment (Kramer et al. 1995, 3).

See selection: habitat selection.

syn. habitat choice (Kramer et al. 1995, 3)

Comments: When we do not know that role of an organism's active habitat selection in relation to other processes, we should use "habitat correlation" or "habitat use" (Kramer et al. 1995, 3). An animal's movements range from simple locomotor, or settlement responses, to behaviorally sophisticated decisions, *q.v.*, concerning its allocation of time to different parts of its familiar home range. Organisms move only slightly to thousands of kilometers when selecting their habitats, and they may move over a broad range of environmental complexity.

♦ **habituation** *n.*

1. A person's making a behavior habitual; forming an action into a habit (*Oxford English Dictionary* 1972, entries from 1449).
2. A person's becoming accustomed to something (*Oxford English Dictionary* 1972, entries from 1816).
3. A mechanism that lowers the responsiveness of each specific behavioral pathway of an individual animal to its particular integrated input; distinguished from adaptation (Gould 1982, 84).
4. An animal's becoming dependent on a drug (Immelmann and Beer 1989, 127).

See learning: habituation.

syn. adaptation (not a preferred synonym according to Verplanck 1957)

cf. ³adaptation

♦ **habitudinal segregation** *n.* The separation and consequent reproductive isolation of organism populations that occupy different habitats (Lincoln et al. 1985).

♦ **habitus** See habit.

♦ **Haeckel's law** See law: biogenetic law.

♦ **Haeckel's laws of heredity** See -genesis: biogenesis.

♦ **haematobium** *n.* An organism that lives in blood (Lincoln et al. 1985).

cf. parasitism: haemoparasitism; -troph-: hemotroph

♦ **haematocryal** See -therm: poikilotherm.

♦ **haematocytozoon** See -zoon: haematocytozoon.

♦ **haematogenesis** See -genesis: haematogenesis.

♦ **haemotophage** See -vore: sanguivore.

♦ **haematotherm** See -therm: haematotherm.

♦ **haematozoon** See zoon: haematozoon.

♦ **haemocoelous viviparity** See -parity: viviparity: haemocoelous viviparity.

♦ **haemoglobin** See molecule: hemoglobin.

♦ **haemoparasitism** See parasitism: haemoparasitism.

♦ **haemophage** See -vore: sanguivore.

♦ **haemopoiesis** See -genesis: haematogenesis.

♦ **haemotroph** See -troph-: haemotroph.

♦ **Hagedoorn effect** See effect: hagedoorn effect.

♦ **hair pulling** *n.* A child's grabbing and pulling on another child's hair during a serious quarrel (Heymer 1977, 81).

♦ **Haldane map distance** See distance: Haldane map distance.

♦ **Haldane's evolutionary unit** See darwin.

♦ **Haldane's exclamation** *n.* Supposedly, when the great British evolutionist J.B.S.

Haldane was asked if he would give his life for his brother, he responded: "Not for one, but for three brothers ... or nine cousins" (Daly and Wilson 1983, 30).

Comment: He seems to have meant that on the average his genome is genetically equivalent to two full brothers or eight full cousins so, sociobiologically speaking, he should be willing to give up his life for more than two (at the least three) full brothers or more than eight (at the least nine) full cousins.

♦ **Haldane's rule** See law: Haldane's rule.

♦ **half sib, half-sib** *n.* Organisms that share only one parent (Lincoln et al. 1985).

♦ **half sister** See sister: half sister.

♦ **hali-, halo-** *combining form* Of, or relating to, salt (Michaelis 1963). Before vowels, hal-.

[Greek, *hals, halos* salt, the sea]

♦ **halibios** See -bios: halibios.

♦ **halicole** See -cole: halicole.

♦ **-haline** *combining form* Referring to salt.

euryhaline *adj.* Referring to an organism that is tolerant of a wide range of salinities (Lincoln et al. 1985).
syn. eurycladous, polyhalophilic (Lincoln et al. 1985)

holeuryhaline *adj.* Referring to an organism that freely inhabits fresh, sea, and brackish water or that has established populations in all three environments (Lincoln et al. 1985).

oligohaline *adj.* Referring to an organism that tolerates only a moderate range of salinities (Lincoln et al. 1985).

polystenohaline *adj.* Referring to an organism that inhabits only oceanic waters of relatively constant high salinity (Lincoln et al. 1985).

stenohaline *adj.* Referring to an organism that tolerates only a narrow range of salinities (Lincoln et al. 1985).

♦ **haliplankton** See plankton: haliplankton.

♦ **halobenthos** See [2]community: benthos: halobenthos.

♦ **halobiont** See -biont: halobiont.

♦ **halobios** See -bios: halobios.

♦ **halocline** See -cline: halocline.

♦ **halodrymium** See [2]community: halodrymium.

♦ **halolimnetic, halolimnic** *adj.* Referring to a marine organism that is adapted to live in fresh water (Lincoln et al. 1985).

♦ **halophile** See [1]-phile: halophile.

♦ **halophobe** See -phobe: halophobe.

♦ **haloplankton** See plankton: haloplankton.

♦ **halorgadophile** See [1]-phile: halorgadophile.

♦ **halosphere** See -sphere: halosphere.

♦ **haloxenic** See -xenic: haloxenic.

♦ **hamabiosis** See -biosis: hamabiosis.

♦ **hamation** See display: wing display: hamation.

♦ **Hamilton's condition for evolution of altruism** See rule: Hamilton's rule.

♦ **Hamilton's inclusive-fitness theory** See [3]theory: kin selection theory.

♦ **Hamilton's rule** See rule: Hamilton's rule.

♦ **hand axe** See tool: hand axe.

♦ **hand patting** *n.* In human babies beginning at 9 months of age: an individual's begging movements involving smacking or patting hands together (Heymer 1977, 81).

♦ **hand-rearing** See fostering.

♦ **handedness** *n.*

1. For example, in some rodent and primate species including Humans: an individual's preference for using one front hand or paw, rather than its other, for manipulating certain objects (Michaelis 1963; Lewin 1986, 115; Deneberg 1981 in Güntürkün et al. 1988, 602).
syn. foot preference, footedness (Güntürkün et al. 1988, 602)
Note: Vallortigara et al. in Leary (1996) reported handedness in the European and cane toads.

2. For example, in the Goldfinch, several parrot species: an individual's preference for using one foot, rather than its other, for manipulation of certain objects (Friedman and Davies 1938, etc. in Güntürkün et al. 1988, 602).
syn. foot preference, footedness (Güntürkün et al. 1988, 602)

3. In spider wasps (Pompilidae): an individual female's tendency to lay eggs on either the right or left side of host insects (called left-handed and right-handed in Day 1988, 15).

Comments: Trilobite fossils from the Burgess Shale often have bite marks on the right sides of their posterior regions (Babcock and Robison in Browne 1993, C1). This trilobite species might have shown "left-handedness" in that it tended to turn to the left when attacked by a crustacean; the crustacean was "right-handed" in that it attacked the left sides of trilobites; or both.

♦ **handicap principle** See hypothesis: handicap principle.

♦ **Hansen system** See sampling technique.

♦ **haphazard sample** See sample: haphazard sample.

♦ **haplo-** *prefix* Simple, single.

♦ **haplobiont** See biont: haplobiont.

♦ **haplodiploid** See -ploid: haplodiploid.

♦ **haplodiploidy** See parthenogenesis: arrhenotoky; -ploidy: haplodiploidy.

♦ **haplodiplomeiosis** See meiosis: haplodiplomeiosis.

h – m

♦ **haplogenesis** See [2]evolution: haplogenesis.

♦ **haploid effect** See effect: haploid effect.

♦ **haploid parthenogenesis** See parthenogenesis: generative parthenogenesis.

♦ **haploidy** See -ploidy: haploidy.

♦ **haplometrosis** See -metrosis: haplometrosis.

♦ **haplont, haplophase** See -plont: haplont.

♦ **haplosis** n. Establishment of an organism's gametic (haploid) chromosome number, usually by meiosis (Lincoln et al. 1985).

♦ **haplotype** See -type, type: haplotype.

♦ **haptephobia** See phobia (table).

♦ **haptobenthos** See [2]community: benthos: haptobenthos.

♦ **haptotropism** See tropism: stereotropism.

♦ **haras, harras** See [2]group: haras.

♦ **hard inheritance** See inheritance: hard inheritance.

♦ **hard science** See study of: science: hard science.

♦ **hard selection** See selection: hard selection.

♦ **hardware** n. The electrical and other parts that comprise a computer which are distinguished from software (programs) (Lincoln et al. 1985).

♦ **hardwoods** See [2]community: deciduous forest.

♦ **Hardy-Weinberg equilibrium** n. Maintenance of more or less constant allele frequencies in a population through successive generations (Lincoln et al. 1985).
syn. genetic equilibrium (Lincoln et al. 1985)

♦ **Hardy-Weinberg law** See law: Hardy-Weinberg law.

♦ **harem** n.
1. The part of a Mohammedan dwelling house appropriated to women, constructed so as to secure the utmost seclusion and privacy (*Oxford English Dictionary* 1972, entries from 1634).
2. The occupants of a human harem collectively; the female members of a Mohammedan family; especially, the wives and concubines collectively of a Turk, Persian, or Indian Mussulman (*Oxford English Dictionary* 1972, entries from 1781).
3. For example, in some parrots and antelope species; many fish, lizard, and monkey species (including Patas Monkeys, Geladas, Hamadryas Baboons, Langurs); Zebras; Elephant Seals: a stable and lasting group of females guarded by a male that prevents other males from mating with them (Wilson 1975, 137, 585; Immelmann and Beer 1989, 128).

Comment: Some workers feel that "harem" is a sexist term and prefer the synonyms "one-male social unit" (Gowaty 1982), "single-male group" (Immelmann and Beer 1989, 195), and "unimale society."
[Arabic *ḥaram*, prohibited or unlawful, that which a man defends and fights for, as his family, a sacred place, sanctuary, enclosure; the women's part of the house; wives, women…]

♦ **harem polygyny** See mating system: polygyny.

♦ **harm avoidance** See need: harm avoidance.

♦ **harmonious development of the type** n. The concept that transitional species (*e.g., Australopithecus* spp.) should have traits that are at approximately the same level of derivedness from a common ancestor (Mayr 1982, 613).
Comment: This concept was a major dogma of idealistic morphology (Mayr 1982, 613).

♦ **harmosis** n. An organisms' total response to a stimulus, comprising both reaction and adaptation (Lincoln et al. 1985).
cf. response, taxis
photoharmosis, photoharmose n. An organism's response to light (Lincoln et al. 1985).
cf. taxis: phototaxis

♦ **harpactophage** See -phage: harpactophage.

♦ **hart** See animal names: hart.

♦ **hatch** See [2]group: hatch.

♦ **hauling grounds** n. The place where male pinnipeds (*e.g.,* sea lions or elephant seals) congregate on shore (Wilson 1975, 297).

♦ **hawk-dove game, hawk-dove-assessor game, hawk-dove-bourgeois game** See game (table).

♦ **hawking** See feeding: hawking.

♦ **head** See [2]group: head.

♦ **head flagging** n. For example, in the Black-Headed Gull, some duck species: a bird's turning its head away from another bird, normally during aggressive interactions (Manley 1960 and Chance 1962 in Heymer 1977, 88).
syn. looking away (Heymer 1977, 88)
cf. display

♦ **head lunge** See behavior: head lunge.

♦ **head mimicry** See mimicry: automimicry: body-self mimicry: head mimicry.

♦ **head-neck dipping** See behavior: head-neck dipping.

♦ **head nodding** See behavior: head nodding.

♦ **head scratching** See behavior: head scratching.

♦ **head-up-tail-up behavior** See behavior: head-up-tail-up behavior.

♦ **heart-weight rule** See law: Hesse's rule.
♦ **heat** See estrus.
♦ **"heat defense"** See behavior: defensive behavior: "heat defense."
♦ **heath barren** See ²community: heath barren.
♦ **Hebb-Williams test** See test: Hebb-Williams test.
♦ **Hebb's rule** See rule: Hebb's rule.
♦ **hebetic** *adj.* Referring to adolescence (Lincoln et al. 1985).
syn. juvenile (Lincoln et al. 1985)
♦ **hedonic** *n.* Referring to factors that stimulate sexual activity (Lincoln et al. 1985).
♦ **hekistoplankton** See plankton: hekistoplankton.
♦ **helcotropism** See tropism: geotropism.
♦ **heleoplankton** See plankton: heleoplankton.
♦ **helio-** *combining form* Sun; of the Sun (Michaelis 1963). Before vowels, "heli-."
[Greek *hēlios*, the Sun]
♦ **heliophile** See ¹-phile: heliophile.
♦ **heliophobe** See -phobe: heliophobe.
♦ **heliophytium** See ²community: heliophytium.
♦ **heliotaxis** See taxis: heliotaxis.
♦ **heliotherm** See therm: heliotherm.
♦ **helioxerophile** See ¹-phile: helioxerophile.
♦ **helium** See ²community: helium.
♦ **helminthology** See study of: helminthology.
♦ **helobius** See uliginous.
♦ **helodium** See ²community: helodium.
♦ **helodrium** See ²community: helodrium.
♦ **helohylium** See ²community: helohylium.
♦ **helohylophile** See ¹-phile: helohylophile.
♦ **helolochmophile** See ¹-phile: helolochmophile.
♦ **helolochmium** See ²community: helolochmium.
♦ **helophile** See ¹-phile: helophile.
♦ **helorgadium** See ²community: helodium.
♦ **helorgadophile** See ¹-phile: helorgadophile.
♦ **helotism** See parasitism: social parasitism: slavery.
♦ **helper** *n.*
1. A person who helps or assists; an auxiliary (*Oxford English Dictionary* 1972, entries from 1300).
2. For example, in the African Wild Dog, Human, Red Fox, Scrub Jay: an individual that performs parent-like behavior toward young or potential young (*e.g.,* in nest building) that are not its own (Skutch 1935, 1961 in Brown 1987a, 301; Henry 1993, 44–45).
Notes: This definition excludes "flocking" and "alarm calling" in winter flocks. In this definition, "helping" is not necessar-

ily beneficial toward conspecific animals (Brown 1987a, 301).
3. An animal that performs parent-like behavior toward young that are not its own offspring, typically in company with the young's parents; thus, helping excludes brood parasitism and brood capture (Brown 1987a, 300–302).
Note: A helper may, or may not, benefit its helped young or its parents; helpers may be altruistic, cooperative, or selfish.
syn. alloparent, attendant, associate, auxiliary, secondary, supernumerary (Brown 1987a, 300–302, who questions the need for all these synonyms and discusses meanings of some of them)
♦ **helping behavior, help, helping** See behavior: helping behavior.
♦ **hemerophile** See ¹-phile: hemerophile.
♦ **hemi-** *prefix* Half. Before vowels, "hem-."
[Greek *hemi*, half]
♦ **hemialloploid** See -ploid: hemialloploid.
♦ **hemiautoploid** See -ploid: hemiautoploid.
♦ **hemichimonophile** See ¹-phile: hemichimonophile.
♦ **hemiclone** See clone: hemiclone.
♦ **hemiendobiotic** See -biotic: hemiendobiotic.
♦ **hemigamy** See -gamy: hemigamy.
♦ **hemikaryon** See -karyon: hemikaryon.
♦ **hemikaryotic** See -karyotic.
♦ **hemilectic** See -tropic: hemitropic.
♦ **hemimetabolous metamorphosis, hemimetaboly, hemimetamorphy** See metamorphosis: hemimetabolous metamorphosis.
♦ **hemiparasite** See parasite: hemiparasite.
♦ **hemipenis** See organ: copulatory organ: hemipenis.
♦ **hemiphile** See ¹-phile: hemiphile.
♦ **hemiplankton** See plankton: hemiplankton.
♦ **hemiploid** See -ploid: hemiploid.
♦ **hemitropic** See -tropic: hemitropic.
♦ **hemizygoid parthenogenesis** See parthenogenesis: hemizygoid parthenogenesis.
♦ **hemizygous** See -zygous: hemizygous.
♦ **hemlock-hardwoods** See ²community: deciduous forest.
♦ **hemoglobin** See molecule: hemoglobin.
♦ **hemophobia** See phobia (table).
♦ **HENNIG-86** See program: HENNIG-86.
♦ **Hennig's progression rule** See law: progression rule.
♦ **Hennigian systematics** See study of: systematics: phylogenetic systematics.
♦ **hepaticology** See study of: hepaticology.
♦ **herbalist** *n.*
1. A person who deals in herbs, especially medicinal ones; one who prepares or

administers herbal remedies (*Oxford English Dictionary* 1972, entries from 1592; Michaelis 1963).

2. A person who is knowledgeable about herbs or plants; a collector or writer on plants; a botanist; now used to refer to early botanical writers (*Oxford English Dictionary* 1972, entries from 1594).

v.i. herbalize (archaic)

♦ **herbicole** See -cole: herbicole.

♦ **herbivore** See -vore: herbivore.

♦ **herbosa** See ²community: herbosa.

♦ **hercogamous** See -gamous: hercogamous.

♦ **herd** See ²group: herd.

♦ **herd instinct** See drive: bonding drive.

♦ **¹herding** *n.*

1. Individual animal's going into, or forming, a herd (*Oxford English Dictionary* 1972, entries from 1392).

2. A person's joining a band, or company, of others; becoming part of a faction or a party; an individual's being one of the "common herd" or crowd (*Oxford English Dictionary* 1972, entries from 1400).

cf. harem

♦ **²herding** *n.*

1. A person's tending (sheep or cattle) (*Oxford English Dictionary* 1972, entries from 1400).

2. For example, in bovids, cervids (mammals): a male's gathering females during "harem" formation and retention (Darling 1937 in Heymer 1977, 86).

3. An individual ant's causing an ant-cow (aphid) to move to a particular place (Ross 1985).

♦ **hereditary** *adj.*

1. Referring to a trait that is transmitted from parents to offspring (*Oxford English Dictionary* 1972, entries from 1597).

2. Referring to a trait that has a genetic basis (Lincoln et al. 1985).

♦ **heredity** *n.*

1. An offspring's inheriting the nature and characteristics of its parents and ancestors in general; the tendency of like organisms to beget like organisms (*Oxford English Dictionary* 1972, entries from Spencer 1863).

2. The mechanism of transmission of specific characters from parent to offspring (Lincoln et al. 1985).

[Latin *herre, heredis,* heir]

♦ **¹heritability** *n.*

1. A trait's capacity of being inherited (Lincoln et al. 1985).

2. A measure of the degree of resemblance of an offspring to its parents due to genes they hold in common (Wade 1992b, 149).

3. Additive genetic variance, *q.v.*

See ²heritability: heritability in the narrow sense.

♦ **²heritability, h^2** *n.* The fraction of the observed variance of a trait within a population caused by genetic variance (differences in heredity); $h_B^2 = V_G/V_P = V_G/(V_G + V_E)$, where V_G is the trait's variance due to an organism's genetics, V_P is the trait's phenotypic variance, and V_E is the trait's variance due to the organism's environment (Luch 1937 in Feldman 1992, 151; Wilson 1975, 68, 586).

syn. h^2 (Hartl 1987, 228)

Comments: "To my knowledge the word 'heritability' first appears in *Animal Breeding Plans* by J.L. Lush (1937)" (Feldman 1992, 151). A heritability estimate is specific to a real population with a particular history of natural selection and a specific environment in which the population exists. A heritability score of 1 means that all of the variation has a genetic basis; a score of 0 means that all of the variation is due to the environment (Wilson 1975, 68, 586). Feldman (1992) also discusses "cultural heritability," "genetic heritability," and "realized heritability."

heritability in the broad sense, h_B^2, coefficient of genetic determination (CGD) *n.* "The proportion that genetic variance contributes to the total phenotypic variance" of a group of organisms (Dewsbury 1978, 121).

syn. broad-sense heritability (Feldman 1992, 154), degree of genetic determination (Falconer 1960 in Feldman 1992, 154)

heritability in the narrow sense, h^2, true heritability *n.* Heritability measured as the fraction of total phenotypic variance only from additive genetic effects; $h^2 = V_A/V_T$, where V_A is additive genetic variance and V_T is total, or phenotypic, variance in a group (Falconer 1960, 135; Dewsbury 1978, 122).

syn. heritability (Falconer 1960, 135), narrow-sense heritability (Feldman 1992, 154)

Comment: Heritability in the narrow sense excludes genetic variation due to dominance and epistasis; this heritability is estimated by the variance in breeding values (Wade 1992b, 150).

heritability of a character, h^2 *n.* "The proportion of the total variance in the phenotype of a given character that is attributable to the average effects of genes in a particular environment" (Wilson 1975, 68).

realized heritability *n.* Heritability, h^2 = (mean S)/(mean R), that is estimated from empirical data (Hartl 1987, 228).

Comments: S = the selection differential, R = the response to selection (Hartl 1987, 228).

◆ **heritage** *n*. Characters, or parts, of organisms that no longer have function, or have reduced function or misfunction, but are still extant, *e.g.*, in Humans, the appendix, vulnerable sacroiliac joint, and poorly built sinuses (Gregory 1913, 1936 in Mayr 1982, 590).
cf. habitus

◆ **herkogamous** See -gamous: hercogamous.

◆ **herkogamy** See -gamy: herkogamy.

◆ **hermaphrodite** *n*.

1. In some mammals, including Humans: an individual with parts of both sexes that are really, or apparently, combined (*Oxford English Dictionary* 1972, entries from 1398).

2. In animals, including many mollusc and worm species: an individual that normally has both male and female sexual organs (*Oxford English Dictionary* 1972, entries from 1727).

3. In the majority of flowering-plant species: a single flower, or plant, that bears both stamens and pistils (*Oxford English Dictionary* 1972, entries from 1727).
syn. cosexual (for plants Lloyd 1980 in Policansky 1987), monoclinous flower

4. In many organism species: an individual that has both female and male sex organs simultaneously or sequentially (Wilson 1975, 586).
Note: Some hermaphrodites, such as many species of plants (Willson 1985, 55), can self-fertilize.

5. A single flower (in contrast to an individual plant) that has both female and male parts (Policansky 1987, 468).
Note: With regard to plants, Policansky (1987, 468) restricts the meaning of "hermaphrodite" to this definition, thus, to him "sequential hermaphroditism" does not make sense for a plant. He prefers the term "sequential cosexuality" (instead of "sequential hermaphroditism") for both plants and animals.

adj. hermaphrodite, hermaphroditic
cf. chimera; gonochorism; hermaphrodism; -oecism: dioecism; mosaic
[Greek *Hermes* + *Aphrodite*]

consecutive hermaphrodite *n*. In some organism species: an individual that is either a protogynous, or protandrous, hermaphrodite (Lincoln et al. 1985).
syn. sequential hermaphrodite (Lincoln et al. 1985)
cf. hermaphrodite: simultaneous hermaphrodite
Comment: In some fish species, consecutive hermaphroditism occurs in which small individuals of a group are functional females and the largest female undergoes a

sex change to become a male when the current male of a group dies or is removed (Wittenberger 1981, 621).

▶ **protandrous hermaphrodite** *n*.

1. For example, in the Anemone Fish, Slipper-Shell (*Crepidula* sp.): a hermaphrodite that is first a male then a female (Barnes 1974, 362; Policansky 1987, 468).

2. In some plant species, an individual whose male flowers open before its female flowers (Policansky 1987, 468).
adj. sex reversal

▶ **protogynous hermaphrodite** *n*.

1. For example, in parrot fish, wrasses (fish): a hermaphrodite that is first a female then a male (Policansky 1987, 468).
syn. metandric hermaphrodite (as an adjective in Lincoln et al. 1985)

2. In some plant species, *e.g.*, some aroid species: an individual plant whose female flowers open before its male flowers (Policansky 1987, 468).
adj. protogynous hermaphrodism, sex reversal
cf. gyny: protogyny

simultaneous hermaphrodite *n*. For example, in many plant species; ctenophores, earthworms, slugs; snails; some parasite, zooplankter, scale-insect, serranid-bass, and tetrapod-vertebrate species; a few hydra species: a hermaphrodite that is functionally both a male and a female at the same time (Barnes 1974, 94, 141; Thornhill and Alcock 1983; E.A. Fischer, personal communication)
syn. synchronous hermaphrodite (Lincoln et al. 1985)

◆ **hermaphroditism** *n*. An organism's being a hermaphrodite, *q.v.*
syn. androgyny, bisexualism

consecutive hermaphroditism *n*. An organism's being a consecutive hermaphrodite, *q.v.*
syn. gender change, sequential cosexuality (Policansky 1987, 468); consecutive sexuality, sequential hermaphrodism, sex reversal (Lincoln et al. 1985)

progestin-induced hermaphroditism *n*. In mammals: masculization of a female's appearance due to an experimenter's administering progestins to her to help her maintain her pregnancy (Dewsbury 1978, 231).
cf. adrenogenital syndrome, testicular feminization

pseudohermaphrodism, pseudogonochorism *n*. In consecutive hermaphrodites: the condition in which either male or female gonads are normally found

h – m

and a hermaphrodite gonad occurs at sex reversal (Lincoln et al. 1985).

simultaneous hermaphroditism *n.* An organism's being a simultaneous hermaphrodite, *q.v.* (Lincoln et al. 1985). *syn.* synchronous hermaphroditism (Lincoln et al. 1985)

♦ **heronry** See [2]group: heronry.

♦ **herpesian** *adj.* Referring to amphibians and reptiles (Lincoln et al. 1985).

♦ **herpetogeny** See -geny: herpetogeny.

♦ **herpetology** See study of: herpetology.

♦ **herpism** *n.* Creeping locomotion of protistans employing pseudopodia (Lincoln et al. 1985).

♦ **herpobenthos** See [2]community: benthos: herpobenthos.

♦ **herpon** *n.* "Crawling organisms" (Lincoln et al. 1985).

♦ **hertz (Hz, hz)** *n.* A unit of electromagnetic wave frequency equal to 1 cycle per second (Michaelis 1963).
[after Heinrich Rudolph Hertz, German physicist]

♦ **Herrnstein's principle of quantitative hedonism** See principle: Herrnstein's principle of quantitative hedonism.

♦ **hesmosis** See swarming: budding.

♦ **Hesse's rule** See law: Hesse's rule.

♦ **heteracme** See -gamy: dichogamy.

♦ **heterauxesis** See auxesis: heterauxesis.

♦ **hetereocism** See -oecism: heteroecism.

♦ **hetero-** *combining form* Other; different; opposed to homo- (Michaelis 1963). Before vowels, "heter-."
[Greek *hetero* < *heteros*, other]

♦ **heteroallelic** *adj.* Referring to genes that have mutations at different mutational sites (Lincoln et al. 1985).
cf. homoallelic

♦ **heterobathmy of characters** *n.* The mosaic distribution of relatively ancestral and relatively derived characters in related species and species groups (Lincoln et al. 1985).

♦ **heterocannibalism** See cannibalism: heterocannibalism.

♦ **heterochromatic coloration, heterochrome coloration** See coloration: gynochromatypic coloration.

♦ **heterochromatism** *n.* Color change (Lincoln et al. 1985).

♦ **heterochromosome** See chromosome: allochromosome.

♦ [1]**heterochrony** *n.* The appearance of a particular fauna, or flora, in two different regions at different times (Lincoln et al. 1985).

♦ [2]**heterochrony** *n.*
1. A change in timing of one organ's development relative to others in an individual organism [coined by Haeckel 1905 in Gould 1992, 161].

Note: Haeckel considered heterochrony to be a disruption and falsification of his biogenetic law, *q.v.* (Gould 1992, 161).
2. Phyletic change in developmental onset, or timing, so that the appearance of, or rate of development of, a feature in a descendant's ontogeny is either accelerated or retarded, relative to the appearance or rate of development of the same feature in an ancestor's ontogeny (De Beer 1930 9, 24, in Gould 1992, 160; Gould 1977, 222, 482).
Note: Haeckel's definition of heterochrony became obsolete, but his term lived on due to De Beer's definition (Gould 1992, 165).
adj. heterochronic, heterochronous
cf. -genesis: cenogenesis; -morphosis: peramorphosis

SOME KINDS OF HETEROCHRONY

I.	acceleration
II.	hypermorphosis
III.	merostasis
IV.	neoteny
	A. hysterotely
	B. metathetely
V.	paedomorphosis
VI.	progenesis
	A. prothetely
VII.	retardation

acceleration *n.* "A speeding up of development in ontogeny (relative to any criterion of standardization), so that a feature appears earlier in the ontogeny of a descendant than it did in an ancestor" (De Beer 1930, 37–38 in Gould 1977, 226, 479); contrasted with retardation.

hypermorphosis *n.* Evolutionary extension of an organism's ontogeny beyond its ancestral termination (usually to larger body sizes and increased complexity of differentiating organs) that results in recapitulation because ancestral adult stages are now intermediate stages of a lengthened descendant ontogeny (De Beer 1930, 37–38 in Gould 1977, 228, 482; Futuyma, 1986, 416).

merostasis, partial paedomorphosis *n.* "Possession of a small number of juvenile characters in an otherwise adult organism" (Lincoln et al. 1985).

neoteny, neoteinia, neotenia, neoteiny *n.*
1. Paedomorphosis (evolutionary retention of formerly juvenile characters by adult descendants) that is produced by retardation of somatic development (Kollmann 1885 and Giard 1887 in Gould 1977, 227–228, 483).
syn. superlarvation (Lincoln et al. 1985)

2. Evolution in which an animal's bodily development is slowed relative to its development of sexual maturity, resulting in reproduction practiced by animals that resemble juvenile stages of their ancestral forms (Dawkins 1982, 291), *e.g.,* in salamanders (that breed in their tadpole stage).
adj. neotene, neoteinic, neotenous
n. neotene
syn. palaeogenesis (Lincoln et al. 1985)
cf. axolotl (Appendix 1); -genesis: paedogenesis; [2]heterochrony: progenesis; -morphosis: paedomorphosis
Comments: Some major steps in evolution (*e.g.,* the origin of the vertebrates and Humans) may have occurred through neoteny (Gould, 1977; Dawkins 1982, 291). Teddy bears and several cartoon heroes, including Mickey Mouse and Pogo, have undergone a process similar to neoteny (Morris et al. 1995, 1697). De Beer (in Gould 1977, 227) uses "neoteny" to describe both acceleration of gonads and retardation of somatic characters. Futuyma (1986, 554) classifies "neoteny" as a kind of "heterochronic evolution."
[coined by Kollmann (1905 in Gould 1977, 356)]
▶ **hysterotely** *n.* In some insect species: neoteny in which a species retains larval characteristics in its pupal or adult stages, or pupal characteristics in adults (Singh-Pruthi 1924, etc. in Gould 1977, 304; Lincoln et al. 1985).
syn. metathetely (Gould 1977, 304; Lincoln et al. 1985), neoteny (Gould 1977, 304)
▶ **metathetely** *n.* In some insect species: neoteny in which a species retains some juvenile characteristics in adults that have undergone a normal number of molts (Lincoln et al. 1985).
paedomorphosis *n.* Retention of juvenile characters of ancestral forms by adults, or later ontogenetic stages of their descendants (Garstang 1928, 62, in Gould 1977, 227, 484).
syn. paedomorphism, pedomorphism, superlarvation (Lincoln et al. 1985)
cf. -genesis: paedogenesis; [2]heterochrony: merostasis
Comment: Gould (1977, 227) uses "paedomorphosis" to specify the common result of both progenesis and neoteny — the appearance of ancestors' youthful characters in later ontogenetic stages of their descendants. This is faithful to Garstang's (1928, 62) meaning.
[New Latin < Greek *pais, paidos,* child + *morphosis,* shaped like]

progenesis *n.* For example, in some parasitic-crustacean and salamander species; possibly in the extinct Isorophid Echinoderms: paedomorphosis (evolutionary retention of formerly juvenile characters by adult descendants) that is produced by precocious sexual maturation of an organism still in a morphologically juvenile stage (Futuyma 1986, 418–419; Giard 1887 in Gould 1977, 228, 485).
See -genesis: pedogenesis.
syn. pedogenesis, paedogenesis (confusing synonyms, Lincoln et al. 1985)
cf. neoteny
▶ **prothetely** *n.* In some insect species: a kind of progenesis that has accelerated sexual maturation and results in an adult with larval, or pupal, characters (Lincoln et al. 1985).
retardation *n.* A decrease in an organism's speed of ontogenetic development so that a given feature appears later in a descendent's ontogeny than its ancestor's ontogeny (Lincoln et al. 1985); contrasted with acceleration.
♦ **heterodichogamy** See -gamy: heterodichogamy.
♦ **heterodynamic** *adj.*
1. Referring to a life cycle comprised of two or more distinct forms, or stages, that exhibit dissimilar biological activity (Lincoln et al. 1985).
2. Referring to a gene that simultaneously effects different developmental processes (Lincoln et al. 1985).
cf. homodynamic
♦ **heterodynamic hybrid** See hybrid: poikilodynamic hybrid.
♦ **heteroecious parasite** See parasite: heteroxenous parasite.
♦ **heterofacial** *adj.* Having regional character differentiation (Lincoln et al. 1985).
♦ **heterogameon** See [2]species: heterogameon.
♦ **heterogamete** See gamete: heterogamete.
♦ **heterogametic** See -gametic: heterogametic.
♦ **heterogamet** See sex determination.
♦ **heterogametic sex** See sex: heterogametic sex.
♦ **heterogamy** See -gamy: heterogamy.
♦ **heterogen** See -genetic: heterogenetic.
♦ **heterogeneous nuclear RNA (hnRNA)** See nucleic acid: ribonucleic acid: heterogeneous nuclear RNA.
♦ **heterogeneous summation** See law: law of heterogeneous summation.
♦ **heterogenesis** See -genesis: heterogenesis.
♦ **heterogenetic** See -genetic: heterogenetic.

♦ **heterogenetic parasite** See parasite: heterogenetic parasite.

♦ **heterogenic** See -genic: heterogenic.

♦ **heterogonic life cycle** See life cycle: heterogonic life cycle.

♦ **heterogony** *n.* The false belief that seeds of one species of plant could germinate into another species of plant (Mayr 1982, 254). See -gamy: heterogamy; mating: disassortative mating; parthenogenesis: heterogony.
syn. allometry (in some cases) (Lincoln et al. 1985)
Comment: Aristotle, Theophrastus, and others believed in heterogony (Mayr 1982, 254).

♦ **heterograft, xenograft** *n.* Tissue transplanted from an organism to a heterospecific organism (Lincoln et al. 1985).

♦ **heterogynistic character** See character: heterogynistic character.

♦ **heterogynous** See -gynous: heterogynous.

♦ **heterogyny** See gyny: heterogyny.

♦ **heteromesogamic species** See ²species: heteromesogamic species.

♦ **heterometabolic metamorphosis** See metamorphosis: hemimetabolic metamorphosis, paurometabolous metamorphosis.

♦ **heteromorphic, heteromorphous** See -morphic: heteromorphic.

♦ **heteromorphic coloration** See coloration: gynochromatypic coloration.

♦ **heteromorphosis** See -morphosis: heteromorphosis.

♦ **heteromorphous coloration** See coloration: gynochromatypic coloration.

♦ **heteroparthenogenesis** See parthenogenesis: heteroparthenogenesis.

♦ **heterophage** See -phage: heterophage.

♦ **heterophasic life cycle** See cycle: life cycle: heterophasic life cycle.

♦ **heterophyadic** *adj.* Referring to a genetic polymorphism (Lincoln et al. 1985).

♦ **heteroplanobios** See -bios: heteroplanobios.

♦ **heteroploidy** See -ploidy: heteroploidy.

♦ **heteroselection** See selection: heteroselection.

♦ **heterosexual** See sexual: heterosexual.

♦ **heterosexual pseudocopulation** See copulation: pseudocopulation: heterosexual pseudocopulation.

♦ **heterosis** *n.*
1. The superiority of the offspring of a cross between two stocks to either of the stocks (Paul 1992a, 166).
   Note: Whaley (1944) protested the growing use of "heterosis" as a synonym for "hybrid vigor;" he felt that "hybrid vigor should denote the manifest effects of heterosis" (Paul 1992a, 169).
   syn. hybrid vigor (Paul 1992a, 166)

2. The superiority of a heterozygote to homozygotes (*e.g.,* AA′ to either homozygote AA or A′A′) (Paul 1992a, 166).
syn. heterozygote advantage, overdominance
cf. luxuriance
Comment: Heterosis has two meanings that are frequency confused or simply fused (Lincoln et al. 1985; Paul 1992a, 166). It has been a theory-laden concept since it was named by Shull (Paul 1992a, 169).
[coined by Shull, who considered it free from any hypothesis regarding its cause (1914 in Paul 1992a, 166)]

♦ **heterospecific** *adj.* Belonging to another species (Lincoln et al. 1985).
cf. conspecific

♦ **heterosymbiosis** See symbiosis: heterosymbiosis.

♦ **heterotelergone** See chemical-releasing stimulus: semiochemical: allelochemic: allomone.

♦ **heterotherm** See -therm: heterotherm.

♦ **heterotopic** *adj.* Referring to an organism that occurs in a wide variety of habitats (Lincoln et al. 1985).

♦ **heterotopy** *n.* Change in differentiation place within an organism (*e.g.,* reproductive organs' developing from endoderm, or ectoderm, in an ancestral line and later from mesoderm in a derived line); an obsolete term; contrasted with ²heterochrony (Haeckel 1905 in Gould 1992, 160).
cf. -genesis: cenogenesis

♦ **heterotroph** See -troph-: heterotroph.

♦ **heteroxenous parasite** See parasite: heteroxenous parasite.

♦ **heteroxeny** See -xeny: heteroxeny.

♦ **heterozon** See zone: heterozone.

♦ **heterozygosity** See -zygosity: heterozygosity.

♦ **heterozygote** See zygote: heterozygote.

♦ **heterozygote superiority** *n.* Balanced polymorphism in which the heterozygote has greater fitness than either homozygote (Wilson 1975, 71).
cf. heterosis

♦ **heterozygous** See -zygous: heterozygous.

♦ **heterozygous lethal (mutation)** See ²mutation: heterozygous lethal (mutation).

♦ **hi-fidelity mimicry** See mimicry: specialized mimicry.

♦ **hibernaculum** *n.* An animal's domicile in which it overwinters or hibernates (*Oxford English Dictionary* 1972, entries from White 1789).
syn. winter-quarters (Lincoln et al. 1985)

♦ **hibernal** *adj.* Referring to winter (Lincoln et al. 1985).
syn. hiemal (Lincoln et al. 1985)
cf. aestival, vernal

♦ **hibernation** See dormancy: hibernation.

◆ **hibernator** *n.* A hibernating animal (Lincoln et al. 1985).

◆ **hidden load** See genetic load: hidden load.

◆ **hider** *n.* In many ungulate species: a young that leaves its mother and chooses a lying-out place where it spends a greater part of its day away from its mother.
cf. follower, period: lying out period
Comment: "Hider" is on a continuum with "follower" (Immelmann and Beer 1989, 177).

◆ **hidroplankton** See plankton: hidroplankton.

◆ **hiemal** See hibernal.

◆ **hierarchical model of behavior** See ⁴model: tinbergen's hierarchical model (of behavior regulation).

◆ **hierarchical promiscuity** See mating system: promiscuity.

◆ **hierarchy** *n.*
1. A group of persons or things arranged in successive orders or classes, each of which is subject to, or dependent on, the one above it (*Oxford English Dictionary* 1972, entries from 1643; Michaelis 1963).
2. A series of systematic groupings in graded order: kingdoms, subkingdoms, phyla, subphyla, etc. (*Oxford English Dictionary* 1972, entries from 1643).
See hierarchy: dominance hierarchy.
cf. ²group

aggregational hierarchy *n.* An ordered arrangement of units made for convenience in which the lower units are not compounded by any interaction into emerging new higher level units as a whole (*e.g.,* the Linnaean hierarchy of taxonomic categories) (Mayr 1982, 65–66).

central hierarchy *n.* In *Cynocephalus* baboons, the Rhesus Macaque, and some species of New World monkeys: an alliance of males that forms a dominance hierarchy in which males call on each other for support in disputes and, thus, together dominate their troop (Rowell 1969a in Wilson 1975, 598; Immelmann and Beer 1989, 41).

constitutive hierarchy *n.* A hierarchy in which the members of a lower level are combined into new units that have unitary functions and emergent properties (*e.g.,* tissues organized into organs) (Mayr 1982, 65).

despotism *n.* The simplest form of dominance hierarchy, comprised of a despot and subordinates that are not of distinctive rank (Wilson 1975, 279).

dominance hierarchy *n.*
1. Any group of persons or things arranged in successive orders or classes, each of which is subject to, or dependent on, the one above it (Michaelis 1963).

2. In some crab, crayfish, cockroach, bumble-bee, lizard, rodent, and primate species; paper wasps; Platyfish; Domestic Fowl; dairy cattle; Reindeer; social-canids: the physical domination of some members of a group by other members, in relatively orderly and long-lasting patterns; except for the highest and lowest ranking individuals, a given member dominates one or more of its companions and is dominated in turn by one or more of the others (Huber 1802, Schjelderup-Ebbe 1922, 1923, 1935 in Wilson 1975, 281, 282, 444, 583).
Note: A hierarchy is initiated and sustained by hostile behavior, albeit sometimes of a subtle and indirect nature (Wilson 1975, 281, 282, 444, 583).
3. The dominance of one member of a group (of two or more members) over another, as measured by superiority in aggressive encounters and order of access to food, mates, resting sites, and other objects that promote survivorship and reproductive fitness (Wilson 1975, 11).
4. A dominance order, especially in birds (Wilson 1975, 281, 283, 591).
5. In chickens: a hierarchy involving a despot, or alpha bird, that has priority of access over all others in its flock; a second ranking animal, the beta, has priority of access over all others except the alpha; etc. (Dewsbury 1978, 93).
Note: This hierarchy gives dominant animals priority of access to food, water, roosts, mates, and other resources (Dewsbury 1978, 93).
6. In monogamous species, especially some bird species: dominance in which a male of a pair is dominant except during the beginning of his species breeding season, when his mate is dominant (Immelmann and Beer 1989, 242).
Note: At least in earlier literature, the highest ranking animal of a hierarchy is designated as the alpha animal; the lowest ranking, the omega animal (Immelmann and Beer 1989, 242).
syn. bunt order, dominance order, dominance system, hierarchy, hook order, peck order (usually refers to birds, now more likely to be used in a colloquial rather than a scientific context, Immelmann and Beer 1989, 216), peck-right hierarchy, pecking order (Immelmann and Beer 1989, 216), rank order, ranking order
cf. hierarchy, rank
[peck order, coined by Schjelderup-Ebbe 1922 in Heymer 1977, 81]
[Late Latin *hieraticus* < Greek rule of a hierarch (sacred ruler)]

h – m

7. "A hierarchical form of social organization based on overt or implied aggression. A status system based on agonistic interactions" (Wittenberger 1981, 621).

▶ **absolute dominance hierarchy** *n.* A dominance hierarchy in which rank order remains the same wherever the group goes and whatever the circumstance (Wilson 1975, 280).

▶ **biological-rank ordering** *n.* A rank order between (among) members of different species that live in the same area and exploit the same resources (*e.g.,* a watering place or a choice food source) (Hediger 1941 in Immelmann and Beer 1989, 243).

▶ **bunt order** *n.* "Social order or rank in de-horned cattle as determined by aggressive bunting [head butting] of one individual by another" (Woodbury 1941, 419).

▶ **hook order** *n.* In some species of horned mammals: a social dominance hierarchy, *q.v.,* that is established by aggressive use of horns (Lincoln et al. 1985).

▶ **linear dominance hierarchy** *n.* A dominance hierarchy in which an alpha animal dominates over all of the group, a beta dominates over all but the alpha, a gamma dominates over all but the first two, etc. (Dewsbury 1978, 93).

▶ **relative dominance hierarchy** *n.* A dominance hierarchy that changes when a social group moves to a new location or other circumstances change (Wilson 1975, 280).
cf. hierarchy: dominance hierarchy: absolute dominance hierarchy

▶ **teat order** *n.* For example, in the Domestic Cat and Domestic Pig: a dominance order that involves young's competing for teat positions that, once established, are maintained until weaning (Wilson 1975, 288).

▶ **territorial hierarchy** *n.* A dominance hierarchy in which higher ranking animals obtain more spacious territories while lower ranking ones obtain less spacious ones (Heymer 1977, 174).

Linnaean hierarchy *n.* A method of classifying organisms into hierarchical taxa such as Domains, Kingdoms, Phyla, Classes, Orders, Family, Genera, and Species (Minelli 1999, 462).
cf. ²group (table)
Comment: Some researchers are developing a rankless classification system (Minelli 1999, 462).

♦ **hierarchy of needs** *n.* A system in which human needs in lower levels must be satisfied before much attention is devoted to the needs in higher levels; these needs are in order of lower to higher level: hunger and sleep, safety, belonging to a group and receiving love, self-esteem, self-actualization and creativity (Maslow 1954, 1972 in Wilson 1975, 550).
cf. need

♦ **high diversity** See diversity: high diversity.

♦ **high eusociality** See sociality: eusociality.

♦ **high-fidelity mimicry** See mimicry: specialized mimicry.

♦ **high-order conditioning** See learning: conditioned learning: high-order conditioning.

♦ **high-resolution electrophoresis** See method: electrophoresis: high-resolution electrophoresis.

♦ **high-throughput DNA sequencing** See method: high-throughput DNA sequencing.

♦ **higher** See derived.

♦ **highly repeated DNA** See nucleic acid: deoxyribonucleic acid: highly repeated DNA.

♦ **hilltopping** *n.* For example, in some fly and butterfly species: an individual's flying to the top of a hill, or mountain, and flying around the top where mates congregate (Alcock 1987).

♦ **hind** See animal names: hind.

♦ **hindbrain** See organ: brain: hindbrain.

♦ **Hinny** See hybrid: Hinny.

♦ **hip thrusting** *n.* A person's moving hips forward and backward, performed as part of mocking and threat behavior during dancing (Heymer 1977, 89).
cf. buttocks display

♦ **histogenesis, histogeny** See -genesis: histogenesis.

♦ **histology** See study of: histology.

♦ **histone** *n.* One of several kinds of proteins that are associated with DNA into molecular aggregates (nucleosomes) which seem to differ in function (Mayr 1982, 575); small, DNA-binding proteins, rich in basic amino acids and classified according to their relative amounts of lysine and arginine (King and Stansfield 1985).

♦ **histophile** See ¹-phile: histophile.

♦ **historical biogeography** See study of: geography: biogeography: historical biogeography.

♦ **historical cause of phenotypic variation** *n.* Phenotypic variation "from deviant cytoplasmic traits transmitted over a period of two or more generations without benefit of special instructions from the hereditary nucleic acids" (Wilson 1975, 68).

♦ **historical mimicry** See mimicry: historical mimicry.

◆ **historical zoogeography** See study of: geography: historical zoogeography.

◆ **historicism** *n.* One's "belief that an adequate understanding of the nature of any phenomenon and an adequate assessment of its value are to be gained through considering it in terms of the place which it occupied and the role it played within a process of development" (Mandelbaum 1971, 42, in Mayr 1982, 129).

◆ **histozoic** *adj.* Referring to an organism that lives within the tissues of another organism (Lincoln et al. 1985).
syn. parasitic (Lincoln et al. 1985)

◆ **histozoite** See -zoite: histozoite.

◆ **hit theory** See ³theory: hit theory.

◆ **hitchhiker hypothesis** See hypothesis: hypotheses of the evolution and maintenance of sex: hitchhiker hypothesis.

◆ **hitchhiking effect** See effect: hitchhiking effect.

◆ **hive, bee hive** *n.* A human-made container in which a bee colony lives, distinguished from "colony" and "nest" (Michener 1974, 372).
cf. ²group: hive

◆ **hive aura, hive odor** See odor: hive aura.

◆ **hoarding** *n.* A person's amassing, or putting away, something for preservation, security, or future use, especially money or wealth (*Oxford English Dictionary* 1972, entries from 1000).

food hoarding *n.* For example, in Burying Beetles, Dung Beetles, Harvester Ants; some carabid and rove beetle species; many bee, spider, wasp species; many bird and mammal species, including Humans: an individual's gathering and storing food reserves for later use (Heymer 1977, 81; Agren et al. 1989, 28; Immelmann and Beer 1989, 110; Vander Wall 1990, 369–372; Clarke and Kramer 1994, 299).
syn. food caching, food storing (Vander Wall 1990, 1)
cf. behavior: caching; food hiding
Comment: Food hoarding appears to be an animal's means of avoiding starvation during food-shortage periods; it may also be a means for an animal to gain more food from a temporarily rich patch than it could immediately consume (Clarke and Kramer 1994, 299).

▶ **larder hoarding** *n.* An animal's hoarding food in one to a few dense aggregations to which it makes repeated visits to add food; contrasted with scatter hoarding (Clarke and Kramer 1994, 299).
Comments: Larder hoarding, in its strictest sense, is an animal's concentration of all of its cached food in one site (*e.g.,* in

Honey Bees) (Vander Wall 1990, 3). Larder hoarding is on a continuum with scatter hoarding. Most larder-hoarding species store food in several discrete, closely spaced sites.

▶ **scatter hoarding** *n.* An animal's hoarding food in many widely dispersed, small caches to which it typically makes only a single visit with food; contrasted with larder hoarding (Vander Wall 1990, 4; Clarke and Kramer 1994, 299).
Comments: Scatter hoarders include Clark's Nutcracker, Fox Squirrels, Green Acouchis, Marsh Tits, and White-Footed Mice. Many species store their food in both larders and scattered surface caches. These species include some species of chipmunks, flying squirrels, kangaroo rats, and wood mice; the Red-Headed Woodpecker, Red Fox, and White-Footed Mouse (Vander Wall 1990, 3–4). In the Eastern Chipmunk (a rodent), scatter hoarding appears to reduce pilferage of hoarded food by individuals unable to defend a larder and to increase the rate at which subordinate individuals can sequester food from ephemeral patches (Clarke and Kramer 1994, 299).
[coined by Morris (1962) to describe captive Green Acouchis' tendencies to space their buried food items widely (Vander Wall 1990, 3)]

◆ **Hofacker-and-Sadler's law** See law: Hofacker-and-Sadler's law.

◆ **hol-** *combining form* Var. of holo-, *q.v.*

◆ **holandric character, holandry** See character: holandric character.

◆ **hold-bottom** *n.* In the stump-tailed monkey: one individual's pulling another onto its lap, sitting behind it, and clasping its hips, a behavior commonly shown during reconciliation (de Waal 1989, 153–154, 163–165).

◆ **hold-up** See ³lactation: milk ejection (comments).

◆ **holeuryhaline** See -haline: holeuryhaline.

◆ **holism** *n.*
1. The philosophical theory that a material object, especially a living organism, has a reality other and greater than the sum of its constituent parts (Michaelis 1963).
2. The whole is more than the some of its parts combined with vitalism, *q.v.* (Smuts 1926 in Mayr 1982, 66).
3. Definition 2 plus the concept that all physical and biological entities form a single, unified, interacting system (Lincoln et al. 1985).
syn. gestalt (Lincoln et al. 1985)
cf. reductionism
[HOL- + -ISM]

h – m

♦ **holo-** *combining form* Whole, wholly, complete, entire (Michaelis 1963). Also, before vowels, "hol-."
[Greek *holos*, whole]

♦ **holobenthic** See [2]benthic: holobenthic.

♦ **holocoen** See [1]system: ecosystem.

♦ **holocyclic** See cyclic: holocyclic.

♦ **holocyclic parthenogenesis** See parthenogenesis: holocyclic parthenogenesis.

♦ **holoendemic** See [1]endemic: schizoendemic: holoendemic.

♦ **hologamodeme** See deme: gamodeme: hologamodeme.

♦ **hologamy** See -gamy: hologamy.

♦ **hologenesis, theory of** See [2]evolution: hologenesis.

♦ **hological method** See method: hological method.

♦ **hologynous character, hologyny** See character: hologyny.

♦ **holometabolous** See metamorphosis: holometabolous metamorphosis.

♦ **holometabolous development, holometabolic development, holometabolism, holometaboly, holometamorphosis, holometamorphy** See metamorphosis: holometabolous metamorphosis.

♦ **holomorphosis** See -morphosis: holomorphosis.

♦ **holomorphospecies** See [2]species: holomorphospecies.

♦ **holoparasite** See parasite: holoparasite.

♦ **holopelagic** See pelagic: holopelagic.

♦ **holophyletic group** See [2]group: monophyletic group (def. 2).

♦ **holoplankton** See plankton: holoplankton.

♦ **holoschisis** See mitosis: amitosis.

♦ **holoschizoendemic** See [1]endemic: schizoendemic: holoschizoendemic.

♦ **holotroph** See -troph-: holotroph.

♦ **holozoic** See -zoic: holozoic.

♦ **home** *n.*
1. A person's dwelling place, house, abode; a family's fixed residence or household; one's seat of domestic life and interests; a person's own house (*Oxford English Dictionary* 1972, entries from 950).
2. An individual animal's dwelling or resting place (*Oxford English Dictionary* 1972, entries from 1774).
3. An individual animal's resting and breeding place (Lincoln et al. 1985).
cf. hotel, range, territory

♦ **home range, home realm** See [3]range: home range.

♦ **home site** *n.* The location of an animal's domicile or resting place that it uses regularly (Lincoln et al. 1985).

♦ **homeo-, homoeo-, homoio-** *combining form* "Like; similar" (Michaelis 1963).
[< Greek *homoios*, similar]

♦ **homeobox** *n.* In most eukaryotes examined: a short region of a gene (*e.g.*, a homeotic-selector gene) that is comprised of about 180 nucleotide pairs and encodes for a sequence of 60 amino acids (Gould 1985c; Alberts et al. 1989, 937–938).
cf. gene: homeobox gene, homeotic homeobox gene, *Hox* cluster
Comments: Homeoboxes are known from many animal phyla and are found in Humans as well as many other species. Homeoboxes are remarkably similar among species. It seems likely that, as an increasingly complex animal body plan evolved, altered homeobox-containing genes were added to evolutionary lines (Alberts et al. 1989, 937–938). Different functions for homeoboxes have been proposed, *e.g.*, producing proteins that bind to DNA sequences that act as enhancers and silencers of gene expression, including the expression of other homeobox-containing genes (Gould 1985c; Alberts et al. 1989, 937). Researchers think that homeoboxes play wide roles in controlling development (Campbell 1990, 391). In frogs, a homeobox-containing gene controls "posteriorness;" in Humans, a homeobox-containing gene turns on genes for antibody formation. Many different classes of regulatory genes share similar homeoboxes, and the orginal homeobox predates the origin of animals (Erwin et al. 1997, 133).
[Greek *homeo*, homeotic complex; box, a discrete unit of tight composition]

♦ **homeochronous** See homochronous.

♦ **homeomorph, homeomorphic, homeomorphy, homoeomorphy** See morphy: homeomorphy.

♦ **homeostasis** *n.*
1. The tendency for the characteristics of a physiological, or morphological, system to be held constant (Cannon 1932 in Bradshaw 1965, 117).
Note: Bradshaw (1965, 117) indicates that discussions of the semantic problems with "homeostasis" are presented by Waddington (1957, 1961) and Lewontin (1957).
2. An organism's tendency "to maintain a uniform and beneficial physiological stability within and between its parts" (Michaelis 1963).
adj. homeostatic
syn. organic equilibrium
cf. canalization (developmental homeostasis); milieu intérieur; plasticity; principle: Bernard's principle

genetic homeostasis *n.*
1. Evolving populations' automatic resistance to selection that proceeds at a rate

fast enough to make marked decreases in genetic variability (Lerner 1954; Mayr 1963).

2. A population's tendency to move to genetic equilibrium and resist sudden changes in its genetic composition (Lincoln et al. 1985).

syn. genetic inertia (Lincoln et al. 1985)

social homeostasis *n.*

1. In social organisms: an individual's, or society's, maintenance of its colony population, caste proportions, and nest environment at a constant level (Emerson 1956a in Wilson 1975, 11).

2. In social animals: an individual's maintenance of steady states at the level of its society either by control of its nest microclimate or by regulation of its population density, behavior, and physiology of its group members as a whole (Wilson 1975, 595).

thermal homeostasis *n.* In birds and mammals: homeostasis involving an individual's maintaining a relatively constant body temperature (McFarland 1985, 259).

cf. -therm: homeotherm

♦ **homeotely** See ²evolution: homeotely.

♦ **homeotherm** See -therm: homeotherm.

♦ **homeothermic** *adj.* Referring to a homeotherm, *q.v.*

♦ **homeotic complex** *n.* A group of genes that code for homeotic mutations (Gould 1985c).

♦ **homeotic-homeobox gene,** *Hox* **gene** *n.* See gene: homeotic-homeobox gene.

♦ **homeotic mutation** See ²mutation: homeotic mutation.

♦ **homing** *n.* In many animal species, including many wasp species; ant, bee, and salmonid-fish species; the Homing Pigeon; Humans: an individual's capability to return, or actually returning, to its nest (or home, as in Humans) (*Oxford English Dictionary* 1972, entries from 1765; Miller 1954 in Halvorsen and Stabell 1990, 1089).

v.i. home

syn. homing behavior (Halvorsen and Stabell 1990, 1089)

cf. behavior, navigation, ortstreue

♦ **homiogamy** See -gamy: homiogamy.

♦ **homiotherm, homiothermic** See -therm: homotherm.

♦ **homo-** *combining form* Similar, alike, like, same; opposed to -hetero (Michaelis 1963) [Greek *homo-* < *homos*, same]

♦ **homoallelic** *adj.* Referring to genes that have mutations at the same mutational sites (Lincoln et al. 1985).

cf. heteroallelic

♦ **homochemy** *n.* A mimic's imitation of a model's emitted odors; contrasted with

homochromy, homoelectry, homokinemy, homomorphy, homophony, homophoty, homothermy, and homotopy (Pasteur 1982, 170).

Comments: For homochemy, homochromy, etc., Pasteur (1982, 170) gives a table that indicates the nature of mimicking signals from a dupe's standpoint. Many kinds of mimicry include more than one kind of imitation (Pasteur 1982, 174).

♦ **homochrome mimicry, homoeochromatic mimicry** See mimicry: androchromatic mimicry.

♦ **homochromy** *n.* A mimic's imitation of a model's emitted light; contrasted with homochemy, homoelectry, homokinemy, homomorphy, homophony, homophoty, homothermy, and homotopy (Pasteur 1982, 170).

♦ **homochronous, homeochronous, simultaneous** *adj.* Referring to phenomena that occur at the same time or age in successive generations (Lincoln et al. 1985).

♦ **homodichogamy** See -gamy: autoallogamy.

♦ **homodynamic** See -dynamic: homodynamic.

♦ **homodynamous** *adj.* "Metamerically homologous" (Lincoln et al. 1985).

♦ **homoecism** See xeny: monoxeny.

♦ **homoelectry** *n.* A mimic's imitation of a model's emitted conductibility or emitted electricity; contrasted with homochemy, homochromy, homokinemy, homomorphy, homophony, homophoty, homothermy, and homotopy (Pasteur 1982, 170).

Comment: Homoelectry is not yet known in nature (Pasteur 1982, 174).

♦ **homoeochromatic mimicry** See mimicry: androchromatypic mimicry.

♦ **homoeologous chromosomes** See chromosome: homoeologous chromosomes.

♦ **homoeomorphy** See -morphy: homoeomorphy.

♦ **homoeotherm** See -therm: homoeotherm.

♦ **homoeotic mutation** See ²mutation: homeotic mutation.

♦ **homogamete** See gamete: homogamete.

♦ **homogametic** See -gametic: homogametic.

♦ **homogametic sex** See ¹sex: homogametic sex.

♦ **homogamety** See -gamety: homogamety.

♦ **homogamy** See breeding: inbreeding; -gamy: homogamy; mating: assortative mating.

♦ **homogen** See -genetic: homogenetic.

♦ **homogeneon** See ²species: homogeneon.

♦ **homogenesis** See -genesis: homogenesis.

♦ **homogenetic** See -genetic: homogenetic.

♦ **homogenetic parasite** See parasite: homogenetic parasite.

♦ **homogenous** See -genous: homogenous.

♦ **homogeny** See homolog: homogeny.

♦ **homogonic life cycle** See cycle: life cycle: homogonic life cycle.

♦ **homoio-** See homeo-.

♦ **homoiogamy** See -gamy: homoiogamy.

♦ **homoiohydric** *adj*. Referring to protistans, fungi, and plants that can compensate for short-term fluctuations in water supply and rate of evaporation so that their water content can be maintained at a given level more or less independently of ambient humidity (Lincoln et al. 1985).
cf. poikilohydric

♦ **homoiolog, homoiologue, homology** See homolog.

♦ **homoiotherm, homoiothermic** See -therm: homotherm.

♦ **homokinemy** *n*. A mimic's imitation of a model's moves and postures; contrasted with homochemy, homochromy, homoelectry, homomorphy, homophony, homophoty, homothermy, and homotopy (Pasteur 1982, 170).

♦ **homolog, homologue, homology** *n*.
1. One animal species' trait that is similar to another's trait because both of the traits are construction variants based on one of God's blueprints (Owen 1843 in Hailman 1976, 181).
2. The same organ in different animals under every variety of form and function (called "homologue" by Owen 1843 in Donoghue 1992, 170).
   Note: Owen cited Saint-Hilaire, who noted its presence in even earlier German literature (*e.g.*, Goethe's writing) (Donoghue 1992, 170).
3. An individual's character (trait) that has a common ancestry with another character of the same individual, or heterospecific individual, but does not necessarily retain similarity of structure, function, or behavior with it; contrasted with analog and homoplasy (*e.g.*, horse forelegs and bird wings are homologs) (Darwin 1859 in Mayr 1982, 45; Campbell and Hodos 1970, 358, in Hailman 1976, 186; Lincoln et al. 1985).
   Notes: This concept is commonly called "homology" today, and this is a widely used definition. Donoghue (1992) discusses changes in the meaning of homology and difficulties and controversies related to this term. These include (1) the implication of both constancy and change in this concept, (2) the argument that a certain definition of homology is a

nonoperational concept because it is circular to define homology with regard to ancestry if it is then to be used as evidence of ancestry, (3) the distinction between homology and nonhomology, and (4) the use of homology in reference to ultrastructural features of organisms, cells, species and higher taxa, and behaviors, in addition to organs.
adj. homolog, homologous
syn. homoiologue, homoiolog, homology (Lincoln et al. 1985)
cf. analog; character: apomorph: synapomorph: homologous synapomorph: homolog: homogeny; homoplasy
4. A character that is derived directly from another character; both are considered to be homologs of one another (Wiley 1981, 9).
5. One of three, or more, characters that are part of a character transformation series in which one character gave rise to the next in a linear sequence (Wiley 1981, 9).
6. A homologous synapomorph (Wiley 1981, 9–10), *q.v.*
7. A derived trait that characterizes a monophyletic group; a synapomorphy (Patterson 1982 in Donoghue 1992, 170).
   Note: Using this definition, "homology" becomes "the relation through which we discover" monophyly resulting in the "taxic" view of homology (Patterson 1982 in Donoghue 1992, 170).
8. A structure that shares "a set of developmental constraints, caused by locally acting self-regulatory mechanisms of organ differentiation" with a structure from another individual or within the same individual (Wagner 1989, 62, in Donoghue 1992, 170).
See chromosome: homologous chromosome; homolog: homogeny, sequence homology; homology.
adj. homologous
[Greek *homologos*, agreeing < *homos*, same + *logos*, measure, a proportion < *legein*, to speak]

homogeny *n*. A homolog that is morphologically similar to another homolog due to inheritance from a common ancestor; contrasted with homoplasy as a division of homolog [coined by E.R. Lankester 1870 in Donoghue 1992, 171].
Comment: "Homogeny" did not become a widely used term; "homology" is used instead (Donoghue 1992, 171).
See homolog.

homotype *n*. An organ that is the same as another organ within an organism's body; contrasted with homolog (def. 2) (Owen 1848 in Donoghue 1992, 170).

iterative homology *pl. n.* "Entities traceable to the same element of a precursor which causes the repetition of parts in a larger whole" (Ghiselin 1976 in Hailman 1976, 195).
Comments: Iterative homology includes homonomy (repeated units within an individual, *e.g.,* hair) and serial homology (Donoghue 1992, 176). Biologists debate whether iterative homology is "true homology."

partial homology *n.* An organ that is intermediate in characteristics between two other organs, *e.g.,* the phylloclade of a monocot that is borne in the position of a branch but is similar in development, symmetry, and internal anatomy to a leaf (Sattler 1984, 1988 in Donoghue 1992, 170).
Comment: Patterson (1987 in Donoghue 1992, 170) considers the concept of partial homology to be incompatible with evolutionary views that hold that organs are either homologous or not homologous, but not partially homologous.

phylogenetic homology *n.* A behavioral homolog that is transmitted genetically (Wickler 1961, 1965 in Heymer 1977, 88).

tradition homology *n.* For example, in Humans: a behavioral homolog that is transmitted culturally (Wickler 1961, 1965 in Heymer 1977, 88).

♦ **homologous** See homolog.

♦ **homologous chromosome** See chromosome: homologous chromosome.

♦ **homologous synapomorph, homologous synapomorphy** See character: apomorph: synapomorph: homologous synapomorph.

♦ **homology** *n.*

1. Formal resemblance between species A and B such that the relations among certain parts of A are similar to the relations among corresponding parts of B; a resemblance considered to be evidence of evolutionary relatedness (Bateson 1979, 242).

2. "A correspondence between two or more characteristics of organisms that is caused by continuity of information" (Van Valen 1982 in Donoghue 1992, 177).

3. "An equivalence relation on a set of structures, partitioning the set into classes whose members share certain invariant internal relationships and are transformable one into the other while preserving the invariance" (Goodwin 1982, 51, in Donoghue 1992, 178).

See homolog.
adj. homologous

Comments: The concept of homology is considered to be among the highly important principles in comparative biology. The history of the meanings of "homology can be interpreted as a series of responses to challenges brought on by underlying conceptual changes" (Donoghue 1992). Developmentally oriented evolutionary biologists seek a causal understanding and definition of "homology" that differs profoundly from a phylogenetic systematic definition (Futuyma 1993, 1153–1154). The meaning of "homology" can be very puzzling, because phylogenetically homologous characters can have different genetic bases and ontogenies (*e.g.,* cartilage in different vertebrate classes), and ontogenetically indistinguisable, perhaps causally identical, character states can originate by convergent and parallel evolution. Hailman (1976) gives more definitions of "homology" and discusses how authors have confounded the definitions of "analogy," "homogeny," "homology," and "homoplasy." Fitch (1976) has organized others to appeal against a confusing genetics definition of "homology" (*syn.* sequence homology, *q.v.*). Since 1972, he and others have been arguing over its usage, yet they still do not totally agree on it (Lewin 1987a). "Homology" and a kind of homology (*e.g.,* serial homology) can refer to both a condition and a trait; I do not include both condition and trait definitions for each kind of homology. The kinds of homology listed below are grouped together because they share the word "homology," not because they are necessary divisions of homology as defined above.
[Greek *homologos*, agreeing < *homos*, same + *logos*, measure, a proportion < *legein*, to speak]

latent homology *n.* The condition in which two or more related lineages share a character that is absent from their ancestral stock although it is presumed to have had the potential to develop the character (Lincoln et al. 1985).

paramorphic homology *n.* Two distinct species' having polypeptide variants whose ancestral forms were polymorphic alleles in their ancestral species (Lincoln et al. 1985).
syn. polymorphism-dependent homology (Lincoln et al. 1985)

parhomology *n.* Characters' apparent similarity of structure or function (Lincoln et al. 1985).

polymorphism-dependent homology See homolog: paramorphic homology.

sequence homology *n.* The degree of matching of DNA bases of two organisms;

the higher the degree, the greater the genealogical relationship between them (Gould 1988).

syn. homology (Lewin 1987a), percent identity (Kimelberg 1987), sequence similarity (Gould 1988 and other workers quoted by him), similarity (Niesbach-Klösgen et al. 1987, 223)

serial homology *n.* In annelids, arthropods, and chordates: an individual's having a series of homologous structures, frequently functionally differentiated (Lincoln et al. 1985).

cf. homology: iterative homology; metamerism

◆ **homomorphy** *n.* A mimic's imitation of a model's body outline; contrasted with homochemy, homochromy, homoelectry, homokinemy, homophony, homophoty, homothermy, and homotopy (Pasteur 1982, 170).

Comment: Homomorphy and homokinemy may involve hearing and touch organs in potential dupes such as fish lateral lines and arachnid trichobothria which enable their bearers to identify solid objects and distant movements (Pasteur 1982, 174).

◆ **homophony** *n.* A mimic's imitation of a model's emitted sound; contrasted with homochemy, homochromy, homoelectry, homokinemy, homomorphy, homophoty, homothermy, and homotopy (Pasteur 1982, 170).

◆ **homophoty** *n.* A mimic's imitation of a model's emitted light; contrasted with homochemy, homochromy, homoelectry, homokinemy, homomorphy, homophony, homothermy, and homotopy (Pasteur 1982, 170).

◆ **homophylic** *adj.* Referring to organisms' traits that show resemblance due to common ancestry (Lincoln et al. 1985).

cf. homoplasy

◆ **homoplasia, homoplasis, homoplast, homoplastic similarity, homoplasty** See homoplasy.

◆ **homoplasy** *n.*
1. A character's resemblance to another due to parallel, or convergent, evolution, rather than common ancestry; opposed to homogeny (Campbell and Hodos 1970, 359 in Hailman 1976, 187; Lincoln et al. 1985).
2. Unrelated species' sharing nucleotides at a particular site by chance alone, not because they share a recent common ancestor (Maley and Marshall 1998, 505).

syn. homoplasis (Lankester 1870 in *Oxford English Dictionary* 1972), homoplastic similarity, nonhomologous similarity, homoplasia, homoplast, homoplasty (Lincoln et al. 1985)

cf. analog, homogeny, homolog

Comments: Hailman (1976) gives more definitions and discusses how authors have confounded the definitions of "analog," "homogeny," "homolog," and "homoplasy." Homoplasies can produce erroneous dendrograms (Maley and Marshall 1998, 505). [coined by Lankester (1870 in the *Oxford English Dictionary* 1972)]

◆ **homorophyletic** See -phyletic: homorophyletic.

◆ **homoselection** See -selection: homoselection.

◆ **homosexual** See -sexual: homosexual.

◆ **homosexual behavior** See behavior: homosexual behavior.

◆ **homosexual copulation** See copulation: homosexual copulation.

◆ **homosexual courtship** See courtship: homosexual courtship.

◆ **homosexual mating** See mating: homosexual mating.

◆ **homosexual mounting** See copulation: pseudocopulation: homosexual pseudocopulation.

◆ **homosexual pseudocopulation** See copulation: pseudocopulation: homosexual pseudocopulation.

◆ **homosymbiosis** See symbiosis: homosymbiosis.

◆ **homotelergone** See chemical-releasing stimulus: semiochemical: pheromone.

◆ **homotene** See homotenous.

◆ **homotenous** *adj.* Referring to a "primitive" form or condition (Lincoln et al. 1985). *n.* homotene *cf.* ancestral

◆ **homotherm** See -therm: homotherm.

◆ **homothermy** *n.* A mimic's imitation of a model's temperature; contrasted with homochemy, homochromy, homoelectry, homokinemy, homomorphy, homophony, homophoty, and homotopy (Pasteur 1982, 170).

◆ **homotopy** *n.* A mimic's imitation of a model's inhabited, or visited, spots; contrasted with homochemy, homochromy, homoelectry, homokinemy, homomorphy, homophony, homophoty, and homothermy (Pasteur 1982, 170).

◆ **homotropic** See -tropic: homotropic.

◆ **homotype** See homolog: homotype.

◆ **homotypy** *n.* A mimetic imitation's eliciting the same reaction from a dupe as the model (Pasteur 1982, 184).

abstract homotypy *n.* Homotypy in which the model is an abstraction of a general type of organism, or organ, that is not identifiable at the species level or at all; contrasted with concrete homotypy (Pasteur 1982, 184).

concrete homotypy *n.* Homotypy in which the model is an actual species or a cluster of similar species; contrasted with abstract homotypy (Pasteur 1982, 184).

♦ **homoxenic** See -xenic: homoxenic.

♦ **homozone** See zone: homozone.

♦ **homozygosity** See -zygosity: homozygosity.

♦ **homozygote** See zygote: homozygote.

♦ **homozygous** See -zygous: homozygous.

♦ **homozygous lethal (mutation)** See ²mutation: homozygous lethal (mutation).

♦ **honest signal** See signal: honest signal.

♦ **honey** *n.* A sweet yellowish through brownish viscid fluid derived from nectar, honeydew, or both, collected by some bee taxa; *e.g.*, species of Honey Bees, some bumble-bee and stingless-bee species (Michener 1974, 320, 335–344).

Comment: Honey is comprised of simple sugars derived from more complex sugars that bees have digested; further, honey is more concentrated than honeydew and nectar as a result of evaporation of their water (Michener 1974, 372).

♦ **honeydew** *n.*

1. A sugar-rich fluid derived from plant phloem that is passed as excrement through the anus of certain homopteran insect species; distinguished from honey and nectar (*Oxford English Dictionary* 1972, entries from 1577 as honey-dew; Wilson 1975, 423, 586; Borror et al. 1989, 797).

   Notes: Honeydew is produced by some kermesid and eriococcid species, many leafhopper species, planthoppers, and aphids (Borror et al. 1989, 328, 333, 337, 343, 346; Batra 1993b, 125) and from the surface of some galls (Torre-Bueno 1978). In some species, it is rich in amino acids; aphid honeydew is comprised of their wastes and consumed excess sap and sugars (Borror et al. 1989, 45). In arid regions, honeydew of the Tamarisk Manna Scale solidifies on leaves and accumulates as thick layers called "manna" (a sweet sugarlike material), which is believed to be the manna mentioned in *The Bible* (Borror et al. 1989, 343). Some ant species tend and protect honeydew-producing insects such as "ant cattle," *q.v.*, and feed on their honeydew (Ebbers and Barrows 1980 and references therein). Paper wasps and yellowjackets defend kermesids from which they obtain honeydew (Barrows 1979b). Bumble bees sometimes and Honey Bees frequently consume honeydew (Batra 1993b, 125).

2. A sweet fluid exuded by leaves of some plant species during warm weather (Michaelis 1963).

Note: This form of honeydew may be a plant's normal extrafloral nectar or sugary fluid caused by a parasite such as a *Monolinia* fungus. This fungus causes blueberry leaves to produce such a fluid, and bees and other insects that consume this honeydew transmit the fungus' spores (Batra and Batra 1985, 228; Batra 1987, 57).

♦ **honeydew gland** See gland: honeydew gland.

♦ **honeypot** *n.* In some bumble-bee species: a wax vessel made for holding nectar (Wilson 1975, 430).

♦ **honing** *n.* An animal's sharpening its teeth (Lincoln et al. 1985).

♦ **hook order** See hierarchy: dominance hierarchy: hook order.

♦ **hop-scratching** See scratching: hop-scratching.

♦ **hopeful monster** *n.*

1. A mutation in a gene(s) that affects development (Goldschmidt 1940, 310 in Dietrich 1992, 198).

2. An individual with a small genetic change that affects its established pathways of ordinary sexual and embryological development, resulting in a markedly different phenotype that is more adapted to its environment than conspecific individuals that do not have this mutation (Goldschmidt 1940 in Gould 1984).

   Notes: Hopeful monsters may be the bases of saltational evolution. "The decisive step in evolution, the first step toward macroevolution, the step from one species to another, requires another evolutionary method [that is, the origin of hopeful monsters] than that of sheer accumulation of micromutations" (Goldschmidt 1940, 183, in Mayr 1982, 562).

3. A coordinated, novel structure of highly improbable origin within the normal spectrum of mutations (Goldschmidt 1940 in Levinton 1983, 103).

   cf. ²mutation: macromutation; ³mutation: systemic mutation

   Comments: The concept of macromutation that most commonly appears in contemporary literature is at least in part based on Goldschmidt's concept of macromutations and their roles in developmental systems. Gould (1982, 88–89) and Dawkins (1983, 414–415) discuss legitimate and illegitimate forms of macromutation (Dietrich 1992, 199).

♦ **Hopkin's bioclimatic law** See law: Hopkin's bioclimatic law.

♦ **horary** *adj.* Lasting one to two hours (Lincoln et al. 1985).

   cf. ephemeral

h – m

◆ **horiodimorphism** See morphism: dimorphism: horiodimorphism.

◆ **horizontal DNA transfer** *n*. DNA's movement from one species to another coexisting species; contrasted with horizontal gene transfer and vertical DNA transfer (Stebbins and Ayala 1985, 75).
cf. transposable element: transposon
Comment: Horizontal DNA transfer can be between highly unrelated species (Stebbins and Ayala 1985, 75).

◆ **horizontal gene transfer** *n*. The movement of a gene from one species to another coexisting species; contrasted with horizontal DNA transfer (Stebbins and Ayala 1985, 75).
cf. transposable element: transposon
Comment: Horizontal gene transfer has been common and intense in Prokaryotes (Koonin and Galperin 1997, 761). Presumed major events in horizontal gene transfer occurred when eukaryotes first evolved, *i.e.,* when a prokaryote acquired another prokaryote which evolved into a mitochondrion and when the eukaryotic plant ancestor first acquired a prokaryote that evolved into a chloroplast (Strickberger 2000, 181).

◆ **hormesis** *n*. The stimulus afforded by an organism's exposure to nontoxic concentrations of a potentially toxic substance (Lincoln et al. 1985).

◆ **hormonal imprinting** See learning: imprinting: hormonal imprinting.

◆ **hormone** *n*.
1. Any substance secreted by an individual animal's endocrine gland into its blood or lymph that affects the physiological activity of its other organs (Starling 1905 in Nordlund 1981, 496; Wilson 1975, 586).
Notes: Hormones are found in the majority, if not all, of multicellular-animal species. This definition is appropriate for many of the more derived species. Hormones are classified by their general function — "developmental hormone," "primer hormone," and "releaser hormone" — and by their chemistry — amino-acid derivatives and lipid, or steroid, hormones (Curtis 1983, 777; Campbell 1996, 914). Hormones can influence an individual's nervous system and, through it, behavior. Hormones can be produced by cells that are not part of endocrine glands (Kolata 1982, 1383). A hormone may move between endocrine glands via nerve cells in some kinds of insects (Chapman 1971). Barrington (1979) reviews hormones of both invertebrate and vertebrates. Humans have more than 50 known hormones (Campbell 1996, 914). Below, I list only a few

of the hundreds of the known kinds of animal hormones.
2. A chemical signal that coordinates the parts of an organism (Campbell 1996, 751).
Notes: Plant hormones are abscissic acid, auxin, cytokinins, ethylene, and gibberellins (Campbell 1996, 753). Each kind of plant hormone has a multiplicity of effects, depending on its action site and concentration and a plant's developmental stage. Two or more kinds of hormone often work in concert effecting a specific plant response.
cf. endocrine disrupter, environmental estrogen, neuropeptide Y, peptide YY
[Greek *hermōn* < *hormaein*, to excite]

SOME CLASSIFICATIONS OF HORMONES

I. Animal hormones
  A. amino-acid derived hormones
    1. amine hormones
      a. catecholamine hormones
        (1) epinephrine (*syn.* adrenalin)
        (2) norepinephrine
      b. melatonin
      c. thyroxine (*syn.* T_4)
      d. triiodothyronine (*syn.* T_3)
    2. glycoproteins
      a. follicle-stimulating hormone (FSH)
      b. luteinizing hormone (LH)
      c. thyroid-stimulating hormone (TSH)
    3. peptide hormones
      a. adrenocorticotropic hormone (ACTH) (*syn.* corticotrophin)
      b. antidiuretic hormone (ADH) (*syn.* vasopressin)
      c. brain hormone (BH)
      d. calcitonin
      e. oxytocin
      f. parathyroid hormone (PTH)
      g. thymosin
    4. protein hormones
      a. insulin
      b. insulin-like growth factor-I (IGF-I)
      c. glucagon
      d. growth hormone (GH)
      e. prolactin (PRL)
  B. steroid hormones (*syn.* lipid hormones)
    1. corticosteroid hormones
      a. glucocorticoids
        (1) corticosterone
        (2) cortisol
        (3) cortisone
      b. mineralocorticoids
        (1) aldosterone

    c. sex hormones
      (1) androgens
        (a) testosterone
        (b) dihydroepiandro-
            sterone
      (2) estrogens (*syn.* oestrogens)
        (a) estradiol
      (3) progesterones (*syn.*
        progestins)
  2. ecdysteroid hormones
    a. ecdysone
    b. ecdysterone
  3. prostaglandins (PGS; about 16
    kinds)
    a. prostaglandin E (PGE)
    b. prostaglandin F (PGF)
II. Plant hormones
  A. abscissic acid
  B. auxin
  C. cytokinins
  D. ethylene
  E. gibberellins

---

abscissic acid (ABA) (ab SIS ik) *n.* A plant hormone (produced in green fruit, leaves, and stems) that has the main functions of closing stomata during water stress, counteracting breaking of dormancy, and inhibiting growth (Campbell 1996, 753). [abscissic < Latin *ab*, loss, and *caedere*, to cut; after leaf abscission of deciduous trees; however, no one has demonstrated a clear-cut role of ABA in leaf abscission (Campbell 1996, 758)]

adrenaline *n.*
1. A hormone that is produced by the medulla of an adrenal gland, increases blood sugar concentration, raises blood pressure and heartbeat rate, and increases muscular power and resistance to fatigue (Curtis 1983, 1087).
2. A neurotransmitter that works across synaptic junctions (Curtis 1983, 1087).
syn. epinephrine (Curtis 1983, 1087)

adrenocorticotropic hormone (ACTH) *n.* In vertebrates: a hormone produced by a pituitary gland that influences activities of adrenal cortices (Immelmann and Beer 1989, 221).
syn. corticotropin

androgen *n.* In vertebrates: one of several sex hormones (*e.g.,* testosterone, androsterone) that controls the appearance and development of masculine characteristics (Curtis 1983, 777).
cf. hormone: estrogen
Comment: In mammals, androgens are 19-carbon steroids with methyl groups at C-10 and C-13 and are secreted by the Leydig cells of the testes (Wilson 1975, 251).

angiotensin *n.* In vertebrates: a hormone with two main effects — causing an animal's vascular system to maintain normal blood pressure and circulation and causing it to drink (McFarland 1985, 269).
cf. hormone: renin

antidiuretic hormone (ADH) See hormone: vasopressin.

antihormone *n.* A synthetic substance that antagonizes the effect of a hormone; for example, antiandrogen (*e.g.,* cyproterone, cyproteronacetate) or antiestrogen (*e.g.,* Tamoxifen®) (Dewsbury 1978, 226; Immelmann and Beer 1989, 17).
cf. hormone

auxin (AUK sin) *n.*
1. A plant hormone (now known to be indoleacetic acid, IAA) that causes an oat seedling to bend toward light (= exhibit phototropism) [coined by F.W. Went in Campbell 1996, 752)].
2. Any chemical substance that promotes the elongation of coleoptiles (Campbell 1996, 754).
3. The plant hormone indoleacetic acid (produced in apical-bud meristems, seed embryos, and young leaves) that has the functions of stimulating branching, differentiation, root growth, and stem elongation; apical dominance; gravitropism; and phototropism (Campbell 1996, 753).
Comment: Researchers have made synthetic auxins.
[Greek *auxein*, to increase]

auxins *n.* A class of plant hormones, including indoleacetic acid (Campbell 1996, G-4).

catecholamine *n.* A hormone (*e.g.,* epinephrine, norepinephrine) that is derived from tyrosine and secreted primarily by the adrenal-gland medullas (Wilson 1975, 253).

corpus-allatum hormones See hormone: juvenile hormones.

corticosteroid *n.* One of several hormones produced by adrenal-gland cortices, including corticosterone, cortisol, and cortisone (Wilson 1975, 253).

corticotropin See hormone: adrenocorticotropic hormone.

corticotropin-releasing factor (CRF) *n.* In Humans: a brain hormone that orchestrates many of a person's responses to stress, including alertness, anxiety, blood pressure, breathing, and heart-rate changes (Anonymous 1996g, A18).
Comment: Researchers discovered CRF in 1981 (Anonymous 1996g, A18).

crustecdysone *n.* In lobsters and decapod crabs: molting hormone which in females attracts males as if it were a sex attractant (Wilson 1975, 224).

h–m

Comment: This is the same substance as "ecdysterone" in insects (Barrington 1979, 235).

cytokinins (Sigh toh Kigh nins) *n.* Plant hormones (synthesized in roots) that have the main functions of affecting root growth and differentiation; stimulating cell division and growth, flowering, and germination; and delaying senescence (Campbell 1996, 753). Zeatin is one of several known cytokinins.
[after cytokinesis (= plant cell division) (Campbell 1996, 755)]

dihydroepiandrosterone (DHEA) *n.* A human androgen hormone produced in adrenal glands (Marano 1997, C1).
Comments: DHEA is involved with the onset of puberty (Marano 1997, C1).

ecdysteroid *n.* In many arthropod species: a hormone that mediates molting, *e.g.,* ecdysone and ecdysterone, which are two important insect ecdysteroids (Barrington 1979, 234–235).

egg-laying hormone (ELH) *n.* In *Aplysia* snails: a chemical that acts as a hormone when it mediates coordinated behavior patterns of egg extraction and deposition and also as a neurotransmitter (Scheller and Axel 1981 in McFarland 1985, 27).

endorphin *n.* A member of a group of peptide hormones that are liberated by proteolytic cleavage from a 29-kilodalton prohormone called pro-opiocortin (King and Stansfield 1985).
Comment: Beta-endorphins occur in the intermediate lobes of pituitary glands and are potent analgesics (pain suppressors) (King and Stansfield 1985).
[endorphin, a contraction of *endo*genous m*orphine*, because an endorphin is produced within an animal's body and attaches to the same neuron receptors that bind morphine and produces similar physiological effects (King and Stansfield 1985)]

environmental hormone See chemical-releasing stimulus: semiochemical: allelochemic: allomone: exocrine substance.

epinephrine See hormone: adrenaline.

estrodiol *n.* In vertebrates: a kind of estrogen.

estrogen, oestrogen *n.* In vertebrates: one of a group of sex hormones (Wilson 1975, 253; Curtis 1983, 780).
cf. hormone: premarin, estrone
Comments: Estrogens are found principally in ovaries but also in adrenals, placentae, testes, and even spermatozoans (Wilson 1975, 253). Receptors are found in about 300 different tissues, including brain, bone, and liver. Estrogens have many effects including mediation of egg development, mediation of female-sex characteristic development, maintenance of female reproductive behavior, and promotion of low levels of aggression and male fertility (Wilson 1975, 253; Curtis 1983, 780). Estrogen is the number one prescription drug in the U.S., being used to reduce menopausal symptoms and bone loss (osteoporosis) (Wallis 1995, 46, 49). Estrogen treatment can increase a woman's risk of breast cancer (Brown 1995, A1). Estrogen stimulates breast growth and other puberty changes in women (Angier 1997, C3).

estrone *n.* A form of human estrogen manufactured by fat cells (Wallis 1995, 50).

ethylene *n.* A gaseous plant hormone (produced in ripening-fruit tissues, senescent flowers and leaves, and stem nodes) that has that main functions of opposing some auxin effects, promoting fruit ripening, and promoting or inhibiting development and growth of flowers, leaves, and roots, depending on the species (Campbell 1996, 753).

follicle-stimulating hormone (FSH) *n.* In vertebrates: a gonadotropic hormone that activates growth of ovarian follicles and estrogen secretion (Immelmann and Beer 1989, 108).
Comments: In Humans, FSH is released by a pituitary gland, and it travels through a bloodstream to ovaries, which it prompts to release estrogen (Angier 1997, C3).

gibberellins (JIB ur EL ins) *n.* Plant hormones (produced in embryos, apical-bud and root meristems, and young leaves) that have the main functions of affecting root differentiation and growth; promoting bud and seed germination, leaf growth, and stem elongation; and stimulating flowering and fruit development (Campbell 1996, 753).
Comment: Scientists have identified more than 70 different gibberellins (Campbell 1996, 757).
[after the fungus genus *Gibberella,* which causes abnormal elongation of rice seedlings (= foolish-seedling disease) (Campbell 1996, 757)]

glandotropic hormones, glandotrophic hormones *n.* In vertebrates: hormones produced by pituitary glands that influence activities of other endocrine glands, including thyroids, adrenal cortices, and gonads (Immelmann and Beer 1989, 221).

glucagon-like factor-1 *n.* A mammalian brain hormone that suppresses appetite (Anonymous 1996g, A18).

Comments: Corticotropin-releasing factor, leptin, and urocortin are other brain hormones that suppress appetite (Anonymous 1996g, A18).

gonadotrophic hormone *n.* In vertebrates: one of a group of hormones (follicle-stimulating hormone, luteinizing hormone, interstitial-cell-stimulating hormone) produced by pituitary glands that stimulates growth and activity of testes and ovaries (*e.g.,* sex-hormone secretion) (McFarland 1985, 293).
syn. gonadotropin (Immelmann and Beer 1989, 120)

gonadotropin-releasing hormone (GNRH) *n.* In the Human: a hormone, released by the hypothalamus, that stimulates the pituitary gland to release Latinizing hormone and follicle-stimulating hormone (Angier 1997, C3).

insulin *n.* In vertebrates: a hormone that controls the availability of glucose to cells (McFarland 1985, 271).

insulinlike-growth factor-I (IGF-I) *n.* In the Human: a protein hormone (Barinaga 1998, 475).
Comment: Men with high blood levels of IGF-1 are about four times more likely to develop prostate cancer than are men with the lowest level (Chan et al. 1998, 563).

juvenile hormones (JHs) *pl. n.* In insects: hormones secreted by corpora allata that regulate metamorphosis and reproductive development (Barrington 1979, 243).
syn. corpus-allatum hormones (Barrington 1979, 243)

lactogenic hormone See hormone: prolactin.

leptin *n.* In laboratory mice: a fat-cell-produced hormone (Kolata 1995, A20).
Comments: Leptin administration to mice causes them to lose weight, even if they are lean mice (Kolata 1995, A20). Humans produce a hormone that is similar to leptin. Leptin's principal role may be to prepare an animal to respond to a long hunger periods, including decreased body temperature, diminished procreation (in either sex), and secretion of substances that help stress survival (Suplee 1996b, A2). Leptin also initiates puberty in female mice and possibly Humans (Angier 1997, C1). Corticotropin-releasing factor, glucagon-like factor-1, and urocortin are other brain hormones that suppress appetite (Anonymous 1996g, A18).

luteinizing hormone (LH) *n.* For example, in the Human: a hormone, released by the pituitary gland, that stimulates ovaries and causes them to release estrogen (Angier 1997, C3).

neurohormone, neurohumor *n.* In many animal species: a secretion of a neuroendocrine cell, *q.v.* (Immelmann and Beer 1989, 202).

neuropeptide Y (NPY) *n.* A protein that signals a rat's and Human's brain to cause it to eat (Anonymous 1996g, A18).
cf. hormone: peptide YY

norepinephrine *n.* In vertebrates: a catecholamine hormone secreted by adrenal-gland medullas (Wilson 1975, 253).
Comment: Norepinephrine coupled with the parasympathetic nervous system has generally different and sometimes opposite effects of epinephrine (Wilson 1975, 253).

oxytocin *n.* In vertebrates: a hormone produced by a posterior pituitary gland that induces uterine contractions of parturition and milk ejection during suckling (Immelmann and Beer 1989, 222).

peptide YY *n.* A protein that signals a rat's brain to cause it to eat (Anonymous 1996g, A18).
cf. hormone: neuropeptide Y

premarin *n.* A form of estrogen made from pregnant mare's urine (Wallis 1995, 48).
[*preg*nant *mare*'s ur*ine*]

progesterone *n.* A progestin, *q.v.* (Immelmann and Beer 1989, 233).

progestin *n.* In vertebrates: one of a group of female sex hormones produced in ovaries and placentas in mammals and in analogous tissues in other vertebrates that regulate reproductive processes usually later than estrogens (Immelmann and Beer 1989, 233).

prolactin *n.* In vertebrates: a hormone produced by a pituitary gland that induces mammary development and milk production in mammals, promotes crop milk production in pigeons and doves, and may be involved in brood-patch development and sustaining incubation in some bird species (Immelmann and Beer 1989, 221, 235).
syn. lactogenic hormone (Immelmann and Beer 1989, 221, 235).

prolactin-release inhibiting factor *n.* A releasing hormone, *q.v.*, that stops release of prolactin (Immelmann and Beer 1989, 251).

prostaglandins (PGs) *pl. n.* A group of about 16 animal hormones that are released from most cell types into interstitial fluid and function as local regulators affecting nearby cells in various ways (Campbell 1996, 915).
Comments: Prostoglandin E (PGE) causes muscles to relax which dilates blood vessels and promotes oxygenation of blood

h – m

(Campbell 1996, 915). Prostaglandin F (PGF) signals muscles to contract, which constricts blood vessels and reduces blood flow through lungs. Prostaglandins, secreted by placental cells, cause chemical changes in nearby uterine muscles, making them more excitable, thereby helping to induce labor during human childbirth. Various prostaglandins help to induce fever and inflammation and intensify the sensation of pain.

releasing hormone *n.* Substances produced by hypothalamus cells and conveyed to an animal's anterior lobe of its pituitary gland via its hypothalamohypophysial portal system; these substances influence secretion of pituitary hormones, which in turn affect release of hormones from other endocrine glands (Immelmann and Beer 1989, 251).

renin *n.* In vertebrates: a hormone produced by kidneys that stimulates production of another hormone, angiotensin (McFarland 1985, 269).

sex hormone *n.* A hormone that regulates development and the process of reproduction; sex hormones include androgens, estrogens, and progestins, *q.v.* (Curtis 1983, 777; Immelmann and Beer 1989, 267).

sociohormone See chemical-releasing stimulus: semiochemical: pheromone.

steroid hormone *n.* A hormone that is a steroid (a fat-soluble organic compound with four linked carbon rings), including cortisol and other glucocorticoids, aldosterone, estrogens, progesterone, and testosterone (Curtis 1983, 792).

testosterone *n.* In vertebrates: a hormone made in testes that has various effects, including increasing aggressiveness, supporting spermatogenesis, and developing and maintaining male sex characteristics (Wilson 1975, 251–252; Curtis 1983, 792).
cf. hormone: androgen

thymosin *n.* One member of a family of hormones that communicates between a person's brain and immune system (Robinson 1987, 74).

thyrotropic hormone *n.* A hormone produced by a vertebrate's pituitary gland that influences thyroid-gland activities (Immelmann and Beer 1989, 221).

urocortin *n.* In Humans, rats, possibly other mammals: a brain hormone that may cause an individual to lose its appetite when it is under stress or in danger, when survival can depend on fighting or fleeing rather than feeding (Anonymous 1996g, A18).

Comments: Wylie Vale, Paul Sawchenko, and other scientists identified urocorin in 1995. Corticotropin-releasing factor, glucagon-like factor-1, and leptin are other brain hormones that suppress appetite (Anonymous 1996g, A18).

vasopressin, antidiuretic hormone *n.* In mammals: a hormone that elevates blood pressure and promotes water reabsorption by kidneys (Dewsbury 1978, 221).

♦ **hormone system** See ²system: hormone system.

♦ **Horn's principle** See principle: Horn's principle.

♦ **horodimorphism** See -morphism: dimorphism: horiodimorphism.

♦ **horos** *combining form*
[Greek *horos*, boundary, landmark, limit, rule, standard]

♦ **horotely** See ²evolution: horotely.

♦ **hospicidal** *adj.* Referring to a larval cuckoo bee that is highly modified having elongate, sickle-shaped mandibles used to destroy its host larva or egg [coined by Rozen 1989, 2].
ant. nonhospicidal [coined by Rozen 1989, 2]
Comment: In a nonhospicidal cuckoo bee, a mother bee destroys an immature host when she oviposits on the host's provisions (Rozen 1989, 2)
[Latin *hospes*, host; *caedo*, to cut down, to kill]

♦ **hospitalism** *n.* Disturbed emotional and bodily development of people who have grown up in orphanages without normal maternal affection (Heymer 1977, 89).
cf. behavior: abnormal behavior
Comment: Similar behavior occurs in Rhesus Monkeys (Spitz 1945 in Heymer 1977, 89).

♦ **hospitator** *n.* An organism that offers refuge, or acts as a host, to another organism (the hospitate) (Lincoln et al. 1985).
v.i. hospitate
cf. host

♦ **hospite** *n.* An organism that takes refuge in or inhabits another organism, the hospitator (Lincoln et al. 1985).
cf. inquiline

♦ **¹host** See ²group: host.

♦ **²host** *n.* An individual animal that receives a tissue graft (Lincoln et al. 1985).

♦ **³host** *n.*
1. A man who lodges and entertains another in his house; the correlative of guest (*Oxford English Dictionary* 1972, entries from 1303).
2. An individual organism that habitually has a relationship with a parasite or commensal (*Oxford English Dictionary* 1972, entries from 1857); the inhabited

symbiont in an endosymbiosis or an exhabited symbiont in an exosymbiosis (Lincoln et al. 1985).

3. An individual organism that provides food or shelter for another organism (Lincoln et al. 1985).

cf. hospitator

carrier host *n.* An animal that carries a phoretic animal (Heymer 1977, 130).

definitive host, primary host *n.* A host in which a parasite attains sexual maturity (Lincoln et al. 1985).

intermediate host, secondary host *n.* A host occupied by a parasite's juvenile stages, prior to its occupying its definitive host, and in which it is likely to undergo asexual reproduction (Lincoln et al. 1985).

paratenic host, transfer host, transport host *n.* A host that is not essential for a parasite's completing its life cycle but is used as a temporary habitat or as a means of reaching its definitive host (Lincoln et al. 1985).

cf. symbiosis: phoresy

primary host See [3]host: definitive host.

reservoir host, reservoir *n.* A host that carries a pathogen without detriment to itself and that serves as a source of infection for other hosts (Lincoln et al. 1985).

secondary host See [3]host: intermediate host.

transfer host, transport host See [3]host: paratenic host.

♦ **host discrimination** *n.*

1. An individual parasitic wasp's ability "to distinguish unparasitized from parasitized hosts and to lay eggs only in the former" (Lenteren et al. 1978, 71, who discuss problems with this definition).

2. A parasitic wasp's ability to distribute her eggs in a "nonrandom way" among her hosts (Lenteren et al. 1978, 71, who discuss virtues of this definition).

♦ **host imprinting** See learning: imprinting: host imprinting.

♦ **host list** See [1]range: host range.

♦ **host race, hostic race** See [1]race: host race.

♦ **host range** See [1]range: host range.

♦ **host specificity** See specificity.

♦ **hotel** *n.* The part of an organism's habitat that it uses for resting, breeding, and feeding (Lincoln et al. 1985).

cf. home, range, territory

♦ **hotspot model** See [4]model: hotspot model.

♦ **housekeeping enzyme** See enzyme: housekeeping enzyme.

♦ **house-keeping gene** See gene: house-keeping gene.

♦ **hover** See [2]group: hover.

♦ **howl** See animal sounds: howl.

♦ *Hox* **cluster** See gene: *hox* cluster.

♦ *Hox* **gene** See gene: homeotic-homeobox gene.

♦ **huddle** See [2]group: huddle.

♦ **huddling, huddling behavior** See behavior: contact behavior.

♦ **hula display** See display: hula display.

♦ **human behavior** See behavior: human behavior.

♦ **human carrying capacity** See carrying capacity: human carrying capacity.

♦ **human culture** See [5]culture: human culture.

♦ **human ethology** See study of: [3]ethology: human ethology.

♦ **human-female "pheromone"** See chemical-releasing stimulus: semiochemical: pheromone: human-female "pheromone."

♦ **Human Genome Project** See project: Human Genome Project.

♦ **human imprinting** See learning: imprinting: human imprinting.

♦ **human-male "pheromone"** See chemical-releasing stimulus: semiochemical: pheromone: human-male "pheromone."

♦ **human perspiration odor** See odor: human underarm odor.

♦ **human reciprocity** See altruism: reciprocal altruism.

♦ **human underarm odor** See odor: human underarm odor.

♦ **human vaginal odor** See odor: human vaginal odor.

♦ **humanology** See study of: humanology.

♦ **humbleness** See abasement.

♦ **humicole** See -cole: humicole.

♦ **humped posture** See posture: humped posture.

♦ **Humpty-Dumptyism** *n.* A person's taking a word, or term, from common usage and redefining it to mean only what he wants it to mean (Beach 1978, 1979).
[from Carroll (1932, 184, in Beach 1978, 1979): Humpty Dumpty said rather scornfully, "When I use a word, it means just what I choose it to mean — neither more nor less."]

♦ **hunting** *n.* An animal's searching for, or pursuing, prey (Michaelis 1963).

mock hunting *n.* An animal's showing prey-capture movements in the absence of prey (*e.g.,* a cat's playing with a wool ball, a young lion's playing with its siblings) (Heymer 1977, 94).

♦ **hunting camouflage** See camouflage: hunting camouflage.

♦ **hurtle** See [2]group: hurtle.

♦ **husbandry** *n.* A person's tillage and cultivation of soil to grow plant crops; husbandry can also include rearing livestock and poultry, even Honey Bees and silk-worms (*Oxford English Dictionary* 1972, entries from 1380).

syn. agriculture, farming (*Oxford English Dictionary* 1972)

♦ **huske** See ²group: huske.

♦ **hybrid** *n.*

1. Often an offspring of a cross between heterospecific individuals; also an offspring of a cross between conspecific populations, strains, or subspecies (*Oxford English Dictionary* 1972, entries from 1601; King and Stansfield 1985; Immelmann and Beer 1989, 137).

   syn. cross (King and Stansfield 1985)

   cf. cross, hybridization

   Note: In taxonomy, a hybrid is usually considered to be from an interspecific cross (Lincoln et al. 1985).

   syn. crossbreed, halfbreed, mongrel (*Oxford English Dictionary* 1972)

2. A community comprised of taxa from two or more separate, distinct communities (Lincoln et al. 1985).

   Comments: Taxonomists indicate a hybrid between two species with an x between the generic and species names: *Ambrosia* x *helenae* Rouleau is a hybrid of *A. trifida* and *A. artemisiifolia* (ragweeds). Taxonomists indicate a cross between two genera with an X before the new genus name: X *Cupressocyparis leylandii* (a conifer in Appendix 1). The plus in front of +*Laburnocytisus adamii* means that it is a graft hybrid of *Cytisus purpureus* and *Laburnum anagyroides.* The plus sign and combining the two parent-plant genus names comprise the designation for a graft chimera outlined by the International Code of Botanical Nomenclature (Geneve 1991, 34).

anisogonous hybrid *n.* A hybrid that does not show equal expression of both of its parental characters (Lincoln et al. 1985).
   cf. hybrid: isogonous hybrid

cryptohybrid *n.* A hybrid that shows unexpected phenotypic traits (Lincoln et al. 1985).

heterodynamic hybrid See hybrid: segregate.

Hinny *n.* A hybrid between a stallion (Horse) and sheass (Donkey) (Michaelis 1963).
   [L. *hinnus* < Greek *ginnos*]

interspecific hybrid See ²species: interspecies.

intervarietal hybrid *n.* A cross between two conspecific varieties (Lincoln et al. 1985).

isogonous hybrid, isogonic hybrid *n.* A hybrid that shows equal expression of parental characters (Lincoln et al. 1985).
   cf. hybrid: anisogonous hybrid

Ligon *n.* A hybrid between a male Lion and a female Tiger produced in a zoo (Hoffmeister 1967, 46).
   syn. Liger (Wilson 1975, 9)
   cf. hybrid: Tigon

Mule *n.*
1. A hybrid between an Ass (Donkey) and a Horse, especially a jackass and a mare (Michaelis 1963).
   cf. hybrid: Hinny
2. Any hybrid or cross, especially a sterile one (*e.g.,* a cross between a canary and related bird) (Michaelis 1963).
[Old French *mul* < Latin *mulus*]

multihybrid *n.* A hybrid individual that is heterozygous at more than one locus (Lincoln et al. 1985).

numerical hybrid *n.* A hybrid that is derived from parental gametes with differing chromosome numbers (Lincoln et al. 1985).

permanent hybrid, permanent heterozygosity *n.* An organism in which heterozygosity is fixed due to balanced, or partially balanced, lethal factors (Lincoln et al. 1985).

poikilodynamic hybrid See hybrid: segregate.

polyploid hybrid See ploidy: allopolyploidy.

pseudohybrid *n.* A hybrid that shows phenotypic traits of only one of its parents (Lincoln et al. 1985).

reciprocal hybrids *pl. n.* A hybrid whose father is species A and mother is species B or whose father is B and mother is A (*e.g.,* a Mule and Hinny) (*Oxford English Dictionary* 1972).
   syn. reciprocal crosses (Lincoln et al. 1985)

segregate *n.* A hybrid that exhibits predominantly only one of its parent's phenotype (Lincoln et al. 1985).
   syn. heterodynamic hybrid, poikilodynamic hybrid (Lincoln et al. 1985)

sesquireciprocal hybrid *n.* A hybrid crossed with one of its parental types (Lincoln et al. 1985).

Tigon *n.* A hybrid between a male Tiger and a female Lion produced in a zoo (Hoffmeister 1967, 46)
   syn. Tiglon (Wilson 1975, 9)
   cf. hybrid: Ligon

♦ **hybrid breakdown** *n.* "Reproductive failure of F₂ hybrids (Lincoln et al. 1985).

♦ **hybrid cline** See cline: hybrid cline.

♦ **hybrid inviability** *n.* Failure of F₁ hybrids to reach sexual maturity (Lincoln et al. 1985).

♦ **hybrid speciation** See speciation: hybrid speciation.

♦ **hybrid vigor** See heterosis.

◆ **hybrid zone** See zone: hybrid zone.
◆ **hybridization** (*Brit.* **hybridisation**) *n.*
Any crossing of individuals of different genetic composition that typically belong to
separate species and result in hybrid offspring (Lincoln et al. 1985; King and Stansfield
1985).
v.i., v.t. hybridize
syn. cross (King and Stansfield 1985)
cf. hybrid
introgressive hybridization See introgression.
nucleic-acid hybridization *n.* Base
pairing of single-stranded nucleic acid from
different nucleic acid molecules (Campbell
et al. 1999, 369).
sympatric hybridization *n.* "The occasional production of hybrids between
two well-defined sympatric species" (Lincoln et al. 1985).
◆ **hybridogenesis** See parthenogenesis:
hybridogenesis.
◆ **hydrarch succession, hydrosere** See
²succession: hydrarch succession.
◆ **hydraulic model of behavior** See
⁴model: Lorenz's hydraulic model.
◆ **hydric** *adj.* Referring to water; wet (Lincoln et al. 1985).
cf. hydroid
◆ **-hydro-, -hydr-** *combining form* Water,
of, related to, or resembling water" (Michaelis 1963).
[Greek *hydro-* < *hydrōr*, water]
◆ **hydrobiology** See study of: biology: hydrobiology.
◆ **hydrobios** See -bios: hydrobios.
◆ **hydrochore** See -chore: hydrochore.
◆ **hydrocole** See -cole: hydrocole.
◆ **hydrogen hypothesis** See hypothesis:
hydrogen hypothesis.
◆ **hydrogenosome** See organelle: hydrogenosome.
◆ **hydrograph** See graph: hydrograph.
◆ **hydroharmose** *n.* An organism's total
reaction and adaptation (harmosis) to water
(Lincoln et al. 1985).
cf. harmosis
◆ **hydroid** *adj.* Wet, watery (Lincoln et al.
1985).
cf. hydric
◆ **hydrologic cycle** See cycle: hydrologic
cycle.
◆ **hydrolysis** *n., pl.* **hydrolyses**
1. Action between the ions of water (H⁺ and
OH⁻) and those of a salt to form an acid or
a base that changes that pH of the solution
when the acid and base are sufficiently
different in strength (Michaelis 1963).
2. Decomposition of a compound by water
resulting in products that each contain
part of the water (Michaelis 1963).

◆ **hydromegatherm** See -therm: hydromegatherm.
◆ **hydromorphic** See -morphic: hydromorphic.
◆ **hydromorphosis** See -morphosis: hydromorphosis.
◆ **hydrophile** See ¹-phile: hydrophile.
◆ **hydrophobe** See -phobe: hydrophobe.
◆ **hydrophytium** See habitat: bog.
◆ **hydrosere** See ²succession: hydrach succession.
◆ **hydrosphere** See -sphere: hydrosphere.
◆ **hydrotaxis** See taxis: hydrotaxis.
◆ **hydrothermal vent** See ²community:
hydrothermal-vent community.
◆ **hydrothermograph** See instrument:
hydrothermograph.
◆ **hydrotribophile** See ¹-phile: hydrotribophile.
◆ **hyemal** See hiemal.
◆ **hyetal** *adj.* Referring to rain or precipitation (Lincoln et al. 1985).
◆ **hygienic behavior** See behavior: hygienic behavior.
◆ **hygric** *adj.* Referring to moisture, moist
conditions, or humid conditions (Lincoln et
al. 1985).
◆ **hygro-** *combining form* Wet; humid;
damp; moist (Michaelis 1963). Before vowels, hygr-.
[Greek *hygros* wet, moist]
◆ **hygrocole** See -cole: hygrocole.
◆ **hygrodrymium** See ²community: rain
forest.
◆ **hygrokinesis** See -kinesis: hygrokinesis.
◆ **hygroklinokinesis** See -kinesis: hygroklinokinesis.
◆ **hygromorphic** See -morphic: hygromorphic.
◆ **hygroorthokinesis** See -kinesis: hygroorthokinesis.
◆ **hygropetric, hygropetrobiontic,
hygropetrobiotic, petrimadicolous**
adj. Referring to an organism that lives in
the surface film of water on rocks (Lincoln
et al. 1985).
◆ **hygropetrobios** See -bios: hygropetrobios.
◆ **hygrophile** See ²-phile: hygrophile.
◆ **hygrophobe** See -phobe: hygrophobe.
◆ **hygrophorbium** See ²community:
hygophorbium.
◆ **hygrosphagnium** See ²community:
hygrosphagnium.
◆ **hygrotaxis** See taxis: hygrotaxis.
◆ **hyla-, hylo-, hyl-** *combining form* Wood,
or referring to wood (Michaelis 1963).
[Greek *hylē*, wood]
◆ **hylacole, hylocole** See -cole: hylacole.
◆ **hylium, hylion** See ²community: forest.
◆ **hylodium** See ²community: hylodium.

♦ **hylodophile** See [1]-phile: hylodophile.

♦ **hylogamy** See -gamy: hylogamy.

♦ **hylophage** See -vore: lignivore.

♦ **hylophile** See [1]-phile: hylophile.

♦ **hylotomous** *adj.* Referring to wood-cutting insects (Lincoln et al. 1985).

♦ **hyper-** *combining form*
1. Over; above; excessive; more than (effect of quantity); higher (position) (Michaelis 1963).
2. In medicine, denoting an abnormal state of excess (Michaelis 1963).
[Greek *hyper-* < *hyper*, above]

♦ **hyperallobiosphere** See -sphere: biosphere: eubiosphere: allobiosphere: hyperallobiosphere.

♦ **hyperanisogamic** See -gamic: hyperanisogamic.

♦ **hyperbenthos** See [2]community: benthos: hyperbenthos.

♦ **hypercapnia** *n.* An animal's sensitivity to an overdose of carbon dioxide which can be lethal (Browne 1996b, C10).

♦ **hypercyesis** See superfoetation.

♦ **hyperdispersed distribution** See distribution: overdispersed distribution.

♦ **hyperdiversity** See diversity: hyperdiversity.

♦ **hypergamesis** *n.* Utilization of excess sperm by a female, or female gamete, for nourishment (Lincoln et al. 1985).

♦ **hypermetabolous metamorphosis** See metamorphosis: hypermetabolous metamorphosis.

♦ **hypermetamorphic metamorphosis** See metamorphosis: hypermetamorphic metamorphosis.

♦ **hypermorph** See -morph: hypermorph.

♦ **hypermorphic allele** See allele: hypermorphic allele.

♦ **hypermorphosis** See [2]heterochrony: hypermorphosis.

♦ **hyperparasitism** See parasitism: hyperparasitism.

♦ **hyperphagia** See -phagia: hyperphagia.

♦ **hyperploidy** See -ploidy: hyperploidy.

♦ **hypersexuality** See sexuality: hypersexuality.

♦ **hypersociality** See sociality: hypersociality.

♦ **hyperteley** See [2]evolution: hyperteley.

♦ **hyperthermia** See thermia: hyperthermia.

♦ **hypertrophy** *n.*
1. Enlargement of an organism's organ due to excessive nutrition; an organ's excessive growth, or development (*Oxford English Dictionary* 1972, entries from 1834).
2. The marked evolutionary development of an organism's trait (*e.g.,* mental hypertrophy) (Wilson 1975, 548).

3. An animal's excessively frequent production of a particular behavior (*e.g.,* hypersexuality in domestic animals) (Immelmann and Beer 1989, 137).
v.i, v.t. hypertrophy
cf. behavioral deficit

♦ **hypervariable marker** See marker: hypervariable marker.

♦ **hyphalomyraplankton** See plankton: hyphalomyraplankton.

♦ **hyphydrogamic, hyphydrogamous** See -gamous: hyphydrogamous.

♦ **hyphydrophile** See [2]-phile: hydrophile: hyphydrophile.

♦ **hypnody** *n.* A prolonged resting period, or diapause, during an organism's development (Lincoln et al. 1985).
adj. hypnodic

♦ **hypnophobia** See phobia (table).

♦ **hypnoplasy** *n.* Arrested development that results in failure to reach normal size (Lincoln et al. 1985).

♦ **hypnosis** *n.* An animal's tonic immobility induced by its restraint by, or its close proximity to, a predator (Immelmann and Beer 1989, 138).
cf. playing dead

♦ **hypnote** *n.* A dormant organism (Lincoln et al. 1985).
cf. dormancy

♦ **hypnotic reflex** See playing dead.

♦ **hypo-, hyp-** *combining form*
1. "Under; beneath" (Michaelis 1963).
2. "Less than" (Michaelis 1963).
3. Medically, "denoting a lack of or deficiency in;" opposed to hyper- (Michaelis 1963).
[Greek < *hypo*, under]

♦ **hypoallobiosphere** See -sphere: biosphere: eubiosphere: allobiosphere: hyperallobiosphere.

♦ **hypobenthos** See [2]community: -benthos: hypobenthos.

♦ **hypobiosis** See -biosis: hypobiosis.

♦ **hypobiotic** See -biotic: hypobiotic.

♦ **hypocaloric** *adj.* Referring to a habitat, or condition, of sparse or reduced food supply (Lincoln et al. 1985).

♦ **hypodispersed distribution** See distribution: underdispersed distribution (def. 1).

♦ **hypogenesis** See -genesis: hypogenesis.

♦ **hypogenous** See -genous: hypogenous.

♦ **hypohydrogamic** See -gamous: hyphydrogamous.

♦ **hypolictic** *adj.* Referring to an organism that lives beneath rocks or stones (Lincoln et al. 1985).
cf. -cole: lapidocole

♦ **hypomorph** See -morph: hypomorph.

- **hypomorphic allele** See allele: hypomorphic allele.
- **hyponeuston** See neuston: hyponeuston.
- **hypopharyngeal gland** See gland: hypopharyngeal gland.
- **hypophloeodal** *adj.* Referring to an organism that lives, or occurs, immediately below bark surface (Lincoln et al. 1985).
- **hypophyllous** *adj.* Referring to an organism that grows on the lower surfaces of leaves (Lincoln et al. 1985).
  cf. -cole: folicole
- **hypophysis** See gland: pituitary gland.
- **hypoplankton** See plankton: hypoplankton.
- **hypoplasia** *n.* Underdevelopment or lack of normal growth (Lincoln et al. 1985).
- **hypoploidy** See -ploidy: hypoploidy.
- **hypostasis** *n.* One gene's (the hypostatic gene) being masked by the expression of another gene (the epistatic gene) at a different locus (Lincoln et al. 1985).
  cf. epistasis
- **hypostatic gene** See gene: hypostatic gene.
- **hypothalamohypophysial system** See ²system: hypothalamohypophysial system.
- **hypothalamus** See organ: brain: hypothalamus.
- **hypothermia** See thermia: hypothermia.
- **hypothesis** *n.*
  1. A proposition that is stated without any reference to its correspondence with fact, merely as a basis for reasoning or argument or as a premise from which to draw a conclusion (*Oxford English Dictionary* 1972, entries from 1656).
  2. A scientific supposition or conjecture made to account for known facts, from which one can draw a conclusion that is in accord with known facts and that serves as a starting point for further investigation (*Oxford English Dictionary* 1972, entries from 1646).
  3. A metaphysical speculation without empirical substance (Dobzhansky et al. 1977, 485).
  Note: This is a definition used during Darwin's time (Dobzhansky et al. 1977, 485). "Darwin expressed distaste and even contempt for empirically untestable hypotheses."
  adj. hypothetical
  v.i., v.t. hypothesize
  syn. supposition (*Oxford English Dictionary* 1972), theory (but some workers, *e.g.,* Moore, 1984, do not recommend this synonymy in science)
  cf. axiom; concept; ⁴model; principle; selection: species selection; ³theory

Comments: A hypothesis is given as a statement, not as a question: "The brilliant plumage of peacocks attracts peahens." Important hypotheses that are not refuted by exhaustive testing can become theories that have the status of facts: the theory of evolution by natural selection. Many hypotheses are given under "theory" in this book because their names contain the words theorem or theory.
[New Latin < Greek foundation, supposition < *hypothitenai*; to put under, *hypo*, under + *tithenia*, to put]

African-origin hypothesis of human mitochondrial DNA evolution *n.* The hypothesis that all mitochondrial DNA types in contemporary Humans (*Homo sapiens sapiens*) stem from a common female ancestor (mitochondrial Eve, *q.v.*) present in an African population about 200,000 years ago (Cann et al. 1987, 31; Vigilant et al. 1991, 1503).
syn. African origin hypothesis (Wilson et al. 1991, 365), Eve hypothesis (Gibbons 1992b, 873), Eve theory (Thorne and Wolpoff 1992, 76), mitochondrial-Eve hypothesis (Lewin 1987c), Mother-Eve hypothesis (Wainscoat 1987, 13), single African origin hypothesis (Templeton 1991, 737)
cf. mitochondrial Eve
Comments: Wolpoff and Thorne (1991) call this hypothesis "the most controversial theory about human origins ever proposed" because it suggests that Modern Humans (wherever they evolved) replaced, rather than mixed with, indigenous Archaic Humans. Some studies support this hypothesis (Lewin 1991; Vigilant et al. 1991), but other analyses do not (Hedges et al. 1991, 738; Maddison 1991; Barinaga 1992a, 686; Templeton 1992, 737). Humans are known to inherit their mtDNA through the mothers but might also inherit a small amount of mtDNA from their fathers (Ross 1991, 30). *Homo s. sapiens* may have arisen sometime from 100,000 to 200,000 years ago (Allan C. Wilson in Lewin 1987c; Maryellen Ruvolo in Gibbons 1993a).

age-and-area hypothesis *n.* The area occupied by a species is proportional to its evolutionary age (Lincoln et al. 1985).
cf. dominance (ecological): faunal dominance

aggression-frustration hypothesis *n.* Aggression regularly follows frustration, and its strength is directly related to the degree of frustration; however, depression, rather than aggression, might result from frustration (Heymer 1977, 70).

allelic-replacement hypothesis *n.* The hypothesis that major steps in social evolution occur via new alleles' replacing earlier ones in a species (West-Eberhard 1987, 38).

cf. hypothesis: "no-allelic-replacement hypothesis"

alternative-adaptation hypothesis *n*. The hypothesis that "novel traits originate and become elaborated as stable alternative phenotypes or morphs within species, prior to reproductive isolation and speciation, when they come to characterize distinctive new lineages;" that is, "drastic innovation can begin not with the branching of a phylogenetic tree but with the bifurcation of a developmental or behavioral program ("epigenetic divergence") giving rise to intraspecific alternative adaptations" (West-Eberhard 1986, 1388).

alternative hypothesis (H_1) *n*. A hypothesis of nonequal relationship between phenomena; *e.g.,* feeding behavior of a test group takes a different amount of time, takes a longer time, or takes a shorter time than that of the control group (Lehner 1979, 227).

syn. experimental hypothesis (Sokal and Rohlf 1969, 557)

cf. hypothesis: null hypothesis

"ancestral-brain hypothesis" *n*. The hypothesis that human ancestors were "primative" in a general way; all their body parts, including their brains, were ancestral in their characteristics (Washburn 1985, 5); contrasted with the "brain-first hypothesis."

Comment: This hypothesis was a major one regarding human evolution at the time of the Raymond Dart's report of the Taung Child in 1925 (Washburn 1985, 5).

angiosperm-origin hypotheses

cf. Darwin's abominable mystery; Plantae: *Amborella trichopoda* (Appendix 1).

Comments: Angiosperms might have originated 200 Ma (Waters 1991, 79). Molecular data do not support the anthophyte hypothesis, or the piperaloid hypothesis, and suggest that the dicot lineage and gymnosperm lineage split about 290 Ma in the late Carboniferous Period and that *Amborella trichopoda* is the sibling group of the rest of the flowering plants (Kenrick 1999, 359; Qui et al. 1999, 404; Soltis et al. 1999, 402).

▶ **anthophyte hypothesis** *n*. The hypothesis that flowering plants (Anthophyta) are most closely related to the extinct Bennettitales and still-extant Gnetales (Crane et al. 1995 in Qui et al. 1999, 404; Kenrick 1999, 359).

▶ **"piperaloid hypothesis"** *n*. The hypothesis that angiosperms originated from a small, rhizomatous perennial which resembles a member of the Piperales and

has secondary growth and diminuative reproductive organs arranged cymosely and subtended by a bract-bracteole complex (Taylor and Hickey 1990, 704; Monastersky 1992, 74); contrasts with "magnolioid hypothesis" and "shrub hypothesis."

syn. lowly-origin hypothesis (Monastersky 1992, 74)

Comments: Taylor and Hickey (1990) found a fossil, possibly in the Piperales, that dates to 120 Ma and thus supports this hypothesis. The oldest fossil of a magnolia-like flower dates to 100 Ma.

▶ **"magnolioid hypothesis"** *n*. The hypothesis that angiosperms originated from a small tree, or shrub, with pinnately veined, simple leaves and moderate to large flowers with numerous reproductive parts (Cronquist 1988 in Taylor and Hickey 1990, 702).

▶ **"shrub hypothesis"** *n*. The hypothesis that angiosperms originated from a shrub with spirally arranged, simple leaves and woody tissues formed from a single vascular cylinder (Stebbins 1974 in Strickberger 1990, 265).

Anthropoidea-origin hypotheses

▶ **"Adapidae hypothesis"** *n*. The lemur-like adaptids give rise to Anthropoidea (Culotta 1995, 1851).

Comment: Adaptoids are extinct (Simons 1995, 1886).

▶ **"Anthropoidea hypothesis"** *n*. The Anthropoidea is a early taxon that originated soon after the first primates appeared (Culotta 1995, 1851).

▶ **"Omomyidae hypothesis"** *n*. The diminutive Omomyidae is the ancestor of Anthropoidea (Culotta 1995, 1851).

Comment: A living descendent of omomyids is *Tarsius* (Simons 1995, 1886).

battered-Earth hypothesis *n*. Two major impacts struck the Earth at the end of the Cretaceous (Mesozoic Era), about 64.4 million years ago, and are responsible for the extinction of many species, including dinosaurs; contrasted with single-impact hypothesis (Fastovsky 1990 in Kerr 1992, 160).

Comments: The Chicxulub and Manson Craters were proposed as sites of impacts that occurred about 64.4 million years ago. More refined dating of the Manson Crater revealed that it formed *ca.* 73.8 million years ago; thus, it was not part of a multiple impact at the end of the Cretaceous, and the battered-Earth hypothesis is not supported (Izett et al. 1993, 729).

beacon hypothesis See hypothesis: firefly-flash-synchronization hypotheses: beacon hypothesis.

bee-language hypothesis See hypothesis: dance-"language" hypothesis.

behavior hypothesis See hypothesis: population-limiting hypotheses: behavior hypothesis.

big-mother hypothesis *n*. In mammals: Compared to a smaller mother, a bigger mother is often a better mother because she might produce a greater number of surviving offspring; produce a larger baby with greater chances of survival; produce more milk, better milk, or both, which would allow her baby to grow faster; carry and defend her young better; or a combination of these activities [coined by Ralls 1976, 268].

Comments: Ralls formulated this hypothesis to suggest why females of many mammal species are larger than conspecific males. In bats, bigger mothers might be better able to fly with and nourish large fetuses, and occasionally young, than smaller mothers (Myers 1978, 709).

bird-origin hypotheses

▸ **crocodylomorph hypothesis** *n*. The hypothesis that birds evolved from crocodylomorphs (Wellnhofer 1994, 301).

▸ **pseudosuchian hypothesis** *n*. The hypothesis that birds evolved from pseudosuchian archosaurs (Heilmann 1926 in Wellnhofer 1994, 300).

▸ **theropod hypothesis** *n*. The hypothesis that birds evolved from theropod dinosaurs (Huxley 1870, Nopcsa 1907, and Ostrum 1969, 1975 in Wellnhofer 1994, 299).

Comments: This is probably the most widely accepted bird-origin hypothesis today (Wellnhofer 1994, 301). Gauthier (1986 in Elzanowski and Wellnohofer 1992) suggested that birds evolved from coelurosaur theropods, while Elzanowski and Wellnohofer (1992, 823) suggest that birds evolved from archaeornithoidid dinosaurs.

***Blitzkrieg* hypothesis** *n*. A computer model that includes the ideas that early American big-game hunters entered North America in Alaska; multiplied rapidly under conditions of unlimited food supply, low incidence of human disease, and an efficient social organization; spread southward and eastward; and extinguished large-game populations, eventually causing these species to go extinct (Mosimann and Martin 1975 in Burney 1993, 534).

Comments: This hypothesis is supported by evidence that the hunters arrived later in southern and eastern parts of the continent, where the large mammals went extinct later than in northern parts and there was only

a brief period of overlap between the very early hunter groups and last mammoths. However, other factors, such as human activities besides hunting and climatic change, may have played roles in extinction of big-game species (Burney 1993, 534–535).

[German *Blitzkrieg*, lightning war, warfare]

"brain-first hypothesis" *n*. The hypothesis that brain development occurred long before other body parts characteristic of Humans (Washburn 1985, 5); contrasted with the "ancestral-brain hypothesis."

syn. brain-first theory (Washburn 1985, 5)

Comment: This hypothesis was a major one regarding human evolution at the time of the Raymond Dart's report of the Taung Child in 1925 (Washburn 1985, 5).

catastrophism *n*.

1. The hypothesis that biological and geological phenomena were caused by sudden, violent natural disturbances (catastrophes, upheavals) rather than by continuous and uniform processes that had a major impact on organic evolution (Huxley 1869 in *Oxford English Dictionary* 1972).

2. The views held by catastrophists or directionists, *q.v.* (Mayr 1982).

syn. convulsionism, directionism, *q.v.*

cf. phyletic gradualism, progressionism, uniformitarianism

Comments: Cuvier used the milder term "revolution" rather than "catastrophe" to refer to the presumed upheavals. Catastrophism is supported by many data (Mayr 1982, 365, 375–378).

[coined by Whewell (1832 in Mayr 1982)]

Chitty hypothesis See hypothesis: population-limiting hypotheses: chitty hypothesis.

climatic hypothesis of human emergence *n*. The hypothesis that climate affected the evolution of Humans because (1) from 5 to 6 million years ago, a cooling and drying of global climate caused African grassland to expand and rain forests to contract; (2) at least one species of tree-dwelling ape left its shrinking forest habitat and foraged on a savanna; (3) this ape became bipedal and evolved into the first hominid (*Australopithecus*); (4) between the late Miocene cooling and about 3 million years ago, the climate of most of tropical Africa fluctuated from mildly warm-moist and mildly cool-dry states, and savanna woodlands proliferated in the warm-moist times and shrank and became fragmented during cool-moist times; (5) two lines of hominids (*Australopithecus* and *Homo*) emerged during the major cooling and drying that set in about 2.8 million years ago; (6) *A. robustus* went extinct; and (7) about 1 million years ago during a major cold spell, *Homo erectus* evolved from *H. habilis* and eventually migrated out of Africa, perhaps using better land connections between Africa and central Asia that formed when ice sheets formed and sea levels lowered (E.S. Vrba in Stevens 1993, C1, C18).

Comments: David Pilbeam (in Stevens 1993, C18) considers this hypothesis to be plausible, but more climatic and fossil data are needed.

compression hypothesis of interspecific competition *n*. As more species are packed into a community, the habitats occupied by particular species shrink, but acceptable food items within the occupied habitat are not changed (MacArthur and Wilson 1967 in Wilson 1975, 277).

conflict hypothesis *n*. The hypothesis that a genome is like a "society, of which the genes are the members, sometimes in conflict and sometimes in harmony" (Haig 1993, 498).

continental drift *n*.
1. The displacement of portions of the Earth's crust relative to an arbitrarily chosen portion (Wegener 1966, 147).
2. The hypothesis that Earth's landmasses were once continuous and then broke apart and drifted away from each other (Lightman and Gingerich 1992, 692).

syn. drift theory, Taylor-Wegener theory (Wegener 1966, 4); hypothesis of continental drift (Sutton 1970, 7; Hallam 1975, 9); continental displacement, drift hypothesis (Hallam 1975, 13); displacement theory, epeirophoresis theory (Lincoln et al. 1985)
cf. ²theory: theory of plate tectonics
Comments: The hypothesis of continental drift as proposed by Wegener was originally rejected by most scientists, but it is now widely accepted by scientists (Hallam 1975, 17; Stanley 1989, 192–198). Wegener (1966, 1) reports that the idea of continental drift came to him in 1910 when he was studying a map of the world; he later found that some aspects of the continental-drift concept were published decades earlier. Wegener's original data that support continental drift include the geometric fit of continental margins, matching mountain chains on opposite continents, corresponding rock succession, similar ancient climatic conditions, and identical species on continents now widely separated by ocean (Wegener 1915 in Erickson 1992, 11).

[from a translation of German *Verschiebung*, which is more precisely translated as continental displacement (Hallam 1975, 13)]

dance-"language" hypothesis, dance-language hypothesis *n*. The hypothesis that the Honey Bee, *Apis mellifera*, recruits its workers to food sources primarily by distance and direction communication transmitted by symbolic dances (Johnson 1967; Wenner 1967, 1971, 27; Wenner et al. 1967, 1969; Wenner and Johnson 1967; Johnson and Wenner 1970; Michener 1973, 175).
syn. bee-language hypothesis (Wenner 1971, 1)
cf. hypothesis: locale-odor hypothesis
Comment: This hypothesis appears to be true based on many studies of Frisch and others (Gould 1976, 211).

dinosaur-extinction hypotheses
Comments: Some paleontologists believe that dinosaurs as a group did not go extinct; instead, the dinosaur lineage is still extant as birds. Thus, to these paleontologists, the following hypotheses relate to nonavian dinosaurs, not all dinosaurs. Below I list hypotheses of varying possibilities concerning the reason for nonavian dinosaur extinction.

▶ **"clumsy hypothesis"** *n*. The hypothesis that dinosaurs became too clumsy to breed or perhaps even move (Lambert 1983, 208).

▶ **"egg-predation hypothesis"** *n*. The hypothesis that small mammals ate dinosaur eggs and eventually extinguished them (Lambert 1983, 208).

▶ **"glutton hypothesis"** *n*. The hypothesis that carnivorous dinosaurs ate all the herbivorous ones, causing all dinosaurs to die out (Lambert 1983, 208).
Comment: There is no good evidence for this hypothesis (Lambert 1983, 208).

▶ **"impact hypothesis-1"** *n*. The hypothesis that an asteroid hit the Earth at the end of the Cretaceous and put much dust into its atmosphere; the dust reflected sunlight, and the Earth cooled, which markedly reduced plant growth, causing starvation in herbivorous dinosaurs followed by carnivores (Gore 1989, 672).
syn. asteroid hypothesis (1978)
Comments: This hypothesis was formulated by Luis and Walter Alvarez, Frank Asaro, and Helen Michel (Kerr 1989, B3). The asteroid may have been about 6 miles across. The impact also might have caused acid precipitation. Evidence that supports this hypothesis includes an iridium-rich layer in about 50 localities on Earth at the KT boundary (= Cretaceous-

Tertiary boundary, the Mesozoic-Cenozoic boundry). Iridium is rare on Earth but common in some meteorites; therefore, Alvarez et al. assumed that the iridium came from an asteroid. Evidence and suppositions that do not support this hypothesis include: (1) Many dinosaur species went extinct over millions of years before the supposed time of impact, and (2) the Earth might have become very hot after the dust settled, because moisture in the atmosphere stopped heat from escaping (Lambert 1983, 209). There might have been a worldwide forest fire, too, because there is a lot of soot in the iridium layer.

▶ **"impact hypothesis-2"** *n*. The hypothesis that a group of comets hit the Earth (Gore 1989, 692).

Comment: These different comet groups explain multiple stages in the K-T extinctions.

▶ **"magnetic-wave hypothesis-1"** *n*. The hypothesis that the Earth switched magnetic poles and caused deadly radiation that killed dinosaurs (Lambert 1983, 208).

▶ **"magnetic-wave hypothesis-2"** *n*. The hypothesis that the Earth switched magnetic poles which somehow chilled the Earth and killed off dinosaurs (Lambert 1983, 208).

▶ **"mammal-competition hypothesis"** *n*. The hypothesis that mammals ate plants used by dinosaurs, causing their starvation (Sloan in Gore 1989, 691).

▶ **"oxygen hypothesis"** *n*. The hypothesis that a reduction in atmosphere caused dinosaur extinction because oxygen was insufficient for dinosaurs due to their inefficient respiratory systems (Associated Press, 1993, A3).

Comments: The oxygen level in air trapped in 120-million-year-old amber dropped from 35% to 28% in 300,000 to 500,000 years (G. Landis et al. in Associated Press, 1993, A3). The oxygen-rich air resulted from volcanic activity that produced carbon dioxide, which was converted to oxygen by plants. Dinosaurs primarily evolved during a period of high atmospheric oxygen, according to the Pele hypothesis. Although most(?) dinosaurs went extinct during the time of low oxygen, the same trend is not apparent in other taxa (M. Norell in Associated Press 1993, A3).

▶ **"pathogen hypothesis"** *n*. The hypothesis that pathogens killed off dinosaurs (Lambert 1983, 208).

▶ **Pele hypothesis** *n*. The hypothesis that dinosaurs primarily evolved during a pe-

riod of high atmospheric oxygen of about 35% (Associated Press, 1993, A3).

cf. hypothesis: dinosaur-extinction hypotheses: "oxygen hypothesis"

[after Pele, a Polynesian goddess of volcanoes]

▶ **"poison-angiosperm hypothesis"** *n*. The hypothesis that newly evolved angiosperms poisoned herbivorous dinosaurs (Lambert 1983, 208).

▶ **racial-senescence hypothesis** *n*. The hypothesis that just as individual organisms are born, grow old, and die, so do populations, species, and other taxa; dinosaur lineages simply got old, and the last living members were not competitive for this reason (Strickberger 1990, 355; Fastovsky 1996, 422).

Comments: There is no evidence for this hypothesis in dinosaurs or other taxa (Fastovsky 1996, 422). Taxa probably go extinct when they can no longer adapt to environmental changes .

▶ **slow-climate-change hypothesis** *n*. The hypothesis that the climate was cool during the late Cretaceous (Steven Stanley in Gore 1989, 692).

Comments: Seas were receding from continental lowlands. Over 50% of the species in many animal groups went extinct around the end of the Cretaceous (Strickberger 1990, Table 17-2). There may have been long-term cooling at the end of the Cretaceous (Steven Stanley in Gore 1989, 692).

▶ **"supernova hypothesis"** *n*. The hypothesis that cosmic rays from a supernova deformed and killed unborn dinosaurs (Lambert 1983, 208).

▶ **"volcanism hypothesis"** *n*. The hypothesis that volcanoes erupted and spewed much dust into the air, which darkened the skies and caused the Earth to cool, or could have heated the Earth with a greenhouse effect. This change caused extinction of dinosaurs and other organisms (Gore 1989).

Comment: This hypothesis is not supported by many paleontologists (Kerr 1989, B3).

directed-mutation hypothesis *n*.

1. The hypothesis that a bacterium that is attacked by a particular virus can somehow acquire the genetic ability to resist further attacks as a result of the initial attack (called "acquired hereditary immunity" by Luria and Delbrück 1943 in Lenski and Mittler 1993a, 188).

2. The hypothesis that bacteria living in an unfavorable environment are able to choose which mutations to produce that

enable them to adapt to this environment; contrasted with the hypothesis of random mutation, *q.v.* (Cairns et al. 1988, 142; Gillis 1991, 202).

Notes: Some experiments might support this hypothesis (Cairns et al. 1988), and other experiments do not support this hypothesis (Hall 1990, 5; Mittler and Lenski 1990; Lenski and Mittler 1993a, 188). This hypothesis is further discussed by Rainey and Moxon (1993), Watson (1993), Hurst and Grafen (1993), and Lenski and Mittler (1993b).

3. The hypothesis that a selective substrate "directs" mutations to particular regions of a bacterium's genome (Shapiro 1995, 374).

Note: This hypothesis remains unsubstantiated (Shapiro 1995, 374).

syn. hypothesis of directed mutation

disinhibition hypothesis *n.* All displacement behavior can be explained by two contrasting behaviors' stopping one another, with the result that a third behavior is allowed to occur (Eibl-Eibesfeldt, Tinbergen in Heymer 1977, 61).

cf. behavior: displacement behavior

dominance hypothesis *n.* In primates, females are merely passive objects of male competition (Smuts 1987, 38).

eat-and-run hypothesis *n.* Ruminant artiodactyls are better able to avoid predation than nonruminant herbivores because they do not have to chew their food thoroughly at their feeding sites and thus can spend less time exposed to predators (Young 1950, 744; Colbert 1955, 400).

[coined by Janis 1976, 758]

ecological-saturation hypothesis *n.* The hypothesis that the number of niches stays roughly constant in a particular habitat due to interspecific competition for a finite supply of resources; as new species evolve in this habitat, others are excluded by competition and become extinct; contrasted with the expanding-resources hypothesis (Raup 1972; May 1974, 1981; Gould 1981 in Strong et al. 1984, 39).

Comment: Available data firmly reject this hypothesis with regard to the evolution of phytophagous-insect communities (Strong et al. 1984, 43). Data on insect families support this hypothesis (Labandeira and Sepkoski 1993, 312).

"ecosystem-diversity-productivity hypothesis" *n.* Ecosystems with more species are more productive than ones with fewer species (Hector et al. 1999, 1123).

Comments: A study of eight European field sites with artificial grassland communities supports this hypothesis. The sites had

different numbers of plant species, and there was an overall log-linear reduction of average above-ground biomass with loss of species (Hector et al. 1999, 1123).

effect of volition See hypothesis: volition.

endosymbiosis hypothesis *n.*

1. The hypothesis that the first eukaryotic cell evolved by physically incorporating prokaryotic organisms into its cytoplasm, and these prokaryotes became mitochondria and chloroplasts (Schimper 1883, etc. in Dodson 1979, 652; Portier 1918, Wallin 1927, Schanderl 1948 in Margulis 1981, xvii; Margulis 1970, 178; Strickberger 1990, 157).

Notes: Alberts et al. (1989, 19, 398–399) hypothesize the following. The first eukaryotic cell evolved from an anaerobic organism without mitochondria and chloroplasts that established a stable endosymbiotic relation with bacteria. These bacteria with oxidative-phosphorylation systems became mitochondria. This ancestral line evolved into other Protistans, Animalia, Fungi, and Plantae. The Plantae ancestor obtained oxygen-evolving, photosynthetic, endosymbiotic bacteria of different taxa which became chloroplasts.

2. The hypothesis that the three classes of organelles (mitochondria; basal bodies, flagella, and cilia; photosynthetic plastids) of eukaryotic cells evolved from free-living prokaryotic ancestors by a series of endosymbiotic relationships (Margulis 1985, 78; Lincoln et al. 1985).

syn. endosymbiont hypothesis (Alberts et al. 1989, 398–399), endosymbiosis theory (Dodson 1979, 653), endosymbiotic hypothesis (Curtis 1983, 414)

cf. hypothesis: serial-endosymbiosis hypothesis

Comments: This hypothesis is contrasted with the autogenous hypothesis (Campbell 1987, 554) and the "genome-duplication hypothesis" (Raff and Mahler 1972, 579; Strickberger 1990, 158). The endosymbiosis hypothesis is currently favored by many biologists. The aerobic bacterium *Paracoccus denitrificans* appears to be similar to an ancestral mitochondrion symbiont because of its many mitochondrion-like enzymatic, membranous, and respiratory features (John and Whatley 1975, 495).

▶ **serial-endosymbiosis hypothesis** *n.* The hypothesis that mitochondria are direct descendants of a taxon of bacterial endosymbiont that became established at an early stage in a nucleus-containing, but amitochondriate, host cell; contrasts with "simultaneous-endosymbiosis hypothesis" (Gray et al. 1991, 1476).

▸ **"simultaneous-endosymbiosis hypothesis"** *n.* The hypothesis that the first mitochondrion and nucleus arose in a common ancestor of all extant eukaryotes at essentially the same time; contrasted with the serial-endosymbiosis hypothesis (Gray et al. 1991, 1476).

cf. hypothesis: hydrogen hypothesis
Comment: The fact that nuclear genomes are evolutionary chimeras that have substantial contributions from both archaean and bacterial progenitors supports this hypothesis (Gray et al. 1991, 1479). Informational genes are largely of archaean origin; whereas, operational genes appear to have come primarily from bacteria. Morever, the bacterial component of a nuclear genome appears to be considerably greater than that usually attributed to specific gene transfer from an evolving mitochondrial genome, and includes genes that have nothing to do with mitochondrial biogenesis and function. Other lines of evidence are consistent with the hypothesis as well.

escalation hypothesis *n.* Through evolutionary time, organisms are faced with increasing risks due to predation by their enemies, and they evolve increasing abilities to cope with these risks; organisms' relative rates of adaptation are at best continually maintained rather than improved (Vermeij in Geary 1988, 1257).

cf. hypothesis: Red-Queen hypothesis
Eve hypothesis, Eve theory See hypothesis: African-origin hypothesis of human mitochondrial DNA evolution.

face-to-face-mating hypothesis *n.* The hypothesis that face-to-face mating in Humans has great psychological importance and is "a major cause of human society" (Washburn 1985, 7).

Comment: Ape behavior does not seem to support this hypothesis. Face-to-face mating is common in Gibbons, Orangutans, Bonobos, and Gorillas (Washburn 1985, 7).

falsifiable hypothesis *n.* A hypothesis that can be demonstrated to be false if, indeed, it is false (Popper in Mayr 1982, 26).

cf. method: popperian method
firefly-flash-synchronization hypotheses

▸ **beacon hypothesis** *n.* In some Southeast Asian firefly species: a tree with synchronously flashing male fireflies provides "a sufficiently bright and large beacon, perhaps enhanced by reflections, that attracts fireflies that wander out into the clear over water" and provides "enough

opportunities for mating to compensate for the (assumed) long flight required of the mated females for egg dispersal" (Buck and Buck 1966, 563).

Comment: Buck and Buck (1978) indicate that data support their individual-selection "lek hypothesis" of firefly flash synchronization rather than their group-selection beacon hypothesis.
[Possibly coined by Buck and Buck (1976, 79)]

▸ **"lek hypothesis"** *n.* In some Southeast Asian firefly species: Each male increases his own fitness by grouping with other conspecific males on trees and synchronously flashing which attracts mates (Buck and Buck 1978, 476–478).

Comment: Buck and Buck (1966, 1978) also discuss physiological hypotheses of firefly synchronous flashing.

"first-cell hypothesis" *n.* The first cell was similar to a mycoplasma (Norstog and Meyerriecks 1983, 38; Alberts et al. 1989, 10).

Gaia hypothesis *n.*

1. The hypothesis that "life, or the biosphere, regulates or maintains the climate and the atmospheric composition at an optimum for itself" (1972 definition in Lovelock 1990).
[coined by Margulis and Lovelock (1974, 471)]

2. "The concept that the atmosphere of the Earth flows in a closed system controlled by and for the biosphere" (Lovelock and Margulis 1974, 93).
Note: In addition, Lovelock and Margulis (1974, 93) suggest that the physical and chemical state of the Earth's surface "is in homeostasis at an optimum set by the contemporary biota."

3. The hypothesis that Earth's biota created an atmosphere with a composition, acidity, redox potential, and temperature history that differ greatly from that of other terrestrial planets (Mars and Venus) and has actively maintained these conditions with a stability akin to the self-regulating physiologies of living organisms (Lovelock and Margulis 1974 and Margulis and Lovelock 1974 in Barlow and Volk 1992, 686).

syn. Gaia, Gaia concept (Barlow and Volk 1992, 686)
Comments: There are many distinct versions of the controversial Gaia hypothesis. Early definitions have been criticized for being teleological (Kerr 1988), but Lovelock (1990) wrote "neither Lynn Margulis nor I have ever proposed a teleological hypothesis." Lovelock (1990 in Barlow and Volk

h–m

1992, 687) modified his original Gaia hypothesis by indicating that the stability was not always homeostatic but punctuated homeostatic. After a disturbance, Earth becomes relatively stable again, albeit sometimes at an altogether new chemical, or thermal, state. Margulis (1990 in Barlow and Volk 1992, 687) proposed that the stability is more homeorrhetic than fixed from outside set points. She considers Gaia to be an ecosystem, not an individual (Margulis 1993). Kirchner (1989 in Resnik 1992, 573) distinguished among five versions of the Gaia hypothesis. He considers many versions to be false, untestable, or internally contradictory. Resnik (1992) analyzes three of these hypotheses and two others. Gaia hypotheses have generated fruitful research (Resnik 1992; Slobodkin 1993). [coined by Margulis and Lovelock (1974, 471) after *Gaia*, the common society to which the ancient Greeks believed all organisms on Earth belonged; *Gaia* is roughly translated as Mother Earth and is also considered to be the Greek goddess of the Earth (Michaelis 1963)]

Gause's hypothesis See principle: gause's principle.

genetic-behavioral-polymorphism hypothesis See hypothesis: population-limiting hypotheses: genetic-behavioral-polymorphism hypothesis.

Geoffroyism *n*. Étienne Geoffroy Saint-Hilaire's hypothesis that an organism's environment causes a direct induction of its organic change (Mayr 1982, 687).

cf. inheritance of acquired characters, Lamarckism

Comments: Lamarck rejected "Geoffroyism," but the neo-Lamarckians at the end of the 18th century held such direct induction in high esteem (Mayr 1982, 362). "Geoffroyism" was widely believed by biologists almost to the end of the 19th century (Mayr 1982, 687).

Goldschmidt's-speciation hypothesis *n*. Speciation results from systematic mutations (Goldschmidt 1940).

Comment: Opposition to this hypothesis led Mayr to propose the widely accepted hypothesis that geographic isolation is a major factor that influences speciation (Mayr 1982, 381).

grandmother hypothesis *n*. The hypothesis that evolution selected for increased longevity in women because older women shared food with their daughters and grandchildren (Suplee 1998, A3).

Comments: Menopause might have evolved because it was not beneficial for older women to reproduce (Suplee 1998, A3). Jonathan Marks suggests that there are other reasons

for menopause besides natural selection, including a woman's running out of eggs. In Chimpanzees, Gibbons, and Macaques, females remain fertile throughout almost all of their lives.

green-beard effect *n*. The hypothesis that a gene might be responsible for both the expression of a specific trait and the recognition of it in other conspecific organisms (Lewin 1984).

[This term was coined by Dawkins (Lewin 1984) and put simply in his words: "I have a green beard and I will be altruistic to anyone else with a green beard."]

handicap principle *n*. The hypothesis that an attribute (*e.g.,* large tail or large antlers) of a male animal is a handicap that he shows to a female demonstrating that he can survive and has "good genes" despite his handicap (Zahavi 1975, 207; 1977; Spencer and Masters 1992, 299).

syn. handicap hypothesis (Kirkpatrick 1986, 222)

Comments: This concept suggests that males with handicaps that survive to mate are compensated by being fitter in other respects, thus their handicaps are advertisements for good genes. Females that mate with such males are expected to gain generally better genes for their offspring. The handicap principle is supported by some models (mathematical representations) but not others (Kirkpatrick 1986; Spencer and Masters 1992, 299).

hydrogen hypothesis *n*. The hypothesis that metabolic syntrophy was the driving force for an association between a hydrogen-producing eubacterial symbiont (assumed to be an α-Proteobacterium) and a hydrogen-requiring archaean (the host), which led to the evolution of the first eukaryote (Martin and Müller 1998 in Gray et al. 1991, 1480).

cf. hypothesis: "simultaneous-endosymbiosis hypothesis"

Comments: This hypothesis suggests the simultaneous origin of the eukaryote ancestor and its mitochondrion with a major bacterial contribution to the eukaryote's nucleus from the same α-bacterial genome whose reduction is postulated to result later in mitochondrial genomes (Gray et al. 1991, 1480). A similar hypothesis involves a δ-Proteobacterium symbiont (Moreira and López-García in Gray et al. 1991, 1480).

hypothesis of biogenesis See law: biogenetic law.

hypothesis of directed mutation, directed-mutation hypothesis *n*.

1. The hypothesis that a bacterium that is attacked by a particular virus somehow

can acquire the genetic ability to resist further attacks as a result of the initial attack, called "acquired hereditary immunity" by Luria and Delbrück (1943 in Lenski and Mittler 1993a, 188).

2. The hypothesis that bacteria living in an unfavorable environment are able to choose which mutations to produce that enable them to adapt to this environment; contrasted with the hypothesis of random mutation, *q.v.* (Cairns et al. 1988, 142; Gillis 1991, 202).

Comments: Some experiments might support this hypothesis (Cairns et al. 1988), and some experiments do not support this hypothesis (Hall 1990, 5; Mittler and Lenski 1990; Lenski and Mittler 1993a, 188). This hypothesis is further discussed by Rainey and Moxon (1993), Watson (1993), Hurst and Grafen (1993), and Lenski and Mittler (1993b).

hypotheses of divergent evolution of animal genitalia

Comment: Eberhard (1985) discusses evidence for and against these hypotheses.

▶ **"genitalic-recognition hypothesis"** *n.* Females determine species identity of a potential mate based on species-specific genitalic stimuli and avoid fertilization if stimuli are not appropriate (Stephenson 1930, etc. in Eberhard 1985, 27).

▶ **lock-and-key hypothesis** *n.* Females avoid fertilization by heterospecific males due to their evolution of complicated genitalia that permit insemination by only conspecific males; a male has the "key" that fits a female's "lock," and "each new speciation event necessitates a new lock and a new key" (Dufour 1844 in Mayr 1963, 63; Eberhard 1985, 19–20, who cites evidence for and against this hypothesis).

Comment: Tatochila butterflies do not support this hypothesis (Porter and Shapiro 1990, 107).

▶ **mechanical-conflict-of-interest hypothesis** *n.* A male's genitalia damage a female's genitalia, thus preventing her further copulations, and female genitalia evolve protection against this damage (Wing 1982 in Eberhard 1985, 29, who indicates that the status of this hypothesis is unclear).

▶ **"pleiotropism hypothesis"** *n.* Genitalia are pleiotropically affected by many genes, any change in a species' genetic constitution may result in an incidental change in its genitalia structure, and because genitalia of many species are internal structures, they are less subject to corrective influences of natural selection,

provided their basic function of gamete transfer is not impaired (Mayr 1963, 104, in Eberhard 1985, 27–28).

syn. pleiotropism argument, pleiotropism theory (Eberhard 1985, 28)

Comments: Arnold (1973) modified this hypothesis; concrete data that support, or do not support, this hypothesis are very limited (Eberhard 1985, 28).

▶ **sexual-selection-by-female-choice hypothesis** *n.* Male genitalia function as "internal courtship devices" that increase the likelihood that females will actually use a given male's sperm to fertilize her eggs rather than those of another male; the diversity of male genitalic form is the result of runaway sexual selection (Eberhard 1985, 14–15, 182; 1992, 1774, who offers data that support this hypothesis).

▶ **sexual-selection-by-male-conflict hypothesis** *n.* Sexual selection in the form of direct male-male competition is responsible for rapid divergence of genitalic structure (Eberhard 1985, 183, who rejects this hypothesis).

hypothesis of nondifference See hypothesis: hypothesis of nondifference.

hypothesis of nonspecificity *n.* There are no distinct major gene groups that exclusively affect one class of characters, such as morphological, physiological, or ethological ones (Lincoln et al. 1985).

hypothesis of punctuated equilibrium *n.* The hypothesis that an organism lineage (equilibrium) changes gradually but also shows rapid, episodic times of change, even speciation (punctuations) during geological time (Eldredge and Gould 1972, 84; Gould 1980; Eldredge 1985; Futuyma 1986, 401–403).

syn. theory of punctuated equilibrium (Futuyma 1986, 402)

cf. ²evolution: punctuational evolution; punctuated equilibrium

Comments: The hypothesis of punctuated equilibrium is contrasted with Charles Darwin's gradualism, but Darwin's views overlapped with those of proponents of this hypothesis (Rhodes 1983, 269; Arthur 1984, 119). Hull (1984) writes that this hypothesis proposes that "evolutionary development is a good deal more "punctuational" than advocates of the modern theory of evolution have been willing to admit and contains an important nonadaptive phase." Gould and Eldredge (1986, 1993) and Futuyma (1986) discuss some of the controversy related to this hypothesis. Dawkins (1986), Hecht and Hoffmann (1986), and Levinton (1988) critique this hypothesis. Gould and Eldredge (1986, 145) conclude that, "Punctuated equilibrium is now quite acceptable, rather smaller in scope than once suspected, and a good, comfortable part (or perhaps, at best, a mild extension) of neo-Darwinism, the Modern Synthesis, or whatever orthodoxy is called." Brown (1987a) indicates that the original data of Eldredge and Gould (1972) do not support their hypothesis. Bryozoan fossils as well as other data support this hypothesis (Jackson and Cheetham 1990, 579), which requires further testing. Gould and Eldredge (1993) explain how this hypothesis is supported by four classes of evidence: cases from particular organism groups, cases from entire faunas, inductions about environments in which punctuated equilibria should and should not prevail, and tests with living organisms.

hypothesis of random mutation *n.* A bacterium has some probability of spontaneously mutating from a viral-sensitive to a viral-resistant state, even in the absence of the virus (Luria and Delbrück 1943 in Lenski and Mittler 1993a, 188); contrasted with the directed-mutation hypothesis.
Comments: This hypothesis was formulated to account for the appearance of bacteria resistant to infection by viruses and has experimental support (Lenski and Mittler 1993a, 188). It is an alternative hypothesis to the hypothesis of directed mutation, *q.v.*

hypotheses of species organization *pl. n.* Hypotheses concerning community equilibrium, organization, and stability.
cf. hypotheses of species richness
▸ **"equilibrium hypothesis of species organization"** *n.* The hypothesis that communities, such as coral reefs and tropical rain forests, are stabilized, ordered systems of species resulting from natural selection; when a factor(s) disturbs such a community, it quickly returns to its original state; contrasted with nonequilibrium hypothesis of species organization (Connell 1978, 1306–1307).
▸ **"nonequilibrium hypothesis of species organization"** *n.* The hypothesis that equilibrium in species organization is seldom attained, species assemblages seldom reach an ordered state, and communities of competing species are not highly organized by diffuse coevolution into systems in which optimal strategies produce highly efficient associations whose species composition is stabilized; contrasted with equilibrium hypothesis of species organization (Connell 1978, 1306–1307).

hypotheses of species richness *n.* Hypotheses that attempt to explain why certain communities, including coral reefs and tropical rain forests, have high numbers of species.
Comments: Connell (1978) discussed six such hypotheses which fall into two categories: nonequilibrium hypotheses (equal-chance hypothesis, gradual-change hypothesis, and intermediate-disturbance hypothesis) and equilibrium hypotheses (circular-networks hypothesis, compensatory-mortality hypothesis, and niche-diversification hypothesis). On Earth, in general, all of these hypotheses could be true, and for a particular location more than one of these hypotheses could be true (Connell 1978, 1309).

SOME HYPOTHESES OF SPECIES RICHNESS

I. "equilibrium hypotheses of species richness"
   A. circular-networks hypothesis
   B. compensatory-mortality hypothesis
   C. niche-diversification hypothesis
II. "nonequilibrium hypotheses of species richness"
   A. equal-chance hypothesis
   B. gradual-change hypothesis
   C. intermediate-disturbance hypothesis
   D. recruitment-limitation hypothesis

▸ **circular-networks hypothesis** *n.* An equilibrium hypothesis of species richness, *q.v.*: Circular competitive hierarchies, in which organisms eliminate others locally, maintain high species richness in an area in general (Connell 1978, 1307).
Comment: Coral-reef research did not support this hypothesis (Connell 1976 in Connell 1978, 1307).
▸ **compensatory-mortality hypothesis** *n.* An equilibrium hypothesis of species richness, *q.v.*: High species richness occurs in a community in which competitive elimination of species is prevented indefinitely because mortality is greatest on the species with the highest competitive ability, or, if all species have approximately equal competitive abilities, mortality is the greatest in the commonest species (Connell 1978, 1307).
Comment: This hypothesis generally does not seem to account for high diversity in coral reefs and mixed tropical rain forests (Connell 1978, 1307).
▸ **equal-chance hypothesis** *n.* A nonequilibrium hypothesis of species richness, *q.v.*: Species richness is high in a

community because each of its species has an equal chance to colonize empty spaces, hold them against invaders, and survive the vicissitudes of physical extremes and natural enemies (Connell 1978, 1306).

Comment: Certain guilds of coral-reef fish might support this hypothesis (Sale 1977 in Connell 1978, 1306).

▶ **"equilibrium hypothesis of species richness"** *n.* The hypothesis that communities, such as coral reefs and tropical rain forests, are stabilized, ordered systems of species; when a factor(s) disturbs such a community, it quickly returns to its original high number of species (inferred from Connell 1978, 1306–1307); contrasted with nonequilibrium hypothesis of species richness.

▶ **gradual-change hypothesis** *n.* A nonequilibrium hypothesis of species richness, *q.v.*: Climate changes (from seasonal through millennial and longer) change species abundances and compositions in a particular location over time (inferred from Connell 1978, 1306, who indicates that Hutchinson 1941 suggested this hypothesis).

Comment: Data from forests and phytoplankton support this hypothesis (Connell 1978, 1306).

▶ **intermediate-disturbance hypothesis** *n.*

1. A nonequilibrium hypothesis of species richness, *q.v.*: Species richness is highest during intermediate states of disturbance frequency, disturbance size, time after a disturbance occurred, or a combination of these factors (Connell 1978, 1303, fig. 1).

Note: Studies of an Australian coral reef and rain forests in Australia, Nigeria, and Uganda support this hypothesis.

2. The hypothesis that many species can coexist in the same area because localized disturbances, such as light gaps, *q.v.*, promote the coexistence of species that have different competitive and dispersal abilities and resource-use strategies (Hubbell et al. 1999, 554).

Notes: This is a more restrictive definition of the intermediate-disturbance hypothesis. In this rendition, the hypothesis predicts that there is a greater species richness in light gaps than in mature forest and that species richness will be greater in gaps collectively than in their forest matrix because gaps provide more diverse conditions and resources in ecological time

(Chazdon et al. 1999, 1459a; Hubbell 1999, 1459a; Hubbell et al. 1999, 554). A study of tree-fall gaps in a Panamanian tropical rainforest supports the recruitment-limitation hypothesis, *q.v.*, rather than this hypothesis (Hubbell et al. 1999, 554).

▶ **niche-diversification hypothesis** *n.* An equilibrium hypothesis of species richness, *q.v.*: Many specialized species can coexist in a habitat with a range of attribute variations (Connell 1978, 1306).

Comment: Data from motile organisms might support this hypothesis (Schoener 1974 in Connell 1978, 1306).

▶ **"nonequilibrium hypothesis of species richness"** *n.* The hypothesis that equilibrium in species organization in a community is seldom attained and that species richness is highest when a factor(s) causes the community to be in a state of disturbance (inferred from Connell 1978, 1306-1307); contrasted with equilibrium hypothesis of species richness.

▶ **recruitment-limitation hypothesis** *n.* A nonequilibrium hypothesis of species richness, *q.v.*: Many species can coexist in the same area because many species cannot dominate the area because they simply do not colonize all available subareas; these species have low local abundance, have poor dispersal ability, do not arrive in a subarea by chance, or a combination of these factors (Tilman 1999a, 495).

Comments: There are several versions of the general recruitment-limitation hypothesis explained by Tilman (1999a, 496).

hypotheses of the beginning of the universe

Comment: These hypotheses are also called theories and models.

▶ **big-bang cosmology, big-bang model, big-bang theory** *n.* In the large-scale average, the universe is expanding in a nearly homogeneous way from a dense early state (Slipher 1922 and Hubble 1929 in Peebles 1994, 53); contrasted with the big-bang hypothesis.

Comment: At present, there are no fundamental challenges to this cosmology, although there are certainly unresolved issues within the theory itself (Peebles et al. 1994, 53).

▶ **big-bang hypothesis** *n.* The hypothesis that the universe was once a small sphere of concentrated energy and matter (energy-matter combination), this sphere exploded, hydrogen and some helium formed when the temperature decreased, protogalaxies formed about 100 million

h–m

years after the blast, and so forth (Ralph Alpher, George Gamow, and Ralph Herman in Strickberger 1990, 66; Sullivan 1997, 1275); contrasted with the steady-state hypothesis.

Comments: This hypothesis was first proposed by a Belgian priest, Georges Lemaître, in 1927 (Sawyer 1992, A1). Many lines of evidence support this hypothesis, including microwave glow thought to result from the original blast, Doppler light shifts that indicate stars and galaxies are moving apart, a background temperature of 3 K, 10% of the universe being helium, and faint "hot spots" in the microwave sky that are 300,000-year-old relics of the big blast detected by the Cosmic Background Explorer Satellite (Strickberger 1990, 66; Powell 1992, 17). [coined by Fred Hoyle, English cosmologist, who used "big bang" to disparage the hypothesis (Peebles et al. 1994, 53). Some people want to rename this hypothesis because "big bang" has sexual connotations (Allen 1993, B1)]

▶ **oscillating-big-bang hypothesis** *n.* A hypothesis regarding the beginning of our Universe that states it started with a big blast and then has been alternating between expanding (big bangs) and contracting (big crunches) perhaps *ad infinitum* (Strickberger 1990, 66); contrasted with the big-bang hypothesis.

▶ **quasi-steady-state model** *n.* A modification of the steady-state hypothesis that states that things including large amounts of matter are produced in discreet explosions called creation events, the creation events occur throughout the universe in creation centers (regions that already contain dense matter and have strong gravity), a series of large creation events occurred 10 to 15 billion years ago and caused the expansion of our part of the universe, and small creation events have continued to occur and produce energetic objects such as quasars and radio galaxies (Hoyle et al. 1993 in Croswell 1993, 14).

Comment: The quasi-steady-state model is meant to account for the microwave background, which is not explained by the steady-state hypothesis (Croswell 1993, 14).

▶ **steady-state hypothesis** *n.* The hypothesis that the universe is expanding to some extent, new galaxies appear in gaps among older ones, and when hydrogen becomes helium new hydrogen is replenished from an unknown source (proposed by Fred Hoyle, Hermann

Bondi, and Thomas Gold 1948 in Strickberger 1990, 66); contrasted with the big-bang hypothesis and quasi-steady-state model.

Comment: This hypothesis proposes that the universe had no beginning and thus is infinitely old (Croswell 1993, 14). Most astronomers do not support this hypothesis (Sullivan 1997, 1275).

hypotheses of the evolution and maintenance of sex

Comments: Bell (1982) discusses data that support and do not support many of these hypotheses. Zinder (1985, 7–12) hypothesizes how sex evolved in prokaryotes. "Why did sex evolve and come to predominate in species" remains unanswered. Sex involves a large set of complex phenomena, and there is a great variety of manifestations of sex; "each of these manifestations may well require separate, though integrated, explanations" (Mooney 1993, 113). There are 20 hypotheses regarding the origin and evolution of sex (Kondrashov 1993 in Judson and Normark 1996, 41).

▶ **best-man hypothesis** *n.* Sexual reproduction maximizes an individual's number of successful offspring by producing genetically variable offspring; some variants are more successful than others in particular environments (Williams 1966, 129; Emlen 1973, 54–55; Williams 1975 in Bell 1982, 104–122).

Comments: Herder (1784–1791 in Mayr 1982, 706) suggested that sexual reproduction causes offspring variability. The best-man hypothesis is similar to the variation hypothesis, *q.v.,* and Williams' hypothesis (Williams 1966a in Wilson 1975, 316). [coined by Bell 1982, 105]

▶ **Crow-Kimura hypothesis, Crow-Kimura model** *n.* "Entire populations evolve faster when they reproduce by sex, and as a result they will prevail over otherwise comparable asexual populations" (Crow and Kimura 1965 in Wilson 1975, 315–316).

▶ **hitchhiker hypothesis** *n.* Sexual reproduction can cause increased recombination when a recombination allele "hitches a ride" with a high-fitness chromosome it has created and, thus, increases in frequency by autoselection [coined by Bell 1982, 125, after Maynard Smith's phraseology].

▶ **inertia hypothesis** *n.* Sex persists in slowly reproducing species due to phylogenetic inertia; *i.e.,* it is very difficult for natural selection to eliminate sex once it evolved in ancestral lines (Williams in Maranto and Brownlee 1984).

▶ **"parasite hypothesis of the evolution and maintenance of sex"** *n.*

1. The hypothesis that parasites that are short lived and rapidly evolving compared to hosts they attack are an evolutionary factor sufficiently general to account for the evolution of sex (Hamilton 1980, 282).

2. The hypothesis that the coevolution between parasites and their hosts provides the necessary changing environment to give a twofold advantage to the host's having sexual reproduction; sexual reproduction gives the host's offspring the possibility of rapid change in the face of its population's losing its resistance to certain kinds of parasites (Hamilton 1980; Dawkins 1986, 136–137).

syn. parasite-avoidance hypothesis (Angier 2000, A18)

▶ **ratchet hypothesis, ratchet mechanism** *n.* Sexual reproduction facilitates the elimination of unfavorable mutations in a group of individuals in uniform environments that do not change in time (Muller 1964 in Bell 1982, 101–104). [coined by Bell 1982, 101]

▶ **Red-Queen hypothesis** *n.*

1. The hypothesis that evolution favors sexual reproduction because continual adaptation through meiosis and gene recombination is necessary for a species to remain extant when its environment changes (Bell 1982, 157).

2. The hypothesis that antagonistic biotic interactions, especially those between parasite and host species are a sufficient evolutionary force to counterbalance the supposed inefficiency of sexual reproduction (Ladle 1992, 405).

Notes: This hypothesis combines parts of the Red-Queen hypothesis (def. 1) above and the Red-Queen hypothesis (def. 1) on p. 357. Some studies provide strong empirical evidence in support of this hypothesis (Ladle 1992, 405).

syn. Red Queen's hypothesis (Boyce 1990, 263)

cf. hypothesis: Red-Queen hypothesis [coined by Bell (1982, 157) due to its generic similarity to Van Valen's Red-Queen hypothesis. This name is derived from *Alice in Wonderland* (Lewis Carrol): "It takes all the running you can do, to keep in the same place," the Red Queen.]

▶ **rejuvenescence hypothesis** *n.* The hypothesis that populations naturally age, and sex rejuvenates them (Maupas 1889 in Mooney 1993, 111).

Comment: Weismann favored his variation hypothesis and strongly attacked the rejuvenescence hypothesis (Mooney 1993, 111–112). The notion that asexual populations age has been rejected, and the idea of rejuvenation has been replaced with "nuclear repair" (*e.g.,* in the repair hypothesis).

▶ **repair hypothesis** *n.* The recombination of genes resulting from sexual reproduction repairs genetic damage and outcrossing yields new gene combinations that mask mutations; the genetic variation found in offspring of sexually reproducing organisms is an important byproduct of the selection for repair but not the principal driver of the evolution of sex (Bernstein et al. 1985).

Comment: The repair hypothesis has been criticized as being too broadly applicable because all organisms "need" to repair DNA, yet they do not all have meiotic sex (Sterns 1987 in Mooney 1993, 113).

▶ **tangled-bank hypothesis** *n.* Sexual conspecific organisms are more fit in an environment comprised of slightly different subniches; if a clone line invades such an environment, its offspring quickly inhabit all subniches to which it is genetically suited but occupies less suitable ones only tenuously; if sexual organisms invade the environment, their offspring can displace clones living in marginal subniches, and clonal and sexual forms compete, with the latter taking over most of the subniches (Bell 1982, 127–142).

Comment: Bell (1982, 388) states that the tangled-bank hypothesis is "by far the most broadly supported of all the empirically vulnerable theories which attempt to identify the function of sex." [coined by Bell 1982, 131, from "tangled bank" in the last paragraph of Darwin (1859), which expresses the complexity of natural habitats]

▶ **variation hypothesis** *n.* The object of sex "is to create those individual differences which form the material out of which natural selection produces new species" (Weismann 1891, 1, in Mooney 1993, 110).

Comment: This hypothesis has been modified, resulting in more recent hypotheses (*e.g.,* the best-man hypothesis, *q.v.*).

▶ **Vicar-of-Bray hypothesis** *n.* Sexual reproduction speeds up evolution by fostering the spread of adaptive mutations in groups of individuals, and sex is adaptive in a uniform, but changing, environment, because it facilitates the rapid fixation of favorable mutations (Weissmann 1889, 281,

Guenther 1904, Fisher, 1930, 137, Muller 1932, 120–123, all in Bell 1982, 91–101).

syn. Fisher-Muller theory (Bell 1982) [named by Bell 1982, 92, after the Vicar of Bray, an English cleric noted for his ability to change his religion whenever a new monarch ascended the British throne]

▸ **Williams' hypothesis** See hypothesis: hypotheses of the evolution and maintenance of sex: best-man hypothesis (comment).

hypothesis of the origin of continents

n. The world's land area was originally united in a single primordial supercontinent called Pangea (Wegener 1915 in Hallam 1975, 13).

cf. hypothesis: continental drift

hypotheses of the origin of eukaryotic organelles

▸ **autogenous hypothesis** *n.* The hypothesis that the first eukaryotic cell evolved by the specialization of internal membranes derived orginally from the plasma membrane of a prokaryote; the endomembrane system (consisting of the nuclear envelope, endoplasmic reticulum, Golgi, and organelles (*e.g.,* lysosomes) that are bound by a single membrane are differentiated products of invaginated membranes; and mitochondria and chloroplasts acquired their double-membrane status by secondary invagination of more elaborate membrane folding (Campbell 1990, 558, illustration).

▸ **"genome-duplication hypothesis"** *n.* The hypothesis that eukaryotic organelles arose from genetic duplications that each became enclosed in a separate set of membranes; mitochondria originated in a one-celled, eukaryotic, plantlike ancestor; and lines from this ancestor became fungi, plants, and animals (Raff and Mahler 1975 in (Strickberger 1990, 158).

hypotheses of the origin of insect wings

Comments: Some of these hypotheses have been called theories. Kulkalova-Peck (1978) also discusses hypotheses related to modification of evolving insect wings and how a group of hypotheses might account for the origin of insect wings. Kingsolver and Koehl (1994) review many hypotheses.

▸ **flying-fish hypothesis of wing origin** *n.* Wings evolved from gill plates used by ancestral pterygote insects to break falls and glide to other pools of water by their flapping (Kukalova-Peck in Evans 1985, 37).

▸ **flying-squirrel hypothesis of insect wing origin** See hypothesis: hypotheses of the origin of insect wings: paranotal hypothesis of insect wing origin.

▸ **gill-cover hypothesis of insect wing origin** *n.* Insect wings evolved from gill covers (Woodworth 1906 in Kukalova-Peck 1978, 70).

▸ **paranotal hypothesis of insect wing origin** *n.* Insect wings evolved from "rigid, doubled, lobe-like expansions of thoracic terga, the paranotal lobes" (Kukalova-Peck 1978, 59); wings evolved from lateral flanges on thoraxes of terrestrial, arboreal ancestral pterygote insects that were used for gliding from tree to tree to the ground (Evans 1984, 38).

syn. flying-squirrel hypothesis of wing origin (Müller 1873–1875 in Kulkalova-Peck 1978, 59)

Comment: Kukalova-Peck (1978) suggests that this hypothesis is not valid.

▸ **spiracular-flap hypothesis of insect wing origin** *n.* Insect wings evolved from integumental evaginations which originally served as spiracular flaps (Bocharova-Messner 1971 in Kulkalova-Peck 1978, 66).

▸ **stylus hypothesis of insect wing origin** *n.* Insect wings evolved from coxal styli of Apterygota (Wigglesworth 1973, 1976 in Kulkalova-Peck 1978, 67).

▸ **thermoregulation hypothesis of insect wing origin** *n.* Insect wings evolved from thoracic lobes that an insect used to heat its thoracic muscles (Walley 1979 in Lewin 1985b; Douglas 1981; Kingsolver and Koehl 1985).

hypotheses regarding the disappearance of Neanderthals, etc.

▸ **Neanderthal-as-progenitors hypothesis** *n.* The hypothesis that Neanderthal Person was not replaced by *Homo sapiens sapiens* but evolved into them (Schwalbe, Hrdlicka, and Weidenreich in Brace and Montagu 1977, 194; Culotta 1991, 376).

▸ **Neanderthal-extinction hypothesis** *n.* The hypothesis that Neanderthal Person disappeared because *Homo sapiens sapiens* killed them off or out-competed them.

syn. complete-replacement hypothesis (Wolpoff 1989, 102)

▸ **Neanderthal-interbreeding hypothesis** *n.* The hypothesis that Neanderthal Person disappeared because *Homo sapiens sapiens* interbred with them.

hypotheses regarding the origin of the Cambrian explosion

Comments: The following hypotheses assume that there really was a Cambrian explosion. Its occurrence is open to debate; see Cambrian explosion.

▸ **Bekner-Marshall hypothesis** *n.* The hypothesis that atmospheric oxygen became high enough to form a protective ozone layer and provide adequate oxygen for respiration in new forms resulting in the Cambrian Explosion (references in Gould 1981b, 305–306; Kerr 1993a, 1274).

▸ **"complexity-level hypothesis"** *n.* The hypothesis that animals reached a certain level of complexity (that of modern worms) that enabled them to exploit many open niches and evolve into many new forms, resulting in the Cambrian Explosion (references in Gould 1981b, 305–306; Kerr 1993a, 1274).

▸ **"continental-drift hypothesis"** *n.* The hypothesis that geologically sudden shifts in latitude and climate due to movement of tectonic plates enabled many species to evolve quickly, resulting in the Cambrian Explosion (references in Gould 1981b, 305–306; Kerr 1993a, 1274).

▸ **"fecal hypothesis"** *n.* The early surface-dwelling, plankton-eating, gut-carrying animals produced fecal pellets that dropped quickly to the ocean floor; this gave surface bacteria less food and their populations shrank; they consumed less oxygen, and more oxygen became available to multicellular creatures which fueled the evolution of more energetic forms which resulted in the Cambrian Explosion (Graham Logan in Oliwenstein 1996b, 43).

▸ **"log-phase hypothesis"** *n.* The hypothesis that ordinal and familial diversity increased as a sigmoid function over time, and the log phase occurred during the Cambrian Explosion (references in Gould 1981b, 305–306; Kerr 1993a, 1274).

▸ **"plentiful-food hypothesis"** *n.* The hypothesis that an increase in nutrients entered the oceans (due to increased erosion on land, when ocean circulation speeds up and pumps up unused nutrients from the deep sea, or both) and caused some organisms to flourish and out-compete and out-reproduce others; the flourishing organisms caused many others to go extinct, leaving many open niches; many new, more productive organisms filled the niches, resulting in the Cambrian Explosion (Ronald Martin in Oliwenstein 1996b, 43).

▸ **"predator hypothesis"** *n.* The hypothesis that newly evolved predators decreased the numbers of dominant prey species and thereby allowed new species to evolve, resulting in the Cambrian Explosion (references in Gould 1981b, 305–306; Kerr 1993a, 1274).

▸ **"volcano hypothesis"** *n.* The hypothesis that submarine volcanoes erupted and changed Earth's climate by producing carbon dioxide, which caused a greenhouse effect resulting in the Cambrian Explosion (Geerat Vermeij in Oliwenstein 1996b, 43).

hypotheses regarding the Permian mass extinction

▸ **"carbon-dioxide hypothesis"** *n.* The hypothesis that a "rapid overturn of deep anoxic oceans, which introduced toxic CO_2 and, perhaps, H_2S into surficial environments" in part caused the Permian mass extinction (Knoll et al. 1996, 452). *Comments:* The overturn perhaps took a few centuries (Browne 1996b, C1, C10). The carbon dioxide may have originated from plant material that settled in deep ocean water. Evidence in support of this hypothesis is the types of calcium-carbonate sediments laid down at the end of the Permian period (Knoll et al. 1996, 453). Contemporaneous Siberian volcanism (= the Siberian Traps eruption) possibly added additional carbon dioxide to the atmosphere.

▸ **"Pangea hypothesis"** *n.* The hypothesis that Pangea's coming together and causing a warm continental climate caused the Permian mass extinction (references in Gould 1981b).

▸ **"poisonous-gas hypothesis"** *n.* The hypothesis that poisonous gases released by a gigantic volcanic eruption (the Siberians Traps eruption) coupled with the climatic effects of a large-scale release of volcanic carbon dioxide caused the Permian mass extinction (Browne 1996b, C1)

▸ **"salt hypothesis"** *n.* The hypothesis that a drop in ocean salt concentrations caused the Permian mass extinction (references in Gould 1981b).

▸ **"supernovae hypothesis"** *n.* The hypothesis that supernovae explosions changed Earth's environment in a way that killed many organisms and caused the Permian mass extinction (references in Gould 1981b).

hypotheses regarding the site of first life on Earth

▸ **agreeable hypothesis** *n.* The hypothesis that life first evolved in a rich broth of organic molecules in a warm pond of fresh to salt water (= a primordial soup) (Darwin 1871 in Nash 1993, 71). *syn.* "warm-little-pond hypothesis" *Comment:* Some biologists are now favoring hypotheses that life originated under conditions that were more like a hot pressure cooker than a nice, warm pond (K. Stetter in Nash 1993, 71).

h–m

▶ **"hot-spring hypothesis"** *n.* The hypothesis that life first evolved in hot springs.

▶ **"hydrothermal-vent hypothesis"** *n.* The hypothesis that life first evolved in very hot areas associated with hydrothermal vents under the sea (Nash 1993). *syn.* hot-world hypothesis (Nash 1993, 71)

▶ **"ocean-foam hypothesis"** *n.* The hypothesis that life first evolved in ocean foam (Louis Lerman in Nash 1993, 73).

▶ **"volcano hypothesis"** *n.* The hypothesis that life first evolved in warm to hot pools of water associated with volcanoes, lava flows, or both.

▶ **"warm-rock-space hypothesis"** *n.* The hypothesis that life first evolved in water-filled spaces in warm (to hot) subterranean and suboceanic rocks (J.A. Baross in Broad 1993a, C14).

information-center hypothesis *n.* In some bird species: "Individuals may track a food supply that varies unpredictably in time and space by nesting colonially or roosting communally" (Fisher 1954 in Waltz 1987, 48; Ward and Zahavi 1973).
Comments: Depending on the bird species, tests of this hypothesis have refuted it, not refuted it, or have produced equivocal results (de Grot 1980, etc. in Waltz 1987, 48; Götmark 1990, 487). Bird colonies might have evolved for other reasons besides tracking food supplies (Heeb and Richner 1994; Clode 1994).

"introns-early hypothesis" *n.* The hypothesis that introns occurred in organisms that are ancestral to both prokaryotes and eukaryotes, but more prokaryotic lines lost their introns due to selection for streamlined DNA replication and improved transcription efficiency; contrasts with the introns-late hypothesis (Doolittle and Daniels 1985 in Strickberger 1996, 167).

"introns-late hypothesis" *n.* The hypothesis that introns arose in eukaryotes after they evolutionarily separated from prokaryotes; contrasts with the introns-early hypothesis (Stoltzfus et al. 1994 in Strickberger 1996, 167).

land-bridge hypothesis *n.* The supposition that continent-sized land bridges existed during the Cenozoic but have since disappeared, leaving no trace of their existence (Lincoln et al. 1985).
cf. hypothesis: continental drift

larval-imagery hypothesis *n.* The hypothesis that lepidopteran wings have color patterns that resemble lepidopteran larvae and these patterns help to protect their owners after they are discovered by avian predators (Grant and Miller 1995, 47).

"late-season–mimic-scarcity hypothesis" *n.* The hypothesis that specialized mimics of aculeate Hymenoptera are absent, or fewer in number, from mid-through late summer because natural selection has caused them not to fly at this time because it is when large numbers of young, inexperienced birds are still learning to avoid hymenopteran models and are highly likely to eat models (Waldbauer and Sheldon 1971, 371).
Comment: Studies in two Illinois habitats and one Michigan habitat support this hypothesis by showing that the majority of these mimics fly in spring and early summer (Waldbauer and Sheldon 1971; Waldbauer et al. 1977; Waldbauer and LaBerge 1985).

"lek hypothesis" See hypothesis: firefly-flash-synchronization hypotheses: "lek hypothesis."

locale-odor hypothesis *n.*
1. The hypothesis that although, Honey Bees, *Apis mellifera*, communicate relatively precise direction information in their waggle dances, local cues (including locality odor, food odor, hive odor, and bee odor at a feeding site), dictate the location(s) at which recruits settle to feed (Johnson 1967, 847).
2. The hypothesis that although, Honey Bees, *Apis mellifera*, communicate food-distance information in their dances, they use other information obtained after they leave their hives (including hive-mate odor and other-bee odor) in the process of orienting to a particular food site (Wenner 1967, 849).
3. The hypothesis that Honey Bees, *Apis mellifera*, indicate the distance as well as the direction of food sites, but hive mates do not use the direction information to the accuracy reported by Frisch or the distance information; instead, newcomers to a food source are guided by local scents (Frisch 1967, 1073; Wenner and Johnson 1967, 1076; Wenner et al. 1969; Gould, 1976; Gould and Gould 1988, 73–74).
syn. locale-odor theory
cf. hypothesis: dance-"language" hypothesis
Comments: This hypothesis and the dance-"language" hypothesis are not mutually exclusive. The locale-odor hypothesis appears to be true under some experimental conditions; however, Honey Bees do often use distance and direction communication as well (Gould et al. 1970; Michener 1973, 1975).
[coined by Gould (1976)]

lock-and-key hypothesis See hypothesis: hypotheses of divergent evolution of animal genitalia: lock-and-key hypothesis.

Ludwig hypothesis, Ludwig effect, Ludwig's theorem, Ludwig's theory *n.* New genotypes can be added to a gene pool if they can utilize new components of their environment and thereby occupy a new subniche, even if they are competitively inferior in their ancestral niche (Lincoln et al. 1985).
cf. niche width-variation hypothesis

macromutationism *n.* The hypothesis that a new taxon can evolve essentially instantaneously via a major genetic mutation that simultaneously establishes both reproductive isolation and new adaptations (Lincoln et al. 1985).
cf. [2]mutation: macromutation

maximum-homology hypothesis, red-king hypothesis *n.* The optimum reconstruction of a phylogenetic tree is one that maximizes identity due to common ancestry, as indicated by homologous genetic coding sites (Lincoln et al. 1985).

Medawar-Williams hypothesis of senescence *n.* "Selection for genes postponing mortality will be most intense at the ages of greatest reproductive value" (Wilson 1975, 339).
Comment: Attributes that can enhance fitness early in the life cycle of an organism may have degenerative consequences later (*e.g.,* the structural modification of a male salmon's jaw).

minimum-interaction hypothesis *n.* Evolution reduces the genetic risk associated with reciprocal translocations (Imai 1986 in Howard 1988).

mitochondrial-Eve hypothesis See hypothesis: African-origin hypothesis of human mitochondrial DNA evolution.

molecular-clock hypothesis *n.* The hypothesis that random, neutral point mutations occur at a sufficiently regular rate in DNA (Zuckerkandl and Pauling 1965, 149; Strickberger 1990, 237).
syn. clock hypothesis (Lewin 1988, 561), evolutionary clock (Strickberger 1990, 237), molecular clock (Vawter and Brown 1986, 194), molecular clock hypothesis (Li and Tanimura 1987, 93; Zuckerkandl 1987, 34), molecular clock model of evolution (Lincoln et al. 1985), molecular evolutionary clock (Zuckerkandl and Pauling 1965, 149), molecular evolutionary clock hypothesis (Zuckerkandl 1987, 34)
Comments: Zuckerkandl conceived the idea of a molecular clock in 1960 (Lewin 1988, 561), and Zuckerkandl and Pauling (1962, 200) discuss this idea. Researchers have used molecular clocks to estimate dates of divergences in the evolution of particular taxa. Molecular clocks "tick at different

speeds;" therefore, researchers are proposing and applying sophisticated statistical methods to deal with clock idiosyncrasies (Strauss 1999, 1437). The speed of these clocks is related to the kind of informational molecule (*e.g.,* mitochondrial DNA and nuclear DNA), molecular sites on a molecule, and taxon (Fitch and Langley 1976 in Strickberger 1990, 238–239; Vawter and Brown 1986, 194; Lewin 1988, 561; Li and Tanimura 1987, 93; Li et al. 1990, 6703). For example, the gene that encodes the enzyme Cu,Zn superoxide dismutase (SOD) and the one that encodes for glycerol-3-phosphate dehydrogenase have a variable evolution rates depending on the taxon measured (Ayala 1997 in Strauss 1999, 1437). Although many analyses of mitochondrial and nuclear DNA suggest divergence dates that match those of fossils, there are some serious discrepancies (Mayr 1982, 577; Fitch 1976, Goodman 1982a in Ruse 1984; Stebbins and Ayala 1985, 77; Lewin 1988; Strauss 1999, 1437), including the time of the Cambrian explosion (= evolution of many animal phyla) and emergences of the more derived bird and mammal orders.

Moon-origin hypotheses
▸ **capture hypothesis** *n.* The hypothesis that the Moon formed independently from the Earth and was gravitationally captured by it (Drake 1990, 128).
syn. capture theory (Drake 1990, 128)
▸ **co-accretion hypothesis** *n.* The hypothesis that the Moon formed simultaneously with the Earth from material in orbit about the Earth (Drake 1990, 128).
syn. co-accretion theory (Drake 1990, 128), double-planet theory (Kerr 1984, 1061)
Comment: This hypothesis was proposed by Ruskol (1972 in Newsom and Taylor 1989, 29).
▸ **fission hypothesis** *n.* The hypothesis that the Moon is derived from the Earth's mantle after core formation through rotational instability (Drake 1990, 128).
syn. rotational fission (Kerr 1984, 1060)
▸ **giant-impact hypothesis** *n.* The hypothesis that the Moon was formed by collision of a Mars-sized object with the Earth late in the Earth's accretionary stages (Boss 1986a,b, 341; Taylor 1987, 469; Drake 1990, 128).
syn. collision-ejection hypothesis (Drake 1990, 128), impact hypothesis, single impact hypothesis (Newsom and Taylor 1989, 29), large-impact hypothesis (Kerr 1984, 1060)
Comments: This hypothesis has become the paradigm for lunar origin (Drake

h—m

1990, 128). It was first proposed by W. Hartmann and D. Davis in 1975 (Kerr 1984, 1060).

Mother-Eve hypothesis See hypothesis: African-origin hypothesis of human mitochondrial DNA evolution.

multiple-factor hypothesis *n.* The true hypothesis that proposes polygeny (multiple-factor inheritance), *q.v.,* and refutes the one-gene-one-character hypothesis, *q.v.,* of early Mendelism (Mayr 1982, 792).

natal-homing hypothesis See hypothesis: turtle-homing hypotheses: natal-homing hypothesis.

nexus hypothesis *n.* Each phenotypic character of an organism is likely to be influenced by more than one gene, and, conversely, most genes affect more than one character (Lincoln et al. 1985).

"no-allelic-replacement hypothesis" *n.* The hypothesis that major steps in social evolution occur via pleiotropic effects and contextual shifts rather than allelic replacement (West-Eberhard 1987, 38). *cf.* hypothesis: allelic-replacement hypothesis

nomogenesis *n.* The hypothesis that external environmental factors can produce heritable adaptations in all individuals of the same species that experience these factors (Lincoln et al. 1985). See -genesis: abiogenesis.

null hypothesis (H_0) *n.*
1. The hypothesis that one quantitative sample is not different than another; that is, one sample is statistically equal to another; *e.g.,* feeding behavior of the control group does not take a statistically different amount of time than that of the test group (mean feeding time of group 1 = mean feeding time of group 2) (Lehner 1979, 227), or that a coefficient of association = 0, meaning that there is no statistical association between two variables (Siegel 1956; Lincoln et al. 1985); contrasted with experimental hypotheses, H_1, H_2, and H_3.
   Note: In testing this null hypothesis, one determines if the data in question depart from those expected on the basis of chance alone (Futuyma 1998, 297, who keeps this kind of null hypothesis in quotes).
   syn. hypothesis of nondifference
   cf. hypothesis: alternative hypothesis
2. The basic, or default, hypothesis of a group of hypotheses (Futuyma 1998, 297).
   Notes: This newer use of "null hypothesis" is analogous to def. 1, and an example of this kind of null hypothesis

is random genetic drift in a population (Futuyma 1998, 297). Alleles at all loci are potentially subject to random genetic drift but not necessarily natural selection at a particular time; therefore, random genetic drift, rather than natural selection, is the "null hypothesis" of evolutionary change.

one-gene–one-character hypothesis, unit-character theory *n.* The hypothesis that one gene affects only one of an organism's traits (Mayr 1982, 774, 786, 792); contrasted with the one-gene-one-enzyme hypothesis and one-gene-one-polypeptide hypothesis.
Comment: The "multiple-factor hypothesis," which proposes that several, if not many, genes may affect or modify a single character, led to the abandonment of the "one-gene–one-character hypothesis" (Mayr 1982, 774, 786, 792).

one-gene–one-enzyme hypothesis *n.* The hypothesis that one gene dictates the production of a specific enzyme (Beadle and Tatum in Campbell 1987, 325); contrasted with the one-gene–one-character and one-gene–one-polypeptide hypotheses.

one-gene–one-polypeptide hypothesis *n.* The hypothesis that one gene dictates the production of a specific polypeptide (Beadle and Tatum in Campbell 1987, 325); contrasted with the one-gene–one-character and one-gene–one-enzyme hypotheses.
Comments: This hypothesis replaced the one-gene–one-enzyme hypothesis after biologists learned more about proteins (Campbell 1987, 326). It is now known that one gene encodes one RNA molecule or one set of closely related polypeptide chains (protein isoforms) (Alberts et al. 1989, 486, 592).

optimal-outbreeding hypothesis *n.* As the degree of genetic similarity between mates decreases, the costs associated with inbreeding should decrease as well (Keane 1990, 264).
cf. breeding: outbreeding: maximum outbreeding; law: Knight-Darwin law

panspermia hypothesis *n.* The hypothesis that "life did not originate on Earth but was introduced from elsewhere in the universe" (Arrhenius 1920s, Brooks and Shaw 1973, Hoyle and Wichramasinghe 1978, Crick 1981 in Strickberger 1990, 104; Lincoln et al. 1985).
syn. panspermia (Lincoln et al. 1985)
Comments: Life may have come to Earth as a space-resistant spore or other means (Strickberger 1990, 104). This hypothesis was fostered by the chemist Arrhenius in

the early 20th century and is still supported by some scientists today.

Parker-Baker-Smith model of the evolution of sex *n*. In the primeval situation, each diploid organism produced a similar number of equally well-provisioned haploid gametes and shed them into the sea; a mutant parent then provisioned its gametes slightly better than any of the others; selection favored these better nourished gametes over smaller ones; the extra fitness that a larger gamete (female) would gain from doubling its amount of nutrients was presumably less than the extra fitness a smaller one (male) would gain from going from almost nothing to the female amount; selection then favored individuals that produced smaller, more mobile gametes that sought larger ones more than those that produced larger gametes that combined with other larger ones because of their difference in the benefit of combining with a large gamete; intermediate-sized gametes did not compete for large gametes as well as small ones; finally, disruptive selection led to evolution of sperm (small gametes) and eggs (large gametes) (Parker et al. 1972 in Thornhill and Alcock 1983, 56–57; Dawkins 1986, 133–134).

periodic-extinction hypothesis *n*. The hypothesis that in the last 600 million years mass extinctions have occurred approximately every 26 million years (Raup and Sepkoski 1984; etc.).

cf. hypothesis: hypotheses regarding mass extinctions

Comments: There are different renditions of this hypothesis implied in the literature. The validity of this hypothesis has been debated. Benton's (1995) quantitative analysis of the fossil record, focusing on the pattern of diversification and seven major extinctions, indicates that the timing of major extinctions is not periodic.

perpetual-intervention hypothesis *n*. New species occur as a result of continuous special creation by God (Lyell in Mayr 1982, 882).

pet hypothesis *n*. A hypothesis that a researcher favors.

Comments: "Pet hypothesis" has a negative connotation when it implies that its owner is biased in alternative hypotheses because he favors his pet one. Scientists should cautiously, ethically, and rigorously use the method of multiple working hypotheses, *q.v.*

pleiotropy hypothesis See hypothesis: hypotheses of divergent evolution of animal genitalia.

polymorphic behavior hypothesis See hypothesis: population-limiting hypotheses: polymorphic behavior hypothesis.

population-limiting hypotheses

▸ **behavior hypothesis** *n*. Mutual interactions involving spacing behavior prevent unlimited increase in population density, and this spacing behavior is not an inherited trait (Lincoln et al. 1985).

▸ **polymorphic-behavior hypothesis, Chitty hypothesis, genetic-behavioral polymorphism hypothesis** *n*. Spacing behavior limits population density, and individual differences in spacing behavior have a genetic basis and respond to rapid natural selection (Lincoln et al. 1985).

▸ **self-regulation hypothesis** *n*. An indefinite increase in population density is prevented by a qualitative change in a population (Lincoln et al. 1985).

▸ **social-subordination hypothesis** *n*. As population density increases, increased intraspecific competition leads to increased aggression, resulting in subordinate individuals' being forced to disperse to suboptimal habitats (Lincoln et al. 1985).

▸ **stress hypothesis** *n*. Mutual interactions in an increasing population lead to physiological changes that are phenotypic in origin and eventually reduce birth rate and increase death rate (Lincoln et al. 1985).

"pregnancy-sickness hypothesis" *n*. The hypothesis that maternal nausea due to pregnancy has evolved to minimize maternal ingestion of dietary teratogens (Profit 1988, 1992 in Haig 1993, 510).

proterogenesis *n*. The hypothesis that new evolutionary features appear suddenly in ancestral juvenile stages and are displaced slowly forwards by neoteny towards the adult stage of descendants (Lincoln et al. 1985).

cf. -genesis: biogenesis

red-king hypothesis See hypothesis: maximum-homology hypothesis.

Red-Queen hypothesis *n*.

1. The hypothesis that a species evolves as quickly as possible to remain extant because its biotic environment is continually deteriorating due to continual evolution of other species (*e.g.,* competitors, predators, and parasites) with which it interacts (Van Valen 1973, 17; Lewin 1985a, 399; Futuyma 1986, 362; Clarke et al. 1994, 4821).

Notes: Van Valen (1973, 17) formulated this hypothesis to account for the fact that, within most taxa, the probability of extinction of a genus, or a family, is independent of its prior duration. He suggests that an improvement in the fitness of one species represents a deterioration in the environment of other species in its community (Strickberger 1990, 443). This hypothesis emphasizes biotic environmental change rather than abiotic environmental change. Michael Rosenzweig conceived of the Red-Queen hypothesis independently of Van Valen (Lewin 1985a) and called this idea the "rat race" because he was interested in predator-prey interactions. Stenseth and Maynard Smith (1984) reformulated this hypothesis by including the idea of evolutionary lag and dismissing Van Valen's zero-sum assumption. They contrast their Red-Queen hypothesis with the stationary hypothesis, *q.v.* Hoffman and Kitchell (1984 in Lewin 1985a, 400) offer some support for the Red-Queen hypothesis. It may not be possible to test the Red-Queen hypothesis with regard to community evolution because of the difficulty in determining if and when periods of physical environmental constancy occurred in the past. Clarke et al. (1994, 4821) indicate that a laboratory study of two competing virus populations supports the Red-Queen hypothesis as it can be extrapolated to nonorganisms.

syn. red queen hypothesis (Lincoln et al. 1985), Red Queen model (Lewin 1985a)

2. The hypothesis that both a predator and its prey continually coevolve ("run faster and faster"), the predator evolves better prey catching, and the prey evolves better predator evasion (Abrahamson et al. 1989).

cf. hypothesis: hypotheses of the evolution and maintenance of sex: Red-Queen hypothesis; law: law of extinction

[coined by Van Valen (1973, 17) as Red Queen's hypothesis. This name is derived from *Alice in Wonderland* (Lewis Carrol): "It takes all the running you can do, to keep in the same place," the Red Queen.]

regional-continuity hypothesis *n.*

1. The hypothesis that modern Humans (*Homo sapiens sapiens*) evolved in different geographic locations at the same time and interbred to form a single species; contrasted with the Noah's-ark model (Gibbons 1992b, 874).

syn. multiregional model (Stringer and Andrews 1988, 1263)

2. The hypothesis that all populations of modern Humans trace back to when Humans first left Africa at least a million

years ago, through an interconnected web of ancient lineages in which the genetic contributions to all living peoples varied regionally and temporally; contrasted with the Noah's-ark model and called "multiregional evolution" in Thorne and Wolpoff (1992, 76).

research hypothesis *n.* A hypothesis about behavior that an investigator perceives to be the true situation (Lehner 1979, 57).

RNA-world hypothesis *n.* The hypothesis that, in the origin of life on Earth, RNA came first about 4 billion years ago; after efficient protein synthesis catalyzed by RNA evolved, DNA took over the primary genetic function, and proteins became the major catalysts, while RNA remained primarily as the intermediary connecting DNA and proteins (Alberts et al. 1989, 10; Campbell 1990, 516; Gibbons 1992c, 31; Illangasekare et al. 1995, 643; Doebler 2000, 19).

Comments: Data that support this hypothesis include RNA having both genetic and catalytic properties and scientists have produced RNAs in a test tube without life from RNA monomers. RNA molecules (ribozymes) catalyze reactions in living cells (*e.g.,* removing introns from mRNA), and RNAs catalyze synthesis of new RNA (rRNA, tRNA, and mRNA) (Cech 1986; Campbell 1990, 516). One kind of RNA molecule catalyzes its own acylation when offered an activated amino acid, showing how RNA self-catalysis could have translated linear nucleotide sequences into amino-acid sequences (Illangasekare et al. 1995). Other related hypotheses are that in the evolution of life on Earth, either peptides or proteins formed and reproduced before RNA (Doebler 2000, 15).

saturation-presaturation-dispersal hypothesis *n.* Individuals dispersing from a population at its peak are qualitatively different from those emigrating from a declining population or one at low density, and the latter are typically in better condition with higher survivorship and greater fecundity (Lincoln et al. 1985).

self-centered-DNA hypothesis See hypothesis: selfish-DNA hypothesis.

selfish-DNA hypothesis *n.* Transposons (selfish DNA) simply spread on their own accord from chromosome to chromosome, making more copies of themselves; these transposons persist because they have no significant effect on the organisms in which they reside (Doolittle et al. in Gould 1981a, 7).

syn. self-centered DNA hypothesis (Gould 1981a, 14)

cf. gene: selfish gene; nucleic acid: deoxyribonucleic acid: selfish DNA

Comment: The transposons are in effect "ultimate parasites" (Doolittle et al. in Gould 1981a, 7).

selfish-herd hypothesis *n.* The hypothesis that an animal herd results from its individual members' moving toward the central area of the group; because predators tend to take the first individuals that they encounter, there is a great advantage for each animal to be protected in the central region of a herd (Galton 1871, Williams 1964, 1966a, Hamilton 1971a in Wilson 1975, 38).

syn. selfish herd, selfish-herd theory (Parrish 1989, 1048)

Comment: Parrish's (1989, 1048) work with fish does not support this hypothesis.

serial-endosymbiosis hypothesis *n.* The hypothesis that three classes of organelles of eukaryotic cells (undulipodia, mitochondria, and plastids) originated from bacterial symbionts; first, a host prokaryotic cell acquired motile bacterial symbionts (spirochetes) that became undulipodia, and this combination became a mastigote; second, the mastigote acquired symbionts that became mitochondria; third, the one-celled algae and plants acquired photosynthetic symbionts that became chloroplasts (Margulis 1992, 150–151).

syn. serial-endosymbiosis theory, theory of symbiosis (Margulis 1992, 150)

cf. hypothesis: endosymbiosis hypothesis

Comment: This hypothesis suggests that all animals have at least three kinds of ancestors, all plant cells have at least four, and all are chimeras (Margulis 1992, 150–151).

"sexual-attraction hypothesis" *n.* The hypothesis that continuous sexual attraction, as opposed to seasonal sexual attraction, holds human society together (Washburn 1985, 7).

Comment: Ape behavior does not seem to support this hypothesis. Orangutans are the most continually sexually active nonhuman apes, and Gorillas are the least sexually active ones, yet Gorillas are the most social (Washburn 1985, 7).

sexual-selection hypothesis *n.* In certain animal species, males regularly invade new groups and kill other males' young, inseminate the young's mothers, and thereby eliminate other males' genes from the population and increase their own genes in the population (Hrdy 1974 in Brown 1996).

Comments: Studies consistent with the hypothesis include Hrdy (1974), Sääl (1982),

Pusey (in Brown 1996, 175), and Emlen (all in Brown 1996). Studies inconsistent with this hypothesis include Sussman et al. (1995) and Hoogland (in Brown 1996).

sexy-son hypothesis *n.* Female animals that mate with males with attractive physical attributes, but poor paternal care, will have fewer surviving offspring, but their sons, having inherited genes for attractiveness (sexiness), will be extremely attractive to the next generation of females; thus, such females of more sexy sons will have higher fitnesses than females with less sexy sons [proposed by and coined by Weatherhead and Robertson 1979].

Comments: This hypothesis is theoretically rejected by Kirkpatrick (1985) and not supported by data of Alatalo and Lundberg (1986, 1454). Diploid models support this hypothesis (Spencer and Masters 1992, 299).

single-impact hypothesis *n.* One major impact of a huge asteroid or comet that struck the Earth about 65 million years ago is responsible for extinction of many species, including dinosaurs, at the end of the Mesozoic Era (Fastovsky 1990 in Kerr 1992, 160); contrasted with battered-Earth hypothesis.

Comments: The Chicxulub crater, perhaps as large as 300 kilometers in diameter, in the Yucatán Peninsula might have been caused by this impact (Kerr 1992, 160; 1993b, 1518; Sharpton et al. 1993). This impact may have caused global wildfires, acid precipitation, and up to 10 years of darkness (Morell 1993c, 1519).

size-fitness-correlation hypothesis *n.* In parasitic wasps: the hypothesis that within a brood from one host individual, a mother's offspring sex ratio is influenced, through evolution, by the sizes of her offspring that can develop from their host; *e.g.,* if daughters benefit by being large, a mother wasp should adjust her sex ratio with regard to host size so that she will have large daughters (Charnov et al. 1981).

Comment: Other factors besides maternal control of egg sex ratio (*e.g.,* sex-biased larval mortality) may be involved in determining the sex ratio of emergent adult progeny (Charnov et al. 1981). [coined by Barash (1983, 300)]

social-cohesion hypothesis *n.* An animal species' social interaction prior to emigration is the major factor that determines its dispersal patterns rather than its agonistic interactions at emigration time (Lincoln et al. 1985).

social-facilitation hypothesis See hypothesis: turtle-homing hypotheses: social-facilitation hypothesis.

species selection *n.*

1. The hypothesis that entire species are selected in the same sense as organisms; *i.e.,* they function as interactors (Eldredge and Gould 1972; Stanley 1975 in Hull 1992, 186).
   Note: This has been a highly controversial topic in evolutionary biology during the past two decades (Hull 1992, 186).
2. "The theory that some evolutionary change takes place by a form of natural selection at the level of species or lineages" (Dawkins 1982, 294).
3. The hypothesis (embraced by some punctualists) that a species with a certain attribute will be more likely to persist and give rise to new species than one lacking the attribute (Stebbins and Ayala 1985, 81).

cf. speciation: geographical speciation, saltational speciation

Comment: Vrba and Gould (1986) expanded the argument for species selection.

spinster hypothesis *n.* The probability of a low-fertility female paper wasp's establishing and bringing a nest through to maturity may be so low that it is more profitable, as measured by inclusive fitness, for these high-risk individuals to subordinate themselves to female relatives in foundress associations (West 1967 in Wilson 1975, 290).

stability-time hypothesis *n.* Where physiological stresses have been historically low, biologically accommodated communities have evolved; as a gradient of environmental stress increases, a community gradually changes to a physically controlled one until the stress conditions exceed organisms' adaptive abilities and an abiotic condition is reached; and the number of species present diminishes continuously along the stress gradient (Lincoln et al. 1985).

stationary hypothesis *n.* The hypothesis that organisms do not evolve during periods of absolutely no perturbations in their physical environments, but as soon as their environments change the organisms are likely to evolve; periods of stasis would be punctuated by periods of change, including speciation and extinction; and the balance between them would determine the overall shift in species diversity in a community (Lewin 1985a).

syn. Stationary hypothesis, Stationary model (Lewin 1985a, 399)

Comments: This hypothesis is contrasted with the Red-Queen hypothesis. Wei and Kennett (in Lewin 1985a) present data in support of the stationary hypothesis.

statistical hypothesis *n.* A null hypothesis or an alternative hypothesis, *q.v.* (Lehner 1979, 227).

stress hypothesis See hypothesis: population-limiting hypotheses: stress hypothesis.

tabula rasa *n.* The false hypothesis that a "higher" animal's nervous system is not genetically programmed for any behavior and all of its behaviors are the result of learning (Wilson 1975, 156; John Locke in Mayr 1982, 80).

See *tabula rasa.*

Comment: This hypothesis was a main tenet of American behaviorism.

Thayer's countershading hypothesis *n.* The little-tested hypothesis that a prey animal's countershading diminishes its three-dimensional shape and improves its background matching, thereby making it more difficult to be seen by predators (A.H. Thayer 1896, 124–125; G.H. Thayer 1909, 14; Kiltie 1989, 542).

Comment: Work with Gray Squirrels suggests that "the degree of shadow obliteration achieved by countershading is imperfect and hence of questionable value in deterring predators" (Kiltie 1989, 543).

[after A.H. Thayer]

toxin hypothesis *n.* The hypothesis that human allergy evolved as an active immunological defense against toxins; thus, it is not a malfunctioning of one's immune system (Profit 1991, 27); contrasted with the helminth hypothesis (Profit 1991, 41–42).

Comment: Data that support this hypothesis include toxins are ubiquitous, chemical correlates of toxicity frequently trigger allergy, most known allergens appear to be toxic substances or carrier proteins that bind well to low-molecular-weight toxins, and allergy symptoms have attributes of an evolved mechanism (Profit 1991, 27–28).

[coined by M. Profit (1991, 27)]

transportation hypothesis *n.* Charles Darwin's hypothesis proposed to account for inheritance of acquired characters: At any stage of an organism's life cycle, its cells may throw off gemmules "which circulate freely throughout the system, and when supplied with proper nutriment multiply by self division, subsequently becoming developed into cells like those from which they were derived" (Darwin 1868 in Mayr 1982, 694).

cf. ³theory: theory of pangenesis

Trivers'-parental-investment hypothesis *n.* "The sex whose typical [average-per-offspring] parental investment is greater than that of the opposite sex will become a limiting resource for that sex" (Trivers 1972).

Trivers-Willard hypothesis, Trivers-Willard principle *n.*

1. In polygynous mammals, healthy females should bias their offspring sex ratio towards sons because their sons will mate more successfully and produce more grandchildren than sickly, small sons (Trivers and Willard 1973 in Wilson 1975, 317–318).
2. In polygynous mammals, the healthier females should have more sons per female than less healthy ones, and more healthy females should produce more healthy sons that are more likely to acquire many mates than less healthy sons; on the other hand, a less healthy female should produce daughters rather than competitively disadvantaged sons because receptive females are competed for by males; small or sickly sons are likely to have low fitnesses (Thornhill and Alcock 1983, 72).

Comment: Thornhill and Alcock (1983, 72) suggest that this hypothesis could apply to insects in certain situations.

true hypothesis *n.* A hypothesis that has not been falsified by all attempts; a hypothesis that is true beyond all reasonable doubt (Moore 1984, 475).

"turnover-pulse hypothesis" *n.* The hypothesis that evolution is concentrated in punctuational bursts at times of world-wide climatic pulsing (Vrba 1985 in Gould and Eldredge 1993, 225).

Comment: A climatic pulse, about 2.5 million years ago, may have stimulated the origin of the genus *Homo* (Gould and Eldredge 1993, 225)

turtle-homing hypotheses

▸ **natal-homing hypothesis** *n.* In marine turtles, a female returns to her natal beach to lay her eggs (Meyland et al. 1990, 724).

Comment: Data on mitochondrial DNA frequencies among green turtles support this hypothesis (Meyland et al. 1990, 724).

▸ **social-facilitation hypothesis** *n.* In marine turtles, a virgin, mature female randomly encounters experienced females in foraging areas; she then follows these females to a nesting beach and, after having a "favorable" nesting experience, fixes on this site for her future nesting (Hendrickson 1958 and Owens et al. 1982 in Meyland et al. 1990, 724–725).

uncorrelated-asymmetry hypothesis *n.* The hypothesis, based on game theory, that any differences in the attributes of two opponents can be used arbitrarily as the means for settling disputes (Maynard Smith and Parker 1976 in Thornhill and Alcock 1983, 220); *e.g.,* when ownership of a territory is correlated with resource-holding power or fighting ability, it could be used as a cue that resolves fights over ownership of a resource.

uniform-convergence hypothesis *n.* Convergent evolution in all parts of a phylogenetic tree takes place at a uniform rate (Lincoln et al. 1985).

uniform-species-substance hypothesis *n.* The hypothesis that a uniform species substance, perhaps in cytoplasm, is transmitted from generation to generation (Mayr 1982, 786).

Comments: This hypothesis attempts to explain continuous variation independently of discontinuous Mendelian genetics. The theory of particulate inheritance gradually replaced this hypothesis (Mayr 1982, 786).

unit-character theory See hypothesis: one-gene-one-character hypothesis.

Upper Devonian-impact hypothesis *n.* The hypothesis that an asteroid impact caused a mass extinction of marine species in the Upper Devonian (McLaren 1970 in Claeys et al. 1992, 1102).

cf. Upper-Devonian-Mass Extinction

Comment: Glass spherules found near the Frasnian-Famennian boundary in Belgium support this hypothesis (Claeys et al. 1992, 1102).

vigilance hypothesis *n.* The hypothesis that there has been natural selection against hallucinations during human sleep that compromise a person's vigilance of external stimuli that might occur during sleep (Donald Symons 1993 in Achenbach 1995).

Comments: This hypothesis attempts to explain why Humans usually do not taste, smell, or feel pain or tactile sensations during dreaming (Symons 1993 in Achenbach 1995, F5). This may be because people evolved vigilance for odors of fires, wolves, and other potentially harmful things that can occur during sleep. The vast majority of "auditory" sensations that occur in dreams usually are speech and other human vocalizations that are generated by mechanisms of speech production, not by those of auditory perception. A sleeping person's audition is not used during dreaming but is free to receive sounds, such as those of a crying baby.

volition, effect of volition, theory of volition *n.* The hypothesis that animal adaptations arise from animals' slow willing of these adaptations (Mayr 1982, 357).

Comment: This idea was erroneously ascribed to Lamarck by other biologists,

h–m

including Charles Darwin. "In part, the misunderstanding was caused by the mistranslation of the word *besoin* into 'want' instead of 'need' and a neglect of Lamarck's carefully developed chain of causation from needs to efforts to physiological excitations to the stimulation of growth to the production of structures" (Mayr 1982, 357).

working hypothesis *n.* A hypothesis that serves as the basis for future experimentation (Lincoln et al. 1985).

zoo hypothesis *n.* Extraterrestrial intelligent life is not interacting with us because it "set us aside as part of a wilderness area or zoo" (Ball 1973, 347).

♦ **hypothetical construct** See variable: intervening variable.

♦ **hypothetico-deductive method** See method: hypothetico-deductive method.

♦ **hypotrophic** *n.* See -troph-: hypotroph.

♦ **hysteresis mechanism** See mechanism: hysteresis mechanism.

♦ **hysterogenic** See -genic: hysterogenic.

♦ **hysterotely** See [2]heterochrony: neoteny: hysterotely.

i

♦ *I* See opportunity for selection.

♦ **ichnite, ichnolite** See fossil: ichnite.

♦ **ichnofauna** See fauna: ichnofauna.

♦ **ichnoflora** See flora: ichnoflora.

♦ **ichnofossil** See fossil: trace fossil.

♦ **ichnolite** See fossil: ichnite.

♦ **ichnology** See study of: ichnology.

♦ **ichnospecies** See ²species: ichnospecies.

♦ **ichthyofauna** See fauna: ichthyofauna.

♦ **ichthyology** See study of: ichthyology.

♦ **ichthyophage** See -phage: ichthyophage.

♦ **ICZN** See International Code of Zoological Nomenclature.

♦ **idea** *n.*
   1. A person's picture, or notion, of anything conceived by his mind; a conception (*Oxford English Dictionary* 1972, entries from 1612).
   2. A person's conception for which no reality exists; an imagined or fanciful conception (*Oxford English Dictionary* 1972, entries from 1588).
   3. Any product of a person's mental apprehension, or activity, that is in his mind as an object of knowledge or thought; a belief; conception; thought; way of thinking (*Oxford English Dictionary* 1972, entries from 1645).
   Note: This is a more widely used definition.
   cf. concept
   Comment: Idea has many other meanings not listed here.

♦ **ideal character** See character: ideal character.

♦ **ideal-dominance distribution** See distribution: ideal-dominance distribution.

♦ **ideal free distribution** See distribution: ideal free distribution.

♦ **ideal trait** See character: ideal trait.

♦ **idealistic morphology** See -morphology: idealistic morphology.

♦ **identical twin** *n.* See twin: fraternal twin.

♦ ¹**identification** *n.* A person's placing an investigated individual organism into one of the classes of an already existing taxonomic classification (Mayr 1982, 147).
   syn. determination
   cf. classification

♦ ²**identification** *n.* An animal's connection, or relationship, to a group or family (Heymer 1977, 90).
   Comment: This behavior may promote group cohesion if coupled with social consciousness (Heymer 1977, 90).

♦ **identified-neuron chauvinist** *n.* One who claims that the way to understand animal behavior is to study animal neurophysiology (Dawkins 1986, 98).
   Comment: Identified-neuron chauvinists debate with those who hold a black-box view of behavior, *q.v.* (Dawkins 1986, 98).

♦ **identifior** *n.* A signal component that specifies a certain place and time (Marler 1961 in Wilson 1975, 217).
   cf. appraisor

♦ **ideology** *n., pl.* **ideologies** The ideas, or manner of thinking, characteristic of an individual or group; especially, the ideas and objectives that influence a whole group or national culture, shaping especially their political and social procedure (Michaelis 1963).
   See study of: ideology.

♦ **idioadaptation** See ²evolution: allomorphosis.

♦ **idiobiology** See study of: biology: idiobiology.

♦ **idiochromosome** See chromosome: idiochromosome.

♦ **idioecology** See study of: autecology.

♦ **idiogamy** See -gamy: idiogamy.

♦ **idiogram** See -type: karyotype.

♦ **idioplasm** See germ plasm.

♦ **idiotaxonomy** See study of: idiotaxonomy.

h – m

♦ **idiotherm** See -therm: idiotherm.

♦ **idiotypic** *adj.* Referring to sex or sexual reproduction (Lincoln et al. 1985).
syn. sexual

♦ **IGS** See external-nontranscribed spacers.

♦ **illegitimate receiver** See eavesdropper.

♦ **illegitimate signaler** See signaler: illegitimate signaler.

♦ **illusion** *n.* A person's false perception; a perception that does not correspond with an objective situation (Storz 1973, 123).

anticipatory visual illusion *n.* A person's initially confusing his searched-for object with its model, *e.g.,* an insect collector's at first perceiving a real leaf as a sought-after leaf butterfly (Krizek 1984, 4). *Comments:* A person who is searching for a particular thing may also confuse it with an object that is not a *bona fide* model of this thing. For example, I have searched for bears in a forest and at first mistaken stumps for them or have searched for the woodland lily and initially mistaken a distant orange daylily for this species. I am not aware of a term for this kind of visual illusion. However, this seems to be a more general case than described by Krizek (1984). Perhaps his illusion should be called "anticipatory-mimic-visual illusion" and the more general case be called "anticipatory visual illusion."

"confusion illusion" *n.* A person's erroneously concluding that another person's research is sophisticated and important after hearing the person give a complicated, confusing, unclear seminar of the research (personal observation).

"irrelevance illusion" *n.* A person's erroneously concluding that a research area is irrelevant, unimportant, or both because the person has erroneous knowledge of the area, limited knowledge of the area, no or little intrinsic interest in the area, or a combination of these things (personal observation).

Moon illusion *n.* A person's perceiving that the Moon seems to be much larger when it is near the Earth's horizon than when it is seen higher in the sky, although the Moon is really the same size (Storz 1973, 173).

Muller-Lyer illusion *n.* An optical illusion in which lines of the same length appear to be of different lengths to a human observer (Immelmann and Beer 1989, 216).

proofreader's illusion *n.* A person's "failure to detect an error in spelling, punctuation, construction, etc.," because he is familiar with the subject matter; a person's "tendency to see things 'as they ought to be' rather than as the are" (Hinsie and Campbell 1976, 378).

♦ **illustrator** See communication: nonverbal communication: illustrator.

♦ **imagination** *n.* In insects: the final developmental stage that produces an imago (Lincoln et al. 1985).

♦ **imago** *n.*
1. In Insects: an adult that is sexually developed (Torre-Bueno 1978).
2. In Termites: usually an adult primary reproductive, rather than other kind of adult (Wilson 1975, 586).

♦ **imitation, imitative behavior** See learning: social imitative learning.

♦ **imitative foraging** See foraging: imitative foraging.

♦ **imitative resemblances** See mimicry.

♦ **immature** *adj.*
1. Referring to the developmental stages of an organism preceding attainment of sexual maturity (Lincoln et al. 1985).
2. Referring to an incompletely differentiated system (Lincoln et al. 1985).
cf. animal names

♦ **immature behavior** See behavior: immature behavior.

♦ **immigrant** *n.* Referring to an adventive, group, or species, that arrived in a new area without deliberate agency of Humans, even though it may have been transported accidently by Humans (Frank and McCoy 1990, 4).
cf. adventive, endemic, indigenous, introduced, precinctive
[Latin *immigrantem* < present participle of *immigrare*, one who, or that which, migrates into a country as a settler]

♦ **immigration** *n.* Movement of an individual or group into a new population or geographical region (Lincoln et al. 1985).
n. immigrant
cf. emigration

♦ **immigration curve** See curve: immigration curve.

♦ **immigration rate** See rate: immigration rate.

♦ **immobility response** See response: immobility response.

♦ **immoral** See moral: immoral.

♦ **immunology** See study of: immunology.

♦ **impaling** *n.* In Shrikes (birds): an individual's pushing a prey onto a thorn (Watson 1910 in Heymer 1977, 33).
v.t. impale

♦ **imperfect** See sexual: unisexual.

♦ **implantation** *n.*
1. A mammal embryo's embedding into the wall of a uterus prior to fetal development (Curtis 1983, 846).
2. An investigator's placing a hormone deposit, usually in capsule form, under an animal's skin or elsewhere in a hormone-

manipulation experiment (Immelmann and Beer 1989, 139).

♦ **imposed altruism** See altruism: imposed altruism.

♦ **impregnation** See insemination.

♦ **imprinting** See learning: imprinting.

♦ ***in copulo*** *adj.* In the act of copulating, or having sexual intercourse or coition (Torre-Bueno 1978).
cf. copulation
Comment: Torre-Bueno (1978) says that the correct form for an act of copulation is "*in copulo*" not "*in copula;*" however, the latter term is frequently used.

♦ ***in situ* hybridization** See method: *in situ* hybridization.

♦ **in-the-head variable** See variable: intervening variable.

♦ ***in vitro* mutagenesis** See method: *in vitro* mutagenesis.

♦ **inadequate stimulus** See ²stimulus: inadequate stimulus.

♦ **inborn behavior** See behavior: inborn behavior.

♦ **inbreeding** See breeding: inbreeding.

♦ **inbreeding coefficient** See coefficient: coefficient of inbreeding.

♦ **inbreeding depression** *n.* Reduction of fitness and vigor of a normally outbreeding population due to increased homozygosity as a result of inbreeding (Lincoln et al. 1985).

♦ **inbreeding polygyny** See mating system: polygyny: inbreeding polygyny.

♦ **incasement theory** See ³theory: encasement theory.

♦ **incentive** *n.*
1. Something that arouses a person's feeling or incites one to action; an exciting cause or motive; an incitement, provocation, spur (*Oxford English Dictionary* 1972, entries from 1432).
2. An individual animal's arousal produced by its associating a stimulus with reinforcement (incentive-motivational effects), not receiving reinforcement, or prevention of reaching a goal (Immelmann and Beer 1989, 141).
Note: The meaning of "incentive" varies among authorities (Immelmann and Beer 1989, 141).
adj. incentive
syn. appetite (in some contexts Immelmann and Beer 1989, 141)

♦ **incest** *n.* "The crime of sexual intercourse or cohabitation between persons related within the degrees within which marriage is prohibited; sexual commerce of near kindred" (*Oxford English Dictionary* 1972, entries from 1225).
[Latin *incestus, incestum,* impure, unchaste]

♦ **incest avoidance, incest taboo** *n.* For example, possibly in some insect species; in some bird and primate species including Humans: an individual's tendency to avoid mating with very close kin (Wilson 1975, 78–79).
Comment: "Incest avoidance" occurs, for example, in actively mobile animals because their young and parents disperse widely and are unlikely to meet again, because one sex leaves a family group as juveniles, or both (Immelmann and Beer 1989, 141).
cf. social inhibition

♦ **incidental** See male: incidental.

♦ **incidental learning** See learning: latent learning.

♦ **incipient** *adj.* Referring to the initial stage in the development of a structure or event (Lincoln et al. 1985).

♦ **incipient species** See ²species: incipient species.

♦ **incipient stages of useful structures, the problem of** *n.* The puzzle of how beginning stages of adapted traits could have evolved (Gould 1985a).

♦ **incitin** See chemical-releasing stimulus (table).

♦ **inciting** *n.* In female ducks: an individual's orienting toward her mate, or prospective mate, following him and simultaneously threatening another conspecific male (Lorenz 1941 in Heymer 1977 86).

♦ **inclinometer** See instrument: inclinometer.

♦ **included niche** See niche: included niche.

♦ **inclusive fitness** See fitness: inclusive fitness.

♦ **incompatibility load** See genetic load: incompatibility load.

♦ **incomplete block design** See experimental design: incomplete block design.

♦ **incomplete dominance** See dominance (genetic): incomplete dominance.

♦ **incross** See cross: incross.

♦ **incubation** *n.* In birds: an individual's sitting on its egg(s) which warms it sufficiently for its embryonic development to proceed (Immelmann and Beer 1989, 142).
syn. brooding
cf. brooding
Comment: Immelmann and Beer (1989, 142) suggest that, for precision, use of "brooding" be restricted to parental care after hatching and "incubation" be restricted to parental care of eggs and hatching eggs.

♦ **independent** *adj.*
1. Referring to a variable's not being influenced or controlled by another variable (Lincoln et al. 1985).

h – m

2. Referring to a datum whose value is not affected by the magnitude of other variables in its set.
n. independence
cf. dependent

♦ **independent assortment, law of** See law: Mendel's laws: Mendel's second law.

♦ **independent samples** *n.* Pairs, or groups, of data in which the selection of one group in no way affects selection of other groups (Lincoln et al. 1985).
cf. independent

♦ **independent variable** See variable: independent variable.

♦ **index** *n., pl.* **indexes, indices**
1. A numerical expression of the ratio between one dimension, or magnitude, and another (Michaelis 1963).
2. A number that summarizes one or more characteristics of a data set, *e.g.,* a diversity index that indicates the number of species and their relative abundances in a community.
cf. diversity
[Latin *index,* forefinger, sign]

F_{RT} *n.* The proportion of heterozygosity reduction of a taxon's regional aggregates relative to its total combined population; $F_{RT} = (H_T - H_R)/H_T$ (Hartl and Clark 1997, 117).

F_{SR} *n.* The fixation index of a taxon's subpopulations relative to regional aggregates; $F_{SR} = (H_R - H_S)/H_R$ (Hartl and Clark 1997, 117).

F_{ST} *n.* The fixation index of a taxon's subpopulations relative to the total area it occupies; $F_{ST} = (H_T - H_S)/H_T$ (Hartl and Clark 1997, 117).
Comments: F_{ST} has a theoretical minimum of 0 (indicating no genetic divergence between, or among, subpopulations) and a maximum of 1 (indicating fixation of alternative alleles in different subpopulations) (Hartl and Clark 1997, 118–119). This index is usually much less than 1. F_{ST} values of 0–0.05 indicate little genetic differentiation; 0.06–0.15, moderate differentiation; 0.016–0.25, great differentiation; and over 0.25, very great differentiation.

index of similarity See coefficient: coefficient of community.

Lincoln index *n.* An index that can be used to estimate a species' local population size: $\hat{N} = an/r$, where $\hat{N}$ is the estimate of the number of individuals in the population (*N*), *a* is the total number of marked individuals, *n* is the total number of individuals in the second captured sample, and *r* is the total number of recaptured individuals (Southwood 1978, 97).

Comment: Southwood (1978, 92–99) discusses different Lincoln-index-type methods, assumptions of these methods, and other mark-recapture methods.

species-diversity index *n.* An index that indicates the number of species in an area or both this number and the species' relative abundances.
Comments: Researchers have invented a large number of diversity indices (Southwood 1978, 421). Indices that combine the number of species and their relative abundances in a sample, such as α, D, and H, summarize most of the biological information on diversity of samples (Krebs 1985, 524).

▸ **α-diversity index** *n.* The number of species in a range of habitats in a geographical areas (*e.g.,* an island) (Southwood 1978, 420).

▸ **β-diversity index** *n.* A measure of the rate and extent of change in species along a gradient, from one habitat to others (Southwood 1978, 420).
Comments: Lawton et al. (1998, 73) use a modified β-diversity index that quantifies species replacement.

▸ **diversity index** *n.* An index of taxon diversity (taxon number or taxon number and taxon abundances) (Southwood 1978, 420).
cf. diversity
Comments: Whittaker (1972 in Southwood 1978, 420) classified organism diversity indices as α-diversity, β-diversity, and τ-diversity. Southwood (1978, 420–437) discusses these and many other diversity indices.

▸ **Shannon-Weaver function (*H*)** *n.*

$$H = -\sum_{i=1}^{S} (p_i)(\log_2 p_i)$$

where *H* = the information content of a sample (in bits per individual) = an index of species diversity, *S* = the number of species, and p_i = the proportion of the total sample belonging to the *i*th species (Krebs 1985, 521).
syn. Shannon-Weiner function [a misleading synonym (Krebs 1985, 521)]
cf. Shannon-Weaver Formula
Comments: This index indicates how difficult it is to predict correctly the species of the next individual collected in a series of collected ones (Krebs 1985, 521). *H* is a measure of uncertainty; diversity of a sample is positively corrected with *H*.
[After Shannon and Weaver, who independently derived this function (Krebs 1985, 521)]

▸ **Simpson's index of diversity (D)** *n.*

$$D = 1 - \sum_{i=1}^{S}(p_i)^2$$

where p_i = the proportion of $i = 1$ individuals of species *i* in a focal community, and *S* is its total species number (Krebs 1985, 523).
Comments: D gives relatively little weight to rare species and more weight to the common ones. It ranges in value from 0 (low diversity) to a maximum of $(1 - 1/S)$.
▸ **τ-diversity index** *n.* The number of species in a range of habitats in a geographical area (e.g., an island) (Southwood 1978, 420).
Comment: τ-Diversity is a consequence of the α-diversity of the habitats together with the extent of the β-diversity between them (Southwood 1978, 420).
species richness *n.* "The absolute number of species in an assemblage or community" (Lincoln et al. 1985).
syn. $S(n)$ (Colwell and Coddington 1994, 104), S_T (Southwood 1978, 420).
♦ **index species** See ²species: index species.
♦ **indicate** *v.t.* To support an experimental hypothesis by rejecting a complementary null hypothesis at a "stipulated *p*" of, *e.g.,* ≤0.05, as in: The analysis indicates a significant difference between groups.
♦ **indicator** *n.* An organism, species, or community that is characteristic of a particular habitat or indicative of a particular set of environmental conditions (Lincoln et al. 1985).
♦ **indicator response** See response: indicator response.
♦ **indicator species** See ²species: indicator species.
♦ **indifferent species** See ²species: companion species, companion species.
♦ **indigenous** *adj.* Referring to a group or species that is native, or aboriginal, to a particular geographic area; contrasted with a group or species that is considered to have immigrated from an outside area (as autochthonous in Tillyard 1926, etc. in Frank and McCoy 1990, 2).
syn. autochthonous, endemic (erroneously), native (Frank and McCoy 1990, 4)
cf. adventive, autochthonous, endemic, immigrant, introduced, precinctive
Comment: Frank and McCoy (1990, 4) prefer indigenous to "autochthonous," or "native," because "indigenous" is used more widely in biology than "autochthonous," and they prefer "indigenous" to "native" because "native" has subsidiary English meanings.
[Latin *indigenus*, native]

♦ **indigenous species** See ²species: native species.
♦ **indirect competition** See competition: indirect competition.
♦ **indirect fitness** See fitness: indirect fitness.
♦ **indirect hyperparasitism** See parasitism: hyperparasitism: indirect hyperparasitism.
♦ **indirect life cycle** See life cycle: indirect life cycle.
♦ **indirect mimicry** See mimicry: indirect mimicry.
♦ **indirect role** See role: indirect role.
♦ **indirect selection** See selection: indirect selection.
♦ **indirect spermatophore transmission** *n.* In some myriapod, spider, mite, and salamander species; scorpions; Whipscorpions; Springtails; Apterygote Insects: a male's depositing a spermatophore on a substrate and inducing a female to pick it up in a way that effects fertilization (Immelmann and Beer 1989, 143).
♦ **indirect tradition** See tradition: indirect tradition.
♦ **individual** *n.* Any physically distinct organism that has genetic uniqueness, including an identical twin, triplet, etc. or a member of a clone with genetically identical individuals (Wilson 1975, 8).
Comment: Hull (1992) discusses the meaning of "individual."
bion, biont See biont.
morphont *n.* A morphological individual (*e.g.,* one of many single organisms that makes up a Portuguese Man-of-War); contrasted with biont (Haeckel in (Hull 1992, 184).
recognized individual *n.* An individual that is classified by conspecifics according to its kinship (Wilson 1987, 10).
syn. cue bearer
recognizing individual, discriminating individual *n.* An individual that responds to appropriate cues by distinguishing kin from nonkin (Wilson 1987, 12).
♦ **individual courtship flight** See flight: courtship flight: individual courtship flight.
♦ **individual distance** See distance: individual distance.
♦ **individual fitness** See fitness (def. 2).
♦ **individual mimicry** See mimicry: individual mimicry.
♦ **individual observation** See observation: individual observation.
♦ **individual odor** See odor: individual odor.
♦ **individual pheromone** See chemical-releasing stimulus: semiochemical: pheromone: individual pheromone.

h – m

♦ **individual psychology** See study of: psychology: individual psychology.

♦ **individual recognition** See recognition: individual recognition.

♦ **individual selection** See selection (table).

♦ **individual territory** See territory: individual territory.

♦ **individual variation** See variation: continuous variation.

♦ **individual variety** See variety: individual variety.

♦ **individualistic model** See [4]model: individualistic model.

♦ **individualized group** See [2]group: individualized group.

♦ **individuation** *n.* The development of a single functional unit (*e.g.,* an individual or colony) (Lincoln et al. 1985).

♦ **induced ovulator** See ovulation.

♦ **induced plant response to herbivory** See response: induced plant response to herbivory.

♦ [1]**induction** *n.*
1. One's inferring, or aiming at, a generality (principle, law) from observation of particular instances of a phenomenon (Michaelis 1963).
2. A conclusion reached by induction (Michaelis 1963).
3. One's relating clues about a phenomenon to construct a hypothesis (Moore 1984, 478).
See inductivism.
cf. deduction

♦ [2]**induction** *n.*
1. One tissue's influencing the development, differentiation, growth, or a combination of these traits of another tissue during ontogeny (Curtis 1983, 858; Immelmann and Beer 1989, 145).
syn. embryonic induction (Curtis 1983)
2. Specific genetic, environmental, or both, factors giving rise to a new trait; *e.g.,* behavior that would not occur without these factors during an animal's ontogeny, or an animal's preference that it establishes during imprinting (Immelmann and Beer 1989, 145).
3. One group of embryonic cell's influencing another adjacent group's development (Campbell 1996, G-11).
Note: Induction occurs by physical contact between cell groups, chemical signals, or both (Campbell 1996, 983).
cf. behavior: allelomimetic behavior (mood induction); facilitation

♦ **inductive method** See method: inductive method.

♦ **inductive statistics** See study of: statistics: inductive statistics.

♦ **inductivism** *n.* The assertion "that a scientist can arrive at objective, unbiased conclusions only by simply recording, measuring, and describing what he encounters without having any prior hypotheses or preconceived expectations" (Bacon in Mayr 1982, 28).
syn. induction, Baconian inductivism, true Baconian method
cf. induction
Comment: This scientific approach is considered a sterile one now (Bacon in Mayr 1982, 28).

♦ **industrial psychology** See study of: psychology: industrial psychology.

♦ **inert** *adj.* "Inactive, physiologically quiescent" (Lincoln et al. 1985).

♦ **inertia hypothesis** See hypothesis: hypotheses of the evolution and maintenance of sex: inertia hypothesis.

♦ **inertia, phylogenetic** See phylogenetic inertia.

♦ **infant killing, infanticide** See -cide: infanticide.

♦ **infantile behavior, infantilism, infantlike behavior** See behavior: infantile behavior.

♦ **infauna** See [2]community: benthos: endobenthos.

♦ **infavoidance** See need: infavoidance.

♦ **inferential statistics** See study of: statistics: inferential statistics.

♦ **inferior colliculus** See organ: brain: inferior colliculus.

♦ **inferior olive** See organ: brain: inferior olive.

♦ **infestation** *n.*
1. A parasite's colonization, utilization, or both, of a host (Michaelis 1963).
2. A host's being colonized, utilized, or both, by parasites (Michaelis 1963).
3. An environment's being colonized, utilized, or both, by pests (Lincoln et al. 1985).

♦ **infibulation** *n.* A person's stitching a girl's genital lips together so that only urine and menstrual blood can pass through the remaining hole (Dugger 1996, 1).
cf. circumcision
Comments: Infibulation can result in highly painful intercourse and childbirth (Dugger 1996, 9).

♦ **inflow theory** See [3]theory: inflow theory.

♦ **influent** *n.* Any organism that influences another organism or group of organisms (Lincoln et al. 1985).

♦ **information** *n.*
1. A person's acquired, or derived, knowledge; facts; data (Michaelis 1963).
Note: Some workers use "information" with regard to nonhuman animals (Michaelis 1963).

2. In cybernetics, a signal's, or message's, property by which it quantitatively conveys something unpredictable and meaningful to a receiver, usually measured in bits (Michaelis 1963; Immelmann and Beer 1989, 146).
Note: This sense is used in animal-communication studies.
3. In development and genetics, instructions, or specifications, for development of a morphological structure or behavioral mechanism (Immelmann and Beer 1989, 147).
cf. communication, message, signal
Comments: Researchers disagree regarding the definition of information (McGregor 1997, 754).

♦ **information-center hypothesis** See hypothesis: information-center hypothesis.
♦ **information theory** See study of: mathematics: cybernetics.
♦ **information transfer** *n.*
1. Technically, an increase in an animal's ability to predict what is going to happen next as a result of a given event; the magnitude of one animal's influence on another's behavior (Dawkins 1986, 106, 108).
Note: If there is no influence, no information transfer has occurred (Dawkins 1986, 106, 108).
2. Colloquially, animals' giving information to each other; in this kind of information transfer, a human observer does not increase his ability to predict what the animals will do next (Dawkins 1986, 110).
Comment: The differences in these two meanings have led to no end of misunderstanding (Dawkins and Krebs 1978, van Rhijn 1980, Hinde 1981 in Dawkins 1986, 110). "Definitions have had an enormous impact on the way people see animal communication. They have affected what we regard as 'communication' and 'signals' and whether we see communication as involving any sort of transfer of information" (Dawkins 1986, 110).

♦ **infra-** *prefix* "Below, beneath; on the lower part" (Michaelis 1963).
cf. supra-
[Latin *infra*]
♦ **infradian rhythm** See rhythm: infradian rhythm.
♦ **infradispersed distribution** See distribution: underdispersed distribution.
♦ **infragenetic mutation** See ²mutation: gene mutation.
♦ **infraneuston** See neuston: hyponeuston.
♦ **infrared-stimulated-luminescence dating** See method: dating method: infrared-stimulated-luminescence dating.

♦ **infrasound** See ¹sound: infrasound.
♦ **infraspecific** *adj.* Within a species.
♦ **infrequent worker** See caste: worker: infrequent worker.
♦ **ingesta** *n.* An organism's total intake of substances into its body (Lincoln et al. 1985).
♦ **ingestion** *n.* Taking in food; feeding.
♦ **ingestive behavior** See behavior: ingestive behavior.
♦ **ingratitude** See ²group: ingratitude.
♦ **inhabited symbiont** See symbiont: inhabited symbiont.
♦ **inhabiting symbiont** See symbiont: endosymbiont.
♦ **inherent mimicry** See mimicry: inherent mimicry.
♦ **inheritance** *n.*
1. In Humans: an attitude, concept, idea, property, or quality that one obtains from his predecessor(s); heritage (Michaelis 1963).
2. In organisms, genetic information transmitted to a descendant from its ancestor(s), parent(s), or both (Lincoln et al. 1985).
cf. law: Mendel's laws; ³theory: Mendel's theory of inheritance
3. In some vertebrate species and Humans: learned information that individuals pass down from generation to generation.
cf. inheritance: cultural inheritance, tradition

A CLASSIFICATION OF INHERITANCE

I. Genetic inheritance
  A. blending inheritance
  B. cytoplasmic inheritance
  C. delayed Mendelian inheritance
  D. hard inheritance
  E. inheritance of acquired characters
  F. maternal inheritance
  G. Mendelian inheritance
  H. multifactorial inheritance
  I. multiple-gene inheritance
  J. qualitative inheritance
II. Nongenetic inheritance (= cultural inheritance)

blending inheritance *n.*
1. Gene fusion during fertilization (Mayr 1982, 644, 779).
Notes: This concept opposes the idea that genes remain separate entities in zygotes and individual organisms (theory of particulate inheritance). Early workers proposed blending inheritance due to first-generation (F_1) hybrids' having most of their characters intermediate between their two parental species and second-generation hybrids' being much

more variable, decreasing the obvious-ness of Mendelian segregation. Nägeli was one of the few post-Darwinian biologists to accept a theory of pure blending inheritance. Darwin appears to have accepted a mixed theory comprised of blending inheritance and particulate inheritance (Mayr 1982, 644, 779).

2. Inheritance in which parental traits appear to blend in their offspring, with no apparent segregation in subsequent generations (Lincoln et al. 1985).

syn. blending concept of heredity, blood theory of heredity, paint-pot theory of heredity (Mayr 1982, 644, 779)

cf. [3]theory: Mendel's theory of particulate inheritance, theory of pure blending inheritance

chromosomal inheritance See inheritance: Mendelian inheritance.

cytoplasmic inheritance *n.* Non-Mendelian heredity that involves replication and transmission of extrachromosomal genetic information found in organelles (*e.g.,* mitochondria, chloroplasts) or in intracellular parasites (*e.g.,* viruses) (Mayr 1982, 732; King and Stansfield 1985).

cf. inheritance: maternal inheritance, Mendelian inheritance; [1]system: inheritance-of-acquired-characteristics system

delayed Mendelian inheritance *n.* A phenotypic character's being shown one generation after the gene(s) involved entered a family line (*e.g.,* coiling direction in *Limnaea* snails) (Mayr 1982, 789).

hard inheritance *n.* Inheritance in which genetic material is constant and changeable only through a sudden and radical alteration — a mutation; contrasted with soft inheritance (Mayr 1982, 584, 667).

inheritance of acquired characters *n.*

1. An imprecise term that refers to the belief that an organism's environment or its use vs. disuse of particular parts, or both, affects the genetically heritable qualities of its characters (Mayr 1982, 687).

cf. inheritance: soft inheritance

2. An organism's obtaining a character during its lifetime that it transmits to its offspring (Lenski and Mittler 1993a, 189).

Notes: A temperate virus can convert a bacterium into a lysogen, wherein the virus is integrated into the host's chromosome (Lenski and Mittler 1993a, 189). In this state, the virus synthesizes a protein that confers immunity to further infections and immunity is inherited.

cf. character: acquired character; Lamarckism; inheritance: soft inheritance; system: inheritance-of-acquired-characteristics system

Comment: Inheritance of acquired characters is an exception to strict neo-Darwinism (Lenski and Mittler 1993a, 189). Inheritance of acquired characters also usually includes "the postulate of modifiability of genetic material by general climatic and other environmental conditions (Geoffroyism) or by nutrition directly, without peripheral (phenotypic) characters necessarily serving as intermediaries" (Mayr 1982, 687).

maternal inheritance, cytoplasmic inheritance, matrilinear inheritance, maternal transmission *n.* Inheritance that is determined by cytoplasmic factors (*e.g.,* mitochondria, chloroplasts) that are carried by female gametes (Lincoln et al. 1985).

Mendelian inheritance, chromosomal inheritance *n.* Inheritance of traits through nuclear chromosomes (Lincoln et al. 1985).

cf. inheritance: cytoplasmic inheritance

multifactorial inheritance, multigenic inheritance, polyfactorial inheritance, polygenic inheritance *n.* Inheritance involving phenotypic characters controlled by the integrated action of multiple independent genes (Lincoln et al. 1985).

multiple-gene inheritance *n.* Inheritance of a trait that is controlled by more than one gene (Lincoln et al. 1985).

particulate inheritance See [3]theory: Mendel's theory of particulate inheritance.

polyfactorial inheritance, polygenic inheritance See inheritance: multifactorial inheritance.

polygenic inheritance See inheritance: multifactorial inheritance.

qualitative inheritance *n.* The inheritance of phenotypic characters that show discontinuous variation and are mainly controlled by only one, or a few, genes; contrasted with quantitative inheritance (Lincoln et al. 1985).

quantitative inheritance *n.* The inheritance of phenotypic characters that show continuous variation and are controlled by several interacting genes; contrasted with qualitative inheritance (Lincoln et al. 1985).

soft inheritance *n.* "Inheritance during which nuclear genetic material is not constant from generation to generation but may be modified by the effects of an organism's environment, by use or disuse, or other factors;" contrasted with hard inheritance (Mayr 1982, 959).

cf. hypothesis: directed-mutation hypothesis; inheritance: inheritance of acquired characters; Lamarckism

Comment: The final disproof of soft inheritance came in the 1950s when molecular geneticists demonstrated that information acquired by proteins cannot be transmitted back to nucleic acids (Mayr 1982, 552, 793).

♦ **inheritance-of-acquired-characteristics system** See ¹system: inheritance-of-acquired-characteristics system.

♦ **inheritance of acquired characters** See inheritance: inheritance of acquired characters.

♦ **inheritance, theory of** See ³theory: Mendel's theory of inheritance.

♦ **inhibition** *n.*

1. Something's (especially a physiological process) being prevented, hindered, or checked (*Oxford English Dictionary* 1972, entries from 1621).
2. A checking or restraining; especially a person's self-imposed restriction on his own behavior (Michaelis 1963).
3. Blockage of a behavior through certain endogenous, or exogenous, stimuli or a simultaneous activation of another behavior that is incompatible with this first behavior (Heymer 1977, 85; Immelmann and Beer 1989, 148).
   cf. behavior: displacement behavior
4. A temporary reduction in activity of a neuron, neuron group, or effector organ due to influences from other neurons or neuron groups (Heymer 1977, 85; Immelmann and Beer 1989, 148).

behavioral inhibition *n.* Blockage of the performance of a particular behavior due to specific external, or internal, stimuli or by simultaneous activation of another behavior tendency incompatible with the first behavior (Immelmann and Beer 1989, 148).
cf. behavior: displacement behavior

disinhibition *n.* Removal of the inhibition of a top-priority activity on a second-priority activity which allows this latter activity to be displayed (McFarland 1985, 389).
cf. hypothesis: disinhibition hypothesis

killing inhibition *n.* Inhibition of serious injury, or rival killing, in intraspecific fights by appeasement gestures, submissive postures, and other behaviors (Heymer 1977, 174).

lateral inhibition *n.* A process in an animal's visual system that enhances its perception of edges of retinal images (Gould 1982, 101–102; Immelmann and Beer 1989, 245).

neural inhibition *n.* Activity suppression of a neuron, neuron group, or an effector organ by influences from other neurons or neuron groups (Immelmann and Beer 1989, 148).

social inhibition *n.* One animal's suppressing a specific behavior of another (*e.g.,* a wolf's appeasement behavior's inhibiting aggression of a conspecific) (Immelmann and Beer 1989, 276).
Comment: A form of social inhibition is psychological castration, *q.v.*

♦ **inhibition of killing** See inhibition: killing inhibition.

♦ **inhibitory pheromone** See chemical-releasing stimulus: semiochemical: pheromone: inhibitory pheromone.

♦ **initial friction** *n.* The phenomenon that the strongest possible and complete expression of a behavior pattern is reached only after it has passed through lower stages (Lorenz 1939 in Heymer 1977, 28).
[term expropriated from the physical sciences]

♦ **injurious fight** See fight: damaging fight.

♦ **injury-feigning display** See display: injury-feigning display.

♦ **innate** *adj.*

1. In genetics, referring to differences in genetic character between two members of the same species that have been raised in the same environment (*e.g.,* a morphological or behavioral character) (Verplanck 1957).
   Notes: This is a widely used definition. Blue or brown eyes are not "innate," but the difference in eye color between two persons is "innate." "Innate" cannot be properly applied to behavior as synonymous with "unlearned" or "inborn." Where "innate behavior" has appeared in past literature, it should be read as "unlearned behavior" or "species-specific behavior" (Verplanck 1957).
2. Referring to a trait that is "inborn, hereditary, genetically inherited" (Immelmann and Beer 1989, 148–149).
3. Referring to a trait that is ontogenetically developed without any environmental, or experiential, shaping influence (*e.g.,* an unlearned behavior) (Immelmann and Beer 1989, 149).
cf. behavior: innate behavior, instinct

♦ **innate behavior** See behavior: innate behavior, instinctive behavior.

♦ **innate-behavior pattern** See pattern: fixed-action pattern.

♦ **innate-language-acquisition device** *n.* In Humans: the innate combination of the deep structure, drive to learn language, its predictable learning-time course, and linguistic phoneme categories (Chomsky in Gould 1982, 538).
cf. deep structure (of language)

♦ **innate releasing mechanism (IRM)** See mechanism: releasing mechanism: innate releasing mechanism.

h – m

♦ **innate releasing mechanism modified by experience** See mechanism: releasing mechanism: innate releasing mechanism: innate releasing mechanism modified by experience.

♦ **innate releasing schema** *n.* The central (internal) correlate of stimuli, effective in eliciting a behavior, that enables an animal to show selective responsiveness to stimuli (von Uexküll 1957 in Immelmann and Beer 1989, 150).
Comment: The innate releasing schema, now an obsolete concept, is the historical antecedent of the innate releasing mechanism (Immelmann and Beer 1989, 150).

♦ **Innendienst** *n.* In social insects: work inside a nest (Wilson 1975, 412).
cf. Außendienst
[German *innen*, inside, within; *dienst*, duty]

♦ **innovation** *n.* A significant, new adaptation (Ruse 1993, 55).
See behavior: innovative behavior.
Comments: Some biologists do not accept the concept of innovation (Ruse 1993, 55). Other biologists support ideas including the possibility that an innovation opens the way to occupying a new niche or seizing an already occupied one, innovations might be mainsprings of macroevolution, and a paradigmatic example of innovations might be endothermal homeothermy (Ruse 1993, 55).

♦ **inorganic** *adj.*
1. Referring to something that does not have the organized anatomical structure of an organism (Michaelis 1963).
2. Referring to something that is not characterized by life processes (Michaelis 1963).
3. In chemistry, of, pertaining to, or designating compounds that lack carbon, except for carbonates and cyanides (Michaelis 1963).

♦ **input load** See load: input load.

♦ **inquiline** *n.*
1. In Humans: a sojourner or lodger; an indweller (*Oxford English Dictionary* 1972, entries from 1641).
Note: This is an obsolete definition.
2. In animals: an individual that lives in the nest, or abode, of another (*e.g.,* "cuckooflies" in insect galls or certain kinds of insects in ant nests) (*Oxford English Dictionary* 1972, entries from 1879).
Note: The "cuckoo-flies" are parasitic wasps.
adj. inquilinous
cf. guest; hospite; parasitism: social parasitism: inquilinism
Comments: Inquilines may take food of their hosts; some are obligate social parasites of their hosts (Wilson 1975, 354, 587).

♦ **inquilinism** See parasitism: social parasitism: inquilinism.

♦ **insect society** See ²society: insect society.

♦ **insect-wing origination, hypotheses of** See hypothesis: hypotheses ("theories") of the origin of insect wings.

♦ **insectivore** See -vore: insectivore.

♦ **insemination** *n.* An animal's sperm introduction into another's genital tract by copulation or indirect spermatophore transmission (Lincoln et al. 1985; Immelmann and Beer 1989, 150).
syn. impregnation

traumatic insemination *n.* In bed bugs, strepsipteran insects: a male's piercing the body wall of a female with his aedeagus and injecting sperm into her haemocoel (Thornhill and Alcock 1983, 326).

♦ **insertion sequence (IS)** See transposable element: insertion sequence.

♦ **insessorial** See -orial: insessorial.

♦ **insight** *n.*
1. A person's understanding, or the faculty to understand, the "hidden nature" of something (*Oxford English Dictionary* 1972, entries from 1580).
2. An individual animal's apprehension of relationships among stimuli or events (Köhler 1925 in McFarland 1985, 347); in problem solving, an individual's sudden grasp of problem relationships leading to its solution; "aha" experience (Storz 1973, 128).
cf. behavior: innovative behavior; learning: insight learning

♦ **insight learning** See learning: insight learning.

♦ **insolation** *n.* "Exposure to solar radiation" (Lincoln et al. 1985).

♦ **inspection behavior** See behavior: inspection behavior.

♦ **instability** See ²stability: instability.

♦ **instantaneous evolution** See ²evolution: instantaneous evolution.

♦ **instantaneous-growth rate** See rate: instantaneous-growth rate.

♦ **instantaneous sampling** See sampling technique: instantaneous sampling.

♦ **instantaneous speciation** See speciation: instantaneous speciation.

♦ **instar** *n.*
1. An individual arthropod "between two successive molts embracing a portion of the somatic growing phase, eclosion from the egg being purposely equated with a molt" (Carlson 1983).
2. The stage of an arthropod between two successive molts (Michaelis 1963).
cf. stadium, stage
[Latin *instar*, form, likeness]

first instar *n*. The larval instar between eclosion from an egg and molting into a second instar.

syn. first larval instar (if pupae, adults, or both are considered to be instars)

Comment: "Second instars," "third instars," etc. occur between the first and last molts of a species.

last instar *n*. The larval instar between a molt and pupation.

syn. final instar, last larval instar (if pupae, adults, or both are considered instars)

♦ **instinct** *n*.
1. An individual animal's innate propensity that manifests itself in "acts that appear to be rational, but are performed without conscious design or intentional adaptation of means to ends; also the faculty supposed to be involved in this operation (formerly often regarded as a kind of intuitive knowledge)" (*Oxford English Dictionary*, 1971, first entry 1595).
    Note: The *Oxford English Dictionary* (1972) indicates that instincts are to be found especially in less derived animals.
2. A source of forces that govern animal behavior and are designed by God in such a way as to make behavior adaptable (Descartes in McFarland 1985, 362).
3. An individual animal's motivational force of behavior (Hutcheson 1728 in McFarland 1987, 309).
4. One of an individual animal's irrational and compelling sources of conduct that orients it toward its goals; there are many instincts, most with corresponding emotions, such as: flight and fear, repulsion and disgust, curiosity and wonder, and pugnacity and anger (McDougall 1908 in McFarland 1987, 309).
5. In animals: "a hierarchically organized nervous mechanism which is susceptible to certain priming, releasing, and directing impulses of internal as well as external origin, and which responds to these impulses by coordinated movements that contribute to the maintenance of the individual and the species" (tentative definition of Tinbergen 1951, 112, who later modified his views about the theory to which this definition relates).
6. In animals: an inborn, behavioral mechanism that manifests itself in an ordered movement sequence (= fixed-action pattern) that is activated by specific stimulation of a releasing mechanism that controls discharge of an associated drive (Immelmann and Beer 1989, 151).
    Note: When "instinct" is currently used, it tends to connote this meaning (Immelmann and Beer 1989, 151).

cf. behavior: innate behavior, instinctive behavior

Comments: Instincts are often named based on their function (*e.g.*, "aggressive instinct" or "reproductive instinct"). Lorenz (1981, 211–212) warns that such naming does not explain instincts; rather, it challenges physiological analysis. According to Immelmann and Beer (1989, 151), "instinct" "is by far the most controversial term in ethology. It was and is used and understood differently by different scientists." Some so-called instincts have been found to have learned components.

[Latin *instinctus*, instigation, impulse]

closed instinct *n*. An instinct that appears in completely functional form and is closed to modification by learning; instincts include fixed-action patterns of some authors (Alcock 1979).

cf. learning: open-instinct learning

Comment: One might consider a closed instinct to be an instinct in the strict sense.

♦ **instinct model** See [4]model: instinct model.

♦ **instinct-training interlocking** *n*. The reciprocity between an animal's genome and experience in its behavioral development resulting in the intermeshing of innate and acquired behavioral components (Immelmann and Beer 1989, 153).

[introduced by Lorenz (1937) as drive-training interlocking (Immelmann and Beer 1989, 153]

♦ **instinctive** *adj*.
1. Referring to "instinct," *q.v.*
2. Often referring to a behavior that is "genetically inherited and independent of experience in its ontogeny."

Note: Like "innate," the term "instinctive" "is often confounded by the mischief of ambiguity," with its "genetic and ontogenetic senses assumed to imply one another despite their lack of logical entailment. Both words should be handled with care…" (Immelmann and Beer 1989, 152).

syn. innate (according to some workers, but others distinguish between "innate" and "instinctive")

cf. behavior: innate behavior, instinctive behavior; innate; instinct

♦ **instinctive activity** See activity: instinctive activity.

♦ **instinctive-behavior pattern** See behavior pattern: instinctive-behavior pattern.

♦ **instinctive movement** See pattern: fixed-action pattern.

♦ **instrument** *n*.
1. A tool, or implement, especially one used for exacting work (Michaelis 1963).
2. An apparatus used for measuring or recording (Michaelis 1963).

bat detector, ultrasound detector *n.*
An instrument that translates bat sounds
(audible squeaks and squeals and ultrasounds) into audible and distinguishable
patterns (Allen 1996, 639).
Comment: A bat detector is invaluable in
surveys of bat species and learning about
their echolocation (Allen 1996, 639).

**C/N-continuous-flow-isotope-ratio
mass spectrometer (CFIRMS)** *n.* An
instrument used to obtain carbon and nitrogen concentrations and their stable isotope
ratios (Ben-David et al. 1998, 48).

evaporation pan *n.* A broad, shallow, water-filled pan of standard size used to obtain
an estimate of evaporation (Allaby 1994)
Comment: The U.S. Weather Bureau uses a
class A pan, 122 centimeters in diameter
and 25 centimeters deep (Allaby 1994).
One estimates evaporation by monitoring
the amount of water in an evaporation pan.

hygrothermograph *n.* An instrument
that records both air temperature and humidity on separate traces (Allaby 1994).
syn. thermo-hygrograph (Allaby 1994)

inclinometer *n.* An instrument used to
measure elevation of one point compared
to another (Ben-David et al. 1998, 49).

lysimeter *n.*
1. An instrument that determines the solubility of a substance (Michaelis 1963).
2. An instrument that directly estimates
   evapotranspiration from a block of veg-
   etated soil (Allaby 1994).
[Greek *lysis*, a loosening + METER]

radio transmitter *n.* An instrument that
sends radio signals to a receiver (Allen
1996, 639).
Comments: Researchers use radio transmitters for tracking animals, including bats,
bears, and Cougars, in their natural environments (Allen 1996, 639).

rain gauge *n.* A gauge used for measuring the amount of rainfall (Allaby 1994).
Comments: A rain gauge is usually made of
copper or polyester (Allaby 1994). It has a
tapering funnel of standard dimensions that
allows the rainwater to collect in an enclosed
bottle, or cylinder, for subsequent measurement. A researcher places a rain gauge
in open ground with its funnel rim up to 30
centimeters above the ground surface.
Manufacturers calibrate some rain gauges
to give direct readings of rainfall amount.
Other rain gauges require researchers to
calculate rainfall from the depth of water in
the funnel and its dimensions.

recording rain gauge *n.* A rain gauge
that measures the amount of precipitation
by determining its weight (inferred from
Adams et al. 1994, 3).

Comment: This gauge can record how
much precipitation falls and when it falls
(Adams et al. 1994, 3).
standard rain gauge *n.* A rain gauge
that collects rain and snow (inferred from
Adams et al. 1994, 3).
water-level recorder *n.* An instrument
that measures water level in a stream over
time (inferred from Adams et al. 1994, 3).

♦ **instrumental communication** See
communication: instrumental communication.

♦ **instrumental conditioning, instrumental learning** See learning: conditioning: trial-and-error conditioning.

♦ **insular** *adj.*
1. Of, like, or referring to islands (Michaelis
   1963).
2. Referring to an organism that dwells on
   an island (Michaelis 1963).
3. Referring to an organism that has a restricted, or limited, habitat or range (Lincoln et al. 1985).
cf. precinctive

♦ **insular endemic** See ²species: endemic
species: insular endemic.

♦ **insulin** See hormone: insulin.

♦ **integration, law of** See oligomerization.

♦ **intelligence** *n.*
1. A person's faculty of understanding; perceiving and comprehending meaning;
   mental quickness; active intellect (*Oxford English Dictionary* 1972, entries from
   1390; Michaelis 1963).
2. An animal's capacity to adjust its behavior
   in accordance with changing conditions
   (Romanes 1882 in McFarland 1985, 505).
3. An individual animal's associating stimuli
   (Thorndike 1911, 20–23).
4. A person's ability to adapt to new situations and to learn from experience
   (Michaelis 1963).
5. A person's inherent ability to seize the
   essential factors of a complex matter
   (Michaelis 1963).
6. An animal's learning ability (Wilson 1975,
   473).
   Note: This general definition is a commonly used by animal behaviorists.
7. In more derived primates: an individual's
   ability to show reasoning or insight learning (Wilson 1975, 381).
cf. awareness and associated terms, learning
Comments: Possibly, the majority view of
intelligence is "a collection of capacities,
including perceptiveness, imagination,
memory, mental and behavioral flexibility,
problem-solving ability, and ability to assess
situations in the light of past experience"
(Immelmann and Beer 1989, 153). Howard

Gardner (in Page 1986, H2) hypothesizes that people have seven intelligences, including: "bodily-kinesthetic intelligence," "linguistic intelligence," "logical-mathematical intelligence," "musical intelligence," "perception-of-self intelligence," and "spatial intelligence." Many philosophers and researchers adamantly deny that "intelligence" exists (Page 1986, H3). "Intelligence" has so many connotations that Jensen (1998, in Plomin 1999, C27) suggested that this word be avoided in scientific discussion.

[Old French < Latin *intelligentia* < *intelligens, -entis,* present participle of *intelligere,* to understand, perceive]

artificial intelligence   *n.* A computer's capability to solve a real-world problem (McCarthy 1952 in Tierney 1983).

emotional intelligence   *n.* A person's being socially graceful and having the ability to read his own feelings, control his own impulses and anger, calm himself down, and maintain a resolve and hope in the face of setbacks (Goleman 1995, 6).

Comments: A person with both a high IQ and emotional intelligence has a competitive edge over others without these traits (Goleman 1995, 6).

[P. Salovey and J. Mayer (1990) formulated the first systematic definition of emotional intelligence (Goleman 1995, 6).]

general-cognitive ability (g)   *n.* Operationally, the composite score created by factor analysis that represents what diverse measures of human cognitive abilities, such as information-processing speed, memory, spatial ability, and verbal ability, have in common (Plomin 1999, C25, C27).

syn. general intelligence

Comments: Tests of spatial ability moderately correlate with tests of verbal ability, and memory tests more modestly correlate with spatial and verbal tests (Plomin 1999, C25, C27). "Factor analysis (an unrotated first principal component) creates a composite that weights each test by its overall correlation with all other tests." General-cognitive ability appears to be heritable and is a highly controversial quantitative trait.

♦ **intelligence quotient (IQ, IQ., I.Q., *IQ., I.Q.*)**   *n.* In Humans: a person's "intellectual age" divided by his chronological age multiplied by 100; IQ usually does not exceed 160 (Michaelis 1963).

Note: What a particular IQ test actually measures is debatable; IQ tests that measure a person's naive and native, education-independent, culture-free intelligence might be impossible to design (Gould 1982, 530–531).

cf. test: IQ test

♦ **intense competition**   See competition: intense competition.

♦ **intensity**   *n.* Usually, a measure of the amplitude, or frequency, of an organism's self-caused movement; a measure of an organism's tendency to show a behavior; *e.g.,* the intensity of incubation based on the strength of an extraneous stimulus that is required to interrupt it (Hinde 1970, 204).

♦ **intensity of sexual selection**   *n.* The amount of sexual selection on males with sexual selection defined as the variance in number of mates per male (Wade and Arnold 1980, 446).

♦ **intent, intention**   *n.*
1. A person's specific aim or purpose; plan; proposal (*Oxford English Dictionary* 1972, entries from 1225; Michaelis 1963).
2. In African Elephants, possibly Pere David's Deer and Red Deer: an individual's aim (*e.g.,* intention to fight), which it may communicate to another individual (Poole 1989, 140).

cf. behavior: intentional behavior, intentionality

♦ **intention movement**   See behavior: intention movement.

♦ **intentional behavior**   See behavior: intentional behavior.

♦ **intentional system**   See ²system: intentional system.

♦ **intentionality**   *n.*
1. In Humans, possibly other vertebrates: one's "planning, or undertaking, action in pursuit of subjectively held aims or goals" (Immelmann and Beer 1989, 153).
   Note: Poole's (1989) data indicate that African elephants show intentionality.
2. Philosophically, "propositional attitudes: all mental states that refer to a content of some sort — that are about something" (*e.g.,* believing, wanting, hoping, knowing, and understanding) (Immelmann and Beer 1989, 154).

cf. behavior: intentional behavior; deceit; intention

♦ **inter-**   *prefix*
1. With each other; together (Michaelis 1963).
2. Mutual; mutually (Michaelis 1963).
3. Between (the units signified) (Michaelis 1963).
4. Occurring or situated between (Michaelis 1963).

[Latin *inter-* < inter, between, among]

♦ ¹**interaction**   *n.*
1. An encounter between, or among, animals in which each responds to the other's presence (Immelmann and Beer 1989, 155).
2. An interrelationship between an animal and its environment that usually involves

its environment's affecting the animal more than vice versa (Immelmann and Beer 1989, 155).

♦ ²**interaction** *n.* In statistics, the joint effects of two or more independent variables over and above that due to their independent effects on a dependent variable(s) and which causes a portion of the variance in an analysis of variance (ANOVA) (Sokal and Rohlf 1969, 344).

Comment: In a case where the independent variables *W* and *X* affect the dependent variable *Y*, there is an "interaction" if the magnitude of *Y* is affected by how *W* and *X* vary with regard to one another; *e.g.,* if both *W* and *X* increase, then *Y* decreases; if *W* increases while *X* decreases, then *Y* increases; or if *W* decreases while *X* increases, then *Y* decreases.

♦ **interbreeding** See breeding: interbreeding.

♦ **intercalated association, intercalated association neuron, intercalated commissural neuron, intercalated neuron** See cell: neuron: interneuron.

♦ **intercistronic complementation** See complementation: intercistronic complementation.

♦ **intercompensation** *n.* When the environment changes to relieve pressure of a density-dependent population's controlling factor(s) and a population increases in size until it reaches a second equilibrium level where another factor halts it (Wilson 1975, 90); for example, if the leading factor, say food shortage, is eliminated, then a second factor (*e.g.,* disease) takes over.

Comment: "This compensation follows a sequence that is peculiar to each species" (Wilson 1975, 587).

♦ **intercross** See cross: intercross.

♦ **interdemic selection** See selection: interdemic selection.

♦ **interdependent association** See symbiosis: mutualism.

♦ **interdigital gland** See gland: interdigital gland.

♦ **interest in participation with a copulating pair** *n.* A nonpaired, conspecific animal's being attracted to a courting, or copulating, pair and participating in sexual behavior with it; *e.g.,* in swans when a second female simultaneously performs courtship movements with the paired female, or in some kinds of beetles when a third (or even more) male simultaneously attempts to copulate with the same female, sometimes resulting in the males holding one another's dorsums (Petzold 1964 in Heymer 1977, 119; Barrows and Gordh 1978, 341).

cf. copulation

♦ **interference** *n.*
1. For example, in Oystercatchers (birds): one individual's reversible and more-or-less immediate decrease in food-intake rate due to an increase in density of conspecific individuals (Ens and Goss-Custard 1984, 217).

Notes: A focal individual's interactions with conspecifics, not food depletion, reduces the food-intake rate (Ens and Goss-Custard 1984, 217). "Interference" may be extended to any situation in which the cost of foraging per prey obtained is increased by the presence of the other individuals. Examples of interference mechanisms are alerting prey to the presence of predators, increased costs of foraging due to simultaneous pursuits of the same prey individuals, increased local movements as a result of contacts with conspecifics, and prey stealing (= kleptoparasitism) (Ens and Goss-Custard 1984, 217).

2. One population's limiting the access of another population to a shared resource (Lincoln et al. 1985).

cf. competition: contest competition

♦ **interference competition** See competition: contest competition.

♦ **interfertile** *adj.* Referring to organisms that are capable of interbreeding (Lincoln et al. 1985).

♦ **intergenic** See -genic: intergenic.

♦ **intergenic spacers** See external-nontranscribed spacers.

♦ **intergenerational conflict** See ³conflict: intergenerational conflict.

♦ **intergradation** *n.* A transitional, or intermediate, form of something (*e.g.,* a population) (Michaelis 1963).

n. intergrade

primary intergradation *n.* The occurrence of intermediates developed *in situ* in a zone of contact between two phenotypically distinct populations (Lincoln et al. 1985).

secondary intergradation *n.* Hybridization that occurs between two phenotypically distinct populations along a zone of secondary contact (suture zone) (Lincoln et al. 1985).

♦ **intergradation zone** See zone: intergradation zone.

♦ **intergrade** See -grade: intergrade.

♦ **intergroup dominance** See dominance (social): intergroup dominance.

♦ **intermediate DNA** See nucleic acid: deoxyribonucleic acid: intermediate DNA.

♦ **intermediate host** See ³host: intermediate host.

♦ **intermittent drift** See drift: genetic drift.

♦ **internal balance** See [3]adaptation: [1]evolutionary adaptation: coadaptation.

♦ **internal chronometer, internal clock** See clock: endogenous clock.

♦ **internal selection** See selection: internal selection.

♦ **internal stimulus, internal stimulation** See [2]stimulus: internal stimulus.

♦ **internal validity** See validity: internal validity.

♦ **International Code of Zoological Nomenclature (ICZN)** *n.* A set of rules for naming animals (Minelli 1999, 462).
Comments: Researchers first established the ICZN in 1961 (Minelli 1999, 462). It replaced the *Règles internationales de la nomenclature zoologique,* which replaced the Series of Propositions for Rendering the Nomenclature of Zoology Uniform and Permanent (*syn.* Strickland Code). Workers published the fourth edition of the ICZN in 2000.

♦ **Internet, Net** *n.* A network of computer networks, basically comprised of computers and cables through which packets of information move (Anderson 1994, 900; www. w3.org/People/Berners-Lee, 2000).
Comments: Vint Cerf and Bob Khan first defined the "Internet Protocol" (IP) in 1973 by which packets are sent from one computer to another until they reach their destinations (www.w3.org/People/Berners-Lee, 2000). The Internet began running in 1983; it delivers information from a source to its destination often in under 1 second.

♦ **interneuron, internuncial neuron** See cell: neuron: interneuron.

♦ **interobserver reliability** See reliability: interobserver reliability.

♦ **interoception** See modality: chemoreception.

♦ **interoceptive senses** See modality: interoceptive senses.

♦ **interoceptor** See [2]receptor: interoceptor.

♦ **interrenal glands** See gland: interrenal glands.

♦ **interrupted swimming** See swimming: interrupted swimming.

♦ **intersegmental glands** See gland: intersegmental glands.

♦ **intersex** See sex: intersex.

♦ **intersexual selection** See selection: epigamic selection.

♦ **interspecies** See [2]species: interspecies.

♦ **interspecies tradition** See tradition: interspecies tradition.

♦ **interspecific** See -specific: interspecific.

♦ **interspecific aggression** See aggression: interspecific aggression.

♦ **interspecific communication** See communication: interspecific communication.

♦ **interspecific competition** See competition: interspecific competition.

♦ **interspecific dominance** See dominance (ecological): interspecific dominance.

♦ **interspecific fostering** See fostering: interspecific fostering.

♦ **interspecific hybrid** See [2]species: interspecies.

♦ **interspecific rearing** See fostering.

♦ **interspecific releaser** See [2]stimulus: releaser: interspecific releaser.

♦ **interspecific selection** See selection: interspecific selection.

♦ **interspecific symbiosis** See symbiosis: interspecific symbiosis.

♦ **interspecific tertiary hyperparasitism** See parasitism: hyperparasitism: interspecific tertiary hyperparasitism.

♦ **interstitial fauna** See fauna: meiofauna.

♦ **interval scale** See scales of measurement: interval scale.

♦ **intervarietal hybrid** See hybrid: intervarietal hybrid.

♦ **intervening variable** See variable: intervening variable.

♦ **intimidation** *n.*
1. A person's intimidating someone, or making someone else afraid; one's condition of being intimidated; one's use of threats or violence to force, or restrain, another person from a particular action or to interfere with his free exercise of his political, or social, rights (*Oxford English Dictionary* 1972, entries from 1658).
2. An individual animal's effective threat against another that can involve visual signals, vocal signals, scentmarking, the mere presence of a dominant animal in a dominance hierarchy, or a combination of these things (Immelmann and Beer 1989, 157).
cf. behavior: threat.

♦ **intra-** *prefix* Within; inside of (Michaelis 1963). [Latin *intra,* within]

♦ **intracistronic complementation** See complementation: intracistronic complementation.

♦ **intrademic selection** See selection: intrademic selection.

♦ **intrageneric isolation** See isolation: ecological isolation.

♦ **intragenic** See -genic: intragenic.

♦ **intragenic mutation** See [2]mutation: gene mutation.

♦ **intragenic recombination** See gene conversion.

♦ **intralaminar nucleus** See organ: brain: intralaminar nucleus.

♦ **intraneous** *n.* Referring to individuals of a species that occur near the center of their species' distribution (Lincoln et al. 1985).
cf. extraneous

h–m

- **intraobserver reliability** See reliability: intraobserver reliability.
- **intrapopulational variant** See variety.
- **intrasexual aggression** See aggression: intrasexual aggression.
  intrasexual selection See selection: epigamic selection.
- **intrasexual territory** See territory: intrasexual territory.
- **intraspecific** See -specific: intraspecific.
- **intraspecific aggression** See aggression: intraspecific aggression.
- **intraspecific behavior** See behavior: intraspecific behavior.
- **intraspecific competition** See competition: intraspecific competition.
- **intraspecific mimicry** See mimicry: automimicry.
- **intraspecific-nest parasitism** See parasitism: nest parasitism.
- **intraspecific symbiosis** See symbiosis: intraspecific symbiosis.
- **intraspecific tertiary hyperparasitism** See parasitism: hyperparasitism: intraspecific tertiary hyperparasitism.
- **intrinsic character** See character: intrinsic character.
- **intrinsic competitive ability** n. "The ability of an individual to inflict costs on an opponent during an aggressive interaction" (Wittenberger 1981, 617).
- **intrinsic isolating mechanism** See isolating mechanism: intrinsic isolating mechanism.
- **intrinsic primary social dominance** See dominance (social): intrinsic primary social dominance.
- **intrinsic rate of increase (r)** See rate: r.
- **intrinsic selection pressure** See selection pressure: intrinsic selection pressure.
- **introduced** adj. Referring to an adventive species that arrived in a new area due to the deliberate agency of a Human(s) (e.g., a biocontrol agent, crop plant, farm animal, or ornamental plant) (Frank and McCoy 1990, 4).
  syn. exotic, immigrant (Sailer 1978 in Frank and McCoy 1990, 4, who prefer not to use these synonyms)
  cf. adventive, endemic, immigrant, indigenous, precinctive
- **introgression, introgressive hybridization** n. The incorporation of the genes of one species into the gene pool of another species as a result of successful hybridization and backcrossing [coined by Anderson 1949 in Mayr 1982, 284].
  cf. leakage
- **intromission** n.
  1. A male's inserting his copulatory organ into a female's genital orifice and genital chamber.

Note: The Oxford English Dictionary (1972) includes a reference on "organs of intromission" from 1836–1839.
  2. For example, in rats: a male's briefly inserting his penis into a female's vagina (Dewsbury 1978, 77).
  cf. coition, coitus, copulation, copulation series, ejaculation, sexual intercourse, mating [Latin intromissus, past participle of intromittere < intro-, within + mittere, to send]
- **intromittent organ** See organ: copulatory organ: aedeagus, penis.
- **intron** n. In Archaea, Eukarya: a portion of a split gene (comprised of DNA) that is included in the gene's RNA transcript (heterogeneous nuclear RNA) but is excised in the process of making mature messenger RNA from the heterogeneous nuclear RNA; contrasted with exon (Curtis 1983, 340; King and Stansfield 1985).
  cf. gene: split gene; hypothesis: "introns-early hypothesis," "introns-late hypothesis;" nucleic acid: deoxyribonucleic acid: junk DNA
  Comments: Different kinds of split genes have from 1 to about 30 introns (King and Stansfield 1985). Genes in some introns appear to function in ribosome assembly and controlling chromosomes (Nowak 1994). An intron gene (XIST) seems to stop activity of X-chromosomes. Gene regulation by introns might have enabled the evolution of more complicated genomes in eukaryotes and their evolution into multicellular organisms (Mattick 1994 in Strickberger 1996, 169).
  [from intervening sequences of DNA, Curtis (1983, 338); coined by Walter Gilbert (Travis 1993)]
- **introspection** n.
  1. A person's looking within, or into, his own mind; his examination of his own thoughts, feelings, or mental state (Oxford English Dictionary 1972, entries from 1695).
  2. A person's attending to the content of his own experience which can be an important means of understanding behavior of other persons (Immelmann and Beer 1989, 158).
  Note: Introspection as it is related from a Human to a nonhuman is generally excluded as part of an objective study of behavior of nonhumans (Immelmann and Beer 1989, 158).
  cf. study of: psychology: behaviorism
- **intrusion** See ²group: intrusion.
- **intuition** n. In Humans: the ability to know something without conscious reasoning (Vogel 1997, 1269; Bechara et al. 1997, 1293).

cf. organ: brain: ventromedial prefrontal cortex

Comment: Many cognitive psychologists think intuition may be based on memories of past emotions (Vogel 1997, 1269).

♦ **invasibility** *n.* "The relative freedom with which an immigrant species can invade an established community" (Lincoln et al. 1985).

♦ **invasion** *n.* The mass movement, or encroachment, of organisms from one area into another (Lincoln et al. 1985).
n. invader

♦ **inventive behavioral combinations**
See behavior: innovative behavior.

♦ **inventory** *n.* A list of things, with the description and quantity of each (Michaelis 1963).

All Taxon Biological Inventory (ATBI)
n. A survey of all species in a particular area (Colwell and Coddington 1994, 103; http://www. discoverlife.org, 1999).
Comment: Discover Life in America is making a comprehensive study of all organisms of the Great Smokey Mountain National Park, U.S., which has about 100,000 species (http://www.discoverlife. org, 1999).

Arthropods of La Selva Inventory (ALAS Inventory) *n.* An effort to survey all arthropod species of La Selva Biological Station, Costa Rica (Colwell and Coddington 1994, 115; Yoon 1995, C4).

♦ **inversion** *n.* A chromosome segment that has been turned through 180° with the result that the segment's gene sequence is reversed with respect to that of the rest of the chromosome (King and Stansfield 1985).
Comments: "Inversions arise when a chromosome breaks in two places and reattaches with the orientation of the middle segment reversed" (Wittenberger 1981, 617). "Inversions" are "paracentric inversions" (= "heterobrachial inversions") or "pericentric inversions" (= "homobrachial inversions") (Mayr 1982, 603, 769; King and Stansfield 1985).

♦ **inversion polymorphism** See -morphism: polymorphism: inversion polymorphism.

♦ **inversion race** See ¹race: inversion race.

♦ **invertebrate zoology** See study of: zoology: invertebrate zoology.

♦ **investigating behavior** See behavior: investigating behavior.

♦ **investment** *n.*
1. A person's purchasing something from which he expects to obtain interest or profit (especially property, stocks, or shares) in order to hold them for accrual of dividends, interest, or profits (*Oxford English Dictionary* 1972, entries from 1613 as a verb).

2. An individual organism's using its resources for a particular function that precludes their being used for other functions.

altruism investment *n.* Any investment by an organism in its relative(s) that increases the relative's chance of surviving (and hence reproductive success) at the cost of the organism's ability to invest in more of its own offspring (Dawkins 1977).

parental investment *n.* A parent's investment in "an individual offspring that increases the offspring's chance of surviving (and hence reproductive success) at the cost of the parent's ability to invest in other offspring" [coined by Trivers 1972, 139].
Comments: Parental investment includes any investment that benefits offspring (*e.g.,* metabolic investment in primary sex cells and feeding and guarding young); parental investment does not include effort expended in finding and obtaining a mate (Trivers 1972, 139). McFarland (1985, 130) suggests that this definition is useful but "suffers from the implication that past investment can influence future behavior." The metaphors of sociobiology (*e.g.,* "investment," "cost," "benefit," and "tradeoff") revive Charles Darwin's analogy between evolutionary biology and economics (Immelmann and Beer 1989, 215).

▸ **maternal investment** *n.* A mother organism's investment in her offspring.

▸ **nonshareable parental investment** *n.* A form of parental investment that parents cannot give to one offspring without reducing their ability to rear other offspring (Wittenberger 1981, 618).
cf. investment: shareable parental investment

▸ **paternal investment** *n.* A father's investment in his offspring.

▸ **shareable parental investment** *n.* "A form of parental investment that can be given to one concurrent offspring without reducing the amount of parental care that can be given to other concurrent offspring" (Wittenberger 1981, 621; Lazarus and Inglis 1986, 1791).
cf. investment: nonshareable parental investment

♦ **inviability** *n.* A measure of the number of individuals that fail to survive in one cohort, or group, relative to another (Lincoln et al. 1985).

♦ **invitation signal** See signal: invitation signal.

♦ **invitation to mate** *n.* A female animal's behavior and body postures that lead to mating (Heymer 1977, 127).

♦ **inviting mimicry** See mimicry: inviting mimicry.

h – m

♦ **ion** *n*. An electrically charged atom, molecule, or radical (Michaelis 1963).
cf. molecule
Comment: The dissolution of an electrolyte or the action of electric fields, high temperatures, various forms of radiation, or other means produces an ion by adding, or removing, an electron(s) (Michaelis 1963).
[Greek *ion*, neuter of *iōn*, present participle of *ienai*, to go]

♦ **iota link** See link: iota link.

♦ **IQ, IQ., I.Q., *IQ., I.Q.*** See intelligence quotient.

♦ **IQ test** See test: IQ test.

♦ **IRM** See mechanism: innate releasing mechanism.

♦ **"iron-pyrite aggregation"** See organized structure: "iron-pyrite aggregation."

♦ **irrelevant behavior** See behavior: displacement behavior.

♦ **irreversibility** *n*. An animal's marked persistence and stability of object preference established during its imprinting (Immelmann and Beer 1989, 158).

♦ *Comment:* Not all imprinting is irreversible (Immelmann and Beer 1989, 158).

♦ **irreversibility law or rule** See law: Dollo's law.

♦ **irritability** See sensitivity.

♦ **irruption** *n*. An irregular, abrupt increase in population size, or density, typically associated with favorable environmental changes and often resulting in a population's mass movement (Lincoln et al. 1985).

♦ **IS** See transposable element: insertion sequence.

♦ **isauxesis** See -auxesis: heterauxesis: isauxesis.

♦ **island biogeography, theory of** See [3]theory: island biogeography, theory of.

♦ **island model** See [4]model: island model.

♦ **island-model migration** See gene flow: island-model migration.

♦ **island tameness** *n*. An animal's showing no fear of Humans upon first encounter with us, at least until they have deleterious experiences with us (*e.g.,* in Elephants) (Gould 1982, 284).

♦ **iso-, is-** *combining form* "Equal; the same; identical" (Michaelis 1963).

♦ [Greek *isos*, equal]

♦ **isoalleles** See allele: isoalleles.

♦ **isochromic mimicry, isomorphous mimicry** See mimicry: androchromatic mimicry.

♦ **isocoenoses** See [2]community: isocoenoses.

♦ **isocommunity** See [2]community: isocommunity.

♦ **isodemic line** See deme: isodeme.

♦ **isogametic** See -gametic: isogametic.

♦ **isogamous** See -gamous: isogamous.

♦ **isogamy** See -gamy: isogamy.

♦ **isogene** See gene: isogene.

♦ **isogeneic** See isogenic: isogeneic.

♦ **isogenesis** See -genesis: isogenesis.

♦ **isogenic** See -genic: isogenic.

♦ **isogenous** See -genous: isogenous.

♦ **isogonous hybrid** See hybrid: isogonous hybrid.

♦ **isolate** *n*. A population, or group of populations, that is geographically, ecologically, or otherwise separated from other populations and in which interbreeding occurs (Lincoln et al. 1985).

♦ **isolating mechanism** See mechanism: isolating mechanism.

♦ **isolation** *n*. Two or more populations' being separated so that they are prevented from interbreeding by extrinsic (premating) mechanisms, including geographical and temporal barriers, and species discrimination or intrinsic (postmating) mechanisms, including gamete death, gamete incompatibility, hybrid inviability, and hybrid sterility (Lincoln et al. 1985).

ecological isolation *n*.
1. Isolation due to an ecological barrier (Lincoln et al. 1985).
2. "A premating isolation" (Lincoln et al. 1985).
syn. intragenetic isolation.

ethological isolation *n*. Isolation due to a behavioral mechanism(s).

geographic isolation, geographical isolation *n*. Isolation due to geographical landmarks, such as a body of water or mountain, that prevents movement of individuals (Mayr 1982, 565).

▸ **minigeographical isolation** *n*. Isolation of domestic animals by Humans to breed new strains (Mayr 1982, 414).

intrageneric isolation See isolation: ecological isolation.

mechanical isolation, mechanical incompatibility *n*. Isolation that results from structural incompatibility of male and female reproductive organs that prevents pollination or copulation (Lincoln et al. 1985).
cf. hypothesis: hypotheses of divergent evolution of animal genitalia

misogamy, sexual isolation See isolation: reproductive isolation.

physiological isolation See isolation: reproductive isolation.

reproductive isolation *n*. Isolation due to behavior and physiological barriers that prevent fertilization between species (Mayr 1982, 565) or incipient species (Immelmann and Beer 1989, 253).

syn. misogamy, sexual isolation (Immelmann and Beer 1989, 253), physiological isolation (Romanes in Mayr 1982, 565)

▶ **premating isolation** *n.* Reproductive isolation due to extrinsic factors that are effective before mating or fertilization, including ecological, ethological, geographical, mechanical, and temporal isolating mechanisms (Lincoln et al. 1985).

▶ **postmating isolation** *n.* Reproductive isolation due to extrinsic factors that are effective after mating or fertilization, including all postmating and prezygotic isolating mechanisms (Lincoln et al. 1985).

▶ **sexual isolation** *n.* Reproductive isolation due to an animal's not mating, or mating infrequently, with heterospecific individuals.
Comments: Sexual isolation appears to be the most important and widespread form of reproductive isolation in animals.

spatial isolation *n.* Isolation due to spatial separation of populations that leads to their reproductive isolation (Lincoln et al. 1985).

♦ **isolation concept of species** See ²species (def. 18).

♦ **isolation experiment** See experiment: deprivation experiment.

♦ **isomerogamy** See -gamy: isomerogamy.

♦ **isometric growth, isometry** See -auxesis: heterauxesis: isauxesis.

♦ **isomorphous mimicry** See mimicry: androchromatypic mimicry.

♦ **isoparthenogenesis** See parthenogenesis: isoparthenogenesis.

♦ **isophage** See -phage: isophage.

♦ **isophenogamy** See -gamy: isophenogamy.

♦ **isophenous, isophenic** *adj.*
1. Referring to organisms' sharing similar phenotypic effects (Lincoln et al. 1985).
2. Referring to something's producing similar phenotypic effects (Lincoln et al. 1985).

♦ **isostasy** *n.* "Essential permanence of continents and oceans in their present positions" (Michener 1987, 449).
cf. hypothesis: continental drift; plate tectonics
Comment: This concept is not currently accepted by geologists.

♦ **isosymbiosis** See symbiosis: isosymbiosis.

♦ **isotelic** *adj.* Referring to factors that produce, or tend to produce, the same effect (Lincoln et al. 1985).
syn. homoplastic (Lincoln et al. 1985)

♦ **isozoic** See -zoic: isozoic.

♦ **issue** See feces.

♦ *Istwert* *n.* The actual parameter value in a feedback circuit, *q.v.*
[German]

♦ **Isua Rocks** See formation: Isua Rocks.

♦ **iterative adaptation** See ²evolution: iterative evolution.

♦ **iterative homology** See homolog: iterative homology.

♦ **itero-** *combining form* Repeating (Michaelis 1963).
[Latin *iteratus*, past participle of *iterare*, to repeat < *iterum*, again]

♦ **iteroparity** See -parity: iteroparity.

♦ **iteroparous** See -parous: iteroparous.

h—m

j

♦ **J** See unit of measure: joule.

♦ **J-shaped distribution** See distribution: J-shaped distribution.

♦ **jackknife** See statistical test: jackknife.

♦ **jackknife-after-bootstrap** See statistical test: jackknife-after-bootstrap.

♦ **Jack Pine plain, Jack Pine barren** See habitat: Jack Pine plain.

♦ **Jacobson's organ** See organ: Jacobson's organ.

♦ **James-Lange theory** See [3]theory: James-Lange theory.

♦ **Janus word** See contranym.

♦ **Japan Prize** See award: Japan Prize.

♦ **jargon** *n.*
1. "The inarticulate utterance of birds, or a vocal sound resembling it; twittering, chattering" (*Oxford English Dictionary* 1972, entries from 1386).
2. "A jingle or assonance of rimes" (*Oxford English Dictionary* 1972, entries from 1570).
3. "Unintelligible or meaningless talk or writing; nonsense, gibberish. (Often a term of contempt for something the speaker does not understand)" (*Oxford English Dictionary* 1972, entries from 1340).
4. "A conventional method of writing or conversing by means of symbols otherwise meaningless; a cipher, or other system of signs having an arbitrary meaning" (*Oxford English Dictionary* 1972, entries from 1594).
5. Specialized or technical language of a profession (Gowaty 1983).

v.i. jargon

cf. language

[Old French *jargon*, warbling of birds, prattle, chatter, talk]

♦ **jealousy** *n.*
1. A person's solicitude, or anxiety, for the preservation or well-being of something; his "vigilance in guarding a possession for loss or damage" (*Oxford English Dictionary* 1972, entries from 1387).
2. A person's state of mind that arises from his suspicion, apprehension, or knowledge of rivalry (*Oxford English Dictionary* 1972, entries from 1303).
3. In Olive Baboons: an individual's stopping a friend, *q.v.*, from associating with a potential rival (Smuts 1987, 43).

Comment: We do not know what "emotional feelings" might be involved in baboon, or any other nonhuman, jealousy.

♦ **jerk dance** See dance: jerk dance.

♦ **joint nesting** See nesting: joint nesting.

♦ **joking behavior** See behavior: joking behavior.

♦ **jordanon** *n.* A microspecies, race, or other infraspecific group (Lincoln et al. 1985).

cf. [1]race, [2]species

♦ **Jordan's law** See law: Jordan's law.

♦ **jostling dance, jostling run** See dance: jostling dance.

♦ **joule** See unit of measure: joule.

♦ **joy** See facial expression: joy.

♦ **jug** See [2]group: jug.

♦ **jumping gene** See gene: mobile genetic element.

♦ **jungle** See [2]community: jungle.

♦ **junk DNA** See nucleic acid: deoxyribonucleic acid: junk DNA.

♦ ***Jurassic Park*** *n.*
1. A science-fiction novel (sci-fi thriller) about a deranged millionaire and a reckless scientist who bring ravenous dinosaurs back to life by cloning DNA from blood-sucking insects encased in amber (Crichton 1990; Morell 1992b, 1861).
2. The 1993 movie based on the book *Jurassic Park.*

♦ **just-so story** *n.* A hypothetical account of why a particular trait of an organism is adaptive (Thornhill and Alcock 1983, 12).

Comment: Stephen Jay Gould uses this term, which is derived from Rudyard Kipling, to refer to such accounts in a humorous, possibly deprecating, way; logical just-so stories can be considered to be working hypotheses (Thornhill and Alcock 1983, 12).

♦ **justification** See confabulation.

♦ **juvenile** See animal names: juvenile.

♦ **juvenile adaptation** See ³adaptation: ²evolutionary adaptation: juvenile adaptation.

♦ **juvenile behavior** See behavior: juvenile behavior.

♦ **juvenile characteristic** See character: juvenile characteristic.

♦ **juvenile coloration** See dress: juvenile dress.

♦ **juvenile dress** See dress: juvenile dress.

♦ **juvenile group** See ²group: juvenile.

♦ **juvenile hormones (JHs)** See hormone: juvenile hormones, JHs.

♦ **juvenile song** See animal sounds: song: juvenile song.

♦ **K** See carrying capacity.

♦ **K extinction** See ²extinction: K extinction.

♦ **K-r spectrum** *n.* A linear continuum of reproductive strategies, with K and r species representing the extremes (Lincoln et al. 1985).

♦ **K-selected species** See ²species: K-selected species.

♦ **K selection, K selection** See selection: *K* selection.

♦ **K strategist, K-strategist** See ²species: K-selected species.

♦ **K-strategy** See strategy: K-strategy.

♦ **K value** See carrying capacity.

♦ **kairomone** See chemical-releasing stimulus: semiochemical: kairomone.

♦ **kakogenesis** See -genesis: cacogenesis.

♦ **kakogenic** See -genic: dysgenic.

♦ **kamikaze sperm** See gamete: sperm: kamikaze sperm.

♦ **Kant's dictum** *n.* "Only so much genuine science can be found in any branch of the natural sciences as it contains mathematics" (Mayr 1982, 39).
Comment: Mayr (1982, 39) disputes this view.

♦ **kappa factor** See factor: kappa factor.

♦ **karyallogamy** See -gamy: karyallogamy.

♦ **-karyo, -kary, -caryo** *combining form* Referring to a cell nucleus, especially to changes that take place within it (*Oxford English Dictionary* 1972).
[Greek *karyon*, nut, kernel]

♦ **karyogamete** *n.* A gamete's nucleus (Lincoln et al. 1985).

♦ **karyogamy** See -gamy: karyogamy.

♦ **karyogenetic** See -genetic: karyogenetic.

♦ **karyogram** See -gram: karyogram.

♦ **karyokinesis** See kinesis: karyokinesis; mitosis.

♦ **karyological** *adj.* Referring to a cell's nucleus, especially its chromosomes (Lincoln et al. 1985).

♦ **karyology** See study of: cytology.

♦ **karyomite** See chromosome.

♦ **karyomitosis** See mitosis.

♦ **-karyon**
cf. -karyo
hemikaryon *n.* A haploid cell (Lincoln et al. 1985).
adj. hemikaryotic
synkaryon *n.* A zygote nucleus that is formed from the fusion of gametic nuclei (Lincoln et al. 1985).

♦ **-karyote**
adj. -otic
eukaryote, eucaryote *n.*
1. A cell with its chromosomes in a distinct nucleus separated from cytoplasm by a double membrane and defined cytoplasmic organelles (Curtis 1983, 84).
2. An organism that is comprised of one, or more, eukaryotic cells (Lincoln et al. 1985).
adj. eucaryotic, eukaryotic
cf. -karyote: prokaryote
[Greek *eu*, true + *karyon*, kernel in reference to a cell nucleus]
prokaryote, procaryote *n.*
1. A cell with a single chromosome and no distinct nucleus separated from cytoplasm by a double membrane (Mayr 1982, 959; Curtis 1983, 84).
2. An organism (*e.g.,* bacterium or cyanobacterium) that is comprised of a prokaryotic cell (Lincoln et al. 1985).
adj. prokaryotic, protokaryotic, protocaryotic
cf. -karyote: eukaryote
[Greek *pro*, before + *karyon*, kernel, in reference to no cell nucleus]

♦ **karyotype** See -type, type: karyotype.

♦ **karyotypic orthoselection** See selection: orthoselection.

♦ **Kasper-Hauser** *n.* An animal that is reared under conditions of severe experience deprivation that denies it the opportunity to learn things necessary for its normal behavioral development (Immelmann and Beer 1989, 163).

cf. behavior: abnormal behavior; experiment: deprivation experiment; hospitalism [after Kasper Hauser, a foundling child who suddenly appeared in Nuremberg in 1882 (Immelmann and Beer 1989, 163)]

♦ **Kasper-Hauser experiment** See experiment: deprivation experiment.

♦ **kata-** See cata-.

♦ **katabolism** See metabolism: catabolism.

♦ **katadromous** See -dromous: catadromous.

♦ **katagenesis** See ²evolution: catagenesis.

♦ **katharobic** *adj.* Referring to an organism that inhabits pure water or water of very low organic content (Lincoln et al. 1985).

♦ **keeping-up-with-the-Joneses** See learning: social-imitative learning.

♦ **kennel** See ²group: kennel.

♦ **kenophobia** See phobia (table).

♦ **keratinophile** See ¹-phile: keratinophile.

♦ **keraunophobia** See phobia (table).

♦ **key character** See character: key character.

♦ **key gene** See gene: oligogene.

♦ **key stimulus** See ²stimulus: releaser.

♦ **keystone mutualist** See symbiont: keystone mutualist.

♦ **keystone predator**, **keystone species** See ²species: keystone species.

♦ **kidnapping** *n.* For example, in White-Winged Choughs (birds): one group's "convincing" a recently fledged young to join it by displaying to it (Heinsohn 1991, 1098). See parasitism: social parasitism: slavery.

♦ **killing, killing behavior** See behavior: killing behavior.

♦ **killing frost** See frost: killing frost.

♦ **killing inhibition** See inhibition: killing inhibition.

♦ **kilocron** See unit of measure: cron: kilocron.

♦ **kin** *n.*
1. A person's relatives by blood, collectively; his family; kinfolk (Michaelis 1963).
2. An organism's genetic relatives (Michaelis 1963).
3. *adj.* Related by blood; consanguineous (Michaelis 1963).
4. Similar in one or more ways; kindred; alike (Michaelis 1963).
[Old English *cyn.* Akin to kind.]

collateral kin *n.* Kin that lie on different branches of a genealogical tree below a particular bifurcation (*e.g.*, siblings) (Brown 1987a, 298).
cf. kin: lateral kin

lateral kin *n.* Kin that lie on the same branch of a genealogical tree (*e.g.*, parents and their offspring) (Brown 1987a, 298).
cf. kin: collateral kin

♦ **kin altruism** See altruism: kin altruism.

♦ **kin recognition** See recognition: kin recognition.

♦ **kin selection, kin-selection theory** See selection: kin selection; ³theory: kin selection.

♦ **kindergarten** See ²group: kindergarten.

♦ **kindle** See ²group: kindle.

♦ **kinematic graph** See graph: kinematic graph.

♦ **kinesics** See study of: kinesics.

♦ **-kinesis** *combining form* A movement, motion (Michaelis 1963).
[Greek *kinesis*, motion]

akinesis See behavior: defensive behavior; playing dead.

cytokinesis *n.* Cytoplasm division that usually follows mitosis or meiosis (Lincoln et al. 1985).

karyokinesis *n.* Nuclear division (Lincoln et al. 1985).
See mitosis.
cf. -kinesis: cytokinesis

psychokinesis (PK) *n.* A person's direct psychological effect upon a physical object that is not explainable by muscular or glandular response (Storz 1973, 213).

telekinesis *n.* Causation of changes in the physical environment by unusual means (*e.g.*, a table's moving due to a human spiritualistic seance) (Storz 1973, 273).

♦ **kinesis** *n.* An organism's change in rate of random movement (linear or angular velocity) as a result of stimulus intensity, but not stimulus direction; kinesis is distinguished from taxis, *q.v.* (Immelmann and Beer 1989, 164).
adj. kinetic
syn. phobotaxis (not an appropriate synonym as explained by Immelmann and Beer 1989, 307)
cf. taxis
Comments: Two principle forms of "kinesis" are "klinokinesis" and "orthokinesis" (Immelmann and Beer 1989, 164). Fraenkel and Gunn (1961) describe kineses in addition to those listed below.

barokinesis *n.* Kinesis in response to barometric pressure (Lincoln et al. 1985).

chemokinesis *n.* Kinesis in response to chemical stimulus (Lincoln et al. 1985).

hygrokinesis *n.* Kinesis in response to a humidity change (Lincoln et al. 1985).

hygroklinokinesis *n.* Klinokinesis in response to humidity (Lincoln et al. 1985).

hygroorthokinesis *n.* Orthokinesis in response to humidity (Lincoln et al. 1985).

h–m

klinokinesis *n.* For example, in a planarian species: kinesis in which an individual's turning rate and magnitude vary with stimulus quality. When stimulation is adverse, for example, an individual changes its direction repeatedly and through wide angles; when stimulation is more favorable, it continues in a particular direction until stimulation becomes less favorable, when it then changes its direction again (Fraenkel and Gunn 1961, 18, 45; Immelmann and Beer 1989, 164).

syn. avoiding reaction, trial-and-error behavior (Immelmann and Beer 1989, 164)

Comments: In re-examining "kinesis," Benhamou and Bovet (1989, 376) replace the spatio-temporal concept of rate of change of direction in "klinotaxis" with a purely spatial concept of sinuosity that expresses the amount of turning associated with a given path length. This redefinition is meant to separate the effects of "orthokinesis" (a purely temporal measure of speed of travel along a track) and those of "klinokinesis" (redefined as a purely spatial concept) and, therefore, lead to better understanding and use of these concepts (Budenberg 1991, 156). However, Budenberg's analyses of insect tracks indicates that Fraenkel and Gunn's (1940) definition of "klinokinesis" should not be replaced with the spatio-temporal concept.

orthokinesis *n.* For example, in woodlice: kinesis in which an individual's locomotion rate varies with stimulus quality. When stimulation is adverse, it moves ahead quickly; when stimulation is more favorable, it slows down; and when stimulation is favorable, it stops its forward movement (Gunn et al. 1937 in Fraenkel and Gunn 1961, 15; Fraenkel and Gunn 1961, 18; Immelmann and Beer 1989, 164).

cf. kinesis: klinokinesis

photokinesis *n.* In a bacterium species: kinesis in response to light (Engelmann 1882, 1883 in Fraenkel and Gunn 1961, 17).

statokinesis *n.* Movement associated with an organism's maintaining its equilibrium (Lincoln et al. 1985).

Comment: Telekinesis is probably not a real phenomenon (Storz 1973, 273).

♦ **kinesophobia** See phobia (table).

♦ **kinesthesia, kinesthetic sense** See modality: kinesthesia.

♦ **kinesthetic learning** See learning: kinesthetic learning.

♦ **kinetics** See study of: physics: kinetics.

♦ **kinetophile** See ²species: kinetophile.

♦ **king** See caste: king.

♦ **kingram** See -gram: kingram.

♦ **kinopsis** *n.* For example, in large-eyed ant species: a form of communication in which visual perception of another individual in motion excites and attracts conspecifics (Wilson 1975, 202, 587; Lincoln et al. 1985).

♦ **kinship** *n.*
1. A person's relation to another by descent; consanguinity (*Oxford English Dictionary* 1972, entries from 1833).
2. Two or more individual organisms' possession of a recent common ancestor; being genetically related (Wilson 1975, 587).

Comment: "Kinship" is precisely measured by the "coefficient of kinship" and "coefficient of relationship," *q.v.* (Wilson 1975, 587).

♦ **kinship component** See fitness: indirect fitness.

♦ **kinship selection** See selection: kin selection.

♦ **kip** See ²group: kip.

♦ **Kirbyan mimicry** See mimicry: Kirbyan mimicry.

♦ **kiss** *n.*
1. A person's touch, or caress, of some object with his lips (while compressing and separating them) (*Oxford English Dictionary* 1972, entries from *ca.* 1000). *Note:* A kiss can be a token of affection, a greeting, or an act of reverence (*Oxford English Dictionary* 1972). *syn.* osculation (*Oxford English Dictionary* 1972)
2. Two Prairie Dogs' touching lips with their mouths open and teeth exposed (Wilson 1975, 473).
3. One Stump-Tailed Macaque's placing its lips on another's mouth, often with some licking and smelling (de Waal 1989, 161). *Note:* Kissing also occurs in Bonobos and Chimpanzees (de Waal 1989, 40, 199).

[Old English *cyssan*]

cloacal kiss *n.* In birds: a male's and female's bringing their cloacal glands together during copulation (Dewsbury 1978, 77).

cf. copulation, mating

French kiss *n.* In Bonobos, Humans: open-mouth kissing with one primate's putting its tongue into the mouth of the other (de Waal 1989, 199).

♦ **kiss feeding** See feeding: kiss feeding.

♦ **kiss-squeak** See animal sounds: kiss-squeak.

♦ **klado-** See clado-.

♦ **kladogenesis** See -genesis: cladogenesis.

♦ **kleptobiosis** See symbiosis: cleptobiosis.

♦ **kleptoparasitism** See parasitism: cleptoparasitism.

♦ **klinokinesis** See -kinesis: klinokinesis.

♦ **klinotaxis** See taxis: klinotaxis.

♦ **kneading** See ³treading.

♦ **knee-jerk reflex** See reflex: knee-jerk reflex.

♦ **knephoplankton** See plankton: knephoplankton.

♦ **Knight-Darwin law** See law: Knight-Darwin law.

♦ **knob** See ²group: knob.

♦ **knophobia** See -phobia: cynophobia.

♦ **knot** See ²group: knot.

♦ **koino-** See coeno-.

♦ **kollaplankton** See plankton: kollaplankton.

♦ **koprophagous** See -phagous: coprophagous.

♦ **kormogene association** See ²group: association: kormogene association.

♦ **kormogene society** See ²society: kormo-gene society.

♦ **Koschevnikov's gland** See gland: Koschevnikov's gland.

♦ **Krebs cycle** See biochemical pathway: Krebs cycle.

♦ **kremastoplankton** See plankton: kremastoplankton.

♦ **Krogh's law** See law: Krogh's law.

♦ **Kromismus** See cannibalism: cronism.

♦ **krummholz** See ²community: krummholz.

♦ **!Kung** *n.* The Bushmen of Southern Africa (Wilson 1975, 287).
[Bushman *!kung*, human beings]

♦ **kybernetics** See study of: mathematics: cybernetics.

♦ **kymatology** See study of: kymatology.

I

♦ **L1** See transposable element: L1.

♦ **L-litter** See litter.

♦ *La Niña* *n., pl.* ***Las Niñas*** Above-average surface cooling of water and high air pressure in the South Pacific; contrasted with El Niño (Suplee 1997b, A1).
syn. El Viejo (Suplee 1997b, A1)
Comments: Las Niñas often follow Los Niños (Suplee 1997b, A1).

♦ **lability, evolutionary lability** *n.* In evolution, "the ease and speed with which particular categories of traits evolve;" *e.g.,* territorial behavior is usually highly labile, and maternal behavior much less so (Wilson 1975, 587).

♦ **labor** See ²group: labor.

♦ **labyrinth** *n.*
1. In vertebrates: part of an animal' inner ear, consisting of semicircular canals and otolith organs (saccules and utricles) and containing mechanoreceptors that register an animal's body position: tilting, turning, acceleration, and deceleration (Immelmann and Beer 1989, 169).
syn. membranous labyrinth, vestibular organ (Immelmann and Beer 1989, 169)
2. A maze used in behavioral experiments (Immelmann and Beer 1989, 169).

♦ **¹lactation** *n.* The period during which milk is produced by a mammal (Michaelis 1963).

♦ **²lactation** *n.* A female mammal's feeding a young at her breast (Michaelis 1963).

♦ **³lactation** *n.* Milk formation and secretion in mammals (Cowie et al. 1951).
cf. milking, suckling
Comment: Cowie et al. (1951) clarify terminology used in lactational physiology.

milk ejection *n.* The effect of alveolar contraction reflexively elicited by a suckling stimulus in mammary glands distended with secretion (Cowie et al. 1951).
cf. nursing, suckling

PARTS OF LACTATION[a]
I. milk secretion
II. milk removal
A. passive withdrawal
B. milk ejection

[a] Cowie et al. (1951)

Comments: Milk ejection involves the movement of some alveolar secretion into the sinus, or cistern, thereby producing a rise in intramammary pressure (called "let-down" or "draught"). Milk ejection may, or may not, involve milk passage to the exterior of a mammary gland (Cowie et al. 1951, who propose that "let-down," "hold-up," and "draught" no longer be used).

milk secretion *n.* Synthesis of milk by alveolar-epithelium cells and milk passage from these cells' cytoplasm into the alveolar lumen (Cowie et al. 1951).

passive withdrawal *n.* "The removal of sinus (or cistern) milk which at low intramammary pressure can be brought about by the action of suckling or milking or by cannulation without the intervention of the special contractile tissue of the mammary glands" (Cowie et al. 1951).

♦ **lactogenic hormone** See hormone: prolactin.

♦ **ladder of nature** See *scala naturae.*

♦ **lag-growth phase** See phase: lag growth phase.

♦ **lag load** See ²evolutionary lag.

♦ **lagena** *n.* In frogs and birds: an organ in an animal's inner ear that seems to be involved with balance rather than hearing (Smith and Takasaka 1971; D.B. Quine, personal communication).

♦ **Lagerstätten effect** See effect: Lagerstätten effect.

♦ **lalophobia** See phobia (table).

♦ **Lamarck's first law** See law: Lamarck's first law.

♦ **Lamarck's second law** See law: Lamarck's second law.

♦ **Lamarckian** *n.* A person who adheres to Lamarckism, *q.v.* (*Oxford English Dictionary* 1972, entries from 1846 as an adjective).

adj. Lamarckian

neo-Lamarckian *n.* One of a group of biologists who were most influential in the 1880s and 1890s and attacked with great hostility Weismann's revolutionary rejections of soft inheritance (Mayr 1982, 540, 701, 882).

Comment: Orthodox Darwinians of the time continued to accept Darwin's occasional reliance on the effects of use and disuse, *q.v.* (Mayr 1982, 540, 701, 882). [coined by Packard 1884 in Mayr 1982]

♦ **Lamarckism** *n.*

1. Lamarck's views of evolution which include the idea that an individual's efforts to satisfy needs play an important role in modifying the individual, an organ is strengthened by use and weakened by disuse (the principle of use and disuse, Lamarck's first law, *q.v.*), and acquired characters are inheritable (Lamarck's second law, *q.v.*) (Mayr 1982, 354).

2. The evolutionary mechanism of inheritance of acquired characteristics popularized through Lamarck's influence (Bowler 1992, 188).

Notes: Biologists have included a number of distinct biological processes within the general framework of Lamarckism; all depend upon the assumption that characters acquired by adult organisms can somehow be transmitted to their offspring and ultimately can be incorporated into their species' genetic makeup (Bowler, 1992, 188).

syn. autogenesis, Lamarckian evolution, the theory of soft inheritance (Mayr 1982, 359) *cf.* Lysenkoism; Michurinism; ¹system: inheritance-of-acquired-characteristics system

Comments: "Relatively few of Lamarck's ideas were entirely new." The "popularity of Lamarckian ideas eventually became an impediment" that "helped to delay for some 75 years after 1859 the general acceptance of the Darwinian explanatory model and of hard inheritance" (Mayr 1982, 360). Lamarckism was popular with many biologists and was a major rival of Darwinism before the rise of genetics (Bowler, 1992, 188). "The orthodox view today is that this theory is completely wrong," being incompatible with the central dogma of genetics (Dawkins 1982, 289). [after Jean Baptiste de Lamarck, 1744–1829, French biologist]

♦ **lambda link** See link: lambda link.

♦ **land-bridge hypothesis** See hypothesis: land-bridge hypothesis.

♦ **Landau index** *n.* A measurement of the strength of a dominance hierarchy (Landau 1965 in Wilson 1975, 294).

♦ **landmark** *n.* A natural object used by an animal that employs topographical orientation (*e.g.*, a solitary tree or group of trees used by Honey Bees or a corner gas station used by a person) (Heymer 1977, 109).

♦ **landmark-based mental map** See cognitive map.

♦ **landmark learning** See learning: landmark learning.

♦ **language** *n.*

1. "The whole body of words and of methods of combination of words used by a nation, people, or race; a 'tongue'" (*Oxford English Dictionary* 1972, entries from 1290).

2. A person's method of transferring his thoughts, feelings, wants, etc. in ways besides using words (*Oxford English Dictionary* 1972, entries from 1606).

3. In nonhuman animals: inarticulate sounds (*Oxford English Dictionary* 1972, entries from 1601).
   cf. jargon

4. In Humans: "words or the method of expressing them for the expression of thought" (*Oxford English Dictionary* 1972, entries from 1599).

5. In Humans: a noninstinctive method of communicating ideas, emotions, and desires by means of a system of voluntarily produced symbols (Sapir 1921, 8).

6. In Humans: a set of plurisituational comsigns restricted in the ways in which they may be combined; behavior having five characteristics: (1) plurality of signs, (2) signs that have a common meaning to a number of interpreters, (3) signs (comsigns) that are producible by members of the interpreter family and have the same meaning to both producers and interpreters, (4) signs that are relatively constant in their meaning in all situations in which they appear (plurisituational signs), and (5) signs that are interconnected in some ways and not in others, making a variety of complex sign processes (Morris 1946).

7. In Humans: behavior that has these 13 design features: a "vocal-auditory channel," "rapid fading" ("transitoriness"), "broadcast transmission" and "directional reception," "interchangeability" (a speaker of language can reproduce any linguistic message he can understand), total feedback, specialization (serves no function

h–m

except as a message), "semanticity," "arbitrariness," "discreetness," "displacement," "productivity," and "traditional transmission" (Hockett 1960).

Note: These traits exist on continua (Hockett and Altmann 1968).

8. "A communications system that permits the exchange of new, unanticipated information" (Lieberman 1973, 59).

9. In Humans: language (def. 8) that "makes use of 'encoded' speech to achieve a rapid transfer of information" (Lieberman 1973, 59).

syn. true language (Lieberman 1973, 59)

10. Nonhuman animal communication, called "animal language" or "animal speech," primarily in nonscientific writing (Immelmann and Beer 1989, 169).

syn. communication (of some authors)

cf. dance-language communication, metacommunication, semiotic

Comments: According to Gardener and Gardener (1969), "language" cannot be distinguished from "communication" in chimpanzees. Human "vocalizations that may have abstract meanings are subject to inflections and variations of syntax such as may distinguish between tenses and between moods" (Medawar and Medawar 1982). "Language," like many other kinds of behavior, is on a continuum being present from nonhuman animals to Humans (McFarland 1987, 336).

Ameslan *n.* American Sign Language that involves gestures of one's hand and arm and that has been taught to Chimpanzees (Gardner and Gardner 1971 in McFarland 1985, 492).

cf. Yerkish

[portmanteau word from *American Sign Language*]

paralanguage *n.* Signals separate from words used to supplement or to modify human language (Wilson 1975, 556).

sign language *n.* Human communication using nonverbal means, largely hand movements (*e.g.,* used by Plains Indians to communicate with tribes that speak other languages or as communication by deaf persons) (Michaelis 1963).

♦ **Lansing effects** See effect: Lansing effects.

♦ **lapidocole** See -cole: lapidocole.

♦ **larder** *n.*

1. A room, or cupboard, where a person stores food articles (Michaelis 1963).

2. "The provisions of a household" (Michaelis 1963).

[Anglo-French larder, Old French *lardier* < Medieval Latin *lardarium*, originally a storehouse for bacon < *lardum*, lard]

3. A particular location where an animal (*e.g.,* Eastern Chipmunk or Honey Bee) stores many food items (inferred from Vander Wall 1990, 3; Clarke and Kramer 1994, 299).

cf. cache; hoarding: larder hoarding

♦ **larva** See animal names: larva; caste: reproductive: nymph.

♦ **larviparity** See parity: larviparity.

♦ **larviparous** See -parous: larviparous.

♦ **larvivore** See -vore: larvivore.

♦ **Lascaux Cave, Lascaux II** See place: Lascaux Cave.

♦ **Lasker Medical Research Awards** See award: Lasker Medical Research Awards.

♦ **last instar, last larval instar** See instar: last instar.

♦ **late-season-mimic-scarcity hypothesis** See hypothesis: late-season-mimic-scarcity hypothesis.

♦ **latebricole** See -cole: latebricole.

♦ **latency** *n.*

1. The time interval between the occurrence of a stimulus and an animal's response to it (*e.g.,* its behavior or nerve impulse) (Immelmann and Beer 1989, 171).

syn. reaction time (Immelmann and Beer 1989, 171)

2. In learning, the time interval between a conditional and an unconditional stimulus or between an animal's operant response and its reinforcement (Immelmann and Beer 1989, 171).

♦ **latent allele** See allele: recessive allele.

♦ **latent homology** See homology: latent homology.

♦ **latent learning** See learning: latent learning.

♦ **latent period** See time: reaction time.

♦ **lateral inhibition** See inhibition: lateral inhibition.

♦ **lateral kin** See kin: lateral kin.

♦ **Latin-square design** See experimental design: Latin-square design.

♦ **laugh, laughing, laughter** See animal sounds: laugh.

♦ **laughter** See ²group: laughter.

♦ **laurophile** See ¹-phile: laurophile.

♦ *Lausen* See grooming: *Lausen.*

♦ **law** *n.*

1. In science and philosophy, a theoretical principle deduced from particular facts, applicable to a defined group or class of phenomena, and expressible by the statement that a particular phenomenon always occurs if certain conditions are present (*Oxford English Dictionary* 1972, entries from 1665).

syn. law of nature, natural law (*Oxford English Dictionary* 1972) dictum, concept, hypothesis, rule, universal law (according to some biological writers, Lincoln et al. 1985)

2. Any generally accepted rule, procedure, or principle that governs a specified area of conduct, field of activity, body of knowledge, and so forth (Michaelis 1963).

3. In mathematics, a rule, or formula, that governs a function or operation (Michaelis 1963).

cf. axiom, doctrine, dogma, effect, hypothesis, principle, rule, theorem, theory, truism
Comments: Physics and chemistry have laws, but not biology, because so-called "biological laws" probably always have exceptions (Mayr 1982, 37–45, 846). The term "law" is often used loosely to refer to a biological hypothesis, or concept, by many authors, as is evident from entries in this book. It seems preferable to use the terms "concept," "generality," "generalization," "principle," or "rule" instead of "law" in biology. Below, I list terms with "law" in them, regardless of their status as actual laws.
[Old English *layu* < Old Norse *lag*, something laid or fixed]

all-or-none law *n.* In physiology, once a specific stimulus threshold has been reached, any stimulus can release no more than a certain transmittable excitation (action potential) which represents its maximum-response value ("all"); below this value, no excitation transmission occurs ("none") (Bowditch 1871 in Heymer 1977, 27).

Allee effect, Allee principle, Allee law *n.* The density of a population of organisms varies according to the spatial distribution of its individuals (degree of aggregation), and both overcrowding and undercrowding may be suboptimal (Lincoln et al. 1985).
cf. density dependence

Allen's law See rule: Allen's rule.

Arber's law *n.* Any structure disappearing from a phylogenetic lineage during evolution is never regained by descendants of that line (Lincoln et al. 1985).
cf. law: Dollo's law

Baker's law *n.* In plants, autogamy is a prerequisite for establishment of long-distance migrants (Baker 1955 in Faegri and van der Pijl 1979, 136).

biogenetic law, biogenic law, biogenesis *n.* "Ontogeny is a concise and compressed recapitulation of phylogeny, conditioned by laws of heredity and adaptation" (Haeckel 1866 in Mayr 1982, 474); during its ontogeny, an organism passes through the morphological stages of its ancestors (Haeckel 1866 in Mayr 1982, 117).

syn. fundamental law of biogenesis, Haeckel's law, law of recapitulation, ontogeny recapitulates phylogeny, recapitulation, theory of recapitulation (Mayr 1982, 117, 215, 474)
cf. -genesis: biogenesis; [1]heterochrony; repetition
Comment: Haeckel based his law on the Meckel-Serrès law, *q.v.*, and Müller (1864) independently came to a similar conclusion. The "biogenetic law" is not true but was immensely popular and successful from about 1870 to 1900. "It led to a splendid flowering of comparative embryology and was responsible for many spectacular discoveries…" (Mayr 1982, 474).

Brook's law See law: Dyar's law.

capacity laws *pl. n.* Concepts of learning that Tolman related to different kinds of learning in many kinds of organisms (listed in Hinde 1970, 576–577).

Castles' law See law: Hardy-Weinberg law.

Cope's law, Cope's rule *n.* In some cnidarian, brachiopod, mollusc, echinoderm, and vertebrate species: the widespread tendency of certain groups to evolve toward species of larger physical sizes through geological time (Cope 1885a–c; Stanley 1973, 1; Wilson 1975, 347).
syn. Depéret's rule (Simpson 1953, 252, in Stanley 1973, 2)
Comment: Depéret (1907), rather than Cope, concisely formulated this rule according, to Stanley (1973, 1–2), who also explains that Cope's rule is "more fruitfully viewed as describing evolution from small size rather than toward large size." Many phyletic lines have evolved smaller species, rather than larger ones.

Depéret's rule See law: Cope's rule.

disharmony law *n.* The logarithm of a dimension of any part of an animal is proportional to the logarithm of its body length (Lincoln et al. 1985).

Dollo's law (of irreversibility), Dollo's rule *n.* Structures that have been lost in evolution can never be reacquired exactly in the same way (Mayr 1982, 609).
syn. irreversibility rule, law of irreversibility, uniquely-evolved-character concept (Mayr 1982, 609; Lincoln et al. 1985)
cf. law: Arber's law

Dyar's law, Dyar's rule, Brook's law *n.* For example, in lepidopterans, hymenopterans, coleopterans, and other kinds of insects: head width increases in a regular geometric progression through a series of larval instars, (Lincoln et al. 1985).

Endangered Species Act (ESA) *n.* A law passed by the U.S. in 1973 to protect endangered species, subspecies, and their isolated populations from becoming extinct (Kenworthy 1995a, A3).

Comments: Biologists, conservationists, developers, politicians, and others have debated the exact intent and power of this law. Some people have criticized the ESA, saying that it ignores human economic needs in favor of absolute protection of often obscure organisms (Kenworthy 1995a, A3; 1995b, A19).

Fechner's Law *n*. $S = k \log I$, where S is an animal's subjective perception of the magnitude of a stimulus, I is stimulus intensity, and k is a constant (McFarland 1985, 206).

cf. law: Weber's law
[after Gustav Fechner, 1801–1887]

Fraser-Darling law See effect: Fraser-Darling effect.

fundamental law of biogenesis See law: biogenetic law.

Galton's law of ancestral inheritance *n*. The genetic contribution of an ancestor of a sexually reproducing, diploid organism is reduced by 50% each generation; that is, an offspring obtains half of its genes from each parent, one fourth of its genes from each grandparent, one eighth of its genes from each great grandparent, etc. (Galton in Mayr 1982, 784–785).

syn. Galton's law, law of ancestral heredity (Lincoln et al. 1985)

cf. r

Comments: This rule also contains these ideas which are weaknesses: it was purely descriptive and does not really provide any causal explanation; it does not permit predictions; it proposes blending inheritance, which does not explain the production of homozygous recessive individuals from heterozygous grandparents. "Galton's law" had to be abandoned before "Mendelism" was universally accepted (Galton in Mayr 1982, 784–785).

Galton's law of filial regression *n*. Offspring of superior genotypes tend to revert to the average genotype for their species (Lincoln et al. 1985).

syn. filial regression (Lincoln et al. 1985)

Gloger's law *n*.

1. Animal colors are darker in warm climates and lighter in cold ones (Emlen 1973, 59).

   Note: This does not appear to be a true generality (Emlen 1973, 59).

2. Among homeothermic animals, populations that live in warm, humid areas are more heavily pigmented than populations in cool, dry areas; pigments are typically black in the former areas and red and yellow in the latter areas and

generally reduced in cool areas (Lincoln et al. 1985).

syn. Gloger's rule (Emlen 1973, 59; Lincoln et al. 1985); pigmentation rule (Lincoln et al. 1985)

Haeckel's law, Haeckel's laws of heredity See law: biogenetic law.

Haldane's law, Haldane's rule *n*. In most cases of interspecific hybridization, offspring of one sex are frequently absent, rare, intersexual, or sterile, and this sex is the heterogametic one (Lincoln et al. 1985).

Hardy-Weinberg law *n*. A population is not evolving (that is, its allele frequencies do not change from generation to generation) unless the allele frequency is affected by genetic drift (due to chance regarding which organisms mate, remain in a population, or both), immigration (resulting in gene flow), mutation, natural selection, or nonrandom mating (Hardy 1908 and Weinberg 1908 in Hartl 1980, 94).

syn. Castle's law, genetic equilibrium, square law (Lincoln et al. 1985), Hardy-Weinberg-equilibrium principle (Mayr 1982, 553), Hardy-Weinberg model (Hartl 1980, 94), Hardy-Weinberg theorem (Campbell 1996, 419)

Comments: This principle demonstrates that meiosis and recombination alone do not alter gene frequencies (Lincoln et al. 1985). [after the scientists G.H. Hardy and W. Weinberg, who each derived this Hardy-Weinberg equation independently and published papers on it in 1908 (Hartl 1980, 94)]

Hofacker-and-Sadler's law *n*. An older male parent crossed with a younger female produces more sons than a younger male crossed with an older female (Lincoln et al. 1985).

Hopkin's bioclimatic law *n*. The generalization that in temperate North America, life-history events tend to occur with an average delay of 4 days for each degree of latitude northward, 5° of longitude eastward, and 122 meters of altitude in spring and early summer, and corresponding earliness occurs in late summer and autumn (Lincoln et al. 1985).

irreversibility law, irreversibility rule See law: Dollo's law.

Jordan's law, Wagner's law *n*.

1. The closest relatives of a species are found immediately adjacent to it and isolated from it by a geographical barrier (Lincoln et al. 1985).

2. In fish: conspecific individuals develop more vertebrae in a cold environment than in a warm one (Lincoln et al. 1985).

Knight-Darwin law *n*. The concept that animals tend not to interbreed (Lincoln et al. 1985).

cf. outbreeding: maximum outbreeding; hypothesis: optimal-outbreeding hypothesis

Krogh's law *n*. The rate of a biological process, or event, is directly correlated with temperature (Lincoln et al. 1985).

Lamarck's first law *n*. An organ is strengthened by use and weakened by disuse; unused organs tend to disappear (Mayr 1982, 354).

syn. principle of use and disuse (Mayr 1982, 354)

cf. Lamarckism

Comment: Persons from Hippocrates (*ca.* 460–370 BC) to Lamarck (19th century) believed in this "law," and belief in it is part of present folklore (Mayr 1982, 304, 355).

Lamarck's second law *n*. Acquired characters are passed on through inheritance (Mayr 1982, 354).

cf. Lamarckism

law of acceleration *n*. The sequence of development of organs and structures in an organism's ontogeny is directly related to their importance to it (Lincoln et al. 1985).

law of conservation of energy See law: laws of thermodynamics: first law of thermodynamics.

law of constant extinction See law: law of extinction.

law of contiguity *n*. An animal may associate two or more events that occur close enough together (Storz 1973, 60).

Comment: According to many hypotheses, contiguity is the basis on which learning takes place (Storz 1973, 60).

law of development See law: von Baer's laws.

law of dominance See law: Mendel's laws: Mendel's third law.

law of effect *n*. Reinforcement has a substantial influence on learning (Thorndike 1898 in Dewsbury 1978, 22).

law of energy dispersal See law: laws of thermodynamics: second law of thermodynamics.

law of extinction *n*. The rule that the probability of extinction within a taxon remained constant through evolutionary time (Van Valen 1973, 1).

syn. law of constant extinction (Benton 1985, 52); Van Valen's law

cf. hypothesis: Red-Queen hypothesis

Comments: Exceptions to this rule are discussed by Van Valen (1973, 10). His explanation for this rule is that various species within a community maintain constant ecological relationships relative to each other, and these relationships themselves are

evolving and cause a relatively constant extinction rate within a taxon.

law of heterogeneous summation *n*. Independent and heterogeneous features of a stimulus situation are additive in their effects upon an animal's behavior (Seitz 1940 in Tinbergen 1951, 81).

[Seitz (1940) coined this as *Reizsummenregel,* which is German for law of heterogeneous summation. Tinbergen (1951, 81) translated this as "heterogeneous summation."]

law of independent assortment See law: Mendel's laws: Mendel's second law.

law of integration See oligomerization.

law of irreversibility See law: Dollo's law.

law of parsimony *n*. The logical principle that one should assume no more causes, or forces, than are necessary to account for the facts (Hamilton 1837, Morgan 1890 in the *Oxford English Dictionary* 1972).

cf. Morgan's canon, Ockham's razor

Comments: One hypothesis is more parsimonious than another if it postulates fewer processes, entities, or events of some specified type than does its rival. A more parsimonious hypothesis is normally said to be simpler. The principle of phylogenetic parsimony is "the idea that derived similarities are evidence of relationship and ancestral similarities are not" (Sober 1992b, 249).

law of recapitulation See law: biogenetic law.

law of relativity of ecological valency *n*. A species' range of tolerance is not constant throughout its range but varies with variations in environmental factors (Lincoln et al. 1985).

law of segregation See law: Mendel's laws.

law of the minimum *n*. A species' minimal requirement of any particular factor is the ultimate determinate in controlling its distribution or survival (Lincoln et al. 1985).

syn. Liebig's law (Lincoln et al. 1985)

law of the unspecialized *n*. The phenomenon that unspecialized taxa, rather than specialized ones, of all geological periods generally have been capable of adapting to new conditions in geological time (Cope 1896, 173–179 in Stanley 1972, 11).

law of tolerance *n*. An organism's distribution is limited by its tolerance to the fluctuation of a single factor (Lincoln et al. 1985).

laws of thermodynamics
▸ **first law of thermodynamics** *n*. "Energy can be changed from one form to another, but it cannot be created or destroyed" (Curtis 1983, 158).

syn. law of conservation of energy (Odum et al. 1988, 13)

Comment: The total energy of any system plus its surroundings remains constant despite change in the energy's form (Curtis 1983, 158).

▸ **second law of thermodynamics** *n.* "In all energy exchanges and conversions, if no energy leaves or enters the system under study, the potential energy of the final state will always be less than the potential energy of the initial state" (Curtis 1983, 160).

syn. law of energy dispersal (Odum et al. 1988, 13), time's arrow (Curtis 1983, 160)

Liebig's law See law: law of the minimum.

loi de balancement *n.* The concept that "the amount of material available during development is limited so that, if one structure is enlarged, another one has to be reduced in order to maintain an exact equilibrium" (Geoffroy 1818 in Mayr 1982, 463).

syn. struggle of parts (Roux in Mayr 1982, 463)

Meckel-Serrès law *n.* The concept of the Meckel-Serrès school that there is a parallelism between the stages of animal ontogeny and the stages of the *scala naturae* (Mayr 1982, 472).

cf. biogenesis

[Named after Meckel's work (1821) and Serrès' work (1860) in Mayr (1982, 472).]

Comment: "The Meckel-Serrès law was often grossly misrepresented" in post-Darwinian days.

Mendel's laws, Mendel's three laws *pl. n.* Gregor Mendel's three principles of genetics: his first, second, and third laws (Mayr 1982, 721).

cf. Mendel's theory of inheritance

▸ **Mendel's first law** *n.* In sexual organisms, two members of an allele pair, or pair of homologous chromosomes, separate during gamete formation and each gamete receives only one member of a pair (Lincoln et al. 1985).

syn. law of segregation (Lincoln et al. 1985)

▸ **Mendel's second law** *n.* Random distribution of alleles to gametes results from random orientation of chromosomes during meiosis (Lincoln et al. 1985).

syn. law of independent assortment (Lincoln et al. 1985)

▸ **Mendel's third law** *n.* A dominant allele may control a phenotypic trait rather than a recessive allele(s) (Lincoln et al. 1985).

syn. law of dominance (Lincoln et al. 1985)

Meyrick's law See law: Dollo's law.

Murphy's law *n.* Bills of birds in populations that inhabit islands are longer than those in conspecific populations on the mainland (Lincoln et al. 1985).

natural law *n.* A regularity of nature that persons may use as a binding principal that offers "a cosmic backing for the transition from *is* to *ought*" (Collins 1959 in Futuyma 1986, 8).

cf. law

pigmentation rule See law: Golger's rule.

Red-Queen law See hypothesis: Red-Queen hypothesis.

Rensch's laws, Rensch's rules *n.*

1. In cold climates, populations of mammals have larger litters and birds have larger clutches than conspecific populations in warmer climates (Lincoln et al. 1985).

2. Birds have shorter wings and mammals have shorter fur in warmer climates compared to colder ones (Lincoln et al. 1985).

syn. wing rule (Lincoln et al. 1985)

3. Races of land snails in colder climates have brown shells and those in warmer climates have white shells (Lincoln et al. 1985).

4. Thickness of land-snail shells is positively associated with strong sunlight and arid conditions (Lincoln et al. 1985).

cf. rule: ecogeographical rule

Schwerdtfeger's law of compensation and moderation *n.* A limiting factor seldom operates in isolation but is influenced by the simultaneous action of other factors (Lincoln et al. 1985).

square law See law: Hardy-Weinberg law.

Van Valen's law See law: law of extinction.

von Baer's laws *pl. n.* Four main concepts proposed by von Baer (1828) to take the place of the idea of a parallelism between ontogeny and level of organization in animals (*scala naturae*): (1) the more general characters of the large group of animals to which an embryo belongs appear earlier in its development than the more special characters; (2) the less general kinds of animals develop from the more general kinds, until the most derived appear; (3) every embryo of a given animal kind becomes separated from all other states of other definite forms instead of passing through them; and (4) fundamentally, the embryo of a higher form resembles the embryo of another animal kind, but not its adult (Mayr 1982, 473–475).

syn. law of development

cf. -genesis: biogenesis

Wallace's law *n.* "Every species has come into existence coincident both in space and time with a pre-existing closely allied species" (Wallace 1855 in Mayr 1982, 419).

Weber's law *n.* ΔI divided by I equals a constant (Weber's function), where I = the magnitude of a standard stimulus and ΔI = the increment in stimulus intensity required to produce a just noticeable difference in intensity (Weber 1934 in McFarland 1985, 206).

cf. law: Fechner's law; study of: physics: psychophysics, physiology: psychophysics

Comment: In general, "Weber's law" holds true for intermediate (normal) stimulus-intensity ranges but tends to break down at range extremes (McFarland 1985, 207).

[after E.H. Weber, German physiologist]

wing rule See law: Rensch's laws.

Yarrell's law *n.* The strength of the influence of a domestic breed, or geographical variety, in cross-breeding is positively correlated with its age (in generations) (Mayr 1982, 692).

[coined by Darwin after William Yarrell, one of his animal-breeder friends from whom he might have obtained this generalization; Darwin admitted that this phenomenon does not always occur (Mayr 1982, 692)]

Yerkes-Dodson law *n.* An animal's "optimum motivation for a learning task decreases with increased complexity of the task" (Hinde 1970, 220).

♦ *Lcyc* See gene: *Lcyc*.

♦ **LD$_{50}$, LD50** *n.* The lethal dose of a toxin that causes 50% of a test group of organisms to die (Ware 1983, 284).

Comment: LD$_{50}$ is expressed as milligrams of toxicants ingested by a mammal per kilogram of body weight (Ware 1983, 284). An LD$_{50}$ for Honey Bees is expressed as micrograms per bee (Johansen 1977, 182).

♦ **leader** *n.* An animal that practices leadership (*q.v.*).

♦ **leadership** *n.*

1. A person's dignity, position, or office of a leader, especially of a political party; also, his ability to lead (*Oxford English Dictionary* 1972, entries from 1834).

2. An individual animal's leading other members of its society when its group progresses from one place to another (Wilson 1975, 311, 588); an individual's ability to influence the movement pattern of its group as it goes from place to place (Dewsbury 1978, 96).

3. For example, in the Chimpanzee: an animal's initiating its group movement (Wilson 1975, 546).

Comment: In many primate troops, females are leaders (Dewsbury 1978, 96).

♦ **leaf litter** See litter.

♦ **leakage** *n.* The natural incorporation of genes of one species into the gene pool of another species (Mayr 1982, 284).

♦ **leaky gene** See allele: hypomorphic allele.

♦ **learned helplessness** See learning: learned helplessness.

♦ **learning** *n.*

1. A person's receiving instruction or acquiring knowledge (*Oxford English Dictionary* 1972, entries from year 897).

2. In animals: "a central nervous process causing more or less lasting changes in the innate behavioural mechanism under the influence of the outer world" (provisional definition of Tinbergen 1951, 142, made with his indication of the difficulty of defining learning).

Note: This definition is restrictive because it excludes "animals" without nerves (*e.g.*, protozoan). Most, if not all, animal species learn.

3. In animals: any systematic behavioral change whether or not it is adaptive, desirable for certain purposes, or in accordance with any other such criteria (Bush and Mosteller 1955, 4, in Dyal and Corning 1973).

4. In animals: the process that "manifests itself by adaptive changes in individual behavior as a result of experience (Thorpe 1956, 55, in Dyal and Corning 1973, 3).

Note: This is a widely used concept of learning. Mpitsos et al. (1978, 497), Lorenz (1981 in Immelmann and Beer 1989, 171), and Alcock (1975, 66–67) give similar definitions.

5. In animals: "the organization of behaviour as a result of individual experience" (Thorpe 1956, 55, in Dyal and Corning 1973, 4).

6. A relatively permanent increase in an individual animal's "response strength that is based on previous reinforcement and that can be made specific to one of two or more arbitrarily selected stimulus situations" (Miller 1967 in Dyal and Corning 1973, 4).

Note: This definition excludes most instances of habituation, sensitization, and pseudoconditioning.

7. A system's having an output (response) to a given test input stimulus pattern that is a function of the system's total previous input-output pattern of which the test input was a part (Eisenstein 1967 in Dyal and Corning 1973, 4).

8. An individual animal's behavioral change that results after its association of particular stimuli and responses, *i.e.*, associative learning (Mpitsos, 1978, 497).

9. In animals: very broadly, "a change in behavior that occurs as the result of practice" (Dewsbury 1978, 320).

10. An individual animal's relatively permanent changes other "than those due to maturation or senescence, fatigue, or sensory adaptation (themselves all hard to define)" (Hinde 1982, 175).

syn. learned behavior

Comments: Dyal and Corning (1973), Dewsbury (1978), and Hinde (1982) discuss problems with defining learning. Dyal and Corning (1973, 5) suggest that it could be useful to drop the term "learning" and replace it with "behavioral plasticity," "nervous-system plasticity," or "behavioral modification" based on experience. Separating "learning" from "physiological maturation" can be difficult. "Learning" includes an animal's habit formation due to "conditioning," information acquisition, and information storage in memory; learning mechanisms in different kinds of animals are likely to be profoundly different (Immelmann and Beer 1989, 171–172). Different workers often use different classifications of learning (Thorpe 1956 in Hinde 1970, 577–583; Thorpe 1963; Alcock 1984). Below, I give definitions of some of the named kinds of learning. Some of these named kinds are not discreet from other kinds.

alternation learning *n.* Learning in which an animal must alternate (*e.g.,* left, right, left, right, etc.) its response in successive trials in order to select the correct sample stimulus (Dewsbury 1978, 331).

associative learning See learning: conditioned learning.

autoinstruction See learning: programmed learning.

automated teaching See learning: programmed learning.

avoidance learning *n.* An animal's learning to keep away from a particular stimulus (Dewsbury 1978, 325).

▶ **active avoidance learning** *n.* For example, in some frog and toad species: avoidance learning in which an animal must make an appropriate response if it is not to be exposed to an aversive stimulus (Dewsbury 1978, 325).

▶ **passive-avoidance learning** *n.* For example, in a land-snail and a toad species: avoidance learning in which an individual stops showing a previously shown response due to an aversive stimulus (Dewsbury 1978, 327).

becoming accustomed, habit formation *n.* An animal's associating certain key stimuli with a "multitude of other stimulus configurations" and then responding to these key stimuli only when they occur with the stimulus configurations [German *angewöhnen,* to become accustomed to something that, during the process, becomes more and more indispensable (Lorenz 1981, 272)].

cf. learning: habituation

behaviorally silent learning *n.* Learning during which an animal shows no outward behavior changes that indicate learning is taking place (Dickinson 1980 in McFarland 1985, 351).

concept learning *n.* An animal's learning to identify and use the attributes that objects, or situations, have in common (Storz 1973, 55).

conditioned learning, conditional learning, conditioning *n.* For example, in Annelida, Arthropoda, Cnidaria, Platyhelminthes, Vertebrata: "The process of acquisition by an animal of the capacity to respond to a given stimulus with the reflex reaction proper to another stimulus (the reinforcement) when the two stimuli are applied concurrently for a number of times" (1948 Round Table Conference on nomenclature in animal behavior in Thorpe 1963, 80; Dyal and Corning 1973).

syn. associative learning

cf. association

Comments: According to Verplanck (1957), there are as many kinds of conditioning as there are sets of conditioning operations. Perhaps conditioning can be subdivided into "classical conditioning" and "operant conditioning;" however, drive-reduction theorists argue that both kinds of conditioning inevitably occur whichever procedure one follows and that the only distinction that can be made is in terms of which one of several responses is observed. Other workers prefer to admit that conditioned responses of both types can be observed in the same animal at the same time as a function of the same reinforcing stimulus. Conditioning also refers to the procedure used to elicit a subject's conditioned response. Scientists disagree as to whether conditioning occurs in Protozoa (Dyal and Corning 1973, 117). Dyal and Corning (1973, 15–21) describe "alpha conditioning," "associative backward conditioning," "beta conditioning," "classical conditioning," "delta conditioning," and "differential conditioning." Below, are definitions for only some of the many kinds of "conditioning" that have been described.

▶ **approximation conditioning** *n*. Operant conditioning of an unusual performance that is taught to a subject in steps; *e.g.,* to train a dog to jump up against a wall, an experimenter may first reinforce a head turn, then one or more steps toward the wall, and so on (Verplanck 1957).
Comment: The timing of the reinforcements is critical: they must follow a reinforced response within a second or so (Verplanck 1957).

▶ **avoidance conditioning** *n*.
1. Conditioning in which a subject's response prevents an experimenter from administering a negatively reinforcing stimulus which would otherwise occur shortly after the onset of a conditioned stimulus (CS) (Verplanck 1957).
Note: Some workers classify "avoidance conditioning" as operant conditioning under intermittent reinforcement (Verplanck 1957).
2. An animal's coming to avoid a stimulus that it initially found neutral, or even sought, due to its associated aversive experience (fright, pain, nausea) with the stimulus (Immelmann and Beer 1989, 25).
syn. avoidance learning (Roper and Redston 1987, 739)

active-avoidance conditioning *n*. Avoidance conditioning in which an animal learns to make a response toward an avoided stimulus (Dyal and Corning 1973, 13).

passive-avoidance conditioning *n*. Avoidance conditioning in which an animal learns to inhibit making a response toward an avoided stimulus (Dyal and Corning 1973, 13).

▶ **classical conditioning** See Pavlovian conditioning.

▶ **counterconditioning** *n*. Conditioning in which a subject learns to give a second and conflicting response to a conditioned, or discriminative, stimulus of a response that is reinforced (Verplanck 1957).

▶ **escape conditioning** *n*. Operant conditioning in which a subject's successive responses repeatedly terminate a negative reinforcement (Verplanck 1957).
cf. behavior: escape behavior

▶ **higher-order conditioning** *n*. Conditioning in which a stimulus-response correlation of a classical-conditioning situation serves as an unconditioned stimulus for a new second-order conditioned response and so on (Verplanck 1957).

▶ **instrumental conditioning, instrumental learning** See learning: conditioning: trial-and-error conditioning.

▶ **negative conditioning, negative reinforcement** *n*. An investigator's use of aversive stimulation, such as electric shock, to change an animal's behavior (Immelmann and Beer 1989, 198).
cf. reinforcement

▶ **operant conditioning** *n*.
1. Laboratory conditioning "in which an operant response (*q.v.*) is followed on a proportion of occasions by a reinforcer (*q.v.*)" (Hinde 1982, 280).
2. Laboratory conditioning in which an animal actively associates a new behavior with reduction of one of its needs or some other sought-after circumstance (Immelmann and Beer 1989, 56).
syn. learning by result, trial-and-error learning, shaping (Immelmann and Beer 1989, 57)
Comments: Most cases of training are operant conditioning. Learning similar to operant conditioning occurs in nature (Immelmann and Beer 1989, 56).

▶ **Pavlovian conditioning** *n*. For example, in many animal species, including Humans and rats: conditioning in which an organism elicits a conditional response (which was at first an unconditional response) toward a conditional stimulus (Dewsbury 1978, 322; MacQueen et al. 1989, 83).
syn. classical conditioning (Hinde 1982, 278), conditioned reflex II, conditioned response II, operant conditioning, respondent conditioning, type-S conditioning, type-1 conditioning (only some of these may be precise synonyms according to Verplanck 1957)
cf. autoshaping
Comment: In a typical classical-conditioning experiment, an investigator, or circumstance, presents an unconditional stimulus (UCS) to an organism, which then shows an unconditional response (UCR) to this stimulus. Then, the organism is presented with a second stimulus, the potential conditional stimulus (CS) that does not reliably elicit the UCR, at about the same time as the UCS for a number of trials. After these paired presentations, the organism shows a conditional response (CR) toward the CS. The CR is identical, or similar, to the UCR.
[after Ivan I. Pavlov, Russian scientist credited with discovery of this kind of conditioning]

▶ **perceptual learning** *n*. Learning in which an organism perceives stimuli as more distinct, or differentiated, through experiencing these stimuli repeatedly; contrasted with associative learning (Zentall 1993, 834).

h – m

▶ **preimaginal conditioning** *n*. An individual's retaining learning through metamorphosis from a larva into an adult (Barron and Corbet 1999, 621).

Comment: Barron and Corbet (1999) did not find preimaginal conditioning in *Drosophila melanogaster*.

▶ **pseudoconditioning** *n*.

1. In some annelid and cephalopod species; many vertebrate species: "the strengthening of a response to a previously neutral stimulus through the repeated elicitation of the response by another stimulus without paired presentation of the two stimuli" (Marquis née Kimble 1961 in Dyal and Corning 1973, 9).

2. An animal's increased probability of eliciting a particular response due to a neutral stimulus "as a consequence of repeated elicitation of its response by the stimulus it is already associated with, but without the contingent relationship between the two stimuli that obtains in true conditioning" (Immelmann and Beer 1989, 236).

syn. sensitization (Storz 1973, 211)

cf. learning: conditioning: sensitization; sensitization

Comment: Dyal and Corning (1973, 10–11) describe "pseudoconditioning type S," "pseudoconditioning type R," and "pseudoconditioning type SR."

▶ **rapid food-avoidance conditioning** *n*. For example, in some bird, rat, slug, and toad species; Humans: an individual's waiting a specific period after eating a new food and, if it becomes sick after eating it, never eating it again (Gould 1982, 264).

▶ **sensitization, alpha conditioning** *n*. The increase in the strength of a reflex that was originally evoked by a conditional stimulus through its conjunction with an unconditional stimulus and response (Dyal and Corning 1973, 9).

cf. learning: conditioning: pseudoconditioning

Comment: "Sensitization" differs from conventional conditioning in that the strengthened response is appropriate to the conditional stimulus, not the unconditional one (Hilgard and Marquis née Kimble 1961 in Dyal and Corning 1973, 9).

▶ **trial-and-error learning** *n*. An animal's "development of an association, as the result of reinforcement during appetitive behaviour, between a stimulus or a situation and an independent motor action as an item in that behavior when both stimulus and motor action precede the rein-forcement and the motor action is not the inevitable inherited response to the reinforcement" (Thorpe 1951 in Thorpe 1963, 88); *e.g.,* a toad's learning to avoid eating a toxic millipede after experiencing it (Alcock 1975, 67).

syn. conditioned reflex I, conditioned response I, instrumental conditioning, instrumental learning (Spence 1956 in Dyal and Corning 1973, 22), operant conditioning (Immelmann and Beer 1989, 57)

Comment: Dyal and Corning (1973, 22–29) describe 12 kinds of instrumental conditioning: "controlled appetitive instrumental conditioning," "controlled aversive instrumental conditioning," "free appetitive instrumental conditioning," "free aversive instrumental conditioning," "controlled appetitive differential instrumental conditioning," "controlled aversive differential instrumental conditioning," "free appetitive differential instrumental conditioning," "free aversive differential instrumental conditioning," "controlled appetitive choice learning," "controlled aversive choice learning," "free appetitive choice learning," and "free aversive choice learning."

contextual learning *n*. Learning that affects sound usages and comprehension; contrasted with vocal learning (Janik and Slater 1997, 59).

Comment: Species that show contextual learning include the Bottlenose Dolphin, Domestic Cat, Domestic Dog, Guinea Pig, Human, and Sea Lion; many bird species; other primate species and rats (many references in Janik and Slater 1997, 59).

discrimination learning *n*. An animal's learning to tell the difference between two stimuli (McFarland 1985, 214).

cf. learning: conditioning: trial-and-error learning

Comment: Dyal and Corning (1973, 22) divide "discrimination learning" into differential "conditioning" (= "successive discrimination learning") and "choice learning."

empathic learning *n*.

1. An animal's learning to avoid a particular object because it has either observed a second animal learn to avoid the object or has established a social relationship with an animal that avoids the object (Klopfer 1957, 61).

2. "Unrewarded learning that occurs when one animal watches the activities of another" (Lincoln et al. 1985).

cf. social imitation

escape learning *n*. An animal's learning to remove itself from a noxious, or unpleasant, situation (*e.g.,* a human child's learning

to run away from a yard where a vicious dog has tried to attack him) (Storz 1973, 90).
cf. learning: conditioning: avoidance conditioning

exploratory learning *n*. An animal's learning new information about its environment that it tends to use primarily later in its life (*e.g.*, a mouse's memorizing landmarks around its nest and later using this knowledge to take the fastest route back to its nest when a predator threatens it) (Alcock 1979, 79).
syn. exposure learning (Hinde 1982, 279)

facilitation *n*. The "most primitive" kind of learning in which an animal improves its behavior by performing it (Lorenz 1981, 263); *e.g.*, a young chicken's improving its pecking aim (Hess 1956 in Lorenz 1981, 264).
cf. facilitation

facultative learning *n*. Learning that is not essential to an animal's life, *e.g.*, individual recognition, learning derived from exploration and play behavior, and many of the "extraneous" things that Humans learn (Immelmann and Beer 1989, 172).
cf. learning: obligate learning
Comment: The distinction between facultative and obligate learning is not hard and fast (Immelmann and Beer 1989, 172).

habituation *n*.
1. For example, in some fungus species, single nerve fibers, Protozoa, and species in all animal phyla: an organism's learning not to respond to stimuli that are not significant in its life (Humphrey 1933 in Thorpe 1963, 60).
2. The waning of the probability, or amplitude, of an individual's unconditioned response as a function of repeated presentation of an effective stimulus (Spencer 1966 in Dyal and Corning 1973, 7).
3. In multicellular animals: an individual's showing a waning response with respect to a repeated stimulus, or a sequential series of similar stimuli (Lorenz 1981, 265).
Note: Habituation is thought to involve both peripheral and central nervous systems in animals with more complex nervous systems (Lorenz 1981, 265).
syn. adaptation (not a preferred synonym according to Verplanck 1957), stimulus adaptation (Lorenz 1981, 265)
cf. ³adaptation: ¹physiological adaptation: sensory adaptation; habituation; learning: conditioning: becoming accustomed; sensitization; stimulus generalization
Comments: For example, a *Hydra* shows "habituation" to turbulent water of a brook (Lorenz 1981, 265). Some workers consider "habituation" to be a simple form of learn-

ing; others argue that no new association is formed during "habituation" (Immelmann and Beer 1989, 127). Thompson and Spencer (1966 in Dyal and Corning 1973, 6–8) advance nine criteria as constituting an operational definition of habituation.
▶ **dishabituation** *n*. An animal's sudden recovery of its response that was just previously a habituated one (Dewsbury 1978, 322).

imitation See learning: social-imitative learning.

imprinting *n*.
1. Something's impressing, or fixing, itself in a person's mind, or memory; formerly often, a person's impressing something on his own mind; considering, or remembering, something carefully (*Oxford English Dictionary* 1972, verb entries from 1374).
2. An individual animal's learning to respond to a sign stimulus (= releaser) only at a particular time (= sensitive period) of its life and often retaining this response throughout its life (*e.g.*, a Jackdaw's imprinting on other Jackdaws as social companions, a young precocial bird's imprinting on its mother, or a mother goat's imprinting on the odor of her young) (Spalding 1872, Heinroth 1910, 1911 in McFarland 1985, 366; Lorenz 1935 and Hess 1973 in Heymer 1977, 131).
v.t. imprint
syn. Prägung (Lorenz 1935 in Tinbergen 1951, 150)
cf. gene imprinting
Comments: "There is little that can be said by way of defining imprinting that cannot be challenged or queried" (Sluckin 1967, 15). Gould (1982, 63) suggests that animals may stop responding to releasers of objects upon which they have imprinted after they start responding to a suite of characters of these objects. Through the years, workers have debated about whether imprinting is irreversible, whether it is learning, etc. McFarland (1985) suggests that imprinting is not a unique kind of learning without components of other kinds of learning. Immelmann and Beer (1989, 140) write that boundaries between imprinting and other kinds of learning are now considered to be less distinct.
▶ **classical imprinting** *n*.
1. Lorenz's original definition of imprinting which is similar to imprinting (def. 2).
2. A young precocial bird's acquisition of a more or less specific attachment to an object without its obtaining a conventional reward such as food or water in the process (Sluckin 1965, 36).

h–m

▸ **egg imprinting** *n*. In birds parasitized by some cuckoos and Cowbirds: a parent's imprinting on its own eggs and thus discriminating them from parasite eggs (Gould 1982, 267).

▸ **erroneous imprinting** *n*. An animal's imprinting on an object other than its natural object, *e.g.*, rarely in ducks and geese, when a female whose nesting is disturbed rears the young of the mother of another species that laid eggs in her nest or under experimental conditions when a young animals imprints on an unnatural object (Immelmann and Beer 1989, 90).
syn. malimprinting (Immelmann and Beer 1989, 90)

filial imprinting *n*. In some mammal species, precocial-bird species: a young's imprinting on its parent, or foster parent, which becomes less important to the imprinter as it reaches adulthood; this imprinting often involves a young's following its imprintee (McFarland 1985, 372; Immelmann and Beer 1989, 104).
syn. following-response imprinting (Immelmann and Beer 1989, 104), parent imprinting (Gould 1982, 61)
Comment: Dog puppies and some birds imprint on Humans as social partners.

▸ **food imprinting** *n*. For example, in some snake, tortoise, and turtle species; some insect species with herbivorous larvae; the Domestic Chicken and European Polecat: a young animal's rapidly forming a specific food preference, which may persist from short to long periods (Immelmann and Beer 1989, 110).
Comment: Because this behavior may not last long, using "imprinting" in its name is questionable (Immelmann and Beer 1989).

▸ **habitat imprinting, environmental imprinting** *n*. An animal's imprinting on a specific habitat type while young (Immelmann and Beer 1989, 125).

▸ **hormonal imprinting** *n*. An animal's hormone-mediated developmental change that occurs early in its development is very persistent and is relatively impervious to later reversal (*e.g.*, determination of whether an animal develops male, or female, sexual behavior) (Immelmann and Beer 1989, 140).
cf. learning: imprinting: neural imprinting

▸ **host imprinting** *n*. In some cuckoo species, the African Widowbird: a female's strong tendency to parasitize the same bird species that reared her (Immelmann and Beer 1989, 136).
Comment: The appropriateness of "imprinting" in this term is not fully understood (Immelmann and Beer 1989, 136).

▸ **human imprinting** *n*. A hand-reared animal's imprinting on a human caretaker (Immelmann and Beer 1989, 90), *e.g.*, a Mockingbird's imprinting on a person's hand (personal observation).
cf. imprinting: erroneous imprinting

▸ **locality imprinting** See imprinting: place imprinting.

▸ **neural imprinting** *n*. A developmental change in an animal's central nervous system that occurs early in its development, is very persistent, and is relatively impervious to later reversal (Immelmann and Beer 1989, 140).
cf. learning: imprinting: hormonal imprinting

▸ **object imprinting** *n*. An animal's imprinting on a biologically inadequate object (*e.g.*, a member of the wrong species or a moving box) (Lorenz 1935 in Heymer 1977, 123).
cf. response: following response

▸ **offspring imprinting** *n*. For example, in the Domestic Sheep, Goat, Herring Gull, Ring Dove: a parent's imprinting on its young (Gould 1982, 61).

▸ **olfactory imprinting** *n*. An animal's place imprinting involving chemoperception (Immelmann and Beer 1989, 222).

▸ **parent imprinting** See learning: imprinting: filial imprinting.

▸ **place imprinting, home imprinting, locality imprinting, site imprinting** *n*. In fish and bird species, possibly Humans: imprinting on a particular geographical region using phenomena including chemoperception and stellar orientation (Immelmann and Beer 1989, 222).

▸ **sexual imprinting** *n*. For example, in the Domestic Fowl, Domestic Pigeon, Zebra Finch; ducks: imprinting on a conspecific, or other object, as an object selected as a mate (Lorenz 1935 in Heymer 1977, 131; Immelmann 1972 in Gould 1982, 267; Immelmann and Beer 1989, 267).

▸ **song imprinting** *n*. An animal's imprinting on its species-specific song (Immelmann and Beer 1989, 140, 282).
cf. learning: song learning

▸ **species imprinting** *n*. An animal's imprinting on its own species (Gould 1982, 61).

insight learning *n*.
1. For example, in the Chimpanzee and Human, possibly in the Honey Bee: an individual's producing "a new adaptive response as a result of insight," its apprehension of relationships between (among) particular things (Thorpe 1963, 121; Hinde 1970, 582; Gould and Gould 1988, 221).

2. An animal's mastering a new task, or solving a new "problem," without the help of pre-programmed innate responses and without previously having attempted the task or the solution to the problem (Heymer 1977, 60); an animal's learning in which it associates stimuli that are within its "mind;" a "mental trial-and-error" learning in which solutions to a problem "come in a flash."

syn. innovation (Immelmann and Beer 1989, 151)

cf. aha experience, behavior: innovative behavior, insight

Comments: "Insight learning" derives from Gestalt psychology, especially Köhler's (1925) study of ape intelligence (Immelmann and Beer 1989, 151). Immelmann and Beer (1989, 150) report that "insight learning" is a little-used term today because it carries a connotation of consciousness; "innovative behavior" (= "innovation") is used instead. Mackintosh (in McFarland 1987, 344) questions Köhler's interpretations of his chimpanzee experiments.

instinct See instinct.

instrumental learning See learning: trial-and-error learning.

kinesthetic learning *n.* Learning in which an animal remembers to perform specific movements or place its body parts in specific positions (Immelmann and Beer 1989, 165).

cf. method: putting-through method; modality: kinesthesia

landmark learning *n.* For example, in the Honey Bee, Human; some digger-wasp species: an individual's learning the position and other attributes of a landmark (Tinbergen and Kruyt 1938, Frisch 1967, etc. in Gould 1987, 35).

latent learning *n.*

1. An animal's "association of indifferent stimuli or situations without patent reward" (Konorski in Thorpe 1963, 101).
2. An animal's being exposed to a situation with no immediate beneficial, or harmful, contingency and later being able to use this experience to obtain a reward or avoid hurt; *e.g.,* rats with prior access to a maze master it more quickly than rats new to the maze when an investigator reinforces their arrival at a goal box (Immelmann and Beer 1989, 171).

syn. incidental learning (Storz 1973, 126) [coined by Blodgett in Thorpe (1963, 101)]

learned helplessness *n.* An animal's learning that it can do nothing to improve a situation (Maier and Seligman 1976 in McFarland 1985, 351).

localized learning *n.* An animal's often species-specific tendency to learn some things much more readily than others (Tinbergen 1951, 145).

obligate learning *n.* Learning that is essential to an animal's life (*e.g.,* imprinting or learning about predator defense and obtaining food) (Immelmann and Beer 1989, 172).

cf. learning: obligate learning

Comment: The distinction between "facultative learning" and "obligate learning" is not hard and fast (Immelmann and Beer 1989, 172).

observational learning See learning: empathic learning.

oddity learning *n.* An animal's learning to select an odd symbol from a group of symbols (McFarland 1985, 509).

cf. test: oddity test

prenatal learning *n.* For example, in some mammal species; the Domestic Fowl, Guillemot (bird): an animal's learning that occurs prior to its birth or hatching (Immelmann and Beer 1989, 231).

programmed learning, programmed instruction, autoinstruction, automated teaching *n.* A person's self-instructed learning with use of textbooks, mechanical apparatuses, various kinds of audiovisual devises, etc. (Storz 1973, 21).

reversal learning *n.* Learning in which a conditioned response is caused to be wrong such that an animal's previously incorrect response becomes a correct one due to experimental design (Bittermann 1965 in Heymer 1977, 184).

cf. test: reversal test

sensitization See learning: conditioning: sensitization.

song learning *n.* For example, in the Swamp Sparrow, White-Throated Sparrow: an individual's learning its species-specific communication during a sensitive period (Alcock 1979, 69; Gould 1982, 268–274).

social-imitative learning, social imitation, imitative learning, imitation *n.* In some vertebrate species: an animal's learning in which it modifies its behavior by mimicking other animals, usually conspecifics (*e.g.,* sweet-potato washing and placer mining of wheat grains in Japanese Macaques) (Kawai, 1965a in Wilson 1975, 170), mobbing a particular kind of bird by other kinds of birds, loss of island tameness, hammering or stabbing shellfish by Oyster-Catchers (birds), or cooking traditions in Humans (Gould 283–287).

syn. aping, cultural learning (Gould 1982, 285), keeping-up-with-the-Joneses, modeling, social learning (Storz 1973, 124), monkey-see-monkey-do behavior

cf. behavior: allelomimetic behavior; cultural exchange; Imo (Animalia, Appendix 1); innovation; social facilitation

Comments: Social imitative learning is common in Humans, and besides cooking traditions it includes dressing and speaking styles. Children, especially teenagers, are very style conscious. For example, recently, in parts of the U.S., children wanted specific styles of backpacks and wore them in specific ways, using both straps over their shoulders and keeping them low so they touched their buttocks (Salmon 1997, A1).

▸ **mate-choice copying** *n.* Social imitative learning in which an animal chooses a mate that it sees a conspecific of its same sex choose (*e.g.*, in the Black Grouse) (Alatalo in Dugatkin and Godin 1998, 60) and Guppy (Dugatkin 1998, 323; Dugatkin and Godin 1998, 60).

trial-and-error learning See learning: conditioning: trial-and-error learning.

true imitation *n.* For example, in some primate species: an individual's "copying of a novel or otherwise improbable act or utterance, or some act for which there is clearly no instinctive tendency" (Thorpe 1963, 135); *e.g.*, learning a particular song dialect by a bird or learning to wash potatoes by Japanese Macaques (Wilson 1975, 51).

verbal learning See study of: verbal learning.

vocal learning *n.* In some bird and mammal species: an individual's learning a sound(s); its modifying a vocalization(s) as a result of experience with another individual's vocalization; contrasted with contextual learning (Janik and Slater 1997, 59).

Comment: Species that show vocal learning include the Harbor Seal, Human, Humpbacked Whale, Oscine Songbirds, and some hummingbird and parrot species (references in Janik and Slater 1997). These authors also report the possibility of vocal learning in some bat species and other primate, seal, and whale species and discuss hypotheses regarding functions of vocal learning.

vocal mimicry, vocal mocking *n.* For example, in some parrot species; the Bottlenose Dolphin, Harbor Seal, Human; Mockingbirds: social-imitative learning in which an individual imitates vocalizations of another species (Immelmann and Beer 1989, 139; Janik and Slater 1997, 62, 64).

syn. vocal imitation (Janik and Slater 1997, 64)

cf. mimicry: vocal mimicry

♦ **learning capacity** *n.* An animal's ranges and limits of what, when, and how it can learn, thus the kinds of experience it can register and store; the kinds of behavior it can acquire; the associations of stimuli and responses it can make; the amount of learned information that it can store at any one time; its ability to retain effects of experience, unlearn associations, and replace associations with new ones (Immelmann and Beer 1989, 172–173).

♦ **learning disposition** *n.* An organism's predetermined, presumably genetically based, learning capacity that governs its potential for information uptake and analysis (Buchholtz 1973, Rensch 1973 in Heymer 1977, 110; Immelmann and Beer 1989, 173–174).

cf. dimension of preparedness, equipotentiality assumption

♦ **learning-program schedule** *n.* In some birds: a young male's imitating only the song of his father (Nicholai 1959 in Heymer 1977, 111).

♦ **learning set** *n.* An animal's improvement over a series of discrimination problems (Harlow 1949 in McFarland 1985, 508).

♦ **learning theory** See ³theory: learning theory.

♦ **learning to learn** *n.* An animal's generalized improvement in its ability to solve a class of problems that results from working on some of the particular problems from the class (Storz 1973, 147).

♦ **lease** See ²group: lease.

♦ **leash** See ²group: leash.

♦ *Lebensspur* See fossil: trace fossil.

♦ **lecithotrophic** See -trophic: lecithotrophic.

♦ **-lectic** *combining term* Referring to choosing, gathering, or both.

combining form for nouns -lecty

cf. -phage: monophage-oligophage-polyphage continuum; -tropy

Comments: Workers use "-lectic" (syn. tropic) to signify animal species' relationships to flowers, and ²-phile, *q.v.*, to signify flower relationships to kinds of animals. Loew (1884) originally coined "-tropy" for this usage; Robertson used "-lecty" (Faegri and van der Pijl 1979, 45). Researchers define "-lectic" with regard to number of taxa a bee species uses for pollen (Michener 1974; Faegri and van der Pijl 1979; Nichols 1989; O'Toole and Raw 1991) and oil (Faegri and van der Pijl 1979). The table below is a suggestion for designating kinds of bee-flower "-lectic" associations.

[Greek *lectos*, choose, gather]

eulectic See -tropic: eutropic.

hemilectic See -tropic: hemitropic.

A Classification of "-Lectic" with Regard to the Number of Plant Species, Genera, and Families a Bee Species Uses for Pollen[a]

Kind of -lectic	Number of Species Used for Pollen	Number of Genera Used for Pollen	Number of Families Used for Pollen
monolectic	1	1	1
oligolectic	2–10	2–3	2–3
polylectic	>10	>4	>4

[a] "-Lectic" is classified by the number of species, genera, or families used for pollen.

monolectic *adj.*
1. Referring to a flower-visiting animal species (*e.g.*, a bee) that obtains one kind of resource (*e.g.*, pollen or nectar) from only one plant species or a group of closely related plant species (Faegri and van der Pijl 1979, 46–48).
syn. monotropic (Faegri and van der Pijl 1979, 46–48)
2. Referring to a bee that obtains pollen from only one plant species (Nichols 1989).
cf. -lectic: oligolectic, polylectic; -phage: monophage
Comments: Researchers determine "-lecty" with regard to kind of resource obtained, number of plant sources used, and taxonomic relatedness of the sources. For example, a *Centris* bee species is monolectic with regard to *Calceolaria* for oil, but polylectic with regard to other species for pollen and nectar (Faegri and van der Pijl 1979, 46). Bees usually use many plant species as nectar sources; therefore, researchers do not use monolectic, oligolectic, and polylectic with regard to nectar collecting. Because "monolecty" grades into "polylecty," for precision it is useful to consider a "monolectic species" to be one that uses one species of plant or one genus; an "oligolectic species" one that uses two through ten plant species, two through three genera, or both; and a "polylectic species," one that uses over ten plant species, over three genera, or both.

oligolectic *adj.*
1. Referring to a bee species that gathers pollen from only a few plant species or genera (Michener 1974, 373).
2. Referring to a flower-visiting animal species (*e.g.*, a bee) that obtains one kind of resource (*e.g.*, pollen) from a group of closely related taxa (Faegri and van der Pijl 1979, 46–48).

3. Referring to a bee species that gathers pollen from only a few closely plant species or genera (O'Toole and Raw 1991, 185).
syn. oligotropic (Faegri and van der Pijl 1979, 46–48)
cf. -lectic: monolectic, polylectic
Comment: See the comments for monolectic.

polylectic *adj.*
1. Referring to a bee species that gathers pollen from many plant species or genera (Michener 1974, 373).
2. Referring to a flower-visiting animal species (*e.g.*, a bee) that obtains one kind of resource (*e.g.*, pollen) from many plant taxa (Faegri and van der Pijl 1979, 46–48).
syn. polytropic (Faegri and van der Pijl 1979, 46–48)
cf. lectic: monolectic, oligolectic
Comment: See the comments for monolectic.

♦ **Lee-Boot effect** See effect: Lee-Boot effect.

♦ **leimocole** See -cole: leimocole.

♦ **leisure behavior** See behavior: leisure behavior.

♦ **leghemoglobin** See molecule: leghemoglobin.

♦ ¹**lek** *n.*
1. In Grouse (Capercaillies): individuals gathering or congregating (*Oxford English Dictionary* 1972, entries from Darwin 1871).
2. In some animal species: individuals' "collective performance" at a lek site (Hjorth 1970, 201).
3. In some animal species: a cluster of displaying males that females visit only for mating purposes (Wittenberger 1981, 618; Gosling and Petri 1990, 272).
participle lekking
cf. incidental; mating system: pure-dominance polygyny; ⁴model: hotspot model; regular
[Swedish *leka*, to play, game, sport]

♦ ²**lek** *n.* In many animal species: a communal display area, often used for many seasons, where males congregate and attract and court females that come to the males as part of their mate-seeking behavior (Selous 1907 in Heymer 1977, 113; Armstrong 1947; Lack 1968; Wilson 1975, 331, 468, 588; Emlen 1976; Stiles and Wolf 1979, 1; Gosling 1987, 620; Cockburn and Lee, 1988, 43; Immelmann and Beer 1989, 174).
See mating system: polygyny: pure: dominance polygyny.
past participle lekking
syn. arena (Hjorth, 1970, 201), court, male-courtship arena, territorial breeding ground (Leuthold 1966, 219)

h–m

cf. court; mating system: pure-dominance polygyny; territory: pairing territory

Comments: Leks occur in many insect species (*e.g.,* Hawaiian Picture-Wing *Drosophila,* some bee species, some empidid-fly species), Bull Grogs, some mammal species (Antechinus, Hammer-Headed Bats, Uganda Kob, Topi, Lechwe), and many bird species (*e.g.,* Sage Grouse, some hummingbirds, Ruffs, Manakins, Wild Turkeys, Bowerbirds, pheasants, some birds-of-paradise). Copulation may occur in, or out of, lek sites, depending on species.

booming field *n.* The lek site of heath hens (Wilson 1975, 169).

classical lek *n.* A lek with densely clustered male display sites (Phillips 1990, 555).

dispersed lek *n.* For example, in some carpenter bee species, some bird species: a lek with dispersed male display sites (Marshall and Alcock 1981; Phillips 1990, 555).

syn. exploded lek (Phillips 1990, 555)

true lek *n.* A lek that is removed from the nesting and feeding areas of an animal species (Mayr 1935, Armstrong 1947, Lack 1968 in Wilson 1975, 331).

syn. arena (Wilson 1975, 331)

♦ ³**lek** *v.i.* In grouse: to congregate (*Oxford English Dictionary* 1972, entries from 1884).

♦ **"lek hypothesis"** See hypothesis: firefly-flash-synchronization hypotheses: "lek hypothesis."

♦ **lek-mating system** See mating system: polygyny: pure: dominance polygyny.

♦ **lentic, lenitic** *n.* Referring to habitats with static, calm, or slow-moving water (Lincoln et al. 1985).

cf. lotic

♦ **lepe** See ²group: lepe.

♦ **lepidophage** See -phage: lepidophage.

♦ **lepidopterid** *adj.* Referring to moth- and butterfly-pollinated flowers (Lincoln et al. 1985).

cf. ²-phile: lepidopterophile

♦ **lepidopterophile** See ²-phile: lepidopterophile.

♦ **leptin** See hormone: leptin.

♦ **leptokurtic distribution** See distribution: leptokurtic distribution.

♦ **leptology** See study of: leptology.

♦ **leptophenic population** See population: leptophenic population.

♦ **Lequesne's-uniquely-derived-character concept** See uniquely derived-character concept.

♦ **lesion** See ablation.

♦ **lestobiosis** See symbiosis: lestobiosis.

♦ **let-down** See ³lactation: milk ejection (comments).

♦ **lethal character** See character: lethal character.

♦ **lethal-equivalent value** *n.* "The mean number of harmful genes carried by each member of a population multiplied by the mean probability that each gene will cause genetic death when homozygous" (Lincoln et al. 1985).

♦ **lethal trait** See character: lethal character.

♦ **lethargy** *n.*

1. A person's torpor, inertness, or apathy (*Oxford English Dictionary* 1972, entries from 1380).

2. An individual animal's showing a slow, or relatively slow, response to an external stimulus (Lincoln et al. 1985).

adj. lethargic

♦ **leucocytic pyrogen** See pyrogen: leucocytic pyrogen.

♦ **leucoplast** See organelle: amyloplast.

♦ **level** *n.*

1. A relative place, degree, condition, or stage; a plane, such as a high level of development (Michaelis 1963).

2. A position in the vertical dimension (*e.g.,* the level in the canopy of a tropical rain forest) (Michaelis 1963).

3. In statistics, a category of an independent variable (= main effect), *e.g.,* a category of nutrient enhancement such as nitrogen, phosphorous, or potassium.

cf. effect: main effect

♦ **leverage advantage** *n.* For example, in gulls: a winning individual's incurring a fitness cost in a conflict that is less than its opponent would incur if it were to win [coined by Hand (1986, 207)].

♦ **leverage cost** See cost: leverage cost.

♦ **leveret** See animal names: leveret.

♦ **Lex-Heinze mood** *n.* A female animal's resistance to mating (Heymer 1977, 111).

cf. coy

♦ **library** *n.* A collection of books, magazines, pamphlets, etc.; especially, one arranged to facilitate reference (Michaelis 1963).

Comment: In gene libraries, each book is a clone of DNA molecules (Campbell et al. 1999, 370).

[Old French *librarie* < Latin *librarium* < *liber, libri,* book]

cDNA library *n.* A genomic library of copy DNA molecules that a researcher produces by cloning cDNA that corresponds to individual mRNA species from an organism's cell(s) (Rieger et al. 1991; Campbell et al. 1999, 370).

Comments: A researcher constructs a cDNA library by synthesizing a double-stranded DNA copy from the mRNA population and integrates these cDNA molecules into a restriction-enzyme site of a cloning vector (Rieger et al. 1991). This library represents the mRNA sequences expressed in the cell type from which it is derived. A researcher uses this library for the study of tissue, or stage-specific, gene expressions.

genomic library *n.*

1. A collection of recombinant DNA molecules that represent an individual organism's genome (King and Stansfield 1985).

2. A gene bank that consists of a set of independent clones that, statistically, contains the entire genome of an organism among the recombinant DNA molecules (Rieger et al. 1991).

Note: The recombinant DNA molecules are recombinant-plasmid clones (= "books"), or phage clones (= "books"), each carrying copies of a particular segment from one particular genome (Campbell et al. 1999, 370). A researcher constructs and screens a complete genome library as a starting point for isolating specific genes (Rieger et al. 1991). The genome library of eukaryotes includes both single-copy and repetitious DNA sequences.

syn. genome library (Rieger et al. 1991), genomic bank (King and Stansfield 1985)

♦ **lichenicole** See -cole: lichenicole.

♦ **lichenology** See study of: lichenology.

♦ **lichenophile** See ¹-phile: lichenophile.

♦ **lie** *n.*

1. A person's act or instance of lying; a false statement made with intent to deceive; a criminal falsehood (*Oxford English Dictionary* 1971, entries from 900).

2. False information given by an individual animal, not necessarily given with use of a subjective consciousness (*e.g.,* a bird's giving a hawk signal that causes other birds to flee and leave all the food for the liar) (Dawkins 1976, 68).

cf. deception

♦ **Liebig's law (of the minimum)** See law: law of the minimum.

♦ **life** *n.*

1. An individual animal's condition or attribute of living or being alive; opposed to death (*Oxford English Dictionary* 1972, entries from *ca.* 1200).

cf. living

Note: Mayr (1982, 53) states that life cannot be defined because "there is no special substance, object, or force that can be identified with life. The process of living, *q.v.,* however, can be defined (Mayr 1982, 53).

2. Living organisms, collectively (Michaelis 1963).

Notes: Tiny bits of elemental carbon in 3800-million-year-old and older rocks of

western Greenland have carbon isotopic compositions that suggest the presence of life on Earth at this time (Holland 1997, 38).

[Old English *lif*]

♦ **life cycle** See cycle: life cycle.

♦ **life expectancy** *n.* The mean time period that an organism is expected to survive at any given point in its life cycle (Lincoln et al. 1985).

♦ **life history** *n.*

1. All stages through which an organism passes between its birth and death (Willson 1983, 2).

2. The significant features of an organism's life cycle with particular reference to strategies that influence its survival and reproduction (Lincoln et al. 1985).

Comment: White et al. (2000, 878) define "aggregative life history" and "circumnatal life history" with regard to Fruit Flies (Tephritidae).

♦ **life-history pattern** *n.* "The way resources are allocated to survival, growth, and reproduction at each age throughout a typical individual's life" (Wittenberger 1981, 618).

cf. principle: principle of allocation

♦ **life-history-strategy game** See game (table).

♦ **life-history theory** See ³theory: life-history theory.

♦ **life-long pair bond** See bond: life-long pair bond.

♦ **life span** *n.* The maximum, or mean, duration of the life of an individual or group of organisms (Lincoln et al. 1985).

See longevity.

syn. longevity (Lincoln et al. 1985)

♦ **life table** *n.* A table that shows the proportion of like-aged individuals surviving to each age and the reproductive rate of surviving individuals at each age (Wittenberger 1981, 618).

cf. population structure

♦ **lifetime monogamy** See mating system: monogamy: lifetime monogamy.

♦ **Liger, Ligon** See hybrid: Ligon.

♦ **light** *n.*

1. In Humans: radiant energy that emanates from the sun, very hot bodies, fire, etc. and can been seen with one's eyes (*Oxford English Dictionary* 1972, entries from *ca.* 1000).

2. In many animal species: the sensation produced by stimulation of an individual's vision organs and brain visual centers (Michaelis 1963).

syn. luminous energy (Michaelis 1963).

Comments: Humans see light from violet to red. Some bee and other arthropod species

h–m

and some hummingbird species also see ultraviolet. Organisms perceive infrared as heat. Some snake species use their pit organs to detect prey by their infrared emission in darkness.

visible light *n*. The part of solar radiation from wavelengths 380 to 780 nm (Lincoln et al. 1985).

♦ **light adaptation** See [3]adaptation: [1]physiological adaptation: light adaptation; [2]physiological adaptation (comment).

♦ **light gap** *n*. An area of increased sunlight that results from a tree's falling in forest (Hubbell et al. 1999, 554).

♦ **light stress** See stress: light stress.

♦ **light trap** See trap: light trap.

♦ **light year** See unit of measure: year: light year.

♦ **lignicole** See -cole: lignicole.

♦ **ligniperdous** *adj*. Referring to organisms that destroy wood (Lincoln et al. 1985).

♦ **lignivore** See -vore: lignivore.

♦ **lignophile** See [1]-phile: lignophile.

♦ **Ligon** See hybrid: Ligon.

♦ **ligualation** *n*. An adult bee's holding a droplet of nectar, or honey water, on the upper surface of its proboscis while extending and retracting it, thus alternately spreading the liquid as a film and letting it contract into a droplet again [coined by Barrows 1976a, 116].
syn. tongue-lashing behavior (in Honey Bees, Winston 1987, 120)
Comment: A bee might use this behavior to cool itself or its nest.

♦ **lilapsophobia** See phobia (table).

♦ *Lim*1 **gene** See gene: *Lim*1 gene.

♦ **limacology** See study of: limacology.

♦ **limicole** See -cole: limicole.

♦ **limiphage, limophage** See -vore: limivore.

♦ **limit cycle** See cycle: limit cycle.

♦ **limiting factor** See factor: limiting factor.

♦ **limivore** See -vore: limivore.

♦ **limnetic** *n*. A stickleback species that feeds in the water column, not on the bottom of a lake; contrasted with benthic (Weiner 1995, 31).

♦ **limnicole** See -cole: limnicole.

♦ **limnium** See [2]community: limnium.

♦ **limnobiology** See study of: biology: limnobiology.

♦ **limnobiont** See biont: limnobiont.

♦ **limnobios** See biosis: limnobiosis.

♦ **limnodophile** See [1]-phile: limnodophile.

♦ **limnology** See study of: limnology.

♦ **limnophile** See [1]-phile: limnophile.

♦ **limnoplankton** See plankton: limnoplankton.

♦ **limophage** See -vore: limivore.

♦ **limp posture** See posture: limp posture.

♦ **Lincoln index** See index: Lincoln index.

♦ **line** See [2]group: line.

♦ **LINE** See nucleotide element: long-interspersed nucleotide element.

♦ **lineage, line** *n*. A line of common descent (Lincoln et al. 1985).

♦ **lineage group** See [2]group: lineage group.

♦ **lineal** *adj*. Referring to a relationship among a series of causes, or arguments, such that the sequence does not come back to the starting point (Bateson 1979, 242). *ant*. recursive

♦ **linear curve** See curve: linear curve.

♦ **linear-dominance hierarchy** See hierarchy: linear-dominance hierarchy.

♦ **linear regression curve** See curve: regression curve: linear regression curve.

♦ **link** *n*. A connecting part, either in a material or immaterial sense; a thing that serves to establish or maintain a connection; "a member of a series or succession; a means of connection or communication" (*Oxford English Dictionary* 1972, entries from before 1584).

gamma link *n*. A dichotomously branching link in a food chain that involves two predators feeding on one prey species (Lincoln et al. 1985).

iota link *n*. "A direct unbranching link in food web, between a prey species and its unique predator or parasite" (Lincoln et al. 1985).

lambda link *n*. A dichotomously branching link in a food web involving one predator that feeds on two prey species (Lincoln et al. 1985).

♦ **linkage** *n*.
1. Genes' being located on the same chromosome (Wittenberger 1981, 618).
   Note: The closeness of linkage depends on how near to each other genes are on the same chromosome (Wittenberger 1981, 618).
2. "The association of certain genes owing to their location on the same chromosome" (Mayr 1982, 958).

sex linkage *n*. The location of a gene, or character, on a sex chromosome (Lincoln et al. 1985). *adj*. sex-linked

♦ **linkage disequilibrium** *n*. The association of alleles at different loci at frequencies higher than expected from random combinations of their frequencies in gametes, or individual organisms, of a population (Dawkins 1982, 289; Rieger et al. 1991).
syn. epistatic selection, gametic excess, gametic phase unbalance (Rieger et al. 1991), gametic disequilibrium (Dawkins 1982, 289)

Comments: Researchers often regard linkage disequilibrium as evidence for selection (Rieger et al. 1991). Causes of linkage disequilibrium are epistatic selection, genetic drift due to small population size, genetic hitchhiking, and population mixing or migration. Linkage disequilibrium should be nearly zero in a large, random-mating population without selection.

♦ **linkage group** See ²group: linkage group.
♦ **linkage map** See map: linkage map.
♦ **linked character** See character: linked character.
♦ **linked extinction** See ²extinction: linked extinction.
♦ **Linnaean hierarchy** See hierarchy: Linnaean hierarchy.
♦ **Linnaean species, Linnaeon** See ²species: Linnaean species.
♦ **lip smacking, lipsmacking** See animal sounds: lip smacking.
♦ **lip-, lipo-** *combining form* Fat; fatty (Michaelis 1963).
[Greek *lipos*, fat]
♦ **lipid** See molecule: lipid.
♦ **lipogenesis** See -genesis: lipogenesis.
♦ **lipopalingenesis** See -genesis: lipopalingenesis.
♦ **liposome** See organized structure: liposome.
♦ **lipoxenous** See -xenous: lipoxenous.
♦ **lipoxeny** See -xeny: lipoxeny.
♦ **liptocoenosis** See -coenosis: necrocoenosis.
♦ **lithodomous** See -domous: lithodomous.
♦ **lithology** See study of: lithology.
♦ **lithophage** See -phage: lithophage.
♦ **lithophile** See ¹-phile: lithophile.
♦ **lithosere** See ²succession: lithosere.
♦ **lithosphere** See -sphere: lithosphere.
♦ **lithotomous** See -domous: lithodomous.
♦ **lithotroph** See ²succession: lithotroph.
♦ **litter** *n.* The accumulation of plant material on the ground, including branches, leaves, reproductive parts, and twigs (Wimmer et al. 1993, 2184; Allaby 1994).
See ²group: litter.
syn. ground litter (Wimmer et al. 1993, 2184), leaf litter (Reardon 1995, 4), L-layer (Allaby 1994)
♦ **litterfall** *n.* The amount of litter, *q.v.*, that falls to the ground per unit time (inferred from DeLucia et al. 1999, 1177, 1178).
♦ **"live-plant shredders"** See ²group: functional feeding group: "live-plant shredders."
♦ **living** *n.*
1. In Humans: one's being alive, animate; not dead (*Oxford English Dictionary* 1972, entries from before 1325; Michaelis 1963).
2. An object's possessing these attributes: complexity and organization, chemical

uniqueness, quality, variability leading to individual uniqueness, inherited genetic programming acted upon by natural selection, and indeterminacy (Mayr 1982, 53–59).
cf. life, organism (Appendix 1)
Comments: Curtis (1983, 18–19) provides a suite of characteristics that distinguish living things (= organisms) from nonliving ones which is similar to Mayr's (1982) list; in addition to his list, she includes homeostasis, growth and development, transformation of environmental energy from one form to another, and response to stimuli. Viruses seem to possess all of these characteristics except homeostasis and transformation of environmental energy from one form to another; therefore, some (perhaps many) researchers would not classify them as organisms.
♦ **living fossil** See fossil: living fossil.
♦ **Lizzie-Borden evolution, Lizzie-Borden syndrome** See ²evolution: Lizzie-Borden evolution.
♦ **Lloyd-Morgan's canon** See Morgan's canon.
♦ **load** See genetic load.
♦ **lobtail** See behavior: lobtail.
♦ **local-character species** See ²species: local-character species.
♦ **local enhancement** *n.* An animal's directing its attention toward some object, or place, in its environment due to its observing how others are directing their attention with regard to this thing (*e.g.*, an animal's finding its way to a food source more quickly if it sees others looking for the source) (Immelmann and Beer 1989, 175).
Comment: In "local enhancement," each animal finds its own way by trial and error, even though it finds its goal more readily by watching other animals; "local enhancement" is distinguished from "facilitation," *q.v.* (Immelmann and Beer 1989, 175).
♦ **local-mate competition (LMC)** See competition: local-mate competition.
♦ **locale-odor hypothesis, local-odor theory** See hypothesis: locale-odor hypothesis.
♦ **locality** *n.* The geographic position of an individual, population, or collection (Lincoln et al. 1985).
♦ **locality imprinting** See learning: imprinting: place imprinting.
♦ **localization** *n.* An animal's ability to discriminate where a stimulus comes from or the area of its body that is being stimulated (Storz 1973, 149).
♦ **localized learning** See learning: localized learning.
♦ **lochmium** See ²community: lochmium.

◆ **lochmocole** See -cole: lochmocole.

◆ **lochmodium** See ²community: lochmodium.

◆ **lochmodophile** See ¹-phile: lochmodophile.

◆ **lochmophile** See ¹-phile: lochmophile.

◆ **lock** *n*. For example, in the Domestic Dog, Golden Mouse, Wolf: a mechanical tie between a male's penis and a female's vagina during copulation (Dewsbury 1978, 79).

◆ **lock-and-key hypothesis** See hypothesis: hypotheses of divergent evolution of animal genitalia: lock-and-key hypothesis.

◆ **locking antlers** *n*. In deer: agonistic behavior in males that fight with lowered heads and interlocked antlers (Heymer 1977, 77).

◆ **locomotion** *n*. An animal's self-caused movement from one place to another (*Oxford English Dictionary* 1972, entries from 1646).
cf. drifting, movement, saltation
Comments: "Locomotion" is characterized by an animal's rhythmic movements of its extremities and limbs — *e.g.*, crawling, running, flying, and swimming in insect species; stepping, trotting, cantering, galloping, swimming, and flying in some mammal species; slow galloping (or hopping) in lagomorph species; trotting and loping in canids; walking, trotting, and running in Humans (Heymer 1977, 77). Position-changing movements (*e.g.*, tentacle waving of some cnidarian species and arthropod species) of sedentary animals and drifting are not included in locomotion (Immelmann and Beer 1989, 176). Some authors occasionally call arthropod crawling "walking."

bipedal locomotion *n*. An animal's locomotion that involves using only its two hindlegs.
cf. biped
Comments: Many animal species, including birds, Chimpanzees, and Humans, show bipedal locomotion. The Bonobo walks on its legs when carrying food and other objects (de Waal and Lanting 1997, 24, illustration).

cross-gait *n*. For example, in ungulates: locomotion in which the front and rear leg of each side of an animal move in opposition to each other; progression of an anterior limb occurs when a corresponding rear leg is on the ground (*e.g.*, in ungulates) (Heymer 1977, 106).
cf. pacing
Comment: Humans have a cross-gait-like locomotion (Heymer 1977, 106).

digitigrade locomotion *n*. Locomotion characteristic of digitigrades, *q.v.* (Eisenberg 1981, 508).

flight See flight.

pacing *n*. A mammal's locomotion in which both legs on one side of its body are moved forward simultaneously (Heymer 1977, 128).
cf. cross-gait
Comment: Pacing is the usual form of locomotion in some antelopes, Elephants, Giraffes, and *Tylopodes* (Heymer 1977, 128).

pinnigrade locomotion *n*. Locomotion by swimming using flippers as paddles (Lincoln et al. 1985).

plantigrade locomotion *n*. Locomotion characteristic of plantigrades, *q.v.* (Eisenberg 1981, 508).

richochetal locomotion *n*. Locomotion characteristic of saltigrades, q.v. (Eisenberg 1981, 508).

swimming See swimming.

◆ **locomotory stimulant** See chemical-releasing stimulus: stimulant: locomotory stimulant.

◆ **locus** See gene locus.

◆ **lodge** *n*. For example, in the Beaver, Otter: an individual's den or lair (*Oxford English Dictionary* 1972, entries from 1567). See ²group: lodge.

◆ **loess** See statistical test: regression: loess.

◆ **loft** See ²group: loft.

◆ **lofting** See display: wing display: lofting.

◆ **log-growth phase** See phase: log-growth phase.

◆ **log-linear curve** See curve: species-accumulation curve: non-asymptotic-species-accumulation curve: log-linear curve.

◆ **log-log curve** See curve: species-accumulation curve: non-asymptotic-species-accumulation curve: log-log curve.

◆ **log-normal curve** See curve: rank-abundance curve: log-normal curve.

◆ **log-normal distribution** See distribution: log-normal distribution.

◆ **"log-phase hypothesis"** See hypothesis: hypotheses regarding the origin of the Cambrian explosion: "log-phase hypothesis."

◆ **logical behaviorism** See behaviorism.

◆ **logical division** See downward classification.

◆ **logical prediction** *n*. Prediction that can be made because individual observations conform with a theory or scientific law (Mayr 1982, 57).
cf. temporal prediction

◆ **logistic** *adj*. Of or pertaining to arithmetical calculation. [Medieval Latin *logisticus* < Greek *logistikos* < *logiszesthai*, to reckon < *logos*, word, calculation]

◆ **logistic growth** *n*.
1. "Growth, especially in the number of organisms constituting a population, that

slows steadily as the entity approaches its maximum size" (Wilson 1975, 588).
2. Growth that follows a sigmoid course with initial gradual growth (lag growth phase), rapid (exponential growth phase) growth, and then slow growth (stationary growth phase) as a population, or system, reaches its maximum size or carrying capacity (Lincoln et al. 1985).
cf. exponential growth

♦ **logistic-population-growth curve** See curve: logistic-population-growth curve.

♦ **-logous**

orthologous *adj.*
1. Referring to homologous structures of organisms determined by a common ancestral gene present in their most recent common ancestor; contrasted with paralogous (Lincoln et al. 1985).
2. Referring to genes that arose from the same common ancestral gene; contrasted with paralogous (Gorr and Kleinschmidt 1993, 76).
Note: Examples of orthologous genes are the gene for the amphibian juvenile form of β-hemoglobin and the gene for β-hemoglobin found in adult boney fish and fleshy finned fish (including the coelacanth) (Gorr and Kleinschmidt 1993, 75–76). These genes apparently evolved from a β-hemoglobin gene present in vertebrates about 400 million years ago.

paralogous *adj.*
1. Referring to homologous structures that arose by gene duplication and evolved in parallel within a single line of descent; contrasted with orthologous (Lincoln et al. 1985).
2. Referring to genes in a group of related organisms that did not arise from the same common ancestral gene; contrasted with orthologous (Gorr and Kleinschmidt 1993, 76).
Note: Examples of paralogous genes are the gene for the amphibian adult form of b-hemoglobin and the gene for b-hemoglobin found in adult boney fish and fleshy finned fish (including the coelacanth) (Gorr and Kleinschmidt 1993, 75–76). The fish genes apparently evolved from the juvenile form of a b-hemoglobin gene present in vertebrates about 400 million years ago. The amphibian gene apparently evolved from the adult form of a b-hemoglobin gene present in vertebrates about this same time.

♦ *loi de balancement* See law: *loi de balancement.*

♦ **lone wolf** *n.*
1. A single individual of a social species (*e.g.,* an old bull elephant) (Immelmann

and Beer 1989, 176); a single individual that leaves its social group for a brief to an extended period (Immelmann and Beer 1989, 273).
2. A solitary animal (Immelmann and Beer 1989, 176).

♦ **long-interspersed nucleotide element** See nucleotide element: long interspersed nucleotide element.

♦ **long-term memory** See memory: long-term memory.

♦ **long-term memory potentiation (LTP)** *n.* A functional change at certain synapses that seems to be directly related to memory storage and learning (Campbell 1996, 1022).
Comment: An LTP is enhanced by a postsynaptic cell to an action potential (Campbell 1996, 1022).

♦ **long-term optimality** See optimality: long-term optimality.

♦ **long-term pair bond** See bond: long-term pair bond.

♦ **long-term potentiation** *n.* A process that makes nerve cells more readily respond the next time that an animal uses them (Lemonick 1999, 58).
Comment: This process is important in learning and memory processes (Lemonick 1999, 58).

♦ **longevity** *n.* "The duration through geological times of a species and all of its descendants" (Wilson 1992, 363).
See lifespan.

♦ **longitudinal study** *n.* A research method that involves following one subject over an extended period (Storz 1973, 149).

♦ **loose school** See ²group: school: loose school.

♦ **lophium** See ²community: lophium.

♦ **lophophile** See ¹-phile: lophophile.

♦ **lophophore** See organ: lophophore.

♦ **lordosis** *n.*
1. A person's anterior curvature of his spine that causes convexity in front and abnormal hollow in one's back (*Oxford English Dictionary* 1972, entries from 1704).
2. In many mammal species including rats, pigs: a stereotyped posture that a female assumes as a male mounts her, consisting of her arching her back, spreading her legs sideways and extending them, lifting her tail, and pointing her head forward and downward (Dewsbury 1978, 77, 136; Forsyth 1985, 28; Immelmann and Beer 1989, 176).
syn. lordosis reflex (Immelmann and Beer 1989, 176)
adj. lordotic
[New Latin < Greek *lordōsis* < *loros*, bent backward]

♦ **Lorenz's hydraulic model, Lorenz's hydraulic model of behavior, Lorenz's hydraulic model of behavior regulation** See ⁴model: Lorenz's hydraulic model.

♦ *Los Niños* See *El Niño*.

♦ **lost call** See animal sounds: call: distress call.

♦ **lot** See ²group: lot.

♦ **lotic** *adj.* Referring to habitats of fast-running water such as rivers and streams (Lincoln et al. 1985).
cf. lentic

♦ **low-cost altruism** See altruism: low-cost altruism.

♦ **low diversity** See diversity: low diversity.

♦ **lower** See ancestral.

♦ **LTP** See long-term memory potentiation.

♦ **lucibufagin** See molecule: lucibufagin.

♦ **lucicole** See -cole: lucicole.

♦ **luciferase** See enzyme: luciferase.

♦ **luciferin** See molecule: luciferin.

♦ **luciferous** *adj.* Referring to an object that brings, produces, or emits light (Michaelis 1963).
cf. bioluminescent

♦ **lucifugal, lucipetal, luciphilous, photophilic** *adj.* Referring to an organism that thrives in open, well-lit habitats; contrasted with lucifugous (Lincoln et al. 1985).
n. luciphile, luciphily

♦ **lucifugous, photophobic** *adj.* Referring to an organism that is intolerant of light; contrasted with lucifugal (Lincoln et al. 1985).
cf. -phobe: photophobe

♦ **luciphile** See ¹-phile: luciphile.

♦ **luciphilous** See lucifugal.

♦ **Ludwig effect, Ludwig hypothesis, Ludwig's theorem, Ludwig's theory** See hypothesis: Ludwig hypothesis.

♦ **luminescence** See bioluminescence.

♦ **lumper** *n.* One of "the majority of taxonomists who prefers a rather large, comprehensive taxa, as being better able to express relationship and reducing the burden on the memory" (Mayr 1982, 240).
ant. splitter

♦ **lunar rhythm** See rhythm: lunar rhythm.

♦ **luring song** See animal sounds: song: luring song.

♦ **lute** See ²group: group: lute.

♦ **luteinizing hormone (LH)** See hormone: luteinizing hormone.

♦ **luticole** See -cole: luticole.

♦ **luxuriance** *n.* The part of hybrid vigor that is nonadaptive and does not increase an organism's fitness (Lincoln et al. 1985).

♦ **l.y.** See unit of measure: year: light year.

♦ **Lyellian survivorship curve** See curve: Lyellian survivorship curve.

♦ **lygophile** See ¹-phile: lygophile.

♦ **lying-out period** See period: lying-out period.

♦ **Lysenkoism** *n.* The doctrine of T.D. Lysenko and his followers that rejects the gene theory of inheritance in favor of inheritance of acquired characters that are responses to environmental factors and assimilated into an organism's genome and transmitted to its offspring (Michaelis 1963; Lincoln et al. 1985).
syn. neo-Lamarckism (Lincoln et al. 1985)
cf. Lamarckism; Michurinism; ¹system: inheritance-of-acquired-characteristics system
Comment: Lysenkoism displaced genetics altogether in the Soviet Union for several decades (Bowler, 1992, 188).
[after T.D. Lysenko, Soviet biologist]

♦ **lysimeter** See meter: lysimeter.

◆ **Ma** See units of measure: Ma.

◆ **macarthur (ma)** See units of measure: macarthur.

◆ **MacArthur-Hamilton-Leigh principle** See principle: MacArthur-Hamilton-Leigh principle.

◆ **MacArthur Prize** See award: MacArthur Prize.

◆ **MacClade** See program: MacClade.

◆ **Mach band** *n.* A narrow strip along the edge of a border between light and dark that seems brighter, or darker, than larger areas to which it belongs to a person due to lateral inhibition in his nervous system (Gould 1982, 101).

◆ **machine** *n.* Any combination of interrelated parts that use, or apply, energy to do work; especially, a mechanical unit with, or without, a driving motor, made to do a particular kind of work (Michaelis 1963).

squeeze machine *n.* A machine with a headrest, hydraulic controls, and inflatable cushioned sides that puts pressure on a person's body (Raver 1997, C1).

Comments: Autistic people reduce their anxiety and relax by using this machine (Raver 1997, C1). Temple Grandin, an autistic scientist who is hypersensitive to touch, invented this machine.

◆ **machismo** See display: machismo.

◆ **macraner** See male: macraner.

◆ **macro-, macr-** *combining form*
1. Pathologically, enlarged or overdeveloped (Michaelis 1963).
2. Large, or long, in size, or duration; great (Michaelis 1963).
[Greek *makros*, large]

◆ **macrobenthos** See [2]community: benthos: macrobenthos.

◆ **macrobiota** See biota: macrobiota.

◆ **macrobiotic** See biotic: macrobiotic.

◆ **macrodichopatric** See -patric: dichopatric.

◆ **macroevolution** See [2]evolution: macroevolution.

◆ **macroevolutionist** See evolutionist: macroevolutionist.

◆ **macrogamete** See gamete: macrogamete.

◆ **macrogamy** See -gamy: hologamy.

◆ **macrogenesis** See [2]evolution: punctuational evolution, saltational evolution.

◆ **macrogyne** See caste: queen: macrogyne.

◆ **macromorphology** See morphology: macromorphology.

◆ **macromutation** See [2]mutation: macromutation.

◆ **macromutationism** See hypothesis: macromutationism.

◆ **macroparapatric** See parapatric: macroparapatric.

◆ **macrophage** See -phage: macrophage.

◆ **macrophylogenesis** See [2]evolution: macroevolution.

◆ **macrophytophage** See -phage: macrophytophage.

◆ **macroplankton** See plankton: macroplankton.

◆ **macrosomatic** See -somatic: macrosomatic.

◆ **macrospecies** See [2]species: macrospecies.

◆ **macrosymbiont** See symbiot: macrosymbiont.

◆ **macrotaxonomy** See taxonomy: beta taxonomy.

◆ **macrotherm** See therm: megatherm.

◆ **macrothermophile** See [1]-phile: thermophile: macrothermophile.

◆ **macrothermophyte** See therm: megatherm.

◆ **macrozoogeography** See study of: geography: macrozoogeography.

◆ **magnet effect** See effect: magnet effect.

◆ **magnetic-field orientation** See taxis: geomagnetotaxis.

◆ **magnetite** See mineral: magnetites.

◆ **magnetoreception** See modality: magnetoreception.

h–m

♦ **magnification process** See process: magnification process.

♦ **main effect** See effect: main effect.

♦ **maintenance** *n.* Organism survival activities such as water regulation, movement, avoidance of predation and disease, baseline (nongrowth) metabolism, and resistance to competitors (Willson 1983, 5).
cf. principle: principle of allocation

♦ **maiosis** See meiosis.

♦ **major community** See ²community: major community.

♦ **major cost of sex** See cost: major cost of sex.

♦ **major-domo stroll** See display: major-domo stroll.

♦ **major gene** See gene: oligogene.

♦ **major-histocompatibility-complex genes** See gene: major-histocompatibility-complex genes.

♦ **major pollinator** See pollinator: major pollinator.

♦ **major worker** See caste: worker: major worker.

♦ **make anosmic** *v.t.* To cause an organism to lose its sense of olfaction [coined by Cheal, personal communication 1985].
cf. anosmize

♦ **malacogamous** See ²-phile: malacophile.

♦ **malacology** See study of: malacology.

♦ **malacophile** See ²-phile: malacophile.

♦ **Malaise trap** See trap: Malaise trap.

♦ **malaxation** See mastication.

♦ **male** *n.* In bisexual, or dioecious, species: a sperm-producing individual (Lincoln et al. 1985).
cf. caste: ergatandromorph; female

complemental male See male: parasitic male.

drone *n.* A male bee, especially a Honey Bee (Michaelis 1963).
cf. caste: queen

ergatomorphic male, ergatoid male See caste: ergatandromorph.

incidental *n.*
1. In many lekking and territorial species including some bee, grasshopper, dragonfly, and bird species including ruffs: a male that does not have a territory and invades the territory of a conspecific male (Hogan-Warburg 1966 in Wilson 1975, 319; Barrows 1976b).
2. An animal that visits a lek but does not successfully hold a place on a lek, *q.v.* (Robel 1967 in Hjorth 1970).
syn. interloper, satellite male (Immelmann and Beer 1989, 258)
cf. male: regular
Comment: "Satellite male" originally referred only to male ruffs (birds) (Immelmann and Beer 1989, 258).

macraner *n.* A male of the larger form in an ant species with two sizes of males; contrasted with micraner (Hölldobler and Wilson 1990, 640).

micraner *n.* A male of the smaller form in an ant species with two sizes of males; contrasted with macraner (Hölldobler and Wilson 1990, 640).

nanander, nannander *n.* A dwarf male (Lincoln et al. 1985).

parasitic male, complemental male *n.* For example, in some angler fish species: a tiny male that lives as a parasite attached to its mate's body and typically has well-developed reproductive organs but an otherwise degenerate body form (Gould 1982b).

peripheral male *n.* In many animal species: a smaller male that uses an alternative mating tactic that parasitizes the mating effort of a larger, dominant male (van den Berghe et al. 1989, 875).
Comments: A peripheral male can reduce the fitness of a dominant male (van den Berghe et al. 1989, 875). Some incidentals are peripheral males.

regular, territory-holding male *n.* A male animal that is well established on a lek, *q.v.* (Robel 1967 in Hjorth 1970).
cf. incidental, lek; male: incidental

satellite male See male: incidental.

♦ **male-assistance monogamy** See mating system: monogamy: male-assistance monogamy.

♦ **male care** See care: male care.

♦ **male mate choice** See mate choice: male mate choice.

♦ **male parthenogenesis** See parthenogenesis: androgenesis.

♦ **malimprinting** See learning: imprinting: malimprinting.

♦ **mammalogy, mammalology** See study of: mammalogy.

♦ **management guild** See guild: management guild.

♦ **mandibular gland** See gland: mandibular gland.

♦ **manduction** See mastication.

♦ **mangrove** See ²community: mangrove.

♦ **manipulation** *n.*
1. A person's operating on, or manipulating, another person or thing with dexterity, especially with disparaging implication; unfair management or treatment (*Oxford English Dictionary* 1972, entries from 1828).
2. An individual animal's altering the behavior of another animal to its own advantage (Krebs and Dawkins 1984 in Dawkins 1986, 129).
Comment: Signaling is sometimes construed as a form of manipulation (Immelmann and Beer 1989, 179).

parental manipulation *n.* A parent's adjusting or manipulating its parental investment, particularly by reducing the reproduction of certain progeny in the interests of increasing the parent's own inclusive fitness via other offspring (Brown 1987a, 305).

Comment: "Variance enhancement," *q.v.*, is very similar to "parental manipulation" but differs from it mainly by emphasizing variance and deemphasizing "adroit or skillful management; fraudulent or deceptive treatment," which Alexander (1974) implies in the concept of "parental manipulation" (Brown 1987a, 305).

♦ **manipulative experiment** See experiment: manipulative experiment.

♦ **Mann-Whitney-Wilcoxon-U statistic** See statistical test: multiple-comparisons test: planned-multiple-comparisons procedure: Mann-Whitney-Wilcoxon-U statistic.

♦ **manometabolous metamorphosis** See metamorphosis: manometabolous metamorphosis.

♦ **map** *n.*
1. A representation of any region (*e.g.,* of the Earth's surface) drawn on a plane surface; a chart (Michaelis 1963).
2. Anything that resembles a map in form, purpose, or both (Michaelis 1963).

chromosome map, cytogenetic map See map: genetic map.

genetic map *n.*
1. The representation of the genetic distance that separates nonallelic gene loci in a linkage structure (= the genes of one linkage group or chromosome) that is deduced from genetic-recombination experiments (King and Stansfield 1985, Rieger et al. 1991).
   Note: This map is measured in centimorgans (cM) over small distances; 1 cM is equivalent to a 1% chance of recombination (Jasny 1991).
   syn. chromosome map (Rieger et al. 1991), cytogenetic map, linkage map (King and Stansfield 1985), genetic linkage map (Jasny 1991)
2. The arrangement of mutational sites (= alleles) of a particular gene that is deduced from genetic-recombination experiments (King and Stansfield 1985, Rieger et al. 1991).
   syn. fine-structure map, gene map (Rieger et al. 1991)

Comments: Genetic maps for most species are linear and unbranching (Rieger et al. 1991). They are circular for bacteria and bacteriophages. Thomas Hunt Morgan's student, Alfred H. Sturtevant, developed the method for constructing genetic maps (Campbell et al. 1999, 267).

physical map *n.* A map of the linear order of genes that is determined by methods other than genetic recombination and whose distances between landmarks (*e.g.,* clones, restriction endonuclease sites, or specific loci) are expressed in kilobases (King and Stansfield 1985; Jasny 1991).

species-density map *n.* A map with isobars that shows numbers of species of a selected taxon (taxa) in a particular part of the world to all of the world (inferred from Colwell and Coddington 1994, 102).

transcript map *n.* A map of transcripts on well-defined DNA segments (inferred from the definition of transcript mapping in Rieger et al. 1991).
Comment: Transcription mapping permits the fine mapping of RNA transcripts with the same level of accuracy as the mapping of restriction endonuclease cleavage sites on DNA (Rieger et al. 1991).

♦ **MAP** See action pattern: modal-action pattern.

♦ **MAPMAKER 3.0** See program: MAPMAKER 3.0.

♦ **maquis** See habitat: maquis.

♦ **marginal species** See ²species: marginal species.

♦ **marisma** See habitat: mudflat: marisma.

♦ **marker** *n.* A gene, molecule, or DNA sequence used to locate a gene(s) on an individual's genome (Rensberger 1995, A10; Tanksley 1993, 205–209).
Comment: Probes and markers are used in DNA fingerprinting and locating genes in genomes.

hypervariable marker *n.* A genetic marker that involves many rare alleles and enables a researcher to identify most of the individuals in a population as genetically distinct individuals (inferred from Hamilton and Rand 1996, 249, 255).

randomly amplified polymorphic DNA marker (RAPD marker) (RAP-id marker) *n.* A marker used for measuring molecular divergence of DNA (inferred from Hartl and Clark 1997, 327).

restriction-fragment-length polymorphism marker (RFLP marker) (RIF-lip marker) *n.* A restriction-fragment-length polymorphism used as a marker in genetics (Campbell et al. 1999, 373).
Comment: Researchers use RFLPs as markers for variants in a population (Campbell et al. 1999, 373).

variable-number-of-tandem-repeats polymorphism marker (VNTR marker) *n.* A marker that involves variable-number-of-tandem-repeats-polymorphic DNA, *q.v.,* and is used for measuring molecular divergence of DNA (inferred from Hamilton and Rand 1996, 249; Hartl and Clark 1997, 327, 398).

♦ **marking** See behavior: marking behavior.

♦ **marking drumming** See animal sounds: marking drumming: drumming; behavior: marking behavior: marking drumming.

♦ **Markoff chain** See chain: Markoff chain.

♦ **Markoff process** See process: Markoff process.

♦ **marriage** *n.*
  1. A man's and woman's being husband and wife; a relation between married persons; spousehood; wedlock (*Oxford English Dictionary* 1972, entries from 1297).
  2. In nonhuman animals (*e.g.,* the Kittiwake Gull, Mongolian Gerbil): pair bonding (Wilson 1975, 331).
  See bond: pair bond.

♦ **Marsden's effect** See effect: Marsden's effect.

♦ **marsh** See habitat: marsh.

♦ **mashing** See behavior: mashing.

♦ **masking experiment** See experiment: masking experiment.

♦ **masking gene** See gene: masking gene.

♦ **mass communication** See communication: mass communication.

♦ **mass extinction** See ²extinction: mass extinction.

♦ **mass provisioning** See provisioning: mass provisioning.

♦ **mass selection** See selection: artificial selection.

♦ **mast** *n.* The fruit of oak, beech, and other trees when it is used as swine food (Michaelis 1963).
  cf. masting
  [Old English *mæst*, mast, fodder]

♦ **mast year** See unit of measure: year: mast year.

♦ **master limiting factor** See factor: master limiting factor.

♦ **mastication, manduction, malaxation** *n.* Chewing (Lincoln et al. 1985).

♦ **masting** *n.* A tree species' producing a very large fruit crop in a particular year subsequent to its producing no to a small crop in the preceding year(s) (Kricher and Morrison 1988, 302).
  cf. year: mast year
  Comments: Masting trees include the American Beech, Black Walnut, and White Pine; Hickories and Oaks; and some dipterocarp, fir, and spruce species (Harlow and Harrar 1958; Kricher and Morrison 1988, 302–303, 308; Blundell 1999, 32). American Beech masts every 2–3 years, and its masting is synchronous over large areas. Populations of mast consumers (*e.g.,* the Blue Jay, Eastern Chipmunk, Eastern Gray Squirrel, White-Footed Mouse, and White-Tailed Deer) may increase temporarily after masting years (Kricher and Morrison 1988, 302; Ostfeld et al. 1996, 234). Masting is probably a tree species' adaptation to seed predation because there are often more fruits produced in mast years than their predators can consume. Food shortages after mast years might cause some bird species to migrate to new locations. Oak masting can increase the number of mammals used as hosts by the tick *Ixodes scapularis,* which transmits Lyme disease (Line 1996, C1; Ostfeld et al. 1996, 324).

♦ **masturbation** *n.* In some primate species including male Chimpanzees, juvenile male and female Orangutans, male and female Humans: an individual's self-stimulation of its genital organs, usually with its hands, often to orgasm in Humans (Hinsie and Campbell, 1976; Smith 1984, 623, 633).
  syn. onanism (not preferred by Storz 1973, 183), self-abuse, self-pollution (*Oxford English Dictionary*, entries from 1857 for Humans)
  n. masturbator
  v.i., v.t. masturbate
  cf. genito-genital rubbing; onanism; *Retrojection*: sexual *Retrojection*.
  Comments: Mutual homosexual and heterosexual masturbation also occurs in Humans (Smith 1984, 636).
  [origin unclear, possibly *mastiturbāri*, virile member + *turba*, disturbance (*Oxford English Dictionary* 1972)]

♦ **matching experiment** See experiment: matching experiment.

♦ **mate choice** *n.* In many species, including some *Drosophila,* finch species; the Bullhead Finch, Bullhead Fish, Human, Pika: an individual's preference of one conspecific, over others, as a copulatory partner or gene donor to its offspring (Darwin 1871, 211; Fisher 1958, 158–159; Bateson 1983, ix; Bisazza and Marconato 1988, 1352; Clayton 1988, 1589; Brandt 1989, 118; Noor 1996, 1205).
  syn. mate preference (Noor 1996, 1205)
  cf. species discrimination
  Comment: To avoid confusion Bateson (1983, ix) does not use "mate selection" as a synonym of "mate choice."

female copying *n.* Female mate choice in which a female imitates the mate choice of another female (Shuster and Wade 1991, 1071).

female mate choice *n.* For example, in some bird, *Drosophila,* other insect species; Humans: intersexual selection in which females choose particular males as mates over other males (Dewsbury 1978, 300; Thornhill and Alcock 1983; Latimer and

Sippel 1987, 887; Alatalo et al. 1990, 244; Noor 1996, 1205).

syn. female choice, female preference (Alatalo 1990, 244)

cf. selection: epigamic selection

male mate choice *n.* For example, in some *Drosophila,* salamander species; Humans: a male's preference for a particular kind of female vs. another kind (Verrell, 1988, 1086; Noor 1996, 1205).

♦ **mate competition** See competition: mate competition.

♦ **mate conflict** See [3]conflict: mate conflict.

♦ **mate desertion** *n.* In many amphibian, fish, mammal, reptile species; some bird species: an individual's leaving its mate and its offspring to be reared by its abandoned mate (Dewsbury 1978, 302; Beissinger and Snyder 1987, 477; Lazarus 1990, 672).

syn. desertion (Dewsbury 1978, 302)

Comment: This behavior is expected to occur if the abandoner can increase its fitness by leaving (Dewsbury 1978, 302).

ambisexual mate desertion *n.* In some bird species: mate desertion by either sex (Kendeigh 1941, etc. in Beissinger and Snyder 1987, 477).

♦ **mate guarding, mate-guarding behavior** See behavior: mate-guarding behavior.

♦ **mate-guarding monogamy** See mating system: monogamy: mate-guarding monogamy.

♦ **mate recognition** See recognition: mate recognition.

♦ **materialism** *n.* Democritus', Thomas Hobbes', and others' doctrine that there are no supernatural, or immaterial, forces that affect organisms — only physico-chemical forces are in play; however, organisms have many characteristics that are without parallel in inanimate objects (Mayr 1982, 52; McFarland 1985, 361).

cf. associationism, rationalism

♦ **maternal** *adj.* Referring to, or derived from, a female parent.

cf. paternal

♦ **maternal behavior** See care: parental care: maternal care.

♦ **maternal-benefit polyandry** See mating system: polyandry: maternal-benefit polyandry.

♦ **maternal care** See care: maternal care.

♦ **maternal family** See family: maternal family.

♦ **maternal half sister** See sister: half sister: maternal half sister.

♦ **maternal inheritance** See inheritance: maternal inheritance.

♦ **maternal investment** See investment: parental investment: maternal investment.

♦ **maternal sex determination** See sex determination: maternal sex determination.

♦ **mathematical model** See [4]model: mathematical model.

♦ **mathematical population ecology** See study of: ecology: mathematical population ecology.

♦ **mathematical population genetics** See study of: genetics: mathematical population genetics.

♦ **mathematics** See study of: mathematics.

♦ **matinal** *adj.* Referring to the morning (Lincoln et al. 1985).

cf. crepuscular, diurnal, nocturnal, pomeridanus

♦ **mating** *n.*

1. In Humans: marrying (*Oxford English Dictionary* 1972, entries from 1621).
2. In animals, especially birds: a person's pairing individuals for breeding (*Oxford English Dictionary* 1972, entry from 1875).
3. For example, in many uni- or multicellular animals: individual's merging, or practicing, hologamy (Heymer 1977, 103).
4. For example, in gametes: a pair's fusing together.

See copulation for a discussion of mating synonyms.

v.i., v.t. mate

syn. associating, copulating, conjugating, consorting, courting, inseminating

cf. behavior: reproductive behavior; breeding; copulation

[Middle Low German < *gemate* < *ge-*, together + *mat*, meat, food; probably akin to MEAT]

assortative mating *n.* In many animal species: nonrandom pairing of individuals that resemble each other in one or more phenotypic traits; the tendency of individuals to choose mates that resemble themselves (Dawkins 1982, 284).

syn. assortative pair formation (Immelmann and Beer 1989), endogamy, homogamous mating, homogamy, positive assortative mating

cf. learning: imprinting: sexual imprinting; mating: disassortative mating, negative assortative mating (Wilson 1975, 80, 579)

Comment: Immelmann and Beer (1989, 21) summarize possible benefits of assortative mating.

▸ **assortative mating by size** *n.* A correlation between the body sizes of mating males and females in a population in the absence of inbreeding (Lewontin et al. 1968 in Crespi 1989, 981, who discusses the five sources of potential ambiguity of this definition).

▸ **negative assortative mating** *n.* The tendency of individuals to choose mates that do not resemble themselves (Dawkins 1982, 284).

h – m

disassortative mating *n*. The nonrandom pairing of individuals that differ from each other in one or more phenotypic traits (Wallace 1968 in Wilson 1975, 80, 319, 582).
syn. exogamy, heterogamous mating (Lincoln et al. 1985)
cf. mating: assortative mating and its synonyms, random mating

homosexual mating See copulation: homosexual copulation.

panmixia, panmixis See mating: random mating; mixis: panmixus.

positive assortative mating See mating: assortative mating.

random mating *n*. Mating within a population without regard to genotypes, or phenotypes, of mates (Lincoln et al. 1985).
syn. panmixia, panmixis (Lincoln et al. 1985)
See mixis: panmixus.

sib mating, adelphogamy *n*. Mating of siblings (Lincoln et al. 1985).

♦ **mating continuum** *n*. An aggregate of interbreeding individuals that systematically exchange genes (Lincoln et al. 1985).

♦ **mating dance** See dance: mating dance.

♦ **mating effort** See effort: reproductive effort: mating effort.

♦ **mating group** See ²group: mating group.

♦ **mating march** *n*.
1. In mammals, especially ungulates: a male's following a female closely, or in direct bodily contact, before copulation; the male moves stiffly, hence the term "march" (Immelmann and Beer 1989, 180).
2. In species with indirect spermatophore transmission (*e.g.*, Pseudoscorpions, Scorpions): a male's and female's grasping one another with their chelate pedipalps (pincers) and running back and forth until the male deposits his spermatophore and the female gets it into her genital opening (Immelmann and Beer 1989, 180).

♦ **mating mistake** *n*. For example, in Fruit Flies (Tephritidae): an individual's attempt to mate with any insect other than a conspecific individual of the opposite sex (White et al. 2000, 892).
cf. behavior: homosexual behavior

♦ **mating pheromone** See chemical-releasing stimulus: semiochemical: pheromone: mating pheromone.

♦ **mating plug** *n*. In some insect, mammal, reptile species: an object left in a female's genital chamber after mating by a male that may stop her sperm leakage, further copulations, or both (Wilson 1975, 321).
syn. copulatory plug (Dickinson and Rutowski 1989, 154)

Comments: A mating plug can be a secretion from a male's accessory glands (in insects such as some Lepidoptera and dytiscid water beetles), part of a male's abdomen (*e.g.*, in a ceratopogonid fly, the Honey Bee), coagulated seminal fluid (in some bats, hedgehogs, marsupials, and rats), and so forth (Wilson 1975, 321). In a checkerspot butterfly, a mating plug impedes copulation (Dickinson and Rutowski, 1989, 154).

♦ **mating position** *n*. The spatial relationship of a conspecific male and female during copulation (Lavigne and Holland 1969, 3).
cf. posture: wheel posture

end-to-end mating position *n*. For example, in lepidopterans, some robber-fly species: a mating position in which a male and female rest, or fly, with their anterior ends in opposite directions (Lavigne and Holland 1969, 3; 32, illustration).

male-beside-female mating position *n*. For example, in some robber-fly species: a mating position in which a male's longitudinal axis is about 30% with regard to his mate's longitudinal axis (Lavigne and Holland 1969, 4; 28, illustration).

male-over-female mating position *n*. For example, the Japanese Beetle; Praying Mantids; many bee, ladybird species; some robber-fly species: a mating position in which a male perches on his mate's dorsum (Lavigne and Holland 1969, 3, 37, illustration; Barrows and Gordh, 1978, 341).

♦ **mating site** *n*. A place where members of the opposite sex copulate (Thornhill and Alcock 1983, 182).
syn. encounter site, rendezvous place (Barrows 1976a)

♦ **mating stimulant** See chemical-releasing stimulus: stimulant: mating stimulant.

♦ **mating system** *n*.
1. Mating patterns in an animal population, including extent of inbreeding, possible pair-bonding, individual's numbers of simultaneous mates, and distribution of parental care within a mated pair (Lincoln et al. 1985; Immelmann and Beer 1989, 180).
Notes: Emlen and Oring (1977) classify mating systems of birds and mammals; Thornhill and Alcock (1983) classify male mating systems in insects; and Kirkendall (1983) classifies mating systems of Ambrosia and Bark Beetles.
2. An archaebacterium gene-transfer system involving a donor and recipient individual (Rosenshine et al. 1989, 1387).
3. Mating patterns in plants (Hamilton and Rand 1996, 255).
cf. breeding, lek

SOME CLASSIFICATIONS OF MATING SYSTEMS

I. Classification by focal organisms based on number of mates and other criteria, including number of serial mates, number of simultaneous mates, extent of inbreeding, and amount of maternal and paternal care given to offspring
  A. monogamy
    1. anonymous monogamy
    2. female-enforced monogamy
    3. male-assistance monogamy
    4. mate-assistance monogamy
    5. mate-guarding monogamy
    6. monogyny
    7. perennial monogamy (= lifetime monogamy)
    8. serial monogamy (*syn.* serial polygamy)
  B. polygamy
    1. bigamy
      a. opportunistic bigamy
      b. usurpatory bigamy
    2. oligogyny
    3. contemporaneous polygamy
    4. facultative polygamy
    5. polybrachygamy (= promiscuity)
      a. broadcast promiscuity
      b. overlap promiscuity
    6. serial polygamy
    7. simultaneous polygamy
    8. polyandry
      a. convenience polyandry
      b. fraternal polyandry
      c. genetic-benefit polyandry
      d. mate-defense polyandry
      e. maternal-benefit polyandry (= prostitution polyandry)
      f. serial polyandry (*syn.* serial monogamy)
      g. simultaneous polyandry
      h. sperm-replenishment polyandry
    9. polygynandry
   10. polygyny
      a. female-defense polygyny
      b. harem polygyny
      c. inbreeding polygyny
      d. lek polygyny
      e. pure-dominance polygyny
      f. resource-defense polygyny
      g. scramble-competition polygyny
        (1) explosive-mating assemblage polygyny
        (2) prolonged-searching polygyny
      h. simultaneous polygyny
      i. territorial polygyny
   11. polygyny-polyandry

II. Classification by number of genders
  A. bipolar mating system
  B. multipolar mating system
III. Classification by actual, or supposed, influencing ecological factors
  A. brachygamy
  B. colonial mating system
  C. inbreeding polygyny
  D. tachygamy

bigamy *n.* For example, in the Tropical House Wren: a male's copulating with two females (Freed 1986, 1895).
Comment: Freed (1986) describes "opportunistic bigamy" and "usurpatory bigamy."

bipolar mating system *n.* A mating system that involves two genders, either of gametes alone or both individual organisms and gametes (Bell 1982, 503).
cf. mating system: multipolar mating system

brachygamy *n.* In some Ambrosia Beetle and Bark Beetle species: a mating system related to a gallery system made by one female in which pair bonds are of variable duration but do not usually last during an oviposition period that a female takes to lay her entire brood [Greek *brachy*, short; *gamos*, marriage, coined by Kirkendall (1983, after Selander 1972)].

colonial mating system *n.* In some Ambrosia Beetle and Bark Beetle species: a mating system related to a gallery system made by more than one female in which more than one males mate with several to many females [coined by Kirkendall 1983].

monarsenous mating system See mating system: polygamy.

monogamy *n.*
1. A mating system in which individuals have only one mate during the production of at least one brood, and both parents almost always rear young together (Wilson 1975, 315, 327, 589; Immelmann and Beer 1989, 189; Alcock 1993, 447).
Notes: Monogamy occurs in animals including some beetle and crab species; 90% of bird species; some mammal species (*e.g.,* some rodent species, including Agouti; some ungulate species, including Dikdik and Klipspringer; some primate species, including five New World monkey species, Gibbons, Marmosets, and Siamangs; Humans; many mustelid species; most canid species, including Foxes, Jackals, and Wolves); Desert Woodlice; and the Crown Shrimp. In many bird species, males are "monogamous" in the sense that they

each pair-bond with one mate and give parental assistance, but they are also polygynous because they obtain extrapair copulations (Alcock 1993, 451).

2. A mating system in which "breeding adults of both sexes pair with only one mate at a time, with the duration of pair bonding lasting for a substantial fraction of the breeding season" (Wittenberger 1981, 618).

3. A mating system in which a male ordinarily copulates (and has reproductive success) with only one female during his entire life or during an entire breeding season if the male is of a long-lived species (Thornhill and Alcock 1983, 81).

4. For example, in some halictine-bee species, possibly many hymenopteran species, and the Screwworm Fly: a female insect's mating only once in her lifetime (Barrows 1976a; Alcock et al. 1978).
   cf. bonding

5. In some Ambrosia Beetle and Bark Beetle species: a mating system related to a gallery system made by one female in which a male stays during the entire oviposition period that his mate takes to lay one brood of eggs (Kirkendall 1983).

Note: Monogamy may be "permanent monogamy" or "serial monogamy" [coined by Kirkendall 1983].

Comment: There are few purely monogamous bird species (Ford 1983 and McKinney et al. 1984 in Freed 1986, 1894).

[French *monogamie* < Late Latin *monogamia* < Greek < *monos*, single + *gamos*, marriage]

▸ **anonymous monogamy** *n.* In the Swallows, Swifts; the White Stork: monogamy that is maintained due to partners' being attached to the same nesting spot (*Ortstreue*) after returning from separate parts of their overwintering region (Immelmann and Beer 1989, 189).

▸ **female-enforced monogamy** *n.* Monogamy that occurs because a dominant female stops subordinate females from copulating with her mate (*e.g.,* in wolves) (Alcock 1993, 447).

▸ **lifetime monogamy** See mating system: perennial monogamy.

▸ **male-assistance monogamy** *n.* Monogamy in which a male remains with his mate which elevates her reproductive output (Thornhill and Alcock 1983, 234).

▸ **mate-assistance monogamy** *n.* Monogamy in which an animal remains with its mate and elevates its reproductive effort (*e.g.,* in many bird species) (Alcock 1993, 447).

▸ **mate-guarding monogamy** *n.* Monogamy in which a male remains with his mate and prevents her from copulating with other males (Thornhill and Alcock 1983, 234).

▸ **monandry** *n.* A female's tendency to mate with only a single male.
cf. mating system: monogamy: monogyny
[< MONO- + Greek *andros*, man]

▸ **monogyny** *n.* A male's tendency to mate with only a single female (Wilson 1975, 589).
cf. -gyny: monogyny; mating system: monogamy: monandry
[MONO- + Greek *gynē*, woman]

▸ **perennial monogamy, life-time monogamy** *n.* For example, the Florida Scrub Jay, Mongolian Gerbil; Swans; some Humans: monogamy in which mates are pair-bonded throughout their lives (Wilson 1975, 327, 454).
cf. mating system: monogamy: serial monogamy

▸ **serial monogamy** *n.* For example, some bird and mammal species, the American Lobster, some Humans: an individual's having only one mate at a time in a series of mates over its lifetime (Wittenberger 1979 and Wittenberger and Tilson 1980 in Cowan and Atema 1990, 1120; Cowan and Atema 1990, 1199).
cf. mating system: monogamy: perennial monogamy; polygamy: serial polygamy
Comment: "Serial monogamy" is a confusing term. Some workers consider it to be synonymous with serial polygamy (Davies 1991, 281).

monothelius mating system See mating system: polyandry.

multipolar mating system *n.* A mating system that involves three or more genders of isogametes (Bell 1982, 510).
cf. mating system: bipolar mating system

oligogyny *n.* The mating of one male with a few females (Lincoln et al. 1985).

overlap promiscuity See mating system: polybrachygamy: overlap promiscuity.

panogamy *n.* A mating system in which female rats change mates during a copulation series (M.K. McClintock, personal communication)

polyandry *n.*

1. For example, in the Dunnock, Honey Bee, Jaçana, Tasmanian Waterhen; some Humans: a female's having more than one male as a mate (Wilson 1975, 327; Alcock 1993, 456) who may cooperate in raising brood (Wilson 1975, 592).
Note: For example, Honey Bee queens mate with an average with 17 drones

(Adams et al. 1977 in Frumhoff and Schneider 1987, 255). Animals that are both polyandrous and polygynous include giant water bugs, many fishes with parental care, and the tinamou (bird) (Alcock 1993, 456).

2. A mating system in which a female copulates and has reproductive success with more than one male during her entire life, or an entire breeding season, whether or not she is pair-bonded to any of her mates (Thornhill and Alcock 1983, 81).

3. "A mating system characterized by a behavioral bond between one breeding female and two or more males breeding with her" (Brown 1987a, 305).

syn. monothelious mating system (Lincoln et al. 1985)

cf. mating system: polygynandry, polygyny

▸ **convenience polyandry** *n.* In some insect species: polyandry in which a female avoids the costs of trying to prevent superfluous copulations by simply allowing males to mate with her instead of resisting their matings (Thornhill and Alcock 1983, 462).

▸ **fraternal polyandry** *n.* A rare form of marriage, not uncommon in Tibet, in which two or more brothers marry the same woman simultaneously; all brothers are sexual mates of the woman, and her children consider all of the brothers as their fathers (Goldstein 1987).

▸ **genetic-benefit polyandry** *n.* In some insect species: polyandry in which a female replaces sperm of a genetically inferior mate with those of a genetically superior one, or in which she adds sperm from a genetically different mate to her sperm supply, thus increasing the genetic variability of her offspring (Thornhill and Alcock 1983, 462).

▸ **mate-defense polyandry** *n.* For example, in the Red-Necked Phalarope: polyandry in which a female guards her male mates from other females (Alcock 1993, 459).

▸ **maternal-benefit polyandry, prostitution polyandry** *n.* In some insect species: polyandry that involves a female's obtaining nutrition or other material resources, reduction of the risk of predation, reduction of competition for a resource, protection from other sexually active males, or a combination of these things (Thornhill and Alcock 1983, 462).

▸ **serial polyandry** *n.* Polyandry in which a female has pair-bonded mates one after another (Wilson 1975, 327).

cf. mating system: polyandry: simultaneous polyandry; polygyny; polyandry

▸ **simultaneous polyandry** *n.* Polyandry in which a female has more than one pair-bonded mate at about the same time (Wilson 1975, 327).

cf. mating system: polyandry: serial polyandry

▸ **sperm-replenishment polyandry** *n.* In some insect species: polyandry in which a female mates with more than one male, enabling her to add to an inadequate, or depleted, sperm supply (Thornhill and Alcock 1983, 462).

polybrachygamy *n.*

1. Nonrandom polygamy without pair bonds (Selander 1972 in Wilson 1975, 327).

2. In the Chimpanzee, Crotophaginine Birds: a mating system in which "males and females do not form long-term pair associations and individuals of at least one sex frequently mate with more than one individual of the opposite sex each reproductive season" (Wilson 1975, 451, 546; Wittenberger 1981, 620).

syn. promiscuity (Wilson 1975, 327)

Comments: "Polybrachygamy" does not necessarily imply that animals concerned do not discriminate among potential mates. The boundaries among "promiscuity," "polygamy," and mating within "one-male mating systems" ("harems") are blurred (Immelmann and Beer 1989, 235).

▸ **broadcast promiscuity** *n.* For example, in some sponge species: sexual reproduction in which male and female gametes are released into their parents' surrounding medium, where fertilization takes place; ordinarily, "gamete release occurs in some coordinated manner and may involve large numbers of individuals of both sexes" (Wittenberger 1981, 613).

▸ **hierarchical promiscuity** *n.* A mating system that occurs in multi-male social groups where a male's sexual access to receptive females is affected by his status in his social hierarchy (Wittenberger 1981, 616).

▸ **overlap promiscuity** *n.* A mating system in which individual males and females live in overlapping home ranges and come together only briefly for the purpose of mating (Wittenberger 1981, 619).

polygamy *n.* For example, in the Three-Spined Stickleback; many mammal, weaverbird species: an individual's having more than one mate as part of its normal life cycle (Wilson 1975, 327), all of which may cooperate in raising the brood (Wilson 1975, 592).

syn. monarsenous mating system (Lincoln et al. 1985)

cf. mating system: polyandry, polybrachygamy, polygynandry, polygyny

h–m

▶ **contemporaneous polygamy** See polygyny: simultaneous polygamy.

▶ **facultative polygamy** *n.* In some songbird species: polygamy in which a male has two, or even more, mates as circumstances allow (Immelmann and Beer 1989, 226).

▶ **serial polygamy** *n.* Polygamy in which an animal has pair-bonded mates one after another rather than simultaneously (Wilson 1975, 327).
cf. mating system: monogamy: serial monogamy; polyandry: serial polyandry; polygamy: simultaneous polygamy; polygyny: serial polygyny

▶ **simultaneous polygamy** *n.* Polygamy in which an animal has more than one pair-bonded mate at about the same time (Wilson 1975, 327).
cf. mating system: polyandry: simultaneous polyandry; polygamy: serial polygamy; polygyny: serial polygyny

polygynandry *n.* For example, in some bird species: a mating system with behavioral bonds between two or more breeding females and two or more males breeding with them (Brown 1987a, 305).
cf. mating system: polyandry, polygamy

polygyny *n.*
1. A male's mating with two or more females (Wilson 1975, 327, 592).
2. A male's mating with more than one female with which he cooperates to some extent in rearing young (Wilson 1975, 592).
3. A mating system in which a male copulates and has reproductive success with more than one female, whether or not he is pair-bonded to any of his mates (Thornhill and Alcock 1983, 81, 85).
Note: This is the most common form of mating system in animals (Thornhill and Alcock 1983, 81, 85).
4. A mating system of a population or species, in which a reasonable percentage (5–10%) of mature males copulate and have reproductive success with more than one female (Thornhill and Alcock 1983, 82).
cf. mating system: polyandry, polygamy, polygynandry

▶ **female-defense polygyny** *n.* For example, in some insect species, a horseshoe-crab species, the Wood Frog: polygyny in which some males prevent others from gaining access to mates by defending groups of females or a series of individual females (Thornhill and Alcock 1983, 234; Alcock 1993, 460–462).

▶ **harem polygyny** *n.*
1. Polygyny in which one adult male controls exclusive sexual access to a social group of two or more breeding females, usually by means of aggressive behavior (Wilson 1975, 327; Baird and Liley 1989, 817).
syn. "one-male-social-unit polygyny"
2. In some Ambrosia Beetle and Bark Beetle species: a mating system related to a gallery system made by more than one female in which a male mates with more than one female to which he is pair-bonded simultaneously or sequentially; this includes bigyny (Kirkendall 1983).
cf. harem.

▶ **inbreeding polygyny** *n.* In Ambrosia and Bark Beetles: a mating system related to a gallery system made by a mother beetle in which her few sons fertilize many daughters (Kirkendall 1983).
cf. mating system: polygyny: harem polygyny; tachygamy

▶ **lek polygyny** *n.* For example, in the Black Grouse, Fallow Deer, Great Snipe, Hammer-Headed Bat, White-Bearded Manakin: polygyny which involves a lek, a group of small territories that each contains a male but no group of females or any resource of value to a potential female mate except for a male's genes (Alcock 1993, 470–471).
cf. lek
Comments: In lek polygyny, females visit males' territories for the sole purpose of becoming inseminated (Alcock 1993, 470).

▶ **mate-defense polyandry** *n.* For example, in the Red-Necked Phalarope: polyandry in which a female guards her male mates from other females (Alcock 1993, 459).

▶ **pure-dominance polygyny** *n.* In some insect species: a polygynous mating system in which some males gain access to mates by excluding others from particular "symbolic" mating territories that are preferred by discriminating females; in this system, males defend a perch on a landmark site or a waiting site on the periphery of a dispersed resource area (Stiles and Wolf 1979, 1).
syn. lek, lek-mating system (Stiles and Wolf 1979, 1; Phillips 1990, 555), lek polygeny
cf. lek

▶ **resource-defense polygyny** *n.* For example, in the African Antelope, Black-Winged Damselfly, Impala, Orange-Rumped Honey Guide, Vicuña: polygyny in which some males prevent others from gaining access to mates by defending resources that attract receptive females; in this system, males defend resources as they occur *in situ* or resources that they collect and (might) prepare for females (Thornhill and Alcock 1983, 234; Alcock 1993, 464–465).

▸ **scramble-competition polygyny** *n.* In the Thirteen-Lined Ground Squirrel, a *Photinus* firefly species: polygyny in which males make no effort to defend individual exclusive mating territories, but instead attempt to outrace their competitors to receptive females (Thornhill and Alcock 1983, 234).

explosive-mating-assemblage polygyny *n.* In the Wood Frog, a horseshoe-crab species, some insect species: polygyny in which receptive females are abundant for a very brief time during which mating is frequent (Thornhill and Alcock 1983, 234; Alcock 1993, 469).

prolonged-searching polygyny *n.* In some insect species: polygyny in which receptive females are "evenly distributed" or male competitors cause a high rate of interferences in a male's mate acquisition (Thornhill and Alcock 1983, 234).

▸ **serial polygyny** *n.* Polygyny in which a male has pair-bonded mates one after another (Wilson 1975, 327).
cf. mating system: polyandry; serial polyandry; polygamy: serial polygamy; polygyny: simultaneous polygyny

▸ **simultaneous polygyny** *n.* Polygyny in which a male has more than one pair-bonded mate at about the same time (Wilson 1975, 327).
syn. contemporaneous polygamy (Palmer 1955)
cf. mating system: polyandry; simultaneous polyandry; polygamy: simultaneous polygamy; polygyny: serial polygyny

▸ **territorial polygyny** *n.* For example, in the Red-Winged Blackbird: a "mating system in which males defend resource-based territories and commonly form pair bonds with more than one concurrent mate" (Wittenberger 1981, 622).

polygyny-polyandry *n.* In some vertebrates: a "mating system in which males are regularly polygynous and females are regularly polyandrous within the same population" (Wittenberger 1981, 619).

promiscuity See mating system: polybrachygamy.

tachygamy *n.* In some Ambrosia Beetle and Bark Beetle species: a mating system related to a gallery system made by one female "in which no lasting pair bond is formed — males that stay at all after copulation leave before oviposition gets underway."
[Greek, *tachys*, rapid, swift; *gamos*, marriage, union; coined by Kirkendall (1983)]

♦ **mating territory** See territory: mating territory.

♦ **mating type** See -type, type: mating type.

♦ **matorral** See habitat: maquis.

♦ **matri-** *combining form* Mother (Michaelis 1963).
cf. patri-
[Latin *mater, matris*, mother]

♦ **matriarchy** See ²group: matriarchy.

♦ **matricliny** *n.* An offspring's having a greater inherited resemblance to its mother than its father (Lincoln et al. 1985).
adj. matriclinic, matriclinal, matriclinous, matrilinear, metroclinal (Lincoln et al. 1985)
cf. patricliny

♦ **matrifocal** *adj.* Referring to a society in which most of the activities and personal relationships are centered on mothers (Wilson 1975, 317, 588).
Comment: Most animal societies are matrifocal (Wilson 1975, 317, 588).

♦ **matrilineal** *adj.*
1. Referring to a maternal line (Michaelis 1963).
2. Referring to things passed from a mother to her offspring (*e.g.,* access to a territory, or status, within a dominance system) (Wilson 1975, 588).
syn. maternal
cf. patrilineal

♦ **matrilineal inheritance** See inheritance: matrilineal inheritance.

♦ **matrix** *n., pl.* **matrices, matrixes**
1. In mathematics, a rectangular array of numbers, or terms, enclosed within parentheses or double vertical bars, to facilitate study of relationships (Michaelis 1963).
2. A table of information.
[Latin *matrix*, womb, breeding animal < *mater, matris*, mother]

amino-acid-substitution matrix *n.* A table that compares the number of amino-acid substitutions between sets of selected taxa (inferred from Doolittle 1996, 471, 474).

▸ **BLOSUM substitution matrix, BLOSUM table** *n.* An amino-acid-substitution matrix (Henikoff and Henikoff 1993 in Doolittle 1996, 474).

▸ **Dayhoff-PAM-250 matrix, PAM-250 matrix** *n.* An amino-acid-substitution matrix (Schwartz and Dayhoff 1978 in Doolittle 1996, 474).

▸ **Gonnet-Cohen-Benner matrix (GCB matrix)** *n.* An amino-acid-substitution matrix (Gonnet et al. 1992 in Doolittle 1996, 474).

h–m

♦ **matrix mutation** See [2]mutation: matrix mutation.

♦ **matromorphic** See -morphic: matromorphic.

♦ **matrone** See chemical-releasing stimulus: matrone.

♦ **maturation** *n.*

1. A person's, or his faculties', becoming fully grown or developed (*Oxford English Dictionary* 1972, entries from 1616).

2. An individual organism's becoming fully grown or developed (*Oxford English Dictionary* 1972, entries from 1664); attaining sexual maturity (Lincoln et al. 1985).

3. An individual animal's seasonal recurrence of reproductive behavior (*e.g.,* nest building or singing) that are separate from initial growth of its underlying neural mechanisms (Tinbergen 1951 in Immelmann and Beer 1989, 181).

4. Gamete differentiation (Lincoln et al. 1985).

5. The increasing complexity, or precision, of an individual animal's behavior that occurs during its growth to sexual maturity and is not learned from prior experience (*e.g.,* a bird's flying without practice) (Heymer 1977, 143; Immelmann and Beer 1989, 180–181).

♦ **maxillation** *n.* A yellowjacket wasps' chewing up a prey insect before presenting it to a larva (Akre 1995, 25).

♦ **maximum-homology hypothesis** See hypothesis: maximum homology hypothesis.

♦ **maximum-likelihood method** See method: maximum-likelihood method.

♦ **maximum population-growth rate** See rate: maximum population-growth rate.

♦ **maximum-power principle** See principle: maximum-power principle.

♦ **maze** *n.* A system of adjoining passageways used to study animals' learning capabilities (Immelmann and Beer 1989, 181).

T-maze *n.* A T-shaped maze.

Y-maze *n.* A Y-shaped maze.

Y-maze globe *n.* An apparatus used to measure turning tendency in some insect species; a Y-maze globe is comprised of six pieces of a thin, flat material that is joined at four points to make four Y-junctions (Hassenstein 1959 in Hinde 1970, 204).

Comment: An insect suspended by its dorsum holds onto the globe with its legs (Hassenstein 1959 in Hinde 1970, 204).

♦ **maze experiment** See experiment: maze experiment.

♦ **Mb** See megabase.

♦ **MCH-dependent odors** See odor: MCH-dependent odors.

♦ **meadow** See habitat: meadow.

♦ **mean, sample mean, $\bar{X}$** *n.* The arithmetic average of a sample (Siegel 1956, xvi). *syn.* arithmetic mean, average, $\bar{Y}$ (Sokal and Rohlf 1969, 42), $\bar{x}$
cf. mode, parameter
Comment: Sokal and Rohlf (1969, 43–44) also define "geometric mean" and "harmonic mean."

trimmed mean ($\bar{x}\{p\}$) *n.* The mean of a central part of a dataset; *e.g.,* $\bar{x}\{0.25\}$ is the average of the middle 50% of a dataset ordered from lowest to highest value (Efron and Tibshirani 1991, 390).

Comment: A trimmed mean can be substantially more accurate than an $\bar{x}$ from a dataset with a long-tailed probability distribution (Efron and Tibshirani 1991, 390).

♦ **mean body temperature** See temperature: mean body temperature.

♦ **mean phenotypic value** See value: mean phenotypic value.

♦ **mean r** See factor: dilution factor.

♦ **mean radiant temperature (T_r)** See temperature: mean radiant temperature.

♦ **meaning** See analysis: message-meaning analysis.

♦ **measurement** *n.*

1. A person's using an instrument to measure anything (Michaelis 1963).

2. A person's assigning numbers to observations and observations into categories in a scientific investigation (Lehner 1979, 109).

See scales of measurement.

♦ **measurement error** See error: measurement error.

♦ **mechanical communication** See communication: mechanical communication.

♦ **mechanical-conflict-of-interest hypothesis** See hypothesis: hypotheses of divergent evolution of animal genitalia: mechanical-conflict-of-interest hypothesis.

♦ **mechanical incompatibility, mechanical isolation** See isolation: mechanical isolation.

♦ **mechanical-physical theory of evolution** See [3]theory: mechanical-physical theory of evolution.

♦ **mechanics** See study of: physics: mechanics.

♦ **mechanism** *n.* The process or technique by which something works or produces an action or effect (Michaelis 1963).

See [3]theory: mechanistic theory.

acquired releasing mechanism (ARM) See mechanism: releasing mechanism.

central excitatory mechanism (CEM) *n.* A mechanism that is receptive to sensory stimuli and hormone influences and dispatches impulses to neural circuits of a behavior pattern (Beach 1942 in Tinbergen 1951, 123).

compensatory mechanism *n*. A postulated central (cerebral) mechanism that can adjust for interim changes in the external environment to direct an animal's behavioral responses (Frisch 1965 in Heymer 1977, 191); *e.g.*, in scout bees that change their dances with the Sun's movement in ways that convey correct information about direction of a new nesting site.

hysteresis mechanism *n*. An animal's mechanism that causes a delay between its receiving a stimulus and responding (Toates and Oatley 1970 in McFarland 1985, 458).

isolating mechanism *n*. "Any agent that hinders the interbreeding of groups of individuals" (Dobzhansky 1937, 230 in Mayr 1982, 274); an agent that causes isolation, *q.v.*

Comment: Brown (1975, 404–405) lists some kinds of isolating mechanisms which are not included here.

▸ **behavioral isolating mechanism** See mechanism: isolating mechanism: sexual isolating mechanism.

▸ **biological isolating mechanism** *n*. "Biological properties of individuals which prevent the interbreeding of populations that are actually or potentially sympatric" (Mayr 1963, 91, in Mayr 1982, 274), *e.g.*, an incompatible behavior during courtship or a physiological barrier that prevents zygote formation if insemination occurs.

cf. mechanism: isolating mechanism: geographical isolating mechanism

Comment: "An occasional individual in an otherwise perfectly good species may hybridize. In other words, isolating mechanisms can only provide the integrity of populations, but not of every last single individual" (Mayr 1963, 91, in Mayr 1982, 274).

▸ **ethological isolating mechanism** See mechanism: isolating mechanism: sexual isolating mechanism.

▸ **extrinsic isolating mechanism** *n*. Any environmental barrier that isolates potentially interbreeding populations (Lincoln et al. 1985).

cf. mechanism: isolating mechanism: intrinsic isolating mechanism

▸ **geographical isolating mechanism** *n*. A physical part of the environment that prevents the interbreeding of populations (*e.g.*, a mountain chain, river, or ocean) (Mayr 1982, 274).

See isolating mechanism.

syn. locomotor isolating mechanism (Brown 1975, 404)

▸ **intrinsic isolating mechanism** *n*. Any of an organism's genetically determined factors that prevents interbreeding between individuals of sympatric species (Lincoln et al. 1985).

cf. mechanism: isolating mechanism: extrinsic isolating mechanism

▸ **postzygotic isolating mechanism** *n*. An isolating mechanism that prevents interbreeding between two or more populations after zygote formation (Lincoln et al. 1985).

syn. post-mating isolating mechanism (Brown 1975, 404)

▸ **prezygotic isolating mechanism** *n*. An isolating mechanism that prevents interbreeding between two or more populations before fertilization or zygote formation (Lincoln et al. 1985).

syn. pre-mating isolating mechanism (Brown 1975, 404)

▸ **sexual isolating mechanism, behavioral isolating mechanism, ethological isolating mechanism** *n*. An organism's behavior that prevents heterospecific interbreeding (Brown 1975, 404).

syn. behavioral isolating mechanism (Brown 1975, 404)

releasing mechanism (RM) *n*.

1. All of an animal's postulated central-nervous-system mechanisms, not including motor mechanisms, that play an important part in releasing one of its responses (Heymer 1977, 27).

2. All of an animal's sensory and neural components contributing to its filtering incoming stimuli, thus determining that only relevant stimuli trigger its behavior (Immelmann and Beer 1989, 251).

Comment: Schleidt (1962 in Heymer 1977, 27) notes that "releasing mechanism" should be used whenever one does not know whether a connection between a stimulus and response is originated by a phylogenetic or ontogenetic adaptation.

▸ **acquired releasing mechanism (ARM)** *n*. A releasing mechanism that an organism obtains in adapting to a certain situation during its ontogeny (Heymer 1977, 57); this involves its learning releasing stimuli (Immelmann and Beer 1989, 251).

cf. mechanism: releasing mechanism: innate releasing mechanism

Comment: This term should be used only if experiments demonstrate that the response in question cannot be released by an innate releasing mechanism (Schleidt 1962 in Heymer 1977, 57).

▸ **innate releasing mechanism (IRM)** *n*.

1. An animal's hypothesized "special sensory mechanism" that releases its

h–m

reactions in response to a "certain set of sign stimuli" and "is responsible for its susceptibility to such a very special combination of sign stimuli" [coined by Tinbergen 1951, 42, which is his free translation of their German term *das angeborene auslösende Schema* of von Uexküll and Lorenz].

2. A hypothesized locus in an animal's central nervous system on which sign stimuli act to release fixed-action patterns (Lorenz 1950; Tinbergen 1950 in McFarland 1985, 364).

cf. acquired mechanism: acquired releasing mechanism

Comments: With further research, workers found that behavioral control stems not so much from actions of IRMs but from filtering processes that occur along afferent pathways (Dewsbury 1978, 21). This term should be used only if the presence of an IRM has been demonstrated in individuals that have been isolated from the sign stimulus in question (Heymer 1977, 19). "IRM" is now used as a term less freely (Immelmann and Beer 1989, 251).

innate releasing mechanism modified by experience (IRME) *n*. An IRM that is modified by ontogenetic adaptation to a certain situation by habituation, sensitization, learning, or a combination of these factors (Schleidt 1962 in Heymer 1977, 57).

♦ **mechanist** *n*. One who holds that natural phenomena can be explained by the laws of chemistry and physics (*Oxford English Dictionary* 1972, entry from 1668; Mayr 1982, 835).

cf. ³theory: mechanistic theory

atomistic mechanist *n*. One who believes that manifestations of life originate through the assembly of appropriate configurations of atoms as did Lucretius (99–55 BC), a follower of Epicurus (Mayr 1982, 90).

♦ **mechanistic theory** See ³theory: mechanistic theory.

♦ **mechanistic world picture, mechanistic world view** *n*. Galileo's view that developed from and is similar to essentialism, *q.v.*, that was spread by Descartes in particular (Mayr 1982, 38, 97).

♦ **mechanoreception** See modality: mechanoreception.

♦ **Meckel-Serrès law** See law: Meckel-Serrès law.

♦ **Medawar-Williams hypothesis of senescence** See hypothesis: Medawar-Williams hypothesis of senescence.

♦ **meddling** *n*.

1. A person's taking part, dealing, managing; now only in a bad sense, a person's taking part officiously in other's affairs; interfering (*Oxford English Dictionary* 1972, entries from 1374).

2. An individual organism's interfering with its rival's activities (*e.g.,* a barnacle's growing over and destroying its rival or a male salamander's disrupting the courtship of another male) (Huntingford and Turner 1987, 6).

cf. aggression, competition

♦ **media worker** See caste: worker: media worker.

♦ **medical psychology** See study of: psychology: medical psychology.

♦ **medroxyprogesterone acetate, Depo-Provera**® See drug: medroxyprogesterone.

♦ **mega- (M)** *prefix* Large, great, greater than usual, used to denote a unit × 10⁶ (Lincoln et al. 1985).

♦ **megabase (Mb)** *n*. 1,000,000 base pairs of a DNA molecule (Goffeau et al. 1996, 566).

♦ **megaevolution** See ²evolution: macroevolution.

♦ **megafauna** See fauna: megafauna.

♦ **megaflora** See flora: megaflora.

♦ **megagamete** See gamete: megagamete.

♦ **megagametogenesis** See -genesis: megagametogenesis.

♦ **megaloplankton** See plankton: megaloplankton.

♦ **megaplankton** See plankton: megaplankton.

♦ **megistotherm** See therm: megatherm.

♦ **meiny, meinie** See ²group: meiny.

♦ **meio-** *prefix* Smaller, less than (Lincoln et al. 1985).

♦ **meiobenthos** See ²community: benthos: meiobenthos.

♦ **meiofauna** See fauna: meiofauna.

♦ **meiosis, maiosis, miosis, gametogenesis, meiotic division, prokaryogamety, reduction division** *n*. Two successive divisions of a diploid nucleus that precede the formation of haploid gametes, or meiospores, each of which contains one of each pair of the homologous chromosomes of their parent cell (Lincoln et al. 1985).

adj. meiotic

cf. mitosis, sexual reproduction

gametic meiosis *n*. In most animal species: meiosis that immediately precedes gametogenesis (Margulis et al. 1985, 73).

zygotic meiosis *n*. In many fungus species: meiosis that immediately follows zygote formation (Margulis et al. 1985, 73).

♦ **meiotic cost** See cost: meiotic cost.

♦ **meiotic division** See meiosis.

♦ **meiotic drive** See drive: meiotic drive.

♦ **meiotic-drive load** See genetic load: meiotic-drive load.

◆ **meiotic-drive locus** See allele: meiotic-drive locus.

◆ **meiotic parthenogenesis** See parthenogenesis: automictic parthenogenesis.

◆ **meiotic sex** See sexual reproduction: eukaryotic sex.

◆ **meiotic thelytoky** See parthenogenesis: automixis.

◆ **melangeophile** See ¹-phile: melangeophile.

◆ **melanophore** *n.* For example, in Octopi, Squid, some fish, frog, and toad species: a cell that contains melanin and expands and contracts under neurendocrine control (Curtis 1983, 786).

◆ **meliphage** See -vore: melivore.

◆ **meliphagous, mellisugous, mellisugent, mellivorous** *adj.* Referring to an organism that feeds on honey (Lincoln et al. 1985).
cf. -vore: mellivore

◆ **melittology** See study of: melittology.

◆ **melittophile** See ²-phile: melittophile.

◆ **melliferous** *adj.* Honey producing.

◆ **mellissophobia** See phobia (table).

◆ **mellivore** See -vore: mellivore.

◆ **melotope** *n.* Bioacoustical neologism: a habitat that has "specific characteristics that affect sound transmission within it" (Immelmann and Beer 1989, 182).
cf. acoustic window

◆ **membrane potential** See potential: membrane potential.

◆ **membranous droplet** See organized structure: membranous droplet.

◆ **membranous labyrinth** See labyrinth: membranous labyrinth.

◆ **meme** *n.* A replicator, or unit, of cultural transmission; a unit of imitation (Dawkins 1977, 206); "a unit of cultural inheritance, hypothesized as analogous to the particulate gene, and as naturally selected by virtue of its 'phenotypic' consequences on its own survival and replication in the cultural environment" (Dawkins 1982, 290).
cf. gene
Comments: Some researchers dismiss the concept of meme (S.J. Gould, H.A. Orr); others value this concept (S. Blackmore, D. Dennett) (Kher 1999, 53).
[coined by Dawkins (1977), Latin *imitatio, -onis,* imitation]

◆ **meme pool** See pool: meme pool.

◆ **memory** *n.*
1. A person's faculty for remembering something; his capacity for retaining, perpetuating, or reviving a thought of things past (*Oxford English Dictionary* 1972, entries from 1340).
2. A person's act, or instance, of remembering; a representation in his memory; a

recollection (*Oxford English Dictionary* 1972, entries from 1842).
3. In animals, except Mesozoa and Porifera: an individual's capacity to store experience-derived information in and subsequently retrieve information from its nervous system (Foffa 1968, etc. in Heymer 1977, 74).
cf. Genetic memory, long-term memory potentiation

episodic memory *n.* A person's memory of past situations (inferred from Hall 1998, 30).

explicit memory *n.* For example, in Humans: a consciously remembered memory; contrasted with implicit memory (Lemonick 1999, 56–57).
syn. declarative memory (Lemonick 1999, 57)
Comment: Explicit memory contains subsystems that handle faces, names, parts of speech, shapes, sounds, and textures.

fact memory *n.* A person's memory of information, such as a phone number or another person's face; contrasted with skill memory (Campbell 1996, 1021).

flashbulb memory *n.* A person's recall of defining moments in his life (*e.g.,* what he was doing when he heard about the shooting of President J.F. Kennedy) (Hall 1998, 30).

implicit memory *n.* For example, in the Human: memory that one retains from an earlier practice session without consciously remembering it; contrasted with explicit memory (Lemonick 1999, 56–57).
syn. nondeclarative memory (Lemonick 1999, 57)
Comment: For example, the person H.M., who cannot form most kinds of new memories, shows implicit memory (Brenda Milner (1962) in Lemonick 1999, 56–57). Associative learning and habituation are parts of implicit memory.

long-term memory *n.* For example, the Honey Bee, Human; *Drosophila:* memory that lasts long enough so that it is not obliterated due to a disturbance (Gould 1982, 261); memory stored for days to years or even a lifetime; contrasted with short-term memory (Immelmann and Beer 1989, 183).
Comment: About 100 genes are involved in long-term memory in *Drosophila* (Tim Tully in Hall 1998, 33).

priming *n.* A person's unconscious recall of short visual, or auditory, cues (Hall 1998, 30).

short-term memory *n.* For example, in the Honey Bee, Human; *Drosophila:* memory that can be obliterated within a

h – m

few minutes of learning due to a disturbance (Gould 1982, 261); memory stored for a few seconds to up to a few hours; contrasted with long-term memory (Immelmann and Beer 1989, 183).

cf. memory: working memory

skill memory *n.* A person's memory of motor activities that he learns by repetition without consciously remembering specific information (*e.g.*, tying shoes, riding a bicycle, walking) (Campbell 1996, 1021–1022).

Comments: Skill memories are difficult to unlearn (Campbell 1996, 1022).

topographic memory *n.* Memory of locations, or sites, that is part of migratory, or homing, behavior (Heymer 1977, 127).

working memory *n.* "A brain system that provides temporary storage and manipulation of the information necessary for such complex cognitive tasks as language comprehension, learning, and reasoning" (Baddeley 1992, 556).

cf. memory: short-term memory

Comments: Working memory can be divided into three subcomponents: the "central executive," "visuospatial sketch pad," and "phonological loop," which are defined by Baddeley (1992). The concept of "working memory" has increasingly replaced the concept of "short-term memory" (Baddeley 1992, 256).

♦ **memory system** See system: memory system.

♦ **memory trace** See -gram: engram.

♦ **menagerie** See ²group: menagerie.

♦ **Mendel's laws** See law: Mendel's laws.

♦ **Mendel's theory of inheritance** See ³theory: Mendel's theory of inheritance.

♦ **Mendel's theory of particulate inheritance** See inheritance.

♦ **Mendel's three laws** See law: Mendel's laws.

♦ **Mendelian** *n.*
1. One who accepts Gregor Johann Mendel's theory of inheritance, *q.v.* (*Oxford English Dictionary* 1972, entries from 1901; Mayr 1982, 716).
2. One who accepts genetic findings from 1900 to 1915 (Mayr 1982, 716).
   Note: Because Mendel did not promote these new findings, then he, himself, was not a Mendelian in this sense (Mayr 1982, 716).
3. An evolutionist who uses "the particulate-ness (discontinuity) of genetic factors as evidence for the importance of saltational processes in evolution, particularly in the origin of species" Mayr 1982, 541).

Note: "Mendelians" believed in typological species and were concerned with frequencies of genes in closed gene pools. By and large, they ignored the existence of populations, problems of multiplication of species, origin of higher taxa, origin of evolutionary novelties, and ultimate causations for genetic change in natural populations (Mayr 1982, 541–542). "Evolutionists" were mainly in two camps after 1900: the "Mendelians" and the "naturalists," *q.v.* (Mayr and Provine, 1980 in Mayr 1982, 541). From 1900 to 1906 the Mendelians opposed the biometricians, *q.v.* (Mayr 1982, 778).

syn. experimental geneticist
cf. Mendelianism

♦ **Mendelian character** See character: Mendelian character.

♦ **Mendelian chromosome theory** See ³theory: chromosome theory.

♦ **Mendelian cross** See cross: Mendelian cross.

♦ **Mendelian genetics** See study of: genetics: Mendelian genetics.

♦ **Mendelian inheritance** See ³theory: Mendel's theory of inheritance, inheritance.

♦ **Mendelian population** See ²population: Mendelian population.

♦ **Mendelian ratio** See ratio: Mendelian ratio.

♦ **Mendelian trait** See character: Mendelian character.

♦ **Mendelianism** *n.*
1. A period from 1900 to about 1909 which "was preoccupied with evolutionary controversies and doubts as to the universal validity of Mendelian inheritance" (Mayr 1982, 731–732).
   Note: This period was dominated by Bateson, de Vries, and Johannsen, who have often been designated as "the early Mendelians." "Mendelism" conveys different meanings to different persons, depending on what aspect of Mendelism one wants to emphasize (Mayr 1982, 731–732).
2. To geneticists, a period in which particulate inheritance was confirmed and hardness of inheritance emphasized (Mayr 1982, 731–732).
3. To evolutionists, "a period in which utterly erroneous ideas about evolution and speciation were promulgated by leading geneticists and during which mutation pressure was considered far more important than selection, ideas which resulted in the alienation of the naturalists" (Mayr 1982, 731–732).

♦ **Mendelism** *n.* Inheritance in accordance with the chromosome theory of heredity (Lincoln et al. 1985).

syn. particulate inheritance (Lincoln et al. 1985)

cf. inheritance

♦ **menopause** *n.* The period in a woman's life when menstruation ceases; contrasted with perimenopause and surgical menopause (Wallis 1995, 50).

cf. post-menopausal zest

Comments: The average age of menopause in the U.S. is 51 (Wallis 1995). Hot flashes, night sweats, vaginal dryness, and other symptoms of estrogen "withdrawal" occur around menopause when a woman's ovaries produce less and less estrogen.

perimenopause *n.* The transitional period between a woman's "fertile period" and her menopause (Wallis 1995, 49).

Comments: In the U.S., perimenopause occurs in women from age 35 through their early 50s (Brody 1998, C7). Symptoms of perimenopause include menstrual irregularity, drier skin, dropping estrogen levels in blood, hair's becoming more brittle and sparser under arms and between legs, hot flashes, loss of concentration, mood swings, vaginal dryness, and possible loss of libido (Wallis 1995, 49; Brody 1998, C7).

surgical menopause *n.* Human menopause caused by removal of a woman's uterus and ovaries (Wallis 1995, 50).

♦ **mensurative experiment** See experiment: mensurative experiment.

♦ **mental** *adj.*
1. Of or pertaining to a person's mind or intellect (*Oxford English Dictionary* 1972, entries from *ca.* 1425).
2. Referring to something that is carried on in a person's mind (*Oxford English Dictionary* 1972, entries from 1526); especially without the aid of written symbols: *mental* calculations (Michaelis 1963).
3. Relating to a person's mind as an object of study; concerned with phenomena of the mind (*Oxford English Dictionary* 1972, entries from *ca.* 1820).

[French < Late Latin *mentalis* < Latin *mens, mentis,* mind]

♦ **mental chronometry** See -metry: mental chronometry.

♦ **mental experience** *n.* An animal's possible "thoughts about objects and events that are remote in time and space from the immediate flux of sensations" (Griffin 1976, 5).

syn. mental image (Griffin 1976, 5)

Comment: Griffin (1976) does not state that nonhuman animals definitely have mental experiences.

♦ **mental gland** See gland: mental gland.

♦ **mental image** See mental experience.

♦ **mental map** See cognitive map.

♦ **mentality** *n.*
1. That which is the nature of a person's mind or mental action (*Oxford English Dictionary* 1972, entries from 1691); mental faculties or experiences (Michaelis 1963).
2. A person's "intellectual quality, intellectuality" (*Oxford English Dictionary* 1972, entries from 1856); intellectual capacity or power (Michaelis 1963).
3. An animal's having a declarative representation, *q.v.* (McFarland 1985, 356).

cf. mind

Comment: McFarland does not state that nonhumans definitely have mentalities.

♦ **mercy call** See animal sounds: call: mercy call.

♦ **merdicole** See -cole: fimicole.

♦ **merdivore** See -vore: merdivore.

♦ **meristic variable** See variable: discontinuous variable.

♦ ***Merkwelt*** See -welt: *Merkwelt.*

♦ **mero-, mer-** *prefix* Part, incomplete, partial (Michaelis 1963).

[Greek *meros,* a part, division]

♦ **merogamodeme** See deme: merogamodeme.

♦ **merogamy** See -gamy: merogamy.

♦ **merohermaphroditic population** See population: merohermaphroditic population.

♦ **merohyponeuston** See neuston: merohyponeuston.

♦ **merological method** See method: merological method.

♦ **meromorphosis** See -morphosis: meromorphosis.

♦ **meroneuston** See neuston: meroneuston.

♦ **meroparasitism** See parasitism: meroparasitism.

♦ **meropelagic** See pelagic: meropelagic.

♦ **meroplankton** See plankton: meroplankton.

♦ **merostasis** See ^{2}heterochrony: merostasis.

♦ **Mertensian mimicry** See mimicry: Emsleyian mimicry.

♦ **mesarch succession, mesosere** See ^{2}succession: mesarch succession.

♦ **meso-, mes-** *combining form*
1. Situated in the middle (Michaelis 1963).
2. "Intermediate in size or degree" (Michaelis 1963).

cf. pro-, meta-

[Greek *mesos,* middle]

♦ **mesobenthos** See ^{2}community: benthos: mesobenthos.

♦ **mesobiota** See biota: mesobiota.

♦ **mesochthonophile** See ^{1}-phile: mesochthonophile.

♦ **mesoderm** See organ: -derm: mesoderm.

h – m

♦ **mesoglea** [MES o GLEE ah] *n*. The middle layer of a cnidarian's body wall (Barnes 1974, 88).
Comments: Mesogleas range from a thin, noncellular membrane to a thick, fibrous, jellylike, mucoid material with, or without, wandering amoebacytes (Barnes 1974, 88).

♦ **mesokurtic distribution** See distribution: mesokurtic distribution.

mesolithion *n*. Organisms inhabiting rock cavities (Lincoln et al. 1985).

♦ **mesolithophage** See -phage: lithophage (comment).

♦ **mesology** See study of: ecology.

♦ **mesophenic population** See ²population: mesophenic population.

♦ **mesophile** See ¹-phile: mesophile.

♦ **mesoplankton** See plankton: mesoplankton.

♦ **mesoplastic** See morphic: mesomorphic.

♦ **mesopredator release** *n*. An increase in the number of smaller predators that results from removal of larger predators in a particular area (Crooks and Soulé 1999, 563; Sæther, 1999, 510).
Comment: For example in Western U.S., removal of large predators (Coyotes) results in an increase of smaller predators (Domestic Cat, Gray Fox, O'possum, Raccoon, and Striped Skunk), which in turn results in a reduction in numbers of scrub birds (Crooks and Soulé 1999, 563; Sæther, 1999, 510).

♦ **mesopsammon** *n*. Organisms living in interstitial spaces of a sandy sediment (Lincoln et al. 1985).
syn. psammon (Lincoln et al. 1985)

♦ **mesosoma** *n*. In insects: the middle of the three major divisions of an individual's body (Wilson 1975, 588).
cf. metasoma, prosoma
Comment: In most insects, "mesosoma" is the strict equivalent of a thorax, but in more derived Hymenoptera it includes the propodeum, the first segment of the abdomen fused to the thorax (Michener 1944 in Michener 1974, 6).

♦ **mesosphere** See -sphere: mesosphere.

♦ **mesotherm, mesothermophyte** See therm: mesotherm.

♦ **mesothermophile** See ¹-phile: thermophile: mesothermophile.

♦ **mesotroph** See -troph-: mesotroph.

♦ **message** *n*.
1. A communication that is transmitted through a human messenger or other agency to a Human; an oral or written communication from one person to another (*Oxford English Dictionary* 1972, entries from 1297).

2. The type of information that is transmitted from one individual organism to another during communication (Storz 1973, 169).
cf. information, signal

♦ **message categories** *pl. n*. The 12 groups into which 10 to 50 displays of vertebrate species are classified (Smith 1969a in Wilson 1975, 217).

♦ **message-meaning analysis** See analysis: message-meaning analysis.

♦ **message set** *n*. A group of species-specific communicatory signals, typically ranging from 15 to 45 signals (Dewsbury 1978, 99).

♦ **messenger RNA** See nucleic acid: ribonucleic acid: messenger RNA.

♦ **meta-, met-** *prefix*
1. "Changed in place or form; reversed; altered" (Michaelis 1963).
2. Anatomically, zoologically, "behind; after; on the farther side of; later; often equivalent to post- or dorsi" (Michaelis 1963).
3. "With; alongside" (Michaelis 1963).
4. "Beyond; over; transcending" (Michaelis 1963).
cf. pro-, meso-
[Greek *meta*, after, beside, with]

♦ **metabiosis** See symbiosis: metabiosis.

♦ **metabolism** *n*.
1. A cell's, or organism's, producing living matter from nutritive matter (*Oxford English Dictionary* 1972, entries from 1878); all of a living organism's synthetic and degradative biochemical processes (Lincoln et al. 1985).
2. Ecological energetics: respiration, *q.v.* (Lincoln et al. 1985).
See metastasis.
cf. metamorphosis

anabolism *n*. Metabolism that involves an organism's manufacture of complex substances from less complex ones which results in energy utilization (Lincoln et al. 1985).

bradymetabolism *n*. The low basal-metabolism levels of reptiles and other nonavian and nonmammalian animals relative to those of birds and mammals of the same body size and at the same tissue temperature; contrasted with tachymetabolism (Bligh and Johnson 1973, 944).
syn. cold bloodedness (an unsatisfactory synonym that is falling into disuse, Bligh and Johnson 1973, 944)
Comments: "Bradymetablism" differs from "poikilothermy" because the latter signifies conformity of body and ambient temperatures and not all bradymetabolic species are temperature conformers, *q.v.*;

however, some are ectothermic temperature regulators (Bligh and Johnson 1973, 944). [Greek *bradus*, slow, sluggish; *metabole*, change]

catabolism, katabolism *n*. Metabolism that involves an organism's degradation of complex substances into less complex ones which results in energy liberation (Lincoln et al. 1985).

tachymetabolism *n*. The high basal-metabolism levels of birds and mammals compared to those of reptiles and nonavian and nonmammalian animals of the same body weight and at the same tissue temperature; contrasted with bradymetabolism (Bligh and Johnson 1973, 954).
syn. warm bloodedness (Bligh and Johnson 1973, 944)
[Greek *takhus*, fast; *metabole*, change]

♦ **metabolous** *adj*. Referring to a pattern of development that includes a metamorphosis (Lincoln et al. 1985).

♦ **-metabolous** See metamorphosis.

♦ **metacentric chromosome** See chromosome: metacentric chromosome.

♦ **metachrosis** *n*. An animal's ability to change color (Lincoln et al. 1985).
cf. color change, coloration

♦ **metacommunication** See communication: metacommunication.

♦ **metacyclic** *adj*. Referring to a parasite's life-cycle state that infects its definitive host, *q.v.* (Lincoln et al. 1985).

♦ **metafemale** See female: metafemale.

♦ **metagamic** See -gamic: metagamic.

♦ **metagamic sex determination** See sex determination: metagamic sex determination.

♦ **metagamy** See -gamy: metagamy.

♦ **metagenesis** See alternation of generations, -genesis: diagenesis.

♦ **metahermaphroditic population** See ¹population: metahermaphroditic population.

♦ **metamere** *n*. A segment, somite (Michaelis 1963); one of a group of an animal's repeated body parts (*e.g.*, an insect's thoracic and abdominal segments or a vertebrate's pair of spinal nerves).
Comments: Metameres occur in phyla including Annelida, Arthropoda, Chordata, Onychophora, and Pogonophora.
[Greek *meta*, with; *mere*, part, division]

♦ **metamerism** *n*. In some cnidarian species, some platyhelminth species, Annelids, Arthropods, and Chordates: an individual's having its body divided into a linear series of similar parts, segments, or metameres; metamerism is primarily a mesodermal phenomenon — body-wall musculature and sometimes coeloms are the primary seg-

mental divisions (Barnes 1974, 63, 233, 434; Strickberger 1990, 281).
cf. homology: serial homology
Comments: Metamerism is involved with a tendency for secondary reduction in forms that adapt to new conditions; metamerism may have evolved in chordates as an adaptation for undulatory swimming and in annelids as an adaptation for burrowing (Barnes 1974, 63, 233, 434). Strickberger (1990, 281) discusses hypotheses regarding the origin of metamerism.

♦ **metamorphosis** *n*.
1. A change in an individual organism, or its parts, during which it develops from an immature to an adult; especially in insects, a change, or series of changes, that certain insects undergo that results in complete alterations of their forms and habits (*Oxford English Dictionary* 1972, entries from 1665).
2. An organism, or other thing, that is metamorphosed (Michaelis, 1963).
adj. metamorphic
cf. animal names
Comment: Especially dramatic examples of metamorphosis include a tadpole's developing into an adult toad or frog and a caterpillar's developing into a moth or butterfly. Insect metamorphosis can be classified primarily as "ametabolous metamorphosis," "hemimetabolous metamorphosis," "paurometabolous metamorphosis," "holometabolous metamorphosis," and "hypermetabolous metamorphosis."
[Latin < Greek *metamorphōsis*, *metamorphoein*, to transform < *meta-*, beyond + *morphe-*, form]

ametabolous metamorphosis, ametabolous development, ametabolic development, ametabolism, ametaboly, anamorphosis, gradual metamorphosis *n*. In Bristletails, Silverfish, Springtails (Arthropoda): metamorphosis in which immatures gradually develop adult characters through a series of molts (Lincoln et al. 1985).

anamorphosis See metamorphosis: ametabolous metamorphosis.

hemimetabolous metamorphosis *n*. For example, in Damselflies, Dragonflies, Mayflies, Stoneflies: metamorphosis in which aquatic immatures eventually develop external wing structures and gradually develop adult characters through a series of molts and a final molt in which an immature transforms into a aerial adult that looks markedly different than an immature, showing a distinct naiad (larval) and adult stage (Borror et al. 1989, 797).

syn. hemimetabolous development, hemimetabolic development, hemimetabolism, hemimetaboly, hemimetamorphy, heterometabolic metamorphosis, incomplete metamorphosis, simple metamorphosis (Lincoln et al. 1985)

heremetabolic metamorphosis *n*. Metamorphosis with a "resting" phase at the end of a larval (nymphal) stage (Lincoln et al. 1985).

heterometabolic metamorphosis See metamorphosis: hemimetabolous metamorphosis, paurometabolous metamorphosis.

holometabolous metamorphosis *n*. For example, in Coleoptera, Diptera, Hymenoptera, Lepidoptera, Neuroptera: metamorphosis with a distinct larval, pupal, and adult stage, in which wings develop inside the body of a larva (Borror et al. 1989, 67–69). *syn*. complete metamorphosis, holometabolous development, holometabolic development, holometabolism, holometaboly, holometamorphosis, holometamorphy (Lincoln et al. 1985)

hypermetabolous metamorphosis *n*. For example, in *Coelioxys* bees, Meloidae, Strepsiptera: holometabolous metamorphosis in which a first instar is active, seeks out a host, or shows other specific activities but as a later instar is more sedentary, primarily feeding and metamorphosing (Borror et al. 1989, 69). *syn*. hypermetabolous development, hypermetabolic development, hypermetabolism, hypermetaboly, hypermetamorphy (Lincoln et al. 1985)

manometabolous metamorphosis *n*. In some insect species: metamorphosis in which a minor or very gradual change occurs without a resting stage (Lincoln et al. 1985).

paurometabolous metamorphosis *n*. For example, in Hemiptera, Orthoptera: metamorphosis in which immatures eventually develop external wing structures and gradually develop adult characters through a series of molts, showing a distinct larval (or nymphal) and adult stage, with adults not appearing dramatically different than larvae (Borror et al. 1989, 66–67). *syn*. heterometabolic metamorphosis, incomplete metamorphosis, paurometabolous development, paurometabolic development, paurometabolism, paurometaboly, paurometamorphy, paurometabolic metamorphosis, simple metamorphosis (Lincoln et al. 1985)

♦ **metamps** See ²species: metamps.
♦ **metandric hermaphrodite** See hermaphrodite: protogynous hermaphrodite.

♦ **metanotal gland** See gland: metanotal gland.
♦ **metaphysical behaviorism** See behaviorism: metaphysical behaviorism.
♦ **metaphysical Darwinism** See Darwinism.
♦ **metaphysical philosophy** See study of: philosophy: metaphysical philosophy.
♦ **metaphysics** See philosophy: metaphysics.
♦ **metaplasia** See -plasia: metaplasia.
♦ **metapleural gland** See gland: metapleural gland.
♦ **metapopulation** See ¹population: metapopulation.
♦ **metapsychology** See study of: psychology: metapsychology.
♦ **metasoma** *n*. The posterior of the three principal divisions of an insect's body (Wilson 1975, 588).
cf. prosoma, mesosoma
Comment: In most insect groups, "metasoma" is the strict equivalent of the abdomen; in more derived Hymenoptera, it is composed of all but the first segment ("propodeum") which is fused with the thorax and has essentially become part of the "mesosoma" (Michener 1944 in Michener 1974, 6).
♦ **metastasis** *n*.
1. Transference of a bodily function, pain, disease, morbific matter, etc. from one part, or organ, to another (*Oxford English Dictionary* 1972, entries from 1586).
2. An organism's transferring chemical compounds into other compounds in assimilation (*Oxford English Dictionary* 1972, entries from 1875).
3. Transportation of pathogens in a host's body (Lincoln et al. 1985).
Comment: Some workers restrict "metastasis" to an organism's change of nonliving matter to living matter; some synonymize "metastasis" with "metabolism," *q.v.* (*Oxford English Dictionary* 1972).
♦ **metastitism** See cannibalism: metastitism.
♦ **metatarsal gland** See gland: metatarsal gland.
♦ **metathetely** See ²heterochrony: neoteny: hysterotely, metathetely.
♦ **metatroph** See -troph-: metatroph.
♦ **metazoan** See -zoan: metazoan.
♦ **meteorite** *n*. A recovered fragment of a meteoroid that survived passage through Earth's atmosphere (Mitton 1993).
Comments: People usually name meteorites after places where they fall (Mitton 1993).
carbonaceous chondrite *n*. A stony meteorite that bears carbon (Mitton 1993).
Comments: Some carbonaceous chondrites contain abiotically produced hydrocarbon compounds (Browne 1996c, D20).

A CLASSIFICATION OF METEORITES[a]

I. iron (siderite) meteorite
II. stony-iron meteorite (*syn.* lithosiderite, siderolite)
III. stony meteorite (*syn.* aerolite, stone)
   A. achondrites
   B. chondrite
     1. carbonaceous chondrite
     2. enstatite chondrite
     3. ordinary chondrite

---

[a] Mitton (1993).

meteorite ALH 84001, meteorite Allen Hills 84001 *n.* A meteorite from Mars found in Allen Hills, Antarctica, in 1984 and which might contain nanobacterium fossils about 3.5 million years old (Gibbs and Powell 1996, 20, 22; Mckay et al. 1996, 924; Wilford 1996b, illustrations).
Comments: This meteorite landed in Antarctica about 13,000 years ago and may contain Martian fossils of small organisms based on four lines of data presented by Mckay et al. (1996); other researchers dispute Mckay's interpretation. This 4.2-pound meteorite formed about 4.5 million years ago and contains gas that is similar to the current Martian atmosphere (Wilford 1996b, D20). It also contains macroscopic black and white circles embedded in a dark matrix; the circles might be microbe fossils.
meteorite EETA 79001, Elephant Moraine 79001 *n.* A meteorite that crystallized 175 million years ago and was ejected from Mars, presumably by an asteroid impact, 600,000 years ago, and that might contain traces of Martian life (Wilford 1996c, A12).
Comment: This meteorite is considered the "Rosetta stone" of the meteoritic fragments found in Elephant Moraine, Antarctica, and is thought to be Martian because it contains trapped gases that are almost identical in composition to atmospheric samples gathered by the Viking Spacecraft in 1976 (Wilford 1996c, A12). Meteorite 79001 has isotope ratios the match those contained in the oldest Earth fossils (Sawyer 1996, A3).
Murchison meteorite *n.* A 200-pound meteorite, from the asteroid belt, that contains amino acids, carbonate compounds, many other kinds of organic compounds, and water (Gibbs and Powell 1996, 20, 22).
Comments: The Murchison meteorite fell on Murchison, Australia, in 1969 (Zimmer 1995c, 6). This meteorite indicates that organic compounds can be made in space abiotically. Pieces of this meteorite are at the Australian National University, Canberra, and the Smithsonian Institution, U.S.

♦ **meteorology** See study of: meteorology.
♦ **methanotroph** See -troph-: methanotroph.
♦ **method** *n.* The procedures, or techniques, used in, or characteristic of, a particular field of knowledge, thought, practice, etc. (Michaelis 1963).
See -metry.

advocacy method of developing science *n.* A method that involves "author X posing a hypothesis to account for a certain phenomenon, selecting and arranging his evidence in the most persuasive manner possible. Author Y then rebuts X in part or in whole, raising a second hypothesis and arguing his case with equal conviction. Verbal skill now becomes a significant factor. Perhaps at this stage author Z appears as an *amicus curiae*, siding with one or the other or concluding that both have pieces of the truth that can be put together to form a third hypothesis — and so forth *seriatim* through many journals and over years of time" (Wilson 1975, 28).
Comment: Often the advocacy method muddles through to the answer, but at its worse it leads to schools of thought that encapsulate logic for a full generation. This method "has been pursued remorselessly by many writers in the reconstruction of human social evolution" (Wilson 1975, 28).
behavioristic method *n.* The objective method (Tinbergen 1951, 6).
brain stimulation *n.* An investigator's use of microelectrodes with weak electrical current to stimulate specific parts of an animal's brain, elicit particular behaviors, or change motivation and thereby help demonstrate the roles of these brain parts in behavioral regulation (*e.g.,* in some insects, crustaceans, birds, and mammals) (Immelmann and Beer 1989, 35).
character-data methods *n.* Methods used for constructing dendrograms that consist of taxa that share derived characters (Mayr and Ashlock 1991, 275, 291–312).
chemical-degradation method *n.* A method used for determining the DNA sequences of cDNAs (Niesbach-Klösgen et al. 1987, 214).
chromosome walking *n.* A technique for making physical maps of chromosomes that uses cloned gene sequences and probes (Campbell et al. 1999, 376).
[after "walking along" (= sequencing) a chromosome starting at a known gene or other sequence (Campbell et al. 1999, 377)]
cladistic method See study of: systematics: cladistic systematics.

h – m

comparative method *n.* An experimental method that assumes one or more factors that could logically cause a certain effect if their occurrence is significantly correlated with this effect; *e.g.,* if two factors are males that can identify females which are about to become receptive (and receptive females are rare) and these factors are significantly correlated with the presence of males guarding females that are about to become receptive, then the two factors are likely to cause the mate guarding (Thornhill and Alcock 1983, 24).
cf. method: observational method
Comment: Aristotle founded this method (Mayr 1982, 88); G.J. Romanes is generally credited with formalizing its use in studying animal behavior (Drickamer and Vessey 1982, 18).

conditioned-suppression technique *n.* A method used to ascertain a Domestic Pigeon's odor discrimination in which an investigator trains the bird to perform a learned operant task (such as key pecking) until the bird reaches a stable performance level and then presents a stimulus and delivers an electric shock to the bird a short time later; the bird soon ceases key pecking when the stimulus is repeated, even though the investigator continues to shock it after the stimulus is presented. The the investigator then varies stimulus parameters to determine whether or not the bird's responding is suppressed and, thereby, determines its sensory capabilities (Henton et al. 1966 in Dewsbury 1978, 176–177).
cf. method: tracking

cross-fostering See fostering: cross-fostering.

CT sereolithography *n.* A technique that scans the outer body of a vertebrate and produces a replica of its bones (Menon 1995, 57).

dating method *n.* A technique used for determining the age of things including fossils, geological strata, meteorites, minerals, rocks, and subfossils.

▸ **infrared-stimulated-luminescence dating** *n.* A dating method that dates feldspar (Holden 1997a, 1268).

▸ **radioactive dating** *n.* A dating method that estimates the age of a stratum and its associated fossil(s) from the amount of a radioactive isotope and its disintegration product in the stratum, fossil(s), or both (Strickberger 2000, 94).
syn. radiodating (Michaelis 1963)
Comments: Radioactive isotopes used for dating include carbon-14, potassium-40, rubidium-87, samarium-147, thorium-232, and uranium-232, -234, and -235 (Strickberger 2000, 95).

▸ **radiocarbon dating** *n.* A radioactive dating method that uses carbon-14, which disintegrates into nitrogen-14.
Comment: Carbon-14 has a half-life of 5730 years and is used for dating objects less than 50,000 years old, such as charcoal in a Pleistocene campfire (Strickberger 2000, 95).

▸ **radiodating** See method: dating method: radioactive dating.

▸ **thermoluminescence (TL)** *n.* A dating method that determines the age of quartz (Holden 1997a, 1268).
Comments: This method gauges how long quartz-containing rock has been buried from the number of electrons trapped in defects of quartz crystals (Holden 1997a, 1268). These electrons build up at a regular rate but are wiped out by sunlight. The quartz in windblown sand, associated with other objects, can be used to date these objects.

deductive method *n.*
1. Reasoning from the general to the particular using a mathematical logic in structure of thought (Descartes in Mayr 1982, 38, 40).
2. Reasoning from the general to the particular; also, reasoning from stated premises to conclusions formally or necessarily implied by such premises (Michaelis 1963).
3. Formulation of theories, or hypotheses, from which singular testable statements (predictions) are deduced (Lincoln et al. 1985).
syn. deduction
cf. method: inductive method

delayed-reaction method, delayed-response method *n.* A method originally used to test an animal's ability to retain an "idea," it employs a multiple-choice apparatus with one door indicated by either a conditioned or unconditioned stimulus. The animal is not allowed to react while the stimulus is present, but only at a particular time after the stimulus is removed, and the animal's maximum reaction delay is used as an indication of its memory (Tinbergen 1951, 9).
Comment: This method has been criticized (Tinbergen 1951, 9).

dendrochronology, tree-ring chronology *n.* A method of dating using annual tree rings (Lincoln et al. 1985).

detrended-correlation analysis (Decorana, DCA) *n.* A method that graphs relative positions of sites with known organism compositions in ordination space; the graphed site positions reflect differences in community structure

(Hill 1996a, Hill and Gauch 1980 in Didham 1997, 59).

dideoxy-chain-termination method *n*. A method used for determining the DNA sequences of cDNAs (Niesbach-Klösgen et al. 1987, 214).

discriminant-function analysis *n*. A statistical method that classifies observations into one of two or more groups using a new variable that is computed from one or more continuous independent variables or classification variables that alone are not adequate to group the observations (Sokal and Rohlf 1969, 488–489; Helwig and Council, 1979, 181).
cf. analysis: discriminant-function analysis

distance methods *n*. Methods used for determining the similarity, or distance, of taxa from each other (Mayr and Ashlock 1991, 275, 284–291).

DNA fingerprinting *n*. A DNA-analysis technique used to ascertain an animal's possible genetic relationship to a presumed parent or sibling (Lewin 1989a, 1550) or to match a person's, or other organism's, DNA with a DNA sample (Campbell et al. 1999, 382).
Comment: This technique produces a DNA fingerprint, DNA bands resulting from electrophoresis (Campbell et al. 1999, 382, illustration). This technique uses microsatellite analysis, or RFLP analysis by Southern blotting, and requires samples as small as only about 1000 cells. Lewin (1989a) indicates some problems with this technique.

DNA-microarray assay *n*. A technique for detecting and measuring the expression of thousands of genes at one time (Campbell et al. 1999, 379).

electrophoresis *n*. A technique that separates mixtures of organic molecules, based on their different rates of travel in an electric field (Lincoln et al. 1985).
▸ **gel electrophoresis** *n*. Electrophoresis used to separate kinds of nucleic acids or proteins (Campbell et al. 1999, 372–373).
▸ **high-resolution electrophoresis** *n*. Electrophoresis used to separate proteins in a strong electric field and stain them for identification (Wilson 1975, 71).
Comment: This process has revealed a far larger amount of genetic polymorphism in organisms than geneticists had once believed possible (Wilson 1975, 71).
▸ **starch-gel electrophoresis** *n*. Electrophoresis used to separate kinds of proteins, first employed by Markert in evolutionary studies (Mayr 1982, 576).
Comments: A researcher separates the proteins by placing them in a gel and exposing them to an electric field. Depending on their sizes and electrical qualities, the kinds of proteins move different distances in the gel and thus can be separated. This technique is a boon to evolutionary studies because it enables one to ascertain the level of heterozygosity of individuals and populations without having to do breeding experiments. This technique also has weaknesses, as discussed by Mayr (1982, 576).

electroporation *n*. A method in which a researcher gives brief electrical pulses to a solution containing cells which introduce DNA into the cells (Campbell et al. 1999, 370).
Comment: Electroporation creates temporary holes in a cell's plasma membrane, through which DNA can enter (Campbell et al. 1999, 370).

evolutionary method See study of: systematics.

factor analysis *n*. A statistical method that, in biology, aims to resolve complex relationships into the interaction of fewer and simpler factors and to isolate and identify causal factors behind biological correlations (Sokal and Rohlf 1969, 542).
cf. analysis: principle-components analysis

gas-chromatography–mass-spectrometric analytical method, GC–mass-spec, GC–mass-spectrometric analytical method *n*. A means of detecting the amount of diflubenzuron in, or on, leaves without having to purify this chemical (Wimmer et al. 1993, 2184, 2192).
syn. GC-MS (Harrahy 1994, 519)
Comment: This method uses heat-induced breakdown of diflubenzuron during gas chromatography and mass spectrometry by using deuterated diflubenzuron as an internal standard (Wimmer et al. 1993). The great sensitivity and selectivity of mass spectrometry permits rapid analysis of complex leaf extracts without derivatization, or purification, of the pesticide.

high-throughput DNA sequencing *n*. A fast, relatively inexpensive means of sequencing DNA which first sequences DNA and then leaves discovery of its function for future study (Wood 1999, 26).
Comment: This method costs about $2000 to sequence a single gene (Wood 1999, 26).

hological method *n*. A study of a system's totality that concentrates on the whole's operation rather than specific contributions of component parts (Lincoln et al. 1985).
cf. method: merological method

hypothetico-deductive method *n.* One's testing a hypothesis "by determining whether the deductions drawn from it conform to observation" (Mayr 1981, 29; Futuyma, 1986, 6).

Comment: On the Origin of Species... (Darwin 1859) is a classical work using this method.

***in situ* hybridization** *n.* A technique in which a researcher base-pairs a radioactive probe with complementary sequences in denatured DNA of intact chromosomes on a microscope slide (Campbell et al. 1999, 374).

Comment: Researchers use this technique for locating genes on chromosomes (Campbell et al. 1999, 374).

***in vitro* mutagenesis** *n.* A technique that can be used to introduce specific changes into the sequence of a cloned gene (Campbell et al. 1999, 379).

inductive method, induction *n.* "A scientific method involving the formulation of universal statements, such as hypotheses or theories, from singular statements, such as empirical observations or experimental results" (Lincoln et al. 1985).

cf. method: deductive method

maximum-likelihood method *n.* A method used for producing a molecular phylogeny (Adachi and Hasegawa 1996 in Tomitani et al. 1999, 161).

merological method *n.* A study of the component parts of a system made to derive a concept of the operation of its whole (Lincoln et al. 1985).

cf. method: hological method

method of multiple working hypotheses *n.* A procedure used in the hypothetico-deductive method that generates and tests all logical, pertinent hypotheses related to an experiment (Uzzell 1984).

Comment: This method, advocated by T.C. Chamberlin, is "useful in encouraging an open mind and broadening inquiry" (Uzzell 1984).

multiple-choice method *n.* A method used for studying memory and perception in which an animal is trained in a choice chamber to choose, for example, one of several doors marked with a signal (*e.g.,* light bulb). After the animal has mastered its task, it is prevented from exercising its choice as long as the signal is on; after a time, the animal is again allowed to choose, and the longest period which elapses with a subsequently correct response serves as a measure of performance (Heymer 1977, 193).

Comment: Hunter (1913) developed this method, and its reliability has been questioned by Maier and Schneirla (1935) and Tinbergen (1951) (Heymer 1977, 193).

neighbor-joining method *n.* A method used for reconstructing phylogenetic trees (Saitou and Nei 1987 in Tomitani et al. 1999, 161).

nonparametric methods for estimating species richness from samples *Comment:* These methods include S_1^*, Chao-1; S_2^*, Chao-2; S_3^*, Jackknife-1; S_4^*, Jackknife-2; S_5^*, Bootstrap; S_6^*, Chao and Lee-1; S_7^*, Chao and Lee-2; and Michaelis-Menten equation (Colwell and Coddington 1994, 109–111). These methods vary in how well they estimate species richness with regard to different sample sizes.

northern blotting *n.* A technique similar to Southern blotting, *q.v.,* that subjects whole mRNA molecules to hybridization analysis, typically to determine whether a particular gene has been transcribed and how much of its mRNA is present (Rieger et al. 1991; (Campbell et al. 1999, 375).

syn. northern transfer (Rieger et al. 1991)

cf. method: western blotting

Comment: In northern blotting, a researcher transfers mRNA from an agarose gel to a nitrocellulose filter on which it can be hybridized to a complementary radioactive, single-stranded DNA or an RNA probe (Rieger et al. 1991).

nucleic-acid hybridization *n.* A technique in which a researcher base-pairs single-stranded nucleic acid from different nucleic acid molecules (Campbell et al. 1999, 369).

observational method *n.* A method promoted by C.L. Morgan involving the use of only data that are gathered by direct experimentation and observation to make generalizations and develop theories (Drickamer and Vessey 1982, 19).

cf. method: comparative method

omnispective method See study of: systematics.

phenetic method See study of: systematics.

polymerase-chain reaction (PCR) *n.* A method used to amplify minute DNA samples (even one DNA molecule) *in vitro* for DNA sequence analysis (Alberts et al. 1989, 269; Morell 1992a, 1860; Campbell et al. 1999, 371).

Comments: Kary Mullis conceived this method in 1983 for which he won the Japan Prize and the Nobel Prize in 1993 (Dwyer 1993, 8; Appenzeller 1993). This method enables workers to extract genes from fossils, positively identify human remains from bones, identify genes in developing fetuses, determine whether someone is infected with the HIV virus, test for chlamydia (a venereal disease that can cause infertility) from a urine sample, and determine if a man really raped someone (resulting in freeing men wrongly accused of rape). PCR using chromosomal DNA does not uniquely identify a person, but it can be used to calculate a probability that two people will match in this DNA coincidentally (Cohen 1995, 22). This gene-copying technique has earned hundreds of millions of dollars for its patent holder, Hoffmann-LaRoche (Milstein 1994, 655).

Popperian method *n.* A scientific method that involves the formulation of general hypotheses from which singular statements (predictions) are deduced that can be tested; a hypothesis may be falsified, or corroborated, but not proved (Lincoln et al. 1985).
syn. deductive method (Lincoln et al. 1985)
cf. method: deductive method

principle-components analysis *n.* A branch of factor analysis that calculates principle axes through hyperellipsoids and removes the greatest, second greatest, and successively smaller sources of variation as each new axis is calculated (Sokal and Rohlf 1969, 532).

putting-through method *n.* A method that one uses to teach an animal (*e.g.*, the Chimpanzee) specific body movements or placement of its body parts (Immelmann and Beer 1989, 165).

site-directed mutagenesis *n.* A method that produces a mutation in a specific amino-acid coding site (Amato 1993).
Comments: This method is used to reorder amino acids to understand protein function and produce enzymes that are more stable under industrial conditions. Further, it raises the prospect of redesigning proteins to have entirely new functions (Amato 1993). Michael Smith developed this technique in the early 1980s and won the Nobel Prize for it in 1993.

Southern blotting *n.* A technique that reveals whether a particular sequence is present in a DNA sample and which restriction fragments contain the sequence (Rieger et al. 1991; Campbell et al. 1999, 373).
syn. genome blotting, Southern-blot technique, Southern transfer (Rieger et al. 1991), Southern-blot hybridization (Niesbach-Klösgen et al. 1987, 214), Southern hybridization (Campbell et al. 1999, 373)
cf. method: northern blotting, western blotting
Comment: In Southern blotting, a researcher transfers single-stranded, restricted DNA fragments, separated in an agarose gel, to a nitrocellulose filter (or other binding matrix); analyzes the fragments by

hybridization to radioactive or biotinylated single-stranded, DNA; and detects the hybrids by autoradiography or color change, respectively (Rieger et al. 1991).
[After E.M. Southern, who developed this technique in 1975 (Campbell et al. 1999, 375)]

subcloning *n.* A technique that identifies a specific part of a gene, such as its promoter, or removes unwanted DNA segments from a clone (Niesbach-Klösgen et al. 1987, 214; Rieger et al. 1991).

synthetic method See study of: systematics.

"3ST" method *n.* A method used to test the branching order of phylogenetic trees (Kimura 1981 in Niesbach-Klösgen et al. 1987, 214).

titration method *n.* An experimental method involving a model (dummy), the features of which are varied in ways that enable an investigator to assess the relative importance of the model's traits (Baerends and Kruijt 1973 in Lehner 1979, 96–97).

tracking *n.* A method used to ascertain an animal's (*e.g.*, a pigeon's) sensory ability that involves two pecking keys, a varying stimulus, and an apparatus that records the bird's response to the stimulus (Blough 1961 in Dewsbury 1978, 178).
cf. method: conditioned-suppression technique
[named after a subject's own "tracking" of its threshold to a stimulus by causing a graph to be plotted automatically by experimental apparatus]

tree-ring chronology See method: dendrochronology.

true Baconian method See inductivism.

two-way indicator-species analysis (TWINSPAN) *n.* A polythetic, divisive method of classification used to cluster sites of similar species composition and identify indicator species characteristic of different sites or groups of sites (Hill 1996b in Didham 1997, 59).

western blotting *n.* A technique that identifies particular protein species (Rieger et al. 1991).
cf. method: northern blotting, Southern blotting
Comment: In western blotting, a researcher transfers proteins from a polyacrylamide gel onto a suitable immobilizing matrix (*e.g.*, a nitrocellulose sheet) and probes the proteins with a specific antibody to identify the proteins (Rieger et al. 1991).

Winkler method *n.* A means of separating soil organisms from leaf litter (Didham 1997, 58).

Comment: In this method, one places a litter sample into a coarse mesh bag which is then suspended inside a large sealed cloth bag. As the litter dries, invertebrates that are sensitive to desiccation move downward through the mesh bag and fall into a jar of alcohol beneath it.

♦ **methodological behaviorism** See behaviorism.

♦ **metoecious** See -oecism: heteroecism.

♦ **metoxenous** See -xenous: metoxenous.

♦ **-metrics**
cf. -metry, study of: -metry
   biometrics *n.* Use of statistical methods to study organisms (Sokal and Rohlf 1969, 2).
   psychometrics *n.* Use of quantitative methods to study human psychology.
   zoometrics, zoometry *n.* Use of statistical methods to study animals (Lincoln et al. 1985).

♦ **metroclinal** See matricliny.

♦ **metromorphy** See -morphy: metromorphy.

♦ **-metrosis**
adj. -metrotic
   haplometrosis, monometrosis *n.* In many social-animal species, including most ant species: a single fertile female's, or queen's, founding a colony (Lincoln et al. 1985; Rissing and Pollock 1987, 975).
   pleometrosis *n.* In many social-animal species, including some ant species: more than one fertile females', or queens', founding a colony (Rissing and Pollock 1987, 975; Hölldobler and Wilson 1990, 641).

♦ **-metry** *combining form* "The process, science, or art of measuring" (Michaelis 1963). See study of: -metry.
cf. method, metrics
[Greek *metria* < *metron*, a measure]
   allometry See -auxesis: heterauxesis.
   biotelemetry *n.* A method for marking and locating animals involving attaching radio transmitters to them (Lehner 1979, 199–203).
syn. radio tracking (Lehner 1979, 199–203)
   chronometry See study of: -metry: chronometry.
   ethometry *n.* The quantitative and function analysis of releasing mechanisms and key stimuli [coined by Jander (1968 in Heymer 1977, 65)].
   geochronometry See study of: geochronology.
   mental chronometry *n.* A technique used, for example, on pigeons and Humans to obtain an objective measure of intangible phenomena such as mental representations and which uses the time required to solve a spatial problem as an index of processes involved (Posner 1978 in McFarland 1985, 498).

morphometry, morphometrics *n.*
1. The art, or process, of measuring an object's external form (*Oxford English Dictionary* 1972, entries from 1856).
2. The methods for the description and statistical analysis of shape variation within and among organism samples and the analysis of shape change as a result of growth, experimental treatment, or evolution (Rohlf and Marcus 1993, 129).
adj. morphometric
Comment: Rohlf and Marcus (1993) also define traditional morphometrics (= multivariate morphometrics) and "new morphometrics."

phenometry *n.* "The quantitative measurement of plant growth, mass, and leaf area" (Lincoln et al. 1985).

photogrammetry *n.* The use of photography as a field technique for quantitative ecological analysis (*e.g.,* in measurements of density taken from photographs) (Lincoln et al. 1985).

sociometry *n.* Quantitative description of social relationships within groups, behaviors of group members, and group organizations (Storz 1973, 261; Immelmann and Beer 1989, 280).
cf. -gram: sociogram
   zoometry See -metrics: zoometrics.

♦ **meute** See ²group: meute.

♦ **mew call** See animal sounds: call: mew call.

♦ **mews** See ²group: mews.

♦ **Meyrick's law** See law: Dollo's law.

♦ **MHC genes** See gene: major-histocompatibility-complex genes.

♦ **micraner** See male: micraner.

♦ **micro-, micr-** *combining form*
1. Very small, minute, short (Michaelis 1963).
2. Enlarging or magnifying size or volume (Michaelis 1963).
3. Pathologically, abnormally small or underdeveloped (Michaelis 1963).
4. Of a science depending on, using, or requiring a microscope (*e.g.,* microbiology) (Michaelis 1963).
5. In the metric system and in technical usage, one millionth of a specific unit (*e.g.,* a microvolt or micron) (Michaelis 1963).
cf. macro-
[Greek *mikros*, small]

♦ **microaerophile** See -¹phile: microaerophile.

♦ **microbenthos** See ²community: benthos: microbenthos.

♦ **microbiology** See study of: biology: microbiology.

♦ **microbiota** See biota: microbiota.

♦ **microbivore** See -vore: microbivore.

♦ **microconsumer** See decomposer.

♦ **microcosm** See ¹system: ecosystem.

♦ **microdichopatric** See -patric: dicho-patric: microdichopatric.

♦ **microdichopatric population** See ¹population: microdichopatric population.

♦ **microendemic** See ²endemic: micro-endemic.

♦ **microevolution** See ²evolution: micro-evolution.

♦ **microfauna** See fauna: microfauna.

♦ **microflora** See flora: microflora.

♦ **microgamete** See gamete: microgamete.

♦ **microgametogenesis** See -genesis: microgametogenesis.

♦ **microgamy** See -gamy: merogamy.

♦ **microgyne** See caste: queen: microgyne.

♦ **microhabitat** See habitat: microhabitat.

♦ **micromelittophile** See -²phile: micro-melittophile.

♦ **micromorphology** See morphology: micromorphology.

♦ **micromutation** See ²mutation: gene mu-tation.

♦ **micromyiophile** See -²phile: micro-myiophile.

♦ **microorganism** See organism: micro-organism (Appendix 1, Part 1).

♦ **micropaleontology** See study of: pale-ontology: micropaleontology.

♦ **microparapatric** See -patric: parapatric: microparapatric.

♦ **microparapatric population** See ¹population: microparapatric population.

♦ **microparasite** See parasite: micropara-site.

♦ **microphage, microphagous, micro-phagy** See -phage: microphage.

♦ **microphile** See -¹phile: microphile.

♦ **microphytophage** See -phage: micro-phytophage.

♦ **microplankton** See plankton: micro-plankton.

♦ **micropredation** See predation: micro-predation.

♦ **micropredator** See predator: micro-predator.

♦ **microsatellite DNA** see nucleic acid: deoxyribonucleic acid: satellite DNA: microsatellite DNA.

♦ **microsere** See ²succession: microsere.

♦ **microsphere** See organized structure: microsphere.

♦ **microsomatic** See -somatic: microsomatic.

♦ **microsomia** See dwarfism.

♦ **microspecies** See ²species: microspecies.

♦ **microsubspecies** See ²species: micro-subspecies.

♦ **microsymbiont** See symbiont: micro-symbiont.

♦ **microtaxonomy** See study of: tax-onomy: microtaxonomy.

♦ **microterritory** See territory: micro-ter-ritory.

♦ **microtherm** See therm: microtherm.

♦ **microthermophile** See -¹phile: micro-thermophile.

♦ **microzoogeography** See geography: microzoogeography.

♦ **microzoon** See -zoon: microzoon.

♦ **microzoophile** See ²-phile: microzoo-phile.

♦ **microzoophobous** *adj.* Referring to or-ganisms that repel small animals (Lincoln et al. 1985).

♦ **mictic** *adj.* In genetics, referring to fe-males that produce offspring of both sexes by apomixis, *q.v.,* or amphigony, *q.v.* (Lin-coln et al. 1985).

♦ **mictium** See ²community: mictium.

♦ **midbrain** See organ: brain: midbrain.

♦ **midden** *n.*

1. "A dunghill, manure-heap, refuse-heap" (*Oxford English Dictionary* 1972, entries from *ca.* 1375).

2. A kitchen midden; a prehistoric refuse heap, or mound, that contains shells, human and animal bones, stone imple-ments, etc. (*Oxford English Dictionary* 1972, entries from 1851; Michaelis 1963). *syn.* shellheap, shellmound (*Oxford En-glish Dictionary* 1972)

3. An underground mass of sticks and other detritus produced by a pack rat and per-meated with its urine (Holden 1992, 155). *Note:* Pack-rat middens over 40,000 years old are known (Holden 1992, 155). Pack rats deposit many kinds of objects in their nests. Their urine evaporates in a dry climate and crystallizes, gradually envel-oping the objects and forming a large, hard clump. Midden study reveals information about past plant communities, cosmic-ray intensity, and carbon dioxide concen-tration, and that certain human groups disappeared due to their degrading their environments. Hydraxes in Africa and the Middle East also make middens that last for millennia (Jaroff 1992, 61).

[Danish *køtten* + *mødding*, dunghill]

♦ **middle bevy** See ²group: middle bevy.

♦ **middle-repetitive DNA** See nucleic acid: deoxyribonucleic acid: middle-repeti-tive DNA.

♦ **midocean ridge** *n.* A ridge that is usu-ally submarine and centrally located within an ocean basin (Stanley 1989, 178). *Comment:* Midocean ridges are evidence for seafloor spreading; part of the Mid-Atlantic Ridge is exposed on Iceland (Stanley 1989, 179, 182).

h—m

♦ **migrant selection** See selection: migrant selection.

♦ **migration** *n.*

1. In many animal species, including lacustrine plankton; some butterfly, locust, bird, bat, and antelope species; salmon; eels: an organism's, or organism group's, active movement from one habitat, or location, to another (Lincoln et al. 1985), excluding its passive travel by wind or water.
   Note: Migrations may be one way or round trip (Immelmann and Beer 1989, 184).

2. A population's periodic or seasonal movement, typically of relatively long distance, from one area, stratum, or climate to another (Lincoln et al. 1985).

3. Any general movement of a population, or individual, that affects its distribution or range (Lincoln et al. 1985).

See gene flow: island-model migration, river-model migration.

cf. dispersion; ²group: migration; movement

ontogenetic migration *n.* An animal's occupying different habitats at different stages of its development (Lincoln et al. 1985).

♦ **migration principle** See principle: migration principle.

♦ **migratory restlessness** *n.* A captive bird's migratory activity that is expressed mainly as hopping, wing flapping, and wing fluttering due to its being unable to fly (Berthold 1971 in Immelmann and Beer 1989, 185).

syn. Zugunruhe (Immelmann and Beer 1989, 185)

♦ **milieu** *n.* An organism's, or population's, characteristic environs or surroundings (Lincoln et al. 1985).

♦ *milieu intérieur* See principle: Bernard's dictum.

♦ **military psychology** See study of: psychology: military psychology.

♦ **milk call** See animal sounds: call: milk call.

♦ **milk ejection** See ³lactation: milk ejection.

♦ **milk removal** See ³lactation (table).

♦ **milk secretion** See ³lactation: milk secretion.

♦ **milking** *n.*

1. In Humans: one's extracting milk by handling the teats of a cow, ewe, goat, etc., rarely a woman (*Oxford English Dictionary* 1972, entries from *ca.* 1000).
   See suckling.

2. In Cows, formerly in women: an individual's giving, or yielding, milk (*Oxford English Dictionary* 1972, entries from 971).
   cf. lactation: milk ejection; nurse; suckle

3. A milking machine's taking milk from a mammary gland (Cowie et al. 1951).
   cf. ³suckling

4. In many ant species: an individual's causing an ant-cow (*e.g.,* an aphid) to secrete honeydew by stroking the ant-cow with antennae (Ross 1985, 51).

v.i., v.t. milk

cf. honeydew

♦ **milking stimulus** See stimulus: milking stimulus.

♦ **millennial** *adj.* Referring to any period of 1000 years (Michaelis 1963).

cf. secular

[Latin *mille*, thousand]

♦ **millicron** See unit of measure cron: millicron.

♦ **millimacarthur (mma)** See units of measure: macarthur: millimacarthur.

♦ **mimesis** *n.* For example, in the Ghost Crab, some tiger-beetle, spider species: an animal's imitation of the appearance of inanimate objects such as sand (Wickler 1968, 238; personal observation).

See mimicry: mimesis.

syn. mimicry (according to some workers)

cf. camouflage, mimicry

phytomimesis *n.* An animal's imitation of objects of plant origin (Wickler 1986, 238).

Comment: This is a confusing term (Wickler 1986, 238).

zoomimesis *n.* An animal's imitation of objects of animal origin (Wickler 1986, 238).

Comment: This is a confusing term (Wickler 1986, 238).

♦ **mimetic behavior** See behavior: mimetic behavior.

♦ **mimetic polymorphism** See mimicry: mimetic polymorphism.

♦ **mimetic resemblances** See mimicry.

♦ **mimic** *n.*

1. An organism that transmits a signal to a receiver organism that is not advantageous to the receiver (Wickler 1968, 239).
   Note: This definition seems to break down when one considers Mertensian mimics and toxic Müllerian mimics.

2. An organism that plagiarizes a model (Pasteur 1982, 169).

See gene: mimic.

v.t. mimic

Comment: For kinds of mimics, see mimicry.

threat mimic See display: bared-teeth display.

♦ **mimic gene** See gene: mimic.

♦ **"mimic-not-perceived-by-dupe mimicry"** See mimicry: "mimic-not-perceived-by-dupe mimicry."

♦ **"mimic-perceived-by-dupe mimicry"** See mimicry: "mimic-perceived-by-dupe mimicry."

◆ **mimicry** *n.*

1. "Resemblance in external appearance, shapes and colours between members of widely distinct [taxonomic] families" (Bates in Wickler 1968, 46).

   Notes: Bates (1861) gives a wide concept of mimicry which he called "deceptive analogies," "mimetic resemblances," "mimetic analogies," and "imitative resemblances" (Pasteur 1982, 173). Pasteur (1982, 172) indicates which different phenomena that many workers include in "mimicry" from 1817 to 1978.

2. An unprotected animal species' "imitation of a protected animal species" (Wickler 1968, 46).

3. "The close resemblance of one organism to another which, because it is unpalatable and conspicuous, is recognized and avoided by some predators at some times;" the unpalatable and conspicuous organism is the model and the organism which resembles it is the mimic (Wickler 1968, 46).

4. One organism's sending a false (counterfeit) signal to another organism (*e.g.,* a wolf spider's looking like sand or a swallowtail caterpillar's looking like a bird dropping) (Wickler 1968, 239, 242; personal observation).

   Note: There are transitional stages between mimicry and camouflage and mimicry and general standard signal transmission (Wickler 1968, 239, 242).

5. An organism's (the mimic's) simulating the signal properties of "a second living organism (the model) which are perceived as signals of interest by a third living organism (the operator), such that the mimic gains in fitness as a result of the operator's identifying it as an example of the model" (Vane-Wright 1976, 1980 in Powell and Jones 1983, 311).

6. "The process whereby the sensory systems of one animal (operator) are unable to discriminate consistently a second organism or parts thereof (mimic) from either another organism or the physical environment (the models), thereby increasing the fitness of the mimic" (Wiens 1978 in Powell and Jones 1983, 311).

7. "A type of convergent evolution in which one or more species have converged in morphology, chemistry, or behavior, on a part of itself or themselves (automimicry), another living entity, or an inanimate object as a result of selection by an animal" (Powell and Jones 1983, 312).

8. A biological system with two or more protagonists that perform three roles: model, mimic, and dupe; the model is a living, or nonliving, agent that emits signals; the mimic is an organism that plagiarizes the model; and the dupe is the mimic's animal enemy, or victim, that perceives the model and is thus deceived by the mimic's similar signals (Pasteur 1982, 169).

syn. biological mimicry, mimesis, unconscious biological mimicry, Batesian mimicry (restricted synonym) (Pasteur 1982, 169); deceptive resemblances, deceptive analogies, mimetic resemblances, mimetic analogies, imitative resemblances (Bates 1861 in Pasteur 1982)

cf. behavior: allelomimetic behavior, cryptic behavior; camouflage; display: distraction display: rodent run; dupe; mimic; ¹system: mimicry system: model-mimic-dupe-tripartite system

Comments: Different workers include from few to all of these things in the concept of mimicry: eucrypsis, cryptic mimesis, phaneric mimesis, nonprotective homotypy, non-Müllerian homotypy, protective homotypy, and Müllerian homotypy (Pasteur 1982, 172). Pasteur (1982) gives a comprehensive review of mimicry, modifies many early definitions of kinds of mimicry, gives a classification of mimicry, and coins many new names for kinds of mimicry. He emphasizes that certain kinds of so-called "mimicry" (*e.g.,* "song mimicry" or "social mimicry") are not necessarily parts of mimicry systems. Many of the kinds of mimicry are defined below.

[Latin *mimicus* < Greek *mimikos* < *mimos*, mime]

adjustable mimicry *n.* Mimicry that is changeable at a certain stage in an organism's life cycle, *e.g.,* endothermic vertebrates that become white in winter or orthopterans that are green in the rainy season and straw-colored in the dry season; contrasted with developmental mimicry, *q.v.* (Pasteur 1982, 177).

▸ *ad hoc* **mimicry** *n.* Adjustable mimicry in which a mimic's coloration blends in with its background; contrasted with circadian mimicry and seasonal mimicry (Pasteur 1982, 177).

   fast *ad hoc* mimicry *n.* Adjustable mimicry in which a mimic's coloration quickly changes to match its background (*e.g.,* a cephalopod's changing its color in less than a second) (Pasteur 1982, 177).

   slow *ad hoc* mimicry *n.* Adjustable mimicry in which a mimic's coloration slowly changes to match its background (*e.g.,* a flatfish's taking a few hours to change color or some homomorphic arthropods' taking a few days to several weeks) (Pasteur 1982, 177).

h–m

A CLASSIFICATION OF MIMICRY
BASED ON PASTEUR (1982)

I. Classification by function
  A. nonprotective mimicry
    1. aggressive mimicry (*syn.* offensive mimicry, Peckhammian mimicry according to some authors)
    2. aggressive reproductive mimicry
    3. commensalist mimicry
    4. mutualistic mimicry
    5. parasitic mimicry, including weed-parasitic mimicry
    6. reproductive mimicry
    7. reproductive mutualistic mimicry
  B. protective mimicry (*syn.* defensive mimicry)
    1. arithmetic mimicry (*syn.* Müllerian mimicry)
    2. Batesian mimicry
    3. Emsleyian mimicry
    4. Gilbertian mimicry
II. Classification by changeability
  A. adjustable mimicry
    1. "*ad hoc* mimicry"
      a. "fast *ad hoc* mimicry"
      b. "slow *ad hoc* mimicry"
    2. "circadian mimicry"
    3. "seasonal mimicry"
  B. developmental mimicry
III. Classification by number of individual mimics involved
  A. collective mimicry (*syn.* population mimicry)
  B. individual mimicry
  C. social mimicry
IV. Classification by directness
  A. direct mimicry
  B. indirect mimicry
V. Classification with regard to inherentness
  A. adventitious mimicry
  B. inherent mimicry
VI. Classification by perception, or nonperception, of mimic by dupe
  A. "mimic-not-perceived-by-dupe mimicry"
    1. eucrypsis
      a. acoustic eucrypsis
      b. tactile eucrypsis
    2. mimesis
      a. cryptic mimesis
        (1) aggressive cryptic mimesis (*syn.* Peckhammian mimicry in some cases)
        (2) "dead-grass mimesis"
        (3) "dead-stick mimesis"
        (4) self-mimesis (*syn.* thanatosis)
        (5) "soil mimesis"
        (6) "stone mimesis"
      b. phaneric mimesis
    3. crypsis (*syn.* eucrypsis plus cryptic mimesis)
  B. "mimic-perceived-by-dupe mimicry"
    1. "model-agreeable-to-dupe mimicry"
      a. Aristotelian mimicry
      b. Bakerian mimicry
      c. Batesian-Poultonian mimicry
      d. Batesian-Wallacian mimicry
      e. Dodsonian mimicry
      f. Kirbyan mimicry
      g. Nicolaian mimicry
      h. Pouyannian mimicry
      i. Vavilovian mimicry
      j. Wasmannian mimicry
      k. Wicklerian mimicry
      l. Wicklerian-Barlowian mimicry
      m. Wicklerian-Eisnerian mimicry
    2. "model-disagreeable-to-dupe mimicry"
      a. Batesian mimicry
      b. Browerian mimicry
      c. Emsleyan mimicry
      d. Gilbertian mimicry
      e. Wicklerian-Guthrian mimicry
VII. Classification by concreteness of the model, "virtual-model mimicry"
  A. "definable-model-virtual-model mimicry"
  B. "nondefinable-model-virtual-model mimicry"

▶ **"circadian mimicry"** *n.* Adjustable mimicry that changes daily (*e.g.,* when homochromy, homotypy, or both, change every night and change back every morning); contrasted with seasonal mimicry (Pasteur 1982, 177).

▶ **"seasonal mimicry"** *n.* Adjustable mimicry that changes with season (*e.g.,* orthopterans that are green during the rainy season and straw-colored during the dry season); contrasted with circadian mimicry (Pasteur 1982, 177).

adventitious mimicry *n.* Mimicry involving a mimic's use of material taken from its environment, as occurs in some species of aquatic molluscs, arthropods, and moths and flatfish; contrasted with inherent mimicry (Pasteur 1982, 181).

aggressive mimicry *n.*

1. Resemblance of a prey species (bumble bee) by its predator (a robber fly *Laphria* sp.) that may enable the predator to capture prey more readily (Brower et al. 1960 in Waldbauer and Sheldon 1971, 380; Wickler 1968, 122-123).

2. For example, in angler fish, a catfish species, the alligator snapping-turtle:

mimicry in which part of a predator mimics food of its prey; this mimicking part attracts prey to the predator (Wickler 1968, 124–128).

3. Mimicry in which a mimic is a predator, parasitoid, or parasite and is camouflaged or has an attractive, or neutral, appearance to its potential victim (Pasteur 1982, 175).

syn. Peckhammian mimicry [after E.C. Peckham; Pasteur (1982, 176) says this synonym is unacceptable because "aggressive mimicry was known and understood as such long before Peckham's publications"]
cf. femme fatale
Comment: Aggressive mimicry is contrasted with aggressive reproductive mimicry, mutualistic mimicry, parasitic mimicry, reproductive mimicry, reproductive mutualistic mimicry, and commensalist mimicry.

aggressive reproductive mimicry *n.* Mimicry employed in both aggression and reproduction (Pasteur 1982, 176).
Comment: Aggressive reproductive mimicry is contrasted with aggressive mimicry, mutualistic mimicry, parasitic mimicry, reproductive mimicry, reproductive mutualistic mimicry, and commensalist mimicry.

allomimetic behavior See mood induction.

androchromatypic mimicry *n.* In some butterfly and dragonfly species: mimicry in which mature adult females have mimicking color patterns of mature adult conspecific males [androchromatypic coined by Hilton 1987, 222].
syn. andromorphic mimicry (Johnson 1964 in Hilton 1987, 221), homochrome mimicry, homoeochromatic mimicry, isochromatic mimicry, isomorphous mimicry (Hilton 1987)
cf. coloration: gynochromatypic coloration

Aristotelian mimicry *n.* Mimicry in which a noninjured brooding, or nourishing, bird makes itself look injured and easy to catch, which deflects a predator away from its young (Aristotle and Edmunds 1974 in Pasteur 1982, 190).
[after Aristotle, in *Historia Animalium*]

automimicry *n.*

1. Intraspecific mimicry of one sex by the other or of one life stage by another; *e.g.,* in some monkey species, males imitate female sexual signals, which they appear to employ in appeasement rituals, and male mouth-breeder fish *Haplochromis* have egg-mimic spots on their anal fins that attract females prior to fertilization of eggs held in the females' mouths (Wickler 1962, 1967, 1969 in Wilson 1975, 229, 579).

2. An individual's replication of patterns present in conspecific individuals (often of the other sex or of different age classes) at homologous sites, or elsewhere, on its body (*e.g.,* a Pronghorn Antelope's pointed, hooked ear that looks like a horn) (Guthrie and Petocz 1970, 585).
See mimicry: Browerian mimicry.
syn. conspecific mimicry (Little 1983, 300), intraspecific mimicry
cf. mimicry: automimicry, body self-mimicry

▶ **body self-mimicry** *n.* One body part's acting as a copy of some other part of an animal's anatomy, *e.g.,* genital-region mimicry by chest coloration in female Gelada Baboons, possible buttocks imitation by female human breasts, and possible penis imitation by a male Human's swollen nose tips (Morris 1977, 239–244).
Comment: Body self-mimicry may be used in social deceit (Weldon 1985, personal communication).
syn. body self-copying

head mimicry *n.* A hypothetical type of mimicry in which an animal's posterior end resembles its anterior end, possibly as a defensive adaptation; contrasted with tail mimicry (Wickler 1968, 74).
cf. false head; mimicry: "definable-model-virtual-model mimicry"

tail mimicry *n.* A hypothetical type of mimicry in which an animal's anterior end resembles its posterior end, possibly as a defensive adaptation; contrasted with head mimicry (Wickler 1968, 74).
cf. mimicry: virtual-model mimicry: nondefinable-model-virtual-model mimicry

▶ **female mimicry** *n.* For example, in some mollusc, insect, and vertebrate species, including Humans: mimicry in which a male animal shows behavior (and sometimes other characters) that are similar to those of a conspecific female (Weldon and Burghardt 1984, 92; Thornhill and Alcock 1983, 372).
syn. transvestism (not preferred)
cf. transvestism
Comments: Subordinate males of the fish *Polycentrus schomburgkii* imitate female color change and behavior as they approach territorial males (Barlow 1967 in Wilson 1975); male Hamadryas Baboons present their rumps to other males (Wilson 1975, 229). In a hangingfly species, some males mimic females by flying to calling rival males, perching close to them, and stealing a nuptial gift if possible (Thornhill and Alcock 1983, 372).

▶ **weapon automimicry** *n*. An animal's imitating the defensive parts of its own, or another, species (Guthrie and Petocz 1970; Weldon 1985, personal communication).

Bakerian mimicry *n*. Mimicry of nectiferous male flowers by conspecific nectarless female flowers (*e.g.,* in the Caricaceae) (Baker 1976 in Pasteur 1982, 190).

Batesian mimicry *n*. Mimicry in which an edible species mimics a less edible or inedible one (Pasteur 1982, 185) (*e.g.,* flies' mimicking stinging Hymenoptera) (Wickler 1968, 12–13); contrasted with Müllerian mimicry, *q.v.*

syn. Bates theory of mimicry; cryptic mimicry (Barlow and Wiens 1977, 161 in Pasteur 1982, 185)

cf. mimicry: floral mimicry: deceptive floral mimicry

Comments: In Batesian mimicry, models are not necessarily more common than mimics, mimics do not necessarily look highly similar to their models, models are not always aposematic or noxious, and models and mimics can be plants as well as animals (Pasteur 1982, 185). Further, in Batesian mimicry, mimics and models are of different edibility depending on the species and dupe involved. This relationship ranges from a model's being slightly inedible and a mimic's being slightly less inedible to a dupe to the model's being totally inedible and a mimic's being almost as inedible to a dupe (Pasteur 1982, 193). There are species which are transitional between "Batesian mimics" and "Müllerian mimics" (Rothschild 1963, 159). Further, a single species can be either a "Batesian mimic" or "Müllerian mimic" depending on its interactions with other organisms, but see "comments" under Müllerian mimicry. [after Henry Walter Bates, English naturalist, who first described this phenomenon in the Amazon]

▶ **"classical Batesian mimicry"** *n*. Batesian mimicry in which a mimetic species does not outnumber its model and appears contemporaneously and sympatrically with its model (Poulton 1890 and Rettenmeyer 1970 in Waldbauer and Sheldon 1971, 379).

Comment: "Classical Batesian mimicry" contrasts with "Waldbauerian-Batesian mimicry." Many insect species are probably classical Batesian mimics.

▶ **"Waldbauerian-Batesian mimicry"** *n*. Batesian mimicry in which a mimetic species is sympatric with its model but is more frequent than it earlier in the warm season (Waldbauer and Sheldon 1971, 379).

Comment: "Waldbauerian-Batesian mimicry" contrasts with "classical Batesian mimicry." Many species of insects appear to be asynchronous Batesian mimics and support the Waldbauer-Sheldon-late-season-mimic-scarcity hypothesis, *q.v.* (Waldbauer 1985).

[after G.P. Waldbauer, entomologist, and Henry W. Bates, naturalist]

Batesian-Poultonian mimicry *n*. Mimicry of a predator by its prey (*e.g.,* mimicry of a common fossorial wasp by an uncommon grasshopper) (Bates 1961 and Poulton 1890 in Pasteur 1982, 189).

[after Henry W. Bates and E.B. Poulton]

Batesian-Wallacian mimicry *n*. Mimicry in which a predaceous-arthropod species resembles its prey species (*e.g.,* a bolas spider's emitting an odor that mimics the female pheromones of two moth species that it attracts and eats) (Eberhard 1977 in Pasteur 1982, 188).

Comment: There are also behavioral, acoustic, and other chemical examples and possibly ultrasonic examples of Batesian-Wallacian mimicry (Pasteur 1982, 188).

behavioral mimicry *n*. Mimicry of a model species' behavior, *e.g.,* a False Cleaner Fish's mimicking the conspicuous movements of cleaner fish, some beetles' and silverfish's mimicking antennal movements of their ant hosts which results in their being fed by the ants, or young burrowing owls' making a call that sounds like a rattlesnake's rattling (Immelmann and Beer 1989, 30).

cf. mimicry: cleaner mimicry, Wicklerian-Eisnerian mimicry

syn. ethomimicry (Heymer 1977, 118; Immelmann and Beer 1989, 93)

Browerian mimicry *n*. Automimicry in which conspecific individuals mimic each other (*e.g.,* in Monarch Butterflies, individuals that are more palatable look like individuals that are less palatable) (Brower, 1969, Brower and Brower, 1972 in Pasteur 1982, 186).

syn. automimicry (Brower 1970; Matthews and Matthews 1978, 332)

cf. mimicry: automimicry

Comment: An individual of the same, or different, sex can be a Browerian mimic (Pasteur 1982, 186).

[after Lincoln P. and Jane V.Z. Brower, biologists who discovered this phenomenon]

chemical mimicry *n*. One species' mimicking the odor of another organism (inferred from Schiestl et al. 1999, 421).

cf. mimicry: Pouyannian mimicry

classical Batesian mimicry See mimicry: Batesian mimicry: classical Batesian mimicry.

cleaner mimicry *n*. Behavioral mimicry that allows a false cleaner fish to move near and bite off parts of fins of a fish waiting to be cleaned instead of cleaning it (Wickler 1968 162–176).

cf. mimicry: behavioral mimicry, Wicklerian Eisnerian mimicry

collective mimicry *n*. Mimicry in which more than one individual conspecific organism performs mimicry which is protective or of unclear significance (Pasteur 1982, 176) (*e.g.,* mimicry of a flower inflorescence by a group of fulgorid homopterans) (Heymer 1977, 118); contrasted with individual mimicry and social mimicry (Pasteur 1982, 176).

commensalist mimicry *n*. In some ant-ectosymbiont species: an individual's imitating the emitted chemicals of its ant host that causes its host to tolerate its presence (Pasteur 1982, 176).

Comment: Commensalist mimicry is contrasted with aggressive mimicry, aggressive reproductive mimicry, mutualistic mimicry, parasitic mimicry, reproductive mimicry, and reproductive mutualistic mimicry.

crypsis *n*. Mimicry that is comprised of both eucrypsis and cryptic mimesis (Pasteur 1982, 183).

developmental mimicry *n*. Mimicry that is present at a certain stage in an organism's life cycle and is not changeable (*e.g.,* a very young sturgeon's resembling a drifting twig which is not found in other stages); contrasted with adjustable mimicry (Pasteur 1982, 177).

direct mimicry *n*. Mimicry directly due to a mimic's own trait(s); contrasted with indirect mimicry (Pasteur 1982, 179).

Comment: "Direct mimicry" is a more common kind of mimicry than "indirect mimicry."

Dodsonian mimicry *n*. Mimicry of the flower of one species by that of another species (*e.g.,* mimicry of *Asclepias curassavica* and *Lantana camara*, which have nectar, by an *Epidendrum* orchid that does not have nectar) (Boyden 1980 in Pasteur 1982, 187).

[after C.H. Dodson, biologist]

egg mimicry See mimicry: Gilbertian mimicry, Wicklerian Barlowian mimicry.

Emsleyan mimicry *n*. Mimicry in which a moderately noxious model is mimicked by a species whose defense is fatal to an assailant that misses its attack (*e.g.,* mimicry of less dangerous coral snakes by lethal coral snakes) (Mertens 1956 in Wickler 1968, 111–121; Emsley 1966 in Pasteur 1982, 186).

syn. Mertensian mimicry [after R. Mertens, but Pasteur (1982, 175) explains that Mertens did not allude to this kind of mimicry]

[after Michael G. Emsley, biologist, who discovered this phenomenon]

ethomimicry See mimicry: behavioral mimicry.

eucrypsis *n*. In many species: mimicry in which a mimic resembles its general background, not particular things in it, based on similar coloration (homochromy) alone; contrasted with mimesis (Pasteur 1982, 182).

Comment: Although eucrypsis is found in many animal species, it is represented in plants by only one case: homochromy of pollinia of certain orchid species with hummingbird beaks (Pasteur 1982, 183).

▸ **acoustic eucrypsis** *n*. Eucrypsis in which the silence achieved by cats and owls when hunting enables them to blend into their silent environment (Pasteur 1982, 184).

▸ **tactile eucrypsis** *n*. A possible kind of eucrypsis in which an inquiline's touching its ant, or termite, host helps the inquiline to blend into its host's colony (Pasteur 1982, 184).

fast-evader mimicry *n*. In some butterfly species: mimicry of the colors of a species adept at evading birds by fast erratic flight by another species that may, or may not, have similar evading flight (Marden 1992, 60).

Comments: These colors indicate to bird predators that the models are difficult to catch. Mimicry analogous to "Batesian mimicry" and "Müllerian mimicry" occurs in "fast-evader mimicry" (Marden 1992, 54).

female mimicry See mimicry: automimicry: female mimicry.

floral mimicry *n*. Mimicry in which one kind of flower resembles another kind that possibly results in the mimic's attracting pollinators and increasing its fitness (Little 1983, 295).

cf. mimicry: Pouyannian mimicry

Comments: Many kinds of floral mimicry are discussed by Endler (1981 in Little 1983, 295). Little (1983, 295) classifies "floral mimicry" as a kind of "plant mimicry." Floral mimicries may be the ends of a continuum of phenomena (Little 1983, 296).

▸ **deceptive floral mimicry** *n*. Floral mimicry in which one flower provides no pollinator reward(s) but receives visits from either naive pollinators or pollinators specializing on a similar rewarding flower (Little 1980 in Powell and Jones 1983, 312).

syn. Batesian mimicry (Brown and Kodric-Brown 1979 in Powell and Jones 1983, 312, but Little 1983, 297, discusses the problem with this synonym), food-deception mimicry, food-source mimicry,

h–m

inviting mimicry (Vane-Wright 1976 in Powell and Jones 1983, 312).

cf. mimicry: floral mimicry: nondeceptive floral mimicry

▸ **nondeceptive floral mimicry** *n.* Floral mimicry in which two or more species of flowers provide rewards for one or more common pollinating species (Little 1980 in Powell and Jones 1983, 312).

syn. advertising mimicry (Proctor and Yeo 1972 in Little 1983, 296), floral mutualism, Müllerian mimicry (Little 1983, 296–297)

cf. mimicry: floral mimicry: deceptive floral mimicry

Comment: The problem with using "Müllerian mimicry" as a synonym is discussed by Little (1983, 297).

food mimicry *n.*

1. Mimicry in which plants produce morphological structures that resemble morphology, odor, or both, of food, *e.g.,* nectar-droplet-like features that attract pollinating insects to Grass-of-Parnassus and some orchids and prey to sundews and rotten-meat odors that attract pollinators to certain aroids (Wickler 1968, 147–151).

2. For example, in the Swordtail Characin: mimicry of food by part of an animal's body that attracts a mate (Wickler 1968, 219–200).

Gilbertian mimicry *n.* Protective mimicry in which the model and dupe are conspecific (*e.g., Passiflora* stipules that mimic *Heliconius* butterfly eggs near the point of hatching) (Gilbert 1975 in Pasteur 1982, 186).

syn. egg mimicry (Gilbert 1983, 278; Sbordoni and Forestiero 1985, 191)

Comment: Because larvae of most *Heliconius* species are cannibalistic, *Heliconius* females avoid ovipositing on plants bearing such eggs (Pasteur 1982, 186)

[after L.E. Gilbert, biologist, who discovered this phenomenon]

hi-fidelity mimicry See mimicry: specialized mimicry.

historical mimicry *n.* For example, in some fly and a mantispid species: mimicry in which a model gains protection from a predator due to its experience with the mimic's model(s) in the previous summer (Rothschild 1981 in Waldbauer and LaBerge 1985, 101).

cf. mimicry: specialized mimicry, vague mimicry

homochrome mimicry, homoeochromatic mimicry See mimicry: androchromatic mimicry.

indirect mimicry *n.* Mimicry that is performed through a dummy or in parasitic species through a host that the parasite alters, causing the host to become a mimic; contrasted with direct mimicry (Pasteur 1982, 176).

Comments: Indirect mimicry is rare among organisms. Indirect mimicry through a dummy may be carried out by West African *Cyclosa* and *Uloborus* spiders which add conspicuous dummy spiders to their webs which may distract predators from actual spiders (Edmunds 1974 in Pasteur 1982, 179).

individual mimicry *n.* Mimicry in which an individual organism fully performs mimicry without involvement of other members of its species; contrasted with collective mimicry and social mimicry (Pasteur 1982, 179).

inherent mimicry *n.* Mimicry that is intrinsic, developing ontogenetically in an organism; contrasted with adventitious mimicry (Pasteur 1982, 180).

intraspecific mimicry See mimicry: automimicry.

isochromic mimicry, isomorphous mimicry See mimicry: androchromatic mimicry.

Kirbyan mimicry *n.* Mimicry of host-bird eggs, young, or both by brood parasites (Kirby and Spence 1923 in Pasteur 1982, 188).

[after W. Kirby]

Mertensian mimicry See mimicry: Emsleyian mimicry.

mimesis *n.* Mimicry in which a mimic resembles its general background based on similar coloration (homochromy), morphology (homomorphy), and if necessary the model's moves and postures (homokinemy); contrasted with eucrypsis (Pasteur 1982, 182).

▸ **cryptic mimesis** *n.* Mimicry in which the model is a dominant element of a mimic's environment, either inanimate (*e.g.,* leaves, sticks, twigs, thick inflorescences and large flowers, bark, grass, lianas, seaweeds, soil, and pebbles) or animal (*e.g.,* gregarious, or colonial, animals that dupes do not react towards); contrasted with phaneric mimesis (Pasteur 1982, 182).

syn. cryptic mimicry (Pasteur 1982, 183)

Comments: In plants, cryptic mimesis includes dead-grass mimicry, dead-stick mimicry, soil mimicry obtained by sticky succulent plants that catch dust and sand, and stone mimicry in five plant families (Pasteur 1982, 183).

aggressive cryptic mimesis *n.* For example, in some species of antlike spiders and a vulture species that resembles a buzzard: cryptic mimesis in which a predator resembles an individual of a gregarious-animal species that is innocuous to the predator's prey and the predator loses itself in its model-species flock (Pasteur 1982, 184).

syn. Peckhammian mimicry (in some cases, Pasteur 1982, 184)

phaneric mimesis *n.* Mimicry in which the model is an isolated, conspicuous inanimate element of a mimic's environment; contrasted with cryptic mimesis (Pasteur 1982, 183).

Comment: Models include an animal dropping, flower bud, mushroom, insect carcass, and part of the mimic's own body (Pasteur 1982, 183).

self-mimesis *n.* For example, in many animal species: mimicry in which a mimic looks like itself but looks dead, which can involve suddenly ceasing to move or feigning death sometimes after abruptly turning upside down (Pasteur 1982, 184).

syn. thanatosis (Pasteur 1982, 184)
cf. playing dead

mimetic polymorphism *n.* Mimicry in which the mimic is one species with different morphological forms that resemble models of more than one species (*e.g.,* the swallowtail butterfly *Papilio dardanus,* which resembles other butterfly species) (Wickler 1968, 19–33).

"mimic-not-perceived-by-dupe mimicry" *n.* Mimicry in which a dupe does not see a mimic (Pasteur 1982, 183).

"mimic-perceived-by-dupe mimicry" *n.* Mimicry in which a dupe sees a mimic (Pasteur 1982, 183).

"model-agreeable-to-dupe mimicry" *n.* Mimicry in which a dupe finds a model not unpleasant (Pasteur 1982, 187).

"model-disagreeable-to-dupe mimicry" *n.* Mimicry in which a dupe finds a model unpleasant in some way (Pasteur 1982, 185).

molecular mimicry *n.* A parasite's making eclipsed antigens (antigenic determinants that resemble its hosts' antigenic determinants to such a degree that they do not elicit host antibodies against them) (Damian 1964, 129; Pasteur 1982, 1978; King and Stansfield 1985).

Müllerian mimicry, Muellerian mimicry *n.*

1. Mimicry between, or among, two or more distasteful, or toxic, species that reduces predation for each species because, in this resemblance situation, predators learn to avoid only one color pattern rather than several (*e.g.,* in the tiger-stripe butterfly complex in South America) (Wickler 1968, 78–85; Matthews and Matthews 1978, 330; Wittenberger 1981, 618); contrasted with Batesian mimicry, *q.v.*

2. Mimicry in which both mimic and model are unpalatable for all potential predators (Müller 1878, 1879, in Pasteur 1982, 185); contrasted with Batesian mimicry, *q.v.*

syn. arithmetic mimicry (Pasteur 1982, 173)
cf. mimicry; mimicry: Batesian mimicry, floral mimicry: nondeceptive floral mimicry, mimicry ring, reproductive mimicry; Müllerian association

Comment: Strictly speaking, Müllerian mimicry is not mimicry because both model and mimic are sending nondeceitful signals to a perceiver (Pasteur 1982, 185); therefore, some workers forsake the term "Müllerian mimicry" for "Müllerian resemblance" and "Müllerian convergence."
[after Fritz Müller, German zoologist]

mutualistic mimicry *n.* Mimicry in which the mimic and dupe help one another (Pasteur 1982, 176).

Comment: Intermediate cases between mutualistic and aggressive mimicry occur (Pasteur 1982, 176). Mutualistic mimicry is contrasted with aggressive mimicry, aggressive-reproductive mimicry, mutualistic mimicry, parasitic mimicry, reproductive mimicry, reproductive mutualistic mimicry, and commensalist mimicry.

Nicolaian mimicry *n.* Mimicry of the vocalization of its estrildid-finch host by a male Widowbird (Nicolai 1964 in Pasteur 1982, 190).

Comment: Nicolaian mimicry conditions a female Widowbird to copulate. A similar kind of mimicry is also found in photomimetic fireflies (Pasteur 1982, 190).
[after J. Nicolai]

nonprotective mimicry *n.* Mimicry that does not protect a mimic from its dupe; contrasted with protective mimicry (Pasteur 1982, 175).

parasitic mimicry *n.* Mimicry used by a parasite in promoting its reproduction; contrasted with aggressive mimicry, aggressive reproductive mimicry, mutualistic mimicry, reproductive mimicry, reproductive mutualistic mimicry, and commensalist mimicry (Pasteur 1982, 176).

Comment: Parasitic mimicry has a gamut of intermediate cases, from aggressive reproductive mimicry to reproductive mimicry (Pasteur 1982, 176).

Peckhammian mimicry See mimicry: aggressive mimicry.

Pouyannian mimicry *n.* An orchid flower's mimicking a female bee, or wasp, which causes a male bee, or wasp, to

h – m

pseudocopulate with the flower (Pouyanne 1917 in Pasteur 1982, 188).

cf. copulation: pseudocopulation

Comments: Male bees of *Andrena nigro-aenea* pseudocopulate with flowers of the orchid *Ophrys sphegodes sphegodes* (Schiestl et al. 1997, 2881). These flowers have different odor bouquets of aldehydes and alkanes which elicit pseudocopulation. The male bees often transfer pollinia during their pseudocopulation bouts (Schiestl et al. 1999, 421). The orchid produces the same compounds and in similar relative proportions as do females of this bee species.

[after M. Pouyanne]

protective mimicry *n.* Mimicry in which a potential victim is camouflaged, or has a formidable appearance, that foils its predators or other potential enemies; contrasted with nonprotective mimicry (Pasteur 1982, 175).

syn. defensive mimicry (Pasteur 1982, 175)

Comment: Protective mimicry includes many kinds of mimicry such as "arithmetic mimicry" and "Batesian mimicry."

rank mimicry *n.* In some bird and primate species: a lower ranking animal's mimicking the behavior of a higher ranking animal (Immelmann and Beer 1989, 241).

reproductive mimicry *n.* Mimicry used in propagation of the mimicking species without assaulting, or harming, dupes (Pasteur 1972 in Pasteur 1982, 176).

Comment: Reproductive mimicry is contrasted with aggressive mimicry, aggressive reproductive mimicry, mutualistic mimicry, parasitic mimicry, reproductive mutualistic mimicry, and commensalist mimicry.

reproductive mutualistic mimicry *n.* Mimicry with both reproductive and mutualistic components (*e.g.,* in ant ectosymbionts) (Pasteur 1982, 176).

Comment: Reproductive mutualistic mimicry is contrasted with aggressive mimicry, aggressive reproductive mimicry, mutualistic mimicry, parasitic mimicry, reproductive mimicry, and commensalist mimicry.

social mimicry *n.*

1. Mimicry of conciliatory and contact signaling that serves to diminish hostility among species forming mixed flocks of birds (Moynihan 1968 in Wilson 1975, 360).
2. Batesian mimicry in which a group of conspecific animals is distasteful to one or more predators and serves as a nucleus for fishes or birds of another species that tend to look and behave like their models; contrasted with collective mimicry and individual mimicry (Pasteur 1982, 179).

specialized mimicry *n.* For example, in some syrphid-fly and robber-fly species: mimicry in which a mimic has a suite of morphological and behavioral adaptations that mimic salient features of its model(s) (given as "specialized mimic" in Waldbauer and Sheldon 1971, 371).

syn. high-fidelity mimicry (given as "high-fidelity mimic" in Waldbauer and LaBerge 1985, 103; "specialized high-fidelity mimic" in Waldbauer 1988, S107)

cf. mimicry: historical mimicry, vague mimicry

Comment: High-fidelity mimicry is contrasted with low-fidelity mimicry (given as "low fidelity" mimic in Waldbauer and LaBerge 1985, 108).

transformational mimicry *n.* For example, in some species of swallowtail butterflies: The phenomenon in which different instars imitate entirely different models (Matthews and Matthews 1978, 332).

syn. I-led-three-lives syndrome (Matthews and Matthews 1978, 332)

vague mimicry *n.* For example, in some syrphid-fly and robber-fly species: mimicry in which a mimic does not have a suite of morphological and behavioral adaptations that mimic salient features of its model, or models (given as "vague mimic" in Waldbauer and Sheldon 1971, 371).

cf. mimicry: historical mimicry, specialized mimicry

Vavilovian mimicry *n.* Mimicry of seeds of one crop by another (*e.g.,* mimicry of wheat seeds by rye) (Vavilov 1951 in Pasteur 1982, 187).

[after N.I. Vavilov]

"virtual-model mimicry" *n.* Mimicry, not of an actual species, but of a character found in models that we can specify in general terms and those that we cannot (Pasteur 1982, 191).

▶ **"definable-model-virtual-model mimicry"** *n.* Mimicry involving a character of more than one species but not mimicry of any particular species, *e.g.,* mimicry of eyes as seen in eye spots of many insect species which may frighten, deflect, or both, predators (Blest 1957, etc. in Pasteur 1982, 191); mimicry of young snakes by caterpillars; mimicry of an insect's head by its rear end and vice versa; mimicry of poisonous snakes by harmless ones or legless lizards; mimicry of prey food by outgrowths on a predatory fish; mimicry of prey food by the tail of a venomous snake; the simulation of flowers by trap-leaves of certain terrestrial carnivorous plants; and mimicry of filamentous algae by extensions on trap vesicles of *Utricularia*; contrasted with "nondefinable-model-virtual-model mimicry (Pasteur 1982, 191).

cf. mimicry: aggressive mimicry;" head mimicry

▶ **"nondefinable-model-virtual-model mimicry"** *n*. Mimicry in which the mimic resembles a generalized character not found in any particular species; contrasted with "nondefinable-model-virtual-model mimicry" (Pasteur 1982, 192).

cf. mimicry: tail mimicry

vocal mimicry *n*. In some parrot species, Myna Birds, Mockingbirds (mimic thrushes): an individual's adding the sounds of other species (including crickets and Humans) to its vocal repertoire (Immelmann and Beer 1989, 324).

cf. learning: vocal mimicry, vocal mocking
Comment: A captive Mockingbird incorporated the chirping of house crickets into its vocal repertoire (personal observation).

Waldbauerian-Batesian mimicry See mimicry: Batesian mimicry: Waldbauerian-Batesian mimicry.

Wasmannian mimicry *n*. Mimicry of an ant by an ant ectosymbiont that may benefit the dupe without damaging it (Wasmann 1925 in Pasteur 1982, 189); *e.g.,* the mimics are some salticid-spider and staphylinid-beetle species that resemble ants in size, coloration, body structure, and behavior (Wickler 1968, 99; Rettenmeyer 1970).

[after E. Wasmann]

Wicklerian-Barlowian mimicry *n*. Mimicry of eggs by a male mouth-brooding cichlid species (Wickler 1962 and Barlow 1967 in Pasteur 1982, 190).

syn. egg mimicry (Wickler 1968, 222)
Comments: The eggs are on a male's anal fin, and a female with a mouth full of eggs approaches his fin to collect the "eggs." When she gets close, he injects his sperm into her mouth and thereby fertilizes her eggs (Pasteur 1982, 190). Another kind of Wicklerian-Barlowian mimicry is also found in fish.

[after W. Wickler and G.W. Barlow]

Wicklerian-Eisnerian mimicry *n*. Mimicry in which the model is not offensive to a dupe, *e.g.,* mimicry of a cleaner fish by the Saber-Toothed Blenny, which approaches a fish to be cleaned and bites off some of its flesh rather than cleaning it (Wickler 1963 in Pasteur 1982, 187), or mimicry of its prey by the larva of a lacewing by covering itself with woolly wax from Woolly Aphids (Eisner et al. 1978).

cf. mimicry: behavioral mimicry, cleaner mimicry

Wicklerian-Guthrian mimicry *n*. Social self-mimicry in which the muzzle of the Forest Baboon appears to mimic its overcolored genitals (Jouventin 1975 in Pasteur 1982, 186).

Comments: It is controversial whether this phenomenon is actually mimicry (Pasteur 1982, 187).

Wicklerian mimicry *n*. Mimicry of female genitalia by a male or vice versa (Wickler 1967 in Pasteur 1982, 191).

Comments: The monstrous red buttocks of a male Hamadryas Baboon mimic a female's estrous genitalia. A male Hamadryas Baboon may display his buttocks to a bellicose dominant male as a submission and appeasement gesture (Wickler 1967 in Pasteur 1982, 191). The pseudopenis of a female hyena resembles a male's penis and may have evolved to allow a female to participate in the conciliatory communication within her pack (Wickler 1967 in Pasteur 1982, 191).

[after Wolfgang Wickler, biologist]

♦ **mimicry ring** *n*. A group of several organisms that are mimics of several other organisms (Wickler 1968, 78).

Batesian mimicry ring *n*. A group of several similar-appearing Batesian mimics that are mimics of a group of several similar-appearing Müllerian mimics (Wickler 1968, 80; Waldbauer and LaBerge 1985).

Müllerian mimicry ring *n*. A mimicry ring comprised of animals that are Müllerian mimics of one another, *e.g.,* certain South American butterflies or North American yellowjackets (wasps) (Wickler 1968, 78–80; Waldbauer and LaBerge 1985).

syn. Müllerian mimicry club (Wickler 1968, 78–80)

♦ **mimicry system** See ²system: mimicry system.

♦ **mimetic analogies** See mimicry.

♦ **mimotype** See -type, type: mimotype.

♦ **mind** *n*.

1. A person's "thought, purpose, intention" (*Oxford English Dictionary* 1972, entries from 971).

2. The seat of a person's "consciousness, thoughts, volitions, and feelings; a system of cognitive and emotional phenomena and powers that constitutes the subjective being of a person; the incorporeal subject of the psychical faculties, the spiritual part of a human being;" a person's soul as distinguished from his body (*Oxford English Dictionary* 1972, entries from *ca.* 1340).

3. A person's cognitive, or intellectual, powers, as distinguished from his will or emotions; often contrasted with heart (*Oxford English Dictionary* 1972, entries from *ca.* 1340).

h–m

4. The aggregate of processes originating in, or associated with, a person's brain, involving conscious and subconscious thought, interpretation of perceptions, insight, imagination, etc. (Michaelis 1963); a person's ability to monitor his own mental states (Gallup 1982, 243).

5. In animals: "that which has mental experiences," *q.v.* (Griffin 1976, 5).

6. Organism activity that involves behavioral pattern formation, appetency, irritability, tropisms, taxis, as well as an ability to learn (Brownlee 1981, iv).

cf. mentality; organ: brain
[Old English *gemynd*, mind]

♦ **mineral** *n.*

1. A naturally occurring, homogeneous substance formed by inorganic processes and having a characteristic set of physical properties, a definite and limited range of chemical composition, and a molecular structure, usually expressed in crystalline form (Michaelis 1963).

2. Inorganic material; distinguished from that produced by organisms (Michaelis 1963).

3. Any of various natural substances such as an element (*e.g.,* gold or silver), a mixture of inorganic compounds (*e.g.,* granite or hornblende), or an organic derivative (*e.g.,* coal or petroleum) (Morris 1982).

magnetite *n.* A massive, isometric black iron oxide, Fe_3O_4, that is strongly magnetic; when it has polarity, it is called lodestone (Michaelis 1963).

cf. taxis: geomagnetotaxis
Comment: Magnetite may be used in geomagnetotaxis in some animals (*e.g.,* the Homing Pigeon, Honey Bee, and Human) or orientation in some kinds of bacteria. Chitons (Mollusca) have magnetite in their scrapers which they use to remove algae from substrates (Baker 1987, 691; D.B. Quine, personal communication). Magnetite is found in human brains (Anonymous 1992, 25).

iron pyrite *n.* Metallic, pale yellow iron disulfide, FeS_2 (Michaelis 1963).

syn. fool's gold, pyrite (Michaelis 1963)
Comment: Organized structures that preceded life on Earth might have been "iron-pyrite aggregations," *q.v.*
[Latin *pyrites* < Greek *pyritēs*, flint < *pyritē s* (lithos), fire (stone) < *pyr*, fire]

Vivanite *n.* An iron-phosphate mineral (Guthrie 1990, 79, illustration; 80).

Comments: Vivanite is whitish gray when unoxidized and blue when oxidized. It occurs as crystals and dust on mammal bone fossils and mummies and as layers in pond sediments (Guthrie 1990, 79).

♦ **minerotroph** See -troph-: minerotroph.

♦ **minigeographical isolation** See isolation: minigeographical isolation.

♦ **minima** See caste: worker: minor worker.

♦ **minimum, law of** See law: law of the minimum.

♦ **minimum-interaction hypothesis** See hypothesis: minimum-interaction hypothesis.

♦ **minisatellite DNA** See nucleic acid: deoxyribonucleic acid: satellite DNA: minisatellite DNA.

♦ **minor community** See ^{2}community: minor community.

♦ **minor pollinator** See pollinator: minor pollinator.

♦ **minor worker** See caste: worker: minor worker.

♦ **miosis** See meiosis.

♦ **misogamy** See isolation: reproductive isolation.

♦ **missense mutation** See ^{2}mutation: missense mutation.

♦ **mist net** See trap: mist net.

♦ **mitochondrial-Eve hypothesis** See hypothesis: African-origin hypothesis of human mitochondrial DNA evolution.

♦ **mitochondrial genome** See genome: mitochondrial genome.

♦ **mitochondrion** See organelle: mitochondrion.

♦ **mitoschisis** See mitosis.

♦ **mitosis** *n.*

1. The division of a cell nucleus (Mayr 1982, 958).

2. In eukaryotes, most archaean species: cell division in which a somatic cell gives rise to daughter cells, each of which has a complete set of all its chromosomes; the ordinary cell division of bodily growth (Dawkins 1982, 290).

3. The result, or process, of mitosis.

syn. cytodiaeresis, karyokinesis, karyomitosis, mitoschisis (Lincoln et al. 1985)
cf. binary fission, meiosis
[Greek *mitos*, thread; coined by Flemming (1882 in Mayr 1982, 674)]

amitosis *n.* A nucleus' division by simple construction into two portions which often forms dissimilar daughter nuclei (Lincoln et al. 1985).

syn. holoschisis (Lincoln et al. 1985)

endomitosis *n.*

1. Chromosomal replication within a nucleus of a cell that does not subsequently divide, resulting in polyploidy (Lincoln et al. 1985).

2. A reproductive system in which there is a premeiotic doubling of chromosomes followed by meiosis (Bernstein et al. 1985, 1279).

♦ **mitotic parthenogenesis** See parthenogenesis: mitotic parthenogenesis.

♦ **mixed colony** See ²group: colony: mixed colony.

♦ **mixed dominance-subordination relationship** See relationship: mixed dominance-subordination relationship.

♦ **mixed dyadic social relationship** See relationship: mixed dyadic social relationship.

♦ **mixed evolutionarily stable strategy** See strategy: evolutionarily stable strategy: mixed evolutionarily stable strategy.

♦ **mixed forest** See ²community: mixed forest.

♦ **mixed group** See ²group: mixed group.

♦ **mixed motivation** See motivation: mixed motivation.

♦ **mixed nest** See nest: mixed nest.

♦ **mixed singer** *n.*
1. For example, in European Firecrests and Goldcrests (birds): an individual that produces the songs, or song parts, of other species (usually closely related ones) in addition to its species-typical song(s) (Immelmann and Beer 1989, 186–187).
2. In dialect-forming birds: an individual that includes song elements of another conspecific dialect in its own song (Immelmann and Beer 1989, 187).
cf. mimic: vocal mimic

♦ **mixed-species group** See ²group: mixed group.

♦ **mixis** *n.*
1. Genetic material rearrangement through meiosis, syngamy, or usually both, almost always resulting in the production of one or more new organisms that differ genetically from one another and from their parents (Bell 1982, 509).
syn. sex, sexuality (Bell 1982, 509)
2. Fusion of gametes, karyogamy, and karyomixis (Lincoln et al. 1985).
See -mixis: amixis, amphimixis.
cf. sex, sexual reproduction

♦ **-mixis, -mixia**
adj. mictic
allomixis See fertilization: cross-fertilization.
amixis *n.* Absence of meiosis and fertilization in all stages in an organism's life cycle (Margulis et al. 1985, 73).
cf. parthenogenesis
amphimixis *n.*
1. "True sexual reproduction; the union of male and female gametes," which "may be either autogamy (inbreeding) or allogamy (outbreeding)" (Lincoln et al. 1985).
2. Syngamy (contact, or fusion of gametes) or karyogamy (fusion of gamete nuclei) that leads to fertilization and forms an individual with two different parents (Margulis et al. 1985, 73).

n. amphimict
syn. mixis (Margulis et al. 1985, 73)
apomixis See parthenogenesis: apomixis.
automixis *n.*
1. Karyogamy of nuclei, or syngamy of cells, that derive from the same parent (Margulis et al. 1985, 73).
See -gamy: autogamy.
syn. autogamy, extreme inbreeding, selfing (Margulis et al. 1985, 73)
2. Obligatory self-fertilization by autogamy, pedogamy, or automictic parthenogenesis (Lincoln et al. 1985).
3. "Reproduction by single cells that are derived from a single parent by meiosis and restoration of ploidy (Judson and Normark 1996, 41).
endomixis *n.* Self-fertilization in which male and female nuclei from one individual fuse (Lincoln et al. 1985).
cf. parthenogenesis
exomixis *n.* Fusion of gametes from different sources (Lincoln et al. 1985).
syn. xenomixis (Lincoln et al. 1985)
holoschisis See mitosis: amitosis.
panmixis *n.* Random breeding among members of a small population or deme (Wilson 1975, 9).
See mating: random mating.
syn. pangamy, panmixia, random mating (Lincoln et al. 1985)
Comment: Panmixis is an important simplifying assumption in population biology (Wilson 1975, 9).
parthenomixis *n.* "The fusion of two female nuclei produced within a single gamete or gametangium" (Lincoln et al. 1985).
xenomixis See -mixis: exomixis.

♦ **mixoploidy** See -ploidy: mixoploidy.
♦ **mixotroph** See -troph-: mixotroph.
♦ **mma** See macarthur: millimacarthur.
♦ **mnemotaxis** See taxis: mnemotaxis.
♦ **mob** See ²group: animal group.
♦ **mobbing** *n.*
1. In many bird species, including the Florida Scrub Jay, Grey-Breasted Jay, and Western Scrub Jay: behavior in which conspecific, and sometimes heterospecific, individuals gather around a predator and vocalize loudly or perform conspicuous visual displays (Hinde 1952, Altmann 1956, Cully and Ligon 1976 in Francis et al. 1989, 795).
2. In many bird species; the American Ground Squirrel, Chimpanzee, Ring-Tailed Lemur; Baboons: a joint assault by an animal group (of one or more species) "on a predator too formidable to be handled by a single individual" that the mobbers attempt to disable, or at least

drive from the vicinity, even though the predator is not engaged in an attack on the group (Wilson 1975, 46–47, 454, 589).

v.t. mob

Comments: All explanations of avian "mobbing" include the idea that it is somehow antipredator in nature (Curio 1978 in Francis et al. 1989, 795). Francis et al. (1989, 798) carefully define "mobbing" in Florida Scrub Jays. "Mobbing" can involve a mobber's changing its position frequently; stereotyped wing or tail, or both, movements in birds; and loud calls, often with a broad frequency spectrum and transients (Heymer 1977, 85; Curio 1978 in Immelmann and Beer 1989, 187). Perhaps "mobbing" should be applied to the defensive behavior of some species of social insects and nonsocial insects (*e.g., Bembix* wasps) (Matthews and Matthews 1978, 351).

♦ **mobile genetic element** See transposable genetic element.

♦ **mobility** See vagility.

adj. mobile

♦ **mock hunting** See play.

♦ **mocking** *n.*
1. A person's holding something up to ridicule; deriding; assailing with scornful words or gestures (*Oxford English Dictionary* 1972, verb entries from *ca.* 1450).
2. A person's using, or uttering, ridicule; acting or speaking in derision; jeering, scoffing; flouting (*Oxford English Dictionary* 1972, verb entries from 1450).
3. A person's deceiving or imposing upon; deluding, befooling; tantalizing, disappointing (*Oxford English Dictionary* 1972, verb entries from *ca.* 1470).
4. A person's behavior directed at another person who deviates from a group norm that places social pressure on him to readopt group-appropriate behavior (*e.g.,* in Bushmen, sticking out one's tongue, vulva presentation, and pubic presentation) (Eibl-Eibesfeldt 1973 in Heymer 1977, 167).

v.i, v.t. mock

syn. mockery (Michaelis 1963), ridiculing (Heymer 1977, 167)

cf. display: genital display; joking; mimicry: vocal mimicry

[Old French *mocquer*, to mock]

♦ **modal-action pattern (MAP)** See action pattern: modal-action pattern.

♦ **modality** *n.*
1. A kind of sense; sensory means (Hinde 1970, 87).
   syn. sensory modality (Hinde 1970, 87)
2. The state, or quality, of a stimulus or sensation (Lincoln et al. 1985).
3. Stimuli (light, sound, smell, taste, touch, magnetism) that an animal perceives via

its particular kind of sense receptor and that give rise to particular sensations (*e.g.,* vision, audition, olfaction, and gustation) (Immelmann and Beer 1989, 188).

cf. receptor, perception, sensation

audition See modality: mechanoreception: audition.

chemoreception *n.* An organism's capability of identifying chemicals and detecting their concentrations (McFarland 1985, 187).

▸ **exteroception** *n.* An organism's chemoreception of chemicals in its external environment; contrasted with interoception (McFarland 1985, 187).

▸ **gustation** *n.*
1. The action, or facility, of tasting; the sense of taste (*Oxford English Dictionary* 1972, entries from 1599).
   adj. gustative, gustatory
2. Chemoreception of chemicals by direct contact (McFarland 1985, 189).
syn. gust, taste (Michaelis 1963)
Comments: The basic tastes in mammals are acid (sour), bitter, salt, and sweet (McFarland 1985, 189). Some researchers consider umami, *q.v.,* to be the Human's fifth taste (Willoughby 1998, C3).
[Latin *gustus,* taste]

▸ **interoception** *n.* An organism's chemoreception of internal chemicals (*e.g.,* carbon dioxide, nutrients, and hormones); contrasted with exteroception (McFarland 1985, 187).

▸ **olfaction** *n.*
1. The action of smell; sense of smell (*Oxford English Dictionary* 1972, entries from 1846).
2. Chemoreception of airborne substances (McFarland 1985, 188).
syn. smell
[Latin *olfactus,* past participle of *olfacere,* to smell]

▸ **osmoreception** *n.* An organism's detection of its cell dehydration and shrinkage due to water loss (McFarland 1985, 267).

electroreception, electrical sensitivity *n.* For example, in Dogfish, Gymnotid Fish, Mormyrid Fish, Sharks: an individual's perception of electric current (McFarland 1985, 226).

interoceptive senses, organic senses, visceral senses *n.* An animal's perceptions of sensations that arise within its body (*e.g.,* hunger, thirst, visceral pain, nausea, and cramps) (Storz 1973, 134).

kinesthesia *n.* An animal's perception of its own movement and body-part positions (Immelmann and Beer 1989, 164–165).

syn. muscle sense, position sense, proprioception (Storz 1973, 140), space orientation (Heymer 1977, 125)

cf. -ceptor: prioprioceptor; -ference: reafference; learning: kinesthetic learning; method: putting-through method

magnetoreception *n.* For example, in the Domestic Pigeon, Honey Bee, Human; some bacterium species: an individual's perception of magnetic waves (Baker 1987, 691).

cf. magnetite; taxis: geomagnetotaxis

mechanoreception *n.* An organism's perception of sound, vibration, or pressure (McFarland 1985, 191).

▸ **audition** *n.* A person's faculty of hearing (*Oxford English Dictionary* 1972, entries from 1599); sound perception.

syn. hearing

Comment: "Audition" is now commonly used for hearing in nonhuman animals as well. These animals hear from 0.1 (Domestic Pigeons) to 100,000 hertz (some bat species) (Gould 1982, 115).

binaural hearing *n.* A vertebrate's use of both of its ears to locate the direction of sound sources (Gregory and Zangwill 1987, 88).

▸ **touch** *n.* A person's means of perceiving an object after touching it with a body part (*Oxford English Dictionary* 1972, entries from 1394).

Comment: "Touch" is now often used to refer to a sense in nonhuman animals as well.

reception *n.* An organism's receiving a stimulus(i) (McFarland 1985, 187).

cf. modality

Comment: McFarland (1985, chap. 12–13) reviews "chemoreception" (including "gustation" and "olfaction"), "mechanoreception" (including "hearing"), "touch reception," and "thermoreception."

thermoreception *n.* Probably in most animal species: an individual's perception of heat (infrared radiation) (McFarland 1985, 190).

vision *n.*
1. A person's seeing with his "bodily eye" (*Oxford English Dictionary* 1972, entries from 1491).
2. An animal's perception of electromagnetic radiation (Gould 1982, 114).

cf. Purkinje shift

Comment: Vision is "monochromatic," "dichromatic," and "trichromatic," depending on the species. The Honey Bee sees from ultraviolet to near red; some hummingbird species see from ultraviolet to far red; pitviper snakes perceive infrared from a distance with their pit organs (Gould 1982).

▸ **binocular vision** *n.* In some invertebrate and vertebrate species, especially predators: vision in which visual fields of both eyes overlap, enabling more accurate depth perception and distance judgement compared to monocular vision (McFarland 1985, 204; Gould 1982).

cf. stereopsis

▸ **monocular vision** *n.* In many vertebrate species: vision in which visual fields of each eye do not overlap (McFarland 1985, 204).

♦ ¹**model** *n.*
1. An object, usually built in miniature and to scale, that represents something to be made or already existing (Michaelis 1963).
   syn. scale model (Michaelis 1963)
2. An object that an investigator uses to represent particular stimuli in a behavioral experiment (*e.g.,* a "hawk model," "goose model," or "stickleback model") (Tinbergen 1948 and 1951 in Hinde 1970, 60, 66).

♦ ²**model** *n.* An artificial, often abstract representation of specific characteristics of stimuli that is used in studying factors that elicit fixed-action patterns or other specific behaviors, *e.g.,* an imitation of part of another animal's body, colored wooden balls or cubes, tape-recorded vocalizations or artificially generated sounds, and artificial scents (Heymer 1977, 31; Dewsbury 1978, 17; Immelmann and Beer 1989, 293).

syn. dummy, stimulus model, surrogate (Immelmann and Beer 1989, 293)

♦ ³**model** *n.*
1. A living, or nonliving, agent that emits perceptible stimuli or signals in a mimicry system, *q.v.* (Pasteur 1982, 169).
2. An organism that is unpalatable, or noxious, and is imitated in some way by its mimic(s) (Lincoln et al. 1985).

cf. mimicry

natural-stimulus model *n.* A parasitic animal's imitating another species' character that has a specific releaser function in the parasite's host; *e.g.,* the mouth markings of Widowbird chicks which are exact copies of those of young of Widowbirds' estrildine-finch hosts (Immelmann and Beer 1989, 293).

predator model *n.* A model that resembles a potential predator of an animal species and is used to study specific stimuli that enable its enemy recognition (Tinbergen 1948 in Heymer 1977, 67).

♦ ⁴**model** *n.*
1. A simple, formal statement of a hypothesis, often expressed mathematically (Bell 1982, 82).
   cf. hypothesis; model: mathematical model

2. A mathematical representation of a phenomenon (Lincoln et al. 1985) which may be a tested, or untested, hypothesis.

3. A group of related hypotheses concerning a phenomenon.

syn. theoretical model

cf. axiom, concept, hypothesis, law, principle, supposition, ³theory

autocatalysis model *n.* The core of a theory of the origin of human sociality: When the earliest hominids became bipedal as part of their terrestrial adaptation, their hands were freed, their manufacture and handling of artifacts were made easier, and their intelligence grew as part of the improvement of their tool-using habit; this was followed by an expansion of their entire materials-based culture, perfected cooperation during hunting, increased intelligence, still more sophisticated tool use, a shift to big-game hunting, and so forth (Wilson 1975, 567–569).

Boorman-Levitt model of group selection *n.* A model of group selection that involves "marginal populations derived from one large, stable population, and altruist genes that do not influence extinction rates until the marginal populations have reached demographic carrying capacity" (Boorman and Levitt 1972, 1973a in Wilson 1973, 1975, 112).

bounded-rationality model of human behavior *n.* A model that examines the psychological limits of Humans in complex situations by simplifying complexity and, thus, accounts for their irrational behavior (Simon in Tierney 1983).

cf. ⁴model: classical model of human rationality

Brock-Riffenburgh model *n.* A model that suggests that by clumping into schools, fish are found less often by predators (Wilson 1975, 135).

classical model of human rationality *n.* Human behavior involving application of a consistent set of values (utility function) over a vast array of choices in order to pick the best possible one (Simon in Tierney 1983).

cf. ⁴model: bounded rationality model of human behavior

deterministic model *n.* A model in which all relationships are fixed (*i.e.,* not probabilistic) so that a particular input produces one exact prediction of the output (Lincoln et al. 1985).

cf. ⁴model: stochastic model

drive model *n.* For example, in a cichlid fish species: a postulated endogenous aggressive drive, a process in the central nervous system that continuously produces energy of an aggressive nature independently of external stimuli (Lorenz 1950, 1963 in Heymer 1977, 179).

Comment: Plack (1973) and Reyer (1975 in Heymer 1977, 179) do not support the existence of a drive model.

dual-inheritance model *n.* The hypothesis that the forces that bring about changes in gene frequencies (drift, migration, mutation, natural selection) have analogs within the realm of cultural evolution (inferred from Dugatkin and Godin 1998, 323).

energy model of behavior *n.* Konrad Lorenz's hypothesis to explain the self-exhausting characteristic of fixed-action patterns (FAPs): Each FAP has its own reservoir of "action-specific energy" (ASE), the amount of ASE in its reservoir increases steadily as an animal refrains from displaying the particular FAP, the ASE level is depleted by repeated occurrence of the FAP, and the ease of occurrence of the FAP is related to characteristics of sign stimuli and the level of ASE at a point in time (Dewsbury 1978, 19).

Gadgil-Bossert model *n.* A mathematical model of life-history evolution that indicates when the optimal strategy is to breed iteroparously, or semelparously, and proposes that the reproductive effort in iteroparous species should increase steadily with age (Gadgil and Bossert 1970 in Wilson 1975, 96–98, 563).

gestalt model *n.* A hypothetical system in which a common odor is created by pooling the recognition pheromones of some, or all, of a group of individuals; conspecifics classify group members as kin, or nonkin, according to the degree to which they possess the group odor (Wilson 1987, 10).

cf. ⁴model: individualistic model

hotspot model *n.* The hypothesis that lekking males clump at places that females visit in greatest numbers (Lill 1976; etc. in Gosling and Petrie 1990, 272).

cf. lek

Comment: Topi data of Gosling and Petrie (1990, 283) support this model rather than alternative hypotheses regarding lek initiation.

individualistic model *n.* A hypothetical system in which an individual judges other conspecific individuals to be kin or nonkin according to whether they possess certain alleles that encode a particular recognition pheromone (Wilson 1987, 11; Hölldobler and Wilson 1990, 639).

cf. ⁴model: gestalt model

instinct model *n.* A theoretical model of instinctive motivation, represented graphically by a diagram that is analogous to a hydraulic system (*e.g.,* Lorenz's hydraulic model and Tinbergen's hierarchical model of behavior) (Lorenz 1950 and Tinbergen 1971 in Immelmann and Beer 1989, 152).

Comment: Because arguments against the concept of instinct apply to these models as well, they are of questionable utility (Immelmann and Beer 1989, 152).

island model *n.* "A population divided into many very small demes and affected by genetic drift that restricts genetic variation within individual demes but increases it between them" (Wright 1943 in Wilson 1975, 78).

Lorenz's hydraulic model, Lorenz's hydraulic model of behavior, Lorenz's hydraulic model of behavior regulation *n.* Konrad Lorenz's model (somewhat analogous to a flush toilet) relating to action-specific energy, *q.v.,* sign stimuli, and innate releasing mechanisms (Lorenz 1950 in Dewsbury 1978).

syn. cistern model of behavior (Dawkins 1986, 70–72), flush-toilet model

cf. ⁴model: instinct model

Comment: The utility of this model is questionable (Hinde 1970 in Dewsbury 1978, 20).

mathematical model *n.* A symbolic representation of a number of hypotheses, or assumptions, about a system in the form of an equation, or set of equations, that is used to describe, or predict, the behavior of a system (Lincoln et al. 1985).

cf. ⁴model

Orians-Verner model of polygyny *n.* For example, in Red-Wing Blackbirds: a graphical model that suggests why it will be advantageous for some females to join a male's "harem" instead of becoming the sole partner of a male in a poor territory (Verner 1965 and Orians 1969 in Wilson 1975, 328; Davies 1989, 226).

Comment: There are now different versions of this model (Davies 1989, 226).

parametric model *n.* A model that is an equation with one or more parameters, *q.v.* (inferred from Colwell and Coddington 1994, 109–111).

probabilistic model See ⁴model: stochastic model.

reflex model of behavior *n.* Animal behavior should be seen as a series of reflexes with an animal's reacting to a series of external stimuli. Coordinated behavior is a result of a chain of these reflexes in which performance of the first small step in the movement stimulates an animal's sense organs to activate muscles to produce the next step, etc.; animals react to environmental stimuli rather than being spontaneously active (Dawkins 1986, 69).

Comment: This view of behavior was modified by Lorenz's work on instinct (Dawkins 1986, 69).

regression model of evolution *n.* Groundwater and other limnic species evolved from populations of marine littoral species stranded by periodic regressions of sea level (Lincoln et al. 1985).

stationary model See hypothesis: stationary model.

stochastic model, probabilistic model *n.* A model in which a given input produces a range of possible outcomes due to chance alone (Lincoln et al. 1985).

cf. ⁴model: deterministic model

Tinbergen's hierarchical model (of behavior regulation) *n.* Tinbergen's model designed to account for the structure of behavior over long periods: Action-specific energy is not completely specific to individual fixed-action patterns; instincts are organized hierarchically, with the final repository of energy being a higher-level instinctive center; and energy flows downward through various inhibitory blocks to progressively finer levels of appetitive behavior and consummatory acts and particular muscular activities down to the most indivisible motor units (Tinbergen 1951, 122–127; Dewsbury 1978, 21, 206).

syn. hierarchical model of behavior (Dewsbury 1978, 21, 206)

Comment: The utility of this model is questionable (Dewsbury 1978, 21).

♦ **"model-agreeable-to-dupe mimicry"** See mimicry: "model-agreeable-to-dupe mimicry."

♦ **"model-disagreeable-to-dupe mimicry"** See mimicry: "model-disagreeable-to-dupe mimicry."

♦ **"model-mimic-dupe-tripartite system"** See ¹system: "model-mimic-dupe-tripartite system."

♦ **"model-vs.-mimic-dupe-bipolar mimicry system"** See ¹system: mimicry system: "model-vs.-mimic-dupe-bipolar mimicry system."

♦ **modern** See derived.

♦ **modern character** See character: modern character.

♦ **modern ethology** See study of: ³ethology: modern ethology.

♦ **Modern Synthesis, the Modern Synthesis, the modern synthetic theory of evolution** See evolutionary synthesis.

♦ **modern theory of the gene** See ³theory: modern theory of the gene.

♦ **modes of evolution** *pl. n.* Phyletic evolution, quantum evolution, and speciation, *q.v.* (Simpson 1944, 197–217 in Gould 1994, 6766).

cf. tempos of evolution

h–m

♦ **modification** *n.* An environmentally, or experimentally, induced change in the life of an individual organism (Immelmann and Beer 1989, 188).

♦ **modifier gene** See gene: modifier gene.

♦ **modular society** See ²society: modular society.

♦ **modular unit** *n.* A member of a society (Lincoln et al. 1985).

♦ **modulation** *n.* In telecommunications, "the process whereby some characteristic of a carrier wave is varied in accordance with another wave;" the result of this process (Michaelis 1963).

amplitude modulation *n.* Variation in a signal's intensity (*e.g.,* sound intensity) (Immelmann and Beer 1989, 13).

frequency modulation *n.* "Variation of the wavelength of energy transmitted in wave form," including sound (Immelmann and Beer 1989, 113).

♦ **modulatory communication** See communication: modulatory communication.

♦ **molar behaviorism, molarism** See behaviorism: purposive behaviorism.

♦ **mole** See unit of measure: mole.

♦ **molecular behaviorism** See behaviorism: molecular behaviorism.

♦ **molecular biology** See study of: biology: molecular biology.

♦ **molecular clock (model of evolution), molecular evolutionary clock** See hypothesis: molecular clock (model of evolution).

♦ **molecular ethology** See study of: ³ethology: molecular ethology.

♦ **molecular evolution** See study of: evolution: molecular evolution.

♦ **molecular fossil** See fossil: molecular fossil.

♦ **molecular genetics** See study of: genetics: molecular genetics.

♦ **molecular mimicry** See mimicry: molecular mimicry.

♦ **molecular paleontology** See study of: paleontology: molecular paleontology.

♦ **molecular systematics** See study of: systematics: molecular systematics.

♦ **molecular theory of evolution** See ²theory: molecular theory of evolution.

♦ **molecular theory of inheritance** See ²theory: molecular theory of inheritance.

♦ **molecule** *n.* One or more atoms that constitute the smallest part of an element, or compound, that can exist separately without losing its chemical properties (Michaelis 1963).

cf. acid; chemical-releasing stimulus: semiochemical: pheromone; drug; enzyme; ion; mineral; nucleic acid; neurotransmitter; ribozyme

Comment: A few of the thousands of kinds of molecules are below.

[French *molécule* < New Latin *molecula,* diminutive of Latin *moles,* mass]

accessory pigment *n.* A molecule that traps light wavelengths to which chlorophyll *a* is not sensitive (Campbell 1996, 542).

Comments: Accessory pigments include carotenoids; chlorophyll *b, c,* and *d;* phycobilins; and xanthophylls (Campbell 1996, 542–543).

▸ **chlorophyll *b*** *n.* A form of chlorophyll, $C_{55}H_{70}O_6N_4Mg$ (Michaelis 1963).

syn. chlorophyll-B (Michaelis 1963)

Comment: Chlorophyll *b* occurs in Prochlorophyta and Chlorophyta and evidently occurred in the common ancestor of chloroplasts, Cyanobacteria, and Prochlorophyta (Tomitani et al. 1999, 161). This chlorophyll is yellow-green (Michaelis 1963) and is an accessory pigment (Campbell 1996, 542).

▸ **phycobilin** *n.* An accessory pigment (Campbell 1996, 542).

Comment: Phycobilins occur in photosynthetic prokaryotes (Cyanobacteria) and eukaryans (Glaucocystophyta, Rhodophyta) and evidently occurred in the common ancestor of chloroplasts, Cyanobacteria, and Prochlorophyta (Tomitani et al. 1999, 161).

adenosine phosphate *n.* One of three molecules in which the nucleotide adenosine is attached through its ribose group to one to three phosphoric acid molecules (King and Stansfield 1985).

See acid, -saccharide.

Comments: The adenosine phosphates are adenosine monophosphate (AMP), adenosine diphosphate (ADP), and adenosine triphosphate (ATP). These compounds are interconvertible.

▸ **adenosine diphosphate (ADP)** *n.* An adenosine phosphate with two phosphoric acid molecules (King and Stansfield 1985).

▸ **adenosine monophosphate (AMP)** *n.* An adenosine phosphate with one phosphoric acid molecule (King and Stansfield 1985).

Comment: Cyclic adenosine monophosphate (cAMP) is a chemical messenger within and between cells (Starr and Taggart 1989, 59).

▸ **adenosine triphosphate (ATP)** *n.* An adenosine phosphate with three phosphoric acid molecules (King and Stansfield 1985).

Comment: When ATP is hydrolyzed, it yields energy that is used to drive a multitude of biological processes, including bioluminescense, muscle contraction, photosynthesis, and biosynthesis of proteins, nucleic acids, polysaccharides, and lipids(King and Stansfield 1985).

carbon dioxide (CO$_2$) *n.* A molecule comprised of one carbon and two oxygen atoms.
Comments: Photosynthetic plants use CO$_2$ as a carbon source for making other compounds; many organisms produce CO$_2$ as a byproduct of respiration. Earth's atmosphere is about 0.04% CO$_2$. The pre-industrial concentration of CO$_2$ in the Earth's atmosphere was about 280 microliters per liter (= parts per million); the current concentration is 360 microliters per liter; and the predicted concentration in 2050 is 560 microliters per liter (DeLucia et al. 1999, 1177, 1179).

chitin [KIGH tin] *n.* A nitrogenous polysaccharide formed primarily of *N*-acetyl-lucosamine units (Borror et al. 1989, 793).
Comments: Chitin is found in the integuments of arthropods and in some fungi. Chitin increases the flexibility of arthropod integuments, and sclerotins (proteins) give integuments rigidity.

chlorofluorocarbon (CFC) *n.* A chemically inert chlorinated hydrocarbon treated with hydrogen fluoride (Allaby 1994).
cf. ozone hole
Comments: Chlorofluorocarbons include CFC-12 (CCl$_2$F^2) and CFC-21 (CCl$_2$F) (Allaby 1994). Industry is using them, and has used them widely, in aerosol propellants, fire extinguishers, foam-plastic manufacturing, refrigerants, and solvents. Chlorofluorocarbons are greenhouse gases and also provide chlorine, which destroys ozone and causes ozone holes. Hydrochlorofluorocarbons and other chemicals are being used to replace CFCs (Rensberger 1993a, A18).

chlorophyll *n.* For example, in Cyanobacteria, most plants, some protistans (Algae, Euglenoids): a nitrogenous molecule, comprised of a hydrocarbon tail and porphyrin ring, found in chloroplasts and essential in carbohydrate production in oxygenic photosynthesis (Michaelis 1963; Campbell 1996, 207, illustration)
Comments: All Algae have chlorophyll *a* and other kinds of chlorophyll used as accessory pigments, along with carotenoids, phycobilins, and xanthophylls, depending on the species (Campbell 1996, 542–543).
[Greek *chlōros*, green + *phyllon*, leaf]

▸ **chlorophyll *a*** *n.* The most abundant form of chlorophyll, C$_{55}$H$_{72}$O$_5$N$_4$Mg (Michaelis 1963).
syn. chlorophyll-A (Michaelis 1963)
Comment: This chlorophyll is blue-green (Michaelis 1963).

▸ **chlorophyll *b*** See molecule: accessory pigment: chlorophyll *b*.

***cis*-9,10-octadecenoamide** *n.* A fatty acid found in cerebrospinal fluid of sleep-deprived cats (Cravatt et al. 1995, 1506).
Comments: Synthetic *cis*-9,10-octadecenoamide induces an apparently normal state of sleep in rats after it is injected into them (Cravatt et al. 1995, 1506). This chemical and other fatty-acid primary amides are natural constituents of the cerebrospinal fluid of cats, rats, and Humans.

greenhouse gas *n.* An atmospheric gas that increases the temperature of Earth's lower atmosphere.
cf. effect: greenhouse effect; global warming
Comments: Greenhouse gases include carbon dioxide, halocarbons, methane, nitrous oxide, ozone, and sulfates (Allaby 1994). We might be increasing the amount of greenhouse gases in the atmosphere which is resulting in global warming.

hemoglobin (*Brit.* **haemoglobin**) *n.* A protein molecule comprised of a globin (protein part) and a heme (iron-containing part) (Michaelis 1963).
Comments: Hemoglobin, in some form, occurs in many taxa, including some bacteria, chironomid flies, cyanobacteria, plants, and yeast, and in Vertebrates (Blakeslee 1999, D2). In Vertebrates, hemoglobin regulates blood pH, serves as an NO-activated deoxygenase, and transports CO$_2$ and O$_2$ (Imai 1999, 437; Minning et al. 1999, 501). About 2 billion years ago, flavohemoglobin helped bacteria detoxify nitric oxide (Blakeslee 1999, D2; Imai 1999, 437). Iron in the hemoglobin catalyzed the conversion of nitric oxide into consumable byproducts. About 1.5 billion years ago, in Round Worms (Achelminthes), hemoglobin bound with unwanted oxygen and used nitric oxide to metabolize oxygen into nitrate and water. From about 245 million years to the present, hemoglobin has carried oxygen from a mammal's lungs to its tissues.

leghemoglobin (*Brit.* **leghaemoglobin**) *n.* A hemoglobin, *q.v.*, found in the root nodules of legumes (Imai 1999, 437).
Comments: Leghemoglobin, which has a very high oxygen affinity, keeps the symbiotic nitrogen-fixing bacteria in root nodules aerobic and protects the nitrogen-fixing enzyme system from oxidation (Imai 1999, 437).

lipid *n.* An organic molecule that is variably soluble in an organic solvent such as alcohol and barely soluble in water (*e.g.,* a carotenoid, fat, oil, phospholipid, sterol, or wax) (King and Stansfield 1985; Strickberger 1990, 113).

syn. lipide (Michaelis 1963)

[LIP(O)- + ID(E)]

lucibufagin *n.* A steroid produced by some firefly species (T. Eisner in Fountain 1999, D5).

Comments: Lucibufagin is a vertebrate heart stimulant. This compound kills pet Australian Bearded Dragons, which did not evolve with fireflies and eat them (Fountain 1999, D5).

luciferin *n.* A molecule in bioluminescent organisms that produces cool light when it is oxidized (McDermott 1948, 17).

cf. firefly, glowworm (Appendix 1); enzyme: luciferase

[named by Raphael DuBois, French physiologist (1886) in McDermott 1948, 17]

methycyclohexanol *n.* An alcohol that smells like the odor of tennis shoes in July (Hall 1998, 32).

murein *n.* A peptidoglycan with chains of sugars cross-linked with short polypeptides, some of which contain D-amino acids (Strickberger 1990, 153).

cf. peptidoglycan

Comment: Murein is found in cell wells of most eubacterian species but not in those of Archaebacteria (Strickberger 1990, 153).

***N*-methyl-D-aspartate (NMDA)** *n.* In vertebrates, a molecule found at the end of dendrites (Lemonick 1999, 57).

Comment: NMDA functions in learning and memory formation (Lemonick 1999, 57).

nitric oxide *n.* A molecule comprised of one nitrogen and one oxygen atom.

Comments: Nitric oxide molecules break apart in about 5 seconds after forming (Rensberger 1992b, A3). In Humans, immune-system cells emit nitric oxide used to kill bacteria and cancer cells, and brain cells use nitric oxide to relay impulses to other brain cells. In rats and dogs, nitric oxide causes muscles of a penis to relax, resulting in its erection.

nitrous oxide (N_2O) *n.* A molecule comprised of two nitrogen and one oxygen atom.

Comments: Nitrous oxide, a greenhouse gas, is about 200 times more efficient as carbon dioxide at trapping heat radiation in Earth's atmosphere (Suplee 1997c, A3). This gas comes from soil and ocean bacteria, fertilizer decomposition, some industrial processes, and an unknown atmospheric source.

octanol *n.* An alcohol that smells like licorice (Hall 1998, 32).

ozone (O_3) *n.* A molecule comprised of three oxygen molecules (Strickberger 2000, 170).

cf. ozone hole

Comments: Stratospheric ozone that, in the stratosphere 15–50 km (9–31 miles) above the ground, protects Earth's life from damage from ultraviolet light (Strickberger 2000, 170). Ground ozone, which forms near the ground due to pollution, is deleterious to life.

peptidoglycan *n.* A compound in bacterial cell walls made up of polymers of modified sugars cross-linked by short polypeptides that vary from species to species (Campbell 1990, 524).

cf. murein

polycyclic aromatic hydrocarbon (PAH) *n.* A molecule comprised of hexagons of carbon and hydrogen atoms linked in various arrangements (Zimmer 1995c, 76).

Comments: Gasoline engines make PAHs and they are found in some meteorites, giving them the scent of outer space, a smell like a musty attic (Zimmer 1995c, 76). A PAH exposed to light can give off an electron. PAHs might have supplied energy for early cells.

polypeptide *n.* A molecule that is composed of two or more amino acids linked together by chemical bonds; a polymer of amino acids (Michaelis 1963).

Comment: A dipeptide is a polypeptide made up of two linked amino acids.

porphyrin *n.* A molecule in a class of organic compounds in which four pyrrole nuclei are connected in a ring structure that is usually associated with metals (*e.g.,* iron or magnesium) (King and Stansfield 1985).

syn. porphyrin ring (Strickberger 1990, 112)

Comments: Porphyrins form parts of chlorophyll, cytochrome, and hemoglobin molecules.

protein *n.* A molecule composed of one or more polypeptide chains twisted, wound, and folded upon themselves to form a macromolecule with a definite three-dimensional shape (= conformation) (Curtis 1983, 78).

cf. proteinoid

Comment: Heterotrophs obtain proteins from autotrophs.

[German < Greek *protetos*, primary < *protos*, first; so called because proteins are a main constituent of living matter]

proteinoid, thermal protein *n.* A synthetic polymer, from 4000 to 10,000 daltons in molecular weight, of amino acids linked together by peptide and other bonds [named by Fox and his colleagues in Strickberger 1990, 115].

cf. protein

Comment: Strickberger (1990, 115–116; 1996, 124, 610) lists traits of proteins and proteinoids. Some proteinoids show protein-like properties such as enzyme activity, color-test reactions, hormonal activity, and nonrandom amino-acid sequences. Researchers make proteinoids by heating dry mixtures of amino acids.

pyrophosphate *n.* A molecule formed from two orthophosphate molecules which contains a high-energy phosphate bond (Strickberger 1990, 114).

saccharide A sugar molecule.

▸ **disaccharide** *n.* A compound formed by the polymerization of two monosaccharide units (*e.g.,* maltose) (Strickberger 1990, 113).

▸ **monosaccharide** *n.* A simple sugar that cannot be decomposed by hydrolysis (*e.g.,* glucose and fructose) (Michaelis 1963).

▸ **polysaccharide** *n.* A carbohydrate formed by polymerization of more than two monosaccharide units (*e.g.,* starch, cellulose, and glycogen) (King and Stansfield 1985).

water (H_2O) *n.* A molecule comprised of two hydrogen and one oxygen atom.

Comments: Water is essential to all known life on Earth (Campbell et al. 1999, 37; Sawyer 1999a, A1). Extraterrestrial microscopic bubbles of water locked inside halite crystals in two meteorites are evidently about 4.5 billion years old. One of the meteorites fell in Monahans, TX, and the other fell in Morocco in 1998.

♦ **molluscivore** See vore: molluscivore.

♦ ***MOLPHY*** See program: *MOLPHY.*

♦ **molt** (*Brit.* **moult**) *n.*

1. Part (feather, antler, skin) of an animal that it casts off in preparation for its replacement by new growth (Michaelis 1963).

2. In arthropods: an animal's castoff skin, comprised of many integumental plates and membranes, that results from molting (Wilson 1975, 589).
Note: An exuvium is usually considered to be one of these plates.
syn. exuviae (preferred), exuvium (not preferred)

3. *v.t.* To produce a molt (Michaelis 1963)
syn. ecdyse (Michaelis 1963)
cf. ecdysis

fright molt *n.* In some bird species: an individual's losing part, or all, of its feathers during severe distress and restraint (Dathe 1955 and Dementiev 1958 in Heymer 1977, 158).

♦ **moltinism** (*Brit.* **moultinism**) See -morphism: polymorphism: moltinism.

♦ **momentary behavior** See behavioral event.

♦ **monacmic** See -acmic: monacmic.

♦ **monadaptive** *adj.* Referring to a character that possesses only one function (Wilson 1975, 22).

♦ **monandry** See mating system: monogamy: monandry.

♦ **monarchistic dominance, monarchy** See dominance (social): monarchistic dominance.

♦ **monarsenous** See mating system: polygamous.

♦ **monaxenic** See -xenic: monaxenic.

♦ **monecious, monoecious** See -oecism: monoecism.

♦ **monestric species, monoestric species** See [2]species: monestric species.

♦ **monestrous, monestrus** See estrus: monestrus.

♦ **monistic evolution, theory of** See [2]theory: monistic evolution, theory of.

♦ **monkey-see-monkey-do behavior** See learning: social-imitative learning.

♦ **mono-, mon-** *combining form* "Single; one" (Michaelis 1963).
[Greek < *monos,* single, one, alone]

♦ **monochronic** See -chronic: monochronic.

♦ **monoclimax** See climax: monoclimax.

♦ **monoclinous flower** See hermaphrodite.

♦ **monocular vision** See modality: vision: monocular vision.

♦ **monocylic** See cylic: monocylic.

♦ **monocyclic parthenogenesis** See parthenogenesis: monocyclic parthenogenesis.

♦ **monodomic, monodomous** See -domous: monodomous.

♦ **monoecism** See -oecism: monoecism.

♦ **monofactorial** See -genic: monogenic.

♦ **monogamous** See -gamous: monogamous.

♦ **monogamy** See mating system: monogamy.

♦ **monogenesis** See parthenogenesis: asexual reproduction.

♦ **monogenetic** See -genetic: monogenetic.

♦ **monogenetic parasite** See parasite: monogenetic parasite.

♦ **monogenic** See -genic: monogenic.

♦ **monogenous** See -genous: monogenous.

♦ **monogony** See parthenogenesis: asexual reproduction.

♦ **monogynopaedium** See -paedium: monogynopaedium.

♦ **monogyny** See mating system: monogyny.

♦ **monohybrid cross** See cross: monohybrid cross.

♦ **monolectic** See -lectic: monolectic.

♦ **monolepsis** *n.* The condition in which only one parent's characters are transmitted to its progeny (Lincoln et al. 1985).

h–m

♦ **monometrosis** See -metrosis: haplotrosis.

♦ **monomorphic** See -morphic: monomorphic.

♦ **monomorphism** See -morphism: monomorphism.

♦ **monomorphous life cycle** See life cycle: monomorphous life cycle.

♦ **monophage** See -phage: monophage.

♦ **monophage-oligophage-polyphage continuum** *n.* A spectrum of how many and what kinds of food that a species consumes (May and Ahmad 1983).
cf. -lectic
Comment: In insects, for example, there is lack of agreement as to what constitutes a "monophage," "oligophage," or "polyphage," and these terms and their subdivisions have been defined differently for different insect taxa (May and Ahmad 1983).

♦ **monophasic allometry** See -auxesis: heterauxesis: monophasic allometry.

♦ **monophenic population** See [1]population: monophenic population.

♦ **monopheny** *n.* The occurrence of only one phenotype in a population (Lincoln et al. 1985).

♦ **monophile** See [2]-phile: monophile.

♦ **monophyletic** See phyletic: monophyletic.

♦ **monophyletic group** See [2]group: monophyletic group.

♦ **monophylistic** *adj.* Referring to phylogenetic relationships between taxa based on both monophyly and genotypic similarity (Lincoln et al. 1985).
n. monophylist

♦ **monophyly** See [2]group: monophyletic group.

♦ **monoplanetic** See -planetic: monoplanetic.

♦ **monoploid** See -ploidy: monoploidy.

♦ **monosaccharide** See -saccharide: monosaccharide.

♦ **monosexual** See sexual: monosexual.

♦ **monosomic** *n.* An individual organism (aneuploid) that is missing an autosome from its chromosome set (Mayr 1982, 759).
adj. monosomic
syn. monosomy, monosome (Lincoln et al. 1985)

♦ **monospermy** See spermy: monospermy.

♦ **monosynaptic reflex** See reflex: monosynaptic reflex.

♦ **monothelius mating system** See mating system: polyandrous mating system.

♦ **monotokous, monotoky** See -toky: monotoky.

♦ **monoton plankton** See plankton: monoton plankton.

♦ **monotroph** See -phage: monophage.

♦ **monotrophic** See -phagous: monotrophic.

♦ **monotypic** See -typic: monotypic.

♦ **monotypic evolution** See [2]evolution: monotypic evolution.

♦ **monotypic species** See [2]species: monotypic species.

♦ **monovoltine** See voltine: monovoltine.

♦ **monovular** See -ovular: monovular.

♦ **monoxenous** See -xeny: monoxeny.

♦ **monoxyny** See -xeny: monoxyny.

♦ **monozygotic** See -zygotic: monozygotic.

♦ **monstrosity** *n.* Any abnormal, malformed, or markedly aberrant individual of a species (Lincoln et al. 1985).
cf. mutation

♦ **monticole** See -cole.

♦ **mood** *n.* An animal's preliminary state of change, or readiness, for action, necessary for the performance of an instinctive behavior (Lincoln et al. 1985).

♦ **mood induction** See behavior: allelomimetic behavior.

♦ **moon illusion** See illusion: moon illusion.

♦ **mooning** See display: buttocks display.

♦ **moral** *adj.* Referring to an act by which a person hurts himself but helps another person(s) (Alexander 1987, 12).
Comment: "Moral" and "immoral" are difficult to define precisely (Alexander 1987, 12).
immoral *adj.* Referring to an act by which a person helps himself but hurts another person(s) (Alexander 1987, 12).
cf. moral

♦ **moral philosophy** See study of: philosophy: moral philosophy.

♦ **moral system** See [1]system: moral system.

♦ **moralistic aggression** See aggression: moralistic aggression.

♦ **morality** See [1]system: moral system.

♦ **morass** See habitat: bog.

♦ **mores, mune** See [2]group: mores.

♦ **Morgan school** *n.* A group of geneticists led by T.H. Morgan that started dominating genetics in 1910 and was occupied primarily with pure genetic problems such as the nature of genes and their arrangements on chromosomes (Mayr 1982, 732).

♦ **Morgan's canon** *n.*
1. "In no case may we interpret an action as the outcome of the exercise of a higher physical faculty, if it can be interpreted as the outcome of the exercise of one which stands lower in the psychological scale" (Morgan 1894); that is, one should interpret data using the most parsimonious explanation (Dewsbury 1978, 10).
2. "Behavior patterns of an animal should not be described in terms of anthropomorphic or higher psychic activity such as love, gentleness, deceit, and so forth,

but instead interpreted exclusively by the simplest mechanisms known to work" (Wilson 1975, 30).

See law: law of parsimony.

syn. law of parsimony, (Lloyd) Morgan's canon (Dewsbury 1978, 10)

cf. Ockham's razor

Comments: This canon went too far in simplifying explanations of behavior. Other behaviorists rationally argued that one should explain behavioral mechanisms as correctly, not as simply, as possible (Wilson 1975, 30). Bateson (1991, 836) remarks that "slavish obedience to such a maxim ... tends to sterilize imagination."

[after C. Lloyd Morgan, animal behaviorist]

♦ **moribund** *adj.* Referring to an organism in the state of dying or being close to death (Lincoln et al. 1985).

n. moribundity

adv. moribundly

[Latin *moribundus* < *mori*, to die]

♦ **morph** *n.*

1. An individual of a polymorphic group (Lincoln et al. 1985).
2. Any phenotypic, or genetic, variant (Lincoln et al. 1985).
3. Any local population of a polymorphic species exhibiting distinctive morphology or behavior (Lincoln et al. 1985).

adj. morph (Lincoln et al. 1985)

syn. form

cf. character, -morphy

♦ **-morph-, -morphic, -morphy** *combining form* Having the form, or shape, of (Michaelis 1963).

[Greek *morphē*, form]

♦ **-morph**

cf. -morphic

antimorph See gene: antimorph.

apomorph, apomorphy, autapomorph, nonhomologous apomorph, pseudoapomorph See character: apomorph.

ergatandromorph, ergatomorphic male, ergatoid male See caste: ergatandromorph.

genomorph *n.* A polyphyletic taxon of generic, or subgeneric, rank that contains a group of superficially similar, but not closely related species (Lincoln et al. 1985).

See ²species: genomorph.

gynandromorph, gynander, mosaic, sex mosaic *n.* In some insect species, Humans: an individual that is a sexual mosaic with genotypic and phenotypic male and female parts (*e.g.*, sexual organs) (Lincoln et al. 1985).

n. gynandrism, gynandromorphism

▶ **bilateral gynandromorph** *n.* In some butterfly and moth species: the most common kind of gynandromorph, which is female on one side of its body and male on the other side (Lincoln et al. 1985; Sbordoni and Forestiero 1985, 44–45).

homeomorph See -morphy: homeomorphy.

hygromorph *n.* An organism that is structurally adapted for life in a moist habitat (given as an adjective in Lincoln et al. 1985).

hypermorph See allele: hypermorphic allele.

hypomorph See allele: hypomorphic allele.

paramorph *n.* Any phenotypic variant that lacks accurate definition (Lincoln et al. 1985).

pleomorph See -morphic: pleomorphic.

plesiomorph, plesiomorphy See character: plesiomorph.

polymorph *n.* One of the two or more forms (morphs) of a polymorphic species (Lincoln et al. 1985).

pseudoapomorph See character: apomorph: nonhomologous apomorph.

stasimorph, stasimorphy See character: stasimorphic character.

synapomorph, homologous synapomorph See character: synapomorph.

♦ **-morphic**

cf. -morph

andromorphic *adj.* Referring to a character that morphologically resembles a male (Lincoln et al. 1985).

n. andromorphy

cf. -morphic: gynomorphic

apomorphic *adj.* Referring to an apomorphic character, *q.v.*

caenomorphic See -morphism: caenomorphism.

dimorphic *adj.* Referring to dimorphism, *q.v.*

ergatomorphic, ergatoid *adj.* Referring to a worker of a social insect species (Lincoln et al. 1985).

gynomorphic *adj.* Referring to a character that morphologically resembles a female (Lincoln et al. 1985).

n. gynomorphy

cf. -morphic: andromorphic

heteromorphic, heteromorphous *adj.*

1. Referring to an organism that has different forms at different times or stages of its life cycle (Lincoln et al. 1985).
2. Referring to a plant species that has an alternation of vegetatively dissimilar generations (Lincoln et al. 1985).

homeomorphic See morphy: homeomorphy.

h—m

hydromorphic *adj.* Referring to an organism that is adapted for aquatic life (Lincoln et al. 1985).

matromorphic *adj.* An organism's resembling its female parent (Lincoln et al. 1985).
cf. -morphic: patromorphic

monomorphic *adj.*
1. Referring to monomorphism, *q.v.*
2. Referring to a population with respect to a particular allele that is present in all of its members (Lincoln et al. 1985).

patromorphic *adj.* Referring to an organism's resembling its male parent (Lincoln et al. 1985).
cf. -morphic: matromorphic

pleomorphic, pleiomorphic *adj.*
1. Referring to a species, or group, that assumes different shapes or forms (Lincoln et al. 1985).
2. Referring to an organism that assumes different shapes or forms during its life cycle (Lincoln et al. 1985).
adj. pleomorphous
n. pleomorph, pleomorphism

polymorphic *adj.* Referring to polymorphism, *q.v.*

protomorphic See primitive.

stasimorphic *adj.*
1. Referring to a deviant form that arose as a result of arrested development (Lincoln et al. 1985).
2. Referring to a stasimorphic character, *q.v.* (Lincoln et al. 1985).
cf. -morphic: apomorphic

stenomorphic *adj.* Referring to an organism that is small due to its cramped habitat conditions (Lincoln et al. 1985).

zoomorphic *adj.*
1. Pertaining to, or produced by, animal activities (Lincoln et al. 1985).
syn. zoomorphosis (Lincoln et al. 1985)
2. "Having the form of an animal" (Lincoln et al. 1985).
cf. zoogenic, -morphism: zoomorphism

◆ **-morphism**
cf. -morph

anthropomorphism *n.*
1. A person's attributing human form and attributes to the Deity (*Oxford English Dictionary* 1972, entries from 1753).
2. A person's attributing human characteristics to nonhuman organisms, deities, or inanimate objects (*Oxford English Dictionary* 1972, entry from 1858; Michaelis 1963).
v.i. anthropomorphize
syn. humanization (Lincoln et al. 1985)
cf. Morgan's canon; -morphism: zoomorphism; zoocentric
[Greek *anthrōpomorphos* < *anthrōpos*, man + *morphē*, form, shape]

balanced genetic polymorphism See balanced polymorphism.

behavioral polymorphism See polyethism under "p."

caenomorphism *n.* A living organism's change from a complex to a simpler form (Lincoln et al. 1985).
adj. caenomorphic

dimorphism *n.*
1. The presence of two forms of an allele, or trait (*e.g.,* physical structure or behavior) in a group of conspecific animals; *e.g.,* in some ant species, two different forms existing in the same colony "including two size classes, not connected by intermediates" (Wilson 1975, 582) or oak-catkin and oak-twig mimics in larvae of the same moth species (Greene 1989, 643).
2. A taxon's having two morphs or marked sexual dimorphism (Lincoln et al. 1985).
See character: dimorphism.
adj. ditypic
syn. ditypism
cf. -morphism: polymorphism
Comment: "Dimorphism" can be classified into many of the same categories as is "polymorphism:" age dimorphism, balanced dimorphism, etc.
[Greek *dimorphos* < *di-,* two + *morphē,* form]

▸ **age dimorphism** *n.* Dimorphism with regard to age (Immelmann and Beer 1989, 8).
cf. -morphism: polymorphism: age polymorphism

▸ **chromosomal dimorphism** *n.* Dimorphism with regard to chromosome type, or structure (*e.g.,* in sex chromosomes) (Lincoln et al. 1985).

▸ **horiodimorphism, horodimorphism** *n.* An organism's showing a seasonal, or periodic, alternation of form (Lincoln et al. 1985).
cf. cyclic: monocyclic

▸ **sexual dimorphism** *n.* Dimorphism in appearance (size, shape, coloration), physiology, and behavior with regard to sex beyond the basic functional portions of a species' sex organs (= primary sex characteristics) (Wilson 1975, 334, 595; Immelmann and Beer 1989, 267).
adj. sexually dimorphic
syn. antigeny (Lincoln et al. 1985)
cf. sex dichromatism

ditypism See -morphism: dimorphism.

gynandromorphism *n.* The occurrence of male and female characteristics (*e.g.,* sexual organs) in the same individual (Michaelis 1963; Wilson 1975, 585).
n. gynander, gynandromorph
syn. gynandrism (Lincoln et al. 1985)

cf. parthenogenesis: androgynous; -morph: gynandromorph

[Greek of doubtful sex < *gynē*, woman + *anēr, andros,* man]

horiodimorphism See -morphism: dimorphism: horiodimorphism.

monomorphism *n.*

1. In many social-insect species: the existence of only one worker subcaste (Wilson 1975, 589).
   cf. -morphism: polymorphism

2. A population's, or taxon's, having no genetically fixed discontinuous variation and, thus, comprising a single morph (Lincoln et al. 1985).

3. A species' having no differences in appearance between its sexes, no division into castes, or no behavioral-role segregation (Immelmann and Beer 1989, 189).
   cf. -morphism: dimorphism, polymorphism

polymorphism *n.* The co-occurrence of more than two different forms of an allele or trait (*e.g.,* structure or behavior) in a group of conspecific organisms (West-Eberhard 1989, 251); *e.g.,* a spectrum of worker sizes in some ant species (Smith 1965, 3) or different color patterns in a garter-snake species (Brodie 1990, 45).
syn. morphism (Lincoln et al. 1985)
cf. -morphism: dimorphism, genetic polymorphism, monomorphism
Comment: Some investigators consider the presence of only two different forms of a trait to constitute "polymorphism;" thus, they make "dimorphism" a subtype of "polymorphism."

▸ **age polymorphism** *n.* Polymorphism with regard to distinct developmental stages (Immelmann and Beer 1989, 8).
   cf. -morphism: dimorphism: age dimorphism

▸ **balanced polymorphism, balanced genetic polymorphism** *n.*

1. Polymorphism in which alleles on the same locus coexist at high frequencies during natural selection for both of them; balanced polymorphism involves heterozygote superiority, frequency-dependent selection, disruptive selection, a spatially heterogeneous environment with migration, cyclical selection, and counteracting selection at different levels (Wilson 1975, 71).
   cf. mimetic polymorphism

2. Polymorphism in which genetically distinct forms are more or less permanent components of their population, maintained by selection in favor of diversity (Lincoln et al. 1985).

▸ **behavioral polymorphism** See polyethism.

▸ **chromosomal polymorphism** *n.* Polymorphism with regard to chromosome types, or structure (*e.g.,* chromosome inversions) (Lincoln et al. 1985).

▸ **cryptic polymorphism** *n.* Genetic polymorphism that is controlled by a recessive gene, such that the different gene forms are not phenotypically distinguishable (Lincoln et al. 1985).

▸ **genetic polymorphism, polymorphism** *n.* Co-occurrence of more than two alleles at the same locus in a population at frequencies that cannot be accounted by recurrent mutation alone (Lincoln et al. 1985).

▸ **inversion polymorphism** *n.* "Polymorphism in chromosomes of both inverted and uninverted forms within a population at frequencies higher than could be maintained by chromosome breakage alone" (Wittenberger 1981, 617).

▸ **mimetic polymorphism** See mimicry: mimetic polymorphism.

▸ **moltinism** (*Brit.* **moultinism**) *n.* Polymorphism in which different strains of a species have a different number of larval stages (Lincoln et al. 1985).

▸ **parallel polymorphism** See morphism: tautomorphism.

▸ **phenotypic polymorphism** *n.* Polymorphism with regard to phenotypes in a population (Lincoln et al. 1985).
   syn. nongenetic polymorphism (Immelmann and Beer 1989, 227)
   Comment: Phenotypic polymorphism differs from other forms of intraspecific variability in that discontinuities separate the different forms; there are no intermediate forms (Immelmann and Beer 1989, 227).

▸ **pleiomorphism, pleomorphism** *n.* Polymorphism at different stages in an animal's life cycle (Lincoln et al. 1985).

▸ **pseudoseasonal polymorphism** *n.* Plants' having phenotype polymorphism according to time of year at which they germinate and flower that is caused by local fluctuations in temperature, season durations, and other climatic factors (Lincoln et al. 1985).

▸ **restriction-fragment-length polymorphisms (RFLPs)** (RIF-lips) *n.* Polymorphisms in a noncoding DNA sequence on homologous chromosomes that result in different restriction-length fragments (Campbell et al. 1999, 373).
   cf. enzyme: restriction enzyme
   Comment: Researchers use RFLPs as markers of variants in a population (Campbell et al. 1999, 373).

h–m

► **transient polymorphism** *n.*

1. Two alleles' (on the same locus) coexisting at high frequencies during the long time that it takes one to replace the other by natural selection (Wilson 1975, 71).

2. Genetic polymorphism that occurs while one form is in the process of replacing another due to natural selection (Lincoln et al. 1985).

protomorphic *adj.* "Primordial; primitive; original" (Lincoln et al. 1985).

tautomorphism *n.* Superficial resemblance of morphs belonging to two or more different polymorphic species (Lincoln et al. 1985).

n. tautomorph

zoomorphism *n.* A person's attributing characteristics of nonhuman animals to Humans, especially aspects of social life that are ascribed more to human biological constraints and imperatives than to ethical, or rational, considerations (Immelmann and Beer 1989, 331).

v.t. zoomorphize

cf. -genic: zoogenic; -morphic: zoomorphic; -morphism: anthropomorphism; teleological

♦ **morphism** See -morphism: polymorphism.

♦ **morphocline** See cline: morphocline.

♦ **morphogenesis** See -genesis: morphogenesis.

♦ **morphogenetic** See -genetic: morphogenetic.

♦ **morphogeny** See -genesis: morphogenesis.

♦ **morphological species concept** See ²species.

♦ **morphological support** *n.* An animal's physical feature that is used in its behavioral display (*e.g.,* a Robin's red breast, a Cockatoo's crest, a Peacock's tail, wing eyespots of many moth species, or the swollen, brightly colored buttocks of receptive female monkeys of many species) (Immelmann and Beer 1989, 189).

♦ **morphological transformation series** See cline: morphocline.

♦ **morphology** *n.* An organism's form and structure, especially its external features (Lincoln et al. 1985).

See study of: morphology.

adj. morphologic, morphological

cf. -morph

endomorphology *n.* An organism's internal morphology; anatomy; contrasted with exomorphology (Lincoln et al. 1985).

exomorphology *n.* An organism's external morphology; contrasted with endomorphology (Lincoln et al. 1985).

functional morphology *n.* One's interpretation of the function of an organism, or organ system by reference to its shape, form, and structure (Lincoln et al. 1985).

idealistic morphology *n.* The explanation of form as a product of an underlying essence or archetype (Mayr 1982, 455).

macromorphology *n.* "Gross external morphology;" contrasted with micromorphology (Lincoln et al. 1985).

micromorphology *n.* Microscopic surface topography, or structure, typically studied with a scanning electron microscope; contrasted with macromorphology (Lincoln et al. 1985).

phylogenetic morphology *n.* "The derivation of form from that of a common ancestor or quite often the tracing back of form to that of the reconstructed common ancestor" (Mayr 1982, 455).

♦ **morphology of growth** See study of: morphology: morphology of growth.

♦ **morphometrics** See -metry: morphometrics.

♦ **morphon, morphone** *n., pl.* **morphontes** (badly formed, *Oxford English Dictionary* 1972)

1. "A morphological individual, element, or factor" (*Oxford English Dictionary* 1972, entries from 1873).

[coined by E. Haeckel]

2. "A definitively formed individual" (Lincoln et al. 1985).

♦ **morphont** See individual: morphont.

♦ **morphosis** *n.*

1. The manner, or order, of development of a plant or plant organ (*Oxford English Dictionary* 1972, entries from 1857).

2. An animal's morbid formation; an organic disease (*Oxford English Dictionary* 1972, entry from 1891).

3. Variation in an individual's morphogenesis caused by environmental changes (Lincoln et al. 1985).

cf. metamorphosis

♦ **-morphosis** *adj.* -morphic.

allomorphosis See ²evolution: allomorphosis.

aromorphosis See ²evolution: aromorphosis.

cyclomorphosis, cyclogeny *n.* Cyclical changes in form (*e.g.,* an organism's seasonal change in morphology) (Lincoln et al. 1985).

epimorphosis *n.*

1. In some arthropod species: development in which all larval forms are suppressed, or passed within an egg prior to hatching, and the juvenile hatches with adult morphology (Lincoln et al. 1985).

2. An organism's regeneration of a structure (Lincoln et al. 1985).
See reproduction (def. 1, note).

gerontomorphosis *n*. Evolutionary change that results from adaptations and modifications of adult structures (Lincoln et al. 1985).

heteromorphosis, xenomorphosis *n*. An organism part's being replaced by a structurally different replacement part following injury (Lincoln et al. 1985).

holomorphosis *n*. An organism's regrowth of an entire part that was previously lost (Lincoln et al. 1985).

hydromorphosis *n*. Structural modification of an organism in response to water or wet habitat conditions (Lincoln et al. 1985).

hypermorphosis See [2]heterochrony: hypermorphosis.

meromorphosis *n*. An organism's partial regeneration of its lost part (Lincoln et al. 1985).

neomorphosis *n*. The regeneration of a structure that differs morphologically from the original structure (Lincoln et al. 1985).

paedomorphosis See [2]heterochrony: paedomorphosis.

peramorphosis *n*. The morphological consequence of hypermorphosis that results in a more exaggerated shape of an evolutionary descendent compared to its evolutionary ancestor (*e.g.,* monstrously large, branched antlers in the extinct Irish elk) (Alberch et al. 1979 in Futuyma 1986, 416).

xenomorphosis See -morphosis: heteromorphosis.

♦ **morphospecies** See [2]species: morphospecies.

♦ **morphotype** See -type, type: morphotype.

♦ **-morphy** See -morph, morphic.

homeomorphy, homoeomorphy *n*.
1. An organism's having the same, or very similar, morphology as another organism as a result of convergent evolution rather than common ancestry (Lincoln et al. 1985).
2. "The state of being structurally similar but phylogenetically distinct" (Lincoln et al. 1985).
adj. homeomorphic
n. homeomorph

metromorphy *n*. The condition of an organism's resembling its mother (Lincoln et al. 1985).

♦ **mortality** *n*.
1. An organism's being mortal, or subject to death (*Oxford English Dictionary* 1972, entries from 1340).
2. Loss of human life on a large scale; abnormal human death frequency due to pestilence or war (*Oxford English Dictionary* 1972, entries from 1400).
3. The number of human deaths in a particular period, or area, from a particular cause; death frequency; death rate (*Oxford English Dictionary* 1972, entries from 1645).
4. A population's death rate expressed as a percentage or fraction (Lincoln et al. 1985).
syn. mortality rate (Lincoln et al. 1985)
cf. natality

concomitant mortality *n*. "Death during exposure to a harmful substance or stimulus" (Lincoln et al. 1985).
cf. mortality: consequential mortality, delayed mortality

consequential mortality *n*. Death during a recovery period that follows exposure to a harmful substance or stimulus (Lincoln et al. 1985).
cf. mortality: concomitant mortality, delayed mortality

delayed-density-dependent mortality *n*. For example, in paramecia, possibly the Bobwhite Quail, the Mule Deer: a population's mortality that is related to its past density of individuals (Lack 1954, 118).

delayed mortality *n*. An animal's death that occurs after a recovery period following exposure to a harmful substance or stimulus, when the animal appears to be healthy (Lincoln et al. 1985).
cf. mortality: consequential mortality

density-dependent mortality *n*. For example, in some barnacle, flour-beetle, trichogrammatid-wasp, waterflea species; probably the Bobwhite Quail; the Alpine Swift, Common Swift, Guinea Pig, Whitefish: a population's mortality that is related to its recent density of individuals (Lack 1954, 116, 121–122).
syn. direct density-dependent mortality, ecological death

♦ **mortality selection** See selection: mortality selection.

♦ **mortification** See abasement.

♦ **mosaic** See chimera.

♦ **mosaic evolution** See [2]evolution: mosaic evolution.

♦ **moss** See habitat: bog.

♦ **mother** *n*.
1. A female organism that has borne an offspring; female parent (Michaelis 1963).
2. A female person who adopts a child, or who otherwise has a maternal relationship with another person (Michaelis 1963).
3. "Anything that creates, nurtures, or protects something else" (Michaelis 1963).
See alloparent.
cf. father, aunt, uncle

h–m

allomother *n.*
1. In some long-lived mammal species, some primate species; the African Elephant: a female that shows maternal behavior (retrieving, guarding, cleaning, etc.) toward a young that is not her own (Lancaster 1971, Douglas-Hamilton and Douglas-Hamilton 1975, etc. in Lee 1987, 278; Immelmann and Beer 1989, 11).
   adj. allomaternal
   syn. aunt (used in earlier literature)
   cf. parent: alloparent
2. *v.i.* In some primate species: to show the behavior of an aunt (Hrdy 1977, 199).
   cf. aunt

play mother *n.* For example, in some monkey species: A female nonrelative that interacts with a young animal in such a way that she, rather than the young animal or its parents, benefits (C.M. Crockett 1984, personal communication).
syn. allomother (suggested synonym from description in Immelmann and Beer 1989, 11)
cf. parent: alloparent

surrogate mother *n.* An artificial mother model, or dummy, used in deprivation experiments with monkeys (Immelmann and Beer 1989, 300).
syn. mother surrogate
cf. surrogate

♦ **mother-child relationship** See relationship: mother-child relationship.

♦ **mother-Eve hypothesis** See hypothesis: African-origin hypothesis of human mitochondrial DNA evolution.

♦ **mother-infant attachment** See attachment: mother-infant attachment.

♦ **mother surrogate** See mother: surrogate mother.

♦ **motif song** See animal sounds: song: functional song.

♦ **motility** *adj.* An organism's having the capacity to move or show movement (Lincoln et al. 1985).
adj. motile

♦ **motivated altruism** See altruism (def. 1).

♦ **motivating stimulus** See [2]stimulus: motivating stimulus.

♦ **motivation** *n.*
1. A factor that induces a person to act in a certain way; a person's desire, fear, or other emotion or a consideration of reason that influences, or tends to influence, his volition (*Oxford English Dictionary* 1972, entries from 1412).
2. A person's contemplated result, or desired object, that tends to influence his

volition (*Oxford English Dictionary* 1972, entries from 1412).
3. An individual animal's "behavioral mood, drive; action readiness or tendency; compulsion or urge to behave in a certain way" (Immelmann and Beer 1989, 190).
4. An individual animal's "total complex of proximate causal factors governing" what it is doing or about to do (Immelmann and Beer 1989, 190).
5. Control of an animal's behavior (Immelmann and Beer 1989, 190).
syn. drive (Hinde 1970, 200)
cf. drive
Comments: Definitions of "drive" and "motivation" are controversial. "Motivation" has a broad range of definitions. Immelmann and Beer (1989, 191) discuss problems with some usages of "motivation" and conclude that "motivation" is a useful term for covering the "subject of proximate causal control of behavior, without commitment to a specific conception or theory" and that "motivation" is now commonly defined in this way.

mixed motivation *n.* An animal's expression of a behavior of more than one behavioral tendency or motivation (*e.g.,* a threat movement that consists of elements of attack and escape behavior) (Immelmann and Beer 1989, 186).
cf. conflict

♦ **motivational analysis** See analysis: motivational analysis.

♦ **motivational change** *n.*
1. An animal's temporary and reversible change in behavior; "more permanent changes are ascribed to learning or ageing" (Hinde 1982, 46–47).
2. An animal's change in the underlying causal control of a behavior in accordance with a change in the behavior's function through evolution (Immelmann and Beer 1989, 192).
cf. emancipation

♦ **motivational competition** See competition: motivational competition.

♦ **motivational conflict** See [2]conflict: motivational conflict.

♦ **motivational energy** See energy: action-specific energy.

♦ **motivational impulse** See energy: action-specific energy.

♦ **motivational state** *n.* An animal's combined physiological and perceptual state represented in its brain, including factors relevant to its incipient and current behavior (McFarland 1985, 279; Immelmann and Beer 1989, 193).
syn. drive (Wilson 1975, 153)
cf. drive

Comment: McFarland (1985) discriminates between "motivational state" and "drive."

♦ **motivational system** See ²system: motivational system.

♦ **motor cortex** See organ: brain: motor cortex.

♦ **motor pattern** See pattern: motor pattern.

♦ **motor program** See program: motor program.

♦ *moule intérieure* *n.* The theory that the first individual of all fundamental species is formed from organic molecules that combined spontaneously, and this primary being becomes the epigenetic inner form that guarantees the permanence of the species. This permanence is incessantly challenged by the "circumstances" which induce the production of varieties; however, the inner form prevents variations from transgressing certain limits and, in this respect, the *moule intérieure* plays a role similar to Aristotle's *eidos* (Buffon in Mayr 1982, 333).

♦ **moultinism** See -morphism: polymorphism: moltinism.

♦ **mount bout** See bout: mount bout.

♦ **mounting, mounting behavior** See behavior: mounting, mounting behavior.

♦ **mounting attempt** *n.* In male Rhesus monkeys raised in isolation: a male's copulation attempt with a female in estrus that is not made in a fashion that leads to copulation (Heymer 1977, 32).
cf. behavior: abnormal behavior

♦ **mouse pounce, stiff-legged jump** *n.* In canids and felids: an individual's pouncing attack on a mouse or other small prey animal (Heymer 1977, 115).

♦ **Mousterian Tool Industry** See tool industry: Mousterian Tool Industry.

♦ **mouth breeder** *n.* A species that shows mouth brooding, *q.v.* (*e.g.,* some cichlid-fish and anabantid-fish species) (Immelmann and Beer 1989, 195).

♦ **mouth-brooding** See brooding: mouth-brooding.

♦ **mouth clashing** See fight: mouth clashing.

♦ **mouth fighting** See fight: mouth fighting.

♦ **mouth pulling** See fight: mouth pulling.

♦ **mouth-to-mouth feeding** See feeding: mouth-to-mouth feeding.

♦ ¹**movement** *n.* A person's emptying his, bowels, or the result of this behavior — feces (*Oxford English Dictionary* 1972, entry from 1891).
cf. elimination: defecation, egestion

♦ ²**movement** *n.*
1. An animal's motor pattern or locomotion (Immelmann and Beer 1989, 195).

2. An animal's change in location (*e.g.,* migratory movement) (Immelmann and Beer 1989, 195).
cf. behavior; kinesis: statokinesis; locomotion; migration

abdominal up-and-down movement *n.* In some dragonfly and damselfly species: a male's moving his abdomen up over his head and then straight back (Heymer 1977, 19).
Comments: This behavior is part of wing grooming in lestid damselflies. In calopterygid damselflies, this movement is a rhythmic oscillation used for signaling (Heymer 1977, 19).

comfort movement *n.* An animal's body-maintenance movement (*e.g.,* preening, shaking, stretching, and washing) (Dewsbury 1978, 262).

demonstrative movement *n.* An animal's communicatory movement that attracts the attention of other animals, particularly in courtship and displays against rivals (*e.g.,* pseudo-preening in which mandarin duck drakes display their particularly striking speculum feathers) (Heymer 1977, 50).

down-up movement *n.* In ducks: a drake's dipping his bill in water for an instant and then immediately thrusting his head high out of the water without lifting his breast which is deep in the water (Lorenz 1941, 1966 in Heymer 1977, 19).

instinctive movement See action pattern: fixed-action pattern.

multipurpose movement *n.* A behavior that is ordinarily simple and brief and appears in more than one functional system (*e.g.,* locomotion motor patterns that occur in foraging, territorial defense, migration, etc.) (Immelmann and Beer 1989, 195).
cf. action: transitional action
[introduced as *Mehrzweckbewegung* by Lorenz (1937 in Immelmann and Beer 1989, 195)]

rapid eye movement *n.* A person's eyes' moving actively across his visual field behind closed lids during rapid-eye-movement sleep (Campbell 1996, 1018).

rapid leaf movement *n.* For example, in *Mimosa* spp., the Venus Flytrap: a leaf's closing quickly after being touched in a specific way (Campbell 1996, 762).
Comments: The function of rapid leaf movements in *Mimosa* spp. (= Sensitive Plants) remains unknown (Campbell 1996, 762). These movements might reduce surface areas and help a plant to conserve water, or they might more prominently expose protective thorns that discourage herbivores, or both. Venus Flytrap leaves entrap prey by closing.

h – m

sleep movement *n.* For example, in many species of Fabaceae (= Legumes), the Prayer Plant (*Maranta bicolor*): a leaf's changing it position on a plant with regard to time of day (Campbell 1996, 762–763, illustrations).

Comments: In many legume species, leaves lower themselves in the evening and rise in the morning (Campbell 1996, 762). A Prayer Plant raises its leaves in the evening and lowers them in the morning (personal observation). Water-stressed Prayer Plants do not markedly raise their leaves in the evening.

♦ **mRNA** See nucleic acid: ribonucleic acid.

♦ **mucivore** See -vore: mucivore.

♦ **mudflat** See habitat: mudflat.

♦ **Muellerian association** See Müllerian association.

♦ **mulada** See ²group: mulada.

♦ **mule** See hybrid: mule.

♦ **Muller-Lyer illusion** See illusion: Muller-Lyer illusion.

♦ **Müllerian association, Muellerian association** *n.* An assemblage of different species in a particular place, all of which display similar aposematic coloration (Lincoln et al. 1985).

cf. mimicry: Müllerian mimicry

♦ **Müllerian mimicry, Muellerian mimicry** See mimicry: Müllerian mimicry.

♦ **Müllerian mimicry club, Muellerian mimicry club** See mimicry ring: Müllerian mimicry ring.

♦ **Muller's ratchet** *n.*
1. The idea that, without recombination, genetic errors continually accumulate in a finite population through chance by the random loss of the least mutated genome. Like a ratchet wheel that moves in only one direction, this process cannot be reversed; once mutants replace beneficial genes, the latter do not return without recombination to replace them, while the deleterious genes continue to accumulate (Morell 1994, 171–172).
2. A mechanism of an asexual population that prevents any of the population's other lines from attaining a mutation load smaller than that already existing in its least-loaded line (Lincoln et al. 1985).

[after H.J. Muller who proposed this concept in 1964 (Morell 1994, 171)]

♦ **multi-, mult-** *combining form*
1. Much; many; consisting of many (Michaelis 1963).
2. Having more than two (or more than one) (Michaelis 1963).
3. Many times over (Michaelis 1963).
4. In medicine: affecting many (Michaelis 1963).

[Latin *multus*, much]

♦ **multicolonial** See colonial: multicolonial.

♦ **multidimensional species concept** See ²species (def. 27).

♦ **multifactorial** See -genic: polygenic.

♦ **multifactorial inheritance** See inheritance: multifactorial inheritance.

♦ **multigene** See gene: multigene.

♦ **multigenic** See -genic: polygenic.

♦ **multihybrid** See hybrid: multihybrid.

♦ **multimale group** See ²group: multimale group.

♦ **multimale society** See ²society: multimale society.

♦ **multiparity, multiparous** See -parity: multiparity.

♦ **multiple alleles** See allele: multiple alleles.

♦ **multiple allelism** *n.* The phenomenon in which more than two alleles occur at a single locus (Mayr 1982, 754).

Comments: Sturtevant (1913) first explained this phenomenon, and it refuted Bateson's "presence-absence theory of gene action" (Mayr 1982, 754). Mouse coat color and human blood groups are caused by "multiple allelism."

♦ **multiple-choice method** See method: multiple-choice method.

♦ **multiple-choice test** See test: choice test.

♦ **multiple-comparisons test** See statistical test: multiple-comparisons test.

♦ **multiparasitism** See parasitism: multiple parasitism.

♦ **multiple-factor hypothesis, multiple-factor theory** See hypothesis: multiple-factor hypothesis.

♦ **multiple factors** See gene: polygene.

♦ **multiple-gene effects, multiple-gene inheritance** See pleiotropy.

♦ **multiple pathways** *pl. n.* The phenomenon that natural selection, which is always opportunistic, makes use of that part of the variation of an organism that leads most easily to the needed adaptation (Mayr 1982, 590).

♦ **multiplier effect** See effect: multiplier effect.

♦ **multiploidy** See -ploidy: multiploidy.

♦ **multipolar mating system** See mating system: multipolar mating system.

♦ **multipurpose movement** See movement: multipurpose movement.

♦ **multistate character** See character: multistate character.

♦ **multivariable analysis, multivariate analysis** See analysis: multivariate analysis.

♦ **multivoltine** See voltine: multivoltine.

♦ **mummy** *n.* A desiccated, mammal corpse preserved in northern permafrost (Guthrie 1990, 22).

See fossil: mummy.

Comments: There are many prehistoric mummies (= fossils) (Guthrie 1990). In cold mummification, the water from a corpse crystallizes and forms ice lenses around the corpse, which shrinks and shrivels. This differs from freeze drying, in which water sublimates into surrounding air, and the corpse retains its size and shape.

♦ **mune** See ²group: mores.

♦ **mural** See rupestral.

♦ **Murchison meteorite** See meteorite: Murchison meteorite.

♦ **murder** See behavior: killing behavior; group: murder.

♦ **murein** See molecule: murein.

♦ **muriphobia** See phobia (table).

♦ **murmuration** See animal sounds: murmuration; ²group: murmuration.

♦ **Murphy's law** See law: Murphy's law.

♦ **muscicole** See -cole: muscicole.

♦ **muscology** See study of: muscology.

♦ **musophobia** See phobia (table).

♦ **must, musth** *n.*

1. In male elephants and camels: an individual's being "must" — in a state of dangerous frenzy that occurs at irregular intervals (*Oxford English Dictionary* 1972, entries from 1871); a rutlike state displayed by health adult male elephants (Schulte and Rasmussen 1999, 1265).
2. An elephant in must (*Oxford English Dictionary* 1972).
3. *adj.* Referring to an animal in must (Michaelis 1963).

v.i. to go must

Comment: Male elephants over 14 years old become "must" — exceptionally aggressive and sexually active while secreting large quantities of temporal-gland liquid (Wilson 1975, 498; Poole 1989, 140).

[Hindu *mast*, drunk, lustful < Persian]

♦ **muster** See ²group: muster.

♦ **mutable gene** See gene: mutable gene.

♦ **mutant** *n.* An organism character, or gene, that has undergone a mutational change (Lincoln et al. 1985).

♦ **mutant type** See -type, type: mutant type.

♦ **¹mutation** See ²group: mutation.

♦ **²mutation** *n.*

1. A kind of change (contrasted with a variation) that results in a new species suddenly from its parental stock; also, the species that results from this process (*Oxford English Dictionary* 1972, entries from 1894). See speciation: saltational speciation.
2. Any drastic change of organism form (Mayr 1963, 168 in Mayr 1982, 742).

Note: This usage goes back to at least the middle of the 17th century; it was first used both for discontinuous variation

---

TWO CLASSIFICATIONS OF MUTATION

I. Classification by how many genes are involved
   A. one gene — point mutation, micromutation
   B. a few genes — micromutation
   C. many genes — chromosomal mutation (*syn.* matric mutation), macromutation (*syn.* rogue, sport)
II. Classification by effect on organism's evolution
   A. advantageous mutation
   B. deleterious mutation
   C. neutral mutation

---

and changes in fossils (Mayr 1963, 168 in Mayr 1982, 742).

3. The smallest distinguishable change in a phyletic series (Waagen 1867 in Mayr 1982, 742).
4. A sudden event (phenotypic change) by which a species originates (de Vries 1901, 1903 in Mayr 1982, 742).

Note: Most of de Vries' mutations in *Oenothera* are now known to be due to chromosomal rearrangements including polyploidy, not gene mutations (Mayr 1982, 743). de Vries used "mutation" to designate a saltation (Mayr 1982, 403).

syn. saltation

5. "Any discontinuous change in the genetic constitution of an organism" (Wilson 1975, 589).
6. A change in the phenotype of a species (Mayr 1982, 742).
7. A change along a very narrow portion of the nucleic acid sequence of a chromosome; a point mutation Wilson 1975, 589).
8. "A discontinuous change of a gene" (Mayr 1982, 794); any error in gene replication (Mayr 1982, 804).

Note: This definition does not apply to all cases of mutation (Mayr 1982, 804).

9. A change in the base-pair sequence of DNA that results from substitution, addition, deletion, or rearrangement of the standard base pairs of DNA (Bernstein et al. 1985, 1278).

syn. sport

cf. group: mutation; saltation

Comment: Mutations do not generally alter the physical regularity of a DNA molecule, and they are replicable and, thus, can be inherited (Bernstein et al. 1985, 1278).

[Old French < Latin *mutatio, -onis* < *mutare*, to change, coined by W. Bateson]

adaptive mutation See ²mutation: directed mutation.

h – m

back mutation See ²mutation: reverse mutation.

Cairnsian mutation *n*. A mutation that occurs with a higher probability when it is advantageous than when it is neutral [coined by Hall 1990, 6].
cf. hypothesis: directed-mutation hypothesis; ²mutation: directed mutation

chromosomal mutation *n*. A chromosomal rearrangement that has a genetic effect; rearrangements include results of unequal crossing overs, inversions, Robertsonian rearrangements, translocations, and polyploidy (Mayr 1982, 769).
syn. chromosome mutation, matrix mutation (Mayr 1982, 804)

directed mutation *n*.
1. A mutation that is nonrandom with respect to normal environmental fluctuations and to effects on a gene product and thus on fitness; contrasted with random mutation (Cairns et al. 1988, 142; Hall 1990, 5).
2. A mutation that "occurs at a higher rate specifically when (and even because) it is advantageous to the organism, whereas comparable increases in rate do not occur either (i) in the same environment for similar mutations that are not advantageous or (ii) for the same mutation in similar environments where it is not advantageous" (Lenski and Mittler 1993a, 188).
3. A selection-induced change in a bacterium's DNA that enables it to grow on a new medium (usually called "adaptive mutation" by Galitski and Roth 1995, 421) (Radicella et al. 1995, 418; Shapiro 1995, 374).
cf. hypothesis: hypothesis of directed mutation; ²mutation: adaptive mutation, Cairnsian mutation
Comments: Adaptive mutation is known in *Escherichia coli*, strain FC40. In the phenomenon of adaptive mutation, bacterial cells change their DNA, using biochemical systems, in response to physiological inputs (Shapiro 1995, 374). This ability is highly evolutionarily significant. Two genetic systems that evidently show adaptive mutation are the lacI-Z33 and the araB-lacZ (Shapiro 1995, 373–374). In the lacI-Z33 system, additional specific molecules participate in selection-induced mutations that restore a bacterium's growth on lactose. Bacterial genetic change is often multicellular. "DNA rearrangements can occur in one cell and be transferred to another before a clone of 'mutant' bacteria proliferates on selective medium. In the lacI-Z33 system, the mutation results from selection-induced recombination activities and plasmid-transfer functions, and in the araB-lacZ system the mutation results from factors other than selective substrates" (Shapiro 1995, 374).
syn. adaptive mutation (Galitski and Roth 1995, 421; Radicella et al. (1995, 418)

forward mutation *n*. Mutation of a gene from its normal, or wild-type, to a mutant state (Lincoln et al. 1985).
cf. back mutation

frameshift mutation *n*. One of a class of mutations that arises from the insertion, or deletion, of one nucleotide, or any number of nucleotides other than three or multiples of three, into or from DNA (Brenner et al. 1961, Crick et al. 1961 in Rieger et al. 1991; Shapiro 1995, 373).
Comments: A frameshift mutation displaces the starting point of genetic transcription of the genetic code, and the resulting mRNA is misread by the translation process from the pinto of the nucleotide(s) addition or deletion (Rieger et al. 1991).
syn. reading mutation, sign mutation (Rieger et al. 1991)

gene mutation, genovariation, infragenic mutation, micromutation, point mutation, transgenation *n*. A heritable change in a single gene (Lincoln et al. 1985).
See ²mutation

heterozygous lethal, heterozygous lethal mutation *n*. A mutation that causes premature death of an organism when it occurs on one of its homologous chromosomes; contrasted with homozygous lethal (King and Stansfield 1985).

homeotic mutation *n*.
1. A mutation that causes one part of an organism's body to develop in a manner appropriate to another part (*e.g.,* the homeotic mutation 'antennopedia' in *Drosophila* flies causes a leg to grow where an antenna normally does) (Dawkins 1982, 288; Arms and Camp 1995, 252, illustration).
   Note: In *Drosophila*, homeotic mutations result from mutations of genes that contain homeoboxes (DNA sequences encoding parts of transcription factors that control development in multicellular organisms) (Arms and Camp 1995, 252).
2. A structure's mutation from one state to another that exists in a series of states, such as the mutation of an insect wing into a haltere (Lincoln et al. 1985).
syn. homoeotic mutation
cf. homeobox-containing gene
[coined by W. Bateson in Gould (1985c)]

homoeotic mutation See ²mutation: homeotic mutation.

homozygous lethal (mutation) *n.* A mutation that causes the premature death of an organism when it occurs on both of its homologous chromosomes; contrasted with heterozygous lethal (King and Stansfield 1985).

intragenic mutation See ²mutation: gene mutation.

lacI-Z33 mutation *n.* A fusion of lacI (encoding repressor) and lacZ (encoding β-galactosidase) which prevents bacterial growth on lactose medium because it includes a frameshift mutation that blocks lacZ translation (Shapiro 1995, 373).
Comments: This a popular system in which to study adaptive mutation (Shapiro 1995, 373).

macromutation *n.*
1. "A heritable change in the genetic material that produces a large effect usually associated with speciation" (traced back to Bateson 1894 and de Vries 1910 in Dietrich 1992, 194).
   Notes: Richard Goldschmidt might have been the first person to use "macromutation;" the concept of a macromutation as a speciation event is usually traced back to de Vries' mutation theory (Dietrich 1992, 195).
2. The "large, easily recognizable mutations of genetic experiments (Carter 1951, 95 in Dietrich 1992, 199).
   Note: This is a redefinition of "macromutation" in terms of degree of phenotypic effect; Carter (1951) only weakly associates macromutation with speciation (Dietrich 1992, 199).
3. An individual that is a highly different variation within a species (Mayr 1982, 428) (*e.g.,* the Ancon Sheep and Turnspit Dog) (Mayr 1982, 738).
4. "Any mutation that results in a major adaptive shift, a change in the way of life that opens a new adaptive zone" (Bush 1982, 125, in Dietrich 1992, 201).
5. The sum of those processes that explain the character-state transitions that diagnose differences of major taxonomic rank; these processes include competition, predation, and habitat change (Levinton 1983, 104).
6. The simultaneous mutation of many different characters (Lincoln et al. 1985).
syn. rogue, sport
cf. hopeful monster; hypothesis: macromutationism; ³mutation: systemic mutation
Comments: There is a lack of consensus regarding the connection of macromutations to speciation events and the relative impor-

tance of this connection as a possible mode of speciation (Dietrich 1992, 201).

matrix mutation See ²mutation: chromosomal mutation.

micromutation See ²mutation: gene mutation.

missense mutation *n.* A point mutation within a codon that results in the incorporation of a different amino acid into part of the polypeptide chain that the codon codes for and produces an inactive, or unstable, enzyme (Lincoln et al. 1985).
cf. ²mutation: nonsense mutation, samesense mutation

neutral mutation *n.* A mutation that has no selective advantage, or disadvantage, compared with its allele (Dawkins 1982, 291).
cf. ³theory: neutral theory
Comment: In theory, a neutral mutation can become predominant in a population (Dawkins 1982, 291).

nonsense mutation *n.* A mutation that converts a sense codon into a chain-terminating codon, or vice versa, resulting in abnormally short, or long, polypeptides usually with altered function; contrasted with missense mutation and samesense mutation (King and Stansfield 1985).
See allele: neutral allele.

numerical mutation *n.* Any change in chromosome number, either by polyploidy or aneuploidy (Lincoln et al. 1985).

paramutation *n.* In Corn: a phenomenon in which one allele influences the expression of another allele that is at the same locus when the two are combined in a heterozygote (King and Stansfield 1985).

quasi-normal mutation *n.* A mutation that causes death of less than 10% of the mutants in a population (Lincoln et al. 1985).
cf. ²mutation: semilethal mutation, subvital mutation

random mutation *n.* A mutation whose initial occurrence is independent of its specific value (advantageous, neutral, or deleterious) to the organism that has it (Lenski and Mittler 1993a, 188).
Comments: "A fundamental tenet of evolutionary biology is that mutations are random events." This "does not mean that mutation rates are unaffected by environmental factors or that all portions of a genome are equally susceptible to mutation" (Lenski and Mittler 1993a, 188).

reaper (*rpr*) *n.* A mutation in a *Drosophila* embryo that blocks normal cell death and causes the embryo to die (White et al. 1994, 677).
Comment: The reaper locus has been cloned; it appears to encode a small peptide that is

h – m

expressed in cells destined to undergo apoptosis (programmed cell death) (White et al. 1994, 677).

regulatory mutation *n.* A mutation in a regulatory gene, *q.v.*, that affects the rates at which gene products are produced, although the products themselves may be unaffected (Strickberger 1990, 187–188).

Comment: A regulatory mutation of the bithorax locus in *Drosophila* may cause an extra set of wings (Strickberger 1990, 191). Regulatory mutations might cause changes in shape of fish and other organisms.

reverse mutation *n.* A change in genetic material back to its previous condition (Mayr 1982, 552).

syn. back mutation

samesense mutation *n.* A point mutation within a codon that produces no phenotypic change because the mutant codon codes for the same amino acid as the original codon (Lincoln et al. 1985).

cf. ²mutation: missense mutation, nonsense mutation

semilethal mutation *n.* A mutation that causes death of more than 50%, but not all of individuals that carry it (King and Stansfield 1985).

cf. ²mutation: quasi-normal mutation, subvital mutation

sport See ²mutation: macromutation.

subvital mutation *n.* A mutation, or gene, that significantly lowers viability but causes the death before maturity of less than 50% of individuals that carry it; contrasted with semilethal mutation (King and Stansfield 1985).

syn. sublethal gene

cf. ²mutation: quasi-normal mutation, semilethal mutation

transgenation See ²mutation: gene mutation.

♦ **³mutation** *n.* The process by which a new "species" originates (de Vries 1901, 1903 in Mayr 1982, 742).

Note: de Vries considered mutation to be an evolutionary phenomenon, but later geneticists considered it to be primarily a genetic phenomenon (Mayr 1982, 742).

systemic mutation *n.*

1. A complete repatterning of a chromosome that might produce a new chemical system that has a definite and completely divergent action upon an organism's development, "an action which can be conceived as surpassing the combined actions of numerous individual changes by establishing a completely new chemical system;" such a repatterning that could bridge the gap

to form a new species (Goldschmidt 1940, 203 in Dietrich 1992, 197).

2. "A catastrophic mutation that would lead to the origin of an entirely new type of organism" (Lincoln et al. 1985).

syn. systematic mutation (Lincoln et al. 1985)

cf. hopeful monster; ²mutation: macromutation

♦ **⁴mutation** *n.*

1. A new species that results from a sudden change in its parental stock (*Oxford English Dictionary* 1972, entries from 1894).

2. An individual, species, etc. that results from a sudden, transmissible variation, especially as the result of new modifications, or combinations, of genes and chromosomes (Michaelis 1963).

syn. mutant

♦ **mutation frequency** *n.* "The proportion of mutant alleles in a population" (Lincoln et al. 1985).

♦ **mutation pressure** *n.*

1. The increase in frequency of one allele and simultaneous decrease in frequency of another allele due to mutation at the gene locus in question (Wilson 1975 64).

2. "Evolution by different mutation rates alone" (Wilson 1975, 589).

Comments: Mutation pressure is usually about 10^{-4} per organism (or cell) (Wilson 1975 64). The evolutionary synthesis "led to a decline of the concept of 'mutation pressure,' and its replacement by a heightened confidence in the powers of natural selection, combined with a new realization of the immensity of genetic variation in natural populations" (Mayr 1982, 567).

♦ **mutational load** See genetic load: mutational load.

♦ **mutationism** *n.*

1. de Vries' (1901, 1903) concept that "speciation must be due to the spontaneous origin of new species by the sudden production of a discontinuous variant" (Mayr 1982, 546).

2. The concept that evolution, in general, and speciation, in particular, are the results of drastic and sudden mutations (Lincoln et al. 1985).

syn. de Vriesianism (Lincoln et al. 1985), de Vries' mutation theory (Mayr 1982, 546)

cf. ²mutation: macromutation; saltation

♦ **mute** See ²group: mute.

♦ **mutilous** *adj.* Referring to an organism without defensive structures; harmless (Lincoln et al. 1985).

♦ **muton** See gene: muton.

♦ **mutual courtship** See courtship: mutual courtship.

♦ **mutualism** See cooperation; symbiosis: mutualism.

♦ **mutualistic mimicry** See mimicry: mutualistic mimicry.

♦ **M-V linkage** *n.* Linkage between morphological-character-determining and variability-controlling genes (Lincoln et al. 1985).

♦ **My, M.Y.** See unit of measure: Ma.

♦ **mycetology** See study of: mycetology.

♦ **mycetophage** See -vore: fungivore.

♦ **mychogamy** See -gamy: mychogamy.

♦ **mycobiont** See -biont: mycobiont.

♦ **mycobiota** See -biota: mycobiota.

♦ **mycorrhiza** *n., pl.* **mycorrhizae** A mutualistic association between a plant root and a fungus (Campbell 1996, 723).
cf. symbiosis
Comments: The fungus secretes a growth factor that stimulates the root to grow and branch (Campbell 1996, 584, 723). Depending on the species involved, the fungus either sheaths the root and extends its hyphae among its cortex cells or invades the root cells themselves. The fungus mycelium provides the root with a greatly increased surface area through which it gains minerals, especially phosphate, and water. The fungus is more efficient at absorbing these materials and secretes an acid that increases the solubility of some minerals. The root obtains minerals and water from the fungus, and the fungus obtains photosynthetic products from the plant. The fungus also protects the plant against certain soil pathogens. Over 95% of vascular plants may have mycorrhizae if they contact appropriate fungus species. Plants without their mycorrhizae can be malnourished and show decreased growth compared to conspecific plants with their mycorrhizae. Ascomycota, Basiodiomycota, and Zygomycota form mycorrhizae. Half of all species of mushroom-forming basiodiomyces live as mycorrhizae with Birches, Oaks, and Pines. Mycorrhizal genera include *Albatrellus, Amanita, Boletopsis, Boletus, Canthrellus, Catathelasma, Chroogomphus, Clavaridelphus, Clitopilus, Cortinarius, Cystoderma, Elaphomyces, Entoloma, Gautieria, Gomphidius, Gyroporus, Hebeloma, Hydnangium, Hydnum, Hygrophorus, Inocybe, Lactarius, Leccinum, Leucocortinarius, Limacella, Melanogaster, Myxocybe, Paxillus, Phylloporus, Pisolithus, Porphyrellus, Ramaria, Rhizopogon, Rizites, Russula, Sarcosphaera, Sepultria, Suillus, Terfezia, Tricholoma, Tuber,* and *Tylopilus* (Pacioni 1981).
[Greek *mycos*, fungus + *rhiza*, root]

♦ **mycotroph** See -troph-: mycotroph.

♦ **myiophile** See ²-phile: myiophile.

♦ **Myr** See unit of measure: Ma.

♦ **Myr B.P.** *n.* 1 million years before the present (Futuyma 1986, 328).

♦ **myrmeco-, myrmec-** *combining form* Referring to an ant; loosely, a termite (Lincoln et al. 1985).
[Greek *myrmēx, myrmēckos*, ant]

♦ **myrmecobromous** See myrmecotropic.

♦ **myrmecochore** See -chore: myrmecochore.

♦ **myrmecochorous** See -chorous: myrmecochorous.

♦ **myrmecoclepty** See symbiosis: myrmecoclepty.

♦ **myrmecocole** See -cole.

♦ **myrmecodomatium** See domatium: myrmecodomatium.

♦ **myrmecodomus** *adj.* Referring to a plant that affords shelter for ants (Lincoln et al. 1985).

♦ **myrmecology** See study of: myrmecology.

♦ **myrmecophage** See -phage: myrmecophage.

♦ **myrmecophile** See ²-phile: myrmecophile.

♦ **myrmecophobia** See phobia (table).

♦ **myrmecophobe** See -phobe: myrmecophobe.

♦ **myrmecophyte** *n.* An angiospermous plant that has an obligatory mutualistic relationship with ants (Wilson 1975, 589).
syn. myrmecoxenous plant (Lincoln et al. 1985)

♦ **myrmecosymbiosis** See symbiosis: myrmecosymbiosis.

♦ **myrmecotroph** See -troph-: myrmecotroph.

♦ **myrmecotropic, myrmecobromous** *adj.* Referring to plants and animals that provide food for ants (Lincoln et al. 1985).

♦ **myrmecoxenous, myrmecoxeny** See -xeny: myrmecoxeny.

♦ **mysophobia** See phobia (table).

♦ **myxotrophic** See -trophic: myxotrophic.

h–m

n

♦ **n, N, *n*, *N*** See population size.

♦ ***N*-acetylaspartylglutamate** See neurotransmitter: *N*-acetylaspartylglutamate.

♦ ***N*-methyl-D-aspartate receptor** See ²receptor: *N*-methyl-D-aspartate receptor.

♦ **NAAG** See neurotransmitter: *N*-acetylaspartylglutamate.

♦ **naive, naïve** *adj.*

1. Referring to a person who has an unaffected, or simple, nature that lacks worldly experience; candid; artless (*Oxford English Dictionary* 1972, entries from 1654); referring to an uninstructed person (Michaelis 1963).

2. Referring to an animal that has no previous experience with infection from a particular parasite (Lincoln et al. 1985) or with a particular kind of food (*e.g.,* a fledgling bird that has not eaten a certain insect species).

[French feminine of *naïf* < Latin *nativus,* natural, inborn]

♦ **namaphile** See ¹-phile: namaphile.

♦ **namatium** See ²community: namatium.

♦ **name** *n.* A word, or group of words, by which a class, concept, organism, or other thing is distinctly referred to; especially, the proper appellation of a person or family (Michaelis 1963).

Comments: According to a Chinese proverb: The beginning of knowledge is knowing the correct name for an object.

common name *n.* A colloquial, or vernacular, name of a taxon (Lincoln et al. 1985); a nonscientific name of a taxon; contrasted with scientific name.

Comments: Common names for species and subspecies are important for many reasons, including: governmental organizations and publishers often require both a common and scientific name for an organism to increase the usefulness of a publication, for convenience, students and others

often prefer to use common names instead of a scientific one for an organism, and conservation efforts benefit from use of common names (Miller 1992, 3–4). For example, public support for butterfly conservation efforts increased markedly when conservationists referred to species by their common names. People base common names on almost any information that is relevant to an organism, including its behavior, collection site, color, feel, geographical range, medical use, shape, size, speed, and toxicity; the person after whom a researcher named a taxon; and who named the taxon. People (*e.g.,* early American colonists) have named many organisms after their resemblance to ones in Europe, even if they were taxonomically unrelated (*e.g.,* buttercup, robin). Many common names are derived from scientific ones: apoid from Apoidea; apid, Apidae; apine, Apinae; drosophila, *Drosophila*; and hominid, Hominidae.

Organism names can be confusing because there are millions of species and so many of the named organisms have published common and scientific names as well as local common names. Further, these organisms may have one or more published common or scientific names, and their common names may be in more than one language. For example, in regard to English names alone, the butterfly *Phyciodes tharos tharos* has 12 published common names (Miller 1992, 87). Its three other subspecies have a total of eight common names. The butterfly *Coenonympha tullia tullia* has five published common names, and its 21 subspecies have a total of 46 common names (Miller 1992, 104–105). Some scientific groups have standardized many common names. For example, the Entomological Society of America (ESA) compiles a standardized list for common names of insects (Committee on Common Names of Insects 1982). According to the ESA, words of insect common names are not capitalized unless they start a sentence or phrase in a table or are proper nouns: *e.g.,* some butterflies are viceroy, Florida purple-wing, hackberry butterfly, tawny emperor, Creole pearly-eye, and Schaus' Swallowtail. Miller (1992) indicated preferred common names for North American butterflies; however, she capitalized all separate parts of common names of butterflies: *e.g.,* Arctic Ringlet, McIsacc's Ringlet, Ochraceous Quaker, and Small Ringlet. Other books and other publications use Miller's style, and I do as well throughout this book. Researchers have not yet given most living organisms common and scientific names. Many species are going extinct before scientists discover them during the current Human-caused mass extinction; thus, unfortunately, these undiscovered species will never be named, let alone be known to science.

scientific name *n.* A formal Latin or Latinized name of a taxon; contrasted with common name (Lincoln et al. 1985).

cf. ²group (table for taxonomic names)

Comments: Scientific names are the major names that scientists use for designating taxa in their publications and are understood by many scientists worldwide. Although many species have more than one scientific name, scientific names are generally more stable than common ones (Justice and Bell 1968, xii). Current researchers publish scientific names in English letters regardless of the principal language of a publication. The Code of Botanical Nomenclature and the Code of Zoological Nomenclature fix the rules for coining, publishing, and revising scientific names, but these Codes do not fix rules for common names. Workers have heated arguments about coinage of new common names and which already coined common names should be used for a particular taxon (Miller 1992, 3). A full subspecific name is comprised of a generic name, a specific name (= specific epithet), subspecific name, taxonomic author(s), and year that a taxon is formally described in a publication: *Anartia jatrophae guantanamo* Munroe 1942 (= the White Peacock, a butterfly). A full specific name is comprised of the same parts except that it lacks a subspecific name: *Calephelis borealis* (Grote and Robinson) 1866 (the Northern Metalmark, a butterfly) and *Calgopogon pulchellus* (Salisb.) R. Br. (the Grass Pink, an orchid).

Taxonomic names above the level of subspecies are capitalized, but not italicized: *e.g.,* Archaea (domain), Animalia (kingdom), Chordata (phylum), Mammalia (order), Insecta (group). Specific and subspecific names are italicized, but not capitalized. Generic names are both italicized and capitalized. Depending on the style of a scientific publication, an author will give only genus, species, and subspecies name; these three names and authors; or all five parts of scientific names. Publications that are not taxonomic ones usually do not include a taxon's year.

In animal names, the taxonomic author(s) (= authority[ies]) is in parentheses if a subsequent researcher moved the species into a genus different than the one into which the first author(s) placed it. In plant names, researchers designate both the original author(s) and the author(s) who placed the species into a new genus. The original author(s) is in parentheses before the subsequent author. Researchers usually abbreviate authors' names, but it would seem less confusing to nonspecialists if the researchers fully wrote out these names as in Opler and Krizek (1984) and other books. The author names of the abovementioned orchid are Salisb. (= Richard Anthony Salisbury) and R. Br. (= Robert Brown). Schuler (1983) usually gives the full author name(s) for each plant species in his book. Named taxa often have more than one scientific name: an earlier published one(s) and a more (most) recently published one. Researchers should use the more (most) recently published name for a taxon from a taxonomic treatise that specifically deals with this taxon (unless the treatise is erroneous).

trinomial *n.* A subspecies' scientific name comprising its generic, specific, and subspecific names in that order; *e.g.,* the British race of the White Wagtail, *Motacilla alba* var. *lugubris*, or the Lowland Gorilla (*Gorilla gorilla gorilla*) (Mayr 1982, 289).

Comments: The "var." is often dropped when a trinomial is written. The first author to employ trinomials routinely was Schlegel (1844) (Mayr 1982, 289).

♦ **nanander** See male: nanander.

♦ **nanism, dwarfism, microsomia** *n.* An organism's being stunted or smaller than normal, or having restricted growth (Lincoln et al. 1985).

adj. nanoid

♦ **nanitic worker** See caste: worker: nanitic worker.

♦ **nannander** See male: nanander.

n–r

♦ **nano-, nanno-, nan-** *combining form*
1. Exceedingly, or abnormally, small; dwarf (Michaelis 1963).
2. A metric unit of measure multiplied by 10^{-9} (Michaelis 1963).
[Greek *nānos*, dwarf]

♦ **nanoid** See nanism.

♦ **nanoplankton** See plankton: nanoplankton.

♦ **naptonuon** See nuon: naptonuon.

♦ **narcotized startle** See startle: narcotized startle.

♦ **narcotropism** See tropism: narcotropism.

♦ **narrow-sense heritability** See heritability: heritability in the narrow sense.

♦ **Nasanov gland** See gland: Nasanov gland.

♦ **Nash equilibrium** See equilibrium: Nash equilibrium.

♦ **nasus** See organ: nasus.

♦ **nasute soldier** See caste: nasute soldier.

♦ **natal dispersal** See dispersal: natal dispersal.

♦ **natal-homing hypothesis** See hypothesis: turtle-homing hypotheses: natal-homing hypothesis.

♦ **natality** *n.*
1. Birth rate in a population (*Oxford English Dictionary* 1972, entries from 1880).
2. Production of new individuals by birth, germination, or fission (Lincoln et al. 1985).
syn. birth rate, natality rate (Lincoln et al. 1985)
cf. mortality
Comment: Natality is calculated per female or head of a population (Lincoln et al. 1985).

♦ **natatorial** See -orial: natatorial.

♦ **National Medal of Science** See award: National Medal of Science.

♦ **National Medal of Technology** See award: National Medal of Technology.

♦ **native species** See [2]species: native species.

♦ **natural balance** See balance of nature.

♦ **natural birth control** See effect: Bruce effect.

♦ **natural experiment** See experiment: natural experiment.

♦ **natural extinction** See [2]extinction: natural extinction.

♦ **natural group** See [2]group: natural group.

♦ **natural history** *n.*
1. A work that deals with the properties of natural objects or organisms; a scientific account of any subject written on similar lines (*Oxford English Dictionary* 1972, entries from 1567).
2. Our sum of knowledge that relates to natural objects, etc., of a place, or the characteristics of a class of persons or things (*Oxford English Dictionary* 1972, entries from 1593).
3. The systematic study of all natural objects including organisms, rocks, and minerals (*Oxford English Dictionary* 1972, entries from 1662).
Note: "Natural history" is often now restricted to animals, and it frequently deals with them from a popular, rather than scientific, standpoint (*Oxford English Dictionary* 1972).
4. The study of the entire universe (Diderot 1751 in Goldensohn 1999, 6).
5. An early stage in the development of a branch of science that involves labeling of phenomena and concepts and syntheses concerning tedious cross-referencing of differing sets of definitions and metaphors erected by the more imaginative thinkers (Wilson 1975, 574).
6. A pejorative term used for a study that is primarily observational and inconclusive compared to a conclusive, experimental one (Mayr 1982, 142–143).
7. Biology; ecology (Goldensohn 1999, 6).

♦ **natural language processing** See study of: natural language processing.

♦ **natural law** See law; law: natural law.

♦ **natural period** See period: natural period.

♦ **natural philosopher** *n.* One who studies the proximate causes by which divine religious laws manifest themselves in nature (Mayr 1982, 103).
cf. natural theology; study of: philosophy: natural philosophy

♦ **natural philosophy** See study of: philosophy: natural philosophy.

♦ **natural predator** See predator: natural predator.

♦ **natural selection** See selection: natural selection.

♦ **natural selection, law of** See [3]theory: Fisher's fundamental theorem.

♦ **natural-stimulus model** See [3]model: natural-stimulus model.

♦ **natural system** See [1]system: natural system.

♦ **natural theology, physcio-theology** *n.* The theory that the harmony and perfection of nature are evidence of God's existence and that the analysis of nature is the way to understand God's plan (Mayr 1982, 92, 103; C.K. Starr, personal communication).
cf. natural philosopher

♦ **naturalism** *n.*
1. A philosophical world view and Humans' relation to the world that admits, or assumes, only natural forces, rather than supernatural or spiritual (*Oxford English Dictionary* 1972, entries from 1750).

2. One of the six main components of uniformitarianism that involved all changes in the Earth's geology after Creation which are due to (1) secondary (or nonsupernatural) causes or (2) secondary causes with occasional divine interventions from the Creator (Mayr 1982, 376–378).

♦ **natural science** See study of: science: natural science (Michaelis 1963).

♦ **naturalist** *n.*
1. An adherent of, or believer in, naturalism, *q.v.* (*Oxford English Dictionary* 1972, entries from 1587).
2. A biologist (usually a zoologist, botanist, or paleontologist) who works with whole organisms in natural environments, is interested in ultimate (evolutionary) causations, and is particularly concerned with problems of diversity (Mayr 1982, 540–542).
Comment: Some post-Darwinian naturalists had erroneous ideas on the nature of inheritance and variation (Mayr 1982, 540–542).

♦ **naturalistic fallacy** See fallacy: naturalistic fallacy.

♦ **naturalistic observation** See observation: naturalistic observation.

♦ **nature vs. nurture** *n.* The amount of phenotypic variation that an organism possesses that is related to its genetic makeup (nature) compared to its environmental influences (nurture) [coined by Francis Galton (Mayr 1982, 781)].
syn. environment versus heredity
cf. -genesis: epigenesis; twin: identical twin
Comment: Ascertaining the relative magnitudes of the effects of nature and nurture on a particular behavior may not be possible.

♦ *Naturphilosophie* See study of: philosophy: natural philosophy.

♦ **navigation** See taxis: navigation.

♦ **NBS** See organization: National Biological Service.

♦ **nealogy** See study of: nealogy.

♦ **neanic** *adj.*
1. Referring to the larval phase that precedes an adult (Lincoln et al. 1985).
2. Referring to a character that first appeared, in evolution, in early ontogenetic stages (Lincoln et al. 1985).
cf. ephebic

♦ **neck bite** See bite: neck bite.

♦ **neck-over-neck diving** See diving: neck-over-neck diving.

♦ **necro-, necr-** *combining form* "Corpse; the dead; death" (Michaelis 1963).
[Greek *neckros*, corpse]

♦ **necrocoenosis** See -coenosis: necrocoenosis.

♦ **necrocoleopterophile** See ²-phile: necrocoleopterophile.

♦ **necrogenous** See -genous: necrogenous.

♦ **necrogeography** See study of: geography.

♦ **necrology** See study of: necrology.

♦ **necrophage** See -phage: necrophage.

♦ **necrophile** See ¹-phile: necrophile.

♦ **necrophilia** See -philia: necrophilia.

♦ **necrophobia** See phobia (table).

♦ **necrophoresis** *n.* In some social-insect species: an individual's transporting a dead member of its colony away from its nest (Wilson 1975, 589).
Comment: This is a highly developed, stereotyped behavior in ants (Wilson 1975, 589).

♦ **necrophoric behavior** See behavior: necrophoric behavior.

♦ **necrophytophage** See -phage: necrophytophage.

♦ **necrotrophic symbiosis** See symbiosis: necrotrophic symbiosis.

♦ **necrotype** See -type, type: necrotype.

♦ **nectar** *n.* A plant-produced sugary solution that is consumed by many kinds of animals, many of which are pollinators (Michaelis 1963).
Comments: Plants produce nectar in extrafloral and floral nectaries. Nectar contains amino acids, antioxidants, lipids, sugars, and other compounds (Buchmann and Nabhan 1996, 250). It varies greatly in sugar concentration. Nectar is the main raw material from which bees make honey, *q.v.*
[Greek *nektar*, the drink of the gods]

♦ **nectar corridor** *n.* A series of different plants that offer abundant nectar seasonally along an annual migration pathway (*e.g.,* the bat-pollinated cacti and century plants in Mexico and southwestern U.S.) (Buchmann and Nabhan 1996, 250).

♦ **nectar flow** *n.* A time of nectar production in a particular habitat (Johansen 1977, 181).

♦ **nectar robber** See robber.

♦ **nectariferous** *adj.* Referring to a nectar-producing plant (Lincoln et al. 1985).

♦ **nectarivore** See -vore: nectarivore.

♦ **nectary** *n.* A specialized area of a plant that produces nectar, *q.v.*
See nectary: floral nectary.

extrafloral nectary *n.* A nectary on a plant's vegetative part (*e.g.,* a leaf or stem).

floral nectary *n.* An area of a flower that produces nectar.
syn. nectary (Buchmann and Nabhan 1996, 250)
Comments: Floral nectaries are often at the base of an innermost floral tube where nectar forms a pool (Buchmann and Nabhan 1996, 250).

n–r

◆ **nectism** See swimming: nectism.

◆ **nectobenthic** See [2]benthic: nectobenthic.

◆ **nectobenthos** See [2]community: benthos: nectobenthos.

◆ **nectogenic** See -genic: nectogenic.

◆ **necton** See nekton.

◆ **need** *n.*

1. Something that a person requires; a necessity (*Oxford English Dictionary* 1972, entries from 900).

2. A construct representing a force in a person's brain that directs and organizes his perception, thinking, and action, so as to change an existing, unsatisfying situation (based on Murray in Wolman 1973, 250).

3. An individual organism's lacking, wanting, or requiring something that if present would benefit it "by facilitating its behavior or satisfying a tension" (Wolman 1973, 250).

v.i., v.t. need

syn. drive (Murray in Wolman 1973, 250)

cf. drive

Comments: "Requirement" seems to sound less anthropomorphic than "need." Some common essential needs in Humans follow (Murray in Wolman 1973, 250).

abasement *n.* One's admitting inferiority, blame, error; accepting punishment or criticism; passively submitting to external forces.

achievement *n.* One's independent mastering of objects, other persons, and ideas to increase self-esteem by successful exercise of one's talent.

affiliation *n.* One's drawing near, cooperating with, and remaining near to another person who is seen as similar to oneself and a friend.

cf. affiliation

aggression *n.* One's opposing, fighting, injuring, or punishing another person.

autonomy *n.* One's independence, unattachedness, and unrestrictedness.

counteraction *n.* One's overcoming failure and weakness through resumed action and repression of fear.

dependence *n.* One's defending oneself against assault, criticism, and blame and justifying or concealing a failure, error, or humiliation.

deference *n.* One's supporting, praising, honoring, or admiring a superior and conforming to custom.

dominance *n.* One's influencing, or controlling, others' behavior.

exhibition *n.* One's impressing others.

harm avoidance *n.* One's avoiding pain, injury, illness, and death.

infavoidance *n.* One's avoiding humiliation.

nurturance *n.* One's supporting, protecting, comforting, healing, and gratifying the needs of helpless persons.

order *n.* One's organizing, balancing, and arranging objects in one's environment.

play *n.* One's seeking enjoyment and relaxation without further purpose.

rejection *n.* One's separating oneself from a negative object.

sentience *n.* One's seeking and enjoying sensuous impressions.

sex *n.* One's forming and pursuing an erotic relationship.

succorance *n.* One's being gratified by an allied object and always having support and protection.

understanding *n.* One's questioning, answering, speculating, formulating, analyzing, and generalizing.

◆ **negation** *n.*

1. A person's denying or making a statement that involves the use of "no," "not," "never," etc. (*Oxford English Dictionary* 1972, entries from 1530).

2. A person's rejecting or negating with use of nonverbal communication, *e.g.,* throwing his head up and back with eyes closed (in Greece, Turkey, many Arab countries of the Near East) or shaking his head laterally (Western Europe) (Heymer 1977, 190).

◆ **negative assortative mating** See mating: negative assortative mating.

◆ **negative-binomial distribution** See distribution: binomial distribution: negative-binomial distribution.

◆ **negative conditioning** See learning: conditioned learning: negative conditioning.

◆ **negative eugenics** See eugenics: negative eugenics.

◆ **negative evidence** *n.*

1. Not finding a significant statistical effect. *Note:* This is an inconclusive result because a different avenue of investigation could reveal an effect in question.

2. Not finding a particular phenomenon. *Note:* This is an inconclusive result because further work could discover the phenomenon.

◆ **negative example** *n.* An example based on negative evidence.

◆ **negative feedback** See feedback: negative feedback.

◆ **negative photophobotaxis** See taxis: photophobotaxis: negative photophobotaxis.

◆ **negative reinforcement** See reinforcement: negative reinforcement.

◆ **negative reinforcer** See reinforcer: negative reinforcer.

◆ **negative rheotaxis** See taxis: rheotaxis: negative rheotaxis.

♦ **negative skewness** See skewness: negative skewness.

♦ **negative thermotropism** See -tropism: thermotropism: negative thermotropism.

♦ **negentropy** See entropy: negentropy.

♦ **negotiation** *n*.

1. A person's interaction, or making a treaty, with another person to obtain, or bring about, some result, especially in affairs of state (*Oxford English Dictionary* 1972, entries from 1579).

2. A person's action or business of negotiating, or making terms, with another person (*Oxford English Dictionary* 1972, entries from 1614).

3. Individual animals' behavior that involves their "attempting to control" each other's behavior, adapting one animal's behavior to that of the other animal, or both (Hinde 1985, 989).

♦ **neidioplankton** See plankton: neidioplankton.

♦ **neighbor-joining method** See method: neighbor-joining method.

♦ **neighbor-modulated fitness** See fitness: inclusive fitness.

♦ **neighbor recognition** See recognition: neighbor recognition.

♦ **neighborhood** *n*. In population genetics, the area around an individual that contains the source(s) of gametes that produced this individual (Lincoln et al. 1985).

♦ **neighboringly sympatric** See -patric: sympatric: neighboringly sympatric.

♦ **neighboringly sympatric population** See ¹population: neighboringly sympatric population.

♦ **nekrophytophage** See -phage: necrophytophage.

♦ **nektobenthic** See ²benthic: nectobenthic.

♦ **nektobenthos** See ²community: benthos: nectobenthos.

♦ **nektogenic** See -genic: nectogenic.

♦ **nekton** *n*. Marine animals of open water that are capable of swimming freely, relatively independently of currents and waves, and that range in size from microorganisms to whales, typically 20 mm to 20 m long (Michaelis 1963; Lincoln et al. 1985).
syn. necton
cf. plankton
Comments: Because Lincoln et al. (1985) call nekton "pelagic organisms," they imply that nekton are both marine and freshwater organisms. They indicate that nekton are divided into "centimeter nekton," "decimeter nekton," and "meter nekton." Ali and Lord (1980) refer to "nektonic organisms" and "nectonic invertebrates" which they collected from freshwater ponds.

[New Latin < Greek *nēktos*, neuter *nēkton*, swimming]

♦ **nematodology, nematology** See study of: nematology.

♦ **Nemesis** *n*.

1. "In Greek mythology, the goddess of retributive justice or vengeance" (Michaelis 1963).

2. A hypothesized companion star of the sun that might be responsible, at least in part, for mass extinctions on Earth (Strickberger 1990, 355).

[Greek *Nemesis*]

♦ **nemoricole** See -cole: nemoricole.

♦ **nemorose, nemericolous** *adj*. Referring to an organism that lives in open woodlands (Lincoln et al. 1985).

♦ **neo-** *combining form* "New; recent;" a modern, or modified, form of (Michaelis 1963). Also ne- before vowels.

[Greek *neos*, new]

♦ **neo-Adansonism** See study of: systematics: phenetic systematics.

♦ **neobiogenesis** See -genesis: neobiogenesis.

♦ **neobiogeography** See study of: geography: biogeography: neobiogeography.

♦ **neobiont** See biont: neobiont.

♦ **neo-Darwinian evolution, Neo-Darwinian theory** See evolutionary synthesis.

♦ **neo-Darwinism, neo-Darwinist** See Darwinism: neo-Darwinism.

♦ **neoendemic** See ¹endemic: schizoendemic: neoendemic.

♦ **neofemale** See female: neofemale.

♦ **neogamy** See -gamy: neogamy.

♦ **neoichnology** See study of: neoichnology.

♦ **neo-Lamarckian** See Lamarckian: neo-Lamarckian.

♦ **neo-Lamarckian evolution** See ²evolution: neo-Lamarckian evolution.

♦ **neologism** *n*. A new term, or expression, that is often disapproved of because of its novelty or barbarousness (Lincoln et al. 1985).

♦ **neomorph, neomorphic allele** See allele: neomorphic allele.

♦ **neomorphosis** See -morphosis: neomorphosis.

♦ **neontology** See study of: neontology.

♦ **neoschizoendemic** See ¹endemic: schizoendemic: neoschizoendemic.

♦ **neoteinic** *adj*. Referring to being a supplementary reproductive termite (*e.g.*, neoteinic reproductive) (Wilson 1975, 589). See caste: neoteinic.

♦ **neoteny, neoteinia, neotenia, neoteiny** See ¹heterochrony: neoteny.

n–r

♦ **nepotism** *n.*

1. A Pope's, or other ecclesiastic's, showing special favors to nephews or other relatives in conferring offices; unfair preferment of nephews or other relatives to other qualified persons (*Oxford English Dictionary* 1972, entries from 1670).
2. For example, in some primate species, including Humans; a sweat-bee and a paper-wasp species; a prairie-dog species; the Acorn Woodpecker, Belding's Ground Squirrel, and Lion: an individual's favoring, or investing in, its kin over nonkin, or certain categories of kin over other categories (Alexander 1974 in Hoogland 1986, 263; Wilson 1987, 11; Fletcher 1987, 23).
3. An individual animal's (sociobiological) altruism towards close kin, including offspring (Immelmann and Beer 1989, 13).

cf. altruism; altruism: kin altruism, parental altruism

Comment: Alexander (1974 in Fletcher 1987, 23) also includes favoring offspring and assisting mates as part of "nepotism." A few authors use "nepotism" in preference to "altruism" where it is appropriate to do so. [Latin *nepotismo*, bestowal of patronage upon the "nephew" of Vatican officials in Renaissance Rome]

♦ **neritopelagic** See pelagic: neritopelagic.
♦ **nerve impulse** See potential: action potential.
♦ **nerve net** *n.*

1. An anatomically dispersed system "of neurons, so connected that excitation can spread through some considerable number of neurons in any direction and diffusely, bypassing incomplete cuts" (Bullock and Horridge 1965, 288 in Brown 1975, 470).
2. For example, in Cnidaria, Echinodermata, Hemichordata: a group of neurons, often occurring in a sheet, which have a low degree of organization and structural diversity (Brown 1975, 470).

cf. ¹system: nervous system

♦ **nerve spike** See potential: action potential.
♦ **nervicole** See -cole: nervicole.
♦ **nervous system** See ²system: nervous system.
♦ **nest** *n.*

1. The structure made, or a place selected, by a female bird in which to lay her eggs and that serves as a shelter for her unfledged young (*Oxford English Dictionary* 1972, entries from 950).
   Note: This concept of a "nest" is now used for many kinds of animals, including some insect, fish, and reptile species, and most bird and probably all mammal species.
2. A place where a person rests or resides; a lodging, shelter, bed, etc., especially of a secluded, or comfortable nature; a snug retreat (*Oxford English Dictionary* 1972, entries from 1000).
3. In some insect and mammal species: the place, or structure, used by a female as an abode or lair, in which she deposits her eggs, spawn, or young (*Oxford English Dictionary* 1972, entries from 1386).
   Note: This concept of nest is now known to be relevant to many kinds of animals in addition to insects and mammals. Social animals have nests occupied by colonies.

See ²group: nest.

v.i., v.t. nest

Comment: Some animals also incubate their eggs, keep their young, or spend significant parts of their lives in other activities (or a combination of these activities) in their nests (Immelmann and Beer 1989, 199).

bubble nest *n.* In Gouramis; the Paradise Fish and Siamese Fighting Fish: a nest of bubbles made by a male that is used as a place in which a female lays her eggs and where young fry might rest (Immelmann and Beer 1989, 200).

compound nest *n.* A nest that contains colonies of two or more species of social insects, "up to the point where the galleries of the nest run together and the adults sometimes intermingle but in which the broods of the species are still kept separate" (Wilson 1975, 354, 581).

formicarium, formicary *n.*

1. An ant's nest (Hölldobler and Wilson 1990, 638).
2. An ant mound, or an artificial nest used to house ants in a laboratory (Hölldobler and Wilson 1990, 638).

mixed nest *n.* A nest containing a mixed colony, *q.v.* (Wilson 1975, 589).

sleep nest *n.* In many songbird species, the Chimpanzee: a nest used for sleeping rather than for rearing young (Immelmann and Beer 1989, 273).

termitarium *n.*

1. A termite nest (Wilson 1975, 597).
2. An artificial nest used in the laboratory to house termites (Wilson 1975, 597).

wasp nest *n.* A structure used and modified (*e.g.,* a hollow stem) or completely built (*e.g.,* a burrow in the ground or a paper nest) by one or more conspecific individuals (Balduf 1954; Evans and West Eberhard 1970; Spradbery 1973).

Comment: Some parts of vespid-wasp nests follow.

▸ **cell** *n.* An elongated, paper chamber, with a hexagonal cross-section, in which a female vespid wasp lays an egg and a larva develops into an adult (Balduf 1954, 445; Evans and West Eberhard 1970, 134–136; Torre-Bueno 1978).

Comment: "Worker cells" and "queen cells" are illustrated by Spradbery (1973, 99).

▸ **comb** *n.* A group of attached cells of a vespid-wasp nest (Balduf 1954, 445; Evans and West Eberhard 1970, 164; Spradbery 1973, 99).
See comb.

▸ **cupola, dormer** *n.* In the bald-faced hornet: a dome shaped sheath that workers build over the top of their nest (Balduf 1954, 446).

▸ **envelope** *n.* The shell of paper around the comb(s) of a vespid-wasp nest (Balduf 1954, 445; Evans and West Eberhard 1970, 166; Spradbery 1973, 99).

▸ **pedicel** *n.* The central supporting pillar at the top of a vespid-wasp nest (Evans and West Eberhard 1970, 181).
syn. main nest pillar (Spradbery 1973, 99), suspensorium (Balduf 1954, 450)
cf. pedicel

♦ **nest aggregation** See aggregation.

♦ **nest-brooding** See brooding: nest-brooding.

♦ **nest-cleaning behavior** See behavior: nest-cleaning behavior.

♦ **nest commensalism** See symbiosis: commensalism.

♦ **nest odor** See odor: nest odor.

♦ **nest parasitism** See parasitism: nest parasitism.

♦ **nest provisioning** See provisioning: nest provisioning.

♦ **nest relief** *n.* In some monogamous fish species, many bird species: one pair member's taking over egg incubation or chick brooding from the other (Immelmann and Beer 1989, 201).

♦ **nesting** *n.* A species' nest-use characteristics.

joint nesting *adj.* In many bird species, including the Groove-Billed Ani: referring to a nesting system in which two or more conspecific females normally lay their eggs in the same nest (Skutch 1959, 302, in Brown 1987a, 21, 304).
cf. nesting: separate nesting
Comment: The term "joint nesting" is independent of a species' breeding system (Brown 1987a, 304).

▸ **separate nesting** *adj.* In many bird species including the Red-Fronted Woodpecker: referring to a nesting system in plural breeding species in which each breeding female has her own nest (Brown 1987a, 21–22, 306).
cf. nesting: joint nesting

♦ **nesting association** *n.*
1. In some bird species: members' of one species nesting close to those of another species (Immelmann and Beer 1989, 200–201).
2. A bird's nesting in a termite, or ant, nest (Immelmann and Beer 1989, 201), *e.g.,* the Indian Ant Woodpecker's nesting in an ant nest (Linsenmaier 1972, 357).
3. An insect's living in a bird nest (Immelmann and Beer 1989, 201).
4. A bird's using a building as a nesting site (Immelmann and Beer 1989, 201).

♦ **nesting symbol** *n.* In many bird species: a piece, or bunch, of nesting material (*e.g.,* grass blades, twigs, straw, leaves, or pebbles) that a male presents, or shows, to a female, often with accompanying body or head movements, prior to or during mating (Immelmann and Beer 1989, 201).

♦ **nestling** *n.* A young bird that is not yet old enough to leave its nest (*Oxford English Dictionary* 1972, entries from 1399), contrasted with a fledgling, a young bird that has recently left its nest.

nidicolous nestling *n.* In many bird species (*e.g.,* Passeriforms): an altricial chick that remains in its nest for some time after hatching and receives food and warmth from its parents while it is in its nest (Immelmann and Beer 1989, 203).
cf. -cial: altricial; nestling: nidifugous nestling

nidifugous nestling *n.* In many bird species (*e.g.,* Galliforms): a precocial chick that hatches at a relatively advanced developmental stage and can actively follow its parent, or parents, shortly after hatching and feed itself (Immelmann and Beer 1989, 203).
cf. -cial: altricial; nestling: nidicolous nestling

♦ **Net** See Internet.

♦ **net assimilation** See ²production: net primary productivity.

♦ **net expected benefit** See benefit: expected benefit: net expected benefit.

♦ **net primary productivity, net primary production** See ²production: net primary productivity.

♦ **net reproductive rate (R_o)** See rate: net reproductive rate.

♦ **neural imprinting** See learning: imprinting: neural imprinting.

♦ **neural inhibition** See inhibition: neural inhibition.

♦ **neuroembryology** See study of: embryology.

♦ **neuroendocrine cell** See cell: neuroendocrine cell.

◆ **neurendocrine system, neuroendo-crine system** See [2]system: neurendocrine system.

◆ **neuroendocrinology** See study of: endocrinology.

◆ **neuroethology** See study of: [3]ethology: neuroethology.

◆ **neurohormone** See hormone: neurohormone.

◆ **neurohumor** See hormone: neurohormone.

◆ **neurohypophysis** See gland: pituitary gland.

◆ **neuron** See cell: neuron.

◆ **neuronal adaptation** See [3]adaptation: [1]physiological adaptation: sensory adaptation.

◆ **neuropeptide Y** See hormone: neuropeptide Y.

◆ **neurophysiology** See study of: physiology: neurophysiology.

◆ **neurosecretion** See secretion: neurosecretion.

◆ **neurotransmitter, neural transmitter** *n*. A substance that is released upon stimulation of a neuron at a nerve ending (with a synapse) and that aids in stimulus transmission by means of changes in the cell's membrane permeability (von Muralt 1939 in Heymer 1977, 26).

syn. transmitter substance (Immelmann and Beer 1989, 318)

Comment: Bradford (in Gregory and Zangwill 1987, 554) lists 48 neurotransmitters, putative neurotransmitters, and neuroactive peptides in six systems.

acetylcholine *n*. In vertebrates: a neurotransmitter secreted from central nervous systems, peripheral nervous systems, and neuromuscular junctions that excites skeletal muscles and both excites and inhibits other sites (Campbell 1999, 975).

aspartate *n*. In vertebrates: an amino-acid neurotransmitter secreted from central nervous systems that is excitatory (Campbell 1999, 975).

dopamine *n*. In vertebrates: a biogenic-amine neurotransmitter secreted from central nervous systems and peripheral nervous systems that is excitatory at most sites and inhibitory at other sites (Campbell 1999, 975).

gamma-aminobutyric acid (GABA) *n*. In invertebrates and vertebrates: an amino-acid neurotransmitter secreted from vertebrate central nervous systems and invertebrate neuromuscular junctions that is inhibitory (Campbell 1999, 975).

glutamate *n*. In invertebrates and vertebrates: an amino-acid neurotransmitter secreted from vertebrate central nervous sys-

tems and invertebrate neuromuscular junctions that is excitatory (Campbell 1999, 975).

glycine *n*. In vertebrates: an amino-acid neurotransmitter secreted from central nervous systems that is inhibitory (Campbell 1999, 975).

***met*-enkephalin** *n*. In vertebrates: a neuropeptide neurotransmitter secreted from central nervous systems that is generally inhibitory (Campbell 1999, 975).

***N*-acetylaspartylglutamate (NAAG)** *n*. In invertebrates and vertebrates: an acidic dipeptide neurotransmitter (Neale et al., 2000, 443).

Comments: Researchers found NAAG in some species of cockroaches, frogs, planarians, and sea anemones; the Brown Rat, Carp, Cattle, Domestic Chicken, Domestic Horse, Domestic Pig, Domestic Rabbit, Goldfish, Guinea Pig, and Human: (Blakely and Coyle 1998, 46–47). NAAG is the most prevalent and widely distributed peptide neurotransmitter in mammalian nervous systems (Neale et al., 2000, 443).

norepinephrine *n*. In vertebrates: a biogenic-amine neurotransmitter secreted from central nervous systems and peripheral nervous systems that both excites and inhibits receptor sites (Campbell 1999, 975).

serotonin *n*. For example, in vertebrates: a biogenic-amine neurotransmitter secreted from central nervous systems that is generally inhibitory (Campbell 1999, 975).

substance P *n*. In vertebrates: a neuropeptide neurotransmitter secreted from central nervous systems and peripheral nervous systems that is excitatory (Campbell 1999, 975).

◆ **neuston, neustont** *n*. Small- to medium-sized organisms that live on or under the surface films of water (Lincoln et al. 1985); *e.g.,* the Portuguese Man-of-War or marine and freshwater water-striders (insects) (Baker et al. 1966).

adj. neustonic

cf. pleuston

epineuston, supraneuston *n*. Neuston that live on the surface films of water (Lincoln et al. 1985).

adj. epineustic

hyponeuston, infraneuston *n*. Neuston that live immediately below a water surface (Lincoln et al. 1985).

▸ **merohyponeuston** *n*. Organisms that spend only part of their life cycles in the hyponeuston (Lincoln et al. 1985).

▸ **planktohyponeuston** *n*. Plankton that congregate immediately below a water surface at night and migrate downward during the day (Lincoln et al. 1985).

meroneuston, meroneustont *n.* Organisms that spend only part of their life cycles in the neuston (Lincoln et al. 1985).

phytoneuston, phytoneustont *n.* A plant in neuston (Lincoln et al. 1985).

planktohyponeuston See neuston: hyponeuston.

supraneuston See neuston: epineuston.

♦ **neustonology** See study of: neustonology.

♦ **neuter** *adj.* Referring to an organism that is of neither sex or with imperfectly developed sex organs (Lincoln et al. 1985).

♦ **neutral allele** See allele: neutral allele.

♦ **neutral association** See neutralism.

♦ **neutral equilibrium** See equilibrium: neutral equilibrium.

♦ **neutral mutation** See ²mutation: neutral mutation.

♦ **neutral object** See object: neutral object.

♦ **neutral theory** See ³theory: neutral theory.

♦ **neutralism, neutral association** *n.*
1. An association between two species in which the population dynamics of each are unaffected by the presence of the other population (Lincoln et al. 1985).
2. "The absence of any interaction between two associated populations" (Lincoln et al. 1985).
See symbiosis: parasymbiosis; ³theory: the neutral theory.

♦ **neutron star** See star: neutron star.

♦ **new synthesis** See evolutionary synthesis.

♦ **new systematics** See study of: systematics: population systematics.

♦ **nexus hypothesis** See hypothesis: nexus hypothesis.

♦ **nibble preening** See preening: nibble preening.

♦ **niche** (NITCH, NICH, NEESH) *n.*
1. A shallow, ornamental recess, or hollow, formed in a building wall, usually made for a statue or other object (*Oxford English Dictionary* 1972, entries from 1611); figurative: a place, or position, adapted to the character or capabilities, or suited to the merits of, a person or thing (*Oxford English Dictionary* 1972, entries from 1726).
2. An organism's place in nature (Johnson 1910 in Griesemer 1992, 232).
3. An organism's place, or role, in an environment (Grinnell and Swarth 1913 and Grinnell 1914, 1917, 1924, 1928 in Griesemer 1992, 232).
Notes: Many textbooks call this kind of niche concept the "habitat niche" and Elton's idea of niche (def. 5), the "functional niche;" however, both definitions consider both habitat and function (Griesemer 1992, 235). Colwell (1992,

241) calls Grinnell's and Elton's niche concepts the "environmental niche concept."
syn. ecologic niche, environmental niche (Grinnell in Griesemer 1992, 233); habitat niche (Griesemer 1992, 235).
4. A subdivision of a habitat (Grinnell 1917 in Krebs 1985, 255).
5. An organism's place, or role, in a community (Elton, 1924, 1927, 1946 in Griesemer 1992, 232).
syn. functional niche (Griesemer 1992, 235)
6. A unit whose limits are set by factors such as food supply, shelter (including refugia from enemies and safe breeding places), competition, parasitism, temperature, humidity, rainfall, insolation, and soil characteristics (Grinnell 1917, 1928 in Griesemer 1992, 232).
7. The multidimensional space that represents the total range of conditions within which a species can function and which it could occupy in the absence of competitors or other interacting species (Hutchinson 1958 in Krebs 1985, 255; Lincoln et al. 1985); "the space framed by the limits of each environmental parameter within which the species in question can exist and reproduce" (Wilson 1975, 25); "the range of each environmental variable, such as temperature, humidity, and food items, within which a species can exist and reproduce" (Wilson 1975, 590); the fundamental niche.
Notes: Hutchinson described the fundamental niche in terms of two or more variables; four or more variables make an *n*-dimensional hypervolume (Krebs 1985, 255). Hutchinson credits Robert MacArthur for the concept of fundamental niche (Griesemer 1992, 238). This concept views niche as an attribute of a population, or species, in relation to its environment.
syn. ecospace, preinteractive niche, prospective ecospace (Lincoln et al. 1985); fundamental niche (Griesemer 1992, 238–239); niche (Krebs 1985, 255); population niche concept [coined by Colwell 1992, 242]
8. Loosely, the microhabitat (physical space) occupied by a species (Lincoln et al. 1985).
cf. biospace; niche: realized niche; principle: competitive-exclusion principle
Comments: Grinnell indicated that niches are unique, thus only one species can occupy a particular niche; Elton, in viewing the concept of niche from a functional perspective, suggested that the same niche could be filled by distinct (but often closely related) species in different communities (Griesemer

n–r

1992, 235). A species' niche can change in ecological and evolutionary time when it adds and deletes variables, changes ranges of one or more variables, or both. Niche has been an important organizing idea in ecology, although criticized by some workers. Modern niche theory focuses on species' actual resource-utilization distributions rather than their permissive ranges of environmental conditions (Levins, 1966, 1968; MacArthur and Levins, 1967; etc. in Griesemer 1992). Colwell (1992) lists many references that review the niche concept; the literature does not agree on many issues. The pronunciation NITCH is common in the U.S. NEESH is a French pronunciation used by some people in the U.S.

[probably from French < *nicher*, to nest, ultimately < Latin *nidus*, nest; introduced by Joseph Grinnell, ecologist (1913 in Griesemer 1992, 233)]

ecospace See niche (def. 7).

empty niche See niche: vacant niche.

fundamental niche See niche (def. 7).

included niche *n.* A niche of a more specialized species that occurs entirely within the niche of a more generalized species (Lincoln et al. 1985).

postinteractive niche See niche: realized niche.

preferred niche *n.* The portion of a Hutchsonian hyperspace in which an organism's fitness is maximum and to which laboratory animals usually also move if given a choice along a series of environmental gradients (Wilson 1975, 25).

Comment: An organism's preferred niche can differ from it realized niche, *q.v.* (Wilson 1975, 25).

preinteractive niche, prospective ecospace See niche (def. 7).

realized niche *n.* "The range of all environmental conditions, including microclimatic regimes and resource characteristics, actually occupied or utilized by a species;" a subset of fundamental niche (Wittenberger 1981, 620).

syn. biospace, postinteractive niche, realized ecospace (Lincoln et al. 1985)

Comment: A species' realized niche is the part of its fundamental niche that does not overlap with that of other species plus the overlapping part in which the species can persist by excluding its competitors (Griesemer 1992, 238).

spatial niche *n.* The spatial parameters of a realized niche (Lincoln et al. 1985).

vacant niche *n.* A role in a community that is available to, but not occupied by, a species (Grinnell 1914; Taylor 1916; etc. in Colwell 1992, 242).

syn. empty niche (Griesemer 1992, 238)

Comment: The hypothesis that a vacant niche (apart from the kind resulting from a recent extinction of a population or species) exists in nature is extremely difficult to test because it is virtually unfalsifiable even in the best cases; nonetheless, popular writers and some recent scientific papers consider vacant niches to occur in nature (Colwell 1992, 245).

♦ **niche breadth** *n.* The upper and lower limits of a given niche parameter (one axis of a niche hyperspace), or the range of a parameter in which a species can function (Lincoln et al. 1985).

Comment: Gould (1982, 338) suggests that "niche width" represents an organism's degree of specialization.

♦ **niche diversification** See character displacement.

♦ **niche-diversification hypothesis** See hypothesis: hypotheses of species richness: niche-diversification hypothesis.

♦ **niche expansion** See ecological release.

♦ **niche overlap** *n.*

1. Direct competition for a given resource by two or more species (Lincoln et al. 1985).

2. Co-occurrence of two or more niches along all, or part, of the same resource axis of a niche (Lincoln et al. 1985).

♦ **niche size, habitat tolerance** *n.* The relative size of a species' realized niche in a species group, community, or habitat (Lincoln et al. 1985).

Comment: "Niche size" is usually estimated in terms of species distribution and tolerance to variables (Lincoln et al. 1985).

♦ **Nicolaian mimicry** See mimicry: Nicolaian mimicry.

♦ **nide** See ²group: nide.

♦ **nidicole** See -cole: nidicole.

♦ **nidicolous** *adj.* Referring to an organism that lives in a nest, *e.g.,* a young animal, especially an altricial bird that remains in its nest for a prolonged period after birth (Lincoln et al. 1985).

cf. -cole: nidicole; nestling: nicolous nestling; nidifuge; nidifugous

♦ **nidicolous nestling** See nestling: nidicolous nestling.

♦ **nidification** *n.* "Nest building" (Lincoln et al. 1985).

♦ **nidifuge** *n.* A young animal, especially a bird, that leaves its nest soon after birth (Lincoln et al. 1985).

adj. nidifugous

cf. nidicole; nestling: nidifugous nestling; nidicolous

♦ **nidifugous** *n.* Referring to an animal that follows its parents out of its nest soon after it is born or hatched (*e.g.,* precocial animals) (Immelmann and Beer 1989, 203, 229).

cf. nestling: nidifugous nestling; nidicolous; nidifuge

♦ **nidus** *n.*
1. "A nest domicile, breeding place, or point of origin" (Lincoln et al. 1985).
2. The focus, or primary site, of an infection (Lincoln et al. 1985).

♦ **night dance** See dance: night dance.

♦ **niphic, nivil, niveal** *adj.* Referring to snow (Lincoln et al. 1985).

♦ **nitric oxide** See molecule: nitric oxide.

♦ **nitrogen cycle** See cycle: biogeochemical cycle: nitrogen cycle.

♦ **nitrophile** See [1]-phile: nitrophile.

♦ **nival, niveal** See niphic.

♦ **nivicole** See -cole: nivicole.

♦ **NMDA** See *N*-methyl-D-aspartate

♦ **NMDA receptor** See [2]receptor: *N*-methyl-D-aspartate receptor.

♦ **NMDA receptor 2B** See [2]receptor: NMDA receptor 2B.

♦ **"no-allelic-replacement hypothesis"** See hypothesis: "no-allelic-replacement hypothesis."

♦ **Nobel Prize** See award: Nobel Prize.

♦ **nocturnae** *adj.* Referring to night-flying insects (Lincoln et al. 1985).
cf. diurnae

♦ **nocturnal** *adj.*
1. Referring to an animal that is active during darkness hours (*Oxford English Dictionary* 1972, entries from 1726).
2. Referring to a phenomenon than lasts only one night (Lincoln et al. 1985).
3. Referring to a phenomenon that occurs during the night, as distinct from daytime (Bligh and Johnson 1973, 951).
cf. crepuscular, diurnal, matinal, pomeridanus [Latin adj. *nocturnus* < *nox,* night]

♦ **nod swimming** See swimming: nod swimming.

♦ **noise** *n.* Background activity in a channel that is unrelated to the signal in question (Dewsbury 1978, 99).
See [1]system: communication system.

♦ **nomad** *n.*
1. "A wandering organism" (Lincoln et al. 1985).
2. "A pasture plant" (Lincoln et al. 1985).

♦ **nomadic phase** See phase: nomadic phase.

♦ **nomadism** *n.*
1. An organism's wandering from place to place, usually within a well-defined territory (Lincoln et al. 1985).
2. For example, in some fish species, army ants: "The relatively frequent movement by an entire society from one nest site or

home range to another" (Wilson 1975, 441, 590).
adj. nomadic

♦ **nomenspecies** See [2]species: nomenspecies.

♦ **nominal fallacy** See fallacy: nominal fallacy.

♦ **nominal scale** See scales of measurement: nominal scale.

♦ **nominal variable** See variable: qualitative variable.

♦ **nominalism** *n.*
1. A philosophical doctrine (opposed to essentialism = realism) that universals, or abstract concepts, exist only as names and without a basis in reality (*Oxford English Dictionary* 1972, entries from 1836; Mayr 1982, 308).
cf. conceptualism
2. A school of scholastic philosophy, that stresses that only individuals, not essences, exist, bracketed together into classes by names (Mayr 1982, 92).
See scholasticism: nominalist.

♦ **nominalistic species concept** See [2]species (def. 10).

♦ **nominalization** *n.* A verb, or other part of speech, that a person uses as a noun: development of indices for community diversity made possible some useful comparative studies.
Comment: Lanciani (1998) decribes how nominalizations unnecessarily complicate scientific writing.

♦ **nomocole** See -cole: nomocole.

♦ **nomogenesis** See -genesis: abiogenesis; hypothesis: nomogenesis.

♦ **nomogram** See -gram: nomogram.

♦ **nomophile** See [1]-phile: nomophile.

♦ **nonadaptive character** See character: nonadaptive character.

♦ **nonadaptive evolution** See [2]evolution: nonadaptive evolution.

♦ **nonadaptive trait** See character: nonadaptive trait.

♦ **nonadaptive zone** See adaptive zone: nonadaptive zone.

♦ **nonasymptotic-species-accumulation curve** See curve: species-accumulation curve: nonasymptotic-species-accumulation curve.

♦ **nonautogenous** See -genous: nonautogenous.

♦ **noncommunicative stridulation** See stridulation.

♦ **noncorrection procedure** See procedure: noncorrection procedure.

♦ **non-Darwinian evolution** See [2]evolution: non-Darwinian evolution.

♦ **nondeceptive floral mimicry** See mimicry: floral mimicry: nondeceptive floral mimicry.

n – r

♦ **"nondefinable-model-virtual-model mimicry"** See mimicry: "virtual-model mimicry:" "nondefinable-model-virtual-model mimicry."

♦ **nondimensional species concept** See ²species (def. 28).

♦ **nondisjunction** *n.* For example, in *Drosophila,* Humans: failure of two homologous chromosomes of a pair to go to opposite poles during a cell's first meiotic division; as a result, one daughter cell has both chromosomes and the other has neither (Mayr 1982, 581, 59).

♦ **nonepigamic behavior** See behavior: epigamic behavior: nonepigamic behavior.

♦ **"nonequilibrium hypothesis of species organization"** See hypothesis: hypotheses of species organization: "nonequilibrium hypothesis of species organization."

♦ **"nonequilibrium hypothesis of species richness"** See hypothesis: hypotheses of species richness: "nonequilibrium hypothesis of species richness."

♦ **nonexclusive-occupancy territory** See territory: nonexclusive-occupancy territory.

♦ **nonfeeding territory** See territory: nonfeeding territory.

♦ **nongenetic polymorphism** See -morphism: polymorphism: phenotypic polymorphism.

♦ **nongenetic sex determination** See sex determination: nongenetic sex determination.

♦ **nonharmonious character transformation** See ²evolution: mosaic evolution; principle: principle of mosaic evolution.

♦ **nonhomologous** *adj.* Referring to chromosomes, or chromosome segments, comprised of dissimilar gene sequences that do not become intimately associated during meiosis (Lincoln et al. 1985).

♦ **nonhomologous apomorph** See morph: apomorph: nonhomologous apomorph.

♦ **nonhomologous similarity** See homoplasy.

♦ **nonhospicidal** See hospicidal: nonhospicidal.

♦ **nonkin-group selection** See selection (table).

♦ **nonmating territory** See territory: nonmating territory.

♦ **nonMendelian** *adj.* Referring to any extrachromosomal hereditary factor (Lincoln et al. 1985).

♦ **nonmotile** See sessile.

♦ **nonobject-bound tradition** See tradition: nonobject-bound tradition.

♦ **nonparametric methods for estimating species richness from samples** See method: nonparametric methods for estimating species richness from samples.

♦ **nonparametric test** See statistical test: nonparametric test.

♦ **nonprotective mimicry** See mimicry: nonprotective mimicry.

♦ **nonrandom** *adj.* Referring to an event that has an *a priori* probability of occurrence of zero or one (unity) (Lincoln et al. 1985). *cf.* random

♦ **nonrearing territory** See territory: nonrearing territory.

♦ **nonreciprocal recombination** See ²theory: transmutation.

♦ **nonsense mutation** See ²mutation: nonsense mutation.

♦ **nonsequential patrolling** See patrol.

♦ **nonsexual selection** See selection: nonsexual selection.

♦ **nonshareable parental investment** See investment: parental investment: nonshareable parental investment.

♦ **nonshivering obligatory thermogenesis (NST(O))** See thermogenesis: nonshivering obligatory thermogenesis.

♦ **nonshivering thermogenesis (NST)** See thermogenesis: nonshivering thermogenesis.

♦ **nonshivering thermoregulatory thermogenesis (NST(T))** See thermogenesis: nonshivering thermoregulatory thermogenesis.

♦ **nonsocial** See solitary.

♦ **nonthermal sweating** See sweating: nonthermal sweating.

♦ **nontranslated spacers** See external-nontranscribed spacers.

♦ **nonverbal communication** See communication: nonverbal communication.

♦ **nonviable** *adj.* Referring to an organism that is incapable of normal development or survival (Lincoln et al. 1985).

♦ **nonworker** See caste: worker: nonworker.

♦ **noosphere** See -sphere: biosphere: noosphere.

♦ **norepinephrine** See hormone: norepinephrine.

♦ **norm of reaction** *n.* "Heritable phenotypic variation among individuals of a single genotype, elicited by variation in environmental conditions" (Gause 1947 and Gupta and Lewontin 1982 in Gordon 1992, 256). *Comment:* "Norm of reaction" has a meaning similar to "phenotypic plasticity," *q.v.*

♦ **normal curve** See curve: normal curve.

♦ **normal demographic distribution** See distribution: normal demographic distribution.

♦ **normal distribution** See distribution: normal distribution.

♦ **normalizing selection** See selection: stabilizing selection.

♦ **normothermy** See -thermy: cenothermy.

♦ **northern blotting** See method: northern blotting.

♦ **northern-evergreen forest** See [2]community: northern-evergreen forest.

♦ **northern hardwoods** See [2]community: northern hardwoods.

♦ **nose-up posture** See display: upward stretch.

♦ **nosogenic** See -genic: nosogenic.

♦ **nosography** See study of: nosology.

♦ **nosology** See study of: nosology.

♦ **noterophile, noterophilous** See [1]-phile: mesophile.

♦ **nothocline** See cline: nothocline.

♦ **nothogamy** See -gamy: nothogamy.

♦ **notify** *v.i., v.t.*
  1. In Humans: "to make known, publish, proclaim; to intimate, give notice of, announce" (*Oxford English Dictionary* 1972, entries from 1386).
  2. In primates (*e.g.,* male Hamadryas Baboons): to turn one's anal region toward a conspecific (Sigg and Falett 1985, 981).
  cf. display: buttocks display

♦ **notifying** See greeting: notifying.

♦ **notochord** See organ: notochord.

♦ **novelty seeking** *n.* In Humans: the personality trait of tending to search for new experiences (Angier 1996, A1).
  cf. drive: curiosity drive, exploratory drive
  Comments: A person with a high novelty-seeking quotient tends to be extroverted, impulsive, extravagant, quick-tempered, excitable, and exploratory (Angier 1996, A1). A novelty seeker tends to have a particular variant of a gene that allows his brain to respond to dopamine. This gene encodes for the D4-dopamine receptor, one of several kinds of brain receptors known to play a role in dopamine reception.

♦ **NPY** See hormone: neuropeptide Y.

♦ **NR2B** See [2]receptor: NMDA receptor 2B.

♦ **NST** See thermogenesis: nonshivering thermogenesis.

♦ **NST(O)** See thermogenesis: nonshivering obligatory thermogenesis.

♦ **NST(T)** See thermogenesis: nonshivering thermoregultory thermogenesis.

♦ **nth order** See order: *n*th order.

♦ **NTS** See external-nontranscribed spacers.

♦ **nucivore** See -vore: nucivore.

♦ **nuclear family** See [1]family: nuclear family.

♦ **nuclear species** See [2]species: nuclear species.

♦ **nucleic acid** *n.*
  1. An acid found in cell nuclei (*Oxford English Dictionary* 1972, entries from 1893).
  2. An acid comprised of bases and a backbone of sugar and phosphate molecules or peptides (King and Stansfield 1985).

cf. acid, molecule
[*nucleus* + IC]

deoxyribonucleic acid (DNA) *n.*
  1. A nucleic acid comprised of the bases adenine, cytosine, guanine, and thymine, which project from a backbone of deoxyribose and phosphate molecules (King and Stansfield 1985).
  2. The basic hereditary material of organisms (Watson and Crick 1953; Mayr 1982, 809–817; Wilson 1975, 582; King and Stansfield 1985).

cf. gene
Comments: The chemical name for DNA has an alleged 207,000 letters (Ash 1976, 108) and is published in its shortened form in *Nature* (9 April 1981). DNA is the molecular carrier of genetic information within cells, has a double-helix structure, *q.v.,* and can self-replicate as well as code for RNA synthesis (King and Stansfield 1985). The great bulk of an organism's DNA is located within its chromosome(s). DNA molecules can have molecular weights of over 1×10^8 daltons (King and Stansfield 1985). Base-pair numbers for selected organisms are given under "gene."

▸ **anonymous DNA** *n.* "A length of DNA of unknown gene content" (Jasney 1991).

▸ **chloroplast DNA (cpDNA)** *n.* DNA found in chloroplasts (Hamilton 1999, 129).
  Comment: This DNA is maternally inherited in most angiosperm species and disperses only through seeds (Hamilton 1999, 129).

▸ **copy DNA (cDNA)** *n.* Complementary DNA produced from an RNA template by the action of a reverse transcriptase (an RNA-dependent DNA polymerase), which may or may not have introns, depending on its processing (King and Stansfield 1985; Campbell et al. 1999, 369).
  syn. complementary DNA
  Comment: If one processes the RNA template to remove its introns, its corresponding cDNA is not identical to the gene from which the RNA was transcribed (King and Stansfield 1985).

▸ **highly repeated DNA** *n.* DNA that contains short and simple sequences repeated hundreds of thousands, or millions, of times (Gould 1981a, 7).
  syn. satellite DNA (Gould 1981a, 7)
  Comment: About 5% of human DNA falls into this class (Gould 1981a, 7).

▸ **intermediate DNA** See nucleic acid: deoxyribonucleic acid: middle-repetitive DNA.

▸ **junk DNA** *n.* "DNA with no real function" (Nowak 1994, 608).

n–r

Some Kinds of DNA

I. anonymous DNA
II. copy DNA (*abbr.* cDNA)
III. mitochondrial DNA (*abbr.* mit DNA, mtDNA)
IV. nonfunctional DNA
V. nuclear DNA
VI. parasitic DNA
VII. polite DNA
VIII. repetitious DNA (*syn.* repeated DNA, repetitive DNA)
   A. highly repeated DNA (*syn.* satellite DNA)
     1. satellite DNA
     2. interspersed repeated DNA
   B. middle-repetitive DNA (*syn.* intermediate DNA)
IX. satellite DNA
X. selfish DNA (*syn.* junk DNA, self-centered DNA, silent DNA)
   A. DNA in general
   B. nonfunctional DNA
   C. parasitic DNA
   D. satellite DNA
   E. spacer DNA
   F. transposable elements

See nucleic acid: deoxyribonucleic acid: selfish DNA.

Comments: DNA originally considered to be junk DNA in Humans is now known to be a complex mix of different kinds of DNA including regulatory elements (Nowak 1994). Some repetitive sequences of junk DNA seem to have crucial functions in maintaining a genome's structure. Researchers previously classified introns as junk DNA, but the *lin*-4 gene of *Caenorhabditis elegans* which occurs in an intron regulates expression of its *lin*-14 by blocking its ability to make a protein (Lee et al. and J. Mattick in Nowak 1994, 609–610). The 3'UTR is a so-called 3' untranslated region which lies at the end of each gene's mRNA; however, a mutation in a 3'UTR region can suppress its gene's action (Nowak 1994, 610). Introns also encode for small nucleolar RNAs (snoRNAs), which might play a role in ribosome assembly.

▶ **middle-repetitive DNA** *n.* DNA comprised of sequences of ten to a few hundred copies of particular DNA segments; the copies are often widely dispersed on several chromosomes (Gould 1981a, 7).
syn. intermediate DNA
Comment: Some 15 to 30% of human and fruit-fly genomes are comprised of middle-repetitive DNA (Gould 1981a, 7).

▶ **mitochondrial DNA (mtDNA)** *n.* DNA of a mitochondrion which consists of a circular duplex usually with 5 to 10 copies per organelle (King and Stansfield 1985). *abbr.* mit DNA (Reid 1980, 620)
Comments: The mitochondrial genetic code differs slightly from the universal genetic code. Mitochondrial DNA encodes a limited number of RNAs and proteins that are essential for formation of a functional mitochondrion (Gray et al. 1991, 1476). Most of the genetic information for a mitochondrion's biogenesis and function resides in that nuclear genome of the cell that produces the mitochondrion. Many species have mtDNAs that map as circles; some have those that map as linear molecules. Mitochondrial genomes size ranges from <6 kilobase pairs (= kbp) in *Plasmodium falciparum* to >200 kbp in land plants. Only egg cells contribute significant numbers of mitochondria to zygotes; thus, mtDNA is primarily maternally inherited (King and Stansfield 1985). A sperm's mitochondrion can enter an egg, and sperm-contributed mtDNA can recombine with that from an egg; this might explain why some people have two different versions of mtDNA in their cells (Strauss 1999, 1438).

▶ **nonfunctional DNA** *n.* DNA, especially a pseudogene, that is not known to code for proteins, or functional RNAs, but may in fact have a function (Zuckerkandl et al. 1989, 504).

▶ **parasitic DNA** *n.* A sequence of DNA (an intron) that is not transcribed into messenger RNA but is excised during transcription (Mayr 1982, 579).
cf. exon
Comment: Orgel and Crick (1980) hypothesize that this DNA is parasitic because an organism is helpless to prevent its replication and accumulation. "Although valid arguments in favor of this hypothesis exist, it is intuitively distasteful to a Darwinian. Surely natural selection, a Darwinian would say, should be able to come up with a defense mechanism against such an expensive type of parasitism" (Mayr 1982, 579).

▶ **polite DNA** *n.* "DNA, that, without being crucially involved in function, is subject to constraints of conformity and, through its base composition, respects a function for which it is not required;" DNA with no crucial function that does not disrupt the action of functional DNA (Zuckerkandl 1986, 12).

▶ **recombinant DNA** *n.* DNA from two different sources (*e.g.,* two different species) (Campbell et al. 1999, 364).

► **repetitious DNA, repeated DNA, repetitive DNA** *n*. DNA of eukaryotes that has nucleotide sequences serially repeated many times (Lincoln et al. 1985).

Comment: Repetitious DNA is also found in prokaryotes (Doolittle and Sapienza 1980, 603).

► **satellite DNA** *n*.

1. Eukaryote DNA that differs sufficiently in its base composition from a cell's main DNA so as to be separable by centrifugation (isopycnic-CsCl-gradient centrifugation) (Lincoln et al. 1985; King and Stansfield 1985).
   See nucleic acid: deoxyribonucleic acid: highly repeated DNA.
2. DNA composed of tandem repeats of a simple sequence (Alberts et al. 1989, 604).

Comments: Satellite DNAs from chromosomes are either lighter (rich in A + T) or heavier (rich in G + C) than the majority of DNA and are usually highly repetitious (King and Stansfield 1985). Satellite DNA sequences usually are not transcribed and are most often located in the heterochromatin associated with the centromeric regions of chromosomes (Alberts et al. 1989, 604). In some mammal species, a single kind of satellite DNA constitutes 10% or more of DNA. Satellite-DNA sequences seem have changed very rapidly and even shifted their positions on chromosomes. Because satellite DNA has no known function, it has been classified as an extreme form of selfish DNA. Satellite DNA might function in maintaining the survival of a chromosome (Nowak 1994).
[satellite, after the minor component of DNA that was separable from the bulk of a cell's DNA (Alberts et al. 1989, 604)]

microsatellite DNA *n*. Satellite DNA that is roughly 10 through 100 base pairs long, has repeating units of only a few base pairs, and is longer than minisatellite DNA (Nowak 1994; Campbell et al. 1999, 382).

Comments: Minisatellite DNAs are found throughout human genes and are highly variable from one person to another (Nowak 1994; Campbell et al. 1999, 382). Defects in these DNAs might cause cancer.

minisatellite DNA *n*. Satellite DNA that is shorter than satellite DNA (Nowak 1994).

Comments: Defects in minisatellite DNAs might cause cancer (Nowak 1994).

► **selfish DNA** *n*.

1. A DNA sequence that spreads within an organism and makes no contribution to its phenotype, except that it is a slight burden to the cell that contains it (Orgel and Crick 1980, 605).
2. DNA in general (Doolittle and Sapienza 1980, 601).

   Note: This concept of selfish DNA is based on the idea that organisms are simply DNA's way of producing more DNA (Doolittle and Sapienza 1980, 601). This idea is also stated in terms of genes: Organisms are simply genes' ways of producing more genes.
3. Repetitive DNA that has increased, or is increasing, in frequency within an organism and whose increase is not stopped because the organism which hosts it cannot stop it, possibly because it has no appreciable effect on its host (Gould 1981a, 13).

syn. junk DNA (King and Stansfield 1985), silent DNA (Mayr 1982, 826; Gould 1985d)
cf. chromosome: supernumerary chromosome; gene machine; nucleic acid: deoxyribonucleic acid: junk DNA, parasitic DNA; gene: selfish gene; nuon; transposable element

Comments: Östergren formulated the concept (but not the name) of selfish DNA in about 1945 (Cavalier-Smith 1980, 618). Because "selfish" implies that the DNA should be doing something besides acting for itself, it might be better to call this DNA "self-centered DNA" (Gould 1981a, 13). Selfish DNA replicates, moves, and inserts into new places in chromosomes and performs no known functions for the organisms containing it (Gould 1981a). King and Stansfield (1985) also suggest the "selfish DNA" is functionless, and that according to selfish-DNA theory, an eukaryotic organism is merely a throwaway survival machine used by selfish DNA to replicate itself. "Spacer DNA," "satellite DNA," and other kinds of repetitive DNA may be examples of "selfish DNA." Brosius and Gould (1992, 10706) write that ambiguous and even derogatory names such as "pseudogene" and "junk DNA" "do not reflect the significance of retroposed sequences as large valuable assets for the future evolvability of species; and as a result, it is more difficult to contemplate their significance, impact, and function."

► **silent DNA** See nuclei acid: deoxyribonucleic acid: selfish DNA.

► **simple-tandem repeats (STRs)** *n*. The repeated short base-pair sequences within a strand of microsatellite DNA (Campbell et al. 1999, 382).

► **3′ untranslated region (3′UTR)** *n*. The final protein coding portion of a gene that

n–r

is transcribed into RNA but not translated into protein (Nowak 1994, 608).

Comment: Researchers once classified 3'UTRs as junk DNAs, *q.v.*, but they are now known to contain sequences that regulate gene activity (Nowak 1994, 608).

peptide-nucleic acid (PNA) *n.* An artificially produced nucleic acid composed of bases linked together with peptides (Zimmer 1995d, 38, illustration).

Comments: Peter Nielsen and colleagues created PNA in 1991 (Zimmer 1995d, 38). PNA peptide bonds are much stronger that ribose-phosphate bonds of DNA and RNA. PNA might have preceded these other nucleic acids in the origin of life on Earth. Prelife might have gone through the states of a PNA world, RNA world, and DNA world, *q.v.*

ribonucleic acid (RNA) *n.* A nucleic acid comprised of the bases adenine, cytosine, guanine, and uracil, which project from a backbone of ribose and phosphate molecules (King and Stansfield 1985).

cf. deoxyribonucleic acid, DNA

Comments: Types of RNA are heterogeneous nuclear RNA (hnRNA); messenger RNA (mRNA); ribosomal RNA (rRNA); small nucleolar RNA (snoRNA); and transfer RNA (tRNA). RNA is the genetic material of many viruses (King and Stansfield 1985). RNAs are single stranded and have lower molecular weights than DNAs.

▸ **18S-ribosomal RNA (18S-rRNA)** *n.* A structural RNA of ribosomes (Maley and Marshall 1998, 505).

Comments: Since 1988, researchers have produced many dendrograms based on 18S-rRNA which might not be as accurate as originally thought (Maley and Marshall 1998, 505). Therefore, researchers might use amino-acid sequences of proteins more in future phylogenetic studies.

▸ **heterogeneous nuclear RNA (hnRNA)** *n.* RNA that is immature messenger RNA or has unknown functions (Nowak 1994).

▸ **ribosomal RNA (rRNA)** *n.* The structural RNA of ribosomes (Lincoln et al. 1985).

♦ **nucleic-acid hybridization** See method: nucleic-acid hybridization.

♦ **nucleic-acid probe** See [1]probe: nucleic-acid probe.

♦ **nuclein** *n.* The phosphorus-rich compound in cell nuclei [named by Miescher (Mayr 1982, 679, 959)].

♦ **nucleoprotein world** *n.* The situation that exists today in which nucleic acids and proteins cooperate in production of new proteins and nucleic acids; contrasted with the RNA world (inferred from Illangasekare et al. 1995, 646).

♦ **nucleosome** *n.* A molecular aggregate comprised of DNA and various proteins, particularly histones (Mayr 1982, 575).

♦ **nucleotide** *n.* A biochemical molecule (comprised of a purine or pyrimidine base, a pentose, and a phosphoric-acid group) that is the basic building block of the polynucleotides DNA and RNA (King and Stansfield 1985).

♦ **nucleotide element** *n.* A repeated sequence of DNA (Zuckerkandl et al. 1989, 504).

long-interspersed nucleotide element (LINE) *n.* A nucleotide element, *q.v.*, of from 20,000 to 50,000 base pairs (Singer 1982 in Zuckerkandl et al. 1989, 504); contrasted with short-interspersed nucleotide element.

short-interspersed nucleotide element (SINE) *n.* A nucleotide element, *q.v.*, of from 150 to 300 base pairs (Singer 1982 in Zuckerkandl et al. 1989, 504); contrasted with long-interspersed nucleotide element.

Comments: Alu sequences (discovered by Houck et al. 1979) represent the great majority of SINEs in Humans (Zuckerkandl et al. 1989, 505, who describe their characteristics). These sequences are variable in structures, numbers, locations, orientations, and transcriptional regulation.

♦ **nucleus** See organelle: nucleus.

♦ **nucleus species** See [2]species: nuclear species.

♦ **nuisance variable** See variable: nuisance variable.

♦ **null hypothesis** See hypothesis: null hypothesis.

♦ **nullipara** See -para: nullipara.

♦ **number** *n.* A specific quantity, or place, in a sequence, usually designated by one or a series of symbols or words called "numerals" (Michaelis 1963).

grail number *n.* The total number of species on Earth (Yoon 1995, C4).

[Old French *graal* < Medieval Latin, *gradalis*]

♦ **number concept** See counting.

♦ **number of replicates** See sample size.

♦ **numeracy** *n.* In the Cottontop Tamarin, Human, Rhesus Monkey: an individual's ability to assess the amount, or bulk, of an object (M. Hauser in Williams 1997, 29).

Comments: Human infants less than 1 year old display numeracy (Williams 1997, 30).

♦ **numerical character** See character: quantitative character.

♦ **numerical hybrid** See hybrid: numerical hybrid.

♦ **numerical mutation** See [2]mutation: numerical mutation.

♦ **numerical phenetics** See study of: taxonomy: numerical taxonomy.

♦ **numerical response** See response: numerical response.

♦ **numerical taxonomy (NT)** See study: taxonomy: numerical taxonomy.

♦ **nuon** *n*. Any stretch of a nucleic-acid sequence that may be identifiable by any criterion; a nuon can be a gene, intergenic region, exon, intron, promoter, enhancer, terminator, pseudogene, short-interspersed element, long-interspersed element, or any other retroelement, transposon, or telomer of a few to thousands of nucleotides long [*nu*cleic acid sequence + ON; coined by Brosius and Gould 1992, 10706].

Comment: Genes are always nuons, but not all nuons are genes (Brosius and Gould 1992, 10707, who give actual and possible examples of kinds of nuons; they also coin "aptogene," "potogene," "potomass," "retro-naptonuon," "retro-potonuon," "retro-xaptonuon," "retro-potogene," "xaptogene," "xaptoprotein").

aptonuon *n*. A nuon with an adaptive function resulting from natural selection [ad*apt*ive + O + NUON; coined by Brosius and Gould 1992, 10709].

naptonuon *n*. A potonuon that never acquires a function in a genome (so far as we know) [*n*onad*apt*ive + O + NUON; coined by Brosius and Gould 1992, 10706].

potonuon *n*. A nuon that has the potential of becoming functional in the course of evolution [*pot*ential + O + NUON; coined by Brosius and Gould 1992, 10706].

xaptonuon *n*. A potonuon that has been coopted into a variant, or novel, function in a genome by exaptation [e*xapt*ation + O + NUON; coined by Brosius and Gould 1992, 10706].

♦ **nuptial dress, nuptial coloration** See dress: advertising dress.

♦ **nuptial flight** See flight: nuptial flight.

♦ **nuptial gift** *n*. For example, in some empidid-fly and scorpionfly species, the European Nursery-Web Spider, the moth *Utetheisa ornatrix*: an object that a male presents to a female as a way of eliciting insemination (Thornhill and Alcock 1983, 190; Austad and Thornhill 1991, 44).

Comments: Depending on the species, empidid males present females with fresh insects, pieces of dried insects, insect parts wrapped in silk, and empty silken "balloons" (Kessel 1955). Male European Nursery-Web Spiders present females with insects wrapped in silk (Austad and Thornhill 1991, 44). A male *Utetheisa ornatrix* gives a female protective alkaloids during copulation (González et al. in Yoon 1999a, D3).

♦ **nursery group** See ²group: creche.

♦ **nursing** *n*.

1. An individual's taking care of (the sick, injured, or infirm) (given as a verb in Michaelis 1963).
2. A female mammal's feeding (an infant) at her breast; suckling (Michaelis 1963). See suckling. *cf*. lactation
3. An individual's bringing up, or nourishing, another (Michaelis 1963).
4. An individual's taking nourishment from a breast (Michaelis 1963).

See sucking: welcome sucking.

substitute nursing *n*. In Cattle: a calf's stereotyped sucking on metal rings, metal chains, or even other calves when it has been fed from buckets rather than being allowed usual nursing (Heymer 1977, 206).

♦ **nursing position, nursing posture** See posture: nursing posture.

♦ **nurturance** See need: nurturance.

♦ **nutricism** See symbiosis: nutricism.

♦ **nutrition** *n*. An organism's action, or process, of supplying, or receiving, nourishment (*Oxford English Dictionary* 1972, entries from 1615); this involves ingestion, digestion, assimilation, or a combination of these activities (Lincoln et al. 1985). *adj*. nutritive

♦ **nurturance, nurturing behavior** See behavior: nurturance.

♦ **nychthemeron** *n*. A period of 24 hours, consisting of a day and a night (Bligh and Johnson 1973, 951). *cf*. period: nyctiperiod

♦ **nyctipelagic** See -pelagic: nyctipelagic.

♦ **nyctiperiod** See period: nyctiperiod.

♦ **nyctophobia** See phobia (table).

♦ **nymph** See animal names: nymph.

♦ **nymphiparity** See parity: nymphiparity.

n–r

O

♦ **object** *n.*
1. Anything that a person actually apprehends, or can potentially apprehend, with his senses; especially, touch and vision; a material thing; a thing, or body, on which an individual makes an observation (*Oxford English Dictionary* 1972, entries from 1398; Michaelis 1963).
2. That to which a person directs his action, thought, or feeling; that to which one does something; that on which something acts or operates (*Oxford English Dictionary* 1972, entries from 1586).

Comment: Animal behaviorists consider objects from the viewpoints of both Humans and nonhuman animals.

neutral object *n.* An object other than the focal one that an animal directs its action toward in a conflict situation (Immelmann and Beer 1989, 203).
syn. substitute object (according to some authors, Immelmann and Beer 1989, 299)
cf. behavior: displacement behavior

substitute object *n.* An object that can elicit an animal's consummatory behavior in the absence of appropriate stimuli and without necessarily fulfilling its biological "need" that underlies this behavior (Heymer 1977, 63), *e.g.,* a rat's treating its own tail as nesting material when it is deprived of nesting material (Immelmann and Beer 1989, 203).

♦ **object-bound tradition** See tradition: object-bound tradition.
♦ **object imprinting** See learning: imprinting: object imprinting.
♦ **object play** See play: object play.
♦ **object taxis** See taxis: object taxis.
♦ **objective self-awareness** See awareness: self-awareness: objective self-awareness.
♦ **obligate** *adj.*
1. Something that is necessarily as it is (*Oxford English Dictionary* 1972, entries

from 1887); "essential; necessary; unable to exist in any other state, mode, or relationship" (Lincoln et al. 1983).
2. Referring to an organism that is restricted to one mode of life (*e.g.,* an obligate parasite) (Michaelis 1963).
cf. facultative

♦ **obligate anaerobe** See aerobe: anaerobe: obligate anaerobe.
♦ **obligate apogamy** See -gamy: apogamy: parthenoapogamy.
♦ **obligate gamete** See gamete: obligate gamete.
♦ **obligate hyperparasitism** See parasitism: hyperparasitism: obligate hyperparasitism.
♦ **obligate-schooling species** See ²species: obligate-schooling species.
♦ **obligate siblicide** See -cide: siblicide: obligate siblicide.
♦ **obligate thermophile** See ¹-phile: thermophile: obligate thermophile.
♦ **obligatory parasitism** See parasitism: obligatory parasitism.
♦ **oblique crossing over** See crossing over: unequal crossing over.
♦ **obliterative coloration, obliterative shading** See camouflage: background imitation: countershading.
♦ **observability sample** See sample: observability sample.
♦ **observation** *n.* A person's action, or act, of observing something scientifically; especially, his careful noting of a phenomenon in regard to its cause or effect, or of phenomena with regard to their mutual relations as they are observed in nature (contrasted with "experiment"); also, a record of this action (*Oxford English Dictionary* 1972, entries from 1559).

descriptive observation *n.* Naturalistic observation with the objective of documenting the natural history of a group, or population, of animals with an emphasis on behavior (Lehner 1979, 60).

cf. experimental manipulation

individual observation *n.* An observation, or measurement, taken on the smallest sampling unit of a population (Sokal and Rohlf 1969, 8).

syn. item

cf. sample of observations, population, variable

naturalistic observation *n.* Scientific investigation involving studying animal behavior "as it occurs naturally, with as little human intrusion as possible" (Lehner 1979, 59).

cf. experimental manipulation

reconnaissance observation *n.* Intense observation of an animal species' behavior before one poses questions and constructs hypotheses about its behavior [coined by Lehner 1979, 24].

♦ **observational learning** See learning: emphatic learning.

♦ **observational method** See method: observational method.

♦ **observed frequency** *n.* The actual number of sample values that belong to a class of a frequency distribution (Lincoln et al. 1985).

♦ **Observer** See program: Observer, The.

♦ **observer error** See error: observer error.

♦ **obsolescence** *n.* A phase, or stage, of population reduction that leads to a species' extinction or the loss of one of its characters (Lincoln et al. 1985).

adj. obsolete

♦ **obstinacy** See ²group: obstinacy.

♦ **Occam's razor** See Ockham's razor.

♦ **occasional** See frequency (table).

♦ **occasional species** See ²species: occasional species.

♦ **oceanic** See pelagic.

♦ **oceanodromous** See -dromous: oceanodromous.

♦ **oceanography** See study of: oceanography.

♦ **oceanophile** See ¹-phile: oceanophile.

♦ **ochthium** See ²community: ochthium.

♦ **ochthophile** See ¹-phile: ochthophile.

♦ **Ockham's razor, Occam's razor** *n.* *Pluralitas non est ponenda sine necessitate* (plurality must not be posited without necessity) (Ockham in Jeffreys and Berger 1992, 64).

cf. law: law of parsimony; Morgan's canon; simplicity

Comments: Lincoln et al. (1985) indicate that Ockham's razor is the idea that the simplest sufficient hypothesis that explains known relevant facts is the preferred one even if other hypotheses are plausible. Jeffreys and

Berger (1992, 64) list other restatements of Ockham's razor: "Entities should not be multiplied without necessity; it is vain to do with more what can be done with less; an explanation of the facts should be no more complicated than necessary ... among competing hypotheses, favor the simplest one." These authors state, "Ockham's razor has proved to be an effective device for trimming away unprofitable lines of inquiry, and scientists use it every day, even when they do not cite it explicitly."

[after William of Ockham, 1300?–1349?, English Franciscan and scholastic philosopher]

♦ **octa-** *prefix* Eight, eightfold (Michaelis 1963).

♦ **oddity effect** See effect: oddity effect.

♦ **oddity learning** See learning: oddity learning.

♦ **oddity problem** See test: oddity test.

♦ **oddity test** See test: oddity test.

♦ **odor** (*Brit.* **odour**) *n.*

1. A substance's quality that renders it perceptible to one's sense of smell (olfaction) (*Oxford English Dictionary* 1972, entries from 1300).

   syn. scent, smell; sometimes a sweet scent, pleasing scent; fragrance (*Oxford English Dictionary* 1972)

2. A substance that emits a sweet smell or scent; a perfume; especially incense, spice, ointment, etc. (*Oxford English Dictionary* 1972, entries from 1388).

cf. chemical-releasing stimulus

clan-specific odor *n.* A group odor arising from the odors of individual male sugar gliders produced by their frontal glands which are rubbed against sternal glands of females of their group (Müller-Schwarze 1974, 319).

colony odor See odor: group odor.

environmental odor *n.* Odor(s) from an animal's environment that it carries (*e.g.,* odors of an ant's nesting substrate and foods) (Crosland 1989, 912).

gestalt odor *n.* For example, in an ant species: odor of a group of nest mates that results from a complete transfer of pheromones among them (Crosland 1989, 920).

group odor *n.* For example, in some rodent species, social Hymenoptera, Marsupial Flying Phalangers: a group-specific smell caused by reciprocal scent marking, or odor exchange, in a group of closely associated conspecific animals (Heymer 1977, 80; Immelmann and Beer 1989, 123).

syn. colony odor (Immelmann and Beer 1989, 123)

cf. chemical-releasing substance: semiochemical: pheromone: surface pheromone; odor: nest odor, species odor

n–r

Comment: Group odors enable animals to tell whether another animal is a member of its group (Immelmann and Beer 1989, 123). Hölldobler and Wilson (1990, 643) refer to a "group odor" that is on the bodies of social insects and peculiar to a particular colony as "colony odor." Individuals recognize their nest mates by colony odor.

hive odor, hive aura *n.* The nest odor of the Honey Bee (Wilson 1975, 212, 589).
cf. odor: nest odor
Comment: Hive aura may be the same substance as Honey Bee footprint substance (Wilson 1975, 212, 589).

human underarm odor, perspiration odor *n.* Odor composed of 3-methyl-2-hexanoic acid and other components (Booth 1990).
Comment: 3-methyl-2-hexanoic acid is produced by bacteria (lipohilic diphtheroids) that feed on apocrine-gland secretions (Booth 1990).

human vaginal odor *n.* An odor comprised of at least 30 compounds derived from several glands and tissues of vaginas (Forsyth 1985, 30).
cf. chemical-releasing stimulus: semiochemical: pheromone: human-female pheromone
Comment: In Rhesus Monkeys, vaginal secretions of a volatile fatty acid increase near the midpoint of a female's estrus cycle and are attractive to male Rhesus Monkeys (Curtis 1983, 805).

individual odor See chemical-releasing stimulus: pheromone: individual pheromone.

MCH-dependent odors *n.* For example, in Domestic Mice and Humans: body odors that are influenced by major-histocompatibility-complex genes (Richardson 1996, 26).
Comments: Women tend to prefer odors of men whose major-histocompatibility-complex genes are dissimilar compared to theirs (Richardson 1996, 26). Choosing MCH-dissimilar mates might have functions including increasing fertility, producing hardier offspring, and reducing the risk of genetic disease.

nest odor *n.* In social insects: a nest's distinctive odor by which its inhabitants are able to distinguish their nest from those of other societies or at least from their surrounding environment (Wilson 1975, 589).
cf. odor: colony odor
Comments: Honey Bees and some ants can orient toward their nests by means of their nest odors. Nest odor may be the same as the colony odor in some cases (Wilson 1975, 589).

species odor *n.* Odor on a social insect's body that is peculiar to a species (Hölldobler and Wilson 1990, 643).
Comment: A species odor might be merely the distinctive component of a larger mixture that comprises a colony odor (Hölldobler and Wilson 1990, 643).

♦ **odor-conditioned anemotaxis** See taxis: anemotaxis.

♦ **odor plume, odor signal** See signal: odor signal.

♦ **odor signature** *n.* The individual odor deposited by an animal (Wilson 1975, 205).

♦ **odor trail** *n.* A chemical trace (trail pheromone or trial substance) laid down by one insect and followed by another (Hölldobler and Wilson 1990, 641).

exploratory trail *n.* In Army Ants: an odor trail that advance workers of a foraging group lay more or less continuously (Hölldobler and Wilson 1990, 638).
Comment: This is a common kind of army-ant communication (Hölldobler and Wilson 1990, 638).

recruitment trail *n.* In ants: an odor trail laid down by scout workers and used to recruit nest mates to a food find, a desirable new nest site, a breach in their nest wall, or some other place where the assistance of many workers is needed; opposed to exploratory trail (Hölldobler and Wilson 1990, 642).

♦ **oecad** See ecad.

♦ **-oecism, -oecy**
adj. -oecious, -oecius, -oekous
cf. -cole, ¹-phile

amphioecism *n.* A population's, or species', showing broad variable tolerance of habitat and environmental conditions, reflected in the presence of clines and a subspecies in the case of a species (given as an adjective in Lincoln et al. 1985).
adj. amphioecious, amphitopic
cf. -oecism: euryoecism, stenoecism

coenomonoecism See -oecism: trimonoecism.

dioecism *n.*
1. A species' having male and female reproductive organs on different individuals (Lincoln et al. 1985).
See sexual: unisexual.
adj. bisexual, dioecious, diecious, dioicious, gonochoristic, unisexual
syn. diecism, dioecy, dioicism, bisexuality, unisexuality (Lincoln et al. 1985)
cf. bisexual; -oecism: monoecism, trioecism
2. A plant species' having unisexual flowers of each sex on separate plants or in separate parts of the same inflorescence (Fernald 1950, 1573).

adj. dioecious, dioecious, dioicious
cf. bisexual; -oecism: monoecism
[Greek *dis*, twice + *oikos*, house]

euryoecism, euroky *n.* An organism's being tolerant of a wide range of habitats and environmental conditions (Lincoln et al. 1985).
adj. euryoekous, euryokous
cf. -oecism: amphioecism; stenoecism

heteroecism, heterecism, heteroxeny, metoxeny *n.* In unisexual organisms: production of male and female gametes by different individuals (Lincoln et al. 1985). See -xeny: pleioxeny.
adj. heterecious, heteroecious, metoecious

homoecism, host specificity, monoxeny See -xeny: monoxeny.

metoxeny *n.*
1. A parasite's not being host specific (*i.e.,* occupying two or more different hosts at different stages of its life cycle) (Lincoln et al. 1985).
   adj. heterecious, heteroecious, metoecious
   cf. -oecism: homoecism
2. In unisexual organisms: production of male and female gametes by different individuals (Lincoln et al. 1985).

monoecism, monecy, monoecy *n.*
1. An individual organism's having both male and female reproductive organs (Lincoln et al. 1985).
   adj. bisexual, hermaphrodite, hermaphroditic
   syn. bisexuality, hermaphrodism (Lincoln et al. 1985)
2. A plant species' having separate male and female flowers (unisexual) on the same individual (Fernald 1950, 1578; Lincoln et al. 1985).
   Note: "Hermaphroditic" refers to a plants' having both stamens and carpels in the same flower (Fernald 1950, 1576).
   syn. autoicy, ambisexuality (Lincoln et al. 1985)
   cf. -oecism: dioecism, trioecism
3. A plant species' having unisexual flowers of each sex on separate plants or in separate parts of the same inflorescence (Fernald 1950, 1573).
   adj. dioecious, dioecious, dioicious
   cf. bisexual; -oecism: monoecism

polygamodioecism *n.* A polygamous plant species' being mostly dioecious (having separate female and male flowers on different individuals or in different parts of the same inflorescence on a single individual) and partly monoecious (having separate male and female flowers [unisexual] on the same individual); contrasted with polygamomonoecism (Fernald 1950, 1580).

syn. polygamodioecious (Strausbaugh and Core 1978, 626)

polygamomonoecism *n.* A polygamous plant species' being partly dioecious (having separate female and male flowers on different individuals or in different parts of the same inflorescence on a single individual) and mostly monoecious (having separate male and female flowers [unisexual] on the same individual); contrasted with polygamodioecism (Fernald 1950, 1580).

syn. polygamo-monoecious (Strausbaugh and Core 1978, 626)

stenoecism, stenoecy *n.* An organism's being restricted to a narrow range of habitats and environmental conditions (given as an adjective in Lincoln et al. 1985).
adj. stenoecic, stenecious, stenoecious, stenoekous
cf. -oecism: amphioecious, euryoecious

synoecy *n.* An organism's producing both male and female gametes (Lincoln et al. 1985). See symbiosis: synoecy.
adj. synoecius

trimonoecism, coenomonoecism *n.* An individual plant's bearing male, female, and hermaphrodite flowers (Lincoln et al. 1985).

trioecism, trioecy, trioiky *n.* A plant species' having male, female, and hermaphrodite flowers on different individuals (Lincoln et al. 1985).
adj. trioikous
cf. -oecism: dioecism, monoecism

♦ **oecology** See study of: ecology.
♦ **oecoparasitism** See parasitism: ecoparasitism.
♦ **oecotrophobiosis** See trophallaxis.
♦ **oekospecies** See ²species: ecospecies.
♦ **oestrus** See estrus.
♦ **offensive threat** See behavior: threat behavior: offensive threat.
♦ **offspring** *n.* The progeny of an organism (Michaelis 1963)
syn. descendant; unit of direct fitness (West Eberhard 1975, 6)
♦ **offspring equivalent, unit of inclusive fitness** *n.* The fraction of genes that (1) a focal individual organism shares with a genetic relative (not its offspring) through common descent, and (2) is mathematically converted into a fraction of the genes that the focal individual shares with one of its own offspring (West Eberhard 1975, 6).
Comments: For example, a mother diplo-diploid organism shares one half of her genes with one of her offspring. She shares an average of one fourth of her genes with her niece or nephew through common descent.

n-r

Thus, on the average, her niece or nephew is equivalent to one half of her offspring; that is, it equals about one half of an offspring from a genetic perspective and can be counted as one half of an "offspring equivalent."

◆ **offspring imprinting** See learning: imprinting: offspring imprinting.

◆ **OGMP** See program: Organelle Genome Megasequencing Program.

◆ **oike** See habitat.

◆ **oikology** See study of: ecology.

◆ **oikos** See habitat.

◆ **oikosite** *n.* A commensal, or parasitic, organism that is attached to another organism (Lincoln et al. 1985).

◆ **oil droplets** *pl. n.* For example, in reptiles and birds: colored droplets of oil found in front of rods, or cones, in a retina which enhance color vision by biasing the wavelengths of light reaching a photoreceptor, thereby enabling an animal to resolve more colors than would otherwise be possible (D.B. Quine, personal communication).

◆ **old field** See habitat: field: old field.

◆ **Oldowan Stone Industry** See tool industry: Oldowan Stone Industry.

◆ **oleaginous** *adj.* Producing, or containing, oil (Lincoln et al. 1985).

◆ **olfaction** See modality: chemoreception: olfaction.

◆ **olfactory imprinting** See learning: imprinting: olfactory imprinting.

◆ **oligo-, olig-** *combining form* "Small; few; scanty" (Michaelis 1963).
[Greek *oligos*, few]

◆ **oligogene** See gene: oligogene.

◆ **oligogenic** See -genic: oligogenic.

◆ **oligogyny** See -gyny: oligogyny.

◆ **oligohaline** See haline: oligohaline.

◆ **oligolectic** See -lectic: oligolectic.

◆ **oligomerization, law of integration** *n.* An evolutionary trend involving a reduction in the number of segments of a body or structure (Lincoln et al. 1985).

◆ **oligonitrophil** See [1]-phile: oligonitrophile.

◆ **oligophage** See -phage: oligophage.

◆ **oligophile** See [2]-phile: oligophile.

◆ **oligopyrene sperm** See sperm: oligopyrene sperm.

◆ **oligotherm** See -therm: oligotherm.

◆ **oligotoky** See -toky: oligotoky.

◆ **oligotroph** See -phage: oligophage.

◆ **oligotypic** See -typic: oligotypic.

◆ **oligoxeny** See -xeny: oligoxeny.

◆ **oligyny** See -gyny: oligyny; mating system: oligyny.

◆ **olisthium** See [2]community: olisthium.

◆ **-ology** *combining form* "Discourse, study of" (Lincoln et al. 1985).
See study of.

◆ **ombrophile** See [1,2]-phile: ombrophile.

◆ **ombrotroph** See -troph-: ombrotroph.

◆ **omega, ω** *n.* The lowest ranking animal in a dominance hierarchy (Wilson 1975, 279).
cf. alpha, beta
[ω, the last letter of the Greek alphabet]

◆ **omni-** *combining form* "All, totally" (Michaelis 1963).
[Latin *omnis*, all]

◆ **omnicole** See -cole: omnicole.

◆ **omnispective method** See study of: systematics: omnispective systematics.

◆ **omnivore** See -vore: omnivore.

◆ **on-the-dot sampling** See sampling technique: instantaneous sampling.

◆ **onanism** *n.* A man's withdrawing his penis from a woman's vagina just prior to his ejaculation as a contraception practice (Storz 1973, 183).
adj. onanistic
n. onanist
syn. masturbation, self-abuse (*Oxford English Dictionary* 1972, entries from 1727; not preferred synonyms, Storz 1973, 183)
cf. coitus interruptus, masturbation
[after Onan, who violated Jewish law by "spilling his seed" rather than inseminating his brother's widow, *The Bible*, 38, 9]

◆ **oncogenic** See -genic: oncogenic.

◆ **one-gene–one-character hypothesis** See hypothesis: one-gene–one-character hypothesis.

◆ **one-gene–one-enzyme hypothesis** See hypothesis: one-gene–one-enzyme hypothesis.

◆ **one-gene–one-polypeptide hypothesis** See hypothesis: one-gene–one-polypeptide hypothesis.

◆ **one-male unit** *n.* For example, in the Anubis Baboon: the basic social element consisting of a mature male and the harem of females permanently associated with him (Devore, Hall in Wilson 1975, 534).
cf. group: band, harem

◆ **one-sided test, one-tailed test** See statistical test: one-tailed test.

◆ **one-zero sampling** See sampling technique: one-zero sampling.

◆ **onomatography** See -graphy: onomatography.

◆ **onto-, ont-** *combining form* "Being; existence" (Michaelis 1963).
[Greek *ōn, ōntos*, present participle of *einai*, to be]

◆ **ontocline** See cline: ontocline.

◆ **ontogenesis** See -genesis: ontogenesis.

◆ **ontogenesis of behavior** See -genesis: ontogenesis of genesis.

◆ **ontogenetic migration** See migration: ontogenetic migration.

◆ **ontogenetic ritualization**. See ritualization: ontogenetic ritualization.

♦ **ontogeny** See -genesis: ontogenesis.

♦ **ontogeny recapitulates phylogeny**
See -genesis: biogenesis.

♦ **ontology** See study of: philosophy: metaphysics: ontology.

♦ **oo-** *combining form* Egg; referring to eggs
(Michaelis 1963).
[Greek, *ōon*, egg]

♦ **ooapogamy** See parthenogenesis: ooapogamy.

♦ **oocyte** *n.* The cell that produces ova (eggs)
by meiosis (Lincoln et al. 1985).
cf. gamete, spermatocyte

♦ **oogenesis** See -genesis: oogenesis.

♦ **oology** See study of: oology.

♦ **oophage** See -phage: oophage.

♦ **oophagy** See -phagy: oophagy.

♦ **oophiophage** See -phage: oophiophage.

♦ **oosphere** See gamete: oosphere.

♦ **oospore** See gamete: oospore.

♦ **open anonymous group** See ²society:
open society: anonymous group.

♦ **open behavioral program** See program: open behavioral program.

♦ **open-brooding** See brooding: open-brooding.

♦ **open community** See ²community: open
community.

♦ **open field** *n.* A large open box, often made
of wood and painted flat gray and having a
floor marked off in equal-sized squares used
to study behavior such as exploration and
"emotionality" (Dewsbury 1978, 62–63).
Comment: After an animal is introduced into an
open field, data such as the number of squares
entered, number of fecal boli deposited,
frequency of sniffing, and other behaviors
can be recorded (Dewsbury 1978, 62–63).

♦ **open-field test** See test: open-field test.

♦ **open genetic program** See program:
genetic program: open genetic program.

♦ **open-group species** See ²species: open-group species.

♦ **open-mouth laughing** See facial expression: open-mouth laughing.

♦ **open pollination** See breeding: outbreeding.

♦ **open population** See ¹population: open
population.

♦ **operant behavior** See behavior: operant behavior.

♦ **operant conditioning** See learning:
conditioning: operant conditioning.

♦ **operational definition** See definition:
operational definition.

♦ **operational taxonomic unit (OTU)** *n.*
1. In numerical taxonomy: the unit of classification (Mayr 1982, 223).
2. In numerical taxonomy: any item, individual, or convenient group used for comparison or analysis (Lincoln et al. 1985).

♦ **operationalism** *n.* One's developing a
repeatable method for measuring, or quantifying, a concept (*e.g.,* foliage density,
human attitudes, or sweetness) (Diamond
1987).
v.t. operationalize
adj. operational

♦ **operator** *n.*
1. In the operon model of bacterial molecular genetics, a location on a gene
where a repressor substance binds and
thereby controls the functioning of adjacent cistrons (Wittenberger 1981, 619;
King and Stansfield 1985).
2. A target organism of a signal (*e.g.,* a
predator that is deceived by mimicry)
(Lincoln et al. 1985).
syn. signal receiver (Lincoln et al. 1985)

♦ **operator gene** See gene: operator gene.

♦ **operon** *n.* In bacteria: adjacent cistrons
that function together under an operator
gene's control (King and Stansfield 1985).

♦ **ophiciophobia** See phobia (table).

♦ **ophidiophobia** See phobia (table).

♦ **ophiology** See study of: ophiology.

♦ **ophiophage** See phage: ophiophage.

♦ **ophiophobia** See phobia (table).

♦ **ophiotoxicology** See study of: ophiotoxicology.

♦ **opium** See ²community: opium.

♦ **opophile** See ¹-phile: opophile.

♦ **opportunistic** *adj.* Referring to an organism that has the ability to exploit newly
available habitats or resources (Lincoln et
al. 1985).

♦ **opportunistic species** See ²species: opportunistic species.

♦ **opportunity for selection (*I*)** *n.* $V/\bar{w}^2$,
where *V* is a population's variance in fitness
and $\bar{w}$ is the population's mean fitness (Crow
1958 in Futuyma 1986, 527).
Comment: I represents only the opportunity for natural selection; selection can
actually occur only if *V* has a genetic
component (Futuyma 1986, 527). A high *I*
suggests a high potential for selection.

♦ **optimal** *adj.*
1. Referring to levels of environmental factors that are best suited for growth and
reproduction of an organism (Lincoln et
al. 1985).
syn. most favorable (Lincoln et al. 1985)
2. Referring to an organism's character that
may not be perfect but, under prevailing
conditions, helps to promote its fitness
(McFarland 1985, 427).
cf. optimality, pessimal

♦ **optimal breeding** See breeding: optimal outbreeding.

♦ **optimal decision making** See decision making: optimal decision making.

n–r

♦ **optimal-diet theory** See [3]theory: optimal-diet theory.

♦ **optimal-foraging theory (OFT)** See [3]theory: optimal-foraging theory (OFT).

♦ **optimal inbreeding** See inbreeding: optimal inbreeding.

♦ **optimal mix of castes** *n*. The ratio of castes that can achieve the maximum rate of production of virgin queens and males while a colony is at, or near, its maximum size (Wilson 1975, 305).

♦ **optimal outbreeding** See breeding: optimal outbreeding.

♦ **optimal-outbreeding hypothesis** See hypothesis: optimal-outbreeding hypothesis.

♦ **optimal yield** *n*. The highest rate of increase that a population can sustain in a given environment (Wilson 1975, 590).

♦ **optimality** *n*. An organism's obtaining as much of its required resources that it can or increasing its fitness as much as it can, given its environmental constraints.

cf. benefit; cost; design; [3]theory: optimality theory; utility

Comment: An organism's actual optimality is unlikely to be as great as its theoretical optimality.

long-term optimality *n*.

1. An organism's leaving as many viable offspring as it can in its lifetime [coined by Dawkins 1986, 21].
   syn. goal function (McFarland and Houston 1981 in Dewsbury 1978, 31)

2. A state that an animal species approaches when natural selection has been in operation for a certain length of time in the same environment (Maynard Smith 1978, Dawkins 1982 in Dawkins 1986, 23).

Comment: The linkage between "long-term optimality" and "short-term optimality" in organisms is variable. Extant species are optimal, subject to provisions such as time-lag effects (Maynard Smith 1978, Dawkins 1982 in Dawkins 1986, 23).

short-term optimality *n*. An organism's optimizing, *q.v.*, a function in its day-to-day life, such as the amount of energy that it collects per unit time [coined by Dawkins 1986, 21].
syn. goal function (McFarland and Houston 1981 in Dewsbury 1978, 31)

♦ **optimality theory** See [3]theory: optimality theory.

♦ **optimization principle of evolution** See compromise of evolution.

♦ **optimization theory** See [3]theory: optimality theory.

♦ **optimon** *n*. The unit of natural selection for whose benefit adaptation may be said to exist (Dawkins 1982, 291).

Comment: An optimon may be a gene, individual organism, or group of organisms (Dawkins 1982, 291).

♦ **optimum** *n*.

1. The value taken by an independent variable when a related dependent variable is maximized or minimized; *e.g.*, optimum values of fecundity, gamete size, or parental care may be paired with optimum fitness (Bell 1982, 510).

2. The conditions that are most favorable for growth and reproduction of an organism or maintenance of a system (Lincoln et al. 1985).

cf. optimal, pessimum

♦ **optimum-permissible trait** See character: optimum-permissible character.

♦ **optokinetic response** See reflex: optomotor reflex.

♦ **optomotor reflex** See reflex: optomotor reflex.

♦ **oral-scent gland** See gland: oral-scent gland.

♦ **oral trophallaxis** See trophallaxis: stomodeal trophallaxis.

♦ **orchard** See [2]community: orchard; habitat: orchard.

♦ **[1]order** See [2]group: A Taxonomic Classification of Groups (table).

♦ **[2]order** *n*.

1. Rank, grade, or class (*Oxford English Dictionary* 1972, entries from 1563).

2. Sequence, disposition, arrangement; arranged or regulated condition (*Oxford English Dictionary* 1972, entries from 1320).

3. Formal disposition or array; regular, methodological, or harmonious arrangement of the position of things in any space, or area, that compose any group or body (*Oxford English Dictionary* 1972, entries from 1374).

4. A method according to which things act or events occur; the fixed arrangement in the existing constitution of things; a natural, moral, or spiritual system in which things proceed according to definite laws; in phrases such as "order of nature, things, and the world; "moral order," "spiritual order," etc. (*Oxford English Dictionary* 1972, entries from 1340).

v.t. order

♦ **[3]order** *n*.

1. The degree of mathematical complexity in analytical form, geometrical form, an equation, expression, operator, or the like, as denoted by an ordinal number (first, second, third, ... , *n*th) (*Oxford English Dictionary* 1972, entries from 1706).

2. The number of rows and columns in mathematical matrices (Lincoln et al. 1985).

♦ ⁴**order** *n.* The relationship between behavioral events with regard to preceding ones.

See need: order.

v.t. order

Comment: Order is used to describe Markoff chains and transition matrices.

first order *n.* The situation in which the probability of a particular behavior's occurring is dependent on only the behavior that immediately precedes it (Lehner 1979, 273).

*n***th order** *n.* The situation in which the probability of a particular behavior's occurring is dependent on the *n* (= the specified number of) behaviors that precede it (Lehner 1979, 273).

second order *n.* The situation in which the probability of a particular behavior's occurring is dependent on only the two behaviors that immediately precede it (Lehner 1979, 273).

zeroth order *n.* The situation in which behavioral events are independent of one another; *i.e.,* the probability of a particular behavior's occurring is not related to any behaviors that precede it (Lehner 1979, 273).

♦ **ordered variable** See variable: ranked variable.

♦ **ordinal scale** See scales of measurement: ordinal scale.

♦ **ordinate** *n.* The vertical, or *y*-axis, of a graph (Lincoln et al. 1985).

cf. abscissa

♦ **ordination** *n.*

1. A mathematical process by which organism communities are ordered along a gradient (Lincoln et al. 1985).

2. A numerical-analysis method used for summarizing similarities between communities, or OTUs, by representing them as points in a multidimensional space in such a way that the interpoint distances are inversely related to their similarities (Lincoln et al. 1985).

♦ **orgadium** See ²community: orgadium.

♦ **orgadocole** See -cole: orgadocole.

♦ **orgadophile** See ¹-phile: orgadophile.

♦ **organ** *n.* An organism's functionally distinct structure composed of tissue, or tissues, that is part of one of its physiological systems (*e.g.,* brain, heart, liver, sense organ, or skin) (Immelmann and Beer 1989, 208).

cf. gland, organelle, spermatheca

Comments: Barnes (1974) describes many of the scores of kinds of organs in invertebrates; Chapman (1971) describes those in insects. Some of the many kinds of organs are defined below.

[Greek *organon,* tool]

brain *n.*

1. A convoluted mass of "nervous substance" contained in a vertebrate's skull; an analogous, but less developed, organ in an invertebrate (*Oxford English Dictionary* 1971, entries from 1000).

Notes: An adult human brain can store up to 100 trillion bits of information and have up to 100 billion neurons (Dowling 1997b, 60–63).

2. The principal ganglion of an invertebrate (Michaelis 1963).

cf. H.M. (Appendix 1), memory, system: memory system

Comments: Vertebrate brains contain centers for complex integration of homeostasis, movements, perceptions, and, in Humans, at least, emotions and intellect (Campbell 1996, 1012).

▸ **amygdala** *n.* In a vertebrate brain: part of the cerebrum that seems to be crucial in forming and triggering recall of a specific subclass of memories tied to strong emotions, especially fear (Lemonick 1999, 57).

▸ **basal ganglia** *n.* In a vertebrate brain: areas in a cerebrum which are important centers for motor coordination (Campbell 1996, 1016, illustration; 1017).

Comments: Degeneration of cells that enter the basal ganglia occurs in Parkinson's disease (Campbell 1996, 1017). In Huntington's disease, patients show destruction of their basal ganglia. They have perfectly functioning explicit memory, but they cannot learn new motor skills (Lemonick 1999, 56).

▸ **brain stem** *n.* In a vertebrate brain: the midbrain and the hindbrain (Campbell 1996, 1015, illustration; 1016).

▸ **cerebellum** *n.* In a vertebrate brain: part of a hindbrain that is a major center for coordinating movement (Campbell 1996, 1014).

▸ **corpus callosum** *n.* In a vertebrate brain: a thick band of fibers (cerebral white matter) that connects the two hemispheres of the cerebrum (Campbell 1996, 1016, illustration; 1017).

▸ **diencephalon** *n.* In a vertebrate brain: a hypothalamus and thalamus; contrasted with telencephalon (Campbell 1996, 1014).

▸ **forebrain** *n.* In a vertebrate brain: the telencephalon (which contains the cerebrum) and the diencephalon (which contains the hypothalamus and thalamus); contrasted with midbrain and hindbrain (Campbell 1996, 1014; 1015, illustration; 1016).

▸ **gray matter** *n.* In a vertebrate brain: the cerebral cortex, which is the outer gray area; contrasted with white matter (Campbell 1996, 1012).

n–r

SOME PARTS OF VERTEBRATE BRAINS

I. forebrain
  A. telencephalon
    1. cerebrum
      a. amygdala
      b. basal ganglia
      c. cerebral cortex
      d. corpus callosum
      e. frontal lobe
        (1) frontal association area
        (2) motor cortex
        (3) speech center
        (4) supplementary motor area
      f. hippocampus
      g. left cerebral hemisphere
      h. olfactory bulb
      i. occipital lobe
        (1) "vision center"
        (2) visual association area
      j. parietal lobe
        (1) "reading center"
        (2) somatosensory
            association area
        (3) "speech center"
        (4) "taste center"
      k. pituitary gland
      l. right cerebral hemisphere
      m. temporal lobe
  B. diencephalon
    1. hypothalamus
      a. hunger-regulating center
      b. pleasure center
      c. suprachiasmatic nucleus
      d. thirst-regulating center
    2. thalamus
      a. intralaminar nucleus
II. brain stem
  A. midbrain
    1. optic lobe
      a. inferior colliculus
      b. reticular formation
      c. superior colliculus
  B. hindbrain
    1. cerebellum
      a. inferior olive
    2. medulla oblongata
    3. pons

▸ **hindbrain** *n*. In a vertebrate brain: the cerebellum, medulla obligata, and pons; contrasted with forebrain and midbrain (Campbell 1996, 1014; 1015, illustration; 1016).

▸ **hypothalamus** *n*. In a vertebrate brain: the ventral part of a diencephalon (Campbell 1996, 1017).
Comments: A hypothalamus is a major center for integrating an animal's endocrine and nervous systems (Immelmann and Beer 1989, 138, 222). Hypothalamus functions include regulation of eating, drinking, reproduction, and general drive; hormone secretion by an individual's pituitary gland and, thus, indirect regulation of its gonadal, thyroid, and adrenocortical functions; and playing a role in the fight-or-flight response, pleasure, mating behavior, and sexual response (Campbell 1996, 1017). The hypothalamus secretes antidiuretic hormone and oxytocin. In mammals, the hypothalamus regulates body temperature. In Humans, the hypothalamus is also associated with pleasure, pain, and anger (Curtis 1983, 819).

▸ **inferior colliculus** *n*. In a vertebrate brain: an auditory center; contrasted with superior colliculus (Campbell 1996, 1017).

▸ **inferior olive** *n*. In a vertebrate brain: a small knot of cells in the cerebellum that acts as the pacemaker for cerebellum cells (discovered by Rodolfo Llinas, in Hilts 1997, C5).

▸ **intralaminar nucleus** *n*. In a vertebrate brain: a group of cells in a thalamus that might set the rhythm for conscious thinking (discovered by Rodolfo Llinas, in Hilts 1997, C5).

▸ **midbrain** *n*. In a vertebrate brain: the optic lobe; contrasted with forebrain and hindbrain (Campbell 1996, 1014; 1015, illustration).
Comments: The midbrain contains centers for the integration and receipt of several types of sensory information (Campbell 1996, 1017). It also sends coded sensory infromation along neurons to specific regions of the forebrain.

▸ **motor cortex** *n*. In a vertebrate brain: an area of the frontal lobe that sends commands to skeletal muscles, signaling appropriate responses to sensory stimuli (Campbell 1996, 1017; 1018, illustration).

▸ **reticular formation** *n*. In vertebrates: a functional neuron system localized in the core of the medulla, midbrain, and pons that regulates arousal states (Campbell 1996, 1017).
Comments: A reticular formation is in the central gray core of a brain (Immelmann and Beer 1989, 20). It receives input indiscriminately from all of the animal's sensory systems and projects this input's pooled activation, together with probable inherent excitation flux, as a background of neural facilitation to more specific areas of the animal's cerebellum, where it makes connections between stimuli and responses.

▸ **somatosensory cortex** *n*. In a vertebrate brain: an area of the parietal lobe that receives and partially integrates signals from pain, pressure, temperature, and touch receptors throughout a body (Campbell 1996, 1017; 1018, illustration).

▶ **superior colliculus** *n.* In a vertebrate brain: a visual center; contrasted with inferior colliculus (Campbell 1996, 1017).

▶ **supplementary motor area** *n.* In Humans: a part of a brain in the left frontal lobe that is related to laughter (Browne 1998, 1).

▶ **suprachiasmatic nucleus** *n.* In a vertebrate brain: part of a hypothalamus that functions as a biological clock (Campbell 1996, 1017).

▶ **telencephalon** *n.* In a vertebrate brain: a cerebrum; contrasted with diencephalon (Campbell 1996, 1014).

▶ **thalamus** *n.* In a vertebrate brain: part of a diencephalon which is a major relay station for sensory information on its way to a cerebrum (Campbell 1996, 1017).

▶ **ventromedial prefrontal cortex** *n.* In Humans: the area of a brain that might store information about past rewards and punishments and triggers the nonconscious emotional responses that normal people may register as intuition or a "hunch" (Antonio Damasio in Vogel 1997, 1269).
cf. intuition

▶ **white matter** *n.* In a vertebrate brain: axons in the inner region with myelin sheaths that give them a white appearance; contrasted with gray matter (Campbell 1996, 1012).

copulatory organ *n.* An organ that is directly involved in an animal's coition.
Comment: Only some kinds of copulatory organs are given below.

▶ **aedeagus** *n.* A male insect's entire copulatory organ; the distal part of a phallus; a penis plus parameres (Borror et al. 1989).
syn. intromittent organ, penis (Barnes 1974, 632), but in some species a penis is considered to be part of an aedeagus
Comment: A "phallus" is the intromittent organ in Cyclorrhaphan Flies and consists of a short "basiphallus" and a usually very elongate "distiphallus" (White et al. 2000, 896).

▶ **bursa copulatrix** *n., pl.* **bursae copulatrices** In many insect species: a female's pouch that receives an aedeagus during copulation (Borror et al. 1989, 792).
syn. genital pouch (in flies, Torre-Bueno 1978)

▶ **cloaca** See gland: cloacal gland.

▶ **hemipenis** *n.* In lizards and snakes: one of a male's two independent copulatory organs (Dewsbury 1978, 77; Tokarz 1988, 1518).

Comment: A male *Anolis sagrei* can use either of his hemipenes in copulation (Tokarz 1988, 1518); alternating the use of his hemipenes is a behavioral means of increasing sperm transfer (Tokarz and Slowinski 1990, 374).

▶ **intromittent organ** See organ: copulatory organ: aedeagus, penis.

▶ **penis** *n., pl.* **penes** For example, in the Tailed Frog; all flightless-bird, duck, mammal species; many arthropod species; some platyhelminth, gastropod, reptile species: a male's copulatory organ (*Oxford English Dictionary* 1972, entries from 1693; Barnes 1974; Wilson 1975, 443; Curtis 1983, 840).
syn. intromittent organ (in males and hermaphrodites of some species)
cf. organ: copulatory organ: aedeagus, vagina
[Latin, *penis* = *cauda*, tail; now as defined above]

▶ **vagina** *n.* In mammals: a female's muscular tube that leads from the cervix of her uterus to outside of her body, receives a penis during copulation, and acts as a birth canal (Curtis 1983, 842).

-derm *suffix* Skin.
[Greek *derma*, skin]

▶ **ectoderm** (EK toe durm) *n.* The outer of an animal embryo's three primary germ layers that develops into its sense organs, skin, and nervous system; contrasted with endoderm and mesoderm (Michaelis 1963).

▶ **endoderm** (EN doe durm) *n.* The inner of an animal embryo's three primary germ layers that develops into its digestive and respiratory systems; contrasted with ectoderm and mesoderm (Michaelis 1963).
syn. endoblast, entoblast, entoderm (Michaelis 1963)

▶ **mesoderm** (ME sew durm) *n.* The middle of the three primary germ layers in an animal embryo that develops into its skeletal and muscular systems, heart, kidneys, etc.; contrasted with ectoderm and endoderm (Michaelis 1963).

flower *n.*
1. A plant's simple to complex cluster of petals, usually brightly colored and with an outer envelope of green sepals enclosing reproductive parts (Michaelis 1963).
2. Any plant's reproductive structure, whether or not it is surrounded by a calyx and petals (Michaelis 1963).
3. An angiosperm structure with one or more of these parts: calyx, petals, stamens, staminodes, and carpels.
Note: Flowers are often structures that produce pollen, seeds, or both. They vary in

many features including color, numbers of parts, scent, shape, and size. Mutant flowers are sometimes nonreproductive ones, including horticultural varieties such as many kinds of double flowers.

[Old French *flour, flor* < Latin *flos, floris,* flower]

▸ **pollen flower** *n.* A flower with one or more anthers that produces pollen but no seeds (inferred from Farrar 1995)

syn. male flower

Comments: Pollen flowers are borne on plants with seed flowers or on plants with only pollen flowers, depending on the species. Pollen flowers can have nonfunctional carpels, as in Sugar Maples and other species (Farrar 1995, 134).

▸ **seed flower** *n.* A flower with one or more carpels that can produce seeds (inferred from Farrar 1995).

syn. female flower

Comments: Seed flowers are borne on plants with pollen flowers or on plants with only seed flowers, depending on the species. Seed flowers can have nonfunctional anthers, as in Sugar Maples and other species (Farrar 1995, 134).

Jacobson's organ *n.* In most vertebrate species: a chemoreceptor area used for detecting chemicals related to aggregation, courtship, feeding, and trailing conspecifics and prey in snakes; identification of prey in a scincid lizard; and pheromone detection in mammals (Halpern 1983 and Wysocki et al. 1986 in Graves and Halpern 1990, 692; Blakeslee 1993, C3).

syn. vomeronasal organ (Immelmann and Beer 1989, 107), VNO (Weiss 1998, A9)

cf. flehmen

Comments: For example, in Indian Elephants, ducts that lead to the Jacobson's organ are in the roof of an individual's mouth (Schulte and Rasmussen 1999, 1269). In Humans, this organ is a pair of tiny pits in a person's nose, and researchers are uncertain whether or not this organ is functional (Weiss 1998, A9).

nasus *n.* In some nasutitermitine-termite species: a snout-like organ of soldiers used to eject poisonous, or sticky, fluid at intruders (Wilson 1975, 589).

[Latin *nasus,* nose]

notochord *n.* In Chordata: a flexible rod of cells (Curtis 1983, 553–556; Bucksbaum et al. 1987, chap. 22).

Comments: A notochord occurs in the the dorsal area of a hemichordate, urochordate, cephalochordate, and vertebrate embryo; in hemichorate, urochordate, and cephalochordate larvae; and in cephalochordate adults (Curtis 1983, 553–556;

Bucksbaum et al. 1987, chap. 22). This organ is composed of large, fluid-filled cells encased in fairly stiff, fibrous tissue, and it provides skeletal support (Campbell et al. 1999, 630). In most vertebrate species, an adult has a complex, jointed skeleton and only remnants of its embryonic notochord as the gelatinous material of disks between vertebrae.

[Greek *noto,* back; Latin *chorda,* gut, cord]

lophophore *n.* In Phyla Brachiopoda, Bryozoa, Entoprocta, Phoronida: the food-catching organ that is a circular or horseshoe-shaped fold that encircles the mouth and bears numerous ciliated tentacles (Barnes 1974, 693, illustration).

Comment: The food-catching organ of Pterobranchia (phylum Hemichordata) resembles a lophophore (Barnes 1974, 805).

pharyngeal gill slits *n.* In Acorn Worms (Hemichordata: Enteropneusta): one of a group of openings in the branchial region of an individual that may function in gas exchange.

Comments: Hemichordate and chordate gill slits probably evolved originally as feeding mechanisms in which the animals strained small particles out of the water passing through their pharyngeal clefts (Barnes 1974, 801–802). Urochordates and cephalochordates still use gill slits for feeding.

placenta *n.* In Eutherian mammals: an organ that provides for fetal anchorage, fetal nourishment, and elimination of the fetal waste products and produces progestins; formed by the union of a fetus' and its mother's membranes (Wilson 1975, 591; Immelmann and Beer 1989, 223).

pollinium *n., pl.* **pollinia** In some milkweed and many orchid species: a specialized pollen mass that can attach to a pollinator by its sticky region (Lawrence 1951, 434; 674; 766, illustrations).

relict, relict organ, evolutionary vestige *n.* An organ that has become functionless (vestigial) yet persists, usually in a fragmentary or greatly reduced form (*e.g.,* pelvic girdles of some snake and whale species and embryonic gill slits in many terrestrial-vertebrate species) (Immelmann and Beer 1989, 252).

cf. behavior: relict behavior

sense organ *n.* An organ that is specialized for receiving sensory stimulation and includes an arrangement of sense cells, nerve cells, and supporting tissue (Immelmann and Beer 1989, 263).

cf. [2]receptor

spermatheca *n., pl.* **spermathecae** In females of many arthropod species: a small pouch attached to a female's common oviduct that stores sperm after mating (*Oxford English Dictionary* 1972, entries from 1826).

Comments: Fruit Flies (Tephritidae) have two, three, or four spermathecae per individual, depending on the taxon (White et al. 2000, 901). In the Honey Bee, *Apis mellifera*, viable sperm are stored in a queen's spermatheca for up to 4 years.

[Greek *sperma*, seed + *theke*, box, coffin, vault]

target organ *n.* An organ upon which a hormone acts and has one or more of these effects: a temporary or lasting change in an animal's ontogenesis, seasonal fluctuation, and sensory, neural, or glandular activity (Immelmann and Beer 1989, 306).

vomeronasal organ See organ: Jacobson's organ.

♦ **organelle** *n.*
1. A cell's complex cytoplasmic structure of particular morphology and function (King and Stansfield 1985).
   Note: This definition is incomplete because some organelles have a simple cytoplasmic structure, some have many variations on a basic structural theme, and some have more than one function (G.B. Chapman, personal communication).
2. Any of a variety of cytoplasmic structures that actively modify the components of a cell, *e.g.,* by participating in the synthesis of a cellular product or structure, its assembly, or both (G.B. Chapman, personal communication).

cf. organ, gland

Comments: Campbell (1987, G-14) indicates that only eukaryotes have organelles; however, if one classifies chromatophores and thylakoids as organelles, organelles are also found in prokaryotes (G.B. Chapman, personal communication). Other kinds of organelles are chloroplasts, Golgi complexes, lysosomes, microbodies, mitochondria, nuclei, plastids, and ribosomes (King and Stansfield 1985; G.B. Chapman, personal communication).

[Greek *organon*, instrument, tool]

amyloplast, leucoplast *n.* A colorless plastid that stores starch, particularly in roots and tubers (Campbell 1996, 135).

chloroplast *n.* A plastid that contains accessory pigments, chlorophyll, enzymes, and other molecules and functions in photosynthesis (Campbell 1996, 135–136, illustration).

Comments: Chloroplasts occur in Algae and Plantae (Campbell 1996, 136). This organelle is comprised of inner and outer membranes, stroma, and thylakoids stacked into grana and is about 2 by 5 micrometers.

chromoplast *n.* A plastid that contains pigments and gives flowers and fruits their orange and yellow colors (Campbell 1996, 135).

hydrogenosome *n.* In Archezoa (*e.g., Plagiopyla* spp.): an organelle that generates ATP anaerobically and produces hydrogen as a reduced end-product of its energy metabolism (Gray et al. 1998, 1479).

Comment: Hydrogenosomes and mitochondria may have a common evolutionary origin (Gray et al. 1998, 1480; Vogel 1998, 1633).

mitochondrion *n., pl.* **mitochondria** In Eukaryotes: an organelle that functions in respiration (Campbell 1996, 126).

syn. chrondriosome (Michaelis 1963)

cf. nucleic acid: deoxyribonucleic acid: mitochondrial DNA

Comments: The energy-producing Krebs cycle and electron transport occur in a mitochondrion's matrix and inner membrane, respectively (Campbell 1996, 170). *Anaplasma, Ehrlichia,* and *Rickettsia* may be the very close eubacterial relatives of mitochondria (Gray et al. 1998, 1476).

[New Latin < Greek *mitos*, thread + *chondros*, cartilage, granule]

nucleus *n.* In Eukaryans: the region of a eukaryotic cell that is enclosed by a double membrane and contains one or more chromosomes (Michaelis 1963).

[Contraction of Latin *nuclueus*, diminutive of *nux, nucis*, nut]

plastid *n.* Any of the autonomous, covalently closed, circular (or rarely linear), double-stranded, and self-replicating DNA molecules found in most bacterial species and some eukaryan species (Rieger et al. 1991).

Comment: Plastids found in plants include chloroplasts, chromoplasts, elaioplasts, and leucoplasts (King and Stansfield 1985).

♦ **Organelle Genome Database** See database: Organelle Genome Database.

♦ **Organelle Genome Megasequencing Program** See program: Organelle Genome Megasequencing Program.

♦ **organic** *adj.* Pertaining to, or derived from, living organisms or to compounds containing carbon as an essential component (Lincoln et al. 1985).

cf. inorganic

[Greek *organon*, instrument, tool]

♦ **organic selection** See effect: Baldwin effect.

♦ **organicism** *n.*
1. "The doctrine that organic structure is merely the result of an inherent property in matter to adapt itself to circumstances" (*Oxford English Dictionary* 1972, entry from 1883).
2. The concept that living processes are the result of the activity of all organs

considered as an autonomous, integrated system (Michaelis 1963).

3. Sociologically, the concept that a society is an organism (Michaelis 1963).

n. organicist

♦ **organism** See Appendix 1.

♦ **organism behavior** See behavior: organism behavior.

♦ **organismal theory, organismal theory of multicellularity** See ³theory: organismal theory.

♦ **organismic theory** See ³theory: organismic theory.

♦ **organized society** See ²society: organized society.

♦ **organized structure** *n.* A naturally occurring, or artificial, group of molecules with particular relative positions due to forces such as hydrogen bonding, ionization, solubility, adhesion, and surface tension (Strickberger 1990, 117).

cf. cell: protocell

Comment: Organized structures might be predecessors of protocells, *q.v.* (Strickberger 1990, 117).

"clay aggregation" *n.* An organized structure that forms on clay (Cairns-Smith in Nash 1993, 73).

Comments: This organized structure has positively charged nucleic acids associated with negatively charged proteins associated with positively charged clay.

coacervate *n.* An organized structure that is a droplet formed from dispersed colloidal particles that separate spontaneously from solution because of special conditions such as acidity and temperature; contrasted with membranous droplet and microsphere (Oparin 1924 in Strickberger 1990, 118).

Comments: Coacervates, depending on the kinds, have certain properties, including a simple but persistent organization, stability in solution, and size increase (Strickberger 1990, 119).

"iron-pyrite aggregation" *n.* An organized structure that forms on iron pyrite (Wächtershäuser in Nash 1993, 74).

Comments: This organized structure has positively charged nucleic acids associated with negatively charged proteins associated with positively charged iron pyrite.

liposome *n.* An organized structure that is a sphere, the surface of which is made of a double layer of lipid molecules (Zimmer 1995c, 76, illustration).

membranous droplet, vesicle *n.* An organized structure composed of lipids, polypeptides, or other molecules, or a combination of these molecules, and produced by the mechanical agitation of molecular films on liquid surfaces or even spontane-

ously; contrasted with coacervate and microsphere (Deamer 1986 in Strickberger 1990, 117).

Comment: Membranous droplets allow certain processes to occur that are necessary for life.

microsphere *n.* A sphere formed from thermally produced proteinoids that are boiled in water and allowed to cool; contrasted with membranous droplet and coacervate (Fox et al. 1974 in Strickberger 1990, 119).

Comment: Microspheres have certain characteristics that suggest they evolved into protocells (Strickberger 1990, 119).

quasi-cell *n.* A spheroidal organized structure comprised of a membrane of a double layer of lipids (= a liposome) with RNA inside (David Deamer et al. in Zimmer 1995c, 78, illustration).

♦ **organogenesis** See -genesis: organogenesis.

♦ **organology** See study of: organology.

♦ **organophyly** *n.* The phylogeny of organs and structures (Lincoln et al. 1985).

♦ **organotroph** See -troph-: organotroph.

♦ **orgasm** *n.*

1. A person's immoderate, or violent, excitement of feeling; rage, fury; a paroxysm of excitement or rage (*Oxford English Dictionary* 1972, entries from 1646).

2. An animal's excitement or violent action in an organ or part, accompanied with turgescence; specifically, the height of venereal excitement in coition of some animal species (*Oxford English Dictionary* 1972, entries from 1684).

Comments: In Humans, "male sexual orgasm" involves muscle contractions that propel sperm out through a man's urethra and associated sensations (Curtis 1983, 841). "Female sexual orgasm" involves the dropping of a woman's cervix down into the upper portion of her vagina, where semen tends to form a pool; this orgasm may also produce contractions in her oviducts that propel sperm upward and associated sensations (Curtis 1983, 847). About 30% of women typically experience orgasms from intercourse alone (Hrdy 1999, 851). In Primates, "female sexual orgasm" also occurs in the Stump-Tailed Monkey and Gorillas (Smith 1984, 621; de Waal 1989, 152).

[Latin *orgasmus*, to swell as with moisture, be excited or eager]

♦ **-orial**

cursorial, running *adj.* Referring to a trait, or animal, that is adapted for running (Lincoln et al. 1985).

fossorial, digging *adj.* Referring to a trait, or animal, that is adapted for digging (Lincoln et al. 1985).

grallitorial *adj.* Referring to a trait, or animal, that is adapted for wading (Lincoln et al. 1985).

insessorial *adj.* Referring to a trait, or animal, that is adapted for perching (Lincoln et al. 1985).

natatorial, natant, natatorious, natatory *adj.* Referring to a trait, or animal, that is adapted for swimming (Lincoln et al. 1985).

raptorial, raptatory, preying *adj.* Referring to a trait, or animal, that is adapted for seizing prey (Lincoln et al. 1985).

rasorial *adj.* Referring to a trait, or animal, that is adapted for scratching, or scraping, the ground (Lincoln et al. 1985).

repugnatorial *adj.* Referring to a trait, or animal, that is adapted for defense or offense (Lincoln et al. 1985).

saltatorial, saltatory *adj.* Referring to a trait, or animal, that is adapted for leaping or bounding locomotion (Lincoln et al. 1985).

scansorial *adj.* Referring to a trait, or animal, that is adapted for climbing (Lincoln et al. 1985).

♦ **Orians effect** See effect: Orians effect.

♦ **Orians-Verner model of polygyny** See [4]model: Orians-Verner model of polygyny.

♦ **orientating reflex, orientating response** See response: orientating response.

♦ **orientation** *n.* An organism's ability to direct its body position and locomotion with respect to object locations and environmental forces (Immelmann and Beer 1989, 208). See taxis.

♦ **orientation flight** See flight: orientation flight.

♦ **orientation response** See response: orientation response.

♦ **origin of insect wings, hypotheses of** See hypothesis: hypotheses of the origin of insect wings.

♦ **original** See -morphic: protomorphic.

♦ **ornithic** *adj.* Referring to birds.

♦ **ornithichite** See fossil: ornithichite.

♦ **ornithischian pelvis** See pelvis: ornithichian pelvis.

♦ **ornithogenic** See -genic: ornithogenic.

♦ **ornithology** See study of: ornithology.

♦ **ornithophile** See [2]-phile: ornithophile.

♦ **ornithophobia** See phobia (table).

♦ **orogenesis, orogeny** See -genesis: orogenesis.

♦ **orohylile community** See [2]community: orohylile community.

♦ **orophile** See [1]-phile: orophile.

♦ **orophytium community** See [2]community: orophytium community.

♦ **orothamnic community** See [2]community: orothamnic community.

♦ **ortet** *n.* The original organism from which a clone originates (Lincoln et al. 1985).
cf. genet, ramet

♦ **ortho-** *combining form*
1. "Straight; upright; in line" (Michaelis 1963).
2. "At right angles; perpendicular" (Michaelis 1963).
3. "Correct; proper; right" (Michaelis 1963).
[Greek *orthos*, straight]

♦ **orthogamy** See -gamy: orthogamy.

♦ **orthogenesis** See -genesis: orthogenesis.

♦ **orthographic error** See error: orthographic error.

♦ **orthography** See -graphy: orthography.

♦ **orthokinesis** See kinesis: orthokinesis.

♦ **orthologous** See -logous: orthologous.

♦ **orthoselection** See selection: orthoselection.

♦ **orthotopic** *adj.* Referring to an organism in its normal habitat (Lincoln et al. 1985).

♦ **ortstreue, *Ortstreue*** *n.*
1. In some mammal species (including bats and the Gray Whale); many migratory-bird and fish species, the monarch butterfly: an individual's tendency to return to a place used by its ancestors for reproducing, feeding, or resting where it performs the same activity (Wilson 1975, 1968).
2. An animal's propensity to return to its birth place, first-breeding place, or any other previously occupied place (Immelmann and Beer 1989, 104).
syn. site fidelity (C.K. Starr, personal communication)
cf. homing
[German *Ort*, site + S + *treue*, faithful, loyal, constant]

♦ **oryctocoenosis** See [2]community: oryctocoenosis.

♦ **oscillating big-bang hypothesis** See hypothesis: hypotheses of the beginning of the universe: oscillating big-bang hypothesis.

♦ **osmallaxis** *n.* Exchange of odor between (among) members of a species [coined by Müller-Schwarze (1974, 319)].
cf. trophallaxis

accidental osmallaxis *n.* Osmallaxis that results from interactions of conspecific animals that are not intentional social marking [coined by Müller-Schwarze (1974, 319)].

stereotyped osmallaxis *n.* For example, the Dall's Sheep, Sugar Glider, Vicuña; rats; rabbits: osmallaxis performed as social communication [coined by Müller-Schwarze (1974, 319)].

♦ **osmoconformer** *n.* An organism whose body fluid is of the same osmotic concentration as its surrounding medium (Lincoln et al. 1985).
cf. osmoregulator

n–r

♦ **osmophile** See [1]-phile: osmophile.

♦ **osmoreception** See modality: chemoreception: osmoreception.

♦ **osmoregulator** *n.* An organism that maintains its body osmotic concentration at a level independent of its surrounding medium (Lincoln et al. 1985).
cf. osmoconformer

♦ **osmotaxis** See taxis: osmotaxis.

♦ **osmotroph** See -troph-: osmotroph.

♦ **osphresiology** See study of: osphresiology.

♦ **ostentation** See [2]group: ostentation.

♦ **osteology** See study of: osteology.

♦ **OTU** See operational taxonomic unit.

♦ **outbreak, outbreak year** See wasp year.

♦ **outbreeding, outcrossing** See breeding: outbreeding.

♦ **outflow theory** See [3]theory: outflow theory.

♦ **outlaw gene** See gene: outlaw gene.

♦ **outlaw sperm** See sperm: outlaw sperm.

♦ **ova** See gamete: ovum.

♦ **ovariectomy** See castration: ovariectomy.

♦ **ovariicole** See -cole: ovariicole.

♦ **ovariole** *n.* One of a group of egg tubes that, together, form the ovary of a female insect (Wilson 1975, 590).

♦ **ovary** See gonad: ovary.

♦ **overdispersed distribution** See distribution: overdispersed distribution.

♦ **overdispersion** See distribution: overdispersed distribution.

♦ **overdominance** See dominance (genetic): overdominance.

♦ **overflow activity** See activity: vacuum activity.

♦ **overgrazing** *n.* Grazing that exceeds a community's recovery capacity and thus reduces its available forage crop or causes undesirable changes in its composition (Lincoln et al. 1985).

♦ **overlap promiscuity** See mating system: polybrachygamy: overlap promiscuity.

♦ **overreactive startle** See startle: overreactive startle.

♦ **oversampling** See sampling: oversampling.

♦ **overstepping** *n.* A descendent organism's development through the ontogenetic stages and beyond the final adult stage of its ancestor (Lincoln et al. 1985).

♦ **ovicide** See -cide: ovicide.

♦ **ovigerous** See gravid.

♦ **oviparity** See parity: oviparity.

♦ **oviposition** *n.* In many animal species: the act, or process, of laying eggs (Lincoln et al. 1985).
v.t. oviposit

♦ **ovipositional stimulant** See chemical-releasing stimulus: ovipositional stimulant.

♦ **ovism** See [3]theory: encasement theory.

♦ **ovogenesis** See -genesis: oogenesis.

♦ **ovovivipary** See -parity: ovoviviparity.

♦ **-ovular** *combining form* Referring to eggs.
  monovular *adj.* Referring to an individual animal that produces a single egg and usually, unless twinning takes place, a single offspring during each of its reproductive cycles (Eisenberg 1981, 507).
  polyovular *adj.* Referring to an individual animal that produces several eggs during each of its reproductive cycles (Eisenberg 1981, 507).

♦ **ovulation** *n.*
  1. In female mammals: formation and development of ovules, or ova, and especially their discharge from an ovary (*Oxford English Dictionary* 1972, entries from 1853).
  2. In female mammals: an individual's release of a ripe ovum from its egg follicle (= enveloping capsule) (Immelmann and Beer 1989, 210).
  Comment: "Ovulation" is rarely used for development and laying eggs in oviparous animals (*Oxford English Dictionary* 1972).

♦ **ovulator** *n.* An animal in which ovulation occurs.
  induced ovulator *n.* In some mammal species including shrews, rabbits, ground squirrels, mink, ferrets: a species in which follicles rupture and release ova after a female has copulated (Immelmann and Beer 1989, 210).
  cf. ovulator: spontaneous ovulator
  spontaneous ovulator, automatic ovulator *n.* In the majority of mammal species, including cats, primates, and rats: a species in which follicles rupture and release ova spontaneously as part the species' ovarian cycle (Immelmann and Beer 1989, 210).
  cf. ovulator: induced ovulator

♦ **ovum** See gamete: egg.

♦ **oxidation** *n.*
  1. In chemistry, something's combining with oxygen (Michaelis 1963).
  2. In chemistry, an atom's, or molecule's, losing, or being deprived of, valence elections (Michaelis 1963);
  3. In chemistry, a compound's gain of oxygen, loss of hydrogen, or loss of electrons (King and Stansfield 1985).
  4. In chemistry, a compound's, or atom's, partial or complete loss of electrons to oxygen or some other substance (Campbell 1987, 184).
  cf. reduction
  Comments: Oxidation can also be the state resulting from these processes. Oxidation increases the valence of an atom or group of

atoms by the loss of elections (Michaelis 1963). Ferrous iron is oxidized to ferric iron. An oxidized atom gives elections (or takes protons).

♦ **oxidizing agent** See agent: oxidizing agent.

♦ **oxodic community, oxodium** See ²community: oxodium.

♦ **oxy-** *combining form* Oxygen; of, or containing, oxygen or one of its compounds (Michaelis 1963).

♦ **oxybiont** See biont: oxybiont.

♦ **oxybiotic, oxybiontic**   See aerobic.

♦ **oxygenotaxis** See taxis: oxygenotaxis.

♦ **oxygeophile** See ¹-phile: oxygeophile.

♦ **oxylium** See ²community: oxodium.

♦ **oxylophile, oxylyphile** See ¹-phile: oxylophile.

♦ **oxyphile** See ¹-phile: oxyphile.

♦ **oxyphobe** See -phobe: oxyphobe.

♦ **oxytaxis** See taxis: oxytaxis.

♦ **oxytocin** See hormone: oxytocin.

♦ **ozone** See molecule: ozone.

♦ **ozone hole** *n.* An area of depleted ozone in the Earth's ozone layer which is mostly in the stratosphere, 15 to 50 kilometers above the Earth's surface (Rensberger 1993a, A19).

cf. molecule: ozone

Comments: The protective ozone layer absorbs most of the ultraviolet light that comes from the Sun (Rensberger 1993a, A18–19). The ozone layer over Antarctica tends to be completely depleted in a 1- to 2-kilometer-thick layer centered 17 kilometers above the Earth. Evidence of the first ozone hole was discovered in the 1970s. Human-made chlorofluorocarbons (CFCs), methyl chloroform, and other chemicals produce ozone holes. In 1995, CFCs were responsible for 75% of the damage to the ozone layer; methyl chloroform, 12.5%; and other chemicals, 12.5% (Cheng 1995, 16 July, 21). In 1995, due to the Montreal Protocol, methyl chloroform decreased in Earth's atmosphere. Sulfuric-acid particles from volcanoes might also cause ozone depletion. Human-made ozone holes are expected to disappear by 2050 due to reduced use of hole-producing chemicals (Rensberger 1993a, A18).

♦ **ozone layer** *n.* A concentration of ozone in Earth's stratosphere at from about 15 to 30 miles altitude; distinguished from ozone near ground level (Gardener 1992, 41; Rensberger 1993a, A18).

cf. molecule: ozone; ozone hole

Comment: The ozone concentration is about 1 molecule per 100,000 molecules of air at 22 miles altitude. The amount of ozone naturally varies from place to place in the ozone layer. This protective shield reduces that amount of deleterious ultraviolet-B that impinges on organisms. Without this process, only a few kinds of organisms could live on Earth.

♦ **ozonosphere** See -sphere: stratosphere: ozonosphere.

n–r

♦ **p, P, *p*, *P*** *n*. In statistics, a probability.
Comment: There are four main kinds of p's that relate to experimental design. Statisticians and others often call each of them simply "p," which can certainly be confusing.

"data-set-choice p" *n*. A person's probability of drawing a particular set of data from a population with random, independent data.
Comment: For example, if a person randomly picks data from a population of 1,000,000, with 100,000 individuals who are Smiths, and gets 10 Smiths in a row, then the probability of obtaining the "data-set-choice p" is 0.1^{10} or 0.000,000,000,1.

"datum-choice p" *n*. A person's probability of drawing a particular datum, or kind of datum, from a population of random, independent data.
Comments: For example, in a population of 1,000,000, if 100,000 individuals are Smiths, then the probability ("datum-choice p") of drawing one Smith from the population is 100,000/1,000,000, or 0.1. The probability of drawing a non-Smith is q, which is 1 – p, or 0.9.

"found p," "calculated p" *n*. A person's probability of obtaining a particular sample selected from a population of random, independent data plus the probabilities of obtaining all rarer samples from this population if the relevant null hypothesis were true (Siegel 1956, 11).
Comment: The "found p" is the sum of the "data-set-choice p" for the selected sample and the "data-set-choice p's" for all rarer samples from its population.

"stipulated p" *n*. The probability value at which a person is willing to reject his *true* null hypothesis.
syn. α, alpha, alpha level, p level, probability level, significance level (Siegel 1956, 8)

Comment: "Stipulated p" is often set at 0.05 by statisticians and biologists. If he uses 0.05, a person is willing to make a type-1 error 5% of the time; that is, he is willing to be wrong (*i.e.*, reject a true null hypothesis) 5 times out of 100. If "found p" ≤ "stipulated p," a person rejects his null hypothesis.

♦ **P element** See transposable element: P element.

♦ **p level** See p: "stipulated p."

♦ **pace** See ²group: pace.

♦ **pachy-** *combining form* "Thick, massive" (Michaelis 1963).
[Greek *pachys*, thick]

♦ **pachytene** *n*. The state during prophase of meiosis during which homologous chromosomes are completely paired (Mayr 1982, 775, 959).

♦ **pacing** See locomotion: pacing.

♦ **pack** See ²group: pack.

♦ **pack dating** See dating: pack dating.

♦ **-paedic** *adj*. Referring to a child (Brown 1956)
[Greek *pais, paidos*, child]

psilopaedic *adj*. Referring to a bird that is featherless when it hatches (Lincoln et al. 1985).

ptilopaedic *adj*. Referring to a bird that is covered with down when it hatches (Lincoln et al. 1985).

♦ **paedium** *n*. "An assemblage of young animals" (Lincoln et al. 1985).

♦ **-paedium**

gynopaedium *n*. A family group whose mother remains with her offspring for some time (Lincoln et al. 1985).

monogynopaedium *n*. An assemblage of one mother and her immediate progeny (Lincoln et al. 1985).

patrogynopaedium *n*. A family group in which both parents remain with their offspring for some time (Lincoln et al. 1985).

patropaedium *n*. A family group whose father remains with his offspring for some time (Lincoln et al. 1985).

polygynopaedium *n*.
1. An association of mothers and daughters, each of which is reproducing parthenogenetically (Deegener 1918).
2. A family group in which both parents remain with their offspring for some time (Lincoln et al. 1985).

sympaedium *n*. An assemblage of young animals that are playing together (Lincoln et al. 1985).

synchronopaedium *n*. A group of young animals of about the same age but of different parents (Lincoln et al. 1985).

♦ **paedogamous autogamy** See -gamy: autogamy: paedogamous autogamy.

♦ **paedogamy** See -gamy: paedogamy.

♦ **paedogenesis** See -genesis: paedogenesis.

♦ **paedomorphosis** See ²heterochrony: paedomorphosis.

♦ **paedoparthenogenesis** See parthenogenesis: paedoparthenogenesis.

♦ **paedophage** See -phage: paedophage.

♦ **paedophile** See -¹phile: pedophile.

♦ **paedophilia, paedophilia erotica** See -philia: pedophilia.

♦ **pagophile** See -¹phile: pagophile.

♦ **PAH** See molecule: polycyclic aromatic hydrocarbon.

♦ **pain** *n*.
1. A person's primary condition of sensation, or consciousness, the opposite of pleasure; his sensation when feeling hurt in body or mind (*Oxford English Dictionary* 1972, entries from 1300).
   syn. distress, suffering (*Oxford English Dictionary* 1972)
2. A person's bodily suffering; a distressing sensation such as soreness, usually of a part of his body (*Oxford English Dictionary* 1972, entries from 1377).
3. A nonhuman animal's "severe, perhaps transient state, recognizable by positive signs such as screaming and struggling" (Dawkins 1980, 25).
   Note: It is controversial whether nonhuman animals feel pain (Bateson 1991, 828).
4. Human "unpleasant sensory or emotional experience associated with actual or potential tissue damage, or described in terms of such damage" (The International Association for the Study of Pain and Iggo, 1984 in Bateson 1991, 828).
cf. animal suffering; general emergency reaction; stress; syndrome: fight syndrome, flight syndrome, general adaptation syndrome
v.t. pain

Comment: Some authors consider "pain" to be part of "suffering," not synonymous with it. Bateson (1991, 827) separates "stress" and "anxiety" from pain. He discusses difficulties of assessing pain in Humans and other animals and the possibility of invertebrates such as octopods' and insects' feeling pain (Wigglesworth 1980, 8).

♦ **paint-pot theory of heredity** See inheritance: blending inheritance.

♦ ¹**pair** *n*. Two associated things; a set of two things (*Oxford English Dictionary* 1972, entries from 1278).

consort pair *n*. In some primate species: a male and female with a consort relationship, *q.v.* (Immelmann and Beer 1989, 58).

contact pair *n*. In some fish species: any two fish that have established, or are in the process of establishing, a dominance-subordination relationship (Braddock 1945, 178).

♦ ²**pair** *v.i.*
1. In Humans: to unite in love, marriage, or both (*Oxford English Dictionary* 1972, entries from 1673).
2. In animals: to mate or copulate, *q.v.* (*Oxford English Dictionary* 1972, entries from 1702; Immelmann and Beer 1989, 212).
3. In animals: to show pair formation, *q.v.* (Immelmann and Beer 1989, 212).
4. In animals: to form a pair bond, *q.v.* (Immelmann and Beer 1989, 212).

♦ **pair bond, pair bonding** See bond: pair bond, marriage.

♦ **pair duet** See animal sounds: song: duet: pair duet.

♦ **pair formation** *n*. For example, in many duck and geese species; Humans: behavioral interactions (*e.g.,* courtship, mating, threats, appeasement, infantile behavior) of potential mates that leads to their pair formation (Heymer 1977, 127; Immelmann and Beer 1989, 212).
See pair bonding.
syn. mating foreplay (Heymer 1977, 127)
cf. bonding, courtship

♦ **paired limiting factor** See factor: paired limiting factor.

♦ **pairing, pairing behavior** See copulation.

♦ **pairing bite** See bite: neck bite.

♦ **pairing territory** See territory: pairing territory.

♦ **pairwise coevolution** See ²evolution: coevolution: pairwise coevolution.

♦ **pairwise comparison** See statistical test: multiple-comparisons test: planned-multiple-comparisons procedure: pairwise comparison.

♦ **paka-paka** *n.* Alternate flashing of differently colored lights that causes tension in Humans (WuDunn 1997, A3).
cf. ²stimulus: photostimulation
Comments: Paka-paka with light from 10 through 30 hertz can cause human seizures (WuDunn 1997, A3).
[Japanese *paka-paka* (WuDunn 1997, A3)]

♦ **palaeo-, palae-, paleo-, pale-** *combining form*
1. "Ancient, old" (Michaelis 1963).
2. "Primitive" (Michaelis 1963).
[Greek *palaios*, old, ancient]

♦ **palaeoagrostology, paleoagrostology** See study of: paleoagrostology.

♦ **palaeoalgology, paleoalgology** See study of: paleoalgology,

♦ **palaeoautecology, paleoautecology** See study of: ecology: autoecology: paleoautecology.

♦ **palaeobiocoenosis** See ²community: paleobiocoenosis.

♦ **palaeobiogeography, paleobiogeography** See study of: geography: paleobiogeography.

♦ **palaeobiology, paleobiology** See study of: biology: paleobiology.

♦ **palaeobotany, paleobotany** See study of: botany: paleobotany.

♦ **paleoclimatology, palaeoclimatology** See study of: climatology: paleoclimatology.

♦ **palaeocommunity, paleocommunity** See ²community: paleocommunity.

♦ **palaeodendrology, paleodendrology** See study of: dendrology: paleodendrology.

♦ **palaeoecology, paleoecology** See study of: ecology: paleoecology,

♦ **palaeoendemic, paleoendemic** See ¹endemic: paleoendemic.

♦ **palaeoethology, paleoethology** See study of: ³ethology: palaeoethology.

♦ **palaeogenesis, paleogenesis** See neoteny.

♦ **palaeogenetics, paleogenetics** See study of: genetics: paleogenetics.

♦ **palaeoichnology, paleoichnology** See study of: ichnology: paleoichnology.

♦ **palaeolimnology, paleolimnology** See study of: limnology: paleolimnology,

♦ **palaeontography, paleontography** See -graphy: paleontography.

♦ **palaeontology, paleontology** See study of: paleontology.

♦ **palaeopalynology, paleopalynology** See study of: palynology: paleopalynology.

♦ **palaeophytogeography, paleophytogeography** See study of: geography: phytogeography: paleophytogeography.

♦ **palaeophytology, paleophytology** See study of: botany: paleobotany.

♦ **palaeornithology** See study of: ornithology: paleornithology.

♦ **palaeospecies, paleospecies** See ²species: chronospecies.

♦ **palaeosynecology, paleosynecology** See study of: ecology: synecology: paleosynecology.

♦ **palaeothanatocoenosis, paleothanatocoenosis** See ²community: palaeothanatocoenosis.

♦ **palaeozoology, paleozoology** See study of: zoology: paleozoology.

♦ **Paley's watch** *n.* The best known of William Paley's (1743–1805) arguments for the existence of God: A watch is too complicated, and too functional, to have come about by accident; it carries its own evidence of having been purposefully designed. This "argument seems to apply *a fortiori* to a living body, which is even more complicated than a watch" (Dawkins 1982, 292).

♦ **palichnology** See study of: ichnology: palaeoichnology.

♦ **palingenesis** See -genesis: palingenesis.

♦ **palingenetic** See -genetic: palingenetic.

♦ **palmigrade** See -grade: plantigrade.

♦ **palpation** *n.* An insect's touching with its labial or maxillary palps, which can serve as a sensory probe or a tactile signal to another insect (Wilson 1975, 590).

♦ **paludal** See uliginous.

♦ **paludicole** See -cole: paludicole.

♦ **palustrine** See uliginous.

♦ **palynology** See study of: palynology.

♦ **pan-** *combining form*
1. "All, every; the whole" (Michaelis 1963).
2. "Comprising, including or applying to all" (Michaelis 1963).
[Greek neutral of *pas*, all]

♦ **panchreston** *n.* "A word or "concept" covering a wide range of different phenomena and loaded with a different meaning for each user; a word that attempts to "explain" everything but explains nothing (*e.g.,* trophallaxis, drive, instinct, aggression, approach-withdrawal, or altruism) (Hardin 1956 in Wilson 1975, 29).
[Greek *pan*, all + *chres* + *-to*, useful]

♦ **panclimax** See climax: panclimax.

♦ **pandemic** *n.* A disease that reaches epidemic proportions simultaneously in many parts of the world (Lincoln et al. 1985).
See ubiquitous.

♦ **panformation** See panclimax.

♦ **pangamy** See -gamy: pangamy.

♦ **"Pangea hypothesis"** See hypothesis: hypotheses regarding the Permian mass extinction: "Pangea hypothesis."

♦ **pangen** *n.* de Vries' "genetic unit" that migrates from a cell nucleus to its cytoplasm and thus determines a cell's character (Mayr 1982, 704).

cf. gene; [3]theory: de Vries' theory of pangens, theory of pangenesis

♦ **pangenesis, hypothesis of, theory of** See [2]theory: theory of pangenesis.

♦ **Panglossian paradigm** See adaptionist program.

♦ **panmictic index** See coefficient: coefficient of relatedness.

♦ **panmictic population** See [1]population: panmictic population.

♦ **panmixis** See mating: random mating; -mixis: panmixis.

♦ **panogamy** See mating system: panogamy.

♦ **panne** See habitat: dune slack.

♦ **panphytophage** See -phage: panphytophage.

♦ **panselectionism** *n.* The view that natural selection is the most likely explanation for most traits of organisms; contrasted with pluralism (Gould and Lewontin 1979, 589).

♦ **panspermia hypothesis** See hypothesis: panspermia hypothesis.

♦ **panthalassic** *adj.* Pertaining to marine organisms, or species, that live in both neritic and oceanic waters (Lincoln et al. 1985).

♦ **panting** *n.*
1. A person's breathing hard or spasmodically, as when out of breath; his drawing quick, labored breaths due to exertion or agitation; his gasping for breath (*Oxford English Dictionary* 1972, verb entries from 1440).
2. A person's longing, or wishing, with breathless eagerness; his grasping with desire; his yearning (*Oxford English Dictionary* 1972, verb entries from 1560).
3. For example, in anatid, corvid, and predatory birds: an individual's breathing fast with its beak wide open in response to overheating (Heymer 1977, 85).
4. For example, in canid mammals: an individual's breathing fast with its mouth open in response to overheating (Heymer 1977, 85).

v.i. pant

♦ **pantomictic plankton** See plankton: pantomictic plankton.

♦ **pantophage** See -phage: pantophage.

♦ **pap** See food: pap.

♦ **-para** *combining form*
multipara *n.* A female mammal that has gone through more than one cycle of pregnancy and parturition; contrasted with nullipara and primapara (Hurnik et al. 1995).
adj. multiparous
cf. -parity: multiparity
nullipara *n.*
1. A female primate without young during a particular breeding season (Hrdy 1977, 199).

2. A female mammal that has not given birth to viable offspring; contrasted with multipara and primapara (Hurnik et al. 1995).
adj. nulliparous
cf. -aparity

♦ **para-, par-** *prefix*
1. "Beside; nearby; along with" (Michaelis 1963).
2. "Beyond; aside from; amiss" (Michaelis 1963).
3. Medically, (a) a functionally disordered or diseased condition; (b) in an accessory, or secondary, capacity; (c) similar to but not identical with a true condition or form (Michaelis 1963).

[Greek *para*, beside]

♦ **parabiosis** See -biosis: parabiosis.

♦ **parabiosphere** See -sphere: biosphere: eubiosphere: parabiosphere.

♦ **paracentric inversion** *n.* A chromosomal inversion, *q.v.*, in which the centromere is not located on the inverted chromosome section (Mayr 1982, 769).
cf. pericentric inversion

♦ **parachoric** *adj.* Referring to a nonparasitic organism that lives within the body of another organism (Lincoln et al. 1985).
cf. symbiosis

♦ **parachuting** See ballooning.

♦ **paraclimax** See climax: paraclimax.

♦ **paraclonal society** See [2]society: paraclonal society.

♦ **paracme** See -acme: paracme.

♦ **paracrine signaling** See signaling: paracrine signaling.

♦ **parade** See [2]group: parade.

♦ **parade formation** *n.* An orderly arrangement of individual animals in space (*e.g.*, in dolphins, groups move in parade formation during silent navigation in clear water) (Pilleri and Knuckey 1969 in Wilson 1975, 478).

♦ **paradigm** *n.*
1. "Pattern, exemplar, example" (*Oxford English Dictionary* 1972, entries from 1483).
2. "An archetype or outstanding example of type" (Lincoln et al. 1985).
3. "A generally accepted way of explaining a scientific phenomenon of first magnitude" (Kuhn 1970 in Moore 1984, 478); an archetypal solution to a problem, such as Ptolemy's theory that the sun revolves around the Earth (van Gelder 1996, B7).
Comments: Thomas S. Kuhn's (1923–1996) thesis is that science is not a steady, cumulative acquisition of knowledge (van Gelder 1996, B7). Instead, it is "a series of peaceful interludes punctuated by intellectually violent revolutions." In those revolutions, one

n–r

"conconceptual world view is replaced by another." Kuhn argued that a typical scientist is not an objective free thinker and skeptic; rather, he is a somewhat conservative individual who accepts what he was taught and applies his knowledge to solving the problems that come before him. He tends to solve a problem in a way that extends the scope of an already present paradigm. During a scientific revolution, a person develops a new paradigm that cannot build on the one that precedes it, but supplants it. Kuhn defined "paradigm" in 22 distinct ways in his book, *The Structure of Scientific Revolutions* (1962) (Gleick 1996, 25, illustration).

[Latin *paradigma* < Greek *paradeigma*, pattern < *para-*, beside + *deiknynai*, to show]

♦ **paradoxical sleep** See sleep: stage-1 sleep.

♦ **paragamy** See -gamy: paragamy.

♦ **paragenesis** See -genesis: paragenesis.

♦ **paragenetic** See -genetic: paragenetic.

♦ **parahomology** See homology: parahomology.

♦ **paralanguage** See language: paralanguage.

♦ **parallel evolution** See ²evolution: parallel evolution.

♦ **parallel polymorphism** See -morphism: polymorphism.

♦ **parallelism** See ²evolution: parallel evolution.

♦ **paralocal character species** See ²species: paralocal character species.

♦ **paralocal species** See ²species: paralocal species.

♦ **paralogous** See -logous: paralogous.

♦ **parameter** *n.*

1. A mathematical quantity that is constant (distinguished from ordinary variables) in a particular case but varies between cases (*Oxford English Dictionary* 1972, entries from 1852).

   Notes: A parameter's value characterizes the other variable, or variables, in a system of expressions, functions, etc. (Michaelis 1963). In a particular kind of investigation, one holds a parameter constant in a model in which other quantities are being varied to study their relationships, and one might also change the parameter's value in other versions of the same model (Wilson 1975, 590).

2. "Any variable property that exerts an effect upon a system" (Wilson 1975, 590).

3. A characteristic of a distribution of a variable, or population, such as a mean or variance (Lincoln et al. 1985).

cf. statistic

[New Latin < Greek *para*, beside + *metron*, measure]

♦ **parametric model** See ⁴model: parametric model.

♦ **parametric test** See test: statistical test: parametric test.

♦ **paramorph** See -morph: paramorph.

♦ **paramorphic homology** See homology: paramorphic homology.

♦ **paramutation** See ²mutation: paramutation.

♦ **paramutualism** See symbiosis: mutualism: paramutualism.

♦ **paranotal hypothesis of insect wing origin** See hypothesis: hypotheses of origin of insect wings: paranotal hypothesis of insect wing origin.

♦ **parapatric** See -patric: parapatric.

♦ **parapatric speciation** See speciation: parapatric speciation.

♦ **paraphilia** See -philia: paraphilia.

♦ **paraphyletic group** See ²group: paraphyletic group.

♦ **paraphyly** See -phyly: paraphyly.

♦ **parapsychology** See study of: psychology: parapsychology.

♦ **pararegional species** See ²species: local pararegional species.

♦ **parasexual** See sexual: parasexual.

♦ **parasite** *n.*

1. A person who frequents rich tables and earns his welcome by flattery; one who obtains the hospitality, patronage, or favor of the wealthy, or powerful, by obsequiousness and flattery; a hanger-on from interested motives; a toady (*Oxford English Dictionary* 1972, entries from 1539).

2. An individual organism that lives in, or upon, another living organism (= host) and draws its nutriment directly from it (*Oxford English Dictionary* 1972, entries from 1727); a symbiont that is metabolically dependent on another living organism for completion of its life cycle and is typically detrimental to its host to a greater, or lesser, extent (Lincoln et al. 1985).

3. A symbiont that obtains more time, energy, metabolites, other resources, or a combination of these resources from another organism than it gives to it, *e.g.,* animal and plant parasites (including Dodders, Louseworts, Mistletoes, and Rust and Smut Fungi) (Lewin 1982).

See ²group: functional feeding group: "parasites;" parasitoid; -zoite: histozoite.

adj. parasitic

v.t. parasitize

syn. biotroph, guest, guest-fly (*Oxford English Dictionary* 1972, entries from 1864), paratroph (Lincoln et al. 1985)

cf. commensal, parasitoid, symbiont, -xenic, -xeny, and parasitism, because many kinds of parasites are classified with regard to kinds of parasitism

Comments: "Parasite" is sometimes inaccurately used to refer to an epiphytic organism that does not draw nutriments from its symbionts (*Oxford English Dictionary* 1972). It is sometimes very difficult, if not impossible, to determine whether a symbiont is a parasite (Lewin 1982). Types of parasites are classified by the kinds of parasitism, *q.v.*, that they show.

[Latin *parasitus* Greek *parasitos*, literally, one who eats at another's table < *para*, beside + *sitos*, food]

allohospitalic parasite *n.* One of two or more parasitic species that occur only on different host species (Lincoln et al. 1985).
cf. parasite: synhospitalic parasite

digenetic parasite *n.* A parasite with two hosts (Bell 1982, 507).
cf. parasite: homogenetic, monogenetic parasite

ectoparasite, ectosite, epiparasite, episite, exoparasite *n.* A parasite that practices ectoparasitism, *q.v.*
cf. parasite: endoparasite

endoparasite, endosite, entoparasite *n.* A parasite that practices endoparasitism, *q.v.* (Lincoln et al. 1985).
cf. ectoparasite

epiparasite, episite See parasite: ectoparasite.

euryadaptive parasite *n.* A parasite that tolerates a wide range of host species (Lincoln et al. 1985).
cf. euryxenous parasite

exoparasite See parasite: ecotoparasite.

heteroecious parasite See parasite: heteroxenous parasite.

heterogenetic parasite *n.* A parasite with a complex life cycle (Bell 1982, 507).

heteroxenous parasite, heteroecious parasite, metoxenous parasite *n.* A parasite that occupies more than one host during its life cycle (Lincoln et al. 1985). See parasite: pleioxenous parasite.

homogenetic parasite *n.* A parasite with a single host (Bell 1982, 507).
cf. parasite: digenetic parasite, monogenetic parasite

metoxenous parasite See parasite: heteroxenous parasite.

monogenetic parasite *n.* A parasite with a single host (Bell 1982, 507).
cf. parasite: digenetic parasite, homogenetic parasite

pleioxenous parasite, polyxenic parasite *n.* A parasite that is not host specific or has several hosts during its life cycle (Lincoln et al. 1985).
n. pleioxeny

syn. heteroxenous parasite (inferred from Lincoln et al. 1985)

prey parasite *n.* A cleptoparasitic organism.

proovigenic parasite *n.* A parasite that ecloses with a full complement of mature, or nearly mature, eggs (*e.g.,* some parasitic-wasp species) (Hassel and Waage 1984).
cf. parasite: synovigenic parasite

symparasite *n.* Any of several competing parasites that use the same host individual (Lincoln et al. 1985).

synhospitalic parasite *n.* One of two or more parasitic species that occur on the same host species (Lincoln et al. 1985).
cf. parasite: allohospitalic parasite

synovigenic parasite *n.* A parasite that ecloses with no, or only a partial complement of, mature or nearly mature eggs (*e.g.,* some parasitic wasp species) (Hassel and Waage 1984).
cf. parasite: proovigenic parasite

xenoparasite *n.* A parasite that infests an organism that is not its normal host (Lincoln et al. 1985).

zooparasite *n.* "A parasitic animal" (Lincoln et al. 1985).

♦ **parasite hypothesis** See hypothesis: hypotheses of the evolution and maintenance of sex: parasite hypothesis.

♦ **parasite pollination** See pollination: parasite pollination.

♦ **parasitic castration** See castration: parasitic castration.

♦ **parasitic chromosome** See chromosome: parasitic chromosome.

♦ **parasitic DNA** See nucleic acid: deoxyribonucleic acid: parasitic DNA.

♦ **parasitic mate** See mate: parasitic mate.

♦ **parasitic mimicry** See mimicry: parasitic mimicry.

♦ **parasitism** *n.*
1. Obligatory symbiosis between individuals of two different species in which the parasite is metabolically dependent on its host and the host is typically adversely affected, even killed (Lincoln et al. 1985). *Note:* There are numerous microorganism, insect, and other parasite species in which the parasites kill their hosts. This phenomenon is sometimes called "parasitoidism," *q.v.*
cf. parasitoid
2. An organism's being a parasite.
adj. parasitic
syn. parasitoidism (in some cases), parasitosis (Lincoln et al. 1985)
cf. parasite, parasitoid, symbiont
Comments: Parasites often have a complex of body-structure, physiological, reproductive, life-history, and behavioral adaptations

related to their modes of life (Immelmann and Beer 1989, 213).

accidental parasitism *n.* Parasitism of an organism that is not a parasite's normal host (Lincoln et al. 1985).

adelphoparasitism *n.* In some cuckoo-bee species and their bee hosts: parasitism in which a parasite utilizes a closely taxonomically related host species (Lincoln et al. 1985).
See parasitism: hyperparasitism: adelphoparasitism.
cf. parasitism: alloparasitism

allohyperparasitism See parasitism: hyperparasitism: tertiary hyperparasitism: interspecific hyperparasitism

alloparasitism *n.* Parasitism in which a parasite utilizes an taxonomically unrelated host organism (Lincoln et al. 1985).
cf. parasitism: adelphoparasitism

autohyperparasitism See parasitism: hyperparasitism: tertiary hyperparasitism: intraspecific hyperparasitism.

autoparasitism See parasitism: hyperparasitism.

brood parasitism *n.* For example, in the Cowbird, European Cuckoo, Longnose Gar, Widow Bird; in a minnow species: parasitism in which a parasite species inserts its egg(s) into the nest of another species, with the result that the host partly, or fully, rears the parasite's brood as if it were its own (Wilson 1975, 353, 580; Immelmann and Beer 1989, 36; Baba et al. 1990, 777).
cf. parasitism: nest parasitism; social parasitism: facultative social parasitism and obligatory social parasitism (comment)

brood rearing See parasitism: social parasitism: slavery.

cleptobiosis See parasitism: cleptoparasitism.

cleptoparasitism, kleptoparasitism *n.* For example, in some spider and stingless-bee species; the Frigate Bird and Hyena; terns: parasitism in which an individual seeks out and takes the prey, or stored food, of another usually heterospecific individual; this frequently involves only female animals that take food for their young (Wilson 1975, 580; Heymer 1977, 43; Griswold and Meikle 1987; Jackson and Griswold 1979; Immelmann and Beer 1989, 167).
syn. food pilfering, prey stealing, piracy (Griswold and Meikle 1990, 8); cleptobiosis, prey parasitism (Heymer 1977, 43)
cf. parasitism: social parasitism: trophic parasitism; robber; symbiosis: cleptobiosis
[Greek *kleptēs*, thief + PARASITISM].

drone parasitism *n.* An alien drone's (of the Africanized Honey Bee) being taken care of by a worker(s) in a colony of the European Honey Bee (Rinderer 1986, 224).

dulosis See parasitism: social parasitism: slavery.

ecoparasitism, oecoparasitism *n.* Parasitism in which a parasite is restricted to a specific host or to a small group of related host species (Lincoln et al. 1985).

ectoparasitism *n.* For example, in some chalcidoid-wasp species, many species of plant-feeding arthropods: a parasite's living on the external surface of its host and obtaining nutrients by feeding on host tissues, fluids, or both (Wittenberger 1981, 615).
cf. parasitism: endoparasitism, hyperparasitism: ectophagous hyperparasitism

endoparasitism, entoparasitism *n.* In many parasite species: parasitism in which a parasite lives within the organs, or tissues, of its host (Lincoln et al. 1985).
n. endoparasite, endocite
cf. parasitism: ectoparasitism, hyperparasitism: endophagous hyperparasitism

ethoparasitism See parasitism: social parasitism: ethoparasitism.

exoparasitism See parasitism: ectoparasitism.

facultative parasitism *n.* Parasitism in which a species can reproduce with, or without, using a particular host species, or even any host, depending on the species.
cf. parasitism: obligatory parasitism, hyperparasitism: facultative hyperparasitism

false parasitism See parasitism: pseudoparasitism.

food pilfering See parasitism: cleptoparasitism.

Futterparasitismus See parasitism: social parasitism: trophic parasitism.

haemoparasitism *n.* Parasitism in which a parasite inhabits its host's blood (Lincoln et al. 1985).
Comment: There are many kinds of haemoparasites; *e.g.,* the protistan *Lankesterella* in amphibians and birds; *Plasmodium* in Humans and other primates; different fluke species in blood vessels of Cattle, Humans, and sheep; the nematode *Angiostrongylus cantonensis* in rodent blood vessels; and the nematode *Dirofilaria immitis* (Heartworm) in canid hearts (Nobel et al. 1989, 4, 101, 171, 324, 339).
cf. -troph-: hemotroph

helotism See parasitism: social parasitism: slavery.

hemiparasitism, meroparasitism *n.* Parasitism in which a partial, or facultative, parasite can survive in the absence of its host (Lincoln et al. 1985).

holoparasitism See parasitism: obligatory parasitism.

hyperparasitism *n.*
1. Parasitism in which an individual host is parasitized by two or more species of primary parasites or by one species more than once [coined by Friske 1910 in Shepard and Gale 1977].
2. Parasitism in which an individual host is parasitized by two or more species of primary parasites (Smith 1916 in Shepard and Gale 1977).
3. In many beetle, fly, and hymenopteran species: parasitism in which an individual parasite (secondary parasite) feeds on another individual parasite (primary parasite) (Sullivan 1987, 49).

Notes: This is a widely used definition. Depending on the species, a primary parasite consumes an arthropod phytophage, predator, or scavenger.
See parasitism: superparasitism (a confusing synonym).
syn. autoparasitism (Lincoln et al. 1985), superparasitism (confusing synonym, Lincoln et al. 1985)
cf. parasitism: primary parasitism

▶ **adelphoparasitism** *n.* In some aphelinid-wasp species: hyperparasitism in which a male larva feeds on a conspecific female larva (Sullivan 1987, 60).
Comment: This is the usual mode of male feeding in some species (Sullivan 1987, 60).

▶ **direct hyperparasitism** *n.* Hyperparasitism in which a mother parasite oviposits in, or on, a primary parasite (Sullivan 1987, 49).

▶ **ectophagous hyperparasitism** *n.* Hyperparasitism in which individuals of a species live outside their hosts (Sullivan 1987, 49).

▶ **endophagous hyperparasitism** *n.* Hyperparasitism in which individuals of a species live inside their hosts (Sullivan 1987, 49).

▶ **facultative hyperparasitism** *n.* Hyperparasitism in which an individual of a parasite species can develop either as a primary or secondary parasite (Sullivan 1987, 49).

▶ **indirect hyperparasitism** *n.* Hyperparasitism in which a secondary parasite feeds on a primary parasite's host rather than consuming it directly (Sullivan 1987, 49).

▶ **obligate hyperparasitism** *n.* Hyperparasitism in which an individual of a secondary-parasite species can develop on, or in, an individual of a primary-parasite species (Sullivan 1987, 49).

▶ **tertiary hyperparasitism** *n.* In aphid hyperparasites: a hyperparasite's feeding on another hyperparasite (Sullivan 1987, 55).

inquilinism See parasitism: social parasitism.

interspecific-tertiary hyperparasitism, allohyperparasitism *n.* Tertiary hyperparasitism that involves members of the same species (Sullivan 1987, 55).

intraspecific-nest parasitism See parasitism: nest parasitism.

intraspecific-tertiary hyperparasitism, autohyperparasitism *n.* Tertiary hyperparasitism that involves members of different species (Sullivan 1987, 55).

kidnapping See parasitism: social parasitism: slavery.

kleptoparasitism See parasitism: social parasitism: trophic parasitism: cleptoparasitism.

meroparasitism See parasitism: hemiparasitism.

multiparasitism *n.*
1. Parasitism in which heterospecific insect parasites feed on the same individual insect host (Torre-Bueno 1978).
2. The occurrence of supernumerary individuals of more than one parasitoid species in, or on, the same individual insect host (Propp and Morgan 1983b, 1232).

cf. parasitism: superparasitism

nest parasitism *n.*
1. Symbiosis between two termite species in which a colony of one species lives in, and feeds on, the nest walls of the other species (host) (Lincoln et al. 1985).
2. In many waterfowl species; a swallow species: a female's laying one or more eggs, in another conspecific female's nest, which may hatch into nestlings that are taken care off by one or more foster parents (Møller 1987, 247; Lank et al. 1989, 77–78).

syn. intraspecific-nest parasitism (Møller 1987, 247)
cf. parasitism: brood parasitism

obligatory parasitism, obligate parasitism, holoparasitism *n.* Parasitism in which a parasite species cannot survive without using a particular host species (Lincoln et al. 1985).
cf. parasitism: facultative parasitism

oecoparasitism See parasitism: ecoparasitism.

partial parasitism See parasitism: semiparasitism.

permanent parasitism See parasitism: social parasitism: inquilinism.

piracy See parasitism: cleptoparasitism.

n–r

prey parasitism, prey stealing See parasitism: cleptoparasitism.

primary parasitism *n*. For example, in chalcidoid wasps: parasitism of a previously nonparasitized host by a particular species (Hooker and Barrows 1989, 460).

pseudoparasitism *n*. A free-living organism's chance entry and survival in the body of another free-living organism (Lincoln et al. 1985).
syn. false parasitism (Lincoln et al. 1985)

semiparasitism *n*.
1. An organism's deriving only part of its nourishment from its host (Lincoln et al. 1985).
2. An organism's living for only part of its life as a parasite (Lincoln et al. 1985).
syn. partial parasitism (Lincoln et al. 1985)

social parasitism *n*. In many animal species: parasitism in which an individual takes food, or some other resource, from another either conspecific or heterospecific individual animal of a social species.

▸ **ethoparasitism** *n*. For example, in some beetle species: social parasitism in which a parasite lives in an ant colony and is fed by the ants (Heymer 1977, 65).

▸ **facultative social parasitism** *n*. For example, in a bumble-bee species, some yellowjacket species: social parasitism in which a queen of a species that can have an independent nest of her own offspring usurps a nest of another species whose workers rear her brood (Akre et al. 1981, 36; Fisher 1987b, 1628).
Comment: "Obligatory social parasitism" occurs in primitively eusocial bee species that produce no workers (Michener 1974, 225).

▸ **inquilinism** *n*. In some insect species: social parasitism in which a parasite spends its entire life cycle in the nests of its host species; the parasite either has no workers or a few workers with degenerate behavior (Wilson 1975, 354, 587).
syn. permanent parasitism (of some authors) (Wilson 1975, 354, 587)
cf. inquiline
[Latin *inquilinus*, lodger < *in-*, in + *colere*, to dwell]

▸ **prey parasitism** See parasitism: cleptoparasitism.

▸ **slave making** See parasitism: social parasitism: slavery.

▸ **slavery** *n*.
1. Severe human toil "like that of a slave; heavy labour, hard work, drudgery" (*Oxford English Dictionary*, 1972, entries from 1551).
2. A human slave's condition; a person's being a slave; servitude; bondage (*Ox-*

ford English Dictionary, 1972, entries from 1604).
3. Human slaves' existing as a class in a community; a person's keeping slaves as a practice or institution (*Oxford English Dictionary* 1972, entries from 1728).
4. Social parasitism in which a group of slave-making ants raid a host-species colony, overwhelm its adults, capture its brood, and transport all, or part, of its brood to the parasite's nest; the adults that arise from the host's brood become workers in the slave-maker's colony and may even participate in future slave raids (Huber 1810 in Wilson 1975, 368; Regnier and Wilson 1971; Alloway 1979, 202).
syn. brood raiding (Rissing and Pollack, 1987, 975); dulosis, helotism (Lincoln et al. 1985), kidnapping (Heinsohn 1991, 1099), slave-making
cf. propaganda substance, scouting, slavery
Comment: Some slave-makers require slaves to feed them because they cannot feed themselves (Heymer 1977, 165).

▸ **temporary social parasitism** *n*.
1. In some ant, bee, and wasp species: social parasitism in which one species spends part of its life cycle as a parasite in another species' society and the remainder of its time as a nonparasite (Wilson 1975, 354).
2. In some ant, bee, and wasp species: parasitism in which a queen enters an alien nest (usually belonging to another species), kills or renders infertile the resident queen, and takes her place; the colony's population then becomes increasingly dominated by the offspring of the parasite queen as the host workers die off from natural causes (Wilson 1975, 354, 597).

▸ **trophic parasitism** *n*. Social parasitism in which one species steals food from another, and one or both of these species are social (Wilson 1975, 354).
syn. Futterparasitismus (Wilson 1975, 354)
cf. parasitism: social parasitism: ethoparasitism
Comment: Some cases of cleptoparasitism are trophic parasitism.
[German *Futter*, food + *parasitismus*, parasitism]

▸ **xenobiosis** *n*. In some ant species: social parasitism in which a member of one ant species lives in the nests of another ant species and moves freely among its hosts, obtaining food from hosts by regurgitation or other means but still keeping their brood separate (*e.g.*, the "shampoo ant," which lives on regurgitated food from the ant *Myrmica brevinodis*) (Wilson 1975, 354, 362, 598). *cf.* -biosis

superparasitism *n.*

1. For example, in chalcidoid wasps: parasitism in which an individual host insect is parasitized by two or more species of primary parasites or by one species more than once [coined by Friske 1910 in Shepard and Gale 1977, 315].

   Note: Although Friske (1910) called both of these kinds of parasitism "superparasitism," today the first kind of parasitism is usually called "multiparasitism," and the second kind is called "superparasitism."

2. For example, in many insect species parasitized by wasps: parasitism of an individual host insect by more than one mother of the same species (Smith 1916 in Shepard and Gale 1977, 315; Hooker and Barrows 1992).

3. The occurrence of more than one individual of a single parasitoid species on, or in, the same individual host insect (Propp and Morgan, 1983a, 561).

[coined by Smith 1916 in Shepard and Gale 1977, 315]

Note: This is the main definition of "superparasitism" used today.

n. superparasite

syn. hyperparasitism (Lincoln et al. 1985, a confusing synonym)

cf. parasitism: hyperparasitism, multiparasitism, primary parasitism

temporary parasitism *n.* Parasitism in which a parasite contacts its host only for feeding (Lincoln et al. 1985).

trophoparasitism *n.* Parasitism in which an organism is an obligate parasite for part of its life cycle (Lincoln et al. 1985).

water parasitism *n.* Parasitism in which a parasite derives only water from its host (Lincoln et al. 1985).

xenoparasitism *n.* Parasitism in which a parasite infests an organism that is not its usual host (Lincoln et al. 1985).

cf. parasitism: social parasitism: temporary social parasitism

♦ **parasitocoenosis** *n.* The total parasite load of a host organism (Lincoln et al. 1985).

♦ **parasitoid** *n.* An organism, intermediate between a parasite and predator, which is usually a hymenopteran species whose larvae feed within, or upon, living bodies of another species, eventually causing deaths of their hosts (Lincoln et al. 1985).

Comments: Many workers lump "parasitoids" with "parasites" and call them all "parasites" (Hooker and Barrows 1992). To have consistent terminology, some workers use the terms "parasitoidism" and "parasitoidize" (Propp and Morgan 1985, 515).

♦ **parasitology** See study of: parasitology.

♦ **parasocial route to eusociality** *n.* A hypothesized pathway of evolution of eusociality in insects (Wilson 1975, 448).

cf. sociality, subsocial route to eusociality

Comment: Birds show a similar tendency (Wilson 1975, 448).

♦ **parasociality** See sociality: parasociality.

♦ **parasymbiosis** See symbiosis: parasymbiosis.

♦ **parasympathetic nervous system** See ²system: nervous system: parasympathetic nervous system.

♦ **paratelic character** See character: paratelic character.

♦ **paratenic host** See host: paratenic host.

♦ **paratonic, aitiogenic** *adj.* Referring to movement or growth stimulated, enhanced, or retarded by an external stimulus (Lincoln et al. 1985).

See -genic: aitiogenic.

♦ **paratrepsis** See display: distraction display.

♦ **paratroph** See -troph-: paratroph.

♦ **parent** *n.*

1. A mother or father (*Oxford English Dictionary*, 1972, entries from 1450).

2. A surrogate mother or father (*Oxford English Dictionary*, 1972, entries from 1526).

3. "A progenitor; forefather" (*Oxford English Dictionary*, 1972, entries from 1413).

4. An organism that generates another organism (*Oxford English Dictionary*, 1972, entries from 1774).

cf. mother, father, aunt, uncle

alloparent *n.* For example, in the Human, Macaques: an individual that assists a parent in care of its young and may or may not be a genetic relative of the young (Wilson 1975, 349, 578; Small 1990, 39). See helper.

syn. "baby-sitter" (Dewsbury 1978, 305)

cf. mother: allomother, aunt, uncle

♦ **parent clinger** *n.* For example, in the Koala; bats, lemurs, monkeys, sloths: a suckling young that continuously holds onto and is passively carried by its mother during the first days to weeks after its birth [coined by Hassenstein 1973 in Heymer 1977, 175].

cf. parent hugger

♦ **parent hugger** *n.* For example, in apes, including Humans: a suckling young that continuously holds onto and is actively carried by its mother during the first days to weeks after its birth (Wickler 1969 in Heymer 1977, 175).

cf. parent-clinger

♦ **parent imprinting** See learning: imprinting: filial imprinting.

n–r

♦ **parent-offspring conflict** See ³conflict: parent-offspring conflict.

♦ **parent-offspring recognition** See recognition: parent-offspring recognition.

♦ **parentage, coefficient of** See coefficient: coefficient of relatedness.

♦ **parental aggression** See aggression: parental aggression.

♦ **parental altruism** See altruism: parental altruism.

♦ **parental behavior** See behavior: parental behavior.

♦ **parental care** See care: parental care.

♦ **parental companion** See companion: parental companion.

♦ **parental disciplinary aggression** See aggression: parental disciplinary aggression.

♦ **parental effort** See effort: effort: parental effort.

♦ **parental facilitation** See ²facilitation: parental facilitation.

♦ **parental investment** See investment: parental investment.

♦ **parental manipulation** See manipulation: parental manipulation.

♦ **parity, -pary** *n.*
1. Medically, an organism's fitness or ability to bear offspring (Michaelis 1963).
2. The number of times that a particular female has given birth (Lincoln et al. 1985).
3. The number of offspring that a female produces at a birth (Lincoln et al. 1985).
adj. -parous
cf. -para
[Latin *parere*, to bear + ITY]

♦ **-parity** *combining form*
adj. -parous
cf. care: parental care

biparity *n.*
1. An organism's production of two offspring per brood (Lincoln et al. 1985).
syn. ditokous (Lincoln et al. 1985)
2. Referring to an organism that has produced only one previous brood (Lincoln et al. 1985).

iteroparity *n.*
1. In many organism species: an organism's reproducing in more than one bout during its lifetime (Wittenberger 1981, 347).
2. An organism's production of offspring in successive groups (Wilson 1975, 587).
cf. parity: semelparity
Comment: "Iteroparity" is on a continuum with "semelparity."

larviparity *n.* An animal's producing eggs that hatch internally and releasing free-living larvae (Lincoln et al. 1985).

multiparity *n.*
1. A female mammal's having more than one young more or less contemporaneously, or producing several offspring within a single brood (Lincoln et al. 1985).
2. A female mammal's having more than one brood of one or more offspring (Wilson 1975, 350).
syn. pluriparity (as the adjective "pluriparous") in Lincoln et al. (1985)

oviparity *n.* In many animal species: a mother's reproducing by laying eggs that hatch outside her body (Lincoln et al. 1985).
cf. -ovular: polyovular; -parity: oviovoviviparity, ovoviparity, viviparity
Comments: Oviparous taxa include most insect and reptile species, many fish species, all bird species, echidnas, and platypuses.
[New Latin < Latin *oviparous*, laying eggs < *ovum*, egg + *parus* < *parere*, to bring forth]

ovoviparity *n.*
1. In a minority of extant and extinct reptile species: a mother's caring for her eggs that are within her body cavity and develop within their shells, or shell membranes, and derive all of their nutrition from their yolks (Inger 1973, 183).
Notes: In reptiles, depending on the species, mothers retain their eggs for from a few days through most of their embryonic developmental period, with their eggs hatching soon after laying (Inger 1973, 183).
2. For example, in many fish genera including *Chalamydoselachus, Latimeria, Scyllium, Sebastes, Xenopoecilus*: a mother's caring for eggs within her body cavity, where they grow into embryos or larvae, using their yolks for food (Balon 1975, 856).
Notes: In ovoviparity, fertilization is internal and is usually facilitated by a father's intromittent organ (Balon 1975, 856). Mothers release hatched embryos or larvae, depending on the species.
3. For example, in many cockroach species: a mother's producing fully formed eggs that are retained and hatched inside her reproductive tract and released as free-living offspring (Lincoln et al. 1985).

oviovoviviparity *n.* In many fish genera including *Chimera, Coelurichthys, Heterodontis, Pantodon, Rhincodon, Scyliorhinus*: a mother's caring for eggs within her body cavity which are internally fertilized and expelled at the moment of fertilization, at the beginning of cleavage, or at any time during their embryological development (Balon 1975, 855).

pluriparity See -parity: multiparity.

primaparity *n.* An animal's having her first "group" of one or more offspring.

pupiparity See -parity: nymphiparity.

scissiparity *n.* Multicellular organisms' reproducing by fission (Lincoln et al. 1985).

semelparity *n.* For example, in the Antechinus (Marsupial Mouse); bamboos, mayflies; some palm-tree species: an organism's reproducing in one bout during its lifetime (Wittenberger 1981, 347).
adj. semelparous
syn. big-bang reproduction (Gadgil and Bossert 1970 in Wittenberger 1981, 347), monocarpy (in plants)
cf. parity: iteroparity
Comment: "Semelparity" is on a continuum with "iteroparity."
[after Semele, one of the many human, mortal lovers of the chief Greek god, Zeus, who was incinerated in a dramatic encounter with her in all his glory (Willson 1983, 14; Cockburn and Lee, 1989, 44)]

semioviparity *n.* Reproduction intermediate between oviparity and viviparity (Lincoln et al. 1985).

sexuparity *n.* An organism's producing male and female offspring either by sexual reproduction or parthenogenesis (Lincoln et al. 1985).

tomiparity *n.* Reproduction by fission (Lincoln et al. 1985).

uniparity *n.*
1. For example, in tsetse flies, many mammal species: an animal's producing one offspring in a single brood (Lincoln et al. 1985).
2. An animal's having only one brood in its lifetime (Lincoln et al. 1985).
cf. parity: semelparity

virginiparity *n.* For example, in some mantid, walkingstick, and wasp species; Whiptailed Lizards: reproduction only by parthenogenesis (Lincoln et al. 1985).

viviparity *n.*
1. In a few extant reptile species, including the Common European Viper; *Denisonia* snakes; some skink species: a mother's caring for her embryos that develop within her body cavity and obtain nutrition from her by means of gas exchange with her through a temporary organ (Inger 1973, 183).
2. In many fish genera, including *Carcharias, Galeocerdo, Gambusia, Goodea, Mustelus, Poecilia, Prionace, Xiphophorus*: reproduction in which a mother provides partial or entire nutrition to her embryo(s) via a special absorptive organ (Balon 1975, 856).
Notes: Carcharias taurus seems to produce one young at a time, which is free within its mother's body cavity and feeds on eggs gradually produced by her (Balon 1975, 856). In some other sharks, young develop in their mother's uterus and receive nutrients from her blood through her placenta (Campbell 1996, 637). The poeciliids have unchorionated eggs (Balon 1975, 856).
3. In placental mammals: reproduction in which young are nourished by a placenta and are born (expelled from their mothers) as animate entities not in eggs; these young are neither incubated as eggs nor hatched within their mothers (Lincoln et al. 1985).
adj. viviparous, zoogonous
syn. zoogony (Lincoln et al. 1985)
cf. -parity: larviparity, oviparity, oviovoviviparity, ovoviviparity
[Latin *viviparus* < *vivus*, alive + *parere*, to bring forth]
  ▸ **haemocoelous viviparity** *n.* Viviparity in which eggs are retained and commence development in their mother's haemocoel (Lincoln et al. 1985).

♦ **Parker-Smith model of evolution of sex** See hypothesis: Parker-Smith model of evolution of sex

♦ **parliament** See ²group: parliament.

♦ **-parous** *combining form* Producing. See -parity.

♦ **parsimony, law of** See Morgan's canon.

♦ **parthenoapogamy** See -gamy: apogamy: parthenoapogamy.

♦ **parthenocaryogamy** See -gamy: parthenokaryogamy.

♦ **parthenogamete** See gamete: parthenogamete.

♦ **parthenogamy** See parthenogenesis: automictic parthenogenesis.

♦ **parthenogenesis** *n.*
1. For example, in many species of microorganisms, Rotifers, Plants, Algae, and Insects; Whip-Tailed Lizards: reproduction without concourse of opposite sexes or union of sexual elements (*Oxford English Dictionary* 1972, entries from 1849); asexual reproduction (Lincoln et al. 1985).
syn. monogony (Lincoln et al. 1985)
2. Reproduction from unfertilized eggs, seeds, spores, or other organism parts (Michaelis 1963; Bernstein et al. 1985, 1279).
Note: "Sexual parthenogenesis," *q.v.*, can produce offspring that are genetically different than their parents.
3. Reproduction from cells other than ovules (*e.g.*, in some species of plants) (Michaelis 1963).
4. The absence of mixis; asexuality (Bell 1982, 501).
syn. apomictic reproduction (in the broad sense), apomixis (in the broad sense),

n–r

geneagenesis, gynecogeny (Lincoln et al. 1985), virgin birth

cf. rule: Dzierzon's rule; mixis; -mixis: amixis, apomixis, endomixis, parthenomixis; polyembryony

[Greek *parthenos*, virgin + GENESIS]

ameiotic parthenogenesis *n.* "Parthenogenesis in which meiosis has been suppressed so that neither chromosome reduction nor any corresponding phenomenon occurs" (Lincoln et al. 1985).

ameiotic thelytoky See parthenogenesis: apomixis.

ampherotoky, amphoterotoky See parthenogenesis: amphitoky.

amphitoky *n.*
1. For example, in some chalcidoid wasps: "parthenogenesis in which unfertilized eggs develop into both sexes" (Gordh in Krombein et al. 1979, 746).
   syn. deuterotoky (Lincoln et al. 1985)
2. Parthenogenesis in which both male and female progeny are produced (Lincoln et al. 1985).
syn. ampherotoky, amphoterotoky, deuterotoky, gametotoky (Lincoln et al. 1985)

androcyclic parthenogenesis *n.* Parthenogenesis in which a series of parthenogenetic generations is followed by the production, by a portion of a population only, of males as a sexual generation (Lincoln et al. 1985).
Comment: This occurs in species that also exhibit cyclic parthenogenesis (Lincoln et al. 1985).

androgenesis *n.*
1. The development of an egg that carries only paternal chromosomes (Lincoln et al. 1985).
   See -genesis: patrigenesis.
   syn. male parthenogenesis, patrigenesis (Lincoln et al. 1985)
2. Production of a haploid plant by germination of a pollen grain within an anther (Lincoln et al. 1985).
cf. parthenogenesis: gynogenesis

anholocyclic parthenogenesis *n.* In some aphid species: parthenogenesis by obligate thelytoky (*e.g.,* without a sexual generation) (Bell 1982, 502).

apomixis *n.*
1. For example, in some plant species: asexual parthenogenesis in which offspring arise through agamospermy (formation of a seed from an ovule without its fertilization) or from cells other than ovules (King and Stansfield 1985).
2. Thelytoky in which there is no meiosis and chromosome number is not reduced (Gordh in Krombein et al. 1979, 746).

syn. ameiotic thelytoky (Lincoln et al. 1985)
3. Reproduction from eggs without meiosis and syngamy (Bell 1982, 502).
See parthenogenesis.
Comment: Each apomictic offspring is genetically identical to its parent (unless mutations occur) (Judson and Normark 1996, 41).

arrhenotoky *n.* In almost all species of Hymenoptera; some species of other arthropods: sexual parthenogenesis in which hemizygous males arise from unfertilized eggs.
syn. haplodiploidy (Lincoln et al. 1985)
cf. parthenogenesis: hemizygoid parthenogenesis
Comment: "Arrhenotoky is a means whereby lethal and deleterious genes can be relatively rapidly eliminated from a population and superior genotypes can be relatively rapidly selected" (Gordh in Krombein et al. 1979, 746).

artificial parthenogenesis *n.* Parthenogenesis resulting from activation of an egg by artificial treatment (Lincoln et al. 1985).

asexual meiotic parthenogenesis *n.* Asexual parthenogenesis in which offspring arise from cells produced by meiosis (Thornhill and Alcock 1983, 34–39).

asexual reproduction, asexual parthenogenesis, monogenesis, monogony *n.* Reproduction without the formation of gametes and in the absence of any sexual process (Lincoln et al. 1985).

automictic parthenogenesis, meiotic parthenogenesis, parthenogamy *n.* Parthenogenesis in which meiosis is preserved and diploidy reinstated either by the fusion of haploid nuclei within a single gamete or by the formation of a restitution nucleus (Lincoln et al. 1985).

automixis *n.*
1. For example, in some chalcidoid wasps: sexual parthenogenesis that has reduction divisions of chromosome numbers, with diploidy maintained in several ways (Gordh in Krombein et al. 1979, 746).
   syn. meiotic thelytoky
2. Obligatory self-fertilization by autogamy, *q.v.* (Lincoln et al. 1985).

autoparthenogenesis *n.* Development of an unfertilized egg that has been activated by a chemical or physical stimulus (Lincoln et al. 1985).

cyclic parthenogenesis *n.*
1. Reproduction in which one or more thelytokous generations is followed by a single arrhenotokous, or amphimictic, generation; usually with a more or less

regular and predictable alternation of kinds of generations (Bell 1982, 504).

2. "Reproduction by a series of parthenogenetic generations alternating with a single sexually reproducing generation" (Lincoln et al. 1985).

syn. holocyclic parthenogenesis, monocyclic parthenogenesis (Lincoln et al. 1985)

deuterotoky See parthenogenesis: amphitoky.

diploid parthenogenesis See -gamy: apogamy: parthenapogamy.

facultative parthenogenesis *n.* Development of some eggs of an organism if they are not fertilized (Lincoln et al. 1985).

fission *n.* In protistans: asexual reproduction in which an organism splits into two parts (= binary fission) or more parts (= multiple fission) (Lincoln et al. 1985).

▸ **schizogony** *n.* A type of multiple fission in which a parent organism (schizont) divides and forms many new individuals (merozoites) (Lincoln et al. 1985).

gametotoky See parthenogenesis: amphitoky.

generative parthenogenesis, haploid parthenogenesis *n.* Development of a haploid organism from a female gamete that has undergone meiotic reduction division but has not been fertilized (Lincoln et al. 1985).

gynogenesis *n.* In some fish and plant species: parthenogenesis in which an egg is activated by pseudogamy (penetration of a sperm without its genome's contributing genetic information to a zygote) (Bell 1982, 511; Licht and Bogart 1987); or, in some plant species, by pollen (Lincoln et al. 1985).

syn. pseudogamy (Lincoln et al. 1985)

cf. parthenogenesis: androgenesis

haplodiploidy *n.* See parthenogenesis: arrhenotoky.

haploid parthenogenesis See parthenogenesis: generative parthenogenesis.

hemizygoid parthenogenesis *n.* "Development of an individual from an unfertilized haploid egg" (Lincoln et al. 1985).

cf. parthenogenesis: arrhenotoky

heterogony *n.* Parthenogenesis in the absence of karyogamy (nuclear fusion) that is stimulated by sperm of another species (Margulis et al. 1985, 73).

heteroparthenogenesis *n.* Parthenogenesis that produces either offspring that reproduce parthenogenetically or offspring that reproduce sexually (Lincoln et al. 1985).

holocyclic parthenogenesis See parthenogenesis: cyclic parthenogenesis.

hybridogenesis *n.* Thelytoky in which a sperm genome, donated by a male of a related amphimictic species, is received and expressed by a hybridogenetic zygote but is not transmitted to this zygote's progeny (Bell 1982, 508).

male parthenogenesis See androgenesis.

meiotic parthenogenesis See parthenogenesis: automictic parthenogenesis.

meiotic thelytoky See parthenogenesis: automixis.

mitotic parthenogenesis *n.* In at least 12 orders of insects: asexual parthenogenesis in which offspring arise from diploid cells (Thornhill and Alcock 1983, 35).

monocyclic parthenogenesis See parthenogenesis: cyclic parthenogenesis.

monogenesis, monogeny See parthenogenesis: vegetative reproduction.

obligate apogamy See parthenogenesis: diploid parthenogenesis.

ooapogamy *n.* Parthenogenetic development of an individual from an unfertilized female gamete (Lincoln et al. 1985).

paedogenesis, pedogenesis See -genesis: paedogenesis.

paedoparthenogenesis *n.* Parthenogenesis in which young arise from larvae (Lincoln et al. 1985).

parthenapogamy, parthenogamy See -gamy: apogamy: parthenogamy.

patrigenesis See parthenogenesis: androgenesis.

polycyclic parthenogenesis *n.* Reproduction involving two or more cycles of alternating parthenogenetic and sexually reproducing generations per year (Lincoln et al. 1985).

cf. parthenogenesis: cyclic parthenogenesis

pseudogamy See parthenogenesis: gynogenesis.

sexual meiotic parthenogenesis *n.* Sexual parthenogenesis in which offspring arise from cells produced by meiosis (Thornhill and Alcock 1983, 34–39).

sexual parthenogenesis *n.* For example, in at least a few species in all orders of insects: parthenogenesis in which offspring arise from unfertilized eggs produced by meiosis and are thus genetically different than their mother (Thornhill and Alcock 1983, 34–35).

cf. parthenogenesis: asexual parthenogenesis

Comment: A special kind of sexual parthenogenesis may occur in some kinds of aphids (Chapman 1971; Thornhill and Alcock 1983, 34–35).

somatic parthenogenesis See parthenogenesis: diploid parthenogenesis.

thelytoky *n.*

1. For example, in a walkingstick species: parthenogenesis in which males are unknown, or rare, and females produce females by various asexual mechanisms (Gordh in Krombein et al. 1979, 746). *Notes:* "Cytologically, diploidy is maintained by apomixis and automixis" (Gordh in Krombein et al. 1979, 746). Thelytoky is classified as "ameiotic thelytoky" (= apomixis) and "meiotic thelytoky" (= automixis).
2. Parthenogenesis in which diploid individuals are formed by karyogamy of an egg with its own female pronucleus (Margulis et al. 1985, 73).

true parthenogenesis *n.* Parthenogenesis in which sperm play no part in egg activation (Licht and Bogart 1987). *cf.* parthenogenesis: gynogenesis

tychoparthenogenesis *n.*

1. Asexual parthenogenesis that occurs in normally sexually reproducing organisms (Lincoln et al. 1985).
2. Occasional parthenogenesis in a species (Margulis et al. 1985, 73).

tychothelotoky *n.* An organism's occasional production of parthenogenetic daughters (Bell 1982, 513).

uniparental reproduction *n.* Reproduction from a single parent as occurs in self-fertilizing hermaphrodism, thelytoky, and vegetative reproduction (Lincoln et al. 1985).

vegetative reproduction *n.* Asexual parthenogenesis in which offspring arise from parental somatic cells (Bernstein et al. 1985, 1279). See parthenogenesis: asexual reproduction. *syn.* asexual reproduction, monogenesis, monogeny, vegetative division, vegetative propagation (Lincoln et al. 1985)

zygoid parthenogenesis *n.* Parthenogenesis involving an "egg that remains diploid or undergoes diploidization during its development" (Lincoln et al. 1985).

♦ **parthenogenetic reproduction** See reproduction: parthenogenetic reproduction.

♦ **parthenogenome, parthenote** *n.* An organism that is produced by parthenogenesis (Lincoln et al. 1985).

♦ **parthenokaryogamy** See -gamy: parthenocaryogamy.

♦ **parthenomixis** See -mixis: parthenomixis.

♦ **parthenote** See parthenogenome.

♦ **partial dormancy** See dormancy: partial dormancy.

♦ **partial ethogram** See -gram: ethogram: partial ethogram.

♦ **partial homology** See homolog: partial homology.

♦ **partial parasitism** See parasitism: partial parasitism.

♦ **partial preference** See preference: partial preference.

♦ **partial reinforcement** See reinforcement: partial reinforcement.

♦ **partial selection** See selection: partial selection.

♦ **partially balanced incomplete block** See experiment: incomplete block: partially balanced incomplete block.

♦ **partially claustral colony founding** *n.* An ant queen's founding her colony by isolating herself in a chamber but still occasionally leaving to forage for part of her food supply (Wilson 1975, 591).

♦ **particulate-organic matter** *n.* Bits of organic material found in an aqueous solution (inferred from Cummins and Merritt in Merritt and Cummins 1984, 61).

coarse-particulate-organic matter (CPOM) *n.* Bits of organic material greater than 3 microns in diameter and found in an aqueous solution; contrasted with fine particulate organic matter (FPOM) (Cummins and Merritt in Merritt and Cummins 1984, 61).

fine-particulate-organic matter (FPOM) *n.* Bits of organic material less than 3 microns in diameter and found in an aqueous solution; contrasted with course particulate organic matter (CPOM) (Cummins and Merritt in Merritt and Cummins 1984, 61). *Comment:* Cummins and Merritt in Merritt and Cummins (1984) do not tell us where to place organic matter that is 3 microns in diameter.

♦ **partner bonding** See bonding: partner bonding.

♦ **partner effect** See effect: group effect.

♦ **parturition** *n.*

1. Chiefly technically, a woman's giving birth; childbirth (*Oxford English Dictionary* 1972, entries from 1646).
2. An individual animal's giving birth (Lincoln et al. 1985).

♦ **party** See ²group: party.

♦ **-pary** See -parity.

♦ **passive-avoidance conditioning** See learning: avoidance conditioning: passive-avoidance conditioning.

♦ **passive-avoidance learning** See learning: avoidance learning: passive-avoidance learning.

♦ **passive nuclear species** See ²species: passive nuclear species.

♦ **passive process** *n.* A system or process that does not require metabolic energy expenditure (Lincoln et al. 1985). *cf.* active process

◆ **passive replicator** See replicator (comment).

◆ **passive withdrawal** See [3]lactation: passive withdrawal.

◆ **patabiont** See -biont: patabiont.

◆ **patacole** See -cole: patacole.

◆ **patch** *n.*
1. In ecology, "a resource clump or habitat island that is essentially homogeneous internally but differs in some important way from surrounding areas" (Wittenberger 1981, 619). *Comments:* In some areas of ecology, "patch" can be almost synonymous with "habitat" because patchiness refers to environmental heterogeneity on any scale (Wiens 1976 in Kramer et al. 1995, 3).
2. In behavioral ecology, a delimited, relatively homogeneous part of an organism's habitat, or microhabitat, that differs, especially in resource availability, from the rest of its habitat (Stephens and Krebs 1986 in Kramer et al. 1995, 3).

◆ **patchy distribution** See distribution: contiguous distribution.

◆ **patchy environment** See environment: patchy environment.

◆ **paternal** *adj.* Referring to a character of, or one derived from, a male parent (Lincoln et al. 1985).
See patrilineal.
cf. maternal

◆ **paternal behavior** See behavior: paternal behavior.

◆ **paternal care** See care: parental care: paternal care.

◆ **paternal family** See [1]family: paternal family.

◆ **paternal half sister** See sister: half sister: paternal half sister.

◆ **paternal investment** See investment: paternal investment.

◆ **paternal sex determination** See sex determination: paternal sex determination.

◆ **path analysis** *n.* A graphical analysis used to determine an inbreeding coefficient (Wilson 1975, 591).

◆ **patho-, path-, pathy** *prefix* "Suffering, feeling, associated with disease" (Lincoln et al. 1985).
[Greek *pathos*, suffering]

◆ **pathogenesis** See -genesis: pathogenesis.

◆ **pathogenic** See -genic: nosogenic.

◆ **pathology** See study of: pathology.

◆ **pathway** *n.* A route, or course, along which something moves; a path (Michaelis 1963).
cf. biochemical pathway
cyclic-AMP pathway *n.* A signaling system in the nervous system of *Aplysia californica* (Hall 1998, 30).

◆ **patocole** See -cole: patocole.

◆ **patoxeny** See -xeny: patoxeny.

◆ **patri-** *combining form* Father (Michaelis 1963).
cf. matri-
[Latin *pater, patris*, father]

◆ **patriarchy** See [2]group: patriarchy.

◆ **-patric** *combining form* Referring to an organism's geographical area.
n. -patry
[Greek *patris*, fatherland, native country]
allopatric *adj.* Referring to populations, species, or other taxa that occupy different and disjunct geographical areas (Lincoln et al. 1985).
n. allopatry
dichopatric, macrodichopatric *adj.* Referring to populations or species that occupy different geographical areas and do not have gene flow (Lincoln et al. 1985).
n. dichopatry
macrodichopatric. See -patric: dichopatric.
microdichopatric, allotopic *adj.* Referring to populations that coexist in the same locality but occupy different microhabitats (Lincoln et al. 1985).
parapatric *adj.* "Referring to two species' having contiguous geographic ranges but no (or only minimal) interbreeding in the zone of contact" (Mayr 1982, 275, 959).
n. parapatry
syn. macroparapatric (Lincoln et al. 1985)
▸ **macroparapatric** See -patric: parapatric.
▸ **microparapatric** *adj.* Referring to a population that coexists in the same locality and in adjacent but not overlapping macrohabitats so that gene flow between them is possible (Lincoln et al. 1985).
sympatric *adj.*
1. Referring to populations or species whose geographical ranges at least partially overlap (Wilson 1975, 596; Immelmann and Beer 1989, 302).
2. Referring to populations, species, or taxa, that occupy the same geographical areas (Lincoln et al. 1985).
cf. -patric: allopatric
▸ **biotically sympatric** *adj.* Referring to sympatry in which groups occupy the same habitats (Lincoln et al. 1985).
▸ **neighboringly sympatric** *adj.* Referring to sympatry in which groups occupy different habitats (Lincoln et al. 1985).

◆ **patricliny, patrocliny** *n.* An offspring's having a greater inherited resemblance to its father than to its mother (Lincoln et al. 1985).
adj. patriclinic, patraclinal, patroclinous
cf. matricliny

◆ **patrigenesis** See -genesis: patrigenesis.

◆ **patrilineal** *adj.*
1. Referring to a paternal line (Michaelis 1963).

n–r

2. Referring to things passed from a father to his offspring (*e.g.,* access to a territory or status within a dominance system) (Wilson 1975, 588).
syn. paternal
cf. matrilineal

♦ **patristic affinity** See affinity: patristic affinity.

♦ **patristic character** See character: patristic character.

♦ **patrocliny** See patricliny.

♦ **patroendemic** See ¹endemic: patroendemic.

♦ **patrogenesis** See -genesis: patrogenesis.

♦ **patrogynopaedium** See -paedium: patrogynopaedium.

♦ **patrol** *v.i.*

1. In Humans: to make the rounds in a camp or garrison; to go on patrol; to act as patrol; to reconnoiter as a patrol; to traverse on duty a particular beat, or district, as a constable or patrolman (*Oxford English Dictionary* 1972, entries from 1691).

2. *v.t.* In Humans: to go over or around (a camp, garrison, town, harbor, etc.) to watch, guard, or protect; to perambulate or traverse (a beat or district) as a constable or patrolman; to traverse leisurely in all directions (*Oxford English Dictionary* 1972, entries from 1765).

3. *v.i.* For example, in males of some bee and wasp species: to fly repeatedly, while not feeding, among landmarks that may be locations where an individual is likely to find a mate (Barrows 1976a, 105).
Note: "Continuous patrolling," "non-sequential patrolling," "food-related patrolling," "sequential patrolling," and "territorial patrolling" have been defined (Barrows 1976a, 108).
syn. fly along a mating path (Haas 1960)

4. For example, in a worker Honey Bee: to investigate the interior of her nest that can lead to her defensive response to a nest intruder (Wilson 1975, 591).

5. In ants: to investigate a nest interior and outer surface (Hölldobler and Wilson 1990, 641).

♦ **patromorphic** See -morphic: patromorphic.

♦ **patropaedium** See -paedium: patropaedium.

♦ **pattern** *n.* A complex of integrated parts that function as a whole, *e.g.,* a behavior pattern (Michaelis 1963) or an ecological community pattern (Cale et al. 1989, 660).
cf. ²movement; pattern: fixed-action pattern

behavior pattern, behavioral pattern, behavior *n.*

1. An organism's set of responses that are statistically organized in time; that is, associated with and manifesting some degree of stereotypy in the temporal sequence in which they occur (Verplanck 1957).
Note: "Behavior pattern" is applied whenever one has analyzed behavior into relatively large units, that is, units composed of a number of responses. Behavior chains, instincts, and the activities that are the empirical bases of particular drives are all labeled behavior patterns, which renders the term of very limited usefulness for theory (Verplanck 1957). In common English, "behavior pattern" can encompass an appreciable number of behaviors, such as the behavior pattern of a 5-year-old person (Michaelis 1963).

2. A major type of behavior (*e.g.,* walking, swimming, diving, or flying) (Chaplin 1968; Lehner 1979, 65).
Note: This is a commonly used definition.
syn. action system (Lincoln et al. 1985)
cf. behavior, classification of (table); behavioral act: component part of a behavioral act
Comment: In definitions in this book, I generally use "behavior," rather than "behavior pattern," to mean behavior of different complexities.

fixed-action pattern (FAP) *n.*

1. An animal's movement that has a generally constant form, need not be learned, generally belongs to a functional system, and is characteristic of its species (Lorenz 1932, 1937; Lorenz and Tinbergen 1939; Tinbergen 1942 in Hinde 1982, 43–44, 47; Lorenz 1939, Schleidt 1974 in Heymer 1977, 62; Immelmann and Beer 1989, 106).
Note: The English translation of "*Erbokoordination*" as "fixed-action pattern" has caused confusion. Yawning may be the best example of a stereotyped action pattern in Humans (Provine and Hamernik 1986); the escape response of *Tritonia* sea slugs may be a classic FAP (Gould 1982, 79).
syn. innate behavior pattern, stereotyped-action pattern (Provine and Hamernik 1986)
cf. modal-action pattern

2. A concept roughly equivalent to consummatory act, *q.v.,* that was important in ethological thinking (Dewsbury 1978, 15).
Note: Different workers have stressed different defining attributes of fixed-action patterns: for example, (1) stereotypy; (2) complexity; (3) display by all members of a species, or at least by all members of a particular group; (4) elicitation by simple but highly specific

stimuli; (5) self-exhaustiveness; (6) triggerability; and (7) occurrence independently of experience (Dewsbury 1978, 15). With further research, investigators found more variability in FAPs than previously expected causing them to be renamed "motor patterns," "action patterns," or "modal-action patterns" (Dewsbury 1978, 21).

3. An animal's "particular behavioral response that is always given in the same stereotyped manner whenever it is evoked by a stimulus situation;" "FAP" is sometimes equated with "instinct" (Wittenberger 1981, 615).

syn. instinctive movement (Immelmann and Beer 1989, 151)

cf. action pattern: modal-action pattern; activity: instinctive activity; program: motor program; stereotypy

Comments: Dawkins (1986, 76–78) gives reasons why the term "fixed-action pattern" should not be used. Gould (1982, 63) indicates that fixed-action patterns are "real." Barlow (1968) proposed that "instinctive activity," "instinctive movement," and "fixed-action pattern" be replaced by the more neutral term "modal-action pattern," because the former terms are so loaded with different conceptual meanings (Immelmann and Beer 1989, 152).

[German *Erbkoordination*, inherited movement coordination]

instinctive-behavior pattern *n.* A behavior sequence involving innate pathways of an animal's central nervous system (Lorenz 1932, 1973 in Heymer 1977, 178). *cf.* behavior: innate behavior, behavior: instinctive behavior, learning

instinctive movement See pattern: fixed-action pattern.

modal-action pattern (MAP) *n.* A behavior that is recognizable, describable at least in statistical terms, indivisible into smaller units, and "widely distributed in similar form throughout an interbreeding population" (Barlow 1977 in Hinde 1982, 47). *cf.* pattern: fixed-action pattern

Comment: Barlow suggests that this term replace "fixed-action pattern."

motor pattern *n.* An animal's spatiotemporal change in its body configuration (*e.g.,* foot flexing, fin fanning, or head nodding) (Immelmann and Beer 1989, 194). *cf.* behavior, Eshkol-Wachmann movement notation; pattern: fixed-action pattern

♦ **pattern of foraging movements** See mode: foraging mode.

♦ **pattern variegation** *n.* For example, in picotee-type petunias and some *Coleus* cultivars: color variegation of a plant part that

is due to developmental differences of cells that are essentially the same genetically (Geneve 1991, 35). *cf.* chimera

♦ **pauling** See unit of measure: macarthur: millimacarthur.

♦ **PAUP** See program: Phylogenetic Analysis Using Parsimony.

♦ **paurometabolous** See metamorphosis: paurometabolous.

♦ **Pavlovian conditioning** See learning: conditioning: Pavlovian conditioning.

♦ **Pavlov's reflex theory** See ³theory: chain-reflex theory.

♦ **paw drumming** See animal sounds: drumming: paw drumming.

♦ **payoff, E(A,B)** *n.* The expected change of fitness of an individual that adopts a strategy A against an opponent that adopts B (Maynard Smith 1982, 204). *cf.* ³theory: game theory

♦ **payoff matrix** *n.* A table that specifies expected gains to a contestant that adopts different strategies in view of which strategy, or strategies, its opponent, or opponents, adopts (Maynard Smith 1982, 204). *cf.* theory: game theory

Comment: Payoff is often measured in terms of fitness but could be measured in other standard units (Maynard Smith 1982, 204).

♦ **PCR** See method: polymerase chain reaction.

♦ **peck** See ²group: peck.

♦ **peck distance** See distance: peck distance.

♦ **peck order, pecking order, peck-right order** See hierarchy: dominance hierarchy.

♦ **Peckhammian mimicry** See mimicry: aggressive mimicry.

♦ **peculiar character** See character: diagnostic character.

♦ **pederosis** See -philia: pedophilia.

♦ **pedicel** *n.*
1. A stalk or supporting part (Michaelis 1963).
2. The "waist" of an ant; made up of either one segment (the petiole) or two segments (the petiole plus the postpetiole) (Wilson 1975, 591).

See nest: wasp nest: pedicel.

syn. petiole (C.K. Starr, personal communication)

[New Latin *pedicellus*, diminutive of Latin *pediculus*, diminutive of *pes, pedis,* foot]

♦ **pedigree** *n.* A formal list, or register, of ancestry, or genealogy, of an individual or family group (Lincoln et al. 1985).

♦ **pediophile** See ¹-phile: pediophile.

♦ **pedo-** See paedo-.

♦ **pedogamy** See -gamy: paedogamy.

n-r

- **pedogenesis** See -genesis: pedogenesis.
- **pedology** See study of: pedology.
- **pedon** *pl. n.* Organisms that live on, or in, the substrate of an aquatic habitat (Lincoln et al. 1985).
  adj. pedonic
- **pedophobia** See phobia (table).
- **pedosphere** See -sphere: biosphere: parabiosphere: pedosphere.
- **pee** See elimination: urination.
- **peep** See ²group: peep.
- **pelagial** See pelagic.
- **pelagic** *adj.*
  1. Referring to the water column of the sea or lake (Michaelis 1963).
  2. Referring to organisms that inhabit open waters of a lake, ocean, or deep pond (Michaelis 1963).
  syn. oceanic (Michaelis 1963), pelagial (Lincoln et al. 1985)
  cf. benthic
  [Latin *pelagicus* < Greek *pelagikos* < *pelagos*, the sea]
- **-pelagic**
  holopelagic *adj.* Referring to aquatic organisms that remain pelagic throughout their life cycles (Lincoln et al. 1985).
  meropelagic *adj.* Referring to aquatic organisms that are only temporary members of the pelagic community (Lincoln et al. 1985).
  cf. pelagic: holopelagic
  neritopelagic *adj.* Referring to an organism that inhabits shallow coastal waters over a continental shelf (Lincoln et al. 1985).
  nyctipelagic *adj.* Referring to organisms that migrate into surface waters at night (Lincoln et al. 1985).
- **pelagium** See ²community: pelagium.
- **pelagophile** See ¹-phile: pelagophile.
- **pelasgic** *adj.* "Moving from place to place" (Lincoln et al. 1985).
- **pelochthium** See ²community: pelochthium.
- **pelochthophile** See ¹-phile: pelochthophile.
- **pelophile** See ¹-phile: pelophile.
- **pelvis** *n.* In bony vertebrates: the part of a skeleton that forms a bony girdle joining the lower, or hind limbs, to the body (Michaelis 1963).
  Comments: In Humans, the pelvis is composed of two innominate bones and the sacrum (Michaelis 1963).
  [Latin *pelvis*, basin]
  ornithischian pelvis *n.* In many dinosaur species: a pelvis in which the longitudinal axes of the ischium and posterior part of the pubis are approximately parallel to one another; distinguished from saurischian pelvis (Strickberger 1996, 400, illustration).

- **saurischian pelvis** *n.* In birds, many dinosaur species: a pelvis in which the longitudinal axes of the ischium and pubis are widely separated at their apices; distinguished from ornithischian pelvis (Strickberger 1996, 400, illustration).
- **pen** See animal names (table).
- **pene-, paene** *prefix* "Almost, nearly" (Brown 1956).
  [Latin, almost]
- **penecontemporary** *adj.* Referring to phenomena that are almost contemporary or occur at about the same time (Lincoln et al. 1985).
- **penetrance** *n.* A phenotypic character's not being shown, although the organism in question has the genetic material responsible for this character; *e.g.*, Humans with the gene(s) responsible for schizophrenia might not show the manifestations of this disease (Mayr 1982, 826).
- **penis** See organ: copulatory organ: penis.
- **penta-, pent-** *combining form* Five, five-fold (Lincoln et al. 1985).
  [Greek *pente*, five]
- **peptide-nucleic acid (PNA)** See nucleic acid: peptide-nucleic acid, PNA.
- **peptidoglycan** See molecule: peptidoglycan.
- **per-** *prefix*
  1. "Through; throughout" (Michaelis 1963).
  2. "Thoroughly; completely" (Michaelis 1963).
  3. "Away" (Michaelis 1963).
  4. "Very" (Michaelis 1963).
  [Latin *per*, through, by means of]
- **percent rate** See rate: percent rate.
- **perception** *n.*
  1. A person's being cognizant, or being aware, of objects in general (*Oxford English Dictionary* 1972, entries from 1611).
  Note: "Perception" is sometimes distinguished from "volition" and sometimes synonymized with "consciousness" (*Oxford English Dictionary* 1972).
  2. A person's being cognizant, or being aware, of a sensible or quasi-sensible object; loosely, personal observation, especially sight (*Oxford English Dictionary* 1972, entries from 1704).
  3. A person's faculty of perceiving something (*Oxford English Dictionary* 1972, entries from 1678).
  4. The result, or product, of a person's perceiving; percept (*Oxford English Dictionary* 1972, entries from 1690).
  5. An individual organism's interpretation of sensory information in light of its experience and unconscious inference (McFarland 1985, 205).

6. Psychologically speaking, some degree of a person's interpretation, or categorization, of stimuli; distinguished from sensation (Immelmann and Beer 1989, 216). *cf.* awareness, consciousness, modality, sensation

depth perception *n.* An animal's perceiving distance away from itself (Storz 1973, 68).

extraocular-light perception *n.* An organism's perceiving light with an organ other than its eye, or eyes, *e.g.,* an amoeba's perceiving light with its body, a fly maggot's perceiving light with its anterior light receptor, or a reptile's perceiving light with its pineal gland (Taylor and Adler 1978; D.B. Quine, personal communication).

gestalt perception *n.* In many animal species: an individual's capacity to apprehend specific stimulus combinations, not only the totality of their component characteristics, but also the particular relational structure among their components; perception of a whole's being more than the sum of its parts; *e.g.,* songbirds with long, varied note sequences in their songs base their reactions to the whole acoustic pattern of an artificial song, rather than on component traits taken separately (Immelmann and Beer 1989, 119). *cf.* stimulus: configurational stimulus

♦ **perceptual psychology** See study of: psychology: perceptual psychology.

♦ **perceptual signs, perceptual stimulus** See ²stimulus: releaser.

♦ **percussive foraging** See foraging: percussive foraging.

♦ **¹perennial** *n.* A plant that remains alive for a number of years, especially a herbaceous plant that dies down to a root and resprouts the next year (*Oxford English Dictionary* 1972, entries from 1672). *cf.* annual, biennial
Comment: Some perennial species also die down to a bulb, corm, or rhizome.

♦ **²perennial** *adj.*
1. Referring to something that occurs throughout a year, or lasts through successive years (*e.g.,* a spring or stream that flows all year) (*Oxford English Dictionary* 1972, entries from 1703).
2. Referring to a structure that grows continuously from a persistent pulp, such as a rodent's front tooth (*Oxford English Dictionary* 1972).
3. Referring to an insect that lives for more than one year (*Oxford English Dictionary* 1972).
4. Referring to an insect colony that lives for more than one year (*Oxford English Dictionary* 1972).

♦ **perennial monogamy** See mating system: monogamy: perennial monogamy.

♦ **pergelicole** See -cole: pergelicole.

♦ **perhalicole** See -cole: perhalicole.

♦ **peri-** *prefix*
1. "Around; encircling" (Michaelis 1963).
2. "Situated near; adjoining" (Michaelis 1963). [Greek *peri,* around]

♦ **pericentric inversion** See inversion (comments).

♦ **periclinal chimera** See chimera: periclinal chimera.

♦ **perigenesis** See -genesis: perigenesis.

♦ **perimenopause** See menopause: perimenopause.

♦ **period** *n.*
1. A course, or extent, of time (*Oxford English Dictionary* 1972, entries from 1413).
2. One of the larger divisions of geological time, usually subordinate to an era and superior to an epoch (*Oxford English Dictionary* 1972, entries from 1833; Lincoln et al. 1985)

activity period *n.* In hibernating animals: the "time between emergence from and entrance into hibernation" (Carpenter 1952a, 238).

critical period See period: sensitive period.

free-running period See period: natural period.

latent period See time: reaction time.

lying-out period *n.* In ungulates, including many gazelle species; the oryx, kudu, and red deer: a period soon after its birth and after suckling when a calf, or fawn, actively leaves its mother and lies down in cover where it remains "couched and concealed" until its mother calls it for its next feeding (Immelmann and Beer 1989, 176–177). *cf.* follower, hider

natural period, free-running period *n.* The fundamental periodicity of a biological clock when it is not entrained to a *zeitgeber* (Lincoln et al. 1985).

nyctiperiod *n.* "A period of darkness; a dark phase in a light-dark cycle" (Lincoln et al. 1985).

phenocritical period *n.* The phase during an organism's development at which a gene's expression is most easily affected by externally applied factors (Lincoln et al. 1985).

photoperiod *n.* The light phase of a light-dark cycle (Lincoln et al. 1985); day length (Immelmann and Beer 1989, 219). *cf.* period: nyctiperiod

refractory period *n.* The time duration that follows excitation of a system and during which a subsequent stimulus of equal magnitude elicits no response or one of reduced amplitude (Heymer 1977, 143).

cf. time: reaction time

sensitive period *n.*

1. For example, in the Domestic Chicken, Japanese Monkey, Norway Rat, Rhesus Monkey; an ant species; ducks; sheep; goats: the restricted period during which an individual becomes imprinted on a particular stimulus (Lorenz 1935 in Champalbert and Lachaud 1990, 850; Dewsbury 1978, 144; Champalbert and Lachaud 1990, 850).
   Notes: "Sensitive period" is preferred to "critical period" by many workers because the former carries less of an implication of suddenness and inflexibility with regard to the onset and termination of the period of maximal susceptibility (Dewsbury 1978, 144). Because "sensitive period" also suggests a demarcated time span, "sensitive phase" seems to be a better term than "sensitive period," but the latter is now preferred in literature (Immelmann and Beer 1989, 263–264). In Zebra Finches, the sensitive period for song learning occurs from 25 to 30 days after hatching and ends about 50 days later (Nash 1997, 55). In Humans, one's ability to learn a second language is from birth up to the age of 6 years.
   syn. critical period, sensitive phase (Dewsbury 1978, 144)
   cf. learning
2. A period during which a person's brain requires certain types of input in order to create and stabilize certain long-lasting structures (Nash 1997, 55).
   Notes: Children exposed to physical abuse in early life develop brains that are exquisitely tuned to danger (Bruce Perry in Nash 1997, 55). Mothers who are disengaged, irritable, impatient, or a combination of these things tend to have sad babies, suggesting that a mother's temperament affects her child's brain development (Geraldine Dawson et al. in Nash 1997, 55).

sensitive phase See period: sensitive period.

♦ **periodic plankton** See plankton: periodic plankton.

♦ **periodicity** *n.*
1. The periodic, or rhythmic, occurrence of an event (Lincoln et al. 1985).
2. The duration of a single phase of an oscillation (Lincoln et al. 1985).
   See rhythm.

♦ **peripatric speciation** See speciation: peripatric speciation.

♦ **peripheral** *adj.* Referring to the surface of an animal's body, *e.g.,* peripheral stimuli as distinguished from central, or internal, stimuli (Immelmann and Beer 1989, 217).

♦ **peripheral male** See male: peripheral male.

♦ **peripheral nervous system** See ²system: nervous system: peripheral nervous system.

♦ **peripheral stimulus, peripheral stimulation** See ²stimulus: external stimulus.

♦ **peripheral stimulus filtering, peripheral filtering** See stimulus filtering: peripheral stimulus filtering.

♦ **periphyton** See ²community: periphyton.

♦ **perittogamy** See -gamy: perittogamy.

♦ **permanent heterozygosity, permanent hybrid** See hybrid: permanent hybrid.

♦ **permanent hybrid, permanent heterozygosity** See hybrid: permanent hybrid, permanent heterozygosity.

♦ **permanent parasitism** See parasitism: social parasitism: inquilinism.

♦ **permeability** *n.* A society's degree of acceptance of, or tendency to accept, immigrants from other conspecific societies (Wilson 1975, 17, 591).

♦ **perpetual-intervention hypothesis** See hypothesis: perpetual intervention hypothesis.

♦ **persistence** *n.* An allele's number of generations that it persists in a population before its elimination (Lincoln et al. 1985).

♦ **persistent dance** See dance: persistent dance.

♦ **personal fitness** See fitness (def. 2, 4); fitness: inclusive fitness.

♦ **personal recognition** See recognition: individual recognition.

♦ **personality** *n.*
1. The quality, character, or fact of one's being a person as distinct from a thing; the quality, or principle, that makes a being personal (*Oxford English Dictionary* 1972, entries from 1380).
2. A personal being, a person (*Oxford English Dictionary* 1972, entries from 1678).
3. The quality, or assemblage of qualities, that makes one person distinct from other persons (*Oxford English Dictionary* 1972, entries from 1795).
4. An individual organism's distinctive suite of behavioral traits (Wilson 1975, 294, 549).

♦ **personnel psychology** See study of: psychology: personnel psychology.

♦ **persuasion signal** See signal: persuasion signal.

♦ **perturbation, disturbance** *n*. Any departure of a biological system from its steady state (Lincoln et al. 1985).

♦ **perverse subsidy** *n*. A subsidy (= a pecuniary aid directly granted by a government to a party and deemed beneficial to the public) that harms our environment (James et al. 1999, 324).
Comments: Perverse subsidies include those for agricultural production, commercial fishing, energy use, road transportation, and water consumption (James et al. 1999, 324). For example, agricultural subsidies, which often harm our environment, average $82,500 km^{-2} in the European Union and $16,100 km^{-2} in the U.S. In contrast, these countries spend on average less than $2000 km^{-2} on their national parks and nature preserves.

♦ *perversum confusum* See ²group: *perversum confusum*.

♦ *perversum simplex n*. An attempted mating of one male with another (Lincoln et al. 1985).
cf. copulation: homosexual copulation

♦ **pessimal** *adj*.
1. Referring to environmental-factor levels that are close to an organism's tolerance level (Lincoln et al. 1985).
2. Referring to a least favorable condition for an organism (Lincoln et al. 1985).

♦ **pessimum** *n*. Conditions outside those for an organism's optimum growth and development or maintenance of a system (Lincoln et al. 1985).

♦ **pet** *n*.
1. Any animal that is domesticated, or tamed, and kept as a favorite or treated with indulgence or fondness; especially, a lamb, or kid, taken into one's house and reared by hand (*Oxford English Dictionary* 1972, entries from 1539).
Note: Undomesticated, or untamed, animals (*e.g.*, fish, hermit crabs, or the Madagascar Hissing Cockroach) are, of course, now kept as pets as well.
2. An artificially reared plant (*Oxford English Dictionary* 1972, entries from 1842).
3. "An indulged (and, usually, spoiled) child" (*Oxford English Dictionary* 1972, entries from 1508).
4. Any person who is indulged, fondled, and treated with special kindness or favor; darling; favorite (*Oxford English Dictionary* 1972, entries from 1755).
cf. animal: companion animal

♦ **Petit effect** See rare-male-mating advantage.

♦ **petric** *adj*. Referring to rock-living communities or species (Lincoln et al. 1985).
cf. ²community: petrochthium, petrodium, phellium

♦ **petricole** See -cole: petricole.
♦ **petrimadicole** See -cole: petrimadicole.
♦ **petrimadicolous** See hygropetric.
♦ **petrium** See ²community: petrium.
♦ **petrobiont** See -biont: petrobiont.
♦ **petrochthium** See ²community: petrochthium.
♦ **petrochthophile** See ¹-phile: petrochthophile.
♦ **petrocole** See -cole: petrocole.
♦ **petrodium** See ²community: petrodium.
♦ **petrodophile** See ¹-phile: petrodophile.
♦ **petrology** See study of: lithology.
♦ **petrophile** See ¹-phile: petrophile.
♦ **pH** See unit of measure: pH.
♦ **phaenology** See study of: phenology.
♦ **phaenotype** See -type, type: phenotype.
♦ **phage** See -phage: bacteriophage.
♦ **-phage** *combining form* One who, or that which, eats or consumes (Michaelis 1963).
adj. -phagous
n. -phagy
cf. -biont, cannibalism, -cole, parasite, -phage, -phagy, ¹-phile, -troph, -vore, zoan, zoite
[Greek *phagein*, to eat]

A CLASSIFICATION OF KINDS OF PHAGES[a]

  I. monophage
    A. generic monophage
    B. specific monophage
    C. subgeneric monophage
  II. oligophage
    A. disjunctive oligophage
    B. sequential oligophage
    C. systematic oligophage
  III. polyphage

[a] Otte and Joern (1977 in May and Ahmad 1983).

acridophage *n*. An organism that feeds on grasshoppers (*e.g.*, a Coyote) (Lincoln et al. 1985).

algophage *n*. An organism that feeds on algae (*e.g.*, a snail species) (Lincoln et al. 1985).

autophage *n*. A young, precocial bird that can locate and secure its own food (Lincoln et al. 1985).

bacteriophage *n*.
1. An ultramicroscopic, filter-passing agent that has the power of destroying bacteria and of inducing bacterial mutation (Michaelis 1963); a virus that infects bacteria (Campbell 1987, G-2).
syn. phage (Campbell 1987, G-2)
2. An organism that feeds on bacteria (*e.g.*, a protozoan species or the Mud Carp) (Lincoln et al. 1985; Scholtissek 1992, 2).

biophage *n*. An organism that consumes, or destroys, other living organisms (*e.g.,* the Chinese Mantis) (Lincoln et al. 1985).

carpophage *n*. An organism that feeds on fruit or seeds (*e.g.,* a weevil species) (Lincoln et al. 1985).

cerophage *n*. An organism that feeds on wax (*e.g.,* the Wax Moth) (Lincoln et al. 1985).

coprophage See -vore: meridovore.

creophage *n*. An organism that feeds on animals (*e.g.,* a carnivorous-plant species) (Lincoln et al. 1985).

cytophage *n*. An organism that feeds on cells (*e.g.,* a malaria *Plasmodium*) (Lincoln et al. 1985).

dendrophage See -vore: lignivore.

detriophage *n*. An organism that feeds on detritus (*e.g.,* a caddisfly species) (Lincoln et al. 1985).

ectoophage *n*. An insect larva that hatches from an egg deposited on, or near, a supply of host eggs and feeds upon them (*e.g.,* a larva of a chalcidoid-wasp species) (Lincoln et al. 1985).

ectophage *n*. An organism that feeds on the outside of its food source (*e.g.,* a parrotfish that feeds on coral) (Lincoln et al. 1985).

endolithophage See -phage: lithophage.

endoophage, ectoophage *n*. An insect larva that hatches from an egg deposited within a host egg and feeds upon the contents of this single egg (Lincoln et al. 1985).

endophage See -phage: entophage.

entomophage See -vore: insectivore.

entophage, endophage *n*. An organism that feeds from within a food source (*e.g.,* a gall-making insect) (Lincoln et al. 1985).

epilithophage See -phage: lithophage.

euryphage, pleophage, plurivore, polyphage *n*. An organism that utilizes, or tolerates, a wide variety of foods or food species (*e.g.,* the Black Bear) (Lincoln et al. 1985).

galliphage See -vore: gallivore.

geophage *n*. An organism that feeds on soil (*e.g.,* an earthworm species) (Lincoln et al. 1985).

gerontophage *n*. For example, in some gall-midge species; ten spider species; a beetle species: an individual that feeds on its own mother (Gould 1976, 24; Lincoln et al. 1985; Seibt and Wickler 1987, 1903).

Comment: In the beetle *Micromalthus debilis*, some parthenogenetic mothers give birth to a single son. This larva attaches to his mother's skin for about 5 days, inserts his head into her genital aperture, and devours her (Gould 1976, 24).

haematophage, haemophage See -vore: sanguivore.

harpactophage *n*. A predator (Lincoln et al. 1985).

cf. predator

heterophage *n*. A parasite that uses a wide variety of hosts (*e.g.,* the Gypsy Moth) (Lincoln et al. 1985).

hylophage See -vore: lignivore.

ichthyophage See -vore: piscivore.

isophage *n*. An organism that feeds on selected foods or hosts but not a single species (*e.g.,* a chalcidoid-wasp species) (Lincoln et al. 1985).

lepidophage *n*. An organism that feeds on external scales of fish (*e.g.,* a false-cleaner-fish species) (Lincoln et al. 1985).

limiphage, limophage See -vore: limivore.

lithophage *n*.
1. An organism that erodes, or bores into, rock (Lincoln et al. 1985).
   Note: "Lithophage" is divided into "epilithophage," "mesolithophage," and "endolithophage," based on the degree that an organism penetrates rock (Lincoln et al. 1985).
2. An organism that consumes small stones (*e.g.,* the Blue Jay and Domestic Chicken, and other birds that use stones in their gizzards) (Lincoln et al. 1985).

macrophage *n*. An organism that feeds on relatively large food particles or prey (*e.g.,* the Lion) (Lincoln et al. 1985).

macrophytophage *n*. An organism that feeds on only material of more derived plants (*e.g.,* the Baltimore Checkerspot, a butterfly) (Lincoln et al. 1985).

cf. -vore: herbivore

meliphage See -vore: melivore.

mesolithophage See -phage: lithophage.

microphage *n*. An organism that feeds on relatively minute particles or on very small prey (*e.g.,* a caddisfly species) (Lincoln et al. 1985).

monophage *n*. An organism that uses only one kind of food or feeds upon a single organism species or genus (*e.g.,* the Pipevine Swallowtail, a butterfly) (Torre Bueno 1978; Lincoln et al. 1985).

syn. monotroph, univore (Lincoln et al. 1985)
Comment: See comment under monophage-oligophage-polyphage continuum.

mycetophage, mycophage See -vore: fungivore.

myrmecophage *n*. An organism that feeds on ants or termites (*e.g.,* the Aardvark) (Lincoln et al. 1985).

necrophage *n*. An organism that feeds on dead organisms (*e.g.,* a carrion-beetle species) (Lincoln et al. 1985).

necrophytophage, nekrophytophage
n. An organism that feeds on dead plants (*e.g.,* a fungus species) (Lincoln et al. 1985). *cf.* -vore: herbivore

oligophage *n.*
1. An insect that feeds on a group of related plant orders, a single order, or even a single genus (Folsom and Wardle 1934 in Torre-Bueno 1978).
2. An organism that uses only a few food species, typically species restricted to a single genus, family, or order (*e.g.,* the Black Swallowtail, a butterfly) (Lincoln et al. 1985).
3. An organism that requires only a small nutrient supply or that is restricted to a narrow range of nutrients (Lincoln et al. 1985).
4. An insect that visits only a small variety of plant species (Lincoln et al. 1985).
See monophage-oligophage-polyphage continuum.
syn. oligotroph (Lincoln et al. 1985)

oophage *n.* An organism that feeds on eggs (*e.g.,* a snake species) (Lincoln et al. 1985).

ophiophage *n.* An organism that feeds on snakes (*e.g.,* the roadrunner bird) (Lincoln et al. 1985).

paedophage *n.* An organism that feeds on embryos, young of other species (*e.g.,* a chalcidoid-wasp species that parasitizes moth eggs) (Lincoln et al. 1985).

panphytophage *n.* An organism that feeds on both less and more derived plant species (*e.g.,* the Mountain Goat) (Lincoln et al. 1985).

pantophage *n.* An organism that feeds on many kinds of organic material (*e.g.,* the Black Bear) (Lincoln et al. 1985).
syn. omnivore (Lincoln et al. 1985)

phycophage See -phage: algophage.

phyllophage *n.* An organism that feeds on leaves (*e.g.,* the Koala) (Lincoln et al. 1985).
syn. folivore (Lincoln et al. 1985)

phytophage *n.* An organism that feeds on living and dead plant material (*e.g.,* a sowbug species) (Lincoln et al. 1985).
syn. herbivore, phytophile (Lincoln et al. 1985)

plasmophage *n.* An organism that feeds on body fluid (Lincoln et al. 1985).

pleophage See -phage: polyphage.

pollenophage *n.* An organism that feeds on pollen (*e.g.,* a bee species) (Lincoln et al. 1985).

polyphage *n.* An organism that feeds on a wide variety of foods or food species (Lincoln et al. 1985).
See monophage-oligophage-polyphage continuum.

syn. euryphage, pleophage, plurivore (Lincoln et al. 1985)

rhizophage See -vore: radicivore.

rypophag *n.* An organism that feeds on putrid matter or refuse (*e.g.,* a bot-fly species) (Lincoln et al. 1985).

saprophage *n.* An organism that feeds on dead, or decaying, matter (*e.g.,* a fungus species) (Lincoln et al. 1985).
cf. -phage: saprophytophage, zoosaprophage; -troph-: saprotroph

saprophytophage *n.* An organism that feeds on decomposing plant material (*e.g.,* a springtail [insect] species) (Lincoln et al. 1985).

sarcophage See -vore: carnivore.

scatophage See -vore: merdivore.

stenophage *n.* An organism that uses, or is tolerant of, only a limited variety of foods or food species (Lincoln et al. 1985).
cf. -phage: euryphage

sycophage *n.* An organism that feeds on figs (*e.g.,* a monkey species) (Lincoln et al. 1985).

tecnophage *n.* An organism that feeds on its own eggs (Lincoln et al. 1985).

telmophage *n.* An organism that feeds on blood from a pool produced by tissue laceration (*e.g.,* a gnat) (Lincoln et al. 1985).
syn. pool feeder (Lincoln et al. 1985)

xylophage See -vore: lignivore.

zoosaprophage *n.* An organism that feeds on decaying animal matter (*e.g.,* the Polar Bear) (Lincoln et al. 1985).
cf. -phage: saprophage, saprophytophage, zoosaprophage; -troph-: saprotroph; -vore: zoosuccivore

♦ **-phagia** *combining form* The consumption, or eating, of (Michaelis 1963).
adj. -phagous
cf. cannibalism, -phage
[Greek *-phagia* < *phagein,* to eat]

aphagia *n.* An organism's lack of eating (Hinde 1970, 269; Dewsbury 1978, 213).
cf. -phagia: hyperphagia
Comment: In rats, aphagia and hyperphagia can be caused by lesions of different parts of their brains (Dewsbury 1978, 213).

hyperphagia *n.* An organism's overeating that can lead to obesity (Hinde 1970, 269; Dewsbury 1978, 213).
cf. -phagia: aphagia

placentophagia *n.* In almost all species of placental mammals: a mother's ingestion of her afterbirth (placenta, fluids, and membranes), usually occurring only at delivery of young (Tinklepaugh and Hartman 1930; M.B. Kristal 1980, personal communication).
syn. placentophagy (Lehrman in Young 1961)
Comment: This term is sometimes misspelled as "placentaphagia."

◆ **phagostimulant** See chemical-releasing stimulant: phagostimulant.

◆ **phagotroph** See -troph-: phagotroph.

◆ **-phagy** *combining form* The consumption, or eating, of (Michaelis 1963).
adj. -phagous
cf. phage
[Greek *-phagia* < *phagein*, to eat]
autocoprophagy *n.* For example, in Rabbits and their kin: an organism's consumption of its own feces (Lincoln et al. 1985).
syn. refection (Lincoln et al. 1985)
autophagy *adj.* For example, in precocial bird species: a young individual's capability of locating and securing its own food (Lincoln et al. 1985).
matriphagy *n.* A young animal's consuming its mother (Theodore Evans in Anonymous 1995b, 34, illustration).
Comments: Aristotle noted spider matriphagy. In an Australian social-spider species, *Doaea ergandros*, the spiderlings suck nutrient-rich blood from their unresisting mother's leg joints and eventually dissolve her insides with their digestive juices which they consume.
oophagy *n.* For example, in some social-insect species: an individual's eating a conspecific egg (Free et al. 1969 in Fisher and Pomeroy 1990, 801).
cf. cannibalism: egg cannibalism; -cide: infanticide

◆ **phalaenophile** See ¹-phile: phalaenophile.

◆ **phalanx** See ²group: phalanx.

◆ **phallic threat** See display: genital display: genital presentation.

◆ **phallocarp** *n.* A decorative penis sheath used by men in the highlands of New Guinea (Diamond 1996, 85).
Comments: Phallocarps are up to 2.3 feet long; are often bright red or yellow; are variously decorated at their tips with fur, leaves, or a forked ornament; and vary in erection angle (Diamond 1996, 85).

◆ **phaneric coloration** See coloration: phaneric coloration.

◆ **phaneric mimesis** See mimicry: mimesis: phaneric mimesis.

◆ **phanerogenic** See -genic: phanerogenic.

◆ **phaoplankton** See plankton: phaoplankton.

◆ **pharmacognosy** *n.* Human knowledge of drugs, especially their origins, structures, and chemical constitutions (Michaelis 1963).
[New Latin *pharmacognosia* < Greek *pharmakon*, drug + *gnōsis*, knowledge]
zoopharmacognosy *n.* One species' using a substance(s) produced by another as a defense against parasites (Gibbons 1992a, 921; Rodriguez and Wrangham 1992 in Clayton and Wolfe 1993, 60).

syn. self-medicating behavior, self-medication (Clayton and Wolfe 1993, 60)
cf. anting
Comment: An example of zoopharmacognosy is a Tobacco Hornworm's using nicotine in its food against bacteria; possible other examples include a Chimpanzee's using juice of *Vernonia amygdalina* against protozoans and helminths, a Spanish-Dancer Nudibranch's using a sponge against fungi, a Kodiak Bear's using roots of *Ligusticum* spp. against viruses and bacteria, and some birds using ants against mites and lice (Clayton and Wolfe 1993).

◆ **pharmacology** See study of: pharmacology.

◆ **pharotaxis** See taxis: pharotaxis.

◆ **pharyngeal gill slits** See organ: pharyngeal gill slits.

◆ **phase** *n.*
1. A characteristic developmental, or behavioral, stage of an organism (Lincoln et al. 1985).
2. A basic unit in cytology (Lincoln et al. 1985).
diplophase *n.* The diploid state of an organism's life cycle (Lincoln et al. 1985).
syn. diplont, zygophase (Lincoln et al. 1985)
escort phase *n.* For example, in the Locust Leafminer Beetle: a long, passive phase in which a male remains with his mate while she feeds, rests, and moves about [coined by Kirkendall 1984, 909].
exponential-growth phase See phase: log-growth phase.
gamophase, haplophase See haplont.
lag-growth phase *n.* A period in which little, or no, population growth occurs and that precedes its exponential growth phase (Lincoln et al. 1985).
log-growth phase, logarithmic-growth phase *n.* The period in which maximum population growth occurs, with its number of individuals doubling per unit of time (Lincoln et al. 1985).
syn. exponential-growth phase (Lincoln et al. 1985)
cf. phase: lag-growth phase
nomadic phase *n.* In Army Ants: the period of a colony's activity cycle during which it forages more actively for food and moves frequently from one bivouac site to another; the queen does not lay eggs, and most brood individuals are larvae (Wilson 1975, 426, 590).
cf. phase: statary phase
sensitive phase See period: sensitive period.
socialization phase *n.* In some social animal species: the certain stage of an animal's development during which it has

specific social contacts important for its socialization, *q.v.* (Immelmann and Beer 1989, 276).

statary phase *n.* A relatively quiescent phase in a social insect colony (Lincoln et al. 1985); *e.g.,* in Army Ants, during a statary phase, a colony does not move from site to site, its queen lays eggs, and the bulk of its brood is eggs and pupae (Wilson 1975, 426, 596).

cf. phase: nomadic phase

stationary-growth phase *n.* A period of little, or no, population growth that follows an exponential growth period (Lincoln et al. 1985).

cf. lag-growth phase

zygophase See phase: diplophase.

♦ **phase sequence** *n.* A connection of a number of cell assemblies, *q.v.* (Immelmann and Beer 1989, 41).

♦ **phase specificity** *n.* An environmental influence's having an especially lasting effect on an animal's behavioral development (*e.g.,* socialization, imprinting) when this influence occurs during an animal's sensitive period, *q.v.* (Immelmann and Beer 1989, 218).

♦ **phatic** *adj.* Referring to communication that establishes and maintains contact between animals (Sebeok 1962 in Wilson 1975, 217).

cf. cognitive

♦ **phellium** See ²community: phellium.

♦ **phellophile** See ¹-phile: phellophile.

♦ **phen-, pheno-** *combining form* Referring to something that is apparent; originally used in chemistry, now used in biology as well (*Oxford English Dictionary* 1972, entry from 1841; Brown 1956).

[Greek *phaino,* appear, shine, derived from to come to light, cause to appear, show, *Oxford English Dictionary* 1972]

♦ **phene** See character: phene.

♦ **phenetic** *adj.* Referring to overall similarity of taxonomic groups based on many characters selected without regard to their evolutionary history, thus including character states arising from common ancestry, parallel evolution, and convergent evolution (Lincoln et al. 1985).

♦ **phenetic method** See method: phenetic method.

♦ **phenetic systematics, phenetics** See study of: systematics: phenetic systematics.

♦ **phengophile** See ¹-phile: phengophile.

♦ **phengophobe** See -phobe: phengophobe.

♦ **phenocline** See cline: phenocline.

♦ **phenocopy** See -type, type: phenotype: phenocopy.

♦ **phenocritic** *adj.* Referring to a phase during an organism's development at which

two different genotypes begin to diverge morphologically and physiologically (Lincoln et al. 1985).

♦ **phenocritical period** See period: phenocritical period.

♦ **phenodeme** See deme: phenodeme.

♦ **phenodeviant** *n.*

1. A scarce aberrant individual that appears regularly in a population because of the segregation of certain unusual combinations of individually common genes, *e.g.,* a *Drosophila* with a pseudotumor or a missing or defective crossvein, a chicken with crooked toes, or a mammal with diabetes (Lerner 1954 and Milkman 1970 in Wilson 1975, 72, 591).

2. An individual that deviates phenotypically from its population norm due to special gene combinations such as excessive homozygosity (Lincoln et al. 1985).

♦ **phenogenesis** See -genesis: phenogenesis.

♦ **phenogenetics** See -genetics: phenogenetics.

♦ **phenogram** See -gram: phenogram.

♦ **phenology** *n.* The timing of a recurring natural phenomenon (Immelmann and Beer 1989, 218), *e.g.,* the flight of adults of an insect species or the flowering of a plant species (Waldbauer and LaBerge 1985). See study of: phenology.

adj. phenological

Comment: When "phenology" is used in a term such as "flowering phenology," it means flowering time of a plant with regard to one or more ecological factors such as climate, drought, and photoperiod.

♦ **phenome** See -type, type: phenotype.

♦ **phenomenology** See study of: phenology.

♦ **phenomenon** *n., pl.* **phenomena**

1. A thing that appears, or is perceived or observed; an individual fact, occurrence, or change that is perceived by any of one's senses or mind: applied chiefly to a fact, or occurrence whose cause, or explanation, is in question (*Oxford English Dictionary* 1972, entries from 1639).

2. Philosophically, that which one's mind, or sense, directly notes; an immediate object of perception, distinguished from substance or a thing in itself; opposed to noumenon (*Oxford English Dictionary* 1972, entries from 1788).

[Late Latin *phaenomenon* < Greek *phainomenon* < present participle of *phainesthal,* to appear, passive of *phainein,* to show]

epiphenomenon *n.*

1. A causally inconsequential and incidental, or secondary, effect compared to a primary one (*e.g.,* mentality according

n–r

to mechanistic materialists) (Immelmann and Beer 1989, 89).

2. An overt concomitant of a physiological state or process (*e.g.,* urination, defecation, trembling, feather ruffling, or hair erection) due to arousal, or activation, of an animal's autonomic nervous system in situations of danger or threat (Immelmann and Beer 1989, 89–90).

syn. Epiphänomen (Immelmann and Beer 1989, 89)

Comment: An epiphenomenon might evolve into a behavior with a particular function (Immelmann and Beer 1989, 89–90).

phenomenon of minimum population size of social species *n.* A conspecific animal group's not showing certain adaptive behavior, or a frequency of this behavior, until it reaches a certain size; *e.g.,* in blue-crowned hanging parrots, a group's emitting a full repertory of vocalizations, and in Weaverbirds, a male group's attracting mates (Buckley 1967 and Collias and Collias 1969 in Wilson 1975, 117).

retrorecognition phenomenon *n.* The process that involves: (1) observing a fact of nature in the context of an existing explanatory framework; (2) not logically explaining this fact in the existing framework but still unquestioning and ignoring, or accepting, this fact as a given property of the world, or simply postulating it to be true; and (3) advancing a new theory in which the observed fact now has a compelling and reasoned explanation and, at the same time, is retroactively recognized as an anomaly in the context of the old theory (Lightman and Gingerich 1992, 693–694).

cf. fact: fact-in-itself, reasoned fact

Comment: Examples of retrorecognition phenomena are the continental-fit problem and the adaptation-of-organism problem (Lightman and Gingerich 1992).

suitor phenomenon See behavior: mate guarding behavior: precopulatory mate guarding.

♦ **phenometry** See -metry: phenometry.

♦ **phenon** See -type, type: phenotype: phenon.

♦ **phenotype** See -type, type: phenotype.

♦ **phenotype matching** *n.* For example, in a sweat-bee species: an individual's identifying a conspecific as kin, or non-kin, by comparing characteristics of its own or those of kin which it has experienced with this conspecific's characteristics (Sherman and Holmes 1985; Immelmann and Beer 1989, 218).

cf. allele: recognition allele, template

♦ **phenotype switching** *n.* For example, in several microorganism species: a revers-

ible, high-frequency phenotype change, "including oscillating expression (phase variation) or antigenic diversification of surface macromolecules" (Rosengarten and Wish 1989, 315).

♦ **phenotypic flexibility** *n.* An organism's capacity to function in a range of environments which may include plastic and stabile responses (Thoday 1953 in Bradshaw 1965, 116).

See developmental homeostasis.

cf. plasticity

♦ **phenotypic plasticity, plasticity** See plasticity: phenotypic plasticity.

♦ **phenotypic polymorphism** See -morphism: polymorphism: phenotypic polymorphism.

♦ **phenotypic selection** See selection: phenotypic selection.

♦ **phenotypic sex determination** See sex determination: phenotypic sex determination.

♦ **phenotypic variance** See variance: phenotypic variance.

♦ **pheromone** See chemical-releasing stimulus: semiochemical: pheromone.

♦ **¹-phile** *combining form* In organisms: thriving in, or living in, a particular habitat (Lincoln et al. 1985).

adj. -philous

n. -phily

cf. -cole, -philo, -vore, -zoite

[Greek *-philos,* loving < *phileein,* to love]

acarophile *n.* An organism that lives in association with mites (Lincoln et al. 1985).

acidophile *n.* An organism that lives in an acidic environment (*e.g.,* bog plants) (Lincoln et al. 1985).

acrodendrophile *n.* An organism that lives in tree tops (*e.g.,* thousands of species of canopy organisms in tree canopies in the tropics) (Lincoln et al. 1985).

cf. -cole: alsocole, arboricole, arbusticole, fruticole, hylacole, nemoricole, silvicole, thamnocole; ¹-phile: aiphyllophile, dendrophile, halorgadophile, hydrophile, hylodophile, orgadophile

actophile *n.* An organism that lives on rocky seashores (*e.g.,* some algal, bivalve, seastar, and urchin species) (Lincoln et al. 1985).

cf. -cole: calcosaxicole, petrimadicole, petrocole, rupicole; ¹-phile: calciphile, chasmophile, lithophile, petrochthophile, petrodophile, petrophile, phellophile

aerohygrophile *n.* An organism that lives in high atmospheric humidity (*e.g.,* a cloud-forest organism) (Lincoln et al. 1985).

cf. -cole: deserticole, siccocole, xerocole; ¹-phile: cheradophile, chersophile, helohylophile, hylodophile, hydrophile, lochnodophile, subxerophile, syrtidophile, telmatophile, xerohylophile, xerophile

aerophile *n.* An organism that lives in exposed windy habitats (*e.g.*, a plant on a mountain slope) (Lincoln et al. 1985).
cf.-chore: aerochore

agrophile *n.* An organism that lives in cultivated soil (*e.g.*, a crop species) (Lincoln et al. 1985).

aigialophile *n.* An organism that lives in beaches (*e.g.*, a Coquina Clam) (Lincoln et al. 1985).
cf. -cole: aigicole, amanthicole, ammocole, amnicole, arenicole, argillicole, limicole, luticole, thinicole; [1]-phile: amathophile, ammochtophile, ammophile, anemophile, argillophile, cheradophile, enaulophile, pelochthophile, pelophile, phellophile, psamathophile, psammophile, spiladophile, syrtidophile, thinophile

aiphyllophile *n.* An organism that lives in evergreen woodlands (*e.g.*, the plant Goldthread) (Lincoln et al. 1985).
cf. -cole: alsocole, arboricole, arbusticole, fruticole, hylacole, nemoricole, silvicole, thamnocole; [1]-phile: acrodendrophile, dendrophile, halorgadophile, hydrophile, hylodophile, orgadophile

aithalophile *n.* An organism that lives in evergreen thickets (*e.g.*, a moss species) (Lincoln et al. 1985).

aletophile *n.* An organism that lives on roadside verges or beside railway tracks (*e.g.*, the red clover) (Lincoln et al. 1985).

alkaliphile *n.* An organism that lives in alkaline habitats (*e.g.*, a sedge species) (Lincoln et al. 1985).

alsophile *n.* An organism that lives in woody groves (*e.g.*, a violet species) (Lincoln et al. 1985).
cf. -cole: hylacole

amathophile *n.* An organism that lives in sandy plains (*e.g.*, a cactus species) (Lincoln et al. 1985).
cf. -cole: aigicole, amanthicole, ammocole, amnicole, arenicole, argillicole, limicole, luticole, thinicole; [1]-phile: aigialophile, ammochtophile, ammophile, anemophile, argillophile, cheradophile, enaulophile, pelochthophile, pelophile, phellophile, psamathophile, psammophile, spiladophile, syrtidophile, thinophile

ammochtophile *n.* An organism that lives on, or in, sandy banks (*e.g.*, the Sandbar Willow) (Lincoln et al. 1985).
cf. -cole: aigicole, amanthicole, ammocole, amnicole, arenicole, argillicole, limicole, luticole, thinicole; [1]-phile: aigialophile, amathophile, ammophile, anemophile, argillophile, cheradophile, enaulophile, pelochthophile, pelophile, phellophile,

psamathophile, psammophile, spiladophile, syrtidophile, thinophile

ammophile *n.* An organism that lives on or in sand or in sandy habitats (*e.g.*, a beach grass, *Ammophila breviligulata*) (Lincoln et al. 1985).
cf. -cole: aigicole, amanthicole, ammocole, amnicole, arenicole, argillicole, limicole, luticole, thinicole; [1]-phile: aigialophile, amathophile, ammochtophile, anemophile, argillophile, cheradophile, enaulophile, pelochthophile, pelophile, phellophile, psamathophile, psammophile, spiladophile, syrtidophile, thinophile

ancophile *n.* An organism that lives in canyon forests (*e.g.*, a sycamore species) (Lincoln et al. 1985).

androphile, anthropophile *n.* An organism that lives in proximity to Humans (*e.g.*, a cockroach species) (Lincoln et al. 1985).

anemophile *n.*
1. An organism that lives in sand draws (*e.g.*, a digger-wasp species) (Lincoln et al. 1985).
2. An organism that is dispersed by wind (Lincoln et al. 1985).
adj. anemochorous
cf. -cole: aigicole, amanthicole, ammocole, amnicole, arenicole, argillicole, limicole, luticole, thinicole; -chore: anemochore; [1]-phile: aigialophile, amathophile, ammochtophile, ammophile, argillophile, cheradophile, enaulophile, pelochthophile, pelophile, phellophile, psamathophile, psammophile, spiladophile, syrtidophile, thinophile

anheliophile *n.* An organism that lives in diffuse sunlight (*e.g.*, a fern species) (Lincoln et al. 1985).
cf. -cole: lucicole, umbraticole; [1]-phile: heliophile, helioxeriophile, lygophile, phengophile, skotophile, sciophile

anthophile *n.* An organism that is attracted to, or feeds, on flowers (Lincoln et al. 1985).
cf. pollinator

anthropophile See [1]-phile: androphile.

argillophile *n.* An organism that lives in clay or mud (*e.g.*, a Mudskipper fish) (Lincoln et al. 1985).
cf. calcifuge; -cole: aigicole, amanthicole, ammocole, amnicole, arenicole, argillicole, calcicole, gelicole, geocole, halicole, humicole, limicole, luticole, pergelicole, perhalicole, silicole, terricole, thinicole; [1]-phile: aigialophile, amathophile, ammochtophile, ammophile, anemophile, cheradophile, enaulophile, geophile, gypsophile, nitrophile, pelochthophile, pelophile, psamathophile, psammophile, spiladophile, syrtidophile, thinophile

barophile *n.* An organism that lives under high hydrostatic, or atmospheric, pressure (*e.g.,* a deep-sea fish) (Lincoln et al. 1985).

basophile *n.* An organism that lives in alkaline habitats (*e.g.,* a crucifer species) (plant) (Lincoln et al. 1985).

bathophile, bathyphile *n.* An organism that lives in lowlands or in deep sea (*e.g.,* a mangrove species) (Lincoln et al. 1985).

cf. [1]-phile: pontophile

biophile *n.* An organism that lives on other living organisms (*e.g.,* an epiphytic plant) (Lincoln et al. 1985).

See [3]phile: biophile.

bryophile *n.* An organism that lives in habitats rich in mosses and liverworts (*e.g.,* a mite species) (Lincoln et al. 1985).

cf. -cole: bryocole, muscicole, sphagnicole; [1]-phile: sphagnophile

calciphile *n.*

1. An organism that lives in environments rich in calcium salts (*e.g.,* a plant species) (Lincoln et al. 1985).
2. An animal that bores into calcareous shells (Lincoln et al. 1985).

cf. -cole: calcosaxicole, petrimadicole, petrocole, rupicole; [1]-phile: actophile, chasmophile, lithophile, petrochthophile, petrodophile, petrophile, phellophile

carboxyphile *n.* An organism that lives in carbon-dioxide-rich habitats (*e.g.,* an alga species) (Lincoln et al. 1985).

chalicophile *n.* An organism that lives on gravel slides (*e.g.,* a rock crawler (insect) (Lincoln et al. 1985).

chasmophile *n.* An organism that lives in rock crevices and fissures (*e.g.,* a fern species) (Lincoln et al. 1985).

cf. -cole: calcosaxicole, petrimadicole, petrocole, rupicole; [1]-phile: actophile, calciphile, lithophile, petrochthophile, petrodophile, petrophile, phellophile

cheiropterophile See [2]-phile: chiropterophile.

cheradophile *n.* An organism that lives on wet sandbars (*e.g.,* a tiger- beetle species) (Lincoln et al. 1985).

cf. -cole: aigicole, amanthicole, ammocole, amnicole, arenicole, argillicole, limicole, luticole, thinicole; [1]-phile: aigialophile, amathophile, ammochtophile, ammophile, anemophile, argillophile, chersophile, enaulophile, pelochthophile, pelophile, phellophile, psamathophile, psammophile, spiladophile, syrtidophile, thinophile

chersophile *n.* An organism that lives in dry wasteland habitats (*e.g.,* a sagebrush species) (Lincoln et al. 1985).

cf. -cole: deserticole, siccocole, xerocole; [1]-phile: aerohygrophile, cheradophile, helohylophile, hylodophile, hydrophile, lochnodophile, subxerophile, syrtidophile, telmatophile, xerohylophile, xerophile

chianophile *n.* An organism that lives under prolonged snow cover (*e.g.,* a lichen species) (Lincoln et al. 1985).

chimonophile *n.*

1. An organism that grows during the (cold) winter (*e.g.,* the Common Chickweed, a winter annual) (Lincoln et al. 1985).
2. A plant that shows maximum development during the winter (Lincoln et al. 1985).

chionophile *n.* An organism that lives in snow-covered habitats (*e.g.,* a rhododendron species) (Lincoln et al. 1985).

chledophile, chomophile *n.* An organism that lives in wasteland habitats and rubbish heaps (*e.g.,* the Great Mullein, a plant) (Lincoln et al. 1985).

cf. ruderal

chrymosymphile *n.* A lepidopterous larva that benefits from interactions with ants (Lincoln et al. 1985).

cf. [1]-phile: myrmecophile

circumneutrophile *n.* An organism that lives in conditions of about neutral pH (Lincoln et al. 1985).

coniophile *n.* An organism that lives on substrates enriched by dust containing excreta (Lincoln et al. 1985).

conophorophile *n.* An organism that lives in coniferous forests (*e.g.,* sawfly species that consumes conifers) (Lincoln et al. 1985).

coprophile *n.* An organism that lives on dung or feces (*e.g.,* a burying-beetle species) (Lincoln et al. 1985).

cf. -zoite: coprozoite

coryphile, coryphophile *n.* An organism that lives in alpine meadows (*e.g.,* a butterfly species) (Lincoln et al. 1985).

cf. -cole: caespiticole, leimocole, nomocole, poocole, pratinicole, psicole; [1]-phile: helolochmophile, nomophile, poophile, psilophile, telmatophile

cremnophile *n.* An organism that lives on cliffs (*e.g.,* a *Saxifraga* plant species) (Lincoln et al. 1985).

crenophile *n.* An organism that lives in, or near, a spring (*e.g.,* Watercress) (Lincoln et al. 1985).

crymophile *n.* An organism that lives in polar habitats (*e.g.,* a penguin species) (Lincoln et al. 1985).

cryophile *n.* An organism that lives in a cold environment (*e.g.,* a boreal-plant species or an ant species) (Wheeler 1930, 11; Lincoln et al. 1985).

syn. frigophile

cf. -tropism: thermotropism: negative thermotropism

dendrophile *n.*
1. An organism that lives in trees (*e.g.,* the howler monkey) (Lincoln et al. 1985). *syn.* dendrocole
2. An organism that lives in orchards (*e.g.,* the Honey Bee) (Lincoln et al. 1985).
cf. -cole: alsocole, arboricole, arbusticole, fruticole, hylacole, nemoricole, silvicole, thamnocole; [1]-phile: acrodendrophile, aiphyllophile, halorgadophile, hydrophile, hylodophile, orgadophile

drimophile *n.* An organism that lives in salt basins or alkaline plains (*e.g.,* the alkali bee) (Lincoln et al. 1985).

enaulophile *n.* An organism that lives in sand dunes (*e.g.,* a grass species) (Lincoln et al. 1985).
cf. -cole: aigicole, amanthicole, ammocole, amnicole, arenicole, argillicole, limicole, luticole, thinicole; [1]-phile: aigialophile, amathophile, ammochtophile, ammophile, anemophile, argillophile, cheradophile, pelochthophile, pelophile, phellophile, psamathophile, psammophile, spiladophile, syrtidophile, thinophile

eremophile *n.* An organism that lives in deserts (*e.g.,* the Gila Monster) (Lincoln et al. 1985).

euhydrophile *n.* An organism that lives submerged in fresh water (*e.g.,* a *Ceratophyllum* (plant) species) (Lincoln et al. 1985).

eurotophile *n.* An organism that lives in, or on, leaf mold (*e.g.,* a millepede species) (Lincoln et al. 1985).

eurythermophile See [1]-phile: thermophile: eurythermophile.

eutroglophile See [1]-phile: troglophile.

frigophile See [1]-phile: cryophile.

geophile *n.* An organism that lives in soil (*e.g.,* a springtail [hexapod] species) (Lincoln et al. 1985).
cf. calcifuge; -cole: aigicole, amanthicole, ammocole, amnicole, arenicole, argillicole, calcicole, gelicole, geocole, halicole, humicole, limicole, luticole, pergelicole, perhalicole, silicole, terricole, thinicole; [1]-phile: aigialophile, amathophile, ammochtophile, ammophile, anemophile, argillophile, cheradophile, enaulophile, geophile, gypsophile, nitrophile, pelochthophile, pelophile, psamathophile, psammophile, spiladophile, syrtidophile, thinophile

gypsophile *n.* An organism that lives on chalk, or gypsum-rich, soil (*e.g., Gypsophila,* Baby's-Breath plant) (Lincoln et al. 1985).
cf. calcifuge; -cole: aigicole, amanthicole, ammocole, amnicole, arenicole, argillicole, calcicole, gelicole, geocole, halicole, humicole, limicole, luticole, pergelicole, perhalicole, silicole, terricole, thinicole; [1]-phile: aigialophile, amathophile, ammoch-

tophile, ammophile, anemophile, argillophile, cheradophile, enaulophile, geophile, gypsophile, nitrophile, pelochthophile, pelophile, psamathophile, psammophile

halophile *n.* An organism that lives in saline habitats (*e.g.,* a starfish species) (Lincoln et al. 1985).

heliophile *n.* An organism that lives under high light intensity (*e.g.,* a euphorb plant species) (Lincoln et al. 1985).
cf. -cole: lucicole, umbraticole; [1]-phile: anheliophile, helioxeriophile, lygophile, phengophile, skotophile, sciophile

helioxerophile *n.* An organism that lives under high light intensity and drought (*e.g.,* the Joshua tree) (Lincoln et al. 1985).
cf. -cole: lucicole, umbraticole; [1]-phile: anheliophile, heliophile, lygophile, phengophile, skotophile, sciophile

helohylophile *n.* An organism that lives in wet, or swampy, forests (*e.g.,* the Bald Cypress, a tree) (Lincoln et al. 1985).
cf. -cole: deserticole, siccocole, xerocole; [1]-phile: aerohygrophile, cheradophile, chersophile, hylodophile, hydrophile, lochnodophile, subxerophile, syrtidophile, telmatophile, xerohylophile, xerophile

helolochmophile *n.* An organism that lives in meadow thickets (*e.g.,* the American Plum) (Lincoln et al. 1985).
cf. -cole: caespiticole, leimocole, nomocole, poocole, pratinicole, psicole; [1]-phile: coryphile, nomophile, poophile, psilophile, telmatophile

helophile *n.* An organism that lives in marshes (*e.g.,* cattail plant species) (Lincoln et al. 1985).

helorgadophile *n.* An organism that lives in swampy woodlands (*e.g.,* the Tupelo tree) (Lincoln et al. 1985).
cf. -cole: alsocole, arboricole, arbusticole, fruticole, hylacole, nemoricole, silvicole, thamnocole; [1]-phile: acrodendrophile, aiphyllophile, dendrophile, hydrophile, hylodophile, orgadophile

hemerophile *n.* An organism that lives in human-influenced, or cultivated, habitats (*e.g.,* a dandelion plant species) (Lincoln et al. 1985).

hemichimonophile *n.*
1. An organism that lives under cold conditions (*e.g.,* the Polar Bear) (Lincoln et al. 1985).
2. A plant species that thrives under cold conditions and begins growth even during frost (Lincoln et al. 1985).

histophile *n.* An organism that inhabits living host tissue (*e.g.,* a filarial-worm species) (given as an adjective in Lincoln et al. 1985).
syn. parasite (Lincoln et al. 1985)

n–r

hydrophile *n.* An organism that lives in wet, or aquatic, habitats (*e.g.,* mayfly species) (Lincoln et al. 1985).
cf. -cole: alsocole, arboricole, arbusticole, fruticole, hylacole, nemoricole, silvicole, thamnocole; [1]-phile: acrodendrophile, aiphyllophile, dendrophile, halorgadophile, hylodophile, orgadophile

hydrotribophile *n.* An organism that lives in badlands (*e.g.,* a sunflower species) (Lincoln et al. 1985).

hygrophile *n.* An organism that thrives in moist habitats (*e.g.,* an earthworm species) (Lincoln et al. 1985).
cf. [1]-phile: hydrophile

hylophile *n.*
1. An organism that lives in forests (*e.g.,* the Spotted Owl) (Lincoln et al. 1985).
2. An organism that lives, or thrives, in forests (Lincoln et al. 1985).

hylodophile *n.* An organism that lives in dry, open woodlands (*e.g.,* the Fragrant Sumac) (Lincoln et al. 1985).
cf. -cole: alsocole, arboricole, arbusticole, fruticole, hylacole, nemoricole, silvicole, thamnocole; [1]-phile: acrodendrophile, aiphyllophile, dendrophile, halorgadophile, orgadophile

keratinophile *n.* An organism that lives on horny (keratin rich) substrates (Lincoln et al. 1985).

kinetophile See [2]species: kinetophile.

laurophile *n.* An organism that lives in sewers and drains (*e.g.,* the Norway Rat) (Lincoln et al. 1985).

lichenophile *n.*
1. An organism that lives on lichens (*e.g.,* a bagworm species) (Lincoln et al. 1985).
2. An organism that thrives on lichens or has an affinity for lichen-rich habitats (Lincoln et al. 1985).
cf. -cole: lichenicole

lignophile *n.* An organism that lives on, or in, wood (*e.g.,* a fungus species) (Lincoln et al. 1985).
cf. -cole: lignicole; [1]-phile: proxylophile, xylophile

limnodophile *n.* An organism that lives in salt marshes (*e.g.,* a salt-marsh grass species) (Lincoln et al. 1985).
cf. -cole: paludicole, pontohalicole, telmicole; [1]-phile: helophile

limnophile *n.* An organism that lives in lakes or ponds (*e.g.,* the Leopard Frog) (Lincoln et al. 1985).
cf. -cole: limnicole

lithophile *n.* An organism that lives in rocky, or stony, habitats (*e.g.,* the horned lizard or the centipede *Lithobius forficatus*) (Lincoln et al. 1985).

cf. -cole: calcosaxicole, petrimadicole, petrocole, rupicole; [1]-phile: actophile, calciphile, chasmophile, petrochthophile, petrodophile, petrophile, phellophile

lochmodophile *n.* An organism that lives in dry thickets (*e.g.,* the Osage-Orange tree) (Lincoln et al. 1985).
cf. -cole: deserticole, siccocole, xerocole; [1]-phile: aerohygrophile, cheradophile, chersophile, helohylophile, hylodophile, hydrophile, subxerophile, syrtidophile, telmatophile, xerohylophile, xerophile

lochmophile *n.* An organism that lives in thickets (*e.g.,* a greenbrier species) (Lincoln et al. 1985).

lophophile *n.* An organism that lives on hilltops (*e.g.,* the White Oak) (Lincoln et al. 1985).

luciphile, photophile *n.* An organism that lives in open, well-lit habitats (*e.g.,* a euphorb species) (Lincoln et al. 1985).
cf. lucifugous, lucipetal

lygophile *n.* An organism that lives in dark, or shaded, habitats (*e.g.,* an orchid species) (Lincoln et al. 1985).
cf. -cole: lucicole, umbraticole; [1]-phile: anheliophile, heliophile, helioxeriophile, phengophile, skotophile, sciophile

macrothermophile See [1]-phile: thermophile: macrothermophile.

melangeophile *n.* An organism that lives in, or on, black loam (*e.g.,* some earthworm species) (Lincoln et al. 1985).

mellitophile *n.* An organism that spends part of its life cycle in association with bees (*e.g.,* the Greater Wax Moth) (Lincoln et al. 1985).
See [2]-phile (pollinated by): mellitophile

mesochthonophile *n.* An organism that lives in midlands (Lincoln et al. 1985).

mesophile, noterophile *n.* An organism that lives under intermediate, or moderate, environmental conditions (Lincoln et al. 1985).
Comment: "Mesophile" is sometimes restricted to an organism that lives under moderate moisture or temperature (Lincoln et al. 1985).

mesothermophile See [1]-phile: thermophile: mesothermophile.

microaerophile *n.* An organism that lives in a free-oxygen concentration significantly less than that of the atmosphere (Lincoln et al. 1985).

microphile, microphil *n.* An organism that lives within a narrow temperature range (*e.g.,* a tropical-orchid species) (Lincoln et al. 1985).

microthermophile See [1]-phile: thermophile: microthermophile.

myrmecophile *n.* "An organism that must spend at least part of its life cycle with ant colonies" (*e.g.,* some species of aphids, beetles, butterflies, or silverfish) (Wilson 1975, 356, 589).

See also ²-phile (pollinated by): myrmecophile below.

cf. ¹-phile: chrymosymphile

namaphile *n.* An organism that lives in brooks and streams (*e.g.,* a caddisfly species) (Lincoln et al. 1985).

necrophile, necrophage, saprophyte *n.* An organism that eats dead organic matter (Lincoln et al. 1985).

nitrophile *n.* An organism that lives in soil rich in nitrogenous compounds (Lincoln et al. 1985).

cf. calcifuge; -cole: aigicole, amanthicole, ammocole, amnicole, arenicole, argillicole, calcicole, gelicole, geocole, halicole, humicole, limicole, luticole, pergelicole, perhalicole, silicole, terricole, thinicole; ¹-phile: aigialophile, amathophile, ammochtophile, ammophile, anemophile, argillophile, cheradophile, enaulophile, geophile, gypsophile, nitrophile, pelochthophile, pelophile, psamathophile, psammophile, spiladophile, syrtidophile, thinophile

nomophile *n.* An organism that lives in pastures (*e.g.,* a thistle species) (Lincoln et al. 1985).

cf. -cole: caespiticole, leimocole, nomocole, poocole, pratinicole, psicole; ¹-phile: coryphile, helolochmophile, poophile, psilophile, telmatophile

noterophile See ¹-phile: mesophile.

oceanophile *n.* An organism that lives in oceans (*e.g.,* the Beluga Whale) (Lincoln et al. 1985).

ochthophile *n.* An organism that thrives on banks (Lincoln et al. 1985).

oligonitrophile *n.* An organism that lives in low-nitrogen habitats; a pitcherplant species (Lincoln et al. 1985).

ombrophile, ombrophil, pluviophile *n.* An organism that lives in habitats with abundant rain (*e.g.,* a rainforest lichen) (Lincoln et al. 1985).

opophile *n.* An organism that lives on, or feeds on, sap (*e.g.,* a sap-beetle species) (Lincoln et al. 1985).

cf. -vore: phytosuccivore

orgadophile *n.* An organism that lives in open woodlands (*e.g.,* a blackberry species) (Lincoln et al. 1985).

cf. -cole: alsocole, arboricole, arbusticole, fruticole, hylacole, nemoricole, silvicole, thamnocole; ¹-phile: acrodendrophile, aiphyllophile, dendrophile, halorgadophile, hylodophile

ornithocoprophile *n.* An organism that lives in habitats rich in bird droppings (Lincoln et al. 1985).

orophile *n.* An organism that lives in subalpine, or mountainous, regions (*e.g.,* the Colorado Spruce) (Lincoln et al. 1985).

osmophile *n.* An organism that lives in a medium of high osmotic concentration (*e.g.,* a brine-shrimp species) (Lincoln et al. 1985).

oxygeophile *n.* An organism that lives in humus-rich habitats (*e.g.,* a trillium species) (Lincoln et al. 1985).

oxylophile *n.* An organism that lives in humus or a humus-rich habitat (*e.g.,* the Skunk Cabbage) (Lincoln et al. 1985).

adj. oxylophilous, oxylophilus, oxylyphilus

oxyphile *n.* An organism that lives in humus or humus-marsh habitats (Lincoln et al. 1985).

paedophile See pedophile.

pagophile *n.* An organism that lives in foothills (*e.g.,* a sagebrush species) (Lincoln et al. 1985).

pediophile *n.* An organism that lives in uplands (*e.g.,* the Ponderosa Pine) (Lincoln et al. 1985).

pedophile (*Brit.* **paedophile**) *n.* A person who engages in pedophila, *q.v.* (Hinsie and Campbell 1976, 552; Leary 1998, A11).

pelagophile *n.* An organism that lives in open ocean surface waters (*e.g.,* the Portuguese Man-of-War, a jellyfish) (Lincoln et al. 1985).

pelochthophile *n.* An organism that lives on mud banks (*e.g.,* a pygmy-mole-cricket species) (Lincoln et al. 1985).

cf. -cole: aigicole, amanthicole, ammocole, amnicole, arenicole, argillicole, limicole, luticole, thinicole; ¹-phile: aigialophile, amathophile, ammochtophile, ammophile, anemophile, argillophile, cheradophile, enaulophile, pelophile, phellophile, psamathophile, psammophile, spiladophile, syrtidophile, thinophile

pelophile *n.* An organism that lives in clay-rich habitats (*e.g.,* the Shale-Barren Evening Primrose) (Lincoln et al. 1985).

petrochthophile *n.* An organism that lives on rock banks (*e.g.,* a stonecrop species) (Lincoln et al. 1985).

cf. -cole: calcosaxicole, petrimadicole, petrocole, rupicole; ¹-phile: actophile, calciphile, chasmophile, lithophile, petrodophile, petrophile, phellophile

petrodophile *n.* An organism that lives in boulder fields (Lincoln et al. 1985).

cf. -cole: calcosaxicole, petrimadicole, petrocole, rupicole; ¹-phile: actophile, calciphile, chasmophile, lithophile, petrochthophile, petrophile, phellophile

n – r

petrophile *n*. An organism that lives on rocks or in rocky habitats (*e.g.*, a lichen species) (Lincoln et al. 1985).
cf. -cole: calcosaxicole, petrimadicole, petrocole, rupicole; [1]-phile: actophile, calciphile, chasmophile, lithophile, petrochthophile, petrodophile, phellophile

phellophile *n*. An organism that lives on rocky, or stony, ground (*e.g.*, Silversword, a plant) (Lincoln et al. 1985).
cf. -cole: aigicole, amanthicole, ammocole, amnicole, arenicole, argillicole, limicole, luticole, thinicole; [1]-phile: aigialophile, amathophile, ammochtophile, ammophile, anemophile, argillophile, cheradophile, enaulophile, pelochthophile, pelophile, petrophile, psamathophile, psammophile, spiladophile, syrtidophile, thinophile

phengophile *n*. An organism that lives in light (*e.g.*, a green-plant species) (Lincoln et al. 1985).
cf. -cole: lucicole, umbraticole; [1]-phile: anheliophile, heliophile, helioxeriophile, lygophile, skotophile, sciophile

photophile See [1]-phile: luciphile.

phretophile *n*. An organism that lives in water tanks (*e.g.*, a mosquito species) (Lincoln et al. 1985).

phycophile *n*. An organism that lives on algae or in algae-rich habitats (*e.g.*, a periwinkle-snail species) (Lincoln et al. 1985).

phytophile See -cole: planticole; -phage: phytophage.

planktophile *n*. An organism that lives on plankton (Lincoln et al. 1985).
cf. -troph-: planktotroph

pluviophile See [1]-phile: ombrophile.

poleophile *n*. An organism that lives in urban habitats (*e.g.*, the American Cockroach) (Lincoln et al. 1985).

polyhalophile *n*. An organism that lives in a wide range of salinities (*e.g.*, a salmon species) (Lincoln et al. 1985).

pontophile *n*. An organism that lives in the deep sea (*e.g.*, a deep-sea fish) (Lincoln et al. 1985).
cf. -[1]phile: bathophile

poophile *n*. An organism that lives in meadows (*e.g.*, a meadowrue species) (Lincoln et al. 1985).
cf. -cole: caespiticole, leimocole, nomocole, poocole, pratinicole, psicole; [1]-phile: coryphile, helolochmophile, nomophile, psilophile, telmatophile

potamophile *n*. An organism that lives in rivers (*e.g.*, the hippopotamus) (Lincoln et al. 1985).

psamathophile *n*. An organism that lives in the strandline of a sandy seashores (*e.g.*, the Ghost Crab) (Lincoln et al. 1985).

cf. -cole: aigicole, amanthicole, ammocole, amnicole, arenicole, argillicole, limicole, luticole, thinicole; [1]-phile: aigialophile, amathophile, ammochtophile, ammophile, anemophile, argillophile, cheradophile, enaulophile, pelochthophile, pelophile, phellophile, psammophile, spiladophile, syrtidophile, thinophile

psammophile *n*. An organism that lives in sandy habits (*e.g.*, the Sunset Shell, *Psammobia vespertina*) (Lincoln et al. 1985).
cf. -cole: aigicole, amanthicole, ammocole, amnicole, arenicole, argillicole, limicole, luticole, thinicole; [1]-phile: aigialophile, amathophile, ammochtophile, ammophile, anemophile, argillophile, cheradophile, enaulophile, pelochthophile, pelophile, phellophile, psamathophile, spiladophile, syrtidophile, thinophile

pseudosymphile See symbiont: symphile: pseudosymphile.

pseudoxerophile *n*. An organism that lives in subxeric habitats (*e.g.*, a plant species that exhibits less moisture sensitivity than a true xerophyte) (Lincoln et al. 1985).

psilophile *n*. An organism that lives in prairies or savannas (*e.g.*, the Western Meadowlark) (Lincoln et al. 1985).
cf. -cole: caespiticole, leimocole, nomocole, poocole, pratinicole, psicole; [1]-phile: coryphile, helolochmophile, nomophile, poophile, telmatophile

psychrophile, psychrophil *n*. An organism that lives at low temperatures (*e.g.*, a species of mountain springtail, a hexapod) (Lincoln et al. 1985).

ptenophyllophile *n*. An organism that lives in deciduous forests (*e.g.*, the Cardinal, a bird) (Lincoln et al. 1985).

ptenothalophile *n*. An organism that lives deciduous thickets (*e.g.*, a grape species) (Lincoln et al. 1985).

pyrophile *n*. An organism that lives in ground that has recently been scorched by fire (*e.g.*, the Fireweed (Lincoln et al. 1985).

pyroxylophile *n*. An organism that lives on burnt wood (Lincoln et al. 1985).
cf. -cole: lignicole; [1]-phile: lignophile, xylophile

rheophile *n*. An organism that lives in running water (*e.g.*, a black-fly species) (Lincoln et al. 1985).
cf. rheophobe

rhizophile *n*. An organism that grows on roots (Lincoln et al. 1985).

rhoophile *n*. An organism that lives in creeks (*e.g.*, the Muskrat) (Lincoln et al. 1985).

rhyacophile *n*. An organism that lives in torrents (Lincoln et al. 1985).

rugophile *n.* An organism that lives on rough surfaces or surface depressions (*e.g.,* a lichen species) (Lincoln et al. 1985).

saprophile *n.* An organism that lives in humus-rich substrates or on decaying organic matter (*e.g.,* the Dog Stinkhorn, a fungus) (Lincoln et al. 1985).
cf. scavenger; -zoite: saprozoite

sathrophile *n.* An organism that lives on humus or on decaying organic matter (*e.g.,* a puffball species of fungus) (Lincoln et al. 1985).
cf. scavenger; -zoite: saprozoite

saurophile See -chore: saurochore.

sciophile, heliophobe, skiophile, umbrophile *n.* An organism that lives in shaded places, or in habitats of low light intensity (*e.g.,* a moss species) (Lincoln et al. 1985).
cf. -cole: lucicole, umbraticole; [1]-phile: anheliophile, heliophile, helioxeriophile, lygophile, phengophile, skotophile

scotophile See -phile: skotophile.

serpentiophile *n.* An organism that lives in a serpentine-rich habitat (Lincoln et al. 1985).

siderophile *n.* An organism that lives in an iron-rich medium (Lincoln et al. 1985).

skiophile See -cole: umbraticole, [1]-phile: sciophile.

skotophile *n.* An organism that lives in darkness or in darkened places (*e.g.,* a soil-mite species) (Lincoln et al. 1985).
cf. -cole: lucicole, umbraticole; [1]-phile: anheliophile, heliophile, helioxeriophile, lygophile, phengophile, sciophile

spermophile *n.* An organism that feeds on seeds (*e.g.,* a weevil beetle species or the goldfinch) (Lincoln et al. 1985).

sphagnophile *n.* An organism that lives on sphagnum moss or in sphagnum-rich habitats (*e.g.,* a blueberry species) (Lincoln et al. 1985).
cf. -cole: bryocole, muscicole, sphagnicole; [1]-phile: bryophile

spiladophile, stygophile *n.* An organism that lives on clay or in clay-rich habitats (Lincoln et al. 1985).
cf. -cole: aigicole, amanthicole, ammocole, amnicole, arenicole, argillicole, limicole, luticole, thinicole; [1]-phile: aigialophile, amathophile, ammochtophile, ammophile, anemophile, argillophile, cheradophile, enaulophile, pelochthophile, pelophile, phellophile, psamathophile, psammophile, syrtidophile, thinophile

stasophile *n.* An organism that lives in stagnant water (*e.g.,* an alga species) (Lincoln et al. 1985).

stenothermophile See [1]-phile: thermophile: stenothermophile.

sterrhophile *n.* An organism that lives in moorland (*e.g.,* a heather species) (Lincoln et al. 1985).

stygophile See [1]-phile: spiladophile.

subhydrophile *n.* An organism that lives in habitats that experience periodic freshwater inundations (*e.g.,* a crayfish species) (Lincoln et al. 1985).

substratohygrophile *n.* An organism that lives on moist substrates (*e.g.,* a liverwort species) (Lincoln et al. 1985).

subxerophile *n.* An organism that lives in moderately dry situations (*e.g.,* Lowrie's Aster) (Lincoln et al. 1985).
cf. -cole: deserticole, siccocole, xerocole; [1]-phile: aerohygrophile, cheradophile, chersophile, helohylophile, hylodophile, hydrophile, lochnodophile, syrtidophile, telmatophile, xerohylophile, xerophile

symphile See symbiont: symphile.

syrtidophile *n.* An organism that lives on dry sand bars (*e.g.,* a wolf-spider species) (Lincoln et al. 1985).
cf. -cole: aigicole, amanthicole, ammocole, amnicole, arenicole, argillicole, limicole, luticole, thinicole; [1]-phile: aigialophile, amathophile, ammochtophile, ammophile, anemophile, argillophile, cheradophile, enaulophile, pelochthophile, pelophile, phellophile, psamathophile, psammophile, spiladophile, thinophile

taphrophile *n.* An organism that lives in ditches (*e.g.,* a rush species) (Lincoln et al. 1985).

telmatophile *n.* An organism that lives in wet meadows (*e.g.,* a pitcherplant species) (Lincoln et al. 1985).
cf. -cole: caespiticole, leimocole, nomocole, poocole, pratinicole, psicole; [1]-phile: coryphile, helolochmophile, nomophile, poophile, psilophile

termitophile *n.* An organism that must spend at least part of its live cycle with a termite colony (Hölldobler and Wilson 1990, 644).

thalassophile *n.* An organism that lives in the sea (*e.g.,* the Hammerhead Shark) (Lincoln et al. 1985).

thermophile *n.* An organism that lives in warm conditions (*e.g.,* a microorganism that has a growth optimum above 45°C) (Lincoln et al. 1985).
Comment: A heat-stable enzyme derived from a microbe from Yellowstone National Park drives the polymerase chain reaction (Milstein 1994, 655).
▶ **eurythermophile** *n.* An organism that is tolerant of a wide range of relatively high temperatures (Lincoln et al. 1985).
cf. [1]-phile: thermophile: stenothermophile, -therm: eurytherm

▶ **facultative thermophile** *n*. A thermophile that requires temperature from 50-65°C for optimum growth but is capable of growth at lower temperatures (Lincoln et al. 1985).

▶ **hyperthermophile** *n*. A microorganism that can grow at temperatures up to 104°C (220°F) and withstand much higher ones, even up to 357°C (700°F) (Broad 1993b, C1, C15).

Comments: Several species of hyperthermophiles are known from hydrothermal-vent communities; their enzymes, which are active at high temperatures, are used in biotechnology (Broad 1993b, C1). K.O. Stetter and his colleagues discovered hyperthermophiles in Icelandic hot springs in 1991 (Broad 1993b, C1).

▶ **macrothermophile** *n*. An organism that lives in the tropics or other warm habitats (*e.g.,* the Queen Palm tree) (Lincoln et al. 1985).

▶ **mesothermophile** *n*. An organism that lives in temperate regions (*e.g.,* the Tulip Tree) (Lincoln et al. 1985).

▶ **microthermophile** *n*. An organism that lives in boreal regions (*e.g.,* the Wolverine); contrasted with macrothermophile (Lincoln et al. 1985).

▶ **obligate thermophile** *n*. A thermophile requiring temperatures from 65–70°C for optimum growth, and is usually unable to grow at temperatures below 40°C (Lincoln et al. 1985).

▶ **stenothermophile** *n*. An organism that is tolerant of only a narrow range of high temperatures (Lincoln et al. 1985).

cf. [1]-phile: thermophile: eurythermophile, -therm: stenotherm

thinophile *n*. An organism that lives on sand dunes (*e.g.,* a grass species) (Lincoln et al. 1985).

cf. -cole: aigicole, amanthicole, ammocole, amnicole, arenicole, argillicole, limicole, luticole, thinicole; [1]-phile: aigialophile, amathophile, ammochtophile, ammophile, anemophile, argillophile, cheradophile, enaulophile, pelochthophile, pelophile, phellophile, psamathophile, psammophile, spiladophile, syrtidophile

thiophile *n*. An organism that lives in sulfur-rich habitats (*e.g.,* a bacterium species) (Lincoln et al. 1985).

cf. -troph: autotroph: chemoautotroph

tiphophile *n*. An organism that lives in ponds (*e.g.,* a turtle species) (Lincoln et al. 1985).

troglophile, eutroglobiont *n*. An animal that is frequently found in underground caves, or passages, but not confined to them (Lincoln et al. 1985).

cf. -cole: cavernicole; stygobie; -xene: trogloxene

▶ **eutroglophile** *n*. A troglophile that reproduces in a cave habitat (Lincoln et al. 1985).

▶ **subtroglophile** *n*. A troglophile that does not reproduce in a cave habitat (Lincoln et al. 1985).

tropophile *n*. An organism that lives in an environment that undergoes marked periodic fluctuations of light, temperature, and moisture (*e.g.,* a mountain-plant species) (Lincoln et al. 1985).

turfophile *n*. An organism that lives in bogs (*e.g.,* a dragonfly species) (Lincoln et al. 1985).

umbrophile See [1]-phile: sciophile.

urophile *n*. An organism that lives in habitats rich in ammonia (Lincoln et al. 1985).

xerohylophile *n*. An organism that lives in dry forests (Lincoln et al. 1985).

cf. -cole: deserticole, siccocole, xerocole; [1]-phile: aerohygrophile, cheradophile, chersophile, helohylophile, hylodophile, hydrophile, lochnodophile, subxerophile, syrtidophile, telmatophile, xerophile

xerophile *n*. An organism that lives dry habitats (*e.g.,* the Creosote Bush) (Lincoln et al. 1985).

cf. -cole: deserticole, siccocole, xerocole; [1]-phile: aerohygrophile, cheradophile, chersophile, helohylophile, hylodophile, hydrophile, lochnodophile, subxerophile, syrtidophile, telmatophile, xerohylophile

xeropoophile *n*. An organism that lives in heathland (*e.g.,* a bumble-bee species) (Lincoln et al. 1985).

xylophile *n*. An organism that lives on, or in, wood (*e.g.,* a carpenter bee *Xylocopa virginica*) (Lincoln et al. 1985).

cf. -cole: lignicole; [1]-phile: lignophile, proxylophile

zodiophile, zoidiophile See [1]-phile: zoophile.

zoophile, zoidiophile, zodiophile *n*. An organism that has an affinity for animals (Lincoln et al. 1985).

◆ [2]-**phile** *combining form* Referring to a plant that is regularly pollinated by a particular agent (*e.g.,* birds, insects, water, or wind).

adj. -gamous, -philous

n. -phily

cf. -lectic; -philo; pollination; -tropic

Comments: The combining form "-gamy" has also been used to indicate kinds of pollination; Faegri and van der Pijl (1979, 34) prefer "-phily." A designation of a kind of "-phile" (*e.g.,* cantharophile) to a plant species does not necessarily indicate that *only* one kind of agent (in this case, only beetles) pollinates it. [Greek -*philos*, loving < *phileein*, to love]

allophile *n.*

1. A plant species that has no morphological floral adaptations for guiding floral visitors and can be utilized by short-tongued visitors (Faegri and van der Pijl 1979, 48).
2. A plant species that is pollinated by an agent to which is it not evolutionarily adapted (Lincoln et al. 1985).

cf. ²-phile: euphile, hemiphile

anemophile, aerophile *n.* A plant species that is regularly pollinated by wind (*e.g.*, species of Ashes, Conifers, Elms, Filberts, Grasses, Marijuana, Meadowrues, Nettles, Oaks, Potomogetons, Ragweeds, Rushes, Sedges, and Walnuts) (Faegri and van der Pijl 1979, 34–35).

autophile *n.* A plant species that is self-pollinated (*e.g., Lobelia dortmanna*) (Faegri and van der Pijl 1979, 40).

cf. -gamy: autogamy

cantharophile *n.* A plant species regularly pollinated by beetles (*e.g., Amorphophallus* spp., *Calycanthus occidentalis, Degenerea* spp., *Magnolia* spp., *Nymphaea citrina, Uvularia perfoliata, Victoria amazonica*, or some species of fagaceous subtropical and tropical trees) (Faegri and van der Pijl 1979, 99–102, 170–171).

Comments: Cantharophiles often have protein- or lipid-rich flowers (Buchmann and Nabhan 1996, 243).

chiropterophile *n.* A plant species that is regularly pollinated by bats (Faegri and van der Pijl 1979, 129–133).

Comments: Bats that feed from flowers include species in the genera *Cynopterus, Eidolon, Eonyctris, Leptonycteris, Macroglossus*, and *Megachiroptera*; chiropterophilous plants include *Agave schotitii, Carnegiea gigantea, Crescentia, Eucalyptus, Kigelia, Marcgravia, Musa*, and *Parkia clappertoniana* (Proctor and Yeo 1972, 330–337; Faegri and van der Pijl 1979, 123–129).

dientiophile *n.* A plant pollinated by two insect species and having two kinds of flowers, each adapted to each insect species (Lincoln et al. 1985).

dientomophile *n.* A plant species that is regularly pollinated by two insect species (Lincoln et al. 1985).

drosophile, drosphile *n.* A plant species that is regularly pollinated by dew (Lincoln et al. 1985).

entomophile *n.* A plant species that is regularly pollinated by insects (Faegri and van der Pijl 1979).

Comments: Thousands of species of plants are entomophiles. Many kinds of insects pollinate, including ants, beetles, bees, flies, thrips, and wasps.

euphile *n.* A species that has morphological floral adaptations for attracting and guiding specialized floral visitors (*e.g.*, the Trumpet Vine, *Campsis radicans*) (Faegri and van der Pijl 1979, 48; Lincoln et al. 1985).

cf. ²-phile: allophile, hemiphile

hemiphile *n.* A plant species with morphological floral adaptations that allow floral visitors with intermediate degrees of specialization to obtain its floral resources (Faegri and van der Pijl 1979, 48).

cf. ²-phile: allophile, euphile

hydroanemophile *n.* (Lincoln et al. 1985).

hydrophile *n.* A plant species that is regularly pollinated by fresh, or salt, water (Faegri and van der Pijl 1979, 40).

Comment: Hydrophiles might be rare because pollen is usually damaged by immersion in water (Buchmann and Nabhan 1996, 248).

▸ **ephydrophile** *n.* A hydrophile that is pollinated on the surface of water (*e.g., Callitriche autumnalis, Vallisneria spiralis*, or species of *Ruppia*) [coined by Delpino, 1868–1875, in Faegri and van der Pijl 1979, 40, 169].

▸ **hyphydrophile** *n.* A hydrophile that is pollinated in water (*e.g., Callitriche hamulata* or species of *Ceratophyllum, Halophila*, or *Najas*) [coined by Delpino, 1868–1875, in Faegri and van der Pijl 1979, 40].

lepidopterophile *n.* A plant species that is regularly pollinated by Lepidoptera (Lincoln et al. 1985).

cf. ²-phile: phalaenophile, psychophile, sphingophile

malacophile *n.* A plant species that is regularly pollinated by snails and slugs (*e.g.*, possibly a species of *Rohdea*) (Faegri and van der Pijl 1979, 119).

Comments: The existence of malacophily is moot (Faegri and van der Pijl 1979, 119).

melittophile *n.* A plant species that is regularly pollinated by bees (*e.g.*, thousands of species including *Astragalus depressus, Coronilla emerus, Cytisus scoparius*, and *Orchis maculata* and species of *Aconitum, Bartsia, Cattleya, Centrosema, Delphinium, Galeopsis, Linaria, Melampyrum, Pedicularis, Prunus, Rosa, Rubus, Salvia, Thunbergia*, or *Trifolium*) (Faegri and van der Pijl 1979, 110–115, 168, 180–184, 188–197).

micromelittophile *n.* A plant species that is regularly pollinated by small bees (Lincoln et al. 1985).

micromyiophile *n.* A plant species that is regularly pollinated by small flies (Lincoln et al. 1985).

cf. ²-phile: myiophile

n–r

microzoophile *n.* A plant species that is regularly pollinated by small animals (Lincoln et al. 1985).

monophile *n.* A plant that is pollinated by only one, or a few closely related, species (Faegri and van der Pijl 1979, 48).
cf. -lectic: monolectic; ²-phile: oligophile, polyphile

myiophile, myophile *n.* A plant species that is regularly pollinated by flies (*e.g., Hedera helix, Theobroma cacao,* or species in *Aruncus, Circaea, Ilex, Mentha,* or *Veronica*) (Faegri and van der Pijl 1979, 102–106).
syn. myiophile (Lincoln et al. 1985)
cf. ²-phile: sapromyophile

myrmecophile *n.* A plant species that is regularly pollinated by ants (*e.g., Glaux maritima, Orthocarpus pusillus,* or *Polygonum cascadense*) (Dahl and Hadac 1940, etc. in Faegri and van der Pijl 1979, 110).
cf. pollination: ant pollination

necrocoleopterophile *n.* A plant species that is regularly pollinated by carrion beetles (Lincoln et al. 1985).

oligophile *n.* A plant that is pollinated by a few related taxa (Faegri and van der Pijl 1979, 48).
cf. -lectic: oligolectic; ²-phile: monophile, polyphile

ombrophile *n.* A plant that is pollinated by rain (*e.g.,* a flower that fills with rain water and whose pollen floats up in the water to its stigma) (Hagerup 1950 in Faegri and van der Pijl 1979, 41).
cf. pollination: rain pollination
Comment: The significance of rain pollination is questionable (Daumann 1970 in Faegri and van der Pijl 1979, 41).

ornithophile *n.* A plant species that is regularly pollinated by birds (Faegri and van der Pijl 1979, 123).
Comments: Birds involved in "ornithophily" include the Brush-Tongued Honeyparrot (Lorikeet), Bulbuls, Honeyeaters, Hummingbirds, Sugarbirds (Honeycreepers, Quits), and Sunbirds; ornithophilous plants include *Salvia splendens* and species in *Aloe, Bombax, Columnea, Cuphea, Erythrina, Eucalyptus, Fuchsia, Heliconia, Iris, Lobelia, Malvaviscus, Mimulus, Monarda, Poinsettia, Puya, Quassia, Rhododendron, Spathodea,* and *Tropaeolum* (Faegri and van der Pijl 1979, 123–129, 184). Ornithophiles usually have colorful, large, and sturdily built flowers whose ovules are not easily damaged by birds' probing bills and feet (Buchmann and Nabhan 1996, 251).

phalaenophile *n.* A plant species that is regularly, or exclusively, pollinated by moths; *e.g., Calonyction bona nox, Cestrum nocturnum, Lonicera periclymenum, Pancratium maritimum,* or species in *Impatiens* and *Yucca* (Faegri and van der Pijl 1979, 117–119, 175).
cf. ²-phile: lepidopterophile, psychophile, sphingophile
Comments: Angraecum sesquipedale, which has a nectar spur up to 30 centimeters long, is pollinated by a hawkmoth *Xanthopan morgani* f. *praedicta,* which has a tongue about 30 centimeters long (Faegri and van der Pijl 1979, 118). *Angraecum longicalcar* has a nectar spur that is nearly 40 centimeters long (Kritsky, 1991, 209).

polyphile *n.* A plant that is pollinated by many taxa (Faegri and van der Pijl 1979, 48).
cf. -lectic: polylectic; ²-phile: monophile, oligophile

protozoophile *n.* A plant species that is regularly pollinated by protozoa and microscopic animals (Lincoln et al. 1985).

psychophile *n.*
1. A plant species that is regularly pollinated by diurnal Lepidoptera (Lincoln et al. 1985).
2. A plant species that is regularly pollinated by butterflies (*e.g., Nigritella nigra;* or species in *Buddleia, Lantana,* or *Phlox*) (Faegri and van der Pijl 1979, 115, 203; Lazri and Barrows 1984; Venables and Barrows 1985).
cf. ²-phile: lepidopterophile, phalaenophile, sphingophile

sapromyophile, sapromyiophile *n.* A plant species that is regularly pollinated by flies that are attracted to carrion and feces (*e.g., Arum maculatum, A. nigrum, Asarum vulgare,* or species in *Arisaema, Aristolochia, Ceropegia,* or *Darlingtonia*) (Faegri and van der Pijl 1979, 103–105, 173).
syn. sapromyiophile (Lincoln et al. 1985)

sphingophile *n.* A plant species that is regularly pollinated by sphingids (hawkmoths) or other nocturnal Lepidoptera (Lincoln et al. 1985).
cf. ²-phile: phalaenophile, psychophile

zoophile *n.* A plant species that is regularly pollinated by animals (Lincoln et al. 1985).

♦ ³-**phile** *combining form*
biophile *n.* An element (*e.g.,* calcium, iron, or sulfur) that is concentrated in a living organism [coined by V.M. Goldschmidt, geochemist (1937) in Fyfe (1996, 448)].
See ¹phile: biophile.
Comment: Most new minerals in the marine sedimentary environment are biocatalyzed (H. Lowenstam in Fyfe 1996, 448).

♦ **-philia** *combining form*
1. A tendency toward (*e.g., hemophila*) (Michaelis 1963).
2. An excessive affection, or fondness for (*e.g., necrophila*) (Michaelis 1963).

syn. phily (Michaelis 1963)

[< Greek *-philia*, loving < *phileein*, to love]

beetlephilia *n.* A person's affection for beetles (inferred from Evans and Bellamy 1996, 163).

biophilia *n.*
1. A person's "innate tendency to focus on life and lifelike processes" [coined by Wilson 1984, 1].
2. A person's positive attitude toward natural diversity and nature itself (Buchmann and Nabhan 1996, 243).

coprophilia *n.* A person's love of feces or filth (Hinsie and Campbell 1975, 163).

necrophilia *n.* A person's reaching orgasm with a dead male or female Human (Hinsie and Campbell 1975, 490).

paraphilia *n.* A person's sexual perversion (*i.e.,* any sexual practice that deviates from the normal or any abnormal means of achieving genital orgasm) (Hinsie and Campbell 1975, 543, 562).

Comments: Perversions include coprophilia, exhibitionism, fetishism, masochism, necrophilia, sadism, transvestism, urolagnia, and voyeurism (Hinsie and Campbell 1975, 562).

pedophilia (*Brit.* **paedophilia**) *n.* A person's love of children, implying, or indicating, a sexual attraction towards or sexual interactions with them (Hinsie and Campbell 1975, 552).

syn. paedophilia erotica (Kraft-Ebling), pederosis (Forel) (both in Hinsie and Campbell 1975, 552)

cf. drug: triptorelin

urophilia *n.* A person's pathologic love for, or interest in, urine (Hinsie and Campbell 1975, 801).

syn. urolagnia (Hinsie and Campbell 1975, 801)

♦ **philo-, philo** *combining form* "Loving; fond of" (Michaelis 1963).

cf. ^{1,2}-phile

[< Greek *-philos*, loving < *phileein*, to love]

♦ **philopatry** *n.* An animal's tendency to remain at certain places or at least to return to them for feeding and resting (Wilson 1975, 337, 340, 591).

♦ **philoprogenetive, prolific** *adj.* Referring to an organism that produces large numbers of offspring (Lincoln et al. 1985).

♦ **philosophical psychology** See study of: psychology: philosophical psychology.

♦ **philosophy** See study of: philosophy.

♦ **philosophy of the mind** See study of: philosophy: philosophy of the mind.

♦ **-phobe** *combining form* "One who fears or has an aversion to" (Michaelis 1963).

adj. phobic

cf. ^{1}-phile, phobia, phobic

[Late Latin < Greek *phobos*, fearing < *phobeesthai*, to fear]

calciphobe, basifuge, calcifuge *n.* A plant that is intolerant of soils rich in calcium salts (Lincoln et al. 1985).

chianophobe *n.* An organism that is intolerant of prolonged snow cover (Lincoln et al. 1985).

chionophobe *n.* An organism that is intolerant of snow-covered habitats (Lincoln et al. 1985).

halophobe *n.* An organism that is intolerant of saline habitats (*e.g.,* the Orange tree) (Lincoln et al. 1985).

heliophobe, skiophile, umbrophile *n.* An organism that is intolerant of high light intensity (Lincoln et al. 1985).

cf. ^{1}-phile: heliophile

hydrophobe *n.* An organism that is intolerant of water or wet conditions (*e.g.,* the White Oak, which is killed by long-term standing water over its roots) (Lincoln et al. 1985).

adj. hydrophobic

cf. ^{1}-phile: hydrophile

hygrophobe *n.* An organism that is intolerant of moist conditions (Lincoln et al. 1985).

myrmecophobe *n.* An organism that repels ants, or termites (*e.g.,* a *Formica* ant species that drives away the ant *Prenolepis imparis*) (Lincoln et al. 1985).

cf. -^{1}phile: myrmecophile

oxyphobe, basophile *n.* An organism that thrives in alkaline habitats or is intolerant of acidic conditions (Lincoln et al. 1985).

phengophobe *n.* An organism that thrives in, or has an affinity for, light (Lincoln et al. 1985).

photophobe *n.* An organism that is intolerant of, or avoids, full light (*e.g.,* a scorpion species) (Lincoln et al. 1985).

adj. lucifugous, photopathic, photophobic, photophobous, photophygous (Lincoln et al. 1985)

cf. ^{1}-phile: photophile

pluviophobe *n.* An organism that is intolerant of abundant rainfall (*e.g.,* a cactus species that rots under conditions that are too moist) (Lincoln et al. 1985).

poleophobe *n.* An organism that is intolerant of urban habitats (*e.g.,* a tree species that cannot grow well in urban air pollution) (Lincoln et al. 1985).

pyrophobe *n.*
1. An organism that is intolerant of soil conditions produced by fire (Lincoln et al. 1985).
2. A plant that cannot re-establish itself following a fire (Lincoln et al. 1985)

rheophobe *n.* An organism that is intolerant of too much water (*e.g.,* a cactus species) (Lincoln et al. 1985).
cf. -phobe: hydrophobe; -¹phile: rheophile

thermophobe *n.* An organism that is intolerant of high temperatures (*e.g.,* the African violet) (Lincoln et al. 1985).

xerophobe *n.* An organism that is intolerant of dry conditions (*e.g.,* the elephant-ear plant) (Lincoln et al. 1985).

♦ **phobia** *n.* (See table next page.)
1. A person's compulsive and persistent fear of any specified type of object, stimulus, or situation (Michaelis 1963).
2. Any strong aversion or dislike (Michaelis 1963).
3. A morbid fear associated with morbid anxiety (Hinsie and Campbell 1976 569).
cf. -phobe, phobic
[Late Latin < Greek *phobos,* fear]

neophobia *n.* For example, in cats, Humans, rats: an individual's avoidance of a novel object (Barnett 1963 in Dewsbury 1978, 63; Bradshaw 1986, 614).

♦ **-phobic** *combining form*
1. Intolerant of (Lincoln et al. 1985).
2. Lacking affinity for (Lincoln et al. 1985).
cf. -phobe, ¹-phile.

♦ **phobo-** *combining form*
1. Referring to evident fear (Lincoln et al. 1985).
2. Referring to negation of an operative prefix (*e.g.,* in phobophototaxis, which means negative phototaxis) (Lincoln et al. 1985).
cf. -phobe

♦ **phobotaxis** See taxis: phobotaxis.

♦ **phonological loop** See working memory (comments).

♦ **phonophobia** See phobia (table).

♦ **phonotaxis** See taxis: phonotaxis.

♦ **phoresy** See symbiosis: phoresy.

♦ **phoretic behavior** See behavior: phoretic behavior.

♦ **phorophyte** See epiphyte.

♦ **phosphene** *n.* One of the colorful patterns that one sees when one closes one's eyes and gently presses against one's eyeball (Holden 1999b, 1845).
Comment: The direct stimulation of one's visual cortex from pressed eyeballs produces phosphenes (Holden 1999b, 1845).

♦ **phosphorescence** *n.*
1. A material's emitting light without sensible heat; also, the light so emitted (Michaelis 1963).

2. A mineral substance's continuing to shine in darkness after exposure to light; distinguished from fluorescence (Michaelis 1963)
adj. phosphorescent
cf. bioluminescence

♦ **photoautotroph** See -troph-: photoautotroph.

♦ **photocliny** *n.* An organism's response to incident-light direction (Lincoln et al. 1985).

♦ **photodynamics** See study of: dynamics: photodynamics.

♦ **photogenic** See -genic: photogenic.

♦ **photogrammetry** See study of: -metry: photogrammetry.

♦ **photoharmosis** See -harmosis: photoharmosis.

♦ **photoheterotroph** See -troph-: photoorganotroph.

♦ **photohoramotaxis, photohorotaxis** See taxis: photohoramotaxis.

♦ **photokinesis** See kinesis: photokinesis.

♦ **photolithotroph** See -troph-: photolithotroph.

♦ **photoorganotroph** See -troph-: photoorganotroph.

♦ **photopathic** See -phobe: photophobe.

♦ **photoperiod** See period: photoperiod.

♦ **photophile** See ¹-phile: photophile.

♦ **photophobe** See -phobe: photophobe.

♦ **photophobotaxis** See taxis: photophobotaxis.

♦ **photophobous, photophygous** See -phobe: photophobe.

♦ **photoreceptor** See receptor: photoreceptor.

♦ **photosynthesis** See biochemical pathway: photosynthesis.

♦ **"photosystem-I photosynthesis"** See biochemical pathway: "photosystem-I photosynthesis."

♦ **"photosystem-II photosynthesis"** See biochemical pathway: "photosystem-II photosynthesis."

♦ **phototaxis** See taxis: phototaxis.

♦ **phototonicity** *n.* An organism's being sensitive to light (Lincoln et al. 1985).
adj. photonic, phototonus

♦ **phototroph** See -troph-: phototroph.

♦ **phototropism** See -tropism: phototropism.

♦ **phototropotaxis** See taxis: phototaxis: phototropotaxis.

♦ **phragmosis** *n.* For example, in some ant, bee, termite, and frog species: an animal's using part of its body to block its burrow (Halliday and Adler 1986, 38, 148; Hölldobler and Wilson 1990, 641).
adj. phragmotic

♦ **Phrap** See program: Phrap.

Some Human Phobias[a]

Name	Stimulus
acrophobia (*syn.* altophobia,[b] hypsophobia,[b] hypsiphobia[b])	high places
aerophobia[c] (*syn.* aviatophobia)	flying
agoraphobia (*syn.* cenophobia,[b] kenophobia[b])	open spaces
aichinophobia	sharp objects
ailurophobia	cats
androphobia	men
anemophobia	cyclones[c]
anthrophobia	people
apeirophobia	infinity
apiphobia (*syn.* apiophobia, mellissophobia)	bees[c]
arachneophobia (*syn.* arachnophobia)	spiders[c]
astraphobia lightning[b]	thunderstorms
astrophobia	stars
autophobia	self, being alone
ballistrophobia	missiles
bathophobia	depths
batrachophobia	frogs or reptiles[c]
brontophobia (*syn.* keraunophobia)	thunderstorms[c]
carsinomaphobia[b] (*syn.* cancer phobia,[e] carcinophobia,[b] carcinomatophobia,[b] cancerphobia,[b] cancerophobia[b])	cancer
chionophobia	snow
claustrophobia (*syn.* clesiophobia, cleithrophobia, clithrophobia)	confined spaces
cynophobia (knophobia[b])	dogs
demophobia	crowds
emetophobia (*syn.* emitophobia)	vomiting[c]
equinophobia (*syn.* hippophobia)	Horses[c]
erythrophobia	red
gamophobia	marriage
haptephobia	being touched
hemophobia	blood
hypnophobia	sleep
kinesophobia	movement
lalophobia	speaking
lilapsophobia	hurricanes and tornadoes[c]
musophobia (*syn.* muriphobia[b])	mice
myrmecophobia	ants[c]
mysophobia	contamination
necrophobia[d] (*syn.* death phobia[e])	dead bodies or death[d]
neophobia	new things
nyctophbia	night, darkness
ophidiophobia	reptiles[a]
ophidiophobia (*syn.* ophiophobia, ophiciophobia, herpetophobia)	snakes[c]
ornithophobia	birds[c]
pedophobia	infants or children
phonophobia	noise
psychrophobia	cold
pyrophobia	fire
sociophobia	social situations
taphephobia	being burned alive
thalassophobia	ocean or sea
thanatophobia	death
toxicophobia	poison
zoophobia	animals

[a] Michaelis (1963).
[b] Ash (1996, 47).
[c] The entire line is from Ash (1996, 47), who indicates that there is no scientific name for fear of, or aversion to, rats.
[d] Morris (1982).
[e] Hinsie and Campbell (1976). These authors also list and discuss these phobias for which they give common names: bathroom phobia (*syn.* toilet phobia), bug phobia, doorknob phobia, hypochondriacal phobia, impregnation phobia, insect phobia, landscape phobia (*syn.* agoraphobia), light-and-shadow phobia, live-burial phobia, poisoning phobia, school phobia, street phobia, traumatic phobia, and vehicle phobia.

n – r

♦ **phreatobiology** See study of: biology: phreatobiology.

♦ **phreatocole** See -cole: phreatocole.

♦ **Phred** See program: Phred.

♦ **phretium** See ²community: phretium.

♦ **phretophile** See ¹-phile: phretophile.

♦ **phycobilin** See molecule: accessory pigment: phycobilin.

♦ **phycobiont** See -biont: phycobiont.

♦ **phycocoenology** See study of: phycocoenology.

♦ **phycology** See study of: phycology.

♦ **phycophage** See -phage: phycophage.

♦ **phycophile** See ¹-phile: phycophile.

♦ **phylacobiosis** See -biosis: parabiosis.

♦ **-phyletic, phylic** *adj.* Of, or referring to, a phylum, genetic type, or strain (Michaelis 1963).
n. -phyly
syn. racial (Michaelis 1963)
cf. ²group: diphyletic group, holophyletic group, paraphyletic group, polyphyletic group
[Greek *phyletikos* < *phyletēs*, tribesman < *phylē*]

diphyletic *adj.* Referring to a group that is derived from two distinct ancestral lineages; referring to a group that is comprised of descendants of two different ancestors; contrasted with monophyletic and polyphyletic (Lincoln et al. 1985).
n. diphyly

homorophyletic *adj.* Referring to a group of taxa that share similar characters that have been inherited from the same ancestral species (Lincoln et al. 1985).
n. homorophyly

monophyletic *adj.* Referring to the process of a taxon's descending from a common ancestor; contrasted with diphyletic and polyphyletic (Hennig 1950 in Mayr 1982, 228).
cf. ²group: monophyletic group

polyphyletic *adj.* Referring to a taxon that is derived from two or more distinct ancestral lineages; contrasted with diphyletic and monophyletic (Lincoln et al. 1985).

♦ **phyletic evolution** See ²evolution: phyletic evolution.

♦ **phyletic extinction** See ²extinction: phyletic extinction.

♦ **phyletic gradualism** See gradualism: phyletic gradualism.

♦ **phyletic group** See ²group: phyletic group.

♦ **phyletic inertia** See phylogenetic inertia.

♦ **phyletic speciation** See speciation: phyletic speciation.

♦ **phyletic tree** See -gram: phylogenetic tree.

♦ **phyletic trogloxene** See -xene: trogloxene.

♦ **PHYLIP** See program: Phylogenetic Inference Package.

♦ **phylistics** *n.* Concepts of phylogenetic relationship based on both cladogenesis and evolutionary divergence (Lincoln et al. 1985).
adj. phylistic
n. phylist

♦ **phylktioplankton** See plankton: phylktioplankton.

♦ **phyllobiology** See study of: biology: phyllobiology.

♦ **phyllophage** See -phage: phyllophage.

♦ **phyllosphere** See -sphere: phyllosphere.

♦ **phylo-, phyl-** *combining form* "Tribe; race; species" (Michaelis 1963).
[Greek *phylē*, *phylon*, tribe]

♦ **PhyloCode** *n.* A classificatory system that names plants according to their clades rather than using the currently used Linnaean classification system (Brown 1999b, 991).
Comment: In the PhyloCode, for example, the herb *Prunella vulgaris* and hundreds of other plants would simply be called *vulgaris*, each with a tag to its clade in a master directory with phylogenetic data (Brown 1999b, 991).

♦ **phylogenesis** See -genesis: phylogenesis.

♦ **phylogenetic** See genetic: phylogenetic.

♦ **phylogenetic adaptation** See ³adaptation: ²evolutionary adaptation (def. 4).

♦ **Phylogenetic Analysis Using Parsimony (PAUP)** See program: Phylogenetic Analysis Using Parsimony (PAUP).

♦ **phylogenetic group** See ²group: phyletic group.

♦ **phylogenetic homology** See homolog: phylogenetic homology.

♦ **phylogenetic inertia** *n.*
1. The deeper properties of a population (including preadaptation and basic genetic mechanisms) that determine the extent to which its evolution can be deflected in one direction or another, as well as the amount by which its rate of evolution can be speeded or slowed (Wilson 1975, 32–33, 592).
2. Underlying genetic mechanisms and preadaptations that determine the direction and rate of evolutionary change (Lincoln et al. 1985).
syn. phyletic constraint, phyletic inertia (Gould and Lewontin 1979, 594)
cf. ³adaptation: ²evolutionary adaptation: preadaptation; prime mover
Comments: "High phylogenetic inertia" implies resistance to evolutionary change; "low phylogenetic inertia," a relatively high degree of lability (Wilson 1975, 32–33, 592).

Evolutionists invoke phylogenetic inertia to explain why Humans are not optimally designed for upright posture, why no molluscs fly in the air, and why no insects are as big as elephants (Gould and Lewontin 1979, 594).

♦ **Phylogenetic Inference Package** See program: Phylogenetic Inference Package.

♦ **phylogenetic morphology** See morphology: phylogenetic morphology.

♦ **phylogenetic noise** *n.* One's misinterpretation of phylogenetic relationships due to convergent evolution (Lincoln et al. 1985).

♦ **phylogenetic rates of evolution** See rate: phylogenetic rates of evolution.

♦ **phylogenetic relationship** *n.* Organisms' affinity based on recency of common ancestry; evolutionary relationship (Lincoln et al. 1985).

♦ **phylogenetic relict** See ²species: relict species.

♦ **phylogenetic scale** See *scala naturae*.

♦ **phylogenetic species concept** See ²species.

♦ **phylogenetic systematics** See study of: systematics: phylogenetic systematics.

♦ **phylogenetic tree** See -gram: phylogenetic tree.

♦ **phylogeny** *n.*
1. The origin and evolution of phyla, tribes, or species; ancestral evolution of an organism type; distinguished from ontogeny (*Oxford English Dictionary* 1972, entries from Darwin 1872).
2. The evolutionary history of a particular group of organisms (Wilson 1975, 591).
3. "The evolutionary history of a lineage, conventionally (though not ideally) depicted as a sequence of successive adult stages" (Gould 1977, 484).
adj. phylogenetic
syn. phylogenesis (Lincoln et al. 1985)
cf. -genesis, ontogeny
[German *phylogenie* < Greek *phylon* tribe + *-geneia*, birth, origin]

♦ **phylogerontic period** See acme: paracme.

♦ **phylogram** See -gram: phylogenetic tree.

♦ **phylon** *n.* Line of descent (Lincoln et al. 1985).
See ²group: a taxonomic classification of groups: phylum (table).

♦ **phylum** See ²group: a taxonomic classification of groups (table).

♦ **-phyly** See -phyletic.

♦ **physical anthropology** See study of: anthropology: physical anthropology.

♦ **physical environment** *n.* The abiotic component (structural and chemical factors) of an ecosystem (Lincoln et al. 1985).

♦ **physical map** See map: physical map.

♦ **physical polyethism** See polyethism: physical polyethism.

♦ **physical psychology** See study of: psychology: physical psychology.

♦ **physical science** See study of: science: physical science (Michaelis 1963).

♦ **physicalism** *n.*
1. The false idea that inheritance is due to forces or excitations, rather than to "concrete materials supplied by maternal egg cells and paternal pollen cells" (Mayr 1982, 79, 720–721).
Note: His, Loeb, Bateson, and Johannsen were physicalists.
cf. corpuscularist, preformationist
2. Application of methodology of the physical sciences (chemistry and physics) to biology, *e.g.*, one's expecting a law to be equally valid for all similar sets of phenomena.
cf. law
Comment: Because so-called "biological laws" usually have exceptions, it is more appropriate to designate "generalities," instead of "laws," in biology (Mayr 1982, 846).

♦ **physically controlled community** See ²community: physically controlled community.

♦ **physico-theology** See natural theology.

♦ **physics** See study of: physics.

♦ **physics envy** *n.* The belief that employing the mathematical methodology of the physical sciences will enhance the reputation of one's own field (Paul Bickart in Hitt 1992).

♦ **physiogenesis** See -genesis: physiogenesis.

♦ **physiognomic dominant** See ²species: physiognomic dominant.

♦ **physiographic climax** See ²community: climax: physiographic climax.

♦ **physiological clock** See clock: physiological clock.

♦ **physiological genetics** See study of: genetics: physiological genetics.

♦ **physiological isolation** See isolation: reproductive isolation.

♦ **physiological longevity** *n.* "The maximum life span of an organism that dies of old age" (Lincoln et al. 1985).

♦ **physiological psychology** See study of: psychology: physiological psychology.

♦ **physiological race, physiological form** *n.* A race characterized by physiological properties (Lincoln et al. 1985).

♦ **physiological specialization** *n.* Evolution of a species into two or more physiological races (Lincoln et al. 1985).

♦ **physiological system** See ²system: physiological system.

n–r

◆ **physiological temperature regulation** See temperature regulation: physiological temperature regulation.

◆ **physiology** See study of: physiology.

◆ **physogastry** *n.* For example, in termite queens, queens of some ant species: excessive enlargement of an abdomen (Lincoln et al. 1985).

◆ **phytalfauna** See fauna: phytalfauna.

◆ **-phytium** *combining form* Referring to a plant community. See ²community.

◆ **phyto-, -phyt-** *combining form* "Plant, or related to vegetation" (Michaelis 1963). [Greek *phyton*, plant]

◆ **phytoalexin** *n.* In more derived-plant species: an antifungal agent produced by a plant that it helps to protect (Müller and Börger 1941 in Harborne 1982, 239–240).

◆ **phytobenthos** See ²community: benthos: phytobenthos.

◆ **phytobiology** See study of: biology: botany.

◆ **phytobiont** See biont: phytobiont; -cole: planticole.

◆ **phytochemical ecology** See study of: ecology: chemical ecology.

◆ **phytochemistry** See study of: taxonomy: phytochemistry.

◆ **phytocoenology** See study of: phytocoenology.

◆ **phytocenose** See -coenosis: phytocenosis.

◆ **phytoecology** See study of: ecology: phytoecology.

◆ **phytoedaphon** See ²community: phytoedaphon.

◆ **phytogamy** See -gamy: phytogamy.

◆ **phytogenesis** See -genesis: phytogenesis.

◆ **phytogenic** See -genic: phytogenic.

◆ **phytogeogenesis** See -genesis: phytogenesis.

◆ **phytogeography** See study of: geography: biogeography: phytogeography.

◆ **phytography** See -graphy: phytography.

◆ **phytology** See study of: botany.

◆ **phytome** See ²community: phytome.

◆ **phytomimesis** See mimesis: phytomimesis.

◆ **phytoneuston** See neuston: phytoneuston.

◆ **phytopalaeontology** See study of: botany: paleobotany.

◆ **phytopathology** See study of: pathology: phytopathology.

◆ **phytophage** See -phage: phytophage.

◆ **phytophenology** See study of: phenology: phytophenology.

◆ **phytophile** See ¹-phile: phytophile.

◆ **phytoplanktivore** See vore: molluscivore.

◆ **phytoplankton** See plankton: phytoplankton.

◆ **phytosociology** See study of: phytocoenology.

◆ **phytostrote** *n.* An organism that migrates, or disperses, through the agency of plants (Lincoln et al. 1985).

◆ **phytosuccivore** See -vore: phytosuccivore.

◆ **phytotelmic** *adj.* Referring to an organism that inhabits small water pools within, or upon, plants (Lincoln et al. 1985).

◆ **phytoteratology** See study of: teratology: phytoteratology.

◆ **phytotopography** See study of: geography: phytotopography.

◆ **phytotroph** See -troph-: autotroph.

◆ **pick** See tool: pick.

◆ **picoplankton** See plankton: picoplankton.

◆ **pigmentation rule** See law: Gloger's rule.

◆ **piloerection** *n.* A mammal's erecting its fur, or hair, in connection with threatening another animal (Heymer 1977, 81) or physiological arousal (Immelmann and Beer 1989, 221).
cf. ruffling

◆ **pinnacles of social evolution** See ²group: pinnacles of social evolution.

◆ **pinnigrade** See -grade: pinnigrade.

◆ **pinnigrade locomotion** See locomotion: pinnigrade locomotion.

◆ **pioneer** See ²community: pioneer; ²species: pioneer.

◆ **piping** See animal sounds: piping.

◆ **piracy** See parasitism: cleptoparasitism.

◆ **pirate** See robber.

◆ **piscicole** See -cole: piscicole.

◆ **piscivore** See -vore: piscivore.

◆ **pitfall trap** See trap: pitfall trap.

◆ **pituitary gland** See gland: pituitary gland.

◆ **PK** See kinesis: psychokinesis.

◆ **place** *n.* A particular point, or portion of space, especially that part of space occupied by, or belonging to, a thing under consideration; a definite locality or location (Michaelis 1963).
See behavior: place.
Comment: A few of the thousands of places of biological interest are below.

Appalachian Trail *n.* A 2158-mile-long hiking trail in eastern U.S. in the Appalachians from Paxter Park Mountain, ME, southwest to Amicolola Falls, GA (Robberson 1996, B3).
syn. Appalachian National Scenic Trail (*Reader's Digest* 1996, map 5)
Comments: From approximately northeast to southwest, this trail is in the White

Mountains, ME; Blue Ridge Mountains (Green Mountains), NH; The Blue Ridge Mountains (The Berkshires), MA; Connecticut; Kittatinny Mountain, NJ; New York; Delaware; Pennsylvania; Blue Ridge Mountains (South Mountain), MD; Shenandoah National Park, VA; Great Smoky Mountains, NC; Tennessee; and Georgia).

Barro Colorado Island (BCI) *n.* The Smithsonian Tropical Research Institute's primary site for study of lowland moist tropical forest (www.si.edu/organiza/centers/stri/about.htm).

Comments: BCI is a 1500-ha island that has been under continuous study since 1923 (www.si.edu/organiza/centers/stri/about.htm). The Institute is also the custodian of five adjacent mainland peninsulas, which, with the island, form the 5400-ha Barro Colorado Nature Monument.

Burgess Shale *n.* Shale in British Columbia, Canada, with beautifully preserved fossils of soft-bodied fauna from the Late Early Cambrian, similar to those from Chengjiang (Erwin et al. 1997, 132–133; illustration, 128–129).

Chauvet Grotto *n.* A cave in Vallon-Pont-d'Arc, France, with Aurignacian depictions of animals and Humans that are between 30,300 and 32,400 years old, making these human paintings the oldest known (Thomas, 1995, A30; Folger 1996, 70–71, photograph).

Comments: Jean-Marie Chauvet and friends discovered the paintings in this grotto in 1994 (Thomas, 1995, A30). There are more than 300 images, including those of birds, bison, butterflies, mammoths, rhinoceroses, and the Sorcerer. The paintings were dated from pieces of charcoal on the ground, samples of pigment from paintings, and carbon in torch marks on cave walls. Before the discovery of Chauvet Grotto, anthropologists thought the most ancient cave paintings were in Cosquer Grotto, near Marseille, France, which date from 27,110 years ago.

Chengjiang, Yunnan Province, China *n.* A location with beautifully preserved fossils of soft-bodied fauna from the Late Early Cambrian, including a number of forms from extinct phyla (Erwin et al. 1997, 132; illustration, 128–129).

cf. Burgess Shale

Chicxulub Crater *n.* A 180-kilometer-wide crater in the Yucatán Peninsula caused by a meteorite that fell to Earth about 65 million years ago at the end of the Cretaceous Period (Sharpton et al. 1993, 1564; Kerr 1997a, 1265).

cf. hypothesis: single-impact hypothesis

Comments: The meteorite impact apparently killed off many species of unicellular organisms (Kerr 1997a, 1265). The effect of this meteorite on Earth's life is not fully understood (Stanley 1989, 509). It might have caused an impact winter that killed off the last dinosaurs and many marine groups; however, their ecosystems were already deteriorating, perhaps due to long-term climatic changes.

During Yuriakh *n.* An archeological site in Siberia with artifacts that are perhaps as old as 260,000 years (Holden 1997a, 1268, map; Waters et al. 1997, 1281).

Ediacara Hills *n.* Hills in Southern Australia that harbor a large number of Ediacaran fossils (Wright 1997, 54).

Everglades, The *n.* A wet grassland that extends from Lake Okeechobee in south-central Florida to its southern shores (Michaelis 1963; Kritcher and Morrison 1988, 87).

Comments: The Everglades is 50 to 75 miles wide and 100 miles long and essentially a very wide, shallow river that flows very slowly from the north to the south (Michaelis 1963; Kritcher and Morrison 1988, 87). Most of the area is a meadow with islands of trees and other woody plants; hammocks (*q.v.*) of subtropical forest occur in the higher and drier areas. Everglades National Park is the southern area of the original Everglades. People have caused many environmental disasters in the Everglades and are now trying to restore many parts of this habitat (Associated Press, 1999, A21).

Gunflint Formation *n.* A Canadian geological formation that dates to 2 billion years ago is found on the north shore of Lake Superior and contains microscopic fossils (Tyler and Barghoorn 1954 in Ravven 1990, 98).

Lascaux Cave *n.* Limestone caverns of Lascaux, France, that contain magnificent Pleistocene paintings from about 17,000 years ago (Milner 1990, 266).

Comments: In 1963, authorities closed this cave to tourists because they produced moisture that enabled patches of algae to grow on the paintings (Milner 1990, 266–267). In 1984, France opened Lascaux II for tourists. This is a painstaking replica of the original Lascaux Cave.

Mistaken Point *n.* A formation of sandstone and siltstone in southeast Newfoundland that juts into the Atlantic and harbors Ediacaran fossils (Wright 1997, 54).

Monte Verde, Chile *n.* The oldest accepted site of human habitation in the Americas (Gibbons 1997, 1256–1257).

n–r

cf. Clovis People, Monte-Verde People (Appendix 1)

Old Crossman Pit *n.* A fossil site in Sayreville, NJ, with many plant species preserved as possibly billions of charcoal and other fossils (Yoon 1999c, D20).

Pilbara Sediments *n.* Slightly metamorphosed sedimentary rocks in Western Australia that are about 3250 million years old and contain elemental carbon that may be remains of early life on Earth (Holland 1997, 38).

Comment: The structures that might be the oldest stromatolite fossils are from the Pilbara Shield, Australia, in rocks 3.4 to 3.5 billion years old (Stanley 1989, 255–256, illustration).

♦ **place imprinting** See learning: imprinting: place imprinting.

♦ **placenta** See organ: placenta.

♦ **placentophagia** See -phagia: placentophagia.

♦ **placer mining** *n.* In the Japanese Macaque: an individual's method of separating wheat grains from sand by throwing a wheat-sand mixture into water and scooping the floating wheat from the water surface (Wilson 1975, 171).

cf. tradition

♦ **plagio-, plagi** *combining form* "Oblique; slanting" (Michaelis 1963).

[Greek *plagios*, oblique]

♦ **plagioclimax** See ²community: climax: plagioclimax.

♦ **plagiosere** See ²succession: plagiosere.

♦ **planet** *n.*
1. One of the celestial bodies that revolves around the sun (Michaelis 1963).
   Notes: From the sun outwards, the planets are Mercury and Venus (the inferior planets); Earth; Mars, Jupiter, Saturn, Uranus, Neptune, and Pluto (the superior planets) (Michaelis 1963). Asteroids (= minor planets) are between Mars and Jupiter.
2. A large celestial body that revolves around a star (Svitil 2000, March, 29).

Comments: As of January 2000, researchers had found 29 planets that revolve around sunlike stars, including three in a solar system (Svitil 2000, 29). Two of these planets may have temperatures from 108 through 130°F, which might be conducive to life. Researchers found planets around pulsars starting in 1991.

[Old French *planeta* < Late Latin *planeta* < Greek *planētēs*, wander < *planaesthai*, to wander]

Planet of 51 Pegasi *n.* A planet that rotates around the star 51 Pegasi in the constellation Pegasus, 50 light-years from Earth (Mayor and Queloz 1995 in Ganz 1997, 1257–1258).

Comments: This is the first planet that orbits a sun-like star found by researchers (Svitil 2000, 29).

♦ **-planetic** *adj.*
1. Referring to a species that has motile, or swimming, stages (Lincoln et al. 1985).
2. Motile (Lincoln et al. 1985).

n. plantism

diplanetic *adj.* Referring to an organism that has two motile stages during its life cycle (Lincoln et al. 1985).

monoplanetic *adj.* Referring to an organism that has a single motile stage during its life cycle (Lincoln et al. 1985).

polyplanetic *adj.* Referring to an organism that has more than two motile stages and intervening resting stages during its life cycle (Lincoln et al. 1985).

♦ **plankter, plankt, planktont** See plankton.

♦ **planktobiont** See biont: planktobiont.

♦ **planktohyponeuston** See neuston: hyponeuston: planktohyponeuston.

♦ **plankton** *pl. n., sing.* **plankter, plankt, planktont**
1. Aquatic organisms that drift, or float, with currents, waves, and so forth; are unable to influence their own courses; and range in size from microorganisms to jellyfish; distinguished from benthos (modified from Michaelis 1963).
   Comments: Michaelis (1963) defines plankton only as marine organisms. Ali and Lord (1980) refer to "planktonic invertebrates" which they collected from freshwater ponds. Schindler et al. 1997, 249) refer to plankton that they studied in freshwater lakes.
2. Aquatic, or aerial, organisms that are unable to maintain their positions, or distributions, independent of water, or air, movement (Lincoln et al. 1985).

adj. planktonic

cf. ²community: benthos; nekton; pleuston; seston; -xene: planktoxene

Comments: A plankter is an individual planktonic organism (*e.g.,* a zooplankter or phytoplankter) (Lincoln et al. 1985). Plankton are classified by size, type of organism, habitat, and other characteristics (Lincoln et al. 1985).

[German < Greek neuter of *planktos*, drifting, wandering]

aeroplankton, aerial plankton *pl. n.* Plankton that are freely suspended in air and dispersed by wind (Lincoln et al. 1985).

collaplankton See plankton: kollaplankton.

euplankton See plankton: holoplankton.

haliplankton, haloplankton *pl. n.* Marine, or inland saltwater, plankton (Lincoln et al. 1985).

hekistoplankton *pl. n.* Flagellated nano-plankton (Lincoln et al. 1985).

heleoplankton *pl. n.* Plankton of small ponds and marshy habitats (Lincoln et al. 1985).

hemiplankton See plankton: mero-plankton.

hidroplankton *pl. n.* Plankton that achieve buoyancy using surface secretions (Lincoln et al. 1985).

holoplankton, euplankton *pl. n.* Plankton that are permanent members of a plankton community (Lincoln et al. 1985). *cf.* plankton: meroplankton

hyphalomyraplankton, hyphal-myroplankton *pl. n.*
1. Plankton of brackish water (Lincoln et al. 1985).
2. Floating organisms of saline waters (Lincoln et al. 1985).

hypoplankton *pl. n.* Plankton that live close to a sea bed or lake floor (Lincoln et al. 1985).

knephoplankton *pl. n.* Plankton of the twilight zone at from about 30 to 500 meters deep (Lincoln et al. 1985). *cf.* plankton: phaoplankton, skotoplankton

kollaplankton, collaplankton *pl. n.* Plankton that are buoyant due to their gelatinous envelopes (Lincoln et al. 1985).

kremastoplankton *pl. n.* Plankton that possess modified appendages, or surface structures, that reduce their rates of sinking (Lincoln et al. 1985).

limnoplankton *pl. n.* Plankton of fresh-water lakes and ponds (Lincoln et al. 1985).

macroplankton *pl. n.* Plankton that are 20–200 millimeters in diameter (Lincoln et al. 1985).

megaloplankton *pl. n.* Plankton that are larger than 10 millimeters in diameter (Lincoln et al. 1985).

megaplankton *pl. n.* Plankton that are 200–2000 millimeters in diameter (Lincoln et al. 1985).

meroplankton *pl. n.* Organisms that are temporary members of a planktonic com-munity (Lincoln et al. 1985). *cf.* plankton: holoplankton

mesoplankton *pl. n.* Plankton that are 0.2–20 millimeters in diameter (Lincoln et al. 1985).

microplankton *pl. n.* Plankton that are 20–200 micrometers in diameter (Lincoln et al. 1985).

monoton plankton *pl. n.* Plankton very heavily dominated by one species (Lincoln et al. 1985). *cf.* plankton: pantomictic plankton, polymictic plankton, and prevalent plank-ton

morphoplankton *pl. n.* Plankton that are rendered buoyant by their anatomical specializations, such as oil droplets or gas vesicles, or in which their rate of sinking is reduced by their structural features or diminutive body size (Lincoln et al. 1985).

nanoplankton *pl. n.* Plankton that have a body diameter of 2–20 micrometers (Lin-coln et al. 1985).

neidioplankton *pl. n.* Plankton with some form of swimming apparatuses (Lin-coln et al. 1985).

pantomictic plankton *pl. n.* Plankton that are not dominated by any particular species (Lincoln et al. 1985). *cf.* plankton: monoton plankton, polymictic plankton, prevalent plankton

periodic plankton *pl. n.* Organisms that enter plankton only at certain times of the day, month, or season (Lincoln et al. 1985).

phaoplankton, phaeoplankton *pl. n.* Surface plankton of the upper photic zone (Lincoln et al. 1985). *cf.* plankton: knephoplankton, scotoplank-ton

phylktioplankton *pl. n.* Plankton that are rendered buoyant by hydrostatic means (Lincoln et al. 1985).

phytoplankton *pl. n.* Photosynthetic plankton (Curtis 1983, 1103). [Greek *phyton*, plant + PLANKTON]

picoplankton *pl. n.* Planktonic organisms from 0.2–2.0 micrometers in diameter (Lin-coln et al. 1985).

planktophyte *pl. n.* A planktonic plant; an individual phytoplankton (Lincoln et al. 1985).

polymictic plankton *pl. n.* Plankton that have several dominant or abundant species (Lincoln et al. 1985). *cf.* plankton: monoton plankton, panto-mictic plankton, prevalent plankton

potamoplankton *pl. n.* Plankton of slow-moving streams and rivers (Lincoln et al. 1985).

prevalent plankton *pl. n.* Plankton dominated by one particular species that comprises at least half of the total plankton of the area (Lincoln et al. 1985). *cf.* plankton: monton plankton, pantomictic plankton, polymictic plankton

pseudoplankton, tychoplankton *pl. n.* Organisms that are not normally plankton, but occur accidentally with plankton (Lin-coln et al. 1985).

rheoplankton *pl. n.* Plankton associated with running water (Lincoln et al. 1985).

saproplankton *pl. n.* Plankton that in-habit water rich in decaying organic matter (Lincoln et al. 1985).

n–r

skotoplankton *pl. n.* Plankton that live in darkness below the photic zone (Lincoln et al. 1985).
cf. plankton: knephoplankton, phaoplankton

stagnoplankton *pl. n.* Plankton and floating vegetation of stagnant water bodies (Lincoln et al. 1985).

tychoplankton See plankton: pseudoplankton.

zooplankton *pl. n.* Nonphotosynthetic plankton (Curtis 1983, 1110).
[Greek *zoe*, life + PLANKTON]

♦ **planktont** See plankter.

♦ **planktophile** See ¹-phile: planktophile.

♦ **planktophyte** See plankton: planktophyte.

♦ **planktotroph** See -troph-: planktotroph.

♦ **planktoxene** See -xene: planktoxene.

♦ **planned multiple-comparisons procedure** See statistical test: multiple-comparisons test: planned multiple-comparisons procedure.

♦ **planogamete** See gamete: planogamete.

♦ **planomenon, ephaptomenon, rhizomenon** *n.* "All free living organisms" (*i.e.,* those not attached to a substrate) (Lincoln et al. 1985).

♦ **planont** See gamete: planont.

♦ **plant biogeography** See study of: geography: biogeography: phytogeography.

♦ **plant-defense guild** See guild.

♦ **"plant piercers"** See ²group: functional feeding group: "plant piercers."

♦ **plant-pollinator landscape** *n.* Long-term mutualistic evolution of pollinators and their floral hosts (Buchmann and Nabhan 1996, 251).
[coined as "plant/pollinator landscape" by J.L. Bronstein (1994) in Buchmann and Nabhan (1996, 251)]

♦ **plant sociology** See study of: plant sociology.

♦ **plant teratology** See study of: teratology: plant teratology.

♦ **planticole** See -cole: planticole.

♦ **plantigrade** See -grade: plantigrade.

♦ **plantigrade locomotion** See locomotion: plantigrade locomotion.

♦ **planula** *n., pl.* **planulae** [plan New la, plan NEW lee] A solid group of cells similar to embryonic stages of Porifera and Cnidaria (Strickberger 1996, 609).
See animal names: planula.
adj. planular, planulate
[New Latin < Latin, diminutive of *planus*, flat]

♦ **-plasia, -plasis, -plasy** *combining form* "Growth; development; formative action" (Michaelis 1963).
adj. -plastic

[Greek *plasis*, molding < *plassein*, to mold, form]

anaplasia, anaplasis *n.*
1. The progressive phase in an organism's ontogeny prior to its attaining maturity (Lincoln et al. 1985).
2. An evolutionary state characterized by increasing vigor and diversification of organisms (Lincoln et al. 1985).
See dedifferentiation.

cataplasia, cataplasis *n.* An evolutionary state characterized by decreasing vigor (Lincoln et al. 1985).

metaplasia *n.*
1. An evolutionary state characterized by maximum vigor and diversification of organisms (Lincoln et al. 1985).
syn. metaplasis (Lincoln et al. 1985)
2. One type of tissue's changing into another (Lincoln et al. 1985).

♦ **plasmagene** See gene: plasmagene.

♦ **plasmagynogamous fertilization** See fertilization: plasmagynogamous fertilization.

♦ **plasmagynogamy** See -gamy: plasmagynogamy.

♦ **plasmaheterogamous fertilization** See fertilization: plasmaheterogamous fertilization.

♦ **plasmaheterogamy** See -gamy: plasmaheterogamy.

♦ **plasmaisogamous fertilization** See fertilization: plasmaisogamous fertilization.

♦ **plasmaisogamy** See -gamy: plasmaisogamy.

♦ **plasmatogamy** See -gamy: plasmogamy.

♦ **plasmid** *n.* An autonomous, extrachromosomal, covalently closed circular (rarely linear), double-stranded, self-replicating DNA molecule found in most bacterial species and in some eukaryote species (Lederberg 1952 in Rieger et al. 1991).
Comments: Linear plasmids occur in organisms including maize, *Streptomyces*, and yeast (Rieger et al. 1991). In maize, linear plasmids are in mitochondria. Plasmids often confer an evolutionary advantage to their hosts (*e.g.,* antibiotic resistance or colicin production) (King and Stansfield 1985).

episome *n.* A bacterial plasmid that can exist in two states within a cell, either independently in its cytoplasm or, following insertion, as an integral part of its chromosomes (Thompson 931 and Jacob and Wollmann 1958 in Rieger et al. 1991).

F plasmid *n.* A conjugative, circular plasmid that mediates mating in *Escherichia coli* (Rieger et al. 1991).
syn. F factor, F sex factor (Rieger et al. 1991)
Comments: An F plasmid can transfer itself from one bacterial strain to another and can

mobilize portions of its hosts' chromosomes (Rieger et al. 1991). Adaptive mutation in *Escherichia coli* involves F plasmids (Galitski and Roth 1995, 421).

Ti plasmid *n.* A plasmid in the bacterium *Agrobacterium tumefaciens* which causes plant tumors called "crown galls" (Campbell et al. 1999, 384).

Comments: Researchers use genetically engineered Ti plasmids as vectors that do not cause plant tumors to introduce new genes into plants (Campbell et al. 1999, 384).

[Ti, after tumor inducing (Campbell et al. 1999, 384)]

♦ **plasmodium** *n.* In true slime molds (Myxomycetales): a life-cycle stage in which a mass of tissue, with multiple nuclei but no distinct cell boundaries, groups and spreads by nuclear division and accretion of cytoplasm (Wilson 1975, 591).

cf. pseudoplasmodium

♦ **plasmogamy** See -gamy: plasmogamy.

♦ **plasmon** See -type, type: plasmon.

syn. plasmotype (Lincoln et al. 1985)

♦ **plasmophage** See -phage: plasmophage.

♦ **plasmotype** See -type, type: plasmon.

♦ **plastic strain** *n.* An irreversible physical, or chemical, change in a living organism produced by stress (Lincoln et al. 1985).

♦ **plasticity** *n.*

1. An organism's genetic and environmentally determined variation (Salisbury 1940 in Bradshaw 1965, 116).

   Note: Salisbury's usage is ambiguous because "his subsequent examples were all of environmentally determined modifications" (Bradshaw 1965, 116).

2. An organism's capacity to vary morphologically and physiologically in response to environmental changes; this does not include variation that is "directly genetic in origin" (Bradshaw 1965, 116).

   Notes: Because these changes are physiological in origin, all plasticity is fundamentally physiological (Bradshaw 1965, 116). "Plasticity" does not have any implications concerning the adaptive value of changes that occur, although many types of plasticity may have important adaptive effects.

3. An organism's capacity to vary morphologically, physiologically, and behaviorally in response to environmental changes (Lincoln et al. 1985).

See plasticity: phenotypic plasticity, phenotypic plasticity: plasticity.

phenotypic plasticity *n.*

1. Heritable phenotypic variation among conspecific individuals (that are not necessarily genetically identical with respect to the focal character) elicited by variation in environmental conditions (*e.g.,* morphological variation within cichlid-fish populations) (Gause 1947, Gupta and Lewontin 1982 in Gordon 1992, 256).

   Note: This definition is similar to that for "norm of reaction," *q.v.,* which stipulates that the conspecifics are of the same genotype.

2. A single genotype's ability to produce more than one alternative form of morphology, physiological state, behavior, or a combination of these things, in response to environmental conditions (West-Eberhard 1989, 249).

   syn. plasticity (West-Eberhard 1989)

   cf. norm of reaction

3. In quantitative genetics, differences between populations in their gene-environment interactions (Sterns 1982 and Via and Land 1985 in Gordon 1992, 257).

   Note: This kind of phenotypic plasticity may not contribute to interindividual differences within a population or generation; rather, it causes one generation as a whole, on the average, to look different from the preceding one (Gordon 1992, 257–258).

4. Morphological changes in successive generations (Lynch 1984 in Gordon 1992, 257).

5. The capacity for marked variation in an organism's phenotype as a result of environmental influences on its genotype during its development (Lincoln et al. 1985).

   cf. developmental homeostasis

6. Variation among individuals within a population in response to environmental conditions (Meyer 1987 in Gordon 1992, 258).

7. In quantitative genetics, the difference in mean value of a phenotypic character from one population to another when it is caused by different reaction norms (Gordon 1992, 257).

Comments: Darwin (1881, 6) speculated about phenotypic plasticity prior to the naming of this phenomenon (Gordon 1992, 255). West-Eberhard (1989, 250) and Gordon (1992, 262) note that there is a bewildering and inconsistent profusion of terms for classifying kinds of "phenotypic plasticity." Gordon (1992) discusses how to measure phenotypic plasticity, its relationship to evolution, and how the diverse meanings of "phenotypic plasticity" can obfuscate evolutionary discussion.

▶ **flexibility** *n.* Phenotypic plasticity that involves reversible change in an organism

n–r

(Bradshaw 1965 and Smith-Gill 1983 in Gordon 1992, 256).

▶ **plasticity** *n.* Phenotypic plasticity that involves irreversible change in an organism (Bradshaw 1965, 116; Smith-Gill 1983 in Gordon 1992, 256).

♦ **plasticity of the phenotype** See phenotypic plasticity.

♦ **plastid** See organelle: plastid.

♦ **plastidule** *n.* "A subcellular building block, conceived by Haeckel as the basic unit of life" (Lincoln et al. 1985).

♦ **plastodeme** See -deme: plastodeme.

♦ **plastoecodeme** See -deme: plastoecodeme.

♦ **plastogamy** See -gamy: plastogamy.

♦ **plastogene** See gene: plastogene.

♦ **plate tectonics** See ²theory: theory of plate tectonics.

♦ **platykurtic distribution** See distribution: platykurtic distribution.

♦ **play, playing** *n.*

1. An individual organism's active bodily exercise; brisk and vigorous action of its body, limbs, or both, as in fencing, leaping, swimming, and clapping hands (*Oxford English Dictionary* 1972, entries from 725).
   Note: This is an obsolete definition (*Oxford English Dictionary* 1972).

2. For example, in the Capercaillie (bird): a cock's gestures that attract a hen (*Oxford English Dictionary* 1972, entry from 1875).

3. A person's action of lightly and briskly wielding or plying (as a weapon in a contest) (*Oxford English Dictionary* 1972, entries from Beowulf).

4. A person's exercise, or action, for amusement or diversion (*Oxford English Dictionary* 1972, entries from *ca.* 1000).

5. A person's carrying on, or playing, a game (*Oxford English Dictionary* 1972, entries from *ca.* 1450).

6. In young of some bird and mammal species, adults of some carnivore, rodent, primate, dolphin, and whale species: "A set of pleasurable activities, frequently but not always social in nature, that imitate the serious activities of life without consuming serious goals" (Wilson 1975, 164) (*e.g.,* subsong of some bird species or inventive acrobatics of primates) (Immelmann and Beer 1989, 223).

7. In some animal species: "any activity that is exaggerated or discrepant, divertive, oriented, marked by novel motor patterns, or combinations of such patterns, and that appears to the observer to have no immediate function" (Darling 1937, Loizos 1966, 1967, Hutt 1966 in Wilson 1975, 165).

8. In some animal species: "any behavior that involves probing, manipulation, experimentation, learning, and the control of one's own body as well as the behavior of others, and that also, essentially, serves the function of developing and perfecting future adaptive responses to the physical and social environment" (Groos 1898, Lorenz 1950, 1956, Klopfer 1970 in Wilson 1975, 1965).
   Note: Three forms of basic play are "creative improvisation," "play-fighting," and "object manipulation" (Fagan, 1983, 72).
   cf. exploration: pure exploration; problem solving: pure problem solving

9. In some animal species: behavior that has these structures: (1) sequences of acts shown by adults may be reordered, (2) individual acts of a sequence may be exaggerated, (3) certain acts from within a sequence may be repeated many times, (4) a normal sequence of movements may be terminated earlier than usual as a result of the introduction of irrelevant activities, (5) some movements may be both exaggerated and repeated, and (6) individual movements within a total sequence may never be completed (Loizos 1966 in Dewsbury 1978, 64).

See need: play.

v.i., v.t. play

syn. play behavior

Comments: "Play" may allow young animals to practice and perfect motor patterns and social skills that will be important in their adult lives (Dewsbury 1978, 64). "Play" is a category of behavior that defies precise definition (Immelmann and Beer 1989, 223).

foreplay *n.* Precopulatory behavior (Immelmann and Beer 1989, 62).
Comment: Using "play" in this term is perhaps ill advised because precopulatory behavior is not necessarily a form of play (Immelmann and Beer 1989, 62).

functional play *n.* For example, in young children and young of some species of nonhuman mammals: either undirected (*e.g.,* kicking, crawling, jabbering) or directed behavior toward simple objects (*e.g.,* touching or grasping) involving exercising muscles, sense organs, and coordination (Tembrock 1961 in Heymer 1977, 72).

mock hunting *n.* An animal's play that shows prey-capture movements and frequent manipulation of a surrogate prey object (*e.g.,* in a cat that plays with a wool ball or a young lion that plays with its sibling) (Heymer 1977, 94).
syn. hunting games (Heymer 1977, 94)

object play *n.* Play in which a lone animal interacts with an inanimate object, toys, or a heterospecific (*e.g.,* a cat's playing with a mouse) (Immelmann and Beer 1989, 224).

social play *n.* Play involving two or more conspecifics, or heterospecifics that treat one another as conspecifics (*e.g.,* a dog's and cat's "wrestling") (Immelmann and Beer 1989, 224).

♦ **play dialect**   See dialect: play dialect.

♦ **play face**   See facial expression: play signal.

♦ **play fighting**   *n.* For example, in kittens, puppies, the Rhesus Monkey: play with frequent agonistic behavior (Wilson 1975, 294).

Comments: Unlike true conflicts, animals' social play does not decide which animal wins a disputed resource. "Play-fights" are not injurious nor do they lead to prolonged mutual avoidance by the players (Fagan, 1983, 72).

♦ **play mother**   See mother: play mother.

♦ **play signal**   See facial expression: play signal.

♦ **play song**   See animal sounds: song: subsong.

♦ **playing a game**   *n.* In game theory, an organism's interacting with other organisms, its environment, or both, in ways that relate to promotion of its fitness (Maynard Smith 1982, 2) (*e.g.,* in bacteria or primates) (Axelrod and Hamilton 1981).

♦ **playing dead, playing O'possum**   See behavior: defensive behavior: playing dead.

♦ **pleio-, pleo-**   *combining form* "More" (Lincoln et al. 1985; Brown 1956).
[Greek *pleon,* more]

♦ **pleiocyclic**   See cyclic: pleiocyclic.

♦ **pleiomorphic**   See -morphic: pleiomorphic.

♦ **pleiomorphism**   See -morphism: polymorphism: pleiomorphism.

♦ **pleiotropic gene**   See gene: pleiotropic gene.

♦ **pleiotropism argument, pleiotropism hypothesis, pleiotropism theory**   See hypothesis: hypothesis of divergent evolution of animal genitalia: pleiotropism hypothesis.

♦ **pleiotropy**   *n.*
1. The same gene's, or set of genes', control of more than one of an organism's "seemingly unrelated" phenotypic characteristic, such as a single gene's effect on both red-blood cell morphology or a person's general health (Wilson 1975, 591).
2. The phenomenon whereby a change at one genetic locus can bring about a variety of seemingly unrelated phenotypic changes (Dawkins 1982, 292).

syn. multiple-gene effects, multiple-gene inheritance, pleiotropism (Lincoln et al. 1985) *cf.* epistasis; hypothesis: one-gene-one-character hypothesis
Comment: "Pleiotropy" is probably the rule rather than the exception in gene control of traits (Dawkins 1982, 292).
[Greek *pleiōn,* more + *trophē,* nourishment]

♦ **pleioxenous parasite**   See parasite: pleioxenous parasite.

♦ **pleioxeny**   See -xeny: pleioxeny.

♦ **plenitude, principle of**   See principle: plenitude, principle of.

♦ **"plentiful-food hypothesis"**   See hypothesis: hypotheses regarding the origin of the Cambrian explosion: "plentiful-food hypothesis."

♦ **pleometrosis**   See -metrosis: pleometrosis.

♦ **pleomorph, pleomorphic, pleiomorphic, pleomorphism, pleomorphous**   See -morphic: pleomorphic.

♦ **pleophage**   See -phage: pleophage.

♦ **pleophyletic group**   See ²group: polyphyletic group.

♦ **pleotroph**   See phage: pleophage.

♦ **plesio-**   *combining form* Near.
[Greek *plesios* near]

♦ **plesiobiosis**   See -biosis: plesiobiosis.

♦ **plesiomorph**   See -morph: plesiomorph.

♦ **plethodontid receptivity factor (PRF)**   See chemical-releasing stimulus: semiochemical: pheromone: plethodontid receptivity factor.

♦ **pleuriilignosa**   See ²community: pleuriilignosa.

♦ **pleuston, pleustont**   *pl. n.* Aquatic organisms that remain permanently at a water surface by their own buoyancy, normally positioned partly in water and partly in air (*e.g., Lemna* or *Wolffia* plant species) (Lincoln et al. 1985).
adj. pleustonic
cf. nekton, neuston, plankton

♦ **-ploid**   *combining form* "Having a (specified) number of chromosomes" (Michaelis 1963).
n. -ploidy
[Greek *-ploos,* fold + OID]

♦ **ploidy**   *n.* The number of sets of homologous chromosomes present in an organism, somatic cell, or gamete (Lincoln et al. 1985).
adj. ploid
cf. -ploid, -somy

♦ **-ploidy**   *combining form*
cf. -ploid, ploidy
allodiploidy   *n.* An individual's, or somatic cell's, having one set of chromosomes from a parent of one species and one set from a parent of another species (Bell 1982, 501).
cf. -ploidy: allopolyploidy

n–r

alloploidy *n.* An individual's, or somatic cell's, having one or more sets of chromosomes from a parent of one species and another set, or sets, from a parent of another species (Bell 1982, 501).
n. allopolyploid, polyploid hybrid
cf. -ploidy: allodiploidy, allopolyploidy, panalloploidy

allopolyploidy *n.* An individual's, or somatic cell's, having more than one set of chromosomes from a parent of one species and more than one set from a parent of another species (Bell 1982, 501).
syn. allopolyploid, polyploid hybrid
cf. -ploidy: allodiploidy

allotetraploidy *n.* Tetraploidy resulting from the doubling of the chromosomes of an allodiploid (Mayr 1982, 957).

aneuploidy *n.*
1. An individual's, or somatic cell's, having one or more whole chromosomes more or less than its species' regular chromosome complement (Bell 1982, 502).
2. An individual's, or somatic cell's, having a chromosome complement that is not a simple multiple of its species' basic haploid number (Bell 1982, 502).
cf. -ploidy: heteroploidy, hyperploidy, hypoploidy; -somy: trisomy

autodiploidy *n.* Diploidy in which both sets of chromosomes are from conspecific parents (Bell 1982, 502).
cf. -ploidy: autoploidy, autopolyploidy

autoploidy *n.* An individual's, or somatic cell's, having two or more sets of chromosomes all derived from conspecifics (Bell 1982, 502).
cf. -ploidy: autodiploidy, autopolyploidy, panautoploidy

autopolyploidy *n.* Polyploidy in which more than one set of chromosomes is from each of two conspecific parents (Bell 1982, 502).
cf. -ploidy: autodiploidy, autopolyploidy

cryptic polyploidy, semicryptic polyploidy *n.* An organism's showing little, or no, phenotypic manifestation of differences in its ploidy levels (Lincoln et al. 1985).

diplodiploidy, diplo-diploidy *n.* Reproduction in which both males and females arise from diploid zygotes arising from fertilized eggs.
cf. -ploidy: haplodiploidy

diploidy, synkaryotoky *n.* In most sexually reproducing species: a cell's, or an organism's, having a chromosome complement consisting of two copies (called homologs) of each chromosome (Lincoln et al. 1985).
cf. -ploidy: autodiploidy, haploidy

Comment: A diploid cell, or organism, usually arises as the result of the union of two sex cells, each bearing just one copy of each chromosome. Thus, the two homologs in each chromosome pair in a diploid cell are of separate origin, one derived from its mother and the other from its father (Wilson 1975, 582).

dysploidy *n.* An individual's, or somatic cell's, having a chromosome complement that bears no obvious relationship to the its polyploid series (Lincoln et al. 1985).

euploidy *n.* Polyploidy with an exact multiple of a basic chromosome number (Lincoln et al. 1985).

haplodiploidy, haplo-diploidy *n.* A genetic system in which haploid unfertilized eggs usually produce males and diploid fertilized eggs usually produce females (Wilson 1975, 415).
See sex determination: haplodiploidy.
adj. haplodiploid
cf. -ploidy: diplodiploidy; parthenogenesis: arrhenotoky
Comment: In haplodiploidy, males have no fathers or sons, but they have grandfathers.

haploidy *n.* For example, in most sex cells, most male Hymenoptera and other kinds of haplodiploid animals: an individual's, or cell's, having only one set of chromosomes, none which are homologous (Wilson 1975, 585).
adj., n. haploid
syn. hemiploidy (Lincoln et al. 1985)
cf. -ploidy: diploidy, monoploidy, pentaploidy, polyploidy, tetraploidy, triploidy; parthenogenesis: arrhenotoky

hemialloploidy *n.* An organism's being "a polyploid hybrid (allopolyploid) having chromosome sets derived from two species that are not fully intersterile" (Lincoln et al. 1985).

hemiautoploidy *n.* An organism's being "a polyploid derived from intraspecific hybrids or by the differentiation of the chromosome sets of successful panautoploids" (Lincoln et al. 1985).

hemiploidy See -ploidy: haploidy.

heteroploidy *n.* A population's having aneuploid, diploid, and euploid individuals (Lincoln et al. 1985).
See -ploidy: aneuploidy.

hyperploidy *n.* An individual's, or somatic cell's, having one or more chromosomes, or chromosome segments, in addition to its species' usual chromosome complement (Lincoln et al. 1985).
cf. -ploidy: aneuploidy, hypoploidy

hypoploidy *n.* An individual's, or somatic cell's, having one or more chromosomes,

or chromosome segments, less than its species' usual chromosome complement (Lincoln et al. 1985).
cf. -ploidy: aneuploidy, hyperploidy

mixoploidy *n.* In mosaic and chimaera individuals: different cells' or tissues' having different chromosome numbers (Lincoln et al. 1985).

monoploidy *n.* A somatic cell's, or individual's, having one set of chromosomes rather than the two sets that are typical of its species (Lincoln et al. 1985).
cf. -ploidy: diploidy, haploidy, pentaploidy, tetraploidy, triploidy; -somy: monosomy, trisomy

multiploidy *n.* In some succulent plant species with crassulacean acid metabolism: an individual's having a simultaneous presence of three, or more, integer multiples of its diploid DNA complement in all of its tissues (De Rocher et al. 1990, 99).

panalloploidy *n.* "A true alloploid" (Lincoln et al. 1985).

panautoploidy *n.* "A true autoploid" (Lincoln et al. 1985).

pentaploidy *n.* An individual's, or somatic cell's, having five sets of homologous chromosomes rather than its species' more usual one, or two, chromosomes (Bell 1982, 511).
cf. -ploidy: diploidy, haploidy, monoploidy, tetraploidy, triploidy

polyploidy *n.* An individual's, or somatic cell's, having more than two sets of homologous chromosomes (Mayr 1982, 769).
cf. ²evolution: instantaneous evolution; speciation: instantaneous speciation
Comment: About 47% of all flowering plants (Stebbins and Ayala 1985, 74) and some vertebrate species, including 12 genera of frogs and toads, are polyploid (Müller 1925 and other references in Ptacek et al. 1994, 898).

pseudopolyploidy *n.*
1. Chromosomal number doubling without a corresponding increase in genetic material (Lincoln et al. 1985).
2. "A numerical relationship of chromosome sets in groups of related species that leads to the erroneous inference that they are polyploids" (Lincoln et al. 1985).

semicryptic polyploidy See -ploidy: cryptic polyploidy.

tetraploidy *n.* An individual's, or somatic cell's, having four sets of homologous chromosomes rather than its species' usual one, or two, chromosomes (Bell 1982, 511).
cf. -ploidy: diploidy, haploidy, monoploidy, pentaploidy, triploidy

triploidy *n.* An individual's, or somatic cell's, having three sets of homologous chromosomes rather than its species' usual one, or two, chromosomes (Bell 1982, 511).
cf. -ploidy: diploidy, haploidy, monoploidy, pentaploidy, tetraploidy

♦ **-plont**
diplont *n.*
1. The diploid stage of an organism's life cycle (Lincoln et al. 1985).
syn. diplophase, zygophase. (Lincoln et al. 1985)
2. An organism with a life cycle in which the direct products of meiosis act as gametes (Lincoln et al. 1985).

haplont *n.*
1. The haploid phase of an organism's life cycle (Lincoln et al. 1985).
syn. gamophase, haplophase (Lincoln et al. 1985)
2. An organism that is diploid as a zygote and haploid in its other phase (Lincoln et al. 1985).

♦ **plot effect** See effect: cage effect (comments).
♦ **plump** See ²group: plump.
♦ **plural breeding** See breeding: plural breeding.
♦ **pluralism** *n.*
1. The view that, although natural selection is the most important evolutionary mechanism, other mechanisms including other evolutionary ones, as well, are responsible for organisms' attributes; contrasted with panselectionism (Gould and Lewontin 1979, 589).
2. "In modern Darwinian jargon, the belief that evolution is driven by many agencies, not just natural selection" (Dawkins 1982, 292).
Comment: Charles Darwin held this view (Gould and Lewontin 1979, 589).
♦ **pluriparity** See -parity: pluriparity.
♦ **plurivore** See -vore: plurivore.
♦ **pluviophile** See ¹-phile: pluviophile.
♦ **pluviophobe** See -phobe: pluviophobe.
♦ **PMS** See syndrome: premenstrual syndrome.
♦ **PMS dysphoria disorder** See syndrome: premenstrual-stress-dysphoria disorder.
♦ **PNA** See nucleic acid: peptide-nucleic acid, PNA.
♦ **PNA world** See world: PNA world.
♦ **pneumotaxis** See taxis: pneumotaxis.
♦ **pnoium** See ²community: pnoium.
♦ **POC, POC theory** See ³conflict: parent-offspring conflict.
♦ **pod** See ²group: pod.
♦ **poecilandry** *n.* A species' having more than one form of male; contrasted with poecilogyny (Lincoln et al. 1985).
adj. poecilandric

♦ **poecilo-, poikilo-** *combining form*
"Various, variable" (Lincoln et al. 1985).
[Greek *poikilos*, variegated]

♦ **poecilogony, poikilogony** *n.*
1. Development in which adult stages of related taxa are virtually identical but juvenile stages are highly divergent (Lincoln et al. 1985).
   adj. poecilogonic
2. Intraspecific variation in duration of an organisms' ontogenetic stages induced by environmental factors (Lincoln et al. 1985).

♦ **poecilogyny** See -gyny: poecilogyny.

♦ **poetic** *adj.* Referring to human communication that evokes complex, personal emotional images, allusory in nature, triggering memories and impulses based upon past associations that can be spelled out only with great difficulty when messages are kept exclusively cognitive in nature (Sebeok 1962 in Wilson 1975, 217).
   cf. cognitive

♦ **poikilodynamic hybrid** See hybrid: poikilodynamic hybrid.

♦ **poikilogony** See poecilogony.

♦ **poikilotherm** See -therm: poikilotherm.

♦ **point mutation** See ²mutation: gene mutation.

♦ **pointer** See dog: pointer.

♦ **"poisonous-gas hypothesis"** See hypothesis: hypotheses regarding the Permian mass extinction: "poisonous-gas hypothesis."

♦ **Poisson distribution** See distribution: Poisson distribution.

♦ **Poisson-lognormal curve** See curve: rank-abundance curve: Poisson-lognormal curve.

♦ **Poisson process** *n.* "A random event in time" (Lincoln et al. 1985).

♦ **poium** See ²community: poium.

♦ **polar** *adj.* Referring to Earth's Frigid Zones, which are within the northern and southern polar circles (Arctic Circle, 66°33′ north, and Antarctic Circle, 66°33′ south); contrasted with temperate, tropical, subtropical (Lincoln et al. 1985).
   Comment: The northern part of Alaska, Canada, and Siberia are in the polar region.

♦ **polar body** *n.* A minute cell that contains a set of sister chromatids and very little cytoplasm and is formed during meiosis in oogenesis (Curtis 1983, 1103).

♦ **polarity** *n.*
1. The direction differentiation, or trend, within a system (Lincoln et al. 1985).
2. The direction of evolution, or change, within a morphocline or transformation series (Lincoln et al. 1985).

♦ **poleophile** See ¹-phile: poleophile.

♦ **poleophobe** See -phobe: poleophobe.

♦ **poleotolerant** *n.* Referring to an organism that is tolerant of urban habitats (Lincoln et al. 1985).

♦ **pollen flower** See organ: flower: pollen flower.

♦ **pollen storer** *n.* A bumble-bee species that stores pollen in its abandoned cocoons (Wilson 1975, 430, 592).
   cf. pouch maker
   Comment: From time to time, an adult female removes pollen from the cocoons and places it into a larval cell in the form of a liquid mixture of pollen and honey (Wilson 1975, 430, 592).

♦ **pollenophage** See -phage: pollenophage.

♦ **pollination, pollenation** *n.* An agent's transferring pollen from an anther to a flower's receptive area (stigma); loosely meaning fertilization of a seed plant (*Oxford English Dictionary* 1972, entries from 1885; Lincoln et al. 1985).
   syn. fertilization (Lincoln et al. 1985)
   cf. fertilization, ²-phile
   Comments: "Pollination" is sometimes misspelled as "pollinization" or "pollenization." Pollination agents include water, wind, and organisms. Some other kinds of "pollination" that are not defined under "pollination" include "bat pollination," "bee pollination," "beetle pollination," "bird pollination," "butterfly pollination," "fly pollination," and "moth pollination." See ²-phile for examples of these kinds of pollination. Pollen collecting by insects does not always result in pollination; *e.g.,* sweat bees can collect corn pollen without pollinating it (personal observation). Basal lineages of modern insect pollinators originated during the Jurassic, probably as generalists on gymnospermous seed plants (Labandeira 1998, 58).
   [Latin *pollen*, fine dust]

   abiotic pollination *n.* Pollination by agents other than organisms (*e.g.,* wind and water) (Faegri and van der Pijl 1979, 34).

   ant pollination *n.* Pollination by ants (Dahl and Hadac 1940, etc. in Faegri and van der Pijl 1979, 110).
   cf. ²-phile: myrmecophile

   biotic pollination *n.* Pollination by organisms (*e.g.,* insects, birds, or lemurs) (Faegri and van der Pijl 1979, 42).

   buzz pollination *n.* In many bee species: an individual's shaking pollen out of poricidal anthers by rapidly contracting her pterothoracic flight muscles and causing the anthers to vibrate (Orsono-Mesa 1947, 465; McGregor 1976, 359, illustration; Buchmann and Cane 1989, 289).
   syn. sonication (Buchmann and Cane 1989, 289; Batra, 1993a); vibratile pollination

(Buchmann 1978 in Buchmann and Cane 1989, 293) vibration (*vibración*) (Orsono-Mesa 1947, 465)

Comments: Bumble bees and other kinds of bees (but not Honey Bees) buzz pollinate some ericaceous plants (*e.g.,* Blueberries, Cranberries, and Huckleberries) and solanaceous plants (Deadly Nightshades, Potatoes, and Tomatoes) (Buchmann and Nabhan 1996, 243). Bees also buzz pollinate flowers with nonporicidal anthers (Houston 1991, 96).

[after the buzzing sound that a bee makes during this kind of pollination]

caprification *n.* Fig-flower pollination by insects (Lincoln et al. 1985).

cross-pollination *n.* Pollination of a flower with pollen from a conspecific plant (Lincoln et al. 1985).

See fertilization: cross-fertilization; -gamy: allogamy.

cf. pollination: exendotrophy, self-pollination, sib pollination

ethodynamic pollination *n.* Pollination in which a pollinator loads pollen from a ripe anther, carries the pollen on its body, moves to a stigma, and directly transfers pollen on the stigma (Galil 1973 in Faegri and van der Pijl 1979, 43).

cf. pollination: topocentric pollination

Comment: This occurs in a few flower-animal associations (*e.g.,* in yucca and yucca-moth species) (Faegri and van der Pijl 1979, 43).

exendotrophy *n.* A flower's being pollinated by pollen from another flower of the same, or different, plant (Lincoln et al. 1985).

cf. pollination: cross-pollination, self-pollination, sib pollination

gravity pollination *n.* Pollination when pollen falls on a flower's stigma from its own anthers or those of another flower (Faegri and van der Pijl 1979, 137).

hand pollination *n.* A person's pollinating a plant, often a crop plant, using a small paint brush or other object (Buchmann and Nabhan 1996, 247).

mess-and-soil pollination *n.* Beetle pollination accompanied by blundering around on a flower, chewing on its floral parts, eating its pollen, and defecating on it (Buchmann and Nabhan 1996, 249).

parasite pollination *n.* Pollination of a flower by an insect species that eats its developing ovules (*e.g.,* in Yucca and Yucca Moths) (Kerner 1898, Stirton 1976 in Faegri and van der Pijl 1979, 72–73).

rain pollination *n.* Pollination of a flower by rain (Hagerup 1950 in Faegri and van der Pijl 1979, 41).

cf. ²phile-: ombrophile

rendezvous pollination *n.* Pollination that is ensured, or increased, by insect copulation on a flower (Faegri and van der Pijl 1979, 75).

cf. ²phile-: hydrophile

self-pollination *n.* Pollination of a flower with its own pollen or with pollen from another flower of the same plant (Lincoln et al. 1985).

syn. selfing

cf. pollination: cross-pollination, exendotrophy, sib pollination

sib pollination, adelphogamy *n.* Pollination of a flower with pollen from another flower of the same plant (Lincoln et al. 1985).

cf. pollination: cross-pollination, exendotrophy, self-pollination; sib mating

sonication See pollination: buzz pollination.

topocentric pollination *n.* A frequent kind of pollination in which a pollinator unwittingly contacts pollen and stigmas (Galil 1973 in Faegri and van der Pijl 1979, 43).

cf. pollination: ethodynamic pollination

vibration, vibratile pollination, *vibración* See pollination: buzz pollination.

water pollination *n.* Pollination of a flower by water's moving pollen to its stigma (Haumann-Merck 1912 in Faegri and van der Pijl 1979, 40–41).

cf. ²-phile: hydrophile

wind pollination *n.* Pollination of a flower by wind's moving pollen to its stigma (Faegri and van der Pijl 1979, 34).

cf. ²-phile: anemophile

♦ **pollination ecology** See study of: ecology: pollination ecology.

♦ **pollination syndrome** See syndrome: pollination syndrome.

♦ **pollinator** *n.* An agent (*e.g.,* bird, insect, mammal, water, or wind) that effects pollination of a flower.

cf. ²-phile for flowers adapted to different kinds of pollinators

Comments: The most common animal pollinators are beetles, followed by hymenopterans, lepidopterans, dipterans, birds, thrips, bats, and other mammals (Buchmann and Nabhan 1996, 274). Bird pollinators include Honeycreepers, Hummingbirds, and Sunbirds. Mammal pollinators include Flying Foxes, Honey Possums, and Sugar Gliders.

major pollinator *n.* A pollinator that is well adapted to a particular flower species and effects much of its pollination; *e.g.,* bats are major pollinators of *Ceiba pentandra* (Baker et al. 1971 in Faegri and van der Pijl 1979, 43).

n–r

minor pollinator *n.* A pollinator that tends to visit a flower after its major pollinator(s) has foraged on the flower and takes what resources remain; *e.g.,* moths, hummingbirds, and bees are minor pollinators of *Ceiba pentandra* (Baker et al. 1971 in Faegri and van der Pijl 1979, 43, 206).

◆ **pollinia, pollinium** See organ: pollinium.

◆ **pollinivore** See -vore: pollinivore.

◆ **poly-** *combining form*
1. "Many; several; much" (Michaelis 1963).
2. "Excessive; abnormal" (Michaelis 1963). [Greek *polys,* much, many]

◆ **polyacmic** See -acmic: polyacmic.

◆ **polyandry** See mating system: polyandry.

◆ **polyandry, degree of** See degree of polyandry.

◆ **polybrachygamy** See mating system: polybrachygamy.

◆ **polycalic** See -domous: polydomous.

◆ **polychotomy** *n.* An unresolved taxon evolutionary radiation (Gray et al. 1991, 1980).

◆ **polyclimax** See climax: polyclimax.

◆ **polycyclic** See cyclic: polycyclic.

◆ **polycyclic aromatic hydrocarbon (PAH)** See molecule: polycyclic aromatic hydrocarbon.

◆ **polycyclic parthenogenesis** See parthenogenesis: polycyclic parthenogenesis.

◆ **polydemic** *adj.* Referring to a species that occurs in different areas (Lincoln et al. 1985).

◆ **polydomous** See -domous: polydomous.

◆ **polyembryony** *n.* For example, in some species of parasitic wasps: reproduction in which an egg laid in a host gives rise to undifferentiated cells that each develop into a genetically identical sibling (Thornhill and Alcock 1983, 37).
cf. parthenogenesis

◆ **polyergistic** See -genic: polygenic.

◆ **polyestric species, polyoestric species** See ²species: polyestric species.

◆ **polyestrous** See estrous: polyestrous.

◆ **polyestrous species, polyoestrous species** See ²species: polyestrous species.

◆ **polyethism** *n.*
1. Age- or caste-related division of labor among members of a society (Wilson 1975, 299, 592).
   Note: An individual animal might show more than one role, or only one role, during its life (Immelmann and Beer 1989, 225).
2. An animal's establishing preferences, especially regarding ecological conditions (*e.g.,* attachment to a specific habitat type) (Immelmann and Beer 1989, 225).

syn. behavioral polymorphism (Immelmann and Beer 1989, 227)
cf. learning: imprinting: habitat imprinting; -morphism: dimorphism, polymorphism

age polyethism *n.* For example, in the Honey Bee and many other social-insects species; the Naked Mole Rat: society members' changing their labor roles as they age (Wilson 1971a in Wilson 1975, 299, 577; Jarvis 1981).
syn. temporal polyethism (Wilson 1971, 470)

caste polyethism *n.* For example, in social-insect species, the Naked Mole Rat: polyethism with regard to caste (Jarvis 1981).

physical polyethism See polyethism: caste polyethism.

temporal polyethism See polyethism: age polyethism.

◆ **polyfactorial** See -genic: polygenic.

◆ **polyfactorial inheritance, polygenic inheritance** See inheritance: multifactorial inheritance.

◆ **polygamy** See mating system: polygamy.

◆ **polygene** See gene: polygene.

◆ **polygenetic** See -genic: polygenic.

◆ **polygenic** See -genic: polygenic.

◆ **polygenic character** See character: polygenic character.

◆ **polygenic inheritance** See inheritance: multifactorial inheritance.

◆ **polygeny** See inheritance: multifactorial inheritance.

◆ **polygoneutism** See -voltine: multivoltine.

◆ **polygynandry** See mating system: polygynandry.

◆ **polygynopaedium** See -paedium: polygynopaedium.

◆ **polygyny** See -gyny; mating system: polygamy: polygyny.

◆ **polygyny, degree of** See degree of polygyny.

◆ **polygyny-polyandry** See mating system: polygyny-polyandry.

◆ **polygyny threshold** See model: Orians-Verner model of polygyny.

◆ **polyhalophile** See ¹-phile: polyhalophile.

◆ **polylectic** See -lectic: polylectic.

◆ **polymerase chain reaction** See method: polymerase chain reaction.

◆ **polymerization** *n.* In chemistry, the formation of a molecule composed of many similar, or identical, molecular subunits (given as a polymer in Raven and Johnson 1989, G-18).
Comment: Polymerization makes larger molecules used by life. Most polymerization involves removing water molecules from

monomers that join; that is, it involves condensation.
[from polymer < Greek *polymeros*, manifold < *polys*, many + *meros*, part]

◆ **polymictic plankton** See plankton: polymictic plankton.

◆ **polymorph** See -morph: polymorph.

◆ **polymorphic** See -morphic: polymorphic.

◆ **polymorphic-behavior hypothesis** See hypothesis: population-limiting hypotheses: polymorphic-behavior hypothesis.

◆ **polymorphism** See -morphism: polymorphism.

◆ **polymorphism-dependent homology** See homology: paramorphic homology.

◆ **polyoestrus** See estrus: polyestrus.

◆ **polyovular** See -ovular: polyovular.

◆ **polyoxybiont** See -biont: polyoxybiont.

◆ **polypeptide** See molecule: polypeptide.

◆ **polyphage** See -phage: polyphage.

◆ **polyphasic species** See ²species: polyphasic species.

◆ **polyphasic taxonomy** See study of: taxonomy: polyphasic taxonomy.

◆ **polyphasy** See ²species: polyphasic species.

◆ **polyphenic population** See population: polyphenic population.

◆ **polypheny** See population: polyphenic population.

◆ **polyphile** See ²-phile: polyphile.

◆ **polyphyletic group** See ²group: polyphyletic group.

◆ **polyplanetic** See -planetic: polyplanetic.

◆ **polyploid hybrid** See -ploidy: allopolyploidy.

◆ **polyploid variety** See -ploidy: autopolyploidy.

◆ **polyploidy** See -ploidy: polyploidy.

◆ **polysaccharide** See molecule: -saccharide: polysaccharide.

◆ **polysomy** See -somy: polysomy.

◆ **polyspermy** See -spermy: polyspermy.

◆ **polystenohaline** See -haline: polystenohaline.

◆ **polytaxis** *n.* Discontinuous change (Lincoln et al. 1985).
cf. taxis

◆ **polytene chromosome** See chromosome: polytene chromosome.

◆ **polytherm, polythermic** See -therm: polytherm.

◆ **polytoky** See -toky: polytoky.

◆ **polytopy** *n.* Character's arising independently in two or more separate localities or geographical areas (Lincoln et al. 1985).
adj. polytopic
syn. polytopism (Lincoln et al. 1985)

◆ **polytroph** See -troph-: polytroph.

◆ **polytrophy** See -trophy: polytrophy.

◆ **polytypic evolution** See ²evolution: polytypic evolution.

◆ **polytypic species** See ²species: polytypic species.

◆ **polyvoltine** See -voltine: polyvoltine.

◆ **polyxenic parasite** See parasite: pleioxenous parasite.

◆ **polyxeny** See -xeny: pleioxeny.

◆ **polyzoic** See -zoic: polyzoic.

◆ **pomeridanus** *adj.* Referring to the afternoon (Lincoln et al. 1985).
cf. crepuscular, diurnal, matinal, nocturnal

◆ **pomology** See study of: pomology.

◆ **pontic** *adj.* Referring to the deep sea (Lincoln et al. 1985).

◆ **pontium** See ²community: pontium.

◆ **pontohalicole** See -cole: pontohalicole.

◆ **pontophile** See ¹-phile: pontophile.

◆ **poocole** See -cole: poocole.

◆ **pool** *n.* "Any combining of efforts or resources: a typists' pool" (Michaelis 1963).
cf. punctuated equilibrium
gene pool *n.* All the genes in a population (Wilson 1975, 585).
meme pool *n.* All of the memes in a population (Dawkins 1977).
cf. gene pool

◆ **pool feeder** See -phage: telmophage.

◆ **poop** See elimination: defecation; feces.

◆ **poophile** See ¹-phile: poophile.

◆ **Popperian method** See method: Popperian method.

◆ ¹**population** *n.*
1. The total number of persons that inhabit a country, town, or other area (*Oxford English Dictionary* 1972, entries from 1612).
2. A set of organisms that belong to the same species and occupy a clearly delimited area at the same time (*Oxford English Dictionary* 1972, entries from 1803; Wilson 1975, 9).
3. In sexually reproducing organisms: a set of organisms capable of freely interbreeding with one another under natural conditions (Wilson 1975, 9).
See society: modular society: discontinuous-modular society.
cf. deme, -patric, group
balanced hermaphrodite population *n.* A population composed solely of hermaphrodites; contrasted with unbalanced hermaphrodite population (Lincoln et al. 1985).
biotically sympatric population *n.* A population that occupies the same habitat within the same geographical area; contrasted with neighboringly sympatric population (Lincoln et al. 1985).
buffered populations *pl. n.* Populations that interact in such a way as to

maintain each others' density within certain limits (Lincoln et al. 1985).

closed population *n.* A population without input of new genetic variation, except that from mutation; contrasted with open population (Lincoln et al. 1985).

diphenic population *n.* A population with two discontinuous phenotypes that lack genetic foundation (Lincoln et al. 1985).
n. diphenism, dipheny
cf. ¹population: leptophenic population, mesophenic population, monophenic population, polyphenic population

equilibrium population *n.*
1. "A population in which gene frequencies are at equilibrium" (Lincoln et al. 1985).
2. A population with a stable size in which death and emigration rates are balanced by birth and immigration rates (Lincoln et al. 1985).

evolutionarily stable population *n.* A population whose "genetic composition is restored after a disturbance, provided the disturbance is not too large" (Maynard Smith 1982, 204).
Comment: An evolutionarily stable population can be genetically monomorphic or polymorphic (Maynard Smith 1982, 204).

floater population *n.* Part of a population of a territorial species comprised of individuals without territories that live perilous vagabonds' existences along the margins of their preferred habitats (*e.g.,* a lemming "march") (Wilson 1975, 83).

leptophenic population *n.* A population with a narrow range of continuous phenotypic variation (Lincoln et al. 1985).
n. leptophenism, leptopheny
cf. ¹population: diphenic population, mesophenic population, monophenic population, polyphenic population

Mendelian population, panmictic unit *n.* A group of interbreeding individuals that share a common gene pool (Lincoln et al. 1985).

merohermaphroditic population, metahermaphroditic population *n.* A population with equal numbers of hermaphrodite and unisexual individuals (Lincoln et al. 1985).

mesophenic population *n.* A population with only a moderately broad range of continuous phenotypic variation (Lincoln et al. 1985).
n. mesophenism, mesopheny
cf. ¹population: diphenic population, leptophenic population, monophenic population, polyphenic population

metahermaphroditic population See ¹population: merohermaphroditic population.

metapopulation *n.* A set of populations that belong to the same species and exist at the same time; by definition, each population occupies a different area (Levin 1970 in Alvarez-Buylla and García-Barrios 1993, 201; Wilson 1975, 588).

microdichopatric population *n.* A population that coexists in the same locality as one or more other populations but occupies different microhabitats (Lincoln et al. 1985).

microparapatric population *n.* A population that coexists in the same locality as, and has gene flow with, one or more other populations and occupies different adjacent microhabitats (Lincoln et al. 1985).

monomorphic population *n.* A population that consists almost exclusively of a single morph, with other morphs' being extremely rare (Campbell 1990, 447).

monophenic population *n.* A population with a single phenotype (Lincoln et al. 1985).
n. monophenism, monopheny
cf. ¹population: diphenic population, leptophenic population, mesophenic population, polyphenic population

neighboringly sympatric population *n.* A population that inhabits different habitats within the same geographical area (Lincoln et al. 1985); contrasted with biotically sympatric population.

open population *n.* A population that is freely exposed to gene exchange with other populations; contrasted with closed population (Lincoln et al. 1985).

panmictic population *n.* A population in which mating is completely random (Wilson 1975, 590).
syn. deme (Lincoln et al. 1985)

polyphenic population *n.*
1. A population with extensive phenotypic variation that lacks genetic fixation (Lincoln et al. 1985).
2. A population with several discontinuous phenotypes (phenes) that lack genetic foundation (Lincoln et al. 1985).
3. A population with environmentally cued alternative phenotypes (West-Eberhard 1989, 251).
n. polyphenism, polypheny
cf. ¹population: diphenic population, leptophenic population, mesophenic population, monophenic population

unbalanced hermaphrodite population *n.* A population composed of a majority of hermaphrodites with varying degrees of maleness, or femaleness, and a minority of pure male and female individuals; contrasted with balanced hermaphrodite population (Lincoln et al. 1985).

♦ ²**population** *n.*
1. Statistically, the "totality of individual observations about which inferences are to be made, existing anywhere in the world or at least within a definitely specified sampling area limited in space and time" (Sokal and Rohlf 1969, 9).
2. Statistically, the "whole group of items or individuals under investigation" (Lincoln et al. 1985).

cf. observation, sample of observations, variable

universe *n.* A subset of a particular population (Sokal and Rohlf 1969, 9).

♦ **population biology** See study of: biology: population biology.

♦ **population bomb** *n.* Earth's huge human population and its impact on this planet (inferred from Myers 1999, 36).
Comment: Paul Ehrlich published his controversial book *The Population Bomb* in 1968.

♦ **population bottleneck** *n.* A sharp decrease in size of a population with consequent reduction in its gene-pool size (Lincoln et al. 1985).
cf. bottleneck effect

♦ **population cage** *n.* A type of cage invented by Teissier and l'Héritier, in which *Drosophila* populations of various sizes and various genetic heterogeneities could be continued for many generations without the input of new alien genes (Mayr 1982, 574).
Comment: This was a major technological advance for the studies of *Drosophila* population biology (Mayr 1982, 574).

♦ **population cycle** *n.* A more or less regular abundance fluctuation of a species' population (Lincoln et al. 1985).

♦ **population density** See density: population density.

♦ **population dynamics** See study of: dynamics: population dynamics.

♦ **population ecology** See study of: ecology: population ecology.

♦ **population fitness** See fitness: population fitness.

♦ **population genetics** See study of: genetics.

♦ **population growth** See growth: population growth.

♦ **population-limiting hypotheses** See hypothesis: population-limiting hypotheses.

♦ **population-niche concept** See niche (def. 7).

♦ **population outbreak** See year: wasp year.

♦ **population regulation** See regulation: population regulation.

♦ **population size (n, N, *n*, *N*)** *n.* The number of items, or individuals, in a finite population (Lincoln et al. 1985).

♦ **population structure** *n.* Population composition according to age, genetic makeup, and sex of individuals (Lincoln et al. 1985).
cf. life table

♦ **population systematics** See study of: systematics: population systematics.

♦ **population thinking** *n.* A school of thought that asserts that there is no "typical" individual organism, and mean values measuring individuals' traits are abstractions (Mayr 1982, 46, 682).
cf. essentialism, typological thinking
Comments: This school emphasizes that "every individual in a sexually reproducing species is uniquely different from all others, with much individuality even existing in uniparentally reproducing ones." Confusion about organismal variation results from the failure to apply population thinking consistently (Mayr 1982, 46, 682).

♦ **population variance, σ²** See variance: population variance.

♦ **porphyrin** See molecule: porphyrin.

♦ **position effect** See effect: position effect.

♦ **positive assortative mating** See mating: assortative mating.

♦ **positive binomial distribution** See distribution: binomial distribution: positive binomial distribution.

♦ **positive chemotaxis** See taxis: chemotaxis: positive chemotaxis.

♦ **positive eugenics** See eugenics: positive eugenics.

♦ **positive feedback** See feedback: positive feedback.

♦ **positive phototaxis** See taxis: phototaxis: prophototaxis.

♦ **positive reinforcement** See reinforcement: positive reinforcement.

♦ **positive reinforcer** See reinforcer: positive reinforcer.

♦ **positive rheotaxis** See taxis: rheotaxis: positive rheotaxis.

♦ **positive skewness** See skewness: positive skewness.

♦ **post-** *prefix*
1. "After in time or order; following" (Michaelis 1963).
2. "Chiefly in scientific terms, after in position; behind" (Michaelis 1963).
[Latin *post*, behind, after]

♦ **postadaptation, postaption** See ³adaptation: ¹,²evolutionary adaptation.

♦ **postclimax** See ²community: climax: postclimax.

♦ **postclisere** See ²succession: postclisere.

♦ **postcopulatory behavior, postcopulatory play** See behavior: copulatory behavior: postcopulatory behavior.

♦ **postcopulatory mate guarding** See behavior: mate guarding: postcopulatory mate guarding.

♦ **postcornual-skin gland** See gland: postcornual-skin gland.

♦ **post-Darwinism** *n.* The attempt to produce "a theory that can predict particular biological events in ecological and evolutionary time" (Wilson 1975, 64).

Comments: "The ultimate success of an event cannot be predicted...;" "Post-Darwinism" is a new phase of evolutionary thinking dating from about 1960, taking up where "The Modern Synthesis" left off in some areas of thought (Wilson 1975, 64).

♦ **poster coloration** See coloration: poster coloration.

♦ **poster signal** See signal: poster signal.

♦ **postinteractive niche** See niche: realized niche.

♦ **postlinguistic signal** See signal: postlinguistic signal.

♦ **postmating isolation** See isolation: postmating isolation.

♦ **post-menopausal zest** *n.* A woman's period of well-being when perimenopause ends and menopause starts (Wallis 1995, 51).

syn. pits to peak phenomenon (Sheehy in Wallis 1995, 51)

[coined by Margaret Mead (Wallis 1995, 51)]

♦ **postpartum** *adj.* Referring to a phenomenon that occurs after parturition (Lincoln et al. 1985).

♦ **posture** *n.* "The relative disposition of parts of anything;" especially, a person's position and carriage of his limbs and body as a whole; "attitude; pose" (*Oxford English Dictionary* 1972, entries from 1606).

cf. behavior, display, sleep position

cleaning-invitation posture *n.* The posture of a cleaning customer (fish) that communicates to a cleaner that the customer is ready to be cleaned (Heymer 1977, 136).

cut-off posture *n.* A submissive posture that decreases, or stops, aggressive stimuli from an opponent; *e.g.,* a primate's looking away from, rather than directly at, an opponent (Hinde 1970, 681).

humped posture *n.* In some cat species: a threat display involving an individual's raising its back into a hump (Lorenz 1951 in Heymer 1977, 48).

limp posture *n.* In carnivores and rodents: a passive posture or "transport catalepsy" involving a slightly bent back and hanging head and legs shown by a young when it is carried by the back of its neck by its parent (Leyhausen 1965, 1975 in Heymer 1977, 176).

cf. neck bite

nose-up posture See display: upward stretch.

nursing posture, nursing position *n.* In mammals: a young's body position that it adopts during suckling and is adjusted to its mother's body position within certain limits and its species characteristics (Immelmann and Beer 1989, 204–205).

obelisk position *n.* In some dragonfly species: the posture of a perching individual that holds its abdomen nearly straight up (Dunkle 1989, 10).

Comment: This posture helps to reduce overheating by reducing the amount of sunlight that falls on a dragonfly's abdomen (Dunkle 1989, 10).

submissive postures *n.* Human ritualized submissive behavior, such as greetings, bowing, tipping one's hat, falling to one's knees, or falling to the ground (Heymer 1977, 51).

cf. submission

tail posture See display: tail display.

wheel posture *n.* In dragonflies and damselflies: a copulation posture in which a male holds a female by the rear part of her head, or her thorax, while she bends her abdomen downward and forward while receiving sperm from his secondary genitalia on his second abdominal segment (Heymer 1966, 1973 in Heymer 1977, 137; Dunkle 1989, 11, 51; Dunkle 1990, 25).

syn. wheel position (Dunkle 1989, 11)

Comment: The abdomen of a male and female form a heart shape when some species are in the wheel posture.

♦ **postventitious** *adj.* "Delayed in development" (Lincoln et al. 1985).

♦ **postzygotic-isolating mechanism** See mechanism: isolating mechanism: postzygotic-isolating mechanism.

♦ **potamic** *adj.* Referring to rivers or transport by river currents (Lincoln et al. 1985).

♦ **potamincole** See -cole: potamincole.

♦ **potamium** See ²community: potamium.

♦ **potamodromous** See -dromous: potamodromous.

♦ **potamology** See study of: potamology.

♦ **potamon** *n.* The lower reaches of rivers and streams (Lincoln et al. 1985).

adj. potamous

cf. rhithron

♦ **potamophile** See ¹-phile: potamophile.

♦ **potamoplankton** See plankton: potamoplankton.

♦ **potential, potentiality** *n.*

1. A physiological difference or change of electrical state in a part of an organism (*e.g.,* nerve membrane) (Immelmann and Beer 1989, 229).

2. An animal's tendency; its probability of showing a particular behavior (Immelmann and Beer 1989, 229).

action potential *n.*

1. A transient electric potential change across a membrane (Curtis 1983, 1086). *Note:* In a nerve cell, an action potential results in transmission of a nerve impulse, and in a muscle cell it results in contraction (Curtis 1983, 1086).

2. A transient, self-regenerating electrical event that is propagated down a neuron membrane when it is depolarized (Gould 1982, 74); firing of a neuron (Immelmann and Beer 1989, 199).
syn. nerve impulse (Immelmann and Beer 1989, 2), nerve spike

3. "The probability of occurrence of a specific behavior pattern" (Immelmann and Beer 1989, 2).
Note: This is a somewhat obsolete ethological term (Immelmann and Beer 1989, 2).

generator potential *n.* A graded electrical potential of receptor cells that results from environmental energy that is converted into a graded electric potential (McFarland 1985, 169).

membrane potential *n.* The electrical difference between a nerve fiber's inside and outside (Immelmann and Beer 1989, 229).

residual reproductive potential *n.* The general "currency" that animals spend at particular times in their lives (Daly and Wilson 1983).

resting potential *n.* The electric potential across the membrane of an inactive, living neuron (McFarland 1985, 167).

resource-holding potential (RHP) *n.*

1. An individual organism's potential for obtaining, or retaining, a resource based on its fighting ability [coined by Parker 1974 in Maynard Smith 1976, 44].

2. An individual organism's potential for obtaining, or retaining, a resource based on one or more relevant attributes, which include fighting ability, physical position with regard to the resource, posturing, size, strength, timing of its relevant behavior, vocalization quality, and weaponry (Parker and Rubenstein 1981, 221, 223–224).

specific-action potential, specific-action potentiality (SAP) *n.* An animal's readiness to show a particular behavior (Immelmann and Beer 1989, 2).
cf. energy: action-specific energy
Comment: Specific-action potential has been replaced by "behavioral tendency," "response tendency," or "response probability" by most writers (Immelmann and Beer 1989, 2).

♦ **potonuon** See nuon: potonuon.
♦ **pouch-brooding** See brooding: pouch-brooding.
♦ **pouch maker** *n.* A bumble-bee species that builds special wax pouches adjacent to groups of larvae and fills them with pollen (Wilson 1975, 430, 592).
cf. pollen storer
♦ **pout face** See facial expression: pout face.
♦ **pouting, pouting behavior** See behavior: pouting.
♦ **Pouyannian mimicry** See mimicry: Pouyannian mimicry.
♦ **power, power of a statistical test** *n.* The probability of rejecting a null hypothesis when it is, in fact, false; power = 1 – probability of a type-II error, *q.v.* (= one's nonrejection of a false null hypothesis) (Siegel 1956, 10; Thomas and Juanes 1996, 856, who discuss the importance of statistical power analysis in animal behavior and list many computer programs that calculate and websites that relate to statistical power).
syn. statistical power (Hedrick 1999, 317)
Comment: Johnson (1996, 860) discusses the overestimation of the role statistical power analysis in science.
♦ **power grip** See grip: power grip.
♦ **power set** *n.* The theoretical upper limit of a combinatorial message which is the set of all possible combinations of its subsets (Wilson 1975, 188).
♦ **praecocial** See -cial: precocial.
♦ **pragmatics** See study of: study of signals: pragmatics.
♦ **prairie** See habitat: prairie.
♦ **pratal** *adj.* Referring to grassland and meadowland (Lincoln et al. 1985).
♦ **pratinicole** See -cole: pratinicole.
♦ **pratum** See ²community: pratum.
♦ **pre-, prae-** *prefix*
1. "Before in time or order; prior to; preceding" (Michaelis 1963).
2. "Before in position; anterior" (Michaelis 1963).
3. "Preliminary to; preparing for" (Michaelis 1963).
[Latin *prae*, before]
♦ **preadaptation** See ³adaptation: ²evolutionary adaptation: preadaptation.
♦ **prebiological, prebios, prebiotic** See biotic: prebiotic.
♦ **precaudal gland** See gland: precaudal gland.
♦ **preceptive behavior** See behavior: soliciting behavior.
♦ **precinctive** *adj.* Referring to a group of organisms confined to a geographic area under discussion (Sharp 1900, 91, in Frank and McCoy 1990, 1).

n–r

cf. authochtonous; [1]endemic; immigrant; indigenous; insular; introduced; [2]species: endemic species

[Latin *praecinctus*, present participle of *praecingere*, to grid, encircle]

♦ **precipice** See habitat: cliff.

♦ **precipitation** *n.*
1. In meteorology, moisture deposition (as drizzle, hail, rain, sleet, snow, etc.) from Earth's atmosphere upon its surface (Michaelis 1963; Allaby 1994).
2. In meteorology, the amount of water (as rain, snow, etc.) deposited on Earth's surface from its atmosphere (Michaelis 1963).
3. Deposition of dust or other substances (*e.g.,* pollution) from air (Allaby 1994).

acid precipitation *n.*
1. Precipitation with a pH less than 5.6 (Krebs 1985, 694).
2. Precipitation with a pH less than about 5.0, which is the value produced when naturally occurring carbon dioxide, sulphate, and nitrogen oxides dissolve into water droplets in clouds (Allaby 1994).

See acid deposition.

syn. acid rain (a confusing synonym because precipitation also includes hail, mist, sleet, and snow)

Comments: Acid precipitation may be caused naturally (*e.g.,* by gases or aerosols ejected by volcanoes) or anthropogenically (*e.g.,* from burning fuels) (Allaby 1994). Acid precipitation effects on plants, soils, and surface waters are complex, their severity depending on the form of deposition and the pH and natural buffering of soil and water into which the acid precipitation falls. Acid rain washes rapidly from plant surfaces but can affect soil. Acid mist tends to coat leaves and is more harmful than acid rain. Acid precipitation has reduced, or eliminated, fish populations in thousands of lakes (Krebs 1985, 695). Starting in the 1980s, countries in North America and Northern Europe have been enacting regulations to reduce smokestack emissions of sulfur in an effort to curb acid precipitation and its harmful effect on our environment (Yoon 1999b, A22). People knew little about what, if any, biological recovery had occurred by 1999.

♦ **precision** *n.* In statistics, "the closeness of repeated measurements of the same quantity" (Lincoln et al. 1985).

♦ **precision grip** See grip: precision grip.

♦ **preclimax** See [2]community: climax: preclimax.

♦ **preclisere** See [2]succession: preclisere.

♦ **precluding of bonds** *n.* Humans' finding it difficult to become peers in a certain endeavor because of past encounters with one another in a situation in which they were not peers, such as professors' and their former students' becoming academic colleagues or siblings' becoming spouses (Tiger and Fox 1971 in Wilson 1975, 79).

♦ **precocial** See -cial: precocial.

♦ **precocious** *adj.* Referring to an event that occurs usually early in an organism's development (Lincoln et al. 1985).

♦ **precoital behavior** See behavior: precoital behavior.

♦ **precopulatory behavior** See courtship.

♦ **precopulatory chasing, precopulatory driving** *n.* In Bovidae without display circling: a bull's chasing a cow for some distance in a straight line and then striking her with his outstretched forelegs (foreleg kicking) (Heymer 1977, 176).
Comment: Similar behavior occurs in doves and pigeons (Heymer 1977, 176).

♦ **precopulatory mate guarding** See behavior: mate guarding: precopulatory mate guarding.

♦ **predation** *n.*
1. Human "plundering or pillaging; depredation" (*Oxford English Dictionary* 1972, entries from 1460).
2. In many animal species: an individual's relatively quick killing and at least partial consumption of its prey (Lincoln et al. 1985).

cf. parasitism, predator

group predation *n.* For example, in Army Ants, Driver Ants, Pelicans, Wolves, Humans (*e.g.,* Pygmies): a conspecific group's cooperating in hunting and retrieving of living prey (Wilson 1975, 585).

micropredation *n.* A small predator's partial consumption of a large prey (Lincoln et al. 1985).

♦ **predator** *n.* An organism that relatively quickly kills and eats another organism (*e.g.,* an ambush bug that kills and eats a bee or a Lion that kills and eats an antelope) (Wilson 1975, 592).

See [2]group: functional feeding group: predators.

syn. harpactophage (Lincoln et al. 1985)
cf. forager, parasite, predation

ambush predator *n.* A predator that awaits its prey and ambushes it rather than stalking it (*e.g.,* a praying mantid) (Heymer 1977, 109).

syn. sit-and-wait predator (a confusing synonym)
Comment: Some researchers do not recommend using this term.

keystone predator See [2]species: keystone species.

A Classification of Predators[a]

I. actively searching predator
II. ambush predator (*syn.* sit-and-wait predator)
III. "clinger"
IV. "cruising predator" (*syn.* actively foraging predator, cruising forager)
V. "fast predator"
VI. flier
VII. intensive predator (intensive forager)
VIII. mobile predator
IX. percher
X. pursuer (*syn.* type-I predator)
XI. saltatory predator
XII. "search-and-chase predator"
XIII. "search-and-destroy predator"
XIV. searcher (*syn.* type-II predator)
XV. sit-and-wait predator (*syn.* ambush predator)
XVI. "slow predator"
XVII. sprinter
XVIII. stationary predator
XIX. stayer
XX. "swimmer"
XXI. widely foraging predator

---

[a] This table reflects how many researchers have classified predators and synonymized their names. Predator terms have been in a state of confusion.

micropredator *n.* An organism that feeds on a prey that is larger than itself and to which it temporarily attaches (*e.g.,* a tick) (Lincoln et al. 1985).
Comment: Many biologists would consider this tick to be a "parasite," not a "predator."

♦ **"predator hypothesis"** See hypothesis: hypotheses regarding the origin of the Cambrian explosion: "predator hypothesis."

♦ **predator warning call** See call: predator warning call.

♦ **predatory aggression** See aggression: predatory aggression.

♦ **predatory drive** See drive: predatory drive.

♦ **predatory model** See [3]model: predatory model.

♦ **prediction** *n.* See logical prediction, temporal prediction.

♦ **prediction error** See error: prediction error.

♦ **predominant species** See [2]species: predominant species.

♦ **preening** See grooming.

♦ **preference** *n.*
1. A person's liking, or estimation, of one thing before, or above, another; prior favor or choice (*Oxford English Dictionary* 1972, entries from 1656).

2. A consuming organism's likelihood of choosing one resource item (= component) over others from a set in which each item type occurs with equal frequency (Johnson 1980, 66).
Notes: In theory, a researcher can rank the items from least preferred through most preferred (Johnson 1980, 66).
3. An individual organism's, or species', being drawn to specific environmental features, such as temperature ranges, humidity ranges, light ranges, foods, or habitats (Immelmann and Beer 1989, 231).
4. An individual animal's choosing a particular conspecific individual over another in a social interaction (*e.g.,* mate choice) (Immelmann and Beer 1989, 231).

partial preference *n.* In some predatory- and parasitic-animal species: an individual's preying upon, or parasitizing, a less profitable prey item when it has the option of utilizing a more profitable one (Krebs and McCleery 1984 in Putters and Assem 1988, 933).

♦ **preferential species** See [2]species: preferential species.

♦ **preferred ambient temperature (T_r)** See temperature: preferred ambient temperature.

♦ **preferred body temperature (T_r)** See temperature: preferred ambient temperature.

♦ **preferred niche** See niche: preferred niche.

♦ **preformationism** See [3]theory: encasement theory, theory of preformation.

♦ **pregnancy** *n.*
1. A mother Human's being with child; gestation (*Oxford English Dictionary* 1972, entries from 1598).
2. A mother mammal's carrying a growing fetus in her uterus.
adj. pregnant
cf. gravid

pseudopregnancy *n.* In some mammal species: a nonpregnant female's having physiological changes associated with pregnancy, such as cessation of ovarian cycling (estrus), due to copulation stimulation that gives rise to her hormonal changes (Immelmann and Beer 1989, 237–238).

♦ **prehensile** *adj.* Referring to an animal's organ that is adapted for gripping (*e.g.,* a monkey's tail or a Black Rhinoceros' upper lip) (Lincoln et al. 1985; Robinson and Challinor 1995, 222).

♦ **preimaginal conditioning** See learning: conditioning: preimaginal conditioning.

♦ **preinteractive niche, prospective ecospace** See niche (def. 7).

n–r

♦ **prelinguistic signal** See signal: prelinguistic signal.

♦ **preparedness** See learning disposition.

♦ **premarin** See hormone: premarin.

♦ **premating isolation** See isolation: premating isolation.

♦ **premenstrual-stress-dysphoria disorder** See syndrome: premenstrual-stress-dysphoria disorder.

♦ **premenstrual syndrome** See syndrome: premenstrual syndrome.

♦ **prenatal learning** See learning: prenatal learning.

♦ **preprogrammed behavior** See behavior: instinct.

♦ **prepubertal, prepuberal** *adj.* Referring to an animal's condition that occurs, or exists, before its puberty (Lincoln et al. 1985).
cf. puberty

♦ **presaturation dispersal** See dispersal: presaturation dispersal.

♦ **prescient forager** See forager: prescient forager.

♦ **prescriptor** *n.* A signal component that designates the appropriate action for the responder to follow (Marler 1961 in Wilson 1975, 217).
cf. appraisor

♦ **presentation** See display: presentation.

♦ **presentation time** See time: presentation time.

♦ **presenting** See behavior: soliciting behavior: presenting.

♦ **preservation** *n.* Human maintenance of individual organisms, populations, or species by planned management and breeding programs (Lincoln et al. 1985).
cf. conservation

♦ **presociality** See sociality: presociality.

♦ **prestige custom** *n.* A person's demonstration of his economic wealth with things such as a large, attractive automobile or house; hospitality; potlatch in Kawikutl Indians (Heymer 1977, 132).

♦ **prevailing climax** See ²community: climax: prevailing climax.

♦ **prevalent plankton** See plankton: prevalent plankton.

♦ **prevernal** *adj.* Referring to early spring (Lincoln et al. 1985).
cf. vernal

♦ **prey** *n.* An organism that is killed relatively quickly and partly, or entirely, consumed by a predator (Lincoln et al. 1985).
cf. predation, predator
Comment: Many researchers include pollen grains, seeds, seedlings, and spores as prey.

♦ **prey-catching behavior** See behavior: prey-catching behavior.

♦ **prey parasite** See parasite: prey parasite.

♦ **prey parasitism** See parasitism: cleptoparasitism.

♦ **prey set** *n.* "The total range of prey items utilized by a predator" (Lincoln et al. 1985).

♦ **prey stealing** See parasitism: cleptoparasitism.

♦ **prezygotic-isolating mechanism** See mechanism: isolating mechanism: prezygotic-isolating mechanism.

♦ **PRF** See chemical-releasing stimulus: semiochemical: pheromone: plethodonitid receptivity factor.

♦ **primaparity** See -parity: primaparity.

♦ **primary** See primordial.

♦ **primary consumer** *n.* "A heterotrophic organism that feeds directly on a primary producer" (Lincoln et al. 1985).

♦ **primary-defensive behavior** See behavior: defensive behavior: primary-defensive behavior.

♦ **primary discontinuity** See discontinuity: primary discontinuity.

♦ **primary drive** See drive: primary drive.

♦ **primary host** See ³host: definitive host.

♦ **primary instinct** See behavior: instinctive behavior: primary instinct.

♦ **primary intergradation** See zone: intergradation zone.

♦ **primary monogyny** See -gyny: monogyny (comment).

♦ **primary parasitism** See parasitism: primary parasitism.

♦ **primary producer** See producer.

♦ **primary production, primary productivity** See production.

♦ **primary queen** See caste: queen: primary queen.

♦ **primary reproductive** See caste: primary reproductive.

♦ **primary-sex ratio** See ratio: sex ratio: primary-sex ratio.

♦ **primary-sexual character, primary-sexual characteristic** See character: primary sexual character.

♦ **primary-social dominance** See dominance (social): primary-social dominance.

♦ **primary speciation** See speciation: primary speciation.

♦ **primatology** See study of: primatology.

♦ **prime mover (of social evolution)** *n.* An ultimate factor, or influence (phylogenetic inertia or environmental pressure) that determines the direction and velocity of evolutionary change in social behavior and, thus, determines the determinants of social organization (Wilson 1975, 32, 592).

♦ **primer pheromone** See chemical-releasing stimulus: semiochemical: pheromone: primer pheromone.

♦ **primeval** *adj.* Referring to the earliest times or the beginning of the universe (Lincoln et al. 1985).

♦ **priming** See memory: priming.

♦ **priming stimulus** See ²stimulus: priming stimulus.

♦ **primitive** See ancestral.

♦ **primitive character** See character: ancestral character, plesiomorph.

♦ **primitive eusociality** See sociality: primitive eusociality.

♦ **primitive similarity, symplesiomorphy** *n.* Taxa's common possession of homologous "primitive" characters or character states (Lincoln et al. 1985).
cf. character: plesiomorph: symplesiomorph

♦ **primitive species** See ²species: primitive species.

♦ **primitive trait** See character: ancestral character, plesiomorph.

♦ **primordial** See primitive.

♦ **principle** *n.*
1. A fundamental truth, or proposition, on which many others depend; a primary truth that comprehends, or forms the basis of, many subordinate truths; a general tenet, or statement, that forms the ground for, or a ground for, or is held to be essential to, a system of thought or belief; a fundamental assumption that forms the basis of a chain of reasoning (*Oxford English Dictionary* 1972, entries from 1380).
2. In physics, a highly inclusive theorem, or law, that admits numerous special applications, or is exemplified in a number of cases (*Oxford English Dictionary* 1972, entries from 1710).
3. "An established mode of action or operation in natural phenomena: the *principle* of relativity" (Michaelis 1963).
4. A generalization useful as a guide to understanding or study (Mayr 1982, 37–45).
Notes: This definition seems appropriate for wide use. Principles in biology are sometimes referred to as "laws," but this is not a preferred use of "law" (Mayr 1982, 37–45).
syn. concept, generalization, rule (Mayr 1982, 37–45)
5. "The intellectual framework of a theory or a generalized way of looking at a phenomenon" (Moore 1984, 475).
syn. concept, theory
cf. axiom, dictum, doctrine, dogma, effect, law, rule, theorem, theory, truism
Comments: See "law" for kinds of biological dictums, laws, principles, rules, etc. Below, I define some of the many terms that contain the word "principle;" all of these terms do not necessarily represent (biological) principles.
[Latin *principium*, a beginning]

Allee effect, Allee principle, Allee law *n.* The density of an organism population varies according to the spatial distribution of its individuals (degree of aggregation), and both overcrowding and undercrowding may be suboptimal (Lincoln et al. 1985).

anthropic principle *n.* Our understanding of the world is a function of our abilities to understand the world (Barrow and Tipler 1986 in Ruse 1993, 59).

August-Krogh principle *n.* "For a large number of problems there will be some animal of choice, or few such animals, on which it can be most conveniently studied" (Krogh 1929 in Lehner 1979, 21); in other words, particular animal species (compared to alternative species) are the better, or best, ones for testing certain hypotheses.
[named after August Krogh by Krebs (1975 in Lehner 1979, 21)]

Bateman's principle *n.*
1. In *Drosophila melanogaster*, male reproductive success is positively correlated with number of copulations (Bateman 1948, 364; Oberhauser 1988, 1384).
Note: Oberhauser's (1988, 1384) work with monarch butterflies supports this principle; she also discusses exceptions to it.
2. Reproductive success is more variable in the sex (usually males) that makes the smaller investment in gametes (Bateman 1948, 364).
3. Reproductive success is more variable in the sex (usually males) that makes the smaller parental investment, *q.v.* (Trivers 1972, 138; Wilson 1975, 324–325).
syn. Bateman effect (Wilson 1975, 327)
cf. bearing-and-caring

Bernard's dictum, Bernard's principle, *milieu intérieur n.* "The stability of the internal environment [of an organism] is the condition for free and independent Life" (Bernard 1859 in McFarland 1985, 238).
[after Claude Bernard (1813–1878), physiologist]

bilateral-primary-linkage principle *n.* In Athapaskans Dogrib Indians: in times of famine, local bands coalesce with temporarily better off ones (Helm 1968 in Wilson 1975, 554).

cicada principle *n.* In Periodical Cicadas: the hypothesis that an individual decreases its chance of being eaten by being part of a large group of conspecific and heterospecific cicadas; this is because a predator (glutton) can become totally satiated from consuming part of such a group,

p–r

but many more group members escape predation in a larger group compared to a smaller one when the number of predators remains the same (Beamer 1931 in Lloyd and Dybas 1966a, 134; Lloyd and Dybas 1966b, 469; Brown 1975, 143).

Comments: "The cicada principle" is a special case of "the gluttony principle," *q.v.* Research with 13-year Periodical Cicadas supports this hypothesis (Williams et al. 1993, 1143).

competitive exclusion, competitive-exclusion principle See principle: Gause's principle.

constant-environment principle *n.* Any population allowed to reproduce itself in a constant environment will attain a stable age distribution (Wilson 1975, 92).

Gause's principle *n.* No two species have the same ecological niche (Gause 1932; Gause 1934 in Odum 1969, 231); "complete competitors cannot coexist" (Hardin 1960 in Krebs 1985, 254).

syn. competitive-exclusion principle (Krebs 1985, 254); Gause's hypothesis, Gause's law; Gause's rule; Grinnell's axiom (Lincoln et al. 1985); Volterra-Gause principle (Hutchinson 1957 in Griesemer 1992, 234)

Comments: This concept was first clearly demonstrated experimentally by Gause (1934 in Odum 1969, 231). It was never formally defined by Gause, and was suggested in Darwin's writing and indicated in writing by Grinnell (1904) and Monard (1920); thus, the "competitive-exclusion principle" might be a preferable name (Krebs 1985, 254). Closely related species can have very similar niches (Odum 1969, 231). Hutchinson (1957, 418, in Griesemer 1992, 237) tried to rescue this principle from apparent circularity "by reformulating it as an empirical claim: realized niches do not intersect." In 1960, Slobodkin referred to Gause's principle as "Gause's Axiom" to emphasize its immunity to test (Slobodkin 1995, 384).

[after G. F. Gause, Russian ecologist]

gluttony principle *n.* A prey organism decreases its chance of being eaten by being part of a large group of conspecific organisms because a predator (glutton) can become totally satiated from consuming part of such a group, but many more group members escape predation in a larger group compared to a smaller one when the number of predators remains the same (Brown 1975, 143).

cf. principle: cicada principle

handicap principle See hypothesis: handicap principle.

Hardy-Weinberg law, Hardy-Weinberg equilibrium, Hardy-Weinberg principle *n.* Allele frequencies in a population do not change from generation to generation if there is no natural selection and pairing of mates is random (Wilson 1975, 63), demonstrating that meiosis and recombination do not alter gene frequencies (Lincoln et al. 1985).

syn. genetic equilibrium (Lincoln et al. 1985)

Comment: "Two alleles (*a* and *d*) will remain at the same frequency in a population from generation to generation unless their frequency is affected by immigration, mutation, selection, nonrandom mating, or errors of sampling" (Hardy 1908; Provine 1971 in Mayr 1982, 554).

Herrnstein's principle of quantitative hedonism *n.* Pigeons trained at two disks, one located to the left and the other to the right, will try one as opposed to the other in precise proportion to the percentage of times each disk reinforces them with food when they peck the disk (Herrnstein 1971a in Wilson 1975, 266).

Horn's principle *n.* The concept that, in some animal species, food-supply variability in space and time is the prime factor for colonial roosting and nesting (Horn 1968; Wilson 1975, 52–53, 525).

Comment: When food is more or less evenly distributed and can be defended economically, it is energetically most efficient to have exclusive territories, but when it occurs in unpredictable patches, it is most efficient to collapse territories to roosting, or nest, sites and forage as a group (Horn 1968; Wilson 1975, 52–53, 525).

MacArthur-Hamilton-Leigh principle *n.* In bisexual species, a parent ideally does not produce equal numbers of each sex; rather, it should make equal investments of time, energy, and other resources in its group of daughters and group of sons (MacArthur 1965, Hamilton 1967, Leigh 1970 in Wilson 1975, 317).

Comment: This principle supersedes Fisher's principle.

maximum-power principle *n.* A system design that survives is organized to bring in energy as quickly as possible and use it to feed back into itself to bring in more energy (Odum et al. 1988, 31).

migration principle *n.* The intensity of the programmed migratory activity of a species is inversely related to the stability of its preferred habitat (Wilson 1975, 104).

principle of allocation *n.*

1. A speculative proposition stating that animals' major requirements differ greatly in the amounts of time and energy that

are profitable to devote to them measured in the currency of genetic fitness (Wilson 1975, 143, 324).

Note: As a rule, these requirements descend in immediacy as food, antipredation, and reproduction (Wilson 1975, 143, 324).

2. Every organism allocates its resources to various essential activities, which can be categorized as maintenance, growth, and reproduction (Willson 1983, 5).

principle of antithesis *n.* Animal signals that are opposite in meaning are often conveyed by expressions, or postures, that are opposites, *e.g.,* postures of a friendly and angry dog or facial expressions of a happy and angry person (Darwin 1872 in McFarland 1985, 398).

principle of commonality *n.* A character state that has the widest distribution among the taxa that comprise a more derived taxon is considered to be the most ancestral (Lincoln et al. 1985).

principle of composition *n.* All homologous structures are composed of the same kinds of elements (*e.g.,* individual hand bones in primates) (Mayr 1982, 462). See principle: principle of connections.

principle of connections *n.* When one is in doubt as to the homology of structures in widely different organisms, consider the relative positions and dependency of the structures; *e.g.,* "the humerus will always lie between the shoulder articulation and the bones of the lower arm (radius and ulna)" (Mayr 1982, 462).

Comment: "The principle of composition" is auxiliary to "the principle of connections." These two principles are used for decisions on homology at present. "The entire modern method of establishing homologies throughout the vertebrate series or throughout the arthropods is ultimately based on Geoffroy's method [these principles]" (Mayr 1982, 462).

principle of correlation of parts *n.* A generalization attributed to Cuvier that proposes that each organ of an organism's body is functionally related to every other organ, and the harmony and well-being of the organism results from their cooperation (Mayr 1982, 460).

principle of expendable surplus *n.* An organism has more of a particular thing (*e.g.,* offspring or resources) than is necessary for optimal fitness promotion in a given environment; for example, sexually reproducing species would not increase their fitnesses for an appreciable period by becoming parthenogenetic (Mayr 1982, 599)

principle of isostasy *n.* The concept that "continents float in hydrodynamic

equilibrium on a substratum of denser material;" contrasts with the theory of plate tectonics (Hallam 1975, 11).

principle of metabolic conservation *n.* The generalization that a species tends to eliminate the biochemical steps required for the synthesis of a kind of nutrient compound (vitamin) that is readily available in its diet, thus enzymatic protein and energy can be diverted for other, more urgent functions (Wilson 1975, 161).

principle of mosaic evolution *n.* Different components of the phenotypes of a species may evolve at highly unequal rates (Dollo 1888, Lamarck 1809, 58, Abel 1924, 21, in Mayr 1982, 579, 613).

syn. mosaic evolution, mosaic-inheritance theory (Mayr 1982, 579, 613); nonharmonious-character transformation (Lincoln et al. 1985)

["mosaic evolution" coined by de Beer (1954 in Mayr 1982, 613)]

principle of plenitude *n.*

1. The idea (doctrine) that God in "the ampleness of His mind had surely created any creature that was possible;" that is, all possible creatures exist (Mayr 1982, 319, 326, 347).

Note: Extinction of a species violates this doctrine, which was adhered to by most of the leading thinkers in the 17th and 18th centuries including Leibniz (Mayr 1982, 319, 326, 347).

2. An ecosystem is delicately balanced and full of exactly the right number of species (Bakker 1985, 72).

syn. plenitude (Mayr 1982, 326)

[Old French < Latin *plenitudo* < *plenus*, full]

principle of progressive development *n.* Chambers (1844) idea that "(1) fauna of the world evolved through geological time, and (2) that the changes were slow and gradual and in no way correlated with any catastrophic events in the environment" (Mayr 1982, 382–383).

cf. gradualism, punctuated equilibrium

principle of reproductive success *n.* Per unit time, all things being equal, a parthenogenetic species can generate twice as many female reproductives as a sexually reproducing species that wastes half of its zygotes in producing males (Mayr 1982, 599).

cf. cost: cost of sex

Comments: The principle of reproductive success suggests that natural selection should favor parthenogenesis rather than sexual reproduction, but it is believed that the latter enables environmental tracking with genetically variable offspring and this accounts for the predominance of sexually

n–r

reproducing species (Williams 1975, Maynard Smith in Mayr 1982, 599). For other possible benefits of sex, see hypothesis: hypotheses of the evolution and maintenance of sex.

principle of stimulus relevance *n.* The associated strength of a cue with its consequences (reinforcer) depends partly upon the nature of the consequences (Capretta 1961 in McFarland 1985, 337).

principle of stringency *n.* A speculative proposition that animals' time-energy budgets evolve so as to fit their times of greatest stringency, periods of food shortage (Wilson 1975, 142).

principle of use and disuse See law: Lamarck's first law.

progressive holism *n.* The idea that an understanding of systemic process in nature can help improve the way in which human societies interact with their natural environments (Jamison 1993, 497).

reafference principle *n.* A concept that attempts to explain the regulation and interaction of internal signals and sensory signals in directing and coordinating an animal's movements: An animal's central nervous system (processing unit) stores an efference copy, which fixes a reference value (*Sollwert*) of parameters required to execute the animal's movement. This information guides its response until the reafference from its motor unit to the processing unit indicates accordance with its *Sollwert*; if, during its movement execution, the animal encounters a disturbance, the animal reexamines its efference copy in relation to the new information and corrects its movements until it reaches its goal (von Holst and Mittelstaedt 1950 in Heymer 1977, 141; Immelmann and Beer 1989, 243).

cf.-ference: afference, efference, reafference; [3]theory: inflow theory, outflow theory
Comment: This is an extension of outflow theory.

social-tolerance principle *n.* A species' social tolerance has evolved to fit its optimal population density and optimal population structure (Lidicker 1965, Eisenberg 1967, Christian 1970 in Wilson 1975, 102).

survival of the stable *n.* A principle that indicates that some groups of atoms (*e.g.,* the Matterhorn, a raindrop, a banana, or an elephant) last long enough on Earth to be observed and named (Dawkins 1977, 13).
Comment: Darwin's "survival of the fittest" is a special case of this principle (Dawkins 1977, 13).

xenophobia principle, xenophobic principle *n.* "The strongest evoker of aggressive response in animals is the sight of a stranger, especially a territorial intruder" (Klopman 1968 and Southwick 1969 in Wilson 1975, 249, 286).

you-first principle *n.* "An individual is safest by allowing other individuals to take the risk of entering a hazardous or unfamiliar area first" (Wittenberger 1981, 622).

♦ **priority** *n.* "The most direct, obvious objective indicated by an animal's behavior; the proximal objective that *appears* to be directing, or 'motivating,' the animal" (Hand 1986).

♦ **prisoner's-dilemma game** See game (table).

♦ [1]**pro-** *prefix*
1. Forward; to, or toward, the front from a position behind; forth (Michaelis 1963).
2. "Forth from its place; away" (Michaelis 1963).
3. "To the front of; forward and down" (Michaelis 1963).
4. "Forward in time or direction" (Michaelis 1963).
5. "In front of" (Michaelis 1963).
6. "In behalf of" (Michaelis 1963).
7. "In place of" (Michaelis 1963).
8. "In favor of" (Michaelis 1963).
[Greek *pro-* < *pro*, before, forward, for]

♦ [2]**pro-** *prefix*
1. "Prior; occurring earlier in time" (Michaelis 1963).
2. "Situated in front; forward; before" (Michaelis 1963).
[Greek *pro-* < *pro*, before, in front]

♦ **proaposematic coloration** See coloration: proaposematic coloration.

♦ **probabilistic model** See [4]model: stochastic model.

♦ **probability, likelihood** *n.* The statistical chance that a particular event will occur (Lincoln et al. 1985).
cf. potential: specific-action potential
Comment: The probability of an impossible event is zero, and that of an inevitable event is 1 (Lincoln et al. 1985).

transition probability *n.* The estimated probability that one particular behavior (*e.g.,* A) will be followed by another particular behavior (*e.g.,* B) (Immelmann and Beer 1989, 317–318).
cf. analysis: stochastic analysis

♦ **probability level** See p: "stipulated p."

♦ [1]**probe** *n.* In surgery, an instrument for exploring cavities, wounds, etc. (Michaelis 1963).
[Late Latin *proba*, proof < Latin *probare*]

♦ ²**probe** *n*. A short sequence of DNA that is designed to recognize only one specific other sequence of DNA (= a marker) (Rensberger 1995, A10).
Comment: Probes are used in DNA fingerprinting.

nucleic-acid probe *n*. A short, single-stranded nucleic acid that is complementary to another strand of nucleic acid (Campbell et al. 1999, 368).
Comment: This probe is used in identifying a gene (Campbell et al. 1999, 368).

♦ **probiogenesis** See -genesis: biogenesis: probiogenesis.

♦ **problem solving, reasoning** *n*. For example, in primates: an individual's finding the solution to a problem (*e.g.,* recognizing oddity or matching to a sample) by using inference, or insight, as opposed to trial-and-error learning (Immelmann and Beer 1989, 233).
cf. learning

pure-problem solving *n*. An animal's working out a solution to a difficult circumstance without entailing pleasurable learning of rules and variations in doing so (Bruner 1968 in Wilson 1975; Wilson 1975, 165).

♦ **procaryotic** See prokaryotic.

♦ **procedural representation** See representation: procedural representation.

♦ **procedure** *n*.
1. A manner of proceeding, or acting, in any course of action (Michaelis 1963).
2. "A course of action; a proceeding" (Michaelis 1963).
cf. method, program

correction procedure *n*. In a study of simultaneous-discrimination learning in primates: an investigator's presenting a subject with a stimulus configuration identical to that of the previous trial (*i.e.,* the stimulus position remains the same) after the subject makes an incorrect choice (Dewsbury 1978, 328).

guidance procedure *n*. In a study of simultaneous-discrimination learning in primates: an investigator's forcing a subject to make a correct choice after the subject makes an incorrect choice because the investigator gives the subject no alternative choice (Dewsbury 1978, 328).

noncorrection procedure *n*. In a study of simultaneous-discrimination learning in primates: an investigator's ending a trial with a subject's incorrect choice (Dewsbury 1978, 328).

♦ **proceptive behavior** See behavior: proceptive behavior.

♦ **proceptivity, female** See behavior: proceptive behavior.

♦ **process** *n*.
1. A course, or method, of operations for producing something (Michaelis 1963).
2. A series of continuous actions that bring about a given result, condition, or both (Michaelis 1963).
3. "A forward movement; progressive or continuous proceeding; passage; advance; course" (Michaelis 1963).

magnification process *n*. An establishment of dominance hierarchies in which an individual's repeatedly successful encounters with others increase the probability of its success in later encounters and make its contest with a timid animal still more of a mismatch (Chase 1973, 1974 in Wilson 1975, 295).

Markoff process *n*.
1. An event occurrence's being influenced to some degree by the nature of immediately preceding events, such that the event's occurrence is different than chance alone (Lincoln et al. 1985; Immelmann and Beer 1989, 295).
cf. order
2. A stochastic process in which future development is determined only by its present state and is independent of this state's developmental mode (Lincoln et al. 1985).
cf. chain: Markoff chain

random process, stochastic process *n*. A process governed in part by a chance, or random, mechanism (Lincoln et al. 1985).

vacancy-chain process *n*.
1. The process by which a series of people sequentially use the same resource; *e.g.,* person-1 obtains an apartment, he leaves it, then person-2 rents it, he leaves it, then person-3 rents it, etc. (Chase et al. 1988, 1265).
2. The process by which a series of hermit crabs uses the same shell sequentially (Chase et al. 1988, 1265).

♦ **process of intercompensation** *n*. "The operation of only one or a small number of control factors at a time, with other mechanisms coming into play only if the primary ones are removed by an amelioration of environmental conditions" (Wilson 1975, 25).

♦ **procession** See ²group: procession.

♦ **prochosium** See ²community: prochosium.

♦ **prochronism** *n*. The general truth: "Organisms carry, in their forms, evidences of their past growth" (Bateson 1979, 242).
Comment: "Prochronism" is to "ontogeny" as "homology" is to "phylogeny" (Bateson 1979, 242).

♦ **proclimax** See ²community: climax: proclimax.

◆ **procrypsis** See -crypsis: procrypsis.
◆ **proctodeal food** See food: proctodeal food.
◆ **proctodeal trophallaxis** See trophallaxis: proctodeal trophallaxis.
◆ **producer** *n.*
  1. An autotrophic organism, usually a photosynthesizer, that contributes to the net primary productivity of a community (Curtis 1983, 1104).
     syn. autotroph, primary producer (Lincoln et al. 1985)
     cf. -troph-: autotroph
  2. An organism that synthesizes complex organic substances from simple inorganic ones (Lincoln et al. 1985).
     syn. autotroph, primary producer (Lincoln et al. 1985)
     cf. -troph-: autotroph
  3. An individual, or species, that obtains a resource through making a behavioral investment; contrasted with scrounger [coined by Barnard and Sibly 1981, 543].
  Notes: For examples of producers and scroungers, see scrounger below.
◆ ¹**production** *n.* The process by which two or more ingredients are combined to form a new product; *e.g.,* plants combine soil nutrients, water, carbon dioxide, and sunlight to form organic matter during photosynthesis (Odum et al. 1988, 29).
  Comment: Odum et al. (1988, 29) also define "gross production" and "net production."
◆ ²**production, rate of production, primary production, productivity, basic productivity, primary productivity** *n.* The rate at which energy is stored by photosynthetic and chemosynthetic activity of producer organisms (chiefly green plants) in the form of organic substances, which can be used as food materials in an ecological system, community, or part thereof; this rate is often measured as grams per square meter per day (Odum 1969, 68, 71).
  gross-primary productivity, gross primary production, total photosynthesis, total assimilation *n.* The total productivity (measured as the total rate of photosynthesis) including the organic matter used up in respiration during a measurement period (Odum 1969, 68).
  net-biome production *n.* Net-primary productivity and the losses of carbon that result from fires, heterotrophic respiration, insect-induced mortality, logging, and other anthropogenic and natural disturbances (Bolin et al. 1999: 1851).

net-primary productivity, net-primary production, apparent photosynthesis, net assimilation *n.* The total productivity (measured as the total rate of storage of organic matter in plant tissues) in excess of plant respiratory utilization during a measurement period (Odum 1969, 68).
  abbr. NNP (Delucia et al. 1999, 1177)
secondary productivity, secondary production *n.* The rates of energy storage at consumer and decomposer trophic levels, which become less and less at successively higher levels (Odum 1969, 68).
  Comment: There is only one kind of "secondary production," because consumers use food that is already produced, with appropriate losses, and convert it to different tissues by one overall process. The total energy flow at heterotrophic levels, which is analogous to gross production of autotrophs, should be designated as "assimilation," not "production" (Odum 1969, 69).
◆ **productivity** See production.
◆ **proepisematic** See -episematic: proepisematic.
◆ **profundal** *adj.* Referring to a deep zone of a lake below the level of effective light penetration and hence green plants (Lincoln et al. 1985).
◆ **progalvanotaxis** See taxis: galvanotaxis: progalvanotaxis.
◆ **progamete** See gamete: progamete.
◆ **progamic sex determination** See sex determination: progamic sex determination.
◆ **progamous** See -gamous: progamous.
◆ **progenesis** See -genesis: paedogenesis; heterochrony: progenesis.
◆ **progenitor** *n.* An ancestor (Lincoln et al. 1985).
◆ **progeny** *n.* Offspring of a single mating or an asexually reproducing organism (Lincoln et al. 1985).
  cf. seed
◆ **progesterone** See hormone: progestin.
◆ **progestin-induced hermaphroditism** See hermaphrodism: progestin-induced hermaphroditism.
◆ **progestins** See hormone: progestins.
◆ **program** *n.* A definite plan, or scheme, of any intended proceedings; an outline, or abstract, of something to be done, whether or not in writing (*Oxford English Dictionary* 1972, entries from 1837).
  Comment: Taxonomic programs include BIOSYS-1, HENNIG-86, MacClade, PAUP, and PHYLIP (Shaffer et al. 1991, 287–288).
behavioral program, program *n.* The genetically and individually stored

information that structures an animal's central nervous system, or is temporarily stored there, and that governs the manner and flexibility with which the animal's behavior adjusts to its environment (Immelmann and Beer 1989, 233–235).

cf. strategy

Comment: Immelmann and Beer (1989) discuss conceptual differences between "computer program" and "behavioral program."

[derived from computer-science language]

▸ **closed behavioral program** *n.* In less-derived invertebrates: a behavioral program that is primarily governed by genetically stored information (Immelmann and Beer 1989, 234).

▸ **open behavioral program** *n.* In more-derived invertebrates, vertebrates: a behavioral program that is primarily governed by individually acquired information (Immelmann and Beer 1989, 234).

CLUSTAL W *n.* A computer program used for aligning amino-acid sequences (Thompson et al. 1994 in Tomitani et al. 1999, 162).

digital organism (DO) *n.* A computer program that self-replicates, mutates, and adapts by a process similar to natural selection (Lenski et al. 1999, 661).

Comments: Digital organisms compete for central-processing-unit (CPU) time, which is the fuel needed for their replications. A DO mutates at random and evolves in a defined computational environment. Each DO has a genome length measured as the number of sequential instructions in its program, and so forth. Digital organisms offer opportunities to test generalizations about living systems that may extend beyond the organic life that biologists usually study, including possible extraterrestrial life (Lenski et al. 1999, 661–662).

▸ **complex digital organism** *n.* A digital organism that is selected to perform mathematical operations that accelerate replication through a set of defined "metabolic" rewards; contrasted with simple digital organism (Lenski et al. 1999, 661).

▸ **simple digital organism** *n.* A digital organism that is selected solely for rapid replication; compared to a complex DO, it does not perform mathematical operations; contrasted with complex digital organism (Lenski et al. 1999, 661).

DNAML Computer Program *n.* A computer program that uses the maximum likelihood method for producing phylogenetic trees (Niesbach-Klösgen et al. 1987, 214).

Fungal Mitochondrial Genome Project (FMGP) *n.* A comprehensive effect to sequence the mtDNAs of fungi (Gray et al. 1991, 1477); http://megasun.bch.umontreal.ca/People/lang/PMGP/).

genetic program *n.* An organism's genes that influence particular, or all of its, phenotypic traits (Mayr 1982, 44).

▸ **closed genetic program** *n.* An organism's genetic program that controls one of its behaviors and is reasonably resistant to any changes during the organism's life (Mayr 1982, 612).

▸ **open genetic program** *n.* An organism's genetic program that controls behaviors (*e.g.,* those involved in choosing food or habitats) that have flexibility and allow incorporation of learning (Mayr 1982, 612).

HENNIG-86 *n.* A computer program developed by J.S. Farris and used to produce parsimonius phylogenetic trees (Mayr and Ashlock 1991, 320).

syn. hennig86 (Shaffer et al. 1991, 288)

[named after Willi Hennig, a German entomologist and pioneer in the development of cladistic analysis]

MacClade *n.* A phylogenetic analysis program for the Apple Macintosh that displays phylogenetic trees on the screen (Maddison and Maddison 1998, history page).

Comments: In 1986, David and Wayne Maddison finished version 1 of MacClade, which evolved into version 3.07 by 1999. It contains many tools for the analysis and manipulation of phylogenies and character evolution (Mayr and Ashlock 1991, 321; http://phylogeny.arizona.edu/macclade.html.

MAPMAKER 3.0 *n.* A computer program used for making maps of gene loci (Lander et al. 1987 in Bradshaw et al. 1995, 763).

MOLPHY *n.* A computer program used for performing a maximum-likelihood analysis (Adachi and Hasegawa 1996 in Tomitani et al. 1999, 162).

motor program *n.* A self-contained unit of behavior of which fixed-action patterns are a conceptually crucial special case (Gould 1982, 163).

neighbor joining *n.* A computer program that tries to find the most parsimonious evolutionary tree (one with the minimum number of mutations) by continually pairing sequences so that it minimizes the total amount of change in a tree (Barinaga 1992a, 687; Gibbons 1992b, 874).

Observer, The *n.* A software system for collection and analysis of observational data (Noldus 1991, 415; www.diva.nl/noldus).

Organelle Genome Megasequencing Program (OGMP) *n.* A comprehensive effort to sequence the mtDNAs of Protista (Gray et al. 1999, 1477; http://megasun.bch.umontreal.ca/OGMP/).

n–r

Phrap *n.* A computer program that assembles the gene sequences that emerge from the program Phred into overlapping sequences (Wade 1998, C5).
cf. program: Phred
Comments: Other computer programs used in gene sequencing are Blast, Genscan, and RepeatMasker (Wade 1998, C5).

Phred *n.* A computer program that scans the output of a gene-sequencing machine and calls the order of the bases along with the level of confidence that can be placed in each call (Wade 1998, C5).
cf. program: Phrap

Phylogenetic Analysis Using Parsimony (PAUP) *n.* A computer program developed by David L. Swofford and used to produce the most parsimonious dendrogram of taxa from a data set (Shaffer et al. 1991, 287–288).
Comments: PAUP has undergone several revisions. It is based on the assumption that the most parsimonious tree is the most likely to mimic best what happened during evolution (Gibbons 1992b, 874). Some systematists debate whether PAUP always produces the most parsimonious tree.

Phylogenetic Inference Package (PHYLIP) *n.* A package of computer programs developed by J. Felsenstein that is used to produce phylogenetic trees (Mayr and Ashlock 1991, 320).
[*Phyl*ogenetic *I*nference *P*ackage]

Swarm *n.* A software package for multiagent simulation of complex systems being developed at The Santa Fe Institute (www.santafe.edu/projects/swarm).
Comments: Swarm is intended to be a useful tool for researchers in a variety of disciplines, especially artificial life (www.santafe.edu/projects/swarm). Swarm's basic architecture is the simulation of collections of concurrently interacting agents. This architecture implements a large variety of agent-based models.

Tinbergian research program *n.* Niko Tinbergen's four main questions about animal behavior as a group: (1) why does the animal show a behavior with regard to the immediate causation of the behavior (function, physiology), (2) why does an animal show a behavior at a particular time in its life with regard to the developmental causation of the behavior, (3) why does an animal show a behavior with regard to short-term adaptation to its environment (ecology), and (4) why does an animal show a behavior with regard to its long-term adaptation to its environment (evolution) (Dewsbury 1978, 5–7; Wiley 1996, 752).

syn. four why's of animal behavior (Dewsbury 1978, 5–7)

Wagner 78 *n.* A computer program that uses the Wagner algorithm to construct a dendrogram (Wagner tree) (Kluge and Farris 1969 in Wiley 1981, 176).
cf. procedure: distance-Wagner cluster analysis
Comments: Wagner's original method uses apomorphic characters to link subtaxa based on their relative degrees of apomorphy. Wiley (1981, 177–192) describes how the Wagner algorithm works.
[after H. Warren Wagner, Jr., botanist]

♦ **programmed behavior** See behavior: instinct.
♦ **programmed cell death** See death: apoptosis.
♦ **programmed instruction** See learning: programmed learning.
♦ **programmed learning** See learning: programmed learning.
♦ **progress** *n.*
1. Evolutionary advancement "to better and better conditions, continuous improvement" (*Oxford English Dictionary* 1933 in McKinney 1987).
   Notes: This is considered to be an incorrect definition from a scientific standpoint. Nonevolutionarily, "progress" implies direction, if not advancement toward a goal, but neither direction, nor goal, is provided by evolutionary mechanisms (Futuyma 1986, 8).
   syn. advancement, improvement, mechanical enhancement (*Oxford English Dictionary*)
2. An increase in efficiency, or effectiveness, of certain adaptations within lineages (Futuyma 1993, 1153).
cf. Darwin's admonition; *scala naturae*
Comment: Gould (1985 in McKinney 1987) suggests that "progress" be replaced with the more objective, less offensive "improvement." McKinney argues that "progress" is not offensive as an evolutionary term. Dawkins discusses progress with regard to evolutionary scaling and the adjectives "higher" and "lower." Progress continues to concern evolutionists, especially those interested in macroevolution, and it continues to be controversial (Ruse 1993, 59).

absolute progress *n.* Evolutionary "improvement" up a scale of fixed value; a directed evolutionary change toward that which is better; contrasted with comparative progress (Ruse 1993, 55).
Comment: Absolute progress is a controversial concept. Enthusiasts for absolute progress generally think that Humans come out on top (Ruse 1993, 55–58).

comparative progress *n.* Adaptive advance of one line of organisms over others; contrasted with absolute progress (Ruse 1993, 55).
Comment: Comparative progress is a controversial concept. At the micro-level, biologists generally agree that it occurs, but there is much debate about its precise nature and extent (Ruse 1993, 55, 58). The arms race, *q.v.*, is classified as a type of comparative progress.

♦ **progressionism** *n.* The assertion made by some of Cuvier's followers that "a brand new creation had taken place after each catastrophe, and that each succeeding creation reflected the changed conditions of the world" (Mayr 1982, 374).
Comment: In a way, progressionism is a creationist reshaping of the *scala naturae*, *q.v.* (Mayr 1982, 374).
[This assertion was designated as "progressionism" by Rudwick (1972) (Bolwer 1976 in Mayr 1982, 374).]

♦ **progressive development, principle of** See principle: principle of progressive development.

♦ **progressive deviation, principle of** See law: von-Baer's laws.

♦ **progressive provisioning** See provisioning: progressive provisioning.

♦ **progressive selection** See selection, directional selection.

♦ **progressive succession** See ²community: ²succession (def. 1).

♦ **progressive uniformitarianism** See ³theory: uniformitarianism: progressive uniformitarianism.

♦ **progyny** See gyny: protogyny.

♦ **prohydrotaxis** See taxis: positive hydrotaxis.

♦ **project** *n.* A research undertaking (Morris 1982).

Biological Dynamics of Forest Fragment Project (BDFFP) *n.* A collaborative research effort between Brazil's National Institute for Amazonian Research (INPA) and the Museum of Natural History, Smithsonian Institution, that is undertaking an experimental study of the process of habitat fragmentation in the Amazon Basin (Didham 1997, 57; 1999, www.si.edu/ biodiversity/bdffp.htm).
Comment: Researchers initiated this Project in 1979, and study sites are near Manaus, Brazil (Hamilton 1999, 129).

Deep Green Project *n.* A 5-year research project that ended in 1999 and had the aim of mapping the entire evolutionary tree of plants (Brown 1999b, 990; Weiss 1999, A5; www.ucjeps.herb.berkeley.edu/bryolab/greenplantpage.html.).

Comment: This project involved over 200 scientists from 12 countries (Weiss 1999, A5).

Human Genome Project *n.* An international effort, officially begun in 1990, to map the entire human genome, producing genetic maps, physical maps, and a complete nucleotide sequence maps of human chromosomes (Lander 1996, 536; Campbell et al. 1999, 376).
cf. gene (table)
Comment: The Human Genome Project expects to complete sequencing the entire human genome in 2005 (Lander 1996, 536).

♦ **prokaryogamety** See meiosis.

♦ **prokaryote** See -karyote: prokaryote.

♦ **prokaryotic sex** See sexual reproduction: prokaryotic sex.

♦ **prolactin** See hormone: prolactin.

♦ **prolactin-release inhibiting factor** See hormone: prolactin-release inhibiting factor.

♦ **proliferation** *n.* "Multiplication; growth; cell division; reproduction" (Lincoln et al. 1985).

♦ **prolonged-searching polygyny** See mating system: polygyny: scramble-competition polygyny: prolonged-searching polygyny.

♦ **promiscuity** *n.*
1. The condition of being promiscuous; indiscriminate mixture, confusion; promiscuousness (*Oxford English Dictionary* 1971, entries from 1849).
2. Promiscuous sexual union, as among some races of low civilization (*Oxford English Dictionary* 1971, entries from 1865).
See mating system: polybrachygamy.
adj. promiscuous.
[French *promiscuité*]

♦ **promiscuous mating effort** See effort: reproductive effort: mating effort: promiscuous mating effort.

♦ **promunturium** See ²community: promunturium.

♦ **pronking** See display: pronking.

♦ **proovigenic parasite** See -parasite: proovigenic parasite.

♦ **propaganda pheromone** See chemical-releasing substance: pheromone: propaganda pheromone.

♦ **propaganda substance** See chemical-releasing stimulus: propaganda substance.

♦ **propagation** *n.*
1. "Vegetative increase" (Lincoln et al. 1985).
2. Sexual, or asexual, multiplication (Lincoln et al. 1985).
v.i. propagate

♦ **propagule** *n.*
1. "Any kind of reproductive particle" that may be a part of either sexual or asexual reproduction (Dawkins 1982, 293).

n–r

syn. diaspore, propagulum (Lincoln et al. 1985)

2. The minimum number of individuals of a species required for colonization of a new, or isolated, habitat (Lincoln et al. 1985).

cf. seed

♦ **property** *n.* An attribute, or quality, that belongs to a person or thing; in earlier uses, sometimes a special or distinctive quality, a peculiarity; in later uses, often a quality, or characteristic, in general, without reference to its essentialness or distinctiveness (*Oxford English Dictionary* 1972, entries from 1303); any attribute, feature, or character (Lincoln et al. 1985).

♦ **prophototaxis** See taxis: phototaxis: prophototaxis.

♦ **proportion rule** See law: Allen's rule.

♦ **proprioceptor** See ²receptor: proprioceptor.

♦ **pros-, pro-** *prefix* Referring to a positive condition (Lincoln et al. 1985).

♦ **prosaerotaxis** See taxis: aerotaxis: prosaerotaxis.

♦ **proschairlimnetic** *adj.* Referring to organisms occasionally found in plankton in fresh water (Lincoln et al. 1985).

♦ **proschemotaxis** See taxis: chemotaxis: proschemotaxis.

♦ **prosody** *n.* A person's nonverbal communication including tone, tempo, rhythm, loudness, pacing, and other aspects of voice that modify the meaning of words (Wilson 1975, 556).

[Greek *proso,* to + *ōidē,* a song]

♦ **prosopagnosia** *n.* A human brain disorder that stops a person from naming someone from facial features alone whether or not the features change substantially over the years (Axelrod and Hamilton 1981).

♦ **prospective** *adj.*

1. Referring to something that looks to, or has regard for, the future; operative with regard to the future (*Oxford English Dictionary* 1972, entries from 1800).

2. Referring to something that looks forward or that a person looks forward to; prospective; expected, hoped for; future (*Oxford English Dictionary* 1972, entries from 1829); "potential; possible; probable" (Lincoln et al. 1985).

cf. realized

♦ **prospective ecospace** See niche: fundamental niche.

♦ **prosphototaxis** See taxis: positive phototaxis.

♦ **prostitution** *n.*

1. A woman's offering of her body to indiscriminate lewdness for hire (especially as a practice or institution); whoredom, harlotry (*Oxford English Dictionary* 1972, entries from 1553).

Note: Zuckerman and Yerkes first used this "loaded term" in connection with nonhuman primates in the 1930s (de Waal 1989, 210).

2. A person's devotion to an unworthy or base use; degradation, debasement, corruption (*Oxford English Dictionary* 1972, entries from 1647).

syn. venality (Michaelis 1963)

cf. mating system: maternal-benefit polyandry

[Latin, *prostitutus,* past participle of *prostituere* < *pro-,* forward + *statuere,* to place]

♦ **prostitution polyandry** See mating system: polyandry: maternal-benefit polyandry.

♦ **protandrous hermaphrodite** See hermaphrodite: protandrous hermaphrodite.

♦ **protandry** *n.*

1. In some species of solitary bees, butterflies, mayflies, and mosquitoes: males' tending to mature sooner and emerge from their natal nests sooner than females (Thornhill and Alcock 1983, 97).

2. A hermaphroditic organism's assuming a functional male condition during its development before reversal to a functional female state (Lincoln et al. 1985).

syn. protandrism, proterandry (Lincoln et al. 1985)

cf. protogyny

♦ **protean behavior** See behavior: escape behavior: protean behavior.

♦ **protean display** See display: protean display.

♦ **protected threat** See behavior: threat behavior: protected threat.

♦ **protective call** See animal sounds: call: alarm call.

♦ **protective coloration** See coloration: protective coloration.

♦ **protective mimicry** See mimicry: protective mimicry.

♦ **protein** See molecule: protein.

♦ **proteinoid** See molecule: proteinoid.

♦ **protelean** *adj.* Referring to an organism that is parasitic when immature and free living as an adult (*e.g.,* a rhipiphorid beetle) (Lincoln et al. 1985).

♦ **proterandry** See protandry.

♦ **proterogenesis** See hypothesis: proterogenesis.

♦ **proterogyny** See protogyny: proterogyny.

♦ **prothetely** See ²heterochrony: progenesis: prothetely.

♦ **Protist Image Database** See database: Protist Image Database.

♦ **protistology** See study of: protistology.

♦ **proto-** *prefix* First, original (Lincoln et al. 1985).

♦ **protobiont** See -biont: protobiont.

♦ **protobios** See -bios: protobios.

♦ **protocaryotic** See prokaryotic.

♦ **protocell** See cell: protocell.

♦ **protocooperation** See symbiosis: proto-cooperation.

♦ **protogamy** See -gamy: protogamy.

♦ **protogene** See gene: protogene.

♦ **protogenesis** See -genesis: protogenesis.

♦ **protogenic** See -genic: protogenic.

♦ **protogynous hermaphrodite** See her-maphrodite: protogynous hermaphrodite.

♦ **protogyny** See -gyny: protogyny.

♦ **protokaryotic** See -karyote: prokaryote.

♦ **protomitochondrial genome** See ge-nome: protomitochondrial genome.

♦ **protomorphic** See -morphic: proto-morphic.

♦ **protoplasm** *n.*
 1. The fluid in living cells considered to be the ultimate building material of all living things and the real agent of all physiologi-cal processes (Mayr 1982, 654).
  Note: From about 1835 to 1935, many biologists held the idea that protoplasm was such an agent.
  syn. sarcode (Dujardin 1935 in Mayr 1982, 654)
 2. The part of a cell, outside its nucleus, containing cell fluid and many kinds of organelles (Mayr 1982, 654).
 syn. protoplasm theory, protoplasm theory of life (Mayr 1982, 654)
 Comment: "Cytoplasm" is replacing the term "protoplasm" (Mayr 1982, 654).
 [coined by Purkinje (1839 in Mayr 1982, 654)]

♦ **protosexual** See sexual: protosexual.

♦ **protospecies** See ²species: protospecies.

♦ **prototroph** See -troph-: prototroph.

♦ **protozoology** See study of: zoology: pro-tozoology.

♦ **protozoophile** See ²-phile: protozoophile.

♦ **protroph** See -troph-: prototroph.

♦ **prove** *v.t.* To test extensively and elegantly, but not falsify, a biological phenomenon (Moore 1984, 475).
 syn. prove beyond all reasonable doubt

♦ **provenance** *n.* A place of origin (Lincoln et al. 1985).

♦ **provisioning** *n.*
 1. A person's providing food or another thing (Michaelis 1963).
 2. A person's taking measures or making preparations for something in advance (Michaelis 1963).

brood provisioning *n.* For example, in some bee, wasp, and other insect species: a parent's providing conditions that pro-mote its young's development before its

egg hatches, including establishing and maintaining sheltering burrows, nests, or cocoons and gathering a food supply, or a mother's laying the eggs in the vicinity of enough food for her hatchling (Immel-mann and Beer 1989, 37).
syn. provisioning (Lincoln et al. 1985)

food provisioning *n.* One animal's feed-ing another, including a parent's feeding its young, a male's feeding a female as part of courtship (courtship feeding), and colony members' feeding one another (trophal-laxis) (Immelmann and Beer 1989, 110).
cf. feeding: courtship feeding

mass provisioning *n.* For example, in Burying Beetles, Dung Beetles, many bee and wasp species: a parent's provisioning a larva with one large meal (all that it eats before it pupates) before it hatches from its egg; contrasted with progressive provision-ing (Wilson 1975, 588; Vander Wall 1990, 5).
Comment: Mass provisioning and progres-sive provisioning are on a continuum (Vander Wall 1990, 5).

nest provisioning *n.* For example, in many vertebrate species: progressive pro-visioning in which a parent returns to its nest regularly and feeds its developing offspring (Lincoln et al. 1985).

progressive provisioning *n.* For ex-ample, in paper wasps, yellowjackets (wasps), bumble bees, the Honey Bee: an individual's providing a larva with many small, sequential meals (Wilson 1975, 592).
Comments: In progressive provisioning an adult actively feeds a developing larvae, either by regurgitating small quantities of partially digested food or by placing the larvae on recently delivered food items (Vander Wall 1990, 5).

♦ **proxemics** See study of: proxemics.

♦ **proximate causation of behavior** See causation: proximate causation: proximate causation of behavior.

♦ **proximate cause, causation** See cau-sation: proximate causation; stimulus: *zeit-geber.*

♦ **proximate factor** See factor: ultimate factor; causation: ultimate causation; stimu-lus: *zeitgeber.*

♦ **proximate function of behavior** See function: proximate function of behavior.

♦ **proxylophile** See ¹-phile: proxylophile.

♦ **psamathium** See ²community: psama-thium.

♦ **psamathophile** See ¹-phile: psamatho-phile.

♦ **psammobiont** See -cole: arenicole.

♦ **psammofauna** See fauna: psammofauna.

♦ **psammogenous** See -genous: psammo-genous.

n–r

- **psammolittoral** *adj.* Referring to a sandy shore (Lincoln et al. 1985).
- **psammon** *n.*
  1. Organisms that live in, or move through, sand (Lincoln et al. 1985).
     cf. -cole: amnicole, arenicole
  2. Interstitial biota (Lincoln et al. 1985).
     syn. mesopsammon (Lincoln et al. 1985)
     cf. benthos
- **psammophile** See [1]-phile: psammophile.
- **psammosere** See [2]succession: psammo-sere.
- **psephonecrocoenosis** See -coenosis: psephonecrocoenosis.
- **pseudaposematic coloration** See coloration: pseudaposematic coloration.
- **pseudepisematic character** See character: episematic character: pseudepisematic character.
- **pseudergate** See caste: pseudergate.
- **pseudo-, pseud-** *combining form*
  1. False; pretended (Michaelis 1963).
  2. Counterfeit; not genuine (Michaelis 1963).
  3. Closely resembling (Michaelis 1963).
  4. Illusory; apparent (Michaelis 1963).
  5. Abnormal; erratic (Michaelis 1963).
  [Greek *pseudēs*, lie]
- **pseudoallele** See allele: pseudoallele.
- **pseudoallelism** *n.* For example, in some microorganism and *Drosophila* species, maize: the phenomenon in which genes (pseudoalleles) that are closely adjacent to one another on the same chromosome behave like alleles in certain genetic crosses but not in others (Mayr 1982, 755, 798–799).
- **pseudoanisogamety** See -gamety: pseudoanisogamety.
- **pseudoapomorph** See -morph: pseudo-apomorph.
- **pseudocaste** See caste: pseudocaste.
- **pseudocelom** See coelom: pseudocoelom.
- **pseudoclonal** See clone: pseudoclone.
- **pseudoclonal society** See [2]society: pseudoclonal society.
- **pseudocoelom, pseudocoelome** See coelom: pseudocoelom.
- **pseudoconditioning** See learning: conditioning: pseudoconditioning.
- **pseudocopulation** See copulation: pseudocopulation.
- **pseudocopulatory behavior** See copulation: pseudocopulation.
- **pseudoepisematic character** See character: episematic character: pseudo-episematic character.
- **pseudoextinction** See extinction: phyletic extinction, pseudoextinction.
- **pseudogamy** See -gamy: pseudogamy; parthenogenesis: gynogenesis.
- **pseudogene** See gene: pseudogene.

- **pseudogonochorism, pseudohermaphroditism** See hermaphrodism: pseudohermaphrodism.
- **pseudohybrid** See hybrid: pseudohybrid.
- **pseudomixis** See -gamy: pseudogamy.
- **pseudoparasitism** See parasitism: pseudoparasitism.
- **pseudoperiphyton** See [2]community: periphyton: pseudoperiphyton.
- **pseudopheromone** See chemical-releasing stimulus: semiochemical: allelochemic: pseudopheromone.
- **pseudoplankton** See plankton: pseudoplankton.
- **pseudoplasmodium** *n.* "The motile, sluglike organism formed by the aggregation of the amebas of cellular slime molds" (Wilson 1975, 593).
- **pseudopolygyny** See -gyny: pseudopolygyny.
- **pseudopolyploid** See -ploid: -polyploid: pseudopolyploid.
- **pseudopregnancy** See pregnancy: pseudopregnancy.
- **pseudopsychology** See study of: psychology: pseudopsychology.
- **pseudoreciprocity** See altruism: reciprocal altruism: pseudoreciprocity.
- **pseudoreplication** See replication: pseudoreplication.
- **pseudoseasonal polymorphism** See -morphism: polymorphism: pseudoseasonal polymorphism.
- **pseudosematic character** See character: pseudepisematic character.
- **pseudosexual behavior** See behavior: sexual behavior: pseudosexual behavior.
- **pseudosound** See [1]sound: pseudosound.
- **pseudospecies** See [2]species: pseudospecies.
- **pseudosymphile** See symbiont: symphile: pseudosymphile.
- **pseudotroglobiont** See -biont: pseudotroglobiont.
- **pseudovicar** See [2]species: pseudovicar.
- **pseudovicarism** See vicarism: pseudovicarism.
- **pseudoxerophile** See [1]-phile: pseudoxerophile.
- **psi, ψ** See degree of satisfaction.
- **psicole** See -cole: psicole.
- **psilicole** See -cole: psilicole.
- **psilile** *n.* An organism that lives in savanna, or prairie, communities or habitats (Lincoln et al. 1985).
  adj. psilic
- **psilium** See [2]community: psilium.
- **psilopaedic** See -paedic: psilopaedic.
- **psilophile** See [1]-phile: psilophile.
- **psittosis** See [2]group: psittosis.

♦ **psolidomeiomology** See study of: psolidomeiomology.

♦ **psyche** *n.*
1. Human soul, mind, or intelligence (Michaelis 1963).
2. In psychoanalysis, the aggregate of a person's mental components, including both his conscious and unconscious states, and often regarded as an entity functioning apart from, or independently of, his body (Michaelis 1963).
[Greek *psychē*, soul, breath of life < *psychein*, to breathe]

♦ **Psyche** *n.* In Greek and Roman mythology, a maiden who, after many tribulations caused by Aphrodite's jealousy, is united with her lover, Eros, and accorded a place among the gods as a personification of the soul (Michaelis 1963).

♦ **psychiatry** *n.* The branch of medicine that deals with diagnosis, treatment, and prevention of human mental disorders (Michaelis 1963); in particular, with disorders that arise from physical and organic causes (Collocot and Dobson 1974).
adj. psychiatric
n. psychiatrist

♦ **psychic** *n.*
1. A person who is sensitive to extrasensory phenomena (Michaelis 1963).
2. "A spiritualistic medium" (Michaelis 1963).
3. *adj.* Referring to minds or souls (Michaelis 1963).
4. "Mental, as distinguished from physical and physiological" (Michaelis 1963).
5. Referring to, or designating, human mental phenomena that are, or appear to be, independent of normal sensory stimuli, such as clairvoyance, telepathy, and extrasensory perception (Michaelis 1963).
6. Referring to a person's being sensitive to mental, or occult, phenomena (Michaelis 1963).
syn. psychical (Michaelis 1963)
[Greek < *psychē*; see psycho-]

♦ **psychism** *n.* The doctrine that living matter has attributes not recognized in nonliving matter; the distinctive attributes or "mentality" of living organisms (Collocot and Dobson 1974).

♦ **psycho-, psych-, psychy-** *combining form* "Mind; soul; spirit" (Michaelis 1963). [Greek *psychikos* < *psychē*, soul, breath of life < *psychein*, to breathe]

♦ **psychoacoustics** See study of: acoustics.

♦ **psychoanalysis** See analysis: psychoanalysis.

♦ **psychobiology** See study of: biology: psychobiology.

♦ **psychochemical** *n.* A drug, or compound, capable of acting directly on a brain and affecting consciousness and behavior (Michaelis 1963).
adj. psychochemical (Michaelis 1963)

♦ **psychodynamics** See study of: dynamics: psychodynamics.

♦ **psychogenesis** See -genesis: psychogenesis.

♦ **psychogenetic** See -genetic: psychogenetic.

♦ **psychogenetics** See study of: genetics: psychogenetics.

♦ **psychogenic** See -genic: psychogenic.

♦ **psychogenic social regulation** See regulation: social regulation: psychogenic social regulation.

♦ **psychognosis** See study of: psychognosis.

♦ **psychokinesis** See kinesis: psychokinesis.

♦ **psychological castration** See castration: psychological castration.

♦ **psychology** *n. pl.* **psychologies**
1. A person's "aggregate of the emotions, traits, and behavior patterns regarded as characteristic of an individual, type, period, group, particular experience, etc." (Michaelis 1963).
2. "A work on psychology" (Michaelis 1963). See study of: psychology, for many branches of psychology and psychological schools of thought.
abbr. psych., psychol.
adj. psychological, psychologic
adv. psychologically

faculty psychology *n.* "The concept that a person has a mind composed of such distinguishable powers or faculties as intellect, will, memory, and feeling" (Storz 1973, 96).

♦ **psychology of personality** See study of: psychology.

♦ **psychometry, psychometrics** See study of: -metry: psychometry.

♦ **psychomotor** *adj.* In physiology, of, or referring to, muscular movements that result from, or are caused by, mental processes (Michaelis 1963).

♦ **psychopathology** See study of: pathology: psychopathology.

♦ **psychopharmacological drug** See drug: psychopharmacological drug.

♦ **psychopharmacology** See study of: pharmacology: psychopharmacology.

♦ **psychophile** See ²-phile: psychophile.

♦ **psychophysics** See study of: physics: psychophysics.

♦ **psychophysiology** See study of: physiology: psychophysiology.

♦ **psychric** *adj.* Referring to low temperatures or cold habitats (Lincoln et al. 1985).

♦ **psychro-** *combining form* "Cold" (Michaelis 1963).

[Greek *psychros*, cold]

♦ **psychrophile** See [1]-phile: psychrophile.

♦ **psychrophobia** See phobia (table).

♦ **ptenophyllium** See [2]community: ptenophyllium.

♦ **ptenophyllophile** See [1]-phile: ptenophyllophile.

♦ **ptenothalium** See [2]community: ptenothalium.

♦ **ptenothalophile** See [1]-phile: ptenothalophile.

♦ **pteridology** See study of: pteridology.

♦ **ptilopaedic** See -paedic: ptilopaedic.

♦ **puberty** *n.*

1. A person's state, or condition, of becoming functionally capable of producing offspring, which is characterized by growth of hair on one's pubes and facial hair in males (*Oxford English Dictionary* 1972, entries 1382).

2. In Humans, the period during which one becomes physiologically capable of reproduction (Michaelis 1963).

Comments: Human puberty appears to begin around 6 years of age (McClintock and Herdt in Marano 1997, C1) and involves a whole series of physical and psychosocial events. A landmark event of puberty is maturation of sexual organs (Marano 1997, C6). In the U.S., at age 8, 48.3% of the black girls and 14.7% of the white girls had begun developing breasts, pubic hair, or both (based on a study of 17,000 girls ages 3 through 12, with 9.6% being black; Marcia E. Herman-Giddens et al. in Associated Press 1997, A2). On the average, menstruation occurred at 12.16 years in Blacks and 12.88 years in whites. "Puberty" is rarely used to refer to a plant that reaches the stage of flowering or bearing fruit (*Oxford English Dictionary* 1972).

cf. sexual attraction, sexuality

[Latin *pubertas* < *pubes, puberis*, an adult]

♦ **pubic presentation** See display: genital display.

♦ **puddle, puddling** See behavior: puddling.

♦ **puddling** See [2]group: puddling.

♦ **Pulitzer Prize** See award: Pulitzer Prize.

♦ **pulsar** *n.* A stellar source of radio waves, characterized by the rapid frequency and regularity of the wave bursts; a rotating neutron star about 10 kilometers in diameter (Mitton 1993).

Comment: Depending on its type, a pulsar emits wave bursts every 1 millisecond to every 4 seconds (Mitton 1993).

♦ **punctuated equilibrium** *n.* An organism lineage that changes gradually (while it is in an equilibrium) but also shows rapid, episodic times of speciation (= punctuations) during geological time (Eldredge and Gould 1972, 84; Futuyma 1986, 401–402).

cf. evolution: Falconer-style evolution, punctuational evolution; gradualism: phyletic gradualism; hypothesis: hypothesis of punctuated equilibrium

Comments: To explain punctuated equilibria, Eldredge and Gould (1972) proposed that most evolutionary change occurs rapidly in small, localized populations in concert with the acquisition of reproductive isolation. A new form that has achieved sexual isolation expands from its site of origin into the range of its unchanged parent species. The new form then becomes sufficiently abundant and widespread to be recovered in the fossil record. Gould (1982c, 84) suggests that punctuated equilibrium is the predominant mode and tempo of evolutionary change.

[coined as "punctuated equilibria" by Eldredge and Gould (1972, 84)]

♦ **punctuational evolution** See [2]evolution: punctuational evolution.

♦ **Punnett's theory of mimicry** See mimicry, mimetic polymorphism.

♦ **pupa** See animal names: pupa; caste: reproductive: nymph.

♦ **pupigerous** *adj.* Referring to something that contains a pupa (Lincoln et al. 1985).

♦ **pupillary response** See response: pupillary response.

♦ **pupiparity** See -parity: pupiparity.

♦ **pupivore** See -vore: pupivore.

♦ **pure blending inheritance, theory of** See [3]theory: theory of pure-blending inheritance.

♦ **pure dominance polygyny** See mating system: polygyny: pure-dominance polygyny.

♦ **pure epigamic selection** See selection: epigamic selection: pure-epigamic selection.

♦ **pure evolutionarily stable strategy** See strategy: evolutionarily stable strategy: pure evolutionarily stable strategy.

♦ **pure exploration** See exploration: pure exploration.

♦ **pure line, pure strain** *n.*

1. A genetically uniform (*i.e.,* a homozygous) population (Mayr 1982, 959).

2. Descendants of a single homozygous parent produced by inbreeding or self-fertilization (Lincoln et al. 1985).

syn. pure strain (Lincoln et al. 1985)

♦ **pure mathematics** See study of: mathematics: pure mathematics.

♦ **pure problem solving** See problem solving: pure problem solving.

◆ **purifying selection** See selection: purifying selection.

◆ **Purkinje shift** *n.* A person's eye's spectral sensitivity changes with light intensity; *e.g.,* red appears to fade before blue as light intensity diminishes (Purkinje 1825 in McFarland 1985, 202).
[after Johannes Evangelista Purkinje, Czech physiologist]

◆ **purposive behavior** See behavior: purposive behavior.

◆ **purposive behaviorism** See behaviorism: purposive behaviorism.

◆ **purposiveness** *n.* The controversial concept that a nonhuman animal has a mental idea, or conception, of a goal for its behavior (Thorpe 1963, 5).
cf. behaviorism, directiveness, intent

◆ **pursuit-deterrent signal** See signal: pursuit-deterrent signal.

◆ **pushup** See display: pushup.

◆ **putting-through method** See method: putting-through method.

◆ **puzzle box** *n.* Any cagelike container used to investigate an animal's learning, memory, other cognitive abilities, or a combination of these attributes (*e.g.,* a maze, *q.v.*), Skinner box, or Wisconsin-general-test apparatus (Immelmann and Beer 1989, 240).

Skinner box *n.* A puzzle box used in operant-conditioning studies, consisting of a receptacle for an experimental animal, a "manipulandum" (a disc at which a pigeon can peck, a lever that a mouse or rat can press, etc.), a "reward dispenser" that delivers food or water as reinforcement, and sometimes a means for presenting a particular stimulus (Immelmann and Beer 1989, 271).
[after B.F. Skinner (1938) who refined the Skinner box originated by Thorndike (1911) (Immelmann and Beer 1989, 271)]

Wisconsin-general-test apparatus *n.* A puzzle box developed by Harlow (1949) and comprised of a cage from which an animal can reach out to uncover one of two or more containers presented to it on a sliding tray; one container has a reward item and the other(s) does not; the containers have differently marked lids; and one observes the animal from behind a one-way screen (Immelmann and Beer 1989, 328).

◆ **pyramid of biomass** See ecological pyramid: pyramid of biomass.

◆ **pyramid of energy** See ecological pyramid: pyramid of energy.

◆ **pyramid of numbers** See ecological pyramid: pyramid of numbers.

◆ **pyramidal system** See ²system: pyramidal system.

◆ **pyric** *adj.* Referring to fire (Lincoln et al. 1985).

◆ **pyrite** See mineral: iron pyrite.

◆ **pyrium** See ²community: pyrium.

◆ **pyrogen** *n.* A substance that causes fever when introduced into, or released in, an animal's body (Bligh and Johnson 1973, 952).

bacterial pyrogen *n.* Any pyrogen from a bacterium (*e.g.,* an endotoxin) (Bligh and Johnson 1973, 952).

endogenous pyrogen *n.* A heat-labile substance formed in an animal's body tissues and which, when released, causes fever by acting upon its central nervous system (Bligh and Johnson 1973, 952).
Comment: Cells exposed to an endotoxin may produce and release endogenous pyrogens (Bligh and Johnson 1973, 952).

endotoxic pyrogen *n.* A heat-stable pyrogen from the cell wall of a Gram-negative bacterium (Bligh and Johnson 1973, 946).
Comment: All examined endotoxins contain lippolysaccharides of high molecular weight (Bligh and Johnson 1973, 946).

leucocytic pyrogen *n.* An endogenous pyrogen released from leucocytes under experimental conditions (Bligh and Johnson 1973, 952).

◆ **pyrophile** See ¹-phile: pyrophile.

◆ **pyrophobe** See -phobe: pyrophobe.

◆ **pyrophobia** See phobia (table).

◆ **pyrophosphate** See molecule: pyrophosphate.

◆ **pyroxylophile** See ¹-phile: pyroxylophile.

◆ **pythmic** *adj.* Referring to lake bottoms (Lincoln et al. 1985).

n – r

q

♦ **Q-K ratio** *n*. The ratio of pheromone emitted (*Q*) to the threshold concentration at which a receiving animal responds (*K*); *Q* is measured as the number of molecules released in a burst or in number of molecules per unit time, and *K* is measured in molecules per unit of volume (Bossert and Wilson 1963 in Wilson 1975, 185, 235).

♦ **QS** See sleep: active sleep.

♦ **QTL** See gene locus: quantitative-trait locus.

♦ **quack, quacking** See animal sounds: quack.

♦ **quadrivoltine** See voltine: quadrivoltine.

♦ **qualitative** *adj*. "Descriptive; non-numerical;" contrasts with quantitative (Lincoln et al. 1985).

♦ **qualitative character** See character: qualitative character.

♦ **qualitative inheritance** See inheritance: qualitative inheritance.

♦ **qualitative multistate character** See character: qualitative multistate character.

♦ **qualitative variable** See variable: qualitative variable.

♦ **quality** *n*. "In evolutionary biology, the extent to which a particular mate, habitat, resource, or context can enhance an individual's inclusive fitness" (Wittenberger 1981, 620).

♦ **quality of complexness** *n*. A compilation of certain aspects of an animal's internal and external milieu (including hunger, time of day, weather conditions, available prey, and estimated prey-capture probability) into a unitary consideration, or whole complex, that permits the animal to behave differently under different conditions (Volkelt 1914, 1937 in Heymer 1977, 101).

♦ **quantitative** *adj*. "Numerical; based on counts, measurements, ratios, or other val-

ues;" contrasts with qualitative (Lincoln et al. 1985).

♦ **quantitative character** See character: quantitative character.

♦ **quantitative ethology** See study of: ethology: [3]quantitative ethology.

♦ **quantitative genetics** See study of: genetics: quantitative genetics

♦ **quantitative inheritance** See inheritance: quantitative inheritance.

♦ **quantitative-multistate character** See character: quantitative-multistate character.

♦ **quantitative-trait locus** See gene locus: quantitative-trait locus.

♦ **quantitative variable** See variable: quantitative variable.

♦ **quantum evolution** See [2]evolution: quantum evolution.

♦ **quantum speciation** See speciation: saltation.

♦ **quartet, quartette** See [2]group: quartet.

♦ **quasi-cell** See organized structure: quasi-cell.

♦ **quasi-normal mutation** See [2]mutation: quasi-normal mutation.

♦ **quasi-sociality** See sociality: quasi-sociality.

♦ **quasi-steady-state model** See hypothesis: hypotheses of the beginning of the Universe: quasi-steady-state model.

♦ **queen** *n*.

1. A woman who is the chief ruler of a state and has the same rank and position as a king (*Oxford English Dictionary* 1972, entries from 825).

2. A king's wife or consort (*Oxford English Dictionary* 1972, entries from *ca.* 893).

3. A woman of preeminence, or authority, in a specified sphere (*Oxford English Dictionary* 1972, entries from 1552).

See animal names: queen; caste: queen.

cf. female

◆ **queen control** *n.* In social insects: a queen's inhibitory influence on reproductive activities of workers and other queens (Hölldobler and Wilson 1990, 642).

◆ **queen substance** See chemical-releasing stimulus: semiochemical: pheromone: queen pheromone.

◆ **queen-worker conflict** See [3]conflict: queen-worker conflict.

◆ **queenright** *adj.* Referring to a social-insect colony with a functional queen (Wilson 1975, 593).
ant. orphan, queenless (C.K. Starr, personal communication)

◆ **quiescence** *n.* An organism's temporary resting phase characterized by its reduced activity, inactivity, or developmental cessation (Lincoln et al. 1985).
adj. quiescent

◆ **quiet sleep** See sleep: quiet sleep.

◆ **quinarian** *n.* A taxonomist who thinks that all taxa in a classification should have five units (Mayr 1982, 241, 846).

Comment: This belief is based on the idea that biology is made "scientific" by making it quantitative or by making it obey definite "laws" (Mayr 1982, 241, 846).

◆ **quinarianism** *n.* The taxonomic beliefs of quinarians, *q.v.*

◆ **quinone** See toxin: quinone.

◆ **quiver dance** See dance: bee dance: quiver dance.

◆ **quivering** *n.*
1. A person's shaking, trembling, or vibrating with a slight but rapid vibration, often by persons under the influence of an emotion or other thing (*Oxford English Dictionary* 1972, entries from 1490 as a verb).
2. A person's trembling his voice (*Oxford English Dictionary* 1972, entry from 1875).
3. A male stickleback fish's striking a female with rapid quivering movements of his snout which causes her spawning (Pelkwijk and Tinbergen 1937 in Heymer 1977, 1958).

♦ *r* See rate: *r*; coefficient: coefficient of relatedness.

♦ **r̄** See factor: dilution factor.

♦ *r, r* **value** See rate: *r* value.

♦ **r extinction** See ²extinction: r extinction.

♦ **r-K spectrum** *n*. A linear continuum of reproductive and life-history strategies with r-selected species at one extreme and K-selected species at the other (Lincoln et al. 1985).

♦ **r-selected species** See ²species: r-selected species.

♦ *r* **selection, r selection** See selection: *r* selection.

♦ **r strategist, R-strategist** See ²species: r strategist.

♦ **r strategy** See strategy: r strategy.

♦ **r value** See rate: r value.

♦ **rabbitry** See ²group: rabbitry.

♦ **rabble** See ²group: rabble.

♦ ¹**race** *n*.

1. A conspecific group of organisms that is characterized by a common character or characters (*Oxford English Dictionary* 1972, entries from 1500).
   See ²species: subspecies.
2. A group of organisms that have descended from a common ancestor (*Oxford English Dictionary* 1972, entries from 1549).
3. An animal breed or stock; a particular species variety (*Oxford English Dictionary* 1972, entries from 1580).
4. One of the large divisions of organisms (*Oxford English Dictionary* 1972, entries from 1580).
5. A genus, species; kind of organism (*Oxford English Dictionary* 1972, entries from 1596).

See stock.

cf. ²group

Comments: Taxonomically, a "race" is now a subdivision of a "species." Some biologists prefer not to use "race" due to sociopolitical connotations of this word. Instead, some workers use "population" or "strain."

biological race, race *n*.

1. An intraspecific group of organisms that is characterized by conspicuous biological properties (Lincoln et al. 1985).
2. One of two or more sympatric conspecific populations that differs biologically, but not morphologically, from others and does not interbreed with another such population due to different food or host preferences or behavior cycles (Lincoln et al. 1985).

climatic race *n*. A plant phenotype that evolved under a particular climatic regime, *e.g.,* members of a species of pine that live at different altitudes of the Himalayas and have different degrees of frost tolerance due to microevolution (Mayr 1982, 560). See -type: ecotype.

ecological race *n*. An intraspecific category characterized by conspicuous ecological properties (Lincoln et al. 1985).

geographical race See ²species: subspecies.

host race, hostic race *n*. A genetic race of a parasite species that occurs on, or in, a particular host species (Lincoln et al. 1985).

inversion race *n*. A naturally occurring population that is characterized by a particular chromosome inversion or inversions (Lincoln et al. 1985).

physiological race *n*. An intraspecific category characterized by conspicuous physiological properties (Lincoln et al. 1985).

substrate race *n*. A local race with a characteristic coloration resembling its substratum's coloration (Lincoln et al. 1985)

♦ ²**race** *n*. In some animal species, including Humans: an individual's running (*Oxford English Dictionary* 1972, entries from 1325).

arms race *n.* "The sequence of mutual counteradaptations of two coevolving organisms, such as a predator and its prey" (Lincoln et al. 1985).

cf. [2]evolution: coevolution; hypothesis: Red-Queen hypothesis

Comment: In an arms race, which is a type of comparative progress, *q.v.*, organisms compete and evolve, throwing up methods of attack and defense in a way that is analogous to human weapon development (Ruse 1993, 55).

♦ **raciation** *n.* The evolution of new races within a species (Lincoln et al. 1985).

♦ **radar** *n.* An electronic device that locates objects by beaming radiofrequency impulses that are reflected back from objects and determines their distance by measuring the time elapsed between impulse transmission and reception (Michaelis 1963).

[portmanteau word < *ra*dio *de*tection *a*nd *r*anging]

♦ **radar angel** See angel: radar angel.

♦ **radar tag** *n.* A radar transmitter that enables a person to determine the location of an object near, or carrying, the transmitter (Caldwell 1997, 42).

Comments: The smallest radar tag is a diode attached to 3-inch-long superfine aluminum wire (Caldwell 1997, 42, illustration). Jens Roland and other researchers use this tag for butterflies.

♦ **radicicole** See -cole: radicicole.

♦ **radicivore** See -vore: radicivore.

♦ **radio tracking** See biotelemetry.

♦ **radio transmitter** See instrument: radio transmitter.

♦ **radioactive dating** See method: dating method: radioactive dating.

♦ **radiocarbon dating** See method: dating method: radiocarbon dating.

♦ **radiodating** See method: dating method: radioactive dating.

♦ **raft** See [2]group: raft.

♦ **rafter** See [2]group: rafter.

♦ **rafting** *n.* Dispersal of terrestrial organisms, sediment, rock, or other material across water on floating objects (Lincoln et al. 1985).

♦ **rag** See [2]group: rag.

♦ **rage copulation** See copulation: rage copulation.

♦ **rain forest** See [2]community: rain forest.

♦ **rain pollination** See pollination: rain pollination.

♦ **rain shadow** *n.* The region on the downwind side of a mountain that receives little rain because the winds rise as they pass over the mountain, cooling and dropping most of their moisture before they reach the other side (Stanley 1989, 665).

Comment: A rain shadow can also be many miles downwind from a mountain group, such as the rain shadow in the Green Ridge State Forest, MD, caused by the higher mountains of the Appalachian Plateau about 20 miles west.

♦ **rain washoff** *n.* A pesticide's removal due to rain.

♦ **rainfastness** *n.* A measure of a pesticide's "stickiness" to a substrate with regard to removal by rain.

♦ **rake** See [2]group: rake.

♦ **ramage** See [2]group: ramage.

♦ **ramet** *n.* A member, or modular member, of a clone that may follow an independent existence if separated from its parent (Lincoln et al. 1985).

cf. genet, ortet

♦ **ramicole** See -cole: ramicole.

♦ **ramification** See [2]group: ramification.

♦ ***Rammelzeit*** See estrus: *Rammelzeit.*

♦ **random** *adj.*
1. In statistics, referring to a datum in a sample of data, all of which have an equal chance of being drawn from the sample (Sokal and Rohlf 1969, 73).
2. Referring to a phenomenon that is haphazard, having no recognizable pattern (Lincoln et al. 1985).

cf. nonrandom; random sequence of events; sample: haphazard sample, random sample

♦ **random distribution** See distribution: random distribution.

♦ **random effect** See effect: main effect: random effect.

♦ **random error** See error: random error.

♦ **random factor** See effect: main effect: random effect.

♦ **random fixation** See fixation: random fixation.

♦ **random genetic drift** See genetic drift: random genetic drift.

♦ **random mating** See mating: random mating.

♦ **random mutation** See [2]mutation: random mutation.

♦ **random-number table** *n.* A table of numbers in which the probability of any number's occurring at any one time is constant and independent of all other tabled values (Hodgman 1963, 237–243).

syn. random numbers (Lincoln et al. 1985)

♦ **random process** See process: random process.

♦ **random sample** See sample: random sample.

♦ **random sequence of events** *n.* A limited set of events in which there is no way that one can predict which kind of event within the set will occur next, and the system obeys the regularities of probability (Bateson 1979, 242).

cf. stochastic sequence of events

♦ **random variable** See variable: random variable.

♦ **randomized-block design** See experimental design: randomized-block design.

♦ **randomly amplified polymorphic DNA marker** See marker: randomly amplified polymorphic DNA marker.

♦ ¹**range** *n.* "A row, line, or series of things" (*Oxford English Dictionary* 1972, entries from 1511).

geologic range *n.* The time span between the base of the period, or epoch, of a taxon's first appearance and the upper boundary of the interval of its last appearance (Flessa et al. 1975, 75).

Comments: A taxon's geological range is an estimate of its time on Earth because of imperfections of the fossil record. For example, the arthropod order Agnostida has a geological range of 135 Myr from the Lower Cambrian through the Upper Ordovician (Flessa et al. 1975, 76).

host range, host list *n.* All the host species that a kind of parasite infects (Lincoln et al. 1985).

tolerated-ambient-temperature range *n.* The ambient-temperature range within which an animal can maintain its body core temperature by means of autonomic thermoregulatory processes, within certain limits for its species or the individual under consideration (Bligh and Johnson 1973, 958).

♦ ²**range** *n.*
1. The extent to which variation in something is possible; the limits between which something may vary in amount or degree (*Oxford English Dictionary* 1972, entry from 1818).
2. Statistically, a measure of variation within a data set, given by the difference between minimum and maximum values (Sokal and Rohlf 1969, 49).

♦ ³**range** *n.*
1. The geographical area that a particular species occupies; the period that a species has been extant (*Oxford English Dictionary* 1972, entries from 1856).
2. The variation in depth occupied by a marine animal (*Oxford English Dictionary* 1972, entries from 1856).
3. The area that an individual animal occupies or has occupied.
cf. dominion, territory

activity range *n.* "The area covered by an animal in the course of its day to day existence" (Carpenter 1952b, 246).
cf. range: home range; territory

home range *n.*
1. An area that an animal learns thoroughly and patrols regularly (Seton 1909, Burt 1943 in Wilson 1975, 256, 586).

Note: If part of a "home range" is defended, this part is a "territory" (Wilson 1975, 256, 586).
2. The area within which an animal confines its activities during all, or part, of its annual cycle (Wittenberger 1981, 616).
3. The area that an animal habitually travels in the course of its normal activities; a home range may be undefended and convey no privileged access to resources (Dewsbury 1978, 96).
4. The area in which an animal normally lives, exclusive of its migrations, emigrations, dispersal movements, or unusual erratic wanderings (Brown 1987a, 299).
syn. home realm (Lincoln et al. 1985)
cf. territory
Comments: Territories of other animals may, or may not, be in a focal animal's "home range." This term has been misused (Brown 1987a, 299). The home range of a grizzly bear can be up to 1600 km² (Weithrich, 1966, 493).

total range *n.* "The entire area covered by an individual animal in its lifetime" (Goin and Goin 1962 in Wilson 1975, 256).
cf. territory

♦ **rangale** See ²group: rangle.

♦ **ranivore** See -vore: ranivore.

♦ ¹**rank** *n.* A class of animals, or things, in a scale of comparison; "hence relative status or position, place" (*Oxford English Dictionary* 1972, entries from 1605).
cf. hierarchy: dominance hierarchy

basic rank *n.* The dominance relationship between two monkeys that is unaffected by influence of kin or other group members (Kawai 1958 in Wilson 1975, 294; Immelmann and Beer 1989, 26).
cf. rank: dependent rank

dependent rank *n.*
1. In monkeys: the ranking outcome resulting from interaction of two individuals that is influenced by their kinship (Kawai 1958 in Wilson 1975, 294).
2. For example, in the Jackdaw (bird), primates: the rank status of an individual that is influenced by its social group (Immelmann and Beer 1989, 26–27).
cf. rank: basic rank

dominance rank *n.* An animal's location within a dominance hierarchy (J.L. Hand, personal communication).
syn. rank, status (J.L. Hand, personal communication)
Comment: In Humans, dominance rank can be acquired by some fortuitous factor, such as birth or ethnic group, or through specific achievement or significant contribution to one's group (Storz 1973, 267).

♦ ²**rank, taxonomic level, taxonomic rank** *n.* A taxon's position in a classification hierarchy (Lincoln et al. 1985).

♦ **rank-abundance curve** See curve: rank-abundance curve.

♦ **rank correlation** See correlation: rank correlation.

♦ **rank demonstration** *n.* An animal's behavior that maintains and strengthens its position in a dominance hierarchy (Heymer 1977, 140).

♦ **rank mimicry** See mimicry: rank mimicry.

♦ **rank order, ranking order** See hierarchy: dominance hierarchy.

♦ **ranked variable** See variable: ranked variable.

♦ *Ranz* See estrus: *Ranz.*

♦ **rapacious, ravenous, voracious** *adj.* Referring to an organism that feeds on prey (Lincoln et al. 1985).

♦ **RAPD marker** See marker: randomly amplified polymorphic DNA marker.

♦ **rape** *n.*
1. A person's act of taking anything by force; violent seizure (of goods), robbery (*Oxford English Dictionary* 1971, entries from 1400).
2. A person's act of carrying away a person, especially a woman, by force (*Oxford English Dictionary* 1972, entries from 1400).
3. A person's violation of, or ravishing, a woman (*Oxford English Dictionary* 1972, entries from 1481).

Note: "Rape" is now used to refer to children and men as well.

v.t. rape

cf. copulation: forced copulation; mating: homosexual mating

Comment: Many behaviorists prefer to use "forced copulation" rather than "rape" for nonhuman animals because "rape" has so many emotional connotations (Wittenberger 1981, 615; Gowaty 1982); thus, I give definitions of "rape" and "forced copulation" in nonhuman animals under "forced copulation."

[Latin *rapère*, to rape]

♦ **rapid eye movement** See movement: rapid eye movement.

♦ **rapid-eye-movement sleep** See sleep: stage-1 sleep.

♦ **rapid food-avoidance conditioning** See learning: conditioned learning: rapid food-avoidance conditioning.

♦ **rapid leaf movement** See movement: rapid-leaf movement.

♦ **rapping behavior** See behavior: rapping behavior.

♦ **raptor** *n.* A predatory two-legged dinosaur whose hallmark feature is large, sharp claws presumably used for attacking and rending prey (Holden 1997b, 1063).

♦ **raptorial** See -orial: raptorial.

♦ **rare** See frequency: rare.

♦ **rare-male mating advantage** *n.* For example, in several *Drosophila* species, *Tribolium castaneum*, *Nasonia vitripennis*, the Two-Spot Ladybird, and Guppy: uncommon males with certain traits are more likely to elicit mating from females than common males with other traits (Petit 1951 in Lichtenberger et al. 1987, 203; Ehrman and Probber 1978, 216; O'Donald and Majerus 1988, 571; Salceda and Anderson 1988, 9870). *syn.* minority effect, Petit effect (Lichtenberger et al. 1987, 203), rare-male advantage (Thornhill and Alcock 1983, 78), rare-male effect (Partridge 1988, 525)

Comments: "Mate competition" among rare males and "female-mate choice" are likely to have different importances in "rare-male mating advantage" among species (Thornhill and Alcock 1983, 78; Partridge 1988, 535). Most studies of "rare-male mating advantage," thus far, have been laboratory ones; this phenomenon is likely to have extremely limited evolutionary significance (Partridge 1988, 536).

♦ **rarity** *n.*
1. "The quality of being rare, uncommon, or infrequent; infrequency" (Michaelis 1963).
2. A wide array of organism spatial and temporal patterns of abundance, from being sparsely populated with a wide geographic range to being a point endemic with a dense local population (Kunin and Gaston 1993, 298).

Comment: Species have been classified as "rare" based on their range extents; number of sites, or grid squares, inhabited; habitat, or niche, restrictions; edge-of-range uncommonness; frequency in censused areas; and population densities where found (Kunin and Gaston 1993, 298).

cf. frequency

♦ **rasorial** See -orial: rasorial.

♦ *Rassenkreis, Rassenkreise* See ²species: polytypic species.

♦ **rat race** See hypothesis: Red-Queen hypothesis.

♦ **ratchet hypothesis, ratchet mechanism** See hypothesis: hypotheses of the evolution and maintenance of sex: ratchet hypothesis.

♦ **rate** *n.*
1. The "measure of a thing by its relation to a standard" (*e.g.,* 50 miles per hour in an automobile) (Michaelis 1963).
2. A proportional, or comparative, amount or degree: a high rate of speed (*e.g.,* 90

n–r

miles per hour in an automobile) (Michaelis 1963).

evolutionary rate *n.* "The speed of evolutionary change" (Lincoln et al. 1985).

extinction rate *n.* The number of species in a particular habitat, or area, that becomes extinct per unit time; contrasted with percent extinction and total extinction rate (Lincoln et al. 1985).

immigration rate *n.* The number of new species that arrive in a given area, or habitat, per unit time (Lincoln et al. 1985).

instantaneous growth rate *n.* A population's growth rate at a point in time (Lincoln et al. 1985).

maximum-population-growth rate *n.* The greatest growth rate that a population can achieve in a particular environment (Wilson 1975, 82).
syn. optimum yield (Lincoln et al. 1985)
cf. rate: r, r_{max}
Comment: This rate occurs at a population size that is one half of its carrying capacity (Wilson 1975, 82).

net reproductive rate (R_o) *n.* The average number of female offspring produced by each female during her entire lifetime (Wilson 1975, 90, 590).

percent extinction *n.* The number of families that go extinct divided by the total number of families in existence during a particular stratigraphic stage; contrasted with extinction rate and total extinction rate (Benton 1995, 55–56).
Comment: Measures of percent extinction can be usually high when species richnesses are low (Benton 1995, 55–56).

phylogenetic rates of evolution *n.* Rates at which a single character, or complexes of characters, evolves within individual lineages (Simpson 1944, 1953 in Futuyma 1986, 397).
cf. rate: taxonomic rates of evolution

r, r value *n.*
1. ($b - d$) in the equation of population growth per unit time: $dN/dt = bN - dN = (b - d)N = rN$.
   Note: r for Humans is up to 0.03 per year; Rhesus monkeys, 0.16 per year; and Norway rats, 0.015 per day (Wilson 1975, 81).
2. "The fraction by which a population is growing in each instant of time" (Wilson 1975, 587).
syn. biotic potential, intrinsic rate of increase, Malthusian parameter, rate of natural increase (Lincoln et al. 1985); little r
Note: Some workers reserve the terms "biotic potential" or "intrinsic rate of increase" for "r_{max}"

cf. coefficient of relatedness, r; rate: maximum population-growth rate; r_{max}

r_{max} *n.* The theoretical maximum intrinsic rate of increase that a population can achieve in an optimum environment that is physically ideal, with abundant space and resources, freedom from predators and competitors, etc. (Wilson 1975, 81).
cf. maximum-population-growth rate, r

rate of natural increase See rate: r.

taxonomic rates of evolution *n.* "Rates at which taxa with different characteristics replace one another" (Simpson 1944, 1953 in Futuyma 1986, 397).
cf. rate: phylogenetic rates of evolution

total-extinction rate *n.* The number of families that went extinct in relation to the duration of a particular stratigraphic stage; contrasted with extinction rate and percent extinction rate (Benton 1995, 56).
Comment: Using total-extinction rate can diminish or enhance the magnitudes of extinction events compared to using other measures of extinction rate (Benton 1995, 56).

♦ **ratio** *n.*
1. The relation between two similar magnitudes with respect to quantity, determined by the number of times one contains the other, integrally or fractionally (*Oxford English Dictionary* 1972, entries from 1660); *e.g.*, the ratio of 3 to 7 is expressed as 3:7 or 3/7 (Michaelis 1963).
2. "Relation of degree, number, etc.; proportion; rate" (Michaelis 1963).

fecundity ratio *n.* The ratio of pregnant, or ovigerous, females, or other specified group, to the total number of females in a population (Lincoln et al. 1985).

fertility ratio *n.* The ratio of offspring to adult females in a population (Lincoln et al. 1985).

Mendelian ratio *n.* The expected allele ratio in offspring generations according to Mendel's law of segregation (Lincoln et al. 1985).

sex ratio *n.*
1. The ratio of males to females in a group of conspecific organisms (*e.g.*, 1:1, 4:1) (Lincoln et al. 1985).
2. The number of males per 100 females (Lincoln et al. 1985).

▸ **primary sex ratio** *n.*
1. The sex ratio of just fertilized eggs (Wittenberger 1981, 620).
2. The number of males per 100 females at zygote formation (Lincoln et al. 1985).

▸ **secondary sex ratio** *n.* "Sex ratio of offspring at the time they become independent of their parents" (Wittenberger 1981, 621).

▶ **spanandrous sex ratio** *n*. A strongly female-biased sex ratio that is associated with inbreeding species (Balachowsky 1949 in Hamilton 1967).

▶ **tertiary sex ratio** *n*. "The number of males per 100 females at sexual maturity" (Lincoln et al. 1985).

♦ **ratio cline** See cline: ratio cline.

♦ **ratio scale** See scales of measurement.

♦ **rationalism** *n*.
1. A view, opposed to empiricism or sensationalism, that regards reason, rather than sense, as the foundation of our certainty of knowledge (*Oxford English Dictionary* 1972, entries from 1857).
2. The view, dating from Plato, that human behavior is a result of rational and voluntary processes, with individuals' being free to choose whatever course of action their reasoning dictates (McFarland, 361).
cf. associationism, materialism

♦ ***Rauschzeit*** See estrus: *Rauschzeit*.

♦ **raw data** See data: raw data.

♦ **ray** *n*.
1. "A line of propagation of any form of radiant energy" (Michaelis 1963).
2. "A stream of particles spontaneously emitted by a radioactive substance" (Michaelis 1963).

cosmic ray *n*. An extremely energetic elementary particle (nucleus) that travels through the universe at practically the speed of light (Mitton 1993).
syn. cosmic radiation, cosmic wave
Comments: V.F. Hess first discovered cosmic rays during a balloon flight in 1912 (Mitton 1993). Cosmic rays are the only particles we can detect that have traversed our galaxy. Very high-energy cosmic rays might come from quasars and active galactic nuclei. Lower-energy cosmic rays are generated within our galaxy in pulsars and supernova explosions and remnants. Solar flares produce very low-energy cosmic rays which increase in intensity at times of maximum solar activity. Cosmic rays may have provided energy used in synthesis of organic molecules before the origin of life on Earth (Strickberger 1996, 115).

♦ **Rayleigh test** See statistical test: Rayleigh test.

♦ **re-** *prefix*
1. Back: rebound, remit (Michaelis 1963).
2. "Again; anew; again and again" (Michaelis 1963).
[Latin *re-*, *red-*, back, again]

♦ **reaction** *n*.
1. A system's, or organ's, response to an external stimulus (*Oxford English Dictionary* 1972, entries from 1896).

2. An organism's activity change in response to a stimulus (Lincoln et al. 1985).
cf. response

alarm reaction (A-R) *n*. All of an organism's nonspecific responses to a sudden exposure to stimuli to which it is not adapted (Seyle 1950; Lincoln et al. 1985).

avoiding reaction See kinesis: klinokinesis.

chain reaction *n*. An organism's sequence of instinctive movements (Spencer 1899, Loef 1905 in Heymer 1977, 96).

disgust reaction *n*. In some mammal species, including Humans: an individual's contracting its pharyngeal musculature, salivating, and having feelings that resemble those accompanying vomiting (von Holst and von St. Paul 1960 in Heymer 1977, 60).
Comment: "Disgust" is an anthropomorphic word when used to describe nonhuman animals.

general emergency reaction *n*. Part of an animal's general adaptation syndrome that involves rapid changes in its body (*e.g.*, taking in extra oxygen by breathing deeply and releasing sugar from its liver) that occur after it is suddenly faced with a situation in which it has to take some sort of rapid action, such as fleeing or fighting (Cannon 1929 in Dawkins 1980, 56).
cf. syndrome: general adaptation syndrome

shoulder reaction *n*. In Humans and other anthropoid apes: an ape's leaning slightly forward, pulling in its head toward its shoulders, and pulling up its shoulders as a fright reaction due to a noise (Spindler 1958 in Heymer 1977, 160).

spook reaction *n*. In the fish *Astatotilapia strigigena*: panic of adults caused by models of conspecifics without the sheen of a normal healthy fish's epithelium [coined by Seitz 1940 in Heymer 1977, 77].

trance reaction See playing dead.

♦ **reaction chain** See chain: action chain.

♦ **reaction norm** *n*. An animal's normal reaction pattern with regard to a directional stimulus that involves the animal's continuing to move independently of the stimuli (Heymer 1977, 142).

♦ **reaction time** See time: reaction time.

♦ **reaction type** See -type, type: phenotype.

♦ **reafference** See -ference: reafference.

♦ **reafference principle** See principle: reference principle.

♦ **real theorizing** *n*. The postulational-deductive process used in scientific discovery (Wilson 1975, 27).

♦ **realism** *n*.
1. The philosophical doctrine that adheres to objective, or absolute, existence of

universals (*Oxford English Dictionary* 1972, entries from 1838).

Notes: This doctrine was chiefly expounded by Thomas Aquinas and is opposed to "nominalism" and "conceptualism" (*Oxford English Dictionary* 1972). "Realism" was the dominant philosophy of the scholastics that involves determining truth by logic, not by observing from experiments (Mayr 1982, 92).

2. The philosophical belief in the real existence of matter as the object of perception (natural realism), which is opposed to idealism; also, the view that the physical world has independent reality and is not ultimately reducible to universal mind or spirit (*Oxford English Dictionary* 1972, entries from 1836).

3. A person's inclination, or attachment, to what is real; tendency to regard things as they really are; any view or system, contrasted with "idealism" (*Oxford English Dictionary* 1972, entries from 1817).

4. A person's attribution of objective existence to a subjective conception (*Oxford English Dictionary* 1972); the philosophical doctrine, opposed to nominalism, that abstract concepts have objective existence and are more real than concrete objects (Michaelis 1963).

5. The doctrine that things have reality apart from our conscious perception of them; opposed to idealism (Michaelis 1963).

n. realist

[Old French < Late Latin *realis* < Latin *res*, thing]

♦ **realized** *adj.* Actual (Lincoln et al. 1985).
cf. prospective

♦ **realized heritability** See [2]heritability (comments).

♦ **realized niche** See niche: realized niche.

♦ **rearing territory** See territory: rearing territory.

♦ **reasoning** See problem solving.

♦ **recapitulation** See -genesis: biogenesis.

♦ **receiver** *n.* An individual organism whose behavior is altered by a signal that it receives (Dewsbury 1978, 99).
See [1]system: communication system.
cf. signaler

♦ **recent** *adj.* Still existing; extant (Lincoln et al. 1985).
See extant.
cf. extinct, fossil

♦ **reception** See modality.

♦ **receptive field** *n.* The area from which any unit, in a sensory system, receives input that changes its activity (Immelmann and Beer 1989, 244).

♦ **receptivity, female receptivity** *n.* In many animal species: responses of a female

toward a mate that include those necessary and sufficient for fertile copulation, including active cooperation in intromission (Beach 1976 in Dewsbury 1978, 237; Hinde 1982, 155).
cf. attractivity, proceptivity

♦ [1]**receptor** *n.* A molecule with a highly specific affinity for a substance (*e.g.,* a hormone) that binds to it (*Oxford English Dictionary* 1972, entries from 1900; Immelmann and Beer 1989, 245).

♦ [2]**receptor** *n.*

1. The terminal structure of an afferent neuron that is specialized to receive external and internal stimuli and transmit them to a vertebrate's spinal cord and brain (Michaelis 1963)

2. An organism's means of detecting a kind of stimulus, comprised of parts of one or more cells (McFarland 1985, chap. 11).
cf. modality, sense organ
Comment: McFarland (1985, chap. 11) reviews kinds of animal receptors.

AMPA receptor *n.* A common synaptic connection between brain cells which plays a role in learning and memory (Gary Lynch in Hall 1998, 49).
cf. drug: ampakine

chemoreceptor *n.* A receptor that perceives chemical stimuli, especially olfactory and gustatory ones (*e.g.,* the taste-and-smell receptor or blood-glucose-concentration receptor) (Heymer 1977, 48).

endoreceptor, enteroceptor See [3]receptor: interoceptor.

exteroceptor *n.* An animal's receptor (*e.g.,* eye, ear, touch receptor) that responds to external stimuli (Michaelis 1963).
[Latin *exterus*, external + (re)ceptor]

interoceptor *n.* In vertebrates: a sensory nerve ending that responds to stimuli originating in an animal's viscera, or other internal organs (Michaelis 1963), and registers variables (*e.g.,* blood pressure, body temperature, degree of stomach fullness, hydration state, oxygen debt, or blood acidity) (Immelmann and Beer 1989, 156).
syn. endoreceptor (Immelmann and Beer 1989, 156), enteroceptor
cf. [3]receptor:proprioceptor
[*inte*rnal + O + re*ceptor*]

NMDA receptor 2B (NR2B) *n.* A component of a vertebrate's MNDA receptor (Lemonick 1999, 58; Tang et al. 1999, 64).
Comment: This component is very active in young mice compared to older ones (Lemonick 1999, 58). It is found primarily in an animal's forebrain and hippocampus, where explicit, long-term memories are formed.

N-methyl-D-aspartate receptor (NMDA receptor) *n.* In mammals, a receptor that turns on after receiving an electrical discharge from its own cell and *N*-methyl-D-aspartate (Lemonick 1999, 57–58; Tang et al. 1999, 64).
Comment: NMDA receptors function in learning and memory formation and sensitizing an animal's brain to drugs, including amphetamines, cocaine, and heroin (Lemonick 1999, 57–58). These receptors are nearly identical in cats, mice, rats, Humans, and other mammals.

photoceptor *n.* A sensory receptor with pigments that are chemically modified by light and give rise to an electrical potential when stimulated by light (McFarland 1985, 169; Immelmann and Beer 1989, 220).

proprioceptor *n.*
1. In vertebrates: an interoceptor in an individual's muscles, tendons, and joint tissues that registers its body position, movement, and strain (Immelmann and Beer 1989, 156).
2. In insects: an interoceptor in a stretch receptor, or chordotonal organ, that registers its body position, movement, and strain (Chapman 1971, 618).
[New Latin < Latin *propri(us)*, one's own + O + re*ceptor*]

sensory receptor *n.* A specialized nerve cell responsible for transducing and transmitting information (McFarland 1985, 169).

thermoreceptor, temperature sensor *n.* A neuronal structure that is differentially sensitive to temperature and responds to a maintained temperature with a characteristic sustained impulse frequency (Bligh and Johnson 1973, 956).
Comment: A thermoreceptor may respond weakly to strong nonthermal stimuli (Bligh and Johnson 1973, 956).

♦ **recessive allele** See allele: recessive allele.
♦ **recessiveness** *n.* In diploid organisms: an allele's not showing its effects if its corresponding dominant allele is also present (Dawkins 1982, 285).
cf. allele: recessive allele; dominance (genetic): incomplete dominance
♦ **reciprocal altruism** See altruism: reciprocal altruism.
♦ **reciprocal cross** See cross: reciprocal cross.
♦ **reciprocal feeding** See trophallaxis.
♦ **reciprocal genes** See gene: complementary gene.
♦ **reciprocal hybrid** See hybrid: reciprocal hybrid.
♦ **reciprocally altruistic behavior, reciprocation** See altruism: reciprocal altruism.

♦ **reciprocity** *n.* In some cooperatively breeding bird species (*e.g.,* the bee-eater, Pied Kingfisher, and possibly the White-Winged Chough): a mature bird's forming a bond, or alliance, with a young bird because of its future value as a helper (Ligon 1983 in Heinsohn 1991, 1097, 1099).
See altruism: reciprocal altruism.

♦ **recognition** *n.*
1. A person's perceiving that some thing, person, etc. is the same as one he has previously known; a person's mental process of identifying what he has previously has known; the result of these processes (*Oxford English Dictionary* 1972, entries from 1798).
2. An individual animal's ability to discriminate one, or a group of, conspecific or heterospecific organism from one another, or other groups.
3. An individual animal's differential treatment of different conspecific, or heterospecific, organisms due to perceived differences between, or among, them (Cheney and Seyfarth 1986, 1722).
4. An individual animal's ability to differentiate between, or among, stimulus categories that are properties of things, including places and individual animals; distinguished from mere sensory discrimination (Immelmann and Beer 1989, 246).
Comment: "Recognition" can be a misleading term if one takes it to imply "consciousness" (Mayr 1963 in Immelmann and Beer 1989, 287).

group recognition *n.* An animal's ability to discriminate a particular conspecific, or group of conspecifics, from a group of other conspecifics, *e.g., Dialictus zephyrus* (a sweat bee) or the Honey Bee, which discriminate between colony members and other conspecifics, or Starfish-Eating Shrimp or desert sowbugs, which discriminate their mates from other conspecifics (Wilson 1975, 205).
cf. recognition: individual recognition

individual recognition *n.* In possibly some woodlouse (isopod) species; in some prawn, bird, and mammal species, including the Black-Tailed Deer, Domestic Dog, European Rabbit, *Lasioglossum zephrum* (a sweat-bee), Sugar Glider: an individual's ability to discriminate other animals (usually conspecifics) as individuals (*e.g.,* distinguishing individuals A, B, and C from each other) (Beer 1970 in Dewsbury 1978, 160; Barrows 1975a).
syn. personal recognition (Wilson 1975, 517), truly personal recognition
cf. recognition: group recognition

n–r

kin recognition *n.*

1. An animal's ability to discriminate genetic kin from nonkin (using individual, group recognition, phenotype matching, or combination of these abilities) (Immelmann and Beer 1989, 165).

   Note: This kind of kin recognition has been reported in a desert woodlouse (isopod), a sweat bee (*Lasioglossum zephrum*), the Honey Bee, and more derived vertebrates (including some bird species, the Belding's Ground Squirrel, Humans, and some other mammal species); however, Grafen (1991c) argues that only one species has been convincingly shown to have a recognition system that works specifically for kin.

2. An animal's "differential treatment of conspecifics as a function of their genetic relatedness" (Sherman and Holmes 1985, 437; Cheney and Seyfarth 1986, 1722).

3. An animal's ability to discriminate genetic kin from nonkin by using genetic similarity detection (Grafen 1991a,b).

   Comment: Grafen (1990) explains how certain previous studies may have confused "kin recognition" with "species recognition" or "group recognition." A series of papers discusses the meaning of "kin selection" and related topics (Blaustein et al. 1991; Grafen 1991a,b,c; Byers and Bekoff 1991; Stuart 1991).

mate recognition *n.* For example, in Starfish-Eating Shrimp, desert sowbugs; some bird and mammal species: an individual's discrimination of its mate from other conspecifics which can involve either individual recognition or group recognition (Wilson 1975, 205).

cf. system: specific-mate-recognition system

neighbor recognition *n.* An animal's differential treatment of a conspecific relative, or nonrelative, as a function of its familiarity with it (Caley and Boutin 1987, 60).

parent-offspring recognition *n.* A parent's differential treatment of its offspring, or vice versa, based on individually distinctive cues (Medwin and Beecher 1986, 1627).

personal recognition See recognition: individual recognition.

species recognition *n.* An animal's determination of another animal as a conspecific by its exchanging species-specific stimuli and responses with the other animal, particularly prior to mating (Lincoln et al. 1985; Immelmann and Beer 1989, 287).

♦ **recognition allele** See allele: recognition allele.

♦ **recognition concept of species, recognition-species concept** See ²species (def. 33).

♦ **recognition gene** See gene: recognition gene.

♦ **recognition label** See discriminator.

♦ **recognition pheromone** See chemical-releasing stimulus: semiochemical: pheromone: recognition pheromone.

♦ **recognized individual** See individual: recognized individual.

♦ **recognizing individual** See individual: recognizing individual.

♦ **recombinant DNA** See nucleic acid: deoxyribonucleic acid: recombinant DNA.

♦ **recombination** *n.* In sexually reproducing organisms: occurrence of progeny with combinations of genes other than those of their parents due to independent reassortment, crossing over, or both, during meiosis and zygote formation (King and Stansfield 1985).

cf. gene conversion

♦ **recombination frequency, recombination fraction** *n.*

1. The number of recombinants vs. the total number of progeny of an organism (Lincoln et al. 1985).

2. The mean frequency with which two particular alleles are separated by recombination (Lincoln et al. 1985).

♦ **recombinational load** See genetic load: recombinational load.

♦ **recombinational repair** See repair: recombinational repair.

♦ **recon** See gene: recon.

♦ **reconnaissance observation** See observation: reconnaissance observation.

♦ **recording error** See observer error: recording error.

♦ ¹**recruitment** *n.*

1. A reinforcement to a group of persons (*Oxford English Dictionary* 1972, entries from 1824).

2. The act, or process, of recruiting a military force or a class of persons (*Oxford English Dictionary* 1972, entries from 1843).

3. The influx of new organism members by reproduction, or immigration, into a population (Lincoln et al. 1985).

♦ ²**recruitment** *n.* In ants, bees, termites, wasps: "a special form of assembly by which members of a society are directed to some point in space where work is required" (Wilson 1975, 211, 213, 594).

syn. alarm behavior (sometimes, Wilson 1965, 1065); recruitment behavior

tandem running *n.* In some ant species: a type of chemical recruitment communication in which one follower runs closely behind a leader, frequently contacting her abdomen with her antennae (Hingston 1929, Wilson 1959a, Hölldobler 1971a, in Wilson 1975, 55, 596).

♦ **recruitment limitation** *n.* A species' failure to colonize (= recruit in) all sites favorable for its growth and survival (Hubbell et al. 1999, 554).
cf. hypothesis: hypotheses of species richness: recruitment-limitation hypothesis

♦ **recruitment-limitation hypothesis** See hypothesis: hypotheses of species richness: recruitment-limitation hypothesis.

♦ **recruitment pheromone** See chemical-releasing stimulus: semiochemical: pheromone: recruitment pheromone.

♦ **recruitment trail** See odor trail: recruitment trail.

♦ **rectilinear evolution** See ²evolution: rectilinear evolution.

♦ **recurrent selection** See selection: recurrent selection.

♦ **recursive** *adj.* Referring to a relationship among a series of causes, or arguments, such that the sequence comes back to the starting point (Bateson 1979, 242).
cf. lineal

♦ **recursiveness** *n.*
1. In linguistics, the possibility of repeating a pattern without limit in constructing a sentence; *e.g.,* Dick went to the store, Tom went to the store, etc. (Immelmann and Beer 1989, 246).
2. In some bird species: an individual's singing repeated syllable insertions (Immelmann and Beer 1989, 246).

♦ **red-king hypothesis** See hypothesis: maximum-homology hypothesis.

♦ **Red List** *n.* A list of organisms that are endangered to some degree and which is a guideline for policy makers (Stevens 1996b, C1).

♦ **Red-Queen hypothesis, red-queen model** See hypothesis: Red-Queen hypothesis; hypotheses of the evolution and maintenance of sex: Red-Queen hypothesis.

♦ **redirected activity, redirected behavior, redirection** See behavior: displacement behavior: redirected behavior.

♦ **redirected aggression** See aggression: redirected aggression.

♦ **reducing agent** See agent: reducing agent.

♦ **¹reduction** *n.*
1. In chemistry, a process that deprives a compound of oxygen (Michaelis 1963).
2. In chemistry, an atom's or molecule's (1) gaining one or more valence electrons, (2) ceasing to share its electrons with a more electronegative atom or compounds, or (3) decreasing its positive valence (Michaelis 1963).

3. In chemistry, a compound's loss of oxygen, gain of hydrogen, or gain of electrons (King and Stansfield 1985).
v. reduce
Comment: Reduction is also the state resulting from these processes.
["Reduction" is derived from the electrical effects of adding one or more electrons to an atom or compound. When an electron is added to a cation, the electron reduces the amount of the cation's positive charge. Thus, adding an electron to a cation is called "reduction" (Campbell 1990, 183).]

♦ **²reduction** *n.* An organism line's producing, or evolving, simpler structures from more complex ones (Lincoln et al. 1985).
cf. elaboration; ²evolution

♦ **reduction division, reduction-division** *n.* "One of the two meiotic divisions, usually the first one, in which the number of chromosomes is halved" (Mayr 1982, 748, 959).
Comment: This phenomenon was anticipated by Galton (1876 in Mayr 1982, 696) and first given a full and completely correct description by Hertwig (1890 in Mayr 1982, 761–762).

♦ **reduction of polymorphy** See effect: Hagedoorn effect.

♦ **reductionism** *n.*
1. Assertion that "the theories and laws formulated in one field of science (usually a more complex field or one higher in the hierarchy) can be shown to be special cases of theories and laws formulated in some other branch of science" (Mayr 1982, 62).
2. The theory that all complex phenomena can be explained by reducing them to their simplest possible terms and that all biological phenomena can be explained by simple physical laws (Lincoln et al. 1985).
3. One's showing that an animal's behavior is consequent on its "physiological mechanisms, which in turn can be understood in terms of their physics and chemistry" (Immelmann and Beer 1989, 247).
cf. holism
Comments: Mayr (1982, 62) is not aware of any biological theory that has ever been reduced to a physicochemical theory. Bateson (1979, 244) remarks that reductionism becomes a vice when it goes beyond one's finding the simplest, most economical, and (usually) most elegant explanation that will account for particular data and it is accompanied by an overly strong insistence that the simplest explanation is the only one.

constitutive reductionism *n.* Assertions that "the material composition of organisms is exactly the same as that found

n–r

in the inorganic world" and "that none of the events and processes encountered in the world of living organisms is in any conflict with the physico-chemical phenomena at the level of atoms and molecules" (Mayr 1982, 60).

syn. reduction (Mayr 1982, 59, who notes that this term has been used in at least three different ways)

cf. emergence (Mayr 1982, 835)

Comment: Constitutive reductionism is accepted by modern biologists (Mayr 1982, 60).

evolutionary reductionism *n.* Humans' developing principles regarding biological evolution [evidently coined by Alexander 1987, 14].

explanatory reductionism *n.* A form of analysis asserting that one cannot understand an entire system "until one has dissected it into its components, and again these components into theirs, down to the lowest hierarchical level of integration" (Mayr 1982, 60).

♦ **redundancy** *n.*
1. In information theory, the occurrence of more signs than are necessary to encode a particular message; the ratio of the difference between the most efficient coding and the observed coding to the most efficient coding (Wilson 1975, 200; Immelmann and Beer 1989, 247).
2. Multiple mapping of sensory input in a vertebrate's brain that enables it to have considerable recovery capacity from neural damage (Immelmann and Beer 1989, 247).

♦ **redundant** *adj.* Referring to the third base pair of a triplet in genetic code which usually can be any of the four possible base pairs without changing the kind of amino acid coded for by the triplet (Gould 1985c).

♦ **redundant character** See character: redundant character.

♦ **refaunation** *n.* The reintroduction of an organism (*e.g.,* a symbiotic gut microorganism) into a host from which it had been removed (Lincoln et al. 1985).

♦ **refection** See -phagy: autocoprophagy.

♦ **reference person** *n.* A person for whom another person shows consistent preference despite opportunities offered for frequent contact with other persons (*e.g.,* a child's reference person is usually its mother) (Heymer 1977, 44).

♦ **referent** *n.* An organism that is chosen by a conspecific as a standard by which it classifies strangers as kin or nonkin; a referent can range from being the organism itself to its entire group (Wilson 1987, 12).

cf. allele: recognition allele; phenotype matching; recognition; template

♦ **reflex** *n.*
1. An involuntary action of a muscle, gland, or other organ caused by the excitation of sensory nerve that is transmitted to a nerve center and thence "reflected" along an efferent nerve to the organ in question (*Oxford English Dictionary* 1972, entries from 1833).
2. An involuntary response made to a stimulus that proceeds in a clearly prescribed fashion; a reflex is comprised of numerous reflex arcs, each comprised of a receptor organ that responds to a stimulus, an afferent pathway that conducts an action potential from the receptor to synapses and ganglion cells of a central nervous system, and an efferent pathway that carries an impulse to an effector organ that processes an appropriate response (Verplanck 1957).
3. The simplest form of reaction to a stimulation that is normally automatic, involuntary, and stereotyped (Storz 1973, 235; McFarland 1985, 239).

syn. reflex action (*Oxford English Dictionary* 1972), reflex arc (some authors), reflex behavior (McFarland 1985)
4. "A simple stimulus-response circuit by which a particular input is inevitably followed, with little or no intervening processing, by a unitary output, *e.g.,* a human knee-jerk reflex" (Gould 1982, 72).

cf. reflex center

Babinski reflex *n.* In adult apes, human infants: an individual's raising its big toe and fanning out its other toes when the side of its foot is stimulated (Dewsbury 1978, 136; Gregory and Zangwill 1987, 68).

[after Joseph Babinski, 1857–1932, French neurologist (Gregory and Zangwill 1987)]

chain reflex *n.* One reflex in an individual animal's reflex sequence that is caused by one reflex's activating the next (Immelmann and Beer 1989, 2).

cf. [3]theory: chain-reflex theory

clamping reflex See reflex: grasp reflex.

clasp reflex See reflex: grasp reflex.

conditional reflex, conditional response, conditioned reflex, conditioned response See learning: conditioning: Pavlovian conditioning.

delayed reflex *n.* An animal's reflex that follows a stimulus after a time lag (*e.g.,* a tongue strike at prey of a toad or frog) and is not corrected after it begins (T-phenomenon; Hinsche 1935 in Heymer 1977; Heymer 1977, 143).

diving reflex *n.* A person's showing an abrupt heart-rate drop, concomitant lower blood pressure, and constriction of peripheral blood vessels that send blood to his brain, ceasing his automatic urge to breath, and changing the brain metabolism, which allows him to go without oxygen for several minutes; this occurs in infants to adults (Gould 1982, 513).

expulsion reflex *n*. In frogs and toads: an individual's opening its mouth wide and moving its forelegs forward from the side in a wiping movement after eating an oversized, or unpalatable, object (Hering 1896, etc., in Heymer 1977, 202).

syn. wiping reflex (Heymer 1977, 202)

grasp reflex, grasping reflex *n*.

1. A premature, or newborn, human infant's closing his fingers about an object, such as an adult's finger, in an orderly sequence, with great strength in some individuals (Prechtl 1955 and Hassenstein 1973 in Heymer 1977, 82).

2. For example, in batrachians, many arboreal-mammal species: an animal's reflex flexion of its fingers, or toes, and adductive gripping of its arms or legs in response to a touch on the palms of its hand or soles of its feet (Immelmann and Beer 1989, 45).

syn. clamping reflex (Heymer 1977, 82); clasp reflex (Immelmann and Beer 1989, 45)

knee-jerk reflex *n*. A person's extending his lower leg after being tapped in his lower knee area (Gould 1982, 72).

lordosis reflex See lordosis.

monosynaptic reflex *n*. An animal's reflex that involves one afferent and one efferent nerve and a single synapse between them in the individual's lower spinal cord (*e.g.*, a human knee jerk) (Immelmann and Beer 1989, 248).

optomotor reflex *n*. In many animal species, *e.g.*, the Siamese fighting fish, praying mantids, rats: an individual's attempting to maintain its entire body, or part of its body, in a constant position with regard to a moving environment; *e.g.*, an individual's attempting to stay in a constant position with regard to stripes on the walls of a revolving cylinder (optomotor apparatus) in which it is placed (McFarland 1985, 250).

syn. optokinetic response (Dewsbury 1978, 176)

orientating reflex See response: orientating response.

rooting reflex *n*. A newborn human baby's turning his head and forming his mouth when his cheek is stimulated (Dewsbury 1978, 136).

cf. reflex: Babinski reflex

startle reflex *n*. A person's involuntary body twitching due to a surprising stimulus (McFarland 1985, 239).

unconditioned reflex *n*. An unlearned reflex (*e.g.*, pupillary reflex, knee jerk, salivary reflex, or clasp reflex) (Immelmann and Beer 1989, 248).

cf. learning: conditioning: Pavlovian conditioning

Unken reflex *n*. In some amphibian species: a defensive posture in which an animal's body is arched inward with its head and tail lifted upward (Halliday and Alder 1986, 34, 148).

Comments: Some very toxic frogs show Unken reflexes, closing their eyes and bending their heads and legs back over their bodies which exposes the bright colors on their bellies and foot bottoms (Brodie 1989, 96). Western newts (*Taricha* spp.) expose the red coloration of the undersides of their hind feet, tail, and venter.

[after the Unke, the German name for the European fire-bellied toad (Brodie 1989, 96)]

wiping reflex See reflex: expulsion reflex.

♦ **reflex action, reflex arc, reflex behavior** See reflex.

♦ **reflex bleeding** *n*. For example, in many lady-beetle species, lycid beetles, an ithomiine-butterfly species: an individual's facile bleeding when it is irritated that indicates to a potential predator that it is toxic, distasteful, or both (Matthews and Matthews 1978, 334).

♦ **reflex center** *n*. All of the synapses and ganglion cells in an animal's central nervous system that are parts of reflex arcs that make up a reflex (Heymer 1977, 143).

cf. reflex

♦ **reflex model of behavior** See [4]model: reflex model of behavior.

♦ **reflex theory** See [3]theory: reflex theory.

♦ **reflexology** *n*. The view that a human brain is an "input-output machine" that makes very complex responses to the world but is essentially responding to its demands (Hilts 1997, C5).

Comment: In contrast to reflexology, Rodolfo Llinas views a human brain as a prediction machine (Hilts 1997, C5). It makes elaborate mental maps that are reliable enough to enable a Human to predict what might lie ahead in space and time.

♦ **refractory period** See period: refractory period.

♦ **refuge** *n*. An area in which prey may escape from, or avoid, a predator (Lincoln et al. 1985).

See refugium.

♦ **refugium** *n., pl.* **refugia**

1. An area that has escaped major climatic changes typical of its geographic region

as a whole and acts as a refuge for biota previously more widely distributed (Lincoln et al. 1985).

2. An isolated habitat that retains environmental conditions that were once widespread (Lincoln et al. 1985).

syn. refuge (Lincoln et al. 1985)

♦ **regeneration** See reproduction (def. 1, note).

cf. -morphosis: epimorphosis

♦ **regional-character species** See ²species: characteristic species: regional- character species.

♦ **regional climax** See climax: regional climax.

♦ **regional dialect** See dialect: regional dialect.

♦ **regression coefficient** See coefficient: regression coefficient.

♦ **regression curve** See curve: regression curve.

♦ **regression line** *n.* The line that is the graphic representation of a regression equation involving one dependent variable and one independent variable (Lincoln et al. 1985).

cf. statistical test: regression

♦ **regression model of evolution** See ⁴model: regression model of evolution.

♦ **regression test** See statistical test: correlation test, regression test

♦ **regressive, retrogressive** *adj.* Referring to dedifferentiation, or reversal, to a simpler state or form (Lincoln et al. 1985).

n. retrogression

♦ **regressive character** See character: regressive character.

♦ **regressive evolution** See ²evolution: regressive evolution.

♦ **regular** See male: regular.

♦ **regular altruism** See altruism (def. 2).

♦ **regular distribution**. See distribution: uniform distribution.

♦ **regulation** *n.*

1. The coordination of units to achieve the maintenance of one or more physical, or biological, variables at a constant level; the result of regulation is homeostasis (Wilson 1975, 11).

2. The processes by which an embryo maintains its normal development (Lincoln et al. 1985).

See regulation: population regulation.

biogenic-social regulation See regulation: social regulation: biogenic-social regulation.

population regulation, regulation *n.* Control of population density by density-dependent factors (Lincoln et al. 1985).

cf. population-limiting hypothesis

self-regulation hypothesis See hypothesis: population-limiting hypotheses.

social regulation *n.* The control of behavior in social groups.

▸ **biogenic-social regulation** *n.* In social insects: social regulation by biological factors such as temperature, humidity, and animals' nutritional needs, combined with their signal mechanisms that convey information about such factors among members of a group.

[coined by Schneirla (1949 in Immelmann and Beer 1989, 33, who indicate this term is a little used now)]

▸ **psychogenic-social regulation** *n.* In more derived vertebrates: social regulation by psychological factors, such as individual attachment, emotion investment, and dominance-subordinance relationships.

[coined by Schneirla (1949 in Immelmann and Beer 1989, 238)]

♦ **regulator** *n.* An animal with certain aspects of its bodily condition that tend to be relatively independent of its environmental conditions (*e.g.*, a homeotherm) (McFarland 1985, 284).

See communication: nonverbal communication: regulator.

cf. conformer; homeostasis; therm: homeotherm

♦ **regulator gene** See gene: regulator gene.

♦ **regulatory gene** See gene: regulatory gene.

♦ **regulon** *n.* A noncontiguous group of genes that is controlled by the same regulator gene (King and Stansfield 1985).

cf. gene

♦ **reinforcement** *n.*

1. An investigator's use of reinforcement, *q.v.*, in an experiment; this involves using partial, positive, or negative reinforcement (Immelmann and Beer 1989, 248–249).

2. Whatever an investigator has to add to a stimulus-response contiguity to effect their association (Immelmann and Beer 1989, 249).

cf. extinction; learning: conditioning

negative reinforcement *n.*

1. An investigator's withholding something from, or causing aversive stimulation of, an animal (Immelmann and Beer 1989, 248).

2. An investigator's punishing an animal to suppress its response (Immelmann and Beer 1989, 248).

partial reinforcement *n.* An investigator's rewarding an animal for only a proportion of its correct responses in a learning experiment (Immelmann and Beer 1989, 216).

positive reinforcement *n.* An investigator's presenting an animal with a "genuine" reward, such as food for a hungry animal, water for a thirsty one, or electrical stimulation in an animal's brain pleasure center (Immelmann and Beer 1989, 248).

secondary reinforcement *n.* An investigator's presenting an animal with a stimulus that is a sign of a reward, or a means of obtaining a reward; *e.g.,* a monkey is given tokens that it can use to obtain food from a dispenser (Immelmann and Beer 1989, 248).

♦ **reinforcement schedule** *n.* The distribution of reward with respect to when an animal shows its responses in a learning experiment (Immelmann and Beer 1989, 249).

continuous reinforcement schedule *n.* A reinforcement schedule that rewards an animal for every correct response that it makes (Immelmann and Beer 1989, 249).

fixed-interval reinforcement schedule *n.* A reinforcement schedule that rewards an animal after it makes a correct response at a set period (Immelmann and Beer 1989, 249).

fixed-ratio reinforcement schedule *n.* A reinforcement schedule that rewards an animal after it makes a certain set number of responses (Immelmann and Beer 1989, 249).

variable-interval reinforcement schedule *n.* A reinforcement schedule that rewards an animal after it makes a correct response and at time intervals that vary randomly within a certain limit (Immelmann and Beer 1989, 249).

variable-ratio reinforcement schedule *n.* A reinforcement schedule that gives rewards at random times with respect to how many responses a animal has made (Immelmann and Beer 1989, 249).

♦ **reinforcer, reinforcement** *n.* An event (*e.g.,* reward) that is predictably followed by an animal's response and can be shown to increase the future probability of its response (Storz 1973, 236; Hinde 1982, 281). *cf.* learning: conditioning: negative conditioning

negative reinforcer *n.* A reinforcer that discourages an animal to approach a stimulus that it associates with this reinforcer (McFarland 1985, 323).

positive reinforcer *n.* A reinforcer that encourages an animal to approach a stimulus that it associates with this reinforcer (McFarland 1985, 323)

♦ **reinforcing selection** See selection: reinforcing selection

♦ **rejection** See need: rejection.

♦ **rejungent species** See ²species: ring species.

♦ ¹**relationship** *n.*
1. Kinship (*Oxford English Dictionary* 1972, entries from 1880).
2. The degree of overall phenotypic similarity between (among) taxa (Lincoln et al. 1985).
3. The degree of genotypic similarity that is inferred from phenotypic similarity (Lincoln et al. 1985).
4. "Recency of common ancestry" (Lincoln et al. 1985).

♦ ²**relationship** *n.* A spatial, social, or both, association between, or among, animals over a sustained period (Immelmann and Beer 1989, 249).

consort relationship *n.* For example, in the Chimpanzee, Barbary Ape, Rhesus Macaque; baboons: a close relationship of a male and female during her estrus that involves their moving and feeding together, much sexual activity, and frequent mutual grooming (Immelmann and Beer 1989, 58). *cf.* pair: consort pair
Comment: Consort relationships last from a few hours to a few days (Immelmann and Beer 1989, 58).

social relationship *n.* Two or more animals' interacting regularly, or periodically, over some extended period that involves their being able to recognize each other as particular individuals and remembering the results of past encounters (Hinde and Stevenson-Hinde 1976 in Hand 1986, 209).

▶ **dominance-subordination relationship** *n.* A social relationship in which "one dyad member accepts (generally follows) the 'rule' that it will defer at points of conflict" (Hand 1986, 209). *Comment:* Dominance-subordination relationships may be pure or mixed (Hand 1986, 209).

▶ **dyadic relationship** *n.* A relationship between two animals.

▶ **egalitarian-social relationship** *n.* A social relationship in which neither member of a social dyad is socially dominant (consistently wins) when the pair has social conflicts; their win-loss ratio approaches 50/50 [coined by Hand (1986, 209)].

stimulus-response relationship *n.* The temporal and quantitative relationship between a specific stimulus' presentation and appearance of an animals' specific behavior (Immelmann and Beer 1989, 294).

♦ **relationship, coefficient of** See coefficient: coefficient of relationship.

♦ **relative abundance** See abundance: relative abundance.

◆ **relative crowding coefficient** See co-efficient: relative crowding coefficient.

◆ **relative density** See density: relative density.

◆ **relative dominance** See dominance (ecological): relative dominance.

◆ **relative-dominance hierarchy** See hierarchy: dominance hierarchy: relative-dominance hierarchy.

◆ **relative frequency** See frequency: relative frequency.

◆ **relative sexuality** See sexuality: relative sexuality.

◆ **relaxed open-mouth display** See display: relaxed open-mouth display.

◆ **release call** See animal sounds: call: release call.

◆ **releaser** See ²stimulus: releaser.

◆ **releaser pheromone** See chemical-releasing stimulus: semiochemical: pheromone: releaser pheromone.

◆ **releasing factor** See hormone: releasing hormone.

◆ **releasing hormone** See hormone: releasing hormone.

◆ **releasing mechanism (RM)** See mechanism: releasing mechanism.

◆ **releasing stimulus** See stimulus: releaser.

◆ **reliability** *n.*
1. The consistency of a measurement test (Storz 1972, 237).
2. Reproducibility of measurements (Lehner 1979, 130).
3. "An indication of the size of chance errors" (Lincoln et al. 1985).

  interobserver reliability *n.* Reliability of more than one investigator in obtaining very similar data from observing the same behavior (Lehner 1979, 130–132).

  intraobserver reliability, self-reliability *n.* Reliability of the same investigator to obtain very similar data again when observing the same behavior (Lehner 1979, 130).

◆ **relic, relict species** See ²species: relict species.

◆ **relict (organ)** See organ: relict.

◆ **relief ceremony** See ceremony: relief ceremony.

◆ **REM sleep** See sleep: stage-1 sleep.

◆ **rendezvous pollination** See pollination: rendezvous pollination.

◆ **renewability** *n.* "The ability of a resource to become replenished during a specified time interval following depletion" (Wittenberger 1981, 620).

◆ **renin** See hormone: renin.

◆ **Rensch's laws** See law: Rensch's laws.

◆ **repair** *n.* Changing a DNA sequence back to its usual (normal) state (Bernstein et al. 1985, 1278).

recombinational repair *n.* Repair of DNA damage (*e.g.,* double-strand breaks and cross-links) that can occur only during recombination of homologous chromosomes.

◆ **repair enzyme** See enzyme: repair enzyme.

◆ **repair hypothesis** See hypothesis: hypotheses of the evolution and maintenance of sex: repair hypothesis.

◆ **repeated bites** See bite: repeated bites.

◆ **repeated DNA** See DNA: repetitive DNA.

◆ **repertoire** *n.* "A stock of dramatic or musical pieces which a company or player is accustomed or prepared to perform; one's stock of parts, tunes, songs, etc." (*Oxford English Dictionary* 1972, entries from 1947); also, such pieces collectively (Michaelis 1963).
syn. repertory (Michaelis 1963)
[French *répertorie* < Late Latin *repertorium,* inventory]

◆ **repertory** *n.*
1. "An index, list, catalogue, or calendar;" obsolete definition (*Oxford English Dictionary* 1972, entries from 1552).
2. "A storehouse, magazine, or repository, where something may be found" (*Oxford English Dictionary* 1972, entries from 1552); a collection of these things gathered together (Michaelis 1963).
See -gram: ethogram; repertory: vocal repertory.
syn. repertoire, *q.v.* (*Oxford English Dictionary* 1972)
[Late Latin *repertorium,* inventory]

  behavioral repertory *n.* "A list of the behavioral acts performed by an individual, caste, or species, sometimes with a specification of relative frequencies of acts" (Hölldobler and Wilson 1990, 636).

  vocal repertory *n.* An inventory of the sounds produced by an individual animal or species (Immelmann and Beer 1989, 325).
  cf. -gram: ethogram
  Comment: Primates (especially Humans, Chimpanzees, and New World monkeys), whales, and birds have the most complex vocal repertories (Immelmann and Beer 1989, 325).

◆ **repetition** *n.* The close resemblance of early developmental stages in related animals, with subsequent divergence in later stages by their acquisition of distinctive characters (Lincoln et al. 1985).
cf. law: biogenetic law

◆ **repetitious DNA, repetitive DNA** See DNA: repetitive DNA.

◆ **repium** See ²community: repium.

◆ **replacement queen** See caste: queen: replacement queen.

♦ **replete** *adj.* "Fully nourished; well fed" (Lincoln et al. 1985).
See caste: replete.
[< Old French *replet* < Latin *repletus*, past participle of *replere*, to fill again]

♦ **replicability** *n.* In statistics, "the idea that replicate experimental units must be extremely similar if not identical at the beginning (premanipulation period) of an experiment" (Hurlbert 1984, 201).
syn. reproducibility (Abbott 1966 in Hurlbert 1984, 201)

♦ **replicate** *n.* In statistics, a repeated experiment (Lincoln et al. 1985).

♦ **replication** *n.*
  1. In statistics, one's repeating an experiment to obtain more information for estimating experimental error (Lincoln et al. 1985).
  2. "Production of identical copies of information from existing units" (Lincoln et al. 1985).
  v.t. replicate

pseudoreplication *n.* In statistics, one's "use of inferential statistics to test for treatment effects with data from experiments where either treatments are not replicated (though samples may be) or replicates are not statistically independent (Hurlbert 1984, 187).
Comment: Hurlbert (1984) describes "implicit pseudoreplication," "sacrificial pseudoreplication," "simple pseudoreplication," and "temporal pseudoreplication."

♦ **replicator** *n.* "Any entity in the universe of which copies are made" Dawkins (1982, chap. 5).
Comment: Dawkins (1982) discusses "active replicator," "dead-end replicator," "germ-line replicator," and "passive replicator."

♦ **representation** *n.* An animal's internal mental image of an object(s) for which it is searching or for complex spatial, or social, situations (Kummer 1982 in McFarland 1985, 354).
cf. search image

declarative representation *n.* A representation of a desired goal or object (*e.g.,* food for a rat) (Dickinson 1980 in McFarland 1985, 354).

procedural representation *n.* A representation of a set of instructions that lead an animal automatically to a desired object without its envisioning the object (*e.g.,* a maze path for a rat in search of food) (Dickinson 1980 in McFarland 1985, 354).

♦ **representative sample** See sample: representative sample.

♦ **reproducibility** See replicability.

♦ **reproduction** *n.*
  1. In some animal species: an individual's regrowing a part that has been removed (*Oxford English Dictionary* 1972, entries from 1727).
  Note: This process is now usually called "regeneration" (*Oxford English Dictionary* 1972).
  2. In organisms: an individual's producing offspring by some form of generation; the generative production of new individuals by, or from, existing ones; also, an individual's power of reproducing in this way (*Oxford English Dictionary* 1972, entries from 1785).
  3. An organism's reproductive activities including mate acquisition, gamete production, and parental care (Willson 1983, 5).
See -genesis.
cf. copulation, fertilization, sexual reproduction

parthenogenetic reproduction See parthenogenesis.

uniparental reproduction *n.* Reproduction from a single parent (*e.g.,* vegetative reproduction, thelytoky, and self-fertilizing hermaphroditism) (Lincoln et al. 1985).
cf. parthenogenesis

♦ **reproduction curve** See curve: reproduction curve.

♦ **reproductive** *n.* An individual that has offspring.

primary reproductive *n.* In termites: the colony-founding type of queen or male derived from a winged adult (Wilson 1975, 592).

♦ **reproductive altruism** See altruism: reproductive altruism.

♦ **reproductive behavior** See behavior: reproductive behavior.

♦ **reproductive community** See [1]population: Mendelian population.

♦ **reproductive conflict** See [3]conflict: reproductive conflict.

♦ **reproductive effort** See effort: reproductive effort.

♦ **reproductive-isolating mechanism** See mechanism: isolating mechanism.

♦ **reproductive isolation** See isolation: reproductive isolation.

♦ **reproductive mimicry** See mimicry: reproductive mimicry.

♦ **reproductive-mutualistic mimicry** See mimicry: reproductive-mutualistic mimicry.

♦ **reproductive pheromone** See chemical-releasing stimulus: semiochemical: pheromone: reproductive pheromone.

♦ **reproductive potential** *n.* The number of offspring produced by one female in a population (Lincoln et al. 1985).

♦ **reproductive stage** See stage; reproductive stage.

♦ **reproductive success** See success: reproductive success.

- **reproductive success, principle of** See principle: principle of reproductive success.
- **reproductive system** See ²system: reproductive system.
- **reproductive value** See value: reproductive value.
- **reproductivity effect** See effect: reproductivity effect.
- **repugnatorial** See -orial: repugnatorial.
- **research hypothesis** See hypothesis: research hypothesis.
- **reservoir host** See host: reservoir host.
- **residence time** See time: residence time.
- **resident** See territory holder.
- **resident avifauna** See fauna: avifauna: resident avifauna.
- **residual** See error: prediction error.
- **residual genotype** See -type, type: genotype: background genotype.
- **residual reproductive potential** See potential: residual reproductive potential.
- **residual variance** See variance: residual variance.
- **resistance adaptation** See ³adaptation: ¹physiological adaptation, resistance adaptation.
- **resource** *n.*
  1. In Humans: "A means of supplying some want or deficiency;" a stock, or reserve, upon which one can draw when necessary; now usually plural (*Oxford English Dictionary* 1972, entries from 1611).
  2. In organisms: "any substance, object, or energy source utilized for body maintenance, growth, or reproduction" (Wittenberger 1981, 620).
  3. In organisms: something that is consumed, or occupied, during growth, maintenance, or reproduction (Ehrlich and Roughgarden 1986, 630, in Abrams 1992, 282).
     Note: This is a broad definition that includes inadvertently ingested nondigestible material as a resource (Abrams 1992, 283).
  4. In organisms: something that potentially influences individual fitness (Wiens 1984, 401, in Abrams 1992, 283).
     Note: This definition would include (1) a nonpreferred food as a resource in an environment that contains sufficient amounts of preferred foods, and (2) foods in general even when they are superabundant and have no fitness effects (Abrams 1992, 283).
  5. In organisms: a thing that increases survival, or reproduction, of an individual or population (Abrams 1992, 282).
  syn. Ressourcen (German)
  Comment: Abrams (1992) discusses some kinds of resources and what distinguishes one resource from another.

coarse-grained resource *n.* A resource distributed in such a way that a consumer encounters it in a proportion different from its actual one; contrasted with fine-grained resource (Lincoln et al. 1985).

fine-grained resource *n.* A resource distributed in such a way that a consumer encounters it in a proportion similar to its actual one; contrasted with coarse-grained resource (Lincoln et al. 1985).

spatiotemporally clumped resource *n.* A resource that is clumped in space and whose availability at any given location varies temporally on a short-term basis (Wittenberger 1981, 621).

- **resource availability** *n.* "The extent to which a given resource can be obtained or utilized by a given individual or species" (Wittenberger 1981, 620).
- **resource axes** See bionomic axes.
- **resource budget** See budget: resource budget.
- **resource competition** See competition: resource competition.
- **resource-defense polygyny** See mating system: polygyny: resource-defense polygyny.
- **resource-holding potential** See potential: resource-holding potential.
- **respect of possession** See behavior: respectful behavior.
- **respectful behavior** See behavior: respectful behavior.
- **respiration** *n.* An organism's taking in oxygen from the air and giving off carbon dioxide and other oxidation products (Michaelis 1963).
  [Latin *respiratio, -onis*]
  Comment: Some definitions of respiration are synonymous with breathing, *q.v.*

aerial respiration See breathing: air breathing.

aquatic respiration See breathing: water breathing.

aquatic-surface respiration (ASR) *n.* For example, in many fish species and larvae of some amphibian species: an animal's selectively inspiring a well-oxygenated surface layer of water while being in a body of poorly oxygenated water (Carter and Beadle 1931 in Kramer and Mehegan 1981, 300; Kramer 1987, 84).
syn. surface skimming, surfacing (Lewis 1970 in Kramer and Mehegan 1981, 300), breathing the surface film (Gee et al. 1978 in Kramer and Mehegan 1981, 300)
Comment: A Guppy engaged in ASR usually moves slowly along the water surface with its head contacting the surface and jaws open just beneath the surface. At intervals it may drop below the surface and

feed, court, or swim about for a while before darting back and recontacting the surface (Kramer and Mehegan 1981, 301). [coined by Kramer and Mehegan (1981, 300)]

♦ **respondent behavior** See behavior: respondent behavior.

♦ **response** *n.* Any change in an organism's behavior, or other attribute, that is caused by a stimulus (Lincoln et al. 1985).
cf. behavior, reflex, taxis

conditional reflex or response, conditioned reflex or response See learning: conditioning: Pavlovian conditioning.

conditioned emotional response *n.* The interruption, or diminution, of ongoing behavior as a result of an intruding stimulus (Estes and Skinner 1941 in Hinde 1985, 986).

evolutionary response See response: genetic response.

following response See behavior, following behavior.

functional response *n.*
1. An increase in enemy density due to increased host availability [coined by Solomon 1949, 16].
   Note: "Host" means "host" or "prey," and enemy means "parasite" or "predator" in this definition.
2. An increase in predation rate in response to an increase in prey density (*e.g.,* in some small-mammal species feeding on sawflies) (Holling 1959, 303).
3. The phenomenon in which "as local populations of the host species increase in numbers, its enemies are able to encounter and to strike individuals at higher frequency" (Holling 1959 in Wilson 1975, 85).
4. A change in rate of predation by an individual predator in response to change in density of its prey (Lincoln et al. 1985; Ruxton and Gurney 1994, 537).

▸ **type-1 functional response** *n.* A functional response in which the predation rate is directly proportional to prey density (Holling 1959, 315, illustration, 317).
Comments: In this functional response, the predator is assumed to search at random and have a constant prey-searching rate for all prey densities (Holling 1959, 315).

▸ **type-2 functional response** *n.* A functional response in which the number of prey attacked per predator increases very rapidly with initial increase in prey density and thereafter increases more slowly, approaching a certain fixed level, with the searching rate becoming progressively less with an increase in prey density (Holling 1959, 316, illustration; 317).

▸ **type-3 functional response** *n.* A functional response in which the searching rate first increases with an increase in prey density and then decreases (*e.g.,* in small mammals that feed on sawflies) (Holling 1959, 315, illustration, 317).

genetic response *n.* The between-generation genetic change that results from phenotypic selection, *q.v.* (Fisher 1930a and others in Endler 1992, 222).
syn. evolutionary response (a confusing synonym Endler 1992, 223)
Comment: The process of natural selection is divided into the sequential subprocesses of phenotypic selection and genetic response by quantitative geneticists and animal and plant breeders (Endler 1992, 222).

graded response *n.* A response that varies in proportion to the strength of the stimulus that evokes it (Immelmann and Beer 1989, 120).

immobility response *n.* An animal's crouching, freezing, or both, due to aversive stimuli (Hinde 1970, 349).
cf. fear, playing dead

indicator response *n.* Movement patterns that occur only in a few species within a family (Seitz 1940–1943 in Heymer 1977, 91).

induced plant response to herbivory *n.* A plant's immune-like response to damage by a herbivore that reduces a herbivore's performance, preference, or both, on that plant (Agrawal 1998, 1201).
Comments: Researchers reported induced plant responses in over 100 herbivore-plant system (Agrawal 1998, 1201). It seems likely that induced plant response to herbivory evolved in natural communities, but researchers have yet to establish this directly. So far, Agrawal's (1998) clear-cut experiments demonstrate that this response occurs in the field but not in natural communities. His experimental communities in California were mixtures of alien and native species. The aliens included his two radish species and the butterfly *Pieris rapae,* all from Europe, with the butterfly also occurring in Asia. The herbivore-plant responses might be quantitatively different in the natural communities of these organisms.

numerical response *n.*
1. An increase in predator number due to increased prey availability and predator survivorship and reproduction [coined by Solomon 1949, 16].

2. A change in the density of a predator population with a change in prey density (Holling 1959, 303; Lincoln et al. 1985).

operant response *n.* An animal's response that acts on its environment and produces an event that affects the animal's subsequent probability of the response (Hinde 1982, 280–281).
cf. reinforcer

optokinetic response See reflex: optomotor reflex.

orientating response, orientating reflex, startle response *n.* An animal's reaction to a novel stimulus that consists of its scrutinizing it attentively, increasing its arousal, and being ready to take flight or defensive action (Immelmann and Beer 1989, 209–210).
cf. response: orientation response
Comment: This term derives from the Russian scientist E.N. Sokolov (Immelmann and Beer 1989, 209–210).

orientation response *n.* In freely moving animals and phytoflagellates: movement from one site to another (Heymer 1977, 126).
cf. response: orientating response

pupillary response *n.*
1. A vertebrate animal's expanding and contracting its pupils in response to different light intensities (Heymer 1977, 135).
2. A person's dilating his pupil in response to pictures that interest him or contracting his pupil in response to an unpleasant stimulus, when no light-intensity change occurs (Heymer 1977, 135).

ritualized response See behavior: ritualized behavior.

skin-conductance response (SCR) *n.* In Humans: skin perspiring that accompanies emotional changes (Vogel 1997, Science 275, 1269).

startle response See response: orientating response.

unconditional reflex or response, unconditioned reflex or response See learning: conditioning: Pavlovian conditioning.

♦ **response tendency** See potential: specific-action potential(ity).

♦ **response to selection (*R*)** *n.* The difference in mean phenotype between the progeny generation (μ') and the previous generation (μ), symbolized as $R = \mu' - \mu$ (Hartl 1987, 228).
cf. equation: prediction equation

♦ **Ressourcen** See resource.

♦ **restibilic** *adj.* "Perennial" (Lincoln et al. 1985).

♦ **resting-level activity** See activity: spontaneous activity.

♦ **resting potential** See potential: resting potential.

♦ **restrictase** See enzyme: restriction enzyme.

♦ **restriction** *n.* A bacterium's using restriction enzyme for protection against intruding DNA from other organisms by cutting up this DNA (Campbell et al. 1999, 366).
cf. enzyme: restriction enzyme

♦ **restriction endonuclease** See enzyme: restriction enzyme.

♦ **restriction enzyme** See enzyme: restriction enzyme.

♦ **restriction fragment** *n.* A part of a DNA molecule that a restriction enzyme(s) has cut out of it (Campbell et al. 1999, 366).

♦ **restriction-fragment-length polymorphism marker** See marker: restriction-fragment-length polymorphism marker.

♦ **restriction site** See site: restriction site.

♦ **retardation** See [2]heterochrony: retardation.

♦ **reticular activating system** See [2]system: reticular activating system.

♦ **reticular formation** See organ: brain: reticular formation.

♦ **reticulate evolution** See [2]evolution: reticulate evolution.

♦ **retinue** See court.

♦ **retrieving** *n.*
1. In some dog strains: an individual's finding, or discovering, again (game that has been temporarily lost); especially to flush or set up (partridges) a second time (*Oxford English Dictionary* 1972, verb entries from 1410).
2. In some dog strains, Humans: an individual's finding and bringing in a bird that has been wounded or killed (*Oxford English Dictionary* 1972, verb entries from 1856).
3. In some rodent species: a mother's recovering her young that are outside her nest (Immelmann and Beer 1989, 254).
4. In some species of ground-nesting birds: an individual's recovering its egg that is outside its nest (Immelmann and Beer 1989, 254).
cf. behavior: egg-rolling behavior
cf. care: parental care: retrieval

♦ **retro** *prefix* Back; backward (Michaelis 1963).
[Latin *retro*, back, backward]

♦ **retrofection** *n.* Reinfection of a host by parasite larvae that hatch on its surface and re-enter its body (Lincoln et al. 1985).

♦ **retrograde evolution** See [2]evolution: retrograde evolution.

♦ **retrogressive** See regressive.

♦ **retrogressive succession** See ²succession: retrogressive succession.

♦ ***Retrojecktion*** *n*. An animal's using its own body, or parts of it, as a substitute object (*e.g.,* a young mammal's sucking its fingers, paws, or other body parts) when deprived of access to its mother's nipples (Immelmann and Beer 1989, 254).
[German, which has no precise English equivalent]

sexual *Retrojecktion* *n*. In animals that are kept alone for long periods: an individual's handling its own genitalia which enables it to derive "sexual gratification" (Immelmann and Beer 1989, 254). *cf.* masturbation

♦ **retrorecognition phenomenon** See phenomenon: retrorecognition phenomenon.

♦ **return-benefit altruism** See altruism: return-benefit altruism.

♦ **reversal learning** See learning: reversal learning.

♦ **reversal problem** See test: reversal test.

♦ **reversal test** See test: reversal test.

♦ **reverse mutation** See ²mutation: reverse mutation.

♦ **reverse sexual dimorphism** See sex role reversal.

♦ **reversion** See atavism.

♦ **revertant** *n*. An organism that bears a revertant allele (Lincoln et al. 1985). See allele: revertant.

♦ **revivescence** *n*. An organism's emergence from hibernation or some other quiescent state (Lincoln et al. 1985).

♦ **revolution** *n*. An instance of great change, or alteration, in affairs or in some particular thing (*Oxford English Dictionary* 1972, entries from 1450).

first-Darwinian revolution *n*. Biologists' accepting the theory of common descent (all organisms evolved from a common ancestor); distinguished from second-Darwinian revolution (Mayr 1982, 117).

genetic revolution *n*. A large amount of genetic change that is generated in a species within a relatively short evolutionary period (Mayr 1982, 885).
syn. genetic transilience (Templeton 1980 in Mayr 1982, 885)
Comment: Stabilization of the genetic change takes a few to thousands of generations (Mayr in Hapgood 1984).

scientific revolution *n*.
1. One of the highly drastic changes in the conceptual framework of a science (Mayr 1972 in 1982, 882).
2. The phenomenon of one paradigm's replacing another; the old paradigm ceases to be a useful and acceptable

way of explaining the phenomenon to which it applies (Kuhn 1970 in Moore 1984, 478).
Comments: Mayr (1972 in 1982, 882) points out the so-called "Darwinian revolutions" occurred over 200 years, which are too many years to be considered revolutions. Also, there was no crisis situation in the 1850s, and there was no replacement of one paradigm by another. Further, Mayr (1982, 857) says, "I cannot think of a single case in biology where there was a drastic replacement of paradigms between two periods of 'normal science.'" Kuhn (1962 in Mayr 1982) gives criteria for identifying a scientific revolution.

second-Darwinian revolution *n*. Biologists' accepting natural selection as the only direction-giving factor in evolution that was completed during the evolutionary synthesis, *ca.* 1936–1947; distinguished from first-Darwinian revolution (Mayr 1982, 117).

♦ **reward** *n*. An object (*e.g.,* food, water) or opportunity (*e.g.,* a chance to see outside its cage) that an animal finds favorable in the context of conditioning, *q.v.* (Immelmann and Beer 1989, 248).

floral reward *n*. An attractant (*e.g.,* floral oils, food bodies, nectar, or pollen) in a flower that invites and lengthens visits of a floral visitor (Buchmann and Nabhan 1996, 246).

♦ **RFLP marker** See marker: restriction-fragment-length polymorphism marker.

♦ **RFLPs** See -morphism: polymorphism: restriction-fragment-length polymorphisms.

♦ **rheo-** *combining form* "Current or flow, as of water or electricity" (Michaelis 1963). [Greek *rheos*, a current]

♦ **rheogameon** See ²species: polytypic species.

♦ **rheology** See study of: limnology: rheology; study of: rheology.

♦ **rheophile** See ¹-phile: rheophile.

♦ **rheophobe** See -phobe: rheophobe.

♦ **rheoplankton** See plankton: rheoplankton.

♦ **rheopositive** *adj*. Referring to an organism's showing behavioral, or other responses, to air or water currents (Lincoln et al. 1985).

♦ **rheoreceptive** *adj*. Referring to an organism's being sensitive to air, or water, currents (Lincoln et al. 1985).

♦ **rheotaxis** See taxis: rheotaxis.

♦ **rheotrophic** *adj.* Referring to an organism that obtains its nutrients largely from percolating, or running, water (Lincoln et al. 1985).

♦ **rheoxene** See -xene: rheoxene.

♦ **rhexigenous** See -genous: rhexigenous.

♦ **rhicochetal locomotion** See locomotion: rhicochetal locomotion.
♦ **rhipium** See ²group: rhipium.
♦ **rhithron** *n.* "The upper reaches of a river" (Lincoln et al. 1985).
adj. rhithrous
cf. potamous
♦ **rhizo-, rhiz-** *combining form* Root; referring to a root or roots (Michaelis 1963). [Greek *rhiza*, root]
♦ **rhizobenthos** See ²community: benthos: rhizobenthos.
♦ **rhizophage, rhizophagy, rhizophagous** See -phage: rhizobenthos.
♦ **rhizophile, rhizophagy, rhizophilous** See ¹-phile: rhizophile.
♦ **rhizoplane** See -sphere: rhizosphere.
♦ **rhizosphere** See -sphere: rhizosphere.
♦ **rhoium** See ²community: rhoium.
♦ **rhoophile, rhoophily, rhoophilus, rhoophilous** See ¹-phile: rhoophile.
♦ **RHP** See potential: resource-holding potential.
♦ **rhyacium** See ²community: rhyacium.
♦ **rhyacophile, rhyacophilous, rhyacophilous** See ¹-phile: rhyacophile.
♦ **rhypophage, rhypophagy, rhypophagous** See -phage: rypophage.
♦ **rhysium** See ²community: rhysium.
♦ **rhythm, periodicity** *n.* A periodic change in an organism's physical environment, such as diurnal, tidal, and seasonal cycles (Immelmann and Beer 1989, 217). See rhythm: biological rhythm.
cf. dormancy: hibernation
biological rhythm, biorhythm, biorhythmicity *n.* A regular periodicity shown by organisms, their behaviors, and their physiological activity and underlying physiological regulation (Immelmann and Beer 1989, 33).
syn. biological clock, clock, periodicity, rhythm, rhythmicity
cf. clock
circadian rhythm *n.* In many animal species: a rhythm that recurs about every 1 day (24.1 hours) (Wilson 1975, 580).
syn. diurnal rhythm (Immelmann and Beer 1989, 44)
cf. time synchronizer
[Latin *circa*, around + *dies*, day]
circalunadian rhythm *n.* A usually bimodal rhythm with a persistent periodicity of about 1 lunar day (24.8 hours) (Lincoln et al. 1985).
circamonthly rhythm See rhythm: lunar rhythm.
circannual rhythm *n.* An organism's endogenous rhythm that is usually slightly less than 1 year (365.2 solar days) (McFarland 1985, 291).

Comments: Circannual rhythms include yearly migration bouts of many bird species, cold-season hibernation of many vertebrate species, and fluctuations in reproduction and other activities of many animal species (Immelmann and Beer 1989, 15). Humans and woodchucks are among the many mammals that show circannual rhythms (Line 1997, C2).
syn. annual rhythm, annual periodicity (Immelmann and Beer 1989, 15)
circasemiannual rhythm *n.* A rhythm with two cycles within about 1 year (Lincoln et al. 1985).
diurnal rhythm See rhythm: circadian rhythm.
endogenous rhythm *n.* An internal rhythm that programs an organism's behavior in synchrony with an exogenous temporal period, particularly an approximately 24-hour or 365-day period (McFarland 1985, 291).
cf. zeitgeber
infradian rhythm *n.* A rhythm with a period less than 24.1 hours (Lincoln et al. 1985).
cf. rhythm: ultradian rhythm
lunar rhythm *n.* For example, in the Palolo Worm and Jamaican Fruit Bat: a rhythm of about 1 lunar (= synodic) period of 29.5 days (McFarland 1985, 303).
syn. circamonthly rhythm (Lincoln et al. 1985)
tidal rhythm *n.* In many marine animals: a rhythm of about 15 days (half a lunar cycle) (McFarland 1985, 303).
ultradian rhythm *n.* A rhythm that slightly exceeds 24 hours (Lincoln et al. 1985).
cf. rhythm: circadian rhythm, infradian rhythm
♦ **rhythmicity** See rhythm.
♦ **ribonucleic acid (RNA)** See nucleic acid: ribonucleic acid.
♦ **ribosome** *n.* In eukaryotes and prokaryotes: an intracellular structure that facilitates the specific coupling of tRNA anticodons with mRNA codons during protein synthesis (Campbell 1996, 308).
Comments: A ribosome is made up of a large subunit and a small subunit (Campbell 1996, 308–309). Each subunit is an aggregate of numerous proteins and rRNA.
♦ **ribozyme** *n.* RNA that cuts and joins pre-existing RNA and can carry out some of the more important subreactions of RNA replication (Orgel 1994, 78–79).
cf. world: RNA world
Comments: Thomas R. Cech and Sidney Altman independently discovered the first known ribozymes in 1983 (Orgel 1994, 78), for which they won the 1989 Nobel Prize in

Chemistry (Hart 1996, 318). Researchers are working on how to use ribozymes to correct genetic codes.

♦ **richness** See ²group: richness; species richness.

♦ **Ricker Mount, Ricker Specimen Mount** *n.* A shallow, glass- or plastic-covered box that contains cotton or polyethylene fiber and is used to hold biological specimens for display and study [after Clarence B. Ricker, a businessman from New Jersey who invented this mount to protect his butterfly specimens in about 1900 (Ehler 1999, 10)].

Comments: Museums hold biological specimens in many kinds of containers including drawers of the styles of Cornell University and the National Museum of Natural History (pinned and other arthropods), in cabinets, jars of preservative (fishes, snakes, etc.), and open boxes (fossils, minerals, rocks, etc.).

♦ **ridiculing** See mocking.

♦ **ring angel** See angel: ring angel.

♦ **ring species** See ²species: ring species.

♦ **riparian, riparial, riparious** *adj.* Referring to living or situated on river, or stream, banks (Lincoln et al. 1985).

♦ **riparious** See riparian.

♦ **ripicole, ripicolous** See -cole: ripicole.

♦ **risk** *n.*
1. A cost that involves a context in which a focal organism could suffer bodily harm (Wittenberger 1981, 621). See cost.
2. Food variance when food rewards are normally distributed (Stephens and Paton 1986, 1659).
cf. cost

♦ **risk-averse preference** *n.* A foraging animal's preference for a constant reward rather than a varying one (Stephens and Paton 1986, 1659).

♦ **risk-prone, risk-loving, risk-preferring preference** *n.* For example, in the Grey Jay: a foraging individual's preference for a varying reward rather than a constant one (Stephens and Paton 1986, 1659; Ha et al. 1990, 91).

♦ **risk-sensitive foraging behavior** See foraging: risk-sensitive foraging behavior.

♦ **ritual** *n.*
1. A person's prescribed form, or method, for performing a religious or other devotional service (*Oxford English Dictionary* 1972, entries from 1649).
2. Human ritual observances; ceremonial acts (*Oxford English Dictionary* 1972, entries from 1656).
3. In animal's symbolic behavior that results from ritualization, *q.v.* (Huxley 1923 and Tinbergen 1952 in McFarland 1985, 392–393).
See behavior: ritualized behavior.
syn. symbolic action (Heymer 1977, 170)

♦ **challenge ritual** *n.*
1. In the blue wildebeest: a male's making the rounds to all of his territorial neighbors and making a territorial advertisement display to each (Estes 1969 in Wilson 1975, 491).
2. In the Bontebok, *Damaliscus dorcas:* a territorial maintenance behavior comprised of at least 30 elements (David 1973 in Heymer 1977, 86).

♦ **ritualization** *n.*
1. The evolutionary process by which a behavior changes to become a display or signal used in communication, or at least becomes increasingly effective as a signal (Wilson 1975, 224–230, 594; Immelmann and Beer 1989, 255).
2. A nondisplay behavior's evolutionary transformation into a display behavior, exclusive of the process involved (Immelmann and Beer 1989, 255).
cf. behavior: ritualized behavior, emancipation, ritual, typical intensity
Comment: Many kinds of behaviors have become ritualized, including chemical production, excretion, flight, food exchange, predation, respiration, and secretion (Huxley 1914, 1923, 1966 in Heymer 1977, 148; Wilson 1975, 224–230, 594; Immelmann and Beer 1989, 255).

♦ **ontogenetic ritualization** *n.* Rigidly stereotyped behavior's arising from highly variable behavior of young animals (*e.g.,* a definitive song's arising from subsong) (Immelmann and Beer 1989, 256).

♦ **ritualized behavior, ritualized response** See behavior: ritualized behavior.

♦ **ritualized display** See display: ritualized display.

♦ **ritualized fight** See fight: ritualized fight.

♦ **ritualized grooming** See grooming: ritualized grooming.

♦ **riverain** *adj.* Referring to a river bank or the general vicinity of a river (Lincoln et al. 1985).

♦ **riverine** *adj.*
1. Referring to a river (Lincoln et al. 1985).
2. Formed by river action (Lincoln et al. 1985).

♦ **r_m, r_{max}** See rate: r_m, r_{max}.

♦ **RNA** See nucleic acid: ribonucleic acid.

♦ **RNA world** See world: RNA world.

♦ **RNA-world hypothesis** See hypothesis: RNA-world hypothesis.

♦ **robber** *n.*
1. A person who practices, or commits, robbery; a depredator; plunderer; despoiler

n–r

(*Oxford English Dictionary* 1972, entries from *ca.* 1175).

2. In some bee, other insect, and anthophilous-bird species: an individual that makes holes in a flower corolla through which is takes nectar, pollen, or both (Barrows 1976c, 132; Inouye 1980).
Notes: Barrows (1976c) categorizes robbers as "nectar-foraging robbers," "nectar-foraging-perforating robbers," and "pollen-foraging robbers." Inouye (1980) further clarifies terminology related to "flower robbers."
syn. flower perforator, thief (Faegri and van der Pijl 1979, 68, 190), nectar robber (Buchmann and Nabhan 1996, 250)
cf. dysteleology, thief

3. In some bee species: an individual that takes provisions of a heterospecific, or conspecific, individual (Michener 1974, 230–232; Winston 1987, 115–116).
syn. cleptoparasite, kleptoparasite (Wilson 1975, 580)

4. In the Black-Headed Gull: an individual that takes the food of a heterospecific (*e.g.,* egret, Godwit, Stilt) individual (Amat and Aquilera 1990, 70).
v.i., v.t. rob
syn. kleptoparasite, pirate (Amat and Aquilera 1990, 70)
cf. egg robbing; parasitism: social parasitism: trophic parasitism: cleptoparasitism

♦ **robustness** *n.* A statistical method's ability to produce a fair result, despite data deviations from the premises upon which the method is based (Lincoln et al. 1985).

♦ **rock-scissors-paper game** See game (table).

♦ **rocky coast** See habitat: coast: rocky coast.

♦ **rodent run** See display: distraction display.

♦ **rogue** *v.t.* To remove rogues (undesirable weeds, diseased plants, or genetic variants) from a population (Lincoln et al. 1985).
See ²mutation: macromutation.

♦ **role** *n.*
1. For example, in social-mammal species: a pattern of behavior displayed by certain members of a society, consisting of communication and activities that influence other individuals directly or indirectly, or both (Wilson 1975, 298, 594).
Note: The same animal can have more than one role (Wilson 1975, 298, 594).
2. In anthropology, the combination of rights and duties imposed on the incumbents of certain positions within a society (Hinde 1982, 282).
cf. institution

direct role *n.* A society subgroups' behavior, or set of behaviors, that benefits

other subgroups and, therefore, its society as a whole (Wilson 1975, 309).

indirect role *n.* A society subgroup's behavior that benefits only it and is neutral, or even destructive, to other subgroups (Wilson 1975, 310).

♦ *Rolligkeit* See estrus: *Rolligkeit.*

♦ **romantic scientist** See scientist: romantic scientist.

♦ **rookery** See ²group: rookery.

♦ **roost** See ²group: roost

♦ **root zone** See -sphere: rhizosphere.

♦ **rooted-phylogenetic tree** See -gram: rooted-phylogenetic tree.

♦ **rooting reflex** See reflex: rooting reflex.

♦ **rootless phylogenetic tree** See -gram: rootless phylogenetic tree.

♦ **Ropartz effect** See effect: Ropartz effect.

♦ **Rosa's theory of hologenesis** See ²evolution: hologenesis.

♦ **rose diagram** See graph: rose diagram.

♦ *Rossigkeit* See estrus, *Rossigkeit.*

♦ **round dance** See dance: bee dance: round dance.

♦ **rout, route** See ²group: rout.

♦ **Roux-Weismann theory** See ²theory: germplasm theory.

♦ **royal cell** *n.*
1. In the Honey Bee: a large, pitted, waxen cell constructed by the workers that is used for rearing an immature queen (Wilson 1975, 594).
2. In some termite species: a special cell in which a queen lives (Wilson 1975, 594).

♦ **royal jelly** *n.* In Honey Bees: a material supplied by workers to female larvae in royal cells that is necessary for the larvae's transformation into queens (Wilson 1975, 594).
Comment: Royal jelly is secreted primarily by a bee's hypopharyngeal gland and consists of a rich mixture of nutrient substances, many with complex chemical structures (Wilson 1975, 594).

♦ **royal substance** See substance: royal substance.

♦ *rpr* See mutation: reaper.

♦ **rRNA** See nucleic acid: ribonucleic acid: rRNA.

♦ **rubicole** See -cole: rubicole.

♦ **ruderal** *n.*
1. A plant that grows on, or among, stone rubbish; a plant that is characteristic of rubbish heaps (*Oxford English Dictionary* 1972, entries from 1858).
2. A plant, or other organism, that inhabits an area of rubbish, debris, or disturbance (Lincoln et al. 1985).
See ²species: r-selected species.
adj. ruderal
cf. ¹-phile: chledophile
[Latin *rūdera*, broken stone]

♦ **rudimentary character** See character: rudimentary character.

♦ **ruffling** *n.*

1. In some bird species: an individual's setting up, stiffening (its feathers), especially "as a sign of anger" (*Oxford English Dictionary* 1972, entries from 1643).
2. In some bird species: an individual's erecting its feathers in connection with threatening another animal (Heymer 1977, 81).

syn. shuffling (Heymer 1977, 81)

cf. piloerection

♦ **rugophile, rugophily, rugophilic** See ¹-phile: rugophile.

♦ **rule** *n.*

1. A principle that regulates a procedure, or method, necessary to be observed in the pursuit, or study, of some art or science (*Oxford English Dictionary* 1972, entries from 1387).
2. A standard of discrimination or estimation; a criterion, test, canon (*Oxford English Dictionary* 1972, entries from 1382).
3. A fact (or the statement of one) that is usually true, that is normally the case (*Oxford English Dictionary* 1972, entries from 1300); a statement that is a biological generality with some exception; contrasted with law (Mayr 1982, 37–45).

syn. concept, generalization, principle (Mayr 1982, 37–45)

cf. axiom, effect, doctrine, dogma, hypothesis, law, principle, theorem, ²theory, truism [Old French *reule* < Latin *regula*, ruler, rule < *regere*, to lead straight, direct]

Comment: The following entries contain terms with "rule" in them; the same definition of "rule" does not relate to all of them.

Allen's rule *n.*

1. Certain species and other taxa show an "enlargement of peripheral parts under high temperature or toward the Tropics; hence southward in North America;" these parts include mammal ears, feet, horns, and pelage and bird bills, claws, and tails (Allen 1877 in Allen 1906, 382; Scholander 1955, 174).

   Note: Allen (1906, 382) indicates that this rule holds only for certain taxa; Scholander (1955, 174) concludes that it is not well supported overall by data. Emlen 1973, 59) states that Allen's rule holds for some mammal groups when Arctic and temperate regions are compared but not when temperate and tropical regions are compared.
2. Extremities in Arctic mammal species tend to be shorter than mammal species in warmer regions (Emlen 1973, 59).

syn. Allen's law, proportion rule (Lincoln et al. 1985)

cf. rule: Bergmann's rule, ecogeographical rule

[after J.A. Allen, biologist]

Aschoff's rule *n.* The direction and rate of an organism's drift from a precise 24-hour cycle under experimental conditions are functions of the light intensity and of whether the organism is normally diurnal or nocturnal (Dewsbury 1978, 66).

Comment: This useful rule has exceptions (Dewsbury 1978, 66).

Bergmann's rule *n.*

1. Within a species or among closely related species, individuals tend to be smaller in warmer environments than in colder ones (Bergmann 1847 in Scholander 1955, 174; Allen 1906, 379; Lindsey 1966, 456).
2. Species of homeothermic vertebrates tend to be smaller in warmer environments than in colder ones (Emlen 1973, 59).
3. Geographically variable species of homeothermic vertebrates are generally represented by larger forms in the colder parts of their ranges than in the warmer parts of their ranges (Lincoln et al. 1985).

syn. Bergmann's law (Emlen 1973, 59); size rule (Lincoln et al. 1985)

cf. law: Allen's law

Comments: Some biologists consider Bergmann's rule to be an empirical generalization without a physiological basis (Lindsey 1966, 456), while others suggest that size increase is due to selection for a smaller surface-to-volume ratio, which minimizes heat loss (Emlen 1973, 59). Within some homeotherm species, Bergmann's rule may hold because there appears to be a positive correlation between large size and higher latitude (Lindsey 1966, 463). However, among homeotherm species, Bergmann's rule is not well supported by data (Scholander 1955, 174). In a study of 12,503 poikilotherm species, Lindsey (1966, 464) found that the proportion of species with large adult size tends to increase from the equator towards the poles in Fish and Amphibians. This trend holds for freshwater fish, deep-sea fish, Anurans, Urodeles, and marine neritic fish, arranged roughly in order of decreasing clarity of this trend. In general, this rule applies not only within these groups of families but also within families. In Reptiles, this rule holds weakly among Snakes but not among Lizards or nonmarine turtles.

[after C. Bergmann]

climatic rule See rule: ecogeographical rule.

n–r

decision rule *n*. An animal's response to conspecifics based on their possession of stimuli that identify them as kin or nonkin (Wilson 1987, 10).

Dzierzon's rule, Dzierzon's theory *n*. Drone Honey Bees arise from unfertilized eggs and are thus haploid; proposed by the beekeeper and Catholic priest Johann Dzierzon in 1845 (Mayr 1982, 661).

syn. arrhenotoky (Lincoln et al. 1985)
cf. parthenogenesis
Comment: "Dzierzon's rule" probably applies to all Hymenoptera that produce males; some hymenopteran species also sometimes produce diploid males.

ecogeographical rule *n*. Any generalization that describes a trend of geographical variation correlated with environmental conditions (Lincoln et al. 1985).

syn. climatic rule (Lincoln et al. 1985)
cf. law: Gloger's law, Hopkin's bioclimatic law, Rensch's laws; rule: Allen's rule, Bergmann's rule, Hesse's rule

ecological-efficiency rule *n*. Approximately 10% of the biomass of one trophic level is transformed into biomass of organisms in an adjacent trophic level that eat members of the former (Wilson 1975, 267).

Emery's rule *n*. "Species of social parasites are very similar to their host species and therefore presumably closely related to them phylogenetically" (Emery 1909 in Wilson 1975, 372, 583).

energetic-equivalence rule *n*. The rule that large animals have predictably lower population densities than small ones (Mohr 1940 and Damuth 1981 in Cotgreave 1993, 244).

Comment: Recent work shows that this rule is rarely true at the scale of local communities (Cotgreave 1993, 244).

Hamilton's rule *n*.
1. A rule that indicates when an individual should be altruistic to its kin; expressed as $k > 1/r$, where k = the benefit in fitness gain to the altruist's genes due to the altruist's helping its relative, with $k = b/c$, where b = benefit to the relative and c = cost to the helper; $1 = 1.0$; and r = the fraction of genes that an altruist and the relative it helps have in common due to descent from the same ancestor (Hamilton 1963).
   Note: Hamilton's rule is often written as $k(r) > 1$. An extreme hypothetical example of altruism is that a person should sacrifice his life to save drowning full siblings only if he can save more than two of them. This is because for full siblings, mean $r = 1/2$; k should be more than two full siblings (Brown 1975, 203).

2. Animals are selected to perform actions for which $rb - c > 0$, where r is the coefficient of relationship between the altruist and beneficiary, b is the benefit to the altruist in indirect fitness, *q.v.*, and c is the cost to the altruist in direct fitness, *q.v.* (Grafen 1984, 69).
   Note: Grafen (1984, 70) explains why the above form of the rule is preferable to other forms such as $b/c > 1/r$; Brown (1987a) discusses this subject further.
   [This definition was coined by Grafen (1982) as "Hamilton's rule."]

heart-weight rule See law: Hesse's rule.

Hebb's rule *n*. Learning and memory are based on modifications of synaptic strength among neurons that are simultaneously active (Hebb 1949 in Tang et al. 1999, 63).

Hesse's rule, "heart-weight rule" *n*. In some mammal species: the ratio of body weight to heart weight is larger in populations found in cold regions compared to those in warmer regions due to individual's maintaining a greater temperature differential between their bodies and environments in colder regions (Lincoln et al. 1985).

pigmentation rule See law: Gloger's rule.

proportion rule See rule: Allen's rule.

size rule See rule: Bergmann's rule.

Watson's rule See principle: principle of mosaic evolution.

wing rule See law: Rensch's rule.

♦ **run-scratching** See scratching: run-scratching.

♦ **runaway-sexual selection** See selection: runaway-sexual selection.

♦ **running on the spot** *n*. In some lizard species: in dominance fighting, a loser's laying flat on the ground and making running movements with its legs without touching the ground (Kitzler 1942 in Heymer 1977, 177).

syn. treading (Heymer 1977, 177)

♦ **rupestral** *adj*.
1. Referring to walls or rocks (Lincoln et al. 1985).
2. Referring to an organism's living on walls or rocks (Lincoln et al. 1985).

syn. mural, rupestrine (Lincoln et al. 1985)

♦ **rupestrine** See rupestral.

♦ **rupicole** See -cole: rupicole.

♦ **rut** *n*.
1. For example, in deer, goats, sheep: annual, or periodic, sexual excitement (*Oxford English Dictionary* 1972, entries from 1410; Immelmann and Beer 1989, 257).
   syn. estrus, heat (Immelmann and Beer 1989, 257)
2. A period during which rut occurs (Michaelis 1963).

3. The roaring, or uproar, made by an animal in rut (Michaelis 1963).

Note: "Rut" is sometimes synonymized with "courtship" in male mammals with the connotation of physiological readiness for mating (Immelmann and Beer 1989, 62).

cf. behavior: rutting behavior; courtship, estrus, foreplay

[Middle French < Latin *rugitus*, bellowing < *rugire*, to roar]

♦ **rutting behavior** See behavior: rutting behavior.

♦ **rypophage, ryophage, rhyophagy, ryophagy, rhypophagous, ryophagous, rypophagous** See -phage: rypophage.

S

◆ **s** See standard deviation.

◆ *s* See coefficient: selection coefficient.

◆ **s²** See variance: sample variance.

◆ **s-selected species** See ²species: s-selected species.

◆ **s strategist, s-strategist** See strategist: s strategist.

◆ **sabulicole** See -cole: sabulicole.

◆ **-saccharide** See molecule: -saccharide.

◆ **sadness** See facial expression: sadness.

◆ **safe** See ²group: safe.

◆ **salamandrin** See toxin: salamandrin.

◆ **saliva spreading** See behavior: saliva spreading.

◆ **salsugine** *n.*
  1. An organism that lives in habitats inundated by salt, or brackish, water (Lincoln et al. 1985).
  2. *adj.* Referring to a salsugine or its habitat (Lincoln et al. 1985).
  cf. -cole: halicole

◆ **"salt hypothesis"** See hypothesis: hypotheses regarding the Permian mass extinction: "salt hypothesis."

◆ **salt marsh** See habitat: marsh: salt marsh.

◆ **saltant** *n.* An organism that exhibits a saltation, *q.v.* (Lincoln et al. 1985).

◆ **¹saltation** *n.*
  1. A drastic and sudden mutational change (Lincoln et al. 1985).
  2. "An abrupt evolutionary change" (Lincoln et al. 1985).
  See speciation: saltatorial speciation.
  cf. ²evolution: macroevolution; mutant

◆ **²saltation** *n.* An animal's movement by leaping or bounding (Lincoln et al. 1985).
  adj. saltatorial, saltatory

◆ **saltatorial** See -orial: saltatorial.

◆ **saltatorial evolution** See ²evolution: saltatorial evolution.

◆ **saltatorial speciation** See speciation: saltatorial speciation.

◆ **saltigrade** See -grade: saltigrade.

◆ **samesense mutation** See ²mutation: samesense mutation.

◆ **sample** *n.*
  1. A subset of a population (Lincoln et al. 1985).
  2. A representative part of a larger unit used to study the properties of a whole unit (Lincoln et al. 1985).
  cf. sample: sample of observations

haphazard sample *n.* A sample taken on some arbitrary basis, generally for convenience; *e.g.,* samples may be taken before lunch and after supper or when the study animals are thought to be most active (Lehner 1979, 111).
  cf. sample: random sample
  Comment: A haphazard sample is assumed to approximate a random one in certain studies (C.K. Starr, personal communication).

observability sample *n.* A census of the number of individuals visible to an observer during some regularly scheduled period (Lehner 1979, 114).

random sample *n.*
  1. An item drawn from a population in such a way that all possible items each have the same probability of being selected (Lehner 1979, 111).
  2. A set of random samples (Lincoln et al. 1985).
  cf. sample: haphazard sample
  Comments: Random, independent samples are required by many statistical tests. "Random sample" is often mistakenly used to mean "haphazard sample," *q.v.* (C.K. Starr, personal communication).

representative sample *n.* A subset of a population that is typical of the population as a whole (Lincoln et al. 1985).

sample of observations *n.* "A collection of individual observations selected by a specified procedure" (Sokal and Rohlf 1969, 8).
  cf. observation, variable

♦ **sample mean** See mean.
♦ **sample size (N, n, *N*, *n*)** *n*.
1. The total number of independent, random data used in a statistical test (Siegel 1956, xvi).
2. The number of observations in a data set (Lincoln et al. 1985).
syn. number of replicates (Lincoln et al. 1985)
Comment: "Sample size" is more often used in connection with experiments, and "number of replicates" is more often used in connection with surveys (Chew 1984, 2).
♦ **sample variance (s²)** See variance: sample variance.
♦ **sampling** *n*. Taking a sample, *q.v.* (Lincoln et al. 1985).
oversampling *n*. One's taking a scientific sample(s) of an organism that is large enough to lower its population size and interfere with its future sampling (inferred from Perry 1995, 27).
♦ **sampling error** See error: sampling error.
♦ **sampling technique** *n*. A procedure used to obtain data.
ad libitum sampling *n*. A sampling technique in which one employs no restraints in sampling behavior; when using this technique, one usually records the more easily observed behaviors (Altmann 1974 in Lehner 1979, 113).
all-occurrences sampling *n*. A sampling technique that attempts to record as many acts of a specified type of behavior as possible during a specified sampling period (Lehner 1979, 117).
syn. event-sampling (Hutt and Hutt 1974 in Lehner 1979, 117), complete record (Slater 1978 in Lehner 1979, 117)
Comment: A "sociometric matrix" can be considered to be a type of "all-occurrences sampling" (Lehner 1979, 115).
focal-animal sampling *n*.
1. A sampling technique in which an observer focuses on an individual animal, pair, or small group of animals during a sampling period; this focal "group" receives the investigator's highest priority in recording behavior, but behavior of other interacting animals may be recorded as well (Lehner 1979, 116).
2. A sampling technique in which an observer focuses on only one animal in a group and its possible interactions with other group members (Immelmann and Beer 1989, 108).
instantaneous sampling *n*. A special type of one-zero sampling in which an observer scores an animal's behavior at predetermined points in time (Lehner 1979, 122).

syn. on-the-dot sampling (Slater 1978 in Lehner 1979, 123), point sampling (Dunbar 1976 in Lehner 1979, 123), time-sampling (Hutt and Hutt 1974 in Lehner 1979, 122).
one-zero sampling *n*. A sampling technique in which an observer scores the occurrence of a particular behavior as "1" and its nonoccurrence as a "0" during a short sampling period (Lehner 1979, 121).
syn. time-sampling (Hutt and Hutt 1974 in Lehner 1979, 121), the Hansen system (Fienberg 1972 in Lehner 1979, 121).
scan sampling *n*. A type of instantaneous sampling in which several animals are viewed at predetermined points in time and their behavior states are scored (Lehner 1979, 123).
sequence sampling *n*. A sampling technique that focuses on chains of behavior, performed by a single animal or a dyad (Lehner 1979, 118–121).
♦ **sandbathing** *n*. For example, in the Bannertail Kangaroo Rat: an individual's flexing and extending its body and sliding forward on its sides and ventrum which scent marks its territory (Randall 1987, 426).
♦ **sanguicole** See -cole: sanguicole.
♦ **sanguinivore** See -vore: sanguinivore.
♦ **sansourire** See habitat: mudflat: marisma.
♦ **saprium** See ²community: saprium.
♦ **sapro-** *combining form*
1. "Decomposition or putrefaction" (Michaelis 1963).
2. "Saprophytic" (Michaelis 1963).
[Greek *sapros*, rotten]
♦ **saprobe** *n*.
1. A saprophytic organism (Lincoln et al. 1985).
2. An organism that thrives in water rich in organic matter (Lincoln et al. 1985).
cf. -cole: cadavericole
♦ **saprobiont** See biont: saprobiont.
♦ **saprogenic** See -genic: saprogenic.
♦ **sapromyiophile, sapromyophile** See ²-phile: sapromyophile.
♦ **saprophage** See -biont: saprobiont.
♦ **saprophile** See ¹-phile: saprophile.
♦ **saprophytophage** See -phage: saprophytophage.
♦ **saproplankton** See plankton: saproplankton.
♦ **saprotroph** See -troph-: saprotroph.
♦ **saproxylobios** See -bios: saproxylobios.
♦ **saprozoic, saprozoite** See -zoite: saprozoite.
♦ **sarco-, sarc-** *combining form* Flesh; of, or related to, flesh (Michaelis 1963).
[Greek *sarx, sarkos*, flesh]
♦ **sarcode** See protoplasm.
♦ **sarcophage, sarcophagous, sarcophagy** See -phage: sarcophage.

S–Z

♦ **satellite DNA** See nucleic acid: deoxyribonucleic acid: satellite DNA.

♦ **satellite male** See male: incidental.

♦ **sathrophile, sathrophilous, sathrophily** See [1]-phile: sathrophile.

♦ **satisfice** *v.i.* Humans' inventing rules of thumb to find solutions that are good enough if not perfect (Simon in Tierney 1983).

♦ **saturation dispersal** See dispersal: saturation dispersal.

♦ **saturation level** See saturation point.

♦ **saturation mutagenesis** mutagenesis: saturation mutagenesis.

♦ **saturation point** *n.* The maximum wild density of grown individuals attained by a species (Leopold 1933 in Pulliam and Haddad 1994, 141).
syn. saturation level, saturation density (Pulliam and Haddad 1994, 141–142)
cf. carrying capacity
Comment: The saturation point of the Bobwhite Quail is about one bird per acre (Pulliam and Haddad 1994, 141). According to Leopold (1933), the saturation point is species specific and does not vary from place to place; however, the species' carrying capacity is habitat dependent.

♦ **saturation-presaturation-dispersal hypothesis** See hypothesis: saturation-presaturation-dispersal hypothesis.

♦ **saurischian pelvis** See pelvis: saurichian pelvis.

♦ **saurochore, saurochorous, saurochory** See -chore: saurochore.

♦ **saurophile, saurophilous, saurophily** See [1]-phile: saurophile.

♦ **savanna** See habitat: savanna.

♦ **savings** *n.* The amount of material that a subject retains after a rest period after it has mastered a specific task (Storz 1973, 241).
syn. relearning (Storz 1973, 241)

♦ **sawt** See [2]group: sawt.

♦ **saxatile** *adj.* Referring to an organism that lives, or grows, among rocks (Lincoln et al. 1985).
cf. -cole: saxicole

♦ **saxicole, saxicoline, saxicolous** See -cole: saxicole.

♦ **saxifrage, saxifragous** *n.* An organism that lives in rock crevices (Lincoln et al. 1985).
adj. saxifragous

♦ **saxigenous** See -cole: saxicole.

♦ *scala naturae* *n.* The continuum from the least perfect atom of matter up to "the most perfect organism — *Homo sapiens*" (Mayr 1982, 201, 242, 305).
syn. chain of being, ladder of nature (Lincoln et al. 1985), phylogenetic scale, The Great Chain of Being (Mayr 1982, 305)
Comments: The *scala naturae* was promoted by Aristotle: Nature passes from inanimate objects through plants to animals in an unbroken sequence. Although Cuvier thoroughly discredited this concept in the beginning of the 19th century, it has remained remarkably persistent, even in much biological writing (Mayr 1982, 201, 242, 305; Dawkins 1992, 263).

♦ **scalar** *adj.* Referring to something completely described by a single number (Lincoln et al. 1985).

♦ **scales of measurement** *n.* The various levels of resolution (or precision) of measurement.

  interval scale *n.* Measurements having attributes of those on an ordinal scale and, in addition, known differences between measurements (Lehner 1979, 109–110).

  nominal scale *n.* Measurement in which observations are classified into predetermined, mutually exclusive, qualitatively different categories (*e.g.,* behaviors A, B, and C) (Lehner 1979, 109).

  ordinal scale *n.* Measurement in which observations are classified into predetermined, mutually exclusive, qualitatively different categories (*e.g.,* behaviors A, B, and C), and the categories are quantitatively ordered with respect to each other (*e.g.,* behavior A > B > C) (Lehner 1979, 109).

  ratio scale *n.* Measurements having the same characteristics as those on an interval scale, and in addition, a known zero point is known (Lehner 1979, 110).

♦ **scamming** See behavior: mashing.

♦ **scan sampling** See sampling technique: scan sampling.

♦ **scansorial, scandent** See -orial: scansorial.

♦ **scarfskin** See cuticle.

♦ **scat** *n.* A fecal dropping (Lincoln et al. 1985).

♦ **scato-, skato-** *combining form* Referring to dung or feces (Brown 1956).
[Greek *skor, skatos,* dung]

♦ **scatology** See study of: scatology.

♦ **scatophage, scatophagous, scatophagy** See -phage: scatophage.

♦ **scatter diagram, scatter plot** See graph: scatter plot.

♦ **scavenger** *n.* An organism that feeds on carrion or organic refuse (Lincoln et al. 1985).
cf. [1]-phile: saprophile, sathrophile

♦ **scenopoetic axes** *pl. n.* Axes that relate to resources (such as light, temperature, and humidity), which have upper and lower tolerance values for a species (niche breadth), but do not involve competitive interactions with neighboring species in the context of a multidimensional hyperspace (= niche) (Lincoln et al. 1985).
cf. bionomic axes

♦ **scent mark, scent post** *n.* For example, in many insect and mammal species, including some bee species, ants, canids, felids, rodents, ungulates: a place where an individual deposited a communicatory odor (*e.g.,* a pheromone, allomone, or acquired smell) (Wilson 1975, 187).
v.t. mark, scent mark
cf. behavior: marking behavior; drumming: marking drumming

♦ **scent marking** See behavior: marking behavior.

♦ **scent post** See scent mark.

♦ **Schankler-immigration event** *n.* A sudden replacement of a parent species by its immigrant daughter species as indicated in the fossil record (Bakker 1985, 75).
cf. ²evolution: Lizzie-Borden evolution
[coined by Bakker (1985) after work of D. Schankler]

♦ **schizoendemic** See ¹endemic: schizoendemic.

♦ **schizogamy** See -gamy: schizogamy.

♦ **schizogenesis** See -genesis: schizogenesis.

♦ **schizogony, schizont** See parthenogenesis: fission: schizogony.

♦ **Scholasticism, scholasticism** *n.*
1. The doctrines of the Schoolmen; the predominant theological and philosophical teaching during AD 1000 to 1500 based upon the authority of Christian fathers and Aristotle and his commentators (*Oxford English Dictionary* 1972, entries from 1756).
2. A philosophical view that promoted determining the truth by logic, not by observation or experiment; two scholastic schools of thought are realism and nominalism, *q.v.* (Mayr 1982, 92).

♦ **school** See ²group: school.

♦ **school psychology** See study of: psychology: school psychology.

♦ *Schreckstoff, Schreckstoffe* See chemical-releasing stimulus: semiochemical: pheromone: alarm pheromone.

♦ **Schwerdtfeger's law of compensation and moderation** See law: Schwerdtfeger's law of compensation and moderation.

♦ *Schwirrlauf* See dance: bee dance: buzzing run.

♦ ¹**science** *n.*
1. "Knowledge acquired by study; acquaintance, mastery, of any department of learning" (*Oxford English Dictionary* 1972, entries from 1289).
2. The state, or fact, of knowing; knowledge, or cognizance, of something specified or implied; "also with wider, reference, knowledge" (more or less extensive) as a personal attribute (*Oxford English Dictionary* 1972, entries from *ca.* 1340); systematic knowledge in general (Michaelis 1963).
3. "The kind of knowledge, of intellectual activity, of which the various 'sciences' are examples" (*Oxford English Dictionary* 1972, entries from 1387).
4. Any department of knowledge in which the results of investigation have been logically arranged and systematized in the form of hypotheses and general laws subject to verification (Michaelis 1963).
5. "Knowledge of facts, phenomena, laws, and proximate causes, gained and verified by exact observation, organized experiment, and ordered thinking" (Michaelis 1963).
   Note: This definition seems appropriate for wide use when one means "science" as it is often used today. For biology, it seems appropriate to substitute "concepts" for "laws" in this definition.
6. An orderly presentation of facts, reasonings, doctrines, and beliefs concerning some subject or group of subjects: the science of theology (Michaelis 1963).
7. "Systematic knowledge in general" (Michaelis 1963).
8. "A self-correcting method of seeking the truth" (Alexander 1987, 201).
See study of: science.
cf. doctrine; law; principle; study of: philosophy: natural philosophy
Comment: Science is predictive when enough is known about a particular subject (Diamond 1987).
[Old French < Latin *scientia* < *sciens, -entis,* present participle of *scrire,* to know]
Big Science *n.* A large, expensive scientific project undertaken by a group of scientists and associates (*e.g.,* the Human Genome Project, which will cost over $1 billion U.S.) (inferred from Wade 1998, C5).

♦ ²**science** *n.* "Expertness, skill, or proficiency resulting from knowledge" (Michaelis 1963).

♦ **scientific breakthrough** *n.* A rare scientific discovery that profoundly changes the practice, or interpretation, of science or its implications for society (Bloom 1997, 2029).

♦ **scientific Darwinism** See Darwinism.

♦ **scientific revolution** See revolution: scientific revolution.

♦ **scientist** *n.* One who studies one or more areas of science (*Oxford English Dictionary* 1972, entries from 1840).
classic scientist *n.* A scientist who concentrates on the perfection of something

that already exists; he tends to work over a subject exhaustively and to defend the *status quo*; contrasted with romantic scientist (Ostwald 1909 in Mayr 1982, 831).

Comment: Firstborn children who become scientists tend to be classical ones and defend the existing paradigms during scientific revolutions (Sulloway 1982 in Mayr 1982, 831).

romantic scientist *n.* A scientist who bubbles over with ideas that have to be dealt with quickly to make room for the next ones; contrasted with classic scientist (Ostwald 1909 in Mayr 1982, 831).

Comment: Some of a romantic scientist's ideas may be superbly innovative; others are invalid, if not silly. The romantic usually does not hesitate to abandon his less successful ideas (Ostwald 1909 in Mayr 1982, 831). Later-borns who become scientists tend to be revolutionaries who propose unorthodox theories (Sulloway 1982 in Mayr 1982, 831).

♦ **scientist-day** See unit of measure: day: scientist-day.

♦ **scientist-hour** See unit of measure: scientist-hour.

♦ **scientist-year** See unit of measure: year: scientist-year.

♦ **scientometrics** See study of: -metry: scientometrics.

♦ **sciophile, sciophilic, sciophilous, sciophily** See ¹-phile: sciophile.

♦ **sciophytium** See ²community: sciophytium.

♦ **scissipary** See -pary: scissipary.

♦ **sclerite** *n.* A portion of an insect's body wall bounded by sutures (*Oxford English Dictionary* 1972, entries from 1861; Wilson 1975, 594).

♦ **scolite** See fossil: scolite.

♦ **scopa** *n., pl.* **scopae** A bee's group of specialized hairs used for transporting pollen (Michener 1974, 9–10, illustration; 143; O'Toole and Raw 1991, 186).

cf. corbicula

Comments: Colletid bees carry pollen in their crops. Fideliid and Megachilid bees have scopae on the undersides of their metasomas. The other bee families have scopae on the hind legs and sometimes on the sides of metasomas (O'Toole and Raw 1991, 186).

♦ **Scopes Trial** *n.* The 1925 court trial of John T. Scopes, who was charged with violating Tennessee's new law (the Butler Act) that forbade teaching human evolution or "any theory that denies the … Divine Creation of man" in public schools (Milner 1990, 397).

syn. Monkey Trial (Milner 1990, 397)

Comments: Scopes was a part-time substitute teacher whom the Civil Liberties Union urged to provoke a test case to challenge the Butler Act and notify authorities of his intentions (Milner 1990, 397). The fundamentalist politician William Jennings Bryan won the case by proving that Scopes broke the law. The liberal lawyer Clarence Darrow defended Scopes. The public perceived that Darrow had won by backing Bryan into an intellectual corner. Tennessee repealed the Butler Act in 1967.

♦ **Scopes II** *n.* The court trial concerning the suit against the Arkansas Board of Education by Reverend Bill McLean and the American Civil Liberties Union to enjoin the Board from enforcing Act 590, which required equal time for teaching the theories of creation science and evolution (Milner 1990, 399).

Comments: In 1982, the court ruled that "creation science" did not qualify as an alternative scientific explanation or theory and that Act 590 was an attempt to establish religion in a state-supported school in violation of the First Amendment of the U.S. Constitution (Milner 1990, 399). In that year, Mississippi passed its own similar law by an overwhelming majority.

♦ **scopulus** *adj.* Referring to crags and steep overhanging cliffs (Lincoln et al. 1985).

♦ **score-keeping mutualism** See symbiosis: mutualism: score-keeping mutualism.

♦ **scoto-, scot-** See skoto-.

♦ **scotophile, scotophilic, scotophilous, scotophily** See ¹-phile: scotophile.

♦ **scotothaxis** See taxis: scototaxis.

♦ **scout** See caste: scout.

♦ **scout dance** See dance: bee dance: scout dance.

♦ **scouting** *n.*
  1. A person's spying, or watching, in order to gain information (*Oxford English Dictionary* 1972, entries from 1553).
  2. A slave-maker ant's emerging from her nest and wandering about in search of a host-species colony (Alloway 1979, 203).
  3. A Honey Bee workers' searching for a nesting site for her swarm.

v.i. scout

cf. bee: scout

♦ **SCR** See reponse: skin-conductance response.

♦ **scramble competition** See competition: scramble competition.

♦ **scramble-competition polygyny** See mating system: polygyny: scramble-competition polygyny.

♦ **scrapers** See ²group: functional feeding group: scrapers.

♦ **scratching** *n.*

1. In some animal species: an individual's superficially wounding itself, or another, by dragging its claws or finger nails over skin; an individual's superficially wounding skin with any hard, pointed object (*Oxford English Dictionary* 1972, entries from 1474).

2. In some animal species: an individual's using its claws or nails as offensive weapons (*Oxford English Dictionary* 1972, entries from 1589).
   cf. behavior: defensive behavior

3. In some animal species: an individual's scraping lightly (a part of its body with its claws or finger nails, *e.g.,* to relieve itching) (*Oxford English Dictionary* 1972, entries from 1530).

4. In some animal species: an individual's removing earth, etc., with its claws or nails (*Oxford English Dictionary* 1972, entries from 1520).
   Comment: The *Oxford English Dictionary* (1972) also defines "scratching out" and "scratching up."

   hop-scratching *n.* A bird's springing forward with both legs and simultaneously pulling its claws backwards over a substrate (Wickler 1970 in Heymer 1977, 153).

   run-scratching, alternating scratching *n.* A bird's scratching a substrate sequentially with alternate legs (Wickler 1970 in Heymer 1977, 153).
   syn. alternating scratching (Heymer 1977, 153)

♦ **scratching over the wing** See grooming: scratching over the wing.

♦ **scream** See animal sounds: scream.

♦ **screech** See animal sounds: screech.

♦ **scrounger** *n.* An organism that obtains resources by parasitizing the behavioral efforts of a producer, thereby reducing its costs of resource gain; contrasted with producer [coined by Barnard and Sibly 1981, 543].
cf. parasitism: brood parasitism, cleptoparasitism, social parasitism
Comments: Examples of scroungers include a kleptoparastic Frigate Bird and Skua that take a host bird's prey, a Brown-Headed Cowbird and cuckoo that exploit the nest-building behavior and parental care of another bird species, a dominant female Anis that exploits communal nest-building and incubation of a subordinant, and a kleptogamist ("sneaky-rutter") male Red Deer that usurps matings from a "harem owner" while he is defending his harem against another competitor (Barnard and Silby 1981, 543).

♦ **scrumping** See behavior: mashing.

♦ **SCUBA, scuba** See self-contained underwater breathing apparatus.

♦ **SD** See standard deviation.

♦ **SE** See standard error.

♦ **search engine** *n.* A means of finding information on the World Wide Web, *q.v.* (Lawrence and Giles 1999, 107).
Comments: Search engines include Alta Vista, Excite, Hotbot, Infoseek, Lycos, NetScape Navigator, Northern Light, Yahoo.

♦ **search image, searching image** *n.* An animal's internal image of, or temporary predisposition to search for, a particular prey type or search for prey in a particular kind of location (von Uexküll 1934 in McFarland 1985, 213; Immelmann and Beer 1989, 259–260).
cf. representation; selection: apostatic selection

♦ **searching automatism** *n.* In Humans and other mammal young: an individual's repeatedly moving its head from side to side spontaneously or after receiving a tactile stimulus in its mouth region; the animal's movement stops after it receives a nipple in its mouth and closes its lips around its opening (Heymer 1977, 170).

♦ **season** *n.*

1. The time of the year when a plant flourishes, blooms, fruits, etc. (*Oxford English Dictionary* 1972, entries from *ca.* 1300).

2. A period of the year, which is naturally divided by the Earth's changing its position with regard to the sun and marked by day-length changes, particular conditions of weather, temperature, etc.; especially, spring, summer, winter, and fall in temperate climates and the rainy and dry seasons in tropical climates (*Oxford English Dictionary* 1972, entries from 1340).

3. The time of the year when an animal is in heat, pairs, breeds, migrates, is killed for food or hunted, etc. (*Oxford English Dictionary* 1972, entries from *ca.* 1400).
cf. aestival, autumnal, hibernal, serotinal, vernal

   dormant season *n.* The period from the first autumnal killing frost through the last vernal killing frost; contrasted with growing season.

   growing season *n.*

   1. In seasonal climates, the period of rapid plant growth; contrasted with dormant season (Allaby 1994).

   2. In Great Britain, the period when the mean temperature exceeds 6°C based on a daily, weekly, or monthly mean of ground, or air, temperature; contrasted with dormant season (Allaby 1994).

   3. In the U.S., the period between the last vernal and first autumnal killing frosts; the number of frost-free days; contrasted with dormant season (Allaby 1994).

S–Z

♦ **seasonal** *adj.* Referring to an organism that shows periodicity related to season (Lincoln et al. 1985).

♦ **seasonal mimicry** See mimicry: seasonal mimicry.

♦ **second Darwinian revolution** See revolution: second Darwinian revolution.

♦ **second instar** See instar: second instar.

♦ **second law of thermodynamics** See law: laws of thermodynamics: second law of thermodynamics.

♦ **second order** See order: second order.

♦ **second-order Markoff chain** See chain: Markoff chain.

♦ **secondary** See helper.

♦ **secondary consumer** See consumer: secondary consumer.

♦ **secondary contact** *n.* Reestablishment of contact between populations after a separation period (Lincoln et al. 1985).

♦ **secondary defensive behavior** See behavior: defensive behavior: behavior: secondary defensive behavior.

♦ **secondary discontinuity** See discontinuity: secondary discontinuity.

♦ **secondary drive** See drive: secondary drive.

♦ **secondary host** See ³host: secondary host.

♦ **secondary instinct** See instinct: instinctive behavior: secondary instinct.

♦ **secondary intergradation** See intergradation: secondary intergradation.

♦ **secondary monogyny** See -gyny: monogyny (comments).

♦ **secondary neoteinic** See caste: secondary neoteinic.

♦ **secondary parasitism** See parasitism: superparasitism.

♦ **secondary productivity, secondary production** See production: secondary productivity.

♦ **secondary reinforcement** See reinforcement: secondary reinforcement.

♦ **secondary robber** See thief.

♦ **secondary sex ratio** See sex ratio: secondary sex ratio.

♦ **secondary sexual character, secondary sexual characteristic** See character: secondary sexual character.

♦ **secondary social dominance** See social dominance: secondary social dominance.

♦ **secondary speciation** See speciation: secondary speciation.

♦ **secreta** See secretion.

♦ **secretion** *n.*
1. A organism gland's, or some analogous organ's, extracting a material from blood, or sap, and transforming it into a particular substance used for a body function or excreted as waste (*Oxford English Dictio-*

nary 1972, entries from 1646); an organism's, or cell's, producing and releasing a secretion (Lincoln et al. 1985).
2. The product of secretion (*Oxford English Dictionary* 1972, entries from 1732); any substance, or product, elaborated and released by a cell, or gland, and that performs a specific function (Lincoln et al. 1985).
syn. secreta (Lincoln et al. 1985)
cf. excretion

neurosecretion *n.*
1. A central-nervous-system cell's producing a physiologically regulative substance that affects target cells an appreciable distance from its secretion site (Immelmann and Beer 1989, 202, 260).
2. The product of neurosecretion of (def. 1).

♦ **sectorial chimera** See chimera: sectorial chimera.

♦ **secular** *adj.*
1. Referring to 100-year periods (Lincoln et al. 1985).
2. Referring to a process that persists for an indefinitely long time (Lincoln et al. 1985).
cf. millennial

♦ **sedentary** *adj.* Referring to an organism that is attached to a substrate or not free living (Lincoln et al. 1985).
cf. sessile

♦ **sedge** See ²group: sedge.

♦ **sedge meadow** See habitat: meadow: sedge meadow.

♦ **seed** *n.*
1. The fertilized ovule of a phanerogam (seed-bearing plant) (*Oxford English Dictionary* 1972, entries from *ca.* 1000); the fertilized, ripened ovule of a flowering plant (Lincoln et al. 1985).
2. Nontechnically, the spore of a cryptogram or a small fruit (*Oxford English Dictionary* 1972, entries from *ca.* 1000).
3. Semen (*Oxford English Dictionary* 1972, entries from *ca.* 1290).
Note: This is now a rare usage.
4. Offspring, progeny (*Oxford English Dictionary* 1972, entries from *ca.* 825).
Note: This is now a rare usage.
5. The ova of a lobster or silkworm moth; oyster spat (*Oxford English Dictionary* 1972, entries from 1662).
6. Any propagative plant, or animal, structure (Lincoln et al. 1985).
cf. gamete, propagule

♦ **seed bank** *n.* The total buried reserve of all the seeds of different ages from all plant species in a particular area (Buchmann and Nabhan 1996, 253).

♦ **seed flower** See organ: flower: seed flower.

♦ **seed shadow** *n*. A plant's seed-dispersal area containing seeds that drop from the plant and those that come directly from the plant and are dropped by seed-dispersal agents (Buchmann and Nabhan 1996, 253).

♦ **segrate** See hybrid: segrate.

♦ **segregation, law of** See law: law of segregation.

♦ **segregation distorter** See gene: segregation distorter.

♦ **segregation-distorter allele** See allele: segregation-distorter allele.

♦ **segregational load** See genetic load: segregational load.

♦ **seismic signal** See signal: seismic signal.

♦ **seismotaxis** See taxis: seismotaxis.

♦ **select** *v.t.* To favor a gene, mutation, or trait by natural selection: "Certain mutations occur more frequently when they are selected" (Shapiro 1995, 373).
syn. favor, select for (*e.g.,* natural selection *selects for* black coats in young Gorillas)
cf. selection: natural selection

♦ **selection** *n*. (See table next page.)
1. The process by which one or more kinds of things, from a group comprised of them and similar things, become more or less frequent with time [suggested by Strickberger's (1990, 120) discussion of selection of protocells].
Notes: This a very broad definition which is not meant to be synonymous with "natural selection," although "selection" is a synonym for "natural selection." Things that are subjected to selection, in the broad sense, include organic compounds, biochemical pathways, genes and traits programmed by genes, individual organisms, species, higher taxa, cultural traits in some primate species, and perhaps protocells, *q.v.* "Selection" can be defined as to whether it is biological or nonbiological, genetic or non-genetic, related or not to mate selection, or is from within or outside an organism; the size of organism group involved; ecological influences; gene-frequency distributions within a population; order; time; and so forth. Different kinds of selection are defined both as a process and a result of a process. Biological selection may have evolved from chemical selection (Strickberger 1990, 121).
2. The result of this process.
cf. fitness, habitat selection

apostatic selection *n*. Decreased natural selection against a rare-phenotype prey, increased natural selection for a common-phenotype prey, or both, because its predator has a search image for a more common phenotype; *e.g.,* individuals of some species of polymorphic insects and snails may

be favored by apostatic selection (Clarke 1962 in Mayr 1982, 593).
Comment: Apostatic selection can help to maintain several morphological forms of the same species of prey (Matthews and Matthews 1978, 330).

artificial selection *n*. Humans' selection and (repeated) breeding of desired phenotypes of a species to obtain individuals with particular characteristics for agricultural, economic, scientific, or other reasons; *e.g.,* particular dog behaviors, particular animal coat colors, increased milk or egg production, increased aggressiveness in Siamese fighting fish, or exaggerated behavior such as tumbling in pigeons (Lincoln et al. 1985; Immelmann and Beer 1989, 262).
syn. mass selection, recurrent selection, selective breeding (Lincoln et al. 1985)
cf. character: domestication character; [2]heterochrony: neoteny
Comment: Humans used a process similar to artificial selection in developing cuter, more baby-like cartoon characters and teddy bears (Morris et al. 1995, 1697).

autoselection *n*. Selection of genes based on their transmission efficiency and without reference to their phenotypic effects (Bell 1982, 503).

catastrophic selection *n*. Natural selection for a particular phenotype during an environmental catastrophe, or emergency, when thousands to millions of other phenotypes of a species succumb (Lewis 1962 in Mayr 1982, 600).
Comment: "Catastrophic selection" is a very important evolutionary process (Lewis 1962 in Mayr 1982, 600).

centrifugal selection See selection: disruptive selection.

counteracting selection *n*.
1. Natural selection that favors a trait during one stage of an organism's life but not at another (Wilson 1975, 71).
2. Selection for genes at some levels (individual, family, and population) but against the same genes at other levels (Wilson 1975, 22, 581).
cf. selection: reinforcing selection

cultural selection *n*. For example, in the Human, Japanese Macaque: nongenetic selection in which learned information or concepts are transmitted that "lead to directional evolutionary change in the social structure of a society" (Wittenberger 1981, 614); contrasted with genetic selection.
syn. nongenetic selection
cf. [3]culture; learning: social imitation; tradition

cyclical selection *n*. Natural selection in which a species is subjected to regular

SOME CLASSIFICATIONS
OF BIOLOGICAL SELECTION[a]

I. genetic selection (*syn.* genic selection, hard selection)
  A. artificial selection
    1. individual selection
    2. truncation selection
  B. natural selection
    1. classification by size of group
      a. individual selection
        (1) sexual selection
          (a) intersexual selection (*syn.* amphiclexis, epigamic selection, sexual selection)
          (b) intrasexual selection (*syn.* sexual selection)
          (c) mutual-sexual selection
          (d) runaway-sexual selection
        (2) nonsexual selection
      b. group selection
        (1) family selection
        (2) kin selection
          (a) kin selection (indirect kin selection alone)
          (b) kin selection (indirect + direct kin selection)
        (3) nonkin-group selection
        (4) demic selection
          (a) interdemic selection
          (b) intrademic selection
        (5) population selection
          (a) interpopulation selection (*syn.* intergroup selection, *Gruppenauslese*)
          (b) intrapopulation selection
        (6) trait-group selection
        (7) specific selection (*syn.* species selection with one species per "group")
          (a) interspecific selection (*syn.* species selection)
          (b) intraspecific selection
         (8) species-group selection (more than one species per "group")
          (a) intergroup selection
          (b) intragroup selection
    2. classification by ecological factor
      a. b selection
      b. d selection
      c. K selection
      d. r selection
    3. classification by resultant trait distribution
      a. directional selection (*syn.* dynamic selection)
      b. disruptive selection (*syn.* centrifugal selection)
      c. stabilizing selection (*syn.* centripetal selection, normalizing selection)
    4. classification regarding the same gene in different types of organism groups — reinforcing selection
    5. classification by order
      a. first-order selection
      b. second-order selection
    6. classification by time
      a. acyclical selection
      b. age-specific selection
      c. continual selection
      d. cyclical selection
      e. endocyclic selection
    7. classification by location of selective pressure
      a. external selection (outside an organism)
      b. internal selection (inside an organism)
    8. some other kinds of selection
      a. a selection
      b. allaesthethic selection
      c. alpha selection
      d. apostatic selection
      e. automatic selection
      f. balancing selection
      g. catastrophic selection (*syn.* catastrophic-mortality selection)
      h. climatic selection
      i. competitive selection
      j. density-dependent selection
      k. habitat selection
      l. migrational selection
      m. organic selection (*syn.* Baldwin effect)
      n. physiological selection
      o. purifying selection
      p. threshold selection
      q. truncation selection
II. nongenetic selection (*syn.* cultural selection)

[a] Endler (1992) divides natural selection into nonsexual selection and sexual selection, with mortality selection and phenotypic selection as divisions of nonsexual selection.

alternation of selective pressures that favor one and then the other of two alleles, resulting in balanced polymorphism (Wilson 1975, 71).

direct selection *n*.

1. Natural selection that affects gene frequencies as a consequence of individuals' producing direct descendants at different rates (Wittenberger 1981, 614).
2. "Natural selection (including sexual selection) based on direct fitness" (Brown 1987a, 299).

cf. selection: indirect selection

directional selection *n*. Natural selection's disproportionate elimination of a group of individuals in one end of a normal (or other kind of population distribution) (Wilson 1975, 67; Mayr 1982, 587–588).

syn. directive selection, dynamic selection, progressive selection (Lincoln et al. 1985)
cf. selection: disruptive selection, stabilizing selection
Comments: "Directional selection" is a principal process by which "directional evolution" takes place. "Directional evolution" can lead to "speciation" (Wilson 1975, 67; Mayr 1982, 587–588).

directive selection See selection: directional selection.

disruptive selection *n*. Natural selection that causes a disproportionate elimination of average, or near average, individuals, with a consequent increase in population variance (Wilson 1975, 67).

syn. centrifugal selection, diversifying selection (Mayr 1982, 588)
cf. selection: directional selection, stabilizing selection.
Comment: "Disruptive selection" is a cause of balanced polymorphism (Wilson 1975, 71). This kind of selection acted (and may still be acting) upon species with mimetic or other forms of polymorphism (Wilson 1975, 582).

double selection *n*. Natural selection involving both male competition and female choice and female competition and male choice (C.R. Darwin in Parker, 1989).

dynamic selection See selection: directional selection.

epigamic selection *n*. The part of sexual selection that involves individuals' choosing members of the opposite sex as mates; contrasted with intrasexual selection (Wilson 1975, 318, 595; Thornhill and Alcock 1983, 74–75).

syn. amphiclexis, intersexual selection, sexual selection (Lincoln et al. 1985); female choice (because most proposed cases of epigamic selection involve females'

choosing among males as mates; Spencer and Masters 1992, 295)
cf. selection: intrasexual selection, sexual selection (def. 2)
Comments: Epigamic selection can occur before, during, and after copulation (which does not necessarily involve insemination) (Thornhill and Alcock 1983, 74–75). It is difficult to demonstrate in the field (Spencer and Masters 1992, 295). Cronin (1992, 292) describes how "intersexual selection" is a terminological curiosity. It implies that males and females are competing for the privilege of being the sex that mates.
[epigamic, Huxley's (1938) term for sexual selection, *q.v.* (Wilson 1975, 318)]

▶ **pure epigamic selection, true epigamic selection** *n*. Epigamic selection that is not mixed with another kind(s) of selection, especially intrasexual selection, which often occurs with it (Wilson 1975, 318–319).

frequency-dependent selection *n*.

1. Natural selection in which the least frequent of two alleles is selected for, and the two alleles strike a balance at some frequency intermediate to their previous ones, *e.g.*, when a parasite, or predator, repeatedly shifts its preference to attack a disproportionate number of prey individuals belonging to the more common type (Moment 1962 and Owen 1963 in Wilson 1975, 71).
2. Natural selection in which a phenotype's fitness depends on the frequency of alternative phenotypes in the population (Wittenberger 1981, 615).

cf. selection: frequency-dependent sexual selection
Comment: "Frequency-dependent selection" is typical of "disruptive selection" and "sexual selection" but sometimes characterizes other forms of "natural selection" as well (Wittenberger 1981, 615).

frequency-dependent sexual selection *n*. Selection in which the mating success of an individual male varies with the frequency of his morph in his male population (Partridge and Hill 1984 in Partridge 1988, 525).

gamete selection *n*. Natural selection that results in differential mortality of gametes, or their precursors, during the period between the reductional division of meiosis and zygote formation (Wilson 1975, 64).

cf. drive: meiotic drive

genetic selection, genic selection *n*. Natural selection that causes the genetic composition of a gene pool to change within, or between, generations (Wittenberger 1981, 616).

S–Z

group selection *n.*

1. Natural selection that acts at the group level, rather than at the individual-organism level (Darwin 1859 in Wilson 1975, 106; Wynne-Edwards 1962 in Mayr 1982, 595; Wilson 1983; Brown 1987a, 300; Immelmann and Beer 1989, 123). *Notes:* Wynne-Edwards (1962) argued that group selection has been of the first importance in the evolution of social behavior, especially behavior that regulates population sizes (Immelmann and Beer 1989, 123). He tended to define groups as discrete populations (= multigenerational demes), isolated from each other except for a trickle of dispersers; these populations persisted indefinitely unless they were driven extinct by a selfish genotypes (Wilson 1992, 145). The credibility of this kind of group selection was seriously undermined by William's (1966) classic book. In the 1960s and 1970s, group selection in any form was rejected by most biologists, but in the 1970s a revised form of group selection (involving trait groups) arose that is accepted by many of today's evolutionary biologists. The importance of group selection depends largely on how one defines the group. The problem is aggravated by the fact that many models that are developed as alternatives to group selection (involving multigenerational demes) actually include a component of group selection (involving trait groups) (Wilson 1992, 146). Much of the evidence that Wynne-Edwards offered in support of group selection has been reinterpreted as resulting from kin selection (Immelmann and Beer 1989, 123). Some supporters of group selection claim that certain phenomena can be explained by individual selection, *e.g.,* characters of entire populations including aberrant sex ratios, mutation rates, dispersal distances and other factors that favor inbreeding or outbreeding, and degrees of sexual dimorphism (Mayr 1982, 595).

2. Natural selection that operates on a unit of two or more members of a lineage group; in the broad sense, group selection includes both kin and nonkin selection (Wilson 1975, 585).

3. Natural selection in which an individual organism sacrifices its reproductive output (direct fitness) in helping its group (Alcock 1979).

4. Natural selection in which the relative frequencies of genetically different groups (species, populations, or sub-units within a population) change through time (Wittenberger 1981, 616).

5. A hypothetical process of natural selection among groups of organisms that is often invoked to explain the evolution of altruism (Dawkins 1982, 288).

syn. kin selection (which is a confusing synonym, Dawkins 1982, 288)

cf. display: epideictic display

Comment: "The study of group selection has a turbulent history that has not yet subsided" (Wilson 1992, 145).

habitat selection *n.* Natural selection exerted by a habitat on the organisms that disperse to it (Bazzaz 1991 in Kramer et al. 1995, 3).

See habitat selection.

heteroselection *n.* Natural selection that favors heterozygotes (Lincoln et al. 1985).

cf. selection: homoselection

homoselection *n.* Natural selection that favors homozygotes (Lincoln et al. 1985).

cf. selection: heteroselection

indirect selection *n.*

1. Natural selection that involves effects of a focal organism (donor) on its non-descendent kin (Brown 1987a, 303). *syn.* kin selection (Wittenberger 1981, 617) [Brown (1987a, 303) discusses problems with this synonym]

2. Natural selection in which an individual helps its relatives, or other genetically similar individuals, to reproduce (Immelmann and Beer 1989, 106).

See selection: epigamic selection, kin selection.

cf. behavior: altruistic behavior; selection: direct selection, epigamic selection, kin selection

individual selection *n.* Artificial selection (*q.v.*) on individual organisms based soley on their own individual pehontypic values (Hartl 1987, 226, 257–258).

cf. selection: truncation selection

interdeme selection, interdemic selection *n.*

1. Natural selection at the level of entire breeding populations (demes) (Wilson 1975, 587).

2. Demes' differential successes due to their genetic properties (Brown 1987a, 304).

cf. selection: intrademic selection

Comment: "Interdeme selection" is one of the extreme forms of "group selection," to be contrasted with "kin selection" (Wilson 1975, 587).

internal selection *n.* Natural selection caused by an organism's internal factors, such as physiology and development (Mayr 1982, 589).

Comment: This is a confusing concept because it is impossible to partition selection into internal and external components (Mayr 1982, 589).

intersexual selection See selection: epigamic selection.

interspecific selection *n.* Natural selection caused by interactions between (among) species (Daly and Wilson 1983).

intrademic selection *n.* Natural selection within a deme (Brown 1987a, 304).
cf. selection: interdemic selection

intrasexual selection *n.* The part of sexual selection that involves same-sex individuals' competing among themselves for mates; contrasted with epigamic selection (Wilson 1975, 595; Thornhill and Alcock 1983, 74–75).
syn. male-male competition [because most proposed examples of intrasexual selection involve males (Spencer and Masters 1992, 294)], sexual selection (Lincoln et al. 1985).

***K* selection** *n.* Natural selection that favors a K-selected species, *q.v.* (MacArthur and Wilson 1967; Wilson 1975, 100, 587).
cf. selection: *r* selection
Comments: Traits thought to be favored by "*K* selection," including large size, long life, and a small number of intensively cared-for offspring, result in a stable population size. Parry (1981) reviews the definitions of "*K* selection" and "*r* selection" that are used by many authors, and he questions the usefulness of these concepts.

kin selection *n.*
1. Natural selection that involves effects of a focal organism (donor) on *both* descendent and nondescendent kin (Maynard Smith 1964 in Brown 1987a, 303). [coined by Maynard Smith 1964]
2. Natural selection within a population that increases genes that cause an organism to help its relatives (exclusive of its own offspring, according to many authors) to survive, reproduce, or both (Wilson 1975, 587; Dawkins, 1982, 289; Immelmann and Beer 1989, 165).
   cf. selection: group selection (in the strict sense), interdemic selection
3. Natural selection for activities that lower an individual's own reproduction but raise a relative's fitness (Alcock 1979, 9).
4. Selection for genes that promote altruistic behavior (McFarland 1985, 140).
5. Natural selection on an actor's genetic trait that affects the genotypic fitness of one or more individuals that are genetically related to the actor in a nonrandom way at the loci determining the trait (Michod 1982, 40).

See selection: indirect selection.
syn. kinship selection (Lincoln et al. 1985); kin-selection theory (Immelmann and Beer 1989, 166)
cf. selection: group selection; [3]theory: kin-selection theory
Comments: Wilson (1975, 587) considers "kin selection" to be one of the extreme forms of "group selection," but McFarland (1985, 128) considers Wilson's classification to be erroneous. "Kin selection" is especially important in that it suggests a biological basis for "altruism," so long as the beneficiary is related to the altruist (Barash 1982, 392). Although many biologists consider "kin selection" to be a reality, Dawkins (1982, 289) considers it to be a hypothetical concept.

mass selection See selection: artificial selection.

migrant selection *n.* Natural selection caused by variation in genotypes' tendencies to emigrate (Wilson 1975, 104); *e.g.,* if new populations are founded more consistently by individuals with gene *A* as opposed to those bearing gene *a*, gene *A* is said to be favored by migrant selection (Wilson 1975, 589).

mortality selection *n.* The effects of consistent phenotypic-specific mortality (Fisher 1930 etc. in Endler 1992, 221).
syn. natural selection (a confusing synonym, Endler 1992, 221)

mutual-sexual selection *n.* Sexual selection in which both sexes affect evolution of characters of the opposite sex (Huxley 1914 in Gherardi 1984, 389).

natural selection *n.*
1. The "preservation of favourable variations and the rejection of injurious variations" (Darwin 1859, 80–81).
   Notes: Darwin coined this term to signify the analogy between artificial selection and that done by nature apart from Humans (Hodge 1992, 213). Others objected to this term because they thought it implied conscious choice in organisms that become modified (Darwin 1959, 164–165, in Hodge 1992, 216). Alfred Russell Wallace urged Darwin to use "survival of the fittest" instead of "natural selection" because he felt the former term is free from metaphorical elements that have caused people to misunderstand "natural selection" (Hodge 1992, 217).
2. A process that causes a change in the relative frequency of particular genotypes in a population due their phenotypes' having different numbers of offspring (Wilson 1975, 67).

S–Z

Notes: Under natural conditions, this differential reproduction can stem from many causes, including genotypes' different abilities in direct competition with other genotypes; differential survival under the onslaught of parasites, predators, and changes in their physical environments; variable reproductive competence; variable ability to penetrate new habitats; and so forth. "Natural selection" is "the agent that molds virtually all of the characteristics of species" (Wilson 1975, 67).

3. A process that results in (1) a difference in trait frequency distribution among age classes, or life-history stages, in a population to an extent beyond that expected from growth and development alone, (2) a difference in trait distribution of all offspring from that of all parents beyond that expected due to phenotypic variation and inheritance alone, or (3) both (Endler 1986 in Endler 1992, 220).

Notes: Based on Endler (1986), Cooke (1986) defines natural selection as a process that causes a trait's frequency to differ between parents and offspring, not just because the trait is variable and has a genetic basis, but also because it causes an individual with it to have a different fitness than an individual without it (Endler 1986 in Cooke 1986). Cooke (1986) writes that this definition effectively eliminates the charge that "natural selection" is a tautology. Endler (1986) also presents many other definitions of "natural selection."

4. The result of the process of natural selection from gene-frequency changes to evolution of new taxa.

See selection: mortality selection, nonsexual selection, phenotypic selection.

syn. evolution (Endler 1992, 220), selection (of many authors), selection theory, survival of the fittest (Hodge 1992, 217)

cf. fitness; selection: artificial selection; survival of the fittest

Comments: The metaphorical terminology "natural selection" is on the borderline of being felicitous and unfortunate because it was strenuously resisted by the majority of Darwin's contemporaries. "They tended to personify whatever did the selecting and to insist that there was no real difference between selection by Nature and creation by God" Mayr (1982, 842). "Survival of the fittest" is a misleading synonym because, usually, it is a group of *fitter* individuals within a population that survive and reproduce; it is not often that just *one* individual

survives and reproduces. "Many evolutionary biologists are in favor of natural selection regardless of what it actually means" (Cooke 1986). "Natural selection" is used to designate both a process and a result of a process; this increases the likelihood of confusion and circular arguments regarding this term (Immelmann and Beer 1989, 260). Philosophers and scientists debate whether natural selection is a law, principle, force, cause, agent, or a combination of these things (Hodge 1992, 218).

[coined by C.R. Darwin (1859)]

nonsexual selection *n.* Natural selection minus sexual selection, *q.v.* (Endler 1992, 221).

syn. natural selection (a confusing synonym; Endler 1992, 221)

Comments: Darwin made a careful distinction between natural selection and sexual selection; Endler (1992, 221–222) and others include sexual selection as part of natural selection.

normalizing selection See selection: stabilizing selection.

organic selection See effect: Baldwin effect.

orthoselection *n.* Sustained natural selection on the members of a lineage over a long period, causing continued evolution in a given direction (Dawkins 1982, 291).

cf. selection: directional selection

Comment: "Orthoselection can create an appearance of 'momentum' or 'inertia' in evolutionary trends" (Dawkins 1982, 291).

partial selection *n.* Natural selection whose selection coefficient is between 0 and 1 (Lincoln et al. 1985).

phenotypic selection *n.* The within-generation change in trait distribution among cohorts (or the difference between the actual number of mates and the effective number of mates in sexual selection) (Endler 1992, 222).

syn. natural selection (confusing synonym; Endler 1992, 220)

pure epigamic selection See selection: epigamic selection.

r selection *n.*

1. Natural selection that favors an r-selected species (Wilson 1975, 99).

2. Selection that favors rapid rates of population increase, especially prominent in species that specialize in colonizing short-lived environments or undergo large fluctuation in population size (Lincoln et al. 1985).

cf. selection: *K* selection; ²species: r-selected species

recurrent selection See selection: artificial selection.

reinforcing selection *n.* Natural selection on two or more levels of organization (*e.g.,* on the individual, family, or population) in such a way that certain genes are favored at all levels and their spread through the population is accelerated (Lincoln et al. 1985).

runaway sexual selection *n.* Natural selection in which females choose males with particular secondary sexual characteristics over generations, causing these characteristics to become more and more elaborate with evolutionary time (*e.g.,* magnificent plumage in birds-of-paradise or peacocks or huge antlers in the extinct Irish elk) (Fisher 1958, 152).

Comment: Some mathematic models support the concept of runaway sexual selection (Spencer and Masters 1992, 297). Chimpanzee and human penile sizes might be causes of runaway sexual selection (Diamond 1996, 85).

sexual selection *n.*

1. "The process by which an individual gains reproductive advantage by being more attractive to individuals of the other sex" (Darwin 1840, 1859, 1871 in Mayr 1982, 596); selection in which there is "differential ability of individuals of different genetic types to acquire mates;" contrasted with nonsexual selection (Wilson 1975, 318, 595).
2. The "struggle between males for possession of females" that relates to the males' offspring numbers; this struggle includes physical combat, searching skill, and ability to attract females from a distance (Darwin 1859, 75, in Thornhill and Alcock 1983, 52–53).
   cf. selection: epigamic selection
3. Natural selection that involves competition among members of the same sex for mates (intrasexual selection) and selection of mates of one sex by the other sex (intersexual selection = epigamic selection) (Thornhill and Alcock 1983, 53).

See selection: epigamic selection, intrasexual selection.

cf. selection: nonsexual selection

Comments: Darwin (1859) originated the concept of sexual selection and originally described it as different from natural selection; the question being addressed determines whether or not sexual selection can be construed as a subclass of natural selection (Spencer and Masters 1992, 294, 296). Leading experts on sexual selection standardly contrast "adaptive sexual selection" (*syn.* good-sense sexual selection, Wallacean sexual selection) with "maladap-

tive sexual selection (*syn.* Fisherian sexual selection, good-taste sexual selection) (Cronin 1992, 292). Many field studies that at first seem to be examples of sexual selection are better described as nonsexual selection after further study; thus, sexual selection should be invoked only after adequate study of a mating system (Spencer and Masters 1992, 296, 301).

social selection *n.* Natural selection that is caused by members of a group in competition for food, dominance, reproductive rights, access to mates, and other resources. *Comment:* "Social selection" is likely to cause a coevolutionary spiral in that new mutations can change a social environment which produces new social pressures which favor further new mutations which produce new social pressures and so on (Williams 1966, 184, in Thornhill and Alcock 1983, 53).

species selection See hypothesis: species selection.

stabilizing selection *n.*

1. Natural selection that causes disproportionate elimination of extreme phenotypes in population, a consequent reduction of population variance, and stabilization of phenotypes around their original trait means (Wilson 1975, 67, 596).
   Note: For example, stabilizing selection can cause the proportions of individuals belonging to different age groups to remain constant for generation after generation (Wilson 1975, 595).
2. "A form of natural selection that maintains the same mean and variance among phenotypes each generation" (Wittenberger 1981, 622)

syn. centripetal selection, normalizing selection (Lincoln et al. 1985)
cf. selection: directional selection, disruptive selection

tandem selection *n.* Natural selection for a second trait once the primary one has reached a certain level of refinement (Lincoln et al. 1985).

truncation selection *n.*

1. An episode of natural selection that favors individuals of a single generation that are heterozygous for a certain fraction of their genetic loci (Wallace 1968; Wilson 1975, 72).
2. Natural selection in which a population is divided sharply into two groups, above and below a certain threshold, with those below it removed (Crow 1992, 134).
3. Artificial selection (*q.v.*) in which a person breeds only individuals which

S–Z

have a focal character state only from a tail region of a frequency distribution of this character (Hartl 1987, 226–227, illustration).

Note: Researchers performed truncation selection for seed weight in edible beans (Hartl 1987, 227).

variability selection *n.* Natural selection caused by large degrees of environmental variability (R. Potts 1998 in Selig 1999, 6).

Comment: Three distinctive areas of human evolution each coincide with a different period of increased environmental oscillation: bipedal walking, stone-tool making, and increased brain size (Selig 1999, 6).

♦ **selection coefficient, coefficient of selection** See coefficient: selection coefficient.

♦ **selection differential (S)** *n.* The difference in mean phenotype between the selected parents (μ_s) and the entire parental population (μ), designated as $S = \mu_s - \mu$ (Hartl 1987, 228).

♦ **selection pressure** *n.*

1. "Any feature of the environment that results in natural selectiong, food shortage, the activity of a predator, or competition from other members of the same sex for a mate can cause individuals of different genetic types to survive to different average ages, to reproduce at different rates, or both" (Wilson 1975, 594); evolution caused by natural selection (Immelmann and Beer 1989, 261).

2. Natural-selection intensity measured as the alteration of a population's genetic composition from generation to generation (Lincoln et al. 1985).

extrinsic selection pressure *n.* Selection pressure from outside an organisms' population (Immelmann and Beer 1989, 262).

intrinsic selection pressure *n.* Selection pressure from inside an organisms' population (Immelmann and Beer 1989, 262).

♦ **selective advantage** *n.* Increased fitness of one genotype relative to a competing one (Lincoln et al. 1985).
cf. selective disadvantage

♦ **selective attention** See attention: selective attention.

♦ **selective breeding** See selection: artificial selection.

♦ **selective disadvantage** *n.* Decreased fitness of one genotype relative to a competing one (Lincoln et al. 1985).
cf. selective advantage

♦ **selective environment** See environment: selective environment.

♦ **selective fruit abortion** See abortion: selective fruit abortion.

♦ **selective species** See ²species: selective species.

♦ **selective value** See adaptive value: fitness: genetic fitness.

♦ **self** *n.*

1. Philosophically, "that which in a person is intrinsically he (in contradistinction to what is adventitious);" a person's ego (often identified with his soul, or mind, as opposed to his body); a permanent subject of successive and varying states of consciousness (*Oxford English Dictionary* 1972, entries from *ca.* 1674).

2. What a person "is at a particular time in a particular aspect or relation;" his "nature, character (or sometimes) physical constitution or appearance; considered as different at different times;" chiefly with a qualifying adjective (his) former, later, self (*Oxford English Dictionary* 1972, entries from 1697).

3. An individual person known, or considered, as the subject of his own consciousness (Michaelis 1963).

4. An individual animal's characteristics only at the time when it is self-aware or self-conscious (Bunge 1984).

cf. awareness: self-awareness
Comment: The presence of self-awareness in nonhuman animals is controversial.

♦ **self-abuse** See masturbation.

♦ **self-awareness** See awareness: self-awareness.

♦ **self-centered DNA** See nucleic acid: deoxyribonucleic acid: selfish DNA.

♦ **self-centered DNA hypothesis** See hypothesis: self-centered DNA hypothesis.

♦ **self-concept** *n.* A person's unique way of perceiving himself (Storz 1973, 247).

♦ **self-consciousness** See consciousness: self-consciousness.

♦ **self-contained underwater breathing apparatus (SCUBA, scuba), Aqualung, aqualung** *n.* A breathing apparatus equipped with compressed-air tanks that is used underwater by divers (Michaelis 1963).

♦ **self-fertilization** See fertilization: self-fertilization.

♦ **self-grooming** See grooming: self-grooming.

♦ **self-incompatibility** See self-sterility.

♦ **self-marking** See behavior: marking behavior; automarking.

♦ **self-medicating behavior, self-medication** See pharmacognosy: zoopharmacognosy.

♦ **self-mimesis** See mimicry: mimesis: cryptic mimesis: self-mimesis.

♦ **self-pollination** See pollination: self-pollination.

♦ **self-pollution** See masturbation.

♦ **self-regulation hypothesis** See hypothesis: population-limiting hypotheses.

♦ **self-reliability** See reliability: intraobserver reliability.

♦ **self-sterility, self-incompatibility** n. In hermaphroditic organisms: an individual's inability to produce viable offspring by self-fertilization (Lincoln et al. 1985).

♦ **self-sterility gene** See gene: self-sterility gene.

♦ **selfing** See -mixis: automixis.

♦ **selfish behavior, selfishness** See behavior: selfish behavior.

♦ **selfish DNA** See nucleic acid: deoxyribonucleic acid: selfish DNA.

♦ **selfish DNA hypothesis** See hypothesis: selfish DNA hypothesis.

♦ **selfish gene** See gene: selfish gene.

♦ **selfish-herd hypothesis** See hypothesis: selfish-herd hypothesis.

♦ **selfish social behavior** See behavior: social behavior: selfish social behavior.

♦ **semantic** n.
1. Of, or pertaining to, meaning (Michaelis 1963).
2. Of, or relating to, semantics (Michaelis 1963).
[Greek *sēmantikos* < *sēmainein*, to signify]

♦ **semantic memory** See memory: fact memory.

♦ **semantics** pl. n. (construed as singular)
1. In logic, the relations between signs, or symbols, and what they signify, or denote (Michaelis 1963).
   See study of: semantics.
   syn. semasiology, semiotics (Michaelis 1963)
2. Loosely, verbal trickery (Michaelis 1963).

♦ **semantide** See character: semantide.

♦ **semantization** n. Any evolutionary change that adds to the communicative function of a behavior (Wickler 1967a in Wilson 1975, 224).

♦ **sematectonic communication** See communication: sematectonic communication.

♦ **semelparity** See -parity: semelparity.

♦ **semi-** *prefix*
1. "Not fully; partially; partly" (Michaelis 1963).
2. "Exactly half" (Michaelis 1963).
3. Occurring twice (in a specified period) (Michaelis 1963).
[Latin *semi*]

♦ **semi-cryptic polyploidy** See -ploidy: semi-cryptic polyploidy.

♦ **semidiurnal** See diurnal: semidiurnal.

♦ **semidominance** See dominance: semidominance.

♦ **semigamy** See -gamy: semigamy.

♦ **semigeographical speciation** See speciation: semigeographical speciation.

♦ **semilethal mutation** See ²mutation: semilethal mutation.

♦ **seminatural community** See ²community: seminatural community.

♦ **semiochemical** See chemical-releasing stimulus: semiochemical.

♦ **semiotics, semeiotics** n.
1. The doctrine of signs (John Locke, 17th century in Sebeok, 1965).
2. Patterned communications in all modalities (Mead 1964 in Sebeok 1968a, 1006).
   cf. communication
3. Semantics; the relation between signs or symbols and what they signify or denote (Michaelis 1963).
   See study of: semiotics.
See semantics
adj. semeiotic, semiotic, semiotical
syn. semasiology (Michaelis 1963)
[Greek *sēmeiōtikos* < *sēmeion*, diminutive of *sēma*, mark]

♦ **semioviparity, semioviparou** See -parity: semioviparity.

♦ **semiparasitism** See parasitism: semiparasitism.

♦ **semiprecocial** See -cial: precocial: semiprecocial.

♦ **semisocial** See social: semisocial.

♦ **semispecies** See ²species: allospecies.

♦ **sender** n. An individual organism that emits a signal (Dewsbury 1978, 99).
cf. signaler; system: communication system

♦ **senescence** n.
1. An individual organism's growing old; aging process (*Oxford English Dictionary* 1972, entries from 1695; Lincoln et al. 1985).
2. An individual organism's post-maturation decline in survivorship and fecundity that accompanies advancing age (Rose and Charlesworth 1980, 141, in Alexander 1987, 43).
3. An individual organism's phenotypic manifestation of deleterious effects that accumulate in its old age (Willson 1983, 32).
adj. senescent
Comment: Alexander (1987, chap. 1) discusses hypotheses of senescence.

♦ **senility** n.
1. A person's being senile; old age or the mental and physical infirmity due to old age (*Oxford English Dictionary* 1972, entries from 1791).
2. An individual organism's degenerative condition of old age (Lincoln et al. 1985).
adj. senile
syn. post-reproductive (Lincoln et al. 1985)

S–Z

♦ ¹**sensation** *n.*
1. Operation, or function, of any of a person's senses; perception by means of a person's senses; now commonly restricted to the subjective element in any operation of a person's senses, a physical feeling that a person considers apart from the resulting perception of an object (*Oxford English Dictionary* 1972, entries from 1642).
2. An individual organism's detecting stimulation (Immelmann and Beer 1989, 216); distinguished from perception.
cf. modality, perception
[Latin *sensus*, sense]

♦ ²**sensation** *n.*
1. A person's psychical affectation or a state of consciousness, due to and related to a particular condition of some portion of his body or a particular impression received by one of his sense organs (*Oxford English Dictionary* 1972, entries from 1615).
2. A person's mental feeling; an emotion; now chiefly a characteristic feeling that arises in a particular circumstance (*Oxford English Dictionary* 1972, entries from 1755).
cf. emotion

♦ ³**sensation** *n.* A person's excited, or violent, emotion (*Oxford English Dictionary* 1972, entries from 1808).

♦ ⁴**sensation** *n.* A stimulus detected by an individual organism's sense organ from which (often in combination with other sensations) it acquires a perception (McFarland 1985, 205).

♦ **sense** See modality.

♦ **sense organ** See organ: sense organ.

♦ ¹**sensibility** *n.*
1. A person's tissue's, or organ's, readiness to respond to sensory stimuli; sensitiveness (*Oxford English Dictionary* 1972, entries from *ca.* 1400).
2. For example, in plants, measuring instruments: an object's aptness to be affected by external influences; sensitiveness (*Oxford English Dictionary* 1972, entries from 1662).
3. Philosophically, a person's power, or faculty, of feeling; capacity of sensation, as distinguished from cognition and will (*Oxford English Dictionary* 1972, entries from 1838).
See sensitivity.

♦ ²**sensibility** *n.*
1. A person's mental perception, awareness of something (*Oxford English Dictionary* 1972, entries from *ca.* 1412).
2. A person's quickness and acuteness of apprehension or feeling; his quality of being easily and strongly affected by emotional influences; sensitiveness (*Oxford English Dictionary* 1972, entries from 1711).

♦ ³**sensibility** *n.* A host organism's degree of reaction to a parasite's infection (Lincoln et al. 1985).

♦ **sensitive period, sensitive phase** See period: sensitive period.

♦ **sensitivity** *n.*
1. An individual animal's (or an individual's part's) having sensation or sensuous perception (*Oxford English Dictionary* 1972, adjective entries from *ca.* 1400).
syn. sensitiveness (Michaelis 1963)
2. An individual organism's having sensation; formerly, often having sense or perception, but not reason (*Oxford English Dictionary* 1972, adjective entries from 1555).
3. An individual animal's having an intense perception or sensation (*Oxford English Dictionary* 1972, adjective entries from 1849).
4. The degree of acuteness with which an organism's organ discriminates sensations (Michaelis 1963).
5. An individual organism's ability to perceive and respond to stimuli (Heymer 1977, 144; Lincoln et al. 1985); this involves an organism's reacting to stimuli in an adaptive and reversible way (Immelmann and Beer 1989, 159).
syn. irritability, sensibility (Lincoln et al. 1985)
Note: "Sensitivity" (often called "irritability") is a fundamental characteristic of life, from single cells to multicellular organisms (Immelmann and Beer 1989, 159).
adj. irritable, sensitive
cf. sensibility

spectral sensitivity *n.* An animal's response threshold to a particular color (light frequency) (Immelmann and Beer 1989, 288).

♦ **sensitization** *n.*
1. An individual organism's being made more sensitive (to something) (*Oxford English Dictionary* 1972, entries from 1880).
2. An individual animal's lowering its response threshold to key stimuli as it responds to these stimuli (Lorenz 1981, 264); an animal's generalized alerting reaction that counteracts habituation (Gould 1982, 78).
cf. learning: habituation
3. An individual animal's showing one stimulus-response connection due to repeated activation of another (Immelmann and Beer 1989, 237).
4. An individual animal's covert motivation increase due to a threshold change (Immelmann and Beer 1989, 232).

cf. learning: conditioning: pseudoconditioning, sensitization; stimulus: priming stimulus

◆ **sensitizative phase** See period: sensitive period (comment).

◆ **sensorial** See sensory.

◆ **sensory** *adj.*
1. Referring to something that is of, or pertaining to, sensation (*Oxford English Dictionary* 1972, entries from 1749).
2. Referring to an organ, or system, that conveys, or procures, sense impulses (*Oxford English Dictionary* 1972, entries from 1799).
3. Referring to a sensorium, an entire sensory apparatus (Michaelis 1963).
syn. sensorial (Michaelis 1963)
[Late Latin *sensorium*, sensorium < *sensus*, perception < *sentire*, to feel]

◆ **sensory adaptation** See [3]adaptation: [1]physiological adaptation: sensory adaptation; [2]physiological adaptation (comment).

◆ **sensory coding** *n.* An individual animal's transformation of a stimulus effect into a neural-transmission pattern (Immelmann and Beer 1989, 264).

◆ **sensory deprivation** *n.* An animal's naturally, or artificially, being deprived of certain stimuli, which can cause abnormal development under some situations (Storz 1973, 251).

◆ **sensory modality** See modality.

◆ **sensory physiology** See study of: physiology.

◆ **sensory receptor** See receptor: sensory receptor.

◆ **sensory system** See [2]system: sensory system.

◆ **sentience** *n.* A person's condition, or quality, of being sentient; conscious; susceptible to sensation (*Oxford English Dictionary* 1971, entries from 1839).
See need: sentience.

◆ **sentinel behavior** See behavior: sentinel behavior.

◆ **separate nesting** See nesting: separate nesting.

◆ **separation syndrome, separation trauma** See syndrome: separation syndrome.

◆ **sepicole** See -cole: sepicole.

◆ **sequence analysis** See analysis: sequence analysis.

◆ **sequence homology** See homology: sequence homology.

◆ **sequence sampling** See sampling technique: sequence sampling.

◆ **sequential hermaphrodite** See hermaphrodite: sequential hermaphrodite.

◆ **sequential mutualism** See symbiosis: mutualism: sequential mutualism.

◆ **sequential patrolling** See patrol.

◆ **seral** *adj.*
1. Referring to a sere (Lincoln et al. 1985).
2. Developmental (Lincoln et al. 1985).

◆ **seratonin** See toxin: seratonin.

◆ **serclimax** See [2]community: climax: serclimax.

◆ **sere** *n.* A stage in community succession (Lincoln et al. 1985).
cf. succession: sere

◆ **serial-endosymbiosis hypothesis, serial-endosymbiosis theory** See hypothesis: endosymbiosis hypothesis: serial-endosymbiosis hypothesis.

◆ **serial homology** See homology: serial homology.

◆ **serial monogamy** See mating system: monogamy: serial monogamy.

◆ **serial polyandry** See mating system: serial polyandry.

◆ **serial polygamy** See mating system: serial polygamy.

◆ **serial polygyny** See mating system: polygyny: serial polygyny.

◆ **series** *n.* For example, in rats: an individual's complete group of intromissions that ends with an ejaculation (Dewsbury 1978, 77).

◆ **seritonergic synapse** See synapse: seritonergic synapse.

◆ **serodeme** See deme: serodeme.

◆ **serological type** See -type, type: serotype.

◆ **serology** See study of: serology.

◆ **serotaxonomy** See study of: taxonomy: serotaxonomy.

◆ **serotype** See -type, type: serotype.

◆ **serpentinophile, serpentinophilous** See [1]-phile: serpentinophile.

◆ **serule** See [2]succession: microsere.

◆ **server** *n.* A computer that supplies information to the World Wide Web (inferred from Lawrence and Giles 1999, 107).

◆ **sesquireciprocal hybrid** See hybrid: sesquireciprocal hybrid.

◆ **sessile** *adj.* In some animal species: referring to an individual, or species, that is firmly attached, permanently attached, or both, to one spot; not ambulatory (*e.g.,* a sponge, mussel, barnacle, or scale insect) (*Oxford English Dictionary* 1972, entries from 1860; Michaelis 1963).
syn. nonmotile (Lincoln et al. 1985)
cf. sedentary
Comment: Some kinds of cells (*Oxford English Dictionary* 1972) and nonanimal organisms are also described as sessile (Wittenberger 1981, 621).
[Latin *sessilis*, sitting down, stunted < *sessus*, past participle of *sedere* to sit]

◆ **Sewall-Wright effect** See drift: random drift.

S–Z

♦ **Sewall-Wright's theorem** See [3]theory: Sewall-Wright's theorem.

♦ [1]**sex** *n.*

1. Either of two divisions of organisms distinguished as males or females; males or females (of a species, etc., especially Humans) viewed collectively (*Oxford English Dictionary* 1972, entries from 1382).
2. The character of being male or female (Michaelis 1963).
3. One of the two genders of bipolar amphimicts (Bell 1982, 512).

See need: sex.

adj. sexual

syn. gender (Bell 1982, 512)

cf. display: advertisement: sex advertisement; gender; hypothesis: hypotheses of the evolution and maintenance of sex; mixis; sexual reproduction

heterogametic sex *n.* The sex that produces gametes that contain unlike sex chromosomes (*e.g.,* female lepidopterans, male flies, female birds, or male mammals); contrasted with homogametic sex (King and Stansfield 1985).

homogametic sex *n.* The sex that produces gametes that contain like sex chromosomes (*e.g.,* male lepidopterans, female flies, male birds, or female mammals); contrasted with heterogametic sex (King and Stansfield 1985).

♦ [2]**sex** *n.* The sum of all structural, functional, and behavioral characteristics that distinguish males and females (*Oxford English Dictionary* 1972, entries from 1631).

♦ [3]**sex** *n.* The activity, or phenomenon, of life that is concerned with sexual desire or reproduction (Michaelis 1963).

trans-kingdom sex *n.* Transferral of genetic information between individuals of different taxonomic kingdoms (Heinemann and Sprague 1989; Booth et al. 1989).

Comments: In the laboratory, the bacterium *Escherichia coli* transfers DNA to yeast (Booth et al. 1989). A crown-gall-tumor bacterium transfers genetic material to injured roots cells of certain plants.

cf. transposable element

viral sex *n.* Exchange of genetic information between two viruses (Goudsmit in Ryan 1997, 17).

cf. transformation

Comment: Viral sex between HIV-1A and an unknown strain might have given rise to HIV-1E (Goudsmit in Ryan 1997, 17).

♦ **sex, hypotheses of the maintenance of** See hypothesis: hypotheses of evolution and the maintenance of sex.

♦ **sex, hypothesis of the origin of** See hypothesis: Parker-Baker-Smith model of the evolution of sex.

♦ **sex, meaning of** *n.* Sexual reproduction is a way of producing great offspring variation characteristic of biological populations; this process creates individual differences from which natural selection produces new species (Weismann 1886 in Mayr 1982, 705).

cf. hypothesis: hypotheses of the evolution and maintenance of sex

Comments: The idea that sex produces great variability goes back to Herder (1784–1791 in Mayr 1982, 706). This is just one of several explanations for the evolution of sex.

♦ **sex attractant** See chemical-releasing stimulus: semiochemical: pheromone: sex pheromone.

♦ **sex change** See hermaphroditism.

♦ **sex chromosome** See chromosome: sex chromosome.

♦ **sex-conditioned gene** See gene: sex-influenced gene.

♦ **sex cycle** *n.* In diploid organisms: "the recurrent series of processes that lead to genetic recombination, such as the alternation of karyogamy and meiosis" (Lincoln et al. 1985).

♦ **sex determination** *n.* The means, or process, by which an individual organism's sex is determined (Wilson 1975, 594).

syn. sex-determining mechanism (Uzzell 1984)

cf. -gamety

environmental sex determination See sex determination: phenotypic sex determination.

genotypic sex determination *n.* Sex determination primarily by the balance of genetic factors in an individual's zygote or spore (Lincoln et al. 1985).

haplodiploidy, haplo-diploidy *n.* For example, in some rotifer, mite, tick, thrips, homopteran, and beetle species; in most hymenopteran species: sex determination in which males are derived from haploid (unfertilized) eggs and females from diploid, usually fertilized eggs (Wilson 1975, 585).

See -ploidy: haplodiploidy.

cf. rule: Dzierzon's rule

heterogamety *n.* Sex determination based on an individual's having heterozygous sex chromosomes (*e.g.,* in female lepidopterans, female birds, and male Humans) (Lincoln et al. 1985).

cf. [1]sex: heterogametic sex

homogamety *n.* Sex determination based on an individual's having homozygous sex chromosomes (*e.g.,* in male lepidopterans, male birds, and female Humans) (Lincoln et al. 1985).

See -gamety: homogamety.

cf. [1]sex: homogametic sex

maternal sex determination *n.* Sex determination by a mother's gametic idiotype (Lincoln et al. 1985).

metagamic sex determination *n.* Sex determination largely by external environmental influences (Lincoln et al. 1985).

paternal sex determination *n.* Sex determination by a father's gametic idiotype (Lincoln et al. 1985).

phenotypic sex determination, environmental sex determination *n.* Sex determination by external environmental factors, not directly by syngamy or karyogamy (Lincoln et al. 1985).

progamic sex determination *n.* Sex determination in an egg prior to its fertilization (Lincoln et al. 1985).

syngamic sex determination *n.* Sex determination as a result of gamete fusion and karyogamy (Lincoln et al. 1985).

♦ **sex-determining mechanism** See sex determination.

♦ **sex dichromatism** *n.* Sexual dimorphism, *q.v.*, in body color (Immelmann and Beer 1989, 267).

♦ **sex differentiation** *n.* The process of diverse development of characters in males and females (Lincoln et al. 1985).

♦ **sex digamety** See -gamety: sex digamety.

♦ **sex drive** See drive: sex drive.

♦ **sex hormone** See hormone: sex hormone.

♦ **sex-influenced gene** See gene: sex-influenced gene.

♦ **sex-linkage** *n.* A gene's, or character's, being on a sex chromosome (Lincoln et al. 1985).

♦ **sex-linked gene** See gene: sex-linked gene.

♦ **sex-limited character** See character: sex-limited character.

♦ **sex-limited gene** See gene: sex-limited gene.

♦ **sex monogamety** See -gamety: sex monogamety.

♦ **sex mosaic** See morph: gynandromorph.

♦ **sex pheromone** See chemical-releasing stimulus: pheromone: sex pheromone.

♦ **sex plurigamety** See -gamety: polygamety.

♦ **sex polygamety** See -gamety: sex polygamety.

♦ **sex ratio** See ratio: sex ratio.

♦ **sex-ratio factor** See factor: sex-ratio factor.

♦ **sex reversal** See hermaphroditism: consecutive hermaphroditism.

♦ **sex-role reversal** *n.* An animal of one sex's ordinarily showing behavior that is usually shown by members of the opposite sex in its taxonomic group (Wilson 1975,

326, 456); *e.g.*, male seahorses take care of fry, male jacanas take care of chicks, or females of a bark-beetle species approach males for mating (Thornhill and Alcock 1983, 442–443).
See hermaphroditism: consecutive hermaphroditism.
cf. eonism; mimicry: automimicry, female mimicry; transvestism
Comment: Sex role reversal is not considered to be abnormal behavior in these animals.

♦ **sexology** See study of: sexology.

♦ **sexual, idiotypic** *adj.* Referring to sex, sex-related characters, or the process of sexual reproduction (Lincoln et al. 1985).
See idiotypic.

♦ **-sexual**
asexual, amictic *adj.* Referring to reproduction without sex (Bell 1982, 502).
syn. zelotypic
cf. sexual reproduction
bisexual *n.*
1. A person with qualities of both sexes (Hinsie and Campbell 1976).
   See hermaphrodite.
   syn. hermaphrodite, intersexual (Lincoln et al. 1985)
2. A person who is erotically attracted to both sexes (Michaelis 1963).
   cf. sexual: monosexual
3. *adj.* Referring to a plant that has both carpels and stamens (Fernald 1950, 1571).
4. Referring to an organism species that has both males and females (Wilson 1975, 286).
   syn. digenic, unisexual (Lincoln et al. 1985)
   cf. sexual: homosexual
cosexual See hermaphrodite.
eusexual *adj.* Referring to organisms that display a regular alternation of karyogamy and meiosis (Lincoln et al. 1985).
cf. sexual: parasexual, protosexual
heterosexual *n.* A person who has a propensity for another person of the opposite sex (*Oxford English Dictionary* 1971, entries from 1901).
adj. heterosexual
cf. sexual: bisexual, homosexual
homosexual *n.* A person who has a propensity for another person of the same sex (*Oxford English Dictionary* 1971, entries from 1897).
adj. homosexual
cf. behavior: homosexual behavior; sexual: bisexual, heterosexual
intersexual See sexual: bisexual.
monosexual *n.* A person who has a propensity for only one sex (*Oxford English Dictionary* 1971).
cf. sexual: bisexual

parasexual *adj.* Referring to organisms that achieve genetic recombination by means other than regular alternation of karyogamy and meiosis (Lincoln et al. 1985).

cf. sexual: eusexual, protosexual

protosexual *adj.* Referring to organisms that achieve genetic recombination by conjugation, transduction, or lysogenation (Lincoln et al. 1985).

cf. sexual: eusexual, parasexual

transexual *n.* A person who "changes" his or her sex by acting like a person of the opposite sex, dressing in clothes of the opposite sex, electrolysis, hormone treatments, surgery, or a combination of these activities (inferred from Jacobs 1998).

Comments: About 1500 people in the U.S. surgically alter their genitals each year, and countless others live in various states of gender nonconformity (Jacobs 1998, 50–51). Surgeons can remove a man's penis and testicles and create a vagina at a cost of more than $12,000 U.S. Surgeons can create an artificial penis from forearm tissue, but the procedure of creating male genitalia for a female using parts of her body is far from perfected. A person's emotional and social aspects of his or her gender disphoria and sex change can be highly complex.

unisexual *adj.*

1. Referring to a population, or generation, composed of individuals of one sex only (Lincoln et al. 1985).
2. Referring to an individual that has either male or female reproductive organs and produces gametes of only one sex (Lincoln et al. 1985).
   syn. diclinous, dioecious, gonochoristic (Lincoln et al. 1985)
3. Referring to a species that has only females (Wilson 1975, 286).
4. Referring to a flower that possesses only male, or female, reproductive organs; contrasted with bisexual (Lincoln et al. 1985).
   syn. imperfect (Lincoln et al. 1985)
5. Referring to a species in which the number of females greatly outnumbers males (*e.g.,* the mole salamander) (Licht and Bogart 1987).

See sexual: bisexual.

cf. -oecism

♦ **sexual aggression** See aggression: sexual aggression.

♦ **sexual attraction** *n.* A person's attraction to another person with sexual arousal (inferred from Marano 1997, C1).

cf. behavior: sexual behavior; puberty; sexuality

Comments: Sexual attraction may first occur in Humans at 9 years of age (Marano 1997, C1). The mean age of first-recalled sexual attraction is 9.6 years in men and 10.1 years in women (G. Herdt inferred from Marano 1997, C6). Sexual desire for another person first occurred at the mean age of 11.2 for males and 11.9 for females, and the first sexual activity occurred at a mean age of 13.1 for males and 15.2 for females.

♦ **"sexual-attraction hypothesis"** See hypothesis: "sexual-attraction hypothesis."

♦ **sexual bargaining** *n.* A male Chimpanzee's obtaining a copulation undisturbed by other Chimpanzees by first grooming dominant males (de Waal 1989, 82).

♦ **sexual behavior** See behavior: sexual behavior.

♦ **sexual bonding** See bond: pair bond.

♦ **sexual companion** See companion: sexual companion.

♦ **sexual cannibalism** See cannibalism: sexual cannibalism.

♦ **sexual characteristic** See character: sexual characteristic.

♦ **sexual checking** *n.* A male animal's examining the reproductive receptivity of a female; *e.g.,* in the Whiptail Wallaby, this involves a male's sniffing at a female's tail and sometimes lifting her tail and pawing and licking her cloaca (Kaufmann 1974a in Wilson 1975, 472).

♦ **sexual competition** See competition: sexual competition.

♦ **sexual conflict** See [3]conflict: reproductive conflict.

♦ **sexual dimorphism** See -morphism: dimorphism: horiodimorphism.

♦ **sexual imprinting** See learning: imprinting: sexual imprinting.

♦ **sexual isolating mechanism** See mechanism: isolating mechanism: sexual isolating mechanism.

♦ **sexual isolation** See isolation: reproductive isolation.

♦ **sexual meiotic parthenogenesis** See parthenogenesis: sexual meiotic parthenogenesis.

♦ **sexual mounting** See behavior: mounting.

♦ **sexual parthenogenesis** See parthenogenesis: sexual parthenogenesis.

♦ **sexual reluctance** See coyness.

♦ **sexual reproduction** *n.*

1. In most eukaryotic organisms: reproduction involving meiosis and fertilization (Curtis 1983, 1106) and associated behaviors (Policansky 1987).
   Note: This is a common definition of sexual reproduction.

2. "Any process that unites genes (DNA) in an individual cell or organism from more than a single source" (Margulis et al. 1985, 73).

3. A process that always involves "at least one autopoietic parent that forms an individual with a genetic constitution that differs from both of the parents" (given as "sex" in Margulis et al. 1985, 69).

4. In most eukaryotic organisms: reproduction in which an individual creates offspring that are genetically different than itself using the process of meiosis (Thornhill and Alcock 1983, 33).

syn. digeny, eusexual reproduction, sex (Lincoln et al. 1985)

cf. gender, mixis, reproduction, sex

Comments: "Sexual reproduction" includes the typical kind of reproduction in which gametes fuse to form zygotes which develop into mature organisms as well as "polyembryony," "hermaphroditism" ("selfing"), and "sexual parthenogenesis" (Thornhill and Alcock 1983, 38). Bell (1982) catalogs many kinds of "sexual reproduction."

eukaryotic sex, meiotic sex *n.* Sex that leads to the alternation of haploid and diploid cells and requires fertilization and meiosis (Margulis et al. 1985, 69).

prokaryotic sex *n.*

1. "Any horizontal transmission of DNA" (given as "sex" in Zinder 1985, 8).

2. Sex that does not lead to the alternation of haploid and diploid cells and requires fertilization and meiosis; prokaryotic sex may involve genetic recombination of autopoietic entities (such as bacterial cells) and non-autopoietic entities (such as plasmids and viruses) (Margulis et al. 1985, 69).

Note: "Prokaryotic" sex probably evolved in the Archean Eon more than 3000 million years ago (Margulis et al. 1985, 69).

♦ **sexual *Retrojecktion*** See *Retrojecktion: sexual Retrojecktion.*

♦ **sexual selection** See selection: sexual selection.

♦ **sexual-selection-by-female-choice hypothesis** See hypothesis: hypotheses of divergent evolution of animal genitalia: sexual-selection-by-female-choice hypothesis.

♦ **sexual-selection-by-male-conflict hypothesis** See hypothesis: hypotheses of divergent evolution of animal genitalia: sexual-selection-by-male-conflict hypothesis.

♦ **sexual skin** *n.* In many species of Old World monkeys: a female's hairless skin area around her genital orifice that becomes reddened and swollen around the time of her estrus (Wilson 1975, 155, 229–231).

Comments: The sexual skin produces sex pheromones in Rhesus Macaques and possibly releases male copulatory behavior in primates that have it. "Males of Hamadryas Baboons and other species possess a permanently colored rump, which they present when greeting and appeasing other males" (Wilson 1975, 155, 229–231).

♦ **sexuality** *n.*

1. An individual animal's being sexual or having sexual characters (*Oxford English Dictionary* 1972, entries from 1800). *Note:* This kind of "sexuality" now refers to nonanimal organisms as well.

2. A person's having sexual powers, or capability of sexual feeling (*Oxford English Dictionary* 1972, entries from 1879).

3. An individual organism's having an appearance of a particular sex or being distinguished by sex (*Oxford English Dictionary* 1972, entries from 1908; Michaelis 1963).

4. A person's sexual activity and feelings, especially after maturation of his or her sexual organs (inferred from Marano 1997, C6).

cf. puberty, sexual attraction

hypersexuality *n.* Excessively frequent sexual behavior, *e.g.,* in some species of domestic animals, wild animals subjected to prolonged social isolation, or some animals with certain brain lesions (*e.g.,* a cat with a damaged amygdala) (Immelmann and Beer 1989, 137).

relative sexuality *n.* A gamete's being able to act as a male or a female when mated to other gametes (Lincoln et al. 1985).

♦ **sexuparity** See -parity: sexuparity.

♦ **sexy-son hypothesis** See hypothesis: sexy-son hypothesis.

♦ **shacking** See behavior: mashing.

♦ **shag time** See estrus: *Rammelzeit.*

♦ **shaking** *n.*

1. A person's trembling his body, limbs, due to physical infirmity or disease; quivering with emotion; shivering with cold; quaking with fear (*Oxford English Dictionary* 1972, verb entries from *ca.* 1100).

2. A person's grasping and seizing (a person) roughly to and fro, especially as a punishment or in a struggle (*Oxford English Dictionary* 1972, verb entries from *ca.* 1300).

3. An individual animal's vibrating its body (to throw off water, wet snow, dust, etc. or to remove stiffness due to repose); figuratively, bestirring itself, arousing itself to activity; also with complement: to shake itself free, loose, awake, sober; and with the construction from (*Oxford English Dictionary* 1972, verb entries from 1390).

S–Z

4. A person's clasping and moving another person's hand to and fro, as a customary salutation or expression of friendly feeling; shaking hands with (another) (*Oxford English Dictionary* 1972, verb entries from 1535).

5. An individual animal's "worrying" (mangling, or killing, by biting, vibrating, or tearing with its teeth) an opponent or prey (*Oxford English Dictionary* 1972, verb entries from 1565).

6. A person's moving a person, or thing, up and down, from side to side, or both rapidly (*Oxford English Dictionary* 1972, verb entries from 1581).

7. A person's moving his limbs, or body, from side to side, up and down, or both, often due to nervous agitation due to fear or horror; also a state of this shaking (*Oxford English Dictionary* 1972, verb entries from 1624).
   n. the shakes

8. A person's rousing up (an animal) to activity; shaking up a horse (*Oxford English Dictionary* 1972, entries from 1853).

9. In many bird and mammal species: a comfort movement that often occurs when an animal leaves water or finishes dust bathing (Heymer 1977, 164).

◆ **shaking dance** See dance: bee dance: shaking dance.

◆ **shaking to death** *n.* In the Hedgehog, Tasmanian Devil; many canid, mustelid, and viverrid species: an individual's vigorously shaking its head from side to side while holding a prey in its jaws (Immelmann and Beer 1989, 268).

◆ **shale barren** See habitat: shale barren.

◆ **shale-barren opening** See habitat: shale barren: shale-barren opening.

◆ **shale-barren woodland** See habitat: shale barren: shale-barren woodland.

◆ **sham pecking** See behavior: displacement behavior: sham pecking.

◆ **sham preening** See behavior: displacement behavior: sham preening.

◆ **shame** See abasement.

◆ **Shannon-Weaver formula** *n.*

1. An equation used to calculate the amount of potential information in a message (Shannon and Weaver 1949 in Wilson 1975, 194, 195).

2. See index: species-diversity index: Shannon-Weiner Function.
   syn. Shannon-Weiner formula (erroneous synonym), Shannon-Weaver Index (Lincoln et al. 1985)

◆ **Shannon-Weaver function, Shannon-Weiner function** See index: species-diversity index: Shannon-Weaver function.

◆ **shaping** *n.* An investigator's, or animal trainer's, using operant conditioning to cause an animal to perform behavior not originally in its repertory (*e.g.,* bicycle riding by bears, ball balancing by seals, waltzing by elephants, or ping-pong playing by pigeons) (Immelmann and Beer 1989, 268–269).

autoshaping *n.* An animal's undergoing Pavlovian conditioning: behaving toward a stimulus (*e.g.,* light) as if it were another stimulus (*e.g.,* food) because it was exposed to both stimuli simultaneously for a number of pairings (Immelmann and Beer 1989, 24).
   v.t. autoshape
   cf. learning: conditioning: Pavlovian conditioning
   Comment: For example, when a pigeon is simultaneously presented with a lighted key (button) and grain of food, it continually pecks at the key as if it were a grain (McFarland 1985, 324). It does not peck at its key "in order" to obtain an actual grain, as it would if it were showing instrumental conditioning. It just persists in pecking at its key.

◆ **Shapiro-Wilk Statistic** See program: Shapiro-Wilk Statistic.

◆ **shareable parental investment** See investment: shareable parental investment.

◆ **Shelford's law of tolerance** See law: law of tolerance.

◆ **shell-searching behavior** See behavior: shell-searching behavior.

◆ **shelter-building behavior** See behavior: shelter-building behavior.

◆ **shelter-seeking behavior** See behavior: shelter-seeking behavior.

◆ **shivering thermogenesis** See thermogenesis: shivering thermogenesis.

◆ **shoal** See ²group: shoal.

◆ **shock disease** *n.* In Snowshoe Hares: "a hormone-mediated idiopathic hypoglycemia that can be identified by liver damage and disturbances in several aspects of carbohydrate metabolism" that occurs during population explosions (Green et al. 1939 in Wilson 1975, 83).
   Comment: Similar diseases occur in rats and the cockroach *Nauphoeta cinerea* (Wilson 1975, 83).

◆ **short-interspersed nucleotide element** See nucleotide element: short interspersed nucleotide element.

◆ **short-term memory** See memory: short-term memory.

◆ **short-term optimality** See optimality: short-term optimality.

◆ **shoulder enhancement** *n.* The tendency of a person, especially a man, to enlarge his shoulders with hair erection, clothes, and other materials (Leyhausen 1969 in Heymer 1977, 159).

♦ **shoulder reaction** See reaction: shoulder reaction.

♦ **show** *v.t.* To demonstrate a phenomenon by revealing that it occurs in a statistically significant way.
syn. indicate

♦ **showing respect** *n.* In Apes: an individual's making a gesture that communicates respect to an object or recipient, *e.g.*, a Human's making such a gesture toward a guest of high rank or a Chimpanzee's laying a slightly folded hand palm on its head as a greeting to a dominant conspecific (Kortlandt 1958, Eibl-Eibesfeldt 1970 in Heymer 1977, 22).
cf. behavior: respectful behavior

♦ **showing-the-nest behavior** See behavior: showing-the-nest behavior.

♦ **shredder** See group: functional feeding group: "dead-plant shredder," "live-plant shredder."

♦ **shrewdness** See ²group: shrewdness.

♦ **shuffling** See ruffling.

♦ **shyness** See learning: conditioning: avoidance conditioning.

♦ **sib** See sibling.

♦ **sib mating** See mating: sib mating.

♦ **sib pollination** See pollination: sib pollination.

♦ **sibicide, siblicide** See -cide: sibicide.

♦ **sibling** *n.*
1. A brother or sister (Michaelis 1963).
2. An organism that is genetically closely related to another one, especially one that is a brother or sister (Lincoln et al. 1985).
syn. sib (Lincoln et al. 1985)

♦ **sibling companion** See companion: sibling companion.

♦ **sibling species** See ²species: sibling species.

♦ **sibship** *n.* The relationship between (among) siblings (Lincoln et al. 1985).

♦ **siccadeserta** *n.* "Dry sand deserts" (Lincoln et al. 1985).

♦ **siccocole, siccicole, siccole, siccocolus** See -cole: siccocole.

♦ **sickle dance** See dance: bee dance: sickle dance.

♦ **siderophile** See ¹-phile: siderophile.

♦ **sideways building** *n.* A individual swan's moving only its neck and head while gathering nesting material that it has previously laid by its nest (Petzold 1964 in Heymer 1977, 188).

♦ **σ²** See variance: population variance.

♦ **sigmoid curve** See curve: sigmoid curve.

♦ **sign stimulus** See ²stimulus: sign stimulus.

♦ **sign vehicle** *n.* The effective part of a signal, *q.v.*; the part that affects a recipient's behavior (Immelmann and Beer 1989, 270).
syn. sign stimulus (in some cases, Immelmann and Beer 1989, 270)
Comment: A "sign vehicle" can include stimuli that act over an extended period and produce a physiological change in an animal (from Morris' [1924] communication theory; Immelmann and Beer 1989, 270).

♦ **signal** *n.*
1. A sign, or notice, that one hears or sees, given especially to warn, direct, or convey information (*Oxford English Dictionary* 1972, entries from 1598).
2. In communication theory, "the physical embodiment of a message" (Michaelis 1963).
3. Any organism behavior that conveys information from one individual to another, whether or not it serves other functions as well (Wilson 1975, 594).
4. The behavior emitted by a sender in a communication system, *q.v.* (Dewsbury 1978, 99).
See alarm; behavior: appeasement; chemical-releasing stimulus: semiochemical; facial expression: play signal; social signal.
n. signaling
v.t. signal
cf. animal sounds: call; display; message; melotope; sign vehicle
Comments: "A signal specially modified in the course of evolution to convey information is called a display" (Wilson 1975, 594). Dawkins and Krebs (1978 in Dawkins 1986, 126) claim that some animal signals do not convey information. Researchers disagree regarding the definition of signal (McGregor 1997, 754).

aggressive signal *n.* For example, in the African Elephant, Human: a sender's signal that he, or she, will be aggressive if challenged by a receiver (Poole 1989, 140).
Comment: A male African Elephant announces his intention to fight with glandular secretions, urine marking, and vocalizing (Poole 1989, 140).

alarm signal *n.* An individual's visual, vocal, or chemical signal that it gives during periods of potential danger and that can alert other individuals to danger (Wittenberger 1981, 612); *e.g.*, White-Tailed Deer tail flag, snort, stamp the ground with forefeet, and release metatarsal-gland pheromones as parts of their alarm signals (LaGory 1987, 20).

analog signal, analogic signal *n.* A signal that varies in intensity, frequency, or both, as a function of an individual's underlying motivational state, thereby

S–Z

transmitting quantitative information about such variables as the sender's mood, distance of the target, and so forth (*e.g.*, the variation in Guppy body markings that correlates with a fish's sexual-arousal level) (Sebeok 1962 in Wilson 1975, 178, 585; Immelmann and Beer 1989, 121).
syn. graded signal (Wilson 1975, 178)

appeasing signal See behavior: appeasement.

assessment signal *n.* A signal that is necessarily correlated with an animal's fighting ability, or resource-holding potential, and is therefore "honest;" *e.g.*, the roaring of Red Deer stags where it appears that stags with a low fighting ability simply cannot roar at a high rate (Clutton-Brock et al. 1979; Maynard Smith and Harper 1988 in Dawkins and Guilford 1991, 866).

chemosignal See chemical-releasing stimulus: semiochemical.

composite signal *n.* A signal made up of two or more simpler signals (Wilson 1975, 581).

conventional signal *n.* A signal that indicates a signaler's membership in a particular category of signalers and is not necessarily a reliable indicator of signaler quality, *e.g.*, the plumage "badges of status" of many bird species (Roper 1986; Maynard Smith and Harper 1988 in Dawkins and Guilford 1991, 865–866).

counterfeit signal *n.* A deceptive signal (Wickler 1968, 241).
cf. deception; lie; signal: deceitful signal

deceitful signal *n.* A signal that is incorrectly identified by its receiver, thus one that may enhance the sender's fitness and decrease the receiver's fitness as a result of its misinterpretation (Weldon 1985, personal communication); *e.g.*, the signal of a female of a predatory firefly (= "femme fatale") that lures heterospecific male fireflies to her which she then eats (Alcock 1983, 449) or an animal's exaggerating its body size which gives it an advantage over a rival (McFarland 1985, 114).
syn. dishonest signal (Dawkins and Guilford 1991, 865)
cf. deception; femme fatale; signal: honest signal
Comment: Dawkins and Guilford (1991, 865) argue "that dishonest signals are likely to be a widespread component of signaling systems concerned with quality advertisement and that previous discussions have neglected a vital evolutionary consideration, the cost to the receiver of eliciting and evaluating honest signals."

digital signal *n.* A signal that is "turned either on or off, without significant inter-

mediate gradation" (Sebeok 1962 in Wilson 1975, 178; Wilson 1975, 582).
syn. discrete signal (Wilson 1975, 178)
cf. analog or graded signal
Comment: A digital signal is most perfectly represented in the act of simple recognition, particularly during courtship (Sebeok 1962 in Wilson 1975, 178; Wilson 1975, 582).

direct signal See signal: digital signal.

graded signal See signal: analog signal.

honest signal *n.*

1. A signal whose level is related to perfect information regarding what a "receiver wants to know" (Zahavi 1975, 1987 in Dawkins and Guilford 1991, 865, 870).
   Notes: Some biologists think that this kind of honest signal is unlikely to occur (Dawkins and Guilford 1991, 870).

2. In the Human, elephants, probably many ungulate species: a signal that is correctly identified by its receiver and, thus, one that enhances the receiver's fitness (Weldon 1985, personal communication; Poole 1989, 151).
   Note: In Humans, facial "beauty" in both sexes, men's muscles, and women's body fat appear be honest signals (Diamond 1996, 82).

3. A signal that is uncheatable because it is maintained by receiver cost (Dawkins and Guilford 1991, 870).
   Note: Future research is needed to ascertain whether this is a reasonable definition of "honest signal."

4. A signal that is a necessary and perfectly reliable indication of a quality that gives the best estimate of outcome (*e.g.*, resource-holding potential) (Dawkins and Guilford 1991, 870).

5. A signal that is a necessary, but not an entirely reliable, indication of a quality that gives the best estimate of outcome (Dawkins and Guilford 1991, 870).
   Note: Not all investigators agree with this definition.

6. One of a group of signals in a conventional system that are worth paying attention to on the average (Dawkins and Guilford 1991, 871).
   Note: This definition trivializes the concept of "honesty" because it could then be used to describe a system in which cheating and "mimicry" are rife (Dawkins and Guilford 1991, 871).

syn. legitimate signal (Weldon, personal communication)
cf. signal: deceitful signal; signaler: deceitful signaler
Comment: Definitions 1, 3, 4, 5, and 6 are based on my perceptions of definitions of

"honesty" by Dawkins and Guilford (1991) that I extrapolated to the term "honest signal."

invitation signal *n.* An individual ant's signal that causes a conspecific individual to join in performing an act, such as following an odor trail or cooperating in adult transport (Hölldobler and Wilson 1990, 639).

legitimate signal See signal: honest signal.

odor signal, odour signal *n.* A signal comprised of airborne molecules that a particular animal can perceive (Murlis 1986).
syn. odor plume (Murlis 1986)
cf. chemical-releasing stimulus: pheromone
Comment: Murlis (1986) gives definitions of many factors related to "odor signals."

persuasion signal *n.* signal employed by individuals with "conflicting interests" that enables them to gain some advantage (Wittenberger 1981, 619).
Comment: "Persuasion signals" are often given during aggressive interactions or "courtship" (Wittenberger 1981, 619).

play signal See facial expression: play signal.

poster signal *n.* In some bird species: a color patch on wings, the tail, or both shown by an individual in flight (*e.g.,* the large white wing patches of mocking birds) (Wilson 1975, 213).
cf. speculum
Comment: Poster patches may be used as assembling signals (Wilson 1975, 213).

postlinguistic signal *n.* A paralanguage signal that has been in service after the evolutionary origin of true language (human verbal communication) (Wilson 1975, 556).
cf. paralanguage; signal: prelinguistic signal

prelinguistic signal *n.* A paralanguage signal that was in service before the evolutionary origin of true language (human verbal communication) (Wilson 1975, 556).
cf. paralanguage; signal: postlinguistic signal

pursuit-deterrent signal *n.* A prey individual's communicating its alertness and ability to escape predator's attack to the predator (Leal 1999, 521).
Comment: The lizard *Anolis cristatellus* uses a pushup display as a pursuit-deterrent signal (Leal 1999, 521).

seismic signal *n.* In the Blind Mole Rate, White-Lipped Frog: an animal-produced, substrate-borne vibration used in communication (Rado et al. 1987, 1249).

status signal *n.* An animal's trait (morphology, behavior, body-part posture, etc.) that communicates socially relevant information to conspecifics such as social rank

or mating experience, *e.g.,* the silver-gray markings of a dominant male gorilla in a troop, larger song repertoires in older males with more breeding experience of some songbird species, or the wing-patch size of male Red-Winged Blackbirds (Schenkel 1947 in Wilson 1975, 280; Immelmann and Beer 1989, 290).
syn. status sign (Wilson 1975, 280)

♦ **signal-detection theory** See [3]theory: signal-detection theory.

♦ **signal entropy** *n.* The information in a signal, which is partly or totally transmitted (Wilson 1975, 195).

♦ **signal interceptor** See eavesdropper.

♦ **signal jump** *n.* In many pomacentrid- and some blennid- and tripterygiid-fish species: an individual's courtship behavior that involves its swimming upward with its rostral end pointing slightly downward, then dropping rapidly downwards and thus inscribing an arc (Heymer 1977, 165).
syn. diving display (Spanier 1970 in Heymer 1977)

♦ **signal receiver** See operator.

♦ **signal stimulus** See stimulus: releaser.

♦ **signaler, signaller** *n.* An organism that produces a signal, *q.v.*
cf. signal

deceitful signaler *n.* An organism that sends a deceitful signal, *q.v.*
syn. cheater, illegitimate signaler
cf. signal: deceitful signal

honest signaler *n.* An organism that sends an honest signal, *q.v.*
syn. legitimate signaler

♦ **signaling** *n.* One cell's affecting itself, or another cell, via a chemical messenger that it produces.

endocrine signaling *n.* A cell's releasing a chemical that affects a distant target cell(Campbell 1996, 912–917).

paracrine signaling *n.* A cell's releasing a chemical messenger (= a local regulator) into interstitial fluid, thereby affecting a nearby target cell (Campbell 1996, 913).
Comments: Local regulators include carbon monoxide, growth factors, histamine, interleukins, neurotransmitters, nitric oxide, and prostaglandins. Histamine and interleukins coordinate a person's body defenses (Campbell 1996, 915). Carbon monoxide and nitric oxide can act as both neurotransmitters and paracrine signals.

synaptic signaling *n.* A neuron's dispatching its transmitter into a synapse, a junction with a target cell (Campbell 1996, 914).

♦ **signaling behavior** See display.

♦ **signature sequence** *n.* A taxon's specific base sequence in ribosomal RNA, or

S–Z

other nucleic acid, that a researcher compared to the specific base sequence of another taxon at a comparable location in its nucleic acid (Campbell 1996, 508).

♦ **significance level** See p: "stipulated p."

♦ **significance test** See statistical test: significance test.

♦ **significant** See statistically significant.

♦ **Signor-Lipps effect** See effect: Signor-Lipps effect.

♦ **silent allele** See allele: silent allele.

♦ **silent bared-tooth display** See display: bared-teeth display: silent bared-tooth display.

♦ **silent DNA** See nucleic acid: deoxyribonucleic acid: silent DNA.

♦ **silicole, silicolous** See -cole: silicole.

♦ **silvicole, silvicolous** See -cole: silvicole.

♦ **silvics** See study of: silvics.

♦ **simple digital organism** See program: simple organism: complex digital organism.

♦ **simplicity** See parsimony.

♦ **Simpson's index of diversity** See index: species-diversity index: Simpson's index of diversity.

♦ **simulation** *n*. A model's imitation, or mimicking, of a system's behavior (Lincoln et al. 1985).

♦ **simultaneous choice test** See test: choice test.

♦ **"simultaneous-endosymbiosis hypothesis** "See hypothesis: endosymbiosis hypothesis: "simultaneous-endosymbiosis hypothesis."

♦ **simultaneous hermaphrodism** See hermaphrodism: simultaneous hermaphrodism.

♦ **simultaneous hermaphrodite** See hermaphrodite: simultaneous hermaphrodite.

♦ **simultaneous polyandry** See mating system: simultaneous polyandry.

♦ **simultaneous polygamy** See mating system: simultaneous polygamy.

♦ **simultaneous polygyny** See mating system: simultaneous polygyny.

♦ **SINE** See nucleotide element: short interspersed nucleotide element.

♦ **singing** See animal sounds: song.

♦ **single African-origin hypothesis** See hypothesis: African-origin hypothesis of human mitochondrial DNA evolution.

♦ **single-impact hypothesis** See hypothesis: single-impact hypothesis.

♦ **single-male group** See harem.

♦ **single-tail test** See statistical test: one-tailed test.

♦ **singular** See ²group: singular.

♦ **singular breeding** See breeding: singular breeding.

♦ **sink habitat** See habitat: sink habitat.

♦ **sire** *n*. In animal breeding, the male parent (Lincoln et al. 1985).
cf. dam

♦ **sister** *n*. A female that has one or both of the same parents as one or more other individuals of either sex (Michaelis 1963).
Comments: Due to terminological problems regarding haplodiploid animals, Page and Laidlaw (1988), classify sisters with a genetic-pairing terminology that includes "super sister," "full sister," "super half sister," "paternal half sister," and "maternal half sister." Their classification differs from a commonly used physical-pairing terminology (which I do not include here); their paper gives further clarification.

full sister *n*. In diploid species: a female that shares the same parents with another female and thus has a pedigree coefficient of relationship, *G*, of 0.5 with her (Michaelis 1963; Page and Laidlaw 1988, 944).
syn. whole sister (Page and Laidlaw 1988, 944)
Comments: G is 0.5 because the sisters share an average of 0.25 of their maternal and paternal genomes, for a total average of 0.5. A full sister in a haplodiploid species is called a "super sister," *q.v.*

half sister *n*. A female that shares only one parent with another female (Michaelis 1963; Page and Laidlaw 1988, 944).

▸ **maternal half sister** *n*. In diplodiploid and haplodiploid species: a female that shares only her mother with another female and thus has a *G* of 0.25 with her (Page and Laidlaw 1988, 944).
Comment: G is 0.25 because these sisters share an average of 0.25 of their maternal genome and share no paternal genes.

▸ **paternal half sister** *n*.
1. In diplodiploid species: a female that shares only her father with another female and thus has a *G* of 0.25 with her (Page and Laidlaw 1988, 944).
Note: G is 0.25 because sisters share an average of 0.25 of their paternal genome and share no maternal genes.
2. In haplodiploid species: a female that shares only the same drone mother with another female but has a different father and thus has a *G* of 0.25 with her (Page and Laidlaw 1988, 945).
Note: The fathers of these paternal half sisters are different genomes from the same drone mother, but on the average such fathers have only 0.25 of their genes in common. Therefore, these daughters have 0.25 of their genes in common via their fathers and no genes in common via their mothers.
cf. sister: half sister: super half sister

▶ **super half sister** *n*. In haplodiploid species: a female that has the same father with another female and thus a *G* of 0.5 with her sister (Page and Laidlaw 1988, 944).

Comment: This *G* occurs because their haploid father gives each daughter an essentially identical set of genes.

▶ **super sister** *n*. In haplodiploid species: a female that shares the same mother and same replicated genome from a common drone (and thus the same mother of this drone) with another female and thus has a pedigree coefficient of relationship, *G*, of 0.75 with her sister (Page and Laidlaw 1988, 944).

♦ **sister group** See ²group: sister group.

♦ **sister species** See ²species: sister species.

♦ **sister-species group** See ²group: sister group.

♦ **sisyphean genotype** See -type, type: genotype: sisyphean genotype.

♦ **site** *n*. "Place; position; habitat" (Lincoln et al. 1985).

restriction site *n*. The place on a DNA molecule where a restriction enzyme cuts the molecule (Campbell et al. 1999, 366).

♦ **site-directed mutagenesis** See method: site-directed mutagenesis.

♦ **site tenacity** See *Ortstreue*.

♦ **sitting** See ²group: sitting.

♦ **situation analysis** See analysis: situation analysis.

♦ **size-fitness-correlation hypothesis** See hypothesis: size-fitness- correlation hypothesis.

♦ **size rule** See rule: Bergmann's rule.

♦ **skato-** See scato-.

♦ **skatobios** See -bios: skatobios.

♦ **skein** See ²group: skein.

♦ **skewness** *n*. Statistically, the degree of asymmetry, or departure from symmetry, of a frequency distribution (Sokal and Rohlf 1969, 113).

Comment: In a skewed distribution, the mean and median do not coincide (Sokal and Rohlf 1969, 113).

negative skewness *n*. A unimodal curve's having a long thin tail on its left side (Lincoln et al. 1985).

syn. left-skewed curve (Sokal and Rohlf 1969, 113)

positive skewness *n*. A unimodal curve's having a long thin tail on its right side (Lincoln et al. 1985).

syn. right-skewed curve (Sokal and Rohlf 1969, 113)

♦ **skill memory** See memory: skill memory.

♦ **skin-conductance response** See response: skin-conductance response.

♦ **Skinner box** See puzzle box: Skinner box.

♦ **skiophile, skiophilous, sciophilous, skiophilic** See ¹-phile: umbrophile.

♦ **skoto-, skot-, scoto-, scot-** *combining form* "Darkness" (Michaelis 1963). [Greek *skotos*, darkness]

♦ **skotophile, skotophilous, skotophilus, skotophilic, skotophily** See ¹-phile: skotophile.

♦ **skotophilic** *adj*. Referring to the dark phase of a light-dark cycle (Lincoln et al. 1985).

cf. ¹-phile: skotophile

♦ **skotoplankton** See plankton: skotoplankton.

♦ **skototaxis, scototaxis, skototactic** See taxis: skototaxis.

♦ **skulk** See ²group: skulk.

♦ **skull** See ²group: skull.

♦ **slack** See habitat: dune slack.

♦ **slave-making** See parasitism: social parasitism: slavery.

♦ **slavery** See parasitism: social parasitism; symbiosis: slavery.

♦ **sleeking** *n*. A bird's depressing its feathers against its skin which reduces heat retention by its feathers (Immelmann and Beer 1989, 101).

cf. feather ruffling

♦ **sleep** *n*.

1. In many animal species: an individual's unconscious state, or condition, that it regularly and naturally assumes, during which its nervous-system activity is almost, or entirely, suspended and recuperation of its powers occurs; slumber; repose (*Oxford English Dictionary* 1972, entries from *ca*. 825).

Note: This historical definition is erroneous because an animal's nervous system is very active during sleep.

2. In many animal species: a period, or occasion, of slumber (*Oxford English Dictionary* 1972, entries from *ca*. 1200).

3. In many plant species: an individual's closing its flowers, leaves, or both, especially at night (*Oxford English Dictionary* 1972, entries from 1757).

cf. movement: rapid eaf movement, sleep movement

4. For example, in some amphibian, bird, fish, insect, mammal, mollusc, and reptile species: a period of prolonged inactivity generally associated with organization by a daily, or tidal, rhythm; an elevation of thresholds for various responses; and occurrence in a species-specific (usually safe) site and in a species-specific posture (Dewsbury 1978, 58).

cf. brain waves

5. A resting animal's active process characterized by a cyclic succession of different psychophysiological phenomena (Kelly 1991, 793).

Comments: Kelly (1991) presents a helpful classification of kinds of sleep, and I include his definitions below, along with more historical ones. The details of sleep vary so much among species "that a comprehensive definition covering all cases remains elusive" (Immelmann and Beer 1989, 271). The hypothalamous mediates sleep in Humans, and probably also in birds and other mammals (Campbell 1996, 1018). Why Humans need sleep remains puzzling (Patricia Churchland in Campbell 1996, 777–778). Sleep might restore basic neurotransmitters because neurons are highly active during sleep. Several centers in a person's cerebrum and brain stem control arousal and sleep (Campbell 1996, 1018).

A CLASSIFICATION OF SLEEP[a]

I. stage-1 sleep (active sleep, AS, desynchronized sleep, paradoxical sleep, rapid-eye-movement sleep, REM sleep)
II. slow-wave sleep
   A. stage-2 sleep
   B. stage-3 sleep (delta sleep)
   C. stage-4 sleep (delta sleep)

[a] After Kelley (1991).

active sleep (AS) See sleep: stage-1 sleep.
AS See sleep: stage-1 sleep.
deep sleep (DS) *n.* Human sleep in which delta brain waves predominate and periods of a desynchronized EEG reminiscent of wakefulness occur (Campbell 1996, 1018).
delta sleep See sleep: stage-3 and -4 sleep.
desynchronized sleep See sleep: stage-1 sleep.
paradoxical sleep See sleep: stage-1 sleep.
quiet sleep *n.* Sleep in which brain waves are relatively long and of high amplitude; contrasted with active sleep (Immelmann and Beer 1989, 272).
rapid-eye-movement sleep See sleep: stage-1 sleep.
REM sleep See sleep: stage-1 sleep.
slow-wave sleep See sleep: slow-wave sleep (table).
stage-1 sleep *n.*
1. Animal sleep in which brain waves are relatively short and of low amplitude; contrasted with quiet sleep (Immelmann and Beer 1989, 272).

2. In birds and mammals: animal sleep during which its brain waves are desynchronized and have a low-voltage, fast-activity pattern that is similar to that of its waking state (Kelly 1991, 794).

syn. active sleep, AS, desynchronized sleep, paradoxical sleep, rapid-eye-movement sleep, REM sleep (Kelly 1991, 794)

Comments: Stage-1 sleep is not definitely known in amphibians and reptiles (Kelly 1991, 794–797). In Humans, stage-1 sleep begins *in utero*, and a person's stage-1 sleep generally declines with age. This is the deepest sleep stage. A person's brain consumes more oxygen during stage-1 sleep than during intense mental or physical exercise during his waking state. Most human sleepers who awaken from stage-1 sleep recall dreaming, whereas less than half who awaken from slow-wave sleep recall dreaming. In human males, penile erection frequently occurs during stage-1 sleep and usually bears little relationship to dream content.

[Stage-1 sleep, after the fact that a person goes into this kind of sleep during a sleeping bout after a period of nonsleep. He then goes into stage-2, -3, and -4 sleep (Kelly 1991, 794, fig. 51-2). Desynchronized sleep, after the brainwave forms of an animal in stage-1 sleep. Rapid-eye-movement sleep, after that fact that an animal in stage-1 sleep shows slow, rolling eye movements with bursts of rapid eye movements.]

stage-2, -3, -4 sleep See sleep (table).
♦ **sleep movement** See movement: sleep movement.
♦ **sleep nest** See nest: sleep nest.
♦ **sleeping position** *n.* An animal's body position that it adopts during sleep and which varies among species (Immelmann and Beer 1989, 273).
cf. posture
♦ **sleuth** See [2]group: sleuth.
♦ **sliding** *n.*
1. A person's passing from one place, or point, to another with a smooth and continuous movement, especially through air or water or along a surface (*Oxford English Dictionary* 1972, verb entries from *ca.* 950).
2. An individual animal's slipping, losing its foothold (*Oxford English Dictionary* 1972, entries from *ca.* 1225).
3. In many reptile species, etc.: an individual's gliding or crawling, now rarely used (*Oxford English Dictionary* 1972, entries from *ca.* 1300).
Note: The "slider" is a basking turtle that "slides into water at least sign of danger" (Conant 1958, 59).

4. A person's moving, going, or proceeding unperceived, quietly, or stealthily; stealing, slinking, creeping, or slipping away, into or out of a place, etc. (*Oxford English Dictionary* 1972, entries from 1382).

5. A person's causing (something) to move in a smooth, gliding motion; pushing (something) over a level surface (*Oxford English Dictionary* 1972, entries from *ca*. 1537).

6. In some penguin species: an individual's sliding on its ventral side, propelling itself forward with its feet (Heymer 1977, 156).
syn. tobogganing (Prevost 1961 in Heymer 1977)

♦ **SLiME** See system: subsurface lithoautotrophic microbial ecosystem.

♦ **"slow *ad hoc* mimicry"** See mimicry: "*ad hoc* mimicry: slow *ad hoc* mimicry."

♦ **slow-wave sleep** See sleep: slow-wave sleep (table).

♦ **smack** See ²group: smack.

♦ **small fry** See animal names: small fry.

♦ **smatch** See animal sounds: smatch.

♦ **smiling** See facial expression: smiling.

♦ **smog** *n*. A combination of smoke and fog, especially as seen in highly populated industrial and manufacturing areas (Michaelis 1963).
[Blend of SM(OKE) and (F)OG]

♦ **smoothing** *n*. In statistics, a person's averaging data in time, or space, in order to compensate for random errors, or variations, on a scale smaller than that presumed significant to a given problem (Lincoln et al. 1985).

♦ **smuth** See ²group: fluther.

♦ **S(n)** See index: species richness.

♦ **sneak copulation** See copulation: sneak copulation.

♦ **sniff yawning** See yawning: sniff yawning.

♦ **snood** *n*. A turkey's pendulous skin over its beak (Zuk 1984, 32).
Comments: In wild turkeys, females prefer males with longer snoods as mates, and males are less likely to engage in aggression with other males with larger snoods (Richard Buchholz in Anonymous 1996f, 34, illustration).
[possibly from a fillet, band, or ribbon for confining a female person's hair < Old English *snōd*]

♦ **snoring** See behavior: snore.

♦ **snort** See animal sounds: snort.

♦ **snuffling** *n*.

1. A person's drawing air into his nostrils to smell something; smelling at a thing (*Oxford English Dictionary* 1972, entries from *ca*. 1600).

2. In adult Wolves: an individual's probing "its nose and mouth around the lip area of another in an apparent attempt to learn whether it has eaten recently" (Mech 1970 in Wilson 1975, 227).
Comments: The *Oxford English Dictionary* (1972) gives many more definitions for human snuffling.

♦ **social** See sociality.

♦ **social attraction** *n*. A mutual attraction between conspecific social animals, or between heterospecific animals, especially in mixed groups, *q.v.* (Immelmann and Beer 1989, 274).

♦ **social behavior** See behavior: social behavior.

♦ **social bucket** *n*. An individual ant's carrying liquid food between her mandibles and sharing it with her nest mates by mouth-to-mouth contact (Hölldobler and Wilson 1990, 643).
cf. trophallaxis

♦ **social buffering** See buffering: social buffering.

♦ **social-cohesion hypothesis** See hypothesis: social-cohesion hypothesis.

♦ **social commensalism** See symbiosis: commensalism.

♦ **social companion** See companion: social companion.

♦ **social conflict** See ¹conflict.

♦ **social conventions** *n*. An individual animal's device by which it curtails its own fitness and thereby increases its group survival (Wynne-Edwards 1962 in Wilson 1975, 109).
cf. display: epideictic display; selection: group selection
Comment: This is a controversial concept, especially as it might relate to nonhuman animals.

♦ **social Darwinism** See Darwinism: social Darwinism.

♦ **social deprivation** See experiment: isolation experiment.

♦ **social distance** See distance: individual distance.

♦ **social dominance** See dominance: social dominance.

♦ **social donorism** See altruism (def. 2).

♦ **social drift** See drift: social drift.

♦ **social drive** See drive: bonding drive.

♦ **social dynamics** *n*. All of the ways that a species-characteristic social structure is maintained or varied (Immelmann and Beer 1989, 274).

♦ **social facilitation** See ²facilitation: social facilitation.

♦ **social-facilitation hypothesis** See hypothesis: turtle-homing hypotheses: social-facilitation hypothesis.

♦ **social grooming** See grooming: social grooming.

♦ **social group** See ²group: social group.

♦ **social hierarchy** See dominance hierarchy.

♦ **social homeostasis** See homeostasis: social homeostasis.

♦ **social hormone** See chemical-releasing stimulus: semiochemical: pheromone.

♦ **social imitative learning, social imitation** See learning: social imitation.

♦ **social inhibition** See inhibition: social inhibition.

♦ **social mimicry** See mimicry: social mimicry.

♦ **social mutualism** See symbiosis: mutualism.

♦ **social organization, class A to class E levels** *n.* A classification used for African antelopes and buffaloes (Jarman 1974 in Wilson 1975, 485–486).

♦ **social parasitism** See parasitism: social parasitism.

♦ **social play** See play: social play.

♦ **social psychology** See study of: psychology: social psychology.

♦ **social regulation** See regulation: social regulation.

♦ **social releaser** See stimulus: releaser: social releaser.

♦ **social selection** See selection: social selection.

♦ **social stimulation** See ²stimulus: social stimulation.

♦ **social stomach** See crop (comment).

♦ **social-subordination hypothesis** See hypothesis: population-limiting hypotheses: social-subordination hypothesis.

♦ **social system** See ¹system: social system.

♦ **social tendency** See drive: bonding drive.

♦ **social-tolerance principle** See principle: social-tolerance principle.

♦ **social tonus** *n.* In group-living mammals: mild tension due to continual stimulation of social encounters (Immelmann and Beer 1989, 278).

♦ **social tool use** See tool use: social tool use.

♦ **social vigilance** See vigilance: social vigilance.

♦ **sociality** *n.*
1. A person's state, or quality, of being social; a person's social intercourse, or companionship, with his fellows, or the enjoyment of this (*Oxford English Dictionary* 1972, entries from 1649).
2. In many animal species: an individual's living in an enduring society (Michaelis 1963; Immelmann and Beer 1989, 274).
3. In some plant species: an individual's growing compactly or in clumps (Michaelis 1963).
4. In some invertebrate and vertebrate species: sociobiologically, a species' having one or more of the following behaviors that occur in a conspecific group: cooperation in caring for young; reproductive division of labor, with more or less sterile individuals working on behalf of individuals engaged in reproduction; and at least two generations of life stages that are capable of contributing to colony labor and that live together (Wilson 1975, 398).
5. "The combined properties and processes of social existence" (Wilson 1975, 398, 595).
6. Broadly, behavior with interactions between conspecifics, or between heterospecific symbiotic animals, in which communication is similar to conspecific communication (*e.g.,* aggression, courtship, or parental care) (Dewsbury 1978, 90; Immelmann and Beer 1989, 274).

adj. social

cf. social, society, solitary

Comments: Classification of kinds of sociality is in a state of flux (Costa 1997, 152–153).

A CLASSIFICATION OF INSECT SOCIALITY[a]

I. solitary
II. presocial
   A. subsocial
   B. parasocial
      1. communal
      2. quasi-social
      3. semisocial
III. eusocial

[a] Costa (1997, 152).

communal sociality *n.* For example, in some bee and wasp species: sociality in which individuals cooperate in nest construction but rear brood separately (Michener 1974, 372; Wilson 1975, 398). *Comment:* Batra (1966, 375) describes this as "colonial" or "communal behavior."

eusociality *n.*
1. Nesting behavior in which a nest-founding female lives long enough to cooperate in division of labor with a group of her mature daughters (used as "eusocial" in Batra 1966, 375).
2. In the Snapping Shrimp; some wasp and bee species: sociality with all of these traits: cooperation in young care; reproductive division of labor, with more or less sterile individuals working

ROUTES TO EUSOCIALITY IN INSECTS[a]

DEGREE OF SOCIALITY	COOPERATIVE BROOD CARE	REPRODUCTIVE CASTES	OVERLAP BETWEEN GENERATIONS
Parasocial route to eusociality			
solitary (presocial)	–	–	–
communal (presocial)	–	–	–
quasi-social (presocial)	+	–	–
semisocial (presocial)	+	+	–
metasocial (presocial)	+	–	+
eusocial (primitive and truly [= "superorganismal"])	+	+	+
Subsocial route to eusociality			
solitary (presocial)	–	–	–
primitively subsocial (presocial)	–	–	–
intermediate subsocial I (presocial)	–	–	+
intermediate subsocial II (presocial)	+	–	+
eusocial (primitively and truly)	+	+	+

[a] After Wilson (1975, 398). Costa (1997, 152) gives a traditional classification of insect sociality which is similar to this classification. A plus (+) indicates the presence of a character; a minus (–), its absence.

on behalf of individuals engaged in reproduction; and overlapping of at least two generations of life stages capable of contributing to colony labor (Wilson 1975, 398; Duffy 1996; Weiss 1996a, A1).
syn. primitive eusociality (Michener 1974, 373)

3. For example, in some vespid-wasp, ant, and termite species; possibly the Naked Mole Rat; the Honey Bee: eusociality (def. 2) with the additional characteristic that queens are externally morphologically distinguishable from female workers (Wilson 1975, 381, 398, 584; Jarvis 1981).
Note: Primitively eusocial bee species have no to little food exchange among adults, compared to highly eusocial species which have extensive food exchange among adults (Michener 1974, 372).
syn. high eusociality, hypersociality (Batra 1966, 1977), true eusociality, true sociality [Greek *eu*, good + Latin *socius*, companion; coined as "hypersocial" by Batra (1977, 291)]

hypersociality See sociality: eusociality.
parasociality *n.* For example, in presocial insect species: sociality in which individuals have one or two of the three traits of eusociality (def. 2) (Wilson 1975, 592).
presociality *n.* Sociality in which there is any degree of social behavior beyond sexual behavior yet short of eusociality

(Wilson 1975, 398); presociality includes "parasociality" and "subsociality."
Comment: "Presociality" is a problematic term because it suggests no sociality, that presocial animals are evolving toward eusociality, or both (Costa 1997, 152–153).
primitive eusociality See sociality: eusociality.
quasi-sociality *n.* In some presocial insect species including bees: sociality in which same-generation conspecific individuals use the same composite nests and cooperate in brood care but do not exhibit caste differentiation or overlap of generations (Michener 1974, 373; Wilson 1975, 398, 593).
semisociality *n.* In some insect species: sociality in which same-generation, conspecific individuals cooperate in brood care and have a reproductive division of labor; *i.e.,* some individuals are primarily egg layers and some are primarily workers (Michener 1958 in Batra 1966, 375; Michener 1974, 374; Wilson 1975, 398, 594).
subsociality *n.* In some insect species including some pentatomids, silphids, passalids: parental care of young (Batra 1966, 375; Wilson 1975, 346, 398, 428, 596).
Comment: "Subsociality" is a problematic term because it suggests no sociality, that subsocial animals are evolving toward eusociality, or both (Costa 1997, 152–153).
♦ [1]**socialization** *n.*
1. A person's "development of those patterns of social behavior basic to every

normal human being" (Mead 1963 in Wilson 1975, 159).

cf. enculturation

2. A process by which a child learns the different behaviors that are expected from him and accepted by his society (Storz 1973, 261).

3. A person's acquisition of social traits (Clausen 1968, Williams 1972 in Wilson 1975, 159).

4. An individual animal's total behavior modification due to its interaction with other members of its society, including its parents; this includes morphogenetic socialization, learning of species-characteristic behavior, and enculturation (Poirier 1972 in Wilson 1975, 159, 595).

5. In some animal species: an individual's learning behavior that is "strongly influenced by the particularities of its social experience" (Wilson 1975, 12–13, 159; Immelmann and Beer 1989, 276).

Note: This process becomes increasingly prominent as one moves upward phylogenetically into more intelligent species (Wilson 1975, 12–13, 159; Immelmann and Beer 1989, 276).

♦ **²socialization** *n.* An individual wild animal's being tamed so that it forms a primary social relationship with a person (*e.g.*, in dog and wolf pups) (Frank and Frank 1982, 95).

♦ **socialization phase** See phase: socialization phase.

♦ **socies** See ²community: socies.

♦ **¹society** *n.* In ecology, a group of organisms living together under the same physiographic conditions and influenced and characterized by a principal species (Michaelis 1963).

♦ **²society** *n.*

1. A person's association with his fellows, especially in a friendly, or intimate, manner; companionship; fellowship (*Oxford English Dictionary* 1972, entries from 1531).

2. An individual animal's living in association, in company, or in intercourse with conspecifics; the system, or mode of life, adapted by a conspecific group for the purpose of harmonious coexistence, mutual benefit, defense, etc. (*Oxford English Dictionary* 1972, entries from 1553 for Humans, 1794 for nonhumans).

3. The system of community life in which people form a continuous and regulatory association for their mutual benefit and protection (Michaelis 1963).

4. The body of people that compose such a community (def. 3); also, all people collectively, regarded as having certain common characteristics and relationships (Michaelis 1963).

5. A cooperating group of conspecific individual organisms (Alverdes 1927, Allee 1931, Darling 1938 in Wilson 1975, 8, 595), drawn together, kept together, or both, by social attraction (Immelmann and Beer 1989, 122).

6. An aggregation of socially intercommunicating, conspecific individual organisms that is delimited from other such societies by limited or no communication with them (Altmann, 1965 in Wilson, 1975, 8).

7. A group of conspecific individual organisms organized in a cooperative manner, extending beyond mere sexual activity (Wilson 1975, 7, 595).

8. For example, in some species of colonial invertebrates: a group of physically united zooids that might even be considered a superorganism or an organism (Wilson 1975, 383).

9. For example, in some insect species, the Naked Mole Rat: a eusocial group with reproductive and sterile castes that might even be considered a superorganism (Wilson 1975, 383).

cf. ²group, sociality

[Old French *societe* < Latin *societas, -tatis* < *socius*, friend]

Some Classifications of Societies

I. Classification by effect of demographic variables
   A. casual society (fusion-fission society)
   B. demographic society

II. Classification by acceptance of newcomers
   A. closed society
   B. open society

III. Level of sociality (see sociality)

IV. Classification by attention structure
   A. acentric society
   B. centripetal society

V. Classification with regard to conflict among members
   A. despotic society
   B. egalitarian society

VI. Level of organization
   A. disorganized society
   B. organized society

acentric society *n.* For example, in Gibbons, Langurs, the Pata: an attention structure (*q.v.*) comprised of a society in which the females and young separate from males in times of inter- and intraspecific aggression; contrasted with centripetal society (Wilson 1975, 517).

allelarkean society *n.* An independent, dense, nonnomadic, and civilized human society (Lincoln et al. 1985).

autarkean society *n.* A simple, independent, nomadic, or sparsely distributed human society (Lincoln et al. 1985).

casual society, causal group *n.* In some primate species: a temporary group formed by individuals within a society which is unstable, being open to new members and losing old ones at a high rate (*e.g.,* feeding groups of monkeys within a troop and groups of playing children). *cf.* ²society: demographic society (Wilson 1975, 133, 580).

centripetal society *n.* An attention structure (*q.v.*) in which the members of society predominantly watch a dominant male and shift their positions according to his approach or departure, and adjust their aggressive behavior toward others according to his responses (Chance 1967, Chance and Jolly 1970 in Wilson 1975, 517).

closed society *n.*
1. A society that communicates relatively little with nearby societies of the same species and seldom if ever accepts immigrants (Dice 1952, 276; Wilson 1975, 17).
2. In some ant, bee, and termite species; many rodent species: a society in which members do not tolerate the presence of nonsociety members and thus do not allow interchange of members with other societies (Immelmann and Beer 1989, 122).
syn. closed anonymous group (Heymer 1977, 29)
cf. society: open society

demographic society *n.* "A society that is stable enough through time, usually owing to its being relatively closed to newcomers, for the demographic processes of birth and death to play a significant role in its composition" (Wilson 1975, 133, 582). *cf.* ²society: casual society

egalitarian society *n.* A society in which "benefits are divided roughly equally or in proportion to the risk or effort taken" (Vehrencamp 1983, 667, in Hand 1986).

euclonal society *n.* A society comprised of genotypically identical modular growth units (ramets) that can follow an independent existence if separated from their parent organism (Lincoln et al. 1985). *cf.* ²society: paraclonal society, pseudoclonal society

insect society *n.*
1. "In the strict sense, a colony of eusocial insects" (ants, termites, eusocial wasps, or eusocial bees) (Wilson 1971, 465).

2. In the broad sense, any group of presocial, or eusocial, insects (Wilson 1971, 465).

kormogene society *n.* An organism colony in which different individuals are organically connected to one another (Lincoln et al. 1985).

modular society *n.* A society with repeated members or modular units (Lincoln et al. 1985).

▶ **continuous modular society** *n.* A society with physically united modular units (Lincoln et al. 1985).
syn. colony (Lincoln et al. 1985)

▶ **discontinuous modular society** *n.* A society with physically separated modular units (Lincoln et al. 1985).
syn. aggregation, population (Lincoln et al. 1985)

multimale society *n.* A society in which more than two males have approximately equal rank (Wilson 1975, 291). *cf.* ²society: unimale society

open society *n.* Especially in mammals that migrate in groups for long distances: a society in which members tolerate the presence of conspecifics, or heterospecifics, that were not originally members of their society and thus allow interchange, which might be continuous, of members with other societies (Immelmann and Beer 1989, 122). *cf.* ²group; ²society: closed society

▶ **anonymous group** *n.* For example, in some bird and insect species, many rodent species: an open group whose members do not recognize each other as individuals but are socially attracted to one another (Heymer 1977, 29; Immelmann and Beer 1989, 16, 122).
syn. open anonymous group (Immelmann and Beer 1989, 16)
cf. aggregation, individualized group

▶ **individualized group** *n.* In many primate species, some carnivore species: an open group whose members recognize each other as individuals (Immelmann and Beer 1989, 15).
cf. ²group: anonymous group

organized society *n.* A group of conspecific organisms with these five attributes: a complex communication system, division of labor based on specialization, cohesion (members' tendency to stay together in time and space), permanence of group members (little emigration and immigration), and a resistance to immigration of nonsociety members (Eisenberg 1965 in Dewsbury 1978, 91).

paraclonal society *n.* A society comprised of genotypically identical modular units that cannot regenerate, or follow, an independent existence if separated from their parent organism (Lincoln et al. 1985). *cf.* ²society: euclonal society, pseudoclonal society

pseudoclonal society *n.* A society comprised of nongenotypically identical modular units that are so closely related and coordinated that they can be regarded as ecologically and functionally equivalent to a clone (Lincoln et al. 1985). *cf.* ²society: euclonal society, paraclonal society

synhesma *n.* "A swarming society" (Lincoln et al. 1985).
See synhesma.

unimale society *n.* A society with one highest ranking male; contrasted with multimale society.

♦ **socio-** *combining form*
1. "Society; social" (Michaelis 1963).
2. "Sociology; sociological" (Michaelis 1963).
[French < Latin *socius*, companion]

♦ **sociobiological-selfish behavior, sociobiological selfishness** See behavior: selfish behavior.

♦ **sociobiology** See study of: biology: behavioral ecology, sociology.

♦ **sociocline** See cline: sociocline.

♦ **sociogenesis** See -genesis: sociogenesis.

♦ **sociogenetics** See study of: genetics: sociogenetics.

♦ **sociogram** See -gram: sociogram.

♦ **sociohormone** See chemical-releasing substance: semiochemical: pheromone.

♦ **sociology** See study of: sociology.

♦ **sociometric matrix** See experimental design, sampling technique.

♦ **sociometry** See -metry: sociometry.

♦ **sociophobia** See phobia (table).

♦ **sociotomy** See swarming: budding.

♦ **sodefrin** See chemical-releasing stimulus: semiochemical: pheromone: sodefrin.

♦ **sodomy** *n.*
1. "Unnatural" human sexual intercourse, especially between males, and including human-nonhuman animal intercourse (bestiality) (*Oxford English Dictionary* 1972, entries from 1297).
2. An act of sodomy (def. 1) (*Oxford English Dictionary* 1972, entries from 1593).
3. Human male penile insertion into the rectum of another person of either sex (Storz 1973, 263).
cf. copulation: homosexual copulation: pseudocopulation; fellatio
Comment: Laws of a number of states of the U.S. also include oral-genital acts and bestiality as sodomy (Storz 1973, 263).
[Old French *sodomie* < Late Latin Sodom, to whose people sodomy was imputed]

♦ **soft inheritance** See inheritance: soft inheritance.

♦ **soft science** See study of: science: soft science.

♦ **soft selection** See selection (table).

♦ **software** *n.* Computer programs (Lincoln et al. 1985).
cf. hardware, program

♦ **soil** *n.*
1. Finely divided rock mixed with organic matter, constituting the portion of Earth's surface in which plants grow (Michaelis 1963).
Note: Soil contains many kinds of organisms including bacteria, algae, arthropods, salamanders, and small mammals.
2. In engineering geology, all unconsolidated materials above bedrock; the regolith (Bates and Jackson 1984).
[Old French *soile, sueil* < Latin *solium*, a seat, mistaken for *solum*, the ground]

♦ **soil climax** See ²community: climax: edaphic climax.

♦ **soil erosion** *n.* The process by which soil is carried away from its source over time.
Comment: Natural erosion in forests can be high during and after storms. Human activities increase soil erosion in many places.

♦ **solar emjoule** See emergy.

♦ **solar mass** *n.* The mass of the sun; 1.9891 × 10³⁰ kg (Mitton 1993, 421).

♦ **solar transformity** *n.* The solar energy required to make 1 joule of some type of energy; expressed as emjoules per joule (sej/J) (Odum et al. 1988, 28).
cf. emergy

♦ **soldier** See caste: soldier.

♦ **soliciting behavior** See behavior: soliciting behavior.

♦ **soliciting song** See animal sounds: soliciting song.

♦ **solipsism** *n.*
1. In philosophy, a person's belief that his, or her, self is the only thing that exists; therefore, his, or her, reality is subjective (Michaelis 1963).
2. In philosophy, the belief that the only consciousness that a person knows about is his own.
Comment: "Solipsism" is contrasted with "objectivism." Bateson (1991, 828) states, "Although it is logically defensible, most people would treat hard-line solipsism as an absurd stance. Certainly any committee that made a case for the treatment of animals on this basis would be treated with derision by the media and with contempt by the public at large."
[Latin *solus*, alone + *ipse*, self]

♦ **solitary** *adj.*
1. Referring to a bee species in which females live alone in their nests and

ordinarily do not interact with other adults, neither their offspring nor others of their own generation (Michener 1974, 38, 374).

2. Referring to a bee species in which females maintain their nests alone but may still be present in their nests when their progeny mature (Michener 1974, 38).

3. Referring to an animal "species in which individuals form no enduring groups or pair bonds" and live most of their lives away from conspecifics; an animal species that is not social (Immelmann and Beer 1989, 281).
Note: Most animal species are solitary.
syn. nonsocial
cf. social, sociality

4. Referring to a female parasitic wasp that lays only one egg on, or into, a particular host individual.
cf. gregarious

♦ **Sollwert** *n.* The predetermined reference value (= efference copy) in an animal's feedback circuit, *q.v.* (Immelmann and Beer 1989, 243).
cf. principle: reafference principle
[German *Sollwert*, should-be value]

♦ **soma** *n.* The parts of an individual organism's body that are mortal and that "work for the preservation of genes in the germ-line" (Dawkins 1982, 287).
cf. germ-line
[Greek *sōma*, body]

♦ **somatic** *adj.*
1. Of, or referring to, the human body; bodily; corporal; physical (*Oxford English Dictionary* 1972, entries from 1775).
2. Referring to the mortal part of an individual organism's body, as opposed to its germ-line (Dawkins 1982, 294).

♦ **-somatic** *adj.*
macrosomatic *adj.* In many mammal species: referring to a species that possesses a highly developed sense of smell (Immelmann and Beer 1989, 178).
cf. -somatic: microsomatic
microsomatic *adj.* In many mammal species, especially primates: referring to a species that possesses a poorly developed sense of smell (Immelmann and Beer 1989, 184).
cf. -somatic: macrosomatic

♦ **somatic character** See character: somatic character.
♦ **somatic effort** See effort: somatic effort.
♦ **somatic-mutation theory** See [3]theory: somatic-mutation theory.
♦ **somatic nervous system** See [2]system: nervous system: somatic nervous system.
♦ **somatic parthenogenesis** See parthenogenesis: diploid parthenogenesis.

♦ **somatogamy** See -gamy: pseudogamy.
♦ **somatogenesis** See -genesis: somatogenesis.
♦ **somatogenic variations** *pl. n.* The various reactions of an individual organism's somatic tissues, or of its body as a whole, to external influences and the effects of use or disuse (Lincoln et al. 1985).
♦ **somatopsychology** See study of: psychology: somatopsychology.
♦ **somatosensory cortex** See organ: brain: somatosensory cortex.
♦ **somewhat common** See frequency (table).
♦ **-somy**
monosomy *n.* An organism's condition of lacking one chromosome of one pair (*e.g.,* an XO *Drosophila*) (King and Stansfield 1985).
polysomy *n.* An organism's condition of having extra chromosomes of some of its complementary pairs (*e.g.,* an XXX *Drosophila*) (King and Stansfield 1985).
trisomy *n.* An organism's condition of having one extra chromosome of one pair (*e.g.,* three chromosome-21s in Humans) (King and Stansfield 1985).
cf. ploidy: aneuploidy

♦ **sonagram** See -gram: sonagram.
♦ **Sonagraph**® See -graph: Sonagraph®.
♦ **song** See animal sounds: song.
♦ **song flight** See flight: courtship flight.
♦ **song imprinting** See learning: imprinting: song imprinting.
♦ **song print** See template.
♦ **sonic, soniferous** *adj.* "Capable of producing sounds" (Lincoln et al. 1985).
♦ **sonication** See pollination: buzz pollination.
♦ **Sorcerer, the** *n.* An Aurignacian cave painting of a black creature that stands upright and has a bison torso and human legs (Thomas, 1995, A30).
Comments: The Sorcerer is found in Chauvet Grotto, and similar images are found in Three Brothers Caves in the Pyrenees and Gabillou Caves in Dordogne, France (Thomas, 1995, A30). The Sorcerer might be a mythical being, force, or spirit.
♦ **sord** See [2]group: sord.
♦ **sort** See [2]group: sort.
♦ **[1]sound** *n.*
1. Any of a class of waves consisting of mechanical disturbances, such as varying pressure or alternating movement, in an elastic system, especially in air (Michaelis 1963); vibration of molecules in air, water, earth, etc. (Gould 1982, 114).
2. An instance of sound (def. 1) (Michaelis 1963).
3. Noise; sound lacking pitch and caused by vibrations of different and dissonant frequencies (Michaelis 1963).

S–Z

adj. sonic

cf. animal sounds

infrasound *n.* Sound (def. 1 above) less than about 25 cycles per second (hertz) and below human hearing range (Michaelis 1963).

adj. infrasonic, subsonic

Comments: D.B. Quine (personal communication) indicates that infrasounds are those below 16 hertz, although Humans can perceive sounds at 1 hertz under certain conditions. McFarland (1985, 254) states that infrasound is that below 10 hertz, and that domestic pigeons can hear infrasound as low as 0.06 hertz.

pseudosound *n.* Sound that results from local air pressure changes which animals distinguish from infrasounds when infrasounds are used for long-distance communication (D.B. Quine, personal communication).

ultrasound *n.* Sound (def. 1) above human hearing threshold, generally considered to be over 20,000 cycles per second, although this upper threshold varies with person and age (Michaelis 1963; D.B. Quine, personal communication).

adj. ultrasonic

Comment: Some bat species, shrews, and many rodent species communicate with ultrasounds.

♦ ²**sound** *n.* Stimulation produced by sound of about 20 to 20,000 cycles per second that is audible to Humans (Michaelis 1963).

♦ ³**sound** *v.i.*

1. To produce sound (Michaelis 1963).
2. "To cause to give forth sound" (Michaelis 1963).
3. "To utter audibly; pronounce" (Michaelis 1963).

♦ ⁴**sound** *v.i.*

1. To test, or examine, by sound (Michaelis 1963).
2. In surgery: to search, or examine, with a sound (Michaelis 1963).

♦ ⁵**sound** *v.i.* In harpooned whales: to dive down suddenly and deeply (Michaelis 1963).

♦ **sound spectrograph** See -graph: sound spectrograph.

♦ **sounder** See ²group: sounder.

♦ **source habitat** See habitat: source habitat.

♦ **Southern-blot hybridization, Southern-blot technique, Southern blotting, Southern hybridization, Southern transfer** See method: Southern blotting.

♦ **southern oscillation** *n.* A fluctuation of the intertropical atmospheric circulation, in particular in the Indian and Pacific Oceans, in which air moves between the southeast-

ern Pacific subtropical high and the Indonesian equatorial low (Allaby 1994).

Comments: The temperature difference between the Indian and Pacific Oceans drives the southern oscillation (Allaby 1994). Its general effect is a high pressure over the Pacific Ocean and a low one over the Indian Ocean, and vice versa. Southern oscillations are strongly linked to Los Niños.

♦ **sowse** See ²group: sowse.

♦ **space psychology** See study of: psychology: space psychology.

♦ **spaced training** See training: spaced training.

♦ **spacer DNA** See nucleic acid: deoxyribonucleic acid: spacer DNA.

♦ **spacing behavior** See behavior: spacing behavior.

♦ **span** See ²group: span.

♦ **spanandrous sex ratio** See ratio: sex ratio: spanandrous sex ratio.

♦ **spanogyny** See -gyny: spanogyny.

♦ **spasmodic dance** See dance: bee dance: spasmodic dance.

♦ **spatial bonding** See bond: spatial bonding.

♦ **spatial distribution** See distribution: spatial distribution.

♦ **spatial facilitation** See ¹facilitation: spatial facilitation.

♦ **spatial isolation** See isolation: spatial isolation.

♦ **spatial niche** See niche: spatial niche.

♦ **spatial summation** See ¹facilitation: spatial facilitation.

♦ **spatiotemporal structure** *n.* A person's analysis and testing of an animal's behavior in relation to its time and place of occurrence (Heymer 1977, 141).

cf. motivational analysis

♦ **spatiotemporal territory** See territory: spatiotemporal territory.

♦ **spatiotemporally clumped resource** See resource: spatiotemporally clumped resource.

♦ **spawn** *n.* Mycelium of fungi (*Oxford English Dictionary* 1972, entries from 1731). See animal names: spawn; gamete: egg; group: spawn.

v.i., v.t. spawn

♦ **special creation, Special Creation, creationism, doctrine of special creation** *n.* The belief that all organisms were specially created by God (Lincoln et al. 1985).

cf. creation science, creationist, evolution

♦ **specialist species** See ²species: specialist species.

♦ **specialization** *n.* An organism's "evolutionary adaptation to a particular mode of life or habitat" (Lincoln et al. 1985).

cf. stereotypy

Comment: Behavioral stereotypy can be part of a species' specialization (Klopfer 1969, 35).

♦ **specialized** *adj.* Referring to an organism that has a narrow tolerance range for one or more ecological conditions (King and Stansfield 1985).

See ²species: specialized species.

cf. character: specialized character; generalized; ²species: specialist species; topic: stenotopic.

♦ **specialized character** See character: specialized character.

♦ **specialized mimicry** See mimicry: specialized mimicry.

♦ **specialized species** See ²species: specialized species.

♦ **speciation** *n.*
1. Genetic diversification of populations and the multiplication of species (Wilson 1975, 595).
2. One phyletic line's branching into two separate lines that are reproductively isolated where secondary contacts between them might occur (Brown 1975, 447).
3. Many coexisting species' evolving within an evolutionary grouping (Carson 1989, 872).

syn. specification (Darwin 1876 in Fleming 1959, 225)

cf. adaptive radiation; ²evolution: divergent evolution

Comment: "Speciation" is difficult to define due to the difficulty of defining "species," *q.v.* (Immelmann and Beer 1989, 285).

allochronic speciation *n.* Speciation that is influenced by divergence of breeding time of individuals within a species (Mayr 1982, 604).

Comment: This is one of three more frequently suggested mechanisms of sympatric speciation, *q.v.* (Mayr 1982, 604).

allopatric speciation *n.*
1. A concept of speciation with these tenets: new species arise by lineage splitting and developing rapidly, a small subpopulation of an ancestral form gives rise to a new species, and the new species originates in a very small part of the ancestral species' geographic range — in an isolated area at the periphery of its range (Eldredge and Gould 1972, 96).

syn. allopatric theory of speciation, (Mayr's theory of) geographic speciation, geographical speciation

cf. speciation: sympatric speciation, geographic speciation; gradualism: phyletic gradualism
2. Speciation in which populations that are isolated by a geographical barrier become new species; contrasted with parapatric speciation and sympatric speciation (Campbell 1990, 465).

Comments: Building up of intrinsic isolating mechanisms occurs in species that may, or may not, have geographical or other physical barriers that promote "allopatric speciation" rather than "sympatric speciation." Although "allopatric speciation" is widely accepted as an important kind of speciation, the relative importances of alternative processes, such as "instantaneous speciation" and "sympatric speciation," are still controversial (Mayr 1982, 565). For example, pupfish allopatric speciation occurred in Death Valley, U.S., where a great inland lake began to dry up about 10,000 years ago (Campbell 1990, 466). Now, usually a single pupfish species occurs in each of the small isolated springs that remain.

divergence speciation *n.* Speciation in which two populations gradually adapt to disparate environments, accumulating differences in genotype and phenotype frequencies; reproductive barriers between the populations evolve coincidentally; and the populations become different species (Alan Templeton in Campbell 1990, 472).

geographic speciation *n.* The phenomenon that species have similar analogs in different geographical areas (Mayr 1982, 411).

Comment: "Actually, the concept of geographic speciation was by no means entirely novel in 1837 when it occurred to Darwin" (Mayr 1982, 411).

hybrid speciation *n.* Speciation as a direct result of interspecific hybridization (Lincoln et al. 1985).

instantaneous speciation *n.* Speciation that occurs in one generation through drastic mutation such as polyploidy (Mayr 1982, 404; Stebbins 1950, Grant 1971 in Mayr 1982, 603).

cf. effect: founder effect; ²evolution: instantaneous evolution; genetic drift; ploidy: polyploidy

Comments: Speciation by polyploidy is common in plants but relatively rare in animals, probably because it disrupts sex determination and dosage compensation in those animals with chromosomal sex determination (Müller 1925 and other references in Ptacek et al. 1994, 898). Frogs and toads (Anura) have more polyploid species than other vertebrate taxa of comparable level. These anurans are in 12 genera and nine families. The tetraploid Gray Treefrog, *Hyla versicolor*, has at least three separate, independent origins from

S–Z

diploid ancestors. Hugo De Vries discovered instantaneous speciation (sympatric speciation by autopolyploidy) in the evening primrose *Oenothera lamarckiana* in the 1920s (Mayr 1982, 404, 603).

parapatric speciation *n.*
1. "Speciation in which initial segregation and differentiation of diverging populations takes place in disjunction but complete reproductive isolation is attained after range adjustment so that populations become separate but contiguous" (Lincoln et al. 1985).
2. Theoretical speciation that occurs in two contiguous populations that have slow gene flow that does not overcome the populations' gene-pool divergences (Campbell 1990, 465).

Comment: Researchers have not reported definite cases of parapatric speciation (Campbell 1990, 465).

"peak-shift speciation" *n.* Speciation that results from a peak shift, *q.v.* (Campbell 1990, 473).

peripatric speciation *n.* The concept that evolutionarily more important events that pertain to speciation take place in peripherally isolated populations of a present species (Hapgood 1984).

phyletic speciation *n.* One species' gradually evolving into a new one (Mayr 1982, 295).
syn. successional speciation (Lincoln et al. 1985)
cf. ²evolution: phyletic evolution
Comment: Phyletic speciation eventually leads to replacement of an ancestral species by a derived species (Lincoln et al. 1985).

polyploidy *n.* An individual's, or somatic cell's, having more than two sets of homologous chromosomes (Mayr 1982, 769).
cf. ²evolution: instantaneous evolution; speciation: instantaneous speciation
Comments: About 47% of all flowering plants (Stebbins and Ayala 1985, 74) and some vertebrate species, including 12 genera of frogs and toads, are polyploid (Müller 1925 and other references in Ptacek et al. 1994, 898).

primary speciation *n.* Speciation by splitting of a single lineage into two or more separate evolutionary lineages (Lincoln et al. 1985).
cf. secondary speciation

quantum speciation See speciation: saltational speciation.

saltational speciation, saltatory speciation, saltation, saltationism *n.*
1. Evolution of new species by macromutations rather than by gradual changes in populations (Mayr 1982, 431, 544).

2. Rapid speciation caused by a combination of natural selection and genetic drift and involving some degree of spatial isolation (Lincoln et al. 1985).

syn. macrogenesis, mutation (in one sense), quantum speciation, saltatory evolution, saltational evolution (Lincoln et al. 1985)
cf. atavism; hypothesis: hypothesis of punctuated equilibrium; mutation; ²evolution: quantum evolution; phyletic gradualism
Comment: There are so many gaps in the fossil record that some geneticists and paleontologists believed that species arose by "saltation" alone as late as the 1940s, and some paleontologists believe that even today (Mayr 1982, 431, 544).

secondary speciation *n.* Interspecific hybridization between distinct evolutionary lineages (Lincoln et al. 1985).
cf. speciation: primary speciation

semigeographical speciation *n.* Speciation along lines of secondary intergradation in zones of marked ecological transition (Lincoln et al. 1985).

stasimorphic speciation *n.* Speciation without morphological differentiation (Lincoln et al. 1985).

stasipatric speciation *n.* A species' (with a wide and essentially continuous range) undergoing fragmentation into a number of descendent races and species (as a result of chromosomal rearrangements) that spread out from their point of origin until their expansion is arrested and the narrow hybridization zone, which has been undergoing a secular movement across the terrain, becomes stabilized (White et al. 1967 in White 1973, 379).

successional speciation See speciation: phyletic speciation.

sympatric speciation *n.*
1. Speciation that occurs in the same geographic locality; contrasted with allopatric speciation and parapatric speciation (Brown 1975, 447; Immelmann and Beer 1989, 285).
cf. speciation: instantaneous speciation
2. Speciation in which a subpopulation in the midst of its parent population becomes a new species (Campbell 1990, 465).

Comments: Biologists have had vigorous debates regarding whether "sympatric speciation" actually occurs (Brown 1975, 447; Immelmann and Beer 1989, 285). It may have occurred in many plant species and fewer animal species (Campbell 1990, 470). It can occur in a single generation. Hugo De Vries discovered sympatric speciation by autopolyploidy (a kind of instantaneous speciation) in the evening primrose

Oenothera lamarckiana in the 1920s. Up to 50% of plant species may have resulted from allopolyploid sympatric speciation, including Cotton, Oats, Potatoes, Tobacco, and Wheat.

♦ ¹**species** *n., pl.* **species**

1. A class composed of individuals that have some common qualities or characters; frequently, a subdivision of a larger class or genus (*Oxford English Dictionary* 1972, entries from 1630).

2. A distinct class, sort, or kind of something that is specially mentioned or indicated (*Oxford English Dictionary* 1972, entries from 1561).

[Latin *species*, form, kind; doublet of SPICE]

♦ ²**species, biological species** *n., pl.* **species**

1. In Greek philosophy, a division of a genus, regardless of the rank of the genus (Mayr 1982, 255).

2. "What is similar and of a single origin" (Saint Augustine in Mayr 1982, 255).

3. A well-defined unit of nature that is constant and sharply separated from other such units; a unit of creation (Mayr 1982, 255).
   Note: This is an herbalist-species concept (Mayr 1982, 255).

4. A group of organisms characterized by having the same unchanging essence (*eidos*), being separated from all other species by a sharp discontinuity, being constant through time, and being severely limited to variation within the group (Mayr 1982, 260).
   Notes: This is an "essentialist-species concept." "Species, thus, were simply defined as groups of *similar* individuals that are *different* from individuals belonging to other species. Individuals, according to this concept, do not stand in any special relation to each other; they are merely expressions of the same *eidos*." This species concept was used by Christian fundamentalists and it was accepted almost unanimously in the post-Linnaean period (Mayr 1982, 256, 260).

5. A group of organisms that have "the distinguishing features that perpetuate themselves in propagation from seed" (Ray 1686 in Mayr 1982, 256–257).
   Notes: "Thus, no matter what variations occur in the individuals or the species, if they spring from the seed of one and the same plant, they are accidental variations and not such as to distinguish a species. …Animals likewise that differ specially preserve their distinct species permanently; one species never springs from the seed of another nor vice versa" (Ray

1686 in Mayr 1982, 256–257). This definition is "a splendid compromise between the practical experience of the naturalist, who can observe in nature what belongs to a species, and the essentialist definition, which demands an underlying shared essence. …Ray's definition was enthusiastically adopted by generations of naturalists. It had the additional advantage that it fitted so well with the creationist dogma" (Mayr 1982, 257).

6. All the individuals that descend from each other, or from a common parentage, and those that resemble them as much as they do each other (Mayr 1982, 365).

7. "The sum of the individuals that are united by common descent" (von Baer 1828 in Mayr 1982, 257).
   Note: "Common descent" means blood relationship here rather than evolutionary heritage (Mayr 1982, 257).

8. A group of organisms that have well-defined and completely constant characters through time (Mayr 1982, 259, 642).
   Notes: This is a "typological-species concept" that was held by Linnaeus during much of his professional life. Later he proposed that perhaps genera had been created by God and that species were the product of hybridization of these genera. This brought Linnaeus attacks from all sides (Mayr 1982, 259, 642). "Species mongers," who use the typological-species concept, created thousands of synonymous taxonomic names. One such person named no less than 14 "species" of House Sparrows in one European village (Mayr 1982, 263). Darwin favored this species concept until 1835 (Mayr 1982, 265).

9. "A constant succession of similar individuals that can reproduce together" (Buffon 1749, 385, in Mayr 1982, 262).
   Notes: A species is constant and invariable with regard to characters of its members. Further species should be delineated by their "habits, temperaments, and instincts" (Buffon 1749, 385, in Mayr 1982, 262). Buffon's ideas eventually led to the biological-species concept.

10. Essentially an individual organism because only individuals exist, but species, which are comprised of more than one individual, and other taxa are human-made constructs (Mayr 1982, 264–265).
    Note: Lamarck used this "nominalistic-species concept" in his earlier work and it remained popular among botanists throughout the 19th century. Some biologists have favored this concept in the 20th century as well (Mayr 1982, 264–265).

S–Z

11. A group of organisms that "remains at large with constant characters, together with other beings of very near structure ..." (Darwin in Mayr 1982, 266).

Note: Darwin further wrote that species may be true ones yet scarcely differ in any external character. Ideas from Darwin's notebooks are similar to those entailed by the modern biological-species concept (Mayr 1982, 266) but later (1856), Darwin wrote that "species" is an undefinable term (Mayr 1982, 267). In *On the Origin of Species...* (1859), Darwin gave the impression that he considered species as something purely arbitrary and invented merely for the convenience of taxonomists (Mayr 1982, 268–269).

12. "A group of individuals which reproduce their like within definite limits of variation, and which are not connected with their nearest allied species by insensible variations" (Wallace in Mayr 1982, 270).

Notes: This "morphological-species concept," which is similar to the essentialist-species concept, raises every isolated geographical race to the rank of a separate species. Nonetheless, this definition was adopted by the majority of taxonomists from the mid-1800s to the early 1900s, as well as many experimental biologists (Mayr 1982, 270). Evolutionary considerations inspired ornithologists to consider species as units separated by discontinuities and subspecies as units with some intermediates (American Ornithologists' Union 1886 in Stevens 1992, 305).

13. A species that is delineated by morphological characters alone, without consideration of other biological factors (Lincoln et al. 1985; Jackson and Cheetham 1990, 579).

syn. morphological-species concept, morphospecies (Lincoln et al. 1985)

14. Any discontinuous variant (De Vries in Mayr 1982, 546).

15. A set of genetically related ecotypes (phenotypes resulting from particular habitat conditions) (Turesson 1922 in Mayr 1982, 277).

Note: Turesson seemed to consider a species to be a "mosaic of ecotypes rather than as a set of variable populations" (Mayr 1982, 277).

See ²species: ecospecies.

syn. ecospecies (introduced by Turesson 1922 in Mayr 1982, 277)

16. A species composed of a number of subspecies (Rensch 1922 in Mayr 1982, 290).

Note: This concept was originally called "*Rassenkreise*" by Rensch (1922) and later called "polytypic species" by Huxley (in Mayr 1982, 290).

See ²species: polytypic species.

cf. ²species: superspecies

17. A group "of actually or potentially interbreeding natural populations which are reproductively isolated from other such groups" (Mayr 1942, 120, in Mayr 1982, 273).

Notes: This is "Mayr's original biological-species concept." Jordan (1896, 1905) and Poulton (1903) were the first authors to define the "biological species" (Mayr 1982, 272).

18. A group of actually interbreeding populations reproductively isolated from other such groups (Mayr 1969 in Mayr 1982, 273).

syn. isolation concept of species (Paterson 1981, 113; 1985, 24)

cf. the recognition concept of species (def. 33)

Notes: This is "Mayr's second biological-species concept." By removing "potentially," he avoided the pragmatic difficulties in evaluating "potential" interbreeding; however, this widely used definition does not relate to parthenogenetically reproducing organisms. In fact, such organisms might not even have species (Ghiselin (1987 in Williams 1992, 319). Starting in the late 1930s, many botanists did not accept the biological-species concept as the unified-species concept because plant data indicated that there could be more than one allowable species concept (Stevens 1992, 306).

19. In sexually reproducing organisms: a population, or set of populations, within which the individuals are capable of freely interbreeding under *natural conditions* (Wilson 1975, 595). 20. "A lineage (an ancestral-descendant sequence of populations) evolving separately from others and with its own unitary evolutionary role and tendencies" (Simpson 1961, 153 in Mayr 1982, 294).

Notes: This is the "evolutionary-species concept." "The vulnerability of this definition is, of course, that it applies equally to most incipient species, such as geographically isolated subspecies. ...Also, what exactly is meant by 'unitary evolutionary role?' Simpson's definition is that of a phyletic lineage, but not of a species. Furthermore, this definition does not tell us at all how to delimit a sequence of species taxa in time." There are other problems with this definition, as well (Mayr 1982, 294).

21. The basic lower unit of classification in biological taxonomy of the Linnaean hierarchy, consisting of a population, or series of populations, of closely related and similar organisms (Wilson 1975, 595).
Note: This is one of the "taxonomic-species concepts."

22. A minimum monophyletic group (Eldredge and Cracraft 1980).
Note: This is one of the "taxonomic-species concepts."

23. "A lineage ... which occupies an adaptive zone minimally different from that of any other lineage in its range" (Van Valen 1976, 233, in Mayr 1982, 275).
Notes: This is an "ecological-species concept." This definition "reflects the principle of competitive exclusion, but is not very practical as a species definition because it is often exceedingly difficult to discover the 'minimal' niche difference between two species, as demonstrated by much ecological research" (Mayr 1982, 275).

24. A diagnosable cluster of individuals within which there is a parental pattern of ancestry and descent, beyond which there is not, and which exhibits a pattern of phylogenetic ancestry and descent among units of like kinds (Eldredge and Cracraft 1980, fig. 3.1).
Note: This definition applies to both sexually and parthenogenetically reproducing organisms.

25. Two or more phenotypes within a population designated as a species (Mayr 1982, 254).
Note: This is the polytypic-species concept.
See ²species: polytypic species.
syn. Rassenkreise (Mayr 1982, 254)

26. A gene pool whose genes reproduce asexually during DNA replication and generate phenotypes (organisms) that can reproduce sexually, producing new combinations of genes (Lincoln et al. 1985).

27. A multi-population system that might have several distinct morphospecies, without phenotypic intermediates, that are replacing each other geographically (Lincoln et al. 1985).
syn. multidimensional-species concept, polytypic-species concept (Lincoln et al. 1985)

28. One of two or more sympatric populations that do not interbreed (Lincoln et al. 1985).
syn. nondimensional-species concept (Lincoln et al. 1985)

29. A taxon of the rank of species below the major rank of genus; a basic unit of biological classification; the lowest principle category of zoological classification (Lincoln et al. 1985).

30. A series of populations that is separated from other series of populations by sharp discontinuities in variation pattern (Raven 1986 in Howard 1988).

31. "An irreducible cluster of organisms, within which there is a parental pattern of ancestry and descent, and which is diagnosably distinct from other such clusters" (Cracraft 1987, 341, in Williams 1992, 320).

32. "The most inclusive population of individuals having the potential for phenotypic cohesion through intrinsic cohesion mechanisms" (Templeton 1989, 12, in Williams 1992, 319); a genetically and demographically cohesive populational unit (Templeton in Carson 1989, 872).
Note: This is called the "cohesion-species concept" (Williams 1992, 319). The cohesion mechanisms include isolating mechanisms, common descent, and natural selection.

33. The "most inclusive population of individual biparental organisms which share a common fertilization system," which includes a specific-mate recognition system (Paterson 1981, 113; 1985, 21, 24).
Notes: This is Paterson's "recognition concept of species," called "recognition-species concept" by Campbell et al. (1999, 450). Paterson (1985) explains why this concept is preferable to the widely used biological-species concept. The recognition-species concept stresses the fact that sexual selection results in reproductive barriers which are adaptations that enhance reproductive success within a single population, not safeguards against interbreeding with other populations. Therefore, a population's reproductive isolation from other populations is a spin-off. Researchers contrast the recognition-species concept, in particular, with the biological-species concept, which emphasizes reproductive isolation mechanisms.

abbr. sp. (for one species), spp. (for more than one species)

cf. endemic, race

Comments: Scientists have been trying to understand the complicated nature of "species" for centuries. Mayr remarked (1982, 252), "Even today several papers on the species problem are published each year, and they reveal almost as much difference of opinion as existed one hundred years ago." Carson (1989, 872) wrote that defining "species" is evolutionary biologists' most

S–Z

knotty problem. Stevens (1992, 303) essentially agrees and discusses species concepts, focusing on those of from 1750 to 1965. According to Dupré (1992, 312), two major theoretical issues regarding the species concept are (1) is a species a natural kind, an individual entity, a set of individual entities, or what; and (2) what criteria should be used for assigning an individual organism to a species — morphology, reproductive characteristics, phylogeny, or what? A natural kind is an entity that exists in nature independently of Humans' discovering it, naming it, or both. Williams (1992, 318–319) explains that none of the presently proposed "species" definitions is completely satisfactory for reasons including a lack of knowledge about groups "we intuitively recognize as species" and species' roles in nature and perhaps the main source of difficulty — we have been seeking "a unitary definition for a pluralistic concept." She also discusses "E-species" (the basic unit of evolution), "T-species" (the basic unit of taxonomy), and their relationships. There are about 1 million named species on Earth. Many species still await discovery, including possibly millions in deep-sea communties (Broad 1999, D5). Below are listed some kinds of species that have been designated and defined. Many of these defined kinds of species are not totally discreet from others.

active-nuclear species *n*. A bird species in a mixed species flock that seeks out other species and follows them (Wilson 1975, 359).
See ²species: nuclear species, passive-nuclear species.

advanced species *n*. A species that has recently evolved (Wilson 1975, 132).
See Darwin's admonition.
syn. derived species (Lincoln et al. 1985)

allochronic species *n*. A species that does not occur in the same geological time horizon as another species to which it is compared (Lincoln et al. 1985).
cf. ²species: synchronic species

allospecies *n*.
1. One of a group of similar species that comprise a superspecies (Lincoln et al. 1985).
2. A population that is intermediate between a race and biological species and exhibits partial interbreeding and intergradation and weak reproductive isolation (Lincoln et al. 1985).
syn. incipient species, semispecies (Lincoln et al. 1985)

aphanic species See ²species: aphanic species.

association element *n*. A constant species of an (ecological) association (Lincoln et al. 1985).

attendant species *n*. A species of bird that regularly joins a particu lar type of mixed flock but is not as attractive to other birds as passive-nuclear species (Wilson 1975, 359).
cf. ²species: nuclear species
Comment: "Attendant species" are less consistent members in a mixed flock than "nuclear species" (Wilson 1975, 359).

biological species, biospecies *n*. "A species defined primarily on biological characters" (Lincoln et al. 1985).
cf. ²species (def. 17, 18); ²species: morphospecies

biological-species concept See ²species (def. 17, 18).

biospecies See ²species: biological species.

buffer species *n*. An organism that acts as an alternative food supply for a predator, thereby buffering its effect on its usual prey (Lincoln et al. 1985).

C-selected species, C strategist, C-strategist *n*. Within the C-S-R triangle of ecological strategies, a species typically with large body size, rapid growth, relatively long life span, and relatively efficient dispersal that devotes only a small portion of its metabolic energy to offspring, or propagule, production (Lincoln et al. 1985).
syn. competitive species (Lincoln et al. 1985)
cf. ²species: species, K-selected species, r-selected species, s-selected species

cenospecies See ²species: coenospecies.

characteristic species *n*.
1. A plant species this is almost always found in a particular community and is used in delimitation of this community (Lincoln et al. 1985).
   syn. guide species, index species, indicator species (Lincoln et al. 1985)
2. A bird, mammal, or plant species that is found in a particular community and is used in delimitation of this community (as indicator species in Kritcher and Morrison 1988, 10).
cf. taxon: indicator taxon; ²species: indicator species
syn. dominant species, indicator species (Kritcher and Morrison 1988, 10)
Comments: Communities are defined by different numbers of characteristic species (Kritcher and Morrison 1988, 10–91). Not all species are characteristic species. Some species are characteristic species only when they occur with certain other species.

▸ **regional-character species** *n.*
1. A characteristic species that is limited to the range of distribution of the community to which it belongs (Lincoln et al. 1985).
2. A species that is a distinguishing feature in the whole range where it occurs together with a particular vegetation unit (Lincoln et al. 1985).

chronospecies, palaeospecies, paleospecies *n.*
1. A species that is represented in more than one geological time horizon (Lincoln et al. 1985).
2. The successive species that replace others in a phyletic lineage and are given ancestor and descendant status according to the geological time sequence (Lincoln et al. 1985).

climax species *n.* A plant species of a climax community (Lincoln et al. 1985).
Comment: This definition could be extended to any kind of organism as well.

coarse-grained species *n.* One of a group of coexisting related species that are specialized and morphologically dissimilar and, therefore, use different food sources in their habitats (Lincoln et al. 1985).
cf. ²species: fine-grained species

coenospecies, cenospecies *n.* A group of ecospecies that are capable of limited genetic exchange by forming fertile hybrids (Lincoln et al. 1985).

cohesion-species concept See ²species (def. 32).

collective species See ²species: superspecies.

common ancestor See ancestor: common ancestor.

companion species *n.* A species that is not restricted to any particular vegetation unit (Lincoln et al. 1985).
syn. indifferent species (Lincoln et al. 1985)

competitive species See ²species: c-selected species.

compilospecies *n.* A species, or species complex, that hybridizes with various related species and, in different areas, takes on some of the characteristics of each of these species (Lincoln et al. 1985).

composite species *n.* A species that is represented by a group of specimens obtained from two or more localities and not all of the same geological age (Lincoln et al. 1985).
See ²species: polytypic species.
cf. ²species: physiognomic dominant, polytypic species

consociation dominant See ²species: physiognomic species.

constant species *n.*
1. A species that is invariably present in a particular community (Lincoln et al. 1985).
2. A species that occurs in at least 50% of samples taken from a given community (Lincoln et al. 1985).

cryptic species See ²species: sibling species.

differential species *n.* A plant species that can be used as an indicator of a particular vegetational type because it has relatively high fidelity to this type (Lincoln et al. 1985).

directive species *n.* A species that attracts a predator, of which it is not a usual prey item, to an area rich in the predator's prey species (Lincoln et al. 1985).

derived species See ²species: advanced species.

disjunct species *n.* A species of plant that lives in separate areas that are sufficiently isolated so that a current natural dispersal from one to the other area seems impossible [coined by Candolle in Mayr 1982, 444].

dominant species *n.* A species that exerts considerable influence upon a community due to its size, abundance, or coverage (Lincoln et al. 1985).
See ²species: characteristic species.
▸ **subdominant** *n.* A species that may appear to be more abundant at certain times of the year than a true dominant species in a climax (Lincoln et al. 1985).
▸ **subdominule** *n.* "A subdominant of a microcommunity" (Lincoln et al. 1985).

E-species See ²species (comments).

ecdemic species *n.* A foreign, non-native species (Lincoln et al. 1985).
cf. ²species: endemic species, native species

ecological equivalent *n.* One of a group of species that functions in a similar way in different regions or habitats (Grinnell 1924 and Elton 1927 in Colwell 1992, 242).
Comment: According to the environmental niche concept, ecological equivalents fill the same niche in different places; according to the population niche concept, ecological equivalents have remarkably similar niches (Colwell 1992, 242, who discusses these ideas further).

ecological-species concept See ²species (def. 23).

ecospecies, oekospecies *n.* A group of populations, or ecotypes, that has the capacity for free genetic-material exchange without loss of fertility or vigor, but having a lesser capacity for such exchange with members of other ecospecies group (Lincoln et al. 1985).

S–Z

See ²species (def. 15).

Comment: "Ecospecies" is similar to "biological species" (Lincoln et al. 1985).

edge species *n.* A species found predominantly, or commonly, in the marginal zone of a community (Lincoln et al. 1985).

endangered species *n.* A species that is faced with possible immediate extinction; contrasted with species of special concern and threatened species (Lincoln et al. 1985).

Comment: U.S. endangered species include many bird, fish, invertebrate, and mammal species, including about 20 butterfly species (Claiborne 1997a, A14). Hotbeds of endangered species include California, Florida, Hawai'i, and Madagascar.

endemic species *n.* A species that is native to, and restricted to, a particular geographical region (Lincoln et al. 1985).

cf. ¹endemic; ²species: ecdemic species, native species

energy maximizer *n.* A species that consumes all available energy regardless of its cost in time (Schoener 1971 in Wilson 1975, 143).

Comment: "Energy maximizer" is on a continuum with "time minimizer," *e.g.,* most opportunistic species (Schoener 1971 in Wilson 1975, 143).

ergatandrous species *n.* A social-insect species that has worker-like males (Lincoln et al. 1985).

cf. caste

espèces jumelles See ²species: sibling species.

ethospecies *n.* A species that is distinguished from others primarily by behavioral traits (Emerson 1959 in Heymer 1977, 65; Lincoln et al. 1985).

cf. ²species: physiological species; -type: ethotype

evolutionary-species concept See ²species def. 20.

exclusive species, faithful species *n.* A species that is totally, or almost totally, confined to one particular community; a species of high fidelity (Lincoln et al. 1985).

cf. ²species: exclusive species, indifferent species, preferential species, strange species

exploitably vulnerable species *n.* A species (*e.g., Ilex verticillata,* Winterberry) that could become endangered if people continue to harvest it in the wild (Galitzki 1996, C9).

facultative schooling species *n.* In some fish species: a species that occasionally forms schools (Breder 1967 in Partridge 1982, 299).

cf. species: obligate schooling species

faithful species See ²species: exclusive species.

fine-grained species *n.* One of a group of coexisting related species that are morphologically similar and use the same foods, but not in equal quantities or proportions (Lincoln et al. 1985).

cf. ²species: coarse-grained species

form species, morphospecies *n.* One of a group of species that has similar morphological characters but is not confirmed to have a common ancestry with this group (Lincoln et al. 1985).

See ²species (def. 13).

fugitive species *n.* An extreme case of an r strategist that is consistently wiped out of the places it colonizes and survives only by its ability to disperse and fill new places at a high rate (Hutchinson 1951 in Wilson 1975, 99).

generalist, generalist species *n.* A species with a broad habitat range or food preference (*e.g.,* the black bear or Humans) (Lincoln et al. 1985); contrasted with specialist species and specialized species.

Comments: Flessa et al. (1975, 72–73) discuss that "generalized" and "specialized" have meanings precise enough to express their respective concepts only within certain well-circumscribed contexts. Measurements of specialization of fossil species are based on their morphologies. Some paleoecologists consider morphological simplicity to indicate generalization and morphological complexity to indicate specialization.

genospecies *n.* An aggregate of interbreeding populations of mutually interfertile forms that contribute to a common gene pool (Lincoln et al. 1985).

geographic race See ²species: subspecies.

group-living species *n.* A species that normally lives in groups at any time of the year, *e.g.,* communally breeding species or species that breed colonially without helpers (Alexander 1974 in Brown 1987a, 300).

cf. breeding: communal breeding

guide species See ²species: characteristic species.

heterogameon *n.* A species comprised of distinct populations that produce morphologically stable populations when selfed but different types of viable and fertile offspring when crossed (Lincoln et al. 1985).

heteromesogamic species *n.* A species in which different individuals employ different modes of fertilization (Lincoln et al. 1985).

holomorphospecies *n.* A species based on the collective morphological characters of fossil assemblages from throughout the geographical range of a given stratigraphic level (Lincoln et al. 1985).

homogeneon *n.* A genetically and morphologically homogeneous species (Lincoln et al. 1985).

ichnospecies *n.* "A subordinate taxon in the classification of traces" (Lincoln et al. 1985).

incipient species See ²species: allospecies.

index species, indicator species See ²species: characteristic species.

indicator species *n.* A species that a researcher uses to monitor biodiversity changes in a particular region (inferred from Lawton et al. 1998, 73).
cf. taxon: indicator taxon

indifferent species, companion species *n.* A plant species that shows no obvious affinity for any particular community (Lincoln et al. 1985).
cf. ²species: exclusive species, preferential species, selective species, strange species

indigenous species See ²species: native species.

interspecies *n.* A hybrid of two species (Lincoln et al. 1985).
syn. interspecific hybrid

isolation concept of species See ²species (def. 18).

K-selected species, K strategist, K-strategist *n.* A species that is characteristic of a relatively constant, or predictable, environment, typically with slow development, relatively high competitive ability, late reproduction, large body size, and iteroparity (Lincoln et al. 1985).
cf. ²species: C-selected species, r-selected species, s-selected species

keystone species *n.*
1. A predator that greatly modifies its community's species composition and physical appearance and determines the community's integrity and unaltered persistence through time by its activities and abundance (Paine 1969 in Mills et al. 1992, 219).
2. A species whose presence is crucial in maintaining the organization and diversity of its ecological community and has an exceptional importance in its community (Paine 1966, 1967 and other authors in Mills et al. 1992, 219).
3. A species whose removal is expected to result in the disappearance of at least half of a considered species assemblage (Mills et al. 1992, 221).
cf. symbiont: keystone mutualist
Comments: "Keystone species" has been applied to many species at many trophic levels (Mills et al. 1992). Five kinds of keystone species can be delineated and described: keystone predator, keystone

prey, keystone plant, keystone link, and keystone modifier. Nonetheless, "keystone species" is broadly applied, poorly defined, and nonspecific in meaning. Further, "the type of community structure implied by the keystone-species concept is largely undemonstrated in nature." Therefore, Mills et al. (1992) recommend abandoning the concept of "keystone species." They advocate the study of interaction strengths and subsequent application of the results into species management plans and policy decisions.
[coined by Paine (1969 in Mills et al. 1992, 219)]

kinetophile *n.* An opportunistic species (Lincoln et al. 1985).

Linnaean species, Linnaeon, macrospecies, taxonomic species *n.* A large, polymorphic species (Lincoln et al. 1985).
cf. ²species: taxospecies

local-character species *n.*
1. A species that is characteristic, *q.v.*, over only the part of its total range where its vegetation unit and species occur in common (Lincoln et al. 1985).
2. A species faithful, *q.v.*, to only part of the range of a single vegetation unit (Lincoln et al. 1985).

macrospecies See ²species: Linnaean species.

marginal species *n.* "A plant species that occurs on the edge of a habitat or community" (Lincoln et al. 1985).

metamps *n.* "Different forms of the same species" (Lincoln et al. 1985).

microspecies, jordanon *n.*
1. A microgeographic group or other infrasubspecific group (Lincoln et al. 1985).
2. A part of a species aggregate (Lincoln et al. 1985).

microsubspecies, microgeographic races *n.* Small subpopulations that comprise a subspecies (Lincoln et al. 1985).

monestric species, monoestric species *n.* A species in which estrus occurs only once per year per female; contrasted with polyestric species (Heymer 1977, 47).

monotypic species *n.* A species with no subspecies or geographical variants; contrasted with polytypic species (Lincoln et al. 1985).

morphospecies See ²species (def. 13).

multidimensional-species concept See ²species (def. 27).

native species, indigenous species *n.* A species that naturally occurs in a particular geographic region (Lincoln et al. 1985).

S–Z

cf. ¹endemic; ²species: ecdemic species, endemic species

nomenspecies See ²species: typological species.

nominalistic-species concept See ²species (def. 10).

nondimensional-species concept See ²species (def. 28).

nuclear species *n.* In mixed-species bird flocks: a species that contributes significantly to the formation and cohesion of a particular kind of mixed-species flock (Wilson 1975, 359).
syn. nucleus species (Wittenberger 1981, 619)
See ²species: active-nuclear species, passive-nuclear species.
cf. ²species: attendant species
Comment: A "nuclear species" may, or may not, actually lead the other birds, but its flock is not likely to exist without it (Wilson 1975, 359).

nucleus species See ²species: nuclear species.

obligate schooling species *n.* In some fish species: a species that nearly always forms schools (Breder 1967 in Partridge 1982, 299).
cf. ²species: facultative schooling species

occasional species *n.* A species that may be found from time to time in, but is not a permanent member of, a particular habitat or community (Lincoln et al. 1985).

oekospecies See ²species: ecospecies.

open-group species *n.* A social species in which individuals move relatively freely among demes (*e.g.,* some ant species or the Chimpanzee) (Wilson 1975, 10).

opportunistic species See ²species: kinetophile, r-selected species.

paleospecies, palaeospecies See ²species: chronospecies.

paralocal species, pararegional species *n.* A local character species that has a geographical range only partly overlapping that of the vegetation unit of which it is a characteristic species (Lincoln et al. 1985).

passive nuclear species *n.* A nuclear species in a mixed-bird flock that attracts other bird species (Wilson 1975, 359).
cf. ²species: active nuclear species, nuclear species

phylogenetic relict See ²species: relict species.

physiognomic dominant, consociation dominant *n.* The dominant species of a consociation that has the life form characteristic of that formation (Lincoln et al. 1985).

physiological species *n.* A species (*e.g.,* a firefly) that is delineated from a sibling species by its behavior (Barber 1951, 1).
cf. ²species: ethospecies

pioneer, pioneer species *n.* The first species to colonize, or recolonize, a barren, or disturbed, area, and which thereby starts a new succession (Lincoln et al. 1985).
cf. ²community: pioneer

polyestric species, polyoestric species *n.* A species in which estrus occurs more than once per year per female (Heymer 1977, 47).
cf. ²species: monestric species

polyestrus species, polyoestrous species *n.* A species that has a succession of breeding periods in one sexual season (Lincoln et al. 1985).

polyphasic species *n.* A species that shows marked morphological variation within a given habitat (Lincoln et al. 1985).
n. polyphasy

polytypic species, composite species, *Rassenkreis*, rheogameon *n.* A species consisting of several subspecies or geographic variants (Lincoln et al. 1985).
See ²species (def. 25); species: composite species.
cf. ²species: monotypic species

polytypic-species concept See ²species (def. 25, 27).

predominant species *n.*
1. A species that is of outstanding abundance, or importance, in a community (Lincoln et al. 1985).
2. "A species that is present in all, or nearly all, of the associations of a formation" (Lincoln et al. 1985).

preferential species *n.* A plant species found in a variety of communities but showing greater abundance in one particular community (Lincoln et al. 1985).
cf. ²species: exclusive species, indifferent species, selective species, strange species

primitive species *n.* A species that originated a relatively long time ago and has undergone little recent evolution (Wilson 1975, 132).
cf. Darwin's admonition; ²species: advanced species

protospecies *n.* "An ancestral species" (Lincoln et al. 1985).

pseudospecies *n.* A culture, or nation, of Humans that is essentially reproductively isolated and, thus, is similar to a biological species (Erikson 1966 in Heymer 1977, 134).

pseudovicar *n.* One of two or more unrelated, or distantly related, but ecologically equivalent, species that tend to be mutually exclusive because they occupy different geographical areas (Lincoln et al. 1985).

r-selected species, r strategist, r-strategist *n.*

1. An opportunistic species, *i.e.*, one that can increase its population size rapidly, witnessing little intraspecific competition (MacArthur and Wilson 1967 in Wilson 1975, 99).

2. A species that relies upon a high intrinsic rate of increase (*r*) that makes use of a fluctuating environment and ephemeral resources (MacArthur and Wilson 1967 in Wilson 1975, 99).

syn. opportunistic species, ruderal species (Lincoln et al. 1985)

cf. ²species: C-selected species, K-selected species, kinetophile; s-selected species; strategist: r strategist

Comment: Characteristics of "r strategists" often include high fecundity, small size, adaptations for long-distance dispersal, fast maturity, no pair bonding, and no parental care (Immelmann and Beer 1989, 256). "Pure r strategists" and "pure K strategists" are at the extremes of a continuum, most real cases lying somewhere between. "Ecologists enjoy a curious love/hate relationship with the r/K concept, often pretending to disapprove of it while finding it indispensable" (Dawkins 1983, 293).

Rassenkreise See ²species (def. 16, 25); ²species: polytypic species.

recognition concept of species, recognition-species concept See ²species (def. 33).

regional-character species See ²species: characteristic species: regional-character species.

rejungent species See ²species: ring species.

relic See ²species: relict species.

relict species *n.*

1. A species that persists in a given area as a survivor from an earlier period or type (Michaelis 1963).

2. A persistent species of a formerly widespread biota that remains in certain isolated areas or habitats (Lincoln et al. 1985), *e.g., Macrotermes darwiniensis* (Morell 1992b, 1861).

   syn. relic (Lincoln et al. 1985)

3. An extant "archaic form" in an otherwise extinct taxon (Lincoln et al. 1985).

syn. phylogenetic relict (Lincoln et al. 1985)

rheogameon See ²species: polytypic species.

ring species, rejungent species *n.* A species comprised of a graded series of populations along an extensive cline that curves around on itself so that the populations at the extremes overlap but are unable to interbreed successfully (Lincoln et al. 1985).

ruderal species See ²species: r-selected species.

s-selected species, s strategist, s-strategist *n.* Within the C-S-R triangle of ecological strategies, a species typically with small body size, slow growth, long to very long life span, low dispersal ability, strong physiological tolerance to environmental stress, and devotion of a small proportion of its metabolic energy to offspring, or propagule, production; a stress-tolerant species (Lincoln et al. 1985).

cf. ²species: C-selected species, K-selected species, r-selected species

selective species *n.* A plant species usually found in one particular community, but also occasionally in other communities as well (Lincoln et al. 1985).

cf. species: exclusive species, indifferent species, preferential species, strange species

semispecies See ²species: allospecies.

sibling species *n.* "Reproductively isolated but morphologically identical or nearly identical species" (Mayr 1982, 959); *e.g.,* a member of a species complex of *Anopheles* mosquitoes or *Microtus subarvalis* and *M. arvalis* (Mayr 1982, 871).

syn. aphanic species, cryptic species, *espèces jumelles,* twin species (Lincoln et al. 1985)

Comments: McDermott (1912 in Lloyd 1989, 52) suggested the concept of sibling species.

sister species *n.* One of the two species that results from the splitting of a cladistic line (Hennig 1950 in Mayr 1982, 226).

specialist species *n.* A species that has a very narrow habitat range or food preference (Lincoln et al. 1985); contrasted with generalist species and specialized species. *q.v.*

specialized species *n.* A species that has a relatively low potential for further evolutionary change (King and Stansfield 1985); contrasted with specialist species.

strange species *n.* A rare, or accidental, plant member of a community, probably a relict of an earlier community (Lincoln et al. 1985).

cf. ²species: exclusive species, indifferent species, preferential species, selective species

subspecies *n.*

1. In organisms, "a geographically defined aggregate of local populations which differ taxonomically from other subdivisions of species" (Mayr 1940, 259, etc., in O'Brien and Mayr 1991, 1188).

   Note: Avise and Ball (in press in O'Brien and Mayr 1991, 1188) urge that the evidence for the designation of the

S–Z

"biological-species concept subspecies" come from the concordant distribution of multiple, independent, genetically based traits. O'Brien and Mayr (1991, 1188) offer guidelines for classifying "subspecies."

2. In plants: a baby species, or species of small magnitude, that is distinguished by less obvious, or significant, morphological features than are more obvious species within the same genus (Lawrence 1957, 54).

3. In plants: a major morphological variation of a species that has a geographic distribution of its own which is distinct from the area occupied by other subspecies of the same species (Lawrence 1957, 55).
syn. geographical race

4. In plants, a group of individuals with geographic, ecologic, and morphologic characters that are suspected to be counterparts of the ecotype (a biologically significant element determinable only after analysis by slow and tedious experimental techniques) (Lawrence 1957, 55).

5. In plants, a group of intraspecific individuals of major morphological distinction with, or without, disjunctive distribution (Englerian School in Lawrence 1957, 55).

6. In plants, a horticultural element (Link et al. in Lawrence 1957, 55).

7. In plants, any group of individuals below the rank of species (Weatherby in Lawrence 1957, 55).

8. In plants, an unproved, but suspected ecotype (Lawrence 1957, 55).

9. In organisms, a portion of a species that is separated from other such groups by distance and geographic barriers that prevent the exchange of individuals, as opposed to the genetically based "intrinsic isolating mechanisms" that hold species apart (Wilson 1975, 10, 596). See ²group.
syn. geographic race (Wilson 1975, 596)
cf. ²species: microsubspecies
Comments: "Subspecies" show every conceivable degree of differentiation from other subspecies, and they are often given scientific names; *e.g.,* in *Gorilla gorilla beringei,* the last name indicates the subspecies (Wilson 1975, 10, 596). Botanical nomenclatural rules provide for the infraspecific categories of subspecies, varietas, subvarietas, forma, forma biologica, forma specialis, and individuum in the order of increasing division of subspecies (Lawrence 1957, 54).

successful species *n.* A species that is still extant (Garrod and Horwood 1984, 367).

successional species *pl. n.* Successive morphospecies within a distinct evolutionary lineage that represent the nodal points of a phyletic evolutionary trend that are sufficiently distinct from one another to justify giving them species names (Lincoln et al. 1985).

superfluent species *n.* An animal species that is of equal significance to a subdominant plant species in its community (Lincoln et al. 1985).

superspecies *n.*
1. A group of similar species (called allospecies) found in the same geographic region, *e.g.,* the *Rana-pipiens* group of frog species in North America (Mayr 1982, 291).
2. "A cluster of incipient species (semispecies)" (Lincoln et al. 1985).
syn. collective species (Lincoln et al. 1985)
cf. ²group: species swarm; ²species (def. 21, 22); ²species: allospecies; species aggregation; species-group
Comment: Mayr (1982, 291) renamed *Artenkreis* of Rensch (1929) "superspecies." *Artenkreis* is singular; *Artenkreise,* plural.

supertramp *n.* A species that has a wide geographical range and a very efficient means of dispersal but which tends to be excluded by competition from species-rich faunas (Lincoln et al. 1985).
cf. ²species: tramp species

synchronic species *n.* A species that occurs within the same geological time horizon as another species (Lincoln et al. 1985).
cf. species: allochronic species

T-species See ²species (comments).

taxonomic species See ²species: Linnaean species.

taxospecies *n.* "A species based on overall similarity, determined by numerical taxonomic methods" (Lincoln et al. 1985).

time minimizer *n.* An animal species that minimizes the time required to harvest the available energy and devotes its remaining time to other activities, including defending its food supply and reproducing (Schoener 1971 in Wilson 1975, 143).

tramp species, tramp *n.* A species with a wide geographical range and an efficient means of dispersal (Lincoln et al. 1985).
cf. ²species: supertramp

twin species *n.* A pair of sibling species, *q.v.* (Lincoln et al. 1985).
See ²species: sibling species.

type species *n.* A species within a genus that a researcher designates as the one that rightfully holds an existing generic name when that researcher, or another one, revises the genus' (or subgenus') limits (White et al. 2000, 904).

typological species, nomenspecies *n.* A species that is defined on the characters of its type specimen (Lincoln et al. 1985).

vicarious species, vicar, vicariad *n.* A species in a group of closely related, ecologically equivalent species that tend to occupy mutually exclusive areas (Lincoln et al. 1985).

♦ **species-abundance curve** See curve: species-abundance curve.

♦ **species-accumulation curve** See curve: species-accumulation curve.

♦ **species aggregate, collective group** *n.* A group of morphologically similar species that are difficult to tell apart (Lincoln et al. 1985).
cf. ²species: microspecies, superspecies

♦ **species-area curve** See curve: species-area curve.

♦ **species category** *n.* The class comprised of species taxa (Mayr 1982, 254).

♦ **species-characteristic, species typical** *adj.* Referring to a trait that is specific to more than one species (Immelmann and Beer 1989, 287).
cf. character: species-specific character

♦ **species concept** See ²species.

♦ **species-density map** See map: species-density map.

♦ **species diversity** See diversity (def. 2).

♦ **species equilibrium** See equilibrium: species equilibrium.

♦ **species-flock** *n.* A group of several ecologically diverse and closely related species that have evolved within a single macrohabitat (*e.g.,* a particular lake basin) (Lincoln et al. 1985).

♦ **species-group** See ²group: species-group.

♦ **species imprinting** See learning: imprinting: species imprinting.

♦ **species odor** See odor: species odor.

♦ **species packing** *n.* The number of species that a given hyperspace, or hyperspace axis, can contain in a multidimensional hyperspace (Lincoln et al. 1985).

♦ **species pairs** See ²species: vicarious species.

♦ **species recognition** See recognition: species recognition.

♦ **species richness** See index: species richness.

♦ **species selection** See hypothesis: species selection.

♦ **species-specific behavior** See behavior: instinctive behavior.

♦ **species-specific character** See character: species-specific character.

♦ **species swarm** *n.* A large number of closely related species that occur together in a specific geographical area and are derived from the same ancestral stock (Lincoln et al. 1985).
cf. ²species: superspecies

♦ **species typical** See species characteristic.

♦ **speciesism** *n.* A person's prejudice, or an attitude, against other species [coined by Singer 1975, 7].

♦ **specific** *adj.* Referring to a particular individual, situation, quality, trait, effect, or species; free from ambiguity (Lincoln et al. 1985).
syn. peculiar, unambiguous (Lincoln et al. 1985)
cf. character: specific character

♦ **specific-action potential** See potential: specific-action potential.

♦ **specific character** See character.

♦ **specific-mate-recognition system** See ¹system: specific-mate recognition system.

♦ **specific modifier** See gene: specific modifier.

♦ **specific nerve energies, doctrine of** *n.* A particular nerve always produces the same type of sensation regardless of its mode of stimulation (Müller 1827 in McFarland 1985, 166).

♦ **specific searching image** See search image.

♦ **specification** See speciation.

♦ **specification error** See error: specification error.

♦ **specificity, host specificity** *n.*
1. A parasite's being restricted to a given host species (Lincoln et al. 1985).
2. An adult parasite's being restricted to a particular host species (Lincoln et al. 1985).

♦ **spectrogram** See -gram: sonagram.

♦ **spectrum analyzer** See -graph: sound spectrograph.

♦ **speculum** *n.* In many bird species, *e.g.,* the Blue Teal: a lustrous mark on a wing (*Oxford English Dictionary* 1972, entries from 1804).
cf. signal: poster signal
Comment: "Speculum" now also refers to white, or colored, patches on body surfaces of other kinds of animals besides birds that contrast with background color, *e.g.,* the white area around and below the tail base of many social-mammal species. German-speaking investigators tend to use "speculum" to refer to mammals; English-speaking investigators, to birds (Immelmann and Beer 1989, 288).

♦ **speleology** See study of: speleology.

S–Z

◆ **sperm** See gamete: sperm.

◆ **sperm competition** See competition: sperm competition.

◆ **sperm displacement** *n*. In the Deer Mouse, Dung Fly and some other insect species: a tactic in sperm competition in which sperm from a female's last mate predominate in fertilizing her eggs (Lefevre and Jonsson 1962 in Parker 1984, 10; Smith 1979; Dewsbury 1985 in Schwagmeyer and Foltz 1990, 156; Immelmann and Beer 1989, 289).

syn. sperm precedence (according to some authors; a confusing synonym)
cf. sperm precedence

Comments: In insects, "sperm displacement" may occur due to how the sperm are positioned in a female's spermatheca or due to a male's exercising sperm removal. In Deermice, "sperm displacement" is related to the interval between different male's copulations (Schwagmeyer and Foltz 1990, 156).

◆ **sperm effect** See effect: sperm effect.

◆ **sperm plug** *n*. In some insect, spider, and snake species; many mammal species; an acanthocephalan-worm species: a plug that a male leaves in a female's vagina, bursa copulatrix, or other sexual organ after mating as a tactic in sperm competition (Parker 1970 in Parker 1984, 15; Immelmann and Beer 1989, 61).

cf. sphragis, tie

◆ **sperm precedence** *n*. For example, in the Deermouse; the Thirteen-Lined Ground Squirrel; a spider-mite species; rabbits: predomination of sperm from a female's first mate in fertilizing her eggs (Immelmann and Beer 1989, 289).

syn. first-male advantage (Schwagmeyer and Foltz 1990, 156), sperm displacement (according to some authors; a confusing synonym)
cf. sperm displacement

Comment: In Deermice, *e.g.,* sperm precedence is related to the interval between different male's copulations and time in diel cycle (Dewsbury 1985 and 1988 in Schwagmeyer and Foltz 1990, 156).

◆ **sperm removal** *n*. A tactic in sperm competition in which a male takes sperm from a previous male's copulation out of a female's spermatheca before he copulates with her; *e.g.,* in a damselfly, a male scrapes out sperm with his aedeagus (penis) (Waage 1979).

◆ **sperm-replenishment polyandry** See mating system: polyandry: sperm-replenishment polyandry.

◆ **spermalege** *n*. In a bed-bug species: a highly specialized tissue mass beneath the cuticle of a female that receives sperm (after traumatic insemination, *q.v.*) and directs it to circulatory channels that eventually carry it to ova (Thornhill and Alcock 1983, 327).

◆ **spermateleosis** See -genesis: spermiogenesis.

◆ **spermatheca, spermathecae** See organ: spermatheca.

◆ **spermatism** See [3]theory: encasement theory: spermism.

◆ **spermatize** See fertilize, impregnate.

◆ **spermatocyte** See -cyte: spermatocyte.

◆ **spermatogamete** See gamete: sperm.

◆ **spermatogenesis** See -genesis: spermatogenesis.

◆ **spermatophore** *n*. In Butterflies, Cephalopods, Leeches, Millepedes, Newts, Onychophorans, Springtails, Silverfish (insects); some katydid and cricket species: a body of sperm and associated materials produced by a male or hermaphrodite (Barnes 1974; 311, 428, 669; Oberhauser 1988, 1384; Immelmann and Beer 1989, 289).

Comment: "Spermatophores" are stalked or unstalked and are placed in an environment by a male and picked up by a female, placed on a female by a male, or directly transferred from one individual to the vagina, bursa copulatrix, etc., of another.

◆ **spermatosome** See gamete: sperm.

◆ **spermatozoon, spermatozoan, spermatozoid, spermatozooid** See gamete: sperm.

◆ **spermiogenesis** See -genesis: spermiogenesis.

◆ **spermism, spermist** See [3]theory: encasement theory.

◆ **spermium** See gamete: sperm.

◆ **spermogenesis** See -genesis: spermatogenesis.

◆ **spermology** See study of: spermology.

◆ **spermophile, spermophilic, spermophily** See [1]-phile: spermophile.

◆ **-spermy** *combining form* Referring to sperm.

dispermy *n*. Penetration of an ovum by two sperm during fertilization (Lincoln et al. 1985).

monospermy *n*. enetration of an ovum by one sperm during normal fertilization (Lincoln et al. 1985).

polyspermy *n*. Penetration of an ovum by more than two sperm during fertilization (Lincoln et al. 1985).

◆ **sphagnicole, sphagnicolous** See -cole: sphagnicole.

◆ **sphagniherbosa** See [2]community: sphagniherbosa.

◆ **sphagnophile, sphagnophilous, sphagnophily** See [1]-phile: sphagnophile.

◆ **sphecology** See study of: sphecology.

♦ **-sphere** *combining form*
1. "Denoting an enveloping spherical mass" (Michaelis 1963).
2. "A sphere-shaped body" (Michaelis 1963).
3. "Denoting a spherical form" (Michaelis 1963).
cf. biome, biotope
[Greek *sphaira*, ball, sphere]

biosphere, ecosphere, ecumene *n.* The part of our Earth that supports life, divided into biocycles, biochores, and biotopes and into the eubiosphere and the parabiosphere (Lincoln et al. 1985).
Comments: Diverse bacterial communities occur as deep as 4.2 kilometers and at a temperature of 110°C (Fyfe 1996, 448).

  ▸ **eubiosphere** *n.* The part of the biosphere in which organisms' physiological processes can occur (Lincoln et al. 1985). *syn.* biogeosphere, true biosphere (Lincoln et al. 1985)

    allobiosphere *n.* Part of the eubiosphere in which heterotrophic organisms occur but into which organic food must be transported because primary production does not occur in it (Lincoln et al. 1985).

      hyperallobiosphere *n.* The upper allobiosphere comprised of the highest parts of major mountain ranges, typically at altitudes greater than 5500 meters (Lincoln et al. 1985).

      hypoallobiosphere *n.* "The lower allobiosphere, including ocean depths below the euphotic zone and hypogean" habitats (Lincoln et al. 1985).

    autobiosphere *n.* Part of the eubiosphere in which energy is fixed by photosynthesis (Lincoln et al. 1985).

  ▸ **noosphere** *n.* The part of the biosphere that is altered, or influenced, by human activities (Lincoln et al. 1985).

  ▸ **parabiosphere** *n.* The space surrounding the eubiosphere in which conditions are too extreme for active metabolism but in which resistant life-cycle stages may exist (Lincoln et al. 1985).

  ▸ **pedosphere** *n.* Part of the biosphere comprised of soil and soil organisms (Lincoln et al. 1985).

ecosphere, ecumene See -sphere: biosphere.

hydrosphere *n.* The Earth's water mass, including atmospheric, surface, and subsurface waters (Lincoln et al. 1985).

lithosphere *n.* The Earth's rigid crustal plates (Lincoln et al. 1985).

mesosphere *n.* The outer layer of the Earth's atmosphere above the stratosphere where temperature decreases with increasing altitude, typically with efficient vertical convection currents (Lincoln et al. 1985).

phyllosphere *n.*
1. A leaf microhabitat (Lincoln et al. 1985).
2. The immediate surroundings of plant leaves that are influenced by the presence of leaves (Lincoln et al. 1985).
cf. -sphere: rhizosphere

rhizosphere, root zone *n.* The soil immediately surrounding plant roots that is influenced structurally, or biologically, by the roots' presence (Lincoln et al. 1985).
cf. -sphere: phyllosphere

  ▸ **rhizoplane** *n.* "The root-surface component of a rhizosphere" (Lincoln et al. 1985).

stratosphere *n.*
1. The nearly uniform cold-ocean water masses in high latitudes and near-bottom waters of middle and low latitudes; ocean water below the thermocline (Lincoln et al. 1985).
2. The atmosphere layer above the Earth's troposphere, 15–50 kilometers above its surface, in which temperature ceases to fall with increasing altitude, typically without strong convection currents (Lincoln et al. 1985).

  ▸ **ozonosphere** *n.* The lower layer of the Earth's stratosphere (def. 2), about 15–30 kilometers above its surface in which the absorption of ultraviolet solar radiation produces ozone (Lincoln et al. 1985).

♦ **spheres of dominance** *n.* The phenomenon in which an individual animal may be dominant in some contexts, but not in others (Hand 1985, 211).

♦ **sphingophile, sphingophilous, sphingophily** See ²-phile: sphingophile.

♦ **sphragis** *n.* In some water-beetle and butterfly species: an external plug that a male applies to his mate's genital opening to prevent other males from mating with her (Thornhill and Alcock 1983, 340).
cf. sperm plug

♦ **spiladophile, spiladophilus, spiladophily** See ¹-phile: spiladophile.

♦ **spinning** See behavior: spinning.

♦ **spinster hypothesis** See hypothesis: spinster hypothesis.

♦ **spiracular-flap hypothesis of insect wing origin** See hypothesis: hypotheses of origin of insect wings: spiracular-flap hypothesis of insect wing origin.

♦ **spiral swimming** See swimming: spiral swimming.

♦ **spite, spiteful behavior** See behavior: spiteful behavior.

♦ **spitting** *n.*
1. A person's ejecting saliva (at, or on, a person, or thing) as an expression of

S–Z

hatred or contempt (*Oxford English Dictionary* 1972, verb entries from 975).

2. For example, in the lion and leopard: an individual's ejecting saliva "when angry" (*Oxford English Dictionary* 1972, verb entries from 1668).

water spitting *n.* In some fish species: an individual's squirting water droplets at prey above the surface of water (*e.g.,* in the archer fish, *Colisa lalia* and *Toxotes jaculatrix*), which is a kind of tool use (Lüling 1969, 1973, etc. in Heymer 1977, 43).

cf. tool use.

♦ **Spjøtvoll method** See statistical test: multiple-comparisons test: planned-multiple-comparisons procedure: Spjotvoll method.

♦ **splendiferin** See chemical-releasing stimulus: semiochemical: pheromone: splendiferin.

♦ **split gene** See gene: split gene.

♦ **splitter** *n.* A taxonomist who considers "even relatively minor differences as justifying the recognition of new [species,] genera, families, and higher taxa" (Mayr 1982, 240).

ant. lumper

♦ **sponge carrying** See behavior: sponge carrying.

♦ **spongicole, spongicolous** See -cole: spongicole.

♦ **spontaneous** *adj.* Referring to a phenomenon that develops, or occurs, without apparent external stimulus or cause (Lincoln et al. 1985).

♦ **spontaneous activity** See activity: spontaneous activity.

♦ **spontaneous behavior** See behavior: spontaneous behavior.

♦ **spontaneous generation** See -genesis: abiogenesis.

♦ **spontaneous ovulator** See ovulator: spontaneous ovulator.

♦ **spook reaction** See reaction: spook reaction.

♦ **sporadic** See frequency (table).

cf. abundant, common, rare

♦ **sporivore** See -vore: sporivore.

♦ **sporophyte** *n.* In many plant species: the diploid phase of a life cycle (Mayr 1982, 959).

cf. -plont: diplont

♦ **sport** See [4]mutation.

♦ **spring** See [2]group: spring.

♦ **squabble** See [1]conflict (def. 2).

♦ **square law** See law: Hardy-Weinberg law.

♦ **S_T** See index: species richness.

♦ **[1]stability** *n.*

1. A thing's "ability to remain in the same relative place, or position, in spite of disturbing influences; capacity for resistance to displacement; condition of being

in stabile equilibrium, tendency to recover its original position after displacement" (*Oxford English Dictionary* 1972, entries from 1542).

2. A population's ability to withstand perturbations without marked changes in its composition (Lincoln et al. 1985).

♦ **[2]stability** *n.* The condition in which an organism does not show variation in a particular trait or group of traits; any condition where there is a lack of plasticity, *q.v.* (Bradshaw 1965, 117).

instability, lack of stability *n.* An organism's variation that is not genetic in origin and has no observed environmental cause (although one might exist) (Mather 1953 in Bradshaw 1965, 116–117).

Comment: Some kinds of instability might be due to developmental errors not connected with any environmental influences (Bradshaw 1965, 117).

♦ **stability-time hypothesis** See hypothesis: stability-time hypothesis.

♦ **stabilizing selection** See selection: stabilizing selection.

♦ **stable equilibrium** See equilibrium: stable equilibrium.

♦ **stable linkage disequilibrium** *n.* The linkage and strong interaction of polygenes that produce particularly favorable or unfavorable combinations.

Comment: When there is tight linkage disequilibria, entire chromosomes can act as genetic units that underline a trait (Franklin and Lewontin 1970 in Wilson 1975, 70).

♦ **stadium** *n., pl.* **stadia**

1. The interval between larval molts (Torre-Bueno 1978).

2. An instar, *q.v.* (Torre-Bueno 1978).

3. Any period in an insect's development (Torre-Bueno 1978).

syn. stage (Torre-Bueno 1978)

♦ **stage, state** *n.* An organism's distinguishable growth, or developmental, phase (Lincoln et al. 1985).

See stadium.

cf. instar

ergonomic stage *n.* In Ants: the period of a colony life cycle that follows the founding stage, has relatively rapid colony growth in which only workers are produced, and is succeeded by the reproductive stage (Hölldobler and Wilson 1990, 638).

founding stage *n.* In Ants: the earliest period of a colony's life cycle in which a newly fecundated queen raises her first worker brood and which is succeeded by the ergonomic stage (Hölldobler and Wilson 1990, 638).

reproductive stage *n.* In Ants: the period of a colony life cycle that follows the ergonomic stage and in which males and virgin queens are produced (Hölldobler and Wilson 1990, 642).

♦ **stage-1 sleep, stage-2 sleep, stage-3 sleep, stage-4 sleep** See sleep (table) and definitions.

♦ **stagnicole, stagnicolous** See -cole: stagnicole.

♦ **stagnoplankton** See plankton: stagno-plankton.

♦ **Stairway to Heaven** *n.* A pathway that leads cattle to a slaughter corridor (Temple Grandin in Raver 1997, C1).

♦ **stamp collecting, stamp-collecting** *n.* A taxonomist's allegedly simply placing taxa in unambiguous, easily determined places in a process similar to affixing stamps in preassigned places in a stamp book (Gould 1987).

n. stamp collector
v.t. stamp collect
cf. bird-watching
Comment: This is a derogatory term that suggests taxonomy is not "real science," but Gould (1987) discusses why this term is inappropriate.

♦ **stamping ground** *n.*
1. A place where Horses, or other animals, gather (Michaelis 1963).
2. For example, in the Blue Wildebeest: a place where an individual displays stamping, striking the ground with its feet (Wilson 1975, 491).

♦ **stand** See ²group: stand.

♦ **standard deviation (s, SD)** *n.* In statistics, a measure of variation within a data set; the square root of the set's variance (Lincoln et al. 1985).

cf. standard error, variance
Comment: For example, the standard deviation of 1, 2, 3, 4, 5 is computed by subtracting each value from the mean (= 3) of this data set, squaring the differences, summing these squares, dividing the sum by $N-1$ (the sample size minus 1), and taking the square root of the results of this division. The SD is 1.58.

♦ **standard error (SE)** *n.* In statistics, a standard deviation of more than one mean (Lincoln et al. 1985).

cf. standard deviation, variance
Comment: The estimated standard error using data from a single mean is the standard deviation (of the mean's data set) divided by the square root of the set's sample size (Lincoln et al. 1985).

♦ **star** *n.* A self-luminous celestial body (exclusive of comets, meteors, and nebulae) (Michaelis 1963).

cf. pulsar, quasar
Comments: In general, stars are luminous due to their nuclear fusion processes (Mitton

1993). Remnants of stars that are no longer luminous are also called stars, *e.g.,* a neutron star (which is a radio-wave-emitting pulsar). Stars are predominantly hydrogen, with helium as their other major constituent. The sun, a typical star in many ways, is 94% hydrogen, 5.9% helium, and less than 0.1% carbon, oxygen, and other elements.

51 Pegasi *n.* A normal star in the constellation Pegasus with one known planet which Swiss astronomers found in 1995 (Wilford 1999a, A1).

neutron star *n.* A star with a mass from about 1.5 to 3.0 solar masses that collapsed under gravity to such an extent that it consists almost entirely of neutrons (Mitton 1993). *Comment:* Neutron stars are about 10 kilometers in diameter and have a density of 10^{17} kg/m³ (Mitton 1993). They are probably comprised of matter of a density second only to black holes (Sawyer 1997, A3).

Sun, sun *n.* The central star of the Solar System (Mitton 1993).

Comments: Generally speaking, the sun is a medium-sized star of medium brightness (Mitton 1993). Specifically, it is a dwarf star of spectral type G2 with a surface temperature of 5700 K.

Upsilon Andromedae *n.* A solar-like star, 44 light years from Earth, with three large planets (Wilford 1999a, A1).

Comments: This star is the first sun-like one found to have multiple planets (Wilford 1999a, A1). The large gaseous planets are not likely to harbor life; however, their possible moons are possible sites for life.

♦ **starch-gel electrophoresis** See method: electrophoresis: starch-gel electrophoresis.

♦ **startle** *n.*
1. A person's start, or shock, of surprise or alarm; a person's being startled (*Oxford English Dictionary* 1972, entries from 1714).
2. An individual animal's sudden involuntary movement due to a particular stimulus (surprise, alarm, acute pain, etc.) (*Oxford English Dictionary* 1972, entries from 1530), *e.g.,* in a fish whose tank is rapped with a hard object or over which a shadow passes (Geiger et al. 1985, 11; R.A. Drummond, personal communication).

v.t. startle

narcotized startle *n.* A startle in which fish appear frightened but move more slowly than usual (*e.g.,* in control fish) (Geiger et al. 1985, 11; R.A. Drummond, personal communication).

Comment: If the fish are fully narcotized, they do not show startle responses to stimuli which ordinarily release them (Geiger et al. 1985, 11; R.A. Drummond, personal communication).

overreactive startle *n.* A startle in which fish appear extremely frightened and form a dense school or swim rapidly (Geiger et al. 1985, 11; R.A. Drummond, personal communication).

♦ **startle reflex** See reflex: startle reflex.

♦ **startle response** See response: startle response.

♦ **stasigenesis** See [2]evolution: stasigenesis.

♦ **stasimorph, stasimorphic character** See character: stasimorph.

♦ **stasimorphic speciation** See speciation: stasimorphic speciation.

♦ **stasipatric speciation** See speciation: stasipatric speciation.

♦ **stasis** *n.* "A period during which no evolutionary change takes place" (Dawkins 1982, 294).

cf. gradualism

typostasis *n.* A relatively static phase in evolutionary history in which few new forms are produced (Lincoln et al. 1985).

cf. -genesis: typogenesis; typolysis

♦ **stasium** See [2]community: stasium.

♦ **stasophile, stasophilous, stasophily** See [1]-phile: stasophile.

♦ **statary phase** See phase: statary phase.

♦ **state** *n.*

1. A particular expression, or condition, of a character (Lincoln et al. 1985).
2. A stage in a plant's life cycle (Lincoln et al. 1985).

See behavior state.

central excitatory state *n.* Changes in nerve excitability that outlast the stimulation of an afferent nerve by a few milliseconds (Sherrington 1906, etc. in Hinde 1970, 268–269).

syn. central motive state

Comment: "Central excitatory state" was later broadened to account for similar effects in more complex responses where time courses are different. This concept becomes less useful in more complicated analyses of neurophysiology (Hinde 1970, 269).

♦ **statics** See study of: physics: statics.

♦ **station** *n.*

1. The precise location of an organism, or species, in its habitat (Lincoln et al. 1985).
2. "A circumscribed area of uniform environmental conditions and vegetation; the typical habitat of a community" (Lincoln et al. 1985).

♦ **stationary-growth phase** See phase: stationary-growth phase.

♦ **stationary hypothesis, Stationary model** See hypothesis: stationary hypothesis.

♦ **stationary model** See hypothesis: stationary hypothesis.

♦ **statistic** *n.* Any function of a sample, often used as an estimate of a corresponding parameter of the population from which the sample is drawn and which is usually denoted with Roman letters (Lincoln et al. 1985).

cf. parameter

♦ **statistical hypothesis** See hypothesis: statistical hypothesis.

♦ **statistical inference** *n.* The process of predicting, or estimating, population parameters on the basis of sample data (Lincoln et al. 1985).

syn. inductive statistics (Lincoln et al. 1985)

♦ **statistical power** See power.

♦ **statistical test** *n.* A method used to draw an inference from data or to attempt to reject a null hypothesis (Lincoln et al. 1985).

cf. statistical design

Comment: Some of the hundreds of statistical tests are defined below.

SOME CLASSIFICATIONS OF STATISTICAL TESTS

I. Classification by conventionality
   A. conventional statistical test
   B. unconventional statistical test

II. Classification by data distribution
   A. nonparametric statistical test
   B. parametric statistical test

III. Classification by level of measurement
   A. interval statistical test
   B. nominal statistical test
   C. ordinal statistical test

IV. Classification by data independence
   A. dependent-data statistical test
   B. independent-data statistical test

V. Classification by number of null hypotheses tested
   A. one-tailed statistical test
   B. two-tailed statistical test

VI. Classification by number of samples
   A. one-sample statistical test
   B. two-sample statistical test
   C. k-sample statistical test

VII. Classification by kind of null hypothesis tested
   A. statistical test for the null hypothesis that a correlation coefficient = 0
   B. statistical test for the null hypothesis that a regression coefficient = 0
   C. statistical test for the null hypothesis that two or more centers of distribution are equal
   D. statistical test for the null hypothesis that two or more means are equal

analysis of covariance (ANCOVA) *n.*
A statistical method that is used to describe
the relationship between a continuous
dependent variable and one or more nominal independent variables and that controls for the effect of one or more continuous independent variables (which are called
covariates) (Kleinbaum and Kupper 1978,
11).

analysis of variance (ANOVA) *n.* A
statistical technique used to assess how
two or more nominal independent variables affect a continuous dependent variable; loosely speaking, ANOVA is usually
concerned with comparisons between, or
among, two or more sample means
(Kleinbaum and Kupper 1978, 11, 244).
Comments: Kinds of ANOVA include
model-1 ANOVA, model-2 ANOVA, nested
ANOVA, 1-way ANOVA, 2-way ANOVA,
nonparametric ANOVA, parametric
ANOVA, and combinations of some of
these kinds (Siegel 1956, chap. 7, 8; Sokal
and Rohlf 1969; Kleinbaum and Kupper
1978). ANOVA can be regarded as a special case of regression, *q.v.* (Kleinbaum
and Kupper 1978, 245).
▶ *t* **test, Student's-*t* test, *t*-test** *n.* ANOVA
used to detect a possible difference between means of two samples; a one-way
analysis of variance with two data sets
(Sokal and Rohlf 1969, 220).
Comment: Other "*t* tests" are used to
look for a possible difference between a
single specimen and a sample or the
possible difference between groups of
paired comparisons (Sokal and Rohlf
1969, 233, 332).
["*t* test" is named after a complicated
mathematical distribution, the *t* distribution, also known as the Student's distribution. "Student" derives from the pseud-
onym of W.S. Gosset, who first described
this distribution (Sokal and Rohlf 1969,
144).]

**before-after-control-impact-pairs
analysis** *n.* A statistical test that compares values in a control and test group
before and after the time of treatment in the
test group (Steart-Oaten et al. 1992 in Perry
1995, 26).

bootstrap *n.* A data-based, computer-
based simulation method used for estimating descriptive statistics and hypothesis
testing (Efron and Tibshirani 1991, 390;
1993, 221).
Comments: This method uses new data sets
drawn from an original one to estimate
statistics, including multimodality of a population, 95% confidence intervals, quantitative differences between two samples, and

standard errors (Efron and Tibshirani 1993,
1, 4, 56, 225–227, 247, 390). Efron introduced the bootstrap in 1979, and the power
of modern computers makes its use fea-
sible. Confidence intervals of a bootstrap
are not formal confidence intervals of other
statistical tests and should not be confused
with them (White et al. 2000, 881).
[From the phrase "to pull oneself up by
one's bootstrap." This is widely thought to
be based on one of the exploits found in
the 18th-century *Adventures of Baron
Munchausen* by Rudolph Erich Raspe. In
one adventure, the Baron fell to the bottom of a deep lake. Just when it looked like
all was lost, he picked himself up by his
own bootstraps.]

correlation test *n.* A statistical method
that determines the association between
two or more variables (*e.g.,* human height
and weight) (Sokal and Rohlf 1969, 495,
549).
syn. regression (but Sokal and Rohlf 1969,
495, indicate that this synonym confounds
matters)
cf. statistical test: regression test
Comments: Kinds of correlation tests include: the parametric test the Pearson's
product-moment correlation coefficient, and
nonparametric tests (contingency coefficient, C; Spearman rank correlation coefficient, rho, ρ; Kendall partial rank correlation coefficient, tau, τ; and Kendall coefficient of concordance, W) (Siegel 1956,
chap. 9). Galton originated the concept of
correlation (Mayr 1982, 697).

cross-validation *n.* A standard statistical test for estimating a prediction error
(Efron and Tibshirani 1991, 237).

discriminate analysis *n.* A statistical
method that determines how one or more
independent variables can be used to discriminate among different categories of
nominal dependent variables (Kleinbaum
and Kupper 1978, 11).

factor analysis *n.* A statistical multivariable method that aims to explain relationships among several difficult to interpret,
correlated variables in terms of a few conceptually meaningful, relatively independent, new composite variables called "factors" (Kleinbaum and Kupper 1978, 11,
376); *e.g.,* among a number of measures of
a capacity such as intelligence or a type of
behavior such as courtship (Immelmann
and Beer 1989, 99).
Comment: Factor analysis can be regarded
as a special case of regression, *q.v.*
(Kleinbaum and Kupper 1978, 245).

Fisher's exact test *n.* A test of statistical
independence of two properties, each

S–Z

occurring in two different states; *e.g.,* moth color (dark and light) and survivorship (high and low) (Sokal and Rohlf 1969, 586).

syn. Fisher exact probability test (Siegel 1956, 96).

jackknife *n.* A data-based, computer-based simulation method for estimating descriptive statistics (Efron and Tibshirani 1993, 141).

Comments: The jackknife was the original computer-based method for estimating biases and standard errors of estimates (Efron and Tibshirani 1993, 141–145). Maurice Quenouille proposed the jackknife estimate of bias in the mid-1950s. This method uses $n-1$ samples from an original sample with n observations; each of these samples is the same as the original one except that it has one different observation removed. For example, in a data set with $i = 1, 2, 3, \ldots n$, the first jackknife sample contains all observations but the first in the series; the second sample, all but the second in the series; etc. The jackknife method is an approximation of the bootstrap method, *q.v.*

jackknife-after-bootstrap *n.* A jackknife method used to measure the variance of a standard error obtained from a bootstrap, $var(\hat{se}_B)$ (Efron and Tibshirani 1993, 275).

multiple-comparisons test *n.* A statistical procedure that looks for possible statistical differences among groups of data, *e.g.,* Scheffé's test, which looks for all possible comparisons among pairs of means, or groups of means (Day and Quinn 1989, 436).

Comment: Day and Quinn (1989) describe many multiple-comparisons tests, including the Mann-Whitney-U statistic, Spjøtvoll method, Behrens-Fisher-*t* test, Welch's test, Studentized-Range statistic (Q), least-significant-difference test (LSD test), Scheffé's test, Bonferroni method, Tukey's method, Duncan's-multiple-range new test, Waller-Duncan-*k*-ratio test, and Dunnett's test.

▶ **Dunnett's test** *n.* A parametric test that compares treatments with a control (Dunnett 1955 in Day and Quinn 1989, 438).

cf. statistical test: multiple comparisons test: planned multiple comparisons test: Steel's test

▶ **planned-multiple-comparisons procedure (PMCP)** *n.* Specific comparisons, out of a set of all possible comparisons, that one decides to make the onset of an experiment (Day and Quinn 1989, 435).

cf. statistical test: multiple-comparisons test

Comments: Ideally planned comparisons should be orthogonal; *i.e.,* they should test completely separate hypotheses and thus provide independent pieces of information. Planned-multiple-comparisons procedures include the Behrens-Fisher-*t* test, Cochran's approximate-*t* test, Fligner-Policello test, Mann-Whitney-Wilcoxon-*U* statistic, Spjøtvoll method, Steel-Dwass test, Steel-Dwass-Ryan test, and Welch's test; contrasts and pairwise comparisons.

Behrens-Fisher-*t* test *n.* A parametric PMCP that is robust to unequal variances (Day and Quinn 1989, 435).

Cochran's approximate-*t* test *n.* A parametric PMCP (Day and Quinn 1989, 435).

contrast *n.* A parametric planned comparison method that compares combinations of means with one another; *e.g.,* an average of two means is compared with a third mean (Day and Quinn 1989, 435).

Fligner-Policello test *n.* A robust version of the Mann-Whitney-U test that requires only that the distributions under each treatment are symmetrical (Day and Quinn 1989, 435).

Mann-Whitney-Wilcoxon-*U* statistic *n.* A nonparametric PMCP; perhaps the simplest pairwise test to run (Day and Quinn 1989, 435).

pairwise comparison *n.* A planned comparison of means that compares two means (Day and Quinn 1989, 435).

cf. statistical test: multiple-comparisons test: planned-multiple-comparisons procedure: contrast

Spjøtvoll method *n.* A parametric PMCP that is best for contrasts (Day and Quinn 1989, 435).

Steel-Dwass test *n.* A nonparametric, instantaneous PMCP that uses pairwise rankings (Day and Quinn 1989, 438).

Steel-Dwass-Ryan test *n.* A step-wise *ad hoc* version of the Steel-Dwass test (Day and Quinn 1989, 438).

Welch's test *n.* A parametric PMCP (Day and Quinn 1989, 435).

▶ **Steel's test** *n.* A nonparametric test that compares treatments with a control (Steel 1959 in Day and Quinn 1989, 438).

cf. statistical test: multiple-comparisons test: planned-multiple-comparisons test: Dunnett's test

▶ **unplanned-multiple-comparisons procedure (UMCP)** *n.* A procedure that compares all possible groups within an experiment; contrasted with planned-multiple-comparisons test.

Comments: This procedure is used to look for any possible differences between groups. UMCPs can be inappropriate because (1) an incorrect type of error rate was used, (2) conclusions drawn using these tests are based on inconsistent criteria, (3) means pronounced different at the significance level of 5% by one test may not be significantly different at this level using another test, and (4) commonly used tests are not robust to violations of assumptions. Different methods have been designed for different purposes, and the best method to use for a particular experiment depends on the precise objectives of an experiment (Day and Quinn 1989, 434). Most UMCPs can be simply modified for unequal sample sizes (Day and Quinn 1989, 436). Unplanned-multiple-comparisons procedures include the Bonferroni method, Bryant-Paulson generalization of Dunn-Sidák method, C method, GH method, joint-rank Ryan test, least-significant-difference test, Nemenyi (joint-rank) test, Student-Newman-Keuls method (SNK), T3 method, Tukey's test, and Waller-Duncan *k*-ratio test, as well as Duncan's multiple-range (-new) test, nonparametric UMCPs, Peritz's test, protected-unplanned multiple-comparisons tests, Ryan's procedure, Ryan's *Q* test (Ryan's *F* test), Scheffé's test (Scheffé test), Shaffer's test, simultaneous tests, stepwise tests, and Welsh's step-up (GAPA test).

nonparametric test *n.* A test with these assumptions about the data it analyzes: data are independent, the variable under study has an underlying continuity, and, depending on the test, data can be measured in a nominal, ordinal, or interval scale (Siegel 1956, 31); contrasted with parametric test.
Comment: Examples of nonparametric tests are the Chi-square test, Fisher-exact-probability test, Mann-Whitney-U test, and Wilcoxon-matched-pairs-signed-ranks test, as well as nonparametric correlation and nonparametric regression (Siegel 1956).

one-tailed test, one-sided test, single-tail test *n.* A test in which the critical region comprises values on only one side of the mean (the left or right tail) (Lincoln et al. 1985).

parametric test *n.* A statistical test with these assumptions about the data it analyzes: data are independent and drawn from normally distributed populations, these populations have the same variances (or in special cases, a known ratio of variances), variables are measured in at least an inter-

val scale, and, in analysis of variance tests, population means are linear combinations of effects due to columns, rows, or both (Siegel 1956, 19); contrasted with nonparametric test.
Comment: Examples of parametric tests are the Pearson's product-moment correlation coefficient, parametric analysis of variance, and parametric regression.

Rayleigh test *n.* One of a number of tests used in circular statistics to determine an event's mean direction, or time, and to determine whether it is statistically different from another event's mean (D.B. Quine, personal communication).

regression, regression test *n.* Statistically, an estimation of the relationship between one or more dependent variables and one or more independent variables by expressing the dependent variable, or variables, as a linear, or more complex, function of the others (Kleinbaum and Kupper 1978, 34; Lincoln et al. 1985).
See [2]evolution: regressive evolution.
syn. least-squares regression (Efron and Tibshirani 1991, 392)
cf. statistical test: correlation
Comments: The concept of regression was first developed by Galton (Mayr 1982, 697). Regression-dependent variables are continuous, and independent variables can be any combination of measurement scales (nominal, ordinal, or interval) (Kleinbaum and Kupper 1978, 245). Types of regression include curvilinear regression, linear regression, multivariate regression, nonparametric regression, parametric regression, partial regression, polynomial regression, stepwise regression, and univariate techniques. Analysis of variance (ANOVA), analysis of covariance (ANCOVA), and discriminate analysis can be looked upon as special cases of regression (Kleinbaum and Kupper 1978, 245). Loess, generalized additive models, and CART (classification and regression trees) are computer algorithms that expand the scope of regression (Efron and Tibshirani 1991, 394).

▶ **loess** *n.* A computer-based, curve-fitting method that fits a series of local regression curves for different values of compliance, in each case using only data points near the compliance value of interest; this is contrasted to a regression that fits only one line, or one smooth curve, to a data plot (Efron and Tibshirani 1991, 392).

significance test *n.* A test that provides a criterion for deciding whether a difference between expected and observed values is

S–Z

small enough to be attributed reasonably to chance alone (Lincoln et al. 1985).

single-tail test See test: one-tailed test.

Student's-*t* test See test: analysis of variance: *t* test.

test of independence *n.* A statistical test that determines whether two or more properties (each different states) are unrelated to one another (= independent of one another) in an experiment (Sokal and Rohlf 1969, 585–606).

Comment: Tests of independence include the Chi-square test, Fisher's exact test, and G-test.

two-tailed test, two-sided test *n.* A test in which the critical region includes values on both sides of the mean of the sampling distribution (both right and left tails) (Sokal and Rohlf 1969, 80).
cf. statistical test: one-tailed test

♦ **statistically significant, significant** *adj.* Referring to the outcome of a statistical test that has a "calculated p" < a "stipulated p", *q.v.* (Sokal and Rohlf 1969, 161).
cf. study of: statistics: Bayesian statistics
Comments: If a test is statistically significant, a null hypothesis is rejected, and a person finds a difference between groups, a difference between a correlation coefficient of 0 and a found correlation coefficient, or other difference. Some statisticians consider rejection of a null hypothesis to be a controversial method. The occurrence of a particular phenomenon might be statistically significant in an experiment, but may not be biologically significant to the organism in question (Thomas and Juanes 1996, 856). Also a biologically significant phenomenon might not show statistical significance in an experiment. Statistical and biological significance can be associated through the use of statistical power analysis. Johnson (1996, 860) reports that researchers' "blind reliance" on the "stipulated p" of 0.05 (= the 5% level) can prevent one's making biologically relevant considerations of experiments.

♦ **statistics** *pl. n.*
1. "Quantitative data, pertaining to any subject or group, especially when systematically gathered and collated" (Michaelis 1963).
2. Procedures used to organize data, summarize data, draw inferences from data, or a combination of these activities.
See study of: statistics.
cf. statistical inference, statistical test

♦ **statokinesis** See kinesis: statokinesis.
♦ **status** See [1]rank: dominance rank.
♦ **status competition** See competition: status competition.

♦ **status sign, status signal** See signal: status signal.
♦ **staurogamy** See -gamy: staurogamy.
♦ **steady-state hypothesis** See hypothesis: hypotheses of the beginning of the universe: steady-state hypothesis.
♦ **Steel-Dwass test** See statistical test: multiple-comparisons test: planned-multiple-comparisons procedure: Steel-Dwass test.
♦ **Steel's test** See statistical test: Steel's test.
♦ **steganochamaephytium** See [2]community: steganochamaephytium.
♦ **stellar black hole** See black hole: stellar black hole.
♦ **steno-, sten-** *combining form* "Tight; narrow; contracted" (Michaelis 1963).
cf. eury-
[Greek *stenos*, narrow]
♦ **stenobaric** *adj.* Referring to an organism that is tolerant of a narrow range of atmospheric, or hydrostatic, pressure (Lincoln et al. 1985).
cf. eurybaric
♦ **stenobathic** *adj.* Referring to an organism that is tolerant of a narrow range of depth (Lincoln et al. 1985).
cf. eurybathic
♦ **stenobenthic** See [2]benthic: stenobenthic.
♦ **stenobiont** See biont: stenobiont.
♦ **stenochore, stenochoric, stenochorous** See -chore: stenochore.
♦ **stenocoenose** *adj.* Referring to an organism with a restricted distribution (Lincoln et al. 1985).
cf. eurycoenose
stenoecism, stenoecy, stenoecic, stenecious, stenoecious, stenoekous See -ecism: stenoecious.
♦ **stenograph** See -graph: stenograph.
♦ **stenohaline** See haline: stenohaline.
♦ **stenohydric** *adj.* Referring to an organism that is tolerant of a narrow range of moisture levels or humidity (Lincoln et al. 1985).
cf. euryhydric
♦ **stenohygrobia** See fauna: stenohygrobia.
♦ **stenoionic** *adj.* Referring to an organism that is tolerant of a narrow range of pH (Lincoln et al. 1985).
cf. euryionic
♦ **stenolume, stenophobic** *adj.* Referring to an organism that is tolerant of a narrow range of light intensity (Lincoln et al. 1985).
cf. eurylume
♦ **stenomorphic** See -morphic: stenomorphic.
♦ **stenophage, stenophagous, stenophagy** See -phage: stenophage.
♦ **stenophobic** See stenolume.
♦ **stenoplastic** *adj.* Referring to an organism that exhibits only a limited developmen-

tal response (phenotypic variation) to different environmental conditions (Lincoln et al. 1985).

cf. euryplastic

♦ **stenosubstratic** *adj.* Referring to an organism that is tolerant of a narrow range of substrate types (Lincoln et al. 1985).

cf. eurysubstratic

♦ **stenotherm, stenothermic, stenothermal, stenothermous** See therm: stenotherm.

♦ **stenothermophile, stenothermophilous, stenothermophily** See [1]-phile: stenothermophile.

♦ **stenotopic, stenotopy** See -topy: stenotopy.

♦ **stenotropism, stenotropic** See -tropism: stenotropism.

♦ **stenoxenous** See -xenous: stenoxenous.

♦ **step-fathering** *n.* In an anemone fish: care of eggs in a brooding father's territory by a conspecific male that obtains the territory after removal of the brooding father (Yanagisawa and Ochi 1986, 1769).

♦ **steppe** See habitat: steppe.

♦ **stercoraceous** *adj.* Referring to a merdicole, *q.v.* (Lincoln et al. 1985).

♦ **stereopsis** *n.* An animal's seeing things in three dimensions (Storz 1973, 267).

♦ **stereotaxis, stereotactic, stereotaxy** See taxis: stereotaxis.

♦ **stereotropism, stereotropic** See -tropism: stereotropism.

♦ **stereotype, stereotypic behavior, stereotyped motor acts** See stereotypy.

♦ **stereotyped-action pattern** See pattern: fixed-action pattern.

♦ **stereotyped osmallaxis** See osmallaxis: stereotyped osmallaxis.

♦ **stereotypy** *n., pl.* **stereotypies**
1. An animal's frequent repetition of behavior that serves no obvious purpose (Russell 1934 in *Oxford English Dictionary* 1989).
2. An animal's "normal" uniform repetition of a kind of behavior (Maier and Schneirla 1964, 164; Immelmann and Beer 1989, 291).
   cf. pattern: fixed-action pattern
3. An animal's "abnormal" repetition of a kind of behavior (Hediger 1934 in Meyer-Holzapfel 1968, 479; Mason and Green 1962 in Mitchell 1970, 222); *e.g.,* flying back and forth, head turning, pacing back and forth, rocking back and forth, or vocalizing (Holzapfel 1939, Hediger 1950 in Hinde 1970, 556; Fox 1974, 73; Immelmann and Beer 1989, 291).

syn. behavior stereotype, stereotype (Hinde 1970, 556), stereotyped motor acts (Spinelli and Markowitz 1985), stereotypic behavior (Line 1987, 856)

cf. behavior: abnormal behavior; captivity degeneration; specialization

Comments: Stereotypies in caged canaries are spot picking (Dawkins 1980, 80); in Rhesus monkeys, crouching, thumb and toe sucking, self-clasping, rocking, etc. (Holzapfel 1939, Hediger 1950 in Hinde 1970, 556; Mason and Green 1962 in Mitchell 1970, 222; Fox 1974, 73; Immelmann and Beer 1989, 291); and in young multihandicapped children, eye gouging, hand flapping, mouthing, rocking, self-stimulation with objects, saliva play, and nonfunctional motor movements (Sisson et al. 1988, 509–510). In captive animals, a stereotypy often consists of early stages of a behavior sequence that has become fixed and modified (Hinde 1970, 556). A stereotypy can be due an animal's stimulus deprivation or confinement in an impoverished environment (Dawkins 1980, 80; Immelmann and Beer 1989, 291).

♦ **sterility** *n.*
1. An organism's being unable to produce viable propagules or reproduce sexually (Lincoln et al. 1985).
2. Something's being free from contamination by microorganisms (Lincoln et al. 1985).

adj. sterile

♦ **sterility factor** See factor: sterility factor.

♦ **sternal gland** See gland: sternal gland.

♦ **steroid hormone** See hormone: steroid hormone.

♦ **sterrhium** See [2]community: sterrhium.

♦ **sterrhophile, sterrhophilus, sterrhophily** See [1]-phile: sterrhophile.

♦ **sterric** *adj.* Referring to a heath community or heathland habitat (Lincoln et al. 1985).

♦ **sticking out one's tongue** See display: sticking out one's tongue.

♦ **stigmergy, stigmergic communication** See communication: sematectonic communication.

♦ **stigmergy theory** See [3]theory: stigmergy theory.

♦ **stimulant** See chemical-releasing stimulus: stimulant.

♦ **stimulation** See [2]stimulus.

♦ **[1]stimulus** *n., pl.* **stimuli** Any effect of a stimulus, *q.v.,* on an organism (Immelmann and Beer 1989, 291).

♦ **[2]stimulus** *n., pl.* **stimuli**
1. Physiologically, "something that acts as a 'goad' or 'spur' to a languid bodily organ; an agent that stimulates, increases, or quickens organic activity" (*Oxford English Dictionary* 1971, entries from 1614).
2. Physiologically, "something that excites an organ or tissue to a specific activity or

S–Z

function; a material agency that produces a reaction in an organism" (*Oxford English Dictionary* 1971, entries from 1793).

3. Anything that rouses a person's mind or spirits; an incentive (Michaelis 1963).
4. Any internal, or external, condition, or change, that produces a reaction, or change, in an organism (Lincoln et al. 1985; Immelmann and Beer 1989, 292).

See ²stimulus: sign stimulus.

syn. stimulation (Immelmann and Beer 1989, 292)

cf. chemical-releasing stimulus, stimulation
Comments: Hinde (1970, 58) indicates that "stimulus" does not necessarily imply an identifiable response on the part of an organism and that he does not use it to refer to "hypothetical events postulated to account for changes in behaviour." Tinbergen (1951, 36) indicates that "stimulus" is a term that is open to criticisms.
[Latin *stimulus*]

Two Classifications of Stimuli

I. Classification by their importance to a receiver
  A. irrelevant stimulus
  B. relevant stimulus
    1. minor stimulus
    2. major stimulus
      (*syn.* sign stimulus, releaser)
    3. pseudoreleaser
    4. supernormal stimulus
      (*syn.* superoptimal stimulus)
II. Classification by their adequacy in eliciting a response from a receiver
  A. adequate
  B. inadequate

▸ **adequate stimulus** *n.* A stimulus to which a particular receptor is most sensitive (Heymer 1977, 144; Immelmann and Beer 1989, 294).
cf. ²stimulus: inadequate stimulus
▸ **inadequate stimulus** *n.* A stimulus to which a particular receptor responds at a relatively high threshold in contrast to an adequate stimulus (Heymer 1977, 144).
chemical-releasing stimulus See chemical-releasing stimulus.
conditional stimulus, conditioned stimulus (CS) *n.* In Pavlovian conditioning, a stimulus that elicits a conditional response (CR), *q.v.* (Dewsbury 1978, 322).
cf. learning: conditioning: Pavlovian conditioning; ²stimulus: unconditional stimulus
configurational stimulus *n.* The particular relational structure among stimulus components (Immelmann and Beer 1989, 119).
cf. perception: gestalt perception
directing stimulus *n.* A stimulus that directs an animal's activity in relation to the spatial arrangement of its surroundings (Tinbergen 1951, 82).
cf. ²stimulus: releasing stimulus
external stimulus, external stimulation *n.* A stimulus from outside an animal's body (Immelmann and Beer 1989, 96).
syn. peripheral stimulation (Immelmann and Beer 1989, 96)
cf. ²stimulus: internal stimulus; ¹receptor: exteroceptor
internal stimulus, internal stimulation *n.* A stimulus from inside an animal's body (Immelmann and Beer 1989, 156).
cf. ²stimulus: external stimulus; ¹receptor: interoceptor
key stimulus See ²stimulus: releaser.
milking stimulus *n.* The sum of stimuli applied to a lactating mammal by suckling (Cowie et al. 1951).
motivating stimulus See ²stimulus: priming stimulus.
perceptual stimulus See ²stimulus: releaser.
peripheral stimulus, peripheral stimulation See ²stimulus: external stimulus.
photostimulation *n.* A stimulus comprised of paka-paka, *q.v.*, and flash (a strong light beam).
Comments: Over 700 people (mostly children) in Japan were hospitalized as a result of photostimulation from an animated television show (WuDunn 1997, A3). These people had one or more of these symptoms: dizziness, losing consciousness, seizures, and vomiting blood. Some video games also cause photostimulation. It might be optically stimulated epilepsy in some cases.
priming stimulus *n.* A stimulus that does not have an immediate overt effect on an animal's behavior, but increases the animal's response tendency to subsequent stimuli (Immelmann and Beer 1989, 232).
cf. sensitization
releaser *n.*

1. Characters exhibited by an individual of a given animal species that activate existing releasing mechanisms in conspecifics and elicit certain chains of instinctive behavior patterns (Lorenz 1932, 1935 in McFarland 1985, 364).
2. A single stimulus, or one out of a very few such crucial stimuli, by which an animal distinguishes key objects such as

an enemy, potential mate, and a suitable nesting place (Uexküll 1939; Wilson 1975, 595).

3. "A specific stimulus responsible for evoking a particular fixed-action pattern" (Wittenberger 1981, 620).

4. A relatively simple yet specific stimulus that elicits a fixed-action pattern; *e.g.,* the red breast feathers of a European Robin in another male's territory are a sign stimulus that releases the latter's aggression (Dewsbury 1978, 17).

5. "A stimulus that serves to initiate instinctive activity or mood" (Lincoln et al. 1985).

6. "A sign stimulus used in communication" (Lincoln et al. 1985).

syn. key stimulus, perceptual signs (Tinbergen 1951, 36), perceptual stimulus (Russell 1943 in Heymer 1977, 156), releasing stimulus, sign stimulus (in many usages, Immelmann and Beer 1989, 250), signal stimulus (Russell 1943)

cf. ²stimulus: sign stimulus

Comments: A "releaser" can be a structure, movement, sound, or odor of a conspecific animal (Dewsbury 1978, 18). In Humans, "releasers" include uniforms, signs of rank, emotional displays (Heymer 1977, 35). One person's "yawn" can be a "releaser" for a yawn in another person; a "yawn" can be considered to be a stereotyped-action pattern (Provine and Hamernik 1986). Immelmann and Beer (1989, 270) distinguish between "sign stimulus" and "releaser."

▶ **interspecific releaser** *n.* A releaser used in communication between species, such as symbionts (Heymer 1977, 35); *e.g.,* alarm calls of songbirds, warning tail beating of a goby that causes its symbiotic shrimp to withdraw into its hole, pheromonal mimicry to ant-larval pheromone by inquilines (Immelmann and Beer 1989, 156).

cf. chemical-releasing stimulus: semiochemical: allelochemic: allomone

▶ **social releaser** *n.* A releaser that elicits innate behavior from a conspecific [translated from *Auslöser* (releaser) of Lorenz (1935 in by Tinbergen 1948) by Tinbergen (1951)].

▶ **supernormal stimulus, supernormal releaser, super-releaser** *n.*

1. A stimulus that has one or more exaggerated attributes and tends to elicit a greater response than the more usual stimuli in its category; *e.g.,* Ringed Plovers (birds) prefer white eggs with black spots to their own eggs with less contrasting spotting, Oystercatchers (birds) prefer five-egg clutches to their usual three-egg clutches, and host birds give young cuckoos which have very large gaping mouths more attention than their own young (Hailman in Wilson 1975, 253).

Note: "Supernormal stimuli" were discovered by Koehler and Zagarus (1937 in Heymer 1977, 183).

2. A releaser that is preferred to a natural "appropriate" stimulus when an animal may choose between a supernormal stimulus and a natural one (Immelmann and Beer 1989, 300).

syn. supernormal-sign stimulus (Tinbergen 1951, 44), superoptimal stimulus (which is little used now because if one considers "optimal" to be the best possible, then "super" is a superfluous prefix; Immelmann and Beer 1989, 300)

sign stimulus *n.* A stimulus that has not evolved as a signal but is used as a signal by a receiver, *e.g.,* branches of a tree with a color that makes it easier for birds to recognize the tree as a nesting site (Immelmann and Beer 1989, 270).

syn. key stimulus [*Schlüsselreiz,* preferred by German ethologists because it suggests a key that unlocks the "gate" of a relevant releasing mechanism (Immelmann and Beer 1989, 270)]

cf. stimulus; stimulus: releaser

Comment: "Sign stimulus" is sometimes distinguished from "releaser" which is a stimulus that evolved as a signal and has adaptive significance for both a sender and receiver (Immelmann and Beer 1989, 250) (*e.g.,* a kairomone).

signal stimulus See ²stimulus: releaser.

social stimulation *n.* A stimulus that affects a focal organism and emanates from a conspecific (Immelmann and Beer 1989, 277, 292).

stress See stress.

subliminal stimulus *n.* A stimulus that is below an organism's threshold level or insufficient to elicit an organism's response (Lincoln et al. 1985).

subthreshold stimulus *n.* A stimulus whose intensity, or other characteristics, is not sufficient to produce a perceptible effect on an animal (Immelmann and Beer 1989, 312).

supernormal stimulus, supernormal releaser, supernormal-sign stimulus, superoptimal stimulus, super-releaser See ²stimulus: releaser: supernormal stimulus.

unconditional stimulus, unconditioned stimulus (UCS) *n.* In Pavlovian conditioning, a stimulus that elicits an unconditional response (UCR) (Dewsbury 1978, 322).

cf. learning: conditioning: Pavlovian conditioning; [2]stimulus: conditional stimulus

zeitgeber, Zeitgeber *n.* In many animal species: a periodic environmental stimulus to which an animal's biological rhythm is entrained (Dewsbury 1978, 65), *e.g.,* the light-dark cycle or lengthening of days in spring (Immelmann and Beer 1989, 330). *syn.* proximate causation, *q.v.,* proximate factor, *q.v.,* synchronizer, *q.v.* (Lincoln et al. 1985)

cf. factor

Comment: The term "*zeitgeber*" is usually restricted to include stimuli in an animal's external environment (Immelmann and Beer 1989, 330).

[coined by Aschoff (1960 in McFarland 1985, 291–292) from German *Zeitgeber,* time giver]

♦ **stimulus adaptation** See learning: habituation.

♦ **stimulus filtering** *n.*
1. An organism's selective neural transmission of energy input from its sensory receptors to its brain or from its lower-level brain centers to its higher-level brain centers (Wittenberger 1981, 622).
2. An organism's selectively responding to some stimuli and its ignoring (filtering out) other stimuli that impinge upon it (McFarland 1985, 210).

peripheral stimulus filtering, peripheral filtering *n.* Stimulus filtering, *q.v.,* in an animal's sense organs (Immelmann and Beer 1989, 292).

♦ **stimulus generalization, generalization** *n.* An animal's first learning to react in a particular way to a particular stimulus and then reacting in the same way toward a similar stimulus; *e.g.,* a toad learns that a black-and-yellow bumble bee is noxious when it tries to eat it due to its sting and then avoids all similar looking insects, even perfectly edible, nonstinging ones (Immelmann and Beer 1989, 292).

cf. learning: habituation

♦ **stimulus modality** See modality.

♦ **stimulus-response psychology** See study of: psychology: stimulus-response psychology.

♦ **stimulus-response relationship** See relationship: stimulus-response relationship.

♦ **stimulus specificity** *n.* A receptor's, or behavior's, characteristic selective responsiveness to a stimulus; *e.g.,* different vertebrate retinal photoreceptor cells have different spectral sensitivities (Immelmann and Beer 1989, 294).

cf. [2]stimulus: adequate stimulus

♦ **stimulus threshold** See threshold: stimulus threshold.

♦ **sting pheromone** See chemical-releasing stimulus: semiochemical: alarm pheromone.

♦ **stink fight** *n.* In males of the Ring-Tailed Lemur: an individual's aggressive behavior that is directed toward another and involves use of odors from pheromone glands (Jolly 1966 in Wilson 1975, 531).

♦ **"stipulated p"** See p: "stipulated p."

♦ **stirp** See germ plasm.

♦ **stochastic** *adj.*
1. Of, referring to, or characterized by conjecture; conjectural (Michaelis 1963).
2. Referring to the process of selecting from among a set of theoretically possible alternatives those elements, or factors, whose combinations will most closely approximate a desired result (Michaelis 1963).
3. In mathematics, referring to a model that "takes into account variations in outcome that are due to chance alone" (Wilson 1975, 596).

syn. probabilistic

cf. deterministic

[Greek *stochastikos* < *stochazesthai,* to guess at < *stochos,* mark, aim; *stochazein,* to shoot with a bow at a target; *i.e.,* scatter events in a partially random manner some of which achieve a preferred outcome (Bateson 1979, 245)]

♦ **stochastic analysis** See analysis: stochastic analysis.

♦ **stochastic-genetic cause of phenotypic variation** *n.* Variation based "on developmental deviations caused by somatic mutations within the lifetimes of particular organisms" (Wilson 1975, 68)

♦ **stochastic model** See [4]model: stochastic model.

♦ **stochastic process** See process: random process.

♦ **stochastic series of events** *n.* A limited set of events that has a random component and a selective process so that "only certain outcomes of the random are allowed to endure" (Bateson 1979, 245).

cf. random sequence of events

♦ **stochastic theory of mass behavior** See [3]theory: stochastic theory of mass behavior.

♦ **stochastic variability** See [3]theory: stochastic theory.

♦ **stock** *n.*
1. A race, stem, or lineage (Lincoln et al. 1985).
2. An organism population (Lincoln et al. 1985).

cf. population, race

♦ **Stoic** *n.*
1. A member of a school of Greek philosophy founded by Zeno about 308 BC,

holding that wisdom lies in being superior to passion, joy, grief, etc. and in unperturbed submission to the divine will (Michaelis 1963).

2. One who supports pantheistic ideas and believes in a designed world created for the benefit of Humans; one who believes "the object of philosophy [is] to understand the order of the world" (Mayr 1982, 90).

[Latin *Stoicus* < Greek *Stoikos* < *Stoa* < *Poikileē,* Painted Porch, the colonnade at Athens where Zeno taught]

♦ **stomach-brooding** See brooding: stomach-brooding.

♦ **stomodeal food** See food: stomodeal food.

♦ **stomodeal trophallaxis** See trophallaxis: stomodeal trophallaxis.

♦ **stomping** *n.* An animal's treading heavily (Michaelis 1963).
cf. grunt

♦ **story telling** *n.* A pejorative term for one's making seemingly unreasonable hypotheses about adaptiveness of organisms' traits (Thornhill and Alcock 1983, 12).
cf. just-so story

♦ **stotting** See display: stotting.

♦ **straight run** *n.* In the Honey Bee: a middle run made by a worker during her waggle dance and the element that contains most of the symbolical information concerning the location of a target outside her hive; "a dancing bee makes a straight run, then loops back to the left, or right, then makes another straight run, then loops back in the opposite direction, and so on — the three basic movements together form the characteristic figure-eight pattern of a waggle dance" (Wilson 1975, 596).
cf. dance

♦ **strain** *n.* An inbred stock of a species produced by artificial selection, *e.g.,* ebony and vestigial winged *Drosophila* (Immelmann and Beer 1989, 296).

♦ **strange species** See ²species: strange species.

♦ **stratagem** See strategy.

♦ **strategist** *n.* An organism that uses a particular kind of strategy, *q.v.*
See ²species: C-selected species, K-selected species, r-selected species, and s-selected species.

♦ **strategy** *n.*
1. A commander-in-chief's science and art (of projecting and directing the larger military movements and operations of a campaign) (*Oxford English Dictionary* 1972, entries from 1810); distinguished from tactics (Michaelis 1963).
2. A person's use of a stratagem (device for obtaining an advantage, a trick) or arti-

fice, as in business or politics (Michaelis 1963).
3. A person's plan, or technique, for achieving some end (Michaelis 1963).
4. A genetically determined life history, or behavior program (composed of tactics), that dictates an organism's action, or course of development, throughout its life (Dawkins 1977; 1983, 294; Gross 1984, 58).
Notes: Strategies evolve through changes in their tactics (Gross 1984, 58). "Strategies" may also be determined by an individual's experience (Maynard Smith 1982, 10). Examples of strategies are age of first reproduction, growth form, and relative numbers of sons and daughters.
5. In game theory, a complete specification (set of rules) of what a contestant (animal or theoretical party) will do in every situation in which it might find itself (Dawkins 1982, 294).
6. An animal's "behavioural phenotype; *i.e.,* it is a specification of what an individual will do in any situation in which it may find itself" (Maynard Smith 1982, 10).
7. An organism's complex of tactics that it employs in "attempting to reach, or actually reaching," a biological goal (*e.g.,* growing, maintaining itself, or reproducing) (inferred from Wootton 1984, 3).
8. The plan, method, or structure used by an individual organism, or organism group, to meet a particular set of conditions (Lincoln et al. 1985); roughly, an animal's "way of life" (Immelmann and Beer 1989, 296).
syn. behavioral pattern (a confusing synonym; Dawkins 1982, 294), stratagem (Wilson 1975, 346), tactic (a confusing synonym; Lincoln et al. 1985; Wootton 1984, 1). "Strategy is a much abused but ... trendy synonym for an animal's behaviour pattern'" (Dawkins 1982, 294). Wootton (1984, 1) distinguishes between "strategy" and "tactic." Chapleau et al. (1988) suggest "pattern" and "option" for "strategy" which they feel do not imply purposefulness. Dawkins (1986, 113) suggests that "program" or "phenotype" might be a better word to describe the concepts labeled by "strategy."
cf. ¹tactic; tactics; ³theory: game theory
Comments: Nonhuman animals are generally not believed to use "strategies" as results of conscious reasoning (Dawkins 1977, 1983, 294; Maynard Smith 1979). Zimmerman and Hicks (1985) also discuss the usefulness of the term "strategy."

adaptive strategy *n.* The sum total of a species' adaptations to its environment (Lincoln et al. 1985).

S–Z

"best-of-a-bad-situation strategy" *n.* An organism's strategy that does not yield the highest fitness of all alternative strategies in its population, but is the best one that the organism can use in a particular constrained situation (*e.g.,* sneak copulation by males that are too small to use physical aggression to obtain a copulation) (inferred from Gross 1984, 58).

syn. best of a bad situation, BBS (Gross 1984, 58)

bonanza strategy *n.* For example, in some dung-beetle species: an individual's exploiting food sources (*e.g.,* dung) that are very rich but at the same time scattered and ephemeral (Wilson 1971, 1975, 49).

C strategy See ²species: C-selected species.

Concorde strategy See effect: Concorde effect.

conditional strategy *n.*

1. In game theory, a theoretical strategy that is composed of more than one tactic that are genetically facultative, provide unequal fitness, and reflect individual competitive ability; contrasted with mixed strategy and pure strategy (Maynard Smith 1982; Gross 1984, 58–59, illustration).
   Note: In organisms, it involves a biont's performing one action if its interactor performs one particular action and performing an alternative action if its interactor performs another particular action; *e.g.,* retaliate if an opponent attacks and do not retaliate if the opponent does not attack (Maynard Smith 1982).
2. A strategy that involves alternative behavior patterns that an animal employs after ascertaining what competitors and other conspecifics are doing (Thornhill and Alcock 1983, 287).
   Comments: These alternatives are not always all shown by one particular animal, and they do not necessarily result in equal reproductive gains. All animals of the same sex in a species are likely to have the genetic basis for all "strategies" within a conditional strategy. For example, if there is a suitable space available, a mate-searching male bee would be territorial; if the space were not available, he would engage in nonterritorial scramble competition for mates (Thornhill and Alcock 1983, 287).

domestic-bliss strategy *n.* A female's examining conspecific males and trying to spot signs of fidelity and domesticity in advance (Dawkins 1977).

evolutionary-stable strategy (ESS) *n.* "A strategy such that, if all the members of a population adopt it, then no mutant strategy could invade the population under the influence of natural selection" (Maynard Smith and Price 1973 in Maynard Smith 1982, 2, 10).

syn. unbeatable strategy (Hamilton 1967 in Maynard Smith 1982, 2)

Comments: Depending on the circumstances, a population, rather than individuals within it, may be an ESS (Maynard Smith 1982, 204). For example, an ESS could involve (1) all animals' of a population always showing behavior A, (2) a certain proportion's always showing behavior A and a certain proportion's showing B, (3) all animals' showing A X% of the time and showing B Y% of the time, (4) all animals' showing A under environmental condition 1 and showing B under condition 2, and so forth. The ESS appears to be an empirical fact, not an hypothesis to be tested.

[coined by Maynard Smith and Price (1973)]

▸ **mixed-evolutionary-stable strategy (mixed ESS)** *n.*

1. An ESS with a strategy polymorphism in which bionts, or a certain proportion of organisms, use one action in one situation, and another action in another situation, depending on what other organisms of its population are doing (Maynard Smith 1982, 15).
   Note: Mixed ESSs may involve conditional strategies (Maynard Smith 1982, 15).
2. An ESS in which an animal has a genetically programmed ability to spend a set proportion of its time in two distinct behaviors with its switch point set genetically and in which both alternative behaviors result in the same reproductive success (Thornhill and Alcock 1983, 291).

cf. strategy: mixed strategy

Comment: This strategy might occur in insect reproduction (Thornhill and Alcock 1983, 291).

▸ **pure-evolutionary-stable strategy (pure ESS)** *n.* An ESS in which all organisms of a population use the same actions in encounters with others (Maynard Smith 1982, 15).

cf. strategy: pure strategy.

K strategy See ²species: K-selected species.

mixed strategy *n.* In game theory, a theoretical strategy that is composed of more than one tactic; the tactics are genetically facultative or stochastic, may or may not reflect individual competitive ability, and provide equal fitness; contrasted with conditional strategy and pure strategy (Gross 1984, 58–59, illustration).

Comments: Theoretically, a species can show a mixed strategy at the individual level in which a single individual has more than one strategy or at the population level in which different individuals each show one strategy or more than one possible strategies (Gross 1984, 59, illustration).

pure strategy *n.* In game theory, a theoretical strategy that is composed of one tactic that is genetically fixed; contrasted with conditional strategy and mixed strategy (Gross 1984, 58–59, illustration).

r strategy See ²species: r-selected species.

reproductive strategy *n.* An organism's complex of reproductive tactics that it employs in attempting to leave, or actually leaving, offspring (inferred from Wootton 1984, 3).

Comment: For example, in a fish, such traits include age of first reproduction, age-specific fecundity, degree of parity, clutch size, gamete nature and size, organization of reproductive behavior, sex change (in some species), and timing of its reproductive season (Wootton 1984, 3).

s strategy See ²species: s-selected species.

truly separate strategy *n.* In some cricket species, possibly in some fig-wasps species: one of two alternative strategies that is linked causally to genes different from those that cause the other strategy (Hamilton 1979, Cade 1981 in Thornhill and Alcock 1983, 291).

Comment: Both members of a pair of these alternative strategies confer the same average reproductive success (Hamilton 1979, Cade 1981 in Thornhill and Alcock 1983, 291).

unbeatable strategy See strategy: evolutionary stable strategy.

wolf-in-sheep's-clothing strategy *n.* In a green lacewing: a larva's disguising itself as an aphid by plucking some of the waxy wool from aphid bodies and applying this material to its own dorsum which protects the larva from ants that ordinarily shepherd the aphids.

[coined by Eisner et al. (1978, 790); wolf refers to the lacewing larva's being a "wolf," the predator of aphids, also known as ant cows, and "sheep's clothing" refers to the covering of the herded animals, the aphids]

♦ **stratification** *n.*
1. "Organization into horizontal strata" (Lincoln et al. 1985).
2. Vertical structuring of a community, or habitat, into superimposed horizontal layers (Lincoln et al. 1985).

3. The grouping of individuals of a community, or habitat, into height classes (Lincoln et al. 1985).

♦ **stratigraphic paleontology** See study of: paleontology: stratigraphic paleontology.

♦ **stratocladistic** See cladistic.

♦ **stratosphere** See -sphere: stratosphere.

♦ **stratospheric ozone** See molecule: ozone.

♦ **stream-gauging station** *n.* A location with instruments used for measuring stream flow and water qualities.

Comment: In the Fernow Experimental Station, water from a watershed is channeled into a weir pond and over a V-notch so flow can be measured. The weirs are 120-degree, sharp-crested, V-notch weirs (Adams et al. 1994, 3, illustration).

♦ ¹**stress** *n.*
1. Any aversive stimulus (Wood-Gush et al. 1959 in Dawkins 1980, 59).
2. An environmental factor that restricts growth and reproduction of an organism or population (Lincoln et al. 1985).
3. A factor that disturbs a system's equilibrium (Lincoln et al. 1985).

♦ ²**stress** *n.*
1. A person's "hardship, straights, adversity, affliction" (a now obsolete definition; *Oxford English Dictionary* 1972, entries from 1303).
2. An individual organism's being "hard pressed" (a now obsolete definition; *Oxford English Dictionary* 1972).
3. A person's, or his part's, witnessing strained exertion, strong effort (a now rare definition; *Oxford English Dictionary* 1972, entries from 1690); strain upon a person's organ or mind (*Oxford English Dictionary* 1972, entries from 1843); a person's emotional, or intellectual, strain or tension (Michaelis 1963).
4. Lack of entropy; a condition that arises when an organism's external environment, or internal sickness, makes excessive or contradictory demands on its ability to adjust; thus, "the organism lacks and needs *flexibility*, having used up its available uncommitted alternatives" (Bateson 1979, 245).
5. A series of physiological changes, such as release of hormones, that goes on within an animal's body when it is subjected to injury, temperature extremes, etc. (Dawkins 1980, 55).
6. The whole general-adaptation syndrome, *q.v.*, or the latter part of it (Dawkins 1980, 60).

cf. animal suffering; general-emergency reaction; pain; syndrome: fight-or-flight syndrome, general-adaptation syndrome

Comment: The definition of "stress" (= "animal stress") is moot, and this controversy has led to some dubious conclusions about what factors cause "stress" (Dawkins 1980, 61).

light stress *n.* In rats: reaction to a bright light in the center of a cage that is measured as defecation frequency, displacement grooming, ambulation, etc. and can be inhibited, or facilitated, with different drugs (Ryall 1958 in Heymer 1977, 111).

thermal stress *n.* Any change in an organism's thermal relationship with its environment which, if uncompensated by a temperature-regulatory response, disturbs the organism's thermal equilibrium (Bligh and Johnson 1973, 957).

♦ **stress avoidance** *n.* An organism's stress resistance, *q.v.*, by avoiding thermodynamic equilibrium with stress either by a physical barrier that insulates its living tissue from the stress or by a steady-state exclusion of stress (*e.g.*, a chemical, or metabolic, barrier) (Lincoln et al. 1985).
cf. stress tolerance

♦ **stress hypothesis** See hypothesis: population-limiting hypotheses: stress hypothesis.

♦ **stress resistance** *n.* An organism's ability to survive exposure to an unfavorable environmental factor, quantifiable as the stress necessary to produce a specific strain (Lincoln et al. 1985).

♦ **stress syndrome** See syndrome: general adaptive syndrome.

♦ **stressor** See stress.

♦ **stretching syndrome** See syndrome: stretching syndrome.

♦ **strict endemic** See [1]endemic: strict endemic.

♦ **stridulation** See animal sounds: stridulation.

♦ **string** See [2]group: string.

♦ **stringency, principle of** See principle: principle of stringency.

♦ **stripes display** See display: stripes display.

♦ **stromatolite** *n.* A knobby intertidal structure made of banded sediment (carbonate mud) that has been constructed by motile bacteria and cyanobacterial colonies as well as these organisms themselves (Campbell 1987, 512–513; Stanley 1989, 94).
cf. fossil: stromatolite fossil
Comments: Stromatolites are made from sediments that stick to the jellylike coats of these microbes. They migrate out of one layer of sediment and form a new one above. As this happens repeatedly, a layered mat becomes thicker (Campbell 1987, 512–513). Stromatolites are almost exclu-

sively found in supratidal and high intertidal settings which are periodically hot and dry and where few grazing marine animals can survive (Stanley 1989, 96). Stromatolites also grow in subtidal channels where tidal currents are very strong and where few animals can live.
[Greek *stroma*, bed; *lithos*, rock]

♦ **strophogenesis** See -genesis: strophogenesis.

♦ **strophotaxis, strophism, strophotactic** See taxis: strophotaxis.

♦ **structural activities** See activity: structural activities.

♦ **structural botany** See study of: botany: structural botany.

♦ **structural chemistry** See study of: chemistry: structural chemistry.

♦ **structural gene** See gene: structural gene.

♦ **structural geology** See study of: geology: structural geology.

♦ **structuralism** *n.*
1. A person's tendency to emphasize structure of something rather than its function (Michaelis 1963); the doctrine that structure rather than function is important (Hughes and Lambert 1984).
2. The policy of concentrating upon structure in different fields including linguistics and biology (Michaelis 1963).
cf. study of: botany: structural botany, geology: structural geology, chemistry: structural chemistry
3. The ideas that natural selection results in a kind of structure (form) that is associated with one particular function, a new kind of function might materialize from this structure, evolution of new structures does not follow all possible paths dictated by external forces, and constraints are imposed both by starting points and by rules of transformation specified by the nature of organic materials; structure determines function (Mayr 1982, 463).
cf. creationism, functionalism
Comments: Structuralism attempts to understand the principles of organization that represent the conceptual basis within which we can properly think about evolution (D.M. Lambert, personal communication). According to one structuralist viewpoint, neither the *elements* nor the *whole* should be the focus of attention but the relationship between them. Structuralists include Georges Cuvier and Richard Owen (Gould 1986, 18).

♦ **struggle for existence** *n.*
1. Interactions among organisms including competition and symbiosis; "I use the term Struggle for Existence in a large and metaphorical sense, including dependence of one being on another" (Darwin 1859, 62).

2. Population members' competition for a limited, vital resource that results in natural selection (Lincoln et al. 1985).
cf. Darwinism

♦ **stubborness** See ²group: stubborness.

♦ **stuck** *n.* In *Drosophila*: a mutation in which a male has difficulty disengaging after a 20-minute copulation, which is usual for his species (Dewsbury 1978, 126).
See ²group: stuck.

♦ **stud** See ²group: stud.

♦ **Student's-*t* test** See statistical test: *t* test.

♦ **study of**
cf. modes of evolution
Comments: "Study of" in the following definitions usually implies the *scientific* study of. There are many ways to classify disciplines and their branches. Here, I generally classify them by main words, *e.g.*, biology, ecology, or geography. Many of these terms end with "-logy," a Greek suffix that denotes knowledge of, science, and study of. Algology, humanology, and mammalogy, for example, are hybrid words derived from both Latin and Greek. Brown (1956, 35) advises coinage of purebred words rather than such "mongrelizations" when possible. Bartholomew (1987, 15) writes, "Composing a definition for a field in biology can be a sterilizing effort because definitions delimit, and biology is functionally indivisible." Therefore, he defined physiological ecology operationally without confining it.

acoustics *n.* The study of sound and hearing (*Oxford English Dictionary* 1972, entries from 1683).

▸ **bioacoustics** *pl. n.* (construed as singular) The study of nonhuman-animal sounds, sound production, and relationships between the characteristics of a nonhuman animal's sounds and its environment (Immelmann and Beer 1989, 32–33).

▸ **psychoacoustics** *pl. n.* (construed as singular) Acoustics with reference to the physiological basis and effects of sound (Michaelis 1963).

adaptive correlation *n.* The study of adaptive correlation, *q.v.* (Immelmann and Beer 1989, 5).

aetiology See study of: etiology.

agrostology See study of: graminology.

aktology *n.* The study of shallow inshore ecosystems (Lincoln et al. 1985).

algology See study of: phycology.

American behaviorism See study of: psychology: behaviorism.

animal behavior See study of: behavior.

animal psychology See study of: psychology: animal psychology.

anthecology, autoecology See study of: ecology.

anthropology *n.*
1. The study of "mankind" in the widest sense (*Oxford English Dictionary* 1972, entries from 1593); embracing human physiology and psychology and their mutual bearing (*Oxford English Dictionary* 1972, entries from 1706).
2. The study of "man" from zoological, evolutionary, and historical standpoints (*Oxford English Dictionary* 1972, entries from 1861).
3. The study of humankind as a natural and social entity (Immelmann and Beer 1989, 16, 137).
Comment: Some areas of anthropology overlap with human ethology (Immelmann and Beer 1989, 16, 137).

▸ **cultural anthropology** *n.* Anthropology of human institutions, including language, religion, and art (Immelmann and Beer 1989, 16).

▸ **demography** *n.* The "branch of anthropology that deals with life-conditions of communities of people, as shown by statistics of birth, death, etc." (*Oxford English Dictionary* 1972, entries from 1880).
See study of: demography.

physical anthropology *n.* Anthropology of human racial physique, technology, and accommodation to and exploitation of our natural environment (Immelmann and Beer 1989, 16).

apiculture *n.* The scientific study of Honey Bees and their management for beeswax, commercial rental for pollination services, increased honey production, and queen bees (Buchmann and Nabhan 1996, 242).

apidology See study of: mellitology.

applied physics See study of: physics: applied physics.

autobiology See study of: biology: idiobiology.

autoecology See study of: ecology: autoecology.

bacteriology *n.* The study of bacteria from medical and biological perspectives (Michaelis 1963).

behavior, animal behavior *n.* The study of animal behavior, *q.v.*
cf. behavior: animal behavior; study of: ethology; biology: behavioral biology, psychobiology, sociobiology; psychology; psychology: animal psychology, behaviorism
Comment: "Modern behavior" includes "behaviorism" and ethological and physiological approaches (McFarland 1985, 9).

behavioral biology See study of: biology: behavioral biology.

behavioral science *n.* All of the empirically based investigations of behavior, including ethology, comparative psychology,

S–Z

human psychology, human ethology, various divisions of sociology and anthropology, and behavioral physiology (Immelmann and Beer 1989, 31).

behavioral sciences *pl. n.* The sciences that study human and nonhuman-animal behavior by means of naturalistic observation; includes behavior, psychology, sociology, and social anthropology (Chaplin 1968).

behaviorism See study of: psychology: behaviorism.

biocenology See study of: biocoenology.

biocenotics See study of: biocoenology.

biochronology *n.* Dating of biological events using biostratigraphic, or paleontological, methods (Lincoln et al. 1985).

bioclimatology See study of: climatology: bioclimatology.

biocoenology, biocenology, biocenotics, biocoenotics, cenobiology, coenobiology *n.* The qualitative and quantitative study of organism communities (Lincoln et al. 1985).

biocoenotics See study of: biocoenology.

biodiversity studies *n.* "The systematic examination of the full array of different kinds of organisms together with the technology by which the diversity can be maintained and used for the benefit of humanity" (Ehrlich and Wilson 1991, 758).

bioecology See study of: ecology.

biogeny See study of: evolution.

biohydrology *n.* "The study of interactions between organisms and the water cycle" (Lincoln et al. 1985).
cf. hydrobiology

biological statistics See study of: -metry: biometry.

biology *n.*
1. The study of life [independently coined by Lamarck (1802), Burdach (1800), and Treviranus (1802), according to Mayr (1982, 108, 656)].
2. The division of physical science that deals with organism morphology, physiology, origin, and distribution; sometimes in a narrower sense, physiology alone (*Oxford English Dictionary* 1972, entries from 1819).
See biology.
cf. study of: botany, phycology, zoology
Comments: Biology now includes the study of any aspect of organisms. Alexander (1987, 7) states that the central question of biology is "What is the evolutionary background, or adaptive function, of organism traits?"

▶ **aerobiology** *n.* "The study of airborne organisms" (Lincoln et al. 1985).

▶ **autobiology** See study of: biology: idiobiology.

▶ **behavioral biology** *n.* The "study of all aspects of behavior, including neurophysiology, ethology, comparative psychology, sociobiology, and behavioral ecology" (Wilson 1975).
cf. study of: behavior, behavioral sciences, behaviorism

▶ **cell biology** *n.* An umbrella term for biological phenomena related to cells that occur both inside and in the immediate environment outside cells (*e.g.,* cellular architecture, embryonic development, or neurobiological and immunobiological events) (Hoffman 1992, 34).

▶ **evolutionary biology** *n.* Biological disciplines that deal with the evolutionary process, the characteristics of organism populations, ecology, behavior, and systematics (Wilson 1975, 584).
cf. study of: behavior, ethology, evolution, psychology
Comment: Evolutionary biology (including sociobiology) looks at behavior in terms of its past evolution and does not address problems of present-day causation (McFarland 1985, 257).

▶ **evolutionary developmental biology** *n.* The study of regions of genomes under strong developmental constraints and how gene-expression changes can result in major morphological changes, using phylogenetic comparisons of development [coined by Raff and Popdi in Honeycutt 1997, 37].

▶ **exobiology** *n.* The study of the origin, evolution, and distribution of life in our universe (http://exobio.ucsd.edu/, 2000).
cf. organization: NASA Specialized Center of Research and Training in Exobiology (Appendix 2).

▶ **geobiology** *n.* "The study of the biosphere" (Lincoln et al. 1985).

▶ **hydrobiology** *n.* The study of organisms in aquatic habitats (Lincoln et al. 1985).
cf. study of: biohydrology

▶ **idiobiology, autobiology** *n.* The biology of individual organisms (Lincoln et al. 1985).
cf. study of: ecology: idioecology

▶ **microbiology** *n.* "The study of microorganisms" (Lincoln et al. 1985).

▶ **molecular biology** *n.*
1. A branch of biology that involves studies such as elucidation of structures of molecules (*e.g.,* DNA) and metabolic pathways (*e.g.,* the citric-acid cycle) (Mayr 1982, 122–124).
2. "Branch of biology concerned with explaining biological phenomena in molecular terms" (King and Stansfield 1985).
Note: "Molecular biologists often use the techniques of physical chemistry to

investigate genetic problems" (King and Stansfield 1985).

3. A branch of biology concerned with the structure of DNA, RNA, and proteins and of processes such as regulation of the levels of gene products, replication, transcription, and translation (Futuyma 1998, 625).

▸ **paleobiology, palaeobiology** *n.*

1. The application of theoretical ecology to the study of extinct organisms (Gould 1981a, 13).
   Note: Paleobiology of today is essentially present-day paleoecology, *q.v.*
2. The study of extinct-organism biology (Lincoln et al. 1985).
   cf. study of: paleontology

▸ **paleoecology, palaeoecology** *n.*

1. A largely atheoretical subject of the 1940s and 1950s concerned primarily with recording the kinds of rock types that given taxa preferred and seeking correlations between environment and fossil morphology, largely in the interest of taxonomic refinement (Gould 1981b, 296).
   Note: Starting in the 1970s, paleoecology transformed by incorporating theoretical ecology (Gould 1981b, 296).
2. "The study of fossil communities" (Lincoln et al. 1985).

▸ **phreatobiology** *n.* "The study of ground-water organisms" (Lincoln et al. 1985).

▸ **phyllobiology** *n.* "The study of leaves" (Lincoln et al. 1985).

▸ **phytobiology** See study of: botany.

▸ **population biology** *n.* The study of temporal and spatial distributions of members of populations and their relationships to all relevant biotic and abiotic factors in their environments (Lincoln et al. 1985; Immelmann and Beer 1989, 227).
   Comment: "Population biology" overlaps in particular with "ecology."

▸ **psychobiology** *n.*

1. The study of human minds in relation to anatomy, physiology, and nervous systems (Michaelis 1963).
2. The biological study of psychology; biological aspects of psychology (Michaelis 1963; Delbridge 1981).
3. A vast, highly complex, hybrid field that studies the biological foundations of behavior, including analyses of behavior in terms of biochemical and physiological factors (Storz 1973, 214).
   syn. biopsychology (Michaelis 1963)
   cf. study of: psychology: physiological psychology

▸ **sociobiology** *n.*

1. The systematic study of the biological basis of all social behavior (Wilson 1975, 3).
2. The "study of the adaptive significance of animal behavior" (Wittenberger 1981, 621).
   cf. study of: biology: evolutionary biology; ecology: behavior ecology; ethology
3. "The comparative, evolutionary study of the biological bases of social behavior, integrating traditional ethological approaches with new advances in population biology, behavioral ecology, demography, and life history theory" (Hrdy et al. 1996).

cf. study of: biology: evolutionary biology; ecology: behavior ecology; ethology.
Comments: McFarland (1985, 357) includes "sociobiology" as part of "evolutionary biology." Alexander (1987, 5) discusses some of the problems with the term "sociobiology" and points out that this term has become "a target of derogation and ridicule by nonbiologists, and this has almost certainly delayed the acceptance and use of evolutionary principles by human-oriented scholars." The sociobiological synthesis profoundly transformed the study of animal behavior and large parts of biological anthropology, and more recently it has influenced other fields including art, botany, development, economics, genetics, law, molecular biology, neurobiology, philosophy, political science, psychology, reproduction, and women's studies (Hrdy et al. 1996).

▸ **symbiology** *n.* "The study of symbioses" (Lincoln et al. 1985).

▸ **zoo biology** *n.* The science of phenomena that occur in zoological gardens and — "in the widest sense — are of biological significance" (Hediger 1941 in Immelmann and Beer 1989, 331).
Comment: Zoo biology contains parts of applied ethology, human medicine, pathology, taxonomy, veterinary medicine, and zoological systematics (Immelmann and Beer 1989, 331).

biomechanics *n.*

1. The study of the mechanics of living-organism structures (Lincoln et al. 1985).
2. The mathematical study of the mechanical, or physical, properties of biological materials (*e.g.,* the hydrostatic "water skeletons" of various aquatic animals and the resiliency and stiffness of wood used as cantilever beams) (Buchmann and Nabhan 1996, 242).

biometeorology See study of: meteorology: biometeorology.

biometry See study of: -metry: biometry.
biophysics See study of: physics: biophysics.
biopsychology See study of: biology: psychobiology.
biospeleology, biospeology *n.* "The study of subterranean life" (Lincoln et al. 1985).
biostatistics See study of: -metry: biometry.
biostratigraphy See study of: stratigraphic paleontology.
biostratinomy *n.* "The study of the relationship between fossils and their environment" (Lincoln et al. 1985).
biosystematics See study of: systematics.
botany *n.* The study of plants (*Oxford English Dictionary* 1972, entries from 1696). *syn.* phytology (Lincoln et al. 1985)
Comment: Botany includes bryology, ethnobotany, geobotany, dendrology, graminology, lichenology, paleobotany, phytoecology, phytography, pteridology, silvics, and structural botany.
[Greek *botanikos*, of herbs]
▶ **ethnobotany** *n.* The study of the use of plants by groups of Humans (Lincoln et al. 1985).
▶ **geobotany, phytogeography, plant biogeography** See study of: geography: phytogeography.
▶ **paleobotany, palaeobotany, paleophytology, palaeophytology, phytopaleontology, phytopalaeontology** *n.* "The study of plant life of the geological past" (Lincoln et al. 1985).
Comments: Paleobotany usually includes formal classification and naming of plant fossils and research on their chemical residues, paleoecology, and phylogeny (Buchmann and Nabhan 1996, 251).
▶ **phytography** *n.* "Descriptive botany" (Lincoln et al. 1985).
▶ **structural botany** *n.* The study of the structure and organization of plants (*Oxford English Dictionary* 1972, entries from 1835).
botrology *n.* The science of organizing items, or concepts, into groups or clusters (Lincoln et al. 1985).
bryology, hepaticology, muscology *n.* The study of mosses and liverworts (Lincoln et al. 1985).
caliology *n.* The study of burrows, nests, hives, tubes, and other domiciles made by animals (Lincoln et al. 1985).
carcinology *n.* The study of crustaceans (Lincoln et al. 1985).
cenobiology See study of: biocoenology.
cetology *n.* The study of whales and their kin (Lincoln et al. 1985).

chemistry *n.* The study of elements and their combinations and various phenomena that accompany their exposure to diverse physical conditions (*Oxford English Dictionary* 1972, entries from 1646).
Comment: Chemistry has many divisions, including agricultural chemistry, analytical chemistry, applied chemistry (= practical chemistry), biochemistry, biogeochemistry, inorganic chemistry, organic chemistry, and physical chemistry.
▶ **biochemistry** *n.* A branch of chemistry that deals with processes and physical properties of organisms (Michaelis 1963).
▶ **biogeochemistry** *n.* The study of mineral cycling and organism-substrate relationships (Lincoln et al. 1985).
▶ **structural chemistry** *n.* The study of the position of atoms in compounds (*Oxford English Dictionary* 1972, entry from 1907).
chorology See study of: biogeography.
chronistics *n.* "The study of time relationships of objects or events" (Lincoln et al. 1985).
chronostratigraphy *n.* "The study of age relationships of rock strata" (Lincoln et al. 1985).
climatology *n.* The study of climate (Lincoln et al. 1985).
cf. meteorology
▶ **bioclimatology, ecoclimatology** *n.* The study of biota in relation to climate (Lincoln et al. 1985).
▶ **dendroclimatology** *n.* The study of tree annual growth rings to determine past climatic conditions (Lincoln et al. 1985).
▶ **paleoclimatology, palaeoclimatology** *n.* The study of climates of past geological times (Lincoln et al. 1985).
clinology *n.* "The study of the decline of organisms or groups after acme" (Lincoln et al. 1985).
coenobiology See study of: biocoenology.
cognitive psychology See study of: psychology: cognitive psychology.
comparative psychology See study of: psychology: comparative psychology.
computer vision *n.* A field of study devoted to discovering algorithms, data representations, and computer architectures that embody the principles underlying visual capabilities (Aloimonos and Rosenfeld 1991, 1249).
conchology *n.* "The study of shells" (Lincoln et al. 1985).
cf. study of: malacology; study of: -metry: conchometry
conchometry See -metry: conchometry.
coprology *n.* The study of feces (Lincoln et al. 1985).

ctetology *n.* "The study of acquired characters" (Lincoln et al. 1985).

cybernetics See study of: mathematics.

cycleology *n.* The study of cyclic phenomena in geology and paleontology (Lincoln et al. 1985).

cytology *n.* "The study of the structure and function of living cells" (Lincoln et al. 1985).

demography *n.* The study of populations, especially their growth rates, age structures, and processes that determine these properties, including immigration, emigration, migration, and kinship structure (Lincoln et al. 1985; Immelmann and Beer 1989, 71).
See study of: anthropology: demography.
▶ **adaptive demography** *n.* The age distribution of the sexes and castes that occurs in populations of social animals and that is influenced by selection on the reproductive individuals (Wilson 1975, 15).

dendrochronology See method: dendrochronology.

dendroclimatology See study of: climatology.

dendrology *n.* The study of trees and shrubs (Lincoln et al. 1985).
cf. study of: botany, silvics
▶ **paleodendrology, palaeodendrology** *n.* "The study of fossil trees" (Lincoln et al. 1985).

dermatoglyphics *n.* The study of fingerprints (Brown 1995, A3).
Comments: Francis Galton (1822–1911) is the probable originator of dermatoglyphics (Brown 1995, A3). The chance of a match of fingerprints from two people is much rarer than that of DNA fingerprints.

development See study of: embryology.

dynamics Also, see study of: physics: dynamics.
▶ **photodynamics** *n.* The study of light effects on living organisms and their metabolic processes (Lincoln et al. 1985).
▶ **population dynamics** *n.*
1. The study of changes within populations and factors that cause, or influence, these changes (Lincoln et al. 1985).
2. "The study of populations as functioning systems" (Lincoln et al. 1985).
▶ **psychodynamics** *pl. n.* (construed as singular) "The study of mental processes in action" (Michaelis 1963).
▶ **syndynamics** *pl. n.* (construed as singular) "The study of successional changes in plant communities" (Lincoln et al. 1985).

dysteleology *n.* The study of apparently functionless, rudimentary organs to sustain the doctrine of dysteleology, *q.v.* (*Oxford English Dictionary* 1972, entries from 1874).

earth-system science *n.* The scientific study of interactions between the solid Earth (from its crust to core) and its atmosphere, biosphere, and hydrosphere (Fyfe 1996, 448).

ecoclimatology See study of: climatology: bioclimatology.

ecography See study of: ecology: ecography.

ecological sociology See study of: ecology: synecology.

ecology, oecology *n.*
1. The study of organisms' interactions with their environments, including both their abiotic (= physical) and biotic components; their habits and modes of life, etc. (*Oxford English Dictionary* 1972, entries from 1873; Mayr 1982, 121).
syn. bioecology, bionomics, bionomy, mesology, oikology (Lincoln et al. 1985)
Note: Haeckel (1866) proposed "ecology" to label the science dealing with the "household of nature" (Mayr 1982, 121). Ecology also includes organism reactions to factors in their environments.
2. The study of human populations in terms of physical environment, spatial distribution, and cultural characteristics (Michaelis 1963).
3. The study of relationships among organisms and their past, present, and future environments (Ecological Society of America, http://esa.sdsc.edu, 1999).
Note: These relationships include interactions among species, organization of biological communities, physiological responses of individuals, processing of energy and matter in ecosystems, structure and dynamics of populations (Ecological Society of America, http://esa.sdsc.edu, 1999).
Comments: Behavior ecology can be classified as part of sociobiology. Population biology overlaps with ecology (Immelmann and Beer 1989, 83). "Ecology" has many metaphors including: (1) the well-being of bio-habitats existing outside of Humans; (2) cleansing ourselves of all mental pollution; (3) housing, *e.g.,* black housing and white housing; and (4) the circumstances, pressures, and requirements placed on people in their work environments (Wali 1995). Wali (1995) lists many terms that contain ec-, eco-, ecologic, ecological, ecology, and environmental, and he discusses some of them.
[Greek *oikos*, home + -OLOGY]
▶ **anthecology** *n.* The study of pollination and the relationships between insects and flowers (Lincoln et al. 1985).
cf. study of: pollination ecology

▸ **autecology, idioecology** *n*. The ecology of individual organisms or species (Lincoln et al. 1985).

cf. study of: ecology: synecology

paleoautecology, palaeoautecology *n*. The study of the ecology of individual fossil species or groups (Lincoln et al. 1985).

cf. paleosynecology

▸ **behavioral ecology** *n*. The study of how behavior enables animals to adapt to their environments.

cf. study of: psychology: behavioral ecology.

▸ **chemical ecology, ecological biochemistry, phytochemical ecology** *n*. The study of the biochemistry of organism interactions (Harborne 1982, v).

▸ **community ecology** *n*. Ecology that focuses on parts of ecosystems, namely communities of interacting populations (Taylor 1992, 52).

▸ **ecosystem ecology, systems ecology** *n*. Ecology that focuses on measurement of nutrient and energy flows in an ecosystem, in contrast to experiments on well-controlled parts of the system (Taylor 1992, 56).

▸ **genecology** *n*. The study of intraspecific variation and genetic composition in relation to the environment (Lincoln et al. 1985).

▸ **geoecology** See study of: geology: geoecology.

▸ **geographical ecology** *n*. The study of geographic variation of adaptations of organisms to their environments (Mayr 1982, 454).

▸ **ecography** *n*. "Descriptive ecology" (Lincoln et al. 1985).

▸ **oral ecology** *n*. The ecology of some 400 microoganism species (mostly bacterial species but also protozoans and yeast) that ordinarily inhabit human mouths (Stevens 1996a, 314).

Comments: These bacteria evolved with Humans, and most species are not harmful (Stevens 1996a, 314–315). Many species help to fend off disease-producing bacteria. For example, some beneficial bacteria produce proprionic and butyric acids, which kill bacteria that cause human intestinal problems. Humans obtain different main communities of mouth microorganisms just after they are born, when their baby teeth start appearing, and during puberty. Climax communities occur in adults.

▸ **paleoecology, palaeoecology** *n*.

1. A largely atheoretical subject of the 1940s and 1950s concerned primarily with recording the kinds of rock types that given taxa preferred and seeking correlations between environment and fossil morphology, largely in the interest of taxonomic refinement (Gould 1981b, 296).

Note: Starting in the 1970s, paleoecology transformed by incorporating theoretical ecology (Gould 1981b, 296).

2. "The study of fossil communities" (Lincoln et al. 1985).

▸ **phytochemical ecology** See study of: ecology: chemical ecology.

▸ **phytoecology, phytosociology, plant sociology** *n*. Plant ecology (Lincoln et al. 1985).

cf. study of: botany, phytocoenology.

▸ **pollination ecology** *n*. The study of the ecological and evolutionary relationships in pollination (Buchmann and Nabhan 1996, 252).

cf. study of: anthecology

▸ **population ecology** *n*. The ecology of organism populations, including their genetics and demography (Immelmann and Beer 1989, 83).

mathematical population ecology *n*. The quantitative study of population ecology which started in 1930 with Fisher's *The Genetical Theory of Natural Selection*, Wright's (1931) *Evolution in Mendelian Populations*, and Haldane's (1932) *Causes of Evolution* (Wilson 1975, 63).

▸ **restoration ecology** *n*. The science and process of using organisms to restore a human-degraded site to its former conditions, *e.g.*, reforestation of mine tailings or timberland (Buchmann and Nabhan 1996, 253).

▸ **synecology, ecological sociology** *n*. The ecology of organism groups, populations, communities, systems, or a combination of these subjects (Lincoln et al. 1985).

cf. study of: ecology: autecology.

paleosynecology, palaeosynecology *n*. The "study of the ecology of fossil populations or communities" (Lincoln et al. 1985).

cf. study of: ecology: autecology, paleoautecology

▸ **systems ecology** See study of: ecology: ecosystem ecology.

▸ **zooecology** *n*. Animal ecology.

edaphology, pedology, soil science *n*. The study of soils, particularly with reference to biota and human use of land for plant cultivation (Lincoln et al. 1985).

embryology *n*. The study of multicellular organism development (Lincoln et al. 1985).

▶ **behavioral embryology** *n.* The study of behavioral development before an animal is born or hatched from its egg (Immelmann and Beer 1989, 29).

▶ **neuroembryology of behavior** *n.* The study of the joint development of an animal's behavior and its central nervous system made to elucidate this system's control of behavior (Immelmann and Beer 1989, 29).

endocrinology *n.* The study of endocrine glands and their products (*i.e.,* hormones); sometimes considered to be a branch of medicine (Michaelis 1963). [ENDO- + Greek *krinein*, to separate + -LOGY]

▶ **behavioral endocrinology, ethoendocrinology** *n.* The study of endocrinology of behavior and effects of behavior on hormone secretion (Immelmann and Beer 1989, 29).
syn. psychoneuroendocrinology (Dewsbury 1978, 218).

▶ **neuroendocrinology** *n.* The study of hormone secretion of central-nervous-system cells and its function in metabolism, growth, and behavioral regulation (Immelmann and Beer 1989, 202).

energetics *pl. n.* (construed as singular) The study of energy (*Oxford English Dictionary* 1972, entries from 1855).

▶ **bioenergetics** *pl. n.* (construed as singular).
 1. The study of energy relations, exchanges, etc., in living matter (Michaelis 1963).
 2. The study of organisms' caloric requirements with regard to their sizes, kinds of metabolism, and activity patterns (Wilson 1975, 142).

entomology *n.*
 1. The study of insects, including their physiology, distribution, and classification (*Oxford English Dictionary* 1972, entries from 1766); the study of insects only (Torre-Bueno 1978).
 2. Broadly, the study of insects and their arthropod kin, including Chilopods, Diplopods, and nonmarine arachnids and crustaceans.
 3. "The study of insects as one of the best ways to revel in the joys of discovery" (Evans 1985, 27).
adj. entomological
[*entomo* < Greek *entoma*, insects (originally neutral pl. of *entomos*, cut up , *en-* in + *temnein* to cut; with reference to insects' segmented body structure) + LOGY]

entomotomy *n.* The study of insect anatomy (Lincoln et al. 1985).

epigenetics See study of: genetics: epigenetics.

epiontology *n.* "The study of the developmental history of plant distributions" (Lincoln et al. 1985).

epistemology *n.* The study of how organisms, or organism groups, "know, think, and decide" (Bateson 1979, 242).
cf. study of: philosophy: epistemology

ergonomics *n.* The quantitative study of work, performance, and efficiency (Lincoln et al. 1985).

ethnobotany See study of: botany: ethnobotany.

ethnology, ethnography *n.* The study of character, history, and culture of human groups (Lincoln et al. 1985).

ethnozoology See study of: zoology: ethnozoology.

ethography See study of: ethnology.

¹**ethology** *n.*
 1. A portrayal of character by mimic gestures; mimicry (an obsolete definition; *Oxford English Dictionary* 1971, entries from 1656).
 2. The science of ethics; also, a treatise on manners or morals (an obsolete definition; *Oxford English Dictionary* 1971, entries from 1678).
 3. The science of character formation (*Oxford English Dictionary* 1971, entries from 1340).

²**ethology** *n.*
 1. Studies of the relationships among organized beings in families, societies, aggregations, and communities (I.G. Saint-Hilaire 1854 in Gherardi 1984, 377).
 2. "The science of the general conditions of existence which pertain to organisms as individuals and at the same time regulate their relations to other organisms and the inorganic environment" (Wheeler 1902 in Gherardi 1984, 386).
syn. ecology (Wheeler 1902 in Gherardi 1984)
Comment: "Ethology mixes physiology, psychology, morphology, and phylogenetics" (Wheeler in Gherardi 1984, 387).

³**ethology** *n.*
 1. "The objective study of behavior" (Tinbergen 1951, 1).
 2. "The biological study of behaviour" (Tinbergen 1963 in Gherardi 1983, 98).
 3. "The evolutionary study of behavior" (Lorenz in Alexander 1987, 6).
 Note: "Ethology" became associated with Lorenz's particular views of physiology, development, and inheritance of behavior and, as a result, eventually became obscure and less frequently used (Alexander 1987, 6).

S–Z

4. A nebulous term that refers to classical ethology, *q.v.*, modern studies in animal behavior, or a discipline that is some combination of these fields.

Note: "Today, ethology is in a dynamic empirical stage in which it is often not easily distinguished from other branches of the study of animal behavior..." (Dewsbury 1978, 22).

5. "The science of animal behavior" that includes "the scientific investigation of behavior by all means and by all kinds of people" (Barnett 1981, 633).

6. The study of animal behaviors that deals with their causation (including external stimulation, internal physiological mechanisms and states), development (including ontogeny and genetics), ecology (including physiological and learning adaptations), and evolution (including origins and modifications) (Gould 1982, preface; McFarland 1985, 4–5, 357; Immelmann and Beer 1989, 92).

Note: This definition seems appropriate for wide use. The boundaries of ethology and other sciences are ill defined (Immelmann and Beer 1989, 92).

syn. classical ethology (according to some authors)

cf. study of: phylogeny: behavioral phylogeny, classical ethology

Comment: Based on surveys that I undertook at a national meeting of the Animal Behavior Society (1988) and at a meeting of the International Ethological Congress (1989), scientists who study animal behavior vary appreciably in their conceptions of what constitutes "ethology." The subjects that they included in "ethology" ranged from only those investigated by European ethologists to all aspects of animal behavior.

[Latin adjective *ethologia*, character + -LOGY < Greek *ethōs*, character; coined by I.G. Saint-Hilaire (1854), meaning ecology; used for the study of behavior by Dollo (1895 in Heymer 1977, 64). Gherardi (1984, 377) suggests that Mill (1843) used this term before Saint-Hilaire.]

▶ **applied ethology** *n.* Ethology of animal species that are of direct practical interest to Humans, emphasizing practical applications of ethology; applied ethology is comprised of veterinary ethology and part of zoo biology (Immelmann and Beer 1989, 19).

cf. study of: biology: zoo biology

▶ **classical ethology** *n.*

1. "The application of orthodox biological methods to the problems of behavior" (Lorenz 1960a in Lehner 1979, 3).

2. "Study of whole patterns of animal behavior under natural conditions, in ways that emphasize(d) the functions and the evolutionary history of the patterns" (Eisner and Wilson 1975, 1, in Lehner 1979, 2).

3. The objective study of animals, including Humans, from a biological standpoint, emphasizing species-typical behavior and its function and evolution; ethology involves a holistic approach to the study of normal behavior, embodying a detailed analysis of all phases of an animal's behavior (Heymer 1977, 64).

4. The objective study of behavior; the biological study of behavior; the study of instinct; or study of behavior by individuals who like their animals (depending on the author in Dewsbury 1978, 13).

5. A branch of the study of animal behavior that was started in Europe in the 1930s by zoologists and promoted studying behaviors that are meaningful in an animal's natural habitat, beginning an analysis with descriptive studies, studying a wide range of species and behaviors, comparing similar behaviors in closely related species (comparative method), and disparaging the exclusive use of domesticated species (Burghardt 1973 in Dewsbury 1978, 13).

adj. ethological

syn. ethology, European ethology, European traditionalism, Lorenz-Tinbergen school of behavior

cf. study of: behavior; psychology: behaviorism, comparative psychology

Comment: The original ethology group placed a strong emphasis on the study of instinct (Immelmann and Beer 1989, 93).

▶ **cognitive ethology** *n.* The comparative study of animal cognition (Smith 1999, A17).

[coined by Donald R. Griffin]

▶ **comparative ethology** *n.* A branch of ethology that places comparisons of evolutionary, ecological, and morphological aspects of species in the foreground; formulated and explained most original ethological concepts; and recognized and demonstrated that fixed-action patterns are species-specific characters useful in phylogenetic analyses (Immelmann and Beer 1989, 52).

cf. study of: psychology: comparative psychology

Comment: "Comparative ethology" is little used today because a comparative perspective is now used in many other branches of biology (Immelmann and Beer 1989, 52).

▸ **cultural ethology** *n.* A branch of ethology concerned with the intellectual and material products of Humans and their culture in relation to their development, ecological relevance, dependence on innate behavior, and relation to corresponding situations in nonhuman animals (Koenig 1970, 1975 in Heymer 1977, 106).

▸ **European ethology** See study of: ethology: classical ethology.

▸ **human ethology** *n.* A branch of ethology that tries to elucidate the phylogenetic bases of human behavior and determine, with the help of a comparative study of cultures, the innate components of complex human behaviors (Eibl-Eibesfeldt and Hass 1966, etc. in Heymer 1977, 89).
Comments: In contrast to psychology, "human ethology" examines human behavior by observation, not questioning persons; concentrates on humanity, rather than individuals; compares Humans with nonhuman primates, but especially compares human cultures; and examines ontogeny of human behavior. "Human ethology" overlaps with some areas of "anthropology" (Immelmann and Beer 1989, 137).

▸ **modern ethology** *n.* The study of the mechanisms and evolution of behavior, using in particular the conceptual tools: motor programs, drives, programmed learning, and releasers based on physiological-feature detectors (Gould 1982, 65).

▸ **molecular ethology** *n.* The study of biochemical, anatomical, and physiological routes between genes and behavior (Gould's 1974 term in Dewsbury 1978, 124).

▸ **neuroethology** *n.* The study of processes in sense organs and central nervous systems that underlie animals' performances and control of their behaviors (Immelmann and Beer 1989, 202).

▸ **paleoethology, palaeoethology** *n.* The study of the behavior of extinct animals (Immelmann and Beer 1989, 213).
Comment: This is a little-used term (Immelmann and Beer 1989, 213).

▸ **quantitative ethology** *n.* The quantitative recording and subsequent analysis of behavior (Immelmann and Beer 1989, 240).
cf. -metry: sociometry

▸ **veterinary ethology, zoo biology** See study of: applied ethology.

etiology, aetiology *n.* The study of the origins, or causes, of phenomena (Lincoln et al. 1985).
See etiology.

etymology *n.* The branch of linguistics concerned with the origins of words (*Oxford English Dictionary* 1972, entries from 1646).
See etymology.

euthenics *n.* The branch of science that attempts to improve Humans by maintaining, or improving, their environments (Storz 1973, 91).
cf. eugenics

evolution *n.* The study of evolution, *q.v.*
cf. study of: biology: evolutionary biology

▸ **biogeny** *n.* The study of evolution including both ontogeny and phylogeny (Lincoln et al. 1985).

▸ **macroevolution** *n.* "The study of evolutionary changes that take place over a very large time-scale (Dawkins 1982, 289).
cf. evolution: macroevolution.

▸ **molecular evolution** *n.* The study of the evolution of molecules.

classical molecular evolution *n.* The study of molecular evolution emphasizing biochemical and molecular-biology technology to solve traditional evolutionary problems (Levin 1984, 454).
Comment: Interest in molecules themselves and molecular biological and biochemical implications of the evolution of molecules are secondary to molecular biological and biochemical processes in this study (Levin 1984, 454).

neoclassical molecular evolution *n.* The study of molecular evolution emphasizing molecular biological and biochemical processes (Levin 1984, 458).

evolutionary medicine *n.* Medical practices that are based on an understanding of the evolution of Humans and their diseases, *e.g.*, not lowering fever when it is not too high because it is a person's natural means of fighting pathogens (Bull 1995, 1296).
syn. Darwinian medicine (Bull 1995, 1296)

evolutionary physiology See study of: physiology: evolutionary physiology.

exobiology See study of: biology: exobiology.

fluviology *n.* "The study of rivers" (Lincoln et al. 1985).

functionalism See study of: psychology: functionalism.

genealogy, geneology *n.* The study of ancestral relationships and lineages" (Lincoln et al. 1985).
See genealogy.

genecology See study of: ecology: genecology.

geneology See study of: genealogy.

general physics See study of: physics: general physics.

S—Z

genesiology *n.* "The study of reproduction" (Lincoln et al. 1985).

genetics *pl. n.* (construed as singular) The study of heredity (Mayr 1982, 732; King and Stansfield 1985)
See genetics.
Comments: Bateson (1906 in Mayr 1982, 732) proposed "genetics" to designate studies and theories of inheritance of his time. Genetics, as a science, was born in 1900 (Mayr 1982, 781).
[genesis < Greek *genesis*, descent; on analogy with *synthetic, antithetic,* etc.]

▶ **bean-bag genetics** *n.* Genetic studies that ignore gene interaction (Mayr 1982, 558).
[coined by Mayr (1959d) based on the fact that "genetics textbooks in the 1940s and 1950s suggested laboratory exercises in which genes were represented by beans of several colors, placed in a bag, mixed and reassembled for each generation, according to certain experimental specifications" (Mayr 1982, 558)]

▶ **behavior genetics, behavioral genetics** *n.* Genetics of inheritance of behaviors and dispositions (Wilson 1975, 349; Immelmann and Beer 1989, 118).
syn. psychogenetics (Hall 1951 in Fuller and Thompson 1960), genetic psychology, psychological genetics
Comment: Immelmann and Beer (1989, 238) restrict "psychogenetics," *q.v.,* to the genetic study of human behavior.
[Fuller and Thompson (1960) appear to have coined "behavior genetics."]

▶ **classical genetics** See study of: genetics: Mendelian genetics.

▶ **cytogenetics** *n.* Genetics that integrates morphology and functions of chromosomes with genetic theory (Mayr 1982, 775).

▶ **developmental genetics, physiological genetics** *n.* Genetics of the translation of genetic programs into phenotypes from zygotes to adults (Mayr 1982, 630).

▶ **ecological genetics** *n.* The genetics of populations of living organisms investigated in both the field and laboratory (Mayr 1982, 553).
[coined by Ford (1964 in Mayr 1982, 553) to distinguish this branch of genetics from mathematical population genetics, *q.v.*]

▶ **epigenetics** *n.* "The study of the causal mechanisms of development" (Lincoln et al. 1985).

▶ **evolutionary genetics** *n.* The genetics of sources of genetic variability and changes in gene frequencies in populations (Mayr 1982, 592).

▶ **Mendelian genetics, classical genetics** *n.* Genetics that uses selective breeding to determine distribution patterns of hereditary characteristics between parent and descendent generations (Immelmann and Beer 1989, 118).

▶ **molecular genetics** *n.* Genetics of subjects such as the biochemical natures of genes, genetic codes and the means by which information is brought forth and utilized in cells, chromosomal and gene mutations, and replication mechanisms that take place during mitosis and meiosis (Immelmann and Beer 1989, 118).
cf. study of: biology: molecular biology

▶ **paleogenetics, palaeogenetics** *n.* Genetics of fossil organisms (Lincoln et al. 1985).

▶ **phenogenetics** *n.* The study of genotypic expression in the development of a phenotype (Lincoln et al. 1985).

▶ **physiological genetics** See study of: genetics: developmental genetics.

▶ **population genetics** *n.* The study of genetic changes and factors that affect them within and among populations (Mayr 1982, 630; Hartl 1987, ix; Immelmann and Beer 1989, 118, 228).
cf. study of: biology: population biology
Comments: Population genetics combines observation with theory in its attempt to study populations (Hartl 1987, ix). It cuts across many diverse disciplines including anthropology, conservation, ecology, evolutionary biology, genetics, mathematics, molecular biology, natural history, organism breeding, sociology, statistics, systematics, and wildlife management. Some of the subjects of population genetics are fitness, gene flow, genetic drift, the Hardy-Weinberg principle and neutral theory, heritability, hybrids, inbreeding, kinds of evolution, levels of population structure, mating, migration, mutations, nucleic-acid evolution, ploidy, polymorphism, quantitative traits, selection, selection limits, transposable genetic elements, and X- and Y-linked genes. Population genetics attempts to relate many of its subjects to one another mathematically in predictive models (Mayr 1982, 630; Immelmann and Beer 1989, 118, 228).

classical population genetics *n.* Population genetics that emphasizes biochemical and molecular-biology technology to solve traditional evolutionary problems and that emphasizes evolution of molecules alone (Levin 1984, 454).

mathematical population genetics *n.* Population genetics that involves

studying hypothetical populations using mathematics (Mayr 1982, 553).
cf. study of: genetics: ecological genetics

▶ **psychogenetics** *n.* Genetics of human behavior (Immelmann and Beer 1989, 238).
cf. study of: genetics: behavior genetics
Comment: "Psychogenetics" is an infrequently used term in psychopathology literature (Immelmann and Beer 1989).

▶ **quantitative genetics** *n.* The scientific study of polygenic traits (= quantitative traits), *q.v.* (Hartl 1987, 216).

▶ **sociogenetics** *n.* Genetics of social behavior and caste systems (Hölldobler and Wilson 1990, 643).

▶ **transmission genetics** *n.* Genetics of the origin of new genetic programs, originating from mutation and recombination, and their transfer to an organism's next generation (Mayr 1982, 630).

▶ **zoogenetics** *n.* "Animal genetics" (Lincoln et al. 1985).

geobiology See study of: biology: geobiology.

geobotany See study of: geography: phytogeography.

geochronology, geological chronology, geochrony *n.* The dating of events and study of time in relation to the Earth's history as revealed by geological data (Lincoln et al. 1985).

▶ **geochronometry** *n.* Quantitative geochronology, usually measuring time in years (Lincoln et al. 1985).

geocosmology *n.* Study of the Earth's origin and geological history (Lincoln et al. 1985).

geoecology See study of: geology: geoecology.

geography *n.* The study of the Earth's surface, treating its form and physical features, natural and political divisions, climate, production, population, etc. (*Oxford English Dictionary* 1972, entries from 1542).
Comments: Main divisions of "geography" include "biogeography," "mathematical geography," "physical geography," and "political geography." "Subterranean geography" is "geology," *q.v.*

▶ **biogeography, chorology, geonemy** *n.* study of the geographical distributions of organisms, their habitats, and the historical and biological factors that produced them (Lincoln et al. 1985).

ecological biogeography *n.* Biogeography concerning the distributions and habits of organisms (Lincoln et al. 1985).

historical biogeography *n.* Biogeography concerning historical and bio-logical factors that produced organisms (Lincoln et al. 1985).

necrogeography, thanogeography *n.* Biogeography of dead organisms (Lincoln et al. 1985).

neobiogeography *n.* Biogeography of modern organisms (Lincoln et al. 1985).
cf. study of: geography: paleobiogeography

paleobiogeography, palaeobiogeography *n.* The biogeography of extinct organisms (Lincoln et al. 1985).

phylogeography *n.* The study of the principles and processes that govern geographic distributions and genealogical lineages both within and between species (Avise 1994 in Larson 1995, 115).

phytogeography, geobotany, plant biogeography *n.* Plant biogeography (Lincoln et al. 1985).

paleophytogeography, palaeophytogeography *n.* "The study of plant distributions through geological time" (Lincoln et al. 1985).

subterranean geography See study of: geology.

thanogeography See study of: geography: necrogeography.

vicariance biogeography *n.* The study of the biogeography of vicarious species, *q.v.* (Lincoln et al. 1985); "the theory and practice of biogeographic analysis based on recognition of speciation events (C.K. Starr, personal communication).

zoogeography *n.* The study of the distribution of animals and their communities (Lincoln et al. 1985).

ecological zoogeography *n.* "The study of distributions of animal species with regard to their life conditions and dispersal potential" (Lincoln et al. 1985).

historical zoogeography *n.* The study of the distribution of animal species and higher taxa with regard to their past history and the palaeogeographical evolution of the region where they occur (Lincoln et al. 1985).

macrozoogeography *n.* "The study of dispersion of higher taxa over wide areas" (Lincoln et al. 1985).

microzoogeography *n.* "The study of dispersion of geographical races, species, or genera over small areas" (Lincoln et al. 1985).

▶ **ecological geography** *n.* The study of the effect of ecological (environmental) factors on organism geographical distribution (Mayr 1982, 454).

cf. study of: geography: biogeography, ecological biogeography

geological chronology See study of: geochronology.

geology *n.*

1. The study of the Earth's crust, including its strata, relationships among strata, and historical changes in strata (*Oxford English Dictionary* 1972, entries from 1795).
2. The study of the Earth's origin and structure, processes and forces that have shaped it, and physical and organic history, especially as evidenced by rocks and rock formations (Michaelis 1963).

cf. study of: geography

▸ **geoecology** *n.* "Environmental geology" (Lincoln et al. 1985).

▸ **structural geology** *n.* The study of the formation of rocks of the Earth's crust (*Oxford English Dictionary* 1972, entry from 1882).

syn. geotectic geology (*Oxford English Dictionary* 1972)

genomics *n.* The study of organisms in terms of their entire genomes (Koonin and Glaperin 1997, 757).

geonomy See study of: geography: biogeography.

geratology *n.* "The study of the decline and senescence of populations" (Lincoln et al. 1985).

gerontology *n.* The study of senescence and the processes and effects of organism aging (Lincoln et al. 1985).
Comment: "Human gerontology" embraces "anthropology," "geriatrics," "psychology," and "sociology" (Chaplin 1968).

Gestalt psychology See study of: psychology: Gestalt psychology.

graminology, agrostology *n.* "The study of grasses" (Lincoln et al. 1985).

▸ **paleoagrostology, palaeoagrostology** *n.* "The study of fossil grasses" (Lincoln et al. 1985).

hard science: See study of: science: hard science.

helminthology *n.* "The study of parasitic flatworms and roundworms" (Lincoln et al. 1985).

hepaticology See study of: bryology.

herpetology *n.* The study of reptiles and amphibians (Lincoln et al. 1985).

histology *n.* The branch of anatomy dealing with tissue structure (Lincoln et al. 1985).

humanology *n.* The "study of human behavior from an evolutionary (biological and cultural) perspective" (Wittenberger 1981, 616).

hydrobiology See study of: biology: hydrobiology.

ichnology *n.* The study of traces of extant and extinct organisms (Lincoln et al. 1985).

▸ **neoichnology** *n.* The study of tracks, burrows, and other traces of extant organisms (Lincoln et al. 1985).

▸ **paleoichnology, palaeoichnology, palichnology** *n.* "The study of trace fossils" (Lincoln et al. 1985).

ichthyology *n.* The study of fish (vertebrates) (Michaelis 1963).

ideology *n.*

1. The study of the origin, evolution, and expression of human ideas (*Oxford English Dictionary* 1972, entries from 1796; Michaelis 1963).
2. The study of how ideas are expressed in language (*Oxford English Dictionary* 1972, entry from 1886).

idioecology See study of: autecology.

immunology *n.* The study of organism immunity to disease, serology, immunochemistry, immunogenetics, hypersensitivity, and immunopathology (King and Stansfield 1985).

karyology *n.* The study of cell nuclei, especially chromosome structure (Lincoln et al. 1985)
Comment: "Karyology" is divided into "alpha karyology," "beta karyology," "gamma karyology," "delta karyology," "epsilon karyology," and "zeta karyology" (Lincoln et al. 1985).

kinesics *n.* The study of human body language, including gesture, movement, posture, and orientation in social interaction (Immelmann and Beer 1989, 204).

kinetics See study of: physics: kinetics.

kybernetics See study of: cybernetics.

kymatology *n.* "The study of waves and wave motion" (Lincoln et al. 1985).

leptology *n.* "The study of minute structures or particles" (Lincoln et al. 1985).

lichenology *n.* "The study of lichens" (Lincoln et al. 1985).

limacology *n.* "The study of slugs" (Lincoln et al. 1985).

limnobiology See study of: limnology.

limnology, limnobiology *n.* "The study of lakes, ponds, and other standing waters, and their associated biota" (Lincoln et al. 1985).

▸ **paleolimnology, palaeolimnology** *n.* The "study of the geological history and development of inland waters" (Lincoln et al. 1985).

▸ **rheology** *n.* The part of limnology devoted to running water (Lincoln et al. 1985).
See study of: rheology.

lithology, petrology *n*. The study of rocks and rock-forming processes (Lincoln et al. 1985).

malacology *n*. The study of molluscs (*Oxford English Dictionary* 1972, entries from 1836).

cf. study of: conchology, limacology

mathematics *pl. n.* (construed as singular)

1. Originally, the collective name for geometry, arithmetic, and certain physical sciences (including astronomy and optics) that involve geometrical reasoning (*Oxford English Dictionary* 1972, entries from 1581).
2. In modern usage, the abstract science (pure mathematics) that deductively investigates the conclusions implicit in the elementary conceptions of spatial and numerical relations, and includes as its main divisions geometry, arithmetic, and algebra (*Oxford English Dictionary* 1972).
3. In a wider, modern sense, mathematics including branches of physical or other research that consist of the application of abstract mathematics to concrete data (*Oxford English Dictionary* 1972).
4. "The study of quantity, form, arrangement, and magnitude; especially, the methods and processes for disclosing, by rigorous concepts and self-consistent symbols, the properties and relations of quantities and magnitudes, whether in the abstract, pure mathematics, or in their practical connections, applied mathematics" (Michaelis 1963).

▸ **cybernetics, kybernetics** *n*.
1. A multidisciplinary research approach to the structure and functions of data- or information-processing systems; cybernetics is divided into the fundamental disciplines of "control theory," "information theory," and "algorithm and automata theory" (Storz 1973, 62).
2. In animals and machines: the study of control and communication in systems, including information systems and feedback control systems (Lincoln et al. 1985; Immelmann and Beer 1989, 66).

control theory *n*. A part of cybernetics that deals with automatic regulation of machines, and organisms (Immelmann and Beer 1989, 59).

information theory *n*.
1. The statistical and mathematical study of the encoding, decoding, transmission, storage, retrieval, etc. of information and of computers, telecommunication channels, and other information-processing systems (Michaelis 1963).
2. A branch of probability theory, or cybernetics, that deals with the likelihood of transmission of messages, accurate within specified limits, when the units of information comprising the message are subject to probabilities of transmission failure, random disturbance, etc., and with the measurement of information content of systems (Lincoln et al. 1985).

Note: Information theory employs a quantitative measure of information and does not reflect a message's qualitative value (= meaning) (Lincoln et al. 1985; Immelmann and Beer 1989, 147).

topology *n*. A branch of mathematics that ignores quantities and deals only with the formal relations between, or among, components, especially those that can be represented geometrically (*e.g.*, body-surface characters that remain unchanged under quantitative distortion) (Bateson 1979, 242).

mechanics See study of: physics: mechanics.

mellitology *n*. "The study of bees" (Lincoln et al. 1985).

syn. apidology (not a preferred synonym; C.K. Starr, personal communication)

mesology See study of: ecology.

metaphysics See study of: philosophy.

meteorology *n*. The study of motions of Earth's atmosphere, especially with a view to forecasting weather (*Oxford English Dictionary* 1972, entries from 1620); "the study of weather or local atmospheric conditions" (Lincoln et al. 1985).

cf. climatology

biometeorology *n*. The study of effects of atmospheric conditions on biota (Lincoln et al. 1985).

-metry See also -metrics, -metry.

▸ **biometry, biological statistics, biostatistics** *n*. Broadly, "the application of statistical methods to the solution of biological problems" (Sokal and Rohlf 1969, 2).

syn. biometrics (Lincoln et al. 1985)
cf. -metry: biometry
["biometry" coined by Weldon (in Sokal and Rohlf 1969, 4)]

▸ **chronometry** *n*. "The science of precise time measurement" (Lincoln et al. 1985).

▸ **conchometry** *n*. The study of mollusc-shell morphometrics (Lincoln et al. 1985).

cf. study of: conchology

▸ **geochronometry** See study of: geochronology.

▸ **psychometry, psychometrics** *n.*
1. "The science of the measurement of psychophysical processes, especially of their interrelationships" and durations in time (Michaelis 1963) and which attempts to avoid subjective-judgement effects (Collocot and Dobson 1974).
2. The measurement of intelligence (Michaelis 1963; Collocot and Dobson 1974).

cf. -metry: psychometrics: mental chronometry

▸ **scientometrics** *n.* The field that uses quantitative methods to analyze the process and development of science (Garfield, 1987).

morphology *n.* The scientific study of the form of organisms and the structures, homologies, and metamorphoses that govern, or influence, this form (*Oxford English Dictionary* 1972, entries from Goethe 1817; Mayr 1982, 455).

▸ **evolutionary morphology** *n.* Morphology that "views form either as response to environmental needs (Lamarck-type explanations) or as adaptation produced by selection pressures" (Mayr 1982, 455).

cf. evolutionary-morphology school

▸ **functional morphology** *n.*
1. The description of structures in terms of their functions (Mayr 1982, 455).
2. Interpretation of the function of an organism, or organ system, by reference to its shape, form, and structure (Lincoln et al. 1985).

▸ **morphology of growth** *n.* The study of growth, "including all processes of growth and development that can be formulated mathematically, particularly allometric growth" (Mayr 1982, 455).

▸ **synmorphology** *n.* The study of a plant community's floristic composition, minimal area, and extent in time and space (Lincoln et al. 1985).

muscology See study of: bryology.

mycology, mycetology *n.* "The study of fungi" (Lincoln et al. 1985).

myrmecology *n.* "The study of ants" (Lincoln et al. 1985).

[*myrmeco* < Greek *myrmēx, myrmēkos,* ant + -LOGY]

natural history See natural history.

natural language processing *n.* "The study of mathematical and computational modeling of various aspects of language and the development of a wide range of systems" (Joshi 1991, 1242).

Comment: These aspects and systems include spoken language systems that integrate speech and natural language; cooperative interfaces to databases and knowledge bases that model aspects of human-human interaction; multilingual interfaces; machine translation; and message-understanding systems, among others (Joshi 1991, 1242).

nealogy *n.* "The study of young and immature organisms" (Lincoln et al. 1985).

necrology *n.* The study of decomposition, fossilization, and other processes that affect organism remains after death (Lincoln et al. 1985).

nematology, nematodology *n.* "The study of nematodes" (Lincoln et al. 1985).

neoichnology See study of: ichnology: neoichnology.

neontology *n.* The study of extant species (Mayr 1982, 292).

cf. study of: paleontology.

neurobiology See study of biology: neurobiology.

neuroembryology of behavior See study of: embryology.

neuroendocrinology See study of: endocrinology.

neustonology *n.* "The study of neuston" (Lincoln et al. 1985).

nosography, nosology See study of: nosology.

oceanography *n.* The study of oceans (Lincoln et al. 1985).

oology *n.* "The study of eggs" (Lincoln et al. 1985).

ophiology *n.* "The study of snakes" (Lincoln et al. 1985).

ophiotoxicology See study of: toxicology: ophiotoxicology.

organology *n.* "The study of organs systems and structures" (Lincoln et al. 1985).

ornithology *n.* "The study of birds" (*Oxford English Dictionary* 1972, entries from 1676).

[Greek *ornis, ornithos,* bird + -LOGY]

▸ **paleornithology, palaeornithology** *n.* "The study of fossil birds" (Lincoln et al. 1985).

osphresiology *n.* "The study of the sense of smell" (Lincoln et al. 1985).

osteology *n.* "The study of the structure and development of bones" (Lincoln et al. 1985).

paleoagrostology, palaeoagrostology See study of: graminology: paleoagrostology.

paleoalgology, palaeoalgology See study of: phycology: paleophycology.

paleoautecology, palaeoautecology See study of: ecology: autecology: paleoautecology.

paleobiogeography, palaeobiogeography See study of: geography: biogeography: paleobiogeography.

paleobiology, palaeobiology See study of: biology: paleobiology.

paleobotany, palaeobotany See study of: botany: paleobotany.

paleoclimatology, palaeoclimatology See study of: climatology: paleoclimatology.

paleodendrology, palaeodendrology See study of: dendrology: paleodendrology.

paleoecology, palaeoecology See study of: ecology: paleoecology.

paleogenetics, palaeogenetics See study of: genetics: paleogenetics.

paleoichnology, palaeoichnology See study of: ichnology: paleoichnology.

paleolimnology, palaeolimnology See study of: limnology: paleolimnology.

paleontology, palaeontology *n.* The study of extinct organisms; a branch of geology, or biology, that treats of fossil organisms, often extinct animals (paleozoology) (*Oxford English Dictionary* 1972, entries from 1838).

cf. study of: biology: paleobiology; ichnology

▸ **micropaleontology** *n.* The study of microscopic fossils (those from a few microns to 2 centimeters long) that must be studied by light, or electron, microscopy (Lincoln et al. 1985).

▸ **molecular paleontology** *n.* Paleontology involving analyzing DNA from fossils (Morell 1992b, 1861).

▸ **stratigraphic paleontology, biostratigraphy** *n.* "The study and classification of rock strata based on their fossil content" (Lincoln et al. 1985).

paleopalynology, palaeopalynology See study of: palynology: paleopalynology.

paleophytogeography, palaeophytogeography See geography: phytogeography: paleophytogeography.

paleophytology, palaeophytology See study of: botany: paleobotany.

paleornithology, palaeornithology See study of: ornithology: paleornithology.

paleosynecology, palaeosynecology See study of: ecology: synecology: paleosynecology.

paleozoology See study of: paleontology.

palynology *n.* The study of pollen and spores of extinct and extant organisms (Lincoln et al. 1985).

▸ **paleopalynology, palaeopalynology** *n.* "The study of fossil spores and pollen" (Lincoln et al. 1985).

parasitology *n.* The study of parasites and parasitism, considered to be branches of biology and medicine (*Oxford English Dictionary* 1972, entries from 1882).

pathology, nosography, nosology *n.* "The study of diseases" (Lincoln et al. 1985).

▸ **phytopathology** *n.* The study of plant diseases (Lincoln et al. 1985).

▸ **psychopathology** *n.* The study of human mental problems (Michaelis 1963).

pedology See study of: edaphology.

perceptual psychology See study of: psychology: perceptual psychology.

petrology See study of: lithology.

pharmacology *n.* "The science of the nature, preparation, administration, and effects of drugs" (Michaelis 1963).

▸ **psychopharmacology** *n.* The branch of pharmacology that deals with drugs that affect human minds (Michaelis 1963).

phenology *n.* "The study of periodic biological phenomena such as migrations, breeding, emergence of insect adults, and flowering, in relation to seasonal changes, and climatic and other ecological factors (Michaelis 1963).

See phenology.

syn. phaenology (Lincoln et al. 1985)

Comment: Lincoln et al. 1985 synonymize "phenology" and "phenomenology," but Michaelis (1963) indicates that "phenology" is a contraction of phenomenology, with a restricted meaning.

▸ **phytophenology** *n.* "The study of periodic phenomena of plants, such as flowering and leafing" (Lincoln et al. 1985).

phenomenology See study of: phenology.

philology *n.*

1. The scientific study of written records, chiefly of literary works, undertaken to set up accurate texts and determine their meaning (Michaelis 1963).
2. "Linguistics, especially comparative and historical" (Michaelis 1963).
3. Literary study, especially classical scholarship (Michaelis 1963; Morris 1982).
4. Loosely, etymology (Michaelis 1963).

[Old French *filsofie, philosophie* < Latin *philosophia* < Greek < *philosophos* < lover of wisdom]

philosophical psychology See study of: psychology: philosophical psychology.

philosophy *n.*

1. "The love, study, or pursuit of wisdom, or of knowledge of things and their causes, whether theoretical or practical" (*Oxford English Dictionary* 1972, entries from 1340).
2. The more advanced knowledge, or study, to which, in medieval universities, the seven liberal arts were recognized as introductory; it includes the

S–Z

three branches of natural, moral, and metaphysical philosophies, commonly called the "three philosophies" (*Oxford English Dictionary* 1972, entries from 1387); sciences as formerly studied in universities (Michaelis 1963).

3. "Sometimes used especially of knowledge obtained by natural reason, in contrast with revealed knowledge" (*Oxford English Dictionary* 1972, entries from 1388).

4. With *of:* "the study of the general principles *of* some particular branch of knowledge, experience, or activity; also less properly, of those *of* any subject or phenomenon" (*Oxford English Dictionary* 1972, entries from 1713).

5. A particular system of ideas that relates to the general scheme of the universe; "a philosophical system or theory" (*Oxford English Dictionary* 1972, entries from 1390); a treatise on a philosophical system (Michaelis 1963).

6. One's system that one forms for conducting his life; "the mental attitude, or conduct, of a philosopher; serenity under disturbing influences, or circumstances; resignation; calmness of temper" (*Oxford English Dictionary* 1972, entries from 1771); practical wisdom; fortitude, as in enduring reverses and suffering (Michaelis 1963).

7. Inquiry into the more comprehensive principles of reality in general, or of some limited sector of it such as human knowledge or human values (Michaelis 1963).

8. General laws that furnish the rational explanation of anything (Michaelis 1963).

9. Reasoned science; a scientific system; an archaic definition (Michaelis 1963).

See philosophy: metaphysical, moral, and natural philosophy.

[Old French *filsofie, philosophie* < Latin *philosophia* < Greek < *philosophos* < lover of wisdom]

▸ **epistemology** *n.* The critical investigation of the nature, grounds, limits, criteria, or validity of human knowledge (Michaelis 1963).

cf. epistemology; study of: epistemology [Greek *epistēmē,* knowledge + -OLGY].

▸ **metaphysical philosophy** *n.* Knowledge, or study, that deals with ultimate reality, or with the most general causes and principles of things (*Oxford English Dictionary* 1972, entries from 1794).

syn. philosophy (*Oxford English Dictionary* 1972)

▸ **metaphysics** *pl. n.* (construed as singular) The investigation of principles of

reality transcending those of any particular science, traditionally including cosmology and ontology (Michaelis 1963). [Medieval Latin *metaphysica* < Medieval Greek < *ta meta ta physika,* the (works) after the physics referring to Aristotle's ontological treatises which came after his *Physics*]

cosmology *n.* "The general philosophy of the universe considered as a totality of parts and phenomena subject to laws" (Michaelis 1963).

cf. study of: philosophy: teleology

ontology *n.* "Philosophical theory of reality, including a consideration of the universal and necessary characteristics of all existences" (Michaelis 1963). [New Latin *ontologia* < Greek *ōn, ontos,* being + LOGY]

▸ **moral philosophy** *n.* The knowledge, or study, of principles of human action or conduct; ethics (*Oxford English Dictionary* 1972, entries from ca. 1400).

syn. philosophy (*Oxford English Dictionary* 1972)

▸ **natural philosophy,** *Naturphilosophie n.* The knowledge, or study, of nature or of natural objects and phenomena; now usually called science (*Oxford English Dictionary* 1972, entries from 1297).

syn. philosophy, science (*Oxford English Dictionary* 1972)

Comments: This branch of philosophy opposed reductionism and the mechanization of Newtonianism (Mayr 1982, 387–388). Some the naturalists who believed in *Naturphilosophie* explained nature in terms of creation and design; others stressed quality, development, uniqueness, and usually a finalistic component. All of the best-known representatives of this philosophy were essentialists, unable to develop a theory of common descent. Much of the naturalists' literature is "fantastic if not ludicrous" (Mayr 1982, 387–388).

▸ **philosophy of the mind** See study of: psychology: philosophical psychology.

▸ **teleology** *n.*

1. The study of a creative design in the processes of nature (Plato, Aristotle, the Stoics in Mayr 1982, 47).

Note: "Teleological" refers to four concepts: "teleonomic activities," "teleomatic processes," "adapted systems," and "cosmic teleology" (Mayr 1982, 48–50).

2. The doctrine, or study, of ends or final causes, especially related to design or purpose in nature; also transferred to such design exhibited in natural objects or phenomena (*Oxford English*

Dictionary 1972, entries from 1728); "the branch of cosmology that treats of final causes; finalism" (Michaelis 1963). See teleology.

photodynamics See study of: dynamics: photodynamics.

phycocoenology *n.* "The study of algal communities" (Lincoln et al. 1985).

phycology, algology *n.* "The study of algae" (Lincoln et al. 1985).

▸ **paleoalgology, palaeoalgology** *n.* "The study of fossil algae" (Lincoln et al. 1985).

phylogenetic systematics See study of: systematics: phylogenetic systematics.

phylogeny *n.* The history, or study of, evolution or genealogical development in phyla, tribes, or species (*Oxford English Dictionary* 1972, entries from 1875).

▸ **behavioral phylogeny** *n.* The study of the evolutionary origin and descent of behaviors (Immelmann and Beer 1989, 30).
cf. ethology: comparative ethology

physicalism *n.* The notion that species are under the direct control of physical variables and that low-diversity communities reflect extreme values of a controlling variable, *e.g.*, hypersalinity or low temperature (Gould 1981b, 297).

physics *pl. n.* (construed as singular)
1. Originally, the study of all nature; natural science in general; especially the Aristotelian system of natural science (*Oxford English Dictionary* 1972, entries from 1589).
   See study of: physiology (def. 1).
2. Now, the study of the properties of matter and energy, or of the action of different forms of energy on matter in general (excluding chemistry, which deals with the different forms of matter, and biology, which deals with vital energy) (*Oxford English Dictionary* 1972, entries from 1715).
3. "The science that treats of the laws governing motion, matter, and energy under conditions susceptible to precise observation, generally considered" distinct from chemistry and biological sciences (Michaelis 1963).
Comment: Two main divisions of physics are "general physics" and "applied physics." Many persons now tend to restrict the term "physics" to "general physics."
[Old French *fisique* < Latin *physica* < Greek *physikē* (*epistēmē*), (the knowledge) of nature < *physics*, nature < *phyein*, to produce]

▸ **applied physics** *pl. n.* (construed as singular) Physics dealing with special phenomena, including astronomy, me-

teorology, terrestrial magnetism, etc. (*Oxford English Dictionary* 1972).

▸ **biophysics** *pl. n.* (construed as singular) "The application of physics to the study of living organisms and systems" (Lincoln et al. 1985).

▸ **dynamics** *pl. n.* (construed as singular)
1. "The branch of physics that treats of the motion of bodies and the effects of forces in producing motion, including kinetics" (Michaelis 1963).
2. "The science that treats of the action of forces, whether producing equilibrium or motion: in this sense including both statics and kinetics" (Michaelis 1963).
cf. study of: physics: kinetics, statics

▸ **general physics** *pl. n.* (construed as singular) Physics of the general phenomena of inorganic nature (dynamics, molecular physics, physics of ether, etc.) (*Oxford English Dictionary* 1972).

▸ **kinetics** *pl. n.* (construed as singular) The branch of physics that deals with the effect of forces in the production, or modification, of motions of bodies (Michaelis 1963).

▸ **mechanics** *pl. n.* (construed as singular) "The branch of physics that treats of motion and the action of forces on material bodies, including statics, kinetics, and kinematics" (Michaelis 1963).

▸ **metaphysics** See study of: philosophy.

▸ **psychophysics** See study of: physiology: psychophysics.

▸ **statics** *pl. n.* (construed as singular) "The branch of mechanics dealing with bodies at rest and with the interaction of forces in equilibrium" (Michaelis 1963).

physiological psychology See study of: psychology: physiological psychology.

physiology *n.*
1. Originally, the study and description of natural objects; natural science or philosophy (*Oxford English Dictionary* 1972, entries from 1654).
   cf. study of: biology, physics (def. 1)
2. The study of the functions of organisms (*Oxford English Dictionary* 1972, entries from 1597) and of their individual organs, tissues, and cells (Wilson 1975, 591); broadly including metabolism, motor mechanisms, hormonal regulation, and stimulus reception, transmission, and processing (Immelmann and Beer 1989, 221).
Comments: "Physiology" includes types listed below and "endocrinology" and "cellular physiology" (Immelmann and Beer 1989, 221). "In its broadest sense physiology also encompasses most of molecular biology and biochemistry" (Wilson 1975, 591).

S–Z

[French *physiologie* < Latin *physiologia* < Greek, natural philosophy < *physiologos*, speaker on nature < *physis*, nature + *logos*, word]

▸ **animal physiology, zoodynamics, zoonomy** *n.* Physiology of animals (Lincoln et al. 1985).

▸ **behavioral physiology** *n.* Physiology of behavior which includes neuroethology and behavioral endocrinology (Immelmann and Beer 1989, 31).

▸ **ecological physiology** *n.* "The study of physiological adaptations that improve an organism's survival and permit it to exploit extreme environments (Feder et al. 1987, i).
syn. ecologically relevant physiology, physiological ecology (used interchangeably by Feder et al. 1987, ix); environmental physiology (Bennett 1987, 1)

▸ **ecophysiology** *n.* The physiology of organisms' adaptations to their environments (Lincoln et al. 1985).

▸ **evolutionary physiology** *n.* The scientific study that integrates functional morphology and physiology with mainsteam evolutionary biology (Kingsolver et al. 1995, 396).

▸ **neurophysiology** *n.* The physiology of nervous systems, especially the physiological processes by which they function (Wilson 1975, 590).
[New Latin < Greek *neuron*, sinew + PHYSIOLOGY]

▸ **psychophysics** *pl. n.* (construed as singular)
1. A branch of physiology that is concerned with human judgments of sensation magnitudes vs. actual magnitudes (Storz 1973, 227).
2. The study of relationships between stimulus magnitude and stimulus intensity experienced by an animal (Immelmann and Beer 1989, 238).

▸ **psychophysiology** *n.* Physiology of human minds (Michaelis 1963).

▸ **sensory physiology** *n.* The physiology of sensory organs and the ways in which they receive stimuli from an animal's environment and transmit them to its nervous system (Wilson 1975, 594).

phytobiology See study of: botany.

phytochemistry See study of: systematics.

phytocoenology, phytosociology *n.* "The study of plant communities" (Lincoln et al. 1985).
syn. phytocenology (Küchler 1964, 1)
Comment: "Phytosociology" is generally recognized as being an unfortunate term both etymologically and philosophically;

"phytoceonology" is replacing it (Küchler 1964, 1).

phytoecology See study of: ecology: phytoecology.

phytogeography See study of: geography: phytogeography.

phytography See study of: botany: phytography.

phytology See study of: botany.

phytopaleontology, phytopalaeontology See study of: botany: paleobotany.

phytopathology See study of: pathology: phytopathology.

phytophenomenology See study of: phenology: phytophenomenology.

phytosociology See study of: ecology: phytoecology; phytocoenology.

phytoteratology See study of: teratology: phytoteratology.

phytotopography *n.* The study of vegetation of a particular region (Lincoln et al. 1985).

plant biogeography See study of: geography: biogeography: phytobiogeography.

plant sociology See study of: ecology: plant ecology.

plant teratology See study of: teratology: plant teratology.

pomology *n.* "The study and practice of fruit cultivation" (Lincoln et al. 1985).

population biology See study of: biology: population biology.

population dynamics See study of: dynamics: population dynamics.

population genetics See study of: genetics: population genetics.

potamology *n.* "The study of rivers" (Lincoln et al. 1985).

primatology *n.* The study of primates, including prosimians, monkeys, and apes (including Humans) (Immelmann and Beer 1989, 232).

protistology *n.* "The study of protistans" (Lincoln et al. 1985).

protozoology See study of: zoology: protozoology.

proxemics *n.* The systematic study of space as a specialized component of human culture (Hall 1966 in Wilson 1975, 259).

psolidomeiomology *n.* The study of scale-row reduction in snakes (Lincoln et al. 1985).

psychiatry *n.* The branch of medicine that deals with the diagnosis, treatment, and prevention of human mental disorders (*Oxford English Dictionary* 1972, entries from 1846; Michaelis 1963).

psychoacoustics See study of: acoustics: psychoacoustics.

psychobiology See study of: biology: psychobiology.

psychodynamics See study of: dynamics: psychodynamics.

psychognosis *n.* "The study of mental states" (Michaelis 1963).

psychology *n.*

1. The study of the nature, functions, and phenomena of human minds (formerly also of souls) (*Oxford English Dictionary* 1989, entries from 1653).

2. The study of a human mind as an entity and its relationship to a physical body, based on one's observations of a person's behavior and activity aroused by specific stimuli (*Oxford English Dictionary* 1989, entries from 1895).

3. The study of an individual, or a selected group of individuals, when he interacts with his environment or in a particular social context (*Oxford English Dictionary* 1989, entries from 1895).

4. The study of human minds in any of their aspects, operations, power, or functions (Michaelis 1963; Storz 1973, 216); the study of human nature (Delbridge 1981).

5. The study of human mental phenomena, especially those associated with consciousness, behavior, and the problems of adjustment to one's environment (Michaelis 1963).

6. The study of human and nonhuman animal behavior (Chaplin 1968; Delbridge 1981); the study of "the behavior of organisms" (Storz 1973, 216).

7. The study of the relationships between antecedent events, or conditions, and the behavior of organisms (Marx and Hillix 1973, 440.

    Note: Marx and Hillix (1973) consider this to be a rough, initial definition of psychology, and they give other definitions of psychology.

8. The study of human and animal behavior as it may elucidate human behavior (Gould 1982, 6).

See psychology.

cf. study of: animal behavior; biology: psychobiology; [2,3]ethology; behaviorism

Comments: An individual animal is the primary unit of study in psychology (Chaplin 1968). Psychology is primarily concerned with proximate cause of behavior and rarely uses rigorous argument based upon evolutionary theory (McFarland 1985, 357). There are many divisions of psychology with great conceptual and methodological differences; this has caused some authorities to write of the sciences of psychology. Areas of psychology overlap with sociol-

ogy, anthropology, and ethology (Immelmann and Beer 1989, 238–239).

▸ **abnormal psychology** *n.* The description, explanation, and treatment of human neuroses, psychoses, and other mental and behavioral disorders (Immelmann and Beer 1989, 238).

▸ **animal psychology** *n.*

1. A diverse mixture of subdisciplines including comparative learning and learning theory, behavioral development, physiological processes that underlie behavior (hormone effects on behavior, neural correlates of behavior, brain chemistry), psychopharmacology, and behavior genetics (Drickamer and Vessey 1982, 24).

2. Study of some kinds of animal behavior that historically stems from human psychology, especially the old school of associationism (parts of comparative psychology and animal cognition) (Immelmann and Beer 1989, 14).

3. Study of animal "personality" characteristics, including individuality, subjectivity (insofar as this can be ascertained), and pathological (= abnormal) behavior (Immelmann and Beer 1989, 14).

syn. comparative animal psychology, comparative psychology (Storz 1973, 15)

cf. study of: animal behavior, [1,2]ethology; psychology: behaviorism, comparative psychology

Comments: "Animal psychology" has been replaced by "ethology" to a large extent. "Animal psychology" is used in a very general and loose way in most popular writing about animal behavior (Immelmann and Beer 1989, 15).

classical animal psychology *n.* A branch of animal psychology that originated within psychology and deals with the evolution of mind and higher mental processes; contrasted with classical ethology (Dewsbury 1978, 13).

cf. study of: psychology: behaviorism

▸ **applied psychology** *n.* The practical use of psychological knowledge that includes divisions such as clinical, educational, industrial, and vocational psychology (Storz 1973, 17–18).

cf. study of: psychology: consumer psychology, counseling psychology, engineering psychology, industrial psychology, personnel psychology

▸ **behavioral ecology** *n.* An area of psychology that focuses on the transaction between an animal's entire ecological environment and its behavior (Storz 1973, 28–29).

cf. study of: ecology: behavioral ecology

▸**behaviorism, Behaviorism** (*Brit.* **Behaviourism**) *n.*

1. Narrowly, the study of learning and related phenomena, often based on white laboratory rats (Verplanck 1957).
2. The study of animal behavior (including human behavior) based on objectively observable and quantifiable behaviors, not on sensation, emotion, mind, consciousness, will, imagery, etc. (Verplanck 1957; Watson 1913 in Chaplin 1968; Storz 1973, 29).
   Note: This is a more widely held idea of "behaviorism."
3. The study of how past events affect an animal's behavior, with the basic tenet that behavior consists of an animal's responses (reflexes), reactions, or adjustments to stimuli, or complexes of stimuli, rather than instincts (Drickamer and Vessey 1982, 23).

syn. American behaviorism
cf. behaviorism; study of: psychology: comparative psychology, stimulus-response psychology

Comments: John B. Watson (1913) is credited for formulating "behaviorism" in his paper "Psychology as a behaviorist views it" (Dewsbury 1978, 23). Early behaviorists include I.P. Pavlov, J.B. Watson, B.F. Skinner, E.L. Thorndike (Chaplin 1968). Friedman (1967, 138) notes that "many areas originally declared off limits by certain varieties of behaviorism — thought, dreams, perception, physiology, consciousness — have been reclaimed by simply renaming them thinking behavior, dreaming behavior, perceptual behavior, physiological behavior, and conscious behavior. This takes, perhaps, undue advantage of the elasticity of the term 'behavior' while allowing the users to remain behaviorists in good standing."

▸**clinical psychology** *n.* The study and treatment of problems of human mental health (Storz 1973, 47).

▸**cognitive psychology** *n.*

1. The study of mental activity that divides it into cognitive (perceiving, thinking, knowing), affective (feelings, emotions), and conative (acting, doing, striving) activity (Storz 1973, 49).
2. The study of processes such as perception, attention, memory, thinking, and problem solving (Levin 1987).

cf. cognition

▸**community psychology** *n.* An area of psychology that deals with human individual-behavioral adaptation in community settings and the need for change of functioning of components of a community or society (Storz 1973, 53).

▸**comparative psychology** *n.*

1. The study of "mind and intelligences as developed in man and animals" (*Oxford English Dictionary* 1989, entries from 1748); the comparative study of "minds" of animals (Immelmann and Beer 1989, 53).
2. Analysis, or dissection, of complex behavioral structures, comparison of all structures analyzed, and classification of analyses (Romanes 1882 in Gherardi 1984, 381).
3. The study of similarities and differences in the behavior of different animal species which includes ethology with studies originally centered in Europe and animal psychology with studies centered in the U.S.; comparative psychology (E.R. Rees in Storz 1973, 53).
4. The attempt to make comparisons across species in order to develop principles regarding animal behavior (Dewsbury in Drickamer and Vessey 1982, 25); the study of "how and why organisms do the things they do" whose paradigm is the synthetic theory of evolution (Demarest 1987, 148).
5. The science that deals mainly with observable behavior and tends to eschew mentalistic terms such as mind, feeling, and intention; compared with ethology, comparative psychology tends to emphasize learning and the physiological basis of behavior, and is concerned with aspects of behavior that are relevant to Humans and cannot be experimentally investigated in Humans for ethical and methodological reasons (Dewsbury 1968).

cf. study of: psychology: animal psychology, behaviorism

Comments: Ratner and Denny (1964, 1, in Dewsbury 1968) state: "The field of comparative psychology includes a wide array of facts, theories, and methods that are generally oriented around the general topic of behavior of organisms (general psychology) and the specific topic of comparative analysis." Waters (1960, 2, in Dewsbury 1968) states: "An adequate description of the nature of comparative psychology must include a statement of its subject matter, its methods, its problems and aims, and its relations with other branches of science." According to Immelmann and Beer (1989, 53), "comparative psychology" has recently broadened and turned more in the direction of ethological questions and methodology, making it impossible to draw a sharp line between these two branches of science. Finally, there is much recent discussion (*e.g.,* in the *Comparative Psychology Newsletter*) about the meaning of comparative psychology.

[Flourens (1864) first officially used this term, according to Gherardi (1984, 381).]

► **consumer psychology** *n.* "A branch of applied psychology concerned with questions about the optimal means of making goods and services available, providing information about them, developing and testing methods for promoting interest in their acquisition, and investigating how they might be consumed, with maximum satisfaction and benefit to the customer" (Storz 1973, 59).

► **counseling psychology** *n.* An area of applied psychology that deals with providing persons with information and guidance with problems of educational planning, choosing and getting jobs, and adjusting to personal and marital requirements (Storz 1973, 61).

► **developmental psychology** *n.*

1. The study of the growth and changes of an organism's behavior from its conception to death; originally, this field concentrated on child development, but its scope now includes an organism's entire life (Storz 1973, 69).

2. The study of how behavioral and cognitive capacities develop in children and young nonhuman animals and how they are affected by aging (Immelmann and Beer 1989, 238).

► **differential psychology** *n.* The study of the psychological differences between, or among, individuals and groups (Storz 1973, 71).

► **dynamic psychology** *n.* Psychology involved with "human nature," as well as consciousness (Storz 1973, 77).

► **education psychology** *n.* Psychology of human learners, learning processes, and instructional strategies as interactive aspects of education (Storz 1973, 78).

► **engineering psychology** *n.* "A branch of applied psychology concerned with the discovery and application of information about human behavior to the design of machines, tools, jobs, and environments so they best match human abilities and limitations" (Storz 1973, 86).

► **experimental psychology** *n.*

1. The experimental study of a person's responses to stimuli (*Oxford English Dictionary* 1989, entry from 1953).

2. "The use of experimental methods in psychological research" (Storz 1973, 94).

3. The study of psychophysics, learning principles, and the nature of motivation that uses laboratory experiments (Immelmann and Beer 1989, 238).

► **faculty psychology** See study of: psychology: faculty psychology.

► **forensic psychology** *n.* Psychology concerned with the reliability of legal evidence (Storz 1973, 99).

► **functionalism** *n.*

1. Psychology that emphasizes mental acts, or processes, with regard to Humans' adaptation to their environments; functionalism stresses objective testing of hypotheses as opposed to subjective introspection (Storz 1973, 103).

2. The study of the functions of minds and how they operate, in contrast to their anatomy (Drickamer and Vessey 1982, 23).

syn. functional psychology (Storz 1973, 103–104)

cf. study of: psychology: behaviorism, structuralism

Comment: Functionalists include W. James, J. Dewey, J.R. Angell, and H. Carr (Chaplin 1968).

► **genetic psychology** *n.* Psychology dealing with the role of genetic influences on behavior, whether innate behaviors exist and how they can be identified and studied, the development of intelligence, etc. (Storz 1973, 106).

► **gestalt psychology, *Gestalt* psychology** *n.*

1. A school of psychology that holds that a person's perceptions, reactions, etc. are gestalts, *q.v.* (*Oxford English Dictionary* 1989, entries from 1890).

2. A branch of psychology that attributes animal's behavioral responses to recognition of the sum of all elements of a configuration in contrast to simple cues; *e.g.,* Honey Bees can be taught to distinguish between more- and less-segmented stimuli (Köhler 1964 in Heymer 1977, 77; Storz 1973, 107).

Note: "Gestalt psychology" permits some degree of analysis of the phenomenological variety but not at the molecular level because this would destroy the unitary quality that is undergoing analysis; holds that conscious experience cannot be meaningfully resolved into structuristic elements; holds that behavior cannot be reduced to combinations of reflexes, or conditioned responses, and still retain its uniqueness and meaningfulness; objects to treating an animal's nervous system "as a static, machinelike structure capable only of responding piecemeal to incoming stimuli;" etc. (Chaplin 1968).

3. A branch of psychology that believes animals gain insight into problems through their innate tendencies to perceive situations as a whole (McFarland 1985, 346).

cf. insight

[German *Gestalt,* form]

▸ **individual psychology** *n.* A school of psychology that holds that important conflicts often occur between an individual person and his environment, rather then within the individual (Storz 1973, 126).

▸ **industrial psychology** *n.* A branch of applied psychology that studies human work including job analysis, personnel selection and training, morale, management-employee relations, working conditions and efficiency, design of machines and work spaces, etc. (Storz 1973, 127).

▸ **medical psychology** *n.*

1. Psychiatry; "the branch of medicine concerned with psychic disorders" (Storz 1973, 159).
   Note: This term is used in the U.K. and in some European countries (Storz 1973, 159).

2. "The application of psychology to all phases of medical research, treatment, and prevention" (Storz 1973, 159).
   Note: This definition is used in the U.S. (Storz 1973, 159).

▸ **metapsychology** *n.* A study that "goes beyond empirical psychology to consider philosophical issues such as the nature of human mind and will and the mind-body problem" (Storz 1973, 170).

▸ **military psychology** *n.* The study of the psychology of armed forces including training programs, evaluating morale, analyzing activity and performance in miliary jobs, exploring the social systems of people in organizations, and designing automated human-machine systems (Storz 1973, 172).

▸ **parapsychology** *n.* The study of human extrasensory perception, psychokinesis, and related topics such as survival after bodily death (Storz 1973, 191).

▸ **perceptual psychology** *n.* The study of human reception of external stimuli, including subjective, compared to physical, measurements of these stimuli (Drickamer and Vessey 1982, 22).

▸ **personnel psychology** *n.* A branch of applied psychology that specializes in selecting, evaluating, and training personnel and in morale and relations of management to employees (Storz 1973, 200).

▸ **philosophical psychology, philosophy of the mind** *n.* The branch of psychology concerned with conceptual and speculative problems and that attempts to analyze and clarify fundamental concepts involved in scientific explanation of mental events and processes, "namely mind, consciousness, self, soul, memory, will, cognition, and perception" (Storz 1973, 201).

▸ **physiological psychology** *n.* The study of physiological mechanism that underlay behavior (Storz 1973, 214).

▸ **pseudopsychology** *n.* "Subprofessional teaching and counsel, or outright quackery, in some field where psychology operates" (Storz 1973, 211).

▸ **psychology of personality** *n.* The study of personality-type variations and how they arise, "especially in light of psychogenic theories such as those of psychoanalysis" (Immelmann and Beer 1989, 238).

▸ **school psychology** *n.* The branch of psychology that specializes in development and adjustments of children to their school situations (Storz 1973, 245).

▸ **social psychology** *n.* The study of a person's behavior and interactions with his social group (*Oxford English Dictionary* 1989, entries from 1908; Immelmann and Beer 1989, 238).

▸ **somatospsychology** *n.* The study of the psychological impact of physiological disease on human disability (Storz 1973, 263).

▸ **space psychology** *n.* The psychological study of Humans during space flight (Storz 1973, 263).

▸ **stimulus-response psychology** *n.* A school of psychological thought that contends that specified responses always occur following stimulation (Storz 1973, 267).

▸ **structuralism** *n.* Psychology that emphasizes conscious contents of minds as a proper subject for study rather than mental acts and processes as environmental adaptations (Chaplin 1968).
cf. study of: psychology: functionalism

▸ **topological psychology** *n.* An attempt to represent the structure and dynamics of Humans and their environments by means of mathematical concepts (Storz 1973, 279).

psychometrics, psychometry See study of: -metry: psychometrics.

psychopathology See study of: pathology: psychopathology.

psychopharmacology See study of: pharmacology: psychopharmacology.

psychophysics See study of: physics: psychophysics.

psychophysiology See study of: physiology: psychophysiology.

pteridology *n.* "The study of ferns" (Lincoln et al. 1985).

rheology *n.* "The study of fluid movement" (Lincoln et al. 1985).
See study of: limnology: rheology.

scatology *n.* "The study of feces" (Lincoln et al. 1985).

science *n.*
1. "A particular branch of knowledge or study; a recognized department of knowledge" (*Oxford English Dictionary* 1972, entries from *ca.* 1348); any knowledge department whose investigation results have been logically arranged and systematized in the form of hypotheses and general laws subject to verification (Michaelis 1963).
2. A branch of study that is "concerned either with a connected body of demonstrated truths, or with observed facts that are systematically classified and more or less colligated by being brought under general laws, and which includes trustworthy methods for the discovery of new truth within its own domain" (*Oxford English Dictionary* 1972, entries from 1725); "an orderly presentation of facts, reasonings, doctrines, and beliefs concerning some subject or group of subjects: the *science* of theology" (Michaelis 1963).
3. In modern use, a natural, or physical, science, thus restricted to branches of study that relate to material-universe phenomena and their laws, or principles, sometimes with exclusion of pure mathematics (*Oxford English Dictionary* 1972, entries from 1867).
See study of: philosophy: natural philosophy; science.
  ▸ **hard science, real science** *n.* Science that includes obtaining its facts from controlled repeatable experiments with all independent variables controlled by an investigator, especially in a laboratory, and obtaining precise numerical values when quantitative data are gathered; hard science includes much of chemistry, physics, and molecular biology; contrasted with soft science (Diamond 1987).
  ▸ **natural science** *pl. n.*
    1. Sciences collectively that deal with the physical universe.
    2. *sing. n.* Biology, chemistry, physics, or a combination of these sciences (Michaelis 1963).

  ▸ **physical science** *n.* Any of the sciences (*e.g.,* astronomy, chemistry, geology, physics, etc.) concerned with structure, properties, and energy relations of matter apart from the phenomena of life (Michaelis 1963).
  ▸ **soft science** *n.*
    1. Pejoratively, a science that is not as rigorous as a hard science (Diamond 1987).
    2. A science that deals with phenomena that are more complicated than those dealt with by hard sciences and, therefore, cannot necessarily involve experiments that are as well controlled as those of hard sciences; nonetheless, a soft science can predict a phenomenon's course with a certain probability (Diamond 1987).
    Note: Complicated "soft sciences" include "animal behavior," "anthropology," "ecology," "economics," "evolution," "psychology," and "sociology." The so-called "soft sciences" are the harder (= more difficult) ones according to Diamond (1987).
    cf. natural history, operationalism

scientometrics See study of: -metry: scientometrics.

semantics *pl. n.* (construed as singular) In linguistics, the study of the meanings of speech forms, especially of the development and changes in meaning of words and word groups (Michaelis 1963).
cf. communication

semiosystematics See systematics: semiosystematics.

semiotics, semeiotics *pl. n.* (construed as singular)
1. The science of language (Carnap 1942 in Sebeok, 1965).
2. The science of signs, having the divisions of semantics, syntactics, and pragmatics, each which can be pure, descriptive, or applied (Morris 1946, 353).
3. Analysis of communication in the broadest sense (Pierce, Morris, Carnap, and Mead in Wilson 1975, 201).
4. The scientific study of communication (Wilson 1975, 594).
cf. communication
adj. semeiotic, semiotic, semiotical
[Greek *sēmeiōtikos* < *sēmeion*, diminutive of *sēma*, mark]
  ▸ **zoosemiotics** *n.*
    1. The "discipline within the behavioral sciences, one which has crystallized at the intersection of semiotics, the science of signs, and ethology, the biological study of behavior" (Sebeok 1963, 1965).

2. The scientific study of animal communication, involving an evolutionary emphasis and logical, analytic techniques associated with human-oriented semiotics (Wilson 1975, 201).
[ZOO + SEMIOTICS, *q.v.*]

serology *n.* "The study of antigens and antibodies" (Lincoln et al. 1985).

sexology *n.* The study of sex and sexual behavior (Lincoln et al. 1985).

silvics *n.* "The study of forest trees" (Lincoln et al. 1985).
adj. silvical

sociobiology See study of: biology: sociobiology.

sociology *n.*
1. The study of origin, history, and constitution of human society; social science (*Oxford English Dictionary* 1972, entries from 1843), at all levels of complexity (Wilson 1975, 4, 595).
Note: "Sociology" differs from "sociobiology" chiefly due to its structural and nongenetic (nonevolutionary) approach (Wilson 1975, 4, 595).
2. The study of plant communities or of collectively organized organism groups (Lincoln et al. 1985).

▸ **ecological sociology** See ecology: synecology.

▸ **phytosociology** See study of: ecology: phytoecology; phytocoenology.

speleology, speology, spelaeology *n.* "The study of caves" (Lincoln et al. 1985).

spermology *n.* "The study of seeds" (Lincoln et al. 1985).

sphecology *n.* The study of Wasps (Lincoln et al. 1985).

statics See study of: physics: statics.

statistics *pl. n.* (construed as singular)
1. "The science that deals with the collection, tabulations, and systematic classification of quantitative data, especially of occurrence as a basis for inference and induction" (Michaelis 1963); constructing probability models to relate data with underlying phenomena being studied is also typically a crucial component of statistics, and testing hypotheses encompasses only about 10% of statistics (J.O. Berger, personal communication).
syn. statistic (Michaelis 1963).
2. "The scientific study of numerical data based on natural phenomena" (Sokal and Rohlf 1969, 2).
3. A branch of mathematics that deals with generalization, estimation, and precision in situations where uncertainty exists, dealing particularly with populations, variation, and reduction of data (Lincoln et al. 1985).

cf. statistics; study of: -metry: biometry
[German *statistik* < Medieval Latin *statisticus*, statesmanlike, ultimately from Latin *status*, condition, state]

▸ **Bayesian statistics** *n.* Unconventional statistics that indicate the probability of a null hypothesis' being correct; Bayesian statisticians argue that "calculated p's" yielded by conventional statistics can result in incorrect inferences about data (Berger and Berry 1988).
[after T. Bayes (1763), who proposed basic ideas of "Bayesian statistics" (Berger and Berry 1988, 159)]

▸ **circular statistics** *n.* Statistics that deal with biological rhythms with regard to a 24-hour clock and animal orientation with regard to 360 degrees (Batschelet 1965; D.B. Quine, personal communication).

▸ **"confidence interval statistics"** *n.* Unconventional statistics that indicate if a null hypothesis is false by examining overlapping and non-overlapping 95% confidence intervals (Jones 1984).

▸ **"conventional statistics"** *n.* Statistics based on "calculated p's" and related probabilities (J.O. Berger, personal communication); contrasted with unconventional statistics.
cf. unconventional statistics
Comment: The majority of today's animal behaviorists use conventional statistics. Bayesian statisticians question whether one may (sensibly) reject null hypotheses with "calculated p's" (J.O. Berger, personal communication).

▸ **descriptive statistics** *n.* Statistics that help to organize and summarize data; contrasted with inferential statistics (Sokal and Rohlf 1969, 40).

▸ **inductive statistics** See statistical inference.

▸ **inferential statistics** *n.* Statistics that help to test hypotheses and generalize from data; distinguished from descriptive statistics (Siegel 1956, 1).

▸ **"unconventional statistics"** *n.* Statistics that make inferences from information other than "calculated p's" (*e.g.*, Bayesian statistics); contrasted with conventional statistics.

stratigraphic paleontology, stratigraphic palaeontology See study of: paleontology: stratigraphic paleontology.

study of signals *n.* The study of signals which has three main branches: pragmatics, semantics, and syntactics.

▸ **pragmatics** *n.* The study of the significance that communicatory signs have for their communicants (Smith 1969a in Wilson 1975, 219).

▶ **semantics** *n.* The study of the message content of signs in communication (Cherry 1957, Smith 1969a in Wilson 1975, 218).

▶ **syntactics** *n.* The study of signals as physical phenomena (Cherry 1957, Smith 1969a in Wilson 1975, 218).

symbiology See study of: biology: symbiology.

synchorology *n.* The study of the occurrence, distribution, and classification of plant communities into the following regional units: region, province, sector, subsector, district, and subdistrict (Lincoln et al. 1985).

syndynamics See study of: dynamics: syndynamics.

synecology See study of: ecology: synecology.

synmorphology See study of: morphology: synmorphology.

synpiontology *n.* "The study of ancient patterns of distribution and migration of plant communities" (Lincoln et al. 1985).

synsystematics See study of: systematics: synsystematics.

systematics *n.*
1. "The scientific study of the kinds and diversity of organisms and of any and all relationships among them" (Simpson 1961 in Mayr 1982, 145, 245–246).
2. Classification of organisms into hierarchical series of groups that emphasizes their phylogenetic interrelationships (Lincoln et al. 1985).
See classification; study of: taxonomy.
syn. taxonomy (in some cases and especially in the first half of the 20th century) (Lincoln et al. 1985)

▶ **biosystematics** *n.*
1. Systematics of genetic and ecological variation in natural populations (Mayr 1982, 277).
2. "The study of evolution and the classification of living organisms, based on biological information at the population level such as genetic variability, hybridization, breeding strategies, competition and local adaptations" (Lincoln et al. 1985).
syn. biosystematy, genonomy (Lincoln et al. 1985), genecology (Turesson 1922 in Mayr 1982, 277)

▶ **cladistic systematics, cladistics** *n.*
1. Systematics that bases classification exclusively on genealogy (the branching pattern of phylogeny) in which parental species split into two sister (= daughter) species, which must be given the same categorical rank, and the ancestral species, together with all of its descendants, must be included in a single holophyletic taxon (Hennig 1950 in Mayr 1982, 226).
2. A classificatory method that employs phylogenetic hypotheses as the basis for classification and uses recency of common ancestry alone as the criterion for grouping taxa, rather than data on phenetic similarity or divergence (Lincoln et al. 1985; Campbell 1990, 491).
syn. cladism, cladistic method, cladistics (Lincoln et al. 1985)
cf. study of: taxonomy: numerical taxonomy
Comment: Cladistics tries to eliminate subjectivity and arbitrariness from classifying by developing a virtually automatic method (Mayr 1982, 226).
[Greek *clados*, branch]

▶ **ecosystematics** *n.* The integrated study of the ontology, evolution, and systematics of ecosystems (Lincoln et al. 1985).

▶ **ethosystematics** *n.* Systematics based on behavioral characters (Whitmann 1899, 1919 and Heinroth 1911 in Heymer 1977, 65).
cf. ²species: ethospecies; -type: ethotype

▶ **evolutionary systematics, evolutionary method, synthetic method** *n.* The branch of systematics in which classification is based on "hypothetical reconstructions of evolutionary history incorporating both cladistic data (on the sequence of branching events) and morphological divergence data" (Lincoln et al. 1985).

▶ **Hennigian systematics** See study of: systematics: phylogenetic systematics.

▶ **molecular systematics** *n.*
1. Systematics using biochemical characters and techniques such as electrophoresis, chromatography, serology, and DNA hybridization (Lincoln et al. 1985).
2. The study of evolutionary relationships using comparative molecular data (Lincoln et al. 1985).

▶ **omnispective systematics, omnispective method** *n.* The branch of systematics in which classification is "based on intuitive and pragmatic considerations using weighted phenotypic similarity as the criterion of relationship, with evolutionary history taken into consideration, but without full phylogenetic analysis" (Lincoln et al. 1985).

▶ **phenetic systematics, neo-Adansonism, phenetic method, phenetics** *n.* The branch of systematics that classifies organisms using the criteria of overall morphological, anatomical, physiological, or biochemical similarity or difference,

S–Z

with all characters equally weighted and without regard to phylogenetic history (Lincoln et al. 1985).

cf.-gram: dendrogram: phenogram; study of: taxonomy: numerical taxonomy

Comments: This method does not attempt to sort homologous characters from analogous ones (Campbell 1990, 491). It arranges species, or other taxa, in "phenograms." Phenetic systematics has few proponents today, but its emphasis on multiple quantitative comparisons with the help of computers had an important impact on the development of presently used phylogenetic methods.

[phenetic from Greek *phainein*, to appear + TYPE]

▶ **phylogenetic systematics, Hennigian systematics** *n.* The branch of systematics that classifies organisms based on their phylogenetic relationships, not considering morphological resemblances between them as a simple measure of phylogenetic relationship, but dividing these resemblances into concepts of convergence, symplesiomorphy, and synapomorphy and using only synapomorphy to establish relationships (Lincoln et al. 1985).

▶ **population systematics** *n.* The study of classification performed in view of evolutionary thinking, more particularly population thinking (Mayr 1982, 247).

syn. "new systematics (which is a less correct term for this concept; Mayr 1982, 561).

Comments: "Population systematics" studies biological properties of organisms rather than the static characters of dead specimens. The greatest number of kinds of characters — physiological, biochemical, and behavioral, as well as morphological ones — are used in classification. Population systematics were first practiced by Rensch (1929, 1933, 1934) (Mayr 1982, 277).

[coined by Huxley, 1940]

▶ **semiosystematics** *n.* A branch of systematics that gives special attention to animal signals used in courtship and reproductive isolation (coined by Lloyd 1989, 51).

Comments: Semiosystematics is attentive to and may even favor mating signals over morphological traits in classifying animals (Lloyd 1989, 52). Frank A. McDermott (1912) may have made the first formal taxonomic decision based on mating signals by taking a firefly species out of synonymy with another species because he believed that its mating signals were

different from the latter's ones (Lloyd 1989, 52). Researchers have used mating signals for classifying crickets and other firefly species.

▶ **synsystematics** *n.* "The classification of plant communities" (Lincoln et al. 1985).

taxonomy *n.*

1. Classification, especially in relation to its general rules or principles (*Oxford English Dictionary* 1972, entries from 1813).

2. The science of classification (*Oxford English Dictionary* 1972, entries from 1813), especially of organisms (Wilson 1975, 597).

3. "The theory and practice of delimiting kinds of organisms and classifying them" (Simpson 1961 and Mayr 1969 in Mayr 1982, 146).

See classification; study of: systematics.

syn. systematics (in some cases) (Lincoln et al. 1985)

cf. study of: taxonomy: idiotaxonomy

Comment: Some consider taxonomy be something of an art, and taxonomists to use judgment with "intuitive perception" and "instinct" (Stevens 1992, 303).

[French *taxonomie* < Greek *taxis*, arrangement + *nomos*, law; coined by Augustin-Pyramus de Candolle (1778–1841), Swiss-French botanist (Margulis 1981, 16)]

▶ **alpha taxonomy** *n.* Taxonomy involving describing and naming species (Mayr 1982, 145); contrasted with beta taxonomy, gamma taxonomy.

▶ **beta taxonomy** *n.* Taxonomy involving classifying species into a natural system of lesser and higher categories (Mayr 1982, 145); contrasted with alpha taxonomy, gamma taxonomy, microtaxonomy.

syn. macrotaxonomy (Lincoln et al. 1985)

▶ **chemotaxonomy** *n.* Taxonomy that uses biochemical characters and techniques (Lincoln et al. 1985).

▶ **cytotaxonomy** *n.* Taxonomy based on cell structure, especially chromosome structure (Lincoln et al. 1985).

▶ **eclectic taxonomy** See study of: taxonomy: evolutionary taxonomy.

▶ **empirical taxonomy** *n.* Classification of organisms based on observed phenotypic similarity (Lincoln et al. 1985).

▶ **evolutionary taxonomy, eclectic taxonomy** *n.* Taxonomy in which phylograms, showing branching of organism phyletic lines but also subsequent divergence into taxa, are produced (Mayr 1982, 233).

Comment: Because this methodology uses newly developed methods such as certain numerical methods of phenetics and ancestral-derived partitioning of characteristics of cladistics, it is also called "eclectic taxonomy" (Mayr 1982, 233).

▸ **gamma taxonomy** *n*. Taxonomy that involves evolutionary interpretations (Mayr 1982, 145); contrasted with alpha taxonomy, beta taxonomy.

▸ **idiotaxonomy** *n*. Taxonomy of individuals, populations, species, and higher taxa; traditional taxonomy (Lincoln et al. 1985).
cf. study of: taxonomy: syntaxonomy

▸ **microtaxonomy** *n*.
1. The subfield of taxonomy that deals with the methods and principles by which kinds ("species") of organisms are recognized and delimited (Mayr 1982, 146).
2. The scientific study of species (Mayr 1982, 297).
3. The branch of taxonomy concerned with the classification of species, varieties, and populations (Lincoln et al. 1985).
cf. study of: taxonomy: macrotaxonomy

▸ **numerical taxonomy (NT)** *n*.
1. A classification process, originating in the 1960s, that attempts to eliminate human subjective evaluation of taxonomic characters by using quantitative methods and computer analysis (Mayr 1982, 222).
2. Taxonomy based on the numerical comparison of large numbers of equally weighted characters, scored consistently for all groups under consideration, in which individuals are grouped solely on the basis of observable similarities (Lincoln et al. 1985).
syn. numerical phenetics, taxometrics, taximetrics, taxonometrics (Lincoln et al. 1985)
Comments: In early NT, characters were given equal weight; in later NT, characters were given "phyletic weights." The species was replaced with the "operational taxonomic unit" as the unit of classification. Because taxonomists have used numerical methods for generations, the "numerical taxonomy" of Sokal and Sneath (1963) is called "numerical phenetics" (Mayr 1982, 222).

▸ **phytochemistry** *n*. "Chemotaxonomy of plants" (Lincoln et al. 1985).

▸ **polyphasic taxonomy** *n*. Taxonomy of a given set of organisms by a number of different techniques, applied successively or simultaneously (Lincoln et al. 1985).

▸ **serotaxonomy** *n*. "Taxonomy based on serological characters" (Lincoln et al. 1985).

▸ **syntaxonomy** *n*. "The classification and nomenclature of plant communities" (Lincoln et al. 1985).
cf. study of: taxonomy: ideotaxonomy

▸ **zootaxy** *n*. Animal taxonomy (Lincoln et al. 1985).

teleology See study of: philosophy: teleology.

teleonomy *n*.
1. The scientific study of "programmed purposiveness (Osche 1973) in living organisms and in certain manmade machines" (Immelmann and Beer 1989, 308).
2. "The science of adaptation" (Dawkins 1982, 295).
cf. study of: philosophy: teleology; teleology
Comments: "In effect, teleonomy is teleology made respectable by Darwin. ...This book is an essay in teleonomy. Teleonomy is a new science that will presumably be involved with questions of units of selection and costs and other constraints on perfection" (Dawkins 1982, 295). Leading neo-Darwinians attempted to replace the term "teleology" with "teleonomy" to avoid the twin specters of natural theology and vitalism (Lennox 1992, 331).

teratology *n*. "The study of malformations and monstrosities" (Lincoln et al. 1985).

▸ **phytoteratology, plant teratology** *n*. The study of plant malformations and monstrosities (Lincoln et al. 1985).

thanatology *n*. The study of the circumstances surrounding human death (Storz 1973, 275).

thermodynamics *n*. The study of energy transformations (Curtis 1983, 157).

thremmatology *n*. The study of organism breeding in domestic situations (Lincoln et al. 1985).

tocology *n*. The science of parturition (Ryan 1828 in Miller 1984, 119).

tokology *n*. The study of the mechanics and fitness maximization of organisms' reproduction (Miller 984, 119).

toxicology *n*. The study of the origin, nature, properties, etc., of poisons (Michaelis 1963).

▸ **ophiotoxicology** *n*. The toxicology of snake venoms (Lincoln et al. 1985).

tree-ring chronology See delete: study of: method: dendrochronology.

trophology *n*. "The study of nutrition" (Lincoln et al. 1985).

typology *n*. "The study of types, as in systems of classification" (Michaelis 1963).

verbal learning *n*. The study of how human verbal associations are learned and

S–Z

remembered in specific experimental situations, *e.g.,* paired-associate learning, serial learning, and transfer (Storz 1973, 282).

xenology *n.* "The study of host-parasite relationships" (Lincoln et al. 1985).

zoodynamics See study of: physiology: animal physiology.

zooecology See study of: ecology: zooecology.

zoology *n.* The study of animals (Lincoln et al. 1985); zoology includes cenozoology, cetology, cryptozoology, entomology, ethnozoology, helminthology, herpetology, ichthyology, invertebrate zoology, limacology, malacology, myrmecology, nematodology, ophiology, ornithology, paleozoology, primatology, protozoology, sphecology, and vertebrate zoology.

▶ **cenozoology** *n.* "The study of extant animals" (Lincoln et al. 1985).

▶ **cryptozoology** *n.* The study of cryptozoa, small terrestrial animals that inhabit crevices and live under stones and in soil and litter (Lincoln et al. 1985).

▶ **ethnozoology** *n.* The study of the use of animals and animal produces by groups of Humans (Lincoln et al. 1985).

▶ **invertebrate zoology** *n.* The study of invertebrate animals which includes all phyla except Chordata (Wilson 1975, 587). *cf.* study of: zoology: vertebrate zoology

▶ **paleozoology, palaeozoology** *n.* The study of animals of the geological past (Lincoln et al. 1985).

▶ **protozoology** *n.* "The study of Protozoa" (Lincoln et al. 1985).

▶ **vertebrate zoology** *n.* Zoology of animals with vertebrae in the phylum Chordata. *cf.* study of: zoology: invertebrate zoology

zoonomy See study of: physiology: animal physiology.

zoosemiotics See study of: semiotics: zoosemiotics.

◆ **stygobie** See -cole: cavernicole.

◆ **stygon** See biotope: stygon.

◆ **stygophile, stygophilic, stygophily** See [1]-phile: stygophile.

◆ **stygoxenous** See -xene: trogloxene.

◆ **stylus hypothesis of insect wing origin** See hypothesis: hypotheses of origin of insect wings: stylus hypothesis of insect wing origin.

◆ **sub-** *prefix*

1. "Under; beneath; below" (Michaelis 1963).
2. Anatomically, "situated under or beneath, or on the ventral side of" (Michaelis 1963).
3. Chiefly scientifically, "almost; nearly; slightly; imperfectly" (Michaelis 1963).

4. "Lower in rank or grade; secondary; subordinate" (Michaelis 1963).
5. "Forming a subdivision" (Michaelis 1963). Also, suc- before a c; suf- before an f; sug- before a g; sup- before a p; sur- before a r; sus- before a c, p, or t.
   cf. super-
   [Latin *sub- < sub,* under]

◆ **subadult** See animal names: adult: subadult.

◆ **subaerial** *adj.* Referring to something that occurs immediately above the ground's surface (Lincoln et al. 1985).

◆ **subaquatic** *adj.* "Not completely aquatic" (Lincoln et al. 1985).

◆ **subclimax** See [2]community: climax.

◆ **subcloning** See method: subcloning.

◆ **subdominant species** See [2]species: dominant species: subdominant species.

◆ **subdominule** See [2]species: dominant: subdominule.

◆ **subfossil** See fossil: subfossil.

◆ **subgeocole, subgeocolous** See -cole: subgeocole.

◆ **subgregarious** See gregarious: subgregarious.

◆ **subhydrophile, subhydrophilous, subhydrophily** See [1]-phile: subhydrophile.

◆ **subjective self-awareness** See awareness: self-awareness: subjective self-awareness.

◆ **sublethal** *adj.* Referring to a factor that does not cause an organism's death by its direct action but by possibly modifying its behavior, physiology, reproduction, or life cycle and showing cumulative effects (Lincoln et al. 1985).

◆ **sublethal gene** See [2]mutation: subvital mutation.

◆ **sublimation** See behavior: displacement behavior: sublimation.

◆ **subliminal stimulus** See [2]stimulus: subliminal stimulus.

◆ **submission** *n.*

1. A person's submitting to an authority, a conquering, or ruling, power; a person's yielding to the claims of another, or surrendering to his will or government; a person's condition of having submitted; an instance of submission (*Oxford English Dictionary* 1972, entries from 1482).
2. An individual animal's showing defensive behaviors that tend to prevent further attack of another contestant, *e.g.,* a dog's or wolf's rolling over on its back and remaining motionless or an ape's assuming a female copulation stance (Darwin 1872 in Heymer 1977, 51).
3. An individual animal's "act of yielding or surrendering" that reflects "resignation"

and relates to individual "temperament" or "inclination" and which occurs when an animal loses a fight and halts further attack (Hand 1986, 216).

syn. catasematic behavior (Mertens 1946 in Heymer 1977, 51), submissive behavior, submissive display

cf. display: submissive display; dominance; posture: submissive postures; subordination

♦ **submissive coloration** See coloration: submissive coloration.

♦ **submissive display** See display: submissive display.

♦ **submissive posture** See posture: submissive posture.

♦ **subordination** *n.* An animal's consistently losing, or deferring (= changing its apparent priority and doing something other than what it had been doing or had been seeking to do), in conflict encounters in a particular dyadic relationship (Hand 1986, 205).

cf. dominance, submission

Comment: "Subordination" implies nothing about an individual's "temperament" or "inclination" (J.L. Hand, personal communication).

♦ **subsere** See ²succession: subsere.

♦ **subsocial** See sociality: subsociality.

♦ **subsocial route to sociality in insects** *n.* One of two hypothesized evolutionary paths to eusociality in insects (Wilson 1975, 398)

See sociality (Wilson 1975, 398).

♦ **subsong** See animal sounds: song: subsong.

♦ **subspecies** See ²species: subspecies.

♦ **substance** *n.* A kind of matter of specific chemical composition (*Oxford English Dictionary* 1972, entries from 1732).

cf. chemical-releasing substance: semiochemical: allelochemic: allomone: propaganda substance

adoption substance *n.* A secretion that a social parasite presents to its host insect that induces it to accept the parasite as a colony member (Hölldobler and Wilson 1990, 635).

allelopathic substance See chemical-releasing substance: allelochemic: allomone: allelopathic substance.

appeasement substance *n.* A secretion that a social parasite presents to its host insect that reduces the host's aggression and aids the parasite's acceptance by the host colony (Hölldobler and Wilson 1990, 635).

exocrine substance See chemical-releasing stimulus: semiochemical: allelochemic: allomone: exocrine substance.

propaganda substance See chemical-releasing substance: semiochemic: allelochemic: allomone: propaganda substance.

royal substance, queen substance See chemical-releasing substance: semiochemical: pheromone: queen pheromone.

symphylic substance *n.* A chemical used by social-insect-colony parasites (*e.g.*, beetles) that influences their host's behavior (Dawkins 1982, 294).

trail-marking substance See chemical-releasing stimulus: semiochemical: pheromone: trail-marking pheromone.

transmitter substance See neurotransmitter.

♦ **substitute activity** See behavior: displacement behavior.

♦ **substitute object** See object: substitute object.

♦ **substitutional load** See genetic load: substitutional load.

♦ ¹**substrate, substratum** *n., pl.* **substrates, substrata**

1. "The matter on which a fungus or other plant grows" (*Oxford English Dictionary* 1972, entries from 1876).
   Note: Fungi are no longer classified as plants (Strickberger 1990, 266).
2. "The sediment, surface, or medium to which an organism is attracted or upon which it grows" (Lincoln et al. 1985).

♦ ²**substrate, substratum** *n., pl.* **substrates, substrata** A substance acted upon by an enzyme (Lincoln et al. 1985).

♦ **substrate race** See race: substrate race.

♦ **substratohygrophile, substratohygrophilous, substratohygrophily** See ¹-phile: substratohygrophile.

♦ **substratum** See substrate.

♦ **substratum stridulation** See stridulation: substratum stridulation.

♦ **subsurface lithoautotrophic microbial ecosystem** See system: subsurface lithoautotrophic microbial ecosystem.

♦ **subterranean, subterrestria** *n.* Referring to an organism that lives, or occurs, beneath the Earth's surface (Lincoln et al. 1985).

♦ **subthreshold stimulus** See stimulus: subthreshold stimulus.

♦ **subtle cheater** See cheater: subtle cheater.

♦ **subtraction** *n.* "Loss of a heredity factor" (Lincoln et al. 1985).

♦ **subtroglophile** See ¹-phile: subtroglophile.

♦ **subtrogloxene** See -xene: subtrogloxene.

♦ **subtropical** *adj.*

1. Referring to the regions adjacent to the Earth's Torrid Zone (Michaelis 1963).

S–Z

2. Having characteristics intermediate between, or common to, both Torrid and
Temperate Zones (Michaelis 1963).
3. Referring to the zone of the Earth between 23°27′ and 34.0° in either hemisphere; contrasted with temperate, tropical, and subtropical (Lincoln et al. 1985).
Comment: Southern parts of California, Arizona, New Mexico, Texas, Oklahoma, Arkansas, Mississippi, Alabama, Georgia, South
Carolina, and North Carolina; the northern
islands of Hawai'i; and all of Louisiana and
Florida are in the subtropical zone.

♦ **subvital mutation** See [4]mutation:
subvital mutation.

♦ **subxerophile** See [1]-phile: subxerophile.

♦ **subzone** See zone: def. 4 (notes).

♦ **success** *n.*
1. A person's prosperous achievement of
something attempted; a person's attaining an object according to his desire, now
often with particular reference to attaining wealth or position (*Oxford English
Dictionary* 1972, entries from 1586).
2. The longevity of a species and all of its
descendants through geological time
(Wilson 1990 in Ruse 1993, 58).
cf. dominance (ecological)

reproductive success (RS) *n.* An individual organism's number of surviving offspring (Wilson 1975, 325).
cf. principle: principle of reproductive success

♦ [1]**succession** *n.*
1. A set of persons, or things, that succeed
in the places of others (*Oxford English
Dictionary* 1972, entries from 1647).
2. The continued sequence in a definite
order of species, types, etc.; specifically,
the decent in uninterrupted series of
forms modified by evolution or development (*Oxford English Dictionary* 1972,
entries from Darwin 1834).

♦ [2]**succession** *n.*
1. The gradual and predictable process of
progressive community change and replacement leading toward a stable climax
community (Lincoln et al. 1985).
syn. biotic succession, ecological succession, progressive succession, sere (Lincoln et al. 1985)
2. The continuous colonization and extinction of species populations in a particular
location (Lincoln et al. 1985).
syn. sere (with regard to plants) (Lincoln
et al. 1985)
3. "The chronological distribution of organisms within an area" (Lincoln et al. 1985).
4. The geological, ecological, or seasonal
sequence of species within a habitat or
community (Lincoln et al. 1985).
cf. climax

adsere *n.* The stage of succession that
precedes its subclimax stage (Lincoln et al.
1985).

aquatosere *n.* A succession that starts in
a wet habitat and leads to an aquatic climax
(Lincoln et al. 1985).

camnium *n.* A succession caused by cultivation (Lincoln et al. 1985).

clisere *n.* A succession that results from a
major climatic change (Lincoln et al. 1985).

geosere *n.* A succession that starts on clay
(Lincoln et al. 1985).
See [2]community: geosere.

hydrarch succession, hydrosere *n.*
A succession that begins in a habitat with
abundant water (Lincoln et al. 1985).
cf. [2]succession: mesarch succession, xerarch
succession

lithosere *n.* A succession that originates
on an exposed rock surface (Lincoln et al.
1985).

mesarch succession, mesosere *n.*
1. A succession that begins in a habitat
with a moderate amount of water (Lincoln et al. 1985).
2. A succession within a microhabitat that
often fails to reach a stable climax
(Lincoln et al. 1985).
syn. serule (Lincoln et al. 1985)
cf. [2]succession: hydrarch succession, xerarch
succession

microsere, serule *n.* A succession within
a microhabitat that often fails to reach a
stable climax (Lincoln et al. 1985).

plagiosere *n.* A succession that is deflected from its natural course by continuous human interference (Lincoln et al.
1985).

postclisere *n.* A succession that arises in
a more mesic condition than is present in its
habitat today; contrasted with preclisere
(Lincoln et al. 1985).

preclisere *n.* A succession that arises
under more xeric conditions than those
present in a particular habitat today; contrasted with postclisere (Lincoln et al. 1985).

progressive succession See [2]commu-
nity: succession (def. 1).

psammosere *n.* A succession that starts
in unconsolidated sand (Lincoln et al. 1985).

retrogressive succession *n.* A succession that recedes from a climax as a result
of an external influence, or interference,
such as human activity (Lincoln et al. 1985).

sere *n.*
1. A stage in community succession (Lincoln et al. 1985).
2. A succession of plant communities in a
given habitat leading to a particular
climax association (Lincoln et al. 1985).
See succession (def. 2).

Comment: Seres are classified by habitat or major influencing factors.

serule See ²succession: microsere.

subsere *n.*
1. A succession on a denuded area (Lincoln et al. 1985).
2. A succession arrested by edaphic, or biotic factors (Lincoln et al. 1985).
syn. secondary sere (Lincoln et al. 1985)
cf. ²community: subclimax

therium, therion *n.* A plant succession caused by animal activity (Lincoln et al. 1985).

xerarch succession, xerosere *n.* A succession that begins in a dry habitat (Lincoln et al. 1985).
cf. ²succession: hydrarch succession, mesarch succession

xerosium *n.* "An ecological succession on drained and dried soils" (Lincoln et al. 1985).

♦ **successional speciation** See speciation: successional speciation.

♦ **successional species** See ²species: successional species.

♦ **successive choice test** See test: choice test.

♦ **succinite** See fossil: succinite.

♦ **succorance** See need: succorance.

♦ **sucking** See ³suckling.

♦ ¹**suckling** *n.*
1. A human infant that is still feeding at its mother's breast or is unweaned (*Oxford English Dictionary* 1972, entries from *ca.* 1440).
2. A young mammal that is still feeding at its mother's breast, especially a calf (*Oxford English Dictionary* 1972, entries from 1530).

♦ ²**suckling** *n.*
1. A woman's nursing a child at her breast (*Oxford English Dictionary* 1972, entries from 1408).
2. An adult mammal's allowing, or causing, a young to take nourishment from her breast (Michaelis 1963).
syn. nursing (Michaelis 1963)
3. An individual animal's bringing up, or nourishing, another (Michaelis 1963).
See nursing.
[Middle English < *sucklen*, probably back formation < SUCKLING]

allosuckling *n.* For example, in the Fallow Deer, Free-Tailed Bat, Lion, Northern Elephant Seal: a mother's giving milk to a young other than her own (Schaller 1972, etc. in Birgersson et al. 1991, 326).
Comment: "Allosuckling" has been interpreted as cooperation between related individuals or as an effect of living under exceptionally crowded conditions (Birgersson et al. 1991, 326).

♦ ³**suckling** *n.* A young mammal's attempting to take, or taking, milk from a mammary gland (Cowie et al. 1951).
v.t. v.i. suckle
syn. sucking (with regard to Humans, *Oxford English Dictionary* 1972)
cf. lactation, milking
Comments: Suckling includes behaviors such as stimulating a mammary gland and sucking milk. To reduce confusion, "nursing" should be used to refer to a lactating female's actions, and "suckling" should refer to the actions of a young that is aiming to obtain milk from a mammary gland. "Nursing" and "suckling" should not be used interchangeably, and "sucking" and "suckling" should not be used interchangeably (Cowie et al. 1951; D. Fraser, personal communication).

♦ **suckling stimulus** See ²stimulus: suckling stimulus.

♦ **suicide** See -cide: suicide.

♦ **suitor phenomenon** See mate guarding: precopulatory mate guarding.

♦ **suffering** See animal suffering.

♦ **sum of squares** *n.* In statistics, the sum of the squared deviations from a mean in an analysis of variance (Lincoln et al. 1985).

♦ **summer solstice** See solstice: summer solstice.

♦ **Sun** See star: Sun.

♦ **sun-compass orientation** See taxis: menotaxis: sun-compass taxis.

♦ **sun-dapple territory** See territory: sun-dapple territory.

♦ **super-** *prefix*
1. "Above in position; over" (Michaelis 1963).
2. Anatomically and zoologically, "situated above, or on the dorsal side of" (Michaelis 1963).
3. "Above or beyond; more than" (Michaelis 1963).
4. "Excessively" (Michaelis 1963).
5. "Greater than, or superior, to others in its class" (Michaelis 1963).
6. "Extra; additional" (Michaelis 1963).
cf. sub-
[Latin *super-* < *super*, above, beyond]

♦ **super beneficiary** See beneficiary.

♦ **super-donor** See donor: super-donor.

♦ **super half sister** See sister: half sister: super half sister.

♦ **super-releaser** See ²stimulus: releaser: supernormal stimulus.

♦ **super sister** See sister: super sister.

♦ **superbout** See bout: superbout.

♦ **supercolony** See colony: supercolony.

♦ **superdispersed distribution** See distribution: contiguous distribution.

♦ **superdonor** See donor: superdonor.

♦ **superfecundation** *n.* Fertilization of separate ova in a single woman by sperm of different males that is followed by plural birth of half siblings (Smith 1984, 616).

Comment: There are more than a dozen reported cases of superfecundation reported since the early 1800s (Smith 1984, 616).

♦ **superfetation** See superfoetation.

♦ **superfluent species** See ²species: superfluent species.

♦ **superfoetation, superfetation** *n.* A plant ovary's fertilization by two or more kinds of pollen (Lincoln et al. 1985).

syn. hypercyesis

♦ **supergene** See gene: supergene.

♦ **superior colliculus** See organ: brain: superior colliculus.

♦ **superlarvation** See ²heterochrony: paedomorphosis.

♦ **supernormal stimulus, supernormal releaser** See ²stimulus: releaser: supernormal stimulus.

♦ **supernova** *n., pl.* **supernovae** A catastrophic stellar explosion (Mitton 1993).

Comments: A supernova occurs when a star of more than 1.8 solar masses contracts after its nuclear fuel is exhausted (Mitton 1993). Then it explodes, blowing off much of its mass into outer space. A supernova can release so much energy that it alone can outshine an entire galaxy of billions of stars. Much of its energy goes into kinetic energy that blows material out into space and is carried off by neutrinos. People see about one supernova in the Milky Way per 200 years. Researchers detect about 50 supernovae in other galaxies each year.

♦ **"supernovae hypothesis"** See hypothesis: hypotheses regarding the Permian mass extinction: "supernovae hypothesis."

♦ **supernumerary** See helper.

♦ **supernumerary chromosome** See chromosome: supernumerary chromosome.

♦ **superoptimal stimulus** See ²stimulus: releaser: supernormal stimulus.

♦ **superorganism** See organism: superorganism (Appendix 1, Part 1).

♦ **superparasitism** See parasitism: superparasitism.

♦ **superpheromone** See chemical-releasing stimulus: semiochemical: pheromone: superpheromone.

♦ **superposition** *n.* In some fish species: the change in rhythm of one kind of fin caused by the dominant rhythm of another kind of fin (Tinbergen 1951, 73).

♦ **super-releaser** See ²stimulus: releaser: supernormal stimulus.

♦ **supersedure queen** See caste: queen: supersedure queen.

♦ **superstitious behavior** See behavior: superstitious behavior.

♦ **superspecies** See ²species: superspecies.

♦ **supertramp** See ²species: supertramp.

♦ **supination** See display: wing display: supination.

♦ **supine test** See test: supine test.

♦ **supplementary reproductive** See reproductive: supplementary reproductive.

♦ **supporting cell** See cell: glial cell.

♦ **suppressor gene** See gene: suppressor gene.

♦ **supra-** *prefix* "Above; beyond" (Michaelis 1963).

cf. infra-

[Latin *supra-*]

♦ **suprabenthos, suprabenthic** See ²community: benthos: suprabenthos.

♦ **suprachiasmatic nucleus** See organ: brain: suprachiasmatic nucleus.

♦ **supraneuston** See neuston: epineuston.

♦ **supraorganism** See organism: superorganism (Appendix 1, Part 1).

♦ **suprapelos** See ²community: suprapelos.

♦ **suprapsammon** See ²community: suprapsammon.

♦ **supraterraneous** *adj.* Referring to something that occurs at, or above, ground level (Lincoln et al. 1985).

♦ **surface pheromone** See chemical-releasing stimulus: semiochemical: pheromone: surface pheromone.

♦ **surgical menopause** See menopause: surgical menopause.

♦ **surprise** See facial expression: surprise.

♦ **surrogate** *n.* A dummy or stimulus model (Immelmann and Beer 1989, 300).

cf. dummy, model, mother: surrogate mother

♦ **surrogate mother** See mother: surrogate mother.

♦ **survival machine** See gene machine.

♦ **survival of the fittest** *n.*

1. The idea that animal life entails a direct and continuous competitive struggle — "nature red in tooth and claw" (Dewsbury 1978, 90).

   Note: This is a misconception of the phrase held by 19th-century social philosophers (Dewsbury 1978, 90).

2. Herbert Spencer's epitomization of Darwinian theory (Mayr 1982, 588).

   Note: "Survival" should mean "reproducing." This epitomization sounds deterministic, but in reality natural selection is a statistical rather than an all-or-none phenomenon. "Even though philosophers may still refer to 'survival of the fittest,' biologists no longer use such deterministic language" (Mayr 1982, 588). "When, at the urging of his friends, Darwin adopted Spencer's term 'survival of the fittest,' he jumped from the frying pan into the fire, because this new metaphor suggested circular reasoning" (Mayr 1982, 842).

Furthermore, it is usually not the fittest (only one) organism in a population that reproduces, but a group of organisms that are *fitter* than others.

3. The differential and greater success of better adapted genotypes (Lincoln et al. 1985).

See selection: natural selection.
cf. fittest, survival of the luckiest

♦ **survival of the luckiest** *n.* The concept that selectively neutral, or very nearly neutral, mutants are often fixed in populations due to random genetic drift; contrasted with survival of the fittest (Kimura 1992, 225).
cf. [3]theory: the neutral theory
Comment: Kimura (1989 in Kimura 1992, 230) proposed this term to emphasize the importance of good fortune vs. natural selection for a mutant's success in evolution.

♦ **survival of the stable** See principle: survival of the stable.

♦ **survival theory** See [3]theory: refugium theory.

♦ **survival value** See adaptive significance.

♦ **survivorship** *n.* The proportion of individuals from a particular cohort that survive at a given time, typically presented as a survivorship curve (Lincoln et al. 1985).

age-specific survivorship *n.* The probability that an individual of age *x* will be alive, and presumably be able to reproduce, at a later time, *x* + 1 (Willson 1983, 16).

♦ **survivorship curve** See curve: survivorship curve.

♦ **survivorship schedule** *n.* The number of individuals of a population that survive to each particular age (Wilson 1975, 90).

♦ **suscept** *n.* An organism that is susceptible to, or harbors, a disease organism (Lincoln et al. 1985).

♦ **susceptibility** *n.* An organism's being prone to influence by a stimulus or to infection by parasites or pathogens (Lincoln et al. 1985).
adj. susceptible

♦ **suspended animation** See biosis: anabiosis.

♦ **suspension feeders** See [2]group: functional feeding group: filtering collectors.

♦ **suspension feeding** See feeding: suspension feeding.

♦ **Sutton-Boveri-chromosome theory, Sutton-Boveri-chromosome theory of inheritance** See [3]theory: Sutton-Boveri chromosome theory.

♦ **suture zone** See zone: suture zone.

♦ **swale** See habitat: swale.

♦ **swamp, swampland** See habitat: swamp.

♦ **swarm** *v.t.*
1. For example, in the Honey Bee: to form a new colony by a queen and a large number of workers that suddenly depart from their parental nest and fly to some exposed site where they cluster while scout workers fly in search of a suitable new nest cavity (Wilson 1975, 596).
2. *v.i.* For example, in ants and termites: to make a mass exodus of reproductive forms from nests at the beginning of a nuptial flight (Wilson 1975, 596).
See [2]group: swarm.

♦ **SWARM** See program: SWARM.

♦ [1]**swarming** *n.* In social insects: colony reproduction in which the queen(s) and a number of workers separate and found a new colony (Hölldobler and Wilson 1990, 643).

budding *n.* In some social-insect species, *e.g.,* the Honey Bee: swarming in which the queen, or queens, and workers leave their main parental nest (Hölldobler and Wilson 1990, 643).
syn. hesmosis (in ants), sociotomy (in termites) (Wilson 1975, 139, 580).

fission *n.* For example, in Army Ants: swarming in which major portions of a colony separate, each with one or more queens and workers (Hölldobler and Wilson 1990, 643).
Comment: Wilson 1975 (581) originally included swarming and budding as divisions of "colony fission."

hesmosis, sociotomy See [1]swarming: budding.

♦ [2]**swarming** *n.* For example, in ants and termites: the mass exodus of reproductive forms from their nest at the beginning of a nuptial flight (Hölldobler and Wilson 1990, 643).

♦ **sweating** *n.* In some mammal species: an individual's exuding, or excreting, moisture from pores of its skin (Michaelis 1963).
adj. sweaty
adv. sweatily
n. sweatiness
syn. perspiring
[Old English *swǣtan* < *swāt*, sweat]

nonthermal sweating *n.* Sweat glands responding to a nonthermal stimulus, *e.g.,* a frightening experience for a Human (Bligh and Johnson 1973, 954).

thermal sweating *n.* Sweat glands responding to a thermal stimulus (Bligh and Johnson 1973, 954).

♦ **swimming** *n.* For example, in some bird, insect, protozoan, insect, species; many amphibian, crustacean, fish, mammal species: an individual's propelling itself through water with organized bodily movements

S–Z

(*Oxford English Dictionary* 1972, entries from 1000; Michaelis 1963; Ricci 1990, 1050).

coquette swimming See swimming: nod swimming.

corkscrew swimming *n*. In some fish species: swimming in which an individual circles on a horizontal plane in water, revolving or spiraling around an imaginary central point (Geiger et al. 1985, 10; R.A. Drummond, personal communication).

interrupted swimming *n*. In some fish species: swimming in which an individual has bouts of normal swimming, or rest periods, punctuated with bouts of strenuous, hyperactive swimming (Geiger et al. 1985, 10; R.A. Drummond, personal communication).

nectism *n*. For example, in *Paramecia* spp.: swimming by means of cilia (Lincoln et al. 1985).

nod swimming *n*.
1. In several dabbling-duck species: an epigamic behavior (Heymer 1977, 123). *syn.* coquette swimming (Heinroth 1910, Lorenz 1941 in Heymer 1977)
2. In some blenniid-fish species: a behavior that shows an individual's conflict between progressing forward or retreating (Wickler 1963 in Heymer 1977, 123).

spiral swimming *n*. In some fish species: corkscrew swimming in which an individual moves vertically up and down (Geiger et al. 1985, 10; R.A. Drummond, personal communication).

tail-down swimming *n*. In fish: swimming in which an individual has its head up, often near the surface of water, and its body axes perpendicular or at a steep angle with regard to water surface (Geiger et al. 1985, 10; R.A. Drummond, personal communication).

tail-up swimming *n*. In some fish species: swimming in which an individual has its head down with its longitudinal body axis perpendicular or at steep angles with regard to a horizontal substrate (Geiger et al. 1985, 10; R.A. Drummond, personal communication).

undulation swimming *n*. In some amphibian, reptile, and mammal species: an individual's swimming with its legs held against its body and propelling itself in a snake-like movement (Dubost 1965 in Heymer 1977, 202).

uninterrupted swimming *n*. In some fish species: swimming in which an individual moves continuously, or hyperactively, for long periods (Geiger et al. 1985, 11; R.A. Drummond, personal communication).

♦ **switch gene** See gene: switch gene.

♦ **sycophage, sycophagous, sycophagy** See -phage: sycophage.

♦ **sylvicole, sylvicolous** See -cole: sylvicole,

♦ **symbiogenesis** See -genesis: biogenesis.

♦ **symbiology** See study of: biology: symbiology.

♦ **symbiont, symbiote, symbion** *n*. An organism that lives in close association with another species (*Oxford English Dictionary* 1972, entries from 1887).
adj. symbiotic
syn. commensal (*Oxford English Dictionary* 1972)
cf. symbiosis

ectosymbiont, exosymbiont *n*. One of two symbionts that are both external to each other; contrasted with endosymbiont (Lincoln et al. 1985).

endosymbiont, inhabiting symbiont *n*. A symbiont that lives inside another symbiont; contrasted with ectosymbiont (Lincoln et al. 1985).

exhabited symbiont *n*. A symbiont that can be regarded as the host in a symbiosis in which neither symbiont is contained within the other; contrasted with exhabiting symbiont (Lincoln et al. 1985).

exhabiting symbiont *n*. A symbiont that lives upon another symbiont in a symbiosis in which neither symbiont is contained within the other; contrasted with exhabited symbiont (Lincoln et al. 1985).

exosymbiont See symbiont: ectosymbiont.

inhabited symbiont *n*. A symbiont in whose body another symbiont lives (Lincoln et al. 1985).

inhabiting symbiont See symbiont: endosymbiont.

keystone mutualist *n*. An organism (*e.g.,* a neotropical fig) that, if removed from its community, can cause a cascade of linked extinctions because of its importance to other organisms (Buchmann and Nabhan 1996, 248).
cf. ²species: keystone species

macrosymbiont *n*. A symbiont that is the larger of two partners in a symbiosis; contrasted with microsymbiont (Lincoln et al. 1985).

microsymbiont *n*. A symbiont that is the smaller of two partners in a symbiosis; contrasted with macrosymbiont (Lincoln et al. 1985).

symphile *n*. A symbiont, particularly a solitary insect or other kind of arthropod, that is accepted to some extent by an insect colony and "communicates with it amicably" (Wilson 1975, 596).
adj. symphilic

n. symphily

cf. symbiont: synechthran, synoekete

Comment: Most symphiles are licked, fed, or transported to host brood chambers, or receive a combination of these actions (Wilson 1975, 596).

▸ **pseudosymphile** *n.* A predator, or parasite, of a social insect that obtains nourishment from trophallactic secretions of this insect (Lincoln et al. 1985).

synechthran *n.* A symbiont that is usually a scavenger, parasite, or predator and is treated with hostility by its host colony (Hölldobler and Wilson 1990, 644).

cf. symbiont: symphile, synoekete; symbiosis: synectry

synoekete *n.* A symbiont that is treated with indifference by its host colony (Hölldobler and Wilson 1990, 644).

cf. symbiont: symphile, synechthran

◆ **symbiosis** *n.*

1. Heterospecific individuals, usually of two plant species, sometimes an animal and a plant species, that live attached to each other, or one as the "tenant" of the other, and contribute to each other's support; distinguished from parasitism (*Oxford English Dictionary* 1972, entries from 1877).
   syn. commensalism, consortism (*Oxford English Dictionary* 1972)

2. Heterospecific organisms' being closely associated with one another, with either, or both, species benefiting, or not benefiting, from their associations; symbiosis includes parasitism, commensalism, and mutualism (de Bary 1879 in Lewin 1982; Committee on Terminology of the American Society of Parasitologists in Goff 1982, 255).
   Note: This definition seems appropriate for wide use. Definitions 1 and 2 were confounded as early as 1893 (Pound in Boucher 1992, 208).

3. Different species' having presumed, or actual, mutually beneficial, close associations (Herwig 1883 in Lewin 1982).

4. Symbiosis in which the growth and survival of populations of two species "is benefitted and neither can survive under natural conditions without the other" (Odum 1969, 225).

5. Dissimilar organisms' (or their parts') living "together in an intimate association (generally implying a physiological interaction)" (Goff, 1982, 255).

syn. mutualism (a confusing synonym, Immelmann and Beer 1989, 301), *Symbiose*

cf. aposymbiotic, -biosis, parachoric, parasitism, predation

Comments: Hertig et al. (1937 in Goff 1982, 255) trace the etymology of "symbiosis."

A CLASSIFICATION OF SYMBIOSIS

I. Nonsocial symbiosis
  A. commensalism
  B. mutualism
  C. parasitism (including nectar robbing, nectar thieving, pollen robbing, pollen thieving)
    1. Classification by necessity
      a. facultative symbiosis
      b. obligate symbiosis (*syn.* obligatory symbiosis)
    2. Classification by order of parasitization of the host
      a. primary parasitism
      b. secondary parasitism
      c. tertiary parasitism
      d. quaternary parasitism, etc.
    3. Classification by number of parasitisms per host individual and number of parasite species involved
      a. single parasitism
      b. multiple parasitism
        (1) superparasitism
        (2) hyperparasitism
    4. predation[a]
II. Social symbiosis
  A. social commensalism
    1. mixed-group commensalism[b]
    2. nest commensalism
    3. plesiobiosis
  B. social mutualism
    1. mixed-group mutualism[b]
    2. parabiosis
    3. trophobiosis
  C. social parasitism
    1. inquilinism
    2. slavery (= dulosis, helotism)
    3. temporary social parasitism
    4. trophic parasitism
    5. xenobiosis

---

[a] Predation is sometimes classified as symbiosis when a predator uses only one to a few prey species.
[b] Group includes flocks, herds, and schools.

"Predation," especially when a predator uses only one to a few prey species, is sometimes classified as a kind of "symbiosis;" *e.g.,* some weevil species are "seed predators" that feed on only one to a few plant species. Boucher (1992) discusses some of the confusion involved in definitions of symbiosis and related terms due to reasons including: (1) combining definitions within and between terms, and (2) not specifying whether relationships are beneficial to individuals or populations. "Symbiosis," "protocooperation," "obligacy,"

S–Z

"facilitation," and "altruism" are used with partial to total overlap in meaning with "cooperation" and "mutualism;" the battle for clarity of these terms is far from won (Boucher 1992, 208, 211).

[Greek *symbiosis*, living together; coined by Anton DeBary to designate close association between two species, whether or not it is beneficial (1879 in Boucher 1992, 208)]

acarophytism, acarophytium *n*. "Symbiosis between plants and mites" (Lincoln et al. 1985).

anisosymbiosis *n*. Symbiosis in which one partner (the macrosymbiont) is larger than the other (microsymbiont) (Lincoln et al. 1985).

cf. symbiosis: isosymbiosis

antipathetic symbiosis *n*. Symbiosis that is either advantageous, or obligatory, to one of the symbionts (Lincoln et al. 1985).

beneficence *n*. Mutualism, commensalism, or both, in organism-organism pairs (Hunter and Aarssen 1988, 34).

biotrophic symbiosis *n*. Symbiosis in which one symbiont uses its living partner as a food source (Lincoln et al. 1985).

cf. symbiosis: necrotrophic symbiosis

byproduct mutualism See symbiosis: mutualism: byproduct mutualism.

calobiosis *n*. In some social-insect species: one species' living in the nest of and at the expense of another species (Lincoln et al. 1985).

cf. symbiosis: nest commensalism

cleaning symbiosis *n*. Symbiosis involving a cleaner, *q.v.*, and its host (Immelmann and Beer 1989, 46).

cleptobiosis, kleptobiosis *n*. In some social-insect species: symbiosis in which one species of social organism steals food from the stores of another social species but does not live, or nest, close to it (Lincoln et al. 1985).

cf. parasitism: cleptoparasitism

commensalism *n*. "Symbiosis in which members of one species are benefitted while those of the other species are neither benefitted nor harmed" (Wilson 1975, 354, 581); *e.g.,* dispersal of seeds on mammals' fur, owls' and chickadees' using woodpecker's holes, or animals' living in social insects' nests.

Comment: The lines between "commensalism," "mutualism," and "parasitism" are often hazy (Immelmann and Beer 1989, 50).

▶ **ectocommensalism** *n*. Commensalism involving one organism's living on its host's external surface; contrasted with endocommensalism (Lincoln et al. 1985).

n. ectocommensal

▶ **nest commensalism** *n*. Commensalism in which one species lives in the nests of another, scavenging on refuse or preying on the scavengers, in either case without harming or benefiting the hosts (*e.g.,* in some millipedes, beetles, or other arthropods that live with army ants) (Wilson 1975, 354).

cf. symbiosis: calobiosis

▶ **social commensalism** *n*. Commensalism involving a social species and another species which may be social or nonsocial (Wilson 1975, 354–356).

Comment: This includes "plesiobiosis," "nest commensalism," "and "mixed-group symbiosis" (mixed flocking," "mixed herding," and "mixed schooling") (Wilson 1975, 354–356).

▶ **synecthry** *n*. "Commensalism in which the participants (synecthrans) display mutual dislike" (Lincoln et al. 1985).

▶ **synoecy** *n*. A symbiosis between a social-insect colony and a tolerated guest organism (= synoekete) (Lincoln et al. 1985).

See -oecism: synoecy.

consortism *n*. Symbiosis between two partners (consorts of a consortium) (Lincoln et al. 1985).

ectosymbiosis, exosymbiosis *n*. Symbiosis in which neither symbiont lives within the body of the other; contrasted with endosymbiosis (Lincoln et al. 1985).

endosymbiosis *n*. Symbiosis in which an endosymbiont (inhabiting symbiont) lives within the body of another (inhabited symbiont); contrasted with ectosymbiosis (Lincoln et al. 1985).

epibiosis *n*. Symbiosis in which one organism lives upon the outer surface of another (Lincoln et al. 1985).

exosymbiosis See symbiosis: ectosymbiosis.

generational mutualism See symbiosis: mutualism: generational mutualism.

heterosymbiosis, interspecific symbiosis *n*. Symbiosis between two heterospecific organisms; contrasted with homosymbiosis (Wilson 1975, 596).

homosymbiosis, intraspecific symbiosis *n*. Symbiosis between two conspecific organisms; contrasted with heterosymbiosis (Lincoln et al. 1985).

interspecific symbiosis See symbiosis: heterosymbiosis.

intraspecific symbiosis See symbiosis: homosymbiosis.

isosymbiosis *n*. Symbiosis in which both symbionts are of the same size; contrasted with anisosymbiosis (Lincoln et al. 1985).

kleptobiosis See symbiosis: cleptobiosis.

lestobiosis *n.* Symbiosis in which a colony of a social-insect species with small individuals nests in the walls of a social-insect species with larger individuals and enters its chambers, where the former preys on the latter's brood or robs its food stores (Wilson 1975, 588).
syn. synclopia (Lincoln et al. 1985)
cf. symbiosis: kleptobiosis

metabiosis, metabios *n.* Symbiosis between two organisms in which one organism inhabits an environment prepared by the other (Lincoln et al. 1985).

mutualism *n.*
1. For example, in many plant species and their animal pollinators, anemone fish and anemones, cleaner fish and their hosts: temporary, or permanent, symbiosis in which both species benefit (*Oxford English Dictionary* 1972, entries from 1876; Lincoln et al. 1985).
   Note: In view of the general definitions for "symbiosis," "symbiosis" and "true symbiosis" are confusing synonyms for "mutualism."
   syn. symbiosis, true symbiosis synergy; interdependent association, synergism, synergy (Lincoln et al. 1985)
   cf. ²evolution: coevolution; cooperation; parasitism; symbiosis; symbiosis: nutricism
2. In many organism species: interactions between two conspecific organisms in which both gain in inclusive fitness; four kinds of cooperation can be distinguished, including reciprocal altruism (West Eberhard 1975, 19).
   syn. cooperation (West Eberhard 1975, 19)
3. A behavioral interaction from which all participating organisms derive a net benefit (Wittenberger 1981, 613).
   Note: Benefit is measured in terms of direct fitness (Wittenberger 1981, 613).
4. Mutualism between two or more conspecific individuals, or species, in which both parties benefit in terms of direct fitness (Brown 1987a, 305).
syn. cooperation (Brown 1987a, 305)
cf. cooperation; ⁴facilitation; parasitism; predation; symbiont: keystone mutualist
Comments: "Symbiosis," "protocooperation," "obligacy," "facilitation," and "altruism" are used with partial to total overlap in meaning with cooperation and mutualism (Boucher 1992, 208, 211). Further, Boucher (1992) indicates how to measure mutualism.
[coined by Belgian zoologist Pierre Van Beneden (1873 in Boucher 1992, 209)]
▸ **byproduct mutualism** *n.* Mutualism in which a neighbor's benefit is an inciden-

tal byproduct of a donor's action that it took for its own benefit (Brown 1987a, 298).
▸ **generational mutualism** *n.* Mutualism between generations (*e.g.,* between parents and their offspring) (Brown 1987a, 299).
▸ **parabiosis** *n.* In some ant species: species' (= parabionts) nesting in close association, defending their nests jointly, foraging together, and possibly even sharing food, but not rearing their offspring together (Wilson 1975, 354, 590).
▸ **paramutualism** *n.* Facultative symbiosis between two species (Lincoln et al. 1985).
▸ **score-keeping mutualism** *n.* Mutualism in which two participants repeatedly exchange favors while each adjusts its behavior in response to partial, or total, reciprocation by the other (Brown 1987a, 306).
Comment: Score-keeping mutualism is found in the games tit-for-tat and judge (Brown 1987a, 306).
▸ **sequential mutualism** *n.* The interactions of a group of plants that grow in the same area and overlap in their blooming periods and their coadapted pollinators that pollinate them as the season progresses (Buchmann and Nabhan 1996, 253–254).
▸ **social mutualism** *n.* Mutualism between a social species and another species that may or may not be social (*e.g.,* mixed-group symbiosis, trophobiosis, and parabiosis) (Wilson 1975, 354, 356).
▸ **syndiacony** *n.* Mutualism between ants and plants (Lincoln et al. 1985).
adj. syndiaconic.
▸ **trophobiosis** *n.* Social mutualism in which an individual of one species (= trophobiont) yields food to a heterospecific individual in exchange for protection from parasites, predators, or inclement weather; *e.g.,* in ants that receive honeydew from trophobionts (aphids and other homopterans, or the caterpillars of certain lycaenid and riodinid butterflies) (Wilson 1975, 354, 597; Hölldobler and Wilson 1990, 644).

myrmecoclepty *n.* Symbiosis between ant species in which the guest species steals food directly from the host species (Lincoln et al. 1985).

myrmecosymbiosis *n.* Symbiosis between an ant species and its host plant (Lincoln et al. 1985).

necrotrophic symbiosis *n.* Symbiosis between two living organisms in which one continues to use the other as a food even after its complete, or partial, death has occurred (Lincoln et al. 1985).

cf. symbiosis: biotrophic symbiosis

neutralism See symbiosis: parasymbiosis.

nutricism *n*. Symbiosis in which only one member of a partnership benefits (Lincoln et al. 1985).

cf. symbiosis: mutualism

parasitism See parasitism.

parasymbiosis *n*.

1. Symbiosis without benefit, or detriment, to either partner (Lincoln et al. 1985). *syn*. neutralism (Lincoln et al. 1985)
2. Association of an organism (the parasymbiont) with an existing symbiont, *e.g.,* a lichenicolous fungus with an alga (Lincoln et al. 1985).

phoresy, phoresis, phoretic behavior *n*. Symbiosis in which one organism is merely transported on the body of a heterospecific individual, *e.g.,* a pseudoscorpion's being carried by an insect (Lincoln et al. 1985) or a carrion-eating nematode's being carried under the wings of a carrion beetle (Immelmann and Beer 1989, 132).

syn. hitchhiking (Immelmann and Beer 1989, 132)

cf. host: carrier host

protocooperation *n*. Symbiosis in which populations of two species benefit by their association but relations are not obligatory (Odum 1969, 225).

See cooperation.

score-keeping mutualism See symbiosis: mutualism: score-keeping mutualism.

slavery See parasitism: social parasitism: slavery.

synclerobiosis *n*. Symbiosis between two species of social insects that usually inhabit separate nesting sites (Lincoln et al. 1985).

syndiacony See symbiosis: mutualism: syndiacony.

synectry *n*. "Commensalism in which the participants (syecthrans) display mutual dislike" (Lincoln et al. 1985).

♦ **symbiotroph, symbiotrophy, symbiotrophic** See -troph-: symbiotroph.

♦ **symbolic action** See ritual.

♦ **symbolic behavior** See behavior: symbolic behavior.

♦ **symbolic territory** See territory: symbolic territory.

♦ **symmetric game** See game.

♦ **sympaedium** See -paedium: sympaedium.

♦ **symparasite** See parasite: symparasite.

♦ **sympathetic induction** See effect: group effect.

♦ **sympathetic nervous system** See ²system: nervous system: sympathetic nervous system.

♦ **sympatric** See -patric: sympatric.

♦ **sympatric hybridization** See hybridization: sympatric hybridization.

♦ **sympatric speciation** See speciation: sympatric speciation.

♦ **symphile** See symbiont.

♦ **symphylic substance** See substance: symphylic substance.

♦ **symplesiomorph** See character: plesiomorph: symplesiomorph.

♦ **symplesiotypy** See -typy: symplesiotypy.

♦ **syn-** *prefix* "With; together; associated with or accompanying" (Michaelis 1963). Also, sy- before sc, sp, st, and z; syl- before l; sym- before b, p, and m; sys- before s. [Latin < Greek *syn*, together]

♦ **synanthropic** See anthropic: synanthropic.

♦ **synapomorph, synapomorphy** See character: apomorph: synapomorph.

♦ **synaposematic coloration** See coloration: synaposematic coloration.

♦ **synapotypy** See -typy: synapotypy.

♦ **synapse** *n*. A communicating junction between two neurons (*Oxford English Dictionary* 1972, entries from 1899), between a receptor cell and a neuron, or between a neuron and an effector-organ cell (Immelmann and Beer 1989, 302).

adrenergic synapse *n*. A synapse whose neurotransmitter is adrenalin (= epinephrin) (Immelmann and Beer 1989, 302).

cholinergic synapse *n*. A synapse whose neurotransmitter is acetylcholine (Immelmann and Beer 1989, 302).

electronic synapse *n*. A synapse across which an electrical charge is directly transmitted (Immelmann and Beer 1989, 302).

seritonergic synapse *n*. A synapse whose neurotransmitter is serotonin (Immelmann and Beer 1989, 302).

♦ **synapsis** *n*. Homologous chromosomes' pairing during the first division of meiosis (Mayr 1982, 762, 960).

♦ **synaptic signaling** See signaling: synaptic signaling.

♦ **syncheimadium** See ²group: syncheimadium.

♦ **synchorology** See study of: synchorology.

♦ **synchronic** See chronic: synchronic.

♦ **synchronic species** See ²species: synchronic species.

♦ **synchronopaedium** See -paedium: synchronopaedium.

♦ **synchronous hermaphrodite** See hermaphrodite: synchronous hermaphrodite.

♦ **synclerobiosis** See symbiosis: synclerobiosis.

♦ **synclopia** See biosis: lestobiosis.

♦ **syncrypsis** See crypsis: syncrypsis.

♦ **syndiacony** See symbiosis: syndiacony.

♦ **syndrome** *n.*

1. The concurrence of several symptoms in a human disease; "a set of such concurrent symptoms" (*Oxford English Dictionary* 1972, entries from 1541).
    syn. deprivation syndrome (in some cases, Immelmann and Beer 1989, 303)
    cf. behavior syndrome

2. An aggregate, or set, of concurrent symptoms that indicate the presence and nature of a disease (Michaelis 1963).
    Comment: This concept now relates to nonhuman animals as well (Immelmann and Beer 1989, 303).
    [New Latin < Greek *syndrome* < *syn-*, together + *dramein*, to run]

adrenogenital syndrome *n.* A hereditary disease in which a female mammal's adrenal gland secretes androgens instead of cortisone (Dewsbury 1978, 231).
cf. hermaphroditism: progestin-induced hermaphroditism, testicular feminization

behavior syndrome See behavior syndrome.

chronic-fatique syndrome *n.* In Humans, a combination of symptoms including persistent fatigue for at least 6 months, impaired concentration and short-term memory, trouble sleeping, and muscle and joint pain (Cheng 1995, C9).
Comment: This is the most widely accepted definition of chronic-fatigue syndrome (Cheng 1995, C9).

contiguous-gene-deletion syndrome *n.* A human syndrome that involves a group of contiguous genes in which the absence of one of the genes of the group causes a particular symptom (*e.g.,* in Williams syndrome) (Blakeslee 1996, C3.)

deprivation syndrome *n.* In primates and some other social-animal species: all of an animal's "deficits and distortions in its behavioral development that can result from its social isolation in early life, including apathy, motor restlessness, stereotypies, and incapacity for normal social interaction (Immelmann and Beer 1989, 73).

fight-or-flight syndrome *n.* In mammals: an individual's condition that results from its secretion of adrenaline and noradrenalin under a stress condition and that involves its soon showing combat or escape behavior (Wilson 1965, 253; Immelmann and Beer 1989, 297).
syn. fight or flight (Wilson 1965, 253)

general-adaptation syndrome (GAS) *n.* In some vertebrate species: The physiological and behavioral changes undergone by an originally normal individual that was subjected to prolonged stress; the GAS involves the stage of alarm (general-emergency reaction), stage of resistance, and stage of exhaustion (Selye 1956 in Wilson 1975).
[coined by Selye: general (the pattern of responses was believed to be the same whatever the precise nature of danger) + adaptation (at least during first stages, an animal's body is adapting to a new condition) + syndrome (many different organs are affected) (Dawkins 1980, 57)]

pollination syndrome *n.*

1. A flower's suite of morphological characters and rewards that may accurately predict its kinds of floral visitors (P. Knuth and H. Müller in Buchmann and Nabhan 1996, 252).

2. A morphologically convergent adaptive trend exhibited by both the floral features of a pollinated plant and the mouthpart structure and other flower-interactive features of its pollinator (Proctor et al. 1996 in Labandeira 1998, 57).

premenstrual-stress-dysphoria disorder, PMS dysphoria disorder *n.* In some women of reproductive age: a more severe derangement of mood that interferes with daily life and requires medical treatment and often psychotherapy (Squires 1995, A1).
cf. premenstrual stress
Comment: Sufferers of this disorder feel extremely "on edge" and experience markedly depressed moods, marked anxiety, and decreased interest in activities (American Psychiatric Association's *Diagnostic and Statistical Manual IV* in Squires 1995, A1).

premenstrual syndrome (PMS) *n.* In some women of reproductive age: a combination of various physical and emotional symptoms that occurs 7–14 days before menstruation (Clayman 1989, 818).
syn. the curse (Helen Gurley Brown in Mansfield 1995, B6)
cf. premenstrual stress dysphoria disorder
Comments: PMS affects more than 90% of fertile women at some time in their lives, and in some women it is so severe that their work and social relationships are seriously disrupted (Clayman 1989, 818). Symptoms include backache, breast tenderness, depression, fluid retention, headache, irritability, lower abdominal pain, and tension. Doctors prescribe some antidepressant drugs that inhibit the reuptake of the mood-altering brain chemical serotonin for PMS and premenstrual stress dysphoria disorder (Squires 1995, A1).

separation syndrome, separation trauma *n.* In some primate species: a young individual's traumatic behavior,

S–Z

including apathy, restlessness, compulsive movements, inability to form social attachments, and other deficits in its social responsiveness, caused by removal of its mother to which it is bonded (Immelmann and Beer 1989, 73, 265).

stretching syndrome *n*. Stretching and yawning behavior (Tembrock 1964, etc. in Heymer 1977, 100).

cf. behavior: comfort behavior; yawning

Williams syndrome *n*. A human disease syndrome with these symptoms: extraordinary language skills for people with their IQ; extreme friendliness; often elfin facial features, hernias, hoarse voice, joint abnormalities, and premature skin aging; poor visual-construction cognition; supravalvular aortic stenosis (a heart defect); usually mental retardation (with an average IQ of 60), which causes trouble crossing a street, making change, and tying shoes without help (Blakeslee 1996, C3, illustration).

Comment: The genes related to Williams syndrome are on chromosome-7. Some people show partial Williams syndrome in that they have only some of the symptoms listed above (Blakeslee 1996, C3).

young-male syndrome *n*. Younger human males' tending to be more competitive and to take risks, which leads to more violence than among older males (Wilson and Daly 1985, 59).

♦ **syndynamics** See study of: dynamics.

♦ **synechthran** See symbiont: synechthran.

♦ **synecology** See study of: ecology.

♦ **synecthry** See symbiosis: synecthry.

♦ **synergism, synergy** *n*.

1. Cooperative action of two or more agencies such that the total is greater than the sum of the component actions (Lincoln et al. 1989), *e.g.,* two hormones that have a joint effect on an animal that neither has alone (Immelmann and Beer 1989, 303).
   cf. antagonism
2. The cooperative action of two microorganisms that effects a change that would not occur, or would take place at a slower rate, in axenic culture (Lincoln et al. 1989).
See symbiosis: mutualism.

♦ **synethogametism** *n*. "Gametic compatibility;" gametes' abilities to fuse (Lincoln et al. 1989).

asynethogametism, aethogametism *n*. "Gametic incompatibility; the inability of two apparently compatible gametes to fuse" (Lincoln et al. 1989).

♦ **syngameon** See ²species: semispecies.

♦ **syngamete** See gamete: syngamete.

♦ **syngamic sex determination** See sex determination.

♦ **syngamodeme** See deme: gamodeme: syngamodeme.

♦ **syngamy** See -gamy: syngamy.

♦ **syngenesis** See -genesis: syngenesis.

♦ **syngenic** See -genic: syngenic.

♦ **syngeographs** See geographical equivalents.

♦ **syngony** *n*. Production of both male and female gametes in the same gonad (Lincoln et al. 1989).
adj. syngonic

♦ **synhesma** See ²group: swarm.

♦ **synhospitalic parasite** See parasite: synhospitalic parasite.

♦ **synkaryon** *n*. A zygote nucleus formed from the fusion of gametic nuclei (Lincoln et al. 1989).

♦ **synkaryotoky** See -ploidy: diploidy.

♦ **synmorphology** See study of: synmorphology.

♦ **synoecete, synecete** See symbiont: synoekete.

♦ **synoecius** See -oecious: synoecius.

♦ **synoecy** See: -oecism: synoecy.

♦ **synoekete** See symbiont: synoekete.

♦ **synomone** See chemical-releasing stimulus: semiochemical: allelochemic: synomone.

♦ **synpiontology** See study of: synpiontology.

♦ **synsystematics** See study of: systematics: synsystematics.

♦ **syntactics** See study of: semiotics (def. 3); study of signals: syntactics.

♦ **syntax** *n*.

1. Grammar rules that govern how words are combined to form phrases, clauses, and sentences (*Oxford English Dictionary* 1972, entries from 1613; Immelmann and Beer 1989, 304).
2. An individual animal's combining and ordering its signals, especially where combining particular signals may convey different information or produce different effects than when the signals are used in isolation (Immelmann and Beer 1989, 304).

♦ **syntaxonomy** See study of: taxonomy: syntaxonomy.

♦ **syntechny** See ²evolution: convergent evolution.

♦ **synthesia** See -thesia: synthesia.

♦ **synthetic lethal** See chromosome: synthetic lethal; gene: synthetic lethal.

♦ **synthetic method** See study of: systematics: evolutionary systematics.

♦ **synthetic theory of classification** See study of: systematics: evolutionary systematics.

♦ **synthetic theory of evolution** See ²evolutionary synthesis.

♦ **syntroph** See -troph-: syntroph.

♦ **synxenic** See -xenic: synxenic.

♦ **synzoochore, synzoochorous** See -chore: synzoochore.

♦ **syringograde** See -grade: syringograde.

♦ **syrtidium** See ²community: syrtidium.

♦ **syrtidophile, syrtidophilus, syrtidophily** See -¹phile: syrtidophile.

♦ **¹system** *n*.
  1. A set, or assemblage, of things connected, associated, or interdependent, so as to form a complex unity; a whole that is composed of parts in orderly arrangement according to some plan or scheme (*Oxford English Dictionary* 1972, entries from *ca*. 1638).
  2. Scientifically and technically, a group, set, or aggregate of natural, or artificial, things from a connected, or complex, whole (*Oxford English Dictionary* 1972, entries from 1830).

adapted system *n*. A kind of teleology that asserts that organism structures were designed by God to perform particular functions (Mayr 1982, 49).
cf. teleology

alarm-defense system *n*. Defensive behavior that also functions as an alarm-signaling device within a colony; *e.g.*, certain ant species use chemical defensive secretions that double as alarm pheromones (Regnier and Wilson 1968, 955; 1969, 893; Wilson and Regnier 1971, 279).

alarm-recruitment system *n*. A communication system that rallies others to some particular place to aid in the defense of their society, *e.g.*, the odor trail system of "lower" termites that is used to recruit colony members to the vicinity of intruders and breaks in their nest wall (Wilson 1975, 48).

attachment-behavior system *n*. A control system that presumably integrates attachment behavior (Hinde 1982, 228).

breeding system *n*. The mode, pattern, and extent to which individuals interbreed with others from the same or different taxa (Lincoln et al. 1985).
cf. mating system

central-place system *n*. For example, in species of nest-using molluscs, annelids, arthropods, and vertebrates: a form of spatial organization in which one or more individuals, or other social units, focuses its activities around a single central location such as a den, nest site, breeding colony, or roosting site (Wittenberger 1981, 613).

communication system *n*. A means of conveying information that involves seven essential components: a sender, receiver, channel, noise, context, signal, and code (Sebeok 1965; Klopfer and Hatch 1968 in Dewsbury 1978, 99).

complex system *n*. A system in which "the whole is more than the sum of the parts, not in an ultimate, metaphysical sense but in the important pragmatic sense that, given the properties of the parts and the laws of their interaction, it is not a trivial matter to infer the properties of the whole" (Simon 1962 in Mayr 1982, 53); *e.g.*, the Solar System, a cell nucleus, a cell, an organ system, an individual organism, a society, or an ecosystem.

dominance system See hierarchy: dominance hierarchy.

ecosystem *n*.
  1. An area of nature that includes living organisms and nonliving substances that interact and produce an exchange of materials between its living and nonliving parts (Odum 1969, 10).
  2. All of the organisms of a particular habitat plus the physical environment in which they live (Wilson 1975, 583).
  3. The entire biological and physical content of a biotope (Lincoln et al. 1985).
syn. ecological system (Odum 1969, 10)
Comment: "Bioinert body" (Vernadsky 1944), "biosystem" (Thienemann 1939), "holocoen" (Friederichs 1930), and "microcosm" (Forbes 1887) are terms that basically express the concept of "ecosystem" (Odum 1969, 10).

Free Air CO₂ Enrichment System (FACE System) *n*. A 30-meter-diameter, circular plenum that delivers air to an array of 32 vertical pipes, with adjustable ports every 50 centimeters, to organisms within the circle; contrasted with closed-growth chambers and open-top growth chambers (DeLucia et al. 1999, 1177–1178, illustration).
Comments: For example, the FACE System can deliver ambient air and CO₂-enriched air to a 14-meter tall forest of Loblolly Pines (DeLucia et al. 1999, 1177).

genetic system *n*. The organization of a species' genetic material and reproductive strategy (Lincoln et al. 1985).

inheritance-of-acquired-characteristics system (IAC system) *n*. The phenomenon in which all, or a large proportion of, individual organisms, or cell cultures, exhibit new traits that they pass on to succeeding generations after they are briefly exposed to a chemical, or physical, treatment under conditions that allow little or no growth, thereby ruling out selection of mutants (Landman 1991, 2).
cf. Lamarckism
Comments: Landman (1991) describes IAC systems in organisms including bacteria, protozoa, *Drosophila*, and mice. He also describes "extranucleic IAC systems,"

S–Z

"epinucleic IAC systems," and "nucleic IAC systems."

mating system See mating system.

memory system *n.* For example, in Humans, one of many systems that enables us to remember past experiences and skills, *e.g.*, a memory system for riding a bicycle, one for drawing, and one for typing words (Lemonick 1999, 56).

Comment: Memory systems work in concert, enabling us to have our complex behavior.

mimicry system *n.* An ecological relationship that includes two or more protagonists that perform three roles: being a model, a mimic, or a dupe (Pasteur 1982, 169).

See "mimicry" for kinds of mimics involved in mimicry systems.

Comment: In organismic mimicry systems, there is a model (a living, or material, agent that emits perceptible stimuli or signals), a mimic (an organism that plagiarizes the model), and a dupe (an animal enemy, or victim, of the mimic that senses the model's signals and is thus deceived by the mimic's similar signals) (Pasteur 1982, 169).

▸ **bipolar mimicry system** *n.* A mimicry system in which two of the protagonists belong to one species and the third protagonist belongs to another species (Vane-Wright 1976 in Pasteur 1982, 178).

▸ **conjunct mimicry system** *n.* A mimicry system in which all three protagonists belong to the same species (Vane-Wright 1976 in Pasteur 1982, 178).

▸ **disjunct mimicry system** *n.* A mimicry system in which each protagonist is a different species (Vane-Wright 1976 in Pasteur 1982, 178).

▸ **model-mimic-dupe-tripartite system** *n.* A system comprised of a model, mimic, and dupe (Vane-Wright 1976 in Pasteur 1982, 178).

Comments: In this system the model, mimic, and dupe may all be the same species, or one to all of them may be different species; in some cases the model is not an organism, but rather a thing such as a rock, sand, and a replica of an organism.

▸ **model-vs.-mimic-dupe bipolar mimicry system** *n.* A model-mimic-dupe tripartite system in which two of the three parts are the same species (Vane-Wright 1976 in Pasteur 1982, 178).

▸ **"S_1R/S_2 mimicry system"** *n.* A model-mimic-dupe tripartite system in which the model and dupe are the same species and the mimic is a second species (Vane-Wright 1976 in Pasteur 1982, 178).

SOME MIMICRY SYSTEMS[a]

I. camouflage mimicry system (model not of interest to a dupe)
  A. cryptic mimesis mimicry system
  B. eucrypsis-mimicry system
  C. mimesis mimicry system
  D. phaneric-mimesis mimicry system
II. homotype mimicry system (model of interest to dupe)
  A. actual-model mimicry system
    1. model-forbidding-to-dupe mimicry system
      a. arithmetic mimicry system
      b. Batesian mimicry system
      c. Browerian mimicry system (automimicry)
      d. Gilbertian mimicry system
      e. Müllerian mimicry system
      f. Wicklerian-Guthrian mimicry system
    2. model-agreeable-to-dupe mimicry system
      a. Aristotelian mimicry system
      b. Bakerian mimicry system
      c. Batesian-Pouyannian mimicry system
      d. Batesian-Wallacian mimicry system
      e. Dodsonian mimicry system
      f. Kirbyan mimicry system
      g. Nicolaian mimicry system
      h. Pouyannian mimicry system
      i. Vavilovian mimicry system
      j. Wasmannian mimicry system
      k. Wicklerian mimicry system
      l. Wicklerian-Barlowian mimicry system
      m. Wicklerian-Eisnerian mimicry system
  B. virtual-model mimicry system
    1. definable-model mimicry system
    2. nondefinable-model mimicry system

---

[a] Pasteur (1982).

▸ **"S_1S_2/R mimicry system"** *n.* A model-mimic-dupe tripartite system in which the model and mimic are one species and the dupe is a second species (Vane-Wright 1976 in Pasteur 1982, 178).

Comment: This system is "automimicry" when the model and mimic are different conspecific individuals and "self-mimicry" when the model and mimic are the same individual.

▸ **"S_1/S_2R mimicry system"** *n.* A model-mimic-dupe tripartite system in which the model is one species and the mimic and dupe are a second species (Vane-Wright 1976 in Pasteur 1982, 178).

Comments: In S_1/S_2R, the slash separates the two species involved. S_1 = the first sender (= model); S_2 = second sender (= mimic); R = receiver (= dupe).

moral system, morality *n.* A person's system of ethics or normative conduct; the question of how agreements, or contracts, about right and wrong are generated and maintained within human societies and why they differ (Alexander 1987, 12).

natural system *n.* The hidden bond of taxonomic connection sought by naturalists and eventually designated "descent" by Darwin (Mayr 1982, 437).

specific-mate-recognition system *n.* "The subset of adaptations, which are involved in signaling between mating partners or their cells" (Paterson 1981, 113, 1985, 24).

Comment: This system relates to the "recognition concept of species," *q.v.*

social system *n.* In some animal species: a species' spatial distributions of its individuals and their associations within their groups, including associations between sexes, extrafamilial social relationships, and intragroup relationships (Immelmann and Beer 1989, 277).

subsurface lithoautotrophic microbial ecosystem (SLiME) *n.* An ecosystem in Earth's crust in which autotrophic bacteria (acetogens) live in basalt, use hydrogen gas for energy, derive carbon from inorganic carbon dioxide, and excrete simple organic compounds consumed by other kinds of bacteria (Fredrickson and Onstoot 1996, 72).

cf. -troph-: acetogen

[SLiME, from *s*ubsurface *li*thoautotrophic *m*icrobial *e*cosystem]

♦ ²**system** *n.*

1. An animal's set of organs, or parts, of the same or similar structure, or that have the same function (*Oxford English Dictionary* 1972, entries from 1740); *e.g.,* endocrine system, circulatory system (blood-vascular system), lymphatic system, nervous system, reproductive system, or respiratory system.

   Note: "System" is now commonly extended to sets of behaviors.

   syn. physiological system (Immelmann and Beer 1989, 221)

2. In more derived plants: each of the primary groups of tissues (*Oxford English Dictionary* 1972, entries from 1875).

action system See pattern: behavior pattern.

active-sensory system *n.* An individual animal's perceptual system that involves its actively emitting energy and perceiving environmental objects on the basis of energy-pattern alterations that return to it, *e.g.,* an echolocation system in bats, oilbirds, or porpoises (Dewsbury 1978, 186).

behavioral system See cycle: functional cycle.

endocrine system *n.* For example, in arthropods, vertebrates: a system comprised of endocrine glands and hormones that affects an animal's behavior and other biological features (Curtis 1983, 667).

cf. hormone; ²system: neurendocrine system

hypothalamohypophysial system *n.* In more derived animals: an "interface" system between an individual animal's central nervous system and its endocrine system (Immelmann and Beer 1989, 222).

intentional system *n.* An animal's system "whose behavior can be (at least sometimes) explained and predicted by relying on ascriptions to the system of beliefs and desires (and other intentionally characterized features) — what I will call intentions here, meaning to include hopes, fears, intentions, perceptions, expectations, etc." (Dennett 1978 in McFarland 1985, 502).

motivational system *n.* The causal regulation that underlies an individual animal's particular kind of behavior (Immelmann and Beer 1989, 193).

Comment: "Motivational system" supersedes the concept of a "specific drive" for each kind of behavior (Immelmann and Beer 1989, 193).

nervous system *n.* In Cnidaria and more derived animal phyla: an organized constellation of nerve cells and associated nonnervous cells (Bullock 1977 in McFarland 1985, 174).

cf. nerve net; ²system: neurendocrine system; receptor

neurendocrine system, neuroendocrine system *n.* In animals: an individual's nervous system plus its endocrine system.

DIVISIONS OF A VERTEBRATE NERVOUS SYSTEM[a]

I. central nervous system
II. peripheral nervous system
   A. sensory nervous system
   B. motor nervous system
      1. somatic nervous system
      2. autonomic nervous system
         a. sympathetic nervous system
         b. parasympathetic nervous system

[a] Curtis (1983, 763).

S–Z

▶ **autonomic nervous system** *n.* In vertebrates: an individual's parasympathetic and sympathetic nervous systems; contrasted with somatic nervous system (McFarland 1985, 181).

Comment: These two nervous systems mediate control over an animal's blood vessels, certain glands, heart, and blood vessels (McFarland 1985, 181). Autonomic activity can cause defecation, feather ruffling, hair erection, and urination, depending on the taxon (Immelmann and Beer 1989, 23–24).

parasympathetic nervous system *n.* Part of an individual's autonomic nervous system that is anatomically distinct from its sympathetic nervous system (McFarland 1985, 524).

syn. cholinergic nervous system (American Academy of Allergy, Asthma, and Immunology 1998, S-10)

Comments: An animals parasympathetic nervous system serves a recuperative function after stress by restoring its blood supply to normal, relaxing curbs on its digestion, and counteracting other effects of its sympathetic nervous system (McFarland 1985, 524).

sympathetic nervous system *n.* Part of an autonomic nervous system with nerve pathways that become active under stress, or exertion and have an emergency function (McFarland 1985, 524).

Comments: The sympathetic nervous system increases blood supply to an animal's brain, heart, lungs, and muscles; increases its heart rate; and reduces blood supply to its intestines and peripheral body parts. In Humans, this system causes emotional arousal which includes increased heart rate, sweating, and changes in peripheral blood circulation, causing a person's face to become pale or flushed (McFarland 1985, 524).

▶ **central nervous system (CNS)** *n.* In vertebrates: a brain plus spinal cord; in Platyhelminths: the anterior ganglion and two nerve cords (McFarland 1985, 176); in annelids: ganglia and a ventral nerve cord; in arthropods: a brain, ganglia, and ventral nerve cord (Barnes 1974).

cf. brain, peripheral nervous system

Comment: The CNS coordinates activities of individual organs, processes incoming messages from sense organs, and in accordance with sensory and stored information, regulates an animal's behavior as a whole (Immelmann and Beer 1989, 42).

▶ **extrapyramidal system** *n.* In vertebrates: all non-reflex motor pathways not included in an individual's pyramidal system, *q.v.* (McFarland 1985, 242).

▶ **peripheral nervous system** *n.* In vertebrates: the part of an individual's nervous system that includes sensory (afferent) and motor (efferent) pathways that carry information to and from its central nervous system (Curtis 1983, 763–767).

▶ **pyramidal system, corticospinal system** *n.* In Mammals, except Monotremes: an individual's most important pathway involved in its voluntary movement control; it begins in the individual's motor cortex and proceeds through its midbrain and brainstem to its spinal cord (McFarland 1985, 242).

cf. ²system: nervous system: extrapyramidal system

▶ **somatic nervous system** *n.* In Vertebrates: the part of an individual's motor nervous system that carries sensory information from its central nervous system and its commands to skeletal muscles that cause bodily movement (McFarland 1985, 180).

physiological system See ²system.

reproductive system *n.* In organisms: a means by which an individual produces offspring, including automixis, selfing, outcrossing, panmixia, endomitosis, apomixis, and vegetative reproduction (Bernstein et al. 1985, 1279).

See parthenogenesis.

reticular-activating system *n.* In more-derived animals: a system in the brain stem whose activity is related to an individual's state of arousal (Hinde 1982, 281).

sensory system *n.* A system of receptors, sense organs, and sensory pathways to an animal's central nervous system and the receiving, or projecting, areas in its central nervous system that serve a particular modality (*e.g.,* the visual system of a vertebrate) (Immelmann and Beer 1989, 264).

♦ **³system** *n.* An organism's entire body that is taken as a functional whole (Michaelis 1963).

♦ **systematic mutation, systemic mutation** See ³mutation: systemic mutation.

♦ **systematics** See study of: systematics.

♦ **systematization** *n.* Production of a natural classification of organisms based on phylogenetic relationships (Lincoln et al. 1989).

♦ **systemic** *n.* Referring to a compound that is absorbed and translocated throughout an organism (Ware 1983, 287).

adj. systemic

♦ **systemic fitness** See fitness: inclusive fitness.

♦ **systemic mutation** See [3]mutation: systemic mutation.

♦ **systems ecology** See study of: ecology: ecosystem ecology.

t

♦ **t** See temperature.

♦ t_a See temperature: ambient temperature.

♦ T_r See temperature: mean radiant temperature.

♦ **t allele** See allele: t allele.

♦ **T-maze** See maze: T-maze.

♦ **T-species** See ²species (Comments).

♦ *tabula rasa* *n., pl.* ***tabulae rasae***

1. A tablet on which the writing is erased and, thus, is ready to be written on again (*Oxford English Dictionary* 1972, entries from 1535).
   syn. clean slate (Michaelis 1963)

2. A human mind before its being exposed to experience (*Oxford English Dictionary* 1972, entries from 1662, quote from Aristotle; Michaelis 1963).
   See hypothesis: *tabula rasa*.
   cf. learning theory
   [Latin scraped tablet, an empty or clean tablet, a clean slate]

♦ **tachy-** *combining form* Speed; swiftness (Michaelis 1963).
   [Greek *tachys*, swift]

♦ **tachyauxesis** See auxesis: heterauxesis: tachyauxesis.

♦ **tachygamy** See mating system: tachygamy.

♦ **tachygen** *n.* A trait that appears suddenly in phylogeny (Lincoln et al. 1985).

♦ **tachygenesis** See ²evolution: tachygenesis.

♦ **tachymetabolism** See metabolism: tachymetabolism.

♦ **tachytely** See ²evolution: tachytely.

♦ ¹**tactic** *n.*

1. An organism's "behavior pattern within a species' repertoire that can potentially be employed in a specified context to achieve a particular goal;" contrasted with strategy, *q.v.* (Wittenberger 1981, 622).

2. An organism's set of coadapted traits (Lincoln et al. 1985).

3. One of an organism's traits that is part of a suite of traits that compose a strategy, *q.v.* (Wootton 1984, 12; Gross 1984, 58).
   syn. strategy (a confusing synonym; Lincoln et al. 1985; Wootton 1984, 1, who, with others, distinguishes between "strategy" and "tactic")

♦ ²**tactic** *adj.* Pertaining to a taxis (Lincoln et al. 1985).
   cf. strategy, tactics

♦ **tactical deception** See deception: tactical deception.

♦ **tactics** *pl. n.* (construed as singular) The art, or science, of deploying military, or naval, forces in battle, and of employing warlike maneuvers (*Oxford English Dictionary* 1972, entries from 1626).
   syn. tactic (sometimes, *Oxford English Dictionary* 1972)
   cf. tactic

♦ **tactile eucrypsis** See mimicry: tactile eucrypsis.

♦ **tactism** See taxis.

♦ **tagmosis** *n.*

1. The condition of specialization of the entire set of limbs on an arthropod body (Flessa et al. 1975, 72).

2. An arthropod's body organization into segment groups of more or less united segments that form distinct trunk sections or tagmata (Snodgrass in Torre-Bueno 1978; Nichols 1989).
   Comment: Flessa et al. (1975, 72) measure tagmosis in arthropods using the Brillouin Expression.

♦ **taphephobia** See phobia (table).

♦ **tail bite** See bite: tail bite.

♦ **tail-down swimming** See swimming: tail-down swimming.

♦ **tail flagging** *n.* For example, in the White-Tailed Deer: an individual's lifting its tail and exposing its white, erected hairs on its rump and ventral surface of its tail (LaGory 1987, 20).

Comment: This is part of its alarm behavior (LaGory 1987, 20).

cf. behavior: alarm behavior; display; signal: alarm signal

♦ **tail mimicry** See mimicry: tail mimicry.

♦ **tail posture** See display: tail display.

♦ **tail-up swimming** See swimming: tail-up swimming.

♦ **tameness** *n.* A wild animal's showing no fleeing tendency with regard to a Human and possibly staying in his vicinity, approaching him for food, or both (Immelmann and Beer 1989, 305).

♦ **taming** *n.*
1. A person's domesticating another animal, individual, variety, or species (Michaelis 1963).
2. In agriculture, a person's bringing a taxon under cultivation; a person's producing a taxon by cultivation (Michaelis 1963).
3. A wild animal's losing its tendencies to flee from Humans ((Hale 1969, 22, in Dewsbury 1978, 267).
4. A wild animal's progressively reducing and eliminating its initial fleeing tendencies and negative reactions toward a Human during interactions with him and possibly showing staying in his vicinity, approaching him for food, or both (Immelmann and Beer 1989, 305).

adv. tamely
n. tamer, tameness
v.t. tame.
syn. imprinting (in some cases, but this is a confusing synonym; Immelmann and Beer 1989, 305)
cf. domestication, socialization
[Old English *tam*]

♦ **tandem calling** See calling: tandem calling.

♦ **tandem evolution** See ²evolution: tandem evolution.

♦ **tandem running** See recruitment: tandem running.

♦ **tandem selection** See selection: tandem selection.

♦ **tangled-bank hypothesis** See hypothesis: hypotheses of the evolution and maintenance of sex: tangled-bank hypothesis.

♦ **taphrium** See ²community: taphrium.

♦ **taphrophile** See -¹phile: taphrophile.

♦ **target organ** See organ: target organ.

♦ **target theory of mutation** See ²theory: hit theory.

♦ **tarsal organ** See gland: tarsal organ.

♦ **taste aversion** *n.* An animal's learning to avoid reingesting a food that causes noxious, or poisonous, effects on it (*e.g.,* a potential predator's learning to avoid arthropod mimicry models) (Immelmann and Beer 1989, 306).

♦ **tattooing** See animal sounds: drumming.

♦ **tautology** *n.*
1. A statement, with repeated parts, that is necessarily true by virtue of its structure: Either it is raining or it is not raining (*Oxford English Dictionary* 1972, entries from 1659; Michaelis 1963).
2. An aggregate of linked propositions (whose truth is not claimed) in which the validity of the *links* between them cannot be doubted (*e.g.,* Euclidean geometry) (Bateson 1979, 245).

cf. survival of the fittest
[Late Latin *tautologia* < Greek *tauto*, the same + *logos*, discourse]

♦ **tautomorphism** See -morphism: tautomorphism.

♦ **taximetrics** See study of: taxonomy: numerical taxonomy.

♦ **taxis** *n.*
1. In motile organisms: an individual's reaction to a particular stimulus by moving toward it (*Oxford English Dictionary* 1971, entries from 1904).
2. In organisms and some individual cells: an involuntary movement made in response to an external stimulus (Michaelis 1963).
   cf. reflex
3. An organism's maintaining its body position, changing its body position, or both, with regard to stimulus direction (Kühn 1919, Fraenkel and Gunn 1961, Schöne 1973 in Immelmann and Beer 1989, 306).
   Note: This definition seems appropriate for wide use.
4. Broadly, taxis (def. 3) plus kinesis, *q.v.* (Heymer 1977, 125).
5. Ethologically speaking, an oriented (directed) movement that occurs along with a fixed-action pattern (Dewsbury 1978, 19).

syn. orientation, tactism (Lincoln et al. 1985); tropism (not favored recently)
cf. Erbkoordination, kinesis, modality, orientation, *Taxiskomponente,* tocotropism, tropism

Comments: Many workers now use "taxis" with regard to animals or animal-like organisms; "tropism" with regard to plant or plant-like organisms. Some workers (Heymer 1977, 172) use "taxis" with regard to freely moving organisms; "tropism," sessile organisms. A taxis may, or may not, involve an organism's locomotion toward, or away from, a stimulus. If an organism moves toward a particular stimulus, it shows a positive taxis, *e.g.,* prosaerotaxis, positive phototaxis (= prophototaxis). If it moves away from a particular stimulus, it shows a negative taxis, *e.g.,* negative phototaxis.
[Greek *taxis*, arrangement (Tinbergen 1951, 87, who translates *Tasikomponente* as taxis)]

S–Z

SOME CLASSIFICATIONS OF TAXES

I. Classification based on temporal aspects of stimulus reception
- A. klinotaxis (*syn.* phobotaxis, scanning orientation, trial-and-error orientation)
- B. tropotaxis (tropic orientation)

II. Classification based on kind of stimulus and stimulus reception[a]
- A. aerotaxis (including prosaerotaxis)
- B. allelothetic orientation
- C. anemotaxis
  1. anemomenotaxis
  2. odor-conditioned anemotaxis
- D. argotaxis
- E. astrotaxis (*syn.* star orientation)
- F. barotaxis
- G. biotaxis
- H. celestial-body taxis
- I. chemotaxis (including tropochemotaxis) (*syn.* chemiotaxis)
- J. clinotaxis (klinotaxis)
- K. diaphototaxis
- L. galvanotaxis (*syn.* electrotaxis, topogalvanotaxis)
- M. geomagnetotaxis (*syn.* magnetic-compass taxis)
- N. geotaxis
- O. hydrotaxis (including aphydrotaxis)
- P. idiothetic orientation
- Q. local orientation
- R. menotaxis (*syn.* compass orientation, compass taxis, constant-angle taxis, direction taxis)
  1. light-compass taxis (*syn.* light-compass orientation, light-compass reaction)
  2. star-compass taxis
- S. navigation (including true navigation)
- T. object taxis
- U. oxygenotaxis (*syn.* oxytaxis)
- V. phobotaxis
- W. phonotaxis
- X. phototaxis (*syn.* heliotaxis, phototropotaxis)
  1. aphototaxis
  2. dorsal-light reaction
  3. photohoramotaxis (*syn.* photohorotaxis) s
  4. photophobotaxi
  5. polarized-light taxis
  6. ultraviolet-light taxis
  7. ventral-light reaction
- Y. pneumatotaxis (*syn.* pneumotaxis)
- Z. remote orientation
- AA. rheotaxis
- BB. scototaxis (*syn.* skototaxis)
- CC. seismotaxis (*syn.* vibrotaxis)
- DD. sonotaxis
- EE. strophotaxis
- FF. telotaxis (*syn.* goal orientation, landmark orientation, pharotaxis, pilotage)
- GG. thermotaxis (including apothermotaxis)
- HH. thigmotaxis (*syn.* stereotaxis)
- II. tonotaxis (*syn.* osmotaxis)
- JJ. topographical orientation (*syn.* mnemotaxis, sometimes)
- KK. traumotaxis (*syn.* tramatotaxis)
- LL. trophotaxis

III. Some other kinds of taxes
- A. primary orientation
- B. secondary orientation

[a] Positive and negative kinds of taxes are not given in this table; some are defined below.

aerotaxis *n.* Taxis with regard to an air-liquid interface or a concentration gradient of dissolved oxygen (Lincoln et al. 1985).
▸ **prosaerotaxis** *n.* Positive aerotaxis (Lincoln et al. 1985).

allelothetic orientation *n.* Taxis with regard to external (environmental) stimuli (Mittelstaedt 1973 in Heymer 1977, 125). *cf.* taxis: idiothetic orientation

anemotaxis *n.* Taxis with regard to wind (Lincoln et al. 1985).
▸ **anemomenotaxis** *n.* Taxis across wind (Heymer 1977, 172).
▸ **odor-conditioned anemotaxis** *n.* Anemotaxis initiated or enhanced by an odor (*e.g.,* in some species of insects that can respond to pheromones or host-plant odors) (May and Ahmad 1983, 183).

aphydrotaxis See taxis: hydrotaxis: aphydrotaxis.

apothermotaxis *n.* No taxis with regard to temperature change (Lincoln et al. 1985). *cf.* taxis: thermotaxis

argotaxis *n.* Passive movement with regard to water-surface tension (Lincoln et al. 1985).

astrotaxis, star orientation *n.* Menotaxis + distance taxis. *cf.* taxis: menotaxis: star-compass taxis

barotaxis *n.* Taxis with regard to a pressure stimulus (Lincoln et al. 1985).

biotaxis *n.* Taxis with regard to a biological stimulus (Lincoln et al. 1985).

celestial-body taxis *n.* Taxis with regard to the Sun or stars (Lincoln et al. 1985). *cf.* taxis: astrotaxis, menotaxis

chemotaxis, chemiotaxis *n.* Taxis with regard to chemicals (Lincoln et al. 1985).

▸ **proschemotaxis** *n*. Positive chemotaxis (Lincoln et al. 1985).

▸ **tropochemotaxis** *n*. Tropotaxis with regard to chemicals.

clinotaxis *n*. Taxis with regard to a stimulus gradient (Lincoln et al. 1985).

compass orientation, compass taxis See taxis: menotaxis.

constant-angle taxis See taxis: constant-angle taxis.

diaphototaxis *n*. Taxis at a right angle with regard to incident light direction (Lincoln et al. 1985).

direction taxis See taxis: menotaxis.

dorsal-light reaction *n*. In many aquatic-animal species: taxis in which an individual keeps its dorsum toward a light source (Tinbergen 1951, 93).

cf. taxis: ventral-light reaction

[possibly coined by Tinbergen 1951]

electrotaxis See taxis: galvanotaxis.

galvanotaxis, electrotaxis, topogalvanotaxis *n*. Taxis with regard to electrical stimuli (Lincoln et al. 1985).

▸ **prosgalvanotaxis** *n*. Positive galvanotaxis (Lincoln et al. 1985).

geomagnetotaxis, magnetic-compass taxis *n*. For example, some termite and bird species, a ladybird beetle, the Honey Bee, possibly the Human, Monarch Butterfly, Loggerhead Turtle: taxis with regard to Earth's magnetism (Middendorf 1855 in Baker 1987, 691; Schneider 1961, Becker 1971, Merkel and Wiltschko 1965, Lindauer 1973 in Heymer 1977, 113; Baker 1980, 555; 1987, 691; Seachrist 1994, 661; Zimmer 1996b, May, 50).

cf. modality: magnetoreception

geotaxis *n*. In many kinds of animals, *Volvox* species (algae): Taxis with regard to gravity (Lincoln et al. 1985).

heliotaxis See taxis: phototaxis.

hydrotaxis *n*. Taxis with regard to water or moisture (Lincoln et al. 1985).

cf. taxis: hygrotaxis

▸ **aphototaxis** *n*. No taxis with regard to light (Lincoln et al. 1985).

hygrotaxis *n*. Taxis with regard to moisture (Lincoln et al. 1985).

cf. taxis: hydrotaxis

idiothetic orientation *n*. Taxis with regard to internal stimuli (Mittelstaedt 1973 in Heymer 1977, 125).

cf. taxis: allelothetic orientation

klinotaxis *n*. For example, in *Euglena*, maggots of some fly species: taxis (of an organism with only one sense organ for a particular kind of stimulus) involving side-to-side or spinning movements used to detect changes in intensity of the stimulus in time and space (Fraenkel and Gunn 1940, 58).

syn. phobotaxis, scanning orientation, trial-and-error orientation (Tinbergen 1951, 89)

cf. taxis: tropotaxis

[phobotaxis coined by Kühn (1919 in Tinbergen 1951, 89)]

landmark orientation See taxis: telotaxis.

light-compass taxis, light-compass orientation, light-compass reaction See taxis: menotaxis.

local orientation *n*. Taxis toward a goal that is continuously within an individual's sight or other perception (Lincoln et al. 1985, 125).

cf. taxis: remote orientation

magnetic-compass taxis See taxis: geomagnetotaxis.

menotaxis *n*. For example, in the Honey Bee and Starling: taxis with regard to a particular compass direction without use of nearby landmarks (McFarland 1985, 251).

syn. compass orientation (McFarland 1985, 245), compass taxis, direction taxis, constant-angle taxis

▸ **light-compass taxis, light-compass orientation, light-compass reaction** *n*. For example, in some ant species: menotaxis with regard to light direction (Tinbergen 1951, 95).

▸ **star-compass taxis** *n*. For example, in the Indigo Bunting: menotaxis with regard to stars, other than the Sun (Sauer and Sauer 1955, Emlen and Emlen 1966 in McFarland 1985, 254).

▸ **sun-compass taxis** *n*. For example, in the Domestic Pigeon, Honey Bee: menotaxis regarding the Sun (McFarland 1985, 254).

mnemotaxis *n*. Taxis based on an animal's memory of a path to a destination (Immelmann and Beer 1989, 307).

See taxis: topographical orientation.

navigation *n*. For example, in the starling: a complex form of spatial orientation involving topographical orientation, compass orientation, and true navigation (McFarland 1985, 251).

cf. homing

▸ **true navigation** *n*. An animal's ability to orient toward a goal, such as a home or breeding area, without using landmarks and without regard to the goal's direction (McFarland 1985, 251).

object taxis *n*. Taxis with regard to an object (Lincoln et al. 1985).

osmotaxis See taxis: tonotaxis.

oxygenotaxis, oxytaxis *n*. Taxis with regard to oxygen (Lincoln et al. 1985).

pharotaxis See taxis: telotaxis.

phobotaxis *n*. Taxis away from a stimulus (Lincoln et al. 1985).

See kinesis.

S–Z

phonotaxis *n*. In many animal species: taxis with regard to sound, used for danger, enemy, and mate detection and auditory communication (Bailey et al. 1988, 33).

photohoramotaxis, photohorotaxis *n*. Taxis with regard to color or a light pattern (Lincoln et al. 1985).

photophobotaxis *n*. Taxis with regard to a temporal change in light intensity (Lincoln et al. 1985).

▶ **negative photophobotaxis** *n*. Taxis away from an increase in light intensity (Lincoln et al. 1985).

▶ **positive photophobotaxis** *n*. Taxis toward a decrease in light intensity (Lincoln et al. 1985).

phototaxis, heliotaxis, phototropotaxis *n*. Taxis with regard to light (Lincoln et al. 1985).

cf. taxis: diaphototaxis, dorsal-light reaction, photohoramotaxis, polarized-light taxis, scototaxis, ultraviolet-light taxis, ventral-light reaction

▶ **phototropotaxis** *n*. For example, in some planarian species: tropotaxis with regard to light (Immelmann and Beer 1989, 209).

▶ **prophototaxis, prosphototaxis** *n*. Positive phototaxis (Lincoln et al. 1985).

phototropotaxis See taxis: phototaxis.

pilotage See taxis: topographical orientation.

pneumatotaxis, pneumotaxis *n*. Taxis with regard to dissolved carbon dioxide or other gas (Lincoln et al. 1985).

polarized-light taxis *n*. For example, in the Honey Bee: taxis with regard to polarized light (McFarland 1985, 257).

polytaxis See polytaxis.

positive phototaxis See taxis: phototaxis: positive phototaxis.

positive rheotaxis See taxis: rheotaxis: positive rheotaxis.

primary orientation *n*. Taxis that controls body posture (Heymer 1977, 125).

cf. taxis: secondary orientation

prophototaxis See taxis: phototaxis.

prosaerotaxis See taxis: aerotaxis: prosaerotaxis.

proschemotaxis See taxis: chemotaxis: proschemotaxis.

prosgalvanotaxis See taxis: galvanotaxis: prosgalvanotaxis.

remote orientation *n*. Taxis toward a goal that is not continuously within an individual's sight or other perception even if the goal is relatively close (Lincoln et al. 1985, 125).

cf. taxis: local orientation

rheotaxis *n*. Water-current, or air-current, taxis (Lincoln et al. 1985).

▶ **negative rheotaxis** *n*. Rheotaxis against a current.

▶ **positive rheotaxis** *n*. Rheotaxis with a current.

scototaxis, skototaxis *n*. Taxis with regard to a dark place; distinguished from negative phototaxis (Lincoln et al. 1985).

cf. negative phototaxis

secondary orientation *n*. Taxis that controls an individual's response to an environmental stimulus (Heymer 1977, 125).

cf. taxis: primary orientation

seismotaxis, vibrotaxis *n*. Taxis with regard to vibration or physical shock (Lincoln et al. 1985).

skototaxis See taxis: scototaxis.

sonotaxis *n*. Taxis with regard to sound.

space orientation See modality: kinesthesia.

stereotaxis See taxis: thigmotaxis.

strophotaxis, strophism *n*. An organism's twisting movement in response to an external stimulus (Lincoln et al. 1985).

sun-compass taxis See taxis: menotaxis.

telotaxis *n*. In many arthropod, cephalopod, and vertebrate species: taxis with regard to a single stimulus (object) without reference to other sources of stimulation (Lincoln et al. 1985)

syn. landmark orientation, goal orientation, pharotaxis (Tinbergen and Kruyt 1983 in Tinbergen 1951, 95)

Comments: Telotaxis can involve an animal's locomotion toward a "goal stimulus" which can be a learned one (*e.g.,* a wasp's going to her burrow). Telotaxis is achieved by use of vision, echolocation, infrared sensitivity, and electric-field sensitivity, depending on the species (Immelmann and Beer 1989, 209).

thermotaxis *n*. Taxis with regard to heat or infrared electromagnetic waves (Lincoln et al. 1985).

thigmotaxis, stereotaxis *n*. Taxis with regard to continuous contact with a solid surface (Lincoln et al. 1985).

tonotaxis, osmotaxis *n*. Taxis with regard to continuous contact with an osmotic pressure or osmotic stimulus (Lincoln et al. 1985).

topogalvanotaxis See taxis: galvanotaxis.

topographical orientation *n*. For example, in some fossorial wasp and bee species, Humans: taxis with regard to familiar landmarks.

syn. mnemotaxis (presently unpreferred; coined by Kühn 1919 in Tinbergen 1951, 97), pilotage (McFarland 1985, 251)

cf. taxis: mnemotaxis, navigation

topotaxis *n.* Taxis with regard to spatial differences in stimulus intensity, especially with reference to stimulus source (Lincoln et al. 1985).

traumataxis, traumotaxis *n.* Organism, cell, or organelle movement in response to injury (Lincoln et al. 1985).

trophotaxis *n.* Taxis with regard to food (Lincoln et al. 1985).

tropochemotaxis See taxis: chemotaxis.

tropophototaxis See taxis: phototaxis.

tropotaxis *n.* In many organism species: taxis in which an individual simultaneously detects a particular kind of stimulus from more than one part of its body, and thus moves without side-to-side, or spinning, motions with regard to this stimulus (*e.g.,* a cat's directly moving away from light). [possibly coined by Kühn (1919 in Tinbergen 1951, 11)].
syn. tropic orientation
cf. taxis: klinotaxis

ultraviolet-light taxis *n.* For example, in the Honey Bee; some *Colias* (butterfly), hummingbird species: taxis with regard to ultraviolet light (McFarland 1985, 257).

ventral-light reaction *n.* In many aquatic-animal species: Taxis in which an individual keeps its venter toward a light source (Tinbergen 1951, 93).
cf. taxis: ventral-light reaction

vibrotaxis See taxis: seismotaxis.

♦ **-taxis, taxy** *combining form* Order; disposition; arrangement (Michaelis 1963).
[Greek *taxis*, arrangement]

xerotaxis *n.* "The condition of a plant succession that remains unaffected by drought" (Lincoln et al. 1985).

zygotaxis, zygotactism *n.* "The mutual attraction between male and female gametes" (Lincoln et al. 1985).
cf. tropism: gamotropism

♦ **taxis component** *n.* A stimulus-dependent taxis in a behavior (Immelmann and Beer 1989, 308).

♦ *Taxiskomponente* *n.* An organism's sequence of reactions to external stimuli that continuously corrects its movement direction in relation to the spatial properties of the organism's environment (Tinbergen 1951, 87).
cf. Erbkoorination (Tinbergen 1951, 87)
Comment: Taxiskomponente is one of two main components of an organism's oriented movement (Tinbergen 1951, 87).
[possibly coined by Kühn (1919 in Tinbergen 1951, 87); Tinbergen translated this term as taxis]

♦ **taxocene** *n.* A group of species that belong to a particular supraspecific taxon and occur together in the same association (Lincoln et al. 1985).

♦ **taxometrics** See study of: numerical taxonomy.

♦ **taxon** *n., pl.* **taxa**
1. A group of organisms sufficiently distinct to be worthy of being named and assigned a definite category within a classification.
[first proposed by Meyer-Abich (1926, 126–137, in Mayr 1982, 207, 869, 960)]
2. Any classificatory division of organisms (*Oxford English Dictionary* 1972, entries from 1826); any organism group that represents a particular unit of classification (*e.g.,* all the members of a given subspecies, species, genus, etc.) (Wilson 1975, 597).
syn. category (which is used erroneously as such, Mayr 1982), group (*Oxford English Dictionary* 1972)
cf. ²group
Comments: "Group" is used to designate an assemblage of organisms that are not designated to another taxonomic category: Group Insecta (Borror et al. 1989, 147). H.J. Lam presented the term "taxon" at the Utrecht symposium on nomenclature in 1948 (Lawrence 1957, 53). Workers voted at the International Botanical Congress in 1950 to use "taxon" wherever appropriate.

indicator taxon *n.* A taxon that a researcher uses to determine the biodiversity of a particular regions (inferred from Lawton et al. 1998, 72).
See ²species: characteristic species.
cf. ²species: indicator species
Comments: It is necessary to study a wide range of taxa, with very different ecologies and life histories, of a particular habitat to monitor changes in biodiversity (Lawton et al. 1998, 72).

♦ **taxon cycle** See cycle: taxon cycle.

♦ **taxonometrics** See study of: taxonomy: numerical taxonomy.

♦ **taxonomic character** See character: species character.

♦ **taxonomic groups** See ²group: taxonomic groups.

♦ **taxonomic level, taxonomic rank** See ²rank.

♦ **taxonomic species** See ²species: Linnaean species.

♦ **taxonomy** See study of: taxonomy.

♦ **taxospecies** See ²species: taxospecies.

♦ **-taxy** *suffix* Directed response of a motile organism.
See taxis.

♦ **teaching** *n.* A person's imparting instruction or knowledge; "the occupation or function of a teacher" (*Oxford English Dictionary* 1972, entries from *ca.* 1175).
syn. tutoring (Michaelis 1963)

S–Z

Comment: Teaching is known only in Chimpanzees and Humans. In Chimpanzees, a sign-language-trained individual attempted to mold and influence the signing performance of a younger companion in a lab (Fouts et al. 1983 in Boesch 1991, 530), and mothers influence their infants' attempts to crack nuts in nature (Boesch 1991, 530). All functions attributed to human tutoring by Wood et al. (1976 in Boesch 1991) seem to be performed by these mothers.
[Old English *tǣcan*, teach]

♦ **team** See ²group: team.

♦ **teat order** See dominance: teat order

♦ **tecnophage** See -phage: tecnophage.

♦ **teem** See ²group: teem.

♦ **teeth chattering** *n.* For example, in Stump-Tailed Monkeys: a monkey's rapidly opening and closing its mouth with bared teeth (de Waal 1989, 154).
cf. animal sounds; display: bared-teeth display

♦ **teeth grinding** See behavior: threat behavior: teeth grinding.

♦ **tegulicole** See -cole: tegulicole.

♦ **telegamic** *adj.* Referring to attracting a mate from a distance (Lincoln et al. 1985).

♦ **telekinesis** See -kinesis: telekinesis.

♦ **telencephalon** See organ: brain: telencephalon.

♦ **teleology** *n.*
1. The study of a creative design in the processes of nature (Plato, Aristotle, the Stoics in Mayr 1982, 47).
   Note: "Teleological" refers to four concepts: "teleonomic activities," "teleomatic processes," "adapted systems," and "cosmic teleology" (Mayr 1982, 48–50).
2. The doctrine, or study, of ends or final causes, especially related to design or purpose in nature; also transferred to such design exhibited in natural objects or phenomena (*Oxford English Dictionary* 1972, entries from 1728).
3. The doctrine that phenomena of life and development can be fully explained only by the action of design and purpose and not by mechanical causes (Michaelis 1963).
   Note: This doctrine has been abandoned by today's biologists.
4. A person's explaining nature, including nonhuman animals, in terms of their utility or purpose, especially divine purpose (Michaelis 1963).
   Notes: Most of today's biologists are not teleological in this way; however, many biologists use "shorthand" statements that sound teleological: A sand wasp uses a pebble *to* tamp her burrow closed, or a syrphid fly mimics a yellowjacket *in order to* avoid predation.

5. A person's assumption that a nonhuman animal "has a conscious purpose underlying its behavior," when in fact it might not have this purpose (Dewsbury 1978, 6).
   Note: Most of today's animal behaviorists do not make this assumption in their studies.
   See study of: teleology.
   cf. behaviorism: purposeful behaviorism; teleonomy
   Comments: Lennox (1992) discusses the Aristotelian model of teleology, external teleology, immanent teleology (= internal teleology), and the Platonic model of teleology.
   [New Latin *teleologia* < Greek *telos, teleos,* end + *logos,* discourse; coined by Christian Wolff (1728) to refer to explanations of final causes, an idea known from ancient Greece (Lennox 1992, 324)]

cosmic teleology *n.* A kind of teleology asserting that the universe has a purpose (Aristotle in Mayr 1982, 50).

dysteleology *n.*
1. Denial of purposelessness or "final causes" in nature; opposed to teleology (*Oxford English Dictionary* 1972, entries from 1874).
2. Frustration of function, such as a bee's taking floral nectar through an artificial hole that it made rather than through a natural floral orifice that would be conducive to cross-pollination (Lincoln et al. 1985).
   See study of: dysteleology.
   cf. robber, thief

♦ **teleomatic processes** *n.* Processes, particularly ones relating to inanimate objects, in which a definite end is reached strictly as a consequence of physical laws (*e.g.,* a falling rock reaches the ground) (Mayr 1982, 49).

♦ **teleonomic activities** *n.* The processes of an organism's individual development (ontogeny) and its seemingly goal-directed behaviors that are underlain by a genetic program (Pittendrigh 1958 in Mayr 1982, 48).

♦ **teleonomy** See study of: teleonomy.

♦ **telmatium** See ²community: telmatium.

♦ **telmatophile** See -¹phile: telmatophile.

♦ **telmicole** See -cole: telmicole.

♦ **telmophage** See -phage: telmophage.

♦ **telocentric chromosome** See chromosome: telocentric chromosome.

♦ **telotaxis** See taxis: telotaxis.

♦ **temperate** *adj.*
1. Referring to the Temperate Zone of the Earth, between 23°27′ north and the northern polar circle (= Arctic Circle, 66°33′ north) and 23°27′ south and the southern polar circle (Antarctic Circle, 66°33′ south) (Lincoln et al. 1985).
2. Referring to a climate with long, warm summers and short, cold winters;

contrasted with polar, subtropical, tropical (Lincoln et al. 1985).

Comment: Much of the contiguous U.S. and Alaska are in the temperate zone.

♦ **temperate rain forest** See ²community: rain forest: temperate rain forest.

♦ **temperature, t., temp** *n.* The degree of heat of an organism (Michaelis 1963).

Comment: Bligh and Johnson (1973) define kinds of temperatures in addition to the ones listed below.

[Latin *temperatura*, due measure < *temperatus*]

ambient temperature, dry-bulb temperature (t_a) *n.* The average temperature of a gaseous or liquid environment (usually air or water) surrounding a body, as measured outside the thermal and hydrodynamic boundary layers that overlay a body (Bligh and Johnson 1973, 954).

Comment: A research measures ambient temperature with a thermometer shielded from light (Bligh and Johnson 1973, 955).

core temperature, deep-body temperature *n.* The average temperature of an organism's tissues that are below tissues directly affected by a temperature-gradient change through its peripheral tissues (Bligh and Johnson 1973, 955).

Comments: Researchers cannot measure core temperature accurately, and they generally represent it by a specific core temperature (*e.g.,* rectal temperature) (Bligh and Johnson 1973, 955).

effective temperature (T_{eff}) *n.* An arbitrary index that combines in a single value the effect of air movement, humidity, and temperature on a person's feeling of warmth or cold (Bligh and Johnson 1973, 955).

mean body temperature (T_b) *n.* The sum of the products of the heat capacity and temperature of all an organism's body's tissues divided by its total heat capacity (Bligh and Johnson 1973, 955).

Comments: Researchers cannot measure mean body temperature precisely in a living organism (Bligh and Johnson 1973, 955).

mean radiant temperature (T_r) *n.* The temperature of an imaginary isothermal "black" enclosure in which a solid body, or occupant, would exchange the same amount of heat by radiation as in the actual nonuniform enclosure (Bligh and Johnson 1973, 955).

preferred ambient temperature *n.* The ambient-temperature range associated with specified air movement, humidity, and radiation intensity from which an unrestrained animal does not seek to move to a warmer, or colder, environment (Bligh and Johnson 1973, 955).

preferred body temperature *n.* The core-temperature range within which an ectothermal animal maintains itself by behavioral means (Bligh and Johnson 1973, 955).

♦ **temperature messenger** See caste: worker: temperature messenger.

♦ **temperature regulation** *n.* An organism's maintaining its body temperature within a restricted range under conditions involving variable external, internal, or both, heat loads; contrasted with temperature conformity (Bligh and Johnson 1973, 955).

autonomic temperature regulation *n.* An organism's regulating its body temperature by autonomic (= involuntary) responses to cold and heat which modify its rates of heat loss (*e.g.,* by basal-metabolism variations, shivering, sweating, thermal tachypnea, vasomotor-tone variations) and heat production (Bligh and Johnson 1973, 955).

behavioral temperature regulation *n.*

1. An organism's regulating its body temperature by changing its behavior (inferred from Bligh and Johnson 1973, 956).

2. An animal's regulating its body temperature by complex behavior controlled by its central nervous system (Bligh and Johnson 1973, 956).

Note: These behaviors include skeletal-muscle responses to temperature that modify rates of heat loss or production, *e.g.,* adjusting clothing (Humans), change in body position, change in feather positions (birds), change in fur positions (mammals), exercise, selecting an environment that reduces thermal stress, or a combination of these things (Bligh and Johnson 1973, 955).

chemical temperature regulation *n.* An animal's regulating its body temperature by changing its heat production; an obsolete term (Bligh and Johnson 1973, 956).

Comment: An animal can achieve chemical temperature regulation by changing its metabolic rate, involuntary muscle movements (*e.g.,* shivering), nonshivering thermogenesis, and voluntary muscle movements (Bligh and Johnson 1973, 956).

physiological temperature regulation *n.* An organism's regulating its body temperature by autonomic responses, behavioral responses, or both (Bligh and Johnson 1973, 955).

♦ **temperature sensor** See ²receptor: thermoreceptor.

♦ **temperature survival limit** *n.* The environmental temperature above or below

S–Z

which an animal cannot maintain its thermal balance for a long period (Bligh and Johnson 1973, 956).

lower temperature survival limit *n.* The environmental temperature below which an animal cannot maintain its thermal balance for a long period and at which it becomes progressively hypothermic (Bligh and Johnson 1973, 956).

upper temperature survival limit *n.* The environmental temperature above which an animal cannot maintain its thermal balance for a long period and at which it becomes progressively hyperthermic (Bligh and Johnson 1973, 956).

♦ **template** *n.*

1. In many songbird species: information on a male bird's adult species-specific song that he memorizes and uses as a model against which he matches his own song production during his vocal development (Immelmann and Beer 1989, 309). *syn.* sound print (Immelmann and Beer 1989, 309)

2. In some invertebrate and many vertebrate species: an individual's memory representation of a set of cues by which it recognizes conspecifics as kin or nonkin (Wilson 1987, 12).
cf. ⁴model: gestalt model, individualistic model

♦ **temporal facilitation** See ¹facilitation: temporal facilitation.

♦ **temporal gland** See gland: temporal gland.

♦ **temporal polyethism** See polyethism: temporal polyethism.

♦ **temporal prediction** *n.* Prediction from the present to the future that is infrequently impossible in many biological situations (Mayr 1982, 57).
cf. logical prediction

♦ **temporal summation** See ¹facilitation: temporal facilitation.

♦ **temporary altruism** See altruism: temporary altruism.

♦ **temporary parasite** See parasite: temporary parasite.

♦ **temporary social parasite** See parasite, parasitism: temporary social parasite.

♦ **tempos of evolution** *pl. n.* Bradytely, horotely, and tachytely, *q.v.* (Simpson 1944 in Gould 1994, 6766).
cf. modes of evolution

♦ **tendency** *n.* An animal's likeliness to behave in a particular way due to one or more probable causal factors (Hinde 1970, 360). See potential: action potential.
Comment: Blurton-Jones (1968 in Hinde 1970, 361) suggests that "tendency" be used for observed likelihood of a behavior's oc-

currence, and "potential tendency" for theoretical likelihood of a behavior's occurrence.

♦ **teneral** See callow.

♦ **teratology** See study of: teratology.

♦ **terminal extinction** See ²extinction: terminal extinction.

♦ **termitarium** See nest: termitarium.

♦ **termiticole** See -cole: termiticole.

♦ **termitophile** See -¹phile: termitophile.

♦ **terricole** See -cole: terricole.

♦ **terrier** See dog: terrier.

♦ **territorial aggression** See aggression: territorial aggression.

♦ **territorial behavior** See territoriality.

♦ **territorial call** See animal sounds: call: territorial call.

♦ **territorial-demarkation substance** See chemical-releasing stimulus: semiochemical: pheromone: territory pheromone.

♦ **territorial flight** See flight: courtship flight.

♦ **territorial hegemony** *n.* In some reptile species: territoriality in which a defended area is inhabited by a tyrant male, a few subordinate males, females, and offspring (Wilson 1975, 444–445).
[Greek *bēgemonia* < *bēgeesthai*, to lead]

♦ **territorial hierarchy** See hierarchy: dominance hierarchy: territorial hierarchy.

♦ **territorial marking** See behavior: marking behavior.

♦ **territorial patrolling** See patrol.

♦ **territorial polygyny** See mating system: territorial polygyny.

♦ **territorial song** See animal sounds: song: territorial song.

♦ **territoriality** *n.* An animal's having one or more territories, *q.v.*
cf. individual distance, spacing behavior
polyterritoriality *n.* In the pied flycatcher: a male's having two nest-site territories (Breiehagen and Slagsvold 1988, 604).

♦ **territory** *n.*

1. An area defended by a male bird (Aristotle and Pliny in Wilson 1975, 260).

2. The land, or country, that belongs to a human ruler or state (*Oxford English Dictionary* 1972, entries from 1494).

3. In the nightingale: a "freehold" (Olina 1622, 1678 in Wilson 1975, 260).

4. In some anemone, coral, crab, spider, insect, salamander, and frog species; many fish, reptile, bird, and mammal species: an area occupied more or less exclusively by an individual, or group of individuals, by the territory holders's repulsion of other animals through overt defense or advertisement (Altum 1867, Eliot 1920, Noble 1939, Brown 1964, Wilson 1971b in Wilson 1975, 256, 442, 444, 564, 597; Robertson 1986, 763; Jaeger 1988, 307).

5. "Any defended area" of an animal (Nobel 1939 in Dewsbury 1978, 93).
6. "An area in which the resident [animal] enjoys priority of access to limited resources that he or she does not enjoy in other areas" (Kaufmann 1971 in Dewsbury 1978, 94).
7. A fixed area from which an individual animal excludes rival intruders by some combination of advertisement, threat, and attack (Brown 1975, 61).

cf. dominion, individual distance, range
Comments: Brown (1975, 60–61) discusses problems with some "territory" definitions. "Dominance hierarchies" and "territories" have intergrading characteristics (Dewsbury 1978, 94). An animal may defend a particular area in different degrees depending on time of year. The original territory concept is tending to fall more and more into obsolescence; instead, workers are describing area inhabitants' degrees of exclusiveness in using an area's resources (Immelmann and Beer 1989, 311). "Territoriality" serves many different functions. Some species exclude both conspecific and heterospecific individuals from their territories (Barrows 1976b, 1983; Immelmann and Beer 1989, 310).
[Latin *territorium* < *terra*, earth; Moffat (1903) introduced the English word "territory" into the scientific literature (Wilson 1975, 260)]

communal territory See territory: group territory.

defended territory *n.* For example, in some bee species, stickleback fish, the Uganda kob: a territory held by overt defense (Alcock et al. 1978; Dewsbury 1978, 95).
Comment: "Defended territory" grades into "undefended territory."

display territory *n.* A space used by one or more individuals of the same sex (usually males) solely for courtship and mating (Immelmann and Beer 1989, 76).
cf. ²lek; territory: mating territory

exclusive-occupancy territory *n.* A territory that is defended against all conspecifics except during a short mating period (Dewsbury 1978, 95).
Comment: "Exclusive-occupancy territory" grades into nonexclusive-occupancy territory."

feeding territory *n.* For example, in some diurnal insectivorous songbirds: a territory in which an individual, or conspecific group, feeds (Dewsbury 1978, 96); in some species, feeding territories also include breeding sites (Immelmann and Beer 1989, 310).

SOME CLASSIFICATIONS OF TERRITORIES

I. Classification with regard to activities[a]
  A. type-A territory (using for shelter, courting, mating, nesting, gathering most food)
  B. type-B territory (breeding, gathering the minority of food)
  C. type-C territory (defending small area around nest)
  D. type-D territory (pairing, mating)
  E. type-E territory (using for shelter, roosting)
II. Classification of type-A and -B territories by type of defense
  A. absolute-defense territory
  B. spatiotemporal-defense territory
III. Classification by constancy of a territory's position
  A. fixed territory
  B. floating territory
IV. Classification by species involved
  A. intraspecific territory
  B. interspecific territory
V. Classification with regard to territory size relative to the territory holder
  A. microterritoriality
  B. macroterritoriality
VI. Classification by territory quality
  A. optimal territory
  B. suboptimal territory

---

[a] These activities can vary with species.

Comment: "Feeding territory" grades into "nonfeeding territory."

fixed territory *n.* A territory that includes the same space throughout its tenure (Dewsbury 1978, 96).
Comment: "Fixed territory" grades into "floating territory."

floating territory *n.* For example, in some bee species, the Bitterling (fish): a territory that changes position from time to time (Wilson 1975, 265; Alcock et al. 1978; Barrows 1983).
cf. floater
Comments: The Bitterling lays eggs into the mantle cavities of certain mussels and defends the area around the mussel even when the mussel moves. "Floating territory" grades into "fixed territory."

group territory *n.* In many animal species: a territory occupied by two or more conspecifics (Brown 1987a, 300; Dewsbury 1978, 95; Agren et al. 1989, 28).
syn. communal territory (Immelmann and Beer 1989, 123)
Comments: "Group territory" grades into "individual territory" (Dewsbury 1978, 95). Group territories are usually held all, or

S–Z

most of a, year and are usually all-purpose territories (Brown 1987a, 300). In one-male groups of pinnipeds, Lion prides, and Humans, a group territory is occupied by a mated pair or more than two conspecifics (Dewsbury 1978, 95). In the Caribbean Striped Parrotfish, a group territory is held by nonkin (Clifton 1989, 90).

individual territory *n*. For example, many animal species, including some bee species, Stickleback Fish: a territory occupied by one male, except during invasions by conspecific contestants or entries of potential mates (Dewsbury 1978, 95).
Comment: This "individual territory" grades into "group territory" (Dewsbury 1978, 95).

interspecific territory *n*. In some frog species, salamander, lizard, and mammal species (including rodents and shrews); many bird species (including woodpeckers, Hummingbirds, Honey-Eaters, Warblers, Shrikes, Goldcrests, Corvids): a territory that an individual defends against a heterospecific individual that uses the same resources (Immelmann and Beer 1989, 157).

intrasexual territory *n*. In Ferrets: a territory defended by an individual against conspecifics of the same sex (Powell 1979 in Clapperton 1989, 436).

mating territory *n*. For example, in lekking species (including some orchid-bee species, Grouse, Ruffs, and the Uganda Kob) and other territorial animals: a territory in which breeding occurs (Dewsbury 1978, 96).
cf. territory: display territory
Comment: "Mating territory" grades into "nonmating territory."

microterritory *n*. In males of a sweat bee species: a territory that is large enough to contain only one individual — its holder. [coined by Barrows (1976, 377)]

nonexclusive-occupancy territory *n*. A territory that is defended against lower-ranking individuals, as in some lizards, or against nongroup members, as in the Steller's jay and bicolored ant birds (Dewsbury 1978, 95).
Comment: "Nonexclusive-occupancy territory" grades into "exclusive occupancy territory."

nonfeeding territory *n*. In most shorebird species (*e.g.,* Gulls and Ibises); some bee species: a territory that ordinarily does not contain food (Dewsbury 1978, 96).
Comment: "Nonfeeding territory" grades into "feeding territory" (Dewsbury 1978, 96).

nonmating territory *n*. A territory in which certain activities besides mating occur (Dewsbury 1978, 96).
Comment: "Nonmating territory" grades into "mating territory."

nonrearing territory *n*. A territory (*e.g.,* a lek) in which young are not reared (Dewsbury 1978, 96).
Comment: "Nonrearing territory" grades into "rearing territory."

pairing territory *n*. For example, in some gull species, lekking species: a site where pair formation occurs that is away from a nesting site and where other reproductive activities occur (Immelmann and Beer 1989, 212).
cf. lek

rearing territory *n*. For example, in colonial shorebirds, sticklebacks: a territory in which young are reared. "Rearing territory" grades into "nonrearing territory" (Dewsbury 1978, 96).

sun-dapple territory *n*. For example, in male Speckled-Wood Butterflies: a territory in a forest whose boundaries are the perimeter of a sun dapple (Davies 1978).

symbolic territory *n*. For example, in males of some hangingfly species, some bee species: a territory that does not contain emerging receptive females or resources that attract nesting or foraging females (Thornhill and Alcock 1983, 112).

type-A territory *n*. "A large defended area within which sheltering, courtship, mating, nesting, and most food gathering occur" (Mayr 1935, Nice 1941, Armstrong 1947, Hinde 1956 in Wilson 1975, 265).
cf. ³range

type-B territory *n*. "A large defended area within which all breeding activities occur but which is not the primary source of food" (Wilson 1975, 265).

type-C territory *n*. A small defended area around an animal's nest (Wilson 1975, 265).

type-D territory *n*. A territory used exclusively for pairing, mating, or both (Wilson 1975, 265).

type-E territory *n*. A territory used for roosting or as a shelter (Wilson 1975, 265).

undefended territory *n*. For example, in the Whiptail Wallaby and coati: a territory held by behavior other than overt defense (Dewsbury 1978, 95).
Comment: "Undefended territory" grades into "defended territory" (Dewsbury 1978, 95).

♦ **territory game** See game (table).

♦ **territory holder** *n*. In many animal species: an individual that is able to obtain and retain a territory compared to a conspecific that cannot do so (Alcock et al. 1978; Barrows 1983; Caro 1989, 54).
syn. regular; resident (Caro 1989, 54)
cf. floater, nonresident, nonterritorial animal

♦ **territory-holding male** See male: regular, territory-holding male.

♦ **territory marking** See behavior: marking behavior: territory marking.

♦ **territory pheromone** See chemical-releasing stimulus: semiochemical: pheromone: territory pheromone.

♦ **tertiary hyperparasitism** See parasitism: hyperparasitism: tertiary hyperparasitism.

♦ **tertiary parasite** See parasite: tertiary parasite.

♦ **tertiary parasitism** See parasitism: superparasitism.

♦ **tertiary sex ratio** See sex ratio: tertiary sex ratio.

♦ **test** *n*. A means by which the existence, quality, or genuineness of anything can be determined; examination, trial, proof (*Oxford English Dictionary* 1972, entries from 1651).
See statistical test.
cf. experiment, experimental design

choice test *n*. A test used to determine an animal's social preferences and investigate its learning and discrimination capacities and stimulus directing and releasing effects; in a choice test, an animal is expected to choose between, or among, two or more objects (Immelmann and Beer 1989, 44).

▸ **multiple-choice test** *n*. A choice test in which an animal chooses among three or more objects (Immelmann and Beer 1989, 44).

▸ **simultaneous-choice test** *n*. A choice test in which choice objects are presented at the same time (Immelmann and Beer 1989, 44).

▸ **successive-choice test** *n*. A choice test in which choice objects are presented one after another (Immelmann and Beer 1989, 44).

Corrositex *n*. An *in vitro* test used as a substitute for the traditional rabbit-skin test (Roush 1996, Science 274,168).
Comments: The Corrositex gauges corrosivity of a product according to the time required for a chemical sample to break through a skinlike-protein membrane (Roush 1996, Science 274, 168).

discrimination test *n*. A test used for determining the perceptual, or learning, abilities of animals in which animals are given choices between stimuli; the animals are rewarded for correct responses and sometimes punished for incorrect responses (Buchholtz 1973, Rensch 1973 in Heymer 1977, 195).

Draize test *n*. A test of a chemical product that involves putting the chemical in the eye of a laboratory rabbit (Roush 1996, 168).
cf. Three R's

Hebb-Williams test *n*. A series of maze-learning problems designed to assess rat "intelligence" (Denenberg and Morton 1962 in Dewsbury 1978, 333).

IQ test *n*. A test that measures IQ (intelligence quotient) by assessing a person's memory, arithmetic and reasoning power, language ability, and concept-formation ability.
cf. intelligence quotient
Comment: The exact meaning of "IQ" is debatable, but IQ scores are relatively stable with a person's age and are correlated with a young person's future academic performance (developed by Albert Benet 1905 in Gould 1982, 529).

mental chronometry *n*. A test used to obtain an objective measure of intangible phenomena such as mental representations which uses the time required to solve a spatial problem as an index of processes involved (Posner 1978 in McFarland 1985, 498).

oddity test *n*. A test used in learning experimentation in which an animal is expected to choose an odd symbol from a group of symbols (McFarland 1985, 509).
cf. learning: oddity learning; test: reversal test

open-field test *n*. A standard method used by experimental psychologists to measure an animal's "general emotionality" as indicated, for example, by its amount of urination or defecation, and to measure the animal's locomotor activity in relation to its previous environmental conditions or experience (Immelmann and Beer 1989, 86, 207).

reversal test *n*. A test used in learning experimentation in which an animal is expected to choose a solution to a problem which was previously the wrong solution (McFarland 1985, 509).
cf. learning: reversal learning, test: oddity test

supine test *n*. A behavioral test in which two anesthetized stimulus animals (*e.g.,* hamsters) are placed adjacent to one another, on their backs (supine position), in order to examine the preference of an unanesthetized test animal for one of these stimulus animals (Landauer et al. 1978, 615; M.R. Landauer, personal communication 1985).
Comment: In this test, the test animal's responses can be evaluated without the influences of reciprocal behavioral interactions from the stimulus animals.
[coined by Landauer et al. 1978, 615; M.R. Landauer, personal communication 1985]

S–Z

traditional rabbit-skin test *n.* A test used to determine the corrosivity of a chemical product by drizzling a sample on the shaved back of a laboratory rabbit (Roush 1996, 168).
cf. test: Corrositex

♦ **test of independence** See statistical test: test of independence.

♦ **Test-Operate-Test-Exit (TOTE)** *n.* A control unit in a hierarchy conception of how behavior is functionally organized; the unit consists of a feedback cycle in which a mismatch (test) between what an animal's action is aimed at and what it achieves leads to its repeating the action (operate) and another comparison between aim and result (test), which, if it shows a match, leads to the animal's moving on (exit) to the next operation in a sequence (Miller et al. 1960 in Immelmann and Beer 1989, 315).

♦ **testicular feminization** *n.* In mammals: a disease in which a genetic male produces functional amounts of androgens, but his target tissues cannot respond to them because he has an enzyme deficit (Dewsbury 1978, 231).
cf. hermaphroditism: progestin-induced hermaphroditism; syndrome: adrenogenital syndrome

♦ **testis, testes** See gonad.

♦ **testosterone** See hormone: testosterone.

♦ **tetany** *n.*
1. A human tetanoid affection that is characterized by intermittent muscular spasms (*Oxford English Dictionary* 1972, entries from 1890).
2. For example, in poisoned fish: a long period of muscle contraction that may last as long as 24 hours, or intermittent bouts of muscle contraction (Geiger et al. 1985, 12; R.A. Drummond, personal communication).

♦ **tetranucleotide theory of the structure of DNA** See ²theory: tetranucleotide theory of the structure of DNA.

♦ **tetraploidy** See -ploidy: tetraploidy.

♦ **tetrodoxin** See toxin: tetrodoxin.

♦ **thalamus** See organ: brain: thalamus.

♦ **thalassium** See ²community: thalassium.

♦ **thalasson** See biotope: thalasson.

♦ **thalassophile** See -¹phile: thalassophile.

♦ **thalassophobia** See phobia (table).

♦ **thamnocole** See -cole: thamnocole.

♦ **thamnophile** See -¹phile: thamnophile.

♦ **thanatology** See study of: thanatology.

♦ **thanatophobia** See phobia (table).

♦ **thanatosis** See playing dead.

♦ **thanogeography** See study of: geography: thanogeography.

♦ **Thayer's countershading hypothesis** See hypothesis: Thayer's countershading hypothesis.

♦ **theistic evolution** *n.* The idea that God oversaw and guided the millions of years of evolution that culminated with Humans (Rosin 1999, A22).
cf. evolution

♦ **thelygenic** See -genic: thelygenic.

♦ **thelygeny** See -geny: thelygeny.

♦ **thelytoky** See parthenogenesis: thelytoky.

♦ **theorem** *n.*
1. A universal, or general, proposition, or statement, not self-evident (as apposed to an axiom), but demonstrable by argument (in the strict sense by necessary reasoning); "a demonstrable theoretical judgement" (*Oxford English Dictionary* 1972, entries from 1551).
2. In mathematics, a proposition that sets forth something to be proved; a proposition that has been proved or assumed to be true; a rule, or statement, of relations formulated in symbols; or in general, any proposition susceptible of proof (Michaelis 1963).
syn. axiom (in some cases)
cf. axiom, doctrine, dogma, law, principle, rule, theory, truism
[Late Latin *theorema* < Greek *theōrēma*, sight, theory < *theōreein*, to look at]

♦ **¹theory** *n.* A person's conception, or mental scheme, of something to be done, or the method of doing it; a systematic statement of rules, or principles, to be followed (*Oxford English Dictionary* 1972, entries from 1597).
[Late Latin *theoria* < Greek *theoria*, view, speculation < *theoreein*, to look at]

♦ **²theory** *n.*
1. A scheme, or system, of ideas, or statements, held as an explanation, or account, of a group of facts or phenomena; a hypothesis that has been confirmed, or established, by observation, or experiment that is propounded, or accepted, as accounting for the known facts; a statement of what one holds to be the general laws, principles, or causes of something known or observed (*Oxford English Dictionary* 1972, entries from 1638); a closely reasoned set of propositions, derived from and supported by established evidence and intended to serve as an explanation for a group of phenomena: the quantum *theory* (Michaelis 1963).
2. Abstractly (without an article), systematic conception, or statement, of the principles of something; abstract knowledge or its formulation, often used to imply more or less supported hypotheses; distinguished from, or opposed to,

practice; in theory, according to theory, theoretically (opposed to in practice, or in fact) (*Oxford English Dictionary* 1972).

3. An integrated group of principles that underlie a science or its practical applications: the atomic *theory* (Michaelis 1963).

4. An arrangement of results, or a body of theorems, that presents a systematic view of some subject: the *theory* of functions (Michaelis 1963).

5. "The grandest synthesis of a large and important body of information about some related group of natural phenomena" (Moore 1984, 474): The Modern Synthesis.

6. "A body of knowledge and explanatory concepts that seeks to increase our understanding of ('explain') a major phenomenon of nature" (Moore 1984, 474).

Note: This kind of theory is not falsifiable (Moore 1984, 474).

♦ ³**theory** *n.*

1. Loosely or generally, a hypothesis proposed as an explanation; hence, a mere hypothesis, speculation, conjecture; an idea, or set of ideas, about something; a person's individual view or notion (*Oxford English Dictionary* 1972, entries from 1792); pejoratively, a dubious notion (Moore 1984, 474).

syn. doctrine, hypothesis (in some cases) (Michaelis 1963)

cf. darwinize

2. Scientifically, a hypothesis that is elevated to a theory because it has not been falsified after many attempts with different kinds of experiments and observations, *e.g.,* Darwin's theory of evolution meaning a body of interconnected statements about natural selection and other processes that are thought to cause evolution (Curtis 1983, 1108; Futuyma 1986, 15).

Notes: This definition seems appropriate for wide use in scientific writing. Some scientists make a strong distinction between "theory" and "hypothesis." Futuyma (1986, 15) states that "organisms have descended with modifications from common ancestors is a historical reality" and, thus, is a fact, not a theory.

cf. axiom; doctrine; dogma; hypothesis; law; principle; rule: Dzierzon's rule; theorem; truism

Comments: A term is listed below because the words "theory" or "theorem" are in its name, in its definition, or both. The same definition of "theory" does not relate to all of the following terms. The theories listed below vary in how well researchers have tested them.

approach-withdrawal theory *n.* When animals begin their lives, they are capable of only forced movements; intense stimulation causes their withdrawal, and mild stimulation causes their approach (Immelmann and Beer 1989, 19–20).

syn. theory of biphasic processes (Immelmann and Beer 1989, 20)

Comment: By interacting with its environment, an animal builds "on this base qualitatively distinct levels of behavioral capacity, such as discrimination, motivation control, and hypothesis formation, which can be distinguished in comparisons of different kinds of animals as well as in stages of individual behavioral development" (Immelmann and Beer 1989, 19–20). Schneirla (1965) formulated this theory (Immelmann and Beer 1989).

Aristotle's theory of inheritance See *eidos.*

autogenetic theory *n.* The hypothesis that "organic evolution has a "built-in capacity for or drive toward increasing perfection" (Mayr 1982, 360–361).

syn. Lamarckian evolution (in part), Lamarckism (in part) (Mayr 1982, 360–361)

Comment: This is one of the six major parts of Lamarckism in the broad sense. "Autogenetic theory" includes "orthogenesis" (Eimer), "aristogenesis" (Osborn), and the "omega principle" (Teilhard de Chardin) (Mayr 1982, 360–361).

balance theories *pl. n.* "Theories concerned with the relation between the attitudes and perceptions of one person towards another person and an object" (Hinde 1982, 277).

biological theory of evolution See evolutionary synthesis.

blood theory of inheritance See blending inheritance.

breakage-fusion theory *n.* The hypothesis proposed by Janssens and Morgan that sister chromatids can break and refuse during meiosis, that is, cross over (Mayr 1982, 767–768).

Comment: This hypothesis has been accepted after refuting theories were rejected (Mayr 1982, 767–768).

Castles' theory *n.* The hypothesis that "contamination" of one genetic factor (*e.g.,* white color of an albino guinea-pig parent) by another (black color of a guinea-pig parent) occurs during meiosis (Mayr 1982, 785–786).

syn. contamination theory (Mayr 1982, 785)

Comment: This was the last hypothesis of "soft inheritance" proposed by a respectable geneticist, and Castle later abandoned it (Mayr 1982, 785–786).

S–Z

cell theory, cell theory of multicellarity *n.*

1. A theory that interprets the developmental relationship of cells to organisms and has the principal tenets: (a) all living substance is concentrated in cells; (b) the cells in a organism's body are all individuals of the same morphological rank; (c) tissue cells are morphologically and physiologically elementary individuals, units of structure and function; (d) an organism's body is an aggregate of cells which are its "building stones;" and (e) a body's action is the sum of the many special actions performed by many kinds of collaborating cells (Schleiden 1938, Schwann 1939, Heidenhain 1907, Sharp 1926 in Kaplan and Hagemann 1991, 693–694).
2. The true hypothesis that all living eukaryotic organisms consist of cells and cell products (Mayr 1982, 653).

cf. ³theory: organismal theory

central dogma of evolutionary theory *n.* Natural selection has a pervasive role in shaping all classes of characters (Wilson 1975, 22).
Comment: Some characters might not be shaped by natural selection.

chain reaction See ³theory: chain-reflex theory.

chain-reflex theory *n.* The hypothesis that all behavior can be explained as a sequence of conditioned and unconditioned reflexes and that complex behavioral responses are simply chain reactions of behavior (Bechterew 1913, Pavlov 1927 in Heymer 1977, 97).
syn. chain reaction, reflex theory (Heymer 1977, 97); Pavlov's reflex theory (Tinbergen 1951; 101)
cf. cyclic-reflex hypothesis
Comment: This hypothesis appears to be a "grotesque simplification" (Tinbergen 1951, 101).

chromosome theory See ³theory: Sutton-Boveri chromosome theory.

classical gene-chromosome theory *n.* Theory that prevailed up to about 1960 and includes these ideas: (1) a chromosome is somewhat like a string of beads, each bead representing a different gene; (2) each gene is a discrete corpuscle, completely constant from generation to generation (except for very rare mutations), being independent of neighboring genes and having no effect on them (except for rare position effects); (3) each gene controls, or affects, a character, mutates independently of other genes, and can be separated from its nearest neighbor

on a chromosome by crossing over; (4) mutation is a slight modification of a gene molecule, resulting in a new allele; and (5) crossing over is a purely mechanical breaking of the "string of beads," followed by a new fusion with the corresponding "piece of string" of a homologous chromosome (Mayr 1982, 795–796).

conflict theory of display *n.* An investigatory approach that recognizes that behavior involves conflict and analyzes behavior in terms of underlying incompatible tendencies (Tinbergen 1959, 1962, Baerends 1975 in McFarland 1985, 401)

contamination theory See ³theory: Castles' theory.

control theory See study of: cybernetics.

cytoplasmic inheritance theory *n.* A hypothesis from the late 19th and early 20th centuries that a special "species substance" in an organism's cytoplasm causes continuous variation in organisms (Mayr 1982, 786–790).
cf. factor: kappa factor, sex-ratio factor, sterility factor
Comment: This hypothesis is incorrect because most of an organism's characters are programmed by its nuclear DNA (Mayr 1982, 786–790).

Darwin's theory of gemmules, Darwin's theory of pangenesis See ³theory: theory of pangenesis.

de Vries' genetic theory, de Vries' theory of pangens *n.* Hugo de Vries' (1889) concepts of genetics that concerned pangens as the basis of heredity; some of his ideas were similar to modern genetic knowledge; other ideas have been shown to be erroneous (Mayr 1982, 708–709).

de Vries' mutation theory See mutationism.

displacement theory See ³theory: continental drift.

encasement theory *n.*

1. A now refuted hypothesis that an embryo, or "germ" (gamete), is a development, or an expansion, of a preexisting form that contains all of the rudiments of the parts of a future organism; it is not brought into existence by fecundation alone (Bonnet 1762 in *Oxford English Dictionary* 1972).
2. A now-discarded doctrine that an organism's bodily complexity is totally mapped out in its egg (ovism, believed by an ovist preformationist) or sperm (spermism, spermist preformationist) (Mayr 1982, 12, 106).

Note: This idea was later "resurrected as the genetic program" (Mayr 1982, 12, 106).

Encasement theory is opposed to epigenesis.

syn. incasement theory, preformation (theory) (Lincoln et al. 1985); preformationism (Mayr 1982, 959); theory of evolution (now obsolete synonym; *Oxford English Dictionary* 1972); theory of preformation

cf. corpuscularist; -genesis: epigenesis; [1]evolution; physicalism; [3]theory: classical-gene chromosome theory

epeirophoresis theory See hypothesis: continental drift.

exchange-and-interdependence theories *pl. n.* Theories of interpersonal behavior that involve the view that this behavior depends on the rewards, or expectations of rewards, that can result from this behavior (Hinde 1982, 278–279).

Comment: These theories stress the effects of each individual of a dyadic relationship on each other (Hinde 1982, 278–279).

Fisher's fundamental theorem *n.* The rate of increase in fitness of any organism at any time is equal to its genetic variance in fitness a that time (McFarland 1985, 112).

[after Ronald Aylmer Fisher, 1890–1962, geneticist and statistician]

Fisher's theory of 1:1 sex ratios *n.*
1. Under outbreeding conditions and assuming that the cost of producing a son or daughter is equal, the optimal sex ratio of a population is 1:1 (Fisher 1930 in Wilson 1975, 317).
2. Assume that in an outbreeding population of diploid, bisexual organisms, both sexes are, on the average, of equal value for the genetic perpetuation of a parent, and also assume that sons and daughters are equally costly for a parent to make, then, if the sex ratio were male biased, one would expect selection for individual females that produced more daughters; if the sex ratio were female biased, one would expect the opposite, until the equilibrium of a 1:1-sex ratio occurs (Fisher 1930, 158–159).

syn. Fisher effect, Fisher's principle (Wilson 1975, 317); Fisherian sexual selection, Fisher's law, Fisher's rule, Fisher's sex-ratio theory

cf. principle: handicap principle; selection: sexual selection

Comment: Sex ratios of close to 1:1 predominate in most animal species. Trivers and Hare (1976) discuss sex ratios related to unequal costs of producing sons and daughters.

[after Ronald Aylmer Fisher]

flying-squirrel theory See hypothesis: paranotal hypothesis of insect wing origin.

free-cell-formation theory *n.* Schleiden's (1838, 1842) hypothesis about cell formation that includes the ideas that (1) a cell's nucleus forms by crystallization from granular material within a cell's contents, and (2) the nucleus grows and eventually forms a new cell around itself with the outer nuclear membrane's becoming the new cell wall (Mayr 1982, 656).

Comment: By the 1860s, this theory was refuted (Mayr 1982, 656).

game theory *n.*
1. A means of analyzing human economic and military behavior that assumes human players who use the same, or different, strategies behave rationally and according to some criterion of self-interest (von Neumann and Morgenstern 1953 in Maynard Smith 1982, 1–2; Dawkins 1982, 287).
2. A means of analyzing organism behavior that (1) considers strategies, *q.v.*, that individuals use, (2) assumes that organisms behave in ways that promote their fitnesses, and (3) considers environmental constraints on organisms and possibly arising population dynamics and equilibria (Lewontin 1961, Slobodkin and Rapoport 1974 in Maynard Smith 1982, 2).

syn. theory of games

cf. game, strategy

Comment: Game theory requires a single scale of measure of game outcomes and assumes that human behavior is rational. Because these criteria are not met well in human economics, game theory has been more fruitful in measuring how nonhuman animals should behave in order to maximize their fitnesses (Dawkins 1982, 287).

gemmule theory See [3]theory: theory of pangenesis.

germ-line theory, germ-plasm theory *n.* The true hypothesis that there are two types of cells, germ-plasm and somatic cells, and germ plasm is potentially immortal through transmission from generation to generation (Lincoln et al. 1985).

Hamilton's inclusive fitness theory See [3]theory: kin-selection theory.

hit theory, target theory *n.* Muller's hypothesis, that if genes are well-defined corpuscles of definite size, bombarding with ionizing radiation (electrons or short-wave rays) would "hit" the genes and produce mutations (Mayr 1982, 803).

Comment: Testing the "hit theory" gave mixed results (Mayr 1982, 803).

S–Z

inflow theory *n*. In vertebrates: an individual's brain distinguishes between the movement of its retinal images that are independent of the individual and movement due to its eyeballs, because receptors in its extraocular muscles send messages to its brain comparator whenever its eyes move (Sherrington 1918 in McFarland 1985, 249).

cf. ³theory: outflow theory; principle: reafference principle

information theory See study of: mathematics: cybernetics.

James-Lange theory *n*. The concept that a person's subjective emotional feelings are generated by his sensory receptors involved in his emotional reaction; *e.g.,* a frightening stimulus elicits certain behavioral and psychological changes, and these give rise to his subjective experience of fear (McFarland 1985, 524).

[after William James (1890) and Carl Lange, contemporary psychologists who proposed similar explanations of emotional experience]

kin-selection theory, theory of kin selection *n*. Hamilton's (1964) ideas regarding the evolution of altruism; altruism can evolve because an altruistic organism helps relatives that share genes with it due to common descent and, thus, the altruist promotes these shared genes (which are identical to its own) by helping (Maynard Smith 1964 in West Eberhard 1975, 2).

syn. kinship theory (C.K. Starr, personal communication)

cf. fitness: inclusive fitness; rule: Hamilton's rule; selection: kin selection

learning theory *n*. The hypothesis that a person's brain at the time of his birth is a *tabula rasa, q.v.* (Montagu 1962, Skinner 1971 in Heymer 1977, 116).

life-history theory *n*. Theory that analyzes "the trade-offs that organisms face in making decisions that affect their fecundity and mortality schedules;" most such trade-offs are concerned with either current vs. future reproduction or with offspring size vs. their number (Clark 1993, 205).

local-odor theory See hypothesis: locale-odor hypothesis.

MacArthur-Wilson theory of island biogeography See ³theory: theory of island biogeography.

marginal-value theorem *n*. The optimal rule that a foraging animal should depart from a food patch when its instantaneous feeding rate falls to the average expected for its habitat (Charnov 1973,

1976, Parker and Stuart 1976 in Giraldeau and Kramer 1982, 1036).

Comment: This theorem predicts that an animal's time spent within patches will increase as the average travel time between them increases (Giraldeau and Kramer 1982, 1036).

Mayr's theory of geographical speciation See speciation: Mayr's theory of geographical speciation.

mechanical-physical theory of evolution *n*. Nägeli's hypothesis that the activity of idioplasm (genetic material) is due to differential states of excitation of different groups of molecules within its strands (Mayr 1982, 671).

mechanistic theory, mechanism *n*. The hypothesis (or belief) that natural phenomena can be explained by the laws of chemistry and physics (Mayr 1982, 835).

Comments: Important controversies in the history of biology are "mechanism" vs. "vitalism" and "mechanism" vs. "teleology" (Mayr 1982, 835). Some biologists (*e.g.,* Nägeli [1884]) considered "mechanistic to be synonymous with "scientific" (Mayr 1982, 851–852). Gherardi (1984) discusses "mechanism" with regard to the history of ethology.

Mendel's theory of inheritance *n*.
1. The concepts that: (a) only one particle (which we now call a gene) not many identical, determinants of a given unit character, is transmitted to a germ cell; (b) particles underlying a certain trait exist in sets (now called alleles) and these particles do not change one another while they are associated in heterozygote organisms; and (c) one particle that affects a character is obtained from each parent (Mayr 1982, 718, 721).
2. "Non-blending inheritance by means of pairs of discrete hereditary factors (now identified with genes), one member of each pair coming from each parent" (Dawkins 1982, 290).

Note: "The main theoretical alternative is 'blending inheritance.' In Mendelian inheritance, genes may blend in their effects on a body, but they themselves do not blend, and they are passed on intact to future generations" (Dawkins 1982, 290).

syn. Mendelian inheritance

cf. inheritance; laws: Mendel's laws

Mendel's theory of particulate inheritance *n*. The hypothesis that genetic factors transmitted to an individual from its parents do not fuse after fertilization but retain their integrity throughout its entire life cycle (Mayr 1982, 781).

syn. theory of particulate inheritance, particulate concept of heredity (Mayr 1982, 781)

cf. blending concept; [3]theory: Sutton-Boveri chromosome theory

Comment: This theory has been widely accepted since the late 19th century (Mayr 1982, 781).

modern synthetic theory of evolution See evolutionary synthesis.

modern theory of the gene *n.* Goldschmidt's (1955) hypothesis that: (1) localized genes do not exist, instead there is "a definite molecular pattern in a definite section of a chromosome; (2) any change of pattern (position effect in the widest sense) changes the action of the chromosomal part and thus appears as a mutant;" and (3) a chromosome, as a whole, is a molecular "field" and genes are discrete, or even overlapping, sections of this field (Mayr 1982, 799).

cf. gene

Comment: The "modern theory of the gene" was proposed to replace the "corpuscular theory" of genes; however, the "modern theory of the gene" was not widely accepted and is now refuted (Mayr 1982, 799).

molecular theory of evolution *n.* A modification of Darwin's theory of evolution as it relates to genetic-level changes (Mayr 1982, 584).

Comment: "At this time it is still uncertain whether some of the recent discoveries of molecular genetics (repetitive DNA, gene splicing, wandering genes) do or do not require any revision of the synthetic theory of evolution" (Mayr 1982, 584).

molecular theory of inheritance *n.* The hypothesis that specific molecules are hereditary factors (Mayr 1982, 774).

Comment: This idea was first proposed in the 1880s by Weismann, de Vries, and others and later promoted by Muller in the 1940s (Mayr 1982, 774).

mosaic inheritance theory See principle: principle of mosaic evolution.

mutation theory See mutationism.

Neo-Darwinian theory See evolutionary synthesis.

neutral theory, the *n.*
1. The hypothesis that "most of the mutant genes that are detected only by the chemical techniques of molecular genetics are selectively neutral; that is, they are adaptively neither more or less advantageous than the genes they replace; at the molecular level most evolutionary changes are caused by the 'random drift' of selectively equivalent mutant genes" Kimura (1978, 98).

Note: This idea can be traced to Kimura (1968) as indicated in Kimura 1978, 98.
2. The hypothesis "that 'most' of the observed sequence variation in DNA and proteins both within and between species is due to the random fixation of nearly neutral alleles by genetic drift" (Gillespie 1984).

Note: Neutral genes are presumably eliminated from populations by chance, not natural selection (Stebbins and Ayala 1985, 76).
3. The hypothesis "that the great majority of evolutionary changes at the molecular (DNA) level do not result from Darwinian natural selection acting on advantageous mutants but rather from random fixation of selectively neutral or very nearly neutral mutants through random genetic drift, which is caused by random sampling of gametes in finite populations" (Kimura 1992, 225).

Note: Selectively neutral means selectively equivalent — neither advantageous nor disadvantageous (Kimura 1992, 225).

syn. neutral gene theory of molecular evolution (King and Stansfield 1985), neutral mutation hypothesis (Li and Tanimura 1987, 93), neutral mutation random drift hypothesis (Kimura 1985, 42), neutral theory of molecular evolution (Kimura 1979; Kimura 1987, 24), neutralism (Curtis 1983, 905; Kimura 1992), non-Darwinian evolution (King and Jukes 1969, Crow and Kimura 1970 in Mayr 1982, 593, who says this is a misleading term), random-walk evolution (Mayr 1982, 593)

cf. drift: genetic drift; [2]evolution: non-adaptive evolution, non-Darwinian evolution; [4]model: stepping-stone model; random walk; selectionist view; [3]theory: non-Darwinian evolution

Comment: Motoo Kimura was in the first group of biologists to embrace the "neutral theory" in the late 1960s and has been its main architect in its present form. This hypothesis is now invoked as routinely as natural selection was a few years back to account for evolutionary DNA changes because a large number of DNA sequences have been published (Gillespie 1984, 732–733). Kimura (1991) summarizes data that support the "neutral theory."

[coined as the "neutral theory" and "neutral mutation random drift hypothesis" (Kimura 1985, 41–42)]

non-Darwinian evolution *n.* A theory of evolution different from that of Charles R. Darwin's theory, *e.g.*, (1) random-walk evolution, (2) Lamarckism,

S–Z

(3) mutationism, and (4) orthogenesis (Mayr 1982, 593).

Note: The term "non-Darwinian evolution" is confusing because different authors use it to refer to one or more of the above concepts of evolution.

See ²evolution: non-Darwinian evolution.

optimal-diet theory *n.* The concept that predators select prey types in a manner that maximizes their net rate of energy intake (Lincoln et al. 1985).

optimal-foraging theory (OFT) *n.* The concept that natural selection favors a strategy in which a predator, or other kind of feeder, utilizes food in a manner that optimizes its net energy gain per unit feeding time (Lincoln et al. 1985).

cf. ⁴model: optimality model

optimality theory, optimization theory *n.*

1. The hypothesis that an organism should optimize (maximize its fitness or correlated characteristics) as much as possible given its genetic and environmental constraints (Wilson 1975, 135; Barash 1981, 47–63).

   syn. compromises of evolution (Mayr 1982, 589)

2. A theory that predicts that natural selection should result in optimal traits, those with benefit-to-cost ratios that are greater than those associated with alternative traits that have arisen over the history of a particular species (Thornhill and Alcock 1983, 16).

Note: Many researchers indicate that optimality theory does not hypothesize, or predict, optimization (D.L. Kramer, personal communication). Rather, a researcher assumes optimization as a basis for predicting particular traits or tactics. Critics, not proponents, of optimality theory often assert that it predicts optimization.

cf. function: cost function, goal function

Comments: An "optimal trait" is not necessarily the one that gives the highest possible benefit-to-cost ratio. Such a ratio is unlikely to be obtained due to numerous constraints on organisms. Gross (1984, 56–57) explains the cost-benefit function in optimality theory.

organismal theory, organismal theory of multicellularity *n.* A theory that interprets the developmental relationship of cells to organisms and has the principal tenets: (1) ontogenesis is a function primarily of an organism as a whole, and consists in the growth and progressive internal differentiation of a single protoplasmic individual which often, but not always, involves the septation of the living mass into subordinate semi-independent parts — the cells; (2) because septation is rarely complete, all parts remain connected, and the whole continues to act as a unit; (3) development is not primarily the establishment of an association of multiplying elementary units to form a new whole, but rather the resolution of one whole into newly formed parts; and (4) development should be thought of not as a multiplication and cooperation of cells, but rather as a differentiation of protoplasm (Sharp 1926 in Kaplan and Hagemann 1991, 694).

cf. ³theory: cell theory

Comment: The organismal theory considers unicellular protozoa to be homologous not with individual cells of an organism but to an organism as a whole (Kaplan and Hagemann 1991, 694).

organismic theory *n.* The theory that rejects human mind-body dualism and conceives a Human as a psychobiological entity endowed with certain capacities (Storz 1973, 187).

outflow theory *n.* An animal's brain distinguishes between the movement of its retinal images that are independent of the animal and movement due to its eyeballs, because instructions to its eye muscles that move its eyeballs are accompanied by parallel signals to a comparator in its brain (von Hemholtz 1867 in McFarland 1985, 249).

cf. principle: reafference principle; ³theory: inflow theory

pangenesis, panspermy See ²theory: theory of pangenesis.

Pavlov's reflex theory See ³theory: chain-reflex theory.

POC theory See ³conflict: parent-offspring conflict (comment).

reflex theory See ³theory: chain-reflex theory.

selection theory See selection: natural selection.

Sewall-Wright's theorem *n.* In the absence of any other evolutionary force besides drift, fixation and loss each proceed at a rate of about $1/(4N)$ per locus per generation, where N is population size (Wilson 1975, 65).

signal-detection theory *n.* The concept that a person's sensitivity to a stimulus depends not only upon stimulus intensity, but also upon his experience, expectation, and motivation (Storz 1973, 255).

somatic-mutation theory *n.* The hypothesis that aging is the result of an accumulation of mutations in somatic tissue (Lincoln et al. 1985).

stigmergy theory *n.* The hypothesis that, with regard to social insects, once a nest builder has identified what construction step its nest (or nest section) is in, it "needs" only to make one decision about whether to continue building as before or to switch to the next step of construction (Grassé 1959 in Downing and Jeanne 1988, 1729).
Comment: Solitary nest-building insects support this hypothesis, but it requires modification for a paper-wasp species (Downing and Jeanne 1988, 1729).

stochastic theory of mass behavior *n.* The theory that transition probabilities in individual social insects' behavior are programmed to produce optimal response of their colonies, these probabilities are determined by selection at the colony level, and they represent a sensitive adaptation to particular environmental conditions in which their species have existed during recent evolutionary time (Hölldobler and Wilson 1990, 643).

Sutton-Boveri chromosome theory, Sutton-Boveri chromosome theory of inheritance *n.* The hypothesis that genes are located on chromosomes, each of which has its particular set of genes (Boveri 1891; Sutton 1902 in Mayr 1982, 749).
syn. chromosomal theory of heredity (Mayr 1982, 749)
Comment: This hypothesis has become widely accepted since about 1930.
[Sutton's professor Wilson named this hypothesis in 1928 (Mayr 1982, 749).]

synthetic theory See evolutionary synthesis.

target theory See ³theory: hit theory.

tetranucleotide theory of the structure of DNA *n.* The hypothesis that a DNA molecule is composed of four nucleotides and has a total molecular weight of about 1500 (Mayr 1982, 815, 819).
cf. DNA

theory of biphasic processes See ³theory: approach-withdrawal theory.

theory of continuity of germ plasm *n.* Weissmann's (1885) hypothesis that the "germ track" is separate from the body (soma) track from the very beginning and thus acquired somatic characters are not transmitted to genetic material (Mayr 1982, 700).
cf. Lamarckism

theory of evolution *n.*
1. Charles R. Darwin's original statement of how (organic) evolution, *q.v.*, occurs (Stebbins and Ayala 1985, 72).
2. The accepted hypothesis that organismic evolution has occurred (Brownlee 1981).
cf. Darwinism: Darwin's own theory

3. "A modified, expanded version of Darwinism that took shape in the 1930s and 1940s" (Stebbins and Ayala 1985, 72).
Note: This version was first known as "Neo-Darwinian theory" (= "Neo-Darwinism") and later the "synthetic theory of evolution," *q.v.* (Stebbins and Ayala 1985, 72).
4. "A body of interconnected statements about natural selection and the other processes that are thought to cause evolution" (Futuyma 1986, 15).
syn. encasement theory (obsolete; *Oxford English Dictionary* 1972)
cf. ²evolution; ³theory: encasement theory

theory of games See ³theory: game theory.

theory of genetic equivalence of all chromosomes *n.* The false hypothesis that every chromosome in an organism's genome has the same effects on all given traits (Mayr 1982, 751).
Comment: McClung's (1901) finding sex chromosome's refuted this hypothesis (Mayr 1982, 751).

theory of island biogeography, MacArthur-Wilson theory of island biogeography *n.* The concept that the number of species that inhabit an island is a function of the island's area and distance from the mainland and is determined by the relationship between immigration (greater on larger and nearer islands) and extinction (greater on smaller islands) (Lincoln et al. 1985).
cf. curve: species-area curve; hypothesis: nonequilibrium hypothesis

theory of monistic evolution *n.* The view that evolutionary changes can be explained by single factors or events (Lincoln et al. 1985).
cf. evolutionary synthesis

theory of pangenesis *n.*
1. The hypothesis that all parts of an organism's body contribute genetic material (gemmules) to its reproductive organs, and particularly to its gametes (Mayr 1982, 959).
Note: This idea dates back to Anaxagoras (*ca.* 500–428 BC) and "had representatives at least to the end of the 19th century, including Charles R. Darwin." This hypothesis also holds that there is an alternation between the "formation of the body (phenotype, soma) and through it the formation of seed stuff (sperm; genotype) which then directly through growth is converted again into the body of the next generation... ." This concept was essentially maintained until first challenged by Galton and

S—Z

Weismann in the 1870s and 1880s (Mayr 1982, 635–636).

2. A false hypothesis that somatic cells contain particles (gemmules) that can be influenced by an organism's environment and activity of organs that contain them; these particles can move to sex cells to influence the course of heredity (Darwin 1868; Lincoln et al. 1985).

syn. Darwin's theory of gemmules, gemmule theory, pangenesis, panspermy (Lincoln et al. 1985)

cf. hypothesis: transportation hypothesis

theory of parent-offspring conflict, POC theory See ³conflict: parent-offspring conflict (comment).

theory of particulate inheritance See ³theory: Mendel's theory of particulate inheritance.

theory of plate tectonics *n.* The hypothesis that continents sit on a series of plates, and slow, convective flows within Earth's mantle force neighboring plates apart and carry the continents piggyback (Lightman and Gingerich 1992, 692).

cf. hypothesis: continental drift

Comments: This hypothesis, proposed in this form in the 1960s, is now widely accepted by scientists. In 1756, Theodor Christoph Lilienthal concluded that Earth's surface tore apart after the Biblical flood based on the shape of coastlines; around 1800, Alexander von Humboldt proposed that lands bordering the Atlantic were once joined; in 1838, Thomas Dick mentioned the possibility that continents moved apart; in 1858, Antonio Snider-Pellegrini claimed that continents were once joined based on fossil evidence; in 1881, Reverend Osmond Fisher published a mechanism to explain the good fit of continents; in 1908, Frank Bursley Taylor proposed that the moon came so close to the Earth that its gravity dragged continents toward the Equator; in 1912, Alfred Wegener argued for an ancient continuity of landmasses based on geological, fossil, and other evidence (Sullivan 1974, 2–3; Goodacre 1991, 261; Lightman and Gingerich 1992, 692).

theory of preformation See ³theory: encasement theory.

theory of pure blending inheritance, exclusively blending inheritance *n.*
A now-refuted hypothesis, subscribed to by Nägeli and a few other biologists, that during fertilization maternal and paternal idioplasms blend, owing to the fusing of homologous strings of micelles into a single strand (Mayr 1982, 723, 780).

theory of symbiosis See hypothesis: serial-endosymbiosis theory.

theory of synthetic evolution See evolutionary synthesis.

theory of volition See hypothesis: volition.

transmutation, theory of transmutation *n.* The hypothesis that all animals have developed from a common original form (Mayr 1982, 388).

tropism theory, Loeb's theory, Loeb's-tropism theory *n.* The hypothesis that all behavior can be explained as a sequence of tropisms (taxes) (Tinbergen 1951, 101).

syn. theory of tropisms

cf. ³theory: chain-reflex theory

Comment: This hypothesis appears to be a "grotesque simplification" (Tinbergen 1951, 101).

uniformitarianism *n.*

1. The doctrine, or principle, held by the uniformian school of geologists; the theory of uniformity in the action and processes of inorganic nature; opposed to catastrophism (*Oxford English Dictionary* 1972, entries from 1865).

2. The doctrine, or hypothesis, that the evolution of the Earth and life on it are affected by secondary natural causes and possibly original direct divine intervention or occasional divine interventions; the same causes have operated throughout the history of the Earth; these causes have always operated at the same intensity; configurational causes have been the same during all geological periods; many changes were gradual, but some were rather cataclysmic; and changes did not occur in a directional manner, instead the world always changed in a steady-state condition, at most changing cyclically (Mayr 1982, 365, 375–378).

3. The hypothesis that natural geological processes that affect the Earth are still operating at essentially the same rate and intensity as they have throughout geological time (Lincoln et al. 1985).

syn. actualism; principle of uniformity (Lincoln et al. 1985)

cf. catastrophism, naturalism, special creation

[coined by Whewell (1832) according to Mayr (1982)]

▶ **progressive uniformitarianism** *n.*
Uniformitarianism that involves organisms' changing toward a goal of increased perfection (Lamarck in Mayr 1982, 358).

Weismann's dissection theory *n.*
Weismann's (1892) now-refuted hypothesis

that, during cell division, daughter cells may receive different kinds and numbers of biophores (particles that relate to particular traits) and so forth (Mayr 1982, 703).

Weismann's genetic theory *n.* Weismann's (1883, 1885) hypothesis that includes two dominant insights: all genetic material is contained in a cell's nucleus and acquired characters are not inherited in any form (Mayr 1982, 699).

Comment: The second part of this hypothesis was vigorously attacked by neo-Lamarckians and orthodox Darwinians (Mayr 1982, 701).

"Wilson's caste theorem" *n.* Insect castes tend to proliferate in evolution until there is one for each task in the colony (Wilson 1975, 399).

Zuckerman's theory *n.* The hypothesis that binding force of primate sociality is sexual attraction (Zuckerman 1932 in Wilson 1975, 514).

Comment: This hypothesis dominated thought in primate sociobiology from about 1932 to 1959 and is no longer supported (Wilson 1975, 514).

♦ **therium** See ²succession: therium.

♦ **-therm, thermo** *combining form* "Heat; of, related to, or caused by heat" (Michaelis 1963).

adj. -thermal, -thermic

cf. -cryo

[Greek *therme, thermos,* heat, warmth]

allotherm See -therm: poikilotherm.

autotherm See -therm: homeotherm.

cold-blooded animal See -therm: poikilotherm.

ectotherm *n.* An animal whose body temperature depends on its behaviorally and autonomically regulated heat uptake from its environment; contrasted with endotherm (inferred from "ectothermy" in Bligh and Johnson 1973, 946).

See -therm: poikilotherm.

[Greek *ektos,* outside; *therme,* heat]

endotherm *n.* An animal whose body temperature depends on its high (= tackymetabolic) and controlled rate of heat production; contrasted with ectotherm (inferred from "endothermy" in Bligh and Johnson 1973, 946).

See -therm: homeotherm, poikilotherm.

[Greek *ektos,* outside; *therme,* heat]

eurytherm *n.* An organism that is tolerant of a wide range of ambient temperatures; contrasted with stenotherm (Lincoln et al. 1985).

cf. ¹-phile: thermophile: eurythermophile

[Greek *eurus,* wide; *therme,* heat]

haematotherm See -therm: homeotherm.

hekistotherm *n.*

1. A plant that requires a temperature of less than 10°C in the warmest month of the year and typically occurs in regions with a mean annual temperature below 0°C (Lincoln et al. 1985).

2. An organism that lives above the tree line in heavy snow (Lincoln et al. 1985).

heliotherm *n.* An ectothermic animal that regulates its core temperature by changing its exposure to the Sun (inferred from "heliothermy" in Bligh and Johnson 1973, 948).

[Greek *helios,* sun; *therme,* heat]

heterotherm *n.*

1. A tachymetabolic animal whose core-temperature variation, either nychthermerally or seasonally, exceeds that which defines homeothermy (inferred from heterothermy in Bligh and Johnson 1973, 948).

2. An animal that can appreciably elevate its body temperature metabolically at certain times but also allows its body temperature to fluctuate with the environmental temperature (*e.g.,* Honey Bees, bumble bees, or some hawkmoth species) (Heinrich 1979, 39, 1990, 55).

cf. -therm: homeotherm, poikilotherm

Comments: "Heterotherm" is an arbitrary term in two senses: (1) "hetero" means "different from" rather than "a wide range," and (2) the distinction that is being made between thermostable and rather less thermally stable species depends on an arbitrary division (Bligh and Johnson 1973, 948). Some plants, including many aroids species and a lotus species, raise their inflorescence temperatures by metabolic heating (Darnton 1996; Raskin et al. 1987; Skubatz et al. 1990). This heating increases floral odors, warms pollinators, or both, depending on the species.

[Greek *hetero,* different; *therme,* heat]

homeotherm, homotherm, homoeotherm, homoiotherm *n.* A tachy-metabolic animal that maintains its core temperature, either nychthemerally or seasonally, within arbitrarily defined limits (±2°C), when it is active, despite much larger variations in ambient temperature; contrasted with poikilotherm (Bligh and Johnson 1973, 958).

syn. autotherm, endotherm, haematotherm, idiotherm, warm-blooded animal (Lincoln et al. 1985; a poor and misleading synonym, according to Curis 1983, 736), temperature regulator (Bligh and Johnson 1973, 954)

Comments: Birds and mammals are homeotherms. Some species (*e.g.,* hummingbirds and bears) let their temperatures drop to save energy. Researchers debate whether dinosaurs were homeotherms (Morell 1996a).

[Greek *homoio*, like, similar; *therme*, heat]

hydromegatherm *n.* An organism that inhabits a warm, wet environment (*e.g.,* a lowland, tropical rain forest) (Lincoln et al. 1985).

idiotherm See -therm: homeotherm.

megatherm, macrotherm, macro-thermophyte, megistotherm *n.* A plant that lives in warm habitats and requires a minimum temperature of 18°C in the coldest month (Lincoln et al. 1985).

mesotherm, mesothermophyte *n.* A plant that lives under intermediate temperatures with a minimum of 22°C in the warmest month and a range from 6 through 18°C in the coldest month (Lincoln et al. 1985).

microtherm *n.* A plant that lives in a relatively cold habitat with a minimum temperature of 6°C in the coldest month and a range from 10 through 22°C in the warmest month (Lincoln et al. 1985).

oligotherm *n.* An organism that tolerates relatively low temperatures (Lincoln et al. 1985).

cf. -therm: polytherm

poikilotherm [PWAH kill low therm] *n.* An animal whose core temperature fluctuates with environmental temperature; that is, it is proportional to ambient temperature; contrasted with homeotherm (= temperature regulator) (Bligh and Johnson 1973, 952; Curtis 1983, 736; Lincoln et al. 1985).

adj. ectothermal, haematocryal, poikilo-thermic (Lincoln et al. 1985)

syn. cold-blooded animal (a poor and misleading synonym; Curtis 1983, 736), ectotherm, heterotherm (in some cases; Lincoln et al. 1985); temperature conformer (the preferred term for this phenomenon according to Bligh and Johnson 1973, 952, who discuss why "cold-blooded" and "warm-blooded" are confusing terms).

Comments: Fish, amphibians, terrestrial reptiles, and most invertebrates are poikilotherms (Curtis 1983, 736–737). "Poikilotherm" should, perhaps, be "poecilotherm" or "pecilotherm" to conform with other uses of this root in biology (Bligh and Johnson 1973, 952).

[Greek *poikilos*, changeable, diversified, variegated; *therme*, heat]

polytherm *n.* An organism that tolerates relatively high temperatures (Lincoln et al. 1985).

cf. -therm: oligotherm

stenotherm *n.*

1. An organism that occurs naturally within a narrow range of environmental temperatures and which, singly or collectively, is intolerant of, or accommodates ineffectually, wide changes in its usual thermal environment; contrasted with eurytherm (inferred from "stenothermy" in Bligh and Johnson 1973, 954).

2. An organism that tolerates only a narrow range of environmental temperatures; contrasted with eurytherm (Lincoln et al. 1985).

[Greek *stenos*, narrow; *therme*, heat]

warm-blooded animal See -therm: homeotherm.

♦ **thermal homeostasis** See homeostasis: thermal homeostasis.

♦ **thermal strain** *n.* Any change in an organism's physiological state caused by thermal stress, *q.v.* (Bligh and Johnson 1973, 957).

♦ **thermal stress** See ²strees: thermal stress.

♦ **thermal sweating** See sweating: thermal sweating.

♦ **thermatology** See study of: thermatology.

♦ **-thermia** *combining form* "Heat; of, related to, or caused by heat" (Michaelis 1963).

cf. -therm, -thermy

[Greek *therme*, *thermos*, heat, warmth]

hyperthermia *n.* A temperature-regulating animal's condition when its core temperature is more than 1 standard deviation above its species' mean temperature in resting conditions in a thermoneutral environment; contrasted with cenothermy (Bligh and Johnson 1973, 948).

[Greek *hyper*, over; *therme*, heat]

hypothermia *n.* A temperature-regulating animal's condition when its core temperature is more than 1 standard deviation below its species' mean temperature in resting conditions in a thermoneutral environment; contrasted with cenothermy (Bligh and Johnson 1973, 948).

[Greek *hypo*, below; *therme*, heat]

♦ **thermium** See ²community: thermium.

♦ **thermodynamics** See study of: thermodynamics.

♦ **thermogenesis** *n.* An organism's producing heat (*e.g.,* in birds, mammals, and the Voodoo-lily).

nonshivering obligatory thermo-genesis (NST(O)) *n.* The component of nonshivering thermogenesis that is independent of short-term ambient-temperature changes (Bligh and Johnson 1973, 957).

Comments: NST(O) corresponds to basal metabolic rate or standard metabolic rate. NST(O) is unaffected by short-term exposure to cold; however, acclimatization to sustained cold, or heat, stress, may change NST(O) (Bligh and Johnson 1973, 957).

nonshivering thermogenesis (NST) *n.*

1. An animal's increasing its heat-production rate during cold exposure due to processes that do not involve contractions of voluntary muscles; contrasted with shivering thermogenesis (Bligh and Johnson 1973, 957).
2. Nonshivering thermoregulatory thermogenesis, *q.v.*

Note: "Nonshivering thermogenesis" usually refers to "nonshivering thermoregulatory thermogenesis" (Bligh and Johnson 1973, 957).

nonshivering thermoregulatory thermogenesis (NST(T)) *n.* An animal's increase in nonshivering thermogenesis which occurs when it is acutely exposed to cold (Bligh and Johnson 1973, 957).

syn. nonshivering thermogenesis (a confusing synonym)

Comments: NST(O) corresponds to basal metabolic rate or standard metabolic rate. NST(O) is unaffected by short-term exposure to cold; however, acclimatization to sustained cold, or heat, stress, may change NST(O) (Bligh and Johnson 1973, 957).

shivering thermogenesis *n.*

1. An animal's increasing its heat-production rate during cold exposure by contracting its voluntary muscles; contrasted with nonshivering thermogenesis (Bligh and Johnson 1973, 957).

Comments: Shivering thermogenesis progresses, as its intensity increases, from thermoregulatory muscle tone, to microvibrations, to clonic contractions of both flexor and extensor muscles (Bligh and Johnson 1973, 958).

♦ **thermoluminescence (TL)** See method: dating method: thermoluminescence.

♦ **thermoneutral zone (TNZ)** See zone: thermoneutral zone.

♦ **thermophile** See [1]-phile: thermophile.

♦ **thermophobe** See -phobe: thermophobe.

♦ **thermopreferendum** *n.* The thermal conditions that an individual organism, or a species, selects for its ambient environment in natural, or experimental, circumstances (Bligh and Johnson 1973, 958).

♦ **thermoreception** See modality: thermoreception.

♦ **thermoreceptor** See [2]receptor: thermoreceptor.

♦ **thermoregulation hypothesis of insect wing origin** See hypothesis: hypotheses of origin of insect wings: thermoregulation hypothesis of insect wing origin.

♦ **thermotropism** See -tropism: thermotropism.

♦ **-thermy** *combining form* "Heat; of, related to, or caused by heat" (Michaelis 1963).

cf. -therm.

Comments: "Thermy" refers to the temperature condition of many of the kinds of "-therms" given above (*e.g.,* heterothermy, homeothermy, poikilothermy). Below, I give terms that specifically relate to a temperature condition, not a kind of "-therm."

[Greek *therme, thermos,* heat, warmth]

cenothermy, euthermy, normothermy *n.* A temperature-regulating animal's condition of being within ±1 SD of the mean core temperature of its species measured under resting conditions in a thermoneutral environment (Bligh and Johnson 1973, 944).

[Greek *ceno,* common; *therme,* heat]

cryothermy *n.* A supercooled organism's thermal status (Bligh and Johnson 1973, 945).

Comment: The temperature of a supercooled organism's body mass is below the freezing point of its tissue (Bligh and Johnson 1973, 945).

[Greek *kruos,* cold; *therme,* heat]

♦ **therodrymium** See [2]community: therodrymium.

♦ **-thesia** *combining form*

alliesthesia *n.*

1. An animal's changed sensation for a given peripheral stimulus that results from stimulation of an internal sensor(s) (Bligh and Johnson 1973, 943).
2. An animal's thermal sensation's dependence on both its core and skin temperatures (Bligh and Johnson 1973, 943).

[Greek *alloioo,* to alter; *aisthesia,* sensation]

chromesthesia, colored hearing *n.* Sound's evoking a person's images of colors (Storz 1973, 272).

kinesthesia See modality: kinesthesia.

synthesia *n.* A person's producing a subjective sensation of a modality, other than the one that is stimulated, by linking his perceptions of his stimulated modality with images from another modality (*e.g.,* chromesthesia, *q.v.*) (Storz 1973, 272).

♦ **theta waves** See brain waves: theta waves.

♦ **thicket** See [2]community: thicket, habitat: thicket.

S–Z

♦ **thief** *n., pl.* **thieves** In some bee, other insect, and anthophilous-bird species: an individual that takes nectar, pollen, or both, through holes in a flower corolla that another animal made (Barrows 1976c, 132; Inouye 1980).

syn. secondary robber (Morse 1978, 191)

cf. dysteleology, robber

♦ **thinicole** See -cole: thinicole.

♦ **thinking** *n.*

1. A person's forming (something) in his mind; a person's conceiving (a thought, etc.); a person's having a notion, idea, etc., in his mind (*Oxford English Dictionary* 1972, verb entries from *ca.* 888).

2. A person's exercising his mind, especially understanding, in any active way; a person's forming connected ideas of any kind; a person's having, or making, a train of ideas pass through his mind; meditating, cogitating (*Oxford English Dictionary* 1972, verb entries from *ca.* 1000).

3. A person's calling to mind; considering, reflecting upon, remembering, bearing in mind (*Oxford English Dictionary* 1972, verb entries from *ca.* 1020).

4. A person's forming, or having, an idea of (a real, or imaginary, thing, action, or circumstance) in his mind; imagining, conceiving, fancying, picturing (*Oxford English Dictionary* 1972, verb entries from *ca.* 1200).

5. A person's thinking of himself of something in the way of a plan or purpose; finding out, or hitting upon, a way to do something by mental effort; contriving, devising, planning, plotting (*Oxford English Dictionary* 1972, verb entries from *ca.* 1330).

6. A person's conceiving, or entertaining, the notion of doing something; meditating, contemplating, intending, purposing, meaning, having (in mind), having thoughts (of) (*Oxford English Dictionary* 1972, verb entries from Beowulf).

7. A person's behavior from which one can infer that he is using symbolic representations of events and objects and manipulating ideas, images, or concepts, or a combination of these things (Storz 1973, 276).

8. In more derived vertebrate species, some more derived invertebrate species: inferential reasoning (Immelmann and Beer 1989, 311).

See behavior: symbolic behavior.

cf. cognition, intelligence

Comment: The *Oxford English Dictionary* (1972) gives many more definitions of thinking (in verb form).

♦ **thinophile** See [1]-phile: thinophile.

♦ **thiophile** See [1]-phile: thiophile.

♦ **thirst drive** See drive: thirst drive.

♦ **Thomism** *n.* The system of dogmatic theology of St. Thomas Aquinas and his followers that formed the basis of 13th-century scholasticism (Michaelis 1963).

Comment: This was the dominant philosophy of "scholasticism" (Mayr 1982, 92).

♦ **threat, threat behavior** See behavior: threat.

♦ **threat face** See facial expression: threat face.

♦ **threat gape** See facial expression: threat yawn.

♦ **threat greeting** See greeting: threat greeting.

♦ **threat mimic** See display: bared-teeth display.

♦ **threat yawn** See facial expression: threat yawn.

♦ **threatened** *adj.* Referring to a species, or population, that could soon face extinction if not monitored and saved (Cooper 1984 in Norden et al. 1984, 9).

cf. endangered, extant, extinct, extirpated

♦ **Three R's** *n.* Reduction, refinement, and replacement of animals used in education, research, and testing (Roush 1996, Science 274, 168).

Comment: Implementing the Three R's is increasing in some parts of the world (Roush 1996, Science 274, 168).

♦ **"3ST" method** See method: "3ST" method.

♦ **thremmatology** See study of: thremmatology.

♦ **threshold** *n.*

1. The minimum strength of a stimulus necessary to elicit an animal's response (Immelmann and Beer 1989, 312).

2. An animal's lowest perceivable level of a stimulus (*e.g.,* the threshold of a specific tone) (Immelmann and Beer 1989, 312).

syn. stimulus threshold

cf. [2]stimulus: subthreshold stimulus

♦ **threshold selection** See selection (table).

♦ **throughfall** *n.*

1. Precipitation that is first intercepted by vegetation and then falls to the ground (Allaby 1994).

2. The water that drips, or falls, through a tree canopy and reaches the forest floor.

Comment: Most throughfall sinks into soil and is used by plants and becomes stream flow and groundwater.

♦ **thwarting** *n.*

1. Something's successfully opposing; preventing (a person, etc.) from accomplishing a purpose; preventing the accomplishment of (a purpose); foiling, frustrating, balking, defeating (*Oxford English Dictionary* 1972, verb entries from 1581).

2. Cessation, or frustration, of an individual's ongoing behavior due to the animal's not receiving a requisite stimulus (Immelmann and Beer 1989, 313).

3. Failure of an individual's behavior to achieve the stimulus that is necessary for the completion of its activated behavior sequence (Immelmann and Beer 1989, 313).

♦ **thymosin** See hormone: thymosin.

♦ **thyrotropic hormone** See hormone: thyrotropic hormone.

♦ **Ti plasmid** See plasmid: Ti plasmid.

♦ **tic** *n.* A persistent muscle movement, such as, in Humans, mouth twitching, lip licking, eye blinking, throat clearing, face making, neck turning, or shoulder shrugging (Storz 1973, 278).

Comment: A person with a tic might not be aware of it (Storz 1973, 278).

♦ **tidal freshwater marsh** See habitat: marsh: tidal freshwater marsh.

♦ **tidal rhythm** See rhythm: tidal rhythm.

♦ **tidbitting** See display: tidbitting.

♦ **tie** *n.* In some mammalian carnivore species: joining of a male and female by their sex organs for up to an hour, or more, after copulation (Immelmann and Beer 1989, 61).

cf. copulation: postcopulatory behavior

♦ **tiered hierarchy of social groupings** *n.* For example, in elephants: a system of organization comprised of individuals, family units, and kinship groups (Wilson 1975, 494).

♦ **Tiglon, Tigon** See hybrid: Tigon.

♦ **time** *n.*

1. A limited stretch, or space, of continued existence, as the interval between two successive events or acts; or the period through which an action, condition, or state continues; a finite portion of "time" (in its infinite sense): a long time, a short time, some time, for a time (*Oxford English Dictionary* 1972, entries from *ca.* 893); "the general concept, relation, or fact of continuous existence, capable of division into measurable portions, and comprising the past, present, and future" (Michaelis 1963).

2. A period of duration (*Oxford English Dictionary* 1972, entries from *ca.* 1000); a definite portion of duration; especially, a definite, specific, or appointed moment, hour, day, season, year, etc.: "The time is 3:33 p.m.," or "Autumn is my favorite time" (Michaelis 1963).

ecological time *n.* A period from about 10^5 to 10^9 seconds during which large sequences of organismic responses occur but during which extensive evolutionary changes usually do not occur within a species (Wilson 1975, 145).

evolutionary time *n.* A period approximately longer than 10^6 seconds during which extensive evolutionary change occurs (Wilson 1975, 145).

generation time *n.*

1. The mean duration of an organism's life cycle from its birth to its reproduction (Lincoln et al. 1985).

2. The average period from the reproduction of an organism's parental generation to reproduction of its first filial generation (Lincoln et al. 1985).

presentation time *n.* The minimum time that an organism takes to respond to a stimulus after it is presented with the stimulus (Lincoln et al. 1985).

reaction time (RT), latent period, utilization time *n.* The time between a stimulus' impinging on an organism and its response to the stimulus (Lincoln et al. 1985).

cf. period: refractory period

DURATIONS OF KINDS OF BIOLOGICAL TIME[a]

KIND OF BIOLOGICAL TIME	APPROXIMATE DURATION (SECONDS)
biochemical-response time	up to 10^3
instinctive-and-reflex-response time	$10^{0.5}$–10^5
learning-response time	$10^{0.7}$–$10^{5.5}$
hormonal-response time	$10^{1.5}$–10^6
reproductive-life-cycle-response time	10^2–$10^{7.5}$
morphogenetic-response time	$10^{2.5}$–10^8
demographic-response time (*syn.* ecological time)	10^5–10^9
evolutionary time	$>10^6$
species-replacement time	$>10^{10}$
species-group-replacement time	$>10^{13}$

[a] Wilson (1975, 145).

S–Z

residence time *n*. The time that a given substance remains within a system (Lincoln et al. 1985).

utilization time See time: reaction time.

♦ **time budget** See budget: time budget.

♦ **time-energy budget** *n*. A summary of the amounts of time and energy allotted by an animal to its various activities (Wilson 1975, 142, 597).

♦ **time minimizer** See ²species: time minimizer.

♦ **time-sampling** See sampling technique: instantaneous sampling.

♦ **time sense** See clock: biological clock.

♦ **time-sharing** *n*. The phenomenon in which the onset and duration of an activity B are under the control of factors normally controlling activity A, in which B occurs due to disinhibition by A, and after a period A reestablishes itself by competition (McFarland 1974 in McFarland 1985, 464; Immelmann and Beer 1989, 314).
cf. experiment: masking experiment

♦ **time-stability hypothesis** See hypothesis: stability-time hypothesis.

♦ **time's arrow** See law: laws of thermodynamics: second law of thermodynamics.

♦ **Tinbergian research program** See program: Tinbergian research program.

♦ **Tinbergen's hierarchical model, Tinbergen's hierarchical model of behavior regulation** See model: Tinbergen's hierarchical model.

♦ **tiphicole** See -cole: tiphicole.

♦ **tiphium** See ²community: tiphium.

♦ **tiphophile** See ¹-phile: tiphophile.

♦ **tissue** *n*.
1. The substance, structure, or texture of which an organism's body or any part, or organ, of it is composed; especially, any one of various structures, each composed of similar cells or modifications of cells, that make up the organism (*Oxford English Dictionary* 1972, entries from 1831).
2. In multicellular organisms: a mass, or sheet, of cells of the same, or similar, histology (*e.g.*, muscle tissue, skeletal tissue, epithelial tissue, or glandular tissue) that form a part of an organ (Immelmann and Beer 1989, 314).

♦ **tit for tat** *n*. "Retaliation in kind; blow for blow" (Michaelis 1963). See game: tit-for-tat game.
[Possibly an alternative of tip for tap; < Middle French *tant pour tant*, tit for tat]

♦ **tit-for-tat game** See game (table).

♦ **titration method** See method: titration method.

♦ **TL** See method: dating method: thermoluminescence.

♦ **TNZ** See zone: thermoneutral zone.

♦ **tobacco juice** See enteric discharge.

♦ **tobogganing** See sliding.

♦ **tocotropism** See -tropism: tocotropism.

♦ **toe clip** *v.t.* To cut one or more toes off a research animal so that it can later be identified in laboratory cultures or in the field (Townsend et al. 1984, 422).
Comment: Research animals that are toe-clipped include amphibians (Townsend et al. 1984, 422), lizards, and rodents.

♦ **toft** See ²group: toft.

♦ **tok** See ²group: tok.

♦ **tokogenetic relationship** *n*. The genetic relationship between individuals (Lincoln et al. 1985).

♦ **-toky** *combining form*
adj. -otokous
See -parity.
cf. atokous

arrhenotoky See parthenogenesis: arrhenotoky.

ditotoky *n*. An organism's production of two offspring per brood (Lincoln et al. 1985).

epitoky *n*. In some species of polychaete worms: formation of a reproductive individual (epitoke) that differs in a varying number of secondary sexual characteristics from a nonsexual form (atoke) (Barnes 1974, 275).

gametoky See parthenogenesis: amphitoky.

monotoky *n*.
1. An organism's having only one offspring per brood (*e.g.*, in tsetse flies) (Lincoln et al. 1985).
2. A plant's fruiting only once during its life cycle (*e.g.*, in *Agave* species or some tropical-tree species) (Lincoln et al. 1985).
syn. monocarpy, uniparity, monotocy (Lincoln et al. 1985)
cf. parity: semelparity

oligotoky *n*. An organism's having only a few offspring per brood (Lincoln et al. 1985).
adj. oligotokous, oligotocous (Lincoln et al. 1985)

polytoky *n*.
1. An organism's having many offspring per brood (Lincoln et al. 1985).
2. A plant's fruiting many times during its life cycle (Lincoln et al. 1985).
syn. caulocarpy, multiparity, olytocy (Lincoln et al. 1985)

♦ **TOL** See Database: Tree of Life.

♦ **tolerated ambient temperature range** See ²range: tolerated ambient temperature range.

♦ **tomiparity** See parity: tomiparity.

♦ **tongue flicking** *n*.
1. A person's brief extension of his tongue, often accompanied by licking movements,

occurring during flirtation, and that may represent ritualized licking (Heymer 1977, 206).

2. An individual reptile's extending and vibrating its tongue which, *e.g.,* in lizards, transfers chemicals from an individual's external environment to its vomeronasal organ (Graves and Halpern 1989 in Graves and Halpern 1990, 692).

♦ **tongue-lashing behavior** See liqualation.

♦ **tonic immobility** See playing dead.

♦ **tonotaxis** See taxis: tonotaxis.

♦ **tool** *n.* An object employed in tool use, *q.v.*
  blade *n.* A bifacially shaped tool used for cutting (Lewin 1993, 141).
  Comments: Homo s. sapiens made very sharp, thin blades (Lewin 1993, 141).
  cleaver *n.* A bifacially shaped tool used for chopping (Lewin 1993, 141).
  hand axe *n.* A bifacially shaped tool used for pounding and cutting food and other materials (Lewin 1993, 141; 142, illustration).
  Comments: Homo habilis made crude hand axes in its Oldowan Tool Industry (Lewin 1993, 141–145). *Homo erectus* and *H. sapiens* made refined teardrop-shaped hand axes, thought to characterize a new technology, the Acheulian Tool Industry. Some researchers describe Acheulian hand axes as Oldowan tools with the addition of large bifaces. Humans used Acheulian hand axes as axes and possibly also as heavy-duty knives and prey-killing projectiles. Humans used hand axes on bone, hide, meat, and wood, based on microwear studies.
  pick *n.* A bifacially shaped tool used for poking and breaking material apart (Lewin 1993, 141).

♦ **tool industry** *n.* A tool-making technology of a particular hominid group (inferred from Michaelis 1963).
  Acheulean Tool Industry *n.* The tool industry of *Homo erectus* (Michaelis 1963). *syn.* Acheulian (Michaelis 1963)
  [Acheulean, after St. Acheul, France, where researchers found tools of *H. erectus* (Michaelis 1963)]
  Aurignacian Tool Industry *n.* The tool industry of *Homo sapiens sapiens* (Michaelis 1963).
  [Aurignacian, after Aurignac, a town in Haute-Garonne, France, where researchers found tools of *H. s. sapiens* (Michaelis 1963)]
  Mousterian Tool Industry *n.* The tool industry of *Homo sapiens neanderthalensis* (Brace and Montagu 1977, 198).
  [Mousterian, after Le Moustier, France, near where researchers first discovered

bones of this species (Brace and Montagu 1977, 198)]
  Oldowan Stone Industry *n.* The tool industry of *Homo habilis* which includes production of crude stone axes (Lewin 1993, 141–145).
  [Oldowan, after the Oldovai Gorge, Africa, where researchers found fossils of *H. habilis* and its tools (Brace and Montagu 1977, 26)]

♦ **tool making** *n.* In Chimpanzees, *Corvus moneduloides* (a crow species), and Humans: an animal's modifying an object which it uses as a tool (Goodall 1965 in Heymer 1977, 200).
  Comments: A Chimpanzee can fashion twigs to be used for termite catching and chew a handful of leaves into a spongy mass before dipping the mass into water and squeezing the water from the mass into its mouth (Goodall 1965 in Heymer 1977, 200). The crow *Corvus moneduloides* appears to have the most complex tool-making behavior, next to Humans. This bird appears to make a tool kit that includes a hook made from a branchlet and lockpick-like, zig-zag probe (= a stepped-cut tool) from a *Pandanus* leaf (Browne 1996a, C1, illustrations). It uses these tools for removing insects from cavities and crevices.

♦ **tool use, tool using** *n.*
1. An animal's using an object, or another living organism, as a means of achieving an advantage by extending its range of movement or increasing its efficiency (Hall 1963 in Pierce 1986); *e.g.,* a primate's using a young in agonistic buffering, an ant's using other ants that form a living bridge, or a Weaver Ant's using a conspecific larva to glue leaves together.
2. An animal's using an external object as a functional extension of its mouth, beak, hand, claw, etc. in attaining an immediate goal (Lawick-Goodall 1970 in Pierce 1986).
3. An animal's manipulating an inanimate object, not internally manufactured, resulting in the animal's improving its efficiency in altering the form, or position, of some separate object (Alcock 1972 in Pierce 1986); *e.g., Ammophila* wasps pound shut their nest entrances with a small pebble held in their mandibles (Peckham 1904 in Heymer 1977, 200), Archer Fish spit drops of water at arthropods, knocking them into the water where they are eaten, and Egyptian Vultures pick up rocks in their beaks and hurl them at Ostrich eggs in order to break open the shells.

S–Z

4. An animal's using an unattached environmental object to alter more efficiently the form, position, or condition of another object, another organism, or itself; this involves carrying its tool during, or just prior to, its use and being responsible for its tool's proper and effective orientation (Beck 1980 in Pierce 1986); *e.g.,* a Egyptian Vulture's opening an egg by dropping stones on it, a primate's throwing its feces at an intruder, a primate's using a conspecific young in agonistic buffering, and a female water strider's using a copulating male to repel other courting males, but not a Herring Gull's opening a mussel by dropping it on rocks or a primate's aiming its defecation at an intruder.

5. An animal's using an external object as a functional extension of its body enabling it to attain an immediate goal (Lawick-Goodall 1970 in McFarland 1985, 510, who lists many examples, some which are in disagreement with other workers).

6. An animal's "active external manipulation of a moveable or structurally modified inanimate environmental object, not internally manufactured for this use, which, when oriented effectively, alters more efficiently the form, position, or condition of another object, another organism, or the user itself (Pierce 1986, 96).

   Comment: This kind of tool use includes an ant's using an absorbent, or adsorbent, material to carry liquid food to her nest (Feller 1987, 1466; Barber et al. 1989), an ant's dropping small pebbles on potential competitors, an ant-lion's throwing sand at a prey, an assassin bug's using termite carton as camouflage material, a tortoise beetle larva's using a shield of dried feces and exuviae that repels ants, and a tree cricket's using a leaf with a small hole gnawed in it as a sound baffle (Pierce 1986, 96).

7. An animal's manipulating an object (an inanimate object or a conspecific individual) that it uses in caring for its body or obtaining and preparing its food (Immelmann and Beer 1989, 314).

cf. tool making

Comments: In the above definitions, an animal's using parts of its own body for altering its environment, or other activity, is not a kind of tool use. For example, a primate's using its finger nail to manipulate objects or a spider's using its web to catch prey are not examples of "tool use" based on these definitions. Bottlenose Dolphins, Cactus Finches, Humans, and Sea Otters also use tools. Captive Bottlenose Dolphins use tools, and wild ones might also use tools (references in Smolker et al., 1997, 463). A captive Bottlenose Dolphin rubbed pieces of broken tile along its tank walls and dislodged pieces of seaweed, which it ate; another Dolphin in the same tank copied the first one. When two other captive Dolphins attempted to dislodge a moray eel from its crevice, one of the Dolphins searched for and killed a spiny Scorpion Fish, took it to the crevice, and poked at the eel with the fish; the eel abandoned the crevice, and the Dolphin caught it. Several Dolphins, generally females, have shown possible tool use in nature (Smolker et al. 1997, 454). Each held a sponge in its rostrum and probably used it to dig into the sandy ocean bottom as a means of uncovering food.

social-tool use *n.* Tool use (def. 1) in which an animal uses a conspecific animal as a tool (Kummer 1982 in Pierce 1986).

♦ **tooth** *n., pl.* **teeth**

1. In boney vertebrate species: a hard oral structure, consisting chiefly of dentine covered on the outer surface of its crown with enamel and a root leading into a pulp cavity richly supplied with blood vessels and nerves (Michaelis 1963).

   Comments: A vertebrate uses its teeth for defense, chewing food, offense, seizing food, or a combination of these activities. Humans normally have 32 adult teeth.

2. In invertebrates: a hard calcareous, or chitinous, body of the oral or gastric region (Michaelis 1963).

3. A small tooth-like projection (*e.g.,* on the margin of a leaf) (Michaelis 1963).

4. In bivalve molluscs: a process near a hinge (Michaelis 1963).

collateral adj. dental

[Old English tōth]

canine, canine tooth *n.*

1. A pointed mammalian tooth situated between incisors and premolars (Michaelis 1963; Strickberger 1996, 422, illustration).

   syn. cuspid (Michaelis 1963)

2. A pointed reptilian tooth, situated between incisors and cheek teeth (Benton 1991, 24, illustration).

Comments: Two canines normally occur in a mammal's lower jaw, and two occur in its upper jaw.

[Latin *caninus* < *canis*, dog]

cheek tooth *n.* A pointed reptilian tooth situated posterior to canine teeth (Benton 1991, 24–25, illustrations).

Comment: Cheek teeth evolved into molars and premolars (Strickberger 1996, 418).

deciduous tooth *n*. One of the first set of a mammal's teeth that is replaced by a permanent tooth (Strickberger 1996, 418).
syn. milk tooth (Strickberger 1996, 418)
incisor, incisor tooth *n*.
1. A mammalian front tooth adapted for cutting (Michaelis 1963; Strickberger 1996, 422, illustration).
2. A pointed reptilian tooth, situated between incisors and cheek teeth (Benton 1991, 24).
Comments: Humans normally have eight incisors, four in each jaw (Michaelis 1963). [New Latin]
labyrinthine tooth, labyrinthodont tooth *n*. A vertebrate tooth with complex (labyrinthine), internal patterns of infolding (Benton 1990, 41, illustration).
Comments: Labyrinthine teeth occurred in Eusthenopteron (Sarcopterygii: Osteolepiformes) and early tetrapods (Benton 1990).
molar, molar tooth *n*. A mammalian tooth adapted for grinding, situated posterior to premolar teeth (Michaelis 1963; Strickberger 1996, 422, illustration).
Comments: Humans normally have 12 molars, six in each jaw (Michaelis 1963). [Latin *molaris* < *mola*, mill]
premolar, premolar tooth *n*. A mammalian tooth adapted for grinding, situated posterior canine teeth and anterior to molar teeth (Michaelis 1963; Strickberger 1996, 422, illustration).
syn. bicuspid
Comments: Humans normally have eight premolars, four in each jaw (Michaelis 1963). [Latin *molaris* < *mola*, mill]
tribosphenic tooth, tritubercular tooth *n*. A mammalian upper-molar tooth whose three cusps are aligned along its anterior-posterior axis; contrasted with a tribosphenic tooth (Strickberger 1996, 423, illustration).
Comments: Tribosphenic teeth occur in Group Theria, which includes the subclasses Eutheria, Metatheria, and Pantotheria (Strickberger 1996, 422).
tricodont tooth *n*. A mammalian molar tooth whose three cusps are aligned along its anterior-posterior axis; contrasted with a tribosphenic tooth (Strickberger 1996, 422, illustration).
Comments: Tricodont teeth occur in Subclass Prototheria (Strickberger 1996, 422).
♦ **top-down control** See ³control: top-down control.
♦ **-topic** See -topy.
♦ **topo-, top-, -topic, -topy** *combining form* "A place or region; regional" (Michaelis 1963). [Greek *topos*, place]

♦ **topocentric pollination** See pollination: topocentric pollination.
♦ **topocline** See cline: topocline.
♦ **topodeme** See deme: topodeme.
♦ **topogalvanotaxis** See taxis: topogalvanotaxis.
♦ **topogamodeme** See deme: topogamodeme.
♦ **topographic memory** See memory: topographic memory.
♦ **topological psychology** See study of: psychology: topological psychology.
♦ **topology** See study of: mathematics: topology.
♦ **topotaxis** See taxis: topotaxis.
♦ **-topy** *combining form*
adj. -topic
See topo-.
amphitopy *n*. A population's, or species', showing a broad, variable tolerance of habitat and environmental conditions, reflected in its having clines and subspecies (Lincoln et al. 1985).
eurytopy, generalization *n*.
1. An organism's being tolerant of a wide range of habitats (Lincoln et al. 1985).
2. An organism's being physiologically tolerant (Lincoln et al. 1985).
3. An organism's having a wide geographic distribution (Lincoln et al. 1985).
stenotopy, specialization *n*.
1. An organism's being tolerant of a narrow range of habitats (Lincoln et al. 1985).
2. An organism's being physiologically intolerant (Lincoln et al. 1985).
3. An organism's having a narrow geographic distribution (Lincoln et al. 1985).
♦ **torpidity, torpor** See dormancy.
♦ **torrenticole** See -cole: torrenticole.
♦ **total assimilation** See production: gross-primary productivity.
♦ **total environment** See environment: total environment.
♦ **total extinction rate** See rate: total extinction rate.
♦ **total load** See genetic load: total load.
♦ **total phenotypic variance of a trait (V_P)** See variance: total phenotypic variance of a trait.
♦ **total photosynthesis** See ²production: gross-primary productivity.
♦ **total range** See ³range: total range.
♦ **TOTE** See Test-Operate-Test-Exit.
♦ **totipotent** *adj*. In ants: referring to an individual (*e.g.*, a founding queen) that is capable of performing all essential tasks in her colony (Hölldobler and Wilson 1990, 644).
♦ **touch** See modality: mechanoreception: touch.
♦ **tower** See ²group: tower.

S—Z

♦ **town** See ²group: town.
♦ **Townes-style Malaise trap** See trap:
Malaise trap: Townes-style Malaise trap.
♦ **toxicology** See study of: toxicology.
♦ **toxicophobia** See phobia (table).
♦ **toxin** *n.*
1. Any of a class of more or less poisonous
compounds produced by an organism
that can be a disease-causative agent
(Michaelis 1963).
2. Any toxic matter generated by a living, or
dead, organism (Michaelis 1963).
Comments: I list and define only a few of
the hundreds of animal toxins below.
[TOX(IC) + -IN; Middle Latin *toxicus*, poisoned, poisonous]
adrenalin *n.* A toxin produced by toads
(Brodie 1989, 88).
Comments: Adrenalin acts as a local irritant, causes numbing, and stimulates heart
rate (Brodie 1989, 88).
bufotoxin *n.* A toxin produced by toads
(Brodie 1989, 88).
Comments: Bufotoxin acts as a local irritant, causes numbing, and stimulates heart
rate (Brodie 1989, 88). The Colorado River
toad produces the highly hallucinogenic
toxin O-methy-bufotenin.
bufotenin *n.* A toxin produced by toads
(Brodie 1989, 88).
Comments: Bufotenin acts as a local irritant, causes numbing, and stimulates heart
rate (Brodie 1989, 88).
cantharidin *n.* A compound produced
by blister beetles and soldier beetles that
blisters human skin and is a severe irritant
of a person's urinary tract when taken
internally (Borror et al. 1989, 446; Brodie
1989, 62).
Comments: Cantharidin is the active component from dried bodies of the Spanishfly,
a soldier beetle (Borror et al. 1989, 8, 11).
People have used it to treat urogenital-
system problems.
formic acid *n.* A component of the
venom of some ant species, the defensive
sprays of some beetle species, and the
ejecta and the urticating hairs of some
caterpillar species (Brodie 1989, 56, 58,
63).
quinone *n.* A component of defensive
sprays of some darkling-beetle species
(bombardier beetles) (Brodie 1989, 63).
salamandarin *n.* A component of the
defensive spray of fire salamanders (Brodie
1989, 103).
seratonin *n.* A component of the skin
secretion of some frog species that repels
predators (Brodie 1989, 91).
tetrodoxin *n.* A neurotoxin found in
some species of frogs, octopuses, sala-

manders, snails, and fish (Filefishes, Puff-
ers, Spikefishes, the Ocean Sunfish, Trig-
gerfishes, and Trunkfishes) (Brodie 1989,
63).
venom *n.* A toxin that an animal secretes
and usually transmits by a bite or sting
(Robinson and Challinor 1995, 227).
Comment: Venom-producing animals include the Platypus, a shrew species, Centipedes, Jellyfish, Sea Anemonies, and Scorpions, as well as many ant, bee, lizard,
snake, and wasp species (Robinson and
Challinor 1995, 227). In the Platypus, only
adult males are venomous.
♦ ¹**trace** See ²group: trace.
♦ ²**trace, trace fossil** See fossil: trace fossil.
♦ **tracking** See environmental tracking;
method: tracking.
♦ **trade-off** *n.* A direct comparison of the
costs and benefits of a particular trait
(McFarland 1985, 432).
♦ **tradition** *n.*
1. That which people hand down; a statement, belief, knowledge, custom, doctrine, practice, etc. that is transmitted
(especially orally) from generation to
generation (*Oxford English Dictionary*
1972, entries from *ca.* 1380).
2. Persons' transmitting, or handing down,
information from one to another or from
generation to generation; transmitting
statements, beliefs, rules, customs, or
the like, especially orally or by practice
without writing, chiefly in the phrase "by
tradition" (*Oxford English Dictionary*
1972, entries from 1591).
3. In some animal species: an individual's
creation of a specific form of behavior
that is passed from generation to generation by learning (Wilson 1975, 168), *e.g.,*
dialects in bird songs, traditional migration routes of some toad and bird species
and reindeer, traditional calving places
of reindeer, traditional game trails of
deer and other mammals, milk stealing
by the blue tit (bird), oyster feeding of
the oystercatcher (McFarland 1985, 515),
or placer mining and sweet-potato washing in the Japanese Macaque (= Red-
Faced Macaque) (Kawai 1965 in Heymer
1977, 175).
cf. cultural exchange, culture, learning: social imitative learning
Comment: Tradition is an "ultimate refinement in environmental tracking" (Wilson
1975, 168).
direct tradition *n.* Tradition in which a
neophyte learns (*e.g.,* by imitation from an
instructor that is usually an adult) (Immelmann and Beer 1989, 316).

indirect tradition *n*. Tradition in which a neophyte is placed in a situation where particular information can be learned without an experienced animal; *e.g.,* in many insect species, a mother lays eggs on plants of choice, and her offspring lay their eggs on the same plant species (Immelmann and Beer 1989, 316).

interspecies tradition *n*. One species' acquiring behavior of another (*e.g.,* a mockingbird's mimicking vocalizations of other bird species) (Immelmann and Beer 1989, 317).

nonobject-bound tradition *n*. Tradition in which information about a specific object (*e.g.,* location and quality of food) is transmitted even in the absence of the object by some form of symbolic representation (*e.g.,* dance communication of Honey Bees) (Immelmann and Beer 1989, 316).

object-bound tradition *n*. Tradition in which experienced and inexperienced conspecific animals gather with the object (*e.g.,* food) about which they are to acquire knowledge or learn a treatment (Immelmann and Beer 1989, 316).

♦ **tradition drift** See drift: tradition drift.

♦ **traditional homology** See homolog: traditional homology.

♦ **traditional rabbit-skin test** See test: traditional-rabbit-skin test.

♦ **trail communication** See communication: trail communication.

♦ **trail kairomone** See chemical-releasing stimulus: semiochemical: allelochemic: allomone: kairomone: trail kairomone.

♦ **trail-marking substance** See substance: trail-marking substance.

♦ **trail pheromone, trail substance** See chemical-releasing stimulus: pheromone: trail pheromone.

♦ **train** See ²group: train.

♦ **training** *n*. A broad term for learning procedures that are very different from one another and are controlled by humans (Immelmann and Beer 1989, 317).
syn. conditioning (in many cases, Immelmann and Beer 1989, 317)
cf. shaping

spaced training *n*. Training in which a researcher paces learning sessions of a laboratory animal with rest intervals (Hall 1998, 32).
cf. memory
Comments: In 1885, the German psychologist Hermann Ebbinghaus used himself as a sole subject in a spaced-training experiment and found that he could memorize a list of nonsense syllables much better if he had a rest interval between his learning sessions compared to having no rest inter-

val (Hall 1998, 32). *Drosophila* that undergo spaced training develop robust long-term memories that last at least 1 week (Tim Tully in Hall 1998, 32).

♦ **trait** See character.

♦ **trait group** See ²group: trait group.

♦ **tramp, tramp species** See ²species: tramp species.

♦ **trampling zone** See zone: trampling zone.

♦ **trance reaction** See playing dead.

♦ **trans-** *prefix*
1. "Across; beyond; through; on the other side of" (Michaelis 1963).
2. "Through and through; changing completely" (Michaelis 1963).
3. "Surpassing; transcending; beyond" (Michaelis 1963).
4. Anatomically, "across; transversely" (Michaelis 1963).
[Latin *trans*, across, beyond, over]

♦ **transcription** *n*. Formation of an RNA molecule upon a DNA template by complementary base pairing that is mediated by RNA polymerase (King and Stansfield 1985).
cf. translation

♦ **transduction** *n*. Gene transfer between prokaryotes by a virus (Campbell 1996, 504).
cf. ¹transformation

♦ **transfer-brooding** See brooding: transfer-brooding.

♦ **transfer host, transport host** See ³host: paratenic host.

♦ **transfer of training** *n*. The influence of an organism's past learning on its new learning (Storz 1973, 280).

♦ **transfer ribonucleic acid (tRNA)** See nucleic acid: ribonucleic acid (comment).

♦ **transform** *v.t.* One's performing a mathematical procedure on a quantitative datum (*e.g.,* taking its arcsine square root, log, or square root).
Comment: A primary reason to transform data in a set is to change the data's variance to meet the assumption of equality of variances of a statistical test. For example, taking the log of data in a set yields a new set of logs with a lower variance than the original set, and these variances might not be significantly different.

♦ ¹**transformation** *n*.
1. An organism's change due to its acquisition of foreign genetic material (Lincoln et al. 1985).
2. Genetic material's moving from one organism individual to another (Bonner 1988, 17).
3. A prokaryote's obtaining genes from its surrounding environment (Campbell 1996, 504).
cf. transduction

Note: This enables considerable gene transfer among bacteria (Campbell 1996, 504).

♦ ²**transformation** See ²evolution: monotypic evolution, phyletic evolution.

♦ **transformational mimicry** See mimicry: transformational mimicry.

♦ **transforming principle** *n*. The heredity information (later identified as DNA) Griffith (1928) found that living, avirulent pneumococci "picked up" from heat-killed virulent pneumococci after both kinds of pneumococci were injected into mice (Mayr 1982, 818).

♦ **transgenation** See ²mutation: gene mutation.

♦ **transgenic** *adj*. Referring to an organism to which a researcher has added a novel gene(s) to its genome (*e.g.*, a transgenetic mouse) (Cohen 1995b, 1715).

♦ **transgenosis** *n*. Transfer of genes from one species to an unrelated one, in which they are maintained and manifested phenotypically (Lincoln et al. 1985).
cf. transposable element: transposon

♦ **transient polymorphism** See morphism: polymorphism.

♦ **transition matrix** *n*. A means of analyzing Markoff chains (Lehner 1979, 274).
Comment: Kinds of transition matrices include "first-order-transition matrix," "second-order-transition matrix," and "*n*th-order-transition matrix."

♦ **transition probability** See probability: transition probability.

♦ **transitional action** See action: transitional action.

♦ **transitional adaptive zone** See zone: adaptive zone: transitional adaptive zone.

♦ **transitional load** See load: transitional load.

♦ **trans-kingdom sex** See ³sex: trans-kingdom sex.

♦ **translation** *n*. A protein's formation that is directed by a specific messenger RNA molecule (King and Stansfield 1985).

♦ **translational load** See load: translational load.

♦ **translocation** *n*. A chromosomal aberration that results in a change in a chromosomal-segment position within an organism's genome but does not change its total number of genes (King and Stansfield 1985).
Comment: Kinds of translocations include "insertional translocation," "intrachromosomal translocation" (= a "shift"), "interchromosomal translocation," "nonreciprocal translocation" (= "aneucentric translocation"), and "reciprocal translocation" (King and Stansfield 1985).

♦ **transmission genetics** See study of: genetics: transmission genetics.

♦ **transmitter substance** See neurotransmitter.

♦ **transmutation** See ³theory: transmutation.

♦ **transparency** See camouflage: background imitation: transparency.

♦ **transpecific evolution** See ²evolution: transpecific evolution.

♦ **transpiration** *n*. A plant's loss of water vapor through its stomata (Curtis 1983, 614, 1109).
[Latin *trans*, across + *spirare*, to breath]
evapotranspiration *n*. The sum of transpiration and water evaporation from soil and water bodies in a particular place over time (Lincoln et al. 1985; Odum et al. 1988, 13).

♦ **transport** *n*. Something's carrying, or conveying, a thing, or a person; conveyance (*Oxford English Dictionary* 1972, entries from 1611).
adult transport *n*. In many social-insect species: one adult's carrying, or dragging, another adult, usually during colony emigration (Hölldobler and Wilson 1990, 635).
chain transport *n*. In some ant species: one worker's relaying food to another in the course of transporting it back to their nest (Hölldobler and Wilson 1990, 636).
group transport *n*. In Ants: two or more workers' coordinated transport of a food item (Hölldobler and Wilson 1990, 638).

♦ **transport host** See host: paratenic host.

♦ **transport of young** *n*. A parent animal's carrying its young either inside or outside its body (Immelmann and Beer 1989, 318).
cf. carrying in, clinging young

♦ **transportation hypothesis** See hypothesis: transportation hypothesis.

♦ **transposable element** *n*. A member of a class of DNA sequences that can move from one chromosomal site to another (King and Stansfield 1985).
See transposable element: transposon.
syn. controlling element (McClintock in King and Stansfield 1985), transposon (a confusing synonym; King and Stansfield 1985)
cf. nucleic acid: deoxyribonucleic acid: selfish DNA
Comments: Transposable elements are classified into families based on their structure, kinds of genes, and model of movement (King and Stansfield 1985). They seem to make up at least 10% of more-derived eukaryote genomes. As transposable elements move in and out of chromosomes, they can change a gene from a coding to a

noncoding form, and vice versa, and duplicate, lose, merge, and move genes (Arms and Camp 1995, 200). Transposable elements can affect gene regulation and thus cause mutations. They might function in formation of new species. Movements of these elements require a transposase and a resolvase that recognize short nucleotide sequences that are repeated in inverted order at both ends of an element. Transposable elements have also been classified as selfish DNA (Alberts et al. 1989, 605–606). Purugganan (1993) discusses transposable elements that can act as introns, and he lists kinds of transposable-element-intron alleles. Some bacterial strains transmit plasmids with drug resistance and have received transposon insertions from other bacteria (Strickberger 1996, 217). Such horizontal DNA transmission might occur between different species and phyla. The Singatoxin gene moves from one bacterium to another via a virus and causes human hemorrhagic colitis and hemolytic-uremic syndrome (Hilts 1996, C1). Barbara McClintock received the Nobel Prize in 1983 for her work with transposable elements in maize which she started in the 1940s. Some of the many kinds of transposable elements are defined below.

Ac-Ds *n.* A transposable element in maize (Alberts et al. 1989, 105).

Comments: Arms and Camp (1995, 199) describe *Ac* and *Ds* as different transposable elements. *Ds* (a disabler gene) stops purple coloration of a corn seed. *Ac* (an activator gene) apparently enables *Ds* to move around within a genome.

Alu *n.* One of a family of possible transposable elements in *Homo sapiens* of about 300 nucleotide pairs and derived from a 7SL-RNA gene that constitutes about 5% of human DNA (Alberts et al. 1989, 605, 608).

Comments: Alu sequences (discovered by Houck et al. 1979) represent the great majority of short interspersed nucleotide elements in Humans (Zuckerkandl et al. 1989, 505). These sequences are variable in structures, numbers, locations, orientations, and transcriptional regulation. *Alus* might be unusually mobile pseudogenes (Alberts et al. 1989, 605, 608), or some kinds might have a function (Zuckerkandl et al. 1989, 505).

[*Alu* refers to a tetranucleotide present in most *Alu* sequences, AGCT, which is specifically cut by a restriction enzyme called *Alu* (Zuckerkandl et al. 1989, 505).]

cin4 *n.* A transposable element in maize (Alberts et al. 1989, 105).

copia element *n.* A transposable element in *Drosophila* that is a member of a family of closely related base sequences that code for abundant mRNAs (King and Stansfield 1985).

syn. Copia (Alberts et al. 1989, 105)

Comment: Several *Drosophila* mutations are known that result from insertions of copia-like elements (King and Stansfield 1985).

insertion sequence (IS) *n.* A small transposable element in *Escherichia coli* that usually encodes proteins for its own transposition and catalyzes the transposition of a sequence it flanks (King and Stansfield 1985).

syn. jumping gene (King and Stansfield 1985)

Comment: Insertion sequences can produce lethal mutations (King and Stansfield 1985).

lambda, λ *n.* A bacteriophage that splices its genes into the DNA of *Escherichia coli* (Arms and Camp 1995, 199).

L1 *n.* A transposable element in *Homo sapiens* that constitutes about 4% of a genome's mass (Alberts et al. 1989, 105).

mariner transposons *pl. n.* A group of transposable elements that were first found in insects (Grady 1996, C1).

Comments: David Hartl found the first mariner transposon (Grady 1996, C1). Some mariner transposons occur in Humans and might have entered our genome from insects via malaria mosquitoes, directly from other insects, or from other organisms. One mariner transposon could be responsible for unequal crossing over near the myelin gene on chromosome-17 in Humans. The resultant duplication can lead to Charcot-Marie-Tooth syndrome which is a disease involving degeneration of hand and feet nerves (and named after its three discoverers). The resultant deletion can lead to heredity neuropathy (HNPP).

P element *n.* A transposable element in *Drosophila melanogaster* that is responsible for one type of hybrid dysgenesis (King and Stansfield 1985).

Comment: a DNA molecule that carries a P element can be microinjected into a *Drosophila* embryo, and the P element may integrate into the germ-line chromosomes and be transmitted to the progeny of injected flies (King and Stansfield 1985).

P transposon *n.* A transposon in *Drosophila melanogaster* that probably entered *D. willistoni* in the 1940s (Daniels et al. 1990 in Strickberger 1996, 217).

retrotransposon *n.* A transposon that resembles a retrovirus (Travis 1992, 884).

S–Z

Tam3 *n.* A transposable element in *Antirrhinum* (Alberts et al. 1989, 105).

tn3 *n.* A transposable element in *Escherichia coli* (Alberts et al. 1989, 105).

transposon *n.*

1. A gene that is repetitive within a cell and can change positions on a chromosome (*e.g.,* in some bacteria) (Gould 1981a, 7).

2. A kind of transposable element that is immediately flanked by inverted repeat sequences that in turn are immediately flanked by direct repeat sequences (King and Stansfield 1985).

Note: Transposons usually have genes in addition to those needed for their insertion into chromosomes (King and Stansfield 1985).

See transposable element.

syn. jumping gene, mobile genetic element, transposable element (a confusing synonym), transposable gene (confusing synonym) (King and Stansfield 1985)

Comments: The gene for leghemoglobin, used in nitrogen fixation in legumes, might have been translocated to a legume from an animal by a virus (Lewin 1981, 636). Circumstantial evidence indicates that a mite carried transposable elements called "P elements," *q.v.,* between *Drosophila* species (Travis 1992, 885). Transposons can alter gene expression that might be favored by natural selection; they might cause gene-regulation changes that lead to speciation.

ty element *n.* A transposable element in the yeast *Saccharomyces cerevisiae* (King and Stansfield 1985).

syn. Ty (Alberts et al. 1989, 105)

[ty, *t*ransposon-*y*east]

♦ **transposon** See transposable element: transposon.

♦ **transreplication** See gene conversion.

♦ **transvestism** *n.*

1. A person's dressing himself, or herself, in garments of the opposite sex (*Oxford English Dictionary* 1972, entries from 1652).

2. A person's compulsive need to dress in garments appropriate for members of the opposite sex (Michaelis 1963).

cf. eonism; mimicry: automimicry: female mimicry

[TRANS- + Latin, *vestilus,* past participle of *vestire,* to clothe]

♦ **trap** *n.* A device used to catch game, or other animals, or collect seeds (based on Michaelis 1963).

Comment: Many factors affect the sizes and compositions of trap samples including age, population size, sex, and species of

---

SOME KINDS OF TRAPS

I. attractive trap
  A. bait trap
  B. host-plant trap
  C. shelter trap
  D. vertebrate-host trap
II. attractive-interceptive trap
  A. light trap
  B. sticky trap
  C. water trap
III. interceptive trap
  A. aquatic trap
  B. emergence trap
  C. flight trap
  D. pitfall trap

---

organisms; geographical location; habitat type; microhabitat conditions; moon phase; time of day and year; trap design, emptying times, kind, and size; and weather (Southwood 1978, chap. 7; Handley and Kalko 1993).

[Old English *treppe, trœppe*]

light trap *n.* A trap that uses ultraviolet, visual light, or both, in attracting animals (Southwood 1978, 253).

Comment: Light traps are probably the most widely used insect traps and include the Haufe-Burgess visual trap, Manitoba trap (= canopy trap), New Jersey trap, Pennsylvanian trap, Robinson trap, Rothamsted trap, and Texas trap (Southwood 1978, 253).

Malaise trap *n.* A trap used for catching arthropods made of gauze, a supporting frame, and a collecting head that collects living or dead specimens (Steyskal 1981).

Comments: Malaise traps catch primarily certain taxa of flying insects, but also collect some taxa that walk into them (Barrows 1986). These traps come in different colors, mesh sizes, overall sizes, and shapes. Some are used with baits such as CO_2.

[after René Malaise, a Swedish entomologist who published a description of a Malise trap in 1937 (Steyskal 1981, 225)]

▸ **Cornell-style Malaise trap** *n.* A pyramid-shaped Malaise trap with four openings in its lower area, a square cross-section at its base, and a collection head at its apex (Matthews and Matthews 1983, figure 1).

▸ **Townes-style Malaise trap** *n.* A Malaise trap with a rectangular cross-section at its base and a collection head at one or both of its top corners (Townes 1962, 253; Matthews and Matthews 1983, fig. 1; Darling and Packer 1988, 787, figures).

[after Henry Townes, an American entomologist who modified Malaise's original trap design]

mist net *n*. A net made of fine, strong
thread, used for trapping flying bats and
birds of many species (Allen 1996, 639).

Comments: Researchers often place these
nets in vertebrate flyways (Allen 1996, 639,
641). Some people use these nets for
capturing these animals for food.

pitfall trap *n*. A steep-sided pit in the
ground that captures primarily cursorial
animals (Southwood 1978, 247).

Comments: A pitfall trap primarily comprises a container that is put into the group
so that its mouth is level with the soil
surface (Southwood 1978, 247). Researchers modify pitfall traps by adding baits,
collecting fluids, precipitation guards, and
timing devices that segregate catches from
different periods and adjusting trap size.
Studies on mammals with the use of pitfall
traps include using arrays of traps, drift
fences, transects of traps, and traps of
different sizes (Handley and Kalko 1993,
19; Handley and Varn 1993, 285; Kalko
and Handley 1993, 3).

♦ **traplining** *n*. In some insect and bird
species: an individual's daily following a
"trapline" of blooming plants in a set order
(Buchmann and Nabhan 1996, 255).

Comment: These animals appear to be
familiar with the exact locations and statuses of these plants (Buchmann and Nabhan
1996, 256).

♦ **traumatic insemination** See insemination: traumatic insemination.

♦ **traumatotaxis, traumotaxis** See taxis:
traumatotaxis.

♦ **tread** See copulate.

♦ **¹treading** *n*.

1. A person's stepping on; pacing, or walking, on (the ground, etc.); walking (in a
place); hence, to go about (in a place,
etc.) (*Oxford English Dictionary* 1972,
verb entries from *ca*. 700).

2. A person's stepping on (something in his
way); a person's accidentally, or intentionally, putting his foot down upon;
especially so as to press upon (*Oxford
English Dictionary* 1972, intransitive verb
entries from 1384).

3. A person's stepping, or walking, upon
something with pressure so as to crush,
beat down, injure, or destroy it; trample
(*Oxford English Dictionary* 1972, entries
from *ca*. 825).

♦ **²treading** *n*. In male birds of some species: copulating (*Oxford English Dictionary*
1972, entries from *ca*. 1250).

cf. copulation

♦ **³treading, kneading** *n*. In many mammal species: an young's using either its
front or hind paw to elicit milk from a teat
during nursing (Heymer 1977, 116).

♦ **treadling** See running on the spot.

♦ **treatment, treatment effect** See effect: main effect.

♦ **tree** See -gram: dendrogram; organism
(Appendix 1, Part 3).

♦ **TreeBase** See database: TreeBase.

♦ **Tree of Life** See Database: Tree of Life.

♦ **tree-ring chronology** See method:
dendrochronology.

♦ **trekking** See behavior: trekking.

♦ **trembling dance** See dance: bee dance:
quiver dance.

♦ **trend** *n*. An evolutionary path up to improved adaptation (*e.g.*, a growth form,
growth rate, or potential for greater habitat
choice) (Ruse 1993, 56).

♦ **trends** See -genesis: orthogenesis.

♦ **treptic behavior, treptics** See behavior: treptic behavior.

♦ **triadic awareness** See awareness: self-
awareness: triadic awareness.

♦ **trial** *n*. One run of part of an entire
experiment; *e.g.*, in a learning experiment,
exposure of one of a group of animals to a
test and control situation can be considered
a trial.

♦ **trial-and-error behavior** See kinesis:
klinokinesis.

♦ **trial-and-error learning** See learning:
trial-and-error learning.

♦ **trial-and-error orientation** See taxis:
phobotaxis.

♦ **tribe** See ²group: tribe.

♦ **trigenetic** See -genetic: trigenetic.

♦ **trigenic** See -genic: trigenic.

♦ **triggered** behavior See behavior: triggered behavior.

♦ **trimmed mean** See mean: trimmed
mean.

♦ **trimonoecism** See -oecism: trimonoecism.

♦ **trinomial** See name: trinomial.

♦ **trio** *n*. For example, in *Hyla* frogs: three
calling individuals that alternate sounds
(Wilson 1975, 443).
See ²group: trio.
cf. animal sounds: duet, quartet; group:
chorus

♦ **trioecism** See -oecism: trioecism.

♦ **trip** See ²group: trip.

♦ **triploidy** See ploidy: triploidy.

♦ **¹tripping** *n*.

1. A person's moving quickly and lightly
(*Oxford English Dictionary* 1972, entries
from 1567).

2. A person's stumbling, erroring, sinning
(*Oxford English Dictionary* 1972, entries
from 1577).

S–Z

♦ ²**tripping** *n.*
1. A bee's causing a flower's anthers to hit it very rapidly when it visits a flower, *e.g.,* when Honey Bees visit some legume species such as alfalfa (Bohart 1971, etc. in Faegri and van der Pijl 1979, 162, 183).
2. The "explosive" anther movement of a flower due to a bee's movements inside it (Faegri and van der Pijl 1979, 162, 183).

♦ ³**tripping** *n.* A provocative, violent behavior, often observed in children, in which one child suddenly holds out his leg and deliberately causes another to stumble over it (Heymer 1977, 39).

♦ **triptorelin** See drug: medroxyprogesterone.

♦ **trisomy** See -somy: trisomy.

♦ **triumph ceremony** See ceremony: triumph ceremony.

♦ **Trivers parental-investment hypothesis** See hypothesis: Trivers' parental-investment hypothesis.

♦ **Trivers-Willard hypothesis, Trivers-Willard principle** See hypothesis: Trivers-Willard hypothesis.

♦ **trivial altercation** *n.* A heated argument between, or among, people that arises for seemingly insignificant reasons (Wilson and Daly 1985, 59).

♦ **trixeny** See -xeny: trixeny.

♦ **tRNA** See nucleic acid: ribonucleic acid (comment).

♦ **troglodyte** *n.* A subterranean organism or a cave dweller (Lincoln et al. 1985). *cf.* ¹-phile: troglophile

♦ **troglon** See biotope.

♦ **troglophile** See ¹-phile: troglophile.

♦ **trogloxene** See -xene: trogloxene.

♦ **trogloxenous** See -xene: trogloxene.

♦ **troop** See ²group: troop.

♦ **-troph-, tropho-, -trophy** *combining form* Referring to nutrition and its processes (Michaelis 1963).
adj. -trophic
Comment: "-Trophs" are classified in many ways, including whether an organism makes all of its own organic compounds used in metabolism, uses organic compounds obtained from other organisms, or both; the

---

**SOME CLASSIFICATIONS OF
FEEDING TYPES (-TROPHS)**

I. Classification by kinds of required organic compounds that an organism manufactures
   A. autotroph (*syn.* anauxotroph, holophyte, holotroph, mesotroph, phytotroph, primary producer, prototroph, protroph)

1. chemoautotroph (*syn.* chemotroph, chemosynthetic autotroph)
2. photoautotroph (*syn.* phototroph, photosynthetic autotroph)
   B. heterotroph (*syn.* allotroph, organotroph)
   C. mixotroph

II. Classification by an organism's energy source
   A. amphitroph
   B. autotroph
   C. chemolithotroph
   D. heterotroph
   E. lithotroph
   F. mixotroph

III. Classification by an organism's energy source and electron-transfer donors
   A. chemoorganotroph
   B. photolithotroph
   C. photoorganotroph (*syn.* photoheterotroph)

IV. Classification by an organism's food type
   A. acidotroph
   B. anauxotroph
   C. biotroph
   D. caecotroph
   E. hemotroph (*syn.* haemotroph)
   F. methanotroph
   G. minerotroph
   H. monotroph
   I. myxotroph
   J. oligotroph
   K. planktotroph
   L. polytroph
   M. prototroph
   N. saprotroph

V. Classification by an organism's food location
   A. ombrotroph
   B. rheotroph

VI. Classification by an organism's location of feeding on its host
   A. dermatotroph
   B. ectendotroph
   C. ectotroph
   D. endotroph

VII. Classification by an organism's symbiotic relationship
   A. mycotroph
   B. myrmecotroph
   C. paratroph (*syn.* parasite)
   D. symbiotroph
   E. syntroph

VIII. Classification by other criteria
   A. atroph
   B. cteinotroph
   C. metatroph
   D. osmotroph

kind of energy source an organism uses; the kind of energy source and electron-transfer donors an organism uses; the kind of food an organism consumes; where in a habitat an organism obtains its food; and where an organism feeds on its host.

[Greek *trophikos* < *trophē*, nourishment < *trephein*, to feed, nourish]

acidotroph *n.* An organism that feeds on acidic food or substrates (Lincoln et al. 1985).

allotroph See -troph-: heterotroph.

amphitroph *n.* An organism that is a phototroph during daylight hours and a chemotroph during dark hours (Lincoln et al. 1985).

anauxotroph See -troph-: autotroph.

atroph *n.* An organism, or life-cycle stage, that does not feed (*e.g.*, an insect pupa or an adult of some moth species) (Lincoln et al. 1985).

autotroph *n.* An organism that can build all the organic molecules it requires using carbon dioxide (in air or water) and energy from its physical environment (*e.g.*, some single-celled-organism species or green plants) (Curtis 1983, 83; Starr and Taggart 1989, glossary); contrasts with heterotroph and mixotroph.
syn. anauxotroph, holophyte, holotroph, mesotroph, phytotroph, primary producer, prototroph, protroph (Lincoln et al. 1985)
cf. producer
Comments: Some autotrophs live far under Earth's surface (Fredrickson and Onstoot 1996, 69).

▶ **chemoautotroph, chemotroph** *n.* An autotroph that obtains metabolic energy from oxidation of inorganic substances, such as ammonia, ferrous ions, and hydrogen sulfide (Jannasch and Mottl 1985, 718; Lincoln et al. 1985; Campbell 1996, 505).
syn. chemosynthetic autotroph (Starr and Taggart 1989, glossary)
cf. [1]-phile: thiophile
Comment: Some bacterium species are chemoautotrophs, including the archaebacterium *Sulfobolus*, which oxidizes sulfur (Starr and Taggart 1989, 109; Campbell 1996, 505).

▶ **photoautotroph, phototroph** *n.* An autotroph that obtains metabolic energy from sunlight by a photochemical process and uses carbon dioxide as a carbon source (Lincoln et al. 1985; Campbell 1996, 505).
syn. photosynthetic autotroph (Starr and Taggart 1989, glossary)
Comments: Photoautotrophs include cyanobacteria, certain protists, and plants (Campbell 1996, 505).

biotroph *n.* An organism that derives nutrients from a living host; a parasite (Lincoln et al. 1985).
See parasite.

caecotroph *n.* An animal that eats caecotrophs, *q.v.*, not caecum feces (Heymer 1977).
cf. -phage: coprophage

chemoautotroph, chemotroph See -troph-: autotroph: chemoautotroph.

chemoheterotroph See -troph-: chemoorganotroph.

chemolithotroph *n.* An organism (*e.g.*, one of many bacterium species) that uses certain reduced inorganic compounds as energy sources by obtaining energy from oxidation-reduction reactions (Jannasch and Mottl 1985, 718; Lincoln et al. 1985).
cf. -troph-: chemoorganotroph, lithotroph

chemoorganotroph *n.* An organism that obtains energy from oxidation-reduction reactions and uses organic electron donors (Lincoln et al. 1985).
syn. chemoheterotroph
cf. -troph-: chemolithotroph

cteinotroph *n.* A parasite that destroys its host (*e.g.*, the AIDS virus or an ichneumon wasp) (Lincoln et al. 1985).

dermatotroph *n.* An organism that lives in, or on, skin (*e.g.*, follicle and scabies mites) (Lincoln et al. 1985).

ectendotroph *n.* A parasite that feeds from both the exterior and interior of its host (Lincoln et al. 1985).

ectotroph *n.* An organism that obtains nourishment externally without marked penetration into its food source (*e.g.*, many phytophagous-arthropod species) (Lincoln et al. 1985).
cf. -troph-: endotroph

endotroph *n.* An organism that obtains nourishment internally by marked penetration into its food source (*e.g.*, mycorrhizal-fungus species or a plant-boring-arthropod species) (Lincoln et al. 1985).
cf. -troph-: ectotroph

hemotroph, haemotroph *n.* An organism that obtains nutrients from blood; especially, an embryo that feeds from its mother's blood (Lincoln et al. 1985).
cf. parasitism: haemoparasitism

heterotroph *n.* An organism that obtains its carbon and all metabolic energy from organic molecules that have already been assembled by autotrophs (Curtis 1983, 83; Starr and Taggart 1989, glossary); contrasted with autotroph.
syn. allotroph, organotroph (Lincoln et al. 1985); chemoheterotroph (Campbell 1996)
Comment: Some plant species, most bacteria, many protistans, and all fungus and

animal species are heterotrophs (Starr and Taggart 1989, glossary; Campbell 1996, 505).

▸ **chemoheterotroph** *n.* A heterotroph species that must consume organic molecules for both carbon and energy (Campbell 1996, 505).

holotroph See -troph-: autotroph.

hypotroph *n.* An "organism" that uses host-cell substances as food (*e.g.*, a bacterium) (Lincoln et al. 1985).

lecithotroph *n.* An organism's developmental stage that feeds on egg yolk (Lincoln et al. 1985).

lithotroph *n.* An organism (*e.g.*, one of many bacterium species) that uses inorganic compounds as electron donors in its energetic processes (Lincoln et al. 1985). *cf.* -troph-: chemolithotroph, photolithotroph

mesotroph *n.* An organism that is completely autotrophic (Lincoln et al. 1985). See -troph-: autotroph.

metatroph *n.* An organism that uses organic nutrients as sources of both carbon and nitrogen (Lincoln et al. 1985).

methanotroph *n.* An organism that metabolizes methane (*e.g.*, a symbiont in a mytillid-mussel species) (MacDonald et al. 1990).

minerotroph *n.* An organism that is nourished by minerals (Lincoln et al. 1985).

mixotroph *n.* An organism that is both autotrophic and heterotrophic (Lincoln et al. 1985).
Comment: Lincoln et al. (1985) are not clear whether a mixotroph can make all of its required organic compounds as an autotroph can.

monotroph *n.* An organism that uses only one kind of food, such as one plant species (*e.g.*, many insect species) (Lincoln et al. 1985). *cf.* -phagous: monophagous

mycotroph *n.* A plant that lives in symbiosis with a fungus on which it is nutritionally dependent (*e.g.*, many orchid species with mycorrhizal associations) (Lincoln et al. 1985).

myrmecotroph, myrmecobromous organism *n.* An organism that provides food for ants (*e.g.*, many aphid species) (Lincoln et al. 1985). *cf.* [2]-phile: myrmecophile

myxotroph *n.* An organism that obtains nutrients though ingestion of particles (*e.g.*, some caddisfly species) (Lincoln et al. 1985).

oligotroph *n.*
1. An organism that requires only a small nutrient supply or is restricted to a narrow range of nutrients (Lincoln et al. 1985).
2. An insect species that visits only a few plant species (Lincoln et al. 1985).
cf. -lecty: oligolecty

ombrotroph *n.* An organism that obtains nutrients largely from precipitation (Lincoln et al. 1985).

organotroph See -troph-: heterotroph.

osmotroph *n.* An organism capable of absorbing organic nutrients directly from an external medium (*e.g.*, Spanish moss, a bromeliad) (Lincoln et al. 1985).

paratroph, parasite *n.* A organism that feeds in a parasitic manner (Lincoln et al. 1985).
See parasite.

phagotroph *n.* A cell that ingests organic particulate matter (*e.g.*, blood cells that ingest foreign particles) (Lincoln et al. 1985).

photoautotroph See -troph-: autotroph: photoautotroph.

photoheterotroph See -troph-: photoorganotroph.

photolithotroph *n.* An organism that utilizes radiant energy and inorganic electron donors (Lincoln et al. 1985). *cf.* -troph-: chemolithotroph, lithotroph, photoorganotroph

photoorganotroph, photoheterotroph *n.* An organism that uses radiant energy and organic electron donors (Lincoln et al. 1985). *cf.* -troph-: photolithotroph

phototroph See -troph-: autotroph: photoautotroph.

planktotroph *n.* A organism that feeds on plankton (Lincoln et al. 1985). *cf.* [1]-phile: planktophile

polytroph *n.* An organism that feeds on a variety of different food substances or food species (*e.g.*, Bears or Humans) (Lincoln et al. 1985). *cf.* -vore: omnivore

prototroph *n.* An organism that obtains nourishment from only one source (Lincoln et al. 1985). See -troph-: autotroph.

rheotroph *n.* An organism that obtains its nutrients largely from percolating, or running, water (Lincoln et al. 1985).

saprotroph *n.* An organism that obtains nourishment from dead, or decaying, organic matter (*e.g.*, many fungus species) (Lincoln et al. 1985). *syn.* saprobe (Campbell 1996, 505) *cf.* -phage: saprophage, saprophytophage, zoosaprophage

symbiotroph *n.* An organism that obtains nourishment through a symbiotic relationship (*e.g.*, many bee species) (Lincoln et al. 1985).

syntroph *n.*
1. One member of a pair of organisms that is mutually dependent on the other member for food (Lincoln et al. 1985).
2. One member of a pair of organisms that each derive one or more essential nutrients from the other member (Lincoln et al. 1985).

zootroph See -troph-: heterotroph.
♦ **trophallactic appeasement** See behavior: appeasement behavior: trophallactic appeasement.
♦ **trophallaxis** *n.*
1. In social wasps: larval donation of salivary secretions to their adult winged sisters (Rouband 1916 in Wilson 1975, 30).
2. In ants and some species of wasps: colony members' exchange of salivary secretions (Wheeler 1918).
3. In social insects: colony members' reciprocal exchange of food and odors (Wheeler 1928 in Wilson 1975, 30).
   Note: "If we select the broadest definition allowed by Wheeler … trophallaxis must be the equivalent of all of chemical communication in the modern sense" (Wilson 1975, 30).
4. In social insects: colony members' exchange of food, odors, and tactile stimuli (Schneirla 1946).
5. Animal communication in general (LeMasne 1953).
   Note: This definition is so broad that is it not very useful.
6. In social insects: interindividual exchange of liquid food delivered by regurgitation from their crops or as secretions from special glands associated with their alimentary tracts (Wilson 1975, 207–208)
7. In some social-insect species: mutual, or unilateral exchange of alimentary liquid between, or among, colony members and heterospecific organisms (social parasites, if present) (Wilson 1975, 597).
   Note: This definition seems appropriate for wide use. "Trophallaxis" can serve as chemical communication in some species (Immelmann and Beer 1989, 319).
adj. trophallactic
syn. ecotrophobiosis, oecotrophobiosis, reciprocal feeding (Lincoln et al. 1985)
cf. biosis: trophobiosis

anal feeding, anal trophallaxis See trophallaxis: proctodeal trophallaxis.
oral trophallaxis See trophallaxis: stomodeal trophallaxis.
proctodeal trophallaxis, anal feeding, anal trophallaxis *n.* For example, in kalotermitid and rhinotermitid termites: trophallaxis that involves food's originat-

ing in an animal's hindgut and passing out of its anus (Wilson 1975, 207).
cf. -phage: coprophage
Comment: In koalas (*Phascolarctos cinereus*), young feed on a special form of feces from their mothers, appearing to show a kind of trophallaxis (Minchin 1937 in Wilson 1975, 207).

stomodeal trophallaxis *n.* For example, in a few theridiid-spider species, many species of more derived social insects, burying beetles (*Necrophorus*): trophallaxis that involves an individual's regurgitation from its crop or secretions from special glands associated with its alimentary tract (Wilson 1975, 207, 208, and references therein).
syn. oral trophallaxis
♦ **trophic cascade, cascade** *n.* A reciprocal predator-prey effect that alters the abundance, biomass, productivity, or a combination of these characteristics, of a population, community, or trophic level across more than one link in a food web (Pace et al. 1999, 483).
Comments: Trophic cascades can result from bottom-up control, top-down control, or both (Pace et al. 1999, 484–485). In a three-level food chain, abundant top predators directly cause lower abundances of mid-level consumers and indirectly cause a higher abundance of basal producers (Pace et al. 1999, 484–485). Trophic cascades occur in many kinds of ecosystems.
[Greek, trophic, referring to feeding; cascade, referring to a series of operations (Morris 1982)]
♦ **trophic egg** *n.* For example, in some ant species, the burrowing cricket *Anurogryllus muticus*: a usually degenerate, inviable egg that a worker ant feeds to other members of her colony or a mother cricket feeds to her young (Wilson 1975, 207, 414, 597; West and Alexander 1963 in Wilson 1975, 208).
♦ **trophic level** *n.* A species' position in a food chain, "determined by which species it consumes and which consumes it" (Wilson 1975, 597).
♦ **trophic parasitism** See parasitism: trophic parasitism.
♦ **tropho-** *combining form* Also, before vowels, troph-.
See -troph-.
♦ **trophobiont** See -biont: trophobiont.
♦ **trophobiosis** See -biosis: trophobiosis.
♦ **trophogenesis** See -genesis: trophogenesis.
♦ **trophology** See study of: trophology.
♦ **trophoparasitism** See parasitism: trophoparasitism.
♦ **trophotaxis** See taxis: trophotaxis.

S–Z

♦ **-trophy** *combining form* See -troph-.
cf. -topic

eutrophy *adj.* Visitation of particular kinds of flowers only by certain specialized insect pollinators (*e.g.,* orchid-bee species that visits only certain orchid species) (Lincoln et al. 1985).
cf. pollination: exendotrophy

polytrophy *n.* An organism's feeding on a variety of different food substances or species (Lincoln et al. 1985).
adj. polytrophic

♦ **-tropic** *combining form*
See -lectic.

allotropic *adj.* Referring to a floral visitor that is poorly adapted for pollinating a particular species, and uses resources from this species' flowers for only part of its food (Faegri and van der Pijl 1979, 48).

dystropic *adj.* Referring to a floral visitor that is unadapted for pollinating a particular flower species (Faegri and van der Pijl 1979, 48).
Comment: A dystropic visitor may be destructive to a flower, but it may effect pollination (Faegri and van der Pijl 1979, 48).

eutropic *adj.* Referring to a floral visitor that is fully adapted for pollinating a particular species and uses resources from this species' flowers as its main food source (Faegri and van der Pijl 1979, 48).

hemitropic, hemilectic *adj.* Referring to a floral visitor that has an intermediate degree of specialization for pollinating a particular flower species (Faegri and van der Pijl 1979, 48).

homotropic *adj.* Referring to a flower that is fertilized by its own pollen (Lincoln et al. 1985).

♦ **tropical** *adj.*
1. Referring to the Torrid Zone of the Earth between the Tropic of Cancer (23°27′ north) and the Tropic of Capricorn (23°27′ south) (Lincoln et al. 1985).
2. Referring to a climate with high temperature, high rainfall (at least part of the year), and no frosts (or very rare light frosts at night); contrasted with polar, subtropical, temperate (Lincoln et al. 1985).
Comment: The more southern islands of Hawai'i are in the tropics.

♦ **tropical rain forest** See ²community: rain forest: tropical rain forest.

♦ **tropism** *n.*
1. In organisms: an individual's turning itself, or part of itself, in a particular direction (either in the way of growth, bending, or locomotion) in response to a particular external stimulus (*Oxford*

English Dictionary 1972, entries from 1899).
2. In organisms: an individual's "forced movement" due to immediate physical and chemical effects of stimuli upon its protoplasm (Loeb in Dewsbury 1978, 10).
Note: Fraenkel and Gunn (1961, 9–10) discuss Loeb's conceptions of "tropism." Kühn (in Fraenkel and Gunn 1961, 317) called Loeb's tropisms "taxes." "Tropism" is now usually distinguished from "taxis," and Fraenkel and Gunn (1961) use "taxis," instead of "tropism" with regard to animals.
3. In organisms: an individual's involuntary response to an external stimulus; an individual's automatic reaction to a stimulus (Michaelis 1963).
4. In plants, plant-like organisms: an individual's response, such as spatial orientation in growth to stimulus direction; *e.g.,* in water in a transparent container, a seedling moves its leaves toward light and its roots away from light (Harman 1953 in Heymer 1977, 182).
adj. -tropic
syn. orientation (sometimes), taxis (sometimes)
cf. taxis; ³theory: tropism theory
[Greek *tropeē*, a turning]

♦ **-tropism**

eurytropism *n.* An organism's exhibiting a marked response, or adaptation, to changing environmental conditions (Lincoln et al. 1985).
cf. tropism: stenotropism

exotropism *n.* An organism's orientation movement of its lateral organs away from its main axis (Lincoln et al. 1985).
adj. exotropic

gamotropism *n.* Gametes' orientation to one another (Lincoln et al. 1985).
cf. taxis: zygotaxis

geotropism, epitropism, helcotropism *n.* Tropism with regard to gravity (Lincoln et al. 1985).

haptropism See tropism: stereotropism.

narcotropism *n.* An organism's "orientation movements resulting from the effects of narcotics" (Lincoln et al. 1985).

phototropism *n.* Tropism with regard to light.

stenotropism *n.* An organism's exhibiting a limited response, or adaptation, to changing environmental conditions (Lincoln et al. 1985).
cf. tropism: eurytropism

stereotropism, haptropism, thigmotropism *n.* An organism's orientation to a contact stimulus (Lincoln et al. 1985).

thermotropism *n.*
1. An organism's being adapted to warm conditions, as occurs in some ant species (Wheeler 1930, 9).
2. A plant's turning toward a temperature stimulus (Bligh and Johnson 1973, 958).
cf. ¹-phile: thermophile; taxis: thermotaxis
▶ **negative thermotropism, cryophily** *n.* An organism's being adapted to cold conditions, as occurs in some ant species and many other insect species (Wheeler 1930, 9, 11).
cf. ¹-phile: cryophile, thermophile

thigmotropism See tropism: stereotropism.

tocotropism *n.* An organism's orientation shown when it gives birth (Lincoln et al. 1985).

♦ **tropism theory** See ³theory: tropism theory.

♦ **tropodrymium** See ²community: tropodrymium.

♦ **tropoparasite** See parasite: tropoparasite.

♦ **tropophile** See ¹-phile: tropophile.

♦ **tropotaxis** See taxis: tropotaxis.

♦ **-tropous** *combining form*
n. -tropy
cf. -²phile
allotropous *adj.* Referring to an unspecialized insect species that is able to feed on many kinds of flowers (Lincoln et al. 1985).
eutropous *adj.* Referring to a specialized insect species adapted to feed on particular kinds of flowers (Lincoln et al. 1985).

♦ **-tropy** See -tropous.

♦ **true altruism** See altruism (def. 2, 4).

♦ **true-Baconian method** See inductivism.

♦ **true biosphere** See -sphere: biosphere: eubiosphere.

♦ **true communication** See communication.

♦ **true consciousness** See consciousness.

♦ **true extinction** See ²extinction: true extinction.

♦ **true heritability** See heritability.

♦ **true hibernation** See dormancy: hibernation: true hibernation

♦ **true hypothesis** See hypothesis: true hypothesis.

♦ **true imitation** See learning: true imitation.

♦ **true lek** See ²lek, arena.

♦ **true parthenogenesis** See parthenogenesis: true parthenogenesis.

♦ **true socialization** See socialization.

♦ **true symbiosis** See symbiosis: mutualism.

♦ **truism** *n.* A self-evident truth, especially one of slight importance; a statement that is so obviously true that it does not require discussion (*Oxford English Dictionary* 1972, entries from 1708); an obvious, or self-evident, fact (Michaelis 1963).
syn. axiom, bromide, platitude (Michaelis 1963)
cf. axiom, doctrine, dogma, law, principle, rule, theorem, theory
Comment: In careful criticism, a truism is always tautological, either explicitly or implicitly; two truisms are that opium puts one to sleep because of its soporific qualities, and a healthy mind requires a healthy body (Michaelis 1963).

♦ **truly altruistic behavior** See altruism: altruism: true altruism.

♦ **truly eusocial** See sociality: eusociality.

♦ **truly personal recognition** See recognition: individual recognition.

♦ **truly separate strategy** See strategy: truly separate strategy.

♦ **truly social** See eusociality.

♦ **trumpet** See animal sounds: trumpet.

♦ **trumpet call** See animal sounds: call: trumpet call.

♦ **truncation selection** See selection: truncation selection.

♦ **Tryon effect** See effect: Tryon effect.

♦ **tubicole** See -cole: tubicole.

♦ **tumor** *n.*
1. A local swelling on, or in, any human body part, especially from some abnormal autonomous growth of tissue that may or may not become malignant; a neoplasm (Michaelis 1963).
2. An organism's circumscribed, noninflammatory growth arising from its existing tissue but growing independently of the normal rate, or structural development, of such tissue and serving no physiological function (Morris 1982).
3. An organism's abnormally swollen part (Morris 1982).
[Latin *tumor* < *tumore*, to swell]
crown-gall tumor *n.* A tumor found in dicotyledons that results from incorporation of plasmid DNA from *Agrobacterium tumefaciens* (Niesbach-Klösgen et al. 1987, 223).

♦ **turfophile** See ¹-phile: turfophile.

♦ **turmoil** See ²group: turmoil.

♦ **turn** See ²group: turn.

♦ **turnover hypothesis** See ³theory: theory of island biogeography.

♦ **"turnover-pulse hypothesis"** See hypothesis: "turnover-pulse hypothesis."

♦ **turtle-homing hypotheses** See hypothesis: turtle-homing hypotheses.

♦ **tussock** See ²group: tussock.

S–Z

♦ **tutoring** See teaching.
♦ **twin** *n.*
   1. One of two young produced by an organism at the same birth event (Michaelis 1963).
   2. The counterpart, or exact mate, of another (Michaelis 1963).
   3. *v.i.* "To bring forth twins" (Michaelis 1963).
   fraternal twin *n.* A twin whose counterpart originated from a different fertilized ovum.
   identical twin *n.* A twin whose counterpart originated from the same fertilized ovum.
   Comments: Human identical twins are clones that occur about once in every 250 births in the U.S. (Allen 1998, 7, illustration; 8). Behavior and life styles of these identical twins vary with the pair, from being similar to different. In 1998, the majority of researchers believed that nature (genetics) is more important nurture (environment) in determining behaviors of human twins (Allen 1998, 9–10). On the average, human identical twins that are raised together are about 80% similar and those raised separately are about 50% similar in their characters, including health, IQ, and political views.
♦ **twin species** See ²species: sibling species.
♦ **TWINSPAN** See method: two-way indicator-species analysis.
♦ **two-parent family** See ¹family: two-parent family.
♦ **two-sided test, two-tailed test** See test: two-tailed test.
♦ **two-state character** See character: binary character.
♦ **two-way indicator-species analysis** See method: two-way indicator-species analysis.
♦ **tychocaval** See -xene: trogloxene.
♦ **tychoparthenogenesis** See parthenogenesis: tychoparthenogenesis.
♦ **tychoplankton** See plankton: pseudoplankton.
♦ **tychothelotoky** See parthenogenesis: tychothelotoky.
♦ **Tyler Prize** See award: Tyler Prize.
♦ **type** *n.*
   1. A certain general plan of structure that characterizes a group of organisms (*Oxford English Dictionary* 1972, entries from 1850).
   2. A species, or genus, that best exhibits the essential characteristics of its family, or group, and from which the family, or group, is (usually) named (*Oxford English Dictionary* 1972, entries from 1840); a taxonomic group that is considered to be representative of the next higher category in a system of classification: a *type* genus (Michaelis 1963).
   3. An individual that embodies all of the distinctive characters of a species, etc. (*Oxford English Dictionary* 1972, entries from 1840); "an organism whose structural and functional characteristics make it representative of a group, species, class, etc." (Michaelis 1963).
   [Middle French < Latin *typus* < Greek, *typus*, impression, figure, type < *typtein*, to strike]
♦ **-type, type** *combining form* "Representative of form; type" (Michaelis 1963).
   archetype, archtype, architype, arquetype *n.*
   1. One of a limited number of forms, or real essences, represented by organisms (Mayr 1982, 458).
      Note: The concept of archetype was proposed in the 19th century before Darwin published his theory of evolution in 1859 (Mayr 1982, 458).
      syn. type, *Urform* (Mayr 1982, 458)
   2. The hypothetical ancestral type; the earliest common ancestor of a line of descendants (Lincoln et al. 1985).
   See *Bauplan*.
   syn. ancestor, *Bauplan*, praeform (Lincoln et al. 1985)
   behavior type See behavior type.
   cytotype *n.* Any group of conspecific individuals that is defined cytologically (Bell 1982, 504).
   ecophene, ecad, oecophene *n.* All of the naturally occurring phenotypes produced within a given habitat by a single genotype (Lincoln et al. 1985).
   See ecad; -type, type: phenotype: ecophenotype.
   ecotype *n.*
   1. A plant ecological unit that arises as a result of genotypical response of an ecospecies to a particular habitat [coined by Turesson 1922 in Lawrence 1975, 177].
   2. A plant population that is distinguished by often quantitative morphological and physiological characters and is interfertile with other ecotypes of the same ecospecies, but is prevented from freely exchanging genes by ecological barriers (Gregor et al. 1936 in Lawrence 1957, 177).
   3. A plant phyletic unit that is adapted to a particular environment but is capable of producing fully fertile hybrids with other ecotypes of the same ecospecies (Lawrence 1957, 176–177).
   Notes: One ecotype differs from another of the same ecospecies by many

genes, and the ecotype unit is similar to a subspecies (Lawrence 1957, 177). Some ecotypes are equivalents of geographic subspecies, but more than one ecotype may be included in these, particularly ecotypes that are physiologically but not morphologically distinct. Ecotypes remain distinct through selection and isolation (Lincoln et al. 1985).

4. A local plant population that has been selected for by its habitat's edaphic and biotic conditions and expresses this in its phenotype (Mayr 1982, 277, 560, 958).

5. A locally adapted organism population (Lincoln et al. 1985).

syn. ecological race (Lincoln et al. 1985)

6. An infraspecific organism group that has distinctive traits that result from the selective pressures from its local environment (Lincoln et al. 1985); in animals, these traits can be physiological, morphological, behavioral, or a combination of kinds (Immelmann and Beer 1989, 83).

cf. ecad; ¹race: climatic race; ²group: variety; ²species: ecospecies

epigenotype See -type, type: genotype.

ethotype *n.* An individual organism that shows behavior different from other members of its population (Curio 1975 in Heymer 1977, 65).

cf. ²species: ethospecies

genophenes *pl. n.* Different phenotypes of the same genotype of a species (Lincoln et al. 1985).

genotype *n.*

1. "The genetic constitution of a biont, designated with reference either to a single trait or to a set of traits" [coined by Johannsen (1909 in Mayr 1982, 44, 781)].

Note: This term helped to cause wide acceptance of a difference between an organism's genetic and phenotypic traits (Mayr 1982, 44, 781).

2. "The particular combination of alleles carried by an individual at one or more gene loci" (Wittenberger 1981, 616).

3. "The total genetic constitution of an organism" (Mayr 1982, 958).

cf. genome

4. All of the genetic material of a cell, usually referring only to its nuclear material (Lincoln et al. 1985).

5. "All of the individuals sharing the same genetic constitution" (Lincoln et al. 1985).

syn. biotype (Lincoln et al. 1985)

6. "The specimen on which a genus-group taxon is based; the primary

type of the type species" (Lincoln et al. 1985).

syn. generitype, genetype (Lincoln et al. 1985)

cf. -type, type: phenotype

▶ **background genotype** *n.* The part of a genotype that is not primarily responsible for producing its phenotype (Lincoln et al. 1985).

syn. residual genotype

▶ **epigenotype** *n.* The entire developmental complex of gene interactions that produces a phenotype (Lincoln et al. 1985).

▶ **residual genotype** See -type, type: genotype: background genotype.

▶ **sisyphean genotype** *n.* A changing genotype found in a sexually reproducing species that is never completely adapted to a changing environment due to gene recombination in offspring (Williams 1975 in Willson 1983, 50).

[after Sisyphus, a man who was condemned by Zeus to roll a huge boulder up a mountain; the boulder forever escaped and rolled back down just as he got it near the top (Willson 1983, 50)]

haplotype *n.*

1. A set of alloantigens coded by a single allele or by closely linked genes, as occurs with antigens of the major histocompatibility system (King and Stansfield 1985).

syn. pheno-group (King and Stansfield 1985)

2. A set of closely linked genes that tend to be inherited together, as occurs with the A, B, and C loci of the human leukocyte antigen gene complex (HLA gene complex) (King and Stansfield 1985) and chloroplast-genome insertion haplotypes in an Amazonian tree species (Hamilton 1999, 129).

karyotype, caryotype *n.*

1. An individual's, or species, somatic chromosomal complement (King and Stansfield 1985).

2. A photomicrograph of an individual's metaphase chromosomes arranged in a standard sequence (King and Stansfield 1985).

mating type *n.* A group of individuals that have genotypically controlled mating behavior, such that they do not usually mate among themselves but with individuals of a complementary type (Lincoln et al. 1985).

mimotype *n.* A form that resembles another phenotypically, but not genotypically, and that occupies a similar niche in a different geographical area (Lincoln et al. 1985).

S–Z

morphotype *n.*
1. A specimen selected to represent a given intrapopulation variant or morph (Lincoln et al. 1985).
2. A list of morphological characters presumed to be present in an ancestral species (Lincoln et al. 1985).

mutant type *n.* Any type that is derived from a wild type by a mutational change (Lincoln et al. 1985).

necrotype *n.* An extinct organism (Lincoln et al. 1985).

phenotype, phaenotype, phenome, phenon *n.*
1. The statistical mean value of a sample (Johannsen 1909 in Mayr 1982, 782).
2. "The totality of characteristics of an individual [organism]" (Johannsen in Mayr 1982, 44, 959), including its biochemistry, morphology, physiology, and behavior (Dewsbury 1978, 114; Witten-berger 1981, 619), but not including its genotype (West-Eberhard 1989, 250).
3. An organism's observable properties produced by its genotype in conjunction with its environment (King and Stansfield 1985).
   Note: This definition is appropriate for wide use. "Phenotype" may also "include functionally important consequences of gene differences, outside the bodies in which the genes sit" (Dawkins 1982, 292).

syn. delete: phaenotype, phenome, phenon, reaction type (Lincoln et al. 1985)
cf. phenocopy; -type: genotype

▶ **alternative phenotype** *n.* One of two or more "forms of behavior, physiological response, or structure, maintained in the same life stage in a single population and not simultaneously expressed in the same individual" (West-Eberhard 1986, 36; 1989, 250).

allelic-switch-alternative phenotype *n.* An alternative phenotype that depends on the allele(s) present at one or more genetic-switch loci (West-Eberhard 1986, 36; 1989, 251).

combined-switch-alternative phenotype *n.* An alternative phenotype that depends on a combination of allelic and environmental factors (West-Eberhard 1989, 251).

conditional phenotype, condition-sensitive phenotype *n.* The sole alternative phenotype that a particular individual adopts, or one of a group of phenotypes that an individual adopts at a particular time (West-Eberhard 1986, 36; 1989, 251).

▶ **ecophenotype, ecad** *n.* A phenotype that exhibits nongenetic adaptations associated with a particular habitat or to a particular environmental factor (Lincoln et al. 1985).
cf. ²group: variety; -type, type: ecotype

▶ **extended phenotype** *n.* "All effects of a gene upon the world" (Dawkins 1982, 286).
cf. phenotype.
Comment: "The 'effect' of a gene is understood as meaning in comparison with its alleles. The conventional phenotype is the special case in which the effects are regarded as being confined to the individual body in which the gene sits. In practice it is convenient to limit 'extended phenotype' to cases where the effects influence the survival chances of the gene, positively or negatively" (Dawkins 1982, 286).

▶ **phenocopy** *n.*
1. A phenotype that is altered by nutritional factors, or exposure to environmental stress, during its development to a form that imitates one characteristically produced by a specific gene; *e.g.,* rickets (due to a lack of vitamin D) is a phenocopy of vitamin-D-resistant rickets (King and Stansfield 1985).
2. "An environmentally induced phenotypic variation that resembles the effect of a known gene mutation" (Lincoln et al. 1985).

▶ **phenon** *n., pl..* **phena** One of two or more different forms or phenotypes that may be encountered within a single population (Mayr 1982, 870).
syn. variety (of older literature, Mayr 1982, 870)
Comments: This includes an organism's "sexes (when there is sexual dimorphism), age stages, seasonal variants, and individual variants (morphs, and so on)" (Mayr 1982, 870). In numerical taxonomy, "phenon" replaces "taxon."

reaction type See -type, type: phenotype.

serotype, serological type *n.* An infraspecific variant characterized by antigenic structure (Lincoln et al. 1985).

wild type *n.*
1. A gene's, strain's, or species' natural or typical form that one arbitrarily designates as standard, or "normal," for comparison with mutant, or aberrant, forms (Muller in Mayr 1982, 591; Lincoln et al. 1985).
2. An organism's most frequently observed phenotype (King and Stansfield 1985).

♦ **type-I error, type-II error** See error: type-I error, type-II error.

♦ **type-1 restrictase, type-II restrictase, type-III restrictase** See enzyme: restriction enzyme (comment).

♦ **type-1 survivorship curve, type-2 survivorship curve, type-3 survivorship curve** See curve: type-1 survivorship curve, type-2 survivorship curve, type-3 survivorship curve.

♦ **type-A territory, type-B territory, type-C territory, type-D territory, type-E territory** See territory: type-A territory, type-B territory, type-C territory, type-D territory, type-E territory.

♦ **type species** See ²species: type species.

♦ **-typic** *combining form*
See -type.

endophenotypic *adj.* Referring to characters, or components, of a phenotype that are not directly adaptive and may not affect an organism's competitive abilities; contrasted with exophenotypic (Lincoln et al. 1985).

exophenotypic *adj.* Referring to characters, or components, of a phenotype that are adaptive and affect an organism's competitive abilities; contrasted with endophenotypic (Lincoln et al. 1985).

oligotypic *adj.* Referring to a group, or assemblage, largely comprised of one type of organism or species (Lincoln et al. 1985).

♦ **typical intensity** *n.* The constancy of form of a ritualized and emancipated display resulting from a range of stimulus intensities (Morris 1957 in Immelmann and Beer 1989, 319).

♦ **typogenesis** See -genesis: typogenesis.

♦ **typological species** See ²species: typological species.

♦ **typology** See study of: typology.

♦ **typolysis** *n.* A period of evolutionary history preceding the extinction of a form (Lincoln et al. 1985).
cf. -genesis: typogenesis; stasis: typostasis

♦ **typostasis** See -stasis: typostasis.

♦ **-typy**
cf. -type

symplesiotypy *n.* Organisms' common possession of a plesiotypic character state (Lincoln et al. 1985).

synapotypy *n.* Organisms' common possession of a derived (apotypic) character state (Lincoln et al. 1985).

♦ **tyrant** *n.*
1. A human king, or ruler, who exercises his power in an oppressive, unjust, or cruel manner; a despot (*Oxford English Dictionary* 1972, entries from 1227).
2. Any person who exercises power, or authority, oppressively, despotically, or cruelly (*Oxford English Dictionary* 1972, entries from 1290).
3. A person who seizes the sovereign power in a state without legal right; absolute ruler; a usurper (*Oxford English Dictionary* 1972, entries from *ca.* 1300).
4. For example, in some reptile species: a male that dominates other males in a group territory (Wilson 1975, 444).

S–Z

u

♦ **ubiquist, ubiquitist** *n.* An organism that has a world-wide distribution, effect, or influence (Lincoln et al. 1985).

♦ **ubiquitous** *adj.*
1. Referring to a ubiquist, *q.v.*
2. "Widespread, cosmopolitan, ecumenical, pandemic" (Lincoln et al. 1985).

♦ **uliginous, uliginose, helobius, paludal, paludine, paldudinous, paludose, palustrine** *adj.* Referring to an organism that inhabits swampy soil or wet muddy habitats (Lincoln et al. 1985).

♦ **ultimate cause, causation** See causation: ultimate causation.

♦ **ultimate factor** See factor: ultimate factor; causation: ultimate causation.

♦ **ultimate function of behavior** See function: ultimate function of behavior.

♦ **ultradian rhythm** See rhythm: ultradian rhythm.

♦ **ultrasound** See ¹sound: ultrasound.

♦ **ultrasound dectector** See instrument: bat detector.

♦ **umami** *n.* A person's taste of free glutumate (Willoughby 1998, C3).
cf. modality: chemoreception: gustation
Comments: Kukunae Ikeda (1908) described umami as a separate and distinct taste, our fifth taste (in addition to bitter, salty, sour, and sweet) (Willoughby 1998, C3). Umami is the brothiness taste of miso soup or the meatiness taste of sautéed mushrooms. Human mouths have specific receptors that seem to respond only to glutamate. Japanese describe the taste of glutamates as "umami" or "Ajinomoto" (a synonym for glutamate), a glutamate-based food additive. U.S. citizens describe the taste of glutamates as beef bouillon, fishy, hamburger grease, seaweed, and most commonly salty. Monosodium glutamate (MSG) produces umami, which can greatly enhance food flavor, not only changing its

taste but also how one's mouth feels. Many foods, including mushrooms, Parmesan cheese, peas, and tomatoes contain free glutamates that enhance food flavors. [Japanese *umami*, roughly translated as deliciousness (Willoughby 1998, C3)]

♦ **umbraticole** See -cole: umbracticole.

♦ **umbrophile** See ¹-phile: umbrophile.

♦ **umwelt, *Umwelt*** See -welt: umwelt.

♦ **unbalanced hermaphrodite population** See ¹population: unbalanced hermaphrodite population.

♦ **unbeatable strategy** See strategy: unbeatable strategy.

♦ **uncle** *n.*
1. The brother of a person's father or mother; a person's aunt's husband (= uncle-in-law) (*Oxford English Dictionary* 1972, entries from *ca.* 1290).
2. In the African wild dog, Humans, possibly other primates: a male that takes care of an offspring to which he is not related (Wilson 1975, 349).
cf. aunt, male care
Comment: Itani (1959) described paternal care that male Japanese Macaques, of unspecified relationship, gave to young macaques. He may have been describing "uncle behavior."

♦ **unconscious biological mimicry** See mimicry.

♦ **unconditional reflex, unconditional response, unconditioned reflex, unconditional response** See learning: conditioning: Pavlovian conditioning.

♦ **unconditional stimulus, unconditioned stimulus (UCS)** See stimulus: unconditional stimulus.

♦ **unconventional statistics** See study of: statistics: unconventional statistics.

♦ **uncorrelated-asymmetry hypothesis** See hypothesis: uncorrelated-asymmetry hypothesis.

♦ **undefended territory** See territory: undefended territory.

♦ **underdispersed distribution, underdispersion** See distribution: underdispersed distribution.

♦ **understanding** See need: understanding.

♦ **undirected tree** See -gram: unrooted tree.

♦ **undulation swimming** See swimming: undulation swimming.

♦ **unequal crossing over** See crossing over: unequal crossing over.

♦ **unguligrade** See -grade: unguligrade.

♦ **uni-** *combining form* Having, or consisting of, one only (Michaelis 1963).
[Latin *unus*, one]

♦ **unicolonial** See colonial: unicolonial.

♦ **unifactorial** See -genic: monogenic.

♦ **uniform-convergence hypothesis** See hypothesis: uniform-convergence hypothesis.

♦ **uniform distribution** See distribution: uniform distribution.

♦ **uniform-species-substance hypothesis** See hypothesis: uniform-species-substance hypothesis.

♦ **uniformitarian** *n.* One who believes in uniformitarianism, *q.v.* (*Oxford English Dictionary* 1972, entries from 1840).

♦ **uniformitarianism, uniformity, principle of** See ³theory: uniformitarianism.

♦ **unigenesis** See -genesis: unigenesis.

♦ **unilateral display** See courtship: unilateral display.

♦ **unimale society** See ²group: one-male group.

♦ **uninterrupted swimming** See swimming: uninterrupted swimming.

♦ **uniparental reproduction** See parthenogenesis: uniparental reproduction.

♦ **uniparity** See parity: uniparity.

♦ **uniparous** See -parous: uniparous.

♦ **uniquely derived character, uniquely derived character concept** See law: Dollo's law; morph: apomorph.

♦ **unisexual** See sexual: bisexual, unisexual.

♦ **unison-bout singing** See animal sounds: song: unison-bout singing.

♦ **unit-character hypothesis** See hypothesis: one-gene-one-character hypothesis.

♦ **unit of measure**
centimorgan (cM) *n.* One map unit in a genetic map, *q.v.*, that is equivalent to a 1% recombination frequency (Campbell et al. 1999, 267).
[coined by Alfred H. Sturtevant, in honor of his teacher, the U.S. embryologist Thomas Hunt Morgan (Campbell et al. 1999, 267)]
cron *n.* One million years; the basic unit of evolutionary time [coined by Huxley 1957, 454].

▶ **kilocron** *n.* One billion years [coined by Huxley 1957, 454].
cf. unit of measure: Ga

▶ **millicron** *n.* One thousand years; 1 millennium [coined by Huxley 1957, 454].

dalton (d) *n.* A unit equal to the mass of a hydrogen atom (1.67×10^{-24} grams), used for measuring molecular masses (King and Stansfield 1985).

darwin *n.* A unit of measure of rate of evolutionary change that is a factor of *e* (the base of natural logarithms, 2.718) per million years [coined by Haldane 1949 in Futuyma 1986, 398].
syn. Haldane's evolutionary unit (Lincoln et al. 1985)
cf. macarthur
[after the British evolutionist Charles R. Darwin, 1809–1982]

day (d) *n.* Twenty-four hours.

▶ **degree day** *n.* The product of number of days and their mean temperatures (Lincoln et al. 1985); *e.g.*, 70% × 20 days = 1400 degree days.
Comment: Researchers use degree days to measure the duration of an organism's life cycle or its particular growth phase (Lincoln et al. 1985).

▶ **scientist day** *n.* Eight scientist hours, *q.v.* (Lawton et al. 1998, 72).

diameter at breast height (d.b.h.) *n.* The diameter of a plant trunk at 4.5 feet (1.4 meters) from the surface of the ground in which the plant grows (Adams et al., 1994).

Ga *n.* One billion years (Price 1996, 123).
[Latin *giga*, billion; *annum*, year]

joule (J) *n.* Physics, a measure of work: the force of 1 newton acting through a distance of 1 meter and equal to 10 million ergs or 0.737324 ft-lbs.

megajoule (MJ) *n.* One million joules.

Ma *n.*
1. 1 million years (Bowring et al. 1993, 1293).
   Note: Ma, My, and Myr have no periods when they are not at the end of a sentence.
   syn. M.Y. (Flessa et al. 1975, 76), My (Benton 1995, 53), Myr (Futuyma 1986, 328; Horai et al. 1992, 32)
2. 1 million years ago (Benton 1995, 52; Price 196, 123).
[Ma < Latin *mega*, million *annum*, years (Price 1996, 123)]

macarthur (ma) *n.*
1. The rate at which the probability of a discrete event per 500 years is 0.5 (Van Valen 1973, 12).
2. The rate of extinction (Ω) that gives a half-life of 500 years (Van Valen 1973, 12).

S–Z

3. The occurrence of one origin of a discrete event per 1000 years per potential ancestor (Van Valen 1973, 12).

cf. darwin

Comment: Van Valen (1973, 23) gives computational aids for macarthurs, kilomacarthurs, and megamacarthurs. [coined by Van Valen (1973) after Robert H. MacArthur, biologist, who showed the importance of extinction in ecology]

▸ **millimacarthur (mma)** *n.* In molecular evolution, the rate of one substitution per 1,000,000 years (Van Valen 1973, 12). *syn.* pauling (Van Valen 1973, 23)

mole *n.*

1. The amount of an element equivalent to its atomic weight expressed in grams (Curtis 1983, 42).
2. The amount of a substance equivalent to its molecular weight expressed in grams (Curtis 1983, 42).

Comments: A mole contains 6.02×10^{23} molecules, a number referred to as Avogadro's number. A mole of hydrogen atoms is 1 g; a mole of water is 18 g of water molecules.

pH *n.* The negative logarithm (base 10) of the hydrogen ion concentration, $[H^+]$ (in moles per liter), of a solution (Michaelis 1963; Curtis 1983, 42).

syn. pH (Michaelis 1963)

Comment: pH expresses relative acidity and alkalinity, with a pH of 0 being highly acidic, pH of 7 being neutral, and pH of 14 being highly alkaline. Human stomach acid is pH 1–2; Coca Cola, 3; human blood and tears, 7.2; and household ammonia, 12 (Campbell 1996, 49).

[from *p*otential of *H*ydrogen]

PART OF THE pH SCALE

CONCENTRATION OF H^+ IONS (MOL L^{-1})	pH	CONCENTRATION OF OH^- IONS (MOL L^{-1})
$1.0 = 10^0$	0	10^{-14}
$0.1 = 10^{-1}$	1	10^{-13}
$0.000001 = 10^{-6}$	6	10^{-8}
$0.0000000000001 = 10^{-14}$	14	10^0

scientist-hour *n.* One hour of continuous work by a scientist (inferred from Lawton et al. 1998, 72).

year *n.* The time that it takes the Earth to revolve around the Sun once, consisting of 365.25 days and now reckoned as beginning January 1 and ending December 31 and consisting of 365 days in regular years and 366 in leap years (Michaelis 1963).

abbr. y, yr (with or without periods)

▸ **flight year** *n.* The year when a group of birds makes an unpredictable appearance in a particular locality (Kricher and Morrison 1988, 307).

Comments: The unpredictable appearance is due to a mass movement called an "irruption" (Kricher and Morrison 1988, 307). Birds in at least ten seed-eating species and six raptorial species exhibit flight years, which may be caused by food shortages in their breeding areas.

▸ **light year (l.y.)** *n.* The distance that light (or any other electromagnetic radiation) travels in a vacuum in 1 year; 9.4605×10^{12} km; 63,240 astronomical units; 0.30660 parsec (Mitton 1993).

Comment: In 1 year, light in a vacuum travels 5,912,000,000,000 miles (about 6 trillion miles), or 9,460,500,000,000 kilometers.

▸ **mast year** *n.* A year when a tree species shows masting (Kricher and Morrison 1988, 302).

▸ **scientist year** *n.* Two-hundred fifty scientist days, *q.v.* (Lawton et al. 1998, 72).

Comment: Some highly enthusiastic scientists work more than 250 days per year (personal observation).

▸ **wasp year, outbreak, outbreak year, population outbreak** *n.* A year during which adult yellowjackets are very common, as seen in the continental U.S., especially in the northwest (Akre 1995, 26).

Comment: Many people abandon areas where yellowjackets are common, *e.g.,* fishing, hiking, picnic, and log-milling and -staging locations and crop fields.

▸ **water year** *n.* A year that starts when soil of an area is fully charged with moisture (inferred from Adams et al. 1994, 4).

Comment: In the Fernow Experimental Forest, WV, the water year is from May 1 of a first year through April 30 of a second year (Adams et al. 1994, 4).

♦ **unit of natural selection, unit of selection** *n.*

1. A biont (Darwin 1859; Mayr 1982; Thornhill and Alcock, 1983, 9; many other biologists).
2. An individual gene (Fisher 1930; Haldane 1957; Dawkins 1978, 1982).
3. Groups of organisms (Wilson, 1983).

cf. selection: group selection

Comment: Lloyd (1992) discusses units of natural selection.

♦ **universals** *pl. n.* Ethologically speaking, behavior patterns that occur in the same context in all Humans regardless of their culture, or race; each elicits a particular same response, has the same meaning, and is interpreted in the same way across cultures (*e.g.,* crying, laughing, smiling, expressions of mistrust, expressions of sadness, pain, or threat) (Eibl-Eibesfeldt 1973, Morin and Piattelli-Palmarini 1974 in Heymer 1977, 186).

♦ **universe** See ²population: universe.

♦ **univoltine** See -voltine: univoltine.

♦ **univore** See -vore: univore.

♦ **univorous** See -vorous: univorous.

♦ **Unken reflex** See reflex: Unken reflex.

♦ **unkindness** See ²group: unkindness.

♦ **unplanned-multiple-comparisons procedure** See statistical test: multiple-comparisons test: unplanned-multiple-comparisons procedure.

♦ **unrooted phylogenetic tree** See -gram: unrooted phylogenetic tree.

♦ **unstable equilibrium** See equilibrium: unstable equilibrium.

♦ **upper temperature survival limit** See temperature survival limit: upper temperature survival limit.

♦ **Upsilon Andromedae** See star: Upsilon Andromedae.

♦ **upward classification** See grouping.

♦ **upward stretch** See display: upward stretch.

♦ **Urform** See -type: archetype.

♦ **urge** *n.*
1. Something's impelling, or prompting, motive, stimulus, or force; inciting or stimulating; exerting pressure or constraint (*Oxford English Dictionary* 1972, entries from 1645).
2. In animal's "manifest motivation as expressed in behavior;" distinguished from a drive when it is defined as an animal's "latent readiness to act in a particular manner" (Immelmann and Beer 1989, 80).
cf. drive
Comment: The *Oxford English Dictionary* (1972) gives many definitions of the verb "urge" as it relates to Humans.

♦ **urination** See elimination: urination.

♦ **urination ceremony** See ceremony: urination ceremony.

♦ **urine licking** *n.* One animal's licking up the urine of a conspecific, *e.g.,* an alpha male wolf's licking the urine of an alpha female, a female wolf's licking urine of

young, or a male ruminant's licking up a female's urine (Heymer 1977, 83).
cf. Flehmen

♦ **urine marking** See behavior: marking behavior: urine marking.

♦ **urine sampling** *n.* In many mammal species: a male's testing the readiness of a female for mating by approaching her during urination and allowing her urine to run over his nose while he smells it, which may be followed by *Flehmen* (Heymer 1977, 84).

♦ **urine spraying** See behavior: marking behavior: urine spraying.

♦ **urine washing** See behavior: marking behavior: urine washing.

♦ **urocortin** See hormone: urocortin.

♦ **urophile** See ¹-phile: urophile.

♦ **urophilia** See -philia: paraphilia.

♦ **use and disuse, principle of** See law: Lamarck's first law.

♦ **utility** *n.*
1. A useful, advantageous, or profitable thing, feature, etc.; a use; used chiefly in the plural: utilities (*Oxford English Dictionary* 1972, entries from 1483).
2. An economic term used by biologists who study optimality that is roughly equivalent to an organism's "satisfaction" or "pleasure" (Dawkins 1986, 30).
syn. long-term optimality (Rubenstein 1980 in Dawkins 1986, 30)
3. An organism's short-term optimization (*e.g.,* energy intake per unit time) (McFarland and Houston 1981 in Dawkins 1986, 31).
syn. goal function (McFarland and Houston 1981 in Dawkins 1986, 31)
4. An organism's long-term optimization (*e.g.,* energy intake per lifetime) (McFarland and Houston 1981 in Dawkins 1986, 31).
syn. cost function (McFarland and Houston 1981 in Dawkins 1986, 31)
5. A quantity that is maximized in the choice behavior of a rational economic person (McFarland 1985, 442).
6. Benefit, or negative cost, of a nonhuman animal (McFarland 1985, 443).

♦ **utility function** See function: utility function.

♦ **utilization time** See time: utilization time.

S–Z

V

♦ **vacancy-chain process** See process: vacancy-chain process.

♦ **vacant niche** See niche: vacant niche.

♦ **vacuum activity** See activity: vacuum activity.

♦ **vacuum behavior** See behavior: vacuum behavior.

♦ **vagility, mobility** *n*. A tendency of an organism, or population, to change its location, or distribution, with time (Lincoln et al. 1985).

♦ **vagina** See organ: copulatory organ: vagina.

♦ **vaginicole** See -cole: vaginicole.

♦ **vagrant** *adj*. Referring to organisms that are wind blown or move about by their own activity (Lincoln et al. 1985).

♦ **vague mimicry** See mimicry: vague mimicry.

♦ **validity** *n*. How well the behavior units in a person's study and the methods he employs answer a research question or test a research hypothesis (Storz 1973, 282; Lehner 1979, 75).

external validity *n*. Validity with regard to an entire animal population, other species, or other situations (Lehner 1979, 75, personal communication).

internal validity *n*. Validity with regard to a person's experimental sample (Lehner 1979, 75).

♦ **value** *n*. The material, or monetary, worth of a thing; the amount at which it may be estimated in terms of some medium of exchange or other standard of a similar nature (*Oxford English Dictionary* 1972, entries from 1303).

reproductive value (RV) *n*.
1. Demographically, "the standard measure of the contribution of an individual to the next generation" (Wilson 1975, 93).
2. Demographically, "the relative number of female offspring remaining to be born to each female of age x" (Wilson 1975, 93).
3. "An organism's age-specific expectation of reproducing" (references in Willson 1983, 18).
4. A measure of a female's expected offspring number when she is a particular age (Lincoln et al. 1985).

♦ **van Valen's law** See law: law of constant extinction.

♦ **variable** *n*.
1. The actual property measured by individual observations (Sokal and Rohlf 1969, 9). *syn*. character (Sokal and Rohlf 1969, 9)
2. A property with respect to which individual observations in a sample differ in some ascertainable way (Sokal and Rohlf 1969, 11).
cf. observation, parameter, ²population, sample of observations

CLASSIFICATION OF BIOLOGICAL VARIABLES[a]

I. measurement variable
   A. continuous variable
   B. discontinuous variable
II. ranked variable
III. qualitative variable

---

[a] Sokal and Rohlf (1969, 11).

continuous variable *n*. A variable that has all possible values in its range, at least theoretically; *e.g.*, 110, 111, 112, etc. and all values between these numbers, such as 110.1, 110.11, 110.111, etc. (Sokal and Rohlf 1969, 11).

dependent variable *n*. A variable (*e.g.*, a behavioral one) that is influenced, or controlled, by one or more independent variables (Sokal and Rohlf 1969, 406).
syn. parameter (restricted to dependent variable by some workers, used to mean

both dependent and independent variables by other workers)

cf. parameter

discontinuous variable, discrete variable, meristic variable *n*. A variable that has only certain fixed numerical values with no intermediate values between them (*e.g.,* the number of plants in a quadrant or the number of grasshoppers in a field) (Sokal and Rohlf 1969, 11).

in-the-head variable See variable: intervening variable.

independent variable *n*. A variable (*e.g.,* a behavioral one) that influences, or controls, a dependent variable (Sokal and Rohlf 1969, 406).

intervening variable *n*.

1. An animal's internal, directly unobservable psychological process that in turn affects behavior (Tolman 1958 in Lehner 1979, 63).
   syn. in-the-head variable (Lehner 1979, 63)
2. A coefficient in an equation related to behavior that is introduced to express the quantitative relationship between its dependent and independent variables (Immelmann and Beer 1989, 72).
   syn. drive (Immelmann and Beer 1989, 157)
3. A process, state, or entity that accounts for the relationship between dependent and independent variables in an equation related to behavior (Immelmann and Beer 1989, 157).

syn. drive, hypothetical construct (Immelmann and Beer 1989, 157)

nuisance variable *n*. An undesired source of variation that may affect a dependent variable being measured and may bias results of an experiment (Lehner 1979, 63).

qualitative variable, attribute *n*. A variable that indicates a data class, or category, but not magnitude of data (*e.g.,* dead or alive; African, American, Asian, or European) (Sokal and Rohlf 1969, 12).

quantitative variable *n*. A variable that varies in magnitude.

random variable *n*. "A variable that can assume any of a given set of values with assigned probability" (Lincoln et al. 1985).

cf. sample of observations, observation

ranked variable, ordered variable *n*. A variable of an ordered set that has a specific relation to other variables (*e.g.,* first, second, or twentieth) (Sokal and Rohlf 1969, 11).

Comment: Ranked variables are expressed only as ranks (*e.g.,* 1, 2, 3, 4, 5), with no indication of the size of intervals between them (Sokal and Rohlf 1969, 11).

♦ **variable-interval reinforcement schedule** See reinforcement schedule: variable-interval reinforcement schedule.

♦ **variable-number-of-tandem-repeats polymorphism marker (VNTR marker)** See marker: variable-number-of-tandem-repeats-polymorphism marker.

♦ **variable-ratio reinforcement schedule** See reinforcement schedule: variable-ratio reinforcement schedule.

♦ **variance** *n*. In statistics, the dispersion of members of a population about their mean, calculated as a sample mean or population mean.

additive-genetic variance *n*. The fraction of total variance in phenotype among individuals due to differences between these individuals in breeding value; variance in breeding values (Falconer 1960, 135; Wade 1992b, 149).

Comment: This component of total genetic variation is used as a measure of degree of resemblance between a parent and offspring. It enables a prediction of a population's response to natural, or artificial, selection when the response is quantified as the change in the mean value of phenotype distribution. It describes the extent to which phenotypic change is possible by natural selection, and it is, thus, the most frequently used definition of heritability (Wade 1992b, 149–150).

environmental variance (V_E) *n*. In genetics, the statistical variance of a character (in an entire population) that is caused by external environmental factors (Wilson 1975, 68).

genetic variance (V_G) *n*.

1. In genetics, the statistical variance of a character that is caused by genetic factors (Wilson 1975, 68).
2. In genetics, the part of phenotypic variance of individuals in a population that is produced by differences, or changes, in their genetic constitution, such as mutation or recombination (Lincoln et al. 1985).

cf. ¹heritability; variance: phenotypic variance

phenotypic variance *n*. In genetics, the total variation in a particular phenotypic character, comprising both environmental and genetic variance (Lincoln et al. 1985).

cf. ¹heritability; variance: genetic variance

population variance (σ^2) *n*. In statistics, the variance that is calculated using all members of a population; the square of the population's standard deviation. Population variance = $\sigma^2 = \sum(\bar{x} - x_i)^2/N$, when $\bar{x}$ is the population mean of the variable x, x_i is a quantitative measure of a specific variable *x*, and *N* is the population size (Sokal and Rohlf 1969, 54).

residual variance *n.* In statistics, the part of the variability of a dependent variable, in an analysis of variance or regression analysis, that cannot be attributed to a specific source of variation (*e.g.*, an independent variable) (Lincoln et al. 1985).

sample variance (*s²*) *n.* In statistics, the variance that is calculated from a sample from a population. Sample variance = $s^2 = \Sigma(\bar{x} - x_i)^2/(N-1)$; the symbols are the same as for population variance.

Comment: Sample variance is "the most commonly used statistical measure of variation (dispersion) of a trait within a population" (Wilson 1975, 597). The denominator $(N-1)$ is a correction for sampling error, but when *N* is reasonably large, the correction is minor (Hartl 1987, 219).

standardized variance (*I*) *n.* The population variance of a character divided by its mean value squared (Crow 1958 in Cabana and Kramer 1991, 228).

Comments: The *I* of a population is equivalent to its opportunity for selection, *q.v.* (Cabana and Kramer 1991, 228).

♦ **variance enhancement** *n.* The increase in the reproductive-potential variance among members of a parent's brood that results in more brood members' being brought to lie above, or below, their breeding threshold, thus increasing the proportion of brood that becomes breeders or potential helpers, respectively (Brown 1987a, 307).

♦ **variance utilization** *n.* A parent's utilization of aid from its offspring that are unable to breed for reasons not imposed by the parent (Brown 1987a, 299)

♦ **variation** *n.* Divergence among individuals of a group, specifically an individual's difference from conspecifics that cannot be ascribed to a difference in age, sex, life-cycle stage, or a combination of these factors (King and Stansfield 1985).

Comment: Variations of evolutionary significance are gene-controlled phenotypic differences (King and Stansfield 1985).

coarse-grained variation *n.* Variation in an organism's environment with a period greater than its life span (Futuyma 1986, 39).

cf. environment: fine-grained environment; variation: fine-grained variation

continuous-phenotypic variation, continuous variation *n.* The existence of different gradations of phenotypic (and its underlying genotypic) variation within a population (Darwin 1859 in Mayr 1982, 738).

syn. individual differences, individual variation, fluctuating variation (Mayr 1982, 738)

cf. law: Galton's law of ancestral inheritance; variation: discontinuous variation

Comment: Strictly speaking, "continuous variation" is not a separate category from "discontinuous variation," *q.v.*, although it was believed to be by breeders and naturalists well into the first quarter of the 20th century (Mayr 1982, 683).

discontinuous (phenotypic) variation *n.* A striking deviation from a species "type" ("normal individual") that is not connected to it by a series of intermediates (*e.g.*, an albino) (Mayr 1982, 682–683).

cf. variation: continuous variation

Comment: Essentialists considered discontinuous variations to be bases of new species. Discontinuous variation is not really a category separate from continuous variation, contrary to Darwin's belief, and this caused conceptual difficulties for him. An individual discontinuous variation is a member of a species at one extreme of a continuum of variation (Mayr 1982, 682–683).

fine-grained variation *n.* Variation in an organism's environment within a period less than its life span (Futuyma 1986, 39).

cf. environment: coarse-grained environment; variation: coarse-grained variation

geographic variation *n.* The phenomenon that a species may have different, geographically isolated forms or subspecies that are often connected with each other by a graded chain of intermediates (Mayr 1982, 411, 566).

cf. speciation: allopatric speciation, geographical speciation

Comment: "The principle of geographic variation" greatly helped to destroy the "essentialist-species concept" (Mayr 1982, 411, 566).

♦ **variegated-leaf chimera** See chimera: variegated-leaf chimera.

♦ **variety** *n.*

1. A deviation from the species type (Mayr 1982, 288).

Note: Linnaeus "did not distinguish between inheritable and noninheritable varieties, nor between those that refer to individuals and those that represent different populations (such as domestic and geographic races). This confusion continued for two hundred years, and some residues of it can be found even in the contemporary literature" (Linnaeus 1751, 158, in Mayr 1982, 288). Linnaeus is generally credited with having formalized the concept "variety."

2. A plant changed by an accidental cause (Mayr 1982, 640).

Note: Linnaeus wrote that there are as many varieties as there are different plants produced from the seed of the same species. He considered a variety to be what we now call "nongenetic modifications of phenotypes" (Mayr 1982, 640).

3. A nongenetic climatic animal variant, breed of domestic animal, intrapopulation genetic animal variant, or a geographical animal variant (Linnaeus 1751 in Mayr 1982, 640).

4. A population of organisms; a geographically isolated portion of a species; a variant (or aberrant individual) within a population (Darwin in Mayr 1982, 43, 268).

Note: By using "variety" and "form" instead of "population" and "individual," Darwin inadvertently made his writing confusing and confounded two rather different modes of speciation, geographic and sympatric speciation (Darwin in Mayr 1982, 43, 268).

5. A cultivated plant variant (cultivar) or an intrapopulational variant (Mayr 1982, 641). *abbr.* cv

6. An animal geographic race (Mayr 1982, 641).

7. An individual (intrapopulational) variant (Esper in Mayr 1982, 289). *syn.* individual variety (Mayr 1982, 683)

8. An individual or a group of organism individuals that differs from a type species in certain characters and is usually fertile with any other members of its species; a subdivision of a species (Michaelis 1963).

syn. geographical race (animals), intrapopulational variant (plants) (Mayr 1982, 641)
[Medieval French *variété* < Latin *varietas* < *varius*, various]

♦ **Vascular Tropicos Nomenclatural Database** See database: Vascular Tropicos Nomenclatural Database.

♦ **vasopressin** See hormone: vasopressin.

♦ **VAST Nomenclatural Database** See database: Vascular Tropicos Nomenclatural Database.

♦ **Vavilovian mimicry** See mimicry: Vavilovian mimicry.

♦ **vector** *n.* An object that carries something else.
[Latin *vector*, carrier < *vehere*, to carry]

artificial chromosome *n.* A human-made vector that contains the essential parts of an eukaryotic chromosome, including an origin for DNA replication, a centromere, and two telomeres, all with foreign DNA (Campbell et al. 1999, 370).
Comment: An artificial chromosome behaves like a normal chromosome in mitosis and clones the foreign DNA when its cell divides (Campbell et al. 1999, 367). An

artificial chromosome can carry more DNA than a cloning vector, enabling cloning of very long DNA pieces.

cloning vector *n.* A plasmid that is a DNA molecule that carries foreign DNA into a cell and replicates there (Campbell et al. 1999, 367).
Comment: Researchers use cloning vectors for gene cloning (Campbell et al. 1999, 367).

expression vector *n.* A cloning vector that contains the requisite prokaryotic promoter just upstream of a restriction site where an eukaryotic gene is inserted (Campbell et al. 1999, 369).
Comment: Using an expression vector overcomes the problem of getting a bacterium to express a foreign gene that is linked to the promoter (Campbell et al. 1999, 369).

yeast-artificial chromosome (YAC) *n.* A cloning vector that can carry inserted fragments a million base pairs long (Campbell et al. 1999, 377).

♦ **vegetation** *n.* A mosaic of plant communities (Küchler 1964, 1).
cf. -coenosis

♦ **vegetative division, vegetative propagation, vegetative reproduction** See parthenogenesis: vegetative reproduction.

♦ **vehicle** *n.* "Any relatively discrete entity, such as an individual organism, which houses a replicator, and which can be regarded as a machine programmed to preserve and propagate the replicators that ride inside it" (Dawkins 1982, 295).
cf. gene machine

♦ **venom** See toxin: venom.

♦ **ventomedial prefrontal cortex** See organ: brain: ventromedial prefrontal cortex.

♦ **verbal behavior** See behavior: verbal behavior.

♦ **verbal learning** See learning: verbal learning.

♦ **vernal** *adj.*
1. "Belonging to, appearing in, or appropriate to spring" (Michaelis 1963).
2. "Youthful; fresh" (Michaelis 1963).
cf. prevernal
[Latin *vernalis* < *vernus*, belonging to spring < *ver*, spring]

♦ **vernal equinox** See equinox: vernal equinox.

♦ **vertebrate zoology** See study of: zoology.

♦ **vertical evolution** See ²evolution: vertical evolution.

♦ **vertigo** *n.* A person's disordered condition in which he has a sensation of whirling, either of external objects or himself, and tends to lose his equilibrium and consciousness; "swimming" in a person's head;

S–Z

giddiness; dizziness (*Oxford English Dictionary* 1972, entries from 1528); a person's state of confusion, or uneasiness, about spatial position, or movement, which may be accompanied by intense dizziness (Storz 1973, 282).

♦ **very likely** See frequency: very likely.

♦ **vesicle** See organized structure: membranous droplet.

♦ **vestibular organ** See labyrinth.

♦ **vestigial character** See character: vestigial character.

♦ **vestigiofossil** See fossil: trace fossil.

♦ **veterinary ethology** See study of: ethology: applied ethology.

♦ **viability** *n.* An organism's capacity to live, grow, germinate, or develop (Lincoln et al. 1985).
adj. viable

♦ **vibratile pollination, vibration, *vibración*** See pollination: buzz pollination.

♦ **vibration dance, vibratory dance** See dance: bee dance: dorsoventral abdominal vibration.

♦ **vibrational communication** See communication: vibrational communication.

♦ **vibrotaxis** See taxis: vibrotaxis.

♦ **vicar, vicarious species, vicariad** See ²species: vicarious species.

♦ **Vicar-of-Bray hypothesis** See hypothesis: hypotheses of the evolution and maintenance of sex: Vicar-of-Bray hypothesis.

♦ **vicariance biogeography** See study of: geography: biogeography: vicariance biogeography.

♦ **vicarism, vicariation** *n.* The existence of vicars, q.v., that replace each other geographically (Lincoln et al. 1985).
pseudovicarism *n.* The existence of pseudovicars that occupy different geographical areas (Lincoln et al. 1985).

♦ **vigilance** *n.* Especially in prey animals: an animal's alertness; its readiness in detecting events that are potentially of serious concern to it or its companions (Immelmann and Beer 1989, 323).
syn. vigilant behavior (Desportes et al. 1989, 771)
social vigilance *n.* An animal's alertness to signs of possible sexual, or aggressive, interactions with others, or for location signals that ensure that it does not lose contact with its group (Immelmann and Beer 1989, 324).

♦ **vigilant behavior** See vigilance.

♦ **violence** *n.* An animal's behavior that "inflicts some form of injury" to another animal (Southwick 1970, 1).

♦ **viral sex** See ³sex: viral sex.

♦ **Virchow's aphorism** *n. Omnis cella e cella* (Virchow 1855 in Mayr 1982, 658):

every cell [arises] from a pre-existing cell.
Comment: This concept is widely accepted by biologists.

♦ ¹**virgin** *n.*
1. A person of either sex who remains in the state of chastity (*Oxford English Dictionary* 1972, entries from *ca.* 1300).
2. A woman, especially a young one, who remains in a state of inviolate chastity; "an absolutely pure maiden or maid;" an old maid, a spinster (*Oxford English Dictionary* 1972, entries from 1310).
3. In many insect species: a female that produces fertile eggs by parthenogenesis (*Oxford English Dictionary* 1972, entry from 1883), *e.g.,* some walking-stick or aphid species.
4. A female animal that has not copulated (Lincoln et al. 1985).

♦ ²**virgin** *adj.* Pertaining to a native habitat, fauna, or flora essentially unaffected by human activities (Lincoln et al. 1985).

♦ **virgin birth** See parthenogenesis.

♦ **Virginia Twin Registry** *n.* One of the world's larger data banks on human twins (Allen 1998, 22).

♦ **viriniparity, virginiparous** See parity: virginiparity.

♦ **"virtual-model mimicry"** See mimicry: "virtual-model mimicry."

♦ **virus** *n., pl.* **viri, viruses**
1. "Any of a class of filterable, submicroscopic pathenogenetic agents, chiefly protein in composition but often reducible to crystalline form, and typically inert except then in contact with certain living cells" (Michaelis 1963).
syn. filterable virus (Michaelis 1963)
2. Any virulent substance developed within an animal's body and capable of transmitting a specific disease (Michaelis 1963).
3. "Venom, as of a snake" (Michaelis 1963).
4. Any of various submicroscopic pathogens that consist of a core of a single nucleic acid (a capsid) surrounded by a protein coat (envelope), having the ability to replicate only inside a living cell (Morris 1982; Campbell 1996, 330).
Note: Human virally caused diseases include chicken pox, the common cold, influenza, polio.
5. A specific pathogen (Morris 1982).
cf. living
Comments: Some researchers consider viruses to be organisms.
[Latin, *virus*, poison, slime]
arbovirus *n.* An arthropod-borne virus (*e.g.,* the West Nile Arbovirus, which killed birds and Humans in New York in 1999) (Wadler 1999, A25).

CLASSES OF ANIMAL VIRUSES GROUPED BY THEIR TYPE OF NUCLEIC ACID[a]

I. dsDNA virus
   A. adenovirus (respiratory disease, tumors in some nonhuman animals)
   B. herpes virus
      1. Epstein-Barr virus (Burkitt's lymphoma, mononucleosis)
      2. herpes-simplex virus (cold cores)
      3. herpes-simplex II virus (genital sores)
      4. varicella zostervirus (chicken pox, shingles)
   C. papova virus
      1. papilloma virus (cervical cancer, human warts)
      2. polyoma virus (tumors in some nonhuman animals)
   D. pox virus
      1. cowpox virus
      2. smallpox virus
      3. vaccinia virus
II. dsDNA = reovirus
   A. diarrhea viruses
III. ssDNA = parvovirus
   A. roseola virus
IV. ssRNA that can serve as mRNA
   A. picornivirus
      1. enteric viruses
      2. poliovirus
      3. rhinovirus (common cold)
   B. togavirus
      1. encephalitis viruses (encephalitis)
      2. rubella virus
      3. yellow-fever virus (yellow fever)
V. ssRNA that is a template for mRNA
   A. orthomyxovirus
      1. influenza viruses (influenza)
   B. paramyxovirus (measles, mumps)
   C. rhabdovirus (rabies)
VI. ssRNA that is a template for DNA synthesis (retrovirus)
   A. human-immunodeficiency viruses (= HIVs) (acquired immunodeficiency syndrome)
   B. RNA-tumor viruses (leukemia, etc.)

[a] Campbell (1996, 330). The diseases produced by these viruses are in parentheses. The subclasses within each class differ mainly in their capsid structures and in the presence, or absence, of envelopes. dsDNA = double-stranded DNA; ssDNA = single-stranded DNA; ssRNA = single-stranded RNA.

Asian Flu Virus of 1957 *n.* A virus that caused an epidemic in Humans 1957 (Associated Press 1998, 26).
Comments: This virus probably mutated in pigs (Associated Press 1998, 26).

herpsevirus *n.* A virus with an envelope derived from its host's nuclear membrane, a capsid with double-stranded DNA, and reproduction within a host's cell nucleus (Campbell 1996, 331).
Comments: Herpseviruses use a combination of their own and host enzymes to replicate and transcribe their DNA. A herpsevirus can become integrated into a host cell's genome as a "provirus" (Campbell 1996, 331). Herpseviruses cause cold sores and genital sores in Humans.

Hong Kong Flu Virus of 1968 *n.* A virus that caused an epidemic in Humans in 1968 (Associated Press 1998, 26).
Comments: This virus probably mutated in pigs (Associated Press 1998, 26).

human-immunodeficiency virus (HIV) *n.* A retrovirus that causes acquired-immunodeficiency syndrome (AIDS) in Humans (Campbell 1996, 331; 873, illustration).
Comments: A family of HIVs causes AIDS (Goudsmit in Ryan 1997, 17). Researchers first recognized AIDS in the U.S. in 1981 (Campbell 1996, 873–875). Virologists identified its causative agent as the HIV in 1983 and found it in preserved blood samples from Africa and England from 1959. An HIV virus might have been extant first as proto-1B and evolved from simian immunodeficiency viruses (Ryan 1997, 17). HIV-1A and -C are the main lethal strains in Africa; HIV-1B, Australia, Japan, U.S., and Western Europe; HIV-1E, Southeast Asia. Viral sex between HIV-1A and an unknown strain might have given rise to HIV-1E. The HIV readily infects helper-T cells of immune systems (Campbell 1996, 873–875). Factors that make it difficult to eradicate AIDS include the fact that the HIV can become a provirus and that the HIV mutates extremely rapidly. It takes an average of 10 years for an HIV infection to develop into full-blown AIDS in Humans older than infants. Infants can be infected with HIV *in utero.* HIV is transmitted in infected blood, breast milk, cells, and semen. Some antiviral drugs (*e.g.,* AZT, ddC, and ddI) inhibit HIV reverse transcriptase and may extend a patient's life but do not completely eliminate HIV.

JC virus *n.* A human virus first isolated in 1971 from a person whose initials are J.C.
Comments: This virus causes the disease progressive multifocal leukoencephalopathy, which is similar to multiple sclerosis but is much more rapid, killing patients in about 6 months. This virus has seven major types and about 20 subtypes. It evolves very slowly and can be used to trace

dispersal of human populations over thousands of years.

1918 Flu Virus *n*. A virus that killed about 30 million people (including 700,000 in the U.S.) in 1918 (Associated Press 1998, 26).

syn. Spanish flu (Associated Press 1998, 26)
Comments: This virus apparently was a virus mutation that evolved in the U.S. Pigs and U.S. troops who were mobilized for World War I spread it (Associated Press 1998, 26). The epidemic from this virus was the deadliest of all known human epidemics. Flu viruses from swine might be the most virulent for Humans.

provirus *n*. A form of a virus that is integrated into its host cell's genome (Campbell 1996, 331).
Comment: Herpseviruses and human immunodeficiency viruses can become proviruses (Campbell 1996, 331, 874).

retrovirus *n*. An RNA virus that transcribes DNA from an RNA template with a reverse transcriptase (Campbell 1996, 331).
Comment: Retroviruses integrate as proviruses composed of transcribed DNA into a host's genome (Campbell 1996, 331).
[Latin *retro*, backward + VIRUS]

RNA virus *n*. A virus with a capsid composed of RNA (Campbell 1996, 331).
Comment: Some animal viruses and bacterial viruses (= phages) and most plant viruses are RNA viruses (Campbell 1996, 331).

simian-immunodeficiency virus (SIV) *n*. A virus of nonhuman primates that might have evolved into a human immunodeficiency virus, *q.v.* (Ryan 1997, 17).

♦ **viral sex** See sex: viral sex.

♦ **viscosity** *n*. "The slowness of individual dispersal and hence the low rate of gene flow" in a population (Hamilton 1964 in Alexander 1987, 65; Wilson 1975, 597).
syn. population viscosity (Lincoln et al. 1985)

♦ **visible light** See light: visible light.

♦ **vision** See modality: vision.

♦ **visual cliff** *n*. An apparatus used to test for animal visual depth perception (*e.g.,* in chicks, goslings, ducklings, or human infants) (Storz 1973, 283).

♦ **visual imagery** *n*. A person's recreation of an experience seen with his eyes in the absence of the actual experience (*e.g.,* his creating an image of the Pantheon in his mind) (Miyashita 1995, 1719).

♦ **visuospatial-construction cognition** See cognition: visuospatial construction cognition.

♦ **visuospatial sketch pad** See working memory (comments).

♦ **vitalism** *n*. The doctrine that life and its phenomena arose from and are the product of a hypothetical vital force (Michaelis 1963); animal behavior is an ideal product of a supernatural life force, neither requiring nor being accessible to natural explanations (Gherardi 1984, 370).
cf. idealism, materialism
Comment: "By the 1920s or 1930s biologists had almost universally rejected vitalism" (Mayr 1982, 36, 52, 66).
[Old French < Latin *vitalis* < *vita*, life]

extreme vitalism *n*. "Organisms are completely controlled by a sensitive, if not thinking soul" (Mayr 1982, 114).

♦ **viticole** See -cole: viticole.

♦ **viviparity** See -parity: viviparity.

♦ **viviparous** See -parous: viviparous.

♦ **VNTR marker** See marker: variable-number-of-tandem-repeats polymorphism marker.

♦ **vocal dialect** See dialect: vocal dialect.

♦ **vocal imitation** See learning: vocal mimicry.

♦ **vocal matching** *n*. For example, in Bonobos: two animals' making calls that change in quality over the course of an encounter but remain similar to each other (de Waal and Lanting 1997, 24).

♦ **vocal mimicry, vocal mocking** See learning: vocal mimicry.

♦ **vocal-pouch brooding** See brooding: vocal-pouch brooding.

♦ **vocal repertory** See repertory: vocal repertory.

♦ **vocal stimulation** *n*. A person's presenting natural, or artificial, sounds, usually by tape recorder, to test their effects on an animal that is being studied for its auditory responsiveness (Immelmann and Beer 1989, 325).

♦ **vocalization** See animal sounds: vocalization.

♦ **volant** *adj*. Referring to a trait, or animal, that is adapted for flying or gliding (Lincoln et al. 1985).

♦ **"volcano hypothesis"** See hypothesis: hypotheses regarding the origin of the Cambrian explosion: "volcano hypothesis."

♦ **volery** See ²group: flock.

♦ **volition** *n*. A person's (consciously) willing or resolving; a person's decision, or choice, made after due consideration, or deliberation; a resolution or determination (*Oxford English Dictionary* 1972, entries from 1615).
See hypothesis: volition.
[French < Medieval Latin *volitio, -onis* < Latin *vol-*, stem of *velle*, will]

♦ **-voltine** *adj*. Referring to the number of broods or generations, per year or per season, in a species or population (Lincoln et al. 1985).

bivoltine, digoneutic *adj.* "Having two generations or broods per year" (Lincoln et al. 1985).

monovoltine See -voltine: univoltine.

multivoltine, polyvoltine, polygoneutic *adj.* Referring to a species, or population, that has several broods or generations per year (Lincoln et al. 1985). *cf.* cyclic: monocyclic

polyvoltine See -voltine: multivoltine.

quadrivoltine *adj.* Referring to a species that has four broods per year or season (Lincoln et al. 1985).

trivoltine *adj.* Referring to a species that has three broods per year or season (Lincoln et al. 1985).

univoltine, monovoltine, monogoneutic *adj.* Referring to a species or population that has one generation per year (Lincoln et al. 1985).

♦ **vomeronasal organ** See organ: Jacobson's organ.

♦ **von Baer's laws** See law: von Baer's laws.

♦ **-vore** *combining form* Feeding on; referring to an organism that feeds on a particular category of food (Lincoln et al. 1985). *cf.* -biont, -cole, guild, -phage, phagy, ¹-phile, -troph-, -zoite

ambivore *n.* An animal that feeds on grasses and broad leaved plants (*e.g.,* Bison) (Lincoln et al. 1985).

aphidivore *n.* An animal that eats aphids (*e.g.,* a lacewing species) (Lincoln et al. 1985).

apivore *n.* An animal that feeds on bees (*e.g.,* a beewolf wasp) (Lincoln et al. 1985).

calcivore *n.* A plant that lives on limestone or in soils rich in calcium salts (Lincoln et al. 1985).

carnivore, sacrophage *n.* An animal that feeds on animal flesh (*e.g.,* Tiger) (Lincoln et al. 1985).

detritivore, deposit feeder *n.* An animal that feeds on detritus (*e.g.,* a filter-feeding insect such as a caddisfly species) (Lincoln et al. 1985).

erucivore *n.* An animal that feeds on caterpillars (*e.g.,* a cuckoo bird) (Lincoln et al. 1985).

exudativore, gumivore *n.* An animal that feeds on tree gum and other exudates from trees (*e.g.,* Mourning-Cloak butterfly) (Lincoln et al. 1985).

folivore phyllophage *n.* An animal that feeds on foliage (*e.g.,* White-Tailed Deer) (Lincoln et al. 1985).

forbivore *n.* An animal that feeds on forbs (*e.g.,* the Woollybear caterpillar or the Impala) (Lincoln et al. 1985).

frugivore *n.* An animal that feeds on fruit (*e.g.,* a fruit-eating bat) (Lincoln et al. 1985).

fucivore, algicole *n.* An animal that feeds on seaweed (*e.g.,* a periwinkle snail) (Lincoln et al. 1985).

fungivore, mycetophage, mycophage *n.* An animal that feeds on fungus (*e.g.,* a fungus gnat) (Lincoln et al. 1985).

gallivore, galliphage *n.* An animal that feeds on galls (Lincoln et al. 1985).

graminivore *n.* An animal that feeds on grass (*e.g.,* a grasshopper) (Lincoln et al. 1985).

granivore *n.* An animal that feeds on grain or seeds (*e.g.,* Slate-Colored Junco) (Lincoln et al. 1985).

gumivore See -vore: exudativore.

herbivore *n.* An animal that feeds on plants (*e.g.,* European Earwig) (Lincoln et al. 1985). *cf.* -phage: macrophytophage, necrophytophage

insectivore, entomophage *n.* An animal that feeds on insects (*e.g.,* Chinese Mantid) (Lincoln et al. 1985).

larvivore *n.* An animal that feeds on larvae (*e.g.,* Caterpillar Hunter, a beetle) (Lincoln et al. 1985).

lignivore, dendrophage, hylophage, xylophage *n.* An animal that feeds on wood (*e.g.,* Eastern Subterranean Termite) (Lincoln et al. 1985).

limivore, limiphage, limophage *n.* An animal that feeds on mud (Lincoln et al. 1985).

mellivore, meliphage *n.* An animal that feeds on honey (*e.g.,* the Black Bear) (Lincoln et al. 1985).

merdivore, coprobiont, coprophage, scatophage *n.* An animal that feeds on dung or feces (*e.g.,* a dung-rolling beetle) (Lincoln et al. 1985). *cf.* -zoite: coprozoite

microbivore *n.* An animal that feeds on microbes (Lincoln et al. 1985).

molluscivore *n.* An organism that feeds on molluscs (*e.g.,* a cichlid fish) (Seehausen et al. 1997b, 894).

mucivore *n.* An animal that feeds on flies (*e.g.,* Golden-Garden Spider) (Lincoln et al. 1985).

nectarivore *n.* An organism that feeds on nectar (Lincoln et al. 1985).

nucivore *n.* An animal that feeds on nuts (*e.g.,* Eastern Chipmunk) (Lincoln et al. 1985).

S–Z

omnivore *n.*

1. An organism that consumes both animals and plants (*e.g.,* the Black Bear or Human) (Allaby 1994, 277).
2. An organism that indiscriminately feeds on all kinds of food (after Michaelis 1963; Lincoln et al. 1985).

syn. diversivore (Allaby 1994, 277), euryphage, panthophage, pleophage (Lincoln et al. 1985)

cf. -troph: polytroph

phytoplanktivore *n.* An organism that feeds on phytoplankton (= photosynthesizing plankton) (*e.g.,* a cichlid fish) (Seehausen et al. 1997b, 895).

phytosuccivore *n.* An animal that feeds on sap (*e.g.,* Cottony-Cushion Scale, an insect) (Lincoln et al. 1985).

piscivore *n.* An animal that feeds on fish (*e.g.,* the Northern Pike fish) (Lincoln et al. 1985).

syn. ichthyophage (Lincoln et al. 1985)

plurivore *n.* An animal that feeds on many kinds of food or species (*e.g.,* the Black Bear or Gypsy Moth) (Lincoln et al. 1985).

syn. euryphage, pleophage (Lincoln et al. 1985)

pollinivore *n.* A organism that consumes pollen (*e.g.,* a bee, masarid wasp, or syrphid fly) (Labandeira 1998, 57).

Comment: Pollinivory has generally been the precursor to pollination (Labandeira 1998, 57).

pupivore *n.* An animal that feeds on pupae (*e.g.,* an ichneumonid wasp or the White-Footed Mouse) (Lincoln et al. 1985).

radicivore *n.* An animal that feeds on roots (*e.g.,* Japanese Beetle) (Lincoln et al. 1985).

syn. rhizophage (Lincoln et al. 1985)

ranivore *n.* An animal that feeds on frogs (*e.g.,* Bullfrog) (Lincoln et al. 1985).

sanguivore *n.* An animal that feeds on blood (*e.g.,* Vampire Bat) (Lincoln et al. 1985).

syn. haematophage, sanguinivore (Lincoln et al. 1985)

cf. parasite: haemoparasite

sporivore *n.* A organism that consumes spores (*e.g.,* an arthropod) (Labandeira 1998, 57).

Comment: Early sporivory occurs in the Late Silurian through Early Devonian in terrestrial ecosystems (Labandeira 1998, 57).

univore *n.* An animal that feeds on one kind of food or host (*e.g.,* the Columbine Leafminer, a fly) (Lincoln et al. 1985).

zooplanktivore *n.* An organism that feeds on zooplankton (= nonphotosynthesizing plankton) (*e.g.,* a cichlid fish) (Seehausen et al. 1997b, 895).

zoosuccivore, zoosaprophage *n.* An animal that feeds on liquid animal secretions or decaying animal matter (*e.g.,* a rove-beetle species) (Lincoln et al. 1985).

cf. -zoite: saprozoite.

♦ **vulva presentation** See display: genital display.

W

♦ **W** See fitness: genetic fitness.

♦ **W-chromosome** See chromosome: W-chromosome.

♦ **waggle dance** See dance: bee dance: waggle dance.

♦ **Wagner tree** See -gram: Wagner tree.

♦ **Waldbauerian-Batesian mimicry** See mimicry: Batesian mimicry: Waldbauerian-Batesian mimicry.

♦ **walk** See ²group: walk.

♦ **Wallace's hypothesis** See effect: Wallace effect.

♦ **Wallace's law** See law: Wallace's law.

♦ **Wallace's line** *n*. An imaginary line that separates the Oriental and Australian zoogeographical regions and runs to the west of Weber's line, *q.v.*, and southwest between the Philippines and Maluku (Moluccas), between Sulawesi and Borneo, and between Lombok and Bali (Wallace 1876 in Mayr 1982, 449; Lincoln et al. 1985).
Comment: Wallace's line divides a richer continental biota from a poorer oceanic one (C.K. Starr, personal communication).
[after Alfred Russel Wallace, 1823–1913, English naturalist]

♦ **wallowing** See behavior: wallowing.

♦ **war** *n*.
1. "Hostile contention by means of armed forces, carried on between nations, states, or rulers, or between parties of the same nation or state; the employment of armed forces against a foreign power, or against an opposing party in the state" (*Oxford English Dictionary* 1972, entries from 1554).
2. Overt aggression between ant worker groups from different colonies that results in one group's appropriation of territorial space or a nest site (Hölldobler and Wilson 1990, 644).
cf. conflict

♦ **war-of-attrition game** See game: war-of-attrition game.

♦ **ward** See ²group: ward.

♦ **warm-blooded animal** See therm: homoiotherm.

♦ **warning behavior** See behavior: alarm behavior.

♦ **warning coloration** See coloration: aposematic coloration.

♦ **warning substance** See chemical-releasing stimulus: semiochemical: pheromone: alarm pheromone.

♦ **warp** See ²group: warp.

♦ **Warrawoona Formation, Warrawoona Rocks** See formation: Warrawoona Rocks.

♦ **warren** *n*.
1. A piece of land that is enclosed and preserved for breeding game (obsolete definition, *Oxford English Dictionary* 1972, entries from 1377); a place granted by a king for keeping certain animals, including hares, conies, partridge, and pheasants (*e.g.*, a rabbit warren) (Sparkes 1975).
2. A piece of land that is appropriated for rabbit breeding, formerly also hare breeding (*Oxford English Dictionary* 1972, entries from *ca.* 1400).
syn. cony warren, hare warren, rabbit warren (*Oxford English Dictionary* 1972)
3. A human-inhabited building or settlement likened to a rabbit warren; a brothel; a building or cluster of dwellings (especially if partly underground) that is densely populated by poor tenants (*Oxford English Dictionary* 1972, entries from *ca.* 1649).
4. A place where rabbits live and breed in communities (Michaelis 1963).
5. A place in a river used for keeping fish (Sparkes 1975).
See ²group: warren.
[Anglo-French *warenne*, game park, rabbit warren < *warir*, to preserve < Germanic]

S–Z

♦ **Wasmannian mimicry** See mimicry: Wasmannian mimicry.

♦ **wasp nest** See nest: wasp nest.

♦ **wasp year** See unit of measure: year: wasp year.

♦ **waste ground, waste place** *n*. An area that is seriously disturbed, but not cultivated, by Humans (*e.g.,* a roadside, vacant lot, dump, or construction site) (Voss 1972, 20).

♦ **watch** See ²group: watch.

♦ **water** See molecule: water.

♦ **water-level recorder** See instrument: water-level recorder.

♦ **water parasitism** See parasitism: water parasitism.

♦ **water pollination** See pollination: water pollination.

♦ **water spitting** See spitting: water spitting.

♦ **water year** See unit of measure: year: water year.

♦ **watershed** *n*.
1. "The line of separation between two contiguous drainage valleys" (Michaelis 1963).
   syn. divide (in Great Britain) (Bates and Jackson 1984), water parting (Michaelis 1963)
2. The whole region from which a river receives its water supply (Michaelis 1963).
3. "A drainage basin" (Bates and Jackson 1984).
   Note: This definition refers to drainage basins of streams and rivers.
syn. catchment (Griffith and Perry 1992)
Comment: The Fernow Experimental Forest has dozens of watersheds of primary through quaternary streams.
[probably from German *Wasserscheide*, water parting]

♦ **weak altruism** See altruism: weak altruism.

♦ **weak cooperation** See cooperation: weak cooperation.

♦ **weak selfishness** See selfishness: weak selfishness.

♦ **weak spite** See spiteful behavior: weak spite.

♦ ¹**weaning** *n*.
1. A person's accustoming a child or young nonhuman animal to the loss of its mother's milk; a person's causing the termination of suckling by a child or other young nonhuman animal (*Oxford English Dictionary* 1972, entries from *ca.* 960 as a transitive verb).
2. An experimenter's removal of a mother from her young (Martin 1985).
   Note: This is the most frequent use of "wean" in the behavioral literature ac-

cording to Martin (1985), who discusses the dubiousness of considering this mother removal as weaning. Counsilman and Lim (1985) disagree with Martin's definition and vice versa.
[Old English *wenian*, to accustom]

♦ ²**weaning** *n*.
1. A parent animal's decreasing the amount of, or ceasing to give, resources to its offspring (Trivers 1972).
2. A mother animal's breaking her offspring's dependence on her, especially with regard to feeding, which can be a lengthy process in some more derived mammal species (Immelmann and Beer 1989, 237).
cf. ³conflict: parent-offspring conflict, weaning conflict

♦ ³**weaning** *n*. An offspring's decreasing the amount of, or ceasing to take, resources (*e.g.,* food, space, or time) from its parent (Trivers 1972).

♦ ⁴**weaning** *n*.
1. The period during ontogeny when the rate of parental investment (*sensu* Trivers 1972) drops most sharply (Martin 1984, 1985).
2. The period when a mother stops suckling a young (Counsilman and Lim 1985).

♦ **weaning aggression** See aggression: weaning aggression.

♦ **weaning conflict** See ³conflict: weaning conflict.

♦ **weapon automimicry** See mimicry: automimicry: weapon automimicry.

♦ ¹**weaving** *n*.
1. A person's forming or fabricating (a stuff or material) by interlacing yarns, or other filaments of a particular substance, in a continuous web (*Oxford English Dictionary* 1972, verb entries from *ca.* 900).
2. A person's practicing weaving; working with a loom (*Oxford English Dictionary* 1972, verb entries from *ca.* 1000).
3. In many spider and insect species: an individual's spinning (a web or cocoon) (*Oxford English Dictionary* 1972, verb entries from 1538).
   v.i., v.t. weave

♦ ²**weaving** *n*. A person's signaling (to a ship or its occupants) by waving a flag or something used as a substitute (*Oxford English Dictionary* 1972, verb entries from 1593).

♦ ³**weaving** *n*.
1. A person's moving repeatedly from side to side; tossing to and fro; swaying his body alternately from one side to the other; pursuing a devious course; treading his way amid obstructions (*Oxford English Dictionary* 1972, verb entries from 1596).

2. A person's moving (his hand, or something held by it) to and fro, up and down (*Oxford English Dictionary* 1972, verb entries from 1607).

♦ ⁴**weaving** *n*. A person's enmeshing, entangling, or wrapping up as in a net, etc. (*Oxford English Dictionary* 1972, verb entries from 1620).

♦ ⁵**weaving** *n*.
1. A Horse's moving its head, neck, and body restlessly from side to side in its stall (*Oxford English Dictionary* 1972, verb entries from 1831).
2. In gazelles and antelopes: a buck's thrusting his horns into grass, or bushes, and then moving them from side to side parallel to the ground (Walther 1968 in Heymer 1977, 197).

Note: Weaving may be a used for enhanced signaling (Walther 1968 in Heymer 1977, 197).

♦ **Web** See World Wide Web.

♦ **Weber's fraction** See law: Weber's law.

♦ **Weber's line** *n*. An imaginary line that separates the Oriental and Australian zoogeographical regions and runs to the east of Wallace's line, *q.v.*, and south between the Maluku (Moluccas) and Sulawesi and between Timor and the Kei Islands (Wallace 1876 in Mayr 1982, 449; Lincoln et al. 1985).

♦ **weed** *n*. A plant that is in the wrong place, in a person's opinion (Allaby 1994, 408).

Comments: Weeds usually occur opportunistically in environments disturbed by Humans (Allaby 1994, 408). Weeds compete for resources used by cultivated plants, causing much crop reduction in many situations. The "weed" concept can be extended to other kinds of organisms that are in the wrong place from a person's standpoint.

♦ **weighting** *n*.
1. A person's loading (something) with a weight; supplying with additional weight; making weighty (*Oxford English Dictionary* 1972, verb entries from 1747).
2. A person's assigning a usefulness value to a taxonomic character for the purpose of using it in classification (Mayr 1982, 866, 960).

Note: Weighting is based on the assumption that a reliance on certain characters leads to better, more "natural," classifications than use of other characters. Exactly how to determine the weight of a character has remained controversial, although it is traditional to give some characters more weight than others (Mayr 1982, 866, 960).

♦ **weir** *n*.
1. An obstruction, or dam, placed in a stream to raise its water, divert it into a millrace or irrigation ditch, etc. (Michaelis 1963).

2. An aperture of a weir (def. 1) used to determine the quantity of water flowing through it (Michaelis 1963).
3. A series of wattled enclosures in a stream used to catch fish (Michaelis 1963).
[Old English *wer* > *werian*, to dam up]

♦ **Weismann's dissection theory** See ³theory: Weismann's dissection theory.

♦ **Weismann's genetic theory** See ³theory: Weismann's genetic theory.

♦ **Weismannism** See central dogma of genetics, germplasm theory.

♦ **Welch's test** See statistical test: multiple-comparisons test: planned-multiple-comparisons procedure: Welch's test.

♦ **-welt** *combining form*
Merkwelt *n*. An animal species' perceptual world; contrasted with umwelt (von Uexküll 1921 in Tinbergen 1951, 16).
umwelt, Umwelt *n*. The total sensory input of an animal (Wilson 1975, 597).
cf. environment, -welt: *Merkwelt*
Comment: Each species has its own distinctive *Umwelt* (Wilson 1975, 597).
[coined by Jacob von Uexküll (1909); German *Umwelt*, world around us, (social) surroundings (Dewsbury 1978, 14)]

♦ **Western blotting** See method: Western blotting.

♦ **wet prairie** See habitat: prairie: wet prairie.

♦ **whale wallow** See habitat: bay.

♦ **wheel position, wheel posture** See posture: wheel posture.

♦ **whisp** See ²group: whisp.

♦ **whistle** See animal sounds: whistle.

♦ **white matter** See organ: brain: white matter.

♦ **Whitten effect** See effect: Whitten effect.

♦ **whole-animal view of behavior** See black-box view of behavior.

♦ **Wicklerian-Barlowian mimicry** See mimicry: Wicklerian-Barlowian mimicry.

♦ **Wicklerian-Eisnerian mimicry** See mimicry: Wicklerian-Eisnerian mimicry.

♦ **Wicklerian-Guthrian mimicry** See mimicry: Wicklerian-Guthrian mimicry.

♦ **Wicklerian mimicry** See mimicry: Wicklerian mimicry.

♦ **widespread** See ubiquitous.

♦ **wiggling call** See animal sounds: call: wiggling call.

♦ **wildlife** *n*.
1. "Wild animals, trees, and other plants collectively" (Michaelis 1963).
2. "Any undomesticated organism" (Allaby 1994).

Note: "Wildlife" is sometimes restricted to wild animals, excluding plants (Allaby 1994). It might be appropriate to extend this term

S–Z

to all wild organisms from the standpoint of wildlife conservation.

cf. biota, flora, fauna

♦ **wild type** See -type, type: wild type.

♦ **wild-type allele** See allele: wild-type allele.

♦ **Williams' hypothesis** See hypothesis: hypotheses of the evolution and maintenance of sex: best-man hypothesis.

♦ **Williams syndrome** See syndrome: Williams syndrome.

♦ **Wilson's caste theorem** See [3]theory: Wilson's caste theorem.

♦ **Wilson's disease** See disease Wilson's disease.

♦ **wilting point** *n.* The percentage of water remaining in soil after a specified test plant has wilted under defined conditions, so that is will not recover unless it is watered (Allaby 1994).

syn. permanent wilting percentage, permanent wilting point, wilting coefficient (Allaby 1994)

Comment: If a plant is wilted too long due to lack of water, it can die (USDA Forest Service, 1993).

♦ **wind pollination** See pollination: wind pollination.

♦ **wind rose** See graph: wind rose.

♦ **wing** See [2]group: wing.

♦ **wing display** See display: wing display.

♦ **wing fanning** *n.* In the Honey Bee: while in front of her nest entrance, a worker's (*sterzelnde* bee's) raising her abdomen, extending her scent organ, and fanning her wings, thereby making a pheromone plume that helps her hive mates to find their nest entrance (Frisch 1965 in Heymer 1977, 69). See display: wing fanning.

cf. bee: fanning bee

♦ **wing rule** See law: Rensch's laws (def. 2).

♦ **Winkler Method** See method: Winkler Method.

♦ **winter solstice** See soltice: winter solstice.

♦ **wiping reflex** See reflex: expulsion reflex.

♦ **Wisconsin-general-test apparatus (WGTA)** See puzzle box: Wisconsin-general-test apparatus.

♦ **wisp** See [2]group: wisp.

♦ **witch's circle** *n.* A hunters' term for circular tracks resulting from display circling, *q.v.* (Kurt 1968 in Heymer 1977, 127).

♦ **withdrawal** *n.* The act, or process, of drawing, or taking, away; removal (Michaelis 1963).

passive withdrawal *n.* "The removal of sinus (or cistern) milk which at low intramammary pressure can be brought about by the action of suckling or milking or by cannulation without the intervention of the special contractile tissue of the mammary glands" (Cowie et al. 1951).

♦ **wolf-in-sheep's-clothing strategy** See strategy: wolf-in-sheep's-clothing strategy.

♦ **Wolff's theory** See -genesis: epigenesis.

♦ **wood, woods** See [2]community: wood.

♦ **wood choir** See chorus: wood choir.

♦ **woodland** See [2]community: woodland.

♦ **work** *n.*

1. Something that a person is doing or has done; a person's act, deed, proceeding, business (*Oxford English Dictionary* 1972, entries from 971).

2. Any structure, or impression, including a tube, burrow, nest, and track, made by an individual animal (Lincoln et al. 1985).

v.i., v.t. work

♦ **work division** See division of labor.

♦ **worker** *n.*

1. A person "who makes, creates, produces, or contrives;" applied to God and Humans (*Oxford English Dictionary* 1972, entries from the 14th century).

2. A person who works or does work of any kind (*Oxford English Dictionary* 1972, entries from 1382).

3. Domestic Dog, Horse, etc. that works (well) (*Oxford English Dictionary* 1972, entries from 1844).

See caste: worker; Appendix 1, Part 3, bee: worker bee.

♦ **worker sperm** See sperm: worker sperm.

♦ **working hypothesis** See hypothesis: working hypothesis.

♦ **working memory** See memory: working memory.

♦ **world** *n.* A division of existing, or created, things that belong to our Earth (*e.g.,* the animal world) (Michaelis 1963)

"DNA world" *n.* The present situation on Earth in which DNAs work in concert with proteins and RNAs in organism heredity (Zimmer 1995d, 39).

syn. nucleoprotein world (Illangasekare et al. 1995, 646)

"PNA world" *n.* The hypothetical situation on Earth before life evolved about 4 billion years ago when peptide-NA (PNA) was the replicating nucleic acid, rather than RNA or DNA (Peter Nielson et al. in Zimmer 1995d, 39, illustration).

Comments: The PNA world might have evolved into the RNA world, which evolved into the DNA world (Zimmer 1995d, 39). PNA is more stable than DNA, which is more stable than RNA.

"RNA world" *n.*

1. The hypothetical situation on Earth before life evolved about 4 billion years

ago when RNA replicated itself and chemical selection occurred on it (Campbell 1990, 526).

2. The hypothetical situation on Earth when RNA catalyzed all the reactions necessary for a precursor of life's last common ancestor to survive and replicate (Orgel 1994, 78).

cf. hypothesis: RNA-world hypothesis

Note: The RNA world could have occurred if RNA had two properties not evident today: a capacity to replicate without help from proteins and an ability to catalyze every step of protein synthesis (Orgel 1994, 78).

♦ **World Wide Web (WWW), Web** *n.* An imaginary space of information in the form of words, pictures, sounds, and video images that resides in the Internet (a network of computer networks, basically comprised of computers and cables through which packets of information move) (www.w3.org/People/Berners-Lee, 2000).

cf. organization: World Wide Web Consortium (Appendix 2); search engine; server

Comments: The WWW contained about 800 million pages, encompassing about 6 terabytes of text data on about 3 million servers in July 1999 (Lawrence and Giles 1999, 8 July, 107). Connections between information sites on the WWW are hypertext links in the Internet (www.w3.org/People/ Berners-Lee). Tim Berners-Lee, while working at the European Particle Physics Labo-

ratory, European Organization for Nuclear Research (CERN), proposed the idea of the WWW to his administrator in 1989 (www.w3.org/History, 2000). He originally described the WWW as "a global hypertext space in which any network-accessible information could be referred to by a single 'Universal Document Identifier.'" The High-Energy Physics Community used a prototype WWW in 1991. His dream was to make the Web a common information space in which people communicated by sharing information (www.w3.org/People/Berners-Lee, 2000). The Web would be universal, with hypertext links that could point to anything (personal, local, or global) with the information varying from drafts to highly polished documents. Further, the Web would be generally used and become a "realistic mirror (or in fact the primary embodiment)" of the ways people work, play, and socialize.

[World Wide Web, coined by Berners-Lee in 1990 (www.w3.org/People/Berners-Lee). He wanted a term for a global hypertext system that stressed that it is decentralized and allows anything to link to anything else, as does a mathematical graph in the form of a web.]

♦ **Wright's inbreeding coefficient** See coefficient: coefficient of relatedness.

♦ **Wright's principle of maximization of W** See adaptive peak.

♦ **WWW** See World Wide Web.

S–Z

♦ **X̄, x̄** See mean.

♦ **x-axis** See abscissa.

♦ **X-chromosome** See chromosome: X-chromosome.

♦ **x{p}** See mean: trimmed mean.

♦ **xaptonuon** See nuon: xaptonuon.

♦ **xenautogamy** See -gamy: xenautogamy.

♦ **xene, xeno, xen-** *combining form* Guest, stranger, foreigner (*Oxford English Dictionary* 1972).
See -xenic, -xeny.
[Greek *xenos*, stranger]

geoxene *n*. An organism that becomes a temporary, or accidental, member of soil fauna (Lincoln et al. 1985).

rheoxene *n*. An organism that occurs only occasionally in running water (Lincoln et al. 1985).

trogloxene, stygoxene, tychocaval, xenocaval *n*. An organism that is found only occasionally in underground caves or passages (Lincoln et al. 1985).
adj. trogloxenous, stygoxenous
cf. cole: cavernicole; ¹-phile: troglophile

▸ **aphyletic trogloxene** *n*. An organism that is found only occasionally in caves and does not reproduce in them (Lincoln et al. 1985).

▸ **eutrogloxene** *n*. An organism found only occasionally in subterranean caves and passages but is not fully tolerant of cave conditions and does not reproduce in cave habitats (Lincoln et al. 1985).

▸ **phyletic trogloxene** *n*. An organism that is found only occasionally in caves and reproduces in them (Lincoln et al. 1985).

▸ **planktoxene** *n*. An organism that is found only occasionally in plankton (Lincoln et al. 1985).

▸ **subtrogloxene** *n*. An organism that is found only occasionally in subterranean caves and passages and can reproduce in them but is not fully tolerant of cave conditions (Lincoln et al. 1985).

♦ **xenia** *n*.

1. A supposed direct action, or influence, of foreign pollen upon a seed, or fruit, that it pollinates (*Oxford English Dictionary* 1972, entries from 1899).
2. "The direct influence of pollen of one species upon the maternal tissues of another species after hybrid fertilization" (Michaelis 1963)

♦ **-xenic** *combining form* See -xene, -xeny.

alloxenic *adj.* Referring to two or more different parasite species that occur on different host species (Lincoln et al. 1985)

axenic *adj.* Referring to an organism culture that is free from contaminant organisms (Lincoln et al. 1985).
cf. -xenic: dixenic, monoxenic

dixenic *adj.* Referring to an organism culture with two other species (Lincoln et al. 1985).
cf. -xenic: axenic, monoxenic

homoxenic *adj.* Referring to two or more parasite species that occur in the same host species (Lincoln et al. 1985).

monoxenic *adj.* Referring to a mixed culture of an organism with one prey species (Lincoln et al. 1985).

planktoxenic *adj.* Referring to an organism that only occasionally occurs in plankton (Lincoln et al. 1985).

polyxenic See -xenous: pleioxenous.
n. polyxeny

synxenic *adj.* Referring to a culture comprised of two or more organisms under controlled conditions (Lincoln et al. 1985).

♦ **xeno-** See -xene.

♦ **xenobiont** See biont: xenobiont.

♦ **xenobiosis** See parasitism: social parasitism.

♦ **xenocaval** See -xene: trogloxene.

♦ **xenodeme** See deme: xenodeme.

♦ **xenoecy** *n.* An organism that inhabits empty domiciles, or shells, of another species (Lincoln et al. 1985).
adj. xenoecic.

♦ **xenogamy** See fertilization: cross-fertilization; -gamy: allogamy: xenogamy.

♦ **xenogenesis** See -genesis: xenogenesis.

♦ **xenogenic** *adj.* Referring to a character in one species that originated in a member of another species (Lincoln et al. 1985).

♦ **xenogenous** See -genous: xenogenous.

♦ **xenology** See study of: xenology.

♦ **xenomixis** See mixis: xenomixis.

♦ **xenomone** See chemical-releasing stimulus: semiochemical: allelochemic.

♦ **xenomorphosis** See -morphosis: heteromorphosis.

♦ **xenoparasite** See parasite: xenoparasite.

♦ **xenoparasitism** See parasitism: xenoparasitism.

♦ **xenophobia principle** See principle: xenophobia principle.

♦ **-xenous** See -xeny.

♦ **-xeny** *adj.* -xenous. See -xenic, -xene.
dixeny *n.* A parasite's using two host species during its life cycle (Lincoln et al. 1985).
euryxeny *n.* An organism's tolerating a wide range of host species (Lincoln et al. 1985).
cf. -xeny: stenoxeny
heteroxeny See -xeny: pleioxeny.
lipoxeny *n.* A parasite's leaving its host after feeding (Lincoln et al. 1985).
metaxeny See -xeny: pleioxeny.
monoxeny *n.* A parasite's using one host species during its entire life cycle (Lincoln et al. 1985).
syn. homoecism, host specificity
myrmecoxeny *n.* A plant's providing food and shelter for ants or termites (Lincoln et al. 1985).

oligoxeny *n.* A parasite's using a few host species during its life cycle (Lincoln et al. 1985).
patoxeny *n.* An organism's inhabiting forest litter for only part of its normal life cycle (Lincoln et al. 1985).
adj. patacolous, patoxenous
cf. -cole: patacole
pleioxeny, polyxeny, heteroecism, heterecism, metaxeny *n.* A parasite's not being host specific, or having two or more hosts during its life cycle (Lincoln et al. 1985).
adj. heteroxenous, pleioxenous, polyxenic
stenoxeny *n.* An organism's tolerating only a narrow range of host species (Lincoln et al. 1985).
cf. -xeny: euryxeny
trixeny *n.* A parasite's using three host species during a its life cycle (Lincoln et al. 1985).

♦ **xerarch succession, xerosere** See [2]succession: xerarch succession.

♦ **xerocole** See -cole: xerocole.

♦ **xerodrymium** See [2]community: xerodrymium.

♦ **xerohylium** See [2]community: xerohylium.

♦ **xerohylophile** See [1]-phile: xerohylophile.

♦ **xerophile** See [1]-phile: xerophile.

♦ **xerophobe** See -phobe: xerophobe.

♦ **xeropoium** See [2]community: xeropoium.

♦ **xeropoophile** See [1]-phile: xeropoophile.

♦ **xerosium** See [2]succession: xerosium.

♦ **xerotaxis** See -taxis: xerotaxis.

♦ **xerothamnium** See [2]community: xerothamnium.

♦ **xylium** See [2]community: woodland.

♦ **xylophage** See -vore: lignivore.

♦ **xylophile** See [1]-phile: xylophile.

♦ **y, yr** See unit of measurement: year.

♦ **Ȳ** See mean.

♦ *y*-**axis** See ordinate.

♦ **Y-chromosome** See chromosome: Y-chromosome.

♦ **Y-maze** See maze: Y-maze.

♦ **Y-maze globe** See maze: Y-maze globe.

♦ **YAC** See vector: yeast-artificial chromosome.

♦ **Yarrell's law** See law: Yarrell's law.

♦ **yawl** See animal sounds: yowl.

♦ **yawn, yawning** *n*.

1. A person's voluntarily opening his mouth wide, especially to swallow, or devour, something (*Oxford English Dictionary* 1972, entries from *ca*. 725).

2. A person's involuntarily making a prolonged inspiration with his mouth wide open and lower jaw much depressed, as from drowsiness or fatigue (*Oxford English Dictionary* 1972, entries from 1450).

3. A person's behavior characterized by gaping of his mouth which is accompanied by a long inspiration followed by a shorter expiration and often body stretching and sometimes vocalization (Provine 1986, 109).
   syn. gaping, q.v. (*Oxford English Dictionary* 1972, entries from 1660)
   cf. threat gape

n. yawner

v.i. yawn

cf. pattern: fixed-action pattern, stimulus: releaser

Comment: Human yawn-like behavior has been reported in fish, reptiles, birds, and other mammals; the function of yawning is unclear (Provine 1986, 110), but Immelmann and Beer (1989, 49) include yawning under comfort behavior, *q.v.*

[probably a fusion of Old English *geonian*, to yawn, and *ganian*, to gape]

clenched-teeth yawning *n*. Yawning in which a person blocks the gaping component of a yawn by keeping his jaw closed (Provine 1986, 113).

sniff yawning *n*. In females of many bovid species: a cow's raising her head, licking her snout and nostrils, and yawning (with the top of her nose wrinkled, her eyes almost, or completely, closed, and her tongue rolled upwards) after she licks her amnionic fluid, placenta after parturition, and offspring (Schenkel 1972 in Heymer 1977, 148).

threat yawning See facial expression: threat yawn.

♦ **year** See unit of measure: year.

♦ **yeast-artificial chromosome** See vector: yeast-artificial chromosome.

♦ **yellow rain** *n*. The mass defecation of airborne Giant Asian Honey Bees, *Apis dorsata* (Buchmann and Nabhan 1996, 256).

♦ **Yerkes-Dodson law** See law: Yerkes-Dodson law.

♦ **Yerkish** *n*. An artificial grammar taught to a Chimpanzee (Rumbaugh 1977 in McFarland 1985, 493).
cf. language: sign language: Ameslan

♦ **yoke** See ²group: team.

♦ **you-first principle** See principle: you-first principle.

♦ **young-male syndrome** See syndrome: young-male syndrome.

♦ **yowl** See animal sounds: yowl.

Z

♦ **Z-chromosome** See chromosome: Z-chromosome.

♦ **Zajonc effect** See effect: Zajonc effect.

♦ **zeal** See ²group: herd.

♦ *zeitgeber, Zeitgeber* See ²stimulus: *zeitgeber.*

♦ **zelotypic** See sexual: asexual.

♦ **zero-truncated-generalized-inverse-Gaussian-Poisson curve** See curve: rank-abundance curve: zero-truncated-generalized-inverse-Gaussian-Poisson curve.

♦ **zero-truncated-Poisson distribution** See distribution: zero-truncated-Poisson distribution.

♦ **zeroth order** See order: zeroth order.

♦ **zigzag dance** See dance: zigzag dance.

♦ **-zoan** *combining form* Referring to an animal (Michaelis 1963).
[Greek *zōion,* animal]
  metazoan *n.* "A multicellular animal" (Lincoln et al. 1985).

♦ **-zoic** *combining form*
cf. -zoan
  holozoic *adj.* Referring to an organism that ingests complex organic matter (Lincoln et al. 1985).
  isozoic *adj.* Referring to habitats that have similar faunas (Lincoln et al. 1985).
  polyzoic *adj.* Referring to a habitat that contains a wide variety of animal life (Lincoln et al. 1985).

♦ **zoidiogamy** See -gamy: zoidiogamy.

♦ **zoidiophile** See ¹-phile: zoidiophile.

♦ **-zoite** *combining form*
adj. -zoic.
cf. -biont, -cole, -phage, -phagy, ¹-phile, -vore, -zoan
  coprozoite *n.* An animal that lives on, or in, dung or feces (Lincoln et al. 1985).
  cf. -biont: coprobiont; -cole: fimicole; -phage: coprophage, scatophage; ¹-phile: coprophile; -vore: merdivore

histozoite *n.* An organism that lives within the tissues of another organism (given as an adjective in Lincoln et al. 1985).
syn. parasite (in some cases)
cf. -cole: piscicole, planticole

saprozoite *n.* An animal that feeds upon decaying plant, or animal, matter in the form of dissolved organic compounds.
cf. ¹-phile: saprophile, sathrophile; -vore: zoosuccivore

♦ **zone** *n.*
1. An area, tract, or section of the Earth, distinguished from other, or adjacent, areas by some special quality, purpose, or condition (*e.g.,* a mountainous zone) (*Oxford English Dictionary* 1972, entries from 1822; Michaelis 1963).
2. In ecology, a belt, or area, of the Earth that is delimited from others by the character of its organisms, climate, geological formations, etc. (*Oxford English Dictionary* 1972, entries from 1829).
3. One of the five divisions of the Earth's surface, enclosed between two parallels of latitude and named for its prevailing climate: Frigid Zones, North Temperate Zone, South Temperate Zone, and Torrid Zone (Michaelis 1963).
4. In geology, a bed, or stratum, that is usually characterized by one to several index fossil species (= guide fossil species) (Ashby 1986; Stanley 1989, 669).
   Notes: The upper and lower boundaries of a zone are approximately the same ages everywhere that this zone is found on Earth (Stanley 1989, 660). A zone is usually named for a particular species that characterizes it (Stanley 1989, 110). A zone that includes the stratigraphic ranges of more than one species that belong to the same genus may be named for this genus. Some zones are divided into subzones.

S–Z

syn. biostratigraphic zone (Stanley 1989, 660)
cf. formation, member
[Latin *zona* < Greek *zōnē*, girdle]

adaptive zone *n.*
1. The ecological pathways along which taxa evolve (Lincoln et al. 1985).
2. The way of life and organizational level of a higher taxonomic group (Lincoln et al. 1985).
3. The fundamental niche at the species level (Lincoln et al. 1985).
cf. adaptive landscape

▸ **nonadaptive zone** *n.* An intermediate area between two adaptive zones that can be regarded as a relatively unstable ecological zone that must be crossed when a group evolves from one adaptive zone to another (Lincoln et al. 1985).

▸ **transitional adaptive zone** *n.* "An evolutionary pathway across a nonadaptive zone between major adaptive zones, in which the strength of selection and rate of evolution are not significantly greater than in the adaptive zones" (Lincoln et al. 1985).

biodeterioration zone *n.* In a central-place system, "the zone around a core area within which resources have been depleted but not exhausted" (Wittenberger 1981, 613).

heterozone *n.* An organism that is not restricted to a single habitat during its development (Lincoln et al. 1985).
cf. zone: homozone

homozone *n.* An organism that is restricted to a single habitat during its development (Lincoln et al. 1985).
cf. zone: heterozone

hybrid zone, hybrid belt, intergradation zone, suture zone *n.* The overlap zone between adjacent populations, subspecies, or species in which interbreeding occurs (Lincoln et al. 1985).

intergradation zone *n.* A "boundary zone between two adjacent subspecies occupied by a population of intergrades" (Lincoln et al. 1985).
See zone: hybrid zone.

suture zone See zone: hybrid zone.

thermoneutral zone (TNZ) *n.* The ambient-temperature range within which an animal has a minimum metabolic rate and regulates its temperature by nonevaporative physical processes alone (Bligh and Johnson 1973, 958).
Comment: Nonevaporative physical processes include autonomic and behavioral responses that vary the thermal conductance between an organism and its environment, *i.e.,* by changes in body conformation and variations in peripheral vaso-

motor tone and piloerection, but excluding changes in thermal conductance due to additional external insulation (*e.g.,* bedding or clothing) (Bligh and Johnson 1973, 958).

trampling zone *n.* In a central-place system, the area around the core area that has been damaged by the many individuals' frequent movements along or through it (Wittenberger 1981, 622).

zone of thermal comfort *n.* The ambient-temperature range associated with specified air movement, humidity, and mean radiant temperature, within which a human in specified clothing expresses satisfaction with his thermal environment for an indefinite period (Bligh and Johnson 1973, 958).

♦ **zones of intergradation** See intergradation zones.

♦ **zoo** See ²group: zoo.

♦ **zoo-, zo-** *combining form* Animal; of or related to animals, or animal forms" (Michaelis 1963).
cf. -zoan
[Greek *zōion*, animal]

♦ **zoo biology** See study of: ethology: applied ethology; study of: biology.

♦ **zoo hypothesis** See hypothesis: zoo hypothesis.

♦ **zooapocrisis** *n.* The strategy, or response, of animals to environmental factors (Lincoln et al. 1985).

♦ **zoobiotic** See -biotic: zoobiotic.

♦ **zoocentric** *adj.* Referring to a comparison of a nonhuman animal with Humans or their attributes (*e.g.,* designation of the lion as the king of beasts) (Wilson 1975, 504).
cf. anthropocentricity; -morphism: zoomorphism

♦ **zoochore** See -chore: zoochore.

♦ **zoodynamics** See study of: physiology.

♦ **zooecology** See study of: ecology: zooecology.

♦ **zoogamete** See gamete: zoogamete.

♦ **zoogamy** See -gamy: zoogamy.

♦ **zoogenesis** See -genesis: zoogenesis.

♦ **zoogenetics** See study of: genetics: zoogenetics.

♦ **zoogenic, zoogenous** See -genic: zoogenic.

♦ **zoogeny, zoogeogenesis** See -genesis: zoogenesis.

♦ **zoogeography** See study of: geography: biogeography: zoogeography.

♦ **zoogony** See -parity: viviparity.

♦ **zoogygogamete** See gamete: zoogamete.

♦ **zooid** *n.*
1. Any organism, usually very small, capable of spontaneous movement and independent existence (*e.g.,* a spermatozoon or spermatozoid) (Michaelis 1963).

2. Any organism capable of independent existence that is produced other than by sexual reproduction (Michaelis 1963).

3. One of the distinct members of a compound, or colonial, organism, as in a bryozoan (Michaelis 1963).
Note: Kinds of zooids include "autozooid," "gonozooid," and "kenozoid." (Wilson 1975, 394).

4. A free-swimming medusa produced as a stage in the life of a jellyfish (Michaelis 1963).
cf. -biont

♦ **zoological garden, zoological park** See ²group: zoo.

♦ **zoology** See study of: zoology.

♦ **zoometry** See -metrics: zoometrics.

♦ **zoomimesis** See mimesis: zoomimesis.

♦ **zoomorphic** See -morphic: zoomorphic.

♦ **zoomorphism** See -morphism: zoomorphism.

♦ **-zoon** *combining form* Referring to an animal.
Comment: Many biologists now classify most animal-like, one-celled species as protistans, or some other group, not Animalia. Thus, for example, a dermatozoon is not an animal parasite according to many biologists.
[New Latin < Greek *zōion,* animal]
dermatozoon *n.* "An animal parasite of skin" (Lincoln et al. 1985).
ectozoon *n.* An "animal" parasite (Lincoln et al. 1985).
epizoon, epizoite, epizoan *n.* An organism that lives attached to an animal's body (Lincoln et al. 1985).
haematocytozoon *n.* A parasite that lives in a blood cell (Lincoln et al. 1985).
cf. parasitism: haematoparasitism
haematozoon *n.* An "animal" parasite that lives freely in blood (Lincoln et al. 1985).
cf. parasitism: haematoparasitism
microzoon *n.* "A microscopic animal" (Lincoln et al. 1985).

♦ **zoonomy** See study of: physiology: animal physiology.

♦ **zooparasite** See parasite: zooparasite.

♦ **zoopharmacognosy** See pharmacognosy: zoopharmacognosy.

♦ **zoophase** See -plont: diplont.

♦ **zoophile** See ²-phile: zoophile.

♦ **zoophobia** See phobia (table).

♦ **zooplanktivore** See vore: molluscivore.

♦ **zooplankton** See plankton: zooplankton.

♦ **zoosaprophage** See -phage: zoosaprophage.

♦ **zoosemiotics** See study of: semiotics: zoosemiotics.

♦ **zoosuccivore** See -vore: zoosuccivore.

♦ **zootaxy** See study of: taxonomy: zootaxy.

♦ **zootic climax** See ²community: climax: zootic climax.

♦ **zootroph** See -troph-: zootroph.

♦ **Zuckerman's theory** See ³theory: Zuckerman's theory.

♦ *Zugunruhe* See migratory restlessness.

♦ **zygogamy** See -gamy: zygogamy.

♦ **zygogenesis** See -genesis: zygogenesis.

♦ **zygoid parthenogenesis** See parthenogenesis: zygoid parthenogenesis.

♦ **zygosis** See conjugation.

♦ **-zygosity** *combining form*
heterozygosity *n.* The overall statistical, within-locus heterogeneity of alleles averaged over all loci in an individual or population (Dawkins 1982, 288).
cf. -zygosity: homozygosity
homozygosity *n.* The overall statistical, within-locus homogeneity of alleles averaged over all loci in an individual or population (Dawkins 1982, 288).
cf. heterozygosity

♦ **zygotaxis** See taxis: zygotaxis.

♦ **zygote** *n.*
1. A cell, or cell nucleus, that is formed from the conjugation of two such bodies in reproduction; a zygospore, or any germ cell that results from the union of two gametes (*Oxford English Dictionary* 1972, entries from 1891); "the cell created by the union of two gametes (sex cells), in which the gamete nuclei are also fused" (Wilson 1975, 598).
2. The organism developed from a union of two gametes (Michaelis 1963).
3. "The earliest stage of the diploid generation" (Wilson 1975, 598).
adj. zygote
syn. syngamete
[Greek *zygotos* < *zygoein* < *zygon,* yoke]
azygote *n.*
1. A haploid propagule that does not have the capacity for syngamy and develops directly into a new organism (Bell 1982, 503).
2. An organism that is produced by haploid parthenogenesis (*e.g.,* a male ant, bee, or wasp) (Lincoln et al. 1985).
heterozygote *n.* A diploid organism that has different alleles of a given gene on its pair of homologous chromosomes that carry that gene (Wilson 1975, 586).
[coined by Bateson (1901 in Mayr 1982, 733)]
homozygote *n.* A diploid organism that has the same alleles of a given gene on its pair of homologous chromosomes that carry that gene (Wilson 1975, 586).

♦ **-zygotic** *combining form*
dizygotic *n.* Referring to twins derived from two fertilized eggs, each fertilized by

S–Z

a different sperm (King and Stansfield 1985).

syn. fraternal (Lincoln et al. 1985)

monozygotic, monovular, uniovular *n.* Referring to twins derived from a single fertilized egg (Lincoln et al. 1985).

♦ **zygotic meiosis** See meiosis: zygotic meiosis.

♦ **-zygous** *combining form*

allozygous *adj.* "Referring to two genes on the same chromosome locus that are different or at least whose identity is not due to common descent" (Wilson 1975, 578).

cf. -zygous: autozygous

autozygous *adj.* "Referring to two more alleles on the same locus that are identical by common descent" (Wilson 1975, 579).

hemizygous *adj.* Referring to unpaired genes, or chromosome segments, in a diploid cell (Lincoln et al. 1985).

heterozygous *adj.*
1. Referring to a heterozygote, *q.v.*
2. Referring to the overall statistical, within-locus heterogeneity of alleles averaged over all loci in an individual or population (Dawkins 1982, 288).

cf. -zygous: homozygous

homozygous *adj.*
1. Referring to a homozygote, *q.v.*
2. Referring to the overall statistical, within-locus homogeneity of alleles averaged over all loci in an individual or population.

cf. -zygous: heterozygous

Comment: "An organism can be a homozygote with respect to one gene and at the same time a heterozygote with respect to another gene" (Wilson 1975, 586).

♦ **zymogen** *n.* "The enzymatically inactive precursor of a proteolytic enzyme" (King and Stansfield 1985).

Appendix 1. Organisms

This appendix includes hierarchical taxonomic tables of organism groups and describes a few of the millions of organism taxa, including fossils. This appendix has three main parts: 1. Domain Archaea, 2. Domain Bacteria, and 3. Domain Eukarya. Eukarya has four parts: Kingdoms Animalia, Fungi, Plantae, and Protista. For convenience, lichens are included with Fungi, although a few lichens are algal-bacterium associations, not fungal-bacterium ones. Appendix 1 also contains the fossil hoax Piltdown Man (in Animalia: Class Mammalia), made of primate parts. Viruses, which many researchers classify as nonorganisms, are in the main body of this book. A term that is not part of a hierarchy or one that is the first term in a hierarchy (= a first-level term) is preceded by a diamond (♦). Subsequent levels of terms are increasingly indented. Second-, fourth-, and sixth-level terms are not preceded by symbols. Third-level terms are preceded by arrowheads (▶); fifth-level terms, black squares (■); seventh-level terms, open squares (□); eighth-level terms (★); ninth-level terms (●); tenth-level terms (○).

♦ **organism** *n.*
1. An organized body that consists of mutually connected and dependent parts that maintain a common life; the material structure of a living being (*Oxford English Dictionary* 1972, entries from 1842); "any living creature" (Wilson 1975, 590).
2. A dead organism (def. 1).
   cf. -cole, name, -phile, psilile (in the main text)
   Comments: Below are tables of taxonomic hierarchies for organisms and entries for a few of the millions of extinct and extant organisms. These entries contain the common and scientific names of organisms, their taxonomic positions, and some comments about their biologies and other subjects. You should consult taxonomic publications to determine how to identify these taxa, because such information is beyond the scope of this book. The listed species are ones of particular conservation, epidemiological, and research importance and interest. For further information regarding organism relationships, taxa, or both, see Biosis (www.york.biosis.org); Richard Stafursky's *World Species* (www.envirolink. org./species), the *Traité de Zoologie*, the Tree of Life (Maddison and Maddison 1998), the Zoological Record (Record, www. york.biosis.org/zrdocs/zrprod/zro.html), field guides, and other books devoted to the different taxa, many of which I indicate in the entries that follow. King and Stansfield (1985) include many organisms of research interest. I capitalize a common name when it refers to a particular species (*e.g.,* Sugar Maple) or a whole taxonomic unit (*e.g.,* Bumble Bees, meaning all of them). I do not capitalize a common name when I use it to mean only some of a taxon that the name designates (*e.g.,* some bumble bees).
[Greek *organon*, instrument, tool]

epiorganism See organism: superorganism.

microbe *n.* A microscopic organism, especially a disease-causing bacterium (Michaelis 1963).
cf. bacterium, microorganism, virus
syn. germ (Michaelis 1963)
Comments: Microbe is not a technical term (Morris 1982). People sometimes confuse the words "bacillus," "bacterium," "germ," "microbe," "microorganism," and "virus" (Michaelis 1963). "Microbe" at first referred to a protozoan and then to a bacterium. "Germ" originally meant a reproductive cell. Both "germ" and "microbe" are commonly used to refer to a disease-producing bacterium. "Microorganism" was a general term for bacteria, protozoa, and viruses. A bacillus is one of several types of bacteria.
[French < Greek, *mikros*, small + *bios*, life]

microorganism *n.* An organism of microscopic or ultramicroscopic size, commonly including Bacteria, Cyanobacteria, Yeasts, some lichens and fungi, Protistans, Viroids, and Viruses (Lincoln et al. 1985).
cf. microbe; virus (which some biologists do not consider to be an organism)

superorganism *n.*
1. Any society, such as the colony of a eusocial-insect species, that acts as a functional unit, possessing features of organization analogous to the physiological properties of a single organism (Wilson 1975, 383, 596).
2. Any group of organisms that acts as a single functional unit (Lincoln et al. 1985).
syn. epiorganism, supraorganism (Lincoln et al. 1985)

Comment: An insect colony, for example, "is divided into reproductive castes (analogous to gonads) and worker castes (analogous to somatic tissue); it may exchange nutrients by trophallaxis (analogous to the circulatory system), and so forth" (Wilson 1975, 383, 596).
[coined by Herbert Spencer (1862, 316, in Laurent 1990)]

supraorganism See organism: superorganism.

transgenic organism *n.* An organism that contains genes from another species (Campbell et al. 1999, 383).
Comments: The majority of Earth's species are likely to be transgenic organisms.

A CLASSIFICATION OF ORGANISMS: DOMAINS AND KINGDOMS[a]

I.   Domain Archaea, Archeans
II. Domain Eubacteria, Bacteria
III. Domain Eukarya, Eukaryans
     (= Eukaryotes)
     A. Kingdom Animalia, Animals[b]
     B. Kingdom Fungi, Fungi
     C. Kingdom Plantae, Plants
     D. Kingdom Protista, Protistans

[a] I list groups in alphabetical order; classification of the larger taxa of life is in a state of flux.
[b] This classification is based on publications by Carl Woese (1983 in Strickberger 2000, 173), Neil Campbell (1996, 496), Gould (1996, H1); and Curt Suplee (1996c, A3).

♦ **Prokaryota** *n.* One of the two large divisions of life that contains the Domains Archaea and Eubacteria, unicellular organisms without nuclear membranes, and many of the kinds of organelles found in Eukaryota; contrasted with Eukaryota (Woese 1983 in Strickberger 2000, 173; Campbell 1996, 498; Suplee 1996c, A3).
cf. Eukaryota, organism (table)
Comments: Researchers have designated about 4000 living prokaryotic species; there may be as many as 4 million species (Campbell 1996, 499). A prokaryote's DNA is one double-stranded molecule in the form of a ring and concentrated into a region called a "nucleoid." Prokaryotes, in contrast to Eukaryotes, have plasmids (smaller rings of DNA with only a few genes that endow a cell with resistance to antibiotics).
[Greek *pro*, before; karyon, *kernal*, referring to the fact that this taxon does not have a true nucleus enclosed by a membranous nuclear envelope (Campbell 1996, 114)]

PART 1. DOMAIN ARCHAEA

♦ **Domain Archaea** *n.* A domain of life that includes one-celled organisms that have cell walls with no murein (a protein) and produce proteins using a process more similar to that of Eukaryota than Eubacteria; distinguished from Eubacteria and Eukarya (Woese 1983 in Strickberger 2000, 173; Suplee 1996c).
See table (Archaea) in this appendix.
syn. Archaebacteria (Strickberger 2000, 173)
cf. Bacterium, basalt archaean, extreme halophile, extreme thermophile, *Methanococcus jannaschii*, methanogen, subsurface archaean
Comments: Karl Woese described Archaea as a new domain in the 1980s. Archaea live in many habitats, including extreme environments such as boiling water, a highly acidic area, or salt flats, which resemble habitats probably found on Earth about 4 billion years ago. Many Archaea feed on CO_2 and produce methane as a waste product; many feed on H_2S and produce sulfur as a waste product (Suplee 1996c, A3). The photosynthetic types do not use chlorophyll and perform a kind of photosynthesis completely distinct from all other Archaea and Bacteria. The Archaea might share a common most recent ancestor with Bacteria (Doolittle et al. 1996; Morell 1996b).
[Greek *archaio*, ancient; *bacterion*, rod; referring to Archaea's origin from very early cells (Campbell 1996, 499)]

"basalt archaean" *n.* Methane-producing archaeans (= methanogens) that live up to 4500 feet below the Earth's surface and apparently subsist on hydrogen from basalt and water (Todd Stevens and James McKinley in Anonymous 1996b, 23–24, illustration).

eocyte See extreme thermophile.

extreme halophile *n.* An archaean that lives in a salty habitat (Campbell 1996, 509).
Comments: Some species live in very saline places such as the Dead Sea and Great Salt Lake (Campbell 1996, 509). Some halophiles merely tolerate a saline environment; other halophiles require a highly saline environment (10 times saltier than seawater). Some halophiles produce a purple-red scum, colored by their bacteriorhodopsin.
[Greek *halo*, salt + *philos*, lover (Campbell 1996, 509).

extreme thermophile *n.* An archaean that thrives in a hot environment (Campbell 1996, 509).
syn. eocyte

Two Classifications of Archaea

Biological Classification
  I. Extreme Halophiles
 II. Extreme Thermophiles
     A. Subsurface Archaeans
     B. *Sulfolobus*
III. Methanogens
     A. Basalt Bacteria
     B. *Methanococcus jannaschii*

Taxonomic Classification
  I. Kingdom Crenarchaeota
     A. *Acidianus*
     B. *Desulfurococcus*
     C. *Hyperthermus*
     D. *Igneococcus*
     E. *Metallosphaera*
     F. *Pyrobaculum*
     G. *Pyrodictium*
     H. *Pyrolobus*
     I. *Staphylothermus*
     J. *Sulfolobus*
     K. *Thermodiscus*
     L. *Thermofilum*
     M. *Thermoproteus*
     N. *Thermosphaera*
 II. Kingdom Euryarchaeota
     A. Archeoglobales
     B. Halobacteriales
     C. Methanomicrobiales
     D. Methanobacteriales
     E. Methanococcales, including *Methanococcus*
     F. Methanopyrales
     G. Thermoplasmales
     H. Thermococcales
III. Kingdom Korarchaeota

[a] Maddison and Maddison (1998, www.phylogeny. arizona.edu/tree/homepages/recent1998.html)

Comments: The optimal conditions for extreme thermophiles range from 60 through 80°C (Campbell 1996, 509). *Sulfolobus* inhabits hot sulfur springs in Yellowstone National Park, U.S., obtaining its energy by oxidizing sulfur. Another sulfur-metabolizing thermophile lives in 105°C water near deep-sea hydrothermal vents. Extreme thermophiles might be the prokaryotes most closely related to eukaryotes.
[Greek *eos*, dawn + *cyte*, cell; James Lake in Campbell 1996, 509].

Haloarcula sp. *n.* An archaean that lives in crusted salt in Baja California (Sawyer 1999b, A11).
Comment: This species, *Bacillus subtilis*, and *Synechococcus cyanobacteria* can survive prolonged periods in the vacuum of outer space (Sawyer 1999b, A11). A few

percent of *Haloarcula* and *S. cyanobacteria* survived 10,000 kilojoules of ultraviolet radiation after 2 weeks in space. These two species do not form spores.

Methanoccocus jannaschii *n.* A micron-long archaean that flourishes 2 miles deep in the Pacific Ocean and at pressures at least 200 times those at sea level (Suplee 1996c, A3).
Comments: Researchers discovered this archaean in 1982 above a hydrothermal vent chimney (Suplee 1996c, A3). This organism converts carbon dioxide and hydrogen into methane. Geneticists sequenced the some 12 million bases in the genome of this bacterium in 1996 (Bult et al. 1996, 1058; Suplee 1996c, A3). Fully 56%, or 1738, of its genes are unlike genes known in other prokaryotes or eukaryotes. Each of these genes might code for a protein unknown to science.
[*jannaschii*, after Holger Jannasch, a leader of the 1998 expedition that discovered this bacterium (Suplee 1996c, A3)]

methanogen *n.* An archaean that consumes CO_2 and hydrogen and produces methane as a waste product (Campbell 1996, 509; Vogel, 1998, 1633).
Comments: Methanogens are obligate anaerobes (Campbell 1996, 509). Many species live in bogs, marshes, and swamps, where other microbes have consumed all the oxygen, and produce bubbles of methane known as marsh gas. Some methanogen species live in animal guts, playing an important role in the nutrition of Cattle, termites, and other herbivores that subsist mainly on cellulose. A methanogen which engulfed one or more bacteria that made CO_2 and hydrogen as waste products might have been the first eukaryotic cell (Vogel 1998, 1633).

subsurface archaean *n.* An archaean that lives deep in the Earth's crust (Fredrickson and Onstoot 1996, 69, 71, illustration).
cf. Archaea
Comments: Subsurface bacteria live in temperatures as high as 75°C up to 2.8 kilometers below Earth's surface (Fredrickson and Onstoot 1996, 71). Some subsurface bacteria can grow at 140°C. Some fungal and protozoan species also live deep below Earth's surface.

Part 2. Domain Bacteria

♦ **Domain Bacteria** *n.* A domain of life that includes the bacteria, one-celled organisms with muramic acid in their cell walls and protein production that uses a process markedly different than Archaea and Eukaryota; distinguished from Archaea and Eukaryota (Woese 1983 in Strickberger 2000, 173; Suplee 1996c, A3).

syn. Domain Eubacteria

cf. organism (table), Prokaryota (Appendix 1, Part 1)

Comments: Researchers published approved lists of bacterial names (Skerman et al. 1980).

A CLASSIFICATION OF BACTERIA[a]

I. Chlamydias
  A. *Chlamydia trachomatis* (Nongonococcal-Urethritis Bacterium)
II. Cyanobacteria
  A. *Anabaena*
III. Gram-positive Bacteria (including some Gram-negative Bacteria)
  A. *Actinomycetes*
    1. Leoprosy Bacterium
    2. *Streptomyces*
    3. Tuberculosis Bacterium
  B. *Bacillus*
  C. *Clostridium*
  D. Mycoplasmas
    1. *Mycoplasma pnemoniae* (Walking-Pneumonia Bacterium)
IV. Proteobacteria
  A. Chemoautotrophic Proteobacteria
    1. *Rhizobium*
  B. Chemoheterotrophic Proteobacteria
    1. *Escherichia coli*
    2. *Salmonella* spp. (Food-Poisoning Bacteria)
    3. *Salmonella typhi* (Typhoid-Fever Bacterium)
  C. Purple Bacteria
V. Spirochetes
  A. *Borrelia burgdorferi* (Lyme-Disease Bacterium)
  B. *Treponema pallidum* (Syphilis Bacterium)

---

[a] Campbell (1996, 509–512).

Agrobacterium tumefaciens *n.* A bacterium that causes crown gall disease in many dicot species (King and Stansfield 1985).

cf. plasmid: Ti plasmid

α-Proteobacteria *n.* The bacterial taxon that includes *Bradyrhizobium, Paracoccus, Rhodobacter,* and *Rickettsia* (Gray et al. 1999, 1478, 1480).

cf. hypothesis: endosymbiosis hypothesis

Bacillus subtilis *n.* A Gram-positive, rod-shaped, spore-forming bacterium that lives in fresh water and soil (Sawyer 1999b, A11).

Comment: This species readily grows on a chemically defined medium and undergoes transduction and transformation (King and Stansfield 1985). This species, *Haloarcula* sp., and *Synechococcus cyanobacteria* can survive prolonged periods in the vacuum of outer space (Sawyer 1999b, A11). After 6 years in space, *B. subtilis* showed an 80% survival in a dry preparation of spores shielded from radiation either by at least one other layer of spores or a thin layer of glucose and salts. Those exposed to radiation did not survive. Other *Bacillus* are *B. cereus* and *B. megatherium.*

bacterium *n.* An individual, or species, in Domain Eubacteria.

cf. Archaea, Eubacteria

Comments: More than 4000 kinds of eubacteria live in human intestines, forming a complex ecosystem (Anonymous 1997b, 23). These bacteria absorb vitamins, make vitamins, and digest food.

cyanobacterium *n.* An autotrophic bacterium that obtains energy from sunlight and uses oxygen as an electron acceptor in electron transport (Campbell 1996, 508).

syn. blue-green alga (Campbell 1996, 508)

Comments: Cyanobacteria may have appeared on Earth as early as 2.7 billion years ago based on fossil 2α-methylhopanes from the Pilbara Craton, Australia (Brocks et al. 1999, 1033); they transformed life on Earth by producing much free oxygen (Campbell 1996, 508), and they and other prokaryotes evidently produced fossil stromatolites.

dwarf bacterium *n.* A starved bacterium that has shrunk from a healthy size to a few microns, to less than a thousandth of its usual volume, as it has used up its stores (Fredrickson and Onstoot 1996, 73).

syn. ultramicro-bacterium (Fredrickson and Onstoot 1996, 73)

Comments: Dwarf bacteria are common deep in Earth's crust (Fredrickson and Onstoot 1996, 73). These bacteria might divide once per century or so, instead of at the much faster rates of non-starved bacteria.

Epulopiscium *n.* The largest known bacterium (E.R. Angert and N.R. Pace in Associated Press 1992, A4).

Comments: This bacterium has a width of about 0.5 mm and is found in intestines of surgeonfish (Associated Press 1992, A4). A researcher discovered it in 1985 and originally classified it as a protozoan.

Escherichia coli *n.* A chemoheterotrophic proteobacterium that inhabits guts of mammals including Humans (Campbell 1996, 514; Suplee 1997a, A10).

Comments: This is the best understood of all organisms (Campbell 1996, 509). Alien strains cause traveler's diarrhea. There are hundreds of strains of this bacterium (Suplee 1997a, A10). This species commonly causes human urinary-tract infections. Some strains (*e.g.,* O157:H7, which was first identified in 1982) dissolve cell membranes of various animal tissues and cause up to 20,000 human infections and about 250 human deaths in the U.S. annually. Humans contract O157:H7 from contaminated beef, milk, and water.

Gram-negative bacterium *n.* A bacterium that has a cell wall with less peptidoglycan than a Gram-positive bacterium and peptidoglycan located in a periplasmic gel between a plasma membrane and an outer membrane; contrasted with Gram-positive bacterium (Campbell 1996, 501).
Comments: Gram staining works with many eubacterian species. It results in a red dyeing of Gram-negative bacteria and violet dyeing of Gram-positive bacteria (Campbell 1996, 501).
[Gram, after Hans Christian Gram, a Danish physician, who developed Gram staining in the late 1800s (Campbell 1996, 501)]

Gram-positive bacterium *n.* A bacterium that has a cell wall with a large amount of peptidoglycan compared to Gram-negative bacteria located outside a plasma membrane; contrasted with Gram-negative bacterium (Campbell 1996, 501).

Hemophilus influenzae *n.* A bacterium (Wade 1995, C1).
Comments: This bacterium does not cause influenza; it colonizes human tissues, where in its virulent form it can cause earaches and meningitis (Wade 1995, C1, C9, illustration). Researchers obtained the first full DNA sequence of a living cell from this species. It has 1,830,137 base pairs and 1743 genes.

Mycoplasma *n.* A bacterial genus in Phylum Aphragmabacteria (Norstog and Meyerriecks 1983, 38, illustration; King and Stansfield 1985).
Comments: Mycoplasmas are smaller and less complex than typical bacteria, and each has a plasma membrane enclosing its DNA, RNA, enzymes, and numerous ribosomes (Norstog and Meyerriecks 1983, 38). Mycoplasmas do not digest their own food but are usually parasites that live on predigested food. Mycoplasmas can live freely in a nutrient broth. They are the simplest organisms capable of independent life (King and Stansfield 1985), and the only prokaryotes that lack cell walls (Campbell 1990, 533, illustration). *Myco-*

plasma pneumonia causes one type of bacterial pneumonia.

prochlorophyte *n.* A prokaryote that performs oxygenic photosynthesis using chlorophyll *b,* as do Green Algae and Land Plants (Tomitani et al. 1999, 161).

purple nonsulfur bacterium *n.* A bacterium that uses an electron-transport system that is a hybrid of photosynthetic and respiratory equipment (Campbell 1996, 508).

Rickettsia prowazekii *n.* The causative agent of epidemic louse-borne typhus (Gray et al. 1999, 1476).
Comment: This species has the most mitochondria-like eubacterial genome known (Gray et al. 1999, 1476).

Streptococcus mitis *n.* A bacterium that lives in human mouths, noses, and throats (Sawyer 1999b, A11).

Synechococcus cyanobacteria *n.* A bacterium that lives in crusted salt in Baja California (Sawyer 1999b, A11, illustration).
cf. Organism: Archaea: *Haloarcula* sp.; gene (table)

PART 3. DOMAIN EUKARYA

♦ **Domain Eukarya** *n.* One of the three domains of life comprised of unicellular and multicellular organisms with nuclear membranes and many kinds of organelles not found in Prokaryota (Domains Archaea and Bacteria); distinguished from Prokaryota, which contains Archaea and Eubacteria (Woese 1983 in Strickberger 2000, 173; Campbell 1996, 498; Suplee 1996c, A3).
cf. Prokaryota
syn. Eucaryota, Eukaryota
Comments: Researchers have named over 1 million extant eukaryan species; there may be as many as 100 million such species (Campbell 1996, 499). A eukaryan's DNA is linear molecules packaged along with proteins into the number of chromosomes that are characteristic of a particular species. The first eukaryan may have arisen on Earth about 2.7 billion years ago, based on fossil steranes, particularly cholestane and its 28- through 30-carbon analogs from the Pilbara Craton, Australia (Brocks et al. 1999, 1033).
[Greek *eu,* true; karyon, *kernal,* referring to a true nucleus enclosed by a membranous nuclear envelope (Campbell 1996, 114)]

Crown Eukaryotes *n.* The eukaryote group composed of Animalia, Fungi, Plantae, and a number of mitochondria-containing protistan groups (Gray et al. 1999, 1480).

▶ **animal** *n.*

1. A sentient living organism typically capable of voluntary motion and sensation, distinguished from a fungus, plant, and protist (*Oxford English Dictionary* 1972, entries from 1398; Michaelis 1963).
2. Nonscientifically, an animal other than a Human; a beast; a brute (*Oxford English Dictionary* 1972, entries from 1600; Michaelis 1963).
3. Colloquially, a domestic quadruped (*e.g.,* Cow, Dog, Horse) (*Oxford English Dictionary* 1972, entries 1600; Michaelis 1963).
4. Colloquially, any vertebrate other than a bird or fish (Michaelis 1963).
   Notes: This is a common definition.
5. Scientifically, a member of the kingdom Animalia; an organism characterized by having eukaryotic cells with mitochondria and nuclear envelopes, heterotrophic nutrition, no chloroplasts or cell walls, multicellularity, and usually fertilization, meiosis, and a nervous system (Curtis 1983, 387; Lincoln et al. 1985).

cf. animal names, -zoan, -zoite, zooid, -zoon (main body); Animalia (Appendix 1)
Comment: Curtis (1983, 387) lists the traits of prokaryotes, protists, fungi, plants, and animals in a useful table.
[Latin *animal*, a living being < *anima*, breath, soul, life]

companion animal *n.* A nonhuman animal that a person keeps for its intrinsic appeal rather than its usefulness; provides for its well-being, at least in part; holds in particular regard (from a subordinate treated as a possession to somewhere between property and a person to a family member with well-prescribed social roles); and lives within intimate association (Veevers 1984 in Veevers 1985, 12; Quackenbush 1985, 395; Young 1985, 298–299).
cf. pet
Comments: Aquarium fish are the most numerous companion animals in the U.S. (Katcher 1985, 403). Companion animals are legally defined in most western countries as property, or chattels, with which owners can do as they please as long as laws are not broken (Fogle 1981, 335). Companion animals can facilitate their human companions' sociability and supplement their human companionship or be an alternative to it (Veevers 1985, 27). Companion animals may contribute to a person's well-being (Voith 1985, 295). Young

(1985, 299) suggests that "companion animal" should not be synonymized with "pet" but should be reserved for "the ancient and emerging companion animal functions now being given more attention."

contact animal *n.* A species that lacks individual distance and may seek bodily contact with conspecifics (Heymer 1977, 102; Immelmann and Beer, 1989, 58).
Comment: "Contact animal" grades into "distance animal," *q.v.* (Immelmann and Beer 1989, 58). Contact animals include some lizard and snake species; most parrot, primate, and rodent species; Catfish, Chelonians, and Eels including Moray Eels, Penguins; and the Domestic Pig, Hippopotamus, Mousebird, and White-Eye Bird.

control animal *n.* In the Capuchin Monkey: a male that maintains vigilance, interposes himself between an intruder and his group, and terminates fights between group members, without exercising aggressive dominance toward other troop members (Bernstein 1966 in Wilson 1975, 282).

density-intolerant animal *n.* An animal species that has low reproductive rates, lives as isolated pairs or families, does not have large population density fluctuations, disperses at the end of each breeding season, and is a distance animal (Immelmann and Beer 1989, 72).

density-tolerant animal *n.* An animal species that has high reproductive rates, maintains these rates even when its population densities approach overcrowding, disperses when crowding exceeds a particular limit, and is generally a contact animal (Immelmann and Beer 1989, 72).

distance animal *n.* For example, in Flamingos, Gulls, Pike, Salmon, Swallows, Trout; most raptor and ungulate species: a species whose members keep particular distances from one another except during copulation and suckling of young (Hediger 1941 in Immelmann and Beer 1989, 76).
cf. animal: contact animal

▶ *Kingdom* **Animalia** *n.* A eukaryotic kingdom comprised of organisms characterized by having heterotrophic nutrition, multicellularity, no chloroplasts or cell walls, nuclear membranes, and usually fertilization, meiosis, and nervous systems; distinguished from Kingdoms Fungi, Plantae, and Protista (Curtis 1983, 387; Lincoln et al. 1985).
cf. animal (above)

Comment: Animals might have first arisen on Earth between 900 and 1000 million years ago (Vermeij 1996, 525; Wray et al. 1996, 568). Tables that show hierarchical classifications of some animal taxa and descriptions of a few of the millions of animal taxa are below.

A Classification of Kingdom Animalia[a]

I. Group Parazoa, animals without true tissues
    A. Phylum Porifera, Sponges
    B. Phylum Placozoa, Placozoans
    C. Phylum Mesozoa, Mesozoans
II. Group Eumetazoa, animals with true tissues
    A. Group Radiata, Radiates
       1. Phylum Cnidaria, Cnidarians (Corals, Hydra, Jellyfish)
         a. Class Anthozoa, Corals
         b. Class Hydrozoa, Hydrozoans
         c. Class Scyphozoa, Scyphozoans
       2. Phylum Ctenophora, Combjellies
    B. Group Bilateria, Bilaterates
       1. Group Acoelomata, Acoelomates
         a. Phylum Gnathostomulida, Gnathostomulids
         b. Phylum Platyhelminthes, Flatworms and kin
            (1) Class Cestoda, Tapeworms
            (2) Class Monogenea, Flukes
            (3) Class Trematoda, Flukes
            (4) Class Turbellaria, Flatworms
         c. Phylum Rhynchocoela
       2. Group Pseudocoelomates, Pseudocoelomates
         a. Phylum Acanthocephala, Spiny-Headed Worms
         b. Phylum Cycliophora, Cyliophorans
         c. Phylum Endoprocta, Endoprocts
         d. Phylum Gastrotricha
         e. Phylum Kinorhyncha
         f. Phylum Loricifera, Loriciferans
         g. Phylum Nematoda (= Aschelminthes), Nematodes (Roundworms and kin)
         h. Phylum Nematomorhpa, Horsehair Worms
         i. Phylum Nemertea (= Rhynchocoela), Nermerteans
         j. Phylum Rotifera, Rotifers
       3. Group Coelomata, Coelomates
         a. Group Protostomia, Protostomes
            (1) Group Protostome Lophophorates

             (a) Phylum Brachiopoda, Brachiopods (= Lamp Shells)
             (b) Phylum Bryozoa, Bryozoans
           (2) Group Ametameria (= Pseudometameria), Ametamerians
             (a) Phylum Echiura
             (b) Phylum Mollusca, Molluscs
                (i) Class Bivalvia, Bivalves (Clams, Oysters, and kin)
                (ii) Class Cephalopoda, Cephalopods (= Nautiluses, Octopuses, Squids)
                (iii) Class Gastropoda, Gastropods (Slugs and Snails)
                (iv) Class Polyplacophora, Chitons
                (v) 4 other classes
             (c) Phylum Pentostomida, Pentostomids
             (d) Phylum Priapulida, Priapulids
             (e) Phylum Sipunculida, Sipunculids
             (f) Phylum Tardigrada, Tardigrades
           (3) Group Metameria, Metamerians
             (a) Phylum Annelida, Annelids
                (i) Class Hirudinea, Leeches
                (ii) Class Oligochaeta, Earthworms
                (iii) Class Polychaeta, Polycheates
             (b) Phylum Arthropoda, Arthropods
             (c) Phylum Onychophora, Onychophorans (= Velvet Worms)
             (d) Phylum Pogonophora, Beard Worms, Pogonophorans

b. Group Deuterostomia,
Deuterostomes
(1) Group Deuterostome
Lophophorates
(a) Phylum Phoronida,
Horseshoe Worms
(2) Group Nonlopho-
phorate Deuterostomia
(a) Phylum
Chaetognatha,[b]
Arrow Worms
(b) Phylum
Echinodermata,
Echinoderms
(i) Class Asteroidea,
Sea Stars
(ii) Class
Concentri-
cycloidea,
Sea Daisies
(iii) Class Crinoidea,
Sea Lilies
(iv) Class
Echinoidea,
Sand Dollars
and Sea Urchins
(v) Class
Holothuroidea,
Sea Cucumbers
(vi) Class
Holothuroidea,
Brittle Stars
(c) Phylum
Hemichordata,
Hemichordates
(i) Class
Pterobranchia,
Pterobranchs
(ii) Class
Enteropneusta,
Acorn Worms
(d) Phylum Chordata,
Chordates
(i) Calcichordata[c]
(ii) Conodonta[c]
(iii) Subphylum
Urochordata,
Urochordates
(iv) Subphylum
Cephalochordata,
Cephalochordates
(= Lancelets,
"Headless
Fishes")
(v) Subphylum
Vertebrata,
Vertebrates

[a] After Strickberger (1996) and others.
[b] Strickberger (1996, 338).
[c] Some researchers place Chaetognatha in Chordata.

A CLASSIFICATION OF PHYLUM ARTHROPODA[a]

I. Subphylum Trilobita, Trilobites
(extinct)
II. Subphylum Chelicerata, Chelicerates
A. Class Merostomata, Horseshoe
Crabs and Eurypterids (extinct)
B. Class Arachnida, Arachnids
(Spiders and their kin)
C. Class Pycnogonida, Seas Spiders
III. Subphylum Crustacea, Crustaceans
A. Class Branchiopoda, Branchiopods
B. Class Branchiura
C. Class Cephalocarida
D. Class Cirripedia
E. Class Copepoda, Copepods
F. Class Malacostraca, Malacostracans
1. Order Isopoda, Isopods
(Pillbugs and Sowbugs)
G. Class Mystacocarida
H. Class Ostracoda, ostracods
I. Class Remipedia
J. Class Tantulocarida
IV. Subphylum Atelocerata, Atelocerates
A. Class Chilopoda, Centipedes
B. Class Diplopoda, Millipedes
C. Class Hexapoda, Hexapods
D. Class Pauropoda, Pauropods
E. Class Symphyla, Symphylans

[a] After Borror et al. (1989).

A CLASSIFICATION OF CLASS ARACHNIDA[a]

I. Order Acari, Mites and Ticks
A. Group I. Opilioacrariformes
B. Group II. Parasitiformes
1. Suborder Holothryina
2. Suborder Mesostigmata
3. Suborder Ixodida, Ticks
C. Group III. Acariformes
1. Suborder Prostigmata
2. Suborder Astigmata
3. Suborder Oribatida
II. Order Amblypygi, Tailless
Whipscorpons, Whipspiders
III. Order Araneae, Spiders
IV. Order Opiliones, Daddylonglegs
(= Harvestmen)
V. Order Palpigrada, Microwhipscorpions
VI. Order Pseudoscorpiones,
Pseudoscorpions
VII. Order Ricinulei, Ricinuleids
VIII. Order Schizomida, Short-Tailed
Whipscorpions
IX. Order Scorpiones, Scorpions
X. Order Solifugae, Windscorpions
XI. Order Uropygi, Whipscorpions

[a] After Borror et al. (1989, 104–105).

SOME FAMILIES OF ORDER ARANEAE, SPIDERS[a]

I. Suborder Orthognatha, Tarantulas and kin
II. Suborder Labidognatha, True Spiders
  A. Section Cribellatae, Hackled-Band Spiders
  B. Section Ecribellatae, Plain-Thread Weavers
    1. Family Agelenidae, Grass Spiders (= Funnel-Web Spiders)
    2. Family Araneidae, Orb Weavers
    3. Family Clubionidae, Two-clawed Hunting Spiders (= Sac Spiders)
    4. Family Hahniidae, Hahniid Sheet-Web Spiders
    5. Family Linyphiidae, Sheet-Web Spiders
      a. *Frontinella pyrmamitela*, Bowl-and-Doily Spider
    6. Family Loxoscelidae, Recluse Spiders
    7. Family Lycosidae, Wolf Spiders (= Ground Spiders)
    8. Family Pholcidae, Long-Legged Spiders (= Cellar Spiders)
    9. Family Pisauridae, Nursery-Web Spiders, Fishing Spiders
      a. *Dolemedes* sp., a fishing spider
    10. Family Salticidae, Jumping Spiders
    11. Family Theridiidae, Comb-Footed Spiders (= Cobweb Spiders, including widows)
    12. Family Thomisidae, Crab Spiders
      a. *Misumena vatia* (Clerck), Crab Spider

[a] After Borror et al. (1989).

A CLASSIFICATION OF CLASS HEXAPODA[a]

I. Group Entognatha, 6859
  A. Order Protura, Proturans, 200
  B. Order Collembola, Springtails, 6000
  C. Order Diplura, Diplurans, 659
II. Group Insecta (= Exognatha), 780,784
  A. Group Apterygota, Wingless Insects
    1. Order Microcoryhia, Bristletails, 250
    2. Order Thysanura, Silverfish, 320
  B. Group Pterygota, Winged insects
    1. Order Blataria, Cockroaches, 4000; 5 families
    2. Order Coleoptera, Beetles, 300,000; 113 families in 4 suborders
      a. Suborder Adephaga
        (1) Family Cantharidae, Soldier Beetles
        (2) Family Carabidae, Ground Beetles
        (3) Family Cicindellidae, Tiger Beetles
        (4) Family Gyrinidae, Whirligig Beetles
      b. Suborder Polyphaga
        (1) Family Bruchidae, Seed Beetles
        (2) Family Buprestidae, Metallic-Woodboring Beetles
        (3) Family Cerambycidae, Long-Horned Beetles
        (4) Family Chrysomelidae, Leaf Beetles
        (5) Family Cleridae, Checkered Beetles
        (6) Family Coccinellidae, Ladybird Beetles (= Ladybugs)
        (7) Family Curculionidae, Weevils (= Snout Beetles)
        (8) Family Dermestidae, Skin Beetles
        (9) Family Elateridae, Click Beetles
        (10) Family Lampyridae, Fireflies (= Lightningbugs)
        (11) Family Lucanidae, Stag Beetles
        (12) Family Lycidae, Net-Winged Beetles
        (13) Family Meloidae, Blister Beetles
        (14) Family Mordellidae, Tumbling-Flower Beetles
        (15) Family Scarabaeidae, Scarab Beetles
        (16) Family Scolytidae, Ambrosia Beetles and Bark Beetles (= Engraver Beetles)
        (17) Family Silphidae, Carrion Beetles
        (18) Family Staphylinidae, Rove Beetles
        (19) Family Tenebrionidae, Darkling Beetles
    3. Order Dermaptera, Earwigs, 1100; 6 families
    4. Order Diptera, Flies, 120,000; 108 families
      a. Suborder Brachycera, Short-Horned Flies
        (1) Family Asilidae, Grass Flies and Robber Flies
        (2) Family Bombyliidae, Bee Flies

(3) Family Dolichopodidae, Long-Legged Flies

(4) Family Drosophilidae, Vinegar Flies (= Fruit Flies)

(5) Family Syrphidae, Flower Flies (= Hover Flies)

(6) Family Tabanidae, Deer Flies and Horse Flies

(7) Superfamily Oestroidea, Oestroid Flies; 5 families

    (a) Family Calliphoridae, Bot Flies

    (b) Family Sarcophagidae, Flesh Flies

    (c) Family Tachinidae, Tachinid Flies

(8) Superfamily Muscoidea, Muscoid Flies; 3 families

    (a) Family Anthomyiidae, Anthomyiid Flies

    (b) Family Muscidae, Muscoid Flies (including House Flies)

b. Suborder Nematocera, Long-Horned Flies

(1) Family Ceratopogonidae, Biting Midges, No-See-Ums, Punkies

(2) Family Chironomidae, Midges

(3) Family Cicidomyiidae, Gall Midges (= Gall Gnats)

(4) Family Culicidae, Mosquitoes

(5) Family Simuliidae, Black Flies (= Buffalo Gnats)

(6) Family Tipulidae, Crane Flies

5. Order Embiidina, Webspinners, 150; 4 families

6. Order Ephemeroptera, Mayflies, 4870

7. Order Grylloblattaria, Rock Crawlers, 20

8. Order Hemiptera, Bugs, 50,000; 42 families

a. Family Coreidae, Leaf-Footed Bugs, Squash Bugs

b. Family Gelastocoridae, Toad Bugs

c. Family Gerridae, Water Striders

d. Family Lygaeidae, Seed Bugs

e. Family Miridae, Leaf Bugs, Plant Bugs

f. Family Pentatomidae, Stink Bugs

g. Family Reduviidae, Ambush Bugs, Assassin Bugs, Thread-Legged Bugs

h. Family Scutelleridae, Shield-Back Bugs

i. Family Tingidae, Lace Bugs

j. Family Veliidae, Broad-Shouldered Water Striders

9. Order Homoptera, Homopterans (Aphids, Cicadas, Hoppers, Psyllids, Scale Insects, Whiteflies), 32,000

a. Family Adelgidae, Pine and Spruce Aphids

b. Family Aphidae, Aphids (= Plantlice)

c. Family Cercopidae, Froghoppers, Spittle Bugs

d. Family Cicadidae, Cicadas

e. Superfamily Coccoidea, Scale Insects

f. Family Fulgoridae, Fulgorid Planthoppers

g. Family Membracidae, Treehoppers

h. Family Phyloxeridae, Phylloxerans

i. Family Psyllidae, Psyllids (= Jumping Plantlice)

10. Order Hymenoptera, Hymenopterans (Ants, Bees, Sawflies, Wasps), 108,000

a. Suborder Symphyta, Horntails and Sawflies, 4 superfamilies

b. Suborder Apocrita, 16 superfamilies

(1) Superfamily Chalcidoidea, Chalcidoid Wasps

(2) Superfamily Cynipoidea

    (a) Family Cynipidae, Gall Wasps

(3) Superfamily Formicoidea

    (a) Family Formicidae, Ants

(4) Superfamily Ichneumonioidea

    (a) Family Braconidae, Braconid Wasps

    (b) Family Ichneumonidae, Ichneumon Wasps

(5) Superfamily Pompiloidea

    (a) Family Pompilidae, Spider Wasps

(6) Superfamily Sphecoidea
  (a) Group
      Spheciformes,
      Sphecid Wasps
  (b) Group Apiformes,
      Bees; 9 families
(7) Superfamily Tiphioidea
  (a) Family Tiphiidae,
      Tiphiid Wasps
  (b) Family Mutillidae,
      Velvet-Ants
(8) Superfamily Vespoidea
  (a) Family Vespidae,
      Vespid Wasps
11. Order Isoptera, termites, 1900; 4 families
12. Order Lepidoptera, Lepidopterans (Butterflies, Moths, Skippers), 112,000; 75 families
  a. Suborder Dacnonypha
  b. Suborder Ditrysia; 17 superfamilies
    (1) Superfamily Bombycoidea
      (a) Family Lasiocampidae, Lappet Moths, Tent Caterpillars, and kin
      (b) Family Saturniidae, Giant Silkworm Moths and Royal Moths
    (2) Superfamily Geometroidea
      (a) Family Geometridae, Cankerworms, Geometers, Measuringworms, and kin
    (3) Superfamily Hesperoidea
      (a) Family Hesperiidae, Skippers
    (4) Superfamily Noctuoidea
      (a) Family Arctiidae, Artiid Moths
      (b) Family Lymantriidae, Lymantriid Moths (Gypsy Moths, Tussock Moths, and kin)
      (c) Family Noctuidae, Noctuid Moths
      (d) Family Notodontidae, Prominent Moths
    (5) Superfamily Papilionidea, Butterflies
      (a) Family Papilionidae, Parnassians and Swallowtails
      (b) Family Pieridae, Orange-Tips,
          Sulphurs, and Whites
      (c) Family Lycaenidae, Blues, Coppers, Hairstreaks, Harvesters, and Metalmarks
      (d) Family Nymphalidae, Brush-Footed Butterflies
      (e) Family Satyridae, Arctics, Satyrs, and Wood Nymphs
    (6) Superfamily Pyraloidea, Grass Moths, Snout Moths, and kin
    (7) Superfamily Sphingoidea
      (a) Family Sphingidae, Sphinx Moths (= Hawkmoths, Hornworms)
    (8) Superfamily Tineoidea
      (a) Family Psychidae, Bagworms
    (9) Superfamily Yponomeutoidea
      (a) Family Yponomeutidae
    (10) Superfamily Zygaenoidea
      (a) Family Limacodidae, Saddleback Caterpillars and Slug Caterpillars
  c. Suborder Exoporia
  d. Suborder Monotrysia
  e. Suborder Zeugloptera
13. Order Mantodea, Mantids, 1500; 1 family: Mantidae
14. Order Mecoptera, Scorpionflies, 480
15. Order Neuroptera, Neuropterans (Alderflies, Antlions, Dobsonflies, Fishflies, Owlflies, Snakeflies), 4670; 13 families
  a. Family Chrysopidae, Common Lacewings, and Green Lacewings
  b. Family Hemerobiidae, Brown Lacewings
  c. Family Mantispidae, Mantidflies
  d. Family Myrmeleontidae, Antlions
16. Order Odonata, Damselflies and Dragonflies, 4870
  a. Suborder Anisoptera, Dragonflies
  b. Suborder Zygoptera, Damselflies

17. Order Orthoptera, Crickets, Grasshoppers, and kin, 12,500; 10 families
    a. Family Acrididae, Short-Horned Grasshoppers
    b. Family Gryllidae, Crickets
    c. Family Tettigoniidae, Long-Horned Grasshoppers
18. Order Phasmida, Timemas and Walkingsticks, 2000
19. Order Phthiraptera, Lice, 5500; 16 families
20. Order Plecoptera, Stoneflies, 1500; 9 families
21. Order Pscoptera, Pscopterans, 2400; 24 families
22. Order Siphonaptera, Fleas, 2300
23. Order Strepsiptera, Twisted-Wing Parasites, 300
24. Order Thysanoptera, Thrips, 4000; 5 families
25. Order Trichoptera, Caddisflies, 7000
26. Order Zoraptera, Zorapterans, 24; 1 family

[a] After Borror et al. (1989, chap. 7). Numbers after orders are numbers of described species worldwide. Most families are not included.

A CLASSIFICATION OF LIVING GROUPS OF PHYLUM CHORDATA, CHORDATES

I. Subphylum Urochordata, urochordates
II. Subphylum Cephalochordata, Cephalochordates (= Lancelets, "Headless Fishes")
III. Subphylum Vertebrata, Vertebrates
   A. Class Acanthodii, Spiny Sharks
   B. Class Agnatha, Jawless Fishes
   C. Class Amphibia, Amphibians (Frogs, Salamanders, Toads, and kin)
   D. Class Aves, Birds
   E. Class Chondrichthyes, Cartilaginous Fishes
   F. Class Mammalia, Mammals
   G. Class Osteichthyes, Boney Fishes
   H. Class Placodermi, Placoderms (= Plated Fishes)
   I. Class Reptilia, Reptiles (Lizards and Snakes)

A CLASSIFICATION OF CLASS AMPHIBIA[a]

I. Subclass Ichthyostegida, Ichthyostegids;[a] Ancestral Labyrinthodonts (extinct)
II. Subclass Labryinthodonta, Labyrinthodonts[a] (extinct)

A. Order Anthracosaura, Anthracosaurs (extinct)
B. Order Temnospondyli, Temnospondyles (Late Paleozoic and Triassic Labyrinthodonts) (extinct)
III. Subclass Lepospondyli, Lepospondyls (Late Paleozoic amphibians with spool-shaped vertebrae) (extinct)
IV. Subclass Lissamphibia (Frogs, Legless Amphibians, Newts, Salamanders, Toads)
   A. Order Proanura, ancestor of frogs and toads (extinct)
   B. Order Anura, Frogs, Toads
   C. Order Apoda, Legless Amphibians
   D. Order Urodela, Newts, Salamanders

[a] After Strickberger (1996).

A CLASSIFICATION OF CLASS REPTILIA[a]

I. Subclass Anapsida, Anapsids
   A. Order Captorhinida (= Cotylosauria), Captorhinids (= Stem Reptiles) (extinct)
   B. Order Mesosauria, Mesosaurs (aquatic freshwater reptiles) (extinct)
   C. Order Testudinata (= Chelonia), Turtles
II. Subclass Diapsida, Diapsids
   A. Infraclass Lepidosauromorpha, Lepidosauromorphs (extinct)
   B. Order Younginifromes (extinct)
   C. Superorder Ichthyopterygia, Ichthyosaurs (extinct)
   D. Superorder Lepidosauria (extinct)
      1. Order Sphenodontida (extinct)
      2. Order Squamata, squamates
         a. Suborder Lacertilia (= Sauria), Lizards
         b. Suborder Serpentes (= Ophidia), Snakes
   E. Superorder Sauropterygia, Sauropterygians (marine Mesozoic reptiles) (extinct)
   F. Infraclass Archosauromorpha, archosauromorphs
      1. Superorder Archosauria, Archosaurs
         a. Order Crocodilia, Alligators, Crocodiles, Ghavials
         b. Order Ornithischia, Ornithiscians, Bird-Hipped Dinosaurs (extinct)
         c. Order Pterosauria, Pterosaurs (extinct)
         d. Order Saurischia, Saurischians, Lizard-Hipped Dinosaurs (extinct)

e. Order Thecodontia,
Thecodonts (extinct)
III. Subclass Parareptilia (extinct)
IV. Subclass Synapsida, Synapsids
(extinct)[c]
  A. Order Pelycosauria, Pelycosaurs
  (extinct)
  B. Order Therapsida, Therapsids,
  Mammal-Like Reptiles (extinct)

---

[a] This classification is based primarily on Kemp
(1982), Benton (1990), and Strickberger (1996).
[b] Benton (1990, 329) includes all Class Aves in
Saurischia. Thus, according to him, only the nonbird
saurischians are extinct.
[c] Benton (1990, 80) includes all Mammalia in
Synapsida. Thus, according to him, only the reptilian synapsids are extinct. Kemp (1988 in Strickberger
1996, 418) distinguishes Synapsida from Sauropsida
(the rest of the Reptilia).

A CLASSIFICATION OF CLASS AVES, BIRDS

I. Subclass Odontoholcae, Toothed
Diving Birds (extinct)
II. Subclass Ornithurae (= Neornithes),
Modern Birds (= True Birds)
  A. Order Aepyornithiformes, Elephant
  Birds (extinct)
  B. Order Apodiformes,
  Hummingbirds, Swifts
  C. Order Apterygiformes, Kiwis
  D. Order Anseriformes, Ducks, Geese,
  Screamers, Swans
  E. Order Casuariiformes, Cassowaries,
  Emu
  F. Order Charadriiformes, Avocets,
  Jacanas, Phalaropes, Plovers,
  Oyster Catchers, Sandpipers,
  Snipes, Stilts
  G. Order Caprimulgiformes,
  Frogmouths, Nightjars
  H. Order Ciconiiformes, Flamingos,
  Herons, Storks
  I. Order Coliiformes, Colies
  J. Order Coraciiformes, Bee Eaters,
  Hoopoe, Hornbills,
  K. Order Columbiformes, Pigeons,
  Sandgrouse
  L. Order Cuculiformes, Cuckoos,
  Turacos
  M. Order Diatrymiformes, *Diatryma*
  and kin (extinct)
  N. Order Dinornithiformes, Moas;
  *Aepyornis, Diatryma,
  Dinornis* (all extinct)
  O. Order Falconiformes, Eagles,
  Hawks, Vultures
  P. Order Galliformes, Curassows,
  Fowl, Grouse, Guinea Fowl,
  Turkeys

Q. Order Gaviiformes, Loons
R. Order Gruiformes, Bustards, Coots,
Cranes, Kagu, Quails, Limpkin,
Mesites, Moorhens, Pinfoots,
Seriamas, Rails, Sun Bittern,
Trumpeters
S. Order Passeriformes, Passerines
(= "Perching Birds:" Bluejays,
Cardinals, Crows, Sparrows, etc.)
T. Order Pelecaniformes, Boobies,
Cormorants, Darters, Frigate Birds,
Gannets, Pelicans, Shaggs, Tropic
Birds
U. Order Piciformes, Barbets,
Toucans, Woodpeckers
V. Order Podicipediformes, Grebes
W. Order Procellariiformes,
Albatrosses, Petrels, Shearwaters
X. Order Psittaciformes, Cockatoos,
Lories, Macaws, Parakeets, Parrots
Y. Order Rheiformes, Rheas
Z. Order Struthioniformes, Ostrich
AA. Order Tinamiformes, Tinamous
BB. Order Sphenisciformes, Penguins
CC. Order Strigiformes, Owls
DD. Order Trogoniformes, Trogons,
Kingfishers, Kookaburras,
Motmots, Rollers, Todies
III. Subclass Sauriurae (= Archaeornithes),
Toothed Reptile-Like Birds
(*Archaeopteryx* and other extinct
genera).

A CLASSIFICATION OF CLASS MAMMALIA[a]

I. Subclass Prototheria (*syn.* Protheria)
  A. Order Docodonta (extinct)
  B. Order Monotremata, Echidnas,
  Platypus, Spiny Anteater
  C. Order Triconodonta (extinct)
  D. Order unnamed (extinct)
    1. Amphilestidae (extinct)
    2. Kuehneotheriidae,
    Kuehneotherium (extinct)
    3. Morganucodontidae,
    Megazostrodon, Morganucodon
    (extinct)
II. Subclass Theria
  A. Infraclass Eutheria
    1. Cohort Edentia
      a. Order Xenarthra, Anteaters,
      Armadillos, Giant Sloth
      (extinct), Glyptodont
      (extinct), Sloths
    2. Cohort Epitheria
      a. Grandorder Anagalida
        (1) Order Lagomorpha,
        Hares, Pikas, Rabbits
          (a) Family Leporidae,
          Hares and Rabbits

(2) Order Macroscelidea, Elephant Shrews

b. Grandorder Arconta
  (1) Order Chiroptera (Bats)
    (a) Family Vespertilionidae, Plainnose Bats
  (2) Order Dermoptera, Flying Lemurs
  (3) Order Scandentia, Tree Shrews
  (4) Order Primates, Apes, including Humans; Monkeys

c. Grandorder Ferae
  (1) Order Carnivora
    (a) Family Felidae, Cats
    (b) Family Canidae, Dogs
    (c) Family Mustelidae, Skunks, Weasels, and kin
    (d) Family Ursidae, Bears
    (e) Family Procyonidae, Coatis and Raccoons
  (2) Order Cimolesta (extinct)
  (3) Order Credonta (extinct)

d. Grandorder Ictopsia (extinct)

e. Grandorder Insectivora, *Asioryctes* (extinct), Hedgehogs, Moles, Shrews, etc.
    (a) Family Soricidae, Shrews
    (b) Family Talpidae, Moles

f. Grandorder Ungulata
  (1) Order Acreodi (extinct)
  (2) Order Arctocyonia (extinct)
  (3) Order Artiodactyla, Even-Toed Ungulates
    (a) Family Bovidae, Bison, Cattle, Goats, Muskox, and Sheep
    (b) Family Cervidae, Deer
    (c) Other families
  (4) Order Astrapotheria (extinct)
  (5) Order Cetacea, Dolphins, Porpoises, Whales, etc.
  (6) Order Condylarthra (extinct)
  (7) Order Desmostylia (extinct)
  (8) Order Dinocerata (extinct)
  (9) Order Embrithopoda (extinct)
  (10) Order Hyracoidea (hyraxes)
  (11) Order Lithopterna (extinct)
  (12) Order Notoungulata (extinct)
  (13) Order Pantodonta (extinct)
  (14) Order Perissodactyla, Odd-Toed Ungulates: Horses, Rhinos, Tapirs, Zebras, etc.)
  (15) Order Proboscidea, Elephants
  (16) Order Pyrotheria (extinct)
  (17) Order Sirenia, Dugongs, Manatees, Sea Cows
  (18) Order Taeniodonta (extinct)
  (19) Order Tillodontia (extinct)
  (20) Order Trigonostylopoidea (extinct)
  (21) Order Tubulidentata, Aardvark
  (22) Order Xenungulata (extinct)

g. Grandorder uncertain
  (1) Order Pholidota, Scaly Anteaters
  (2) Order Rodentia, Rodents
    (a) Family Castoridae, Beavers
    (b) Family Cricetidae, Lemmings, Mice, Rats, and Voles
    (c) Family Erethizontidae, Porcupine
    (d) Family Scuiridae, Squirrels
    (e) Family Zapodidae, Jumping Mice
    (f) Family Castoridae, Beaver

B. Infraclass Metatheria (*syn.* Marsupialia)
  1. Order Dasyurida
  2. Order Diprotodonta, Cucus, Diprotodonts (extinct), Kangaroos, Koala, Possums, Sugar Glider, Wallabies, Wombats, Rat Kangaroos, Tree Kangaroo
  3. Order Marsupicarnivora, Australian Native Cat, Monitos del Monte, Marsupial Mice, Marsupial Moles, O'possums (= Opossums), Numbat, Tasmanian Devil, Tasmanian Pouched Wolf
    a. Family Didelphiidae, O'possums

4. Order Paucituberculata, Rat O'possums and related fossil forms
5. Order Peramelina, Bandicoots
C. Infraclass Pantotheria (extinct)
 1. Order Eupantotheria, Eupanthotheres including *Aegialodon* (extinct)
 2. Order Symmetrodontia, Symmetrodonts including *Spalacotherium* (extinct)

---

[a] Based primarily on Savage and Long (1986). Taxa are in alphabetical order within their next highest taxon. This table omits many lower taxa. References do not agree on many details of mammal classification (Savage and Long 1986, inside cover; Benton 1991a, 330–331; 1991b). A commonly used evolution textbook (Strickberger 1996) does not give a mammalian classification in a tabular form.

ORDER SCANDENITA (TREE SHREWS) AND ORDER PRIMATES (PRIMATES)

I. Order Scandentia, Tree Shrews
II. Order Primates, Primates
 A. Prosimian Gade
  1. Suborder Plesiapiformes, Near Primates (extinct)
   a. Family Paramomyidae
    (1) *Purgatorius* sp.
   b. Family Plesiadapidae
    (1) *Plesiadapis* sp.
  2. Family Adapidae
   a. *Adapis* sp.
   b. *Notharctus* sp.
   c. *Pelycodus* sp.
   d. *Pronycticebus* sp.
   e. *Smilodectes* sp.
  3. Family Omomyidae
   a. *Hemiacodon* sp.
   b. *Omomys* sp.
   c. *Macrotarsius* sp.
   d. *Tetonius* sp.
  4. Suborder Prosimii, Prosimians (= Strepsirhini, "Premonkeys:" Tarsiers, Lorises, Lemurs)
   a. Superfamily Lemuroidea, Lemurs
   b. Superfamily Lorisoidea, Lorises
   c. Superfamily Tarsoidea, Tarsiers
 B. Monkey Grade
  1. Suborder Anthropoidea, Anthropoids (= Haplorhini), Monkeys
   a. Infraorder Platyrrhini, New World Monkeys (Capuchins, Marmosets, Spider Monkeys, Squirrel Monkeys, Tamarins, etc.)

(1) *Biretia* sp. (extinct)
(2) Superfamily Ceboidea
 (a) Family Callitrichidae, Marmosets, Tamarins
 (b) Family Cebidae, Capuchins, Howler Monkeys, Spider Monkeys, Uakaris, etc.
   b. Infraorder Catarrhini, Old World Monkeys (Baboons; Macaques, including Rhesus Macaque; Mandrills; etc.)
    (1) *Aeolopithecus* sp. (extinct)
    (2) *Amphipithecus* sp. (extinct)
    (3) Parapithecidae sp. (extinct)
    (4) Propliopithecidae (extinct)
     (a) *Aegyptopithecus zeuxis* 1965 (= *Propliopithecus* sp., Dawn Monkey (= Dawn Ape, Egyptian Ape)
    (5) Superfamily Cercopithecoidea
     (a) Family Cercopithecidae, Baboons, Colobus, Drills, Geladas, Macaques (including Rhesus Monkeys), Vervet Monkeys (= Grivets)
      (i) *Oreopithecus* sp.
     (b) Family Colobidae, Langurs
C. Nonhuman-Ape Grade
 1. Suborder Anthropoidea, Anthropoids: Apes (including Humans)
  a. Superfamily Hominoidea, Apes (= Hominoids)
   (1) Family Dryopithecidae, Dryopithecids
    (a) *Dryopithecus* sp., Oak Ape (= Forest Ape) (extinct)
   (2) Family Hylobatidae, Gibbons and Siamangs
   (3) Family Pliopithecidae
    (a) *Pliopithecus* sp. (extinct)
   (4) Family Pongidae, Pongids (Orangutan, Gorillas, Chimpanzees)
    (a) *Pongo pygmaeus*, Orangutan
    (b) *Gorilla*, Gorillas, 2 subspp.

(c) *Pan* spp.
Chimpanzees
(= Chimps)
   (i) *Pan troglodytes*,
Chimpanzee
   (ii) *Pan paniscus*,
Bonobo
(= Pigmy
Chimpanzee)
  (5) Family Proconsulidae
   (a) *Proconsul* sp.
(extinct)
  (6) Family Ramapithecidae,
Ramapithecids
   (a) *Gigantopithecus* sp.
   (b) *Kenyapithecus* sp.,
Kenya Ape (extinct)
   (c) *Ramapithecus*,
Rama's Apes, 2 spp.
   (d) *Sivapithecus* spp. (in
the broad sense, this
genus, includes
Ramapithecus)
D. Human-Ape Grade
  1. Family Hominidae, Humans
(= Hominids)

FAMILY HOMINIDAE, HUMANS

  I. *Ardipithecus*
   A. *Ardipithecus ramidus*, Stem Person
  II. *Australopithecus*
   A. *Australopithecus aethiopicus*,
"Aethiopian Person" (*Parapithecus
aethiopicus*)
   B. *Australopithecus afarensis*, Afar
Person
    1. Chad jawbone
    2. Desi (= Ricky)
    3. First Family
    4. Laetoli footprints
    5. Lucy (= Al 288, "Louis")
    6. Little Foot, total skeleton
    7. 53 new specimens
   C. *Australopithecus africanus*,
Southern Ape Person
    1. *A. a. aethiopicus*
    2. *A. a. africanus*
     a. Taung Child (= Taung Boy)
    3. *A. a. tanzaiensis*
    4. *A. a. transvallensis*
     a. Laetoli footprints
    5. Mr. Ples
   D. *Australopithecus anamensis*, Lake
Person (= Anamen Person)
   E. *Australopithecus boisei*, Boise Ape
Person (= *Parapithecus boisei*,
Nutcracker Person, *Zinjanthropus
boisei*, Zinj; possibly part of *A.
robustus*)

    1. Black Skull (= KNM-WT 17000)
(possibly *A. aethiopicus*)
    2. Dear Boy
   F. *Australopithecus bahrelghazalia*
   G. *Australopithecus garhi* Asfaw,
White, Lovejoy, Latimer, Simpson,
Suwa, 1999, "Surprise Ape Person"
   H. *Australopithecus robustus*, Robust
Ape Person (several synonymous
scientific names including
Parapithecus robustus)
    1. Kromdraai Cave fossils
    2. Swartkrans Cave
III. *Homo* (2.33 Ma to present)
   A. Palate (2.33 Ma, oldest known
fossil of *Homo*; Hadar, Ethiopia)
   B. *Homo antecessor*, "Explorer Person"
   C. *Homo erectus* Dubois, "Upright
Person" (= Eugene)
(= *Meganthropus* sp.,
Pithecanthropus erectus)
    1. *Homo erectus pekinensis* Pei
Wei-Chung, 1928, "Beijing
Person" (= *Sinanthropus
pekinensis*, "Peking Person")
     a. Zoukoutien-Cave fossils
    2. "Java Person" (= Java Man)
    3. Poloyo Skull
    4. several skulls
   D. *Homo ergaster*, "Ergaster Person"
(African specimens formerly
classified as *H. erectus*)
    1. Turkana Boy
    2. several skulls
   E. *Homo erectus-sapiens*, Solo Person
   F. *Homo ergaster*, "Ergaster Person"
    1. Much of a skull, a few teeth
   G. *Homo habilis-erectus*, "Handy
Upright Person"
   H. *Homo habilis*, Handy Person
(= Louis, *Telanthropus* sp.)
    1. Ethiopia
    2. Kenya
     a. first fossils
     b. Johnny's Child
     c. 1963 fossils
     d. OH62
    3. South Africa
    4. Zaire
     a. Cindy
     b. George
     c. Tiggy
  I. *Homo rudolfensis*, Rudolf Person
  J. *Homo sapiens*, Archaic Humans
    1. *Homo sapiens neanderthalensis*,
Neanderthal Person (= *H.
neanderthalensis*, Neandertal
Person, Neanderthals)
    2. *Homo sapiens sapiens*, Modern
Person (Humans, Human
Beings, Man, Modern Humans)

a. First-known fossils
b. Cro-Magnon Person
c. Qafeh Skull (Israel)

Adalia bipunctata, Two-Spotted Lady Beetle See Animalia: Phylum Arthropoda: *Adalia bipunctata.*

African Elephant See Animalia: Phylum Chordata: Class Mammalia: Subclass Theria: Infraclass Eutheria: Order Proboscidea: *Loxodonta africana.*

Africanized Bee, Africanized Honey Bee See Animalia: Phylum Arthropoda: Group Apiformes: *Apis mellifera.*

Ailuropoda melanoleuca See Animalia: Class Mammalia: Subclass Theria: Infraclass Eutheria: Order Carnivora: *Ailuropoda melanoleuca.*

Alkali Bee See Animalia: Phylum Arthropoda: Group Apiformes: *Nomia melanderi.*

American Crow See Animalia: Phylum Chordata: Class Aves: *Corvus brachyrhyncos.*

Amphibians See Animalia: Phylum Chordata: Class Amphibia.

Anser anser See Animalia: Phylum Chordata: Class Aves: *Anser anser.*

Ants See Animalia: Phylum Arthropoda: Formicidae.

Apidae, Apid Bees See Animalia: Phylum Arthropoda: Group Apiformes: Apidae.

Archaeornithes See Animalia: Phylum Chordata: Class Aves: Archaeornithes.

Archeopteryx See Animalia: Phylum Chordata: Class Aves: *Archeopteryx.*

Archeornis See Animalia: Phylum Chordata: Class Aves: *Archeornis.*

Archezoa *n.* A group of eukaryotes that lack mitochondria and diverged early from the eukaryote common ancestor (Gray et al. 1999, 1479).
Comments: Archezoa includes *Giardia lamblia* and other diplomonads. Recent findings suggest that Archezoa once had mitochondria but subsequently lost them, and Archezoa is not the earliest group of eukaryotes to diverge from the eukaryote common ancestor (Gray et al. 1999, 1479–1480).

Asian Cerambycid Beetle, Asian Long-Horned Beetle See Animalia: Phylum Arthropoda: *Anoplophora glabripennis.*

Asian Elephant, Asiatic Elephant See Animalia: Phylum Chordata: Class Mammalia: Subclass Theria: Infraclass Eutheria: Order Proboscidea: *Elephas maximus.*

Birds See Animalia: Phylum Chordata: Class Aves.

Blue Babe See Animalia: Phylum Chordata: Class Mammalia: Subclass Theria: Infraclass Eutheria: Order Artiodactyla: Blue Babe.

Bonobo See Animalia: Phylum Chordata: Class Mammalia: Subclass Theria: Infraclass Eutheria: Order Primates: *Pan paniscus.*

Bravo Bee See Animalia: Phylum Arthropoda: Apiformes: *Apis mellifera.*

Brown Tree Snake See Animalia: Phylum Chordata: Class Reptilia: *Boiga irregularis.*

bumble bee See Animalia: Phylum Arthropoda: Group Apiformes: bumble bee

Chimp, Chimpanzee See Animalia: Phylum Chordata: Class Mammalia: Subclass Theria: Infraclass Eutheria: Order Primates: *Pan troglodytes.*

Chondrichthes See Animalia: Phylum Chordata: Class Chondrichthyes.

Clovis People See Animalia: Phylum Chordata: Class Mammalia: Subclass Theria: Infraclass Eutheria: Order Primates: Clovis People.

Cocinellidae See Animalia: Phylum Arthropoda: Order Coleoptera: Cocinellidae.

Coelacanth See Animalia: Phylum Chordata: Class Osteichthyes: *Latimeria chalumnae.*

Coelomata *n.* Animal phyla with coeloms including Arthropoda and Chordata (King and Stansfield 1985).

coelomate *n.* An animal with a coelom, an internal cavity formed in and surrounded by mesoderm; contrasted with acoelomate, pseudocoelomate, and schizocoelomate (King and Stansfield 1985).

Common Crow See Animalia: Phylum Chordata: Class Aves: *Corvus brachyrhyncos.*

Common Langur See Animalia: Phylum Chordata: Class Mammalia: Subclass Theria: Infraclass Eutheria: Order Primates: *Semnopithecus entellus.*

Common Rat See Animalia: Phylum Chordata: Class Mammalia: Subclass Theria: Infraclass Eutheria: Order Rodentia: *Rattus norvegicus.*

companion animal See Animalia: animal: companion animal.

contact animal, contact-type animal See Animalia: animal: contact animal.

control animal See Animalia: animal: control animal.

Corvus brachyrhyncos See Animalia: Phylum Chordata: Class Aves: *Corvus brachyrhyncos.*

crawler See Animalia: Phylum Arthropoda: Apiformes: crawler.

cuckoo See Animalia: Phylum Chordata: Class Aves: Cuculidae.

cuckoo bee See Animalia: Phylum Arthropoda: Group Apiformes: cuckoo bee.

dancing bee See Animalia: Phylum Arthropoda: Group Apiformes: dancing bee.

Darwin's Finches See Animalia: Phylum Chordata: Class Aves: Geospizinae.

Dawn Monkey See Animalia: Phylum Chordata: Class Mammalia: Subclass Theria: Infraclass Eutheria: Order Primates: *Eosimias.*

Dawson's Dawn Man See Animalia: Phylum Chordata: Class Mammalia: Piltdown Man.

density-intolerant animal See Animalia: animal: density-intolerant animal.

density-tolerant animal See Animalia: animal: density-tolerant animal.

Dicyemidae *n.* A family of microscopic parasites of octopuses and squids that have very simple body plans for eukaryan animals (Kobayashi et al. 1999, 762). *Comments:* Researchers classified Dicyemids as mesozoans; however, Kobayashi et al. (1999) conclude that dicyemids are in the Lophotrochozoa and are related to Annelida, Brachiopoda, Mollusca, Playthelminthes, and Nemertea.

distance animal See Animalia: animal: distance animal.

Dominulus Paper Wasp See Animalia: Phylum Arthropoda: *Polistes (Polistes) dominulus.*

Doogie See Animalia: Phylum Chordata: Class Mammalia: Order Rodentia: *Mus musculus.*

Early Toothed Birds See Animalia: Phylum Chordata: Class Aves: Archaeornithes.

Ediacaran [ee dee ACK a ran] *n.* An organism from the Ediacara Epoch (570–543 million years ago), Vendian Subperiod, Sinian Era (Erwin et al. 1997, 128, illustration). *Comments:* Some Ediacarans might have lived in the Varanger Epoch before the Ediacara Epoch (Wright 1997, 56). Ediacaran fossils occur in rocks worldwide (Wright 1997, 53). Fossils of Ediacaran fauna suggest that they had no bone, heads, shells, tails, teeth, or obvious circulatory, digestive, and nervous sys-

tems. Researchers debate whether Ediacarans were animals, lichens, or plants; single or multicellular; and ancestors of other extant phyla or evolutionary dead ends.

[after Ediacara Hills, southern Australia, which harbor a large number of fossils of Ediacaran organisms (Wright 1997, 53)]

- **Pre-Ediacaran metazo** *n.* Metazoa that date before the Ediacara Epoch (Doolittle et al. 1996, 473).

Elk See Animalia: Phylum Chordata: Class Mammalia: Order Artiodactyla: *Cervus canadensis.*

Enantiornithines See Animalia: Phylum Chordata: Class Aves: Enantiornithines.

Eoanthropus dawsonii See Animalia: Phylum Chordata: Class Mammalia: Piltdown Man.

Eosimias See Animalia: Phylum Chordata: Class Mammalia: Subclass Theria: Infraclass Eutheria: Order Primates: *Eosimias.*

Eumetazoa *n.* The animal group that possess digestive cavities, mouths, and organ systems (King and Stansfield 1985).

fanning bee See Animalia: Phylum Arthropoda: Group Apiformes: fanning bee.

Firebrat See Animalia: Phylum Arthropoda: *Thermobia domestica.*

firefly See Animalia: Phylum Arthropoda: Lampyridae.

Fish See Animalia: Phylum Chordata: Class Chondrichthyes, Osteichhthyes.

follower bee See Animalia: Phylum Arthropoda: Group Apiformes: follower bee.

Formosan Subterranean Termite See Animalia: Phylum Arthropoda: *Coptotermes formosanus.*

Galápagos Finche See Animalia: Phylum Chordata: Class Aves: Geospizinae.

Galápagos Tortoise See Animalia: Phylum Chordata: Class Reptilia: *Geochelone elephantophus.*

Geochelone elephantophus See Animalia: Phylum Chordata: Class Reptilia: *Geochelone elephantophus.*

Giant Panda See Animalia: Class Mammalia: Subclass Theria: Infraclass Eutheria: Order Carnivora: *Ailuropoda melanoleuca.*

Grayhound See Animalia: Phylum Chordata: Mammalia: Dog: Greyhound.

Gray Langur See Animalia: Phylum Chordata: Class Mammalia: Subclass Theria: Infraclass Eutheria: Order Primates: *Semnopithecus entellus.*

Greyhound See Animalia: Phylum
Chordata: Class Mammalia: Subclass
Theria: Infraclass Eutheria: Order Car-
nivora: Greyhound.

Greylag Goose See Animalia: Phylum
Chordata: Class Aves: *Anser anser.*

guard bee See Animalia: Phylum Arthropoda: Group Apiformes: follower bee.

H.M. See Animalia: Phylum Chordata:
Class Mammalia: Subclass Theria:
Infraclass Eutheria: Order Primates: H.M.

Hanabi-Ko See Animalia: Phylum
Chordata: Class Mammalia: Subclass
Theria: Infraclass Eutheria: Order Pri-
mates: Koko.

Hanuman Langur See Animalia: Phylum Chordata: Class Mammalia: Subclass Theria: Infraclass Eutheria: Order
Primates: *Semnopithecus entellus.*

***Harmonia axyridis* Pallas, Multicol-
ored-Asian Lady Beetle** See Animalia:
Phylum Arthropoda: *Harmonia axyridis.*

hawk moth, hawkmoth See Animalia:
Phylum Arthropoda: Sphingidae.

Heterocephalus glaber See Animalia:
Phylum Chordata: Class Mammalia: Sub-
class Theria: Infraclass Eutheria: Order
Rodentia: *Heterocephalus glaber.*

Homing Pigeon See Animalia: Phylum
Chordata: Class Aves: *Columba livia.*

Honey Bees See Animalia: Phylum Arthropoda: Group Apiformes: Apidae,
honey bee.

Honey Possum See Animalia: Class
Mammalia: Subclass Theria: Infraclass
Metatheria: *Tarsipes spenserae.*

honeycreeper See Animalia: Phylum
Chordata: Class Aves: Drepanidae.

House Mouse See Animalia: Phylum
Chordata: Class Mammalia: Subclass
Theria: Infraclass Eutheria: Order Ro-
dentia: *Mus musculus.*

Human See Animalia: Phylum Chordata:
Class Mammalia: Subclass Theria:
Infraclass Eutheria: Order Primates:
Homo sapiens sapiens.

hummingbirds See Animalia: Phylum
Chordata: Class Aves: Trochilidae.

Ice Man See Animalia: Phylum Chordata:
Class Mammalia: Subclass Theria:
Infraclass Eutheria: Order Primates: Ice
Man.

Inca Girl See Animalia: Phylum
Chordata: Class Mammalia: Subclass
Theria: Infraclass Eutheria: Order Pri-
mates: Inca Girl.

Jernigan, Joseph Paul See Animalia:
Phylum Chordata: Class Mammalia: Sub-
class Theria: Infraclass Eutheria: Order
Primates: Jernigan, Joseph Paul.

Killer Bee See Animalia: Phylum Arthropoda: Apiformes: *Apis mellifera.*

Koko See Animalia: Chordata: Mammalia: Subclass Theria: Infraclass Eutheria:
Order Primates: Koko.

Laboratory Mouse See Animalia: Phylum Chordata: Class Mammalia: Subclass Theria: Infraclass Eutheria: Order
Rodentia: *Mus musculus.*

Lady Beetles, Ladybirds, Ladybugs
See Animalia: Phylum Arthropoda:
Cocinellidae.

lightningbug See Animalia: Phylum
Arthropoda: Lampyridae.

Mammals See Animalia: Phylum Chordata: Class Mammalia.

Man See Animalia: Phylum Chordata:
Class Mammalia: Subclass Theria:
Infraclass Eutheria: Order Primates:
Homo sapiens sapiens.

Marine Toad See Animalia: Phylum
Chordata: Class Amphibia: *Bufo
marinus.*

Megan See Animalia: Phylum Chordata:
Subclass Theria: Infraclass Eutheria:
Order Artiodactyla: Dolly.

Michael See Animalia: Phylum Chor-
data: Class Mammalia: Subclass Theria:
Infraclass Eutheria: Order Primates:
Michael.

Modern Birds See Animalia: Phylum
Chordata: Class Aves: Neornithes.

Modern Human, Modern Man See
Animalia: Phylum Chordata: Class Mammalia: Subclass Theria: Infraclass
Eutheria: Order Primates: *Homo sapiens
sapiens.*

Monte-Verde People See Animalia: Phylum Chordata: Class Mammalia: Subclass Theria: Infraclass Eutheria: Order
Primates: Monte-Verde People.

Morag See Animalia: Class Mammalia:
Subclass Theria: Infraclass Eutheria:
Order Artiodactyla: Dolly.

Multicolored-Asian Lady Beetle See
Animalia: Phylum Arthropoda: *Harmonia axyridis.*

Naked Mole Rat See Animalia: Phylum
Chordata: Class Mammalia: Subclass
Theria: Infraclass Eutheria: Order Ro-
dentia: *Heterocephalus glaber.*

native bee, non-*Apis* bee See Animalia:
Phylum Arthropoda: Group Apiformes:
pollen bee.

Neornithes See Animalia: Phylum
Chordata: Class Aves: Neornithes.

nerd See Animalia: Phylum Chordata:
Class Mammalia: Subclass Theria:
Infraclass Eutheria: Order Primates: nerd.

Nim Chimpsky See Animalia: Chordata:
Class Mammalia: Subclass Theria:

Infraclass Eutheria: Order Primates: Nim Chimpsky.

Nomia melanderi See Animalia: Phylum Arthropoda: Group Apiformes: *Nomia melanderi.*

Norway Rat See Animalia: Phylum Chordata: Class Mammalia: Subclass Theria: Infraclass Eutheria: Order Rodentia: *Rattus norvegicus.*

nurse bee See Animalia: Phylum Arthropoda: Group Apiformes: nurse bee.

Opposite Birds See Animalia: Phylum Chordata: Class Aves: Enantiornithines.

Orchid Bees See Animalia: Phylum Arthropoda: Group Apiformes: Apidae, Euglossini.

Panda See Animalia: Phylum Chordata: Class Mammalia: Subclass Theria: Infraclass Eutheria: Order Carnivora: *Ailuropoda melanoleuca.*

Parazoa *n.* An animal group that contains organisms with indeterminate shapes and no organs (King and Stansfield 1985).

Phylum Annelida, Annelids, Segmented Worms *n.* A phylum of animals characterized by having a circulatory system, an elongated and segmented body, a head, nephridia, a one-way digestive tract, a well-defined nervous system, and a well-developed coelom (Curtis 1983, 1082).
Comments: Annelida has about 9000 species, including Earthworms.

Phylum Arthropoda, Arthropods *n.* An animal phylum characterized by having an exoskelton and jointed appendages (King and Stansfield 1985).
Comment: This is the largest and most diverse animal phylum with almost 1 million named species and probably millions of unnamed species (King and Stansfield 1985).

▪ ***Adalia bipunctata*, Two-Spotted Lady Beetle** *n.* North American Lady Beetle with a large black spot on each of its orange elytra (Animalia: Phylum Arthropoda: Order Coleoptera: Family Coccinellidae) (Milne and Milne 1997, 580, fig. 146).
Comment: This species can be common in buildings during the cold season, when it diapauses (C. Klass, 1998; www.cce.cornell.edu/act sheets/home/insects).

▪ ***Aedes*** *n.* A mosquito genus with over 700 species (Diptera: Culicidae) (King and Stansfield 1985).
Comments: Aedes aegypti transmits dengue fever and yellow fever to Humans. Researchers have intensively studies this mosquito's resistance to pesticides (King and Stansfield 1985).

▪ ***Anopheles*** *n.* A mosquito genus of about 150 species (King and Stansfield 1985).
Comment: Many *Anopheles* are medically important; *A. albimanus, A. atroparvus, A. pharoensis, A. quadri-maculatus,* and *A. stephensi* transmit malaria (King and Stansfield 1985). Their larval salivary glands have polytene chromosomes.

▪ ***Anoplophora glabripennis* (Motchulsky), Asian Cerambycid Beetle, Asian Long-Horned Beetle** *n.* A tree-consuming, bluish-black, long-horned beetle from China and Korea, with adults up to 3.5 centimeters long (Coleoptera: Cerambycidae) (R.J. Favrin, 1998, www.cfia-acia.agr.ca/english/ppc/science/pps/datasheets/anogla; http://willow.ncfes.umn.edu/pa_ceram/ceramb.htm, illustrations)
syn. Starry-Sky Beetle (in China)
Comments: This beetle entered the U.S. in sewer pipes from China and now infests trees in Illinois and New York (Warrick 1998), where people have undertaken major efforts to control it because it could become a devastating pest in North America. It killed millions of trees in China in 1998.

▪ **Apiformes, Bees** *n.* Insects found throughout the world and characterized by having branched hairs, no strigili, and hind basitari that are broader than other hind tarsi that distinguish them from a closely related group, the sphecid wasps (Hymenoptera: Apoidea) (Michener 1974, 3; Michener et al. 1994, 2–3).
Comments: Bees evolved from a sphecid wasp (Michener et al. 1994, 2–3). Researchers have delineated 11 bee families. Buchmann and Nabhan (1996) define many kinds of bees not in this book and indicate that there are about 40,000 bee species (p. 254). Bees might have originated during the Cretaceous Period, 146–65 million years ago (O'Toole and Raw 1991, 19).

Africanized Honey Bee See Animalia: Phylum Arthropoda: Group Apiformes: *Apis mellifera.*

Apidae, Apid Bees *n.* A family of bees (Apiformes) found

A Classification of Bees (Group Apiformes)[a]

I. Andrenidae, Andrenid Bees (Miner Bees), worldwide except Australia
   A. Andreninae, Adrenine Bees
   B. Panurginae, Panurgine Bees
II. Anthophoridae, Anthophorid Bees (Digger Bees), worldwide
   A. Anthophorinae, Anthophorine Bees
   B. Nomadinae, Nomadine Bees
   C. Xylocopinae, Xylocopinae Bees (Carpenter Bees)
III. Apidae, Apid Bees, worldwide
   A. Apinae, Apine Bees (Honey Bees, Stingless Bees)
   B. Bombinae, Bombine Bees (Bumble Bees)
   C. Euglossinae, Euglossines Bees (Orchid Bees)
IV. Colletidae, Colletid Bees, worldwide
   A. Colletinae, Colletine Bees
   B. Diphaglossinae, Diphaglossine Bees
   C. Hylaeinae, Hylaeine Bees (Yellow-faced Bees)
V. Ctenoplectridae, Ctenoplectrid Bees, Africa through China and Australia
VI. Fideliidae, Fideliid Bees, South Africa
VII. Halictidae, Halictid Bees (Burrowing Bees), worldwide
   A. Halictinae, Halictine Bees (includes Sweat Bees)
   B. Nomiinae, Nomiine Bees
   C. Dufoureinae, Doufoureine Bees
VIII. Megachilidae, Megachilid Bees, worldwide
   A. Lithurginae, Lithurgine Bees
   B. Megachilinae, Megachiline Bees (includes Leaf-Cutter Bees)
IX. Melittidae, Melittid Bees, Africa, Eurasia, North America
X. Oxaeidae, Oxaeid Bees, American subtropics and tropics
XI. Stenotritidae, Stenotritid Bees, Australia

[a] These taxa are in alphabetical order, instead of "systematic order."

throughout most of the world and characterized by having a labrum that is broader than long, a long glossa with a flabellum, three forewing submarginal cells or much-reduced wing venation and a weak marginal cell or an open apex, a scopa (when present) made of marginal hairs on the hind tibia that surround a bare area on its outer surface and forming a corbicula, and no facial fovea or pygidial plate (Michener et al. 1994, 167).

Comment: Apidae comprises the Honey Bees, Orchid Bees, and Stingless Bees (Michener et al. 1994, 167).

***Apis mellifera* Linnaeus, the Honey Bee** *n.* A eusocial bee (Apidae: Apinae) native to Europe, Middle East, and North Africa (O'Toole and Raw 1991, 112).
Comments: There are seven *Apis* spp.(O'Toole and Raw 1991, 113). *Apis mellifera* is the basis of a $US multibillion industry of beeswax, honey, and pollination. It has about 25 races, including Carni-olan, Caucasian, German, and Italian races (King and Stansfield 1985). *Apis mellifera scutellata* (the Africanized Bee, Africanized Honey Bee, Bravo Bee, Killer Bee) is a race, or subspecies, of *Apis mellifera* that is highly defensive (Buchmann and Nabhan 1996, 241). Warwick Kerr, Brazilian geneticist, brought this race to Brazil to use it for breeding a better honey bee for tropical regions. It is now in a large area of South America, Central America, Mexico, and the southern U.S.

crawler *n.* A worker Honey Bee that has been stupefied by a toxin and may be found crawling near her colony's nest entrance (Robinson and Johansen 1978, 16).

cuckoo bee *n.* A bee species that does not collect provisions for its offspring but instead uses those of another bee species (Rozen 1989, 1).
cf. hospicidal; organism: cuckoo, cuckoo wasp; parasitism: clepto-parasitism
Comments: Cuckoo-bee species occur in many of the 11 families of bees (O'Toole and Raw, 1991).

dancing bee *n.* In stingless bees, the Honey Bee: a worker that has returned to her hive and transmits information about exterior resources (food, water, a nesting site) by her movements (Heymer 1977, 161).
cf. dance: bee dance

drone *n.* A male social bee, especially a male Honey Bee or bumble bee (*Oxford English Dictionary* 1972, entries from *ca.* 1000).

cf. caste: king

[Middle English *dronen* < *drone*, male bee]

Euglossini, Orchid Bees *n.* A tribe of apid bees from the American tropics characterized by having bright colors in many species, very long tongues, and very fast flight (O'Toole and Raw 1991, 30).

Comments: Euglossini includes the nonparasitic genera *Eufriesea*, *Euglossa*, and *Eulaema*, and the parasitic genera *Aglae* and *Exaerete*. Male Euglossines collect fragrance from the flowers of some orchid species which they transfer to the tibiae of their hind legs (Orchidales, Encyclopaedia Britannica Online: www.eb.com, 2000). In the process of moving around on the flowers, males of the nonparasitic genera may pick up pollinia, which they may transfer to other flowers, thereby effecting cross-pollination. Males of individual euglossine species are attracted to the odors of only one, or a few, orchid species.

fanning bee *n.* A worker Honey Bee that buzzes her wings and causes water evaporation in her nest which cools it (Heymer 1977, 66).

follower bee *n.* A worker Honey Bee that moves after a dancing bee and may induce her to release a food sample (Heymer 1977, 161).

guard bee *n.* A worker Honey Bee that stands watch at her hive entrance (Michener 1974, 126).

gyne See female: gyne (main part of book).

honey bee *n.* A species, or individual bee, in the genus *Apis* (Michener 1974, 347).

Honey Bees See Animalia: Phylum Arthropoda: Group Apiformes: Apidae.

Killer Bee See Animalia: Phylum Arthropoda: Group Apiformes: *Apis mellifera*.

native bee, non-*Apis* bee See Animalia: Phylum Arthropoda: Group Apiformes: pollen bee.

***Nomia melanderi*, Alkali Bee** *n.* A halictid-bee species, native to western U.S., that nests in alkaline soil (Tepedino and Geer 1991, 4).

Comments: Groups of this bee often inhabit the same patch of soil for many years and can form noisy aggregations of thousands of independently nesting females (Tepedi-no and Geer 1991, 4). People manage this bee for pollination of Alfalfa.

nurse bee *n.* A worker Honey Bee that feeds larvae (Michener 1974, 126).

Orchid Bees See Animalia: Phylum Arthropoda: Group Apiformes: Apidae: Euglossini.

pollen bee *n.* A bee species that is not in the genus *Apis*; contrasted with the Honey Bee, *Apis mellifera*, and other members of the genus *Apis* (inferred from Batra 1994, 591).

syn. native bee, non-*Apis* bee, wild bee (Batra 1994, 591)

Comments: Many researchers are now preferring the name "pollen bee" to designate bees that are not Honey Bees and are better pollinators than Honey Bees for many plant species. Pollen bees collect and feed pollen and nectar to their young; some of them do make and store honey as well (*e.g.,* some bumble bees and meliponine bees). Honey Bees all make honey, and they also collect pollen and nectar. Therefore, the term "pollen bee" is a useful term but not a perfect reflection of the biologies of Pollen Bees. These bees include Bumble Bees, Burrowing Bees, Carpenter Bees, Digger Bees, Leaf-Cutter Bees, Miner Bees, Orchid Bees, Stingless Bees, Sweat Bees, and Yellow-Faced Bees (table above).

scout bee *n.* A worker Honey Bee that flies off from a swarm, investigates possible sites for a new hive, and communicates the location of a site that she has found to her swarm after returning to it (McFarland 1985, 416).

sterzelnde **bee** *n.* A worker Honey Bee that wing fans, *q.v.,* at her nest entrance [coined by Frisch 1965 in Heymer 1977, 69].

Stingless Bees See Animalia: Phylum Arthropoda: Group Apiformes: Apidae.

Trigona prisca **Michener and Grimaldi** *n.* The oldest known bee (O'Toole and Raw 1991, 21). *cf.* bee

Comments: This bee is from amber from the Upper Cretaceous of New Jersey, 96–74 millioin years ago.

wild bee See Animalia: Phylum Arthropoda: Group Apiformes: pollen bee.

worker bee *n.* In Bumble Bees, Honey Bees, Stingless Bees, and other kinds of social bees: a female that lays no eggs, or few eggs relative to a queen, and performs tasks in her nest (Michener 1974, 374).

cf. caste: worker

■ **Arachnida** *n.* A class of Arthropoda that contains Mites, Opilionids, Pseudoscorpions, Scorpions, Spiders, Ticks, and their kin (King and Stansfield 1985).

■ *Biston betularia* **(Linnaeus), Pepper-and-Salt Geometer, Peppered Moth** *n.* A geometrid moth native to Europe and North America (Covell 1984, 356; King and Stansfield 1985). *Comment:* This species shows industrial melanism (H.B.D. Kettlewell in Covell 1984, 356).

■ **Cocinellidae, Lady Beetles, Ladybirds, Ladybugs** *n.* A beetle family characterized by having species with oval, convex, often brightly colored adults, with heads concealed from above by their pronota (Animalia: Phylum Arthropoda: Order Coleoptera).

cf. Animalia: Phylum Arthropoda: *Adalia bipunctata, Harmonia axyridis,* and *Rodolia cardinalis*

Comment: Most Lady Beetles consume other insects, or mites, as larvae and adults; a small minority consume plants (Borror et al. 1989, 441–442).

■ **Coleoptera, Beetles** *n.* An insect order characterized by having chewing mouthparts, elytra, and holometabolous metamorphosis (Borror et al. 1989, 370).

Comment: Coleoptera is the largest insect order, with almost 1,000,0000 described species (Buchmann and Nabhan 1996, 244).

■ *Coptotermes formosanus* **Shiraki, 1909, Formosan Subterranean Termite** *n.* A termite species native to China and Taiwan (Isop-

tera: Rhinotermitidae) (Catalog of Termites of the New World, R. Constantino, 1999, http://www.nb.br/ib/zoo/docente/constant/catal/cat.html).

Comments: This termite is common in China, Guam, Hawaii, and Taiwan. Its colonies have up to 6.8 million workers, its winged adults are attracted to lights, and it is now causing millions of dollars of losses in the southern U.S. (J.B. Benavides, 1999, www.netside.net/~jb/images/formosan.html).

■ **cuckoo wasp** *n.* A wasp species in the Chrysididae (Borror et al. 1989, 723–724).

Notes: Depending on the species, chrysidids feed on bee larvae, sawfly larvae, walkingstick eggs, and wasp larvae (Borror et al. 1989, 723–724).

cf. hospicidal; organism: cuckoo, cuckoo bee; parasitism: cleptoparasitism

■ *Culex pipiens* *n.* A mosquito species that is found in many parts of the Earth (King and Stansfield 1985).

Comment: Researchers have studied this species' giant polytene chromosomes and its insecticide resistance (King and Stansfield 1985).

■ *Dendroctonus frontalis* **Zimmerman, Southern Pine Beetle** *n.* A North American beetle (Coleoptera: Scolytidae).

Comments: The Southern Pine Beetle attacks and kills all pine species in its range (Day 1997; Southern Pine Beetle Information Directory, 1999; www.ento.vt.edu/~salom/SPBinfodirect/spbinfo direct.html).

■ *Drosophia* **spp., Fruit Flies, Pomace Flies** *n.* Flies native to many parts of the world (Diptera: Drosophilidae) (King and Stansfield 1985).

Comments: There are over 900 described Drosophila species in eight subgenera, and this genus is the most studied from the standpoint of cytology and genetics (King and Stansfield 1985). *Drosophila melanogaster* is a well-studied member of this genus. This genus underwent dramatic radiation in Hawai'i.

■ **Ephemeroptera, Mayflies** *n.* An insect order characterized by having terrestrial adults; an abdomen with two or three threadlike caudal filaments; short, bristlelike, and inconspicuous antennae; membranous forewings and hindwings with numerous cells and veins; mouthparts vestigial in adults; and aquatic nymphs (Borror et al. 1989, 175–176).

Comments: Mayfly larvae of most species consume algae and detritus. Some species eat other plants and animal tissue (Edmunds in Merritt and Cummins 1984, 94–125). A few species are predators. Adults do not feed. There are about 2000 described species worldwide. In North America, there are 16 families and dozens of genera. Mayflies can be abundant in some habitats and are food for many organisms, including birds, fish, other arthropods, and salamanders.

▪ **femme fatale,** *femme fatale* *n*. A female *Photuris* firefly that mimics the species-characteristic flash patterns of females of several other species and preys on heterospecific males that perceive her as a mate and approach her (Lloyd 1975 in Dewsbury 1978, 284).

cf. Animalia: Lampyridae; signal: deceitful signal; signaler: deceitful signaler

[French *femme fatale*, a dangerously seductive woman; vamp]

▪ **Formicidae, Ants** *n*. Hymenoptera characterized by having eusociality, one or two petioles per individual, metapleural glands (which make an antibiotic spray), and usually elbowed antennae (Borror et al. 1989, 737; Hilts 1998, A12).

Comments: Researchers found very old ants in amber from New Jersey that is 90 million years old (Hilts 1998, A12). The fossil record shows that ants started flourishing about 50 million years ago. *Sphecomyrma*, a 92-million-year-old fossil, is a possible ant, but it lacks a metapleural gland. Ants are a large part of the biomass of some habitats.

▪ **glowworm** *n*.

1. A wingless bioluminescent adult female or a bioluminescent larval firefly (McDermott 1948, 7).

2. A European beetle (*Lampyris* sp.) with bioluminescent larvae and wingless females (Michaelis 1963).

3. A bioluminescent adult, or larval, firefly (Michaelis 1963).

See Lampyridae.

syn. glowfly (Michaelis 1963)
cf. bioluminescence, firefly, luciferin

▪ *Harmonia axyridis* **Pallas, Multicolored-Asian Lady Beetle** *n*. An Asian Lady Beetle (Animalia: Phylum Arthropoda: Order Coleoptera: Family Coccinellidae).

Comment: Since 1994, this species started becoming common in buildings during the cold season, when it diapauses (C. Klass, 1998, www.cce.cornell.edu/factsheets/home/insects).

▪ **Hymenoptera** *n*. An insect order characterized by four wings with tiny hooks (hamuli) in winged forms, holometabolous metamorphosis, mandibulate mouthparts, and usually five-segmented tarsi and well-developed ovipositors (Borror et al. 1989, 665).

Comments: Hymenoptera contains Ants, Bees, Sawflies, and other wasps and is the second largest insect order (Buchmann and Nabhan 1996, 248).

▪ **Lampyridae, Fireflies, Lightningbugs** *n*. A beetle family characterized by having a pronotum that extends forward over an individual's head; soft, flexible elytra; and light organs in many species (Borror et al. 1989, 432)

See Animalia: femme fatale, glowworm.

cf. Animalia: glowworm; bioluminescence; molicule: luciferin.

Comments: Lightning bug (term in the U.S., Michaelis 1963) and lightningbug (Borror et al. 1989, 432; one word because a lightningbug is a beetle, not a true bug) are synonyms for firefly. A glowworm, *q.v.*, is a bioluminescent female or larval lampyrid.

▪ **Lepidoptera** *n*. An insect order characterized by a small labrum, body and wing scales, small or lacking maxillary palps, usually no mandibles, sucking mouthparts, and well-developed labial palps (Borror et al. 1989, 588–589).

Comment: This order contains butterflies and moths.

▪ *Musca domestica*, **House Fly** *n*. A common household fly (Diptera: Muscidae) (King and Stansfield 1985).

Comment: Researchers have studied pesticide resistence in this Fly (King and Stansfield 1985).

- *Nicrophorous americanus*, **American Burying Beetle** *n*. A large, black and red silphid beetle that feeds on carrion (Line 1996, C1, C4, illustrations).
cf. organism: Arthropoda (Appendix 1, table)
Comments: This beetle was once widespread in the eastern U.S., but had disappeared from all areas except Block Island, RI, by 1947 (Line 1996, C1, C4). It occurs in six states west of the Mississippi River.

- **Noctuidae, Noctuid Moths, Owlet Moths** *n*. Moths characterized by having forewings, each usually with one areole; a Cu vein that appears four-branched; a stout body, usually with many long, narrow scales; metathoracic tympana that point outward or posteriorly; and ocelli in most species (Covell 1984, 77, many plates; Borror et al. 1989, 657, illustrations).
Comments: Noctuidae is the largest moth family, with about 20,000 species worldwide and 2900 species in North America. Noctuids include Armyworm Moths, Bird-Dropping Moths, Borer Moths, Dagger Moths, Darts, Deltoid Noctuids, Flower Moths, Groundlings, *Heliothus zea* (a major pest species), Looper Moths, Midgets, Foresters, Owlets, Pinions, Sallows, Underwing Moths, and Zales (Covell 1984).

- *Papilio aristodemus ponceanus* **Schaus, 1911, Schaus' Swallowtail** *n*. A swallowtail-butterfly subspecies originally native to Miami and the Florida Keys and now found only in the Keys (Opler and Krizek 1984, 47, illustration; Daerr 1999, 39, illustration).
Comments: Habitat loss and pesticides nearly caused this subspecies to go extinct (Daerr 1999, 39). Thomas Emmel saved it from extinction by breeding it and releasing individuals into their natural habitats. About 1200 of these insects were alive in 1999. This is one of the few butterflies known to stop in mid-air and fly backwards as part of predator evasion.
[Latin *papilio*, butterfly; *aristodemus*, after Aristodemus, a king and hero in the war at Messenia, 716 BC; *ponceanus*, after Ponce de Léon, Spanish statesman and explorer; Schaus, after William Schaus, phy-

sician and butterfly enthusiast who came to Miami in 1898 to treat yellow-fever victims and named this butterfly (Opler and Krizek 1984, 47; Daerr 1999, 39)]

- *Polistes (Polistes) dominulus* **Christ, 1791 (= *gallicus* auct., nec L.), Dominulus Paper Wasp** *n*. A paper wasp native to Eurasia and introduced into North America (Hymenoptera: Vespidae) (V.V. Dubatolov, 1999; http://pisum. bionet.nsc.ru/szmn/Hymenop/Vespidae.htm).

- *Rodolia cardinalis* **(Mulsant)** *n*. An Australian Lady Beetle (Animalia: Phylum Arthropoda: Order Coleoptera: Family Coccinellidae) (Borror et al. 1989, 442).
Comment: Entomologists introduced this species into California in the 1880s to control the Cottony-Cushion Scale, which threatened the citrus industry (Borror et al. 1989, 442). This beetle quickly reduced the population sizes of this Scale.

- **Sphingidae, Hawk Moths, Hummingbird Moths, Sphinx Moths** *n*. A moth family characterized by connection of the radius and subcosta veins of hindwings by a cross vein (R_1) about opposite the middle of the discal cell, heavy adult bodies, and a long proboscis in most species (Borror et al. 1989, 652–653).
Comments: Synonymous names for members of the Sphingidae are hawk moth, hawkmoth, sphingid, sphinx moth (Buchmann and Nabhan 1996, 247). Many sphingid species have tongues longer than their bodies. Sphingidae includes the Bumble-Bee Moth, Tobacco Hornworm, and Tomato Hornworm. Hawkmoths are frequent pollinators.
[hawk, after the fast flight of adults in these moths; hummingbird, after hummingbird-like hovering behavior; sphinx, after the larvae of some species that looks like a sphinxes when they rear back the anterior parts of their bodies (Borror et al. 1989, 653)]

- **Syrphidae, Flower Flies, Hover Flies, Syrphid Flies** *n*. A worldwide fly family characterized by having a spurious vein on each forewing between the radius and media veins (Animalia: Phylum Ar-

thropoda: Group Hexapoda: Order Diptera) (Borror et al. 1989, 557).
Comments: Some syrphid species are bee or wasp mimics; many syrphid species are important pollinators (Buchmann and Nabhan 1996, 255).
[Greek Syrphidae, *syrphos*, a kind of fly; English Flower Fly, after the frequent presence of many species on flowers; hover, after the habit of many species to hover over flowers and in territories]

- *Thermobia domestica* **(Packard), Firebrat** *n.* A cosmopolitan silverfish, possibly originally from Eurasia (Thysanura: Lepismatidae) (Milne and Milne 1997, 353).
Comments: This species often lives in building near warm to hot places, such as furnaces, heating pipes, and ovens (Milne and Milne 1997, 353). This insect consumes starchy organic matter, including book bindings and labels, human foods, and wallpaper glue (Borror et al. 1989, 174).

- **Trilobitomorpha, Trilobites** *n.* An arthropod subphylum characterized by usually having hard exoskeletons and three-lobed bodies, each with a central lobe and two lateral lobes (Stanley 1989, 468, figs. 12-2, 12-3, 12-15, 12-16, 13-5; www.ualberta.ca/~kbrett/Trilobites, illustrations).
Comments: Trilobites lived from the Cambrian through the Permian Periods (Stanley 1989, 468). Trilobites in the Burgess Shale have soft parts preserved that are not preserved in fossils from other sites (Gould 1993, 55).

Phylum Aschelminthes, Round Worms *n.* An animal phylum characterized by having a bilaterally symmetrical, cylindrical, elongated, and unsegmented body and a pseudoco-elom (Curtis 1983, 1081).
syn. Nematoda (Curtis 1983, 1081)
Comments: Aschelminthes has about 12,000 named species, and perhaps up to 500,000 total species, including Hookworms, Nematodes, and Vinegar Eels (Curtis 1983, 1081).

- *Caenorhabditis elegans* *n.* A small nematode worm (Animalia: Aschelminthes) (King and Stansfield 1985).
Comments: Researchers use this worm in developmental genetics studies (King and Stansfield 1985;

Campbell 1996, 979, illustration). It is transparent, which enables researchers to see its internal anatomy during its development, ending in 2000 cells with about 302 neurons (Campbell 1996, 979). At 20°C, it has a life cycle of 3.5 days. This worm usually reproduces as a self-fertilizing hermaphrodite and has two X chromosomes per cell, plus five pairs of autosomes. Loss of an X by meiotic nondisjunction leads to the production of a male. Mating hermaphrodites with males provides a method of genetic analysis. This species has about 100,000,000 base pairs and 17,000 genes.
cf. apoptosis

Phylum Chordata, Chordates *n.* An animal phylum with about 52,000 living species characterized by organisms that each have a hollow dorsal nerve cord, gill slits, and a notochord, *q.v.*, some time in its development (King and Stansfield 1985).

- **Class Agnatha, Jawless Fish** *n.* A chordate class of about 45 living species characterized by being aquatic and having developed brains enclosed by cartilage and notochords reduced to tissue between vertebrae, usually having tails, and having no legs or wings (Curtis 1983, 1084).
Comment: Fish are in the Classes Acanthodii, Agnatha, Chondricthyes, Osteichthyes, and Placodermi.

- **Class Amphibia, Amphibians** *n.* A chordate class of about 3000 living species, containing frogs, salamanders, and toads, characterized by having developed brains enclosed by bone, eggs unprotected by embryonic membranes or shells, incomplete double blood circulation, legs, and notochords reduced to tissue between vertebrae; often having tails; and usually having both aquatic and terrestrial stages, gills in young and lungs in adults, and naked skin (Curtis 1983, 1085).

  Acanthostega **Jarvik 1952** *n.* A very early tetrapod genus (Amphibia) thought to be similar to the first amphibian (Zimmer 1995a, illustrations).
  cf. Ichthyostega, Sarcopterygii
  Comments: Fossils of this species date to 360 million years ago (Zimmer 1995, June). Säve-Söderberg and Jarvis found the first

Acanthostega fossils in Greenland in 1933. *Acanthostega* has eight digits on its forelimbs, a pair of prongs on the back of its skull, hips, a scarum, and fins on the dorsal and ventral sides of its tail; probably had an internal gill system; and may have been exclusively aquatic. Researchers also found a tetrapod 5 million years older than *Acanthostega*.

[Greek *akantha*, thorn + *stegos*, roof; spine plate; referring to a pair of prongs on the back of *Acanthostega*'s head (Zimmer 1995a, 123)]

axolotl *n*.

1. *Seridon pisciforme*, Proteidae, Amphibia (*Oxford English Dictionary* 1972, entries from 1786). *Note:* This species comes from Mexican lakes and retains external gills throughout its life.

2. The neotene of salamander species including *A. tigrinum mavortium* (Conant 1958, 213, 259, illustration; Stebbins 1996, 33, illustration, plate 2).

Comment: Other salamander species have neotene individuals (Stebbins 1996).

[Aztec *axolotl*, the common name for *Seridon pisciforme* (*Oxford English Dictionary* 1972)]

[Spanish, *nahautl*, servant of water (Michaelis 1963)]

***Bufo marinus*, Marine Toad** *n*. A large toad originally from Mexico through Brazil (Associated Press 1996c, 29).

Comments: Humans introduced this Toad into Australia, Florida, Hawaii, Philippines, and Puerto Rico as a biocontrol agent for sugar-cane pests in the 1930s (Associated Press 1996c, 29). This Toad is not a good biocontrol agent and has become a pest for such reasons as it eats cat and dog food, kills and sickens dogs with its venom, and scares people.

Ichthyostega [ik thee o STEG AH, IK thee OHS teg gah] *n*. An extinct genus of early terrestrial amphibians characterized by an osteolepiform body shape with a streamlined, broad, flat, fish-like head; bulky body; a small fin on the dorsum of its tail; and unusually massive ribs (Benton 1990, 51–53; Zimmer 1995a, 120).

Comments: Ichtyostega probably lived in water part of the time and waddled when moving on land. It had sharp fangs, suggesting that it ate centipedes, insects, millipedes, scorpions, and rhipidistians. This tetrapod held the distinction of being the earliest known one until J. Clack discovered *Acanthostega* and other workers discovered tetrapods predating *Acanthostega* (Zimmer 1995a, 120). In 1931, Gunner Säve-Söderberg found the first *Ichthyostega* fossil (a skull in the mountain Celsius Bjerg, Greenland), which dates to 360 million years ago (Zimmer 1995a, 121).

[Greek *ichthyos*, fish + *stega*, plate; referring to the roof of this amphibian's skull which is shaped like that of a fish; named by Säve-Söderberg (Zimmer 1995a, 121)]

salamander *n*. An amphibian characterized by clawless toes, gelatinous eggs, and smooth, sometimes slimy, skin (Order Urodela) (Cohn 1996, H1).

syn. lizard (in West Virginia, personal observation); spring lizard (in Appalachian areas, Cohn 1996, H1).

Comments: The greatest salamander diversity occurs in the Southern Appalachian Mountains of the U.S. (Cohn 1996, H6). Most adult salamanders are under 6 inches long, but the Hellbender can grow up to 2 feet long. The Tiger Salamander grows up to 13 inches long and is the world's largest terrestrial salamander. Adults of about 220 species do not have lungs or gills. They absorb oxygen and emit carbon dioxide through their skins and mucous membranes in their mouths. Adults of another 170 species either have gills throughout their lives or develop lungs at maturity.

■ **Class Aves, Birds** *n*. A chordate class, containing extinct toothed birds and about 8000 extant toothless ones, characterized by having bills, complete double blood circulation, developed brains enclosed by bone, eggs protected by embryonic membranes or shells, feathers, homeo-thermy, legs, notochords reduced to tissue between vertebrae, tails, and wings (Curtis 1983, 1085).

Alex *n*. An African Gray Parrot that is a subject of experiments on thinking undertaken by Irene Pepperberg (Smith 1999, A1, illustration).

Comments: Alex was 23 years old in 1999 and could live for at least another 20 years (Smith 1999, A1, A17). Alex identifies 50 different objects, recognizes quantities up to six, and understands bigger and smaller and same and different, according to Pepperberg.

***Anser anser*, Greylag Goose** *n*. A European goose (Anseriformes: Anseridae) (King and Stansfield 1985).

Comment: Researchers, including Konrad Lorenz, studied imprinting and other behavior of this bird (King and Stansfield 1985).

Archeopteryx *n*. An extinct European bird genus from the Jurassic Period (King and Stansfield 1985).

Archeornis *n*. An extinct bird genus from the Jurassic Period, characterized by having a long, feathered tail with many vertebrae, teeth, and three free digits with claws on each wing (King and Stansfield 1985).

Archaeornithes, Early Toothed Birds *n*. Birds, including *Archaeopteryx* spp., that lived in the Jurassic Period and are characterized by having skeletons that resemble some dinosaur taxa and teeth; contrasted with Neornithes (Stanley 1989, 651).

***Columba livia*, Domestic Pigeon, Pigeon, Rock Dove** *n*. An Old World bird with a chunky body; short, rounded tail; and typically bluish-gray plumage with black and white markings (Columbiformes: Columbidae) (Udvardy and Farrand 1994, 545–546, plate 360).

Comments: People often domesticate this bird as a pet and for food and sport; there are several artificially bred strains. The Homing Pigeon is a strain bred for quickly returning to its roost when artificially released at a distance from its home; the Pouter, for displaying its plumage; and the Tumbler, for its ability to perform aerial acrobatics (Brown 1975, 12).

Confuciusornis sanctus *n*. A Cretaceous bird from China with a long boney tail, needle-sharp claws on its wings, and a toothless beak, and dating from 143–132 million years ago (Lian-hai Hou and Alan Feduccia in Zimmer 1996a, 50, photograph).

***Corvus brachyrhyncos*, American Crow, Common Crow** *n*. A North American bird with black plumage, a fan-shaped tail, stocky body, and stout bill (Passeriformes: Corvidae) (Udvardy and Farrand 1994, 635–636, plate 673).

Comments: American Crows are becoming more common in urban areas, where many people consider them to be pests (Brody 1997, C1). Most U.S. states protect Crows, classifying them as game birds without a hunting season. Crows help conspecifics, play, and have complex vocal communication.

Cuculidae, Cuckoos *n*. A nearly worldwide bird family with 143 species characterized by having conspecific females and males with similar plumage, slender bodies with long tails, and zygodactyl feet (Passeriformes) (Peterson 1980, 182, illustrations; Bull and Farrand 1998, 540).

See animal sounds: cuckoo; cuckoo.

Comments: Cuculidae includes Anis, and the Black-Billed Cuckoo, European Cuckoo, Roadrunner, and Yellow-Billed Cuckoo (Bull and Farrand 1998, 540). Some of the Old World species lay their eggs in nests of other bird species that rear their young. American Cuckoos lay their eggs in their own nests. [Old French *cucu, coucou*, imitative word]

Drepanidae, Hawaiian Honey Creepers *n*. A Hawaiian bird family with 23 species characterized by being perching, small, and tree dwelling (Passeriformes) (Buchmann and Nabhan 1996, 247).

Comments: Depanididae is native to forests of the Hawaiian Islands (Buchmann and Nabhan 1996, 247). Some of these species are pollinators. About eight drepanid species are lately extinct (Austin 1971, 183).

Enantiornithines, Opposite Birds *n.* A group of birds that lived in the Cretaceous and characterized by having fat, fleshy tails and foot bones partially fused from their tops down and usually having teeth; contrasted with Archaeornithes and Neornithes (Stanley 1989, 651; Zimmer 1995b, 40–41).

Geospizinae, Darwin's Finches *n.* A finch subfamily that includes the 14 species of Darwin's Finches of the Galápagos Islands, Equador (King and Stansfield 1985).

Comments: These Finches arose from a single ancestral species which appeared on the Islands, as Charles R. Darwin first suggested. These birds have differently shaped and sized bills that correlate with their food types (King and Stansfield 1985).

***Gymnogyps californianus*, California Condor** *n.* An endangered North American predaceous bird characterized by having a bare head, black plumage, and a wingspan up to 9.5 feet, making it the largest North American bird of prey (Falconiformes: Cantharidae) (Peterson 1990, 182; Udvardy and Farrand 1994, 426, plate 328).

Comments: Conservationists captured 27 California Condors for a captive breeding program in the 1980s (Kenworthy, 1996a, A10). In 1999, there were 162 of these Condors, with 113 in captivity, 29 flying free in California, and 20 in Arizona near the Grand Canyon (Whittaker 1999, A7).

Neornithes, Modern Birds *n.* Birds that lived from the late Mesozoic Era through today and characterized by having foot bones that are partially fused from the bottom up and by having no teeth; contrasted with Archaeornites (Stanley 1989, 651; Zimmer 1995b, 40–41).

Sinornis santensis *n.* A fully volant, perching, 135-million-year-old, sparrow-sized bird from China (Barinaga 1992b, 796; Feduccia 1993, 791).

Trochilidae, Hummingbirds *n.* A New World bird family with about 340 species, characterized by being extremely specialized for feeding on nectar from tubu-lar blossoms and having glittering, iridescent plumage (Trochiliformes) (Austin 1971, 118; Udvardy and Farrand 1994, 575, many plates; Buchmann and Nabhan 1996, 248).

Comments: Hummingbirds are the only known birds that can fly backwards. The hover in front of flowers and sip nectar and pick insects off substrates while flying (Udvardy and Farrand 1994, 575). Many species migrate along nectar corridors, some are important pollinators, and some use traplining (Buchmann and Nabhan 1996, 248). Species richness is highest in the Andes Mountains.

- **Class Chondrichthyes, Cartilaginous Fishes** *n.* A chordate class with about 575 living species, including Rays and Sharks, characterized by being aquatic; having developed brains enclosed by cartilage, notochords reduced to tissue between vertebrae, and tails; and having no bones, legs, or wings (Curtis 1983, 1084).

Comment: Fishes are in the Classes Acanthodii, Agnatha, Chondrichthyes, Osteichthyes, and Placodermi. An earlier classification placed all fishes in the Class Pisces (Curtis 1983, 1084). People are over fishing many shark species (Warrick 1996, A2). A fishing ban protects Basking Sharks, Big-Eye Sharks, Sand Sharks, Tiger Sharks, Whale Sharks, White Sharks, and other sharks.

- **Class Mammalia, Mammals** *n.* A chordate class with about 4000 living species characterized by having complete double blood circulation, developed brains enclosed by bone, fur, homeothermy, legs (sometimes arms), live birth (except for Monotremes), milk-producing glands, notochords reduced to tissue between vertebrae, placentas (in Eurtherians), pouches in females (Metatherians), and tails (Curtis 1983, 1085; Strickberger 1996, 417).

Piltdown Man *n.* A fossil forgery composed of the doctored jaw of an Orangutan and the cranium of a Human (Milner 1990, 363; Milner 1996, 101).

Comments: William Dawson and Teilhard de Chardin found Piltdown Man and a bone shaped like a cricket bat in a gravel pit at Piltdown, Sussex, U.K., in 1912, 25 miles from Charles Darwin's home (Milner 1990, 363–364; Milner 1996, 101). Many people, including some reputable scientists, were fooled into thinking that this forgery was a missing link between Nonhuman Apes and Humans. William Dawson proclaimed the skull a new species; researchers at the British Museum authenticated the skull and named it *Eoanthropus dawsonii*, Dawson's Dawn Man. In 1912, a newspaper headline read: "Missing Link Found — Darwin's Theory Proved." In 1952, a newspaper headline read: "Piltdown Ape-Man a Fake — Fossil Hoax Makes Monkeys Out of Scientists." The British Museum's Kenneth Oakley devised a new method for determining whether ancient bones were the same age, using a radioactive fluorine test. Critics used Piltdown Man as an example of the weakness of evolutionary anthropology. Some creationists proclaimed that all evolutionary science was phony. The British Parliament proposed a vote of "no confidence" in the scientific leadership of the British Museum. The motion failed to carry when a member of the Parliament reminded his colleagues that politicians had "enough skeletons in their own closets." Possible creator(s) of the Piltdown hoax include William Dawson, Teilhard de Chardin, and Sir Arthur Conan Doyle; however, Brian Gardiner and Andrew Currant appear to have solved the mystery (Menon 1997, 34). They concluded that Martin A.C. Hinton, Keeper of Zoology at the British Museum from 1936 to 1945, created Piltdown Man. In 1975, workmen found a box of bones in a trunk with the initials M.A.C.H. in a loft in the museum. These bones were dipped in acid and stained with manganese and iron oxides, as were Piltdown's bones. The cricket-bat-shaped bone was a carved elephant femur. Hinton's possible motivation for the hoax was to embar-

rass Arthur Smith Woodward, Keeper of Paleontology at the museum, who would not give him a salary. Hinton became a respected scholar and did not find it appropriate to confess.

□ **Subclass Prototheria** (*syn.* **Protheria**)
  ***Morganucodon* W.G. Kühne 1949** [MOR gan YOOK oh don] *n.* A Triassic mammalian genus (Subclass Protoheria: Family Morganucodontidae) (Benton 1991, 31–33).
  Comments: Morganucodon is known from complete skeletons in caves in South Wales, Bristol Region, Southwestern England (Benton 1991, 31). Members of this genus were 2.5 centimeters long and had complete secondary palates, large eye sockets (suggesting nocturnal behavior), possible fur, specialized teeth (canines, incisors, molars, premolars), and two sets of teeth (suggesting suckling by its young). Many small openings in their snout bones suggest that nerves passed to sensory whiskers.
  [after Morgan + Greek *dont*, tooth; Morgan's tooth (Benton 1991, 141)]

□ **Subclass Theria**
★ **Infraclass Eutheria**
● **Order Artiodactyla, Even-Toed Ungulates**
○ **Blue Babe** *n.* A 36,000-year-old, mummified Pleistocene male Bison unearthed at a placer gold mine just north of Fairbanks, AK, in 1979 (Animalia: Class Mammalia: Subclass Theria: Infraclass Eutheria: Order Artiodactyla: Family Bovidae) (Guthrie 1990, 1).
  [blue, after the blue color of the mineral Vivanite, which dusted and formed crystals on the dried skin of this mummy, Guthrie 1990, 79]

○ ***Cervus canadensis*, Elk, Wapiti** *n.* A large North American deer, possibly the same species as the Old World *C. elaphus* (Animalia: Class Mammalia: Subclass Theria: Infra-class Eutheria: Order Artiodactyla: Family Cervidae) (Burt 1980, 215).

○ **Dolly** *n.* The first Domestic Sheep cloned from a mature sheep cell (Specter and Kolata 1997, A20, photograph).

Comment: The mature cell came from a sheep mammary gland (Specter and Kolata 1997, A20). Megan and Morag, born in 1994, are the first cloned sheep.

cf. clone.

[after the singer Dolly Parton, famous for her mammary cells (Specter and Kolata 1997, A20)]

● **Order Carnivora**

○ *Ailuropoda melanoleuca*, **Giant Panda, Panda** *n.* An endangered Central and Western Chinese bear characterized by a unique wrist bone that functions as a feeding thumb (Lynn Seizer, K and M International, 1955 Midway Drive, Twinsburg, OH 44087) (Class Mammalia: Subclass Theria: Infraclass Eutheria: Order Carnivora).

Comments: Giant Pandas are active primarily at night and eat mostly bamboo shoots, but also other plants and small animals (Lynn Seizer, K and M International). Young are called cubs and are very small and helpless at first. They can care for themselves in 1 year. Adults weigh up to 350 pounds.

○ *Canis familiaris*, **Dog, Domestic Dog** *n.* A canid derived from the Wolf by artificial breeding (King and Stansfield 1985).

Comments: There are more than 100 dog breeds (King and Stansfield 1985). Researchers have extensively studied the genetic control of dog social behavior and other areas of dog biology. The Beagle is a small, short-coated, short-legged, droopy-eared dog bred for rabbit hunting (Brown 1975, 12). The Greyhound is a tall, slender, smooth-coated dog bred for its ability to run rapidly over open country. The Pointer is a smooth-haired, usually white dog with brown spots, bred for its ability to scent and point silently at game birds without making them fly away. The Sheep Dog (= Shepard Dog, Shepherd's Dog)

is a dog bred for shepherding (Brown 1975, 12), *e.g.,* a Collie or an Old English Sheep Dog (Michaelis 1963). The Terrier is a small, active, wiry dog bred for its ability to go down fox and badger burrows and for its persistent pursuit of prey (Michaelis 1963; Brown 1975, 12). The Thievish Dog is a dog bred to be essentially barkless and is used by game poachers who hunt mainly at night.

○ *Canis latrans*, **Coyote** *n.* A large predatory North American mammal found in Alaska, much of Canada, and all of the contiguous U.S.(Canidae) (Animalia: Class Mammalia: Subclass Theria: Infraclass Eutheria: Order Carnivora: Family Canidae) (Whitaker 1996, 684).

Comment: Coyotes and Domestic Dogs interbreed and produce Coydogs (Whitaker 1996, 684). In the western U.S., removal of large predators (Coyotes) results in an increase of smaller predators (Domestic Cat, Gray Fox, O'possum, Raccoon, and Striped Skunk) which in turn results in a reduction in numbers of scrub birds (Sæther, 1999, 510).

[Latin *Canis*, dog; *latrans*, barking, after its barking vocalizations accompanied by howling; Nahautl Indian Coyote, *coyotl*]

○ *Felis catus*, *Felis domesticus*, **Domestic Cat** *n.* The cat frequently kept as a pet by Humans (Animalia: Class Mammalia: Subclass Theria: Infraclass Eutheria: Order Carnivora: Family Felidae).

cf. organization: Allie Cat Allies, People for the Ethical Treatment of Animals (Appendix 2)

Comments: Owned and feral (= free-ranging, stray) Domestic Cats are often common, effective predators of smaller animals, especially in suburbia and adjacent areas (Crooks and Soulé 1999, 563). Feral and pet *F. domesticus* caused 39 native Australian birds and mammals to go extinct, locally extinct, or near extinct in the wild (Mydans 1997a, A1). Each pet Cat kills about 25 native animals per year. Each feral Cat kills about 100 native animals per year.

○ *Felis concolor*, **Catamount, Cougar, Mountain Lion, Panther** *n*. A large predatory cat native to Central, North, and South America (Animalia: Class Mammalia: Subclass Theria: Infraclass Eutheria: Order Carnivora: Family Felidae). *Comments:* There might be as many as 30 Cougar subspecies (Derr 1999, D2; The Florida Panther Society: http://www.atlantic.net/~oldfla/panther/panther.html; Wildlife Rescue Inc.: www.stonehill.org/wrifccqk.htm). Females weigh up to 90 pounds; males, 140 pounds. Stephen O'Brien (in Derr 1999, D2) showed that the Cougar once had a single, unbroken population throughout its former range. The Cougar was the widest-ranging nonhuman mammal in the Americas.

○ *Felis concolor coryi*, **Florida Panther** *n*. A subspecies of *Felis concolor* originally native throughout the southeastern United States (Derr 1999, D2). *Comments:* By 1970, the Florida Panther became a population with its original genes and those from introduced Costa Rican Panthers. For much related information, see the website of the Florida Panther Society.

● **Order Chiroptera, Bats** *n*. A chordate order with about 900 species of mammals characterized by the presence of wings (paper-thin, elastic membranes extending from the sides of an individual's body, legs, and tail that are back and belly skin extensions) and true flight (Reader's Digest Association 1971, 411; Graham 1994, 52; Allen 1996, 639). *cf.* echolocation *Comments:* Bats are the only true flying mammals (Allen 1996, 639). The oldest bat fossil is from about 60 million years ago, and the oldest Megachirop-teran fossil is from about 35 million years ago (Gra-

ham 1994, 10). Researchers debate whether Bats arose from one (shrew-like) or two (shrew-like and flying-lemur-like) ancestral stocks (Graham 1994, 10–11).

○ **Megachiroptera, Megachiropterans, Megabats** *n*. A bat suborder characterized by feeding mainly on fruit and nectar, having no echo location, and having a claw on the first and second finger of each wing (Reader's Digest Association 1971, 411). *Comments:* Megachiroptera contains one family, which includes the Flying Foxes and Old World Fruit-Eating Bats (Reader's Digest Association 1971, 411).

○ **Microchiroptera, Microchiropterans, Microbats** *n*. A bat suborder characterized by feeding mainly on insects, fish, and frogs and less frequently on fruit, in some species; having echo location; and having a claw on the first finger of each wing (Reader's Digest Association 1971, 411). *Comment:* Microchiroptera contains 16 families, including the Bumble-Bee Bat, which weighs less than a penny (Reader's Digest Association 1971, 411).

● **Order Primates, Apes, including Humans; Monkeys**

○ *Chilecebus carrascoensis* *n*. The oldest known New World monkey (Anonymous 1995a, 31, skull photograph). *Comments:* John Flynn and André Wyss found a skull of this species in the Chilean Andes in 1994 (Anonymous 1995a, 31). Its molars and the shape of its ear region indicate that New World monkeys originated in Africa. Primates immigrated into South America before 28 million years ago.

○ **Chimp, Chimpanzee** See Animalia: Chordata: Mammalia: *Pan troglodytes*.

○ **Clovis People** *n*. Possibly the first people who lived in the Americas (Gibbons 1997, 1256). *Comment:* The Clovis People were big-game hunters who came over the Bering Land

Bridge and then swept rapidly through the Americas about 11,500 years before the present (Gibbons 1997, 1256). The Monte-Verde People might have come to the Americas before the Clovis People.

o **Eosimias, Dawn Monkey** *n.* An extinct anthropoid primate genus from China (Beard in Culotta 1995, 1851).

Comments: Eosimias appears to be a transitional primate between Prosimians and more-derived Anthropoidea, lived about 45 million years ago, weighed about 1 ounce, was probably nocturnal, and fed on fruit and insects (Gebo et al. 2000, 276; Wilford 2000b, A1). [Greek *eos*, dawn + Latin *simias*, ape]

o **Gorilla gorilla gorilla, Western Lowland Gorilla** *n.* An endangered ape native to Central Africa characterized by having long arms, massive bodies, and tusk-like canine teeth (Primates: Pongidae).

cf. Organization: Gorilla Foundation (Appendix 2)

Comments: Western Lowland Gorillas live in tropical forests, (Turco and Gruner 1999). They eat bamboo shoots, bark, fruit, leaves, and vines (especially *Galium* sp.) and live up to 35 years in the wild. A "silverback" is an older male Gorilla with gray through white hairs on his back and is the highest ranking male in his group (Immelmann and Beer 1989, 290). Silverbacks can weigh up to 400 pounds. They are usually gentle and peaceful but become aggressive in protecting their troops. Gorillas are the largest living primates.

[New Latin < Greek *Gorilla*, apparently a native name (Michaelis 1963)]

[Greek, *Gorillai*, a tribe of hairy women (Morris 1982)]

[Gorilla, hairy person, a new name given by an explorer from Ancient Carthage (Turco and Gruner 1999)]

o **H.M.** *n.* A person who received drastic brain surgery to cure his severe epilepsy in 1953

(Lemonick 1999, September 13, 56). *Comments:* Surgeons removed parts of H.M.'s temporal lobes, including his hippocampus (Lemonick 1999, 56). This destroyed his ability to form new long-term memories. H.M. has reasonably good short-term memory and shows implicit memory. He has no permanent memory of what happened to him since 1953.

o **Homo sapiens sapiens, Human, Man, Modern Human, Modern Man** *n.* The only member of Hominidae still living on Earth (Primates: Humanoidea); contrasted with prehumans, nonhuman animal, and nonhuman organism.

adj. Humans, Human

cf. Aridopithecus, Australopithecus, Hominidae, *Homo,* Ice Man, prehumans (Appendix 1, table)

Comments: Benton (1995, 52) lists *Homo sapiens* as a species of ape. Some workers call all members of Hominidae "Humans." There are about 6 billion Humans alive on Earth today. [Old French *humain* < Latin *humansus*]

o **Ice Man** *n.* A 5300-year-old Neolithic man preserved in ice in the Schnals Valley, the Alps, Italy, near the province of South Tyrol (Menon 1995, 57, photos of head; Fowler 1997, 13, photo).

syn. Ötzi (Tyrolian name, after Ötztal near Ice Man's discovery site; Fowler 1997, 13)

Comments: Hikers found Ice Man in 1991 when his body emerged from a glacier (Rensberger 1992a, A1; Menon 1995, 57). He and his belongings, including a copper-headed ax, food, and medicines, were very well preserved, making him an extraordinary find. He was about 5 feet, 4 inches tall and around 35 years old. He might have been a herdsman, hunter, prospector, or a shaman overcome by an early snowfall. Ice Man has a severely shrunken brain that indicates he was exposed to drying sun and wind when ice did not cover him.

○ **Imo** *n.* The "monkey genius;" a female Japanese Macaque, *Macaca fuscata*, that started traditions of washing Sweet Potatoes and placer mining of Wheat grains when she was a juvenile (Kawamura 1954 in Wilson 1975, 170).

○ **Inca Girl** *n.* The frozen body of an Inca girl who died on Mount Ampato, Peru, about 500 years ago (Menon 1996, 22–23).
Comments: Johan Reinhard found the girl in 1995. She was between 12 and 14 years old when other Incas killed her on the summit of Mount Ampato as a human sacrifice. They buried her inside a stone platform with ritual offerings. Reinhard found two other bodies in the area.

○ **Java Man** *n.* A population of *Homo erectus* represented by bones from Java (Wilford 1999b, A1).
Comments: Java Man is the first discovered fossil of this species; the Dutch physician Eugène Dubois found them in 1891 (Wilford 1999b, A1, A21). [Java Man, after the location of this fossil; coined by Dubois (Wilford 1999b, A21)]

○ **Jernigan, Joseph Paul; The Visible Man** *n.* A 38-year-old man whose body anatomy is computerized based on photographs and other images of thin slices of his corpse (Dowling 1997a, illustrations).
cf. Animalia: Class Mammalia: Subclass Theria: Infraclass Eutheria: Order Primates: Visible Woman
Comments: Jernigan was a prisoner sentenced to death who willed his body to science (Dowling 1997a). His corpse had a missing appendix, testicle, and tooth. Victor Spitzer, David Whitlock, and colleagues froze his body and cut it into 1878 slices. The CAT scans, MRIs, and photographs comprise 15 gigabytes of information and fill 15,000 floppy disks or 23 CD-ROM discs. Computer users can view Jenigan's virtual image from any angle and at any depth.

○ **Koko, Hanabi-Ko** *n.* A female Lowland Gorilla that is part of a study of American Sign Language, Ameslan (www.koko.org, 2000).
Comments: Koko was born in 1971 in the San Francisco Zoo in California (www.koko.org, 2000). The Gorilla Foundation reports that she has a working vocabulary of over 1000 signs, understands about 2000 words of spoken English, typically constructs statements averaging three to six words, shows an IQ of 70 to 95 based on human IQ tests, communicates in Ameslan with a male Lowland Gorilla (Michael), and has favorite things.
[Japanese *Hanabi-Ko*, Fireworks Child]

○ **!Kung** *n.* The Bushmen of Southern Africa (Wilson 1975, 287).
[Bushman *!kung*, human beings]

○ ***Macaca mulata*, Rhesus Macaque, Rhesus Monkey** *n.* An Old World monkey from India (King and Stansfield 1985).
Comment: This Monkey is an important research primate and the original source of the Rh antigen (King and Stansfield 1985).
[New Latin, *Rhesus* < Greek *Rhēsos*, a mythical king of Thrace]

○ **Michael** *n.* A male Lowland Gorilla that is part of a study of American Sign Language (www.koko.org, 2000).
Comments: Michael was born in 1973 in Cameroon, Africa (www.koko.org, 2000). The Gorilla Foundation reports that he has a working vocabulary of 600 signs, and has favorite things.

○ **Mitochondrial Eve** *n.* A common female ancestor of living *Homo sapiens sapiens* who lived in Africa about 200,000 years ago and whose existence was reported by Cann et al. (1987) based on their study of human mitochondrial DNA (Barinaga 1992, 687).
cf. hypothesis: African-origin hypothesis of human mitochondrial DNA evolution

Comments: Wilson and Cann (1992, 70) suggest that Mitochondrial Eve was one of 10,000 people who lived at her time. Some workers doubt her existence because they conclude that the original work on her was statistically flawed (Barinaga 1992a, 686; Templeton 1992, 737; Hedges et al. 1992, 738).

[mitochondrial, from analysis of human mitochondrial DNA; Eve, after the first woman in *The Bible*]

o **Monte-Verde People** *n.* Possibly the first people who lived in the Americas (Gibbons 1997, 1256).

Comment: The Monte-Verde People, lived in huts with wooden frames and animal-hide roofs at least 12,500 years ago in Chile (Thomas Dillehay in Gibbons 1997, 1256). The Clovis People might have come to the Americas before the Monte-Verde People.

o **nerd** *n.* A person who is socially inept, foolish, ineffectual, or a combination of these characters (Morris 1982).

Comments: Kinds of nerds include "computer nerd," "congressional-staff-tax nerd," "fumbling nerd," "geeky nerd (a somewhat repetitive term)," "outcast nerd," "unsmiling, charmless, uptight nerd," "sexually obsessed nerd," and "stingy nerd" (Frerking 1997, B5). Some people categorize engineers, scientists, or students who major in science as nerds, even though many of these people are the opposite of "nerd" as defined above. *If I Ran a Zoo* by Dr. Seuss (1950) includes a nerd who is a Seuss version of a grumpy professor. The popular movie *Independence Day* portrays a lead scientist as a nerd.

[probably an alteration of NUT, a crazy person (Morris 1982)]

o **Nim Chimpsky** *n.* A Chimpanzee that learned a large number of hand gestures that represent words in American Sign Language and appeared to use signs combined grammatically into sentences (de Waal 1989, 48; Smith 1999, A17).

Comments: This Chimp did not make the "sentences" on his own, but instead was cleverly imitating his teacher, Herbert Terrance, who famously repudiated his own earlier studies of this animal in the 1970s (Smith 1999, A17).

[after Naom Chomsky, linguist at the Massachusetts Institute of Technology]

o *Pan paniscus*, **Bonobo, Pigmy Chimpanzee** *n.* An endangered ape native to Zaire between the Rivers Kasai and Zaire and characterized by having a completely black face as an adult and a lighter build than a Chimpanzee (Primates: Pongidae).

[Greek *Pan*, God of the Hills, Flocks, Forests, Pastures, and Sportsmen; African *chimpanzee*; *paniskos*, diminutive of *Pan*]

o *Pan troglodytes*, **Chimpanzee** *n.* An endangered ape native to Central and West Africa from Senegal through Tanzania (Primates: Pongidae).

cf. Animalia: Chordata: Mammalia: Nim Chimsky

Comments: Chimpanzee subspecies are *Pan troglodytes schweinfurthii* (Eastern Africa), *P. troglodytes troglodytes* (Central Africa), possibly *P. troglodytes vellerosus* (Southern Nigeria and Western Cameroon), and *P. troglodytes verus* (Western Africa) (Stein 1997, A2). Chimpanzees live in deciduous forests, humid evergreen forests, and savannas (Turco and Gruner 1999). They eat flowers, fruits, nuts, seeds, and less often birds and small vertebrates. Chimps live up to 50 years in the wild. They live in social groups, each with an alpha male. They make and use tools, including rocks as nut crushers and sticks as levers, probes, and weapons.

[Greek *Pan*, God of the Hills, Flocks, Forests, Pastures, and Sportsmen; African *chimpanzee*; *troglodytes*, hole dweller]

○ **Poloyo Skull** *n*. A skull of *Homo erectus* from Poloyo, Java (Wilford 1999b, A1).

Comments: This skull is probably from a male, has a high forehead atypical of this species, shows differential development of the two sides of its brain, and has a Broca cap that suggests a potential for higher language processing, although this species may have had only simple verbal ability (Wilford 1999b, A1, A21). Possibly a farmer found this skull and sold it to a collector in 1997. It reappeared in 1999 in the Maxilla and Mandible, a shop in New York City.

○ *Pongo pygmaeus abelii*, **Sumatran Orangutan** *n*. An endangered ape native to Borneo and Sumatra characterized by having grasping feet and hands, long arms, and long reddish-brown fur (Primates: Pongidae) (Turco and Gruner 1999).

Comments: Sumatran Orangutans live in lowland and upland tropical rain forests and peat swamp forests and spend most of their time in trees, where they often locomote by brachiation (Turco and Gruner 1999). They eat primarily fruit (including Breadfruit, Jackfruit, Langsat, Rambutan, and Wild Plum), also bark, flowers, insects and other small invertebrates, leaves, seeds, and shoots, and they live up to 60 years in the wild. A mature male has cheek flaps and a throat pouch used in making "long calls," which advertise his exact location. Such a male may have a long, flowing mustache and beard.

[Malay *orang-utan*, old man of the forest]

○ **Princess** *n*. An Orangutan studied by Biruté Galdikas and Gary Shapiro (Galdikas 1995, 400).

Comments: Shapiro taught Princess more than 30 signs of sign language (Galdikas 1995, 400).

○ *Semnopithecus entellus* **(Dufresne, 1797), Common Langur, Gray Langur, Hanuman Langur** *n*. An Old World

monkey native to India, Kashmir, Nepal, Pakistan, Southern Tibet, and Sri Lanka (Primates: Cercopithecidae) (Constable, 1998, A18; www.selu.com/~bio/PrimateGallery, 1999).

cf. -cide: infanticide

Comments: Millions of Hanuman Langurs once roamed India's forests (Constable, 1998, A18). In 1983, fewer than 200,000 remained. People protect the Tughluqabad Monkeys (= Hanuman Langurs) at the Tughluqabad Ruins around the clock. Many people feed these monkeys as a way of making offerings to the Hindu Monkey God. Status: CITES, Appendix I; U.S. ESA, Endangered.

[Hanuman, after *Hanuman*, the Hindu Monkey God]

○ **silverback** *n*. An older male Gorilla with gray to white hairs on his back that is the highest ranking male in his group (Immelmann and Beer 1989, 290).

○ **Trimates, The** *n*. The founding mothers of the contemporary field of primatology: Dian Fossey, Biruté Marija Filomena Galdikas, and Jane Goodall (Morell 1993a, 420; Galdikas 1995, 65).

Comments: Jane Goodall was the first researcher to begin her field work and studied Chimpanzees. Dian Fossey was second and studied Mountain Gorillas. Biruté Galdikas was the third and studied Orangutans. Galdikas (1995) describes the other two Trimates and their work.

○ **Visible Woman, The** *n*. An anonymous 59-year-old woman whose body anatomy is being computerized based on photographs and other images of thin slices of her corpse (Dowling 1997a, 47).

cf. Animalia: Class Mammalia: Subclass Theria: Infraclass Eutheria: Order Primates: Jernigan, Joseph Paul

Comments: The Visible Woman died of a heart blockage that did not affect the appearance of her body. Victor Spitzer and colleagues froze her body and cut it into 5189 slices.

○ **Washoe** *n*. A research Chimpanzee that learned sign language (Galdikas 1995, 400).

● **Order Proboscidea, Elephants**

○ *Elephas maximus*, **Asian Elephant, Asiatic Elephant** *n*. A large Asian mammal (Animalia: Class Mammalia: Subclass Theria: Infraclass Eutheria: Order Proboscidea: Family Elephantidae).
Comments: China has executed several Elephant poachers in the 1990s (Faison 1998, 9). A pair of Elephant tusks sells for about $10,000 per pair.

○ *Loxodonta africana*, **African Elephant** *n*. A living elephant species native to Africa (Animalia: Class Mammalia: Subclass Theria: Infraclass Eutheria: Order Proboscidea: Family Elephantidae).

● **Order Rodentia, Rodents**

○ *Eomys quercyi n*. An extinct rodent that lived about 26 million years ago with a gliding membrane similar to that of a flying squirrel (Animalia: Class Mammalia: Subclass Theria: Infraclass Eutheria: Order Rodentia) (Burkart Engesser in Anonymous 1996c, 28, illustration).
Comments: A fossil from Enspel, Germany, shows the outline of this rodent's fur and gliding membrane (Anonymous 1996c, 28, illustration).

○ *Heterocephalus glaber*, **Naked Mole Rat** *n*. A social rodent found in Ethiopia, Kenya, and Somalia characterized by having pink, wrinkled skin with sparse individual hairs, a very loose skin that enables it to turn its body inside the skin when crawling in narrow burrows (Animalia: Class Mammalia: Subclass Theria: Infraclass Eutheria: Order Rodentia) (D. Diszek 1996, www. personal.umich.edu/~cberger/ syllabusfolder/animal diversity, illustration).

○ *Mus musculus*, **House Mouse** *n*. An Asian mouse (Animalia: Class Mammalia: Subclass Theria: Infraclass Eutheria: Order Rodentia: Family Musidae) (King and Stansfield 1985; Whitaker 1996, 664–665).
Comments: Researchers have extensively used this Mouse in laboratory studies, including those in behavior and genetics, and have produced many inbreed lines by brother-sister matings (King and Stansfield 1985). Commonly used strains include Albino (A, Ak, BALB, R_m), Agouti (C3H, CBA), Black (black, C_{57}, C_{58}), Dilute Brown (DBA/2), and Dilute Brown Piebald (I). *Doogie* is a transgenic strain of *M. musculus* engineered by scientists to have greater learning ability than ordinary Laboratory Mice (Lemonick 1999, 54; Tang et al. 1999, 64, 68).
[Latin *Mus*, mouse; Sanskrit *musha*, thief; *Doogie*, after the television show *Doogie Howser, M.D.*, about a young man of superior intelligence who is an M.D.]

○ *Rattus norvegicus*, **Brown Rat** *n*. An Asian rodent (Animalia: Class Mammalia: Subclass Theria: Infraclass Eutheria: Order Rodentia: Family Musidae) (Whitaker 1996, 659).
syn. Common Rat, Norway Rat, Sewer Rat, Water Rat (Whitaker 1996, 659)
Comment: This Rat probably originated in Central Asia and spread across Europe in the 16th to 18th centuries. It arrived in North America in boxes of grain brought by Hessian troops in about 1776 (Whitaker 1996, 660). White laboratory rats are albino Brown Rats (King and Stansfield 1985).

○ *Rattus rattus*, **Black Rat** *n*. A Southeast Asian rodent (Animalia: Class Mammalia: Subclass Theria: Infraclass Eutheria: Order Rodentia: Family Musidae) (Whitaker 1996, 662).
syn. Roof Rat, Ship Rat (Whitaker 1996, 662)
Comment: This Rat probably originated in Southeast Asia and across Europe before the Brown Rat's arrival in the 16th century (Whitaker 1996, 662). It arrived in Central and South

America in the mid-16th century, evidently carried there aboard Spanish ships, and in North America in 1609 with the Jamestown colonists.

★ **Infraclass Metatheria, Metatherians** *n*. A therian mammalian infraclass of about 260 species characterized by having bones that support external pouches; short development of young in wombs followed by further development in external pouches (present in most species); and no nerve tissue that connects brain hemispheres (Reader's Digest Association 1971, 409).

syn. Marsupialia, Marsupials

Comments: Australia is the current center of metatherian diversity (Reader's Digest Association 1971, 409).

● **Benjamin** *n*. The Tasmanian Tiger that died in the zoo in Hobart, Australia, in 1936 and may have been the last living member of this species (Bergamini 1964, 94, photo; Drollette 1996, 32).

See Animalia: Class Mammalia: Subclass Theria: Infraclass Metatheria: *Thylacinus cynocephalus*.

● ***Tarsipes spenserae*, Honey Possum** *n*. An Australian meta-therian mammal (Animalia: Class Mammalia: Subclass Theria: Infraclass Metatheria: Order Diprotodontia: Family Tarsipdidae) (Buchmann and Nabhan 1996, 247).

Comments: Honey Possums have long protrusible tongues used for feeding on nectar and pollen and occur in Southwestern Australia (Buchmann and Nabhan 1996, 247). They are excellent pollinators.

● ***Thylacinus cynocephalus*, Tasmanian Tiger** *n*. A possibly still extant, nocturnal, predatory marsupial from Tasmania (and originally Australia and New Guinea) with a wolf-like body shape and head (Animalia: Class Mammalia: Subclass Theria: Infraclass Metatheria: Order Polyprotodonta: Family Dasyuridae) (Drollette 1996, 32–33, illustration).

syn. Tasmanian Wolf (Bergamini 1964, 94); Thylacine (Michaelis 1963); Dog-Faced Dasyurus, O'possum Hyena, Wolf O'possum, Zebra Wolf (www.zoo.utas.edu.au/profiles.html, 2000)

cf. Animalia: Class Mammalia: Subclass Theria: Infraclass Metatheria: Benjamin

Comments: This is the largest known living marsupial carnivore in the 20th century; it was up to 160 cm long, including a 50-centimeters-long tail, and consumed Wallabies and other mammals (Bergamini 1964, 94; E. Guiler, www.zoo.utas.edu.au/thylacine, 2000). Aborigines, vagabonds, and wild Dogs might have killed the majority of Tasmanian Tigers. A few of them might still be alive (Drollette 1996, 32–34).

[Tiger, after brown vertical stripes on its back and tail]

[Latin *thyla*, pouch; *cinus*, dog; *cyno*, wolf; *cephalus*, head; pouched dog with a wolf's head]

■ **Class Osteichthyes, Boney Fish** *n*. A chordate class of about 20,000 living species characterized by being aquatic and having developed brains enclosed by bone, notochords reduced to tissue between vertebrae, tails, and no legs or wings (Curtis 1983, 1084).

Comment: Fishes are in the Classes Acanthodii, Agnatha, Chondrichthyes, Osteichthyes, and Placodermi. An earlier classification placed all fishes in the Class Pisces (Curtis 1983, 1084).

***Lates niloticus*, Nile Perch** *n*. A fish native to the Congo, Lake Chad, and the Volta and Niger River Systems in Africa (Pisces: Snook Family) (*Lycos Encyclopedia of Animals,* http://versaware.animalszone.lycos.com/).

Comments: Lates niloticus is a snook, not a true perch (*Lycos Encyclopedia of Animals,* http://versaware. animalszone. lycos. com/). People introduced this fish into Lake Victoria, Africa, in the 1950s (Vick 1999, A28). It greatly changed the ecosystem of the lake by consuming many native cichlid fishes, which eat algae (American Museum of Natural

History, http://www.gambit.co. za/corporate/fishingafrica/ boards/fishreport/messages/ 542.html, 1999). Now Lake Victoria is frequently choked with algal blooms, and *L. niloticus* mostly feeds on a freshwater shrimp species. The introduced fish originated from Lakes Albert and Turkana, which contain three *Lates* species, and it has never been entirely clear which of these became established in Lake Victoria, or indeed whether the Lake Victoria population is derived from hybridization between *Lates* species (Hauser, 1999, illustrations).

***Latimeria chalumnae* J.L. Smith, 1939, Coelacanth** [SEEL uh kanth] *n*. A South African and South Asian fish (Order Crossopterygii: Suborder Coelacanthini) (Thomson 1991, 13–14, 35, illustration).

Comment: Researchers knew Coelacanths only from Cretaceous fossils (which were probably different species than the living one) until Marjorie Courtenay-Latimer obtained a specimen from a fisherman in 1938 (Thomson 1991, 13–14). People have caught about 200 Coelacanths since 1938. Some people call this fish a living fossil. Crossopterygiians are related to ancestors of Amphibia.

[*Latimeria*, after M. Courtenay-Latimer; *chalumnae*, after the Chalumna River, Cape Province, South Africa, which is near the estuary where a fisherman caught the 1938 specimen).

■ **Class Reptilia, Reptiles** *n*. A chordate class containing crocodiles, lizards, snakes, and turtles, with about 6000 living species and characterized by having eggs enclosed by protective embryonic membranes and shells, developed brains enclosed by bone, incomplete double blood circulation, legs (and sometimes arms), lungs, notochords reduced to tissue between vertebrae, skin usually covered with scales, and tails (Curtis 1983, 1085).

Comments: Reptilia is polyphyletic in some classifications because it contains Classes Aves and Mammalia. In the legal reptile business, the U.S. imported 1.8 million live rep-

tiles in 1997, worth about $US 7 million and exported 9.7 million valued at $13.2 million (Cooper 1999, A23). There is a thriving black market in illegal reptiles.

***Boiga irregularis*, Brown Tree Snake** *n*. A snake native to Australia and Asia (Reptilia).

Comments: This snake feeds on many species of birds and small lizards and mammals (Claiborne, 1997, A6). Adults reach up to 8 feet long. It is orginally from Indonesia, Northern Australia, Papua New Guinea, and the Solomon Islands and first appeared in Guam in the 1950s, where it became abundant and is extinguishing many native animal species due to its predation. This snake is also an important threat to other areas of the world, including Hawai'i.

Galápagos Tortoise See Animalia: Reptilia: *Geochelone elephantophus*.

Geochelone elephantophus (formerly ***Testudo indica, T. nigra***), **Galápagos Tortoise** *n*. A species of large tortoise native to the Galápagos Islands, Equador (Milner 1990, 279).

Comments: Individuals of this species live over 200 years (Milner 1990, 279). Sailors and others have consumed over 200,000 of these Tortoises. Lonesome George is the last Galápagos Tortoise, *Geochelone elephantophus abingdoni*, from Pinta Island and is in a pen on Santa Cruz Island (Milner 1990, 279). No one found a mate of his population for him. Other populations of this species are surviving well due to conservation efforts. Charles R. Darwin first encountered this species on September 17, 1835 (Darwin 1860, 375).

Phylum Ciliophora *n*. A phylum that contains one species, *Symbon pandora*, which has a funnel-like, round mouth and seems to live exclusively on lobster mouthparts (established by Peter Funch and Reinhardt Kristensen in Anonymous 1996b, illustration).

[Greek, *cyclio*, small wheel; *phereum*, to carry (Anonymous 1996b, illustration).

Pigeon See Animalia: Phylum Chordata: Class Aves: *Columba livia*.

Pigmy Chimpanzee See Animalia: Phylum Chordata: Class Mammalia: Subclass Theria: Infraclass Eutheria: Order Primates: *Pan paniscus.*

Pouter See Animalia: Phylum Chordata: Class Aves: *Columba livia.*

Princess See Animalia: Phylum Chordata: Class Mammalia: Subclass Theria: Infraclass Eutheria: Order Primates: Princess.

Reptiles See Animalia: Phylum Chordata: Reptilia.

Rock Dove See Animalia: Phylum Chordata: Class Aves: *Columba livia.*

Rodolia cardinalis See Animalia: Phylum Arthropoda: *Rodolia cardinalis.*

Roof Rat See Animalia: Phylum Chordata: Class Mammalia: Subclass Theria: Infraclass Eutheria: Order Rodentia: *Rattus rattus.*

scout bee See Animalia: Phylum Arthropoda: Group Apiformes: scout bee.

Sewer Rat See Animalia: Phylum Chordata: Class Mammalia: Subclass Theria: Infraclass Eutheria: Order Rodentia: *Rattus norvegicus.*

Sheep Dog See Animalia: Phylum Chordata: Class Mammalia: Subclass Theria: Infraclass Eutheria: Order Carnivora: *Canis familiaris.*

Ship Rat See Animalia: Phylum Chordata: Subclass Theria: Infraclass Eutheria: Order Rodentia: *Rattus rattus.*

silverback See Phylum Chordata: Class Mammalia: Subclass Theria: Infraclass Eutheria: Order Primates: silverback.

Southern Pine Beetle See Animalia: Phylum Arthropoda: *Dendroctonus frontalis.*

Starry-Sky Beetle See Animalia: Phylum Arthropoda: *Anoplophora glabripennis.*

sterzelnde **bee** See Animalia: Phylum Arthropoda: Apiformes: *sterzelnde* bee.

Stingless Bees See Animalia: Phylum Arthropoda: Group Apiformes: Apidae.

Tarsipes spenserae See Animalia: Class Mammalia: Subclass Theria: Infraclass Metatheria: *Tarsipes spenserae.*

Tasmanian Tiger, Tasmanian Wolf See Animalia: Phylum Chordata: Subclass Theria: Infraclass Metatheria: *Thylacinus cynocephalus.*

Thievish Dog See Animalia: Phylum Chordata: Class Mammalia: Subclass Theria: Infraclass Eutheria: Order Carnivora: *Canis familiaris.*

Trigona prisca See Animalia: Phylum Arthropoda: Apiformes: *Trigona prisca.*

Trilobites, Trilobitomorpha See Animalia: Phylum Arthropoda: Trilobitomorpha.

Trimates, The See Animalia: Phylum Chordata: Class Mammalia: Subclass Theria: Infraclass Eutheria: Order Primates: Trimates.

Trochilidae See Animalia: Phylum Chordata: Class Aves: Trochilidae.

Tumbler See Animalia: Phylum Chordata: Class Aves: *Columba livia.*

Two-Spotted Lady Beetle See Animalia: Phylum Arthropoda: *Adalia bipunctata.*

Visible Man, The See Animalia: Phylum Chordata: Class Mammalia: Subclass Theria: Infraclass Eutheria: Order Primates: Jernigan, Joseph Paul.

Visible Woman, The See Animalia: Phylum Chordata: Class Mammalia: Subclass Theria: Infraclass Eutheria: Order Primates: Visible Woman.

Wapiti See Animalia: Phylum Chordata: Class Mammalia: Subclass Theria: Infraclass Eutheria: Order Artiodactyla: *Cervus canadensis.*

Washoe See Animalia: Chordata: Mammalia: *Pan troglodytes:* Washoe.

Water Rat See Animalia: Phylum Chordata: Class Mammalia: Subclass Theria: Infraclass Eutheria: Order Rodentia: *Rattus norvegicus.*

Western Lowland Gorilla See Animalia: Phylum Chordata: Class Mammalia: Subclass Theria: Infraclass Eutheria: Order Primates: *Gorilla gorilla gorilla.*

wild bee See Animalia: Phylum Arthropoda: Group Apiformes: pollen bee.

worker bee See Animalia: Phylum Arthropoda: Group Apiformes: worker bee.

▶ **Kingdom Fungi** *n.* An eukaryan kingdom with six taxonomic divisions and comprised of heterotrophic, usually multicellular organisms with cell walls (often with chitin); distinguished from Kingdoms Animalia, Plantae, and Protista (Curtis 1983, 1078).

Comments: Workers have named about 100,000 fungus species, and they include Mildews, Molds, and Mushrooms (Curtis 1983, 1078). The filamentous forms consist basically of a continuous mycelium which becomes septate (partitioned off) in certain taxa and at certain stages of their life cycles. Reproductive cycles often include both asexual and sexual phases. Most fungus taxa are haploid, with diploid zygotes. Yeasts are unicellular ascomycetes.

[Latin *fungus,* mushroom, akin to SPONGE]

Aeroaquatic Hypomycetes *n.* A fungus taxon that lives in water and produces conidia above a water surface (Dubey 1995, 34).

A CLASSIFICATION OF KINGDOM FUNGI[a]

I. Phylum Acrasiomycota, Cellular Slime Molds
II. Phylum Amastigomycota
    A. Class Ascomycetes, Sac Fungi (including most Yeasts and Molds, Morels, Truffles)
    B. Class Basidiomycetes, Club Fungi: Mushrooms, Shelf Fungi, Bird's-Nest Fungi, Stinkhorns
    C. Class Zygomycetes, Bread Molds and kin
III. Form-Phylum Deutermycota, Imperfect Fungi
IV. Phylum Mastigomycota
    A. Class Chytridiomycetes, Chytrids
    B. Class Oomycetes, Water Molds and kin
V. Phylum Myxomycota, Plasmodial Slime Molds (including Ceratiomyxa, Fuligo Flowers-of-Tan, *Stemonitis*, Wolf's-Milk Fungi)

[a] Curtis (1983, 1078).

Aquatic Hypomycetes *n.* A fungus that lives in water (Dubey 1995, 34).

Armillaria bulbosa, **Honey Mushroom, Shoestring Root Rot** *n.* A European and North American fungus in the Tricholomataceae (Basiodiomycetes) (McKnight and McKnight 1987, plate 15).
Comments: Individuals of this fungus can grow very large. One individual can weigh over 220,000 pounds and occupy over 37 A (Anderson et al. 1992 in Rensberger 1992a, A1). This fungus inhabits hardwood forests throughout Eastern North America and Europe (Rensberger 1992a, A1). In comparison an adult Blue Whale weighs up to 250,000 pounds. *Armillaria ostoyae* has the largest known fungus individuals (K. Russell and T. Shaw in Rensberger 1992a, A1). It occurs in southwestern Washington, covers 1500 A, and is up to 1000 years old. The largest living organism might be *Populus tremuloides* (Plantae), *q.v.*

Entomophaga maimaiga **Humber, Shi and Soper** *n.* A fungus (Zygomeycetes: Entomophthorales) that consumes arthropods, primarily the Gypsy Moth (Valenti 1998, 20).
Comments: This fungus currently causes drastic population declines of Gypsy Moths and evidently negligible impact on related nontarget Lepidoptera in North America (Valenti 1998, 20–21). People originally introduced *E. maimaiga* into the U.S. in 1910 from Japan. It begin to make great impacts on Gypsy Moth populations in the 1980s. It is unclear whether the original strain or a new strain of this fungus is currently affecting Gypsy Moths, how this fungus will affect Gypsy Moth populations in the future, and what indirect effects it will have on other Gypsy Moth consumers.

fungus *n., pl.* **fungi, funguses** A member of Kingdom Fungi, *q.v.*

lichen *n.* A symbiotic association of an alga and a fungus (as in most kinds of lichens) or an alga and a bacterium (Red-Blanket Lichens) (Kritcher and Morrison 1998, 206–207).
Comments: Workers classify lichens as Crustose Lichens, Foliose Lichens, and Fruticose Lichens (Hale 1961, 7–17). Although they are obviously combinations of two species, researchers give them genus and species names, *e.g.*, *Umbilicaria mammulata*. A lichen's fungus holds onto a substrate, and its bacterium, or alga, photosynthesizes, producing carbohydrates used by both parties of its symbiosis (Kritcher and Morrison 1988, 206–207). Researchers debate whether lichen partners are equal partners; one might derive more benefits from the association than the other. Lichens are important in early succession in some habitats.

A CLASSIFICATION OF LICHENS

I. Crustose Lichens
II. Foliose Lichens
III. Fruticose Lichens

Microsporidium **spp.** *n.* A group of amitochondriate eukaryotes, now affiliated with fungi (Gray et al. 1999, 1480).

Terrestrial Geofungus *n.* A fungus that lives in soil (Dubey 1995, 34).

▶ **Kingdom Plantae** *n.* An eukaryan kingdom with organisms that have chloroplasts, fertilization, involuntary motion (in almost all species), meiosis, multicellularity, no nervous systems, photosynthetic nutrition, and rigid cell walls of cellulose and other polysaccharides; distinguished from Kingdoms Animalia, Fungi, and Protista (*Oxford English Dictionary* 1972, entries from 1551; Michaelis 1963; Curtis 1983, 387).
Comment: There are about 265,000 species of living plants, with about 235,000 of them being Flowering Plants (Curtis 1983, 1079).
syn. Metaphyta

plant *n.*

1. An organism with rigid cell walls that is usually autotrophic and without voluntary motion; distinguished from an animal, fungus, and protistan (*Oxford English Dictionary* 1972, entries from 1551; Michaelis 1963); a member of Kingdom Plantae.

2. A herbaceous plant, rather than a bush or tree (*Oxford English Dictionary* 1972, entries from 1551; Michaelis 1963).

3. A member of the Kingdom Plantae; an organism with these characteristics: eukaryotic cells with nuclear envelopes, mitochondria, chloroplasts (in almost all species); cell walls of cellulose and other polysaccharides; fertilization and meiosis; photosynthetic nutrition (in almost all species); multicellularity; and no nervous system (Curtis 1983, 387; Lincoln et al. 1985).
Notes: This definition seems appropriate for wide use.

cf. animal, -cole, ²-phile, -phobe
Comment: Curtis (1983, 387) tabulates the traits of Animals, Fungi, Plants, Prokaryotes, and Protists.
[Old English *plante* < Latin *planta* sprout, slip, cutting, graft]

A CLASSIFICATION OF PLANTS (KINGDOM PLANTAE)[a]

I. Group Nonvascular Plants[b] (*syn.* Bryophytes, in the broad sense)
  A. Division Anthocerophyta (*syn.* Anthocerophytes, Hornworts)
  B. Division Bryophyta (*syn.* Bryophytes, Mosses)
  C. Division Hepatophyta (*syn.* Hepatophytes, Liverworts)
  D. Division Takakiophyta (*syn.* Takakiophytes) (extinct)
II. Group Vascular plants (*syn.* Tracheophyta, Tracheophytes, plants with phloem and xylem)
  A. Group Seedless plants (plant that produce spores, not seeds)
    1. Division Lycophyta (*syn.* Clubmosses, Lycophytes), 7 orders
      a. *Asteroxylon* (extinct)
      b. *Lepidodendron* (extinct)
      c. *Lycopodium*
      d. *Selaginella*
      e. *Zosterophyllum* (extinct)
    2. Division Psilophyta (*syn.* Psilophytes, Wiskferns)
    3. Division Progymnospermophyta (extinct)

4. Division Pterophyta (*syn.* Ferns, Pterophytes), 10 orders
      a. *Pteridium*
    5. Division Rhiniophyta (*syn.* Rhyniophytes) (extinct)
    6. Division Sphenophyta (*syn.* Horsetails, Sphenophytes), 3 orders
      a. *Calamites* (extinct)
      b. *Equisetum*
    7. Division Trimerophytophyta (extinct)
    8. Division Zosterophyllophyta (extinct)
  B. Group Seed Plants (plants that produce seeds)
    1. Division Pteridospermophyta (*syn.* Pteridospermophytes, Seedferns)
    2. Group Gymnospermae (Gymnosperms)
      a. Division Coniferophyta (*syn.* Conifers, Coniferophytes) (Bald Cypress, Cypresses, Dawn Redwood, Douglas Fir, False Cedars, Firs, Hemlocks, Junipers, Larches, Pines, Redwood, Sierra Bigtree, Spruces, True Cedars, Yews), 2 orders
      b. Division Cycadophyta (*syn.* Cycads, Cycadophytes)
      c. Division Ginkgophyta (*syn.* Ginkgos, Ginkgophytes), 2 orders
      d. Division Gnetophyta (*syn.* Gnetae, Gnetophytes), 1 order
    3. Group Angiospermae (Angiosperms)
      a. Division Anthophyta (*syn.* Magnoliophyta, Flowering Plants), 2 classes, 31 orders
        (1) Class Dicotyledonae (*syn.* Dicots, Group Dicotyledonae, Magnoliopsida) (includes Buttercups, Cacti, Daisies, Legumes, Magnolias, Mallows, Maples, Melons, Mustards, Parsleys, Poppies, Potatoes, Roses, Spurges, Water-lilies, Willows)
          (a) Order Asterales, Asters and kin (includes Chrysanthemums, Daisies, Ragweeds)

(b) Order Lamiales, Mints and kin

(c) Order Magnoliales (includes Magnolias, Tuliptree)

(d) Order Piperales (includes Black Pepper, Lizard Tails)

(e) Order Rosales (includes Briars [*Rubus* spp.], Cherries, Hawthornes, Legumes, Peaches, Plums)

(f) Order Umbellales, Parsnips and kin

(g) Order Urticales (includes Elms, Hackberries, Nettles)

(2) Class Monocotyledonae (*syn.* Group Monocotyledonae, Monocots, Class Liliopsida, Liliopsids) (Bromeliads, Irides, Lilies, Orchids, Palms, Rushes, Sedges, etc.)

(a) Order Arales, Aroids

(b) Order Liliales (includes Amaryllids, Irides, Lilies, Narcissus, Squills, Tulips)

(c) Order Orchidales, Orchids

(d) Order Poales, Grasses, Rushes, and Sedges

---

[a] Taylor and Taylor (1993), Campbell (1996, 550). Taxa not listed as extinct are still extant (living). A division is comparable to a phylum as a classificatory unit.

[b] Some moss species have food- and water-conducting tissue which is not xylem or phloem, respectively.

- ***Amborella trichopoda*** *n.* A plant species from New Caledonia (Dicotyledonae: Amborellaceae) (Weiss 1999, A5; Kenrick 1999, 358). *Comments:* This extant plant taxon probably belongs on the earliest branch of the plant evolutionary tree (Brown 1999b, 990, illustration; Weiss 1999, A5).

- **Angiospermae, Angiosperms** [AN gee owe SPERM mee] *n.* A seed-bearing, tracheophyte group characterized by having flowers and fruits, composed of a seed(s) enclosed in an ovary(ies); distinguished from Gymnospermae (Campbell 1999, 567–569, illustration; Strickberger 2000, 314).
Comments: There are about 235,000 living angiosperm species (Campbell 1996, 550, 564). Most species have vessel elements which carry water, rather than tracheids, which are found in conifers. Angiosperms have xylem reinforced by fiber cells that are specialized for support and evolved from tracheids.
[Greek *angeion*, vessel, container; *sperma*, seed; referring to the fact that angiosperm seeds are enclosed in fruits, usually with fleshy tissue]

- ***Angraecum sesquepedale*** See Plantae: Orchidaceae: *Angraecum sesquepedale.*

- **Anthophyta, Flowering Plants** [an THO figh tah, an THOHF fih tah] *n.* The only division in Angiospermae (Campbell 1996, 564, illustration; Strickberger 2000, 314–315).
cf. Plantae: Angiospermae
[Greek *anthos*, flower + plant]

- ***Arabidopsis thaliana*** **(Linnaeus) Gustav Heynhold, Mouse-Ear Cress** [a RAB ih DOP sis THAL lee an ah] *n.* A small, fast-growing annual in the Brassaceae, the Mustard Family (Fernald 1950, 710; Weiss 1995, A3).
syn. Thale Cress (Wood 1999, 26)
Comments: This plant is a major model used for understanding plant genetics (Weiss 1995, A3). It grows from seed to maturity in about 8 weeks and has about 20,000 genes and no "junk DNA." Geneticists are planning to identify all of its genes in 2000 and subsequently learn the function of each (Wood 1999, 26). Fletcher (1995) contains photographs of many of its floral mutants.
[Latin *Arabis* + *opsis*, resembling the genus *Arabis*; *thaliana*, after Johann Thal, who described this species in the 16th century)

- ***Archaefructus liaoningensis*** **Sun, Dilcher, Zheng, and Zhou, 1999** *n.* An early angiosperm represented by a fossil 145 million years ago (Subphylum Magnoliophyta: Class Magnoliopsida: Sub-

class Archaemagnoliidae) (Sun et al. 1999, 1692, 1693, illustration).
Comments: This species is from Western Liaoning Province, Northeast China (Sun et al. 1999, 1692). Its fossil includes carpels that enclose ovules, a major character of Angiospermae, and this fossil evidently represents the oldest know angiosperm. Some researchers have putatively identified fossils from the Triassic, Jurassic, and Cretaceous Periods as angiosperms, which do not have such carpels.
cf. Plantae: *Amborella*
[Greek *archae*, ancient; *fructus*, fruiting; liaoningensis, after the Liaoning Province, China]

- **Atlantic White Cedar** See Plantae: *Chamaecyparis thyoides.*

- ***Bromus tectorum* L., Cheatgrass, Downy Brome** *n.* A Eurasian annual grass species (Anthophyta: Poaceae) (Morrow and Stahlman 1984, 2; http://ianrwww.unl.edu/ianr/jgg/grasses/dbpub.htm, 1999).
Comments: People introduced Cheatgrass into the U.S. in the late 1800s (Kenworthy 1999, A11); it is now an important weed in Canada, Mexico, and the U.S. By 1999, Cheatgrass dominated millions of acres in the western U.S., where it has greatly changed communities by displacing Blue Bunch Wheat Grass, Bottle Brush Squirrel-Tail Grass, Great Basin Wild Rye, Idaho Fescue, and Indian Rice Grass. Cheatgrass is a major weed in alfalfa fields, grass-seed fields, overgrazed rangeland, winter-wheat fields, and other crop lands. Cheatgrass totally changed the fire regime of the area from a 50- to 100-year interval to a 5- to 10-year interval. The more frequent fires do not allow sagebrush and the organisms that depend on it to regenerate, and they increase erosion, because they kill plant roots, which hold soil in place. The Great Basin Restoration Initiative, proposed by the Bureau of Land Management, would attempt to restore native vegetation to areas of the Great Basin now dominated by Cheatgrass.

- **Bryophyta, Bryophytes** [BREYE Oh FIH tah, breye OHF fih tah, breye oh FIGHTS] *n.*

1. A plant division comprised of Hornworts, Liverworts, and Mosses and characterized by having food-conducting tissue that is not phloem, gametangia, water-conducting tissue that is not xylem, waxy cuticles, and zygotes (Campbell 1996, 553; Strickberger 2000, 300).
2. A plant division comprised of Mosses (Campbell 1996, 553).
syn. Mosses
Notes: Hornworts, Liverworts, and Mosses might not be closely related groups (Campbell 1996, 553). Most plant biologists continue to use "bryophyte" to designate all three groups. *Comment:* There are about 10,000 living moss species (Campbell 1996, 550).
[Latin *bryum*, moss + *phyta*, plant]

- **Bucket Orchid** See Plantae: Orchidaceae: *Coryanthes.*

- **bush** *n.* A low, tree-like or thickly branching shrub; distinguished from forb, shrub, and tree (Michaelis 1963).
See bush (main section).
cf. habitat: forest

- **C₃ plant** *n.* A photosynthetic plant species that fixes carbon via ribulose bisphosphate carboxylase (= rubisco, RUBP carboxylase), the Calvin-cycle enzyme that adds CO_2 to birubulose bisphosphate (RuBP); contrasted with C_4 plant (Campbell et al. 1999, 182).
Comment: C_3 plants include Rice, Soybean, and Wheat (Campbell et al. 1999, 182).
[C_3, after the fact that the first organic product of carbon fixation is a three-carbon compound, 3-phosphoglycerate]

- **C₄ plant** *n.* A photosynthetic plant species that prefaces its Calvin cycle with a mode of carbon fixation that forms the four-carbon compound malate as its first product; contrasted with C_3 plant (Campbell et al. 1999, 182).
Comments: C_4 plants include Corn and Sugarcane (Campbell et al. 1999, 183). Such plants have unique leaf anatomy correlated with C_4 photosynthesis.
[C_4, after the fact that the plant makes the four-carbon compound malate as its first product]

- ***Cannabis indica*, Marijuana** *n.* An Asian herbaceous dicot (Cannabinaceae).

Comments: People use this species as a recreational and medicinal drug. In 1999, drought conditions in Maryland and Virginia caused tended, illegal Marijuana plants to be more evident as green plants among browned, drought-stressed ones (Gray 1999, B1). This enabled authorities to find and seize an increased number of these plants. *Cannabis sativa* is another Asian species grown for its fiber and drug component which now grows wild in part of the U.S. (Fernald 1950, 555).
[Greek *Cannabis*, possibly from Persian *Kanab*, *indica*, from India]

■ *Chamaecyparis thyoides* (Linnaeus) L. Britton, E.E. Sterns, and J.F. Poggenberg, Atlantic White Cedar *n*. A North American coniferous, gymnospermous tree (Gymnospermae: Cupressaceae) (Fernald 1950, 58).
[Greek *chamai*, on the ground; *cyparissos*, cypress; *thyoides*, like *Thya* or *Thuja*]

■ Cheatgrass See Plantae: *Bromus tectorum.*

■ Clubmosses See Plantae: Lycophyta.

■ Comet Orchid See Plantae: Orchidaceae: *Angraecum sesquepedale.*

■ Coniferophyta, Coniferophytes, Conifers [CON nih fer O fih tah, CON nih fihr AHF fih tah] *n*. A gymnosperm division characterized by evergreen foliage (in most species), xerophytic leaves (Campbell 1996, 557, illustration; Strickberger 2000, 314).
syn. gnetae (Campbell 1996, 550)
Comment: There are about 550 living conifer species (Campbell 1996, 550). A Bristlecone Pine named Methuselah is more than 4000 years old.
[Latin *conus*, cone + *ferre*, to bear + Latin *phyta*, plant]

■ *Cooksonia* [COOK sowen nee ah] *n*. A plant genus of several extinct species, possibly in the Rhyniophyta, and historically regarded as the oldest vascular plant (Taylor and Taylor 1993, 191–192; Strickberger 2000, 304, illustration).
Comments: Researchers found *Cooksonia* fossils in Africa, Asia, Australia, Europe, North America (Taylor and Taylor 1993, 191).

Cooksonia and other cooksonioids might represent a highly artificial group of plants that include ancestral stock of both bryophytes and vascular plants.

■ Corn See Plantae: *Zea mays.*

■ *Coryanthes* See Plantae: Orchidaceae: *Coryanthes.*

■ cryptogam *n*.
1. A member of the former division of plants Cryptogamia that has no carpels or stamens, propagates by spores, and includes Algae, Bryophyta, Charophyta, Chlorophyta, Fungi, Lycophyta, Progymnospermphyta, Psilophyta, Pterophyta, Rhiniophyta, and Sphenophyta (Michaelis 1963).
2. A plant that lacks true seeds and flowers; contrasted with phanerogam (Michaelis 1963).
adj. cryptogamic, cryptogamous
syn. cryptophyte
[French *cryptogame*, ultimately from Greek *kryptos*, hidden; *gamos*, marriage]

■ Cycadophyta, Cycads, Cycadophytes [SIGH cahd o FIGHT tah, SIGH cahd DOHF fah ta, SIGH cads, Sigh ca do fights] *n*. A gymnosperm division (= phylum) characterized by having large, motile sperm; pinnately compound leaves; separate female and male cones; seeds born on cone scales; and woody stems (Lawrence 1951, 356; Campbell 1996, 560; 561, illustration; Strickberger 2000; 311, 314, illustration).
Comment: There are about 100 living cycad species (Campbell 1996, 550).
[Greek cycad, *kykas, -ados*, the name for an African plant + *phyton*, plant]

■ *Datura n*. A genus in the Solanaceae with species having hawkmoth-pollinated, large, nocturnal, sweetly scented, white flowers (Buchmann and Nabhan 1996, 248).
Comments: Jimsonweed is *Datura stramonium*; researchers performed classical studies of polysomy on this plant (King and Stansfield 1985).
[Arabic *Tatorah*, Hindustani *Dhatura*]

■ Dicotyledonae, Dicots *n*. A plant class in Division Anthophyta found throughout the world and characterized by species that have two cotyledons and stem vascular bundles in rings and usually have floral parts in multiples of fours or fives; netlike, palmate, or pinnate leaf venation; and true secondary growth with vascular cambia; distinguished from Monocotyledonae (Curtis 1983, 1080; chap. 28, illustrations).

Comments: There are about 170,000 extant Dicot species (Curtis 1983, 1080).

■ **Dipterocarpaceae, Dipterocarps** *n.* A plant family found in Africa, India, South America, and Southeast Asia (Angiospermae: Malvales) (Blundell 1999, 32).

Comment: Many dipterocarp species are tall, some growing up to 230 feet; canopy-emergent; and the more common trees in certain Asian rainforests (Buchmann and Nabhan 1996, 2451; Blundell 1999, 32).

■ **Downy Brome** See Plantae: *Bromus tectorum.*

■ ***Eichhornia crassipes* (Mart.) Solms, Water Hyacinth** *n.* A South American, aquatic, floating, perennial herb (Monocotyledonae: Pontederiaceae) (Bell and Taylor 1982, 48).

Comments: This species was, or is, a troublesome, highly invasive weed in bodies of water in many parts of the world, including Florida and Lake Victoria, Africa (Vick 1999, A25). For example, in Lake Victoria it has greatly affected electrical power production, the fishing industry, spread of water-related human disease, and transportation. Two weevil species *Neochetina bruchi* and *N. eichhorniae*, introduced into the lake in 1997, greatly reduced the population size of this plant.

■ **Ferns** See Plantae: Pterophyta.

■ **forb** *n.*

1. A weed, or other herb, that is not a grass (Michaelis 1963).

2. A nongrassy herb; an herb that is not a grass, rush, or sedge; contrasted with herb (Allaby 1994).

Comment: Forb is contrasted with alga and nonseed plants, including clubmoss, horsetail, and moss and the seed plants grass, herb, shrub, and tree.

[Apparently from Greek *phorbe-*, fodder]

■ **Gnetophyta, Gnetophytes** [knee toh FIGHT tah, knee TOHF fah tah] *n.* A gymnosperm division (= phylum) characterized by compound staminate strobili in some species,

double fertilization in *Ephedra* which also occurs in Angiosperms, opposite whorled leaves, and vessels in secondary wood (Lawrence 1951, 368; Campbell 1996, 557, illustration; Strickberger 2000, 314).

syn. Gnetae (Campbell 1996, 550)

Comment: There are about 70 living gnetophyte species (Campbell 1996, 550). Genera include *Ephedra, Gnetum,* and *Welwitschia.*

[New Latin, *gnetum* + Greek, *phyton*]

■ **Golden Aspen** See Plantae: *Populus tremuloides.*

■ **grass** *n.*

1. A species in the Grass Family, Poaceae.

2. Broadly, a grass (Poaceae), rush (Juncaceae), or a sedge (Cyperaceae).

Comment: Grass is contrasted with the nonseed plants including alga, clubmoss, horsetail, and moss and the seed plants forb, herb, shrub, and tree.

■ **Gymnospermae, Gymnosperms** [JIM no SPERM mee, JIM no SPERMS] *n.* A seed-bearing, tracheophyte group characterized by having no enclosed chambers (ovaries) in which seeds develop; gymnosperm seeds are not in fruits and are called "naked;" distinguished from Angiospermae (Campbell 1996, 560; 561, illustration; Strickberger 2000, 307).

Comment: Gymnospermae contains Coniferophyta, Cycadophyta, Ginkogophyta, Gneotophyta, and Pteriodospermophyta. There are about 721 living gymnosperm species (Campbell 1996, 550). Gymnospermae might be a polyphyletic group (Arnold 1948 in Stewart 1983, 229). On the other hand, Gymnospermae might be a monophyletic group derived from a progymnosperm ancestor (Beck 1976 in Stewart 1983, 229). The first gymnosperms appeared in the Devonian Period.

[Greek *gymnos,* naked + *sperm,* seed; referring to seeds which are not enclosed in fruits as in angiosperms]

■ **herb** *n.*

1. A nonwoody plant that withers and dies away after flowering; contrasted with alga and nonseed plants including clubmoss, horsetail, and moss and the seed plants grass, herb, shrub, and tree (Michaelis 1963).

2. A small, nonwoody, seed-bearing plant whose aerial parts all die back to the ground at the end of each growth season (Allaby 1994).

Comments: Definition 1 is more accurate because there are exceptions to definition 2, such as some banana, canna, and heliconia species. Some banana species grow over 15 feet tall; some canna species, over 8 feet tall; and some *Heliconia* species, up to 29 feet tall (Berry and Kress 1991). These plants have trunklike main stems. Many perennial tropical herbs eventually die back to the ground sometime after they flower, not necessarily at the end of a growing season. In annual species, entire plants die at the end of each season.

[Latin *herba*, grass, herbage]

■ **honey plant** *n.* A plant whose nectar or honeydew, or both, is used by Honey Bees for honey production (inferred from Pellett 1978).

    major-honey plant *n.* A kind of honey plant from which Honey Bees frequently store surplus honey, that is, honey that is not quickly consumed by the colony that produces it; contrasted with minor-honey plant (inferred from Pellett 1978).

    Comments: Some kinds of plants are always minor honey plants. Other kinds can be major or minor honey plants, depending on their abundances, genetic strains, geographic locations, weather conditions, or a combination of these factors (Pellett 1979, 12–14).

    minor-honey plant *n.* A kind of honey plant from which Honey Bees do not, or infrequently, store surplus honey, that is, honey that is not quickly consumed by the colony that produces it; contrasted with major-honey plant, *q.v.* (inferred from Pellett 1979).

■ **Horsetails** See Plantae: Sphenophyta.

■ **invasive plant** *n.* An alien plant that thrives in a location in which it is not native, *e.g.,* the Common Dandelion, a Eurasian species that now grows in many other parts of the world.

cf. bioinvasion

Comments: Scores of such plants occur in the U.S. (www.state.va.ur/ ~dcr/dnh/invlist.htm). The Invasive Plants Alert (IPA) is an outreach program funded by the Biodiversity Convention Office, Environment Canada, with in-kind support provided by the Canadian Nature Federation. Its site includes much information on Garlic-Mustard and other northern North American invasive plants (www.infoweb.magi.com/ ~ehaber/alert.html). Invasive plants are a worldwide, multibillion-dollar problem.

■ **Jimsonweed** See *Datura.*

■ **Leyland Cypress** See Plantae: *Cupressocyparis leylandii.*

■ **Lycophyta, Lycophytes, Clubmosses** [LEYE koh FIGH tah, leye KOPH fah tah, LEYE ko Fights] *n.* A seedless tracheophyte group characterized by having microphylls, roots, and sporophylls (leaves specialized for reproduction) (Campbell 1996, 557; Strickberger 2000, 314).

syn. Clubmosses, groundpines (for *Lycopodium*)

Comment: There are about 100 living lycophyte species (Lawrence 1951, 337).

[Greek *lykos*, wolf + plant]

■ *Lygodium microphyllum,* **Old World Climbing Fern** *n.* A climbing fern from China, east India, Southeast Asia, and tropical Africa (Pterophyta: Schizeaceae) (Mirsky 1999, 24).

Comment: This invasive fern is choking out native organisms in Southern Florida (Mirsky 1999, 24). An Australian moth and a Thai sawfly consume this plant.

■ **Maize** See Plantae: *Zea mays.*

■ **major-honey plant** See Plantae: honey plant: major-honey plant.

■ **Marijuana** See Plantae: *Cannabis indica.*

■ *Metasequoia glyptostroboides,* **Dawn Redwood** *n.* A gymnospermous tree (Coniferophyta: Taxodiaceae) from the Szechuan Province, China (Stewart and Rothwell 1993, 432–433).

cf. fossil: living fossil.

Comments: The genus *Metasequoia* is known from fossils as old as *ca.* 70 million years (Stewart and Rothwell 1993, 432–433). Scientists did not know this genus was extant until Miki reported it in 1941.

- **minor-honey plant** See Plantae: honey plant: minor-honey plant.
- **Monarch-of-the-East** See Plantae: *Sauromatum guttatum.*
- **Monotyledonae, Monocots** *n.* A plant class in Division Anthophyta found throughout the world and characterized with species that have one cotyledon, no true secondary growth with vascular cambia, and scattered stem vascular bundles and usually have floral parts in multiples of threes, "parallel" leaf venation, and true secondary growth with vascular cambia; distinguished from Dicotyledonae (Curtis 1983, 1080; chap. 28, illustrations).
  Comments: There are about 65,000 extant Dicot species (Curtis 1983, 1080).
- **Mosses** See Plantae: Bryophyta.
- **Mouse-Ear Cress** See Plantae: *Arabidopsis thalliana.*
- **Old World Climbing Fern** See Plantae: *Lygodium microphyllum.*
- *Oncidium* See Plantae: Orchidaceae: *Oncidium.*
- *Ophrys* See Plantae: Orchidaceae: *Ophrys.*
- **Orchidaceae, Orchids** *n.* A monocot family characterized by having flowers with three sepals, three petals including one modified into a labellum which is often larger and more showy than the other two petals, a column (a fleshy, club-shaped fusion of female and male reproductive organs), and one anther with two to eight pollinia (Phylum Anthophyta: Class Monocotyledon: Subclass Liliidae: Order Orchidales) (Shuttleworth et al. 1970, 4–5; Encyclopaedia Britannica Online, www.eb.com, 2000).
  Comments: Below an orchid's anther is its stigma on its column. A single orchid fruit (= capsule) can contain over a million tiny seeds (Shuttleworth et al. 1970, 4–5; Encyclopaedia Britannica Online, www.eb.com, 2000). Orchidaceae contains over 25,000 described species and over 25,000 artificial hybrids; it might be the largest plant family. Orchid plants are from about 1 inch to over 25 feet tall. Most species grow as nonparasitic epiphytes, others grow in the ground, and one species, *Rhizanthella gardneri,* is evidently a saprophyte that grows and flowers underground.

Most orchid species set seeds after cross-pollination, and many species are pollinated by only a few kinds of insects. Many kinds of orchid flowers have constructions that cause insect floral visitors to contact and pick up pollinia. Bees pollinate *Catasetum* spp., *Coryanthes macrantha, Cycnoches* spp., *Cypripedium* spp., *Gongora* spp., *Ophyrs* spp., *Orchis* spp., and others; gnats and mosquitoes, *Pterosylis* spp.; butterflies, *Epidendrum secundum*; hummingbirds, *Comparettia falcata, Elleanthus capitatus,* and *Laelia milleri*; sphinx moths, *Angraecum sesquipedale*; male wasps, *Chiloglottis* spp., *Ophyrs* spp. Orchids as a group use nectar as their major attractant for pollinators, and they do not provide all of the food that their pollinators use for feeding themselves and their young as do many other plant species. Many orchid species are nectarless and attract pollinators usually by deceptive attractants such as bright colors, fragrances, and pseudopollen.

Some genera readily hybridize and produce viable seeds under artificial conditions, including *Brassavola, Cattelya,* and *Laelia,* which produce X *Brassocattelya,* X *Brassolaeliacattelya* and *Laeliacattelya.* Orchidaceae is economically important as a source of folk medicine and cures, many kinds of horticultural plants, seasoning from leaves, and vanilla (from three species of the genus *Vanilla*). Many orchid species are illustrated in Jay's Internet Orchid Species Encyclopedia and links to this site (www.orchidspecies.com, 2000).
cf. copulation: pseudocopulation [Greek *orchis,* testicle, after some orchid species that have ribbed, thickened stems that resemble human testicles]
Angraecum sesquepedale **Thours, 1822, Star-of-Bethelehem Orchid** *n.* An orchid species from Malagasy Republic charcterized by having white petals, a flower with a star-like form, and a nectar spur almost 12 inches long (Shuttleworth et al. 1970, 152, illustration;www.orchid species.com, 2000).

syn. Comet Orchid and other names (www.orchidspecies.com, 2000)

Comments: Darwin predicted that this orchid is pollinated by a moth with a tongue long enough to reach nectar at the end of its nectar spur (Shuttleworth et al. 1970, 152). Researchers later found the moth *Xanthopan morgani praedicta* (Sphingidae), whose tongue is long enough to reach the nectar of this species, but it has not yet been observed to pollinate it.

Coryanthes Hooker, 1831, Bucket Orchids *n.* An orchid genus from Honduras and Guatemala through Brazil and Peru that comprises 15 species characterized by having extremely complex flowers, each with two faucet glands that fill its bucket-like labellum with liquid (Shuttleworth et al. 1970, 100, illustration; www.orchidspecies.com, 2000).

Comments: Male euglossine bees (*Euglossa* and *Eulaema*) are attracted to the strong odor of the fleshy margins of the *Coryanthes* labellum, which they scratch (Encyclopaedia Britannica Online, www.eb.com, 2000). Some of these bees fall into the liquid. When a bee crawls through the flower's spout-like aperture between its column and labellum, his only escape, he may pick up a pollinium that he may deposit on another flower, effecting cross-pollination. A *Coryanthes* flower stops emitting scent within minutes of its pollinia removal (O'Toole and Raw 1991, 152). Euglossines no longer find it attractive, until its starts emitting scent again in about 1 day when its stigma is receptive. This floral behavior evidently promotes cross-pollination.

Oncidium Swartz, 1800 *n.* An orchid genus from the Caribbean Area, Central America, Florida, and South America and having over 750 species (Shuttleworth et al. 1970, 126, illustrations; www.orchidspecies.com, 2000).

Comments: Male *Centris* bees perceive flowers of some *Oncidium* spp. as territorial intrud-

ers. They strike at the flowers in flight, and pick up pollinia in the process (Encyclopaedia Britannica Online, www.eb.com, 2000).

Ophrys Linnaeus, 1753 *n.* An orchid genus with about 30 species in Eurasia and North Africa and characterized by having metallic-colored, hairy flowers that resemble insects (Shuttleworth et al. 1970, 4–5, illustration; www. orchidspecies.com, 2000).

Comments: The Bee Orchid (*O. apifera*) and Fly Orchid (*O. insectifera*) are common European species (Shuttleworth et al. 1970, 4–5). Male bees and wasps pollinate these *Ophrys* through pseudocopulation, *q.v.*

Vanilla Miller, 1754, Vanilla Orchid *n.* An orchid genus that produces the compound vanillin used as a spice (Encyclopaedia Britannica Online, www.eb.com, 2000; www.orchidspecies.com, 2000).

Comment: Vanilla planifolia is the main species from which people obtain vanillin; people also cultivate *V. pompona* and *V. tahitensis* for this compound (Encyclopaedia Britannica Online, www.eb.com, 2000).

[New Latin < Spanish *vainilla*, diminutive of *vaina*, sheath, pod < Latin *vagina*, sheath; after its pods that contain its seeds]

■ **Paleoclusia chevalieri** *n.* A dicot known from 90-million-year-old fossils from the Old Crossman Pit in New Jersey (Yoon 1999c, D1, illustration).

Comment: This species has resin ducts in its flowers that might have produced a resin used by bees.

■ **phanerogam** *n.*

1. A plant that produces seeds; a seed plant; contrasted with cryptogam (Lawrence 1951, 764).

   Notes: Phanerogams include Anthophyta, Coniferophyta, Cycadophyta, Ginkgophyta, Gnetophyta, and Pteridopermophyta. *syn.* spermatophyte

2. A member of the former division of plants Phanerogamia that includes plants with flowers and seeds; distinguished from cryptogam (Michaelis 1963).

   adj. phanerogamic, phanerogamous

[French *phenérogame*, ultimately from Greek *phaneros*, visible; *gamos*, marriage]

- ***Populus tremuloides*** **André Michaux, Golden Aspen, Quaking Aspen, Trembling Aspen** *n.* A widespread North American, clonal, pioneer, short-lived tree (Dicotyledonae: Salicaceae) (Little 1996, 344–345; Matthews 1998, A3). *Comments:* This species grows often as pure stands in western mountains and in an altitudinal zone below fir-spruce forests covering millions of acres in North America (Little 1996, 344–345; Matthews 1998, A3). Conifers replace this species in succession. It is a major organism in ecosystems that harbor many other organisms. Much plant undergrowth occurs in Golden Aspen forests. This species leaves more water in the ground than conifers and, therefore, allows more water to flow into streams and other bodies of water. Golden Aspens reproduce primarily by cloning. One clone in southern Utah covers 106 acres, contains about 47,000 stems, and weighs about 6000 tons, possibly making it the largest living organism.
  cf. Fungus: *Armillaria*

- **Progymnospermophyta, Progymnosperms** [pro Jim nowe sper mo Figh tah, pro JIM nowe sperm MOHF ih tah] *n.* An extinct plant division characterized by gymnospermic secondary wood and pteriodophytic reproduction (Beck 1969 in Stewart 1983, 216–229, illustrations; Campbell 1996, 562, illustration; Strickberger 2000, 311, illustration).
  [Greek *pro*, before + *gymnos*, bare + *sperm*, seed + *phyton*, plant]

- **Psilophyta, Psilophytes** [seye lo FIGH tah sege LOHF fah tah] *n.* A seedless-tracheophyte group characterized by having lateral sporangia and no leaves or roots (Campbell 1996, 557, illustration; Strickberger 2000, 304).
  Comment: There are about 10 to 13 living psilophyte species (Campbell 1996, 550).
  [Greek *psilos*, bare, smooth + *phyton*, plant]

- **Pterophyta, Ferns, Pterophytes** [ter o FIGH tah, ter AHF fah tah, ter Oh fights] *n.* A seedless-tracheophyte division (or phylum) charac-

terized by having catapulting sporangia often arranged in clusters called sori and megaphylls (Campbell 1996, 558, illustration; Strickberger 2000, 307).
  syn. Class Filicinae (Lawrence 1951, 342), Pteridophyta (Brown and Brown 1984, 1), Pteridophytes (Michaelis 1963)
  Comment: There are about 12,000 living fern species in many genera (Campbell 1996, 550).
  [Greek *pteron*, feather + *phyton*, plant; referring to feathery leaves of some species]

- **Quaking Aspen** See Plantae: *Populus tremuloides.*

- **Rhyniophyta, Rhyniophytes** [reye nee o FIGH tah, reye nee OHF fah ta] *n.* An extinct, seedless, tracheophyte group characterized by having fossils indicating that it was the first to inhabit land, no leaves or roots, and terminal sporangia (Campbell 1996, 556, illustration; Strickberger 2000, 304).
  [after the Rhynie, Aberdeenshire, Scotland, near where some rhyniophyte fossils occur + Greek *phyton*, plant]

- ***Sauromatum guttatum*** **(Wall.) Schott, 1832,** ***S. gutattum*** **var.** ***venosum*** **(Aiton) Engl., 1920,** ***S. venosum*** **(Aiton) Kunth, 1841, Monarch-of-the-East, Voodoolily** *n.* A herbaceous aroid from the Himalaya Mountains characterized by an inflorescence with a maroon spadix, a green spathe with maroon spots, and an odor of cow manure (Monocotyledonae: Araceae) (Brickell and Zuk, 1996, 936; Vascular Tropicos Nomenclatural Database, 1999).
  Comments: Researchers have investigated the heat production of *S. guttatum* inflorescences (Skubatz and Kunkel 1999, 841, and references therein). Its floral odor attracts flies and other merdivorous arthropods, some of which pollinate it. This plant can flower from a corm in a humid atmosphere without soil, and some people grow this species as a curiosity.

- **shrub** *n.*
  1. A low, woody perennial plant with persistent stems and branches springing from its base; contrasted with forb, grass, subshrub, tree, and woody vine (Michaelis 1963).
  [Old English *scrybb*, brushwood]

2. A woody plant that branches below, or near, the ground level into several main stems and has no main trunk, whose branches do not ordinarily die back at the end of each growing season (Allaby 1994).

3. A woody-plant species that grows under 15 feet (inferred from Farrar 1995, viii).

syn. bush (Allaby 1994)

cf. bush

Comment: Brockman (1968, 13) notes that a shrub is usually shorter than 15 feet.

> **subshrub** *n.* A woody plant that is smaller than a shrub, produces wood only at its base, and has abundant growth branching upwards from its base and whose upper stems die back at the end of each growing season (Allaby 1994).

■ **Sphenophyta, Horsetails, Sphenophytes** [SPHEN o FIGH tah, sphen NOHF fah tah] *n.* A seedless-tracheophyte group characterized by having jointed stems, tiny or no leaves in extant species, and terminal sporangia (Campbell 1996, 556, illustration; Strickberger 2000, 306).

Comment: There are about 25 living sphenophyte species all in the genus *Equisetum* (Lawrence 1951, 336). [Greek *sphenos*, wedge; perhaps after the shape of the stroblis in some species]

■ **Star-of-Bethelehem Orchid** See Plantae: Orchidaceae: *Angraecum sesque-pedale.*

■ **Tracheophyta, Vascular Plants** [TRAY key o FIGH tah, TRAY key OHF fah tah] *n.* A plant group characterized by having phloem and xylem (Campbell 1996, 555–556; Strickberger 2000, 303).

Comment: There are about 248,746 living tracheophyte species (Campbell 1996, 550).

[Greek *trachea*, windpipe + plant, referring to fluid-conducting cells]

■ **tree** *n.*

1. A perennial woody plant, usually with a single, self-supporting trunk of considerable height and with branches and foliage growing at some distance above the ground; contrasted with herb, shrub, and woody vine (Michaelis 1963).

2. Any shrub or plant that assumes treelike shape or dimensions (*e.g.,* a banana "tree") (Michaelis 1963).

3. A woody plant that is at least 15 feet tall at maturity and has a well-developed crown and single trunk at least several inches in diameter (Brockman 1968, 13).

4. A woody plant with a single main stem (its trunk) that is unbranched near the ground and whose branches do not ordinarily die back at the end of each growing season (Allaby 1994).

5. A woody perennial plant species that reaches a height of at least 15 feet (4.5 meters) (Farrar 1995, vii).

Comments: Most tree species have trunks that are unbranched near the ground in older plants. Saplings (= young trees) are often branched near the ground. The American basswood and trees that sprout from stumps often have several main trunks.

[Old English *trēow, triow, trēo*]

■ **Trembling Aspen** See Plantae: *Populus tremuloides.*

■ **Triurids** *n.* Very early monocots known from 90-million-year-old fossils from the Old Crossman Pit in New Jersey (Yoon 1999c, D1, illustration).

Comment: Triurids are leafless, without chlorophyll, saprophytic, and possibly ancestors of Monocotyledonae (Yoon 1999c, D20).

cf. Clusia gaudichadii

■ *Vanilla* See Plantae: Orchidaceae: *Vanilla.*

■ **Vanilla Orchids** See Plantae: Orchidaceae: *Vanilla.*

■ **Vascular Plants** See Plantae: Tracheophyta.

■ **Voodoo-lily** See Plantae: *Sauromatum guttatum.*

■ **Water Hyacinth** See Plantae: *Eichornia crassipes.*

■ **X** *Cupressocyparis leylandii* **(A.B. Jackson and Dallimore) Dallimore and A.B. Jackson, Leyland Cypress** *n.* A natural hybrid between *Chamaecyparis nootkatensis* ([D. Don] Spach), Nootka Cypress, and *Cupressus macrocarpa* Hartw., Monterey Cypress, both native of the western U.S. (Choukas-Bradley and Alexander, 1987, 103; White 1995, 62).

Comments: This hybrid occurred in England (White 1995, 63). A person planted hybrid seeds from ovules of Nootka Cypress and pollen of Monterey Cypress in 1888. Offspring are fast growing and relatively disease free. Hybrids from ovules of Monterey Cypress and pollen of Nootka cypress also occur. Their seeds are usually sterile. 'Leighton Green,' 'Haggerston Grey,' and 'Naylor's Blue' are cultivars. People frequently plant Leyland Cypresses in the U.S.

[Greek *Cupressus, kuo,* to produce; *parisos,* equal (-sided tree); *Chamaecyparis, chamae,* small; *cyparis,* cypress (White 1995, 62)]

[Leyland, after C.J. Leyland, owner of Haggerston Hall where a person grew the first clones of this tree]

['Leighton Green' and 'Naylor's Blue,' after Mr. Naylor of Leighton Hall, brother-in-law of C.J. Leyland. 'Haggerston Grey,' after Haggerston Hall]

■ *Zea mays* **L., Corn, Maize** *n.* A cultivated member of the Grass Family used for food for Humans and other organisms.

syn. Indian Corn (King and Stansfield 1985)

Comment: Five commercial varieties of Corn are Dent Corn, Flint Corn, Flour Corn, Popcorn, and Sweet Corn (King and Stansfield 1985).

▶ *Kingdom* **Protista** *n.* A kingdom that includes all unicellular eukaryotes, such as Algae, Amoebae, Flagellated Protozoa, and Multicellular Algae (table below); distinguish from Kingdoms Animalia, Fungi, and Plantae.

syn. Protoctista (King and Stansfield 1985)

cf. plankton: zooplankton

Comment: Protista has about 55,000 living species and about 35,000 known extinct species (Curtis 1983, 1076). Protista is the kingdom with the most phylogenetic diversity within Domain Eucarya; many of the evolutionary relationships among Protistan taxa await discovery (Gray et al. 1999, 1476).

CLASSIFICATIONS OF KINGDOM PROTISTA

Classification 1[a]

I. Phylum Actinopoda, Heliozoans, Radiolarians (aquatic)

II. Phylum Chlorophyta, Green Protists (Algae) (includes *Chlamydomonas, Closterium, Eudorina, Euglena, Gonium, Pandorina, Protococcus, Scenedesmus, Volvox*)

III. Phylum Chrysophyta, Yellow-Green Protists (Algae) (includes *Dinobryon*)

IV. Phylum Ciliophora, Ciliates (includes *Paramecium* and *Stentor*)

V. Phylum Gamophyta, Filamenous Green Protists (Algae) (includes *Spirogyra, Zygnema*)

VI. Phylum Rhizopoda, Amoebas

VII. Phylum Sporozoa, Sporozoans

VIII. Phylum Xanthophyta, Filamenous Green Protists (Algae) (includes *Vacheria*)

IX. Phylum Zoomastigina, Zoomastigotes (includes Choanoflagellates, Trypanosomes)

Classification 2[b]

I. Phylum Apicomplexa (*syn.* Sporozoa, Apicoplexans, Sporozoans)

II. Phylum Antinopoda (*syn.* Radiolaria, Heliozoans, and Radiozoans)

III. Phylum Chlorophyta (*syn.* Green Algae)
  A. *Fritschiella tuberosa*
  B. *Spirogyra*
  C. *Ulva*

IV. Phylum Charophyta (*syn.* Charophytes, Stoneworts)
  A. *Chara*
  B. *Nitella*

V. Phylum Chrysophyta (*syn.* Golden Algae, including Diatoms)

VI. Phylum Ciliophora (*syn.* Ciliates)

VII. Phylum Dinoflagellata (*syn.* Dinoflagellates)

VIII. Phylum Euglenophyta (*syn.* Euglenoids)

IX. Phylum Foraminifera (*syn.* Foraminiiferida, Forams, Foraminiferans)

X. Phylum Phaeophyta (*syn.* Brown Algae)

XI. Phylum Rhizpoda (Naked and Shelled Amoebas)

XII. Phylum Rhodophyta (*syn.* Red Algae)

XIII. Phylum Zoomastigophora (*syn.* Zooflagellates)

XIV. No assigned phylum
  A. "Closed-mitosis flagellate"
  B. "Open-mitosis flagellate"
  C. Protoflagellate
  D. Zooflagellate

---

[a] Rainis and Russell (1986).
[b] Strickberger (1996, 312–313).

alga *n. pl.* **algae** An individual or other group within the Group Algae.

Algae *n.* A protistan group that includes Chlorophyta, Dinoflagellata, Euglenophyta, Gamophyta, Phaecophyta, Rhodophyta, and Xanthophyta (King and Stansfield 1985).
Comment: Algal individuals range from single cells to large kelps.

Amoeba proteus *n.* A rhizopod protistan (King and Stansfield 1985).
Comment: Researchers use this giant protozoan for microsurgical nuclear transplantations (King and Stansfield 1985).

chlorophyte *n.* A eukaryan that performs oxygenic photosynthesis using chlorophyll *b* (Tomitani et al. 1999, 161).
Comments: Chlorophytes include Green Algae and most plants (Tomitani et al. 1999, 161).

Giardia *n.* A genus of protistans with a simple cytoskeleton, flagella, no mitochondria or plastids, and two separate nuclei (Kingdom Archaezoa: Diplomonads) (Brody 1999, D7; Campbell et al. 1999, 524).
Comment: These protistans cause giardiasis in some mammal species, first named beaver fever because researchers traced an outbreak in Humans in Canada to water contaminated with beaver excrement (Brody 1999, 97). *Giardia lamblia* causes giardiasis in Humans (Campbell et al. 1999, 524–525).

Plasmodium *n.* A protistan genus in the Sporozoa (King and Stansfield 1985).
Comment: Plasmodium falciparum causes the most dangerous kind of malaria, subtertian malaria, in Humans (King and Stansfield 1985). *Plasmodium malariae* produces benign quartan malaria; *P. vivax*, benign tertian malaria. *Anopheles* mosquitoes transmit these protistans.

Reclinomonas americana *n.* A flagellated, heterotrophic protozoan (Gray et al. 1999, 1476).
Comment: This species has the most bacteria-like mitochondrial genome known (Gray et al. 1999, 1476).

Trypanosoma *n.* A genus of flagellated protozoa (King and Stansfield 1985).
Comment: Trypanosome species cause African sleeping sickness and Chagas disease in Humans (King and Stansfield 1985).

Appendix 2. Companies, Organizations, and Societies Concerned with Animal Behavior, Animal Welfare, Conservation, Ecology, Evolution, and Related Subjects

Comments: A partial list of these groups is provided below, with descriptions for some of them. Most of these groups have websites replete with information about themselves. The Conservation Directory (published by the National Wildlife Federation and updated annually) is an excellent reference on conservation organzations.

♦ **Alley Cat Allies (ACA)** *n.* An organization, founded in 1990, that advocates and implements humane treatment of feral *Felis catus* (www.alleycat.org, 1999).
   cf. Animalia: *Felis catus* (Appendix 1)

♦ **American Academy of Allergy, Asthma, and Immunology (AAAAI)** *n.* An organization, founded in 1943, that is the largest professional medical-speciality organization, representing allergists, allied health professionals, immunologists, and other physicians with a special interest in allergy (American Academy of Allergy, Asthma, and Immunology 1998, S-1; www.aaaai.org, 1999).

♦ **American Association for the Advancement of Science (AAAS)** [Triple-A-S] *n.* A not-for-profit, professional society, founded in 1848, dedicated to the advancement of scientific and technological excellence across all disciplines and to the public's understanding of science and technology (www.aaas.org, 1999).
   Comments: In 1999, the AAAS had more than 143,000 members who were engineers, science educators, scientists, policymakers, and others dedicated to scientific and technological progress in service to society, as well as many affiliated organizations (www.aaas.org, 1999). According to the Constitution of the AAAS, its mission is to further the work of scientists, facilitate cooperation among them, foster scientific freedom and responsibility, improve the effectiveness of science in the promotion of human welfare, advance education in science, and increase the public's understanding and appreciation of the promise of scientific methods in human progress. AAAS is among the older societies in the U.S. Many of today's most prestigious and influential scientific societies have their historical origins in the AAAS. For example, groups such as the American Chemical Society (founded in 1876), the

American Anthropological Association (1902), and the Botanical Society of America (1906) all grew out of informal gatherings at AAAS annual meetings or from established AAAS Sections.

♦ **American Association of Botanical Gardens and Arboreta (AABGA)**

♦ **American Association of Zoological Parks and Aquariums** *n.* A national association, founded in 1924, that is dedicated to promoting responsible animal care and exhibition by zoos and is a strong advocacy group of preservation of captive wildlife species (Buchmann and Nabhan 1996, 263; www.aza.org, 1999).

♦ **American Bryological and Lichenological Society**

♦ **American Fern Society**

♦ **American Institute of Biological Sciences (AIBS)** *n.* A society, founded in 1947, dedicated to advancing the biological sciences and their application to human welfare, fostering and encouraging research and education in the biological sciences, including agricultural, environmental, and medical sciences (www.aibs.org, 1999).
   Comments: Many biologically oriented societies affiliate with the AIBS.

♦ **American Museum of Natural History (AMNH)**

♦ **American Ornithologists' Union**

♦ **American Phytopathological Society**

♦ **American Society for Gravitational and Space Biology**

♦ **American Society for Photobiology**

♦ **American Society for the Prevention of Cruelty to Animals**

♦ **American Society of Agronomy**

♦ **American Society of Limnology and Oceanography**

♦ **American Society of Mammalogists**

♦ **American Society of Naturalists** *n.* A society, founded in 1883, that advances and diffuses knowledge of organic evolution and other broad biological principles to enhance the conceptual unification of the biological science (www.amnat.org, 1999).
   Comments: The ASN publishes *The American Naturalist.* Since its inception in 1867, *The American Naturalist* has been one of the world's most highly renowned peer-reviewed publications in ecology, evolution, and population and integrative biology

research (www.amnat.org, 1999). This journal addresses topics in community and ecosystem dynamics, evolution of sex and mating systems, organismal adaptation, and genetic aspects of evolution, and in particular emphasizes sophisticated methodologies and innovative theoretical syntheses.

♦ **American Society of Plant Physiologists**

♦ **American Society of Plant Taxonomists**

♦ **American Type Culture Collections**

♦ **American Zoo and Aquarium Association (AZA)** *n.* An organization, founded in 1924, whose main mission is conserving our natural world; it also supports membership excellence in conservation, education, research, and science (www.aza.org, 1998).
Comments: Further, on behalf of its 183 accredited zoos and aquariums throughout North America, AZA promotes education and habitat-protection programs and manages cooperative Species Survival Plans (SSPs) for over 135 endangered species (www.aza.org, 1998).

♦ **Animal Behavior Society (ABS)** *n.* A society, founded in 1964, that promotes and encourages the biological study of animal behavior in the broadest sense, including studies at all levels of organization using both descriptive and experimental methods under natural and controlled conditions. The ABS encourages both research studies and the dissemination of knowledge about animal behavior through publications, educational programs, and activities (www.animalbehavior.org, 1999).
Comments: The ABS encourages both research studies and the dissemination of knowledge about animal behavior through publications, educational programs, and activities (www.animalbehavior.org, 1999).

♦ **Appalachian Mountain Club** *n.* A club, founded in 1876, that promotes the protection, enjoyment, and wise use of the mountains, rivers, and trails of the northeastern U.S. (www.outdoors.org, 1999).
Comment: This club is the oldest conservation and recreation organization in the U.S. and had over 80,000 members in 1999 (www.outdoors.org, 1999). The club believes that the mountains and rivers have an intrinsic worth and also provide recreational opportunity, spiritual renewal, and ecological and economic health for the northeastern U.S. The club encourages people to enjoy and appreciate the natural world because it believes that successful conservation depends on this experience.

♦ **Arizona Sonora Desert Museum** *n.* A museum in the Sonoran Desert of southern Arizona which is the home to such things as the Forgotten Pollinators Public Awareness Campaign, the Desert Alert Program, animal and plant exhibits, and pollinator gardens (Buchmann and Nabhan 1996, 263).

♦ **Association for Tropical Biology**

♦ **Association of College and University Biology Educators**

♦ **Association or Ecosystem Research Centers**

♦ **Association of Southeastern Biologists**

♦ **Association of Systematics Collections**

♦ **Audubon Naturalist Society (ANS)** *n.* A society, founded in 1897, that works "to increase the enjoyment and appreciation of the natural world and to preserve and protect the treasures of the Washington D.C. Metropolitan Region" (www.Audubon Naturalist.org, 1999).

♦ **Avida Artificial Life Group, Avida Group** *n.* A group of students, professors, and other researchers founded in 1993 and interested in the study of the fundamental properties of living and evolving systems (http://www.krl.caltech.edu/avida/).
Comments: The Avida Group is based at the California Institute of Technology (http://www.krl.caltech.edu/avida/). Christoph Adami heads the group and founded it with C. Titus Brown and Charles Ofria with the writing of the initial version of the Avida software.

♦ **AZA** See American Zoo and Aquarium Association.

♦ **Biological Resources Division (BRD) of the U.S. Geological Survey** *n.* Part of the U.S. Geological Survey whose "mission is to work with others to provide the scientific understanding and technologies needed to support the sound management and conservation of the U.S.'s biological resources" (www.biology.usgs.gov, 1999).
Comments: The BRD started the Ornithological Office in the Entomology Division of the U.S. Department of Agriculture in 1885 (www.biology.usgs.gov, 1999). As the years progressed, this organization became the Division of Ornithology and Mammalogy (1886), the Division of Biological Survey (1896), Bureau of Biological Survey (1905) which became part of the U.S. Fish and Wildlife Service in the Department of the Interior (1939), National Biological Service (1993), and the Biological Resources Division in the U.S. Geological Survey (1996). A main goal of the BRD is to make data and other information on the U.S.'s biological

resources more accessible to more people; other goals are explained in its website.
cf. Geologic Division of the U.S. Geological Survey

♦ **Bat Conservation International (BCI)** *n.* An orgnaization that seeks to promote bat conservation by increasing public awareness of their ecological importance (Buchmann and Nabhan 1996, 264).

♦ **Biological Sciences Curriculum Study**

♦ **BioQUEST Curriculum Consortium**

♦ **BIOSIS**

♦ **Botanical Society of America**

♦ **BRD** See Biological Resources Division of the U.S. Geological Survey.

♦ **CITES** See organization: Convention on International Trade in Endangered Species of Wild Fauna and Flora.

♦ **CITES Secretariat (UNEP)** *n.* The regulatory body that controls the trade in endangered species and products derived from their harvest (Buchmann and Nabhan 1996, 264).

♦ **Conservation International**

♦ **Convention on International Trade in Endangered Species of Wild Fauna and Flora, The (CITES)** *n.* A group of 146 countries that ban commercial international trade in an agreed list of endangered species and regulate and monitor trade in others that might become endangered (www.wmc.org.uk/CITES, which includes the Convention Text).
Comments: CITES started to act in 1975 (www.wmc.org.uk/CITES). The international wildlife trade is worth billions of dollars annually and has caused massive declines in the numbers of many species. CITES' aims are major components of *Caring for the Earth, a Strategy for Sustainable Living,* launched in 1991 by UNEP, the United Nations' Environment Programme (UNEP); The World Conservation Union (IUCN); the World Wide Fund for Nature (WWF).

♦ **Cooper Ornithological Society**

♦ **Coordinating Group of Alien Pest Species** *n.* A U.S. group comprised of 14 governmental agencies and private groups that drafted a ten-point "Silent Invasion" action plan to improve alien pest prevention and control programs (Claiborne, 1997, A6).

♦ **Department of Defense (DOD)**

♦ **Ecological Society of America (ESA)** *n.* A non-partisan, not-for-profit society of scientists, founded in 1915, to stimulate sound ecological research, clarify and communicate the science of ecology, and promote the responsible application of ecological knowledge to public issues (www.esa.sdsc.edu).

Comments: ESA's 7600 members conduct research, teach, and work to provide the ecological knowledge necessary to solve environmental problems that include biotechnology, ecological restoration, ecosystem management, habitat alteration and destruction, natural resource management, ozone depletion and global climate change, species extinction and loss of biological diversity, and sustainable ecological systems (www.esa.sdsc.edu).

♦ **Entomological Society of America (ESA)** *n.* A not-for-profit, educational society established in 1889 "to promote the science of entomology in all of its subdisciplines for the advancement of science and the benefit of society both at the national and international levels, to publish and encourage publications pertaining to entomology, and to assure cooperation both locally and globally in all measures leading to these ends" (constitution and bylaws, www.entsoc.org, 1999).
Comments: The ESA has more than 7400 members who are consultants; educators; extensional personnel; researchers and scientists from agricultural departments, colleges and universities, health agencies, private industries, and federal and state governments; and students (www.entsoc.org, 1999).

♦ **Entomological Society of Washington** *n.* A not-for-profit organization, founded in 1884, whose objectives "are to promote the study of entomology in all its aspects and to cultivate mutually advantageous relations among those in any way interested in entomology" (Gurney 1976, 225; Entomological Society of Washington, 1996, 610).
Comments: Members operate this society exclusively for scientific and educational purposes (Entomological Society of Washington 1996, 610). This society's publications include the *Proceedings of the Entomological Society of Washington.*

♦ **Environmental Defense Fund (EDF)** *n.* A not-for-profit organization, founded in 1967, that provides assistance with economic, legal, and scientific issues related to conservation biology and pesticide abuse (Buchmann and Nabhan 1996, 264; www.edf.org, 1999).

♦ **Friends of Dyke Marsh (FODM)** *n.* A not-for-profit organization, founded in 1975, that promotes the preservation of Dyke Marsh Wildlife Preserve in Virginia and its biota (www.dykemarsh.org).
Comment: This preserve has species-rich habitats that include a tidal freshwater marsh along the Potomac River and associated low forest and swamp forest.

◆ **Friends of the National Zoo (FONZ)**
n. "A nonprofit organization of individuals, families, and organizations who are interested in helping to maintain the status of the Smithsonian National Zoological Park as one of the world's great zoos; to foster its use for education, research, and recreation; to increase and improve its facilities and collections; and to advance the welfare of its animals (*Zoogoer* 1999, Jan./Feb., 4).
Comments: FONZ publishes the magazine *Zoogoer.*

◆ **Geologic Division of the U.S. Geological Survey** *n.* Part of the U.S. Geological Survey whose chief function is to undertake geologic and mineral-resource surveys and mapping for the Department of the Interior (www.geology.usgs.gov, 1999). *cf.* Biological Resources Division of the U.S. Geological Survey

◆ **Global Warming International Center (GWIC)** *n.* An international body that disseminates information on global warming science and policy, serving both governmental and non-governmental organizations, as well as industries in more than 120 countries (www.globalwarming.net).
Comments: GWIC sponsors unbiased research supporting the understanding of global warming and its mitigation (www.globalwarming.net). Among its best-known projects are the Global Treeline Projects (GTP), Greenhouse Gas Reduction Benchmark (GHGRB), Himalayan Reforestation Project (HRP), and the Extreme Event Index (EEI). The GWIC sponsors the annual Global Warming International Conference & Expo and the Executive Workshop on Industry Technology and Greenhouse Gas Emission, which facilitates international exchange and provides the most up-to-date, hands-on workshops for corporation and utilities executives. The Global Warming International Conference & Expo occurred in the U.S. in April 2000.

◆ **Gorilla Foundation, The; Koko.org**
n. An organization that promotes teaching Ameslan (American Sign Language) to two Lowland Gorillas, Koko and Michael, *q.v.* (Appendix 1) (www.koko.org, 2000).
Comments: This Foundation supports the Gorilla Language Project (= Project Koko), the longest continuous interspecies communications project of its kind, and promotes international conservation (www.koko.org, 2000).

◆ **Greenpeace**
◆ **Human Genome Organization**
◆ **Institute of Ecosystem Studies**
◆ **Intergovernmental Panel on Climate Change (IPCC)**

◆ **International Association for Landscape Ecology, U.S. Region**
◆ **International Bee Research Association (IBRA)** *n.* A not-for-profit organization, founded in 1949, that is devoted to advancing apicultural education and science worldwide (Buchmann and Nabhan 1996, 264).
Comments: IBRA was formed in 1949 and serves as a central clearinghouse for all published information on bees of the world (with a focus on apiculture and the study of honey bees). Members and visitors can access its world-class library. IBRA sponsors numerous international scientific colloquia and conferences and publishes scientific journals on beekeeping, including *Apicultural Abstracts*, *Bee World*, and the *Journal of Apicultural Research.*

◆ **International Council for Bird Preservation** *n.* An organization that determines the conservation and protection status of bird species worldwide (Buchmann and Nabhan 1996, 265).

◆ **International Society for Ecological Economics**
◆ **International Society for Ecological Modeling**
◆ **International Society for Ecosystem Health**
◆ **International Union for the Conservation of Nature and Natural Resources (IUCN)** *n.* The primary coodinating agency for international conservation efforts, founded in 1948 (Buchmann and Nabhan 1996, 265; www.iuc.org).
Comments: IUCN has many nations as members, publishes directories of environmental specialists, maintains the Red List of organisms that are endangered to some degree and which is a guideline for policy makers (Buchmann and Nabhan 1996, 265; Stevens 1996b, C1).

◆ **Kansas Entomological Society, Central States Entomological Society**
◆ **Lady Bird Johnson Wildflower Center**
◆ **League of Conservation Voters**
◆ **Monell Chemical Senses Center**
◆ **Mycological Society of America**
◆ **NASA Specialized Center of Research and Training in Exobiology (NSCORT)** *n.* A center that conducts research in exobiology (the study of the origin, evolution, and distribution of life in our universe); trains young scientists for research careers in exobiology and related areas; and communicates with scientists, students, and the general public about exobiology (http://exobio.ucsd.edu/, 2000).

Comment: NSCORT is a consortium among the University of California, San Diego (UCSD), The Scripps Research Institute (TSRI), and The Salk Institute (SALK).

♦ **National Academy of Sciences (NAS)** *n.* An organization, created by the U.S. Congress in 1863, to advise the U.S. government in scientific and technical matters (www.nas.org, 1999).

Comments: Sibling organizations of the National Academy of Science are the National Academy of Engineering, the Institute of Medicine, and the National Research Council (www.nas.org, 1999). The academies and the institute are honorary societies that elect new members to their ranks each year. The Institute of Medicine also conducts policy studies on health issues, but the National Research Council conducts the bulk of the institute's science-policy and technical work. These non-profit organizations provide a public service by working outside the framework of government to ensure independent advice on matters of science, technology, and medicine.

♦ **National Association of Biology Teachers**

♦ **National Audubon Society** *n.* The premier birding organization that was founded in 1905 in the U.S. and is active in public educational outreach, scientific research, and wildlife conservation (Buchmann and Nabhan 1996, 265; www.audubon.org, 1999).

Comment: Various states of the U.S. have their own branches of this society.

♦ **National Bioethics Advisory Commission**

♦ **National Geographic Society**

♦ **National Institute for Environmental Renewal (NIER)** *n.* An organization, founded in 1994, to support environmental renewal and economic development throughout the U.S. (www.nier.org, 1999).

♦ **National Institute for the Environment (NIE)** *n.* An institute, proposed to Congress by the NIE Committee, whose goal would be to "to improve the scientific basis for making decisions on environmental issues" (www.nie.org, 1999).

Comments: In October 1999, the CNIE publicly endorsed the report of the National Science Board which recommended implementing nearly all of the activities proposed for a NIE (www.nie.org, 1999). The CNIE suspended its call for the creation of a NIE to work in support of the NSF initiative.

♦ **National Museum of Natural History (NHNH)**

♦ **National Oceanic and Atmospheric Administration (NOAA)**

♦ **National Parks and Conservation Association (NPCA)**

♦ **National Wildlife Federation (NWF)** *n.* A leading advocate for wildlife preservation (Buchmann and Nabhan 1996, 265).

Comment: The NWF publishes *International Wildlife, National Wildlife, Ranger Rick, The Conservation Directory,* and *Your Big Backyard.*

♦ **Natural Resources of Defense Council, Inc. (NRDC)** *n.* An organization that uses both legal and scientific methods to influence and monitor government actions and proposed legislation affecting conservation issues and pesticide abuse (Buchmann and Nabhan 1996, 265).

♦ **Nature Conservancy, The (TNC)** *n.* The premier organization dedicated to the preservation of biota by means of habitat preservation through local, private, and state land acquisition and an extensive array of wildlife refuges and plant-protection sites and founded in 1951 (Buchmann and Nabhan 1996, 265; www.tnc.org, 1999).

Comment: The Nature Conservancy maintains extensive databases on the distribution of rare species in the Americas, especially North America (Buchmann and Nabhan 1996, 265).

♦ **New York Zoological Society** *n.* An organization that leads in wildlife conservation and related research (Buchmann and Nabhan 1996, 266).

♦ **North American Butterfly Association (NABA)** *n.* A not-for-profit organization of North American amateur and professinal lepidopterists who are actively engaged in the conservation and protection of endangered and threatened Lepidoptera and their habitats (Buchmann and Nabhan 1996, 265).

Comment: The NABA "educates the public about the joys of nonconsumptive, recreation 'butterflying'" (Buchmann and Nabhan 1996, 266).

♦ **Orangutan Foundation International (OFI)** *n.* A society that supports the study of wild Orangutans in Central Indonesian Borneo and the repatriation of ex-captive, wild-born Orangutans back to their native forests. OFI has an outreach educational program for children and assists needy people near Camp Leaky, Indonesia (N. Briggs in Turco and Gruner, 1999; www.nss.net/orangutans, 2000).

Comment: The OFI successfully returned over 200 formerly captive Orangutans into the forest and cared for over 60 orphaned Orangutans after the severe fires in Borneo in 1977 (www.nss.net/orangutans, 2000).

♦ **Organization for Tropical Studies (OTS)**

♦ **Organization of Biological Field Stations**

♦ **People for the Ethical Treatment of Animals (PETA)** *n*. An international, not-for-profit organization, founded in 1980, that promotes the animal-rights movement and attempts to eradicate all animal cruelty (Adams 1995, F1; www.peta-online.org, 1999).

Comments: PETA uses tactics including running ads about treatment of animals used in experimentation (naming Universities, labs, etc.), investigating animals used in lab studies, and terrorizes its perceived enemies (Adams 1995, F1, F4).

♦ **Phi Sigma Biological Sciences Honor Society**

♦ **Pollen Bee Foundation**

♦ **Potomac Conservancy, The** *n*. An action-oriented land trust, founded in 1993, that works with landowners to conserve the beauty and integrity of the Potomac River especially from Georgetown (Washington, D.C.) through Harpers Ferry, WV (Richard W. Fox, Conservancy flier, 1998; www.potomac.org).

♦ **Potomac River Basin Consortium, Inc., The**

♦ **Poultry Science Association**

♦ **Rain Forest Action Network (RAN)** *n*. A not-for-profit organization, founded in 1985, that works to protect Earth's rainforests and supports the rights of their inhabitants through education, grassroots organizing, and non-violent direct action (www.ran.org, 1999).

Comments: RAN is member and volunteer based and coordinates its activities with environmental and human-rights organizations around the world. Over 150 Rainforest Action Groups (RAGs) associated with the RAN in 1999. It has played a key role in strengthening the worldwide rainforest conservation movement through supporting activists in tropical countries as well as organizing and mobilizing consumers and community action groups throughout the U.S. (www.ran.org, 1999).

♦ **Rain Forest Alliance** *n*. An international not-for-profit organization, incorporated in 1987 and dedicated to the conservation of tropical forests for the benefit of the global community (www.rainforest-alliance.org, 1999).

Comments: Its mission is to develop and promote economically viable and socially desirable alternatives to the destruction of this endangered, biologically diverse natural resource (www.rainforest-alliance.org, 1999). It pursues this mission through education, research in the social and natural sciences, and the establishment of cooperative partnerships with businesses, governments, and local peoples. This organization had 14,000 members in 1999.

♦ **Save the Manatee Club** *n*. A group in Florida, founded in 1981, that has about 30,000 members who work to stop manatees from going extinct (Booth 1992, 3A; www.savethemanatee.org).

Comment: "Drunken and idiotic boaters" are manatees' greatest enemies (Booth 1992, 3A).

♦ **Search for Extraterrestrial Intelligence (SETI) Institute**

♦ **Sierra Club, The** *n*. An organization, founded in 1892, through which the Sierra Club Foundation advances the preservation and protection of our natural environment by empowering the citizenry, especially democratically based grassroots organizations, with charitable resources to protect our environment (www.sierraclub.org, 1999).

♦ **Sigma Xi** *n*. An international honor society for science and engineering founded in 1886 (www.sigmaxi.org).

Comments: Sigma Xi has more than 80,000 members in over 500 chapters at colleges and universities, industrial research centers, and government laboratories (www.sigmaxi.org). Sigma Xi hosts an array of programs and activities, including publication of the *American Scientist* magazine, a 75-year-old Grants-in-Aid of Research Program, a number of prestigious prizes and awards, and the college of Distinguished Lecturers.

♦ **Smithsonian Tropical Research Institute (STRI)** *n*. A part of the Smithsonian Institution's research program with marine terrestrial research stations in Columbia and Panama and research conducted in many parts of the world (www.si.edu/organiza/centers /stri/about.htm).

Comment: STRI became a separate part of the Smithsonian Institution in 1966. STRI's website provides a wealth of information about it.

♦ **Society for Conservation Biology**

♦ **Society for Ecological Restoration**

♦ **Society for Industrial Microbiology**

♦ **Society for Integrative and Comparative Biology** (originally the American Society of Zoologists [1903–1995])

♦ **Society for Mathematical Biology**

♦ **Society for the Study of Evolution**

♦ **Society of Environmental Toxicology and Chemistry**

♦ **Society of Nematologists**

♦ **Society of Systematic Biologists**

♦ **Society of Toxicology**

♦ **Society of Wetland Scientists**

♦ **Sonoran Arthropod Studies Institute (SASI)** *n*. A not-for-profit oganization that promotes educational outreach and research on the biology of native Sonoran Desert arthropods (Buchmann and Nabhan 1996, 266; www.sasionline.org, 1999).

Comment: SASI offers training workshops for in-service teachers, hosts the annual "Invertebrates in Captivity" Conference, publishes a members' magazine (*Backyard BUGwatching*) and newsletter (*The Instar*), operates The Arthropod Discovery Center in the Tucson Mountains, and sponsors members' events (Buchmann and Nabhan 1996, 266).

♦ **Southern Appalachian Botanical Society**

♦ **The Institute for Genomic Research (TIGR)**

♦ **The Wilderness Society** *n.* A U.S.-based organization that works to protect all U.S. wild public lands (The Wilderness Society flier, 1999).

Comments: This society watches over national forests, parks, and refuges, and wilderness areas (The Wilderness Society flier, 1999).

♦ **TIGR** See The Institute for Genomic Research.

♦ **Torrey Botanical Society**

♦ **U.S. Energy Information Administration**

♦ **U.S. Federation for Culture Collections**

♦ **U.S. Fish and Wildlife Service (FWS)**

♦ **W3 Consortium** See Organization: World Wide Web Consortium.

♦ **W3C** See Organization: World Wide Web Consortium.

♦ **Weed Science Society of America**

♦ **Wildlife Conservation International and New York Zoological Society** *n.* An organization that leads in wildlife conservation and related research (Buchmann and Nabhan 1996, 266).

♦ **Wildlife Defense Fund (WDF)**

♦ **World Conservation Monitoring Centre** *n.* An organization that monitors the global wildlife trade, the status of endangered species, the use of natural resources, and conserved and protected areas worldwide (Buchmann and Nabhan 1996, 267; www.wcmc.org.uk, 1999).

♦ **World Wide Web Consortium (W3C, W3 Consortium)** *n.* An organization,

founded in 1994 by Tim Berners-Lee, as "a neutral open forum where companies and organizations to whom the future of the Web is important come to discuss and to agree on new common computer protocols" (www.w3.org/People/Berners-Lee).

Comments: The W3C has been a center for making decisions by consensus, designing, and raising issues and also as a vantage point from which to view evolution of the World Wide Web. The W3C is based at the Institut National pour la Recherche en Informatique et Automatique (INRIA), France; Keio University, Japan; and Massachusetts Institute of Technology, U.S.

♦ **World Wildlife Fund (WWF)** *n.* A major conservation organization, founded in 1961, with branches throughout the world and active in both management and scientific research within many national parks (Buchmann and Nabhan 1996, 267).

syn. World Wildlife Fund for Nature (Buchmann and Nabhan 1996, 267; www.worldwildlife.org)

♦ **Xerces Society** *n.* An international, not-for-profit organization that focuses on the conservation of insects and other invertebrates, especially butterflies (Buchmann and Nabhan 1996, 267; www.xerces.org).

Comments: The Xerces Society publishes the beautifully illustrated newsletter *Wings*.

♦ **Zero Population Growth (ZPG)** *n.* A U.S. not-for-profit organization, founded in 1968, that works to slow population growth and achieve a sustainable balance between our Earth's people and its resources (www.zpg.org, 1999).

Comments: ZPG seeks to protect our environment and ensure a high quality of life for present and future generations (www.zpg.org, 1999). ZPG's education and advocacy programs aim to influence public policies, attitudes, and behavior on national and global population issues and related concerns. The human population was 1 billion in 1804, 5 billion in 1987, and 6 billion on about October 12, 1999 (www.y6b.org).

REFERENCES

Abele, L.G. and S. Gilchrist. 1977. Homosexual rape and sexual selection in acanthocephalan worms. *Science,* 197: 81–83.

Abrahamson, W.G., T.G. Whitham, and P.W. Price. 1989. Fads in ecology. *BioScience,* 39(5): 321–325.

Abrams, P. 1992. Resource, pp. 282–285 in E.F. Keller and E.A. Lloyd, Eds., *Keywords in Evolutionary Biology,* Harvard University Press, Cambridge, MA, 414 pp.

Adams, L. 1995. What is PETA's beef? *Washington Post,* 28 May, F1, F4.

Adams, M.B., J.N. Kochenderfer, F. Wood, T.R. Angradi, and P. Edwards. 1994. *Forty Years of Hydrometeorological Data from the Fernow Experimental Forest, West Virginia.* General Technical Report NE-184, U.S. Department of Agriculture, Forest Service, Northeastern Forest Experiment Station, 24 pp.

Adams, S. 1996. Sorting look-alike soybeans. *Agricultural Research,* 44(8): 12–13.

Agrawal, A.A. 1998. Induced responses to herbivory and increased plant performance. *Science,* 279: 1201–1202.

Agren, G.Q. et al. 1989. Territory, cooperation and resource priority: hoarding in the Mongolian Gerbil, *Meriones unguiculatus. Animal Behaviour,* 37: 28–32.

Ainsworth, M.D.S. and S.M. Bell. 1970. Attachment, exploration, and separation: illustrated by the behavior of one-year-olds in a strange situation. *Child Development,* 41: 49–67.

Akre, R.D., A. Greene, J.F. MacDonald, P.J. Landolt, and H.G. Davis. 1981. *Yellowjackets of America North of Mexico.* USDA Agricultural Handbook No. 552, 102 pp.

Alatalo, R.V. and A. Lunberg. 1986. The sexy son hypothesis: data from the Pied Flycatcher *Ficedula hypoleuca. Animal Behaviour,* 34: 1454–1462.

Alatalo, R.V., A. Lunberg, and J. Sundberg. 1990. Can female preference explain sexual dichromatism in the Pied Flycatcher, *Ficedula hypoleuca? Animal Behaviour,* 39: 244–252.

Alberts, B., D. Bray, J. Lewis, M. Raff, K. Roberts, and J.D. Watson. 1989. *The Molecular Biology of the Cell,* 2nd ed., Garland Publishing, New York, 1219 pp.

Alcock, J. 1975. *Animal Behavior: An Evolutionary Approach,* Sinauer Associates, Sunderland, MA, 547 pp.

Alcock, J. 1979. *Animal Behavior: An Evolutionary Approach,* 2nd ed., Sinauer Associates, Sunderland, MA, 532 pp.

Alcock, J. 1984. *Animal Behavior: An Evolutionary Approach,* 3rd ed., Sinauer Associates, Sunderland, MA, 596 pp.

Alcock, J. 1987. Leks and hilltopping in insects. *Journal of Natural History,* 21: 319–328.

Alcock, J., E.M. Barrows, G. Gordh, L.J. Hubbard, C. Kirkendall, D.W. Pyle, T.L. Ponder, and F.G. Zalom. 1978. The ecology and evolution of male reproductive behaviour in bees and wasps. *Zoological Journal of the Linnaean Society,* 64(4): 293–326.

Alexander, B. 1994. People of the Amazon fight to save the flooded forest. *Science,* 265: 606–607.

Alexander, R.D. 1987. *The Biology of Moral Systems,* Aldine de Gruyter, New York, 301 pp.

Allaby, M. 1994. *The Concise Oxford Dictionary of Ecology,* Oxford University Press, London, 415 pp.

Allen, A. 1998. Nature and nurture. *Washington Post Magazine,* 11 January, 6–11, 21–25.

Allen, H. 1993. Big bang sucked into black hole: cosmically correct seek to rename theory, *Washington Post,* 16 June, B1, B2.

Allen J.A. 1906. *The Influence of Physical Conditions in the Genesis of Species.* Annual Report of the Board of Regents of the Smithsonian Institution Showing the Operations, Expenditures, and Condition of the Institution for the Year Ending June 30, 1905, U.S. Government Printing Office, Washington, D.C, pp. 375–402 (reprint of Allen, 1877).

Allen, J. 1997. Biospherian viewpoints. *Science,* 275: 1249.

Ali, A. and J. Lord. 1980. Impact of experimental insect growth regulators on some nontarget aquatic invertebrates. *Mosquito News,* 40: 564–571.

Alloway, T.M. 1979. Raiding behaviour of two species of slave-making ants, *Harpagoxenus americanus* (Emery) and *Leptothorax duloticus* Wesson (Hymenoptera: Formicidae). *Animal Behaviour,* 27: 202–210.

Aloimonos, Y. and A. Rosenfeld. 1991. Computer vision. *Science,* 253: 1249–1266.

Aluja, M. and A. Norrbom. 2000. *Fruit Flies (Tephritidae): Phylogeny and Evolution of Behavior,* CRC Press, Boca Raton, FL.

Alvarez-Buylla, E.R. and R. García-Barrios. 1993. Models of patch dynamics in tropical forests. *Trends in Ecology and Evolution,* 8: 201–204.

Alverson, W.S., D.M. Waller, and S.S. Solheim. 1998. Forests too deer: edge effects in northern Wisconsin. *Conservation Biology,* 2: 348–358.

Amat, J. A. and E. Aquilera. 1990. Tactics of Black-Headed Gulls robbing egrets and waders. *Animal Behaviour,* 39: 70–77.

Amato, I. 1993. Chemistry: changing the landscape of the possible. *Science,* 262: 507.

American Academy of Allergy, Asthma, and Immunology. 1998. The Allergy Report 1998. *Discover,* March: S-1–S-31.

Anderson, C. 1994. Cyberspace offers chance to do "virtually" real science. *Science,* 264: 900–901.

Anderson, R.B. 1995. The unnatural moment. *Natural History,* 104: 88–89.

Angier, N. 1996. Variant gene tied to a love of new thrill. *New York Times,* 2 January: A1, B11.

Angier, N. 1997. Chemical tied to fat control could help trigger puberty. *New York Times,* 7 January: C1, C3.

Angier, N. 1998. Study finds signs of elusive pheromones in humans. *New York Times Northeast,* 12 March: A22.

Angier, N. 2000. William Hamilton, 63, dies: an evolutionary biologist. *New York Times Northeast,* 10 March: A18.

Anonymous. 1986. Little-known animal disobeys the rules of chromosomes. *Washington Post,* 5 January: A12.

Anonymous. 1992. Magnets on the brain. *Time,* 139 (20): 25.

Anonymous. 1995a. Whence the monkeys? *Discover,* June: 31.

Anonymous. 1995b. Arachnophila. *Discover,* November: 32.

Anonymous. 1996a. Rock-eating slime. *Discover,* March: 20–21.

Anonymous. 1996b. Life on lobster lips. *Discover,* March: 24.

Anonymous. 1996c. Like a gopher in the sky. *Discover,* June: 28.

Anonymous. 1996d. Gastric groomers. *Discover,* July: 32.

Anonymous. 1996e. Gene may be clue to nature of nurturing. *New York Times,* 26 July: A21.

Anonymous. 1996f. That's some snood. *Discover,* December: 34.

Anonymous. 1996g. Scientists find a brain protein spurs appetite. *New York Times,* 11 July: A18.

Anonymous. 1996h. Tonight's blue moon leaves experts pondering the term. *Washington Post,* 30 June: A7.

Anonymous. 1997a. Questionable life. *Discover,* February: 16.

Anonymous. 1997b. Bac talk. *Discover,* February: 23.

Anonymous. 1999. Once in a blue moon. *Washington Post,* 1 April: B4.

Aoki, S., U. Kurosu, and D. L. Stern. 1991. Aphid soldiers discriminate between soldiers and non-soldiers, rather than between kin and non-kin in *Ceratoglyphin*a *bambusae. Animal Behavior,* 42: 865–866.

Appenzeller, T. 1993. Chemistry: laurels for a late-night brainstorm. *Science,* 262: 506–507.

Arcese, P. 1987. Age, intrusion pressure and defence against floaters by territorial male song sparrows. *Animal Behaviour,* 35: 773–784.

Armstrong, E.A. 1947. *Bird Display and Behaviour: An Introduction to the Study of Bird Psychology,* 2nd ed. Lindsay Drummond, London, 431 pp. (reprinted in 1965 by Dover Publications, New York, 431 pp.).

Arthur, W. 1984. *Mechanisms of Morphological Evolution.* John Wiley & Sons, New York, 275 pp.

Artz, L.B. 1937. Plants of the shale banks of the Massanutten Mountains of Virginia. *Claytonia,* 3: 45–51; 4: 10–15.

Ash, R. 1996. *The Top 10 of Everything 1997.* Dorling Kindersley, Ltd., London, 256 pp.

Aspey, W.P. 1976. Behavioral ecology of the "edge effect" in *Schizocosa crassipes* (Araneae: Lycosidae). *Psyche,* 83: 42–50.

Associated Press. 1992. World's largest bacteria found in surgeonfish: organisms dwarf other known types, defy theories on structure, researchers say. *Washington Post,* 31 May: A4.

Associated Press. 1993. Lack of oxygen blamed for dinosaurs' extinction. *Washington Post,* 28 October: A3.

Associated Press. 1996a. Nature leads to nurture for mice: study finds a good mother gene. *Washington Post,* 27 June: A2.

Associated Press. 1996b. Tyrannosaurus bite was mighty, study finds. *New York Times,* 22 August: A21.

Associated Press. 1996c. Toad imported as solution turns out to be bigger problem. *New York Times,* 22 September: 29.

Associated Press. 1997. American girls are maturing earlier: research indicates some begin puberty by age 8. *Washington Post,* 8 April: A2.

Associated Press. 1998. Alaskan victim of 1918 flu yields sample of killer virus. *New York Times Northeast,* 8 February: 26.

Associated Press. 1999. Everglades restoration schedule is moved up. *New York Times Northeast,* 8 April: A21.

Atkins, A. 1989. The drinking and perching habits of skippers. *Antenna,* 13(3): 103–104.

Atsatt, P.R. and D.J. O'Dowd. 1976. Plant defense guilds. *Science,* 193: 24–29.

Austad, S.N. 1984. Evolution of sperm priority patterns in spiders, pp. 223–249 in R.L. Smith, Ed., *Sperm Competition and the Evolution of Animal Mating Systems.* Academic Press, New York, 687 pp.

Austad, S.N. and R. Thornhill. 1991. This bug's for you. *Natural History,* December: 44–49.

Austin, O.L., Jr. 1971. *Families of Birds.* Golden Press, New York, 200 pp.

Axelrod, R. and W.D. Hamilton. 1981. The evolution of cooperation. *Science,* 211: 1390–1396.

Baba, R, Y. Nagata, and S. Tamagishi. 1990. Brood parasitism and egg robbing among three freshwater fish. *Animal Behaviour,* 40: 776–778.

Baddeley, A. 1992. Working memory. *Science,* 255: 556–559.

Bailey, W.J., R.J. Cunningham, and L. Lebel. 1990. Song power, spectral distribution and female phonotaxis in the bushcricket *Requena verticalis* (Tettigoniidae: Orthoptera): active female choice or passive attraction. *Animal Behaviour,* 40: 33–42.

Baird, T.A., and N.R. Liley. 1989. The evolutionary significance of harem polygyny in the Sand Tilefish, *Malacanthus plumieri*: resource or female defence? *Animal Behaviour,* 38: 817–829.

Baker, C.S. and L.M. Herman. 1985. Whales that go to extremes. *Natural History,* 94(10): 52–61.

Baker, R.R. 1980. Goal orientation by blindfolded humans after long-distance displacement: possible involvement of a magnetic sense. *Science,* 210: 555–557.

Baker, R.R. 1987. Human navigation and magnetoreception: the Manchester experiments do replicate. *Animal Behaviour,* 35: 691–704.

Baker, R.R. 1989. *Human Navigation and Magneto-Reception.* Manchester University Press, Manchester, 305 pp.

Baker, R.R. and M.A. Bellis. 1988. "Kamikaze" sperm in mammals. *Animal Behaviour,* 36: 936–939.

Baker, R.R. and M.A. Bellis. 1989. Elaboration of the kamikaze sperm hypothesis: a reply to Harcourt. *Animal Behaviour,* 37: 865–867.

Balda, R.P. and A.C. Kamil. 1989. A comparative study of cache recovery by three corvid species. *Animal Behaviour,* 38: 486–495.

Balduf, W.V. 1954. Observations on the White-Faced Wasp *Dolichovespula maculata* (Linn) (Vespidae, Hymenoptera). *Annals of the Entomological Society of America,* 47: 445–458.

Ball, J.A. 1973. The zoo hypothesis. *Icarus,* 19: 347–349.

Barash, D.P. 1982. *Sociobiology and Behavior,* 2nd ed. Elsevier, New York, 426 pp.

Barber, H.S. and F.A McDermott. 1951. North American fireflies of the genus *Photuris. Smithsonian Miscellaneous Collections,* Smithsonian Institution, Washington, D.C., 117(1): i–58.

Barber, J.T., E.G. Ellgaard, L.B. Thien, and A.E. Stack. 1989. The use of tools for food transportation by the Imported Fire Ant, *Solenopsis invicta. Animal Behaviour,* 38: 550–552.

Barinaga, M. 1992a. "African Eve" backers beat a retreat. *Science,* 255: 686–687.

Barinaga, M. 1992b. Evolutionists wing it with a new fossil bird. *Science,* 255: 796.

Barinaga, M. 1998. Study suggests new way to gauge prostate cancer risk. *Science,* 279: 475.

Barlow, C. and T. Volk. 1992. Gaia and evolutionary biology. *BioScience,* 42: 686–692.

Barnard, M.M. 1981. The coincidences of mimicries and other misleading coincidences. *American Naturalist,* 117: 372–378.

Barnes, R.D. 1974. *Invertebrate Zoology,* 3rd ed. W.B. Saunders, Philadelphia, PA, 870 pp.

Barnett, S.A. 1958. An analysis of social behaviour in wild rats. *Proceedings of the Zoological Society of London,* 130: 107–152.

Barrington, E.J.W., Ed. 1979. *Hormones and Evolution.* Academic Press, New York, pp. 1–491 (Vol. 1), pp. 492–989 (Vol. 2).

Barron, A.B. and S.A. Corbet. 1999. Preimaginal conditioning in *Drosophila* revisited. *Animal Behaviour,* 58: 621–628.

Barrows, E.M. 1974. Aggregation behavior and response to sodium chloride in females of a solitary bee, *Augochlora pura* (Hymenoptera: Halictidae). *Florida Entomologist,* 57: 189–193.

Barrows, E.M. 1975a. Individually distinctive odors in an invertebrate. *Behavioral Biology,* 15: 57–64.

Barrows, E.M. 1975b. Mating behavior in halictine bees (Hymenoptera: Halictidae): III. Copulatory behavior and olfactory communication. *Insectes Sociaux,* 22: 307–322.

Barrows, E.M. 1976a. Mating behavior in halictine bees (Hymenoptera: Halictidae): I. Patrolling and age-specific behavior in males. *Journal of the Kansas Entomological Society,* 49: 105–119.

Barrows, E.M. 1976b. Mating behavior in halictine bees (Hymenoptera: Halictidae): II. microterritorial and patrolling behavior in ♂ ♂ of *Lasioglossum rohweri. Zeitschrift für Tierpsychologie,* 40: 377–389.

Barrows, E.M. 1976c. Nectar robbing and pollination of *Lantana camara* (Verbenaceae). *Biotropica,* 8: 132–135.

Barrows, E.M. 1979a. Flower biology and arthropod associates of *Lilium philadelphicum. Michigan Botanist,* 18: 109–116.

Barrows, E.M. 1979b. *Polistes* wasps (Hymenoptera: Vespidae) show interference competition with other insects for *Kermes* scale insect (Homoptera: Kermesidae) secretions. *Proceedings of the Entomological Society of Washington,* 81: 570–575.

Barrows, E.M. 1982. Observation, description, and quantification of behavior: a study of praying mantids, pp. 8–20, 242–246 in J.R. and R.W. Matthews, Eds., *Behavioral Biology: Laboratory and Field Investigations with Insects.* Westview Press, Boulder, CO.

Barrows, E.M. 1983. Male territoriality in the carpenter bee *Xylocopa virginica virginica. Animal Behaviour,* 31: 806–813.

Barrows, E.M. and G. Gordh. 1978. Sexual behavior in the Japanese Beetle, *Popillia japonica,* and comparative notes on sexual behavior of other scarabs (Coleoptera: Scarabaeidae). *Behavioral Biology,* 23: 341–354.

Barrows, E.M., G.B. Chapman, J.E. Zenel, and A.S. Blake. 1986. Ultrastructure of the Dufour's gland of the Horn-Faced Bee, *Osmia cornifrons* (Hymenoptera: Megachilidae). *Journal of the Kansas Entomological Society,* 59: 480–493.

Bastock, M., D. Morris, and M. Moynihan. 1953. Some comments on conflict and thwarting in animals. *Behaviour,* 6: 66–84.

Bates, R. L. and J. A. Jackson, Eds. 1984. *Dictionary of Geological Terms,* 3rd ed. Doubleday, New York, 571 pp.

Bateson, G. 1955. A theory of play and fantasy. *Psychiatric Research Reports,* 2: 39–51.

Bateson, P., Ed. 1983. *Mate Choice.* Cambridge University Press, Cambridge, U.K., 462 pp.

Bateson, P. 1991. Assessment of pain in animals. *Animal Behaviour,* 42: 827–839.

Batra, S.W.T. 1966. Nests and social behavior of halictine bees from India. *Indian Journal of Entomology,* 18: 300–318.

Batra, S.W.T. 1987. Deceit and corruption in the blueberry patch. *Natural History,* 96(8): 56–59.

Batra, S.W.T. 1993a. Male-Fertile Potato flowers are selectively buzz-pollinated only by *Bombus terricola* Kirby in upstate New York. *Journal of the Kansas Entomological Society,* 66(2): 252–254.

Batra, S.W.T. 1993b. Opportunistic bumble bees congregate to feed at rare, distant alpine honeydew bonanzas. *Journal of the Kansas Entomological Society,* 66: 125–127.

Batra, S.W.T. 1994. Diversify with pollen bees. *American Bee Journal,* 134: 591–593.

Batra, S.W.T. and L.R. Batra. 1985. Floral mimicry induced by mummy-berry fungus exploits host's pollinators as vectors. *Science,* 228: 1011–1013.

Batschelet, E. 1965. *Statistical Methods for the Analysis of Problems in Animal Orientation and Certain Biological Rhythms.* American Institute of Biological Sciences, Washington, D.C., 57 pp.

Beach, F.A. 1978. Sociobiology and interspecific comparisons of behavior, pp. 116–135 in M.S. Gregory, A. Silver, and D. Sutch, Eds., *Sociobiology and Human Nature.* Jossey-Bass, San Francisco, CA, 326 pp.

Beach, F.A. 1979. Animal models and psychological inference, pp. 98–112 in H.A. Katchadourian, Ed., *Human Sexuality: A Comparative and Developmental Perspective.* University of California Press, Berkeley, 358 pp.

Beani, L. and S. Turillazzi. 1999. Stripes display in hover-wasps (Vespidae: Stenogastrinae): a socially costly status badge. *Animal Behaviour,* 57: 1233–1239.

Beatty, J. 1992a. Fitness: theoretical contexts, pp. 115–119 in E.F. Keller and E.A. Lloyd, Eds., *Keywords in Evolutionary Biology.* Harvard University Press, Cambridge, MA, 414 pp.

Beatty, J. 1992b. Random drift, pp. 273–281 in E.F. Keller and E.A. Lloyd, Eds., *Keywords in Evolutionary Biology.* Harvard University Press, Cambridge, MA, 414 pp.

Bechara, A., H. Damasio, D. Tranel, and A.R. Damasio. 1997. Deciding advantageously before knowing the advantageous strategy. *Science,* 275: 1293–1295.

Becker, H. 1997. Floral gems. *Agricultural Research,* September: 8–13.

Beissinger, S.R. and N.F.R. Synder. 1987. Mate desertion in the snail kite. *Animal Behaviour,* 36: 477–487.

Bell, C.R. and B.J. Taylor. 1982. *Florida Wild Flowers and Roadside Plants.* Laurel Hill Press, Chapel Hill, NC, 308 pp.

Bell, G. 1982. *The Masterpiece of Nature.* University of California Press, Berkeley, 635 pp.

Belluck, P. 1999. Board for Kansas deletes evolution from curriculum. *New York Times Northeast,* 12 August: A1, A13.

Ben-David, M., T.A. Hanley, and D.M. Schell. 1998. Fertilization of terrestrial vegetation by spawning Pacific salmon: the role of flooding and predator activity. *Oikos,* 83: 47–55.

Benhamou, S. and P. Bovet. 1989. How animals use their environment: a new look at kinesis. *Animal Behaviour,* 38: 375–383.

Benton, D. and V. Wastell. 1986. Effects of androstenol on human sexual arousal. *Biological Psychology,* 22: 141–147.

Benton, M.J. 1990. *Vertebrate Palaeontology: Biology and Evolution.* Harper Collins Academic, London, 377 pp.

Benton, M.J. 1991. *The Rise of the Mammals.* Crescent Book, New York, 144 pp.

Benton, M.J. 1995. Diversification and extinction in the history of life. *Science,* 268: 52–58.

Berenbaum, M. 1996. I'm okay — are you o.k.? *American Entomologist,* 42(1): 5–6.

Bergamini, D. 1964. *The Land and Wildlife of Australia.* Time, Inc., New York, 198 pp.

Berger, J.O. and D.A. Berry. 1988. Statistical analysis and the illusion of objectivity. *American Scientist,* 76: 159–165.

Berghe, E.P. van den, F. Wernerus, and R.R. Warner. 1989. Female choice and mating cost of peripheral males. *Animal Behaviour,* 38: 875–884.

Bernstein, H., H.C. Byerly, F.A. Hopf, and R.E. Michod. 1985. Genetic damage, mutation, and the evolution of sex. *Science,* 229: 1277–1281.

Berry, F. and W.J. Kress. 1991. *Heliconia: An Identification Guide.* Smithsonian Institution Press, Washington, D.C. 334 pp.

Birch, L.D. 1957. The meanings of competition. *American Naturalist,* 91: 5–18.

Birkhead, T.R., K.E. Lee, and P. Young. 1988. Sexual cannibalism in the praying mantis *Hierodula membranacea. Behaviour,* 106 (1,2): 112–118.

Bisazza, A. and A. Marconato. 1988. Female mate choice, male-male competition and parental care in the river bullhead, *Cottus gobio* L. (Pisces, Cottidae). *Animal Behaviour,* 36: 1352–1360.

Blackburn, I.M. 1984. Cognitive approaches to clinical psychology, pp. 290–319 in J. Nicholson and H. Beloff, Eds., *Psychology Survey 5.* Allen and Unwin, London. 425 pp.

Blakely, R.D. and J.T. Coyle. 1998. The neurobiology of *N*-acetyaspartylglutamate. *International Review of Neurobiology,* 30: 39–100.

Blakeslee, S. 1993. Human nose may hold an additional organ for a real sixth sense. *New York Times,* 7 September: C3.

Blakeslee, S. 1996. Researchers track down a gene that may govern spatial abilities. *New York Times,* 23 July: C3.

Blakeslee, S. 1999. Thanks to a "horrible worm," new ideas on hemoglobin. *New York Times Northeast,* 5 October: D2.

Bland, K.P., and B.M. Jubilan. 1987. Correlation of flehmen by male Sheep with female behaviour and oestrus. *Animal Behaviour,* 35: 735–738.

Blattner, F.R., G. Plunkett III, C.A. Bloch, N.T. Perna, V. Burland, M. Riley, J. Collado-Vides, J.D. Glasner, C.K. Rode, G. Mayhew et al. 1997. The complete genome sequence of *Escherichia coli* K–12. *Science,* 277: 1453–1462.

Blaustein, A.R., M. Beckoff, J.A. Byers, and T.J. Daniels. 1991. Kin recognition in vertebrates: what do we really know about adaptive value? *Animal Behaviour,* 41: 1079–1083.

Bligh, J. and K.G. Johnson. 1973. Glossary of terms for thermal physiology. *Journal of Applied Physiology,* 35: 941–961.

Bliss, T.V.P. 1999. Young receptors make smart mice. *Nature,* 401: 25–26.

Bloom, F.E. 1997. Breakthroughs 1997 (editorial). *Science,* 378: 2029.

Blundell, A. 1999. Flowering of the forest. *Natural History,* July-August: 30–39.

Boake, C.R.B. 1983. Mating systems and signals in crickets, pp. 28–44 in D.T. Gwynne and G.K. Morris, Eds., *Orthopteran Mating Systems: Sexual Competition in a Diverse Group of Insects.* Westview Press, Boulder, CO, 376 pp.

Boake, C.R.B. 1984. Male displays and female preferences in the courtship of a gregarious cricket. *Animal Behaviour,* 32: 690–967.

Boaz, N.T. 1984. Evolution myths (book review). *BioScience,* 34(1): 54–55.

Bolin, B., J. Canadell, B. Moore III, I. Noble, and W. Steffen. 1999. Effect on the biosphere of elevated atmospheric CO_2. *Science,* 285: 1851–1852.

Bonner, J.T. 1988. *The Evolution of Complexity.* Princeton University Press, Princeton, NJ, 260 pp.

Booth, W. 1990. Scientists follow scent to underarm discovery. *Washington Post,* 28 August: A1, A5.

Booth, W. 1992. Squabble in Margaritaville over the Manatee: conservation club's fate has songwriter Buffett, Florida Audubon Society at odds. *Washington Post,* 3 May: A3.

Booth, W., M. Gladwell, and K. Sawyer. 1989. Crossing distant species. *Washington Post,* 24 July: A2.

Bordas, M.L. 1894. Appareil glandulaire des Hyménoptères (Glandes salivaires, tube digestif, tubes de Malphigi et glandes venimeuses). Thèses Présentées à la Faculté des Sciences de Paris pour Obtenir Le Grade de Docteur ès Sciences Naturelles. Librairie de L'Académie de Médecine, Paris. Série A. No. 217. 362 pp.

Borden, J.H. 1974. Aggregation pheromones in the Scolytidae, pp. 135–160 in M.C. Birch, Ed., *Pheromones.* North Holland, Amsterdam, 495 pp.

Borgia, G. and M.A. Gore. 1986. Feather stealing in the satin bowerbird (*Ptilonorhynchus violacea*): male competition and the quality of display. *Animal Behaviour,* 34: 727–738.

Boroughs, D. 1999. Battle for a wild garden. *International Wildlife,* May-June: 12–21.

Borror, D.J., C.A. Triplehorn, and N.F. Johnson. 1989. *An Introduction to the Study of Insects,* 6th ed. W.B. Saunders, Philadelphia, PA, 875 pp.

Boss, A.P. 1986. The origin of the Moon. *Science,* 231: 341–345

Boucher, D.H. 1992. Mutualism and cooperation, pp. 208–211 in E.F. Keller and E.A. Lloyd, Eds., *Keywords in Evolutionary Biology.* Harvard University Press, Cambridge, MA, 414 pp.

Bourke, A.F.G. 1991. Queen behaviour, reproduction and egg cannibalism in multiple-queen colonies of the ant *Leptothorax acervorum. Animal Behaviour,* 42: 295–310.

Bowers, J.M. and B.K. Alexander. 1967. Mice: individual recognition by olfactory cues. *Science,* 158: 1208–1210.

Bowler, P.J. 1992. Lamarckism, pp. 188–193 in E.F. Keller and E.A. Lloyd, Eds., *Keywords in Evolutionary Biology.* Harvard University Press, Cambridge, MA, 414 pp.

Bowmaker, J.K. and A. Knowles. 1977. The visual pigments and oil droplets of the chicken retina. *Vision Research,* 17: 755–764.

Bowring, S.A., J.P. Grotzinger, C.E. Isachsen, A.H. Knoll, S.M. Pelechaty, and P. Kolosov. 1993. Calibrating rates of early Cambrian evolution. *Science,* 261: 1293–1298.

Boyce, M.S. 1990. The Red Queen visits sage grouse leks. *American Zoologist,* 30: 263–270.

Boyle, H. and E.M. Barrows. 1978. Oviposition and host feeding behavior of *Aphelinus asychis* (Hymenoptera: Chalcidoidea: Aphelinidae) on *Schizaphis graminum* (Homoptera: Aphidae) and some reactions of aphids to this parasite. *Proceedings of the Entomological Society of Washington,* 80: 441–455.

Brace, C.L. and A. Montagu. 1977. *Human Evolution,* 2nd ed. Macmillan, New York, 493 pp.

Bradley, D. 1993. Frog venom cocktail yields a one-handed painkiller. *Science,* 261: 1117.

Bradshaw, A.D. 1965. Evolutionary significance of phenotypic plasticity in plants, pp. 115–155 in E.W. Caspari and J.M Thoday, Eds., *Advances in Genetics,* Vol. 13. Academic Press, New York, 378 pp.

Bradshaw, H.D., Jr., S.M. Wilbert, K.G. Otto, and D.W. Schemske. 1995. Genetic mapping of floral traits associated with reproductive isolation in monkeyflowers (*Mimulus*). *Nature,* 376: 762–765.

Bradshaw, J.W.S. 1986. Mere exposure reduces cats' neophobia to unfamiliar food. *Animal Behaviour,* 34: 613–614.

Bradshaw, J.W.S., P.E. Howse, and R. Baker. 1986. A novel autostimulatory pheromone regulating transport of leaves in *Atta cephalotes. Animal Behaviour,* 34: 234–240.

Bradsher, K. 1999. Daughter's death prompts fight on "date rape" drug. *New York Times Northeast,* 16 October: A8.

Brandon, R.N. 1992. Environment, pp. 81–86 in E.F. Keller and E.A. Lloyd, Eds., *Keywords in Evolutionary Biology.* Harvard University Press, Cambridge, MA, 414 pp.

Brandt, C.A. 1989. Mate choice and reproductive success of pikas. *Animal Behaviour,* 37: 118–132.

Breiehagen, T. and T. Slagsvold. 1988. Male polyterritoriality and female-female aggression in pied flycatchers *Ficedula hypoleuca. Animal Behaviour,* 36: 604–605.

Brémond, J.-C. and T. Aubin. 1990. Responses to distress calls by black-headed gulls, *Larus ridibundus*: the role of non-degraded features. *Animal Behaviour,* 39: 503–511.

Brewer, J.W. 1981. The Dr. Fox effect: studies on the validity of student evaluations. *Bulletin of the Entomological Society of America,* 27: 121–123.

Brickell, C. and J.C. Zuk, Eds. 1996. *The American Horticultural Society A–Z Encyclopedia of Garden Plants.* DK Publishing, New York, 1092 pp.

Broad, W.J. 1993a. A voyage into the abyss: gloom, gold and Godzilla. *New York Times,* 2 November: C1, C12.

Broad, W.J. 1993b. Strange oases in sea depths offer map to riches. *New York Times,* 16 November: C1, C15.

Broad, W.J. 1999. The diverse creatures of the deep may be starving. *New York Times,* 1 June: D5.

Brockman, C.F. 1968. *Trees of North America.* Golden Press, New York, 280 pp.

Brockmann, H. J., and A. Grafen. 1989. Mate conflict and male behaviour in a solitary wasp, *Trypoxylon (Trypargilum) politum* (Hymenoptera: Sphecidae). *Animal Behaviour,* 37: 232–255.

Brodie, E.D., Jr. 1989. *Venomous Animals.* Golden Press, Racine, WI, 160 pp.

Brodie, E.D. III. 1990. Genetics of the garter's getaway. *Natural History,* July: 44–51.

Brody, J.E. 1997. The too-common crow, too close for comfort. *New York Times,* 27 May: C1, C6.

Brody, J.E. 1998. Menopause begins silently, before its symptoms. *New York Times Northeast,* 27 January: C7.

Brody, J.E. 1999. Sly parasite menaces pets and their owners. *New York Times Northeast,* 21 December: D7.

Brosius, J. and S.J. Gould. 1992. On "genomenclature:" a comprehensive (and respectful) taxonomy for pseudogenes and other "junk DNA." *Proceedings of the National Academy of Science USA,* 89: 10706–10710.

Brower, L.P. and J.V.Z. Brower. 1972. Parallelism, convergence, divergence, and the new concept of advergence in the evolution of mimicry. *Connecticut Academy of Arts and Sciences Transactions,* 44: 59–67.

Brown, D. 1995. Arches, loops and whorls make your mark on the world unique. *Washington Post,* 3 July: A3.

Brown, D. 1999a. Microbe's map of migration. *Washington Post,* 9 August 1999: A7.

Brown, G.E., J.-G.J. Godin, and J. Pedersen. 1999. Fin-flicking behaviour: a visual antipredator alarm signal in a characin fish, *Hemigrammus erythrozonus. Animal Behaviour,* 58: 469–475.

Brown, J.L. 1975. *The Evolution of Behavior*. Norton, New York, 761 pp.

Brown, J.L. 1987a. *Helping and Communal Breeding in Birds: Ecology and Evolution*. Princeton University Press, Princeton, NJ, 354 pp.

Brown, J.L., E.R. Brown, S.D. Brown, and D.D. Dow. 1982. Helpers: effects of experimental removal on reproduction success. *Science*, 215: 421–422.

Brown, K.S. 1999b. Deep Green rewrites evolutionary history of plants. *Science:* 285: 990–991.

Brown, M.L. and R.G. Brown. 1984. *Herbaceous Plants of Maryland*. Port City Press, MD, 1127 pp.

Brown, R.G. and M.L. Brown. 1972. *Woody Plants of Maryland*. Port City Press, MD, 347 pp. (1992 reprint).

Brown, R.W. 1956. *Composition of Scientific Words*. Smithsonian Institution Press, Washington, D.C., 882 pp.

Brown, S.M. 1986. Of mantises and myths. *BioScience*, 36(7): 421–423.

Brown, W.L., Jr. 1987b. Punctuated equilibrium excused: the original examples fail to support it. *Biological Journal of the Linnaean Society*, 31: 383–404.

Browne, M.W. 1993. "Handedness" seen in nature, long before hands. *New York Times*, 15 June: C1, C10.

Browne, M.W. 1996a. Second greatest toolmaker? A title crows can crow about. *New York Times Northeast*, 30 January: C1, C8.

Browne, M.W. 1996b. Mass extinction of Permian Era linked to a gas. *New York Times*, 30 July: C1, C10.

Browne, M.W. 1996c. Planetary experts say Mars life is still speculative. *New York Times*, 8 August: D20.

Browne, M.W. 1998. Who needs jokes? Brain has a ticklish spot. *New York Times Northeast*, 10 March: C1.

Brownlee, A. 1981. *Biological Complementariness: A Study of Evolution*. A. Brownlee, Edinburgh. 133 pp.

Brush, G.S., C. Lenk, and J. Smith. 1976. *Vegetation Map of Maryland: The Existing Natural Forests*. Department of Geography and Environmental Engineering, The Johns Hopkins University, Baltimore, MD, 1 p.

Brush, J.S. and P.M. Narins. 1989. Chorus dynamics of a neotropical amphibian assemblage: comparison of computer simulation and natural behavior. *Animal Behaviour*, 37: 33–44.

Buchmann, S.L. and J.H. Cane. 1989. Bees assess pollen returns while sonicating *Solanum* flowers. *Oecologia*, 81: 289–294.

Buchmann, S.L. and G.P. Nabhan. 1996. *The Forgotten Pollinators*. Island Press, Washington, D.C., 292 pp.

Buck, J. and E. Buck. 1966. Biology of synchronous flashing of fireflies. *Nature*, 211: 562–564.

Buck, J. and E. Buck. 1976. Synchronous fireflies. *Scientific American*, 234(5): 74–85.

Buck, J. and E. Buck. 1978. Toward a functional interpretation of synchronous flashing by fireflies. *American Naturalist*, 112: 471–492.

Bugos, G.E. 1993. Hot neologisms (book review). *Science*, 262: 121–122.

Bull, J.J. 1995. (R)evolutionary medicine (book review). *Evolution*, 49: 1296–1298.

Bull, J. and J. Farrand, Jr. 1998. *National Audubon Society Field Guide to North American Birds: Eastern Region*. Alfred A. Knopf, New York, 797 pp.

Bult, C.J. et al. 1996. Complete genome sequence of the methanogenic archaeon, *Methanococcus jannaschii. Science,* 273: 1058–1073.

Bunge, M.A. 1984. *The Mind-Body Problem: A Psychological Approach.* Pergamon Press, New York, 250 pp.

Buren, W.F. 1958. A review of the species of *Crematogaster,* sensu stricto, in North America (Hymenoptera: Formicidae), Part I. *Journal of the New York Entomological Society,* 66:119–134.

Burghardt, G.M. 1970. Defining "communication," in J.W. Johnston, Jr., D.G. Mouton, and A. Turk, Eds., *Advances in Chemoreception.* Vol. 1. *Communication by Chemical Signals.* Appleton-Century-Crofts, New York, 412 pp.

Burghardt, G.M. 1985. Animal awareness: current perceptions and historical perspective. *American Psychologist,* 40: 905–919.

Burghardt, G.M. and H.W. Greene. 1988. Predator simulation and duration of death feigning in neonate hognose snakes. *Animal Behaviour,* 36: 1842–1844.

Burian, R.M. 1992. Adaptation: historical perspectives, pp. 7–12 in E.F. Keller and E.A. Lloyd, Eds., *Keywords in Evolutionary Biology.* Harvard University Press, Cambridge, MA, 414 pp.

Burk, T. 1982. Evolutionary significance of predation on sexually signalling males. *The Florida Entomologist,* 65: 90–104.

Burney, D.A. 1993. Recent animal extinctions: recipes for disaster. *American Scientist,* 81: 530–541.

Burns, J.T., K.M. Cheng, and F. McKinney. 1980. Forced copulation in captive Mallards. I. Fertilization of eggs. *Auk,* 97: 875–879.

Burt, W.H. 1980. *A Field Guide to the Mammals of North America North of Mexico,* 3rd ed. Houghton Mifflin, Boston, MA, 289 pp.

Buschsbaum, R., M. Buschsbaum, J. Pearse, and V. Pearse. 1987. *Animals Without Backbones,* 3rd ed. University of Chicago Press, Chicago, IL, 572 pp.

Butler, C. 1609. *The Feminine Monarchie.* Da Cap Press, New York, 200 pp.

Butler, D. 1999. Venter's *Drosophila* "success" set to boost human genome efforts. *Nature,* 401: 729–730.

Cabana, G. and D.L. Kramer. 1991. Random offspring mortality and variation in parental fitness. *Evolution,* 45: 228–234.

Cairns, J., J. Overbaugh, and S. Miller. 1988. The origin of mutants. *Nature,* 335: 142–145.

Caldwell, M. 1997. The wired butterfly. *Discover,* February: 40–48.

Cale, W.G., G.M. Henebry, and J.A. Yeakley. 1989. Inferring process from pattern in natural communities. *BioScience,* 39: 600–605.

Caley, M.J. and S.A. Boutin. 1987. Sibling and neighbour recognition in wild juvenile muskrats. *Animal Behaviour,* 35: 60–66.

Calhoun, C. and R. Solomon, Eds. 1984. *What Is an Emotion: Classic Readings in Philosophical Psychology.* Oxford University Press, New York, 358 pp.

Calman, W.T. and W.L. Schmitt. 1973. Crustacea, pp. 836–842 in W.E. Preece, Ed., *Encyclopedia Britannica,* Vol. 6. William Benton, Publisher, London, 996 pp.

Campbell, B. and E. Lack, Eds. 1985. *A Dictionary of Birds.* Buteo Books, Vermillion, SD, 670 pp.

Campbell, N.A. 1987. *Biology.* Benjamin/Cummings, Menlo Park, CA, 1101 pp. + appendices.

Campbell, N.A. 1990. *Biology,* 2nd ed. Benjamin/Cummings, Menlo Park, CA, 1165 pp. + appendices.

Campbell, N.A. 1993. *Biology,* 3rd ed. Benjamin/Cummings, Menlo Park, CA, 1190 pp. + appendices.

Campbell, N.A. 1996. *Biology,* 4th ed. Benjamin/Cummings, Menlo Park, CA, 1206 pp. + appendices.

Campbell, N.A., J.B. Reece, and L.G. Mitchell. 1999. *Biology,* 5th ed. Benjamin/Cummings, Menlo Park, CA, 1175 pp. + appendices.

Cann, R.L., M. Stoneking, and A.C. Wilson. 1987. Mitochondrial DNA and human evolution. *Nature,* 325: 31–36.

Carlet, G. 1884. Sur le venin des Hyménoptères et ses organes sécréteurs. *C.R. Academy of Science (Paris),* 7: 403–410.

Carlet, G. 1890. Mémoire sur le venin et l'aiguillon de l'abeille. *Annales des Sciences Naturelles, Zoologie et Biologie Animale,* 9: 1–17.

Caro, T.M. 1986. The functions of stotting: a review of hypotheses. *Animal Behaviour,* 34: 649–662.

Caro, T.M. 1988. Why do Tommies stott? *Natural History,* 97(9): 26–30.

Caro, T.M. 1989. The brotherhood of Cheetahs. *Natural History,* 98(6): 50–59.

Carpenter, C.C. 1952a. Growth and maturity of three species of *Thamnophis* in Michigan. *Copeia,* 4: 237–243.

Carpenter, C.C. 1952b. Comparative ecology of the Common Garter Snake (*Thamnophis s. sertalis*) and Butler's Garter Snake (*Thamnophis butleri*) in mixed populations. *Ecological Monographs,* 22: 235–258.

Carrol, L. 1932. *Alice's Adventures in Wonderland, Through the Looking-Glass, and The Hunting of the Snark,* National Home Library Foundation, Washington, D.C., 262 pp.

Carson, H.L. 1989. Evolution: the pattern or the process (book review). *Science,* 245: 872–873.

Cavalier-Smith, T. 1980. How selfish is DNA? *Nature,* 285:617–618.

Cech, T.R. 1986. RNA as an enzyme. *Scientific American,* 255(5): 64–75.

Champalbert, A., and J.-P. Lachaud. 1990. Existence of a sensitive period during the ontogenesis of social behaviour in a primitive ant. *Animal Behaviour,* 39: 850–859.

Chapin, J.P. 1968. *Dictionary of Psychology.* Dell Publishing, New York, 537 pp.

Chapleau, F, P.H. Johansen, and M. Williamson. 1988. The distinction between pattern and process in evolutionary biology: the use and abuse of the term "strategy." *Oikos,* 53: 136–138.

Chapman, R.F. 1971. *The Insects: Structure and Function.* Elsevier, New York, 819 pp.

Chan, J.M., M.J. Stampfer, E. Giovannucci, P.H. Gann, J. Ma, P. Wilkinson, C.H. Hennekens, and M. Pollak. 1998. Plasma insulin-like growth factor-I and prostate cancer risk: a prospective study. *Science,* 279: 563–566.

Chance, M.R.A. 1956. Social structure of a colony of *Macaca mulatta. British Journal of Animal Behaviour,* 4: 1–13.

Chase, I.D., M. Weissburg, and T.H. DeWitt. 1988. The vacancy chain process: a new mechanism of resource distribution in animals with application to hermit crabs. *Animal Behaviour,* 36: 1265–1274.

Chazdon, R.L., R.K. Colwell, and J.S. Denslow. 1999. Tropical tree richness and resource-based niches. *Science,* 285: 1459 (www.sciencemag.org/cgi/content/full/285/533/1459a).

Cheal, M. 1975. Social olfaction: a review of the ontogeny of olfactory influences on vertebrate behavior. *Behavioral Biology,* 15: 1–25.

Cheney, D.L. and R.M. Seyfarth. 1986. The recognition of social alliances by Vervet Monkeys. *Animal Behaviour,* 34: 1722–1731.

Cheng, V. 1995. Fatigue syndrome is tracked in study. *New York Times,* 18 July: C9.

Chew, V. 1985. Number of replicates in experimental research. *Southwestern Entomologist,* 6: 2–9.

Choate, E.A. 1985. *The Dictionary of American Bird Names,* rev. ed. Harvard Common Press, Boston, MA, 226 pp.

Choukas-Bradley, M. and P. Alexander. 1987. *City of Trees: The Complete Field Guide to the Trees of Washington, D.C.,* rev. ed. The Johns Hopkins University Press, Baltimore, MD, 354 pp.

Chu, E. 1985. Where is the *co-* in coevolution? *BioScience,* 35: 622–623.

Claiborne, W. 1997a. Early El Niño blows hot and cold: weird weather brings good news for some species, ill winds for others. *Washington Post,* 4 July: A14.

Claiborne, W. 1997b. Trouble in paradise? Serpentless Hawaii fears snake invasion. *Washington Post,* 23 August: A1.

Clapperton, B.K. 1989. Scent-marking behaviour of the ferret, *Mustela furo* L. *Animal Behaviour,* 38: 436–446.

Clark, C.W. 1993. Dynamic models of behavior: an extension of life history theory. *Trends in Ecology and Evolution,* 8: 205–209.

Clark, J.S., C. Fastie, G. Hurtt, S.T. Jackson, C. Johnson, G.A. King, M. Lewis, J. Lynch, S. Pacala, C. Prentice, E.W. Schupp, T. Webb III, and P. Wyckoff. 1998. Reid's paradox of rapid plant migration. *BioScience,* 48: 12–24.

Clarke, M.F. and D.L. Kramer. 1994. Scatter hoarding by a larder-hoarding rodent: intraspecific variation in the hoarding behaviour of the Eastern Chipmunk, *Tamias striatus. Animal Behaviour,* 48: 299–308.

Clayman, C.B., Ed. 1989. *The American Medical Association Home Medical Encyclopedia.* Random House, New York, pp. 1–565 (Vol. 1), pp. 566–1184 (Vol. 2).

Clayton, D.H. and N.D. Wolfe. 1993. The adaptive significance of self-medication. *Trends in Ecology and Evolution,* 8: 60–63.

Clayton, N.S. 1988. Song learning and mate choice in estrildid finches raised by two species. *Animal Behaviour,* 36: 1589–1600.

Clewell, D.B. 1985. Sex pheromones, plasmids, and conjugation in *Streptococcus faecalis,* pp. 13–28 in H.O. Halvorson and A. Monroy, Eds., *The Origin and Evolution of Sex.* Alan R. Liss, New York, 345 pp.

Clode, D. 1994. Reply from D. Clode. *Trends in Ecology and Evolution,* 9: 25.

Cockburn, A. and A.K. Lee. 1988. Marsupial femmes fatales. *Natural History,* 97(3): 40–47.

Cohen, J. 1995. Genes and behavior make an appearance in the O.J. trial. *Science,* 268: 22–23.

Cohen, J.E. and D. Tilman. 1996. Biosphere 2 and biodiversity: the lessons so far. *Science,* 274: 1150–1151.

Cohn, D'V. 1999. A purr-fect debut at the National Zoo. *Washington Post,* 30 September: B1, B4.

Cohn, J.P. 1996. Creatures of the damp. *Washington Post,* 11 September: H1, H6.

Colbert, E.H. 1980. *The Evolution of the Vertebrates: A History of the Backboned Animals Through Time.* Wiley, New York, 510 pp.

Collias, N.E. 1944. Aggressive behavior among vertebrate animals. *Physiological Zoology,* 17: 83–123.

Collocot, T.C. and A.B. Dobson, Eds. 1974. *Dictionary of Science and Technology,* rev. ed. W. & R. Chambers, Edinburgh, 1328 pp.

Colmenares, F. 1991. Greeting behaviour between male baboons: oestrous females, rivalry, and negotiation. *Animal Behaviour,* 41: 49–60.

Colwell, R.K. 1992. Niche: a bifurcation in the conceptual lineage of the term, pp. 241–248 in E.F. Keller and E.A. Lloyd, Eds., *Keywords in Evolutionary Biology.* Harvard University Press, Cambridge, MA, 414 pp.

Conant, R. 1958. *A Field Guide to the Reptiles and Amphibians.* Houghton Mifflin, Boston, MA, 366 pp.

Connell, J.H. 1978. Diversity of tropical rain forests and coral reefs. *Science,* 199: 1302–1310.

Conner, D.A. 1982. Dialects versus geographic variation in mammalian vocalizations. *Animal Behaviour,* 30: 297–298.

Connor, R.C. 1986. Pseudo-reciprocity: investing in mutualism. *Animal Behaviour,* 34: 1562–1566.

Conrad, M. 1977. The thermodynamic meaning of ecological efficiency. *American Naturalist,* 111: 99–106.

Constable, P. 1998. And God said, let there be monkeys. *Washington Post,* 21 September: A18.

Constantz, G.D. 1984. Sperm competition in poeciliid fishes, pp. 465–485 in R.L. Smith, Ed., *Sperm Competition and the Evolution of Animal Mating Systems.* Academic Press, New York, 687 pp.

Conte, Y. Le, G. Arnold, J. Trouiller, C. Masson, B. Chappe, and G. Ourisson. 1989. Attraction of the parasitic mite *Varroa* to the drone larvae of Honey Bees by simple aliphatic esters. *Science,* 245: 638–640.

Convey, P. 1989. Post-copulatory guarding strategies in the non-territorial dragonfly *Sympetrum sanguineum* (Müller) (Odonata: Libellulidae). *Animal Behaviour,* 37: 56–63.

Cooke, F. 1986. Demonstrating natural selection (book review). *Science,* 233: 1332.

Cooper, M. 1999. Zoo's tortoises are stolen and on black market, they might be going fast. *New York Times,* 10 September: A23.

Cope, E.D. 1885a. On the evolution of the Vertebrata, progressive and retrogressive, Part 1. *American Naturalist,* 19: 140–148.

Cope, E.D. 1885b. On the evolution of the Vertebrata, progressive and retrogressive, Part 2. *American Naturalist,* 19: 234–247.

Cope, E.D. 1885c. On the evolution of the Vertebrata, progressive and retrogressive, Part 3. *American Naturalist,* 19: 341–353.

Corbet, S.A. 1971. Mandibular gland secretion of larvae of the Flour Moth, *Anagasta kuehniella,* contains an epideictic pheromone and elicits oviposition movements in a hymenopteran parasite. *Nature,* 232: 481–484.

Costa, J.T. 1997. Caterpillars as social insects. *American Scientist,* 85: 150–159.

Cotgreave, P. 1993. The relationship between body size and population abundance in animals. *Trends in Ecology and Evolution,* 8: 244–248.

Coulson, R.N. and J.A. Witter. 1984. *Forest Entomology, Ecology and Management.* Wiley Interscience, New York, 669 pp.

Council of Europe Commission of European Communities. 1987. *Map of the Natural Vegetation of Member Countries of the European Community and the Council of Europe,* 2nd ed. Office for Official Publications of the European Communities, Luxembourg, 80 pp. + 4 maps.

Counsilman, J.J. and L.M. Lim. 1985. The definition of weaning. *Animal Behaviour,* 33: 1023–1024.

Covell, C.V., Jr. 1984. *A Field Guide to the Moths of Eastern North America.* Houghton Mifflin, Boston, MA, 496 pp.

Cowan, D.F. and J. Atema. 1990. Moult staggering and serial monogamy in American lobsters, *Homarus americanus. Animal Behaviour,* 39: 1199–1206.

Cowie, A.T., S.J. Folley, B.A. Cross, G.W. Harris, D. Jacobsohn, and K.C. Richardson. 1951. Terminology for use in lactational physiology. *Nature,* 4271: 421.

Crawford, M.P. 1939. The social psychology of the vertebrates. *Psychological Bulletin,* 36: 407–446.

Crespi, B.J. 1989. Causes of assortative mating in arthropods. *Animal Behaviour,* 38: 980–1000.

Crewe, R.M. and M.S. Blum. 1970. Identification of the alarm pheromones of the ant *Myrmica bevinodis. Journal of Insect Physiology,* 16: 141–146.

Crichton, M. 1990. *Jurassic Park.* Alfred A. Knopf, New York, 402 pp.

Cronin, H. 1992. Sexual selection: historical perspectives, pp. 286–293 in E.F. Keller and E.A. Lloyd, Eds., *Keywords in Evolutionary Biology.* Harvard University Press, Cambridge, MA, 414 pp.

Crook, J. 1983. On attributing consciousness to animals. *Nature,* 303: 11–14.

Crooks, K.R. and M.E. Soulé. 1999. Mesopredator release and avifaunal extinctions in a fragmented system. *Nature,* 400: 563–566.

Crosland, M.W.J. 1989. Kin recognition in the ant *Rhytidoponera confusa.* I. Environmental odour. *Animal Behaviour,* 37: 912–919.

Crosland, M.W.J. 1989. Kin recognition in the ant *Rhytidoponera confusa.* II. Gestalt odour. *Animal Behaviour,* 37: 920–926.

Crossley, S.A. 1986. Courtship sounds and behaviour in the four species of the *Drosophila bipectinata* complex. *Animal Behaviour,* 34: 1146–1159.

Croswell, K. 1993. Return of the steady-state universe. *New Scientist,* 27 February: 14.

Crow, J.F. 1992. Genetic load, pp. 132–136 in E.F. Keller and E.A. Lloyd, Eds., *Keywords in Evolutionary Biology.* Harvard University Press, Cambridge, MA, 414 pp.

Culotta, E. 1995. New finds rekindle debate over anthropoid origins. *Science,* 268: 1851.

Cumming, J., Ed. 1972. *Encyclopedia of Psychology.* Herder and Herder, New York, 311 pp.

Curtis, H. 1983. *Biology,* 4th ed. Worth Publishers, New York, 1159 pp.

Daerr, E.G. 1999. Taking wing. *National Parks,* March/April: 39.

Daly, H.V. 1985. Insect morphometrics. *Annual Review of Entomology,* 30: 415–438.

Daly, M. and M. Wilson. 1983. *Sex, Evolution, and Behavior,* 2nd ed. Willard Grant Press, Boston, MA, 402 pp.

Damuth, J. 1992. Extinction, pp. 106–111 in E.F. Keller and E.A. Lloyd, Eds., *Keywords in Evolutionary Biology.* Harvard University Press, Cambridge, MA, 414 pp.

Darden, L. 1992. Character: historical perspectives, pp. 41–44 in E.F. Keller and E.A. Lloyd, Eds., *Keywords in Evolutionary Biology.* Harvard University Press, Cambridge, MA, 414 pp.

Darling, D.C., and L. Packer. 1988. Effectiveness of Malaise traps in collecting Hymenoptera: the influence of trap design, mesh size, and location. *Canadian Entomologist,* 120: 787–796.

Darling, F.F. 1938. *Bird Flocks and the Breeding Cycle: A Contribution to the Study of Avian Sociality.* Cambridge University Press, Cambridge, U.K., 124 pp.

Darton, J. 1996. A blooming let down at Kew: nothing to sniff at. *New York Times,* 1 August: A4.

Darwin, C.R. (1859) 1876. *On the Origin of Species by Means of Natural Selection or the Preservation of Favoured Races in the Struggle for Life.* John Murray, London, 458 pp.

Darwin, C.R. (1868) 1875. *The Variation in Animals and Plants Under Domestication,* Appleton & Co., New York, 413 pp. (Vol. 1), 446 pp. (Vol. 2).

Darwin, C.R. 1871. *The Descent of Man, and Selection in Relation to Sex.* Appleton & Co., New York, 553 pp.

Darwin, C.R. 1903. *More Letters of Charles Darwin.* Appleton & Co., New York, 494 pp. (Vol. 1), 508 pp. (Vol. 2).

Darwin, C.R. 1959. *On the Origin of Species by Means of Natural Selection or the Preservation of Favoured Races in the Struggle for Life,* a facsimile of the first edition with an introduction by Ernst Mayr. Harvard University Press, Cambridge, MA (reprinted in 1964).

Davies, N. 1978. Territorial defense in the Speckled Wood Butterfly (*Paragoa aegeria*): the resident always wins. *Animal Behaviour,* 26: 138–147.

Davies, N.B. 1989. Sexual conflict and the polygamy threshold. *Animal Behaviour,* 38: 226–234.

Davis, H. and M. Albert. 1986. Numerical discrimination by rats using sequential auditory stimuli. *Animal Learning and Behavior,* 14: 57–59.

Davis, H. and J. Memmott. 1983. Autocontingencies: rats count to three to predict safety from shock. *Animal Learning and Behavior,* 11: 95–100.

Dawkins, M.S. 1980. *Animal Suffering: The Science of Animal Welfare.* Chapman & Hall, London, 149 pp.

Dawkins, M.S. 1986. *Unravelling Animal Behaviour,* Longman Group, Ltd., Essex, 161 pp.

Dawkins, M.S. 1988. Book review. *Animal Behaviour,* 36: 316.

Dawkins, M.S. and T. Guilford. 1991. The corruption of honest signalling. *Animal Behaviour,* 41: 865–873.

Dawkins, R. 1976. *The Selfish Gene.* Oxford University Press, London, 224 pp.

Dawkins, R. 1982. *The Extended Phenotype: The Gene as The Unit of Selection.* W.H. Freeman, San Francisco, CA, 307 pp.

Dawkins, R. 1986. *The Blind Watchmaker.* W.W. Norton, New York, 332 pp.

Dawkins, R. 1992. Progress, pp. 263–272 in E.F. Keller and E.A. Lloyd, Eds., *Keywords in Evolutionary Biology.* Harvard University Press, Cambridge, MA, 414 pp.

Dawkins, R. and H.J. Brockmann. 1980. Do digger wasps commit the Concorde fallacy? *Animal Behaviour,* 28: 892–896.

Day, E.R. 1997. *Southern Pine Beetle Fact Sheet.* Entomology Publication 444-243. Insect Identification Laboratory, Entomology Department, Virginia Technical and State University, Blacksburg, VA.

Day, R.W. and G.P. Quinn. 1989. Comparisons of treatments after an analysis of variance in ecology. *Ecological Monographs,* 59: 433–463.

De Rocher, E.J., K.R. Harkins, D.W. Galbraith, and H.J. Bohnert. 1990. Developmentally regulated systemic endopolyploidy in succulents with small genomes. *Science,* 250: 99–101.

de Waal, F. 1989. *Peacemaking Among Primates.* Harvard University Press, Cambridge, MA, 294 pp.

de Waal, F. and F. Lanting. 1997. Bonobo dialogues. *Natural History,* May: 22–25.

Deag, J.M. and J.H. Crook. 1971. Social behaviour and "agonistic buffering" in the wild Barbary macaque *Macaca sylvana* L. *Folia Primatologia,* 15: 183–200.

DeGrazia, D. and A. Rowan. 2000. Animal pain, suffering, and anxiety (manuscript).

Delbridge, A., Ed. 1981. *The MacQuarie Dictionary.* MacQuarie Library Party, Ltd., South Australia, 2049 pp.

DeLucia, E.H., J.U.G. Hamilton, S.L. Naidu, R.B. Thomas, J.A. Andrews, A. Finzi, M. Lavine, R. Matamala, J.E. Mohan, G.R. Hendrey, and W.H. Schlesinger. 1999. Net primary production of a forest ecosystem with experimental CO_2 enrichment. *Science,* 284: 1177–1179.

Demarest, J. 1987. On changing the framework of psychology: comparative psychology is what general psychology should be. *Teaching of Psychology,* 14: 147–151.

Dempster, W.F. 1997. Biospherian viewpoints. *Science,* 275: 1247–1248.

Dennis, R.L.H. 1984a. Egg-laying sites of the Common Blue Butterfly, *Polyommatus icarus* (Rottemburg) (Lepidoptera: Lycaenidae): the edge effect and beyond the edge. *Entomologist's Gazette,* 35: 85–93.

Dennis, R.L.H. 1984b. The edge effect in butterfly oviposition: batch siting in *Aglais urticae* (L.) (Lepidoptera: Nymphalidae). *Entomologist's Gazette,* 35: 157–173.

Derr, M. 1999. Texas rescue squad comes to aid of Florida Panther. *New York Times Northeast,* 2 November: D2.

Desalle, R., J. Gatesy, W. Wheeler, and D. Grimaldi. 1992. DNA sequences from a fossil termite in Oligo-Miocene amber and their phylogenetic implications. *Science,* 257: 1933–1936.

Desportes, J.-P., N.B. Metcalfe, F. Cezilly, G. Lauvergeon, and C. Kervella. 1989. Tests of the sequential randomness of vigilant behaviour using spectral analysis. *Animal Behaviour,* 38: 771–777.

Dethier, V.G. 1970. Chemical interactions between plants and insects, pp. 83–102 in E.Sondheimer and J.B. Simeone, Eds., *Chemical Ecology.* Academic Press, New York.

Dethier, V.G., L.B. Browne, and C.N. Smith. 1960. The designation of chemicals in terms of the responses they elicit from insects. *Journal of Economic Entomology,* 53: 134–136.

DeVries, P.J. 1990. Enhancement of symbioses between butterfly caterpillars and ants by vibrational communication. *Science,* 248: 1104–1106.

Dewsbury, D.A. 1968. Comparative psychology and comparative psychologists: an assessment. *Journal of Biological Psychology,* 10: 35–38.

Dewsbury, D.A. 1978. *Comparative Animal Behavior.* McGraw-Hill, New York, 452 pp.

Dewsbury, D.A. 1984. Sperm competition in muroid rodents, pp. 547–571 in R.L. Smith, Ed., *Sperm Competition and the Evolution of Animal Mating Systems.* Academic Press, New York, 687 pp.

Diamond, J. 1987. Soft sciences are often harder than hard sciences. *Discover,* 8(8): 34–39.

Diamond, J. 1996. The best ways to sell sex. *Discover,* December: 78–85.

Dice, L.R. 1952. *Natural Communities.* University of Michigan Press, Ann Arbor, 547 pp.

Dickinson, J.L. and R.L. Rutowski. 1989. The function of the mating plug in the Chalcedon Checkerspot Butterfly. *Animal Behaviour,* 38: 154–162.

Didham, R.K. 1997. The influence of edge effects and forest fragmentation on leaf litter invertebrates in Central Amazonia, pp. 55–70 in W.F. Laurance and R.O. Bierregaard, Jr., Eds., *Tropical Forest Remnants. Ecology, Management, and Conservation of Fragmented Communities.* The University of Chicago Press, Chicago, IL, 616 pp.

Dietrich, M.R. 1992. Macromutation, pp. 194–201 in E.F. Keller and E.A. Lloyd, Eds., *Keywords in Evolutionary Biology.* Harvard University Press, Cambridge, MA, 414 pp.

Dobkin, D.S. 1979. Functional and evolutionary relationships of vocal copying phenomena in birds. *Zeitschrift für Tierpsychologie,* 50: 348–363.

Dobzhansky, T. 1962. *Mankind Evolving: The Evolution of the Human Species.* Yale University Press, New Haven, CT, 381 pp.

Dobzhansky, T., F.J. Ayala, G.L. Stebbins, and J.W. Valentine. 1977. *Evolution.* W.H. Freeman, San Francisco, CA, 572 pp.

Doebler, S. 2000. The dawn of the protein era. *BioScience,* 50: 15–20.

Dold, C. 1996. Hormone hell. *Discover,* September: 52–59.

Donoghue, M.J. 1992. Homology, pp. 170–179 in E.F. Keller and E.A. Lloyd, Eds., *Keywords in Evolutionary Biology.* Harvard University Press, Cambridge, MA, 414 pp.

Doolittle, R.F., D.-F. Feng, S. Tsang, G. Cho, and E. Little. 1996. Determining divergence times of the major kingdoms of living organisms with a protein clock. *Science,* 271: 470–477.

Doolittle, W.F. and C. Sapienza. 1980. Selfish genes, the phenotype paradigm and genome evolution. *Nature,* 284: 601–603.

Dorst, J. 1970. *A Field Guide to the Larger Mammals of Africa.* Houghton Mifflin, Boston, MA, 287 pp.

Douglas, M.M. 1981. Thermoregulatory significance of thoracic lobes in the evolution of insect wings. *Science,* 211: 84–86.

Dowling, C.G. 1997a. The visible man. *Life,* February: 33–47.

Dowling, C.G. 1997b. The brain. *Life,* February: 60–63.

Downing, H.A. and R.L. Jeanne. 1988. Nest construction by the paper wasp, *Polistes:* a test of stigmergy theory. *Animal Behaviour,* 36: 1729–1739.

Doyle, J.A. 1978. Origin of angiosperms. *Annual Review of Ecology and Systematics,* 9: 365–392.

Drake, M.J. 1990. Experiment confronts theory. *Nature,* 347: 128–129.

Drever, J. 1974. *A Dictionary of Psychology.* Penguin Books, New York, 320 pp.

Drickamer, L.C. and S.H. Vessey. 1982. *Animal Behavior: Concepts, Processes, and Methods.* Prindle, Weber, and Schmidt, Boston, MA, 619 pp.

Drollette, D. 1996. On the tail of the Tiger. *Scientific American,* October: 32–34.

Dubey, T. 1995. Aquatic fungi, pp. 31–37 in R.C. Reardon, Ed., *Effects of Diflubenzuron on Non-Target Organisms in Broadleaf Forested Watersheds in the Northeast.* USDA Forest Service FHM-NC–0595, U.S. Department of Agriculture, Washington, D.C., 174 pp.

Duckworth, G.E. and H.J. Rose. 1989. Cronus, pp. 497–498 in W.D. Halsey, Ed., *Colliers Encyclopedia,* Vol. 7. Macmillan, New York, 772 pp.

Duffey, S.S. 1976. Arthropod allomones: chemical effronteries and antagonists, pp. 323–394 in *Proceedings of the XV International Congress of Entomology,* August 19–27, 1976. The Entomological Society of America, College Park, MD, 824 pp.

Dufour, L. 1835. Recherches anatomiques et physiologiques sur les orthoptères, les hyménoptères et les névroptères. Mémories Présentées par Divers Savants a l'Académie Royale des Sciences de l'Institut de France, 647 pp.

Dugatkin, L.A. 1998. Genes, copying, and female mate choice: shifting thresholds. *Behavioral Ecology,* 9: 323–327.

Dugatkin, L.A. and J.-G.J. Godin. 1998. How females choose their mates. *Scientific American,* April: 56–61.

Dugger, C.W. 1996. Tug of taboos: African genital rite vs. U.S. law. *New York Times,* 28 December: 1, 9.

Dunham, P.J. and A. Hurshman. 1990. Precopulatory mate guarding in the amphipod, *Gammarus lawrencianus*: effects of social stimulation during the post-copulation interval. *Animal Behaviour,* 39: 976–979.

Dunham, P.J., T. Alexander, and A. Hurshman. 1986. Precopulatory mate guarding in an amphipod, *Gammarus lawrencianus* Bousfield. *Animal Behaviour,* 34: 1680–1686.

Dunkle, S.W. 1989. *Dragonflies of the Florida Peninsula, Bermuda, and the Bahamas.* Scientific Publishers, Gainesville, FL, 155 pp.

Dunkle, S.W. 1990. *Damselflies of Florida, Bermuda, and the Bahamas.* Scientific Publishers, Gainesville, FL, 148 pp.

Dupré, J. 1992. Species: theoretical contexts, pp. 312–317 in E.F. Keller and E.A. Lloyd, Eds., *Keywords in Evolutionary Biology.* Harvard University Press, Cambridge, MA, 414 pp.

Dwyer, J. 1993. The quirky genus who is changing our world. *Parade Magazine,* 10 October: 8–9.

Eastwood, E. 1967. *Radar Ornithology.* Methuen, London, 278 pp.

Ebbers, B.C. and E.M. Barrows. 1980. Individual ants specialize on particular aphid herds (Hymenoptera; Formicidae, Homoptera: Aphididae). *Proceedings of the Entomological Society of Washington,* 82: 405–407.

Eberhard, W. 1985. *Sexual Selection and Animal Genitalia.* Harvard University Press, Cambridge, MA, 244 pp.

Edmunds, P.J. 1996. Ten days under the sea. *Scientific American,* October: 88–95.

Edwards, D., P.A. Selden, J.B. Richardson, and L. Axe. 1995. Coprolites as evidence for plant-animal interaction in Siluro-Devonian terrestrial ecosystems. *Nature,* 377: 329–331.

Edwards, R. 1986. Behavior vs. behaviors. *Sphecos,* 12:2.

Efron, B. and R.J. Tibshirani. 1991. Statistical data analysis in the computer age. *Science,* 253: 390–395.

Efron, B. and R.J. Tibshirani. 1993. *An Introduction to the Bootstrap.* Chapman & Hall, New York, 436 pp.

Ehler, L.E. 1999. A modified "Riker Specimen Mount" for soft-bodied arthropods. *American Entomologist,* 45(1): 10–11.

Ehret, G. and C. Bernecker. 1986. Low-frequency sound communication by mouse pups (*Mus musculus*): wiggling calls release maternal behaviour. *Animal Behaviour,* 34: 821–830.

Ehrlich, P.R. and P.H. Raven. 1964. Butterflies and plants: a study in coevolution. *Evolution,* 18: 586–608.

Ehrlich, P.R. and E.O. Wilson. 1991. Biodiversity studies: science and policy. *Science,* 253: 758–762.

Ehrman, L. and J. Probber. 1978. Rare *Drosophila* males: the mysterious matter of choice. *American Scientist,* 66: 216–222.

Eisenberg, J.F. 1981. *The Mammalian Radiations. An Analysis of Trends in Evolution, Adaptation, and Behavior.* University of Chicago Press, Chicago, IL, 610 pp.

Eisner, T., K. Hicks, M. Eisner, and D.S. Robson. 1978. "Wolf-in-sheep's-clothing" strategy of a predaceous insect larva. *Science,* 199: 790–794.

Eldredge, N. 1985. *Time Frames: The Rethinking of Darwinian Evolution and the Theory of Punctuated Equilibria.* Simon & Schuster, New York, 240 pp.

Eldredge, N. and J. Cracraft. 1980. *Phylogenetic Patterns and the Evolutionary Process: Method and Theory in Comparative Biology.* Columbia University Press, New York, 349 pp.

Eldredge, N. and S.J. Gould. 1972. Punctuated equilibria: an alternative to phyletic gradualism, pp. 82–115 in T.J.M. Schopf, Ed., *Models in Paleobiology.* Freeman, Cooper, and Co., San Francisco, CA, 250 pp.

Elgar, M.A., and D.R. Nash. 1988. Sexual cannibalism in the garden spider *Araneus diadematus. Animal Behaviour,* 36: 1511–1517.

Emlen, J.M. 1973. *Ecology: An Evolutionary Approach.* Addison-Wesley, Reading, MA, 493 pp.

Emlen, S.T. 1976. Lek organization and mating strategies in the bullfrog. *Behavioral Ecology and Sociobiology,* 1: 283–313.

Emlen, S.T. and L.W. Oring. 1977. Ecology, sexual selection, and the evolution of mating systems. *Science,* 197: 215–223.

Endler, J.A. 1992. Natural selection: current usages, pp. 220–224 in E.F. Keller and E.A. Lloyd, Eds., *Keywords in Evolutionary Biology.* Harvard University Press, Cambridge, MA, 414 pp.

Ens, B.J. and J.D. Goss-Custard. 1984. Interference among oystercatchers *Haematopus ostralegus,* feeding on muscles, *Mytilis edulus,* on Exe Estuary. *Journal of Animal Ecology,* 53: 217–231.

Enserink, M. 1999. Biological invaders sweep in. *Science,* 285: 1834–1835.

Entomological Society of Washington. 1996. Bylaws. *Proceedings of the Entomological Society of Washington,* 98: 610–614.

Epstein, R., R.P. Lanza, and B.F. Skinner. 1981. "Self-awareness" in the pigeon. *Science,* 212: 695–696.

Erickson, J. 1992. *Plate Tectonics: Unraveling the Mysteries of Earth.* Facts on File, New York, 197 pp.

Erwin, D., J. Valentine, and D. Jablonski. 1997. The origin of animal body plans. *American Scientist,* 85: 126–137.

Erwin, J. 1979. Strangers in a strange land: abnormal behaviors or abnormal environments?, pp. 1–28 in J. Erwin, T.L. Maple, and G. Mitchell, Eds., *Captivity and Behavior: Primates in Breeding Colonies, Laboratories, and Zoos.* Van Nostrand-Reinhold, New York, 286 pp.

Estep, D.Q. and K.E.M. Bruce. 1981. The concept of rape in non-humans: a critique. *Animal Behaviour,* 29: 1272–1273.

Evans, A.V. and C.L. Bellamy. 1996. *An Inordinate Fondness for Beetles.* Henry Holt & Company, New York, 208 pp.

Evans, D.A and R.W. Matthews. 1976. Comparative courtship behaviour in two species of the parasitic chalcid wasp *Melittobia* (Hymenoptera: Eulophidae). *Animal Behaviour,* 24: 46–51.

Evans, H.E. 1984. *Insect Biology: A Textbook of Entomology.* Addison-Wesley, Reading, MA, 436 pp.

Evans, H.E. 1985. *The Pleasures of Entomology. Portraits of Insects and the People Who Study Them.* Smithsonian Institution Press, Washington, D.C., 238 pp.

Ewing, A.W., and J.A. Miyan. 1986. Sexual selection, sexual isolation and the evolution of song in the *Drosophila repleta* group of species. *Animal Behaviour,* 34: 421–429.

Eysenck, H.J., W. Arnold, and R. Meili, Eds. 1979. *Encyclopedia of Psychology.* The Seabury Press, New York, 1187 pp.

Faegri, K., and L. van der Pijl. 1979. *The Principles of Pollination Ecology,* 3rd rev. ed. Pergamon Press, New York, 244 pp.

Fagen, R.M. 1978. Repertoire analysis, pp. 25–42 in P.W. Colgan, Ed., *Quantitative Ethology.* John Wiley & Sons, New York, 364 pp.

Fagan, R.M. 1983. Horseplay and monkeyshines. *Science,* 83 December: 70–76.

Faiola, A. 1999. Amazon cash crop: Brazil seeks "bioroyalties" from western drug firms. *Washington Post,* 9 July: A21.

Faison, S. 1998. Elephants are friends, China tells its farmers. *New York Times Northeast,* 15 February: 9.

Falconer, D.S. 1960. *Introduction to Quantitative Genetics.* The Ronald Press Company, New York, 365 pp.

Farkas, S.R. and H.H. Shorey. 1974. Mechanisms of orientation to a distant pheromone source, pp. 81–95 in M.C. Birch, Ed., *Pheromones.* North-Holland, London.

Farrar, J.L. 1995. *Trees of the Northern United States and Canada.* Iowa State University Press, Ames, IA, 502 pp.

Feduccia, A. 1993. Evidence from claw geometry indicating arboreal habits of *Archaeopteryx. Science,* 259: 790–793.

Feldman, M.W. 1992. Heritability: some theoretical ambiguities, pp. 151–157 in E.F. Keller and E.A. Lloyd, Eds., *Keywords in Evolutionary Biology.* Harvard University Press, Cambridge, MA, 414 pp.

Fellers, J.H. 1987. Interference and exploitation in a guild of woodland ants. *Ecology,* 68(5): 1466–1478.

Fenton, M.B. 1984. Sperm competition? The case of verspertilionid and rhonolophid bats, pp. 573–587 in R.L. Smith, Ed., *Sperm Competition and the Evolution of Animal Mating Systems.* Academic Press, New York, 687 pp.

Fernald, M.L. 1950. *Gray's Manual of Botany,* 8th ed. American Book Company, New York, 1632 pp.

Ferster, C.B. and B.F. Skinner. 1957. *Schedules of Reinforcement.* Appleton-Century-Crofts, New York, 741 pp.

Fichter, G.S. 1990. *Whales and Other Marine Mammals.* Golden Press, Racine, WI, 160 pp.

Fink, L.S. 1986. Costs and benefits of maternal behaviour in the green lynx spider (Oxyopodidae, *Peucetia viridans*). *Animal Behaviour,* 34: 1051–1060.

Fischman, J. 1993. Forever (in) amber. *Science,* 262: 655–656.

Fisher, R.A. (1930) 1958. *The Genetical Theory of Natural Selection,* 2nd rev. ed. Dover Publications, New York, 291 pp.

Fisher, R.M. 1987a. Queen-worker conflict and social parasitism in bumble bees (Hymenoptera: Apidae). *Animal Behaviour,* 35: 1026–1036.

Fisher, R.M. 1987b. Temporal dynamics of facultative social parasitism in bumble bees (Hymenoptera: Apidae). *Animal Behaviour,* 35: 1628–1636.

Fisher, R.M. and N. Pomeroy. 1990. Sex discrimination and infanticide by queens of the bumble bee *Bombus terrestris* (Hymenoptera: Apidae). *Animal Behaviour,* 39: 801–802.

Fitch, W.M. 1976. Molecular evolutionary clocks, pp. 160–178 in F.J. Ayala, Ed., *Molecular Evolution.* Sinauer, Sunderland, MA, 277 pp.

Fitzgerald, T.D. 1995. Caterpillars roll their own. *Natural History,* April: 30–37.

Fleischmann, R.D., M.D. Adams, O. White, R.A. Clayton, E.F. Kirkness, A.R. Kerlavage, C.J. Bult, J.F. Tomb, B.A. Doughtery, J.M. Merrick et al. 1995. Whole-genome random sequencing and assembly of *Haemophilus influenzae* Rd. *Science,* 269: 496–512.

Fleming, C.A. 1959. Palaeontological evidence for speciation preceded by geographic isolation, pp. 225–241 in G. Leeper, Ed., *The Evolution of Living Organisms. A Symposium to Mark the Century of Darwin's "Origin of Species" and The Royal Society of Victoria,* December 1959, Melbourne. Melbourne University Press, Australia. 459 pp.

Flessa, K.W., K.V. Powers, and J.L. Cisne. 1975. Specialization and evolutionary longevity in the Arthropoda. *Paleobiology,* 1: 71–81.

Fletcher, C. 1995. A garden of mutants. *Discover,* August: 48–53.

Fletcher, D.J.C. 1975. Significance of dorsoventral abdominal vibration among honey-bees (*Apis mellifera* L.). *Nature,* 256:721–723.

Fletcher, D.J.C. 1987. The behavioral analysis of kin recognition: perspectives on methodology and interpretation, pp. 19–54 in D.J.C. Fletcher and C.D. Michener, Eds., *Kin Recognition in Animals.* Wiley, New York, 465 pp.

Fogle, B. 1981. Attachment-euthanasia-grieving, pp. 331–343 in B. Fogle, Ed., *Interrelations Between People and Pets.* Charles C Thomas, Springfield, IL, 352 pp.

Folger, T. 1996. The first masterpieces. *Discover,* January: 70–71.

Formanowicz, D.R., Jr. 1990. The antipredator efficacy of spider leg autotomy. *Animal Behaviour,* 40: 400–409.

Forsyth, A. 1985. Good scents and bad. *Natural History,* 94(11): 24–32.

Fountain, H. 1999. Lizards with big appetite find fireflies a fatal attraction. *New York Times Northeast,* 27 July: D5.

Fowler, B. 1997. Following stone-age footsteps. *New York Times,* 3 August: 13, 20.

Fox, M.W. 1974. *Concepts in Ethology: Animal and Human Behavior.* University of Minnesota Press, Minneapolis, 139 pp.

Fraenkel, G.S., and D.L. Gunn. (1940) 1961. *The Orientation of Animals.* Dover Publications, New York, 376 pp.

Francis, A.M., J.P. Hailman, and G.E. Woolfenden. 1989. Mobbing by Florida Scrub Jays: behaviour, sexual asymmetry, role of helpers and ontogeny. *Animal Behaviour,* 38: 795–816.

Frank, J.H. and E.D. McCoy. 1990. Introduction to attack and defense: behavioral ecology of predators and their prey: endemics and epidemics of shibboleths and other things causing chaos. *Florida Entomologist,* 73: 1–9.

Fraser, C.M., J.D. Gocayne, O. White, M.D. Adams, R.A. Clayton, R.D. Fleischmann, C.J. Bult, A.R. Kerlavage, G. Sutton, J.M. Kelley et al. 1995. The minimal gene complement of *Mycoplasma genitalium. Science,* 270: 397– 403.

Fredrickson, J.K. and T.C. Onstott. 1996. Microbes deep inside the Earth. *Scientific American,* October: 68–73.

Freed, L.A. 1986. Usurpatory and opportunistic bigamy in tropical house wrens. *Animal Behaviour,* 34: 1894–1896.

Frerking, B. 1997. Good news for nerds! Now, trading cards that make it cool to be a brainiac. *Washington Post,* 8 April, B5.

Friedman, N. 1967. *The Social Nature of Psychological Research: The Psychological Experiment as a Social Interaction.* Basic Books, New York, 204 pp.

Frisch, K. von. 1942. Über einen Schreckstoff der Fischhaut und seine biologische Bedeutung. *Zeitschrift für Vergleitchende Physiologie,* 29: 47–145.

Frisch, K. von. 1955. *The Dancing Bees.* Harcourt Brace and Company, New York, 183 pp.

Frisch, K. von. 1967. *The Dance Language and Orientation of Bees.* Belknap Press, Cambridge, MA, 566 pp.

Fristrup, K. 1992. Character: current usages, pp. 45–51 in E.F. Keller and E.A. Lloyd, Eds., *Keywords in Evolutionary Biology.* Harvard University Press, Cambridge, MA, 414 pp.

Frondel, J.W. 1973, pp. 717–718 in W.E. Preece, Ed., *Encyclopedia Britannica*, Vol. 1. William Benton, Publisher, London, 1016 pp.

Frumhoff, P.C. and S. Schneider. 1987. The social consequences of honey bee polyandry: the effects of kinship on worker interactions within colonies. *Animal Behaviour*, 36: 255–262.

Fuller, J.L. and W.R. Thompson. 1960. *Behavior Genetics*. Wiley, New York, 396 pp.

Futuyma, D.J. 1979. *Evolutionary Biology*. Sinauer Associates, Sunderland, MA, 565 pp.

Futuyma, D.J. 1986. *Evolutionary Biology*, 2nd ed. Sinauer Associates, Sunderland, MA, 600 pp.

Futuyma, D.J. 1989. Coevolution: cautious views (book review). *Science*, 245: 991–992.

Futuyma, D.J. 1993. Attention to terms (book review). *Science*, 260: 1153–1154.

Futuyma, D.J. and M. Slatkin, Eds. 1983. *Coevolution.* Sinauer Associates, Sunderland, MA, 555 pp.

Fyfe, W.S. 1996. The biosphere is going deep. *Science*, 273: 448.

Gabriel, T. 1997. Pack dating: for a good time, call a crowd. *New York Times Educational Life*, 5 January: 22–23, 38.

Gadagkar, R. 1993. Can animals be spiteful? *Trends in Ecology and Evolution*, 8: 232–234.

Galdikas, B.M.F. 1995. *Reflections of Eden: My Years with the Orangutans of Borneo.* Little, Brown & Co., New York, 408 pp.

Galitzki, D. 1996. Garden Q. & A. *New York Times*, 26 December: C9.

Gallup, G.G. 1973. Towards an operational definition of self-awareness, pp. 309–313 in R.H. Tuttle, Ed., *Socioecology and Psychology of Primates*. Mouton Publishers, The Hague, 474 pp.

Gallup, G.G. 1979. Self-awareness in primates. *American Scientist*, 67: 417–421.

Gallup, G.G. 1982. Self-awareness and the emergence of mind in primates. *American Journal of Primatology*, 2: 237–248.

Ganz, J. 1997. Is first extrasolar planet a lost world? *Science*, 275:1256–1257.

Gardener, R. and B. Gardener. 1969. Teaching sign language to a chimpanzee. *Science*, 165: 664–672.

Gardner, J.L., Ed. 1992. *Reader's Digest Atlas of the World*. The Reader's Digest Association, Pleasantville, NY, 240 pp.

Garfield, E. 1987. Citation data is subtle stuff. *Current Contents AB and ES*, 18(17): 18–19.

Garreau, J. 1993. Just put on a happy face. *Washington Post*, 9 August: D1, D10.

Gary, N.E. 1962. Chemical mating attractants in the queen honey bee. *Science*, 136: 773–774.

Gary, N.E. 1974. Pheromones that affect the behavior and physiology of honey bees, pp. 200–221 in M.C. Birch, Ed., *Pheromones*. North-Holland, Amsterdam. 495 pp.

Gause, G.F. 1932. Ecology of populations. *Quarterly Review of Biology*, 7: 27–46.

Geary, D.H. 1988. *Animal Behaviour* (book review), 36: 1257–1259.

Gebo, D.L., M. Dagosto, K.C. Beard, T Qi, and J. Wang. 2000. The oldest known anthropoid postcranial fossils and the early evolution of higher primates. *Nature*, 404: 276–278.

Geneve, R. 1991. Attainable chimeras. *American Horticulturist*. 70: 33–36.

Getty, T. 1989. Are dear enemies in a war of attrition? *Animal Behaviour*, 37: 337–339.

Ghent, R.L. and N.E. Gary. 1962. A chemical alarm releaser in Honey Bee stings (*Apis mellifera* L.). *Psyche*, 69: 1–6.

Gherardi, F.C. 1984. A historical reconstruction of some pre-ethological approaches to animal behaviour, *Rivista di Storia della Scienza,* I(3): 355–402.

Gibbons, A. 1992a. Plants of the apes. *Science,* 255: 921.

Gibbons, A. 1992b. Mitochondrial Eve: wounded, but not yet dead. *Science,* 257: 873–875.

Gibbons, A. 1992c. Biologists trace the evolution of molecules. *Science,* 257: 30–31.

Gibbons, A. 1993a. Mitochondrial Eve refuses to die. *Science,* 259: 1249–1250.

Gibbons, A. 1993b. Pleistocene population explosions. *Science,* 262: 27–28.

Gibbons, A. 1997. Monte Verde: blessed but not confirmed. *Science,* 275: 1256–1257.

Gibbs, H.L. 1990. Cultural evolution of male song types in Darwin's medium ground finches. *Animal Behaviour,* 39: 253–263.

Gibbs, W.W. and C.S. Powell. 1996. Bugs in the data? *Scientific American,* October: 20, 22.

Gibson, E.K., Jr., D.S. McKay, K. Thomas-Keprta, and C.S. Romanek. 1996. Evaluating the evidence for past life on Mars. *Science,* 274: 2125.

Gilbert, J., Ed., 1976. *The Complete Aquarist's Guide to Freshwater Tropical Fish.* Eurobook, Ltd., England. 249 pp.

Gilbert, L.E. 1983. Coevolution and mimicry, pp. 263–281 in D.J. Futuyma and M. Slatkin, Eds., *Coevolution.* Sinauer Associates, Sunderland, MA, 555 pp.

Gilbert, L.W. 1976. Postmating female odor in *Heliconius* butterflies: a male contributed antiaphrodisiac? *Science,* 193: 419–420.

Gillam, C. 1999. Kansas eliminates evolution from public school curricula. *Washington Post,* 12 August: A13.

Gillespie, J.H. 1984. The status of the neutral theory (book review). *Science,* 224: 732–733.

Gillis, A.M. 1991. Can organisms direct their evolution? *BioScience,* 41: 202–205.

Gladstone, D.E. 1979. Promiscuity in monogamous colonial birds. *American Naturalist,* 114: 545–557.

Goff, L.J. 1982. Symbiosis and parasitism: another viewpoint. *BioScience,* 32: 255–256.

Goffeau, A. et al. 1996. Life with 6000 genes. *Science,* 274: 546–567.

Goldensohn, E. 1999. What's in the name? *Natural History,* September: 6.

Goldenson, R.M., Ed., 1984. *Longman Dictionary of Psychology and Psychiatry.* Longman, New York. 815 pp.

Goldschmidt, T. 1996. *Darwin's Dreampond: Drama in Lake Victoria.* The MIT Press, Cambridge, MA, 274 pp.

Goldstein, M.C. 1987. When brothers share a wife. *Natural History,* 96(3): 38–49.

Goleman, D. 1995. The decline of the nice-guy quotient. *New York Times,* 10 September: 6.

Good, K. 1995. The Yanomami keep on trekking. *Natural History,* 104: 57–65.

Goodacre, A. 1991. Continental drift. *Nature,* 354: 261.

Goodman, M. 1982. Decoding the pattern of protein evolution. *Progress in Biophysics and Molecular Biology,* 38: 105.

Gordon, D.M. 1992. Phenotypic plasticity, pp. 255–262 in E.F. Keller and E.A. Lloyd, Eds., *Keywords in Evolutionary Biology.* Harvard University Press, Cambridge, MA, 414 pp.

Gore, R. 1989. Extinctions. *National Geographic,* 175(6): 662–699.

Gosling, L.M. 1987. Scent marking in an antelope lek territory. *Animal Behaviour,* 35: 620–622.

Gosling, L.M. and M. Petrie. 1990. Lekking in topi: a consequence of satellite behaviour by small males at hotspots. *Animal Behaviour,* 40: 272–287.

Götmark, F. 1987. White underparts in gulls function as hunting camouflage. *Animal Behaviour,* 35: 1786–1792.

Götmark, F. 1990. A test of the information-centre hypothesis in a colony of Sandwich Terns *Sterna sandvicensis. Animal Behaviour,* 39: 487–495.

Gould, J.L. 1976. The dance-language controversy. *The Quarterly Review of Biology,* 51: 211–244.

Gould, J.L. 1982. *Ethology: The Mechanisms and Evolution of Behavior.* W.W. Norton, New York, 544 pp.

Gould, J.L. 1987. Landmark learning in Honey Bees. *Animal Behaviour,* 35: 26–34.

Gould, J.L. and C.G. Gould. 1988. *The Honey Bee.* Scientific American Library, New York, 239 pp.

Gould, S.J. 1976. The advantages of eating mom. *Natural History,* 85(12): 24–31.

Gould, S.J. 1977. *Ontogeny and Phylogeny.* Harvard University Press, Cambridge, MA, 501 pp.

Gould, S.J. 1980. Is a new and general theory of evolution emerging? *Paleobiology,* 6: 119–130.

Gould, S.J. 1981a. The ultimate parasite. *Natural History,* 90(11): 7–14.

Gould, S.J. 1981b. Palaeontology plus ecology as palaeobiology, pp. 295–317 in R.M. May, Ed., *Theoretical Ecology: Principles and Applications.* Blackwell Scientific, Boston, MA.

Gould, S.J. 1982a. Darwinism and the expansion of evolutionary theory. *Science,* 216: 380–387.

Gould, S.J. 1982b. The oddball human male. *Natural History,* 91(7): 14–22.

Gould, S.J. 1982c. The meaning of punctuated equilibrium and its role in validating a hierarchical approach to macroevolution, pp. 83–104 in R. Milkman, Ed., *Perspectives on Evolution.* Sinauer Associates, Sunderland, MA, 241 pp.

Gould, S.J. 1984. A short way to corn. *Natural History,* 93(3): 12–20.

Gould, S.J. 1985a. Here goes nothing. *Natural History,* 94(7): 12–19.

Gould, S.J. 1985b. Not necessarily a wing. *Natural History,* 94 (10): 12–25.

Gould, S.J. 1985c. Geoffroy and the homeobox. *Natural History,* 94(11): 12–23.

Gould, S.J. 1986. Archetype and adaptation. *Natural History,* 95(10): 16–27.

Gould, S.J. 1987. Hatracks and theories. *Natural History,* 96(3): 12–23.

Gould, S.J. 1988. The heart of terminology. *Natural History,* 97(2): 24–31.

Gould, S.J. 1989. The Horn of Triton. *Natural History,* 99(12): 18–27.

Gould, S.J. 1992. Heterochrony, pp. 158–165 in E.F. Keller and E.A. Lloyd, Eds., *Keywords in Evolutionary Biology.* Harvard University Press, Cambridge, MA, 414 pp.

Gould, S.J., Ed., 1993. *The Book of Life.* W.W. Norton, New York, 256 pp.

Gould, S.J. 1994. Tempo and mode in the macroevolutionary reconstruction of Darwinism. *Proceedings of the National Academy of Sciences USA,* 91: 6764–6771.

Gould, S.J. 1996. Planet of the bacteria. *Washington Post,* 13 November: H1.

Gould, S.J. 1999. Message from a mouse. *Time,* 13 September: 62.

Gould, S.J. and N. Eldredge. 1986. Punctuated equilibrium at the third stage. *Systematic Zoology,* 35: 143–148.

Gould, S.J. and N. Eldredge. 1993. Punctuated equilibrium comes of age. *Nature,* 366: 223–227.

Gould, S.J. and R.C. Lewontin. 1979. The spandrels of San Marco and the Panglossian paradigm: a critique of the adaptationist programme. *Proceedings of the Royal Society London B,* 205: 581–598.

Gould, S.J. and E. Vrba. 1981. Exaptation: a missing term in the science of form. *Paleobiology,* 8: 4–15.

Gowaty, P.A. 1981. The aggression of breeding eastern bluebirds (*Sialia sialis*) toward each other and intra- and inter-specific intruders. *Animal Behaviour,* 29: 1013–1027.

Gowaty, P.A. 1982. Sexual terms in sociobiology: emotionally evocative and, paradoxically, jargon. *Animal Behaviour,* 30: 630–631.

Grady, D. 1996. Tracing a genetic disease to bits of traveling DNA. *New York Times,* 5 March: C1, C5.

Grady, R.M. and J.L. Hoogland. 1986. Why do male black-tailed prairie dogs (*Cynomys ludovicianus*) give a mating call? *Animal Behaviour,* 34: 108–112.

Grafen, A. 1982. How not to measure inclusive fitness. *Nature,* 298: 425–426.

Grafen, A. 1984. Natural selection, kin selection and group selection, pp. 62–84 in J.R. Krebs and N.B. Davies, Eds., *Behavioural Ecology: An Evolutionary Approach,* 2nd ed. Blackwell Scientific, Oxford. 493 pp.

Grafen, A. 1990. Do animals really recognize kin? *Animal Behaviour,* 39: 42–54.

Grafen, A. 1991a. A reply to Blaustein et al. *Animal Behaviour,* 41: 1085–1087.

Grafen, A. 1991b. A reply to Byers and Bekoff. *Animal Behaviour,* 41: 1091–1092.

Grafen, A. 1991c. Kin vision?: A reply to Stuart. *Animal Behaviour,* 41: 1095–1096.

Graham, G.L. 1994. *Bats of the World.* Golden Press, New York, 160 pp.

Grant, V. 1963. *The Origin of Adaptations.* Columbia University Press, New York, 606 pp.

Grassle, J.F. 1985. Hydrothermal vent animals: distribution and biology. *Science,* 229: 713–717.

Graves, B.M., and M. Halpern. 1990. Roles of vomeronasal organ chemoreception in tongue flicking, exploratory and feeding behaviour of the lizard, *Chalcides ocellatus. Animal Behaviour,* 39: 692–698.

Gray, S. 1999. Drought leaves marijuana growers high and dry. *Washington Post,* 28 August: B1.

Gray, M.W., G. Burger, and B.F. Lang. 1999. Mitochondrial evolution. *Science,* 283: 1476–1481.

Greene, E. 1989. A diet-induced developmental polymorphism in a caterpillar. *Science,* 243: 643–646.

Greenwood, P.H., Ed. 1981. *The Haplochrome Fishes of the East African Lakes: Collected Papers on Their Taxonomy, Biology and Evolution.* Cornell University Press, Ithaca, NY, 839 pp.

Greenwood, P.J., P.H. Harvey, and M. Slatkin. 1985. *Evolution: Essays in Honour of John Maynard Smith.* Cambridge University Press, Cambridge, U.K., 328 pp.

Gregory, R.L. and O.L. Zangwell. 1987. *The Oxford Companion to the Mind.* Oxford University Press, New York, 856 pp.

Grene, M., Ed. 1983. *Dimensions of Darwinism: Themes and Counterthemes in Twentieth-Century Evolutionary Theory.* Cambridge University Press, New York, 336 pp.

Griesemer, J.R. 1992. Niche: historical perspectives, pp. 231–240 in E.F. Keller and E.A. Lloyd, Eds., *Keywords in Evolutionary Biology.* Harvard University Press, Cambridge, MA, 414 pp.

Griffin, D.A. 1976. *The Question of Animal Awareness.* Rockefeller University Press, New York, 135 pp.

Griffin, D.A. 1984. *Animal Thinking.* Harvard University Press, Cambridge, MA, 237 pp.

Griffin, D.A. 1985. Bugs, slugs, computers, and consciousness. *American Scientist,* 73: 121–122.

Grishin, A.V., J.L. Weiner, and K.J. Blumer. 1994. Control of adaptation to mating pheromone by G protein b subunits of *Saccharomyces cerevisiae. Genetics,* 138: 1081–1092.

Griswold, C.E. and T. Meikle-Griswold. 1987. *Archaeodictyna ulova,* a new species (Araneae: Dictynidae), a remarkable kleptoparasite of group-living eresid spiders (*Stegodyphus* spp., Araneae: Eresidae). *American Museum Novitates,* 2897: 1–11.

Griswold, C.E. and T.C. Meikle. 1990. Social life in a web. *Natural History,* March: 6–8.

Groskin, H. 1943. The Scarlet Tanager's "anting." *Auk,* 60: 55–58.

Groskin, H. 1950. Additional observations and comments on "anting" by birds. *Auk,* 67: 201–209.

Gugliotta, G. 1999. Evidence found of cannibalism in Ice Age. *New York Times,* 1 October: A3.

Güntürkün, O., S. Kesch, and J.D. Delius. 1988. Absence of footedness in domestic pigeons. *Animal Behaviour,* 36:602–604.

Gurney, A.B. 1976. A short history of the Entomological Society of Washington. *Proceedings of the Entomological Society of Washington,* 78: 225–239.

Guthrie, R.D. 1990. *Frozen Fauna of the Mammoth Steppe: The Story of Blue Babe.* University of Chicago Press, Chicago, IL, 323 pp.

Guthrie, R.D. and R.G. Petocz. 1970. Weapon automimicry among mammals. *American Naturalist,* 104: 585–588.

Gutteridge, A. 1983. *The Barnes and Noble Thesaurus of Biology.* Barnes and Noble Books, New York, 256 pp.

Gwynne, D.T. 1984. Male mating effort, confidence of paternity, and insect sperm competition, pp. 117–149 in R.L. Smith, Ed., *Sperm Competition and the Evolution of Animal Mating Systems.* Academic Press, New York, 687 pp.

Ha, J.C., P.N. Lehner, and S.D. Farley. 1990. Risk-prone foraging behaviour in captive Grey Jays, *Perisorius canadensis. Animal Behaviour,* 39: 91–96.

Haemig, P.D. 1997. Effect of birds on the intensity of ant rain: a terrestrial form of invertebrate drift. *Animal Behaviour,* 54: 89–97.

Haig, D. 1993. Genetic conflicts in human pregnancy. *Quarterly Review of Biology,* 68: 495–532.

Hailman, J.P. 1969. How an instinct is learning. *Scientific American,* 221(6): 98–106.

Hailman, J.P. 1976. Homology: logic, information, and efficiency, pp. 181–198 in R.B. Masterton, W. Hodos, and H. Jerison, Eds., *Evolution, Brain, and Behavior: Persistent Problems.* Lawrence Erlbaum Assoc., Hillsdale, NJ, 276 pp.

Hailman, J.P. 1984. Robert A. Hinde's view of ethology (book review). *Bioscience,* 34(1): 57.

Haldane, J.B.S. 1932. *The Causes of Evolution.* Longman, Green, and Company, London, 234 pp.

Hale, E. 1996. *Arabidopsis* plant's gene map called "our only hope." *The Detroit News,* 29 December: 5A, 11A.

Hale, M.E., Jr. 1961. *Lichen Handbook.* Smithsonian Institution, Washington, D.C., 178 pp.

Hall, B.G. 1990. Spontaneous point mutations that occur more often when advantageous than when neutral. *Genetics,* 126: 5–16.

Hallam, A. 1975. Alfred Wegener and the hypothesis of continental drift, pp. 9–17 in J.T. Wilson, Ed., *Continents and Continents Aground.* W.H. Freeman, San Francisco, CA, 230 pp.

Halliday, T.R. and K. Adler, Eds. 1986. *The Encyclopedia of Reptiles and Amphibians.* Facts on File, New York, 152 pp.

Halvorsen, M. and O.B. Stabell. 1990. Homing behaviour of displaced stream-dwelling Brown Trout. *Animal Behaviour,* 39: 1089–1097.

Hamer, D.H., S. Hu, V.L. Magnuson, N. Hu, and A.M.L. Pattatucci. 1993. A linkage between DNA markers on the X chromosome and male sexual orientation. *Science,* 261: 321–327.

Hamilton, M.B. 1999. Tropical tree gene flow and seed dispersal. *Nature,* 401: 129.

Hamilton, W.D. 1963. The evolution of altruistic behavior. *American Naturalist,* 97: 354–356.

Hamilton, W.D. 1964. The genetical theory of social behaviour, parts I and II. *Journal of Theoretical Biology,* 7: 1–52.

Hamilton, W.D. 1967. Extraordinary sex ratios. *Science,* 156: 477–488.

Hamilton, W.D. 1980. Sex versus non-sex versus parasite. *Oikos,* 35: 282–290.

Hand, J.L. 1986. Resolution of social conflicts: dominance, egalitarianism, spheres of dominance, and game theory. *Quarterly Review of Biology,* 61: 201–220.

Handby, J.P., and J.D. Bygott. 1987. Emigration of subadult lions. *Animal Behaviour,* 35: 161–169.

Hansen, A.J., and S. Rohwer. 1986. Coverable badges and resource defence in birds. *Animal Behaviour,* 34: 69–76.

Hanson, B., Ed., 1997. Cichlid loss: a murky tale. *Science,* 277: 1737.

Hapgood, F. 1984. The importance of being Ernst. *Science,* 84 5(5): 40–46.

Harborne, J.B. 1982. *Introduction to Ecological Biochemistry,* 2nd ed. Academic Press, New York, 278 pp.

Harnad, S. 1985. Bugs, slugs, computers, and consciousness. *American Scientist,* 73: 121–122.

Harrahy, E.A., S.A. Perry, M.J. Wimmer, and W.B. Perry. 1994. The effects of diflubenzuron (Dimilin®) on selected mayflies (Heptageniidae) and stoneflies (Peltoperlidae and Pteronarcyidae). *Environmental Toxicology and Chemistry,* 13: 517–522.

Harré, R. and R. Lamb, Eds. 1983. *The Encyclopedic Dictionary of Psychology.* MIT Press, Cambridge, MA, 718 pp.

Harris, L.D. 1988. Edge effects and conservation of biotic diversity. *Conservation Biology,* 2: 330–332.

Harris, M.A. and J.O. Murie. 1982. Responses to oral gland scents from different males in Columbian ground squirrels. *Animal Behaviour,* 30: 140–148.

Hart, S. 1996. RNA's revising machinery. *BioScience,* 46: 318–321.

Hartl, D.L. 1980. Principles of Population Genetics. Sinauer Associates, Sunderland, MA, 488 pp.

Hasegawa, M. and W.M. Fitch. 1996. Dating the cenancester [sic] of organisms. *Science,* 274: 1750.

Haseman, L. 1912. The evergreen bagworm. *Missouri Agricultural Station Bulletin,* 104: 308–329.

Hastings, J.W. 1989. Bioluminescence, pp. 545–550 in S.P. Parker, Ed. *McGraw-Hill Encyclopedia of Science and Technology,* 6th ed., Vol. 2. McGraw-Hill, New York, 661 pp.

Hauser, L. 1999. *Genetics and Taxonomy of Nile Perch (Lates niloticus) Introduced to Lake Victoria.* http://www .hull.ac.uk /molecol/lates.htm.

Hauser, M.D. 1993. Right hemisphere dominance for the production of facial expression in monkeys. *Science,* 261: 475–477.

Hawkins, C.P. and J.A. MacMahon. 1989. Guilds: the multiple meanings of a concept. *Annual Review of Entomology,* 34: 423–451.

Hazlett, B.A. 1987. Information transfer during shell exchange in the Hermit Crab *Clibanarius antillensis. Animal Behaviour,* 35: 218–226.

Hecht, M.K. and A. Hoffmann. 1986. Why not neo-Darwinism? A critique of paleobiological challenges. *Oxford Surveys in Evolutionary Biology,* 3: 1–47.

Hector, A., B. Schmid, C. Beierkuhnlein, M.C. Caldeira, M. Diemer, P.G. et al. 1999. Plant diversity and productivity experiments in European grasslands. *Science,* 286: 1123–1127.

Hedges, S.B., S. Kumar, K. Tamura, and M. Stoneking. 1992. Human origins and analysis of mitochondrial DNA sequences. *Science,* 255: 737–739.

Hedges, S. B., S. Kumar, K. Tamura, and M. Stoneking. 1991. Human origins and analysis of mitochondrial DNA sequences. *Science,* 255: 737–739.

Heeb, P. and H. Richner. 1994. Seabird colonies and the appeal of the information center hypothesis. *Trends in Ecology and Evolution,* 9: 25.

Hefetz, A., H.M. Fales, and S.W.T. Batra. 1979. Natural polyesters: Dufour's gland macrocyclic lactones form brood cell laminesters in *Colletes* bees. *Science,* 204: 415–417.

Hefetz, A., F. Bergström, and J. Tengö. 1986. Species, individual and kin specific blends in Dufour's gland secretions of halictine bees: chemical evidence. *Journal of Chemical Ecology,* 12: 197–208.

Heinrich, B. 1979. *Bumblebee Economics.* Harvard University Press, Cambridge, MA, 245 pp.

Heinrich, B. 1990. The antifreeze of bees. *Natural History,* July: 52–59.

Heinsohn, R. G. 1991. Kidnapping and reciprocity in cooperatively breeding White-Winged Choughs. *Animal Behaviour,* 41: 1097–1100.

Helwig, J.T. and K.A. Council. 1979. *SAS® User's Guide.* SAS Institute, Cary, NC, 494 pp.

Henry, J.D. 1986. *Red Fox: The Catlike Canine.* Smithsonian Institution Press, Washington, D.C., 174 pp.

Henry, L.K. 1933. Shale-barren flora in Pennsylvania. *Proceedings of the Pennsylvania Academy of Science,* 7: 65–68.

Hermann, H.R., Ed., 1981. *Social Insects,* Vol. II. Academic Press, New York, 491 pp.

Heske, E.J. 1995. Mammalian abudances on forest-farm edge versus forest interiors in southern Illinois: is there an edge effect? *Journal of Mammalogy,* 76: 562–568.

Hews, D.K. 1988. Alarm response in larval Western Toads, *Bufo boreas*: release of larval chemicals by a natural predator and its effect on predator capture efficiency. *Animal Behaviour,* 36: 125–133.

Heymer, A. 1977. *Ethological Dictionary.* Garland Publishing, New York, 238 pp.

Higley, L.G. and D.W. Stanley-Samuelson. 1993. What do you mean, have I read my own paper? *American Entomologist,* 39: 74–75.

Hill, G.E. 1986. The function of distress calls given by tufted titmice (*Parus bicolor*): an experimental approach. *Animal Behaviour,* 34: 590–598.

Hilton, D.F.J. 1987. A terminology for females with color patterns that mimic males. *Entomological News,* 98: 221–223.

Hilts, P.J. 1996. Gene jumps to spread a toxin in meat. *New York Times,* 23 April: C1.

Hilts, P.J. 1997. Listening to the conversation of neurons. *New York Times,* 27 May: C1, C5.

Hilts, P.J. 1998. Fossil shows ants evolved much earlier than thought. *New York Times,* 27 May: C1, C5.

Himmelreich, R., H. Hilbert, H. Plagens, E. Pirkl, B.C. Li, and R. Herrmann. 1996. Complete sequence analysis of the genome of the bacterium *Mycoplasma pneumoniae. Nucleic Acids Research,* 24: 4420–4449.

Hinde, R.A. 1970. *Animal Behaviour,* 2nd ed. McGraw-Hill, New York, 876 pp.

Hinde, R.A. 1974. *Biological Bases of Human Social Behaviour.* McGraw-Hill, New York, 462 pp.

Hinde, R.A. 1982. *Ethology: Its Nature and Relations with Other Sciences.* Oxford University Press, New York, 320 pp.

Hinde, R.A. 1985. Was "The Expression of the Emotions" a misleading phrase? *Animal Behaviour,* 33: 985–992.

Hinsie, L.E. and R.J. Campbell. 1976. *Psychiatric Dictionary,* 4th ed. Oxford University Press, London, 816 pp.

Hirth, L., D. Abadanian, and H.W. Goedde. 1986. Incidence of specific anosmia in Northern Germany. *Human Heredity,* 36: 1–5.

Hitt, J., Ed., 1992. *In a Word: A Dictionary of Words That Don't Exist, But Ought To.* Dell Publishing, New York, 215 pp.

Hjorth, I. 1970. Reproductive behavior in Tetraonidae with specific reference to males. *Viltrevy,* 7: 184–587.

Hockett, C.F. 1960. The origin of speech. *Scientific American,* 203: 89–96.

Hockett, C.F. and S.A. Altmann. 1968. A note on design features, pp. 61–72 in T.A. Sebeok, Ed., *Animal Communication: Techniques of Study and Results of Research.* Hafner Publishing, New York, 203 pp.

Hodge, M.J.S. 1992. Natural selection: historical perspectives, pp. 212–219 in E.F. Keller and E.A. Lloyd, Eds., *Keywords in Evolutionary Biology.* Harvard University Press, Cambridge, MA, 414 pp.

Hodgman, C.D. 1963. *CRC Standard Mathematical Tables,* 12th ed. CRC Press, Cleveland, OH, 525 pp.

Hoffman, M. 1991. How parents make their mark on genes. *Science,* 252: 1250–1251.

Hoffman, M. 1992. Anything goes at the cell biology meeting. *Science,* 255: 34–35.

Hoffmann, A. 1982. Punctuated versus gradual mode of evolution, pp. 411–442 in M.K. Hecht, B. Wallace, and G.T. Prance, Eds., *Evolutionary Biology,* Vol. 15. Plenum Press, New York, 442 pp.

Hoffmeister, D.F. 1967. *Zoo Animals.* Western Publishing, New York, 160 pp.

Hograefe, T. 1984. Substratum-stridulation in the colonial sawfly larvae of *Hemichroa crocea* (Geoff.) (Hymenoptera: Tenthedinidae). *Zoologischer Anzeiger Jena,* 213: 234–241.

Holden, C., Ed., 1992. Pack rats' liquid legacy. *Science,* 225: 154.

Holden, C., Ed., 1993. Random samples. *Science,* 262: 1510–1511.

Holden, C. 1997a. Tooling around: dates show early Siberian settlement. *Science,* 275: 1268.

Holden, C. 1997b. Largest dino claw unearthed. *Science,* 278: 1063.

Holden, C., Ed., 1999a. Random samples. *Science,* 285: 17.

Holden, C. 1999b. Kansas dumps Darwin, raises alarm across the United States. *Science,* 285: 1186–1187.

Holland, H.D. 1997. Evidence for life on Earth more than 3850 years ago. *Science,* 275: 38–39.

Hölldobler, B. and E.O. Wilson. 1990. *The Ants.* The Belknap Press of Harvard University Press, Cambridge, MA, 732 pp.

Honeycutt, R.L. 1992. Naked mole-rats. *American Scientist,* 80: 43–53.

Honeycutt, R.L. 1997. Evolutionary issues (book review). *Science,* 275: 36–37.

Hoogland, J.L. 1986. Nepotism in Prairie Dogs (*Cynomys ludovicianus*) varies with competition but not with kinship. *Animal Behaviour,* 34: 263–270.

Hooker, M.E. and E.M. Barrows. 1989. Clutch sizes and sex ratios in *Pediobius foveolatus* (Hymenoptera: Eulophidae), primary parasites of *Epilachna varivestis* (Coleoptera: Coccinellidae). *Annals of the Entomological Society of America,* 82: 460–465.

Hooker, M.E. and E.M. Barrows. 1992. Clutch size reduction and host discrimination in the superparsitizing gregarious endoparasitic wasp *Pediobius foveolatus* (Hymenoptera: Eulophidae). *Annals of the Entomological Society of America,* 85: 207–213.

Horai, S., Y. Satta, K. Hayasaka, R. Kondo, T. Inoue, T. Ishida, S. Hayashi, and N. Takahata. 1992. Man's place in Hominoidea revealed by mitochondrial DNA genealogy. *Journal of Molecular Evolution,* 35: 32–43.

Horrocks, J.A. and W. Hunte. 1986. Sentinel behaviour in vervet monkeys: who sees whom first? *Animal Behaviour,* 34: 1566–1568.

Houck, L.D. and N.L. Reagan. 1990. Male courtship pheromones increase female receptivity in a plethodontid salamander. *Animal Behaviour,* 39: 729–734.

Houston, T.R. 1991. Two new and unusual species of the bee genus *Leioproctus* Smith (Hymenoptera: Colletidae), with notes on their behaviour. *Record of the Western Australian Museum,* 15: 83–96.

Howard, D.J. 1988. The species problem (book review). *Evolution,* 42: 1111–1112.

Hrdy, S.B. 1977. *The Langurs of Abu: Female and Male Strategies of Reproduction.* Harvard University Press, Cambridge, MA, 198 pp.

Hrdy, S.B. 1996. The evolution of female orgasms: logic please but no atavism. *Animal Behaviour,* 52: 851–852.

Hubbell, S.P. 1999. Response to letters of R.L. Chazdon et al. and R.K. Kobe. *Science,* 285: 1459 (www.sciencemag.org/cgi/content/full/285/5433/1459a).

Hubbell, S.P., R.B. Foster, S.T. O'Brien, K.E. Harms, R. Londi, B. Wechsler, S.J. Wright, and S. Loo de Lao. 1999. Light-gap distributions, recruitment limitation, and tree diversity in a neotropical forest. *Science,* 283: 554–557.

Hughes, A.J., and D.M. Lambert. 1984. Functionalism, structuralism, and "ways of seeing." *Journal of Theoretical Biology,* 111: 787–800.

Hull, D.L. 1984. Evolutionary thinking observed (book review). *Science,* 223: 923–924.

Hull, D.L. 1992. Individual, pp. 180–187 in E.F. Keller and E.A. Lloyd, Eds., *Keywords in Evolutionary Biology.* Harvard University Press, Cambridge, MA, 414 pp.

Hunter, A.F. and L.W. Aarssen. 1988. Plants helping plants. *BioScience,* 38: 34–40.

Hurlbert, S.H. 1984. Pseudoreplication and the design of ecological field experiments. *Ecological Monographs,* 54: 187–211.

Hurnick, J.F., A.B. Webster, and P.B. Siegel. 1995. *Dictionary of Farm Animal Behavior.* Iowa State University, Ames, 200 pp.

Hurst, L.D. and A. Grafen. 1993. Unusual mutational mechanisms and evolution. *Science,* 260: 1959.

Huxley, J. 1957. The three types of evolutionary process. *Nature,* 180: 454–455.

Illangasekare, M., G. Sanchez, T. Nickeles, and M. Yarus. 1995. Amnoacyl-RNA synthesis catalyzed by an RNA. *Science,* 267: 643–647.

Imai, K. 1999. The haemoglobin enzyme. *Nature,* 401: 437.

Immelmann, K., Ed., 1977. *Grzimek's Encyclopedia of Ethology,* English ed. Van Nostrand-Reinhold, New York, 705 pp.

Immelmann, K. and C. Beer. 1989. *A Dictionary of Ethology.* Harvard University Press, Cambridge, MA, 336 pp.

Inouye, D.W. 1980. The terminology of floral larceny. *Ecology,* 61: 1251–1253.

Itani, J. 1959. Paternal care in the Wild Japanese Monkey, *Macaca fuscata fuscata. Primates,* 2: 61–93.

Jackson, J.B.C. and A.H. Cheetham. 1990. Evolutionary significance of morphospecies: a test with cheilostome Bryozoa. *Science,* 248: 579–583.

Jackson, R.R. and C.E. Griswold. 1979. Nest associates of *Phidippus johnsoni* (Araneae, Salticidae). *Journal of Arachnology,* 7: 59–67.

Jacobson, E. 1987. *Biology of Emotions.* Charles C Thomas, Springfield, IL, 211 pp.

Jaeger, R.G. 1988. A comparison of territorial and non-territorial behaviour in two species of salamanders. *Animal Behaviour,* 36: 307–309.

James, A.N., K.J. Gaston, and A. Balmford. 1999. Balancing the Earth's accounts. *Nature,* 401: 323–324.

Jamieson, I.G. and J.L. Craig. 1987. Male-male and female-female courtship and copulation in a communally breeding bird. *Animal Behaviour,* 35: 1251–1252.

Jamison, A. 1993. A tale of two brothers (book review). *Science,* 261: 497–498.

Janik, V.M. and P.J.B. Slater. 1997. Vocal learning in mammals, pp. 59–99 in P.J.B. Slater, J.S. Rosenblatt, C.T. Snowdon, and M. Milinski, Eds., *Advances in the Study of Behavior,* Vol. 26. Academic Press, San Diego, CA, 484 pp.

Janis, C. 1976. The evolutionary strategy of the Equidae and the origins of rumen and cecal digestion. *Evolution,* 30: 757–774.

Jannasch, H.W. and M.J. Mottl. 1985. Geomicrobiology of deep-sea hydrothermal vents. *Science,* 229: 717–725.

Janzen, D.H. 1980. When is it coevolution? *Evolution,* 34: 611–612.

Jaroff, L. 1992. Nature's time capsules. *Time,* 139(14): 61.

Jarvis, J.U.M. 1981. Eusociality in a mammal: cooperative breeding in Naked Mole-Rat colonies. *Science,* 212: 571–573.

Jasny, B.R., coordinator. 1991. Genome maps 1991. *Science,* 254: 1 page.

Jeffreys, W.H. and J.O. Berger. 1992. Ockham's razor and Bayesian analysis. *American Scientist,* 80: 64–72.

Jenssen, T.A., D.L. Marcellini, K.A. Buhlmann, and P.H. Goforth. 1989. Differential infanticide by adult curly-tailed lizards, *Leiocephalus schreibersi. Animal Behaviour,* 38: 1054–1061.

Johansen, C.A. 1977. Pesticides and pollinators. *Annual Review of Entomology,* 22: 177–192.

John, P. and F.R. Whatley. 1975. *Paracoccus denitrificans* and the evolutionary origin of the mitochondrion. *Nature,* 254: 495–498.

Johnson, D.L. 1967. Honey Bees: do they use the direction information contained in their dance maneuver? *Science,* 155: 844–847.

Johnson, D.L. and A.M. Wenner. 1970. Recruitment efficiency in honeybees: studies of the role of olfaction. *Journal of Apicultural Research,* 9: 13–18.

Johnson, F.H. 1966. Introduction, pp. 3–28 in F.H. Johnson and Y. Hareda, Eds. *Bioluminescence in Progress.* Princeton University Press, New Jersey, 650 pp.

Johnson, R. 1972. *Aggression in Man and Animals.* W.B. Saunders, Philadelphia, PA, 253 pp.

Johnson, T.E. 1990. Increased life-span *age-1* mutants in *Caenorhabditis elegans* and lower Gompertz rate of aging. *Science,* 249: 908–912.

Johnsson, J.I. 1996. Statistics and biological sense: a reply to Thomas and Juanes. *Animal Behaviour,* 52: 860.

Jolivet, P., J. Vasconcellos-Neto, and P. Weinstein. 1990. Cycloalexy: a new concept in the larval defense of insects. *Insecta Mundi,* 4: 133–142.

Jones, C.G., R.S. Ostfeld, M.P. Richard, E.M. Schauber, and J.O. Wolff. 1998. Chain reactions linking acorns to gypsy moth outbreaks and lyme disease risk. *Science,* 279: 1023–1026.

Jones, D. 1984. Use, misuse, and role of multiple-comparison procedures in ecological and agricultural entomology. *Environmental Entomology,* 13: 635–649.

Jordon, G. 1997. Seeking flora and fauna in the wilds of the web. *New York Times,* 4 August: D4.

Joshi, A.K. 1991. Natural language processing. *Science,* 253: 1242–1249.

Judson, O.P. and B.B. Normark. Ancient asexual scandals. *Trends in Ecology and Evolution,* 11(2): 41–46.

Justice, W.S. and C.R. Bell. 1968. *Wild Flowers of North Carolina.* University of North Carolina Press, Chapel Hill, 217 pp.

Kaiser, J. 1999. Stemming the tide of invading species. *Science,* 285: 1839–1841.

Kamen, A. 1986a. Creationism case raises issues of faith, "freedom." *Washington Post,* 10 December: A1, A4–5.

Kamen, A. 1986b. Creation science law had secular intent, High Court told. *Washington Post,* 11 December: A13.

Kaneko, T., S. Sato, H. Kotani, A. Tanaka, E. Asamizu, Y. Nakamura, N. Miyajima, M. Hirosawa, M. Sugiura, S. Sasamoto et al. 1996. Sequence analysis of the genome of the unicellular cyanobacterium *Synechocystis* sp., strain PCC6803. II. Sequence determination of the entire genome and assignment of potential protein-coding regions. *DNA Research,* 3: 190–136.

Kaplan, D.R. and W. Hagemann. 1991. The relationship of cell and organism in vascular plants. *BioScience,* 41: 693–703.

Kaplan, E.H. 1982. *A Field Guide to Coral Reefs: Caribbean and Florida.* Houghton Mifflin, Boston, MA, 289 pp.

Kaplan, N.L., R.R. Hudson, and C.H. Langley. 1989. The "hitchhiking effect" revisited. *Genetics,* 123: 887–899.

Karlson, P. and A. Butenandt. 1959. Pheromones (ectohormones) in insects. *Annual Review of Entomology,* 4: 39–58.

Karlson, P. and M. Luscher. 1959. "Pheromones," a new term for a class of biologically active substances. *Nature,* 183: 155–176.

Katcher, A.H. 1985. Physiologic and behavioral responses to companion animals. *Symposium on the Human-Companion Animal Bond,* Veterinary Clinics of North America, Small Animal Practice, 15: 403–410.

Kaufmann, J.H. 1989. The Wood Turtle stomp. *Natural History,* 98(8): 8–13.

Keane, B. 1990. The effect of relatedness on reproductive success and mate choice in the white-footed mouse, *Peromyscus leucopus. Animal Behaviour,* 39: 264–273.

Keener, C.S. 1983. Distribution and biohistory of the endemic flora of the mid-Appalachian shale barrens. *Botanical Review,* 49: 65–115.

Keller, E.F. 1992a. Competition: current usages, pp. 68–73 in E.F. Keller and E.A. Lloyd, Eds., *Keywords in Evolutionary Biology.* Harvard University Press, Cambridge, MA, 414 pp.

Keller, E.F. 1992b. Fitness: reproductive ambiguities, pp. 120–121 in E.F. Keller and E.A. Lloyd, Eds., *Keywords in Evolutionary Biology.* Harvard University Press, Cambridge, MA, 414 pp.

Keller, E.F., and E.A. Lloyd, Eds. 1992. *Keywords in Evolutionary Biology.* Harvard University Press, Cambridge, MA, 414 pp.

Kelley, D.D. 1991. Sleep and dreaming, pp. 792–804 in E.R. Kandel, J.H. Schwartz, and T.M. Jessell, Eds., *Principles of Neural Science,* 3rd ed. Appleton and Lange, Norwalk, CT, 1135 pp.

Kelves, D.J. 1992. Eugenics, pp. 92–94 in E.F. Keller and E.A. Lloyd, Eds., *Keywords in Evolutionary Biology,* Harvard University Press, Cambridge, MA, 414 pp.

Kenrick, P. 1999. The family tree flowers. *Nature,* 402: 358–359.

Kenworthy, T. 1995a. Tour operator was client of Babbitt's ex-law firm; rise in Alaska permits may benefit company. *Washington Post,* 24 January: A3.

Kenworthy, T. 1995b. Justices affirm wide power to protect wildlife habitat. *Washington Post,* 30 June: A1, A19.

Kenworthy, T. 1996a. Condor's return may dispel some distrust of species act: release of spectacular bird set in Arizona. *Washington Post,* 9 August: A1, A10.

Kenworthy, T. 1996b. Condors soar again over Grand Canyon. Decades-long struggle to save endangered bird passes critical milestone. *Washington Post,* 13 December: A1, A35.

Kenworthy, T. 1999. Invader fuels a fiery cycle. *Washington Post,* 15 November: A11.

Kevles, D.J. 1992. Eugenics, pp. 92–94 in E.F. Keller and E.A. Lloyd, Eds., *Keywords in Evolutionary Biology.* Harvard University Press, Cambridge, MA, 414 pp.

Kerr, R.A. 1984. Periodic impacts and extinctions reported. *Science,* 223: 1277–1279.

Kerr, R.A. 1985. Making the Moon from a big splash. *Science,* 226: 1060–1061.

Kerr, R.A. 1988. No longer willful, Gaia becomes respectable. *Science,* 240: 393–395.

Kerr, R.A. 1989. Asteroids, dinosaurs and the big splat. *Washington Post,* 7 May, B3.

Kerr, R.A. 1992. Extinction by a one-two comet punch? *Science,* 255: 160–161.

Kerr, R.A. 1993a. Evolution's big bang gets even more explosive. *Science,* 261: 1274–1275.

Kerr, R.A. 1993b. A bigger death knell for the dinosaurs? *Science,* 261: 1518–1519.

Kerr, R.A. 1997a. Cores document ancient catastrophe. *Science,* 275: 1265.

Kerr, R.A. 1997b. Did a blast of seafloor gas usher in a new age? *Science,* 275: 1267.

Kessel, E.L. 1955. The mating activities of Balloon Flies. *Systematic Zoology,* 4: 997–1004.

Kiltie, R.A. 1989. Testing Thayer's countershading hypothesis: an image processing approach. *Animal Behaviour,* 38: 542–544.

Kimelberg, H.K. 1987. "Homology" controversy. *Science,* 238: 1217.

Kimura, M. 1968. Evolutionary rate at the molecular level. *Nature,* 217: 624–626.

Kimura, M. 1979. The neutral theory of molecular evolution. *Scientific American,* 241(11): 98–126.

Kimura, M. 1985. The neutral theory of molecular evolution. *New Scientist,* 1464: 41–43, 46.

Kimura, M. 1987. Molecular evolutionary clock and neutral theory. *Journal of Molecular Evolution,* 26: 24–33.

Kimura, M. 1991. Some recent data supporting the neutral theory, pp. 3–14 in M. Kimura and N. Takahata, Eds., *New Aspects of the Genetics of Molecular Evolution*. Japan Scientific Societies Press, Tokyo, 322 pp.

Kimura, M. 1992. Neutralism, pp. 225–230 in E.F. Keller and E.A. Lloyd, Eds., *Keywords in Evolutionary Biology*. Harvard University Press, Cambridge, MA, 414 pp.

King, J.A. and N.L. Gurney. 1954. Effect of early social experiences on adult aggressive behavior in C57BL/10 mice. *Journal of Comparative Physiological Psychology,* 47: 326–330.

King, R.C., and W.D. Stansfield. 1985. *A Dictionary of Genetics,* 3rd ed. Oxford University Press, New York. 480 pp.

King's College Sociobiology Group, Cambridge 1982. *Current Problems in Sociobiology*. Cambridge University Press, New York, 394 pp.

Kingsolver, J.G., H.A. Woods, and G. Gilchrist. 1995. Traits related to fitness (book review). *Science,* 267: 396.

Kingsolver, J.G. and M.A.R. Koehl. 1985. Aerodynamics, thermoregulation, and the evolution of insect wings: differential scaling and evolutionary change. *Evolution,* 39: 488–504.

Kinston, J.D., B.D. Marino, and A. Hill. 1994. Isotopic evidence for neogene hominid paleoenvironments in the Kenya Rift Valley. *Science,* 264: 955–959.

Kirkendall, L.R. 1983. The evolution of mating systems in bark and ambrosia beetles (Coleoptera: Scolytidae and Platypodidae). *Zoological Journal of the Linnaean Society,* 77: 293–352.

Kirkendall, L.R. 1984. Long copulations and post-copulatory "escort" behaviour in the Locust Leaf Miner, *Odontota dorsalis* (Coleoptera: Chrysomelidae). *Journal of Natural History,* 18: 905–919.

Kirkendall, L.R. and N.C. Stenseth. 1985. On defining "breeding once." *American Naturalist,* 125: 189–204.

Kirkpatrick, M. 1985. Evolution of female choice and male parental investment in polygynous species: the demise of the "sexy son." *American Naturalist,* 125: 788–810.

Kirkpatrick, M. 1986. The handicap mechanism of sexual selection does not work. *American Naturalist,* 127: 222–240.

Kitcher, P. 1992. Gene: current usages, pp. 128–131 in E.F. Keller and E.A. Lloyd, Eds., *Keywords in Evolutionary Biology*. Harvard University Press, Cambridge, MA, 414 pp.

Kleinbaum, D.G., and L.L. Kupper. 1978. *Applied Regression Analysis and Other Multivariable Methods*. Duxbury Press, North Scituate, MA, 556 pp.

Klopfer, P. 1969. *Habitats and Territories: A Study of the Use of Space by Animals*. Basic Books, New York, 177 pp.

Knoll, A.H., R.K. Bambach, D.E. Canfield, and J.P. Grotzinger. 1996. Comparative Earth history and Late Permian mass extinction. *Science,* 273: 452–457.

Kobayashi, M., H. Furuya, and P.W.H. Holland. 1999. Dicyemids are higher animals. *Nature,* 401: 762.

Kobe, R.K. 1999. Tropical tree richness and resource-based niches. *Science,* 285: 1459 (www.sciencemag.org/cgi/content/full/285/533/1459a).

Kolata, G. 1982. New theory of hormones proposed. *Science,* 215: 1383–1384.

Kolata, G. 1985. Finding biological clocks in fetuses. *Science,* 230: 929–930.

Kolata, G. 1986. What does it mean to be "rare" or "likely?" *Science,* 234: 542.

Kolata, G. 1995. Researchers find hormone causes a loss of weight. *New York Times,* 27 July. A1, A20.

Kolata, G. 1997. Scientist reports first cloning ever of adult mammal. *New York Times,* 23 February: 1, 22.

Koonin, E.V. and M.Y. Glaperin. 1997. Prokaryotic genomes: the emerging paradigm of genome-based microbiology. *Current Opinion in Genetics and Development,* 7: 757–763.

Krajick, K. 1997. The riddle of the Carolina bays. *Smithsonian,* September: 45–55.

Kramer, D.L. 1983. The evolutionary ecology of respiratory mode in fishes: an analysis based on the costs of breathing. *Environmental Biology of Fishes,* 9: 145–158.

Kramer, D.L. 1987. Dissolved oxygen and fish behavior. *Environmental Biology of Fishes,* 18: 81–92.

Kramer, D.L. 1988. The behavioral ecology of air breathing by aquatic animals. *Canadian Journal of Zoology,* 66: 89–94.

Kramer, D.L. and J.P. Mehegan. 1981. Aquatic surface respiration, an adaptive response to hypoxia in the Guppy, *Poecilia reticulata* (Pisces, Poeciliidae). *Environmental Biology of Fishes,* 6: 299–313.

Kramer, D.L. and D.M. Weary. 1991. Exploration versus exploitation: a field study of time allocation to environmental tracking by foraging chipmunks. *Animal Behavior,* 41: 443–449.

Krebs, C.J. 1985. *Ecology: The Experimental Analysis of Distribution and Abundance,* 3rd ed. Harper & Row, New York, 800 pp.

Kricher, J.C. and G. Morrison. 1988. *Eastern Forests.* Houghton Mifflin, Boston, MA, 368 pp.

Kricher, J.C. and G. Morrison. 1993. *A Field Guide to the Ecology of Western Forests.* Houghton Mifflin, Boston, MA, 554 pp.

Kristal, M.B. 1980. Placentophagia: a biobehavioral enigma (or *De gustibus non disputandum est*). *Neuroscience and Biobehavior Review,* 4: 141–150.

Kritsky, G. 1991. Darwin's Madagascan hawk moth prediction. *American Entomologist,* 37: 206–210.

Krizek, G.O. 1984. Entomologists and their visual illusions. *Bulletin of the Entomological Society of America,* 30: 4.

Küchler, A.W. 1964. *Manual to Accompany the Map Potential Natural Vegetation of the Conterminous United States,* Special Publ. No. 36. American Geographical Society, New York, 38 pp + map.

Kukalova-Peck, J. 1978. Origin and evolution of insect wings and their relation to metamorphosis as documented by the fossil record. *Journal of Morphology,* 156: 53–126.

Kukuk, P.F. 1983. Evidence of an antiaphrodisiac in the sweat bee *Lasioglossum* (*Dialictus*) *zephyrum. Science,* 227: 656–657.

Kunin, W.E. and K.J. Gaston. 1993. The biology of rarity: patterns, causes and consequences. *Trends in Ecology and Evolution,* 8: 298–301.

Kutschera, U. and P. Wirtz. 1986. A leech that feeds its young. *Animal Behaviour,* 34: 941–942.

Labandeira, C.C. 1998. How old is the flower and the fly? *Science,* 280: 57–59.

Labandeira, C.C. and J.J. Sepkoski, Jr. 1993. Insect diversity in the fossil record. *Science,* 261: 310–315.

Lack, D. 1954. *The Natural Regulation of Animal Numbers.* Oxford University Press, London. 343 pp.

Ladle, R. J. 1992. Parasites and sex: catching the Red Queen. *Trends in Ecology and Evolution,* 7: 405–408.

LaGory, K.E. 1987. The influence of habitat and group characteristics on the alarm and flight response of White-Tailed Deer. *Animal Behaviour,* 35: 20–25.

Lambert, D. 1983. *A Field Guide to Dinosaurs.* The Diagram Group, Avon Books, New York, 256 pp.

Lambert, D. and the Diagram Group. 1990. *The Dinosaur Data Book.* Avon Books, New York, 320 pp.

Lambert, D.M. and A.J. Hughes. 2000. Keywords and concepts in structuralist and functionalist biology (manuscript).

Lanciani, C.A. 1998. Reader-friendly writing in science. *Bulletin of the Ecological Society of America,* 9(2): 171–172.

Landauer, M.R., E.M. Banks, and C.S. Carter. 1978. Sexual and olfactory preferences of naive and experienced male hamsters. *Animal Behaviour,* 26: 611–621.

Lander, E.S. 1996. The new genomics: global views of biology. *Science,* 274: 536–539.

Landman, O.E. 1991. The inheritance of acquired characteristics. *Annual Review of Genetics,* 25: 1–20.

Lank, D.B., P. Mineau, R.F. Rockwell, and F. Cooke. 1989. Intraspecific nest parasitism and extra-pair copulation in lesser snow geese. *Animal Behaviour,* 37: 74–89.

Lapidus, D.F. 1987. *Dictionary of Geology and Geophysics.* Facts on File, New York, 347 pp.

Larson, A. 1995. Evolutionary revelations (book review). *Science,* 267: 115–116.

Latimer, W. and M. Sippel. 1987. Acoustic cues for female choice and male competition in *Tettigonia cantans. Animal Behaviour,* 35: 887–900.

Laurent, J. 1990. Roots of "superorganism" (letters to the editors). *American Scientist,* 78: 197.

Lavigne, R.J. and F.R. Holland. 1969. *Comparative Behavior of Eleven Species of Wyoming Robber Flies (Diptera: Asilidae).* Agricultural Experiment Station, University of Wyoming, Laramie, 61 pp.

Law, J.H. and F.E. Regnier. 1971. Pheromones. *Annual Review of Biochemistry,* 40: 533–548.

Lawrence, G.H.M. 1951. *Taxonomy of Vascular Plants.* Macmillan Company, New York, 823 pp.

Lawton, J.H., D.E. Bignell, B. Bolton, G.F. Bloemers, P. Eggleton, P.M. Hammond, M. Hodda, R.D. Holts, T.B. Larsen, N.A. Mawdsley, N.E. Stork, D.S. Srivastava, and A.D. Watt. 1998. Biodiversity inventories indicator taxa and effects of habitat modification in tropical forest. *Nature,* 391: 72–76.

LaRue, P., L. Bélander, and J. Huot. 1995. Riparian edge effects on boreal Balsam Fir bird communities. *Canadian Journal of Forest Research,* 25: 555–566.

Lazarus, J. 1990. The logic of mate desertion. *Animal Behaviour,* 39: 672–684.

Lazarus, J. and I. R. Inglis. 1986. Shared and unshared parental investment, parent-offspring conflict and brood size. *Animal Behaviour,* 34: 1791–1804.

Leal, M. 1999. Honest signalling during prey-predator interactions in the lizard *Anolis cristalellus. Animal Behaviour,* 58: 521–526.

Lean, G., D. Hinrichsen, and A. Markham. 1990. *WWF Atlas of the Environment.* World Wildlife Federation, New York, 195 pp.

Leary, W.E. 1996. The toad shows that "handedness" is not exclusive to the higher orders. *New York Times,* 6 February: C10.

Leary, W.E. 1998. New therapy offers promise in treatment of pedophiles. *New York Times Northeast,* 12 February: A11.

Lee, A. and R. Martin. 1990. Life in the slow lane. *Natural History,* August: 34–43.

Lee, P.C. 1987. Allomothering among African Elephants. *Animal Behaviour,* 35: 278–291.

Lehman, H.J. 1991. Developers put the accent on green: nature-friendly plans unite homes, environmentally protected areas. *Washington Post,* 14 December: E1, E4.

Lehner, P.N. 1979. *Handbook of Ethological Methods.* Garland STPM Press, New York, 403 pp.

Lehner, P.N. 1988. Avian Davian behavior. *Wilson Bulletin.* 100: 293–294.

Lemonick, M.D. 1996. Big, fast and vicious. *Time,* 147: 45.

Lemonick, M.D. 1999. Smart genes? *Time,* 13 September: 54–58.

Lemonick, M.D. and A. Dorfman. 1999. Up from the apes. *Time,* 23 August: 50–58.

Lennox, J.G. 1992. Teleology, pp. 324–333 in E.F. Keller and E.A. Lloyd, Eds., *Keywords in Evolutionary Biology.* Harvard University Press, Cambridge, MA, 414 pp.

Lenski, R.E. and J.E. Mittler. 1993a. The directed mutation controversy and neo-Darwinism. *Science,* 259: 188–194.

Lenski, R.E. and J.E. Mittler. 1993b. Unusual mutational mechanisms and evolution. *Science,* 260: 1959–1960.

Lenski, R.E., C. Ofria, T.C. Collier, and C. Adami. 1999. Genome complexity, robustness and genetic interactions in digital organisms. *Nature,* 400: 661–664.

Lenteren, J.C. van, K. Bakker, J.J.M. van Alphen. 1978. How to analyse host discrimination. *Ecological Entomology,* 3: 71–75.

Leuthold, W. 1966. Variations in territorial behavior of Uganda kob *Adenota kob thomasi* (Neumann 1896). *Behaviour,* 27: 215–258.

Levin, B.R. 1984. Science as a way of knowing: molecular evolution. *American Zoologist,* 24: 451–464.

Levin, H. 1987. Successions in psychology (book review). *Science,* 236: 1683–1684.

Levinton, J.S. 1983. Stasis in progress: the empirical basis of macroevolution. *Annual Review of Ecology and Systematics,* 14: 103–137.

Levinton, J.S. 1988. *Genetics, Paleontology, and Macroevolution.* Cambridge University Press, New York, 637 pp.

Lewin, R. 1981. Do jumping genes make evolutionary leaps? *Science,* 213: 634–636.

Lewin, R. 1984. Practice catches theory in kin recognition. *Science,* 223: 1049–1051.

Lewin, R. 1985a. Red queen runs into trouble? *Science,* 227: 399–400.

Lewin, R. 1985b. On the origin of insect wings. *Science,* 230: 428–429.

Lewin, R. 1986. Dexterous early hominids. *Science,* 231: 115.

Lewin, R. 1987a. When does homology mean something else? *Science,* 237: 1570.

Lewin, R. 1987b. Do animals read minds, tell lies? *Science,* 238: 1350–1351.

Lewin, R. 1987c. The unmasking of Mitochondrial Eve. *Science,* 238: 24–26.

Lewin, R. 1988. Molecular clocks turn a quarter century. *Science,* 239: 561–563.

Lewin, R. 1989a. Limits to DNA fingerprinting. *Science,* 243: 1549–1551.

Lewin, R. 1989b. Sources and sinks complicate ecology. *Science,* 243: 477–478.

Lewin, R. 1991. DNA evidence strengthens Eve hypothesis. *New Scientist,* 132(1791): 20.

Lewin, R. 1993. *Human Evolution: An Illustrated Introduction.* Blackwell Scientific, Oxford, 203 pp.

Lewin, R.A. 1982. Symbiosis and parasitism: definitions and evaluations. *BioScience,* 32: 254–259.

Lewis, D.B. and D.M. Gower. 1980. *Biology of Communication.* John Wiley & Sons, New York, 231 pp.

Lewis, E.E. and J.H. Cane. 1990. Stridulation as a primary antipredator defense of a beetle. *Animal Behaviour,* 40:1003–1004.

Lewis-Beck, M.S. 1980. *Applied Regression: An Introduction.* Sage Publications, Beverly Hills, CA, 77 pp.

Lewontin, R.C. 1992. Genotype and phenotype, pp. 137–144 in E.F. Keller and E.A. Lloyd, Eds., *Keywords in Evolutionary Biology.* Harvard University Press, Cambridge, MA, 414 pp.

Li, W.-H. and M. Tanimura. 1987. The molecular clock runs more slowly in man than in apes and monkeys. *Nature,* 326: 93–96.

Li, W.-H., M. Gouy, P.M. Sharp, C. O'hUigin, and Y.-W. Yang. 1990. Molecular phylogeny of Rodentia, Lagomorpha, Primates, Artiodactyla, and Carnivora and molecular clocks. *Proceedings of the National Academy of Sciences USA,* 87: 6703–6707.

Licht, L.E. and J.P. Bogart. 1987. More myths. *BioScience,* 37: 4–5.

Lichtenberger M., J. Lechien, and A. Elens. 1987. Influence of light intensity on rare-male advantage in *Drosophila melanogaster. Behavior Genetics,* 17: 203–210.

Lieberman, P. 1973. On the evolution of language: a unified view. *Cognition,* 2: 59–94.

Liebling, B.A., and P. Shaver. 1973. Evaluation, self-awareness, and task performance. *Journal of Experimental Social Psychology,* 9: 297–306.

Lifjeld, J.T., and T. Slagsvold. 1986. The function of courtship feeding during the incubation in the Pied Flycatcher *Ficedula hypoleuca. Animal Behaviour,* 34: 1441–1453.

Lightman, A. and O. Gingerich. 1992. When do anomalies begin? *Science,* 255: 690–695.

Lincoln, R.J., G.A. Boxshall, and P.F. Clark. 1985. *A Dictionary of Ecology, Evolution, and Systematics.* Cambridge University Press, New York, 298 pp.

Lindauer, M. 1961. *Communication Among Social Bees.* Harvard University Press, Cambridge, MA, 143 pp.

Lindeman, R.L. 1942. The trophic-dynamic aspect of ecology. *Ecology,* 23: 399–418.

Lindsey, C. 1966. Body sizes of poikilotherm vertebrates at different latitudes. *Evolution,* 20: 456–465.

Line, L. 1996. Microcosmic captive breeding project offers new hope for beleaguered beetle. *New York Times,* 17 September: C1, C4.

Line, L. 1997. Woodchucks are in the lab, but their body clocks are wild. *New York Times,* 28 January: C2.

Line, S.W. 1987. Environmental enrichment for laboratory primates. *JAVMA,* 190: 854–859.

Linnaeus, C. (C. von Linné). (1735) 1758. *Caroli Linnaei Systema Naturae, Regnum Animale,* 10th ed. British Museum (Natural History), London. 30 pp.

Linsenmaier, W. 1972. *Insects of the World.* McGraw-Hill, New York, 392 pp.

Lipton, J. 1968. *An Exaltation of Larks or The Venereal Game.* Grossman Publishers, New York, 119 pp.

Little, E.L. 1996 (1980). *The Audubon Society Field Guide to North American Trees: Western Region.* Alfred A. Knopf, New York, 640 pp.

Little, R.J. 1983. A review of floral food deception mimicries with comments on floral mutualism, pp. 294–309 in C.E. Jones and R.J. Little, Eds., *Handbook of Experimental Pollination Biology.* Scientific and Academic Editions, New York, 558 pp.

Lloyd, E.A. 1992. Unit of selection, pp. 334–340 in E.F. Keller and E.A. Lloyd, Eds., *Keywords in Evolutionary Biology*. Harvard University Press, Cambridge, MA, 414 pp.

Lloyd, M. and H.S. Dybas. 1966a. The periodical cicada problem. I. Population ecology. *Evolution,* 20: 133–149.

Lloyd, M. and H.S. Dybas. 1966b. The periodical cicada problem. II. Evolution. *Evolution,* 20: 466–505.

Lorenz, K.Z. 1957. The past twelve years in the comparative study of behavior, pp. 288–310 in C.H. Schiller, Ed., *Instinctive Behavior*. International University Press, New York, 328 pp.

Lorenz, K.Z. 1981. *The Foundations of Ethology*. Springer-Verlag, New York, 380 pp.

Lovelock, J.E. and L. Margulis. 1974. Homeostatic tendencies of the Earth's atmosphere. *Origins of Life,* 5: 93–103.

Lyons, W., Ed., 1980. *Emotion.* Cambridge University Press, Cambridge, U.K., 230 pp.

MacDonald, J.F., R.W. Matthews, and R.S. Jacobson. 1980. Nesting biology of the yellowjacket, *Vespula flavopilosa* (Hymenoptera: Vespidae). *Journal of the Kansas Entomological Society,* 53: 448–458.

MacDonald, I.R., J.F. Reilly II, N.K. Guinasso, Jr., J.M. Brooks, R.S. Carney, W.A. Bryant, and T.J. Bright. 1990. Chemosynthetic mussels at a brine-filled pockmark in the northern Gulf of Mexico. *Science,* 248: 1096–1099.

Mace, R. 1986. Importance of female behaviour in the dawn chorus. *Animal Behaviour,* 34: 621–622.

MacQueen, G., J. Marshall, M. Perdue, S. Siegel, and J. Bienenstock. 1989. Pavlovian conditioning of rat mucosal mast cells to secrete rat mast cell protease II. *Science,* 243: 83–85.

Maddison, D.R. 1991. African origin of human mitochondrial DNA reexamined. *Systematic Zoology,* 40: 355–363.

Maddison, D.R. and W.P. Maddison. 1998. The Tree of Life: a multi-authored, distributed Internet project containing information about phylogeny and biodiversity. Internet address: http://phylogeny.arizona.edu/tree/phylogeny.html

Maienschein, J. 1992. Gene: historical perspectives, pp. 122–127 in E.F. Keller and E.A. Lloyd, Eds., *Keywords in Evolutionary Biology*. Harvard University Press, Cambridge, MA, 414 pp.

Maier, N.R.F., and T.C. Schneirla. 1964. *Principles of Animal Psychology*. Dover Publications, New York, 683 pp.

Malakoff, D. 1997. Thirty Kyotos needed to control warming. *Science,* 279: 2048.

Malakoff, D. 1999. Fighting fire with fire. *Science,* 285: 1841–1843.

Maley, L.E. and C.R. Marshall. 1998. The coming of age of molecular systematics. *Science,* 279: 505–506.

Mann, C. and M. Plummer. 1997. Qualified thumbs up for habitat plan science. *Science,* 278: 2052–2053.

Mann, J. 1991. Home observations of high-risk premature infants and their mothers during the first year of life: a microanalytic behavioral study. Ph.D. dissertation. University of Michigan, Ann Arbor.

Mansfield, S. 1995. Women's right to be in a bad mood. *Washington Post,* 10 June: B1, B6.

Marano, H.E. 1997. Puberty may start at 6 as hormones surge. *New York Times,* 1 July: C1, C6.

Marden, M.H. 1992. Newton's second law of butterflies. *Natural History,* January: 54–61.

Margulis, L. 1981. *Symbiosis in Cell Evolution: Life and Its Environment on the Early Earth*. W.H. Freeman, San Francisco, CA. 419 pp.

Margulis, L. 1992. Symbiosis theory: cells a microbial communities, pp. 149–172 in L. Margulis and L. Olendzenski, Eds., *Environmental Evolution*. MIT Press, Cambridge, MA, 404 pp.

Margulis, L. 1993. Gaia in science. *Science,* 259: 745.

Margulis, L. and J.E. Lovelock. 1974. Biological modulation of the Earth's atmosphere. *Icarus,* 21: 471–489.

Margulis, L., D. Sagan, and L. Olendzenski. 1985. What is sex? pp. 69–85 in H.O. Halvorson and A. Monroy, Eds., *The Origin and Evolution of Sex*. Alan R. Liss, New York, 345 pp.

Marler, P. 1961. The logical analysis of animal communication. *Journal of Theoretical Biology,* 1: 295–317.

Marshall, E. 1996. The Genome Program's conscience. *Science,* 274: 488–490.

Marshall, L. and J. Alcock. 1981. The evolution of the mating system of *Xylocopa varipuncta* (Hymenoptera: Anthophoridae). *Journal of Zoology,* 193: 315–324.

Martin, P. 1984. The meaning of weaning. *Animal Behaviour,* 32: 1257–1259.

Martin, P. 1985. Weaning: a reply to Counsilman and Lim. *Animal Behaviour,* 33: 1024–1026.

Marx, J. 1992. Homeobox genes go evolutionary. *Science,* 255: 399–401.

Marx, M.H. and W.A. Hillix. 1973. *Systems and Theories in Psychology,* 2nd ed. McGraw-Hill, New York, 625 pp.

Maschwitz, U.W. 1964. Alarm substances and alarm behavior in social insects. *Vitamins and Hormones,* 24: 267–290.

Maslow, A.H. 1935. Individual psychology and social behavior of monkeys and apes. *International Journal of Individual Psychology,* 1: 47–59.

Mason, R.T., H.M. Fales, T.H. Jones, L.K. Pannell, J.W. Chinn, and D. Crews. 1989. Sex pheromones in snakes. *Science,* 245: 290–293.

Massey, A. 1988. Sexual interactions in Red-Spotted Newt populations. *Animal Behaviour,* 36: 205–210.

Masters, R.D. 1984. Sociobiology comes of age (book review). *Bioscience,* 34: 54.

Matthews, M. 1998. Conifers, grazing animals are stumping the Aspen. *Washington Post,* 5 October: A3.

Matthews, R.W. and J.R. Matthews. 1978. *Insect Behavior*. John Wiley & Sons, New York, 503 pp.

Maxwell, L.S. 1980. *Florida Fruit*. Lewis S. Maxwell, Publisher, Tampa, FL, 120 pp.

May, M.L. and S. Ahmad. 1983. Host location in the Colorado Potato Beetle: searching mechanisms in relation to oligophagy, pp. 173–199 in S. Ahmad, Ed., *Herbivorous Insects: Host-Seeking Behavior and Mechanisms*. Academic Press, New York, 257 pp.

May, R.M. and M. Robertson. 1980. Just-so stories and cautionary tales. *Nature,* 286: 327–329.

Maynard Smith, J. 1982. *Evolution and the Theory of Games*. Cambridge University Press, Cambridge, U.K., 224 pp.

Mayr, E. 1940. Speciation phenomena in birds. *American Naturalist,* 74: 249–278.

Mayr, E. 1963. *Animal Species and Evolution*. Belknap Press, Cambridge, MA, 672 pp.

Mayr, E. 1969. *Principles of Systematic Zoology*. McGraw-Hill, New York, 428 pp.

Mayr, E. 1982. *The Growth of Biological Thought: Diversity, Evolution, and Inheritance*. Harvard University Press, Cambridge, MA, 974 pp.

Mayr, E. 1993. What was the evolutionary synthesis? *Trends in Ecology and Evolution,* 8: 31–34.

Mayr, E. and P.D. Ashlock. 1991. *Principles of Systematic Zoology,* 2nd ed. McGraw-Hill, New York, 475 pp.

Mayr, H. 1985. *A Guide to Fossils.* Princeton University Press, Princeton, NJ. 256 pp.

McAllister, M.K., B.D. Roitberg, and K.L. Weldon. 1990. Adaptive suicide in Pea Aphids: decisions are cost sensitive. *Animal Behaviour,* 40: 167–175.

McAtee, W.L. 1938. 'Anting' by birds. *Auk,* 55:98–105.

McClendon, J.H. 1975. Efficiency. *Journal of Theoretical Biology,* 49: 213–218.

McClintock, M.K. 1971. Menstrual synchrony and suppression. *Nature,* 229: 244–245.

McDermott, F.A. 1948. *The Fireflies of Delaware with General Notes on Fireflies.* Society of Natural History, Wilmington, DE, 36 pp.

McFarland, D. 1985. *Animal Behavior.* Benjamin/Cummings, Menlo Park, CA, 576 pp.

McFarland, D. 1987. *The Oxford Companion to Animal Behavior.* Oxford University Press, New York, 685 pp.

McGregor, P.K. 1997. Book review (M.D. Hauser's *The Evolution of Communication,* MIT Press, Cambridge, MA, 1996, 760 pp.). *Animal Behaviour,* 54: 754–755.

McIntosh, R. 1992. Competition: historical perspectives, pp. 61–67 in E.F. Keller and E.A. Lloyd, Eds., *Keywords in Evolutionary Biology.* Harvard University Press, Cambridge, MA, 414 pp.

McKay, D.S., E.K. Gibson Jr., K.L. Thomas-Keprta, H. Vali, C.S. Romanek, S.J. Clemett, X.D.F. Chiller, C.R. Maechling, and R.N. Zare. 1996. Possible relic biogenic activity in Martian meteorite ALH84001. *Science,* 273: 924–930.

McKechnie, J.L., Ed., 1979. *Webster's New Universal Unabridged Dictionary,* deluxe 2nd ed. Simon & Schuster, New York, 2129 pp. + appendices.

McKinney, F.K. 1987. "Progress" in evolution. *Science,* 237: 575.

McKinney, F.K., K.M. Cheng, and D.J. Bruggers. 1984. Sperm competition in apparently monogamous birds, pp. 523–545 in R.L. Smith, Ed., *Sperm Competition and the Evolution of Animal Mating Systems.* Academic Press, New York. 687 pp.

Medawar, P.B. and J.S. Medawar. 1983. *Aristotle to Zoos: A Philosophical Dictionary of Biology.* Harvard University Press, Cambridge, MA, 340 pp.

Medvin, M.B., and M.D. Beecher. 1986. Parent-offspring recognition in the Barn Swallow (*Hirundo rustica*). *Animal Behaviour,* 34: 1627–1639.

Menon, S. 1995. The dry Ice Man. *Discover,* January: 57.

Menon, S. 1996. To appease the mountain. *Discover,* January 22–23.

Menon, S. 1997. The Piltdown perp. *Discover,* January: 34.

Merriam, J., M. Ashburner, D.L. Hartl, and F.C. Kafatos. 1991. Toward cloning and mapping the genome of *Drosophila. Science,* 254: 221–225.

Mestel, R. 1996. Secrets in a fly's eye. *Discover,* July 106–114.

Mettler, L.E., T.G. Gregg, and H.E. Schaffer. 1988. *Population Genetics and Evolution,* 2nd ed. Prentice Hall, Englewood Cliffs, NJ. 325 pp.

Meyer-Holzapfel, M. 1968. Abnormal behavior in zoo animals, pp. 476–503 in M.W. Fox, Ed., *Abnormal Behavior in Animals.* W.B. Saunders, Philadelphia, PA, 476 pp.

Meyers, P. 1978. Sexual dimorphism in size of Vespertilionid Bats. *American Naturalist,* 112: 701–711.

Meylan, A.B., B.W. Bowen, and J.C. Avise. 1990. A genetic test of the natal homing versus social facilitation models for Green Turtle migration. *Science,* 248: 724–727.

Michaelis, R.R., Ed., 1963. *Funk & Wagnalls Standard College Dictionary,* text ed. Harcourt, Brace and World, New York, 1606 pp.

Michener, C.D. 1974. *The Social Behavior of the Bees.* Harvard University Press, Cambridge, MA, 404 pp.

Michener, C.D. 1987. Taxonomy, phylogeny and zoogeography (book review). *Evolution,* 41: 449–450.

Michener, C.D. 1988a. Reproduction and castes in social halictine bees, pp. 75–119 in W. Engles, Ed., *Social Insects: An Evolutionary Approach to Castes and Reproduction.* Springer-Verlag, New York, 264 pp.

Michener, C.D. 1988b. Caste in xylocopine bees, pp. 120–144 in W. Engles, Ed., *Social Insects: An Evolutionary Approach to Castes and Reproduction.* Springer-Verlag, New York, 264 pp.

Michod, R.E. 1982. The theory of kin selection. *Annual Review of Ecology and Systematics,* 13: 23–55.

Miller, J. 1999. U.S. to use lab for more study of bioterrorism. *New York Times Northeast,* 22 September: A1, A25.

Mills, L.S., M.E. Soulé, and D.F. Doak. 1993. The keystone-species concept in ecology and conservation. *BioScience,* 43: 219–223.

Milne, L. and M. Milne. (1980) 1997. *The Audubon Society Field Guide to North American Insects and Spiders.* Alfred A. Knopf, New York, 989 pp.

Milner, R. 1990. *The Encyclopedia of Evolution: Humanity's Search for Its Origins.* Facts on File, New York, 481 pp.

Milner, R. 1996. Charles Darwin and associates, ghostbusters. *Scientific American,* October: 96–101.

Milstein, M. 1994. Yellowstone managers eye profits from hot microbes. *Science,* 264: 655.

Minelli, A. 1996. Linnaean categories. *Science,* 274: 1193.

Minelli, A. 1999. The names of animals. *Trends in Ecology and Evolution,* 14: 462–463.

Minning, D.M., A.J. Gow, J. Bonaventura, R. Braun, M. Dewhirst, D.E. Goldberg, and J.S. Stamier. 1999. *Ascaris* haemoglobin is a nitric oxide-activated "deoxygenase." *Nature,* 401: 497–502.

Mirsky, S. 1999. Floral fiend. *Scientific American,* 281(5): 24.

Mitchell, A. 1999. Bonsai flies. *Nature,* 400: 8 July: 115.

Mitchell, G. 1970. Abnormal behavior in primates, pp. 149–195 in L.A. Rosenblum, Ed., *Primate Behavior. Developments in Field and Laboratory Research,* Vol. 1. Academic Press, New York, 400 pp.

Mittler, J.E. and R.E. Lenski. 1990. New data on excisions of Mu from *E. coli* MCS2 cast doubt on directed mutation hypothesis. *Nature,* 344: 173–175.

Mitton, J. 1993. *The Penguin Dictionary of Astronomy.* Penguin Books, New York, 431 pp.

Miyasato, L.E. and M.C. Baker. 1999. Black-Capped Chickadee call dialects along a continuous habitat corridor. *Animal Behaviour,* 57: 1311–1318.

Miyashita, Y. 1995. How the brain creates imagery: projection to primary visual cortex. *Science,* 268: 1719–1720.

Mock, D.W. and L.S. Forbes. 1992. Parent-offspring conflict: a case of arrested development. *Trends in Ecology and Evolution,* 7: 409–413.

Mock, D.W., H. Drummond, and C.H. Stinson. 1990. Avian siblicide. *American Scientist,* 78: 450–459.

Mode, C.J. 1958. A mathematical model of the co-evolution of obligate parasites and their hosts. *Evolution,* 12: 158–165.

Møller, A.P. 1987. Intraspecific nest parasitism and anti-parasite behaviour in swallows, *Hirundo rustica. Animal Behaviour,* 36: 247–254.

Mooney, S.M. 1993. The evolution of sex: variation and rejuvenescence in the 19th century and today. *BioScience,,* 43: 110–113.

Moore, B.P. 1968. Studies on the chemical composition and function of the cephalic gland secretion in Australian termites. *Journal of Insect Physiology,* 14: 529–535.

Moore, J.A. 1984. Science as a way of knowing: evolutionary biology. *American Zoologist,* 24: 467–534.

Moorehead, A. 1969. *Darwin and the Beagle.* Harper & Row, New York, 280 pp.

Morell, V. 1992a. 30-million-year-old DNA boosts an emerging field. *Science,* 257: 1860–1862.

Morell, V. 1992b. Science imitates art imitating science. *Science,* 257: 1861.

Morell, V. 1993a. Called "Trimates," three bold women shaped their field. *Science,* 260: 420–425.

Morell, V. 1993b. Evidence found for a possible "aggression gene." *Science,* 260: 1722–1723.

Morell, V. 1993c. How lethal was the K-T impact? *Science,* 261: 1518–1519.

Morell, V. 1993d. *Archaeopteryx:* early bird catches a can of worms. *Science,* 259: 764–765.

Morell, V. 1994. Rise and fall of the Y chromosome. *Science,* 263: 171–172.

Morell, V. 1996a. A cold, hard look at dinosaurs. *Discover,* December: 98–108.

Morell, V. 1996b. Proteins "clock" the origins of all creatures — great and small. *Science,* 271: 448.

Morgan, C.L. 1894. *Introduction to Comparative Psychology.* Scott, London. 382 pp.

Morris, C. 1946. *Signs, Language and Behavior.* Prentice Hall, Englewood Cliffs, NJ, 365 pp.

Morris, D. 1970. *Patterns of Reproductive Behavior.* J. Cape, London, 528 pp.

Morris, W., Ed., 1982. *The American Heritage Dictionary,* 2nd college ed. Houghton Mifflin, Boston, MA, 1568 pp.

Morris, M.R. and M.J. Ryan. 1996. Sexual difference in signal-receiver coevolution. *Animal Behaviour,* 52: 1017–1024.

Morris, P.H., V. Reddy, and R.C. Bunting. 1995. The survival of the cutest: who's responsible for the evolution of the teddy bear? *Animal Behaviour,* 50: 1697–1700.

Morrow, L.A. and P.W. Stahlman. 1984. The history and distribution of Downy Brome (*Bromus tectorum*) in North America. *North America Weed Science,* 32 (suppl. 1): 2–6.

Morse, D.H. 1978. Size-related foraging differences of bumble bee workers. *Ecological Entomology,* 3: 189–192.

Morse, L.E. 1983. A shale barren on Silurian strata in Maryland. *Castanea,* 48: 206–208.

Morse, R.A. 1963. Swarm orientation in Honeybees. *Science,* 141: 357–358.

Morse, R.A. and M.S. Blum. 1963. Trail marking substance of the Texas leaf-cutting ant: source and potency. *Science,* 140: 1228.

Morse, R.A., N.E. Gary, and T.S.K. Johansson. 1962. Mating of virgin queen honey bees (*Apis mellifera* L.) following mandibular gland extirpation. *Nature,* 194: 605.

Moser, J.C. 1964. Inquiline roach responds to trail-marking substance of leaf-cutting ants. *Science,* 141: 1048–1049.

Moser, J.C., R.C. Brownlee, and R. Silverstein. 1968. Alarm pheromones of the ant *Atta texana. Journal of Insect Physiology,* 14: 529–535.

Mpitsos, G.J., S.D. Collins, and A.D. McClellan. 1978. Learning: a model system for physiological studies. *Science,* 100: 497–506.

Müller-Schwarze, D. 1974. Olfactory recognition of species, groups, individuals and physiological status among mammals, pp. 316–326 in M.C. Birch, Ed., *Pheromones.* North-Holland, London, 495 pp.

Munson, R.H. 1981. Integrated methods of cultivar identification: a case study of selected Ericaceae. Ph.D. dissertation. Cornell University, Ithaca, NY, 265 pp.

Murlis, J. 1986. The structure of odour plumes, pp. 27–48 in T.L. Payne, M.C. Birch, and C.E.J. Kennedy, Eds., *Mechanisms of Insect Olfaction.* Clarendon Press, Oxford, 364 pp.

Murray, H.A. 1938. *Explorations in Personality: A Clinical and Experimental Study of Fifty Men of College Age by Workers of the Harvard Psychological Clinic.* Oxford University Press, New York, 761 pp.

Mydans, S. 1997a. The stray cats of Australia: 9 lives seen as 9 too many. *New York Times,* 28 January: A1, A4.

Mydans, S. 1997b. Indonesia airliner crashes and all 234 aboard may have been killed. *New York Times,* 27 September: A6.

Mydans, S. 1997c. Southeast Asia tourism is smothered by smog. *New York Times,* 2 November: 3A.

Myers, N. 1999. What we must do to counter the biotic holocaust. *International Wildlife,* March/April: 30–39.

Nafus, D.M. and I.H. Schreiner. 1988. Parental care in a tropical nymphalid butterfly *Hypolimnas anomala. Animal Behaviour,* 36: 1425–1431.

Nash, J.M. 1993. How did life begin? *Time,* 11 October: 68–74.

Nash, J.M. 1995. Where do toes come from? *Time,* 31 July: 56–57.

Nash, J.M. 1997. Fertile minds. *Time,* 3 February: 48–56.

Nash, J.M., A. Park, and J. Willwerth. 1995. "Consciousness" may be an evanescent illusion. *Time,* 17 July: 52.

Nason, J.D., E. Allen-Herre, and J.L. Hamrick. 1998. The breeding structure of a tropical keystone plant resource. *Nature,* 391: 685–687.

Neale, J.H., T. Bzdega, and B. Wroblewska. 2000. N-acetylaspartylglutamate: the most abundant peptide neurotransmitter in the mammalian central nervous system. *Journal of Neurochemistry,* 75:443–452.

Nelson, M. 1997. Biospherian viewpoints. *Science,* 275: 1248–1249.

Newell, N.D. 1973. Graptolite, pp. 695–696 in Preece, W.E., Ed., *Encyclopedia Britannica,* Vol. 10. William Benton, Publisher, London, 1133 pp.

Newman, K.S., and Z.T. Halpin. 1988. Individual odours and mate recognition in the Prairie Vole, *Microtus ochrogaster. Animal Behaviour,* 36: 1779–1787.

Newsom, H E. and S.R. Taylor. 1989. Geochemical implications of the formation of the Moon by a single giant impact. *Nature,* 338: 29–34.

Nichols, S.W. 1989. *The Torre-Bueno Glossary of Entomology.* New York Entomological Society, New York, 840 pp.

Nielsen, C. 1995. *Animal Evolution: Interrelationships of the Living Phyla.* Oxford University Press, Oxford, 467 pp.

Niesbach-Klösgen, U., E. Barzen, J. Bernhardt, W. Rohde, Z. Schwarz-Sommer, H.J. Reif, U. Wienand, and H. Saedler. 1987. Chalcone synthase genes in plants: a tool to study evolutionary relationships. *Journal of Molecular Evolution,* 26: 213–225.

Niesenbaum, R.A. 1999. The effects of pollen load size and donor diversity on pollen performance, selective abortion, and progeny vigor in *Mirabilis jalapa* (Nyctaginaceae). *American Journal of Botany,* 86: 261–268.

Noble, E.R., G.A. Noble, G.A. Schad, and A.J. MacInnes. 1989. *Parasitology: The Biology of Animal Parasites,* 6th ed. Lea & Febiger, Philadephia, PA, 574 pp.

Noldus, L.P.J.J. 1991. The Observer: software system for collection and analysis of observational data. *Behavior Research Methods, Instruments, and Computers,* 23: 415–429.

Noor, M.A.F. 1996. Absence of species discrimination in *Drosophila pseudoobscura* and *D. persimilis. Animal Behaviour,* 52: 1205–1206.

Norden, A.W., D.C. Forester, and G.H. Fenwick, Eds. 1984. *Threatened and Endangered Plants and Animals of Maryland.* Special Publ. 84-1. Maryland Natural Heritage Program, Annapolis, MD, 475 pp.

Nordlund, D.A. 1981. A glossary of terms used to describe chemicals that mediate intra- and interspecific interactions, pp. 495–497 in E.R. Mitchell, Ed., *Management of Insect Pests with Semiochemicals.* Plenum Press, New York, 514 pp.

Nordlund, D.A. and W.J. Lewis. 1976. Terminology of chemical releasing stimuli in intraspecific and interspecific interactions. *Journal of Chemical Ecology,* 2: 211–220.

Nordlund, D.A., R.L. Jones, and W.J. Lewis. 1981. *Semiochemicals: Their Role in Pest Control.* John Wiley & Sons, New York, 306 pp.

Norman, A.W. and G. Litwack. 1987. *Hormones.* Academic Press, New York, 806 pp.

Norman, D. 1985. *The Illustrated Encyclopedia of Dinosaurs.* Crescent Books, New York, 208 pp.

Norstog, K. and A.J. Meyerriecks. 1983. *Biology.* Charles E. Merrill Publishing, Columbus, OH, 671 pp.

Norval, R.A. I., H.R. Andrew, and C.E. Yunker. 1989. Pheromone-mediation of host selection in bont ticks (*Amblyomma hebraeum* Koch). *Science,* 243: 364–365.

Norvell, S. 1996. Force of nature. *Washington Post Magazine,* 23 June: 14–18, 26–27.

Nowak, R. 1994. Mining treasures from "junk DNA." *Science,* 263: 608–610.

Nur, U., J.H. Werren, D.G. Eickbush, W.D. Burke, and T.H. Eickbush. 1988. "Selfish" B chromosome that enhances its transmission by eliminating the paternal genome. *Science,* 240: 512–513.

Oberhauser, K.S. 1988. Male monarch butterfly spermatophore mass and mating strategies. *Animal Behaviour,* 36:1384–1388.

O'Brien, S.J. and E. Mayr. 1991. Bureaucratic mischief: recognizing endangered species and subspecies. *Science,* 251: 1187–1187.

O'Donald, P., and M.E.N. Majerus. 1988. Frequency-dependent sexual selection. *Philosophical Transactions of the Royal Society of London B,* 319: 571–586.

O'Dowd, D.J., C.R. Brew, D.C. Christophel, and R.A. Norton. 1991. Mite-plant associations from the Eocene of southern Australia. *Science,* 252: 99–101.

Odum, E.P. 1969. *Fundamentals of Ecology,* 2nd ed. W.B. Saunders, Philadelphia, PA, 546 pp.

Odum, E.P. 1971. *Fundamentals of Ecology,* 3rd ed. W.B. Saunders, Philadelphia, PA, 574 pp.

Odum, H.T., E.C. Odum, M.T. Brown, D. LaHart, C. Bersok, and J. Sendzimir. 1988. *Environmental Systems and Public Policy: A Text on Science, Technology and Society that Unifies Basic Sciences, Environment, Energetics, Economics, Microcomputers, and Public Policy.* Ecological Economics Program, Phelps Lab, University of Florida, Gainesville, 253 pp.

Ogurlu, I. 1996. Habitat use of Red Deer (*Cervus elaphus* L.) in Çatacik Forest. *Turkish Journal of Zoology,* 20: 427–435.

Oliver, J.A. 1955. *The Natural History of North American Amphibians and Reptiles.* Van-Nostrand, New York, 359 pp.

Oliwenstein, L. 1995. Roots. *Discover,* 16 January: 37–38.

Oliwenstein, L. 1996a. Headless. *Discover,* January: 34.

Oliwenstein, L. 1996b. Life's grand explosions. *Discover,* January: 42–43.

Olson, E.C. 1973. Fossil, pp. 649–651 in W.E. Preece, Ed., *Encyclopedia Britannica,* Vol. 9. William Benton, Publisher, London, 1153 pp.

Opler, P.A. and G.O. Krizek. 1984. *Butterflies East of the Great Plains.* The Johns Hopkins University Press, Baltimore, MD. 294 pp.

Orgel, L.E. 1994. The origin of life on the Earth. *Scientific American,* October: 76–83.

Orgel, L.E. and F.H.C. Crick. 1980. Selfish DNA: the ultimate parasite. *Nature,* 284: 604–607.

Ostfeld, R.S., C.G. Jones, and J.O. Wolff. 1996. Of mice and mast: ecological connections in eastern deciduous forests. *BioScience,* 46: 323–330.

O'Toole, C. and A. Raw. 1991. *Bees of The World.* Facts on File, New York, 192 pp.

Oxford English Dictionary. 1972. Compact edition, Vols. 1–2. Oxford University Press, London, 4116 pp.

Oxford English Dictionary. 1989. Second edition, Clarendon Press, Oxford, 1019 pp. (Vol. 2), 1143 pp. (Vol. 3), 1143 pp. (Vol. 6), 1016 pp. (Vol. 12), 1015 pp. (Vol. 16).

Pace, M.L., J.J. Cole, S.R. Carpenter, and J.F. Kitchell. 1999. Trophic cascades revealed in diverse ecosystems. *Trends in Ecology and Evolution,* 14: 483–488.

Packer, C. 1977. Reciprocal altruism in *Papio anubis. Nature,* 265: 441–443.

Page, J. 1986. Does genius come in seven flavors? *Washington Post,* 14 December: H3.

Page, R.E., Jr. and H.H. Laidlaw, Jr. 1988. Full sisters and super sisters: a terminological paradigm. *Animal Behaviour,* 36: 944–945.

Parker, G.A. 1984. Sperm competition and the evolution of animal mating systems, pp. 1–60 in R.L. Smith, Ed., *Sperm Competition and the Evolution of Animal Mating Systems.* Academic Press, New York, 687 pp.

Parker, S.T. 1989. Early hominid mating systems (letter to the editor). *Science,* 246: 195.

Parrish, J.K. 1989. Re-examining the selfish herd: are central fish safer? *Animal Behaviour,* 38: 1048–1053.

Parry, G.D. 1981. The meanings of r- and K-selection. *Oecologia,* 48:260–264.

Partridge, B.L. 1982. Rigid definitions of schooling behaviour are inadequate. *Animal Behaviour,* 30: 298–299.

Partridge, L. 1988. The rare-male effect: what is its evolutionary significance? *Philosophical Transactions of the Royal Society of London B,* 319: 525–539.

Pasteur, G. 1982. A classification of mimicry systems. *Annual Review of Ecology and Systematics,* 12: 169–199.

Paterson, H.E.H. 1981. The continuing search for the unknown and unknowable: a critique of contemporary ideas of speciation. *South African Journal of Science,* 77: 113–119.

Paterson, H.E.H. 1982. Perspective on speciation by reinforcement. *South African Journal of Science,* 78: 53–57.

Paterson, H.E.H. 1985. The recognition concept of species, pp. 21–29 in E.S. Vrba, Ed., *Species and Speciation.* Monograph No. 4, Transvaal Museum, Pretoria, South Africa, 176 pp.

Paul, D. 1992a. Heterosis, pp. 166–169 in E.F. Keller and E.A. Lloyd, Eds., *Keywords in Evolutionary Biology.* Harvard University Press, Cambridge, MA, 414 pp.

Paul, D. 1992b. Fitness: historical perspectives, pp. 112–114 in E.F. Keller and E.A. Lloyd, Eds., *Keywords in Evolutionary Biology.* Harvard University Press, Cambridge, MA, 414 pp.

Peebles, P.J.E., D.N. Schramm, E.L. Turner, and R.G. Kron. 1994. The evolution of the universe. *Scientific American,* October: 53–57.

Pellett, F.C. 1979. *American Honey Plants,* 5th ed. Dadant and Sons, Hamilton, IL, 467 pp.

Pennisi, E. 2000. Ideas fly at gene-finding jamboree. *Science,* 287: 2183–2184.

Pepperberg, I.M. 1987. Evidence of conceptual quantitative abilities in the African Grey Parrot: labeling of cardinal sets. *Ethology,* 75: 37–61.

Perry, S.A. 1995. Macroinvertebrates, pp. 23–30 in R.C. Reardon, Ed., *Effects of Diflubenzuron on Non-Target Organisms in Broadleaf Forested Watersheds in the Northeast,* USDA Forest Service FHM–NC–0595, U.S. Department of Agriculture, Washington, D.C., 174 pp.

Peterson, R.T. 1947. *A Field Guide to the Birds.* Houghton Mifflin, Boston, MA, 290 pp.

Peterson, R.T. 1980. *A Field Guide to the Birds: A Completely New Guide to All the Birds of Eastern and Central North America.* Houghton Mifflin, Boston, MA, 384 pp.

Peterson, R.T. 1990. *A Field Guide to Western Birds,* 3rd ed. Houghton Mifflin, Boston, MA, 432 pp.

Peterson, R.T. and E.L. Chalif. 1973. *A Field Guide to Mexican Birds.* Houghton Mifflin, Boston, MA, 298 pp.

Peterson, R.T., G. Mountfort, and P.A.D. Hollom. 1967. *A Field Guide to the Birds of Britain and Europe,* 2nd ed. Houghton Mifflin, Boston, MA, 344 pp.

Peuhkuri, N. 1997. Size-assortative shoaling in fish: the effect of oddity on foraging behaviour. *Animal Behaviour,* 54: 271–278.

Phillips, A. 1992. An old crab myth shattered. *Washington Post,* 8 July: E1, E3.

Phillips, J.B. 1990. Lek behaviour in birds: do displaying males reduce nest predation? *Animal Behaviour,* 39: 555–565.

Picman, J. and J.-C. Belles-Isles. 1987. Intraspecific egg destruction in marsh wrens: a study of mechanisms preventing filial ovicide. *Animal Behaviour,* 35: 236–246.

Pierce, J.D., Jr. 1986. A review of tool use in insects. *Florida Entomologist,* 69: 95–104.

Pierotti, R. and E.C. Murphy. 1987. Intergenerational conflicts in gulls. *Animal Behaviour,* 35: 435–444.

Pitman, G.B., J.P. Vité, G.W. Kinzer, and A.F. Fentiman, Jr. 1968. Bark beetle attractants: trans-verbenol isolated from *Dendroctonus. Nature,* 218: 168–169.

Platt, R.B. 1951. An ecological study of the mid-Appalachian shale barrens and of the plants endemic to them. *Ecological Monographs,* 21: 269–300.

Plomin, R. 1999. Genetics and general cognitive ability. *Nature,* 402(suppl.): C25–C29.

Plutchik, R. 1980. *Emotion: A Psychological Synthesis.* Harper & Row, New York, 440 pp.

Pockley, P. 1999. Global warming "could kill most coral reefs by 2100." *Nature,* 400: 98.

Poinar, G. and R. Poinar. 1995. *The Quest of Life in Amber.* Addison-Wesley, Reading, MA, 219 pp.

Polis, G.A. 1989. The unkindest sting of all. *Natural History,* 98(7): 34–39.

Pool, R. 1993. Evidence for a homosexuality gene. *Science,* 261: 291–292.

Poole, J.H. 1989. Announcing intent: the aggressive state of musth in African elephants. *Animal Behaviour,* 37: 140–152.

Powell, E.A. and C.E. Jones. 1983. Floral mutualism in *Lupinus bethamii* (Fabaceae) and *Delphinium parryi* (Ranunculaceae), pp. 310–337 in C.E. Jones and R.J. Little, Eds., *Handbook of Experimental Pollination Biology,* Scientific and Academic Editions, New York, 558 pp.

Pribram, K.H. 1967. Emotion: steps toward a neuropsychological theory, pp. 3–40 in D.C. Glass, Ed., *Neurophysiology and Emotion.* The Rockefeller University Press, New York, 234 pp.

Price, P.W. 1996. *Biological Evolution.* Saunders College Publishing, New York, 429 pp.

Proctor, M. and P. Yeo. 1972. *The Pollination of Flowers.* Taplinger Publishing, New York, 418 pp.

Propp, G.D. and P.B. Morgan. 1983a. Superparasitism of House Fly, *Musca domestica* L., pupae by *Spalangia endius* Walker (Hymenoptera: Pteromalidae). *Environmental Entomology,* 12: 561–566.

Propp, G.D. and P.B. Morgan. 1983b. Multiparasitism of House Fly, *Musca domestica* L., pupae by *Spalangia endius* Walker and *Muscidifurax raptor* Girault and Sanders (Hymenoptera: Pteromalidae). *Environmental Entomology,* 12: 1232–1238.

Propp, G.D. and P.B. Morgan. 1985. Effect of host distribution on parasitoidism of House Fly (Diptera: Muscidae) pupae by *Spalangia* spp. and *Muscidifurax raptor* (Hymenoptera: Pteromalidae). *Canadian Entomologist,* 117: 515–542.

Provine, R.R. 1986. Yawning as a stereotyped action pattern and releasing stimulus. *Ethology,* 72: 109–122.

Provine, R.R. and H.B. Hamernik. 1986. Yawning: effects of stimulus interest. *Bulletin of the Psychomonic Society,* 24: 437–438.

Ptacek, M.B. and J. Travis. 1996. Inter-population variation in male mating behaviours in the Sailfin Mollie, *Poecilia latipinna. Animal Behaviour,* 52: 59–71.

Ptacek, M.B. and J. Travis. 1997. Mate choice in the Sailfin Molly, *Poecilia latipinna. Evolution,* 51: 1217–1231.

Ptacek, M.B., H.C. Gerhardt, and R.D. Sage. 1994. Speciation by polypolidy in treefrogs: multiple origins of the tetraploid, *Hyla versicolor. Evolution,* 48: 898–908.

Pulliam, H.R. and N.M. Haddad. 1994. Human population growth and the carrying capacity concept. *Bulletin of the Ecological Society of America,* 75:141–157.

Purugganan, M.D. 1993. Transposable elements as introns: evolutionary connections. *Trends in Ecology and Evolution,* 8: 239–243.

Putters, F.A., and J. van den Assem. 1988. The analysis of partial preferences in a parasitic wasp. *Animal Behaviour,* 36: 933–934.

Quackenbush, J. 1985. The death of a pet: how it can affect owners. *Symposium on the Human-Companion Animal Bond,* Veterinary Clinics of North America, Small Animal Practice, 15(2): 283–470.

Qui, Y.-L., J. Lee, F. Bernasconi-Quadroni, D.E. Soltis, P.S. Soltis, M. Zanis, E.A. Zimmer, Z. Chen, V. Savolainen, and M.W. Chase. 1999. The earliest angiosperms: evidence from mitochondrial, plastic, and nuclear genomes. *Nature,* 402: 407.

Rado, R., N. Levi, H. Hauser, J. Witcher, N. Adler, N. Intrator, Z. Wooburg, and J. Terkel. 1987. Seismic signalling as a means of communication in a subterranean mammal. *Animal Behaviour,* 35: 1249–1266.

Raff, R.A. and H.R. Mahler. 1992. The non symbiotic origin of mitochondria. *Science,* 177: 575–582.

Rainey, P. and R. Moxon. 1993. Unusual mutational mechanisms and evolution. *Science,* 260: 1958.

Rainis, K.G. and B.J. Russell. 1986. *Guide to Microlife.* Franklin Watts, Danbury, CT, 287 pp.

Ralls, K. 1976. Mammals in which females are larger than males. *Quarterly Review of Biology,* 51: 245–276.

Randall, J.A. 1987. Sandbathing as a territorial scent-mark in the Bannertail Kangaroo Rat, *Dipodomys spectabilis. Animal Behaviour,* 35: 426–434.

Randall, J.A. 1989. Individual footdrumming signatures in Banner-Tailed Kangaroo Rats *Dipodomys spectabilis. Animal Behaviour,* 38: 620–630.

Ranta, E., V. Kaitala, and P. Lundberg. 1999. A tale of big game and small bugs. *Science,* 285: 1022–1023.

Raskin, I., A. Ehmann, W.R. Melander, and B.J.D. Meeuse. 1987. Salicylic acid: a natural inducer of heat production in Arum lilies. *Science,* 237: 1601–1608.

Ratner, S.C. 1964. *Comparative Psychology: Research in Animal Behavior.* Dorsey Press, Homewood, IL, 773 pp.

Raup, D.M. and J.J. Sepkoski, Jr. 1986. Periodic extinction of families and genera. *Science,* 231: 833–835.

Raver, A. 1997. Qualities of an animal scientist: cow's eye view and autism. *New York Times,* 5 August: C1, C6.

Ravven, W. 1990. In the beginning. *Discover,* October: 98–102.

Reader's Digest Association. 1971. *Fascinating World of Animals.* Reader's Digest Association, Pleasantville, NY, 428 pp.

Reader's Digest Association. 1996. *Drive America Road Atlas: Northern and Central States.* Reader's Digest Association, Pleasantville, NY.

Reese, K.P. and J.T. Ratti. 1988. *Edge Effect: A Concept Under Scrutiny.* Transactions of the 53rd North American Wildlife and Natural Resources Conference. Wildlife Management Institute, Washington, D.C., pp. 127–136.

Regnier, F.E. and E.O. Wilson. 1968. The alarm-defense system of the ant *Acanthomyops claviger. Journal of Insect Physiology,* 14: 955–970.

Regnier, F.E. and E.O. Wilson. 1969. The alarm-defense system of the ant *Lasius alienus. Journal of Insect Physiology,* 15: 893–898.

Regnier, F.E. and E.O. Wilson. 1971. Chemical communication and "propaganda" in slave-maker ants. *Science,* 172: 267–269.

Reichman, O.J. 1988. Caching behaviour by Eastern Woodrats, *Neotoma floridana,* in relation to food perishability. *Animal Behaviour,* 36: 1525–1532.

Reid, R.A. 1980. Selfish DNA in "Petite" mutants. *Nature,* 285: 620.

Reingold, H.L., Ed., 1963. *Maternal Behavior in Mammals.* Wiley, New York, 349 pp.

Rensberger, B. 1992a. Underground goliath: Michigan mushroom over 1500 years old. *Washington Post,* 2 April: A1.

Rensberger, B. 1992b. Nitric oxide signal causes erection, scientists discover: finding offers hope for treatment of impotence, priapism. *Washington Post,* 17 July: A3.

Rensberger, B. 1993a. After 2000, outlook for the ozone layer looks good. *Washington Post,* 15 April: A1, A18–A19.

Rensberger, B. 1993b. Greenhouse effect seems benign so far: warming most evident at night, in winter. *Washington Post,* 1 June: A1.

Rensberger, B. 1995. How genes work. *Washington Post,* 8 November: H1.

Resnik, D.B. 1992. Gaia: from fanciful notion to research program. *Perspectives in Biology and Medicine,* 35: 572–582.

Rettenmeyer, C. 1970. Insect mimicry. *Annual Review of Entomology,* 15: 43–74.

Reuters. 1992. Reprieve for the London Zoo. *Washington Post,* 8 September: B4.

Reuters. 1996. Six men are charged in smuggling of rare Madagascan reptiles. *New York Times,* 24 August: 9.

Rhodes, F.H.T. 1983. Gradualism, punctuated equilibrium and the *Origin of Species. Nature,* 305: 269–272.

Ricci, N. 1990. The behaviour of ciliated protozoa. *Animal Behaviour,* 40: 1048–1069.

Richards, M.J. and M.A. El Mangoury. 1968. Further experiments on the effects of social factors on the rate of sexual maturation in the desert locust. *Nature,* 219: 865–866.

Richards, R.J. 1992. *Evolution,* pp. 95–105 in E.F. Keller and E.A. Lloyd, Eds., *Keywords in Evolutionary Biology.* Harvard University Press, Cambridge, MA, 414 pp.

Richardson, S. 1995. Battle in the burrow. *Discover,* November: 42.

Richardson, S. 1996. Scent of a man. *Discover,* February: 26–27.

Rieger, R., A. Michaelis, and M.M. Green. 1991. *Glossary of Genetics: Classical and Molecular,* 5th ed. Springer-Verlag, New York, 553 pp.

Rimer, S. 1996. Founder of program to build students' self-respect, and houses, wins "genius" grant. *New York Times,* 18 June: B7.

Rinderer, T.E. 1986. Africanized Bees: the Africanization process and potential range in the United States. *Bulletin of the Entomological Society of America,* 32: 222–227.

Rissing, S.W., and G.B. Pollock. 1987. Queen aggression, pelometrotic advantage and brood raiding in the ant *Veromessor pergandei* (Hymenoptera: Formicidae). *Animal Behaviour,* 35: 975–981.

Robbins, C.S., B. Bruun, and H.S. Zim. 1966. *A Guide to Field Identification: Birds of North America.* Western Publishing, Racine, WI, 340 pp.

Robbins, J. 1996. As snow falls in Tetons, elk temperatures rise. *New York Times,* 22 September: 14.

Robbins, R.K. 1980. The lycaenid "false head" hypothesis: historical review and quantitative analysis. *Journal of the Lepidopterists' Society,* 34: 194–208.

Roberts, L. 1993a. Wetlands trading is a loser's game, say ecologists. *Science,* 260: 1890–1892.

Roberts, L. 1993b. Bringing vanished ecosystems to life. *Science,* 260: 1891.

Robertson, J.G.M. 1986. Male territoriality, fighting and assessment of fighting ability in the Australian frog *Uperoleia rugosa. Animal Behaviour,* 34: 763–772.

Roe, K.E. and R.G. Frederick. 1981. *Dictionary of Theoretical Concepts in Biology.* Scarecrow Press, Methuchen, NJ, 267 pp.

Rohlf, F.J. and L.F. Marcus. 1993. A revolution in morphometrics. *Trends in Ecology and Evolution,* 8: 129–132.

Rohwer, S. 1978. Parent cannibalism of offspring and egg raiding as a courtship strategy. *American Naturalist,* 112: 429–440.

Roland, J. and W.J. Kaupp. 1995. Reduced transmission of Forest Tent Caterpillar (Lepidoptera: Lasiocampidae) nuclear polyhedrosis virus at the forest edge. *Environmental Entomology,* 24: 1175–1178.

Rollmann, S.M., L.D. Houck, and R.C. Feldhoff. 1999. Proteinaceous pheromone affect female receptivity in a terrestrial salamander. *Science,* 285: 1907–1909.

Romanes, G.J. 1897. *Darwin, and After Darwin,* Vol. 1. The Open Court Publishing Company, Chicago, IL, 460 pp.

Roper, T.J. and S. Redston. 1987. Conspicuousness of distasteful prey affects the strength and durability of one-trial avoidance learning. *Animal Behaviour,* 35: 739–747.

Rosenberg, A. 1992. Altruism: theoretical contexts, pp. 19–28 in E.F. Keller and E.A. Lloyd, Eds., *Keywords in Evolutionary Biology.* Harvard University Press, Cambridge, MA, 414 pp.

Rosenfeld, M. 1996. MacArthur names 21 new "geniuses." *Washington Post,* 18 June: B1, B8.

Rosengarten, R. and K.S. Wise. 1989. Phenotypic switching in mycoplasmas: phase variation of diverse surface lipoproteins. *Science,* 247: 315–318.

Rosenshine, I., R. Tchelet, and M. Mevarech. 1989. The mechanism of DNA transfer in the mating system of an archaebacterium. *Science,* 245: 1387–1389.

Rosin, H. 1999. Creationism evolves: Kansas Board targets Darwin. *Washington Post,* 8 August: A1, A22.

Ross, G.N. 1985. The case of the vanishing caterpillar. *Natural History,* 94(11): 48–55.

Ross, P.E. 1991. Crossed lines: Eve's family tree may have a few branches from Adam. *Scientific American,* 265(4): 30, 32.

Rothschild, M. 1963. Is the Buff Ermine (*Spilosoma Lutea* (Huf.)) a mimic of the White Ermine (*Spilosoma Lubricipeda* (L.))? *Proceedings of the Royal Entomological Society of London,* 38: 159–164.

Rothschild, M. 1964. A note on the evolution of defense and repellant odours of insects. *The Entomologist,* December: 276–288.

Roubik, D.W. 1989. *Ecology and Natural History of Tropical Bees.* Cambridge University Press, Cambridge, U.K., 514 pp.

Roughgarden, J. 1979. *Theory of Population Genetics and Evolutionary Ecology: An Introduction.* Macmillan, New York, 634 pp.

Roush, W. 1996. Hunting for animal alternatives. *Science,* 274: 168–171.

Rowan, A.N. 1986. *Of Mice, Models, and Men: A Critical Evaluation of Animal Research.* State University of New York Press, Albany, 323 pp.

Rowell, T.E., R.A. Hinde, and Y. Spencer-Booth. 1964. "Aunt"-infant interaction in captive Rhesus Monkeys. *Animal Behaviour,* 12: 219–229.

Rowland, W.J. 1989. The effects of body size, aggression and nuptial coloration on competition for territories in male Threespine Sticklebacks, *Gasterosteus aculeatus. Animal Behaviour,* 37: 282–289.

Rozen, J.G., Jr. 1989. Morphology and systematic significance of first instars of the cleptoparasitic bee tribe Epeolini (Anthophoridae: Nomadinae). *American Museum Novitates,* No. 2957, 19 pp.

Ruse, M. 1984. Is there a limit to our knowledge of evolution? *BioScience,* 34: 100–104.

Ruse, M. 1992. Darwinism, pp. 74–80 in E.F. Keller and E.A. Lloyd, Eds., *Keywords in Evolutionary Biology.* Harvard University Press, Cambridge, MA, 414 pp.

Ruse, M. 1993. Evolution and progress. *Trends in Ecology and Evolution,* 8: 55–59.

Ryan, F. 1997. The smartest disease. *New York Times Book Review,* 24 August: 17.

Sæther, B.-E. 1999. Top dogs maintain diversity. *Nature,* 4000: 510–511.

Safire, W. 1997. Janus lives. *New York Times Magazine,* 4 May: 22, 24.

Sakaluk, S.K. 1991. Post-copulatory mating guarding in decorated crickets. *Animal Behaviour,* 41: 207–216.

Salceda, V.M. and W.W. Anderson. 1988. Rare male mating advantage in a natural population of *Drosophila pseudoobscura. Proceedings of the National Academy of Science USA,* 85: 9870–9874.

Salmon, J.L. 1997. For students today, life is in the bag: backpacks no longer just a casual purchase. *Washington Post,* 30 August: A1.

Salmon, J.L. and M. Bombardieri. 1999. Drought is worst since Depression. *Washington Post,* 2 August: A1, A7.

Sapir, E. 1921. *An Introduction to the Study of Speech.* Harcourt, Brace & World, New York, 223 pp.

Sattler, H.R. 1990. *The New Illustrated Dinosaur Dictionary.* Lothrop, Lee and Shepard Books, New York, 363 pp.

Savage, R.J.G. and M.R. Long. 1986. *Mammal Evolution: An Illustrated Guide.* Facts on File, New York, 259 pp.

Sawyer, K. 1995. Zoology: the newt's stimulating secretions. *Washington Post,* 20 May, A2.

Sawyer, K. 1996. British team detects new evidence of life on Mars. *Washington Post,* 1 November: A3.

Sawyer, K. 1997. Scientists find first direct evidence of black holes: report describes a rim of super hot gas. *Washington Post,* 14 January: A3.

Sawyer, K. 1999a. A meteoric discovery: extraterrestrial water. *Washington Post,* 27 August: 1, A9.

Sawyer, K. 1999b. Hardy microbes appear able to survive in space. *Washington Post,* 4 October: A11.

Sawyer, L.A., J.M. Hennessy, A.A. Peixoto, E. Rosato, H. Parkinson, R. Costa, and C.P. Kyriacou. 1997. Natural variation in a *Drosophila* clock gene and temperature compensation. *Science,* 278: 2117–2120.

Sbordoni, V. and S. Forestiero. 1984. *Butterflies of the World.* Times Books, New York, 312 pp.

Schemske, D. W. 1982. Limits to specialization and coevolution in plant-animal mutualisms, pp. 67–109 in M.H. Nitecki, Ed., *Coevolution.* The University of Chicago Press, Chicago, IL, 392 pp.

Schiestl, F.P., M. Ayasse, H.F. Paulus, D. Erdmann, and T. Francke. 1997. Variation of floral scent emission and postpollination changes in individual flowers of *Ophrys sphegodes* subsp. *sphegodes. Journal of Chemical Ecology,* 23: 2881–2895.

Schiestl, F.P., M. Ayasse, H.F. Paulus, C. Löfstedt, B.S. Hansson, F. Ibarra, and W. Francke. 1999. Orchid pollination by sexual swindle. *Nature,* 399: 421–422.

Schmidt, K. 1996. Creationists evolve new strategy. *Science,* 273: 420–422.

Schneider, J.M. and Y. Lubin. 1997. Infanticide by males in a spider with suicidal maternal care, *Stegodyphus lineatus* (Eresidae). *Animal Behaviour,* 54: 304–312.

Schneider, S.S., J.A. Stamps, and N.E. Gary. 1986a. The vibration dance of the Honey Bee. I. Communication regulating foraging on two time scales. *Animal Behaviour,* 34: 377–385.

Schneider, S.S., J.A. Stamps, and N.E. Gary. 1986b. The vibration dance of the Honey Bee. II. The effects of foraging success on daily patterns of vibration activity. *Animal Behaviour,* 34: 386–391.

Scholander, P.F. 1955. Evolution of climatic adaptation in homeotherms, pp. 166–180 in J.B. Bresler, Ed., *Readings in Human Ecology.* Addison-Wesley, Reading, MA, 472 pp.

Scholtissek, C. 1992. Cultivating a killer virus. *Natural History,* January: 2,4,6.

Schuler, G. D. et al. 1996. A gene map of the human genome. *Science,* 274: 540–546.

Schuler, S., Ed., 1983. *Guide to Garden Flowers.* Simon & Schuster, New York, 91 pp. + plates.

Schulte, B.A. and L.E.L. Rasmussen. 1999. Signal-receiver interplay in the communication of male condition by Asian Elephants. *Animal Behaviour,* 57: 1265–1274.

Schwagmeyer, P.L. and D.W. Foltz. 1990. Factors affecting the outcome of sperm competition in thirteen-lined ground squirrels. *Animal Behaviour,* 39: 156–162.

Scott, D. 1987. The timing of the sperm effect on female *Drosophila melanogaster* receptivity. *Animal Behaviour,* 35: 142–149.

Scott, D., R.C. Richmond, and D.A Carlson. 1988. Pheromones exchanged during mating: a mechanism for mate assessment in *Drosophila. Animal Behaviour,* 36: 1164–1173.

Scott, J.P. 1956. The analysis of social organization in animals. *Ecology,* 37: 213–221.

Scott, J.P. 1975. *Aggression,* 2nd ed. University of Chicago Press, Chicago, IL, 233 pp.

Scott, J.P. and E. Fredericson. 1951. The causes of fighting in rats and mice. *Physiological Zoology,* 24: 273–309.

Scott, M.P. and J.F.A. Traniello. 1989. Guardians of the underworld. *Natural History,* 98(6): 32–37.

Scott, P.M., Ed., 1974. *The World Atlas of Birds.* Random House, New York, 272 pp.

Scriber, J.M. 1996. Tiger tales: natural history of native North American swallowtails. *American Entomologist,* 42(1): 19–32

Seachrist, L. 1994. Sea turtles master migration with magnetic memories. *Science,* 264: 661–662.

Sebeok, T.A. 1963. Communication among social bees; porpoises and sonar; man and dolphin. *Language,* 39: 448–466.

Sebeok, T.A. 1965. Animal communication. *Science,* 147: 1006–1014.

Sebeok, T.A., Ed., 1968a. *Animal Communication.* Indiana University Press, Scarsborough, Canada, 1128 pp.

Sebeok, T.A., Ed., 1968b. *Animal Communication: Techniques of Study and Results of Research.* Hafner Publishing, New York, 203 pp.

Seehausen, O., J.J.M. van Alphen, and F. Witte. 1997a. Cichlid fish diversity threatened by eutrophication that curbs sexual selection. *Science,* 277: 1808–1811.

Seehausen, O., F. Witte, E.F. Ktaunzi, J. Smits, and N. Bouton. 1997b. Patterns of the remnant cichlid fauna in southern Lake Victoria. *Conservation Biology,* 11: 890–904.

Seibt, U. and W. Wickler. 1987. Gerontophagy versus cannibalism in the social spiders *Stegodyphus mimosarum* Pavesi and *Stegodyphus dumicola* Pocock. *Animal Behaviour,* 35: 1903–1904.

Sekgororane, G.B. and T.G. Dilworth. 1995. Relative abundance, richness, and diversity of small mammals at induced forest edges. *Canadian Journal of Zoology,* 73: 1432–1437.

Selander, R.K. 1972. Sexual selection and dimorphism in birds, pp. 180–230 in B. Campbell, Ed., *Sexual Selection and the Descent of Man.* Aldine Press, Chicago, IL, 378 pp.

Selig, R.O. 1999 Human origins: one man's search for the causes in time. *AnthroNotes,* 22: 1–9.

Sereno, P.C., D.B. Dutheil, M. Iarochene, H.C.E. Larsson, G.H. Lyon, P.M. Magwene, C.A. Sidor, D.J. Varricchio, and J.A. Wilson. 1996. Predatory dinosaurs from the Sahara and Late Cretaceous faunal differentiation. *Science,* 272: 986–991.

Seyfarth, R.M., and D.L. Cheney. 1984. Grooming, alliances and reciprocal altruism in Vervet Monkeys. *Nature,* 308: 541–543.

Seymour-Smith, C. 1986. *Dictionary of Anthropology.* G.K. Hall and Company, Boston, MA, 305 pp.

Shaffer, H.B., J.M. Clark, and F. Kraus. 1991. When molecules and morphology clash: a phylogenetic analysis of the North American ambystomatic salamanders (Caudata: Ambystomatidae). *Systematic Zoology,* 40: 284–303.

Shapiro, J.A. 1995. Adaptive mutation: who's really in the garden? *Science,* 268: 373–374.

Sharpton, V.L., K. Burke, A. Camargo-Zanoguera, S.A. Hall, D.S. Lee, L.E. Marín, G. Suárez-Reynoso, J.M. Quezada-Muñeton, P.D. Spudis, and J. Urrutia-Fucugauchi. 1993. Chicxulub multiring impact basin: size and other characteristics derived from gravity analysis. *Science,* 261: 1564–1567.

Shelton, N. 1985. *Saguaro National Monument, Arizona.* National Park Service, U.S. Department of the Interior, Washington, D.C., 98 pp.

Shepard, M. and G.T. Gale. 1977. Superparasitism of *Epilachna varivestis* (Col.: Coccinellidae) by *Pediobius foveolatus* (Hym.: Eulophidae): influence of temperature and parasitoid-host ratio. *Entomophaga,* 22: 315–321.

Sherman, P.W. and W.G. Holmes. 1985. Kin recognition: issues and evidence. *Fortschritte der Zoologie,* 31: 437–460.

Shields, T. 1996. Busy beavers gnaw on suburban nerves: homeowners take lethal action as property damage grows. *Washington Post,* 14 December: A1, A18.

Shields, W.M. 1982. *Philopatry, Inbreeding, and the Evolution of Sex.* State University of New York Press, Albany, NY, 245 pp.

Shreeve, J. 1996. Sunset on the savanna. *Discover,* July: 116–125.

Shuster, S.M. and M.J. Wade. 1991. Female copying and sexual selection in a marine isopod crustacean, *Parascerceis sculpta. Animal Behaviour,* 42: 1071–1078.

Sibly, R.M., H.M.R. Nott, and D.J. Fletcher. 1990. Splitting behaviour into bouts. *Animal Behaviour,* 39: 63–69.

Siegel, S. 1956. *Nonparametric Statistics for the Behavioral Sciences.* McGraw-Hill, New York, 312 pp.

Sigg, H. and J. Falett. 1985. Experiments on respect of possession and property in Hamadryas Baboons (*Papio hamadryas*). *Animal Behaviour,* 33: 978–984.

Silberhorn, G.M. 1968. The shale baren flora of the Virginias. *The Radford Review,* 22: 111–118.

Simons, M. 1996. New species of early human reported found in Africa. *New York Times,* 23 May: A8.

Simpson, G.G. 1944. *Tempo and Mode in Evolution.* Columbia University Press, New York, 237 pp.

Simpson, G.G. 1953. *The Major Features of Evolution.* Columbia University Press, New York, 434 pp.

Simpson, G.G. 1961. *Principles of Animal Taxonomy.* Columbia University Press, New York, 247 pp.

Singer, P. 1975. *Animal Liberation.* Avon Books, New York, 297 pp.

Sisson, L.A., V.B. Van Hasselt, M. Hersen, and J.C. Aurand. 1988. Tripartite behavioral intervention to reduce stereotypic and disruptive behaviors in young multihandicapped children. *Behavior Therapy,* 19: 503–526.

Sivinski, J. 1984. Sperm in competition, pp. 85–115 in R.L. Smith, Ed., *Sperm Competition and the Evolution of Animal Mating Systems.* Academic Press, New York, 687 pp.

Skerman, V.B.D., V. McGowan, and P.H.A. Sneath. 1980. Approved lists of bacterial names. *International Journal of Systematic Bacteriology,* 30: 225–420.

Skubatz, H., and D.D. Kunkel. 1999. Further studies of the glandular tissue of the *Sauromatum guttatum* (Araceae). *American Journal of Botany,* 86:841–854.

Skubatz, H., T.A. Nelson, A.M. Dong, B.J.D. Meeuse, and A.J. Bendich. 1990. IR thermography of arum lily inflorescences. *Planta (Heidelberg),* 182: 432–436.

Slobodkin, L.B. 1993. Gaia: hoke and substance (book review). *BioScience,* 43: 255–256.

Slobodkin, L.B. 1995. A scientific sermon (book review). *Trends in Ecology and Evolution,* 10: 384–385.

Sluckin, W. 1967. *Imprinting and Early Learning.* Aldine Publishing, Chicago, IL, 147 pp.

Small, M.F. 1989. Aberrant sperm and the evolution of human mating patterns. *Animal Behaviour,* 38: 544–546.

Small, M.F. 1990. Alloparental behaviour in Barbary Macaques, *Macaca sylvanus. Animal Behaviour,* 39: 297–306.

Smith, C.A. and T. Takasaka. 1971. Auditory receptor organs of reptiles, birds, and mammals. *Contributions to Sensory Physiology,* 5: 129–178.

Smith, D. 1999. A thinking bird or just another birdbrain? *New York Times Northeast,* 9 October: A1, A17.

Smith, H.M. 1984. Terminological barbarisms overdue for retirement. *Bioscience,* 34(11): 679.

Smith, M.R. 1965. *House-Infesting Ants of the Eastern United States: Their Recognition, Biology, and Economic Importance.* U.S. Department of Agriculture, Washington, D.C., 105 pp.

Smith, R.L. 1979. Repeated copulation and sperm precedence: paternity assurance for a male brooding water bug. *Science,* 205: 1029–1031.

Smith, R.L. 1984. Human sperm competition, pp. 601–659 in R.L. Smith, Ed., *Sperm Competition and the Evolution of Animal Mating Systems.* Academic Press, New York, 687 pp.

Smolker, R., A. Richards, R. Connor, J. Mann, and P. Berggren. 1997. Sponge carrying by Dolphins (Delphinidae, *Turiops* sp.): a foraging specialization involving tool use? *Ethology,* 103: 454–465.

Smuts, B. 1987. What are friends for? *Natural History,* 96(2): 36–45.

Sober, E. 1992a. Monophyly, pp. 202–207 in E.F. Keller and E.A. Lloyd, Eds., *Keywords in Evolutionary Biology.* Harvard University Press, Cambridge, MA, 414 pp.

Sober, E. 1992b. Parsimony, pp. 249–254 in E.F. Keller and E.A. Lloyd, Eds., *Keywords in Evolutionary Biology.* Harvard University Press, Cambridge, MA, 414 pp.

Sokal, R.R. and F.J. Rohlf. 1969. *Biometry.* W.H. Freeman, San Francisco, CA, 776 pp.

Soltis, P.S., D.E. Soltis, and M.W. Chase. 1999. Angiosperm phylogeny inferred from multiple genes as a tool for comparative biology. *Nature,* 402: 402–404.

Southern, H.N. 1948. Sexual and aggressive behaviors in the Wild Rabbit. *Behaviour,* 1(3,4): 173–194.

Southwick, C., Ed., 1970. *Animal Aggression: Selected Readings.* Van Nostrand-Reinhold, New York, 229 pp.

Southwood, T.R.E. 1978. *Ecological Methods with Particular Reference to the Study of Insect Populations,* 2nd ed. Chapman & Hall, London, 524 pp.

Sparkes, I.G. 1975. *A Dictionary of Collective Nouns and Group Terms.* Gale Research Company, Detroit, MI, 191 pp.

Spears, R.A. 1981. *Slang and Euphemism.* Jonathan David Publishers, Middle Village, NY, 448 pp.

Specter, M. and G. Kolata. 1997. After decades and many missteps, cloning success. *New York Times,* 3 March: A1, A20–23.

Spencer, H.G. and J.C. Masters. 1992. Sexual selection: contemporary debates, pp. 294–301 in E.F. Keller and E.A. Lloyd, Eds., *Keywords in Evolutionary Biology.* Harvard University Press, Cambridge, MA, 414 pp.

Spencer, K.C. 1988. *Chemical Mediation of Coevolution.* Academic Press, New York, 609 pp.

Spinelli, J.S. and H. Markowitz. 1985. Prevention of cage-associated distress. *Lab Animal,* November/December: 19–28.

Spradbery, J.P. 1973. *Wasps: An Account of the Biology and Natural History of Solitary and Social Wasps.* University of Washington Press, Seattle, 408 pp.

Spurr, S.H. and B.V. Barnes. 1980. *Forest Ecology,* 3rd ed. John Wiley & Sons, New York, 687 pp.

Squires, S. 1995. Drug Prozac relieves PMS, study finds: antidepressant works in most severe cases. *Washington Post,* 8 June: A1.

Stanley, S.M. 1973. An explanation for Cope's rule. *Evolution,* 27: 1–26.

Stanley, S.M. 1989. *Earth and Life Through Time,* 2nd ed. W.H. Freeman, New York, 689 pp.

Stanley, S.M. 1993. *Exploring Earth and Life Through Time.* W.H. Freeman, New York, 538 pp.

Starr, C. and R. Taggart. 1984. *Biology: The Unity and Diversity of Life,* 3rd ed. Wadsworth, Belmont, CA. 697 pp.

Stebbins, G.L., Jr. 1951. Cataclysmic evolution. *Scientific American,* 184(4): 54–59.

Stebbins, G.L. and F. J. Ayala. 1985. The evolution of Darwinism. *Scientific American,* 253(1): 72–82.

Stebbins, R.C. 1966. *Field Guide to Western Reptiles and Amphibians.* Houghton Mifflin, Boston, MA, 279 pp.

Steele, E.S. 1911. New or noteworthy plants from the eastern United States. *Contributions of the United States National Herbarium,* 12: 259–374.

Steele, R.H. 1986. Courtship feeding in *Drosophila subobscura.* I. The nutritional significance of courtship feeding. *Animal Behaviour,* 34: 1087–1098.

Steen, E.B. 1973. *Dictionary of Biology.* Barnes and Noble, New York, 637 pp.

Stenseth, N.C. and J. Maynard Smith. 1984. Coevolution in ecosystems: Red Queen evolution or stasis? *Evolution,* 38: 870–880.

Stephens, D.W. and J.R. Krebs. 1986. *Foraging Theory.* Princeton University Press, Princeton, NJ, 247 pp.

Stephens, D.W. and S.R. Paton. 1986. How constant is the constant of risk-aversion? *Animal Behaviour,* 34: 1659–1667.

Stevens, W.K. 1993. Dust in sea mud may link human evolution to climate. *New York Times Northeast,* 14 December: C1, C18.

Stevens, W.K. 1995. More extremes found in weather, pointing to greenhouse gas effect. *New York Times,* 23 May: C4.

Stewart, W.N. 1983. *Paleobotany and the Evolution of Plants.* Cambridge University Press, Cambridge, U.K., 405 pp.

Stewart, W.N. and G.W. Rothwell. 1993. *Paleobotany and the Evolution of Plants.* Cambridge University Press, Cambridge, U.K., 521 pp.

Stevens, J.E. 1996a. It's a jungle in there. *BioScience,* 46: 314–317.

Stevens, P.F. 1992. Species: historical perspectives, pp. 302–311 in E.F. Keller and E.A. Lloyd, Eds., *Keywords in Evolutionary Biology*. Harvard University Press, Cambridge, MA, 414 pp.

Stevens, W.K. 1996b. Fierce debate erupts over degree of peril facing ocean species. *New York Times,* 17 September: C1, C6.

Stiles, F.G. and L.L. Wolf. 1979. Ecology and evolution of lek mating behavior in the long-tailed hermit hummingbird. *Ornithological Monographs,* No. 27, 78 pp.

Stone, R. 1999. Keeping paradise safe for the natives. *Science,* 285: 1837.

Storz, A., Ed., 1973. *Psychology Encyclopedia*. Dushkin Publishing, Guildord, CT, 312 pp.

Strausbaugh, P.D. and E.L. Core. 1978. *Flora of West Virginia,* 2nd ed. Seneca Books, Grantsville, WV, 1079 pp.

Strauss, E. 1999. Can mitochondrial clocks keep time? *Science,* 283: 1435–1438.

Strickberger, M.W. 1990. *Evolution*. Jones and Bartlett Publishers, Boston, MA, 579 pp.

Strickberger, M.W. 1996. Imminent domain: microbe may redefine life's paradigm. *Washington Post,* 23 September: A3.

Strickberger, M.W. 2000. *Evolution,* 3rd ed. Jones and Bartlett Publishers, Boston, MA, 722 pp.

Stringer, C.B. and P. Andrews. 1988. Genetic and fossil evidence for the origin of modern humans. *Science,* 239: 1263–1268.

Stuart, R.J. 1991. Kin recognition as a functional concept. *Animal Behaviour,* 41: 1093–1094.

Stutz, B. 1987. Leaping *Lepus. Natural History,* 96(2): 46–48.

Sullivan, J.J. 1987. Insect hyperparasitism. *Annual Review of Entomology,* 32: 49–70.

Sullivan, W. 1974. *Continents in Motion: The New Earth Debate*. McGraw-Hill, New York, 399 pp.

Sullivan, W.T. III. 1997. The clash of cosmologies (book review). *Science,* 275: 1275.

Sun, G., D.L. Dilcher, S. Zheng, and Z. Zhou. 1998. In search of the first flower: a Jurassic angiosperm, *Archaefructus,* from Northeast China. *Science,* 282: 1692–1693.

Suplee, C. 1996a. "Environmental estrogens" may pose greater risk, study shows. *Washington Post,* 7 June: A4.

Suplee, C. 1996b. Biology: alarm may inhibit pregnancy in famine. *Washington Post,* 22 July: A2.

Suplee, C. 1996c. Imminent domain: microbe may redefine life's paradigm. *Washington Post,* 23 September: A3.

Suplee, C. 1997a. In a family of good bacteria, tiny subset has turned bad. *Washington Post,* 23 August: A10.

Suplee, C. 1997b. *El Niño:* preparing for the worst. Weather conditions ripe for most destructive pattern in history, scientists say. *Washington Post,* 21 September: A1, A16.

Suplee, C. 1997c. Chemistry: a sobering abundance of nitrous oxide. *Washington Post,* 8 December: A3.

Suplee, C. 1998. From ancient grandmas, theory of longevity and menopause. *Washington Post,* 9 February: A3.

Sussman, R. W. 1993. A current controversy in human evolution. *American Anthropologist,* 95: 9–13.

Sustare, B.D. 1978. Systems diagrams, pp. 275–311 in P.W. Colgan, Ed., *Quantitative Ethology*. John Wiley & Sons, New York, 364 pp.

Sutton, J. 1970. *Continental Drift*. Francis Hodgson, U.K., 22 pp.

Svitil, K.A. 2000. Field guide to new planets. *Discover,* March: 49–55.

Tallamy, D.W. 1986. Age specificity of "egg dumping" in *Gargaphia solani* (Hemiptera: Tingidae). *Animal Behaviour,* 34: 599–603.

Tang, Y.-P., E. Shimizu, G.R. Dube, C. Rampon, G.A. Kerchner, M. Zhuo, G. Liu, and J.Z. Tsien. 1999. Genetic enhancement of learning and memory in mice. *Nature,* 401: 63–69.

Tanksley, S.D. 1993. Mapping polygenes. *Annual Review of Genetics,* 27: 205–233.

Tavolga, W.N. 1968. Fishes, pp. 271–288 in T.A. Sebeok, Ed., *Animal Communication: Techniques of Study and Results of Research.* Indiana University Press, Scarsborough, Canada, 1128 pp.

Taylor, D.H. and K. Adler. 1978. The pineal body: site of extraocular perception of celestial cues for orientation in the tiger salamander (*Ambystoma tigrinum*). *Journal of Comparative Physiology,* 124: 357–361.

Taylor, D.W. and L.J. Hickey. 1990. An aptian plant with attached leaves and flowers: implications for angiosperm origin. *Science,* 247: 702–704.

Taylor, P. 1992. Community, pp. 52–60 in E.F. Keller and E.A. Lloyd, Eds., *Keywords in Evolutionary Biology.* Harvard University Press, Cambridge, MA, 414 pp.

Taylor, S.R. 1987. The origin of the Moon. *American Scientist,* 75: 468–477.

Taylor, T.N. and E.L. Taylor. 1993. *The Biology and Evolution of Fossil Plants.* Prentice Hall, Englewood Cliffs, NJ, 982 pp.

Teall, E.N., Ed., 1984. *New Concise Webster's Dictionary.* Modern Promotions Publishers, New York, 370 pp.

Tedford, K., D. Sa, K. Stevens, and M. Tyers. 1997. Regulation of the mating pheromone and invasive growth responses in yeast by two MAP kinase substrates. *Current Biology,* 7:228–238.

Templeton, A.R. 1991. Human origins and analysis of mitochondrial DNA sequences. *Science,* 255: 737.

Tepidino, V.J. and S. Geer. 1991. The hairy Hymenoptera. *American Horticulturist,* May 70(5): 4–5.

Thomas, L. and F. Juanes. 1996. The importance of statistical power analysis: an example from *Animal Behaviour, Animal Behaviour,* 52: 856–859.

Thomas, R.H. and D.W. Zeh. 1984. Sperm transfer and utilization strategies in arachnids: ecological and morphological constraints, pp. 179–221 in R.L. Smith, Ed., *Sperm Competition and the Evolution of Animal Mating Systems.* Academic Press, New York, 687 pp.

Thompson, J.N. 1982. *Interaction and Coevolution.* John Wiley & Sons, New York, 179 pp.

Thomson, J.D., E.A. Herre, J.L. Hamrick, and J.L. Stone. 1991. Genetic mosaics in strangler fig trees: implications for tropical conservation. *Science,* 254: 1214–1216.

Thomson, K.S. 1985. Is paleontology going extinct? *American Scientist,* 73: 570–572.

Thomson, K.S. 1991. *Living Fossil: The Story of The Coelacanth.* W.W. Norton, New York, 252 pp.

Thorne, A.G. and M.H. Wolpoff. 1992. The multiregional evolution of humans. *Scientific American,* 266(4): 76–83.

Thornhill, R. 1980. Rape in *Panorpa* scorpionflies and a general rape hypothesis. *Animal Behaviour,* 28: 52–59.

Thornhill, R. and J. Alcock. 1983. *The Evolution of Insect Mating Systems.* Harvard University Press, Cambridge, MA, 537 pp.

Thornhill, R. and S.W. Gangestad. 1996. Human female copulatory orgasm: a human adaptation or phylogenetic holdover. *Animal Behaviour,* 52: 853–855.

Thorpe, W.H. 1951. The definitions of some terms used in animal behavior studies. *Bulletin of Animal Behaviour,* 9: 1–7.

Thorpe, W.H. 1956. *Learning and Instinct in Animals.* Methuen, London, 493 pp.

Thorpe, W.H. 1963. *Learning and Instinct in Animals,* 2nd ed. Methuen, London, 558 pp.

Thorsteinson, A.J. 1953. The chemotactic responses that determine host specificity in an oligophagous insect (*Plutella maculipennis* (Curt.) Lepidoptera). *Canadian Journal of Zoology,* 31: 52–72.

Thorsteinson, A.J. 1955. The experimental study of the chemotactic basis of host specificity in phytophagous insects. *Canadian Entomologist,* 87: 49–57.

Tierney, P. 1983. Herbert Simon's simple economics. *Science,* 83 November: 82–88.

Tilman, D. 1999a. Diversity by default. *Science,* 283: 495–496.

Tilman, D. 1999b. Diversity and production in European grasslands. *Science,* 286: 1099–2000.

Tinbergen, N. 1951 (1969). *The Study of Instinct.* Oxford University Press, New York, 228 pp.

Tinbergen, N. 1952. Derived activities: their causation, biological significance, origin and emancipation during evolution. *Quarterly Review of Biology,* 27: 1–32.

Tinbergen, N. 1961. *The Herring Gull's World.* Basic Books, New York, 255 pp.

Tinklepaugh, O.L. and C.G. Hartman. 1930. Behavioral aspects of parturition in the monkey (*Macacus rhesus*). *Journal of Comparative Psychology,* 11: 63–98.

Todd, S.C. 1999. A view from Kansas on that evolution debate. *Nature,* 401: 423.

Tokarz, R.R. 1988. Copulatory behaviour of the lizard *Anolis sagrei*: alternation of hemipenis use. *Animal Behaviour,* 36: 1518–1524.

Tokarz, R.R. and J.B. Slowinski. 1990. Alternation of hemipenis use as a behavioural means of increasing sperm transfer in the lizard *Anolis sagrei. Animal Behaviour,* 40: 374–379.

Tomitani, A., K. Okada, H. Miyashita, H.C.P. Matthijs, T. Ohno, and A. Tanaka. 1999. Chlorophyll b and phycobilins in the common ancestor of cyanobacteria and chloroplasts. *Nature,* 400: 159–162.

Torre-Bueno, J.R. de la. 1978. *A Glossary of Entomology.* New York Entomological Society, New York, 336 pp.

Townes, H. 1962. Design for a Malaise trap. *Proceedings of the Entomological Society of Washington,* 64: 253–262.

Townsend, D.S., M.M. Stewart, and F.H. Pough. 1984. Male parental care and its adaptive significance in a neotropical frog. *Animal Behaviour,* 32: 421–431.

Travis, J. 1992. Possible evolutionary role explored for "jumping genes." *Science,* 257: 884–885.

Travis, J. 1993. Medicine: discovery of genes in pieces wins for two biologists. *Science,* 262: 506.

Trivers, R.L. 1971. The evolution of reciprocal altruism. *Quarterly Review of Biology,* 46: 35–57.

Trivers, R.L. 1972. Parental investment and sexual selection, pp. 136–179 in B. Campbell, Ed., *Sexual Selection and the Descent of Man 1871–1971.* Aldine Publishing, Chicago, IL, 378 pp.

Trivers, R.L. 1974. Parent-offspring conflict. *American Zoologist,* 14: 249–264.

Trivers, R.L. 1985. *Social Evolution.* Benjamin/Cummings, Menlo Park, CA. 462 pp.

Trivers, R.L. and H. Hare. 1976. Haplodiploidy and the evolution of the social insects. *Science,* 191: 249–263.

Trowbridge, A.C., Ed., 1962. *Dictionary of Geological Terms.* Doubleday, Garden City, NY, 545 pp.

Turchin, P, A.D. Taylor, and J.D. Reeve. 1999. Dynamical role of predators in population cycles of a forest insect: an experimental test. *Science,* 285: 1068–1071.

Turco, M. and S. Gruner. 1999. *Great Apes Calendar 2000.* Avalance Publishing, Huntington Beach, CA.

Turning Point Project. 1999. Warning: bioinvasion. *New York Times,* 20 September: A14.

Tuttle, R.H., Ed., 1973. *Socioecology and Psychology of Primates.* Mouton Publishers, The Hague, 474 pp.

Udvardy, M.D.F. and J. Farrand, Jr. 1994. *National Audubon Society Field Guide to North American Birds: Western Region,* rev. ed. Alfred A. Knopf, New York, 822 pp.

Uyenoyama, M.K. and M.W. Feldman. 1992. Altruism: some theoretical ambiguities, pp. 34–40 in E.F. Keller and E.A. Lloyd, Eds., *Keywords in Evolutionary Biology.* Harvard University Press, Cambridge, MA, 414 pp.

Uzzell, T. 1984. Sex determination (book review). *Science,* 224: 733–734.

Valenti, M.A. 1998. *Entomophaga maimaiga:* salvation from the Gypsy Moth or fly in the ointment? *American Entomologist,* 44(1): 20–21.

van Gelder, L. 1996. Thomas S. Kuhn, scholar who altered the paradigm of scientific change, dies at 73. *New York Times,* 19 June: B7.

Van Valen, L. 1973. A new evolutionary law. *Evolutionary Theory,* 1: 1–30.

Vander Wall, S.B. 1990. *Food Hoarding in Animals.* University of Chicago Press, Chicago, IL, 445 pp.

Vascular Tropicos Nomenclatural Database. 1999. Missouri Botanical Garden, St. Louis, MO (http://mobot.mobot.org/ Pick/Search/pick.html).

Vawter, L. and W.M. Brown. 1986. Nuclear and mitochondrial DNA comparisons reveal extreme rate variation in the molecular clock. *Science,* 234: 194–196.

Veevers, J.E. 1985. The social meanings of pets: alternative roles for companion animals, pp. 11–28 in M.B. Sussman, Ed., *Pets and the Family.* Haworth Press, New York, 238 pp.

Veiga, J.P. 1990. Infanticide by male and female house sparrows. *Animal Behaviour,* 39: 496–502.

Venning, F.D. 1974. *Cacti.* Western Publishing, Racine, WI. 160 pp.

Vermeij, G.J. 1996. Animal origins. *Science,* 274: 525–526.

Verplanck, W.S. 1957. A glossary of some terms used in the objective study of behavior. *Psychological Review,* 64(suppl.): 1–42.

Verrell, P.A. 1989. Male mate choice for fecund females in a plethodontid salamander. *Animal Behaviour,* 38: 1086–1087.

Vick, K. 1999. Betting on the bugs: machine-like weevils rid Lake Victoria of choking weed. *Washington Post,* 22 September: A25 A28.

Vigilant, L., M. Stoneking, H. Harpending, K. Hawkes, and A.C. Wilson. 1991. African populations and the evolution of human mitochondrial DNA. *Science,* 253: 1503–1507.

Vogel, G. 1997. Scientists probe feelings behind decision-making. *Science,* 275: 1269.

Vogel, G. 1998. Did the first complex cell eat hydrogen? *Science,* 279: 1663–1664.

Voith, V.L. 1985. Attachment of people to companion animals. *Human-Companion Animal Bond,* Veterinary Clinics of North America, Small Animal Practice, 15: 403–410.

von Neumann, J. and O. Morgenstern. 1953. *Theory of Games and Economic Behavior.* Princeton University Press, Princeton, NJ, 641 pp.

Voss, E.G. 1972. *Michigan Flora.* Part 1. *Gymnosperms and Monocots.* Cranbrook Institute of Science, Bloomfield Hills, MI, 488 pp.

Vrba, E. and S.J. Gould. 1986. The hierarchical expansion of sorting and selection. *Paleobiology,* 12: 217–228.

Waage, J.K. 1984. Sperm competition and the evolution of odonate mating systems, pp. 251–290 in R.L. Smith, Ed., *Sperm Competition and the Evolution of Animal Mating Systems.* Academic Press, New York, 687 pp.

Wabnitz, P.A., J.H. Bowie, M.J. Tyler, J.C. Wallace, and B.P. Smith. 1999. Aquatic sex pheromone from a male tree frog. *Nature,* 401: 444–445.

Wade, M.J. 1992a. Epistasis, pp. 87–91 in E.F. Keller and E.A. Lloyd, Eds., *Keywords in Evolutionary Biology.* Harvard University Press, Cambridge, MA, 414 pp.

Wade, M.J. 1992b. Heritability: historical perspectives, pp. 149–150 in E.F. Keller and E.A. Lloyd, Eds., *Keywords in Evolutionary Biology.* Harvard University Press, Cambridge, MA, 414 pp.

Wade, M.J. and S.J. Arnold. 1980. The intensity of sexual selection in relation to male sexual behaviour, female choice and sperm precedence. *Animal Behaviour,* 28: 446–461.

Wade, N. 1995. First sequencing of a cell's DNA defines basis of life. *New York Times,* 1 August: C1, C9.

Wade, N. 1998. The struggle to decipher human genes. *New York Times Northeast,* 10 March: C1, C5.

Wade, N. 1999a. The genome's combative entrepreneur. *New York Times,* 18 May: D1–D2.

Wade, N. 1999b. Count of human genes is put at 140,000, a significant increase. *New York Times,* 23 September: A17.

Wadler, J. 1999. Passionate life in a lab of dead animals. *New York Times Northeast,* 1 October: A25.

Wainscoat, J. 1987. Out of the garden of Eden. *Nature,* 325: 13.

Waldbauer, G.P. and W.E. LaBerge. 1985. Phenological relationships of wasps, bumblebees, their mimics and insectivorous birds in Northern Michigan. *Ecological Entomology,* 10: 99–110.

Waldbauer, G.P. and J.K. Sheldon. 1971. Phenological relationships of some aculeate Hymenoptera, their dipteran mimics, and insectivorous birds. *Evolution,* 25: 371–382.

Waldbauer, G.P., J.G. Sternburg, and C.T. Maier. 1977. Phenological relationships of wasps, bumblebees, their mimics, and insectivorous birds in an Illinois sand area. *Ecology,* 58: 583–591.

Walker, L.C. 1990. *Forests: A Naturalist's Guide to Trees and Forest Ecology.* John Wiley & Sons, New York, 288 pp.

Wali, M.K. 1995. ecoVocabulary: a glossary of our times. *Bulletin of the Ecological Society of America,* 76:106–111.

Waltz, E.C. 1987. A test of the information-centre hypothesis in two colonies of Common Terns, *Sterna hirundo. Animal Behaviour,* 35: 48–59.

Ward, P. and A. Zahavi. 1973. The importance of certain assemblages of birds as "information centers" for food finding. *Ibis,* 115: 517–534.

Warrick, J. 1996. Proposals cast protective net around depleted shark populations. *Washington Post,* 31 December: A2.

Warrick, J. 1998. In U.S., an Asian beetle instills full-bore economic fear. *Washington Post,* 19 October: A3.

Waser, N.M. 1986. Flower constancy: definition, cause, and measurement. *American Naturalist,* 127: 593–603.

Waser, N.M. and M.V. Price. 1983. Optimal and actual outcrossing in plants and the nature of plant-pollinator interaction, pp. 341–359 in C.E. Jones and R.J. Little, Eds., *Handbook of Experimental Pollination Biology*. Scientific and Academic Editions, New York, 558 pp.

Washburn, S.L. 1985. Human evolution after Raymond Dart, pp. 3–18 in P.V. Tobias, Ed., *Hominid Evolution: Past, Present, and Future*. Alan R. Liss, New York.

Wasser, S.K. 1986. Book review. *Animal Behavior,* 34: 623–624.

Waters, M.R., S.L. Forman, and J.M. Pierson. 1997. Diring Yuriakh: a Lower Paleolithic site in central Siberia. *Science,* 275: 1281–1284.

Waters, T. 1991. The flowers of Koonwarra. *Discover,* January: 79.

Watson, D.A. 1993. Unusual mutational mechanisms and evolution. *Science,* 260: 1958–1959.

Watson, J.D. and F.H.C. Crick. 1953. Genetical implications of the structure of deoxyribonucleic acid. *Nature,* 171: 964–967.

Weatherhead, P.J. and R.J. Robertson. 1979. Offspring quality and the polygyny threshold: "the sexy son hypothesis." *American Naturalist,* 113:201–208.

Wegener, A. 1966. *The Origin of Continents and Oceans*. Dover Publications, New York, 246 pp.

Weinrich, J. D. 1980. Toward a sociobiological theory of the emotions, pp. 113–138 in R. Plutchik and H. Kellerman, Eds. *Emotion: Theory, Research, and Experience*. Academic Press, New York, 399 pp.

Weiss, R. 1991. Floral colour changes as cues for pollinators. *Nature,* 354: 227–229.

Weiss, R. 1995. Floral engineering helps decipher nature's code. *Washington Post,* 25 September: A3.

Weiss, R. 1996a. Swimming in high social circles: for Belize shrimp, bee-like cooperation is way of life — and survival. *Washington Post,* 6 June: A1, A17.

Weiss, R. 1996b. Way to hustle! A fish's ultimate thrill. *Washington Post,* 12 August: A3.

Weiss, R. 1996c. Genetics: fruit flies' sex-behavior source. *Washington Post,* 16 December: A2.

Weiss, R. 1998. Sniffing out human pheromones: scientists find proof of "chemistry" between people. *Washington Post,* 12 March: A1, A9.

Weiss, R. 1999. Plant kingdoms' new family tree: project explains flowers, identifies a green "Eve." *Washington Post,* 5 August: A1, A8.

Weissenbach, J. 1996. Landing on the genome. *Science,* 274: 479.

Weldon, P.J. 1980. In defense of "kairomone" as a class of chemical releasing stimuli. *Journal of Chemical Ecology,* 6: 719–725.

Weldon, P.J. and G.M. Burghardt. 1984. Deception divergence and sexual selection. *Zeitschrift für Tierpsychologie,* 65: 89–102.

Welty, J.C. 1966. *The Life of Birds*. W.B. Saunders, Philadelphia, PA, 546 pp.

Welty, J.C. 1975. *Life of Birds*. W.B. Saunders, Philadelphia, PA, 623 pp.

Wenner, A.M. 1967. Honey Bees: do they use the distance information contained in their dance maneuver? *Science,* 155: 844–847.

Wenner, A.M. 1971. *The Bee Language Controversy: An Experience in Science*. Educational Programs Improvement Corp., Boulder, CO. 109 pp.

Wenner, A.M., P.H. Wells, and F.J. Rohlf. 1967. An analysis of the waggle dance and recruitment in honeybees. *Physiological Zoology,* 40: 317–344.

Wenner, A.M., P.H. Wells, and D.L. Johnson. 1969. Honeybee recruitment to food sources: olfaction or language? *Science,* 164: 84–86.

West Eberhard, M.J. 1975. The evolution of social behavior by kin selection. *Quarterly Review of Biology,* 50: 1–33.

West-Eberhard, M.J. 1986. Alternative adaptations, speciation, and phylogeny (a review). *Proceedings of the National Academy of Sciences USA,* 83: 1388–1392.

West-Eberhard, M.J. 1987. Flexible strategy and social evolution, pp. 35–51 in Y. Itô, J.L. Brown, and J. Kikkawa, Eds. *Animal Societies: Theories and Facts.* Japan Scientific Societies Press, Tokyo, 291 pp.

West-Eberhard, M.J. 1989. Phenotypic plasticity and the origins of diversity. *Annual Review of Systematics and Ecology,* 20: 249–278.

West-Eberhard, M.J. 1992. Adaptation: current usages, pp. 13–18 in E.F. Keller and E.A. Lloyd, Eds., *Keywords in Evolutionary Biology.* Harvard University Press, Cambridge, MA, 414 pp.

Westneat, D.F. 1987. Extra-pair copulations in a predominantly monogamous bird: observations of behaviour. *Animal Behaviour,* 35: 865–876.

Westoby, M. 1994. Adaptive thinking and medicine. *Trends in Ecology and Evolution,* 9: 1–2.

Wheeler, M. 1996. The vertical forest. *Discover,* February: 76–82.

Wheeler, W.M. 1930. The ant *Prenolepis imparis* Say. *Annals of the Entomological Society of America,* 33: 1–26.

Wherry, E.T. 1929. Three shale-slope plants in Maryland. *Torreya,* 29: 104–107.

Whewell, W. 1840. *The Philosophy of the Inductive Sciences.* Parker, London, 532 pp. (Vol. 1), 546 pp. (Vol. 2).

Whitaker, B. 1999. Released to the wild, Condors choose a nice peopled retreat. *New York Times Northeast,* 2 October: A7.

Whitaker, J.O., Jr. 1996. *National Audubon Society Field Guide to North American Mammals,* 2nd ed. Alfred A. Knopf, New York, 937 pp.

White, I.M., D.H. Headrick, A.L. Norrbom, and L.E. Carroll. 2000. Glossary, pp. 877–920 in M. Aluja and A. Norrbom, Eds. *Fruit Flies (Tephritidae): Phylogeny and Evolution of Behavior.* CRC Press, Boca Raton, FL.

White, J. 1995. *Forest and Woodland Trees in Britain.* Oxford University Press, London, 217 pp.

White, K., M.E. Grether, J.M. Abrams, L. Young, K. Farrell, and H. Steller. 1994. Genetic control of programmed cell death in *Drosophila. Science,* 264: 677–683.

White, M.J.D. 1973. *Animal Cytology and Evolution,* 3rd ed. Cambridge University Press, Cambridge, U.K., 961 pp.

White, R.E. 1983. *A Field Guide to the Beetles of North America.* Houghton Mifflin, Boston, MA, 368 pp.

Whitehead, H. 1998. Cultural selection and genetic diversity in matrilineal whales. *Science,* 282: 1708–1711.

Whitman, D.W. 1988. Allelochemical interactions among plants, herbivores, and their predators, pp. 11–64 in P. Barbosa and D.L. Letourneau, Eds., *Novel Aspects of Insect-Plant Interactions.* Wiley, New York, 362 pp.

Whittaker, R.H. 1970a. The biochemical ecology of higher plants, pp. 43–70 in E. Sondheimer and J.B. Simeone, Eds. *Chemical Ecology.* Academic Press, New York, 336 pp.

Whittaker, R.H. 1970b. *Communities and Ecosystems.* Macmillan, New York, 385 pp.

Whittaker, R.H. and P.P. Feeny. 1971. Allelochemics: chemical interactions between species. *Science,* 171: 757–770.

Whoriskey, F.G. 1991. Stickleback distraction displays: sexual or foraging deception against egg cannibalism? *Animal Behaviour,* 41: 989–995.

Wickler, W. 1968. *Mimicry in Plants and Animals.* World University Library, London, 255 pp.

Wicklund, R.A. and S. Duval. 1971. Opinion change and performance facilitation as a result of objective self-awareness. *Journal of Experimental Social Psychology,* 7: 319–342.

Wickman, P.-O. 1986. Courtship solicitation by females of the Small Heath Butterfly, *Coenonympha pamphilus* (L.) (Lepidoptera: Satyridae) and their behaviour in relation to male territories before and after copulation. *Animal Behaviour,* 34: 153–157.

Wigglesworth, V.B. 1980. Do insects feel pain? *Antenna,* 4: 8–9.

Wiley, E.O. 1981. *Phylogenetics: The Theory and Practice of Phylogenetic Systematics.* John Wiley & Sons, New York, 439 pp.

Wiley, R.H. 1997. Book review (M.D. Hauser's *The Evolution of Communication,* MIT Press, Cambridge, MA, 1996, 760 pp.). *Animal Behaviour,* 54: 751–754.

Wilford, J.N. 1996a. Fossil is found of a terror that dwarfed *T. rex. New York Times,* 17 May: A16.

Wilford, J.N. 1996b. Replying to skeptics, NASA defends claims about Mars. *New York Times,* 8 August: A1, D20.

Wilford, J.N. 1996c. New traces of past life on Mars. *New York Times,* 1 November: A12.

Wilford, J.N. 1997. New findings suggest massive black holes lurk in the hearts of many galaxies. *New York Times,* 14 January: C1, C7.

Wilford, J.N. 1999a. At long last, another sun with a family of planets. *New York Times,* 16 April: A1, A19.

Wilford, J.N. 1999b. Off a shelf in Manhanttan, a pre-human skull. *New York Times Northeast,* 7 September: A1, A21.

Wilford, J.N. 2000a. In the very distant Universe, objects even older than light. *New York Times Northeast,* 14 January: A1, A20.

Wilford J.N. 2000b. Wee animal called earliest link to lower primates and Humans. *New York Times Northeast,* 16 March: A1, A18.

Williams, G.C. (1966) 1974. *Adaptation and Natural Selection: A Critique of Some Current Evolutionary Thought.* Princeton University Press, Princeton, NJ, 307 pp.

Williams, G.C. and R.M. Nesse. 1991. The dawn of Darwinian medicine. *The Quarterly Review of Biology,* 66: 1–20.

Williams, K.S., K.G. Smith, and F.M. Stephen. 1993. Emergence of 13-yr Periodical Cicadas (Cicadidae: *Magicicada*): phenology, mortality, and predator satiation. *Ecology,* 74: 1143–1152.

Williams, M.B. 1992. Species: current usages, pp. 318–323 in E.F. Keller and E.A. Lloyd, Eds., *Keywords in Evolutionary Biology.* Harvard University Press, Cambridge, MA, 414 pp.

Williams, N. 1997. Selling Darwinism in a citadel of social science. *Science,* 275: 29.

Willoughby, J. 1998. A chemical mystery that excites the taste buds. *New York Times Northeast,* 14 January: C3.

Willson, M.F. 1983. *Plant Reproductive Ecology.* John Wiley & Sons, New York, 282 pp.

Wilson, A.C. and R.L. Cann. 1992. The recent African genesis of humans. *Scientific American,* 266(4): 68–73.

Wilson, A.C., M. Stoneking, and R.L. Cann. 1991. Ancestral geographic states and the peril of parsimony. *Systematic Zoology,* 40: 363–365.

Wilson, D.S. 1983. The group selection controversy: history and current status. *Annual Review of Ecology and Systematics,* 14: 159–187.

Wilson, D.S. 1992. Group selection, pp. 145–148 in E.F. Keller and E.A. Lloyd, Eds., *Keywords in Evolutionary Biology.* Harvard University Press, Cambridge, MA, 414 pp.

Wilson, D.S. and L.A. Dugatkin. 1992. Altruism: contemporary debates, pp. 29–33 in E.F. Keller and E.A. Lloyd, Eds., *Keywords in Evolutionary Biology.* Harvard University Press, Cambridge, MA, 414 pp.

Wilson, E.O. 1965. Chemical communication in the social insects. *Science,* 149: 1064–1071.

Wilson, E.O. 1969. Source and possible nature of the odor trail of fire ants. *Science,* 128: 643–644.

Wilson, E.O. 1971. *The Insect Societies.* The Belknap Press, Cambridge, MA, 548 pp.

Wilson, E.O. 1975. *Sociobiology: The New Synthesis.* Harvard University Press, Cambridge, MA, 697 pp.

Wilson, E.O. 1984. *Biophilia.* Harvard University Press, Cambridge, MA, 157 pp.

Wilson, E.O. 1987. Kin recognition: an introductory synopsis, pp. 7–18 in J.C. Fletcher and C.D. Michener, Eds., *Kin Recognition in Animals.* Wiley, New York, 465 pp.

Wilson, E.O. and F.E. Regnier, Jr. 1971. The evolution of the alarm-defense system in the formicine ants. *American Naturalist,* 105: 279–289.

Wilson, M. and M. Daly. 1985. Competitiveness, risk taking, and violence: the young male syndrome. *Ethology and Sociobiology,* 6: 59–73.

Wimmer, M.J., R.R. Smith, D.L. Wellings, S.R. Toney, D.C. Faber, J.E. Miracle, J.T. Carnes, and A.B. Rutherford. 1993. Persistence of diflubenzuron on Appalachian forest leaves after aerial application of Dimilin. *Journal of Agricultural and Food Chemistry,* 41: 2184–2190.

Winston, M.L. 1987. *The Biology of the Honey Bee.* Harvard University Press, Cambridge, MA, 281 pp.

Winters, J. 1996. Death of a star. *Discover,* July: 126–129.

Wittenberger, J.F. 1981. *Animal Social Behavior.* Duxbury Press, Boston, MA, 722 pp.

Wolman, B.B., Ed., 1973. *Dictionary of Behavioral Science.* Litton Educational Publishing, New York, 478 pp.

Wolpoff, M. and A. Thorne. 1991. The case against Eve. *New Scientist,* 130(1774): 37–41.

Wood, M. 1999. *Arabidopsis.* A model plant genome. *Agricultural Research,* January: 26.

Wray, G.A., J.S. Levinton, and L.H. Shapiro. 1996. Molecular evidence for deep Precambrian divergences among metazoan phyla. *Science,* 274: 568–573.

Wright, K. 1996. What the dinosaurs left us. *Discover,* June: 59–65.

Wright, K. 1997. When life was odd. *Discover,* March: 51–61.

Wright, S. 1948. Genetics of populations. *Encyclopaedia Britannica* (1961 printing), Vol. 10, pp. 111–112.

WuDunn, S. 1997. TV cartoon's flashes send 700 Japanese into seizures. *New York Times Northeast,* 18 December: A3.

Wuethrich, B. 1996. Wayward grizzlies spark debate. *Science,* 274: 493.

Yanagisawa, Y, and H. Ochi. 1986. Step-fathering in the anemonefish *Amphiprion clarkii:* a removal study. *Animal Behaviour,* 34: 1769–1780.

Yoon, C.K. 1995. Monumental inventory of Costa Rican forest's insects is under way. *New York Times,* 11 July: C4.

Yoon, C.K. 1996a. Ecosystem's productivity rises with diversity of its species. *New York Times Northeast,* 5 March: C4.

Yoon, C.K. 1996b. Within nests, egret chicks are natural born killers. *New York Times,* 6 August: C1, C4.

Yoon, C.K. 1996c. Lake Victoria's lightning-fast origin of species. *New York Times,* 27 August: C1, C4.

Yoon, C.K. 1996d. Pronghorn's speed may be legacy of past predators. *New York Times,* 24 December: C1, C6.

Yoon, C.K. 1999a. A moth in search of a lifesaving encounter. *New York Times,* 18 May: D3.

Yoon, C.K. 1999b. Report on acid rain finds good news and bad news. *New York Times Northeast,* 7 October: A22.

Yoon, C.K. 1999c. In tiny fossils, botanists see flowery world. *New York Times Northeast,* 21 December: D1, D6.

Young, J.Z. 1950. *The Life of Vertebrates.* Oxford University Press, New York, 767 pp.

Young, M.S. 1985. The evolution of domestic pets and companion animals. *Human-Companion Animal Bond,* Veterinary Clinics of North America, Small Animal Practice, 15: 297–309.

Young, P.T. 1961a. *Motivation and Emotion.* John Wiley & Sons, New York, 648 pp.

Young, W.C., Ed., 1961b. *Sex and Internal Secretions.* Williams & Wilkins, Baltimore, MD, 1069 pp.

Zahavi, A. 1975. Mate selection: a selection for a handicap. *Journal of Theoretical Biology,* 53: 205–214.

Zahavi, A. 1977. The cost of honesty (further remarks on the handicap principle). *Journal of Theoretical Biology,* 67: 603–605.

Zentall, T.R. 1993. Mechanisms of learning (book review). *Science,* 260: 834.

Zim, H.S. and D.F. Hoffmeister. 1955. *Mammals: A Guide to Familiar American Species.* Golden Press, Racine, WI, 160 pp.

Zim, H.S. and H.H. Shoemaker. 1987. *Fishes: A Guide to Fresh- and Salt-Water Species.* Golden Press, Racine, WI, 160 pp.

Zim, H.S. and H.M. Smith. 1956. *Reptiles and Amphibians. A Guide to Familiar American Species.* Golden Press, Racine, WI, 160 pp.

Zimmer, C. 1995a. Coming onto the land. *Discover,* June: 117–127.

Zimmer, C. 1995b. The descent of birds. *Discover,* October: 40–41.

Zimmer, C. 1995c. First cell. *Discover,* November: 70–78.

Zimmer, C. 1995d. Life takes a backbone. *Discover,* December: 38–39.

Zimmer, C. 1996a. From teeth to beak. *Discover,* January: 50.

Zimmer, C. 1996b. The flight of the butterfly. *Discover,* May: 50.

Zimmer, C. 1996c. An explosion defused? *Discover,* December: 52.

Zimmerman, M. and D.J. Hicks. 1985. Strategy: misuse or insight? *BioScience,* 35: 66.

Zinder, N.D. 1985. The origin of sex: an argument, pp. 7–12 in H.O. Halvorson and A. Monroy, Eds., *The Origin and Evolution of Sex.* Alan R. Liss, New York, 345 pp.

Zuckerkandl, E. 1986. Polite DNA: functional density and functional compatibility in genomes. *Journal of Molecular Evolution,* 24: 12–27.

Zuckerkandl, E. 1987. On the molecular evolutionary clock. *Journal of Molecular Evolution,* 26: 34–46.

Zuckerkandl, E. and L. Pauling. 1962. Molecular disease, evolution, and genic heterogeneity, pp. 189–228 in M. Kasha and B. Pullman, Eds., *Horizons in Biochemistry.* Academic Press, New York, 603 pp.

Zuckerkandl, E. and L. Pauling. 1965. Evolutionary divergence and convergence in proteins, pp. 97–166 in V. Bryson, and H.J. Vogel, Eds., *Evolving Genes and Proteins.* Academic Press, New York, 629 pp.

Zuckerkandl, E., G. Latter, and J. Jurka. 1989. Maintenance of function without selection: *Alu* sequences as "cheap genes." *Journal of Molecular Evolution,* 29: 504–512.

Zuk, M. 1984. A charming resistance to parasites. *Natural History,* 93(4): 28–34.